HANDBOOK OF
Graph Theory

SECOND EDITION

DISCRETE MATHEMATICS

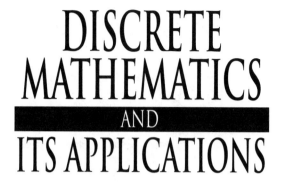

AND

ITS APPLICATIONS

Titles (continued)

Stig F. Mjølsnes, A Multidisciplinary Introduction to Information Security

Jason J. Molitierno, Applications of Combinatorial Matrix Theory to Laplacian Matrices of Graphs

Richard A. Mollin, Advanced Number Theory with Applications

Richard A. Mollin, Algebraic Number Theory, Second Edition

Richard A. Mollin, Codes: The Guide to Secrecy from Ancient to Modern Times

Richard A. Mollin, Fundamental Number Theory with Applications, Second Edition

Richard A. Mollin, An Introduction to Cryptography, Second Edition

Richard A. Mollin, Quadratics

Richard A. Mollin, RSA and Public-Key Cryptography

Carlos J. Moreno and Samuel S. Wagstaff, Jr., Sums of Squares of Integers

Gary L. Mullen and Daniel Panario, Handbook of Finite Fields

Goutam Paul and Subhamoy Maitra, RC4 Stream Cipher and Its Variants

Dingyi Pei, Authentication Codes and Combinatorial Designs

Kenneth H. Rosen, Handbook of Discrete and Combinatorial Mathematics

Douglas R. Shier and K.T. Wallenius, Applied Mathematical Modeling: A Multidisciplinary Approach

Alexander Stanoyevitch, Introduction to Cryptography with Mathematical Foundations and Computer Implementations

Jörn Steuding, Diophantine Analysis

Douglas R. Stinson, Cryptography: Theory and Practice, Third Edition

Roberto Tamassia, Handbook of Graph Drawing and Visualization

Roberto Togneri and Christopher J. deSilva, Fundamentals of Information Theory and Coding Design

W. D. Wallis, Introduction to Combinatorial Designs, Second Edition

W. D. Wallis and J. C. George, Introduction to Combinatorics

Jiacun Wang, Handbook of Finite State Based Models and Applications

Lawrence C. Washington, Elliptic Curves: Number Theory and Cryptography, Second Edition

DISCRETE MATHEMATICS AND ITS APPLICATIONS

Series Editor KENNETH H. ROSEN

HANDBOOK OF
Graph
Theory

SECOND EDITION

Edited by

Jonathan L. Gross

Columbia University
New York, USA

Jay Yellen

Rollins College
Winter Park, Florida, USA

Ping Zhang

Western Michigan University
Kalamazoo, USA

CRC Press
Taylor & Francis Group
Boca Raton   London   New York

CRC Press is an imprint of the
Taylor & Francis Group, an **informa** business

A CHAPMAN & HALL BOOK

CRC Press
Taylor & Francis Group
6000 Broken Sound Parkway NW, Suite 300
Boca Raton, FL 33487-2742

© 2014 by Taylor & Francis Group, LLC
CRC Press is an imprint of Taylor & Francis Group, an Informa business

No claim to original U.S. Government works

Printed on acid-free paper
Version Date: 20130923

International Standard Book Number-13: 978-1-4398-8018-0 (Hardback)

Library of Congress Cataloging-in-Publication Data

Handbook of graph theory. -- Second edition / [edited by] Jonathan L. Gross, Jay Yellen, Ping Zhang.
    pages cm -- (Discrete mathematics and its applications)
  Includes bibliographical references and index.
  ISBN 978-1-4398-8018-0 (hardback)
    1. Graph theory--Handbooks, manuals, etc. I. Gross, Jonathan L., editor of compilation. II. Yellen, Jay, editor of compilation. III. Zhang, Ping.

QA166.H36 2014
511'.5--dc23
                                                     2013038209

Visit the Taylor & Francis Web site at
http://www.taylorandfrancis.com

and the CRC Press Web site at
http://www.crcpress.com

Jonathan dedicates this book to Hadas, Noa, Nili, Tirzah, Eli Chaim, Bezi, Benjamin, Naomi, Rebecca, Abigail, Shayna, Alice, and Ruth.

Jay dedicates this book to Betsey and Tara.

Ping dedicates this book to Gary Chartrand.

CONTENTS

PREFACE

Over the past fifty years, graph theory has been one of the most rapidly growing areas of mathematics. Since 1960, more than 10,000 different authors have published papers classified as graph theory by *Math Reviews*, and for the past decade, more than 1000 graph theory papers have been published each year. Not surprisingly, this Second Edition is about 450 pages longer than the First Edition, which appeared in 2004.

This *Handbook* is intended to provide as comprehensive a view of graph theory as is feasible in a single volume. Many of our chapters survey areas that have large research communities, with hundreds of active mathematicians, and which could be developed into independent handbooks. The 89 contributors to this volume, 31 of whom are new to this edition, collectively represent perhaps as much as 90% or more of the main topics in pure and applied graph theory. Thirteen of the sections in the Second Edition cover newer topics that did not appear in the First Edition.

Format

In order to achieve this kind of comprehensiveness, we challenged our contributors to restrict their expository prose to a bare minimum, by adhering to the ready-reference style of the CRC Handbook series, which emphasizes quick accessibility for the non-expert. We thank the contributors for responding so well to this challenge.

The 13 chapters of the *Handbook* are organized into 65 sections. Within each section, several major topics are presented. For each topic, there are lists of the essential definitions and facts, accompanied by examples, tables, remarks, and in some cases, conjectures and open problems. Each section ends with a bibliography of references tied directly to that section. In many cases, these bibliographies are several pages long, providing extensive guides to the research literature and pointers to monographs.

To ensure that each section be reasonably self-contained, we encouraged contributors to include some definitions that may have appeared in earlier sections. Each contributor was also asked to include a glossary with his or her section. These section glossaries were then merged by the editors into 13 chapter glossaries.

Terminology and Notation

Graph theory has attracted mathematicians and scientists from diverse disciplines and, accordingly, is blessed (and cursed) with a proliferation of terminology and notations. Since the *Handbook* objective is to survey topics for persons whose expertise may be elsewhere, either on other topics, or outside of graph theory, we asked our contributors to tilt toward the general usage in the mathematical community, rather than staying strictly within the idioms of their specialties. But to understand graph theory literature, it helps to accept the legacy of history. As editors, we tried to strike a balance between preserving the notation and terminology that evolved from each area's rich history and our desire create a cohesive, uniform body of material.

Some uniformity of usage came easily. In general, the word *graph* is used inclusively to refer to graphs with directed edges and/or to graphs with multi-edges and self-loops. In most sections, G denotes a graph and V and E denote its vertex- and edge-sets, respectively.

However, some words are used differently by different graph theory communities. For instance, to an algebraic graph theorist, a *Cayley graph* is simple, connected, and undirected, but to a topological graph theorist, it may be non-connected, possibly directed, and have multi-edges and/or self-loops. To some graph theorists, a *clique* is a complete subgraph, maximal under set inclusion, and to others maximality is not required.

Consistency in notation was also problematic. In the literature of graph coloring, the Greek letter χ is used as the chromatic number, but to an algebraic topologist, it means the Euler characteristic.

Notes regarding terminology and notation were added to make these variations explicit, thereby improving cross-chapter compatibility.

Acknowledgments

We would like to thank Bob Stern of CRC Press for his continued enthusiasm and patience during the gestation period and Bob Ross at CRC for providing the final support in bringing the *Handbook* to publication.

Jonathan Gross, Jay Yellen, and Ping Zhang

About the Editors

Jonathan Gross is professor of computer science at Columbia University. His research in topology, graph theory, and cultural sociometry has earned him an Alfred P. Sloan Fellowship, an IBM Postdoctoral Fellowship, and various research grants from the Office of Naval Research, the National Science Foundation, and the Russell Sage Foundation.

Professor Gross is the inventor of the voltage graph, a construct widely used in topological graph theory and in other branches as well. His main current research interest is the genus distribution of graphs. His other recent areas of research publication include computer graphics and knot theory. He has received several awards for outstanding teaching at Columbia University, including the career Great Teacher Award from the Society of Columbia Graduates. He appears on the Columbia Video Network and on the video network of the National Technological University.

Prior to Columbia University, Professor Gross was in the mathematics department at Princeton University. His undergraduate work was at M.I.T., and he wrote his Ph.D. thesis on 3-dimensional topology at Dartmouth College.

His previous books include *Topological Graph Theory*, coauthored with Thomas W. Tucker, *Graph Theory and Its Applications*, coauthored with Jay Yellen, and *Combinatorial Methods with Computer Applications*. Another previous book, *Measuring Culture*, coauthored with Steve Rayner, constructs network-theoretic tools for measuring sociological phenomena.

Jay Yellen is Archibald Granville Bush Professor of Mathematics at Rollins College. He received his B.S. and M.S. in mathematics at Polytechnic University of New York and did his doctoral work in finite group theory at Colorado State University. Dr. Yellen has had regular faculty appointments at Allegheny College, the State University of New York at Fredonia, and the Florida Institute of Technology, where he was chair of Operations Research from 1995 to 1999. He has had visiting appointments at Emory University, Georgia Institute of Technology, Columbia University, and the University of Nottingham, UK.

In addition to his book *Graph Theory and Its Applications*, coauthored with Professor Gross, Professor Yellen has written manuscripts used at IBM for two courses in discrete mathematics within the Principles of Computer Science Series and has contributed two sections to the *Handbook of Discrete and Combinatorial Mathematics*. He also has designed and conducted several summer workshops on creative problem solving for secondary-school mathematics teachers, which were funded by the National Science Foundation and New York State. At Rollins, he has received the Hugh F. McKean Award for Outstanding Teaching, the Student's Choice Professor Award, and the Hugh F. McKean Research Grant Award.

Dr. Yellen has published research articles in character theory of finite groups, graph theory, power-system scheduling, and timetabling. His current research interests include graph theory, discrete optimization, and graph algorithms for software testing and course timetabling.

Ping Zhang is professor of mathematics at Western Michigan University. She wrote her Ph.D. thesis on algebraic combinatorics at Michigan State University. Her previous books, coauthored with Gary Chartrand, include *Graphs & Digraphs* (5th edition), *Mathematical Proofs: A Transition to Advanced Mathematics* (3rd edition), *Chromatic Graph Theory*, *A First Course in Graph Theory* and *Discrete Mathematics*. The first of these was also coauthored with Linda Lesniak and the second coauthored with Albert D. Polimeni. Her research interests are algebraic combinatorics and colorings, distance and convexity, traversability, decompositions, and domination within graph theory.

CONTRIBUTORS

Alfred V. Aho
Columbia University

Brian Alspach
University of Newcastle, Australia

Dan Archdeacon
University of Vermont

David C. Arney
West Point

Camino Balbuena
Universitat Politècnica de Catalunya, Spain

Lowell W. Beineke
Purdue University at Fort Wayne

Jacek Blazewicz
Poznan University of Technology, Poland

Béla Bollobás
University of Memphis
Trinity College, Cambridge, UK

Anthony Bonato
Ryerson University, Canada

Danail Bonchev
Virginia Commonwealth University

Richard B. Borie
University of Alabama

Robert C. Brigham
University of Central Florida

Edmund Burke
University of Stirling, Scotland

Gary Chartrand
Western Michigan University

Jianer Chen
Texas A&M University

Maria Chudnovsky
Columbia University

Fan Chung
University of California, San Diego

Alice M. Dean
Skidmore College

Camil Demetrescu
University of Rome La Sapienza, Italy

A. K. Dewdney
University of Western Ontario, Canada

Dominique de Werra
École Polytechnique Fédérale,
de Lausanne, Switzerland

Emilio Di Giacomo
University of Perugia, Italy

Michael Doob
University of Manitoba, Canada

Ernesto Estrada
University of Strathclyde, Scotland

Josep Fàbrega
Universitat Politècnica de Catalunya, Spain

Ralph J. Faudree
University of Memphis

Irene Finocchi
University of Rome La Sapienza, Italy

Miquel Àngel Fiol
Universitat Politècnica de Catalunya, Spain

Lisa Fleischer
Dartmouth College

Herbert Fleischner
Technical University of Vienna, Austria

Tomaž Pisanski
University of Ljubljana, Slovenia

Michael Plummer
Vanderbilt University

Primož Potočnik
University of Ljubljana, Slovenia

K. B. Reid
California State University, San Marcos

Dana Richards
George Mason University

R. Bruce Richter
University of Waterloo, Canada

Gelasio Salazar
Universidad Autónoma de San Luis Potosí, Mexico

Jay Sethuraman
Columbia University

Anthony Shaheen
California State University, Los Angeles

Douglas R. Shier
Clemson University

David Simchi-Levi
Massachusetts Institute of Technology

Martin Škoviera
Comenius University, Slovakia

Clifford Stein
Columbia University

Paul K. Stockmeyer
The College of William and Mary

Roberto Tamassia
Brown University

Krishnaiyan "KT" Thulasiraman
University of Oklahoma

Craig A. Tovey
Georgia Institute of Technology

Thomas W. Tucker
Colgate University

Zsolt Tuza
University of Veszprém, Hungary

Nikos Vlassis
LCSB, University of Luxembourg

Mark E. Watkins
Syracuse University

Arthur T. White
Western Michigan University

Robin J. Wilson
Pembroke College, Oxford University, UK

Nicholas Wormald
University of Waterloo, Canada

Jay Yellen
Rollins College

Ping Zhang
Western Michigan University

Chapter 1

Introduction to Graphs

Section 1.1
Fundamentals of Graph Theory

Jonathan L. Gross, Columbia University

Jay Yellen, Rollins College

INTRODUCTION

Configurations of nodes and connections occur in a great diversity of applications. They may represent physical networks, such as electrical circuits, roadways, or organic molecules. They are also used in representing less tangible interactions as might occur in ecosystems, sociological relationships, databases, or in the flow of control in a computer program.

1.1.1   Graphs and Digraphs

Any mathematical object involving points and connections between them may be called a *graph*. If all the connections are unidirectional, it is called a *digraph*. Our highly inclusive definition in this initial section of the *Handbook* permits fluent discussion of almost any particular modification of the basic model that has ever been called a graph.

Basic Terminology

DEFINITIONS

D1: A *graph* $G = (V, E)$ consists of two sets V and E.

- The elements of V are called *vertices* (or *nodes*).

- The elements of E are called *edges*.

- Each edge has a set of one or two vertices associated to it, which are called its *endpoints*. An edge is said to *join* its endpoints.

NOTATION: The subscripted notations V_G and E_G (or $V(G)$ and $E(G)$) are used for the vertex- and edge-sets when G is not the only graph under consideration.

D2: If vertex v is an endpoint of edge e, then v is said to be ***incident*** on e, and e is incident on v.

D3: A vertex u is ***adjacent*** to vertex v if they are joined by an edge.

D4: Two adjacent vertices may be called ***neighbors***.

D5: ***Adjacent edges*** are two edges that have an endpoint in common.

D6: A ***proper edge*** is an edge that joins two distinct vertices.

D7: A ***multi-edge*** is a collection of two or more edges having identical endpoints.

D8: A ***simple adjacency*** between vertices occurs when there is exactly one edge between them.

D9: The ***edge-multiplicity*** between a pair of vertices u and v is the number of edges between them.

D10: A ***self-loop*** is an edge that joins a single endpoint to itself.

TERMINOLOGY

- An alternative term for *self-loop* is **loop**. This can be used in contexts in which *loop* has no other meanings.

- In computer science, the term *graph* is commonly used either to mean a graph as defined here, or to mean a computer-represented data structure whose value is a graph.

EXAMPLE

E1: A line drawing of a graph $G = (V, E)$ is shown in Figure 1.1.1. It has vertex-set $V = \{u, v, w, x\}$ and edge-set $E = \{a, b, c, d, e, f\}$. The set $\{a, b\}$ is a multi-edge with endpoints u and v, and edge c is a self-loop.

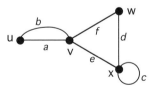

Figure 1.1.1: A graph.

REMARKS

R1: A graph is realized in a plane or in 3-space as a set of points, representing the vertices, and a set of curved or straight line segments, representing the edges. The curvature or length of such a line segment is irrelevant to the meaning. However, if a *direction* is indicated, that is significant.

R2: Occasionally, a graph is *parametrized* so that each edge is regarded as the homeomorphic image of the real interval $[0, 1]$ (except that for a self-loop, the endpoints 0 and 1 have the same image).

Simple Graphs

Most of theoretical graph theory is concerned with *simple* graphs. This is partly because many problems regarding general graphs can be reduced to problems about simple graphs.

DEFINITIONS

D11: A *simple graph* is a graph that has no self-loops or multi-edges.

D12: A *trivial graph* is a graph consisting of one vertex and no edges.

D13: A *null graph* is a graph whose vertex- and edge-sets are empty.

Edge Notation for Simple Adjacencies and for Multi-Edges

NOTATION: An edge joining vertices u and v of a graph may be denoted by the juxtaposition uv if it is the only such edge. Occasionally, the ordered pair (u, v) is used in this situation, instead of uv. To avoid ambiguities when multi-edges exist, or whenever else desired, the edges of a general graph may be given their own names, as in Figure 1.1.1 above.

EXAMPLE

E2: The simple graph shown in Figure 1.1.2 has edge-set $E = \{uv, vw, vx, wx\}$.

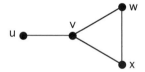

Figure 1.1.2: A simple graph.

General Graphs

Many applications require non-simple graphs as models. Moreover, some non-simple graphs serve an essential role in theoretical constructions, especially in constructing graph drawings (simple and non-simple) on surfaces (see Chapter 7).

TERMINOLOGY NOTE: Although the term "graph" means that self-loops and multi-edges are allowed, sometimes, for emphasis, the term *general graph* is used.

DEFINITIONS

D14: A *loopless graph* is a graph that has no self-loops. (It might have multi-edges.) Sometimes a loopless graph is referred to as a *multigraph*.

D15: The *dipole* D_n is a loopless graph with two vertices and n edges joining them.

D16: The *bouquet* B_n is a graph with one vertex and n self-loops.

EXAMPLES

E3: The loopless graph in Figure 1.1.3 depicts the benzene molecule C_6H_6.

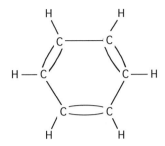

Figure 1.1.3: Graph model for a benzene ring.

E4: The dipole D_3 is shown in Figure 1.1.4.

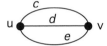

Figure 1.1.4: The loopless graph D_3.

E5: Two graphs with self-loops are shown in Figure 1.1.5.

Figure 1.1.5: The dumbbell graph and the bouquet B_4.

Attributes

Allowing graphs to have additional attributes beyond vertices and edges enables them to serve as mathematical models for a wide variety of applications. Two of the most common additional edge attributes, both described in great detail later in the *Handbook*, are edge *direction* (e.g., Chapters 3 and 11) and edge *weight* (e.g., Chapters 4 and 11). Another common attribute (for edges or vertices) is *color*. Graph coloring is discussed in Chapter 5.

DEFINITIONS

D17: A ***vertex attribute*** is a function from the vertex-set to some set of possible attribute values.

D18: An ***edge attribute*** is a function from the edge-set to some set of possible attribute values.

Digraphs

An edge between two vertices creates a connection in two opposite senses at once. Assigning a direction makes one of these senses *forward* and the other *backward*. Viewing direction as an edge attribute is partly motivated by its impact on computer implementations of graph algorithms. Moreover, from a mathematical perspective, regarding *directed graphs* as augmented graphs makes it easier to view certain results that tend to be established separately for graphs and for digraphs as a single result that applies to both. The attribute of edge direction is developed extensively in Chapter 3 and elsewhere in this *Handbook*.

DEFINITIONS

D19: A ***directed edge*** (or ***arc***) is an edge *e*, one of whose endpoints is designated as the ***tail***, and whose other endpoint is designated as the ***head***. They are denoted *head*(*e*) and *tail*(*e*), respectively.

TERMINOLOGY: A directed edge is said to be ***directed from*** its tail and ***directed to*** its head. (The tail and the head of a directed self-loop are the same vertex.)

NOTATION: In a line drawing, the arrow points toward the head.

D20: A ***multi-arc*** is a set of two or more arcs having the same tail and same head.

D21: A ***digraph*** (or ***directed graph***) is a graph each of whose edges is directed.

D22: A ***simple digraph*** is a digraph with no self-loops and no multi-arcs.

D23: A ***mixed graph*** (or ***partially directed graph***) is a graph that has both undirected and directed edges. In a mixed graph, using the unmodified term *edge* avoids specifying whether the edge is directed or undirected.

D24: The ***underlying graph*** of a directed or partially directed graph *G* is the graph that results from removing all the designations of *head* and *tail* from the directed edges of *G* (i.e., deleting all the edge-directions).

Ordered-Pair Representation of Arcs

NOTATION: In a simple digraph, an arc from vertex *u* to vertex *v* is commonly denoted (*u, v*) (or sometimes *uv*). When multi-arcs are possible, using distinct names is often necessary.

COMPUTATIONAL NOTE: (*A caution to software designers*) From the perspective of object-oriented software design, the ordered-pair representation of arcs in a digraph treats digraphs as a different class of objects from graphs. This could seriously undermine *software reuse*. Large portions of computer code might have to be rewritten in order to adapt an algorithm that was originally designed for a digraph to work on an undirected graph.

The ordered-pair representation could also prove awkward in implementing algorithms for which the graphs or digraphs are *dynamic* structures (i.e., they change during the algorithm). Whenever the direction on a particular edge must be reversed, the associated ordered pair has to be deleted and replaced by its reverse. Even worse, if a directed edge is to become undirected, then an ordered pair must be replaced with an unordered pair. Similarly, the undirected and directed edges of a partially directed graph would require two different types of objects.

EXAMPLES

E6:   The digraph on the left in Figure 1.1.6 has the undirected graph on the right as its underlying graph. The digraph has two multi-arcs: $\{a, b\}$ and $\{f, h\}$.

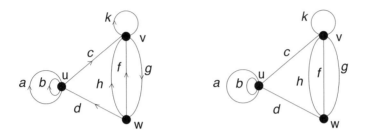

Figure 1.1.6: A digraph and its underlying graph.

E7:   A simple digraph can have one arc in each direction between two vertices, as illustrated in Figure 1.1.7.

Figure 1.1.7: A simple digraph whose underlying graph is not simple.

Vertex-Coloring

When the vertex-set of a graph is partitioned, the cells of the partition are commonly assigned distinct *colors*. This is developed at length in Chapter 5.

DEFINITIONS

D25:   A ***vertex-coloring*** of a graph G is a function from its vertex-set V_G vertices to a set C whose elements are called *colors*.

D26:   A vertex-coloring is ***proper*** if two adjacent vertices are always assigned different colors.

D27:   A graph is c-***colorable*** if it has a proper vertex-coloring with c or fewer colors.

D28:   The (***vertex***) ***chromatic number of a graph*** G, denoted $\chi(G)$, is the smallest number c of colors such that G is c-colorable.

REMARK

R3: Definitions of *edge-coloring*, *c-edge-colorable*, and *edge-chromatic number*, denoted $\chi'(G)$, are obtained by simply replacing the word "vertices" with the word "edges" in the definitions above.

EXAMPLE

E8: The graph G in Figure 1.1.8 is shown with a 3-coloring of its vertex-set. Since it is not 2-colorable, its chromatic number is 3. Also, the graph is easily seen to be 3-edge-colorable and clearly is not 2-edge-colorable; hence, $\chi'(G) = 3$.

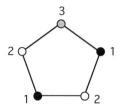

Figure 1.1.8: A graph G with $\chi(G) = \chi'(G) = 3$.

1.1.2   Degree and Distance

Two of the most fundamental notions in graph theory are those of the *degree* of a vertex and the *distance* between two vertices. Distance is developed fully in Chapter 9.

Degree

DEFINITIONS

D29: The *degree* (or *valence*) of a vertex v in a graph G, denoted $deg(v)$, is the number of proper edges incident on v plus twice the number of self-loops. (For simple graphs, of course, the degree is simply the number of neighbors.)

TERMINOLOGY: Applications of graph theory to physical chemistry motivate the use of the term *valence* as an alternative to *degree*. Thus, a vertex of degree d is also called a *d-valent vertex*.

D30: The *degree sequence* of a graph is the sequence formed by arranging the vertex degrees into non-decreasing order.

D31: The *indegree* of a vertex v in a digraph is the number of arcs directed to v; the *outdegree* of vertex v is the number of arcs directed from v. Each self-loop at v counts one toward the indegree of v and one toward the outdegree.

D32: An *isolated vertex* in a graph is a vertex of degree 0.

EXAMPLES

E9: The graph in Figure 1.1.9 has degree sequence $< 0, 1, 1, 4, 6, 6 >$. Vertices u and v both have degree 6.

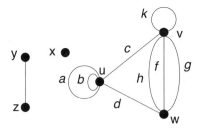

Figure 1.1.9: A graph with degree sequence $< 0, 1, 1, 4, 6, 6 >$.

E10: Figure 1.1.10 below shows the indegrees and outdegrees of a digraph.

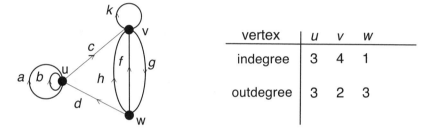

vertex	u	v	w
indegree	3	4	1
outdegree	3	2	3

Figure 1.1.10: The indegrees and outdegrees of the vertices of a digraph.

FACTS

For proofs of the following elementary facts, see [GrYe06, §1.1] or other basic texts.

F1: (Euler) The sum of the degrees of the vertices of a graph is twice the number of edges.

F2: In every graph, the number of vertices having odd degree is an even number.

F3: A non-trivial simple graph G must have at least one pair of vertices whose degrees are equal.

F4: In a digraph, the sum of the indegrees and the sum of the outdegrees both equal the number of edges.

F5: The degree sequence of a graph is a finite, non-decreasing sequence of non-negative integers whose sum is even.

F6: Conversely, any non-decreasing, non-negative sequence of integers whose sum is even is the degree sequence of some graph, but not necessarily of a simple graph.

Walks, Trails, and Paths

DEFINITIONS

D33: A **walk** in a graph G is an alternating sequence of vertices and edges,

$$W = v_0, e_1, v_1, e_1, \ldots, e_n, v_n$$

such that for $j = 1, \ldots, n$, the vertices v_{j-1} and v_j are the endpoints of the edge e_j. If, moreover, the edge e_j is directed from v_{j-1} to v_j, then W is a **directed walk**.

- In a simple graph, a walk may be represented simply by listing a sequence of vertices: $W = v_0, v_1, \ldots, v_n$ such that for $j = 1, \ldots, n$, the vertices v_{j-1} and v_j are adjacent.

- The **initial vertex** is v_0.

- The **final vertex** (or **terminal vertex**) is v_n.

- An **internal vertex** is a vertex that is neither initial nor final.

D34: The **length of a walk** is the number of edges (counting repetitions).

D35: A walk is **closed** if the initial vertex is also the final vertex; otherwise, it is **open**.

D36: A **trail** in a graph is a walk such that no edge occurs more than once.

D37: An **eulerian trail** in a graph G is a walk that contains each edge of G *exactly* once. (See §4.2.)

D38: A **path** in a graph is a trail such that no internal vertex is repeated.

D39: A **cycle** is a closed path of length at least 1.

D40: A **trivial** walk, trail, or path consists of a single vertex and no edges.

EXAMPLE

E11: In the graph shown in Figure 1.1.11, the vertex sequence $\langle u, v, x, v, z \rangle$ represents a walk that is not a trail, and the vertex sequence $\langle u, v, x, y, v, z \rangle$ represents a trail that is not a path.

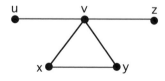

Figure 1.1.11: A graph.

Distance and Connectivity

DEFINITIONS

D41: The *distance between two vertices* in a graph is the length of the shortest walk between them.

D42: The *directed distance from* a vertex u *to* a vertex v in a digraph is the length of the shortest directed walk from u to v.

D43: A graph is *connected* if between every pair of vertices there is a walk.

D44: A digraph is *(weakly) connected* if its underlying graph is connected.

D45: A digraph is *strongly connected* if from each vertex to each other vertex there is a directed walk.

D46: The *eccentricity* of a vertex v in a connected graph is its distance to a vertex farthest from v.

D47: The *radius* of a connected graph is its minimum eccentricity.

D48: The *diameter* of a connected graph is its maximum eccentricity.

EXAMPLE

E12: The digraph shown on the left in Figure 1.1.12 is strongly connected; the digraph on the right is connected but not strongly connected.

Figure 1.1.12: A strongly connected digraph and a weakly connected one.

1.1.3 Basic Structural Concepts

We are concerned with the possible equivalence of two graphs, with the symmetries of an individual graph, and with the possible appearance of one graph within another graph.

Isomorphism

In concept, two graphs are *isomorphic* if they are structurally identical, which means that they correspond in all structural details. A formal vertex-to-vertex and edge-to-edge correspondence is called an *isomorphism*.

DEFINITIONS

D49: An ***isomorphism between two simple graphs*** G and H is a vertex bijection $\phi : V_G \to V_H$ such that for $u, v \in V_G$, the vertex u is adjacent to the vertex v in graph G if and only if $\phi(u)$ is adjacent to $\phi(v)$ in graph H. Implicitly, there is also an edge bijection $E_G \to E_H$ such that $uv \mapsto \phi(u)\phi(v)$.

D50: An ***isomorphism between two general graphs*** G and H is a pair of bijections $\phi_V : V_G \to V_H$ and $\phi_E : E_G \to E_H$ such that for every pair of vertices $u, v \in V_G$, the set of edges in E_G joining u and v is mapped bijectively to the set of edges in E_H joining the vertices $\phi(u)$ and $\phi(v)$.

D51: We say that G and H are ***isomorphic graphs*** and we write $G \cong H$ if there is an isomorphism $G \to H$.

D52: An ***adjacency matrix*** for a simple graph G whose vertices are explicitly ordered $v_1, v_2, \ldots, v_n$ is the $n \times n$ matrix A_G such that

$$A_G(i, j) = \begin{cases} 1 & \text{if } v_i \text{ and } v_j \text{ are adjacent} \\ 0 & \text{otherwise} \end{cases} \tag{1.1.1}$$

D53: A property associated with all graphs is an ***isomorphism invariant*** if it has the same value (or is the same) for any two isomorphic graphs.

EXAMPLES

E13: The two graphs in Figure 1.1.13 are isomorphic under the mapping

$$u_1 \mapsto v_1 \quad u_2 \mapsto v_1 \quad u_3 \mapsto v_4 \quad u_4 \mapsto v_3$$

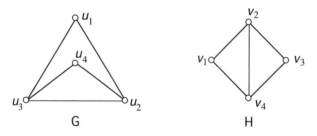

Figure 1.1.13: Two isomorphic graphs.

If one flips vertex u_4 of graph G downward to the bottom and rotates the figure a quarter-turn counterclockwise, then the resulting image of graph G "looks just like" graph H. Their adjacency matrices are:

$$A_G = \begin{array}{c} \\ u_1 \\ u_2 \\ u_3 \\ u_4 \end{array} \begin{array}{cccc} u_1 & u_2 & u_3 & u_4 \\ \begin{pmatrix} 0 & 1 & 1 & 0 \\ 1 & 0 & 1 & 1 \\ 1 & 1 & 0 & 1 \\ 0 & 1 & 1 & 0 \end{pmatrix} \end{array} \qquad A_H = \begin{array}{c} \\ v_1 \\ v_2 \\ v_3 \\ v_4 \end{array} \begin{array}{cccc} v_1 & v_2 & v_3 & v_4 \\ \begin{pmatrix} 0 & 1 & 0 & 1 \\ 1 & 0 & 1 & 1 \\ 0 & 1 & 0 & 1 \\ 1 & 1 & 1 & 0 \end{pmatrix} \end{array}$$

We observe that transposing rows u_3 and u_4 and also transposing columns u_3 and u_4 transforms the matrix A_G into matrix A_H.

E14: The two graphs in Figure 1.1.14 are isomorphic, even if the drawings look quite different. The vertex-labels indicate an isomorphism.

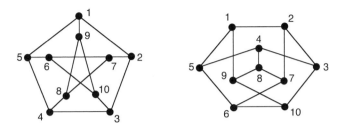

Figure 1.1.14: Two isomorphic graphs that look quite different.

E15: Figure 1.1.15 shows two non-isomorphic graphs with identical degree sequences. (It is easy to show that connectedness is an isomorphism invariant.)

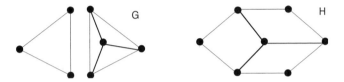

Figure 1.1.15: Two graphs whose degree sequences are both $\langle 2, 2, 2, 3, 3, 3, 3 \rangle$.

FACTS

F7: Considering all possible bijections of the vertex-sets of two n-vertex graphs requires $O(n!)$ steps.

F8: Although some fast heuristics are known (see §2.2), there is no known polynomial-time algorithm for testing graph isomorphism.

F9: The number of vertices, the number of edges, and the degree sequence are all isomorphism invariants. On the other hand, having the same values for all three of these invariants does not imply that two graphs are isomorphic, as illustrated by Example 15.

F10: Each row sum (and column sum) in an adjacency matrix equals the degree of the corresponding vertex.

Automorphisms

The notion of symmetry in a graph is formalized in terms of isomorphisms of the graph to itself.

DEFINITIONS

D54: A **graph automorphism** is an isomorphism of the graph to itself.

D55: The ***orbit of a vertex*** u of a graph G is the set of all vertices $v \in V_G$ such that there is an automorphism ϕ such that $\phi(u) = v$.

D56: The ***orbit of an edge*** d of a graph G is the set of all edges $e \in E_G$ such that there is an automorphism ϕ such that $\phi(d) = e$.

D57: A graph is ***vertex-transitive*** if all the vertices are in the same orbit.

D58: A graph is ***edge-transitive*** if all the edges are in the same orbit.

FACTS

F11: The vertex orbits partition the vertex-set of a graph.

F12: The edge orbits partition the edge-set of a graph.

EXAMPLE

E16: For the graph on the left in Figure 1.1.16, the vertex orbits are $\{u_1, u_4\}$ and $\{u_2, u_3\}$, and the edge orbits are $\{u_1 u_2, u_1 u_3, u_2 u_4, u_3 u_4\}$ and $\{u_2 u_3\}$. The graph on the right is vertex-transitive and edge-transitive.

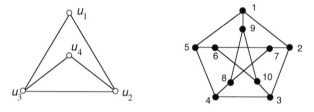

Figure 1.1.16: The graph $K_4 - e$ and the Petersen graph.

Subgraphs

DEFINITIONS

D59: A subgraph of a graph G is a graph H such that $V_H \subset V_G$ and $E_H \subset E_G$. (Usually, any graph isomorphic to a subgraph of G is also said to be a subgraph of G.)

D60: In a graph G, the ***induced subgraph*** on a set of vertices $W = \{w_1, \ldots, w_k\}$, denoted $G(W)$, has W as its vertex-set, and it contains every edge of G whose endpoints are in W. That is,

$$V(G(W)) = W \text{ and } E(G(W)) = \{e \in E(G) \mid \text{the endpoints of edge } e \text{ are in } W\}$$

D61: A subgraph H of a graph G is a ***spanning subgraph*** if $V(H) = V(G)$. (Also, if H is isomorphic to a spanning subgraph of G, we may say that H spans G.)

D62: A ***component*** of a graph G is a connected subgraph H such that no subgraph of G that properly contains H is connected. In other words, a component is a *maximal* connected subgraph.

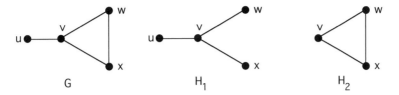

Figure 1.1.17: A spanning subgraph H_1 and an induced subgraph H_2.

EXAMPLE

E17: For the graph G in Figure 1.1.17, H_1 is a spanning subgraph but not an induced subgraph, and H_2 is an induced subgraph but not a spanning subgraph.

FACTS

F13: Let $\phi : G \to H$ be a graph isomorphism, and let J be a subgraph of G. Then the restriction of ϕ to the subgraph J is an isomorphism onto its image $\phi(J)$.

F14: If a graph J is a subgraph of a graph G but not a subgraph of a graph H, then $G \not\cong H$. This is a corollary of Fact F13.

Graph Operations

The operations of adding and deleting vertices and edges of a graph are regarded as *primary operations*, because they are the foundation for other operations, which may be called *secondary operations*.

DEFINITIONS

D63: The operation of **adding the vertex** u to a graph $G = (V, E)$, such that $u \notin V$, yields a new graph with vertex-set $V \cup \{u\}$ and edge-set E, which is denoted $G \cup \{u\}$. (The new vertex u has no neighbors.)

D64: The operation of **deleting the vertex** u from a graph $G = (V, E)$ not only removes the vertex u but also removes every edge of which u is an endpoint. The resulting graph is denoted $G - u$.

D65: The operation of **adding an edge** d (or uv) to a graph $G = (V, E)$ joining the vertices u and v yields a new graph with vertex-set V and edge-set $E \cup \{d\}$ (or $E \cup \{uv\}$), which is denoted $G \cup \{d\}$ (or $G \cup \{uv\}$).

D66: The operation of **deleting an edge** d (or uv) from a graph $G = (V, E)$ removes only that edge. The resulting graph is denoted $G - d$ (or $G - uv$).

D67: A **cut-vertex** (or **cutpoint**) is a vertex whose removal increases the number of components.

D68: A **cut-edge** is an edge whose removal increases the number of components.

D69: The **edge-complement** of a simple graph G is the graph $\overline{G}$ (alternatively denoted G^c) that has the same vertex-set as G, such that uv is an edge of $\overline{G}$ if and only if it is *not* an edge of G.

D70: The *join* (or *suspension*) of two graphs G and H is denoted by $G + H$. It has the following vertex-set and edge-set:

$$V(G + H) = V(G) \cup V(H)$$
$$E(G + H) = E(G) \cup E(H) \cup \{uv \mid u \in V(G) \text{ and } v \in V(H)\}$$

D71: The *cartesian product* (or *product*) of two graphs G and H is denoted by $G \times H$. Its vertex-set and edge-set are as follows:

$$V(G \times H) = V(G) \times V(H)$$
$$E(G \times H) = E(G) \times V(H) \cup V(G) \times E(H)$$

The endpoints of the edge $(d, v) \in E(G) \times V(H)$ are the vertices (x, v) and (y, v), where x and y are the endpoints of edge $d \in E(G)$. The endpoints of the edge $(u, e) \in V(G) \times E(H)$ are the vertices (u, s) and (u, t), where s and t are the endpoints of edge $e \in E(H)$.

D72: The *graph union* of two graphs G and H is the graph $G \cup H$ whose vertex-set and edge-set are the disjoint unions, respectively, of the vertex-sets and the edge-sets of G and H.

D73: The **m-*fold self-union*** mG is the iterated disjoint union $G \cup \cdots \cup G$ of m copies of the graph G.

EXAMPLES

E18: Figure 1.1.18 illustrates the operation of edge-complementation.

Figure 1.1.18: Edge-complementation.

E19: Figure 1.1.19 illustrates the join operation.

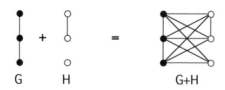

Figure 1.1.19: Join operation.

E20: In Figure 1.1.18, the vertex in the upper left corner of the drawing of the graph G is a cut-vertex, and the edge from that vertex to the center vertex is a cut-edge.

E21: Figure 1.1.20 illustrates the product operation.

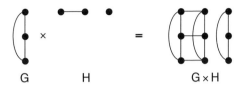

Figure 1.1.20: Cartesian product.

1.1.4 Trees

Trees are important to the structural understanding of graphs and to the algorithmics of information processing, and they play a central role in the design and analysis of connected networks. A standard characterization theorem for trees appears in Chapter 2.

Acyclic Graphs

DEFINITIONS

D74: A *tree* is a connected graph with no cycles (i.e., *acyclic*).

D75: A *forest* is a (not necessarily connected) graph with no cycles.

D76: A *central vertex* in a graph is a vertex whose eccentricity equals the radius of the graph.

D77: The *center* of a graph is the subgraph induced on its set of central vertices.

TERMINOLOGY NOTE: Classically (see Chapter 3), the words *center* and *bicenter* were used to mean the set of central vertices of a tree, when there was only one vertex or two vertices, respectively. (See Fact F15 below.)

EXAMPLE

E22: The graph on the left in Figure 1.1.21 is a tree; the other two graphs are not.

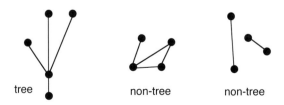

Figure 1.1.21: A tree and two non-trees.

FACT

F15:   The center of a tree is isomorphic to K_1 or to K_2. (See §1.3 for information about the historical context of this fact.)

Trees as Subgraphs

Several different problem-solving algorithms involve growing a tree within a graph, one edge and one vertex at a time. All these techniques are refinements and extensions of the same basic tree-growing scheme given in this section.

DEFINITIONS

TERMINOLOGY: For a given tree T in a graph G, the edges and vertices of T are called **tree edges** and **tree vertices**, and the edges and vertices of G that are not in T are called **non-tree edges** and **non-tree vertices**.

D78:   A **frontier edge** for a given tree T in a graph is a non-tree edge with one endpoint in T and one endpoint not in T.

D79:   A **spanning tree** of a graph G is a spanning subgraph of G that is a tree.

EXAMPLE

E23:   For the graph in Figure 1.1.22, the tree edges of a tree T are drawn in bold. The tree vertices are black, and the non-tree vertices are white. The frontier edges for T, appearing as dashed lines, are edges a, b, c, and d. The plain edges are the non-tree edges that are not frontier edges for T.

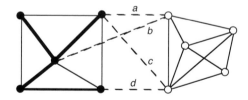

Figure 1.1.22: A tree with frontier edges a, b, c, and d.

Observe that when any one of the frontier edges in Figure 1.1.22 is added to the tree T, the resulting subgraph is still a tree. This property holds in general, and applying it iteratively forms the core of the tree-growing scheme of this section.

FACT

F16:   Let T be a tree in a graph G, and let e be a frontier edge for T. Then the subgraph of G formed by adding edge e to tree T is a tree. (Formally, adding frontier edge e to a tree involves adding a new vertex to current tree T, i.e., its non-tree endpoint.)

Basic Tree-Growing Algorithm

The basic tree-growing scheme uses vertex labels to keep track of the order in which vertices are added to the tree.

TERMINOLOGY NOTE: A ***standard (0-based) vertex-labeling*** of an n-vertex graph is a one-to-one assignment of the integers $0, 1, \ldots, n - 1$ to the vertices of that graph.

Algorithm 1.1.1: Basic Tree-Growing with Vertex Labels

Input: a graph G and a starting vertex $v \in V_G$.
Output: a spanning tree T of $C_G(v)$ and a standard vertex-labeling of $C_G(v)$.

Initialize tree T as vertex v.
Write label 0 on vertex v.
Initialize label counter $i := 1$
While tree T does not yet span component $C_G(v)$
    Choose a frontier edge e for tree T.
    Let w be the endpoint of edge e that lies outside of T.
    Add edge e and vertex w to tree T.
    Write label i on vertex w.
    $i := i + 1$
Return tree T and vertex-labeling of $C_G(v)$.

REMARK

R4: Uniqueness of the Output Tree from Tree-Growing

Without a rule for choosing a frontier edge (including a way to break ties), the output tree from Algorithm 1.1.1 would not be unique (in which case, many computer scientists would hesitate to use the term *algorithm*). The uniqueness of the output depends on some *default priority* based on the ordering of the edges (and vertices) in the data structure chosen to implement the algorithm. The default priority is used whenever no other rule is given and as a way of breaking ties left from other rules.

FACTS

F17: If an execution of the basic tree-growing algorithm starts at vertex v of a graph G, then the subgraph consisting of the labeled vertices and tree edges is a spanning tree of the component $C_G(v)$.

F18: A graph is connected if and only if the basic tree-growing algorithm labels all its vertices.

Prioritizing the Edge Selection

The edge-prioritized tree-growing algorithm, Algorithm 1.1.2, is a refinement of basic tree-growing.

Algorithm 1.1.2: Edge-Prioritized Tree-Growing

Input: a connected graph G, a starting vertex $v \in V_G$,
  and a rule for prioritizing frontier edges.
Output: a spanning tree T and a standard vertex-labeling of V_G.

  Initialize tree T as vertex v.
  Initialize the set of frontier edges for tree T as empty.
  Write label 0 on vertex v.
  Initialize label counter $i := 1$
  While tree T does not yet span G
      Update the set of frontier edges for T.
      Let e be the frontier edge for T of highest priority.
      Let w be the unlabeled endpoint of edge e.
      Add edge e (and vertex w) to tree T.
      Write label i on vertex w.
      $i := i + 1$
  Return tree T with its vertex-labeling.

FACT

F19:   Different rules for prioritizing the frontier edges give rise to different spanning trees: the *depth-first search* tree (*last-in-first-out* priority), the *breadth-first search* tree (*first-in-first-out* priority), the *Prim tree* (*least-cost* priority), and the *Dijkstra tree* (*closest-to-root* priority). See §10.1

References

[Be85] C. Berge, *Graphs*, North-Holland, 1985.

[Bo98] B. Bollobás, *Modern Graph Theory*, Springer, 1998.

[ChLeZh10] G. Chartrand, L. Lesniak, and P. Zhang, *Graphs and Digraphs*, Fifth Edition, CRC Press, 2010.

[GrYe06] J. L. Gross and J. Yellen, *Graph Theory and Its Applications*, Second Edition, CRC Press, 2006.

[Ha94] F. Harary, *Graph Theory*, Perseus reprint, 1994. (First Edition, Addison-Wesley, 1969.)

[ThSw92] K. Thulasiraman and M. N. S. Swamy, *Graphs: Theory and Algorithms*, John Wiley & Sons, 1992.

[Tu00] W. T. Tutte, *Graph Theory*, Cambridge University Press, 2000.

[We01] D. B. West, *Introduction to Graph Theory*, Second Edition, Prentice-Hall, 2001. (First Edition, 1996.)

Section 1.2
Families of Graphs and Digraphs

Lowell W. Beineke, Purdue University at Fort Wayne

INTRODUCTION

Whenever a property of graphs is defined, a family of graphs — those with that property — results. Consequently, we focus on basic families. Along with the definitions of families, we include characterizations where appropriate. [ReWi98] offers a detailed catalog of the members of various graph and digraph families.

1.2.1   Building Blocks

Some simple graphs have as few edges or as many as possible for a given number of vertices. Some multigraphs and general graphs have as few vertices as possible for a given number of edges.

DEFINITIONS

D1: A simple graph is a *complete graph* if every pair of vertices is joined by an edge. The complete graph with n vertices is denoted K_n.

D2: The *empty graph* $\overline{K_n}$ is defined to be the graph with n vertices and no edges.

D3: The *null graph* K_0 is the graph with no vertices or edges.

D4: The *trivial graph* K_1 is the graph with one vertex and no edges.

D5: The *bouquet* B_n is the general graph with one vertex and n self-loops.

D6: The *dipole* D_n is the multigraph with two vertices and n edges.

D7:   A simple digraph is a ***complete digraph*** if between every pair of vertices there is an arc in each direction. The complete digraph with n vertices is denoted $\overset{\leftrightarrow}{K}_n$.

D8:   The ***path graph*** P_n is the n-vertex graph with $n - 1$ edges, all on a single open path. (Quite commonly elsewhere, the subscript of the notation P_n denotes the number of edges.)

D9:   The ***cycle graph*** C_n is the n-vertex graph with n edges, all on a single cycle.

REMARKS

R1:   Although the empty graph may seem to some a "pointless" concept, it is the default initial value in computer representations of graph-valued variables.

R2:   Whereas a "path" and a "cycle" are alternating sequences of vertices and edges, a "path graph" and a "cycle graph" are kinds of graphs.

EXAMPLES

E1:   Figure 1.2.1 shows the complete graph K_4 and the complete digraph $\overset{\leftrightarrow}{K}_4$.

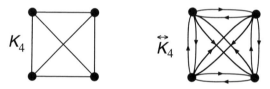

Figure 1.2.1: A complete graph and a complete digraph.

E2:   Figure 1.2.2 shows a path graph and a cycle graph.

Figure 1.2.2: A path graph and a cycle graph.

1.2.2   Symmetry

Graphs with various kinds of symmetry are of particular interest.

Local Symmetry: Regularity

Regularity of a graph is an elementary form of local symmetry.

DEFINITIONS

D10: A graph is ***regular*** if every vertex is of the same degree.

- It is **k-*regular*** if every vertex is of degree k.

D11: A **k-*factor*** of a graph G is a k-regular spanning subgraph.

FACT

F1: All *vertex-transitive* graphs (see §1.1) are regular.

EXAMPLES

E3: For $k = 0, 1, 2, 3$, there is exactly one k-regular simple graph with 4 vertices.

E4: The only regular simple graphs with 5 vertices are the empty graph $\overline{K_5}$ (degree 0), the cycle graph C_5 (degree 2), and the complete graph K_5 (degree 4).

E5: [ReWi98] There are exactly two 3-regular simple graphs with 6 vertices.

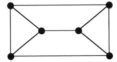

Figure 1.2.3: The two 3-regular simple graphs with 6 vertices.

E6: The disjoint union of the complete graphs K_3 and K_4 is a 2-regular simple 7-vertex graph that is not vertex-transitive. Its edge-complement is a 4-regular connected simple 7-vertex graph that is not vertex-transitive.

E7: Of the five 3-regular connected simple graphs with 8 vertices, two are vertex-transitive.

Figure 1.2.4: The five 3-regular connected simple graphs with 8 vertices.

Global Symmetry: Vertex-Transitivity

Often vertex-transitivity arises from algebra or geometry. See §6.2 for further discussion of Cayley graphs and circulant graphs.

DEFINITIONS

D12: The **Cayley graph** $C(\mathcal{A}, X)$ for a group $\mathcal{A}$ with generating set X has the elements of $\mathcal{A}$ as vertices and has an edge directed from a to ax for every $a \in \mathcal{A}$ and $x \in X$. We assume that vertices are labeled by elements of $\mathcal{A}$ and that edges are labeled by elements of X.

- We note that an involution x gives rise to a pair of oppositely directed edges between a and ax, for each $a \in \mathcal{A}$; sometimes we identify each such pair of directed edges to a single undirected edge labeled x.

D13: A **circulant graph** $\text{Circ}(n; X)$ is defined for a positive integer n and a subset X of the integers $1, 2, \ldots, \lfloor \frac{n}{2} \rfloor$, called the **connections**.

- The vertex set is $\mathbb{Z}_n$, the integers modulo n.

- There is an edge joining two vertices j and k if and only if the difference $|j - k|$ is in the set X. A circulant graph is a special case of a Cayley graph; an involution in the connection set gives rise to a single edge.

D14: The **1-skeleton** (often in graph theory, the **skeleton**) of a k-complex K is the graph consisting of the vertices and the edges of K.

D15: The **d-hypercube graph** Q_d (or **d-cube graph**) is the 1-skeleton of the d-dimensional hypercube $\{(x_1, \ldots, x_n) \mid 0 \le x_j \le 1\}$. This graph has 2^d vertices and is regular of degree d.

D16: The **d-octahedral graph** $\mathcal{O}_d$ is defined recursively:

$$\mathcal{O}_d = \begin{cases} \overline{K_2} & \text{if } n = 1 \\ \mathcal{O}_{d-1} + \overline{K_2} & \text{if } n \ge 2 \end{cases}$$

D17: The **Petersen graph** is the 10-vertex 3-regular graph depicted in Figure 1.2.5.

Figure 1.2.5: The Petersen graph.

EXAMPLES

E8: The **n-simplex** is the convex hull of $n + 1$ affinely independent points in n-dimensional space. Its 1-skeleton is isomorphic to the complete graph K_n.

E9: A **Platonic graph** is the 1-skeleton of one of the five Platonic solids: the tetrahedron, the cube, the octahedron, the dodecahedron, and the icosahedron.

E10: The Petersen graph is vertex-transitive, since there is an automorphism that swaps the pentagram (i.e., the star) with the pentagon. It is not a Cayley graph of either of the two groups of order 10, i.e., of the cyclic group $\mathbb{Z}_{10}$ or of the dihedral group $\mathbb{D}_5$, and thus, not a Cayley graph.

E11: The octahedral graph $\mathcal{O}_d$ is isomorphic to $\overline{dK_2}$.

FACTS

F2: [Hypercube Characterization Theorem] The graph whose vertices are the binary sequences of length d in which two vertices are adjacent if their sequences differ in exactly one place is isomorphic to Q_d.

F3: We can construct the d-dimensional hypercube Q_d recursively, using the cartesian product operation:
$$Q_d = \begin{cases} K_1 & \text{if } d = 0 \\ Q_{d-1} \times K_2 & \text{if } d \geq 1 \end{cases}$$

1.2.3   Integer-Valued Invariants

Some of the most useful graph properties are provided by integer-valued invariants of isomorphism type. Such invariants partition all graphs into an infinite list of subclasses. Often the subclasses with low invariant values are of special interest.

Cycle Rank

The connected graphs of *cycle rank* 0 are of great special interest, since they are the *trees* (see §1.1).

DEFINITION

D18: The **cycle rank** of a connected graph $G = (V, E)$ is the number $|E| - |V| + 1$. (See §6.4 for an interpretation of cycle rank as the rank of a vector space.) More generally, for a graph G with $c(G)$ components, the cycle rank is the number $|E(G)| - |V(G)| + c(G)$.

EXAMPLE

E12: The connected graphs of cycle rank 0 are the trees. The smallest trees are shown in Figure 1.2.6.

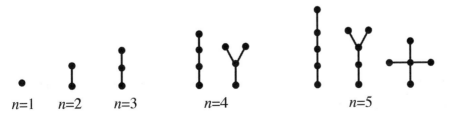

n=1   *n=2*   *n=3*   *n=4*   *n=5*

Figure 1.2.6: The trees with up to five vertices.

FACTS

F4: [Tree Characterization Theorem] The following statements are equivalent for a graph T with n vertices (e.g., see [GrYe06, Theorem 3.1.8]):

- T is a tree (that is, G is connected and has no cycles).
- T is connected and has $n - 1$ edges.
- T has no cycles and has $n - 1$ edges.
- Any two vertices of T are connected by exactly one path.

F5: [Inductive (Recursive) Definition of Trees] Let $\mathcal{T}$ be the family of graphs defined as follows:

(i) $K_1 \in \mathcal{T}$.

(ii) If $T \in \mathcal{T}$ and T' can be obtained by adding a new vertex and joining it to a vertex of T, then $T' \in \mathcal{T}$.

Then $\mathcal{T}$ is the family of all trees.
(Several more classes of recursively defined graphs are presented in §2.4.)

F6: The cycle rank of a graph is the sum of the cycle ranks of its components.

F7: A forest is a graph such that every component is a tree.

Chromatic Number and k-Partite Graphs

In a proper coloring of a graph, no two vertices with the same color are adjacent, and thus, every edges joins vertices in different color classes. The graphs with a proper 2-coloring are of special interest. Graph coloring is covered extensively in §5.1 and §5.2.

DEFINITIONS

D19: A simple graph or multigraph is **bipartite** if its vertices can be partitioned into two sets (called **partite sets**) in such a way that no edge joins two vertices in the same set. (For technical reasons, this includes the graph K_1 in this definition.) If r and s are the orders of the partite sets, then the graph is said to be an r-**by**-s **bipartite graph**.

D20: A **complete bipartite graph** is a simple bipartite graph in which each vertex in one partite set is adjacent to all the vertices in the other partite set. If the two partite sets have cardinalities r and s, then this graph is denoted $K_{r,s}$.

D21: A graph is k-**partite** if its vertices can be partitioned into k sets (called **partite sets**) in such a way that no edge joins two vertices in the same set.

D22: A **complete k-partite graph** is a simple k-partite graph in which two vertices are adjacent if and only if they are in different partite sets. All such graphs are called **complete multipartite graphs**. If the k partite sets have orders $n_1, n_2, \ldots, n_k$, then the graph is denoted $K_{n_1,n_2,\ldots,n_k}$, and if each partite set has order r, then $K_{k(r)}$.

EXAMPLES

E13: Every tree is bipartite.

E14: Every cycle with an even number of vertices is bipartite, and no cycle with an odd number is bipartite.

E15: The complete d-partite graph $K_{d(2)}$ is isomorphic to the d-octahedral graph $\mathcal{O}_d$. The first four complete d-partite graphs are shown in Figure 1.2.7.

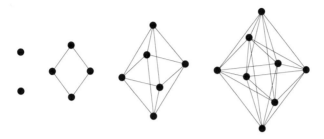

Figure 1.2.7: The complete d-partite graphs $K_{d(2)}$, for $d = 1, \ldots, 4$.

FACTS

F8: [Bipartite Graph Characterization Theorem] A graph is bipartite if and only if the length of each of its cycles is even (e.g., see [GrYe06, Theorem 1.5.4]).

F9: A graph is k-colorable if and only if it is k-partite.

F10: For $k \geq 3$, the problem of deciding whether a graph is k-partite is NP-complete.

K-Connectivity and K-Edge-Connectivity

Graphs can be categorized according to their *connectivity* and their *edge-connectivity*. There are analogues for strong connectedness in digraphs. See §4.1 and §4.7 for extensive coverage of connectivity.

DEFINITIONS

D23: The *(vertex-)connectivity* of a graph G, denoted $\kappa_v(G)$, is the minimum number of vertices whose removal from G leaves a non-connected or trivial graph.

D24: The *edge-connectivity* of a nontrivial graph G, denoted $\kappa_e(G)$ is the minimum number of edges whose removal from G results on a non-connected graph.

NOTATION: The subscripted "G" is often suppressed when the graph G is understood. Elsewhere, the notation κ and λ are used instead of κ_v and κ_e, respectively.

D25: A graph G with connectivity $\kappa_v \geq k \geq 1$ is called k-*connected*. Equivalently, G is k-connected if the removal of $k - 1$ or fewer vertices leaves neither a non-connected graph nor a trivial one.

D26:   A graph G with edge-connectivity $\kappa_e \geq k \geq 1$ is called **k-edge-connected**. That is, the removal of $k - 1$ or fewer edges from a k-edge-connected graph results in a connected graph.

D27:   A digraph is **strongly k-connected** (or **k-strong**) if the result of removing any set of fewer than k vertices is strongly connected and non-trivial.

D28:   A digraph is **strongly k-arc-connected** (or **k-arc-strong**) if the result of removing any set of fewer than k arcs is strongly connected and non-trivial.

Minimum Genus

Graphs can be categorized according to their topological properties.

DEFINITIONS

D29:   The **minimum genus** (or simply the **genus**) of a connected graph G is the smallest number g such that G can be drawn on the orientable surface S_g (see §7.1) without any edge-crossings.

D30:   A graph of genus 0 is **planar**.

1.2.4   Criterion Qualification

A graph family is also specified as the set of all graphs or digraphs that match a stated criterion, e.g., traversibility and various forms of minimality and maximality.

DEFINITIONS

D31:   A graph is **eulerian** if it has a closed walk that contains every edge exactly once. (See §1.3 for the history of eulerian graphs and §4.2 for an extensive discussion.)

D32:   A graph is **hamiltonian** if it has a spanning cycle. (See §1.3 for the history of hamiltonian graphs and §4.5 for an extensive discussion.)

D33:   A k-chromatic graph is **critically k-chromatic** if its chromatic number would decrease if any edge were removed. (See §5.1.)

D34:   A k-connected graph is **critically k-connected** if its connectivity would decrease if any vertex were removed. (See §4.1.)

D35:   A k-edge-connected graph is **critically k-edge-connected** if its edge-connectivity would decrease if any edge were removed. (See §4.1.)

D36:   A **tournament** is a digraph in which there is exactly one arc between each pair of vertices. (See §3.3.)

D37: The *line graph* $L(G)$ of a graph G has the edges of G as its vertices; two vertices of $L(G)$ are adjacent if the edges in G to which they correspond have a common vertex. A graph and its line graph are illustrated in Figure 1.2.8. Also, a graph H is said to be a *line graph* if there exists a graph G such that H is isomorphic to $L(G)$.

Figure 1.2.8: A graph and its line graph.

FACTS

F11: [Line Graph Characterization] The following statements are equivalent:
- G is a line graph.
- [Kr43] The edges of G can be partitioned into complete subgraphs in such a way that no vertex is in more than two.
- [Be70] None of the nine graphs in Figure 1.2.9 is an induced subgraph of G.

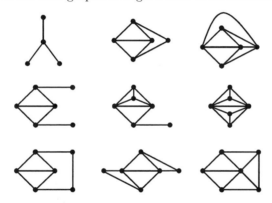

Figure 1.2.9: The nine forbidden induced subgraphs.

F12: A strongly connected tournament contains a directed spanning cycle.

EXAMPLE

E16: The eight tournaments with one to four vertices are shown in Figure 1.2.10.

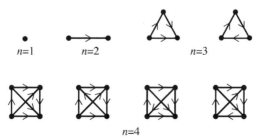

Figure 1.2.10: All tournaments with one to four vertices.

References

[Be70] L. W. Beineke, Characterizations of derived graphs, *J. Combin. Theory* 9 (1970), 129–135.

[GrYe06] J. L. Gross and J. Yellen, *Graph Theory and Its Applications*, Second Edition, CRC Press, 2006.

[Kr43] J. Krausz, Dèmonstation nouvelle d'une théorème de Whitney sur les réseaux, *Mat. Fiz. Lapok* 50 (1943), 75–89.

[ReWi98] R. C. Read and R. J. Wilson, *An Atlas of Graphs*, Oxford University Press, 1998.

Section 1.3
History of Graph Theory

Robin J. Wilson, Pembroke College, Oxford University, UK

INTRODUCTION

Although the first mention of a graph was not until 1878, graph-theoretical ideas can be traced back to 1735 when Leonhard Euler (1707–83) presented his solution of the Königsberg bridges problem. This chapter summarizes some important strands in the development of graph theory since that time. Further information can be found in [BiLlWi98] or [Wi99].

1.3.1   Traversability

The origins of graph theory can be traced back to Euler's work on the Königsberg bridges problem (1735), which subsequently led to the concept of an *eulerian graph*. The study of cycles on polyhedra by the Revd. Thomas Penyngton Kirkman (1806–95) and Sir William Rowan Hamilton (1805–65) led to the concept of a *Hamiltonian graph*.

The Königsberg Bridges Problem

The *Königsberg bridges problem*, pictured in Figure 1.3.1, asks whether there is a continuous walk that crosses each of the seven bridges of Königsberg exactly once — and if so, whether a closed walk can be found. See §4.2 for more extensive discussion of issues concerning *eulerian graphs*.

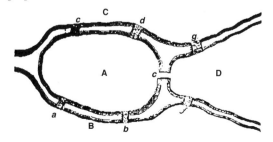

Figure 1.3.1: The seven bridges of Königsberg.

FACTS [BiLIWi98, Chapter 1]

F1: On 26 August 1735 Leonhard Euler gave a lecture on "The solution of a problem relating to the geometry of position" to the Academy of Sciences of St. Petersburg, Russia, proving that there is no such continuous walk across the seven bridges.

F2: In 1736, Euler communicated his solution to several other mathematicians, outlining his views on the nature of the problem and on its situation in the geometry of position [HoWi04].

F3: Euler [Eu:1736] sent his solution of the problem to the *Commentarii Academii Scientiarum Imperialis Petropolitanae* under the title "Solutio problematis ad geometriam ad geometriam situs pertinentis". Although dated 1736, it did not appear until 1741, and was later republished in the new edition of the *Commentarii* in 1752.

F4: Euler's paper is divided into 21 sections, of which 9 are on the Königsberg bridges problem, and the remainder are concerned with general arrangements of bridges and land areas.

F5: Euler did not draw a graph in order to solve the problem, but he reformulated the problem as one of trying to find a sequence of eight letters A, B, C, or D (the land areas) such that the pairs AB and AC are adjacent twice (corresponding to the two bridges between A and B and between A and C), and the pairs AD, BD, and CD are adjacent just once (corresponding to the remaining bridges). He showed by a counting argument that no such sequence exists, thereby proving that the Königsberg bridges problem has no solution.

F6: In discussing the general problem, Euler first observed that the number of bridges written next to the letters A, B, C, etc. together add up to twice the number of bridges. This is the first appearance of what some graph-theorists now call the "handshaking lemma", that the sum of the vertex-degrees in a graph is equal to twice the number of edges.

F7: Euler's main conclusions for the general situation were as follows:

- If there are more than two areas to which an odd number of bridges lead, then such a journey is impossible.

- If the number of bridges is odd for exactly two areas, then the journey is possible if it starts in either of these two areas.

- If, finally, there are no areas to which an odd number of bridges lead, then the required journey can be accomplished starting from any area.

These results correspond to the conditions under which a graph has an eulerian, or semi-eulerian, trail.

F8: Euler noted the converse result, that if the above conditions are satisfied, then a route is possible, and gave a heuristic reason why this should be so, but did not prove it. A valid demonstration did not appear until a related result was proved by C. Hierholzer [Hi:1873] in 1873.

Diagram-Tracing Puzzles

A related area of study was that of *diagram-tracing puzzles*, where one is required to draw a given diagram with the fewest possible number of connected strokes. Such puzzles can be traced back many hundreds of years – for example, there are some early African examples.

FACTS [BiLIWi98, Chapter 1]

F9: In 1809 L. Poinsot [Po:1809] wrote a memoir on polygons and polyhedra in which he posed the following problem:

> Given some points situated at random in space, it is required to arrange a single flexible thread uniting them two by two in all possible ways, so that the two ends of the thread join up and the total length is equal to the sum of all the mutual distances.

Poinsot noted that a solution is possible only when the number of points is odd, and gave a method for finding such an arrangement for each possible value. In modern terminology, the question is concerned with eulerian trails in complete graphs of odd order.

F10: Other diagram-tracing puzzles were posed and solved by T. Clausen [Cl:1844] and J. B. Listing [Li:1847]. The latter appeared in the book *Vorstudien zur Topologie*, the first place that the word "topology" appeared in print.

F11: In 1849, O. Terquem asked for the number of ways of laying out a complete ring of dominoes. This is essentially the problem of determining the number of eulerian tours in the complete graph K_7, and was solved by M. Reiss [Re:1871–3] and later by G. Tarry.

F12: The connection between the Königsberg bridges problem and diagram-tracing puzzles was not recognized until the end of the 19th century. It was pointed out by W. W. Rouse Ball [Ro:1892] in *Mathematical Recreations and Problems*. Rouse Ball seems to have been the first to use the graph in Figure 1.3.2 to solve the problem.

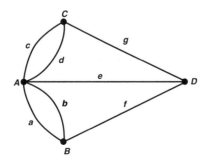

Figure 1.3.2: Rouse Ball's graph of the Königsberg bridges problem.

Hamiltonian Graphs

A type of graph problem that superficially resembles the eulerian problem is that of finding a cycle that passes just once through each vertex of a given graph. Because of Hamilton's influence, such graphs are now called *hamiltonian graphs* (see §4.5), instead of more justly being named after Kirkman, who, prior to Hamilton's consideration of the dodecahedron, as discussed below, considered the more general problem.

FACTS [BiLIWi98, Chapter 2]

F13: An early example of such a problem is the *knight's tour problem*, of finding a succession of knight's moves on a chessboard, visiting each of the 64 squares just once and returning to the starting point. This problem can be dated back many hundreds of years, and systematic solutions were given by Euler [Eu:1759], A.-T. Vandermonde [Va:1771], and others.

F14: In 1856 Kirkman [Ki:1856] wrote a paper investigating those polyhedra for which one can find a cycle passing through all the vertices just once. He proved that every polyhedron with even-sided faces and an odd number of vertices has no such cycle, and gave as an example the polyhedron obtained by "cutting in two the cell of a bee" (see Figure 1.3.3).

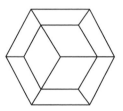

Figure 1.3.3: Kirkman's "cell of a bee" example.

F15: Arising from his work on non-commutative algebra, Hamilton considered cycles passing through all the vertices of a dodecahedron. He subsequently invented a game, called the icosian game (see Figure 1.3.4), in which the player was challenged to find such cycles on a solid dodecahedron, satisfying certain extra conditions.

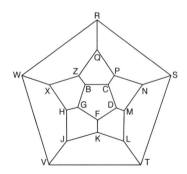

Figure 1.3.4: Hamilton's icosian game.

F16:   In 1884, P. G. Tait asserted that every 3-valent polyhedron has a hamiltonian cycle. This assertion was subsequently disproved by W. T. Tutte [Tu46] in 1946 (see Figure 1.3.5).

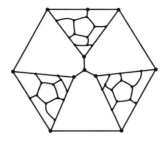

Figure 1.3.5: Tutte's 3-valent non-hamiltonian polyhedron.

F17:   Sufficient conditions for a graph to be hamiltonian were later obtained by G. A. Dirac [Di52], O. Ore [Or60], J. A. Bondy and V. Chvátal [BoCh76], and others.

F18:   Hamiltonian digraphs have also been investigated, by A. Ghouila-Houri (1960), H. Meyniel (1973), and others.

1.3.2   Trees

The concept of a tree, a connected graph without cycles, appeared implicitly in the work of Gustav Kirchhoff (1824–87), who employed graph-theoretical ideas in the calculation of currents in electrical networks. Later, trees were used by Arthur Cayley (1821–95), James Joseph Sylvester (1806–97), Georg Pólya (1887–1985), and others, in connection with the enumeration of certain chemical molecules.

Counting Trees

Enumeration techniques involving trees first arose in connection with a problem in the differential calculus, but they soon came to be fundamental tools in the counting of chemical molecules, as well as providing a fascinating topic of interest in their own right. Enumeration of various kinds of graphs is discussed in §6.3.

FACTS [BiLlWi98, Chapter 3] [PóRe87]

F19:   While working on a problem inspired by some work of Sylvester on "differential transformation and the reversion of serieses", Cayley [Ca:1857] was led to the enumeration of rooted trees.

F20: Cayley's method was to take a rooted tree and remove its root, thereby obtaining a number of smaller rooted trees (see Figure 1.3.6).

Figure 1.3.6: Splitting a rooted tree.

Letting A_n be the number of rooted trees with n branches, Cayley proved that the generating function

$$1 + A_1 x + A_2 x^2 + A_3 x^3 + \ldots$$

is equal to the product

$$(1 - x)^{-1} \cdot (1 - x^2)^{-A_1} \cdot (1 - x^3)^{-A_2} \cdot \ldots$$

Using this equality, he was able to calculate the first few numbers A_n, one at a time.

F21: Around 1870, Sylvester and C. Jordan independently defined the *center/bicenter* and the *centroid/bicentroid* of a tree.

F22: In 1874, Cayley [Ca:1874] found a method for solving the more difficult problem of counting unrooted trees. This method, which he applied to chemical molecules, consisted essentially of starting at the center or centroid of the tree or molecule and working outwards.

F23: In 1889, Cayley [Ca:1889] presented his n^{n-2} formula for the number of labeled trees with n vertices. He explained why the formula holds when $n = 6$, but he did not give a proof in general. The first accepted proof was given by H. Prüfer [Pr18]: his method was to establish a one-to-one correspondence between such labeled trees and sequences of length $n - 2$ formed from the numbers $1, 2, \ldots, n$.

F24: In a fundamental paper of 1937, Pólya [Pó37] combined the classical idea of a generating function with that of a permutation to obtain a powerful theorem that enabled him to enumerate certain types of configuration under the action of a group of symmetries. Some of Pólya's work was anticipated by J. H. Redfield [Re27], but Redfield's paper was obscurely written and had no influence on the development of the subject.

F25: Later results on the enumeration of trees were derived by R. Otter [Ot48] and others. The field of graphical enumeration (see [HaPa73]) was subsequently further developed by F. Harary [Ha55], R. C. Read [Re63], and others.

Chemical Trees

By 1850 it was already known that chemical elements combine in fixed proportions. Chemical formulas such as CH_4 (methane) and C_2H_5OH (ethanol) were known, but it was not understood how the elements combine to form such substances. Around this time, chemical ideas of valency began to be established, particularly when Alexander Crum Brown presented his graphic formulae for representing molecules. Figure 1.3.7 presents his representation of ethanol, the usual drawing, and the corresponding tree graph.

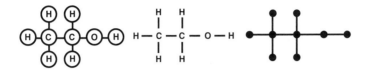

Figure 1.3.7: Representations of ethanol.

FACTS [BiLIWi98, Chapter 4]

F26: Crum Brown's graphic notation explained for the first time the phenomenon of isomerism, whereby there exist pairs of molecules (isomers) with the same chemical formula but different chemical properties. Figure 1.3.8 shows isomers with chemical formula C_4H_{10}.

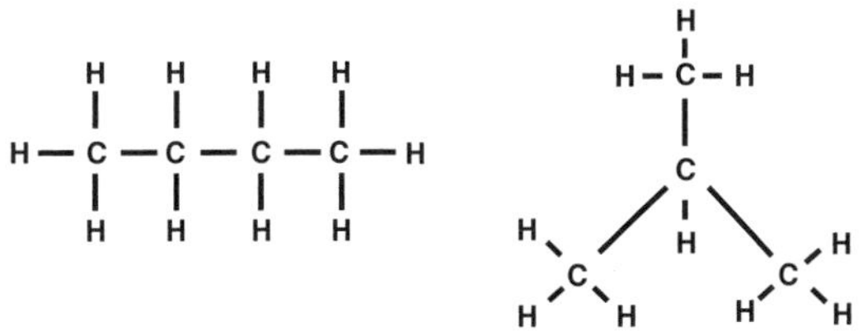

Figure 1.3.8: Two isomers: butane and isobutane.

F27: Cayley [Ca:1874] used tree-counting methods to enumerate paraffins (alkanes) with up to 11 carbon atoms, as well as various other families of molecules; the followiing table gives the number of isomers of alkanes for $n = 1, \ldots, 8$.

Formula	CH_4	C_2H_6	C_3H_8	C_4H_{10}	C_5H_{12}	C_6H_{14}	C_7H_{16}	C_8H_{18}
Number	1	1	1	2	3	5	9	18

F28: W. K. Clifford and Sylvester believed that a connection could be made between chemical atoms and binary quantics in invariant theory, a topic to which Cayley and Sylvester had made significant contributions. In 1878, Sylvester [Sy:1877–8] wrote a short note in *Nature* about this supposed connection, remarking that:

> Every invariant and covariant thus becomes expressible by a *graph* precisely identical with a Kekuléan diagram or chemicograph.

This was the first appearance of the word *graph* in the graph-theoretic sense.

F29: In 1878, Sylvester [Sy:1878] wrote a lengthy article on the graphic approach to chemical molecules and invariant theory in the first volume of the *American Journal of Mathematics*, which he had recently founded.

F30: Little progress was made on the enumeration of isomers until the 1920s and 1930s. A. C. Lunn and J. K. Senior [LuSe29] recognized the importance of permutation groups for this area, and Pólya's above-mentioned paper solved the counting problem for several families of molecules.

1.3.3   Topological Graphs

Euler's polyhedron formula [Eu:1750] was the foundation for topological graph theory, since it holds also for planar graphs. It was later extended to surfaces other than the sphere. In 1930, a fundamental characterization of graphs imbeddable in the sphere was given by Kazimierz Kuratowski (1896–1980), and recent work – notably by Neil Robertson, Paul Seymour, and others – has extended these results to the higher order surfaces.

Euler's Polyhedron Formula

The Greeks were familiar with the five regular solids, but there is no evidence that they knew the simple connection between the numbers V of vertices, E of edges, and F of faces of a polyhedron:

$$V - E + F = 2$$

In the 17th century, René Descartes studied polyhedra, and he obtained results from which Euler's formula could later be derived. However, since Descartes had no concept of an edge, he was unable to make such a deduction.

FACTS [BiLlWi98, Chapter 5] [Cr99] [BeWi09]

F31: The first appearance of the polyhedron formula appeared in a letter, dated 14 November 1750, from Euler to C. Goldbach. Denoting the number of faces, solid angles (vertices) and joints (edges) by $\mathcal{H}$, $\mathcal{S}$, and $\mathcal{A}$, he wrote:

- In every solid enclosed by plane faces the aggregate of the number of faces and the number of solid angles exceeds by two the number of edges, or $\mathcal{H} + \mathcal{S} = \mathcal{A} + 2$.

F32: Euler was unable to prove his formula. In 1752 he attempted a proof by dissection, but it was deficient. The first valid proof was given by A.-M. Legendre [Le:1794] in 1794, using metrical properties of spherical polygons.

F33: In 1813, A.-L. Cauchy [Ca:1813] obtained a proof of Euler's formula by stereographically projecting the polyhedron onto a plane and considering a triangulation of the resulting planar graph.

F34: Around the same time, S.-A.-J. Lhuilier [Lh:1811] gave a topological proof that there are only five regular convex polyhedra, and he anticipated the idea of duality by noting that four of them occur in reciprocal pairs. He also found three types of polyhedra for which Euler's formula fails – those with indentations in their faces, those with an interior cavity, and ring-shaped polyhedra drawn on a torus (that is, polyhedra with a 'tunnel' through them). For such ring-shaped polyhedra, Lhuilier derived the formula

$$V - E + F = 0$$

and extended his discussion to prove that, if g is the number of tunnels in a surface on which a polyhedral map is drawn, then

$$V - E + F = 2 - 2g$$

The number g is now called the **genus of the surface**, and the value of the quantity $2 - 2g$ is called the **Euler characteristic**. (See §7.1.)

F35: In 1861–2, Listing [Li:1861–2] wrote *Der Census räumliche Complexe*, an extensive investigation into complexes, and studied how their topological properties affect the generalization above of Euler's formula. This work proved to be influential in the subsequent development of topology. In particular, H. Poincaré took up Listing's ideas in his papers of 1895–1904 that laid the foundations for algebraic topology.

F36: Poincaré's work was instantly successful, and it appeared in an article by M. Dehn and P. Heegaard [DeHe07] on analysis situs (topology) in the ten-volume *Encyklopädie der Mathematischen Wissenschaften*. His ideas were further developed by O. Veblen [Ve22] in a series of colloquium lectures on analysis situs for the American Mathematical Society in 1916.

Planar Graphs

The study of planar graphs originated in two recreational problems involving the complete graph K_5 and the complete bipartite graph $K_{3,3}$. These graphs (shown in Figure 1.3.9) are the main obstructions to planarity, as was subsequently demonstrated by Kuratowski.

 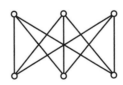

Figure 1.3.9: The Kuratowski graphs K_5 and $K_{3,3}$.

FACTS [BiLIWi98, Chapter 8]

F37: Around the year 1840, A. F. Möbius presented the following puzzle to his students:

> There was once a king with five sons. In his will he stated that after his death the sons should divide the kingdom into five regions so that the boundary of each region should have a frontier line in common with each of the other four regions. Can the terms of the will be satisfied?

This question asks whether one can draw five mutually neighboring regions in the plane. The connection with graph theory can be seen from its dual version, later formulated by H. Tietze:

> The king further stated that the sons should join the five capital cities of his kingdom by roads so that no two roads intersect. Can this be done?

In this dual formulation, the problem is that of deciding whether the graph K_5 is planar.

F38: An old problem, whose origins are obscure, is the *utilities problem*, or *gas–water–electricity problem*, mentioned by H. Dudeney [Du13] in the *Strand Magazine* of 1913:

> The puzzle is to lay on water, gas, and electricity, from W, G, and E, to each of the three houses, A, B, and C, without any pipe crossing another (see Figure 1.3.10).

This problem is that of deciding whether $K_{3,3}$ is planar.

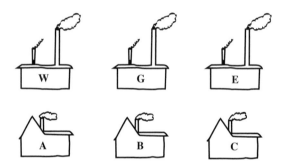

Figure 1.3.10: The gas–water–electricity problem.

F39: In 1930 Kuratowski [Ku30] published a celebrated paper proving that every nonplanar graph has a subgraph homeomorphic to K_5 or $K_{3,3}$; this result was obtained independently by O. Frink and P. A. Smith.

F40: In 1931 H. Whitney [Wh31] discovered an abstract definition of duality that is purely combinatorial and agrees with the geometrical definition of duality for planar graphs. He proved that, with this general definition of duality, a graph is planar if and only if it has an abstract dual. Related results were obtained by S. MacLane and others.

F41: In 1935 Whitney [Wh35] generalized the idea of independence in graphs and vector spaces to the concept of a matroid. The dual of a matroid extends and clarifies the duality of planar graphs, and Tutte [Tu59] used these ideas in the late 1950s to obtain a Kuratowski-type criterion for a matroid to arise from a graph (see §6.6).

Graphs on Higher Surfaces

A graph drawn without crossings on a plane corresponds (by stereographic projection) to a graph similarly drawn on the surface of a sphere. This leads to the idea of graphs drawn on surfaces other than the sphere. The initial work in this area was carried out, in the context of coloring maps, by Percy Heawood (1861–1955) and Lothar Heffter (1862–1962) for orientable surfaces, and by Heinrich Tietze (1880-1964) for non-orientable surfaces, but the basic problems in the area were not solved until Gerhard Ringel and Ted Youngs solved the Heawood conjecture in the 1960s and Neil Robertson and Paul Seymour generalized Kuratowski's theorem to other surfaces in the 1980s.

FACTS [BiLlWi98, Chapter 7; Ri74]

F42: In 1890, Heawood [He:1890] presented an imbedding of the complete graph K_7 on a torus. He also derived a formula for the genus of a surface on which a given complete graph can be imbedded, but his attempted proof of this formula was deficient.

F43: In 1891, L. Heffter [He:1891] investigated the imbedding of complete graphs on orientable surfaces other than the sphere and the torus, and he proved that Heawood's formula is correct for orientable surfaces of low genus and certain other surfaces.

F44: In 1910, H. Tietze [Ti10] extended Heffter's considerations to certain non-orientable surfaces, such as the Möbius band and the projective plane, and stated a corresponding Heawood formula. He was unable to prove it for the Klein bottle, but this case was settled in 1934 by P. Franklin [Fr34], who found that it was an exception to the formula. In 1935, I. N. Kagno [Ka35] proved the formula for surfaces of non-orientable genus 3, 4, and 6.

F45: The Heawood formula for general non-orientable surfaces was proved in 1952 by Ringel. The proof for orientable surfaces proved to be much more difficult, involving 300 pages of consideration of 12 separate cases. Most of these were settled in the mid-1960s, and the proof was completed in 1968 by Ringel and Youngs [RiYo68], using W. Gustin's [Gu63] combinatorial inspiration in 1963 of a *current graph*. Since then, the transformation by J. L. Gross [Gr74] of numerous types of specialized combinatorial current graphs into a unified topological object, with its dualization to a *voltage graph* (see §7.4), has led to simpler solutions (see Gross and T. W. Tucker [GrTu74]).

F46: In a sequence of papers in the 1980s of great mathematical depth, Robertson and Seymour [RoSe85] proved that, for each orientable genus g, the set of "forbidden subgraphs" is finite (see §7.7). However, apart from the sphere, the number of forbidden subgraphs runs into hundreds, even for the torus. For non-orientable surfaces, there is a similar result, and in 1979 H. H. Glover, J. P. Huneke, and C. S. Wang [GlHuWa79] obtained a set of 103 forbidden subgraphs for the projective plane.

1.3.4   Graph Colorings

Early work on colorings concerned the coloring of the countries of a map and, in particular, the celebrated four-color problem. This was first posed by Francis Guthrie in 1852, and a celebrated (incorrect) "proof" by Alfred Bray Kempe appeared in 1879. The four-color theorem was eventually proved by Kenneth Appel and Wolfgang Haken in 1976, building on the earlier work of Kempe, George Birkhoff, Heinrich Heesch, and others, and a simpler proof was subsequently produced by Neil Robertson, Daniel Sanders, Paul Seymour, and Robin Thomas [1997]. Meanwhile, attention had turned to the dual problem of coloring the vertices of a planar graph and of graphs in general.

There was also a parallel development in the coloring of the edges of a graph, starting with a result of Tait [1880], and leading to a fundamental theorem of V. G. Vizing in 1964. As mentioned earlier, the corresponding problem of coloring maps on other surfaces was settled by Ringel and Youngs in 1968.

The Four-Color Problem

Many developments in graph theory can be traced back to attempts to solve the celebrated four-color problem on the coloring of maps.

FACTS [BiLlWi98, Chapter 6] [Wi02]

F47: The earliest known mention of the four-color problem occurs in a letter from A. De Morgan to Hamilton, dated 23 October 1852. De Morgan described how a student had asked him whether every map can be colored with just four colors in such a way that neighbouring countries are colored differently. The student later identified himself as Frederick Guthrie, giving credit for the problem to his brother Francis, who formulated it while coloring the counties of a map of England. Hamilton was not interested in the problem.

F48: De Morgan wrote to various friends, outlining the problem and trying to describe where the difficulty lies. On 10 April 1860, the problem appeared in print, in an unsigned book review in the *Athenaeum*, written by De Morgan. This review was read in the U.S. by C. S. Peirce, who developed a life-long interest in the problem. An earlier printed reference, signed by "F.G.", appeared in the *Athenaeum* in 1854 [McK12].

F49: On 13 June 1878, at a meeting of the London Mathematical Society, Cayley asked whether the problem had been solved. Shortly after, he published a short note describing where the difficulty might lie, and he showed that it is sufficient to restrict one's attention to trivalent maps.

F50: In 1879, Kempe [Ke:1879], a former Cambridge student of Cayley, published a purported proof of the four-color theorem in the *American Journal of Mathematics*, which had recently been founded by Sylvester. Kempe showed that every map must contain a country with at most five neighbours, and he showed how any coloring of the rest of the map can be extended to include such a country. His solution included a new technique, now known as a ***Kempe-chain*** argument, in which the colors in a two-colored section of the map are interchanged. Kempe's proof for a map that contains a digon, triangle, or quadrilateral was correct, but his argument for the pentagon (where he used two simultaneous color-interchanges) was fallacious.

F51: In 1880, Tait [Ta:1878–80] presented "improved proofs" of the four-color theorem, all of them fallacious. Other people interested in the four-color problem at this time were C. L. Dodgson (Lewis Carroll), F. Temple (Bishop of London), and the Victorian educator J. M. Wilson.

F52: In 1890, Heawood [He:1890] published a paper in the *Quarterly Journal of Pure and Applied Mathematics*, pointing out the error in Kempe's proof, salvaging enough to deduce the five-color theorem, and generalizing the problem in various ways, such as for other surfaces (see §1.1.3). Heawood subsequently published another six papers on the problem, the last while he was in his 90th year. Kempe admitted his error, but he was unable to put it right.

F53: During the first half of the 20th century two ideas emerged, each of which finds its origin in Kempe's paper. The first is that of an ***unavoidable set*** — a set of configurations, at least one of which must appear in any map. Unavoidable sets were produced by P. Wernicke [We:1904] (see Figure 1.3.11), by P. Franklin, and by H. Lebesgue.

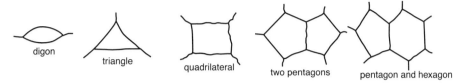

digon   triangle   quadrilateral   two pentagons   pentagon and hexagon

Figure 1.3.11: Wernicke's unavoidable set.

The second is that of a ***reducible configuration*** — a configuration of countries with the property that any coloring of the rest of the map can be extended to the configuration: no such configuration can appear in any counter-example to the four-color theorem. Birkhoff [Bi13] showed that the arrangement of four pentagons in Figure 1.3.12 (known as the ***Birkhoff diamond***) is a reducible configuration.

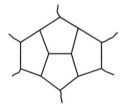

Figure 1.3.12: The Birkhoff diamond.

F54: In 1912, Birkhoff [Bi12] investigated the number of ways of coloring a given map with k colors, and he showed that this is always a polynomial in k, now called the *chromatic polynomial* of the map.

F55: In 1922, Franklin [Fr22] presented further unavoidable sets and reducible configurations, and he deduced that the four-color theorem is true for all maps with up to 25 countries. This number was later increased several times by other authors.

F56: Around 1950 Heesch started to search for an unavoidable set of reducible configurations. Over the next few years, Heesch [He69] produced thousands of reducible configurations.

F57: In 1976, Appel and Haken [ApHa77, ApHaKo77], with the assistance of J. Koch, obtained an unavoidable set of 1482 reducible configurations, thereby proving the four-color theorem. Their solution required substantial use of a computer to test the configurations for reducibility.

F58: Around 1994, Robertson, Sanders, Seymour, and Thomas [RoSaSeTh97] produced a more systematic proof. Using a computer to assist with both the unavoidable set and the reducible configuration parts of the solution, they systematized the Appel–Haken approach, and they obtained an unavoidable set of 633 reducible configurations.

Other Graph Coloring Problems

Arising from work on the four-color problem, progress was being made on other graph problems involving the coloring of edges or vertices.

FACTS [BiLlWi98, Chapter 6] [FiWi77] [JeTo95]

F59:   In his 1879 paper on the coloring of maps, Kempe [Ke:1879] outlined the dual problem of coloring the vertices of a planar graph in such a way that adjacent vertices are colored differently. This dual approach to map-coloring was later taken up by H. Whitney in a fundamental paper of 1932 and by most subsequent workers on the four-color problem.

F60:   In 1880, Tait [Ta:1878–80] proved that the four-color theorem is equivalent to the statement that the edges of every trivalent map can be colored with three colors in such a way that each color appears at every vertex.

F61:   In 1916, D. König [Kö16] proved that the edges of any bipartite graph with maximum degree d can be colored with d colors. (See §11.3.)

F62:   The idea of coloring the vertices of a graph so that adjacent vertices are colored differently developed a life of its own in the 1930s, mainly through the work of Whitney, who wrote his Ph.D. thesis on the coloring of graphs.

F63:   In 1941, L. Brooks [Br41] proved that the chromatic number of any simple graph with maximum degree d is at most $d + 1$, with equality only for odd cycles and odd complete graphs. (See §5.1.)

F64:   In the 1950s, substantial progress on vertex-colorings was made by G. A. Dirac, who introduced the idea of a *critical graph*.

F65:   In 1964, V. G. Vizing [Vi64] proved that the edges of any simple graph with maximum degree d can always be colored with $d + 1$ colors. In the following year, Vizing produced many further results on edge-colorings.

F66:   The concepts of the chromatic number and edge-chromatic number of a graph have been generalized by a number of writers — for example, M. Behzad and others introduced total colorings in the 1960s, and P. Erdős and others introduced list colorings.

Factorization

A graph is *k-regular* if each of its vertices has degree k. Such graphs can sometimes be split into regular subgraphs, each with the same vertex-set as the original graph. A *k-factor* in a graph is a k-regular subgraph that contains all the vertices of the original graph. Fundamental work on factors in graphs was carried out by Julius Petersen [1839–1910] and W. T. Tutte [1914–2002]. (See §5.4.)

FACTS [BiLlWi98, Chapter 10]

F67:   In 1891, Petersen [Pe:1891] wrote a fundamental paper on the factorization of regular graphs, arising from a problem in the theory of invariants. In this paper he proved that if k is even, then any k-regular graph can be split into 2-factors. He also proved that any 3-regular graph possesses a 1-factor, provided that it has not more than two "leaves"; a leaf is a subgraph joined to the rest of the graph by a single edge.

F68: In 1898, Petersen [Pe:1898] produced a trivalent graph with no leaves, now called the **Petersen graph** (see Figure 1.3.13), which cannot be split into three 1-factors; it can, however, be split into a 1-factor (the spokes) and a 2-factor (the pentagon and pentagram).

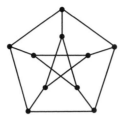

Figure 1.3.13: The Petersen graph.

F69: In 1947, Tutte [Tu47] produced a characterization of graphs that contain a 1-factor. Five years later he extended his result to a characterization of graphs that contain a k-factor, for any k.

1.3.5 Graph Algorithms

Graph theory algorithms can be traced back to the 19th century, when Fleury gave a systematic method for tracing an eulerian graph and G. Tarry showed how to escape from a maze (see §4.2). The 20th century saw algorithmic solutions to such problems as the minimum connector problem, the shortest and longest path problems, and the *Chinese Postman Problem* (see §4.3), as well as to a number of problems arising in operational research. In each of these problems we are given a network, or weighted graph, to each edge (and/or vertex) of which has been assigned a number, such as its length or the time taken to traverse it.

FACTS [Da82] [LLRS85] [LoPl86]

F70: The *Traveling Salesman Problem*, in which a salesman wishes to make a cyclic tour of a number of cities in minimum time or distance, appeared in rudimentary form in 1831. It reappeared in mathematical circles in the early 1930s, at Princeton, and was later popularized at the RAND Corporation. This led to a fundamental paper of G. B. Dantzig, D. R. Fulkerson, and S. M. Johnson [DaFuJo54] that included the solution of a traveling salesman problem with 49 cities. In the 1980s a problem with 2392 cities was settled by Padberg and Rinaldi [PaRi87]. (See §4.6.)

F71: The greedy algorithm for the *minimum connector problem*, in which one seeks a minimum-length spanning tree in a weighted graph, can be traced back to O. Boruvka [Bo26] and was later rediscovered by J. B. Kruskal [Kr56]. A related algorithm, due to V. Jarník (1931), was rediscovered by R. C. Prim (1957). (See §10.1.)

F72: Graph algorithms were developed by D. R. Fulkerson and G. B. Dantzig [FuDa55] for finding the maximum flow of a commodity between two nodes in a capacitated network, and by R. E. Gomory and T. C. Hu [GoHu61] for determining *maximum flows* in multi-terminal networks.

F73: Finding a longest path, or critical path, in an activity network dates from the 1940s and 1950s, with *PERT* (Program Evaluation and Review Technique) used by the U.S. Navy for problems involving the building of submarines and CPM (*Critical Path Method*) developed by the Du Pont de Nemours Company to minimize the total cost of a project. (See §3.2.)

F74: There are several efficient algorithms for finding the shortest path in a given network, of which the best known is due to E. W. Dijkstra [Di59]. (See §10.1.)

F75: The Chinese postman problem, for finding the shortest route that covers each edge of a given weighted graph, was originated by Meigu Guan (Mei-Ku Kwan) [Gu60] in 1960. (See §4.3.)

F76: In matching and assignment problems one wishes to assign people as appropriately as possible to jobs for which they are qualified. This work developed from work of König and from a celebrated result on matching due to Philip Hall [Ha35], later known as the "marriage theorem" [HaVa50]. These investigations led to the subject of polyhedral combinatorics and were combined with the newly emerging study of linear programming. (See §11.3.)

F77: By the late 1960s it became clear that some problems seemed to be more difficult than others, and Edmonds [Ed65] discussed problems for which a polynomial-time algorithm exists. Cook [Co71], Karp [Ka72], and others later developed the concept of NP-completeness. The assignment, transportation, and minimum spanning-tree problems are all in the *polynomial-time class* P, while the traveling salesman and Hamiltonian cycle problems are NP-hard. It is not known whether P = NP. Further information can be found in [GaJo79].

References

[ApHa77] K. Appel and W. Haken, Every planar map is 4-colorable: Part 1, Discharging, *Illinois J. Math.* 21 (1977), 429–490.

[ApHaKo77] K. Appel, W. Haken, and J. Koch, Every planar map is 4-colorable: Part 2, Reducibility, *Illinois J. Math.* 21 (1977), 429–490.

[BeWi09] L. W. Beineke and R. J. Wilson, *Topics in Topological Graph Theory 1736-1936*, Cambridge University Press, 2009.

[BiLlWi98] N. L. Biggs, E. K. Lloyd, and R. J. Wilson (eds.), *Graph Theory 1736-1936*, Oxford University Press, 1998.

[Bi12] G. D. Birkhoff, A determinantal formula for the number of ways of coloring a map, *Ann. of Math.* 14 (1912), 42–46.

[Bi13] G. D. Birkhoff, The reducibility of maps, *Amer. J. Math.* 35 (1913), 115–128.

[BiLe46] G. D. Birkhoff and D. C. Lewis, Chromatic polynomials, *Trans. Amer. Math. Soc.* 60 (1946), 355–451.

[BoCh76] J. A. Bondy and V. Chvátal, A method in graph theory, *Discrete Math.* 15 (1976), 111–136.

[Bo26] O. Boruvka, O jistém problému minimálním, *Acta Soc. Sci. Natur. Moravicae* 3 (1926), 37–58.

[Br41] R. L. Brooks, On colouring the nodes of a network, *Proc. Cambridge Philos. Soc.* 37 (1941), 194–197.

[Ca:1813] A.-L. Cauchy, Recherches sur les polyèdres-premier mémoire, *J. Ecole Polytech.* 9 (Cah. 16) (1813), 68–86.

[Ca:1857] A. Cayley, On the theory of the analytical forms called trees, *Phil. Mag.* (4) 13 (1857), 172–176.

[Ca:1874] A. Cayley, On the mathematical theory of isomers, *Phil. Mag.* (4) 47 (1874), 444–446.

[Ca:1879] A. Cayley, On the colouring of maps, *Proc. Roy. Geog. Soc.* (new Ser.) 1 (1879), 259–261.

[Ca:1889] A. Cayley, A theorem on trees, *Quart. J. Pure Appl. Math.* 23 (1889), 376–378.

[Co71] S. A. Cook, The complexity of theorem-proving procedures, *Proc. 3rd Annual ACM Symp. Theory of Computing*, pp151–158, ACM, New York, 1971.

[Cl:1844] T. Clausen, [Second postscript to] De linearum tertii ordinis proprietatibus, *Astron. Nachr.* 21 (1844), col. 209–216.

[Cr99] P. R. Cromwell, *Polyhedra*, Cambridge University Press, 1999.

[Da82] G. B. Dantzig, Reminiscences about the origins of linear programming, *Oper. Res. Lett.* 1 (1982), 43–48.

[DaFuJo54] G. B. Dantzig, D. R. Fulkerson, and S. M. Johnson, Solution of a large-scale traveling-salesman problem, *Oper. Res.* 2 (1954), 393–410.

[DeHe07] M. Dehn and P. Heegaard, Analysis situs, *Encyklopädie der Mathematischen Wissenschaften* (1907), 153–120.

[DeM:1860] A. De Morgan, A review of the philosophy of discovery, chapters historical and critical, by W. Whewell, D. D., *Athenaeum* No. 1694 (1860), 501–503.

[Di59] E. W. Dijkstra, A note on two problems in connexion with graphs, *Numer. Math.* 1 (1959), 269–271.

[Di52] G. A. Dirac, Some theorems on abstract graphs, *Proc. London Math. Soc.* (3) 2 (1952), 69–81.

[Du13] H. E. Dudeney, Perplexities, *Strand Mag.* 46 (July 1913), 110 and (August 1913), 221.

[Ed65] J. R. Edmonds, Paths, trees and flowers, *Canad. J. Math.* 17 (1965), 449–467.

[Eu:1736] L. Euler, (1736) Solutio problematis ad geometriam situs pertinentis, *Commentarii Academiae Scientiarum Imperialis Petropolitanae* 8 (1752), 128–140.

[Eu:1759] L. Euler, Solution d'une question curieuse qui ne paroit soumise à aucune analyse, *Mem. Acad. Sci. Berlin* 15 (1759), 310–337.

[FiWi77] S. Fiorini and R. J. Wilson, *Edge-Colourings of Graphs*, Pitman, 1977.

[FoFu56] L. R. Ford and D. R. Fulkerson, Maximal flow through a network, *Canad. J. Math.* 8 (1956), 399–404.

[Fr22] P. Franklin, The four color problem, *Amer. J. Math.* 44 (1922), 225–236.

[Fr34] P. Franklin, A six color problem, *J. Math. Phys.* 13 (1934), 363–369.

[FuDa55] D. R. Fulkerson and G. B. Dantzig, Computation of maximum flow in networks, *Naval Research Logistics Quarterly* 2 (1955), 277–283.

[GaJo79] M. R. Garey and D. S. Johnson, *Computers and Intractability: A Guide to the Theory of NP-Completeness*, W. H. Freeman and Co., 1979.

[GlHuWa79] H. H. Glover, J. P. Huneke and C. S. Wang, 103 graphs that are irreducible for the projective plane, *J. Combin. Theory* (B) 27 (1979), 332–370.

[GoHu61] R. E. Gomory and T. C. Hu, Multi-terminal network flows, *SIAM J. Appl. Math.* 9 (1961), 551–556.

[Gr74] J. L. Gross, Voltage graphs, *Discrete Math.* 9 (1974), 239–246.

[GrTu74] J. L. Gross and T. W. Tucker, Quotients of complete graphs: revisiting the Heawood problem, *Pacific J. Math.* 55 (1974), 391–402.

[Gu60] Guan Meigu, Graphic programming using odd or even points, *Acta Math. Sinica* 10 (1962), 263–266; *Chinese Math.* 1 (1962), 273–277.

[Gu63] W. Gustin, Orientable embedding of Cayley graphs, *Bull Amer. Math. Soc.* 69 (1963), 272–275.

[Ha35] P. Hall, On representatives of subsets, *J. London Math. Soc.* 10 (1935), 26–30.

[HaVa50] P. R. Halmos and H. E. Vaughan, The marriage problem, *Amer. J. Math.* 72 (1950), 214–215.

[Ha:1856] W. R. Hamilton, Memorandum respecting a new system of roots of unity, *Phil. Mag.* (4) 12 (1856), 446.

[Ha55] F. Harary, The number of linear, directed, rooted, and connected graphs, *Trans. Amer. Math. Soc.* 78 (1955), 445–463.

[HaPa73] F. Harary and E. M. Palmer, *Graphical Enumeration*, Academic Press, 1973.

[He:1890] P. J. Heawood, Map-colour theorem, *Quart. J. Pure Appl. Math.* 24 (1890), 332–338.

[He:1891] L. Heffter, Über das Problem der Nachbargebiete, *Math. Ann.* 38 (1891), 477–580.

[He69] H. Heesch, Untersuchungen zum Vierfarbenproblem, *B. I. Hochschulscripten*, 810/810a/810b, Bibliographisches Institut, Mannheim-Vienna-Zürich, 1969.

[Hi:1873] C. Hierholzer, Über die Möglichkeit, einen Lineanzug ohne Wiederholung und ohne Unterbrechung zu umfahren, *Math. Ann.* 6 (1873), 30–32.

[HoWi04] B. Hopkins and R. Wilson, The truth about Konigsberg?, *College Math. J.* 35 (2004), 198-207.

[JeTo95] T. R. Jensen and B. Toft, *Graph Coloring Problems*, Wiley–Interscience, 1995.

[Ka35] I. N. Kagno, A note on the Heawood color formula, *J. Math. Phys.* 14 (1935), 228–231.

[Ka72] R. M. Karp, Reducibility among combinatorial problems, 85–103 in *Complexity of Computer Computations* (ed. R. E. Miller and J. W. Thatcher), Plenum Press, 1972.

[Ke:1879] A. B. Kempe, On the geographical problem of four colours, *Amer. J. Math.* 2 (1879), 193–200.

[Ki:1856] T. P. Kirkman, On the representation of polyedra, *Phil. Trans. Roy. Soc. London* 146 (1856), 413–418.

[Kö16] D. König, Über Graphen und ihre Anwendung auf Determinantentheorie und Mengenlehre, *Math. Ann.* 77 (1916), 453–465.

[Kr56] J. B. Kruskal, On the shortest spanning subtree of a graph and the traveling salesman problem, *Proc. Amer. Math. Soc.* 7 (1956), 48–50.

[Ku30] K. Kuratowski, Sur le problème des courbes gauches en topologie, *Fund. Math.* 15 (1930), 271–283.

[LLRS85] E. L. Lawler, J. K. Lenstra, A. H. G. Rinooy Kan, and D. B. Schmoys (eds.), *The Traveling Salesman Problem: A Guided Tour through Combinatorial Optimization*, Wiley, 1985.

[Le:1794] A.-M. Legendre, *Eléments de Géométrie* (1st ed.), Firmin Didot, Paris, 1794.

[Lh:1811] S.-A.-J. Lhuilier, Démonstration immédiate d'un théorème fondamental d'Euler sur les polyhèdres, et exceptions dont ce théorème est susceptible, *Mém. Acad. Imp. Sci. St. Pétersb.* 4 (1811), 271–301.

[Li:1847] J. B. Listing, Vorstudien zur Topologie, *Göttingen Studien* (Abt. 1) Math. Naturwiss. Abh. 1 (1847), 811–875.

[Li:1861–2] J. B. Listing, Der Census räumliche Complexe, *Abh. K. Ges. Wiss. Göttingen Math. Cl.* 10 (1861–2), 97–182.

[LoPl86] L. Lovász and M. D. Plummer, Matching Theory, *Annals of Discrete Mathematics* 29, North-Holland, 1986.

[Lu:1882] E. Lucas, *Récréations Mathématiques*, Vol. 1, Gauthier-Villars, Paris (1882).

[LuSe29] A. C. Lunn and J. K. Senior, Isomerism and configuration, *J. Phys. Chem.* 33 (1929), 1027–1079.

[Ma69] J. Mayer, Le problème des régions voisines sur les surfaces closes orientables, *J. Combin. Theory* 6 (1969), 177–195.

[McK12] B. McKay, A note on the history of the four-colour conjecture, *J. Graph Theory* 72 (2013), 361–363.

[Or60] O. Ore, Note on Hamiltonian circuits, *Amer. Math. Monthly* 67 (1960), 55.

[Ot48] R. Otter, The number of trees, *Ann. of Math.* 49 (1948), 583–599.

[PaRi87] M. W. Padberg and G. Rinaldi, Optimization of a 532-city symmetric traveling salesman problem by branch and cut, *Oper. Res. Lett.* 6 (1987), 1–7.

[Pe:1891] J. Petersen, Die Theorie der regulären Graphs, *Acta Math.* 15 (1891), 193–220.

[Pe:1898] J. Petersen, Sur le théorème de Tait, *Interméd. Math.* 5 (1898), 225–227.

[Po:1809–10] L. Poinsot, Sur les polygones et les polyèdres, *J. Ecole Polytech.* 4 (1809–10) (Cah. 10), 16–48.

[Pó37] G. Pólya, Kombinatorische Anzahlbestimmungen für Gruppen, Graphen und chemische Verbindungen, *Acta Math.* 68 (1937), 145–254.

[PóRe87] G. Pólya and R. C. Read, *Combinatorial Enumeration of Groups, Graphs and Chemical Compounds*, Springer, 1987.

[Pr18] H. Prüfer, Neuer Beweis eines Satzes über Permutationen, *Arch. Math. Phys.* (3) 27 (1918), 142–144.

[Re63] R. C. Read, On the number of self-complementary graphs and digraphs, *J. London Math. Soc.* 38 (1963), 99–104.

[Re27] J. H. Redfield, The theory of group-reduced distributions, *Amer. J. Math.* 49 (1927), 433–455.

[Re:1871–3] M. Reiss, Evaluation du nombre de combinaisons desquelles les 28 dés d'un jeu du domino sont susceptibles d'après la règle de ce jeu, *Ann. Mat. Pura. Appl.* (2) 5 (1871–3), 63–120.

[Ri74] G. Ringel, *Map Color Theorem*, Springer, 1974.

[RiYo68] G. Ringel and J. W. T. Youngs, Solution of the Heawood map-coloring problem, *Proc. Nat. Acad. Sci. U.S.A.* 60 (1968), 438–445.

[RoSe85] N. Robertson and P. D. Seymour, Graph minors — a survey, in *Surveys in Combinatorics 1985* (ed. I. Anderson), London Math. Soc. Lecture Notes Series 103 (1985), Cambridge University Press, 153–171.

[RoSaSeTh97] N. Robertson, D. Sanders, P. Seymour, and R. Thomas, The four-colour theorem, *J. Combin. Theory, Ser. B* 70 (1997), 2–44.

[Ro:1892] W. W. Rouse Ball, *Mathematical Recreations and Problems of Past and Present Times* (later entitled *Mathematical Recreations and Essays*), Macmillan, London, 1892.

[SaStWi88] H. Sachs, M. Stiebitz and R. J. Wilson, An historical note: Euler's Königsberg letters, *J. Graph Theory* 12 (1988), 133–139.

[Sh49] C. E. Shannon, A theorem on coloring the lines of a network, *J. Math. Phys.* 28 (1949), 148–151.

[Sy:1877–8] J. J. Sylvester, Chemistry and algebra, *Nature* 17 (1877–8), 284.

[Sy:1878] J. J. Sylvester, On an application of the new atomic theory to the graphical representation of the invariants and covariants of binary quantics, *Amer. J. Math.* 1 (1878), 64–125.

[Ta:1878–80] P. G. Tait, Remarks on the colouring of maps, *Proc. Roy. Soc. Edinburgh* 10 (1878–80), 729.

[Ti10] H. Tietze, Einige Bemerkungen über das Problem des Kartenfärbens auf einseitigen Flächen, *Jahresber. Deut. Math.-Ver.* 19 (1910), 155–179.

[Tu46] W. T. Tutte, On hamiltonian circuits, *J. London Math. Soc.* 21 (1946), 98–101.

[Tu47] W. T. Tutte, The factorizations of linear graphs, *J. London Math. Soc.* 22 (1947), 107–111.

[Tu59] W. T. Tutte, Matroids and graphs, *Trans. Amer. Math. Soc.* 90 (1959), 527–552.

[Tu70] W. T. Tutte, On chromatic polynomials and the golden ratio, *J. Combin. Theory* 9 (1970), 289–296.

[Va:1771] A.-T. Vandermonde, Remarques sur les problèmes de situation, *Mém. Acad. Sci. (Paris)* (1771), 556–574.

[Ve22] O. Veblen, *Analysis Situs*, Amer. Math. Soc. Colloq. Lect. 1916, New York, 1922.

[Vi64] V. G. Vizing, On an estimate of the chromatic class of a p-graph, *Diskret. Analiz* 3 (1964), 25–30.

[Vi65] V. G. Vizing, The chromatic class of a multigraph, *Diskret. Analiz* 5 (1965), 9–17.

[We:1904] P. Wernicke, Über den kartographischen Vierfarbensatz, *Math. Ann.* 58 (1904), 413–426.

[Wh31] H. Whitney, Non-separable and planar graphs, *Proc. Nat. Acad. Sci. U.S.A.* 17 (1931), 125–127.

[Wh35] H. Whitney, On the abstract properties of linear dependence, *Amer. J. Math.* 57 (1935), 509–533.

[Wi99] R. J. Wilson, Graph Theory, Chapter 17 in *History of Topology* (editor, I. M. James), Elsevier Science, 1999.

[Wi02] R. Wilson, *Four Colors Suffice*, Allen Lane, 2002; Princeton University Press, 2002.

Glossary for Chapter 1

bipartite graph: a graph whose vertices can be partitioned into two sets (called the *partite sets*) in such a way, that no edge joins two vertices in the same set. (For technical reasons, this includes the graph K_1 in this definition.)

bouquet B_n: the general graph with one vertex and n self-loops.

Cayley graph $C(\mathcal{A}, X)$ – for a group $\mathcal{A}$ with generating set X: the digraph whose vertex-set is $\mathcal{A}$ with an edge directed from a to ax for every $a \in \mathcal{A}$ and every $x \in X$. Sometimes two oppositely directed edges corresponding to an involution x are merged into a single undirected edge. (The underlying undirected graph of a Cayley graph is also commonly called a Cayley graph.)

circulant graph $\mathrm{Circ}(n; X)$: a Cayley graph for a cyclic group $\mathbb{Z}_n$.

complete k-partite graph $K_{n_1, n_2, \ldots, n_k}$: a simple k-partite graph such that two vertices are adjacent if and only if they are in different partite sets. All such graphs are called *complete multipartite graphs*.

complete bipartite graph: a simple bipartite graph such that each vertex in one partite set is adjacent to all the vertices in the other partite set. If the two partite sets have cardinalities r and s, then this graph is denoted $K_{r,s}$.

complete digraph $\overleftrightarrow{K}_n$: the simple digraph on n vertices such that between every pair of vertices, there is an arc in both directions.

complete graph K_n: the simple graph with n vertices in which every pair of vertices is joined by an edge.

k-connected graph: a graph such that the result of removing fewer than k vertices is connected and nontrivial, for all possible choices of the vertices.

connectivity of a graph G: the largest number k such that G is k-connected. It is denoted $\kappa(G)$ or $\kappa_V(G)$.

critically k-chromatic graph: a graph of chromatic number k whose chromatic number would decrease if any edge were removed. (See §5.1.)

critically k-connected graph: a graph of connectivity k whose connectivity would decrease if any vertex were removed. (See §4.1.)

critically k-edge-connected graph: a graph of edge-connectivity k whose edge-connectivity would decrease if any edge were removed. (See §4.1.)

cube: see *hypercube*.

cube graph: see *hypercube graph*.

cycle graph C_n: the n-vertex graph with n edges, such that every edge lies on a single cycle.

cycle rank – for a graph $G = (V, E)$ with $c(G)$ components: the number $|E(G)| - |V(G)| + c(G)$.

dipole D_n: the multigraph with two vertices and n edges.

edge-connectivity of a graph G: the largest number k such that G is k-edge-connected. It is denoted $\kappa'(G)$ or $\kappa_E(G)$.

k-edge-connected graph: a graph such that the result of removing fewer than k edges is connected and nontrivial, for all possible choices of the edges.

empty graph $\overline{K_n}$: the graph with n vertices and no edges.

eulerian graph: a graph with a closed walk that contains every edge exactly once. (See §1.3 for the history of eulerian graphs and §4.2 for an extensive discussion.)

k-factor of a graph G: a k-regular subgraph.

genus: see *minimum genus*.

hamiltonian graph: a graph that has a spanning cycle. (See Section 1.3 for the history of hamiltonian graphs and Section 4.6 for an extensive discussion.)

hypercube of dimension d: $\{(x_1, \ldots, x_d) \mid 0 \geq x_j)\}$.

hypercube graph Q_d: the 1-skeleton of a d-dimensional hypercube.

line graph of a graph G: the simple graph $L(G)$ whose vertex-set is the edge-set of G, and in which two vertices are adjacent if the edges in G to which they correspond have a common vertex. Also, a graph H is said *to be a line graph* if there exists a graph G such that H is isomorphic to $L(G)$.

minimum genus (or **genus**) of a connected graph G: the smallest number g such that G can be drawn on the orientable surface S_g (see Section 7.1) without any edge-crossings. Notation: $\gamma_{min}(G)$ or $\gamma(G)$.

null graph K_0: the graph with no vertices and no edges.

octahedral graph $\mathcal{O}_d$: the edge-complement of a 1-factor in K_{2d}.

partite sets: see *k-partite* graph.

p-partite graph: a graph whose vertex-set can be partitioned into p subsets (called the *partite sets*) in such a way that no edge joins two vertices in the same subset.

Petersen graph: a 10-vertex 3-regular graph, commonly depicted as a 5-pointed star inside a pentagon, with a 1-factor joining the vertices of the pentagon to the points of the star.

path graph P_n: the n-vertex graph with $n - 1$ edges, such that every edge lies on a single open path. (Quite commonly elsewhere, the subscript of the notation P_n denotes the number of edges.)

planar graph: a graph of minimum genus 0, i.e., a graph that can be drawn in the sphere or plane with no edge crossings.

platonic graph: the skeleton of any of the five platonic solids.

platonic solid: any of the five regular 3-dimensional polyhedra — tetrahedron, cube, octahedron, dodecaheron, icosahedron.

regular graph: a graph in which every vertex is of the same degree. It is *k-regular* if every vertex is of degree k.

simplex: the convex hull of a set S of affinely independent points in Euclidean space. It is a *k-simplex* if $|S| = k + 1$.

skeleton (or *1-skeleton*) of a k-complex K: the graph consisting of the vertices and the edges of K.

trivial graph K_1: the graph with one vertex and no edges.

Chapter 2

Graph Representation

Section 2.1

Computer Representations of Graphs

Alfred V. Aho, Columbia University

INTRODUCTION

Many problems in science and engineering can be modeled in terms of directed and undirected graphs. The data structures and algorithms used to represent graphs can have a significant impact on the size of problems that can be implemented on a computer and the speed with which they can be solved. This section presents the fundamental representations used in computer programs for graphs and illustrates the tradeoffs among the representations using key algorithms for some of the most common graph problems.

Throughout this section we use the notation $|X|$ to denote the number of elements in a set X. The graphs and digraphs in this section are assumed to be simple.

2.1.1   Basic Representations for Graphs

The two most basic representations for a graph are the adjacency matrix and the adjacency list.

DEFINITIONS

D1: A ***directed graph*** or ***digraph*** $G = (V, E)$ consists of a finite, nonempty set of *vertices* V and a set of *edges* E. Each edge is an ordered pair (v, w) of vertices.

D2: An ***undirected graph*** $G = (V, E)$ consists of a finite, nonempty set of *vertices* V and a set of *edges* E. Each edge is a set $\{v, w\}$ of vertices.

D3: In a directed graph $G = (V, E)$, vertex w is ***adjacent*** to vertex v if (v, w) is an edge in E. The number of vertices adjacent to v is called the ***out-degree*** of v.

D4: In an undirected graph $G = (V, E)$, vertex w is ***adjacent*** to vertex v if $\{v, w\}$ is an edge in E. The number of vertices adjacent to v is called the ***degree*** of v.

D5: A **path** in a directed or undirected graph is a seqence of edges (v_1, v_2), (v_2, v_3),..., (v_{n-1}, v_n). This path is from vertex v_1 to vertex v_n and has length $n - 1$.

D6: A graph $G = (V, E)$ is **dense** when the number of edges is close to $|V|^2$.

D7: A graph $G = (V, E)$ is **sparse** when the number of edges is much less than $|V|^2$.

D8: An **adjacency matrix** representation for a simple graph or digraph $G = (V, E)$ is a $|V| \times |V|$ matrix A, where $A[i, j] = 1$ if there is an edge from vertex i to vertex j; $A[i, j] = 0$ otherwise.

D9: An **adjacency list** representation for a graph or digraph $G = (V, E)$ is an array L of $|V|$ lists, one for each vertex in V. For each vertex i, there is a pointer L_i to a linked list containing all vertices j adjacent to i. A linked list is terminated by a **nil pointer**.

D10: An **incidence matrix** representation for a simple digraph $G = (V, E)$ is a $|V| \times |E|$ matrix I, where

$$I[v, e] = \begin{cases} -1 & \text{if edge } e \text{ is directed } to \text{ vertex } v \\ 1 & \text{if edge } e \text{ is directed } from \text{ vertex } v \\ 0 & \text{otherwise} \end{cases}$$

For an undirected graph, $I[v, e] = 1$ if e is incident on v and 0 otherwise.

EXAMPLES

E1: Figure 2.1.1 shows the adjacency matrix and adjacency list representations of a directed graph.

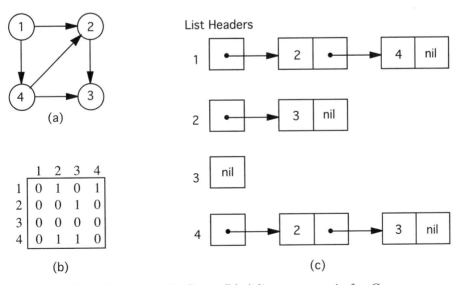

Figure 2.1.1: (a) A directed graph G.   (b) Adjacency matrix for G.
(c) Adjacency list representation for G.

E2: An incidence matrix for the digraph of Figure 2.1.1 is shown below.

$$
I_G \;=\; \begin{array}{c} 1 \\ 2 \\ 3 \\ 4 \end{array}
\begin{pmatrix}
\overset{(1,2)}{1} & \overset{(1,4)}{1} & \overset{(2,3)}{0} & \overset{(4,2)}{0} & \overset{(4,3)}{0} \\
-1 & 0 & 1 & -1 & 0 \\
0 & 0 & -1 & 0 & -1 \\
0 & -1 & 0 & 1 & 1
\end{pmatrix}
$$

FACTS

F1: An adjacency matrix representation for a graph $G = (V, E)$ always takes $O(|V|^2)$ space.

F2: An adjacency list representation for a graph $G = (V, E)$ takes $O(|V| + |E|)$ space.

REMARKS

R1: For a more detailed discussion of graph representations, see [AhHoUl74, AhHoUl83, CoLeRiSt09, Ev79, Ta83].

R2: As a general rule, an adjacency list representation is preferred when a graph is sparse, because it takes space that is linearly proportional to the number of vertices and edges.

R3: When a graph $G = (V, E)$ is dense, both an adjacency matrix and an adjacency list representation require $O(|V|^2)$ space. However, with the adjacency matrix, we can determine whether an edge exists in constant time, whereas with the adjacency list we may need $O(|V|)$ time. For this reason, adjacency matrix representations are often used with dense graphs.

R4: Note that in an adjacency list representation of an undirected graph, an edge i, j appears on two adjacency lists: the list for vertex i and the list for vertex j.

2.1.2   Graph Traversal Algorithms

One of the most fundamental tasks in algorithms involving graphs is visiting the vertices and edges of a graph in a systematic order. Depth-first and breadth-first search are frequently used traversal techniques for both directed and undirected graphs. For both these techniques, the adjacency list representation of a graph works well.

Depth-First Search

ALGORITHM

Depth-first search systematically visits all the vertices of a graph. Initially, all vertices are marked "new". When a vertex is visited, it is marked "old". Depth-first search works by selecting a new vertex v, marking it old, and then calling itself recursively on each of the vertices adjacent to v. The algorithm below is called "depth-first search" because it searches along a path in the forward (deeper) direction looking for new vertices as long as it can.

Algorithm 2.1.1 Depth-First Search

Input: A graph $G = (V, E)$, where $V = \{1, 2, \ldots, n\}$
     and $L[v]$ is a pointer to the list of vertices adjacent to vertex v.
Output: Traversal of all vertices in V in a depth-first order.

procedure $DepthFirstSearch(G)\{$
  for $v := 1$ **to** n **do**
    $mark[v] := new;$
  for $v := 1$ **to** n **do**
    if $mark[v] = new;$ **then**
      $dfs(v);$
$\}$
procedure $dfs(v)\{$
  $mark[v] := old;$
  for each vertex w on $L[v]$ **do**
    if $mark[w] = new$ **then**
      $dfs(w);$
$\}$

DEFINITIONS

During the course of its traversal, depth-first search partitions the graph into a collection of **depth-first trees** that make up a **depth-first forest**. The forest and its trees are determined by the edges, which are partitioned by the search into four sets:

D11: **Tree edges** are those edges (v, w) where w is first encountered by exploring edge (v, w).

D12: **Back edges** are those edges (v, w) that connect a vertex v to an ancestor w in a depth-first tree.

D13: **Forward edges** are those nontree edges (v, w) that connect a vertex v to a proper descendant in a depth-first tree.

D14: **Cross edges** are the remaining edges. They connect vertices that are neither ancestors nor descendants of one another.

FACTS

F3: Depth-first search takes $O(|V| + |E|)$ time on a graph $G = (V, E)$.

F4: If we represent the first visit of a vertex v with a left parenthesis "$(v$" and its last visit by a right parenthesis "$v)$", then the sequence of first and last visits forms an expression in which the parentheses are properly nested.

F5: In a depth-first search of an undirected graph, every edge is either a tree edge or a back edge.

REMARKS

R5: Depth-first search is a fundamental graph algorithm that has been in use since the 1950s. [Ta72, HoTa73] developed several efficient graph algorithms using depth-first search.

R6: Depth-first search forms the basis of many important graph algorithms such as determining the biconnected components of an undirected graph and finding the strongly connected components of a directed graph.

Breadth-First Search

Breadth-first search is another fundamental technique for exploring a graph G. It starts from a specified *source* vertex s from which it constructs a **breadth-first tree** consisting of all vertices of G reachable from s. In the process it computes a breadth-first tree rooted at s such that if a vertex v is reachable from s in G, there is a path in the tree from the root to s. The path in the tree is a shortest path from s to v in G.

ALGORITHM

Breadth-first search uses the abstract data type *queue* to hold vertices as they are being processed. The operation $enqueue(s, Q)$ places vertex s on the back of the queue Q. The operation $dequeue(Q)$ removes the element at the front of the queue Q.

Breadth-first search (Algorithm 2.1.2) visits the vertices of a graph G uniformly across the breadth of the frontier of its search, visiting all vertices at distance d from s, before looking for vertices at distance $d + 1$. In contrast, depth-first search plunges as deeply into the graph along a path as it can before backtracking to visit nodes closer to s.

DEFINITION

D15: Let BFT be the tree with root s, vertices v such that $parent[v]$ is not **nil**, and edges $\{(parent[v], v)|parent[v]$ is not **nil**$\}$. BFT is the **breadth-first tree** constructed by $BreadthFirstSearch(G, s)$.

FACTS

F6: Breadth-first search takes $O(|V| + |E|)$ time on a graph $G = (V, E)$.

F7: $BreadthFirstSearch(G, s)$ computes the length of the shortest path from s to v in $distance[v]$.

Algorithm 2.1.2 Breadth-First Search

Input: A graph $G = (V, E)$, where $V = \{1, 2, \ldots, n\}$, $L[v]$ is a pointer to the list
      of vertices adjacent to vertex v, and where s is a specified source vertex.
Output: A breadth-first tree consisting of root s and all vertices in V
      that are reachable from s.

procedure $BreadthFirstSearch(G, s)$ {
  for $v := 1$ **to** n **do** {
    $mark[v] := new$;
    $distance[v] := \infty$;
    $parent[v] := $ **nil**;
  }
  $mark[s] := visited$;
  $distance[s] := 0$;
  initialize queue Q;
  $enqueue(s, Q)$;
  while Q is not empty **do** {
    $v := dequeue(Q)$;
    for each vertex w on $L[v]$ **do**
      if $mark[w] = visited$ **then** {
        $mark[w] := visited$;
        $distance[w] := distance[w] + 1$;
        $parent[w] := v$;
        $enqueue(w, Q)$;
      }
  }
}

REMARKS

R7: Like depth-first search, breadth-first search has been used since the 1950s. Early applications of breadth-first search included maze searching and routing wires on printed circuit boards.

R8: The ideas found in breadth-first search are the building blocks of many other graph algorithms such as Dijkstra's single-source shortest-paths algorithm and Prim's algorithm for finding minimal spanning trees.

2.1.3 All-Pairs Problems

This section considers two algorithms: one for computing the shortest paths between all pairs of vertices in a directed graph and the other for computing the transitive closure of a directed graph. For both algorithms the adjacency matrix is a natural representation for the graph.

All-Pairs Shortest-Paths Algorithm

Suppose that we have a schedule that tells us the driving time between n cities at a given time of day and that we wish to compute the shortest driving time between all pairs of cities. This is an instance of the *all-pairs shortest-paths* problem. We could iterate through every pair of cities and compute the shortest path between each using a single-source shortest-path algorithm such as Dijkstra's algorithm.

ALGORITHM

An easier way is to use the Floyd–Warshall algorithm below. The natural representation for a graph in the Floyd–Warshall algorithm is an adjacency matrix. Assume that we are given a directed graph $G = (V, E)$ and that the vertices in V are numbered $1, 2, \ldots, n$. Further assume that we are given a matrix $C[i, j]$ that tells us the cost of edge (i, j). If there is no edge $C[i, j]$, then we assume $C[i, j]$ is set to infinity. We assume all other costs are non-negative.

Algorithm 2.1.3 Floyd–Warshall

Input: A directed graph $G = (V, E)$, where $V = \{1, 2, \ldots, n\}$;
    and a cost matrix $C[i, j]$.
Output: Cost matrix $A[1..n, 1..n]$ where $A[i, j]$ is the cost of
    the cheapest path from i to j.

procedure $FloydWarshall(G)$ {
  for $i := 1$ **to** n **do**
    for $j := 1$ **to** n **do**
      $A[i, j] := C[i, j]$;
  for $i := 1$ **to** n **do**
    $A[i, i] := 0$;
  for $k := 1$ **to** n **do**
    for $i := 1$ **to** n **do**
      for $j := 1$ **to** n **do**
        if $A[i, k] + A[k, j] < A[i, j]$ **then**
          $A[i, j] := A[i, k] + A[k, j]$;
}

The Floyd–Warshall algorithm computes a cheapest-cost array A, where $A[i, j]$ gives the cheapest cost of any path from vertex i to vertex j. For the algorithm to work correctly, it is important that there are no negative cost cycles in the graph.

FACT

F8: The Floyd–Warshall algorithm computes the cost matrix of the cheapest paths between all pairs of vertices of a directed graph $G = (V, E)$ in $O(|V|^3)$ time and $O(|V|^2)$ space.

REMARKS

R9: For additional discussion of the Floyd–Warshall algorithm and its variants see [AhHoUl74] and [CoLeRiSt09].

R10: Let $A^k[i, j]$ be the cost of the cheapest path from vertex i to vertex j that does not pass through a vertex numbered higher than k, except possibly for the endpoints. We can prove by induction on k that $A^k[i, j] = min(A^{k-1}, A^{k-1}[i, k] + A^{k-1}[k, j])$. In the next section we see that the Floyd–Warshall algorithm is a special case of Kleene's algorithm.

Transitive Closure

In some problems we may just want to know whether there exists a path from vertex i to vertex j of length one or more in a graph $G = (V, E)$. We call this the problem of computing the **transitive closure** of G. Given a directed graph $G = (V, E)$ with adjacency matrix A, we want to compute a Boolean matrix T such that $T[i, j]$ is 1 if there is a path from i to j of length 1 or more, and 0 otherwise. We call T the *transitive closure* of the adjacency matrix.

The transitive-closure algorithm below is similar to the Floyd–Warshall algorithm except that it uses the Boolean operation **and** to conclude that if there is a path from i to k and one from k to j, then there is a path from i to j.

Algorithm 2.1.4 Transitive Closure

Input: A directed graph $G = (V, E)$, with $V = \{1, 2, \ldots, n\}$
    and adjacency matrix $A[i, j]$.
Output: Boolean transitive-closure matrix $T[1..n, 1..n]$ where $T[i, j]$ is
    1 if there is a path from i to j of length 1 or more, and 0 otherwise.

procedure $TransitiveClosure(G)$ {
  for $i := 1$ **to** n **do**
    for $j := 1$ **to** n **do**
      $T[i, j] := A[i, j]$;
  for $k := 1$ **to** n **do**
    for $i := 1$ **to** n **do**
      for $j := 1$ **to** n **do**
        if $A[i, j] =$ **false then**
          $A[i, j] := A[i, k]$**or**$A[k, j]$;
}

FACT

F9: The algorithm $TransitiveClosure(G)$ computes the transitive closure of G in $O(|V|^3)$ time and $O(|V|^2)$ space.

REMARKS

R11: The transitive-closure algorithm is due to S. Warshall [Wa62].

R12: Let $T^k[i, j] = 1$ if there is a path of length one or more from vertex i to vertex j that does not pass through an intermediate vertex numbered higher than k, except for the endpoints. We can prove by induction on k that

$$C^k[i, j] = C^{k-1}[i, j] \textbf{ or } C^{k-1}[i, k] \textbf{ and } C^{k-1}[k, j]$$

where **and** and **or** are the Boolean *and* and *or* operators. In the next subsection we will see the transitive-closure algorithm is a special case of Kleene's algorithm.

2.1.4   Applications to Pattern Matching

Graphs play a major role in problems arising in the specification and translation of programming languages. A special kind of graph called a ***finite automaton*** is used in language theory to specify and recognize sets of strings called ***regular expressions***. Regular expressions are used to specify the lexical structure of many programming language constructs. They are also widely used in many string-pattern-matching applications.

This section presents an algorithm due to S. C. Kleene to construct representations called ***regular expressions*** for all paths between the vertices of a directed graph.

DEFINITIONS

D16: A ***nondeterministic finite automaton*** (NFA) is a labeled, directed graph $G = (V, E)$ in which

1. one vertex is distinguished as the *start* vertex

2. a set of vertices are distinguished as *final* vertices

3. each edge is labeled by a symbol from a set $\Sigma \cup \{\epsilon\}$ where
   Σ is a finite set of *alphabet symbols*, and
   ϵ is a special symbol denoting the empty string

D17: An NFA G ***accepts*** a string x if there is a path in G from the start vertex to a final vertex whose edge labels spell out x.

D18: The set of strings accepted by an NFA G is called the ***language*** defined by G.

D19: If R and S are sets of strings, then their ***concatenation*** $R \cdot S$ is the set of strings $\{xy | x$ is in r and y is in $S\}$.

D20: Let S be a set of strings. Define $S^0 = \{\epsilon\}$ and $S^i = S \cdot S^{i-1}$ for $i \geq 1$. The ***Kleene closure*** of S, denoted S^*, is defined to be $\cup_{i=0}^{\infty} S^i$.

D21: Let Σ be a finite set of alphabet symbols. The ***regular expressions*** over Σ and the languages they denote are defined recursively as follows:

1. ϕ is a regular expression that denotes the empty set.

2. ϵ is a regular expression that denotes $\{\epsilon\}$.

3. For each a in Σ, a is a regular expression that denotes $\{a\}$.

4. If r and s are regular expressions denoting the languages R and S, then
   $(r + s)$ is a regular expression denoting the language $R \cup S$,
   rs is a regular expression denoting $R \cdot S$, and
   (r^*) is a regular expression denoting R^*.

We can avoid writing many parentheses in a regular expression by adopting the convention that the Kleene closure operator $*$ has higher precedence than concatenation or $+$, and that concatenation has higher precedence than $+$. For example, $((a(b^*)) + c)$ may be written $ab^* + c$. This regular expression denotes the set of strings $\{ab^i | i \geq 0\} \cup \{c\}$.

Kleene's Algorithm

S. C. Kleene presented an algorithm for constructing a regular expression from a non-deterministic finite automaton. This algorithm, shown below, includes the Floyd–Warshall algorithm and the transitive-closure algorithm as special cases.

ALGORITHM

Let $G = (V, E)$ be an NFA in which the vertices are numbered $1, 2, \ldots, n$. Kleene's algorithm (Algorithm 2.1.5) works by constructing a sequence of matrices C^k in which the entry $C^k[i, j]$ is a regular expression for all paths from vertex i to vertex j with no intermediate vertex on the path (except possibly for the endpoints) that is numbered higher than k.

Algorithm 2.1.5 Kleene's Algorithm

Input: A directed graph $G = (V, E)$, where $V = \{1, 2, \ldots, n\}$,
      and a label matrix $L[i, j]$.
Output: Matrix $C[1..n, 1..n]$ where $C[i, j]$ is a regular expression describing
      all paths from i to j.

procedure $Kleene(G)$ {
  for $i := 1$ **to** n **do**
    for $j := 1$ **to** n **do**
      $C^0[i, j] := L[i, j]$;
  for $i := 1$ **to** n **do**
    $C^0[i, i] := \epsilon + C^0[i, i]$;
  for $k := 1$ **to** n **do**
    for $i := 1$ **to** n **do**
      for $j := 1$ **to** n **do**
        $C^k[i, j] := C^{k-1}[i, j] + C^{k-1}[i, k] \cdot (C^{k-1}[k, k])^* \cdot C^{k-1}[k, j]$;
  for $i := 1$ **to** n **do**
    for $j := 1$ **to** n **do**
      $C[i, j] := C^n[i, j]$;
}

REMARKS

R13: Kleene's algorithm appeared in [Kl56].

R14: To prove the correctness of Kleene's algorithm, we can prove by induction on k that $C^k[i, j]$ is the set of path labels of all paths from vertex i to vertex j with no intermediate vertex numbered higher than k, excluding the endpoints. The term $C^{k-1}[i, k]$ in the inner loop represents the labels of all paths from vertex i to vertex k that do not have an intermediate vertex numbered higher than $k-1$. The term $(C^{k-1}[k,k])^*$ represents the labels of all paths that go from vertex k to vertex k zero or more times without passing through an intermediate vertex numbered higher than $k-1$. The term $C^{k-1}[k,j]$ represents the labels of all paths from vertex k to vertex j that do not have an intermediate vertex numbered higher than $k-1$. Thus, the term $C^{k-1}[i,j] \cdot (C^{k-1}[k,k])^* \cdot C^{k-1}[k,j])$ represents the path labels of all paths with the segments: from i to k, from k to k zero or more times, and from k to j with no intermediate vertex numbered higher than $k-1$ on any of the segments.

R15: The Floyd–Warshall algorithm is a special case of Kleene's algorithm with the inner loop replaced by $C^k[i,j] := min(C^{k-1}[i,j], C^{k-1}[i,k] + C^{k-1}[k,j]$. In the Floyd–Warshall algorithm we don't need to consider paths from k to k since we assume the edge costs are non-negative. Also, in the Floyd–Warshall algorithm the operator representing concatenation ($\cdot$) is arithmetic addition.

R16: The transitive-closure algorithm is a special case of Kleene's algorithm with the inner loop replaced by $C^k[i,j] := C^{k-1}[i,j] + C^{k-1}[i,k] \cdot C^{k-1}[k,j]$ where $+$ represents Boolean **or** and $\cdot$ represents Boolean **and**.

R17: Aho, Hopcroft, and Ullman present Kleene's algorithm in the general setting of a closed semiring [AhHoUl74].

R18: One of the key results of formal language theory is that the set of languages defined by NFAs is exactly the same as the set of languages defined by regular expressions. These languages are called *regular sets*.

R19: For applications of finite automata and regular expressions to string pattern matching and compiling see [Ah90, AhSeUl86].

References

[Ah90] A. V. Aho, Algorithms for finding patterns in strings, pp. 255–300 in *Handbook of Theoretical Computer Science A, Algorithms and Complexity*, Ed. J. Van Leeuwen, MIT Press, 1990.

[AhHoUl74] A. V. Aho, J. E. Hopcroft, and J. D. Ullman, *The Design and Analysis of Computer Algorithms*, Addison-Wesley, 1974.

[AhHoUl83] A. V. Aho, J. E. Hopcroft, and J. D. Ullman, *Data Structures and Algorithms*, Addison-Wesley, 1983.

[AhSeUl86] A. V. Aho, R. Sethi, and J. D. Ullman, *Compilers: Principles, Techniques, and Tools*, Addison-Wesley, 1986.

[CoLeRiSt09] T. H. Cormen, C. E. Leiserson, R. L. Rivest, and C. Stein, *Introduction to Algorithms, Third Edition*, MIT Press, 2009.

[Ev79] S. Even, *Graph Algorithms*, Computer Science Press, 1979.

[Fl62] R. W. Floyd, Algorithm 97 (Shortest Path), *Comm. ACM* 5(6) (1962), 345.

[HoTa73] J. E. Hopcroft and R. E. Tarjan, Efficient algorithms for graph manipulation, *Comm. ACM* 16(6) (1973), 372–378.

[Jo77] D. B. Johnson, Efficient algorithms for shortest paths in sparse networks, *J. ACM* 24(1) (1977), 1–13.

[Kl56] S. C. Kleene, Representation of events in nerve nets and finite automata, pp. 3–40 in *Automata Studies*, Eds. C. E. Shannon and J. McCarthy, Princeton Univ. Press, 1985.

[Ta72] R. E. Tarjan, Depth first search and linear graph algorithms, *SIAM J. Comput.* 1(2) (1972), 146–160.

[Ta83] R. E. Tarjan, *Data Structures and Network Algorithms*, Soc. Industrial and Applied Mathematics, 1983.

[Wa62] S. Warshall, A theorem on Boolean matrices, *J. ACM* 9(1) (1962), 11–12.

Section 2.2
Graph Isomorphism

Brendan D. McKay, Australian National University

INTRODUCTION

Isomorphism between graphs and related objects is a fundamental concept in graph theory and its applications to other parts of mathematics. The problem also occupies a central position in complexity theory as a proposed occupant of the region that must exist between the polynomial-time and NP-complete problems if P$\neq$NP. Due to its many practical applications a considerable number of algorithms for graph isomorphism have been proposed.

2.2.1   Isomorphisms and Automorphisms

Informally, two graphs are isomorphic if they are the same except for the names of their vertices and edges. Formally, this relationship is defined by means of bijections between them.

Basic Terminology

DEFINITIONS

D1: Let $G_1 = (V_1, E_1)$ and $G_2 = (V_2, E_2)$ be simple graphs. An *isomorphism* from G_1 to G_2 is a bijection $\phi : V_1 \to V_2$ such that $vw \in E_1$ if and only if $\phi(v)\phi(w) \in E_2$.

D2: A second way to define an isomorphism is that there are two bijections $\phi : V_1 \to V_2$ and $\phi' : E_1 \to E_2$, such that the incidence relation between vertices and edges is preserved. That is, $v \in V_1$ is incident to $e \in E_1$ if and only if $\phi(v)$ is incident to $\phi(e)$. This method is preferred if edges have additional attributes that should be preserved by the mapping. However, we will use the previous definition where it applies.

D3: An isomorphism from a graph to itself is called an *automorphism* or *symmetry*.

D4: The set of automorphisms of a graph G form a group under the operation of composition, called the ***automorphism group*** $\mathrm{Aut}(G)$. The automorphism group of a simple graph is a subgroup of the symmetric group acting on the vertex set of the graph.

EXAMPLE

E1: Figure 2.2.1 shows an isomorphism between two graphs and gives the automorphism group of the first graph.

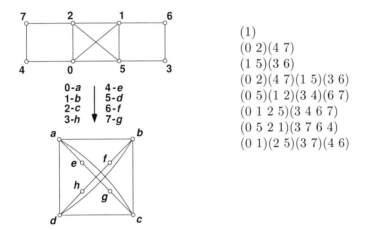

(1)
(0 2)(4 7)
(1 5)(3 6)
(0 2)(4 7)(1 5)(3 6)
(0 5)(1 2)(3 4)(6 7)
(0 1 2 5)(3 4 6 7)
(0 5 2 1)(3 7 6 4)
(0 1)(2 5)(3 7)(4 6)

Figure 2.2.1: An isomorphism between two graphs and the automorphism group of the first graph.

DEFINITION

D5: Closely related to isomorphism is the concept of canonical labeling. Arbitrarily choose one member of each isomorphism class of graphs, and call it the ***canonical form*** of that isomorphism class. Replacing a graph by the canonical form of its isomorphism class is called ***canonical labeling*** or ***canonizing the graph***. Two graphs are isomorphic if and only if their canonical forms are identical, as shown in Figure 2.2.2.

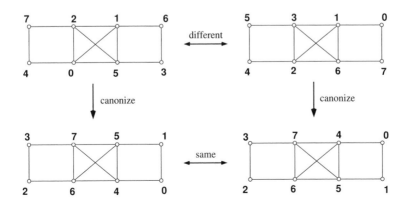

Figure 2.2.2: Isomorphism between graphs becomes equality when they are canonized.

REMARKS

R1: The labeled graph which gives the lexicographically greatest adjacency matrix is an example of an explicitly defined canonical form. In practice more complex definitions are used to assist efficient computation.

R2: Canonical labeling has central importance to practical applications. One task is to determine whether a graph is isomorphic to any graph in a database of graphs. This is best achieved by storing the canonical forms of graphs in the database and comparing them to the canonical form of the new graph. Another task is to remove isomorphs from a large collection of graphs. This is best achieved by applying a sorting algorithm to the canonical forms of the graphs. Both tasks are very expensive if only pair-wise isomorphism testing is available.

Related Isomorphism Problems

Many types of isomorphism problem can be modeled as isomorphism between simple graphs or digraphs.

FACTS

F1: *Vertex colors* that must be preserved by isomorphisms can be modeled by attaching gadgets to the vertices, a different gadget for each color. However, this is such an important generalization that most software can handle vertex colors directly.

F2: *Edge colors* can be modeled using layers, once vertex colors are available. Figure 2.2.3 illustrates one approach. The edge colors are assigned numbers according to the table in the center. The vertices of the original graph are assigned to vertical paths, with the first layer identified by vertex color. Then the original edges with each color c are represented by horizontal edges within those layers where the binary expansion of c has ones.

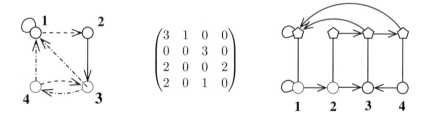

Figure 2.2.3: Modeling of graphs with colored edges.

F3: *Hypergraphs* and other types of incidence structures like *block designs* and *finite geometries* can be represented by bipartite graphs. One color class consists of the vertices of the hypergraph, while the other has a vertex for each of the hyperedges of the hypergraph. An edge of the bipartite graph represents a vertex of the hypergraph and a hyperedge it lies in.

F4: Other types of isomorphism easily modeled by graph isomorphism include equivalence of matrices defined by permutation of rows and columns, Hadamard equivalence, and isotopy (such as for Latin squares) [McPi12b].

2.2.2 Complexity Theory

The problem of determining whether two graphs are isomorphic, called ***GI*** or ***ISO***, has received a great deal of interest from theorists due to its unsolved nature.

FACTS

F5: GI is not known to have a polynomial-time algorithm, nor to be NP-complete. While obviously in NP, its presence in co-NP is also undecided. Indeed, it is considered a prime candidate for the intermediate territory between P and NPC that must exist if P$\neq$NP. One reason for this is that the NP-completeness of GI would imply the collapse of the polynomial-time hierarchy [GoMiWi91].

F6: The fastest proven running time for GI has stood for three decades at $e^{O(\sqrt{n \log n})}$ [BaKaLu83].

F7: On the other hand, many special classes of graph are known to have polynomial-time isomorphism tests. The most general such classes are those defined by a forbidden minor [Po88, Gr10] or by a forbidden topological minor [Gr12]. These classes include many earlier classes, including graphs of bounded degree [Lu82], bounded genus [FiMa80, Mi80], and bounded tree-width [Bo90]. However, very few of these polynomial algorithms are practical.

DEFINITION

D6: A decision problem is called ***isomorphism-complete*** if it is polynomial-time equivalent to GI.

FACTS

F8: All of the isomorphism problems noted in the previous subsection are isomorphism-complete. Many other examples are known, including isomorphism of semigroups and finite automata [Bo78], homeomorphism of 2-complexes [STPi94], and polytope isomorphism [KaSc03].

F9: Isomorphism of linear codes (vector spaces over finite fields) defined by permutation of the coordinate positions, where the codes are presented as generator matrices, is at least as hard as graph isomorphism but might be harder [PeRo97].

F10: Isomorphism of groups given as multiplication tables is at least as easy as GI but might be easier [Bo78]. The best algorithm is very elementary and takes time $n^{O(\log n)}$ [Bo78, Mi78]).

F11: Some problems similar to GI are NP-complete. The best known is the subgraph isomorphism problem: given two graphs, is the first isomorphic to a subgraph of the second? Another is the presence of an automorphism without fixed points, or of such an automorphism of order 2 [Lu81].

DEFINITION

D7: A ***graph invariant*** is a property of graphs that is equal for isomorphic graphs. A ***complete graph invariant***, also called a ***certificate***, is an invariant that always distinguishes between non-isomorphic graphs.

FACTS

F12: Examples of invariants include the degree sequence and the eigenvalue set of the adjacency matrix. However, neither of those invariants is complete. Nevertheless, even incomplete invariants can sometimes be used as a short proof of non-isomorphism.

F13: An example of a complete invariant is a canonical form. However, it is not known if there is a complete invariant computable in polynomial time. In fact, it is not even known if there is a complete invariant checkable in polynomial time (which would place GI in co-NP).

2.2.3   Algorithms

The development of computer programs for graph isomorphism has been such a popular pursuit that already in 1976 it was called a "disease" [ReCo77]. Literally hundreds of algorithms have been published (many wrong).

FACTS

F14: The earliest software appeared in the 1960s. The approach which has been the most successful is the "individualization-refinement" paradigm introduced by Parris and Read [PaRe69] and further developed by Corneil and Gotlieb [CoGo70] and Arlazarov et al. [ArZuUsFa74]. This genre is now represented by the author's **nauty** and other software mentioned below.

REMARK

R3: We will focus our attention on canonical labeling, which is the method used by the most useful modern algorithms.

DEFINITIONS

D8: A key routine is that of ***partition refinement***, which is any process of making a partition finer (i.e., breaking its cells into smaller cells) by detecting combinatorial differences between the vertices. For isomorphism purposes, only properties independent of the numbering of the vertices may be used. This implies that vertices equivalent under the action of an automorphism fixing the input partition cannot be separated.

D9: An ***equitable partition*** is a partition of the vertices of a graph into cells such that, for any two vertices v, w in the same cell, and any cell C, we have that v and w are adjacent to the same number of vertices in C.

EXAMPLE

E2: Figure 2.2.4 shows a graph with an equitable partition of two cells. Each black vertex is adjacent to no black and two white vertices, while each white vertex is adjacent to one white and two black vertices. As this example shows, vertices in the same cell of an equitable partition do not need to be equivalent under the automorphism group of the graph.

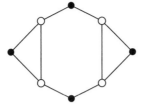

Figure 2.2.4: An equitable partition with two cells.

FACTS

F15: The most well-known partition refinement method splits cells until the partition becomes equitable. Given two cells C_1, C_2, the vertices in C_1 are separated into subcells according to their number of neighbors in C_2. This is repeated for different pairs of cells until no more splitting is possible. Algorithms differ according to the method used to choose the pairs. The fastest algorithm is in [Mc80].

F16: A generalization of this type of refinement, called the *k-th order Weissfeiler–Lehman refinement*, uses partitions of the set of k-tuples of vertices rather than just a partition of the vertices. For some classes of graphs, it is known that there is a fixed k for which this refinement provides the automorphism partition of the graph, which can be used to build a polynomial-time isomorphism algorithm. However, Cai, Füredi, and Immerman showed that no such k is sufficient for all graphs [CaFuIm92].

Search Tree

A partition refinement method is used to define a search tree that is used to find a canonical labeling and the automorphism group.

DEFINITION

D10: Consider a graph and an initial partition (perhaps trivial). Define a *search tree* whose nodes are refined partitions. The root of the tree is the refined initial partition. Any node which is a discrete partition (one with only singleton cells) is a leaf of the tree. Consider any node ν which is not discrete. Choose a non-singleton cell C. Then ν has one child for each $v \in C$, obtained by splitting C into two cells $\{v\}$ and $C-\{v\}$, then refining.

FACTS

F17: The leaves of the search tree are discrete partitions, and thus, lists of the vertices in a definite order. The orders define a set of numberings of the vertices of the graph.The maximum labeled graph, according to lexicographic or other convenient ordering, is a canonical form. Moreover, two numberings that define the same labeled graph yield an automorphism, and all automorphisms can be found in that way.

F18: In practice the search tree may be much too large, so various means are employed to reduce it. One way is to employ automorphisms as they are discovered to prune branches of the tree thus shown to be equivalent to other branches. Another way is to employ invariants computed at the nodes of the tree to perform a type of branch-and-bound. A third method is to use a more powerful refinement procedure.

Software

The first program that could process structurally highly regular graphs and graphs with hundreds of vertices was that of the author, which became known as `nauty` [Mc78, Mc80] and dominated the field from the 1970s until recent years. Now there are several strong competitors.

EXAMPLES

E3: `nauty`, by this author, can find automorphisms groups and canonical forms, of graphs and digraphs. It comes in two forms, with either dense or sparse data structures [McPi12b, McPi13].

E4: `saucy`, by Darga, Liffiton, Sakallah and Markov, computes automorphism groups and is especially efficient for large sparse graphs having many automorphisms that move few vertices [DaLiSaMa04, DaSaMa04].

E5: `Bliss`, by Junttila and Kaski, can also perform canonical labeling and has very dependable performance for highly regular graphs [JuKa07, JuKa11].

E6: `Traces`, by Piperno, introduced an entirely new way of scanning the search tree, using a combination of breadth-first and depth-first search [Pi08, McPi13]. At the time of writing, `Traces` is the most efficient program for processing many classes of very difficult graphs, as well as being highly competitive for easy graphs [McPi13]. Since January 2013, `Traces` has been distributed with `nauty` [McPi12].

E7: Other worthy programs are `conauto` by López-Presa, Fernándes Anta, and Núñez Chiroque [LPFe09, JPFe11], `VSEP` by Stoichev, and `VF` by Cordella, Foggia, Sansone, and Vento.

E8: Packages which contain high-quality graph isomorphism facilities (usually via `nauty`) include Magma, GRAPE, LINK, Sage-combinat, and Macaulay2.

REMARKS

R4: An experimental comparison of `nauty`, `Traces`, `saucy`, `Bliss`, and `conauto` can be found in [McPi13].

R5: All the named programs have exponential running time in the worst case. However, the worst-case graphs are rather difficult to find and most users will see only scaling according to a polynomial of low degree. The state of the art is that the easiest graphs can be handled if they fit into main memory (tens of millions of vertices). The most difficult graphs cause difficulty in the hundreds or thousands of vertices.

References

[ArZuUsFa74] V. L. Arlazarov, I. I. Zuev, A. V. Uskov, and I. A. Faradzev, An algorithm for the reduction of finite non-oriented graphs to canonical form. *Zh. vȳchisl. Mat. mat. Fiz.* 14 (1974) 737–743.

[BaKaLu83] L. Babai, W. M. Kantor, and E. M. Luks, Computational complexity and the classification of finite simple groups. In: Proceedings of the 24th Annual Symposium on the Foundations of Computer Science (1983) 162–171.

[Bo90] H. Bodlaender, Polynomial algorithms for graph isomorphism and chromatic index on partial k-trees. *J. Algorithms* 11 (1990) 631–643.

[Bo78] K. S. Booth, Isomorphism testing for graphs, semigroups, and finite automata are polynomially equivalent problems. *SIAM J. Comput.* 7 (1978) 273–279.

[CaFuIm92] Jin-yi Cai, Martin Fürer, and Neil Immerman, An optimal lower bound on the number of variables for graph identifications. *Combinatorica* 12 (1992) 389–410.

[CoGo70] D. G. Corneil and C. C. Gotlieb, An efficient algorithm for graph isomorphism. *JACM* 17 (1970) 51–64.

[DaLiSaMa04] P. T. Darga, M. H. Liffiton, K. A. Sakallah, and I. L. Markov, Exploiting structure in symmetry detection for CNF. In: Proceedings of the 41st Design Automation Conference (2004), 530–534.

[DaSaMa04] P. T. Darga, K. A. Sakallah, and I. L. Markov, Faster Symmetry Discovery using Sparsity of Symmetries. In: Proceedings of the 45th Design Automation Conference (2004), 149–154.

[FiMa80] I. S. Filotti and J. N. Mayer, A polynomial-time algorithm for determining the isomorphism of graphs of fixed genus. In: Proceedings of the 12th ACM Symposium on Theory of Computing (1980), 236–243.

[GoMiWi91] O. Goldreich, S. Micali, and A. Wigderson, Proofs that yield nothing but their validity, or all languages in NP have zero-knowledge proof systems. *JACM* 38 (1991) 690–728.

[Gr10] M. Grohe, Fixed-point definability and polynomial time on graphs with excluded minors. In: Proceedings of the 25th Annual IEEE Symposium on Logic in Computer Science (2010), 179–188.

[Gr12] M. Grohe, Structural and Logical Approaches to the Graph Isomorphism Problem, In: Proceedings of the 23rd Annual ACM-SIAM Symposium on Discrete Algorithms (2012), 188.

[JuKa07] T. Junttila and P. Kaski, Engineering an efficient canonical labeling tool for large and sparse graphs. In: Proceedings of the 9th Workshop on Algorithm Engineering and Experiments and the 4th Workshop on Analytic Algorithms and Combinatorics (2007), 135–149.

[JuKa11] T. Junttila and P. Kaski, Conflict Propagation and Component Recursion for Canonical Labeling. In: Proceedings of the 1st International ICST Conference on Theory and Practice of Algorithms (2011), 151–162.

[KaSc03] V. Kaibel and A. Schwartz, On the complexity of polytope isomorphism problems. *Graphs and Combinatorics* 19 (2003) 215–230.

[LPFe09] J. L. López-Presa and A. Fernández Anta, Fast algorithm for graph isomorphism testing. In: Proceedings of the 8th International Symposium on Experimental Algorithms (2009), 221–232.

[JPFe11] J. L. López-Presa, A. Fernández Anta, and L. Núñez Chiroque, Conauto-2.0: Fast isomorphism testing and automorphism group computation. Preprint 2011. Available at http://arxiv.org/abs/1108.1060.

[Lu81] A. Lubiw, Some NP-complete problems related to graph isomorphism. *SIAM J. Comput.* 10 (1981) 11–21.

[Lu82] E. Luks, Isomorphism of graphs of bounded valence can be tested in polynomial time. *J. Comp. System Sci.* 25 (1982) 42–65.

[Mc78] B. D. McKay, Computing automorphisms and canonical labelings of graphs. In: Combinatorial Mathematics, Lecture Notes in Mathematics, 686. Springer-Verlag, Berlin (1978), 223–232.

[Mc80] B. D. McKay, Practical graph isomorphism. *Congr. Numer.* 30 (1980) 45–87.

[McPi12] B. D. McKay and A. Piperno, nauty Traces, Software distribution web page. http://cs.anu.edu.au/~bdm/nauty/ and http://pallini.di.uniroma1.it/.

[McPi12b] B. D. McKay and A. Piperno, nauty and Traces User's Guide (Version 2.5). available at [McPi12].

[McPi13] B. D. McKay and A. Piperno, Practical graph isomorphism II, arXiv:1301.1493..

[Mi78] G. L. Miller, On the $n^{\log n}$ isomorphism technique. In: Proceedings of the 10th ACM Symposium on Theory of Computing (1978) 51–58.

[Mi80] G. L. Miller, Isomorphism testing for graphs of bounded genus. In: Proceedings of the 12th ACM Symposium on Theory of Computing (1980), 225–235.

[PaRe69] R. Parris and R. C. Read, A coding procedure for graphs. Scientific Report. UWI/CC 10. University of West Indies Computer Centre, 1969.

[PeRo97] E. Petrank and R. M. Roth, Is code equivalence easy to decide? *IEEE Trans. Inform. Th.* 43 (1997) 1602–1604.

[Pi08] A. Piperno, Search space contraction in canonical labeling of graphs. Preprint 2008–2011. Available at http://arxiv.org/abs/0804.4881.

[Po88] I. N. Ponomarenko, The isomorphism problem for classes of graphs that are invariant with respect to contraction (Russian). *Zap. Nauchn. Sem. Leningrad. Otdel. Mat. Inst. Steklov. (LOMI)* 174 (1988) no. Teor. Slozhn. Vychisl. 3, 147–177.

[ReCo77] R. C. Read and D. G. Corneil, The graph isomorphism disease. *J. Graph Theory* 1 (1977) 339–363.

[STPi94] J. Shawe-Taylor and T. Pisanski, Homeomorphism of 2-complexes is graph isomorphism complete. *SIAM J. Comput.* 23 (1994) 120–132.

Section 2.3
The Reconstruction Problem

Josef Lauri, University of Malta, Malta

INTRODUCTION

In the first volume of the *Journal of Graph Theory* published in 1977, the journal editors wrote, "The foremost currently unsolved problem in Graph Theory is, in our considered opinion, the Reconstruction Conjecture." One might agree or disagree with this assessment, but surely one must admit that this is an inviting problem, made all the more tantalizing by the fact that, although thirty-five years have passed since the editors of the *Journal of Graph Theory* expressed their views, and several researchers have made efforts to attain, at least, some partial reconstruction results, we are still nowhere near any complete solution of this problem. In the first section below we shall give the two main variants of the Reconstruction Problem, and in the subsequent sections we shall expand on these definitions hoping, this way, to provide a panoramic view of the present state of knowledge on the Reconstruction Problem.

In this section all graphs are assumed to be simple.

2.3.1   Two Reconstruction Conjectures

Some classical problems in mathematics are of the following type. If the structure S' is associated with the given structure S, does S' determine S uniquely? In graph theory we ask what knowledge, short of its full incidence relations, is sufficient to determine the graph completely. The structure S would be the graph and S' could be its line-graph, or chromatic polynomial, or spectrum, say. The best known problem of this type in graph theory is the Reconstruction Problem.

Decks and Edge-Decks

We first need to introduce some definitions and notations connected with families of vertex-deleted and edge-deleted subgraphs of a graph.

DEFINITIONS

D1: Let G be a graph (assumed here to have no loops or multiple edges) on n vertices. For any vertex v of G, let $G - v$ denote the **vertex-deleted subgraph** of G that is obtained from G by removing v and all edges incident to v. Similarly, let e be an edge of G. The **edge-deleted subgraph** of G, denoted by $G - e$, is obtained from G by deleting the edge e.

D2: If A is a subset of vertices of G then $G - A$ will denote the graph obtained from G by deleting all vertices in A and any edge incident to at least one of them. If B is a subset of edges of G then $G - B$ will denote that subgraph of G obtained by deleting all the edges in B.

D3: The collection of all vertex-deleted subgraphs of G is called the **deck** of G, while the collection of all edge-deleted subgraphs of G is called the **edge-deck** of G.
NOTATION: The deck of G is denoted by $\mathcal{D}(G)$ and the edge-deck of G is denoted by $\mathcal{ED}(G)$.

Note that the graphs in the deck are unlabeled and if G contains isomorphic vertex-deleted subgraphs, then such subgraphs are repeated in $\mathcal{D}G$ according to the number of isomorphic subgraphs which G contains. The same holds for the edge-deck. Therefore the deck and the edge-deck are multi-sets, rather than sets, of isomorphism types of graphs.

EXAMPLE

E1: Figure 2.3.1 shows a graph and its deck.

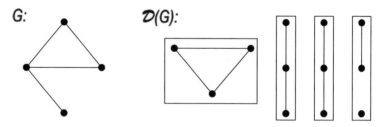

Figure 2.3.1: A graph and its deck.

Reconstructibility

Now suppose that H is another graph with $\mathcal{D}(H) = \mathcal{D}(G)$. The question we are interested in is: must H be isomorphic to G?

DEFINITION

D4: Any graph H with the same deck as G is called a **reconstruction** of G. If every reconstruction of G is isomorphic to G, then G is said to be **reconstructible**.

EXAMPLE

E2: A graph that is not reconstructible is K_2, the graph consisting of just one edge, because if H is the graph consisting of two isolated vertices, then clearly H is a reconstruction of K_2, but not isomorphic to K_2. The Reconstruction Conjecture asserts that these are the only non-reconstructible graphs.

The Reconstruction Conjecture

C1: [Ke57, Ul60] *Every graph with at least three vertices is reconstructible.*

Another way of looking at reconstruction is by saying that a graph G is reconstructible if it can be uniquely (up to isomorphism) determined from $\mathcal{D}(G)$. Note that the problem stated in this way is not about finding an efficient algorithm for reconstructing G from $\mathcal{D}(G)$. In principle, given the deck, one can consider all graphs on n vertices to check which of them have the given deck. The question remains one of uniqueness, that is, whether this search will find only one graph with the given deck. The deck is a collection of isomorphism types with appropriate multiplicities, and the question is whether the isomorphism type of G can be determined uniquely from this collection.

DEFINITIONS

Closely related to the Reconstruction Problem is the *Edge-Reconstruction Problem*.

D5: An *edge-reconstruction* of G is a graph which has the same edge-deck as G, and G is said to be *edge-reconstructible* if every edge-reconstruction of G is isomorphic to it.

The Edge-Reconstruction Conjecture

C2: [Ha64] *Every graph with at least four edges is edge-reconstructible.*

EXAMPLE

E3: The graph $G = K_3 \cup kK_1$ is not edge-reconstructible because, if H is the graph $K_{1,3} \cup (k-1)K_1$, then H is an edge-reconstruction of G that is not isomorphic to it. Also, $G = 2K_2$ is not edge-reconstructible because if $H = P_3 \cup K_1$ (where P_3 is the path on three vertices), then $\mathcal{ED}(G) = \mathcal{ED}(H)$ but $G \not\cong H$.

The Edge-Reconstruction Conjecture asserts that these are the only graphs that are not edge-reconstructible.

REMARK

R1: Two important surveys on the Reconstruction Problem were published in 1977 and 1978 [BoHe77, Na78]. These give the state of knowledge on this problem till that date, together with a complete bibliographic list. Since then, a number of survey or expository articles have been published [El88, Ma88, La87, Bo91]. The two early surveys and the most recent one should be consulted by anyone who intends to do serious work in this field. The book [LaSc03] also contains four chapters on the reconstruction problem.

Comparing Reconstruction with Edge-Reconstruction

Intuition seems to suggest that it is easier to reconstruct a graph from its edge-deck than from its deck; there are generally more graphs in the edge-deck, and edge-deleted subgraphs are generally more nearly like the original graph than vertex-deleted subgraphs. This intuitive notion is borne out by the following theorem of Greenwell which, essentially, says that if the Reconstruction Conjecture is true then so is the Edge-Reconstruction Conjecture.

FACTS

F1: [Gr71] [**Greenwell's Theorem**] Let G be a graph without isolated vertices. The deck of G is edge-reconstructible, that is, $\mathcal{D}(G)$ is uniquely determined from $\mathcal{ED}(G)$. Therefore, if G is reconstructible, then it is also edge-reconstructible.

Therefore, for graphs without isolated vertices, if the Reconstruction Conjecture is true, then so is the Edge-Reconstruction Conjecture. (We shall henceforth tacitly assume, unless otherwise stated, that any graph to be reconstructed has no isolated vertices.) So, if we can prove a result for vertex reconstruction, say that a certain class of graphs is reconstructible, then this result would automatically hold for edge-reconstruction.

Another result which shows that the problem of edge-reconstruction is a special case of reconstruction is the following theorem of Hemminger.

F2: [He69] A graph is edge-reconstructible if and only if its line graph is reconstructible and is not K_3.

However, the Edge-Reconstruction Problem holds its own considerable independent interest because several results are known in edge-reconstruction which have not yet been proved for vertex-reconstruction, and some elegant proof techniques have been developed for edge-reconstruction. We shall consider these techniques in some detail in a later section.

Reconstruction and Graph Symmetries

It is well to emphasize here that at the heart of the difficulty of reconstructing a graph G from its deck is the symmetry of G and of the subgraphs in its deck. As an illustration, let us consider an extreme case. Suppose that the vertices of G are labeled $1, \ldots, n$ and that these labels are preserved on every vertex-deleted subgraph. Then, clearly, the graph G can be uniquely reconstructed by considering any three subgraphs in its deck and 'superimposing' them accordingly. This happens because the labelings have essentially removed all symmetries of G and its subgraphs, and therefore all ambiguities of how these subgraphs are embedded inside G. A later section of this chapter will bring out more clearly the role of the automorphism of G in the edge-reconstruction of G.

But the situation just described does, in fact, occur quite often, and considering how this happens should help one understand where the difficulty of reconstruction lies.

DEFINITION

D6: Let k be a fixed but arbitrary integer. Then the graph G is said to have **property** A_k if, whenever A and B are distinct k-sets of vertices of G, the graphs $G - A$ and $G - B$ are not isomorphic. In other words, if G has n vertices, then any two subgraphs of G induced by different sets of $n - k$ vertices are not isomorphic.

FACTS

F3: If G has property A_{k+1} then it has property A_k and if it has property A_1 then its automorphism group is trivial.

F4: [Ko71, Mü76, Bo90] For a fixed k almost every graph has property A_k, meaning that the proportion of labeled graphs on n vertices which have this property tends to 1 as n goes to ∞.

F5: [My88, Bo90] If a graph G has property A_3 then it can be reconstructed uniquely from any three subgraphs in its deck.

REMARK

R2: Considering why the last fact holds helps to gain insight into the Reconstruction Problem. Let $G - u, G - v, G - w$ be any three subgraphs from the deck of G. Since G also has property A_2, we can identify v in $G - u$ and u in $G - v$ as the only vertices which give $G - u - v \simeq G - v - u$. Also, by property A_3, there is a unique isomorphism from $G - u - v$ to $G - v - u$. This isomorphism labels the two graphs uniquely, and we have the situation of a labeled graph which we described above. By comparing the two graphs $G - u, G - v$ we can then clearly put the vertex u back in $G - u$ and join it to its neighbors in G. The only uncertainty is whether or not u is adjacent to v. But this can be resolved by repeating the above with $G - u$ and $G - w$ instead of $G - v$.

2.3.2 Some Reconstructible Parameters and Classes

Faced with the olympian task of settling either way the Reconstruction Conjectures, most mathematicians have settled on attacking the conjectures partially, by showing that certain graph parameters or graph classes are reconstructible. We shall here present the most important parameters or classes which have been shown to be reconstructible, focusing mainly on results which have been obtained since the surveys mentioned in Section 1 and on those classes and parameters with which we shall not be dealing in the next three sections.

Reconstructible Parameters

DEFINITION

D7: A graph parameter $\mathcal{P}$ is said to be reconstructible (or edge-reconstructible) if, for any graph G with the value p for the parameter, any reconstruction (or edge-reconstruction) of G also has value p for $\mathcal{P}$. Equivalently, $\mathcal{P}$ is reconstructible from $\mathcal{D}(G)$ (or $\mathcal{ED}(G)$) if it is uniquely determined by the deck (or edge-deck).

FACTS

Recall that, by Greenwell's Theorem, for graphs without isolated vertices, a parameter is edge-reconstructible if it is reconstructible. For some of the following facts it is easy to show that the respective parameter is reconstructible.

F6: The number of vertices and edges are both reconstructible and edge-reconstructible. (See [LaSc03].)

F7:   The degree sequence is reconstructible and edge-reconstructible. (See [Bo91, LaSc03].)

F8:   Given a $G - v$ from the deck of G, the degree in G of the missing vertex and the degrees in G of the neighbors of v are reconstructible. (See [LaSc03].)

F9:   Given a $G - e$ in the edge-deck of G, the degrees in G of the vertices with which the missing edge e is incident is edge-reconstructible.

NOTATION

Let H and G be two graphs. Then $\binom{G}{H}$ denotes the number of subgraphs of G isomorphic to H.

FACTS

Perhaps the single most useful result in reconstruction has proven to be this very simple lemma referred to as *Kelly's Lemma*.

F10:   [Ke57] Let G and H be graphs with G having more vertices than H. Then $\binom{G}{H}$ is reconstructible from $\mathcal{D}(G)$. Similarly, if G has at least as many vertices as H and strictly more edges than H, then $\binom{G}{H}$ is reconstructible from $\mathcal{ED}(G)$.

Reconstructible Classes

Recall that graphs which do not have isolated vertices and are reconstructible are, by Greenwell's Theorem, also edge-reconstructible.

Also, one must keep in mind that, when we say that the class of graphs $\mathcal{C}$ is reconstructible, one is only given the deck or the edge-deck, and not the information that the graph to be reconstructed is in $\mathcal{C}$. Reconstruction here usually proceeds in two steps, first determining from the deck that the graph is in $\mathcal{C}$, then using this extra piece of information to prove that G is reconstructible. The following is a more exact definition of these two stages.

DEFINITIONS

D8:   A class $\mathcal{C}$ of graphs is said to be ***recognizable*** or ***edge-recognizable*** if, for any graph $G \in \mathcal{C}$, any reconstruction, or edge-reconstruction, of G is also in $\mathcal{C}$. Equivalently, $\mathcal{C}$ is recognizable or edge-recognizable if it can be determined from $\mathcal{D}(G)$ or $\mathcal{ED}(G)$ whether or not G is in $\mathcal{C}$.

D9:   A graph $G \in \mathcal{C}$ is said to be ***weakly reconstructible*** or ***weakly edge-reconstructible*** if any reconstruction, or edge-reconstruction, of G which is also in $\mathcal{C}$ is isomorphic to G. Equivalently, G is weakly reconstructible or weakly edge-reconstructible if it can be determined uniquely from the deck, or edge-deck, with the extra information that G is in $\mathcal{C}$.

This two-step process was essential in practically all proofs of reconstructibility of the following classes.

FACTS

F11:   Regular graphs are reconstructible. (See [LaSc03].)

F12: Disconnected graphs are reconstructible. (See [LaSc03].)

F13: [Ke57] Trees are reconstructible.

F14: [Bo69b] Separable graphs (that is, graphs with connectivity 1) without vertices of degree 1 are reconstructible.

F15: [Zh88] The reconstruction conjecture is true if all 2-connected graphs are reconstructible.

F16: [Mc77] A computer search has shown that all graphs on nine or fewer vertices are reconstructible.

F17: [FiMa78, FiLa81, La81] Maximal planar graphs are reconstructible.

F18: [Gi76] Outerplanar graphs are reconstructible.

F19: [GoMc81] If all but at most one eigenvalue of G is simple and the corresponding eigenvectors are not orthogonal to the all-1's vector, then G is reconstructible. In particular, if G and its complement share no eigenvalue, then G is reconstructible.

F20: [Yu82] If there exists a subgraph $G - v$ of G none of whose eigenvectors is orthogonal to the all-1's vector, then G is reconstructible.

F21: [Fa94] Planar graphs with minimum degree at least 3 are edge-reconstructible.

F22: [Zh98a, Zh98b] Any graph of minimum degree 4 that triangulates a surface is edge-reconstructible. Any graph that triangulates a surface of characteristic at least 0 is edge-reconstructible. A graph G that triangulates a surface Σ of characteristic $\chi(\Sigma)$ is edge-reconstructible if $|V(G)| \geq -43\chi(\Sigma)$.

F23: [FaWuWa01] Series parallel networks (that is, 2-connected graphs without a subdivision of K_4) are edge-reconstructible.

F24: [Ch71] If a graph has property A_2 then it is reconstructible.

F25: [ElPyXi88] Claw-free graphs are edge-reconstructible.

F26: [MyElHo87] Bidegreed graphs are edge-reconstructible.

REMARKS

R3: A *claw-free graph* is one which has no induced subgraph isomorphic to $K_{1,3}$. This result made essential use of Nash–Williams' Lemma, which we shall discuss below.

R4: A *bidegreed graph* is a graph whose vertices can have only one of two possible degrees (the degrees have to be consecutive numbers, otherwise edge-reconstruction is trivial). The next step after this result would be the edge-reconstruction of tridegreed graphs (again, if the three degrees are not consecutive, then edge-reconstruction is easy). However, even the most elementary instance of this case, that is, degrees equal to 1, 2, and 3, seems to be extremely difficult to tackle [Sc85].

R5: As a next step, after the above series of results, attempting to reconstruct bipartite graphs seems worthwhile. However, no progress has been achieved to date either in reconstructing or edge-reconstructing bipartite graphs, in spite of the fact that, in Bondy and Hemminger's survey, this was even then suggested as a worthwhile problem.

R6: Several of the proofs of the above results involved long arguments very specific to the class of graphs under consideration, although in some cases common techniques began to emerge. But if these results are viewed as a step-wise attempt at solving the Reconstruction Problem for all graphs, then these results would simply be nibbles at a big mountain. What makes these results interesting, really, is the fact that use is made of the properties of such classes, and often new properties have to be unearthed. In Sections 4 and 5 we shall see sets of results which are more general and less specific to particular classes of graphs.

2.3.3   Reconstructing from Less than the Full Deck

The proofs of most of the reconstructibility results given above use much less information than is given by the full deck or edge-deck. This situation is best epitomized by the reconstruction of trees. Trees have been shown to be reconstructible by deleting only their endvertices (vertices of degree 1) [HaPa66], or only their peripheral vertices (vertices at maximum distance from the center of the tree) [Bo69a]. This immediately suggests, for graphs with many endvertices, reconstructibility from only the endvertex-deleted subgraphs.

Endvertex-Reconstruction

DEFINITION

D10: The ***endvertex-deck*** of a graph G is the collection of graphs $G - v$ for all vertices v with degree 1 in G. A graph G is ***endvertex-reconstructible*** if it is uniquely determined by its endvertex-deck.

FACTS

F27: [HaPa66] Trees are endvertex-reconstructible.

One natural question which arises is therefore whether a graph with a sufficiently large number of endvertices is necessarily endvertex-reconstructible. A negative result of Bryant, however, puts paid to any such hopes.

F28: [Br71] For any integer k there is a graph with k endvertices which is not endvertex-reconstructible.

However, this is not the end of the story for endvertex-reconstructibility. Toward the end of Section 5 we shall present a result which indicates that it is the proportion of endvertices in a graph that determines its endvertex-reconstructibility.

Reconstruction Numbers

Again noting that not all graphs in the deck are usually needed for reconstruction, Harary and Plantholt [HaPl85] introduced the definition of reconstruction numbers.

DEFINITIONS

D11: The *reconstruction number* of a graph G, denoted by $\operatorname{rn}(G)$, is the least number of subgraphs in the deck of G which guarantees that G is uniquely determined. The *edge-reconstruction number*, denoted by $\operatorname{ern}(G)$, is analogously defined.

D12: Let $\mathcal{C}$ be a class of graphs. The *class reconstruction number* of a graph G in $\mathcal{C}$, denoted by $\mathcal{C}\operatorname{rn}(G)$, is the minimum number of subgraphs in the deck of G which, together with the information that G is in $\mathcal{C}$, guarantees that G is uniquely determined. The *class edge-reconstruction number*, denoted by $\mathcal{C}\operatorname{ern}(G)$, is analogously defined.

Since almost every graph has property A_3, when discussing above the relationship between reconstruction and symmetries we have already met the first of the following results that answers in the positive a question raised by Harary and Plantholt in their paper.

FACTS

F29: [My88, Bo90] Almost every graph has reconstruction number equal to 3.

The next two results imply that there is no disconnected graph with c vertices in each component and reconstruction number equal to $c + 1$. They also raise the natural question of investigating the gap between 3 and $c + 1$ for the reconstruction number of disconnected graphs. Thus, let G be a disconnected graph consisting of a number of copies of H with $|V(H)| = c$. Determine the number $g = g(c)$ such that if $\operatorname{rn}(G) \geq g$ then $H = K_c$ but there is a G, with $H \neq K_c$, such that $\operatorname{rn}(G) = g - 1$. The last result above shows that $g \leq c$. Also, is there a constant g_0 such that if G is a disconnected graph with $\operatorname{rn}(G) \geq g_0$ then G must be a union of complete graphs?

F30: [My90] A disconnected graph with components not all isomorphic has reconstruction number 3. If all components are isomorphic and have c vertices each, then the reconstruction number can be equal to $c + 2$.

F31: [AsLa02] If the reconstruction number of a disconnected graph is at least $c + 1$ then G must consist of copies of K_c.

F32: [BaBaHo87] If $\mathcal{C}$ is the class of total graphs and G is in $\mathcal{C}$, then $\mathcal{C}\operatorname{rn}(G)$ equals 1.

F33: [My90] The reconstruction number of trees is 3.

Harary and Lauri [HaLa88] have conjectured that if $\mathcal{C}$ is the class of trees and T is a tree then $\mathcal{C}\operatorname{rn}(T)$ is at most 2.

F34: [HaLa87] If $\mathcal{C}$ is the class of maximal planar graphs and G is maximal planar then $\mathcal{C}\operatorname{rn}(G)$ is at most 2. Those maximal planar graphs with class reconstruction number equal to 1 are characterized.

F35: Almost every graph has edge-reconstruction number equal to 2. (See [LaSc03].)

F36: [Mo95] Let G be a disconnected graph. If G contains a pair of nontrivial, non-isomorphic components, then $\mathrm{ern}(G)$ is at most 3. If, furthermore, G is not a forest and contains a component other than K_3 and $K_{1,3}$, then $\mathrm{ern}(G)$ is at most 2. If the components of G are all isomorphic and contain k edges, then the edge-reconstruction number can be as high as $k + 2$.

Remarks analogous to those made above concerning the reconstruction number of disconnected graphs also can be made here.

F37: [Mo93] Every tree with at least 4 edges has edge-reconstruction number at most 3.

FURTHER REMARKS

R7: An intriguing question which has hardly been given any attention is the relationship between $\mathrm{rn}(G)$ and $\mathrm{ern}(G)$. While we have seen results which say that edge-reconstruction is implied by reconstruction, no such relationship seems to exist between these two parameters. In fact, the edge-reconstruction number for a graph could be greater than its reconstruction number.

R8: Reconstruction numbers are also interesting from another point of view. We have seen that lack of symmetry favors reconstruction, but whereas this should imply that high symmetry makes reconstruction more difficult, highly symmetric graphs are regular, and these are trivially reconstructible. Reconstruction numbers seem to put this in a better perspective because, while graphs with property A_3 have reconstruction number 3, it seems [My88] that regular graphs are the candidates for being the graphs with the largest reconstruction number. It should also be pointed out that the reconstruction number of regular graphs is not yet known.

R9: In her thesis [My88], Myrvold calls the reconstruction number the ***ally reconstruction number*** and she also defines another parameter which she calls the ***adversary reconstruction number***. If **A** and **B** are two players, the reconstruction number can be seen as the smallest number of graphs from the deck which **A** can give to **B** such that the latter can determine the graph uniquely; here **A** and **B** are allies. However, we can also ask for the largest number of subgraphs which **A** can give **B** such that **B** cannot determine the graph uniquely; here **A** and **B** are adversaries. The adversary reconstruction number of a graph is equal to 1 plus this last number. In other words, a graph G has adversary reconstruction number k if and only if any k subgraphs from the deck of G determine it uniquely; equivalently, no other graph has these same k subgraphs in its deck. Only partial results have been obtained on the adversary reconstruction number, which seems even more difficult to tackle than the (ally) reconstruction number, and this might perhaps be an area for further interesting research.

Set Reconstruction

In 1964, Harary [Ha64] suggested another way of reconstructing by not using the full deck when he made the following conjecture.

C3: [Set Reconstruction Conjecture] *Any graph G with $n \geq 4$ vertices can be reconstructed uniquely from its set of nonisomorphic subgraphs $G - v$.*

In other words, one is now only given *one* graph from each isomorphism class in the deck, and one does not know how many times each given graph appears in the deck.

DEFINITION

D13:   A graph or a parameter which can be determined from the respective set of non-isomorphic subgraphs is said to be ***set reconstructible***.

FACTS

F38:   [Ma76] The number of edges and the set of degrees of a graph is set reconstructible.

F39:   [Ma76] For every graph in which no vertex of minimum degree lies on a triangle, the degree sequence is set reconstructible.

F40:   [Ma76] The degree sequence of any graph with minimum degree at most 3 is set reconstructible.

F41:   [Ma76] The connectivity of any graph is set reconstructible.

F42:   [Ma76] Disconnected graphs are set reconstructible.

F43:   Separable graphs (that is, graph of connectivity 1) without vertices of degree 1 are set reconstructible.

F44:   [Ma70] Trees are set reconstructible.

F45:   [Gi76] Outerplanar graphs are set reconstructible.

F46:   [ArCo74] Unicyclic graphs (that is, graphs having only one cycle) are set reconstructible.

REMARK

R10:   The idea of set reconstruction can also be applied to edge-reconstruction, that is, only one copy of each isomorphism type in the edge-deck is given. When a parameter or a class of graph is so reconstructible we say that it is ***set edge-reconstructible***. We highlight a few results in set edge-reconstructibility.

FACTS

F47:   [Ma76] The degree sequence of a graph is set edge-reconstructible.

Delorme, Favaron, and Rautenbach have improved this result of Manvel.

F48:   [DeFaRa02] The degree sequence of a graph with at least four edges is uniquely determined by the set of degree sequences of its edge-deleted subgraphs with one well-described class of exceptions. Moreover, the multiset of the degree sequences of the edge-deleted subgraphs determines the degree sequence of the graph.

F49:   [AnDiVe96] If a graph G with at least four edges has at most two non-isomorphic edge-deleted subgraphs, then G is set edge-reconstructible.

Reconstruction from the Characteristic Polynomial Deck

In [Sc79], Schwenk proposed the problem of reconstructing a graph from the characteristic polynomial of each subgraph in the deck, which we shall call the **polynomial deck**. He also showed that the answer to this general problem is in the negative, that is, a graph is not necessarily reconstructible from the polynomial deck. But he suggested a weakening of the problem so that what is required is the reconstruction of the characteristic polynomial of G from its polynomial deck. This problem is still open, and we here limit ourselves to presenting four results from the few which have been obtained.

FACTS

F50:   [Sc79] The characteristic polynomial of any graph is reconstructible up to a constant from the polynomial deck.

F51:   If a subgraph in the deck of G has a characteristic polynomial with repeated roots, then the characteristic polynomial of G is reconstructible from its polynomial deck. (See [LaSc03].)

F52:   [CvLe98] The characteristic polynomial of a tree is reconstructible from its polynomial deck.

F53:   [Sc02] If a graph of order n has at least $n/3$ vertices of degree 1 then its characteristic polynomial is reconstructible from its polynomial deck.

Reconstructing from k-Vertex-Deleted Subgraphs

A k-vertex-deleted subgraph of G is a subgraph obtained from G by deleting k of its vertices and all edges incident to them. We shall have more to say about this mode of reconstruction in Section 5 when we consider k-edge-deleted subgraphs. We here limit ourselves to one result.

F54:   [Ta89] Let $k \geq 3$ be an integer. Then the degree sequences of all sufficiently large graphs are determined by their k-vertex-deleted subgraphs. In particular, this result is true for all graphs on at least $f(k)$ vertices, where $f(k)$ is a certain function which is asymptotic to ke.

2.3.4   Tutte's and Kocay's Results

If Kelly's Lemma were true for all spanning subgraphs of G (that is, subgraphs with the same number of vertices of G) then this would solve the reconstruction problem. In [Tu79], however, Tutte managed to show that Kelly's Lemma can be extended to certain classes of spanning subgraphs, and this has had very important consequences. In [Ko81], Kocay managed to obtain Tutte's results with proofs that were much easier. We shall here present a sketch of Kocay's method. A fuller treatment is given in [Bo91] and [LaSc03].

Kocay's Parameter

DEFINITION

D14: Let G be a graph and $\mathcal{F} = (F_1, F_2, \ldots, F_k)$ a sequence of graphs (we do not exclude the possibility that different F_i could be isomorphic). A **cover** of G by $\mathcal{F}$ is a sequence $\mathcal{G} = (G_1, G_2, \ldots, G_k)$ of subgraphs of G (not necessarily distinct) such that: (i) $G_i \simeq F_i$, $i = 1, \ldots, k$ and (ii) $G = \cup_i G_i$. The number of covers of G by $\mathcal{F}$ is denoted by $c(\mathcal{F}, G)$.

FACTS

[Tu79, Ko81] (See also [Bo91, LaSc03].)

F55: Let G be a graph and let $\mathcal{F} = (F_1, F_2, \ldots, F_k)$ be a sequence of graphs with each $|V(F_i| < |V(G)|$. Let $\kappa(\mathcal{F}, G)$ be the parameter defined by

$$\sum_X c(\mathcal{F}, X) \binom{G}{X}$$

where the summation is taken over all isomorphism types X of graphs such that $|V(X)| = |V(G)|$. Then $\kappa(\mathcal{F}, G)$ is reconstructible.

The following results are then obtained by defining a suitable choice for the F_i.

F56: The number of 1-factors of G is reconstructible.

F57: The number of spanning trees of G is reconstructible.

F58: The number of Hamiltonian cycles of G is reconstructible.

F59: The number of 2-connected spanning subgraphs of G with a specified number of edges is reconstructible.

Characteristic and Chromatic Polynomials

DEFINITIONS

D15: An **elementary graph** is a graph in which every component is either an edge or a cycle. For any graph X, $c(X)$ denotes the number of components of X and $s(X)$ the number of cycles.

FACTS

The proofs of the first two of the following results, due to Sachs and Whitney, respectively, can be found in [Bi93].

F60: Let the characteristic polynomial of G be

$$\lambda^n + a_1 \lambda^{n-1} + a_2 \lambda^{n-2} + \ldots + a_n$$

Then each coefficient a_i is given by

$$a_i = \sum_X (-1)^{c(X)} 2^{s(X)} \binom{G}{X}$$

where the summation extends over all isomorphism types X of elementary graphs on i vertices.

F61: Let the chromatic polynomial of G be

$$b_1 x + b_2 x^2 + \ldots + b_n x^n$$

Then each coefficient b_i is given by

$$b_i = \sum_X (-1)^{|E(X)|} \binom{G}{X}$$

where the summation extends over all isomorphism types X of graphs on n vertices and i components.

From these characterizations of the characteristic and chromatic polynomials together with the previous reconstruction results, the next important results follow. [Tu79, Ko81] (See also [Bo91, LaSc03].)

F62: The characteristic polynomial is reconstructible.

F63: The chromatic polynomial is reconstructible.

2.3.5   Lovász's Method and Nash–Williams's Lemma

It is quite arguable that the deepest and most general results obtained in reconstruction are those which we shall be presenting in this section. In 1972, Lovász [Lo72] published a beautiful two-page paper in which he showed that if a graph has one half more than the largest possible number of edges, then it is edge-reconstructible. This paper made a surprising and elementary use of the inclusion-exclusion principle. In 1977, Müller [Mü77], using the same method as Lovász, obtained a stronger conclusion for edge-reconstructibility. In his survey of 1978 [Na78], Nash–Williams proved a lemma from which Lovász's and Müller's results follow (but still applying the method introduced by Lovász). We shall here give these results. A more extended treatment can be found in [El88] or [Bo91] or [LaSc03].

The Nash–Williams Lemma

DEFINITIONS

In the following, let G and H be graphs which are assumed to share the same vertex-set V, and let X be a subset of the edge-set of G.

D16: A *homomorphism* from G to H is a permutation of V such that any edge of G is mapped into an edge of H. The number of such homomorphisms is denoted by $[H]_G$.

D17: A *homomorphism with forbidden* X is a permutation of V such that all edges in $E(G) - X$ are mapped into edges of H but all edges in X are mapped into non-edges in H. The number of such homomorphisms is denoted by $[H]_{G \backslash X}$.

REMARK

R11: Note that $[H]_{G \backslash X}$ is quite different from $[H]_{G-X}$; the latter counts homomorphisms from $G - X$ to H, that is, where all edges in $E(G) - X$ are mapped into edges of H, but all edges in X can be mapped either into edges or non-edges of H; this is unlike the case of $[H]_{G \backslash X}$ where all the edges of X must be mapped into non-edges of H.

FACTS

F64: [Lo72] Let G, H, and X be as above. Then

$$[H]_{G \setminus X} = \sum_{Y \subset X} (-1)^{|Y|} [H]_{G-X+Y}$$

F65: *Nash–Williams' Lemma* Let G, H and X be as above, and suppose that G and H have the same edge-deck. Then

$$[H]_G = |\text{Aut}(G)| + (-1)^{|X|}([H]_{G \setminus X} - [G]_{G \setminus X})$$

(See [Bo91,LaSc03].)

The following is an important corollary to Nash–Williams's Lemma. (See [Bo91, LaSc03].)

F66: Suppose G, H and X are as in Nash–Williams's Lemma, and assume that $G \not\simeq H$. Then,
  (i) if $|X|$ is odd, then $[H]_{G \setminus X} > 0$;
  (ii) if $|X|$ is even, then $[G]_{G \setminus X} > 0$.

From this, Lovász and Müller's results follow.

F67: [Lo72] Let G be a graph such that $|E(G)| > \binom{n}{2}/2$. Then G is edge-reconstructible.

F68: [Mü77] Let G be a graph such that $2^{|E(G)|-1} > n!$. Then G is edge-reconstructible.

Perhaps the most striking result in edge-reconstruction obtained by these methods is the following.

F69: [Py90] A Hamiltonian graph with a sufficiently large number of vertices is edge-reconstructible.

Structures Other than Graphs

The real power and generality of the above methods appear with the realization that edge-reconstruction can be generalized to the reconstruction, up to some group of isomorphisms, of a combinatorial object or structure from its subobjects, again given up to isomorphism.

DEFINITION

D18: More exactly, define a ***structure*** to be a triple (D, Γ, E) where D is a finite set, Γ is a group of permutations acting on D, and E is a subset of D. By edge-reconstruction we mean here that the subsets $E - x$ are given, up to 'translation' by the group Γ, and the question is whether E can be reconstructed uniquely, again up to action by the group Γ.

Therefore in edge-reconstruction for graphs, D would be the set of all possible $\binom{n}{2}$ edges on n vertices, E would be the edges which define the graph to be reconstructed, and Γ would be the full symmetric group with its induced action on the unordered, distinct pairs of vertices.

An extended treatment of edge-reconstruction seen in this light can be found in [Bo91, LaSc03].

FACTS

We shall here present some results obtained by viewing edge-reconstruction in this more general setting. The first result, although a straightforward application of Nash–Williams' Lemma to structures, gives some life to the problem of reconstructing from the endvertex-deck.

F70: [LaSc03] Let H be a graph with minimum degree 2, and let G be obtained from H by adding k endvertices such that no two have a common neighbor. Then G is endvertex-reconstructible if either $k > V|H|/2$ or $2^{k-1} > \mathrm{Aut}(H)$.

REMARK

R12: This result in conjunction with Bryant's negative result leads to the natural question asking what is the minimum proportion of endvertices required to guarantee endvertex-reconstructibility.

DEFINITION

D19: The following generalizes Müller's result not only to structures, but also in this fashion: instead of removing one edge at a time, k edges at a time are removed. Let us call this k-*edge-reconstruction*.

FACTS

F71: [AlCaKrRo89] Let (D, Γ, E) be a structure such that $2^{|E|-k} > |\Gamma|$. Then the structure is k-edge-reconstructible.

The paper [AlCaKrRo89] contains several other results of this type and should be studied carefully by anyone who is interested in extending the reconstruction of structures in the direction of k-edge-reconstruction. Many of the results in this paper have been extended more recently by Radcliffe and Scott. They consider the reconstruction of a subset of the integers modulo n, Z_n, or of the reals, R, up to translation from the collection of its subsets of a given size, also given up to translation. The following summarizes their important results.

F72: [RaSc98] Suppose p is prime. The every subset of Z_p is reconstructible from the collection of its 3-subsets.

F73: [RaSc98] For arbitrary n, almost all subsets of Z_n are reconstructible from the collections of their 3-subsets.

F74: [RaSc98] For any n, every subset of Z_n is reconstructible from its $9\alpha(n)$-subsets, where $\alpha(n)$ is the number of distinct prime factors of n.

F75: [RaSc99] A locally finite subset of R (that is, a subset which contains only finitely many translates of any given finite set of size at least 2) is reconstructible from its 3-subsets.

Another recent result of this type is the following. Here subsets of R^2 are considered, and any two such subsets are considered isomorphic if one can be transformed into the other by a translation or a rotation by a multiple of 90 degrees.

F76: [Ra02] Any finite subset A of the plane R^2 is uniquely determined by at most 5 of its subsets of cardinality $|A| - 1$, given up to isomorphism; that is, in the terminology of graph reconstruction, A has reconstruction number 5.

The Reconstruction Index of Groups

Looking at edge reconstruction in this guise, that is, as reconstruction of structures (D, Γ, E), has led some authors to focus attention more directly on the permutation group Γ.

DEFINITION

D20: The ***reconstruction index*** $\rho(\Gamma, D)$ of the permutation group Γ acting on D is the smallest t such that for any $E \subset D$ with $|E| \geq t$, the structure (D, Γ, E) is edge-reconstructible.

The Edge-Reconstruction Conjecture therefore states that, if $Y = \{1, 2, \ldots, n\}$ and D is the set of unordered, distinct pairs of Y, and if $S_n^{(2)}$ is the symmetric group of Y acting on these pairs, then $\rho(S_n^{(2)}, D) = 4$.

FACT

The following is one result obtained on the reconstruction index of permutation groups.

F77: [Mn98] The reconstruction index of an abelian group is 4 and the reconstruction index of Hamiltonian groups is 5.

Other works which deal with the reconstruction index of groups are [Ca96, Ma96, Mn87, Mn92, Mn95]

2.3.6 Digraphs

The situation with the reconstruction of digraphs is quite different from that of graphs, for here it has been shown that the conjecture is false.

FACTS

F78: [St77, Ko85] There exists an infinite family of tournaments which are not reconstructible.

A positive partial result has been obtained by Harary and Palmer.

F79: [HaPa67] Tournaments on at least five vertices which are not strongly connected are reconstructible. (See also [BoHe77].)

Some more positive results on the reconstructibility of tournaments can be found in [DeGu90, Gu96, Vi99].

Therefore the problem here should be to investigate which digraphs are reconstructible, or to determine what reconstruction question one should ask for digraphs. In fact, Ramachandran has noted that all non-reconstructible digraphs which have been discovered to date have the property that they would be reconstructible if, with every $D - v$, one is also given the in-degree and the out-degree in D of the missing vertex v.

DEFINITION

D21: [Ra97] A digraph D is said to be *N-reconstructible* if it is uniquely determined by the triples $(D - v_i, \deg_{\text{in}}(v_i), \deg_{\text{out}}(v_i))$, for all vertices v_i of D.

And Ramachandran goes on to make the following conjecture.

C4: [The N-Reconstruction Conjecture for Digraphs] *Every digraph is N-reconstructible.*

No counterexamples to this conjecture are known.

2.3.7   Illegitimate Decks

As we have already said, the reconstruction problem is not about finding an efficient algorithm to determine G from its deck, but the question is one of uniqueness: is there only one graph with the given deck?

However, there is one problem in graph reconstruction which falls naturally within the setting of computational complexity.

DEFINITION

D22: A collection of graphs $G_1, G_2, \ldots, G_n$ each on $n - 1$ vertices is said to be an *illegitimate deck* if there is no graph G having the given collection as deck. The *illegitimate deck problem* is to determine whether or not such a given collection of graphs is indeed the deck of some graph.

FACTS

F80: [Ma82] Determining whether a given collection of graphs is an illegitimate deck is at least as hard as the isomorphism problem.

F81: [HaPlSt82] The graph isomorphism problem is polynomially equivalent to the illegitimate deck problem for regular graphs.

More information about the relationship between the computational complexities of the legitimate deck problem and the graph isomorphism problem can be found in [KöScTo93, KrHe94].

2.3.8   Recent Results

We briefly outline the most important results in reconstruction obtained since the publication of the *Handbook of Graph Theory, First Edition*. Notably, Asciak, Francalanza, Lauri and Myrvold [AsFrLaMy10] have given a survey of open questions regarding reconstruction numbers. At about the same time, Bowler, Brown, and Fenner have commenced a systematic attack on the adversary reconstruction number.

FACTS

F82: [BiKwYu07] The class of planar graphs is recognizable.

F83: [BoBrFe10] There exist pairs of graphs on n vertices with $2\lfloor\frac{1}{3}(n-1)\rfloor$ common cards for every $n \geq 10$.

F84: There exist pairs of trees on n vertices with $2\lfloor\frac{1}{3}(n-5)\rfloor$ cards in common for $n \geq 8$.

F85: [BoBrFeMy11] The maximum number of cards in common between a connected and a disconnected graph on n vertices is $\lfloor\frac{n}{2}\rfloor + 1$. Graphs that attain this upper bound are characterized.

F86: [AsLa10] If G is a connected graph all of whose components are isomorphic and contain k edges, and if G has edge-reconstruction number at least $k + 1$, then the components of G are isomorphic to the star $K_{1,k}$.

CONJECTURES

C5: [BoBrFe10] The largest possible number of cards that two graphs on n vertices can have in common is $2\lfloor\frac{1}{3}(n-1)\rfloor$. Affirming this would imply that any such graph can be reconstructed from any $2\lfloor\frac{1}{3}(n-1)\rfloor + 1$ of its cards.

C6: [BaWe10] They also conjecture that $drn(T) \leq 2$ for all but finitely many trees T.

C7: [AsLa10] For a disconnected graph all of whose components are isomorphic to H, suppose that $ern(G) > 3$. Then H is isomorphic to the star $K_{1,k}$.

REMARK

R13: The reconstruction number and the edge-reconstruction number of all graphs with $n \leq 11$ vertices are given by [RiRa11], along with computational results on reconstruction numbers associated with the removal of more than one vertex at a time.

DEFINITION

D23: The *degree associated reconstruction number* $drn(G)$ of a graph G is the minimum number of cards required to reconstruct G if, along with any card $G - v$, one is also given the degree of the missing vertex v. Note that $drn(G)$ is equivalent to the class reconstruction number of graphs with a given value of m as the number of edges.

FACTS

F87: [Ra06] If G is a connected non-regular graph and kG is the disconnected graph made up of k copies of G then $drn(kG) \leq 1 + darn(G)$. If G is r-regular of order $n > 2$ then $drn(kG) \leq n + 2 - r$.

F88:   [BaWe10] The degree associated reconstruction number equals 2 for all caterpillars, except stars and one 6-vertex example.

References

[AlCaKrRo89] N. Alon, Y. Caro, I. Krasikov, and Y Roditty, Combinatorial reconstruc-
    tion problems, *J. Combin. Theory (Ser. B)* 47:153–161, 1989.

[AnDiVe96] L. D. Andersen, S. Ding, and P. D. Vestergaard, On the set edge-
    reconstruction conjecture, *J. Combin. Math. Combin. Comput.* 20:3–9, 1996.

[ArCo74] E. Arjomandi and D. G. Corneil, Unicyclic graphs satisfy Harary's conjecture,
    Canad. Math. Bull. 17:593–596, 1974.

[AsFrLaMy10] K. J. Asciak, M. A. Francalanza, J. Lauri, and W. Myrvold, A survey
    of some open questions in reconstruction numbers, *Ars Combin.* 97:443–456, 2010.

[AsLa02] K. J. Asciak and J. Lauri, On disconnected graphs with large reconstruction
    number, *Ars Combin.* 62:173–181, 2002.

[AsLa10] K. J. Asciak and J. Lauri, On the edge-reconstruction number of disconnected
    graphs, *Bull. ICA* 63:87-110, 2011.

[BaBaHo87] D. W. Bange, A. E. Barkauskas, and L. H. Host, Class-reconstruction of
    total graphs, *J. Graph Theory* 11:221–230, 1987.

[BaWe10] Degree-associated reconstruction number of graphs, *Discrete Math.*
    310:2600–2612, 2010.

[Bi93] N. L. Biggs, *Algebraic Graph Theory*, Cambridge University Press, 1993.

[BiKwYu07] M. Bilinski, Y. S. Kwon, and X. Yu, On the reconstruction of planar
    graphs, *J. Combin. Theory Ser. B* 97:745–756, 2007.

[Bo90] B. Bollabás, Almost every graph has reconstruction number 3, *J. Graph Theory*
    14:1–4, 1990.

[Bo69a] J. A. Bondy, On Kelly's congruence theorem for trees, *Proc. Camb. Phil. Soc.*
    65:387–397, 1969.

[Bo69b] J. A. Bondy, On Ulam's conjecture for separable graphs, *Pacific J. Math.*
    31:281–288, 1969.

[Bo91] J. A. Bondy, A graph reconstructor's manual, In *Surveys in Combinatorics* (A.
    D. Keedwell, ed.), 221–252. Cambridge University Press, 1991.

[BoHe77] J. A. Bondy and R. L. Hemminger, Graph reconstruction—A survey, *J. Graph
    Theory* 1:227–268, 1977.

[BoBrFe10] A. Bowler, P. Brown, and T. Fenner, Families of pairs of graphs with a
    large number of common cards, *J. Graph Theory* 63:146–163, 2010.

[BoBrFeMy11] A. Bowler, P. Brown, T. Fenner, and W. Myrvold, Recognizing connect-
    edness from vertex-deleted subgraphs, *J. Graph Theory* 67:285–299, 2011.

[Br71] R. M. Bryant, On a conjecture concerning the reconstruction of graphs, *J. Combin. Theory* 11:139–141, 1971.

[Ca96] P. J. Cameron, Stories from the age of reconstruction, *Congr. Num.* 113:31–41, 1996.

[Ch71] P. Z. Chinn, A graph with p points and enough distinct $p-2$-order subgraphs is reconstructible, In *Recent Trends in Graph Theory*, M. Capobianco, J. B. Frechen, and M. Krolik, editors, 71–73. Springer-Verlag, 1971.

[CvLe98] D. M. Cvetković and M. Lepović, Seeking counterexamples to the reconstruction conjecture for the characteristic polynomial of graphs and a positive result, *Bull. Cl. Sci. Math. Nat. Sci. Math.* 23:91–100, 1998.

[DeFaRa02] C. Delorme, O. Favaron, and D. Rautenbach, On the reconstruction of the degree sequence, *Discrete Math.* 259:293–300, 2002.

[DeGu90] D. C. Demaria and C. Guido, On the reconstruction of normal tournaments, *J. Combin. Inform. System Sci.* 15(1–4):301–323, 1990.

[El88] M. N. Ellingham, Recent progress in edge reconstruction, *Congr. Numer.* 62:3–20, 1988.

[ElPyXi88] M. N. Ellingham, L. Pyber, and Y. Xingxing, Claw-free graphs are edge reconstructible, *J. Graph Theory* 12:445–451, 1988.

[Fa94] H. Fan, Edge reconstruction of planar graphs with minimum degree at least three—IV, *Systems Sci. Math. Sci.* 7:218–222, 1994.

[FaWuWa01] H. Fan, Y-L. Wu, and C. K. Wang, On fixed edges and the edge reconstruction of series parallel networks, *Graphs Combin.* 17(2):213–225, 2001.

[FiLa81] S. Fiorini and J. Lauri, The reconstruction of maximal planar graphs, I: Recognition, *J. Combin. Theory (Ser. B)* 30:188–195, 1981.

[FiMa78] S. Fiorini and B. Manvel, A theorem on planar graphs with an application to the reconstruction problem, II, *J. Combin. Inf. Sys. Sci.* 3(3):200–216, 1978.

[Gi76] W. B. Giles, Point deletions of outerplanar blocks, *J. Combin. Theory (Ser. B)* 20:103–116, 1976.

[GoMc81] C. D. Godsil and B. D. McKay, Spectral conditions for reconstructibility of a graph, *J. Combin. Theory (Ser. B)* 30:285–289, 1981.

[Gr71] D. L. Greenwell, Reconstructing graphs, *Proc. Amer. Math. Soc.* 30:431–433, 1971.

[Gu96] C. Guido, A larger class of reconstructible tournaments, *Discrete Math.* 152(1–3):171–184, 1996.

[Ha64] F. Harary, On the reconstruction of a graph from a collection of subgraphs, In M. Fiedler, Editor, *Theory of Graphs and Its Applications* Proc. Symposium Smolenice, 1963, pages 47–52, Academic Press, 1964.

[HaLa87] F. Harary and J. Lauri, The class-reconstruction number of a maximal planar graph, *Graphs and Combinatorics* 3:45–53, 1987.

[HaLa88] F. Harary and J. Lauri, On the class-reconstruction number of trees, *Quart. J. Math. Oxford (2)* 39:47–60, 1988.

[HaPa66] F. Harary and E. M. Palmer, The reconstruction of a tree from its maximal subtrees, *Canad. J. Math.* 18:803–810, 1966.

[HaPa67] F. Harary and E. M. Palmer, On the problem of reconstructing a tournament from subtournaments, *Monatsh. Math.* 71:14–23, 1967.

[HaPl85] F. Harary and M. Plantholt, The graph reconstruction number, *J. Graph Theory* 9:451–454, 1985.

[HaPlSt82] F. Harary, M. Plantholt, and R. Statman, The graph isomorphism problem is polynomially equivalent to the legitimate deck problem for regular graphs, *Caribbean J. Math.* 1(1):15–23, 1982.

[He69] R. L. Hemminger, On reconstructing a graph, *Proc. Amer. Math. Soc.* 20:185–187, 1969.

[Ke57] P. J. Kelly, A congruence theorem for trees, *Pacific J. Math.* 7:961–968, 1957.

[KöScTo93] J. Köbler, U. Schöning, and J. Torán, *The Graph Isomorphism Problem: Its Structural Complexity*, Birkhäuser, 1993.

[Ko81] W. L. Kocay, On reconstructing spanning subgraphs, *Ars Combinatoria* 11:301–313, 1981.

[Ko85] W. L. Kocay, On Stockmeyer's non-reconstructible tournaments, *J. Graph Theory* 9:473–476, 1985.

[Ko71] A. D. Korshunov, Number of nonisomorphic graphs in an n-point graph, *Math. Notes of the Acad. of Sciences of the USSR* 9:155–160, 1971.

[KrLeTh02] I. Krasikov, A. Lev, and B. D. Thatte, Upper bounds on the automorphism group of a graph, *Discrete Math.* 256(1–2):489–493, 2002.

[KrHe94] D. Kratsch and L. A. Hemaspaandra. On the complexity of graph reconstruction, *Math. Systems Theory* 27(3):257–273, 1994.

[La81] J. Lauri, The reconstruction of maximal planar graphs II: Reconstruction, *J. Combin. Theory (Ser. B)* 30:196–214, 1981.

[La87] J. Lauri, Graph reconstruction—some new techniques and new problems, *Ars Combinatoria (Ser B)* 24:35–61, 1987.

[LaSc03] J. Lauri and R. Scapellato, *Topics in Graph Automophisms and Reconstruction* Cambridge University Press, 2003.

[Lo72] L. Lovász, A note on the line reconstruction problem, *J. Combin. Theory (Ser. B)* 13:309–310, 1972.

[Ma82] A. Mansfield, The relationship between the computational complexities of the legitimate deck and isomorphism problems, *Quart. J. Math. Oxford (2)* 33:345–347.

[Ma70] B. Manvel, Reconstruction of trees, *Canad. J. Math.* 22:55–60, 1970.

[Ma76] B. Manvel, On reconstructing graphs from their sets of subgraphs, *J. Combin. Theory (Ser. B)* 21:156–165, 1976.

[Ma88] B. Manvel, Reconstruction of graphs: progress and prospects, *Congr. Numer.* 63:177–187, 1988.

[Ma96] P. Maynard, *On Orbit Reconstruction Problems*, PhD thesis, University of East Anglia, Norwich, 1996.

[Mc77] B. D. McKay, Computer reconstruction of small graphs, *J. Graph Theory* 1:281–283, 1977.

[Mn87] V. B. Mnukhin, Reconstruction of k-orbits of a permutation group, *Math. Notes* 42:975–980, 1987.

[Mn92] V. B. Mnukhin, The k-orbit reconstruction and the orbit algebra, *Acta. Applic. Math.* 29:83–117, 1992.

[Mn95] V. B. Mnukhin, The reconstruction of oriented necklaces, *J. Combin., Inf. & Sys. Sciences* 20(1–4):261–272, 1995.

[Mn98] V. B. Mnukhin, The k-orbit reconstruction for abelian and hamiltonian groups, *Acta. Applic. Math.* 52:149–162, 1998.

[Mo93] R. Molina, The edge reconstruction number of a tree, *Vishwa Int. J. Graph Theory* 2(2):117–130, 1993.

[Mo95] R. Molina, The edge reconstruction number of a disconnected graph, *J. Graph Theory* 19(3):375–384, 1995.

[Mü76] V. Müller, Probabilistic reconstruction from subgraphs, *Comment. Math. Univ. Carolinae* 17:709–719, 1976.

[Mü77] V. Müller, The edge reconstruction hypothesis is true for graphs with more than $n \log_2 n$ edges, *J. Combin. Theory (Ser. B)* 22:281–283, 1977.

[My88] W. J. Myrvold, *Ally and Adversary Reconstruction Problems*, PhD thesis, University of Waterloo, 1988.

[My89] W. J. Myrvold, The ally-reconstruction number of a disconnected graph, *Ars Combinatoria* 28:123–127, 1989.

[My90] W. J. Myrvold, The ally-reconstruction number of a tree with five or more vertices is three, *J. Graph Theory* 14:149–166, 1990.

[MyElHo87] W. J. Myrvold, M. N. Ellingham, and D. G. Hoffman, Bidegreed graphs are edge reconstructible, *J. Graph Theory* 11(3):281–302, 1987.

[Na78] C. St. J. A. Nash–Williams, The reconstruction problem. In L. W. Beineke and R. J. Wilson, editors, *Selected Topics in Graph Theory*, Chapter 8, Academic Press, London, 1978.

[Py90] L. Pyber, The edge-reconstruction of hamiltonian graphs, *J. Graph Theory* 14:173–179, 1990.

[RaSc98] A. J. Radcliff and A. D. Scott, Reconstructing subsets of Z_n, *J. Combin. Theory (Ser. A)* 83(2):169–187, 1998.

[RaSc99] A. J. Radcliff and A. D. Scott, Reconstructing subsets of reals, *Electronic J. Combin.* (1):Research Paper 20, 7pp., 1999.

[Ra97] S. Ramachandran, N-reconstructibility of nonreconstructible tournaments, *Graph Theory Notes of N.Y.* 32:23–29, 1997.

[Ra06] S. Ramachandran, Reconstruction number for Ulam's conjecture, *Ars Combin.* 78:289–296, 2006.

[Ra02] D. Rautenbach, On a reconstruction problem of Harary and Manvel, *J. Combin. Theory (Ser. A)* 99:32–39, 2002.

[RiRa09] D. Rivshin and S. Radziszowski, The vertex and edge graph reconstruction numbers of small graphs, *Australas. J. Combin* 45:175–188, 2009.

[RiRa11] D. Rivshin and S. Radziszowski, Multi-vertex deletion graph number reconstructions, *J. Combin. Math. Combin. Comput.* 78:303–321, 2011.

[Sc85] J. Schönheim, personal communication.

[Sc79] A. J. Schwenk, Spectral reconstruction problems, in *Topics in Graph Theory*, Vol 328 of Annals New York Academy of Science, pages 183–189, New York Academy of Sciences, 1979.

[Sc02] I. Sciriha, Polynomial reconstruction and terminal vertices, *Linear Algebra and its Applications* 356:145–156, 2002.

[St77] P. K. Stockmeyer, The falsity of the reconstruction conjecture for tournaments, *J. Graph Theory* 1:19–25, 1977.

[Ta89] R. Taylor, Reconstructing degree sequences from k-vertex-deleted subgraphs, *Discrete Maths.* 79:207–213, 1989/90.

[Tu79] W. T. Tutte, All the king's horses—a guide to reconstruction, in J. A. Bondy and U. S. R. Murty, editors, *Graph Theory and Related Topics*, Academic Press, 1979.

[Ul60] S. M. Ulam, *A Collection of Mathematical Problems*, Wiley (Interscience), New York, 1960.

[Vi99] P. Vitolo, The reconstruction of simply disconnected tournaments, *J. Combin. Inform. System Sci.* 24(2–4):65–77, 1999.

[Yu82] H. Yuan, An eigenvector condition for reconstructibility, *J. Combin. Theory (Ser. B)* 32:245–256, 1982.

[Zh98a] Y. Zhao, On the edge reconstruction of graphs embedded in surfaces. II, *J. London Math. Soc.* 57:268–274, 1998.

[Zh98b] Y. Zhao, On the edge reconstruction of graphs embedded in surfaces. III, *J. Combin. Theory (Ser. B)* 74:302–310, 1998.

[Zh88] Yang Yong Zhi, The reconstruction conjecture is true if all 2-connected graphs are isomorphic, *J. Graph Theory* 12:237–243, 1988.

Section 2.4
Recursively Constructed Graphs

Richard B. Borie, University of Alabama
R. Gary Parker, Georgia Institute of Technology
Craig A. Tovey, Georgia Institute of Technology

INTRODUCTION

The core idea of recursively constructed graphs is captured in Definition 1, but the substantial literature on the subject has motivated a considerable breadth and variety of notational distinctions.

NOTATION: All graphs in this section are simple, and an edge with endpoints x and y is denoted (x, y).

DEFINITIONS

D1: A *recursively constructed graph class* is defined by a set (usually finite) of primitive or *base graphs*, in addition to one or more operations that compose larger graphs from smaller subgraphs. Each operation involves either fusing specific vertices from each subgraph or adding new edges between specific vertices from each subgraph.

D2: Each graph in a recursive class has a corresponding *decomposition tree* that shows how to build it from base graphs.

REMARK

R1: Graphs in these classes possess a modular structure, so fast algorithms can often be designed to solve hard problems restricted to these classes. The algorithms typically proceed by solving the desired problem on the base graphs, then employ dynamic programming to combine solutions for small subgraphs into a solution for a larger graph. The construction of these algorithms is the subject of Section 10.4.

2.4.1   Some Parameterized Families of Graph Classes

Trees

DEFINITION

D3:   The graph with a single vertex r (and no edges) is a ***tree*** with root r (the sole base graph). Let (G, r) denote a tree with root r. Then $(G_1, r_1) \oplus (G_2, r_2)$ is a tree formed by taking the disjoint union of G_1 and G_2 and adding an edge (r_1, r_2). The root of this new tree is $r = r_1$.

TERMINOLOGY NOTE: Technically, the pairs (G, r) in Definition D3 denote *rooted trees*. However, the specification of distinguished vertices r_1 and r_2 (and hence r) is relevant here only as a vehicle in the recursive construction.

EXAMPLE

E1:   Figure 2.4.1 illustrates the recursive construction of trees.

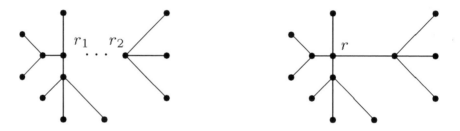

Figure 2.4.1: Recursive construction of a tree.

Series-Parallel Graphs

From a non-recursive perspective, a graph is series-parallel if it has no subgraph homeomorphic to K_4 [Du65]. The graph on the left of Figure 2.4.2 is not series-parallel; the offending subgraph is identified by bold edges. Removal of two edges, as indicated, yields the graph to the right which is series-parallel.

Figure 2.4.2: Non-series-parallel and series-parallel graphs.

Following, we give a recursive definition of this class.

DEFINITION

D4: A *series-parallel graph* with distinguished *terminals* l and r is denoted (G, l, r) and is defined recursively as follows:
- The graph consisting of a single edge (v_1, v_2) is a series-parallel graph (G, l, r) with $l = v_1$ and $r = v_2$.
- The *series operation* $(G_1, l_1, r_1) \odot_s (G_2, l_2, r_2)$ forms a series-parallel graph by identifying r_1 with l_2. The terminals of the new graph are l_1 and r_2.
- The *parallel operation* $(G_1, l_1, r_1) \odot_p (G_2, l_2, r_2)$ forms a series-parallel graph by identifying l_1 with l_2 and r_1 with r_2. The terminals of the new graph are l_1 and r_1.
- The *jackknife operation* $(G_1, l_1, r_1) \odot_j (G_2, l_2, r_2)$ forms a series-parallel graph by identifying r_1 with l_2; the new terminals are l_1 and r_1.

COMPUTATIONAL NOTE: The jackknife operation can also be specified where the new terminals, after composition, are defined to be l_1 and l_2.

EXAMPLE

E2: The three operations defining series-parallel graphs are demonstrated in Figure 2.4.3. The pair-specific composition is on the left; the result is shown to the right. Terminal vertices are circled and labeled.

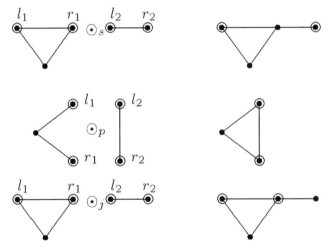

Figure 2.4.3: Composition operations for series-parallel graphs.

k-Trees and Partial k-Trees

DEFINITIONS

D5: The k-vertex complete graph, K_k, is a k-*tree*. A k-tree with $n+1$ vertices $(n \geq k)$ is constructed from a k-tree on n vertices by adding a vertex adjacent to all vertices of one of its K_k subgraphs, and only to those vertices.

D6: A *partial k-tree* is a subgraph of a k-tree.

TERMINOLOGY NOTE: In a given construction of a k-tree, the original K_k subgraph is referred to as its *basis*.

D7: A graph is *chordal* (or *triangulated*) if it contains no induced cycles of length greater than 3.

D8: A graph is *perfect* if every induced subgraph has chromatic number equal to the size of its maximum clique.

FACTS

F1: Trees are 1-trees, and forests are partial 1-trees.

F2: Series-parallel graphs are partial 2-trees.

F3: Any K_k subgraph of a k-tree can act as its basis.

F4: All k-trees are chordal graphs and, hence, perfect (because every chordal graph is perfect).

EXAMPLES

E3: A 3-tree is shown on the left in Figure 2.4.4, and a partial 3-tree is shown to the right.

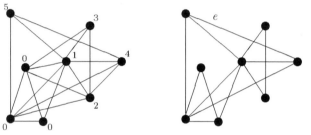

Figure 2.4.4: Construction of a 3-tree and a partial 3-tree.

Demonstrated by the graph to the left in Figure 2.4.4 is the "creation" of a 3-tree following a small number of composition operations starting from the basis given by an initial K_3 identified by vertex labels of 0. At each step, a new (consecutively labeled) vertex is added. Observe that if edge e is eliminated from the graph on the right in Figure 2.4.4, a partial 2-tree is created.

E4: The graph on the left in Figure 2.4.5 is series-parallel; it is a subgraph (and hence a partial 2-tree) of the 2-tree on the right. The dotted edges complete the 2-tree where the construction is verified by the labels on the vertices that are interpreted similarly as for Figure 2.4.4.

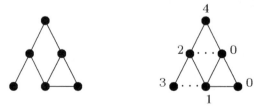

Figure 2.4.5: A series-parallel graph and a 2-tree.

Halin Graphs

DEFINITION

D9: A ***Halin graph*** is a planar graph having the property that its edge set E can be partitioned as $E = \langle T, C \rangle$, where T is a tree with no vertex of degree 2 and C is a cycle including only and all leaves of T.

FACTS

F5: Halin graphs are contained in the class of partial 3-trees.

F6: The set of Halin graphs is not closed under the taking of subgraphs, i.e., some subgraphs of Halin graphs are not Halin graphs.

EXAMPLES

E5: A Halin graph is given in Figure 2.4.6, with the cycle edges drawn on the outer face; their removal yields a tree satisfying the stated degree stipulation.

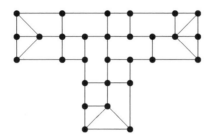

Figure 2.4.6: A Halin graph.

E6: The graph G on the left in Figure 2.4.7 is a 3-tree; vertex labels guide the construction as before. But this graph G is not a Halin graph. However, by removing one edge, we obtain the subgraph G' on the right, which is both a partial 3-tree and a Halin graph. The edges shown in bold form a tree of the Halin graph, and the cycle edges can be easily traced through the leaves of this tree.

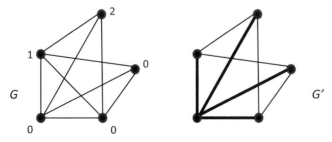

Figure 2.4.7: Non-Halin graph and Halin graph.

Bandwidth-k Graphs

DEFINITION

D10: A graph $G(V, E)$ is a **bandwidth-k graph** if there exists a vertex labeling $h : V \rightarrow \{1, 2, \ldots |V|\}$ such that $\{u, v\} \in E \Rightarrow |h(u) - h(v)| \leq k$. (Bandwidth is discussed in §9.4.)

EXAMPLE

E7: A bandwidth-3 graph is shown to the left in Figure 2.4.8; displayed to the right is a bandwidth-2 graph.

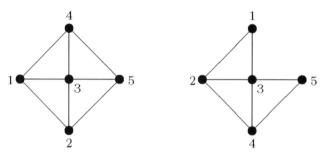

Figure 2.4.8: Bandwidth-3 and bandwidth-2 graphs.

Treewidth-k Graphs

The seminal works by Robertson and Seymour (cf. [RoSe86-a], [RoSe86-b], [RoSe91-a]) are commonly identified as being responsible for motivating the creation of the graph classes in this section. Most notable is the concept of treewidth, which played a key role in the authors' work on graph minors culminating, ultimately, in the proof of Wagner's conjecture, a topic addressed in §2.4.2.

DEFINITIONS

D11: A **tree-decomposition** of a graph $G = (V, E)$ is a pair $(\{X_i \mid i \in I\}, T)$, where $\{X_i \mid i \in I\}$ is a family of subsets of V, and T is a tree with vertex set I such that:

- $\bigcup_{i \in I} X_i = V$,
- for all edges $(x, y) \in E$ there is an element $i \in I$ with $x, y \in X_i$, and
- for all triples $i, j, k \in I$, if j is on the path from i to k in T, then $X_i \cap X_k \subseteq X_j$.

D12: The **width** of a given tree-decomposition is measured as max $_{i \in I}\{|X_i| - 1\}$.

D13: The **treewidth** of a graph G is the minimum width taken over all tree-decompositions of G.

D14: A graph G is a **treewidth-k graph** if it has treewidth no greater than k.

REMARK

R2: Trivially, every graph, G, has a tree-decomposition that is defined by a single vertex (representing G itself). On the other hand, we are interested in tree-decompositions and, hence, their graphs, in which the X_i are small (i.e., graphs with small treewidth).

EXAMPLE

E8: A sample tree-decomposition is shown in Figure 2.4.9. For the stated graph, G, one family of suitable vertex sets can be given by: $X_1 = \{v_1, v_2, v_3\}$, $X_2 = \{v_2, v_7, v_8\}$, $X_3 = \{v_2, v_3, v_7\}$, $X_4 = \{v_3, v_5, v_7\}$, $X_5 = \{v_3, v_4, v_5\}$, and $X_6 = \{v_5, v_6, v_7\}$. An appropriate tree T is shown next and then on the right side of Figure 2.4.9, the relevant subgraphs of G induced by the stated pair $(\{X_i\}, T)$ are displayed. Moreover, the graph G has treewidth 2; in fact, the graph is series-parallel.

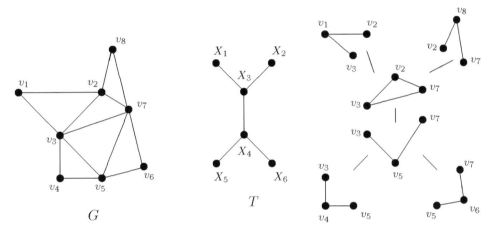

Figure 2.4.9: A sample tree-decomposition.

Pathwidth-k Graphs

DEFINITIONS

D15: A **path-decomposition** is a tree-decomposition whose tree is a path.

NOTATION: A path-decomposition is often denoted simply by a sequence of vertex subsets of V, say $\{X_1, X_2, \ldots, X_t\}$, listed in order defined by their position on the path.

D16: The **width** of a path-decomposition is $\max_{1 \le i \le t}\{|X_i| - 1\}$.

D17: The **pathwidth** of a graph G is the smallest width taken over all path-decompositions of G.

D18: A **pathwidth-k graph** is a graph that has pathwidth no greater than k.

EXAMPLE

E9:   A sample path-decomposition is shown in Figure 2.4.10. The vertex-sets X_i and the first edge occurrences are displayed below the corresponding vertices of T.

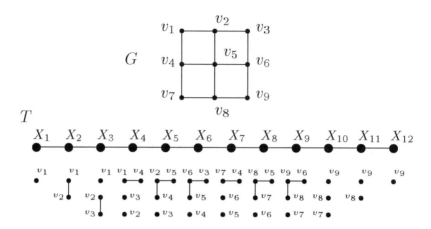

Figure 2.4.10: A path-decomposition.

Branchwidth-k Graphs

DEFINITIONS

D19:   A **branch-decomposition** of a graph $G = (V, E)$ is a pair (T, f), where T is a tree in which every non-leaf vertex has exactly three neighbors and f is a bijection from the leaves of T to E.

D20:   If the degree of every non-leaf vertex in T is *at least* 3, the pair (T, f) is called a **partial branch-decomposition.**

D21:   Let (T, f) be a branch decomposition of a graph $G = (V, E)$. The **order** of an edge e of T is the number of vertices v in V such that there exist leaves l_1 and l_2 of T residing in different components of $T - e$, where $f(l_1)$ and $f(l_2)$ are both incident on v.

D22:   The **width** of a branch decomposition (T, f) is the maximum order of the edges of T.

D23:   The **branchwidth** of G is the minimum width taken over all branch-decompositions of G.

D24:   A graph G is a **branchwidth-k graph** if it has branchwidth no greater than k.

FACTS

F7:   [RoSe91-a] A graph G is branchwidth-0 if and only if every component of G has at most one edge.

F8:   [RoSe91-a] A graph G is branchwidth-1 if and only if every component of G has no more than one vertex with degree greater than or equal to 2.

F9:   [RoSe91-a] A graph G is branchwidth-2 if and only if G has treewidth no greater than 2.

EXAMPLE

E10: A branchwidth-2 graph is shown on the left in Figure 2.4.11 (edges are numbered); its branch-decomposition is given to the right.

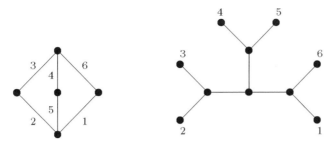

Figure 2.4.11: Branchwidth-2 graph and its branch-decomposition.

k-Terminal Graphs

DEFINITIONS

D25: A k-***terminal graph*** $G = (V, T, E)$ has a vertex set V, an edge set E, and a set of *distinguished terminals* $T = \{t_1, t_2, \ldots, t_{|T|}\} \subseteq V$, where $|T| \leq k$.

D26: A k-***terminal recursively structured class*** $C(B, R)$ is specified by a set B of base graphs and a finite rule set $R = \{f_1, f_2, \ldots, f_n\}$, where each f_i is a *recursive composition operation*.

EXAMPLE

E11: A construction for a 2-terminal graph is shown in Figure 2.4.12. Vertices are labeled in order to clarify how constituent subgraphs compose; terminals are denoted by doubly circled vertices.

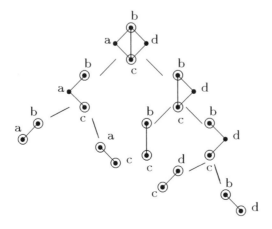

Figure 2.4.12: Recursive construction of a 2-terminal graph.

REMARKS

R3: Typically, for some k, B is the set of connected k-terminal graphs (V, T, E) with $V = T$. But each such base graph is trivially composed of individual edges, so it is reasonable and hence convenient to simply use $C(R)$ to denote $C(B, R)$, where B is a singleton consisting only of edges (i.e., K_2).

R4: The notion of ***composition*** typically permitted in the context of k-terminal graphs can be described in a more formal way. For $1 \leq i \leq m$, let $G_i = (V_i, T_i, E_i)$, such that $V_1, \ldots, V_m$ are mutually disjoint vertex sets. Let $G = (V, T, E)$ as well. Then a ***valid vertex mapping*** is a function $f : \cup_{1 \leq i \leq m} V_i \to V$ such that:

- Vertices from the same G_i remain distinct:

$$v_1 \in V_i, v_2 \in V_i, f(v_1) = f(v_2) \Rightarrow v_1 = v_2$$

- Only (not necessarily all) terminals map to terminals:

$$v \in V_i, f(v) \in T \Rightarrow v \in T_i$$

- Only terminals can merge:

$$v_1 \in V_{i_1}, v_2 \in V_{i_2}, i_1 \neq i_2, f(v_1) = f(v_2) \Rightarrow v_1 \in T_{i_1}, v_2 \in T_{i_2}$$

- Edges are preserved:

$$(\exists i)(\{v_1, v_2\} \in E_i) \Leftrightarrow \{f(v_1), f(v_2)\} \in E$$

NOTATION: If f is a valid vertex mapping, then the corresponding m-ary composition operation (denoted by f) is generally written $f(G_1, \ldots, G_m) = G$.

Cographs

DEFINITION

D27: A ***cograph*** is defined recursively as follows:
- A graph with a single vertex is a cograph.
- If G_1 and G_2 are cographs, then the disjoint union $G_1 \cup G_2$ is a cograph.
- If G_1 and G_2 are cographs, then the cross-product $G_1 \times G_2$ is a cograph, which is formed by taking the union of G_1 and G_2 and adding all edges (v_1, v_2) where v_1 is in G_1 and v_2 is in G_2.

TERMINOLOGY NOTE: Cographs are also referred to as ***complement reducible graphs***.

FACTS

F10: [CoLeBu81] The complement of any cograph is also a cograph.

F11: [CoLeBu81] All cographs are perfect.

EXAMPLE

E12: A cograph construction is demonstrated in Figure 2.4.13. The relevant operations are signified at each node of the decomposition tree (left) for the graph G shown on the right.

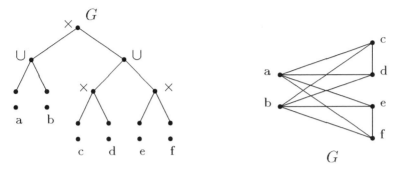

Figure 2.4.13: Cograph construction.

Cliquewidth-k Graphs

The graph parameter *cliquewidth* was introduced in [CoEnRo93] and formed a seminal concept in linking research in graph theory and logic.

DEFINITION

D28: Let [k] denote the set of integers $\{1, 2, \ldots, k\}$. A ***cliquewidth-k graph*** is defined recursively as follows:

- Any graph G with $V(G) = \{v\}$ and $l(v) \in [k]$ is a cliquewidth-k graph.

- If G_1 and G_2 are cliquewidth-k graphs and $i, j \in [k]$ with $i \neq j$, then:

  - The disjoint union $G_1 \cup G_2$ is a cliquewidth-k graph.
  - The graph $(G_1)_{i \times j}$ is a cliquewidth-k graph, where $(G_1)_{i \times j}$ is formed from G_1 by adding all edges (v_1, v_2) such that $l(v_1) = i$ and $l(v_2) = j$.
  - The graph $(G_1)_{i \to j}$ is a cliquewidth-k graph, where $(G_1)_{i \to j}$ is formed from G_1 by switching all vertices with label i to label j.

REMARK

R5: Definition D28 defines the *class* of cliquewidth-k graphs. The cliquewidth of a graph G is the smallest value of k such that G is a cliquewidth-k graph. A cliquewidth decomposition for a graph is a rooted tree such that the root corresponds to G, each leaf corresponds to a labeled, one-vertex graph, and each non-leaf node of the tree is obtained by applying one of the operations $\cup$, $i \times j$, or $i \to j$ to its child or children.

TERMINOLOGY NOTE: In this section, the term *clique* refers to *any* complete subgraph of the graph. In some other sections of this handbook, clique is defined to be a *maximal* subset of pairwise adjacent vertices of the graph.

TERMINOLOGY NOTE: Every tree is a treewidth-1 graph, so treewidth is a measure of how much a graph varies from a tree. Similarly, every clique is a cliquewidth-2 graph, so cliquewidth is a measure of how much a graph varies from a clique. This analogy forms the basis for coining the term *cliquewidth* (cf. [CoOl00]).

EXAMPLE

E13: A cliquewidth-3 construction is given in Figure 2.4.14. As in Example E12, the relevant operations are identified at each node of the decomposition tree (left) for the graph G shown on the right.

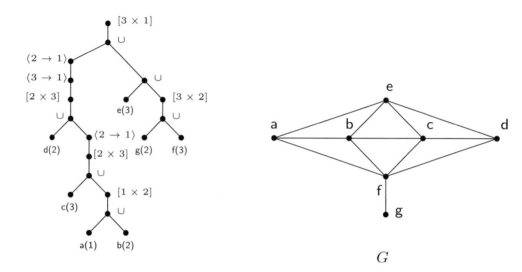

Figure 2.4.14: A cliquewidth-3 graph construction.

k-NLC Graphs

DEFINITION

D29: Let [k] denote the set of integers $\{1, 2, \ldots, k\}$ and let B denote a bipartite graph on $[k] \times [k]$. A **k-NLC (node-label-controlled) graph** is defined recursively as follows:

- Any graph G with $V(G) = \{v\}$ and $l(v) \in [k]$ is a k-NLC graph.

- If G_1 and G_2 are k-NLC graphs and $i, j \in [k]$, then:

    - The join $G_1 \times_B G_2$ is a k-NLC graph, where $G_1 \times_B G_2$ is formed from $G_1 \cup G_2$ by adding all edges (v_1, v_2) where $v_1 \in V_1$, $l(v_1) = i$, $v_2 \in V_2$, $l(v_2) = j$, and (i, j) is an edge in E_B.

    - The graph $(G_1)_{i \to j}$ is a k-NLC graph, which is formed from G_1 by switching all vertices with label i to label j.

EXAMPLE

E14: The same graph G previously shown in Figure 2.4.14 is a 2-NLC graph. In Figure 2.4.15, its decomposition tree has leaves corresponding to the vertices a, b, c, d, e, f, and g with starting labels drawn from the set $k = \{1, 2\}$ as shown. Relevant operations are identified with the internal nodes of the tree, i.e., $(i \rightarrow j)$ for label switching and (i, j) indicating the specific edge from E_B inducing the stated composition.

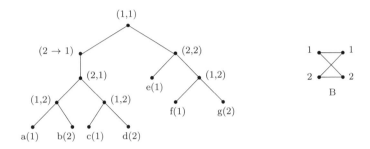

Figure 2.4.15: A 2-NLC graph construction.

k-HB Graphs

The *homogeneous balanced* graphs produce a modular decomposition upon application of a certain decomposition algorithm.

DEFINITION

D30: *k-HB (homogeneous balanced) graphs* are graphs for which there is a particular $O(n^{k+2})$-time top-down decomposition algorithm that constructs a pseudo-cliquewidth-$(k + 2^k)$ balanced decomposition.

REMARKS

R6: Top-down decomposition refers to a recognition algorithm that places a candidate graph at the root of a tree and then decomposes this graph into smaller subgraphs that become its children in the tree, and so on, recursively, until reaching the leaves of the tree.

R7: A pseudo-cliquewidth decomposition is similar to a k-NLC decomposition, except that the vertex labels used at one node in the tree are not enforced at other nodes.

R8: A balanced decomposition of an n-vertex graph is a decomposition tree that has height $O(\log n)$.

R9: The requirement that the decomposition must be balanced is more restrictive, while simultaneously the pseudo-cliquewidth condition is less restrictive. This trade-off yields the class of k-HB graphs. For more details on these matters, see [Jo03], [BoJoRaSp04].

R10: However, k-HB graphs are an ambiguously defined class due to the nondeterministic nature of this decomposition algorithm. On the other hand, the decomposition is guaranteed to succeed for every cliquewidth-k graph despite this nondeterminism, so every cliquewidth-k graph is a k-HB graph.

2.4.2   Equivalences and Characterizations

Relationships between Recursive Classes

A number of equivalences serve to relate many of the recursive graph classes defined in the previous subsection. Several of these are listed below. Unless a specific source is cited, a good general and fairly comprehensive reference for Facts F12 through F19 (and others) is [BrLeSp99].

FACTS

F12: A graph has treewidth at most k if and only if it is a partial k-tree.

F13: Every bandwidth-k graph is a pathwidth-k and thus a treewidth-k graph.

F14: The class of partial k-trees can be defined as a $(k + 1)$-terminal recursive graph class (cf. [WiHe88], [Wi87]).

F15: 1-trees are trees in the usual sense and have treewidth 1.

F16: Trees are series-parallel graphs where only the jackknife operation is used.

F17: Series-parallel graphs in which only the series and parallel operations are used are precisely the 2-terminal series-parallel graphs.

F18: Series-parallel and outerplanar graphs are partial 2-trees and have treewidth 2.

F19: Halin graphs are contained in the class of partial 3-trees; they are also defined as a class of 3-terminal graphs by an appropriate choice of composition operations.

F20: [CoEnRo93] Cographs are precisely the cliquewidth-2 graphs.

F21: [CoRo05] Every treewidth-k graph has cliquewidth at most $3 \cdot 2^{k-1}$.

F22: [RoSe91-a] Every graph of branchwidth at most k has treewidth at most $3k/2$.

F23: [RoSe91-a] Every graph of treewidth at most k has branchwidth at most $k + 1$.

F24: [Wa94] Cographs are exactly the 1-NLC graphs.

F25: [Wa94] Every treewidth-k graph has NLC width at most $2^{k+1} - 1$.

F26: [Jo98] Every cliquewidth-k graph is a k-NLC graph.

F27: [Jo98] Every k-NLC graph is a cliquewidth-$2k$ graph.

F28: [Jo03], [BoJoRaSp04] Every cliquewidth-k graph is a k-HB graph.

Characterizations

Structural characterizations of recursive graph classes are generally stated in terms of forbidden subgraph minors.

DEFINITIONS

D31: An ***edge-extraction*** operation on a graph $G = (V, E)$ removes an edge e leaving a graph, $G - e$, with $V(G - e) = V$ and $E(G - e) = E - \{e\}$.

D32: The operation of ***edge-contraction*** produces a graph with edge-set $E - \{e\}$ but with a vertex-set obtained by replacing ("merging") the vertices defining e in G, thus creating a new single vertex where the latter inherits all of the adjacencies of the pair of replaced vertices, without introducing loops or multiple edges.

D33: A graph H is a ***minor*** of a graph G if and only if it can be obtained from G by a finite sequence of edge-extraction and edge-contraction operations.

REMARKS

R11: A result apparently first conjectured (but unpublished) by K. Wagner asserts the following: Suppose $\mathcal{F}$ is a graph class with the property that if G is in $\mathcal{F}$ and H is contained as a minor in G, then H is in $\mathcal{F}$, i.e., the class $\mathcal{F}$ is closed under minors. Then there exists a finite set $\{H_1, H_2, \ldots, H_k\}$ of graphs, the ***forbidden minors*** such that G is in $\mathcal{F}$ if and only if it contains no minor isomorphic to any member H_i for $1 \leq i \leq k$.

R12: Robertson and Seymour ([RoSe88-b]) confirmed Wagner's conjecture and with their proof established that any graph class $\mathcal{F}$ closed under minors can be recognized in polynomial time. Unfortunately, this outcome, although deep, is an existential one; we do not know the number of forbidden minors or their sizes in an arbitrary case.

R13: The class of partial k-trees is closed under minors and thus, by the Robertson–Seymour results is completely characterized by a finite set of forbidden minors.

R14: The forbidden minors for partial 3-trees are known (see Fact F33 below), but complete lists of explicit minors for partial k-trees are not known for values of $k \geq 4$.

FACTS

F29: [CoLeBu81] Cographs have no induced paths P_4.

F30: Trees are graphs having no K_3 minor.

F31: The class of partial 2-trees is characterized by a single forbidden minor: the complete graph K_4.

F32: The forbidden minors of outerplanar graphs are K_4 and $K_{2,3}$.

F33: The class of partial 3-trees has four forbidden minors: K_5 and the three graphs shown in Figure 2.4.16.

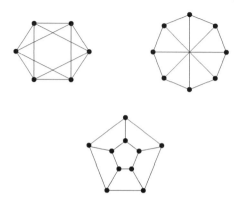

Figure 2.4.16: Forbidden minors of partial 3-trees.

2.4.3   Recognition

In order to solve graph problems on recursive classes and particularly, to do so efficiently, it is necessary that membership in the classes be *quickly recognized*.

REMARKS

R15: Some recognition algorithms are direct and essentially ad hoc. For example, Halin graphs can be recognized by first testing for 3-connectivity and for planarity. Then simply embed the graph in the plane, select a largest cycle of edges that corresponds to a face on the plane embedding, remove these edges and test if the remaining graph is a tree of the stated form (see [CoNaPu83]).

R16: Partial 2-trees or series-parallel graphs are recognizable, unambiguously, by successive application of the following *reduction operations* (cf. [Du65]): replace any vertex of degree 2, say v_j, and its incident edges (v_i, v_j) and (v_j, v_k) by a new edge (v_i, v_k); replace any pair of multiple edges by a single edge; and eliminate any edges incident to a vertex of degree 1 unless only one edge remains. Then a single edge remains, upon an admissible application of these reduction operations, if and only if the original graph is a partial 2-tree; otherwise, the process will stop with either K_4 or a graph with a K_4 minor.

R17: Similar reduction operations have also been described in the case of partial 3-trees (cf. [ArPr86]) as well as for partial 4-trees ([Sa96]).

EXAMPLE

E15: An illustration of a successful reduction sequence is shown in Figure 2.4.17.

Figure 2.4.17: Reduction operations for a partial 2-tree.

Recognition of Recursive Classes

FACTS

F34: Trees can be recognized and their decomposition constructed in linear time.

F35: Series-parallel graphs can be recognized and their decomposition constructed in linear time.

F36: Treewidth-k, pathwidth-k, branchwidth-k, and bandwidth-k graphs can be recognized and their decompositions constructed in $O(n^{k+2})$ time.

COMPUTATIONAL NOTE: For fixed k the polynomial-time algorithms of Fact F36 are practical.

F37: The graph classes of Fact F36 can be recognized in *linear time* for fixed k.

COMPUTATIONAL NOTE: The corresponding algorithms referred to by Fact F37 are not practical because their running times possess enormous hidden constants.

F38: When $k \leq 4$, more practical linear-time recognition algorithms have been found for the graph classes in Fact F36 (cf. [MaTh91] for $k = 3$; [Sa96] when $k = 4$).

F39: When k is part of the problem instance, the recognition problems associated with the graphs of Fact F36 are $\mathcal{NP}$-complete.

F40: Branchwidth can be determined in polynomial time for planar graphs [SeTh94].

F41: Since partial k-trees are characterizable by a finite set of forbidden minors, they are polynomially recognizable (cf. [RoSe88-b]).

COMPUTATIONAL NOTE: Fact F41 was established in the graph minors results of Robertson and Seymour. However, the result is existential rather than constructive and so the actual exhibition of the implied algorithms remains elusive.

F42: [Wi87] Every k-terminal graph is a treewidth-k' graph for some k' that depends upon k and the particular set of recursive composition operations. For example, if m denotes the maximum *arity* of any operation, then $k' \leq km$.

TERMINOLOGY NOTE: The term "arity" refers to the number of operands. For example, a binary operation has arity 2.

F43: [CoPeSt85] Cographs can be recognized and their decomposition constructed in linear time.

F44: [CoHaLaReRo00] Recognition of cliquewidth-k graphs is solvable in polynomial time for $k \leq 3$, but this problem remains open for fixed $k \geq 4$.

F45: [FeRoRoSz09] Recognition of cliquewidth-k graphs is an $\mathcal{NP}$-complete problem for arbitrary k.

F46: [Jo00] Recognition of k-NLC graphs is solvable in polynomial time for $k \leq 2$, but this problem remains open for fixed $k \geq 3$.

F47: [GuWa05] Recognition of k-NLC graphs is $\mathcal{NP}$-complete for arbitrary k.

F48: [Jo03], [BoJoRaSp04] Algorithms for problems defined on k-HB graphs are robust with respect to cliquewidth-k graphs. That is, such an algorithm either determines the correct answer or reports that the decomposition was unsuccessful and hence the input graph is not a cliquewidth-k graph. The $O(n^{k+2})$-time decomposition algorithm for k-HB graphs is guaranteed to succeed for all cliquewidth-k graphs as well as some others.

References

[ArCoPr87] S. Arnborg, D. G. Corneil, and A. Proskurowski, Complexity of finding embeddings in a k-tree, *SIAM J. Algebraic and Discrete Methods* 8 (1987), 277–284.

[ArCoPrSe93] S. Arnborg, B. Courcelle, A. Proskurowski, and D. Sesse, An algebraic theory of graph reductions, *J. ACM* 40 (1993), 1134–1164.

[ArHePr94] S. Arnborg, S. Hedetniemi, and A. Proskurowski (eds.), Efficient algorithms and partial k-trees, special issue of *Discrete Applied Math.* 54 (1994).

[ArPr85] S. Arnborg and A. Proskurowski, Characterization and recognition of partial k-trees, *Congressus Numerantium* 47 (1985), 69–75.

[ArPr86] S. Arnborg and A. Proskurowski, Characterization and recognition of partial 3-Trees, *SIAM J. Algebraic and Discrete Methods* 7 (1986) 305–314.

[ArPrCo90] S. Arnborg, A. Proskurowski, and D. G. Corneil, Forbidden minors characterization of partial 3-trees, *Discrete Math.* 80 (1990), 1–19.

[BaOl98] L. Babel and S. Olariu, On the structure of graphs with few P_4s, *Discrete Applied Math.* 84 (1998), 1–13.

[BePi71] L. W. Beineke and R. E. Pippert, Properties and characterizations of k-trees, *Mathematika* 18 (1971), 141–151.

[BiRoSeTh91] D. Bienstock, N. Robertson, P. D. Seymour, and R. Thomas, Quickly excluding a forest, *J. Combin. Theory Series B* 52 (1991), 274–283.

[Bo88] R. B. Borie, Recursively constructed graph families: membership and linear algorithms, Ph.D. Dissertation, School of Information and Computer Science, Georgia Institute of Technology, 1988.

[Bo90] H. L. Bodlaender, Classes of graphs with bounded treewidth, *Tech. Rep. RUU-CS-86-22*, Department of Computer Science, Utrecht University, The Netherlands, 1990.

[Bo93] H. L. Bodlaender, A tourist guide through treewidth, *Acta Cybernetica* 11 (1993), 1–23.

[Bo96] H. L. Bodlaender, A linear-time algorithm for finding tree decompositions of small treewidth, *SIAM J. Computing* 25 (1996), 1305–1317.

[BoGiHaKl95] H. L. Bodlaender, J. R. Gilbert, H. Hafsteinsson, and T. Kloks, Approximating treewidth, pathwidth, and minimum elimination tree height, *J. Algorithms* 18 (1995), 238–255.

[BoJoRaSp04] R. B. Borie, J. L. Johnson, V. Raghavan, and J. Spinrad, Robust algorithms for some problems on graphs of bounded clique-width, manuscript, presented at *SIAM Conference on Discrete Mathematics* (2004).

[BoKl96] H. L. Bodlaender and T. Kloks, Efficient and constructive algorithms for the pathwidth and treewidth of graphs, *J. Algorithms* 21 (1996), 358–402.

[BoMo93] H. L. Bodlaender and R. H. Möhring, The pathwidth and treewidth of cographs, *SIAM J. Discrete Math.* 6 (1993), 181–186.

[BoPaTo91] R. B. Borie, R. G. Parker, and C. A. Tovey, Deterministic decomposition of recursive graph classes, *SIAM J. Discrete Math.* 4 (1991), 481–501.

[BrLeSp99] A. Brandstädt, V. B. Le, and J. Spinrad, *Graph Classes: A Survey*, SIAM Monographs on Discrete Math. and Applications, SIAM, Philadelphia (1999).

[Co90] B. Courcelle, The monadic second-order logic of graphs I: Recognizable sets of finite graphs, *Inform. and Comput.* 85 (1990), 12–75.

[Co92] B. Courcelle, The monadic second-order logic of graphs III: Tree-decompositions, minors, and complexity issues, *Inform. Théorique Appl.* 26 (1992), 257–286.

[Co95] B. Courcelle, The monadic second-order logic of graphs VIII: Orientations, *Annals of Pure and Applied Logic* 72 (1995), 103–143.

[Co96] B. Courcelle, The monadic second-order logic of graphs X: Linear orderings, *Theoretical Computer Science* 160 (1996), 87–143.

[CoEnRo93] B. Courcelle, J. Engelfriet, and G. Rozenberg, Handle-rewriting hypergraph grammars, *J. of Computer and Systems Sciences* 46 (1993), 218–270.

[CoHaLaReRo00] D. G. Corneil, M. Habib, J. M. Lanlignel, B. Reed, and U. Rotics, Polynomial-time recognition of clique-width ≤ 3 graphs, *Lecture Notes in Computer Science* 1776 (2000), 126–134.

[CoKi83] D. G. Corneil and D. G. Kirkpatrick, Families of recursively defined perfect graphs, *Congressus Numerantium* 39 (1983), 237–246.

[CoLeBu81] D. G. Corneil, H. Lerchs, and L. S. Burlington, Complement reducible graphs, *Discrete Applied Math.* 3 (1981), 163–174.

[CoMo93] B. Courcelle and M. Mosbah, Monadic second-order evaluations on tree-decomposable graphs, *Theoretical Computer Science* 109 (1993), 49–82.

[CoNaPu83] G. Cornuejols, D. Naddef, and W. R. Pulleyblank, Halin graphs and the travelling salesman problem, *Math Programming* 26 (1983), 287–294.

[CoOl00] B. Courcelle and S. Olariu, Upper bounds to the clique-width of graphs, *Discrete Applied Math.* 101 (2000), 77–114.

[CoPeSt84] D. G. Corneil, Y. Perl, and L. K. Stewart, Cographs: recognition, applications and algorithms, *Congressus Numerantium* 43 (1984), 249–258.

[CoPeSt85] D. G. Corneil, Y. Perl, and L. Stewart, A linear recognition algorithm for cographs, *SIAM J. Computing* 14 (1985), 926–934.

[CoRo05] D. G. Corneil and U. Rotics, On the relationship between clique-width and treewidth, *SIAM Journal on Computing* 34 (2005), 825–847.

[Du65] R. J. Duffin, Topology of series-parallel graphs, *J. Math. Anal. Appl.* 10 (1965), 303–318.

[FeRoRoSz09] M. R. Fellows, F. A. Rosamond, U. Rotics, and S. Szeider, Clique-width is $\mathcal{NP}$-complete, *SIAM J. on Discrete Mathematics* 23 (2009), 909–939.

[GoRo99] M. C. Golumbic and U. Rotics, On the clique-width of perfect graph classes, *Lecture Notes in Computer Science* 1665 (1999), 135–147.

[GrSk91] D. Granot and D. Skorin-Kapov, NC Algorithms for recognizing partial 2-trees and 3-trees, *SIAM J. Algebraic and Discrete Methods* 4 (1991), 342–354.

[GuWa05] F. Gurski and E. Wanke, Minimizing NLC-width is $\mathcal{NP}$-complete, *Lecture Notes in Computer Science* 3787 (2005), 69–80.

[HeYe87] X. He and Y. Yesha, Parallel recognition and decomposition of two-terminal series-parallel graphs, *Information and Computing* 75 (1987), 15–38.

[Jo98] O. Johansson, Clique-decomposition, NLC-decomposition, and modular decomposition relationships and results for random graphs, *Congressus Numerantium* 132 (1998), 39–60.

[Jo00] O. Johansson, NLC 2-decomposition in polynomial time, *International Journal of Foundations of Computer Science* 11 (2000), 373–395.

[Jo03] J. Johnson, Polynomial time recognition and optimization algorithms on special classes of graphs, Ph.D. Dissertation, Computer Science, Vanderbilt University, 2003.

[KaIsUe85] Y. Kajitani, A. Ishizuka, and S. A. Ueno, Characterization of the partial k-tree in terms of certain structures, *Proc. ISCAS '85* (1985), 1179–1182.

[Kl94] T. Kloks, Treewidth: computations and approximations, *Lecture Notes in Computer Science* 842 (1994).

[KlKr95] T. Kloks and D. Kratsch, Treewidth of chordal bipartite graphs, *J. Algorithms* 19 (1995), 266–281.

[MaTh91] J. Matoušek and R. Thomas, Algorithms for finding tree-decompositions of graphs, *J. Algorithms* 12 (1991), 1–22.

[Pr93] A. Proskurowski, Graph reduction and techniques for finding minimal forbidden minors, in *Proc. AMS Workshop on Graph Minors*, Seattle 1991, Contemporary Math. AMS 147, Graph Structure Theory (1993), 591–600.

[Re92] B. Reed, Finding approximate separators and computing treewidth quickly, in *Proc. 9th Symposium on Theoretical Aspects of Computer Science* (1992), 221–228.

[Re93] B. Reed, Treewidth and Tangles: A New Connectivity Measure and Some Applications, in *Surveys in Combinatorics* (1997), Cambridge University Press, 87–162.

[Ro74] D. J. Rose, On simple characterization of k-trees, *Discrete Math.* 7 (1974), 317–322.

[RoSe83] N. Robertson and P. D. Seymour, Graph minors. I. Excluding a forest, *J. Combin. Theory Series B* 35 (1983), 39–61.

[RoSe84] N. Robertson and P. D. Seymour, Graph minors. III. Planar treewidth, *J. Combin. Theory Series B* 36 (1984), 49–64.

[RoSe86-a] N. Robertson and P. D. Seymour, Graph minors. II. Algorithmic aspects of treewidth, *J. Algorithms* 7 (1986), 309–322.

[RoSe86-b] N. Robertson and P. D. Seymour, Graph minors. V. Excluding a planar graph, *J. Combin. Theory Series B* 41 (1986), 92–114.

[RoSe86-c] N. Robertson and P. D. Seymour, Graph minors. VI. Disjoint paths across a disc, *J. Combin. Theory Series B* 41 (1986), 115–138.

[RoSe88-a] N. Robertson and P. D. Seymour, Graph minors. VII. Disjoint paths on a surface, *J. Combin. Theory Series B* 45 (1988), 212–254.

[RoSe88-b] N. Robertson and P. D. Seymour, Graph minors. XX. Wagner's conjecture, manuscript (1988), *J. Combin. Theory Series B* 92 (2004), 325–357.

[RoSe90–a] N. Robertson and P. D. Seymour, Graph minors. IV. Treewidth and well-quasi-ordering, *J. Combin. Theory Series B* 48 (1990), 227–254.

[RoSe90-b] N. Robertson and P. D. Seymour, Graph minors. IX. Disjoint crossed paths, *J. Combin. Theory Series B* 49 (1990), 40–77.

[RoSe90-c] N. Robertson and P. D. Seymour, Graph minors. VIII. A Kuratowski theorem for general surfaces, *J. Combin. Theory Series B* 48 (1990), 255–288.

[RoSe91-a] N. Robertson and P. D. Seymour, Graph minors. X. Obstructions to tree-decompositions, *J. Combin. Theory Series B* 52 (1991), 153–190.

[RoSe91-b] N. Robertson and P. D. Seymour, Graph minors. XVI. Excluding a non-planar graph, manuscript (1991), *J. Combin. Theory Series B* 89 (2003), 43–76.

[RoSe92] N. Robertson and P. D. Seymour, Graph minors. XXII. Irrelevant vertices in linkage problems, manuscript (1992), *J. Combin. Theory Series B*, to appear.

[RoSe94] N. Robertson and P. D. Seymour, Graph minors. XI. Distance on a surface, *J. Combin. Theory Series B* 60 (1994), 72–106.

[RoSe95] N. Robertson and P. D. Seymour, Graph minors. XIII. The disjoint paths problem, *J. Combin. Theory Series B* 63 (1995), 65–110.

[RoSeTh94] N. Robertson, P. D. Seymour, and R. Thomas, Quickly excluding a planar graph, *J. Combin. Theory Series B* 62 (1994), 323–348.

[Sa96] D. P. Sanders, On linear recognition of treewidth at most four, *SIAM J. Discrete Math.* 9 (1996), 101–117.

[SaTu90] A. Satyanarayana and L. Tung, A characterization of partial 3-trees, *Networks* 20 (1990), 299–322.

[Sc88] P. Scheffler, What graphs have bounded treewidth?, in *Proc. Fischland Collo-quium on Discrete Math. and Applications*, Rostock Math. Kolloq. (1988).

[SeTh94] P. D. Seymour and R. Thomas, Call routing and the ratcatcher, *Combinator-ica* 14 (1994), 217–241.

[Wa94] E. Wanke, k-NLC graphs and polynomial algorithms, *Discrete Applied Math.* 54 (1994), 251–266.

[Wi87] T. V. Wimer, Linear algorithms on k-terminal graphs, Ph.D. Dissertation, De-partment of Computer Science, Clemson University, 1987.

[WiHe88] T. V. Wimer and S. T. Hedetniemi, *K*-terminal recursive families of graphs, *Congressus Numerantium* 63 (1988), 161–176.

Section 2.5
Structural Graph Theory

Maria Chudnovsky, Columbia University

INTRODUCTION

The goal of this section is to survey some recent results in structural graph theory. One of the greatest achievements of structural graph theory to this day has been the Robertson–Seymour Graph Minor Project, which, in addition to answering a long standing open conjecture of Wagner, revolutionized the field and laid foundation to a large body of research that is being conducted today. For this very reason, numerous excellent surveys on the topic of minors have been written over the years (see, for example, [Di05, Re97, Th99]), and so in this section we concentrate on other aspects of structural graph theory. The bulk of the section is devoted to the area of induced subgraphs, which received a fair amount of attention in the past 10 years, due to the proof of the Strong Perfect Graph Conjecture, a famous open question, posed by Claude Berge in 1961 [Be61], which was finally solved in the early 2000s [ChRoSeTh06]. We conclude the section with a survey of new results on tournament immersion.

DEFINITIONS

All our graphs are finite and simple (except when we explicitly say otherwise). Let us start with some definitions. We omit some very basic and standard definitions here; the reader is referred to [Di05] or [We01] for those. Let G be a graph. We denote its vertex set by $V(G)$, and its edge set by $E(G)$.

D1: The *complement* of G, denoted by G^c, is a graph with vertex set $V(G)$, such that two vertices are adjacent in G^c if and only if they are non-adjacent in G.

D2: A *clique* in G is a set of vertices of G, all pairwise adjacent.

D3: A *stable set* in G is a set of vertices, all pairwise non-adjacent. Thus, $S \subseteq V(G)$ is a stable set in G if and only if it is a clique in G^c.

D4: The largest size of a clique in G, the *clique number* of G, is denoted by $\omega(G)$, and the largest size of a stable set, the *stability number* of G, by $\alpha(G)$.

D5: The *chromatic number* of G, denoted by $\chi(G)$, is the smallest number k for which the vertices of G can be colored with k colors, so that no two adjacent vertices receive the same color.

D6: Given a graph H, we say that H is an *induced subgraph* of G if $V(H) \subseteq V(G)$, and $uv \in E(H)$ if and only if $uv \in E(G)$ for every $u, v \in V(H)$.

D7: Given a graph H, we say that the graph G is an *H-free graph* if no induced subgraph of G is isomorphic to H. For $X \subseteq V(G)$, the subgraph of G induced by X is denoted $G|X$.

D8: A *component* of G is a maximal connected subgraph of G.

D9: We say that the graph G is *anticonnected* if its complement G^c is connected.

D10: An *anticomponent* of G is a maximal anticonnected induced subgraph of G.

2.5.1   Perfect Graphs

Let G be a graph. It follows immediately from the definitions of the clique number and the chromatic number that $\chi(G) \geq \omega(G)$. It is then natural to ask: for what graphs G does equality hold, that is, when is $\chi(G) = \omega(G)$? It turns out that the two parameters are equal in many natural classes of graphs: bipartite graphs (these are graphs whose vertex set is the union of two stable sets), complements of bipartite graphs [Ko16], comparability graphs [Di50] (these are graphs whose vertices are elements of a given partially ordered set P, and two elements are adjacent if and only if they are comparable in P), and many others. However, here is an unfortunate example: take a graph G', and let G be the union of G' with a complete graph on $|V(G')|$ vertices. Clearly $\chi(G) = \omega(G) = |V(G')|$, and yet we have not learned anything about the structure of the subgraph G' of G.

Thus, to get a nice answer, we need to modify the question a little. In 1961 Claude Berge came up with what seems to be right modification, the notion of a "perfect graph" [Be61].

DEFINITIONS

D11: A graph G is a *perfect graph* if $\chi(H) = \omega(H)$ for every induced subgraph H of G. A graph that is not perfect is called *imperfect*.

D12: A *cycle of length* n (where $n \geq 3$ is an integer) is the graph with vertex set $\{v_1, \ldots, v_n\}$, such that v_i is adjacent to v_j if and only if $|i - j| = 1 \bmod n$. We denote this graph by C_n.

EXAMPLES

E1: Bipartite graphs, their complements, comparability graphs, and many other natural classes of graphs are all perfect, but the pathological example we constructed at the end of the last paragraph may not be.

These two examples of imperfect graphs will be important in what follows.

E2: It is easy to see that if $n \geq 5$ is odd, then $\omega(C_n) = 2$ and $\chi(C_n) = 3$, and therefore C_n is imperfect.

E3: Let us next consider C_n^c, again with $n \geq 5$ odd. In this graph, the largest clique has size $\lfloor \frac{n}{2} \rfloor$, and the chromatic number is $\lceil \frac{n}{2} \rceil$ (since the size of the largest stable set is two, and so at most two vertices can be colored with a given color), so C_n^c is also imperfect.

Fact F1 arose as a conjecture of Berge. It is now known as the **Weak Perfect Graph Theorem**.

FACT, FIRST CONJECTURED BY BERGE

F1: (Lovász [Lo72]). A graph G is perfect if and only if G^c is perfect.

DEFINITIONS

D13: Let G be a graph, and H be an induced subgraph of G. We say that H is a **hole** if H is isomorphic to C_n for some integer $n \geq 4$; moreover, H is an **odd hole** if n is odd, and an **even hole** if n is even.

D14: Similarly, H is an **antihole** if H is isomorphic to C_n^c for some integer $n \geq 4$; also H is an **odd antihole** if n is odd, and an **even antihole** if n is even.

D15: Let us say that a graph is a **Berge graph** if G has no odd holes and no odd antiholes (this terminology is due to Chvátal).

Fact F2 also arose as a conjecture of Berge. Since every induced subgraph of a perfect graph is perfect, the "only if" direction of Fact F2 follows immediately. The "if" direction remained open for more than 40 years, until it was proved in the early 2000s. This fact is commonly known as the **Strong Perfect Graph Theorem**.

ANOTHER FACT, ALSO FIRST CONJECTURED BY BERGE

F2: (Chudnovsky, Robertson, Seymour, and Thomas [ChRoSeTh06]). A graph G is perfect if and only if it has no odd holes and no odd antiholes.

Outline of the Proof of the Strong Perfect Graph Theorem

Since its introduction by Berge, the theory of perfect graphs became a very active area of graph theory (largely motivated by the attempts to prove Fact F2), and so we will spend some time here discussing the ideas of that proof. The key idea is to "describe all Berge graphs".

The main part of the proof of Fact F2 is devoted to proving a theorem that says that every Berge graph is either "basic" (meaning that it belongs to a well-understood class of graphs, all of whose members are perfect), or admits a "useful decomposition" (this is a decomposition that cannot occur in an imperfect Berge graph with a minimum number of vertices). Let us call this a "decomposition theorem".

Any theorem of this form would imply Fact F2; let us quickly run through the implication. Assume that Fact F2 is false; then there exists an imperfect Berge graph G with $|V(G)|$ minimum. Apply the theorem to G. Then G is either basic (and therefore perfect, which is a contradiction), or admits a useful decomposition, which is again a contradiction by the definition of usefulness. The fact that a decomposition theorem of this form should exist had been a growing belief in the field for a number of years prior to the proof of Fact F2, and was finally formulated, but not published, as a conjecture by Cornuéjols, Conforti, and Vušković.

Let us now make this more precise.

DEFINITIONS

D16: Given a graph H, its *line graph* $L(H)$ is the graph with vertex set $E(H)$, and $ef \in E(L(H))$ if and only if the edges e and f share an endpoint in H.

D17: We say that G is a *double-split graph* if $V(G)$ can be partitioned into four sets $\{a_1, \ldots, a_m\}$, $\{b_1, \ldots, b_m\}$, $\{c_1, \ldots, c_n\}$, $\{d_1, \ldots, d_n\}$ for some $m, n \geq 2$, such that:
- a_i is adjacent to b_i for $1 \leq i \leq m$, and c_j is nonadjacent to d_j for $1 \leq j \leq n$.
- there are no edges between $\{a_i, b_i\}$ and $\{a_{i'}, b_{i'}\}$ for $1 \leq i < i' \leq m$, and all four edges between $\{c_j, d_j\}$ and $\{c_{j'}, d_{j'}\}$ for $1 \leq j < j' \leq n$.
- there are exactly two edges between $\{a_i, b_i\}$ and $\{c_j, d_j\}$ for $1 \leq i \leq m$ and $1 \leq j \leq n$, and these two edges have no common end.

D18: Let us say that a graph G is a *Berge-basic graph* if either G or G^c is bipartite, the line graph of a bipartite graph, or a double-split graph ("Berge-basic" is an ad hoc definition; we use it to avoid confusion with graphs that we would like to consider basic in other settings). It is not difficult to see (using theorems of König [Ko16]) that all Berge-basic graphs are perfect.

Next let us define the useful decompositions used in [ChRoSeTh06].

D19: A *skew-partition* in a graph G is a partition (A, B) of $V(G)$ such that A is not connected and B is not anticonnected. Skew-partitions were first introduced by Chvátal [Ch85].

D20: A *proper 2-join* (a special case of a decomposition defined by Cornuéjols and Cunningham [CoCu85]) in G is a partition (X_1, X_2) of $V(G)$ such that there exist disjoint nonempty $A_i, B_i \subseteq X_i$ ($i = 1, 2$) satisfying:
- every vertex of A_1 is adjacent to every vertex of A_2, and every vertex of B_1 is adjacent to every vertex of B_2,

- there are no other edges between X_1 and X_2,
- for $i = 1, 2$, every component of $G|X_i$ meets both A_i and B_i, and
- for $i = 1, 2$, if $|A_i| = |B_i| = 1$ and $G|X_i$ is a path joining the members of A_i and B_i, then it has odd length ≥ 3.

D21: If $X \subseteq V(G)$ and $v \in V(G) \setminus X$, we say that v is X-*complete* if v is adjacent to every vertex in X, and that v is X-*anticomplete* if v has no neighbors in X.

D22: If $X, Y \subseteq V(G)$ are disjoint, we say that X is **complete** to Y (or the pair (X, Y) is **complete**) if every vertex in X is Y-complete; and being **anticomplete** to Y is defined similarly.

D23: Finally, a ***proper homogeneous pair*** in G (a slight variation of a decomposition by Chvátal and Sbihi [ChSb87]) is a pair of disjoint nonempty subsets (A, B) of $V(G)$, such that, if A_1, A_2, respectively, denote the sets of all A-complete vertices and all A-anticomplete vertices in $V(G)$, and if B_1, B_2 are defined similarly, then:

- $A_1 \cup A_2 = B_1 \cup B_2 = V(G) \setminus (A \cup B)$ (and in particular, every vertex in A has a neighbor in B and a non-neighbor in B, and vice versa), and
- the four sets $A_1 \cap B_1, A_1 \cap B_2, A_2 \cap B_1, A_2 \cap B_2$ are all nonempty.

REMARK

R1: Note that if G admits a skew-partition then so does G^c, and the same holds for homogeneous pairs. However, a 2-join in G is substantially different from a 2-join in G^c. In fact, Fact F2 uses a slight variant of the skew-partition decomposition (also invariant under taking complements), called a *balanced skew-partition*, but the definition of that is somewhat technical, and we omit it here.

FACT

We can now state the decomposition theorem of [ChRoSeTh06].

F3: [ChRoSeTh06] Let G be a Berge graph. Then either

- G is Berge-basic, or
- G admits a balanced skew-partition, or
- G admits a proper homogeneous pair, or
- one of G, G^c admits a proper 2-join.

REMARKS

R2: In [Ch03, Ch06], the author showed that the proper homogeneous pair decomposition is in fact unnecessary, and Fact F3 remains true if we simply omit it. We will come back to this result later, in §2.5.4, since it contains a tool that has since been very useful in the study of induced subgraphs. The proof of Fact F3 occupies most of [ChRoSeTh06], which is over 150 pages long. Recently, Seymour and the author were able to shorten the proof of Fact F2 by about a third, proving a theorem that is similar to Fact F3 but uses more kinds of decompositions [ChSe09].

R3: Even though Fact F3 provides enough structural information about Berge graphs to prove Fact F2, it is not what we would call a "structure theorem" for Berge (or perfect) graphs. The problem is that while Fact F3 gives a way to break (or decompose) a big Berge graph into smaller pieces, this decomposition cannot be "reversed". We cannot use Fact F3 to prove a theorem that says "every perfect graph can be built from basic pieces by gluing them together via certain operations; and every graph built in this way is perfect". Thus, while we have a *decomposition theorem* for perfect graphs, we do not have a *structure theorem*. The question of explicitly describing the structure of perfect graphs is currently wide open, and any progress on it would be a breakthrough in the area.

2.5.2 Other Decomposition Theorems

This subsection is devoted to more decomposition results for graphs with certain induced subgraphs forbidden. The theorems in this subsection have not, so far, been turned into explicit structure theorems. Of course, a carefully crafted decomposition theorem is often enough to answer a specific question about a certain class of graphs (just as Fact F3 was strong enough to prove Fact F2); decomposition theorems also often have strong algorithmic consequences.

DEFINITION

D24: Let $\mathcal{F}$ be a family of graphs. We say that a graph G is $\mathcal{F}$-*free* if G is F-free for every $F \in \mathcal{F}$.

Defining $\mathcal{C}$ to be the family of all cycles of odd length at least five, and their complements, Fact F3, with the strengthening of [Ch03, Ch06], can be restated as follows:

FACT

F4: Let G be a $\mathcal{C}$-free graph. Then either

- G is Berge-basic, or
- G admits a balanced skew-partition, or
- one of G, G^c admits a proper 2-join.

DEFINITION

D25: A k-*star cutset* in a graph G is a partition (A, B) of $V(G)$ such that A is not connected, and there is a clique $K \subseteq B$, such that $|K| = k$, and every vertex of $B \setminus K$ has a neighbor in K. A 1-star cutset is usually called a *star cutset*.

We observe that every star cutset is a skew-partition, and every skew-partition is a 2-star cutset (but the converse implications are false). In this subsection we also refer to a 2-*join decomposition*; it is a slight variant of the proper 2-joins that we have used so far, but we will not define it exactly.

Next, we discuss a number of decomposition theorems for $\mathcal{F}$-free graphs for various families $\mathcal{F}$. Here we use the word "basic" loosely, to mean that a graph belongs to some well-understood explicitly constructed class of graphs, all of which are $\mathcal{F}$-free for the family $\mathcal{F}$ in question.

DEFINITION

D26: Let $\mathcal{B}$ be the family of all cycles C_n where n is not divisible by 4; the $\mathcal{B}$-free graphs are called ***balanced graphs***.

FACT

F5: Conforti and Rao [CoRa92] proved a decomposition theorem for $\mathcal{B}$-free graphs, where they showed that every such graph is either basic ("strongly balanced") or admits a 2-join, or an "extended star cutset" (this is a variation on the star cutset theme).

This was a ground-breaking result, the first theorem of the kind. In particular it lead to a polynomial-time recognition algorithm for the class of balanced graphs.

Here are more results of a similar flavor.

DEFINITIONS

D27: Let $\mathcal{F}_{odd}$ be the family of cycles C_n with odd $n > 3$, and $\mathcal{F}_{even}$ the family of cycles C_n with even n. An $\mathcal{F}_{odd}$-free graph is called ***odd-hole-free***, and an $\mathcal{F}_{even}$-free graph is called ***even-hole-free***.

FACTS

F6: [CoCoVu04]. Every odd-hole-free graph is either basic, or admits a 2-join or a k-star cutset, for some $k \leq 2$.

In [CoCoKaVu02] a similar result was proved for $\mathcal{F}_{even}$-free graphs. More precisely:

F7: [CoCoKaVu02] Every even-hole-free graph is either basic, or admits a 2-join or admits a k-star cutset, for some $k \leq 3$.

REMARK

R4: Fact F7 was later strengthened in [daSiVu13], where only 2-joins and star cutsets are used (though the list of basic classes had to be extended). Each of these two theorems can be used to design a polynomial-time recognition algorithm for even-hole-free graphs; we will discuss those later, in §2.5.5.

Incidentally, here is another structural property of even-hole-free graphs (which is neither a structure theorem nor a decomposition theorem).

DEFINITION

D28: A vertex v in a graph G is a ***bisimplicial vertex*** if the set of neighbors of v in G is the union of two cliques.

FACT

F8: [AdChHaReSe08] Every non-null even-hole-free graph has a bisimplicial vertex.

Star cutsets (along with other "tightly structured" cutsets) are used in the proof of Fact F8 as a way to reduce the problem to a smaller graph and use inductive arguments.

2.5.3   Structure Theorems

In contrast with the previous subsection, here we survey some results that provide explicit structural descriptions of $\mathcal{F}$-free graphs for some families $\mathcal{F}$. In each of the classes of graphs now considered, the proof of the structure theorem follows a certain outline: first a decomposition theorem is proved, and then it is shown that the decompositions can be reversed and turned into "compositions" (that is, ways to glue two smaller graphs in a class in such a way that the resulting graph is also in the class). One then analyzes the effect of repeatedly performing the composition operations, starting from basic graphs, and an explicit structural description emerges. Theorems 6.2 and 7.3 of [Vu13] provide further examples of decomposition theorems that can be turned into compositions.

Claw-Free Graphs

DEFINITION

D29: A *claw* is the complete bipartite graph $K_{1,3}$, and a graph is *claw-free* if it is $K_{1,3}$-free (in other words, no vertex has three pairwise non-adjacent neighbors).

EXAMPLES

E4: Claw-free graphs are a generalization of line graphs (it is not difficult to see that if $G = L(H)$ for some graph H, then G is claw-free). But there are others. For example, the skeleton of the icosahedron (this is the unique 5-regular planar graph on 12 vertices) and the Schläfli graph (a highly symmetric 27-vertex graph that comes up naturally in the geometry of polytopes) are examples of claw-free graphs that are far from being line graphs.

DEFINITIONS

Another interesting subclass of claw-free graphs is called *circular interval graphs*.

D30: Let Σ be a circle, and let $F_1, \ldots, F_k \subseteq \Sigma$ be homeomorphic to the interval $[0, 1]$, such that no two of $F_1, \ldots, F_k$ share an end-point. Now let $V \subseteq \Sigma$ be finite, and let G be a graph with vertex set V in which for distinct $u, v \in V$, u is adjacent to v if and only if $u, v \in F_i$ for some i. Such a graph G is called a *circular interval graph*. If in addition no three of $F_1, \ldots, F_k$ have union Σ, then G is a *long circular interval graph*.

D31: A *composition of strips* is a generalization of a line-graph: given a graph H, in $L(H)$ every edge of H is replaced by a vertex, and vertices that correspond to edges that share an end are made adjacent; in a composition of strips that corresponds to H, every edge of H is replaced by a member of 1 of 15 prescribed families of graphs (a "strip"), and then certain edges are added between the subgraphs corresponding to edges of H that share an end.

FACT

The main theorem of [ChSe08] gives an explicit description of all claw-free graphs. To state this theorem precisely would take several pages, so let us instead describe it roughly.

F9: [ChSe08] Every connected claw-free graph G with $\alpha(G) \geq 4$ is either a ***a thickening of a long circular interval graph*** (this is a slight generalization of a long circular interval graph), or a ***composition of strips***.

Obviously, all graphs G with $\alpha(G) = 2$ are claw-free, so it remains to construct those claw-free graphs G for which $\alpha(G) = 3$. This turns out to be the most complex part of both the proof and the statement of the main theorem of [ChSe08]; graphs obtained in certain ways from the skeleton of the icosahedron are one class of claw-free graphs with stability number three, but there are many others that we will not describe here.

Quasi-Line Graphs

DEFINITION

D32: A graph is a ***quasi-line graph*** if each of its vertices is bisimplicial.

EXAMPLES

E5: Every line graph is quasi-line.

E6: Every quasi-line graph is claw-free.

E7: The 5-wheel (C_5, together with a vertex complete to its vertex set) is claw-free and not quasi-line.

E8: Long circular interval graphs are quasi-line and not line graphs.

FACT

The structure of quasi-line graphs is described in [ChSe12], and it is much simpler than that of general claw-free graphs. The main result of [ChSe12] states that:

F10: [ChSe12] There are only two kinds of connected quasi-line graphs: thickening of circular interval graphs (again, this is a slight generalization of circular interval graphs), and compositions of strips, with only two kinds of strips permitted.

REMARK

R5: In [ChPl13] an explicit structural description of **perfect claw-free graphs** is given (these were originally studies in [ChSb88] and [MaRe99]).

Bull-Free Graphs

DEFINITIONS

D33: The **bull** is the graph with vertex set $\{v_1, v_2, v_3, v_4, v_5\}$ where $\{v_2, v_3, v_4\}$ is a clique, v_1 is adjacent to v_2, v_4 is adjacent to v_5, and there are no other edges.

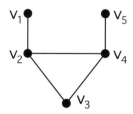

Figure 2.5.1: The graph known as the bull.

D34: A graph G is **bull-free** if no induced subgraph of G is isomorphic to the bull.

REMARK

R6: The structure of bull-free graphs was originally studied in connection with perfect graphs in [ChSb87], but that paper contains only decomposition theorems, and no structure theorem. The structure of general bull-free graphs was described in [Ch12, Ch12a], and we discuss it next.

FACTS

F11: Since the bull is isomorphic to its complement, the class of bull-free graphs is closed under complementation.

F12: [Ch12, Ch12a] There are basically two types of bull-free graphs, $\mathcal{T}_1$, and $\mathcal{T}_2$, which are described in what follows.

DEFINITIONS

D35: Let us first define the **substitution operation.** Given disjoint graphs H_1 and H_2, each with at least two vertices, and $v \in V(H_1)$, we say that H is **obtained from H_1 by substituting H_2 for v**, or **obtained from H_1 and H_2 by substitution** (when the details are not important) if:

- $V(H) = (V(H_1) \cup V(H_2)) \setminus \{v\}$,

- $H|V(H_2) = H_2$,

- $H|(V(H_1) \setminus \{v\}) = H_1 \setminus v$, and

- $u \in V(H_1)$ is adjacent in H to $w \in V(H_2)$ if and only if u is adjacent in H_1 to v.

D36: A graph is **prime** if it is not obtained from smaller graphs by substitution.

FACTS

F13: The bull is a prime graph.

F14: Substituting one bull-free graph for a vertex of another bull-free graph results in yet another bull-free graph.

If follows from Facts F13 and F14 that to understand bull-free graphs, it is enough to construct all prime bull-free graphs.

EXAMPLES

E9: Graphs with no clique of size three are bull-free.

E10: Graphs with no stable set of size three are bull-free.

E11: However, there are prime bull-free graphs that contain both a big clique and a big stable set: let $n \geq 3$ be an integer, and let G be a graph with vertex set $k_1, \ldots, k_n, s_1, \ldots, s_n$ where

- $\{k_1, \ldots, k_n\}$ is a clique,

- $\{s_1, \ldots, s_n\}$ is a stable set,

- for $i, j \in \{1, \ldots, n\}$, s_i is adjacent to k_j if and only if $i + j > n$.

This richness of examples suggests that the structure theorem, if it exists, should be quite complex; and indeed it is. Once again, it would take several pages to state the construction explicitly, so instead we will try to give the reader a flavor of the result.

DEFINITION

D37: A graph F is ***triangle-free*** if $\omega(F) \leq 2$.

INFORMAL DEFINITIONS

D38: Graphs in $\mathcal{T}_1$ consist of a triangle-free induced subgraph F, together with a disjoint union of cliques $K_1, \ldots, K_t$, and the edges between $V(F)$ and $\bigcup_{i=1}^{t} K_i$ are carefully controlled. Thus, even though graphs in $\mathcal{T}_1$ may have both a large clique (one of $K_1, \ldots, K_t$) and a large stable set (in F), these two structures tend to "live" in different parts of the graph.

On the other hand, in the class $\mathcal{T}_2$, large cliques and large stable sets happily intertwine.

D39: Let G be a graph. Let $a, b \in V(G)$ be distinct vertices, and let $A = \{a_1, \ldots, a_n\}$ and $B = \{b_1, \ldots, b_m\}$ be disjoint subsets of $V(G)$ such that $A \cup B = V(G) \setminus \{a, b\}$. Let us now describe the adjacency in G.

- a is complete to A and anticomplete to B.

- b is complete to B and anticomplete to A.

- the adjacency between a and b is not specified.

- If $i, j \in \{1, \ldots, n\}$, and $i < j$, and a_i is adjacent to a_j, then a_i is complete to $\{a_{i+1}, \ldots, a_{j-1}\}$, and a_j is complete to $\{a_1, \ldots, a_{i-1}\}$.

- If $i, j \in \{1, \ldots, m\}$, and $i < j$, and b_i is adjacent to b_j, then b_i is complete to $\{b_{i+1}, \ldots, b_{j-1}\}$, and b_j is complete to $\{b_1, \ldots, b_{i-1}\}$.

- If $p \in \{1, \ldots, n\}$ and $q \in \{1, \ldots, m\}$, and a_p is adjacent to b_q, then a_p is complete to $\{b_{q+1}, \ldots, b_m\}$, and b_q is complete to $\{a_{p+1}, \ldots, a_n\}$.

Under these circumstances we say that G is a 1-***thin graph***.

D40: A variant of this is called a 2-*thin* graph, but we omit the definition here.

D41: Graphs in $\mathcal{T}_2$ are built from 1-thin and 2-thin graphs by gluing them together in prescribed ways.

D42: Another class of bull-free graphs, called $\mathcal{T}_0$, consists of a few sporadic bull-free graphs. We omit the precise definition of $\mathcal{T}_0$ in order to avoid getting too technical.

FACT

To state Fact F15 precisely we would need to define "expansions", but let us regard them in their non-technical sense, as "slight generalizations". The main result of the series [Ch12, Ch12a] is the following:

F15: [Ch12, Ch12a] Let G be a bull-free graph. Then either

- the graph G is obtained from smaller bull-free graphs by substitution, or

- one of the graphs G, G^c is an expansion of a member of $\mathcal{T}_0 \cup \mathcal{T}_1 \cup \mathcal{T}_2$.

Moreover, every graph obtained this way is bull-free.

REMARK

R7: In [ChPe13], a similar description of **perfect bull-free graphs** is obtained.

2.5.4   Trigraphs

In this subsection we describe an object slightly more general than a graph, called a *trigraph*. Trigraphs have proved to be a useful tool in describing structural properties of graphs with certain induced subgraphs excluded. Roughly, trigraphs are used to record the fact that the adjacency between a certain two vertices, say u and v, cannot be determined from the condition that the graph is $\mathcal{F}$-free for the family $\mathcal{F}$ in question. In most cases that means that each of u and v can be replaced by a set of vertices (with certain restrictions on it), say U and V, where the adjacencies between members of U and the members of V are arbitrary. This in turn gives rise to notions such as "thickening" or "expansion" that were briefly mentioned in §2.5.4.

Let us now be more formal.

DEFINITION

D43: A **trigraph** G consists of a finite set $V(G)$, called the **vertex set** of G, and a map $\theta : V(G)^2 \to \{-1, 0, 1\}$, called the **adjacency function**, satisfying:

- for all $v \in V(G)$, $\theta_G(v, v) = 0$

- for all distinct $u, v \in V(G)$, $\theta_G(u, v) = \theta_G(v, u)$

- for all $u \in V(G)$, there exists at most one $v \in V(G) \setminus \{u\}$ such that $\theta_G(u, v) = 0$

The idea is that, while a graph has two kinds of vertex pairs uv with $u \neq v$, adjacent and non-adjacent ones, a trigraph has three kinds: adjacent (for which $\theta(u, v) = 1$), non-adjacent (for which $\theta(u, v) = -1$), and semi-adjacent (for which $\theta(u, v) = 0$). A good way to think of semi-adjacent vertex pairs is as vertex pairs whose adjacency is "undecided". "Deciding" the adjacency of the undecided pairs results in a graph.

A version of trigraphs was first used in [Ch03, Ch06], where the last condition of the present definition was omitted. However, it seems that in order to study families of graphs, the more restricted definition that we use here is both sufficient and much nicer to work with; see [Ch03, Ch06, ChSe08, ChSe12, ChPl13, Ch12, Ch12a, ChPe13, ChKi14, Pe13].

Using Trigraphs: Structure Theorems

Let us now explain the use of trigraphs in a little more detail. Suppose that we are trying to understand the structure of the class of $\mathcal{F}$-free graphs for some family $\mathcal{F}$ (denote this class by $\mathcal{G}_\mathcal{F}$), and we can prove a theorem of the following form: we describe a few classes of $\mathcal{F}$-free graphs, and then say that every graph in $\mathcal{G}_\mathcal{F}$ is obtained from members of these classes by "expanding" certain vertex pairs uv. This means that u and v are replaced by two disjoint sets of new vertices, U and V, respectively, with arbitrary adjacencies between members of U and members of V. In order for this construction to be explicit, we need to provide a description of all pairs $(G, \mathcal{P})$, where G is a graph in $\mathcal{G}_\mathcal{F}$, and where $\mathcal{P}$ is the set of vertex pairs of G that can be expanded. To accomplish that, instead of working with $\mathcal{G}_\mathcal{F}$, we consider the class $\mathcal{T}_\mathcal{F}$ of $\mathcal{F}$-free trigraphs; these are trigraphs that have the property that however the adjacency of the undecided pairs is decided, the resulting graph is $\mathcal{F}$-free. Now we prove a similar theorem for trigraphs in $\mathcal{T}_\mathcal{F}$, where the vertex pairs that can be expanded are precisely the semi-adjacent pairs of the trigraph. This summarizes the way in which trigraphs are used in a purely structural setting.

Using Trigraphs: Algorithms

More recently, trigraphs have been used in the setting of algorithms. Algorithms that are based on decomposition theorems usually work as follows: take a graph G, "break it apart" via a decomposition given by the theorem, construct two new graphs G_1, G_2, where each G_i consist of a "piece" of G, together with a few more vertices recording the information about the remainder of G (let's call this part of G_i the *marker* for G_i); process each of G_1, G_2 separately, and then put the results together to get an answer for G. It turns out that semi-adjacent pairs are a good way to keep track of the information recorded in the markers for G_1 and G_2. Trigraphs were used in this way in [ChTrTrVu14].

2.5.5   Recognition Algorithms

In this subsection, we discuss the question of testing if a given graph G is $\mathcal{F}$-free for a given family $\mathcal{F}$. Obviously this can be done in polynomial time if $\mathcal{F}$ is finite (just run through all subsets X of $V(G)$ of size at most $\max_{F \in \mathcal{F}} |V(F)|$ and check if the subgraph $G|X$ is isomorphic to a member of $\mathcal{F}$), so this question is only of interest when $\mathcal{F}$ is infinite. For brevity, let us say "testing for $\mathcal{F}$" when we mean "testing if a graph is $\mathcal{F}$-free".

After Fact F2 was proved, the following major question in the theory of perfect graphs remained open: given a graph G, test (in polynomial time) whether G is perfect. With Fact F2 in our ammunition bag, this turns into a question of testing if a given graph is Berge (we remind the reader that a graph is Berge if it is $\mathcal{C}$-free, where $\mathcal{C}$ is the family consisting of all odd cycles of length at least five, and their complements). This was done in [ChCoLiSeVu05], and we will describe the algorithm briefly (the remainder of this subsection is closely based on [ChSe07]).

DEFINITION

D44: A *pyramid* is a graph consisting of a triangle $\{b_1, b_2, b_3\}$, called the *base*, a vertex $a \notin \{b_1, b_2, b_3\}$, called the *apex*, and three paths P_1, P_2, P_3, such that for $i, j \in 1, 2, 3$

- the ends of P_i are a and b_i,

- if $i \neq j$ then $V(P_i) \setminus \{a\}$ is disjoint from $V(P_j) \setminus \{a\}$ and the only edge between them is $b_i b_j$, and

- at most one of P_1, P_2, P_3 has length one.

In this case we say that the pyramid is *formed* by the paths P_1, P_2, P_3.

Let $\mathcal{P}$ be the family of all pyramids.

FACTS

F16: It is easy to see that every pyramid contains C_n for some odd $n \geq 5$, and therefore every Berge graph is $\mathcal{P}$-free.

It turns out that:

F17:   [ChCoLiSeVu05]. Testing for $\mathcal{P}$ is relatively easy and can be done in time $O(|V(G)|^9)$.

This is the first step of the algorithm of [ChCoLiSeVu05].

Testing for Pyramids: the Shortest Paths Detector

The idea is as follows. If G contains a pyramid, then it contains a pyramid P with the number of vertices smallest. We are going to "guess" (by trying all possibilities) some of the vertices of P in G, then find shortest paths in G between pairs of vertices that we guessed that were joined by a path in P, and then test whether the subgraph of G formed by the union of these shortest paths is a pyramid. If the answer is "yes", then G contains a pyramid, and we stop. Surprisingly, it turns out, that choosing the shortest paths with a little bit of care, we can guarantee that if the answer is "no", then there is no pyramid in P. We call this general strategy of testing for a family $\mathcal{F}$ a *shortest-paths detector* for $\mathcal{F}$. Let us now be more precise.

DEFINITIONS

D45: For $u, v \in V(G)$ we denote by $d_G(u, v)$ the length of the shortest path of G between u and v.

D46: If P is a pyramid, formed by three paths P_1, P_2, P_3, with apex a and base $\{b_1, b_2, b_3\}$, we say its **frame** is the 10-tuple

$$a, b_1, b_2, b_3, s_1, s_2, s_3, m_1, m_2, m_3$$

where

- for $i = 1, 2, 3$, s_i is the neighbor of a in P_i
- for $i = 1, 2, 3$, $m_i \in V(P_i)$ satisfies $d_{P_i}(a, m_i) - d_{P_i}(m_i, b_i) \in \{0, 1\}$

D47: A pyramid P in G is **optimal** if there is no pyramid P' with $|V(P')| < |V(P)|$.

FACT

F18: [ChCoLiSeVu05] Let P be an optimal pyramid, with frame

$$a, b_1, b_2, b_3, s_1, s_2, s_3, m_1, m_2, m_3$$

Let S_1, T_1 be the subpaths of P_1 from m_1 to s_1, b_1, respectively. Let F be the set of all vertices non-adjacent to each of s_2, s_3, b_2, b_3.

1. Let Q be a path between s_1 and m_1 with interior in F, and with minimum length over all such paths. Then a-s_1-Q-m_1-T_1-b_1 is a path (say P_1'), and P_1', P_2, P_3 form an optimal pyramid.

2. Let Q be a path between m_1 and b_1 with interior in F, and with minimum length over all such paths. Then a-s_1-S_1-m_1-Q-b_1 is a path (say P_1'), and P_1', P_2, P_3 form an optimal pyramid.

Analogous statements hold for P_2, P_3.

F19: Fact F18 can be used to design an algorithm to test for $\mathcal{P}$:

- guess the frame $a, b_1, b_2, b_3, s_1, s_2, s_3, m_1, m_2, m_3$ of an optimal pyramid P of G, by trying all 10-tuples of vertices;

- find shortest paths between m_1 and b_1, and between m_1 and s_1, not containing any neighbors of s_2, s_3, b_2, and b_3; do the same for m_2, b_2, s_2 and m_3, b_3, s_3;

- test if the union of the six shortest paths, together with the vertex a, forms a pyramid.

Now, by Fact F18, the answer is "yes" if and only if G contains a pyramid.

REMARK

R8: The algorithm in [ChCoLiSeVu05] is similar; it was modified a little to bring the running time down to $O(|V(G)|^9)$.

Easily Detectable Configurations; Cleaning; Finding Odd Holes

The next idea in [ChCoLiSeVu05] is to use the shortest-path detector for odd holes. Unfortunately, there does not seem to be a theorem similar to Fact F18 for odd holes, and so, first, the graph needs to be "prepared" for using a shortest-paths detector. The first step is to test for $\mathcal{P}$, and a few other families $\mathcal{F}$ that are easy to test for, and such that every Berge graph is $\mathcal{F}$-free. Now we can assume that the graph in question is $\mathcal{F}$-free for all these $\mathcal{F}$. The next step is applying *cleaning*, a technique first proposed in [CoRa93]. The idea of cleaning is to find, algorithmically, polynomially many subsets $X_1, \ldots, X_k$ of $V(G)$, such that if G contains an odd hole, then for at least one value of $i \in \{1, \ldots, k\}$ the graph $G_i = G \setminus X_i$ contains an odd hole that can be found using a shortest-paths detector. Finally, applying a shortest-paths detector for odd holes to each of $G_1, \ldots, G_k$, we detect an odd hole if and only if G contains one.

REMARKS

R9:   In addition to the algorithm just described, [ChCoLiSeVu05] contains another algorithm to test for Bergeness, which instead of a shortest-paths detector for odd holes, uses a decomposition theorem for odd-hole-free graphs from [CoCoVu04], but we will not describe this algorithm here.

R10:   Both algorithms in [ChCoLiSeVu05] test for Bergeness, and not for the existence of an odd hole in a graph. The complexity of testing if a graph contains an odd hole is still unknown.

Testing for Even Holes

On the other hand, the problem of testing if a graph contains an even hole can be solved in polynomial time. There are three (!) known algorithms. One is due to Conforti, Cornuéjols, Kapoor, and Vušković [CoCoKaVu02a], another to Kawarabayashi, Seymour and the author [ChKaSe05], and a third one to da Silva and Vušković [daSiVu13]. All three algorithms use cleaning; the first algorithm uses a decomposition theorem of [CoCoKaVu02] for even-hole-free graphs, the second one is based on the shortest-paths detector, and the last one is again decomposition based [daSiVu13].

More Algorithms

There are two other kinds of graphs that are somewhat similar to the pyramid, called a *theta* and a *prism*.

DEFINITIONS

D48:   A *theta* is a graph consisting of two non-adjacent vertices s, t and three paths P_1, P_2, P_3, each between s and t, such that the sets $V(P_1) \setminus \{s, t\}$, $V(P_2) \setminus \{s, t\}$, and $V(P_3) \setminus \{s, t\}$ are pairwise disjoint, and the union of every pair of P_1, P_2, P_3 is a hole.

D49:   A *prism* is a graph consisting of two disjoint triangles $\{a_1, a_2, a_3\}$ and $\{b_1, b_2, b_3\}$ and three paths P_1, P_2, P_3, with the following properties:

- for $i = 1, 2, 3$, the ends of P_i are a_i and b_i,

- P_1, P_2, P_3 are pairwise disjoint, and

- for $1 \leq i < j \leq 3$, there are precisely two edges between $V(P_i)$ and $V(P_j)$, namely, $a_i a_j$ and $b_i b_j$.

FACT

F20: Let $\mathcal{T}$ be the family of all thetas, and $\mathcal{P}r$ the family of all prisms. Then every even-hole-free graph is $\mathcal{T} \cup \mathcal{P}r$-free. (This is easy to check.)

In view of Fact F20 prisms and thetas play a similar role for even-hole-free graphs to the one that pyramids play for odd-hole-free graphs. It turns out, however, that:

FACTS

F21: [MaTr05] Unlike with $\mathcal{P}$, the problem of testing for $\mathcal{P}r$ is NP-complete.

F22: [LeLiMaTr09] The problem of testing for $\mathcal{P}r$ is NP-complete even in graphs with clique number two.

On the other hand,

F23: [ChSe10a] Testing for $\mathcal{T}$ can be done in polynomial time.

F24: [MaTr05] Testing for $\mathcal{P} \cup \mathcal{P}r$ can be done in polynomial time.

F25: [ChKa08] Testing for $\mathcal{T} \cup \mathcal{P}r$ can be done in polynomial time.

The Three-in-a-Tree Problem

All the algorithms mentioned so far, except one, use variations on the ideas of cleaning and shortest paths detectors (or decomposition theorems), and that one exception is the algorithm for testing for $\mathcal{T}$. There the approach is different. In order to be able to test for $\mathcal{T}$, a slightly more general problem is studied: given a graph G, and three vertices v_1, v_2, v_3 of G, does there exist an induced subgraph T of G, such that T is a tree and $v_1, v_2, v_3 \in V(T)$? This is the *three-in-a-tree* problem.

It turns out that the answer to this question is "no" if and only if the graph admits a certain structure. This fact is then used to design a polynomial time algorithm for the three-in-a-tree problem. Now, if $\{v_1, v_2, v_3\}$ is a stable set of size three with a common neighbor w in G, the degree of each of v_1, v_2, v_3 in $G \setminus \{w\}$ is one, and the degree of w in G is three, then the answer to the three-in-a-tree problem with input $(G \setminus \{w\}, v_1, v_2, v_3)$ is "yes" if and only if G contains a theta using v_1, v_2, v_3, w.

On the other hand, if $\{v_1, v_2, v_3\}$ is a clique of size three, and no vertex of G has two neighbors in it, then the answer to the three-in-a-tree problem with input (G, v_1, v_2, v_3) is "yes" if and only if G contains a pyramid with base $\{v_1, v_2, v_3\}$. Thus, the algorithm to solve the three-in-a-tree problem can be used, after some pre-processing, to test both for $\mathcal{P}$ and for $\mathcal{T}$ (and this is the only algorithm known to test for $\mathcal{T}$). This result is particularly pleasing from the point of view of a structural graph theorist, because this is one of the few times that a structure (and not just a decomposition) theorem and an algorithm appear together in the study of graphs with forbidden induced subgraphs.

As we have seen, the complexity of testing for $\mathcal{F}$ varies with $\mathcal{F}$: for some families polynomial-time algorithms are known, while for others the problem can be shown to be NP-complete. An interesting open question is: what causes this difference? Can one characterize the families for which testing can be done efficiently?

2.5.6   Erdős–Hajnal Conjecture and χ-Boundedness

As the results surveyed in Subsection 2.5.5 illustrate, structure theorems for $\mathcal{F}$-free graphs tend to be complicated to state, difficult to prove, and hard to use. At the moment, we are nowhere near having a structural conjecture for excluding a general induced subgraph. But what if we lower our sights, and ask whether excluding a general induced subgraph guarantees that the graph has certain special properties that a general graph does not possess? In 1989, Erdős and Hajnal made a beautiful conjecture of this kind; it is now known as the Erdős–Hajnal Conjecture:

CONJECTURE

C1:   [ErHa89] For every graph H, there exists a constant $\delta(H) > 0$, such that every H-free graph G has either a clique or a stable set of size at least $|V(G)|^{\delta(H)}$.

FACTS

In the same paper a partial result in this direction is proved, showing that for every H, H-free graphs behave differently from general graphs. It is a well-known theorem of Erdős that

F26:   [Er47] There exist graphs on n vertices, with no clique or stable set of size larger than $O(\log n)$.

However,

F27:   [ErHa89] For every graph H, there exists a constant $c(H) > 0$, such that every H-free graph G has either a clique or a stable set of size at least $e^{c(H)\sqrt{\log|V(G)|}}$.

DEFINITION

D50:   Let us say that a graph H has the **Erdős–Hajnal property** if there exists a constant $\delta(H) > 0$, such that every H-free graph G has either a clique or a stable set of size at least $|V(G)|^{\delta(H)}$.

FACTS

F28:   Clearly, H has the Erdős–Hajnal property if and only if H^c does.

Very few graphs have been shown to have the Erdős-Hajnal property.

F29:   It is not difficult to show that all graphs on at most four vertices have the Erdős-Hajnal property.

A much more complicated argument is needed to show that:

F30: [ChSa08] The bull has the Erdős–Hajnal property.

In [AlPaSo01] it was shown that:

F31:   [AlPaSo01] If H_1, H_2 have the property, then so does every graph obtained from H_1 and H_2 by substitution.

Thus in order to prove Conjecture C1, it is enough to show that every prime graph has the Erdős–Hajnal property. However, this question is still open for C_5, and for the five-vertex path. No prime graphs on at least six vertices have been shown to have the Erdős–Hajnal property.

Tournaments

Next we introduce another version of Conjecture C1.

DEFINITION

D51: A *tournament* is a directed graph G where for every distinct $u, v \in V(G)$, exactly one of the (ordered) pairs uv and vu belongs to $E(G)$. If $uv \in E(G)$, we say that u is *adjacent to* v.

D52: A tournament is *transitive* if it has no directed cycles (or, equivalently, no directed cycles of length three).

D53: For a tournament T, we denote by $\alpha(T)$ the largest number of vertices in a transitive subtournament of T.

D54: For tournaments S and T, we say that T is *S-free* if no subtournament of T is isomorphic to S.

CONJECTURE

C2: [AlPaSo01] For every tournament S, there exists a constant $\delta(S) > 0$, such that every S-free tournament T satisfies $\alpha(T) \geq |V(T)|^{\delta(H)}$.

FACT

It is also shown that:

F32: [AlPaSo01] Conjectures C1 and C2 are equivalent.

DEFINITION

D55: As with graphs, let us say that a tournament S has the *Erdős–Hajnal property* if there exists $\delta(S) > 0$, such that every S-free tournament T satisfies $\alpha(T) \geq |V(T)|^{\delta(H)}$.

D56: Similarly to graphs, a tournament T is *prime* if there is no $X \subseteq V(T)$ with $1 < |X| < |V(T)|$ such that for every $v \in V(T) \setminus X$, either v is adjacent to every vertex of X, or v is adjacent from every vertex of X.

For some reason, Conjecture C2 seems to be a little more approachable than Conjecture C1:

FACT

F33: [BeChCh14] Unlike in the case of graphs, there is a known infinite family of prime tournaments, all of which have the Erdős–Hajnal property.

We refer the reader to [Ch13] for more information about recent progress on Conjecture C1 and Conjecture C2.

χ-**Boundedness**

Let us now consider another notion, related to Conjecture C1.

DEFINITIONS

D57: A class of graphs $\mathcal{G}$ is **hereditary** if $H \in \mathcal{G}$ for every $G \in \mathcal{G}$ and every induced subgraph H of G.

D58: We say that a hereditary graph $\mathcal{G}$ is χ-**bounded** if there exists a function $f : \mathbb{N} \to \mathbb{N}$ such that $\chi(G) \leq f(\omega(G))$ for every $G \in \mathcal{G}$. In this situation we call f a χ-**bounding function** for $\mathcal{G}$.

EXAMPLES

E12: The class of $\mathcal{F}$-free graphs is hereditary for every family $\mathcal{F}$.

E13: The class of perfect graphs is χ-bounded by the identity function.

E14: Suppose that for some graph H the class of H-free graphs has a χ-bounding function that is a polynomial. Then there exists $t \geq 1$ such that $\chi(G) \leq \omega(G)^t$ for every H-free graph G. Since in every coloring of G, each color class has size at most $\alpha(G)$, it follows that
$$\omega(G)^t \alpha(G) \geq |V(G)|$$
and so G has either a clique or a stable set of size at least $|V(G)|^{1/(t+1)}$, and H has the Erdős–Hajnal property.

FACT

It is tempting to conjecture that the class of H-free graphs is χ-bounded for every H. However, this is false, as shown by the following theorem of Erdős [Er59]:

F34: [Er59] For every pair of integers $k, g > 0$ there exists a graph G with $\chi(G) > k$ and no cycle of length less than g.

Thus in order for the class of H-free graphs to be χ-bounded, H must contain no cycles (otherwise, every graph G with no cycle of length at most $|V(H)|$ is H-free, and has $\omega(G) \leq 2$; and by Fact F34 there exist such graphs with arbitrarily large chromatic numbers). A famous conjecture of Gyárfás and Sumner [Gy75, Su81] states that this necessary condition is in fact sufficient:

CONJECTURE

C3: [Gy75, Su81] For every forest F, the class of F-free graphs is χ-bounded.

This conjecture is still open. Gyárfás [Gy75] proved that it holds when F is a path. Kierstead and Penrice [KiPe90, KiPe94] and Scott [Sc97] made further progress. Some of the theorems in Subsection 2.5.2 and Subsection 2.5.3 also imply χ-boundedness results for certain classes of $\mathcal{F}$-free graphs. We list some of them here.

FACTS

F35: [AdChHaReSe08] For every even-hole-free graph G, $\chi(G) \leq 2\omega(G) - 1$. This follows from Fact F8. Incidentally, the question of whether the class of odd-hole-free graphs is χ-bounded is still open.

F36: [ChSe10] If G is an induced subgraph of a connected claw-free graph G' with $\alpha(G') \geq 3$, then $\chi(G) \leq 2\omega(G)$. This follows from the main result of [ChSe08].

F37: [ChFr07] For every quasi-line graph G, $\chi(G) \leq \frac{3}{2}\omega(G)$. This follows from the main result of [ChSe12].

For more results of this type, see [Vu13].

2.5.7   Well-Quasi-Ordering and Rao's Conjecture

DEFINITION

D59: A **quasi-order** Q consists of a class $E(Q)$ and a transitive reflexive relation which we denote by $\leq$ or $\leq_Q$; and it is a **well-quasi-order** or **wqo** if for every infinite sequence q_i $(i = 1, 2 \ldots)$ of elements of $E(Q)$ there exist $j > i \geq 1$ such that $q_i \leq_Q q_j$.

FACTS
One of the consequences of the Robertson–Seymour graph minor project is that:

F38: [RoSe04] The class of all graphs forms a well-quasi-order under minor containment.

F39: The same is not true for induced subgraphs: the sequence $C_3, C_4, \ldots$ is an infinite sequence of graphs, none of which is an induced subgraph of another.

This is disappointing, but S.B. Rao proposed the following "fix":

DEFINITIONS

D60: Let us say two graphs G, G' are **degree-equivalent** if they have the same vertex set, and for every vertex, its degrees in G and in G' are equal.

D61: A graph H is **Rao-contained** in a graph G if H is isomorphic to an induced subgraph of some graph that is degree-equivalent to G.

FACT
In the early 1980s Rao [Ra81] conjectured the following , which was proved in [ChSe13]:

F40: [ChSe13] In any infinite set of graphs, there exist two, say G and H, such that H is Rao-contained in G.

Outline of the Proof of Fact F40

DEFINITION

D62: A graph is called **split** if its vertex set can be partitioned into a clique and a stable set.

The first part of the proof is a structural result. It states, roughly, that

FACT

F41: For every graph H there exists a constant c_H, such that if G is a graph that does not Rao-contain H, then $V(G) = A \cup B \cup C$ where

- A, B, C are pairwise disjoint,

- $G|A$ is a split graph,

- either every vertex of B has at most c_H neighbors in $G|B$, or every vertex of B has at most c_H non-neighbors in $G|B$,

- $|C| \leq c_H$,

- the edges between A and B are tightly controlled.

Now suppose that the class of all graphs is not a wqo under Rao-containment. Then there exists an infinite sequence $G_1, G_2, \ldots$ such that G_j does not Rao-contain G_i for $1 \leq i < j$. In particular, for all $1 < j$, G_j does not Rao-contain G_1. By the structural result that we just mentioned, that means that all of $G_2, G_3, \ldots$ have the structure described in the previous paragraph. It is therefore enough to prove that the class of graphs with that structure, where $H = G_1$, is a wqo. Because the size of C is bounded by a constant that depends only of G_1, and the edges between A and B are tightly controlled, standard techniques allow us to reduce the problem to proving that the class of pairs (F, J) where F is a split graph and J is a graph with all degrees at most c_{G_1} is a wqo under Rao-containment. This, in turn (using Higman's theorem [Hi52]), reduces to proving that

FACTS

F42: In any infinite set of split graphs, there exist two, say G and H, such that H is Rao-contained in G.

F43: For every $c > 0$, in any infinite set of graphs where every vertex has at most c neighbors, there exist two, say G and H, such that H is Rao-contained in G.

We will not dwell on the proof of Fact F43; let us just mention that it has recently been proven again using different and interesting methods in [Al12]. The proof Fact F42 though developed in a somewhat unexpected direction. It was reduced to proving that a certain family of directed graphs (called *contests*) is a wqo under a certain containment relation, which is very closely related to the well-known concept of *immersion*. This led to a number of new results regarding immersion of directed graphs; we will discuss some of them in the next subsection.

2.5.8   Tournament Immersion

A directed graph H is *immersed* in a directed graph G if the vertices of H are mapped to (distinct) vertices of G, and the (directed) edges of H are mapped to directed paths joining the corresponding pairs of vertices of G, in such a way that the paths are pairwise edge-disjoint.

More precisely,

DEFINITION

D63: Let G, H be directed graphs. A ***weak immersion*** of H in G is a map η such that

- $\eta(v) \in V(G)$ for each $v \in V(H)$

- $\eta(u) \neq \eta(v)$ for distinct $u, v \in V(H)$

- for each edge $e = uv$ of H (this notation means that e is directed from u to v), $\eta(e)$ is a directed path of G from $\eta(u)$ to $\eta(v)$ (paths do not have "repeated" vertices)

- if $e, f \in E(H)$ are distinct, then $\eta(e), \eta(f)$ have no edges in common, although they may share vertices

If in addition we add the condition

- if $v \in V(H)$ and $e \in E(H)$, and e is not incident with v in H, then $\eta(v)$ is not a vertex of the path $\eta(e)$

we call the relation ***strong immersion***. (For undirected graphs the definitions are the same except we use paths instead of directed paths.)

The following is a theorem from [RoSe10], conjectured by Nash-Williams:

FACT, CONJECTURED BY NASH-WILLIAMS

F44: [RoSe10] The class of all graphs is a wqo under weak immersion.

REMARK

R11: It remains open whether the class of all graphs is a wqo under strong immersion (this is another conjecture of Nash-Williams). Robertson and Seymour believe that at one time they had a proof, but they have never written it down [RoSe11].

FACTS

F45: Unfortunately, weak immersion does not provide a wqo of the class of directed graphs. To see this, let D_n be a cycle of length $2n$ and direct its edges alternately clockwise and counterclockwise; then no member of the set $\{D_i : i \geq 2\}$ is weakly immersed in another.

But what about tournaments? The main result of [ChSe11] is that:

F46: [ChSe11] The class of all tournaments is a wqo under strong immersion.

The proof of Fact F46 relies on a new graph parameter, defined in [ChSe11].

DEFINITION

D64: If $k \geq 0$ is an integer, an enumeration $(v_1, \ldots, v_n)$ of the vertex set of a directed graph has **cut-width** at most k if for all $j \in \{1, \ldots, n-1\}$, there are at most k edges uv such that $u \in \{v_1, \ldots, v_j\}$ and $v \in \{v_{j+1}, \ldots, v_n\}$; and a digraph has **cut-width** at most k if there is an enumeration of its vertex set with cut-width at most k.

So if a directed graph has low cut-width, it means that it is in some sense degenerate.

On the Proof of Fact F46

The proof of Fact F46 consists of two ingredients. First it is shown that:

FACT

F47: [ChSe11] For every tournament S there exists an integer c_S, such that if a tournament T does not strongly immerse S, then T has cut-width at most c_S.

Similarly to the proof of Fact F40, the proof of Fact F46 now reduces to proving that for every $c > 0$, the class of tournaments of cut-width at most c is a wqo under strong immersion. This is a much more manageable task than proving Fact F46 directly; the proof can be found in [ChSe11].

2.5.9   Topological Containment in Tournaments

In this subsection we discuss another containment relation on tournaments. Let F, H be directed graphs. Then F is a *subdivision* of H if it can be obtained from H by repeatedly deleting an edge uv, adding a new vertex w, and adding two new edges uw and wv. We say that a directed graph G *topologically contains* a graph H if G has a subgraph isomorphic to a subdivision of H.

It turns out that a theorem similar to Fact F47 exists for topological containment, except "cut-width" needs to be replaced by a more complicated parameter. The following was defined in [FrSe13].

DEFINITIONS

D65: Given a directed graph D, a sequence $W = [W_1, \ldots, W_r]$ of subsets of $V(D)$ is a **path-decomposition** of D if the following conditions are satisfied:

1. $\bigcup_{i=1}^{r} W_i = V(D)$

2. $W_i \cap W_k \subseteq W_j$ for $1 \leq i < j < k \leq r$

3. for each edge $uv \in E(D)$, $u \in W_i, v \in W_j$ for some $i \geq j$

The **width** of a path-decomposition is $\max_{1 \leq i \leq r}(|W_i| - 1)$.

D66: The **path-width** of a directed graph D is the minimum width over all path-decompositions of D.

REMARK

R12: Some readers may be familiar with the concept of path-decomposition and path-width for undirected graphs; this is, of course, the directed analogue.

FACTS

F48: [FrSe13] For every tournament S there exists an integer c_S, such that if a tournament T does not topologically contain S, then T has path-width at most c_S.

F49: [FrSe13] Fact F48 is then used to obtain a polynomial-time algorithm to test if a given tournament topologically contains a fixed tournament S.

REMARKS

R13: Unlike in the case of immersion, topological containment turns out not to be a wqo.

R14: Some of the results mentioned here and in Subsection 2.5.10 are in fact proved for a wider class of directed graphs, called *semi-complete digraphs*, but we will not discuss it here. The reader is referred to [ChSe11] and [FrSe13].

2.5.10   Disjoint Paths Problems in Tournaments

DEFINITIONS

D67: Let $s_1, t_1, \ldots, s_k, t_k$ be vertices of a graph or directed graph G. The k **edge-disjoint paths problem** is the problem of determining whether there exist edge-disjoint paths $P_1, \ldots, P_k$ (directed paths, in the case of a directed graph) such that P_i is from s_i to t_i for $1 \leq i \leq k$.

D68: The **vertex-disjoint paths problem** is defined similarly, except that the paths are required to be vertex disjoint.

FACTS

F50: [RoSe95] For undirected graphs, both problems are solvable in polynomial time for all fixed k; this was one of the highlights of the Graph Minors project of Robertson and Seymour.

The directed version is therefore a natural and important question, but it was shown by Fortune, Hopcroft, and Wyllie [FoHoWy80] that

F51: [FoHoWy80] Without further restrictions on the input G, both the edge-disjoint paths problem and the vertex-disjoint paths problem are NP-complete for directed graphs, even for $k = 2$.

So, it becomes significant to study subclasses of directed graphs for which the problems can be solved in polynomial time.

The Edge-Disjoint Paths Problem

Let us first address the edge-disjoint version of the problem.

FACT

F52: [FrSe13a] The edge-disjoint paths problem in a tournament can be solved in polynomial time for any fixed k.

The idea of the algorithm is as follows. If the tournament in question has high cut-width, then a certain structure is found that can be excised from the graph without changing the answer, and so the problem is reduced to a smaller graph; if the cut-width is low, the problem is solved using dynamic programming. Also in [FrSe13a], this result is extended to directed graphs for which the underlying undirected graph has bounded stability number.

The Vertex-Disjoint Paths Problem

One might hope that a similar approach would work when considering the vertex-disjoint paths problem in tournaments, with path-width replacing cut-width. However, this does not seem to be the case. The main result of [ChScSe13] is that:

FACT

F53: [ChScSe13] The vertex-disjoint paths problem can be solved in polynomial time for any fixed k in tournaments (in fact, in semi-complete digraphs).

But the only tool used there is a version of dynamic programming.

2.5.11   Acknowledgment

The author is very grateful to Irena Penev and Paul Seymour for their careful reading of the manuscript, and many helpful suggestions. We also thank Jonathan Gross for his thoughtful editing.

This work was partially supported by NSF grants DMS-1001091 and IIS-1117631.

References

[AdChHaReSe08] L. Addario-Berry, M. Chudnovsky, F. Havet, B. Reed, P. Seymour, Bisimplicial vertices in even-hole-free graphs, *J. Combin. Theory Ser B* **98** (2008), 1119–1164.

[Al12] C. J. Altomare, A semigroup proof of the bounded degree case of S.B. Rao's Conjecture on degree sequences and a bipartite analogue, *J. Combin. Theory Ser B* **102** (2012), 756–759.

[AlPaSo01] N. Alon, J. Pach, and J. Solymosi, Ramsey-type theorems with forbidden subgraphs, *Combinatorica* **21** (2001), 155–170.

[Be61] C. Berge, Färbung von Graphen, deren sämtliche bzw. deren ungerade Kreise starr sind, *Wiss. Z. Martin-Luther-Univ. Halle-Wittenberg Math.-Natur. Reihe* **10** (1961), 114.

[BeChCh14] E. Berger, K. Choromanski, and M. Chudnovsky, Forcing large transitive subtournaments, *submitted for publication.*

[Ch03] M. Chudnovsky, Berge trigraphs and their applications, PhD thesis, Princeton University, 2003.

[Ch06] M. Chudnovsky, Berge trigraphs, *J. of Graph Theory,* **53** (2006), 1–55.

[Ch12] M. Chudnovsky, The structure of bull-free graphs I — Three-edge-paths with centers and anticenters, *J. Combin. Theory Ser B* **102** (2012), 233–251.

[Ch12a] M. Chudnovsky, The structure of bull-free graphs II and III—a summary, *J. Combin. Theory Ser B* **102** (2012), 252–282.

[Ch13] M. Chudnovsky, The Erdős–Hajnal Conjecture — A Survey, *submitted for publication.*

[ChCoLiSeVu05] M. Chudnovsky, G. Cornuéjols, X. Liu, P. Seymour, and K. Vušković, Recognizing Berge graphs, *Combinatorica* **25** (2005), 143–187.

[ChFr07] M. Chudnovsky and A. O. Fradkin, Coloring quasi-line graphs, *J. Graph Theory* **54** (2007), 41–50.

[ChFrSe12] M. Chudnovsky, A. O. Fradkin, and P. Seymour, Tournament immersion and cutwidth, *J. Combin. Theory Ser. B* **106** (2012), 93–101.

[ChKa08] M. Chudnovsky and R. Kapadia, Detecting a theta or a prism, *SIAM J. Discrete Math.* **22** (2008), 1164–1186.

[ChKaSe05] M. Chudnovsky, K. Kawarabayashi, and P. Seymour, Detecting even holes, *J. Graph Theory* **48** (2005), 85–111.

[ChKi14] M. Chudnovsky and A. King, Optimal anti-thickenings of claw-free graphs, *submitted for publication.*

[ChPe13] M. Chudnovsky and I. Penev, The structure of bull-free perfect graphs, *J. Graph Theory,* to appear.

[ChPl13] M. Chudnovsky and M. Plumettaz, The structure of claw-free perfect graphs, *submitted for publication,* 2012.

[ChRoSeTh06] M. Chudnovsky, N. Robertson, P. Seymour, and R. Thomas, The strong perfect graph theorem, *Annals of Math.* **164** (2006), 51–229.

[ChSa08] M. Chudnovsky and S. Safra, The Erdős–Hajnal conjecture for bull-free graphs, *J. Combin. Theory, Ser. B* **98** (2008), 1301–1310.

[ChScSe13] M. Chudnovsky, A. Scott, and P. Seymour, Disjoint paths in tournaments, *submitted for publication.*

[ChSe07] M. Chudnovsky and P. Seymour, Excluding induced subgraphs, in *Surveys in Combinatorics 2007, London Math. Soc. Lecture Note Series* **346** (2007), 99–119.

[ChSe08] M. Chudnovsky and P. Seymour, Clawfree graphs V — Global structure, *J. Combin. Theory Ser B* **98** (2008), 1373–1410.

[ChSe09] M. Chudnovsky and P. Seymour, Even pairs in Berge graphs, *J. Combin. Theory Ser B* **99** (2009), 370–377.

[ChSe10] M. Chudnovsky and P. Seymour, Clawfree Graphs VI. Coloring claw-free graphs, *J. Combin. Theory Ser B* **100** (2010), 560–572.

[ChSe10a] M. Chudnovsky and P. Seymour, The three-in-a-tree problem, *Combinatorica* **30** (2010), 387–417.

[ChSe11] M. Chudnovsky and P. Seymour, A well-quasi-order for tournaments, *J. Combin. Theory Ser. B*, **101** (2011), 47–53.

[ChSe12] M. Chudnovsky and P. Seymour, Clawfree graphs VII. Quasi-line graphs, *J. Combin. Theory Ser. B*, to appear.

[ChSe13] M. Chudnovsky and P. Seymour, Rao's conjecture on degree sequences, *submitted for publication*.

[ChTrTrVu14] M. Chudnovsky, N. Trotignon, T. Trunck and K. Vušković, Coloring perfect graphs with no balanced skew-partitions, *submitted for publication*.

[Ch85] V. Chvátal, Star-cutsets and perfect graphs, *J. Combin. Theory Ser. B* **39** (1985), 189–199.

[ChSb87] V. Chvátal and N. Sbihi, Bull-free Berge graphs are perfect, *Graphs and Combinatorics* **3** (1987), 127–139.

[ChSb88] V. Chvátal and N. Sbihi, Recognizing claw-free Berge graphs, *J. Combin. Theory Ser. B* **44** (1988), 154–176.

[CoCoKaVu02] M. Conforti, G. Cornuéjols, A. Kapoor, and K. Vušković, Even-hole-free graphs, Part I: Decomposition theorem, *J. Graph Theory* **39** (2002), 6–49.

[CoCoKaVu02a] M. Conforti, G. Cornuéjols, A. Kapoor, and K. Vušković, Even-hole-free graphs, Part II: Recognition algorithm, *J. Graph Theory* **40** (2002), 238–266.

[CoCoVu04] M. Conforti, G. Cornuéjols, and K. Vušković, Decomposition of odd-hole-free graphs by double star cutsets and 2-joins, *Discrete Applied Math.* **141** (2004), 41–91.

[CoRa92] M. Conforti and M. R. Rao, Structural properties and decomposition of linear balanced matrices, *Mathematical Programming* **55** (1992), 129–168.

[CoRa93] M. Conforti and M. R. Rao, Testing balancedness and perfection of linear matrices, *Mathematical Programming* **61** (1993), 1–18.

[CoCu85] G. Cornuéjols and W. H. Cunningham, Compositions for perfect graphs, *Discrete Math.* **55** (1985), 245–254.

[Di05] R. Diestel, *Graph Theory*, 3rd Edition, Springer, 2005.

[Di50] R. P. Dilworth, A decomposition theorem for partially ordered sets, *Annals of Math* **51** (1950), 161–166.

[Er47] P. Erdős, Some remarks on the theory of graphs, *Bull. Amer. Math. Soc.* **53** (1947), 292–294.

[Er59] P. Erdős, Graph theory and probability, *Canad. J. Math* **11** (1959), 34–38.

[ErHa89] P. Erdős and A. Hajnal, Ramsey-type theorems, *Disc. Appl. Math.* **25** (1989), 37-52.

[FoHoWy80] S. Fortune, J. Hopcroft, and J. Wyllie, The directed subgraphs homeomorphism problem, *Theoret. Comput. Sci.* **10** (1980), 111–121.

[FrSe13] A. O. Fradkin and P. Seymour, Tournament pathwidth and topological containment, *J. Combin. Theory Ser B* (2013), to appear.

[FrSe13a] A. O. Fradkin and P. Seymour, Edge-disjoint paths in digraphs with bounded independence number, *submitted for publication.*

[Gy75] A. Gyárfás, On Ramsey covering-numbers, *Coll. Math. Soc. János Bolyai*, in *Infinite and Finite Sets*, North Holland American Elsevier, New York (1975), 10.

[Hi52] G. Higman, Ordering by divisibility in abstract algebras, *Proc. London Math. Soc., 3rd series* 2 (1952), 326–336.

[KiPe90] H. A. Kierstead and S. G. Penrice, Recent results on a conjecture of Gyárfás, *Congr. Numer.* **79** (1990), 182–186.

[KiPe94] H. A. Kierstead and S. G. Penrice, Radius two trees specify χ-bounded classes, *J. Graph Theory* **18** (1994), 119–129.

[Ko16] D. König, Über Graphen und ihre Anwendung auf Determinantentheorie und Mengenlehre, *Math. Ann.* **77** (1916), 453–465.

[LeLiMaTr09] B. Lévêque, D. Lin, F. Maffray, and N. Trotignon, Detecting induced subgraphs, *Discrete Applied Mathematics* **157** (2009), 3540–3551.

[Lo72] L. Lovász, A characterization of perfect graphs, *J. Combin. Theory Ser. B* **13** (1972), 95–98.

[MaRe99] F. Maffray and B. Reed, A description of claw-free perfect graphs, *J. Combin. Theory Ser. B* **75** (1999), 134–156.

[MaTr05] F. Maffray and N. Trotignon, Algorithms for perfectly contractile graphs, *SIAM J. Discrete Math* **19** (2005), 553–574.

[Pe13] I. Penev, Coloring bull-free perfect graphs, *SIAM J. Discrete Math* **26** (2012), 1281–1309.

[Ra81] S. B. Rao, Towards a theory of forcibly hereditary P-graphic sequences, *Combinatorics and Graph Theory, Proc. Int. Conf., Kolkata, India, Feb. 1980, S. B.Rao, ed., Lecture Notes in Mathematics, Springer Verlag*, **885** (1981), 441–458.

[Re97] B. Reed, Tree width and tangles: a new connectivity measure and some applications, in *Surveys in Combinatorics, 1997 (London), London Mathematical Society Lecture Notes Series* **241** (1997), 87–162.

[RoSe95] N. Robertson and P. D. Seymour, Graph minors. XIII. The disjoint paths problem, *J. Combin. Theory Ser. B* **63** (1995), 65–110.

[RoSe04] N. Robertson and P. Seymour, Graph minors. XX. Wagner's conjecture, *J. Combin. Theory Ser. B* **92** (2004), 325–357.

[RoSe10] N. Robertson and P. Seymour, Graph minors. XXIII. Nash-Williams's immersion conjecture, *J. Combin. Theory Ser. B* **100** (2010), 181–205.

[RoSe11] N. Robertson and P. Seymour, *private communication.*

[daSiVu13] M. V. G da Silva and K. Vušković, Decomposition of even-hole-free graphs with star cutsets and 2-joins, *J. Combin. Theory Ser. B* **103** (2013), 144–183.

[Sc97] A. Scott, Induced trees in graphs of large chromatic number, *J. Graph Theory* **24** (1997), 297–311.

[Su81] D. P. Sumner, Subtrees of a graph and chromatic number, in *The Theory and Applications of Graphs*, (G. Chartrand, ed.), John Wiley & Sons, New York (1981), 557–576.

[Th99] R. Thomas, Recent excluded minor theorems for graphs, in *Surveys in Combinatorics, 1999 (Canterbury)*, London Mathematical Society Lecture Notes Series **267** (1999), 201–222.

[Vu13] K. Vušković, The world of hereditary graph classes viewed through Truemper configurations, *Surveys in Combinatorics 2013, London Mathematical Society Lecture Notes Series*, to appear.

[We01] D. B. West, *Introduction to Graph Theory*, Second Edition, Prentice-Hall, 2001.

Glossary for Chapter 2

adjacency list representation – for a graph or digraph $G = (V, E)$: an array L of $|V|$ lists, one for each vertex in V; for each vertex i, there is a pointer L_i to a linked list containing all vertices j adjacent to i.

adjacency matrix representation – of a simple graph or digraph $G = (V, E)$: a $|V| \times |V|$ matrix A, where $A[i, j] = 1$ if there is an edge from vertex i to vertex j, and $A[i, j] = 0$ otherwise.

adversary reconstruction number – of a graph G: the minimum number k such that every choice of k subgraphs from the deck of G determines G uniquely.

all-pairs shortest-paths problem: determining the shortest path between every pair of vertices in a graph.

ally reconstruction number – of a graph G: same as the reconstruction number.

X**-anticomplete vertex** in a graph G, where $X \subseteq V(G)$: a vertex $v \in V(G) \setminus X$ such that v has no neighbors in X. For disjoint subsets $X, Y \subseteq V(G)$, we say that Y is *complete* to X (or that the pair (X, Y) is anticomplete) if every vertex in Y is X-anticomplete.

anticonnected graph: a graph G whose complement G^c is connected.

anticomponent of a graph G: a maximal anticonnected induced subgraph of G.

antihole in a graph G: an induced subgraph H that is isomorphic to the complement C_n^c of a cycle graph C_n, for some integer $n \geq 4$.

___, **odd**: an antihole with an odd number of vertices.

___, **even**: an antihole with an even number of vertices.

automorphism of a graph: an isomorphism from the graph to itself.

automorphism group – of a graph: the group of automorphisms of the graph under the operation of functional composition.

back edge – for a spanning tree in a directed graph: a nontree edge that joins a vertex to a proper ancestor.

balanced graph: a graph all of whose induced cycles are of length divisible by 4.

bandwidth-k **graph**: a graph for which there exists a vertex labeling $h : V \to \{1, 2, \ldots |V|\}$ such that $\{u, v\} \in E \Rightarrow |h(u) - h(v)| \leq k$.

Berge graph: a graph with no odd holes and no odd antiholes (terminology due to Chvátal).

Berge-basic graph: a graph G such that either G or G^c is bipartite, the line graph of a bipartite graph, or a double-split graph.

bidegreed graph: a graph whose vertices have only two possible degrees.

bisimplicial vertex in a graph G: a vertex v whose set of neighbors is the union of two cliques.

branch-decomposition – of a graph $G = (V, E)$: a pair (T, f), where T is a tree in which every non-leaf vertex has exactly three neighbors and f is a bijection from the leaves of T to E.

___, **partial**: a branch-decomposition in which the degree of every non-leaf vertex in T is *at least* 3.

branchwidth-k graph: a graph whose branchwidth is no greater than k.

branchwidth – of a graph G: the minimum width taken over all branch-decompositions of G.

breadth-first search: a systematic method for finding all vertices of a graph that are reachable from a given start vertex, by beginning at the start vertex and then visiting the unvisited vertices in a shortest-distance-from-the-start-vertex order.

breadth-first tree: tree of all vertices reachable from a given start vertex of a graph during a breadth-first search.

bull: the graph with vertex set $\{v_1, v_2, v_3, v_4, v_5\}$ where $\{v_2, v_3, v_4\}$ is a clique, v_1 is adjacent to v_2, v_4 is adjacent to v_5, and there are no other edges.

___, **-free graph**: a graph with no induced subgraph isomorphic to the bull.

canonical form – of a simple graph (for isomorphism testing): an arbitrary labeling of the vertex set of a representative of each isomorphism type of graph. Thus, two graphs are isomorphic if and only if their canonical forms are identical.

CAP: see *color automorphism problem*.

certificate for isomorphism: synonym for a complete invariant.

chordal graph: a graph that contains no induced cycles of length greater than 3.

chromatic number $\chi(G)$ of a graph G: the smallest number k for which the vertices of G can be colored with k colors, so that no two adjacent vertices receive the same color.

circular interval graph: the intersection graph of a set of arcs on the circle.

___, **long**: a circular interval graph for which no three of the arcs cover the circle.

class edge-reconstruction number – of a graph G in a class $\mathcal{C}$: the least number of subgraphs in the edge-deck of G which, together with the information that G is in the class $\mathcal{C}$, guarantees that G is uniquely determined.

class reconstruction number – of a graph G in a class $\mathcal{C}$: the least number of subgraphs in the deck of G which, together with the information that G is in the class $\mathcal{C}$, guarantees that G is uniquely determined.

claw: the complete bipartite graph $K_{1,3}$.

clique in a graph G: a set of vertices of G, all pairwise adjacent.

clique number $\omega(G)$ of a graph G: the largest size of a clique in G.

cliquewidth – of a graph: the minimum number of labels that are sufficient to construct a graph from isolated vertices, while using only the union, module join, and relabeling operations.

cliquewidth-k graph: defined recursively as follows ($[k]$ denotes the set of integers $\{1, 2, \ldots, k\}$):

- Any graph G with $V(G) = \{v\}$ and $l(v) \in [k]$ is a cliquewidth-k graph.
- If G_1 and G_2 are cliquewidth-k graphs and $i, j \in [k]$, then
  1. the disjoint union $G_1 \cup G_2$ is a cliquewidth-k graph.
  2. the graph $(G_1)_{i \times j}$ is a cliquewidth-k graph, where $(G_1)_{i \times j}$ is formed from G_1 by adding all edges (v_1, v_2) such that $l(v_1) = i$ and $l(v_2) = j$.
  3. the graph $(G_1)_{i \to j}$ is a cliquewidth-k graph, where $(G_1)_{i \to j}$ is formed from G_1 by switching all vertices with label i to label j.

cograph: defined recursively as

- A graph with a single vertex is a cograph.
- If G_1 and G_2 are cographs, then the disjoint union $G_1 \cup G_2$ is a cograph.
- If G_1 and G_2 are cographs, then the cross-product $G_1 \times G_2$ is a cograph, which is formed by taking the union of G_1 and G_2 and adding all edges (v_1, v_2) where v_1 is in G_1 and v_2 is in G_2.

color automorphism problem (CAP): the problem of finding a set of generators for the subgroup of color-preserving permutations, within a given permutation group acting on a given colored set.

color class – for a graph: the set of all vertices that are assigned the same color.

coloring – of a graph G: a mapping $\sigma : V_G \to \mathbb{C}$ from its vertex set to a set $\mathbb{C}$ (often a set of integers); alternatively, a partition $\sigma = [C_1, \ldots, C_m]$ of the vertex set into *color classes*.

___, **trivial** – for a graph: a coloring that assigns the same color to every vertex.

color-preserving mapping: a graph mapping such that any two like-colored vertices of the domain are mapped to like-colored vertices in the codomain.

X-complete vertex in a graph G, where $X \subseteq V(G)$: a vertex $v \in V(G) \setminus X$ such that v is adjacent to every vertex in X. For disjoint subsets $X, Y \subseteq V(G)$, we say that Y is complete to X (or that the pair (X, Y) is complete) if every vertex in Y is X-complete.

complement or **edge-complement** G^c of a simple graph G: a graph with vertex set $V(G)$, such that two vertices are adjacent in G^c if and only if they are non-adjacent in G.

component of a graph G: a maximal connected subgraph of G.

composition of strips – a generalization of a line-graph of H: every edge of H is replaced by a member of 1 of 15 prescribed families of graphs (a "strip"), and then certain edges are added between the subgraphs corresponding to edges of H that share an end.

cover of a graph G by $\mathcal{F}$ – for a sequence $\mathcal{F} = (F_1, F_2, \ldots, F_k)$ of graphs (in which different F_i could be isomorphic): a sequence $\mathcal{G} = (G_1, G_2, \ldots, G_k)$ of subgraphs of G (not necessarily distinct) such that (i) $G_i \simeq F_i$, $i = 1, \ldots, k$ and (ii) $G = \cup_i G_i$.; the number of covers of G by $\mathcal{F}$ is denoted by $c(\mathcal{F}, G)$.

cross edge – for a spanning forest in a directed graph: a nontree edge that joins two vertices that are neither ancestors nor descendants of each other.

cutwidth of a graph G: the smallest integer k such that the vertices of G can be arranged in a sequence $v_1, \ldots, v_n$ so that, for every $i = 1, \ldots, n-1$, there are at most k edges with one endpoint in $\{v_1, \ldots, v_i\}$ and the other in $\{v_{i+1}, \ldots, v_n\}$.

cycle (graph) C_n **of length** n (where $n \geq 3$ is an integer: the graph with vertex set $\{v_1, \ldots, v_n\}$, such that v_i is adjacent to v_j if and only if $|i - j| = 1 \bmod n$.

deck – of a graph G: the collection $\mathcal{D}(G)$ of all vertex-deleted subgraphs of the graph G.

degree of a vertex v: the number of vertices adjacent to v.

degree sequence – of a graph G: the sequence of degrees of the vertices of G, written in non-descending order.

degree-equivalent pair of graphs: two graphs with the same degree sequence.

degree vector – of a graph coloring $\sigma = [C_1, \ldots, C_m]$: the vector assignment

$$v \mapsto \vec{deg}_\sigma(v) = [|N(v) \cap C_1|, \ldots, |N(v) \cap C_m|]$$

dense graph $G = (V, E)$: one in which the order of magnitude of $|E|$ is close to $|V|^2$.

depth-first forest: set of depth-first trees formed in a depth-first search of a graph.

depth-first search: a systematic method for visiting all vertices of a graph by beginning at a vertex, picking an unvisited adjacent vertex, and recursively continuing the search from that vertex.

depth-first tree: tree formed by tree edges discovered in a depth-first search of a graph.

disjoint paths problem, edge: the problem of determining for a graph or for a directed graph G with vertices $s_1, t_1, \ldots, s_k, t_k$, whether there exist edge-disjoint paths $P_1, \ldots, P_k$ (directed paths, in the case of a directed graph) such that P_i is from s_i to t_i for $1 \leq i \leq k$.

disjoint paths problem, vertex: the problem of determining for a graph or for a directed graph G with vertices $s_1, t_1, \ldots, s_k, t_k$, whether there exist vertex-disjoint paths $P_1, \ldots, P_k$ (directed paths, in the case of a directed graph) such that P_i is from s_i to t_i for $1 \leq i \leq k$.

double-split graph: a graph G, whose vertex-set $V(G)$ can be partitioned into four sets $\{a_1, \ldots, a_m\}$, $\{b_1, \ldots, b_m\}$, $\{c_1, \ldots, c_n\}$, $\{d_1, \ldots, d_n\}$ for some $m, n \geq 2$, such that

- a_i is adjacent to b_i for $1 \leq i \leq m$, and c_j is non-adjacent to d_j for $1 \leq j \leq n$.

- there are no edges between $\{a_i, b_i\}$ and $\{a_{i'}, b_{i'}\}$ for $1 \leq i < i' \leq m$, and all four edges between $\{c_j, d_j\}$ and $\{c_{j'}, d_{j'}\}$ for $1 \leq j < j' \leq n$.

- there are exactly two edges between $\{a_i, b_i\}$ and $\{c_j, d_j\}$ for $1 \leq i \leq m$ and $1 \leq j \leq n$, and these two edges have no common end.

edge-contraction – of an edge e in a graph $G = (V, E)$: an operation that results in a graph with edge-set $E - \{e\}$ but with a vertex-set obtained by replacing ("merging") the endpoints of e in G, thus creating a new single vertex where the latter inherits all of the adjacencies of the pair of replaced vertices, without introducing loops or multiple edges.

edge-deck – of a graph G: the collection $\mathcal{ED}(G)$ of all edge-deleted subgraphs of G.

edge-deleted subgraph – of a graph G: a graph $G - e$ obtained from G by deleting an edge e; also called *edge-deletion subgraph*.

k**-edge-deleted subgraph** – of a graph G: a subgraph obtained from G by deleting k of its edges.

edge-extraction – on a graph $G = (V, E)$: an operation that removes an edge e leaving the edge-deletion graph $G - e$.

edge-recognizable class: a class $\mathcal{C}$ of graphs such that, for any graph $G \in \mathcal{C}$, every edge-reconstruction of G is also in $\mathcal{C}$.

edge-reconstructible graph: a graph G whose every edge-reconstruction is isomorphic to G.

edge-reconstructible parameter: a graph parameter $\mathcal{P}$ such that, for any graph G with parameter value p, every edge-reconstruction of G also has value p for that parameter.

Edge-Reconstruction Conjecture: the conjecture that every graph on at least four edges is edge-reconstructible.

edge-reconstruction number – of a graph G: the least number of subgraphs in the edge-deck of G which guarantees that G is uniquely determined.

edge-reconstruction of a graph G: a graph H with the same edge-deck as G.

edge-reconstruction problem for a structure (D, Γ, E) – where all the subsets $E - x$ are given, up to action by the group Γ: the question of whether E can be reconstructed from these subsets uniquely, again up to action by the group Γ.

k**-edge-reconstruction problem**: the problem of determining uniquely, up to isomorphism, a graph or a structure from its k-edge-deleted subgraphs or substructures.

elementary graph: a graph in which any component is either an edge or a cycle.

endvertex – of a graph G: a vertex whose degree is 1.

endvertex-deck – of a graph G: the collection of graphs $G - v$ for all endvertices v of G.

endvertex-reconstructible graph: a graph that is uniquely determined by its endvertex deck.

equitable partition: a partition of the vertices of a graph into cells such that, for any two vertices v, w in the same cell, and for any cell C of the partition, the vertices v and w are adjacent to the same number of vertices in C.

Erdős–Hajnal property of a graph H: the property that there exists a constant $\delta(H) > 0$, such that every H-free graph G has either a clique or a stable set of size at least $|V(G)|^{\delta(H)}$.

Floyd–Warshall algorithm: an algorithm to compute the shortest length path (or least cost) between vertex i and vertex j, for all vertices i and j.

forward edge – for a spanning tree in a directed graph: a nontree edge that joins a vertex to a proper descendant.

H-free graph: a graph G such that no induced subgraph of G is isomorphic to H. For a family $\mathcal{F}$ of graphs, a graph G is $\mathcal{F}$-**free** if G is H-free for every $H \in \mathcal{F}$.

GI: an abbreviation for the general decision problem of determining whether two graphs are isomorphic.

graph invariant: a property of graphs that has the same value for any two isomorphic graphs.

___, **complete**: an invariant that assigns different values to any two non-isomorphic graphs.

p-group – for a prime p: a group whose order is a power of the prime p.

Halin graph: a planar graph whose edge set can be partitioned into a spanning tree, with no vertices of degree 2, and a cycle through the leaves of this tree.

k-HB graph: graph that yields a balanced modular decomposition when a certain decomposition algorithm is applied; see Definition D30 in §2.4.

hereditary class of graphs: a class $\mathcal{G}$ such that every induced subgraph of a graph in $\mathcal{G}$ is also in $\mathcal{G}$.

hole in a graph G: an induced subgraph H that is isomorphic to a cycle graph C_n for some integer $n \geq 4$.

___, **odd**: a hole with an odd number of vertices.

___, **even**: a hole with an even number of vertices.

illegitimate deck: a collection of graphs $G_1, G_2, \ldots, G_n$, each on $n - 1$ vertices such that there is no graph G having the given collection as its deck.

illegitimate deck problem: the problem to determine whether or not a given collection of graphs is indeed the deck of some graph.

immersion of a (directed) graph H in a (directed) graph G: a one-to-one mapping $\eta : V(H) \to V(G)$ along with a mapping of the (directed) edges of H to (directed) paths joining the corresponding pairs of vertices of G, in such a way that the paths are pairwise edge-disjoint.

___, **strong**: an immersion such that if $v \in V(H)$ and $e \in E(H)$, and if e is not incident with v in H, then $\eta(v)$ is not a vertex of the path $\eta(e)$.

imperfect graph: a graph that is not perfect.

incidence matrix$_1$ representation – of a simple graph $G = (V, E)$: a $|V| \times |E|$ matrix I, where $I[v, e] = 1$ if e is incident on v and 0 otherwise.

incidence matrix$_2$ representation – of a simple digraph $G = (V, E)$: a $|V| \times |E|$ matrix I, where

$$I[v,e] = \begin{cases} -1 & \text{if edge } e \text{ is directed } to \text{ vertex } v \\ 1 & \text{if edge } e \text{ is directed } from \text{ vertex } v \\ 0 & \text{otherwise} \end{cases}$$

induced subgraph of a graph G: a graph H such that $V(H) \subseteq V(G)$, and $uv \in E(H)$ if and only if $uv \in E(G)$ for every $u, v \in V(H)$. For $X \subseteq V(G)$, the subgraph of G induced by X is denoted $G|X$.

ISO: another abbreviation for the general decision problem of determining whether two graphs are isomorphic.

isomorphic graphs: two graphs G and H, such that there is an isomorphism $G \to H$.

isomorphism of labeled graphs G and H: an isomorphism $\phi : G \to H$, such that for each $v \in V_G$, the vertices v and $\phi(v)$ have the same label.

isomorphism of simple graphs: a vertex bijection that preserves adjacency relationships.

isomorphism-complete problem: a decision problem that is polynomial-time equivalent to GI.

Kleene closure of a set of strings S: the set $S^* = \cup_{i=0}^{\infty} S^i$.

Kleene's algorithm: an algorithm for constructing a regular expression that describes all paths between every pair of vertices in a labeled graph.

labeled graph: a graph whose vertices and/or edges are labeled, possibly with repetitions, using symbols from a finite alphabet.

line graph $L(H)$ of a graph H: the graph with vertex set $E(H)$, such that $ef \in E(L(H))$ if and only if the edges e and f share an endpoint in H.

linear-time algorithm: algorithm that runs in $O(V + E)$ time for input graph $G = (V, E)$.

minor – of a graph G: a graph that can be obtained from G by a finite sequence of *edge-extraction* and *edge-contraction* operations.

module: with respect to a subgraph, a set of vertices that share exactly the same neighbors outside this subgraph.

monomorphism with forbidden X – of simple graphs G and H, where X is a subset of the edges of G: a bijection of V such that if $\{u, v\}$ is an edge in $E(G) - X$ then $\{f(u), f(v)\}$ is also an edge in H, but if $\{u, v\}$ is an edge in X then $\{f(u), f(v)\}$ is not an edge in H. The number of monomorphisms from G to H with forbidden X is denoted by $[H]_{G \backslash X}$.

monomorphism – of simple graphs G and H: a one-to-one function $f : V_G \to V_H$ such that if $\{u, v\}$ is an edge of G, then $\{f(u), f(v)\}$ is an edge of H. The number of monomorphisms from G to H is denoted by $[H]_G$.

nauty: the name of a practical computer program for use in graph isomorphism testing. (The name is a quasi-acronym for "no automorphisms, yes".)

neighborhood – of a vertex v of a graph: the set of all vertices adjacent to v. It is denoted by $N(v)$.

k-NLC (node-label-controlled) graph: defined recursively as follows ($[k]$ denotes the set of integers $\{1, 2, \ldots, k\}$, and B denotes a bipartite graph on $[k] \times [k]$):

- Any graph G with $V(G) = \{v\}$ and $l(v) \in [k]$ is a k-NLC graph.
- If G_1 and G_2 are k-NLC graphs and $i, j \in [k]$, then the join $G_1 \times_B G_2$ is a k-NLC graph, where $G_1 \times_B G_2$ is formed from $G_1 \cup G_2$ by adding all edges (v_1, v_2) where $v_1 \in V_1$, $l(v_1) = i$; $v_2 \in V_2$, $l(v_2) = j$, and (i, j) is an edge in E_B.
- The graph $(G_1)_{i \to j}$ is a k-NLC graph, which is formed from G_1 by switching all vertices with label i to label j.

nondeterministic finite automaton: a directed graph (possibly with multiple edges) between the same pair of vertices, having a distinguished start state, a set of final states, and labels on the edges.

N-reconstructible digraph: a digraph D such that the set of triples $(D - v_i, \deg_{in}(v_i), \deg_{out}(v_i))$, for all vertices v_i of D, is sufficient information to determine D uniquely.

order – of an edge e in T in a branch-decomposition (T, f) of a graph $G = (V, E)$: the number of vertices $v \in V$ such that there exist leaves l_1 and l_2 of T residing in different components of $T - e$, where $f(l_1)$ and $f(l_2)$ are both incident on v.

partial k-tree: subgraph of a k-tree.

partition refinement (for isomorphism testing): any process of making a partition of the vertex sets finer (i.e., breaking cells of partitions into smaller cells) by detecting combinatorial differences between the vertices.

path in a graph: a sequence of edges $(v_1, v_2), (v_2, v_3), \ldots, (v_{n-1}, v_n)$.

path-decomposition: a tree-decomposition whose tree is a path.

pathwidth-k graph: a graph that has pathwidth no greater than k.

pathwidth – of a graph G: the smallest width taken over all path-decompositions of G; measures how closely the graph resembles a path.

perfect graph: a graph G such that $\chi(H) = \omega(H)$, for every induced subgraph H of G.

peripheral vertex of a tree: a vertex that has maximum distance from the center of the tree.

polynomial deck: the collection (multi-set) of the characteristic polynomials of all subgraphs in the deck.

polynomial-time algorithm: an algorithm that runs in $O((V + E)^k)$ time for input graph $G = (V, E)$ for some constant k.

prime graph: a graph that cannot be obtained from smaller graphs by substitution.

prism (graph): a graph consisting of two disjoint triangles $\{a_1, a_2, a_3\}$ and $\{b_1, b_2, b_3\}$ and three paths P_1, P_2, P_3, with the following properties:

- for $i = 1, 2, 3$, the ends of P_i are a_i and b_i,
- P_1, P_2, P_3 are pairwise disjoint, and
- for $1 \leq i < j \leq 3$, there are precisely two edges between $V(P_i)$ and $V(P_j)$, namely, $a_i a_j$ and $b_i b_j$.

property A_k – of a graph G: the property that whenever A and B are distinct k-sets of vertices of G, the graphs $G - A$ and $G - B$ are not isomorphic.

pyramid: a graph consisting of a triangle $\{b_1, b_2, b_3\}$, called the *base*, a vertex $a \notin \{b_1, b_2, b_3\}$, called the *apex*, and three paths P_1, P_2, P_3, such that for $i, j \in 1, 2, 3$

- the ends of P_i are a and b_i,
- if $i \neq j$ then $V(P_i) \setminus \{a\}$ is disjoint from $V(P_j) \setminus \{a\}$ and the only edge between them is $b_i b_j$, and
- at most one of P_1, P_2, P_3 has length one.

In this case we say that the pyramid is *formed* by the paths P_1, P_2, P_3.

___, **frame of**: the 10-tuple
$$a, b_1, b_2, b_3, s_1, s_2, s_3, m_1, m_2, m_3,$$
where

- for $i = 1, 2, 3$, s_i is the neighbor of a in P_i
- for $i = 1, 2, 3$, $m_i \in V(P_i)$ satisfies $d_{P_i}(a, m_i) - d_{P_i}(m_i, b_i) \in \{0, 1\}$.

___, **optimal**: a pyramid P such that there is no pyramid P' with $|V(P')| < |V(P)|$.

quasi-line graph: a graph in which every vertex is bisimplicial.

quasi-order Q consists of a class $E(Q)$ and a transitive, reflexive relation which we denote by $\leq$ or $\leq_Q$.

___, **well** or **wqo**: a quasi-order in which for every infinite sequence q_i $(i = 1, 2 \ldots)$ of elements of $E(Q)$, there exist $j > i \geq 1$ such that $q_i \leq_Q q_j$.

Rao-contained graph in a graph G: a graph H that is isomorphic to an induced subgraph of some graph that is degree-equivalent to G.

recognizable class of graphs: a class $\mathcal{C}$ of graphs such that, for any $G \in \mathcal{C}$, every reconstruction of G is also in C.

reconstructible graph: a graph G whose every reconstruction is isomorphic to G.

reconstructible parameter: a graph parameter $\mathcal{P}$ such that, for any graph G with the value p for that parameter, every reconstruction of G also has parameter value p.

Reconstruction Conjecture: the conjecture that every graph with at least three vertices is reconstructible.

reconstruction index – of a group Γ: the smallest number t such that for any $E \subset D$ with $|E| \geq t$, the structure (D, Γ, E) is edge-reconstructible.

reconstruction number – of a graph G: the least number of subgraphs in the deck of G which guarantees that G is uniquely determined.

reconstruction of a graph G: a graph H with the same deck as G.

recursively constructed graph class: defined by a set (usually finite) of primitive or *base graphs*, in addition to one or more operations that compose larger graphs from smaller subgraphs; each operation involves either fusing specific vertices from each subgraph or adding new edges between specific vertices from each subgraph.

refinement of a graph coloring – an operation that yields a new coloring of the graph: two vertices with the same old color get the same new color if and only if they have the same numbers of neighbors of every old color.

regular expression: a notation for describing a regular set by using the operators union, concatenation, and Kleene closure.

series-parallel graph with distinguished *terminals l and r*, denoted (G, l, r) – defined recursively:

- The graph consisting of a single edge (v_1, v_2) is a series-parallel graph (G, l, r) with $l = v_1$ and $r = v_2$.
- A *series operation* $(G_1, l_1, r_1) \odot_s (G_2, l_2, r_2)$ forms a series-parallel graph by identifying r_1 with l_2. The terminals of the new graph are l_1 and r_2.
- A *parallel operation* $(G_1, l_1, r_1) \odot_p (G_2, l_2, r_2)$ forms a series-parallel graph by identifying l_1 with l_2 and r_1 with r_2. The terminals of the new graph are l_1 and r_1.
- A *jackknife operation* $(G_1, l_1, r_1) \odot_j (G_2, l_2, r_2)$ forms a series-parallel graph by identifying r_1 with l_2; the new terminals are l_1 and r_1.

set edge-reconstructible – graph or a parameter: a graph or a parameter that can be determined from the set of non-isomorphic subgraphs in the edge-deck.

set reconstructible – graph or a parameter: a graph or a parameter that can be determined from the set of non-isomorphic subgraphs in the deck.

skew-partition in a graph G: a partition (A, B) of $V(G)$ such that A is not connected and B is not anticonnected.

sparse graph $G = (V, E)$: one in which the order of magnitude of $|E|$ is $|V|$ or less.

split graph: a graph whose vertex set can be partitioned into a clique and a stable set.

stabilization of a coloring σ: the coloring that results from iterating the refinement process until a stable coloring is obtained. It is denoted σ^*.

stable coloring: a graph coloring that is unchanged by the refinement operation.

stable set or **independent set** in a graph G: a set of vertices, all pairwise non-adjacent.

stability number or **independence number** $\alpha(G)$ of a graph G: the largest size of a stable set.

k-star cutset in a graph G: a partition (A, B) of $V(G)$ such that A is not connected, such that there is a clique $K \subseteq B$, with $|K| = k$, and such that every vertex of $B \setminus K$ has a neighbor in K. A 1-star cutset is usually called a **star cutset**.

structure: a triple (D, Γ, E) where D is a finite set, Γ is a group of permutations acting on D, and E is a subset of D.

substitution, result of of H_2 for $v \in V(H_1)$: the graph H such that
- $V(H) = (V(H_1) \cup V(H_2)) \setminus \{v\}$,
- $H|V(H_2) = H_2$,
- $H|(V(H_1) \setminus \{v\}) = H_1 \setminus v$, and
- $u \in V(H_1)$ is adjacent in H to $w \in V(H_2)$ if and only if u is adjacent in H_1 to v.

k-terminal recursive graph: graph that has at most k special vertices called terminals, and that can be obtained by operations that fuse some of the terminals in its constituent k-terminal subgraphs. (See Definition D25 in §2.4.)

theta (graph): a graph consisting of two non-adjacent vertices s, t and three paths P_1, P_2, P_3, each between s and t, such that the sets $V(P_1) \setminus \{s, t\}$, $V(P_2) \setminus \{s, t\}$, and $V(P_3) \setminus \{s, t\}$ are pairwise disjoint, and the union of every pair of P_1, P_2, P_3 is a hole.

transitive closure of a graph G: a graph G^* that has an edge (i, j) if and only if there is a path of length 1 or more in G from i to j.

tree: a connected graph with no cycles, and sometimes with a designated *root*.

___, **recursively defined**: a graph with a single vertex r as its root r; or, a graph formed by joining the roots of two trees.

k-tree (recursively defined): the complete graph K_k; or, a graph constructed from a k-tree on n vertices by adding a vertex adjacent to all vertices of one of its K_k subgraphs, and only to those vertices.

___, **partial**: a subgraph of a k-tree.

tree-decomposition – of a graph $G = (V, E)$: a pair $(\{X_i \mid i \in I\}, T)$, such that $\{X_i \mid i \in I\}$ is a family of subsets of V and T is a tree with vertex set I such that
- $\bigcup_{i \in I} X_i = V$,
- for all edges $(x, y) \in E$ there is an element $i \in I$ with $x, y \in X_i$,
- for all triples $i, j, k \in I$, if j is on the path from i to k in T, then $X_i \cap X_k \subseteq X_j$.

treewidth – of a graph G: the minimum width taken over all tree-decompositions of G; measures how closely the graph resembles a tree.

treewidth-k graph: a graph whose treewidth is no greater than k.

trigraph: a graph G with a map $\theta : V(G)^2 \to \{-1, 0, 1\}$, called the *adjacency function*, satisfying:
- for all $v \in V(G)$, $\theta_G(v, v) = 0$,
- for all distinct $u, v \in V(G)$, $\theta_G(u, v) = \theta_G(v, u)$,
- for all $u \in V(G)$, there exists at most one $v \in V(G) \setminus \{u\}$ such that $\theta_G(u, v) = 0$.

tournament: a directed graph G where for every distinct $u, v \in V(G)$, exactly one of the (ordered) pairs uv and vu belongs to $E(G)$.

___, **transitive**: a tournament in which there are no directed cycles.

vertex-deleted subgraph – of a graph G: a graph $G - v$ obtained from G by deleting a vertex v and all the edges incident to it; also called a *vertex-deletion subgraph*.

k**-vertex-deleted subgraph** – of a graph G: a subgraph obtained from G by deleting k of its vertices and all the edges incident to them.

weakly edge-reconstructible graph – relative to a class $\mathcal{C}$: a graph $G \in \mathcal{C}$ such that every edge-reconstruction of G which is also in the class $\mathcal{C}$ is isomorphic to G.

weakly reconstructible graph – relative to a class $\mathcal{C}$: a graph $G \in \mathcal{C}$ such that every reconstruction of G which is also in the class $\mathcal{C}$ is isomorphic to G.

width$_1$ – of a branch decomposition (T, f): the maximum order of the edges of T.

width$_2$ – of a tree-decomposition $(\{X_i \mid i \in I\}, T)$: $\max_{i \in I}\{|X_i| - 1\}$.

Chapter 3

Directed Graphs

Section 3.1
Basic Digraph Models and Properties

Jay Yellen, Rollins College

INTRODUCTION

This section extends the basic terminology and properties begun in Chapter 1, and it describes several classical digraph models that preview later sections of the *Handbook*. Many of the basic methods and algorithms for digraphs closely resemble their counterparts for undirected graphs. Some general references for digraphs are [ChLeZh10], [GrYe06], and [We01]. A comprehensive and in-depth reference for digraphs is [BaGu01].

3.1.1   Terminology and Basic Facts

TERMINOLOGY NOTE: The term *arc* is used throughout this section instead of its synonym *directed edge*.

NOTATION: Often, when the digraphs under consideration do not have multi-arcs, an arc that is directed from vertex u to v is represented by the ordered pair (u, v) or by the juxtaposition uv.

TERMINOLOGY: An arc that is directed from vertex u to v is said to have **tail** u and **head** v.

Reachability and Connectivity

DEFINITIONS

D1: In a digraph, a **directed walk** from v_0 to v_n is an alternating sequence
$$W = \langle v_0, e_1, v_1, e_2, ..., v_{n-1}, e_n, v_n \rangle$$
of vertices and arcs, such that $tail(e_i) = v_{i-1}$ and $head(e_i) = v_i$, for $i = 1, ..., n$.

TERMINOLOGY: A directed walk from a vertex x to a vertex y is also called an x-y *directed walk*.

D2: The *length of a directed walk* is the number of arc-steps in the walk sequence.

D3: A *connected digraph* is a digraph whose underlying graph is connected. Elsewhere, the term *weakly connected* is often used to describe such digraphs.

D4: Let u and v be vertices in a digraph G. Then u and v are said to be *mutually reachable* in G if G contains both a directed u-v walk and a directed v-u walk. Every vertex is regarded as reachable from itself (by the trivial walk).

D5: A digraph is *strongly connected* if every two vertices are mutually reachable.

D6: A *strong component* of a digraph G is a maximal strongly connected subdigraph of G. Equivalently, a strong component is a subdigraph induced on a maximal set of mutually reachable vertices.

D7: Let $S_1, S_2, \ldots, S_r$ be the strong components of a digraph G. The *condensation* of G is the simple digraph G^* with vertex-set $V_{G^*} = \{s_1, s_2, \ldots, s_r\}$, such that there is an arc in digraph G^* from vertex s_i to vertex s_j if and only if there is an arc in digraph G from a vertex in component S_i to a vertex in component S_j.

EXAMPLE

E1: Figure 3.1.1 shows a digraph G, its four strong components, S_1, S_2, S_3, S_4, and its condensation G^*. Notice that the vertex-sets of the strong components of G partition the vertex-set of G and that the edge-sets of the strong components do *not* include all the edges of G. This is in sharp contrast to the situation for an undirected graph G, in which the edge-sets of the components of G partition E_G.

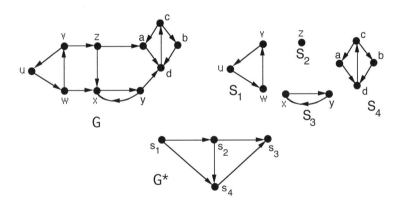

Figure 3.1.1: A digraph, its four strong components, and its condensation.

FACT

F1: Let G be a digraph. Then the mutual-reachability relation is an equivalence relation on V_G, and the strong components of digraph G are the subdigraphs induced on the equivalence classes of this relation.

Measures of Digraph Connectivity

We introduce a few basic measures of the connectedness of a digraph. Connectivity of graphs and digraphs is discussed extensively in §4.1 and §4.7. The concept of an *edge-cut* plays an important role in the study of flows in networks (§11.1) and in certain algebraic properties of a graph or digraph (§6.4).

DEFINITIONS

D8: A *complete digraph* is a simple digraph such that between each pair of its vertices, both (oppositely directed) arcs exist.

D9: A *vertex-cut* in a strongly connected digraph $G = (V, E)$ is a vertex subset $S \subset V$ such that the vertex-deletion subdigraph $G - S$ is not strongly connected, and an *edge-cut* (*arc-cut*) is an arc subset $F \subset E$ such that the arc-deletion subdigraph $G - F$ is not strongly connected.

D10: The *(vertex-) connectivity* of an n-vertex non-complete digraph $G = (V, E)$, denoted $\kappa_v(G)$, is the minimum size of a vertex subset S such that $G - S$ is neither strongly connected nor the trivial digraph. (The connectivity of a complete n-vertex digraph is $n - 1$.)

D11: The *edge-connectivity* of a non-trivial digraph, denoted $\kappa_e(G)$, is the minimum size of an edge subset F such that $G - F$ is not strongly connected.

NOTATION: When the context is clear, the vertex- and edge-connectivity are denoted κ_v and κ_e, respectively. Some other sections of the *Handbook* use the "traditional" κ and λ instead of κ_v and κ_e, respectively.

TERMINOLOGY NOTE: Synonyms for vertex-cut are *cut* and *disconnecting set*. Synonyms for edge-cut are *edge-disconnecting set* (or *arc-disconnecting set*) and *cut-set*.

Directed Trees

DEFINITIONS

D12: A *directed tree* is a digraph whose underlying graph is a tree.

D13: A *rooted tree* is a directed tree having a distinguished vertex r, called the *root*, such that for every other vertex v, there is a directed r-v path.

TERMINOLOGY NOTE: Occasionally encountered synonyms for rooted tree are *out-tree*, *branching*, and *arborescence*.

REMARKS

R1: Since the underlying graph of a rooted tree is acyclic, the directed r-v path is unique.

R2: Designating a root in a directed tree does *not* necessarily make it a rooted tree.

Tree-Growing in a Digraph

Algorithm 3.1.1, shown below, is simply the basic tree-growing algorithm of §1.1 (Algorithm 1.1.1), recast for digraphs. Its output, as in Algorithm 1.1.1, is a rooted tree whose vertices are reachable from the starting vertex. But because the paths to these vertices are directed (i.e., one-way), the vertices in this *output tree* need not be mutually reachable from one another.

DEFINITION

D14: A ***frontier arc*** for a rooted tree T in a digraph is an arc whose tail is in T and whose head is not in T.

Algorithm 3.1.1: Basic Tree-Growing in a Digraph

Input: a digraph G and a starting vertex $v \in V_G$.
Output: a rooted tree T with root v and a standard vertex-labeling of T.

Initialize tree T as vertex v.
Write label 0 on vertex v.
Initialize label counter $i := 1$
While there is at least one frontier arc for tree T
    Choose a frontier arc e for tree T.
    Let w be $head(e)$ (which lies outside of T).
    Add arc e and vertex w to tree T.
    Write label i on vertex w.
    $i := i + 1$
Return tree T and vertex-labeling of T.

COMPUTATIONAL NOTE: We assume that there is some implicit *default priority* for choosing vertices or edges, which is invoked whenever there is more than one frontier arc from which to choose.

EXAMPLE

E2: Figure 3.1.2 shows a digraph and all possible output trees that could result for each of the different starting vertices and each possible default priority. Two opposite extremes for possible output trees are represented here. When the algorithm starts at vertex u, the output tree spans the digraph. The other extreme occurs when the algorithm starts at vertex x (because x has outdegree 0). Notice that any two output trees in Figure 3.1.2 with the same vertex-set have roots that are mutually reachable.

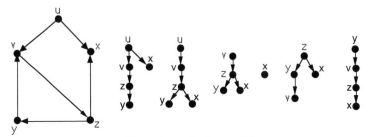

Figure 3.1.2: A digraph and all possible output trees.

FACTS

F2: Let u and v be two vertices of a digraph G. Then u and v are in the same strong component of G if and only if the output trees that result from starting Algorithm 3.1.1 at vertex u and at vertex v have the same vertex-set.

F3: If the digraph G is strongly connected, then the output tree is a spanning rooted tree of G, regardless of the starting vertex.

REMARK

R3: Example E2 above illustrates an important distinction between undirected and directed graphs: whereas tree-growing in an undirected graph provides a simple algorithm to determine the components of the graph, in a digraph this is not the case. Other differences were suggested earlier in Example E1. The use of tree-growing, specifically *depth-first search* (§10.1), in finding the strong components of a digraph is considerably more intricate than its undirected counterpart. For discussions of strong-component-finding algorithms, see, e.g., [BaGe99], [GrYe06, §11.4], and [St93].

Oriented Graphs

DEFINITIONS

D15: An ***oriented graph*** is a digraph obtained by choosing an orientation for each edge of an undirected simple graph. Thus, an oriented graph does not have both oppositely directed arcs between any pair of vertices, which means that an oriented tree is the same as a directed tree.

D16: A ***tournament*** is an oriented complete graph. That is, it has no self-loops, and between every pair of vertices, there is exactly one arc. See §3.3 for extensive coverage of tournaments.

D17: A graph G is ***strongly orientable*** if there exists an assignment of directions to the edge-set of G such that the resulting digraph is strongly connected.

EXAMPLE

E3: Of the three graphs shown in Figure 3.1.3, only the graph G_2 is strongly orientable.

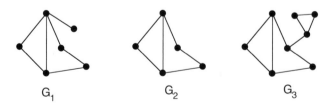

Figure 3.1.3: Only the graph G_2 is strongly orientable.

Notice that G_2 is the only graph in the example that does not have a cut-edge. In fact, the absence of cut-edges is a necessary and sufficient condition for a graph to be strongly orientable. This characterization of strongly orientable graphs was proved by H. E. Robbins in 1939.

FACT

F4: Robbins's Theorem [Ro39] A connected graph G is strongly orientable if and only if G has no cut-edges.

Adjacency Matrix of a Digraph

DEFINITION

D18: The **adjacency matrix of a digraph** $G = (V, E)$, denoted A_G, is given by

$$A_G[u, v] = \begin{cases} \text{the number of arcs from } u \text{ to } v & \text{if } u \neq v \\ \text{the number of self-loops} & \text{if } u = v \end{cases}$$

FACTS

F5: A row-sum in a directed adjacency matrix equals the outdegree of the corresponding vertex, and a column-sum equals the indegree.

F6: Let G be digraph with adjacency matrix A_G. Then the value of the entry $A_G^r[u, v]$ of the rth power of matrix A_G equals the number of directed u-v walks of length r.

EXAMPLE

E4: The adjacency matrix of the digraph in Figure 3.1.4 uses the vertex ordering u, v, w, x. As an illustration of Fact F6, observe that the number of directed w-v walks equals 3, which is the (w, v)-entry of A_G^2.

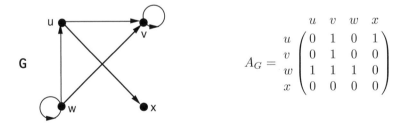

$$A_G = \begin{array}{c} \\ u \\ v \\ w \\ x \end{array} \begin{array}{cccc} u & v & w & x \\ \begin{pmatrix} 0 & 1 & 0 & 1 \\ 0 & 1 & 0 & 0 \\ 1 & 1 & 1 & 0 \\ 0 & 0 & 0 & 0 \end{pmatrix} \end{array}$$

Figure 3.1.4: A digraph and its adjacency matrix.

REMARK

R4: Another matrix representation of a digraph is the *node-arc incidence matrix*, where the columns are labeled by the arcs, and the rows are labeled by the vertices (nodes). Connections between the incidence matrix of a graph or digraph and the structural properties of the graph are explored in §6.4.

3.1.2   A Sampler of Digraph Models

In this subsection, we sample a few of the digraph models. Acyclic digraph models are the focus of §3.2.

Markov Chains and Markov Digraphs

The topic of Markov processes is part of a more general area known as *stochastic processes*, a branch of mathematics and operations research with far-ranging applications and theoretical challenges. The reader may consult any of the standard texts in this subject for a formal presentation of these concepts (e.g., [Ci75], [Wi94]).

DEFINITIONS

D19: A sequence of random variables $\{X_t\}$, $t = 0, 1, 2, \ldots$, is a **(finite) discrete-time Markov chain (DTMC)** on a state-space $S = \{1, 2, \ldots, n\}$ if $X_t \in S$ for all times $t = 0, 1, 2, \ldots$, and the probability distribution of X_{t+1} depends only on the value of X_t. In particular,

$$prob(X_{t+1} = j | X_t = i, X_{t-1} = i_{t-1}, \ldots, X_0 = i_0) = prob(X_{t+1} = j | X_t = i)$$

D20: A **stationary DTMC** satisfies the additional condition that for all states $i, j \in S$ and all times t, the **transition probability** $prob(X_{t+1} = j | X_t = i) = p_{ij}$ is independent of t.

D21: A **Markov digraph** $G = (V, E)$ of a stationary DTMC with state-space S and transition probabilities p_{ij} is a digraph with vertex-set $V = S$, arc-set $E = \{ij | p_{ij} > 0\}$, and to each arc $ij \in E$ is assigned the probability p_{ij}.

D22: The **transition matrix** of a Markov chain is the matrix whose ij^{th} entry is the transition probability p_{ij}.

EXAMPLE

E5: *A Gambler's Problem*:   A gambler starts with \$3 and plays the following game. Two coins are tossed. If both come up heads, then he wins \$3; otherwise, he loses \$1. He plays until either he loses all his money or he reaches a total of at least \$5. Let X_t be the amount of money he has after t plays, with $X_0 = 3$. The state space is $S = \{0, 1, 2, 3, 4, 5\}$, and the sequence $\{X_t\}$ is a discrete-time Markov chain. The transition matrix and Markov digraph for this Markov chain are shown in Figure 3.1.5.

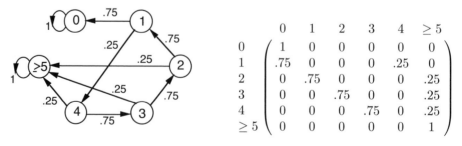

Figure 3.1.5: Gambler's transition matrix and Markov digraph.

Equipment-Replacement Policy

We present a digraph model that can be used to determine a replacement policy that minimizes the net cost of owning and operating a car for a pre-specified number of years.

EXAMPLE

E6: Suppose that today's price for a new car is $16,000, and that the price will increase by $500 for each of the next four years. The projected annual operating cost and resale value of this kind of car are shown in the table below. To simplify the setting, assume that these data do not change for the next five years.

Annual Operating Cost	Resale Value
$600 (for 1st year of car)	$13,000 (for a 1-year-old car)
$900 (for 2nd year of car)	$11,000 (for a 2-year-old car)
$1200 (for 3rd year of car)	$9,000 (for a 3-year-old car)
$1600 (for 4th year of car)	$8,000 (for a 4-year-old car)
$2100 (for 5th year of car)	$6,000 (for a 5-year-old car)

Digraph Model: The digraph has six vertices, labeled 1 through 6, representing the *beginning* of years 1 through 6. The beginning of year 6 signifies the end of the planning period. For each i and j with $i < j$, an arc is drawn from vertex i to vertex j and is assigned a weight c_{ij}, where c_{ij} is the total net cost of purchasing a new car at the beginning of year i and keeping it until the beginning of year j. Thus,

$$c_{ij} = \begin{cases} \text{price of new car at beginning of year } i \\ + \text{ sum of operating costs for years } i, i+1, \ldots, j-1 \\ - \text{ resale value at beginning of year } j \end{cases}$$

Figure 3.1.6 shows the resulting digraph with 7 of its 15 arcs drawn. The arc-weights are in units of $100.

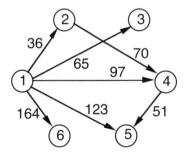

Figure 3.1.6: Part of the digraph model for a car-replacement problem.

The problem of determining the optimal replacement policy is reduced to finding the shortest (least-cost) path from vertex 1 to vertex 6. This is a simple task for ***Dijkstra's algorithm***, even for much larger instances of this kind of problem. Dijkstra's algorithm is discussed in §10.1.

The Digraph of a Relation and the Transitive Closure

Our focus here is on general relations and their transitive closure. Digraphs of *posets* (partially ordered sets) are discussed in §3.2.

DEFINITIONS

D23: A *relation* R on a finite set S is a subset of the cartesian product $S \times S$.

D24: The *digraph representation of a relation* R on a finite set S is the digraph whose vertices correspond to the elements of S, and whose arcs correspond to the ordered pairs in the relation; that is, an arc is drawn from vertex x to vertex y if $(x, y) \in R$.

Conversely, a digraph G induces a relation R on V_D in a natural way, namely, $(x, y) \in R$ if and only if there is an arc in digraph G from vertex x to vertex y.

D25: A *transitive digraph* is a digraph whose corresponding relation is transitive. That is, if there is an arc from vertex x to vertex y and an arc from y to z, then there is an arc from x to z.

D26: The *transitive closure* R^* of a binary relation R is the relation R^* defined by $(x, y) \in R^*$ if and only if there exists a sequence $x = v_0, v_1, v_2, \ldots, v_k = y$ such that $k \geq 1$ and $(v_i, v_{i+1}) \in R$, for $i = 0, 1, \ldots, k-1$. Equivalently, the transitive closure R^* of the relation R is the smallest transitive relation that contains R.

D27: Let G be the digraph representing a relation R. Then the digraph G^* representing the transitive closure R^* of R is called the *transitive closure of the digraph* G. Thus, an arc (x, y), $x \neq y$, is in the transitive closure G^* if and only if there is a directed x-y path in G. Similarly, there is a self-loop in digraph D^* at vertex x if and only if there is a directed cycle in digraph G that contains x.

EXAMPLES

E7: Suppose a relation R on the set $S = \{a, b, c, d\}$ is given by

$$\{(a, a), (a, b), (b, c), (c, b), (c, d)\}$$

Then the digraph G representing the relation R and the transitive closure G^* are as shown in Figure 3.1.7.

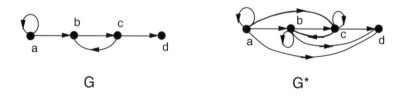

Figure 3.1.7: The digraph G and its transitive closure G^*.

E8: *Transitive Closure in a Paging Network:* Suppose that the arcs of an n-vertex digraph G represent the one-way direct links between specified pairs of nodes in an n-node paging network. Thus, an arc from vertex i to vertex j indicates that a page call can be transmitted from person i to person j.

To send an alert from person i to person j, it is not necessary to have a direct link from i to j. There need only be a directed i-j path. The transitive closure G^* of digraph G specifies all pairs i, j of vertices for which there exists a directed i-j path in G.

Constructing the Transitive Closure of a Digraph: Warshall's Algorithm

Let G be an n-vertex digraph with vertices $v_1, v_2, \ldots, v_n$. A computationally efficient algorithm, due to Warshall [Wa62], constructs a sequence of digraphs,

$$D_0, D_1, \ldots, D_n,$$

such that $D_0 = G$, D_{i-1} is a subgraph of D_i, $i = 1, \ldots, n$, and such that D_n is the transitive closure of D. Digraph D_i is obtained from digraph D_{i-1} by adding to D_{i-1} an arc (v_j, v_k) (if it is not already in D_{i-1}) whenever there is a directed path of length 2 in D_{i-1} from v_j to v_k, having v_i as the internal vertex (see Figure 3.1.8).

A related algorithm of Floyd [Fl62] determines the shortest distance between all pairs of vertices in an edge-weighted digraph.

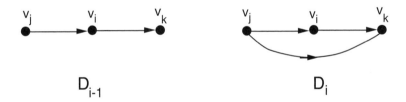

Figure 3.1.8: The arc (v_j, v_k) is added to digraph D_{i-1}.

Algorithm 3.1.2: Warshall's Transitive Closure [Wa62]

Input: an n-vertex digraph D with vertices $v_1, v_2, \ldots, v_n$.
Output: the transitive closure of digraph D.

  Initialize digraph D_0 to be digraph G.
  For $i = 1$ to n
    For $j = 1$ to n
      If (v_j, v_i) is an arc in digraph D_{i-1}
        For $k = 1$ to n
          If (v_i, v_k) is an arc in digraph D_{i-1}
            Add arc (v_j, v_k) to D_{i-1} (if it is not already there).
  Return digraph D_n.

Activity-Scheduling Networks

In large projects, often there are some tasks that cannot start until certain others are completed. Figure 3.1.9 shows a digraph model of the *precedence relationships* among some tasks for building a house. Vertices correspond to tasks. An arc from vertex u to vertex v means that task v cannot start until task u is completed. To simplify the drawing, arcs that are implied by transitivity are not drawn. This digraph is the *cover diagram* of a partial ordering of the tasks. Section 3.2 discusses this model further and introduces a different model in which the tasks are represented by the arcs of a digraph.

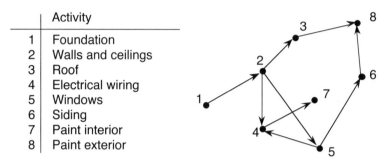

	Activity
1	Foundation
2	Walls and ceilings
3	Roof
4	Electrical wiring
5	Windows
6	Siding
7	Paint interior
8	Paint exterior

Figure 3.1.9: An activity digraph for building a house.

Scheduling the Matches in a Round-Robin Tournament

Suppose that each pair of n teams is to play one match in a tournament. Typically, one would like to schedule the matches so that all matches are completed in a minimum number of days (assume that each team plays at most one match on a given day). If the teams are from different cities, an additional objective is to have an equitable distribution of home and away matches. We preview here a strategy that is discussed in §5.6.

DEFINITIONS

D28: A *compact schedule* for a round-robin tournament is one in which each team plays a match each day.

D29: A team is said to have a *break* if it is either home for two consecutive matches or away for two.

D30: A *proper arc-coloring* of a digraph $G = (V, E)$ is an assignment of colors to the arcs in G so that any two arcs that have an endpoint in common are assigned different colors. Graph coloring is discussed in §5.1 and §5.2, and the related concept of graph factorization is discussed in §5.4.

REMARK

R5: An algorithm for constructing a compact schedule for a n-team round-robin tournament, where n is even, that minimizes the total number of breaks is given in §5.6 (Algorithm 5.6.1). The strategy is based on orienting the edges of a complete graph and then producing a proper arc-coloring so that each color is assigned to exactly $n/2$ arcs.

Flows in Networks

A pipeline network for transporting oil from a single source to a single sink is one proto-type of a network model. Each arc represents a section of pipeline, and the endpoints of an arc correspond to the junctures at the ends of that section. The arc capacity is the maximum amount of oil that can flow through the corresponding section per unit time. A network could just as naturally represent a system of truck routes for transporting commodities from supply points to demand points, or it could represent a network of phone lines from one distribution center to another.

DEFINITIONS

D31: A *cost flow network* $G = (V, E, cap, c, b)$ is a directed graph with vertex-set V, arc-set E, a nonnegative capacity function $cap : E \to N$, a linear cost function $c : E \to Z$, and an integral supply vector $b : V \to Z$ that satisfies $\sum_{w \in V} b(w) = 0$.

D32: An *s-t flow network* $G = (V, E, cap, s, t)$ is a directed graph (typically without the cost and supply functions) with a nonnegative capacity function $cap : E \to N$, that has a distinguished vertex s, called the *source*, with nonzero outdegree, and a distinguished vertex t, called the *sink*, with nonzero indegree.

D33: The *maximum-flow problem* is to determine the maximum flow that can be pushed through an *s-t* network from source s to sink t such that the flow into each intermediate node equals the flow out (*conservation of flow*) and the flow across any arc does not exceed the capacity of that arc. (See §11.1.)

D34: The *minimum-cost-flow problem* is to find an assignment of *flows* on the arcs of the flow network that satisfy the supply and demand (negative supply) requirements at minimum cost. (See §11.2.)

Software Testing and the Chinese Postman Problem

During execution, an application software's flow moves between various *states*, and the *transitions* from one state to another depend on the input. In testing software, one would like to generate input data that forces the program to test all possible transitions.

DEFINITIONS

D35: An *eulerian tour* of a digraph G is a closed directed walk that uses each arc *exactly* once.

D36: A *postman tour* (or *covering walk*) is a closed directed walk that uses each arc *at least* once.

D37: Given a directed edge-weighted graph G, the *Directed Chinese Postman Problem* is to find a minimum-weight postman tour.

Digraph Model: The software's execution flow is modeled as a digraph, where the states of the program are represented by vertices, the transitions are represented by arcs, and each of the arcs is assigned a label indicating the input that forces the corresponding transition. Then the problem of finding an input sequence for which the program invokes all transitions and minimizes the total number of transitions is equivalent to the Directed Chinese Postman Problem, where all arc-weights equal one.

REMARKS

R6: Since certain transitions take more execution time than others, one might want to minimize the total time of execution during the testing (instead of the number of transitions). In that case, each arc is assigned a weight equal to the transition time corresponding to that arc.

R7: Under certain reasonable assumptions, the flow digraph modeling a program's execution can be assumed to be strongly connected, which guarantees the existence of a postman tour.

R8: Eulerian digraphs and graphs, along with algorithms to construct eulerian tours, are discussed in detail in §4.2, and various versions of the Chinese Postman Problem and its algorithms are discussed in §4.3.

Lexical Scanners

The source code of a computer program may be regarded as a string of symbols. A *lexical scanner* must scan these symbols, one at a time, and recognize which symbols go together to form a syntactic *token* or *lexeme*. We now consider a single-purpose scanner whose task is to recognize whether an input string of characters is a valid *identifier* in the C programming language. Such a scanner is a special case of a *finite-state recognizer* and can be modeled by a labeled digraph, as in Figure 3.1.10. One vertex represents the *start* state, in effect before any symbols have been scanned. Another represents the *accept* state, in which the substring of symbols scanned so far forms a valid C identifier. The third vertex is the *reject* state, indicating that the substring has been discarded because it is not a valid C identifier. Each arc label tells what kinds of symbols cause a transition from the tail state to the head state. If the final state after the input string is completely scanned is the accept state, then the string is a valid C identifier.

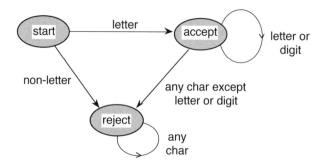

Figure 3.1.10: Finite-state recognizer for identifiers.

3.1.3 Binary Trees

At first glance, a discussion of binary trees does not seem to belong in a section on digraphs. In fact, binary trees are digraphs. In particular, they are special *rooted trees*. Here we describe a few applications.

Rooted Tree Terminology

DEFINITIONS

D38: In a rooted tree, the **depth** or **level** of a vertex v is its distance from the root, that is, the length of the unique path from the root to v. (Thus, the root has depth 0.)

D39: The **height** of a rooted tree is the length of a longest path from the root (which equals the greatest depth in the tree).

D40: If vertex v immediately precedes vertex w on the path from the root to w, then v is the **parent** of w and w is the **child** of v.

D41: A vertex w is called a **descendant** of a vertex v (and v is called an **ancestor** of w), if v is on the unique path from the root to w. If, in addition, $w \neq v$, then w is a **proper descendant** of v (and v is a **proper ancestor** of w).

D42: An **ordered tree** is a rooted tree in which the children of each vertex are assigned a fixed ordering.

D43: A **standard plane representation** of an ordered tree is a standard plane drawing of the tree such that at each level, the left-to-right order of the vertices agrees with their prescribed order.

D44: A **binary tree** is an ordered tree in which each vertex has at most two children, and each child is designated either a **left-child** or a **right-child**.

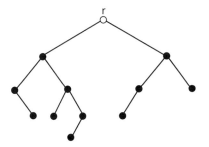

Figure 3.1.11: A binary tree of height 4.

D45: The **left (right) subtree** of a vertex v in a binary tree T is the binary subtree spanning the left (right)-child of v and all of its descendants.

FACT

F7: Every binary tree of height h has at most $2^{h+1} - 1$ vertices.

Binary Search

An entry in a *random-access table* consists of two fields. One field is for the actual data element, and the other one is for the *key*. An entry is found in a random-access table by searching for its key, and the most generally useful implementation of a random-access table uses the following information structure.

DEFINITIONS

D46: A ***binary-search tree*** (BST) is a binary tree, each of whose vertices is assigned a key, such that the key assigned to any vertex v is greater than the key at each vertex in the left subtree of v, and is less than the key at each vertex in the right subtree of v.

D47: A binary tree is ***balanced*** if for every vertex, the number of vertices in its left and right subtrees differ by at most one.

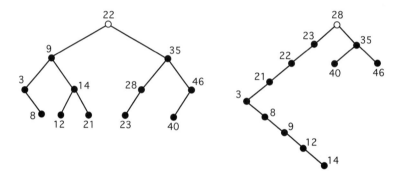

Figure 3.1.12: A balanced binary-search tree and an unbalanced one.

Algorithm 3.1.3: Binary-Search-Tree Search

Input: a binary-search tree T and a target key t.
Output: a vertex v of T such that $key(v) = t$ if t is found,
          or NULL if t is not found.

$v := root(T)$
While $(v \neq \text{NULL})$ and $(t \neq key(v))$
    If $t > key(v)$
            $v := rightchild(v)$
    Else $v := leftchild(v)$
Return v.

COMPUTATIONAL NOTE: Since each comparison of a binary search performed on a binary-search tree moves the search down to the next level, the number of comparisons is at most the height h of the tree plus one. If the tree is balanced, then it is not hard to show that the number of vertices n is between 2^h and 2^{h+1}. Hence, the worst-case performance of the binary search on a perfectly balanced binary-search tree is $O(log_2 n)$. The other extreme occurs when each internal vertex of the binary tree has only one child. Such a binary tree is actually an ordinary linked list, and therefore the performance of the search degenerates to $O(n)$.

References

[BaGe99] S. Baase and A. V. Gelder, *Computer Algorithms: Introduction to Design and Analysis*, Third Edition, Addison-Wesley, 1999.

[BaGu01] J. Bang-Jensen and G. Gutin, *Digraphs. Theory, Algorithms and Applications*, Springer-Verlag, London, 2001.

[ChLeZh10] G. Chartrand, L. Lesniak, and P. Zhang, *Graphs and Digraphs*, Fifth Edition, CRC Press, 2010.

[Ci75] E. Cinlar, *An Introduction to Stochastic Processes*, Prentice-Hall, 1975.

[Fl62] R. Floyd, Algorithm 97: Shortest Path, *Communications of the ACM* 5 (6) (1962), 345.

[GrYe06] J. L. Gross and J. Yellen, *Graph Theory and Its Applications*, Second Edition, CRC Press, 2006.

[Ro39] H. E. Robbins, A theorem on graphs with an application to a problem of traffic control, *Amer. Math. Monthly* 46 (1939), 281–283.

[Ro76] F. Roberts, *Discrete Mathematical Models*, Prentice-Hall, 1976.

[Ro84] F. Roberts, *Applied Combinatorics*, Prentice-Hall, 1985.

[St93] H. J. Straight, *Combinatorics: An Invitation*, Brooks/Cole, 1993.

[ThSw92] K. Thulasiraman and M. N. S. Swamy, *Graphs: Theory and Algorithms*, John Wiley & Sons, 1992.

[Wa62] S. Warshall, A theorem of boolean matrices, *Journal of the ACM* 9 (1962), 11–12.

[We01] D. B. West, *Introduction to Graph Theory*, Second Edition, 2001, Prentice-Hall, (First Edition 1996).

[Wi94] W. L. Winston, *Operations Research: Applications and Algorithms*, Third Edition, Duxbury Press, 1994.

Section 3.2
Directed Acyclic Graphs

Stephen B. Maurer, Swarthmore College

INTRODUCTION

When a digraph has no directed cycles, it is called a directed acyclic graph, or a DAG. While being acyclic may seem to be a stringent condition, it arises quite naturally because vertices often have a natural ordering. For instance, vertices may represent events ordered in time or ordered by hierarchy. This ordering makes results and algorithms for DAGs relatively simple.

3.2.1   Examples and Basic Facts

DEFINITIONS

D1: A digraph is *acyclic* if it has no *directed* cycles.

D2: *DAG* is an acronym for directed acyclic graph.

D3: A *source* in a digraph is a vertex of indegree zero.

D4: A *sink* in a digraph is a vertex of outdegree zero.

D5: A *basis* of a digraph is a minimal set of vertices such that every other vertex can be reached from some vertex in this set by a directed path.

EXAMPLES

E1: *Operations Research.* A large project consists of many smaller tasks with a
precedence relation — some tasks must be completed before certain others can begin.
One graphical representation of such a project has a vertex for each task and an arc from
u to v if task u must be completed before v can begin. For instance, in Figure 3.2.1, the
food must be loaded and the cabin cleaned before passengers are loaded, but luggage
unloading is independent of the timing of cabin activities. This model of a project will
always be a DAG, because if there were a directed cycle, the project could not be done:
every task on the cycle would have to be started before every other one on the cycle.

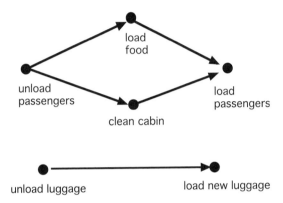

Figure 3.2.1: A digraph of precedence in an airplane stopover.

E2: *Sociology and Sociobiology.* A business (or army, or society, or ant colony) has a
hierarchical dominance structure. The nodes are the employees (soldiers, citizens, ants)
and there is an arc from u to v if u dominates v. If the chain of command is unique,
with a single leader, and if only arcs representing immediate authority are included,
then the result is a *rooted tree*, as in Figure 3.2.2. (Also see §3.2.2.)

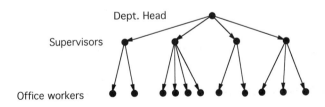

Figure 3.2.2: A corporate hierarchy.

E3: *Computer Software Design.* A large program consists of many subprograms, some
of which can invoke others. Let the nodes of D be the subprograms, and let there be
an arc from u to v if subprogram u can invoke subprogram v. Then this *call graph D*
encapsulates all possible ways control can flow within the program. Must D be a DAG?

No, but each directed cycle represents an indirect recursion and serves as a warning to the designer to ensure against infinite loops. See Figure 3.2.3, where Proc 2 can call itself indirectly. To determine if a digraph is a DAG or not, do a *topological sort* (§3.2.4).

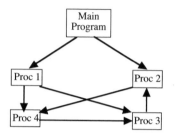

Figure 3.2.3: The call graph of a computer program.

E4: *Ecology.* A *food web* is a digraph in which nodes represent species and in which there is an arc from u to v if species u eats species v. Figure 3.2.4 shows a small food web. In general, food webs are acyclic, because animals tend to eat smaller animals or animals in some way "lower down" in the "food chain." The very fact that phrases like this are used indicates that there is a hierarchy, and thus no directed cycles.

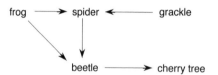

Figure 3.2.4: A small food web.

E5: *Genealogy.* A "family tree" is a digraph, where the orientation is traditionally given not by arrows but by the direction down for later generations. Despite the name, a family tree is usually not a tree, since people commonly marry distant cousins, knowingly or unknowingly. However, it is always a DAG, because if there were a cycle, everyone on it would be older than everyone else on the cycle.

E6: *State Diagrams.* Let the vertices of D be a set of states of some process, and let the arcs represent possible transitions. For instance, the process might be a board game, where the states are the configurations and each arc represents the transition of a single move. Then walks through D represent "histories" that the process/game can follow. If the game can never return to a previous configuration (e.g., as in tic-tac-toe), the state diagram of the game is a DAG.

FACTS

F1: Every DAG has at least one source and at least one sink.

F2: Every DAG has a unique basis, namely, the set of all its sources.

F3: Every subgraph of a DAG is a DAG.

F4: The transitive closure of a DAG is a DAG.

F5: A digraph is a DAG if and only if every walk in it is a path.

F6: A digraph is a DAG if and only if it is possible to order the vertices so that, in the adjacency matrix, all nonzero entries are above the main diagonal. (Topological sort in §3.2.4 finds the ordering.)

F7: The condensation of any digraph is a DAG. Figure 3.2.5 shows a digraph and its condensation.

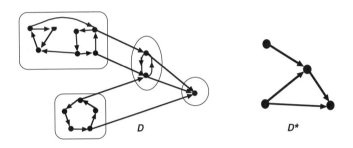

Figure 3.2.5: A digraph and its condensation.

F8: A digraph is a DAG if and only if it is isomorphic to its condensation.

F9: A digraph is strongly connected (unilateral, weakly connected) if and only if its condensation is strongly connected (unilateral, weakly connected).

F10: A DAG is never strongly connected, unless it consists of a single vertex.

F11: A DAG is unilateral if and only if it is a path.

F12: Every undirected graph without self-loops can be given an acyclic orientation, in fact, usually many. Namely, arbitrarily index the vertices as $v_1, v_2, \ldots, v_n$ and direct each edge from its lower indexed end to its higher indexed end.

REMARKS

R1: For more basic information on DAGs, see [Ha94, Ch. 16] and [Ro76, §2.2–2.3].

R2: Most of the acyclic orientations in Fact F12 are arbitrary and uninteresting, but occasionally an acyclic orientation is natural. In a tree, it is natural to orient edges away from a *root*; see §3.2.2. In a bipartite graph, it is natural to direct all edges from one side to the other. Still, most interesting orientations are already imposed by the nature of the problem, and the question is whether they are acyclic.

3.2.2   Rooted Trees

If the underlying graph of a digraph D is a tree, then D is certainly a DAG, because it doesn't even have any undirected cycles. However, the important tree DAGs have further restrictions on their edge directions.

For more on rooted trees, see [GrYe06, §3.2].

DEFINITIONS

D6: A ***directed tree*** is a digraph whose underlying graph is a tree.

D7: A ***rooted tree*** is a directed tree with a distinguished vertex r, called the ***root***, such that for every other vertex v, the unique path from r to v is a directed path from r to v.

CONVENTION: In drawing a rooted tree with the root marked, the arrows are usually omitted because the direction of each arc is always away from the root. In fact, if the direction is always down or left-to-right, as in Figure 3.2.6, it is not even necessary to indicate the root.

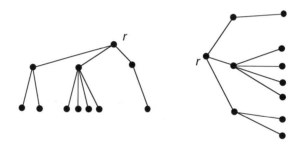

Figure 3.2.6: Two standard ways to draw a rooted tree.

D8: A rooted tree is also called an ***out-tree***. This alternative name is typically used when the arc directions are shown explicitly, for instance, when the tree is a spanning subgraph of a larger digraph.

D9: An ***in-tree*** is an out-tree with all the directions reversed, so that all paths are directed toward the root.

EXAMPLES

Previous Example E2 is about rooted trees. Here are some others.

E7: *Decision trees.* Any branching process leads to a rooted tree, where each node is a decision point, each arc from a node is an allowed decision, and the root is the start. For instance, the stages in a game may be represented this way. Figure 3.2.7 shows the first two moves in a game of tic-tac-toe, one by each player. Each node is represented by the way the board looks just *before* the decision. If we take into account symmetry, the figure is complete through the first two moves.

CONVENTION: In Figure 3.2.7 the two nodes on the bottom level (3rd move) illustrate that different nodes in the tree can represent the same state. While the board looks the same at these two nodes, the two ordered sequences of decisions leading to these nodes are different. Thus in a decision tree, each node represents both a state and the complete history of how it was achieved. Compare with Example 6, where these nodes would be one, and the digraph would not be a tree.

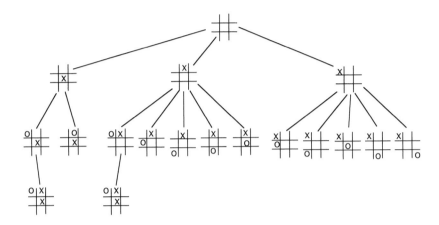

Figure 3.2.7: The first two moves in the tic-tac-toe game tree and a bit of the third level.

E8: *Decomposition trees.* Any decomposition of an object or structure into finer and finer parts can be modeled with a rooted tree. Figure 3.2.8 shows an example of *sentence parsing.*

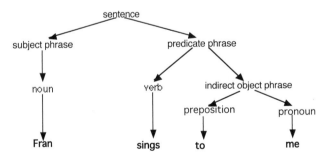

Figure 3.2.8: A sentence parse tree.

FACTS

F13: Every directed tree is a DAG.

F14: A digraph is a rooted tree if and only if its underlying graph is connected, exactly one vertex (the root) has indegree 0, and all others have indegree 1.

DEFINITIONS FOR ROOTED TREES

D10: The **depth** or **level** of a vertex v is its distance from the root, that is, the number of edges in the unique directed path from the root to v.

D11: The **height** of a rooted tree is the greatest depth of a vertex.

D12: If (u, v) is an edge, the u is the **parent** of v and v is the **child** of u.

D13: Vertices having the same parent are **siblings**.

D14: If there is a directed path from vertex u to vertex v, then u is an **ancestor** of v and v is a **descendant** of u.

D15: A **leaf** is a vertex with outdegree 0 (no children).

D16: An **internal vertex** is a vertex that is not a leaf.

D17: An m-**ary tree** is a rooted tree in which every vertex has m or fewer children.

D18: A **complete** m-**ary tree** is an m-ary tree in which every internal vertex has exactly m children and all leaves are at the same level. See Figure 3.2.9.

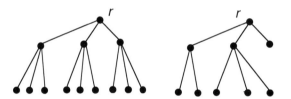

Figure 3.2.9: Complete and incomplete ternary (3-ary) trees.

D19: A **ordered tree** is a rooted tree in which the order of the children at each vertex makes a difference.

D20: A **binary tree** is an ordered 2-ary tree in which, even when a vertex has only one child, it makes a difference whether it is a **left child** or a **right child**.

REMARKS

R3: Trees, rooted trees, ordered trees, and binary trees make finer and finer distinctions, which should only be used if the distinctions are important in the application being modeled. For instance, binary trees are used to model computations with binary operations, as in $3 \times (4/5)$. Since division is *noncommutative* $(4/5 \neq 5/4)$, binary trees are an appropriate model for such computations.

R4: Figure 3.2.10 shows four graphs. As trees they are all the same (that is, *isomorphic*). However, as rooted trees, $G_1 = G_2$ and $G_3 = G_4$, so there are two rooted trees. There are three ordered trees, as G_1 and G_2 are still the same, but G_3, G_4 are different. Finally, as binary trees they are all different. In G_1, vertex c is a right child; in G_2 it is a left child.

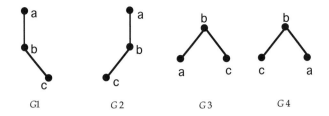

Figure 3.2.10: Four trees: the same and not the same.

FACTS

F15: An m-ary tree has at most m^k vertices at level k.

F16: Let T be an n-vertex m-ary tree of height h. Then

$$h + 1 \le n \le \frac{m^{h+1} - 1}{m - 1}.$$

The lower bound is attained if and only if T is a path. The upper bound is attained if and only if T is a complete m-ary tree.

Spanning Directed Trees

Since every connected graph has a spanning tree, every digraph has a spanning directed tree. In a graph, a spanning tree connects all the vertices, while using the minimum number of edges. However, in a digraph, a spanning directed tree may contain few directed *paths* and thus may allow fewer connections than the whole digraph does. So the more interesting question is whether a digraph has a spanning rooted tree. This question is answered algorithmically by the directed version of *depth first search*; see §10.1 and [GrYe06, §12.1]. It is answered algebraically by the directed matrix tree theorems; see §6.4. Here we simply state two key facts.

FACTS

F17: A digraph D has a spanning tree rooted at v if and only if directed depth first search starting at v finds one.

F18: For every vertex of a digraph D there is a spanning tree rooted at that vertex if and only if D is strongly connected.

Functional Graphs

Closely related structurally to rooted trees, but devised for a different purpose, are functional graphs.

DEFINITIONS

D21: A ***functional graph*** is a digraph in which each vertex has outdegree one.

EXAMPLES

E9: For each function f from a finite domain U to itself, define a digraph D whose vertex set is U and for which (u, v) is an arc if and only if $f(u) = v$. By definition of a function, there is one such v for every $u \in U$. Hence, D is a functional graph (whence the name).

E10: Specifically, consider the doubling function on the positive integers, but consider only the effect on the ones digit. This function is completely described by its effect on the domain $\{0, 1, \ldots, 9\}$. Its functional graph is shown in Figure 3.2.11.

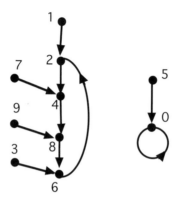

Figure 3.2.11: The functional graph for doubling (mod 10).

FACTS

F19: Let D be a functional graph, and let G be the underlying undirected graph. Then each component of G contains exactly one cycle. In D this cycle is a directed cycle, and the removal of any arc in it turns that component into an in-tree.

3.2.3   DAGs and Posets

There is a very close connection between DAGs and posets. Every DAG represents a poset, and every poset can be represented by DAGs in several ways. For more information, see [Bo00, §7.1–7.2].

DEFINITIONS

D22: A *partial order* is a binary relation $\preceq$ on a set X that is
- *reflexive*: for all $x \in X$, $x \preceq x$;
- *antisymmetric*: for all $x, y \in X$, if $x \preceq y$ and $y \preceq x$, then $x = y$;
- *transitive*: for all $x, y, z \in X$, if $x \preceq y$ and $y \preceq z$, then $x \preceq z$.

D23: A *poset*, or *partially ordered set* $P = (X, \preceq)$ is a pair consisting of a set X, called the *domain*, and a partial order $\preceq$ on X.

D24: Elements x, y of P are ***comparable*** if either $x \preceq y$ or $y \preceq x$.

D25: Element x is ***less than*** element y, written $x \prec y$, if $x \preceq y$ and $x \neq y$.

D26: The ***comparability digraph*** of the poset $P = (X, \preceq)$ is the digraph with vertex set X such that there is an arc from x to y if and only if $x \preceq y$.

D27: The element y ***covers*** the element x in a poset if $x \prec y$ and there is no element z such that $x \prec z \prec y$.

D28: The ***cover graph*** of a poset $P = (X, \preceq)$ is the graph with vertex set X such that x, y are adjacent if and only if one of them covers the other.

D29: A ***Hasse diagram*** of poset P is a straight-line drawing of the cover graph such that the lesser element of each adjacent pair is lower in the drawing.

EXAMPLES

E11: Let $X = \{2, 4, 5, 8, 10, 20\}$ and let $\preceq$ be the *divisibility relation* on X. That is, $x \preceq y$ if and only if y/x is an integer. The comparability digraph and the Hasse diagram for $P = (X, \preceq)$ are as shown in Figure 3.2.12.

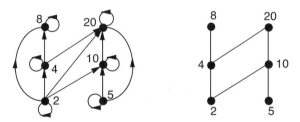

Figure 3.2.12: Comparability digraph and Hasse diagram for a poset.

FACTS

F20: If the loops are deleted, the comparability digraph of any poset is a DAG.

F21: Every Hasse diagram is a DAG if one considers all edges to be directed up (or all down).

F22: Every DAG D represents a poset in the following sense. The domain of P is the vertex set of D, and $x \preceq y$ if there is a directed path from x to y.

TERMINOLOGY NOTE: In passing from DAG D to poset P, null paths are included, so that $x \preceq x$ for all x. Alternatively, we obtain the poset by taking the transitive closure D^* of D. Then $x \prec y$ if and only if (x, y) is an arc of D^*.

3.2.4   Topological Sort and Optimization

In a DAG, the vertices can always be numbered consecutively so that all arcs go from lower to higher numbers. Using this numbering, many optimization problems can be solved by essentially the same algorithm, one that makes a single pass through the vertices in numbered order. For more general digraphs, algorithms for these optimization problems are less efficient or at least more complicated to describe.

DEFINITIONS

D30: A ***linear extension ordering*** of a digraph is a consecutive numbering of the vertices as $v_1, v_2, \ldots, v_n$ so that all arcs go from lower-numbered to higher-numbered vertices.

D31: A ***topological sort***, or ***topsort***, is any algorithm that assigns a linear extension ordering to a digraph when it has one. (This name is traditional, but the relation to topology in the sense understood by topologists is obscure.) A simple topological sort algorithm is shown as Algorithm 3.2.1. See also [RoTe03, §11.6.2].

FACTS

F23: A digraph has a linear extension ordering if and only if it is a DAG.

F24: Topological sort determines if a digraph is a DAG and finds a linear extension ordering if it is.

Algorithm 3.2.1: Topological Sort

Input: a digraph D.
Output: A linear extension ordering if D is a DAG; failure otherwise

   $H := D;\ k := 1$
   while $V_H \neq \phi$ {vertex set of H non-empty}
      $v_k :=$ any vertex in H of indegree zero.
        {If no such vertex exists, **exit**: D is not a DAG}
      $H := H - v_k$ {New H is a DAG if old H was}
      $k := k+1$

REMARK

R5: Because of the close connection between DAGs and posets, this whole discussion of linear extensions and topological sort can just as well be stated in the poset context. For instance, every poset has a linear extension, which may be found by a topological sort. See [GrYe06, pp. 507–510].

Optimization

There are many computational problems about graphs, with important real-world applications, when the graphs have *weights* on their vertices and/or edges. For DAGs, many of these problems can be solved by essentially the same single-pass algorithm. This algorithm is the basic form of the sort of staged algorithm called *dynamic programming* in operations research circles [HiLi10, Ch. 10]. Algorithms 3.2.2 and 3.2.3 provide templates for two versions of this algorithm. The examples that follow fill in the templates by giving specific formulas for updating the functions they compute.

In Algorithm 3.2.2, topsort is done first, and then the function F is computed vertex by vertex in topsort order. In Algorithm 3.2.3, the topsort is done simultaneously with improving F on vertices not yet sorted.

Algorithm 3.2.2: Basic Dynamic Programming, First Version

Input: DAG D with vertices numbered $v_1, v_2, \ldots, v_n$ in topsort order; weights
$\qquad w(v)$ on vertices or $w(v, u)$ on arcs, as needed.
Output: Correct values of desired function F.

$\quad$ Initialize $F(v_1)$
$\quad$ For $k = 2$ to n
$\qquad$ Determine $F(v_k)$ in terms of weights and all $F(v_i)$ with $i < k$.

Algorithm 3.2.3: Basic Dynamic Programming, Second Version

Input: DAG D with n vertices and weights $w(v)$ on vertices or $w(v, u)$ on arcs,
$\qquad$ as needed.
Output: Correct values of desired function F.

$\quad$ Initialize $F(v)$ for all v.
$\quad$ $H := D$
$\quad$ For $k = 1$ to n
$\qquad$ $v_k :=$ a source in H $\quad$ {exists since H is a DAG}
$\qquad$ Update $F(u)$ for all u for which (v_k, u) is an edge in H.
$\qquad$ $H := H - v_k$

EXAMPLES

CONVENTION: Below we assume that each DAG has just one source, the vertex Start. In other words, every other vertex has at least one predecessor. Without this assumption, some formulas below involve operations on empty sets (e.g., sum or max), and each example would need to specify how to correctly interpret those operations. For simplicity we also assume that the DAGs have no multiple edges.

E12: *Project Scheduling.* Consider Figure 3.2.13, which repeats Figure 3.2.1 with the following additions: Start and Finish vertices, a topsort ordering, and times for the tasks as weights on the vertices. Start and Finish, being merely marker vertices, take time 0. Recall that (u, v) is an arc if task u must be completed directly before task v begins, and that these tasks are the steps necessary to complete an airplane stopover. How quickly can the stopover be completed? The bottleneck is the *longest path* from Start to Finish, where the length of a (directed) path is the sum of the weights on its vertices. Dynamic programming can answer this question as follows.

Let

$$F(u) = \text{the length of the longest path (using vertex weights) from Start to } u.$$

Then in Algorithm 3.2.2 use

$$\text{Initialization:} \quad F(v_1) = w(v_1) = 0, \quad (\text{Note: } v_1 = \text{Start})$$
$$\text{Update:} \quad F(v_k) = w(v_k) + \max\{F(v_i) \mid (v_i, v_k) \text{ is an arc}\}.$$

In Algorithm 3.2.3 use

$$\text{Initialization:} \quad \text{For all } v, \ F(v) = w(v),$$
$$\text{Update:} \quad \text{For all } u \text{ such that } (v_k, u) \text{ is an arc, } F(u) = \max\{F(u), \ F(v_k)+w(u)\}.$$

For either algorithm, at termination the desired answer is $F(\text{Finish})$, that is, $F(v_n)$.

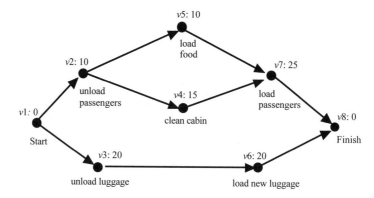

Figure 3.2.13: Airplane stopover as CPM graph.

This method of finding the optimal schedule by iteratively finding the longest path is the essence of the *critical path method*, or *CPM* [HiLi10, §9.8]. This example uses the *activity on node* model, or *AoN*. See Example 13 for the *activity on arc* model, or *AoA*.

E13: *Project Scheduling, second model.* If *edges* represent subtasks, and tasks earlier on directed paths must be completed before those later are begun, then the longest path from the Start to Finish vertex is the shortest time in which the whole project can be completed, where now the length of a path is the sum of the weights on its edges. Let

$$F(u) = \text{the length of the longest path (using edge weights) from Start to } u.$$

Then in Algorithm 3.2.2 use

$$\text{Initialization:} \quad F(v_1) = 0,$$
$$\text{Update:} \quad F(v_k) = \max\{F(v_i) + w(v_i, v_k) \mid (v_i, v_k) \text{ is an arc}\}.$$

In Algorithm 3.2.3 use

Initialization: For all v, $F(v) = 0$,

Update: For all u such that (v_k, u) is an arc,

$$F(u) = \max\{F(u),\ F(v_k) + w(v_k, u)\}.$$

For either algorithm, at termination the desired answer is $F(\text{Finish})$.

E14: *Shortest Paths.* What is the shortest directed path between two vertices u and u', where the length of a path is the sum of the weights on its edges? If a graph represents a road network, and the weights on the edges are the lengths of the road segments (or the travel times, or the toll on that segment), then shortest path means the shortest road distance (or least time, or lowest toll). If the graph is a DAG, and we make u the Start vertex (by eliminating earlier vertices in the topsort if necessary), then dynamic programming finds the shortest path as follows. Let

$$F(v) = \text{the length of the shortest path (using edge weights) from Start to } v.$$

Then in Algorithm 3.2.2 use

Initialization: $F(v_1) = 0$, (Note: $v_1 = u = \text{Start}$)

Update: $F(v_k) = \min\{F(v_i) + w(v_i, v_k) \mid (v_i, v_k) \text{ is an arc}\}.$

In Algorithm 3.2.3 use

Initialization: $F(v_1) = 0$, $F(v) = \infty$ for $v \neq v_1$,

Update: For all v such that (v_k, v) is an arc,

$$F(v) = \min\{F(v),\ F(v_k) + w(v_k, v)\}.$$

For either algorithm, at termination the desired answer is the value of $F(u')$.

E15: What is the shortest directed path between two vertices, where the length of a path is the sum of the weights on its *vertices*? Dynamic programming solves this problem too for DAGs, with a slight change in the formulas in Example 14 (replace edge weights with vertex weights).

E16: *Counting Paths.* How many directed paths are there between a given pair of vertices? If the digraph is a DAG, and the vertices are Start and Finish, let

$$F(u) = \text{the number of directed paths from Start to } u.$$

Then in Algorithm 3.2.2 use

Initialization: $F(v_1) = 1$, $(v_1 = \text{Start})$

Update: $F(v_k) = \sum\{F(v_i) \mid (v_i, v_k) \text{ is an arc}\}.$

In Algorithm 3.2.3 use

Initialization: $F(\text{Start}) = 1$, $F(v) = 0$ for $v \neq \text{Start}$,

Update: For all v such that (v_k, v) is an arc,

$$F(v) = F(v) + F(v_k).$$

For either algorithm, at termination the desired answer is the value of $F(\text{Finish})$.

E17:   *Maximin Paths.* What is the directed path between two vertices for which the minimum edge weight on that path is maximum among all paths between those two vertices? This is called the *maximin path* and that maximum value is called the *maximin value.* In Figure 3.2.14 the maximin path from v_1 to v_6 is $v_1v_3v_4v_6$ and the maximin value is 4. If the edges represent railroad segments, and each edge weight is the weight limit on that railroad segment, then this is the path between the two points over which the heaviest load can be shipped.

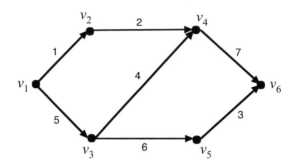

Figure 3.2.14: The maximin path $v_1v_3v_4v_6$ has value 4 and the minimax path $v_1v_3v_5v_6$ has value 6.

If the digraph is a DAG, and the vertices of interest are Start and Finish, let $F(u) = \emptyset$ mean that $F(u)$ is so far undefined, and by convention let $\max\{\emptyset, a\} = \min\{\emptyset, a\} = a$. Our goal is at the end to have

$$F(u) = \text{the maximin value for directed paths from from Start to } u.$$

Then in Algorithm 3.2.2 use

Initialization: $F(v_1) = \emptyset$, $(v_1 = \text{Start})$
     Update: $F(v_k) = \max\{\min\{F(v_i), w(v_i, v_k)\} \mid (v_i, v_k) \text{ is an arc}\}.$

In Algorithm 3.2.3 use

Initialization: For all v, $F(v) = \emptyset$.
     Update: For all v such that (v_k, v) is an arc,
          $F(v) = \max\{F(v), \min\{F(v_k), w(v_k, v)\}\}.$

For either algorithm, at termination the desired answer is the value of $F(\text{Finish})$.

E18:   *Minimax Paths.* What is the directed path between two vertices for which the maximum edge weight on the path is minimum? This *minimax* question is relevant if the graph represents a pipeline network, and each edge weight is the maximum elevation on that segment, because the work necessary to push a fluid through a pipeline route is related to the maximum height to which the fluid must be raised along the way. In Figure 3.2.14 the minimax path from v_1 to v_6 is $v_1v_3v_5v_6$ and the minimax value is 6. Dynamic programming solutions to the minimax problem are found by interchanging the roles of min and max in the algorithms for Example E17.

FACTS

F25: Algorithms 3.2.2–3.2.3 each solve critical path problems and many other optimization and computation problems on DAGs. (See the examples above.)

F26: In project scheduling problems modeled by DAGs, the minimum completion time is the length of the longest path from the Start node to the Finish node.

F27: Any DAG may be augmented to have just one source and one sink (just create a new node named Start adjacent *to* all existing sources, and a new node named Finish adjacent *from* all existing sinks).

References

[Bo00] K. Bogart, *Introductory Combinatorics*, Third Edition, Brooks Cole, 2000.

[GrYe06] J. L. Gross and J. Yellen, *Graph Theory and Its Applications*, Second Edition, CRC Press, 2006.

[Ha94] F. Harary, *Graph Theory*, Perseus reprint, 1994 (First Edition, Addison Wesley, 1969).

[HiLi10] F. Hillier and G. Lieberman, *Introduction to Operations Research*, Ninth Edition, McGraw-Hill, 2010.

[Ro76] F. Roberts, *Discrete Mathematical Models*, Prentice-Hall (or Pearson, Facsimile edition), 1976.

[RoTe03] F. Roberts and B. Tesman, *Applied Combinatorics*, Second Edition, Prentice-Hall, 2003.

Section 3.3

Tournaments

K. B. Reid, California State University San Marcos

INTRODUCTION

Tournaments comprise a large and important class of directed graphs. Application areas in which tournaments arise as models include round-robin tournaments (hence the name), paired-comparison experiments, domination in some animal societies, majority voting, population ecology, and communication networks. Many early results were motivated by applications; more recently, much focus has been on the combinatorial structure of tournaments as a separate area of graph theory. J. W. Moon's excellent monograph [Mo68] contains most of the results on tournaments up to 1968. In large part because of the influence of that work, tournament theory has so flourished during the past 35 years that subsequent surveys covered only a fraction of the results available. However, these surveys remain good sources for results about tournaments and directed graphs related to tournaments (see [HaNoCa65], [HaMo66], [BeWi75], [ReBe79], [Be81], [ZhSo91], [Gu95], [BaGu96], and [Re96]). Much work has been done on generalizations and extensions of tournaments (see [BaGu98]). A good source for digraphs in general, with extensive coverage of tournaments and various generalizations, is the book by Bang-Jensen and Gutin [BaGu01].

3.3.1   Basic Definitions and Examples

NOTATION: An arc from vertex x to vertex y will be denoted (x, y) or by $x \to y$.

DEFINITIONS

D1:   A *tournament* is an oriented complete graph, i.e., there is exactly one arc between every pair of distinct vertices (and no loops).

D2: The *order of a tournament* T is the number of vertices in T. A tournament of order n will be called an n-*tournament*.

D3: A vertex x in a tournament T *dominates* (or *beats*) vertex y in T whenever (x, y) is an arc of T. We also say that y is *dominated* (or beaten) by x.

D4: A vertex that dominates every other vertex in a tournament is called a *transmitter*. A vertex that is dominated by every other vertex in a tournament is called a *receiver*.

D5: The *score* (or *out-degree*) of a vertex v in a tournament T is the number of vertices that v dominates. It is denoted by $d_T^+(v)$. Note that if the tournament T under consideration is clear from the context, then T will be dropped and the score of v will be denoted $d^+(v)$. The *in-score* (or *in-degree*) of a vertex u in a tournament T is the number of vertices that dominate u. It is denoted by $d_T^-(u)$ (or $d^-(u)$).

D6: The *score sequence* (or *score vector*) of an n-tournament T is the ordered n-tuple $(s_1, s_2, \ldots, s_{n-1}, s_n)$, where s_i is the score of vertex v_i, $1 \le i \le n$, and

$$s_1 \le s_2 \le \ldots \le s_{n-1} \le s_n$$

D7: A tournament is *reducible* if its vertex-set can be partitioned into two non-empty subsets V_1 and V_2 such that every vertex in V_1 dominates every vertex in V_2. A tournament that is not reducible is said to be *irreducible.*

D8: The *out-set* of a vertex x in a digraph D, denoted $O(x)$, is the set of all vertices that x dominates, and the *in-set* of x, denoted $I(x)$, is the set of all vertices that dominate x.

TERMINOLOGY: In a digraph D, the out-set of a vertex x is also called the *neighborhood* of x, denoted $N_D^+(x)$ (or $N^+(x)$ if D is understood).

FACTS

F1: There are $2^{\binom{n}{2}}$ different labeled n-tournaments using the same n distinct labels, since for each pair of distinct labels $\{a, b\}$, either the vertex labeled a dominates the vertex labeled b or b dominates a.

F2: [Da54] The number $t(n)$ of non-isomorphic (unlabeled) n-tournaments is given by a rather complicated formula involving a summation over certain partitions of n. Moreover,

$$t(n) > \frac{2^{\binom{n}{2}}}{n!} \quad \text{and} \quad \lim_{n \to \infty} \frac{t(n)}{2^{\binom{n}{2}}/n!} = 1$$

The first few values of $t(n)$ are given by

n	1	2	3	4	5	6	7	8	9	10
$t(n)$	1	1	2	4	12	56	456	6880	191536	9733056

EXAMPLES

E1: Tournaments of orders 1 through 4 are illustrated in Figure 3.3.1.

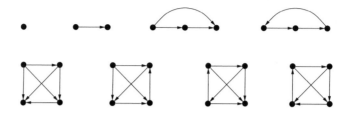

Figure 3.3.1: Tournaments of orders 1 through 4.

E2: *Tournament Scheduling.* To minimize the number of days of an n-tournament, games are scheduled in parallel. If n is even, then at least $n-1$ rounds are needed (since at most $n/2$ of the $n(n-1)/2$ games may be played at once). If n is odd, then at least n rounds are needed. There are several scheduling methods that achieve these compact schedules (see [Mo68]). Round-robin sports tournament scheduling and its relationship to edge-coloring is discussed in §5.6. Another related topic, *factors and factorization*, is discussed in §5.4.

Regular Tournaments

DEFINITIONS

D9: A *regular tournament* is a tournament T in which all scores are the same (i.e., there is an integer s so that $d^+(v) = s$ for all vertices $v \in V(T)$). An *almost regular* (or *near regular*) *tournament* is a tournament T in which $\max\limits_{v \in V(T)} \{|d^+(v) - d^-(v)|\} = 1$.

D10: A *doubly-regular tournament* is a tournament in which all pairs of vertices jointly dominate the same number of vertices (i.e., there is an integer k so that for all distinct pairs of vertices $x, y \in V(T)$, we have $|O(x) \cap O(y)| = k$).

D11: Let G be an abelian group of odd order $n = 2m + 1$ with identity 0. Let S be an m-element subset of $G - \{0\}$ such that for every $x, y \in S$, $x + y \neq 0$. That is, choose exactly one element from each of the m 2-sets of the form $\{x, -x\}$, where x ranges over all $x \in G - \{0\}$. Form the digraph D with vertex-set $V(D) = G$ and arc-set $A(D)$ defined by: arc $(x, y) \in A(D)$ if and only if $y - x \in S$. Then D is called a *rotational tournament* with *symbol set* S and is denoted $R_G(S)$, or simply $R(S)$ if the group G is understood.

D12: Let $G = GF(p^k)$ be the finite field with p^k elements, where p is a prime, $p \equiv 3$ (modulo 4), and k is an odd positive integer, and let S be the set of elements that are multiplicative squares of G (called the *quadratic residues*). Then the rotational tournament $R_G(S)$ is called a *quadratic residue tournament*.

FACTS

F3: The rotational tournament $R_G(S)$, where $|G| = n$, is a regular n-tournament.

F4: [ReBe79] If T is a doubly-regular n-tournament, then T is regular and $n \equiv 3$ (modulo 4). Moreover, there exists a doubly-regular $(4k+3)$-tournament if and only if there exists a $(4k+4)$ by $(4k+4)$ matrix H of $+1$'s and -1's such that $HH^t = (4k+4)I$ and $H + H^t = 2I$, where I is the identity matrix (such an H is called a skew-Hadamard matrix) [ReBr72].

REMARK

R1: Frequently, the group G for the rotational tournament $R_G(S)$ is taken to be Z_n, the integers modulo $n = 2m + 1$.

EXAMPLES

E3: The 9-tournament shown in Figure 3.3.2 is regular since every vertex has score 4, and it is also irreducible. Moreover, it is the rotational tournament $R_G(S)$, where $G = Z_9$ and $S = \{2, 4, 6, 8\}$.

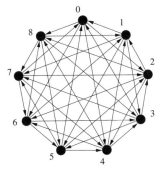

Figure 3.3.2: The regular, rotational tournament $R(\{2, 4, 6, 8\})$.

E4: The regular 7-tournament shown in Figure 3.3.3 is the quadratic residue tournament $R_G(S)$, where $G = GF(7)$ and $S = \{1, 2, 4\}$. Observe that it is irreducible, and it is a doubly-regular tournament since $|O(x) \cap O(y)| = 1$ for all distinct pairs of vertices x and y. The quadratic residue 7-tournament is notorious in tournament theory due to its occurrence as an exception to many results on tournaments.

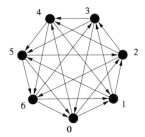

Figure 3.3.3: $R(\{1, 2, 4\})$.

Arc Reversals

Any n-tournament can be transformed into any other n-tournament by a sequence of reversals of arcs.

FACTS

F5: [Ry64] If T and W are two n-tournaments with the same score sequence, then T can be transformed into an isomorphic copy of W by a properly chosen sequence of reversals of arcs in 3-cycles.

F6: [Re73] If T and W are two n-tournaments and k is a fixed integer with $1 \leq k \leq n-1$, then T can be transformed into an isomorphic copy of W by a properly chosen sequence of reversals of arcs in k-paths.

REMARK

R2: C. Thomassen [Th88] extended these results to pairs of tournaments with the same vertex-set. He also described an infinite class of non-tournament digraphs for which the reversal of no arc decreases the total number of cycles. However, little progress has been made on the corresponding tournament problem (see [Re84]).

CONJECTURE

Ádám's Conjecture [Ád64]: Every non-transitive tournament contains at least one arc whose reversal reduces the total number of cycles.

3.3.2 Paths, Cycles, and Connectivity

Paths and cycles are fundamental sub-structures in tournaments and have been well studied in tournament theory. Many more results than given here have been collected by Bang-Jensen and Gutin in their survey [BaGu96] and their book [BaGu01].

NOTATION: All paths in this section are directed unless specified otherwise.

DEFINITIONS

D13: A *hamiltonian path* (or *spanning path*) in a digraph D is a path that includes all vertices of D. A *hamiltonian cycle* (or *spanning cycle*) in a digraph D is a cycle that includes all vertices of D. (Hamiltonian paths and cycles are discussed in §4.5.)

D14: A digraph D is *strong* (or *strongly connected*) if for every pair of distinct vertices x and y of D, there is a path from x to y and a path from y to x.

EXAMPLE

E5: The two tournaments in Figures 3.3.2 and 3.3.3 are strong and irreducible. For example, the hamiltonian cycle in $R(\{2,4,6,8\})$ given by

$$0 \to 2 \to 6 \to 3 \to 5 \to 4 \to 8 \to 1 \to 7 \to 0$$

implies that for every pair of distinct vertices x and y, there is a path from x to y and a path from y to x.

REMARK

R3: Fact F7, below, is perhaps the most fundamental result about tournaments and is used frequently in their study. The first part has several inductive proofs.

FACTS

F7: [Re34] Every tournament contains a hamiltonian path. Moreover, every tournament contains an odd number of hamiltonian paths.

F8: The following four statements are equivalent for any n-tournament T:

(a) T is strong.

(b) T is irreducible.

(c) T contains a hamiltonian cycle [Ca59].

(d) For every vertex x of T and for every integer k, $3 \leq k \leq n$, x is contained in a cycle of length k [Mo68]. (See also [HaMo66].)

F9: [Ga72] A curious fact: the number of n-tournaments containing a unique hamiltonian cycle is equal to the $(2n-6)^{\text{th}}$ Fibonacci number.

F10: [MoMo62] The fraction of labeled n-tournaments that are strong approaches 1 as $n \to \infty$.

F11: There is an $O(n^2)$ algorithm for finding a hamiltonian path in a tournament, and there is an $O(n^2)$ algorithm for finding a hamiltonian cycle in a tournament. (See [BaGu01] and [Ma92]).)

F12: [Vo02] Every arc of a strongly connected n-tournament is contained in a path of length
$$\lceil (n+3)/2 \rceil - 1$$

Condensation and Transitive Tournaments

DEFINITIONS

D15: If T is a tournament with vertex partition $\{V_1, V_2, \ldots, V_k\}$, where each V_i induces a maximal strongly connected sub-tournament of T, then the **condensation tournament** of T, denoted T^*, is the k-tournament with vertex-set $\{u_1, u_2, \ldots, u_k\}$ and in which u_i dominates u_j whenever all of the vertices in V_i dominate all of the vertices in V_j in T.

D16: A tournament T is **transitive** if for all three distinct vertices x, y, and z in T, if x dominates y, and y dominates z, then x dominates z.

EXAMPLE

E6: Consider the 9-tournament T consisting of three vertex-disjoint 3-cycles A_1, A_2, A_3, in which every vertex of A_1 dominates every vertex of A_2 and every vertex of A_3, and every vertex of A_2 dominates every vertex of A_3. The vertex partition of T in Definition D15 is $V(A_1) \cup V(A_2) \cup V(A_3)$, and T^* is the transitive 3-tournament with vertex-set $\{u_1, u_2, u_3\}$, where u_1 dominates u_2 and u_3, and u_2 dominates u_3.

FACTS

F13: [HaNoCa65] The condensation T^* of a tournament T is a transitive tournament.

F14: The following five statements are equivalent for an n-tournament. See [Mo68] for references.

(a) T is transitive.

(b) T contains no cycles.

(c) T contains a unique hamiltonian path.

(d) T has score sequence $(0, 1, 2, 3, \ldots, n-2, n-1)$.

(e) The vertices of T can be labeled $v_1, v_2, v_3, \ldots, v_{n-1}, v_n$ so that v_i dominates v_j if and only if $1 \leq i < j \leq n$ (i.e., T is a complete [linear] order).

F15: Every (2^{n-1})-tournament contains a transitive sub-tournament of order n.

Cycles and Paths in Tournaments

FACTS

F16: [Al67] Every arc in a regular n-tournament, $n \geq 3$, is in cycles of all lengths m, $3 \leq m \leq n$. (See [Th80] for extensions.)

F17: [Ja72] Every arc in an almost regular n-tournament, $n \geq 8$, is in cycles of all lengths m, $4 \leq m \leq n$. (See [Th80] for extensions.)

F18: [AlReRo74] For every arc (x, y) of a regular n-tournament T, where $n \geq 7$, and for every integer m, $3 \leq m \leq n-1$, T contains a path of length m from x to y. (See [Th80] for extensions.)

F19: [Th80] For every arc (x, y) of an almost regular n-tournament T, where $n \geq 10$, and for every integer m, $3 \leq m \leq n-1$, T contains a path of length m from x to y. (See also [GuVo97].)

Hamiltonian Cycles and Kelly's Conjecture

CONJECTURE

Kelly's conjecture (see [Mo68]): The arc-set of a regular n-tournament can be partitioned into $(n-1)/2$ subsets, each of which induces a hamiltonian cycle.

EXAMPLE

E7: The arc-set of the quadratic residue rotational 7-tournament $R(\{1,2,4\})$ can be decomposed into 3 hamiltonian cycles:

$$0 \to 1 \to 2 \to 3 \to 4 \to 5 \to 6 \to 0$$

$$0 \to 2 \to 4 \to 6 \to 1 \to 3 \to 5 \to 0$$

$$0 \to 4 \to 1 \to 5 \to 2 \to 6 \to 3 \to 0$$

REMARK

R4: Kelly's conjecture has stimulated much work in tournament theory. Evidence for the conjecture includes: it is true for $n \leq 9$ (B. Alspach, see [BeTh81]); every n-tournament, $n \geq 5$, contains two arc-disjoint hamiltonian cycles [Zh80]; regular or almost regular n-tournaments contain at least $\lfloor \sqrt{n/1000} \rfloor$ arc-disjoint hamiltonian cycles [Th82]. The best published result to date is the next result. A covering result then follows.

FACTS

F20: [Ha93] Then there exists a positive constant c, $c \geq 2^{-18}$, so that each regular n-tournament contains at least cn arc-disjoint hamiltonian cycles.

F21: [Th85] Each regular n-tournament T contains $12n$ hamiltonian cycles so that each arc of T is in at least one of the cycles.

Higher Connectivity

DEFINITION

D17: D is k-*strong* (or k-*strongly connected*) if for every subset S of $k-1$ or fewer vertices of D, $D - S$ is a strong digraph.

FACTS

F22: [Th80] Every arc in a 3-strong tournament is contained in a hamiltonian cycle. Moreover, this is false for infinitely many 2-strong tournaments. For every pair of distinct vertices x and y in a 4-strong tournament there is a hamiltonian path from x to y and there is a hamiltonian path from y to x. Moreover, this is false for infinitely many 3-strong tournaments.

F23: [FrTh87] If T is a k-strong tournament and B is any set of $k-1$ or fewer arcs of T, then the arc-deletion digraph $T-B$ contains a hamiltonian cycle.

F24: [Th84] There is a function h so that given any k independent arcs, $a_1, a_2, \ldots, a_k$, in an $h(k)$-connected tournament T, there is a hamiltonian cycle in T containing $a_1, a_2, \ldots, a_k$ in cyclic order.

F25: [So93] For any integer m, $3 \le m \le n-3$, every 2-strong n-tournament T, $n \ge 6$, contains two vertex-disjoint cycles of lengths m and $n-m$, unless T is isomorphic to the quadratic residue rotational tournament $RT(1, 2, 4)$. (This result is based on the case $m = 3$, which was established earlier in [Re85]. See also [BaGu00].)

F26: [ChGoLi01] If T is a k-strong n-tournament with $n \ge 8k$, then T contains k vertex-disjoint cycles that use all of the vertices of T.

Anti-Directed Paths

During the last 30 years, researchers have also searched for copies of other orientations of undirected paths and cycles in tournaments. Initially, study focused on oriented paths and cycles that contain no directed path of length 2 (called *anti-directed* paths and cycles), and successes there led to more general results on arbitrary oriented paths and cycles.

TERMINOLOGY: In a digraph, a directed path of length k is sometimes called a k-**path**.

DEFINITION

D18: An ***anti-directed path*** (or ***cycle***) in a digraph D is a sequence of arcs that forms a path or cycle in the underlying graph of D but does not contain a directed path of length 2 in D.

EXAMPLE

E8: Two anti-directed paths and an anti-directed cycle are illustrated in Figure 3.3.4.

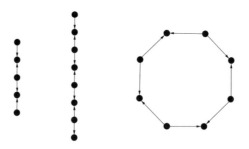

Figure 3.3.4: Two anti-directed paths and an anti-directed cycle.

FACTS

F27: [HaTh00a] Let T be a tournament different from the 3-cycle, the regular 5-tournament, and the quadratic residue rotational tournament $R(1, 2, 4)$. Then T contains every orientation of an undirected hamiltonian path. (This was first proved for anti-directed hamiltonian paths in 1971 [Gr71].)

F28: [Pe84] If n is even and $n \geq 16$, then every n-tournament contains an anti-directed hamiltonian cycle. (The major part of this result was first established for all even $n \geq 50$ [Th73], and then the result was improved to all even $n \geq 28$ [Ro74].)

F29: [Ha00a] Every n-tournament, $n \geq 68$, contains every orientation of a (undirected) hamiltonian cycle except possibly the (directed) hamiltonian cycle when the tournament is reducible. (This was first proved for $n \geq 2^{128}$ [Th86].)

3.3.3 Scores and Score Sequences

Fact F30, below, due to the mathematical sociologist H. G. Landau, is another basic result that is useful in studies on tournaments. Nearly a dozen proofs appear in the literature (see the survey [Re96] and subsequent paper [GrRe99]).

FACTS

F30: [La53] A sequence of n integers $(s_1, s_2, \ldots, s_{n-1}, s_n)$, where $s_1 \leq s_2 \leq \ldots \leq s_{n-1} \leq s_n$, is the score sequence of some n-tournament if and only if

$$\sum_{i=1}^{k} s_i \geq \binom{k}{2}, \text{ for } k = 1, 2, \ldots, n-1, \quad \text{and} \quad \sum_{i=1}^{n} s_i = \binom{n}{2}$$

F31: [HaNoCa65] A sequence of n integers $\langle s_1, s_2, \ldots, s_{n-1}, s_n \rangle$, where $s_1 \leq s_2 \leq \ldots \leq s_{n-1} \leq s_n$, is the score sequence of some strong n-tournament if and only if

$$\sum_{i=1}^{k} s_i > \binom{k}{2}, \text{ for } k = 1, 2, \ldots, n-1, \quad \text{and} \quad \sum_{i=1}^{n} s_i = \binom{n}{2}$$

(See also [HaMo66].)

F32: Let $S = (s_1, s_2, s_3, \ldots, s_{n-1}, s_n)$ be a sequence of $n \geq 2$ nonnegative integers, where $s_1 \leq s_2 \leq \ldots \leq s_{n-1} \leq s_n \leq n - 1$, and let $m = s_n$. S is the score sequence of some n-tournament if and only if the new sequence

$$s_1, s_2, \ldots, s_m, s_{m+1} - 1, s_{m+2} - 1, \ldots, s_{n-1} - 1$$

when arranged in non-decreasing order, is the score sequence of some $(n-1)$-tournament. (See [ReBe79].)

F33: [Av80] A score sequence $S = (s_1, s_2, s_3, \ldots, s_{n-1}, s_n)$ is the score sequence for exactly one n-tournament T if and only if each of the strong components of T is simple, and the simple strong score sequences are (0), $(1, 1, 1)$, $(1, 1, 2, 2)$, and $(2, 2, 2, 2, 2)$. (See also [Te98].)

F34: [BrQi84] Let $S = (s_1, s_2, \ldots, s_n)$ be a score sequence. Every n-tournament with score sequence S has a unique hamiltonian cycle if and only if

$$S = (1, 1, 2, 3, \ldots, n-3, n-2, n-2)$$

F35: [Ya88] and [Ya89] Every non-empty set of nonnegative integers is the set of scores for some tournament.

EXAMPLE

E9: It is easy to verify that the sequence (1, 1, 1, 4, 4, 4) satisfies the conditions of Fact F30, and hence, it is the score sequence for some 6-tournament. In fact, the 6-tournament consisting of two vertex-disjoint 3-cycles, A and B, where every vertex of A dominates every vertex of B, has score sequence (1, 1, 1, 4, 4, 4). Since the sum of the first three scores equals $\binom{3}{2}$, Fact F31 implies that no 6-tournament with score sequence (1, 1, 1, 4, 4, 4) is strong.

REMARKS

R5: The transitive n-tournament is the only n-tournament in which all of the scores are distinct. So, each score occurs with frequency 1. Regular n-tournaments are the only n-tournaments in which all of the scores are the same. So, each score in a regular n-tournament occurs with frequency n. Thus, each of the sets $\{1\}$ and $\{n\}$ is the set of frequencies of scores in some n-tournament. Given a non-empty set F of positive integers, the least possible order of a tournament with a set of score frequencies given by F was explicitly determined in [AlRe78].

R6: *Tournament Rankings.* Given the results of a round-robin competition, one would like to rank the teams or at least pick a clearcut winner. Unfortunately, not one of the many ranking methods that have been proposed is entirely satisfactory. Ranking by the order of a hamiltonian path (whose existence is guaranteed by Fact F7 in the previous subsection) does not work unless the path is unique, which is only the case for transitive tournaments (Fact F14). Ranking by score vector usually results in ties, and a team that beats only a few teams, with those few teams having lots of wins, might deserve a better ranking. This suggests considering the *second-order score vector*, where each team's score is the sum of the out-degrees of the teams it beats. One can continue by defining the nth-order score vectors recursively. There is an asymptotic ranking obtained this way, related to the eigenvalues of the digraph. See [Mo68] for more details and references.

The Second Neighborhood of a Vertex

DEFINITION

D19: Let x be a vertex in a digraph D. The **second neighborhood** of x, denoted $N_D^{++}(x)$, is the set of all vertices of D reachable from x by a 2-path but not a 1-path. That is, $N_D^{++}(x) = [\ \bigcup_{y \in N_D^+(x)} N_D^+(y)\ ] - N_D^+(x)$.

FACTS

F36: [Fi96] Every tournament T contains a vertex x so that $\left|N_T^{++}(x)\right| \geq \left|N_T^+(x)\right|$.

F37: [HaTh00b] If a tournament T contains no transmitter, then there are at least two vertices that satisfy the condition in Fact F36.

CONJECTURE

Seymour's second neighborhood conjecture (see [Fi96]): Every digraph D contains a vertex z for which $\left|N_D^{++}(z)\right| \geq \left|N_D^+(z)\right|$.

3.3.4 Transitivity, Feedback Sets, and Consistent Arcs

In a tournament that represents the outcomes of a paired-comparison experiment (or the results of a round-robin competition or the results of majority voting by an electorate in which there are no ties), there is much interest in attempts to measure the consistency of choices by the subject who made the comparisons (or the consistency of wins among the participants or the consistency of the electorate's choice among the alternatives). Consistency corresponds to a lack of cycles. So, one measure is the largest number of vertices that induce a transitive sub-tournament in the outcome tournament. Another measure is the largest number of arcs of the outcome tournament that do not contain the arcs of a cycle.

DEFINITION

D20: A *feedback set of arcs* in a tournament T is a set S of arcs such that the digraph $T - S$ contains no cycle.

EXAMPLE

E10: Let T be the tournament with vertex-set $\{1, 2, \ldots, n\}$ in which j dominates k whenever $j > k$, except that i dominates $i + 1$, for $i = 1, 2, \ldots, n - 1$. Then the set of arcs $\{(i, i + 1) \mid 1 \leq i \leq n - 1)\}$ is a feedback set of arcs in T, and the smaller set of arcs $\{(i, i + 1) \mid 1 \leq i \leq n - 1, \ i \text{ odd}\}$ is also a feedback set of arcs in T.

Smallest Feedback Sets

Finding a smallest feedback set in an n-tournament T is equivalent to finding a transitive n-tournament (or linear order) W such that $V(W) = V(T)$ and the number of pairs $\{x, y\}$ of distinct vertices in which x dominates y in T but not in W is as small as possible. The latter problem is known as a *Slater problem* in the voting literature involving tournaments. See [ChHuWo96]. Work on the computation of a minimum-weighted feedback set of arcs in an arc-weighted tournament is reported in [ChGuHuWo97] and [ShYu01].

FACTS

F38: A smallest set of arcs in a tournament T whose reversal yields a transitive tournament is a smallest feedback set in T. (See [BaHuIsRoTe95].)

F39: The number of arcs in a smallest feedback set in a tournament T is equal to the number of arcs in a smallest transversal of the cycles in T. (See [BaHuIsRoTe95].)

F40: [BaHuIsRoTe95] If R is a smallest feedback set in a tournament T, then every arc of R is contained in some 3-cycle of T.

F41: [BaHuIsRoTe95] A digraph D is acyclic if and only if its arc set is a smallest feedback set of some tournament.

Acyclic Subdigraphs and Transitive Sub-Tournaments

DEFINITION

D21: A set of arcs in a digraph D is a ***consistent set of arcs*** if it induces an acyclic subdigraph of D.

FACTS

F42: [Sp71/72, Fe83] If $g(n)$ denotes the largest integer so that every n-tournament contains a consistent set of $g(n)$ arcs, then there are positive constants c_1 and c_2 so that $\frac{1}{2}\binom{n}{2} + c_1 n^{3/2} \leq g(n) \leq \frac{1}{2}\binom{n}{2} + c_2 n^{3/2}$. Moreover, values for $n \leq 12$ are as follows ([Re69, Be72]):

N	2	3	4	5	6	7	8	9	10	11	12
$g(n)$	1	2	4	6	9	12	20	24	30	35	44-46

F43: [PaRe70, Ne94, Sa94] If $v(n)$ denotes the largest integer such that every n-tournament contains a transitive sub-tournament with at least $v(n)$ vertices, then

$$v(n) = \begin{cases} 3 & \text{for } 4 \leq n \leq 7 \\ 4 & \text{for } 8 \leq n \leq 13 \\ 5 & \text{for } 14 \leq n \leq 27 \\ 6 & \text{for } 28 \leq n \leq 31 \end{cases}$$

$$\lfloor \log_2(16n/7) \rfloor \leq v(n) \leq 2\lfloor \log_2 n \rfloor + 1 \quad \text{for } 32 \leq n \leq 54$$
$$v(n) \geq \lfloor \log_2(n/55) \rfloor + 7 \quad \text{for } n \geq 55$$

F44: [AoHa98] and [GuGyThWe98] For any tournament score sequence $S = (s_1, s_2, \ldots, s_{n-1}, s_n)$, where $s_1 \leq s_2 \leq s_3 \leq \ldots \leq s_{n-1} \leq s_n$, there exists a tournament T on vertex set $\{1, 2, \ldots, n\}$ so that the score of vertex i is s_i and the sub-tournaments of T on both the even and the odd indexed vertices are transitive, i.e., i dominates j whenever $i > j$ and $i \equiv j \pmod 2$. (See also [BrSh01] for a shorter proof, and see [BaBeHa92] for origins of the result in terms of the so-called cyclic chromatic number of a tournament.)

Arc-Disjoint Cycles

NOTATION: For a given tournament T, $a(T)$ denotes the maximum number of arc-disjoint cycles in T, and $c(T)$ denotes the number of arcs in a smallest feedback set in T. Also, let $a(n) = \max\{a(T)\}$ and $c(n) = \max\{c(T)\}$, where the maxima are taken over all n-tournaments T.

REMARK

R7: Note that the quantity $a(n)$ equals the maximum number of edge-disjoint (undirected) cycles in the complete graph of order n, which has been shown by [ChGeHe71] to equal $\lfloor (n/3)\lfloor (n-1)/2 \rfloor \rfloor$.

FACTS

F45: For any tournament T, $a(T) \leq c(T)$.

F46: [BeKo76] For $n \geq 10$, $a(n) < c(n)$. That is, for each $n \geq 10$, there exists an n-tournament T such that a smallest feedback set in T contains more arcs than in a largest collection of arc-disjoint cycles in T. (See also [Be75] and the discussions in [BaGu01] and [Is95].)

3.3.5 Kings, Oriented Trees, and Reachability

Kings arose in an attempt to determine the "strongest" individuals in certain small animal societies in which there exists a pairwise "pecking" relationship (see work referenced in [La53]). The delightful article by Maurer [Ma80] stimulated early interest in the topic. Extensions of the idea led to new investigations into combinatorial sub-structures in tournaments involving oriented trees and other "reachability sub-structures." The concept of a king appeared independently as "uncovered vertices" in some of the voting theory literature (see §3.3.8 below). Moreover, there is current interest in kings and generalizations in other digraphs, particularly in multi-partite tournaments (e.g., see the discussion in [Re96]).

DEFINITIONS

D22: A *king* in a tournament T is a vertex x such that for every other vertex y, there is a 1-path or a 2-path from x to y in T.

D23: A *serf* in a tournament is a vertex x such that for every other vertex y, there is a 1-path or a 2-path from y to x.

D24: A tournament is *k-stable*, $k \geq 1$, if every vertex is a king and more than k arcs must be reversed in order to reduce the number of kings.

FACTS

F47: [Va52] Every tournament contains a king. In fact, every vertex of maximum score is a king [La53].

F48: [Ma80] For positive integers k and n, there exists an n-tournament with exactly k kings if and only if $1 \leq k \leq n$, $k \neq 2$, and $(k, n) \neq (4, 4)$ (see also [Re82]).

F49: [Re80] For integers $0 \leq b \leq s \leq k \leq n$, there exists an n-tournament with exactly k kings, exactly s serfs, and exactly b vertices that are both kings and serfs if and only if the following four conditions are satisfied:

(1) $k + s - b \leq n$;

(2) $s \neq 2$ and $k \neq 2$;

(3) $n = k = s \neq 4$ or $\{k < n$ and $b < s\}$;

(4) (n, k, s, b) is none of $(n, 4, 3, 2)$, $(5, 4, 1, 0)$, or $(7, 6, 3, 2)$.

F50: [ReBr84] A k-stable tournament must have at least $4k + 3$ vertices. Moreover, the following three statements are equivalent:

(1) There exists a k-stable $(4k + 3)$-tournament.

(2) There exists a $(4k + 4)$ by $(4k + 4)$ skew-symmetric Hadamard matrix.

(3) There exists a doubly regular $(4k + 3)$-tournament.

F51: [Ma80] In almost all tournaments every vertex is a king. (See also [Mo68].) In fact, for each positive integer k, almost all tournaments are k-stable [ReBr84].

REMARKS

R8: Bounds on the least number of vertices that need be adjoined to an n-tournament T to form a new super tournament W so that the set of kings in W is exactly the vertices of T were described in [Re82] and [Wa84]. The least order of a tournament Z in which all vertices of Z are kings and that contains T as a sub-tournament was determined in [Re80]. For other work on kings in tournaments the reader is referred to the references in [Re96].

R9: By definition, a king is the root of a rooted spanning tree of depth at most 2. So, it is natural to consider the existence of other oriented trees in tournaments.

Tournaments Containing Oriented Trees

TERMINOLOGY: An **out-branching** (or **out-tree**) in a digraph is a rooted spanning tree, and an **in-branching** (or **in-tree**) is an out-branching with all the arcs reversed.

CONJECTURE

Sumner's conjecture (see [ReWo83]): Every $(2n - 2)$-tournament contains every orientation of every tree of order n.

EXAMPLE

E11: All of the 8 oriented trees of order 4 are shown in Figure 3.3.5. A copy of each can be found in any 6-tournament.

Figure 3.3.5: The oriented trees of order 4.

REMARK

R10: Note that no integer smaller than $2n - 2$ will suffice in the statement of the conjecture, for a score of at least $n - 1$ is required to accommodate the "out-orientation" of the tree $K_{1,n-1}$; any regular $(2n - 3)$-tournament fails to have a score of $n - 1$. Over the last 20 years several papers reported partial and related results (see the references in [HaTh00b]), all of which support the conjecture.

FACTS

F52: [HaTh00b] If $f(n)$ denotes the least integer m so that every m-tournament contains every orientation of every tree of order n, then $f(n) \leq (7n - 5)/2$. (Earlier efforts yielded $f(n) \leq 12n$ [HaTh91] and then $f(n) \leq \frac{38}{5}n - 6$ [HaTh00b].)

F53: [LuWaPa00] Every n-tournament, $n \geq 800$, contains a spanning rooted 2-tree of depth 2 so that, with at most one exception, all vertices that are not a leaf or the root have out-degree 2.

F54: [Pe02] Each rotational $(2n + 1)$-tournament contains all rooted trees of order $2n + 1$ in which there are at most n branches, each of which is a directed path.

F55: [Ba91] A tournament T contains an out-branching and an in-branching that are arc-disjoint, both rooted at a specified vertex v, if and only if T is strong and for each arc (x, y) of T, the digraph $T - \{(x, y)\}$ contains either an out-branching rooted at v or an in-branching rooted at v. If T is 2-arc-strong, then for every pair of vertices x and y, there is an out-branching rooted at x and an arc disjoint in-branching rooted at y.

Arc-Colorings and Monochromatic Paths

CONJECTURE

P. Erdős conjecture (see [SaSaWo82]): for each positive integer k, there is a least positive integer $s(k)$ so that every arc-colored tournament involving k colors contains a set S of $s(k)$ vertices with the property that for every vertex y not in S, there is a monochromatic path from y to some vertex in S.

REMARK

R11: This conjecture considers reachability in tournaments via monochromatic paths. Since every tournament contains a hamiltonian path, $s(1) = 1$. Fact F56 below implies that $s(2) = 1$. A certain coloring of the 9-tournament that is the lexicographic product of a 3-cycle with a 3-cycle shows that $s(3) > 2$ (see [SaSaWo82]). In particular, Erdős asked if $s(3) = 3$. It is not even known that $s(k)$ is finite for $k \geq 3$. Some progress on this conjecture is included below. (See also [LiSa96] and [Re00] for a relaxation of the problem and several open questions.)

FACTS

F56: [SaSaWo82] If the arcs of a tournament T are colored with two colors, then there exists a vertex x in T so that for every vertex $y \neq x$ in T, there is a monochromatic path from y to x. (See [Re84] for another proof.)

F57: [Sh88] If the arcs of a tournament T are colored with three colors and T does not contain a 3-cycle or a transitive sub-tournament of order 3 whose arcs use all three colors, then there exists a vertex x in T so that for every vertex $y \neq x$ in T, there is a monochromatic path from y to x.

3.3.6   Domination

Issues concerning domination have played an important role in the development of tournament theory. However, exact results on domination numbers of tournaments are scarce. For example, the problem of determining the smallest order of a tournament T with domination number $\gamma(T) = k$ for a given integer k has only some partial results. Bounds are known, some of which are constructive, but the exact value is known only for small values of k. Domination in general (undirected) graphs is discussed in §9.2.

DEFINITIONS

D25: A *dominating set* in a tournament T is a set S of vertices in T such that every vertex not in S is dominated by some vertex in S.

D26: The *domination graph* of a tournament T is an undirected graph G that has the same vertex-set as T, and x is adjacent to y in G whenever $\{x, y\}$ is a dominating set in T.

D27: A *spiked cycle* is a connected (undirected) graph with the property that when all vertices of degree 1 are removed, a cycle results.

D28: The *domination number* of a tournament is the minimum cardinality of a dominating set in T, denoted $\gamma(T)$.

EXAMPLE

E12: Let T denote the transitive n-tournament with vertex-set $\{1, 2, \ldots, n\}$ in which j dominates i whenever $j > i$. Reversal of the arc $(n, 1)$ yields a strong n-tournament W in which vertex 1 can reach vertex n via a 1-path, and 1 can reach vertices $2, 3, \ldots, (n-1)$ via 2-paths (through vertex n). So, 1 is a king in W. Since W has no transmitter, and every vertex in W is dominated by 1 or n, $\{1, n\}$ is a dominating set in W and $\gamma(W) = 2$.

FACTS

F58: [GrSp71] For a positive integer k, let p denote the smallest prime number greater than $k^2 2^{2k-2}$, where $p \equiv 3 \pmod 4$. The domination number of the quadratic residue p-tournament is greater than k. (See also [ReMcHeHe02].)

F59: [FiLuMeRe98] The domination graph of a tournament is either a spiked odd cycle with or without isolated vertices, or a forest of caterpillars. In particular, the domination graph of an n-tournament has at most n edges. Furthermore, any spiked odd cycle with or without isolated vertices is the domination graph of some tournament.

F60: [FiLuMeRe99] A connected graph is the domination graph of a tournament if and only if it is either a spiked odd cycle, a star, or a caterpillar whose spine has positive length and has at least three vertices of degree 1 adjacent to one end of its spine. The tournaments that have a connected domination graph were characterized in [JiLu98].

F61: [SzSz65] If T is an n-tournament with $n \geq 2$, then the domination number of T satisfies $\gamma(T) \leq log_2 n - log_2 log_2 n + 2$.

3.3.7 Tournament Matrices

Some early work on tournament matrices included the results by Brauer and Gentry and by J. W. Moon and N. J. Pullman (see [Mo68] and [BrGe72]); work by H. J. Ryser [Ry64] on tournament matrices with given row and column sum that minimize the number of *upsets*, i.e., the number of 1's above the main diagonal; and work by D. R. Fulkerson [Fu65] that described the tournament matrices with prescribed row sums that minimize and maximize the number of upsets. R. A. Brualdi and Q. Li [BrLi83] continued the upset theme and expanded on the work of Ryser and Fulkerson. These last three references may be thought of as papers on ranking since minimizing upsets gives rise to orderings of the vertices that minimize the number of losses by stronger players to weaker players.

DEFINITION

D29: A **tournament matrix** is a square matrix $M = (m_{ij})$ of 0's and 1's, with 0's on the main diagonal and $m_{ij} + m_{ji} = 1$, for all distinct i and j.

TERMINOLOGY: For a given ordering of the vertices, $v_1, v_2, \ldots, v_n$, of a tournament T, the **adjacency matrix** $M = [m_{ij}]$ of T is the 0-1 matrix given by

$$m_{ij} = \begin{cases} 1 & \text{if } v_i \text{ dominates } v_j \\ 0 & \text{otherwise} \end{cases}$$

Thus, a tournament matrix is the adjacency matrix of some tournament for a given ordering of the vertices.

EXAMPLE

E13: A tournament matrix of order 6 is shown in Figure 3.3.6.

$$
\begin{bmatrix}
0 & 1 & 1 & 1 & 0 & 1 \\
0 & 0 & 1 & 0 & 0 & 0 \\
0 & 0 & 0 & 1 & 0 & 1 \\
0 & 1 & 0 & 0 & 1 & 1 \\
1 & 1 & 1 & 0 & 0 & 0 \\
0 & 1 & 0 & 0 & 1 & 0
\end{bmatrix}
$$

Figure 3.3.6: A tournament matrix.

REMARK

R12: An elementary observation about an n by n tournament matrix M is that $M + M^t + I_n = J_n$, where I_n is the n by n identity matrix, M^t is the transpose matrix of M, and J_n is the n by n matrix of all 1's. Moreover, any adjacency matrix M of T can be obtained from any other adjacency matrix N of T by permuting the order of the vertices used to obtain N, i.e., there is a permutation matrix P such that $M = P^{-1}NP$. Thus, the eigenvalues are the same for all of the tournament matrices corresponding to a particular tournament.

FACTS

F62: [BrGe68] Let $\lambda_1, \lambda_2, \ldots, \lambda_n$ denote the eigenvalues of an n by n tournament matrix A, where $|\lambda_1| \geq |\lambda_2| \geq \ldots \geq |\lambda_n|$. Then $0 < \lambda_1 \leq (n-1)/2$, and $|\lambda_j| \leq \lfloor n(n-1)/2j \rfloor^{1/2}$, $j = 2, 3, \ldots, n$. Moreover, if M is an n by n tournament matrix and λ_M denotes $\max\{|\lambda| : \lambda$ an eigenvalue of $M\}$, then for odd n, $\max\{\lambda_M : M$ an n by n tournament matrix$\}$ is attained by the regular tournament matrices.

F63: [CaGrKiPuMa92] For all $n \geq 3$, each irreducible n by n tournament matrix M has at least three distinct eigenvalues; such a matrix has exactly three distinct eigenvalues if and only if it is a Hadamard tournament matrix (i.e., $M^tM = nI$). There is an irreducible n by n tournament matrix with exactly n distinct eigenvalues.

F64: [Mi95] If A is an n by n tournament matrix, then the rank of A is equal to $(n-1)/2$ if and only if n is odd and $AA^t = 0$. Equality implies that the characteristic of the field divides $(n-1)/2$ (without the hypothesis of regularity). Examples of order n having rank $(n-1)/2$ for $n \equiv 1 \pmod 4$ for fields of characteristic p, where p divides $(n-1)/4$, can be obtained from doubly regular tournament matrices of order $(n-2)$ by adding an $(n-1)$st row of n 0's (and hence, an $(n-1)$st column of all 1's save for the 0 in the $(n-1, n-1)$ position) followed by an nth row of $(n-1)$ consecutive 1's and a 0 in the (n, n) position (and hence, an nth column of n 0's).

F65: [Sh92] A tournament matrix is singular if more than one-fourth of the triples of vertices in the corresponding tournament induce 3-cycles. All tournament matrices realizing a given score sequence are nonsingular if and only if the scores are "sufficiently close" to one another. The spectral radius of a singular n by n tournament matrix is less than or equal to $(1/2)(n-1)$, and equality implies that exactly one-fourth of the triples of vertices in the corresponding tournament induce 3-cycles.

3.3.8   Voting

Work on acyclic digraphs in tournaments, including transitive sub-tournaments, is of interest in voting theory since such structures give a measure of group consistency by the voters. Readers can find a rich source of problems and issues in selected articles in the social choice literature that treats voting theory; particular examples of such literature include the periodicals *Social Choice and Welfare*, *Mathematical Social Sciences*, *Public Choice*, and *The American Journal of Political Science*.

Deciding Who Won

A central issue in voting theory is to pick a "best" alternative (or subset of the alternatives) given that voter preferences have been aggregated. A "best" alternative or subset of alternatives is called a solution and is thought of as the *winners*. Several tournament solutions have been considered in the literature. Each is to be non-empty, invariant under isomorphism, and uniquely the Condorcet winner if there is one in the tournament. Some of these solutions are: the vertices of largest score (the Copeland solution), vertices based on the maximum eigenvalue of the adjacency matrix of the tournament, vertices associated with a Markov method, the Condorcet winner of a transitive tournament that is "closest" to the given tournament (called the *Slater solution*), vertices that are uncovered relative to a certain *covering* relation, vertices that are transmitters of maximal transitive sub-tournaments (the Banks set), and vertices satisfying a special axiomatic formulation (the tournament equilibrium set). These are discussed in detail in J.-F. Laslier's monograph [La97].

DEFINITION

D30: The ***Condorcet winner*** is a candidate (or alternative) x such that for every other candidate (or alternative) y, x is preferred over y by a majority of the voters.

D31: The ***majority digraph*** D of a set of n-tournaments, all with the same vertex-set V, has vertex-set V, and vertex x dominates vertex y in D if and only if x dominates y in a majority of the n-tournaments.

D32: A digraph D is ***induced*** by a set of voters if D is the majority digraph based on a collection of linear orderings of the vertices of D, exactly one for each voter. (The linear orders represent preferences by the voters for the alternatives that are the vertices of D. Different voters might have the same linear order.)

D33: The ***Condorcet paradox*** is that the voters may be consistent in their preferences (i.e., each of their rankings of the n candidates is a linear order), but the amalgamation of voters' preferences using majority rule can result in inconsistencies (i.e., cycles in the majority digraph).

REMARK

R13: In the definition of a majority digraph, the common vertex set may be thought of as a finite set of n "alternatives," and each n-tournament may be thought of as the pairwise preferences of the alternatives by a "voter." So, the resulting majority digraph represents voters' preferences under majority voting. If there are an odd number of voters or there are no ties, then the majority digraph is a tournament.

EXAMPLE

E14: Figure 3.3.7 illustrates the majority tournament T of the set of three transitive 5-tournaments T_1, T_2, and T_3. In the drawings of T_1, T_2, and T_3, the long lines directed downward mean that each vertex dominates exactly the vertices below it. For example, in the second tournament from the left, vertex c dominates exactly vertices a and d, and vertex a is dominated by exactly vertices b, e, and c. The existence of cycles in the majority digraph illustrates the Condorcet paradox. For instance, it shows that a majority of voters prefer a to b, a majority prefer b to c, and yet, a majority prefer c to a.

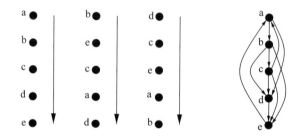

Figure 3.3.7: The majority digraph of 3 tournaments.

Tournaments That Are Majority Digraphs

FACTS

F66: [St59] Every n-tournament (indeed, every oriented graph of order n) is the majority digraph of some collection of $n + 1$ tournaments, for n odd, and of $n + 2$ tournaments, for n even.

NOTATION: Let $m(n)$ denote the smallest integer such that any n-vertex digraph can be induced as a majority digraph by a collection of $m(n)$ or fewer voters, and let $g(n)$ denote the smallest integer such that any n-tournament can be induced as a majority digraph by a collection of $g(n)$ or fewer voters.

F67: For large n, $m(n) > (.55n/logn)$ [St59], and there exists a constant c so that $m(n) < (cn/logn)$ [ErMo64].

F68: [Mo68] The integer $g(n)$ is always odd, $g(3) = g(4) = g(5) = 3$, $g(n+1) \leq g(n)+2$, and $m(n) \leq 2g(n)$.

F69: [Ma99] In contrast to the situation for majority tournaments, for any λ, $1/2 < \lambda \le 1$, there exists an integer n and a labeled n-tournament T, so that for every collection C of transitive tournaments on the same label set as T, there is an arc (u, v) of T such that the proportion of C in which u dominates v is less than λ. In short, T is not the λ-*majority tournament* for any collection of transitive n-tournaments.

Agendas

DEFINITIONS

D34: An *agenda* is an ordered list of alternatives (i.e., an ordered list of the vertices of a majority tournament).

D35: An *amendment procedure* of voting is a sequential voting process in which, given an agenda $(a_1, a_2, \ldots, a_n)$ of alternatives, alternative a_1 is pitted against a_2 in the first vote, then the winner is pitted against a_3 in the second vote, then the winner is pitted against a_4 in the third vote, etc.

D36: Given a majority n-tournament T and an agenda $(a_1, a_2, \ldots, a_n)$ of alternatives given by the vertices of T, the *sincere decision* is the alternative surviving the last vote (i.e., the $(n-1)^{\text{th}}$ vote) in an amendment procedure of voting using majority voting at each stage. It is a function of the agenda and T. The *decision tree* is the spanning, rooted subtree of T, rooted at the sincere decision, induced by the $n - 1$ arcs of T that describe the $n - 1$ votes taken in the amendment procedure using T and $(a_1, a_2, \ldots, a_n)$.

EXAMPLE

E15: Given the agenda (b, e, c, a, d) and the majority tournament shown in Figure 3.3.8, alternative a is the sincere decision. The corresponding decision tree rooted at vertex a is also shown.

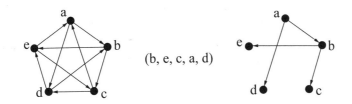

(b, e, c, a, d)

Figure 3.3.8: Majority tournament, agenda, and decision tree.

FACT

F70: [Mi77] For any tournament T, the set of vertices that can be obtained as the sincere decision under an amendment procedure is exactly the set of vertices in the initial strong component of T. (For another proof, see [Re91a].)

Division Trees and Sophisticated Decisions

DEFINITIONS

D37: Given an agenda $(a_1, a_2, \ldots, a_n)$, the **division tree** of $(a_1, a_2, \ldots, a_n)$ is the labeled, balanced, binary, rooted tree on $2^n - 1$ vertices labeled by non-empty subsequences of the agenda $(a_1, a_2, \ldots, a_n)$; the root is labeled $(a_1, a_2, \ldots, a_n)$; and, for $0 \leq j \leq n - 2$, a vertex at level j which is labeled by a subsequence of $(a_1, a_2, \ldots, a_n)$, say $(b_1, b_2, b_3, \ldots, b_{n-j})$, dominates exactly two vertices at level $j + 1$, one labeled $(b_1, b_3, \ldots, b_{n-j})$, and one labeled $(b_2, b_3, \ldots, b_{n-j})$.

D38: Let T be a majority n-tournament and let $(a_1, a_2, \ldots, a_n)$ be an agenda of alternatives given by the vertices of T. The **sophisticated decision** is the anticipated decision at the root of the division tree relative to $(a_1, a_2, \ldots, a_n)$ and T, where the anticipated decision at each vertex at level $n - 2$ of the division tree is the majority choice in T between the two alternatives that make up the ordered pair labeling that vertex in the division tree; and inductively, for $0 \leq j < n - 2$, the anticipated decision at each vertex v of level j in the division tree is the majority choice in T between the anticipated decisions at the two vertices at level $j + 1$ that are dominated by v.

FACTS

F71: [Ba85] The set of vertices in a tournament T that can be obtained as the sophisticated voting decision under an amendment procedure relative to some agenda consists of those vertices of T that are transmitters of maximal transitive sub-tournaments of T.

F72: No alternative is unanimously preferred to the sophisticated voting decision. (Observed in [Mi77] and [Mi80] and proved in [Re91b].)

F73: [Re97] A tournament T admits an agenda for which the sincere voting decision and the sophisticated voting decision are identical if and only if the initial strong component of T is not a 3-cycle. As a result, asymptotically, most tournaments admit such an agenda.

EXAMPLES

E16: The division tree of the agenda (x, y, z) is shown in Figure 3.3.9. Given the majority tournament shown, the anticipated decisions at levels 1 and 0 of the division tree are underlined in the vertex labels. The anticipated decision at the root is y, so y is the sophisticated decision relative to this tournament and agenda. Note that the sincere decision is z, which illustrates Fact F73.

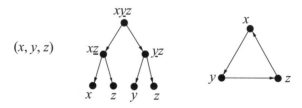

Figure 3.3.9: Agenda, division tree, and majority tournament.

E17: The majority 4-tournament shown in Figure 3.3.10 illustrates the positive case for Fact F73. As before, the anticipated decisions are underlined. For the agenda (y, v, u, x), the sincere decision and the sophisticated decision are both u.

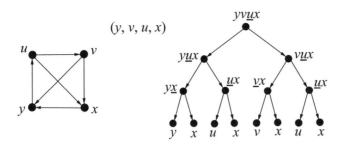

Figure 3.3.10: A majority 4-tournament.

Inductively Determining the Sophisticated Decision

The following result yields an algorithm for determining the sophisticated decision that is much more straightforward than using the definition. (Recall that $I(z)$ denotes the in-set of vertex z.)

FACT

F74: [ShWe84] Let T be a majority n-tournament and let $(a_1, a_2, \ldots, a_n)$ denote an agenda composed of the alternatives that make up the vertices of T. Inductively define the sequence $\langle z_1, z_2, \ldots, z_n \rangle$ as follows: $z_n = a_n$, and for $1 \leq j < n$,

$$
z_j = \begin{cases} a_j & \text{if } a_j \in \bigcap_{i=j+1}^{n} I(z_i) \\ z_{j+1} & \text{otherwise} \end{cases}
$$

Then z_1 is the sophisticated decision.

References

[Ad64] A. Ádám, Problem, 12–18 in *Theory of Graphs and Applications, Proc. Coll. Smolenice*, Czech. Acad. Sci. Publ., 1964.

[Al67] B. Alspach, Cycles of each length in regular tournaments, *Canad. Math. Bull.* 10 (1967), 283–285.

[AlRe78] B. Alspach and K. B. Reid, Degree frequencies in digraphs and tournaments, *J. Graph Theory* 2 (1978), 241–249.

[AlReRo74] B. Alspach, K. B. Reid, and D.P. Roselle, Bypasses in asymmetric digraphs, *J. Combin. Theory B* 17 (1974), 11–18.

[AlTa82] B. Alspach and C. Tabib, A note on the number of 4-circuits in a tournament, *Annals Discrete Math.* 12 (1982), 13–19.

[AoHa98] S. Ao and D. Hanson, Score vectors and tournaments with cyclic chromatic number 1 or 2, *Ars Combin.* 49 (1998), 185–191.

[Av80] P. Avery, Condition of a score sequence to be simple, *J. Graph Theory* 4 (1980), 157–164.

[Ba85] J. S. Banks, Sophisticated voting outcomes and agenda control, *Social Choice and Welfare* 1 (1985), 295–306.

[Ba91] J. Bang-Jensen, Edge-disjoint in- and out-branchings in tournaments and related path problems, *J. Combin. Theory B* 51 (1991), 1–23.

[BaBeHa92] K. S. Bagga, L. W. Beineke, and F. Harary, Two problems on coloring tournaments, Vishrva Internat. *J. Graph Theory* 1 (1992), 83–94.

[BaGu00] J. Bang-Jensen, Y. Guo, and A. Yeo, Complementary cycles containing prescribed vertices in tournaments, *Discrete Math.* 214 (2000), 77–87.

[BaGu01] J. Bang-Jensen and G. Gutin, *Digraphs: Theory, Algorithms, and Applications*, Springer, 2001.

[BaGu96] J. Bang-Jensen and G. Gutin, Paths, trees, and cycles in tournament, *Congressus Numer.* 115 (1996), 131–170.

[BaGu98] J. Bang-Jensen and G. Gutin, Generalizations of tournaments: a survey, *J. Graph Theory* 28 (1998), 171–202.

[BaHuIsRoTe95] J.-P. Barthélémy, O. Hurdy, G. Isaak, F. S. Roberts, and B. Tesman, The reversing number of a digraph, *Discrete Appl. Math.* (1995), 39–76.

[Be72] J. C. Bermond, Ordes à distance minimum d'un tournoi et graphes partiels sans circuits maximal, *Math. Inform. Sci. Humaines* 37 (1972), 5–25.

[Be75] J.-C. Bermond, The circuit hypergraph of a tournament, 165–180 in *Infinite and Finite Sets*, North Holland, 1975.

[Be81] L. W. Beineke, A tour through tournaments or bipartite and ordinary tournaments: a comparative survey, 41–55 in *Combinatorics*; Swansea, 1981, *London Math. Soc. Lecture Notes* 52, Cambridge University Press, Cambridge, 1981.

[BeKo76] J.-C. Bermond and Y. Kodratoff, Une heuristic pour le calcul de l'indice de transitivite d'un tournoi, *R. A. I. R. O. Informatique Theorique* 10 (1976), 83–92.

[BeTh81] J.-C. Bermond and C. Thomassen, Cycles in digraphs: a survey, *J. Graph Theory* 5 (1981) 1–43.

[BeWi75] L. W. Beineke and R. J. Wilson, A survey of recent results on tournaments, 31–48 in *Recent Advances in Graph Theory, Proc. Symp. Prague 1974*, Academia, 1975.

[BrGe68] A. Brauer and I. C. Gentry, On the characteristic roots of tournament matrices, *Bull. Amer. Math. Soc.* 74 (1968), 1133–1135.

[BrGe72] A. Brauer and I. C. Gentry, Some remarks on tournament matrices, *Linear Algebra Appl.* 5 (1972), 311–318.

[BrLi83] R. A. Brualdi and Q. Li, Upsets in round robin tournaments, *J. Combin. Theory B* (1983), 62–77.

[BrQi84] R. A. Brualdi and Q. Li, The interchange graph of tournaments with the same score vector, 129–151 in *Progress in Graph Theory*, Academic Press, 1984.

[BrSh01] R. A. Brualdi and J. Shen, Landau's inequalities for tournament scores and a short proof of a theorem on transitive sub-tournaments, *J. Graph Theory* 38 (2001), 244–254.

[Ca59] P. Camion, Chemins et circuits hamiltonioens des graphes complets, *C. R. Acad. Sci. Paris* 249 (1959), 2151–2152.

[CaGrKiPuMa92] D. de Caen, D. A. Gregory, S. J. Kirkland, N. J. Pullman, and J. S. Maybe, Algebraic multiplicity of the eigenvalues of a tournament matrix, *Linear Algebra Appl.* 169 (1992), 179–193.

[ChGeHe71] G. Chartrand, D. Geller, and S. Hedetniemi, Graphs with forbidden subgraphs, *J. Combin. Theory B* 10 (1971), 12–41.

[ChGoLi01] G. Chen, R. J. Gould, and H. Li, Partitioning vertices of a tournament into independent cycles, *J. Combin. Theory B* 83 (2001), 213–220.

[ChGuHuWo97] I. Charon-Fournier, A. Guénoche, O. Hudry, and F. Woirgard, New results on the computation of median orders, *Discrete Math.* 165/166 (1997), 139–153.

[ChHuWo96] I. Charon-Fournier, O. Hudry, and F. Woirgard, Ordres medians et orders de Slater des tournois, *Math. Inform. Sci. Humaines* 133 (1996), 23–56.

[Da54] R. L. Davis, Structures of dominance relations, *Bull. Math. Biophys.* 16 (1954), 131–140.

[ErMo64] P. Erdős and L. Moser, On the representation of directed graphs as unions of orderings, *Publ. Math. Inst. Hung. Acad. Sci.* 9 (1964), 125–132.

[Fe83] W. Fernandez de la Vega, On the maximum cardinality of a consistent set of arcs in a random tournament, *J. Combin. Theory B* (1983), 328–332.

[Fi96] D. C. Fisher, Squaring a tournament: proof of Dean's conjecture, *J. Graph Theory* 23 (1996), 43–48.

[FiLuMeRe98] D. C. Fisher, J. R. Lundgren, S. Merz, and K. B. Reid, The domination and competition graphs of a tournament, *J. Graph Theory* 29 (1998), 103–110.

[FiLuMeRe99] D. C. Fisher, J. R. Lundgren, S. Merz, and K. B. Reid, Connected domination graphs of tournaments, *J. Comb. Math. and Combin. Comput.* 31 (1999), 169–176.

[FrTh87] P. Fraise and C. Thomassen, Hamiltonian dicycles avoiding prescribed arcs in tournaments, *Graphs and Combin.* 3 (1987), 239–250.

[Fu65] D. R. Fulkerson, Upsets in round robin tournaments, *Canad. J. Math.* 17 (1965), 957–969.

[Ga72] M. R. Gary, On enumerating tournaments that admit exactly one Hamiltonian circuit, *J. Combin. Theory B* 13 (1972), 266–269.

[Gr71] B. Grünbaum, Anti-directed Hamiltonian paths in tournaments, *J. Combin. Theory B* 11 (1971), 249–257.

[GrRe99] J. R. Griggs and K. B. Reid, Landau's Theorem revisited, *Australasian J. Combin.* 20 (1999), 19–24.

[GrSp71] R. L. Graham and J. H. Spencer, A constructive solution to a tournament problem, *Canad. Math. Bull.* 14 (1971), 45–48.

[Gu95] G. Gutin, Cycles and paths in semicomplete multipartite digraphs, theorems, and algorithms: a survey, *J. Graph Theory* 19 (1995), 481–505.

[GuGyThWe98] B. Guiduli, A. Gyárfás, S. Thomassé, and P. Weidl, 2-partition-transitive tournaments, *J. Combin. Theory B* 72 (1998), 181–196.

[GuVo97] Y. Guo and L. Volkman, Bypasses in tournaments, *Discrete Appl. Math.* 79 (1997), 127–135.

[Ha00a] F. Havet, Oriented Hamiltonian cycles in tournaments, *J. Combin. Theory B* (2000), 1–31.

[Ha93] R. Häggkvist, Hamiltonian cycles in oriented graphs, *Combin. Probab. Comput.* 2 (1993), 25–32.

[HaMo66] F. Harary and L. Moser, The theory of round robin tournaments, *Amer. Math. Monthly* 73 (1966), 231–246.

[HaNoCa65] F. Harary, R. Z. Norman, and D. Cartwright, *Structural Models: An Introduction to the Theory of Directed Graphs*, John Wiley & Sons, 1965.

[HaTh00a] F. Havet and S. Thomassé, Oriented Hamiltonian paths in tournaments: a proof of Rosenfeld's conjecture, *J. Combin. Theory B* 78 (2000), 243–273.

[HaTh00b] F. Havet and S. Thomassé, Median orders of tournaments: a tool for the second neighborhood problem and Sumner's conjecture, *J. Graph Theory* 35 (2000), 244–256.

[HaTh91] R. Häggkvist and A. Thomason, Trees in tournaments, *Combinatorica* 11 (1991), 123–130.

[HaTh97] R. Häggkvist and A. Thomason, Oriented Hamiltonian cycles in oriented graphs, 339–353 in *Combinatorics, Geometry, and Probability (Cambridge, 1993)*, Cambridge University Press, 1997.

[Is95] G. Isaak, Tournaments as feedback arc sets, *Electronic J. Combin.* 2 (1995), 19.

[Ja72] O. S. Jakobsen, Cycles and paths in tournaments, *Thesis*, University of Aarhus, 1972.

[JiLu98] G. Jimenez and J. R. Lundgren, Tournaments which yield connected domination graphs, *Congressus Numer.* 131 (1998), 123–133.

[La53] H. G. Landau, On dominance relations and the structure of animal societies. III. The condition for a score structure, *Bull. Math. Biophys.* 15 (1953), 143–148.

[La97] J.-F. Laslier, *Tournament Solutions and Majority Voting*, Springer, 1997.

[LiSa96] V. Linek and B. Sands, A note on paths in edge-colored tournaments, *Ars Comb.* 44 (1996), 225–228.

[LuWaPa00] X. Lu, D.-W. Wang, and J. Pan, Rooted spanning trees in tournaments, *Graphs and Combinatorics* 16 (2000), 411–427.

[Ma80] S. Maurer, The king chicken theorems, *Math. Mag.* 53 (1980), 67–80.

[Ma99] J. Mala, On λ-majority voting paradoxes, *Math. Soc. Sci.* 37 (1999), 39–44.

[Ma92] Y. Manoussakis, A linear-time algorithm for finding Hamiltonian cycles in tournaments, *Discrete Appl. Math.* 36 (1992), 199–201.

[Mi77] N. R. Miller, Graph-theoretical approaches to the theory of voting, *Amer. J. Polit. Sci.* 21 (1977), 769–803.

[Mi80] N. R. Miller, A new solution set for tournaments and majority voting: further graph-theoretical approaches to the theory of voting, *Amer. J. Polit. Sci.* 24 (1980), 68–96.

[Mi95] T. S. Michael, The ranks of tournament matrices, *Amer. Math. Monthly* 102 (1995), 637–639.

[Mo68] J. W. Moon, *Topics on Tournaments*, Holt, Rinehart, and Winston, 1968.

[MoMo62] J. W. Moon and L. Moser, Almost all tournaments are irreducible, *Canad. Math. Bull.* 5 (1962), 61–65.

[Ne94] V. Neumann Lara, A short proof of a theorem of Reid and Parker on tournaments, *Graphs and Combin.* 10 (1994), 363–366.

[PaRe70] E. T. Parker and K. B. Reid, Disproof of a conjecture of Erdös and Moser on tournaments, *J. Combin. Theory* 9 (1970), 225–238.

[Pe02] V. Petrovic, Claws in rotational tournaments, *Graphs and Combin.* (2002), 591–596.

[Pe84] V. Petrovic, Antidirected Hamiltonian circuits in tournaments, 259–269 in *Graph Theory, Novi Sad, 1983*, Univ. Novi Sad, Novi Sad, 1984.

[Re00] K. B. Reid, Monotone reachability in arc-colored tournaments, *Congressus Numer.* 146 (2000), 131–141.

[Re34] L. Rédei, Ein kombinatorischer satz, *Acta Litterarum ac Scientiarum*, Szeged. 7 (1934), 39–43.

[Re69] K. B. Reid, On sets of arcs containing no cycles in a tournament, *Canad. Math. Bull.* 12 (1969), 261–267.

[Re73] K. B. Reid, Equivalence of n-tournaments via k-path reversals, *Discrete Math.* 6 (1973), 263–280.

[Re80] K. B. Reid, Tournaments with prescribed numbers of kings and serfs, *Congressus Num.* 29 (1980), 809–826.

[Re82] K. B. Reid, Every vertex a king, *Discrete Math.* 38 (1982), 93–98.

[Re84] K. B. Reid, Monochromatic reachability, complementary cycles and single arc reversals in tournaments, *Springer-Verlag Lecture Notes in Mathematics* 1073 (1984), 11–21.

[Re85] K. B. Reid, Two complementary cycles in two-connected tournaments, 171–211 in *Cycles in Graphs*, vol. 115 of *North-Holland Math. Stud.*, North-Holland, 1985.

[Re91a] K. B. Reid, Majority tournaments: Sincere and sophisticated voting decisions under amendment procedure, *Math. Soc. Sci.* 21 (1991), 1–19.

[Re91b] K. B. Reid, The relationship between two algorithms for decisions via sophisticated majority voting with an agenda, *Discrete Appl. Math.* 31 (1991), 23–28.

[Re96] K. B. Reid, Tournaments: scores, kings, generalizations and special topics, *Congressus Numer.* 115 (1996), 171–211.

[Re97] K. B. Reid, Equitable agendas: agendas ensuring identical sincere and sophisticated voting decisions, *Social Choice and Welfare* 14 (1997), 363–378.

[ReBe79] K. B. Reid and L. W. Beineke, Tournaments, 169–204 in *Selected Topics in Graph Theory*, Academic Press, London, 1979.

[ReBr72] K. B. Reid and E. Brown, Doubly regular tournaments are equivalent to skew Hadamard matrices, *J. Combin. Theory* 12 (1972), 332–338.

[ReBr84] K. B. Reid and M. F. Bridgland, Stability of kings in tournaments, 117–128 in *Progress in Graph Theory*, Academic Press, 1984.

[ReMcHeHe02] K. B. Reid, A. A. McRae, S. M. Hedetniemi, and S. T. Hedetniemi, Domination and irredundance in tournaments, preprint, 2002.

[ReWo83] K. B. Reid and N. Wormald, Embedding oriented n-trees in tournaments, *Studia Sci. Math. Hungarica* 18 (1983), 377–387.

[Ro74] M. Rosenfeld, Anti-directed Hamiltonian circuits in tournaments, *J. Combin. Theory B* 16 (1974), 234–242.

[Ry64] H. J. Ryser, Matrices of zeros and ones in combinatorial mathematics, 103–124 in *Recent Advances in Matrix Theory*, Univ. Wisconsin Press, 1964.

[Sa94] A. Sánchez-Flores, On tournaments and their largest transitive subtournaments, *Graphs and Combin.* 10 (1994), 367–376.

[SaSaWo82] B. Sands, N. Sauer, and R. Woodrow, On monochromatic paths in edge-colored digraphs, *J. Combin. Theory B* 45 (1982), 108–111.

[Sh88] M. Shen, On monochromatic paths in m-coloured tournaments, *J. Combin. Theory B* (1988), 108–111.

[Sh92] B. L. Shader, On tournament matrices, *Linear Algebra Appl.* 162/164 (1992), 335–368.

[ShWe84] K. A. Shepsle and B. Weingast, Uncovered sets and sophisticated voting outcomes with implications for agenda institutions, *Amer. J. Polit. Sci.* 28 (1984), 49–74.

[ShYu01] I. Sharon (I. Charon-Fournier) and O. Yudri (O. Hudry), The branch and bound method for solving the problem of linear ordering on weighted tournaments (Russian), *Diskretn. Anal. Issled. Oper.* 8 (2001), 73–91.

[So93] Z. M. Song, Complementary cycles of all lengths in tournaments, *J. Combin. Theory B* (1993), 18–25.

[Sp71/72] J. Spencer, Optimal ranking of tournaments, *Networks* 1 (1971/72), 135–138.

[St59] R. Stearns, The voting problem, *Amer. Math. Monthly* 66 (1959), 761–763.

[SzSz65] E. Szekeres and G. Szekeres, On a problem of Schütte and Erdös, *Math. Gaz.* 49 (1965), 290–293.

[Te98] P. Tetali, A characterization of unique tournaments, *J. Combin. Theory B* 72 (1998), 157–159.

[Th73] C. Thomassen, Antidirected Hamiltonian circuits and paths in tournaments, *Math. Ann.* 201 (1973), 231–238.

[Th80] C. Thomassen, Hamiltonian-connected tournaments, *J. Combin. Theory B* 28 (1980), 142–163.

[Th82] C. Thomassen, Edge-disjoint Hamiltonian paths and cycles in tournaments, *Proc. London Math. Soc.* 45 (1982), 151–168.

[Th84] C. Thomassen, Connectivity in tournaments, 305–313 in *Graph Theory and Combinatorics*, Cambridge, 1983, Academic Press, 1984.

[Th85] C. Thomassen, Hamiltonian circuits in regular tournaments, *Annals Discrete Math.* 27 (1985), 159–162.

[Th86] A. Thomason, Paths and cycles in tournaments, *Trans. Amer. Math. Soc.* 296 (1986), 167–180.

[Th88] C. Thomassen, Arc reversals in tournaments, *Discrete Math.* 71 (1988), 73–86.

[Va52] H. E. Vaughan, On well-ordered subsets and maximal elements of ordered sets, *Pacific J. Math.* 2 (1952), 407–412.

[Vo02] L. Volkman, Cycles in multipartite tournaments: results and problems, *Discrete Math.* 245 (2002), 19–53.

[Wa84] K. Wayland, Getting your chickens elected, *Congressus Numer.* 45 (1984), 311–318.

[Ya88] T. X. Yao, Reid's conjecture on score sets in tournaments (in Chinese), *Kexue Tongbao* 33 (1988), 481–484.

[Ya89] T. X. Yao, On Reid conjecture of score sets for tournaments, *Chinese Sci. Bull.* 34 (1989), 804–808.

[Zh80] C. Q. Zhang, Every regular tournament has two arc-disjoint Hamiltonian cycles, *J. Qufu Normal College, Special Issue Oper. Res.* (1980), 70–81.

[ZhSo91] K.-M. Zhang and Z.-S. Song, Cycles in digraphs – a survey, *J. Nanjing Univ., Natural Sci.* 27 (1991) 188–215.

Glossary for Chapter 3

Activity on Arc: a digraph scheduling model in which arcs represent subtasks to be scheduled as part of a large project.

Activity on Node: a digraph scheduling model in which nodes represent subtasks to be scheduled as part of a large project.

adjacency matrix – of a digraph: the $|V| \times |V|$ matrix in which the ij entry is the number of arcs from v_i to v_j.

agenda – in voting: an ordered list of alternatives (i.e., an ordered list of the vertices of a majority tournament).

almost regular tournament (or *near regular*): see *regular tournament*.

amendment procedure – for voting: a sequential voting process in which, given an agenda $(a_1, a_2, \ldots, a_n)$ of alternatives, alternative a_1 is pitted against a_2 in the first vote, then the winner is pitted against a_3 in the second vote, then the winner is pitted against a_4 in the third vote, etc.

ancestor of a vertex v – in a rooted tree: a vertex that lies on the unique path from v to the root; see also *descendant*.

anti-directed cycle – in a digraph D: a sequence of arcs that forms a cycle in the underlying graph of D but does not contain a directed path of length 2 in D.

anti-directed path – in a digraph D: a sequence of arcs that forms a path in the underlying graph of D but does not contain a directed path of length 2 in D.

antisymmetric relation R: one in which, for all x, y, if xRy and yRx, then $x = y$.

AoA: Activity on Arc.

AoN: Activity on Node.

arc: see *directed edge*.

arc-cut: synonym for edge-cut.

basis of a digraph: a minimal set of vertices such that every other vertex can be reached from some vertex in this set by a directed path.

beats: synonym for *dominates*.

binary tree: an ordered tree in which each vertex has at most two children, and each child is designated either a *left-child* or a *right-child*.

   ___, **balanced**: a binary tree such that for every vertex, the number of vertices in its left and right subtrees differ by at most one.

binary-search tree (BST): a binary tree, each of whose vertices is assigned a key, such that the key assigned to any vertex v is greater than the key at each vertex in the left subtree of v, and is less than the key at each vertex in the right subtree of v.

child of vertex v – in a rooted tree: a vertex to which there is an edge from v; see also *parent*.

   ___, **left** – in a binary tree: a child which is designated to be on the left, whether or not there is another child.

   ___, **right** – in a binary tree: a child which is designated to be on the right, whether or not there is another child.

comparability digraph – of a poset $(X, \preceq)$: the digraph with vertex set X such that there is an arc from x to y if and only if $x \preceq y$.

comparable elements – of a poset $(X, \preceq)$: elements x, y such that either $x \preceq y$ or $y \preceq x$.

complete m-ary tree: an m-ary tree in which every internal vertex has exactly m children and all leaves are at the same level.

complete digraph: a simple digraph such that between each pair of its vertices, there is an arc in both directions.

condensation – of a digraph G whose strong components are $S_1, S_2, \ldots, S_r$: a digraph G^* with vertex-set $V_{G^*} = \{s_1, s_2, \ldots, s_r\}$ such that $(s_i, s_j) \in E(G^*)$ if and only if there is an arc in digraph G from a vertex in component S_i to a vertex in component S_j.

condensation of a tournament T: a tournament T^* whose vertex-set $\{u_1, u_2, \ldots, u_k\}$ corresponds to a vertex partition $\{V_1, V_2, \ldots, V_k\}$ of $V(T)$, where each V_i induces a maximal strongly connected sub-tournament of T, and in which vertex u_i dominates u_j whenever all of the vertices in V_i dominate all of the vertices in V_j in T.

Condorcet paradox: the possibility that the voters may be consistent in their preferences (i.e., each of their rankings of the n candidates is a linear order), but the amalgamation of voters' preferences using majority rule can result in inconsistencies (i.e., cycles in the majority digraph).

Condorcet winner – in voting: a candidate (or alternative) x such that for every other candidate (or alternative) y, x is preferred over y by a majority of the voters.

connectivity (or *vertex-connectivity*) – of a non-complete digraph: the minimum size of a vertex subset S such that $G - S$ is neither strongly connected nor the trivial digraph. (The connectivity of a complete n-vertex digraph is $n-1$.) Denoted $\kappa_v(G)$ or $\kappa(G)$. Synonyms for vertex-cut are *cut* and *disconnecting set*.

consistent set of arcs – in a digraph D: a set of arcs that induces an acyclic subdigraph of D.

cost flow network: see *network*.

cover graph – of a poset $(X, \preceq)$: the graph with vertex set X such that x, y are adjacent if and only if one of them covers the other.

covering – in a poset $(X, \preceq)$: the element y *covers* the element x if $x \prec y$ and there is no element z such that $x \prec z \prec y$.

CPM: Critical Path Method.

Critical Path Method: a method for scheduling models where subtasks have fixed times and precedence is known. The whole project is modeled as an AoA or AoN digraph, and a single-pass iterative algorithm is used to find the longest path from start to finish.

DAG: a directed acyclic graph.

decision tree – for a sincere decision corresponding to a given majority n-tournament T and an agenda $(a_1, a_2, \ldots, a_n)$ of alternatives: the spanning, rooted subtree of T, rooted at the sincere decision, induced by the $n-1$ arcs of T which describe the $n-1$ votes taken in the amendment procedure using T and $(a_1, a_2, \ldots, a_n)$.

depth (or *level*) – of a vertex v in a rooted tree: the length (i.e., number of arcs) of the unique directed path from the root to v.

descendant of a vertex v – in a rooted tree: a vertex w such that v is on the unique path from the root to w; the vertex v is called an *ancestor* of w.

digraph: a directed graph.

   ___, **acyclic**: a digraph with no directed cycles, i.e., a directed acyclic graph, a *DAG*.

   ___, **connected**: a digraph whose underlying graph is connected. The term *weakly connected* is also used.

   ___, **representation of a relation** R on a finite set S: the digraph whose vertices correspond to the elements of S, and whose arcs correspond to the ordered pairs in the relation.

   ___, **weak**: short form of *weakly connected* digraph.

   ___, **weakly connected**: a digraph whose underlying graph is connected; synonym for *connected digraph*.

directed acyclic graph: a digraph without directed cycles.

Directed Chinese Postman Problem: to find a minimum-weight postman tour in a given weighted digraph.

directed cycle: a closed directed path.

directed edge (or *arc*): an edge e, one of whose endpoints is designated as the *tail*, and whose other endpoint is designated as the *head*. In a line drawing, the arrow points toward the head.

directed path: a path in a digraph or partial digraph in which all edges are oriented in the same direction.

directed tree: a digraph whose underlying graph is a tree.

directed walk – from v_0 to v_n: an alternating sequence $\langle v_0, e_1, v_1, e_2, ..., v_{n-1}, e_n, v_n \rangle$ of vertices and arcs, such that $tail(e_i) = v_{i-1}$ and $head(e_i) = v_i$, for $i = 1, 2 ..., n$. Also called a v_0-v_n directed walk.

division tree – in voting: see §3.3, Definition D36.

dominating set – in a tournament T: a set S of vertices in T such that every vertex not in S is dominated by some vertex in S.

domination (or *beating*)– a vertex y in a tournament: a property that a vertex x has if there is an arc from x to y.

   ___, **graph** – of a tournament T: an undirected graph G that has the same vertex-set as T, and x is adjacent to y in G whenever $\{x, y\}$ is a dominating set in T.

   ___, **number** – of a tournament: the minimum cardinality of a dominating set in T; denoted $\gamma(T)$.

doubly-regular tournament: see *regular tournament*.

edge-connectivity – of a non-trivial digraph: the minimum size of an edge subset F such that $G - F$ is not strongly connected. Denoted $\kappa_e(G)$ or $\lambda(G)$.

edge-cut (or *arc-cut*) – in a strongly connected digraph: an arc subset whose deletion results in a digraph that is not strongly connected. Synonyms are *edge-disconnecting set*, *arc-disconnecting set*, and *cut-set*.

eulerian tour of a digraph G: a closed directed walk that uses each arc *exactly* once.

feedback set of arcs – in a tournament T: a set S of arcs such that the digraph $T - S$ contains no cycle.

flow network: see *network*.

frontier arc – relative to a rooted tree T in a digraph: an arc whose tail is in T and whose head is not in T.

functional graph: a digraph in which each vertex has outdegree 1.

hamiltonian cycle (or *spanning cycle*) – in a digraph D: a cycle that includes all vertices of D.

hamiltonian path (or *spanning path*) – in a digraph D: a directed path that includes all vertices of D.

Hasse diagram – of a poset: a straight-line drawing of the cover graph such that the lesser element of each adjacent pair is lower in the drawing.

head: see *directed edge.*

height – of a rooted tree: the length of a longest path from the root.

in-branching (or *in-tree*) – in a digraph: a rooted spanning tree with all the arcs reversed.

in-score – of a vertex v in a tournament T: the number of vertices that dominate v (i.e., its *indegree*; denoted $d_T^-(v)$ (or $d^-(v)$ when T is understood).

in-set – of a vertex x in a digraph D: the set of all vertices that dominate x; denoted $I(x)$.

internal vertex – in a tree or rooted tree: a non-leaf.

in-tree: synonym for *in-branching.*

k**-strong tournament**: see *strong tournament.*

king – in a tournament T: a vertex x such that for every other vertex y, there is a 1-path or a 2-path from x to y in T.

leaf – in a rooted tree: a vertex with outdegree 0.

left subtree – of a vertex v in a binary tree: the binary subtree spanning the left-child of v and all of its descendants.

length of a directed walk: the number of arc-steps in the walk sequence.

level of a vertex – in a rooted tree: synonym for depth.

linear extension ordering – of a digraph: a consecutive numbering of the vertices as $v_1, v_2, \ldots, v_n$ so that all arcs go from lower-numbered to higher-numbered vertices.

linear ordering: a consecutive numbering.

m**-ary tree**: see *rooted tree.*

majority digraph D – of a set of n-tournaments, all with the same vertex-set V: a digraph with vertex-set V and such that vertex x dominates vertex y in D if and only if x dominates y in a majority of the n-tournaments.

Markov digraph: a complete digraph with a self-loop at each vertex and whose arcs are assigned probabilities such that the out-probabilities at each vertex sum to one; models a stationary Markov chain.

maximum-flow problem: to determine the maximum flow that can be moved through an s-t network from source s to sink t such that the flow into each intermediate node equals the flow out (*conservation of flow*) and the flow across any arc does not exceed the capacity of that arc.

minimum-cost-flow problem: to find an assignment of *flows* on the arcs of the flow network that satisfy the supply and demand (negative supply) requirements at minimum cost.

mutually reachable vertices – in a digraph G: vertices that have a directed walk from one to the other and vice versa. Every vertex is regarded as mutually reachable with itself (via the trivial walk).

neighborhood: see *out-set.*

network: a digraph $G = (V, E)$ used to model a variety of network flow problems; vertices might have supply or demand, and arcs might have capacities and or flow costs.

___, s-t **flow**: a network $G = (V, E, cap, s, t)$ with a nonnegative capacity function $cap : E \to N$, a distinguished vertex s, called the *source*, with nonzero outdegree, and a distinguished vertex t, called the *sink*, with nonzero indegree.

___, **capacitated cost flow** $G = (V, E, cap, c, b)$: a directed graph with vertex-set V,
   arc-set E, a nonnegative capacity function $cap : E \to N$, a linear cost function
   $c : E \to Z$, and an integral supply vector $b : V \to Z$ that satisfies $\sum_{w \in V} b(w) = 0$.

___, **cost flow**: a network $G = (V, E, cap, c, b)$ with nonnegative capacity function
   $cap : E \to N$, a linear cost function $c : E \to Z$, and an integral supply vector
   $b : V \to Z$ that satisfies $\sum_{w \in V} b(w) = 0$.

order of a tournament: the number of vertices it contains. A tournament of order n
   is an *n-tournament*.

ordered tree: a rooted tree in which the children of each vertex are assigned a fixed
   ordering.

ordering: a linear ordering.

orientation – of a graph: an assignment of directions to its edges, thereby making it a
   digraph.

oriented graph: a digraph obtained by choosing an orientation for each edge of an
   undirected simple graph.

out-branching (or *out-tree*) – in a digraph: synonym for rooted spanning tree.

out-set (or *neighborhood*) – of a vertex x in a digraph D: the set of all vertices that x
   dominates; denoted $O(x)$ or $N^+(x)$ (or with a subscripted "D" if necessary).

out-tree: a rooted tree, especially when the arc directions are shown explicitly.

parent of a vertex w – in a rooted tree: a vertex v that immediately precedes w on
   the path from the root to w; also, w is the *child* of v.

partial order: a binary relation $\preceq$ on a set X that is reflexive, antisymmetric, and
   transitive.

partially ordered set: a pair $(X, \preceq)$ consisting of a set X and a partial order $\preceq$ on
   X.

path in a digraph: a directed path.

___, **k-**: a directed path of length k.

poset: a partially ordered set.

postman tour (or *covering walk*): a closed directed walk that uses each arc *at least*
   once.

proper arc-coloring – of a digraph: an assignment of colors to the arcs such that any
   two arcs that have an endpoint in common are assigned different colors.

receiver – in a tournament: a vertex that is dominated by every other vertex in a
   tournament.

reflexive relation R: one in which, for all x, xRx.

regular tournament: a tournament T in which all scores are the same.

___, **almost** (or *near*): a tournament T in which $\max_{v \in V(T)} \{|d^+(v) - d^-(v)|\} = 1$.

___, **doubly-**: a tournament in which all pairs of vertices jointly dominate the same
   number of vertices (i.e., there is an integer k so that $|O(x) \cap O(y)| = k$, for all
   distinct pairs of vertices x and y in T).

right child – in a binary tree: a child which is designated to be on the right, whether
   or not there is another child.

right subtree – of a vertex v in a binary tree: the binary subtree spanning the right-
   child of v and all of its descendants.

root: see *rooted tree*.

rooted tree: a directed tree having a distinguished vertex r, called the *root*, such that for every other vertex v, there is a directed r-v path. Occasionally encountered synonyms for rooted tree are *out-tree*, *branching*, and *arborescence*.

___, **m-ary**: a rooted tree in which every vertex has m or fewer children; also called an *m-ary tree*.

rotational tournament: denoted $R_G(S)$, or simply $R(S)$ if the group G is understood; see §3.3, Definition D11.

s-t flow network: see *network*.

score of a vertex v in a tournament T: the number of vertices that v dominates (i.e., its *outdegree*). Denoted $d_T^+(v)$ (or $d^+(v)$ when T is understood).

score sequence (or *score vector*) – of an n-tournament: the ordered n-tuple $(s_1, s_2, \ldots, s_{n-1}, s_n)$, where s_i is the score of vertex v_i, $1 \leq i \leq n$, and $s_1 \leq s_2 \leq \ldots \leq s_{n-1} \leq s_n$.

score vector: synonym for *score sequence*.

second neighborhood – of a vertex x in a digraph D: the set of all vertices of D reachable from x by a 2-path but not a 1-path; denoted $N_D^{++}(x)$.

serf – in a tournament T: a vertex x such that for every other vertex y, there is a 1-path or a 2-path from y to x.

siblings – in a rooted tree: children of the same parent.

simple digraph: a digraph with no self-loops and no multi-arcs.

sincere decision – for a given majority n-tournament T and an agenda $(a_1, a_2, \ldots, a_n)$ of alternatives given by the vertices of T: the alternative surviving the last vote (i.e., the $(n-1)^{\text{th}}$ vote) in an amendment procedure of voting using majority voting at each stage.

sink – in a digraph: a vertex of outdegree zero.

sophisticated decision – in voting: see §3.3, Definition D38.

source – in a digraph: a vertex of indegree zero.

spanning subgraph – of a graph or digraph: a subgraph that includes all the vertices of the original graph.

spiked cycle: a connected (undirected) graph with the property that when all vertices of degree 1 are removed, a cycle results.

standard plane representation of an ordered tree: a standard plane drawing of the tree such that at each level, the left-to-right order of the vertices agrees with their prescribed order.

strong component – of a digraph G: maximal strongly connected subdigraph of G.

strong digraph: short form of *strongly connected digraph*.

strong orientation – of a graph: an orientation that results in a strong digraph.

strong tournament: a tournament that is a *strongly connected* digraph.

___, **k-**: a strong tournament such that the removal of any set of $k-1$ or fewer vertices results in a strong digraph.

strongly connected digraph: a digraph in which every two vertices are mutually reachable, i.e., there is a directed path from each of the two vertices to the other.

strongly orientable graph: a graph for which there exists an assignment of directions to the edges such that the resulting digraph is strongly connected.

symbol set – for a rotational tournament: see §3.3, Definition D11.

tail: see *directed edge*.

topological sort or *topsort*: any algorithm that assigns a linear extension ordering to a digraph when it has one.

topsort: short form of *topological sort*.

tournament matrix: a square matrix $M = (m_{ij})$ of 0's and 1's, with 0's on the main diagonal and $m_{ij} + m_{ji} = 1$, for all distinct i and j (i.e., the adjacency matrix of some tournament).

tournament: a simple digraph such that between each pair of vertices there is exactly one arc.

   ___, **irreducible**: a tournament that is not a *reducible tournament*.

   ___, **quadratic residue**: a special rotational tournament; see §3.3, Definition D12.

   ___, **reducible**: a tournament whose vertex-set can be partitioned into two non-empty subsets V_1 and V_2 such that every vertex in V_1 dominates every vertex in V_2.

   ___, k-**stable**: a tournament in which every vertex is a king and more than k arcs must be reversed in order to reduce the number of kings, where $k \geq 1$.

   ___, n-: a tournament of order n, i.e., an n-vertex tournament.

transitive closure – of a graph of digraph D: the smallest supergraph of D that is transitive.

transitive digraph: a digraph in which, if (u, v) and (v, w) are arcs, then so is (u, w).

transitive orientation – of a graph: an orientation that results in a transitive digraph.

transitive relation R: a relation in which, for all x, y, z, if xRy and yRz, then xRz.

transitive tournament: a tournament such that for every set of three distinct vertices x, y, and z, if x dominates y, and y dominates z, then x dominates z.

transmitter – in a tournament: a vertex that dominates every other vertex in a tournament.

unilateral digraph: a digraph in which, for all pairs of vertices u, v, there is a directed path between them in at least one direction.

vertex-cut – in a strongly connected digraph: a vertex subset whose deletion results in a digraph that is not strongly connected.

weights – in a graph or digraph: numbers on the vertices or edges or arcs, often representing something that is to be maximized or minimized.

Chapter 4

Connectivity and Traversability

Section 4.1

Connectivity: Properties and Structure

Camino Balbuena, Universitat Politècnica de Catalunya, Spain
Josep Fàbrega, Universitat Politècnica de Catalunya, Spain
Miquel Àngel Fiol, Universitat Politècnica de Catalunya, Spain

INTRODUCTION

Connectivity is one of the central concepts of graph theory, from both a theoretical and a practical point of view. Its theoretical implications are mainly based on the existence of nice *max-min* characterization results, such as Menger's theorems. In these theorems, one condition which is clearly necessary also turns out to be sufficient. Moreover, these results are closely related to some other key theorems in graph theory: Ford and Fulkerson's theorem about flows and Hall's theorem on perfect matchings. With respect to the applications, the study of connectivity parameters of graphs and digraphs is of great interest in the design of reliable and fault-tolerant interconnection or communication networks.

Since graph connectivity has been so widely studied, we limit ourselves here to the presentation of some of the key results dealing with finite simple graphs and digraphs. For results about infinite graphs and connectivity algorithms the reader can consult, for instance, Aharoni and Diestel [AhDi94], Gibbons [Gi85], Halin [Ha00], Henzinger, Rao, and Gabow [HeRaGa00], Wigderson [Wi92]. For further details, we refer the reader to some of the good textbooks and surveys available on the subject: Berge [Be76], Bermond, Homobono, and Peyrat [BeHoPe89], Frank [Fr90, Fr94, Fr95], Gross and Yellen [GrYe06], Hellwig and Volkmann [HeVo08], Lovász [Lo93], Mader [Ma79], Oellermann [Oe96], Tutte [Tu66].

4.1.1   Connectivity Parameters

In this first subsection the basic notions of connectivity and edge-connectivity of simple graphs and digraphs are reviewed.

NOTATION: Given a graph or digraph G, the vertex-set and edge-set are denoted $V(G)$ and $E(G)$, respectively. Often, when there is no ambiguity, we omit the argument and refer to these sets as V and E.

Preliminaries

DEFINITIONS

D1:   A graph is **connected** if there exists a walk between every pair of its vertices. A graph that is not connected is called **disconnected**.

D2:   The subgraphs of G which are maximal with respect to the property of being connected are called the **components** of G.

D3:   Let $G = (V, E)$ be a graph and $U \subset V$. The **vertex-deletion subgraph** $G - U$ is the graph obtained from G by deleting from G the vertices in U. That is, $G - U$ is the subgraph induced on the vertex subset $V - U$. If $U = \{u\}$, we simply write $G - u$.

D4:   Let $G = (V, E)$ be a graph and $F \subset E$. The **edge-deletion subgraph** $G - F$ is the subgraph obtained from G by deleting from G the edges in F. Thus, $G - F = (V, E - F)$. As in the case of vertex deletion, if $F = \{e\}$, it is customary to write $G - e$ rather than $G - \{e\}$.

D5:   A **disconnecting (vertex-)set** (or **vertex-cut**) of a connected graph G is a vertex subset U such that $G - U$ has at least two different components.

D6:   A vertex v is a **cut-vertex** of a connected graph G if $\{v\}$ is a disconnecting set of G.

D7:   A **disconnecting edge-set** (or **edge-cut**) of a connected graph G is an edge subset F such that $G - F$ has at least two different components.

D8:   An edge e is a **bridge** (or **cut-edge**) of a connected graph G if $\{e\}$ is a disconnecting edge-set of G.

FACTS

F1:   Every nontrivial connected graph contains at least two vertices that are not cut-vertices.

F2:   An edge is a bridge if and only if it lies on no cycle.

Vertex- and Edge-Connectivity

The simplest way of quantifying connectedness of a graph is by means of its parameters *vertex-connectivity* and *edge-connectivity*.

DEFINITIONS

D9: The *(vertex-)connectivity* $\kappa(G)$ of a graph G is the minimum number of vertices whose removal from G leaves a disconnected or a trivial graph.

D10: The *edge-connectivity* $\lambda(G)$ of a nontrivial graph G is the minimum number of edges whose removal from G results in a disconnected graph.

NOTATION: When the context is clear, we suppress the dependence on G and simply use κ and λ.

NOTATION: In some other sections of the *Handbook*, $\kappa_v(G)$ and $\kappa_e(G)$ are used instead of $\kappa(G)$ and $\lambda(G)$.

EXAMPLE

E1: Figure 4.1.1 shows an example of a graph with $\kappa = 2$ and $\lambda = 3$.

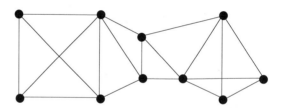

Figure 4.1.1: $\kappa = 2$ and $\lambda = 3$.

FACTS

F3: We have $\kappa = 0$ if and only if G is disconnected or $G = K_1$. If G has order n, then $\kappa = n - 1$ if and only if G is the complete graph K_n. In this case, the removal of $n - 1$ vertices results in the trivial graph K_1. Moreover, if $G \neq K_n$ is a connected graph, then $1 \leq \kappa \leq n - 2$ and there exists a disconnecting set U of κ vertices.

F4: If $G \neq K_1$ we have $\lambda = 0$ if G is disconnected. By convention, we set $\lambda(K_1) = 0$.

F5: If $G \neq K_1$ is connected, then the removal of λ edges results in a disconnected graph with precisely two components.

F6: The parameters κ and λ can be computed in polynomial time.

Relationships Among the Parameters

NOTATION: The **minimum degree** of a graph G is denoted $\delta(G)$. When the context is clear, we simply write δ. (In some other sections of the *Handbook*, the notation $\delta_{min}(G)$ is used.)

FACTS

F7: [Wh32] For any graph, $\kappa \leq \lambda \leq \delta$.

F8: [ChHa68] For all integers a, b, c such that $0 < a \leq b \leq c$, there exists a graph G with $\kappa = a$, $\lambda = b$, and $\delta = c$.

DEFINITIONS

D11: A graph G is **maximally connected** when $\kappa = \lambda = \delta$, and G is **maximally edge-connected** when $\lambda = \delta$.

D12: A graph G with connectivity $\kappa \geq k \geq 1$ is called k-**connected**. Equivalently, G is k-connected if the removal of fewer than k vertices leaves neither a disconnected graph nor a trivial one. Analogously, if $\lambda \geq k \geq 1$, G is said to be k-**edge-connected**.

D13: A connected graph G without cut-vertices ($\kappa > 1$ or $G = K_2$) is called a **block**.

Some Simple Observations

The following facts are simply restatements of the definitions.

FACTS

F9: A nontrivial graph is 1-connected if and only if it is connected.

F10: If G is k-connected, either $G = K_{k+1}$ or it has at least $k + 2$ vertices and $G - U$ is still connected for any $U \subset V$ with $|U| < k$.

F11: A graph G is k-edge-connected if the deletion of fewer than k edges does not disconnect it.

F12: Every block with at least three vertices is 2-connected.

Internally-Disjoint Paths and Whitney's Theorem

DEFINITIONS

D14: An **internal vertex** of a path is a vertex that is neither the initial nor the final vertex of that path.

D15: The paths $P_1, P_2, \ldots, P_k$ joining the vertices u and v are said to be **internally-disjoint** (or **openly-disjoint**) u–v paths if no two paths in the collection have an internal vertex in common. Thus, $V(P_i) \cap V(P_j) = \{u, v\}$ for $i \neq j$.

FACTS

F13: [Wh32] A graph G with order $n \geq 3$ is 2-connected if and only if any two vertices of G are joined by at least two internally-disjoint paths.

F14:   Fact F13 implies that every 2-connected graph is a block.

F15: A graph G with at least three vertices is a block if and only if every two vertices of G lie on a common cycle.

Strong Connectivity in Digraphs

For basic concepts on digraphs, see, for example, the textbooks of Bang-Jensen and Gutin [BaGu01], Chartrand, Lesniak, and Zhang [ChLeZh11], Harary, Norman, and Cartwright [HaNoCa68].

DEFINITIONS

D16: In a digraph G, vertices u and v are **mutually reachable** if G contains both a directed $u-v$ walk and a directed $v-u$ walk.

D17: A digraph G is said to be **strongly connected** if every two vertices u and v are mutually reachable.

D18: For a strongly connected digraph G, the **(vertex) connectivity** $\kappa = \kappa(G)$ is defined as the minimum number of vertices whose removal leaves a non-strongly connected or trivial digraph. Analogously, if G is not trivial, its **edge-connectivity** $\lambda = \lambda(G)$ is the minimum number of directed edges (or **arcs**) whose removal results in a non-strongly connected digraph.

D19: Let G be an undirected graph. The **associated symmetric digraph** G^* is the digraph obtained from G by replacing each edge $uv \in E(G)$ by the two directed edges (u, v) and (v, u) forming a *digon*.

REMARKS

R1: In our context, the interest for studying digraphs is that we can deal with an undirected graph G by considering G^*. In particular, $\kappa(G^*) = \kappa(G)$, and, since a minimum edge-disconnecting set cannot contain digons, we also have $\lambda(G^*) = \lambda(G)$.

NOTATION: The symbols δ^+ and δ^- denote the minimum outdegree and indegree among the vertices of a digraph G. Then, the **minimum degree** of G is defined as $\delta = \min\{\delta^+, \delta^-\}$.

R2: Note that, if G is a strongly connected digraph, then $\delta \geq 1$. The following result, due to Geller and Harary, is the analogue of (and implies) Fact F7.

FACT

F16: [GeHa70] For any digraph, $\kappa \leq \lambda \leq \delta$.

TERMINOLOGY: A digraph G is said to be **maximally connected** when $\kappa = \lambda = \delta$, and G is **maximally edge-connected** when $\lambda = \delta$.

An Application to Interconnection Networks

The interconnection network of a communication or distributed computer system is usually modeled by a (directed) graph in which the vertices represent the switching elements or processors, and the communication links are represented by (directed) edges. *Fault-tolerance* is one of the main factors that have to be taken into account in the design of an interconnection network. See, for instance, the survey of Bermond, Homobono, and Peyrat [BeHoPe89] and the book by Xu [Xu01]. Indeed, it is generally expected that the system be able to work even if several of its elements fail. Thus, it is often required that the (di)graph associated with the interconnection network be sufficiently connected, and, in most cases, a good design requires that this (di)graph has maximum connectivity. Communication networks are discussed in §11.4 of the *Handbook*.

4.1.2   Characterizations

When a graph G is k-connected we need to delete at least k vertices to disconnect it. Clearly, if any pair u, v of vertices can be joined by k internally-disjoint $u-v$ paths, G is k-connected. In fact, it turns out that the converse statement is also true. That is, in a k-connected graph any two vertices can be joined by k internally-disjoint paths. We review in this subsection some key theorems of this type that characterize k-connectedness.

Menger's Theorems

DEFINITION

D20:   Let u and v be two non-adjacent vertices of a connected graph $G \neq K_n$. A $(u|v)$-**disconnecting set** X, or simply $(u|v)$-**set**, is a disconnecting set $X \subset V - \{u, v\}$ whose removal from G leaves u and v in different components.

NOTATION: For any pair of non-adjacent vertices u and v, $\kappa(u|v)$ denotes the minimum number of vertices in a $(u|v)$-set.

NOTATION: For any two vertices u and v, $\kappa(u-v)$ denotes the maximum number of internally-disjoint $u-v$ paths.

FACTS

F17:   For any graph G, $\kappa(G) = \min\{\kappa(u|v) : u, v \in V, \text{nonadjacent}\}$.

F18:   (Menger's theorem) [Me27] For any pair of non-adjacent vertices u and v,

$$\kappa(u-v) = \kappa(u|v)$$

F19:   Although $\kappa(u-v)$ can be arbitrarily smaller than the minimum of the degrees of u and v, Mader proved that every finite graph contains vertices for which equality holds:

F20:   [Ma73] Every connected non-trivial graph contains adjacent vertices u and v for which $\kappa(u-v) = \min\{deg(u), deg(v)\}$.

NOTATION: For any pair of distinct vertices u and v, $\lambda(u|v)$ denotes the minimum number of edges whose removal from G (G non-trivial) leaves u and v in different components and $\lambda(u-v)$ denotes the maximum number of *edge-disjoint* $u-v$ paths.

F21:   For any non-trivial graph G, $\lambda(G) = \min\{\lambda(u|v),\ u, v \in V\}$.

F22:   (Edge-analogue of Menger's theorem) [ElFeSh56, FoFu56] For any pair of vertices u and v,

$$\lambda(u-v) = \lambda(u|v).$$

REMARKS

R3:   Digraph versions of Menger's theorems are the same except that all paths are directed paths.

R4:   The edge form and arc form of Menger's theorem were proved by Ford and Fulkerson [FoFu56] using *network-flow* methods. Network flow is discussed in Chapter 11 of this *Handbook*.

Other Versions and Generalizations of Menger's Theorem

In addition to the ones given below, there exist other versions and generalizations of Menger's theorem; see, for example, Diestel [Di00], Frank [Fr95], and McCuaig [McCu84]. A comprehensive survey about variations of Menger's theorem can be found in Oellermann [Oe12].

DEFINITIONS

D21:   Given $A, B \subset V$, an $A-B$ **path** is a $u-v$ path P with $u \in A$, $v \in B$, $u \neq v$, and any other vertex of P is neither in A nor in B.

D22:   A set $X \subset V$ **separates** A from B (or is $(A|B)$**-separating**) if every $A-B$ path in G contains a vertex of X.

D23:   An A**-path** is an $A-B$ path with $A = B$.

D24:   A subset $X \subset V - A$ **totally separates** A if each component of $G-X$ contains at most one vertex of A (or, equivalently, every A-path between different vertices contains some vertex of X).

D25:   A vertex subset is an **independent set** if no two of its vertices are adjacent.

NOTATION: The maximum number of (internally-)disjoint $A-B$ paths is denoted $\kappa(A-B)$, and the size of a minimum $(A|B)$-separating set is denoted $\kappa(A|B)$.

FACTS

F23:   The minimum number of vertices separating A from B is equal to the maximum number of disjoint $A-B$ paths. That is,

$$\kappa(A-B) = \kappa(A|B).$$

F24:   If A is an independent set, the maximum number of internally-disjoint A-paths is at most the minimum number of vertices in a totally A-separating set, that is,
$$\kappa(A-A) \leq \kappa(A|A).$$

F25: The corresponding Menger-type result does not hold and inequality can be strict. In fact, there exist examples for which $\kappa(A-A) = \kappa(A|A)/2$.

F26:   Gallai [Ga61] conjectured that Fact F25 corresponds to the "extremal" situation and that always $\kappa(A-A) \geq \kappa(A|A)/2$, and Lovász [Lo76] conjectured that $\lambda(A-A) \geq \lambda(A|A)/2$. Both conjectures were proved by Mader.

F27:   [Ma78b, Ma78c] $\kappa(A-A) \geq \kappa(A|A)/2$ and $\lambda(A-A) \geq \lambda(A|A)/2$.

REMARK

R5:   The classical version of Menger's theorem (Fact F18) is easily derived from Fact F23 by taking A and B as the sets of vertices adjacent to u and v, respectively.

Another Menger-Type Theorem

NOTATION: For any pair of vertices u and v, $\kappa_n(u-v)$ denotes the maximum number of internally-disjoint $u-v$ paths of length less than or equal to n. For any pair of non-adjacent vertices u and v, $\kappa_n(u|v)$ denotes the minimum number of vertices of a set $X \subset V - \{u, v\}$ such that every $u-v$ path in $G - X$ has length greater than n.

FACTS

F28:   There are examples for which we have the strict inequality $\kappa_n(u-v) < \kappa_n(u|v)$. However, for $n = d(u,v) \geq 2$ (i.e., for shortest $u-v$ paths), we have $\kappa_n(u-v) = \kappa_n(u|v)$. This Menger-type result is equivalently restated as Fact F29.

F29:   [EnJaSl77, LoNePl78] The maximum number of internally-disjoint shortest $u-v$ paths is equal to the minimum number of vertices (different from u and v) necessary to destroy all shortest $u-v$ paths.

Whitney's Theorem

In a connected graph, there exists a path between any pair of its vertices, and if the graph is 2-connected, then there exist at least two internally-disjoint paths between two distinct vertices (Fact F13). As a corollary of Menger's theorem, we have the remarkable result that this property can be generalized to k-connected graphs, which was independently proved by Whitney. It provides a natural and intrinsic characterization of k-connected graphs.

FACTS

F30: (Whitney's theorem) [Wh32] A non-trivial graph G is k-connected if and only if for each pair u, v of distinct vertices there are at least k internally-disjoint $u-v$ paths (or, alternatively, if and only if every cut-set has at least k vertices).

F31: (Edge version of Whitney's theorem) A nontrivial graph G is k-edge-connected if and only if for each pair u, v of distinct vertices there exist at least k edge-disjoint u–v paths.

F32: (The Fan Lemma) Let G be a k-connected graph ($k \geq 1$). Let $v \in V$ and let $B \subset V$, $|B| \geq k$, $v \notin B$. Then there exist distinct vertices $b_1, b_2, \ldots, b_k$ in B and a v–b_i path P_i for each $i = 1, 2 \ldots, k$, such that the paths $P_1, P_2, \ldots, P_k$ are internally-disjoint (that is, with only vertex v in common) and $V(P_i) \cap B = \{b_i\}$ for $i = 1, 2, \ldots k$.

Other Characterizations

Another interesting characterization of k-connected graphs was independently conjectured by Frank and Maurer. The conjecture was proved by Lovász and by Györi (who worked independently), and it appears as Fact F33. Su proved a characterization of k-edge-connectivity for digraphs (Fact F34).

FACTS

F33: [Lo77, Gy78] A graph G with $n \geq k + 1$ vertices is k-connected if and only if, for any distinct vertices $u_1, u_2, \ldots, u_k$ and any positive integers $n_1, n_2, \ldots, n_k$ such that $n_1 + n_2 + \cdots + n_k = n$, there is a partition $V_1, V_2, \ldots, V_k$ of $V(G)$ such that $u_i \in V_i$, $|V_i| = n_i$, and the induced subgraph $G(V_i)$ is connected, $1 \leq i \leq n$.

F34: [Su97] A digraph G with at least k edges is k-edge-connected if and only if, for any k distinct arcs $e_i = (u_i, v_i)$, $1 \leq i \leq k$, the digraph $G - \{e_1, e_2, \ldots, e_k\}$ contains k edge-disjoint spanning arborescences (rooted trees) $T_1, T_2, \ldots, T_k$ such that T_i is rooted at v_i, $1 \leq i \leq n$.

4.1.3   Structural Connectivity

Here our purpose is to give results about certain configurations that must be present in a k-connected or k-edge-connected graph.

Cycles Containing Prescribed Vertices

The first is a classical result by Dirac, which generalizes Fact F15.

FACTS

F35: [Di60] Let G be a k-connected graph, $k \geq 2$. Then G contains a cycle through any given k vertices.

F36: [WaMe67] Let G be a k-connected graph with $k \geq 3$. Then G has a cycle containing a given set H with $k + 1$ vertices if and only if there is no set $T \subset V - H$ with $|T| = k$ vertices whose removal separates the vertices of H from each other.

The Lovász–Woodall Conjecture

Lovász [Lo74] and Woodall [Wo77] independently conjectured that every k-connected graph has a cycle containing a given set F of k *independent edges* (that is, no two edges have a vertex in common), if and only if F is not an edge-disconnecting set of odd cardinality. Partial results on this conjecture are given in Facts F37 → F39.

FACTS

F37: [Lo74, Lo77, ErGy85, Lo90, Sa96] The *Lovász–Woodall Conjecture* is true for $k = 3, 4, 5$.

F38: [HaTh82] The *Lovász–Woodall Conjecture* is true assuming that G is $(k+1)$-connected (without restriction on the edge set F).

F39: [Ka02] Under the same assumptions of the conjecture, F is either contained in a cycle or in two disjoint cycles.

TERMINOLOGY: A subset of independent edges is also called a **matching**. Matchings are discussed in Section 11.3 of this *Handbook*.

Paths with Prescribed Initial and Final Vertices

Given any two subsets $A, B \subset V$ of k vertices of a k-connected graph, the existence of k disjoint paths P_i $(1 \le i \le k)$ connecting A and B is guaranteed by Menger's theorem. Menger's theorem does not, however, ensure that each of these paths can be so chosen to join a fixed u_i, v_i pair of vertices, $u_i \in A$, $v_i \in B$, $(1 \le i \le k)$. Now we consider the existence of paths with prescribed end-vertices.

DEFINITIONS

D26: A graph G is called k-**linked** if it has at least $2k$ vertices, and for every sequence $u_1, u_2, \ldots, u_k, v_1, v_2, \ldots, v_k$ of $2k$ different vertices, there exists a u_i–v_i path P_i, $i = 1, 2, \ldots, k$, such that the k paths are vertex-disjoint.

D27: A graph is **weakly** k-**linked** if it has at least $2k$ vertices, and for every k pairs of vertices (u_i, v_i), there exists a u_i–v_i path P_i, $1 \le i \le k$, such that the k paths are edge-disjoint.

D28: A graph is said to be k-**parity-linked** if one can find k disjoint paths with prescribed end-vertices and prescribed parities of the lengths.

D29: The **bipartite index** of a graph is the smallest number of vertices whose deletion creates a bipartite graph.

FACTS

F40: A k-linked graph is always $(2k-1)$-connected, but the converse is not true.

F41: [Ju70], [LaMa70] (independently) For each k, there exists an integer $f(k)$ such that if $\kappa \ge f(k)$ then G is k-linked.

F42:   Thomassen [Th80a] and Seymour independently characterized the graphs that are not 2-linked. This is the first problem in the so-called k-**paths problem** that has been solved using the Robertson–Seymour theory [RoSe85].

NOTATION: For $k \geq 1$, $g(k)$ denotes the smallest integer such that every $g(k)$-edge-connected graph G is weakly k-linked.

CONJECTURE

[Th80a] For every integer $k \geq 1$, $g(2k + 1) = g(2k) = 2k + 1$.

FACTS

F43:   [Ok84, Ok85, Ok87] If $k \geq 3$ is odd, $u_1, u_2, \ldots, u_k, v_1, v_2 \ldots, v_k$ are (not necessarily distinct) vertices from a set T with $|T| \leq 6$, and $\lambda(u_i, v_i) \geq k$ $(1 \leq i \leq k)$, then there exists a $u_i - v_i$ path for $1 \leq i \leq k$ such that the k paths are edge-disjoint.

F44:   [Hu91] For every integer $k \geq 1$, $g(2k + 1) \leq 2k + 2$ and $g(2k) \leq 2k + 2$.

F45:   [Ok88, Ok90a] For every integer $k \geq 1$,

  (a) $g(2k + 1) \leq 3k$ and $g(2k + 2) \leq 3k + 2$,

  (b) $g(3k) \leq 4k$ and $g(3k + 2) \leq 4k + 2$.

F46:   [Th01] Every $f(k)$-connected graph (defined in Fact F41) with bipartite index at least $4k - 3$ is k-parity-linked.

F47:   [Su97] Let G be a k-edge-connected digraph, and let (u_1, f_1, v_1), (u_2, f_2, v_2), $\ldots$, (u_k, f_k, v_k) be any k triples, where $u_1, u_2 \ldots, u_k$, $v_1, v_2 \ldots, v_k$ are not necessarily distinct vertices, and $f_1, f_2, \ldots, f_k$ are k distinct arcs, either of the form $f_i = (u_i, t_i)$, $i = 1, \ldots, k$, or $f_i = (t_i, v_i)$, $i = 1, \ldots, k$. Then there exist k edge-disjoint $u_i - v_i$ paths P_i in G such that $f_i \in E(P_i)$, $i = 1, \ldots, k$.

Subgraphs

High connectivity implies a large minimum degree (Fact F7). Conversely, a large minimum degree does not guarantee high connectivity (Fact F8). However, it does ensure the existence of a highly connected subgraph.

FACT

F48:   [Ma72a] Every graph of minimum degree at least $4k$ contains a k-connected subgraph.

REMARK

R6:   In fact, Mader [Ma72a] proved that if the average of the degrees of the vertices of G is at least $4k$, then G contains a k-connected subgraph. Concerning the proof of Fact F48, see also Thomassen [Th88].

4.1.4   Analysis and Synthesis

An interesting question in the study of graph connectivity is to describe how to obtain every k-(edge-)connected graph from a given "simple" one by a succession of elementary operations preserving k-connectedness. A classical result on this topic is Tutte's theorem, which states how to construct all 3-connected graphs, starting with a *wheel graph*. We also consider some relevant results dealing with deletion of edges or vertices. Finally, some facts concerning minimally and critically k-connected graphs, as well as a reference to connectivity augmentation problems, are considered.

Contractions and Splittings

DEFINITIONS

D30: The ***contraction*** of an edge uv consists of the identification of its endpoints u and v (keeping the old adjacencies but removing the self-loop from $u = v$ to itself). Let G be a k-connected graph. An edge of G is said to be k-***contractible*** if its contraction results in a k-connected graph.

D31: The converse operation is called ***splitting***: A vertex w with degree δ is replaced by an edge uv in such a way that some of the vertices adjacent to w are now adjacent to u and the rest are adjacent to v. Moreover, if the new vertices u, v have degrees at least $k = \delta/2 + 1$ we speak about a k-***vertex-splitting***.

D32: For any integer $n \geq 4$, the ***wheel graph*** W_n is the n-vertex graph obtained by joining a vertex to each of the $n - 1$ vertices of the cycle graph C_{n-1}.

FACTS

F49: If G is a k-connected graph, the operations of k-vertex splitting and edge addition always produce a graph that is also (at least) k-connected. In fact, as shown below, for $k = 3$ these operations suffice to derive all 3-connected graphs.

F50: [Th80b] Every 3-connected graph distinct from K_4 has a 3-contractible edge.

F51: [Th81] Every *triangle-free* (no 3-cycles) k-connected graph has a k-contractible edge.

F52: [Tu61] Every 3-connected graph can be obtained from a wheel by a finite sequence of 3-vertex-splittings and edge additions.

REMARK

R7: In general, k-connectedness does not ensure the existence of k-contractible edges.

EXAMPLE

E2:   In Figure 4.1.2, the cube graph Q_3 is synthesized from the wheel graph W_5 in four steps. All but the second step are 3-vertex-splittings.

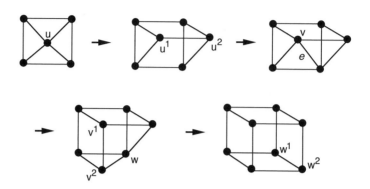

Figure 4.1.2: A 4-step Tutte synthesis of the cube graph Q_3.

REMARKS

R8:   Thomassen used Fact F50 to give a short proof of Kuratowski's theorem on planarity. Fact F50 can also be derived from Tutte's theorem (Fact F52).

R9:   Since Tutte's paper, the distribution of contractible edges in graphs of given connectivity has been extensively studied. For a comprehensive survey of this subject, we refer the reader to Kriesell [Kr02], where the author also considers subgraph contractions (see below).

R10:   Fact F52 is a reformulation of the following proposition [Tu61]: a 3-connected graph is either a wheel, or it contains an edge whose removal leaves a 3-connected subgraph, or it contains a 3-contractible edge that is not in a cycle of length 3.

R11:   Slater [Sl74] gave a similar result for constructing all 4-connected graphs starting from K_5, but in this case three more operations are required. For $k \geq 5$ the problem is still open. However, Lovász [Lo74] and Mader [Ma78a] managed to construct all k-edge-connected pseudographs (loops and multiple edges allowed) for every k even and odd, respectively.

Subgraph Contraction

The contraction of a subgraph is a natural generalization of edge contraction.

DEFINITION

D33:   A connected subgraph H of a k-connected graph G is said to be k-***contractible*** if the contraction of H into a single vertex results in a k-connected graph.

FACTS

F53: [McOt94] Every 3-connected graph on $n \geq 9$ vertices has a 3-contractible path of length two.

F54: [ThTo81] Every 3-connected graph with minimum degree at least four contains a 3-contractible cycle.

F55: [Kr00] Every 3-connected graph of order at least eight has a 3-contractible subgraph of order four.

CONJECTURE

[McOt94] For every n, a 3-connected graph of sufficiently large order has a 3-contractible subgraph of order n.

Edge Deletion

DEFINITION

D34: A subgraph H of a k-edge-connected graph G is said to be ρ-***reducible*** if the graph obtained from G by removing the edges of H is $(k - \rho)$-connected.

FACTS

F56: [Ma74] Every k-connected graph G with minimum degree at least $k+2$ contains a cycle C such that $G - E(C)$ is k-connected.

F57: [Ok88] Let G be a k-edge-connected graph with $k \geq 4$ even. Let $\{u, v\} \subset V$ and $\{e_1, e_2, f\} \subset E$, $e_i \neq f$ $(i = 1, 2)$. Then,

(a) There exists a 2-reducible cycle containing e_1 and e_2, but not f.

(b) There exists a 2-reducible $u-v$ path containing e_1, but not f.

F58: [Ok90b] Let G be a k-edge-connected graph with $k \geq 2$ even. If $\{u_1, v_1, u_2, v_2\}$ are distinct vertices, with edges $e_0 = v_1 v_2$, $e_i = u_i v_i$ $(i = 1, 2)$, and there is no edge-cut with k or $k + 1$ elements containing $\{e_0, e_1, e_2\}$, then there exists a 2-reducible cycle containing $\{e_0, e_1, e_2\}$.

F59: [HuOk92] For each odd $k \geq 3$, there exists a k-edge-connected graph containing two vertices u and v such that every cycle passing through u, v is ρ-reducible with $\rho \geq 3$.

REMARK

R12: For the case of three consecutive edges e_1, e_2, e_3 of a k-connected graph, Okamura [Ok95] also found a nontrivial equivalent reformulation of the condition that no cycle of G containing e_1, e_2, and e_3 is 2-reducible.

Vertex Deletion

FACTS

F60: [ChKaLi72] Every 3-connected graph of minimum degree at least 4 has a vertex v such that $G - v$ is 3-connected.

F61: [Th81] Every $(k + 3)$-connected graph has an induced (chordless) cycle whose deletion results in a k-connected graph.

F62: [Eg87] Every $(k + 2)$-connected triangle-free graph has an induced cycle whose deletion results in a k-connected graph.

REMARK

R13: Fact F61 was conjectured by Lovász, and Thomassen used Fact F51 to prove it.

Products of Graphs

DEFINITIONS

D35: Recall that the ***cartesian product*** of two graphs $G_i = (V_i, E_i)$, $i = 1, 2$, is the graph $G_1 \square G_2$ with vertex set $V_1 \times V_2$, and for which vertices (x_1, x_2) and (y_1, y_2) are adjacent if $x_1 = y_1$ and $x_2 y_2 \in E_2$, or $x_1 y_1 \in E_1$ and $x_2 = y_2$.

D36: The ***Kronecker product*** of two graphs $G_i = (V_i, E_i)$, $i = 1, 2$, is the graph $G_1 \times G_2$ with vertex set $V_1 \times V_2$, and for which vertices (x_1, x_2) and (y_1, y_2) are adjacent if $x_1 y_1 \in E_1$ and $x_2 y_2 \in E_2$.

D37: [BeDeFa84] Let $G_i = (V_i, E_i)$, $i = 1, 2$, be two graphs with the edges of G_1 arbitrarily oriented, in such a way that an oriented edge from x_1 to y_1 is denoted by $e_{x_1 y_1}$. For each arc $e_{x_1 y_1}$, let $\pi_{e_{x_1 y_1}}$ be a permutation of V_2. Then the ***twisted product*** $G_1 * G_2$ has $V_1 \times V_2$ as vertex set, with two vertices (x_1, x_2), (y_1, y_2) being adjacent if and only if either

$$x_1 = y_1 \quad \text{and} \quad x_2 y_2 \in E_2$$

or

$$x_1 y_1 \in E_1 \quad \text{and} \quad y_2 = \pi_{e_{x_1 y_1}}(x_2).$$

D38: [BaDaFiMi09] Given two graphs $G_i = (V_i, E_i)$, $i = 1, 2$, and a non-empty vertex subset $U_1 \subset V_1$, the ***generalized hierarchical product*** $G_1(U_1) \sqcap G_2$ is the graph with vertex set $V_1 \times V_2$, and for which vertices (x_1, x_2) and (y_1, y_2) are adjacent if $x_1 y_1 \in E_1$ and $x_2 = y_2$, or $x_1 = y_1 \in U_1$ and $x_2 y_2 \in E_2$.

FACTS

F63: [XuYa06] For any nontrivial graphs G_1 and G_2,

$$\kappa(G_1 \square G_2) \geq \min\{\kappa(G_1) + \delta(G_2), \kappa(G_2) + \delta(G_1)\}$$

and

$$\lambda(G_1 \square G_2) \geq \min\{\lambda(G_1)|V_2|, \lambda(G_2)|V_1|, \delta(G_1) + \delta(G_2)\}.$$

F64: [Sp08] For any nontrivial graphs G_1 and G_2,

$$\kappa(G_1 \square G_2) = \min\{\kappa(G_1)|V_2|, \kappa(G_2)|V_1|, \delta(G_1) + \delta(G_2)\}.$$

F65: [We62] If G_1 and G_2 are two connected graphs, then $G_1 \times G_2$ is connected if and only if G_1 and G_2 are not both bipartite graphs.

F66:

(a) [MaVu08] $\kappa(K_n \times K_m) = (n-1)(m-1)$ for any $n \geq m \geq 2$ and $n \geq 3$.

(b) [WaWu11] $\kappa(G \times K_n) = \min\{n\kappa(G), (n-1)\delta(G)\}$ for any nontrivial graph G and $n \geq 3$.

F67: [BaGVMa06, BaCeDiGVMa07]

(a) For any nontrivial graphs G_1 and G_2,

$$\min\{\kappa(G_1)|V_2|, (\delta_1 + 1)\kappa(G_2), \delta_1 + \delta_2)\} \leq \kappa(G_1 * G_2) \leq \delta_1 + \delta_2;$$

$$\min\{\lambda(G_1)|V_2|, (\delta_1 + 1)\lambda(G_2), \delta_1 + \delta_2)\} \leq \lambda(G_1 * G_2) \leq \delta_1 + \delta_2,$$

where $\delta_1 + \delta_2$ is the minimum degree of $G_1 * G_2$.

(b) If G_1 and G_2 are maximally connected, then $G_1 * G_2$ is also maximally connected.

(c) For every connected graph G, the graph $G * G$ is maximally connected.

F68: [BaDaFiMi09] The connectivity of the generalized hierarchical product satisfies

$$\kappa(G_1(U_1) \sqcap G_2) \leq \min\{\kappa(G_1)|V_2|, \kappa(U_1|U_1'), \delta(G_1(U_1) \sqcap G_2)\},$$

where $U_1' \subset V_1 - U_1$ and $\delta(G_1(U_1) \sqcap G_2) = \min\{\delta(G_1 - U_1), \delta(G_1(U_1)) + \delta_2\}$.

REMARKS

R14: The graph $G_1 * G_2$ can be viewed as formed by $|V_1|$ disjoint copies of G_2, each oriented edge $x_1 y_1$ indicating that some perfect matching between the copies $G_1^{x_1}$, $G_1^{y_1}$ (respectively generated by the vertices x_1 and y_1 of G_1) is added. Moreover, $K_2 * G$ is a **permutation graph** [ChHa67].

R15: If in Definition D37, $\pi_{e_{x_1 y_1}}$ is the identity permutation for any oriented edge $e_{x_1 y_1}$, the twisted product $G_1 * G_2$ is the cartesian product $G_1 \square G_2$.

R16: If U_1 is consits of only one vertex, then $G_1(U_1) \sqcap G_2$ is the standard hierarchical product [BaCoDaFi09], whereas if $U_1 = V_1$ we obtain the cartesian product $G_1 \square G_2$.

R17: Fact F66(b) was previously proved for G bipartite in [GuVu09].

R18: Regarding Fact F66, the connectivity of Kronecker products by K_2 has been recently studied in [WaYa12].

Minimality and Criticality

A standard technique used to study a certain property $\mathcal{P}$ is to consider those graphs that are edge-minimal or vertex-minimal (critical) with respect to $\mathcal{P}$, in the sense that the removal of any vertex or edge produces a graph for which $\mathcal{P}$ does not hold.

DEFINITIONS

D39:   A graph or digraph G is said to be ***minimally k-connected*** if $\kappa(G) \geq k$ but, for each edge $e \in E$, $\kappa(G - e) < k$. Analogously, G is ***minimally k-edge-connected*** if $\lambda(G) \geq k$, but for each $e \in E$, $\lambda(G - e) < k$.

D40:   A vertex u of a digraph has ***half degree*** k if either $deg^+(u) = k$ or $deg^-(u) = k$.

FACTS

F69:   [Ma71, Ma72b] Every minimally k-connected (or k-edge-connected) graph contains at least $k + 1$ vertices of degree k.

F70:   [Ma72b] Every cycle of a minimally k-connected graph contains a vertex of degree k.

F71:   Every cycle in a k-connected graph G contains either a vertex of degree k or an edge whose removal does not lower the connectivity of G.

F72:   [Ha81] Every minimally k-connected digraph contains at least $k + 1$ vertices of half degree k.

F73:   [Ma02] Every minimally k-connected digraph contains at least $k + 1$ vertices of outdegree k and at least $k + 1$ vertices of indegree k.

REMARKS

R19:   Halin [Ha69, Ha00] proved the existence of a vertex of degree k in every minimally k-connected graph, and the corresponding theorem for minimally k-edge-connected graphs was proved by Lick [Li72]. Both results were then improved by Mader (Fact F69).

R20:   Fact F72, a consequence of Mader's result Fact F73, is due to Hamidoune and is the digraph analogue of (and implies) Mader's theorem (Fact F69) about the existence of vertices of degree k. The existence of at least one vertex of half degree k had been previously asserted by Kameda [Ka74].

Vertex-Minimal Connectivity – Criticality

Maurer and Slater [MaSl77] introduced the general concept of *critically connected* and *critically edge-connected graphs*, graphs whose connectivity decreases when one or more vertices are removed.

DEFINITION

D41: A graph G is called k-**critically** n-**connected**, or an (n, k)-**graph**, if, for each vertex subset U with $|U| \leq k$, we have $\kappa(G - U) = n - |U|$. When $k = 1$, we simply refer to the graph as **critically** n-**connected**.

FACTS

F74: [MaSl77] The only (n, n)-graph is the complete graph K_{n+1}.

F75: The "cocktail party graph" (obtained from K_{2n+2} by removing a 1-factor [perfect matching]) is a $(2n, n)$-graph but not a $(2n, n + 1)$-graph.

F76: [Su88] The complete graph on $k+1$ vertices is the unique k-critically n-connected graph with $n < 2k$.

F77: [Ma77] If G is a $(n, 3)$-graph, then its order is at most $6n^2$. Thus, for each n, there are only finitely many of $(n, 3)$-critical graphs.

REMARKS

R21: An early survey about (n, k)-graphs can be found in [Ma84].

R22: Fact F75 led Slater to conjecture that, apart from K_{n+1}, there is no (n, k)-graph with $k > n/2$, which, after some partial results, was finally proved by Su (Fact F76).

R23: Fact F77 was generalized by Mader to the class of all finite n-connected graphs.

Connectivity Augmentation

We conclude the section by referring the reader to Frank [Fr94] for an in-depth discussion of *connectivity augmentation*. In the *edge-connectivity augmentation problem*, we are given a graph $G = (V, E)$ and a positive integer k, and the goal is to find the smallest set of edges F that we can add to G such that $G' = (V, E \cup F)$ is k-connected. Due to its applicability to the design of fault-tolerant networks, connectivity augmentation has also been widely investigated from an algorithmic point of view. Watanabe and Nakamura [WaNa87] gave the first polynomial-time algorithm solving the edge-connectivity augmentation problem. In the same paper, the authors formulated a necessary and sufficient condition to decide if a given graph G can be made k-connected by adding at most a certain number of edges. The same question for digraphs was solved in [Fr92].

References

[AhDi94] R. Aharoni and R. Diestel, Menger's theorem for a countable source set. *Combin. Probab. Comput.* **3** (1994), 145–156.

[BaGVMa06] C. Balbuena, P. García-Vázquez, and X. Marcote, Reliability of interconnection networks modeled by a product of graphs. *Networks* **48** (2006), 114–120.

[BaCeDiGVMa07] C. Balbuena, M. Cera, A. Diánez, P. García-Vázquez, and X. Marcote, On the edge-connectivity and restricted edge-connectivity of a product of graphs. *Discrete Appl. Math.* **155** (2007), 2444–2455.

[BaGu01] J. Bang-Jensen and G. Gutin, *Digraphs. Theory, Algorithms and Applications*, Springer-Verlag, London, 2001.

[BaCoDaFi09] L. Barrière, F. Comellas, C. Dalfó, and M. A. Fiol, The hierarchical product of graphs. *Discrete Appl. Math.* **157** (2009), 36–48.

[BaDaFiMi09] L. Barrière, C. Dalfó, M. A. Fiol, and M. Mitjana, The generalized hierarchical product of graphs. *Discrete Math.* **309** (2009), 3871–3881.

[Be76] C. Berge, *Graphs and Hypergraphs*, Second Edition, North-Holland Pub. Co., New York, 1976.

[BeDeFa84] J.-C. Bermond, C. Delorme, and G. Farhi, Large graphs with given degree and diameter II. *J. Combin. Theory Ser. B* **36** (1984), 32–48.

[BeHoPe89] J.-C. Bermond, N. Homobono, and C. Peyrat, Large fault-tolerant interconnection networks. *Graphs Combin.* **5** (1989), 107–123.

[ChHa67] G. Chartrand and F. Harary, Planar permutation graphs. *Ann. Inst. H. Poincaré Sect. B* **3** (1967), 433–438.

[ChHa68] G. Chartrand and F. Harary, Graphs with prescribed connectivities, pp. 61–63 in *Theory of Graphs (Proc. Colloq., Tihany, 1966)*, Academic Press, New York, 1968.

[ChKaLi72] G. Chartrand, A. Kaugars, and D. R. Lick, Critically n-connected graphs. *Proc. Amer. Math. Soc.* **32** (1972), 63–68.

[ChLeZh11] G. Chartrand, L. Lesniak, and P. Zhang, *Graphs & Digraphs*, Fifth Edition, Chapman and Hall/CRC, Boca Raton, FL, 2011.

[Di00] R. Diestel, *Graph Theory*, Second Edition, Graduate Texts in Mathematics, Volume 173, Springer-Verlag, New York, 2000.

[Di60] G. A. Dirac, In abstrakten Graphen vorhandene vollständige 4-Graphen und ihre Unterteilungen. *Math. Nachr.* **22** (1960), 61–85.

[Eg87] Y. Egawa, Cycles in k-connected graphs whose deletion results in a $(k-2)$-connected graph. *J. Combin. Theory Ser. B* **42** (1987), 371–377.

[ElFeSh56] P. Elias, A. Feinstein, and C. E. Shannon, A note on the maximum flow through a network. *IRE Trans. Inform. Theory* **IT–2** (1956), 117–119.

[EnJaSl77] R. Entringer, D. Jackson, and P. Slater, Geodetic connectivity of graphs. *IEEE Trans. Circuits and Systems* **24** (1977), 460–463.

[ErGy85] P. L. Erdös and E. Györi, Any four independent edges of a 4-connected graph are contained in a circuit. *Acta Math. Hung.* **46** (1985), 311–313.

[FoFu56] L. R. Ford and D. R. Fulkerson, Maximal flow through a network. *Canad. J. Math.* **8** (1956), 399–404.

[Fr90] A. Frank, Packing paths, circuits, and cuts – a survey, pp. 47–100 in B. Korte, L. Lovász, H-J. Prömel, and A. Schrijver (Eds.), *Paths, Flows and VLSI-Layouts*, Springer, Berlin, 1990.

[Fr92] A. Frank, Augmenting graphs to meet edge-connectivity requirements. *SIAM J. Discrete Math.* **5** (1992), 22–53.

[Fr94] A. Frank, Connectivity augmentation problems in network design, pp. 34–63 in J. R. Birge and K. G. Murty (Eds.), *Mathematical Programming: State of the Art 1994*, The University of Michigan, Ann Arbor, 1994.

[Fr95] A. Frank, Connectivity and network flows, pp. 111–177 in R. Graham, M. Grötschel and L. Lovász (Eds.), *Handbook of Combinatorics*, Elsevier Science B.V., 1995.

[Ga61] T. Gallai, Maximum-Minimum Sätze und verallgemeinerte Faktoren von Graphen. *Acta Math. Sci. Hungar.* **12** (1961), 131–173.

[GeHa70] D. Geller and F. Harary, Connectivity in digraphs. *Lec. Not. Math.* **186**, Springer, Berlin (1970), 105–114.

[Gi85] A. Gibbons, *Algorithmic Graph Theory*, Cambridge University Press, Cambridge, 1996.

[GrYe06] J. L. Gross and J. Yellen, *Graph Theory and Its Applications*, Second Edition, CRC Press, Boca Raton, 2006.

[GuVu09] R. Guji and E. Vumar, A note on the connectivity of Kronecker products of graphs. *Appl. Math. Lett.* **22** (2009), 1360–1363.

[Gy78] E. Györi, On division of graphs to connected subgraphs, pp. 485–494 in *Combinatorics*, North-Holland, Amsterdam, 1978.

[Ha69] R. Halin, A theorem on n-connected graphs. *J. Combin. Theory* **7** (1969), 150–154.

[Ha00] R. Halin, Miscellaneous problems on infinite graphs. *J. Graph Theory* **35** (2000), 128–151.

[Ha81] Y. O. Hamidoune, Quelques problèmes de connexité dans les graphes orientés. *J. Combin. Theory Ser. B* **30** (1981), 1–10.

[HaNoCa68] F. Harary, R. Z. Norman, and D. Cartwright, *Introduction à la Théorie des Graphes Orientés*, Dunod, Paris, 1968.

[HaTh82] R. Häggkvist and C. Thomassen, Circuits through specified edges. *Discrete Math.* **41** (1982), 29–34.

[HeVo08] A. Hellwig and L. Volkmann, Maximally edge-connected and vertex-connected graphs and digraphs: a survey. *Discrete Math.* **308** (2008), 3265–3296.

[HeRaGa00] M. R. Henzinger, S. Rao, and H. Gabow, Computing vertex connectivity: new bounds from old techniques. *J. Algorithms* **34** (2000), 222–250.

[Hu91] A. Huck, A sufficient condition for graphs to be weakly k-linked. *Graphs and Combin.* **7** (1991), 323–351.

[HuOk92] A. Huck and H. Okamura, Counterexamples to a conjecture of Mader about cycles through specified vertices in n-edge-connected graphs. *Graphs Combin.* **8** (1992), 253–258.

[Ju70] H. A. Jung, Eine Verallgemeinerung des n-fachen Zusammenhangs für Graphen. *Math. Ann.* **187** (1970), 95–103.

[Ka74] T. Kameda, Note on Halin's theorem on minimally connected graphs. *J. Combin. Theory Ser. B* **17** (1974), 1–4.

[Ka02] K. Kawarabayashi, One or two disjoint cycles cover independent edges: Lovász-Woodall Conjecture. *J. Combin. Theory Ser. B* **84** (2002), 1–44.

[Kr00] M. Kriesell, Contractible subgraphs in 3-connected graphs. *J. Combin. Theory Ser. B* **80** (2000), 32–48.

[Kr02] M. Kriesell, A survey on contractible edges in graphs of a prescribed vertex connectivity. *Graphs Combin.* **18** (2002), 1–30.

[LaMa70] D. G. Larman and P. Mani, On the existence of certain configurations within graphs and the 1-skeletons of polytopes. *Proc. London Math. Soc* **20** (1970), 144–160.

[Li72] D. R. Lick, Minimally n-line connected graphs. *J. Reine Angew. Math.* **252** (1972), 178–182.

[Lo90] M. V. Lomonosov, Cycles through prescribed elements in a graph. *Algorithms Comb.* **9** (1990), 215–234.

[Lo74] L. Lovász, Problem 5. *Period. Math. Hung.* **4** (1974), 82.

[Lo76] L. Lovász, On some connectivity properties of eulerian graphs. *Acta Math. Acad. Sci. Hung.* **28** (1976), 129–138.

[Lo77] L. Lovász, A homology theory for spanning trees of a graph. *Acta Math. Acad. Sci. Hung.* **30** (1977), 241–251.

[Lo93] L. Lovász, *Combinatorial Problems and Exercises*, 2nd Edition, North-Holland, Amsterdam, 1993.

[LoNePl78] L. Lovász, V. Neumann-Lara, and M. D. Plummer, Mengerian theorems for paths of bounded length. *Period. Math. Hungar.* **9** (1978), 269–276.

[Ma71] W. Mader, Minimale n-fach kantenzusammenhängende Graphen. *Math. Ann. (Basel)* **191** (1971), 21–28.

[Ma72a] W. Mader, Existenz n-fach zusammenhängender Teilgraphen in Graphen genügend grosser Kantendichte. *Abh. Math. Sem. Univ. Hamburg* **37** (1972), 86–97.

[Ma72b] W. Mader, Ecken vom Grad n in minimalen n-fach zusammenhängenden Graphen. *Arch. Math. (Basel)* **23** (1972), 219–224.

[Ma73] W. Mader, Grad und lokaler Zusammenhang in endlichen Graphen. *Math. Ann.* **205** (1973), 9–11.

[Ma74] W. Mader, Kreuzungsfreie a, b-Wege in endlichen Graphen. *Abh. Math. Sem. Univ. Hamburg* **42** (1974), 187–204.

[Ma77] W. Mader, Endlichkeitssätze für k-kritische Graphen. *Math. Ann.* **229** (1977), 143–153.

[Ma78a] W. Mader, A reduction method for edge-connectivity in graphs. *Ann. Discrete Math.* **3** (1978), 145–164.

[Ma78b] W. Mader, Über die Maximalzahl kantendisjunkter A-Wege. *Arch. Math. (Basel)* **30** (1978), 325–336.

[Ma78c] W. Mader, Über die Maximalzahl kreuzungsfreier H-Wege. *Arch. Math. (Basel)* **31** (1978/79), 387–402.

[Ma79] W. Mader, Connectivity and edge-connectivity in finite graphs, pp. 66–95 in *Surveys in Combinatorics, Proc. 7th Br. Comb. Conf.*, Cambridge 1979, Lond. Math. Soc. Lect. Note Ser. 38, 1979.

[Ma84] W. Mader, On k-critically n-connected graphs, pp. 389–398 in *Progress in Graph Theory (Waterloo, Ont., 1982)*, Academic Press, Toronto, ON, 1984.

[Ma02] W. Mader, On vertices of outdegree n in minimally n-connected digraphs, *J. Graph Theory* **39** (2002), 129–144.

[MaVu08] A. Mamut and E. Vumar, Vertex vulnerability parameters of Kronecker products of complete graphs. *Inform. Process. Lett.* **106** (2008), 258–262.

[MaSl77] S. B. Maurer and P. J. Slater, On k-critical, n-connected graphs. *Discrete Math.* **20** (1977), 255–262.

[McCu84] W. D. McCuaig, A simple proof of Menger's Theorem. *J. Graph Theory* **8** (1984), 427–429.

[McOt94] W. D. McCuaig and K. Ota, Contractible triples in 3-connected graphs. *J. Combin. Theory Ser. B* **60** (1994), 308–314.

[Me27] K. Menger, Zur allgeminen Kurventheorie. *Fund. Math.* **10** (1927), 96–115.

[Oe96] O. R. Oellermann, Connectivity and edge-connectivity in graphs: A survey. *Congr. Numerantium* **116** (1996), 231–252.

[Oe12] O. R. Oellermann, Menger's theorem, in L. W. Beineke and R. J. Wilson (Eds.), *Topics in Structural Graph Theory*, Cambridge University Press, 2012.

[Ok84] H. Okamura, Paths and edge-connectivity in graphs. *J. Combin. Theory B* **37** (1984), 151–172.

[Ok85] H. Okamura, Paths and edge-connectivity in graphs, II, pp. 337–352 in J. Akiyama et al. (Eds.), *Number Theory and Combinatorics*, World Scientific Publishing, 1985.

[Ok87] H. Okamura, Paths and edge-connectivity in graphs, III: Six-terminal k paths. *Graphs Combin.* **3** (1987), 159–189.

[Ok88] H. Okamura, Paths in k-edge-connected graphs. *J. Combin. Theory Ser. B* **45** (1988), 345–355.

[Ok90a] H. Okamura, Every $4k$-edge-connected graph is weakly $3k$-linked. *Graphs Combin.* **6** (1990), 179–185.

[Ok90b] H. Okamura, Cycles containing three consecutive edges in $2k$-edge-connected graphs, pp. 549–553 in R. Bodendiek and R. Henn (Eds.), *Topics in Combinatorics and Graph Theory*, Physica-Verlag, Heidelberg, 1990.

[Ok95] H. Okamura, 2-reducible cycles containing three consecutive edges in $(2k+1)$-edge-connected graphs. *Graphs Combin.* **11** (1995), 141–170.

[RoSe85] N. Robertson and P. D. Seymour, Graph minors—a survey, pp. 153–171 in *Surveys in Combinatorics 1985* (Glasgow, 1985), London Math. Soc. Lecture Note Ser. 103, Cambridge Univ. Press, Cambridge, 1985.

[Sa96] D. P. Sanders, On circuits through five edges. *Discrete Math.* **159** (1996), 199–215.

[Sl74] P. J. Slater, A classification of 4-connected graphs. *J. Combin. Theory* **17** (1974), 281–298.

[Sp08] S. Špacapan, Connectivity of cartesian product of graphs. *Appl. Math. Lett.* **21** (2008), 682–685.

[Su88] J. J. Su, Proof of Slater's conjecture on k-critical n-connected graphs. *Kexue Tongbao (English ed.)* **33** (1988), 1675–1678.

[Su97] X.-Y. Su, Some generalizations of Menger's theorem concerning arc-connected digraphs. *Discrete Math.* **175** (1997), 293–296.

[Th80a] C. Thomassen, 2-linked graphs, *European J. Combin.* 1 (1980), 371–378.

[Th80b] C. Thomassen, Planarity and duality of finite and infinite graphs. *J. Combin. Theory Ser. B* **29** (1980), 244–271.

[Th81] C. Thomassen, Non-separating cycles in k-connected graphs. *J. Graph Theory* **5** (1981), 351–354.

[Th88] C. Thomassen, Paths, circuits and subdivisions, pp. 97–131 in L. W. Beineke and R. J. Wilson (Eds.), *Selected Topics in Graph Theory III*, Academic Press, London, 1988.

[Th01] C. Thomassen, The Erdös-Pósa property for odd cycles in graphs of large connectivity. *Combinatorica* **21** (2001), 321–333.

[ThTo81] C. Thomassen and B. Toft, Non-separating induced cycles in graphs. *J. Combin. Theory B* **31** (1981), 199–224.

[Tu61] W. T. Tutte, A theory of 3-connected graphs. *Nederl. Akad. Wetench. Indag. Math.* **23** (1961), 441–455.

[Tu66] W. T. Tutte, *Connectivity in Graphs*, University of Toronto Press, London, 1966.

[WaYa12] W. Wang and Z. Yan, Connectivity of Kronecker products by K_2. *Appl. Math. Lett.* **25** (2012), 172–174.

[WaWu11] Y. Wang and B. Wu, Proof of a conjecture on connectivity of Kronecker product of graphs. *Discrete Math.* **311** (2011), 2563–2565.

[WaMe67] M. E. Watkins and D. M. Mesner, Cycles and connectivity in graphs. *Canad. J. Math.* **19** (1967), 1319–1328.

[WaNa87] T. Watanabe and A. Nakamura, Edge-connectivity augmentation problems. *J. Comput. System Sci.* **35** (1987), 96–144.

[We62] P. M. Weichesel, The Kronecker product of graphs. *Proc. Amer. Math. Soc.* **13** (1962), 47–52.

[Wh32] H. Whitney, Congruent graphs and the connectivity of graphs. *Amer. J. Math.* **54** (1932), 150–168.

[Wi92] A. Wigderson, The complexity of graph connectivity, pp. 112–113 in I. M. Havel and V. Koubek (Ed.), *Lecture Notes in Comput. Sci.*, 629, Proc. Mathematical Foundations of Computer Science 1992, Springer, Berlin, 1992.

[Wo77] D. R. Woodall, Circuits containing specified edges. *J. Combin. Theory Ser. B* **22** (1977), 274–278.

[Xu01] J. M. Xu, *Topological Structure and Analysis of Interconnection Networks*, Kluwer Academic Publishers, Dordrecht/Boston/London, 2001.

[XuYa06] J. M. Xu and C. Yang, Connectivity of Cartesian product graphs. *Discrete Math.* **306** (2006), 159–165.

Section 4.2

Eulerian Graphs

Herbert Fleischner, Technical University of Vienna, Austria

INTRODUCTION

Eulerian graph theory has its roots in the Königsberg Bridges Problem: Four landmasses are being connected by seven bridges as depicted in Figure 4.2.1. The graph theoretical model of this problem is depicted in Figure 4.2.2.

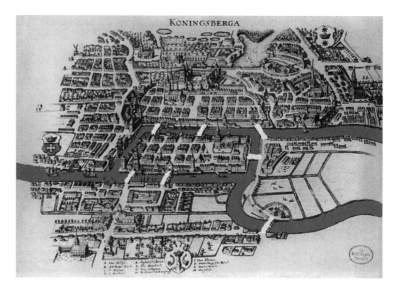

Figure 4.2.1: A map of Königsberg as it was in Euler's days, with highlighted bridges.

QUESTION: Starting at any of the four landmasses, is it possible to perform a walk such that every bridge is crossed once and only once, the walk ending at any of these four landmasses? L. Euler wrote an article on this problem in 1736 [Eu1736]; hence

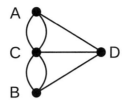

Figure 4.2.2: Graph model of the Königsberg Bridges Problem.

the name eulerian graph. This paper can be viewed as the "birth certificate" for graph theory, in general. For an extensive treatment of eulerian graphs and related topics see [Fl90, Fl91]. The equivalent question asks for a walk in the graph of Figure 4.2.2 such that every edge is traversed precisely once.

4.2.1   Basic Definitions and Characterizations

NOTATION: Throughout this section, a graph, digraph, or mixed graph is denoted $G = (V, E)$, where V is the vertex-set of G and E is the edge-set of G, consisting of undirected edges, directed edges (arcs), or both, respectively.

TERMINOLOGY: Sometimes, for emphasis and to avoid confusion, the adjective "undirected" is used for "graph" or "edge."

DEFINITIONS

D1: An **eulerian tour** in a graph (or digraph) G is a closed walk that uses each edge (or arc) of G *exactly* once, and hence, is a closed trail. An eulerian tour in a mixed graph is a closed trail that uses each edge and each arc exactly once. An **open eulerian trail** is an open trail that uses each edge and/or arc exactly once.

D2: A graph, digraph, or mixed graph that has an eulerian tour is called **eulerian**.

D3: An undirected graph is **even** if every vertex has even degree.

D4: In a digraph, a vertex v is **balanced** if the indegree and outdegree of v are equal. A digraph is balanced if all of its vertices are balanced.

TERMINOLOGY NOTE: In §4.3, the term *symmetric* is used instead of balanced when the indegree and outdegree of v are equal.

D5: A **balanced orientation** of a graph (or mixed graph) G is an assignment of a direction to each edge of the graph (or each undirected edge of the mixed graph) so that the resulting digraph is balanced.

D6: A **cycle decomposition** of a graph (digraph) G is a partition of the edge-set (arc-set) of G such that each partition set forms a cycle (directed cycle).

Some Basic Characterizations

Unless stated otherwise, we assume graphs not to have any self-loops. Note that the existence or non-existence of self-loops has no effect on whether a graph is eulerian.

FACTS

For details of the following facts, see, e.g., [To73, Mc84, Wo90, Fl89, Fl90].

F1: *The Classical Characterization* ([Eu1736], [Hi1873], [Ve12, Ve31])
Let G be a connected graph. The following are equivalent:

(a) G is eulerian.

(b) G is an even graph.

(c) G has a cycle decomposition.

F2: A graph is even if and only if it has a balanced orientation.

F3: A graph is even if and only if it has a decomposition into closed trails.

F4: A graph is even if and only if every edge belongs to an odd number of cycles.

F5: A graph is even if and only if it has an odd number of cycle decompositions.

F6: A connected graph $G = (V, E)$ is eulerian if and only if the number of subsets of E (including the empty set) that induce an acyclic subgraph of G is odd ([Sh79, Fl89, Fl90]).

F7: For a connected digraph D the following are equivalent.

(a) D is eulerian.

(b) D is a balanced digraph.

(c) D has a directed cycle decomposition.

REMARKS

R1: For the classical characterization, Euler ([Eu1736]) showed that Fact F1(a) implies Fact F1(b), while the converse is due to Hierholzer ([Hi1873]). The equivalence of Fact F1(b) and Fact F1(c) is due to Veblen ([Ve12, Ve31]).

R2: By Fact F1, the statements in Facts F2 through F6 can be viewed as alternative characterizations of eulerian graphs.

R3: Note that a connected eulerian digraph is strongly connected.

R4: There is no digraph or mixed graph analog for the characterization expressed in Fact F4.

Characterizations Based on Partition Cuts

DEFINITIONS

D7: Let G be a graph and let $X \subset V(G)$. The **partition-cut** associated with X, denoted $E(X, \overline{X})$, is the set of edges in G with one endpoint in X and one endpoint in $\overline{X} = V(G) - X$. A partition-cut in a digraph or mixed graph is analogously defined.

D8: An **edge-cut**, **arc-cut**, and **mixed-cut** are partition-cuts in a graph, digraph, and mixed graph G, respectively, associated with some $X \subset V(G)$.

D9: The **out-arcs** of an arc-cut (or mixed-cut) $E(X, \overline{X})$ is the subset of directed edges whose tail is in X and is denoted $E^+(X, \overline{X})$. The **in-arcs** of $E(X, \overline{X})$ is the subset of directed edges whose head is in X and is denoted $E^-(X, \overline{X})$.

D10: Let v be a vertex of a graph, digraph, or mixed graph G. The **incidence set** of v, denoted E_v, is the partition-cut $E(X, \overline{X})$, where $X = \{v\}$.

NOTATION: In a digraph, the out-arcs and in-arcs of the incidence set of v are denoted E_v^+ and E_v^-, respectively.

FACTS

F8: A graph G is even if and only if $|E(X, \overline{X})|$ is even for every $X \subset V(G)$.

F9: A connected digraph G is eulerian if and only if $|E^+(X, \overline{X})| = |E^-(X, \overline{X})|$ for every $X \subset V(G)$.

F10: Let G be a connected mixed graph. The following are equivalent:

(a) G is eulerian.

(b) $|E(X, \overline{X})| - ||E^+(X, \overline{X})| - |E^-(X, \overline{X})||$ is nonnegative and even for every $X \subset V(G)$.

(c) G has a cycle decomposition.

REMARKS

R5: While Fact F1(b) and its digraph analogue, Fact F7(b), are (*local*) degree conditions guaranteeing that a graph (undirected or directed) is eulerian, for mixed graphs one needs the *global* condition in Fact F10(b) (which reduces to Facts F8 and F9 for undirected graphs and digraphs).

R6: Although the condition in Fact F10 is impractical from an algorithmic point of view for producing an eulerian tour in a mixed graph G, such a tour can be obtained using *network-flow techniques* by first getting a balanced orientation D_G of G; then any eulerian tour of D_G corresponds to an eulerian tour in G [FoFu62].

4.2.2   Algorithms to Construct Eulerian Tours

We begin with two classical algorithms for constructing an eulerian tour. All three algorithms in this subsection are polynomial-time (see [Fl90]).

Algorithm 4.2.1: Hierholzer's Algorithm [Hi1873]

Input: a connected graph G whose vertices all have even degree.
Output: an eulerian tour T.

 Start at any vertex v, and construct a closed trail T in G.
 While there are edges of G not already in trail T
  Choose any vertex w in T that is incident on an unused edge.
  Starting at vertex w, construct a closed trail D of unused edges.
  Enlarge trail T by splicing trail D into T at vertex w.
 Return T.

COMPUTATIONAL NOTE: A modified *depth-first search* (see §10.1), in which every un-used edge remains in the stack, can be used to construct the closed trails.

EXAMPLE

E1:   The key step in Algorithm 4.2.1 is enlarging a closed trail by combining it with a second closed trail — the *detour*. To illustrate, consider the closed trails, $T = \langle t_1, t_2, t_3, t_4 \rangle$ and $D = \langle d_1, d_2, d_3 \rangle$, in the graph shown in Figure 4.2.3. The closed trail that results when detour D is spliced into trail T at vertex w is given by $T' = \langle t_1, t_2, d_1, d_2, d_3, t_3, t_4 \rangle$. At the next iteration, the trail $\langle e_1, e_2, e_3 \rangle$ is spliced into trail T', resulting in an eulerian tour of the entire graph.

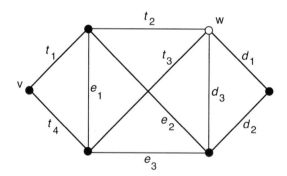

Figure 4.2.3: Splicing $\langle d_1, d_2, d_3 \rangle$ into $\langle t_1, t_2, t_3, t_4 \rangle$ results in $\langle t_1, t_2, d_1, d_2, d_3, t_3, t_4 \rangle$.

REMARKS

R7:   The splicing operation in Hierholzer's algorithm is also called a κ-*absorption* and is discussed later in this section.

R8:   The strategy in Fleury's algorithm, shown below, is to avoid, if possible, traversing a bridge in the subgraph induced on the set of untraversed edges. Fleury's algorithm also appeared in [Lu1894].

Algorithm 4.2.2: Fleury's Algorithm [Fl1883]

Input: Eulerian graph G with q edges and $v_0 \in V(G)$.
Output: Eulerian tour T_q.

    Choose $e_1 = v_0 v_1 \in E_{v_0}$ arbitrarily.
    Let $T_1 = \langle v_0, e_1, v_1 \rangle$
    For $i = 1$ to $q - 1$
        Let $G_i = G - E(T_i)$.
        If $\deg_{G_i}(v_i) = 1$
            Let $e_{i+1} = v_i v_{i+1} \in E(G_i)$.
        Else
            Choose $e_{i+1} = v_i v_{i+1} \in E(G_i)$ that is not a bridge in G_i.
        Extend T_i to $T_{i+1} = \langle v_0, e_1, v_1, \ldots, v_i, e_{i+1}, v_{i+1} \rangle$.

The Splitting and Detachment Operations

The *splitting* and *detachment* operations can serve as the basis for many of the characterizations, constructions, and decompositions discussed in this section.

DEFINITIONS

D11: Let G be a graph with vertex v such that $\deg(v) \geq 3$, and let e_a, e_b be incident on v and w_a, w_b, respectively. The graph $G_{a,b}$ obtained from G by introducing a new vertex $v_{a,b}$, adding new edges e'_a, e'_b joining $v_{a,b}$ and w_a, w_b, respectively, and deleting e_a, e_b is called the **a-b split of G at v**. The operation that produces $G_{a,b}$ is called the **splitting operation** (see Figure 4.2.4.)

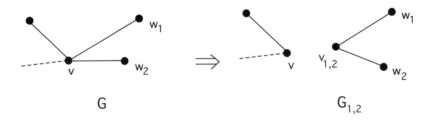

Figure 4.2.4: The splitting operation producing the 1-2 split of G at v.

D12: Let v be a vertex of a graph G with $\deg(v) \geq 2$, and let the edge subsets $E_1(v), E_2(v), \ldots, E_k(v)$, $k \geq 2$, be a partition of the incidence set E_v. Replace v with new vertices $v_1, v_2, \ldots, v_k$, and let v_i, $i = 1, 2, \ldots, k$, be incident on the edges of $E_i(v)$ (without altering any other incidence). The graph H thus obtained is called a **detachment of G at v**. This action is called a **detachment operation at v** (see Figure 4.2.5).

D13: A graph H is a **detachment of G** if it results from a sequence of detachment operations performed at each of the vertices of some vertex subset $W \subseteq V(G)$. For a discussion of detachments of graphs, see [Na79, Na85a, Na85b].

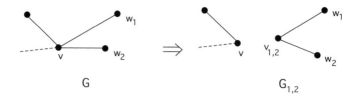

Figure 4.2.5: Graph H is a detachment of G at v.

FACTS

F11: *Splitting Lemma.* Let v be a vertex of a connected, bridgeless graph G with $\deg(v) \geq 4$, and let $e_1, e_2, e_3 \in E_v$.
(a) If v is not a cut-vertex then at least one of the splits $G_{1,2}$ or $G_{1,3}$ is connected and bridgeless.
(b) If v is a cut-vertex and e_1 and e_3 belong to different blocks, then $G_{1,3}$ is connected and bridgeless.

F12: Let v be a vertex in a 2-connected graph G (i.e., no cut-vertices) with $\deg(v) \geq 4$. If neither the 1-2 split $G_{1,2}$ at v nor the 1-3 split $G_{1,3}$ is 2-connected, then $G_{1,2}$ and $G_{1,3}$ have the same cut-vertex x and no other cut-vertices. In this case, both $G_{1,2}$ and $G_{1,3}$ are connected and bridgeless.

F13: A graph is connected if and only if there is a detachment of G that is a tree.

F14: A graph is eulerian if and only if it has a detachment that is a cycle.

REMARKS

R9: The Splitting Lemma (Fact F11) can serve as the basis for many of the results and algorithms mentioned in this section (see, e.g., [Fl90]). It can also be used to restrict, with no loss in generality, various other (solved as well as unsolved) graph theoretical problems to 3-regular graphs. For a short proof of the Splitting Lemma, see [Fl00].

R10: Definitions D11 through D13 and Facts F11 through F14 can be formulated for graphs with self-loops as well. In this case (but also later on) it makes sense to consider an edge e (not just self-loops) as composed of two *half-edges* incident on the respective endpoints of e. Correspondingly, one then considers the splitting operation as involving different half-edges and the sets E_v, $E_i(v)$ as being sets of half-edges.

R11: The splitting operation can be viewed as a special case of the detachment operation, where the partition of the incidence set E_v has exactly two cells, with at least one cell containing exactly two edges.

Algorithm 4.2.3: Splitting Algorithm

Input: Eulerian graph G with q edges and $v_0 \in V(G)$.
Output: Eulerian tour T_q in the form of a detachment of G.

> Initialize $H = G$.
> Choose $e_1 = v_0 v_1 \in E_{v_0}$ arbitrarily.
> Let $T_1 = \langle v_0, e_1, v_1 \rangle$
> For $i = 1$ to q
> > If $\deg_H(v_i) = 2$
> > > Let $e_{i+1} = v_i v_{i+1} \in E_{v_i}(H) - E(T_i)$.
> > Else {apply splitting lemma}
> > > If v_i is not a cut-vertex of H
> > > > Choose $e_{i+1} = v_i v_{i+1} \in E_{v_i}(H) - E(T_i)$ arbitrarily.
> > > Else
> > > > Choose $e_{i+1} = v_i v_{i+1}$ in a different block than e_i.
> > > $H := H_{i,(i+1)}$ {the i-$(i+1)$ split of H at v_i}
> > Extend T_i to $T_{i+1} = \langle v_0, e_1, v_1, \ldots, v_i, e_{i+1}, v_{i+1} \rangle$.

REMARKS

R12: Algorithms 4.2.1–4.2.3 can easily be adapted to construct an eulerian tour in a digraph: all one needs to do is choose e_{i+1} such that v_i is its tail since v_i is the head of e_i.

R13: The difference between the Splitting Algorithm and Fleury's Algorithm lies exclusively in the fact that the intermediate trails T_i, $0 \le i < q$, are stored separately as edge sequences, say, by Fleury's Algorithm, while the Splitting Algorithm retains them as part of the graphs considered. In both cases, however, it is the Splitting Lemma which guarantees the correctness of these algorithms (see [Fl90]). Observe that all even graphs are necessarily bridgeless.

4.2.3 Eulerian-Tour Enumeration and Other Counting Problems

The BEST-Theorem gives an explicit, computationally good formula for the number of eulerian tours in an eulerian digraph. It rests on the *Matrix Tree Theorem* (Fact F15) and can be applied to (undirected) graphs by summing over all balanced orientations of G. The latter, however, grows exponentially large with the number of vertices. We also briefly mention deBruijn (di)graphs because of their relevance to DNA-sequencing and other questions. DeBruijn digraphs are discussed in §4.4.

DEFINITIONS

D14: An *out-tree* in a digraph is a tree having a root of indegree 0 and all other vertices of indegree 1, and an *in-tree* is an out-tree with edges reversed.

D15: Let D be a digraph, $\boldsymbol{A}(D)$ its adjacency matrix with entries $a_{i,j}$, and let λ_i be the number of self-loops at $v_i \in V(D) = \{v_1, \ldots, v_n\}$. The ***Kirchhoff matrix*** $\boldsymbol{A}^*(D)$ with entries $a_{i,j}^*$ is defined by setting

$$a_{i,j}^* = -a_{i,j} \ \text{ if } \ i \neq j, \ a_{i,i}^* = id(v_i) - \lambda_i \ ; \qquad 1 \leq i, j \leq n$$

D16: Let a set $A = \{a_1, \ldots, a_n\}$ be called an ***alphabet*** whose ***letters*** are the elements of A. A k-letter ***word*** over A is an ordered k-tuple whose components are letters. A k-***deBruijn sequence*** over A is a cyclic sequence of letters from A such that every k-letter word over A appears exactly once in this cyclic sequence.

D17: Let $n \geq 2, k \geq 2$. The ***deBruijn graph*** $D_{n,k}$ has as its vertices the $(k-1)$-letter words over an n-letter alphabet A; thus, there are altogether n^{k-1} vertices. For each k-letter word $a_{i_1}, \ldots, a_{i_k}$ in the alphabet A, there is an arc of $D_{n,k}$ that joins the vertex $a_{i_1}, \ldots, a_{i_{k-1}}$ to the vertex $a_{i_2}, \ldots, a_{i_k}$.

TERMINOLOGY: For a matrix A, $A_{i,j}$ denotes the (i, j)-th ***minor***, i.e., the matrix obtained by deleting the i-th row and j-th column from A.

FACTS

F15: *Matrix Tree Theorem.* Given a digraph $D, V(D) = \{v_1, \ldots, v_n\}$, let $\boldsymbol{A}^* = \boldsymbol{A}^*(D)$ be its Kirchhoff matrix. The number of spanning out-trees of D rooted at v_i is $\det \boldsymbol{A}_{i,i}^*$.

F16: In an eulerian digraph D, the number of spanning in-trees rooted at v_i equals the number of spanning out-trees rooted at v_i.

F17: For an eulerian digraph D, $\det \boldsymbol{A}_{i,i}^* = \det \boldsymbol{A}_{j,j}^*, 1 \leq i, j \leq n$.

F18: *BEST-Theorem.* [EhBr51, TuSm41] Let D be an eulerian digraph of order n, and let $v_i \in V(D), a \in E_{v_i}^+$ be chosen arbitrarily. The number of eulerian tours starting at v_i with the traversal of a is

$$\det \boldsymbol{A}_{i,i}^* \prod_{j=1}^n (od(v_j) - 1)!$$

F19: For an eulerian graph G with p vertices and q edges, and chosen $e \in E(G)$, the number $\mathcal{O}_E(G)$ of balanced orientations of G containing a fixed orientation of e satisfies

$$\left(\frac{3}{2}\right)^{q-p} \leq \mathcal{O}_E(G) \leq 2^{q-p}.$$

F20: The deBruijn graph $D_{n,k}$ is an n-regular digraph $(id(v) = od(v) = n$ for every $v \in V(D_{n,k}))$ with n^{k-1} vertices.

F21: There is a 1-1 correspondence between the set of k-deBruijn sequences over an n-letter alphabet and the set of eulerian tours of the deBruijn graph $D_{n,k}$. Consequently, and as an application of the BEST-Theorem, the number of k-deBruijn sequences over an n-letter alphabet is

$$\frac{(n!)^{n^{k-1}}}{n^k}$$

EXAMPLE

E2: The deBruijn graphs $D_{2,3}$ and $D_{2,4}$ are shown in Figure 4.2.6 (see also [ChOe93, p. 220]).

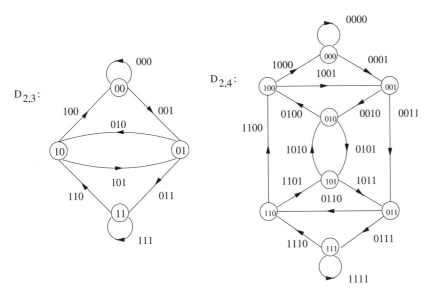

Figure 4.2.6: The deBruijn digraphs $D_{2,3}$ and $D_{2,4}$.

REMARK

R14: DeBruijn graphs are of particular interest in the case $n = 2$, i.e., when the words are binary sequences. The study of the graphs $D_{2,k}$ has been applied in biochemistry when considering the problem of DNA sequencing. These same graphs have also been of interest in telecommunications when one is concerned with the question of **network reliability**. Apart from $D_{2,k}$, Kautz graphs and hypercubes also play an important role because these graphs perform well with respect to diameter and other parameters, although the number of edges is relatively small in comparison to the number of vertices (see, e.g., [Xu02]).

4.2.4 Applications to General Graphs

In this subsection, we introduce some applications of eulerian graph theory to graph theory in general; some of these applications are also relevant in computer science and operations research, for example, the *Chinese Postman Problem* (§4.3). Interestingly, while certain analogues of results in eulerian graph theory hold equally well for general graphs, there are other quite natural analogues that lead to yet unsolved problems.

Covering Walks and Double Tracings

DEFINITIONS

D18: A *covering walk* (or *postman tour*) in an arbitrary graph G is a closed walk containing every edge of G.

D19: A *double tracing* is a closed walk that traverses every edge exactly twice. A double tracing is *bidirectional* if every edge is used once in each of its two directions.

D20: A *retract* or *retracing* in a walk W is a section of the form $v_{i-1}, e_i, v_i, e_{i+1}, v_{i+1}$ such that $e_i = e_{i+1}$ (and thus $v_{i+1} = v_{i-1}$). W is called *retract-free* if it has no retracts.

D21: A double tracing is called *strong* if it is both bidirectional and retract-free.

D22: The *edge-connectivity* of a connected graph G, denoted $\lambda(G)$, is the minimum number of edges whose removal can disconnect G. G is called k-*edge-connected* if $\lambda(G) \geq k$.

FACTS

F22: Let G be a graph with $2k$ vertices of odd degree, $k > 0$. Then G has a decomposition into k open trails whose initial and end vertices are of odd degree in G. Consequently, G has a decomposition into cycles and k paths; and if $k = 1$ and G is connected, then it has an open eulerian trail.

F23: Every connected graph has a bidirectional double tracing. In a tree, every double tracing is bidirectional.

F24: [Sa77] A connected graph has a retract-free double tracing if and only if it has no end-vertices (vertices of degree 1).

F25: [Th85] If G is a graph without 1- and 3-valent vertices, then it has a strong double tracing. Consequently, every 4-edge-connected graph has a strong double tracing.

F26: [Tr66], [Th85] A connected 3-regular graph with $|V(G)| \equiv 0 \bmod 4$ has no strong double tracing.

F27: [Ve75] Let G be a connected graph and $E_0 \subseteq E(G)$. G has a double tracing using every $e \in E(G) - E_0$ twice in the same (not prescribed) direction, and acting bidirectional on E_0, if and only if $G - E_0$ is an even graph. Observe that this implies Fact F23 (taking $E_0 = E(G)$).

REMARKS

R15: The condition for a double tracing to be bidirectional (Definition D19) applies to the case of self-loops if one views edges as composed of two half-edges, which allows a loop to be viewed as being also orientable in two ways.

R16: The double tracings quoted in Facts F23 and F25 can be obtained in polynomial time by reducing the respective problems to problems of finding eulerian tours satisfying certain restrictions, in eulerian digraphs derived from the given graphs by replacing every edge by two oppositely oriented arcs joining the same pair of vertices.

Maze Searching

In the context of this section, a *maze* may be viewed as a connected graph for which one has at each vertex local information only. Tarry's algorithm is just one of several *maze-searching algorithms*. (See [Fl91] for a more extensive study.)

NOTATION: In the description of Algorithm 4.2.4, $e_{in}(v)$, $v \neq v_0$, denotes the edge that was traversed in visiting vertex v for the first time, $\{e_{in}(v_0)\} = \emptyset$, and $E_{left}(v)$ denotes the set of edges that have been already traversed in leaving v.

Algorithm 4.2.4: Tarry's Algorithm [Ta1895]

Input: a connected graph G.
Output: a bidirectional double tracing of G.

    Choose $v_0 \in V(G)$.
    Initialize $i = 0$ and $W = \langle v_0 \rangle$.
    While $(E_{v_i} - E_{left}(v_i) \neq \emptyset)$
        While $([E_{v_i} - E_{left}(v_i)] - \{e_{in}(v_i)\} \neq \emptyset)$
            Choose edge $e_i = v_i v_{i+1} \in [E_{v_i} - E_{left}(v_i)] - \{e_{in}(v_i)\}$.
            $W := W, \langle e_i, v_{i+1} \rangle$ {Extend W to v_{i+1} via edge e_i.}
            $i := i + 1$
        Let $e_i = v_i v_{i+1} = e_{in}(v_i)$
        $W := W, \langle e_i, v_{i+1} \rangle$

FACTS

F28: Tremaux's (maze search) Algorithm also operates with local information only: at vertex v_i reached by the walk W, the number $\lambda_W(e)$ of traversals of every $e \in E_{v_i}$ by W is known. The output of Tremaux's Algorithm is also a bidirectional double tracing.

F29: Applying Tarry's Algorithm to eulerian graphs G, with additionally having the local information $\lambda_W(e)$ of Tremaux's Algorithm and choosing as next edge e_i with minimal $\lambda_W(e_i)$, then the list of the edges according to their second traversal yields an eulerian trail of G (for details of Facts F28 and F29, see [Fl91]).

REMARK

R17: A. S. Fraenkel's Algorithm uses a counter in connection with a modification of Tarry's Algorithm; its outcome is a covering walk using every edge at most twice.

Covers, Double Covers, and Packings

DEFINITIONS

D23: A *cycle cover* of a graph G is a family S of cycles of G such that every edge of G belongs to at least one element of S.

D24: A cycle cover S is a *cycle double cover* (CDC) if every edge of G belongs to exactly two elements of S.

D25: A *cycle packing* in G is a set of edge disjoint cycles in G.

D26: A CDC S is called **orientable** if the elements of S can be cyclically oriented in such a way that every edge e is given opposite orientations in the two elements of S containing e.

CONJECTURES

Cycle Double Cover Conjecture (CDCC): Every bridgeless graph has a CDC.

Oriented Cycle Double Cover Conjecture: Every bridgeless graph has an oriented CDC.

Strong Cycle Double Cover Conjecture: Every bridgeless graph has a CDC containing a prescribed cycle of the graph.

Three Optimization Problems

DEFINITIONS

D27: Let G be a bridgeless edge-weighted graph with weight function $w : E(G) \to R^+$. The **weight of a cycle** C in G, denoted $w(C)$, is given by $w(C) = \sum\limits_{e \in E(C)} w(e)$. The weight of a cycle cover or cycle packing S is $w(S) = \sum\limits_{C \in S} w(C)$.

D28: The **Minimum-Weight Cycle-Cover Problem** (MWCCP) is to find a cycle cover S in G such that $w(S)$ is minimum.

D29: The **Maximum-Weight Cycle-Packing Problem** (MWCPP) is to find a cycle packing S such that $w(S)$ is maximum.

D30: The **Chinese Postman Problem** is to find a minimum-weight covering walk W in G where $w(e)$ is counted as often as e is traversed by W (see §4.3).

FACTS

F30: [Fl86] Let G be a planar, bridgeless graph. Then G has an oriented CDC, and for any given cycle packing S, G has a CDC containing S as a subset. Thus, the Strong Cycle Double Cover Conjecture is true for planar graphs.

F31: A bridgeless graph having a hamiltonian path admits a double cover with at most six even subgraphs ([Tar86]). Later on, it was shown in [HuKo95] that five even subgraphs suffice to double cover cubic graphs having a hamiltonian path.

F32: [FlHa09] Every hypohamiltonian graph has a SCDC.

F33: [FlGu85] The Undirected Chinese Postman Problem and the Maximum-Weight Cycle-Packing Problem are both solvable in polynomial time, and for planar, bridgeless graphs, the Minimum-Weight Cycle-Cover Problem can be solved in polynomial time.

F34: [FlGu85] Let G be an edge-weighted graph with weight function w. If W is a solution of the Undirected Chinese Postman Problem and S a solution of the Maximum-Weight Cycle-Packing Problem, then $w(S) = w(W) - 2w(E_d)$, where $E_d \subset E(G)$ is the set of those edges used twice in W, and $w(E_d) := \sum\limits_{e \in E_d} w(e)$.

F35: [FlGu85] For any planar, connected, bridgeless graph, if S is a solution of the Minimum-Weight Cycle-Cover Problem and W is a solution of the Undirected Chinese Postman Problem, then $w(S) = w(W)$.

F36: For any connected, bridgeless graph G with weight function w, if W is a solution of the Undirected Chinese Postman Problem and S is a solution of the Minimum-Weight Cycle-Cover Problem, then $w(S) \geq w(W)$. The Petersen graph (§1.2) shows that the inequality can be strict ($w(S) = 21$ and $w(W) = 20$, for $w \equiv 1$).

Nowhere-Zero Flows

DEFINITIONS

D31: Let $f: E(D) \to R$ be given for a digraph D. The function f is called a **flow** if for every $v \in V(D)$, $\sum_{a \in E_v^+} f(a) = \sum_{a \in E_v^-} f(a)$.

D32: Let $f: E(G) \to N$ be given for a graph G. Let D be an orientation of G with $a_e \in E(D)$ the directed edge corresponding to $e \in E(G)$, and define $f'(a_e) := f(e)$. Then f is an **integer flow** in the graph G if f' is a flow in the digraph D.

D33: An integer flow f in G is **nowhere-zero** if $f(e) \neq 0$ for each edge $e \in E(G)$.

D34: A **k-flow** is an integer flow f such that $f(e) < k$ for each edge $e \in E(G)$.

CONJECTURE

Nowhere-Zero 5-Flow Conjecture (NZ5FC). Every bridgeless graph has a nowhere-zero 5-flow. [Tu54]

FACTS

F37: [Se81a] Every bridgeless graph has a nowhere-zero 6-flow.

F38: [Tu54] In a plane graph G, a (proper) k-face coloring of G corresponds to a nowhere-zero k-flow, and vice versa.

F39: A 3-regular graph G has a nowhere-zero 4-flow if and only if it is 3-edge-colorable, and it has a nowhere-zero 3-flow if and only if it is bipartite.

F40: [Ja75, Ja79] Every 4-edge-connected graph has a nowhere-zero 4-flow because it contains a spanning eulerian subgraph E^*. Likewise, it has a CDC containing the elements of a cycle decomposition of E^* (in fact, Tutte conjectured that every 4-edge-connected graph has a nowhere-zero 3-flow; see §5.2.2).

F41: To prove or disprove the NZ5FC and CDCC, one can assume without loss of generality that the graphs are 3-regular.

F42: [Se79] Let G be a bridgeless, planar graph, and let $f\colon E(G) \to Z^+$. Then the following two statements are equivalent:

(a) There exists a cycle cover S such that for every edge $e \in E(G)$, e belongs to exactly $f(e)$ elements in S.

(b) For every edge-cut $E_0 \subseteq E(G)$,
$$\sum_{e \in E_0} f(e) \text{ is even and } \frac{1}{2}\sum_{e \in E_0} f(e) \geq \max\{f(e)\colon e \in E_0\}.$$

REMARKS

R18: Double tracings in arbitrary connected graphs are the natural analogue to eulerian tours – Euler was already aware of that. Correspondingly, cycle double covers seem to be the natural analogue to cycle decompositions, yet their existence has been guaranteed so far only for certain classes of graphs, apart from the planar case. See [AlGoZh94], [Zh97], and [Zh12] for a thorough treatment of integer flows and cycle covers.

R19: Nowhere-zero flows can be viewed as eulerian tours in an eulerian multidigraph derived from an appropriate orientation of the given graph G, by replacing every arc a_e (corresponding to $e \in E(G)$) by $f(e)$ arcs with the same head and tail that a_e has.

4.2.5   Various Types of Eulerian Tours and Cycle Decompositions

DEFINITION

D35: Let G be an eulerian digraph and D_0 a subdigraph of G. If for every $v \in V(G)$, an eulerian trail T of G traverses every arc of D_0 incident from v before it traverses any other arc incident from v, then T is called D_0-*favoring*.

FACTS

F43: [Ko56] Let G be a connected graph with vertex-set $V(G) = \{v_1, \ldots, v_n\}$ and having an even number of edges. Then G is eulerian if and only if G is the edge-disjoint union of graphs G_1, G_2 with $\deg_{G_1}(v_i) = \deg_{G_2}(v_i)$, $1 \leq i \leq n$; and if G is the union of two such graphs, then G has an eulerian tour in which the edges of G_1 and G_2 alternate.

F44: [Se81b, FlFr90] A planar even graph G has a decomposition into even cycles if and only if every block of G has an even number of edges.

F45: Let v be an arbitrary vertex of a strongly connected digraph G. Then there exists a spanning in-tree of G with root v.

F46: [EhBr51] Let D' be a spanning in-tree with root v in the eulerian digraph G, and let $D_0 = G - E(D')$. Then there exists a D_0-favoring eulerian tour of G starting and ending at v.

F47: [CaFl95] Let $\{e_1, \ldots, e_m\} \subseteq E(G)$ be an ordered set where G is eulerian. An eulerian tour T of the form $T = \ldots, e_1, \ldots, e_2, \ldots, e_m, \ldots$ exists if the edge-connectivity $\lambda(G) \geq m - 1$; and if $\lambda(G) \geq 2m$, then one can even prescribe the direction in which these m edges are traversed by T.

REMARKS

R20: Fact F43 can be proved using the Splitting Lemma (Fact F11).

R21: Fact F44 is stated for planar graphs, but it can be extended to a more general class of graphs (see [Zh97]).

R22: D_0-favoring eulerian tours are studied in [FlWe89, Fl90]. However, in-trees are a special case of a more general class of digraphs D' for which there is a $(G - E(D'))$-favoring eulerian tour. We restricted Fact F46 to in-trees because of its relevance to enumerating eulerian tours in digraphs (see the BEST-Theorem [Fact F18]).

Incidence-Partition and Transition Systems

DEFINITIONS

D36: For each vertex v in a graph G, let $P(v) = \{E_1(v), \ldots, E_{k_v}(v)\}$, $k_v \geq 1$, be a partition of the incidence set E_v. Then $P(G) = \bigcup_{v \in V} P(v)$ is called an ***incidence-partition system*** of G.

D37: A ***transition system*** of an even graph G, denoted $\tau(G)$, is an incidence-partition system $\tau(G) = \bigcup_{v \in V} P(v)$ such that for every $v \in V(G)$, $|E_i(v)| = 2$ for every cell of the partition $P(v)$. Each cell $E_i(v)$ is called a ***transition***.

D38: An eulerian tour T and a cycle decomposition S give rise to transition systems, denoted τ_T and τ_S, respectively, in a natural way. Each transition in the **eulerian-tour transition system** τ_T is a pair of consecutive edges in the tour T. Similarly, each transition in the **cycle-decomposition transition system** τ_S is a pair of consecutive edges in a cycle $C \in S$.

TERMINOLOGY: A transition in τ_T and a transition in τ_S are referred to as a **transition of T** and a **transition of S**, respectively.

D39: Let $P(G)$ be an incidence-partition system of a graph G. An eulerian tour T is $P(G)$-***orthogonal*** (or ***orthogonal to*** $P(G)$) if no transition of T is a subset of any cell $E_i(v)$ of $P(G)$. $P(G)$-orthogonal cycle decompositions are defined analogously.

D40: A cycle decomposition S and an eulerian tour T are ***orthogonal*** if $\tau_S \cap \tau_T = \emptyset$.

TERMINOLOGY: The term *orthogonal* has been suggested by several authors as describing the underlying concept more accurately than the original term *compatible*.

D41:   An incidence-partition system $P(G)$ satisfies the **cut condition** if for every vertex subset X, the edge-cut $E(X, \overline{X})$ satisfies $|E(X, \overline{X}) \cap E_i(v)| \leq \frac{1}{2}|E(X, \overline{X})|$ for every cell $E_i(v)$ of $P(G)$.

FACTS

F48:   [Ko68] A loopless eulerian graph G has an eulerian tour orthogonal to a given partition system $P(G)$ if and only if $P(G)$ satisfies the cut condition restricted to the edge-cuts E_v, $v \in V(G)$.

F49:   [Fl80] Given a cycle decomposition S of the eulerian graph G with $\deg(v) > 2$ for every $v \in V(G)$, there exists an eulerian tour orthogonal to S.

F50:   [Fl80] Let T be an eulerian tour of the eulerian graph G. If $\deg(v) \equiv 0 \bmod 4$ for every $v \in V(G)$, then there exists a cycle decomposition orthogonal to T.

F51:   [FlFr90] Let G be a planar, even, loopless graph with incidence-partition system $P(G)$. Then G has a $P(G)$-orthogonal cycle decomposition if and only if $P(G)$ satisfies the cut condition.

F52:   [Fl80] Let G be a planar eulerian graph and let T be an eulerian tour of G. If $\deg(v) > 2$ for every $v \in V(G)$, then G has a cycle decomposition orthogonal to T.

EXAMPLE

E3:   The complete graph K_5 in Figure 4.2.7, with transition system $\tau(K_5) = \{\{i, i+1\}, \{i', (i+1)'\}\colon\ 1 \leq i \leq 5$, setting $6 = 1\}$ has no $\tau(K_5)$-orthogonal cycle decomposition, which shows that Fact F51 cannot be generalized to arbitrary non-planar graphs.

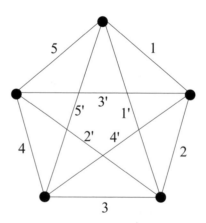

Figure 4.2.7: K_5 having no $\tau(K_5)$-orthogonal cycle decomposition.

REMARKS

R23: To produce a cycle decomposition S orthogonal to a given eulerian tour in a graph with $\deg(v) \equiv 0 \bmod 4$ for every $v \in V(G)$, one can apply a procedure developed by J. Petersen in his celebrated paper [Pe1891]: Color the edges of T alternately *blue* and *red*, and combine a cycle decomposition of the blue even graph with one of the red even graph.

R24: Fact F50 follows from Fact F43 by using the classical characterization (Fact F1).

R25: Fact F49 is basically a special case of Fact F48. We stated it separately because its converse (given an eulerian tour T, there exists a cycle decomposition orthogonal to T) is an open problem known as *Sabidussi's Compatibility Conjecture*. Its relevance to other open problems such as the Cycle Double Cover Conjecture and the Nowhere-Zero 5-Flow Conjecture is discussed in [Fl84, Fl88, Fl01, Fl02].

R26: Facts F48 and F51 show that the existence of eulerian tours satisfying certain restrictions does not necessarily imply the existence of cycle decompositions satisfying the same restrictions: Fact F48 relates to *arbitrary* loopless graphs and uses the cut condition only locally, whereas in Fact F51, the full strength of the cut condition is invoked.

R27: While Facts F51 and F52 have been formulated for planar graphs only, they can be extended to a somewhat more general class of graphs (see [Zh97]).

R28: $P(G)$-orthogonal eulerian tours in digraphs have been studied in [Fl90]. Naturally, due to the appearance of arcs instead of edges, somewhat stronger conditions than the cut condition of Definition D41 are needed to prove the existence of $P(G)$-orthogonal eulerian tours.

R29: Fact F51 can be viewed as a generalization of Fact F42 because one obtains an eulerian planar graph by replacing every $e \in E(G)$ by $f(e)$ parallel edges and by defining the incidence-partition system correspondingly.

Orderings of the Incidence Set, Non-Intersecting Tours, and A-Trails

DEFINITIONS

D42: Given a graph G and a vertex v, a fixed sequence $\langle e_1, e_2, \ldots, e_{\deg(v)} \rangle$ of the edges in the incidence set E_v is called a *positive ordering of* E_v and is denoted $O^+(v)$. If G is imbedded in some surface, one such $O^+(v)$ is given by the counterclockwise cyclic ordering of the edges incident on v.

D43: Let G be an even graph and v a vertex with $\deg(v) \geq 4$ and with a positive ordering of its incident set E_v given by $O^+(v) = \langle e_1, e_2, \ldots, e_{\deg(v)} \rangle$. A transition system $\tau(G)$ is *non-intersecting with respect to* $O^+(v)$ if for any $e_i, e_j, e_k, e_l \in E_v$ with $i < j < k < l$, $\{e_i, e_k\}$ and $\{e_j, e_l\}$ cannot both be transitions of $\tau(G)$. That is,

$$\{e_i, e_k\} \in \tau(G) \Rightarrow \{e_j, e_l\} \notin \tau(G)$$

D44: Let G be an even graph with a given positive ordering $O^+(v)$ for each $v \in V$. A transition system $\tau(G)$ is ***non-intersecting*** if $\tau(G)$ is non-intersecting with respect to $O^+(v)$ for every $v \in V$ with $\deg(v) \geq 4$. An eulerian tour T and a cycle decomposition S are non-intersecting if their corresponding transition systems, τ_T and τ_S, respectively, are non-intersecting.

D45: Let G be an eulerian graph with a given positive ordering $O^+(v)$ for each $v \in V$. An eulerian tour T is an ***A-trail*** if $\{e_i, e_j\} \in \tau_T$ implies $j = i+1$ or $j = i-1$ (modulo $\deg(v)$).

D46: An ***outerplanar graph*** is a graph with an imbedding in the plane such that every vertex appears on the boundary of the exterior face.

D47: A graph (imbedding) ***triangulates*** a surface if every region is 3-sided.

EXAMPLE

E4: An A-trail T in the octahedron, given by the sequence $1, 2, 3, \ldots, 11, 12$, is shown in Figure 4.2.8 below. A cycle decomposition orthogonal to T is given by the sets $\{2, 6, 10\}$, $\{4, 8, 12\}$, $\{1, 11, 9, 7, 5, 3\}$.

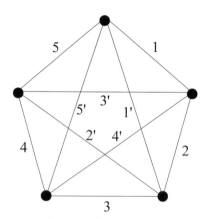

Figure 4.2.8: An A-trail in the octahedron.

FACTS

F53: Given an eulerian graph G and $O^+(v)$ for every $v \in V(G)$, a non-intersecting eulerian tour exists.

F54: In an eulerian graph with $\deg(v) \leq 4$ for every $v \in V(G)$, the concepts of non-intersecting eulerian tour and A-trail are equivalent.

F55: [AnFl95] The decision problem whether a given simple, planar, 3-connected eulerian graph has an A-trail is NP-complete.

F56: [AnFlRe98] Simple, outerplanar, eulerian graphs have A-trails; they can be constructed in polynomial time.

F57: Let G be a simple eulerian graph that triangulates the plane. Suppose that G has maximum degree $\delta_{\max}(G) \leq 8$ with at most one 8-valent vertex, which, if it exists, is adjacent to a 4-valent vertex. Then G has an A-trail.

REMARKS

R30: Facts F48, F49, and F53 can be proved by employing the Splitting Lemma (Fact F11). Consequently, algorithms for constructing eulerian tours that are based on the Splitting Lemma, can be modified so as to yield $P(G)$-orthogonal eulerian tours or non-intersecting eulerian tours.

R31: There is a 1-1 correspondence between transition systems $\tau(G)$ of the even graph G and the decompositions of $E(G)$ into closed trails: traversing edges of G following the given transitions results in closed trails, one at a time; together they form a decomposition into closed trails. Likewise, each of these trails defines a subset of $\tau(G)$ (for a given $\tau(G)$), and since these trails are edge-disjoint, the union of the subsets is $\tau(G)$.

4.2.6 Transforming Eulerian Tours

The Kappa Transformations

The kappa transformations consist of various combinations of splitting, splicing, and reversing closed trails. They form the basis for constructing eulerian tours and for transforming one eulerian tour into another. For a detailed discussion, see, e.g., [Fl90].

DEFINITIONS

D48: The **reverse of a trail** $T = \langle v_0, e_1, v_1, \ldots, e_l, v_l \rangle$ is the trail

$$T^{-1} = \langle v_l, e_l, v_{l-1}, \ldots, e_1, v_0 \rangle$$

D49: Let $T = \langle \ldots, e_i, v_i, e_{i+1}, \ldots, e_j, v_j, e_{j+1}, \ldots \rangle$ be an eulerian tour in a graph G such that $v_i = v_j$, and consequently, $\{e_i, e_{i+1}, e_j, e_{j+1}\} \subseteq E_{v_i}$. The closed subtrail $\langle v_i, e_{i+1}, \ldots, e_j, v_j \rangle$ is called a **segment** of tour T and is denoted $S_{i,j}$.

D50: A **segment reversal** (or κ-**transformation**) is the replacement of one of the segments in an eulerian tour T by its reverse segment. The resulting eulerian tour is denoted $\kappa(T)$. Thus, if tour $T = \langle \ldots, e_i, S_{i,j}, e_{j+1}, \ldots \rangle$, then

$$\kappa(T) = \langle \ldots, e_i, S_{i,j}^{-1}, e_{j+1}, \ldots \rangle$$

D51: Let $T = \langle \ldots, e_i, S_{i,j}, e_{j+1}, \ldots \rangle$ be an eulerian tour of a graph G with segment $S_{i,j}$. The 2-cell partition of $E(G)$ consisting of the edge set of $S_{i,j}$ and the edge set of the ("rest of the way around") segment $S_{j,i} = \langle v_j, e_{j+1}, \ldots, e_i, v_i \rangle$ is called a κ-**detachment** and is denoted $\kappa'(T)$.

D52: Given a trail decomposition of $E(G)$ into closed trails $T_1, \ldots, T_k$, $k \geq 2$, choose trails T_i and T_j such that $v \in V(T_i) \cap V(T_j)$ for some vertex v. Let $e_{m,i}$, $e_{m+1,i} \in E_v$ be consecutive in T_i, and $e_{n,j}$, $e_{n+1,j} \in E_v$ consecutive in T_j (i.e., they are transitions of their respective trails). Thus, we may write $T_i = \langle \ldots, e_{m,i}, v, e_{m+1,i}, \ldots \rangle$ and $T_j = \langle v, e_{n,j}, \ldots, e_{n+1,j}, v \rangle$. A **splice at** v of trail T_j into trail T_i (or the κ-**absorption** at v of T_j by T_i) is either one of the closed trails:

$$\text{splice}(T_i, T_j, v) \ = \langle \ldots, e_{m,i}, T_j, e_{m+1,i} \ldots \rangle$$
$$\text{splice}(T_i, T_j^{-1}, v) = \langle \ldots, e_{m,i}, T_j^{-1}, e_{m+1,i} \ldots \rangle$$

NOTATION: Either one of the closed trails that result from a splice of T_j into T_i is denoted $\kappa''(\{T_i, T_j\})$.

D53: Let T be an eulerian tour in a graph G. An eulerian tour T' is obtained from T by a κ^*-**transformation**, denoted $T' = \kappa^*(T)$, if there exists a κ-detachment $\kappa'(T) = \{S_{i,j}, S_{j,i}\}$ such that $T' = \kappa''\{S_{i,j}, S_{j,i}\}$. That is, $T' = \kappa^*(T) = \kappa''(\kappa'(T))$.

D54: Let T_1 and T_2 be two eulerian tours in a graph G. Tour T_2 is obtained from T_1 by a κ_1-transformation, denoted $T_2 = \kappa_1(T_1)$, if either $T_2 = \kappa(T_1)$ or $T_2 = \kappa^*(T_1)$.

D55: Two eulerian trails, T_1 and T_2, are considered different if their corresponding transition systems are different, i.e., if $\tau_{T_1} \neq \tau_{T_2}$.

REMARK

R32: The various transformations defined above carry over to eulerian digraphs with the added restriction that each transition at a vertex v must comprise an arc incident *to* v and an arc incident *from* v.

FACTS

F58: Let T_1 and T_2 be two different eulerian tours of an eulerian graph G (they exist unless G is a cycle). T_2 can be obtained from T_1 by a sequence of κ-transformations (see [AbKo80], [Sk84], [Fl90]).

F59: Let G be an eulerian graph with a partition system $P(G)$, and suppose that T_1 and T_2 are different $P(G)$-orthogonal eulerian tours. Then T_2 can be obtained from T_1 by a sequence of κ_1-transformations in such a way that any eulerian tour and any trail decomposition S with $|S| = 2$ arising in this sequence are $P(G)$-orthogonal.

F60: Let G be an eulerian graph with a given positive ordering $O^+(v)$ for every $v \in V(G)$, and let T_1 and T_2 be different non-intersecting eulerian tours of G. Then T_2 can be obtained from T_1 by a sequence of κ_1-transformations in such a way that any eulerian tour and any trail decomposition S with $|S| = 2$ arising in this sequence are non-intersecting.

F61: Let T_1 and T_2 be two different eulerian tours in a digraph G. Then tour T_2 can be obtained from T_1 by a sequence of κ_1-transformations.

F62: In 4-regular plane graphs, A-trails (which are non-intersecting eulerian tours in this case) are in 1-1 correspondence with spanning trees in an (easily constructed) auxiliary graph. The κ_1-transformations correspond to the edge-addition and edge-deletion process in transforming one spanning tree into another spanning tree.

EXAMPLES

E5: The complete bipartite graph $K_{2,4}$ with eulerian tour $T = \langle 1, 2, 3, \ldots, 8 \rangle$ (written as edge sequence) is shown in Figure 4.2.9 (a). The transitions at v and w are marked with little arcs. Tour T is transformed into the eulerian tour $T' = \langle 1, 2, 3, 4, 8, 7, 6, 5 \rangle$ by a κ-transformation (segment reversal) at v (see Figure 4.2.9 (b)).

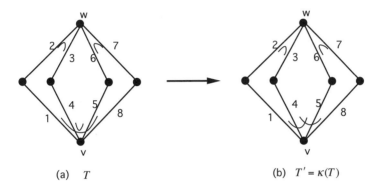

(a) T (b) $T' = \kappa(T)$

Figure 4.2.9: T and T' in Example E5.

E6: The tour T in Example E5 is a non-intersecting eulerian tour. By a κ-detachment at v, one obtains the non-intersecting trail decomposition $S = \{T_1, T_2\}$ with $T_1 = \langle 1, 2, 3, 4 \rangle$ and $T_2 = \langle 5, 6, 7, 8 \rangle$ (written as edge sequences; Figure 4.2.10 (a)). A κ-absorption at w results in $T'' = \langle 1, 2, 7, 8, 5, 6, 3, 4 \rangle$, another non-intersecting eulerian tour (Figure 4.2.10 (b)).

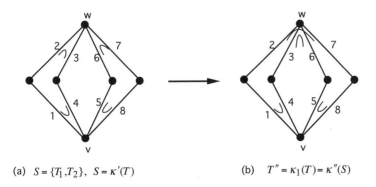

(a) $S = \{T_1, T_2\}$, $S = \kappa'(T)$ (b) $T'' = \kappa_1(T) = \kappa''(S)$

Figure 4.2.10: S and T'' in Example E6.

Splicing the Trails in a Trail Decomposition

We close the section with an eulerian-tour construction by A. Tucker that starts with a closed-trail decomposition and iteratively splices pairs of trails together (i.e., performs κ-absorptions) until there is only one trail left.

Algorithm 4.2.5: Tucker's Algorithm [Tuc76]

Input: eulerian graph G.
Output: eulerian tour T.

    Produce a trail decomposition of G by forming an arbitrary 2-regular detachment H of G.
    Let $W = \{T_1, \ldots, T_k\}$ be the set of components of H.
    While $k \geq 2$
        Choose T_i, T_j with $i \neq j$ such that $V(T_i) \cap V(T_j) \neq \emptyset$.
        Let $v \in T_i \cap T_j$.
        Let $T_{i,j} = \kappa''(T_i, T_j)$ {a κ-absorption at v}.
        $W := W \cup \{T_{i,j}\} - \{T_i, T_j\}$.
        $k := k - 1$

References

[AbKo80] J. Abraham and A. Kotzig, Transformations of Euler tours, *Ann. Discrete Math.* **8** (1980), 65–69.

[AlGoZh94] B. Alspach, L. Goddyn, and C-Q. Zhang, Graphs with the circuit cover property, *Transactions Amer. Math. Soc.* **344**, 1 (1994), 131–154.

[AnFl95] L. D. Andersen and H. Fleischner, The NP-completeness of finding A-trails in eulerian graphs and of finding spanning trees in hypergraphs, *Discrete Appl. Math.* **59** (1995), 203–214.

[AnFlRe98] L. D. Andersen, H. Fleischner, and S. Regner, Algorithms and outerplanar conditions for A-trails in plane eulerian graphs, *Discrete Appl. Math.* **85** (1998), 99–112.

[CaFl95] M-C. Cai and H. Fleischner, An eulerian trail traversing specified edges in given order, *J. Graph Theory* **19** (1995), 137–144.

[ChOe93] G. Chartrand and O. R. Oellermann, *Applied and Algorithmic Graph Theory*, Internat. Series in Pure and Applied Math., McGraw-Hill, 1993.

[EhBr51] T. van Aardenne-Ehrenfest and N. G. de Bruijn, Circuits and trees in oriented linear graphs, *Simon Stevin* **28** (1951), 203–217.

[Eu1736] L. Euler, Solutio problematis ad geometriam situs pertinentis, *Commentarii Academiae Petropolitanae* **8** (1736), 1741, 128–140 = Opera omnia Ser. I, Vol. 7, 1–10.

[Fl1883] Fleury, Deux problemes de geometrie de situation, *Journal de Mathematiques Elementaires* 1883, 257–261.

[Fl80] H. Fleischner, Eulersche Linien und Kreisüberdeckungen, die vorgegebene Durchgänge in den Kanten vermeiden, *J. Combin. Theory* **B**29, 2 (1980), 145–167.

[Fl84] H. Fleischner, Cycle decompositions, 2-coverings, removable cycles, and the 4CD, in *Progress in Graph Theory*, J. A. Bondy and U. S. R. Murty (Eds.), Academic Press, 1984, 233–246.

[Fl86] H. Fleischner, Proof of the strong 2-cover conjecture for planar graphs, *J. Combin. Theory* **B**40, 2 (1986), 229–230.

[Fl88] H. Fleischner, Some blood, sweat, but no tears in eulerian graph theory, *Congressus Numer.* **63** (1988), 8–48.

[Fl89] H. Fleischner, Elementary proofs of (relatively) recent characterizations of eulerian graphs, *First Montreal Conference on Combinatorics and Computer Science*, 1987, *Discrete Appl. Math.* **24** (1989), No. 1–3, 115–119.

[Fl90] H. Fleischner, *Eulerian Graphs and Related Topics, Part 1, Vol. 1*, Ann. Discrete Math **45** North-Holland, Amsterdam, 1990.

[Fl91] H. Fleischner, *Eulerian Graphs and Related Topics, Part 1, Vol. 2*, Ann. Discrete Math **50** North-Holland, Amsterdam, 1991.

[Fl00] H. Fleischner, *Traversing Graphs: The Eulerian and Hamiltonian Theme* in *ARC ROUTING: Theory, Solutions, and Applications*, M. Dror (Ed.), Kluwer Academic Publishers, 2000, 19–87.

[Fl01] H. Fleischner, (Some of) The many uses of eulerian graphs in graph theory (plus some applications), *Discrete Math.* **2**30 (2001), 23–43.

[Fl02] H. Fleischner, Bipartizing matchings and Sabidussi's compatibility conjecture, *Discrete Math.* **2**44 (2002), 77–82.

[FlFr90] H. Fleischner and A. Frank, On circuit decompositions of planar eulerian graphs, *J. Combin. Theory* **B**50 (1990), 245–253.

[FlGu85] H. Fleischner and M. Guan, On the minimum weighted cycle covering problem for planar graphs, *Ars Combinatoria* **2**0 (1985), 61–67.

[FlHa09] H. Fleischner and R. Häggkvist, Circuit double covers in special types of cubic graphs, *Discrete Math.* **3**09 (2009), 5724–5728.

[FlWe89] H. Fleischner and E. Wenger, D_0-favouring eulerian trails in digraphs, *Arch. Math. (Brno)* **2**5 (1989), 55–60.

[FoFu62] L. R. Ford and D. R. Fulkerson, *Flows in Networks*, Princeton University Press, Princeton, NJ, 1962.

[Hi1873] C. Hierholzer, Über die Möglichkeit, einen Linienzug ohne Wiederholung und ohne Unterbrechung zu umfahren, *Math Annalen* **VI** (1873), 30–32.

[HuKo95] A. Huck and M. Kochol, Five cycle double covers of some cubic graphs, *J. Combin. Theory* **B**64 (1995), 119–125.

[Ja75] F. Jaeger, On nowhere-zero flows in multigraphs, *Proceedings of the Fifth British Combinatorial Conference 1975*, Congr. Numer., **XV** (1975), 373–378.

[Ja79] F. Jaeger, Flows and generalized coloring theorems in graphs, *J. Combin. Theory* **B**26 (1979), 205–216.

[Ko56] A. Kotzig, Euler lines and decompositions of a regular graph of even order into two factors of equal orders (in Slovak), *Mat.-Fyz. Časopis Slovensk. Akad.* **6** (1956), No. 3, 133–136.

[Ko68] A. Kotzig, Moves without forbidden transitions in a graph, *Mat.-Fyz. Časopis* **18** (1968), No. 1, 76–80.

[Lu1894] M. É. Lucas, *Récréations Mathématiques IV*, Gauthiers–Villars et fils, Paris, 1894.

[Mc84] T. A. McKee, Recharacterizing eulerian: Intimations of new duality, *Discrete Math.* **51** (1984), 237–242.

[Na79] C. St. J. A. Nash-Williams, Acyclic detachments of graphs, in *Graph Theory and Combinatorics*, R. J. Wilson (Ed.), Proc. Conf. Open University; Milton Keynes, 1978, *Res. Notes in Math.* **34**, Pitman, San Francisco (1979), 87–97.

[Na85a] C. St. J. A. Nash-Williams, Detachment of graphs and generalized Euler trails, in *Surveys in Combinatorics*, I. Anderson (Ed.), 1985, *Math. Soc. Lecture Notes Ser.* **103**, Cambridge University Press, London, (1985), 137–151.

[Na85b] C. St. J. A. Nash-Williams, Connected detachments of graphs and generalized Euler trails, *J. London Math. Soc.* **31** (1985), No. 2, 17–29.

[Pe1891] J. Petersen, Die Theorie der regulären Graphs, *Acta Math.* **15** (1891), 193–220.

[Sa77] G. Sabidussi, Tracing graphs without backtracking, in *Methods of Operations Research XXV, Part 1*, Henn et al. (Eds.), First Symp. on Oper. Res., Univ. Heidelberg, Sept. 1–3, 1976.

[Se79] P. D. Seymour, Sums of circuits, in *Graph Theory and Related Topics*, J. A. Bondy and U. S. R. Murty (Eds.), Academic Press, New York 1979, 341–356.

[Se81a] P. D. Seymour, Nowhere-zero 6-flows, *J. Combin. Theory* **B**30 (1981), 130–135.

[Se81b] P. D. Seymour, Even circuits in planar graphs, *J. Combin. Theory* **B**31 (1981), 327–338.

[Sh79] H. Shank, Some parity results on binary vectors spaces, *Ars. Combin.* **8** (1979), 107–108.

[Sk84] D. K. Skilton, Eulerian chains and segment reversals, in *Graph Theory, Proc. First Southeast Graph Theory Colloq.*, K. M. Koh and H. P. Yap (Eds.), Singapore, May 1983, Lecture Notes in Math. **1073**, Springer, Berlin-New York, 1984, 228–235.

[Ta1895] G. Tarry, Le problème des labyrinthes, *Nouv. Ann. Math.* 14 (1895), 187–190.

[Tar86] M. Tarsi, Semi-duality and the cycle double cover conjecture, *J. Combin. Theory*, **B**41 (1986), 332–340.

[Th85] C. Thomassen, Retracting-free double tracings of graphs, *Ars Combinatoria* **19** (1985), 63–68.

[To73] S. Toida, Properties of a Euler graph, *J. Franklin Inst.* **295** (1973), 343–345.

[Tr66] D. J. Troy, On traversing graphs, *Amer. Math. Monthly* **73** (1966), 497–499.

[Tuc76] A. Tucker, A new applicable proof of the Euler circuit theorem, *Amer. Math. Monthly* **83** (1976), 638–640.

[Tu54] W. T. Tutte, A contribution to the theory of chromatic polynomials, *Canad. J. Math.* **6** (1954), 80–91.

[TuSm41] W. T. Tutte and C. A. B. Smith, On unicursal paths in a network of degree 4, *Amer. Math. Monthly* **48** (1941), 233–237.

[Ve12] O. Veblen, An application of modular equations in analysis situs, *Ann. of Math* (2)14 (1912/13), 86–94.

[Ve31] O. Veblen, Analysis situs, *Amer. Math. Soc. Colloq. Publ.* **5**, Part II (1931), 1–39.

[Ve75] P. D. Vestergaard, Doubly traversed Euler circuits, *Arch. Math.* **26** (1975), 222–224.

[Wo90] D. R. Woodall, A proof of McKee's eulerian-bipartite characterization, *Discrete Math.* **84** (1990), 217–220.

[Xu02] J. Xu, *Topological Structure and Analysis of Interconnection Networks*, Network Theory and Applications **7**, Kluwer Academic Publishers, Dordrecht, 2002.

[Zh97] C-Q. Zhang, *Integer Flows and Cycle Covers of Graphs*, Marcel Dekker Inc., New York, 1997.

[Zh12] C-Q. Zhang, *Circuit Double Cover of Graphs*, London Math. Soc. Lecture Notes Ser. 399, Cambridge University Press, New York, 2012.

Section 4.3
Chinese Postman Problems

R. Gary Parker, Georgia Institute of Technology
Richard B. Borie, University of Alabama

INTRODUCTION

The *Chinese Postman Problem* (CPP) is one of the more celebrated problems in graph optimization. It acts as a useful model in an array of practical contexts such as refuse collection, snow removal, and mail delivery. The basic problem was first posed by the mathematician Guan (or Kwan Mei-Ko) in 1962 [Gu62]; hence the problem was dubbed "Chinese" by Jack Edmonds ([Ed65-a]), based on a suggestion by his supervisor Alan Goldman at the U.S. National Bureau of Standards.

4.3.1   The Basic Problem and Its Variations

DEFINITIONS

D1:   A *postman tour* in a graph G is a closed walk that uses each edge of G *at least* once.

D2:   Given a finite graph, $G = (V, E)$, with edges weighted as $w : E \to R^+$, the *Chinese Postman Problem* seeks a minimum-weight postman tour.

TERMINOLOGY NOTE: The basic problem is sometimes simply referred to as a *postman problem*.

D3:   The *undirected version of CPP (UCPP)* assumes that the instance graph G is an undirected graph.

D4: The ***directed version of CPP (DCPP)*** assumes that the instance graph G is a digraph.

D5: The ***mixed version of CPP (MCPP)*** assumes that the instance graph G is a *mixed graph*, that is, some edges are directed, and some edges are undirected.

FACTS

F1: UCPP and DCPP are polynomial-time solvable. (See §4.3.2 and §4.3.3.)

F2: MCPP is $\mathcal{NP}$-hard. (See §4.3.4.)

The Eulerian Case

DEFINITIONS

D6: An ***eulerian tour*** in a graph (or digraph) G is a closed walk that uses each edge (or arc) of G *exactly* once. An eulerian tour in a mixed graph is a closed walk that uses each edge and each arc exactly once.

D7: A graph, digraph, or mixed graph that has an eulerian tour is called ***eulerian***.

D8: A digraph G is ***strongly connected*** if for every two of its vertices, u and v, there is a directed walk from u to v and one from v to u.

D9: If $G = (V, E)$ is a digraph, a vertex $v \in V$ is ***symmetric*** if the indegree and outdegree of v are equal.

TERMINOLOGY NOTE: If every vertex in a digraph G is symmetric, then G is sometimes referred to as a ***symmetric digraph***. However, in other contexts, "symmetric digraph" is sometimes taken to mean $(x, y) \in E \to (y, x) \in E$ for all $x, y \in V$.

FACTS

F3: A connected graph G is eulerian if and only if every vertex of G has even degree.

F4: A strongly connected digraph G is eulerian if and only if G is symmetric.

F5: If the instance graph G (undirected, directed, or mixed) is eulerian, then CPP is solved by producing an eulerian tour.

REMARK

R1: A characterization of eulerian mixed graphs is given later in the subsection Mixed Postman Problems.

Variations of CPP

DEFINITIONS

D10: *open postman tour*: The postman is required to start and end at distinct vertices of the graph (or digraph).

D11: *not requiring a specified edge*: A specified edge is not *required* to be in an admissible tour but its inclusion is at least permitted.

D12: *a specified edge cannot be duplicated*: A specified edge is required to be present in the postman tour but cannot be duplicated (i.e., cannot be traversed more than once).

D13: *windy postman problem*: Instances of UCPP place no restriction on the direction of traversal along an edge. This does not suggest, however, that in a practical application, the postman need necessarily experience the same "cost" of traversal in both directions (suppose the edge-weight metric that is relevant is not distance but rather time). If one allows edge weights to differ depending upon which direction an edge is traversed, the problem becomes the *windy postman problem*.

D14: *rural postman problem*: This variant, also motivated by practical settings, arises when rather than requiring that all edges or arcs be traversed at least once, only a given subset has to be used. This version derives its name from the apparent case of postal delivery in non-urban settings where, perhaps, the postman may have to traverse every street within a small town or village and then move on to another one but can do so by selecting any of a number of connecting roads (edges) that exist to connect the towns.

D15: *stacker crane problem*: This is the rural postman problem for mixed graphs.

FACTS

F6: The open postman problem remains polynomial-time solvable. If v_1 and v_2 are the pre-specified source and destination vertices, then simply add to G an artificial edge from v_2 to v_1 and assign the new edge a weight of M, where M is sufficiently large. Clearly, in the application of the algorithm, this artificial edge would never be part of a shortest path and hence, would never be duplicated. In the resultant $\overline{G}$, one would simply find an optimal postman tour and then remove the artificial edge from this tour to obtain the desired open postman tour from v_1 to v_2.

F7: For the second and third variations above, where a specified edge is not required or cannot be duplicated, the problem remains polynomial-time solvable (cf. [EdJo73]).

F8: [Wi89] The windy postman problem is $\mathcal{NP}$-hard although solvable in polynomial time if instances are eulerian.

F9: The rural postman problem is $\mathcal{NP}$-hard on both graphs and digraphs even if all edges/arcs have the same weight (see [GaJo79]).

REMARK

R2: It is important to note that the list of extensions presented here is not exhaustive. Additional variations to these sorts of problems in general are often easy to create whether motivated by purely combinatorial interests or ones more pragmatic, stemming from a given practical setting. A good starting place for a sense of the breadth of cases, degree of analysis, and categorization of results is the rather expansive survey in [EiGeLa95-a] and [EiGeLa95-b]. Extensive coverage for the Chinese Postman Problem and its variations may also be found in [Fl91] and [Dr00].

4.3.2 Undirected Postman Problems

The solution posed by Guan, though clever, was not fast in the universally adopted, complexity-theoretic sense (i.e., not polynomial time in the size of the input graph). This flaw was pointed out by Edmonds ([Ed65-a]), who then proposed a polynomial-time algorithm for the problem.

DEFINITIONS

D16: A *matching M* in a graph G is a subset of edges no two of which have a common vertex. (Matchings are discussed in §11.3.)

D17: A matching is *perfect* if every vertex in G is incident to some edge in the matching.

Algorithm 4.3.1: Solution to UCPP

Input: Connected graph G with positive edge weights.
Output: Minimum-weight postman tour in G.
   Let V_O be the set of vertices with odd degree in G.
   For each pair of vertices $x, y \in V_O$
      Find a shortest path P in G between x and y.
   Form a complete graph K on the vertex set V_O with edges weighted by
      the respective shortest-path lengths.
   Find a minimum-weight perfect matching M in K.
   For each edge $e \in M$
      Duplicate the edges in G of the shortest path P corresponding to e.
   Let $\overline{G}$ be the resulting super(multi)graph.
   Produce an eulerian tour in $\overline{G}$.

COMPUTATIONAL NOTE: The shortest path computation in Algorithm 4.3.1 is straightforward and fast. Producing a minimum-weight perfect matching in a graph, although complicated, can be accomplished in polynomial time following the seminal work by Edmonds (cf. [Ed65-b],[Ed65-c]). The implementation of the step relative to traversal-finding in $\overline{G}$ is also easy. The following strategy by Fleury (cf. [Ka67]) can be applied recursively: Given a position in the walk, select the next edge arbitrarily so long as its removal would not disconnect the graph $\overline{G}$, *unless* this is the only choice. (Fleury's algorithm appears in §4.2.)

EXAMPLE

E1:   Consider G on the left in Figure 4.3.1; weights are specified directly on the edges. The set V_O is given by $\{v_1, v_3, v_4, v_5\}$ and the stated shortest paths along with the path lengths result as follows:

Vertex Pair	Path	Length
v_1, v_3	v_1, e_1, v_2, e_2, v_3	3
v_1, v_4	$v_1, e_1, v_2, e_7, v_5, e_4, v_4$	5
v_1, v_5	v_1, e_1, v_2, e_7, v_5	3
v_3, v_4	v_3, e_3, v_4	3
v_3, v_5	v_3, e_2, v_2, e_7, v_5	2
v_4, v_5	v_4, e_4, v_5	2

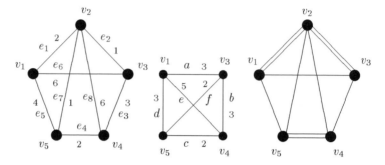

Figure 4.3.1: Application of Algorithm 4.3.1.

An optimal matching in the complete graph K_4, shown in the middle of Figure 4.3.1, consists of edges a and c, having total weight 5. These edges correspond to the paths v_1, e_1, v_2, e_2, v_3 and v_4, e_4, v_5. The respective edges in these paths are duplicated in G, producing the multigraph, $\overline{G}$, shown on the right in Figure 4.3.1. The latter is eulerian, and an eulerian tour (of total weight 30) is given by the walk below:

$$v_1, e_1, v_2, e_2, v_3, e_2, v_2, e_1, v_1, e_6, v_3, e_3, v_4, e_4, v_5, e_7, v_2, e_8, v_4, e_4, v_5, e_5, v_1$$

REMARKS

R3:   The stipulation that edges be weighted by positive values cannot be relaxed, because otherwise negative weight closed walks will occur in G (by simply going back and forth on such an edge), which creates an intractability in the shortest path computation (see [GaJo79]).

R4:   Implementation of the Fleury traversal strategy requires some attention due, largely, to the requirement to test the stated connectivity stipulation. An alternative that relaxes this complication was proposed by Edmonds and Johnson ([EdJo73]).

R5:   Trivially, a necessary condition for a perfect matching to exist in a graph is that the graph possesses an even number of vertices. Since we may take G, in any interesting instance of UCPP, to be connected, there must be a path between every pair of vertices. Hence the complete graph specification is clear and since $|V_O|$ is even, it follows that the perfect matching step of the procedure is well-defined.

R6:   Assuming a correct application of Algorithm 4.3.1, a postman would never traverse an edge more than twice in an optimal walk.

COMPUTATIONAL NOTE: If double-traversing occurred, the supergraph, $\overline{G}$, would have an edge from G duplicated more than once. But this would deny that $\overline{G}$ had been constructed correctly since two of the duplicated copies could be eliminated, leaving a connected graph with the same (even) degree parity everywhere and with smaller weight than $\overline{G}$.

4.3.3   Directed Postman Problems

The strategy for solving the DCPP is analogous to the one used for the undirected case. If the digraph is not symmetric, then a minimum-weight arc duplication produces a symmetric super(multi)digraph. The number of copies of each arc is determined by solving a *circulation* problem.

FACTS

F10:   Since easy to test, we may take G to be strongly connected.

F11:   The multigraph $\overline{G}$ produced from a correct application of Algorithm 4.3.2 below is symmetric; obviously it remains strongly connected.

COMPUTATIONAL NOTE: The circulation problem in Algorithm 4.3.2 is easily solved by standard *network flow techniques* (Chapter 11).

Algorithm 4.3.2: Solution to DCPP

Input: Strongly connected digraph G with positive arc weights.
Output: Minimum-weight postman tour in G.
    If G is symmetric
        Produce an eulerian tour in G.
    Else
        For each vertex v_k, set $b_k = indegree(v_k) - outdegree(v_k)$.
        Solve the following circulation problem:

$$\text{minimize} \quad \sum_{(v_i,v_j)\in E} w_{ij}x_{ij}$$

$$\text{s.t.} \quad \sum_{(v_k,v_j)\in E} x_{kj} - \sum_{(v_i,v_k)\in E} x_{ik} = b_k \text{ for } v_k \in V$$

$$x_{ij} \geq 0$$

        For each pair of vertices v_i, v_j, add x_{ij} copies of arc (v_i, v_j) to G.
        Call the resulting super(multi)digraph $\overline{G}$.
        Produce an eulerian tour in $\overline{G}$.

Producing an Eulerian Tour in a Symmetric (Multi)Digraph

Algorithm 4.3.2 requires the traversal of an eulerian tour in the original digraph (if it is symmetric) or in a symmetric multidigraph. This can be accomplished by applying Algorithm 4.3.3 below (cf. [EhBr51]).

DEFINITION

D18: An *intree* is a connected, acyclic digraph where the outdegree of every vertex is at most 1.

Algorithm 4.3.3: Producing a Tour in an Eulerian Digraph

Input: Eulerian digraph G.
Output: Eulerian tour in G.
    Select any vertex in G and denote it by v^*.
    Form an intree T that spans G and that is rooted at v^*.
    For each vertex w in G, $w \neq v^*$,
        Label the out-arcs from w randomly with consecutive integers
            subject to the restriction that the last (highest) label is given
            to the arc in intree T.
    Label the out-arcs from v^* arbitrarily.
    Starting at vertex v^*, trace an eulerian tour in G by always selecting
        the untraversed out-arc with the smallest label.

COMPUTATIONAL NOTE: An easy way to form an intree T of digraph G is to start with $T = \{v^*\}$ and proceed iteratively: select at each iteration an arc in G that is directed from a vertex in $V(G) - V(T)$ to a vertex in T; repeat until T spans G.

EXAMPLE

E2: Consider the instance digraph in the upper left of Figure 4.3.2. Specified beside each vertex is the respective value for b_k. Solving the explicit circulation problem defined in Algorithm 4.3.2 produces the following outcome: $x_{51} = x_{34} = 1$; $x_{45} = 2$; and $x_{ij} = 0$ elsewhere. Copies of the respective arcs are added, forming the multigraph shown in the upper right of the figure. Applying Algorithm 4.3.3 and selecting (arbitrarily) vertex v_4 as a root, an intree T is constructed and shown at the bottom of Figure 4.3.2. The stated arc-labeling scheme is applied with labels affixed to the arcs in the multidigraph. Starting with vertex v_4 and proceeding in label order produces an eulerian tour specified (unambiguously) by the following vertex sequence:

$$v_4, v_5, v_1, v_3, v_4, v_5, v_1, v_2, v_3, v_4, v_5, v_2, v_4$$

REMARKS

R7: The network flow formulation in Algorithm 4.3.2 is due to Edmonds and Johnson ([EdJo73]).

R8: Trivially, a correct application of Algorithm 4.3.2 may require that an arc be duplicated several times.

R9: In applying Algorithm 4.3.3, the requirement that a tour be traced beginning with vertex v^* and proceeding in label order cannot be casually relaxed. For instance, if one starts from vertex v_3 on the labeled digraph in the upper right in Figure 4.3.2, any tour generated will violate the label ordering.

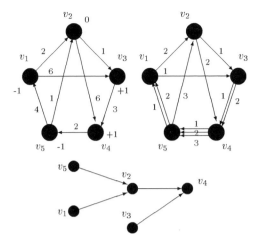

Figure 4.3.2: Applications of Algorithms 4.3.2 and 4.3.3.

4.3.4   Mixed Postman Problems

FACTS

F12:   The mixed postman problem, MCPP, is $\mathcal{NP}$-hard; the reduction is from the 3-satisfiability problem (cf. [Pa76]). Papadimitriou's reduction makes use of a widget (a special subgraph) that has the following property: any optimal postman tour must traverse the widget in one of only two possible ways. The reduction creates one copy of the widget for each variable appearance within the 3-satisfiability instance, and the two possible ways to traverse the widget correspond to this variable having value 0 or 1. The remainder of the reduction provides edges that connect the various copies of the widget to ensure consistency of the values chosen for all the appearances of the same variable. See [Pa76] for further details.

F13:   MCPP remains $\mathcal{NP}$-hard even on planar graphs with no vertex (total) degree exceeding 3 and with all edge weights the same (see [Pa76]).

Deciding if a Mixed Graph Is Eulerian

DEFINITIONS

D19:   The **total degree** of a vertex v in a mixed graph G is the total number of arcs and undirected edges incident on v.

D20:   A mixed graph is **even** if the total degree of each of its vertices is even.

D21:   A vertex in a mixed graph is **symmetric** if its indegree and outdegree are equal.

TERMINOLOGY NOTE: A mixed graph is said to be **symmetric** if all of its vertices are symmetric.

D22: A mixed graph G satisfies the **balance condition** if for every $S \subseteq V(G)$, the difference between the number of arcs from S to $V(G) - S$ and the number of arcs from $V(G) - S$ to S is no greater than the number of undirected edges joining vertices in S and $V(G) - S$ (cf. [FoFu62]).

FACTS

F14: A (strongly) connected, mixed graph G is eulerian if and only if G is even and satisfies the balance condition.

F15: Mixed graphs that are even and symmetric are balanced.

EXAMPLE

E3: Clearly, the even-degree condition is necessary for a mixed graph to be eulerian while symmetry at each vertex is not. The graph in Figure 4.3.3 illustrates. An eulerian tour is specified by the vertex sequence $v_1, v_5, v_6, v_1, v_2, v_3, v_4, v_5, v_3, v_1$.

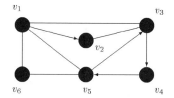

Figure 4.3.3: An eulerian mixed graph.

COMPUTATIONAL NOTE: The (nontrivial) requirement in the mixed-graph case is to create a graph that satisfies the symmetry condition at each vertex or show that this is not possible. That is, we seek to orient some undirected edges in such a way that symmetry is created, albeit artificially. There is an easy network flow formulation that will do this or correctly conclude that no such orientation is possible.

EXAMPLE

E4: The application of Algorithm 4.3.4 on the (mixed) graph in Figure 4.3.3 is illustrated in Figure 4.3.4. In the upper graph in the figure, the values b_k are written beside each vertex. The non-zero variables x_{23} and x_{31} induce the specified orientation for the original, undirected edges (v_2, v_3) and (v_1, v_3), as indicated by the lower graph in the figure. The existing eulerian tour can be found using Algorithm 4.3.3 (see Remark R11 below).

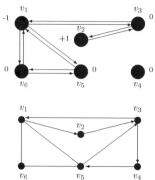

Figure 4.3.4: Application of Algorithm 4.3.4 on the graph in Figure 4.3.3.

Algorithm 4.3.4: Deciding if a Mixed Graph Is Eulerian

Input: An even and strongly connected mixed graph G.
Output: An orientation of some or all of the undirected edges of G that is
            eulerian or a conclusion that no such orientation is possible.
    For each vertex v_k, set $b_k = indegree(v_k) - outdegree(v_k)$.
    Replace each undirected edge in G by a pair of oppositely oriented arcs.
    Let U be the set of these new pairs of arcs.
    Solve the following network flow problem P_s:

$$\text{minimize} \quad \sum_{(v_i,v_j)\in U} x_{ij}$$

$$\text{s.t.} \quad \sum_{(v_k,v_j)\in U} x_{kj} - \sum_{(v_i,v_k)\in U} x_{ik} = b_k \text{ for } v_k \in V$$

$$0 \le x_{ij} \le 1 \text{ for } (v_i, v_j) \in U$$

    If P_s has an admissible solution (i.e., G is eulerian)
        For each undirected edge $\{v_i, v_j\}$
            If $x_{ij} = 1$
                Orient edge $\{v_i, v_j\}$ so that it is directed from v_i to v_j.
            Else if $x_{ji} = 1$
                Orient edge $\{v_i, v_j\}$ so that it is directed from v_j to v_i.
            Else
                Leave edge $\{v_i, v_j\}$ undirected.
    Else (P_s has no admissible solution)
        Conclude that G cannot be made eulerian through edge orientation.

REMARKS

R10:   The second part of the balance condition in Definition D22 asks that for every subset of vertices, a lack of symmetry (a difference between total indegree and total outdegree of vertices in the subset) must be made up for by some or all of the undirected edges joining vertices in the subset to those outside.

R11:   If the orientation produced by Algorithm 4.3.4 results in a graph with all edges directed, then an eulerian tour is produced by employing the strategy described previously in the case of eulerian digraphs (Algorithm 4.3.3). Alternatively, if Algorithm 4.3.4 outputs a graph with some undirected edges remaining, proceed as follows: First, for each connected component induced by the remaining undirected edges, find an eulerian tour of that component, and then orient each edge in the direction it is traversed. Now all the edges in the original mixed graph G have been oriented, so apply Algorithm 4.3.3 to the resulting digraph (cf. [EdJo73]).

The Postman Problem for Mixed Graphs

Since MCPP is $\mathcal{NP}$-hard in general, options are few. We may have to look for special cases that do submit to polynomial-time resolution, or we will simply have to be less ambitious and settle for *approximation algorithms*, i.e., fast procedures that cannot guarantee optimal solutions but that will produce ones that are, in some well-defined sense, reasonably close to optimal. Of course, for instances of manageable size, it might be feasible to resort to exact procedures. However, these approaches are inherently enumerative and will require effort that is exponential in the worst case (cf. [EiGeLa95-a], [EiGeLa95-b]). Such algorithms typically model the MCPP as an integer linear program, and then solve this ILP by adapting standard techniques such as the *cutting-plane method* (cf. [NoPi96], [Dr00]).

DEFINITION

D23:   An algorithm is an ***approximation algorithm*** for a given problem if given any instance of the problem, it finds at least a ***candidate*** solution for the instance.

REMARK

R12:   If the instance for MCPP is at least even but perhaps not symmetric, we can apply Algorithm 4.3.4 in order to test if symmetry at each vertex can be created. If so, the instance is eulerian (by Facts F14 and F15), and we can proceed accordingly. Otherwise, it is not eulerian, and we have to determine if it can be made so through some duplication of edges and/or arcs.

FACT

F16:   A mixed graph G has a postman tour if and only if G is strongly connected.

COMPUTATIONAL NOTE: Testing for strong connectivity in mixed graphs can be done in polynomial time, because each undirected edge can be replaced by a pair of oppositely directed arcs, and then the (polynomial-time) algorithm for digraphs will apply.

REFERENCE NOTE: The graphs that are employed in the remaining figures are either explicitly drawn from or are alluded to in an important paper by Frederickson ([Fr79]).

EXAMPLE

E5:   To illustrate the problematical aspect unique to the mixed postman problem, consider the mixed graph in Figure 4.3.5, part a; all edges are assumed to have weight 1. It is easy to see that no orientation exists for undirected edges that would create symmetry. Now, duplication of two arcs creates symmetry, as shown in part b of the figure; however, the resulting structure is not even so further duplication is required. On the other hand, the multigraph in part c also has only two arcs duplicated but is both symmetric *and* even; clearly, this graph is preferred. Unfortunately, it is not easy to distinguish, in any general way, its selection over the structure of part b.

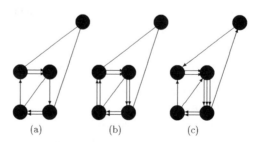

(a)   (b)   (c)

Figure 4.3.5: Interaction between symmetry and even-degree.

COMPUTATIONAL NOTE: It is possible to deal with a few of these complications, albeit in somewhat ad hoc fashion, by employing various network flow formulations; relevant results are discussed in [EdJo73]. It is important to note that the prime contributor to the intractability of the general, mixed postman case is the ambiguity in effectively dealing with the interaction between symmetry and even-degree creation and/or preservation.

Approximation Algorithm ES

The following approximation algorithm combines an easy even-degree-creation phase followed by a more intricate, joint symmetry-producing/even-degree-preserving phase. The details are based on solving a min-cost flow problem which is somewhat complicated and beyond the scope of this section (cf. [EdJo73] and [Fr79]), and so in the statement of Algorithm 4.3.5, the step is simply referenced as "symmetric/even-parity."

Algorithm 4.3.5: Approximation Algorithm ES

Input: Strongly connected, mixed graph G with positive edge/arc weights.
Output: Admissible postman tour.
  Apply the even-degree-creation component of UCPP to the underlying graph
    of mixed graph G.
  Restore orientation to edges as specified in G.
  Let $\overline{G}$ be the resulting super(multi)graph.
  Operating on $\overline{G}$, apply the symmetric/even-parity construction.
  Let $\overline{\overline{G}}$ be the resulting graph.
  Produce an eulerian tour in $\overline{\overline{G}}$.

TERMINOLOGY NOTE: The approximation procedure stated by Algorithm 4.3.5 is sometimes referred to as the ***even-symmetric*** strategy, i.e., ***ES***.

REMARK

R13: Since it cannot guarantee an optimal solution, it is interesting to consider the limit (if any) to *how* poorly Algorithm ES could perform. In fact, this was answered by Frederickson.

FACT

F17: [Fr79] The ratio of the value of a postman solution produced by Algorithm ES to an optimal value cannot exceed 2. Importantly, the value of 2 is approachable as established by Example E6.

EXAMPLE

E6: Consider the mixed graph in Figure 4.3.6, part a, where edge weights are specified on the graph. The even-degree-creation phase of Algorithm ES duplicates the directed edges (considered undirected for the stated step), yielding the multigraph in part b. Operating on this graph to produce symmetry while preserving the even degree condition yields the structure in part c having total edge weight $4 + 12\epsilon$. The eulerian tour in this multigraph is not optimal, however. Had one been less greedy in the even-degree-creation application, duplicating instead the edges with weight 2ϵ, the structure in part d of the figure would have resulted, yielding an optimal multigraph and hence, a correct tour directly. Its weight is $2 + 10\epsilon$.

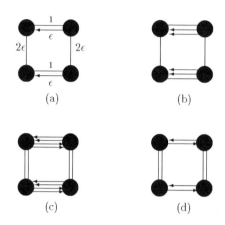

Figure 4.3.6: Application of Approximation Algorithm ES.

Approximate Algorithm SE

A natural alternative approximation is to reverse the strategy proposed by Algorithm ES, yielding the following, *symmetric-even* approach (*SE*) (cf. [Fr79]).

Algorithm 4.3.6: Approximation Algorithm SE

Input: Strongly connected, mixed graph G with positive edge/arc weights.
Output: Admissible postman tour.
    Create symmetry on mixed graph G, and denote resulting mixed graph G^s.
    Let H be the subgraph induced on the undirected edges of G^s.
    Apply the even-degree-creation component of UCPP to H.
    Let $\overline{G}$ be the resulting even-degree, symmetric super(multi)graph of G^s.
    Produce an eulerian tour in $\overline{G}$.

COMPUTATIONAL NOTE: The symmetry-creation construction of step 1 in Algorithm 4.3.6 takes polynomial time (cf. [Fr79]) and employs the same symmetry-creation component of the symmetry/even-parity step in Algorithm 4.3.5. Recall this step is based on solving a min-cost flow problem, but again we omit the details here.

EXAMPLE

E7:   Consider the graph in Figure 4.3.7, part a. A correct application of Algorithm SE produces the multigraph in part b. However, the structure in part c is optimal. The tour obtained using Algorithm SE has weight $4 + 2\epsilon$, whereas the optimal tour has weight $2 + 3\epsilon$.

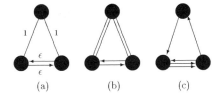

Figure 4.3.7: Application of Approximation Algorithm SE.

Some Performance Bounds

FACT

F18: [Fr79] The ratio of the length of an outcome from Algorithm SE to an optimal tour value will also never exceed 2. Example E7 provides evidence that this value is approachable as well.

EXAMPLE

E8: Proceeding from left to right in Figure 4.3.8 on the respective input instances, it is evident that Algorithm SE solves a worst-case instance for Algorithm ES, while the latter achieves the same outcome on a worst-case instance for Algorithm SE.

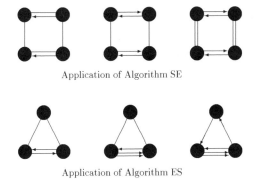

Application of Algorithm SE

Application of Algorithm ES

Figure 4.3.8: Algorithms ES and SE on each other's worst-case instances.

REMARKS

R14: In the worst-case sense, both approximation procedures Algorithm 4.3.5 (ES) and Algorithm 4.3.6 (SE) perform the same. However, Example E8 demonstrates an interesting phenomenon: if each approximation is applied to a worst-case instance of the other, the outcome is that Algorithm ES solves (to optimality) the worst-case instance for Algorithm SE, while the latter, when operating on the worst-case instance for Algorithm ES, produces the optimal outcome.

R15: The outcome of Example E8 motivates an obvious question which is stated loosely as follows: What if Algorithms ES and SE realize their respective worst-case behaviors on different classes of graphs? If this is the case, it is conceivable that they could be employed in a "composite" fashion where each strategy is applied separately and the best outcome is then selected. Since each algorithm runs in polynomial time, the total time required to separately run both algorithms is also polynomial. Preserving polynomial time is meaningful because if this composite strategy is applied, the outcome does indeed yield an improvement in guaranteed performance. The first result of this sort is also due to Frederickson ([Fr79]) where it was shown that applying the two stated heuristics and selecting the best result would never produce a tour having length that when compared to an optimal value yielded a ratio in excess of $\frac{5}{3}$.

R16:   In the same paper ([Fr79]), Frederickson also proposed a separate composite strategy for planar instances. The bound on its performance was shown to be $\frac{3}{2}$.

R17:   When the $\frac{5}{3}$ result in [Fr79] appeared, attempts to create an instance establishing realizability were not fruitful. The closest was a $\frac{3}{2}$-inducing instance shown in Figure 4.3.9; here the optimal tour has weight $4 + 13\epsilon$, but Algorithm ES yields a tour with weight $6 + 15\epsilon$, and Algorithm SE yields a tour with weight $6 + 12\epsilon$. Eventually, Raghavachari and Veerasamy (RaVe99]) were able to employ a modification of the stated Frederickson approximation to obtain a performance ratio bounded by $\frac{3}{2}$. The instance in Figure 4.3.9 establishes tightness.

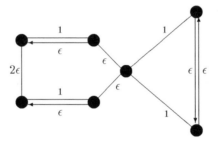

Figure 4.3.9: A worst case for composite use of Algorithms ES and SE.

R18:   Following its attendant proof of intractability, MCPP is not likely to submit to any fast solution. However, as with any other provably $\mathcal{NP}$-hard problem, this attribute does not preclude the existence of nor the value in pursuing special cases which might prove to be quickly solvable. This is certainly the case for MCPP.

COMPUTATIONAL NOTE: If input instances are confined to the class of *recursively structured graphs*, then it is possible to solve MCPP on members of this class and, in fact, by strategies requiring only linear-time effort (cf. [Co90], [ArLaSe91], [BoPaTo91], [BoPaTo92], [CoMo93], and §10.4). Typical recursive graph classes include *trees, series-parallel graphs, Halin graphs, partial k-trees, bandwidth-k graphs, pathwidth-k graphs,* and *treewidth-k graphs.*

4.3.5   Recent Research

All the results presented in the previous subsections of this section were obtained prior to the year 2000. Those results established a strong foundation for what today remains an active research area. Next we briefly summarize the current status of research in this area.

DEFINITIONS

In addition to the (classical) variations of CPP given previously in Definitions D2 to D5 and also in Definitions D10 to D15, many additional variations have been developed to more closely match the constraints and objectives of certain real-world applications. Several such variations are defined below. As before, each variation can be applied to either an undirected or directed or mixed graph. Some of these variations were defined fairly recently, while others were defined long ago but have recently attracted greater attention.

D24: *generalized CPP*: The edges are partitioned into subsets, and at least one edge from each subset must be traversed.

D25: *hierarchical CPP*: Each edge has both a service time and a cruising time. The edges are partitioned into subsets, and a precedence relation ($\prec$) is given between these subsets. If $E_i \prec E_j$ then all edges of E_i must be serviced before any edges of E_j are serviced. (See [EiGeLa95-a], [EiGeLa95-b].)

D26: *time-dependent CPP*: The cost to traverse each edge (x, y) depends on the time t of departure from x, and is given by a function $W_t(x, y)$.

D27: *time-constraint CPP*: Each edge (x, y) must be serviced no earlier than a specified start time $S(x, y)$ and/or no later than a specified finish time $F(x, y)$.

D28: *k-CPP*: Generalizes the CPP to use $k \geq 2$ postmen, and designates a particular *depot* vertex. Every postman's tour must include the depot vertex, and the goal is to minimize the total weight of the k tours.

D29: *min-max k-CPP*: Same scenario as the k-CPP, except here the goal is to minimize the maximum weight incurred by any of the k tours. (See [FrHeKi78].)

D30: *capacitated CPP*: Generalizes the k-CPP such that every postman's vehicle has capacity C, each edge has both a weight and a demand, and the sum of the demands of each tour is at most C.

D31: *capacitated arc routing problem (CARP)*: Generalizes the capacitated CPP such that only a specified subset of edges must be included (as in the rural postman problem). (See [GoWo81], [AsGo95], [Wo08], [CoPr10].)

D32: *generalized routing problem (GRP)*: Generalizes the CARP such that both a required subset of edges and a required subset of vertices may be specified.

REMARKS

R19: Most recent papers in this area typically select one or more ($\mathcal{NP}$-complete) problem variations such as those given above, and then provide one or more of the following types of results:

- Describe real-world applications to motivate the choices made in the definition of the problem.
- Formulate the problem using a precise mathematical programming model, for example, an integer linear program.

- Design an optimal algorithm that is guaranteed to produce an exact solution (but may take exponential time). Often this is accomplished using one of these methods: branch-and-bound, cutting planes, or a hybrid (branch-and-cut).

- Design an approximation algorithm that is guaranteed to produce a solution whose value is always within some known ratio of optimal.

- Design a heuristic algorithm that has no guarantees, but that is expected to usually perform well in practice. This may be accomplished using genetic algorithms, simulated annealing, ant-colony algorithms, or other machine learning or ad hoc approaches.

- Implement two or more proposed algorithms for the same variation, and perform experiments to empirically compare their performance. This may be done using benchmark data or by generating large random instances.

- Analyze the running time or approximation ratio of an algorithm.

- Develop better lower bounds for the value of an optimal solution. This can lead to better performance of exact algorithms such as branch-and-bound, and also enables better comparisons to determine the quality of non-optimal algorithms.

R20: The subsequent list of references includes many papers which together encompass all the preceding problem variations, and also all the preceding kinds of results.

References

[Ah04] D. Ahr, Contributions to Multiple Postmen Problems, Ph.D. Thesis, Heidelberg University (2004).

[AhRe02] D. Ahr and G. Reinelt, New Heuristics and Lower Bounds for the Min-Max k-Chinese Postman Problem, in *Proceedings of the 10th Annual European Symposium on Algorithms*, Springer-Verlag, London (2002), 64–74.

[ArLaSe91] S. Arnborg, J. Lagergren, and D. Seese, Easy Problems for Tree-decomposable Graphs, *Journal of Algorithms* **12** (1991), 308–340.

[AsGo95] A. Assad and B. Golden, Arc Routing Methods and Applications, in *Handbooks in Operations Research and Management Science, Volume 8: Network Routing* (1995), 375–483.

[BaCoLa11] E. Bartolini, J. Cordeau, and G. Laporte, Improved Lower Bounds and Exact Algorithm for the Capacitated Arc Routing Problem, CIRRELT technical report (2011).

[BeBe98] J. Belenguer and E. Benavent, The Capacitated Arc Routing Problem: Valid Inequalities and Facets, *Computational Optimization and Applications* **10** (1998), 165–187.

[BeBe03] J. Belenguer and E. Benavent, A Cutting Plane Algorithm for the Capacitated Arc Routing Problem, *Computers and Operations Research* **30** (2003), 705–728.

[BeBo74] E. L. Beltrami and L. D. Bodin, Network and Vehicle Routing for Municipal Waste Collection, *Networks* **4** (1974), 65–94.

[BoPaTo91] R. B. Borie, R. Parker, and C. Tovey, Algorithms for Recognition of Regular Properties and Decomposition of Recursive Graph Families, *Annals of Operations Research* **33** (1991), 127–149.

[BoPaTo92] R. B. Borie, R. Parker, and C. Tovey, Automatic Generation of Linear-Time Algorithms from Predicate Calculus Descriptions of Problems on Recursively Constructible Graph Families, *Algorithmica* **7** (1992), 555–581.

[Br80] P. Brucker, The Chinese Postman Problem for Mixed Networks, in *Proceedings of the International Workshop on Graphtheoretic Concepts in Computer Science, Lecture Notes in Computer Science* **100**, Springer-Verlag, New York (1980), 354–366.

[ChBeCaCoMo84] C. Christofides, E. Benavent, V. Campos, A. Corberan, and E. Mota, An Optimal Method for the Mixed Postman Problem, in *System Modeling and Optimization, Lecture Notes in Control and Information Science* **59**, Springer, New York (1984).

[Co90] B. Courcelle, The monadic second-order logic of graphs I: Recognizable sets of finite graphs, *Information and Computation* **85** (1990), 12–75.

[CoMo93] B. Courcelle and M. Mosbah, Monadic Second-order Evaluations on Tree-decomposable Graphs, *Theoretical Computer Science* **109** (1993), 49–82.

[CoMoSa06] A. Corberán, E. Mota, and J. Sanchis, A Comparison of Two Different Formulations for Arc Routing Problems on Mixed Graphs, *Computers and Operations Research* **33** (2006), 3384–3402.

[CoPr10] A. Corberán and C. Prins, Recent Results on Arc Routing Problems: An Annotated Bibliography, *Networks* **56** (2010), 50–69.

[CoRoSa03] A. Corberán, A. Romero, and J. Sanchis, The Mixed General Routing Polyhedron, *Mathematical Programming* **96** (2003), 103–137.

[DaKrSr08] P. Damodaran, M. Krishnamurthi, and K. Srihari, Lower Bounds for Hierarchical Chinese Postman Problem, *International Journal of Industrial Engineering* **15** (2008), 36–44.

[De04] J. Degenhardt, An Ant-Algorithm for the Balanced k-Chinese Postmen Problem, in *Operations Research Conference*, Tilburg University (2004).

[Dr00] M. Dror (editor), *Arc Routing: Theory, Solutions and Applications*, Kluwer Academic Publishers, Norwell, MA (2000).

[Ed65-a] J. Edmonds, The Chinese Postman Problem, *Operations Research* **13**, Suppl. 1 (1965), 373.

[Ed65-b] J. Edmonds, Maximum Matching and a Polyhedron with 0,1 Vertices, *J. Research National Bureau of Standards* **69B** (1965), 125–130.

[Ed65-c] J. Edmonds, Paths, Trees, and Flowers, *Canadian J. Mathematics* **17** (1965), 449–467.

[EdJo73] J. Edmonds and E. Johnson, Matching, Euler Tours, and the Chinese Postman, *Mathematical Programming* **5** (1973), 88–124.

[EhBr51] T. van Aardenne-Ehrenfest and N. G. de Bruin, Circuits and Trees in Oriented Linear Graphs, *Simon Stevin* **28** (1951), 203–217.

[EiGeLa95-a] A. Eiselt, M. Gendreau, and G. Laporte, Arc Routing Problems, Part I: The Chinese Postman Problem, *Operations Research* **43** (1995), 231–242.

[EiGeLa95-b] A. Eiselt, M. Gendreau, and G. Laporte, Arc Routing Problems, Part II: The Rural Postman Problem, *Operations Research* **43** (1995), 399–414.

[Fl91] H. Fleischner, Eulerian Graphs and Related Topics, Part 1, Vol. 2, *Annals of Discrete Math* **50**, North-Holland, Amsterdam, (1991).

[FoFu62] L. R. Ford and D. R. Fulkerson, *Flows in Networks*, Princeton University Press, Princeton, NJ (1962).

[Fr79] G. Frederickson, Approximation Algorithms for Some Postman Problems, *Journal of the ACM* **26** (1979), 538–554.

[FrHeKi78] G. Frederickson, M. Hecht, and C. Kim, Approximation Algorithms for Some Routing Problems, *SIAM Journal on Computing* **7** (1978), 178–193.

[GaJo79] M. Garey and D. Johnson, *Computers and Intractability: A Guide to the Theory of $\mathcal{NP}$-Completeness*, W.H. Freeman and Co., New York (1979).

[GhIm00] G. Ghiani and G. Improta, An Algorithm for the Hierarchical Chinese Postman Problem, *Operations Research Letters* **26** (2000), 27–32.

[GoMoPi10] L. Gouveia, M. Mourão, and L. Pinto, Lower Bounds for the Mixed Capacitated Arc Routing Problem, *Computers and Operations Research* **37** (2010), 692–699.

[GoWo81] B. Golden and R. Wong, Capacitated Arc Routing Problems, *Networks* **11** (1981), 305–315.

[GrWi92] M. Grötschel and Z. Win, A Cutting Plane Algorithm for the Windy Postman Problem, *Mathematical Programming* **55** (1992), 339–358.

[Gu62] M. Guan, Graphic Programming Using Even and Odd Points, *Chinese Mathematics* **1** (1962), 273–277.

[Gu84-a] M. Guan, A Survey of the Chinese Postman Problem, *J. Math. Res. and Expos.* **4** (1984), 113–119 (in Chinese).

[Gu84-b] M. Guan, On the Windy Postman Problem, *Discrete Applied Mathematics* **9** (1984), 41–46.

[JiKaZhZh10] H. Jiang, L. Kang, S. Zhang, and F. Zhu, Genetic Algorithm for Mixed Chinese Postman Problem, in *Proceedings of the 5th International Conference on Advances in Computation and Intelligence*, Springer-Verlag, Berlin (2010), 193–199.

[Ka67] A. Kaufmann, *Graphs, Dynamic Programming and Finite Games*, Academic Press, New York (1967).

[KaKo79] C. Kappauf and G. Koehler, The Mixed Postman Problem, *Discrete Applied Mathematics* **1** (1979), 89–103.

[KoVo06] P. Korteweg and T. Volgenant, On the Hierarchical Chinese Postman Problem with Linear Ordered Classes, *European Journal of Operational Research* **169** (2006), 41–52.

[LiZh88] Y. Lin and Y. Zhao, A New Algorithm for the Directed Chinese Postman Problem, *Computers and Operations Research* **15** (1988), 577–584.

[Mi79] E. Mineka, The Chinese Postman Problem for Mixed Networks, *Management Science* **25** (1979), 643–648.

[NoPi96] Y. Norbert and J. C. Picard, An Optimal Algorithm for the Mixed Chinese Postman Problem, *Networks* **27** (1996), 95–108.

[Or74] C. S. Orloff, A Fundamental Problem in Vehicle Routing, *Networks* **4** (1974), 35–64.

[OsMa05] A. Osterhues and F. Mariak, On Variants of the k-Chinese Postman Problem, Dortmund University technical report (2005).

[Pa76] C. H. Papadimitriou, On the Complexity of Edge Traversing, *J. ACM* **23** (1976), 544–554.

[Pe94] W. L. Pearn, Solvable Cases of the k-Person Chinese Postman Problem on Mixed Networks, *Operations Research Letters* **16** (1994), 241–244.

[PeCh99] W. L. Pearn and J. Chou, Improved Solutions for the Chinese Postman Problem on Mixed Networks, *Computers and Operations Research* **26** (1999), 819–827.

[PeLu95] W. L. Pearn and C. M. Lui, Algorithms for the Chinese Postman Problem on Mixed Networks, *Computers and Operations Research* **22** (1995), 479–489.

[Ra93] T. K. Ralphs, On the Mixed Chinese Postman Problem, *Operations Research Letters* **14** (1993), 123–127.

[RaVe99] B. Raghavachari and J. Veerasamy, A $\frac{3}{2}$-Approximation Algorithm for the Mixed Postman Problem, *SIAM J. Discrete Mathematics* **12** (1999), 425–433.

[SunTanHou11-a] J. Sun, G. Tan, and G. Hou, A New Integer Programming Formulation for the Chinese Postman Problem with Time Dependent Travel Times, *World Academy of Science, Engineering and Technology* **76** (2011), 965–969.

[SunTanHou11-b] J. Sun, G. Tan, and G. Hou, Branch-and-Bound Algorithm for the Time Dependent Chinese Postman Problem, in *Proceedings of International Conference on Mechatronic Science, Electric Engineering and Computer* (2011), 949–954.

[Th03] H. Thimbleby, The Directed Chinese Postman Problem, *Software: Practice and Experience* **33** (2003), 1081–1096.

[WaWe02] H. Wang and Y. Wen, Time-Constrained Chinese Postman Problems, *Computers and Mathematics with Applications* **44** (2002), 375–387.

[Wi89] Z. Win, On the Windy Postman Problem on Eulerian Graphs, *Mathematical Programming* **44** (1989), 97–112.

[Wo05] S. Wøhlk, Contributions to Arc Routing, Ph.D. Thesis, University of Southern Denmark (2005).

[Wo08] S. Wøhlk, A Decade of Capacitated Arc Routing, in *The Vehicle Routing Problem: Latest Advances and New Challenges*, Springer, New York (2008), 29–48.

[YaCh02] K. Yaoyuenyong and P. Charnsethikul, A Heuristic Algorithm for the Mixed Chinese Postman Problem, *Optimization and Engineering* **3** (2002), 157–187.

[YuBa11] W. Yu and R. Batta, Chinese Postman Problem, in *Wiley Encyclopedia of Operations Research and Management Science* (2011).

[Zh11] J. Zhang, Modeling and Solution for Multiple Chinese Postman Problems, *Communications in Computer and Information Science* **215** (2011), 520–525.

Section 4.4
DeBruijn Graphs and Sequences

A. K. Dewdney, University of Western Ontario, Canada

INTRODUCTION

N. deBruijn solved the problem of finding a minimum-length binary string that contains as a (contiguous) substring every binary string of a prescribed length k. For this purpose, he prescribed a special directed graph, of in-degree 2 and out-degree 2, now called a deBruijn graph. In this section, we cover the basics of deBruijn graphs, two methods to generate deBruijn sequences, and applications to the generation of pseudorandom numbers and to genetics.

4.4.1   DeBruijn Graph Basics

DeBruijn Sequences

DEFINITIONS

D1: A ***deBruijn sequence of order*** k is a binary string of length $n = 2^k$ in which

- the last bit is considered to be adjacent to the first, and

- every possible binary k-tuple appears exactly once.

Two deBruijn sequences are considered to be the "same sequence" if one can be obtained from the other by a cyclic permutation.

D2: In a string s of length $m > k$, the ***successor of a substring*** t of length k is the k-bit substring t' that begins at the second bit of t. This is understood cyclically within s, so that if needed, the last bit of the successor substring t' is the first bit of string s.

D3: In a string s of length $m > k$, the k-**tour** is the sequence of substrings of length k, starting with the initial substring. Since this is understood cyclically within s, there are m substrings in the tour.

D4: A k-bit string b is said to be obtained from a k-bit string $a = a_1 a_2 a_3 \cdots a_k$ by a **(left) shift operation** if $b_i = a_{i+1}$, for $i = 1, 2, \ldots, k-1$. The bit b_k may be arbitrary.

D5: A left shift $a_1 a_2 \ldots a_k \longrightarrow b_1 b_2 \ldots b_k$ is a **cycle shift** if $b_k = a_1$.

D6: A left shift $a_1 a_2 \ldots a_k \longrightarrow b_1 b_2 \ldots b_k$ is a **deBruijn shift** if $b_k \neq a_1$.

FACTS

F1: An obvious lower bound on the length of a deBruijn sequence of order k is 2^k, since there are 2^k different bitstrings of length k, and since each bit in a sequence starts only one k-bitstring.

F2: The successor of each k-bit substring t in a deBruijn sequence is either a cycle shift or a deBruijn shift of t.

EXAMPLES

E1: 00010111 is a deBruijn sequence of order 3. Its 3-tour is

$$000, 001, 010, 101, 011, 111, 110, 100$$

E2: 0000101101001111 is a deBruijn sequence of order 4.

DeBruijn Graphs

An intuitive approach to the problem of constructing a deBruijn sequence is to construct a graph in which a hamiltonian tour corresponds to such a sequence.

DEFINITIONS

D7: A **deBruijn graph of order** k, denoted by $G(k)$, is a directed graph with 2^k vertices, each labeled with a unique k-bit string. Vertex a is joined to vertex b by an arc if bitstring b is obtainable from bitstring a by either a cycle shift or a deBruijn shift. Additionally, each arc of $G(k)$ is designated as a **cycle-shift arc** or a **deBruijn arc**, according to the shift operation it represents. Each arc is labeled by the first bit of the vertex at which it originates, followed by the label of the vertex at which it terminates.

D8: The **cycle-shift 2-factor** in a deBruijn graph is the 2-factor formed by all of its cycle-shift arcs.

D9: The **deBruijn 2-factor** in a deBruijn graph is the 2-factor formed by all of its deBruijn arcs.

EXAMPLE

E3: Figure 4.4.1 below illustrates the deBruijn graph of order 3.

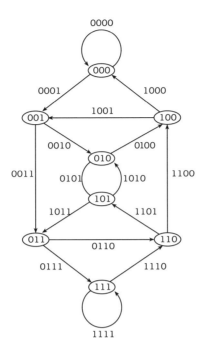

Figure 4.4.1: A deBruijn graph of order 3.

FACTS

F3: The cycle-shift arcs form a directed 2-factor, because the cycle-shift operation acts as a permutation on the bitstrings. Similarly, the deBruijn arcs form a directed 2-factor.

F4: Every vertex of a deBruijn graph has out-degree 2. The first bit of the label on one of the vertices to which it points is 0, and the first bit on the label of the other is 1.

F5: Every vertex of a deBruijn graph has in-degree 2.

F6: Every deBruijn graph is strongly connected.

F7: Every deBruijn graph is hamiltonian.

F8: The hamiltonian (directed) circuits in the deBruijn graph $G(k)$ are in one-to-one correspondence with the deBruijn sequences of order k. The correspondence is realized by listing, in sequence, the first bit of each vertex encountered on a hamiltonian tour.

F9: deBruijn's Theorem [dB46] For each positive integer k, there are $2^{2^{k-1}-k}$ deBruijn sequences of order k.

k	1	2	3	4	5	6	$\ldots$
$2^{2^{k-1}-k}$	1	1	2	16	2048	67108864	$\ldots$

REMARKS

R1: A hamiltonian circuit in a deBruijn graph can be constructed by splicing together the components of its deBruijn 2-factors. However, deBruijn's theorem depends on a more elegant way to construct deBruijn sequences.

R2: Since each component of the deBruijn 2-factor of the deBruijn graph $G(k)$ has cardinality at most k, it follows that the number of components of the deBruijn 2-factor grows exponentially in k.

4.4.2   Generating deBruijn Sequences

An efficient algorithm for constructing a deBruijn sequence of order k is based not on finding a hamiltonian circuit in the deBruijn graph of order k, but rather on the easier task of constructing an Eulerian tour in the deBruijn graph of order $k - 1$. Another interesting method is strictly lexicographic.

FACTS

F10: [Go46]: A strongly connected directed graph in which every vertex has the same indegree as outdegree has an Eulerian tour.

F11: In a deBruijn graph $G(k)$, the k-sequence of arc labels encountered on every directed path of length k originating at a vertex v is the binary string that labels vertex v. (This is an immediate consequence of the specification of the arc labels in the definition of a deBruijn graph.)

F12: The sequence of arc labels encountered on an Eulerian tour of the deBruijn graph of order k is a deBruijn sequence of order $k + 1$.

REMARK

R3: The proof of Fact F12 is not difficult. Since an Eulerian tour of $G(k)$ visits each vertex twice, it follows from Fact F11 that each bitstring label occurs twice in the sequence of arc labels. By Fact F4, one occurrence is followed by a 0 and the other by a 1.

ALGORITHM

A1: To construct a deBruijn sequence of order k, use Fleury's algorithm (quadratic time) to construct an Eulerian tour of the deBruijn graph $G(k - 1)$. Then record the sequence of arc labels on the Eulerian tour. (Fleury's algorithm appears in §4.2.)

EXAMPLE

E4: Figure 4.4.2 illustrates the construction of a deBruijn sequence of order 4 from the deBruijn graph of order 3.

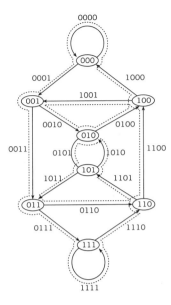

Figure 4.4.2: An Eulerian circuit in $G(3)$.

Necklaces and Lyndon Words

Fredricksen and Kessler [FrKe77] have published a remarkable alternative method for constructing deBruijn sequences.

DEFINITIONS

D10: A *rotation of a binary string* is the result of an iteration of cycle shifts. (Rotation is clearly an equivalence relation.)

D11: An equivalence class under rotation of the binary strings of length n is called a *necklace* of order n.

D12: A *Lyndon word* of order n is a necklace of order n whose rotation class has n binary strings. A Lyndon necklace of length 1 is called *trivial*. We take the lexicographically least element of the equivalence class as representative of the necklace.

FACTS

F13: A necklace representative is a Lyndon word if and only if it is aperiodic, i.e., cannot be written as the concatenation of two or more identical strings.

F14: By an elementary application of Burnside–Polya enumeration, the number of necklaces of order n is

$$\frac{1}{n} \sum_{k:k|n} \phi(k) \cdot 2^{\frac{n}{k}}$$

where $\phi(n)$ is the number of integers in the interval $[1, n]$ that are relatively prime to n.

F15: [FrKe77]: If the (lexicographically least) representatives of all the nontrivial Lyndon words whose lengths divide n are arranged into lexicographic order and concatenated, with the terminal string 10 appended at the end, then the result is a deBruijn sequence of order n that is lexicographically minimum.

REMARK

R4: The number $N(n)$ of necklaces grows exponentially with n. While $N(5) = 8$, we have $N(10) = 108$, and $N(15) = 2192$.

EXAMPLES

E5: Figure 4.4.3 displays five equivalent strings of length 5.

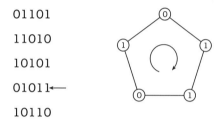

01101

11010

10101

01011←—

10110

Figure 4.4.3: A necklace and its representations.

E6: The only nontrivial Lyndon word of length 2 is 01. We observe that 0110 is a deBruijn sequence of order 2.

E7: The only nontrivial Lyndon words of length 3 are 001 and 011. We observe that 00101110 is a deBruijn sequence of order 3.

E8: We now illustrate Fact F15 for $n = 4$. In lexicographic order, the nontrivial Lyndon words of lengths that divide 4 are

$$0001, 0011, 01, 0111$$

If we now concatenate these words in the order given, we obtain the lexicographically least deBruijn sequence of order 4:

$$0001001101011110$$

4.4.3   Pseudorandom Numbers

For Monte Carlo applications, the numbers produced by ordinary pseudorandom number generators (e.g., congruential generators) are close enough to random not to affect the outcome of the study. But for certain applications called precision Monte Carlo simulation, special sequences must be used. DeBruijn sequences, which already appear somewhat random to the untrained eye, may be made more random by interchanging runs of zeros and ones.

DEFINITIONS

D13: A **run** in a binary sequence is a subsequence of identical bits, and a **maximal run** is a run that is not contained in any longer run.

D14: [Go67] The **Golomb postulates of randomness** for a periodic sequence X are as follows:

- The number of 1's in X differs from the number of 0's by at most unity.

- At least half the runs in X have length 1, at least one-quarter have length 2, at least one-eighth have length 3, etc.

- The bit in position i is correlated to the same degree with adjacent bits ($i+1$ and $i-1$) as it is with ones further away ($i+100$ and $i-100$).

D15: A **run-permuted sequence** is a pseudorandom sequence obtained by the following procedure:

0. Generate a random deBruijn sequence of order n.

1. Randomly permute the maximal runs of 0's.

2. Randomly permute the maximal runs of 1's.

D16: [Ch87] The **randomness of an infinite binary sequence** S is defined to be

$$\lim_{m \to \infty} \frac{s(m)}{m}$$

where $s(m)$ is the minimum number of states in a 2-symbol Turing machine that produces the first m bits of the sequence S.

FACTS

F16: Obviously, every deBruijn sequence can be generated as a run-permuted sequence.

F17: Any deBruijn sequence X of order n satisfies Golomb's first two postulates. First, the number of 1's exactly equals the number of 0's. Second, it is easily shown that over all possible binary subsequences of each length n, exactly half of the runs have that length.

F18: Interchanging (maximal) runs in permuting a deBruijn sequence does not change the number of runs of any length or kind. One therefore obtains a much larger class of sequences that are, by Golombs measure, just as random as the original deBruijn sequence from which the new sequences are generated.

F19: [Je01] The class C_n of run-permuted sequences of order n contains a vanishingly small proportion of deBruijn sequences of order n as n increases.

4.4.4   A Genetics Application

Typically, the short DNA fragments observed in experiments are not sufficient to reconstruct the genome of an organism completely. Because of the time and expense of such experiments, it is desirable to minimize the remaining work. To this end, biologists algorithmically assemble as much of the genome as they can, thereby obtaining longer DNA fragments that are fewer in number. They then perform additional experiments at specific locales in the resulting sequences, in order to extend the reconstruction.

A phenomenon that complicates the stepwise reconstruction of the genome is the natural occurrence of multiple copies of the same substring in a number of DNA sequences acquired by experiment. To help resolve this difficulty, Pvezner, Tang and Waterman [PvTaWa01] have applied modified de Bruijn graphs, in which a repeated k-string in a given sequence s_i results in multiple vertices, and consequently, in multiple paths connecting certain pairs of vertices in the graph. Such a graph need not be connected.

DEFINITIONS

D17: A *DNA sequence* is any finite sequence of the letters A, C, G, T.

D18:   For any set $S = \{s_1, s_2, \ldots, s_n\}$ of DNA sequences, we define the *S-relative deBruijn graph of order* k to have vertices corresponding to all k-substrings from the elements of S, one for each occurrence of a substring. Two such vertices u and v are adjacent if their substrings belong to the same DNA sequence s_i and the last $k - 1$ letters of u coincide with the first $k - 1$ letters of v.

REMARK

R5:   Since Eulerian paths can be found very quickly in connected portions of the S-relative deBruijn graph, partial paths can be produced efficiently for the graph as a whole. These not only (in most cases) recapture the original sequences, but suggest where additional experiments need to be performed to choose between different possible paths through the S-relative deBruijn graph.

References

[COS] Information regarding necklaces, unlabeled necklaces, Lyndon words, deBruijn sequences. Available at www.theory.csc.uvic.ca/inf/neck/Necklaceinfo/html.

[Ch87] G. J. Chaitin, *Algorithmic Information Theory*, Cambridge University Press, 1987.

[dB46] N. G. deBruijn, A combinatorial problem, *Nederl. Akad. Wetensch., Proc.* 49 (1946), 758–764.

[FrKe77] H. Fredricksen and I. Kessler, Lexicographic compositions and deBruijn sequences, *J. Combin. Theory, Ser. A* 22 (1977), 17–30.

[GaJo79] M. R. Garey and D. S. Johnson, *Computers and Intractability: A Guide to the Theory of NP-Completeness*, W. H. Freeman & Co, 1979.

[Go67] S. W. Golomb, *Shift Register Sequences*, Holden-Day, 1967.

[Go46] I. J. Good, Normal recurring decimals, *J. London Math. Soc.* 21 (1946), 167–172.

[GrGeMiLe99] A. J. F. Griffiths, W. M. Gelbart, J. H. Miller, and R. C. Lewontin, *Modern Genetic Analysis*, W. H. Freeman, 1999.

[GrYe06] J. L. Gross and J. Yellen, *Graph Theory and Its Applications*, Second Edition, CRC Press, 2006.

[Ha67] M. Hall Jr., *Combinatorial Theory*, Blaisdell Publishing Co, 1967.

[Je01] C. J. A. Jensen, On the construction of run permuted sequences, pp. 196–203 in *Advances in Cryptology - Eurocrypt '90* (edited by L. B. Damgard), Springer-Verlag, 2001.

[MuPv02] Z. Mulyukov and P. A. Pvezner, Euler-PCR: Finishing experiments for repeat resolution, *Pacific Symposium on Biocomputing* 7 (2002), 199–210.

[PvTaWa01] P. A. Pvezner, H. Tang, and M. S. Waterman, An Eulerian path approach to DNA fragment assembly, *Proc. Natl. Acad. Sci.* 98 (2001).

Section 4.5
Hamiltonian Graphs

Ronald J. Gould, Emory University

INTRODUCTION

Named for Sir William Rowan Hamilton, the hamiltonian problem traces its origins to the 1850s.

4.5.1   History

Characterizing hamiltonian graphs is an NP-complete problem (see [GaJo79]); thus the hamiltonian problem is generally considered to be determining conditions under which a graph contains a hamiltonian cycle. Hamilton exhibited his *Icosian Game* at a meeting in Dublin in 1857. The game involved finding various paths and cycles, including spanning cycles, of the regular dodecahedron. The game was marketed by a wholesale dealer in 1859, but apparently was not a big hit. Perhaps the only profit was Hamilton's, as he sold the game to the dealer for 25 pounds.

Hamilton does not appear to be the first to have considered the question of spanning cycles. In a paper [Ki56] submitted in 1855, Thomas Penyngton Kirkman posed the question: Given the graph of a polyhedron, can one always find a circuit (cycle) that passes through each vertex once and only once. Thus, Kirkman actually asked a more general question than Hamilton. Unfortunately for Kirkman, the term hamiltonian cycle is much too ingrained to be changed now. For a more detailed account of this history see [BiLlWi86].

DEFINITIONS

D1: A graph G is **hamiltonian** if it contains a spanning cycle (**hamiltonian cycle**).

D2: A graph G is **traceable** if it contains a spanning path.

D3: Further, G is **hamiltonian connected** if any pair of vertices are the ends of a spanning path.

4.5.2 The Classic Attacks

There are certain fundamental results that deserve attention, both for their contribution to the overall theory and for their affect on the later development of the area.

The approach taken to developing sufficient conditions for a graph to be hamiltonian usually involved some sort of edge density condition, providing enough edges to ensure the existence of a hamiltonian cycle.

TERMINOLOGY NOTE: The **order** of a graph is the cardinality of its vertex set, and the **size** of a graph is the cardinality of its edge set.

Degrees

NOTATION: The minimum degree of the vertices of a graph G is denoted $\delta_{min}(G)$, and the maximum degree is denoted $\delta_{max}(G)$.

DEFINITIONS

D4: We say a set $X \subseteq V$ is **independent** if there are no edges between vertices in X. The largest cardinality of an independent set in G is called the **independence number** of G and is denoted $ind(G)$.

D5: The k-**degree closure** of G, denoted $C_k(G)$, is the graph obtained by recursively joining pairs of nonadjacent vertices whose degree sum is at least k, until no such pair remains.

D6: For a balanced bipartite graph $G = (X \cup Y, E)$, the **bipartite degree closure** is that graph obtained by joining any nonadjacent pair $x \in X$ and $y \in Y$ whose degree sum is at least $n + 1$.

NOTATION: The following notation has become standard in the area:

$$\sigma_k(G) = min\{\sum_{i=1}^{k} deg\ x_i\ |x_1, \ldots, x_k\ \text{are independent}\}.$$

FACTS

F1: [Di52] If G is a graph of order n such that $\delta_{min}(G) \geq n/2$, then G is hamiltonian.

F2: Let G be a graph of order n.
i. [Or60] If $\sigma_2(G) \geq n$, then G is hamiltonian,
ii. [Or63] If $\sigma_2(G) \geq n + 1$, then G is hamiltonian connected.

EXAMPLE

E1:   Consider two $K_{(p+1)/2}$ with one vertex from each identified (graph on left in Figure 4.5.1). This graph is not hamiltonian, but has order p, $\delta_{min}(G) = (p-1)/2$ and $\sigma_2(G) = p - 1$, illustrating the sharpness of Dirac's Theorem and Ore's Theorem (Fact F2 i). The graph obtained by identifying a pair of vertices from two copies of $K_{(p+2)/2}$ is not hamiltonian connected, has $\delta_{min}(G) = p/2$ and $\sigma_2(G) = p$, showing Fact F2ii is sharp.

 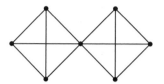

Figure 4.5.1: Illustrating the sharpness of Dirac's and Ore's results.

FACTS

F3:   [Ja80] If G is a d-regular 2-connected graph of order n with $d \geq n/3$, then G is hamiltonian.

F4:   [MoMo63] If $G = (X \cup Y, E)$ is a balanced bipartite graph of order $2n$ $(n \geq 2)$ with $deg\ u + deg\ v \geq n + 1$ for each nonadjacent pair $u \in X$ and $v \in Y$, then G is hamiltonian.

F5:   [BoCh76] Let G have order n. Then
i. $C_k(G)$ is well defined,
ii. G is hamiltonian if, and only if, $C_n(G)$ is hamiltonian,
iii. if $C_{n+1}(G)$ is complete, then G is hamiltonian connected.

F6:   [He91] A balanced bipartite graph is hamiltonian if, and only if, its bipartite closure is hamiltonian.

REMARK

R1:   These closure results provide an interesting relaxation of the degree conditions. The closure is (hopefully) a denser graph, making it easier to find a hamiltonian cycle. However, the number of edges actually added in forming these closures can vary widely. It is easy to construct examples for all possible values from 0 to all the missing edges. Thus, we might receive no help in deciding if the original graph is hamiltonian, or we might conclude trivially that it is (when the closure is the complete graph).

Other Counts

DEFINITIONS

D7:   The **neighborhood** of a vertex x, denoted $N_G(x)$, is the set of all vertices adjacent to x in G. Similarly, $N_G(S)$ denotes the **neighborhood of the set** S and is the collection of all vertices adjacent to some vertex in S.

NOTATION: When the graph in which the neighborhood is defined is clear, the subscript is omitted.

D8: The (***vertex-***)***connectivity*** of a connected graph, denoted $\kappa_v(G)$, is the minimum number of vertices whose removal either disconnects G or reduces it to a 1-vertex graph.

D9: A graph G is k-***connected*** if $\kappa_v(G) \geq k$.

EXAMPLE

E2: The graph $G(r,p)$ is that graph with order p and vertex set $S \cup T \cup U$ where $|S| = |T| = r$ and $|U| = p - 2r$ and where two vertices are adjacent if either belongs to S or both belong to U. Hence, the graphs induced by S, T, and U are K_r, $\bar{K}_r$, and K_{p-2}, respectively. Figure 4.5.2 below shows the graphs $G(1,6)$ and $G(2,5)$.

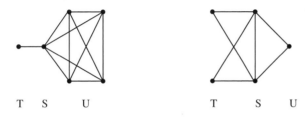

T   S   U   T   S   U

Figure 4.5.2: Graphs $G(1,6)$ and $G(2,5)$.

FACTS

F7: [Or63] If G is a graph of order n and size greater than $\binom{n-1}{2} + 1$, then G is hamiltonian. Furthermore, the only nonhamiltonian graphs with exactly size $\binom{n-1}{2} + 1$ are $G(1,n)$ and $G(2,5)$. In addition, if G has size at least $\binom{n-1}{3}$, then G is hamiltonian connected.

F8: [Fa84] If G is a 2-connected graph of order n such that

$$min\{max(deg\ u, deg\ v) \mid d(u,v) = 2\} \geq n/2,$$

then G is hamiltonian.

F9: [BaBrVeLi89] If G is a 2-connected graph of order n and connectivity $\kappa_v(G)$ such that $\sigma_3(G) \geq n + \kappa_v(G)$, then G is hamiltonian.

F10: [ChEr72] Let G be a graph of order at least 3.
i. If $\kappa_v(G) \geq ind(G)$, then G is hamiltonian.
ii. If $\kappa_v(G) \geq ind(G) - 1$, then G is traceable.
iii If $\kappa_v(G) \geq ind(G) + 1$, then G is hamiltonian connected.

F11: [Wo78] If for any nonempty $S \subseteq V$, $|N(S)| \geq \frac{|S|+n+3}{3}$, then G is hamiltonian.

F12: [Fr86] Let G be a k-connected graph of order n. Suppose there exists some $t \leq k$, such that for every independent set S of vertices with cardinality t we have $|N(S)| \geq \frac{t(n-1)}{t+1}$, then G is hamiltonian.

F13: [BrVe91], [FaGoJaLe92] If G is a 2-connected graph of sufficiently large order n such that $|N(S)| \geq n/2$ for every set S of two distinct vertices of G, then G is hamiltonian.

REMARK

R2: The Petersen graph is the only counterexample for $n \leq 10$ (see [BrVe91]).

Powers and Line Graphs

TERMINOLOGY: A ***circuit*** is a closed walk having no repeated edges (also called a closed trail).

DEFINITIONS

D10: The ***line graph*** $L(G)$ of a graph G is that graph whose vertices can be put into 1-1 correspondence with the edges of G in such a way that two vertices of $L(G)$ are adjacent if, and only if, the corresponding edges of G are incident (have an endpoint in common).

D11: A circuit C such that every edge of G is incident to a vertex of C is called a ***dominating circuit***.

D12: We say that G contains a ***k-system that dominates*** if G contains a collection of k edge disjoint circuits and stars (here stars are $K_{1,n_i}, n_i \geq 3$), such that each edge of G is either contained in one of the circuits or stars, or is adjacent to one of the circuits.

D13: The ***k-th power***, G^k, of a connected graph G is that graph with $V(G^k) = V(G)$ for which $uv \in E(G)$ if, and only if, $1 \leq d_G(u,v) \leq k$.

D14: A ***k-factor*** of a graph G is a k-regular spanning subgraph of G. In particular, a 2-factor is a (vertex-disjoint) union of cycles that covers $V(G)$.

FACTS

F14: [HaNW65] Let G be a graph without isolated vertices. Then $L(G)$ is hamiltonian if and only if $G \simeq K_{1,n}$, for some $n \geq 3$, or G contains a dominating circuit.

F15: [GoHy99] Let G be a graph with no isolated vertices. The graph $L(G)$ contains a 2-factor with k ($k \geq 1$) cycles if, and only if, G contains a k-system that dominates.

F16: [ChWa73] If G is connected with $\delta_{min}(G) \geq 3$, then $L^2(G) = L(L(G))$ is hamiltonian.

F17: [Fl74] If G is a 2-connected graph, then G^2 is hamiltonian.

F18: If G is connected then G^3 is hamiltonian (in fact, hamiltonian-connected) (see [Be78]).

Planar Graphs

FACTS

F19: [Th83] Every 4-connected planar graph is hamiltonian connected (and hence hamiltonian [Tu56]).

F20: [Gr68] Let G be a plane graph of order n with hamiltonian cycle C. If r_i denotes the number of i sided regions interior to C and r'_i the number of i sided regions exterior to C, then

$$\sum_{i=3}^{n} (i-2)(r_i - r'_i) = 0.$$

4.5.3 Extending the Classics

Adding Toughness

DEFINITION

D15: If every vertex cut-set S of G satisfies $t \cdot c(S) \leq |S|$, where $c(S)$ is the number of components of $G - S$, we say that G is t-***tough***. The ***toughness*** of G is the maximum t such that G is t-tough.

FACTS

F21: [Ju78] Let G be a 1-tough graph of order $n \geq 11$ such that $\sigma_2(G) \geq n - 4$. Then G is hamiltonian and this bound is sharp.

F22: [BaMoScVe90] Let G be a 2-tough graph of order n such that $\sigma_3(G) \geq n$. Then G is hamiltonian.

F23: [BrVe90] Let G be a 1-tough graph of order $n \geq 3$ with $\delta_{min}(G) \geq \frac{n + \kappa_v(G) - 2}{3}$. Then G is hamiltonian.

REMARK

R3: Chvátal conjectured that there is a t_0 such that all t_0-tough graphs are hamiltonian. For years $t_0 = 2$ seemed possible. However, in [BaBrLiVe00], examples of $(9/4 - \epsilon)$-tough nonhamiltonian graphs, for arbitrary $\epsilon > 0$, were presented.

More Than Hamiltonian

DEFINITIONS

D16: A graph G of order n is ***pancyclic*** if it contains cycles of all lengths l, $3 \leq l \leq n$.

D17: A bipartite graph G of order n is ***bipancyclic*** if it contains cycles of all possible even lengths from 4 to n.

D18:   A graph of order n is **cycle extendable** if any cycle C of length $m < n$ can be extended to a cycle of length $m + 1$ containing all of $V(C)$. Further, if G is cycle extendable and every vertex is on a triangle, then G is called **fully cycle extendable.**

D19:   A graph is **k-ordered** (**hamiltonian**) if for every ordered sequence of k vertices there is a cycle (hamiltonian cycle) that encounters the vertices of the sequence in the given order.

FACT

F24:   [BrChFaGoLe97] If G is a graph of order n satisfying
(1) $\delta_{min}(G) \geq n/2$ and $n \geq 4k$ or
(2) $\sigma_2(G) \geq n$ and $n \geq 4k$
then G contains a 2-factor with k cycles for each k, $1 \leq k \leq \lceil n/4 \rceil$, and this result is best possible.

EXAMPLE

E3:   To see this result is best possible we need only consider the complete bipartite graph $K_{n/2,n/2}$. The smallest cycle in any 2-factor of this graph is a 4-cycle, hence the bounds on k are sharp.

FACTS

F25:   [Bo77] If G is a hamiltonian graph of order n with $|E(G)| \geq \frac{n^2}{4}$, then either G is pancyclic or $G \simeq K_{n/2,n/2}$.

F26:   [He90] If G has order $n \geq 3$ and $\sigma_2(G) \geq n$, then G is cycle extendable unless G belongs to one of two special classes. Also, if $\sigma_2(G) \geq (4n - 5)/3$, then G is cycle extendable. Further, if $\delta_{min}(G) \geq (n + 1)/2$, then G is fully cycle extendable.

F27:   [He91] If $G = (X \cup Y, E)$ is a balanced bipartite graph of order $2n$ such that for any nonadjacent pair $x \in X$ and $y \in Y$ we have $deg\ x + deg\ y \geq n + 1$, then G is bipancyclic.

F28:   [He91] Let $n \geq 2m \geq 2$. If $G = (X \cup Y, E)$ is a balanced bipartite graph of order $2n$ satisfying $\delta_{min}(G) \geq m$ and $|E(G)| > n^2 - mn + m^2$, then G is bipancyclic.

F29:   [KoSaSz96], [KoSaSz98] There exists a natural number n_0 such that if G has order n and $n \geq n_0$ and $\delta_{min}(G) \geq kn/(k + 1)$, then G contains the k-th power of a hamiltonian cycle.

F30:   [KiSaSe99] Let $k \geq 2$ be an integer and let G be a graph of order $n \geq 11k - 3$. If $\delta_{min}(G) \geq \lceil \frac{n}{2} \rceil + \lfloor \frac{k}{2} \rfloor - 1$, then G is k-ordered hamiltonian.

F31:   [FaGoKoLeScSa03] Let k be an integer with $3 \leq k \leq n/2$ and let G be a graph of order n. If $\deg(u) + \deg(v) \geq n + (3k - 9)/2$ for every pair u, v of nonadjacent vertices of G, then G is k-ordered hamiltonian.

REMARK

R4: The bounds in Facts F30 and F31 are both sharp for the respective values of k. Unexpectedly, the Dirac type bound does not follow from the Ore type bound.

FACTS

F32: [Ha79] Let G be a graph of order $n \geq 4$, n even. If $\sigma_2(G) \geq n+1$, then for any 1-factor F of G, there is a hamiltonian cycle containing F.

F33: [LV72] Let $G = (A \cup B, E)$ be a bipartite graph with $|A| = |B| = n \geq 2$. If for each pair of nonadjacent vertices u, v with $u \in A$ and $v \in B$, $deg\ u + deg\ v > n+1$, then any 1-factor F of G is contained in a hamiltonian cycle of G.

F34: [Ya99] If G is a graph of order $n \geq 4$, n even and $\delta_{min}(G) \geq 2$ and $|E(G)| \geq \frac{(n-1)(n-2)}{2} + 1$, then for any 1-factor F, there is a hamiltonian cycle of G containing F.

F35: [KaYo01] Let G be a graph of order n with $\delta_{min}(G) \geq n/2$ and let d be a positive integer such that $d \leq \frac{n}{4}$. Then, for any vertex subset A with $|A| \leq \frac{n}{2d}$, there is a hamiltonian cycle C such that $d_C(u, v) \geq d$ for any $u, v \in A$.

F36: [SaSe08] There are $\omega, n_0 > 0$ such that if G is a graph of order $n \geq n_0$ with $\delta_{min}(G) \geq n/2$ and d is an arbitrary integer with $3 \leq d \leq \omega n/2$ and $S \subset V(G)$ with $2 \leq |S| = k \leq \omega n/d$, then for every sequence d_i of integers with $3 \leq d_i \leq d$, $1 \leq i \leq k-1$, there is a hamiltonian cycle C of G and an ordering of the vertices of S, $a_1, a_2, \ldots, a_k$ such that the vertices of S are visited in this order on C and we have $|d_C(a_i, a_{i+1}) - d_i| \leq 1$ for all but one $1 \leq i \leq k-1$.

F37: [FaGoJaMa] Let $t \geq 3$ be an integer and let $0 < \epsilon \frac{t}{2}$. For $n \geq \frac{7t^6 \times 10^{10}}{\epsilon^6}$, let G be a graph of order n having $\delta_{min}(G) \geq \frac{n}{2}$ and $\kappa(G)_v \geq 2\lceil \frac{t}{2} \rceil$. For every $X = \{x_1, x_2, \ldots, x_t\} \subseteq V(G)$, there exists a hamiltonian cycle C such that $d_C(x_i, x_j) \geq \left(\frac{1}{t} - \epsilon\right)n$ for all $1 \leq i < j \leq t$. Furthermore, the minimum degree and connectivity conditions are sharp.

F38: [FaGoJaMa] Let $t \geq 3$ be an integer and $\epsilon, \gamma_1, \gamma_2, \ldots, \gamma_t$ positive real numbers having $\sum_{i=1}^{t} \gamma_i = 1$ and $0 < \epsilon < \min\{\frac{\gamma_i}{2}\}$. For $n \geq \frac{7t^6 \times 10^{10}}{\epsilon^6}$, let G be a graph of order n having $\delta_{min}(G) \geq \frac{n+t-1}{2}$ or $\delta_{min}(G) \geq \frac{n}{2}$ and $\kappa_v(G) \geq \frac{3t}{2}$. For every $X = \{x_1, x_2, \ldots, x_t\} \subseteq V(G)$, there exists a hamiltonian cycle C containing the vertices of X in order such that $(\gamma_i - \epsilon)n \leq d_C(x_i, x_{i+1}) \leq (\gamma_i + \epsilon)n$ for all $1 \leq i \leq t$. Furthermore, the minimum degree and connectivity conditions are sharp.

F39: [FaGoJa09] Let G be a graph of order n, and let F be a k-edge forest composed of t paths, where $2 \leq k + t \leq n$. If
(a) $\sigma_2(G) \geq n+k$ when either $n+k$ is odd or $F = P_{k+1} \cup (t-1)K_1$, and
(b) $\sigma_2(G) \geq n+k-1$, otherwise,
then G has a hamiltonian cycle containing F.

4.5.4   More Than One Hamiltonian Cycle

A Second Hamiltonian Cycle

FACTS

F40: Every edge of a 3-regular graph is contained in an even number of hamiltonian cycles. Thus, every 3-regular hamiltonian graph contains a second and, in fact, a third hamiltonian cycle (C.A.B. Smith, see [Tu46]).

F41: [Th98] If G is hamiltonian and m-regular with $m \geq 300$, then G has a second hamiltonian cycle.

F42: [Th97] Let G be a graph with a hamiltonian cycle C. Let A be a vertex set in G such that A contains no two consecutive vertices of C and A is *dominating* in $G - E(C)$ (i.e., $N_{G-E(C)}(A) \supseteq V(G - E(C))$). Then G has a hamiltonian cycle C' such that $C' - A = C - A$ and there is a vertex v in A such that one of the two edges of C' incident with v is in C and the other is not in C.

F43: [HoSt00] For any real number $k \geq 1$, there exists $f(k)$ so that every hamiltonian graph G with $\delta_{max}(G) \geq f(k)$ has at least $\delta_{min}(G) - \lfloor \frac{\delta_{max}(G)}{k} \rfloor + 2$ hamiltonian cycles. In particular, every hamiltonian graph with $\delta_{max}(G) \geq f(\delta_{max}(G)/\delta_{min}(G))$ has a second hamiltonian cycle.

F44: [Ma76], [GrMa76] There exist 4-regular, 4-connected planar graphs that do not have two edge-disjoint hamiltonian cycles.

F45: [Za76], [Ro89] There exist infinitely many examples of 5-connected planar graphs (both regular and nonregular) in which every pair of hamiltonian cycles have common edges.

REMARK

R5: Thomason [Th78] extended Smith's result (Fact F40) to all r-regular graphs where r is odd (in fact, to all graphs in which all vertices have odd degree). Thomassen extended this further (see Fact F41).

Many Hamiltonian Cycles

DEFINITION

D20: A *planar triangulation* of a planar graph is the process of adding edges between pairs of non-adjacent vertices to produce another planar graph, each of whose regions is bounded by a triangle.

FACTS

F46: [KrZe88] If a planar triangulation (except K_3 and K_4) is hamiltonian, then it contains at least four hamiltonian cycles.

F47: [Th96] Let $C : x_1, y_1, x_2, y_2, \ldots, x_n, y_n, x_1$ be a hamiltonian cycle in a bipartite graph G.
(a) If all the vertices $y_1, \ldots, y_n$ have degree at least 3, then G has another hamiltonian cycle containing the edge $x_1 y_1$.
(b) If all the vertices $y_1, \ldots, y_n$ have degree $d > 3$ and if $P_1, P_2, \ldots, P_q$ $(0 \leq q \leq d - 3)$ are paths in C of length 2 of the form $y_{i-1} x_i y_i$, then G has at least $2^{q+1-d}(d-q)!$ hamiltonian cycles containing $P_1 \cup \cdots \cup P_q$.

F48: [FaRoSc85] Let k be a positive integer.
(a) If G is a graph of order $n \geq 60k^2$ such that $\sigma_2(G) \geq n + 2k - 2$, then G contains k edge-disjoint hamiltonian cycles.
(b) If G has order $n \geq 6k$ and size at least $\binom{n-1}{2} + 2k$, then G contains k edge-disjoint hamiltonian cycles.

F49: [Eg93] Let $n, k \geq 2$ be integers with $n \geq 44(k - 1)$. If G is a graph of order n with $\sigma_2(G) \geq n$ and $\delta_{min}(G) \geq 4k - 2$, then G contains k edge disjoint hamiltonian cycles.

Uniquely Hamiltonian Graphs

DEFINITION

D21: A graph is ***uniquely hamiltonian*** if it contains exactly one hamiltonian cycle.

FACTS

F50: [EnSw80] There exist infinitely many uniquely hamiltonian graphs with minimum degree three.

F51: [JaWh89] Any uniquely hamiltonian graph contains a vertex of degree at most $(n + 9)/4$, and if the graph has a unique 2-factor, then it contains a vertex of degree 2.

F52: [BoJa98] Every uniquely hamiltonian graph of order n has a vertex of degree at most $c\log_2(8n) + 3$ where $c = (2 - \log_2 3)^{-1} \approx 2.41$. Furthermore, every uniquely hamiltonian plane graph has at least two vertices of degree less than four.

Products and Hamiltonian Decompositions

DEFINITIONS

D22: A ***hamiltonian decomposition*** is a partitioning of the edge set of G into hamiltonian cycles if G is $2d$-regular or into hamiltonian cycles and a perfect matching if G is $(2d + 1)$-regular.

D23: Each of the following four kinds of product graphs has vertex set $V(G_1) \times V(G_2)$. The ***cartesian product*** $G = G_1 \times G_2$ has edge set

$$E(G) = \{(u_1, u_2)(v_1, v_2) \mid u_1 = v_1 \text{ and } u_2 v_2 \in E(G_2) \text{ or } u_2 = v_2 \text{ and } u_1 v_1 \in E(G_1)\}.$$

The **direct product (or conjunction)** $G = G_1 \cdot G_2$ has edge set

$$E(G) = \{(u_1, u_2)(v_1, v_2) \mid u_1 v_1 \in E(G_1) \text{ and } u_2 v_2 \in E(G_2)\}.$$

The **strong product** $G = G_1 \otimes G_2$ has edge set

$$E(G) = \{(u_1, u_2)(v_1, v_2) \mid u_1 = v_1 \text{ and } u_2 v_2 \in G_2, \text{ or}$$

$$u_2 = v_2 \text{ and } u_1 v_1 \in E(G_1), \text{ or both } u_1 v_1 \in E(G_1) \text{ and } u_2 v_2 \in E(G_2)\}.$$

Finally, the **lexicographic product** (sometimes called composition, tensor or wreath product) $G = G_1[G_2]$ has edge set

$$E(G) = \{(u_1, u_2)(v_1, v_2) \mid u_1 v_1 \in E(G_1), \text{ or } u_1 = v_1 \text{ and } u_2 v_2 \in E(G_2)\}.$$

REMARK

R6: Jackson [Ja79] conjectured that every k-regular graph on at most $2k+1$ vertices is hamiltonian decomposable. Another natural question is: If G_1 and G_2 are hamiltonian decomposable, is the appropriate product of G_1 and G_2 also hamiltonian decomposable?

FACTS

F53: [St91] Let G_1 and G_2 be two graphs that are decomposable into s and t hamiltonian cycles, respectively, with $t \leq s$. Then $G_1 \times G_2$ is hamiltonian decomposable if one of the following holds:
(1) $s \leq 3t$
(2) $t \geq 3$
(3) the order of G_2 is even, or
(4) the order of G_1 is at least $6\lceil s/t \rceil - 3$.

F54: It is easy to see that if G_1 and G_2 are both bipartite, then the direct product $G_1 \cdot G_2$ is disconnected. Hence, the set of hamiltonian decomposable graphs is not closed under the direct product.

F55: [Bo90], [Zh89] Suppose both G_1 and G_2 are hamiltonian decomposable. If at least one of them has odd order, then $G_1 \cdot G_2$ is hamiltonian decomposable.

F56: [FaLi98] The set of hamiltonian decomposable graphs is closed under strong products, that is, if G_1 and G_2 are hamiltonian decomposable, then so is $G_1 \otimes G_2$.

F57: [BaSz81] The lexicographic product of two hamiltonian decomposable graphs is hamiltonian decomposable.

F58: [Kr97] Let H be a subgraph of a graph G. Each of the following conditions is sufficient for the lexicographic product $G[H]$ to be hamiltonian.
(a) G is 1-tough and contains a 2-factor, and $|E(H)| \geq 2$.
(b) G is 2-tough and $|E(H)| \geq 2$.
(c) G is connected and $2k$-regular, and $|V(H)| \geq k$.
(d) G is $(2k + 1)$-regular, connected, and has a 1-factor, and $|V(H)| \geq k + 1$.
(e) G is connected and vertex transitive of degree k, and $|V(H)| \geq k/2$.
(f) G is connected and vertex transitive, and $|E(H)| \geq 2$.
(g) G is cubic and 2-edge connected, and $|V(H)| \geq 2$.
(h) G is 4-regular and connected, and $|V(H)| \geq 2$.

4.5.5   Random Graphs

NOTATION: Throughout this subsection $Pr(X)$ denotes the probability of event X and N denotes the quantity $\binom{n}{2}$.

DEFINITIONS

D24: (*The edge density model*) Suppose that $0 \leq p \leq 1$. Let $G_{n,p}$ denote a graph on n vertices obtained by inserting any of the N possible edges with probability p.

D25: (*The fixed size model*) Suppose that $M = M(n)$ is a prescribed function of n which takes on values in the set of positive integers. Then there are $s = \binom{N}{M}$ different graphs with M edges possible on the vertex set $\{1, 2, \ldots, n\}$. We let $G_{n,M}$ denote one of these graphs chosen uniformly at random with probability $1/s$.

D26: A somewhat different approach is to consider a *graph process* as a sequence $(G_t)_{t=0}^N$ such that

1. each G_t is a graph with vertex-set V,

2. G_t has t edges for $t = 0, 1, \ldots, N$,

3. $G_0 \subset G_1 \subset \ldots$.

D27: If Ω_n is a model of random graphs of order n, we say *almost every graph* in Ω_n has property Q if $Pr(Q) \to 1$ as $n \to \infty$. Note that this is equivalent to saying that the proportion of all labeled graphs of order n that have Q tends to 1 as $n \to \infty$.

D28: The *k-in, l-out random digraph* $D_{k-in,l-out}$ has n vertices and for each vertex v, a set of k arcs into v and l arcs out of v are chosen independently and uniformly at random. The union of these arc subsets is the arc-set of $D_{k-in,l-out}$.

FACTS

F59: [Po76], [Ko76] There exists a constant c such that almost every labeled graph on n vertices and at least $cn\log n$ edges is hamiltonian.

F60: [Ko76], [KoSz83] Suppose $\omega(n) \to \infty$ as $n \to \infty$, and let

$$p = \frac{1}{n}\{log\,n + log(log\,n) + \omega(n)\} \text{ and } M(n) = \left\lfloor \frac{n}{2}\{log\,n + log(log\,n) + \omega(n)\} \right\rfloor.$$

Then almost every $G_{n,p}$ is hamiltonian and almost every $G_{n,M}$ is hamiltonian.

F61: [KoSz83] For $M(n) = n/2\,(log\,n + log(log\,n) + c_n)$

$$lim_{n\to\infty} Pr(G_{n,M} \text{ is hamiltonian}) = \begin{cases} 0 & : \text{ if } c_n \to -\infty \\ e^{-e^{-c}} & : \text{ if } c_n \to c \\ 1 & : \text{ if } c_n \to \infty. \end{cases}$$

F62: [RoWo92], [RoWo94] For every $r \geq 3$, almost all r-regular graphs are hamiltonian.

F63: [CoFr94] Almost all random digraphs $D_{3-in,3-out}$ are hamiltonian.

F64: [CoFr00] Almost all random digraphs $D_{2-in,2-out}$ are hamiltonian, In particular, this implies that G_{4-out}, the underlying graph of $D_{2-in,2-out}$, is hamiltonian. On the other hand, almost all $D_{1-in,2-out}$ and $D_{2-in,1-out}$ are not hamiltonian.

F65: [BoFr09] Almost all random graphs G_{3-out} are hamiltonian.

REMARKS

R7: Considering the probability space of all $N!$ graph processes (with equal probability) allows us to consider when a property "appears" (called the **hitting time**). Erdös and Spencer where the first to conjecture that with probability tending to 1, the very edge that increases the minimum degree to 2 also makes the graph hamiltonian. This was verified by Bollobás [Bo84].

R8: It is natural to ask for a polynomial algorithm which, with probability tending to 1, finds a hamiltonian cycle in $G_{n,M(n)}$. Bollobás, Fenner, and Frieze [BoFeFr85] constructed such an algorithm which is essentially best possible.

4.5.6   Spectral Attacks

NOTATION: Let $A(G)$ be the adjacency matrix of the graph G, let $D(G)$ be the degree matrix of G, let $L(G) = D(G) - A(G)$ be the *Laplacian* of G, and let $Q(G) = D(G) + A(G)$.

REMARK

R9: We extend our concept of a graph by allowing **free edges**, which are edges with only one end vertex. In this case the degree of a vertex counts both the ordinary and free edges incident with the vertex. However, the free edges do not appear in the adjacency matrix.

DEFINITIONS

D29: The **subdivision graph** of G, denoted $S(G)$, is the graph obtained from G by subdividing each edge of G.

D30: Let $C_{2n,l}$ denote the cycle C_{2n} with l free edges added to every second vertex of C_{2n}.

D31: The **eigenvalues of a graph** are the eigenvalues of the adjacency matrix of that graph. For a graph G we denote the eigenvalues of G as $\lambda_1(G) \le \lambda_2(G) \le \ldots \le \lambda_n(G)$.

FACTS

F66: [Mo92] Let G be a k-regular graph of order n. If G is not hamiltonian, then for $i = 1, 2, \ldots, n$, $\lambda_i(L(S(G)) \le \lambda_i(L(C_{2n,k-2}))$.

F67: [vdH95] Let G be a graph of order n and size m. If G is not hamiltonian, then for $i = 1, 2, \ldots, n$, $\lambda_i(L(C_n)) \le \lambda_i(L(G))$ and $\lambda_i(Q(C_n)) \le \lambda_i(Q(G))$.

F68: [KrSu03] If the second largest absolute value of an eigenvalue λ of the adjacency matrix of a d-regular graph satisfies

$$\lambda \leq c \frac{(log\ log\ n)^2}{log\ n\ (log\ log\ log\ n)} d$$

for a constant c and n sufficiently large, then G is hamiltonian.

F69: [BuCh10] Let G be a graph of order n and average degree d, and let $0 = \lambda_1 \leq \lambda_2 \leq \ldots \leq \lambda_n$ be the eigenvalues of the Laplacian of G. If there is a constant c so that

$$|d - \lambda_i| \leq c \frac{(log\ log\ n)^2}{log\ n\ (log\ log\ log\ n)} d$$

for $i \neq 1$ and n sufficiently large, then G is hamiltonian.

F70: [FiNi10] Let G be a graph of order n, and let $\mu(G)$ be the largest eigenvalue of the adjacency matrix of G. Then,
(a) If $\mu(G) \geq n - 2$, then G is traceable unless G is the disjoint union of K_{n-1} and a vertex.
(b) If $\mu(G) > n - 2$, then G is hamiltonian unless G is K_{n-1} with a pendant edge.
(c) If $\mu(\overline{G}) \leq \sqrt{n-1}$, then G is traceable unless G is the disjoint union of K_{n-1} and a vertex.
(d) If $\mu(\overline{G}) \leq \sqrt{n-2}$, then G is hamiltonian unless G is K_{n-1} with a pendant edge.

4.5.7 Forbidden Subgraphs

DEFINITION

D32: A graph G is said to be $\{F_1, F_2, \ldots, F_k\}$-***free*** if G contains no induced subgraph isomorphic to any F_i, $1 \leq i \leq k$.

NOTATION: The graph $N_{i,j,k}$ is a graph which consists of K_3 and vertex-disjoint paths of length i, j, k with one path rooted at each of its three vertices. The graph L consists of two vertex-disjoint copies of K_3 and an edge joining them. The graph P_i is a path with i vertices. The graph $K_{1,3}$ is the four vertex star (also called the claw).

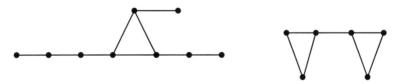

Figure 4.5.3: The graphs $N_{1,2,3}$ and L.

FACTS

F71: [DuGoJa81] If G is a $\{K_{1,3}, N_{1,1,1}\}$-free graph, then
(a) if G is 2-connected, then G is hamiltonian;
(b) if G is connected, then G is traceable.

F72: [BrDrKo00] There exists a linear time algorithm for finding a hamiltonian cycle in a $\{K_{1,3}, N_{1,1,1}\}$-free graph.

F73: [BrVe90] If G is a 2-connected $\{K_{1,3}, P_6\}$-free graph, then G is hamiltonian.

F74: [GoJa82] If G is a 2-connected $\{K_{1,3}, N_{2,0,0}\}$-free graph, then G is hamiltonian.

F75: [Be91] If G is a 2-connected $\{K_{1,3}, N_{2,1,0}\}$-free graph, then G is hamiltonian.

F76: [FaGoRySc95] If G is a 2-connected $\{K_{1,3}, N_{3,0,0}\}$-free graph of order $n \geq 10$, then G is hamiltonian.

REMARK

R10: A natural question is: Are these the only such pairs? This was investigated in [Be91] for all graphs, and in [FaGo97] for graphs of order 10 or more. We now summarize these combined results.

FACTS

F77: [Be91], [FaGo97] Let R and S be connected graphs $(R, S \neq P_3)$ and G a 2-connected graph of order n. Then G is $\{R, S\}$-free implies G is hamiltonian if, and only if, $R = K_{1,3}$ and S is one of the graphs $N_{1,1,1}$, P_6, $N_{2,1,0}$, $N_{2,0,0}$ (or $N_{3,0,0}$ when $n \geq 10$), or a connected induced subgraph of one of these graphs.

F78: [FaGo97] Let R, S be connected graphs $(R, S \neq P_3)$ and let G $(G \neq C_n)$ be a 2-connected graph of order $n \geq 10$. Then G is $\{R, S\}$-free implies G is pancyclic if, and only if, $R = K_{1,3}$ and S is one of P_4, P_5, P_6, $N_{1,0,0}$ or $N_{2,0,0}$.

F79: [GoLuPf04] Let X and Y be connected graphs on at least three vertices such that $X, Y \neq P_3$ and $Y \neq K_{1,3}$. Then the following statements are equivalent:
(a) Every 3-connected $\{X, Y\}$-free graph G is pancyclic.
(b) $X = K_{1,3}$ and Y is a subgraph of one of the graphs from the family $\mathcal{F} = \{P_7, L, N_{4,0,0}, N_{3,1,0}, N_{2,2,0}, N_{2,1,1}\}$.

F80: [FaGo97] It is an easy observation that P_3 is the only nontrivial single graph that when forbidden implies G is hamiltonian.

REMARK

R11: (Claw-free graphs) In each of the forbidden-pair results above, the claw $K_{1,3}$ is one of the two forbidden graphs. This led naturally to the question: Is the claw in every triple of forbidden subgraphs implying hamiltonicity? This was answered negatively in [FaGoJaLe02] where all triples containing no $K_{1,t}$ with $t \geq 3$ for sufficiently large 2-connected graphs were given. Further, in [FaGoJa04] other forbidden triples for sufficiently large graphs were investigated. Brousek [Br02] characterized the collection of all forbidden triples that include the claw and imply hamiltonicity for all 2-connected graphs. In [FaGoJa05], all possible remaining triples implying hamiltonicity for all graphs were given.

DEFINITIONS

D33: For a vertex x such that the induced subgraph $G[N(x)]$ is connected, a ***local completion*** of G at x is the graph obtained by replacing $G[N(x)]$ by a complete subgraph on $V(N(x))$. (Observe that a local completion of a claw-free graph is claw-free.)

D34: The ***claw-free closure*** of G, denoted by $cl(G)$, is that graph obtained by repeatedly finding the local completion of a vertex x until the resulting graph has the property that for every vertex x, $G[N(x)]$ is either non-connected or a complete subgraph.

D35: The ***circumference*** of a graph G, denoted $circum(G)$, is the length of a longest cycle in G.

FACTS

F81: [FaGo97] Let R, S be connected graphs ($R, S \neq P_3$) and G a 2-connected graph of order $n \geq 10$. Then G is $\{R, S\}$-free implies G is cycle extendable if, and only if, $R = K_{1,3}$ and S is one of K_3, P_4, $N_{1,0,0}$ or $N_{2,0,0}$.

F82: [Sh97] If G is a 3-connected $\{K_{1,3}, N_{1,1,1}\}$-free graph, then G is hamiltonian-connected.

F83: [Ry97] Let G be a claw-free graph. Then
(a) the closure $cl(G)$ is well-defined,
(b) there is a triangle-free graph H such that $cl(G) = H$, and
(c) $circum(G) = circum(cl(G))$.

REMARKS

R12: The claw-free closure is different from the degree closure (see [BoCh76]) or any of several other closures that have recently been developed. For more information on closures, see [BrRySc00].

R13: By Fact F83, if G is claw-free, then G is hamiltonian if and only if $cl(G)$ is hamiltonian.

R14: Matthews and Sumner [MaSu84] conjectured that every 4-connected claw-free graph is hamiltonian. At about the same time Thomassen [Th86] conjectured that every 4-connected line graph is hamiltonian. These were shown to be equivalent by Ryjáček [Ry97]. Several other statements are known to be equivalent to these (see [KaVra]).

FACT

F84: [KaVra] Every 5-connected line graph with minimum degree at least 6 is hamiltonian (in fact, hamiltonian connected).

REMARK

R15: This vast area contains far more than can be written here. For more details on hamiltonian graphs the reader should see [Be78], [Bo78], [WiGa84], [Bo95], [CuGa96], [Go91] and [Go03].

References

[BaSz81] Z. Baranyai and Gy. R. Szasz, Hamiltonian decompositions of lexicographic product, *J. Combin. Theory Ser. B* 31 (1981), 253–261.

[BaBrLiVe00] D. Bauer, H. J. Broersma, R. Li, and H. J. Veldman, Not every 2-tough graph is hamiltonian, *Discrete. Appl. Math.* 99 (2000), 317–321.

[BaBrVeLi89] D. Bauer, H. J. Broersma, H. J. Veldman, and R. Li, A generalization of a result of Haggkvist and Nicoghossian, *J. Combin. Theory Ser. B* 47 (1989), no. 2, 237–243.

[BaMoScVe90] D. Bauer, A. Morgana, E. Schmeichel, and H. J. Veldman, Long cycles in graphs with large degree sums, *Discrete Math.* 79, no. 1, (1989/90), 59–70.

[Be91] P. Bedrossian, *Forbidden subgraphs and minimum degree conditions for hamiltonicity*, Ph.D. Thesis, Memphis State University, 1991.

[Be78] J.-C. Bermond, Hamiltonian graphs, in *Selected Topics in Graph Theory*, L. Beineke and R. Wilson, ed., Academic Press, London (1978).

[BiLlWi86] N. L. Biggs, E. K. Lloyd, and R. J. Wilson, *Graph Theory 1736–1936*, Oxford University Press, Oxford (1986).

[Bo84] B. Bollobás, The evolution of sparse random graphs, *Trans. Am. Math. Soc.*, 286 (1984), 257–274.

[Bo77] J. A. Bondy, Pancyclic graphs, *J. Combin. Theory Ser. B* 11 (1977), 80–84.

[Bo78] J. A. Bondy, Hamilton cycles in graphs and digraphs, *Proceedings 9th S.E. Conf. on Combin., Graph Theory and Computing*, in Congr. Numer. XXI (1978), 3–28.

[Bo90] J. Bosak, *Decompositions of Graphs*, Kluwer Academic Publishers (1990).

[Bo95] J. A. Bondy, Basic graph theory - paths and cycles, *Handbook of Combinatorics I*, Elsevier, Amsterdam (1995), 5–110.

[BoCh76] J. A. Bondy and V. Chvátal, A method in graph theory, *Discrete Math.* 15 (1976), no. 2, 111–135.

[BoFr09] T. Bohman and A. Frieze, Hamiltonian cycles in 3-out, *Random Structures Algorithms* 35 (2009), no. 4, 393–417.

[BoFeFr85] B. Bollobás, T. I. Fenner and A. M. Frieze, On matchings and hamiltonian cycles in random graphs, *Proceedings of ACM Symposium on Theory of Computing*, New York (1985), 430–439.

[BoJa98] J. A. Bondy and B. Jackson, Vertices of small degree in uniquely hamiltonian graphs, *J. Combin. Theory Ser. B*, 74 (1998), 265–275.

[Br02] J. Brousek, Forbidden triples for hamiltonicity, *Discrete Math.* 251 (2002), 71–76.

[BrChFaGoLe97] S. Brandt, G. Chen, R. J. Faudree, R.J. Gould, and L. Lesniak, On the number of cycles in a 2-factor, *J. Graph Theory* Vol. 24, no. 2 (1997), 165–173.

[BrDrKo00] A. Brandstädt, F. F. Dragan, and E. Köhler, Linear time algorithms for hamiltonian problems on (claw, net)-free graphs, *SIAM J. Comput.* Vol. 30 (2000), no. 5, 1662–1677.

[BrRySc00] H. Broersma, Z. Ryjáček, and I. Schiermeyer, Closure concepts: a survey, *Graphs and Combinatorics* 16 (2000), 17–48.

[BrVe90] H. J. Broersma and H. J. Veldman, Restrictions on induced subgraphs ensuring hamiltonicity or pancyclicity of $K_{1,3}$-free graphs, *Contemporary Methods in Graph Theory*, R. Bodendiek, ed., BI-Wiss.-Verl., Mannheim-Wien-Zurich (1990), 181–194.

[BrVe91] H. J. Broersma and H. J. Veldman, Long dominating cycles and paths in graphs with large neighborhood unions, *J. Graph Theory* 15 (1991), 20–38.

[BuCh10] S. Butler and F. Chung, Small spectral gap in the combinatorial Laplacian implies hamiltonian, *Ann. Comb.* 13 (2010), 403–412.

[ChWa73] G. Chartrand and C. E. Wall, On the hamiltonian index of a graph, *Studia Sci. Math Hungar.*, 8 (1973), 43–48.

[ChEr72] V. Chvátal and P. Erdös, A note on hamiltonian circuits, *Discrete Math.* 2 (1972), 111–113.

[CoFr94] C. Cooper and A. M. Frieze, Hamilton cycles in a class of random directed graphs, *J. Combin. Theory Ser. B* 62 (1994), no. 1, 151–163.

[CoFr00] C. Cooper and A. M. Frieze, Hamilton cycles in random graphs and directed graphs, *Random Structures Algorithms* 16 (2000), no. 4, 369–401.

[CuGa96] S. J. Curran and J. A. Gallian, Hamiltonian cycles and paths in Cayley graphs and digraphs - a survey, *Discrete Math.* 156 (1996), no. 1-3, 1–18.

[Di52] G. A. Dirac, Some theorems on abstract graphs, *Proc. London Math. Soc.* 2 (1952), 69–81.

[DuGoJa81] D. Duffus, R. J. Gould, and M. S. Jacobson, Forbidden subgraphs and the hamiltonian theme, *The Theory and Applications of Graphs*, ed. by G. Chartrand, J. Alavi, D. Goldsmith, L. Lesniak, and D. Lick (1981), 297–316.

[Eg93] Y. Egawa, Edge-disjoint hamiltonian cycles in graphs of Ore type, *SUT J. Math.* 29 (1993), no. 1, 15–50.

[EnSw80] R. Entringer and H. Swart, Spanning cycles of nearly cubic graphs, *J. Combin. Theory Ser. B* 29 (1980), 303–309.

[Fa84] G. H. Fan, New sufficient condition for cycles in graphs, *J. Combin. Theory Ser B* 37 (1984), 221–227.

[FaGo97] R. J. Faudree and R. J. Gould, Characterizing forbidden pairs for hamiltonian properties, *Discrete Math.* 173 (1997) 45–60.

[FaGoJa04] R. J. Faudree, R. J. Gould, and M. S. Jacobson, Forbidden triples implying hamiltonicity: for all graphs, *Discuss. Math. Graph Theory* 24 (2004), no. 1, 47–54.

[FaGoJa05] R. J. Faudree, R.J. Gould, and M. S. Jacobson, Forbidden triples implying hamiltonicity: for sufficiently large graphs, *Discuss. Math. Graph Theory* 25 (3) (2005), 273–289.

[FaGoJa09] R. J. Faudree, R. J. Gould, and M. S. Jacobson, Pancyclic graphs and linear forests. *Discrete Math.* 309 (2009), no. 5, 1178–1189.

[FaGoJaLe92] R. J. Faudree, R. J. Gould, M. S. Jacobson, and L. Lesniak, On a generalization of Dirac's theorem, *Discrete Math.* 105 (1992), 61–71.

[FaGoJaLe02] R. J. Faudree, R.J. Gould, M. S. Jacobson, and L. Lesniak, Characterizing forbidden clawless triples implying hamiltonian graphs, *Discrete Math.* 249 (2002), no. 1–3, 71–81.

[FaGoJaMa] R. J. Faudree, R. J. Gould, M. S. Jacobson, and C. Magnant, Distributing vertices on hamiltonian cycles, *J. Graph Theory*, to appear.

[FaGoKoLeScSa03] R. J. Faudree, R.J. Gould, A. Kostochka, L. Lesniak, I. Schiermeyer, and A. Saito, Degree Conditions for k-ordered hamiltonian graphs, *J. Graph Theory* 42, no. 3 (2003), 199–210.

[FaGoRySc95] R. J. Faudree, R.J. Gould, Z. Ryjáček, and I. Schiermeyer, Forbidden subgraphs and pancyclicity, *Congr. Numer.* 109 (1995), 13–32.

[FaLi98] C. Fan and J. Liu, Hamiltonian decompositions of strong products, *J. Graph Theory* 29 (1998), no. 1, 45–55.

[FaRoSc85] R. J. Faudree, C. Rousseau, and R. Schelp, Edge disjoint hamiltonian cycles, *Graph Theory with Applications to Algorithms and Computer Science* (Kalamazoo, Mich., 1984), Wiley, New York (1985), 231–249.

[FiNi10] M. Fiedler and V. Nikiforov, Spectral radius and hamiltonicity of graphs, *Linear Algebra and its Apps.* 432 (2010), 2170–2173.

[Fl74] H. Fleischner, The square of every two-connected graph is hamiltonian, *J. Combin. Theory Ser. B* 16 (1974), 29–34.

[Fr86] P. Fraisse, A new sufficient condition for hamiltonian graphs, *J. Graph Theory*, 10 (1986), 405–409.

[GaJo79] M. R. Garey and D. S. Johnson, *Computers and Intractability A Guide to the Theory of NP-Completeness*, Freeman and Co., New York, (1979).

[Go91] R. J. Gould, Updating the hamiltonian problem - a survey, *J. Graph Theory* 15 (1991), no. 2, 121–157.

[Go03] R. J. Gould, Advances on the hamiltonian problem - a survey, *Graphs and Combinatorics*, 19 (2003), no. 1, 7–52.

[GoHy99] R. J. Gould and E. Hynds, A note on cycles in 2-factors of line graphs, *Bull. of the I.C.A.*, 26 (1999), 46–48.

[GoJa82] R. J. Gould and M. S. Jacobson, Forbidden subgraphs and hamiltonian properties of graphs, *Discrete Math.* 42 (1982), 189–196.

[GoLuPf04] R. J. Gould, T. Łuczak, and F. Pfender, Pancyclicity of 3-connected graphs: pairs of forbidden subgraphs, *J. Graph Theory*, 47 (2004), no. 3, 183–202.

[Gr68] E. J. Grinberg, Plane homogeneous graphs of degree three without hamiltonian circuits, *Latvian Math. Yearbook* 4 (1968), 51–58.

[GrMa76] B. Grunbaum and J. Malkevitch, Pairs of edge-disjoint hamiltonian circuits, *Aequationes Math.* 14 (1976), no. 1/2, 191–196.

[Ha79] R. Häggkvist, On *F*-hamiltonian graphs, *Graph theory and related topics* (Proc. Conf., Univ. Waterloo, Waterloo, Ont., 1977), pp. 219–231, Academic Press, New York-London 1979.

[HaNW65] F. Harary and C. St. J. A. Nash-Williams, On eulerian and hamiltonian graphs and line graphs, *Canad. Math. Bull.* (1965), 701–710.

[He90] G. Hendry, Extending cycles in graphs, *Discrete Math.* 85 (1990), no. 1, 59–72.

[He91] G. Hendry, Extending cycles in bipartite graphs, *J. Combin. Theory Ser. B* 51 (1991), no. 2, 292–313.

[HoSt00] P. Horak and L. Stacho, A lower bound on the number of hamiltonian cycles, *Discrete Math.* 222 (2000), no. 1–3, 275–280.

[Ja79] B. Jackson, Edge-disjoint hamiltonian cycles in regular graphs of large degree, *J. London Math. Soc.* (2) 19 (1979), no. 1, 13–16.

[Ja80] B. Jackson, Hamilton cycles in regular 2-connected graphs, *J. Combin. Theory Ser B* 29 (1980), 27–46.

[JaWh89] B. Jackson and R. W. Whitty, A note concerning graphs with unique *f*-factors, *J. Graph Theory* 13 (1989), 577–580.

[Ju78] H. A. Jung, On maximal circuits in finite graphs, *Annals of Discrete Math.* 3 (1978), 129–144.

[KaVra] T. Kaiser and P. Vtána, Hamilton cycles in 5-connected line graphs, *European J. Combin.*, to appear.

[KaYo01] A. Kaneko and K. Yoshimoto, On a hamiltonian cycle in which specified vertices are uniformly distributed, *J. Combin. Theory Ser. B* 81 (2001), no. 1, 100–109.

[Ki56] T. P. Kirkman, On the representation of polyedra, *Philosophical Transactions of the Royal Society London*, 146 (1856), 413–418.

[KiSaSe99] H. S. Kierstead, G. N. Sárközy and S. Selkow, On *k*-ordered hamiltonian graphs, *J. Graph Theory* 32 (1999), no. 1, 17–25.

[Ko76] A. D. Korshunov, Solution of a problem of Erdös and Rényi on Hamilton cycles in non-oriented graphs, *Soviet Mat. Dokl.* 17 (1976), 760–764.

[KoSaSz96] J. Komlós, G. N. Sárközy, and E. Szemerédi, On the square of a hamiltonian cycle in dense graphs, *Random Structures and Algorithms* 9 (1996), no. 1–2, 193–211.

[KoSaSz98] J. Komlós, G. N. Sárközy, and E. Szemerédi, Proof of the Seymour conjecture for large graphs, *Ann. Comb.* 2 (1998), no. 1, 43–60.

[KoSz83] J. Komlós and E. Szermerédi, Limit distribution for the existence of hamiltonian cycles in random graphs, *Discrete Math.* 43 (1983), no. 1, 55–63.

[Kr97] M. Kriesell, A note on hamiltonian cycles in lexicographic products, *J. Autom. Lang. Comb.* 2 (1997), no. 2, 135–138.

[KrSu03] M. Krivelevich and B. Sudakov, Sparse pseudo-random graphs are hamiltonian, *J. Graph Theory* 42 (2003), 17–33.

[KrZe88] J. Kratochvil and D. Zeps, On the number of hamiltonian cycles in triangulations, *J. Graph Theory* 12 (1988), 191–194.

[LV72] M. Las Vergnas, Thesis, University of Paris, 1972.

[Ma76] P. Martin, Cycles hamiltoniens dans les graphes 4-reguliers 4-connexes, *Aequationes Math.* 14 (1976), no.1/2, 37–40.

[MaSu84] M. M. Matthews and D. P. Sumner, Hamiltonian results in $K_{1,3}$-free graphs, *J. Graph Theory* 8 (1984), 139–146.

[Mo92] B. Mohar, A domain monotonicity theorem for graphs and hamiltonicity, *Discrete Appl. Math.* 36 (1992), 169–177.

[MoMo63] J. Moon and L. Moser, On hamiltonian cycles in bipartite graphs, *Isr. J. Math.* 1 (1963), 163–165.

[Or60] O. Ore, A note on hamiltonian circuits, *Amer. Math. Monthly* 67 (1960), 55.

[Or63] O. Ore, Hamiltonian connected graphs, *J. Math. Pures. Appl.* 42 (1963), 21–27.

[Po76] L. Pósa, Hamiltonian circuits in random graphs, *Discrete Math.* 14 (1976), 359–364.

[Ro89] M. Rosenfeld, Pairs of edge disjoint hamiltonian circuits in 5-connected planar graphs, *Aequationes Math.* 38 (1989), no. 1, 50–55.

[RoWo92] R. W. Robinson and N. C. Wormald, Almost all cubic graphs are hamiltonian, *Random Structures Algorithms* 3 (1992), no. 2, 117–125.

[RoWo94] R. W. Robinson and N. C. Wormald, Almost all regular graphs are hamiltonian, *Random Structures Algorithms* 5 (1994), no. 2, 363–374.

[Ry97] Z. Ryjáček, On a closure concept in claw-free graphs, *J. Combin. Theory Ser B* 70 (1997), 217–224.

[SaSe08] G. Sáközy and S. Selkow, Distributing vertices along a hamiltonian cycle in Dirac graphs, *Discrete Math.* 308 (2008), 5757–5770.

[Sh97] F. B. Shepherd, Hamiltonicity in claw-free graphs, *J. Combin. Theory Ser. B* 53 (1991) 173–194.

[St91] R. Stong, Hamilton decompositions of cartesian products of graphs, *Discrete Math.* 90 (1991), 169–190.

[Th78] A. G. Thomason, Hamiltonian cycles and uniquely edge colourable graphs, *Ann. Discrete Math.* 3 (1978), 259–268.

[Th83] C. Thomassen, A theorem on paths in planar graphs, *J. Graph Theory*, 7 (1983), 169–176.

[Th86] C. Thomassen, Reflections on graph theory, *J. Graph Theory* 10 (1986), 309–324.

[Th96] C. Thomassen, On the number of hamiltonian cycles in bipartite graphs, *Combinatorics, Probability and Computing* 5 (1996), no. 4, 437–442.

[Th97] C. Thomassen, Chords of longest cycles in cubic graphs, *J. Combin. Theory Ser. B* 71 (1997), 211–214.

[Th98] C. Thomassen, Independent dominating sets and a second hamiltonian cycle in regular graphs, *J. Combin. Theory Ser. B* 72 (1998), 104–109.

[Tu46] W. T. Tutte, On hamiltonian circuits, *J. London Math. Soc.* 21 (1946), 98–101.

[Tu56] W. T. Tutte, A theorem on planar graphs, *Trans. Amer. Math. Soc.* 82 (1956), 99–116.

[vdH95] J. van den Heuvel, Hamilton cycles and eigenvalues of graphs, *Linear Algebra App.* 226–228 (1995), 723–730.

[WiGa84] D. Witte and J. A. Gallian, A survey – Hamiltonian cycles in Cayley graphs, *Discrete Math.* 51 (1984), 293–304.

[Wo78] D. R. Woodall, A sufficient condition for hamiltonian circuits, *J. Combin. Theory Ser. B* 25, no. 2 (1978), 184–186.

[Ya99] Z. Yang, Note on F-hamiltonian graphs, *Discrete Math.* 196 (1999), 281–286.

[Za76] J. Zaks, Pairs of hamiltonian circuits in 5-connected planar graphs, *J. Combin. Theory, Ser. B* 21 (1976), 116–131.

[Zh89] M. Zhou, Decomposition of some product graphs into 1-factors and hamiltonian cycles, *Ars Combinatoria* 28 (1989), 258–268.

Section 4.6
Traveling Salesman Problems

Gregory Gutin, Royal Holloway, University of London

INTRODUCTION

The Traveling Salesman Problem (TSP) is perhaps the most frequently studied discrete optimization problem. Its popularity is due to the facts that TSP is easy to formulate, difficult to solve, and has a large number of applications. TSP has a number of variations and generalizations extensively studied in the literature [Pu02]. In this section, we consider TSP, the Generalized TSP and the Vehicle Routing Problem.

4.6.1   The Traveling Salesman Problem

K. Menger [Me32] was perhaps the first researcher to consider the Traveling Salesman Problem (TSP). He observed that the problem can be solved by examining all permutations one by one. Realizing that the complete enumeration of all permutations was not possible for graphs with a large number of vertices, he looked at the most natural *nearest neighbor* strategy and pointed out that this *heuristic*, in general, does not produce the shortest route. In fact, as we will see below, the nearest neighbor heuristic will generate the worst possible route for some problem instances. (For an interesting overview of TSP history, see [HoWo85].)

In applications, both the symmetric and asymmetric versions of the TSP are important.

Symmetric and Asymmetric TSP

DEFINITIONS

D1: Symmetric TSP (STSP):
Given a complete (undirected) graph K_n with weights on the edges, find a hamiltonian cycle in K_n of minimum (total) weight.

D2: Asymmetric TSP (ATSP):
Given a complete directed graph $\overleftrightarrow{K}_n$ with weights on the arcs, find a hamiltonian cycle in $\overleftrightarrow{K}_n$ of minimum weight.

D3: The **Euclidean TSP** is the special case of STSP in which the vertices are points in the Euclidean plane and the weight on each edge is the Euclidean distance between its endpoints.

D4: A hamiltonian cycle in K_n or $\overleftrightarrow{K}_n$ is called a **tour**.

NOTATION: Throughout this section, the set $\{1, 2, \ldots, n\}$ denotes the vertices of K_n or $\overleftrightarrow{K}_n$ or any other n-vertex graph under discussion.

NOTATION: By **TSP** we refer to both STSP and ATSP simultaneously.

Matrix Representation of TSP

Every instance of TSP can be associated with the matrix of edge-weights of the corresponding complete graph. Such a matrix is symmetric for STSP and, in general, asymmetric for ATSP.

DEFINITIONS

D5: The **distance** (or **weight**) **matrix** of an instance of STSP is the matrix $D = [d_{ij}]$, where d_{ij} is the weight of the edge between vertices i and j. The **distance matrix** of an instance of ATSP is the matrix $D = [d_{ij}]$, where d_{ij} is the weight of the arc directed from i to j. Accordingly, the diagonal entries d_{ii} are set to zero.

D6: An instance of TSP is said to satisfy the **triangle inequality** if $d_{ij} + d_{jk} \geq d_{ik}$ for all distinct vertices i, j, k.

D7: Metric TSP is the special case of STSP where every instance satisfies the triangle inequality.

Clearly, the Euclidean TSP is a special case of the Metric TSP.

EXAMPLES

E1: An instance of ATSP with distance matrix

$$\begin{pmatrix} 0 & 6 & 5 & 10 \\ 3 & 0 & 3 & 9 \\ 7 & 4 & 0 & 8 \\ 12 & 7 & 5 & 0 \end{pmatrix}$$

is shown in Figure 4.6.1. There are $3! = 6$ tours of total weight 29, 27, 30, 23, 27, and 22. The optimal tour is $(1, 4, 3, 2, 1)$ of weight 22.

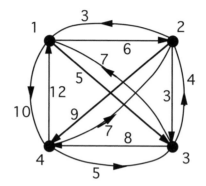

Figure 4.6.1: An instance of ATSP.

E2: An instance of STSP with distance matrix

$$D = \begin{pmatrix} 0 & 10 & 7 & 7 & 11 \\ 10 & 0 & 9 & 6 & 5 \\ 7 & 9 & 0 & 9 & 10 \\ 7 & 6 & 9 & 0 & 6 \\ 11 & 5 & 10 & 6 & 0 \end{pmatrix}$$

is shown in Figure 4.6.2. Since this graph has 5 vertices, there are $4!/2 = 12$ tours. The optimal tour is $(1, 3, 2, 5, 4, 1)$ of weight 34.

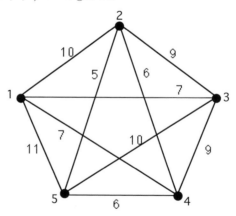

Figure 4.6.2: An instance of STSP.

Algorithmic Complexity

FACTS

F1: The hamiltonian cycle problem on an n-vertex graph G can be transformed into STSP by converting G to an edge-weighted K_n as follows: assign weight 0 to each edge of G, and assign weight 1 to each edge in the edge-complement of G. A similar transformation can be used for digraphs and ATSP.

F2: The previous fact implies that even the Metric TSP is NP-hard.

By replacing the weights 0 by 1 and the weights 1 by nr in the transformation above, we obtain the following result.

F3: [SaGo76] For an arbitrary constant r, unless P = NP, there is no polynomial time algorithm that always produces a tour of total weight at most r times the optimal.

Exact and Approximate Algorithms

DEFINITIONS

D8: An ***exact algorithm*** is an algorithm that always produces an optimal solution.

D9: An ***approximate*** (or ***approximation***) ***algorithm*** is an algorithm that typically makes use of heuristics in reducing its computation but produces solutions that are not necessarily optimal.

NOTATION: Let $\mathcal{A}$ be an approximate algorithm for TSP and I a problem instance. Then $w_{\min}(I)$, $w_{\max}(I)$, $w_{\mathcal{A}}(I)$ denote the weights, respectively, of an optimal tour, a heaviest tour, and a tour produced by algorithm $\mathcal{A}$, for instance I.

D10: The ***Zemel measure*** [Ze81] of an algorithm $\mathcal{A}$, denoted $\rho_z(\mathcal{A})$, is the supremum of $(w_{\mathcal{A}}(I) - w_{\min}(I))/(w_{\max}(I) - w_{\min}(I))$, taken over all TSP instances I for which $w_{\max}(I) \neq w_{\min}(I)$.

FACT

F4: [HaKh01] There is a polynomial time approximate algorithm $\mathcal{A}$ for ATSP with $\rho_z(\mathcal{A}) \leq \frac{1}{2}$, and one for STSP with $\rho_z(\mathcal{A}) \leq \frac{1}{3}$.

The Euclidean TSP

Despite the fact that Euclidean TSP is NP-hard [Pa77,GaGrJo76], there was a feeling among some researchers that the Euclidean TSP is somewhat simpler than the general STSP. This was confirmed by Arora [Ar98] in 1996 (see below). Mitchell [Mi99] independently made a similar discovery a few months later (see [Ar02]).

FACTS

F5: [Ar02] For every $\epsilon > 0$, there is a polynomial time algorithm $\mathcal{A}_\epsilon$ that, for any instance of the Euclidean TSP, finds a tour at most $1 + \epsilon$ times longer than the optimal one.

F6: As of this writing, the fastest algorithm $\mathcal{A}_\epsilon$ has time complexity $O(n \log n + n/\text{poly}(\epsilon))$ [RaSm98].

COMPUTATIONAL NOTE: These $\mathcal{A}_\epsilon$ algorithms have been implemented, but, in their current form, they are not competitive with other TSP heuristics [Ar02, RoCiFa09].

F7: [Tr97] There exists a constant $r > 1$ such that, for the Euclidean TSP in $O(\log n)$-dimensional Euclidean space, the problem of finding a tour that is at most r times longer than the optimal tour is NP-hard.

REMARKS

R1: Arora's result above can be generalized to d-dimensional Euclidean space for any constant d. However, the previous fact limits the scope of this generalization.

R2: Exact algorithms cannot be relied on for applications requiring very fast solutions (online, for example) or ones that involve huge problem instances. Although approximate algorithms forfeit the guarantee of optimality, with good heuristics they can normally produce solutions close to optimal. For many applications this is good enough, since often the data are inexact anyway.

R3: TSP heuristics can be roughly partitioned into two classes: ***construction heuristics*** and ***improvement heuristics***. Both types are discussed below. Other overviews of TSP heuristics can be found in [GoSt85], [JoGuMcYeZhZv02], and [JoMc02].

4.6.2   Exact Algorithms

The NP-hardness results mentioned in the previous subsection indicate that it is rather difficult to solve large instances of TSP to optimality. Nevertheless, there are computer codes that can solve many instances with thousands of vertices within days (on a single-processor computer) [ApBiChCo06]. For a discussion of some TSP software implementing both exact algorithms and heuristics, see [LoPu02].

FACT

F8: The *brute-force* method of explicitly examining all possible TSP tours is impractical for even moderately sized problem instances because there are $(n-1)!/2$ different tours in K_n and $(n-1)!$ different tours in $\overset{\leftrightarrow}{K}_n$.

Integer Programming Approaches

Various methods have been suggested to solve TSP to optimality. They include *Lagrangian relaxation* ([BeLu00]), *dynamic programming* ([PaSt82]), and *branch-and-bound* and *branch-and-cut* (see [BaTo85], [FiLoTo02], and [Na02]). These are all well-known methods in integer programming ([Wo98]). The earliest (and still useful) integer programming formulation of ATSP is due to Dantzig, Fulkerson, and Johnson [DaFuJo54].

DANTZIG, FULKERSON, AND JOHNSON FORMULATION: Define zero-one variables x_{ij} by

$$x_{ij} = \begin{cases} 1, & \text{if the tour traverses arc}(i,j) \\ 0, & \text{otherwise} \end{cases}$$

Let d_{ij} be the weight on arc (i,j). Then ATSP can be expressed as:

$$\min z = \sum_{i=1}^{n} \sum_{j=1}^{n} d_{ij} x_{ij}$$

$$\text{subject to} \quad \sum_{i=1}^{n} x_{ij} = 1, \ j = 1, 2, \ldots, n$$

$$\sum_{j=1}^{n} x_{ij} = 1, \ i = 1, 2, \ldots, n$$

$$\sum_{i \in S} \sum_{j \in S} x_{ij} \leq |S| - 1 \text{ for all } 0 < |S| < n$$

$$x_{ij} = 0 \text{ or } 1, \quad i, j = 1, \ldots, n$$

FACTS

F9: The first set of constraints ensures that a tour must come into vertex j exactly once, and the second set of constraints indicates that a tour must leave every vertex i exactly once. These two sets of constraints ensure that there are two arcs adjacent to each vertex, one in and one out. However, this does not prevent non-hamiltonian cycles. Instead of having one tour, the solution can consist of two of more vertex-disjoint cycles (called ***sub-tours***).

F10: The third set of constraints, called ***sub-tour elimination constraints***, requires that no proper subset of vertices, S, can have a total of $|S|$ arcs.

F11: The formulation without the third set of constraints is an integer programming formulation of the *Assignment Problem* that can be solved in time $O(n^3)$ [Wo98]. A solution of the Assignment Problem is a minimum-weight collection of vertex-disjoint cycles $C_1, \ldots, C_t$ spanning $\overset{\leftrightarrow}{K_n}$. If $t = 1$, then an optimal solution of ATSP has been obtained. Otherwise, one can consider two or more subproblems. For example, for a particular arc $a \in C_i$, one subproblem could add the constraint that arc a be in the solution, and a second subproblem could require that a not be in the solution. This simple idea gives a basis for branch-and-bound algorithms for ATSP.

4.6.3   Construction Heuristics

Approximate algorithms based on construction heuristics build a tour from scratch and stop when one is produced.

Greedy-Type Algorithms

The simplest and most obvious construction heuristic is *nearest neighbor*. The nearest neighbor greedy algorithm constructs a tour by always choosing as the next vertex to visit one that is nearest to the last one visited.

Algorithm 4.6.1: Nearest Neighbor (NN)

Input: $n \times n$ distance matrix $[d_{ij}]$ and a fixed vertex i_1.
Output: TSP tour $(i_1, i_2, \ldots, i_n, i_1)$.

     Initialize $S := \{1, 2, \ldots, n\} - \{i_1\}$.
     For $k = 2, 3, \ldots, n$
         Choose i_k such that $d_{i_{k-1}, i_k} = \min\limits_{s \in S} \{d_{i_{k-1}, s}\}$.
         $S := S - \{i_k\}$.

A second greedy-type algorithm is based on the observation that a vertex-disjoint collection of paths in $\overleftrightarrow{K}_n$ (K_n) can be extended to a tour in $\overleftrightarrow{K}_n$ (K_n).

Algorithm 4.6.2: Greedy Heuristic (GR)

Input: $n \times n$ distance matrix $[d_{ij}]$.
Output: ATSP (STSP) tour as a set S of arcs (edges).

     Set $S = \emptyset$ and $m = n(n-1)$ (for ATSP) or $m = n(n-1)/2$ (for STSP).
     Sort the arcs (edges) $a_1, a_2, \ldots, a_m$ in non-decreasing order of weight.
     For $i = 1, 2, \ldots, m$
         If $S \cup \{a_i\}$ is the arc (edge) set of a collection of vertex-disjoint paths
            or is the arc (edge) set of a tour,
                $S := S \cup \{a_i\}$

EXAMPLE

E3: The performance of Nearest Neighbor on STSP or ATSP can be arbitrarily bad. For the instance of STSP shown in Figure 4.6.3, starting at vertex 1, Algorithm NN moves to vertex 4, then to vertex 3, and on to vertex 2. The resulting tour is $(1, 4, 3, 2, 1)$ with weight 103, whereas, the optimal tour is $(1, 3, 2, 4, 1)$ with weight 6.

COMPUTATIONAL NOTE: Computational experiments in [JoGuMcYeZhZv02] indicate that, in fact, on most real-world problem instances of ATSP, Algorithm NN performs better than Algorithm GR; GR fails completely on one family of instances, where the average GR-tour is more than 2000% above the optimum. Computational experiments for STSP in [JoMc02] show that both GR and NN perform relatively well on Euclidean instances and perform poorly for general STSP. GR appears to perform better than NN for STSP.

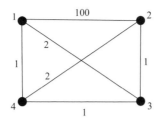

Figure 4.6.3: Performance of Nearest Neighbor can be arbitrarily bad.

Insertion Algorithms

Vertex insertion, another type of construction heuristic, applies to both Symmetric and Asymmetric TSP. For ATSP, the insertion algorithm begins with a cycle of length 2, and in each iteration, inserts a new vertex into the cycle. For STSP, the algorithm begins with a cycle of length 3. The algorithm descriptions and examples that follow are for ATSP, but with the obvious adjustments, they apply equally well to STSP.

DEFINITION

D11: Let $C = (i_1, i_2, \ldots, i_m, i_1)$ be the vertex sequence of a cycle in $\overset{\leftrightarrow}{K}_n$, and let v be a vertex not on C. For any arc (a, b) on cycle C, the **insertion of vertex v at arc** (a, b) is the operation of replacing arc (a, b) with the arcs (a, v) and (v, b) (see Figure 4.6.4). The resulting cycle is denoted $C(a, v, b)$. Thus, if $(a, b) = (i_k, i_{k+1})$, $1 \le k \le m - 1$, then $C(a, v, b) = (i_1, i_2, \ldots, a, v, b, \ldots i_m, i_1)$, and if $(a, b) = (i_m, i_1)$, then $C(a, v, b) = (i_1, i_2, \ldots i_{m-1}, a, v, b)$.

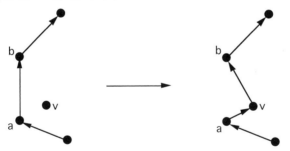

Figure 4.6.4: Insertion of vertex v at arc (a, b).

Algorithm 4.6.3: Vertex Insertion (VI)

Input: $n \times n$ distance matrix $[d_{ij}]$.
Output: TSP tour $(i_1, i_2, \ldots, i_n, i_1)$.

Let i_1 and i_2 be two vertices of $\overset{\leftrightarrow}{K}_n$, chosen by some rule.
Initialize cycle $C = (i_1, i_2, i_1)$.
   For $s = 3, \ldots, n$
      Let i_s be a vertex not on cycle C, chosen by some rule [*].
      Insert vertex i_s at an arc (a^*, b^*) of cycle $C = (i_1, i_2, \ldots, i_{s-1}, i_1)$
      such that the weight of $C(a^*, i_s, b^*)$ is minimum among the
      cycles $C(a, i_s, b)$ for all arcs (a, b) in C.
      $C := C(a^*, i_s, b^*)$.

REMARK

R4:   *Random vertex insertion, nearest vertex insertion,* and *farthest vertex insertion,*
which are defined below, are three different versions of Algorithm VI. Each is determined
by how rule [*] chooses vertex i_s.

NOTATION: Given a vertex v and a cycle C in $\overset{\leftrightarrow}{K}_n$, $d(v, C)$ denotes the distance from v
to C, that is, $d(v, C) = \min\limits_{x \in V(C)} \{d_{vx}\}$.

DEFINITIONS

D12:   The **random vertex insertion** (**RVI**) chooses vertex i_s randomly.

D13:   The **nearest vertex insertion** (**NVI**) chooses vertex i_s so that its distance to
cycle C is a minimum. That is, $d(i_s, C) = \min\limits_{v \notin V(C)} \{d(v, C)\}$.

D14:   The **farthest vertex insertion** (**FVI**) chooses vertex i_s so that its distance to
cycle C is a maximum. That is, $d(i_s, C) = \max\limits_{v \notin V(C)} \{d(v, C)\}$.

COMPUTATIONAL NOTE: The vertex insertion heuristics described above perform quite
well for Euclidean TSP (see [JoMc02]). Computational experiments with Algorithm RVI
for ATSP in [GlGuYeZv01] show that RVI is good only for instances close to Euclidean.

Minimum Spanning Tree Heuristics

There are many more construction heuristics for TSP and especially for STSP (see, e.g.,
[JoGuMcYeZhZv02], [JoMc02], [KaRe04]). For STSP, the heuristics that are often given
in the literature include the *Double Minimum Spanning Tree* (DMST) and *Christofides*
(CH) heuristics (see, e.g., Algorithms 6.4.2 and 6.4.3 in [GrYe06]). Here is pseudocode
for the DMST heuristic. Following the pseudocode is a remark highlighting the difference
between the DMST and CH heuristics.

Algorithm 4.6.4: DMST

Input: $n \times n$ distance matrix $[d_{ij}]$.
Output: STSP tour $(i_1, i_2, \ldots, i_n, i_1)$.

  Find a minimum spanning tree T in K_n.
  Create a eulerian multigraph H by doubling the edge set of T.
  Construct a eulerian tour W of H.
  Let $W = (i_1, i_2, \ldots, i_m, i_1)$ (written as a sequence of vertices).
  For $s = 3, \ldots, n$ (look for *shortcuts*)
    If $i_s = i_t$ for some $t < s$, then delete $i_{t+1}, \ldots, i_s$ from W.

REMARK

R5:   The only difference between the algorithms that use the CH and DMST heuristics
is the way in which the eulerian multigraph H is constructed. In particular, using the
CH heuristic, $E(H) = E(T) \cup M$, where M is a minimum-weight perfect matching in
the subgraph of K_n induced on the odd-degree vertices of T.

DEFINITION

D15: The operations to eliminate repeated vertices in the eulerian tour W in order to obtain an STSP tour are called ***shortcuts***.

COMPUTATIONAL NOTE: Implementing the ordinary shortcuts when using the CH heuristic already produces a relatively good heuristic for the Euclidean TSP [JoMc02]. However, so-called greedy shortcuts (see [JoMc02]) result in a modification of CH, which seems to be one of the best construction heuristics for the Euclidean TSP. Deineko and Tiskin [DeTi10] designed and implemented an algorithm that finds an optimal sequence of shortcuts for the multigraph H of DMST. The resulting heuristic appears to be among the best construction heuristics for STSP.

FACTS

F12: [JoPa85] The algorithm that uses the CH heuristic can be implemented to run in time $O(n^3)$.

F13: [DeTi10] For the Euclidean TSP, the algorithm of Deineko and Tiskin mentioned above runs in time $O(n^2)$ and memory $O(n)$.

Worst Case Analysis of Heuristics

While computational experiments are important in the evaluation of heuristics, they cannot cover all possible families of instances of TSP, and, in particular, they normally do not cover the most difficult instances. Moreover, certain applications may produce families of instances that are much harder than those normally used in computational experiments. For example, such instances can arise when Generalized TSP (discussed later) is transformed into TSP. Thus, theoretical analysis of the worst possible cases is also important in evaluating and comparing TSP heuristics.

FACTS

F14: [GuYeZv02-a] For every $n \geq 3$, there is an instance of ATSP and an instance of STSP with n vertices satisfying the triangle inequality on which NN outputs the unique worst possible tour.

F15: [GuYeZv02-a, GuYe07] For every $n \geq 3$, there is an instance of ATSP and an instance of STSP with n vertices satisfying the triangle inequality on which GR outputs the unique worst possible tour.

F16: [Ru73] Let H be a tour produced by RVI for an instance I_n of STSP with $n \geq 3$ vertices. Then H is not worse than at least $(n-2)!$ tours when n is odd and $(n-2)!/2$ tours when n is even (including H itself).

F17: [GuYeZv02-a] For every $n \geq 2$, $n \neq 6$ and every instance of ATSP with n vertices, RVI computes a tour T that is not worse than at least $(n-2)!$ tours, including T itself.

F18: [PuMaKa03] For the Metric TSP, the DMST algorithm always produces a tour no more than twice as long as the optimal one, while the CH one produces tours never worse than 1.5 times the optimum (see [JoPa85]). However, there are instances for which DMST produces the unique worst possible tour, and there are instances for which CH produces a tour worse than all but at most $\lceil n/2 \rceil!$ tours.

F19: [PaVa84] Given the multigraph H created by the CH heuristic, it is NP-hard to find an optimal sequence of shortcuts even for the Euclidean TSP.

F20: For the Metric TSP, given the multigraph H of DMST, we can find an optimal sequence of shortcuts (a) in time $O(n^3 + 2^d n^2)$ and memory $O(2^d n^2)$ [BuDeWo98] or (b) in time $O(4^d n^2)$ and $O(4^d n)$ [DeTi10], where d equals half the maximum degree of H.

Part (b) of Fact F20 implies Fact F13 as $d \leq 6$ for the Euclidean TSP [DeTi10].

REMARKS

R6: A simplified proof of Fact F14 for ATSP can be found in [GuYeZv02-b]. It is based on a proof of a much more general result for the greedy algorithm in combinatorial optimization (see [GuYe02-a]).

R7: The proof of Fact F15 for ATSP in [GuYeZv02-b] cannot be directly used for STSP. A proof of Fact F15 for STSP is given in [GuYe07].

R8: A proof of Fact F17 is similar to the proof of Fact F16, but is based on a different result that was first proved for n odd by Sarvanov [Sa76], and for n even by Gutin and Yeo [GuYe02-b]. The proofs use decompositions of K_n and $\overleftrightarrow{K}_n$, into hamiltonian cycles, which exist for K_n if and only if n is odd (see, e.g., [Ha69]), and for $\overleftrightarrow{K}_n$ if and only if $n \neq 4$ or 6 [Ti80].

4.6.4   Improvement Heuristics

Approximate algorithms based on improvement heuristics start from a tour (normally obtained using a construction heuristic) and iteratively improve it by changing some parts of it at each iteration. The best known tour improvement procedures are based on *edge exchange*, in which a tour is improved by replacing k of its edges (arcs) with k edges (arcs) not in the solution.

COMPUTATIONAL NOTE: For many combinatorial optimization problems, well-known metaheuristics including *tabu search*, *simulated annealing*, and *genetic algorithms* provide the best tools for producing good quality approximate solutions. This has not been the case for TSP, for which variations of the edge-exchange algorithms of Lin and Kernighan (*Lin–Kernighan local search*) are still state-of-the-art ([JoMc02], [ApBiChCo06], [He09]). They are typically much faster than the exact algorithms, yet often produce solutions very close to the optimal one. Interested readers can find a detailed description of the Lin–Kernighan local search and its generalizations in [ReGl02] and [ApBiChCo06]. Although the Lin–Kernighan local search can be applied only to STSP, ATSP can be transformed into STSP (see, e.g., [JoGuMcYeZhZv02]).

DEFINITIONS

D16: For STSP, the **2-opt** algorithm starts from an initial tour T and tries to improve T by replacing two of its non-adjacent edges with two other edges to form another tour (see Figure 4.6.5). Once an improvement is obtained, it becomes the new T. The procedure is repeated as long as an improvement is possible (or a time limit is exceeded).

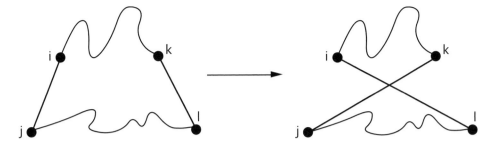

Figure 4.6.5: Edges $\{i, j\}$ and $\{k, l\}$ are replaced by $\{i, l\}$ and $\{j, k\}$.

D17: For $k \geq 3$, the **k-opt** algorithm is the same as 2-opt except that k edges (arcs) are replaced at each iteration.

D18: The **best improvement k-opt** is similar to the k-opt defined above; the difference is that every set of k edges (arcs) is tried for deletion from T and all possibilities of adding k edges (arcs) are considered before the *best* replacement of T is retained (as a replacement for T). The procedure is repeated.

FACT

F21: [PuMaKa03] Although best improvement 2-opt can take exponential time to find a local optimum, any (possibly sub-optimal) tour obtained after $O(n^3 \log n)$ iterations is not worse than $\frac{1}{n-1}$ of all STSP tours. Similar statements hold for the best improvement k-opt, $k \geq 3$.

Exponential Neighborhoods

Best improvement k-opt considers $\Theta(n^k)$ tours that can be obtained from a tour T by replacing edge-exchanges involving exactly k edges (arcs). Thus, it considers only a polynomial number of tours in the *neighborhood* of T. For TSP, one can construct various neighborhoods with an exponential number of tours in which the best tour can be found in polynomial time. In particular, there exist TSP neighborhoods of size $2^{\Theta(n \log n)}$ $(2^{\Theta(n)})$, where the best tour can be found in time $O(n^2)$ $(O(n))$. These neighborhoods are discussed in [AhErOrPu02], [DeWo00], and [GuYeZv02-b].

COMPUTATIONAL NOTE: While having seemingly strong theoretical properties, TSP exponential-neighborhood, local-search algorithms have not shown strong computation properties so far. Perhaps this is due to the fact that it is not the size of the neighborhood that matters, but the total number of tours of TSP that are worse than the best tour in the neighborhood. This may explain why computational experiments show that some exponential-neighborhood, local-search heuristics are worse than the (seemingly much weaker) 3-opt.

4.6.5   The Generalized TSP

The *Generalized TSP* (GTSP) has numerous applications and is one of the most studied extensions of TSP [FiSaTo02].

DEFINITIONS

D19: The ***Generalized Asymmetric Traveling Salesman Problem*** (GATSP): Given a weighted complete digraph $\overset{\leftrightarrow}{K}_n$ and a partition $V_1, \ldots, V_k$ of its vertex-set, find a minimum-weight cycle containing exactly one (at least one) vertex from each set V_i, $i = 1, \ldots, k$.

D20: The sets V_i are called ***clusters***, and a cycle containing exactly one (at least one) vertex from each cluster is called a ***tour***.

D21: The ***Generalized Symmetric Traveling Salesman Problem*** (GSTSP) is formulated similarly with $\overset{\leftrightarrow}{K}_n$ replaced by K_n.

REMARK

R9: Observe that the requirements 'at least one' and 'exactly one' in GATSP and GSTSP coincide when the triangle inequality holds. The 'exactly one' versions of GATSP and GSTSP have received the most attention in the literature, and only these versions are discussed here.

Transforming Generalized TSP to TSP

One of the ways to solve instances of the Generalized TSP is to transform them into TSP instances. The most efficient transformations from GATSP to ATSP and from GSTSP to STSP appear to be the ones given in [NoBe93] and [LaSe99], respectively.

FACTS

F22: In the transformation of [NoBe93], from GATSP into ATSP, the number of vertices remains the same. Weights are modified so that an optimal ATSP tour must visit all the vertices that belong to the same cluster in the original problem before moving on to the next cluster. This is achieved by adding a large positive constant M to the weight of each inter-cluster arc. If the constant is large enough, an optimal tour will contain exactly k such heavy arcs, thus ensuring that no cluster is visited more than once.

F23: In the transformation of [LaSe99], from a GSTSP instance into an STSP instance, we first add a sufficiently large positive constant to every edge-weight, if needed, to ensure that all edge-weights are nonnegative. Then we consider each cluster V_i of cardinality at least 2. For each vertex v_i in such a cluster, we create a copy v_i'. In each such cluster, we form a hamiltonian cycle $C = (v_1, v_1', \ldots, v_t, v_t', v_1)$ and assign weight $-M$ to every edge of the form $v_i v_i'$ and weight $-2M$ to the rest of the edges in cycle C, where M is any constant larger than the sum of n heaviest edges in the GSTSP instance. The weights of the remaining edges within the clusters and between the clusters are inherited from the corresponding weights of the GSTSP instance. Clearly, an optimal tour T of the resulting STSP instance will use all edges of weight $-2M$, all but one of the $(-M)$-weight edges from each cluster of cardinality at least 2, and edges between the clusters. By contracting every vertex v_i and its copy v_i' in T, we obtain an optimal tour of the GSTSP instance.

F24: For the transformations in [LaSe99] and [NoBe93], there is a bijection between *optimal* tours in the original problem and those in the transformed one, making the transformations suitable for exact algorithms.

Exact Algorithms

FACTS

F25: Computational experiments ([BeGuPeYeZv03] and [LaSe99]) have shown that the transformations given in [NoBe93] and [LaSe99] can be used to solve to optimality small to moderate instances of Generalized TSP. However, even small instances require substantial computation because of the corresponding TSP instances' very large weights on some of its edges (arcs).

F26: A successful branch-and-cut algorithm for GSTSP is described and analyzed in [FiSaTo02], and a Lagrangian-based approach for GATSP is given in [NoBe91].

The next result appears to be a major stumbling block for using a standard branch-and-bound for ATSP adapted for GATSP.

F27: [GuYe03] Let $D = (V, A)$ be a digraph and let $V_1, V_2, \ldots, V_k$ be a partition of V. The problem of checking whether D has a 1-regular subdigraph containing exactly one vertex from each $V_1, V_2, \ldots, V_k$ is NP-complete even if $|V_i| \leq 2$ for every $i = 1, 2, \ldots, k$. (A digraph H is 1-regular if the indegree and outdegree of every vertex in H equal 1.)

Approximate Algorithms

Researchers designed, implemented, and tested many GTSP heuristics: metaheuristics, local search algorithms, and construction heuristics. As of this writing, the most powerful GTSP heuristic approach is the use of *memetic* algorithms, which combine genetic algorithms and local search. The following is a general scheme of memetic algorithms.

Algorithm 4.6.5: General Scheme of Memetic Algorithms

1. *Initialize*: Construct the first generation of solutions using construction heuristics.
2. *Improve*: Use a local search procedure to replace each of the first generation solutions by the local optimum. Eliminate duplicate solutions.
3. *Produce next generation*: Use reproduction, crossover, and mutation genetic operators to produce the non-optimized next generation. Each of the genetic operators selects parent solutions from the previous generation. The length of a solution is used as the evaluation function.
4. *Improve next generation*: Use a local search procedure to replace each of the current generation solutions except the reproduced ones by the local optimum. Eliminate duplicate solutions.
5. *Evolve*: Repeat Steps 3–4 until a termination condition is reached.

COMPUTATIONAL NOTE: Memetic algorithms differ in their use of local search and genetic operators. Computational experiments for GTSP show that memetic algorithms with powerful local search provide best results. At the time of this writing, the best GTSP memetic algorithm is by Gutin and Karapetyan [GuKa10]. The algorithm applies to both GATSP and GSTSP. The algorithm can likely be improved if some local search algorithms recently developed in [KaGu] are used. For efficient adaptation of the Lin–Kernighan heuristic for GTSP, see [KaGu11].

COMPUTATIONAL NOTE: The following *Cluster Optimization* heuristic [FiSaTo02] is used in many successful GTSP heuristics. It finds a minimum-weight cycle C having exactly one vertex from each cluster and traversing the clusters in the order $V_{i_1}, V_{i_2}, \ldots, V_{i_k}$ by solving $|V_{i_1}|$ shortest path problems, each associated with a different vertex $v \in V_{i_1}$. In particular, for each $v \in V_{i_1}$, let v' and v'' be two distinct copies of vertex v, and construct an acyclic digraph D_v with vertex-set $\{v'\} \cup V_{i_2} \cup \ldots \cup V_{i_k} \cup \{v''\}$ and whose weighted arcs are defined as follows: for each $y \in V_{i_2}$ and each $x \in V_{i_k}$, there are arcs (v', y) and (x, v'') whose weights equal the weights of the arcs (v, y) and (x, v), respectively, in the original digraph; for each $x \in V_{i_t}$ and $y \in V_{i_{t+1}}$, $t = 2, \ldots, k-1$, there is an arc (x, y) with the same weight as arc (x, y) in the original digraph. The heuristic proceeds by finding a shortest (v', v'')-path in each D_v, and the minimum-weight cycle C will correspond to a smallest one of these shortest paths.

4.6.6   The Vehicle Routing Problem

The *Vehicle Routing Problem* (VRP) was introduced by Dantzig and Ramser [DaRa59]. This problem (including its versions with additional constraints) seems to be the most applicable of all generalizations of TSP. Vehicle routing is the generic name given to a large family of problems involving the distribution of goods, information, services, or people. A particularly important special case of VRP is that of minimizing the total distance traveled by a fleet of vehicles that deliver goods ordered by customers. The vehicles are assumed to have equal capacity Q, and their delivery tours start and end at a central depot.

DEFINITIONS

D22: Given a weighted complete directed or undirected graph on vertices $\{0, 1, \ldots, n\}$, a demand $d_i \geq 0$ for $i = 1, 2, \ldots, n$, and two positive integers Q (vehicle capacity) and k (number of vehicles), a **CVRP tour** is a collection of k cycles $C_1, C_2, \ldots, C_k$ such that

(i) $\bigcup\limits_{j=1}^{k} V(C_j) = \{0, 1, \ldots, n\}$;

(ii) $V(C_j) \cap V(C_l) = \{0\}$, for $j \neq l$; and

(iii) $\sum\limits_{i \in V(C_j)} d_i \leq Q$ for each $j = 1, 2, \ldots, k$.

D23: The **Capacitated Vehicle Routing Problem (CVRP)**: Given a weighted complete directed or undirected graph on vertices $\{0, 1, \ldots, n\}$, a demand $d_i \geq 0$ for $i = 1, 2, \ldots, n$, and two positive integers Q and k, find a CVRP tour for which the total weight of the cycles is minimum.

REMARKS

R10: Practitioners and researchers often consider additional complicating constraints. Some examples are: the total weight of each cycle is limited; each vertex must be visited within a prescribed *time window*; vehicles are allowed to have different capacities; routes of different vehicles cannot cross, etc. [PoKaWa99].

R11: In most research papers, the symmetric CVRP (on K_{n+1}) is considered. Nevertheless, the asymmetric (i.e., 'directed') CVRP version is also of interest [Vi96].

Exact Algorithms

FACTS

F28: The most efficient exact algorithms for symmetric CVRP are those based on branch-and-cut ([BlHo00], [NaRi01], [RaKoPuTr03]).

F29: For the asymmetric version of CVRP, it seems that the state-of-the-art exact algorithms still use branch-and-bound [ToVi01,ToVi02].

F30: Since CVRP has aspects of both TSP and *Bin Packing*, *set-covering* methods can sometimes be applied to CVRP and its generalizations with great success [BrSi01].

COMPUTATIONAL NOTE: The exact algorithms appear to be less powerful for CVRP than they are for TSP. Although they are able to solve some instances with 100 or more vertices, the exact algorithms were unable to solve an instance of symmetric CVRP with as few as 51 vertices [RaKoPuTr03,ToVi02]. Often, practical versions of CVRP have various complicating constraints that cannot be tackled by exact algorithms. Thus, heuristics are of great importance for CVRP.

Heuristics for CVRP

CVRP heuristics fall roughly into two categories: those that produce a CVRP tour relatively quickly, and those that try to produce a near-optimal solution, using a substantial amount of computing, if necessary. The latter kind are mostly metaheuristic-based algorithms. *Tabu search* seems to provide a good tradeoff between the quality of solution and running time ([ErOrSt06], [GeLaPo01], [Ta93], [ToVi02]). [DeFiTo06] uses TSP exponential neighborhoods to great effect.

Fast CVRP heuristics are of great importance, supplying quick and flexible solutions, good starting tours for metaheurisic-based algorithms, and upper bounds for exact algorithms. We close this section with brief descriptions of three classes of fast CVRP heuristics: *savings heuristics*, *insertion heuristics*, and *two-phase heuristics*.

REMARKS

R12: An important difference between TSP and CVRP is that a CVRP heuristic may not produce a feasible solution even if one exists. We illustrate this fact below for the Clarke–Wright savings heuristic.

R13: The descriptions that follow are for the asymmetric CVRP, but they also apply to the symmetric CVRP with *digraph* replaced by *graph* and *arc* replaced by *edge*.

Savings Heuristics

The Clarke–Wright savings heuristic is perhaps the earliest [ClWr64] and best known heuristic for the VRP. Here, we describe a generic savings heuristic, whose concrete implementations may be found in [AlGa91], [ClWr64], and [LaSe01].

NOTATION: (a) For a vertex subset S, $t(S)$ denotes (an approximation of) the weight of an optimal TSP tour of the induced subdigraph on S. (b) The total demand of a vertex subset S is $d(S) = \sum\limits_{i \in S} d_i$.

DEFINITIONS

D24: A **merge** of cycles C_1 and C_2, denoted $\mathrm{merge}(C_1, C_2)$, is a cycle whose vertex set equals $V(C_1) \cup V(C_2)$. The resulting cycle is determined by some prescribed rule. Cycles C_1 and C_2 can be merged only if the total demand of their vertices does not exceed capacity Q (i.e., $d(V(C_1) \cup V(C_2)) \leq Q$).

D25:   Given cycles C_1 and C_2, the **saving** of $\mathrm{merge}(C_1, C_2)$, denoted $s(C_1, C_2)$, is given by $s(C_1, C_2) = t(V(C_1)) + t(V(C_2)) - t(V(C_1) \cup V(C_2))$.

D26:   Let $R = C_1, C_2, \ldots, C_m$ be a collection of m cycles of $\overset{\leftrightarrow}{K}_n$ whose pairwise intersections are vertex 0. The **savings digraph**, $D(R)$, is the weighted digraph on m vertices, labeled $C_1, C_2, \ldots, C_m$, such that arc (C_i, C_j) exists if $d(V(C_1) \cup V(C_2)) \leq Q$, and the weight assigned to arc (C_i, C_j) is the saving $s(C_i, C_j)$.

REMARKS

R14: In the *Clarke–Wright savings algorithm* ([ClWr64], [LaSe01]), the weight of cycle C is used as an estimate of $t(V(C))$. To obtain the exact value of $t(V(C))$, one would have to solve a TSP on the induced subdigraph on $V(C)$, which may be too costly computationally.

R15: The simplest way to merge cycles C_1 and C_2 is the one used in the Clarke–Wright algorithm. If $(i, 0)$ is the arc in C_1 that enters vertex 0, and $(0, j)$ is the arc in C_2 that leaves vertex 0, then these two arcs are deleted and arc (i, j) is added to complete the new cycle. (See Figure 4.6.6.)

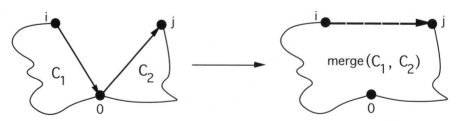

Figure 4.6.6: A Clarke–Wright merge of cycles C_1 and C_2.

Algorithm 4.6.6: Savings Heuristic (SH)

Input: distance matrix $[d_{ij}]$; demands d_i, $i = 1, \ldots, n$; capacity Q;
    and number of vehicles k.
Output: CVRP tour $R = \{C_1, \ldots, C_k\}$.

    Initialize cycle count $m = n$ and cycles $C_i = (0, i, 0)$, $i = 1, \ldots, m$.
    Initialize $R = \{C_1, \ldots, C_m\}$.
    While $m > k$
        Construct savings digraph $D(R)$.
        Construct a matching M in $D(R)$ with $|M| \leq m - k$.
        For each arc (C_i, C_j) in M,
           $R := (R - \{C_i, C_j\}) \cup \text{merge}(C_i, C_j)$
        $m := m - 1$

REMARKS

R16: The easiest way to construct M is to simply choose a pair (C_a, C_b) with maximum saving $s(C_a, C_b)$ as the only arc in M. In some versions in which cycle-merging occurs in parallel, M is built in a greedy manner [LaSe01].

R17: Often, after R is produced by Algorithm SH, each of the cycles in R is improved by some TSP improvement heuristic. For example, in [LaSe01], a CVRP tour found by the Clarke–Wright algorithm is improved by applying the best improvement 3-opt to each of its cycles.

R18: Let $n = 12$, $k = 2$, $Q = 6$, and $d_i = 1$ for each i. Clearly, there is a feasible solution. However, the Clarke–Wright algorithm may first produce six cycles, each containing two vertices different from 0, and then construct three cycles, each containing four vertices different from 0. Now there is no possibility to obtain a feasible solution.

Insertion Heuristics

In CVRP iteration algorithms [LaSe01], we start from k cycles of the form $C_p = (0, i_p, 0)$. The vertices outside of the cycles are inserted one by one in sequential or parallel manner. The word 'parallel' here means that a vertex is inserted in one of the current cycles C_p for which the insertion is most beneficial. In the sequential mode, we start constructing a new cycle only when the previous one cannot be increased because of the capacity constraints. The cost of insertion of a vertex i into a current cycle C_p can be measured by $t(V(C) \cup \{i\}) - t(V(C))$.

REMARKS

R19: An example of such an algorithm is the Christofides–Mingozzi–Toth insertion heuristic [ChMiTo79].

R20: Fisher and Jaikumar [FiJa81] suggested inserting all vertices at the same time. They apply the *Generalized Assignment Problem* to find 'optimal' insertions that do not violate the capacity constraints.

Two-phase Heuristics

The basic idea of *two-phase heuristics* is to partition vertices $\{1, 2, \ldots, n\}$ into k clusters $V_1, \ldots, V_k$ and solve TSP for each of the graphs induced by $V_i \cup \{0\}$, $i = 1, 2, \ldots, k$. Wren and Holliday [WrHo72] suggested a sweeping technique for the Euclidean CVRP, in which the depot 0 and vertices i, $1 \leq i \leq n$ are points on the Euclidean plane, and d_{ij} is the Euclidean distance between i and j.

We introduce a polar coordinate system, in which one of the n vertices, say i, is chosen as the reference point, with polar coordinates $(0, d_{0i})$. Then every vertex $j \neq i$ has coordinates (ϕ_j, d_{0j}), where ϕ_j is the angle between the rays from 0 to i and from 0 to j. The sweeping algorithm in its simplest form is as follows.

Algorithm 4.6.7: Sweeping Heuristic

Input: distance matrix $[d_{ij}]$; polar angles ϕ_i for $1 \leq i \leq n$;
  demands d_i, $i = 1, \ldots, n$; capacity Q; and number of vehicles k.
Output: CVRP tour consisting of cycles C_j, $j = 1, \ldots, k$.

Sort the vertices $i_1, i_2, \ldots, i_n$ such that $\phi_{i_s} \leq \phi_{i_{s+1}}$, $s = 1, \ldots, n-1$.
Initialize $S_j = \emptyset$, $j = 1, \ldots, k$.
Set $j = 1$.
For $s = 1, \ldots, n$
  If $d(S_j \cup \{i_s\}) > Q$
    $j := j + 1$
  $S_j := S_j \cup \{i_s\}$
For $j = 1, \ldots, k$
  Let C_j be a TSP tour for the subgraph induced on $S_j \cup \{0\}$.

REMARK

R21: An extension of this approach to the general CVRP is described in [BrSi95]. Another example of a two-phase heuristic is a truncated branch-and-bound provided in [CHMiTo79].

References

[AhErOrPu02] R. K. Ahuja, O. Ergun, J. B. Orlin, and A. P. Punnen, A survey of very large-scale neighborhood search techniques, *Discrete Appl. Math.* **123** (2002), 75–102.

[AlGa91] K. Altinkemer and B. Gavish, Parallel savings based heuristics for the delivery problem, *Oper. Res.* **39** (1991), 456–469.

[ApBiChCo06] D. Applegate, R. E. Bixby, V. Chvátal, and W. Cook, *The Traveling Salesman Problem: A Computational Study*, Priceton Univ. Press, 2006.

[Ar98] S. Arora, Polynomial-time approximation schemes for Euclidean TSP and other geometric problems, *JACM* **45** (1998), 753–782.

[Ar02] S. Arora, Approximation algorithms for geometric TSP, *The Traveling Salesman Problem and its Variations* (G. Gutin and A. P. Punnen, Eds.), Kluwer, 2002.

[BaTo85] E. Balas and P. Toth, Branch and Bound Methods, In *The Traveling Salesman Problem: A Guided Tour of Combinatorial Optimization* (E. L. Lawler, J. K. Lenstra, A. H. G. Rinnooy Kan, and D. B. Shmoys, Eds.), Wiley, 1985.

[BeLu00] A. Belloni and A. Lucena, A relax and cut algorithm for the traveling salesman problem, *A talk at Intern. Symp. Math. Prog., Atlanta*, 2000.

[BeGuPeYeZv03] D. Ben-Arieh, G. Gutin, M. Penn, A. Yeo, and A. Zverovitch, Transformations of Generalized ATSP into ATSP: experimental and theoretical study, *Oper. Res. Lett.* **31** (2003), 357–365.

[BlHo00] U. Blasum and W. Hochstättler, Application of the branch and cut method to the vehicle routing problem. *Zentrum für Angewandte Informatik*, Köln, Tech. Report zrp2000–386.

[BrSi95] J. Bramel and D. Simchi-Levi, A location based heuristic for general routing problems, *Oper. Res.* **43** (1995), 649–660.

[BrSi01] J. Bramel and D. Simchi-Levi, Set-Covering Based Algorithms for the Capacitated VRP, in *The Vehicle Routing Problem* (P. Toth and D. Vigo, Eds.), SIAM, 2001.

[BuDeWo98] R. E. Burkard, V. G. Deineko, and G. J. Woeginger, The traveling salesman and the PQ-tree. *Math. Oper. Res.* **23** (1998), 613–623.

[ClWr64] G. Clarke and J. W. Wright, Scheduling of vehicles from a central depot to a number of delivery points. *Oper. Res.* **12** (1964), 568–581.

[ChMiTo79] N. Christofides, A. Mingozzi, and P. Toth, The Vehicle Routing Problem, In *Combinatorial Optimization* (A. Mingozzi, P. Toth, and C. Sandi, Eds.), Wiley, 1979.

[DaFuJo54] G. B. Dantzig, D. R. Fulkerson and S. M. Johnson, Solution of large scale traveling salesman problem. *Oper. Res.* **2** (1954), 393–410.

[DaRa59] G. B. Dantzig, and R. H. Ramser, The truck dispatching problem. *Manag. Sci.* **6** (1959), 80–91.

[DeFiTo06] R. De Franceschi, M. Fischetti, and P. Toth, A new ILP-based refinement heuristic for Vehicle Routing Problems, *Math. Prog. B* **105** (2006), 471–499.

[DeTi10] V. G. Deineko and A. Tiskin, Fast minimum-weight double-tree shortcutting for metric TSP: Is the best one good enough? *J. Exp. Algorithmics* **14** (2010), 6:4.6–6:4.16.

[DeWo00] V. G. Deineko and G. J. Woeginger, A study of exponential neighbourhoods for the traveling salesman problem and the quadratic assignment problem, *Math. Program., Ser. A* **87** (2000), 519–542.

[ErOrSt06] O. Ergun, J. B. Orlin, and A. Steele-Feldman, Creating very large scale neighborhoods out of smaller ones by compounding moves: a study on the vehicle routing problem. *J. Heuristics* **12** (2006), 115–140.

[FiJa81] M. L. Fisher and R. Jaikumar, A generalized assignment heuristic for vehicle routing, *Networks* **11** (1981), 109–124.

[FiLoTo02] M. Fischetti, A. Lodi, and P. Toth, Exact Methods for the Asymmetric Traveling Salesman Problem, in *The Traveling Salesman Problem and its Variations* (G. Gutin and A. P. Punnen, Eds.), Kluwer, 2002.

[FiSaTo02] M. Fischetti, J. J. Salazar-González, and P. Toth, The Generalized Traveling Salesman and Orienteering Problem, in *The Traveling Salesman Problem and its Variations* (G. Gutin and A. P. Punnen, Eds.), Kluwer, 2002.

[GaGrJo76] M. R. Garey, R. L. Graham, and D. S. Johnson, Some NP-complete geometric problems, in *Proc. 8th ACM Symp. Theory Comput.* (1976), 10–22.

[GeLaPo01] M. Gendreau, G. Laporte, and J. Y. Potvin, Metaheuristics for the Capacitated VRP, *The Vehicle Routing Problem* (P. Toth and D. Vigo, Eds.), SIAM, 2001.

[GlGuYeZv01] F. Glover, G. Gutin, A. Yeo, and A. Zverovich, Construction heuristics for the asymmetric TSP, *Eur. J. Oper. Res.* **129** (2001), 555–568.

[GoSt85] B. L. Golden and W. R. Stewart, Empirical Analysis of Heuristics, in *The Traveling Salesman Problem: A Guided Tour of Combinatorial Optimization* (E. L. Lawler, J. K. Lenstra, A. H. G. Rinnooy Kan and D. B. Shmoys, Eds.), Wiley, 1985.

[GrYe06] J. L. Gross and J. Yellen, *Graph Theory and Its Applications*, Second Edition, CRC Press, 2006.

[GuKa10] G. Gutin and D. Karapetyan, A memetic algorithm for the generalized traveling salesman problem, *Nat. Comput.* **9** (2010), 47–60.

[GuYe02-a] G. Gutin and A. Yeo, Anti-matroids, *Oper. Res. Lett.* **30** (2002) 97–99.

[GuYe02-b] G. Gutin and A. Yeo, Polynomial approximation algorithms for the TSP and the QAP with factorial domination number, *Discrete Appl. Math.* **119** (2002), 107–116.

[GuYe03] G. Gutin and A. Yeo, Assignment Problem based algorithms are impractical for the Generalized TSP, *Ausralasian J. Combinatorics* **27** (2003), 149–154.

[GuYe07] G. Gutin and A. Yeo, The Greedy Algorithm for the Symmetric TSP, *Algorithmic Oper. Res.* **2** (2007), 33–36.

[GuYeZv02-a] G. Gutin, A. Yeo, and A. Zverovich, Traveling salesman should not be greedy: domination analysis of greedy-type heuristics for the TSP, *Discrete Appl. Math.* **117** (2002), 81–86.

[GuYeZv02-b] G. Gutin, A. Yeo, and A. Zverovitch, Exponential Neighborhoods and Domination Analysis for the TSP, in *The Traveling Salesman Problem and its Variations* (G. Gutin and A. P. Punnen, Eds.), Kluwer, 2002.

[HaKh01] R. Hassin and S. Khuller, z-Approximations, *J. Algorithms* **41** (2001), 429–442.

[Ha69] F. Harary, *Graph Theory*, Addison-Wesley, 1969.

[He09] K. Helsgaun, General k-opt submoves for the Lin–Kernighan TSP heuristic, *Math. Programing Comput.* **1** (2009), 119–163.

[HoWo85] A. J. Hoffman and P. Wolfe, History, in *The Traveling Salesman Problem: A Guided Tour of Combinatorial Optimization* (E. L. Lawler, J. K. Lenstra, A. H. G. Rinnooy Kan and D. B. Shmoys, Eds.), Wiley, 1985.

[JoGuMcYeZhZv02] D. S. Johnson, G. Gutin, L. A. McGeoch, A. Yeo, W. Zhang, and A. Zverovitch, Experimental Analysis of Heuristics for ATSP, in *The Traveling Salesman Problem and its Variations* (G. Gutin and A. P. Punnen, Eds.), Kluwer, Dordrecht, 2002.

[JoMc02] D. S. Johnson and L. A. McGeoch, Experimental Analysis of Heuristics for STSP, in *The Traveling Salesman Problem and its Variations* (G. Gutin and A. P. Punnen, Eds.), Kluwer, 2002.

[JoPa85] D. S. Johnson and C. H. Papadimitriou, Performance guarantees for heuristics, in *The Traveling Salesman Problem: A Guided Tour of Combinatorial Optimization* (E. L. Lawler, J. K. Lenstra, A. H. G. Rinnooy Kan and D. B. Shmoys, Eds.), Wiley, 1985.

[KaRe04] A. Kahng and S. Reda, Match twice and stitch: a new TSP tour construction heuristic, *Oper. Res. Lett.* **32** (2004), 499–509.

[KaGu] D. Karapetyan and G. Gutin, Efficient local search algorithms for known and new neighborhoods for the Generalized Traveling Salesman Problem, submitted, 2011.

[KaGu11] D. Karapetyan and G. Gutin, Lin–Kernighan heuristic adaptations for the Generalized Traveling Salesman Problem. *Europ. J. Oper. Res.* **208** (2011), 221–232.

[LaSe99] G. Laporte and F. Semet, Computational evaluation of a transformation procedure for the symmetric generalized traveling salesman problem, *INFOR* **37** (1999), 114–120.

[LaSe01] G. Laporte and F. Semet, Classical Heuristics for the Capacitated VRP, *The Vehicle Routing Problem* (P. Toth and D. Vigo, Eds.), SIAM, 2001.

[LoPu02] A. Lodi and A. P. Punnen, TSP Software, *The Traveling Salesman Problem and its Variations* (G. Gutin and A. P. Punnen, Eds.), Kluwer, 2002.

[Me32] K. Menger, Das botenproblem, *Ergebnisse Eines Mathematischen Kolloquiums* **2** (1932), 11–12.

[Mi99] J. C. B. Mitchell, Guillotine subdivisions approximate polygonal subdivisions: A simple polynomial time approximation scheme for geometric TSP, k-MST, and related problem, *SIAM J. Comput.* **28** (1999), 1298–1309.

[Na02] D. Naddef, Polyhedral Theory and Branch-and-Cut Algorithms for the Symmetric TSP, *The Traveling Salesman Problem and its Variations* (G. Gutin and A. P. Punnen, Eds.), Kluwer, 2002.

[NaRi01] D. Naddef and G. Rinaldi, Branch-and-Cut Algorithms for the Capacitated VRP, *The Vehicle Routing Problem* (P. Toth and D. Vigo, Eds.), SIAM, 2001.

[NoBe91] C. E. Noon and J. C. Bean, A Lagrangian based approach for the asymmetric generalized traveling salesman problem, *Oper. Res.* **39** (1991), 623–632.

[NoBe93] C. E. Noon and J. C. Bean, An efficient transformation of the generalized traveling salesman problem, *INFOR* **31** (1993), 39–44.

[Pa77] C. H. Papadimitriou, The Euclidean traveling salesman problem is NP-complete, *Theoret. Comput. Sci.* **4** (1977), 237–244.

[PaSt82] C. H. Papadimitriou and K. Steiglitz, *Combinatorial Optimization*, Prentice-Hall, 1982.

[PaVa84] C. H. Papadimitriou and U. V. Vazirani, On two geometric problems related to the travelling salesman problem, *J. Algorithms* **5** (1984), 231–246.

[PoKaWa99] A. Poot, G. Kant, and A. P. M. Wagelmans, A savings based method for real-life vehicle routing problems, Erasmus University Rotterdam, Econometric Institute Report Ei 9938/A.

[Pu02] A. P. Punnen, The Traveling Salesman Problem: Aplications, Formulations and Variations, *The Traveling Salesman Problem and its Variations* (G. Gutin and A. P. Punnen, Eds.), Kluwer, 2002.

[PuMaKa03] A. P. Punnen, F. Margot, and S. Kabadi, TSP heuristics: domination analysis and complexity, *Algorithmica* **35** (2003), 111–127.

[RaKoPuTr03] T. K. Ralphs, L. Kopman, W. R. Pulletblank, and L. E. Trotter, On the capacitated vehicle routing problem, *Math. Program.* **94** (2003), 343–359.

[RaSm98] S. Rao and W. Smith, Approximating geometric graphs via "spanners" and "banyans", *Proc. 30th Ann. ACM Symp. Theory Comput.* (1998), 540–550.

[ReGl02] C. Rego and F. Glover, Local search and metaheuristics, *The Traveling Salesman Problem and its Variations* (G. Gutin and A. P. Punnen, Eds.), Kluwer, 2002.

[RoCiFa09] B. Rodeker, M. V. Cifuentes and L. Favre, An empirical analysis of approximation algorithms for Euclidean TSP, *Proc. Intern. Conf. Sci. Comput., CSC 2009* (2009), 190–196.

[Ru73] V. I. Rublineckii, Estimates of the accuracy of procedures in the Traveling Salesman Problem, *Numerical Mathematics and Computer Technology* no. 4 (1973), 18–23 (in Russian).

[SaGo76] S. Sahni and T. Gonzalez, P-complete approximation problems, *JACM* **23** (1976), 555–565.

[Sa76] V. I. Sarvanov, On the minimization of a linear form on a set of all n-elements cycles, *Vestsi Akad. Navuk BSSR, Ser. Fiz.-Mat. Navuk* no. 4 (1976), 17–21 (in Russian).

[Ta93] E. Taillard, Parallel iterative search methods for vehicle routing problem, *Networks* **23** (1993), 661–673.

[Ti80] T. W. Tillson, A Hamiltonian decomposition of K_{2m}^*, $2m \geq 8$, *J. Combin. Theory, Ser. B* **29** (1980), 68–74.

[ToVi01] P. Toth and D. Vigo, Branch-and-Bound Algorithms for the Capacitated VRP, *The Vehicle Routing Problem* (P. Toth and D. Vigo, Eds.), SIAM, 2001.

[ToVi02] P. Toth and D. Vigo, Models, relaxations and exact approaches for the capacitated vehicle routing problem, *Discrete Appl. Math.* **123** (2002), 487–512.

[Tr97] L. Trevisan, When Hamming meets Euclid: the approximability of geometric TSP and MST, *Proc. 29th ACM Symp. Theory Comput.* (1997), 21–39.

[Vi96] D. Vigo, A heuristic algorithm for the asymmetric capacitated vehicle routing problem, *Europ. J. Oper. Res.* **89** (1996), 108–126.

[Wo98] L. A. Wolsey, *Integer Programming*, Wiley, 1998.

[WrHo72] A. Wren and A. Holliday, Computer scheduling of vehicles from one or more depots to a number of delivery points, *Oper. Res. Quart.* **23** (1972), 333–344.

[Ze81] E. Zemel, Measuring the quality of approximate solutions to zero-one programming problems, *Math. Oper. Res.* **6** (1981), 319–332.

Section 4.7

Further Topics in Connectivity

Camino Balbuena, Universitat Politècnica de Catalunya, Spain
Josep Fàbrega, Universitat Politècnica de Catalunya, Spain
Miquel Àngel Fiol, Universitat Politècnica de Catalunya, Spain

INTRODUCTION

Continuing the study of connectivity, initiated in §4.1 of the *Handbook*, we survey here some (sufficient) conditions under which a graph or digraph has a given connectivity or edge-connectivity. First, we describe results concerning maximal (vertex- or edge-) connectivity. Next, we deal with conditions for having (usually lower) bounds for the connectivity parameters. Finally, some other general connectivity measures, such as one instance of the so-called "conditional connectivity," are considered.

For unexplained terminology concerning connectivity, see §4.1.

4.7.1   High Connectivity

Since connectivity has to do with "connection," intuitively we can expect to find high connectivity when the "edge density" of the graph is large. Different situations in which this seems to be the case are:

(a) Large minimum or average degree.

(b) Small diameter (for given girth).

(c) Small number of vertices (for given degree and girth).

(d) Large number of vertices (for given degree and diameter).

The results in this subsection give several conditions of the above types, under which maximum vertex- or edge-connectivity is attained. An extensive collection of results about maximally edge-connected and vertex-connected graphs and digraphs can be found in the survey by Hellwig and Volkmann [HeVo08b].

Minimum Degree and Diameter

NOTATION: Let $G = (V, E)$ be a graph with order n, minimum degree δ, maximum degree Δ, edge-connectivity λ, and (vertex-)connectivity κ. In some other sections of the *Handbook*, the notations δ_{min}, δ_{max}, κ_e, and κ_v are used instead of δ, Δ, λ, and κ, respectively.

DEFINITIONS

D1: The **girth** g of a graph G with a cycle is the length of its shortest cycle. An acyclic graph has infinite girth.

D2: The **diameter** D of G is $\max\limits_{u,v \in V} \{dist_G(u, v)\}$.

D3: The **clique number** of a graph G, denoted $\omega(G)$, is the maximum number of vertices in a complete subgraph of G.

D4: A (di)graph G is p-**partite** if its vertex-set can be partitioned into p independent (or stable) sets.

FACTS

F1: [Ch66] If $\delta \geq \lfloor n/2 \rfloor$, then G is maximally edge-connected (i.e., $\lambda = \delta$).

F2: [Le74] If for any non-adjacent vertices u and v, $deg(u) + deg(v) \geq n - 1$, then $\lambda = \delta$.

F3: [Pl75] If G is a graph with diameter $D = 2$, then $\lambda = \delta$.

F4: [HeVo08a] For any graph G, $\lambda(G) = \delta(G)$ or $\lambda(\overline{G}) = \delta(\overline{G})$.

F5: [Vo88] If G is bipartite and $\delta \geq \lfloor n/4 \rfloor + 1$, then $\lambda = \delta$.

F6: [Vo89] If G is p-partite ($p \geq 2$) and $n \leq 2 \left\lfloor \frac{p}{p-1}\delta \right\rfloor - 1$, then $\lambda = \delta$.

F7: [ToVo93] If G is p-partite ($p \geq 2$) and $\delta \geq n \frac{2p-3}{2p-1}$, then G is maximally connected (i.e., $\kappa = \delta$).

F8: [DaVo95] If G is p-partite ($p \geq 2$) with clique number $\omega \leq p$ and $n \leq 2 \left\lfloor \frac{p}{p-1}\delta \right\rfloor - 1$, then $\lambda = \delta$.

REMARKS

R1: It is easily shown that Fact F3 $\Rightarrow$ Fact F2 $\Rightarrow$ Fact F1.

R2: Fact F5 is a slight improvement of Fact F6 for $p = 2$.

R3: In addition to Fact F8, the authors in [DaVo95] gave other sufficient conditions for $\lambda = \delta$ that mostly generalize conditions in [PlZn89].

R4: A consequence of Fact F4 is that $\lambda(G) = \delta(G)$ for any self-complementary graph $(G = \overline{G})$.

Degree Sequence

NOTATION: For the next group of results, G is an n-vertex graph with degree sequence $d_1 \geq d_2 \geq \cdots \geq d_n = \delta$. For a vertex u, $N(u)$ denotes the set of vertices adjacent to u.

FACTS

F9: [GoWh78] If the vertex set of G can be partitioned into $\lfloor n/2 \rfloor$ pairs of vertices (u_i, v_i) (and, if n is odd, one "unpaired" vertex w) such that $deg(u_i) + deg(v_i) \geq n$, $i = 1, 2, \ldots, \lfloor n/2 \rfloor$, then $\lambda = \delta$.

F10: [GoEn79] If each vertex u of minimum degree satisfies

$$\sum_{v \in N(u)} deg(v) \geq \begin{cases} \lfloor n/2 \rfloor^2 - \lfloor n/2 \rfloor, & \text{for even } n \text{ or odd } n \leq 15, \\ \lfloor n/2 \rfloor^2 - 7, & \text{for odd } n \geq 15, \end{cases}$$

then $\lambda = \delta$.

F11: [Bo79] Let G be a graph with order $n \geq 2$. If its degree sequence $d_1 \geq d_2 \geq \cdots \geq d_n = \delta$ satisfies $\sum_{i=1}^{k}(d_i + d_{n-i}) \geq kn - 1$ for all k with $1 \leq k \leq \min\{\lfloor n/2 \rfloor - 1, \delta\}$, then $\lambda = \delta$.

F12: [DaVo97] If $\delta \geq \lfloor n/2 \rfloor$ or if $\delta \leq \lfloor n/2 \rfloor - 1$ and $\sum_{i=1}^{k}(d_i + d_{n+i-\delta-1}) \geq k(n-2) + 2\delta - 1$ for some k with $1 \leq k \leq \delta$, then $\lambda = \delta$.

F13: [Vo03] Suppose that G is p-partite ($p \geq 2$) and has order $n \geq 6$ with clique number $\omega \leq p$. Let $\nu = 1$ when n is even and $\nu = 0$ when n is odd. If $\delta \geq \lfloor n/2 \rfloor$ or if $\delta \leq \lfloor n/2 \rfloor - 1$ and $\sum_{i=1}^{\delta+1} d_{n+1-i} \geq (\delta+1)\frac{p-1}{p}\frac{n+1+\nu}{2} - \frac{2\delta+2}{p(n-3+\nu)}$, then $\lambda = \delta$.

REMARKS

R5: Note that Fact F9 implies Fact F1 only when n is even. Fact F10 also implies Fact F1. Moreover, as shown by the examples in [PlZn89], Fact F10 is independent of Fact F2 and Fact F3.

R6: Fact F11 implies Fact F1 when n is even, but in general, as shown in [PlZn89], it is independent of Facts 1, 2, 3 and 10.

R7: Fact F12 is even valid for digraphs, and a theorem of Xu [Xu94] follows easily (see Fact F23). It is easily shown that Fact F12 implies Fact F11.

R8: Fact F13 generalizes results in [Vo88, Vo89], as well as Fact F8. Furthermore, as shown in [HeVo03b], the conditions in Fact F13 also guarantee maximum local edge-connectivity for all pairs u and v of vertices in G; that is, $\lambda(u-v) = \min\{deg(u), deg(v)\}$.

Distance

DEFINITIONS

D5:   The ***distance*** $dist_G(U_1, U_2)$ between two given subsets $U_1, U_2 \subset V(G)$ is the minimum of the distances $dist_G(u_1, u_2)$ for all vertices $u_1 \in U_1$ and $u_2 \in U_2$. (When there is no ambiguity, we omit the subscript G.)

D6:   The ***line graph*** $L(G)$ of a graph G has vertices representing the edges of G, and two vertices are adjacent if and only if the corresponding edges are adjacent (that is, they have one endpoint in common).

FACTS

F14:   Let u_1v_1 and u_2v_2 be edges in a graph G, and let $U_i = \{u_i, v_i\}$, $i = 1, 2$. Then, the distance between the corresponding vertices of $L(G)$ satisfies $d_{L(G)}(u_1v_1, u_2v_2) = d_G(U_1, U_2) + 1$ and thus, the diameters of $L(G)$ and G satisfy $D(L(G)) \leq D(G) + 1$.

F15:   [PlZn89] Let G be a connected graph such that every pair of vertex subsets U_1, U_2 of cardinality two satisfies $dist(U_1, U_2) \leq 2$. Then $\lambda = \delta$.

F16:   [BaCaFaFi96] Let G be a graph with minimum degree δ and line graph $L(G)$. Then,

(a) If $L(G)$ has diameter at most three, then $\lambda = \delta$.

(b) If $L(G)$ has diameter two, then $\kappa = \delta$.

REMARKS

R9:   The sufficient condition given in Fact F15 is slightly weaker than the one given in Fact F3. Furthermore, it suffices to require such a condition on the 2-element subsets that are the endpoints of some edge, as shown in Fact F16(a).

R10:   From the above remark, Fact F16(a) generalizes both Fact F15 and Fact F3 (Plesnik's result).

Super Edge-Connectivity

Here we consider a stronger measure of edge-connectivity.

DEFINITION

D7:   A maximally edge-connected graph is ***super-λ*** if every minimum edge-disconnecting set is ***trivial***, that is, consists of the edges incident on a vertex of minimum degree.

EXAMPLE

E1:   Figure 4.7.1 shows a 3-regular maximally edge-connected graph that is not super-λ. The set $\{e, f, g\}$ is a non-trivial minimum edge-disconnecting set.

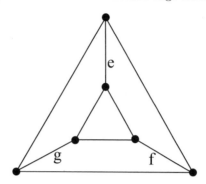

Figure 4.7.1: G is maximally edge-connected but not super-λ.

FACTS

F17:   [Le74] Let $G \neq K_{n/2} \times K_2$. If for any non-adjacent vertices u and v, $deg(u) + deg(v) \geq n$, then G is super-λ.

F18:   If for any non-adjacent vertices u and v, $deg(u) + deg(v) \geq n + 1$, then G is super-λ.

F19:   [Ke72] If $\delta \geq \lfloor n/2 \rfloor + 1$, then G is super-λ.

F20:   [Fi92] If G has diameter two and contains no complete subgraph H on δ vertices with $\deg_G(v) = \delta$ for all $v \in V(H)$, then G is super-λ.

F21:   [So92] Let G be a graph with maximum degree Δ. If $n > 2\delta + \Delta - 1$, then G is super-λ.

REMARKS

R11:   Facts F18 and F19, which are analogues of Facts F2 and F1, are direct consequences of Fact F17.

R12:   Fact F20 can be seen as a refinement of Fact F3 (where only the diameter condition is required) and has Fact F21 as a corollary.

Digraphs

As mentioned in §4.1, since the connectivity parameters of a graph G equal those of its *symmetric digraph* G^* (obtained by replacing each edge of G by a digon), many of the previous results can be generalized to the directed case.

DEFINITIONS

D8:   The ***vertex-connectivity*** of a digraph G, denoted $\kappa(G)$, is the minimum size of a vertex subset whose deletion results in a non-strongly connected or trivial digraph.

D9: The ***edge-connectivity*** of a digraph G, denoted $\lambda(G)$, is the minimum size of an edge subset whose deletion results in a non-strongly connected digraph.

NOTATION: (a) For a vertex $u \in V(G)$, $deg^+(u)$ denotes the *out-degree*, the number of vertices adjacent *from* vertex u, and $deg^-(u)$ denotes the *in-degree*, the number of vertices adjacent *to* vertex u. Let $\delta(u) = \min\{deg^+(u), deg^-(u)\}$.
(b) $\delta^+ = \min_{u \in V}\{deg^+(u)\}$ and $\delta^- = \min_{u \in V}\{deg^-(u)\}$.
(c) $\delta = \min_{u \in V} \delta(u) = \min\{\delta^+, \delta^-\}$.
Similar notations with Δ stand for maximum degrees.

NOTATION: For vertices $u, v \in V(G)$, $\lambda(u - v)$ denotes the maximum number of edge-disjoint directed paths from u to v.

FACTS

F22: [Jo72] If G is a digraph with diameter $D = 2$, then $\lambda = \delta$.

F23: [Xu94] Let G be a digraph of order n. If there are $\lfloor n/2 \rfloor$ pairs of (different) vertices (u_i, v_i) such that $\delta(u_i) + \delta(v_i) \geq n$, $i = 1, 2, \ldots, \lfloor n/2 \rfloor$, then $\lambda = \delta$.

F24: [HeVo03b] Let G be a digraph with diameter at most two. Then, $\lambda(u-v) = \min\{deg^+(u), deg^-(v)\}$ for all pairs u and v of vertices in G.

F25: [HeVo03a] Let G be a strongly connected digraph with edge-connectivity λ and minimum degree δ. If for all maximal pairs of vertex sets X and Y at distance 3 there exists an isolated vertex in the induced subgraph on $X \cup Y$, then $\lambda = \delta$.

F26: [HeVo03b] Let G be a p-partite digraph of order n and minimum degree δ with $p \geq 2$. If $n \leq 2\lfloor (p\delta)/(p-1) \rfloor - 1$, then $\lambda(u-v) = \min\{deg^+(u), deg^-(v)\}$ for all pairs u and v of vertices in G.

F27: [HeVo03b] Let G be a bipartite digraph of order n and minimum degree $\delta \geq 2$ with the bipartition $V' \cup V''$. If $deg(x) + deg(y) \geq (n+1)/2$ for each pair of vertices $x, y \in V'$ and each pair of vertices $x, y \in V''$, then $\lambda(u-v) = \min\{deg^+(u), deg^-(v)\}$ for all pairs u and v of vertices in G.

REMARKS

R13: Notice that Plesník's result (Fact F3) is, in fact, a consequence of the older result of Jolivet (Fact F22). Similarly, Fact F23 generalizes Fact F9.

R14: Fact F23 was improved by Dankelmann and Volkmann in two subsequent papers [DaVo97, DaVo00], where the bipartite case was also considered.

R15: A restatement of Fact F24 states that a digraph with diameter two has maximum local edge-connectivity. Moreover, this obviously implies Jolivet's result (Fact F22) and the corresponding local connectivity result for undirected graphs, proved in [FrOeSw00].

R16: A consequence of Fact F25 is the directed version of Fact F15.

Oriented Graphs

DEFINITIONS

D10: A digraph is **super-λ** if every minimum edge-disconnecting set consists of the edges directed to or from a vertex with minimum degree. A digraph is **super-κ** if every minimum disconnecting set consists of the vertices adjacent to or from a vertex with minimum degree.

D11: An **oriented graph** G (also called an **antisymmetric digraph**) is a digraph such that between any two vertices u, v, there is at most one (directed) edge $((u, v)$ or $(v, u))$.

EXAMPLE

E2: Figure 4.7.2 shows a 2-regular maximally connected digraph G that is not super-κ. If $F = \{x, y\}$, then $G - F$ is not strongly connected (for instance, there is no [directed] path in $G - F$ from u to v) and F is non-trivial (it does not consist of the vertices adjacent to or from a vertex with minimum degree).

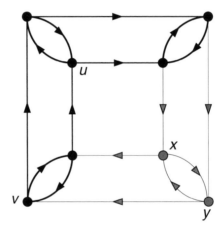

Figure 4.7.2: G is maximally connected but not super-κ.

FACTS

F28: [AyFr70] Let G be an oriented graph with n vertices and minimum degree δ. If $\delta \geq \lfloor (n + 2)/4 \rfloor$, then $\lambda = \delta$.

F29: [Fi92] If G is an oriented graph with n vertices and minimum degree $\delta \geq \lfloor n/4 \rfloor + 1$, then G is super-λ.

F30: [Fi92] If G is an oriented graph with diameter two, then G is super-λ.

REMARKS

R17: Facts F28 and F29 are analogues of Fact F1, whereas Fact F30, similar to Fact F22, is a consequence of Fact F20.

R18: In fact, the sufficient conditions given in [AyFr70] and [Fi92] (Facts F28 and F29) were $\delta^+ + \delta^- \geq \lfloor n/2 \rfloor$ and $\delta^+ + \delta^- \geq \lfloor n/2 \rfloor + 1$, respectively. Furthermore, it is easily shown that Facts F29 and F30 do not imply each other.

R19: Higher connectivity in *tournaments*, which are oriented complete graphs, is discussed in §3.3 of the *Handbook*.

Semigirth

To generalize Jolivet's result (Fact F22) and give new results on superconnectivity, it is relevant to consider a new parameter related to the path structure of the digraph. In our context, this parameter plays a role similar (and is tightly related) to the girth of a graph.

DEFINITIONS

D12: [FaFi89, FiFaEs90] For a given digraph $G = (V, E)$ with diameter D, the **semi-girth**, denoted $\ell(G)$, is the greatest integer ℓ between 1 and D such that for any $u, v \in V$,

(a) if $dist(u, v) < \ell$, the shortest u-v directed walk is unique and there are no u-v directed walks of length $dist(u, v) + 1$.

(b) if $dist(u, v) = \ell$, there is only one shortest u-v directed walk.

D13: A digraph G is a **generalized p-cycle** when it has its vertex set partitioned in p parts cyclically ordered, and vertices in one part are adjacent only to vertices in the next part. Thus, a generalized 2-cycle is the same as a bipartite digraph.

EXAMPLE

E3: Figure 4.7.3 shows a 2-regular digraph for which the semigirth ℓ is equal to its diameter, namely, $\ell = D = 3$.

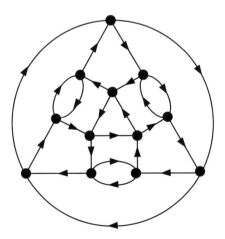

Figure 4.7.3: Semigirth $\ell = D = 3$.

FACTS

F31: [FaFi89] Let G be a digraph with minimum degree $\delta > 1$, diameter D, semigirth ℓ, and connectivities κ and λ.

(a) If $D \leq 2\ell$, then $\lambda = \delta$.

(b) If $D \leq 2\ell - 1$, then G is super-λ and $\kappa = \delta$.

(c) If $D \leq 2\ell - 2$, then G is super-κ.

F32:   [FaFi96a, PeBaGo01] Let G be a generalized p-cycle ($p \geq 2$).

(a) If $D \leq 2\ell + p - 1$, then $\lambda = \delta$.

(b) If $D \leq 2\ell + p - 2$, then G is super-λ and $\kappa = \delta$.

(c) If $D \leq 2\ell + p - 3$, then G is super-κ.

F33: Any bipartite digraph with diameter three is maximally edge-connected.

REMARKS

R20: The main idea in the proof of the results in Fact F31 is that semigirth ℓ measures how far away one can move from or to a given subset F of vertices. For instance, in proving (a), it is shown that if $|F| < \delta$, in any connected component of $G - F$ there are vertices u, v such that $dist(u, F), dist(F, v) \geq \ell$. Hence, any shortest path of length at most $2\ell - 1$ cannot contain a vertex of F. As a conclusion, F cannot be a disconnecting set.

R21: Since any digraph G has semigirth $\ell \geq 1$, Fact F22 is included in Fact F31(a).

R22:   Fact F33 is the analogue for bipartite digraphs of Jolivet's result (Fact F22). In fact, for a bipartite (di)graph, the condition $\delta \geq \lfloor n/4 \rfloor + 1$ implies $D \leq 3$, so that Fact F33 can be also seen as a generalization of Fact F5.

Line Digraphs

DEFINITION

D14: The *line digraph* of a digraph G, denoted $L(G)$, has $V(L(G)) \equiv E(G)$, and a vertex (u, v) is adjacent to a vertex (w, z) if $v = w$ (that is, the *head* of edge (u, v) is the *tail* of edge (w, z) in digraph G). The k-*iterated line digraph*, $L^k(G)$, is defined recursively by $L^k(G) = L(L^{k-1}(G))$.

FACTS

F34:   The order of $L(G)$ equals the size of G, $|V(L(G))| = |E(G)|$, and their minimum degrees coincide, $\delta(L(G)) = \delta(G) = \delta$. Moreover, $\kappa(L(G)) = \lambda(G)$.

F35: If G is d-regular, $d > 1$, has order n, diameter D, and semigirth ℓ, then $L^k(G)$ is also d-regular, has $d^k n$ vertices, diameter $D(L^k(G)) = D(G) + k$, and semigirth $\ell(L^k(G)) = \ell(G) + k$. See the papers [Ai67, ReKuHoLe82, FiYeAl84, FaFi89].

F36: [FaFi89] Let G be a digraph with minimum degree $\delta > 1$, diameter D, and semigirth ℓ.

(a) If $k \geq D - 2\ell$, then $L^k(G)$ is maximally edge-connected.

(b) If $k \geq D - 2\ell + 1$, then $L^k(G)$ is super-λ and maximally connected.

(c) If $k \geq D - 2\ell + 2$, then $L^k(G)$ is super-κ.

REMARK

R23: As shown in Fact F36, the interest of considering k-iterated line digraphs stems from the fact that if k is large enough, Fact F35 guarantees that the conditions of Fact F31 hold.

Girth

For a given girth, high density/connectivity graphs occur when they have a reduced diameter, and also when they have a small number of vertices.

DEFINITION

D15: The same definition for the semigirth (Definition D12) applies for an undirected graph G (considering undirected walks). In this case, it turns out that the semigirth $\ell = \ell(G) = \ell(G^*)$ equals $\lfloor (g-1)/2 \rfloor$ where $g = g(G)$ stands for the girth of G.

FACTS

F37: Let G be a graph with minimum degree $\delta > 1$, diameter D, girth g, and connectivities κ and λ.

(a) [SoNaIm85, SoNaImPe87, FaFi89] If $D \leq \begin{cases} g - 1, & g \text{ odd,} \\ g - 2, & g \text{ even,} \end{cases}$ then $\lambda = \delta$.

(b) [SoNaIm85, SoNaImPe87, FaFi89] If $D \leq \begin{cases} g - 2, & g \text{ odd,} \\ g - 3, & g \text{ even,} \end{cases}$ then G is super-λ and $\kappa = \delta$.

(c) [SoNaIm85, SoNaImPe87, FaFi89] If $D \leq \begin{cases} g - 3, & g \text{ odd,} \\ g - 4, & g \text{ even,} \end{cases}$ then G is super-κ.

(d) [BaCeDiGVMa06] If $D \leq g - 3$, then G is super-κ.

(e) [BaTaMaLi09] If G is regular and $D \leq g - 2$, g odd, then G is super-κ.

(f) [BaMaMo10] If $\delta \geq 3$, $\Delta \leq 3\delta/2 - 1$, and $D \leq g - 2$, g odd, then G is super-κ.

F38: [BaCaFaFi96, CaFa99] Let G be a graph with minimum degree $\delta > 1$, girth g, and connectivities κ and λ. Let $L(G)$ be the line graph of G, with diameter $D(L(G))$. Then,

(a) If $D(L(G)) \leq \begin{cases} g, & g \text{ odd,} \\ g-1, & g \text{ even,} \end{cases}$ then $\lambda = \delta$.

(b) If $D(L(G)) \leq \begin{cases} g-1, & g \text{ odd,} \\ g-2, & g \text{ even,} \end{cases}$ then G is super-λ and $\kappa = \delta$.

(c) If $D(L(G)) \leq \begin{cases} g-2, & g \text{ odd,} \\ g-3, & g \text{ even,} \end{cases}$ then G is super-κ.

F39: [FaFi96a] Any bipartite graph with diameter three is maximally edge-connected.

F40: [KnNi03] For every graph G there is a number $i(G)$ such that $L^k(G)$ is maximally connected when $k \geq i(G)$.

REMARKS

R24: Fact F37 is a simple consequence of Definition D15 and Fact F31.

R25: Fact F39 is the undirected version of Fact F33, which can be seen as Plesník's analogue for the bipartite case.

R26: Fact F40 is based on a result of Hartke and Higgins [HaHi99] about the growth of minimum degree in iterated line graphs. For regular graphs this result is not needed, and in this case $i(G) \leq 5$.

Girth Pair

DEFINITIONS

D16: The **girth pair** (g_1, g_2) of a graph G gives the length g_1 of a shortest odd cycle and the length g_2 of a shortest even cycle.

EXAMPLE

E4: The **Dodecahedron graph** is a cubic graph with girth 5 and a shortest even cycle has length 8. Hence its girth pair is $(5, 8)$.

FACTS

F41: [BaCeDiGVMa07, BaGVMo11] Let G be a graph with minimum degree $\delta \geq 3$, diameter D, girth pair (g, h), odd g and even h with $g + 3 \leq h < \infty$, and connectivities κ and λ.

(a) If $D \leq h - 3$, then $\lambda = \delta$.

(b) If $D \leq h - 4$, then $\kappa = \delta$.

(c) If $D \leq h - 4$ and $\delta \geq 4$, then G is super-κ.

(d) If $D \leq h - 5$ and $\delta = 3$, then G is super-κ.

(e) If $g \geq 5$, $D(L(G)) \leq h - 3$, and the maximum degree of G satisfies $\Delta \leq 2\delta - 3$, then $\kappa = \delta$.

REMARKS

R27: Fact F41 improves Fact F37 for graphs with girth pair (g, h), g odd and $h \geq g+3$ even.

Cages

DEFINITIONS

D17: A (k, g)-***cage*** is a k-regular graph with girth g having the least possible number of vertices.

D18: A 3-connected graph $G = (V, E)$ is said to be ***quasi 4-connected*** if for every vertex-cut $F \subset V$ such that $|F| = 3$, F is the neighborhood of a vertex of degree 3 and $G - F$ has exactly two components.

EXAMPLE

E5: The ***Heawood graph***, shown in Figure 4.7.4, is a $(3, 6)$-cage with order 14 and diameter 3.

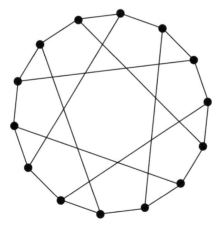

Figure 4.7.4: The Heawood graph.

FACTS

F42: [FuHuRo97] All (k, g)-cages are 2-connected.

F43: [JiMu98, DaRo99] Every (k, g)-cage with $k \geq 3$ is 3-connected.

F44: [MaPeBa02] Every $(3, g)$-cage is superconnected, edge-superconnected, and quasi 4-connected.

F45: [WaXuWa03] Every (k, g)-cage with $k \geq 3$ and odd girth g is maximally edge-connected.

F46: [MaBa04] Every (k, g)-cage with $k \geq 3$ and odd girth g is super-λ.

F47: [LiMiBaMa06] Every (k, g)-cage with $k \geq 3$ and even girth g is super-λ.

F48: [XuWaWa02] Every $(4, g)$-cage is 4-connected.

F49: [MaBaPeFa05] Every (k, g)-cage with $k \geq 4$ and $g \geq 10$ is 4-connected.

F50: [LiMiBa05] Every (k, g)-cage with $k \geq 3$ and odd girth $g \geq 7$ is r-connected with $r \geq \sqrt{k+1}$.

F51: [LiBaMaMi08] Every (k, g)-cage with $k \geq 3$ and even girth $g \geq 6$ is $(r + 1)$-connected, r being the largest integer such that $r^3 + 2r^2 \leq k$.

F52: [MaBaPe07] $(k, 6)$- and $(k, 8)$-cages are maximally connected.

F53: [ArGoMoSe07] $(k, 12)$-cages are maximally connected.

F54: [LuWuLuLi10] Every (k, g)-cage with $k \geq 3$ and odd girth $g \geq 9$ is $\lceil \sqrt{k + \sqrt{k} - 2} \rceil$-connected; and every (k, g)-cage with $k \geq 3$ and even girth $g \geq 10$ is r-connected, where r is the largest integer such that $r(r - 1)^2/4 + 1 + 2r(r - 1) \leq k$.

F55: [BaSa12] Every (k, g)-cage with $k \geq 3$ and odd girth $g \geq 7$ is r-connected with $r \geq \lceil k/2 \rceil$.

CONJECTURE

[FuHuRo97] Every (k, g)-cage is maximally connected.

Large Digraphs

The following results support the intuitive idea that dense (di)graphs have high connectedness.

DEFINITION

D19: For a digraph with maximum degree Δ and diameter D, the **Moore bound**, denoted $n(\Delta, D)$, is given by $n(\Delta, D) = 1 + \Delta + \Delta^2 + \cdots + \Delta^D$.

FACTS

F56: An n-vertex digraph with maximum degree Δ and diameter D has $n \leq n(\Delta, D)$.

F57: [Wa67] The order of a (di)graph with connectivity $\kappa > 1$ and diameter D satisfies $n \geq \kappa(D - 1) + 2$.

F58: [ImSoOk85]

   (a) If $\lambda < \delta$, then $n \leq \lambda\left(n(\Delta, D - 2) + \Delta + 1\right)$.

   (b) If $\kappa < \delta$, then $n \leq \kappa\left(n(\Delta, D - 1) + \Delta\right)$.

F59:

 (a) If $n > (\delta - 1)(n(\Delta, D - 2) + \Delta + 1)$, then $\lambda = \delta$.

 (b) If $n > (\delta - 1)(n(\Delta, D - 1) + \Delta)$, then $\kappa = \delta$.

F60: [Fi93]

 (a) If $\lambda < \delta$, then $n \le \lambda(n(\Delta, D - 2) + 1) + \Delta$.

 (b) If $\kappa < \delta$, then $n \le \kappa(n(\Delta, D - 1) - 1) + \Delta + 1$.

F61: [Xu92, Fi93] Let G be d-regular.

 (a) If $n > d^{D-1} + 2d - 2$, then $\lambda = d$.

 (b) If $n > d^D + 1$, then $\kappa = d$.

F62: [So92, Fi94] Let G be a d-regular digraph, $d \ge 2$, with diameter D.

 (a) If G satisfies either of the following conditions, then G is super-λ.

    (i) $D = 2$ and $n > 3d$.
    (ii) $D \ge 3$ and $n > 2d^{D-1} + d^{D-2} + \cdots + d^2 + 2d$.

 (b) If G satisfies either of the following conditions, then G is super-κ.

    (i) $D = 3$ and $n > 3d^2 + 1$.
    (ii) $D \ge 4$ and $n > 2d^{D-1} + d^{D-2} + \cdots + d^3 + 2d^2 + 1$.

EXAMPLE

E6: Figure 4.7.5 shows a regular digraph for which $n = 6$, $\Delta = \delta = d = 2$, and $D = 2$. Since $n > d^{D-1} + 2d - 2$ and $n > d^D + 1$, Fact F61 guarantees that it is maximally connected ($\kappa = \lambda = d$).

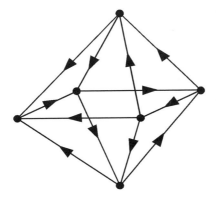

Figure 4.7.5: $\kappa = \lambda = d = 2$.

REMARKS

R28: To our knowledge, Fact F57, due to Watkins, was the first result in which the order n, the diameter D, and the connectivity κ were related (in the undirected case). It follows easily from counting the minimum number of vertices involved in κ internally disjoint u-v paths between a pair of vertices u, v at distance D, as Menger's theorem guarantees.

R29: Similar reasoning gives a lower bound for the number of edges m of a (di)graph with edge-connectivity λ, namely, $m \geq \lambda D$. However, it is not difficult to realize that this is not a very strong result. (The situation seems to depend heavily on the values of λ and D: for $\lambda = 3$ there are constructions giving a lower bound of the order of $\frac{7}{2}D$, whereas for $\lambda = 4$ we have a bound which is "asymptotically optimal," that is, of the order of $4D$.)

R30: If we take into account the connectivity parameters κ or λ, the Moore bound can be refined. Intuitively, a disconnecting set with few vertices or edges is a kind of "bottleneck" that prevents the order from being large, as shown in Fact F58.

R31: Fact F59 is a direct consequence of Fact F58, and Fact F60 is an improvement of Fact F58. Notice that if we set $\kappa = \Delta$ in the upper bound on n of Fact F60(b), we obtain the Moore bound $n(\Delta, D)$.

Large Graphs

Similar results for graphs were derived independently by Esfahanian [Es85], Fiol [Fi93, Fi94], and Soneoka et al. [SoNaImPe87].

DEFINITION

D20: The **_Moore bound_** for an undirected graph with maximum degree Δ and diameter D is given by $n(\Delta, D) = 1 + \Delta + \Delta(\Delta - 1) + \cdots + \Delta(\Delta - 1)^{D-1}$.

FACTS

F63: [SoNaImPe87]

  (a) If $n > (\delta - 1)\left(n(\Delta - 1, D - 2) + 1\right) + \Delta - 1$, then $\lambda = \delta$.

  (b) If $n > (\delta - 1)(\Delta - 1)^{D-1} + 2$, then $\kappa = \delta$.

F64: [So92, Fi94]

  (a) Let $D \geq 2$ and $\delta \geq 2$. If $n > \delta\left(n(\Delta - 1, D - 2) + 1\right) + (\Delta - 1)^{D-1}$, then G is super-λ.

  (b) Let $D \geq 3$, $\delta \geq 3$ and $g \geq 5$. If $n > (\delta - 1)\left(n(\Delta, D - 1) + \Delta\right)$, then $\kappa = \delta$.

4.7.2   Bounded Connectivity

The techniques used for proving the results of the preceding subsection can often be used to derive bounds on the connectivity or edge-connectivity of a (di)graph. In this subsection, we provide some examples.

π-Semigirth

The following definition generalizes semigirth (Definition D12).

DEFINITION

D21: [FaFi89] Let $G = (V, E)$ be a digraph with minimum degree δ and diameter D, and let π be an integer, $0 \leq \pi \leq \delta - 2$. The π-*semigirth* of G, denoted $\ell_\pi(G)$, is the greatest integer ℓ_π between 1 and D such that, for any $u, v \in V$,

(a) if $dist(u, v) < \ell_\pi$, the shortest u-v path is unique and there are at most π distinct u-v walks of length $dist(u, v) + 1$.

(b) if $dist(u, v) = \ell_\pi$, there is only one shortest u-v walk.

FACT

F65: [FaFi89, MaBaPe04] Let G be a connected digraph with minimum degree $\delta > 1$, diameter D, π-semigirth ℓ_π for $0 \leq \pi \leq \delta - 2$, and with k-iterated line digraph $L^k(G)$. Then,

(a) If $D \leq 2\ell_\pi$, then $\lambda \geq \delta - \pi$.

(b) If $D \leq 2\ell_\pi - 1$, then $\kappa \geq \delta - \pi$.

(c) If $D \leq 2\ell_\pi - 1$ and $\pi \leq \lfloor (\delta - 1)/2 \rfloor$, then $\lambda = \delta$.

(d) If $D \leq 2\ell_\pi - 2$, $\ell_0 \geq 2$ and $\pi \leq \lfloor (\delta - 1)/2 \rfloor$, then $\kappa = \delta$.

(e) If $k \geq D - 2\ell_\pi$, then $\lambda(L^k(G)) \geq \delta - \pi$.

(f) If $k \geq D - 2\ell_\pi + 1$, then $\kappa(L^k(G)) \geq \delta - \pi$.

REMARKS

R32: Note that ℓ_0 corresponds to the ordinary semigirth ℓ. Moreover, for $\pi \geq 1$, ℓ_π is well defined even for a digraph with self-loops.

R33: The definition of ℓ_π is restricted to $\pi \leq \delta - 2$ since, otherwise, the above results become irrelevant.

Imbeddings

Here we cite one of the earliest results relating the connectivity of a graph to a topological property of that graph. Other more recent results of this kind can be found in [PluZh98, PluZh02].

DEFINITION

D22: A graph G is said to be ***imbeddable*** in a given surface S if G can be drawn on S without edge crossings.

FACT

F66: [Co73] Let G be any graph embeddable in a oriented surface of genus $g > 0$ (where the genus is, informally, the number of *handles* on its surface [see Chapter 7 of this *Handbook*]). Then, $\kappa \leq \lfloor (5 + \sqrt{1 + 48g})/2 \rfloor$.

Adjacency Spectrum

Given a (di)graph G with some associated matrix A, a natural problem is to study how much can be said about the structure of G from the spectrum of A. This is a major topic in algebraic graph theory, and has been the object of research (see §6.5 of the *Handbook* or the classic textbooks D. Cvetković, M. Dragoš, and H. Sach [CvDoSa95], Biggs [Bi94]).

DEFINITIONS

D23: Given a graph G on n vertices, its ***adjacency matrix*** $A = (a_{uv})$ is the $n \times n$ matrix indexed by the vertices of G with entries $a_{uv} = 1$ if u and v are adjacent and $a_{uv} = 0$ otherwise.

D24: The ***toughness*** t of a graph G is defined as $t = \min_S \{ |S|/c(G - S) \}$, where S runs over all vertex-cuts of G and $c(G - S)$ denotes the number of components of $G - S$.

FACTS

F67: [Al95, Br95] Let G be a connected, non-complete d-regular graph and let λ be the maximum of the absolute values of the eigenvalues of G distinct from d. Then, $t > d/\lambda - 2$.

NOTATION: Given a graph G, let D_2 denote the maximum distance between vertex subsets of G with two vertices. (This parameter is a particular case of the so-called *conditional diameter*, introduced in [BaCaFaFi96].)

F68: [FiGaYe97] Let G be a d-regular graph with $D_2 > 1$ and distinct eigenvalues (of its adjacency matrix A) $\lambda_0 (= d) > \lambda_1 > \cdots > \lambda_r$. Let $P(x) := 2(x - \lambda_r)/(\lambda_1 - \lambda_r) - 1$. Then, $\kappa(G) \geq \min \left\{ d, \left\lceil \frac{2(P(d)^2 - 1)(n-2)}{2(P(d)^2 - 1) + n} \right\rceil \right\}$.

REMARKS

R34: Besides Fact F67, Brouwer [Br96] gave some other interesting examples of results about the connectivity of a graph G in terms of its spectrum.

R35: For other results concerning the toughness of a graph, mainly used in the study of vulnerability of network topologies [BoHaKa81], see, for instance, [ChLi02].

R36: Notice that, from Fact F16(b), if $D_2 = 1$ then G is maximally connected. Otherwise, Fact F68 applies.

Laplacian Spectrum

DEFINITION

D25: Given a graph G, its ***Laplacian matrix*** L is defined as $L = D - A$, where D is the diagonal matrix of the vertex degrees and A is the adjacency matrix of G (see, for instance [Bi94]). The ***Laplacian eigenvalues*** of G are the eigenvalues of its Laplacian matrix.

TERMINOLOGY: The second smallest Laplacian eigenvalue, θ_1, usually denoted by $a = a(G)$, is called the ***algebraic connectivity*** of G because it has some properties which are similar to those satisfied by the connectivity κ.

FACTS

F69: Since the Laplacian matrix L is positive semidefinite, its eigenvalues are all nonnegative, with the first one equal to zero. If G is d-regular with (distinct) eigenvalues $\lambda_0(= d) > \lambda_1 > \cdots \geq \lambda_r$, then its Laplacian eigenvalues are $\theta_0, \theta_1, \ldots, \theta_r$, where $\theta_i = d - \lambda_i$, $i = 1, 2, \ldots, r$.

F70: [Fi73] Let G be a graph with second smallest Laplacian eigenvalue a.

(a) $\kappa \geq a \geq 0$, and $a = 0$ if and only if G is not connected.

(b) For any spanning subgraph H of G we have $a(H) \leq a(G)$.

(c) For any vertex subset U of G we have $a(G - U) \geq a(G) - |U|$.

F71: Let G be a d-regular graph with n vertices, $D_2 > 1$, and Laplacian eigenvalues $\theta_0(= 0) < \theta_1 < \theta_2 < \cdots < \theta_r$. If $d < \frac{n(\theta_r - \theta_1)^2 + 8\theta_1\theta_r(n-1)}{n(\theta_r - \theta_1)^2 + 8\theta_1\theta_r}$, then $\kappa = d$.

REMARK

R37: Fact F71 is just a consequence of Fact F68 in terms of the Laplacian eigenvalues.

4.7.3   Symmetry and Regularity

Boundaries, Fragments, and Atoms

The concepts of *fragment* and *atom* are very useful in the study of connectivity, both in the undirected and the directed case, and, in particular, for (di)graphs with strong symmetries. For graphs, the concept of an atom was introduced independently by Mader [Ma70] and Watkins [Wa70]. The notion of an atom for digraphs was introduced by Chaty [Ch76] and first used extensively by Hamidoune [Ha77, Ha80, Ha81].

Because of the close relationship between a graph G and its corresponding symmetric digraph G^*, we only give the definitions for digraphs. (For undirected graphs, the corresponding definitions are unsigned.)

DEFINITIONS

D26: The *positive boundary* of a vertex subset F in a digraph G, denoted $\partial^+ F$, is the set of vertices that are adjacent *from* F, and the *negative boundary*, $\partial^- F$, is the set of vertices adjacent *to* F.

D27: The *positive edge-boundary* and the *negative edge-boundary*, denoted $\omega^+ F$ and $\omega^- F$, respectively, are given by

$$\omega^+ F = \{(u, v) \in E : u \in F \text{ and } v \in V - F\};$$
$$\omega^- F = \{(u, v) \in E : u \in V - F \text{ and } v \in F\}.$$

D28: Let G be a strongly connected digraph with connectivity κ. A vertex subset F is a *positive fragment* of G if $|\partial^+ F| = \kappa$ and $V - (F \cup \partial^+ F) \neq \emptyset$, and F is a *negative fragment* if $|\partial^- F| = \kappa$ and $V - (\partial^- F \cup F) \neq \emptyset$.

D29: Let G be a digraph with edge-connectivity λ. A vertex subset F is a *positive α-fragment* of G if $|\omega^+ F| = \lambda$, and F is a *negative α-fragment* if $|\omega^- F| = \lambda$.

D30: A vertex u of a positive [negative] α-fragment F is called *interior* if none of the edges adjacent from [to] u belongs to $\omega^+ F$ [$\omega^- F$].

D31: An *atom* is a (positive or negative) fragment of minimum cardinality.

EXAMPLE

E7: For the digraph of Figure 4.7.6, $\kappa = 2$ and F is a positive (respectively, negative) fragment with positive (respectively, negative) boundary $\{u, v\}$ (respectively, $\{z, t\}$). Analogously, $\omega^+ F = \{(x, v), (y, u)\}$ and $\omega^- F = \{(z, x), (t, y)\}$. In this digraph, each single vertex is an atom.

FACT

F72: If $F \cup \partial^+ F \neq V$ [$F \cup \partial^- F \neq V$], then $\partial^+ F$ [$\partial^- F$] is a vertex-cut of G. Similarly, if F is a proper (nonempty) subset of V, then $\omega^+ F$ [$\omega^- F$] is an edge-cut. Using these concepts, we have the following alternative definitions of the connectivity parameters:

$$\kappa = \min\{|\partial^+ F| : F \subset V, F \cup \partial^+ F \neq V \text{ or } |F| = 1\}$$

$$\lambda = \min\{|\omega^+ F| : F \text{ is a nonempty, proper subset of } V\}$$

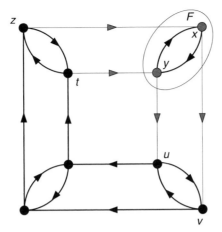

Figure 4.7.6: F is a fragment.

Fragments and Atoms in Undirected Graphs

FACTS

F73: [Wa70, Ma71a] In a connected graph, any two distinct atoms are disjoint.

F74: [Ma71a] Let G be a graph with order n and connectivity κ. Let F_1 and F_2 be distinct minimal fragments of G, with at most $n - 3\kappa/2$ vertices. Then $F_1 \cap F_2 = \emptyset$.

F75: [Ma71a] Let G be a graph with connectivity κ. If T is a disconnecting set with κ vertices and A is an atom, then either $A \subset T$ or $A \cap T = \emptyset$.

REMARKS

R38: To quote a personal communication from Watkins [Wa02]: "It is an amazing coincidence that Prof. Mader and I not only conceived of the notion of 'atom' independently and simultaneously, but we also accorded this notion almost identical names." In fact, Watkins [Wa70] used the term "atomic part," while Mader [Ma70] used the term "kleinstes Glie" (that is, "smallest member"). Then, in a subsequent paper, Mader [Ma71a] mentioned the main result in [Wa70] on atomic parts, and deduced it from his results. Moreover, inspired by Watkins' terminology, he first used the simpler name "atom."

R39: Fact F73 was proved by Watkins for transitive graphs, whereas the general case is due to Mader.

R40: As already mentioned, the seminal papers on atoms are those of Mader [Ma70] and Watkins [Wa70]. Notice that Fact F75 is a generalization of Fact F73, and it is considered as the more important property of an atom.

R41: Results on atoms and the connectivity of infinite graphs can be found in [JuWa77] and [Ha89].

Fragments and Atoms in Digraphs

The results above can be seen as consequences of the corresponding directed versions, which are due to Hamidoune.

FACTS

F76: [Ha77] Let G be a connected digraph with a positive (negative) atom A and a positive (negative) fragment F. Then, either $A \subset F$ or $A \cap F = \emptyset$. In particular, two distinct positive (negative) atoms are disjoint.

F77: [Ha80] If G is a connected digraph with $\lambda < \delta^+$ ($\lambda < \delta^-$), then every positive (negative) α-fragment contains an interior vertex.

REMARKS

R42: Contrary to the case of graphs, where the presence of an atom is always assured, a digraph does not necessarily have an atom with a prescribed sign.

R43: Fact F77 implies Jolivet's theorem (Fact F22).

Graphs with Symmetry

Graphs with high symmetry often have "good" properties, and their study has special relevance to other areas of mathematics. In particular, the results here show that, for connected graphs, high symmetry goes hand in hand with high connectivity. Graph automorphisms and symmetry are discussed in §6.1 and §6.2.

DEFINITION

D32: A (di)graph G is **vertex-transitive** (or **vertex-symmetric**) if for any vertices u, v there is an automorphism of G which maps u into v. Similarly, G is called **edge-transitive** (or **edge-symmetric**) if for any (possibly oriented) edges uv, wz there is an automorphism of G that maps uv into wz.

FACTS

F78: [Ma70, Ma71b] Let G be a vertex-symmetric connected graph with degree $d \geq 3$. Then, $\lambda = d$, $\kappa \geq 2\lfloor d/3 \rfloor + 2$ if $d > 3$, and $\kappa = 3$ otherwise. Furthermore, if G does not contain K_4, then $\kappa = d$.

F79: [Ma70] Let G be an edge-symmetric connected graph with degree d. Then $\kappa = \lambda = d$.

F80: [Ma70, Wa70] Let G be a vertex-transitive graph with an atom A. Then the subgraph $G(A)$ induced on A is also vertex-transitive. Moreover, the set of atoms of G constitutes a partition of $V(G)$.

F81: [Ha77] Let G be a vertex-transitive digraph with a positive (negative) atom A. Then, the induced subdigraph $G(A)$ is also vertex-transitive. Furthermore, the set of positive (negative) atoms of G constitutes a partition of $V(G)$.

F82: [Ha81] Let G be a vertex-symmetric strongly connected digraph with (constant) outdegree d^+. Then $\lambda = d^+$ and $\kappa \geq \frac{1}{2}d^+$. Moreover, if G is an oriented graph, then $\kappa \geq \frac{2}{3}d^+$.

REMARKS

R44: The inequality in Fact F78, which is best possible, is a consequence of Fact F73 and, for $d \not\equiv 2 \mod 3$, it is an improvement of a result of Watkins [Wa70], who showed that $\kappa > (2/3)d$.

R45: From Facts F80 and F81, the order of a (positive or negative) atom of G divides the order of G. Consequently, every connected vertex-transitive (di)graph with a prime number of vertices is maximally connected ($\kappa = \lambda = \delta$). In fact, it is known that such (di)graphs must be Cayley (di)graphs of cyclic groups.

R46: By Fact F76, Hamidoune [Ha77] proved Kameda's result stating that every minimal k-connected digraph has one vertex of out-degree or in-degree k [Ka74], and Hamidoune also proved that every edge-transitive digraph is maximally connected.

Cayley Graphs

The *Cayley graphs* are among the most interesting vertex-symmetric (di)graphs, mainly because of their relationship with group theory (see §6.1 and §6.2). In particular, the study of the connectivity of Cayley graphs has striking connections with some key results in additive number theory, such as the well-known Cauchy–Davenport theorem: If p is a prime number and A, B are two nonempty subsets of the cyclic group Z_p, then either $A + B = Z_p$ or $|A + B| \geq |A| + |B| - 1$.

DEFINITIONS

D33: Let Γ be a finite group with identity element e and generating set $S \subset \Gamma - \{e\}$. The *Cayley digraph* $G = (\Gamma, S)$ has vertices labeled with the elements of Γ, and edges of the form (u, ug) where $g \in S$. In particular, when $S^{-1} = S$ (where $S^{-1} = \{x^{-1} : x \in S\}$) we obtain a symmetric Cayley digraph or, simply, a *Cayley graph*.

D34: If Γ is a cyclic group, then the Cayley graph is called a *circulant graph*.

D35: A generating set S of a group Γ is called *minimal* when any proper subset $S' \subset S$ does not generate Γ.

D36: The *symmetric group* on n elements, denoted Σ_n, is the group of all permutations of the set $\{1, 2, \ldots, n\}$.

D37: Let H be a subgroup of a group G, and let $x \in G$. Then the set $xH = \{xh | h \in H\}$ is a *left coset* of G with respect to H.

FACTS

F83: [Im79] Let S be a generating set of the symmetric group $\Gamma = \Sigma_n$ with $n \geq 5$, such that $xSx^{-1} = S$ for every $x \in \Gamma$. Then, the Cayley digraph (Γ, S) is maximally connected (that is, $\kappa = |S|$).

F84: [Ha84] Let Γ be a finite group with identity e and generating set S. Let A be a positive (respectively, negative) atom of (Γ, S) containing e. Then A is the subgroup of Γ generated by $S \cap A$, and the positive (respectively, negative) atoms of (Γ, S) are the left cosets of Γ with respect to A.

F85: [Ha84] Let Γ be a finite group with a minimal generating set S. Let $S' \subset S^{-1}$. Then, the Cayley digraph $(\Gamma, S \cup S')$ is maximally connected.

F86: [HaSe96] Let Γ be an Abelian group of order n and let S be a generating subset of Γ such that $|S \cup \{0\}| \leq n - 1$. Let D be the diameter of $G = (\Gamma, S)$. Then there is a vertex-cut of size less than $(4n \ln(n/2))/D$ whose deletion separates G into a negative fragment B and a positive fragment $\overline{B}$ such that $|B| = |\overline{B}|$. Moreover, G can be separated into two equal parts of size $|B|$ by deleting less than $(8e/|S|)n^{1-1/|S|} \ln(n/2)$ vertices.

REMARKS

R47: Fact F84, due to Hamidoune, provides a very short proof of Fact F83.

R48: For the case of Cayley graphs, Fact F85 was previously proved by Godsil [Go81]. Subsequently, Akers and Krishnamurthy [AkKr87], Hamidoune, Lladó, and Serra [HaLlSe92], and Alspach [Als92] improved these results by considering Cayley (di)graphs with a **hierarchical** generating set (that is, when the group generated by the first k generators is a proper subgroup of the group generated by the first $k + 1$ for each k).

Circulant Graphs

Because of their circular symmetry, circulant graphs have been proposed as good models for local area network topologies, where they are called *loop networks*. In this context, other good topologies are provided by Cayley graphs of Abelian groups, also called loop networks (see [BeCoHs95], [BoTi84]).

FACTS

F87: [Ha84] Let Γ be the cyclic group Z_n. Let S be the strictly increasing sequence of s integers $(1 =)b_1 < b_2 < \cdots < b_s(< n)$ such that $b_{i+1} - b_i \geq \min\{2, b_i - b_{i-1}\}$ for $i = 2, 3, \ldots, s - 1$. Then the circulant digraph (Γ, S) is maximally connected ($\kappa = s$).

F88: [Ha84] Let Γ be the cyclic group Z_n. Let S be the strictly increasing sequence of s integers $(1 =)b_1 < b_2 < \cdots < b_s(< n/2)$ such that $b_{i+1} - b_i \geq \min\{2, b_i - b_{i-1}\}$ for $i = 2, 3, \ldots, s - 1$, and let $S' \subset -S$, where $-S$ denotes the set of (additive) inverses of the elements in S. Then the circulant digraph $(\Gamma, S \cup S')$ is maximally connected ($\kappa = |S \cup S'|$).

F89: The Cauchy–Davenport theorem is equivalent to stating that, for any generating set $S \subset Z_p$, p prime, the Cayley digraph (Z_p, S) is maximally connected (that is, $\kappa = |S|$).

REMARKS

R49:   The case $S' = -S$ in Fact F88 (that is, for circulant graphs) was proved in [BoFe70] using the "convexity conditions" $b_{i+1} - b_i \geq b_i - b_{i-1}$ (see also [BoTi84]).

R50:   Fact F89, noted by Hamidoune, is a bridge between additive number theory and graph theory. (For a comprehensive survey on the subject, we refer the reader to [Ha96].)

Distance-Regular Graphs

The concept of distance-regularity was introduced by Biggs [Bi71] in the early 1970s, by changing a symmetry-type requirement, that of distance-transitivity, to a regularity-type condition concerning the cardinality of some vertex subsets. Distance-regular graphs have important connections with other branches of mathematics, such as geometry, coding theory, and group theory, as well as with other areas of graph theory. In our context, their high regularity seems also to induce a high degree of connectedness.

DEFINITIONS

D38:   Let G be a regular graph with diameter D and let k be an integer between 1 and D. Graph G is said to be ***distance-regular*** if, for any two vertices u and v with $dist(u, v) = k$, the numbers c_k, a_k, and b_k of vertices that are adjacent to v and whose distance from u is $k - 1$, k, and $k + 1$, respectively, depend only on k.

D39:   An n-vertex k-regular graph G is called $(n, k; a, c)$-***strongly-regular*** if any two adjacent vertices have a common neighbors and any two non-adjacent vertices have c common neighbors.

FACTS

F90:   Let G be a connected graph. Then G is strongly-regular if and only if G is distance-regular of diameter two.

F91:   Every strongly regular graph is maximally edge-connected.

F92:   [BrMe85] Every strongly regular graph is maximally connected and super-κ.

F93:   [BrKo09] Every distance-regular graph is maximally connected. If the degree is at least 3, it is also super-κ.

REMARKS

R51:   Fact F91 is a consequence of Facts F3 and F90.

R52:   Fact F93 was a long standing conjecture of Brouwer [Br96]. The result was previously proved for some families of distance-regular graphs, such as the so-called *odd graphs* O_k (having the k-subsets of a $(2k - 1)$-set as its vertices and adjacencies defined by void intersection); see [Gh92].

R53:   Fact F93 implies a previous conjecture of Godsil [Go81], stated in the context of association schemes, that every distance-regular graph is maximally edge-connected. In fact, this last result was proved by Brouwer and Haemers in [BrHa05].

4.7.4   Generalizations of Connectivity Parameters

The standard connectivity parameters have been generalized in different ways, giving rise to numerous articles; see, for instance, [BaBeLiPi87], [BeOePi02], [Ha83], [Wo73]. Here we will consider several examples, some of which have special relevance to the study of network vulnerability.

Conditional Connectivity

The next two definitions generalize the concept of superconnectivity.

DEFINITIONS

D40: Given a graph $G = (V, E)$ and a nonnegative integer s, a vertex subset $V' \subset V$ is said to be *n-trivial* if it contains the boundary $\partial(H)$ of some subgraph $H \subset G$ with s' vertices, $1 \le s' \le s$. Similarly, an edge subset $E' \subset E$ is said to be *s-trivial* if it contains the edge-boundary $\omega(H)$ of some subgraph $H \subset G$ with s' vertices, $1 \le s' \le s$.

D41: The *conditional connectivity* κ_s of a graph G is the minimum cardinality of a disconnecting set that is not s-trivial. The *conditional edge-connectivity* λ_s of G is the minimum cardinality of a disconnecting edge set that is not s-trivial.

FACT

F94: [FaFi89, FiFaEs90, FaFi94] Let G be a graph with minimum degree $\delta > 1$, diameter D and girth g. Let $\ell = \left\lfloor \frac{g-1}{2} \right\rfloor$.

 (a) If $D \le 2\ell$, then $\lambda_0 = \delta$.

 (b) If $D \le 2\ell - 1$, then $\kappa_0 = \delta$ and $\lambda_1 \ge 2\delta - 2$.

 (c) If $D \le 2\ell - 2$, then $\kappa_1 \ge 2\delta - 2$ and $\lambda_2 \ge 3\delta - 4$.

 (d) If $D \le 2\ell - 3$, then $\kappa_2 \ge 3\delta - 4$.

CONJECTURE

C1: [FaFi94]

 (a) If $D \le 2\ell - s$, then $\lambda_s \ge (s+1)\delta - 2s$.

 (b) If $D \le 2\ell - s - 1$, then $\kappa_s \ge (s+1)\delta - 2s$.

REMARKS

R54: Harary [Ha83] introduced the general concept of conditional connectivity. In our context, the graphs are assumed to be those for which κ_s and λ_s are well-defined.

R55: Note that the conditional connectivities κ_0 and λ_0 correspond to the standard connectivities κ and λ (thus, Fact F94 generalizes Fact F37). If $\kappa_1 > \delta$, then G is super-κ, and if $\lambda_1 > \delta$, then G is super-λ.

R56: The conjecture above was proved to be true for even s (provided that $\delta > 2$ and $\ell > (s+1)/2$) [FaFi96b]. Moreover, if s is large enough in comparison with the minimum degree δ, further improvements of the sufficient conditions were given in [BaCaFaFi97b, Ba99].

Restricted Connectivity

D42: The ***restricted edge-connectivity*** $\lambda' = \lambda'(G)$, introduced by Esfahanian and Hakimi [Es88], is the minimum cardinality over all ***restricted edge-cuts***, i.e., those edge-cuts S such that there are no isolated vertices in $G - S$.

D43: The ***minimum edge-degree*** of G is $\xi = \xi(G) = \min\{deg(u) + deg(v) - 2 : uv \in E\}$.

D44: A graph is said to be λ'-***optimal*** if $\lambda' = \xi$.

D45: Given a graph $G = (V, E)$, an edge set $S \subset E$ is called a k-***restricted-edge-cut*** if $G - S$ is disconnected and every component of $G - S$ has at least k vertices.

D46: The k-***restricted-edge-connectivity*** of G, denoted by $\lambda^{(k)} = \lambda^{(k)}(G)$, is defined as the cardinality of a minimum k-restricted-edge-cut.

D47: The edge-boundary $\omega(F)$ is called a $\lambda^{(k)}$-cut if $|\omega(F)| = \lambda^{(k)}$, and $F \subset V$ is called a k-***fragment*** of G. A minimum k-fragment is called a k-***atom***, and its cardinality is denoted $a_k(G) = a_k$.

D48: The ***minimum*** k-***edge-degree*** of G is $\xi_k = \xi_k(G) = \min\{|\omega(U)| : \emptyset \neq U \subset V(G), |U| = k$ and $G[U]$ is connected$\}$.

D49: A graph with k-restricted edge cuts is said to be $\lambda^{(k)}$-***optimal*** if $\lambda^{(k)} = \xi_k$.

D50: A graph G is ***super*** k-***restricted edge connected***, or ***super-***$\lambda^{(k)}$, if G is $\lambda^{(k)}$-optimal and the deletion of every $\lambda^{(k)}$-cut isolates a component with k vertices; that is, if every k-fragment X has cardinality $|X| \in \{k, n - k\}$.

REMARKS

R57: Esfahanian and Hakimi [Es88] showed that each connected graph of order $n \geq 4$ except a star has restricted edge-cuts and satisfies $\lambda' \leq \xi$.

R58: The restricted edge-connectivity λ' corresponds to the 2-restricted-edge-connectivity $\lambda^{(2)}$ and also to the conditional connectivity λ_1 defined in Definition D41.

R59: Furthermore, the k-restricted-edge-connectivity $\lambda^{(k)}$ corresponds to the conditional connectivity λ_{k-1} given in Definition D41, for any $k \geq 1$.

R60: If G is super-$\lambda^{(k)}$, then $a_k = k$.

FACTS

F95: [BaGVMa06] Every graph with girth g and $\delta \geq 2$ is super-λ and has $\lambda' = \xi$ if its diameter $D \leq g - 2$.

F96: [BaGVMa06] Every graph G with girth g odd and $\delta \geq 2$ is super-λ and has $\lambda' = \xi$ if $D = g - 1$ and either of the following assertions holds.

(i) All pairs u, v of vertices at distance $d(u, v) = g - 1$ are such that neither vertex u nor v lies on a cycle of length g.

(ii) $|N_{(g-1)/2}(u) \cap N_{(g-1)/2}(v)| \geq 3$ for all pairs u, v of vertices at distance $d(u,v) = g - 1$ where $N_h(u) = \{w \in V(G) : d(u,w) = h\}$.

F97: [BaCeDiGVMa05] Every graph with girth g even and $\delta \geq 2$ is super-λ and has $\lambda' = \xi$ if $D = g - 1$ and only $\delta - 1$ vertices are mutually at distance $g - 1$ apart.

F98: [BaLiMi08] Every graph G with girth g, minimum degree $\delta \geq 3$, and diameter D is super-λ' if $D \leq g - 3$ or if the diameter of the line graph satisfies $D(L(G)) \leq g - 3$.

F99: [ZhYu05] Let G be a connected graph with order at least $2(\delta + 1)$ not isomorphic to any $G_{s,\delta}^*$, where $G_{s,\delta}^*$ is the graph obtained from s copies of K_δ by adding a new vertex u and joining u to every other vertex. Then for any $k \leq \delta + 1$, G has k-restricted edge cuts and $\lambda^{(k)} \leq \xi_k$.

F100: [ZhYu07] Let G a connected graph on $n \geq 2k$ vertices. If $deg(u) + deg(v) \geq n + 2k - 3$, for every pair of nonadjacent vertices u and v, then G is $\lambda^{(k)}$-optimal.

F101: [BoUeVo02, XuXu02] For $k = 2, 3$, a graph with $\lambda^{(k)}$-cuts has $\lambda^{(k)} = \xi_k$ if and only if $a_k = k$.

F102: [BaGMMa09] Let G be a graph with $\lambda^{(k)}$-cuts and such that $\lambda^{(k)} \leq \xi_k$. Then G is $\lambda^{(k)}$-optimal if $a_k = k$. Moreover, $a_k = k$ follows if G is $\lambda^{(k)}$-optimal and either of the following conditions holds.

(i) $\delta \geq 2k - 1$.

(ii) $\delta \geq k + 1$ and $g \geq k + 1$.

F103: [BaGMMa09] Let G be a graph with $\lambda^{(k)}$-cuts such that $\lambda^{(k)} \leq \xi_k$ and $\lambda^{(k+1)}$ exists. Then G is super-$\lambda^{(k)}$ if and only if $\lambda^{(k+1)} > \xi_k$.

F104: [ZhLi10] Let G be a graph with $\lambda^{(k)}$-cuts, $k \geq 3$, girth $g \geq 5$, minimum degree $\delta \geq k$, and diameter D. Then G is $\lambda^{(k)}$-optimal if either of the following conditions holds.

(i) $D \leq g - 4$ when g is even, or $D \leq g - 3$ when g is odd;

(ii) $D \leq g - 3$ and $\delta \geq 2k - 3$.

F105: [BaGV10] Let G be a graph with $\lambda^{(k)}$-cuts, girth g, minimum degree $\delta \geq \max\{3, k\}$, and diameter D. Then G is super-$\lambda^{(k)}$ if either of the following conditions holds.

(i) $D \leq g - 4$ when g is even, or $D \leq g - 4$ when g is odd and $\delta \geq k + 1$;

(ii) the diameter of the line graph is $D(L(G)) \leq g - 4$.

Distance Connectivity

Here we consider a generalization of the concepts of connectivity and edge-connectivity of a (di)graph, introduced in [FiFa94] and [BaCaFi96], which takes into account the distance between vertices.

DEFINITIONS

D51: Let $G = (V, E)$ be a digraph. Given $u, v \in V$ such that $(u, v) \notin E$, recall (from §4.1) that a set $S \subset V - \{u, v\}$ is called a $(u|v)$-**set** if there is no u-v path in $G - S$, and $\kappa(u|v)$ is the minimum cardinality of a $(u|v)$-set. Similarly, a given edge-set $T \subset E$ is called a $(u|v)$-**edge-set** for some $u, v \in V$ if there is no u-v path in $G - T$, and $\lambda(u|v)$ is the minimum cardinality of a $(u|v)$-edge-set.

D52: Let $G = (V, E)$ be a digraph with diameter D. Given t, $1 \le t \le D$, the t-**distance connectivity** of G, denoted by $\kappa(t; G) = \kappa(t)$, is defined as $\kappa(t) = \min\{\kappa(u|v) : u, v \in V, \ dist(u, v) \ge t\}$ if $t \ge 2$, and $\kappa(1) = \kappa$, where κ is the standard connectivity of G. Analogously, the t-**distance edge-connectivity** is $\lambda(t; G) = \lambda(t) = \min\{\lambda(u|v) : u, v \in V, \ dist(u, v) \ge t\}$ for $t \ge 1$.

FACTS

F106:

(a) $\kappa = \kappa(1) = \kappa(2) \le \kappa(3) \le \cdots \le \kappa(D)$.

(b) $\lambda = \lambda(1) \le \lambda(2) \le \cdots \le \lambda(D)$.

F107: Let G be a digraph with minimum degree $\delta > 1$ and semigirth ℓ (see Definition D12).

(a) If $\lambda < \delta$ then $D \ge 2\ell + 1$ and $\lambda = \lambda(2\ell + 1)$.

(b) If $\kappa < \delta$ then $D \ge 2\ell$ and $\kappa = \kappa(2\ell)$.

F108:

(a) $\lambda = \delta$ if and only if $D \le 2\ell$ or $\lambda(2\ell + 1) \ge \delta$.

(b) $\kappa = \delta$ if and only if $D \le 2\ell - 1$ or $\kappa(2\ell) \ge \delta$.

F109: Every digraph with distance connectivity $\lambda(3) \ge \delta$ has maximum edge-connectivity.

F110: Let G be an undirected graph with associated symmetric digraph G^*. Since a minimum t-distance disconnecting set of G^* cannot contain digons, $\kappa(t; G^*) = \kappa(t; G)$ and $\lambda(t; G^*) = \lambda(t; G)$.

F111: Let G be an undirected graph with girth g and $\delta > 1$.

(a) If $\lambda < \delta$ then
$$\begin{cases} D \ge g \text{ and } \lambda = \lambda(g), & g \text{ odd}, \\ D \ge g - 1 \text{ and } \lambda = \lambda(g - 1), & g \text{ even}. \end{cases}$$

(b) If $\kappa < \delta$ then
$$\begin{cases} D \ge g - 1 \text{ and } \kappa = \kappa(g - 1), & g \text{ odd}, \\ D \ge g - 2 \text{ and } \kappa = \kappa(g - 2), & g \text{ even}. \end{cases}$$

F112:

(a) $\lambda = \delta$ if and only if $\begin{cases} D \geq g - 1 \text{ or } \lambda(g) \geq \delta, & g \text{ odd}, \\ D \geq g - 2 \text{ or } \lambda(g - 1) \geq \delta, & g \text{ even}. \end{cases}$

(b) $\kappa = \delta$ if and only if $\begin{cases} D \geq g - 2 \text{ or } \kappa(g - 1) \geq \delta, & g \text{ odd}, \\ D \geq g - 3 \text{ or } \kappa(g - 2) \geq \delta, & g \text{ even}. \end{cases}$

F113: Any graph with distance connectivity $\lambda(3) \geq \delta$ has maximum edge-connectivity.

REMARKS

R61: In Fact F108, since $\kappa(t)$ and $\lambda(t)$ are defined only for $t \leq D$, the two sufficient conditions on the diameter and the distance connectivity are complementary to one another.

R62: Since the semigirth of any digraph is at least one, Fact F107(a) implies Fact F109, which complements Jolivet's result (Fact F22).

R63: Fact F111 follows from Fact F107 by considering Fact F110 and $\ell(G^*) = \lfloor (g - 1)/2 \rfloor$.

High Distance Connectivity

DEFINITIONS

D53: Given a vertex u of a digraph G, the **out-** and **in-eccentricity** of u are $ecc^+(u) = \max_{v \in V}\{dist(u, v)\}$ and $ecc^-(u) = \max_{v \in V}\{dist(v, u)\}$, respectively.

D54: For any integer t, $1 \leq t \leq D$, the **minimum t-degree** of a digraph G is $\delta(t) = \min\{\delta^+(t), \delta^-(t)\}$, where $\delta^+(t) = \min_{u \in V}\{deg^+(u) : ecc^+(u) \geq t\}$ and $\delta^-(t) = \min_{u \in V}\{deg^-(u) : ecc^-(u) \geq t\}$.

D55: A connected digraph G with diameter D is said to be **s-geodetic**, for some $1 \leq s \leq D$, if any two vertices of G are joined by at most one path of length less than or equal to s. If $s = D$, the digraph is called **strongly geodetic** (see [BoKoZn68], [PlZn74]).

FACTS

F114: $\delta = \delta(1) = \cdots = \delta(r) \leq \delta(r + 1) \leq \cdots \leq \delta(D)$.

F115: For any t, $1 \leq t \leq D$, $\kappa(t) \leq \lambda(t) \leq \delta(t)$.

TERMINOLOGY: A digraph G is called **maximally t-distance connected** when $\kappa(t) = \lambda(t) = \delta(t)$, and **maximally t-distance edge connected** when $\lambda(t) = \delta(t)$.

F116: If a digraph G is maximally connected, then G is maximally t-distance connected for any $1 \leq t \leq r$.

F117: [BaCaFaFi97a] Let G be an s-geodetic digraph. Then,

(a) $\lambda(t) = \min\{\delta(t), \lambda(2\ell + 1)\}$, for any $t \le 2s + 1$.

(b) $\kappa(t) = \min\{\delta(t), \kappa(2\ell)\}$, for any $t \le 2s$.

F118: [BaCaFaFi97a] Let G be an s-geodetic digraph.

(a) G is maximally t-distance connected for any $t \le 2s$ if $D \le 2\ell - 1$.

(b) G is maximally t-distance edge connected for any $t \le 2s + 1$ if $D \le 2\ell$.

F119: Let G be a graph with girth g and diameter D. Then, for any $1 \le t \le D$,

(a) G is maximally t-distance edge connected if $\begin{cases} D \le g - 1, & g \text{ odd,} \\ D \le g - 2, & g \text{ even.} \end{cases}$

(b) G is maximally t-distance connected if $\begin{cases} D \le g - 2, & g \text{ odd,} \\ D \le g - 3, & g \text{ even.} \end{cases}$

Maximal Connectivity

Instead of looking for minimum disconnecting sets, we can consider those (minimal) disconnecting sets with maximum cardinalities. This leads to considering the following connectivity parameters.

NOTATION: Denote by $\kappa_{\max}$ and $\lambda_{\max}$ the maximum cardinality of a minimal disconnecting (vertex) set and a minimal disconnecting edge set, respectively.

FACTS

F120: $\kappa_{\max} \ge \kappa$ and $\lambda_{\max} \ge \lambda$.

F121: [PeLaHe86] For any non-trivial graph G with order n and maximum degree $\Delta \ne n - 1$ we have $\kappa_{\max} \le \lambda_{\max}$. Furthermore, if G is 2-connected, then $\lambda_{\max} \ge \Delta$.

F122: [PeLaHe86] Let G be an n-vertex graph with minimum degree δ.

(a) If $\delta \ge \lfloor n/2 \rfloor$, then $\lambda_{\max} \ge \delta$.

(b) If $\delta \ge \lfloor (n + i)/2 \rfloor$ for some i with $1 \le i \le n/2$, then $\lambda_{\max} \ge i\lfloor (n - i + 2)/2 \rfloor$.

(c) If $\delta \ge \lfloor (n + i)/2 \rfloor$ for some i with $n/2 < i < n - 2$, then $\lambda_{\max} \ge \lceil n/2 \rceil \cdot \lfloor (i + 1)/2 \rfloor$.

Hamiltonian Connectivity

DEFINITIONS

D56: A graph G is **hamiltonian connected** if between any pair of vertices u, v there is a hamiltonian u-v path in G.

D57: A graph G **k-leaf-connected** if $|V(G)| > k$ and for each subset S of $V(G)$ with $k = |S|$ there exists a spanning tree T with precisely S as the set of endvertices (vertices of degree 1).

FACTS

F123: [GuWa86] Let u and v be non-adjacent vertices of G with $d(u) + d(v) \geq |V(G)| + k - 1$. If $G + uv$ is k-leaf-connected, then G is k-leaf-connected.

F124: [GuWa86] For all natural numbers n, k, $2 \leq k < n - 2$, there are k-leaf-connected graphs with $\lceil (k + 1)n/2 \rceil$ edges (the minimum number of edges that a k-leaf-connected graph on n vertices can have).

REMARK

R64: The generalization of the concept of hamiltonian connectivity (Definition D57) is due to Murty. Notice that G is hamiltonian-connected if and only if G is 2-leaf-connected.

References

[Ai67] M. Aigner, On the linegraph of a directed graph. *Math. Z.* **102** (1967), 56–61.

[AkKr87] S.B. Akers and B. Krishnamurthy, On group graphs and their fault tolerance. *IEEE Trans. Comput.* **36** (1987), 885–888.

[Al95] N. Alon, Tough Ramsey graphs without short cycles. *J. Algebraic Combin.* **4** (1995), 189–195.

[Als92] B. Alspach, Cayley graphs with optimal fault tolerance. *IEEE Trans. Comput* **41** (1992), 1337–1339.

[ArGoMoSe07] G. Araujo, D. González Moreno, J.J. Montellano, and O. Serra, On upper bounds and connectivity of cages. *Australas. J. Combin.* **38** (2007), 221–228.

[AyFr70] J.N. Ayoub and I.T. Frisch, On the smallest-branch cuts in directed graphs. *IEEE Trans. Circuit Theory* **CT-17** (1970), 249–250.

[Ba99] M.C. Balbuena, Extraconnectivity of s-geodetic digraphs and graphs. *Discrete Math.* **195** (1999), 39–52.

[BaBeLiPi87] K.S. Bagga, L.W. Beineke, M.J. Lipman, and R.E. Pippert, On the edge-integrity of graphs. *Congressus Numerantium* **60** (1987), 141–144.

[BaCaFaFi96] M.C. Balbuena, A. Carmona, J. Fàbrega, and M.A. Fiol, On the connectivity and the conditional diameter of graphs and digraphs. *Networks* **28** (1996), 97–105.

[BaCaFaFi97a] M.C. Balbuena, A. Carmona, J. Fàbrega, and M.A. Fiol, On the order and size of *s*-geodetic digraphs with given connectivity. *Discrete Math.* **174** (1997), 19–27.

[BaCaFaFi97b] M.C. Balbuena, A. Carmona, J. Fàbrega, and M.A. Fiol, Extraconnectivity of graphs with large minimum degree and girth. *Discrete Math.* **167/168** (1997), 85–100.

[BaCaFi96] M.C. Balbuena, A. Carmona, and M.A. Fiol, Distance connectivity in graphs and digraphs. *J. Graph Theory* **22** (1996), 281–292.

[BaCeDiGVMa05] C. Balbuena, M. Cera, A. Diánez, P. García-Vázquez, and X. Marcote, Sufficient conditions for λ'-optimality of graphs with small conditional diameter. *Inf. Process. Letters* **95** (2005), 429–434.

[BaCeDiGVMa06] C. Balbuena, M. Cera, A. Diánez, P. García-Vázquez, and X. Marcote, On the restricted connectivity and superconnectivity in graphs with given girth. *Discrete Math.* **307** (2007), 659–667.

[BaCeDiGVMa07] C. Balbuena, M. Cera, A. Diánez, P. García-Vázquez, and X. Marcote, Connectivity of graphs with given girth pair. *Discrete Math.* **307** (2007), 155–162.

[BaGVMa06] C. Balbuena, P. García-Vázquez, and X. Marcote, Sufficient conditions for λ'-optimality in graphs with girth *g*. *J. Graph Theory* **52** (2006), 73–86.

[BaGV10] C. Balbuena and P. García-Vázquez, Edge fault tolerance analysis of super *k*-restricted connected networks. *Appl. Math. Comput.* **216** (2010), 506–513.

[BaGVMo11] C. Balbuena, P. García-Vázquez, and L. P. Montejano, Superconnectivity of graphs with odd girth *g* and even girth *h*. *Discrete Appl. Math.* **159** (2011), 91–99.

[BaGMMa09] C. Balbuena, D. González-Moreno, and X. Marcote, On the 3-restricted edge connectivity of Permutation Graphs. *Discrete Appl. Math.* **157** (2009), 1586–1591.

[BaLiMi08] C. Balbuena, Y. Lin, and M. Miller, Diameter-sufficient conditions for a graph to be super restricted connected. *Discrete Applied Math* **156** (2008), 2827–2834.

[BaMaMo10] C. Balbuena, K. Marshall, and L. P. Montejano, On the connectivity and superconnected graphs with small diameter. *Discrete Appl. Math.* **58** (2010), 397–403.

[BaSa12] C. Balbuena and J. Salas, A new bound for the connectivity of cages, *Appl. Math. Letters* **25** (2012) 1676–1680.

[BaTaMaLi09] C. Balbuena, J. Tang, K. Marshall, and Y. Lin, Superconnectivity of regular graphs with small diameter. *Discrete Appl. Math.* **157** (2009), 1349–1353.

[BeCoHs95] J.-C. Bermond, F. Comellas, and D.F. Hsu, Distributed loop computer networks: a survey. *J. Parallel Distrib. Comput.* **24** (1995) 2–10.

[BeOePi02] L.W. Beineke, O. Oellermann, and R.E. Pippert. The average connectivity of a graph. *Discrete Math.* **252** (2002), 31–45.

[Bi71] N. Biggs, Intersection matrices for linear graphs, in: D.J.A. Welsh (Ed.), *Combinatorial Mathematics and Its Applications*, Academic Press, London, 1971, 15–23.

[Bi94] N. Biggs, *Algebraic Graph Theory*, Second edition, Cambridge University Press, Cambridge, 1994.

[BoFe70] F. Boesch and A.P. Felger, A class of non vulnerable graphs. *Networks* **2** (1970), 261–283.

[BoHaKa81] F. Boesch, F. Harary, and J.A. Kabell, Graphs as models of communication network vulnerability: Connectivity and persistence. *Networks* **11** (1981), 57–63.

[Bo79] B. Bollobás, On graphs with equal edge connectivity and minimum degree. *Discrete Math.* **28** (1979), 321–323.

[BoUeVo02] P. Bonsma, N. Ueffing, and L. Volkmann, Edge-cuts leaving components of order at least three. *Discrete Math.* **256** (2002), 431–439.

[BoKoZn68] J. Bosák, A. Kotzig, and Š. Znám, Strongly geodetic graphs. *J. Combin. Theory* **5** (1968), 170–176.

[BoTi84] F. Boesch and R. Tindell, Circulants and their connectivities. *J. Graph Theory* **8** (1984), 487–499.

[Br95] A.E. Brouwer, Toughness and spectrum of a graph. *Linear Algebra Appl.* **226-228** (1995), 267–271.

[Br96] A.E. Brouwer, Spectrum and connectivity of graphs. *CWI Q.* **9** (1996), 37–40.

[BrHa05] A.E. Brouwer and W.H. Haemers, Eigenvalues and perfect matchings. *Linear Algebr. Appl.* **395** (2005), 155–162.

[BrKo09] A.E. Brouwer and J.H. Koolen, The vertex-connectivity of a distance-regular graph. *European J. Combin.* **30** (2009), no. 3, 668–673.

[BrMe85] A.E. Brouwer and D.M. Mesner, The connectivity of strongly regular graphs. *European J. Combin.* **6** (1985), 215–216.

[CaFa99] A. Carmona and J. Fàbrega, On the superconnectivity and the conditional diameter of graphs and digraphs. *Networks* **34** (1999), 197–205.

[Ch66] G. Chartrand, A graph-theoretic approach to a communications problem. *SIAM J. Appl. Math.* **14** (1966), 778–781.

[Ch76] G. Chaty, On critically and minimally k vertex (arc) strongly connected digraphs, *Proc. Keszthely* (1976), 193–203.

[ChLi02] E. Cheng and M.J. Lipman, Vulnerability issues of star graphs, alternating group graphs and split-stars: Strength and toughness. *Discrete Appl. Math.* **118** (2002), 163–179.

[Co73] R.J. Cook, Heawood's theorem and connectivity. *Mathematica Lond.* **20** (1973), 201–207.

[CvDoSa95] D. Cvetković, M. Dragoš, and H. Sach, *Spectra of Graphs, Theory and Applications*. Third edition, J.A. Barth Verlag, Leipzig, 1995.

[DaRo99] M. Daven and C.A. Rodger, (k, g)-cages are 3-connected. *Discrete Math.* **199** (1999), 207–215.

[DaVo95] P. Dankelmann and L. Volkmann, New sufficient conditions for equality of minimum degree and edge-connectivity. *Ars Comb.* **40** (1995), 270–278.

[DaVo97] P. Dankelmann and L. Volkmann, Degree sequence conditions for maximally edge-connected graphs and digraphs. *J. Graph Theory* **26** (1997), 27–34.

[DaVo00] P. Dankelmann and L. Volkmann, Degree sequence conditions for maximally edge-connected graphs depending on the clique number. *Discrete Math.* **211** (2000), 217–223.

[Es85] A.H. Esfahanian, Lower-bounds on the connectivities of a graph. *J. Graph Theory* **9** (1985), 503–511.

[Es88] A.H. Esfahanian and S.L. Hakimi, On computing a conditional edge-connectivity of a graph, *Inf. Process. Lett.* **27** (1988), 195–199.

[FaFi89] J. Fàbrega and M.A. Fiol, Maximally connected digraphs. *J. Graph Theory* **13** (1989), 657–668.

[FaFi94] J. Fàbrega and M.A. Fiol, Extraconnectivity of graphs with large girth. *Discrete Math.* **127** (1994), 163–170.

[FaFi96a] J. Fàbrega and M.A. Fiol, Bipartite graphs and digraphs with maximum connectivity. *Discrete Appl. Math.* **69** (1996), 269–277.

[FaFi96b] J. Fàbrega and M.A. Fiol, On the extraconnectivity of graphs. *Discrete Math.* **155** (1996), 49–57.

[Fi73] M. Fiedler, Algebraic connectivity of graphs. *Czech. Math. J.* **23** (1973), 298–305.

[Fi92] M.A. Fiol, On super-edge-connected digraphs and bipartite digraphs. *J. Graph Theory* **16** (1992), 545–555.

[Fi93] M.A. Fiol, The connectivity of large digraphs and graphs. *J. Graph Theory* **17** (1993), 31–45.

[Fi94] M.A. Fiol, The superconnectivity of large digraphs and graphs. *Discrete Math.* **124** (1994), 67–78.

[FiFa94] M.A. Fiol and J. Fàbrega, On the distance connectivity of graphs and digraphs. *Discrete Math.* **125** (1994), 169–176.

[FiFaEs90] M.A. Fiol, J. Fàbrega, and M. Escudero, Short paths and connectivity in graphs and digraphs. *Ars Combin.* **29B** (1990), 17–31.

[FiGaYe97] M.A. Fiol, E. Garriga, and J.L.A. Yebra, The alternating polynomials and their relation with the spectra and conditional diameters of graphs. *Discrete Appl. Math.* **167/168** (1997), 297–307.

[FiYeAl84] M.A. Fiol, J.L.A. Yebra, and I. Alegre, Line digraph iterations and the (d, k) digraph problem. *IEEE Trans. Comput.* **C-33** (1984), 400–403.

[FrOeSw00] G. Fricke, O.R. Oellermann, and H.C. Swart, The edge-connectivity, average edge-connectivity and degree conditions, manuscript (2000).

[FuHuRo97] H.L. Fu, K.C. Huang, and C.A. Rodgers, Connectivity of cages. *J. Graph Theory* **24** (1997), 187–191.

[Gh92] A. Ghafoor, Connectivity, persistence and fault diagnosis of interconnection networks based on O_k and $2O_k$ graphs. *Discrete Appl. Math.* **37/38** (1992), 205–226.

[Go81] C. D. Godsil, Equiarboreal graphs. *Combinatorica* **1** (1981), 163–167.

[GoEn79] D.L. Goldsmith and R.C. Entringer, A sufficient condition for equality of edge-connectivity and minimum degree of a graph. *J. Graph Theory* **3** (1979), 251–255.

[GoWh78] D.L. Goldsmith and A.T. White, On graphs with equal edge-connectivity and minimum degree. *Discrete Math.* **23** (1978), 31–36.

[GuWa86] M.A. Gurgel and Y. Wakabayashi, On k-leaf-connected graphs. *J. Combin. Theory Ser. B* **41** (1986), 1–16.

[Ha77] Y.O. Hamidoune, Sur les atomes d'un graphe orienté. *C. R. Acad. Sci. Paris Sér. A* **284** (1977), 1253–1256.

[Ha80] Y.O. Hamidoune, A property of α-fragments of a digraph. *Discrete Math.* **31** (1980), 105–106.

[Ha81] Y.O. Hamidoune, Quelques problèmes de connexité dans les graphes orientés. *J. Combin. Theory Ser. B* **30** (1981), 1–10.

[Ha83] F. Harary, Conditional connectivity. *Networks* **13** (1983), 347–357.

[Ha84] Y.O. Hamidoune, On the connectivity of Cayley digraphs. *European J. Combin.* **5** (1984), 309–312.

[Ha89] Y.O. Hamidoune, Sur les atomes d'un graphe de Cayley infini. *Discrete Math.* **73** (1989), 297–300.

[Ha96] Y.O. Hamidoune, Additive group theory applied to network topology, pages 1–39 in D.-Z. Du et al. (Eds.), *Combinatorial network theory*, Appl. Optim. 1, Kluwer Academic Publishers, Dordrecht, 1996.

[HaLlSe92] Y.O. Hamidoune, A.S. Lladó, and O. Serra, The connectivity of hierarquical Cayley digraphs. *Discrete Appl. Math.* **37/38** (1992), 275–280.

[HaSe96] Y.O. Hamidoune and O. Serra, On small cuts separating Abelian Cayley graphs into two equal parts. *Math. Syst. Theory* **29** (1996), 407–409.

[HaHi99] S.G. Hartke and A.W. Higgins, Maximum degree growth of the iterated line graph. *Electron. J. Combin.* **6** (1999), Research paper 28.

[HeVo03a] A. Hellwig and L. Volkmann, Maximally edge-connected digraphs. *Austral. J. Combin.* **27** (2003), 23–32.

[HeVo03b] A. Hellwig and L. Volkmann, Maximally local-edge-connected graphs and digraphs. *Ars Combin.* **72** (2004), 295–306.

[HeVo08a] A. Hellwig and L. Volkmann, The connectivity of a graph and its complement. *Discrete Appl. Math.* **156** (2008), 3325–3328.

[HeVo08b] A. Hellwig and L. Volkmann, Maximally edge-connected and vertex-connected graphs and digraphs: a survey. *Discrete Math.* **308** (2008), 3265–3296.

[ImSoOk85] M. Imase, T. Soneoka, and K. Okada, Connectivity of regular directed graphs with small diameter. *IEEE Trans. Comput.* **C-34** (1985), 267–273.

[Im79] W. Imrich, On the connectivity of Cayley graphs. *J. Combin. Theory Ser. B* **26** (1979), 323–326.

[JiMu98] T. Jiang and D. Muyabi, Connectivity and separating sets of cages. *J. Graph Theory* **29** (1998), 35–44.

[Jo72] J.L. Jolivet, Sur la connexité des graphes orientés. *C.R. Acad. Sci. Paris* **274A** (1972), 148–150.

[JuWa77] H.A. Jung and M.E. Watkins, On the connectivities of finite and infinite graphs. *Mh. Math.* **83** (1977), 121–131.

[Ka74] T. Kameda, Note on Halin's theorem on minimally connected graphs. *J. Combin. Theory Ser. B* **17** (1974), 1–4.

[Ke72] A.K. Kelmans, Asymptotic formulas for the probability of k-connectedness of random graphs. *Theory Probab. Appl.* **17** (1972), 243–254.

[KnNi03] M. Knor and L. Niepel, Connectivity of iterated line graphs. *Discrete Appl. Math.* **125** (2003), 255-266.

[Le74] L. Lesniak, Results on the edge-connectivity of graphs. *Discrete Math.* **8** (1974), 351–354.

[LiMiBa05] Y. Lin, M. Miller, and C. Balbuena, Improved lower bound for the vertex connectivity of $(\delta; g)$-cages. *Discrete Math.* **299** (2005), 162–171.

[LiMiBaMa06] Y. Lin, M. Miller, C. Balbuena, and X. Marcote, All $(k; g)$-cages are edge-superconnected. *Networks* **47(2)** (2006), 102–110.

[LiBaMaMi08] Y. Lin, C. Balbuena, X. Marcote, and M. Miller, On the connectivity of (k, g)-cages of even girth. *Discrete Math.* 308(15) (2008), 3249–3256.

[LuWuLuLi10] H. Lu, Y. Wu, Q. Lu, and Y. Lin, New improvements on connectivity of cages. *Acta Math. Sinica English Series,* **26(5)** (2010), 1–12.

[Ma70] W. Mader, Über den Zusammenhang symmetrischer Graphen. *Arch. Math. (Basel)* **21** (1970), 331–336.

[Ma71a] W. Mader, Eine Eigenschaft der Atome endlicher Graphen. *Arch. Math. (Basel)* **22** (1971), 333–336.

[Ma71b] W. Mader, Minimale n-fach kantenzusammenhängende Graphen. *Math. Ann.* **191** (1971), 21-28.

[MaBa04] X. Marcote and C. Balbuena, Edge-superconnectivity of cages. *Networks* **43(1)** (2004), 54–59.

[MaBaPe04] X. Marcote, C. Balbuena, and I. Pelayo, Diameter, short paths and super-connectivity in digraphs. *Discrete Math.* **288** (2004), 113–123.

[MaBaPe07] X. Marcote, C. Balbuena, and I. Pelayo, On the connectivity of cages with girth five, six and eight. *Discrete Math.* **307** (2007), 1441–1446.

[MaBaPeFa05] X. Marcote, C. Balbuena, I. Pelayo, and J. Fàbrega, (δ, g)-cages with $g \geq 10$ are 4-connected. *Discrete Math.* **301** (2005), 124–136.

[MaPeBa02] X. Marcote, I. Pelayo, and C. Balbuena, Every cubic cage is quasi 4-connected. *Discrete Math.* **266** (2003), 311–320.

[PeBaGo01] I. Pelayo, C. Balbuena, and J. Gómez, On the connectivity of generalized p-cycles. *Ars Combin.* **58** (2001), 215–231.

[PeLaHe86] K. Peters, R. Laskar, and S. Hedetniemi, Maximal/minimal connectivity in graphs. *Ars Comb.* **21** (1986), 59–70.

[Pl75] J. Plesník, Critical graphs of given diameter. *Acta Fac. Rerum Natur. Univ. Comenian. Math.* **30** (1975), 71–93.

[PlZn74] J. Plesník and Š. Znám, Strongly geodetic directed graphs. *Acta Fac. Rerum Natur. Univ. Comenian., Math. Publ.* **29** (1974), 29–34.

[PlZn89] J. Plesník and Š. Znám, On equality of edge-connectivity and minimum degree of a graph. *Arch. Math. (Brno)* **25** (1989), 19–25.

[PluZh98] M. Plummer and X. Zha, On the connectivity of graphs embedded in surfaces. *J. Combin. Theory Ser. B* **72** (1998), 202–228.

[PluZh02] M. Plummer and X. Zha, On the connectivity of graphs embedded in surfaces II. *Electron. J. Combin.* **9** (2002), no. 1, Research Paper 38, 27 pp.

[ReKuHoLe82] S.M. Reddy, J.G. Kuhl, S.H. Hosseini, and H. Lee, On digraphs with minimum diameter and maximum connectivity, *Proc. 20th Annual Allerton Conference* (1982), 1018–1026.

[So92] T. Soneoka, Super edge-connectivity of dense digraphs and graphs. *Discrete Appl. Math.* **37/38** (1992), 511–523.

[SoNaIm85] T. Soneoka, H. Nakada, and M. Imase, Sufficient conditions for dense graphs to be maximally connected, *Proc. ISCAS85* (1985), 811–814.

[SoNaImPe87] T. Soneoka, H. Nakada, M. Imase, and C. Peyrat, Sufficient conditions for maximally connected dense graphs. *Discrete Math.* **63** (1987), 53–66.

[ToVo93] J. Topp and L. Volkmann, Sufficient conditions for equality of connectivity and minimum degree of a graph. *J. Graph Theory* **17** (1993), 695–700.

[Vo88] L. Volkmann, Bemerkungen zum p-fachen Kantenzusammenhang von Graphen, [Remarks on the p-fold edge connectivity of graphs.] *An. Univ. Bucure ti Mat.* **37** (1988), 75–79.

[Vo89] L. Volkmann, Edge-connectivity in p-partite graphs. *J. Graph Theory* **13** (1989), 1–6.

[Vo03] L. Volkmann, Degree sequence conditions for equal edge-connectivity and minimum degree, depending on the clique number. *J. Graph Theory* **42** (2003), 234–245.

[WaXuWa03] P. Wang, B. Xu, and J. Wang, A note on the edge-connectivity of cages. *Electron. J. Combin.* **10** (2003), Note 2, 4 pp.

[Wa67] M.E. Watkins, A lower bound for the number of vertices of a graph. *Amer. Math. Monthly* **74** (1967), 297.

[Wa70] M.E. Watkins, Connectivity of transitive graphs. *J. Combin. Theory* **8** (1970), 23–29.

[Wa02] M.E. Watkins, Personal communication, 2002.

[Wo73] D.R. Woodall, The binding number of a graph and its Anderson number. *J. Combin. Theory Ser. B* **15** (1973), 225–255.

[Xu92] J.-M. Xu, An inequality relating the order, maximum degree, diameter and connectivity of a strongly connected digraph. *Acta Math. Appl. Sinica* **8** (1992), 144-152.

[Xu94] J.-M. Xu, A sufficient condition for equality of arc-connectivity and minimum degree of a digraph. *Discrete Math.* **133** (1994), 315–318.

[XuWaWa02] B. Xu, P. Wang, and J. Wang, On the connectivity of $(4, g)$-cages. *Ars Combin.* **64** (2002), 181–192.

[XuXu02] Jun-Ming Xu and Ke-Li Xu, On restricted edge-connectivity of graphs. *Discrete Math.* **243** (2002), 291–298.

[ZhLi10] Z. Zhang and Q.H. Liu, Sufficient conditions for a graph to be $\lambda^{(k)}$-optimal with given girth and diameter. *Networks* **55** (2010), 119–124.

[ZhYu05] Z. Zhang and J.J. Yuan, A proof of an inequality concerning k-restricted edge connectivity. *Discrete Math.* **304** (2005), 128–134.

[ZhYu07] Z. Zhang and J.J. Yuan, Degree conditions for restricted-edge-connectivity and isoperimetric-edge-connectivity to be optimal. *Discrete Math.* **307** (2007), 293–298.

Glossary for Chapter 4

A-B **path**: see *path*.

$A|B$ **set**: see *separating set*.

adjacency matrix – of a graph (or digraph) G: the matrix $A = (a_{uv})$ indexed by the vertices of G with entries $a_{uv} = 1$ if uv is an edge (or arc) of G and $a_{uv} = 0$ otherwise.

algebraic connectivity – of a graph G: the smallest non-zero eigenvalue of the Laplacian matrix of G.

antisymmetric digraph (or oriented graph): a digraph such that between any pair of vertices there is at most one directed edge.

approximate (or approximation) algorithm: an algorithm that typically makes use of heuristics in reducing its computation but produces solutions that are not necessarily optimal.

arborescence: synonym for *rooted tree*.

associated symmetric digraph: see *digraph*.

asymmetric TSP (ATSP): see *TSP*.

atom – of a graph G: each minimum component obtained by removing a minimum disconnecting set from G.

balance condition – for a mixed graph G: for every $S \subseteq V(G)$, the difference between the number of arcs from S to $V(G) - S$ and the number of arcs from $V(G) - S$ to S is no greater than the number of undirected edges joining vertices in S and $V(G) - S$.

deBruijn graph of order k: a directed graph with 2^k vertices, each labeled with a unique k-bit string; vertex a is joined to vertex b by an arc if bitstring b is obtainable from bitstring a by either a cycle shift or a deBruijn shift; each arc is labeled by the first bit of the vertex at which it originates, followed by the label of the vertex at which it terminates.

balanced digraph: see *digraph*.

balanced orientation – of a graph (or mixed graph): an assignment of a direction to each edge of the graph (or each undirected edge of the mixed graph) so that the resulting digraph is balanced.

balanced vertex – in a digraph: a vertex whose indegree and outdegree are equal.

bipartite degree closure: bipartite graph of order $2n$ obtained by recursively joining pairs of non-adjacent vertices $x \in X$ and $y \in Y$ whose degree sum is at least $n + 1$, until no such pair remains.

bipartite graph: a graph G with two independent vertex subsets that partition $V(G)$.

bipartite index – of a graph: the smallest number of vertices whose removal leaves a bipartite graph.

block – in a connected graph: a maximal 2-connected subgraph.

boundary – of a vertex subset U: the set of vertices which are at distance one from U.

branching: synonym for *rooted tree*.

bridge – of a connected graph G: an edge whose deletion disconnects G; synonym for *cut-edge*.

(k, g)-**cage**: a regular graph of degree k and girth g with the minimum number of vertices.

Cayley (di)graph – of a group Γ with generating set S: a (di)graph whose vertices are identified with the elements of Γ, and there is an edge uv when $u^{-1}v \in \Gamma$.

Chinese Postman Problem: finding a postman tour of minimum length in a graph where all edges are undirected; see *postman*.

claw-free closure: the graph obtained by repeatedly applying the local completion until it is no longer possible to do.

clique number – of a graph G: the maximum number of vertices in a complete subgraph of G.

component$_1$ – of a graph: a maximal connected subgraph.

component$_2$ – of a digraph: a maximal strongly connected subdigraph.

connected graph: a graph in which there exists a walk between any pair of vertices.

___, **critically** k-: a graph G such that $\kappa(G) \geq k$ but, for each vertex $v \in V$, $\kappa(G - v) < k$.

___, **hamiltonian**: containing a spanning path between any two vertices.

___, k-: a graph with connectivity $\kappa \geq k \geq 1$.

___, k-**edge**-: a graph with edge-connectivity $\lambda \geq k \geq 1$.

contraction: an operation involving the identification (amalgamation) of vertices.

___, **edge**: given an edge uv, identification of its endpoints u and v (keeping the old adjacencies but removing the *self-loop* from $u = v$ to itself).

___, **subgraph**: identification of all the vertices of a given subgraph H by a succession of elementary contractions of the edges of H.

k-**contractible edge**: an edge of a k-connected graph whose contraction results in a k-connected graph.

k-**contractible subgraph**: a subgraph of a k-connected graph whose contraction results in a k-connected graph.

covering walk (or *postman tour*): – in an arbitrary graph G: a closed walk containing every edge of G.

critically k-**connected graph**: see *connected graph*.

cut-edge: synonym for *bridge*.

cut-vertex – of a connected graph: a vertex whose deletion disconnects it.

CVRP tour – in a weighted directed or undirected complete graph with vertex set $\{0, 1, \ldots, n\}$, with a demand $d_i \geq 0$ for $i = 1, 2, \ldots, n$, and two parameters Q and k: a collection of k cycles $C_1, C_2, \ldots, C_k$, which contain all the vertices, pairwise intersect only in vertex 0, and satisfy $\sum_{i \in V(C_j)} d_i \leq Q$ for each $j = 1, 2, \ldots, k$.

cycle cover – of a graph G: a family S of cycles of G such that every edge of G belongs to at least one element of S.

cycle decomposition – of a graph (or digraph) G: a partition of the edge-set (or arc-set) of G such that each partition set forms a cycle (or directed cycle).

cycle double cover (CDC): a cycle cover S such that every edge of G belongs to exactly two elements of S.

cycle extendable: any cycle C of length $m < |V(G)|$ can be extended to a cycle of length $m + 1$ containing all of $V(C)$.

___, **fully**: a graph that is cycle extendable and having any vertex on a triangle.

cycle packing – in a graph G: a set of edge disjoint cycles in G.

cycle shift – $a_1 a_2 \ldots a_k \longrightarrow b_1 b_2 \ldots b_k$: a left shift such that $b_k = a_1$.

deBruijn shift – $a_1 a_2 \ldots a_k \longrightarrow b_1 b_2 \ldots b_k$: a left shift such that $b_k \neq a_1$. **bipancyclic**: a bipartite graph of order $2n$ containing cycles of all even lengths from 4 to $2n$.

k-degree closure: graph obtained by recursively joining pairs of non-adjacent vertices whose degree sum is at least k, until no such pair remains.

degree sequence – of a graph G: the degrees of the vertices of G ordered in non-increasing (or non-decreasing) order.

detachment operation: see §4.2, Definition D**??**.

detachment – of a graph G: a graph that results from a sequence of detachment operations performed at each of the vertices of some vertex subset $W \subseteq V(G)$; used to transform or produce eulerian tours.

diameter – of a (di)graph G: the maximum distance between vertices of G.

digraph: a graph all of whose edges are directed; a *directed graph.*

    ___, **associated symmetric** – of a graph G: the digraph $\overset{\leftrightarrow}{G}$ obtained from G by replacing each edge uv by the two directed edges (u,v) and (v,u) forming a *digon.*

    ___, **balanced**: a digraph whose vertices are all balanced; in §4.3, this is called a *symmetric digraph.*

    ___, **connected**: a digraph whose underlying graph is connected; also called *weakly connected.*

    ___, **critically k-connected**: a digraph G such that $\kappa(G) \geq k$ but, for each vertex $v \in V$, $\kappa(G - v) < k$.

    ___, **strongly connected**: a digraph with a $u-v$ walk for any pair of vertices u, v; also called a *strong digraph.*

    ___, **k-(strongly) connected**: a digraph with connectivity at least k; also called *k-strong* or *k-strongly connected.*

    ___, **symmetric$_1$**: a digraph such that between each pair of distinct vertices, either both (oppositely directed) arcs exist or neither does.

    ___, **symmetric$_2$**: (used in §4.3) a digraph such that indegree equals outdegree for each vertex.

disconnecting edge-set$_1$ – of a graph G: a subset of edges whose removal from G results in a non-connected graph.

disconnecting edge-set$_2$ – of a digraph G: a subset of arcs whose removal from G results in a non-strongly-connected digraph.

disconnecting (vertex-)set – of a graph G: a subset of vertices whose removal from G results in a non-connected graph.

    ___, $(u|v)$- – in a graph G: a disconnecting (vertex-)set whose removal from G leaves u and v in different components.

disconnecting (vertex-)set – of a digraph G: a subset of vertices whose removal from G results in a non-strongly-connected digraph.

t-distance connectivity – of a graph (or digraph) G: the minimum cardinality of a $(u|v)$-set with $dist(u,v) \geq t$.

distance matrix of an instance of TSP: the matrix $D = [d_{ij}]$, where d_{ij} is the weight of the edge between vertices i and j; analogously defined for digraphs.

distance$_1$ – between vertices u, v in a graph (or digraph) G: the length of a shortest path joining u to v.

distance$_2$ – between vertex sets U, W in a graph G: the minimum of the distances between vertices of U and W.

distance-regular graph: a graph whose number of l-walks between any pair of vertices u, v only depends on $dist(u,v)$ and $l \geq 0$.

dominating circuit: A circuit C such that every edge of G is incident to a vertex of C.

double tracing: a closed walk that traverses every edge exactly twice.

___, **bidirectional**: a double tracing that uses every edge once in each of its two directions.

___, **strong**: a double-tracing that is both bidirectional and retract-free.

edge addition: given two non-adjacent vertices u, v of a graph $G = (V, E)$, $G + e = (V, E \cup \{uv\})$ is the graph obtained from G by addition of the edge $e = uv$.

edge contraction: see *contraction*.

edge-boundary – of a vertex subset U: the set of edges having exactly one endpoint in U.

edge-connectivity$_1$ – of a graph: the minimum number of edges whose removal leaves a connected graph; denoted λ (or κ_e).

edge-connectivity$_2$ – of a non-trivial digraph: the minimum number of arcs whose removal leaves a non-strongly connected graph; denoted λ or κ_e.

edge-cut: a *disconnecting edge-set*.

edge-disjoint paths: paths that have no edge in common.

edge-symmetric graph: a graph whose automorphism group acts transitively on its edge set.

Euclidean TSP: see *TSP*.

eulerian graph (or digraph or mixed graph): a graph that has an eulerian tour.

eulerian tour – in a graph (or digraph): a closed walk that uses each edge (or arc) *exactly* once. An eulerian tour in a mixed graph is a closed walk that uses each edge and each arc exactly once.

even graph: an undirected graph whose vertices all have even degree.

exact algorithm: an algorithm that solves a certain optimization problem to optimality.

$\{F_1, F_2, \ldots, F_k\}$**-free**: containing no induced subgraph isomorphic to any F_i, $1 \leq i \leq k$.

1-factor – of a graph G: a 1-regular subgraph H such that $V(H) = V(G)$ (i.e., spans G); thus, a 1-factor is a spanning *matching*.

2-factor of a graph G: the (vertex-disjoint) union of cycles of G that covers $V(G)$.

fragment – of a graph G: a component obtained from G by removing a minimum disconnecting set.

generalized p-cycle: a graph (or digraph) whose vertex set can be partitioned into p subsets, say $V_0, V_1, \ldots, V_{p-1}$, in such a way that every edge uv is of the form $u \in V_i$, $v \in V_{i+1}$ (arithmetic modulo p).

generalized TSP: see *TSP*.

s**-geodetic digraph**: see §4.7, Definition D55.

girth – of a graph G: the length of a shortest cycle.

half degree – of a vertex u in a digraph G: the indegree or outdegree of u in G.

hamiltonian connected: containing a spanning path between any two vertices.

hamiltonian cycle: a spanning cycle.

hamiltonian decomposition: a partitioning of the edge set of G into hamiltonian cycles if G is $2d$-regular or hamiltonian cycles and a perfect matching if G is $(2d+1)$-regular.

hamiltonian graph: a graph containing a hamiltonian cycle.

___, k**-ordered**: a hamiltonian graph such that for every ordered sequence of k vertices there is a hamiltonian cycle that encounters the vertices of the sequence in the given order.

___, **uniquely**: containing exactly one hamiltonian cycle.

imbedding – of a graph G on a surface S: a drawing of G on S with no crossing edges.

in-arcs – of a partition-cut $E(X, \overline{X})$: the subset of arcs whose head is in X; denoted $E^-(X, \overline{X})$.

incidence set – of a vertex v: the partition-cut $E(X, \overline{X})$, where $X = \{v\}$; denoted E_v (the out-arcs and in-arcs of E_v are denoted E_v^+ and E_v^-, respectively).

incidence-partition system: see §4.2, Definition D??.

indegree – of a vertex u in a digraph G: the number of vertices in G adjacent to u.

___, *t*-: the indegree of a vertex v with in-eccentricity at least t.

independence number: cardinality of a largest set of independent vertices.

independent set – of vertices: a subset of pairwise non-adjacent vertices.

independent vertices: vertices with no edges between them.

induced subgraph (or subdigraph) on a set $U \subset V$: the maximal subgraph (or subdigraph) of G with vertex set U; denoted $G(U)$ or $G[U]$.

in-eccentricity – of a vertex u in a digraph G: the maximum of the distances $dist(v, u)$ for all vertices v in G.

interior vertex – of a fragment F: a vertex with no adjacent vertices outside of F.

internally-disjoint $u-v$ **paths**: $u-v$ paths that have pairwise exactly vertices u and v in common.

in-tree: a rooted tree with all arcs reversed.

k-**connected graph (or digraph)**: a graph (or digraph) with connectivity at least k.

k-**edge-connected graph (or digraph)**: a graph (or digraph) with edge-connectivity at least k.

k-**paths problem**: given $2k$ distinct vertices $u_1, u_2, \cdots, u_k$ and $v_1, v_2, \cdots, v_k$ in a graph G, determining whether there exist k mutually edge-disjoint paths $P_1, P_2, \cdots, P_k$ in G such that P_i connects u_i and v_i for $i = 1, 2, \ldots, k$.

k-**system that dominates**: a collection of k edge-disjoint circuits and stars (with at least 3 endvertices), such that each edge of G is either contained in one of the circuits or stars, or is adjacent to one of the circuits.

kappa transformations: various combinations of splitting, splicing, and reversing closed trails; they form the basis for constructing eulerian tours and for transforming one tour into another; see §4.2.6.

Laplacian matrix – of a graph G: the matrix $L = (l_{uv})$ with entries $l_{uv} = deg(u) - 1$ if u, v are adjacent vertices of G, and $l_{uv} = 0$ otherwise.

line digraph – of a digraph G: the digraph whose vertices are the directed edges of G and vertex (u, v) is adjacent to vertex (v, w).

___, *k*-**iterated** – of a digraph G: the digraph obtained by applying recursively k times the line digraph operation on G.

line graph – of a graph G: the graph whose vertices can be put into $1-1$ correspondence with the edges of G in such a way that two vertices of $L(G)$ are adjacent if and only if the corresponding edges of G are incident.

k-**linked graph**: a graph that has at least $2k$ vertices, and for every sequence of $2k$ different vertices, $u_1, u_2, \ldots, u_k, v_1, v_2, \ldots, v_k$, there exists a u_i-v_i path P_i, $i = 1, 2, \ldots, k$, such that the k paths are vertex-disjoint.

k-**parity-linked graph**: a graph in which one can find k disjoint paths with prescribed endvertices and prescribed parities of the lengths.

local completion – at a vertex x of a graph G such that the induced subgraph $G[N(x)]$ is connected: the graph obtained by replacing $G[N(x)]$ by a complete subgraph on $V(N(x))$.

matching: a subset of edges no two of which have a common vertex.

___, **perfect** – in a graph G: a matching such that every vertex in G is incident to some edge in the matching.

maximally connected graph (or digraph): a graph (or digraph) whose connectivity equals its minimum degree.

maximally edge-connected graph (or digraph): a graph (or digraph) whose edge-connectivity equals its minimum degree.

maximum degree – of a (di)graph G: the maximum of the (positive and negative) degrees of the vertices of G.

minimally k-connected graph (or digraph): a graph (or digraph) G such that $\kappa(G) \geq k$ but, for each edge $e \in E$, $\kappa(G - e) < k$.

minimally k-edge-connected graph (or digraph): a graph (or digraph) G such that $\lambda(G) \geq k$, but for each $e \in E$, $\lambda(G - e) < k$.

minimum (t-)degree – of a (di)graph G: the minimum among all t-(out and in)degrees of the vertices of G.

minimum degree – of a (di)graph G: the minimum of the (out and in) degrees of the vertices of G.

minimum degree – of a (di)graph G: the minimum of the (positive and negative) degrees of the vertices of G.

mixed graph: a graph that has both undirected and directed edges.

Moore bound – of a graph G: an upper bound for the maximum number of vertices, given its maximum degree and its diameter.

negative α-fragment – of a digraph G: the subset of edges whose negative edge-boundary is a minimum disconnecting edge set of G.

negative atom – of a digraph G: a negative fragment with minimum cardinality.

negative boundary – of a vertex subset F in a digraph G: the set of vertices in G which are at distance one to F.

negative edge-boundary – of a vertex subset F in a digraph G: the set of edges in G which have only its final vertex in F.

negative fragment – of a digraph G: the subset of vertices whose negative boundary is a minimum disconnecting set of G.

neighborhood$_1$ – of a vertex x: the set of all vertices adjacent to x.

neighborhood$_2$ – of the set S: vertices adjacent to some vertex in S.

out-arcs – of a partition-cut $E(X, \overline{X})$: the subset of arcs whose tail is in X; denoted $E^+(X, \overline{X})$.

outdegree – of a vertex u in a digraph: the number of vertices adjacent from u.

___, t-: the outdegree of a vertex v with out-eccentricity at least t.

out-eccentricity – of a vertex u in a digraph G: the maximum of the distances $dist(u, v)$ for all vertices v in G.

out-tree: synonym for *rooted tree*.

pancyclic: containing cycles of all lengths from 3 to $|V(G)|$.

p-partite graph: a graph whose vertex-set can be partitioned into p independent vertex subsets.

partition-cut – in a graph $G = (V, E)$ associated with $X \subset V(G)$: the set of edges in G with one endpoint in X and one endpoint in $\overline{X} = V(G) - X$; denoted $E(X, \overline{X})$.

path: a simple walk, that is, a walk in which all defining terms are distinct.

___, A−B: given $A, B \subset V$, a u−v path P such that u is the only vertex of P belonging to A and v is the only vertex of P that belongs to B.

___, u-v – in a graph G: a path in G joining u to v; analogously defined for digraphs.

positive α-fragment – of a digraph G: the subset of edges whose positive edge-boundary is a minimum disconnecting edge set of G.

positive atom – of a digraph G: a positive fragment with minimum cardinality.

positive boundary – of a vertex subset F in a digraph G: the set of vertices in G that are at distance one from F.

positive edge-boundary – of a vertex subset F in a digraph G: the set of edges in G which have only its initial vertex in F.

positive fragment – of a digraph G: the set of vertices whose positive boundary is a minimum disconnecting set of G.

postman problem: the class of problems of finding a minimum-length (or minimum-weight) postman tour in a graph under various conditions.

___, **directed**: a postman tour of minimum length in a digraph.

___, **mixed**: a postman tour of minimum length in a mixed graph (with both directed and undirected edges).

___, **rural**: relaxation of the basic postman version (undirected, directed, mixed) where only a subset of edges has to be included at least once.

___, **stacker crane**: a rural postman version of a mixed postman problem but where each directed edge is traversed at least once.

___, **undirected**: a postman tour of minimum length in a graph where all edges are undirected.

___, **windy**: an undirected postman problem where the cost of edge traversal depends on the direction the (undirected) edge is traversed.

postman tour (or *covering walk*): – in an arbitrary graph G: a closed walk containing every edge of G.

k-th power – of a connected graph G: the graph with $V(G^k) = V(G)$ for which $uv \in E(G)$ if and only if $1 \leq d_G(u,v) \leq k$.

quasi 4-connected graph: a 3-connected graph G such that, for each vertex-cut $F \subset V$ with $|F| = 3$, F is the neighborhood of a vertex of degree 3 and $G - F$ has exactly two components.

retract or *retracing* – in a walk W: a section of the walk of the form $v_{i-1}, e_i, v_i, e_{i+1}, v_{i+1}$ such that $e_i = e_{i+1}$ (and $v_{i+1} = v_{i-1}$).

retract-free walk: a walk that has no retracts.

rooted tree: a directed tree having a distinguished vertex r, called the *root*, such that for every other vertex v, there is a directed r-v path. Occasionally encountered synonyms for rooted tree are *out-tree*, *branching*, and *arborescence*.

semigirth ℓ – of a digraph G: the parameter ℓ_π for $\pi = 0$.

semigirth, π- – of a digraph G: the greatest integer such that, for any pair of vertices u, v: (a) if $dist(u,v) < \ell$, the shortest u-v directed walk is unique and there are at most π u-v directed walks of length $dist(u,v) + 1$; (b) if $dist(u,v) = \ell$, there is only one shortest u-v directed walk; denoted ℓ_π.

separating set – of sets $A, B \subset V$: a set $X \subset V$ such that every A–B path in G contains a vertex of X.

(left) shift operation – converting a k-bit string $a = a_1 a_2 a_3 \cdots a_k$ to a k-bit string $b_1 b_2 \ldots b_k$: an operation such that $b_i = a_{i+1}$, for $i = 1, 2, \ldots, k - 1$.

spanning subgraph (or subdigraph) – of a graph G: a subgraph (or subdigraph) that contains all the vertices of G.

splitting operation: see §4.2, Definition D11.

stacker crane problem: a rural postman version of a mixed postman problem but where each directed edge is traversed at least once.

strongly connected digraph: see *digraph*.

strongly regular graph: a connected distance-regular graph with diameter two.

subgraph contraction: see *contraction*.

super-κ graph: a maximally connected graph whose minimum disconnecting sets are the vertices adjacent to some vertex of minimum degree; analogously defined for digraphs.

super-λ graph: a maximally edge-connected graph whose minimum disconnecting edge sets are the edges incident on some vertex of minimum degree; analogously defined for digraphs.

symmetric digraph$_1$: see *digraph*.

symmetric digraph$_2$: see *digraph*.

symmetric digraph$_3$: a digraph which is both vertex- and edge-transitive; used in §4.7.

symmetric graph: a graph which is both vertex- and edge-transitive.

symmetric TSP (STSP) – for a complete (undirected) weighted graph: finding a minimum-weight hamiltonian cycle in K_n.

total degree – of a vertex v in a mixed graph G: the total number of arcs and undirected edges incident on v.

totally separating set: Given $A \subset V$, a subset $X \subset V - A$ totally separates A if each component of $G - X$ has at most one vertex of A.

toughness – of a graph G: maximum t such that $t\, c(S) \leq |S|$, where $c(S)$ is the number of components of $G - S$.

traceable: containing a spanning path.

transition system: see §4.2, Definition D??.

traveling salesman problem: finding a minimum-weight hamiltonian cycle in a weighted complete graph. A non-complete graph is made complete by adding the missing edges and assigning to them a prohibitively large weight; also referred to as the *symmetric TSP*.

triangle inequality – on a weighted digraph (or graph): the condition $d_{ij} + d_{jk} \geq d_{ik}$ for all distinct vertices i, j, k, where d_{ij} is the weight of the arc from vertex i to vertex j (of the edge between i and j).

TSP: *traveling salesman problem*.

___, asymmetric (ATSP): finding a minimum-weight hamiltonian cycle in a weighted complete digraph $\overleftrightarrow{K}_n$ (every pair of vertices has both oppositely directed arcs).

___, Euclidean: the special case of TSP in which the vertices are points in the Euclidean plane and the weight on each edge is the Euclidean distance between its endpoints.

___, generalized – for a weighted complete digraph $\overleftrightarrow{K}_n$ and a partition $V_1, \ldots, V_k$ of its vertices: finding a minimum-weight cycle containing exactly one (at least one) vertex from each set V_i, $i = 1, \ldots, k$.

$(u|v)$-(disconnecting) set: see *disconnecting set*.

$u-v$ path: a path from vertex u to vertex v.

vehicle routing problem, capacitated (CVRP) – in a weighted directed or undirected complete graph on $n + 1$ vertices with a demand $d_i \geq 0$ for $i = 1, 2, \ldots, n$ and with two parameters Q and k: finding a CVRP tour for which the total weight of the cycles is minimum.

(vertex-)connectivity$_1$ – of a graph: the minimum number of vertices whose removal leaves a non-connected or trivial graph; denoted κ or κ_v.

(vertex-)connectivity$_2$ – of a strongly connected digraph: the minimum number of vertices whose removal leaves a non-strongly connected or trivial digraph; denoted κ or κ_v.

vertex splitting: operation in which a vertex w is replaced by an edge uv in such a way that some of the vertices adjacent to w are now adjacent to u and the rest are adjacent to v.

vertex-symmetric graph: a graph whose automorphism group acts transitively on its vertex set; that is, for any pair of vertices u, v, there exists an automorphism of G mapping u to v.

walk: an alternating sequence of vertices and edges such that for each edge, one endpoint precedes and the other succeeds that edge in the sequence.

weakly k-linked graph: a graph that has at least $2k$ vertices, and for every k pairs of vertices (u_i, v_i), there exists a u_i-v_i path P_i, $1 \leq i \leq k$, such that the k paths are edge-disjoint.

weakly connected digraph: a digraph whose underlying graph is connected; also called *connected digraph*.

wheel: a graph consisting of a cycle C and an additional vertex that is adjacent to every vertex of C.

Chapter 5

Colorings and Related Topics

Section 5.1

Graph Coloring

Zsolt Tuza, University of Veszprém, Hungary

INTRODUCTION

§5.1 concentrates on the classical concept of *chromatic number* and on the more recent but closely related concept of *choice number*, mostly in connection with other important graph invariants. Further developments of graph colorings appear in §5.2.

Various problems, some of which are equivalent to colorings (e.g., 1-factorizations) are dealt with in the other sections of this chapter. The book [JeTo95] is a rich source of additional information, where results are organized around more than 200 open problems. On list coloring and related topics, a comprehensive survey can be found in [Tu97] and its update [KrTuVo99]. The recent book [StScToFa12] is devoted to the theory of edge colorings.

TERMINOLOGY NOTE: We consider a *graph* to be without self-loops, to have at least one vertex, and, except within the last subsection, to be finite.

5.1.1   General Concepts

Graph coloring deals with the general and widely applicable concept of partitioning the underlying set of a structure into parts, each of which satisfies a given requirement (e.g., to be an independent set). One of the most famous problems in this area, and even within graph theory, what is now known as the Four Color Theorem (see §5.2.2), has been a driving force of research on graphs for nearly a century.

Proper Vertex-Coloring and Chromatic Number

DEFINITIONS

D1:   A *vertex-coloring* of a graph $G = (V, E)$ is a function

$$\varphi : V \to \mathcal{C}$$

from the set V of vertices to a set $\mathcal{C}$ of *colors*.

D2:   The coloring φ is *proper* if no two adjacent vertices are assigned the same color.

D3:   A *k-coloring* is a vertex-coloring with at most k colors.

NOTATION: If k is a positive integer, we assume (unless specified otherwise explicitly) that $\mathcal{C} = \{1, 2, \dots, k\}$.

D4:   A proper k-coloring φ may also be viewed as a *vertex partition*

$$V_1 \cup \dots \cup V_k = V$$

where the disjoint subsets $V_i = \varphi^{-1}(i)$ are called the *color classes*. (Thus, a k-coloring φ is proper if and only if $\varphi^{-1}(i)$ is an *independent set* for each i.)

D5:   A graph is *k-colorable* if it admits a proper vertex-coloring with at most k colors.

D6:   A graph is *k-chromatic* if it is k-colorable but not $(k-1)$-colorable.

D7:   The *chromatic number* of a graph G, denoted $\chi(G)$, is the smallest nonnegative integer k such that G is k-colorable.

EXAMPLES

E1:   The complete graph K_n on n vertices has chromatic number n.

E2:   A graph G (other than the null graph) has chromatic number

$$\chi(G) = 1 \quad \text{if and only if } G \text{ is edgeless,}$$
$$\chi(G) \le 2 \quad \text{if and only if } G \text{ is bipartite.}$$

In particular, cycles of even length are 2-chromatic, while cycles of odd length are 3-chromatic.

DEFINITION

D8:   The (vertex) *independence number* of a graph G, denoted $\alpha(G)$, is the size of a maximum independent set in G.

FACTS

F1:   An immediate consequence of the definitions is that for every graph G,

$$\chi(G) \ge \frac{|V|}{\alpha(G)}$$

F2:   [Bo88] The lower bound $\chi(G) \geq \frac{|V|}{\alpha(G)}$ is almost tight for almost all graphs, since the *random graph* $\mathbf{G}_{n,p}$ satisfies

$$\chi(\mathbf{G}_{n,p}) = \left(\frac{1}{2} + o(1)\right)\left(\log\frac{1}{1-p}\right)\frac{n}{\log n}$$

for every fixed p ($0 < p < 1$) with probability $1 - o(1)$ as $n \to \infty$.

F3:   Let $f : \mathbb{R}^{\geq 0} \to \mathbb{R}^+$ be a nondecreasing, continuous function with $f(0) = 1$. If $G = (V, E)$ is a graph such that, for every $1 \leq i \leq |V|$, every subgraph of G on i vertices has independence number at least $f(i)$, then

$$\chi(G) \leq \int_0^{|V|} \frac{1}{f(x)}\, dx$$

List Coloring and Choice Number

Many results on the chromatic number can be discussed in the following more general setting.

DEFINITIONS

D9:   A (vertex) *list assignment* L on a graph G associates a set L_v of colors with each vertex v of G. Each L_v is interpreted as the set of allowed colors for vertex v.

D10:   The graph G is *L-colorable* (or *list colorable*, when L is understood from context) if it admits a proper vertex-coloring φ such that $\varphi(v) \in L_v$ for all v.

TERMINOLOGY NOTE: The term "list" is used in the literature for historical reasons only. No particular ordering on the color set L_v is assumed here.

D11:   If $|L_v| = k$ for all $v \in V$, then the list assignment L is called a *k-assignment*.

D12:   A graph G is *k-choosable* if it is L-colorable for every k-assignment L.

D13:   The *choice number* of G, denoted $ch(G)$, is the smallest nonnegative integer k such that G is k-choosable. (In part of the literature, the choice number is called *list chromatic number*, and also the notation $\chi_\ell(G)$ is commonly used for $ch(G)$.)

EXAMPLES

E3:   The complete graph K_n on n vertices has choice number n.

E4:   The choice number may be strictly larger than the chromatic number. In particular, the complete bipartite graphs $K_{3,3}$ and $K_{2,4}$ are 2-colorable but not 2-choosable (the former with the lists $(1,2),(1,3),(2,3)$ in both bipartition subsets, and the latter with lists $(1,2),(3,4)$ and $(1,3),(1,4),(2,3),(2,4)$ in its 2- and 4-element class, respectively). It is true, however, that cycles of even length are 2-choosable, while cycles of odd length have choice number 3 (although the former is not immediate to verify).

FACTS

F4: [ErRuTa79] The complete bipartite graph $K_{n,n}$ has choice number $(1+o(1))\log_2 n$ as $n \to \infty$.

F5: (J. Kahn in [Al93]; [TuVo94]) Asymptotically $ch(\mathbf{G}_{n,p}) = (1 + o(1))\chi(\mathbf{G}_{n,p})$ for every constant p $(0 < p < 1)$ with probability $1 - o(1)$ as $n \to \infty$.

F6: [Kr00] If the edge probability $p = p(n)$ is such that $p(n) \to 0$ as $n \to \infty$ and $p(n) \geq n^{-1/4+\varepsilon}$ for some $\varepsilon > 0$, then $ch(\mathbf{G}_{n,p}) = (1 + o(1))\chi(\mathbf{G}_{n,p})$ with probability $1 - o(1)$.

F7: [AlKrSu99, Vu99] The bound $ch(\mathbf{G}_{n,p}) = \mathcal{O}(\chi(\mathbf{G}_{n,p}))$ holds also for $p(n) > 2/n$, and upper bounds of the form $\mathcal{O}(n\,p(n)/\log(n\,p(n)))$ follow deterministically, too, from conditions (analogous to the expected values of parameters in $\mathbf{G}_{n,p}$) on degrees, pair-degrees, and complementary degrees.

The Hajós Construction

FACTS

F8: [Ha61] Every graph of chromatic number at least k can be constructed from the complete graph K_k by a sequence of operations of the following three types:

(1) Insert new vertices and/or edges.

(2) Having constructed vertex-disjoint graphs G_1 and G_2, select edges $u_i v_i$ in G_i $(i = 1, 2)$, remove $u_1 v_1$ and $u_2 v_2$, identify u_1 with u_2, and insert the new edge $v_1 v_2$.

(3) Identify nonadjacent vertices.

F9: [Gr96] Every graph of choice number at least k can be constructed from any one complete bipartite graph of choice number k by a sequence of the operations (1) and (2) above, and the following third type:

(3') Having constructed a graph $G = (V, E)$ that has an uncolorable list assignment L where $|L_v| \geq k$ for all $v \in V$ and two nonadjacent vertices u, v have the same list in L, then identify u with v.

Lovász's Topological Lower Bound

DEFINITION

D14: The **neighborhood complex** of a graph $G = (V, E)$ is the simplicial complex $\mathcal{N}(G)$ whose vertices are the vertices of G, and such that the set $X \subseteq V$ is a simplex if all the $x \in X$ have a common neighbor in G.

FACT

F10: [Lo78] If the neighborhood complex $\mathcal{N}(G)$ of a graph G is a k-connected topological space, then $\chi(G) \geq k + 3$.

Alon and Tarsi's Graph Polynomial Characterization

DEFINITION

D15: The **graph polynomial**, also called the **edge difference polynomial**, of a graph $G = (V, E)$ with $E \neq \emptyset$ and vertex set $V = \{v_1, \ldots, v_n\}$ is

$$P_G = P_G(x_1, \ldots, x_n) := \prod_{\substack{i < j \\ v_i v_j \in E}} (x_i - x_j)$$

Given a list assignment L on G, we also define

$$Q_i = Q_i(x_i) := \prod_{q \in L_i} (x_i - q)$$

for $i = 1, \ldots, n$.

FACT

F11: [AlTa92] A graph G admits a list L-coloring if and only if its graph polynomial P_G does not belong to the ideal generated by the polynomials Q_i.

List Reduction

FACT

F12: [TuVo97] Let $G = (V, E)$ be a graph, let L be a k-assignment on G ($k \geq 2$), let $X \subset V$ be any subset, and let $F \subseteq E$ be the set of those edges which have at least one vertex in X. If (V, F) is a 2-choosable graph, then one can properly color the set X with colors $\varphi(x) \in L_x$, for all $x \in X$, in such a way that at most one color occurs in the neighborhood of each vertex $v \in V \setminus X$. In particular, $ch(G) \leq ch(G - X) + 1$.

5.1.2   Vertex Degrees

DEFINITIONS

D16: The (Erdős–Hajnal) **coloring number** of G, denoted $col(G)$ (and also called the **Szekeres–Wilf number**) is the smallest positive integer k for which there exists an order $v_1 v_2 \cdots v_n$ of the vertices, such that every v_i has fewer than k neighbors v_j with $j > i$.

D17: A graph is d-**degenerate** if none of its subgraphs has minimum degree larger than d. ($col(G)$ is the smallest integer k such that G is $(k-1)$-degenerate.)

D18: The **clique number** of a graph G, denoted $\omega(G)$, is the size of a maximum clique in G.

FACTS

F13: For every graph G,

$$\omega(G) \leq \chi(G) \leq ch(G) \leq col(G) \leq \Delta(G) + 1$$

F14: (Nordhaus–Gaddum Theorem) [NoGa56] For any graph $G = (V, E)$ and its edge-complement $\overline{G}$,

$$\chi(G) + \chi(\overline{G}) \leq |V| + 1$$

F15: [ErRuTa79] The Nordhaus–Gaddum upper bound also holds for list colorings:

$$ch(G) + ch(\overline{G}) \leq |V| + 1 \quad \text{and} \quad col(G) + col(\overline{G}) \leq |V| + 1$$

F16: [HaSz70] Every graph $G = (V, E)$ with maximum degree Δ admits a proper $(\Delta + 1)$-coloring such that each color class has cardinality $\left\lfloor \frac{|V|}{\Delta+1} \right\rfloor$ or $\left\lceil \frac{|V|}{\Delta+1} \right\rceil$ (called *equitable* coloring). See [KiKoYu09] for a short proof and further references.

F17: [KiKo08] If $d(x) + d(y) \leq 2D + 1$ for every edge xy in G, then G has an equitable $(D + 1)$-coloring.

F18: (Brooks's Theorem) [Br41] If the graph G is connected, then

$$\chi(G) \leq \Delta(G)$$

unless G is complete or $\Delta(G) = 2$ and G is an odd cycle.

F19: [Vi76, ErRuTa79] (analogue to Brooks's theorem for list colorings) If the graph G is connected, then

$$ch(G) \leq \Delta(G)$$

unless G is a complete graph or $\Delta(G) = 2$ and G is an odd cycle. (See also §5.2.1 for generalizations.)

REMARK

R1: Concerning the sequence of inequalities $\omega \leq \chi \leq ch \leq col \leq \Delta + 1$, Brooks's theorem and its analogue characterize equality in $\chi = \Delta + 1$ and $ch = \Delta + 1$. The other end, $\omega = \chi$, is studied in the theory of perfect graphs (see §5.5). So far, the problems [Tu97a] of finding tight conditions for ensuring $\omega = ch$ or $\chi = ch$ are open, except for $ch = 2$.

FACTS

F20: [ErRuTa79] A connected graph is 2-choosable if and only if the sequential removal of degree-1 vertices yields the trivial graph K_1, or an even cycle, or an even cycle plus a degree-2 vertex whose two neighbors are at distance two along the cycle.

F21: [Al00] There exists a sequence of real numbers ϵ_d with $\epsilon_d \to 0$ as $d \to \infty$, such that the inequality $ch(G) \geq (\frac{1}{2} - \epsilon_d) \log_2 d$ holds whenever the graph G on n vertices contains a subgraph of minimum degree d. Equivalently, $ch(G) \geq (\frac{1}{2} - o(1)) \log_2 col(G)$ as $col(G)$ gets large.

5.1.3   Critical Graphs and Uniquely Colorable Graphs

The two extremes are considered here: k-chromatic graphs that are almost $(k-1)$-colorable, and k-chromatic graphs with just one proper k-coloring. List critical graphs are mentioned only briefly, to indicate that they behave quite differently.

DEFINITIONS

D19: For $k \geq 2$, a k-chromatic graph $G = (V, E)$ is k-**critical** if

$$\chi(G - e) = k - 1$$

for every edge $e \in E$; and G is k-**vertex-critical** if

$$\chi(G - v) = k - 1$$

for every vertex $v \in V$.

D20: A graph $G = (V, E)$ is k-**list critical** if there is a $(k-1)$-assignment L for G such that every proper subgraph of G is L-colorable, but G itself is not L-colorable.

FACTS

F22: Every k-vertex-critical graph is connected, without any vertex cutsets inducing a complete subgraph, and contains a k-critical spanning subgraph.

F23: [To78] A graph G is vertex-critical if and only if the complement of each block in the complementary graph $\overline{G}$ is vertex-critical.

F24: For $k = 2$ and $k = 3$, a graph is k-critical if and only if it is k-vertex-critical. The unique 2-critical graph is K_2, and the 3-critical graphs are the odd cycles.

F25: (same as Brooks's Theorem) Every k-critical graph G has minimum degree at least $k - 1$, and if G is $(k-1)$-regular, then either $G = K_k$ or $k = 3$ and G is an odd cycle.

F26: [Mi92] Every k-critical graph $G \neq K_k$ $(k \geq 4)$ contains all trees of k edges as subgraphs.

F27: [Ga63] If G is k-critical, then its vertices of degree $k - 1$ induce a subgraph in which every block is a complete graph or an odd cycle. Conversely, if H is a K_k-free graph of maximum degree at most $k - 1 \geq 3$ where each block is a complete graph or an odd cycle, then there exists a k-critical graph G in which the vertices of degree $k - 1$ induce a subgraph isomorphic to H. (This graph H may also be the null graph.)

F28: [St85] For $k \geq 4$, let F be a connected graph of maximum degree at most $k - 1$, such that each of its blocks is a complete graph or an odd cycle. There exist only finitely many k-critical graphs in which the vertices of degree $k - 1$ induce F and in which the other vertices induce either K_{k-1} or a $(k-2)$-colorable graph.

F29: [GrLo74] A graph H is a proper subgraph of some k-critical graph if and only if H and each of its edge contractions H/e (for $e \in E(H)$) is $(k-1)$-colorable.

F30: [Si72, To72] There exists an infinite sequence of 4-critical graphs on n vertices with minimum degree at least $\frac{1}{2}n^{1/3}$. (It is not known whether the minimum degree can be as large as cn in 4-critical or 5-critical graphs.)

F31: [KoYa12] The number of edges of a k-critical graph, $k \geq 4$, on $n > k$ vertices is at least $\left\lceil \frac{(k+1)(k-2)}{2(k-1)}n - \frac{k(k-3)}{2(k-1)} \right\rceil$. This bound is tight for all $n \geq 6$ if $k = 4$, for all $n \equiv 1$ (mod $(k-1)$) if $k \geq 5$, and also for all $n \equiv 2$ (mod 4) with $n \geq 10$ if $k = 5$. (Classical lower bounds are $\frac{k-1}{2}n + \frac{k-3}{2}$ in [Di57] and $\frac{k-1}{2}n + \frac{k-3}{2(k^2-3)}n$ in [Ga63].)

F32: [Lo73] For the largest possible independence number $\alpha_{n,k}$ in k-critical graphs on n vertices, the difference $n - \alpha_{n,k}$ is at least $\frac{1}{6}kn^{1/(k-2)}$ and, for infinitely many values of n, at most $2kn^{1/(k-2)}$.

F33: [MoRe01] Let k_Δ be the largest integer k' such that $(k'+1)(k'+2) \leq \Delta$. There is an absolute constant Δ_0 such that for any $\Delta \geq \Delta_0$ and any $k < k_\Delta$, if G has maximum degree Δ and if $\chi(G) > \Delta - k$ then G contains a $(\Delta - k + 1)$-chromatic subgraph of maximum degree at most $\Delta + 1$. On the other hand, for every $\Delta \geq 2$ and $k \geq k_\Delta$, there exist arbitrarily large $(\Delta - k + 1)$-critical graphs with maximum degree Δ. Viewing it from the algorithmic side, on graphs with Δ large (fixed), $(\Delta - k)$-colorability can be tested in linear time if $k \leq k_\Delta$, but it is **NP**-hard if $k_\Delta < k < \Delta - 2$. (It is conjectured that the condition $\Delta \geq \Delta_0$ can be omitted.)

F34: [RöTu85] Let $t(k, \ell)$ denote the minimum number of edges in graphs with k vertices and independence number less than ℓ (i.e., the complement of the Turán number, cf. §8.1). If G is $(k + 1)$-critical and has at least $2t(k, \ell) - 1$ vertices, then in order to obtain an ℓ-colorable subgraph, one must delete at least $t(k, \ell)$ edges. This bound is tight. Moreover, for unrestricted $(k+1)$-chromatic graphs, the minimum is $t(k+1, \ell+1)$.

F35: [RöTu85] Let $k \geq 2$, and G be a graph with n vertices and m edges. Suppose that the automorphism group of G acts transitively, either

(a) on the vertices and $\chi(G) \geq 2k + 1$, or

(b) on the edges and $\chi(G) \geq k^2 + 1$.

Then, to obtain a k-colorable subgraph of G, one has to delete at least $\sqrt{n}$ vertices in Case (a) and $\sqrt{m}$ edges in Case (b). (It is not known whether vertex/edge transitive 4-chromatic graphs can always be made bipartite by removing a bounded number of vertices/edges.)

F36: (V. Rödl in [To85]) The number of nonisomorphic 4-critical graphs is at least c^{n^2}, for some constant $c > 1$.

F37: Every k-critical graph G is k-list critical, by taking the list assignment $L_v = \{1, \ldots, k - 1\}$ for all $v \in V(G)$.

F38: [StTuVo09] For every $k \geq 5$, every complete graph on at least k vertices is k-list critical, whereas there are only finitely many 4-list critical complete graphs. In particular, a list critical graph may contain other list critical graphs as (induced) subgraphs, with the same sizes of 'critical lists' (whereas k-critical graphs contain no k-critical proper subgraphs); moreover, a graph may be critical with respect to many different values of list sizes.

OPEN PROBLEMS

P1: Determine tight asymptotic bounds on the maximum number of edges in k-critical graphs on n vertices ($k \geq 4$ fixed, $n \to \infty$).

P2: (J. Nešetřil and V. Rödl) For $k \geq 3$, does there exist a function $f_k(n)$ such that every $(k+1)$-critical graph with at least $f_k(n)$ vertices contains a k-critical subgraph with at least n vertices?

CONJECTURES

C1:   Double-critical graph conjecture (Erdős and Lovász) If $\chi(G - u - v) = \chi(G) - 2$ holds for any two distinct vertices of G, then D is a complete graph.

C2: (Hedetniemi) The equality $\chi(G \times H) = \min(\chi(G), \chi(H))$ holds for any two graphs G and H. (See [Zh98] for a survey. The fractional version of the equality is valid; see §5.2.1.)

FACT

F39: [KeKe54] The function $f_3(n)$ exists, i.e., every 'large' 4-critical graph contains a 'long' odd cycle.

EXAMPLES

E5: The complete join of two disjoint cycles of length $2t + 1$ is a 6-critical graph on $n = 4t + 2$ vertices with $\frac{1}{4}n^2 + n$ edges ([Di52]). More generally, the number of edges in k-critical graphs on n vertices can be as large as $(\frac{1}{2} - c_k)n^2$, where $c_k \to 0$ as $k \to \infty$.

E6: On $n = 8t + 4$ vertices, let V_i ($1 \leq i \leq 4$) be disjoint vertex subsets of cardinality $2t + 1$ each, V_1 and V_2 induce the cycle C_{2t+1}, V_3 and V_4 be independent and completely joined, and let the $V_1 - V_3$ edges and the $V_2 - V_4$ edges form a perfect matching. This graph is 4-critical, with more than $n^2/16$ edges ([To70]).

Uniquely Colorable Graphs

DEFINITION

D21: A graph $G = (V, E)$ is **uniquely k-colorable** if it admits just one proper k-coloring with $k = \chi(G)$ colors (apart from the renumbering of colors).

The uniquely 1-colorable graphs are the edgeless graphs, and therefore we assume below that $k \geq 2$.

EXAMPLES

E7: The uniquely 2-colorable graphs are the connected bipartite graphs.

E8: If $G - v$ is a uniquely k-colorable graph, and v a vertex of degree $k - 1$ whose neighbors are in mutually distinct color classes of $G-v$, then also the graph G is uniquely k-colorable. In particular, uniquely 4-colorable planar graphs are obtained from a planar embedding of K_4 by sequentially inserting new degree-3 vertices inside triangular faces (where the neighbors of each new vertex are the vertices of the corresponding face).

FACTS

F40: [ChGe69] Every uniquely k-colorable graph is $(k - 1)$-connected.

F41: [Bo78] If the graph G is k-colorable on n vertices, with minimum degree larger than $\frac{3k-5}{3k-2}\, n$, then G is uniquely k-colorable. This lower bound is best possible for infinitely many n.

F42: [Bo78] Let the graph G be k-colorable on n vertices, with a k-coloring where the union of any two color classes is connected. If the minimum degree is larger than $\frac{k-2}{k-1}\, n$, then G is uniquely k-colorable. This lower bound is best possible for infinitely many n.

5.1.4   Girth and Clique Number

The results below show that the exclusion of short cycles does not make the chromatic number bounded; but on graphs without large complete subgraphs, the general upper bounds in terms of vertex degrees can be improved.

FACT

F43: [Er59] For a suitably chosen constant c, and for every $k > 1$ and $g > 2$ there exists a non-k-colorable graph of girth g on at most $c\, k^{2g}$ vertices. Moreover, for $k \geq 4$, the maximum girth of k-chromatic graphs on n vertices grows with $\Theta(\frac{\log n}{\log k})$ as $n \to \infty$.

REMARK

R2: The proof in [Er59] is probabilistic. Constructions usually are of much larger size than guaranteed by Erdős's theorem. For g small, some examples are listed below; for unrestricted g, see, e.g., [Lo68, NeRö79]. A fairly small general construction, involving Ramanujan graphs on $\mathcal{O}(k^{4g})$ vertices, can be found in [LuPhSa88].

EXAMPLES

E9: [Zy49] If G_k is a k-chromatic triangle-free graph, let G_{k+1} consist of k vertex-disjoint copies G_k^i of G_k $(i = 1, \ldots, k)$, together with a new independent set X of size $|V(G_k)|^k$. To each k-tuple $(v_1, \ldots, v_k) \in V(G_1) \times \cdots \times V(G_k)$, join a distinct common neighbor in X. This G_{k+1} is $(k + 1)$-chromatic and triangle-free.

E10: [De48] If G_k is a k-chromatic graph of girth at least 6, take an independent set X of cardinality $s = k|V(G_k)|-k+1$, together with $m = \binom{s}{|V(G_k)|}$ vertex-disjoint copies G_k^i of G_k $(i = 1, \ldots, m)$. For each of the m distinct $|V(G_k)|$-tuples $Y_i \subset X$, draw a perfect matching between Y_i and $V(G_k^i)$. The resulting graph G_{k+1} is a $(k+1)$-chromatic graph of girth 6.

E11: [My55] For each vertex v of a k-chromatic triangle-free graph G_k, take a distinct new vertex v' adjacent to all neighbors of v, and join a new vertex w to all of these v'. The graph obtained is $(k+1)$-chromatic and triangle-free.

E12: (Kneser graphs) [Lo78, Bá78] The vertices of the Kneser graph $K(n,k)$ are the k-element subsets of $\{1,\ldots,n\}$ ($n/2 > k > 1$), and two vertices are adjacent if and only if the corresponding two k-sets are disjoint. Then $\chi(K(n,k)) = n - 2k + 2$, and $K(n,k)$ is triangle-free if $n < 3k$. Moreover, a vertex-critical subgraph of $K(n,k)$ is induced by the vertices corresponding to the k-tuples that have no pair of consecutive elements in the cyclic order $\{1,2,\ldots,n\}$ ([Sc78]).

CONJECTURE

C3: [Re97] For every graph G,

$$\chi(G) \leq \frac{1}{2}(\Delta(G) + \omega(G)) + 1$$

FACTS

F44: [Re97] For every m there is a number Δ_m such that, if $\Delta(G) \geq \Delta_m$ and $\omega(G) \leq \Delta(G) + 1 - 2m$, then $\chi(G) \leq \Delta(G) + 1 - m$. Moreover, there is a constant $\epsilon > 0$ such that if $\Delta(G) \geq 3$, then $\chi(G) \leq (1 - \epsilon)(\Delta(G) + 1) + \epsilon\,\omega(G)$.

F45: Conjecture C3 is true for line graphs of multigraphs [KiReVe07] and for several other graph classes [RaSc07], [KiRe08], [KoSc10].

F46: Combining Fact F3 with the known estimates [AjKoSz80] on the Ramsey numbers $R(s,t)$, it follows that if $\omega(G) \leq t$, then for some constant $c = c(t)$, we have

$$\chi(G) \leq c \left(\frac{n}{\log n} \right)^{1-1/t}$$

F47: [BoKo77, Ca78, La78] Let $t \geq 3$. If $\omega(G) \leq t \leq \Delta(G)$, then

$$\chi(G) \leq \frac{t}{t+1}\,(\Delta(G) + 2)$$

F48: [Er67] The maximum ratio in the set

$$\left\{ \frac{\chi(G)}{\omega(G)} \;\middle|\; |V_G| = n \right\}$$

grows with $\Theta(n/\log^2 n)$ as $n \to \infty$.

F49: [Jo96a] For every $r \in \mathbb{N}$ there exists a constant c_r such that if the graph G is K_r-free, then

$$ch(G) \leq \frac{c_r\,\Delta(G) \log\log\Delta(G)}{\log\Delta(G)}$$

F50: [Jo96] If the graph G is triangle-free, then $ch(G) \leq c\,\Delta(G)/\log\Delta(G)$, for some constant c independent of $\Delta(G)$.

F51: [Ki95] If the graph G has girth at least 5, then

$$ch(G) \leq \frac{(1 + \epsilon_{\Delta(G)})\,\Delta(G)}{\log \Delta(G)}$$

where $\epsilon_{\Delta} \to 0$ as $\Delta \to \infty$.

F52: [NoReWu12] If $|V(G)| \leq 2\chi(G) + 1$, then $ch(G) = \chi(G)$. (This is equivalent to $ch = \omega$ on complete multipartite graphs with $n \leq 2\omega + 1$; formerly called Ohba's conjecture.)

F53: [Tu97a] Every induced subgraph H of the square C_n^2 of a cycle ($n \geq 3$) satisfies $ch(H) = \omega(H)$.

F54: [KoMa77] For every integer $\Delta > 3$, there exists a triangle-free graph G with maximum degree Δ and $\chi(G) > \frac{\Delta}{2\log \Delta}$.

F55: [Br02] Let $G = (V, E)$ be a maximal triangle-free d-regular graph, with $d > |V|/3$. Then $\chi(G) \leq 4$; moreover, if G has a nontrivial automorphism, then $\chi(G) \leq 3$.

F56: [Ko78] For every Δ there is a $g = g(\Delta)$ such that $\chi(G) \leq \frac{1}{2}\Delta + 2$ whenever G has girth at least g and maximum degree at most Δ.

F57: [MiSc04] If the cycles in G have at most p distinct even lengths and at most q odd lengths, then $\chi(G) \leq \min\{2p + 3, 2q + 2\}$. The bound is tight for all p and q.

F58: [RaSc01] If every induced cycle of G has length 4 or 5, then $\chi(G) \leq 3$. (This generalizes the fact [Su81] that every triangle-free graph without induced P_6 and C_6 is 3-colorable.) Also, all pairs F_1, F_2 of graphs with the property that every graph without induced F_1 and F_2 is 3-colorable can be characterized ([Ra04]).

CONJECTURE

C4: (Erdős–Faber–Lovász Conjecture) The union of any n edge-disjoint copies of K_n has chromatic number n.

FACT

F59: [Ka92] If G is the edge-disjoint union of n complete graphs of n vertices, then $\chi(G) = n + o(n)$ and also $ch(G) = n + o(n)$ as $n \to \infty$.

The Conjectures of Hadwiger and Hajós

DEFINITIONS

D22: A *contraction of an edge* $e = uv$ in a graph G is obtained by removing the edge e and identifying the vertices u and v, i.e., replacing u and v by a single vertex w adjacent to those vertices that were adjacent to u or v. A graph G is **contractible** to a graph H if H can be obtained from G by a sequence of edge contractions.

D23: A graph H is a ***minor*** of a graph G if a subgraph of G is contractible to H. An H-***minor*** is a minor isomorphic to H.

D24: A ***subdivision of an edge*** $e = uv$ in a graph H is obtained by replacing e with a new vertex w and two new edges uw and wv. A graph G is a ***subdivision*** of a graph H if G can be obtained from H by a sequence of edge subdivisions.

D25: A graph G is said to be an ***odd*** K_p-***expansion*** if G consists of p vertex-disjoint trees, every two of which are joined by an edge, and the vertices of the trees can be two-colored so that the edges of the trees are bichromatic but that the edges between trees are monochromatic.

CONJECTURES

C5: (Hadwiger's Conjecture) [Ha43] Every k-chromatic graph contains a K_k-minor.

C6: (Hajós's Conjecture) For $k \leq 6$, every k-chromatic graph contains a subdivision of the complete graph K_k.

C7: (Gerards and Seymour — "Odd Hadwiger Conjecture") If G contains no odd K_{k+1}-expansion, then $\chi(G) \leq k$.

FACTS

F60: [Di52] Both Hadwiger's and Hajós's conjectures are true for $k = 4$.

F61: [Ca79] If $\chi(G) = 4$, then G contains an *odd-triangle* subdivision of K_4, where each of the four cycles corresponding to the triangles of K_4 has odd length. Moreover, there is a polynomial-time algorithm that either properly colors an input graph G with 3 colors or outputs an odd-triangle subdivision of K_4 in G ([Za98]). (It is not known whether G also contains a subdivision of K_4 where each of the six paths obtained from the edges of K_4 have odd length.)

F62: [Ca79] For every $k \geq 7$ there exist graphs that are k-chromatic but do not contain any subdivision of K_k.

F63: [Wa37] The 4-colorability of all planar graphs implies the 4-colorability of all graphs not contractible to K_5. Hence, for $k = 5$, Hadwiger's conjecture (stated later than Wagner's theorem) is equivalent to the Four Color Theorem on planar graphs (cf. §5.2.2).

F64: [RoSeTh93] For $k = 6$, the validity of Hadwiger's conjecture can be deduced from the Four Color Theorem. For $k = 7$, as a partial result toward the conjecture, it is known that every 7-chromatic graph contains a K_7-minor or a $K_{4,4}$-minor ([KaTo05]). For general k, estimates in terms of connectivity and also of independence number can be found (e.g., see [KaSo07]).

F65: [BoCaEr80, Ko82] For the random graph on n vertices, Hadwiger's conjecture is valid with probability $1 - o(1)$ as $n \to \infty$.

F66: [Ko84] Let $h(G)$ denote the largest number of vertices in a complete graph to which G can be contracted. Then the inequality $h(G) + h(\overline{G}) \leq \lfloor 6n/5 \rfloor$ holds for every graph G on $n \geq 5$ vertices, and the upper bound is tight.

F67: There exists a constant c such that every graph G without a K_k-minor has $col(G) \leq ck \log k$, hence G is $O(k \log k)$-colorable and also $O(k \log k)$-choosable as $k \to \infty$ ([Ko84a], [Th84]). As a lower bound, for every k there exists a K_k-minor-free graph which is not $\lfloor 4(k-2)/3 \rfloor$-choosable ([BaJoWo11]).

F68: [GeGeReSeVe09] There is a constant c such that, for every $k \geq 1$, if a graph G contains no odd K_k-expansion, then $\chi(G) \leq ck\sqrt{\log k}$. (See §5.2.1 for related results.)

5.1.5 Edge-Coloring and χ-Binding Functions

TERMINOLOGY NOTE: In this subsection we shall explicitly use the term *multigraph* in those cases where multiple edges are allowed.

DEFINITIONS

D26: A *proper edge-coloring* of a graph or multigraph G is an assignment of colors to the edges of G, such that all edges incident with the same vertex get distinct colors.

D27: The *chromatic index* of a graph or multigraph G, denoted $\chi'(G)$, is the smallest number of colors in a proper edge-coloring of G.

D28: Definitions D26 and D27 for edge-colorings can be generalized to list edge-colorings in the natural way, as for vertex-colorings. The *edge choice number* (or *list chromatic index* or *list edge chromatic number*) is the minimum list-size that guarantees a list edge-coloring of G; it is denoted by $ch'(G)$ (or by $\chi'_\ell(G)$).

D29: A *proper total coloring* of $G = (V, E)$ is an assignment φ of colors to the vertices and the edges of G, such that φ induces a proper coloring on both V and E, and such that $\varphi(e) \neq \varphi(v)$ whenever $e \in E$ is incident with $v \in V$.

D30: The smallest number of colors in a proper total coloring of G is denoted by $\chi''(G)$; and the analogous quantity for the smallest size of lists is denoted by $ch''(G)$ (or by $\chi''_\ell(G)$).

D31: Let $G = (V, E)$ be a graph or multigraph.
- The *line graph* $L(G)$ of G has as its vertices the edges of G, two of them being adjacent in $L(G)$ if they share a vertex in G.
- The *total graph* $T(G)$ has $V \cup E$ as its vertex set, and its subgraphs induced by V and E are isomorphic to G and $L(G)$, respectively; moreover, $v \in V$ is adjacent to $e \in E$ in $T(G)$ if v is an endpoint of e in G.
- The *square* G^2 has the same vertex set as G; two vertices are adjacent in G^2 if they are at distance 1 or 2 apart in G.

D32: A *χ-binding function* (also called *χ-bounding* in a part of the literature) on a class $\mathcal{G}$ of graphs is a function $f : \mathbb{N} \to \mathbb{N}$ such that $\chi(G) \leq f(\omega(G))$ for all $G \in \mathcal{G}$.

REMARKS

R3: Clearly,
$$\chi'(G) = \chi(L(G)) \quad \text{and} \quad ch'(G) = ch(L(G))$$
and similarly,
$$\chi''(G) = \chi(T(G)) \quad \text{and} \quad ch''(G) = ch(T(G))$$

In general, however, much stronger results are valid for χ' and ch' than for χ and ch, due to the restricted structure of line graphs. (The one-to-one correspondence between edge colorings of G and vertex colorings of $L(G)$ is valid for almost all kinds of coloring; one exception is known to have opposite bijection, namely, *3-consecutive coloring*; see §5.2 for details.)

R4: The cliques in the line graph $L(G)$ (and in the total graph $T(G)$) correspond to the stars (with their centers) and to the — possibly multiple-edged — triangles in the (multi)graph G. Thus,
$$\Delta(G) \le \omega(L(G)) \le \frac{3}{2}\Delta(G)$$

Moreover, $ch'(G) \ge \chi'(G) \ge \Delta(G)$.

R5: The total graph $T(G)$ of $G = (V, E)$ is isomorphic to the square of the bipartite graph $B = B(G)$ whose bipartition classes are V and E, where $v \in V$ and $e \in E$ are adjacent in B if v is an endpoint of e in G.

FACTS

F69: (Vizing's Theorem) [Vi64] If G is a simple graph, then
$$\chi'(G) \le \Delta(G) + 1$$

If a multigraph G has maximum edge multiplicity $\mu(G)$, then
$$\chi'(G) \le \Delta(G) + \mu(G)$$

For unrestricted edge multiplicity the upper bound $\chi'(G) \le \frac{3}{2}\Delta(G)$ follows (first proved in [Sh49]). Thus, $f(\omega) = \omega + 1$ is a χ-binding function on the class of line graphs of simple graphs, and $f(\omega) = 3\omega/2$ is one on the class of all line graphs.

F70: [KiSc83] If the graph G contains no induced $K_{1,3}$ (*claw*) and no induced $K_5 - e$, then $\chi(G) \le \omega(G) + 1$.

F71: (König's Theorem) [Kö16] If G is a bipartite multigraph, then $\chi'(G) = \Delta(G)$.

F72: [MoRe98] There exists a constant C such that $\chi''(G) \le \Delta(G) + C$ holds for every graph G.

F73: [Ga95] If G is a bipartite multigraph, then $ch'(G) = \Delta(G)$.

F74: [ElGo96] Suppose that G is a d-regular multigraph with $\chi'(G) = d$. If G has an odd number of proper edge d-colorings, or if G is planar, then $ch'(G) = d$.

F75: For every multigraph G, $ch''(G) \le ch'(G) + 2$.

F76: [Ka96] For every graph G of maximum degree Δ, $ch'(G) \leq (1 + o(1))\Delta$, where $o(1) \to 0$ as $\Delta \to \infty$.

CONJECTURES

C8: Total Coloring Conjecture   (Vizing [Vi64]; Behzad) For every multigraph G with maximum edge multiplicity $\mu(G)$,

$$\chi''(G) \leq \Delta(G) + \mu(G) + 1$$

In particular, if G is a simple graph, then $\chi''(G) \leq \Delta(G) + 2$.

C9: Overfull Subgraph Conjecture   (A. J. W. Hilton) If $\chi'(G) = \Delta(G) + 1$ and $\Delta(G) > |V(G)|/3$, then there is an "overfull" subgraph H in G, such that

$$|E(H)| > \Delta(H) \left\lfloor \frac{1}{2}|V(H)| \right\rfloor$$

C10:   (Goldberg; Seymour) Every multigraph G with $\chi'(G) \geq \Delta(G) + 2$ has

$$\chi'(G) = \max_{H \subseteq G,\ |V(H)| \geq 2} \left\lceil \frac{|E(H)|}{\lfloor |V(H)|/2 \rfloor} \right\rceil$$

C11: List Coloring Conjecture   (Vizing; Gupta; Albertson and Collins; Bollobás and Harris) For every multigraph G, we have $ch'(G) = \chi'(G)$.

C12:   (Gravier and Maffray) More generally than Conjecture 11, for every claw-free graph G we have $ch(G) = \chi(G)$.

C13:   (Borodin, Kostochka and Woodall; Juvan, Mohar and Škrekovski; Hilton and Johnson) For every graph G we have $ch''(G) = \chi''(G)$.

C14:   (Kostochka and Woodall) More generally than Conjecture 13, $ch(G^2) = \chi(G^2)$ for the square of every graph G.

Snarks

DEFINITIONS

D33: A graph is cyclically k-connected if at least k edges must be deleted in order to leave two components, each containing a cycle.

D34: A ***snark*** (often called a "nontrivial snark") is a 3-regular, cyclically 4-edge-connected graph of girth at least 5, that is *not* edge-3-colorable.

D35: A snark G is ***irreducible*** if every nontrivial edge-cut (i.e., leaving at least two vertices in each component) separates G into 3-edge-colorable components.

REMARKS

R6:   According to the two possible cases provided by Vizing's theorem, a commonly used terminology says that a graph G is of **Class 1** or **Class 2** if its chromatic index is equal to $\Delta(G)$ or $\Delta(G)+1$, respectively. Hence, snarks represent the smallest nontrivial subfamily of Class 2. There has been much effort to construct snarks with various specific properties. Several methods of construction with further references can be found, e.g., in [Ko02] and in the earlier paper [CaMeRuSp98].

R7:   The idea of an irreducible snark is that it cannot be "reduced" to a smaller snark by separating from G a proper subgraph H which is itself not 3-edge-colorable and by restoring 3-regularity.

FACTS

F77:   [NeŠk96] A cubic graph G with chromatic index 4 is an irreducible snark if and only if G is *bicritical*, i.e., $G - \{u, v\}$ is 3-edge-colorable for every pair of distinct vertices $u, v \in V(G)$.

F78:   [ChŠk10] Every irreducible snark G can be factorized into a collection $\{H_1, \ldots, H_n\}$ of cyclically 5-edge-connected irreducible snarks such that G can be reconstructed from them by repeated dot products. Moreover, such a collection is unique up to isomorphism and ordering of the factors.

F79:   [Ko09] For each surface of genus at least 5 there exists a snark that admits a polyhedral embedding on the surface. (This disproves Grünbaum's conjecture for every orientable surface of genus at least 5; see §7.6.)

F80:   [Go85] If the Cycle Double Cover Conjecture (see §7.6.1) is false, then a smallest counterexample is a snark of girth at least 8.

F81:   [Ce84] If the 5-flow conjecture (see §5.2.2) is false, then a smallest counterexample is a cyclically 5-edge-connected snark of girth at least 7.

F82:   [Ko96] For every $g \geq 5$ there exists a cyclically 5-edge-connected snark of girth at least g.

Uniquely Edge-Colorable Graphs

DEFINITION

D36:   A graph $G = (V, E)$ is **uniquely edge k-colorable** if it admits just one proper edge-coloring with $k = \chi'(G)$ colors (apart from the renumbering of colors).

EXAMPLES

E13: The uniquely edge 1-colorable graphs are the matchings. The uniquely edge 2-colorable graphs are the paths and the even cycles. The star graph $K_{1,3}$ and the complete graph K_4 are uniquely 3-edge-colorable. (No complete characterization is available for $k = 3$.)

E14: [Tu76] The graph consisting of two 9-cycles $a_1 a_2 \ldots a_9$ and $b_1 b_2 \ldots b_9$ and the further edges $a_i b_{2i}$ ($1 \le i \le 9$, subscript addition modulo 9) is uniquely 3-edge-colorable. This is the only known triangle-free non-planar example for $k = 3$.

FACTS

F83: [Th78] The star graph $K_{1,k}$ is the only uniquely edge k-colorable graph, for each $k \ge 4$.

F84: If G is 3-regular and uniquely 3-colorable, then the following operation called "Y–Δ replacement" yields again a uniquely 3-colorable 3-regular graph. Let u_1, u_2, u_3 be the neighbors of vertex v. Remove v and insert the new vertices v_1, v_2, v_3 and the new edges $v_1 v_2, v_1 v_3, v_2 v_3$ and $u_i v_i$ for $i = 1, 2, 3$.

Further χ-Bound Graph Classes

CONJECTURES

C15: (Gyárfás [Gy75]; Sumner [Su81]) For every tree T, there exists a χ-binding function on the class of graphs not containing T as an induced subgraph.

C16: (Gyárfás [Gy87]) There exists a χ-binding function for the class of graphs in which every induced cycle has length 3 or 4.

C17: Divisibility conjecture (McDiarmid and Hoàng) If a graph contains no induced odd cycle of length at least 5 (i.e., an **odd hole**), then it can be 2-colored so that no maximum clique is monochromatic. (Equivalently, every induced subgraph of G has such a 2-coloring precisely when the graph contains no odd hole longer than 3.)

FACTS

F85: [KiPe94] A χ-binding function exists on the class of graphs not containing T, if the forbidden induced tree T has radius 2.

F86: [Sc97] For every tree T, there exists a χ-binding function on the class of graphs not containing any subdivision of T as an induced subgraph.

F87: The class of intersection graphs of axis-aligned rectangles in the plane is χ-bound ([AsGr60]), while that of the triangle-free intersection graphs of axis-aligned boxes in $\mathbb{R}^3$ isn't ([Bu65]).

F88: [PaKoKrLaMiTrWa12] The class of intersection graphs of line segments in the plane is not χ-bound; it contains triangle-free graphs with any large chromatic number. (This fact also disproves Scott's conjecture for a large class of non-planar graphs.) On the other hand, the class of complements of those graphs is χ-bound ([PaTö94]).

F89: [BoTh12] If H is a triangle-free graph with diameter 2, then the class of graphs not containing any induced subdivision of H is χ-bound. (Hence Scott's conjecture is true for maximal triangle-free graphs.)

F90: [RaSc01] There exists no linear χ-binding function for the class of graphs in which every induced cycle has length 3 or 4.

REMARKS

R8: The validity of Conjecture C17 would imply by induction that the corresponding class of graphs has χ-binding function $2^{\omega-1}$. While perfect graphs need to exclude both the odd holes and the odd ***antiholes*** (i.e., induced subgraphs that are the complements of odd holes), to make a class χ-bound it would suffice to exclude odd holes. The exclusion of odd antiholes, however, is not sufficient, as shown by the graphs of girth 6; they can have any large chromatic number.

R9: For more details on χ-bound classes of graphs, see [Gy87] and [RaSc04].

5.1.6   Coloring and Orientation

Paths and Cycles

FACTS

F91: [Ga68, Ro67] A graph has $\chi(G) \leq k$ if and only if G admits an orientation without directed paths on more than k vertices.

F92: [Mi62] (*Minty's Condition*) A graph G has chromatic number $\chi(G) \leq k$ if and only if G admits an orientation such that every cycle $C \subseteq G$ has at least $|C|/k$ arcs oriented in each of the two directions around C.

F93: [Tu92] It suffices to assume Minty's Condition above just for the cycles of length $|C| \equiv 1 \pmod{k}$. (This implies the Gallai–Roy Theorem (F91), too.) Moveover, if such an orientation is given, then a proper k-coloring of G can be found in polynomial time. In particular, if the undirected graph G contains no cycles of length 1 modulo k, then G can be properly k-colored in linear time.

F94: [Bo76] In every strongly connected orientation of a k-chromatic graph ($k \geq 2$), there is a directed cycle of length at least k.

Eulerian Subgraphs

DEFINITION

D37: An ***Eulerian spanning subgraph*** of an oriented graph $\vec{G} = (V, \vec{E})$ is a subgraph $\vec{H} = (V, \vec{F})$ of $\vec{G}$ with the same vertex set V, and with $d_H^+(v) = d_H^-(v)$ for all $v \in V$. (Some or all vertices of $\vec{H}$ may be isolated; hence, $\vec{F} = \emptyset$ shows that every $\vec{G}$ has at least one such subgraph.)

FACT

F95: [AlTa92] If the number of Eulerian spanning subgraphs of an oriented graph $\vec{G}$ with an even number of edges differs from the number of those with an odd number of edges, then $\vec{G}$ is L-colorable whenever the list assignment L satisfies $|L_v| > d^+(v)$ for all $v \in V$. (The spanning subgraph condition holds in every bipartite graph; but this class can be handled in a simpler way using kernels; see below.)

Choosability and Orientations with Kernels

Some background and references for the following results can be found in [Tu97].

FACTS

F96: Suppose that in the oriented graph $\vec{G} = (V, \vec{E})$, every induced subgraph $\vec{H}$ contains an *independent* set $Y \subseteq V$ such that from each vertex $v \notin Y$ of $\vec{H}$ there is at least one arc to Y. If L is a list assignment with $|L_v| > d^+(v)$ for all $v \in V$, then G admits a list coloring.

F97: The condition above holds in every bipartite directed graph. A more general class where the required subsets Y (termed ***kernels***) exist is the class of so-called *kernel-perfect graphs* (see §2.5). In those graphs, it suffices to consider orientations without directed 3-cycles.

F98: If every induced subgraph of a graph G has average degree at most $2k$, then G has an orientation with maximum out-degree at most k, and such an orientation can be found in polynomial time (in several ways, via the Kőnig–Hall theorem or by sequential improvements of suboptimal solutions). In particular, if G is bipartite and satisfies the average-degree condition, then $ch(G) \leq k + 1$ and a list coloring can be found efficiently for any given $(k + 1)$-assignment.

More results can be derived by applying this machinery on edge-colorings and planar graphs (see §5.1.5 and §5.2.2).

Acyclic Orientations

DEFINITION

D38: (Cf. [MaTo84] and [Ai88, p. 323]) The ***acyclic orientation game*** starts with an undirected graph. In each round, Player A ('Algy') selects a non-oriented edge e of G, and Player S ('Strategist') orients that edge in one direction, under the condition that no directed cycles may occur. The game is over when the graph G admits just one acyclic orientation that extends the partial orientation obtained so far. The goal of A is to finish the game in as few rounds as possible, while S aims at making the game long.

NOTATION: We denote by $c(G)$ the number of rounds when both A and S play optimally on graph G.

FACTS

F99: [AiTrTu95] If $G = (V, E)$ is a graph on $n \geq 6$ vertices with $c(G) = |E|$, then $|E| \leq \frac{1}{4}n^2$, and for $n \geq 7$ equality holds if and only if G is the complete bipartite graph $K_{\lfloor n/2 \rfloor, \lceil n/2 \rceil}$.

F100: [Pi10] There exists a constant $c > 0$ such that $c(G) \geq cm\frac{\log n}{n}$ holds for all graphs with n vertices and m edges. Moreover, as $n \to \infty$, a general asymptotic upper bound is $c(G) \leq \frac{1}{4}n^2 + o(n^2)$ for all graphs on n vertices.

F101: [AiTrTu95] For every g, there exists a graph $G = (V, E)$ with girth at least g and $c(G) < |E|$.

F102: [AlTu95] For the random graph $\mathbf{G}_{n,p}$ with n vertices and edge probability p, $c(\mathbf{G}_{n,p}) = \Theta(n \log n)$ with probability $1 - o(1)$ as $n \to \infty$ whenever $p > 0$ is fixed. For unrestricted p, a general upper bound is $c(\mathbf{G}_{n,p}) = \mathcal{O}(n \log^3 n)$.

F103: [Pi10] It is **APX-hard** to determine the value $c(G)$; it cannot be approximated within $74/73 - \varepsilon$ for any $\varepsilon > 0$. (Perhaps it is **PSPACE-complete** to determine $c(G)$.)

5.1.7   Colorings of Infinite Graphs

FACTS

F104: [BrEr51] For any $k \in \mathbb{N}$, an infinite graph G has $\chi(G) \leq k$ if and only if every finite subgraph of G is k-colorable.

F105: [Jo94] For any $k \in \mathbb{N}$, an infinite graph G has $ch(G) \leq k$ if and only if every finite subgraph of G is k-choosable.

F106: [Bo77] If $\chi(G) = \infty$, then for every infinite arithmetic progression $A \subseteq \mathbb{N}$, G contains a cycle whose length belongs to A.

F107: [GaKo91] That $\chi(G)$ is well-defined for every graph G is equivalent to the set-theoretic Well-Ordering Theorem, and assuming that every set has a cardinality, it is equivalent to the Axiom of Choice.

F108: [Ko88] The following assertion is consistent: There exists a graph $G = (V, E)$ such that $|V| = \aleph_{\omega+1}$, $\chi(G) = \aleph_1$, and $\chi(H) \leq \aleph_0$ whenever $|V(H)| \leq \aleph_\omega$.

Coloring Euclidean Spaces and Distance Graphs

DEFINITIONS

D39: The **unit distance graph** $\mathcal{U}^n$ has the points of $\mathbb{R}^n$ as its vertices; the edges are the pairs of points whose Euclidean distance is 1.

D40: Given a (finite or infinite) "distance set" $D = \{d_1, d_2, \dots\} \subset \mathbb{N}$, the **distance graph** $G(D)$ has vertex set $\mathbb{Z}$; two vertices $i, j \in \mathbb{Z}$ are adjacent if and only if $|i - j| \in D$.

D41: A **packing coloring** of graph $G = (V, E)$ is a vertex partition $V_1 \cup \cdots \cup V_k = V$ such that, for all $1 \leq i \leq k$ and for any two $u, v \in V_i$ ($u \neq v$), $d(u, v) > i$ holds for their distance. The **packing chromatic number** (also called **broadcast chromatic number**), denoted by $\chi_p(G)$, is the smallest k for which G admits a packing coloring.

FACTS

F109: [Ha45] In the plane, $\chi(\mathcal{U}^2) \leq 7$.

F110: [MoMo61] By a 4-chromatic subgraph on seven vertices, $\chi(\mathcal{U}^2) \geq 4$.

F111: [Co02] In 3-dimensional space, $\chi(\mathcal{U}^3) \leq 15$.

F112: A 5-chromatic subgraph on 9 vertices yields $\chi(\mathcal{U}^3) \geq 5$.

F113: [LaRo72] As $n \to \infty$, $\chi(\mathcal{U}^n)$ is at most $(3 + o(1))^n$.

F114: [FrWi81] As $n \to \infty$, $\chi(\mathcal{U}^n)$ is at least $(1 + o(1)) \left(\frac{6}{5}\right)^n$.

F115: [Sc95] The choice number $ch(\mathcal{U}^n)$ is countably infinite if and only if $n = 2$ or $n = 3$. (Infinity was first observed in [JeTo95].)

F116: For every finite distance set D, $\chi(G(D)) \leq |D| + 1$.

F117: For every D with $|D| = 3$, $\chi(G(D))$ has been determined ([Zh02]). For $k > 3$, however, the chromatic number for k-element distance sets has not been characterized so far.

F118: [EgErSk85] If D is the set of all prime numbers, then $\chi(G(D)) = 4$. Moreover, $\chi(G(D \setminus \{3\})) = 3$ and $\chi(G(D \setminus \{2\})) = 2$.

F119: [Ka01, RuTuVo02, PeSc10] Let $\varepsilon_1 \geq \varepsilon_2 \geq \cdots > 0$ be a sequence of positive reals, and let $\mathcal{D}$ be the family of distance sets $D = \{d_1, d_2, \dots\}$ such that $d_{i+1}/d_i \geq \varepsilon_i$ for all $i \geq 1$. If $\lim_{i \to \infty} \varepsilon_i > 0$, then $\chi(G(D))$ is finite for all those D; and if $\lim_{i \to \infty} \varepsilon_i = 0$, then there exists a $D \in \mathcal{D}$ such that $\chi(G(D)) = \infty$.

F120: [AlKo11] For any finite set $X \subset \mathbb{R}^2$, and for any $k \in \mathbb{N}$, there exists a list k-assignment to the points of $\mathbb{R}^2$ such that any coloring of the plane from those lists contains a monochromatic isometric copy of X.

F121: The packing chromatic number of the square lattice (infinite grid) $G(\mathbb{Z}^2)$ is at least 12 and at most 17 (both bounds attained by using computer programs, the latter published in [SoHo10]); but the triangular lattice and also $G(\mathbb{Z}^2) \square K_2$ — i.e., two layers of the square lattice — have infinite packing chromatic number ([FiRa10] and [FiKlLi09], respectively). The packing chromatic number of the hexagonal lattice is equal to 7, but its Cartesian product with P_6 has $\chi_p = \infty$.

References

[Ai88] M. Aigner, *Combinatorial Search*, Wiley–Teubner, 1988.

[AiTrTu95] M. Aigner, E. Triesch, and Zs. Tuza, Searching for acyclic orientations of graphs, *Discrete Math.* 144 (1995), 3–10.

[AjKoSz80] M. Ajtai, J. Komlós, and E. Szemerédi, A note on Ramsey numbers, *J. Combin. Theory, Ser. A* 29 (1980), 354–360.

[Al93] N. Alon, Restricted colorings of graphs, pages 1–33 in K. Walker (Ed.), *Surveys in Combinatorics*, Proc. 14th British Combinatorial Conference, London Math. Soc. Lecture Notes Series 187, Cambridge University Press, 1993.

[Al00] N. Alon, Degrees and choice numbers, *Random Struct. Alg.* 16 (2000), 364–368.

[AlKo11] N. Alon and A. V. Kostochka, Hypergraph list coloring and Euclidean Ramsey Theory, *Random Struct. Alg.* 39 (2011), 377–390.

[AlKrSu99] N. Alon, M. Krivelevich, and B. Sudakov, List coloring of random and pseudo-random graphs, *Combinatorica* 19 (1999), 453–472.

[AlTa92] N. Alon and M. Tarsi, Colorings and orientations of graphs, *Combinatorica* 12 (1992), 125–134.

[AlTu95] N. Alon and Zs. Tuza, The acyclic orientation game on random graphs, *Random Struct. Alg.* 6 (1995), 261–268.

[AsGr60] E. Asplund and B. Grünbaum, On a colouring problem, *Math. Scand.* 8 (1960), 181–188.

[Bá78] I. Bárány, A short proof of Kneser's conjecture, *J. Combin. Theory, Ser. A* 25 (1978), 325–326.

[BaJoWo11] J. Barát, G. Joret, and D. R. Wood, Disproof of the List Hadwiger Conjecture, *Electron. J. Combin.* 18:1 (2011), #P232.

[Bo76] J. A. Bondy, Diconnected orientations and a conjecture of Las Vergnas, *J. London Math. Soc.* (2) 14 (1976), 277–282.

[Bo77] B. Bollobás, Cycles modulo k, *Bull. London Math. Soc.* 9 (1977), 97–98.

[Bo78] B. Bollobás, Uniquely colorable graphs, *J. Combin. Theory, Ser. B* 25 (1978), 54–61.

[Bo88] B. Bollobás, The chromatic number of random graphs, *Combinatorica* 8 (1988), 49–55.

[BoCaEr80] B. Bollobás, P. A. Catlin, and P. Erdős, Hadwiger's conjecture is true for almost every graph, *Europ. J. Combin.* 1 (1980), 195–199.

[BoKo77] O. V. Borodin and A. V. Kostochka, On an upper bound of a graph's chromatic number, depending on the graph's degree and density, *J. Combin. Theory, Ser. B* 23 (1977), 247–250.

[BoTh12] N. Bousquet and S. Thomassé, Scott's induced subdivision conjecture for maximal triangle-free graphs, *Combin. Probab. Comput.* 21 (2012), 512–514.

[Br02] S. Brandt, A 4-colour problem for dense triangle-free graphs, *Discrete Math.* 251 (2002), 33–46.

[Br41] R. L. Brooks, On colouring the nodes of a network, *Proc. Cambridge Philos. Soc.* 37 1941, 194–197.

[Bu65] J. P. Burling, On coloring problems of families of prototypes, PhD Thesis, University of Colorado, 1965.

[BrEr51] N. G. de Bruijn and P. Erdős, A colour problem for infinite graphs and a problem in the theory of relations, *Nederl. Akad. Wetensch. Proc. Ser. A* 54 (1951), 371–373.

[Ca78] P. A. Catlin, A bound on the chromatic number of a graph, *Discrete Math.* 22 (1978), 81–83.

[Ca79] P. A. Catlin, Hajós' graph-coloring conjecture: variations and counterexamples, *J. Combin. Theory, Ser. B* 26 (1979), 268–274.

[CaMeRuSp98] A. Cavicchioli, M. Meschiari, B. Ruini, and F. Spaggiari, A survey on snarks and new results: products, reducibility and a computer search, *J. Graph Theory* 28 (1998), 57–86.

[Ce84] U. A. Celmins, On cubic graphs that do not have an edge-3-colouring, PhD Thesis, University of Waterloo, Canada, 1984.

[ChGe69] G. Chartrand and D. P. Geller, On uniquely colorable planar graphs, *J. Combin. Theory* 6 (1969), 271–278.

[ChŠk10] M. Chladný and M. Škoviera, Factorisation of snarks, *Electron. J. Combin.* 17 (2010), #R32.

[Co02] D. Coulson, A 15-colouring of 3-space omitting distance one, *Discrete Math.* 256 (2002), 83–90.

[De48] B. Descartes, Solutions to problems in Eureka No. 9, *Eureka* 10 (1948).

[Di52] G. A. Dirac, A property of 4-chromatic graphs and some remarks on critical graphs, *J. London Math. Soc.* 27 (1952), 85–92.

[Di57] G. A. Dirac, A theorem of R. L. Brooks and a conjecture of H. Hadwiger, *Proc. London Math. Soc. (3)* 7 (1957), 161–195.

[EgErSk85] R. B. Eggleton, P. Erdős, and D. K. Skilton, Coloring the real line, *J. Combin. Theory, Ser. B* 39 (1985), 86–100.

[ElGo96] M. N. Ellingham and L. Goddyn, List edge colourings of some 1-factorizable multigraphs, *Combinatorica* 16 (1996), 343–352.

[Er59] P. Erdős, Graph theory and probability, *Canad. J. Math.* 11 (1959), 34–38.

[Er67] P. Erdős, Some remarks on chromatic graphs, *Colloq. Math.* 16 (1967), 253–256.

[ErHa66] P. Erdős and A. Hajnal, On chromatic number of graphs and set-systems, *Acta Math. Acad. Sci. Hungar.* 17 (1966), 61–99.

[ErRuTa79] P. Erdős, A. L. Rubin, and H. Taylor, Choosability in graphs, Proc. West-Coast Conference on Combinatorics, Graph Theory and Computing, Arcata, California, *Congr. Numer.* XXVI (1979), 125–157.

[FiKlLi09] J. Fiala, S. Klavžar, and B. Lidický. The packing chromatic number of infinite product graphs, *Europ. J. Combin.* 30 (2009), 1101–1113.

[FiRa10] A. S. Finbow and D. F. Rall, On the packing chromatic number of some lattices, *Discrete Applied Math.* 158 (2010), 1224–1228.

[FrWi81] P. Frankl and R. M. Wilson, Intersection theorems with geometric consequences, *Combinatorica* 1 (1981), 357–368.

[Ga63] T. Gallai, Kritische Graphen I., *Publ. Math. Inst. Hungar. Acad. Sci.* 8 (1963), 165–192.

[Ga68] T. Gallai, On directed paths and circuits, pages 115–118 in P. Erdős and G. O. H. Katona (Eds.), *Theory of Graphs*, Colloq. Math. Soc. J. Bolyai, Tihany (Hungary), 1966, Academic Press, San Diego, 1968.

[Ga95] F. Galvin, The list chromatic index of a bipartite multigraph, *J. Combin. Theory, Ser. B* 63 (1995), 153–158.

[GaKo91] F. Galvin and P. Komjáth, Graph colorings and the axiom of choice, *Periodica Math. Hungar.* 22 (1991), 71–75.

[GeGeReSeVe09] J. Geelen, B. Gerards, B. Reed, P. Seymour, and A. Vetta, On the odd-minor variant of Hadwiger's conjecture, *J. Combin. Theory, Ser. B* 99 (2009), 20–29.

[Go85] L. Goddyn, A girth requirement for the double cycle cover conjecture, pages 13–26 in *Cycles in Graphs*, North-Holland Math. Stud. 115, 1985.

[Gr96] S. Gravier, A Hajós-like theorem for list coloring, *Discrete Math.* 152 (1996), 299–302.

[GrLo74] D. Greenwell and L. Lovász, Applications of product colouring, *Acta Math. Acad. Sci. Hungar.* 25 (1974), 335–340.

[Gy75] A. Gyárfás, On Ramsey covering-numbers, pages 801–816 in *Infinite and Finite Sets*, Colloq. Math. Soc. J. Bolyai, Vol. 10, Keszthely (Hungary), 1973, North-Holland/American Elsevier, 1975.

[Gy87] A. Gyárfás, Problems from the world surrounding perfect graphs, *Zastos. Mat.* XIX (1987), 413–441.

[Ha43] H. Hadwiger, Über eine Klassifikation der Streckenkomplexe, *Vierteljahrsch. Naturforsch. Ges. Zürich* 88 (1943), 133–142.

[Ha45] H. Hadwiger, Überdeckung des Euklidischen Raumes durch kongruente Mengen, *Portug. Math.* 4 (1945), 238–242.

[HaSz70] A. Hajnal and E. Szemerédi, Proof of a conjecture of Erdős, pages 601–623 in P. Erdős, A. Rényi, and V. T. Sós (Eds.), *Combinatorial Theory and its Applications, Vol. II*, Colloq. Math. Soc. J. Bolyai 4, North-Holland, 1970.

[Ha61] Gy. Hajós, Über eine Konstruktion nicht n-färbbarer Graphen, *Wiss. Z. Martin-Luther-Univ. Halle-Wittenberg Math.-Natur. Reihe*, 10 (1961), 116–117.

[JeTo95] T. R. Jensen and B. Toft, *Graph Coloring Problems*, Wiley-Interscience, 1995.

[Jo96] A. Johansson, An improved upper bound on the choice number for triangle free graphs, Manuscript, January 1996.

[Jo96a] A. Johansson, The choice number of sparse graphs, Preliminary version, April 1996.

[Jo94] P. D. Johnson, The choice number of the plane, *Geombinatorics* 3 (1994), 122–128.

[Ka92] J. Kahn, Coloring nearly-disjoint hypergraphs with $n + o(n)$ colors, *J. Combin. Theory, Ser. A* 59 (1992), 31–39.

[Ka96] J. Kahn, Asymptotically good list-colorings, *J. Combin. Theory, Ser. A* 73 (1996), 1–59.

[Ka01] Y. Katznelson, Chromatic numbers of Cayley graphs on $\mathbb{Z}$ and recurrence, *Combinatorica* 21 (2001), 211–219.

[KaSo07] K. Kawarabayashi and Z. Song, Some remarks on the odd Hadwiger's conjecture, *Combinatorica* 27 (2007), 429–438.

[KaTo05] K. Kawarabayashi and B. Toft, Any 7-chromatic graph has a K_7-minor or a $K_{4,4}$-minor, *Combinatorica* 25 (2005), 327–353.

[KeKe54] J. B. Kelly and L. M. Kelly, Paths and circuits in critical graphs, *Amer. J. Math.* 76 (1954), 786–792.

[KiKo08] H. A. Kierstead and A. V. Kostochka, An Ore-type theorem on equitable coloring, *J. Combin. Theory, Ser. B* 98 (2008), 226–234.

[KiKoYu09] H. A. Kierstead, A. V. Kostochka, and G. Yu, Extremal graph packing problems: Ore-type versus Dirac-type, pages 113–135 in S. Huczynska et al. (Eds.), *Surveys in Combinatorics*, Proc. 22nd British Combinatorial Conference, London Math. Soc. Lecture Notes Series 365, Cambridge University Press, 2009.

[KiPe94] H. A. Kierstead and S. G. Penrice, Radius two trees specify χ-bounded classes, *J. Graph Theory* 18 (1994), 119–129.

[KiSc83] H. A. Kierstead and J. H. Schmerl, Some applications of Vizing's theorem to vertex colorings of graphs, *Discrete Math.* 45 (1983), 277–285.

[Ki95] J. H. Kim, On Brooks' theorem for sparse graphs, *Combin. Probab. Comput.* 4 (1995), 97–132.

[KiReVe07] A. D. King, B. A. Reed, and A. Vetta, An upper bound for the chromatic number of line graphs, *Europ. J. Combin.* 28 (2007), 2182–2187.

[KiRe08] A. D. King and B. A. Reed, Bounding χ in terms of ω and Δ for quasi-line graphs. *J. Graph Theory* (2008), 215–228.

[Ko96] M. Kochol, Snarks without small cycles, *J. Combin. Theory, Ser. B* 67 (1996), 34–47.

[Ko02] M. Kochol, Superposition and constructions of graphs without nowhere-zero *k*-flows, *Europ. J. Combin.* 23 (2002), 281–306.

[Ko09] M. Kochol, Polyhedral embeddings of snarks in orientable surfaces, *Proc. Amer. Math. Soc.* 137 (2009), 1613–1619.

[KoSc10] A. Kohl and I. Schiermeyer, Some results on Reed's Conjecture about ω, Δ, and χ with respect to α, *Discrete Mathematics* 310 (2010), 1429–1438.

[Ko88] P. Komjáth, Consistency results on infinite graphs, *Israel J. Math.* 61 (1988), 285–294.

[Kö16] D. König, Über Graphen und ihre Anwendung auf Determinantentheorie und Mengenlehre, *Math. Ann.* 77 (1916), 453–465.

[Ko78] A. V. Kostochka, Degree, girth and chromatic number, pages 679–696 in A. Hajnal and V. T. Sós (Eds.), *Combinatorics*, Colloq. Math. Soc. J. Bolyai 18, Keszthely (Hungary), 1976, North-Holland, 1978.

[Ko82] A. V. Kostochka, The minimum Hadwiger number for graphs with a given mean degree of vertices, *Metody Diskret. Analiz.* 38 (1982), 37–58. (in Russian)

[Ko84] A. V. Kostochka, On Hadwiger numbers of a graph and its complement, pages 537–545 in A. Hajnal, L. Lovász, and V. T. Sós (Eds.), *Finite and Infinite Sets*, Colloq. Math. Soc. J. Bolyai 37, Eger (Hungary) 1981, North-Holland, 1984.

[Ko84a] A. V. Kostochka, Bounds on the Hadwiger number of graphs by their average degree, *Combinatorica* 4 (1984), 307–316.

[KoMa77] A. V. Kostochka and N. P. Mazurova, An inequality in the theory of graph coloring, *Metody Diskret. Analiz.* 30 (1977), 23–29. (in Russian)

[KoYa12] A. V. Kostochka and M. Yancey, Ore's Conjecture on color-critical graphs is almost true, manuscript, 2012, http://arxiv.org/abs/1209.1050

[KrTuVo99] J. Kratochvíl, Zs. Tuza, and M. Voigt, New trends in the theory of graph colorings: Choosability and list coloring, pages 183–197 in R. L. Graham et al. (Eds.), *Contemporary Trends in Discrete Mathematics*, DIMACS Series in Discrete Mathematics and Theoretical Computer Science 49, Amer. Math. Soc., 1999.

[Kr00] M. Krivelevich, The choice number of dense random graphs, *Combin. Probab. Comput.* 9 (2000), 19–26.

[LaRo72] D. G. Larman and C. A. Rogers, The realization of distances within sets in Euclidean space, *Mathematika* 19 (1972), 1–24.

[La78] J. Lawrence, Covering the vertex set of a graph with subgraphs of smaller degree, *Discrete Math.* 21 (1978), 61–68.

[Lo68] L. Lovász, On chromatic number of finite set-systems, *Acta Math. Acad. Sci. Hungar.* 19 (1968), 59–67.

[Lo73] L. Lovász, Independent sets in critical chromatic graphs, *Studia Sci. Math. Hungar.* 8 (1973), 165–168.

[Lo78] L. Lovász, Kneser's conjecture, chromatic number, and homotopy. *J. Combin. Theory, Ser. A* 25 (1978), 319–324.

[LuPhSa88] A. Lubotzky, R. Phillips, and P. Sarnak, Ramanujan graphs, *Combinatorica* 8 (1988), 261–277.

[MaTo84] U. Manber and M. Tompa, The effect of the number of Hamiltonian paths on the complexity of a vertex coloring problem, pages 220–227 in *Proc. 24th FOCS*, 1984.

[Mi92] P. Mihók, An extension of Brooks' theorem, *Annals of Discrete Math.* 51 (1992), 235–236.

[MiSc04] P. Mihók and I. Schiermeyer, Cycle lengths and chromatic number of graphs, *Discrete Math.* 286 (2004), 147–149.

[Mi62] G. J. Minty, A theorem on n-colouring the points of a linear graph, *Amer. Math. Monthly* 67 (1962), 623–624.

[MoRe98] M. Molloy and B. Reed, A bound on the total chromatic number, *Combinatorica* 18 (1998), 241–280.

[MoRe01] M. Molloy and B. Reed, Colouring graphs when the number of colours is nearly the maximum degree, pages 462–470 in *Proc. STOC'01*, Hersonissos, Greece, July 2001.

[MoMo61] L. Moser and W. Moser, Problem and solution P10, *Canad. J. Math.* 4 (1961), 187–189.

[My55] J. Mycielski, Sur le coloriage des graphes, *Colloq. Math.* 3 (1955), 161–162.

[NeŠk96] R. Nedela and M. Škoviera, Decompositions and reductions of snarks, *J. Graph Theory* 22 (1996), 253–279.

[NeRö79] J. Nešetřil and V. Rödl, A short proof of the existence of highly chromatic hypergraphs without short cycles, *J. Combin. Theory, Ser. B* 27 (1979), 225–227.

[NoReWu12] J. Noel, B. Reed, and H. Wu, A proof of Ohba's conjecture, Manuscript, 2012.

[NoGa56] E. A. Nordhaus and J. W. Gaddum, On complementary graphs, *Amer. Math. Monthly* 63 (1956), 175–177.

[PaTö94] J. Pach and J. Törőcsik, Some geometric applications of Dilworth's theorem, *Discrete Comput. Geom.* 12 (1994), 1–7.

[PaKoKrLaMiTrWa12] A. Pawlik, J. Kozik, T. Krawczyk, M. Lason, P. Micek, W. T. Trotter, and B. Walczak, Triangle-free intersection graphs of line segments with large chromatic number, arXiv:1209.1595v1, 2012.

[PeSc10] Y. Peres and W. Schlag, Two Erdős problems on lacunary sequences: chromatic number and Diophantine approximation, *Bull. London Math. Soc.* 42 (2010), 295–300.

[Pi10] O. Pikhurko, Finding an unknown acyclic orientation of a given graph, *Combin. Probab. Comput.* 19 (2010), 121–131.

[Ra04] B. Randerath, 3-colorability and forbidden subgraphs. I: Characterizing pairs, *Discrete Math.* 276 (2004), 313–325.

[RaSc01] B. Randerath and I. Schiermeyer, Colouring graphs with prescribed induced cycle lengths, *Discuss. Math. Graph Theory* 21 (2001), 267–282.

[RaSc04] B. Randerath and I. Schiermeyer, Vertex colouring and forbidden subgraphs — a survey, *Graphs Combin.* 20 (2004), 1–40.

[RaSc07] B. Randerath and I. Schiermeyer, On Reed's Conjecture about ω, Δ and χ, pages 339–346 in A. Bondy et al. (Eds.), *Graph Theory in Paris*, Proc. Conf. in Memory of Claude Berge, Trends in Mathematics, Birkhäuser, 2007.

[Re97] B. Reed, ω, Δ, and χ, *J. Graph Theory* 27 (1997), 177–212.

[RoSeTh93] N. Robertson, P. D. Seymour, and R. Thomas, Hadwiger's conjecture for K_6-free graphs, *Combinatorica* 13 (1993), 279–361.

[RöTu85] V. Rödl and Zs. Tuza, On color critical graphs, *J. Combin. Theory, Ser. B* 38 (1985), 204–213.

[Ro67] R. Roy, Nombre chromatique et plus longs chemins d'un graphe, *Revue AFIRO* 1 (1967), 127–132.

[RuTuVo02] I. Z. Ruzsa, Zs. Tuza, and M. Voigt, Distance graphs with finite chromatic number, *J. Combin. Theory, Ser. B* 85 (2002), 181–187.

[Sc95] J. H. Schmerl, The list-chromatic number of Euclidean space, *Geombinatorics* 5 (1995), 65–68.

[Sc78] A. Schrijver, Vertex-critical subgraphs of Kneser graphs, *Nieuw Arch. Wisk.* 26 (1978), 454–461.

[Sc97] A. D. Scott, Induced trees in graphs of large chromatic number, *J. Graph Theory* 24 (1997), 297–311.

[Se81] P. D. Seymour, Nowhere-zero 6-flows, *J. Combin. Theory, Ser. B* 30 (1981), 130–135.

[Sh49] C. E. Shannon, A theorem on coloring the lines of a network, *J. Math. Phys.* 28 (1949), 148–151.

[Si72] M. Simonovits, On colour-critical graphs, *Studia Sci. Math. Hungar.* 7 (1972), 67–81.

[SoHo10] R. Soukal and P. Holub, A note on packing chromatic number of the square lattice, *Electronic J. Combinatorics* 17 (2010), #N17.

[St85] M. Stiebitz, Colour-critical graphs with complete major-vertex subgraph, pages 169–181 in H. Sachs (Ed.), *Graphs, Hypergraphs and Applications* (Proc. Conf. Graph Theory, Eyba, GDR, 1984), Teubner-Texte zur Mathematik 73, Teubner, 1985.

[StScToFa12] M. Stiebitz, D. Scheide, B. Toft, and L. M. Favrholdt, *Graph Edge Coloring — Vizing's Theorem and Goldberg's Conjecture*, Wiley, 2012.

[StTuVo09] M. Stiebitz, Zs. Tuza, and M. Voigt, On list critical graphs, *Discrete Math.* 309 (2009), 4931–4941.

[Su81] D. P. Sumner, Subtrees of a graph and the chromatic number, pages 557–576 in *The Theory and Applications of Graphs*, Proc. Conf. Kalamazoo (Michigan), Wiley, 1981.

[Th78] A. G. Thomason, Hamiltonian cycles and uniquely edge colourable graphs, *Annals of Discrete Math.* 3 (1978), 259–268.

[Th84] A. Thomason, An extremal function for contractions of graphs, *Math. Proc. Cambridge Philos. Soc.* 95 (1984), 261–26.

[To70] B. Toft, On the maximal number of edges of critical *k*-chromatic graphs, *Studia Sci. Math. Hungar.* 5 (1970), 461–470.

[To72] B. Toft, Two theorems on critical 4-chromatic graphs, *Studia Sci. Math. Hungar.* 7 (1972), 83–89.

[To78] B. Toft, An investigation of colour-critical graphs with complements of low connectivity, *Annals of Discrete Math.* 3 (1978), 279–287.

[To85] B. Toft, Some problems and results related to subgraphs of colour critical graphs, pages 178–186 in R. Bodendiek, H. Schumacher, and G. Walther (Eds.), *Graphen in Forschung und Unterricht: Festschrift K. Wagner*, Barbara Franzbecker Verlag, 1985.

[Tu54] W. T. Tutte, A contribution to the theory of chromatic polynomials, *Canad. J. Math.* 6 (1954), 80–91.

[Tu70] W. T. Tutte, More about chromatic polynomials and the golden ratio, pages 439–453 in R. K. Guy et al. (Eds.), *Combinatorial Structures and their Applications*, Gordon and Breach, 1970.

[Tu76] W. T. Tutte, Hamiltonian circuits, pages 193–199 in *Colloquio Internazionale sulle Teorie Combinatorie* (Roma, 1973), Tomo I, Atti dei Convegni Lincei, No. 17, Accad. Naz. Lincei, Rome, 1976.

[Tu92] Zs. Tuza, Graph coloring in linear time, *J. Combin. Theory, Ser. B* 55 (1992), 236–243.

[Tu97] Zs. Tuza, Graph colorings with local constraints — a survey, *Discuss. Math. Graph Theory* 17 (1997), 161–228.

[Tu97a] Zs. Tuza, Choice-perfect graphs and Hall numbers, manuscript, 1997; updated version: Choice-perfect graphs in *Discuss. Math. Graph Theory*) 33 (2013), 231–242.

[TuVo94] Zs. Tuza and M. Voigt, Restricted types of graph colorings, in *Kolloquium Kombinatorik* (German Combinatorics Conference), Hamburg, 1994.

[TuVo97] Zs. Tuza and M. Voigt, List colorings and reducibility, *Discrete Applied Math.* 79 (1997), 247–256.

[Vi64] V. G. Vizing, On an estimate of the chromatic class of a *p*-graph, *Metody Diskret. Analiz.* 3 (1964), 9–17. (in Russian)

[Vi76] V. G. Vizing, Coloring the vertices of a graph in prescribed colors, *Metody Diskret. Anal. v Teorii Kodov i Schem*, 29 (1976), 3–10. (in Russian)

[Vu99] V. H. Vu, On some simple degree conditions that guarantee the upper bound on the chromatic (choice) number of random graphs. *J. Graph Theory* 31 (1999), 201–226.

[Wa37] K. Wagner, Über eine Eigenschaft der ebenen Komplexe, *Math. Ann.* 114 (1937), 570–590.

[Za98] W. Zang, Coloring graphs with no odd-K_4, *Discrete Math.* 184 (1998), 205–212.

[Zh02] X. Zhu, Circular chromatic number of distance graphs with distance sets of cardinality 3, *J. Graph Theory* 41 (2002), 195–207.

[Zh98] X. Zhu, A survey on Hedetniemi's conjecture, *Taiwanese J. Math.* 2 (1998), 1–24.

[Zy49] A. A. Zykov, On some problems of linear complexes, *Mat. Sbornik* 24 (1940), 163–188. (in Russian)

Section 5.2
Further Topics in Graph Coloring

Zsolt Tuza, University of Veszprém, Hungary

INTRODUCTION

In this section we consider variants of graph coloring, and also the algorithmic complexity of the problems. Some of the concepts here (e.g., face coloring of planar graphs) may be viewed as equivalents of proper vertex-coloring on restricted classes of graphs, a perspective for which there is an extensive literature. Some other topics here are generalizations in various directions. We also include a new area in the theory of hypergraph coloring, which provides a general framework for many kinds of coloring problems.

From the many interesting variants of coloring, several ones are discussed in detail in [ChZh09], while [Ku04] is a collection of shorter surveys.

The current section applies several concepts introduced in §5.1. Some familiarity with §5.1 is assumed. (Please see the chapter glossary as needed.)

5.2.1   Multicoloring and Fractional Coloring

DEFINITIONS

D1: A *fractional vertex-coloring* of G is a real function $\varphi^* : \mathcal{S} \to \mathbb{R}^{\geq 0}$ on the collection $\mathcal{S}$ of all independent vertex sets in $G = (V, E)$ such that for all $v \in V$

$$\sum_{\{S \in \mathcal{S} | v \in S\}} \varphi^*(S) \geq 1$$

This definition extends naturally to other types of coloring (e.g., fractional edge-coloring), leading in this way to the fractional versions of further graph invariants.

D2: The **fractional chromatic number** of G is

$$\chi^*(G) := \min_{\varphi^*} \sum_{S \in \mathcal{S}} \varphi^*(S)$$

where the minimum is taken over all fractional vertex-colorings φ^* of G. (The *fractional chromatic index* $\chi'^*(G)$ is defined analogously.) Equivalently, the fractional chromatic number of a graph $G = (V, E)$ is definable as the minimum ratio p/q such that there exists a cover of V by p independent sets $S_1, S_2, \ldots, S_p$ (not necessarily distinct), with each $v \in V$ contained in precisely q of them.

D3: For two functions f and g from V to $\mathbb{N}$, with $g(v) \leq f(v)$ for all $v \in V$, the graph $G = (V, E)$ is (f, g)-**choosable** if for every list assignment L with $|L_v| = f(v)$ there can be chosen subsets $C_v \subseteq L_v$ such that $|C_v| = g(v)$ for all $v \in V$ and $C_u \cap C_v = \emptyset$ for all $uv \in E$. The functions f and/or g may be constant; e.g., the terminology "(a, b)-choosable" means that $f(v) = a$ and $g(v) = b$ for all $v \in V$. If $g(v) = 1$ for every vertex v, we simply say that G is f-**choosable**.

EXAMPLE

E1: For every $t \in \mathbb{N}$, the odd cycle C_{2t+1} is $(2t + 1, t)$-choosable ([AlTuVo97]) and its fractional chromatic number is $2 + \frac{1}{t}$.

FACTS

F1: For every graph G, we have

$$\omega(G) \leq \chi^*(G) \leq \chi(G)$$

In particular, if G is a perfect graph, then $\chi^*(G) = \chi(G)$.

F2: For every graph $G = (V, E)$, we have $\chi^*(G) \geq |V|/\alpha(G)$.

F3: For every graph G on n vertices, the value $\chi^*(G) = p/q$ is attained for some $q \leq n^{n/2}$ ([ChGaJo78]), but there exists a constant C ($C > 1.34619$) and infinitely many graphs G for which $q > C^n$ is necessary ([Fi95]).

F4: [ReSe98] For every natural number t, if $\chi^*(G) > 2t$, then some subgraph of G is contractible to K_{t+1}. (Cf. Hadwiger's conjecture in §5.1.4.)

F5: [Lo75] For every graph G,

$$\chi(G) \leq (1 + \log \alpha(G)) \max_{H \subseteq G} \frac{|V(H)|}{\alpha(H)}$$

As a consequence, χ/χ^* is bounded above by $1 + \log n$ (where n is the number of vertices).

F6: [Zh11a] For any two graphs G and H, we have $\chi^*(G \times H) = \min\{\chi^*(G), \chi^*(H)\}$. (Cf. Hedetniemi's conjecture in §5.1.3.)

F7: [KiRe93] For fractional colorings of simple graphs, the Total Coloring Conjecture (cf. §5.1.5) holds true, i.e., $\chi''^*(G) \leq \Delta(G) + 2$ for every graph G.

F8: [Ka00] The edge choice number asymptotically equals the fractional chromatic index, i.e., for every $\varepsilon > 0$ there exists a $k = k(\varepsilon)$ such that $\chi'^*(G) \geq k$ implies $ch'(G) \leq (1+\varepsilon)\chi'^*(G)$, for every multigraph G.

F9: [AlTuVo97] For every graph G, the minimum value of a/b such that G is (a, b)-choosable is equal to $\chi^*(G)$. The theorem can be generalized for list colorings of hypergraphs, too ([MiTuVo99]).

F10: (Brooks's Theorem for unequal lists) [ErRuTa79] Suppose that G is connected, and that at least one of its blocks is neither a complete graph nor an odd cycle. If $f(v)$ is the degree of v for every vertex v, then G is f-choosable.

F11: *For multicoloring:* [TuVo96] Under the same conditions as Fact F10, G is also (mf, m)-choosable for all $m \in \mathbb{N}$.

F12: The List Reduction method (see [TuVo97] in §5.1.1) can be applied for (km, m)-choosability, too.

F13: [TuVo96a] Every 2-choosable graph is $(2m, m)$-choosable, for every $m \in \mathbb{N}$.

OPEN PROBLEMS

P1: [ErRuTa79] Is every (a, b)-choosable graph (am, bm)-choosable for all $m \in \mathbb{N}$?

P2: [ErRuTa79] Given any pair of graphs G and H on the same set of vertices, is $ch(G \cup H) \leq ch(G)ch(H)$?

REMARK

R1: An affirmative answer to Problem 1 would imply an affirmative answer on Problem 2 also. Various particular cases of the former are proved in [TuVo96].

5.2.2   Graphs on Surfaces

Historically, for more than a half century, the theory of graph coloring dealt with face colorings of maps, which can equivalently be interpreted as vertex-colorings of their dual graphs. Via duality, there is a natural correspondence between total colorings of graphs and simultaneously coloring the edges and faces of maps on surfaces. Adjacency of faces of a map means sharing an edge on their boundary, and incidence of a vertex or edge with a face means belonging to its boundary walk.

DEFINITIONS

D4: A *plane graph* is a planar graph together with a given imbedding in the plane.

D5: A *triangulation* (in the plane or on a higher surface) is a graph imbedding in a surface such that all the face boundaries are cycles of length 3.

D6: A graph is *outerplanar* if it has an imbedding in the plane such that all vertices lie on the boundary walk of the exterior face.

FACTS

F14: (Five Color Theorem) [He:1890] Every planar graph is 5-colorable.

F15: (Four Color Theorem) [ApHa77, ApHaKo77] Every planar graph is 4-colorable.

F16: [Ta:1880] A plane triangulation G is 4-colorable if and only if its dual G^* is 3-edge-colorable.

F17: [Bo79] Every planar graph has a proper 5-coloring such that the union of any two color classes induces a forest (termed **acyclic coloring**).

F18: (Grötzsch's Theorem) [Gr59] Every K_3-free planar graph is 3-colorable.

F19: [He:1898] A planar triangulation is 3-colorable if and only if all of its vertices have even degrees.

F20: Every planar graph is 5-choosable ([Th94]), but there exist non-4-choosable planar graphs ([Vo93]).

F21: [KrTu94] Every K_3-free planar graph is 4-choosable.

F22: Every planar graph of girth at least five is 3-choosable ([Th95]), but there exist non-3-choosable K_3-free planar graphs ([Vo95]).

F23: [AlTa92] All bipartite planar graphs are 3-choosable.

F24: [ChGeHe71] Every planar graph has a vertex partition into two classes where each class induces an outerplanar graph.

F25: [He93] The edge set of any planar graph can be partitioned into two outerplanar graphs.

REMARKS

R2: A simpler proof to the Four Color Theorem is given in [RoSaSeTh97], but so far no proof without the extensive use of a computer is known.

R3: A very short proof of Grötzsch's Theorem can be derived from the lower bound $\frac{5n-2}{3}$ on the number of edges in a 4-critical graph of order n ([KoYa12]; see Fact F31 in §5.1.3).

Heawood Number and the Empire Problem

DEFINITION

D7: The **Heawood number** of a closed surface S of Euler characteristic ϵ is

$$H(\epsilon) = \left\lfloor \frac{7 + \sqrt{49 - 24\epsilon}}{2} \right\rfloor$$

More generally, for every natural number m we write

$$H(\epsilon, m) = \left\lfloor \frac{6m + 1 + \sqrt{(6m+1)^2 - 24\epsilon}}{2} \right\rfloor$$

FACTS

F26: [He:1890] If a surface S has Euler characteristic $\epsilon < 2$, then the connected regions of any map drawn on S — or equivalently, any graph imbedded in S — can be properly colored with at most $H(\epsilon)$ colors.

F27: [Fr34] Every graph drawn on the Klein bottle is 6-colorable.

F28: [RiYo68] On any other surface S except the Klein bottle, the maximum chromatic number of graphs imbeddable in S is equal to $H(\epsilon)$.

F29: [Di52] $(-1 \neq \epsilon \neq 1)$, [AlHu79] $(\epsilon = -1$ or $1)$ If S is a surface of Euler characteristic $\epsilon < 2$, other than the Klein bottle, then every $H(\epsilon)$-chromatic graph imbedded in S contains $K_{H(\epsilon)}$ as a subgraph.

F30: [BöMoSt99] With the possible exception of $\epsilon = -1$, every graph G imbedded in a surface S of Euler characteristic ϵ with $ch(G) = H(\epsilon)$ contains the complete graph $K_{H(\epsilon)}$.

F31: [AlMoSa96] As $\epsilon \to \infty$, every graph imbedded in a surface with Euler characteristic ϵ admits an acyclic coloring with at most $O(\epsilon^{4/7})$ colors (and at least $\Omega(\epsilon^{4/7}/\log^{1/7}\epsilon)$ colors are necessary for some graphs).

F32: [He:1890] If each country on a surface of Euler characteristic ϵ consists of at most m connected regions $(m \geq 2)$, then the countries can be colored with at most $H(\epsilon, m)$ colors so that any two neighbor countries are colored differently. (For planar maps this means $6m$ colors.)

F33: [JaRi84] For every $m \geq 2$, there exist planar maps with countries of m regions each, where $6m$ colors are necessary.

F34: $H(\epsilon, m)$ colors are necessary for every $m \geq 2$ on the torus (H. Taylor in [Ga80]), on the projective plane ([JaRi83]) and on the Klein bottle ([JaRi85] for $m \geq 3$ and [Bo89] for $m = 2$).

OPEN PROBLEM

P3: (Empire Problem) For which surfaces S and for which values of m do there exist maps on S with at most m connected regions in each country (here called an *empire* or an *m-pire*), such that $H(\epsilon, m)$ colors are necessary for a proper coloring of all countries?

Nowhere-Zero Flows

DEFINITION

D8: Let $\vec{G} = (V, \vec{E})$ be an oriented multigraph and $k \geq 2$ an integer. A **nowhere-zero** **k-flow** is a function $\phi : E \to \{1, 2, \ldots, k-1\}$ such that

$$\sum_{uv \in \vec{E}} \phi(uv) = \sum_{vw \in \vec{E}} \phi(vw)$$

holds for every vertex $v \in V$.

FACTS

F35: [Tu54] A plane graph G is k-colorable if and only if its planar dual G^* admits an orientation with a nowhere-zero k-flow. (The analogous property holds for a graph imbedded in any orientable surface.)

In particular, the Four Color Theorem is equivalent to the assertion that every planar graph without cut-edges has a nowhere-zero 4-flow, and Grötzsch's theorem asserts (in dual form) that every 4-edge-connected planar graph has a nowhere-zero 3-flow.

F36: [Tu50, Mi67] A 3-regular multigraph is bipartite if and only if it has a nowhere-zero 3-flow, and it is 3-edge-colorable if and only if it has a nowhere-zero 4-flow. In particular, the former assertion generalizes the fact that the *skeleton* (i.e., the graph) of every Eulerian planar triangulation is 3-colorable.

F37: [Se81] Every graph without cut-edges has a nowhere-zero 6-flow.

F38: [LoThWuZh12] Every 6-edge-connected graph has a nowhere-zero 3-flow.

CONJECTURES

C1: 5-flow conjecture [Tu54] Every graph without cut-edges has a nowhere-zero 5-flow.

C2: [Tu54] Every 4-edge-connected graph has a nowhere-zero 3-flow.

Chromatic Polynomials

DEFINITION

D9: The **chromatic polynomial** $P(G, \lambda)$, $\lambda \in \mathbb{N}$, of graph $G = (V, E)$ is the function whose value at λ ($\lambda = 1, 2, 3, \dots$) is the number of proper colorings $\varphi : V \to \{1, \dots, \lambda\}$ of G with at most λ colors. Here, two colorings are counted as different even if they yield the same color classes by renumbering the colors.

EXAMPLE

E2: The chromatic polynomials of the edgeless graph and the complete graph on n vertices are, respectively,

$$P(\overline{K}_n, \lambda) = \lambda^n \quad \text{and} \quad P(K_n, \lambda) = \binom{\lambda}{n} n! = \lambda(\lambda - 1) \cdots (\lambda - n + 1)$$

FACTS

F39: (Deletion-Contraction Formula) For every graph $G = (V, E)$ and every edge $e \in E$, we have $P(G, \lambda) = P(G - e, \lambda) - P(G/e, \lambda)$, where '$-e$' and '$/e$' mean the deletion and contraction of edge e, respectively.

F40: [Bi12] If the graph G has n vertices, then $P(G, \lambda)$ is a polynomial of degree n in λ, with integer coefficients, and $\chi(G)$ is the smallest natural number λ such that $P(G, \lambda) \neq 0$.

F41: (Golden Identity) [Tu70] If G is a planar triangulation on n vertices, then

$$P(G, \tau + 2) = (\tau + 2)\tau^{3n-10} P^2(G, \tau + 1)$$

where $\tau = \dfrac{1 + \sqrt{5}}{2}$ denotes the golden ratio.

REMARK

R4: Although one can deduce from the Golden Identity that $P(G, \tau + 2) > 0$ holds for every planar triangulation ($\tau + 2 = 3.618...$), this does not seem to lead closer to a computer-free proof of the Four Color Theorem. (As a matter of fact, $P(G, \tau + 1)$ is nonzero for every connected graph G.)

5.2.3 Some Further Types of Coloring Problems

Variants of Proper Coloring

We briefly mention some further coloring concepts, most of them with only a few references.

DEFINITIONS

D10: A **Grundy coloring** of a graph is a proper vertex-coloring $\varphi : V \to \mathbb{N}$ such that every vertex v has a neighbor of color i for all $1 \leq i < \varphi(v)$. The **Grundy number** of a graph (also called **Grundy chromatic number** and **online chromatic number**) is the largest number of colors in a Grundy coloring.

D11: An **achromatic coloring** of a graph is a proper vertex-coloring such that each pair of color classes is adjacent by at least one edge. The largest possible number of colors in an achromatic coloring is called the **achromatic number**.

D12: A b-**coloring** of a graph is a proper vertex-coloring such that each color class contains a vertex adjacent to some vertices in every other color class. The b-**chromatic number** of a graph is the largest number of colors in a b-coloring.

D13: A λ-**coloring** (also called a **radio coloring** or an $L(2,1)$-**labeling**) is a vertex-coloring $\varphi : V \to \{0, 1, \ldots, k\}$ (i.e., $k + 1$ colors may be used) such that if $uv \in E$ then $|\varphi(u) - \varphi(v)| \geq 2$, and if vertices u and v have a common neighbor then $\varphi(u) \neq \varphi(v)$. More generally, for integers $r_1 \geq \cdots \geq r_d \geq 1$, an $L(r_1, \ldots, r_d)$-**labeling** is a vertex-coloring $\varphi : V \to \{0, 1, \ldots, k\}$ such that, if two vertices u, v are at distance $d(u, v) = i \leq d$ apart, then $|\varphi(u) - \varphi(v)| \geq r_i$. The difference between the largest and smallest color is called the **span** of the coloring.

D14: Given a set $T \subseteq \{0, 1, 2, \ldots\}$, a T-**coloring** is a mapping $\varphi : V \to \mathbb{N}$ such that $|\varphi(u) - \varphi(v)| \notin T$ for all edges $uv \in E$.

D15: A *circular C-coloring* is an assignment of a real number $\varphi(v)$ (with $0 \le \varphi(v) < C$) to each vertex v, in such a way that if uv is an edge then $(\varphi(v) - \varphi(u)$ modulo $C) \ge 1$. The **circular chromatic number** $\chi_c(G)$ of G (called **star chromatic number** in early papers) is the minimum C for which G has a circular C-coloring.

D16: An **harmonious coloring** of a graph is a partition of the vertex set into independent sets such that the union of any two induces at most one edge.

D17: A **rainbow connection** of a connected graph G is an edge coloring such that any two vertices are joined by a path on which no color occurs more than once. A state-of-the-art account on this subject is given in [LiSu12].

FACTS

F42: The smallest possible number of colors in a Grundy coloring of a graph G is just $\chi(G)$. On the other hand, the algorithmic complexity of finding the Grundy number – the largest number of colors – has not been determined so far. See, e.g., [ChSe79], [HeHeBe82]; and see also [Si83] with a different terminology.

F43: It is NP-hard to determine the achromatic number, even for trees ([CaEd97]), but polynomial-time solvable for trees of bounded degree ([CaEd98]). It is also hard to *approximate* the achromatic number on a general input graph within a factor $2 - \epsilon$, for any $\epsilon > 0$ ([KoKr01]).

F44: [HaReSe12] For any positive integer p there exists a constant Δ_p such that every graph with maximum degree $\Delta \ge \Delta_p$ has an $L(p, 1)$-labeling with span at most Δ^2. (It remains an open problem to prove or disprove that every graph with maximum degree Δ admits an $L(2, 1)$-labeling with span at most Δ^2, for *all* $\Delta \ge 2$, conjectured by Griggs and Yeh [GrYe92].)

F45: For every graph G, the circular chromatic number satisfies $\chi(G) - 1 < \chi_c(G) \le \chi(G)$. In this way, χ_c is a refinement of χ, and χ is an approximation of χ_c. Both bounds are tight: If G is uniquely colorable, then $\chi_c(G) = \chi(G)$ ([Zh96]); if G is k-critical and has girth g, then $\chi_c(G) \le \chi(G) - 1 + 1/s$, where $s = \lfloor (g - 1)/(k - 1) \rfloor$, i.e., for fixed chromatic number, large girth makes a critical graph have χ_c close to $\chi - 1$ ([StZh96]).

F46: The values of circular chromatic number form a dense set: Among the rational numbers r, for any $r \ge 2$ and for any integer g there is a graph with girth at least g and with $\chi_c = r$ ([StZh96]); for any $2 \le r \le 4$ there is a planar graph with $\chi_c = r$ ([Mo97, Zh99]); and for any $2 \le r \le n - 1$ there is a K_n-minor-free graph with $\chi_c = r$ ([LiPaZh03]).

REMARKS

R5: We refer to [IrMa99] and [KrTuVo02] for various estimates and complexity results regarding b-colorings (e.g., hardness on connected bipartite graphs).

R6: For results and further references on λ-colorings, see the surveys [Ye06], [GrKr09], and [Ca11].

R7: For results on T-colorings, see, e.g., [Te93], [Wa96], and [ChLiZh99]. It is worth noting that in the list coloring version of T-coloring, the case $0 \notin T$ leads to interesting questions.

R8: Results on circular colorings are surveyed in [Zh01]. The digraph version of the concept has also been introduced; see, e.g., [BoFiJuKaMo04].

R9: The goal of harmonious coloring investigations is to minimize the number of vertex classes. See [Ed97] for a survey.

Graph Homomorphisms

DEFINITION

D18: A *homomorphism* from a graph $G = (V, E)$ to a graph $H = (X, F)$, where loops may occur in both G and H, is a vertex mapping $\varphi : V \to X$ such that

$$\varphi(u)\varphi(v) \in F \quad \text{for all } uv \in E$$

A homomorphism into the complete graph K_k can be viewed as a proper k-coloring and vice versa. Moreover, a homomorphism into H is also called an H-*coloring of* G.

D19: A *list homomorphism* from $G = (V, E)$ to $H = (X, F)$ is a homomorphism φ with the further constraint that each $v \in V$ has a prescribed list $L_v \subseteq X$ such that $\varphi(v) \in L_v$.

NOTATION: The notation $G \to H$ means that G has at least one H-coloring, and $G \not\to H$ denotes that G is not H-colorable. The concept is extended to digraphs in a natural way (i.e., where uv and $\varphi(u)\varphi(v)$ are *ordered* pairs).

FACT

F47: [NeTa99] Let G_1 and G_2 be graphs such that $\chi(G_2) \geq 3$ and $G_1 \to G_2$ but $G_2 \not\to G_1$. Then there exists a graph G with $G_1 \to G \to G_2$ and $G_2 \not\to G \not\to G_1$.

For complexity results on H-coloring, see §5.2.5. Detailed discussion and references can be found in the book [HeNe04] and in the surveys [He03], [HeNe08], and [He12].

Coloring with Costs

DEFINITIONS

D20: A *cost set* $C = \{c_1, c_2, \dots\}$ associates a cost $c_i > 0$ with each color i. It is assumed without loss of generality that $0 < c_1 < c_2 < \cdots$, and also that $|C| \geq \chi(G)$.

D21: Given a graph $G = (V, E)$ and a cost set C, the *cost of a coloring* $\varphi : V \to \mathbb{N}$ is the sum $\sum_{v \in V} c_{\varphi(v)}$. We denote by $\Sigma_C(G)$ the smallest possible cost of a proper vertex-coloring φ of G. If $C = \mathbb{N}$, this notion simplifies to $\Sigma(G) := \min_\varphi \sum_{v \in V} \varphi(v)$ and usually is called the *chromatic sum* or *color cost* of G.

D22: The *cost chromatic number* of G with respect to a cost set C is the smallest possible number of colors in a minimum-cost coloring. If $C = \mathbb{N}$, this parameter is often called the *strength* or *chromatic strength* of G.

FACTS

F48: [ThErAlMaSc89] If G is connected and has m edges, then

$$\left\lceil \sqrt{8m} \right\rceil \leq \Sigma(G) \leq \left\lfloor \frac{3}{2}(m+1) \right\rfloor$$

F49: [MiMo97] For every finite cost set C there exists a tree T whose cost chromatic number is equal to $|C|$.

F50: [Tu90] For every $s \geq 2$, the minimum number of vertices in a tree of strength s equals $((2+\sqrt{2})^{s-1} - (2-\sqrt{2})^{s-1})/\sqrt{2}$. Moreover, for every $s \geq 3$, there exist precisely two trees of strength s, which are minimal in the sense that every tree of strength at least s is contractible to at least one of them.

F51: [MiMo97] Every tree of maximum degree Δ has cost chromatic number at most $\lceil \Delta/2 \rceil + 1$.

F52: The bound in Fact F51 is tight for every Δ with the cost set $C = \{1, 1.1, 1.11, \dots\}$ ([MiMoSc97]) and also with $C = \mathbb{N}$ ([JiWe99]).

REMARK

R10: Beside the chromatic sum, various notions concerning coloring with costs have been motivated by scheduling problems. In some of them, given numbers of colors have to be assigned to the vertices. For results of this type, see, e.g., [Ma02].

Vertex Ranking

DEFINITION

D23: A *vertex ranking* of graph $G = (V, E)$ is a (necessarily proper) coloring $\varphi : V \to \mathbb{N}$ with the property that for any two vertices u, v of the same color, every u–v path contains some vertex z with $\varphi(z) > \varphi(u)$. The smallest possible number of colors, called *ranking number*, will be denoted by $\chi_r(G)$. In the directed analogue for digraphs, the requirement is put on directed paths only.

EXAMPLE

E3: [BoDeJaKlKrMüTu98] The line graph of K_n ($n \geq 2$) has ranking number $\frac{1}{3}(n^2 + g(n))$, where the function $g(n)$ is defined recursively with $g(1) = -1$, $g(2k) = g(k)$, and $g(2k+1) = g(k+1) + k$ for every natural number k. (No closed formula is available.)

FACTS

F53: For every graph G, $col(G) \leq \chi_r(G)$, and if the ranking number is at most k, then the graph is (km, m)-choosable for every $m \in \mathbb{N}$ ([TuVo96]).

F54: [BoDeJaKlKrMüTu98] If $\chi_r(G) = \chi(G)$, then also $\omega(G) = \chi(G)$. Moreover, $\chi_r(H) = \omega(H)$ holds for every induced subgraph H of G if and only if G contains no P_4 and C_4 as an induced subgraph.

Non-Repetitive (Thue) Colorings

DEFINITION

D24: A ***non-repetitive coloring*** of a graph $G = (V, E)$ is a vertex coloring φ such that, for all $r \geq 1$, no path $v_1 v_2 \ldots v_{2r}$ with an even number of vertices satisfies $\varphi(v_i) = \varphi(v_{r+i})$ for all $1 \leq i \leq r$.

D25: The ***Thue chromatic number*** of G, denoted by $\pi(G)$, is the smallest k such that G has a non-repetitive k-coloring.

FACTS

F55: [Th06] If P_n is a path on $n \geq 4$ vertices, then $\pi(P_n) = 3$. (For cycles, $\pi(C_n) = 4$ for $n = 5, 7, 9, 10, 14, 17$, and $\pi(C_n) = 3$, otherwise [Cu02].)

F56: The value of $\pi(G)$ is finite for graphs of maximum degree at most Δ [AlGrHaRi02] (the largest value is between $c_1 \frac{\Delta^2}{\log \Delta}$ and $c_2 \Delta^2$) and for graphs of treewidth at most w [KüPe08] (then $\pi(G) \leq 4^w$).

F57: [BaWo08] Every graph has a subdivision which is non-repetitively 3-colorable.

CONJECTURE

C3: The value of $\pi(G)$ is bounded on the class of planar graphs.

REMARK

R11: For planar graphs of order n the currently best upper bound seems to be $O(\log n)$ ([DuFrJoWo12]). Further results and conjectures are surveyed in [Gr07].

Partial Colorings and Extensions

Partial colorings and their extensions are of interest in several aspects: from the viewpoint of theory and also as a proof technique. We mention here some concepts with only a few references.

DEFINITIONS

D26: A ***partial coloring*** of graph $G = (V, E)$ on a vertex subset $W \subseteq V$ is a coloring $\varphi_W : W \to \mathbb{N}$.

D27: In the ***Precoloring Extension*** problem, abbreviated ***PrExt***, we are given a graph G, a **color bound** k, and a proper partial coloring φ_W. The question is whether φ_W can be extended to a proper k-coloring of the entire G.

D28:   For a nonnegative integer t, the problem t-***PrExt*** is the restricted version of Precoloring Extension where the given partial coloring uses each color at most t times. (Hence, 0-PrExt is an equivalent formulation of asking whether the graph in question is k-colorable.)

D29:   In an ***on-line coloring*** the vertices $v_1, v_2, \ldots, v_n$ of graph G are received one by one in some unknown order. When v_i appears, we also get the information which its neighbors in $\{v_1, \ldots, v_{i-1}\}$ are. A color has to be assigned to v_i without any information on its adjacencies to the v_j, $i < j \leq n$.

FACTS

F58:   If $G = (V, E)$ is k-colorable, and $t \leq k$ is a positive integer, then G has a proper partial coloring on at least $\frac{t}{k} |V|$ of its vertices.

F59:   [AlGrHa00] If G is a k-colorable graph with n vertices, then for every list t-assignment it has a partial proper list coloring on at least $\left(1 - \left(\frac{k-1}{k}\right)^t\right) n$ vertices.

F60:   If G is a k-choosable graph with n vertices, then it has a partial list coloring on more than $\frac{6}{7} \frac{t}{k} n$ vertices, for every list t-assignment $(1 \leq t < k)$ [Ch99]. Moreover, if G has maximum degree k or the union of lists contains at most k colors, then at least tn/k vertices can be colored from their lists [Ja01]. (It is conjectured in [AlGrHa00] that this lower bound is valid for all G and all t-assignments.)

F61:   (Cf. §5.1.5) If T is a tree of radius 2, then there exists an on-line χ-binding function on the class of graphs not containing T as an induced subgraph ([KiPeTr94]), but the class of induced-P_6-free graphs is not on-line χ-bound ([GyLe91]).

REMARKS

R12:   In several graph classes, efficiently testable necessary and sufficient conditions can be given for the extendability of partial colorings. Details can be found in [HuTu96].

R13:   Results on on-line coloring are surveyed in [Ki98a].

R14:   An on-line version of list coloring is ***paintability***. It can be described as a two-person game played on a graph $G = (V, E)$ where each vertex $v \in V$ has an initial bound $f(v) \in \mathbb{N}$, which corresponds to f-choosability in the off-line setting. In each round, Alice marks a subset of vertices having positive current f-value, and Bob removes an independent subset of the marked ones; the marked but non-removed vertices decrease their f-value by 1 and erase the marking. Alice wins if some f-value decreases to zero in some step; Bob wins if all vertices get removed. Graph G is k-***paintable*** if Bob has a winning strategy on G when $f(v) = k$ holds for all $v \in V$. Every k-choosable graph is k-paintable, but the converse implication is not valid. Some sufficient conditions on list colorability can be extended for paintability, too; e.g., the kernel method is a sequential selection of monochromatic subsets and can easily be interpreted in the on-line scenario. Further major results on list colorings can also be generalized to paintability; see, e.g., the seminal papers [Sc09] and [Sc10].

R15:   An on-line version of Precoloring Extension can be presented as a game where an uncolored graph G and a color bound k are given, and two persons alternately color the vertices so that a proper partial coloring is obtained after each step. The game ends when no legal move is available. In the **Achievement game** the person making the last move wins. In the **Avoidance game** the person making the last move loses. Although these games were introduced two decades ago ([HaTu93]), still very little is known about them.

R16:   The **game chromatic number** $\chi_g(G)$ of graph $G = (V, E)$ is the smallest number k of colors for which the first player can force to color the entire vertex set V, in the game where two players alternately color the vertices properly. (It follows, in particular, that if $|V|$ is odd, then the first player has a winning strategy in the Achievement game, and if $|V|$ is even, then he/she can win the Avoidance game.) It is known, for example, that $\chi_g(T) \leq 4$ for every tree T ([FaKeKiTr93]), and the analogous upper bound $ch_g(T) \leq 4$ remains valid in the list coloring version of the game, too (i.e., where the color of every $v \in V$ has to be chosen from a given list L_v). While the graphs with $\chi_g(G) = 2$ are the star forests, it is already a little more complicated to characterize the graphs with $ch_g(G) = 2$ ([BoSiTu07]). Some strategies for the coloring game to obtain bounds on χ_g are surveyed in [BaGrKiZh07].

Coloring Cubic Graphs with Triple Systems

DEFINITIONS

D30:   A **Steiner triple system** is a pair $\mathcal{S} = (X, \mathcal{B})$ where X is an underlying set called the set of points and $\mathcal{B}$ is a collection of 3-element subsets of X, called blocks, such that any two $x, x' \in X$ occur together in *precisely one* block $B \in \mathcal{B}$. Less restrictively, a **partial triple system** $(X, \mathcal{B})$ with a collection $\mathcal{B}$ of 3-element subsets of X requires that any two $x, x' \in X$ are contained in *at most one* $B \in \mathcal{B}$. Trivial system means $|X| = 3$ and $\mathcal{B} = \{X\}$.

D31:   For a partial triple system $\mathcal{S} = (X, \mathcal{B})$, an **$\mathcal{S}$-coloring** of a cubic (i.e., 3-regular) graph $G = (V, E)$ is an assignment $\phi : E \to X$ such that the colors of any three edges meeting at a vertex form a block of $\mathcal{S}$.

D32:   A **k-line Fano coloring** is one that uses at most k different blocks (lines) as color patterns around the vertices. (The Fano plane is the smallest nontrivial Steiner triple system, which has seven points and seven blocks.)

D33:   The **Cremona-Richmond configuration** is a symmetrical configuration $\mathcal{R}$ of fifteen points and fifteen lines such that each line contains three of the fifteen points and each point lies on three of the fifteen lines. Combinatorially, it can be represented by the partial triple system consisting of fifteen points corresponding to the 2-element subsets of the set $\{1, 2, 3, 4, 5, 6\}$ and the fifteen "perfect matchings," i.e., blocks formed by triples of points that cover all the six elements of $\{1, 2, 3, 4, 5, 6\}$. (The configuration was discovered in the context of real projective geometry in the 19th century and can be realized in the 4-dimensional projective space over R.)

FACTS

F62: [HoŠk04] Every bridgeless cubic graph has an $\mathcal{S}$-coloring for every nontrivial Steiner triple system $\mathcal{S}$.

F63: [MáŠk05] Every bridgeless cubic graph G has a 6-line Fano coloring. Moreover, the bridgeless cubic graphs have a 4-line Fano coloring if and only if they contain three perfect matchings with no edge in common. That is, the existence of 4-line Fano colorings is equivalent to the Fan-Raspaud conjecture.

F64: [KrMáPaRaSeŠk09] A bridgeless cubic graph has a coloring with the blocks of the Cremona-Richmond configuration if and only if it has a family of six perfect matchings such that each edge belongs to precisely two of those matchings. Hence, the bridgeless cubic graphs are $\mathcal{R}$-colorable if and only if Fulkerson's conjecture ([Fu71]) is valid.

Neighbor-Distinguishing Colorings

REMARK

R17:   There are many variations of colorings with the aim to distinguish between vertices: vertex and edge colorings, with or without lists, distinguishing all vertices or only the adjacent ones, assuming or not assuming that the coloring is proper, considering for distinction the set or the multiset of colors in the neighborhood, using abstract colors or natural numbers or elements of a group, etc. For example, it is proved in [GyPa09] and [HoSo10] that if G is a connected graph with $\chi(G) \geq 3$, then exactly $\lceil \log_2 \chi(G) \rceil + 1$ colors are needed as the minimum in a (non-proper) edge coloring such that, for any two adjacent vertices u, v in G, there is a color incident with precisely one of u and v (i.e., the color sets on their stars are distinct). For bipartite graphs, two or three colors suffice in a coloring with this property ([GyHoPaWo08]).

Maximizing the Number of Colors

REMARKS

R18:   Coloring constraints may be of opposite type as well. For example, if for each vertex v the number of distinct colors occurring in the neighborhood of v is bounded from *above*, the parameter of interest is the *largest* possible number of colors which can be achieved on the entire vertex set. Results of this kind are surveyed in [BuTu11]. See also *mixed hypergraphs* and their generalizations in §5.2.4 below.

R19:   A pair of vertex and edge colorings, of interest for a special reason, is as follows. A **3-consecutive vertex coloring** of graph $G = (V, E)$ is a mapping $\varphi : V \to \mathbb{N}$ such that for any 3-vertex path $P = \langle v_1, v_2, v_3 \rangle$ in G, the middle vertex v_2 has the same color as one (or both) of the ends v_1, v_3 of P. Similarly, a **3-consecutive edge coloring** of G is a mapping $\phi : E \to \mathbb{N}$ such that for any three consecutive edges e_1, e_2, e_3 (forming a cycle C_3 or a path P_4 in this order), the middle edge e_2 has the same color as one (or both) of e_1 and e_3. (The conditions force every triangle to be monochromatic in both φ and ϕ.) As indicated already in §5.1, these colorings are very special in the following sense. If G has minimum degree at least two, then there is a bijection between the 3-consecutive *vertex* colorings of G and the 3-consecutive *edge* colorings of its line graph $L(G)$ ([BuSaTuDoPu12]). This is the opposite of the standard correspondence between edge colorings of G and vertex colorings of $L(G)$, which is valid for many kinds of colorings. Since both φ and ϕ are allowed to make the entire G monochromatic, the relevant parameter to consider for 3-consecutive colorings is the *maximum* possible number of colors.

R20: The maximum number of colors — as well as colorability — is of interest in the context of *interval edge coloring*, too. For a graph $G = (V, E)$, an ***interval edge coloring*** is a proper edge coloring $\phi : E \to \mathbb{N}$ such that, for each $v \in V$, the colors on the edges incident with v form an *interval* of consecutive integers. Recent bounds and further references can be found in [KaPe12].

Partitions with Weaker Requirements

REMARKS

R21: There are many papers dealing with vertex- or edge-partitions into parts that are not necessarily independent, but satisfy some weaker properties. Usually it is assumed that the property to be satisfied in each part is *hereditary* or *induced-hereditary*, i.e., if it holds for a graph H then it also holds for all (induced) subgraphs of H. A detailed discussion on the general theory can be found in the survey [BoBrFrMiSe97].

R22: Sometimes conditions are imposed on the vertex degrees in each partition class. For results and references on this kind of problem, see, e.g., the surveys [Wo01] and [BaTuVa10].

5.2.4   Colorings of Hypergraphs

Beside some results on the coloring of finite set systems (*hypergraphs*), here we also mention the basic definitions and a few facts from the recently fast-developing theory of *mixed hypergraph coloring* (and beyond). See the informative monograph [Vo02] for a detailed account, and the survey [BaBuTuVo10] for open problems.

DEFINITIONS

D34: A ***hypergraph*** $\mathcal{H} = (X, \mathcal{F})$ has vertex set X; its edge set $\mathcal{F}$ consists of subsets of X. We assume that $\mathcal{F} \neq \emptyset$ and that $|F| \geq 2$, for all $F \in \mathcal{F}$.

D35: A ***proper vertex k-coloring of a hypergraph*** $\mathcal{H}$ is a mapping $\varphi : X \to \{1, 2, \ldots, k\}$ such that no edge of $\mathcal{H}$ is monochromatic. Equivalently, it is a vertex partition into k classes such that no color class contains any edge.

D36: A ***proper edge-coloring of a hypergraph*** $\mathcal{H}$ is an edge partition such that the edges in the same class are mutually vertex-disjoint.

D37: The **chromatic number** $\chi(\mathcal{H})$ and **chromatic index** $\chi'(\mathcal{H})$ are the smallest numbers of colors in a proper vertex and proper edge-coloring, respectively. **List coloring**, **choice number**, **choice index**, (a,b)-**choosability**, etc., can be defined for hypergraphs analogously.

D38: A hypergraph is r-**uniform** if every edge has precisely r vertices.

D39: The **complete** r-**uniform hypergraph** of order n $(n \geq r)$, denoted $\mathcal{K}_n^r$, has $|X| = n$, and its edge set consists of all the r-element subsets of X.

D40: A **mixed hypergraph** $\mathcal{H} = (X, \mathcal{C}, \mathcal{D})$ has vertex set X, and two types of edges: the C-**edges** in $\mathcal{C}$ and the D-**edges** in $\mathcal{D}$, respectively. It is called a **bi-hypergraph** if $\mathcal{C} = \mathcal{D}$, a C-**hypergraph** if $\mathcal{D} = \emptyset$, and a D-**hypergraph** if $\mathcal{C} = \emptyset$. We shall assume that at least one of $\mathcal{C}$ and $\mathcal{D}$ is nonempty, and also that every (C- and D-) edge has at least two vertices.

D41: A **strict** k-**coloring** of a mixed hypergraph is a vertex-coloring with exactly k colors, such that every C-edge has two vertices with a <u>c</u>ommon color and every D-edge has two vertices with <u>d</u>ifferent colors. (In this way, the D-hypergraphs are just the hypergraphs in the usual sense with respect to proper vertex coloring.)

D42: A mixed hypergraph is said to be **colorable** if it admits at least one strict coloring, and **uncolorable** if it doesn't.

D43: If a mixed hypergraph $\mathcal{H}$ is colorable, then the smallest and largest number of colors in a strict coloring is called the **lower** and **upper chromatic number**, denoted $\chi(\mathcal{H})$ and $\overline{\chi}(\mathcal{H})$, respectively.

D44: A mixed hypergraph $\mathcal{H}$ is said to be **uniquely colorable** if $\chi(\mathcal{H}) = \overline{\chi}(\mathcal{H})$ and $\mathcal{H}$ has only one strict coloring (apart from renaming the colors).

D45: More general models for hypergraph coloring are the **color-bounded** and **stably bounded hypergraphs**. The latter means a six-tuple $\mathcal{H} = (X, \mathcal{F}, s, t, a, b)$, where $(X, \mathcal{F})$ is a hypergraph, while each of the four color-bound functions s, t, a, b assigns positive integers to the edges. A proper vertex coloring of $\mathcal{H}$ requires that for each edge $F \in \mathcal{F}$ the number of distinct colors assigned to the vertices of F is at least $s(F)$ and at most $t(F)$; moreover, the cardinality of the largest monochromatic subset inside F is between $a(F)$ and $b(F)$. The color-bounded hypergraphs are just the four-tuples $\mathcal{H} = (X, \mathcal{F}, s, t)$. The concepts of colorability, upper chromatic number, etc., extend to color-bound and stably bound hypergraphs, too, in a natural way.

EXAMPLES

E4: If $\mathcal{H}$ is a C-hypergraph, then $\chi(\mathcal{H}) = 1$, since the entire vertex set may be colored with the same color, and if $\mathcal{H}$ is a D-hypergraph, then $\overline{\chi}(\mathcal{H}) = |X|$, because its vertices may get mutually distinct colors.

E5: The complete hypergraph (viewed as a D-hypergraph) has $\chi(\mathcal{K}_n^r) = \lceil \frac{n}{r-1} \rceil$, and when viewed as a C-hypergraph, it has $\overline{\chi}(\mathcal{K}_n^r) = r - 1$. If $\mathcal{C} = \mathcal{K}_n^p$, $\mathcal{D} = \mathcal{K}_n^q$, and $n > (p-1)(q-1)$, then $\mathcal{H}$ is uncolorable.

E6: [ErLo75] Let $X = X_1 \cup \cdots \cup X_r$, with $|X_i| = i$ for all $1 \le i \le r$, and let an r-element set be an edge if and only if for some i it contains X_i and has precisely one vertex in every X_j with $i < j \le r$. This r-uniform hypergraph is not 2-colorable, for all $r \ge 2$.

E7: [TuVo00] The following mixed hypergraphs $\mathcal{H} = (X, \mathcal{C}, \mathcal{D})$ are uncolorable: starting from a k-chromatic graph $G = (V, E)$, set $X = V$, $\mathcal{D} = E$, and let a k-subset $Y \subset V$ be a C-edge if the subgraph of G induced by Y has a hamiltonian path.

E8: [KoKü01] The following *planar* hypergraph $\mathcal{H} = (X, \mathcal{C}, \mathcal{D})$ has strict colorings with two and four colors, but not with three colors: $X = \{a, b, c, d, e, f\}$, $\mathcal{C} = \{abe, bce, bcf, cdf\}$, $\mathcal{D} = \{ab, bc, cd, da, af, de\}$.

FACTS

F65: [Er64] For every $r \ge 2$, there exists a non-2-colorable r-uniform hypergraph with fewer than $r^2 2^{r+1}$ edges.

F66: [RaSr00] For sufficiently large values of r, every r-uniform hypergraph with at most $0.7\sqrt{r/\ln r}\, 2^r$ edges is 2-colorable, and efficient algorithms can also be designed to find a proper 2-coloring. (The previous bound $r^{\frac{1}{3}-\varepsilon} 2^r$ is given in [Be78].)

F67: [AlKo11] For every $r, k \in \mathbb{N}$ there is a $d_r(k)$ such that if an r-uniform hypergraph has average degree at least $d_r(k)$ and any two of its edges share at most one vertex, then it is not k-choosable.

F68: (See [Lo68, NeRö79] in §5.1.4.) For every triple of integers $r, k, g \ge 3$, there exists an r-uniform hypergraph with chromatic number at least k and girth at least g.

F69: [KrVu01] For every positive integer r and for every real-valued function $p = p(n)$ with
$$n^{-(r-1)^2/(2r)+\epsilon} \le p \le 0.9 \quad (\epsilon > 0 \text{ fixed})$$
the random r-uniform hypergraph with n vertices and edge probability $p = p(n)$ has
$$ch(\mathbf{H}_{n,p}) = (1 + o(1))\chi(\mathbf{H}_{n,p}) \quad \text{with probability } 1 - o(1)$$
Also, if $p \ge Cn^{1-r}$ for a sufficiently large constant C, then
$$ch(\mathbf{H}_{n,p})/\chi(\mathbf{H}_{n,p}) = (1 + \delta_p)\, r^{1/(r-1)} \quad \text{almost surely}$$
where $\delta_p \to 0$ as $n^{r-1}p \to \infty$.

F70: [KoSt00] If a hypergraph on n vertices is $(k+1)$-chromatic, critical, and does not contain any 2-element edges, then the number of its edges is at least $(k - 3/\sqrt[3]{k})\, n$.

F71: [BuTu09] For every fixed $k \ge 2$, and also for $k = o(n^{1/3})$, the minimum number of edges in a k-uniform C-hypergraph of upper chromatic number $k-1$ is $(1 + o(1))\frac{2}{k}\binom{n}{k-1}$ as $n \to \infty$. In other words, this is the minimum number of edges in a k-uniform hypergraph whose edges cover all vertex partitions with k non-empty classes. For $k = 3$ the exact minimum is $\lceil n(n-2)/3 \rceil$ ([DiLiRaZh06], [ArTe07]) and for $k = 2$ it is $n - 1$ (tree graphs). For $k = n - 2$ the minimum is equal to the smallest number of edges in a graph whose complement has girth at least 5; i.e., it is the complementary value of the Turán function $ex(n, \{C_3, C_4\})$, whose determination is a famous open problem in extremal graph theory.

F72: [TuVo00] The largest possible value of $\overline{\chi}(\mathcal{C}) - \chi(\mathcal{D})$ in *uncolorable* mixed hypergraphs $\mathcal{H} = (X, \mathcal{C}, \mathcal{D})$ with $|X| = n$ vertices is equal to $n - 4$.

F73: [TuVo00] The List Coloring problem can be reduced to the colorability problem of mixed hypergraphs, in the following sense. For every graph $G = (V, E)$ and for every list assignment L on G, there can be constructed (in linear time) a mixed hypergraph $\mathcal{H}$ on $|V| + \left|\bigcup_{v \in V} L_v\right|$ vertices such that $\mathcal{H}$ is colorable if and only if G is L-colorable.

F74: [TuVoZh02] Every colorable mixed hypergraph is the induced subhypergraph of a uniquely colorable mixed hypergraph. More generally, given a colorable mixed hypergraph $\mathcal{H}$ and two of its coloring partitions, say P with t classes and P' with t' classes such that P' is a refinement of P (i.e., every class in P' is a subset of some class in P), there exists a mixed hypergraph $\mathcal{H}'$ with $\chi(\mathcal{H}') = t$ and $\overline{\chi}(\mathcal{H}') = t'$ that contains $\mathcal{H}$ as an induced subhypergraph; moreover, $\mathcal{H}$ has just one coloring with t and also with t' colors, and those two colorings induce P and P' on $\mathcal{H}$.

F75: [JiMuTuVoWe02] For every finite set $S \subset \mathbb{N} \setminus \{1\}$ there exists a mixed hypergraph $\mathcal{H}$ such that $\mathcal{H}$ admits a strict k-coloring if and only if $k \in S$. Moreover, for every $r \geq 3$ and for every $S = S' \cup S''$ with $S' \subset \mathbb{N} \setminus \{1, 2, \ldots r-1\}$, and $S'' = \emptyset$ or $S'' = \{\ell, \ldots, r-1\}$ for some $2 \leq \ell \leq r - 1$, there exists an r-uniform bi-hypergraph $\mathcal{H}$ such that $\mathcal{H}$ admits a strict k-coloring if and only if $k \in S$ ([BuTu08]).

F76: [Kr04] For every finite sequence $(r_1, r_2, \ldots, r_\ell)$ of nonnegative integers with $r_1 = 0$ and $r_\ell > 0$, there exists a mixed hypergraph $\mathcal{H}$ such that $\overline{\chi} = \ell$, and for each $1 \leq k \leq \ell$ there are exactly r_k different strict k-colorings of $\mathcal{H}$ apart from renumbering of colors. (For the cases where $r_k \in \{0, 1\}$ holds for all k, the smallest possible number of vertices is investigated in a series of papers starting with [ZhDiWa12].)

F77: ([KrKrPrVo06]; A. Niculitsa) If each edge of a mixed hypergraph $\mathcal{H}$ induces a subtree in some fixed tree, then the set of values k for which $\mathcal{H}$ has a strict k-coloring consists of *consecutive* integers.

F78: As a function of $k \in \mathbb{N}$, the number of proper colorings $\varphi : X \to \{1, \ldots, k\}$ of a colorable stably bounded hypergraph $\mathcal{H} = (X, \mathcal{F}, \boldsymbol{s}, \boldsymbol{t}, \boldsymbol{a}, \boldsymbol{b})$ with at most k colors is a polynomial of degree $\overline{\chi}(\mathcal{H})$. (The subsets of the color-bounding functions $\boldsymbol{s}, \boldsymbol{t}, \boldsymbol{a}, \boldsymbol{b}$ generate subclasses of stably bounded and mixed hypergraphs, to which the corresponding subclasses of chromatic polynomials are associated in the natural way. Those subclasses of polynomials are partially ordered under inclusion; the Hasse diagram of this poset is determined in [BuTu07].)

REMARKS

R23: Starting with [MiTu97], several papers study the strict colorings of Steiner systems (block designs). It is not known, for example, whether there exists an infinite family of Steiner quadruple systems S such that the bi-hypergraphs with $\mathcal{C} = \mathcal{D} = S$ can have any large upper chromatic number. For a survey, see [MiTuVo03].

R24: The perfectness of mixed hypergraphs is introduced via the comparison of independence number and upper chromatic number, since $\overline{\chi}(\mathcal{H})$ cannot exceed the independence number of $(X, \mathcal{C})$ for any $\mathcal{H} = (X, \mathcal{C}, \mathcal{D})$. The characterization problem of C-perfect hypergraphs, raised in [Vo95a], is still open, even in the 3-uniform case. For subtrees of a tree, C-perfectness is characterized ([BuTu10]), with several algorithmic consequences, as well.

R25: In a very general model of hypergraph coloring, called *pattern hypergraph*, the allowed color partitions can be specified locally for each edge ([DvKáKrPa10]).

Clique Hypergraphs

DEFINITION

D46: The *clique hypergraph* of graph $G = (V, E)$ has the same vertex set V; a subset $H \subseteq V$ of cardinality at least two is considered to be a hyperedge if it induces a maximal complete subgraph in G. Its chromatic number will be denoted by $\chi_C(G)$.

OPEN PROBLEM

P4: [DuSaSaWo91] Does there exist a constant k such that $\chi_C(G) \leq k$ for every *perfect* graph G?

FACTS

F79: [BaGrGyPrSe04] The following upper bounds are valid on $\chi_C(G)$ for every graph G: the domination number plus one, the independence number unless G is a complete graph or $G = C_5$, and $2\sqrt{n}$ (on n vertices).

F80: $\chi_C(G) \leq 2$ holds for comparability graphs ([DuSaSaWo91]), claw-free graphs without induced odd cycles longer than 3, and also the complements of such graphs ([BaGrGyPrSe04]).

F81: [DuKiTr91] For the complements of comparability graphs, $\chi_C(G) \leq 3$.

F82: [MoSk99] If the graph G is planar, then $\chi_C(G) \leq 3$, and the clique hypergraph is 4-choosable also if G is imbeddable in the projective plane.

5.2.5 Algorithmic Complexity

FACTS

F83: Bipartite (i.e., 2-colorable) graphs can be recognized and properly 2-colored in linear time.

F84: The 2-choosable graphs can be recognized in linear time (by the structural characterization theorem in [ErRuTa79]).

F85: [Ma68, FiSa69] The coloring number (see §5.1.2) can be determined in polynomial time.

F86: [Zh11] It is NP-complete to decide whether $\chi(G) < col(G)$ holds for a generic input graph G.

F87: [Ka72] For every $k \geq 3$, it is NP-complete to decide whether $\chi(G) \leq k$. Also, it is NP-complete to decide whether a planar graph of maximum degree 4 is 3-colorable ([GaJoSt76]).

F88: [KoYa12] Let $k \geq 4$ and G a graph. If each subgraph $H \subseteq G$ has fewer than $\frac{(k+1)(k-2)}{2(k-1)}|V(H)| - \frac{k(k-3)}{2(k-1)}$ edges, then G can be properly $(k-1)$-colored in polynomial time, in at most $\mathcal{O}(k^{3.5} n^{6.5} \log n)$ steps. (The condition on the sparseness of subgraphs H is set to ensure that G cannot contain any k-critical subgraph and hence is $(k-1)$-colorable; cf. Fact F31 of §5.1.3.)

F89: [Kö16] Every bipartite graph G can be properly edge-colored in polynomial time with $\Delta(G)$ colors.

F90: [Ho81] ($k = 3$), [LeGa83] ($k \geq 4$) It is NP-complete to decide whether a k-regular graph is edge k-colorable.

F91: [McSa94] For every $k \geq 3$ it is NP-complete to decide whether a k-regular bipartite graph admits a total coloring with $k + 1$ colors.

F92: [BrLo98] For every fixed k, it can be decided in polynomial time whether G admits a vertex partition into k-element sets, each of which is independent in G or in its complement $\overline{G}$.

F93: [GrLoSc84] The chromatic number of perfect graphs can be determined in polynomial time.

F94: [KrKrTuWo01] For a fixed graph H, let $Forb(H)$ be the class of graphs not containing any induced subgraphs isomorphic to H. Determining $\chi(G)$ for the graphs $G \in Forb(H)$ is polynomial if H is an induced subgraph of P_4 or of $P_3 \cup K_1$ (the path of length two plus an isolated vertex), and is NP-hard for any other H.

F95: Although the chromatic number of graphs containing no induced P_5 is NP-hard to compute (as implied by Fact F94), for every fixed k there is a polynomial-time algorithm that decides whether a P_5-free graph is k-colorable, and finds a proper k-coloring if it is ([HoKaLoSaSh10]).

F96: The 3-colorability of graphs without any induced P_6 subgraph can be decided, and a 3-coloring can be found if it exists, in polynomial time ([RaSc04]). On the other hand, deciding the 4-colorability of P_{12}-free graphs and the 5-colorability of P_8-free graphs are NP-complete ([LeRaSc07] and [SgWo01], respectively).

F97: [KrTu02] For every $k \geq 3$, it is NP-complete to decide whether the clique hypergraph of a *perfect* graph G with $\omega(G) = k$ is 2-colorable.

F98: [KrTu02] If G is planar, then the decision of 2-colorability (and the determination of chromatic number) of its clique hypergraph is polynomial-time solvable.

F99: Planar graphs can be properly 5-colored, and also properly 4-colored, in polynomial time. On the other hand ([VeWe92]), it is #P-complete to determine the number of proper 4-colorings of a planar graph.

F100: [Th94] Given a planar graph with a 5-assignment L on its vertices, a proper L-coloring can be found in linear time.

F101: [HeNe90] For every non-bipartite graph H it is NP-complete to decide whether an input graph G is H-colorable.

F102: [FeHeHu03] The graphs H (with or without loops) have been characterized into which the *list homomorphism* (list H-coloring) problem $G \to H$ for a generic input graph G can be decided in polynomial time. (There is a dichotomy between polynomial-time solvability and NP-completeness.) In particular, if all vertices of H have loops, then list H-coloring is solvable in polynomial time if and only if H is an interval graph ([FeHe98]), and if H has no loops, then list H-coloring is solvable in polynomial time if and only if the complement $\overline{H}$ of H is a circular-arc graph ([FeHeHu99]).

F103: [DyGr00] The number of homomorphisms $G \to H$ (where H is fixed and G is the input graph) can be determined in polynomial time if each component of H is a complete graph with loops at all of its vertices, or a complete bipartite graph without loops, or an isolated vertex; and it is #P-complete otherwise.

F104: [HeNeZh96] Suppose that the following property holds for the digraph H: A digraph G is *not* H-colorable if and only if there exists an oriented tree T such that $T \to G$ and $T \not\to H$. Then H-colorability is decidable in polynomial time.

F105: [GuWeWo92] There exist oriented trees T such that it is NP-complete to decide whether $G \to T$.

F106: For every $k \geq 3$, it is Π_2^p-complete to decide whether $ch(G) \leq k$ ([GuTa09]). It remains Π_2^p-complete for $k = 4$ on planar graphs and for $k = 3$ on triangle-free planar graphs ([Gu96]).

F107: [JaSc97] It is NP-complete to decide whether a complete bipartite graph with given lists on its vertices is colorable.

F108: [KrTu94] The List Coloring problem remains NP-complete on the instances satisfying all the following three conditions (also if restricted to planar graphs): every list has at most 3 colors, every color occurs in at most 3 lists, and every vertex has degree at most 3. On the other hand, both the decision and search versions of the problem can be solved in linear time if every list has at most 2 colors, or every color occurs in at most 2 lists, or every vertex has degree at most 2.

F109: [Lo73] For every $k \geq 2$, it is NP-complete to decide whether a hypergraph is k-colorable. It remains NP-complete for $k = 2$ on 3-uniform hypergraphs. For all $k \geq 2$ and $r \geq 2$ it is NP-complete also on r-uniform hypergraphs in which any two edges share at most one vertex, except for $k = r = 2$ ([PhRö84]). Also, deciding 14-colorability of a Steiner triple system is NP-complete.

F110: [TuVoZh02] It is NP-complete to decide whether a mixed hypergraph is colorable, and given a mixed hypergraph $\mathcal{H}$ with a strict coloring, it is coNP-complete to decide whether $\mathcal{H}$ is uniquely colorable.

F111: [Ma11] For every $k \geq 2$, on the clique hypergraphs of graphs it is Σ_2^p-complete to test k-colorability, and Π_3^p-complete to decide whether a given list assignment on the vertices admits a proper list clique coloring. (To decide whether the clique hypergraph of every induced subgraph is k-colorable is also Π_3^p-complete.)

F112: [BuTu09a] If a color-bounded interval hypergraph $\mathcal{H} = (X, \mathcal{F}, \boldsymbol{s}, \boldsymbol{t})$ is colorable, then $\chi(\mathcal{H}) = \max_{F \in \mathcal{F}} \boldsymbol{s}(F)$, and there is a polynomial-time algorithm that transforms any proper vertex coloring of $\mathcal{H}$ to one with $\chi(\mathcal{H})$ colors. (Time complexity results for coloring problems on several functional and structural subclasses of stably bounded hypergraphs — and also on subclasses of mixed hypergraphs — are systematically summarized in [BuTu13]).

F113: On an unrestricted input graph it is NP-hard to determine the chromatic sum, but it is polynomial on trees ([KuSc89]) and also on the line graphs of trees ([GiKu00]).

F114: [Ja97] If the cost set contains at least four colors, then on bipartite graphs it is NP-hard to determine the minimum cost of a proper coloring.

F115: [HuTu93] Precoloring extension on the complements of bipartite graphs, and also on split graphs, is solvable with exactly the same efficiency (in polynomial time) as the Bipartite Matching problem, but it is NP-complete on bipartite graphs. It remains NP-complete on bipartite graphs even if just 3 colors may be used ([Kr93]).

F116: [BiHuTu92] On interval graphs, 1-PrExt (see D28) is solvable in polynomial time, but 2-PrExt is NP-complete. (On unit interval graphs, the unrestricted PrExt problem is NP-complete [Ma06].)

F117: 1-PrExt is polynomial-time solvable on chordal graphs ([Ma07]) but NP-complete on permutation graphs ([Ja97]).

F118: On the class of trees, to decide whether there exists a labeling with span at most λ (i) for $L(r_1, r_2)$-labeling with fixed r_1, r_2 is polynomial-time solvable if r_1 is a multiple of r_2, and NP-complete otherwise [FiGoKr08]; (ii) for $L(r_1, 1)$-labeling is polynomial-time solvable even when r_1 is part of the input [ChKeKuLiYe00]; and (iii) for $L(2, 1, 1)$-labeling is NP-complete [GoLiPa10], despite that the minimum span for $L(2, 1, 1)$ on any tree T is either $\omega(T^3) - 1$ or $\omega(T^3)$ [FiGoKr04].

F119: [GoHeHeHaRa08] It is NP-complete to decide whether a graph admits a packing coloring with 4 colors (see definition in §5.1.7), while the graphs with $\chi_p \leq 3$ can be recognized in polynomial time.

F120: [BoDeJaKlKrMüTu98] It is NP-hard to determine the ranking number of bipartite graphs, and also of complements of bipartite graphs.

F121: [LaYu98] The ranking number of line graphs is NP-hard to compute.

F122: The ranking number can be determined in polynomial time for line graphs of trees ([dToGrSc95]), for graphs contained in chordal graphs of bounded clique size ([BoDeJaKlKrMüTu98]), interval graphs ([AsHe94]), and graphs in which there is only a polynomially bounded number of minimal separators ([BrKlKrMü02]).

F123: [KrTu99] It is NP-complete to decide whether an acyclic, planar directed graph has ranking number at most 3 (while for any k, the *undirected* connected graphs of ranking number at most k can be recognized in constant time [BoDeJaKlKrMüTu98]).

REMARKS

R26: The 5-coloring and 5-list-coloring algorithms on planar graphs are efficient. Although the proofs of the Four Color Theorem also yield polynomial algorithms that 4-color a planar graph, those algorithms are not practical.

R27: Most applications of the Lovász Local Lemma [ErLo75] can be made algorithmic by the method presented in [MoTa10].

R28: Further algorithmic aspects of coloring problems are discussed in the Special Issue on Computational Methods for Graph Coloring and Its Generalizations, *Discrete Applied Math.* 156:2 (2008), and in the survey [GaHe06].

Approximation

DEFINITIONS

D47: Let $r(n) : \mathbb{N} \to \mathbb{R}^+$ be a function. An algorithm is an $r(n)$-***approximation*** for chromatic number if, for every n and every input graph G with n vertices it outputs an integer k such that $\chi(G) \leq k \leq r(n)\chi(G)$. Analogous terminology applies to any minimization problem, e.g., to determine $ch(G)$.

D48: A ***doubly-periodic graph*** is an infinite graph whose vertices are labeled $v_{ij\ell}$ $(i, j \in \mathbb{Z}, \ell \in \{1, \ldots, n\})$; the subgraphs induced by $\{v_{ij1}, v_{ij2}, \ldots, v_{ijn}\}$ — called cells — are isomorphic for all pairs i, j, any other edge joins neighboring cells (i.e., cells (i, j) and (i', j') where $|i - i'| \leq 1$ and $|j - j'| \leq 1$), and both mappings $i \mapsto i+1$ and $j \mapsto j+1$ are automorphisms of G.

FACTS

F124: [Zu07] Unless $\mathsf{P} = \mathsf{NP}$, no polynomial-time $\mathcal{O}(n^{1-\varepsilon})$-approximation exists for $\chi(G)$, with any $\varepsilon > 0$. (Earlier non-approximability results were proved for $\mathcal{O}(n^{1/7-\varepsilon})$ assuming $\mathsf{P} \neq \mathsf{NP}$ [BeGoSu98] and for $\mathcal{O}(n^{1-\varepsilon})$ assuming $\mathsf{ZPP} \neq \mathsf{NP}$ [FeKi98]. The analogue of the latter is proved for uniform hypergraphs in [KrSu98].)

F125: [KhLiSa00] Unless $\mathsf{P} = \mathsf{NP}$, no polynomial-time algorithm can possibly color the k-colorable graphs with $k + 2\lfloor k/3 \rfloor - 1$ colors.

F126: The chromatic number can be approximated in polynomial time within the ratio $\mathcal{O}(n(\log \log n)^2/(\log n)^3)$ ([Ha93]), and also within

$$\max \left\{ \mathcal{O}(n/\log^{m-1} n), \mathcal{O}(\Delta \log \log n / \log n) \right\}$$

for any fixed m ([Pa01]).

F127: [KaTh12] There is a combinatorial algorithm that colors any 3-colorable graph in polynomial time using $\widetilde{\mathcal{O}}(n^{4/11})$ colors, and applying semi-definite programming methods the number of colors can be made $O(n^{0.2038})$. Moreover, there is a randomized polynomial-time algorithm that colors any graph with at most

$$\min \left\{ \mathcal{O}(\Delta^{1-2/k} \log^{1/2} \Delta \log n), \mathcal{O}(n^{1-3/(k+1)} \log^{1/2} n) \right\}$$

colors, where $k = \chi(G) \geq 3$ and $\Delta = \Delta(G)$ ([KaMoSu98]).

F128: Due to the inequalities $c \log col(G) \leq ch(G) \leq col(G)$, the choice number is constant-approximable on classes of graphs with bounded choice number, which is closely related to assuming bounded average degree.

F129: [DuFü97] The difference $n - \chi(G)$ is approximable within $360/289$.

F130: [BoGiHaKl95] The ranking number can be approximated within $\mathcal{O}(\log^2 n)$.

F131: [Ki98] For every k there is an on-line algorithm that properly colors every k-colorable graph with at most $\mathcal{O}(n^{1-1/k!})$ colors. For $k = 3$ and $k = 4$ the bound can be improved to $\mathcal{O}(n^{2/3} \log^{1/3} n)$ and $\mathcal{O}(n^{5/6} \log^{1/6} n)$, respectively.

F132: [Bu84] For every integer $k \geq 3$, there exists a doubly-periodic planar graph G of maximum degree 4 and a properly colored finite subgraph $F \subset G$ such that it is undecidable whether the coloring of F can be extended to a proper k-coloring of G. An analogous result holds for the undecidability of whether a partial homomorphism $F \to H$ can be extended to a homomorphism $G \to H$, whenever H is a finite non-bipartite graph ([DuEmGi98]). For bipartite H, the necessary and sufficient conditions of (un)decidability are not known.

References

[AlGrHa00] M. O. Albertson, S. Grossman, and R. Haas, Partial list colorings, *Discrete Math.* 214 (2000), 235–240.

[AlHu79] M. O. Albertson and J. P. Hutchinson, The three excluded cases of Dirac's map-color theorem, *Ann. New York Acad. Sci.* 318 (1979), 7–17.

[AlGrHaRi02] N. Alon, J. Grytczuk, M. Hałuszczak, and O. Riordan, Nonrepetitive colorings of graphs, *Random Struct. Alg.* 21 (2002), 336–346.

[AlKo11] N. Alon and A. V. Kostochka, Hypergraph list coloring and Euclidean Ramsey Theory, *Random Struct. Alg.* 39 (2011), 377–390.

[AlMoSa96] N. Alon, B. Mohar, and D. P. Sanders, On acyclic colorings of graphs on surfaces, *Israel J. Math.* 94 (1996), 273–283.

[AlTa92] N. Alon and M. Tarsi, Colorings and orientations of graphs, *Combinatorica* 12 (1992), 125–134.

[AlTuVo97] N. Alon, Zs. Tuza, and M. Voigt, Choosability and fractional chromatic numbers, *Discrete Math.* 165/166 (1997), 31–38.

[ApHa77] K. Appel and W. Haken, Every planar map is four colorable. Part I: Discharging, *Illinois J. Math.* 21 (1977), 429–490.

[ApHaKo77] K. Appel, W. Haken, and J. Koch, Every planar map is four colorable. Part II: Reducibility, *Illinois J. Math.* 21 (1977), 491–567.

[ArTe07] J. L. Arocha and J. Tey, The size of minimum 3-trees, *J. Graph Theory* 54 (2007), 103–114.

[AsHe94] B. Asprall and P. Heggernes, Finding minimum height elimination trees for interval graphs in polynomial time, *BIT* 34 (1994), 484–509.

[BaBuTuVo10] G. Bacsó, Cs. Bujtás, Zs. Tuza, and V. Voloshin, New challenges in the theory of hypergraph coloring, pages 45–57 in B. D. Acharya et al. (Eds.), *Advances in Discrete Mathematics and Applications: Mysore, 2008*, Ramanujan Math. Soc. Lecture Notes Ser. Vol. 13, 2010.

[BaGrGyPrSe04] G. Bacsó, S. Gravier, A. Gyárfás, M. Preissmann, and A. Sebő, Coloring the maximal cliques of graphs, *SIAM J. Discr. Math.* 17 (2004), 361–376.

[BaWo08] J. Barát and D. R. Wood, Notes on nonrepetitive graph colouring, *Electron. J. Combin.* 15 (2008), #R99.

[BaGrKiZh07] T. Bartnicki, J. Grytczuk, H. A. Kierstead, and X. Zhu, The map-coloring game, *Amer. Math. Monthly* 114 (2007), 793–803.

[BaTuVa10] C. Bazgan, Zs. Tuza, and D. Vanderpooten, Satisfactory graph partition, variants, and generalizations, *Europ. J. Oper. Res.* 206 (2010), 271–280.

[Be78] J. Beck, On 3-chromatic hypergraphs, *Discrete Math.* 24 (1978), 127–137.

[BeGoSu98] M. Bellare, O. Goldreich, and M. Sudan, Free bits, PCPs and non-approximability - towards tight results, *SIAM J. Comp.* 27 (1998), 804–915.

[Bi12] G. D. Birkhoff, A determinantal formula for the number of ways of coloring a map, *Annals of Math.* 14 (1912), 42–46.

[BiHuTu92] M. Biró, M. Hujter, and Zs. Tuza, Precoloring extension. I. Interval graphs, *Discrete Math.* 100 (1992), 267–279.

[Bo79] O. V. Borodin, On acyclic colorings of planar graphs, *Discrete Math.* 25 (1979), 211–236.

[Bo89] O. V. Borodin, Representing K_{13} as a 2-pire map on the Klein bottle, *J. Reine Angew. Math.* 393 (1989), 132–133.

[BoBrFrMiSe97] M. Borowiecki, I. Broere, M. Frick, P. Mihók, and G. Semanišin, Survey of hereditary properties of graphs, *Discuss. Math. Graph Theory* 17 (1997), 5–50.

[BoDeJaKlKrMüTu98] H. L. Bodlaender, J. S. Deogun, K. Jansen, T. Kloks, D. Kratsch, H. Müller, and Zs. Tuza, Rankings of graphs, *SIAM J. Discr. Math.* 11 (1998), 168–181.

[BoFiJuKaMo04] D. Bokal, G. Fijavž, M. Juvan, P. M. Kayll, and B. Mohar, The circular chromatic number of a digraph, *J. Graph Theory* 46 (2004), 227–240.

[BoGiHaKl95] H. L. Bodlaender, J. R. Gilbert, H. Hafsteinsson, and T. Kloks, Approximating treewidth, pathwidth and minimum elimination tree height, *J. Algorithms* 18 (1995), 238–255.

[BöMoSt99] T. Böhme, B. Mohar, and M. Stiebitz, Dirac's map-color theorem for choosability, *J. Graph Theory* 32 (1999), 327–339.

[BoSiTu07] M. Borowiecki, E. Sidorowicz, and Zs. Tuza, Game list colouring of graphs, *Electron. J. Combin.* 14 (2007), #R26.

[BrKlKrMü02] H. J. Broersma, T. Kloks, D. Kratsch, and H. Müller, A generalisation of AT-free graphs and a generic algorithm for solving triangulation problems, *Algorithmica* 32 (2002), 594–610.

[BrLo98] K. Bryś and Z. Lonc, Clique and anticlique partitions of graphs, *Discrete Math.* 185 (1998), 41–49.

[Bu84] S. A. Burr, Some undecidable problems involving the edge-coloring and vertex-coloring of graphs, *Discrete Math.* 50 (1984), 171–177.

[BuSaTuDoPu12] Cs. Bujtás, E. Sampathkumar, Zs. Tuza, Ch. Dominic, and L. Pushpalatha, When the vertex coloring of a graph is an edge coloring of its line graph — a rare coincidence, *Ars Combinatoria*, to appear.

[BuTu07] Cs. Bujtás and Zs. Tuza, Color-bounded hypergraphs, III: Model comparison, *Appl. Anal. Discrete Math.* 1 (2007), 36–55.

[BuTu08] Cs. Bujtás and Zs. Tuza, Uniform mixed hypergraphs: The possible numbers of colors, *Graphs Combin.* 24 (2008), 1–12.

[BuTu09] Cs. Bujtás and Zs. Tuza, Smallest set-transversals of k-partitions, *Graphs Combin.* 25 (2009), 807–816.

[BuTu09a] Cs. Bujtás and Zs. Tuza, Color-bounded hypergraphs, II: Interval hypergraphs and hypertrees, *Discrete Math.* 309 (2009), 6391–6401.

[BuTu10] Cs. Bujtás and Zs. Tuza, Voloshin's conjecture for C-perfect hypertrees, *Australas. J. Combin.* 48 (2010), 253–267.

[BuTu11] Cs. Bujtás and Zs. Tuza, Maximum number of colors: C-coloring and related problems, *J. Geometry* 101 (2011), 83–97.

[BuTu13] Cs. Bujtás and Zs. Tuza, Color-bounded hypergraphs, VI: Structural and functional jumps in complexity, *Discrete Math.* 313 (2013), 1965–1977.

[CaEd97] N. Cairnie and K. Edwards, Some results on the achromatic number, *J. Graph Theory* 26 (1997), 129–136.

[CaEd98] N. Cairnie and K. Edwards, The achromatic number of bounded degree trees, *Discrete Math.* 188 (1998), 87–97.

[Ca11] T. Calamoneri, The $L(h, k)$-labelling problem: An updated survey and annotated bibliography, *Comp. J.* 54 (2011), 1344–1371.

[ChKeKuLiYe00] G. J. Chang, W. T. Ke, D. Kuo, D. F. Liu, and R. K. Yeh, On $L(d, 1)$-labelings of graphs, *Discrete Math.* 220 (2000), 57–66.

[ChLiZh99] G. J. Chang, D. D.-F. Liu, and X. Zhu, Distance graphs and T-coloring, *J. Combin. Theory, Ser. B* 75 (1999), 259–269.

[Ch99] G. G. Chappell, A lower bound for partial list colorings, *J. Graph Theory* 32 (1999), 390–393.

[ChGeHe71] G. Chartrand, D. P. Geller, and S. T. Hedetniemi, Graphs with forbidden subgraphs, *J. Combin. Theory* 10 (1971), 12–41.

[ChZh09] G. Chartrand and P. Zhang, *Chromatic Graph Theory*, CRC Press, 2009.

[ChSe79] C. A. Christen and S. M. Selkow, Some perfect colouring properties of graphs, *J. Combin. Theory, Ser. B* 27 (1979),49–59.

[ChGaJo78] V. Chvátal, M. R. Garey, and D. S. Johnson, Two results concerning multicoloring, *Annals of Discrete Math.* 2 (1978), 151–154.

[Cu02] J. D. Currie, There are ternary circular square-free words of length n for $n \geq 18$, *Electron. J. Combin.* 9:1 (2002), #N10.

[dToGrSc95] P. de la Torre, R. Greenlaw, and A. A. Schäffer, Optimal edge ranking of trees in polynomial time, *Algorithmica* 13 (1995), 592–618.

[DiLiRaZh06] K. Diao, G. Liu, D. Rautenbach, and P. Zhao, A note on the least number of edges of 3-uniform hypergraphs with upper chromatic number 2, *Discrete Math.* 306 (2006), 670–672.

[Di52] G. A. Dirac, Map colour theorem, *Canad. J. Math.* 4 (1952), 480–490.

[DuKiTr91] D. Duffus, H. A. Kierstead, and W. T. Trotter, Fibres and ordered set coloring, *J. Combin. Theory, Ser. A* 58 (1991), 158–164.

[DuSaSaWo91] D. Duffus, B. Sands, M. Sauer, and R. E. Woodrow, Two-colouring all two-element maximal antichains, *J. Combin. Theory, Ser. A* 57 (1991), 109–116.

[DuFü97] R. Duh and M. Fürer, Approximation of k-set cover by semi-local optimization, pages 256–265 in *29th STOC*, Proc. Ann. ACM Symp. on Theory of Computing, ACM, 1997.

[DuFrJoWo12] V. Dujmović, F. Frati, G. Joret, and D. R. Wood, Nonrepetitive colourings of planar graphs with $O(\log n)$ colours, arXiv:1202.1569v2, 2012.

[DuEmGi98] P. Dukes, H. Emerson, and G. MacGillivray, Undecidable generalized colouring problems, *J. Combin. Math. Combin. Comput.* 26 (1998), 97–112.

[DvKáKrPa10] Z. Dvořák, J. Kára, D. Král', and O. Pangrác, Pattern hypergraphs, *Electron. J. Combin.* 17 (2010), #R15.

[DyGr00] M. Dyer and C. Greenhill, The complexity of counting graph homomorphisms, *Random Struct. Alg.* 17 (2000), 260–289.

[Ed97] K. Edwards, The harmonious chromatic number and the achromatic number, pages 13–47 in *Surveys in Combinatorics*, London Math. Soc. Lecture Note Ser. Vol. 241, London Math. Soc., 1997.

[Er64] P. Erdős, On a combinatorial problem II., *Acta Math. Acad. Sci. Hungar.* 15 (1964), 445–447.

[ErLo75] P. Erdős and L. Lovász, Problems and results on 3-chromatic hypergraphs and some related questions, pages 609–627 in A. Hajnal, R. Rado and V. T. Sós (Eds.), *Infinite and Finite Sets*, Colloq. Math. Soc. J. Bolyai 10, North-Holland, 1975.

[ErRuTa79] P. Erdős, A. L. Rubin, and H. Taylor, Choosability in graphs, Proc. West-Coast Conference on Combinatorics, Graph Theory and Computing, Arcata, California, *Congr. Numer.* XXVI (1979), 125–157.

[FaKeKiTr93] U. Faigle, U. Kern, H. Kierstead, and W. T. Trotter, On the game chromatic number of some classes of graphs, *Ars Combinatoria* 35 (1993), 143–150.

[FeHe98] T. Feder and P. Hell, List homomorphisms to reflexive graphs, *J. Combin. Theory, Ser. B* 72 (1998), 236–250.

[FeHeHu99] T. Feder, P. Hell, and J. Huang, List homomorphisms and circular arc graphs, *Combinatorica* 19 (1999), 487–505.

[FeHeHu03] T. Feder, P. Hell, and J. Huang, Bi-arc graphs and the complexity of list homomorphisms, *J. Graph Theory* 42 (2003), 61–80.

[FeKi98] U. Feige and J. Kilian, Zero knowledge and the chromatic number, *J. Comput. System Sci.* 57 (1998), 187–199.

[FiGoKr04] J. Fiala, P. A. Golovach, and J. Kratochvíl, Elegant distance constrained labelings of trees, pages 58–67 in *Proc. WG 2004*, Lecture Notes in Computer Sci. Vol. 3353, Springer, 2004.

[FiGoKr08] J. Fiala, P. A. Golovach, and J. Kratochvíl, Computational complexity of the distance constrained labeling problem for trees, pages 294–305 in *Proc. ICALP 2008, Part I*, Lecture Notes in Computer Sci. Vol. 5125, Springer, 2008.

[FiSa69] H.-J. Finck and H. Sachs, Über eine von H. S. Wilf angegebene Schranke für die chromatische Zahl endlicher Graphen, *Math. Nachr.* 39 (1969), 373–386.

[Fi95] D. Fisher, Fractional colorings with large denominators, *J. Graph Theory* 20 (1995), 403–409.

[Fr34] P. Franklin, A six-color problem, *J. Math. Phys.* 13 (1934), 363–369.

[Fu71] D. R. Fulkerson, Blocking and anti-blocking pairs of polyhedra, *Math. Progr.* 1 (1971), 168–194.

[Ga80] M. Gardner, Mathematical games, *Sci. Amer.* 242 (2) (1980), 14–21.

[GaHe06] Ph. Galinier and A. Hertz, A survey of local search methods for graph coloring, *Comput. Oper. Res.* 33 (2006), 2547–2562.

[GaJoSt76] M. R. Garey, D. S. Johnson, and L. J. Stockmeyer, Some simplified NP-complete graph problems, *Theor. Computer Sci.* 1 (1976), 237–267.

[GiKu00] K. Giaro and M. Kubale, Edge-chromatic sum of trees and bounded cyclicity graphs, *Inf. Proc. Letters* 75 (2000), 65–69.

[GoHeHeHaRa08] W. Goddard, S. M. Hedetniemi, S. T. Hedetniemi, J. M. Harris, and D. F. Rall, Broadcast chromatic numbers of graphs, *Ars Combinatoria* 86 (2008), 33–49.

[GoLiPa10] P. A. Golovach, B. Lidický, and D. Paulusma, $L(2,1,1)$-labeling is NP-complete for trees, pages 211–221 in *Proc. TAMC 2010*, Lecture Notes in Computer Sci. Vol. 6108, 2010.

[GrKr09] J. R. Griggs and D. Král', Graph labellings with variable weights, a survey, *Discrete Applied Math.* 157 (2009), 2646–2658.

[GrYe92] J. R. Griggs and R. K. Yeh, Labeling graphs with a condition at distance 2, *SIAM J. Discr. Math.* 5 (1992), 586–595.

[GrLoSc84] M. Grötschel, L. Lovász, and A. Schrijver, Polynomial algorithms for perfect graphs, pages 325–356 in *Topics on Perfect Graphs*, North-Holland Math. Stud., 88, North-Holland, Amsterdam, 1984.

[Gr59] H. Grötzsch, Ein Dreifarbensatz für dreikreisfreie Netze auf der Kugel, *Wiss. Z. Martin-Luther-Univ. Halle-Wittenberg Math.-Natur. Reihe.* 8 (1959), 109–120.

[Gr07] J. Grytczuk, Nonrepetitive colorings of graphs—a survey, *Intern. J. Math. & Math. Sci.* Vol. 2007, Article ID 74639, doi:10.1155/2007/74639.

[GuWeWo92] W. Gutjahr, E. Welzl, and G. Woeginger, Polynomial graph colourings, *Discrete Applied Math.* 35 (1992), 29–46.

[Gu96] S. Gutner, The complexity of planar graph choosability, *Discrete Math.* 159 (1996), 119–130.

[GuTa09] S. Gutner and M. Tarsi, Some results on $(a:b)$-choosability, *Discrete Math.* 309 (2009), 2260–2270 (original manuscript in 1997).

[GyLe91] A. Gyárfás and J. Lehel, Effective on-line coloring of P_5-free graphs, *Combinatorica* 11 (1991), 181–184.

[GyHoPaWo08] E. Győri, M. Horňák, C. Palmer, and M. Woźniak, General neighbour-distinguishing index of a graph, *Discrete Math.* 308 (2008), 827–831.

[GyPa09] E. Győri and C. Palmer, A new type of edge-derived vertex colorings, *Discrete Math.* 309 (2009), 6344–6352.

[Ha93] M. M. Halldórsson, A still better performance guarantee for approximate graph coloring, *Inf. Proc. Letters* 45 (1993), 19–23.

[HaReSe12] F. Havet, B. Reed, and J.-S. Sereni, Griggs and Yehs conjecture and $L(p,1)$-labelings, *SIAM J. Discr. Math.* 26 (2012), 145–168.

[HaTu93] F. Harary and Zs. Tuza, Two graph-colouring games, *Bull. Austral. Math. Soc.* 48 (1993), 141–149.

[He:1890] P. J. Heawood, Map colour theorem, *Quart. J. Pure Appl. Math.* 24 (1890), 332–338.

[He:1898] P. J. Heawood, On the four-colour map theorem, *Quart. J. Pure Appl. Math.* 29 (1898), 270–265.

[He93] L. S. Heath, Edge coloring planar graphs with two outerplanar graphs, pages 195–202 in *Proc. 2nd ACM–SIAM Symp. on Discrete Algorithms* (San Francisco, 1991), ACM, 1991.

[He03] P. Hell, Algorithmic aspects of graph homomorphisms, pages 239–276 in *Surveys in Combinatorics*, London Math. Soc. Lecture Note Ser. Vol. 307, Cambridge University Press, 2003.

[He12] P. Hell, Graph partitions with prescribed patterns, Manuscript, 2012.

[HeHeBe82] S. M. Hedetniemi, S. T. Hedetniemi, and T. Beyer, A linear algorithm for the Grundy (coloring) number of a tree, Proc. 13th S-E Conf. on Combinatorics, Graph Theory and Computing, *Congr. Numer.* 36 (1982), 351–363.

[HeNe90] P. Hell and J. Nešetřil, On the complexity of *H*-coloring, *J. Combin. Theory, Ser. B* 48 (1990), 92–110.

[HeNe04] P. Hell and J. Nešetřil, *Graphs and Homomorphisms*, Oxford Univ. Press, 2004.

[HeNe08] P. Hell and J. Nešetřil, Colouring, constraint satisfaction, and complexity, *Comp. Sci. Rev.* 2 (2008), 143–163.

[HeNeZh96] P. Hell, J. Nešetřil, and X. Zhu, Duality and polynomial testing of tree homomorphisms, *Trans. Amer. Math. Soc.*, 348 (1996), 1281–1297.

[HoKaLoSaSh10] C. T. Hoàng, M. Kamiski, V. Lozin, J. Sawada, and X. Shu, Deciding *k*-colorability of P_5-free graphs in polynomial time, *Algorithmica* 57 (2010), 74–81.

[HoŠk04] F. Holroyd and M. Škoviera, Colouring of cubic graphs by Steiner triple systems, *J. Combin. Theory, Ser. B* 91 (2004), 57–66.

[Ho81] I. Holyer, The NP-completeness of edge-coloring, *SIAM J. Computing* 10 (1981), 718–720.

[HoSo10] M. Horňák and R. Soták, General neighbour-distinguishing index via chromatic number, *Discrete Math.* 310 (2010), 1733–1736.

[HuTu93] M. Hujter and Zs. Tuza, Precoloring extension. II. Graph classes related to bipartite graphs, *Acta Math. Univ. Comen.* 62 (1993), 1–11.

[HuTu96] M. Hujter and Zs. Tuza, Precoloring extension. III. Classes of perfect graphs, *Combin. Probab. Comput.* 5 (1996), 35–56.

[IrMa99] R. W. Irving and D. F. Manlove, The b-chromatic number of a graph, *Discrete Applied Math.* 91 (1999), 127–141.

[Ja97] K. Jansen. Optimum cost chromatic partition problem, *Algorithms and Complexity*, Lecture Notes in Computer Sci. Vol. 1203, 25–36, Springer, 1997.

[JaRi83] B. Jackson and G. Ringel, Maps of *m*-pires on the projective plane, *Discrete Math.* 46 (1983), 15–20.

[JaRi84] B. Jackson and G, Ringel, Solution of Heawood's empire problem in the plane, *J. Reine Angew. Math.* 347 (1984), 146–153.

[JaRi85] B. Jackson and G. Ringel, Heawood's empire problem, *J. Combin. Theory, Ser. B* 38 (1985), 168–178.

[JaSc97] K. Jansen and P. Scheffler, Generalized colorings for tree-like graphs, *Discrete Applied Math.* 75 (1997), 135–155.

[Ja01] J. C. M. Janssen, A partial solution of a partial list colouring problem, *Proc. 32nd S-E International Conf. on Combinatorics, Graph Theory and Computing, Congr. Numer.* 152 (2001), 75–79.

[JiMuTuVoWe02] T. Jiang, D. Mubayi, Zs. Tuza, V. Voloshin, and D. B. West, The chromatic spectrum of mixed hypergraphs, *Graphs Combin.* 18 (2002), 309–318.

[JiWe99] T. Jiang and D. B. West, Coloring of trees with minimum sum of colors, *J. Graph Theory* 32 (1999), 354–358.

[Ka72] R. Karp, Reducibility among combinatorial problems, pages 85–104 in R. E. Miller and J. W. Thatcher (Eds.), *Complexity of Computer Computations*, Plenum Press, 1972.

[Ka00] J. Kahn, Asymptotics of the list-chromatic index for multigraphs, *Random Struct. Alg.* 17 (2000), 117–156.

[KaPe12] R. R. Kamalian and P. A. Petrosyan, A note on upper bounds for the maximum span in interval edge-colorings of graphs, *Discrete Math.* 312 (2012), 1393–1399.

[KaMoSu98] D. Karger, R. Motwani, and M. Sudan, Approximate graph coloring by semidefinite programming, *J. ACM* 45 (1998), 246–265.

[KaTh12] K. Kawarabayashi and M. Thorup, Combinatorial coloring of 3-colorable graphs, *Proc. 53rd FOCS* (2012), 68–75.

[KhLiSa00] S. Khanna, N. Linial, and S. Safra, On the hardness of approximating the chromatic number, *Combinatorica* 20 (2000), 393–415.

[Ki98] H. A. Kierstead, On-line coloring *k*-colorable graphs, *Israel J. Math.* 105 (1998), 93–104.

[Ki98a] H. A. Kierstead, Recursive and on-line graph coloring, pages 1233–1269 in *Handbook of Recursive Mathematics*, Vol. 2, North-Holland, 1998.

[KiPeTr94] H. A. Kierstead, S. G. Penrice, and W. T. Trotter, On-line coloring and recursive graph theory, *SIAM J. Discr. Math.* 7 (1994), 72–89.

[KiRe93] K. Kilakos and B. Reed, Fractionally colouring total graphs, *Combinatorica* 4 (1993), 435–440.

[KoKü01] D. Kobler and A. Kündgen, Gaps in the chromatic spectrum of face-constrained plane graphs, *Electron. J. Combin.* 8:1 (2001), #N3.

[Kö16] D. König, Über Graphen und ihre Anwendung auf Determinantentheorie und Mengenlehre, *Math. Ann.* 77 (1916), 453–465.

[KoKr01] G. Kortsarz and R. Krauthgamer, On approximating the achromatic number, *SIAM J. Discr. Math.* 14 (2001), 408–422.

[KoSt00] A. V. Kostochka and M. Stiebitz, On the number of edges in colour-critical graphs and hypergraphs, *Combinatorica* 20 (2000), 521–530.

[KoYa12] A. V. Kostochka and M. Yancey, Ore's Conjecture on color-critical graphs is almost true, manuscript, 2012, http://arxiv.org/abs/1209.1050

[Kr04] D. Král', On feasible sets of mixed hypergraphs, *Electron. J. Combin.* 11 (2004), #R19.

[KrKrPrVo06] D. Král', J. Kratochvíl, A. Proskurowski, and H.-J. Voss, Coloring mixed hypertrees, *Discrete Applied Math.* 154 (2006), 660–672.

[KrKrTuWo01] D. Král', J. Kratochvíl, Zs. Tuza, and G. J. Woeginger, Complexity of coloring graphs without forbidden induced subgraphs, pages 254–262 in A. Brandstädt and V. B. Le (Eds.), *Graph-Theoretic Concepts in Computer Science*, Lecture Notes in Computer Sci. 2204, Springer, 2001.

[KrMáPaRaSeŠk09] D. Král', E. Máčajová, O. Pangrác, A. Raspaud, J.-S. Sereni, and M. Škoviera, Projective, affine, and abelian colorings of cubic graphs, *Europ. J. Combin.* 30 (2009), 53–69.

[Kr93] J. Kratochvíl, Precoloring extension with fixed color bound, *Acta Math. Univ. Comen.* 62 (1993), 139–153.

[KrTu94] J. Kratochvíl and Zs. Tuza, Algorithmic complexity of list colorings, *Discrete Applied Math.* 50 (1994), 297–302.

[KrTu99] J. Kratochvíl and Zs. Tuza, Rankings of directed graphs, *SIAM J. Discr. Math.* 12 (1999), 374–384.

[KrTu02] J. Kratochvíl and Zs. Tuza, On the complexity of bicoloring clique hypergraphs of graphs, *J. Algorithms* 45 (2002), 40–54.

[KrTuVo02] J. Kratochvíl, Zs. Tuza, and M. Voigt, On the b-chromatic number of graphs, pages 310–320 in L. Kučera, Ed., *Graph-Theoretic Concepts in Computer Science*, Lecture Notes in Computer Sci. Vol. 2573, Springer, 2002.

[KrSu98] M. Krivelevich and B. Sudakov, Approximate coloring of uniform hypergraphs, *J. Algorithms* 49 (2003), 2–12.

[KrVu01] M. Krivelevich and V. H. Vu, Choosability in random hypergraphs, *J. Combin. Theory, Ser. B* 83 (2001), 241–257.

[Ku04] M. Kubale (Ed.), *Graph Colorings*, Contemporary Mathematics Vol. 352, Amer. Math. Soc., 2004.

[KüPe08] A. Kündgen and M. J. Pelsmajer, Nonrepetitive colorings of graphs of bounded tree-width, *Discrete Math.* 308 (2008), 4473–4478.

[KuSc89] E. Kubicka and A. J. Schwenk, An introduction to chromatic sum, pages 39–45 in *Proc. 17th ACM Computer Science Conf.*, 1989.

[LaYu98] T. W. Lam and F. L. Yue, Edge ranking of graphs is hard, *Discrete Applied Math.* 85 (1998), 71–86.

[LeGa83] D. Leven and Z. Galil, NP-completeness of finding the chromatic index of regular graphs, *J. Algorithms* 4 (1983), 35–44.

[LeRaSc07] V. B. Le, B. Randerath, and I. Schiermeyer, On the complexity of 4-coloring graphs without long induced paths, *Theor. Computer Sci.* 389 (2007), 330–335.

[LiPaZh03] S.-C. Liaw, Z. Pan, and X. Zhu, Construction of K_n-minor free graphs with given circular chromatic number, *Discrete Math.* 263 (2003), 191–206.

[LiSu12] X. Li and Y. Sun, *Rainbow Connections of Graphs*, SpringerBriefs in Mathematics, 2012.

[Lo73] L. Lovász, Coverings and colourings of hypergraphs, Proc. 4th S–E Conf. on Combinatorics, Graph Theory and Computing, *Congr. Numer.* 8 (1973), 3–12.

[Lo75] L. Lovász, On the ratio of optimal integral and fractional covers, *Discrete Math.* 13 (1975), 383–390.

[LoThWuZh12] L. M. Lovász, C. Thomassen, Y. Wu, and C.-Q. Zhang, Nowhere-zero 3-flows and modulo k-orientations, manuscript, 2012.

[Ma68] D. W. Matula, A min-max theorem for graphs with application to graph coloring, *SIAM Review* 10 (1968), 481–482.

[Ma02] D. Marx, The complexity of tree multicoloring, pages 532–542 in *Proc. MFCS 2002*, Lecture Notes in Computer Sci. Vol. 2420, Springer, 2002.

[Ma06] D. Marx, Precoloring extension on unit interval graphs. *Discrete Applied Math.* 154 (2006), 995–1002.

[Ma07] D. Marx, Precoloring extension on chordal graphs, pages 255–270 in A. Bondy et al. (Eds.), *Graph Theory in Paris*, Proc. Conf. in Memory of Claude Berge, Trends in Mathematics, Birkhäuser, 2007.

[Ma11] D. Marx, Complexity of clique coloring and related problems, *Theor. Computer Sci.* 412 (2011), 3487–3500.

[MáŠk05] E. Máčajová and M. Škoviera, Fano colourings of cubic graphs and the Fulkerson Conjecture, *Theor. Computer Sci.* 349 (2005), 112–120.

[McSa94] C. J. H. McDiarmid and A. Sánchez-Arroyo, Total colouring regular bipartite graphs is NP-hard, *Discrete Math.* 124 (1994), 155–162.

[MiTuVo99] P. Mihók, Zs. Tuza, and M. Voigt, Fractional $\mathcal{P}$-colourings and $\mathcal{P}$-choice-ratio, *Tatra Mountain Math. Publ.* 18 (1999), 69–77.

[Mi67] G. J. Minty, A theorem on three-coloring the edges of a trivalent graph, *J. Combin. Theory* 2 (1967), 164–167.

[MiMo97] J. Mitchem and P. Morriss, On the cost-chromatic number of graphs, *Discrete Math.* 171 (1997), 201–211.

[MiMoSc97] J. Mitchem, P. Morriss, and E. Schmeichel, On the cost chromatic number of outerplanar, planar, and line graphs, *Discuss. Math. Graph Theory* 17 (1997), 229–241.

[MiTu97] L. Milazzo and Zs. Tuza, Upper chromatic number of Steiner triple and quadruple systems, *Discrete Math.* 174 (1997), 247–259.

[MiTuVo03] L. Milazzo, Zs. Tuza, and V. Voloshin, Strict colorings of Steiner triple and quadruple systems: a survey, *Discrete Math.* 261 (2003), 399–411.

[Mo97] D. Moser, The star-chromatic number of planar graphs, *J. Graph Theory* 24 (1997), 33–43.

[MoSk99] B. Mohar and R. Škrekovski, The Grötzsch theorem for the hypergraph of maximal cliques, *Electron. J. Combin.* 6 (1999), #R26.

[MoTa10] R. A. Moser and G. Tardos, A constructive proof of the general Lovász Local Lemma, *J. ACM* 57:2 (2010), #11.

[NeTa99] J. Nešetřil and C. Tardif, Density, pages 229–235 in R. L. Graham et al. (Eds.), *Contemporary Trends in Discrete Mathematics*, DIMACS Series in Discrete Mathematics and Theoretical Computer Science 49, Amer. Math. Soc., 1999.

[Pa01] V. T. Paschos, A note on the approximation ratio of graph-coloring, *Found. Comput. Decision Sci.* 26 (2001), 267–271.

[PhRö84] K. T. Phelps and V. Rödl, On the algorithmic complexity of coloring simple hypergraphs and Steiner triple systems, *Combinatorica* 4 (1984), 79–88.

[RaSc04] B. Randerath and I. Schiermeyer, 3-colorability $\in \mathcal{P}$ for P_6-free graphs, *Discrete Applied Math.* 136 (2004), 299–313.

[RaSr00] J. Radhakrishnan and A. Srinivasan, Improved bounds and algorithms for hypergraph 2-coloring, *Random Struct. Alg.* 16 (2000), 4–32.

[ReSe98] B. Reed and P. Seymour, Fractional colouring and Hadwiger's conjecture, *J. Combin. Theory, Ser. B* 74 (1998), 147–152.

[RiYo68] G. Ringel and J. W. T. Youngs, Solution of the Heawood map-coloring problem, *Proc. Natl. Acad. Sci. USA* 60 (1968), 438–445.

[RoSaSeTh97] N. Robertson, D. P. Sanders, P. D. Seymour, and R. Thomas, The four-colour theorem, *J. Combin. Theory, Ser. B* 70 (1997), 2–44.

[Se81] P. D. Seymour, Nowhere-zero 6-flows, *J. Combin. Theory, Ser. B* 30 (1981), 130–135.

[Sc09] U. Schauz, Mr. Paint and Mrs. Correct, *Electron. J. Combin.* 16 (2009), #R77.

[Sc10] U. Schauz, A paintability version of the Combinatorial Nullstellensatz, and list colorings of k-partite k-uniform hypergraphs, *Electron. J. Combin.* 17 (2010), #R176.

[SgWo01] J. Sgall and G. J. Woeginger, The complexity of coloring graphs without long induced paths, *Acta Cybern.* 15 (2001), 107–111.

[Si83] G. J. Simmons, On the ochromatic number of a graph, Proc. 14th S-E Conf. on Combinatorics, Graph Theory and Computing, *Congr. Numer.* 40 (1983), 339–366.

[StZh96] E. Steffen and X. Zhu, Star chromatic numbers of graphs, *Combinatorica* 16 (1996), 439–448.

[Ta:1880] P. G. Tait, On the colouring of maps, *Proc. Royal Soc. Edinburgh Sect. A* 10 (1878–1880), 501–503.

[Te93] B. A. Tesman, List T-colorings of graphs, *Discrete Applied Math.* 45 (1993), 277–289.

[Th94] C. Thomassen, Every planar graph is 5-choosable, *J. Combin. Theory, Ser. B* 62 (1994), 180–181.

[Th95] C. Thomassen, 3-list-coloring planar graphs of girth 5, *J. Combin. Theory, Ser. B* 64 (1995), 101–107.

[Th06] A. Thue, Über unendliche Zeichenreichen, *Norske Vid. Selsk. Skr., I Mat. Nat. Kl.*, Christiania, 7 (1906), 1–22.

[ThErAlMaSc89] C. Thomassen, P. Erdős, Y. Alavi, P. J. Malde, and A. J. Schwenk, Tight bounds on the chromatic sum of a connected graph, *J. Graph Theory* 13 (1989), 353–357.

[Tu50] W. T. Tutte, On the imbedding of linear graphs in surfaces, *Proc. London Math. Soc. (2)* 51 (1950), 474–483.

[Tu54] W. T. Tutte, A contribution to the theory of chromatic polynomials, *Canad. J. Math.* 6 (1954), 80–91.

[Tu70] W. T. Tutte, More about chromatic polynomials and the golden ratio, pages 439–453 in R. K. Guy et al. (Eds.), *Combinatorial Structures and their Applications*, Gordon and Breach, 1970.

[Tu90] Zs. Tuza, Contractions and minimal k-colorability, *Graphs Combin.* 6 (1990), 51–59.

[TuVo96] Zs. Tuza and M. Voigt, On a conjecture of Erdős, Rubin and Taylor, *Tatra Mountain Math. Publ.* 9 (1996), 69–82.

[TuVo96a] Zs. Tuza and M. Voigt, Every 2-choosable graph is $(2m, m)$-choosable, *J. Graph Theory* 22 (1996), 245–252.

[TuVo00] Zs. Tuza and V. Voloshin, Uncolorable mixed hypergraphs, *Discrete Applied Math.* 99 (2000), 209–227.

[TuVoZh02] Zs. Tuza, V. Voloshin, and H. Zhou, Uniquely colorable mixed hypergraphs, *Discrete Math.* 248 (2002), 221–236.

[VeWe92] D. L. Vertigan and D. J. A. Welsh, The computational complexity of the Tutte plane: the bipartite case, *Combin. Probab. Comput.* 1 (1992), 181–187.

[Vo93] M. Voigt, List colourings of planar graphs, *Discrete Math.* 120 (1993), 215–219.

[Vo95] M. Voigt, A not 3-choosable planar graph without 3-cycles, *Discrete Math.* 146 (1995), 325–328.

[Vo95a] V. I. Voloshin, On the upper chromatic number of a hypergraph, *Australas. J. Combin.* 11 (1995), 25–45.

[Vo02] V. I. Voloshin, *Coloring Mixed Hypergraphs: Theory, Algorithms and Applications*, Amer. Math. Soc., Providence, 2002.

[Wa96] A. O. Waller, An upper bound for list T-coloring, *Bull. London Math. Soc.* 28 (1996), 337–342.

[Wo01] D. R. Woodall, List colourings of graphs, pages 269–301 in J. W. P. Hirschfeld (Ed.), *Surveys in Combinatorics*, Proc. British Combinatorial Conference, London Math. Soc. Lecture Note Series Vol. 288, Cambridge University Press, 2001.

[Ye06] R. K. Yeh, A survey on labeling graphs with a condition at distance two, *Discrete Math.* 306 (2006), 1217–1231.

[Zh96] X. Zhu, Uniquely *H*-colorable graphs with large girth, *J. Graph Theory* 23 (1996), 33–41.

[Zh99] X. Zhu, Planar graphs with circular chromatic numbers between 3 and 4, *J. Combin. Theory, Ser. B* 76 (1999), 170–200.

[Zh01] X. Zhu, Circular chromatic number: a survey, *Discrete Math.* 229 (2001), 371–410.

[Zh11] X. Zhu, Graphs with chromatic numbers strictly less than their colouring numbers, *Ars Math. Contemp.* 4 (2011), 25–27.

[Zh11a] X. Zhu, The fractional version of Hedetniemi's conjecture is true, *Europ. J. Combin.* 32 (2011), 1168–1175.

[ZhDiWa12] P. Zhao, K. Diao, and K. Wang, The smallest one-realization of a given set, *Electron. J. Combin.* 19:1 (2012), #P19.

[Zu07] D. Zuckerman, Linear degree extractors and the inapproximability of Max Clique and Chromatic Number, *Theory of Computing* 3 (2007), 103–128.

Section 5.3

Independence and Cliques

Gregory Gutin, Royal Holloway, University of London

INTRODUCTION

Finding maximum cliques and maximum independent sets are among the most applicable problems in graph theory. We give an overview of both algorithmic and theoretical results on these problems.

5.3.1 Basic Definitions and Applications

In this section, all graphs are *simple*, i.e., they do not have self-loops or multi-edges.

Some Combinatorial Optimization Problems

DEFINITIONS

D1: For a graph G, a set S of vertices is an ***independent set*** if no two vertices in S are adjacent.

D2: The number of vertices in a maximum-size independent set of G is called the ***independence number*** of G and is denoted $ind(G)$.

D3: A ***clique*** in a graph G is a maximal set of mutually adjacent vertices of G. The ***clique number***, denoted $\omega(G)$, is the number of a vertices in a largest clique of G.

D4: A ***vertex cover*** in a graph G is a set S of vertices such that at least one endpoint of every edge of G is in S.

D5: A ***matching*** in a graph G is a set of mutually non-adjacent edges of G.

REMARKS

R1: The ***maximum-clique problem*** (determining $\omega(G)$ for a given graph G) and the ***maximum-independent-set problem*** (determining $ind(G)$) are the main problems considered in this section. Observe that $\omega(G) = ind(\overline{G})$ for any graph G, where $\overline{G}$ denotes the edge-complement graph.

R2: Also considered in this section is the ***minimum-vertex-cover problem*** of finding a vertex cover of minimum cardinality.

R3: Sometimes, we will consider graphs with non-negative weights on their vertices.

EXAMPLE

E1: It is easy to verify for the Peterson graph G shown in Figure 5.3.1 that $\omega(G) = 2$, $ind(G) = 4$, and a minimum vertex cover has size 5.

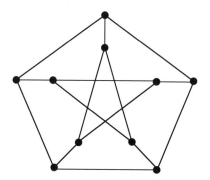

Figure 5.3.1: The Petersen graph.

Vertex-Weighted Graphs

DEFINITIONS

D6: A graph G is ***vertex-weighted*** if every vertex x is assigned a non-negative weight $w(x)$; the graph is denoted (G, w).

D7: The ***weight of a vertex set*** S in a vertex-weighted graph is the sum of the weights of the vertices in S.

NOTATION: The weight of a maximum-weight independent set in G is denoted $ind(G, w)$. The weight of a maximum-weight clique in G is denoted $\omega(G, w)$.

REMARKS

R4: The ***maximum-weight-independent-set problem*** (determining $ind(G, w)$) generalizes the unweighted one: assign weight 1 to every vertex of a graph. A similar remark holds for the ***maximum-weight-clique problem***.

R5: The first three of the following facts are immediate consequences of the definitions.

FACTS

F1:   For every vertex-weighted graph (G, w),

$$ind(G, w) = \omega(\overline{G}, w),$$

which establishes a duality between the two problems.

F2:   A set S of vertices in a graph $G = (V, E)$ is a vertex cover if and only if $V - S$ is an independent set.

F3:   The size of a maximum matching in a graph G is less than or equal to the size of a minimum cover of G.

F4:   [Ko31, Eg31] If G is a bipartite graph, then the maximum size of a matching in G equals the minimum cardinality of a vertex cover of G.

Applications Involving Hamming Distance

There are numerous and varied applications of the combinatorial optimization problems introduced above. Perhaps among the most studied are those from coding theory.

DEFINITIONS

D8:   The **Hamming distance** between a pair

$$u = (u_1, \ldots, u_n), v = (v_1, \ldots, v_n)$$

of binary vectors is the number of indices i for which $u_i \neq v_i$.

D9:   The vertex set of the **Hamming graph** $H(n, d)$ consists of all binary vectors with n coordinates. A pair u, v of vertices in $H(n, d)$ are adjacent if the Hamming distance between them is at least d.

REMARKS

R6:   The Hamming graph is of interest for error-correcting codes. In particular, a set of binary vectors in which any two have Hamming distance at least d forms a binary code that can correct $\lfloor (d-1)/2 \rfloor$ errors [MaSl79].

R7:   A natural question arises: How many vectors with n coordinates can be in a code in which any two vectors have Hamming distance d? It follows from the definitions that this number equals the clique number, $\omega(H(n, d))$.

R8:   For further discussions and results of Hamming distance, see, for example, [BoBu-PaPe99] and [Os01]. Other applications include fault diagnosis [BePe90, HaPaVa93], machine learning [HoSk89, HaJa90], and detecting embedded network structures in linear programs [GuGuMiMa00, GuGuMiZv04].

5.3.2   Integer Programming Formulations

The simplest formulation of the maximum-weight-clique problem is based on the edges of the input graph.

Edge-based formulation: Let $G = (V, E)$ be a vertex-weighted graph with $V = \{v_1 \ldots, v_n\}$ and weights $w_i = w(v_i)$. The maximum weight of a clique can be found by solving the following integer program:

$$\max w \;=\; \sum_{i=1}^{n} w_i x_i$$
$$\text{subject to} \qquad x_i + x_j \leq 1 \quad \forall \{v_i, v_j\} \notin E$$
$$x_i = 0 \text{ or } 1, \quad i = 1, \ldots, n$$

FACTS

F5:   In the edge-based formulation, each feasible solution x corresponds to the clique $S = \{v_i \in V : x_i = 1\}$.

F6:   [NeTr74, NeTr75] Let x be an optimal $(0, \frac{1}{2}, 1)$-valued solution to the linear relaxation of the edge formulation, and let $J = \{j : \; x_j = 1\}$. Then there exists an optimal solution x^* to the edge-based formulation such that $x_j^* = 1$ for every $j \in J$.

REMARKS

R9:   Unfortunately, the result in the last fact appears to be of relatively minor computational value since optimal solutions of the linear relaxation of the edge-based formulation normally have only a small number of integer components, and the gap between optimal solutions of the edge-based formulation and its linear relaxation is usually too large. [BoBuPaPe99]

R10:   Using Fact F1 we can transform the edge-based formulation of the maximum-weight-clique problem to the corresponding formulation of the maximum-weight-independent-set problem, which is equivalent to the following nonlinear optimization problem first studied in [Sh90].

Shor formulation:

$$\min w \;=\; \sum_{i=1}^{n} w_i x_i$$
$$\text{subject to} \qquad x_i x_j = 0 \quad \forall \{v_i, v_j\} \in E$$
$$x_i^2 - x_i = 0 \quad i = 1, \ldots, n$$

COMPUTATIONAL NOTE: Shor [Sh90] reported very good computational results using his formulation.

Two More Formulations of the Maximum-Clique Problem

While the formulations above are relatively straightforward, the following ones obtained by Motzkin and Straus [MoSt65] are less obvious.

DEFINITIONS

D10: The *standard simplex* Δ in R^n is defined as follows:

$$\Delta = \{x \in R^n : x_i \geq 0, \; i = 1, \ldots, n, \; \sum_{i=1}^{n} x_i = 1\}$$

NOTATION: For a graph G with adjacency matrix A_G and $x \in R^n$, let $g(x) = x^T A_G x$.

D11: Let $G = (V, E)$ be a graph with vertices $v_1, \ldots, v_n$ and let $S \subseteq V$ be arbitrary. The *characteristic vector* $x^S \in R^n$ is defined as follows: $x^S = (x_1^S, x_2^S, \ldots, x_n^S)$, where $x_i^S = \frac{1}{|S|}$ if $v_i \in S$ and $x_i^S = 0$, otherwise.

Motzkin–Straus Theorem [MoSt65]: Let G be a graph and let

$$x^* = argmax\{g(x) : \; x \in \Delta\}$$

(i.e., $x^* \in \Delta$ and $g(x^*) \geq g(x) \; \forall x \in \Delta$). Then $\omega(G) = 1/(1 - g(x^*))$. Moreover, a subset S of vertices of G is a maximum clique if and only if $x^S = argmax\{g(x) : \; x \in \Delta\}$.

REMARK

R11: One drawback of the Motzkin–Straus formulation is the fact that some solutions of $argmax\{g(x) : \; x \in \Delta\}$ are not characteristic vectors [PaPh90, PeJa95]. Thus, the following variation of the Motzkin–Straus formulation due to Bomze [Bo97] is of interest.

NOTATION: For a graph G with adjacency matrix A_G, let $f(x) = x^T A_G x + (x^T x)/2$.

Bomze Theorem [Bo97]: Let S be a subset of vertices of a graph G. Then
(a) S is a maximum clique in G if and only if $x^S = argmax\{f(x) : \; x \in \Delta\}$.
(b) S is a clique in G if and only if x^S is a local maximizer of $\{f(x) : \; x \in \Delta\}$.
(c) All local maximizers x of $\{f(x) : \; x \in \Delta\}$ are characteristic vectors.

5.3.3 Hardness Results

The maximum-clique problem is one of the first shown to be NP-hard [Ka72]. Since then many researchers have tried to gain a more precise understanding of the difficulty of the problem (see, e.g., [AuCrGaKaMaPr99, BoBuPaPe99]).

NOTATION: If π is an algorithm for the maximum-independent-set problem, then $\pi(G)$ denotes the independent set produced by π when the input graph is G.

FACTS

F7: [Ha99] Unless P = NP, there can be no polynomial-time algorithm that approximates the maximum clique to within a factor better than $O(n^{1-\epsilon})$, for any $\epsilon > 0$.

The following positive result has a nice (and short) proof based on some ideas of Paul Erdős (see [Ha98]).

F8: [BoHa92] There is a polynomial-time algorithm π such that for any n-vertex graph G, $ind(G)/|\pi(G)| = O(n/\log^2 n)$.

This result was improved by Fiege [Fi04].

F9: [Fi04] There is a polynomial-time algorithm π such that for any n-vertex graph G, $ind(G)/|\pi(G)| = O(n(\log\log n)^2/\log^3 n)$.

REMARK

R12:   Since the approximation in Fact F9 is still weak, it is natural to consider the approximation for graphs for which certain parameters are restricted. One such parameter is the maximum degree, $\Delta(G)$, of a graph G. For an overview of approximation algorithms whose performance is measured in terms of $\Delta(G)$, see [LaTi01].

FACTS

F10: [AlFeWiZu95] Unless P = NP, there exists a constant $\epsilon > 0$ such that there is no polynomial-time algorithm π for which $ind(G)/|\pi(G)| = O(\Delta(G)^\epsilon)$ for every graph G.

The following result is due to Vishvanathan (see [Ha98]).

F11:   There is a polynomial-time algorithm π such that for every graph G,

$$ind(G)/|\pi(G)| = O(\Delta(G) \log\log \Delta(G)/ \log \Delta(G))$$

REMARKS

R13:   There is a simple polynomial-time algorithm for the minimum-vertex-cover problem that provides a 2-approximation (i.e., no worse than twice the optimum): find a maximum matching M in a given graph G and output the vertices of M as a vertex cover of G. For slightly better approximation results, see [AuCrGaKaMaPr99].

R14:   The last remark and Fact F7 are, in a way, at odds with each other. The maximum-clique and the minimum-vertex-cover problems are dual, in a sense (via the maximum-independent-set problem as noted earlier). Nevertheless, while the former cannot be approximated to any good degree, the latter can be. This 'strange' situation is somewhat resolved by the next fact. Observe that a feasible solution of the maximum-clique problem is a set of vertices that induces a complete subgraph.

FACT

F12: [GuVaYe03] Let ψ be any polynomial-time algorithm for the maximum-clique problem and let $p(n)$ be any polynomial function of n. Unless P = NP, there exists an n-vertex graph G such that $\psi(G)$ has fewer vertices than at least $\frac{p(n)-1}{p(n)}$ of the complete subgraphs of G. The analogous fact holds for the minimum-vertex-cover problem.

DEFINITIONS

D12: [ImPa99] The ***exponential time hypothesis*** states that there is an $\epsilon > 0$ such that 3-SAT cannot be solved in time $O(2^{\epsilon n})$, where n is the number of variables.

D13: The ***k-clique problem*** is the problem of deciding whether a graph has a clique with at least k vertices.

D14: ***Parameterized complexity*** studies problems equipped with a small integer parameter k; such problems are called ***parameterized problems***. A parameterized problem is said to be ***fixed-parameter tractable*** if there is an algorithm for solving it on inputs of size n in time $f(k)n^{O(1)}$, where f is an arbitrary function in k only. For more information on parameterized algorithms and complexity, see [DoFe99, FlGr06].

FACTS

F13: It is easy to solve the k-clique problem in time $O(k^2 n^k)$.

F14: [ChHuKaXi06] showed that the k-clique problem cannot be solved in time $n^{o(k)}$ unless the exponential time hypothesis fails.

REMARK

R15: Downey and Fellows [DoFe95] defined a hierarchy of parameterized problems, the *W hierarchy*, that they conjectured are not fixed-parameter tractable; they proved that the k-clique problem is hard for the first level of this hierarchy, W[1]. Thus, according to their conjecture, the k-clique problem is not fixed-parameter tractable. Moreover, this result provides the basis for proofs of W[1]-hardness of many other problems, and thus serves as an analogue of the Cook–Levin theorem for parameterized complexity.

5.3.4   Bounds on Independence and Clique Numbers

Every maximum-clique or independent-set heuristic provides an 'algorithmic' lower bound to the corresponding problem. In this subsection, we consider 'analytical' ones that require only certain parameters of the input graph.

Lower Bounds

Caro and Wei obtained the lower bound given in Fact F15, and Alon and Spencer gave an elegant probabilistic proof of this bound [AlSp92]. Sakai, Togasaki, and Yamazaki generalized Fact F15 to vertex-weighted graphs (Fact F16).

NOTATION: For a vertex v in a graph G, $N(v)$ is the set of vertices adjacent to v and $N[v] = N(v) \cup \{v\}$.

FACTS

F15: [Ca79,We81] Let $G = (V, E)$ be a graph. Then $ind(G) \geq \sum_{v \in V} 1/(deg(v) + 1)$.

F16: [SaToYa03] Every vertex-weighted graph $G = (V, E)$ contains an independence set S of weight at least

$$\max\{\sum_{x \in V} w(x)/(deg(x) + 1), \sum_{x \in V} w(x)^2/[\sum_{y \in N[x]} w(y)]\}.$$

Moreover, a set S with the above property can be found in polynomial time.

Selkow [Se94] improved the Caro-Wei bound using additional information.

F17: [Se94] Let $G = (V, E)$ be a graph. Then

$$ind(G) \geq \sum_{v \in V} \frac{1}{deg(v) + 1}(1 + \max\{0, \frac{deg(v)}{deg(v) + 1} - \sum_{u \in N(v)} \frac{1}{deg(u) + 1}\}).$$

The next few results involve the eigenvalues of the adjacency matrix A_G of G.

DEFINITION

D15: The **Perron root** $\lambda_P(G)$ is the largest eigenvalue of A_G.

FACTS

F18: The adjacency matrix of a connected graph G is irreducible, symmetric, and has all non-negative entries; hence, all its eigenvalues are real (see, e.g., [HoJo85]).

F19: [Wi86] For a connected graph G on n vertices, $\omega(G) \geq \frac{n}{n - \lambda_P(G)}$. (This bound was improved by Budinich [Bu03].)

Upper bounds

FACTS

F20: [Wi67] For a connected graph G, $\omega(G) \leq \lambda_P(G) + 1$. Equality holds if and only if G is complete.

NOTATION: For a connected graph G, let $\Lambda_{-1}(G)$ denote the number of eigenvalues of A_G that do not exceed -1.

F21: [AmHa72] For a connected graph G, $\omega(G) \leq \Lambda_{-1}(G) + 1$. Equality holds if and only if G is complete multipartite.

F22: [Bu03] For a connected graph G on n vertices, $\omega(G) \leq n - \frac{1}{2}rank A_{\overline{G}}$.

F23: Each of the three lower bounds given above can be computed in time $O(n^3)$, and one can find examples that show the bounds are sharp [Bu03].

COMPUTATIONAL NOTE: Budinich [Bu03] tested the lower bounds on a set of 700 random graphs of order 100 and 200. For these graphs, the smaller of the first two upper bounds was almost always better than the third upper bound.

5.3.5 Exact Algorithms

Clique Enumeration

Harary and Ross [HaRo57] initiated an algorithmic and theoretical study of the enumeration of all cliques in a graph. This topic has a variety of applications (see, e.g., [Bo64, HaRo57, PaUn59]). The first significant theoretical result (see Fact F24) is due to Moon and Moser [MoMo65].

FACTS

F24: [MoMo65] The maximum number of cliques in an n-vertex graph equals

$$
\begin{cases}
3^{n/3} \text{ if } n \equiv 0 \ (mod \ 3) \\
4 \cdot 3^{(n-4)/3} \text{ if } n \equiv 1 \ (mod \ 3) \\
2 \cdot 3^{(n-2)/3} \text{ if } n \equiv 2 \ (mod \ 3)
\end{cases}
$$

(Some extensions of this result are discussed in [SaVa02].)

F25: [ToTaTa06] There is an algorithm for listing all cliques of an n-vertex graph in time $O(3^{n/3})$. This is a modification of the backtracking algorithm of Bron and Kerbosch [BrKe73].

DEFINITIONS

D16: We say that an algorithm A for generation of all cliques has **time delay** T if T bounds from above the time spent by A between two consecutive cliques and before the first clique is generated.

D17: Given a graph G, the **arboricity** is the minimum number of edge-disjoint acyclic subgraphs whose union is G.

FACTS

F26: [TsIdAvSh77] There is an algorithm for listing all cliques of a graph $G = (V, E)$ with time delay $O(|V||E|)$.

F27: [ChNi85] There is an algorithm for listing all cliques of a graph $G = (V, E)$ with time delay $O(a|E|)$, where a is the arboricity of G.

F28: [MaUn04] There is an algorithm for listing all cliques in an n-vertex graph with time delay $O(M(n))$, where $M(n)$ is the time required to multiply two $n \times n$ matrices.

F29: [MaUn04] There is an algorithm for listing all cliques in a graph $G = (V, E)$ with time delay $O(\Delta^4)$, where Δ is the maximum degree of G; the algorithm also requires $O(|V||E|)$ time before the first clique is generated.

Maximum-Clique and Maximum-Weight-Clique Algorithms

Clearly, the algorithms mentioned previously can be used (directly or after simple modifications) to find a maximum clique in a graph. However, the maximum-clique problem has attracted more attention than the clique-enumeration problem, and the use of certain procedures has made algorithms for the maximum-clique problem quite fast.

FACTS

In the following three facts, n denotes the number of vertices in the input graph.

F30: [TaTr77] There is an algorithm for solving the maximum-clique problem with time complexity $O(1.2599^n)$.

F31: [Ro86] There is an algorithm for solving the maximum-clique problem with time complexity $O(1.2108^n)$.

F32: [Ro01] There is an algorithm for solving the maximum-clique problem with time complexity $O(1.1889^n)$. This appears to be the fastest (with respect to the worst-case complexity) currently known algorithm for the problem.

COMPUTATIONAL NOTE: Branch-and-cut algorithms were used with great success for several combinatorial optimization problems; see, e.g., Section 4.6. However, for the maximum-clique problem, branch-and-cut algorithms currently remain, in general, inferior to the state-of-the-art branch-and-bound algorithms [RoSm01].

COMPUTATIONAL NOTE: There are several quite efficient branch-and-bound algorithms that use fast coloring heuristics to produce upper and lower bounds. See, e.g., algorithms in [Os02, ToKa07].

R16: Branch-and-bound algorithms for the maximum-weight-clique problem are discussed in [BoBuPaPe99, Os01].

R17: Some algorithms have been compared on random graphs and on special families of graphs inspired by applications. For the maximum-clique problem, frequently occurring special families of graphs are collected in [JoTr96] and explained in [HaPaVa93]. [Os01] introduces a special family of instances for the maximum-weight-clique problem.

COMPUTATIONAL NOTE: There are a few maximum-clique and maximum-weight-clique computer codes freely available for research purposes; see, e.g., [Di,NiOs03].

5.3.6   Heuristics

When the instance of the maximum-weight-clique problem under consideration is of large size or the data is not precise (which is the case in many applications) or a solution has to be obtained very quickly, one should resort to heuristics rather than exact algorithms. Moreover, heuristics form important parts of many exact algorithms.

Construction Heuristics and Local Search

D18: A *construction heuristic* produces a feasible solution without any attempt to improve it.

REMARKS

R18: Construction heuristics are normally very fast and provide quick solutions, and lower or upper bounds. However, their solutions cannot be expected to be of high quality.

R19: The simplest construction heuristic for the maximum-clique problem is to add one vertex at a time to an emerging clique. It is logical to choose in each iteration an eligible vertex of maximum degree [KoRu87]. Alternatively, one may delete vertices from the given graph one by one until a clique is obtained [KoRu87].

R20: Most approximation algorithms for various optimization problems are, in fact, construction heuristics. Examples of construction heuristics used in maximum-clique approximation algorithms may be found in [BoHa92] and [Fi04].

DEFINITIONS

D19: A *local search (LS)* heuristic starts from a feasible solution and in each iteration until termination chooses the next solution from a *neighborhood* of solutions that are in some prescribed sense close to the current solution.

D20: An *improvement LS* (or simply *local improvement*) is a local search that always chooses a solution that is better than the current one and terminates when it cannot find one in the neighborhood.

COMPUTATIONAL NOTE: Local improvement does not appear to perform particularly well for the maximum-clique problem [GuGuMiZv04] as it may terminate at a relatively small clique. Local search algorithms that do not require monotonic improvement are much more flexible in that they can escape from local optima that are non-maximum cliques; perhaps the most flexible among them is briefly discussed below.

Tabu Search

This *metaheuristic* appears to provide a good trade-off between computational time and solution quality.

DEFINITION

D21: *Tabu search* is a local search in which solutions that are worse than the current one can be chosen provided that they are not in any of the so-called *tabu lists*.

REMARKS

R21: Tabu search was introduced independently by Glover [Gl89, Gl90] and Hansen and Jaumard [HaJa90a]. 'Pure' tabu search techniques for the maximum-clique problem were implemented in a number of papers; see, e.g., [FrHeWe89, SoGe96].

COMPUTATIONAL NOTE: Tabu-search algorithms use parameters that have to be fine-tuned in order to achieve good results. This slows down the development and use of tabu-search computer codes. Battiti and Protasi [BaPr01] deal with this issue by adjusting the parameters using an internal learning loop.

R22: For brief descriptions and discussions of other metaheuristics applied to the maximum-clique problem, see [BoBuPaPe99].

References

[AlFeWiZu95] N. Alon, U. Feige, A. Wigderson, and D. Zuckerman, Derandomized graph products. *Computational Complexity* **5** (1995), 60–75.

[AlSp92] N. Alon and J. H. Spencer, *The Probabilistic Method*, Wiley, 1992.

[AmHa72] A. T. Amin and S. L. Hakimi, Upper bounds on the order of a clique of a graph. *SIAM J. Appl. Math.* **22** (1972), 569–573.

[AuCrGaKaMaPr99] G. Ausiello, P. Crescenzi, G. Gambosi, V. Kann, A. Marchetti-Spaccamela, and M. Protasi, *Complexity and Approximation*, Springer, 1999.

[BaPr01] R. Battiti and M. Protasi, Reactive local search for the maximum clique problem. *Algorithmica* **29** (2001), 610–637.

[BePe90] P. Berman and A. Pelc, Distributed fault diagnosis for multiprocessor systems. In *Proc. 20th Annual Int. Symp. Fault-Tolerant Comput.* (Newcastle, UK), 340–346, 1990.

[Bo64] R. E. Bonner, On some clustering techniques. *IBM J. Res. Develop.* **8** (1964), 22–32.

[Bo97] I. M. Bomze, Evolution towards the maximum clique. *J. Global Optim.* **10** (1997), 143–164.

[BoBuPaPe99] I. Bomze, M. Budinich, P. M. Pardalos, and M. Pelillo, The Maximum Clique Problem. In *Handbook of Combinatorial Optimization (Supplement Volume A)* (D.-Z. Du and P. M. Pardalos, eds.), Kluwer, 1999.

[BoHa92] R. Boppana and M. M. Halldórsson, Approximating maximum independent sets by excluding subgraphs. BIT 32 (1992), 180–196.

[BrKe73] C. Bron and J. Kerbosch, Algorithm 457: Finding all cliques of an undirected graph. *Commun. ACM* **16** (1973), 575–577.

[Bu03] M. Budinich, Bounds on the maximum clique of a graph. *Discrete Appl. Math.* **127** (2003), 535–543.

[Ca79] Y. Caro, New results on the independence number. Tech. Report, Tel-Aviv University, 1979.

[ChHuKaXi06] J. Chen, X. Huang, I. A. Kanj, and G. Xia, Strong computational lower bounds via parameterized complexity. *J. Comput. Syst. Sci.* **72** (2006), 1346–1367.

[ChNi85] N. Chiba and T. Nashizeki, Arboricity and subgraph listing algorithms. *SIAM J. Comput.* **14** (1985), 210–223.

[Di] C programs available at ftp://dimacs.rutgers.edu/pub/challenge/graph/solvers/

[DoFe95] R. G. Downey and M. R. Fellows, Fixed-parameter tractability and completeness. II. On completeness for W[1]. *Theor. Comput. Sci.* **141** (1995), 109–131.

[DoFe99] R. G. Downey and M. R. Fellows, *Parameterized Complexity*, Springer, 1999.

[Eg31] E. Egerváry, On combinatorial properties of matrices. *Math. Lapok* **31** (1931), 16–28.

[Fi04] U. Fiege, Approximating maximum clique by removing subgraphs. *SIAM J. Discrete Math.* **18** (2004), 219–225.

[FlGr06] J. Flum and M. Grohe, *Parameterized Complexity Theory*, Springer, 2006.

[FrHeWe89] C. Friden, A. Hertz, and M. de Werra, STABULUS: A technique finding stable sets in large graphs with tabu search. *Computing* **42** (1989), 35–44.

[Gl89] F. Glover, Tabu search - Part I. *ORSA J. Comput.* **1** (1989), 190–260.

[Gl90] F. Glover, Tabu search - Part II. *ORSA J. Comput.* **2** (1990), 4–32.

[GrYe06] J. L. Gross and J. Yellen, *Graph Theory and Its Applications*, Second Edition, CRC Press, 2006.

[GuGuMiMa00] N. Gulpinar, G. Gutin, G. Mitra, and I. Maros, Detecting embedded network structures in linear programs. *Comput. Opt. Appl.* **15** (2000), 235–247.

[GuGuMiZv04] N. Gulpinar, G. Gutin, G. Mitra, and A. Zverovitch, Extracting pure network submatrices in linear programs using signed graphs. *Discrete Appl. Math.* **137** (2004), 359–372.

[GuVaYe03] G. Gutin, A. Vainshtein, and A. Yeo, Domination analysis of combinatorial optimization problems. *Discrete Appl. Math.* **129** (2003), 513–520.

[Ha98] M. Halldórsson, Approximation of independent sets in graphs. In *Proc. of APPROX'98*, 1998, 1–13.

[Ha99] J. Hastad, Clique is hard to approximate within $n^{1-\epsilon}$. *Acta Mathematica* **182** (1999), 105–142.

[HaJa90] M.-H. Han and D. Jang, The use of maximum curvature points for the recognition of partially occluded objects. *Pattern Recognition* **23** (1990), 21–33.

[HaJa90a] P. Hansen and B. Jaumard, Algorithms for the maximum satisfability problem. *Computing* **44** (1990) 279–303.

[HoJo85] R. A. Horn and C. R. Johnson, *Matrix Analysis*, Cambridge Uni. Press, 1985.

[HaPaVa93] J. Hasselberg, P. P. Pandalos, and G. Vairaktarakis, Test case generators and computational results for the maximum clique problem. *J. Global Optim.* **3** (1993), 463–482.

[HaRo57] F. Harary and I. C. Ross, A procedure for clique detection using the group matrix. *Sociometry* **20** (1957), 205–215.

[HoSk89] R. Horaud and T. Skordas, Stereo correspondence through feature grouping and maximal cliques. *IEEE Trans. Pattern. Anal. Machine Intell.* **11** (1989), 1168–1180.

[ImPa99] R. Impagliazzo and R. Paturi, The complexity of k-SAT. *Proc. 14th IEEE Conf. on Computational Complexity* (1999), 237–240.

[JoTr96] D. S. Johnson, and M. Trick (eds.), *Clique, Coloring and Satisfiability: Second DIMACS Implementation Challenge*, DIMACS Vol. 26, AMS, 1996 (see also http://dimacs.rutgers.edu/Volumes/Vol26.html).

[Ka72] R. M. Karp, Reducibility among combinatorial problem. In *Complexity of Computer Computations* (R. E. Miller and J. W. Thatcher, eds.), Plenum Press, 1972.

[Ko31] D. König, Graphen und matrizen. *Math. Lapok* **38** (1931), 116–119.

[KoRu87] R. Kopf and G. Ruhe, A computational study of the weighted independent set problem for general graphs. *Found. Control Eng.* **12** (1987), 167–180.

[LaTi01] H. Y. Lau and H. F. Ting, The greedier the better: an efficient algorithm for approximating maximum independent set. *J. Combin. Optim.* **5** (2001), 411–420.

[MaSl79] J. MacWillimas and N. J. A. Slone, *The Theory of Error Correcting Codes*, North-Holland, 1979.

[MaUn04] K. Makino and T. Uno, New algorithms for enumerating all maximal cliques. *Lect. Notes Comput. Sci.* **3111** (2004), 260–272.

[MoMo65] J. W. Moon and L. Moser, On cliques in graphs, *Israel J. Math.* **3** (1965), 23–28.

[MoSt65] T. S. Motzkin and E. G. Straus, Maxima for graphs and a new proof of a theorem of Turán. *Canad. J. Math.* **17** (1965), 533–540.

[NeTr74] G. L. Nemhauser and L. E. Trotter, Properties of vertex packings and independence system polyhedra. *Math. Prog.* **6** (1974), 48–61.

[NeTr75] G. L. Nemhauser and L. E. Trotter, Vertex packings: Structural properties and algorithms. *Math. Prog.* **8** (1975), 232–248.

[NiOs03] S. Niskanen and P. R. J. Östergard, Cliquer User's Guide, Version 1.0, *Communications Lab., Helsinki Univ. Technology*, Tech. Rep. T48, 2003. The current release of the *t*Cliquer code is available from www.hut.fi/~pat/cliquer.html

[Os01] P. R. Östergard, A new algorithm for the maximum-weight clique problem. *Nordic J. Comput.* **8** (2001), 424–436.

[Os02] P. R. Östergard, A fast algorithm for the maximum clique problem. *Discrete Appl. Math.* **120** (2002), 197–207.

[PaPh90] P. M. Pardalos and A. T. Phillips, A global optimization approach for solving the maximum clique problem. *Int. J. Comput. Math.* **33** (1990), 209–216.

[PaUn59] M. C. Paull and S. H. Unger, Minimizing the number of sates in incompletely specified sequential switching functions. *IRE Trans. Electr. Comput.* **EC-8** (1959), 356–367.

[PeJa95] M. Pelillo and A. Jagota, Feasible and infeasible maxima in a quadratic program for maximum clique. *J. Artif. Neural Networks* **2** (1995), 411–420.

[Ro86] J. M. Robson, Algorithms for maximum independent sets. *J. Algorithms* **7** (1986), 425–440.

[Ro01] J. M. Robson, Finding a maximum independent set in time $O(2^{n/4})$. *Technical Report 1251-01, LaBRI, Universit de Bordeaux I*, 2001.

[RoSm01] F. Rossi and S. Smriglio, A branch-and-cut algorithm for the maximum cardinality stable set problem. *Oper. Res. Lett.* **28** (2001), 63–74.

[SaToYa03] S. Sakai, M. Togasaki, and K. Yamazaki, A note on greedy algorithms for the maximum weighted independent set problem. *Discrete Appl. Math.* **126** (2003), 313–322.

[SaVa02] B. E. Sagan and V. R. Vatter, Maximal independent sets in graphs with at most r cycles. Preprint, 2002.

[Se94] S. M. Selkow, A probabilistic lower bound on the independence number of graphs. *Discrete Math.* **132** (1994), 363–365.

[Sh90] N. Z. Shor, Dual quadratic estimates in polynomial and Boolean programming. In *Computational Methods in Global Optimization* (P. M. Pardalos and J. B. Rosen, eds.), *Ann. Oper. Res.* **25** (1990), 163–168.

[SoGe96] P. Soriano and M. Gendreau, Tabu search algorithms for the maximum clique problem. In [JoTr96], 221–242, 1996.

[TaTr77] R. E. Tarjan and A. E. Trojanowski, Finding a maximum independent set. *SIAM J. Comput.* **13** (1977), 537–546.

[ToKa07] E. Tomita and T. Kameda, An efficient branch-and-bound algorithm for finding a maximum clique with computational experiments. *J. Glob. Optim.* **37** (2007), 95–111.

[ToTaTa06] E. Tomita, A. Tanaka, and H. Takahashi, The worst-case time complexity for generating all maximal cliques and computational experiments. *Theor. Comput. Sci.* **363** (2006), 28–42.

[TsIdAvSh77] S. Tsukiyama, M. Ide, H. Aviyoshi, and I. Shirakawa, A new algorithm for generating all maximum independent sets. *SIAM J. Comput.* **6** (1977), 505–517.

[We81] V. K. Wei, A lower bound on the stability number of a simple graph. Bell Lab. Tech. Memo., No. 81-11217-9, 1981.

[Wi67] H. S. Wilf, The eigenvalues of a graph and its chromatic number. *J. London Math. Soc.* **42** (1967), 330–332.

[Wi86] H. S. Wilf, The spectral bounds for the clique and independent numbers of graphs. *J. Combin. Theory B* **40** (1986), 113–117.

Section 5.4
Factors and Factorization

Michael Plummer, Vanderbilt University

INTRODUCTION

The vast body of work on factors and factorizations has much in common with other areas of graph theory. Indeed, factorization significantly overlaps the topic of *edge-coloring* (cf. §5.1), since any color class of a proper edge-coloring in a graph is just a matching. Moreover, the *hamiltonian cycle problem* (cf. §4.5) can be viewed as the search for a connected 2-factor. Due to space constraints, we will treat factors of *finite undirected* graphs only. Nevertheless, several papers dealing with infinite graph factors and directed graph factors are included in our list of references.

5.4.1   Preliminaries

DEFINITIONS

D1: Given a graph (multigraph, general graph) G, we say that H is a **factor** of G if H is a spanning subgraph of G.

D2: A factor that is n-regular is called an n-**factor**.

D3: A factor defined only in terms of the degrees of its vertices is called a **degree factor**.

D4: A factor described in terms of graph-theoretic properties other than its vertex degrees is called a **component factor**.

D5: If a graph G can be represented as the edge-disjoint union of factors $F_1, F_2, \ldots F_k$, we shall refer to $\{F_1, F_2, \ldots F_k\}$ as a **factorization** of graph G.

FACTS

Today most workers in the field attribute the birth of graph factorization to two theorems of the Danish mathematician Julius Petersen. The analogous result of Bäbler for regular graphs of *odd degree* did not appear until almost fifty years later.

F1: [Pe1891] A 3-regular multigraph with at most one cutedge contains a 1-factor (and hence also a 2-factor).

F2: [Pe1891] Every $2k$-regular multigraph contains a 2-factor (and, hence, it has a factorization into 2-factors).

F3: [Ba38] Every 2-edge-connected $(2k + 1)$-regular multigraph contains a 2-factor.

REMARKS

R1: The names *degree factors* and *component factors* for the two main categories of factors treated in the literature seem to be due to Akiyama and Kano [AkKan85a].

R2: These two main problem categories overlap. For example, finding a 1-factor and finding a factor each component of which is an edge amounts to the same thing.

R3: A thorough survey tracing the descendants of Petersen's factorization results for regular graphs may be found in [Vo95].

5.4.2   1-Factors

The most studied of degree factors are those in which each component is a single edge. We observe that §11.3 applies matchings to assignments. In our complementary approach here, we are principally interested in those properties of 1-factors that most naturally extend to analogous properties of more general factors.

Conditions for a Graph to Have a 1-Factor

DEFINITIONS

D6: A 1-*factor* (or *perfect matching*) of graph G is a set of vertex-disjoint edges in G which together span $V(G)$.

D7: The bipartite graph $K_{1,3}$ is often called a *claw*. A graph containing no $K_{1,3}$ as an induced subgraph is said to be *claw-free*.

D8: Graph G is said to have the *odd-cycle property* if every pair of odd cycles in G either have a vertex in common or are joined by an edge.

D9: The *toughness* of graph G, denoted by tough (G), is defined to be $+\infty$ when G is complete and otherwise to be

$$\min\{|S|/c(G - S)|S \subseteq V(G)\}$$

where the minimum is taken over all subsets $S \subseteq V(G)$ and $c(G-S)$ denotes the number of components of $G - S$.

D10: The ***binding number*** of a graph G, denoted $\mathrm{bind}\,(G)$, is defined to be

$$\min\left\{|N(X)|/|X| \;\middle|\; \emptyset \neq X \subseteq V(G) \text{ and } N(X) \neq V(G)\right\}$$

D11: We call a sequence of non-negative integers $d_1, \ldots, d_n$ ***graphical*** if there exists a graph G of order n the vertices of which have, in some order, degrees $d_1, \ldots, d_n$.

D12: If $|V(G)| \geq 2m+2$ and G has a perfect matching, then G is m-***extendable*** if every matching of size m extends to (i.e., is a subset of) a perfect matching.

D13: Let T be a set of vertices in graph G.

- A T-***path*** is a path beginning and ending in T.
- A T-***path covering*** is a union of vertex-disjoint T-paths covering T.

FACTS

Arguably, the most influential theorem in the study of 1-factors has been the seminal result called *Tutte's 1-factor Theorem*.

F4: [Tu47] **Tutte's 1-Factor Theorem**: A graph G has a 1-factor if and only if for each $S \subseteq V(G)$, $c_o(G-S) \leq |S|$, where $c_o(G-S)$ denotes the number of components of $G-S$ which have an odd number of vertices.

F5: [Pe1891] **Petersen's Theorem**: Every 2-edge-connected 3-regular multigraph has a 1-factor.

F6: [Ple79] If G is an r-regular, $(r-1)$-edge-connected multigraph of even order, then G has a 1-factor not containing any of $r-1$ prescribed edges. (And hence G has a 1-factor containing any prescribed edge.)

F7: [Kai08] If $r \geq 2$ and G is an r-regular, r-edge-connected graph with T a subset of $V(G)$, then every edge of G is contained in a T-path covering.

F8: [Su74, Su76], [La75] If G is a connected claw-free graph of even order, then G has a 1-factor.

F9: [Su76] If G is an n-connected graph of even order, and if G has no induced subgraph isomorphic to the bipartite graph $K_{1,n+1}$, then G has a 1-factor.

F10: [FuHoMc65] If G is r-regular of even order and has the odd-cycle property, then G has a 1-factor.

F11: [Ni78, Ni79] If G is a k-connected graph ($k \geq 4$) of even order and if $\gamma(G) < k(k-2)/4$, then G has a 1-factor. (As in Chapter 7, $\gamma(G)$ denotes the (orientable) genus of G.)

F12: If G is of even order and $\mathrm{tough}\,(G) \geq 1$, then G has a 1-factor. This follows immediately from Tutte's 1-factor Theorem.

F13: [Ku73], [Lo74] There exists a graph G having a 1-factor and degree sequence $d_1, d_2, \ldots, d_n$ if and only if both the sequences $d_1, \ldots, d_n$ and $d_1-1, \ldots, d_n-1$ are graphical.

F14: (essentially due to Anderson [An73]) Let G be a graph of even order. If, for all $X \subseteq V(G)$,

$$|N(X)| \geq \min\left\{|V(G)|, \frac{3}{4}|X| - \frac{2}{3}\right\}$$

then G has a 1-factor. This theorem can be regarded as a binding number result.

F15: [LiGrHo75] If G is a connected graph of even order the automorphism group of which acts transitively on $V(G)$, then G has a 1-factor containing any given edge. Highly symmetric graphs of even order are guaranteed to have 1-factors by this result.

F16: [AlPl11] If $m \geq 0$ and G is a 5-connected even planar triangulation with $|V(G)| \geq 2m + 2$, and M is a matching in G with $|M| \leq m$ such that the distance between any two of its edges is at least 5, then G has a perfect matching that contains M.

REMARKS

R4: A number of sufficient conditions quite similar to that of Anderson above are collected and compared in [Wo90]. A similar condition sufficient for a bipartite graph to have a k-factor (respectively, $[a, b]$-factor (see below)) may be found in [EnOtKan88] (respectively, [Kan90a], [Che93]).

R5: There are now many papers investigating the existence of 1-factors containing or excluding specified edge sets. However, space does not permit us to treat these results and for the case of 1-factors, we direct the interested reader to two survey articles [Pl94, Pl96] and the recent book [YuLi09].

The Number of 1-Factors: Bounds

DEFINITIONS

D14: A graph G is said to be **bicritical** if $G - x - y$ has a 1-factor for every choice of two different vertices x and y. (For further reading on bicritical graphs, see [LoPl86].)

D15: A **hexagonal system** is a 2-connected plane graph in which every face is a hexagon.

D16: A 3-connected cubic plane graph with 12 pentagonal faces and all remaining faces hexagons is called a **fullerene**.

NOTATION: $\Phi(G)$ denotes the number of 1-factors in graph G.

FACTS

F17: Let G be connected and have a unique 1-factor. Then:

(a) [Ko59] G has a cutedge belonging to the 1-factor;

(b) [LoPl86] G contains a vertex of degree $\leq \lfloor \log_2(p + 1) \rfloor$; and

(c) [Hetyei (unpublished)] $|E(G)| \leq (|V(G)|/2)^2$.

F18: If G is k-connected and has a 1-factor, then either

(a) G has at least $k!$ 1-factors, or else

(b) G is bicritical. It seems somewhat counterintuitive that bicritical graphs should be the exception here. It has proven much more difficult to bound $\Phi(G)$ in the bicritical case.

F19: If G is bicritical, then $\Phi(G) \geq |V(G)|/2 + 1$. Study of the *perfect matching polytope* of G, $PM(G)$, (see [LoPl86]) can be utilized to give the bound in this result.

F20: If the graph G is k-connected and contains a 1-factor, and if $|V(G)|$ is sufficiently large, then G has at least $k!$ 1-factors.

F21: The number of perfect matchings in a cubic bridgeless graph is exponential. More specifically:
[EsKaKi KrNo11] If G is cubic and bridgeless, then $\Phi(G) \geq 2^{|V(G)|/3656}$.
(Hence, in particular, this bound holds for fullerenes.)

F22: [AlFr08] If G is a simple graph of even order and degree sequence $d_1, d_2, \ldots, d_n$, then $\Phi(G) \leq \prod_{i=1}^{n} (d_i!)^{1/2d_i}$.

REMARKS

R6: Gabow, Kaplan and Tarjan [GaKaTa99, GaKaTa01] developed an $O(|E| \log^4 |V|)$ algorithm to test whether a graph has a *unique* 1-factor and find it, if it exists.

R7: One can bound $\Phi(G)$ below by the matrix function called a *Pfaffian*. (For details, see [LoPl86; §8.3].) In the case when G is *planar*, the Pfaffian can be used to exactly compute $\Phi(G)$ in polynomial time.

R8: The connectivity of the graph G can also be employed to yield a lower bound on $\Phi(G)$ in some cases.

R9: Computing $\Phi(G)$ when G is a hexagonal system or a fullerene is of interest to chemists. For a popular introduction to the former, see [Ri01]. Hexagonal systems are used to model benzenoid hydrocarbons. As for fullerenes, the literature is vast, but a brief background on the $\Phi(G)$ problem can be found in [Dos02], [KaKrMiSe09]. Some other mathematical questions about fullerenes may be found in [Ma00].

R10: A long-standing open conjecture about fullerenes states that they all contain a hamiltonian cycle. Currently, the best lower bound for a longest cycle in a fullerene of order n is $6n/7 + 2/7$. In [ErKaMi09] this bound is obtained by studying the 2-factors of the fullerene.

1-Factors in Bipartite Graphs

In the special case of bipartite graphs, the story of 1-factors has two principal historical roots, one in a result due to P. Hall [Ha35], and the other in a result due to König [Ko31, Ko33].

DEFINITIONS

D17: A ***vertex cover*** of a graph G is a subset $C \subseteq V(G)$ such that every edge of G has at least one endvertex in C.

D18: The ***vertex-covering number*** of a graph G is the size of any smallest vertex cover in G. Notation: $\tau(G)$.

D19: The ***matching number*** of a graph G is the size of any largest matching in G. Notation: $\nu(G)$.

D20: The ***permanent*** of an $n \times n$ matrix A, denoted per A, is given by

$$\text{per } A = \sum a_{1\pi(1)} a_{2\pi(2)} \cdots a_{n\pi(n)}$$

where the sum extends over all permutations π of the set $\{1, \ldots, n\}$.

FACTS

F23: [Ha35] **Hall's Theorem**: Let G be a bipartite graph with vertex bipartition $V(G) = A \cup B$. Then G has a matching of A into B if and only if $|N(X)| \geq |X|$, for all $X \subseteq A$.

F24: [Fr1912] **Marriage Theorem**: Let G be a bipartite graph with vertex bipartition $V(G) = A \cup B$. Then G has a 1-factor matching A onto B if and only if

(a) $|A| = |B|$ and

(b) $|N(X)| \geq |X|$, for all $X \subseteq A$.

This earlier result of Frobenius is an immediate consequence of Hall's Theorem.

F25: It is clear that in any graph G, the matching number and the vertex-covering number are related by the inequality $\nu(G) \leq \tau(G)$.

F26: [Ko31, Ko33] **König's Theorem**: If G is bipartite, then $\nu(G) = \tau(G)$.

F27: [Ha48] Let G be a simple bipartite graph with bipartition $V(G) = A \cup B$, and assume that each vertex in A has degree at least k. If G has at least one 1-factor, then it has at least $k!$ 1-factors.

F28: Let G be a simple k-regular bipartite graph on $2n$ vertices. Then

$$n! \left(\frac{k}{n}\right)^n \leq \Phi(G) \leq (k!)^{n/k}$$

The first inequality is equivalent to the famous *van der Waerden Conjecture* [Wa26] on permanents, which was proved independently by [Fa81] and [Eg80, Eg81]. The second inequality was proved by [Br73].

F29:   [Sc98b] If G is a k-regular bipartite graph of order $2n$, then

$$\Phi(G) \geq \left(\frac{(k-1)^{k-1}}{k^{k-2}} \right)^n$$

REMARKS

R11:   In fact, it can be shown that Hall's Theorem and König's Theorem are equivalent.

R12:   Since König's Theorem asserts the equality of the maximum of one quantity and the minimum of another, it is often referred to as a *minimax* theorem, especially in the study of linear programming. For an introduction to such ideas within the confines of graph theory, and for the associated *polytopal* ideas, see [LoPl86; Ch. 7 and 12].

The Number of 1-Factors: Exact Counting

DEFINITIONS

D21:   A graph G is a ***threshold graph*** if there is a real-valued assignment of weights to the vertices such that $e = uv$ if and only if $w(u) + w(v) > 0$.

D22:   A graph G is a ***split graph*** if $V(G) = A \cup B$ where A spans a complete graph and B spans an independent set.

D23:   A graph G is ***Pfaffian*** if the edges can be oriented so that every alternating cycle C (i.e., an even cycle C such that $G - V(C)$ contains a 1-factor) has an odd number of edges oriented clockwise.

D24:   If G is a graph drawn in the plane and M is a 1-factor in G, then the ***crossing number of*** M, denoted by $cr(M)$, is the number of pair-wise crossings of edges of M.

REMARK

R13:   As mentioned previously, $\Phi(G)$ can be computed in polynomial time if G is planar.

FACTS

F30:   [OkUeUn10] $\Phi(G)$ can be computed in polynomial time if the graph is a threshold graph.

F31:   [Kas61, Kas63, Kas67, Fi61, TeFi61] If G is Pfaffian, then $\Phi(G)$ can be computed in polynomial time.

F32:   [McRoSeTh97, RoSeTh99, Mc04, Thomas06] There is a polynomial algorithm to decide if a bipartite graph is Pfaffian or not.

F33:   [No08] A graph G is Pfaffian if and only if there exists a drawing of G in the plane such that $cr(M)$ is even for every 1-factor M of G.

F34: [ArLaSe91] $\Phi(G)$ can be computed in polynomial time if G has bounded tree-width (cf. Chapt. 2).

F35: [Va79a] $\Phi(G)$ is #P-complete for bipartite graphs.

F36: [DaLu92] $\Phi(G)$ remains #P-complete even if the bipartite graphs are 3-regular.

F37: [OkUeUn10] Determining $\Phi(G)$ is #P-complete for chordal graphs, split graphs and even chordal bipartite graphs.

F38: [OkUeUn10] The complexity of computing $\Phi(G)$ is presently unsettled for the classes of bipartite permutation graphs, proper interval graphs, interval graphs and bipartite interval graphs.

REMARKS

R14: The problems of counting (and bounding) the number of 1-factors have important applications to the so-called *dimer problem* of crystal physics. Here the underlying (bipartite) graphs are d-dimensional rectangular lattices; the edges in the lattices are called *dimers*. (A perfect matching in such a lattice is called a *dimer tiling*.) We refer the reader to three references which give succinct overviews of the dimer problem: [KeRaSi96, BeSu99, Sc03].

R15: Recently, possible connections between the dimer problem and its relation to string theory and the theory of black holes have come under investigation (cf. [HeVa07].)

5.4.3 Degree Factors

REMARK

R16: The names *degree factors* and *component factors* for the two main categories of factors treated in the literature seem to be due to Akiyama and Kano [AkKan85]. Note that they sometimes overlap, however. For example, a 2-factor is just a factor in which each component is a cycle.

k-factors

DEFINITIONS

D25: A k-***factor*** of a graph G is a k-regular subgraph that spans G.

D26: A graph G is ***hypohamiltonian*** (respectively, ***hypotraceable***) if G does not have a hamiltonian cycle (respectively, path), but $G - v$ does, for all $v \in V(G)$.

D27: For any graph G, let $\sigma_2(G) = \min\{d_G(x) + d_G(y) | x, y \in V(G), x \neq y, xy \notin E(G)\}$.

D28: The *face-width* of a graph G embedded in a surface is the minimum number of times any non-contractible curve in the surface intersects G. (Also called *representativity*; see Section 7.6.)

NOTATION: The face-width of an embedded graph G is denoted $fw(G)$.

D29: The *Euler genus* of a surface Σ is twice the ordinary genus (or handle number) if the surface is orientable and equal to the ordinary genus (or cross-cap number) if the surface is non-orientable.

FACTS

F39: [EnJaKatSa85] If tough $(G) \geq k$, then G has a k-factor. (This was conjectured by Chvátal [Ch73].)

F40: [Nis89] Let G be a graph and k an even non-negative integer. If

$$\kappa(G) \geq \max\{k(k+2)/2, (k+2)\alpha(G)/4\}$$

then G has a k-factor. ($\alpha(G)$ and $\kappa(G)$ are the independence number and the connectivity, respectively.)

NOTATION: We denote the minimum degree in a graph G by $\delta(G)$, and the minimum degree sum, $\deg(u) + \deg(v)$, over all pairs of non-adjacent vertices u and v, by $\sigma_2(G)$.

F41: [IiNi91] Let k be a positive integer and let G be a graph of order $n \geq 4k - 5$ with $\delta(G) \geq k$ and kn even. If $\sigma_2(G) \geq n$, then G has a k-factor. (The condition on $\sigma_2(G)$ is called an *Ore condition* after Ore who first introduced a condition of this type and showed it sufficient for the existence of a hamiltonian cycle.)

F42: [KaYo02] Let G be a graph with $|V(G)| \geq 4k + 1$ and let e be an edge of G. If $\sigma_2(G) \geq |V(G)| + 1$, then G contains a 2-factor F such that F contains edge e and consists of exactly k cycles.

F43: [IiNi97] Let $k \geq 2$ be an integer and let G be a connected graph of order n with $\delta(G) \geq k$. Suppose that kn is even and $n \geq 9k - 1 - 4\sqrt{2(k-1)^2 + 2}$. If $|N_G(u) \cup N_G(v)| \geq (1/2)(n + k - 2)$ for each pair of non-adjacent vertices u and v, then G has a k-factor. (The sufficiency condition here is called a *neighborhood union condition*.)

F44: [Nis92] Let G be a connected graph of order n and let k be an integer ≥ 3 such that kn is even, $n \geq 4k - 3$ and $\delta(G) \geq k$. Then if $\max\{d(u), d(v)\} \geq n/2$, for all pairs of non-adjacent vertices u and v, G has a k-factor.

F45: [FaFaFlLiLi99] If a graph G is claw-free of order n with $\delta(G) \geq 4$, then G has a 2-factor with at most $[6n/(\delta(G) + 2)] - 1$ components. Moreover, there is an $O(n^3)$ algorithm to construct such a 2-factor.

F46: [Ku73] If k is a positive integer and the sequences $d_1, \ldots, d_n$ and $d_1 - k, \ldots, d_n - k$ are both graphical, then $d_1, \ldots, d_n$ can be realized by a graph G which contains a k-factor.

F47: [Kat83] If G is a graph and k is an even non-negative integer, then if $G - v$ has a k-factor for all $v \in V(G)$, G also has a k-factor.

F48: [Sa91] Suppose G is a graph with a 1-factor F and order at least four and let k be a positive integer. Then if $G - \{u, v\}$ has a k-factor for each edge $uv \in F$, G itself has a k-factor.

F49: [Kat83] If G is either hypohamiltonian or hypotraceable, then G has a 2-factor.

F50: [Nis90] Let G be a graph and m a non-negative integer.

(i) If $m \geq 2$ and even, and $\delta(G) \geq m + 1$, then $L(G)$ has a $2m$-factor.

(ii) If G is connected, $|E(G)|$ even, and $\delta(G) \geq m + 2$, then $L(G)$ has a $(2m+1)$-factor.

F51: [Kat90] Let G be a bipartite graph with bipartition $V(G) = X \cup Y$ and k be a positive integer. Suppose that the following three conditions hold:

(i) $|X| = |Y|$;

(ii) $\delta(G) \geq \lceil |X|/2 \rceil \geq k$;

(iii) $|X| \geq 4k - 4\sqrt{k} + 1$ when $|X|$ is odd, and $|X| \geq 4k - 2$ when $|X|$ is even.

Then G has a k-factor.

F52: [Nis91] If $k \geq 2$ is an integer and G is a connected graph with $k|E(G)|$ even and if $\delta(L(G)) \geq (9k + 12)/8$, then $L(G)$ has a k-factor.

F53: [Kan93] Let k be a positive integer and let G be a connected graph of order n with $\delta(G) \geq k$, where kn is even and $n \geq 4k - 3$. If $\sigma_2(G) \geq n$, then G has both a hamiltonian cycle C and a k-factor F, and hence, G has a connected $[k, k + 2]$-factor.

F54: [AlEgFuOtSa11]

(a) Let k and n be integers with $k \geq 2$ and $n \geq 3$, and let G be a k-edge-connected $K_{1,n}$-free graph. If $(k, n) \neq (2, 3)$ and $\delta(G) \geq n - 2 + (n - 1)/(k - 1)$, then G contains a 2-factor.

(b) Let $n \geq 3$ be an integer. Every 2-connected $K_{1,n}$-free graph G with $\delta(G) \geq n$ contains a 2-factor.

(c) Let $n \geq 4$ be an integer. Every $(n - 1)$-connected $K_{1,n}$-free graph has a 2-factor.

(d) A corollary to (a) says that if $n \geq 3$, every n-edge-connected $K_{1,n}$-free graph has a 2-factor.

F55: [KaOz11] (a) If G is a 4-connected graph embedded in a surface of Euler genus g, $\delta(G) \geq 5$, and $fw(G) \geq 4g - 12$, then G has a 2-factor. (b) If G is a 5-connected graph of even order embedded on a surface of Euler genus g and $fw(G) \geq \max\{44g - 117, 5\}$, then G has a 3-factor.

REMARKS

R17: [KatWo87, EgEn89] proved theorems similar to Fact F27, concerning the binding number.

R18: [KlWa73] gives an alternative proof of Fact F32 and also a polynomial algorithm for constructing the graph G containing the k-factor.

R19: Suppose that G is r-regular and has edge-connectivity λ. All values of k for which a multigraph G is guaranteed to have a k-factor are known [BoSaWo85]. Similarly, all such k are known when G is simple [NiesRa98].

R20: Hendry [He84] initiated the study of graphs with *unique* k-factors and his conjecture on the maximum number of edges that such a graph may have was proved by Johann [Joh00a, Joh00b].

R21: There are many results on the existence of 2-factors having special properties such as (a) having a given number of cycles [BrChFaGoLe97]; (b) having cycles of specified lengths [Go01]; and (c) having k cycles containing k specified vertices [EgEn-FaLiSc03]. In [FaFaRy08] there is a complete list of single forbidden subgraphs and pairs of forbidden subgraphs which guarantee that a 2-connected graph contains a 2-factor.

f-factors

Whereas a k-factor is a subgraph with the same degree at every vertex, an f-factor may have a prescription of different degrees.

DEFINITIONS

D30: Let G be a multigraph possibly with loops and f a non-negative, integer-valued function on $V(G)$. Then a spanning subgraph H of G is called an ***f*-factor** of G if $\deg_H(v) = f(v)$, for all $v \in V(G)$.

D31: A set $S \subseteq V(G)$ such that $c_o(G - S) > |S|$ is called a **1-barrier** or **antifactor set**. (Recall that by Tutte's 1-factor Theorem, a graph G with no 1-factor has a 1-barrier.)

NOTATION: Let $e_G(A, B)$ denote the number of edges in graph G joining vertex sets A and B.

FACTS

F56: [Tu54] **Tutte's *f*-Factor Theorem**: The graph G has an f-factor if and only if the following two conditions hold for all disjoint sets $D, S \subseteq V(G)$:

  (i) $f(D) - f(S) + d_{G-D}(S) - q_G(D, S, f) \geq 0$, where $q_G(D, S, f)$ denotes the number of components C of $G - (D \cup S)$ such that $e_G(V(C), S) + f(V(C)) \equiv 1(\bmod 2)$;

  (ii) $f(D) - f(S) + d_{G-D}(S) - q_G(D, S, f) \equiv f(V(G))(\bmod 2)$.

F57: [Tu81] A graph G has an f-factor if and only if it does not have an f-barrier. (An f-barrier is a generalization of a 1-barrier. We omit the details here.)

F58: [Ko01] Let G be a connected graph without multiple edges or loops and let p be an integer such that $0 < p < |V(G)|$. Let f be an integer-valued function on $V(G)$ such that $2 \leq f(v) \leq \deg_G(v)$ for all $v \in V(G)$. If every connected induced subgraph of order p of G has an f-factor, then G has an f-factor, or else $\sum_v f(v)$ is odd.

F59: [KatTs00] Let G be a graph, and let $a \leq b$ be two positive integers such that $\delta(G) \geq \frac{b}{a+b}|V(G)|$ and $|V(G)| > \frac{a+b}{a}(b + a - 3)$. If f is a function from $V(G)$ to $\{a, a+1, \ldots, b\}$ such that $\sum_v f(v)$ is even, then G has an f-factor.

F60: [JaWh89] If G is a 2-edge-connected graph with a unique f-factor F, then some vertex has the same degree in F as in G.

REMARK

R22: There is a procedure for reducing the f-factor problem on a graph G to the 1-factor problem on a larger graph G'. See Chapter 10 of [LoPl86].

[a,b]-factors

DEFINITIONS

D32: Let a and b be integers such that $1 \leq a \leq b$. An $[a, b]$-*factor* of a graph G is a subgraph H such that $a \leq \deg_H(v) \leq b$, for all $v \in V(G)$. (Thus, it is an f-factor such that $a \leq f(v) \leq b$, for all $v \in V(G)$.)

D33: Let f be a function from $V(G)$ to the odd positive integers. A spanning subgraph F of graph G in which $\deg_F(v) \in \{1, 3, \ldots, f(v)\}$ is called a $(1, f)$-*odd-factor* of G.

D34: A $[k, k + 1]$-factor is sometimes called an *almost regular* (or *semiregular*) factor.

D35: A graph G is an $[a, b]$-*graph* if $a \leq \deg(v) \leq b$, for every vertex $v \in V(G)$.

NOTATION: Let $i(G)$ denote the number of isolated vertices of G, and if $S \subseteq V(G)$, let $n(S, j)$ denote the number of vertices of degree j in $G - S$.

FACTS

F61: [HeHeKiLi90] If $1 \leq a < b$, then G has an $[a, b]$-factor if and only if for every set $S \subseteq V(G)$, $\sum_{0 \leq j < a}(a - j)n(S, j) \leq b \cdot |S|$.

F62: [LiLi98] If G is a 2-connected claw-free graph, then G has a connected $[2, 3]$-factor.

F63: [LiZhCh02] If G is a 2-connected claw-free graph containing a k-factor where $k \geq 2$, then G contains a connected $[k, k + 1]$-factor.

F64: [KanSa83] Suppose that k, r, s and t are integers such that $0 \leq k \leq r$ and $1 \leq t$. If $ks \leq rt$, then an $[r, r + s]$-graph has a $[k, k + t]$-factor.

F65: [Nis94] If G is a graph, if a and b are integers such that $1 \leq a < b$, and if $\delta(G) \geq (\alpha(G)/2) + 1$, where $\alpha(G)$ is the independence number, then the line graph $L(G)$ has an $[a,b]$-factor.

F66: [La78, AmKan82] Let $n \geq 2$ be an integer, and let $i(G)$ denote the number of isolated vertices of graph G. The graph G has a $[1,n]$-factor if and only if $i(G-S) \leq n|S|$, for all $S \subset V(G)$. This is an analogue of Tutte's 1-factor theorem.

F67: [YuKan88] Let G be a graph and f a function from $V(G)$ to $\{1,3,\ldots\}$. Then G has a $(1,f)$-odd-factor if and only if $c_o(G - S) \leq \sum_{v \in S} f(v)$, for all $S \subset V(G)$. This is another generalization of Tutte's 1-factor theorem. (See also [KaKa03].) For an n^3 algorithm for finding a $(1,f)$-odd-factor see [KaKa07].)

F68: [Lo70, Tu78] If G is r-regular, then G has a $[k,k+1]$-factor for all k, $0 \leq k \leq r$.

F69: [Th81] If G is an $[r,r+1]$-graph, then G has a $[k,k+1]$-factor for all k, $0 \leq k \leq r$.

REMARKS

R23: A sufficient condition in the spirit of binding number for the existence of an $[a,b]$-factor is given in [Kan90a].

R24: A characterization of graphs with $[a,b]$-factors can be derived immediately from Lovász's (g,f)-factor theorem. (See Fact F70 below.)

R25: [HiWo05] surveyed the area of $[r,r+1]$-factors in $[d,d+s]$-graphs. Somewhat later, Hilton [Hi08, Hi09] introduced an additional parameter into the mix by asking: For which values of d, r and x does a $[d, d+1]$-graph admit an $[r, r+1]$-factorization into x $[r,r+1]$-factors?

(g,f)-factors

DEFINITIONS

D36: Let G be a finite general graph, and let f, g be mappings of $V(G)$ into the non-negative integers. A (g,f)-**factor** of G is a spanning subgraph F such that $g(v) \leq \deg_F(v) \leq f(v)$ for all $v \in V(G)$.

D37: Let G be a finite general graph, and let f, g be mappings of $V(G)$ into the non-negative integers. Graph G is said to have **all** (g,f)-**factors** if and only if G has an h-factor for every h such that $g(v) \leq h(v) \leq f(v)$ for all $v \in V(G)$. Notice that if $f \equiv g \equiv 1$, then a (g,f)-factor (i.e., a $(1,1)$-factor) is just a 1-factor.

FACTS

F70: [Lo70] (g,f)-**factor theorem**: The graph G has a (g,f)-factor if and only if both of the following conditions hold:

(i) $e_G(V(C), S) + f(V(C)) \equiv 1(\bmod\ 2)$;

(ii) $f(D) - g(S) + \deg_{G-D}(S) - \hat{q}_G(D, S, g, f) \geq 0$ for all pairs of disjoint sets $D, S \subseteq V(G)$, where $\hat{q}_G(D, S, g, f)$ denotes the number of components C of $G - (D \cup S)$ having $g(v) = f(v)$ for all $v \in V(C)$.

F71: [La78] let G be a graph and f and g two integer-valued functions defined on $V(G)$ such that $0 \le g(x) \le 1 \le f(x)$. Then G contains a (g, f)-factor if and only if for every subset $X \subseteq V(G)$, the value $f(X)$ is at least equal to the number of connected components C of $G[V - X]$ such that either $C = \{x\}$ and $g(x) = 1$, or $|C|$ is odd and $|C| \ge 3$ and $g(x) = f(x) = 1$ for all $x \in C$.

F72: [EgKan96] Let G be a graph and f and g functions from $V(G)$ to the non-negative integers such that $g(v) \le \deg_G(v), 0 \le f(v)$ and $g(v) < f(v)$, for all $v \in V(G)$. If

$$\frac{g(x)}{\deg_G(x)} \le \frac{f(y)}{\deg_G(y)}$$

for every pair of adjacent vertices x and y in G, then G has a (g, f)-factor.

F73: [Nies98] Let G be a multigraph and let g and f be as in Fact F72 above. Then G has all (g, f)-factors if and only of

$$g(D) - f(S) + d_{G-D}(S) - q_G^*(D, S, g, f) \ge \begin{cases} -1, & \text{if } f \ne g, \\ 0, & \text{if } f = g \end{cases}$$

for all disjoint sets $D, S \subset V(G)$, where $q_G^*(D, S, g, f)$ denotes the number of components C of $G - (D \cup S)$ such that there exists a vertex $v \in V(C)$ with $g(v) < f(v)$ or $e_G(V(C), S) + f(V(C)) \equiv 1(\bmod\ 2)$.

F74: [XuLiTo98] Let $n \ge 3$ be an integer and let G be a $K_{1,n}$-free graph. Let f and g be positive integer-valued functions on $V(G)$, such that $g(v) \le f(v)$, for all $v \in V(G)$. If G has a (g, f)-factor, then G has a *connected* $(g, f + n - 1)$-factor.

F75: [An92] Let G be a graph with the odd-cycle property (D8) and with a k-regular spanning subgraph, and let r be an integer, $1 \le r \le k$. Then G has an r-regular spanning subgraph if either r is even or $|V(G)|$ is even and r is odd.

REMARKS

R26: See also [AnNa98] for more sufficiency conditions for the existence of a (g, f)-factor, [An90, HeHeKiLi90] for simplified existence theorems for such factors, and [Kan84, Kan90b, Li89] for the existence of such a factor having additional properties such as including or excluding prescribed sets of edges.

R27: It is apparently unknown whether there is a polynomial algorithm to test if a graph G has all (g, f)-factors.

R28: For further information on the connections between network flows and graph factors see [FrJu99a, FrJu99b, FrJu99c, FrJu01, KoSt93].

Factors in Random Graphs

There are several popular models of so-called *random* graphs. We will be content to refer to only one of these.

DEFINITIONS

D38: Let $1, \ldots, n$ be a labeling of the vertices and let $\{e_{ij}\}, 1 \leq i < j \leq n$, be an array of independent random variables, where each e_{ij} assumes the value 1 with probability p and 0 with probability $1 - p$. This array determines a **random graph** on $\{1, \ldots, n\}$ where each (ij) is an edge if and only if $e_{ij} = 1$. It is denoted by $G_{n,p}$.

D39: An event E concerning a graph $G \in G_{n,p}$ is said to hold **asymptotically almost surely** (or **a.a.s.**), if $\lim_{n \to \infty}$ Prob $E = 1$.

FACTS

F76: [ErRe66] Let n be even and $p = (1/n)(\log n + w(n))$, with $\lim_{n \to \infty} w(n) = \infty$. Then $G \in G_{n,p}$ has a 1-factor a.a.s.

F77: [ShUp81] Let $p = (1/n)(\log n + (r - 1) \log \log n + w(n))$, with $r \geq 1$ and suppose $\lim_{n \to \infty} w(n) = \infty$. Suppose further that f is a mapping from $V(G)$ into $\{1, \ldots, r\}$ with $\sum_{i=1}^{r} f(x_i)$ even. Then $G \in G_{n,p}$ has an f-factor a.a.s.

REMARK

R29: For several excellent treatments of random graphs, including their factors, see [Ka82, Ka95, JaLuRu00, Bo85, Bo01, MoRe02, ShUp81, ShUp82, JohKaVu08] and Section 8.2 of this *Handbook*.

5.4.4   Component Factors

DEFINITIONS

D40: A **path factor** of graph G is a spanning subgraph of G each component of which is a path.

D41: A P_3-**packing** in a graph G is a subgraph F of G such that each component of F is isomorphic to P_3. If, in addition, F spans G, then F is a P_3-**factor**.

D42: If $\mathcal{H}$ is a set of graphs, then an $\mathcal{H}$-**factor** of a graph G is a spanning subgraph of G each component of which is a member of $\mathcal{H}$.

D43: An F-**factor** is a spanning subgraph in which each component is a single edge or an odd cycle.

D44: A graph G is **well-covered** if every maximal independent set of vertices has the same size.

D45: A graph G is **factor-critical** if $G - x$ contains a 1-factor for every $x \in V(G)$.

D46: Let R be a factor-critical graph with $V(R) = \{x_1, \ldots, x_n\}$, $n \geq 3$. Add new vertices $\{y_1, \ldots, y_n\}$ to R, together with the edges $x_i y_i$, $1 \leq i \leq n$. The resulting graph H is called a **sun**.

FACTS

F78: [AkAvEr80] If G is a simple graph, then G has a path factor if and only if $i(G - S) \leq 2|S|$, for all $S \subseteq V(G)$.

F79: [KaLuYu10] If $i(G-S) \leq |S|/2$ for all $S \subseteq V(G)$, then G contains a $\{K_{1,2}, K_{1,3}, K_5\}$-factor.

F80: [Kane03] Let $c_s(G)$ denote the number of sun components of G. A graph G has a $P_{\geq 3}$-factor if and only if $c_s(G - S) \leq 2|S|$, for all $S \subseteq V(G)$.

F81: [KaKeNi01] Every 2-connected claw-free graph of order n has a P_3-packing which covers at least $\lfloor n/3 \rfloor$ vertices.

F82: [KoZy08]] Every 2-connected cubic graph of order $n > 8$ has a P_3-packing which covers at least $\lceil 9n/11 \rceil$ vertices.

F83: [KaKeNi01] Every 2-connected claw-free graph of order n has a P_3-packing which covers at least $\lfloor n/3 \rfloor$ vertices.

F84: [AnEgKaKawMa02] Let d be a non-negative integer and let G be a claw-free graph with $\delta(G) \geq d$. Then G has a path factor in which all paths have at least $d + 1$ vertices.

F85: [St82] If G is a graph then G has an F-factor if and only if $|N(S)| \geq |S|$, for every independent $S \subseteq V(G)$. (This result can be viewed as a generalization of Hall's Theorem to the non-bipartite case.)

F86: [Mu79, HeKi81a] There is a polynomial algorithm for finding an F-factor or showing that none exists.

F87: [RaVe06] If G is well-covered without isolated vertices, then G has an F-factor.

CONJECTURE

C1: [AkKan85]: Every 3-connected cubic graph of order $3n$ contains a P_3-factor.

REMARK

R30: Graphs that contain, for each edge e, a $P_{\geq k}$-factor, $k \geq 2$, containing e are said to be $P_{\geq k}$-factor covered. Graphs which are $P_{\geq 2}$-factor covered and those which are $P_{\geq 3}$-factor covered are characterized in [ZhHeZh09].

5.4.5 Graph Factorization

Roughly speaking, one could classify graph factorization problems as one of two kinds: those in which the edge set is partitioned, and those in which the vertex set is partitioned.

Edge Partitions

DEFINITIONS

D47: A **1-factorization** of a graph G of even order is a partition of $E(G)$ into edge-disjoint 1-factors.

D48: A k-**linear forest** is a forest in which all components are paths of length at most k.

D49: The k-**linear arboricity** of a graph G is the minimum number of k-linear forests which partition $E(G)$.

D50: A **Hamiltonian partition** of an r-regular graph G is a partition of $E(G)$ in $r/2$ Hamilton cycles when r is even and $(r-1)/2$ Hamilton cycles plus a 1-factor when r is odd.

D51: An H-**decomposition of** G (or simply an H-**decomposition** when G is understood) is a partition of $E(G)$ into edge-disjoint subgraphs all isomorphic to H.

NOTATION: We denote the edge connectivity of graph G by $\lambda(G)$.

CONJECTURES

C2: The 1-Factorization Conjecture: Let G be a simple graph of even order n. If G is regular with $\Delta(G) \geq n/2$, then $\chi'(G) = \Delta(G)$; that is, G has a 1-factorization. (See [Wa97, Ch. 19].)

C3: [AkExHa80]: The linear arboricity of every d-regular graph is $\lceil (d+1)/2 \rceil$.

C4: The Tree-Decomposition Conjecture [BaTh06]: For each tree T there exists a positive integer k_T such that the following holds: if $\lambda(G) \geq k_T$, and $|E(T)|$ divides $|E(G)|$, then there is a T-decomposition of G.

C5: The P_3-packing Conjecture [AkKan85]: Every 3-connected cubic graph on $3n$ vertices has a P_3-factor.

FACTS

F88: [ChHi89],[NiesVo90] If one replaces $(\sqrt{7} - 1)n/2$ by $(\sqrt{7} - 1)/2$, then the 1-Factorization Conjecture becomes true. This result is regarded as the best to date toward the conjecture.

F89: [PlTi91] Let G be a regular multigraph of even order n and multiplicity $\mu(G) \leq r$. Then if $\Delta(G) \geq r(5n/6 + 1)$, $\chi'(G) = \Delta(G)$. This result may be viewed as an extension of Fact F88 to the multigraph case.

F90: Given $\epsilon > 0$, there is a number $N = N(\epsilon)$ such that if G is a simple graph of even order greater than N, and $\Delta \geq (\frac{1}{2} + \epsilon)|V(G)|$, then G is 1-factorizable. This provides evidence in favor of the truth of the 1-Factorization Conjecture for "large" graphs. (See [PeRe97] and Häggkvist (unpublished).)

F91: [ZhZh92] Every k-regular graph of order $2n$ contains at least $\lfloor k/2 \rfloor$ edge-disjoint 1-factors, if $k \geq n$. This result represents another approach to the 1-Factorization Conjecture.

F92: [Kan85] Let a and b be integers such that $0 \leq a \leq b$. Then

(i) a graph G has a $[2a, 2b]$-factorization if and only if G is a $[2am, 2bm]$-graph, for some integer m; and

(ii) every $[8m + 2k, 10n + 2k]$-graph has a $[1, 2]$ factorization.

F93: [YaPaWoTo00] Let G be a multigraph and let g and f be two functions mapping $V(G)$ into the non-negative integers. Let m be a positive integer and ℓ an integer with $0 \leq \ell \leq 3$ and $\ell \equiv m(\bmod 4)$. If G is an $(mg + 2\lfloor m/4 \rfloor + \ell, mf - 2\lfloor m/4 \rfloor - \ell)$ graph, then G is (g, f)-factorizable. (See [Ya95] for other such results.)

F94: [Eg86] Let $k \geq 2$ be an integer.

(i) Every r-regular graph G with $r \geq 4k^2$ has a $[2k, 2k + 1]$-factorization.

(ii) Every $(k^2 - 4k + 2)$-regular graph G has a $[2k - 1, 2k]$-factorization.

NOTATION: We denote the number of components of a graph G by $\omega(G)$.

F95: [Tu61, Na61] Let k be a positive integer.

(i) A connected graph G can be decomposed into k edge-disjoint connected factors if and only if $k(\omega(G - L) - 1) \leq |L|$, for all $L \subseteq E(G)$.

(ii) A graph G contains k edge-disjoint spanning trees if and only if $k(\omega(G - L) - 1) \leq |L|$, for all $L \subseteq E(G)$.

F96: [Po71, Ku74, Gu83] If $\lambda(G) \geq 2k$, then G contains at least k edge-disjoint spanning trees. (See also [Cat92, Pa01].)

F97: [Kr11] If G has a factorization into two spanning trees, then it has such a factorization in which all leaves of both factors have degree at most 8 in G.

F98: [Als80] If G is a connected vertex-transitive graph of order $2p$, where $p \equiv 3$ (mod 4) is a prime, then G has a Hamiltonian partition.

F99: The Tree-Decomposition Conjecture is true when T is the 4-path P_5. For a proof see [Th08a], where it is also proved that if $\lambda(G) \geq 10^{10^{10^{14}}}$, then there is a P_5-decomposition of G if and only if $|E(G)|$ is divisible by 4.

F100: In [Th08b] the same author shows that if $\lambda(G) \geq 171$, has a P_4-decomposition if and only if $|E(G)|$ is divisible by 3.

CONJECTURE

C6: El-Zahár's Conjecture [El84]: If G is a graph with $n = n_1 + \cdots + n_k$ vertices and $\delta(G) \geq \lceil n_1/2 \rceil + \cdots + \lceil n_k/2 \rceil$, then G has a 2-factor in which the cycles have lengths $n_1, \ldots, n_k$, respectively.

El-Zahár himself proved the conjecture true in the case $k = 2$ and [CoHa63] when each $n_i = 3$. Further partial results can be found in [Johs00]. Abbasi [Ab98] proved the conjecture true in the case when $n = |V(G)|$ is sufficiently large.

REMARKS

R31: Perkovic and Reed [PeRe97] make the interesting observation that if the 1-Factorization Conjecture is true, then it follows that for any regular graph G, either G or its complement has a 1-factorization.

R32: It is not difficult to show that the linear arboricity of a d-regular graph is at least the bound given above. It is the inequality in the opposite direction that has proved intractible so far. See [Al88, LiWo98] for further details.

R33: See [PlTi01] for extensions of 1-factorization to the multigraph case.

R34: It is pointed out in [Gu83] that if there are M edge-disjoint spanning trees in a graph, one can find them all in polynomial time using a matroid partition algorithm.

R35: [ShSe92, ElVa99] The complete bipartite graph $K_{m,n}$ can be factored into isomorphic spanning trees if and only if $(m + n - 1)|mn$.

R36: In order for a complete graph K_n to factor into isomorphic spanning trees, n must be even. The area of spanning tree factorization of the complete graph K_{2n} has attracted considerable attention. For an overview of this area, see [Kov11, KoKu09].

R37: In [BaLiDaWi99] the relationship between the problem of finding k edge-disjoint spanning trees in hypercubes and the development of efficient communication algorithms in parallel computing architectures is discussed.

R38: Connected $[a, b]$-factors are studied in [Na10] and connected (f, g)-factors in [ElNaVo02]. The subject of connected factors of various kinds is surveyed in [KoVe05].

R39: The study of factorizations of K_{2n} and $K_{n,n}$ has evolved into an area of combinatorics all its own that includes (in the latter case) the widely studied discipline of latin squares. We refer the interested reader to [Wa97, DeKe74, DeKe91].

R40: A partition of the edge set of a graph G into matchings (not necessarily perfect) is exactly an edge coloring of G. For much on the topic of edge coloring, we refer the reader to [JeTo95].

Vertex Partitions

When does a graph G admit a decomposition of its vertex set $V(G) = V_1 \cup \cdots V_k$ such that the induced subgraphs $G_i = G[V_i]$ have certain specified properties? We present several results of this genre.

DEFINITION

D52: For any graph G, let $\sigma_2(G) = \min\{d_G(x) + d_G(y) | x, y \in V(G), x \neq y, xy \notin E(G)\}$.

FACTS

F101: [Gy78, Lo77] Let G be a k-connected graph and suppose $v_1, \ldots, v_k$ are k distinct vertices of G. Suppose further that $|V(G)| = n = n_1 + \cdots + n_k$ is a partition of $|V(G)| = n$ into k positive parts. Then there exists a subgraph G' of G with the following three properties:

   (i) G' consists exactly of k components.

   (ii) Each of the components contains exactly one of the vertices v_i.

   (iii) The component containing v_i contains exactly n_i vertices.

F102: [EnMa97] Let G be a graph of order n and suppose $n = a_1 + \cdots + a_k$ is a partition of n where each $a_i \geq 2$. Suppose $\delta(G) \geq 3k - 2$. Then given any k distinct vertices $v_1, \ldots, v_k \in V(G)$, $V(G)$ can be partitioned as $V(G) = A_1 \cup \cdots \cup A_k$ such that $|A_i| = a_i$, $v_i \in A_i$ and $\delta(G[A_i]) > 0$, for all $1 \leq i \leq k$.

F103: (a) [Th83] For each pair of positive integers (s, t), there exist positive integers $f(s, t)$ and $g(s, t)$ such that each graph G with $\kappa(G) \geq f(s, t)$ (respectively, $\delta(G) \geq g(s, t)$) admits a partition of its vertex set $V(G) = S \cup T$ such that the induced subgraphs $G[S]$ and $G[T]$ have connectivity (respectively, minimum degree) at least s and t, respectively. (b) [Ha83] Moreover, if $s \geq 3$ and $t \geq 2$, then $f(s, t) \leq 4s + 4t - 13$.

REMARKS

R41: Fact F102 has been used to derive best known error bounds in certain branches of coding theory [CsK81].

R42: A nice survey of vertex partitions into cycles and paths versus $\delta(G)$ and $\sigma_2(G)$ may be found in [En01].

R43: Conditions sufficient to guarantee the existence of an H-factor, for all connected H with $|V(H)| \leq 4$, are summarized and discussed in [EgFuOt08].

Factor Algorithms and Complexity

DEFINITIONS

D53: Let G be an arbitrary graph. A *G-factor* of a graph H is a set $\{G_1, \ldots, G_d\}$ of subgraphs of H such that each subgraph G_i is isomorphic to G and the sets $V(G_i)$ collectively partition $V(G)$.

D54: The *G-factor recognition problem* $FACT(G)$: INSTANCE: A graph H. QUESTION: Does H admit a G-factor?

D55: The **clique partition number** of a graph G is the smallest number $cp(G)$ such that there exists a set of $cp(G)$ cliques in G such that the cliques form a partition of $E(G)$.

D56: A graph G is **chordal** if every cycle in G of length greater than 3 has a chord.

D57: The *H-decomposition Problem:* Given a fixed graph H, can the edge set of an input graph G be partitioned into copies of H?

D58: If graph G admits a partition of its edge set into t isomorphic subgraphs, then we say that G *is divisible by* t. (An obvious necessary condition for G to be divisible by t is that the number of graphs in the partition must divide $|E(G)|$.)

D59: [El88a, El88b] A graph G is *t-rational* if G is divisible by t or if $t \nmid |E(G)|$.

D60: The *t-Rational Recognition Problem* (or the **Isomorphic Factorization Problem**): Given a graph G and a positive integer t, is G a t-rational graph? Note that t and G form the input to the problem. A candidate for subgraph H is not part of the input.

D61: An *δ-separated matching* is one in which, given any two edges e and e' of the matching, the length of a shortest path from a vertex of e to a vertex of e' is at least d. (A 2-separated matching is more often called an *induced* matching.)

D62: A problem is said to be in the class NC if it can be solved in parallel time polynomial in the logarithm of the size of the input by a set of parallel processors the number of which is polynomial in the input size.

FACTS

NOTATION: In the next group of facts, $n = |V(G)|$.

F104: The first polynomial algorithm for matching in an arbitrary graph was formulated by Edmonds [Ed65] and has come to be popularly known as the *blossom algorithm*. Its running time is $O(n^4)$.

F105: The fastest algorithm to date for maximum matching in a general (i.e., not necessarily bipartite) graph with m edges has complexity $O(m\sqrt{n})$ and is due to Micali and Vazirani [MiVa80]. (See also [PeLo88].) (Curiously, a proof of correctness of this algorithm was not published until fourteen years later! (See [Va94]).) Since the Micali–Vazirani algorithm was introduced, two other matching algorithms [GaTa91, Bl90] having the same complexity as Micali–Vazirani have been produced.

F106: Faster matching algorithms exist, however, in certain special cases. If the graph is 3-regular and has no cutedge, then by the classical result of Petersen [Pe1891], the graph must have a 1-factor. In this case, an $O(n \log^4 n)$ algorithm is given in [BiBoDeLu01] for finding a 1-factor. An $O(n)$ algorithm is also given, if, in addition, the graph is planar.

F107: [StVa82] The maximum δ-separated matching problem is NP-complete for $\delta \geq 2$, even when restricted to bipartite graphs of degree 4.

F108: The Gabow, Kaplan, and Tarjan algorithm [GaKaTa99, GaKaTa01] cited above can be modified to test whether a graph has a *unique f-factor* and find it, if it exists, and to check whether a given f-factor is unique, all in polynomial time.

F109: Anstee [An85] gave algorithmic proofs of both the (g, f)-factor theorem and the f-factor theorem and his algorithms either return one of the factors in question or show that none exists, all in $O(n^3)$ time. Note that this complexity bound is independent of the number of edges in the graph and also independent of g and f.

F110: A polynomial algorithm for finding a 2-factor, if one exists, was first found by Edmonds and Johnson [EdJo70]. If one additionally demands that the 2-factor be triangle-free, the problem remains polynomially solvable. (See [CoPu80].) If one demands that the cycle lengths to be disallowed form a non-empty subset of $\{5, 6, \ldots\}$, the problem has been shown to be NP-hard [HeKiKrKr88]. The complexity in the two remaining cases, namely, where only 4-cycles are forbidden or where only triangles and 4-cycles are forbidden, remains unresolved.

F111: The problem of deciding whether or not a graph has a hamiltonian cycle is one of first decision problems proved ([Ka72, Ka75]) to be NP-complete. The problem remains NP-complete, even if the graphs are restricted to be 3-regular and planar [GaJoTa76] or 4- or 5-regular and planar [Pi94].

F112: The answer to the factor recognition problem $FACT(K_1)$ is (trivially) always "yes" and so $FACT(K_1) \in P$. Problem $FACT(K_2)$ is just the question of the existence of a perfect matching in H and hence also lies in P. More generally, if G consists of a disjoint union of copies of K_1 and K_2, then $FACT(G)$ belongs to P.

F113: [KiHe83] If any component of G has more than two vertices, then $FACT(G)$ is NP-complete.

F114: [Kaw02] Let G be a graph of order $4k$ with $\delta(G) \geq 5k/2$. Then G contains a K_4^--factor, where K_4^- denotes the complete graph K_4 with one edge removed.

F115: [ShWaJu88] The problem of determining $cp(G)$ is NP-hard, for the class of K_4-free graphs and for the class of chordal graphs. However, the problem is polynomial for the class of graphs which are both K_4-free and chordal.

F116: [Ho81a] The problem of determining the chromatic index of a graph is NP-complete. If G is bipartite, however, see Fact F117.

F117: König's Edge-Coloring Theorem [Ko16a, Ko16b]: If G is bipartite, then $\chi'(G) = \Delta(G)$. The proof yields an $O(mn)$ algorithm ($m = |E(G)|$) to produce an optimal edge-coloring.

F118: Presently, it seems that either an algorithm of Kapoor and Rizzi [KaRi00] or an algorithm of Schrijver [Sc98a] is best for edge-coloring a bipatite graph G, depending upon the relative sizes of $|V(G)|$ and $\Delta(G)$. If the bipartite graphs involved are regular, then even faster algorithms exist. (See [Ri02].)

F119: [Ho81b] Suppose $n \geq 3$. Then the problem of partitioning $E(G)$ into copies of K_n is NP-complete. Holyer used the above result to prove five other edge partition problems to be NP-complete in the same paper.

F120: [BrLo95] If H has no connected component with three or more edges, then the H-decomposition Problem is polynomial.

F121: [DoTa97] The H-decomposition Problem is NP-complete whenever H contains a connected component with three edges or more. (See also [AlCaYu98].)

F122: [BeHo97] There is a polynomial algorithm that finds a factorization of any given 4-regular graph into two triangle-free 2-factors or else shows that such a factorization does not exist.

F123: [Wo84] If $r > 2t$, then almost all labeled r-regular graphs cannot be factorized into $t \geq 2$ isomorphic subgraphs.

F124: But curiously, there is no known example of a regular non-factorizable graph as in Fact F123 which satisfies the obvious necessary divisibility condition: $t \| E(G) |$.

F125: [ElWo88] Let G be a multigraph and suppose t is an integer such that $t \geq \chi'(G)$. Then G is t-rational.

F126: If G is r-regular and $t \geq r + 1$, then G is t-rational. This follows from Vizing's theorem.

F127: [El88a] Let G be a $2k$-regular graph of even order that contains no 3-cycles or 5-cycles. Then $E(G)$ can be partitioned into $2k$ isomorphic subgraphs. Moreover, this factorization can be constructed in polynomial time.

F128: [ScBi78, HaRoWo78] Given the complete graph K_n, then there exists a graph H such that K_n is the edge-disjoint union of t copies of H if and only if $n(n-1) \equiv 0(\bmod 2t)$.

REMARKS

R44: It is unknown whether or not the problem of finding a maximum matching (or 1-factor) is in the parallel class NC. For a general reference on this subject, see [KaRy98].

R45: [HaWa77] provides some observations about connections between the Isomorphic Factorization Problem and combinatorial designs. Even the subject of 1-factorizations of graphs (that is, where the isomorphs are 1-factors) is an enormous topic unto itself and quickly leads one into the discipline of combinatorial design theory. See the excellent surveys [StGo81, MeRo85] and the encyclopedic volume [Wa97].

R46: The so-called *packing* problems are closely allied to factor and factorization problems. Here instead of searching for a factor of a particular kind in a given graph G, one seeks a subgraph of G of maximum order which admits the factor. See [LoPo90] for a nice survey of the state of the art.

CONJECTURES

C7: [BeHo97]: The problems of recognizing

  (a) which $2n$-regular graphs factor into two triangle-free n-factors, and

  (b) which $2n$-regular graphs factor into n triangle-free 2-factors

are both NP-complete for all $n \geq 3$.

C8: [Hi85]: Let G be a d-regular simple graph of order $2n$ and let $d = p_1 + \cdots + p_r$ be a partition of d. If $d \geq n$, then G has a factorization into edge-disjoint subgraphs $H_1 \cup \cdots \cup H_r$, where H_i is regular of degree p_i. (The author proves the conjecture true in various special cases.)

Subgraph Problems

DEFINITIONS

D63: The *k-regular Subgraph Recognition Problem*: given a graph G, does it contain a k-regular subgraph? (Here we do not require that the k-regular subgraph span G.) If $k = 1$ or 2, clearly the problem takes only polynomial time.

FACTS

F129: [Ga83] Let G be a k-regular graph of order n, and let $v(G)$ denote the minimum number of extra vertices needed to ensure that there exist a $(k+1)$-regular supergraph of G.

  (i) If the graph $\overline{G}$ has a 1-factor, then $v(G) = 0$.

  (ii) If the graph $\overline{G}$ has no 1-factor and if n and k are of opposite parity, then $v(G) = 1$.

  (iii) If the graph $\overline{G}$ has no 1-factor and n and k are of the same parity, then $n < 2k$ and $v(G) = k + 2$.

F130: [Ta84] Every 4-regular simple graph G contains a 3-regular subgraph. (However, the proof here does not provide an algorithm for finding the 3-regular subgraph.)

F131: [Ple84] The k-regular Subgraph Recognition Problem is NP-complete for all $k \geq 3$.

F132: [EgOt99] If G is a graph with $|V(G)| \geq 4k+6$ and $\delta(G) \geq k+2$, then G contains k pairwise vertex-disjoint claws (i.e., copies of $K_{1,3}$). (The claws are *not* considered to be induced.)

REMARKS

R47: There are hundreds of papers in the literature dealing with a wide variety of "subgraph problems" as well as "graph decompositions." The reader is referred to the survey papers [ChGr81, Di90b, Ro90, Be96] and to the books [Bo90, Di90a, CoRo99].

R48: There are indeed infinite analogs of some of the matching and factor theorems for finite graphs. See [Ra49, Ah84a, Ah84b, Ah88, Ah91, AhMaSh92, AhNa84, AhNaSh83, Br71, HoPoSt87, St77, St85a, St85b, St89, Nied91, NiedPo94].

R49: Since the first edition of this *Handbook* was published, several new books and survey articles on graph factors and factorization have appeared. The reader is referred to [AkKan11, KoVe05, Pl07, YuLi09].

References

[Ab98] S. Abbasi, Spanning subgraphs of dense graphs and a combinatorial problem on strings, Ph.D. Thesis, Rutgers University, 1998.

[Ah84a] R. Aharoni, A generalization of Tutte's 1-factor theorem to countable graphs, *J. Combin. Theory Ser. B* 37 (1984), 199–209.

[Ah84b] R. Aharoni, König's duality theorem for infinite bipartite graphs, *J. London Math. Soc.* 29 (1984), 1–12.

[Ah88] R. Aharoni, Matchings in infinite graphs, *J. Combin. Theory Ser. B* 44 (1988), 87–125.

[Ah91] R. Aharoni, Infinite matching theory, *Discrete Math.* 95 (1991), 5–22.

[AhMaSh92] R. Aharoni, M. Magidor and R. Shore, On the strength of König's duality theorem for infinite bipartite graphs, *J. Combin. Theory Ser. B* 54 (1992), 257–290.

[AhNa84] R. Aharoni and C. Nash-Williams, Marriage in infinite societies, *Progress in Graph Theory*, Academic Press, 1984, 71–79.

[AhNaSh83] R. Aharoni, C. Nash-Williams and S. Shelah, A general criterion for the existence of transversals, *Proc. London Math. Soc.* 47 (1983), 43–68.

[AkAvEr80] J. Akiyama, D. Avis and H. Era, On a $\{1,2\}$-factor of a graph, *TRU Math.* 16 (1980), 97–102.

[AkExHa80] J. Akiyama, G. Exoo and F. Harary, Covering and packing in graphs III, cyclic and acyclic invariants, *Math. Slovaca* 30 (1980), 405–417.

[AkKan85] J. Akiyama and M. Kano, Factors and factorizations of graphs – a survey, *J. Graph Theory* 9 (1985), 1–42.

[AkKan11] J. Akiyama and M. Kano, *Factors and Factorizations of Graphs*, Springer, 2011.

[Al88] N. Alon, The linear arboricity of graphs, *Israel J. Math.* 62 (1988), 311–325.

[AlCaYu98] N. Alon, Y. Caro and R. Yuster, Packing and covering dense graphs, *J. Combin. Des.* 6 (1998), 451–472.

[AlEgFuOtSa11] R. Aldred, Y. Egawa, J. Fujisawa, K. Ota and A. Saito, The existence of a 2-factor in $K_{1,n}$-free graphs with large connectivity and large edge-connectivity. *J. Graph Theory* 68 (2011), 77–89.

[AlFr08] N. Alon and S. Friedland, The maximum number of perfect matchings in graphs with a given degree sequence, *Electron. J. Combin.* 15 (2008), N13.

[AlPl11] R. Aldred and M. Plummer, Proximity thresholds for matching extension in planar and projective planar triangulations, *J. Graph Theory* 67 (2011), 38–46.

[Als80] B. Alspach, Hamiltonian partitions of vertex-transitive graphs of order $2p$, *Congr. Numer.* 28 (1980), 217–221.

[AmKan82] A. Amahashi and M. Kano, On factors with given components, *Discrete Math.* 42 (1982), 1–6.

[An73] I. Anderson, Sufficient conditions for matching, *Proc. Edinburg Math. Soc.* 18 (1973), 129–136.

[An85] R. Anstee, An algorithmic proof of Tutte's f-factor theorem, *J. Algorithms* 6 (1985), 112–131.

[An90] R. Anstee, Simplified existence theorems for (g, f)-factors, *Discrete Math.* 27 (1990) 29–38.

[An92] R. Anstee, Matching theory: fractional to integral, *New Zealand J. Math.* 21 (1992), 17–32.

[AnEgKaKawMa02] K. Ando, Y. Egawa, A. Kaneko, K. Kawarabayashi, and H. Matsuda, Path factors in claw-free graphs, *Discrete Math.* 243 (2002), 195–200.

[AnNa98] R. Anstee and Y. Nam, More sufficient conditions for a graph to have factors, *Discrete Math.* 184 (1998), 15–24.

[ArLaSe91] S. Arnborg, J. Lagergren, and D. Seese, Easy problems for tree-decomposable graphs, *J. Algor.* 12 (1991), 308–340.

[AsWo] H. Assiyatun and N. Wormald, 3-star factors in random d-regular graphs, *European J. Combin.* 27 (2006), 1249–1262.

[Ba38] F. Bäbler, Über die Zerlegung regulärer Streckencomplexe ungerader Ordnung, *Comment. Math. Helv.* 10 (1938), 275–287.

[BaLiDaWi99] B. Barden, R Libeskind-Hadas, J. Davis, and W. Williams, On edge-disjoint spanning trees in hypercubes, *Inform. Process. Lett.* 70 (1999), 13–16.

[BaTh06] J. Barát and C. Thomassen, Claw-decompositions and Tutte-orientations, *J. Graph Theory* 52 (2006), 135-146.

[Be96] L. Beineke, Graph decompositions, *Congress. Numer.* 115 (1996), 213–226.

[BeHo97] E. Bertram and P. Horák, Decomposing 4-regular graphs into triangle-free 2-factors, *SIAM J. Discrete Math.* 10 (1997), 309–317.

[BeLa78] C. Berge and M. Las Vergnas, On the existence of subgraphs with degree constraints, *Nederl. Akad. Wetensch. Indag. Math.* 40 (1978), 165–176.

[BeSu99] I. Beichl and F. Sullivan, Approximating the permanent via importance sampling with application to the dimer covering problem, *J. Comput. Phys.* 149 (99), 128–147.

[BiBoDeLu01] T. Biedl, P. Bose, E. Demaine, and A. Lubiw, Efficient algorithms for Petersen's matching theorem, *J. Algorithms* 38 (2001), 110–134.

[Bl90] M. Blum, A new approach to maximum matchings in general graphs, *Languages and Programming* (Proc. of the 17th International Colloquium on Automata) (1990), 586–597.

[Bo85] B. Bollobás, *Random Graphs*, Harcourt Brace Jovanovich, 1985.

[Bo90] J. Bosák, *Decompositions of Graphs*, Kluwer Academic Publishers Group, 1990.

[Bo01] B. Bollobás, *Random Graphs*, Second Edition, Cambridge University Press, 2001.

[BoSaWo85] B. Bollobás, A. Saito and N. Wormald, Regular factors of regular graphs, *J. Graph Theory* 9 (1985), 97–103.

[Br71] R. Brualdi, Strong transfinite versions of König's duality theorem, *Monatsh. Math.* 75 (1971), 106–110.

[Br73] L. Brègman, Certain properties of nonnegative matrices and their permanents, *Dokl. Akad. Nauk SSSR* 211 (1973), 27–30. (Russian); *Soviet Math. Dokl.* 14 (1973), 945–949. (English translation)

[BrChRaGoLe97] S. Brandt, G. Chen, R. Faudree, R. Gould, and L. Lesniak, Degree conditions for 2-factors, J. Graph Theory 24 (1997) 165-173.

[BrLo95] K. Bryś and Z. Lonc, A complete solution of a Holyer problem, *Proc. 4th Twente workshop on Graph and Combinatorial Optimization*, University of Twente, Enschede, The Netherlands, 1995.

[Ca91] M. Cai, $[a, b]$-factorizations of graphs, *J. Graph Theory* 15 (1991), 283–302.

[Cat77] P. Catlin, Embedding subgraphs under extremal degree conditions, *Congr. Numer.* 19 (1977), 139–145.

[Cat92] P. Catlin, Super-Eulerian graphs: a survey, *J. Graph Theory* 16 (1992), 177–196.

[Ch73] V. Chvátal, Tough graphs and hamiltonian circuits, *Discrete Math.* 5 (1973), 215–228.

[Che93] Binding number and minimum degree for $[a, b]$-factors, *Systems Sci. Math. Sci.* 6 (1993), 179–185.

[ChGr81] F. Chung and R. Graham, Recent results in graph decompositions, *London Math. Soc. Lecture Note Ser.* 52 (1981), Cambridge University Press, 103–123.

[ChHi89] A. Chetwynd and A. Hilton, 1-factorizing regular graphs with high degree: an improved bound, *Discrete Math.* 75 (1989), 103–112.

[CoHa63] K. Corrádi and A. Hajnal, On the maximal number of independent circuits in a graph, *Acta Math. Acad. Sci. Hungar.* 14 (1963), 423–439.

[CoPu80] G. Cornuéjols and W. Pulleyblank, Perfect triangle-free 2-matchings, *Combinatorial optimization II (Proc. Conf., Univ. East Anglia, Norwich, 1979)*, Math. Programming Stud. 13 (1980), 1–7.

[CoRo99] C. Colbourn and A. Rosa, *Triple Systems*, Oxford University Press, 1999.

[CsK81] I. Csiszár and J. Körner, Graph decomposition: a new key to coding theorems, *IEEE Trans. Inform. Theory* 27 (1981), 5–12.

[DaLu92] P. Dagum and M. Luby, Approximating the permanent of graphs with large factors, *Theoret. Comput. Sci.* 102 (1992), 283–305.

[DeKe74] J. Dénes and A. Keedwell, *Latin squares and their applications*, Academic Press, 1974.

[DeKe91] J. Dénes and A. Keedwell, *Latin Squares: New developments in the theory and applications*, Ann. Discrete Math. 46, North-Holland, 1991.

[Di90a] R. Diestel, *Graph Decompositions. A Study in Infinite Graph Theory*, Oxford University Press, 1990.

[Di90b] R. Diestel, Decomposing infinite graphs, *Contemporary Methods in Graph Theory*, Bibliographisches Inst., Mannheim, 1990, 261–289.

[Dos02] T. Došlić, On some structural properties of fullerene graphs, *J. Math. Chem.* 31 (2002), 187–195.

[DoTa97] D. Dor and M. Tarsi, Graph decomposition is NP-complete: a complete proof of Holyer's conjecture, *SIAM J. Comput.* 26 (1997), 1166–1187.

[Ed65] J. Edmonds, Paths, trees and flowers, *Canad. J. Math.* 17 (1965), 449–467.

[EdJo70] J. Edmonds and E. Johnson, Matching: a well solved class of integer linear programs, *Combinatorial Structures and their Applications*, Gordon and Breach, 1970, 89–92.

[Eg80] G. Egoryčev, Solution of the van der Waerden problem for permanents, *Sibirsk. Mat. Zh.* 22 (1981), 65–71, 225. (Russian)

[Eg81] G. Egoryčev, The solution of van der Waerden's problem for permanents, *Adv. in Math.* 42 (1981), 299–305.

[Eg86] Y. Egawa, Era's conjecture on $[k, k+1]$-factorizations of regular graphs, *Ars Combin.* 21 (1986), 217–220.

[EgEn89] Y. Egawa and H. Enomoto, Sufficient conditions for the existence of k-factors, *Recent Studies in Graph Theory*, Vishwa International Publications (1989), 96–105.

[EgEnFaLiSc03] Y. Egawa, H. Enomoto, R. Faudree, H. Li and I. Schiermeyer, Two-factors each component of which contains a specified vertex, *J. Graph Theory* 43 (2003), 188–198.

[EgFuOt08] Y. Egawa, S. Fujita and K. Ota, $K_{1,3}$-factors in graphs, *Discrete Math.* 308 (2008), 5965–5973.

[EgKan96] Y. Egawa and M. Kano, Sufficient conditions for graphs to have (g, f)-factors, *Discrete Math.* 151 (1996), 87–90.

[EgOt99] Y. Egawa and K. Ota, Vertex-disjoint claws in graphs, *Discrete Math.* 197/198 (1999), 225–246.

[El84] M. El-Zahár, On circuits in graphs, *Discrete Math.* 50 (1984), 227–230.

[El88a] M. Ellingham, Isomorphic factorizations of regular graphs of even degree, *J. Austral. Math. Soc. Ser. A* 44 (1988), 402–420.

[El88b] M. Ellingham, Isomorphic factorization of r-regular graphs into r parts, *Discrete Math.* 69 (1988), 19–34.

[ElNaVo02] M. Ellingham, Y. Nam, and H.-J. Voss, Connected (g, f)-factors, *J. Graph Theory* 39 (2002), 62–75.

[ElVa99] S. El-Zanati and C.Vanden Eynden, Factorizations of $K_{m,n}$ into spanning trees, *Graphs Combin.* 15 (1999), 287–293.

[ElWo88] M. Ellingham and N. Wormald, Isomorphic factorization of regular graphs and 3-regular multigraphs, *J. London Math. Soc.* 37 (1988), 14–24.

[En01] H. Enomoto, Graph partition problems into cycles and paths, *Discrete Math.* 233 (2001), 93–101.

[EnJaKatSa85] H. Enomoto, B. Jackson, P. Katerinis, and A. Saito, Toughness and the existence of k-factors, *J. Graph Theory* 9 (1985), 87–95.

[EnMa97] H. Enomoto and S. Matsunaga, Graph decompositions without isolated vertices III, *J. Graph Theory* 24 (1997), 155–164.

[EnOtKan88] H. Enomoto, K. Ota, and M. Kano, A sufficient condition for a bipartite graph to have a k-factor, *J. Graph Theory* 12 (1988), 141–151.

[ErKaMi09] R. Erman, F. Kardoš, and J. Miškuf, Long cycles in fullerene graphs, *J. Math Chem.* 46 (2009), 1103–1111.

[ErRe66] P. Erdős and A. Rényi, On the existence of a factor of degree one of a connected random graph, *Acta Math. Acad. Sci. Hungar.* 17 (1966), 359–368.

[EsKaKiKrNo11] L. Esperet, F. Kardoš, A. King, D. Král, and S. Norine, Exponentially many perfect matchings in cubic graphs, *Adv. Math.* 227 (2011), 1646–1664.

[Fa81] D. Falikman, Proof of the van der Waerden conjecture on the permanent of a doubly stochastic matrix, *Mat. Zametki* 29 (1981), 931–938, 957. (Russian) *Math. Notes* 29 (1981), 475–479.

[FaFaFlLiLi99] R. Faudree, O. Favaron, E. Flandrin, H. Li, and Z. Liu, On 2-factors in claw-free graphs, *Discrete Math.* 206 (1999), 131–137.

[FaFaRy08] J. Faudree, R. Faudree, and Z. Ryjáček, Forbidden subgraphs that imply 2-factors, *Discrete Math.* 308 (2008), 1571–1582.

[Fi61] M. Fisher, Statistical mechanics of dimers on a plane lattice, *Phys. Rev.* 124 (1961), 1664–1672.

[Fr1912] G. Frobenius, Über Matrizen aus nicht negativen Elementen, *Sitzungsber. König. Preuss. Akad. Wiss.* 26 (1912), 456–477.

[FrJu99a] C. Fremuth-Paeger and D. Jungnickel, Balanced network flows. I. A unifying framework for design and analysis of matching algorithms, *Networks* 33 (1999), 1–28.

[FrJu99a] C. Fremuth-Paeger and D. Jungnickel, Balanced network flows. I. A unifying framework for design and analysis of matching algorithms, *Networks* 33 (1999), 1–28.

[FrJu99b] C. Fremuth-Paeger and D. Jungnickel, Balanced network flows. II. Simple augmentation algorithms, *Networks* 33 (1999), 29–41.

[FrJu99c] C. Fremuth-Paeger and D. Jungnickel, Balanced network flows. III. Strongly polynomial augmentation algorithms, *Networks* 33 (1999), 43–56.

[FrJu01] C. Fremuth-Paeger and D. Jungnickel, Balanced network flows. IV. Duality and structure theory, *Networks* 37 (2001), 194–201.

[FuHoMc65] D. Fulkerson, A.Hoffman, and M. McAndrew, Some properties of graphs with multiple edges, *Canad. J. Math.* 17 (1965), 166–177.

[Ga83] A. Gardiner, Embedding k-regular graphs in $k+1$-regular graphs, *J. London Math. Soc.* 28 (1983), 393–400.

[GaJoTa76] M. Garey, D. Johnson, and R. Tarjan, The planar hamiltonian circuit problem is NP-complete, *SIAM J. Comput.* 5 (1976), 704–714.

[GaKaTa99] H. Gabow, H. Kaplan, and R. Tarjan, Unique maximum matching algorithms, *Annual ACM Symp. on Theory of Comput.*, ACM New York 1999, 70–78.

[GaKaTa01] H. Gabow, H. Kaplan, and R. Tarjan, Unique maximum matching algorithms, *J. Algorithms* 40 (2001), 159–183.

[GaTa91] H. Gabow and R. Tarjan, Faster scaling algorithms for general graph-matching problems, *J. Assoc. Comput. Mach.* 38 (1991), 815–853.

[Go01] R. Gould, Results on degrees and the structure of 2-factors, *Discrete Math.* 230 (2001), 99–111.

[Gu83] D. Gusfield, Connectivity and edge-disjoint spanning trees, *Inform. Process. Lett.* 16 (1983), 87–89.

[Gy78] E. Györi, On division of graphs to connected subgraphs, *Combinatorics (Proc. Fifth Hungarian Colloq., Keszthely, 1976) I*, North-Holland, 1978, 485–494.

[Ha35] P. Hall, On representatives of subsets, *J. London Math. Soc.* 10 (1935), 26–30.

[Ha48] M. Hall, Distinct representatives of subsets, *Bull. Amer. Math. Soc.* 54 (1948), 922–926.

[Ha83] A. Hajnal, Partition of graphs with condition on the connectivity and minimum degree, *Combinatorica* 3 (1983), 95–99.

[HaRoWo78] F. Harary, R. Robinson, and N. Wormald, Isomorphic factorizations. I. Complete graphs, *Trans. Amer. Math. Soc.* 242 (1978), 243–260.

[HaWa77] F. Harary and W. Wallis, Isomorphic factorizations. II. Combinatorial designs, *Congress. Numer.* XIX (1977), 13–28.

[He84] G. Hendry, Maximum graphs with a unique k-factor, *J. Combin. Theory Ser. B* 37 (1984), 53–63.

[HeHeKiLi90] K. Heinrich, P. Hell, D. Kirkpatrick, and G. Liu, Communication: A simplified existence criterion for $(g < f)$-factors, *Discrete Math.* 85 (1990), 313–317.

[HeKi81a] P. Hell and D. Kirkpatrick, On generalized matching problems, *Inform. Process. Lett.* 12 (1981), 33–35.

[HeKiKrKr88] P. Hell, D. Kirkpatrick, J. Kratochvíl, and I. Kříž, On restricted two-factors, *SIAM J. Discrete Math.* 1 (1988), 472–484.

[HeVa07] J. Heckman and C. Vafa, Crystal melting and black holes, *J. High Energy Phys.* 2007, no. 9.

[Hi85] A. Hilton, Factorizations of regular graphs of high degree, *J. Graph Theory* 9 (1985), 193–196.

[Hi08] A. Hilton, $(r, r+1)$-factorizations of $(d, d+1)$-graphs, *Discrete Math.* 308 (2008), 645–669.

[Hi09] A. Hilton, On the number of $(r, r + 1)$-factors in an $(r, r + 1)$-factorization of a simple graph, *J. Graph Theory* 60 (2009), 257–268.

[HiWo05] A. Hilton and J. Wojciechowski, Semiregular factorization of simple graphs, *AKCE Int. J. Graphs Comb.* 2 (2005), 57–62.

[Ho81a] I. Holyer, The NP-completeness of edge-coloring, *SIAM J. Comput.* 10 (1981), 718–720.

[Ho81b] I. Holyer, The NP-completeness of some edge partition problems, *SIAM J. Comput.* 10 (1981), 713–717.

[Hof03] A. Hoffmann, Regular factors in connected regular graphs, *Ars. Combin.* 68 (2003), 235–242.

[HoPoSt87] M. Holz, K.-P. Podewski and K. Steffens, Injective choice functions, *Lecture Notes in Mathematics*, 1238, Springer-Verlag, 1987.

[IiNi91] T. Iida and T. Nishimura, An Ore-type condition for the existence of k-factors in graphs, *Graphs Combin.* 7 (1991), 353–361.

[IiNi97] T. Iida and T. Nishimura, Neighborhood conditions and k-factors, *Tokyo J. Math.* 20 (1997), 411–418.

[JaLuRu00] S. Janson, T. Łuczak, and A. Rucinski, *Random Graphs*, Wiley Interscience, 2000.

[JaWh89] B. Jackson and R. Whitty, A note concerning graphs with unique f-factors, *J. Graph Theory* 13 (1989), 577–580.

[JeTo95] T. Jensen and B. Toft, *Graph Coloring Problems*, John Wiley, 1995.

[Joh00a] P. Johann, On the structure of graphs with a unique k-factor, *J. Graph Theory* 35 (2000), 227–243.

[Joh00b] P. Johann, On the structure of graphs with a unique k-factor, *Electron. Notes Discrete Math.* 5 (2000), 193–195.

[JohKaVu08] A. Johansson, J. Kahn, and V. Vu, Factors in random graphs, *Random Structures Algorithms* 33 (2008), 1–28.

[Johs00] R. Johansson, On the bipartite case of El-Zahár's conjecture, *Discrete Math.* 219 (2000), 123–134.

[Ka72] R. Karp, Reducibility among combinatorial problems, *Complexity of Computer Computations*, Plenum Press, 1972, 85–103.

[Ka75] R. Karp, On the computational complexity of combinatorial problems, *Networks* 5 (1975), 45–68.

[Ka82] M. Karoński, A review of random graphs, *J. Graph Theory* 6 (1982), 349–389.

[Ka95] M. Karoński, Random Graphs, *Handbook of Combinatorics*, Elsevier (1995), 351–380.

[Kai08] T. Kaiser, Disjoint T-paths in tough graphs, *J. Graph Theory* 50 (2008), 1–10.

[KaKa02] M. Kano and G. Katona, Note: Odd subgraphs and matchings, *Discrete Math.* 250 (2002), 265–272.

[KaKa07] M. Kano and G. Katona, Structure theorem and algorithm on $(1, f)$-odd subgraph, *Discrete Math.* 307 (2007), 1404–1417.

[KaKeNi01] A. Kaneko, A. Kelmans, and T. Nishimura, On packing 3-vertex paths in a graph, *J. Graph Theory* 36 (2001), 175–197.

[KaKrMiSe09] F. Kardoš, D. Král', J. Miškuf, and J.-S. Sereni, Fullerene graphs have exponentially many perfect matchings, *J. Math. Chem.* 46 (2009), 443–447.

[KaLuYu10] Component factors with large components in graphs, *Appl. Math. Lett.* 23 (2010), 385–389.

[KaMaOdOt02] K. Kawarabayashi, H. Matsuda, Y. Oda, and K. Ota, Path factors in cubic graphs, *J. Graph Theory* 39 (2002), 188–193.

[Kan84] M. Kano, Graph factors with given properties, *Graph Theory, Singapore 1983*, Lecture Notes in Math., 1073 (1984) 161–168.

[Kan85] M. Kano, $[a, b]$-factorization of a graph, *J. Graph Theory* 9 (1985), 129–146.

[Kan90a] M. Kano, A sufficient condition for a graph to have $[a, b]$-factors, *Graphs Combin.* 6 (1990), 245–251.

[Kan90b] M. Kano, Sufficient conditions for a graph to have factors, *Discrete Math.* 80 (1990), 159–165.

[Kan93] M. Kano, Current results and problems on factors of graphs, *Combinatorics, Graph Theory, Algorithms and Applications* (Beijing, 1993), World Scientific Publishers, 1994, 93–98.

[Kane03] A. Kaneko, A necessary and sufficient condition for the existence of a path factor every component of which is a path of length at least two, *J. Combin. Theory Ser. B* 88 (2003), 195–218.

[KanSa83] M. Kano and A. Saito, Note: [a, b]-factors of graphs, *Discrete Math.* 47 (1983), 113–116.

[KaOz11] K. Kawarabayashi and K. Ozeki, 2- and 3-factors of graphs on surfaces, *J. Graph Theory* 67 (2011), 306–315.

[KaRi00] A. Kapoor and R. Rizzi, Note: Edge-coloring bipartite graphs, *J. Algorithms* 34 (2000), 390–396.

[KaRy98] M. Karpinski and W. Rytter, *Fast Parallel Algorithms for Graph Matching Problems*, Oxford, 1998.

[Kas61] P. Kasteleyn, The statistics of dimers on a lattice. I. The number of dimer arrangements on a quadratic lattice, *Physica* 21 (1961), 1209–1225.

[Kas63] P. Kasteleyn, Dimer statistics and phase transitions, *J. Math. Phys.* 4 (1963), 287–293.

[Kas67] P. Kasteleyn, Graph theory and crystal physics, *Graph Theory and Theoretical Physics*, Academic Press, 1967, 43–110.

[Kat83] P. Katerinis, Some results on the existence of $2n$-factors in terms of vertex deleted subgraphs, *Ars Combin.* 16-B (1983), 271–277.

[Kat90] P. Katerinis, Minimum degree of bipartite graphs and the existence of k-factors, *Graphs Combin.* 6 (1990), 253–258.

[KatTs00] P. Katerinis and N. Tsikopoulos, Minimum degree and F-factors in graphs, *New Zealand J. Math.* 29 (2000), 33–40.

[KatWo87] P. Katerinis and D. Woodall, Binding numbers of graphs and the existence of k-factors, *Quart. J. Math.* 38 (1987), 221–228.

[Kaw02] K. Kawarabayashi, K_4^--factor in a graph, *J. Graph Theory* 39 (2002), 111–128.

[KaYo02] A. Kaneko and K. Yoshimoto, On a 2-factor with a specified edge in a graph satisfying the Ore condition, *Discrete Math.* 257 (2002), 445–461.

[KeRaSi96] C. Kenyon, D. Randall, and A. Sinclair, Approximating the number of monomer-dimer coverings of a lattice, *J. Statist. Phys.* 83 (1996), 637–659.

[KiHe83] D. Kirkpatrick and P. Hell, On the complexity of general graph factor problems, *SIAM J. Comput.* 12 (1983), 601–609.

[KlWa73] D. Kleitman and D. Wang, Algorithms for constructing graphs and digraphs with given valences and factors, *Discrete Math.* 6 (1973), 78–88.

[Ko16a] D. König, Über Graphen und ihre Andwendung auf Determinantentheorie und Mengenlehre, *Math. Ann.* 77 (1916), 453–465.

[Ko16b] D. König, Graphok és alkalmazásuk a determinánsok és a halmazok elméletére, *Math. Termész. Ért.* 34 (1916), 104–119.

[Ko31] D. König, Graphs and matrices, *Mat. Fiz. Lapok* 38 (1931), 116–119. (Hungarian)

[Ko33] D. König, Über trennende Knotenpunkte in Graphen (nebst. Anwendungen auf Determinanten und Matrizen), *Acta Sci. Math. (Szeged)* 6 (1933), 155–179.

[Ko59] A. Kotzig, On the theory of finite graphs with a linear factor II, *Mat.-Fyz. Časopis Slovensk. Akad. Vied* 9 (1959), 136–159.

[Ko01] K. Kotani, Factors and connected induced subgraphs, *Graphs Combin.* 17 (2001), 511–515.

[KoKu09] P. Kovář and M. Kubesa, Factorizations of complete graphs into spanning trees with all possible maximum degrees, *Lecture Notes in Comput. Sci.* 5874 (2009), 334–344.

[KoSt93] W. Kocay and D. Stone, Balanced network flows, *Bull. Inst. Combin. Appl.* 7 (1993), 17–32.

[Kov11] P. Kovář, Decompositions and factorizations of complete graphs, *Structural analysis of complex networks*, Birkhäuser/Springer, 2011, 169–196.

[KoVe05] M. Kouider and P. Vestergaard, Connected factors in graphs - a survey, *Graphs Combin.* 21 (2005), 1–26.

[KoZy08] A. Kosowski and P. Žyliński, Packing three-vertex paths in 2-connected cubic graphs, *Ars Combin.* 89 (2008), 95–113.

[Kr11] M. Kriesell, Balancing two spanning trees, *Networks* 57 (2011), 351–353.

[Ku73] S. Kundu, The k-factor conjecture is true, *Discrete Math.* 6 (1973), 367–376.

[Ku74] S. Kundu, Bounds on the number of disjoint spanning trees, *J. Combin. Theory Ser. B* 17 (1974), 199–203.

[La75] M. Las Vergnas, A note on matchings in graphs, *Colloque sur la Théorie des Graphes (Paris, 1974)*, Cahiers Centre Étude Rech. Opér., 17 (1975), 257–260.

[La78] M. Las Vergnas, An extension of Tutte's 1-factor theorem, *Discrete Math.* 23 (1978), 241–255.

[Li88] G. Liu, On (g, f)-covered graphs, *Acta Math. Sci. (English ed.)* 8 (1988), 181–184.

[Li89] G. Liu, On $[a, b]$-covered graphs, *J. Combin. Math. Combin. Comput.* 5 (1989), 14–22.

[LiGrHo75] C. Little, D. Grant, and D. Holton, On defect d-matchings in graphs, *Discrete Math.* 13 (1975), 41–54.

[LiLi98] G. Li and Z. Liu, On connected factors in $K_{1,3}$-free graphs, *Acta Math. Appl. Sinica* 14 (1998), 43–47.

[LiWo98] T. Lindquester and N. Wormald, Factorisation of regular graphs into forests of short paths, *Discrete Math.* 186 (1998), 217–226.

[LiZhCh02] G. Li, B. Zhu, and C. Chen, On connected $[k, k + 1]$-factors in claw-free graphs, *Ars Combin.* 62 (2002), 207–219.

[Lo70] L. Lovász, Subgraphs with prescribed valencies, *J. Combin. Theory* 8 (1970), 391–416.

[Lo74] L. Lovász, Valencies of graphs with 1-factors, *Period. Math. Hungar.* 5 (1974), 149–151.

[Lo77] L. Lovász, A homology theory for spanning trees of a graph, *Acta Math. Acad. Sci. Hungar.* 30 (1977), 241–251.

[LoPo90] M. Loebl and S. Poljak, Subgraph packing — a survey, *Topics in Combinatorics and Graph Theory* (Oberwolfach, 1990), Physica, Heidelberg, 1990, 491–503.

[LoPl86] L. Lovász and M. Plummer, *Matching Theory*, North-Holland, 1986. Corrected reprint, AMS Chelsea Publishing, 2009.

[Ma00] J. Malkevitch, Geometrical and combinatorial questions about fullerenes, *Discrete mathematical chemistry*, DIMACS Ser. Discrete Math. Theoret. Comput. Sci. 51 (2000), 261-266.

[Mc04] W. McCuaig, Pólya's permanent problem, *Electron. J. Combin.* 11 (2004), Research Paper 79.

[McRoSeTh97] W. McCuaig, N. Robertson, P. Seymour, and R. Thomas, Permanents, Pfaffian orientations and even directed circuits (Extended abstract), *Proc. 29th Sympos. Theory Comput.* (1997), 402–405.

[MeRo85] E. Mendelsohn and A. Rosa, One-factorizations of the complete graphs — a survey, *J. Graph Theory* 9 (1985), 43–65.

[MiVa80] S. Micali and V. Vazirani, An $O(|V|^{1/2}|E|)$ algorithm for finding maximum matchings in general graphs, *Proc. 21st Ann. Sympos. Found. Comput. Sci. (Syracuse)* (1980), 17–27.

[MoRe02] M. Molloy and B. Reed, *Graph Colouring and the Probabilistic Method*, Springer-Verlag (2002).

[Mu79] J. Mühlbacher, F-factors of graphs: a generalized matching problem, *Inform. Process. Lett.* 8 (1979), 207–214.

[Na10] Y. Nam, Binding numbers and connected factors, *Graphs Combin.* 26 (2010), 805–813.

[Nas61] C. Nash-Williams, Edge-disjoint spanning trees of finite graphs, *J. London Math. Soc.* 36 (1961), 445–450.

[Ni78] T. Nishizeki, Lower bounds on the cardinality of the maximum matchings of graphs, *Proc. Ninth Southeastern Conf. on Combinatorics, Graph Theory and Computing* Utilitas Mathematica (1978), 527–547.

[Ni79] T. Nishizeki, On the relationship between the genus and the cardinality of the maximum matchings of a graph, *Discrete Math.* 25 (1979), 149–156.

[Nied91] F. Niedermeyer, f-optimal factors of infinite graphs, *Discrete Math.* 95 (1991), 231–254.

[NiedPo94] F. Niedermeyer and K.-P. Podewski, Matchable infinite graphs, *J. Combin. Theory Ser. B* 62 (1994), 213–227.

[Nies98] T. Niessen, Note: A characterization of graphs having all (g, f)-factors, *J. Combin. Theory Ser. B* 72 (1998), 152–156.

[NiesRa98] T. Niessen and B. Randerath, Regular factors of simple regular graphs and factor-spectra, *Discrete Math.* 185 (1998), 89–103.

[NiesVo90] T. Niessen and L. Volkmann, Class 1 conditions depending on the minimum degree and the number of vertices of maximum degree, *J. Graph Theory* 14 (1990), 225–246.

[Nis89] T. Nishimura, Independence number, connectivity and degree factors, *SUT J. Math.* 25 (1989), 79–87.

[Nis90] T. Nishimura, Note: Regular factors of line graphs, *Discrete Math.* 85 (1990), 215–219.

[Nis91] T. Nishimura, Regular factors of line graphs. II, *Math. Japon.* 36 (1991), 1033–1040.

[Nis92] T. Nishimura, A degree condition for the existence of k-factors, *J. Graph Theory* 16 (1992), 141–151.

[Nis94] T. Nishimura, Degree factors of line graphs, *Ars Combin.* 38 (1994), 149–159.

[No08] S. Norine, Pfaffian graphs, T-joins and crossing numbers, *Combinatorica* 28 (2008), 89–98.

[OkUeUn10] Y. Okamoto, R. Uehara, and T. Uno, Counting the number of matchings in chordal and chordal bipartite graph classes, *Lecture Notes in Comput. Sci.* 5911 (2010), 296–307.

[Pa01] E. Palmer, On the spanning tree packing number of a graph: a survey, *Discrete Math.* 230 (2001), 13–21.

[PeRe97] L. Perkovic and B. Reed, Edge coloring regular graphs of high degree, *Discrete Math.* 165/166 (1997), 567–578.

[Pe1891] J. Petersen, Die Theorie der regulären Graphen, *Acta Math.* 15 (1891), 193–220.

[PeLo88] P. Peterson and M. Loui, The general maximum matching algorithm of Micali and Vazirani, *Algorithmica* 3 (1988), 511–533.

[Pi94] C. Picouleau, Note: Complexity of the hamiltonian cycle in regular graph problem, *Theoret. Comput. Sci.* 131 (1994), 463–473.

[Pil83] J. Pila, Connected regular graphs without one-factors, *Ars Combin.* 18 (1983), 161–172.

[Pl94] M. Plummer, Extending matchings in graphs: a survey, *Discrete Math.* 127 (1994), 277–292.

[Pl96] M. Plummer, Extending matchings in graphs: an update, *Utilitas Math.* 116 (1996), 3–32.

[Pl07] M. Plummer, Graph factors and factorization: 1985-2003: A survey, *Discrete Math.* 307 (2007), 791–821.

[Ple79] J. Plesník, Remark on matchings in regular graphs, *Acta Fac. Rerum Natur. Univ. Comenian. Math.* 34 (1979), 63–67.

[Ple84] J. Plesník, A note on the complexity of finding regular subgraphs, *Discrete Math.* 49 (1984), 161–167.

[PlTi91] M. Planthold and S. Tipnis, Regular multigraphs of high degree are 1-factoriz-able, *Proc. London Math. Soc.* 44 (1991), 393–400.

[PlTi01] M. Plantholt and S. Tipnis, All regular multigraphs of even order and high degree are 1-factorable, *Electron. J. Combin.* 8 (2001), #R41.

[Po71] V. Polesskiĭ, A certain lower bound for the reliability of information networks, *Problemy Peredači Informacii* 7 (1971), 88–96 (Russian. English Transl. *Problems of Information Transmission* 7 (1971), 172–179 (1973).)

[Ra49] R. Rado, Factorization of even graphs, *Quart. J. Math. Oxford Ser.* 20 (1949), 94–104.

[RaVe06] B. Randerath and P. Vestergaard, Well-covered graphs and factors, *Discrete Appl. Math.* 154 (2006), 1416–1428.

[Ri02] R. Rizzi, Finding 1-factors in bipartite regular graphs and edge-coloring bipartite graphs, *SIAM J. Discrete Math.* 15 (2002), 283–288.

[Ris01] F. Rispoli, Counting perfect matchings in hexagonal systems associated with benzenoids, *Math. Mag.* 74 (2001), 194–200.

[Ro90] C. Rodger, Graph decompositions, *Matematiche (Catania)* 45 (1990), 119–139.

[RobSeTh99] N. Robertson, P. D. Seymour, and R. Thomas, Permanents, Pfaffian orientations, and even directed circuits, *Ann. of Math.* 150 (1999), 929–975.

[Sa91] A. Saito, One-factors and k-factors, *Discrete Math.* 91 (1991), 323–326.

[Sc83] A. Schrijver, Bounds on permanents, and the number of 1-factors and 1-factorizations of bipartite graphs, *Surveys in Combinatorics*, London Math. Soc. Lecture Note Ser. 82 (1983), 107–134.

[ScBi78] J. Schönheim and A. Bialostocki, Decomposition of K_n into subgraphs of prescribed type, *Arch. Math. (Basel)* 31 (1978/79), 105–112.

[Sc98a] A. Schrijver, Bipartite edge colouring in $O(\Delta m)$ time, *SIAM J. Comput.* 28 (1998), 841–846.

[Sc98b] A. Schrijver, Counting 1-factors in regular bipartite graphs, *J. Combin. Theory Ser. B* 72 (1998), 122–135.

[Sc03] A. Schrijver, Matching, edge-colouring, dimers, *Lecture Notes in Computer Sci.* 2880, 13–22.

[ScVa80] A. Schrijver and W. Valiant, On lower bounds for permanents, *Nederl. Akad. Wetensch. indag. Math.* 42 (1980), 425–427.

[ShSe92] Y. Shibata and Y. Seki, The isomorphic factorization of complete bipartite graphs into trees, *Ars Combin.* 33 (1992), 3–25.

[ShUp81] E. Shamir and E. Upfal, On factors in random graphs, *Israel J. Math.* 39 (1981), 296–302.

[ShUp82] E. Shamir and E. Upfal, One-factor in random graphs based on vertex choice, *Discrete Math.* 41 (1982), 281–286.

[ShWaJu88] M. Shaohan, W. Wallis, and W. Ju-Lin, The complexity of the clique partition number problem, Nineteenth Southeastern Conference on Combinatorics, Graph Theory, and Computing (Baton Rouge, LA, 1988). *Congress. Numer.* 67 (1988), 59–66.

[St77] K. Steffens, Matchings in countable graphs, *Canad. J. Math.* 29 (1977), 165–168.

[St82] F. Steinparz, On the existence of *F* factors, *Conference on Graphtheoretic Concepts in Computer Science (7th: 1981: Linz, Austria)*, Hanser, 1982, 61–73.

[St85a] K. Steffens, Maximal tight sets and the Edmonds-Gallai decomposition for matchings, *Combinatorica* 5 (1985), 359–365.

[St85b] K. Steffens, Faktoren in unendlichen Graphen, *Jahresber. Deutsch. Math.-Verein.* 87 (1985), 127–137.

[St89] K. Steffens, The *f*-factors of countable graphs, *Grüne Reihe Preprint Series* 233 Univ. Hannover (1989).

[StGo81] R. Stanton and I. Goulden, Graph factorization, general triple systems, and cyclic triple systems, *Aequationes Math.* 22 (1981), 1–28.

[StVa82] L. Stockmeyer and V. Vazirani, NP-completeness of some generalizations of the maximum matching problem, *Inform. Process. Lett.* 15 (1982), 14–19.

[Su74] D. Sumner, On Tutte's factorization theorem, *Graphs and Combinatorics*, Lecture Notes in Math. Vol. 406, Springer (1974), 350–355.

[Su76] D. Sumner, 1-factors and anti-factor sets, *J. London Math. Soc.* 13 (1976), 351–359.

[Ta84] V. Tashkinov, 3-regular subgraphs of 4-regular graphs, *Mat. Zametki* 36 (1984), 239–259. (Russian) (English transl.: *Math. Notes* 36 (1984), 612–623.)

[TeFi61] H. Temperley and M. Fisher, Dimer problem in statistical mechanics - an exact result, *Phil. Mag.* 6 (1961), 1061–1063.

[Th81] C. Thomassen, A remark on the factor theorems of Lovász and Tutte, *J. Graph Theory* 5 (1981), 441–442.

[Th83] C. Thomassen, Graph decomposition with constraints on the connectivity and minimum degree, *J. Graph Theory* 7 (1983), 165–167.

[Th08a] C. Thomassen, Edge-decompositions of highly connected graphs into paths, *Abh. Math. Semin. Univ. Hambg.* 78 (2008), 17–26.

[Th08b] C. Thomassen, Decompositions of highly connected graphs into paths of length 3, *J. Graph Theory* 58 (2008), 286–292.

[Thomas06] R. Thomas, A survey of Pfaffian orientations of graphs, *International Congress of Mathematicians Vol. III*, Eur. Math. Soc. (2006), 963–984.

[Thomass81] C. Thomassen, A remark on the factor theorems of Lovász and Tutte, *J. Graph Theory* 5 (1981), 441–442.

[Tu47] W. Tutte, The factorization of linear graphs, *J. London Math. Soc.* 22 (1947), 107–111.

[Tu54] W. Tutte, A short proof of the factor theorem for finite graphs, *Canad. J. Math.* 6 (1954), 347–352.

[Tu61] W. Tutte, On the problem of decomposing a graph into n connected factors, *J. Lond. Math. Soc.* 36 (1961), 221–230.

[Tu78] W. Tutte, The subgraph problem, *Advances in Graph Theory (Cambridge Comb. Conf., Trinity College, Cambridge, 1977)*, Ann. Discr. Math. 3 (1978), 289–295.

[Tu81] W. Tutte, Graph factors, *Combinatorica* 1 (1981), 79–97.

[Va79a] L. Valiant, The complexity of computing the permanent, *Theoret. Comput. Sci.* 8 (1979), 189–201.

[Va79b] L. Valiant, The complexity of enumeration and reliability problems, *SIAM J. Comput.* 8 (1979), 410–421.

[Va94] V. Vazirani, A theory of alternating paths and blossoms for proving correctness of the $O(\sqrt{V}E)$ general graph maximum matching algorithm, *Combinatorica* 14 (1994), 71–109.

[Vo95] L. Volkmann, Regular graphs, regular factors, and the impact of Petersen's theorems, *Jahresber. Deutsch. Math.-Verein.* 97 (1995), 19–42.

[Wa26] B. van der Waerden, Problem 45, *Jahresber. Deutsch. Math.-Verein.* 35 (1926), 117.

[Wa97] W. Wallis, *One-Factorizations*, Kluwer Academic Publishers, 1997.

[Wo90] D. Woodall, k-factors and neighbourhoods of independent sets in graphs, *J. London Math. Soc.* 41 (1990), 385–392.

[Wo84] N. Wormald, Isomorphic factorizations. VII. Regular graphs and tournaments, *J. Graph Theory* 8 (1984), 117–122.

[XuLiTo98] B. Xu, Z. Liu, and T. Tokuda, Connected factors in $K_{1,n}$-free graphs containing a (g, f)-factor, *Graphs Combin.* 14 (1998), 393–395.

[Ya95] G. Yan, Some new results on (g, f)-factorizations of graphs, *J. Combin. Math. Combin. Comput.* 18 (1995), 177–185.

[YaPaWoTo00] G. Yan, J. Pan, C. Wong, and T. Tokuda, Decompositions of graphs into (g, f)-factors, *Graphs Combin.* 16 (2000), 117–126.

[YuKan88] C. Yuting and M. Kano, Some results on odd factors of graphs, *J. Graph Theory* 12 (1988), 327–333.

[YuLi09] Q. Yu and G. Liu, *Graph Factors and Matching Extensions*, Higher Education Press, 2009.

[ZhHeZh09] H. Zhang, P. He, and S. Zhou, Characterizations for $P_{\geq 2}$-factor and $P_{\geq 3}$-factor covered graphs, *Discrete Math.* 309 (2009), 2067–2076.

[ZhZh92] C.-Q. Zhang and Y. Zhu, Factorizations of regular graphs, *J. Combin. Theory Ser. B* 56 (1992), 74–89.

Section 5.5
Applications to Timetabling

Edmund Burke, University of Stirling, Scotland
Dominique de Werra, École Polytechnique Fédérale
de Lausanne, Switzerland
Jeffrey Kingston, University of Sydney, Australia

INTRODUCTION

The construction of timetables for educational institutions and other organizations is a rich area of research with strong links to graph theory, especially to node- and edge-coloring, bipartite matching, and network flow problems. A significant amount of recent research has developed powerful hybrids of graph coloring/meta-heuristic methods. The purpose of this section is to demonstrate how graph theory plays a pivotal role in timetabling research today and to provide insight into the close relationship between graph coloring and a range of timetabling problems. We concentrate on four timetabling problems: class-teacher timetabling, university course timetabling, university examination timetabling and sports timetabling, and we illustrate some of the key points that have underpinned graph-theoretical approaches to timetabling over the years. We aim to highlight the role of graph theory in modern timetabling research and to provide some pointers to the relevant literature for the interested reader.

Automated Timetabling: Historical Perspective

The problem of developing computer programs and systems to solve timetabling problems has been addressed by the scientific community for over 40 years. Bardadym in his 1995 survey [Ba96] examines the distribution of educational timetabling publications from 1960 to 1995. This shows a significant growth in educational timetabling research throughout the 1960s and into the 1970s. There is a lowering of interest in the late 1970s, which picks up again in the 1980s and reaches a peak of over 60 published papers in 1995 alone, the year of the 1st International Conference on the Practice and Theory of Automated Timetabling (PATAT) [BuRo96]. In 1996 the European Association of Operational Research Societies Working Group on Automated Timetabling was launched, and today it has over 300 members from more than 60 countries.

REMARKS

R1: The broad definition of the timetabling problem covers a wide variety of important scheduling problems, which include school timetabling, university course timetabling, examination timetabling, sports timetabling, transport timetabling and a wide variety of employee timetabling and rostering problems.

R2: Welsh and Powell [WePo67] observed the relationship between the graph-coloring problem and timetabling in 1967. This relationship has been a significant feature of timetabling research ever since. A broad generation of timetabling algorithms was based upon graph-coloring methods.

R3: It is not our purpose to survey all of these approaches. Carter's 1986 survey paper [Ca86] on examination timetabling provides an excellent review of the early examination timetabling methods, and Carter and Laporte updated this survey paper in 1995 [CaLa96]. There are a number of other timetabling survey papers that cover the field (e.g., [de85-b], [Ba96], [Wr96], [BuJaKiWe97], [CaLa98], [Sc99], [QuBuMcMeLe09], and [Pi10]).

TERMINOLOGY: Throughout this section, *node* is used instead of vertex.

5.5.1 Specification of Timetabling Problems

Timetabling problems are complex and vary widely in structure. Our definition is general enough to cover most cases.

The General Problem

DEFINITION

D1: A *timetabling problem* is a problem with four parameters: T, a finite set of times; R, a finite set of resources; M, a finite set of meetings; and C, a finite set of constraints. The problem is to assign times and resources to the meetings so as to satisfy the constraints as far as possible. The parts of this definition are elaborated below.

Times

Although it is possible to allow arbitrary time intervals for meetings, in practice time is usually discretized by dividing it into a fixed finite set of intervals of equal length.

DEFINITIONS

D2: A *time* t is an element of the set of times T of an instance of the timetabling problem.

D3: A *time slot* is a variable constrained to contain one time.

FACTS

F1: Time slots are occasionally preassigned (fixed to a particular value in advance).

F2: In practice, constraints involving time often use information about the actual time intervals being represented. For example, a constraint could specify that two time slots must contain times whose underlying time intervals are directly adjacent, or that a set of time slots must contain times that are spread fairly uniformly through the week, and so on.

F3: Some timetables recur: they are repeated every week, or every two weeks, etc. School and university course timetables recur. Other timetables are used only once (e.g., examination timetables).

Resources

Meetings contain teachers, rooms, items of special equipment, students (or groups of students), and so on, which we call *resources*.

DEFINITIONS

D4:   A *resource* r is an element of the set of resources R of an instance of the timetabling problem.

D5: A *resource slot* is a variable constrained to contain one resource.

FACT

F4:   Resource slots are often preassigned (fixed to a particular value in advance). Student group slots are usually preassigned.

EXAMPLE

E1: The basic constraint of timetabling, that no resource appear in two meetings that share a time, applies equally to teachers, students, and rooms, and hence, these items are often treated together as part of a meeting. Other constraints may be specialized for different resources. For example, if a meeting contains several times, it may be required that a particular teacher be present in that meeting for all of those times, whereas in filling the room slot it may be acceptable to use a split assignment, that is, to assign different rooms at different times.

Meetings

DEFINITION

D6: A *meeting* m is a named collection of time slots and resource slots. Assigning values to these slots means that all of the assigned resources attend this meeting at all of the assigned times.

EXAMPLES

E2: In examination timetabling, one meeting will usually represent one examination and contain: one time slot, a large number of preassigned students (those students enrolled in the corresponding course), and one or more room slots.

E3: In school timetabling, one meeting will usually represent one subject studied through one week, and will contain some small number of time slots, one preassigned student group slot, one teacher slot (often preassigned) and one room slot.

E4: In staff rostering, one meeting will represent the total staff requirements for one time interval, and will contain one preassigned time and a number of staff slots, not preassigned.

Constraints

Timetabling practitioners have documented dozens of different constraints in the many organizations they have investigated, so it is not possible to give a comprehensive list of constraints in such a general setting. When evaluating constraints against solutions it is convenient to assign a value of 0 to perfectly acceptable outcomes, and to assign progressively higher values to less acceptable outcomes.

DEFINITIONS

D7: Let S be the set of all solutions to a given timetabling problem. A **hard constraint** is a constraint that must be satisfied. Associated with each hard constraint is a binary-valued function $h : S \to \{0, 1\}$, defined for each solution $w \in S$ by

$$h(w) = \begin{cases} 1, & \text{if } w \text{ does not satisfy the constraint} \\ 0, & \text{otherwise} \end{cases}$$

D8: A **feasible** solution is any solution $w \in S$ that satisfies all the hard constraints, i.e., $h(w) = 0$ for all h.

D9: Let S be the set of all solutions to a given timetabling problem. A **soft constraint** is a constraint that it is desirable, but not necessary, to satisfy. Associated with each soft constraint is a function $s : S \to Z^+$. The interpretation is that a solution $w \in S$ for which $s(w)$ is small is preferred.

D10: Let S be the set of all solutions to a given timetabling problem. The **badness function** of that problem is a function $b : S \to Z^+$ that encapsulates in a single number $b(w)$ an overall rating for a solution $w \in S$.

D11: The **completeness constraint** requires that every time slot receive a value.

D12: The **no-clashes constraint** (or **no-conflicts constraint**) requires that each resource not participate in any two meetings that share a time.

D13: The **availability constraint** specifies that a particular resource is only available for a certain subset of the times T. For example, a part-time teacher might be available only on Thursdays and Fridays.

EXAMPLE

E5: Let the hard constraints for a given problem be $h_1, h_2, \ldots, h_n$ and the soft constraints be $s_1, s_2, \ldots, s_m$. A common approach is to choose a badness function that is a weighted sum of these values:

$$b(S) = \sum_{i=1}^{n} v_i h_i(S) + \sum_{j=1}^{m} w_j s_j(S)$$

where the weights v_i and w_j are nonnegative integers chosen to reflect the importance of the corresponding constraints, with the v_i much larger than the w_j.

REMARKS

R4: In university course timetabling, the no-clashes constraint would typically be a hard constraint for lecturers but a soft constraint for students as far as optional courses are concerned (since it is usually impossible to satisfy every student).

R5: When a resource slot is not preassigned, it almost always carries a resource type constraint, which specifies that the value is constrained to some subset of R. For example, a slot may require one English teacher or one science laboratory. Within the basic categories (rooms, teachers, etc.) these subsets are typically not disjoint; for example, some English teachers may also teach history. Preassignment can be viewed as a type constraint that constrains a slot to a subset of size 1.

R6: The availability constraint for a particular resource may also be expressed by creating an artificial meeting that contains just that resource and those times when the resource is to be unavailable for actual meetings.

R7: Examples of other constraints often considered are: each teacher is to have at least one hour free each day; each student is to have a lunch hour; large gaps between classes during any one day should be minimized; walking time between classes is to be minimized; etc.

5.5.2   Class-Teacher Timetabling

Class-teacher timetabling is a special case of the general problem in which each meeting contains one preassigned student-group slot, one preassigned teacher slot, and any number of time slots. We first consider this basic version of the problem, and then generalize it to *school problems* (pre-college), in which students are timetabled in groups rather than individually. School problems are characteristically dominated by hard constraints, since constraint violations that might be acceptable when they affect one individual are unacceptable when they affect an entire student group.

The Basic Class-Teacher Timetabling Problem

DEFINITIONS

D14: The **basic class-teacher timetabling problem** [Go62] is a timetabling problem in which each meeting contains one preassigned student-group slot, one preassigned teacher slot, and one completely unconstrained time slot. The no-clashes constraint is a hard constraint and applies to every resource.

D15: A **proper edge-coloring** in a graph G is a mapping of the edge-set $E(G)$ to a set of colors such that adjacent edges are assigned different colors.

D16: The **edge-chromatic number** of a graph G, denoted $\chi'(G)$, is the minimum number of different colors required for a proper edge-coloring of G.

FACTS

F5: There is no requirement that each student group and teacher meet exactly once, or indeed at most once. We could allow each meeting to contain any number k of unconstrained time slots, since that would be equivalent to having k meetings between the given student group and teacher.

F6: The class-teacher timetabling problem can be modeled as an **edge-coloring problem in a bipartite graph** [Be83, de85-a]. Each student group is represented by a left node, each teacher is represented by a right node, and each meeting m is represented by an edge between the nodes corresponding to the student group and teacher preassigned to m. If a student group and teacher meet k times, there will be k parallel edges between the two corresponding nodes. Assigning a time to a meeting corresponds to assigning a color to the corresponding edge; the no-clashes constraint is equivalent to requiring a proper edge-coloring.

F7: An obvious lower bound on the edge-chromatic number of a graph, and hence, on the number of different times needed to timetable an instance of the basic class-teacher problem, is the maximum vertex degree. König's theorem (Fact F8) asserts that for the basic class-teacher problem, this is an upper bound as well.

NOTATION: The maximum vertex degree in a graph G is denoted $\Delta(G)$. Sometimes, when the context is clear, we use Δ.

F8: [Ko16] Let G be a bipartite graph. Then $\chi'(G) = \Delta$. (See [GrYe06], §9.3, for a proof.)

REMARKS

R8: A timetable using Δ different times can be constructed in low-order polynomial time [Be83]. The algorithm is based on finding maximum matchings in a bipartite graph. Matchings are discussed in §11.3 of the *Handbook*.

R9: The connection between class-teacher timetabling and edge-coloring in a bipartite graph was first made by Csima [Cs65], according to [ScSt80].

Extensions to the Basic Class-Teacher Problem

We give some examples of extensions to the basic class-teacher timetabling problem. These and others are described in [Pi10] and [Po11].

EXAMPLES

E6: Some teachers may be available for only certain subsets of the full set of times. This was the first timetabling problem, identified as such, shown to be NP-complete [EvItSh76]. Allowing some times to be preassigned is essentially the same case, since meetings with preassigned times reduce the availability of the teachers within them.

E7: Multiple time slots within meetings may be constrained to be contiguous. There is an easy reduction from the bin packing problem [GaJo79], where the bins are days, showing that this problem is NP-complete.

E8: Some meetings may be "group meetings" involving several student groups coming together for a large lecture. This problem is NP-complete, but there is a good approximation algorithm [Asde02].

E9: Room slots may be added to the meetings. In most school-timetabling problems, each student group attends some class at every time, and therefore there must be at least as many rooms as there are student groups. In that case, if rooms are not differentiated into different types, each student group can be permanently allocated to some room. If rooms are typed or preassigned, we have an NP-complete problem equivalent to the basic problem with teacher unavailabilities described in Example E6.

Graph Models for Subproblems of the Class-Teacher Problem

It is frequently the case that intractable timetabling problems have tractable subproblems that may be useful to solve within a larger framework. If the subproblem has no solution, then the entire problem is infeasible (and analysis of the model can uncover the deficiency). If the subproblem reveals that there is only one feasible assignment for some slot, then that assignment might as well be made immediately [Go62].

MODELING EXAMPLES

E10: Suppose we need to determine whether a set of meetings can be scheduled to run simultaneously. First we must check that preassignments or other constraints on their time slots do not preclude this. Then we must check that the combined resource slots of all these meetings can be covered by the complete set of resources R. This is trivial if all the resource slots are preassigned (simply check that no resource is used twice), but in general these slots will be constrained to overlapping subsets of R.

Bipartite Graph Model: Each resource slot becomes a left node, each available resource in R becomes a right node, and an edge joins slot s to resource r whenever r is an acceptable resource for slot s. The meetings may run simultaneously if a matching touching every left node exists [CoKi93].

E11: Example E10 generalizes to multiple times in a way that allows us to check whether the resources and times available can cover all the meetings.

Bipartite Graph Model: There is one left node for each possible triple (m, ts, rs), where m is a meeting, ts is a time slot from m, and rs is a resource slot from m. These triples represent indivisible units of demand for one resource at one time. There is one right node for each possible pair (r, t), where r is a resource and t is a time when r is available. These pairs represent indivisible units of supply (of resources). An edge joining a triple to a pair means that the given constraints are not violated by the implied time and resource assignment. For example, if ts is preassigned we would join triples containing it only to pairs containing its preassigned time; if rs requires an English teacher we would join triples containing it only to pairs containing resources r that are teachers whose capabilities include English. Clearly, if there is no matching that touches every triple, then the problem is infeasible.

E12: We may have a partial solution in which some time slots have been assigned times and others have not. We ask whether we can extend this set of time assignments by assigning workable times to all currently unassigned time slots in the set M_r of all meetings containing a particular fixed resource r (e.g., a student group). These time slots must be assigned distinct times, otherwise there will be a clash involving r.

Bipartite Graph Model: The left nodes are the time slots of M_r, and the right nodes are all the times of T. Create an edge between each time slot that is already assigned and the time it has been assigned. For each time slot that has not been assigned, create an edge between it and each of its allowable times. A time is allowable for a time slot if, when the time slot's meeting is added to those meetings that already contain this time, the resulting collection of meetings can run simultaneously. The meetings may be assigned allowable times if there exists a matching in the resulting graph that touches every time slot node [de85-a, CoKi93].

E13: If the times of all meetings are preassigned it may be possible to create models for assigning teachers. For example, suppose that all meetings occupy one time and may be taught by all teachers, but that each teacher is available for a limited set of times and for a limited total number of classes. This problem, which arises in allocating staff to university tutorials, can be modeled as a *network flow problem*.

Network Flow Model: From the source there is one edge directed to a *teacher-node* for each teacher, with capacity equal to the maximum number of classes for that teacher. From each teacher-node there is one edge with capacity 1 for each time that that teacher is available. Each such edge is directed to a *time-node* that represents the set of all meetings assigned that time. This time-node receives edges from all teachers available at that time. From each time-node, there is an edge directed to the sink, with capacity equal to the number of simultaneous classes allowed at that time. A *minimum-cost network flow model* would allow the inclusion of soft constraints such as teacher preferences for certain times.

REMARK

R10: For a discussion of minimum-cost network flow, see, for example, [Pa82] or §11.2 of the *Handbook*.

5.5.3   University Course Timetabling

University course timetabling differs from the basic class-teacher timetabling problem essentially by the fact that each student may in principle choose the courses of his program, and that there are no other classes of students that are given beforehand and that follow exactly the same program.

Basic Model

The following notation will be used for the rest of this subsection.

NOTATION: Let $\mathcal{C} = \{C_1, \ldots, C_n\}$ denote a collection of courses to be offered during the week W, where W is viewed as a set of time periods. We assume that each course C_i consists of c_i one-period lectures, that is, $C_i = \{C_i^1, C_i^2, \ldots, C_i^{c_i}\}$. For each student s_t, let $\overline{S}_t$ be the collection of courses chosen by student s_t.

DEFINITIONS

D17: A *course timetable* is an assignment to each course C_i a set $\overline{C}_i \subset W$ of c_i time periods, one for each of its c_i lectures.

D18: Given a course timetable, a *conflict* occurs if for some student s_t, there exist two courses $C_i, C_j \in \overline{S}_t$ such that $\overline{C}_i \cap \overline{C}_j \neq \emptyset$. In other words, there are two courses chosen by student s_t that have at least one lecture at the same time.

D19: The *university course timetabling problem* is to produce a conflict-free (or *feasible*) course timetable.

REMARKS

R11: For the moment we assume that there are no capacity obstacles (i.e., the classrooms are large enough and a course may accommodate any number of students).

R12: It may occur that with a given set of data, no feasible timetable can be found. In such a case we may need to relax our requirements and consider allowing certain conflicting lectures to occur. The resulting timetabling problem becomes one of minimizing the severity of the conflicts. A measure of the severity of a conflict is the number of students who have elected to take both of these lectures. This is formalized in Definition D20 below.

A Graph Formulation

Our graph model consists of nodes representing lectures, and edges joining pairs of these nodes, where the edges are weighted according the severity of the conflicts they represent.

DEFINITIONS

D20: The *(penalty) weight*, w_{ij}, of a conflict between two lectures C_i^r and C_j^s is the number of students who have to take both of these lectures, i.e.,

$$w_{ij} = |\{t | C_i, C_j \in \overline{S}_t\}|$$

D21:   A *conflict graph* G is an edge-weighted graph defined as follows: for each course C_i, there are c_i nodes,

$$C_i^1, C_i^2, \ldots, C_i^{c_i},$$

representing its lectures. For each pair i, j, $i \neq j$, if $w_{ij} > 0$, then an edge with weight w_{ij} is created between nodes C_i^r and C_j^s for each pair r, s, $r \neq s$. In addition, an edge with weight ∞ is created between nodes C_i^r, C_i^s for each possible pair r, s, representing a *prohibitive penalty* corresponding to two lectures of the same course.

NOTATION: The edge joining nodes x and y is denoted $[x, y]$. This causes no ambiguity here because conflict graphs have no multi-edges. Some other sections of the *Handbook* use (x, y) or xy to denote simple adjacency between x and y.

D22:   In a graph G, a subset of mutually non-adjacent nodes is called a *stable* (or *independent*) set of nodes.

D23:   A *proper node-coloring* of a graph G is an assignment of colors to the nodes of G such that adjacent nodes receive different colors. A *proper node k-coloring* is a proper node-coloring that uses k different colors.

FACTS

F9:   The timetabling problem reduces to finding a *partition* $\mathcal{P}$ of the node-set $V(G)$ into $k = |W|$ subsets, $S_1, \ldots, S_k$, that minimizes the total penalty

$$z(\mathcal{P}) = \sum_{u=1}^{k} (w_{ij} \mid C_i^r, C_j^s \in S_u)$$

F10:   It is easy to see that there is a one-to-one correspondence between feasible (conflict-free) timetables and partitions $\mathcal{P}$ with $z(\mathcal{P}) = 0$: given such a partition, $C_i^r \in S_u$ means that lecture r of course C_i is scheduled at period $u \in W$.

F11:   A partition $\mathcal{P}$ for which $z(\mathcal{P}) = 0$ gives rise to a proper node-coloring, obtained by assigning the same color to each node in one cell of the partition so that different cells get different colors. Conversely, given a proper node-coloring, the node-subsets receiving the same color (called *color classes*) form a partition with $z(\mathcal{P}) = 0$. Thus, there exists a feasible timetable in $k = |W|$ periods if and only if G has a proper node k-coloring.

F12:   Node-coloring models are more general than edge-coloring models: one can always transform an edge-coloring instance into a node-coloring instance in an auxiliary graph, but the converse is not true.

F13:   For some classes of graphs, the determination of the smallest k for which there exists a node k-coloring (the chromatic number) is easy; it is in particular the case for perfect graphs (see [Be83]). But in general the problem is NP-hard. Node-coloring (vertex-coloring) is discussed in detail in §5.1 and §5.2.

EXAMPLE

E14: Figure 5.5.1 gives an example of a university timetabling problem. In the basic model introduced above, we have not mentioned the teachers giving the various courses. We have assumed that all courses are to be taught by different teachers. Should this not be the case, we would simply introduce edges with a prohibitively large weight between lectures (of different courses) that have to be given by the same teacher. This would not change the nature of the problem, which remains a node-coloring problem in a graph or a weighted extension as shown above.

$$
\begin{array}{lll}
C_1 = 3 \text{ lectures} & \overline{S}_1 = C_1, C_2 & w_{12} = 1 \\
C_2 = 2 \text{ lectures} & \overline{S}_2 = C_2, C_3 & w_{13} = 1 \\
C_3 = 2 \text{ lectures} & \overline{S}_3 = C_1, C_3, C_4 & w_{14} = 1 \\
C_4 = 1 \text{ lecture} & \overline{S}_4 = C_3, C_4 & w_{23} = 1 \\
& \overline{S}_5 = C_2 & w_{24} = 0 \\
& & w_{34} = 2
\end{array}
$$

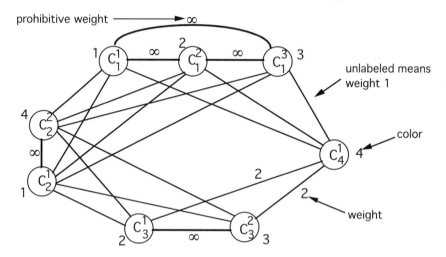

Figure 5.5.1: An example of university timetabling.

$$
\begin{array}{llll}
C_1^1, C_2^1 & : & \text{period 1} \qquad & C_1^3, C_3^2 & : & \text{period 3} \\
C_1^2, C_3^1 & : & \text{period 2} \qquad & C_2^2, C_4^1 & : & \text{period 4}
\end{array}
$$

$$
z(\mathcal{P}) = 1 \quad (edge[C_1^1, C_2^1])
$$

Scheduling Multi-Section Courses

Suppose that a collection of m courses, $\{C_1, C_2, \ldots, C_m\}$, has to be scheduled. Assume, for notational simplicity, that each course consists of a single weekly lecture and that there are exactly h_i sections of course C_i, where $h_1 \geq h_2 \geq \cdots \geq h_m$. The following four-step strategy produces a timetable for all sections of all m courses, *in advance*, that can accommodate any collection of student groups, $\{g_1, g_2, \ldots, g_n\}$, as long as no more than h_i of those groups need course C_i. The strategy is followed by an example illustrating each step on a sample problem.

Step 1: Construct the bipartite graph $G^* = (L^*, R^*, E^*)$, where as a left set, we have $L^* = \{C_1, C_2, \ldots, C_m\}$, the right set $R^* = \{1, 2, \ldots, h_1\}$, and for each $i = 1, 2, \ldots, m$, $[C_i, j] \in E^*$ for $j = 1, 2, \ldots h_i$.

Step 2: Produce a set of *feasible colors* for the sections of each course.

- If Δ is the maximum degree of nodes in G^*, then clearly, $\Delta = \max\{m, h_1\}$.

- From König's theorem (Fact F8), G^* has a proper edge Δ-coloring, which can be constructed easily.

- For this edge-coloring of bipartite graph G^*, let $p(C_i)$, $i = 1, 2, \ldots, m$, denote the set of colors used for the edges incident on node (course) C_i. Observe that $|p(C_i)| = h_i$, $i = 1, 2, \ldots, m$.

Step 3: Given an actual collection $\mathcal{G} = \{g_1, g_2, \ldots, g_n\}$ of student groups, construct a bipartite graph $G^{**} = (L^{**}, R^{**}, E^{**})$, where the left set $L^{**} = \{C_1, C_2, \ldots, C_m\}$, the right set $R^{**} = \{g_1, g_2, \ldots, g_n\}$, and for each pair i, j, $i = 1, 2, \ldots, m$ and $j = 1, 2, \ldots, n$, edge $[C_i, g_j] \in E^{**}$ if and only if student group g_j needs course C_i.

Step 4: Assign the collection $\mathcal{G}$ of student groups to the sections of courses $C_1, C_2, \ldots, C_m$ without changing the time-period of any section.

NOTATION: Let $g(C_i)$ denote the set of student groups needing course C_i.

FACT

F14: [Hä83] Given the bipartite graph G^{**} defined in Step 3, if $deg_{G^{**}}(C_i) = h_i$ for each i, then there is a proper edge Δ-coloring of G^{**} such that the edges incident on node C_i are assigned the feasible colors of C_i obtained from Step 2. (See also [AsDeHa98].)

REMARK

R13: In terms of the timetabling problem, Fact F14 says that if the number of student groups that need course C_i equals the number of sections that have been scheduled for C_i, i.e., $|g(C_i)| = |p(C_i)| (= h_i)$, $i = 1, 2, \ldots, m$, then there exists an assignment of the student groups to sections such that each student group gets the courses it needs and the original set of time-periods for the sections of each course is unchanged.

EXAMPLE

E15: (Step 1) The graph G^* with $m = 5$ and $(h_1, h_2, \ldots, h_5) = (6, 4, 4, 2, 1)$ is shown in Figure 5.5.2.

(Step 2) For the graph in Figure 5.5.2, $\Delta = 6$, and a proper edge 6-coloring using colors $\{a, b, c, d, e, f\}$ is represented by the following matrix whose (i, j)th entry is the color assigned to edge $[C_i, j]$.

$$
\begin{array}{c}
\\ C_1 \\ C_2 \\ C_3 \\ C_4 \\ C_5
\end{array}
\begin{array}{cccccc}
1 & 2 & 3 & 4 & 5 & 6 \\
\left(\begin{array}{cccccc}
a & b & c & d & e & f \\
b & e & d & a & & \\
c & d & a & e & & \\
d & a & & & & \\
e & & & & &
\end{array}\right)
\end{array}
$$

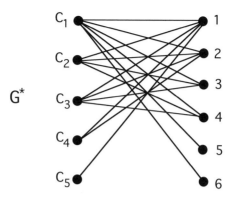

Figure 5.5.2: Bipartite graph G^* for $m = 5$ and $(h_1, h_2, \ldots, h_5) = (6, 4, 4, 2, 1)$.

Thus, the corresponding timetable for the courses is given by the following sets $p(C_i)$ of feasible colors (time-periods) for the sections of course C_i, $i = 1, 2, \ldots, m$:

$$p(C_1) = \{a, b, c, d, e, f\}$$
$$p(C_2) = \{a, b, d, e\}$$
$$p(C_3) = \{a, c, d, e\}$$
$$p(C_4) = \{a, d\}$$
$$p(C_5) = \{e\}$$

(Step 3) The bipartite graph G^{**} shown in Figure 5.5.3 below represents the specific requirements of seven student groups, $g_1, g_2, \ldots, g_7$. For instance, group g_1 needs courses C_1, C_2, and C_4, and group g_6 needs courses C_1 and C_4.

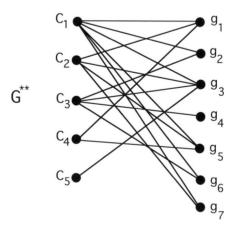

Figure 5.5.3: Bipartite graph G^{**}.

(Step 4) A proper edge 6-coloring for bipartite graph G^{**} in Figure 5.5.3 is represented by the matrix below. Observe that the colors used for the edges incident on a given course-node are precisely that node's feasible colors determined in Step 2.

$$
\begin{array}{c}
\begin{array}{ccccccc}
g_1 & g_2 & g_3 & g_4 & g_5 & g_6 & g_7
\end{array}\\
\begin{array}{c}
C_1\\ C_2\\ C_3\\ C_4\\ C_5
\end{array}
\left(
\begin{array}{ccccccc}
f & e & d &   & a & b & c\\
d &   & b &   & e &   & a\\
  & d & c & a &   & e & \\
a &   &   &   & d &   & \\
  &   & e &   &   &   &
\end{array}
\right)
\end{array}
$$

5.5.4   University Examination Timetabling

Basic Model

Examination timetabling differs from university course timetabling in a number of ways. However, the very core of the problem can be considered to be the same. We have a collection of exams $E_1, \ldots, E_n$ that have to be assigned time slots (periods) and rooms. The number of periods that are available can play a crucial role. In many universities the number of periods extends over a time length of two to four weeks. The constraints that characterize the examination timetabling problem are quite different from constraints that are important in course timetabling.

DEFINITION

D24: Given an examination timetable, a ***conflict*** occurs if two exams taken by the same student are scheduled in the same time period.

FACTS

F15: In examination timetabling it is often desirable (or necessary) to have several exams allocated to the same room. It would, of course, not be very sensible to assign a number of lectures to the same room!

F16: In examination timetabling, it is usually considered desirable to spread exams out over the number of periods so that students do not have exams in succession. On the other hand, for course timetabling it is often considered undesirable to spread the lectures out. Students tend to prefer to have lectures in contiguous blocks. The prototype problem given in Example E16 below has seven exams $(E_1, \ldots, E_7)$ that it has to allocate to five time periods $P_1, \ldots, P_5$. It only attempts to satisfy the constraint that no student can attend more than one examination at the same time. We say that there is a conflict in the timetable if that constraint is not satisfied.

EXAMPLE

E16: For our prototype problem, there are seven exams $(E_1, \ldots, E_7)$ to assign to five time periods $P_1, \ldots, P_5$ such that there are no conflicts. In our graph model, nodes represent examinations, and edges join two nodes whose corresponding exams have at least one student in common. Weights on the edges between two nodes (exams) can represent the number of students who have to take both of those exams. The graph model is shown in Figure 5.5.4 below. Exam 1 only conflicts with Exam 4 (seven students need to take both exams). However, Exam 2 has one, seven, and three students in common with Exam 4, Exam 5, and Exam 7, respectively. This simplified examination-timetabling problem is directly analogous to the node-coloring problem where the colors are represented by the periods and is very similar to the graph-theoretical models discussed earlier. The solution to this simplified problem, shown in Figure 5.5.5, uses all five colors (periods).

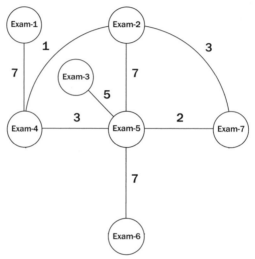

Figure 5.5.4: A graph model for our prototype problem.

Period 1	Period 2	Period 3	Period 4	Period 5
Exam-1	Exam-3	Exam-2	Exam-6	Exam-7
Exam-5	Exam-4			

Figure 5.5.5: A solution to the problem using all the colors (periods).

A More Compact Schedule

Is the solution given in the table above a good solution? If we consider the problem purely as producing a coloring that uses the fewest possible colors, then the answer is clearly no. An alternative coloring can be seen in Figure 5.5.6.

Period 1	Period 2	Period 3	Period 4	Period 5
Exam-1	Exam-3	Exam-2		
Exam-5	Exam-4			
	Exam-6			
	Exam-7			

Figure 5.5.6: A solution to the problem that uses only three colors.

REMARK

R14: The solution in Figure 5.5.6 uses only three colors (periods), rather than five, so it is clearly better in terms of the number of colors used. However, if we consider it as an examination timetabling problem, then a student who has to take Exam 1, Exam 2, and Exam 7 would (almost certainly) consider the solution in Figure 5.5.5 to be better. On the other hand, a university administrator might think that the solution in Figure 5.5.6 is better than the one in Figure 5.5.5 because it gets the exams completed more quickly.

FACTS

F17: The node-coloring problem can be considered to be an underlying model of examination timetabling, but for realistic applications there are a number of other constraints, both hard and soft, that need to be considered. For instance, room capacity is a hard constraint, but avoiding having students take consecutive exams is a soft constraint.

F18: In situations where no proper node-coloring exists, the objective might be to minimize the number of students having conflicts. But a further complication is weighing this consideration against other features of an examination timetable: how spread out the exams are, how many days are used, etc.

DEFINITION

D25: The *quality of a solution* to an examination timetabling problem can be defined as a measure of the level of satisfaction of the soft constraints (provided all the hard constraints are satisfied).

The Breadth and Variation of Exam Timetabling Constraints

In 1996, Burke, Elliman, Ford, and Weare [BuElFoWe96] published a paper that analyzed and discussed the results of a questionnaire completed by examination-timetabling administrators from 56 British universities. The aim of this exercise was to determine the nature of the problem as it occurred in British universities (circa 1995). The questionnaire concentrated upon 13 constraints but also asked administrators to include other constraints thought to be important to their institution. An additional 19 constraints were listed. The 32 constraints demonstrate the breadth and variation of requirements and priorities among British universities.

REMARK

R15: The message for developers of examination-timetabling decision-support software is that if a system is to be generic and widely applicable, then it has to be flexible in designating which soft constraints are important and which are not. It has to allow the user to weight soft constraints according to the needs and requirements of the user's own institution.

Heuristic Methods

Early approaches to solving the examination-timetabling problem [Br64], [Co64] employed heuristic construction methods. As mentioned in the introduction, the analogy with graph coloring was observed by Welsh and Powell in 1967 [WePo67]. This observation has underpinned the development of examination-timetabling methods. The survey paper [Ca86] and its sequel by Carter and Laporte [CaLa96] provide an excellent overview of the development of graph-coloring-based heuristic approaches for the examination-timetabling problem.

FACTS

F19: One of the basic approaches for solving the examination-timetabling problem is to construct the timetable by sequentially placing exams into periods according to some measure (heuristic) of how difficult the exams are to schedule (see [Ca86]). The early approaches mentioned above used this strategy.

F20: The strategy of scheduling the most troublesome exams first corresponds to coloring the nodes in the graph model that are expected to be the most difficult. Examples of four of the most common node-coloring heuristics used in examination-timetabling systems are presented below.

HEURISTICS

H1: Largest Degree: This heuristic takes the nodes with the largest degree (number of edges) and schedules them first. This corresponds to the exams that have the most conflicts with other exams.

H2: Largest Weighted Degree: This heuristic is similar to Largest Degree except that the edges are weighted by the number of students who are involved in the conflict.

H3: Color Degree: Here, we first schedule the exams that have the largest number of conflicts (degree) with the other exams that have already been placed into the timetable.

H4: Saturation Degree: This heuristic chooses first those exams that have the least number of available periods in the timetable that can be selected without violating hard constraints.

REMARKS

R16: While these four heuristics do not form an exhaustive list, they do cover the key node-coloring-based heuristics used in examination-timetabling. Examples of these and other similar approaches include [Br64], [Co64], [WePo67], [Wo68], [Me81], [Me82], and [BuElWe94]. For further discussion, see the survey papers [de85-b], [Ca86], [Ba96], [Wr96], [BuJaKiWe97], [CaLa96], [CaLa98], and [Sc99].

R17: There is an obvious limitation with the simple timetable construction method outlined above (independent of the heuristics used). Exams scheduled early in the process might make certain other exams impossible to schedule later on. This can be addressed by adding a ***backtracking*** component to the process. An algorithm that gets stuck can unschedule or re-schedule exams. Examples of this kind of approach can be found in [CaLaCh94] and [CaLaLe96].

R18: Consistent with the philosophy of scheduling the most troublesome exams first, Carter and his colleagues investigated methods based on finding a ***maximum clique*** of the conflict graph. A maximum clique is a largest subgraph where each node is adjacent to every other node. See [CaJo01] for more details about the role of cliques in examination timetabling.

Two Different Random-Selection Strategies

Burke, Newall, and Weare [BuNeWe98-a] use a random element in the process of selecting the next exam to schedule. This approach produced good results quickly and can be seen as a compromise between the relatively simple coloring-heuristic-based methods discussed earlier and the more complex *meta-heuristic* methods (discussed briefly below), which generally require much more computational time. The two randomization approaches that are considered in [BuNeWe98-a] are described in the next two examples.

(1) A random subset of exams is selected, and the most difficult from within the subset is selected (according to some heuristic).

(2) The x most difficult exams to schedule are selected (according to some heuristic), and then one of those x exams is selected at random.

Hybrid Graph-Coloring/Meta-Heuristic Approaches

Throughout the 1990s, meta-heuristic approaches, such as *simulated annealing, evolutionary methods*, and *tabu search*, were investigated and developed for various timetabling problems. Significant progress has been made by combining the more modern meta-heuristic methods with some of the older graph-coloring-based methods. For a discussion of the advantages and disadvantages of using such approaches for timetabling, see [de85-b], [Ca86], [Ba96], [Wr96], [BuJaKiWe97], [CaLa96], [CaLa98], and [Sc99]. To find out more about the meta-heuristics themselves, see [GlKo03].

EXAMPLES

E17: Dowsland and Thompson [ThDo96-a, ThDo96-b] implemented a ***simulated annealing***/graph-coloring hybrid approach for solving the examination-timetabling problem at the University of Wales Swansea. Their method works in two phases. The first phase satisfies the hard (*binding*) constraints:

(a) all exams to be scheduled within 24 time slots,

(b) no student clashes to be allowed,

(c) certain pairs of exams to be scheduled at the same time,

(d) certain pairs of exams to be scheduled at different times,

(e) certain groups of exams to be scheduled in order,

(f) certain exams to be scheduled within time windows,

(g) no more than 1200 students to be involved in any one session.

The second phase of this simulated annealing approach attempts to optimize the soft constraints of the problem:

(a) minimize the number of exams with over 100 students scheduled after period 10,

(b) minimize the number of occurrences of students having exams in consecutive periods.

E18: Burke, Newall, and Weare in 1998 [BuNeWe98-b] used graph-coloring heuristics (Largest Degree, Color Degree, and Saturation Degree) to construct initial solutions that were then fine-tuned by *memetic algorithms*. Memetic algorithms refer to evolutionary methods (often genetic algorithms) combined with local search (often hill-climbing). The memetic algorithm that they investigated was based upon one that had already been shown to work well on benchmark examination-timetabling problems [BuNeWe96].

E19: Burke and Newall used the heuristics outlined above in conjunction with a *decomposition* approach [BuNe99]. The authors investigated ways of decomposing large problems into smaller subproblems, which were then solved using memetic algorithms. However, the authors noted that the decomposition approach is independent of the method that is used to solve each of the subproblems. Decomposition had been previously addressed by Carter [Ca83].

E20: Di Gaspero and Schaerf [DiSc01] presented an approach, based on the work of Hertz and de Werra [Hede87], that combined graph-coloring heuristics and *tabu search*. They employed weights on the edges to represent the number of students who were involved in the conflicts between the corresponding pairs of exams, and they also employed weights on the nodes to indicate the number of students taking the exams. For a range of benchmark problems, their method was competitive with (and in some cases, better than) state-of-the-art methods in 2001.

REMARK

R19: A potential drawback with the decomposition approach described in Example E19 is that exams can be assigned time slots in earlier subproblems that then lead to the infeasibility of later subproblems. Burke and Newall employed graph coloring heuristics to build the subproblems in order to tackle this difficulty. For the problems they considered in [BuNe99], the approach that used the saturation-degree heuristic along with using a subproblem size of 50 exams for the smaller problems and 100 exams for the larger problems was the most effective one. They also employed a look-ahead approach to try and detect difficulties. It considered two subproblems together and fixed the solution to the ith one only after it had solved the $(i + 1)$th one.

5.5.5   Sports Timetabling

This section focuses on modeling and solving some basic problems occurring in the construction of season schedules for sports leagues. We show how the design of some round-robin tournaments can be modeled as an edge-coloring problem in a digraph. Such a model should then be extended to handle more general constraints that arise when a season schedule involves travel that should be optimized. This *traveling-tournament problem* (TTP) is described in [EaNeTr03]. Instead of discussing the general problem here, we concentrate on a simple model using elementary properties of graphs. We use the terminology of [Be83] for general graphs and that of [de81] for sports scheduling.

DEFINITIONS

D26: A *(single) round-robin tournament* for a set of l teams is a collection of games such that each team plays each other team exactly once. Each game is played in one of the two teams' home city.

D27: If the game between teams i and j is played in the home city of team j, then the game is a *home game*, H, for team j and an *away game*, A, for team i.

D28: Given a sports league consisting of l teams, a *basic sports timetable (schedule)* (for a round-robin tournament) has two components for each pair of teams i and j:

- designating the day on which the game between i and j is played;

- designating the home city for that game.

REMARK

R20: For the rest of this subsection, we assume that the sports league consists of $2n$ teams for some integer n.

A Simple Graph Model

The league of $2n$ teams is identified with the node set of a graph G, and an edge joining node i and node j corresponds to a game between team i and team j. Observe that if a round-robin tournament is to be scheduled, graph G is the complete graph K_{2n}.

NOTATION: An undirected edge between node i and node j is denoted $[i,j]$. A directed edge from i to j is denoted (i,j) and indicates that the game is a home game for team j and an away game for team i.

DEFINITIONS

D29: Let G be the $2n$-node graph representing a league of $2n$ teams. An *oriented d-coloring* of graph G is a proper edge-d-coloring together with an assignment of a direction to each edge. This oriented d-coloring results in a digraph, each of whose arcs is assigned one of the d colors. This arc-colored digraph specifies a sports timetable using d days for the $2n$ teams: the arcs that are assigned color c_k correspond to those games that are scheduled for day k, and the arc (i,j) indicates that the game between teams i and j is a home game for j and an away game for i.

NOTATION: The digraph created from an oriented coloring of a graph G is denoted $\vec{G}$.

TERMINOLOGY: When a round-robin tournament is to be scheduled (i.e., G is K_{2n}), the digraph $\vec{G}$ is a *tournament*. This family of digraphs is covered in detail in §3.3.

D30: Let $\{c_1, c_2, \ldots, c_d\}$ be the colors used for an edge d-coloring of a graph G. For each color c_k, $k = 1, 2, \ldots, d$, the *color class* M_k is the set of edges assigned color c_k. For a given oriented d-coloring of graph G, $\vec{M}_k$ denotes the set of arcs assigned color c_k.

D31: A *factor* of a graph (digraph) G is a subset F of edges (arcs) such that every node of G is incident on exactly one edge (arc) in F.

D32: A *d-factorization* of a graph G is a partition, $\{F_1, F_2, \ldots, F_d\}$, of the edge-set of G such that each F_i is a factor of G. A graph G is *d-factorizable* if there exists a d-factorization of G. A d-factorization $\{\vec{F_1}, \vec{F_2}, \ldots, \vec{F_d}\}$ of the arcs of a digraph is defined analogously.

TERMINOLOGY NOTE: A factor in an undirected graph is also called a *perfect matching* and is actually a 1-factor, where an *r-factor* is an r-regular, spanning subgraph of G. Matchings are discussed in §11.3, regular graphs are introduced in §1.2, and graph factors and factorization are discussed in §5.4.

FACTS

F21: A d-factorization $\{F_1, \ldots, F_d\}$ of graph G induces a proper edge d-coloring of G, obtained by assigning color c_k to each of factor F_k, $k = 1, 2, \ldots, d$. Thus, if a graph G is d-factorizable, then there exists a proper edge d-coloring of G.

F22: Analogous to Fact F21, an oriented d-coloring of a graph G induces a d-factorization of the digraph $\vec{G}$.

F23: The d-day schedule that corresponds to a d-factorization of a graph G has the property that each team plays a game on each of the d days, i.e., no team has a day off.

TERMINOLOGY: Sometimes, an oriented d-coloring, its induced d-factorization of the arc-set of the resulting digraph, and the corresponding schedule will all be regarded as the same thing.

EXAMPLE

E21: Figure 5.5.7 shows an oriented 5-coloring (using colors $1, 2, \ldots, 5$) of a complete graph $G = K_6$ representing a league of $2n = 6$ teams.

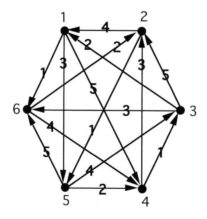

Figure 5.5.7: An oriented 5-coloring of K_6.

The corresponding 5-factorization, $(\vec{F}_1, \ldots, \vec{F}_5)$, of the digraph $\vec{G}$ is shown in table below (Figure 5.5.8). The ith row of the table lists the arcs assigned color i. This factorization specifies the complete 5-day schedule for the six teams. In particular, the arcs in the ith row indicate the games scheduled on the ith day. For instance, on day 2, team 3 plays team 1, and team 1 is at home.

$$
\begin{array}{cccc}
 & \text{game 1} & \text{game 2} & \text{game 3} \\
\vec{F}_1 \text{ (day1)} & (1,6) & (2,5) & (4,3) \\
\vec{F}_2 \text{ (day2)} & (6,2) & (3,1) & (5,4) \\
\vec{F}_3 \text{ (day3)} & (3,6) & (4,2) & (1,5) \\
\vec{F}_4 \text{ (day4)} & (6,4) & (5,3) & (2,1) \\
\vec{F}_5 \text{ (day5)} & (5,6) & (1,4) & (3,2)
\end{array}
$$

$$
\begin{array}{cccc}
 & \text{game 1} & \text{game 2} & \text{game 3} \\
\vec{F}_1 \text{ (day1)} & (1,6) & (2,5) & (4,3) \\
\vec{F}_2 \text{ (day2)} & (6,2) & (3,1) & (5,4) \\
\vec{F}_3 \text{ (day3)} & (3,6) & (4,2) & (1,5) \\
\vec{F}_4 \text{ (day4)} & (6,4) & (5,3) & (2,1) \\
\vec{F}_5 \text{ (day5)} & (5,6) & (1,4) & (3,2)
\end{array}
$$

Figure 5.5.8: A compact schedule.

Observe that each team plays a game on each of the five days, which illustrates Fact F23. This kind of *compact* schedule always exists for a round-robin tournament of $2n$ teams because the complete graph K_{2n} has a $(2n-1)$-factorization, where each factor has n edges.

Profiles, Breaks, and Home-Away Patterns of a Schedule

DEFINITIONS

D33: Let S be a schedule for a league of $2n$ teams. The **home-away pattern** (HAP) associated with S (see [de81]), denoted $H(S)$, is a $2n \times (2n-1)$ array defined by

$$
h_{ik}(S) = \begin{Bmatrix} A \\ H \\ \emptyset \end{Bmatrix} \text{ if team } i \text{ has } \begin{Bmatrix} \text{an away game} \\ \text{a home game} \\ \text{no game} \end{Bmatrix} \text{ on day } k
$$

D34: For a given an $(2n-1)$-day schedule S for a league of $2n$ teams, the **profile** of team i is the ith row of $H(S)$. Thus, the profile is the sequence of H's and A's indicating when team i is home and away for the $2n-1$ days.

D35: For a given schedule S, the profiles of two teams are **complementary** if for each day in the schedule, one of the teams is at home and the other is away.

D36: For a given schedule S, team i has a **break** on day $(k+1)$ if $h_{ik}(S) = h_{i,k+1}(S)$. In other words, the profile of team i has two consecutive H's or two consecutive A's, with the second one falling on day $(k+1)$.

EXAMPLE

E22: Figure 5.5.9 shows the HAP associated with the oriented 5-coloring of Figure 5.5.7 and its corresponding schedule in Figure 5.5.8. The breaks are indicated by underlining. Notice that teams 4 and 5 have complementary profiles.

		days			
	1	2	3	4	5
1	A	H	A	H	A
2	A	H	H	A	H
3	H	A	A	H	A
4	A	H	A	H	H
5	H	A	H	A	A
6	H	A	H	A	H

Figure 5.5.9: The HAP associated with the schedule of Figure 5.5.8.

A Lower Bound on the Number of Breaks

Often in round-robin tournaments, one tries to construct schedules in which for each team, home games and away games alternate as regularly as possible (i.e., the number of breaks is minimized).

DEFINITION

D37: A subset of mutually non-adjacent nodes in a graph G is called a **stable** (or **independent**) set. The **independence number** of G, denoted $\alpha(G)$, is the maximum size of a stable set. Some other sections of the *Handbook* use $ind(G)$ instead of $\alpha(G)$.

FACTS

F24: [de88] Let G be a d-factorizable graph on $2n$ nodes, and let $(\vec{F}_1, \vec{F}_2, \ldots, \vec{F}_d)$ be a d-factorization arising from an oriented d-coloring of G. Then the corresponding schedule has at least $2(n - \alpha(G))$ breaks.

F25: Since the independence number of a complete graph equals 1, any oriented $(2n - 1)$-coloring of K_{2n}, $(\vec{F}_1, \ldots, \vec{F}_{2n-1})$, has at least $2n - 2$ breaks.

REMARK

R21: Fact F25 implies that the schedule given in Figure 5.5.8 has a minimum number of breaks.

Irreducible and Compact Schedules

DEFINITIONS

D38: A schedule is ***irreducible*** if, whenever two teams play against each other, at most one of them has a break on that day.

D39: A schedule is ***compact*** if each team plays one game on each day (i.e., its HAP has no ϕ symbols).

FACTS

F26: A compact d-day schedule corresponds to a d-factorization of the associated graph.

F27: In a compact schedule, if there is a team with an A in its profile for days k and $k+1$, there must be another team with an H in its profile for days k and $k+1$. Thus, in a compact schedule, *breaks occur in pairs*. (In Figure 5.5.9, teams 2 and 3 and teams 4 and 5 are two such pairs.)

F28: By reversing the orientation of some arcs (corresponding to games with a break for each one of its teams), one may always generate from a schedule S an irreducible schedule that does not have more breaks than S. For the rest of this section, we assume (without loss of generality) that the schedules we consider are irreducible.

NOTATION: Fact F29 below uses the following notation. Given a compact schedule S constructed on a d-regular graph, b_i denotes the ith day on which breaks occur, and γ_i is the number of breaks occurring on day b_i (where $2 \le b_1 < b_2 < \cdots < b_p \le d$). In addition, we define $b_0 = 1$ and $b_{p+1} = d + 1$.

F29: Let G be a d-regular graph. Then the following conditions are equivalent:

(1) There exists a compact schedule S constructed on G, where $2 \cdot \gamma_i$ breaks occur on day $b_i, i = 1, \ldots, p$.

(2) The edge-set of G can be partitioned into subsets $E_1, \ldots, E_{p+1}$ such that

    (a) The edge subset E_i induces a $(b_i - b_{i-1})$-regular bipartite graph with vertex bipartition $\{X_i, \overline{X}_i\}$ for $i = 1, \ldots, p+1$.

    (b) $|\overline{X}_{i+1} \cap X_i| = |\overline{X}_i \cap X_{i+1}| = \gamma_i$ for $i = 1, \ldots, p$.

EXAMPLE

E23: For the compact schedule S in Figure 5.5.8 (and its corresponding HAP in Figure 5.5.9), $(b_0, b_1, b_2, b_3) = (1, 3, 5, 6)$ and $\gamma_1 = \gamma_2 = 1$, and it is easy to see that condition (1) of Fact F29 is satisfied.

To show that condition (2) is satisfied, let $E_1 = F_1 \cup F_2$, $E_2 = F_3 \cup F_4$, and $E_3 = F_5$. Then the vertex bipartitions of the induced subgraphs are:

$$\begin{array}{ll} X_1 = \{1, 2, 4\}, & \overline{X}_1 = \{3, 5, 6\} \\ X_2 = \{1, 3, 4\}, & \overline{X}_2 = \{2, 5, 6\} \\ X_3 = \{1, 3, 5\}, & \overline{X}_3 = \{2, 4, 6\} \end{array}$$

It is now straightforward to verify that condition (2) is also satisfied.

Complementarity

Another property of compact schedules that is of interest in practice is *complementarity*.

DEFINITION

D40: A compact schedule S for K_{2n} has the **complementarity property** if the $2n$ teams can be grouped into n disjoint pairs $T_1, \ldots, T_n$ such that the two teams in each T_i have complementary profiles.

FACTS

F30: [de88] If S is a compact schedule (for K_{2n}) such that each team has at most one break, then S has the complementarity property (by Fact F27).

F31: If S is a compact schedule (for K_{2n}) with exactly $2n - 2$ breaks, then S has the complementarity property.

F32: There are compact schedules with the complementarity property where some teams have more than one break.

EXAMPLES

E24: Consider the compact schedule S in Figure 5.5.8 and its corresponding HAP, given in Figure 5.5.9. The three pairs $T_1 = \{1, 6\}$, $T_2 = \{2, 3\}$, and $T_3 = \{4, 5\}$ show that S has the complementarity property.

E25: Figure 5.5.10 shows a 3-factorization of $G = K_4$ that corresponds to an irreducible compact schedule S. Its HAP shows that S does not have the complementarity property (team a has two breaks).

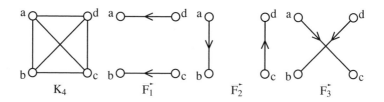

$$\begin{array}{c} & \text{day 1} \quad \text{day 2} \quad \text{day 3} \\ \begin{array}{l} \text{team a} \\ \text{team b} \\ \text{team c} \\ \text{team d} \end{array} & \left(\begin{array}{ccc} H & A & \underline{A} \\ H & \underline{H} & \underline{H} \\ A & \underline{A} & H \\ A & H & A \end{array}\right) \end{array}$$

Figure 5.5.10: An irreducible compact schedule of K_4.

Constructing a Compact Schedule with a Minimum Number of Breaks

We restrict our attention to the most common case, when $G = K_{2n}$. Algorithm 5.5.1 below gives a simple construction that produces an oriented coloring (and hence, a schedule) having exactly $2n - 2$ breaks, which, by Fact F25, is the minimum.

EXAMPLE

E26:   Figure 5.5.11 illustrates Algorithm 5.5.1 for K_6. Observe that the schedule reproduces the oriented 5-coloring given in Figure 5.5.7.

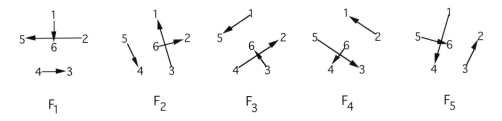

Figure 5.5.11: The 5-day schedule of K_6 produced by Algorithm 5.5.1.

Algorithm 5.5.1: A $(2n - 1)$-Day Schedule of K_{2n} With $2n - 2$ Breaks

Input: Complete graph K_{2n}.
Output: A $(2n - 1)$-day schedule with $2n - 2$ breaks.

Step 1. Construct a $(2n - 1)$-factorization of K_{2n}:
    For $i = 1$ to $2n - 1$
        $F_i = \{[2n, i]\} \cup \{[i + k, i - k] \pmod{2n - 1} : k = 1, 2, \ldots, n - 1\}$
Step 2. Orient the edges:
    For $i = 1$ to $2n - 1$
      If i is odd
          Orient edge $[2n, i]$ as $(i, 2n)$
      Else
          Orient edge $[2n, i]$ as $(2n, i)$
      For $k = 1$ to $n - 1$
        If k is odd
            Orient edge $[i + k, i - k]$ as $(i + k, i - k)$
        Else
            Orient edge $[i + k, i - k]$ as $(i - k, i + k)$

REMARK

R22:   The factorization specified in Step 1 of the algorithm is discussed in [Be83, Chapter 5]. It is called a ***canonical*** factorization [de88].

An Alternate View of the Canonical Factorization

Let $\alpha_n, \alpha_1, \beta_1, \alpha_2, \beta_2, \alpha_3, \ldots, \alpha_{n-1}, \beta_{n-1}, \beta_n$ denote the nodes $1, 2, \ldots, 2n$, respectively, and let $(F_1, \ldots, F_{2n-1})$ be the canonical factorization produced by Algorithm 5.5.1. Consider the partition $(E_1, E_2, \ldots, E_n)$ of the edge-set of K_{2n}, defined by $E_n = F_{2n-1}$ and $E_i = F_{2i-1} \cup F_{2i}$, $i = 1, \ldots, n-1$.

FACTS

F33: For $i = 1, 2, \ldots, n-1$, E_i defines a 2-regular bipartite graph on node sets

$$X_i = \{\alpha_i, \alpha_{i+1}, \ldots, \alpha_n, \beta_1, \beta_2, \ldots, \beta_{i-1}\}$$

$$\text{and} \quad \overline{X}_i = \{\beta_i, \beta_{i+1}, \ldots, \beta_n, \alpha_1, \alpha_2, \ldots, \alpha_{i-1}\}.$$

In addition, $X_i \cap \overline{X}_{i+1} = \alpha_i$ and $\overline{X}_i \cap X_{i+1} = \beta_i$. Thus, by Fact F29, it defines a schedule where nodes α_i and β_i have a simultaneous break on day $b_i = 2i + 1$, $i = 1, \ldots, n-1$.

F34: Fact F33 implies that the canonical factorization produces a compact schedule having exactly $2n - 2$ breaks and satisfying the complementarity property.

EXAMPLE

E27: For K_6, we have $\alpha_1 = 2, \alpha_2 = 4, \alpha_3 = 1, \beta_1 = 3, \beta_2 = 5, \beta_3 = 6$, and from the HAP of Figure 5.5.9, one sees that teams $\alpha_1 = 2$ and $\beta_1 = 3$ have a break on day 3, while teams $\alpha_2 = 4$ and $\beta_2 = 5$ have a break on day 5, and teams $\alpha_3 = 1$ and $\beta_3 = 6$ have no break.

Some Characterization Results

FACTS

F35: Let S_1 and S_2 be two compact schedules for K_{2n}, each with exactly $2n-2$ breaks. If both schedules have the same sequence $b_1, b_2, \ldots, b_{n-1}$ of days where breaks occur in pairs, then their HAPs, $H(S_1)$ and $H(S_2)$, are the same (up to a permutation of rows).

F36: Equivalently, by setting $b_0 = 1$ and $b_n = 2n$, we could start with the sequence $D = (b_1 - b_0, b_2 - b_1, \ldots, b_n - b_{n-1})$, which is the sequence of degrees of the $(b_i - b_{i-1})$-regular bipartite graphs appearing in the partition of the edge-set of K_{2n} defined in Fact F33. (For instance, the schedule of Figure 5.5.8 has $D = (2, 2, 1)$ since $(b_0, b_1, b_2, b_3) = (1, 3, 5, 6)$.)

F37: Given a sequence $D = (d_1, d_2, \ldots, d_n)$ with $d_1 + \cdots + d_n = 2n - 1$, we can reconstruct a unique HAP as follows: for $i \leq n-1$ the profile of α_i starts with an A and has a unique break on day $d_1 + \cdots + d_i + 1$; the profile of α_n starts also with an A and has no break. For each $i \leq n$ the profile of β_i is the complement of the profile of α_i.

EXAMPLE

E28: The HAP in Figure 5.5.12 illustrates Fact F37 for $2n = 6$ and $D = (3, 1, 1)$.

$$
\begin{array}{c c c c c c}
 & \text{day 1} & \text{day 2} & \text{day 3} & \text{day 4} & \text{day 5} \\
\alpha_1 & H & A & H & \underline{H} & A \\
\alpha_2 & H & A & H & A & \underline{A} \\
\alpha_3 & H & A & H & A & H \\
\beta_1 & A & H & A & \underline{A} & H \\
\beta_2 & A & H & A & H & \underline{H} \\
\beta_3 & A & H & A & H & A
\end{array}
$$

Figure 5.5.12: A HAP corresponding to $D = (3, 1, 1)$.

REMARK

R23: A sequence $D = (d_1, \ldots, d_n)$ of positive integers with $d_1 + \cdots + d_n = 2n - 1$ does not in general give a HAP that corresponds to a compact schedule for K_{2n} with $2n - 2$ breaks. For instance, the HAP in Figure 5.5.12 does not correspond to any compact schedule for K_6.

Feasible Sequences

DEFINITION

D41: A sequence $D = (d_1, \ldots, d_n)$ and its corresponding HAP are **feasible** if they correspond to a compact schedule for K_{2n} with $2n - 2$ breaks.

FACTS

F38: [de88] If $D = (d_1, d_2, \ldots, d_n)$ is feasible for K_{2n}, then $\overline{D} = (d_n, d_{n-1}, \ldots, d_1)$ and all sequences obtained by a cyclic permutation of D or $\overline{D}$ are also feasible.

F39: No complete characterization of the feasible sequences has been obtained yet; however, for $n \leq 13$ the feasible sequences have been characterized (see [MiIwMa02]).

F40: Given a sequence $(d_1, d_2, \ldots, d_n)$, we can reconstruct the associated HAP such that: the rows are ordered $\alpha_1, \alpha_2, \ldots, \alpha_n, \beta_1, \beta_2, \ldots, \beta_n$, where α_i and β_i have complementary profiles; α_i and β_i have their break on day $d_1 + \cdots + d_i + 1$ $(i = 1, \ldots, n - 1)$; and the profiles of $\alpha_1, \ldots, \alpha_n$ start with an A.

NOTATION: For a given HAP and any subset T of teams, we define the quantity $\alpha_k(T)$ for each day k by $\alpha_k(T) = \min \{ |\{i \in T | h_{ik} = A\}|, \ |\{i \in T | h_{ik} = H\}| \}$.

F41: [MiIwMa02] If a given HAP is feasible, then for any subset T of teams, $\alpha_k(T)$ is an upper bound on the number of games between teams in T that can be scheduled at period k. Moreover, since all these teams have to play against each other over the $2n - 1$ days, we have

$$\sum_{k=1}^{2n-1} \alpha_k(T) \geq \binom{|T|}{2}$$

F42: [MiIwMa02] Instead of checking explicitly all possible subsets T, it is sufficient to assume that the $2n$ teams are cyclically ordered $(\alpha_1, \alpha_2, \ldots, \alpha_n, \beta_1, \ldots, \beta_n, \alpha_1, \ldots)$, and to examine only subsets T that are intervals of at most n consecutive teams in the cyclic order. Thus, the number of inequalities to check is $\mathrm{O}(n^2)$. Using these observations, the authors were able to eliminate, as infeasible, a number of sequences $D = (d_1, \ldots, d_n)$; it turned out that for $2n \leq 26$, all sequences D that were not eliminated did correspond to feasible HAPs.

CONJECTURE

[MiIwMa02] The inequalities given in Fact F41 are necessary and sufficient conditions for a sequence D to correspond to a feasible HAP.

REMARKS

R24:   There are also season schedules where each pair of teams has to meet several times. The schedule consists of rounds that have to satisfy additional requirements.

R25:   Also, there are often constraints that require more breaks in the schedule (for instance, some teams might have away games or home games on prespecified days), so we cannot use all the properties of schedules with a minimum number of breaks.

R26:   Some references on various types of sports-scheduling problems are given in [EaNeTr03].

R27:   Finally, one should observe that canonical factorizations are not the only factorizations that should be considered for constructing the schedules. There are other types of factorization that are of interest (in particular when a league is divided into several subleagues in which internal games have also to be played). Such constraints are considered in [de82] and [de85-c].

References

[AsDeHa98] A. S. Asratian, T. M. J. Denley, and R. Häggkvist, *Bipartite Graphs and their Applications*, Cambridge University Press, 1998.

[Asde02] A. S. Asratian and D. de Werra, A Generalized Class-Teacher Model for Some Timetabling Problems, *European J. of Operational Research* 143 (2002), 531–542.

[Ba96] V. A. Bardadym, Computer Aided School and University Timetabling: The New Wave, in E. Burke and P. Ross (Eds.), *The Practice and Theory of Automated Timetabling I (PATAT 1995, Edinburgh, Aug/Sept, selected papers) (Lecture Notes in Computer Science, Vol. 1153)*. Springer, (1996), 22–45.

[Be83] C. Berge, *Graphes*, Gauthiers-Villars, Paris, 1983.

[Br64] S. Broder, Final Examination Scheduling, *Communications of the ACM* 7 (1964), 494–498.

[BuCa98] E. Burke and M. Carter (Eds.), *Practice and Theory of Automated Timetabling II (PATAT 1997, Toronto, Canada, August, selected papers) (Lecture Notes in Computer Science, Vol. 1408)*, Springer, 1998.

[BuDe03] E. Burke and P. De Causmaecker (Eds.), *Practice and Theory of Automated Timetabling IV (PATAT 2002, Gent, Belgium, August, selected papers) (Lecture Notes in Computer Science, Vol. 2740)*, Springer, 2003.

[BuElWe94] E. K. Burke, D. G. Elliman, and R. F. Weare, A University Timetabling System Based on Graph Colouring and Constraint Manipulation, *J. of Research on Computing in Education* 27 (1994), 1–18.

[BuElFoWe96] E. K. Burke, D. G. Elliman, P. H. Ford, and R. F. Weare, Examination Timetabling in British Universities: a Survey, in E. Burke and P. Ross (Eds.), *The Practice and Theory of Automated Timetabling I (PATAT 1995, Edinburgh, Aug/Sept, selected papers) (Lecture Notes in Computer Science, Vol. 1153)*, Springer (1996), 76–90.

[BuEr01] E. Burke and W. Erben (Eds.), *Practice and Theory of Automated Timetabling III (PATAT 2000, Konstanz, Germany, August, selected papers) (Lecture Notes in Computer Science, Vol. 2079)*, Springer, 2001.

[BuJaKiWe97] E. K. Burke, K. S. Jackson, J. H. Kingston, and R. F. Weare, Automated Timetabling: The State of the Art, *The Computer J.* 40 (1997), 565–571.

[BuNe99] E. K. Burke and J. P. Newall, A Multi-stage Evolutionary Algorithm for the Timetable Problem, *IEEE Transactions on Evolutionary Computation* 3 (1999), 63–74.

[BuNeWe96] E. K. Burke, J. P. Newall, and R. F. Weare, A Memetic Algorithm for University Exam Timetabling, in E. Burke and P. Ross (Eds.), *The Practice and Theory of Automated Timetabling I (PATAT 1995, Edinburgh, Aug/Sept, selected papers) (Lecture Notes in Computer Science, Vol. 1153)*, Springer (1996), 241–250.

[BuNeWe98-a] E. K. Burke, J. P. Newall, and R. F. Weare, A Simple Heuristically Guided Search for the Timetable Problem, *Proceedings of the International ICSC Symposium on Engineering of Intelligent Systems (EIS'98)*, ICSC-Academic, New York (1998), 575–579.

[BuNeWe98-b] E. K. Burke, J. P. Newall, and R. F. Weare, Initialization Strategies and Diversity in Evolutionary Timetabling, *Evolutionary Computation* 6 (1996), 81–103.

[BuRo96] E. Burke and P. Ross (Eds.), *The Practice and Theory of Automated Timetabling I (PATAT 1995, Edinburgh, Aug/Sept, selected papers) (Lecture Notes in Computer Science, Vol. 1153)*, Springer, 1996.

[Ca83] M. W. Carter, A Decomposition Algorithm for Practical Timetabling Problems, *Working Paper 83-06*, Department of Industrial Engineering, Univ. of Toronto (1983).

[Ca86] M. W. Carter, A Survey of Practical Applications of Examination Timetabling Algorithms, *Operations Research* 34 (1986), 193–201.

[CaJo01] M. W. Carter and D. G. Johnson, Extended Clique Initialization in Examination Timetabling, *J. of the Operations Research Society* 52, No. 5 (2001), 538–544.

[CaLa96] M. W. Carter and G. Laporte, Recent Developments in Practical Examination Timetabling, in E. Burke and P. Ross (Eds.), *The Practice and Theory of Automated Timetabling I (PATAT 1995, Edinburgh, Aug/Sept, selected papers) (Lecture Notes in Computer Science, Vol. 1153)*, Springer (1996), 3–21.

[CaLa98] M. W. Carter and G. Laporte, Recent Developments in Practical Course Timetabling, in E. Burke and M. Carter (Eds.), *Practice and Theory of Automated Timetabling II (PATAT 1997, Toronto, Canada, August, selected papers) (Lecture Notes in Computer Science, Vol. 1408)*, Springer (1998), 3–19.

[CaLaCh94] M. W. Carter, G. Laporte, and J. W. Chinneck, A General Examination Scheduling System, *Interfaces* 11 (1994), 109–120.

[CaLaLe96] M. W. Carter, G. Laporte, and S. Lee, Examination Timetabling: Algorithmic Strategies and Applications, *J. of the Operations Research Society* 47 (1996), 373–383.

[Ch71] P. Z. Chinn, A Graph with p Points and Enough Distinct $p-2$-Order Subgraphs Is Reconstructible, in M. Capobianco, J. B. Frechen, and M. Krolik (Eds.), *Recent Trends in Graph Theory*, Springer (1971), 71–73.

[Co64] A. J. Cole, The Preparation of Examination Timetables Using a Small Store Computer, *The Computer J.* 7 (1964), 117–121.

[CoKi93] T. B. Cooper and J. H. Kingston, The Solution of Real Instances of the Timetabling Problem, *The Computer J.* 36 (1993), 645–653.

[Cs65] J. Csima, *Investigations on a Time-Table Problem*, Phd thesis, School of Graduate Studies, University of Toronto, 1965.

[de81] D. de Werra, Scheduling in Sports, in P. Handesn (Ed.), *Studies on Graphs and Discrete Programming*, North-Holland (1981), 381–395.

[de82] D. de Werra, Minimizing Irregularities in Sports Schedules Using Graph Theory, *Discrete Applied Mathematics* 4 (1982), 217–226.

[de85-a] D. de Werra, An Introduction to Timetabling, *Euro. J. Oper. Res.* 19 (1985), 151–162.

[de85-b] D. de Werra, Graphs, Hyper-graphs and Timetabling, *Methods of Operations Research (Germany, F.R.)* 49 (1985), 201–213.

[de85-c] D. de Werra, On the Multiplication of Divisions: the Use of Graphs for Sports Scheduling, *Networks* 4 (1985), 125–136.

[de88] D. de Werra, Some Models of Graphs for Scheduling Sports Competitions, *Discrete Applied Mathematics* 21 (1988), 47–65.

[de97] D. de Werra, The Combinatorics of Timetabling, *Euro. J. of Oper. Res.* 96 (1997), 504–513.

[DiSc01] L. Di Gaspero and A. Schaerf, Tabu Search Techniques for Examination Timetabling, in E. Burke and W. Erben (Eds.), *Practice and Theory of Automated Timetabling III (PATAT 2000, Konstanz, Germany, August, selected papers) (Lecture Notes in Computer Science, Vol. 2079)*, Springer, 2001.

[EaNeTr03] K. Easton, G. Nemhauser, and M. Trick, Solving the Traveling Tournament Problem: A Combined Integer Programming and Constraint Programming Approach, in E. Burke and P. De Causmaecker (Eds.), *Practice and Theory of Automated Timetabling IV (PATAT 2002, Gent, Belgium, August, selected papers) (Lecture Notes in Computer Science, Vol. 2740)*, Springer, 2003.

[EvItSh76] S. Even, A. Itai, and A. Shamir, On the Complexity of Timetable Multi-commodity Flow Problems, *SIAM J. Comput.* 5 (1976), 691–703.

[GaJo79] M. R. Garey and D. S. Johnson, *Computers and Intractability*, W.H. Freeman and Co., 1979.

[GlKo03] F. Glover and K. Kochenberger, *Handbook of Metaheuristics*, Kluwer, 2003.

[Go62] C. C. Gotlieb, The Construction of Class-Teacher Timetables, *Proceedings of the IFIP Congress* (1962), 73–77.

[GrYe06] J. L. Gross and J. Yellen, *Graph Theory and Its Applications*, Second Edition, CRC Press, 2006.

[Hä83] R. Häggkvist, *Restricted Edge Colorings of Bipartite Graphs*, Manuscript 1983.

[Hede87] A. Hertz and D. de Werra, Using Tabu Search Techniques for Graph Coloring, *Computing* 39 (1987), 345–351.

[Ko16] D. König, Über Graphen und ihre Andwendung auf Determinantentheorie und Mengenlehre, *Math. Ann.* 77 (1916), 453–465.

[Me81] N. K. Mehta, The Application of a Graph Colouring Method to an Examination Scheduling Problem, *Interfaces* 11 (1981), 57–64.

[Me82] N. K. Mehta, A Computer Based Examination Management System, *J. of Educational Technology Systems* 11 (1982), 185–198.

[MiIwMa02] R. Miyashiro, H. Iwasaki, and T. Matsui, *Characterizing Feasible Pattern Sets with a Minimum Number of Breaks*, in E. Burke and P. De Causmaecker (Eds.), *Practice and Theory of Automated Timetabling IV (PATAT 2002, Gent, Belgium, August, selected papers) (Lecture Notes in Computer Science, Vol. 2740)*, Springer, 2003.

[Pa82] C. H. Papadimitriou and K. Steiglitz, *Combinatorial Optimization: Algorithms and Complexity*, Prentice-Hall, 1982.

[Pi10] N. Pillay, An Overview of School Timetabling, *Practice and Theory of Automated Timetabling VIII*, 321–335 (2010).

[Po11] Gerhard Post, Jeffrey H. Kingston, Samad Ahmadi, Sophia Daskalaki, Christos Gogos, Jari Kyngas, Cimmo Nurmi, Nysret Musliu, Nelishia Pillay, Haroldo Santos, and Andrea Schaerf, XHSTT: An XML Archive for High School Timetabling Problems in Different Countries, *Annals of Operations Research* (2011).

[QuBuMcMeLe09] R. Qu, E. K. Burke, B. McCollum, L. T. G. Merlot, and S. Y. Lee, A Survey of Search Methodologies and Automated Approaches for Examination Timetabling, *Journal of Scheduling*, 12(1) (2009), 55–89.

[Sc99] A. Schaerf, A Survey of Automated Timetabling, *Artificial Intelligence Review* 13 (1999), 87–127.

[ScSt80] G. Schmidt and T. Ströhlein, Timetable Construction—an Annotated Bibliography, *The Computer J.* 23 (1980), 307–316.

[ThDo96-a] J. Thompson and K. Dowsland, Variants of Simulated Annealing for the Examination Timetabling Problem, *Annals of Operations Research* 63 (1996), 105–128.

[ThDo96-b] J. Thompson and K. Dowsland, General Cooling Schedules for a Simulated Annealing Based Timetabling System, in E. Burke and P. Ross (Eds.), *The Practice and Theory of Automated Timetabling I (PATAT 1995, Edinburgh, Aug/Sept, selected papers) (Lecture Notes in Computer Science, Vol. 1153)*, Springer (1996), 345–363.

[WePo67] D. J. A. Welsh and M. B. Powell, An Upper Bound for the Chromatic Number of a Graph and its Application to Timetabling Problems, *The Computer J.* 10 (1967), 85–86.

[Wo68] D. C. Wood, A System for Computing University Examination Timetables, *The Computer J.* 11 (1968), 41–47.

[Wr96] A. Wren, Scheduling, Timetabling and Rostering – A Special Relationship?, in E. Burke and P. Ross (Eds.), *The Practice and Theory of Automated Timetabling I (PATAT 1995, Edinburgh, Aug/Sept, selected papers) (Lecture Notes in Computer Science, Vol. 1153)*, Springer (1996), 46–75.

Section 5.6
Graceful Labelings

Joseph A. Gallian, University of Minnesota Duluth

INTRODUCTION

There are more than 1000 papers published on a wide variety of graph labeling methods. Many of these methods trace their origin to one introduced by Rosa [Ro67] in 1967. An extensive survey article on graph labelings that is periodically updated is available online at [Ga11].

DEFINITIONS

D1: Rosa called a function f a β-***valuation of a graph*** G with q edges, if f is an injection from the vertices of G to the set $\{0, 1, \ldots, q\}$ such that, when each edge xy is assigned the label $|f(x) - f(y)|$, the resulting edge labels are distinct.

D2: Golomb [Go72] subsequently called such labelings ***graceful labelings*** and this is now the popular term.

REMARKS

R1: Rosa introduced β-valuations as well as a number of other labelings as tools for decomposing the complete graph into isomorphic subgraphs. In particular, β-valuations originated as a means of attacking the conjecture of Ringel [Ri64] that the complete graph K_{2n+1} can be decomposed into $2n + 1$ subgraphs that are all isomorphic to a given tree with n edges.

R2: Although an unpublished result of Erdős (see [GS80]) says that most graphs are not graceful, many graphs that have some sort of regularity of structure are graceful. Sheppard [Sh76] has shown that there are exactly $q!$ gracefully labeled graphs with q edges. Rosa [Ro67] has identified essentially three reasons why a graph fails to be graceful:

1. G has "too many vertices" and "not enough edges";
2. G has "too many edges"; and
3. G has "the wrong parity."

FACTS

F1: The disjoint union of trees is a case where there are too many vertices for the numbers of edges it has.

F2: An infinite class of graphs that are not graceful for the second reason is given by [BG86].

F3: [Ro67] If every vertex of a given graph has even degree, and if the number of edges is congruent to 1 or 2 (mod 4), then the graph is not graceful. In particular, the cycles C_{4n+1} and C_{4n+2} are not graceful. These are examples of the third possible reason for non-gracefulness.

F4: [Ach82] Every graph can be embedded as an induced subgraph of a graceful graph.

F5: [Ach82] Every connected graph can be embedded as an induced subgraph of a graceful connected graph.

The next three results demonstrate that there is no forbidden subgraph characterization of various particular kinds of graceful graphs.

F6: [ARA08] Every triangle-free graph can be embedded as an induced subgraph of a triangle-free graceful graph.

F7: [ARA08] Every planar graph can be embedded as an induced subgraph of a planar graceful graph.

F8: [ARA08] Every tree can be embedded as an induced subgraph of a graceful tree.

5.6.1   Trees

DEFINITIONS

D3: A *caterpillar* is a tree such that the deletion of all univalent vertices leaves a path.

D4: A *lobster* is a tree such that the deletion of all univalent vertices leaves a caterpillar.

CONJECTURES

A conjecture by Ringel and Kotzig has spurred much research, which Kotzig [HKR82] has characterized as a "disease." A special case of that conjecture was made by Bermond in 1979.

C1: (Graceful Tree Conjecture) [Ringel–Kotzig] All trees are graceful.

C2: [Be79] Lobsters are graceful.

FACTS

F9: [Fa] Trees with at most 35 vertices are graceful.

F10: [Ro67] Caterpillars are graceful.

F11: [HKR82, Zha89, JMW93] Trees with at most 4 end-vertices are graceful.

F12: [Zha89, HH01] Trees with diameter at most 5 are graceful.

F13: [BeS76, PoS82] Rooted trees in which every level contains vertices of the same degree are graceful.

REMARK

R3: Methods for combining graceful trees to yield larger graceful trees are given in [StZa73, KR80, KRT81, KTR79b]. Recursive constructions to create graceful trees are provided in [Rog78, KTR79a].

5.6.2   Cycle-Related Graphs

Cycle-related graphs have been a major focus of attention.

FACTS

F14: [Ro67] The n-cycle C_n is graceful if and only if $n \equiv 0$ or $3 \pmod 4$.

F15: [Fr79] Wheels $W_n = C_n + K_1$ are graceful.

F16: [BNS03] The n-***cone*** (also called the n-***point suspension*** of C_m) $C_m + \overline{K_n}$ is graceful when $m \equiv 0$ or $3 \pmod{12}$. When n is even and m is 2, 6 or 10 $\pmod{12}$ $C_m + \overline{K_n}$ violates Rosa's parity condition for a graceful graph.

F17: [AF84] The ***helm graphs*** obtained from a wheel W_n by attaching a pendant edge at each vertex of the n-cycle are graceful.

F18: [KLG96] The ***web graphs*** obtained by joining the pendant points of a helm to form a cycle and then adding a single pendant edge to each vertex of this outer cycle are graceful.

F19: [MF84b] The **gear graphs** obtained from the wheel W_n by adding a vertex between every pair of adjacent vertices of the n-cycle are graceful.

F20: [LY95] The graphs obtained from a gear graph by attaching one or more pendant edges to each vertex between the vertices of the n-cycle are graceful.

F21: [LY96] The graphs obtained when two or more vertices are inserted between every pair of vertices of the n-cycle of the wheel W_n are graceful.

F22: [DMT80, MF84a, KY85, PP87] The graphs obtained from a cycle by joining two nonconsecutive vertices of the cycle with a path of fixed length at least 3 are graceful.

F23: [KP82, KRY80, GoL92] For $3 \leq p \leq n - r$, the n-cycle with consecutive vertices $v_1, v_2, \ldots, v_n$ to which the r chords $v_1 v_p, v_1 v_{p+1}, \ldots, v_1 v_{p+r-1}$ have been added is graceful.

Cycles that share a common edge or a vertex have received some attention.

F24: [MuAr] Books with n pentagonal pages (i.e., n copies of C_5 with an edge in common) are graceful when n is even and not graceful when n is odd.

F25: [BBG78, BKT78] Graphs, $C_3^{(t)}$, that are the one-point union of t 3-cycles are graceful if and only if $t \equiv 0$ or 1 (mod 4).

F26: [Sh91] Graphs, $C_4^{(t)}$, that are the one-point union of t 4-cycles are graceful for all t.

CONJECTURE

C3: [KRT79] A graph $C_n^{(t)}$ that is the one-point union of t n-cycles is graceful if and only if $nt \equiv 0$ or 3 (mod 4).

FACTS about 1-point unions

F27: (see [Ga11]) Conjecture C3 has been proved for $n = 5, 7, 9, 11, 13$ and when $t = 2$ and n is even.

F28: [BSW75] The one-point union of any two cycles is graceful when the number of edges is congruent to 0 or 3 modulo 4. (The other cases violate the necessary parity condition.)

F29: [SeSe01b] For a variety of choices of points, the one-point union of any number of non-isomorphic complete bipartite graphs is graceful. The question of whether this is true for all choices of the common point is open.

DEFINITIONS

D5: A **block** of a graph is a maximal connected subgraph that has no cut-vertex.

D6: The **block-cutpoint graph** of a graph G is a bipartite graph in which one partite set consists of the cut vertices of G, and the other has a vertex b_i for each block B_i of G.

D7: A C_n**-cactus** is a connected graph all of whose blocks are C_n.

FACTS about cacti

F30: [Mo89] Triangular cacti are graceful if and only the number of blocks of the graph is 0 or 1 (mod 4).

F31: [Sek02] The C_n-cacti whose block-cutpoint graphs are paths are graceful when $n \equiv 0 \pmod 4$ ($n \geq 8$) and when $n \equiv 2 \pmod 4$ and the number of C_n is even.

F32: [Sek02] The graphs obtained from C_m by appending a path P_n to each vertex are graceful.

F33: [Qi, KLG96] The graphs formed by adding a single pendant edge to arbitrarily many vertices of a cycle are graceful.

CONJECTURE

The graphs in the previous two facts are special cases of a long standing conjecture.

C4: [Tr84] All unicyclic graphs except C_n for $n \equiv 1$ or 2 (mod 4) are graceful.

REMARK

R4: For given cycle C_n with $n \equiv 0$ or 3 (mod 4) and a family of trees

$$\mathcal{T} = \{T_1, T_2, \ldots, T_n\},$$

let u_i and $v_i, 1 \leq i \leq n$, be fixed vertices of C_n and T_i, respectively. Figueroa-Centeno, Ichishima, Muntaner-Batle, and Oshima [FMO] provided two construction methods that generate a graceful labeling of the unicyclic graphs obtained from C_n and $\mathcal{T}$ by amalgamating them at each u_i and v_i. Their results encompass all previously known results for unicyclic graphs whose cycle length is 0 or 3 (mod 4) and considerably extend the known classes of graceful unicyclic graphs.

5.6.3 Product-Related Graphs

Graphs that are Cartesian products and related graphs have been the subject of many papers.

FACTS

F34: [AG81, Mah80] Planar grids, $P_m \times P_n$, are graceful.

F35: [FG88] The graphs $C_m \times P_2$ are graceful.

F36: [JR92] The graphs $C_m \times P_n$ are graceful when m and n are even or when $m \equiv 0$ (mod 4).

REMARK

R5: The graphs $C_m \times P_n$ can be viewed as grids on cylinders.

FACTS

F37: [YW92, YW94] The graphs $C_{4n+2} \times P_{4m+3}$, $C_n \times P_2$, and $C_6 \times P_m$ $(m \geq 2)$ are graceful.

F38: [Sin92] The graphs $C_3 \times P_n$ are graceful.

F39: [HuS94] The graphs $C_m \times P_n$ are graceful for all n when m is even and for all n with $3 \leq n \leq 12$ when m is odd.

F40: [JR92] The torus grids $C_m \times C_n$ are graceful when $m \equiv 0 \pmod{4}$ and n is even.

REMARKS

R6: Jungreis and Reid [JR92] also investigated the existence of a stronger form of graceful labeling called an α-labeling (see Section 5.1.7) for graphs of the form $P_m \times P_n, C_m \times P_n$, and $C_m \times C_n$ (see also [Ga11]).

R7: The graceful parity condition is violated for $C_m \times C_n$ when m and n are odd. A complete determination of which torus grids are graceful will most likely involve a large number of cases.

FACTS about prism graphs

F41: [GPW93] The graphs $C_m \times P_2$ with a single vertex deleted or single edge deleted are graceful.

F42: [Ga89] ***Möbius ladders*** obtained from the ladder $P_n \times P_2$ by joining the opposite endpoints of the two copies of P_n are graceful.

F43: [Rop90] The graphs $C_m \times P_2$ with a single pendant edge at each vertex are graceful.

F44: [Rop90] The graphs $C_m \times P_2$ with a single pendant edge at each vertex of one of the m-cycles are graceful.

F45: [Mah80] The graphs $S_{2m} \times P_2$ (***book graphs***) where S_n is the star with $n + 1$ vertices are graceful.

F46: [De80] The books $S_{4m+1} \times P_2$ are graceful.

REMARK

R8: The books $S_{4m+3} \times P_2$ do not satisfy the graceful parity condition.

FACTS

F47: [Mah80] The graphs $(P_n \times P_2) \times P_2$ and $(S_{2m} \times P_2) \times P_2$ are graceful.

F48: [K75, Mah80] The ***n-cubes*** $K_2 \times K_2 \times \cdots \times K_2$ (n copies) are graceful.

F49: [GJ88] The graphs $S_{2m} \times P_n$ (***stacked books***) are graceful.

REMARK

R9: Whether the graphs $S_{2m+1} \times P_n$ are graceful is an open question.

5.6.4   Complete Graphs

Complete graphs and variations have been well studied.

FACTS

F50: [Go72, Si74, BH02] The complete graphs K_n are graceful if and only if $n \leq 4$.

F51: [Ro67, Go72] The bipartite complete graphs $K_{m,n}$ are graceful.

F52: [AM, Gn91, BH02] The graphs $K_{1,m,n}, K_{1,1,m,n}$, and $K_{2,m,n}$ are graceful.

CONJECTURE

C5: [BH02] The graphs $K_{1,m,n}, K_{2,m,n}, K_{1,1,m,n}$, and $K_{m,n}$ are the only complete multipartite graphs that are graceful.

REMARK

R10: Beutner and Harborth [BH02] have verified their above conjecture for graphs with up to 23 vertices via computer.

FACTS

F53: [Jir03, AK] The graphs $K_{m,n}$ with a pendant edge attached to each vertex are graceful.

F54: [SeEl01] The graphs $K_{1,m,n}$ with a pendent edge attached to each vertex are graceful.

F55: [BH02] The graphs K_n with an edge deleted are graceful only if $n \leq 5$.

F56: [BH02] The graphs K_n with two or three edges deleted are graceful only if $n \leq 6$.

F57: [KRY80] A necessary condition for the graphs consisting of m copies of K_n with a vertex in common, $K_n^{(m)}$ $(n > 3)$, to be graceful is that $n = 4$ or 5.

REMARK

R11: The gracefulness of $K_4^{(m)}$ is equivalent to the existence of a $(12m+1, 4, 1)$-perfect difference family, which is known to exist for $m \leq 1000$ (see [HuS94, AB04, WC10], and [GMS10]).

CONJECTURE

C6: [Be79] The graphs $K_4^{(m)}$ are graceful for all $m \geq 4$.

FACTS

F58: [BKT78] The graphs $K_2^{(2)}, K_2^{(3)}$, and $K_5^{(2)}$ are not graceful.

F59: [KRT79, RaPu] The graphs that are the one-point union of t copies of $K_{m,n}$ are graceful.

F60: [SeSe01a] The one-point union of graphs of the form K_{2,m_i} for $i = 1, 2, \ldots, n$, where the union is taken at a vertex from the partite set with exactly two vertices, are graceful if at most two of the m_i are equal.

CONJECTURE

C7: [SeSe01a] The restriction in Fact F60 that at most two of the m_i are equal is not necessary.

FACT

Bermond [Be79] raised the question: "For which m, n, and r is, $B(n, r, m)$, the graph consisting of m copies of K_n with a K_r in common $(n \geq r)$ graceful?

F61: For $r > 1$ the graphs $B(n, r, m)$ are graceful in the following cases:

- $n = 3$, $r = 2$, $m \geq 1$ [KRL79];
- $n = 4$, $r = 2$, $m \geq 1$ [De80];
- $n = 4$, $r = 3$, $m \geq 1$ [Be79, KRL79].

REMARK

R12: Combining results of Bermond and Farhi [BF82] and of Smith and Puget [SmPu10] shows that $B(n, 2, 2)$ is not graceful for $n > 5$.

5.6.5 Disconnected Graphs

For any graph G the graph mG denotes the disjoint union of m copies of G. In 1984 Kotzig [Ko81] investigated the gracefulness of rC_s as well as of graphs that are the disjoint unions of odd cycles. For graphs of the latter kind he gives several necessary conditions. His paper concludes with an elaborate table that summarizes what was then known about the gracefulness of rC_s. When $rs \equiv 1$ or $2 \pmod 4$, these graphs violate the gracefulness parity condition.

FACTS

F62: [Ko73] For $r = 3$ and $4k > 4$, rC_{4k} has a stronger form of graceful labeling called α-labeling (see Section 5.1.7) whereas when $r \geq 2$ and $s = 3$ or 5, rC_s is not graceful.

F63: [He95] The graphs $2C_{2m}$ and graphs obtained by connecting two copies of C_{2m} with an edge are graceful.

F64: [KT76] The graphs mK_n are graceful if and only if $m = 1$ and $n \leq 4$.

REMARKS

R13: Bu and Cao [BuC95] give some sufficient conditions for the gracefulness of graphs of the form $K_{m,n} \cup G$ and they prove that $K_{m,n} \cup P_t$ and the disjoint union of complete bipartite graphs are graceful under some conditions.

R14: In 1985 Frucht and Salinas [FS85] conjectured that $C_s \cup P_n$ is graceful if and only if $s + n \geq 7$ and proved the conjecture for the case that $s = 4$. Between 1985 and 2012 more than a dozen authors proved many special cases. Building on partial results in a paper by Buratti and Traetta [BT], the Frucht–Salinas conjecture was proved by Traetta [Tr] in 2012. He used his result to get a complete solution to the well-known two-table Oberwolfach problem: given an odd number of people and two round tables, determine when is it possible to arrange a series of seatings so that each person sits next to each other person exactly once during the series. The t-table Oberwolfach problem $OP(n_1, n_2, \ldots, n_t)$ asks to arrange a series of meals for an odd number $n = \sum n_i$ of people around t tables of sizes $n_1, n_2, \ldots, n_t$ so that each person sits next to each other exactly once. A solution to $OP(n_1, n_2, \ldots, n_t)$ is a 2–factorization of K_n whose factors consists of t cycles of lengths $n_1, n_2, \ldots, n_t$. The λ–fold Oberwolfach problem $OP_\lambda(n_1, n_2, \ldots, n_t)$ refers to the case where K_n is replaced by λK_n. Traetta used his proof of the Frucht and Salinas conjecture to provide complete solutions to both $OP(2r + 1, 2s)$ and $OP(2r + 1, s, s)$, except possibly for $OP(3, s, s)$. He also gave a complete solution of the general λ-fold Oberwolfach problem $OP_\lambda(r, s)$.

FACTS

F65: [SeYo2] The graphs $K_5 \cup K_{m,n}, K_{m,n} \cup K_{p,q}$ $(m, n, p, q \geq 2), K_{m,n} \cup K_{p,q} \cup K_{r,s}$ $(m, n, p, q, r, s \geq 2, \ (p, q) \neq (2, 2))$, and $pK_{m,n}$ $(m, n \geq 2, (m, n) \neq (2, 2))$ are graceful; the graphs $C_4 \cup K_{1,n}$ $(n \neq 2)$ are not graceful.

F66: [CK4, Kis96] The graphs $C_s \cup K_{1,n}$ are graceful for $s \geq 7$.

F67: [LQW88] The graphs $P_s \cup K_{1,n}$ are graceful.

F68: [AK96] The graphs $C_p \cup C_q$ are graceful if and only if $p + q \equiv 0$ or $3 \pmod 4$.

F69: [Zho93] The graphs $K_m \cup K_n$ $(n > 1, m > 1)$ are graceful when $\{m, n\} = \{4, 2\}$ or $\{5, 2\}$, and only then.

F70: [BD96] The graphs $C_{4t} \cup K_{1,4t-1}$ and $C_{4t+3} \cup K_{1,4t+2}$ are graceful.

REMARK

R15: Subsection 5.6.7 includes numerous families of disconnected graphs that have a stronger form of graceful labelings.

5.6.6   Joins of Graphs

Various joins of graphs have been shown to be graceful.

FACTS

F71: [Ach82] If G is a connected graceful graph, then $G + \overline{K_n}$ is graceful.

F72: [Ba05] If G is a graceful graph of order n and size $n-1$, then $G + \overline{K_n}$ is graceful.

F73: [Re03] The graphs $C_n + \overline{K_2}$ (*double cones*) are graceful for $n = 3, 4, 5, 7, 8, 9, 11$. That $C_n + \overline{K_2}$ is not graceful for $n \equiv 2 \pmod 4$ follows from Rosa's parity condition.

F74: [SeYo4] The join of any two stars and the join of any path and any star are graceful.

F75: [You03] If G is a graceful graph with p vertices and q edges with $p = q + 1$, then $G + K_{1,n}$ is graceful.

F76: [BG86] The graphs $2K_2 + \overline{K_n}$ are not graceful.

F77: [You03, Ma86] The graphs $mK_2 + \overline{K_n}$ are graceful if $m \equiv 0$ or $1 \pmod 4$ and $mK_2 + \overline{K_n}$ is not graceful if n is odd and $m \equiv 2$ or $3 \pmod 4$.

REMARK

R16:   Balakrishnan and Sampathkumar [BS96] ask for which $m \geq 3$ is the graph $mK_2 + \overline{K_n}$ graceful for all n.

5.6.7 α-labelings

In 1966 Rosa [Ro67] defined an α-**labeling** (or α-*valuation*) as a graceful labeling with the additional property that there exists an integer k so that for each edge xy either $f(x) \leq k < f(y)$ or $f(y) \leq k < f(x)$. (Other names for such labelings are **balanced**, **interlaced**, and **strongly graceful**.) It follows that such a k must be the smaller of the two vertex labels that yield the edge labeled 1. Also, a graph with an α-labeling is necessarily bipartite and therefore cannot contain a cycle of odd length.

Graphs with α-labelings have proved to be useful in the development of the theory of graph decompositions. Rosa [Ro67], for instance, has shown that if G is a graph with q edges and has an α-labeling, then for every natural number p, the complete graph K_{2qp+1} can be decomposed into copies of G in such a way that the automorphism group of the decomposition contains the cyclic group of order p. In the same vein El-Zanati and Vanden Eynden [EV96] proved that if G has q edges and admits an α-labeling then $K_{qm,qn}$ can be partitioned into subgraphs isomorphic to G for all positive integers m and n. Although a proof of the graceful tree conjecture has withstood many attempts, examples of trees that do not have α-labelings are easy to construct (see [Ro67]).

FACTS

F78: [Wu] A necessary condition for a bipartite graph with n edges and degree sequence $d_1, d_2, \ldots, d_p$ to have an α-labeling is that the $\gcd(d_1, d_2, \ldots, d_p, n)$ divides $n(n-1)/2$.

F79: [Ko73] Almost all trees have α-labelings.

F80: [Ro67] All paths have α-labelings.

F81: [Ro67] The n-cycle has an α-labeling if and only if $n \equiv 0$ (mod 4).

F82: Other familiar graphs that have α-labelings include the following:
- caterpillars [Ro67],
- the n-cube [Ko65],
- Möbius ladders M_n when n is odd [Pas10],
- books with $4n + 1$ pages [GJ88],
- $C_{2m} \cup C_{2m}$ and $C_{4m} \cup C_{4m} \cup C_{4m}$ for all $m > 1$ [Ko73],
- $C_{4m} \cup C_{4m} \cup C_{4n}$ for all $(m,n) \neq 1,1)$ [EC01],
- $P_n \times Q_n$ [Mah80],
- $K_{1,2k} \times Q_n$ [Mah80],
- $C_{4m} \cup C_{4m} \cup C_{4m} \cup C_{4m}$ [LV87],
- $C_{4m} \cup C_{4n+2} \cup C_{4r+2}, C_{4m} \cup C_{4n} \cup C_{4r}$ when $m + n \leq r$ [AK96],
- $C_{4m} \cup C_{4n} \cup C_{4r} \cup C_{4s}$ when $m \geq n + r + s$ [ACE],
- $C_{4m} \cup C_{4n} \cup C_{4r+2} \cup C_{4s+2}$ when $m \geq n + r + s + 1$ [ACE],
- $((m + 1)^2 + 1)C_4$ for all m [Zhi98],
- $k^2 C_4$ for all k [Zhi98], and
- $(k^2 + k)C_4$ for all k [Zhi98].

F83: [Es97] With the exception $C_4 \cup C_4 \cup C_4$, every 2-regular bipartite graph with 3 components has an α-labeling if and only if the number of edges is a multiple of four.

F84: [AK92] The graphs kC_4 have an α-labeling for $4 \leq k \leq 10$.

F85: [AK92] If kC_4 has an α-labeling then so do $(4k+1)C_4, (5k+1)C_4$, and $(9k+1)C_4$.

F86: [Es02] The graphs $5C_{4k}$ have α-labelings for all k.

F87: [Le] The graphs $C_6 \times P_{2t+1}$ have α-labelings.

F88: [FMB03] If $m \equiv 0$ (mod 4) then the one-point union of 2, 3, or 4 copies of C_m admits an α-labeling, and if $m \equiv 2$ (mod 4) then the one-point union of 2 or 4 copies of C_m admits an α-labeling.

F89: [Zhi98] The connected graphs all of whose blocks are C_{4m} and whose block-cutpoint graph are paths have α-labelings but the connected graphs all of whose blocks are C_m and whose block-cutpoint graph are paths do not have α-labelings when m is odd.

CONJECTURE

C8: The one-point union of n copies of C_m admits an α-labeling if and only if $mn \equiv 0$ (mod 4).

REMARKS

R17:   [EC01] Eshghi and Carter show several families of graphs of the form $C_{4n_1} \cup C_{4n_2} \cup \cdots \cup C_{4n_k}$ have α-labelings.

R18:   Jungreis and Reid [JR92] investigated the existence of α-labelings for graphs of the form $P_m \times P_n, C_m \times P_n$, and $C_m \times C_n$ (see also [Ga11]). Of course, the cases involving C_m with m odd are not bipartite, so there is no α-labeling. The only unresolved cases among these three families are $C_{4m+2} \times P_{2n+1}$ and $C_{4m+2} \times C_{4n+2}$. All other cases result in α-labelings.

FACTS

F90:   [Ro67, Ba3] The graphs $K_{m,n}$ have α-labelings.

F91:   [Sel02] The one-point unions of the following forms have an α-labeling: K_{m,n_1} and K_{m,n_2}; K_{m_1,n_1}, K_{m_2,n_2}, and K_{m_3,n_3} where $m_1 \leq m_2 \leq m_3$ and $n_1 < n_2 < n_3$; $K_{m_1,n}$, $K_{m_2,n}$, and $K_{m_3,n}$ where $m_1 < m_2 < m_3 \leq 2n$.

F92:   [Ba3] For n even the graphs obtained from the wheel W_n by attaching a pendant edge at each vertex have α-labelings.

F93:   [Sn2] Compositions of the form $G[\overline{K_n}]$ have an α-labeling whenever G does. (The *composition* $G_1[G_2]$ is the graph having vertex set $V(G_1) \times V(G_2)$ and edge set $\{(x_1, y_1), (x_2, y_2) | \; x_1 x_2 \in E(G_1) \text{ or } x_1 = x_2 \text{ and } y_1 y_2 \in E(G_2)\}$.)

F94:   [Qi] For n even graphs obtained from an n-cycle by adding one or more pendant edges at some vertices have α-labelings as long as at least one vertex has degree 3 and one vertex has degree 2.

F95:   [SeH11] All gear graphs have an α-labeling (see also [Le]), all graphs obtained by joining an endpoint of a path to a cycle of order $n \equiv 0 \pmod 4$ have an α-labeling, and the graphs obtained by identifying an endpoint of a star S_m ($m \geq 3$) with a vertex of C_{4n} have an α-labeling.

References

[AB04] R. J. R. Abel and M. Buratti, Some progress on $(v, 4, 1)$ difference families and optical orthogonal codes, *J. Combin. Theory Ser. A* **106** (2004), 59–75.

[ACE] J. Abraham, M. Carter, and K. Eshghi, personal communication.

[AK92] J. Abraham and A. Kotzig, Two sequences of 2-regular graceful graphs consisting of 4-gons, *4th Czechoslovakian Symp. on Combinatorics, Graphs, and Complexity* (J. Nesetril and M. Fiedler, eds.), Elsevier, Amsterdam (1992), 1–14.

[AK94] J. Abraham and A. Kotzig, All 2-regular graphs consisting of 4-cycles are graceful, *Discrete Math.* **135** (1994) 1–14.

[AK96] J. Abraham and A. Kotzig, Graceful valuations of 2-regular graphs with two components, *Discrete Math.* **150** (1996) 3–15.

[Ach82] B. D. Acharya, Construction of certain infinite families of graceful graphs from a given graceful graph, *Def. Sci. J.* **32** (1982) 231–236.

[AG81] B. D. Acharya and M. K. Gill, On the index of gracefulness of a graph and the gracefulness of two-dimensional square lattice graphs, *Indian J. Math.* **23** (1981), 81–94.

[ARA08] B. D. Acharya, S. B. Rao, and S. Arumugan, Embeddings and NP-complete problems for graceful graphs, in *Labeling of Discrete Structures and Applications*, Narosa Publishing House, New Delhi, 2008, 57–62.

[AK] S. Amutha and K. M. Kathiresan, Pendant edge extensions of two families of graphs, *Proceed. of the National Seminar on Algebra and Discrete Math.* Univ. Kerala, Thiruvananthapuram, India, 146–150.

[AM] R. Aravamudhan and M. Murugan, Numbering of the vertices of $K_{a,1,b}$, preprint.

[AF84] J. Ayel and O. Favaron, Helms are graceful, in *Progress in Graph Theory* (Waterloo, Ont., 1982), Academic Press, Toronto (1984), 89–92.

[BS96] R. Balakrishnan and R. Sampathkumar, Decompositions of regular graphs into $K_n^c \bigvee 2K_2$, *Discrete Math.* **156** (1996), 19–28.

[Ba3] C. Barrientos, Equitable labelings of corona graphs, *J. Combin. Math. Combin. Comput.* **41** (2002), 139–149.

[Ba4] C. Barrientos, Graceful labelings of chain and corona graphs, *Bull. Inst. Combin. Appl.* **34** (2002) 17–26.

[Ba05] C. Barrientos, Graceful graphs with pendant edges, *Australas. J. Combin.* **33** (2005), 99–107.

[Be79] J. C. Bermond, Graceful graphs, radio antennae and French windmills, *Graph Theory and Combinatorics*, Pitman, London (1979), 18–37.

[BBG78] J. C. Bermond, A. E. Brouwer, and A. Germa, Systemes de triplets et differences associèes, *Problems Combinatories et Thèorie des Graphs*, Colloq. Intern. du Centre National de la Rech. Scient., 260, Editions du Centre Nationale de la Recherche Scientifique, Paris (1978), 35–38.

[BF82] J. C. Bermond and G. Farhi, Sur un probleme combinatoire d'antennes en radioastronomie II, *Annals of Discrete Math.* **12** (1982), 49–53.

[BKT78] J. C. Bermond, A. Kotzig, and J. Turgeon, On a combinatorial problem of antennas in radioastronomy, in *Combinatorics* (A. Hajnal and V. T. Sós, eds.), *Colloq. Math. Soc. János Bolyai* **18** 2 vols., North-Holland, Amsterdam (1978), 135–149.

[BeS76] J. C. Bermond and D. Sotteau, Graph decompositions and G-design, in *Proc. 5th British Combin. Conf., 1975, Congr. Numer.* **XV** (1976), 53–72.

[BH02] D. Beutner and H. Harborth, Graceful labelings of nearly complete graphs, *Result. Math.* **41** (2002), 34–39.

[BD96] V. Bhat-Nayak and U. Deshmukh, Gracefulness of $C_{4t} \cup K_{1,4t-1}$ and $C_{4t+3} \cup K_{1,4t+2}$, *J. Ramanujan Math. Soc.* **11** (1996), 187–190.

[BG86] V. N. Bhat-Nayak and S. K. Gokhale, Validity of Hebbare's conjecture, *Util. Math.* **29** (1986) 49–59.

[BNS03] V. N. Bhat-Nayak and A. Selvam, Gracefulness of n-cone $C_m \vee K_n^c$, *Ars Combin.* **66** (2003), 283–298.

[BSW75] R. Bodendiek, H. Schumacher, and H. Wegner, Über eine spezielle Klasse groziöser Eulerscher Graphen, *Mitt. Math. Gesellsch. Hamburg* **10** (1975), 241–248.

[BuC95] C. Bu and C. Cao, The gracefulness for a class of disconnected graphs, *J. Natural Sci. Heilongjiang Univ.* **12** (1995), 6–8.

[BT] M. Buratti and T. Traetta, *J. Combin. Des.* DOI:10.1002/jcd.21296.

[CK4] S. A. Choudum and S. P. M. Kishore, Graceful labelling of the union of cycles and stars, unpublished.

[De80] C. Delorme, Two sets of graceful graphs, *J. Graph Theory* **4** (1980), 247–250.

[DMT80] C. Delorme, M. Maheo, H. Thuillier, K. M. Koh, and H. K. Teo, Cycles with a chord are graceful, *J. Graph Theory* **4** (1980), 409–415.

[EV96] S. El-Zanati and C. Vanden Eynden, Decompositions of $K_{m,n}$ into cubes, *J. Combin. Designs* **4** (1996), 51–57.

[Es97] K. Eshghi, The existence and construction of α-valuations of 2-regular graphs with 3 components, Ph. D. Thesis, Industrial Engineering Dept., University of Toronto, 1997.

[Es02] K. Eshghi, α-valuations of special classes of quadratic graphs, *Bull. Iranian Math. Soc.* **28** (2002), 29–42.

[EC01] K. Eshghi and M.Carter, Construction of α-valuations of special classes of 2-regular graphs, Topics in Applied and Theoretical Mathematics and Computer Science, *Math. Comput. Sci. Eng.*, WSEAS, Athens (2001), 139–154.

[Fa] W. Fang, A computational approach to the graceful tree conjecture, arXiv:1003.3045v1 [cs.DM].

[FMB03] R. Figueroa-Centeno, R. Ichishima, and F. Muntaner-Batle, Labeling the vertex amalgamation of graphs, *Discuss. Math. Graph Theory* **23** (2003), 129–139.

[FMO] R. Figueroa-Centeno, R Ichishima, F. Muntaner-Batle, and A. Oshima, Gracefully cultivating trees on a cycle, preprint.

[Fr79] R. Frucht, Graceful numbering of wheels and related graphs, *Ann. N.Y. Acad. Sci.* **319** (1979) 219–229.

[FG88] R. Frucht and J. A. Gallian, Labeling prisms, *Ars Combin.* **26** (1988), 69–82.

[FS85] R. Frucht and L. C. Salinas, Graceful numbering of snakes with constraints on the first label, *Ars Combin.* **20** (1985), B, 143–157.

[Ga89] J. A. Gallian, Labeling prisms and prism related graphs, *Congr. Numer.* **59** (1989), 89–100.

[Ga11] J. A. Gallian, Graph labeling, *Electronic J. Combin.* **18** (2011) #DS6
http://www.combinatorics.org/ojs/index.php/eljc/article/view/ds6.

[GJ88] J. A. Gallian and D. S. Jungreis, Labeling books, *Scientia* **1** (1988), 53–57.

[GPW93] J. A. Gallian, J. Prout, and S. Winters, Graceful and harmonious labelings
of prisms and related graphs, *Ars Combin.* **34** (1992), 213–222.

[GMS10] G. Ge, Y. Miao, and X. Sun, Perfect difference families, perfect difference
matrices and related combinatorial structures, *J. Combin. Des.* **18** (2010) 415–
449.

[Gn91] R. B. Gnanajothi, Topics in graph theory, Ph.D. Thesis, Madurai Kamaraj
University, 1991.

[GoL92] C. G. Goh and C. K. Lim, Graceful numberings of cycles with consecutive
chords, 1992, unpublished.

[Go72] S. W. Golomb, How to number a graph, in *Graph Theory and Computing* (R.
C. Read, ed.), Academic Press, New York (1972), 23–37.

[GS80] R. L. Graham and N. J. A. Sloane, On additive bases and harmonious graphs,
SIAM J. Alg. Discrete Math. **1** (1980), 382–404.

[He95] M. He, The gracefulness of the graph $2C_n$, *Neimenggu Daxue Xuebao Ziran
Kexue* **26** (1995), 247–251.

[HH01] P. Hrnčiar and A. Haviar, All trees of diameter five are graceful, *Discrete Math.*
233 (2001), 133–150.

[HKR82] C. Huang, A. Kotzig, and A. Rosa, Further results on tree labellings, *Util.
Math.* **21c** (1982), 31–48.

[HuS94] J. Huang and S. Skiena, Gracefully labeling prisms, *Ars Combin.* **38** (1994)
225–242.

[JMW93] D. J. Jin, F. H. Meng, and J. G. Wang, The gracefulness of trees with diameter
4, *Acta Sci. Natur. Univ. Jilin.* (1993), 17–22.

[Jir03] Jirimutu, On k-gracefulness of r-crown $I_r(K_{1,n})$ $(n \geq 2, r \geq 2)$ for complete
bipartite graph, *J. Inner Mongolia University for Nationalities* **2** (2003), 108–
110.

[JR92] D. Jungreis and M. Reid, Labeling grids, *Ars Combin.* **34** (1992), 167–182.

[KLG96] Q. D. Kang, Z.-H. Liang, Y.-Z. Gao, and G.-H. Yang, On the labeling of some
graphs, *J. Combin. Math. Combin. Comput.* **22** (1996), 193–210.

[Kis96] S. P. Kishore, Graceful labellings of certain disconnected graphs, Ph.D. Thesis,
Indian Institute of Technology, Madras, 1996.

[KP82] K. M. Koh and N. Punnim, On graceful graphs: cycles with 3-consecutive
chords, *Bull. Malaysian Math. Soc.* **5** (1982), 49–63.

[KRL79] K. M. Koh, D. G. Rogers, and C. K. Lim, On graceful graphs: sum of graphs,
Research Report 78, College of Graduate Studies, Nanyang University (1979).

[KR80] K. M. Koh, D. G. Rogers, and T. Tan, Products of graceful trees, *Discrete Math.* **31** (1980), 279–292.

[KRT81] K. M. Koh, D. G. Rogers, and T. Tan, Another class of graceful trees, *J. Austral. Math. Soc. Ser. A* **31** (1981), 226–235.

[KRY80] K. M. Koh, D. G. Rogers, H. K. Teo, and K. Y. Yap, Graceful graphs: some further results and problems, *Congr. Numer.* **29** (1980), 559–571.

[KTR79a] K. M. Koh, T. Tan, and D. R. Rogers, Interlaced trees: a class of graceful trees, *Combinatorial Mathematics, VI* , Proc. Sixth Austral. Conf., Univ. New England, Armidale (1978), Lecture Notes in Math. **748** Springer, Berlin, 1979, 65–78.

[KTR79b] K. M. Koh, T. Tan, and D. G. Rogers, Two theorems on graceful trees, *Discrete Math.* **25** (1979), 141–148.

[KY85] K. M. Koh and K. Y. Yap, Graceful numberings of cycles with a P_3-chord, *Bull. Inst. Math. Acad. Sinica* **12** (1985), 41–48.

[KRT79] K. M. Koh, D. G. Rogers, P. Y. Lee, and C. W. Toh, On graceful graphs V: unions of graphs with one vertex in common, *Nanta Math.* **12** (1979), 133–136.

[Ko65] A. Kotzig, Decompositions of a complete graph into $4k$-gons (in Russian), *Matematický Casopis* **15** (1965), 229–233.

[Ko73] A. Kotzig, On certain vertex valuations of finite graphs, *Util. Math.* **4** (1973), 67–73.

[K75] A. Kotzig, β-valuations of quadratic graphs with isomorphic components, *Util. Math.* **7** (1975), 263–279.

[Ko81] A. Kotzig, Decomposition of complete graphs into isomorphic cubes, *J. Combin. Theory, Series B* **31** (1981), 292–296.

[Ko84] A. Kotzig, Recent results and open problems in graceful graphs, *Congr. Numer.* **44** (1984), 197–219.

[KT76] A. Kotzig and J. Turgeon, β-valuations of regular graphs with complete components, *Colloq. Math. Soc. János Bolyai* **18**, *Combinatorics*, Keszthély, Hungary, 1976.

[LV87] D. R. Lashmi and S. Vangipuram, An α-valuation of quadratic graph $Q(4, 4k)$, *Proc. Nat. Acad. Sci. India* Sec. A, **57** (1987), 576–580.

[Le] P.-S. Lee, On α-labelings of prism graphs and gear graphs, preprint.

[LQW88] S. M. Lee, L. Quach, and S. Wang, On Skolem-gracefulness of graphs which are disjoint union of paths and stars, *Congr. Numer.* **61** (1988), 59–64.

[LWa90] S. M. Lee and P. Wang, On the k-gracefulness of the sequential join of null graphs, *Congr. Numer.* **71** (1990), 243–254.

[LY95] Y. Liu, The gracefulness of the star graph with top sides, *J. Sichuan Normal Univ.* **18** (1995), 52–60.

[LY96] Y. Liu, Crowns graphs Q_{2n} are harmonious graphs, *Hunan Annals Math.* **16** (1996), 125–128.

[MF84a] K. J. Ma and C. J. Feng, About the Bodendiek's conjecture of graceful graph, *J. Math. Research and Exposition* **4** (1984), 15–18.

[MF84b] K. J. Ma and C. J. Feng, On the gracefulness of gear graphs, *Math. Practice Theory* (1984), 72–73.

[Ma86] X. D. Ma, Some classes of graceful graphs, *J. Xinjiang Univ. Nat. Sci.* **3** (1986), 106–107.

[Ma88] X. Ma, A graceful numbering of a class of graphs, *J. Math. Res. and Exposition* (1988), 215–216.

[MLL90] X. Ma, Y. Liu, and W. Liu, Graceful graphs: cycles with $(t-1)$ chords, *Math. Appl.* **9** (1990), suppl., 6–8.

[Mah80] M. Maheo, Strongly graceful graphs, *Discrete Math.* **29** (1980), 39–46.

[Mo89] D. Moulton, Graceful labelings of triangular snakes, *Ars Combin.* **28** (1989), 3–13.

[MuAr] M. Murugan and G. Arumugan, On graceful numberings of nC_5 with a common edge, preprint.

[Pas10] A. Pasotti, Constructions for cyclic Moebius ladder systems, *Discrete Math.* **310** (2010), 3080–3087.

[PoS82] S. Poljak and M. Sûra, An algorithm for graceful labeling of a class of symmetrical trees, *Ars Combin.* **14** (1982), 57–66.

[PP87] N. Punnim and N. Pabhapote, On graceful graphs: cycles with a P_k-chord, $k \geq 4$, *Ars Combin.* **23A** (1987), 225–228.

[Qi] J. Qian, On some conjectures and problems in graceful labelings graphs, unpublished.

[RaPu] I. Rajasingh and P. R. L. Pushpam, On graceful and harmonious labelings of t copies of $K_{m,n}$ and other special graphs, personal communication.

[Re03] T. Redl, Graceful graphs and graceful labelings: Two mathematical formulations and some other new results, *Congr. Numer.* **164** (2003), 17–31.

[Ri64] G. Ringel, Problem 25, in Theory of Graphs and its Applications, *Proc. Symposium Smolenice 1963*, Prague (1964), 162.

[Rog78] D. G. Rogers, A graceful algorithm, *Southeast Asian Bull. Math.* **2** (1978), 42–44.

[Rop90] D. Ropp, Graceful labelings of cycles and prisms with pendant points, *Congress. Numer.* **75** (1990), 218–234.

[Ro67] A. Rosa, On certain valuations of the vertices of a graph, *Theory of Graphs (Internat. Symposium, Rome, July 1966)*, Gordon and Breach, New York, and Dunod Paris (1967), 349–355.

[Sek02] C. Sekar, Studies in graph theory, Ph. D. Thesis, Madurai Kamaraj University, 2002.

[Sel02] P. Selvaraju, New classes of graphs with α-valuation, harmonious and cordial labelings, Ph. D. Thesis, Anna University, 2001. Madurai Kamaraj University, 2002.

[SeH11] M. A. Seoud and E. F. Helmi, Some α-graphs and odd graceful graphs, *Ars Comb.* **101** (2011), 385–404.

[SeYo2] M. A. Seoud and M. Z. Youssef, On gracefulness of disconnected graphs, unpublished.

[SeYo4] M. A. Seoud and M. Z. Youssef, The effect of some operations on labelling of graphs, unpublished.

[SeEl] G. Sethuraman and A. Elumalai, Every graph is a vertex induced subgraph of a graceful graph and elegant graph, preprint.

[SeEl01] G. Sethuraman and A. Elumalai, On graceful graphs: Pendant edge extensions of a family of complete bipartite and complete tripartite graphs, *Indian J. Pure Appl. Math.* **32** (2001), 1283–1296.

[SeEl05] G. Sethuraman and A. Elumalai, Gracefulness of a cycle with parallel P_k-chords, *Australas. J. Combin.* **32** (2005), 205–211.

[SeSe01a] G. Sethuraman and P. Selvaraju, On graceful graphs: one vertex unions of non-isomorphic complete bipartite graphs, *Indian J. Pure Appl. Math.* **32** (2001), 975–980.

[SeSe01b] G. Sethuraman and P. Selvaraju, On graceful graphs I: Union of non-isomorphic complete bipartite graphs with one vertex in common, *J. Combin. Inform. System Sci.* **26** (2001), 23–32.

[Sh91] S. C. Shee, Some results on λ-valuation of graphs involving complete bipartite graphs, *Discrete Math.* **28** (1991), 73–80.

[Sh76] D. A. Sheppard, The factorial representation of major balanced labelled graphs, *Discrete Math.* **15** (1976), 379–388.

[Si74] G. J. Simmons, Synch-sets: a variant of difference sets, *Proc. 5th Southeastern Conference on Combinatorics, Graph Theory and Computing*, Util. Math. Pub. Co., Winnipeg (1974), 625–645.

[Sin92] G. S. Singh, A note on graceful prisms, *Nat. Acad. Sci. Lett.* **15** (1992), 193–194.

[SmPu10] B. M. Smith and J.-F. Puget, Constraint models for graceful graphs, *Constraints* **15** (2010), 64–92.

[Sn2] H. Snevily, New families of graphs that have α-labelings, *Discrete Math.* **170** (1997), 185–194.

[StZa73] R. Stanton and C. Zarnke, Labeling of balanced trees, *Proc. 4th Southeast Conf. Combin., Graph Theory, Comput.* (1973), 479–495.

[Tr] T. Traetta, A complete solution to the two-table Oberwolfach problems, preprint.

[Tr84] M. Truszczyński, Graceful unicyclic graphs, *Demonstatio Mathematica* **17** (1984), 377–387.

[WC10] X. Wang and Y. Chang, Further results on $(v, 4, 1)$-perfect difference families, *Discrete Math.* **310** (2010), 1995–2006.

[Wu] S.-L. Wu, A necessary condition for the existence of an α-labeling, unpublished.

[YW92] Y. C. Yang and X. G. Wang, On the gracefulness of the product $C_n \times P_2$, *J. Math. Research and Exposition* **1** (1992) 143–148.

[YW94] Y. C. Yang and X. G. Wang, On the gracefulness of product graph $C_{4n+2} \times P_{4m+3}$, *Combinatorics, Graph Theory, Algorithms and Applications (Beijing, 1993)*, 425–431, World Sci. Publishing, River Edge, NJ, 1994.

[You03] M. Z. Youssef, New families of graceful graphs, *Ars Combin.* **67** (2003), 303–311.

[YLC] T.-K. Yu, D. T. Lee, and Y.-X. Chen, Graceful and harmonious labelings on 2-cube, 3-cube and 4-cube snakes, preprint.

[Zha89] S. L. Zhao, All trees of diameter four are graceful, *Graph Theory and its Applications: East and West (Jinan, 1986)*, Ann. New York Acad. Sci. **576** (1989), 700–706.

[Zhi98] L. Zhihe, The balanced properties of bipartite graphs with applications, *Ars Combin.* **48** (1998), 283–288.

[Zho93] S. C. Zhou, Gracefulness of the graph $K_m \cup K_n$, *J. Lanzhou Railway Inst.* **12** (1993), 70–72.

Glossary for Chapter 5

achromatic number – of a graph G: largest number of colors in a proper vertex-coloring such that the union of any two color classes induces at least one edge.

almost regular factor (or semiregular factor) – of a graph G: a factor of G of type $[k, k+1]$, for some integer $k \geq 0$.

antifactor set (or 1-barrier) – in a graph G: a set $S \subseteq V(G)$ such that $c_o(G-S) > |S|$.

approximation (or approximate) algorithm: an algorithm that typically makes use of heuristics in reducing its computation but produces solutions that are not necessarily optimal.

$r(n)$**-approximation algorithm** – for χ: for every n and every input graph G with n vertices, the algorithm outputs an integer k such that $\chi(G) \leq k \leq r(n)\chi(G)$ (where $r(n) : \mathbb{N} \to \mathbb{R}^+$ is a given function).

arboricity – of a graph G: the minimum number of edge-disjoint acyclic subgraphs whose union is G.

k**-assignment** – on the vertices (edges) of a graph: a list assignment L where $|L_v| = k$ ($|L_e| = k$) for every vertex v (every edge e).

asymptotically almost surely (or *a.a.s.*): an event E concerning a graph $G \in G_{n,p}$ is said to hold asymptotically almost surely (or a.a.s.) if $\lim_{n \to \infty}$ Prob $E = 1$.

1-barrier (or antifactor set) – in a graph G: a set $S \subseteq V(G)$ such that $c_o(G-S) > |S|$.

β**-valuation** – of a graph G: a synonym for **graceful labeling**.

bicritical graph: a graph G in which $G - x - y$ has a 1-factor for every choice of two different vertices x and $y \in V(G)$.

bi-hypergraph: a mixed hypergraph with $\mathcal{C} = \mathcal{D}$.

binding number bind (G) – of a graph G: defined to be
$$\min \left\{ |N(X)|/|X| \mid \emptyset \neq X \subseteq V(G), \text{ and } N(X) \neq V(G) \right\}$$

bipartite graph: a graph G whose vertex set V can be partitioned into two sets V_1 and V_2 such that every edge of G connects a vertex in V_1 with a vertex in V_2.

block – of a graph: a maximal connected subgraph that has no cut-vertex.

block-cutpoint graph – of a graph G: a bipartite graph in which one partite set consists of the cut vertices of G, and the other has a vertex b_i for each block B_i of G.

book graph: the graph $S_{2m} \times P_2$, where S_{2m} is the star with $2m + 1$ vertices.

___, **stacked**: the graph $S_{2m} \times P_n$.

C_n**-cactus**: a connected graph all of whose blocks are C_n.

caterpillar: a tree such that the deletion of all univalent vertices leaves a path.

choice number – of a graph G: the smallest nonnegative integer k such that G is k-choosable. Denoted by $ch(G)$.

(f, g)**-choosable graph** – for two functions $f, g : V \to \mathbb{N}$: if $|L_v| = f(v)$ for every vertex, then one can choose subsets $C_v \subseteq L_v$ such that $|C_v| = g(v)$ and $C_u \cap C_v = \emptyset$ for all $uv \in E$.

k-choosable graph: L-colorable for every k-assignment L.

f-choosable graph – for a function $f : V \to \mathbb{N}$: L-colorable for every list assignment L with $|L_v| = f(v)$ for all $v \in V$.

chordal graph: a graph in which every circuit of length ≥ 4 has a *chord* (i.e., an edge joining non-consecutive vertices on the circuit).

k-chromatic graph: has precisely k as the smallest number of colors in a proper vertex coloring = k-colorable but not $(k-1)$-colorable.

chromatic index – of a graph G: smallest number of colors in a proper edge-coloring of G; same as the chromatic number of the line graph. Denoted by $\chi'(G)$.

chromatic number – of a graph G: the minimum number of colors in a proper vertex coloring of G. Denoted by $\chi(G)$.

chromatic polynomial – of a graph G: for every natural number k, its value is the number of proper k-colorings of G; denoted $P(G, \lambda)$.

chromatic sum – of a graph G: smallest sum of colors in a proper vertex-coloring with *natural numbers*; denoted $\Sigma(G)$.

Class 1/2: graph G is of Class 1 if its chromatic index is $\Delta(G)$, and of Class 2 if $\chi'(G) = \Delta(G) + 1$.

claw – of a graph G: an induced subgraph of a graph G isomorphic to the bipartite graph $K_{1,3}$.

claw-free graph G: a graph G containing no $K_{1,3}$ as an induced subgraph.

clique cover – of a graph G: a collection of cliques of G that contains every vertex of G.

clique hypergraph – of a graph $G = (V, E)$: the hypergraph on V whose edges are the vertex subsets inducing inclusionwise-maximal complete subgraphs in G, other than isolated vertices. Its chromatic number is denoted $\chi_C(G)$.

clique number – of a graph G, denoted $\omega(G)$: the size of the largest clique in G.

clique partition number – of a graph G: the smallest number $cp(G)$ such that there exists a set of $cp(G)$ cliques in G such that the cliques form a partition of $E(G)$.

clique$_1$ – in a graph G: a subset of vertices in G that are mutually adjacent to one another (caution: non-uniform definition).

clique$_2$ – in a graph G: a maximal mutually adjacent set of vertices (caution: non-uniform definition).

color class – in a vertex (edge) coloring: set consisting of all vertices (edges) having the same color.

color cost: same as chromatic sum.

k-colorable graph: has a proper vertex coloring with at most k colors.

colorable mixed hypergraph: a mixed hypergraph that has at least one strict coloring.

L-colorable – graph G, with respect to list assignment L: if G admits a proper vertex coloring φ such that $\varphi(v) \in L_v$ for all v.

coloring number – of a graph G: smallest integer k such that every subgraph of G contains a vertex of degree less than k; denoted $col(G)$.

H-coloring – of a graph G: homomorphism from G to H.

k-coloring: coloring with at most k colors.

comparability graph: a graph whose edges can be directed so that directed adjacency becomes a transitive relation, that is, whenever there exist directed edges (a, b) and (b, c) there must also exist the directed edge (a, c).

complement – of a graph G: the graph $G^c = (V, E^c)$ which is related to graph $G = (V, E)$ as follows: it has the same vertex set V as G and edges defined by (x, y) is in E^c if and only (x, y) is not in E.

complete r-uniform hypergraph: its edges are all the r-element subsets of the vertex set; denoted $\mathcal{K}_n^r$ (n is the number of vertices).

n-cone: the graph $C_m + \overline{K_n}$ (also called the n-point suspension of C_m).

conflict graph: a graph in which the nodes represent events (e.g., courses, exams) and an edge between two nodes indicates that the two events cannot be scheduled in the same time slot.

construction heuristic: a heuristic that produces a feasible solution without any attempt to improve it.

cost chromatic number – of a graph G, with given cost set C: smallest number of colors in a minimum-cost coloring.

cost set: associates a positive real cost with each color.

k-critical graph: k-chromatic graph whose chromatic number decreases to $k-1$ whenever an edge is deleted.

n-cube: the graph $K_2 \times K_2 \times \cdots \times K_2$ (n copies).

cyclically k-edge-connected graph: a graph in which at least k edges must be deleted in order to leave two components, each containing a cycle.

H-decomposition problem: the problem defined as follows: given a fixed graph H, can the edge-set of an input graph G be partitioned into copies of H?

divisible by t: A graph G which admits a partition of its edge-set into t isomorphic subgraphs is said to be *divisible by t*.

edge k-colorable graph: has a proper edge-coloring with at most k colors.

edge choice number – of a graph G: the choice number of the line graph of G; denoted $ch'(G)$.

edge, C-, D-edge – of a mixed hypergraph: see *strict coloring*.

edge-chromatic number – of a graph G: the minimum number of different colors required for a proper edge-coloring of G; denoted $\chi'(G)$.

edge-coloring: assignment of colors to the edges (each edge gets one color).

___, **proper**: edge-coloring where any two edges sharing a vertex have distinct colors.

exact algorithm: an algorithm that solves a certain optimization problem to optimality.

factor – of a graph G: a spanning subgraph of G.

___, **$(1, f)$-odd-**: a spanning subgraph F of G in which $\deg_F(v) \in \{1, 3, \ldots, f(v)\}$, where f is a function from $V(G)$ to the odd positive integers.

___, **(g, f)-**: a spanning subgraph F of G such that $g(v) \le \deg_F(G) \le f(v)$ for all $v \in V(G)$.

___, **$[a, b]$-**: a factor of G for which $a \le \deg_H(v) \le b$, for all $v \in V(G)$, where a and b are integers such that $1 \le a \le b$.

___, **1-**: a set of vertex-disjoint edges in G which together span $V(G)$.

___, **F-**: a spanning subgraph of G in which each component is a single edge or an odd cycle.

___, **f- –** of multigraph G (possibly with loops): a spanning subgraph H of G such that $\deg_H(v) = f(v)$, for all $v \in V(G)$, where f, a non-negative, integer-valued function on $V(G)$.

___, **G- –** of a graph H: a set $\{G_1, \ldots, G_d\}$ of subgraphs of H such that each G_i is isomorphic to G and such that the sets $V(G_i)$ collectively partition $V(G)$.

___, **k-**: a k-regular spanning subgraph.

G-factor recognition problem $FACT(G)$: defined by
>   INSTANCE: A graph H.
>   QUESTION: Does H admit a G-factor?

factorization – of a graph G: a set of factors $\{F_1, F_2, \ldots F_k\}$ of G such that the edge-disjoint union of factors $F_1, F_2, \ldots F_k$ is $E(G)$.

all (g, f)-factors: a graph G is said to have all (g, f)-factors if and only if G has an h-factor for every h such that $g(v) \leq h(v) \leq f(v)$ for all $v \in V(G)$.

feasible solution: a solution that satisfies all hard constraints.

fractional chromatic number – of a graph G: smallest ratio p/q such that there exist p independent sets that cover each vertex precisely q times; denoted $\chi^*(G)$.

fractional vertex-coloring: function from the family of independent vertex sets to $\mathbb{R}^{\geq 0}$, such that the sum over the sets containing vertex v is at least 1, for each v.

fractional (g, f)-factor – of a graph G: a vector $x = (x_e)$ with $|E(G)|$ real components such that $0 \leq x_e \leq c_e$ and $g(v) \leq \deg_x(v) \leq f(v)$. Here $\deg_x(v) = \sum x_{uv}$, where the sum is over all edges incident with vertex v and c_e is the "capacity" or perhaps the "multiplicity" of edge e.

gear graph: obtained from the wheel W_n by adding a vertex between every pair of adjacent vertices of the n-cycle.

graceful labeling – of a graph G with q edges: an injection f from the vertices of G to the set $\{0, 1, \ldots, q\}$ such that, when each edge xy is assigned the label $|f(x) - f(y)|$, the resulting edge labels are distinct.

___, strongly: synonym for α-labeling.

$[a, b]$-graph: a graph G in which $a \leq \deg(v) \leq b$, for every vertex $v \in V(G)$.

graphical – degree sequence: a sequence of non-negative integers $d_1, \ldots, d_n$ such that there exists a graph G of order n and degrees (in some order) $d_1, \ldots, d_n$.

Grundy number – of a graph G: largest number of colors in a proper vertex-coloring with natural numbers where each vertex v has a neighbor in each color smaller than the color of v.

Hamming distance – between a pair $u = (u_1, \ldots, u_n)$, $v = (v_1, \ldots, v_n)$ of binary vectors: the number of indices i for which $u_i \neq v_i$.

Hamming graph $H(n, d)$: a graph whose vertices are all binary vectors with n coordinates. A pair u, v of vertices in $H(n, d)$ are adjacent if the Hamming distance between them is at least d.

hard constraint: a constraint that must be satisfied.

helm graph: the graph obtained from a wheel by attaching a pendant edge at each vertex of the cycle.

homomorphism – from (di)graph G to H: maps the vertices of G to vertices of H in such a way that the image of every edge is an edge of H. Notation if it exists: $G \to H$; if it does not exist: $G \not\to H$.

hypergraph: a pair $\mathcal{H} = (X, \mathcal{F})$ where X is a set (vertex set) and $\mathcal{F}$ is a set system on X (edge set).

hypohamiltonian – graph G: a graph G which has no Hamilton cycle, but for which $G - v$ does, for all $v \in V(G)$.

hypotraceable – graph G: a graph G which has no Hamilton path, but for which $G - v$ does, for all $v \in V(G)$.

independence number: the number of vertices in a maximum-size independent set of a graph.

independent set – of a graph G: a subset of vertices in G that are mutually non-adjacent.

independent set: a mutually non-adjacent set of vertices.

independent-set cover – of a graph G: a collection of independent sets of G that contains every vertex of G.

intersection graph – of a family F of subsets of a given set: a graph $G(F)$ with a 1-to-1 correspondence between subsets of F and vertices of G such that two vertices of G are adjacent if and only if they correspond to two subsets of F with a non-empty intersection.

interval graph: a graph for which there exists a family F of intervals on a line such that G is an intersection graph, that is, there is a 1-to-1 correspondence between intervals of F and vertices of G such that two vertices of G are adjacent if and only if they correspond to overlapping intervals of F.

proper: an interval graph with the property that there is an interval model F for G in which no interval of F is properly contained within another interval of F.

interval model – for an interval graph G: a family of intervals on the line for which G is an intersection graph.

α-**labeling**: a graceful labeling with the additional property that there exists an integer k so that for each edge xy, either $f(x) \leq k < f(y)$ or $f(y) \leq k < f(x)$. (Other names for α-labelings are **balanced**, **interlaced**, and **strongly graceful**.)

line graph – of a graph G: a graph, denoted $L(G)$, with a vertex for each edge of G and two vertices of $L(G)$ joined by an edge if and only if they correspond to two edges in G with a common endpoint.

k-**linear arboricity** – of a graph G: the minimum number of k-linear forests which partition $E(G)$.

k-**linear forest**: a forest in which all components are paths of length at most k.

list assignment L – on the vertex set of a graph G: associates a set L_v of "allowed" colors with each vertex v of G.

list chromatic index or **list edge chromatic number**: same as edge choice number.

list chromatic number: same as choice number.

list colorable graph: L-colorable with respect to a list assignment L that is understood.

lobster: a tree such that the deletion of all univalent vertices leaves a caterpillar.

local search: a heuristic that starts from a feasible solution and in each iteration, until termination, chooses the next solution from a *neighborhood* of solutions that are, in some prescribed sense, close to the current solution.

___, **improvement**: a local search in which we choose only a solution that is better than the current one and stop if we cannot find one.

lower chromatic number – of a colorable mixed hypergraph: the smallest number of colors in a strict coloring; denoted $\chi(\mathcal{H})$.

matching number of G: the size of a largest matching in G; denoted $\nu(G)$.

maximum clique: the number of vertices in a maximum-size clique of a graph.

mixed hypergraph: has two sets of edges, $\mathcal{C}$ and $\mathcal{D}$. (Cf. *strict coloring*.)

___, C-: mixed hypergraph without D-edges.

___, D-: mixed hypergraph without C-edges. Same as hypergraph.

___, **uncolorable**: does not have any strict coloring.

___, **uniquely colorable**: has just one strict coloring, apart from renaming the colors.

Möbius ladders: obtained from the ladder $P_n \times P_2$ by joining the opposite endpoints of the two copies of P_n.

neighborhood complex – of a graph G: simplicial complex whose vertices are the vertices of G, and the simplexes are the vertex subsets having a common neighbor in G.

nowhere-zero k-**flow** – on an oriented graph: weight function from the edge set to $\{0, 1, \ldots, k-1\}$, such that the in-flow (sum of weights on the edges oriented toward a vertex) equals the out-flow (sum on the out-going edges), at each vertex.

odd-cycle property: the property which states that every pair of odd cycles either has a vertex in common or is joined by an edge.

on-line coloring: receiving the vertices $v_1, \ldots, v_n$ of a graph G one by one, a color has to be assigned to each successive vertex v_i only after its neighbors in $\{v_1, \ldots, v_{i-1}\}$ are known.

oriented d-coloring – of a graph G: a proper edge-d-coloring together with an assignment of a direction to each edge; used to model the schedule of home and away games in sports timetabling.

partial vertex/edge **coloring**: assignment of colors to a subset of the vertices/edges.

partition number – of a graph G, denoted $\rho(G)$: the size of the smallest clique cover of G.

path factor – of a graph G: spanning subgraph of G each component of which is a path.

permanent – of a matrix A: the matrix function defined by
$$\text{per } A = \sum a_{1\pi(1)} a_{2\pi(2)} \cdots a_{n\pi(n)}$$
where the sum extends over all permutations π of the set $\{1, \ldots, n\}$; denoted per A.

Perron root $\lambda(G)$: the largest eigenvalue of A_G.

planar dual – of a plane graph G: the dual vertices are the faces of G, and the endpoints of a dual edge are the faces whose boundary contains the original edge; denoted $G^* = (V^*, E^*)$.

plane graph: planar graph imbedded in the plane.

precoloring extension or **PrExt** problem: asks whether a given partial coloring can be extended to a proper coloring of the entire graph, using at most a given number k of colors.

t-PrExt: Precoloring Extension where each color occurs on at most t vertices in the given partial coloring.

proper vertex-coloring – of a hypergraph $\mathcal{H}$: vertex partition where no partition class contains any edge of $\mathcal{H}$.

random graph $G_{n,p}$: Let $1, \ldots, n$ be a labeling of the vertices and let $\{e_{ij}\}, 1 \leq i < j \leq n$, be an array of independent random variables, where each e_{ij} assumes the value 1 with probability p and 0 with probability $1 - p$. This array determines a random graph on $\{1, \ldots, n\}$ where each (ij) is an edge if and only if $e_{ij} = 1$. This probability space (or random graph) is denoted by $G_{n,p}$.

ranking number – of a graph G: smallest number of colors in a vertex-coloring such that each path with the same color on its endpoints contains a vertex of larger color; denoted $\chi_r(G)$.

t-rational graph G: a graph G which is divisible by t or else $t \nmid |E(G)|$.

t-rational Problem: The problem defined by: Given a graph G and a positive integer t, is G t-rational?

snark: 3-regular graph of chromatic index 4, which is also cyclically 4-edge-connected and has girth at least 5.

soft constraint: a constraint that it is desirable, but not necessary, to satisfy.

square – of a graph G: obtained from G by joining the vertex pairs at distance 2; denoted G^2.

stability number – of a graph G, denoted $\alpha(G)$: the size of the largest independent set in G.

standard simplex: $\Delta = \{x \in R^n : x_i \geq 0, \ i = 1, \ldots, n, \ \sum_{j=1}^n x_j = 1\}$.

strength – of a graph G: cost chromatic number of G where the cost set is $\mathbb{N}$.

strict k-coloring – of a mixed hypergraph: vertex-coloring with exactly k colors, such that every C-edge has two vertices with a <u>c</u>ommon color and every D-edge has two vertices with <u>d</u>ifferent colors.

n-point suspension – of the cycle graph C_m: the graph $C_m + \overline{K_n}$.

Szekeres–Wilf number: same as coloring number.

tabu search: a local search in which solutions that are worse than the current one can be chosen provided that they are not in any of the so-called tabu lists.

timetabling problem: the assignment of times and resources to meetings so as to satisfy a set of constraints as best as possible.

total coloring: assignment of colors to the vertices and to the edges (each vertex and each edge gets one color).

    ___, **proper**: total coloring where no two adjacent or incident vertices/edges have the same color.

total graph – of $G = (V, E)$: its vertex set is $V \cup E$, and $x, y \in V \cup E$ are adjacent if they are incident or adjacent in G; denoted $T(G)$.

toughness – of a graph G: defined to be $+\infty$ when G is complete and otherwise to be

$$\min\{|S|/c(G - S)|S \subseteq V(G)\},$$

where the minimum is taken over all subsets $S \subseteq V(G)$ and $c(G - S)$ denotes the number of components of $G - S$. Denoted $\mathrm{tough}\,(G)$.

triangulation: graph imbedded in a surface, with all faces being cycles of length 3.

r-uniform hypergraph: every edge has precisely r vertices.

uniquely (vertex-)colorable graph: has just one proper coloring with the minimum number of colors, apart from renaming the colors.

uniquely edge-colorable graph: has just one proper edge-coloring with the minimum number of colors, apart from renaming the colors.

upper chromatic number – of a colorable mixed hypergraph: the largest number of colors in a strict coloring; denoted $\overline{\chi}(\mathcal{H})$.

vertex cover – of a graph G: a subset $C \subseteq V(G)$ such that every edge of G has at least one endpoint in C.

k-vertex-critical graph: k-chromatic graph whose chromatic number decreases to $k - 1$ whenever a vertex is deleted.

vertex-coloring: assignment of colors to the vertices (each vertex gets one color).

    ___, **proper**: coloring where no two adjacent vertices have the same color.

vertex-covering number – of a graph G: size of any smallest vertex cover in G; denoted $\tau(G)$. **adjacency matrix** – of a graph G: a 0-1 matrix with a row and a column for each vertex and entry (i, j) is 1 if and only if the ith vertex is adjacent to the jth vertex.

vertex-induced subgraph: a subgrapn $H = (V', E')$ of a graph $G = (V, E)$ with the property that $V' \subseteq V$ and for any two vertices $x, y \in V'$, $(x, y) \in E'$ if and only if $(x, y) \notin E$.

vertex-weighted graph: a graph G in which every vertex x is assigned a non-negative weight $w(x)$.

web graph: the graph obtained by joining the pendant points of a helm graph to form a cycle and then adding a single pendant edge to each vertex of this outer cycle.

Chapter 6

Algebraic Graph Theory

Section 6.1

Automorphisms

Mark E. Watkins, Syracuse University

INTRODUCTION

An automorphism of a graph is a permutation of its vertex set that preserves incidence of vertices and edges. Under composition, the set of automorphisms of a graph forms a group that gives much information about both the local and the global structure of the graph. It may determine the graph's connectivity structure (Section 4.2) and the kinds of surfaces in which it may be embedded (Sections 7.1, 7.5). It is indispensable for counting the number of essentially "distinct" graphs with a variety of different properties (Section 6.3). In this section one will also encounter infinite graphs. It will usually be assumed that the infinite graphs are *locally finite*, i.e., all valences will be presumed to be finite, although they may be arbitrarily large.

6.1.1 The Automorphism Group of a Graph

NOTATION: We need to use the formal notation $\{u, v\}$ (rather than uv) to represent the edge incident with vertices u and v for reasons that become evident in Subsection 6.1.3.

DEFINITIONS

D1: Given a graph X, a permutation α of $V(X)$ is an ***automorphism*** of X if

$$\{u, v\} \in E(X) \Leftrightarrow \{\alpha(u), \alpha(v)\} \in E(X), \text{ for all } u, v \in V(X).$$

D2: The set of all automorphisms of X, together with the operation of composition of functions, forms a subgroup of the symmetric group on $V(X)$ called **the automorphism group** of X.

NOTATION: The automorphism group of X is denoted by $\text{Aut}(X)$. The identity of any group of permutations is denoted by ι.

D3: A graph is called **asymmetric** if ι is its only automorphism.

D4: A **ray** is a one-way infinite path; a **double ray** is a two-way infinite path. (The vertices of a ray admit a natural indexing with $\mathbb{N}$; the vertices of a double ray admit a natural indexing with $\mathbb{Z}$.)

REMARKS

R1: While all vertices in the same *orbit* of $\text{Aut}(X)$ must have the same valence, there exist asymmetric graphs all of whose vertices have the same valence.

R2: If Y is a subgraph of X, even an induced subgraph, then except in special cases there is little relationship between $\text{Aut}(X)$ and $\text{Aut}(Y)$.

R3: When one speaks of $\text{Aut}(X)$ being isomorphic to a group G, it is ambiguous whether the implied isomorphism is between *abstract* groups or between *permutation* groups (see Subsection 6.1.2). In the examples immediately below, the automorphism groups $\text{Aut}(X)$ are abstractly isomorphic to the given groups G.

EXAMPLES

E1: Let $V(K_4) = \{a, b, c, d\}$ and let $X = K_n - \{a, c\}$. Then $\text{Aut}(X) = \{\iota, \alpha, \beta, \alpha\beta\}$ where α interchanges a and c but fixes both b and d, while β fixes a and c but interchanges b and d. Thus $\text{Aut}(X) = \mathbb{Z}_2 \times \mathbb{Z}_2$. Strictly speaking, one should say that $\text{Aut}(X)$ is *isomorphic* to $\mathbb{Z}_2 \times \mathbb{Z}_2$, but for brevity we abuse language in this way throughout this section.

E2: The automorphism group of a circuit of length n is the dihedral group D_n with $2n$ elements.

E3: [Fr37] The automorphism group of the Petersen graph is $\text{Sym}(5)$.

E4: The automorphism group of a double ray is the infinite dihedral group D_∞, but a ray is asymmetric.

E5: The automorphism group of the underlying graph of the regular square tiling of the plane is generated by D_∞ (acting on a major axis) together with a rotation of $\pi/2$ about the origin.

FACTS

F1: For any graph X, the automorphism group of X and that of its complement $\overline{X}$ are identical.

F2: Given any finite tree, either there is a unique vertex or there is a unique edge that is fixed by all automorphisms.

F3: Let the components of X be $X_1, \ldots, X_k$. If no two of the components are isomorphic subgraphs, then $\mathrm{Aut}(X)$ is the direct product $\prod_{i=1}^{k} \mathrm{Aut}(X_i)$. If all of the components are mutually isomorphic subgraphs, then $A(X)$ is the wreath product $\mathrm{Sym}(k) \wr \mathrm{Aut}(X_1)$.

6.1.2   Graphs with Given Group

D5: A group G of permutations of a set S ***acts transitively*** or ***is transitive*** on S if for every $x, y \in S$, there exists $\alpha \in G$ such that $\alpha(x) = y$.

D6: A graph X is said to be ***vertex-transitive*** if $\mathrm{Aut}(X)$ acts transitively on $V(X)$. (Intuitively speaking, a vertex-transitive graph looks the same no matter from what vertex it is viewed.)

D7: A group G of permutations of a set S ***acts doubly transitively*** on S if, for any two ordered pairs of distinct elements $(x_1, x_2), (y_1, y_2) \in S \times S$, there exists $\alpha \in G$ such that $\alpha(x_1) = y_1$ and $\alpha(x_2) = y_2$.

D8: For $i = 1, 2$, let G_i be a group of permutations of the set S_i. We say that G_1 and G_2 are ***isomorphic as permutation groups*** if there exist a group-isomorphism $\Phi : G_1 \to G_2$ and a bijection $f : S_1 \to S_2$ such that

$$f(\alpha(x)) = [\Phi(\alpha)](f(x)) \text{ for all } \alpha \in G_1, \ x \in S_1,$$

i.e., the diagram in Figure 6.1.1 commutes.

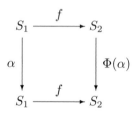

Figure 6.1.1: Isomorphism of permutation groups.

In this case, $|G|$ is the ***order*** of G and $|S|$ is the ***degree*** of G.

D9: An ***edge-isomorphsm*** from a graph X to a graph Y is a bijection $\eta : E(X) \to E(Y)$ such that edges e_1 and e_2 of X are incident with a common vertex of X if and only if $\eta(e_1)$ and $\eta(e_2)$ are incident with a common vertex of Y. An ***edge-isomorphsim*** is an edge-automorphism from a graph to itself.

D10: The set of edge-automorphisms of a graph X, together with the operation of composition of functions, forms a subgroup of the symmetric group on $E(X)$ called the ***edge-group*** of X.

EXAMPLES

E6: Let n denotes any cardinal, finite or infinite. The automorphism group of the complete graph K_n is $\mathrm{Sym}(n)$, and the implied isomorphism is between permutation groups.

E7: For no graph X on n vertices is $\mathrm{Aut}(X)$ ever isomorphic as a permutation group to the alternating group $\mathrm{Alt}(n)$. This is because the permutation group $\mathrm{Alt}(n)$ acts doubly transitively on an n-set. The only graphs on n vertices whose automorphism group acts doubly transitively are K_n and its complement, the edgeless graph $\overline{K}_n$, but the automorphism group of these two graphs is $\mathrm{Sym}(n)$.

E8: The edge-groups of the 3-circuit C_3 and of $K_{1,3}$ are isomorphic as permutation groups to each other and to $\mathrm{Aut}(C_3)$ but are abstractly isomorphic to $\mathrm{Aut}(K_{1,3})$.

FACTS

F4: Every automorphism α of a graph X induces a unique edge-automorphism η_α; namely, if $\{u, v\} \in E(X)$, then $\eta_\alpha(\{u, v\}) = \{\alpha(u), \alpha(v)\}$. The converse is not true.

NOTATION: From here on, unless stated otherwise, one understands the term "isomorphic" between groups to mean "abstractly isomorphic".

F5: [HarPa68] The permutation group induced by $\mathrm{Aut}(X)$ on $E(X)$ is isomorphic to $\mathrm{Aut}(X)$ if and only if X has at most one isolated vertex and K_2 is not a component of X.

F6: Frucht's Theorem [Fr38]: Given any group G, there exist infinitely many connected graphs X such that $\mathrm{Aut}(X)$ is isomorphic to G. Moreover, X may be chosen to be 3-valent [Fr49].

This result was extended by G. Sabidussi [Sa57] as follows:

F7: In addition to having $\mathrm{Aut}(X)$ isomorphic to a given group G, one may further impose that X
- has connectivity κ for any integer $\kappa \geq 1$, *or*
- has chromatic number c for any integer $c \geq 2$ (see Section 5.1), *or*
- be r-valent for any integer $r \geq 3$, *or*
- be spanned by a graph $\hat{Y}$ homeomorphic to a given connected graph Y.

For most of the graphs X constructed by Frucht and Sabidussi, the degree of $\mathrm{Aut}(X)$, namely, $|V(X)|$, is several times as large as its order $|A(X)|$. Consequently, given a group G, there has been interest in seeking the smallest graph X such that $A(X)$ is isomorphic to G. (The situation where the order and degree of $A(X)$ are equal and $A(X)$ acts transitively on $V(X)$ is the subject of Subsection 6.1.6.)

NOTATION: For a finite group G, let $\mu(G)$ denote the least $|V(X)|$ such that $\mathrm{Aut}(X)$ is isomorphic to G.

F8: The asymmetric graph with the fewest edges is obtained from a path of length 5 by adjoining a new edge to a vertex at distance 2 from an end-vertex of the path, yielding a tree on seven vertices. Thus $\mu(\{\iota\}) = 7$.

F9: [Bab74] If G is a nontrivial finite group different from the cyclic groups of orders 3, 4, and 5, then $\mu(G) \le 2|G|$.

F10: $\mu(\mathbb{Z}_3) = 9$; $\mu(\mathbb{Z}_4) = 10$; $\mu(\mathbb{Z}_5) = 15$. (See [Sa67].)

FURTHER READING

The automorphism groups of all the generalized Petersen graphs are presented in detail in [FrGraWa71].

6.1.3   Groups of Graph Products

In this subsection, we use the symbol & to indicate an arbitrary graph product $X\&Y$ of graph X by graph Y, where we define a ***graph product*** of X by Y to be the graph with vertex set $V(X) \times V(Y)$ and whose edge set is determined in a prescribed way by (and only by) the adjacency relations in X and in Y. It has been shown (see [ImIz75]) that there exist exactly 20 graph products that satisfy this definition. One is generally interested in products that are *associative*, in the sense that, for all graphs W, X, Y, the graphs $(W\&X)\&Y$ and $W\&(X\&Y)$ are isomorphic.

DEFINITIONS

The four most commonly used associative graph products are now defined.

D11:   Let Z be a graph product of arbitrary graphs X and Y. Let x_1, x_2 be (not necessarily distinct) vertices of X, and let y_1, y_2 be (not necessarily distinct) vertices of Y. Suppose that $\{(x_1, y_1), (x_2, y_2)\} \in E(Z)$ if and only if

- $[\{x_1, x_2\} \in E(X)$ and $y_1 = y_2]$ or $[x_1 = x_2$ and $\{y_1, y_2\} \in E(Y)]$. Then Z is the ***cartesian product*** of X by Y, and we write $Z = X \square Y$;

- $[\{x_1, x_2\} \in E(X)$ and $y_1 = y_2]$ or $[x_1 = x_2$ and $\{y_1, y_2\} \in E(Y)]$ or $[\{x_1, x_2\} \in E(X)$ and $\{y_1, y_2\} \in E(Y)]$. Then Z is the ***strong product*** of X by Y, and we write $Z = X \boxtimes Y$;

- $\{x_1, x_2\} \in E(X)$ and $\{y_1, y_2\} \in V(Y)$. Then Z is the ***categorical product*** or ***weak product*** of X by Y (also called the ***Kronecker product*** or ***direct product***), and we write $Z = X \times Y$;

- $\{x_1, x_2\} \in E(X)$ or $[x_1 = x_2$ and $\{y_1, y_2\} \in E(Y)]$. Then Z is the ***lexicographic product*** of X by Y, and we write $Z = X[Y]$.

These four products are illustrated in Figure 6.1.2 wherein both X and Y denote the path of length 2.

D12:   A graph X is a ***divisor*** of a graph Z (with respect to a product &) if there exists a graph Y such that $Z = X\&Y$ or $Z = Y\&X$.

D13:   A graph Z is ***prime*** (with respect to a given product &) if Z has no ***proper divisor***, i.e., no divisor other than itself and the graph consisting of a single vertex.

D14:   Graphs X and Y are ***relatively prime*** (with respect to a given product &) if they have no common proper divisor.

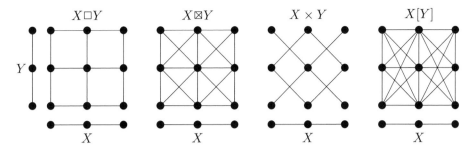

Figure 6.1.2: The four products of the 2-path by the 2-path.

D15: Let G and H be groups of permutations of sets S and T, respectively. We define the **wreath product** $G \wr H$ to be the group of permutations of $S \times T$ with the property that, for each $\pi \in G \wr H$, there exists $\alpha \in G$ and for each $s \in S$ there exists $\beta_s \in H$ such that $\pi(s, t) = (\alpha(s), \beta_{\alpha(s)}(t))$ holds for all $(s, t) \in S \times T$. The group operation is componentwise composition.

GENERAL FACTS

F11: The lexicographic product is the only one of these four products that is not commutative. In fact, if $X[Y] \cong Y[X]$, then either both X and Y are complete, or both are edgeless, or both are powers (with respect to the lexicographic product) of the same graph W. (See [Im69].)

F12: The lexicographic product is the only one of these four products that is self-complementary, in the sense that for any graphs X and Y, we have $\overline{X}[\overline{Y}] \cong \overline{X[Y]}$.

FACTS ABOUT CONNECTEDNESS

F13: The cartesian (respectively, strong) product of two graphs is connected if and only if each factor is connected.

F14: [Weic62] Let X and Y be graphs with at least one edge. Then $X \times Y$ is connected if and only if both X and Y are connected and not both X and Y are bipartite.

F15: The lexicographic product $X[Y]$ is connected if and only if X is connected.

FACTS ABOUT FACTORIZATION

F16: [Sa60] Every connected graph has a unique prime factorization with respect to the cartesian product and with respect to the strong product.

F17: Finite nonbipartite graphs have a prime decomposition into nonbipartite divisors with respect to the categorical product.

F18: [Im72] Any prime decomposition with respect to the lexicographic product can be transformed into any other by transpositions of edgeless or complete divisors. Thus if a graph has such a prime decomposition without such divisors, then it is unique.

FACTS ABOUT AUTOMORPHISM GROUPS OF PRODUCTS

F19: Each of the four product graphs $X \& Y$ named in D11 is vertex-transitive if and only if both X and Y are vertex-transitive.

F20: [Sa60] If X is connected, then $\mathrm{Aut}(X)$ is generated by the automorphisms of its prime divisors with respect to the cartesian product and the transpositions interchanging isomorphic prime divisors.

An important corollary is the following:

F21: Let X be the cartesian product $X = X_1 \square, \cdots, \square X_k$ of pairwise relatively prime connected graphs. Then $\mathrm{Aut}(X)$ is the direct product $\prod_{i=1}^{k} \mathrm{Aut}(X_i)$.

The following notation is needed in order to characterize the group of the lexicographic product.

NOTATION: For a subgraph Y of X, let

$$\partial(Y) = \{x \in V(X) \backslash V(Y) : \{x, y\} \in E(X) \text{ for some } y \in V(Y)\}.$$

If Y consists of a single vertex $Y = \{y\}$, we write simply $\partial(y)$.

NOTATION: We define binary relations $R(X)$ and $S(X)$ on $V(X)$ by

- $(u, v) \in R(X) \Leftrightarrow \partial(u) = \partial(v)$;

- $(u, v) \in S(X) \Leftrightarrow \partial(u) \cup \{u\} = \partial(v) \cup \{v\}$;

- $(u, v) \in \Delta(X) \Leftrightarrow u = v$.

F22: [Sa61] Let X and Y be graphs that are not edgeless. Then $\mathrm{Aut}(X[Y]) = \mathrm{Aut}(X) \wr \mathrm{Aut}(Y)$ if and only if the following two conditions hold:
(i) $R(X) \neq \Delta(X) \Rightarrow Y$ is connected, and
(ii) $S(X) \neq \Delta(X) \Rightarrow \overline{Y}$ is connected.

FURTHER READING

For a comprehensive and up-to-date treatment of all graph products, see [ImKl00].

6.1.4   Transitivity

DEFINITIONS

D16: If G is a group of permutations of a set S and $x \in S$, then the **stabilizer** of x (in G) is the subgroup $G_x = \{\alpha \in G : \alpha(x) = x\}$.

NOTATION: The stabilizer of a vertex $u \in V(X)$ is denoted by $\mathrm{Aut}_u(X)$.

D17: A graph X is **edge-transitive** if, given $e_1, e_2 \in E(X)$, there exists $\alpha \in \mathrm{Aut}(X)$ such that $\alpha(e_1) = \alpha(e_2)$.

D18: A graph X is **arc-transitive** if given ordered pairs (u_1, v_1), (u_2, v_2) of adjacent vertices, there exists $\alpha \in \mathrm{Aut}(X)$ such that $\alpha(u_1) = v_1$ and $\alpha(u_2) = v_2$.

D19: A graph that is vertex-transitive and edge-transitive but not arc-transitive is *half-transitive*.

D20: A graph with constant valence that is edge-transitive but not vertex-transitive is *semisymmetric*.

The following definition is useful when discussing infinite graphs.

D21: Given a graph X, let $d(k)$ denote the number of vertices at distance k from some given vertex, and let a be some real number such that $a > 1$. Then the *growth* of X is defined to be $\mathrm{gr}(X) = \liminf_{k\to\infty}[d(k)/a^k]$.

- If $\mathrm{gr}(X) > 0$, then X has *exponential growth*.

- if $\mathrm{gr}(X) = 0$, then the growth is *subexponential*.

- If $r > 1$ and $0 < \liminf_{k\to\infty}[\sum_{j=0}^{k} d(j)/k^r] < \infty$, then X has *polynomial growth of degree* r.

FACTS

F23: If X is vertex-transitive, then for any $u, v \in V(X)$ we have:

- $\mathrm{Aut}_u(X)$ and $\mathrm{Aut}_v(X)$ are conjugate subgroups of $\mathrm{Aut}(X)$;

- $|\mathrm{Aut}_u(X)| = |\{\alpha \in \mathrm{Aut}(X) : \alpha(u) = v\}|$;

- If X is finite, then $|\mathrm{Aut}(X)| = |\mathrm{Aut}_u(X)| \cdot |V(X)|$.

F24: If a vertex-transitive graph is not connected, then its components are pairwise isomorphic and each is vertex-transitive.

F25: If a graph X is edge-transitive but not vertex-transitive, then it is bipartite. In this case, $\mathrm{Aut}(X)$ induces exactly two orbits in $V(X)$, namely, the two sides of the bipartition.

F26: [Fo67] The smallest semisymmetric graph is 4-valent and has 20 vertices.

F27: If a graph is arc-transitive, then it is both vertex-transitive and edge-transitive.

F28: Tutte's Theorem [Tut66]: Every finite half-transitive graph has even valence.

Finite half-transitive graphs do indeed exist, though they are not plentiful:

F29: [Bouw70] For every positive integer n, there exists a half-transitive graph on $9 \cdot 6^n$ vertices with valence $2(n + 1)$.

F30: [Ho81] The smallest half-transitive graph has 27 vertices and is 4-valent.

F31: [ThWa89] Every infinite half-transitive graph of subexponential growth has even valence. However, there exist half-transitive graphs with exponential growth that have odd valence.

F32: [Tr85] Let X be a connected vertex-transitive, infinite, locally finite graph. Then the following are equivalent:

- X has polynomial growth;

- There is a system $\mathcal{S}$ of imprimitivity of $\mathrm{Aut}(X)$ on $V(X)$ with finite (possibly singleton) blocks such that $\mathrm{Aut}(X)/\mathcal{S}$ is a finitely generated nilpotent-by-finite group and stabilizers in $\mathrm{Aut}(X)/\mathcal{S}$ of vertices in $X/\mathcal{S}$ are finite.

Combining F32 with the result of H. Bass [Bas72] that the rate of growth of a nilpotent group is always polynomial yields the following:

F33:   If the growth of a connected locally finite vertex-transitive graph X is not greater than that of any polynomial, then X has polynomial growth, i.e., if $0 < \liminf_{k\to\infty}[\sum_{i=0}^{k} d(i)/k^r] < \infty$ for some real number r, then r must be a positive integer.

EXAMPLES

E9: $K_{m,n}$ when $m \neq n$ is edge-transitive but not vertex-transitive.

E10:   The graph P_n of the n-sided prism is the cartesian product $C_n \square K_2$. If $n = 3$ or $n \geq 5$, then $\mathrm{Aut}(P_n) = D_n \times \mathbb{Z}_2$. The graph is vertex-transitive but it has two edge-orbits. Thus, although $\mathrm{Aut}(x)$ and the edge-group are (abstractly) isomorphic, they are not isomorphic as permutation groups. However, P_4 is the graph of the 3-dimensional cube; it is arc-transitive and $|\mathrm{Aut}(P_4)| = 48$.

E11:   The lexicographic product of an n-circuit by the edgeless graph on k vertices is arc-transitive.

E12:   Let $D = D_1$ be a double ray, and for $n \in \mathbb{N}$ define $D_{n+1} = D_n \square D$ or $D_{n+1} = D_n \boxtimes D$. Then D_n has polynomial growth of degree n for all $n \in \mathbb{N}$.

E13:   The cartesian, strong, or lexicographic product of the (infinite) r-valent tree for $r \geq 3$ with any finite connected graph has exponential growth.

E14:   An infinite half-transitive graph may be constructed by taking a two-way infinite sequence of copies of Folkman's graph (see F26) with bipartition $\{V_1, V_2\}$ and identifying the vertices in V_2 of the nth copy with the vertices in V_1 in the $(n+1)$st copy for $n \in \mathbb{Z}$ (cf. [ThWa89]). This graph has linear growth.

FACTS

F34: [Gri85] There exist r-valent graphs for small values of r that have subexponential growth but grow faster than any polynomial. (This is called ***intermediate growth***.) They are Cayley graphs (see Section 6.2) of groups with the same rate of growth. These groups are generated by a set of four elements, and all elements have finite order, but these groups are not finitely presentable.

F35: [Se91] Finitely generated groups with intermediate growth cannot act vertex-transitively on connected locally finite graphs of polynomial growth.

FURTHER READING

For a detailed list of conditions for the existence and non-existence of semisymmetric graphs, see [Iv87].

6.1.5 *s*-Regularity and *s*-Transitivity

In the next three subsections we consider some refinements of transitivity of automorphism groups of graphs and begin by reviewing two more notions from the theory of permutation groups.

DEFINITIONS

D22: If G is a group of permutations of a set S, then G **acts semiregularly** if $G_x = \{\iota\}$ for all $x \in S$.

D23: A permutation group **acts regularly** or **is regular** if G acts both transitively and semiregularly.

D24: For $s \geq 0$, an *s-**arc*** in a graph X is a directed walk of length s in which consecutive edges are distinct.

D25: A graph X is *s-**transitive*** if it contains at least one s-arc and $\mathrm{Aut}(X)$ acts transitively on the set of all s-arcs. The terms *1-transitive* and *arc-transitive* are synonymous.

D26: A graph X is *s-**regular*** if it contains at least one s-arc and $\mathrm{Aut}(X)$ acts regularly on the set of all s-arcs.

D27: An *m-**cage*** is a smallest 3-valent graph with girth m.

FACTS

F36: Let G be a group of permutations of a finite set S. Then the following three statements are equivalent:

1. G acts regularly on S.

2. For all $x, y \in S$, there is a unique $\alpha \in G$ such that $\alpha(x) = y$.

3. G acts transitively on S and $|G| = |S|$.

F37: [Tut66] Let X be a connected s-transitive graph with no pendant vertex that is not a circuit. Then

- X is r-transitive for $1 \leq r \leq s$;

- $s \leq \frac{1}{2}\gamma(X) + 1$, where $\gamma(X)$ denotes the girth of X;

- if X is not s-regular, then X is $(s+1)$-transitive.

F38: [Weis74] Let X be a finite $(1 + p^r n)$-valent graph, where p is prime, $r \geq 1$, and $1 \leq n \leq p$. If $\mathrm{Aut}(X)$ contains a subgroup that acts regularly on the s-arcs of X, then $s \leq 7$ and $s \neq 6$. (This result has been generalized to infinite graphs of polynomial growth. See [Se91].)

EXAMPLES

E15: A circuit is s-transitive for all $s \geq 0$.

E16: For all $r \geq 2$, the infinite r-valent tree is s-transitive but not s-regular for all $s \geq 0$.

E17: The graphs of the cube and the dodecahedron are 2-regular.

E18: $K_{1,n}$ for $n \geq 2$ is 2-transitive but not 1-transitive. It has pendant vertices (cf. F37).

E19: K_4 is the unique 3-cage; $K_{3,3}$ is the unique 4-cage.

E20: The Petersen graph is 3-regular. It is the unique 5-cage.

E21: The unique 6-cage is the 4-regular **Heawood graph** H, defined as follows. Let $V(H)$ be the cyclic group $\mathbb{Z}_{14}$. For each $j = 0, \ldots, 6$, let the vertex $2j$ be adjacent to the three vertices $2j - 1$, , $2j + 1$, and $2j + 5$.

6.1.6   Graphical Regular Representations

DEFINITIONS

D28: Given a group G, a graph X such that $\mathrm{Aut}(X)$ is isomorphic to G and acts regularly on $V(X)$ is called a **graphical regular representation**, or **GRR**, of G.

D29: A **generalized dicyclic group** is a (finite or infinite) group G with the following properties:

- G contains an abelian subgroup A of index 2;

- There exists $b \in G \backslash A$ such that $bab^{-1} = a^{-1}$ for all $a \in A$;

- There exists an element $a_0 \in A$ of order $2m$ where $m \geq 2$;

- $b^2 = a_0^m$.

FACTS

F39: The 8-element quaternion group $\mathbb{Q}$ is the smallest generalized dicyclic group.

F40: If X is a GRR of G, then $|V(X)| = |G|$. (In the finite case, this would follow directly from F23.) It will be seen in Section 6.2 that X must be a Cayley graph of G. Of importance is that the automorphism group of every Cayley graph contains a subgroup that acts regularly on its vertex set. If the Cayley graph is a GRR, that subgroup is exactly the whole automorphism group of the graph.

EXAMPLES

E22: K_2 is a GRR of $\mathbb{Z}_2$.

E23: The dihedral group D_n has the presentation

$$\langle a, b \mid a^n = b^2 = (ab)^2 = \iota \rangle.$$

The Cayley graph $\mathrm{Cay}(D_n, S)$ with connection set $S = \{a, a^{-1}, b, ba, ba^3\}$ (see Subsection 6.2.1) is a GRR of D_n when $n \geq 6$. If $3 \leq n \leq 5$, then the vertex-stabilizers have order 2.

FACTS

The first several facts answer questions as to which abstract groups do and which do not admit a GRR.

F41: [Ch64, Sa64, Im70] The only abelian groups that admit a GRR are the elementary abelian groups of order 2^n for $n = 1$ and $n \geq 5$.

F42: [No68, Wa71] No generalized dicyclic group admits a GRR.

F43: The following ten groups do not admit a GRR, namely:

- the dihedral groups D_n, for $n = 3, 4, 5$ [Wa71];

- $\mathrm{Alt}(4)$ [Wa74]);

- $\mathbb{Q} \times \mathbb{Z}_n$ for $n = 2, 3, 4$ [Wa72];

- the nonabelian group of order 27 and exponent 3 [NoWa72b];

- the nonabelian group of order 18 and exponent 6, another group of order 16, and another group of order 24 [NoWa72a, Wa72].

F44: [He76] Every finite solvable group which is non-abelian, non-generalized dicyclic, and not one of the ten groups listed in F43 admits a GRR.

F45: [Go81] Every finite non-solvable group admits a GRR.

F46: [ImWa76] Let G be a finite abelian group. Then there exists a graph X such that $\mathrm{Aut}(X)$ contains a regular subgroup of index ≤ 2 isomorphic to G unless G is one of the seven groups

$$\mathbb{Z}_2^3, \ \mathbb{Z}_2^4, \ \mathbb{Z}_4 \times \mathbb{Z}_2, \ \mathbb{Z}_4 \times \mathbb{Z}_2^2, \ \mathbb{Z}_4^2, \ \mathbb{Z}_3^2, \ \text{and} \ \mathbb{Z}_3^3.$$

F47: [BabGo82] Let G be a nilpotent non-abelian group of odd order g. Let $\mathcal{C}(G)$ be the set of all graphs whose automorphism group contains a regular subgroup isomorphic to G. Then almost all the graphs in $\mathcal{C}(G)$ are GRR's of G, i.e., there are only $o(2^{(g-1)/2})$ exceptions as $g \to \infty$.

We close this subsection with a result about infinite GRR's.

NOTATION: If n is a positive integer, let $[n] = \{j \in \mathbb{Z} : 1 \leq j \leq n\}$.

F48: [Wa76] Let $n \geq 2$, and let $\{G_j : j \in [n]\}$ be a family of pairwise-disjoint groups such that $r(G_j) \leq \aleph_0$, where $r(G_j)$ denotes the cardinality of a smallest generating set for G_j. Then the free product $G = \coprod_{j \in [n]} G_j$ admits a GRR. If $\sum_{j \in [n]} r(G_j)$ is finite, then G admits a locally finite GRR.

6.1.7   Primitivity

DEFINITIONS

D30: Let G be a group of permutations of a set S. A subset $B \subseteq S$ is called a **block** (of imprimitivity with respect to G) if for all $\alpha \in G$, either $\alpha(B) = B$ or $\alpha(B) \cap B = \emptyset$.

- Clearly $\emptyset, S$, and the singleton subsets of V are blocks; they are called **trivial blocks**.

- If G acts transitively and admits no block other than the trivial blocks, then G is **primitive** on S.

- If G acts transitively but admits nontrivial blocks, then G is **imprimitive** on S.

- If G is imprimitive on S and B is a nontrivial block, then the set of images of B under G forms a partition of S, called a **system of imprimitivity**.

D31: A graph is said to be **primitive** if its automorphism group acts as a primitive permutation group on its vertex set.

D32: A **lobe** of a graph is a maximal 2-connected subgraph.

EXAMPLES

E24: An n-circuit $(n \geq 3)$ is primitive if and only if n is prime. On the other hand, if d is a proper divisor of n, then a nontrivial block is obtained by starting at any vertex v and selecting all vertices lying at distance a multiple of d from v. In this case, there exists a system of imprimitivity consisting of d blocks, each of size n/d.

E25: K_n is primitive for all n, but $K_{n,n}$ is never primitive; the two sides of the bipartition form a system of imprimitivity of $V(K_{n,n})$.

E26: The Petersen graph is primitive.

E27: Except for the tetrahedron, the graphs of all the Platonic solids are imprimitive. The blocks of imprimitivity are the pairs of antipodal vertices.

FACTS

F49: The vertex sets of the components of a vertex-transitive graph X are blocks of imprimitivity of $\mathrm{Aut}(X)$. Hence primitive graphs with nonempty edge-sets are connected.

F50: In a primitive graph, the connectivity κ must equal the valence. Otherwise the graph has nontrivial atoms (see §4.2.3), and the family of vertex sets of the atoms is a system of imprimitivity.

F51: [GraWa88] Let X be a finite planar graph. Then X is primitive if and only if it is K_2, K_4, a circuit of prime length, or an edgeless graph.

We state five primitivity results for infinite graphs.

F52: [JuWa89] Let n be a nonnegative integer. There exists an infinite, locally finite, primitive graph with connectivity $\kappa = n$ if and only if $n \neq 2$.

F53: [JuWa77b] Let X be a vertex-transitive graph with connectivity $\kappa(X) = 1$. A necessary and sufficient condition for X to be primitive is that it have no cut-edge and that its lobes themselves be primitive and pairwise isomorphic.

F54: [WaGra04] An inifinite, locally finite, planar graph X is primitive if and only if $\kappa(X) = 1$, X has no cut-edge, and there exists an integer $m \geq 2$ such that every vertex of X is incident with exactly m lobes. Moreover, all of the lobes are isomorphic to K_4 or all are p-circuits for some fixed odd prime p.

F55: [Sm10] Let Δ be an infinite, imprimitive directed graph with connectivity $\kappa(\Delta) = 1$. The associated (undirected) graph of Δ is primitive if and only if it has no cut-edge and, for some odd prime p, the lobes of Δ are pairwise-isomorphic directed p-circuits.

F56: [GoImSeWaWo89] Let X be a locally finite, connected, vertex-transitive graph of polynomial growth. Then X is not primitive.

6.1.8   More Automorphisms of Infinite Graphs

The overall "shape" of an infinite graph can be described effectively with the aid of the concept of the *ends* of the graph. This notion, previously used in describing infinite groups, was first applied to graphs by R. Halin [Hal64]. In this subsection, the symbol X will always denote an infinite graph.

NOTATION: Given a graph X, let $\mathcal{R}(X)$ denote the set of all rays in X.

DEFINITIONS

D33: Let $R_1, R_2 \in \mathcal{R}(X)$. We say that R_1 and R_2 are **end-equivalent** and write $R_1 \sim R_2$ if there exists $R_3 \in \mathcal{R}(X)$ such that both $V(R_3 \cap R_1)$ and $V(R_3 \cap R_2)$ are infinite.

It is not hard to show that $\sim$ is an equivalence relation on the set $\mathcal{R}(X)$.

D34: The equivalence classes with respect to $\sim$ are the **ends** of X.

NOTATION: $\mathcal{E}(X)$ will denote the set of ends of X, and we define the cardinal $\epsilon(X) = |\mathcal{E}(X)|$.

D35: If Y is a subgraph of X and $R \in \mathcal{R}(X)$, then R **is contained in** Y if Y contains a subray of R. If $R_1, R_2 \in \mathcal{R}(X)$, then Y **separates** R_1 and R_2 if these two rays are contained in distinct components of $X - Y$.

D36: An automorphism α of X is a **translation** if it fixes no finite nonempty subgraph, i.e., if $\alpha(Y) = Y$ implies that $V(Y)$ is empty or infinite.

D37: A subgroup $G \leq \text{Aut}(X)$ acts **almost transitively** on $V(X)$ if its action induces only finitely many orbits. We say that X is **almost transitive** if $\text{Aut}(X)$ acts almost transitively.

D38: A ***torsion subgroup*** of an infinite group G is a subgroup all of whose elements have finite order.

FACTS

F57: König's "Unendlichkeitslemma" [K36]: Let $S_1, S_2, \ldots$ be a (countably infinite) sequence of finite, nonempty, pairwise-disjoint subsets of $V(X)$. Suppose that for each positive integer n, each vertex in S_{n+1} is adjacent to some vertex in S_n. Then X contains a ray with vertices $x_1, x_2, \ldots$, where $x_n \in S_n$ for all $n \in \mathbb{N}$.

F58: If a graph Y contains no ray at all, then either Y is finite or is infinite but not locally finite. In this case $\epsilon(Y) = 0$.

F59: [Hal64] Let $R_1, R_2 \in \mathcal{R}(X)$. The following statements are equivalent:

- $R_1 \sim R_2$;

- no finite subgraph of X separates R_1 and R_2;

- X contains infinitely many pairwise-disjoint (finite) $R_1 R_2$-paths.

F60: When X is locally finite, then $\epsilon(X)$ equals the supremum of the number of infinite components of $X - S$ as S ranges over all finite subsets of $V(X)$.

Clearly $\operatorname{Aut}(X)$ induces a group of permutations of $\mathcal{E}(X)$.

F61: Given an automorphism α of a connected locally finite graph, either all orbits of α are infinite (in which case α is a translation) or all orbits of α are finite (although they may be arbitrarily large).

F62: [Hal73] Every translation of a connected locally finite graph fixes some double ray and at most two ends.

F63: [Hal73] If X is connected and locally finite, and if $\operatorname{Aut}(X)$ contains a translation, then $\epsilon(X) = 1$, 2, or ∞.

By combining this result with a theorem of H. A. Jung [Ju81], we obtain the following important classification of almost transitive infinite graphs.

F64: If X is connected, locally finite, and almost transitive, then $\epsilon(X) = 1$, 2 or $2^{\aleph_0}$.

F65: Suppose that X is almost transitive. If X has linear growth, then $\epsilon(X) = 2$ (cf. "strips" below). If X has polynomial growth of degree $d \geq 2$, then $\epsilon(X) = 1$. If X has exponential growth, then $\epsilon(X) = 1$ or $2^{\aleph_0}$.

F66: [SeTr97] If a graph X is almost transitive and has quadratic growth, then $\operatorname{Aut}(X)$ contains a subgroup isomorphic to $\mathbb{Z} \oplus \mathbb{Z}$ that acts on $V(X)$ with finitely many orbits.

F67: [SeTr97] There are only countably many nonisomorphic almost transitive graphs with linear or quadratic growth.

F68: [Se91] Let X be connected, locally finite, and vertex-transitive. Then $\operatorname{Aut}(X)$ is uncountable if and only if it contains a finitely generated subgroup of exponential growth that acts transitively on $V(X)$.

F69: [Hal68] Every end of a graph contains a family of pairwise disjoint rays of maximum cardinality, i.e., if an end of a graph contains arbitrarily large finite families of pairwise disjoint rays, then that end contains an infinite family of pairwise disjoint rays. In this same sense, every graph contains a family of pairwise disjoint double rays of maximum cardinality.

F70: [BabWa80] Suppose that X is connected and locally finite. If $\text{Aut}(X)$ contains a torsion subgroup T that acts almost transitively on $V(X)$, then $\epsilon(X) = 1$. If T acts transitively on $V(X)$ and X is r-valent, then X has connectivity $\kappa(X) \geq \frac{3}{4}(r+1)$ and edge-connectivity $\lambda(X) = r$.

REMARK

R4: In F70 it is not known whether, under the stated hypotheses, the lower bound $\frac{3}{4}(r+1)$ is attainable, although examples X exist for which $\kappa(X) = \frac{4}{5}(r+1)$. More generally, if X is merely locally finite and vertex transitive, then it is known that $\kappa(X) > \frac{2}{3}(r+1)$, and this bound is indeed sharp (see [JuWa77a]).

EXAMPLES

E28: The complete bipartite graph $K_{n,\aleph_0}$, where $n \in \mathbb{N}$, has zero ends, because it contains no rays. Of course, it is not locally finite. Since it has finite diameter, its growth is not defined.

E29: Let D be a double ray, and let Y be any connected graph. If Y is infinite, then the cartesian product $D \square Y$ has exactly one end. However, if Y is a finite, then $D \square Y$ has exactly two ends. In fact, $D \square Y$ is a *strip* (see below).

E30: Let $V(X) = \mathbb{Z} \times \mathbb{Z}$ and let $E(X)$ consist of all edges of the forms $\{(m,n),(m,n+1)\}$ and $\{(m,0),(m+1,0)\}$ for $m, n \in \mathbb{Z}$. Then X has quadratic growth and $\epsilon(X) = \aleph_0$. By F64, X cannot be almost transitive. In fact, $\text{Aut}(X)$ clearly has $\aleph_0$ orbits.

E31: The cartesian, strong, or lexicographic product of the (infinite) r-valent tree for $r \geq 3$ with any finite connected graph has $2^{\aleph_0}$ ends.

E32: There are three so-called *regular tessellations* of the Euclidean plane, namely,

1. six congruent equilateral triangles meeting at each vertex, or

2. four congruent squares meeting at each vertex, or

3. three congruent regular hexagons meeting at each vertex.

Their underlying graphs all have quadratic growth and exactly one end.

E33: The regular tessellations of the hyperbolic plane (e.g., four congruent pentagons meeting at every vertex) also have exactly one end, but their growth is exponential.

Strips

A special class of 2-ended graphs is of some interest.

DEFINITIONS

D39: A connected graph X is called a **strip** if there exists a connected subgraph Y of X and an automorphism $\alpha \in \text{Aut}(X)$ such that $\partial(Y)$ and $Y - \alpha(Y)$ are finite and $\alpha(Y \cup \partial(Y)) \subseteq Y$.

D40: The **infinite connectivity** of X, denoted $\kappa_\infty(X)$, is the minimum cardinality of a set $S \subset V(X)$ such that $X - S$ has at least two infinite components. If no such minimum exists, then $\kappa_\infty(X) = \infty$.

REMARK

R5: In §4.2.3, the notions of *fragment* and *atom* are presented with respect to the connectivity κ of graph. These terms may also be defined with respect to the more restrictive parameter of infinite connectivity with very similar results. In particular, distinct κ_∞-atoms are disjoint. (See [JuWa77a].)

FACTS ABOUT STRIPS

F71: Let X be a connected infinite graph. Then the following statements are equivalent:

- X is a strip.

- X is locally finite and $\text{Aut}(X)$ contains an automorphism with finitely many orbits [JuWa84].

- X is locally finite, $\epsilon(X) = 2$, and $\text{Aut}(X)$ contains a translation [ImSe88].

F72: [ImSe88] Let X be connected, locally finite, and vertex-transitive. Then X has linear growth if and only if X is a strip.

F73: [JuWa84] Suppose that X is connected and that $\text{Aut}(X)$ contains an abelian subgroup H that acts transitively on $V(X)$. Then either

- $\kappa_\infty(X) = \infty$, i.e., $\epsilon(X) = 1$, or

- X is a strip and $H \cong \mathbb{Z} \oplus F$ for some finite abelian group F.

F74: [Wa91] If a strip is edge-transitive, then all vertices have even valence.

F75: [Wa91] Let S be a planar edge-transitive strip with connectivity $\kappa(S) = k \geq 3$. Then $V(S) = (\mathbb{Z} \times \mathbb{Z})/\rho$ where $\rho = \{\{(x, -x), (-x + k, x + k)\} : x \in \mathbb{Z}\}$ and the vertex $(x, y)\rho$ is adjacent to $(x, y \pm 1)\rho$ and $(x \pm 1, y)\rho$ for all $(x, y) \in \mathbb{Z} \times \mathbb{Z}$. (In this case, S admits a regular embedding on an infinitely long circular cylinder obtained by "rolling up" the Euclidean plane at a 45° angle to the major axes.)

The next result says that strips can be found as subgraphs of multi-ended graphs.

F76: [Ju94] Let X be locally finite and suppose that a subgroup $G \leq \text{Aut}(X)$ fixes some 2-subset of $\mathcal{E}(X)$. If G contains a translation, then there is a G-invariant induced subgraph S of X that is a strip (with respect to G), and ∂Y is finite for every component Y of $X - S$.

Automorphisms and Distance

DEFINITIONS

D41: An automorphism $\alpha \in \text{Aut}(X)$ is **bounded** if there exists $M > 0$ such that for all $v \in V(X)$, the distance $d(v, \alpha(v)) < M$.

D42: A path or ray or double ray is said to be **geodetic** if it contains a shortest path joining any two of its vertices. A geodetic double ray is a **geodesic**.

D43: Let R be a ray or a double ray in X. The **straightness** $\sigma(R)$ of R is defined to be

$$\sigma(R) = \liminf_{d_R(u,v)\to\infty} \frac{d(u,v)}{d_R(u,v)},$$

where $u, v \in V(R)$ and $d_R(u, v)$ is the length of the subpath of R joining u and v.

D44: A ray or double ray R is **metric** if $\sigma(R) > 0$.

D45: If $\alpha \in \text{Aut}(X)$, then a ray R is α-**essential** if $\alpha^n(R) \subset R$ for some positive integer n, and α is **of metric type** if there exists a metric α-essential ray.

FACTS

F77: [JuWa84] The set of bounded automorphisms of a graph X forms a normal subgroup of $\text{Aut}(X)$.

The following result extends König's *Unendlichkeitslemma*:

F78: [Wa86] If X has infinite diameter, then for each vertex $u \in V(X)$ and for each end of X, there exists a geodetic ray R belonging to that end that originates at u.

F79: $0 \leq \sigma(R) \leq 1$ holds for any ray or double ray R. If R is a geodesic, then $\sigma(R) = 1$, but not conversely.

F80: [PoWa95] If a translation $\tau \in \text{Aut}(X)$ fixes some metric double ray, then every double ray fixed by a nonzero power of τ is also metric.

F81: [PoWa95] Suppose that a translation $\tau \in \text{Aut}(X)$ fixes some metric double ray D_0. If $\sigma(D_0) < 1$, then some power of τ fixes a metric double ray D_1 such that $\sigma(D_1) > \sigma(D_0)$. If $\sigma(D_0) = 1$, then D_0 is a geodesic.

F82: [JuNi94] If $\tau \in \text{Aut}(X)$ is a translation of metric type, then

$$\sup\{\sigma(R) : R \text{ is } \tau\text{-essential}\} = 1.$$

EXAMPLE

E34: Let $V(X) = \mathbb{Z}_m \times \mathbb{Z}$ where each vertex (x, y) is adjacent to vertices $(x \pm 1, y)$ and $(x, y \pm 1)$, the first coordinate being read modulo m. Let $\tau(x, y) = (x + 1, y + 1)$ for all $(x, y) \in V(X)$. Let D be the double ray with edges of the forms $\{(x, x), (x, x+1)\}$ and $\{(x, x+1), (x+1, x+1)\}$. Then $\sigma(D) = 1/2$, and so D is metric but not geodetic. Note that all rays with vertex set $\{(x, y) : y \in \mathbb{Z}\}$ are τ-essential, as they are fixed by τ^m.

6.1.9   Distinguishability

DEFINITIONS

D46: Given a group G of permutations of a set V and $d \in \mathbb{N}$, we say that the action of G on V is d-***distinguishable*** if there exists a partition of V into d cells (or "color classes") such that the only element of G that fixes (setwise) every cell in the partition is the identity ι. Such a partition is called a ***distinguishing coloring***. The ***distinguishing number*** of this group of permutations is the least number d such that a distinguishing coloring with exactly d cells exists.

D47: A graph X is d-***distinguishable*** if there exists a partition of $V(X)$ into d cells such that the only element of $\mathrm{Aut}(X)$ that fixes (setwise) every cell in the partition is the identity ι. The ***distinguishing number*** of X is the least number d such that X is d-distinguishable.

One may think of the distinguishing number of a graph as the least number of colors in a coloring of its vertex set such that every nonidentity automorphism maps some vertex of one color to some vertex of a different color.

D48: The ***distinguishing chromatic number*** of a graph X is the least number of color classes in a distinguishing partition of X that is at the same time a proper vertex-coloring of X.

NOTATION: The distinguishing number of a graph X is denoted by $D(X)$, and the distinguishing chromatic number of X is denoted by $D_\chi(X)$. Clearly for any graph X with chromatic number $\chi(X)$, one has $D_\chi(X) \geq \max\{\chi(X), D(X)\}$.

FURTHER READING

The main topic of the remainder of the subsection is the distinguishing number of finite and infinite graphs. See [ColHovTre09] for more information concerning the relationship between this parameter and the distinguishing chromatic number. The reader interested in the distinguishing number of a permutation group acting on other combinatorial objects as well as on certain graphs may consult [Chan06] and [ConTuc11].

D49: Given a map M (see §7.6.1), the ***distinguishing number*** of M is the distinguishing number of the group of map-automorphism (see §7.6.5) acting on the vertices of M.

EXAMPLES

E35: Since a graph and its complement have the very same automorphism group, $D(X) = D(\overline{X})$ holds for all graphs X.

E36: $D(K_n) = n$ for all n.

E37: $D(K_{m,n}) = n$ if $m < n$ but $D(K_{n,n}) = n + 1$.

E38: If X is the graph K_{2n} with the edges of a 1-factor deleted, then its complement $\overline{X}$ is the disjoint union of n copies of K_2. Thus $\mathrm{Aut}(X) = \mathrm{Sym}(n) \wr \mathbb{Z}_2$, and so $D(X) = \binom{n}{2}$.

E39: Let $V(C_n) = \{x_0, x_1, \ldots, x_{n-1}\}$ be labeled in the natural way. If $n \geq 6$, then $D(C_n) = 2$ with $S = \{x_0, x_1, x_3\}$ being one color class and its complement being the other. However, for $3 \leq n \leq 5$, every subset of $V(C_n)$ is invariant under some reflection, but a third color can "break" that reflection, in which case $D(C_n) = 3$ (cf. E23).

E40: The vertices of the **Kneser graph** $K(n, k)$ are the k-subsets of an n-set ($1 \leq k \leq n/2$), and two such vertices are adjacent when their corresponding subsets are disjoint. All Kneser graphs $K(n, k)$ have distinguishing number 2 except for the $K(5, 2)$, which is the Petersen graph, and $K(n, 1)$, which is complete [AlBout07]. In particular, the distinguishing number of the Petersen graph is 3 [AlCol96].

E41: Here is an example of a locally finite graph with infinite distinguishing number. Start with a ray R, where $V(R) = \{x_0, x_1, x_2, \ldots\}$ indexed in the natural way. For each $n \in \mathbb{N}$, append to x_n exactly n pendant edges.

FACTS

F83: [AlCol96] For any finite group G there exists a graph X such that $\text{Aut}(X) = G$ and $D(X) = 2$ (cf. F7).

F84: [AlCol96] Let X be a graph and let G be a finite group and let m be the number of groups in a longest ascending chain of proper subgroups of G. If $\text{Aut}(X) = G$, then $D(X) \leq m$.

F85: [AlCol96] If a nontrivial group G has the property that all its subgroups are self-conjugate (this includes abelian groups), and if $\text{Aut}(X) = G$, then $D(X) \leq m$.

F86: [ColTre06] For any finite graph X, $D(X) \leq \Delta(X) + 1$, where Δ indicates the maximum valence. Equality holds if and only if X is of one of the graphs K_n, $K_{n,n}$, or C_5.

An extension of F86 to infinite graphs is the following:

F87: [ImKlTr07] If X is a connected infinite graph with no vertex of valence greater than $\mathbf{m}$, then $D(X) \leq \mathbf{m}$.

F88: [Tuc11] There exist only finitely many maps M on orientable surfaces with $D(M) > 2$. Only four maps exist for which the distinguishing number of the subgroup of orientation-preserving automorphisms equals 3.

Distinguishing Number of Graph Products

NOTATION: The symbol $\mathbf{m}$ will denote any cardinal. If & denotes one of the graph products defined in D11, then the product of $\mathbf{m}$ copies of a graph X will be denoted by $X^{\&\mathbf{m}}$. In particular, $K_2^{\square \mathbf{m}}$ is the hypercube of dimension $\mathbf{m}$, which is conventionally denoted by $Q_{\mathbf{m}}$.

F89: $D(Q_{\mathbf{m}}) = 2$ for all $\mathbf{m} \geq 4$. (For finite $\mathbf{m}$, see [BogCow04]; for transfinite $\mathbf{m}$, see [ImKlTr07].)

F90: [ImKlTr07] If $\mathbf{m}$ and $\mathbf{n}$ are infinite cardinals, then

$$D(K_\mathbf{m} \,\square\, K_\mathbf{m}) = D(K_\mathbf{m} \,\square\, K_{2^\mathbf{m}}) = 2,$$

but if $\mathbf{n} > 2^\mathbf{m}$, then $D(K_\mathbf{m} \,\square\, K_\mathbf{n}) > \mathbf{m}$.

F91: [ImKl06] Let X be any connected graph, and let $k \geq 2$ be any integer. Then $D(X^{\square k}) = 2$ with the three exceptions of Q_2, Q_3, and K_3^2, which have distinguishing number 3.

Analogous results hold for the strong product and the categorical product as follows, but with some restrictions.

F92: [ImKl06] Let X be a connected graph that satisfies the following property: for no two distinct vertices $u, v \in V(X)$ does it hold that every vertex other than u and v is adjacent to either both u and v or to neither u nor v. Then for all $k \geq 2$, one has $D(X^{\boxtimes k}) = 2.$

F93: [ImKl06] Let X be a connected, nonbipartite graph with the property that no two vertices have the same closed neighborhood (see §5.3.4). Then for all $k \geq 2$, one has $D(X^{\times k}) = 2$.

More Distinguishing Number Results for Infinite Graphs

DEFINITIONS

D50: The *n-sphere* of a vertex u is the set of vertices at distance n from u.

D51: [SmTucWa12] A graph X satisfies the **Distinct Spheres Condition** if there exists a vertex $x \in V(X)$ such that every pair y, z of vertices equidistant from x have distinct n-spheres.

FACTS

F94: [SmTucWa12] If a connected denumerable graph satisfies the Distinct Spheres Condition, then it is 2-distinguishable.

F95: [SmTucWa12] A connected denumerable graph X satisfying any of the following conditions also satisfies the Distinct Spheres Condition and hence is 2-distinguishable:

- X is a primitive graph with infinite diameter;

- X is a denumerable connected graph such that the deletion of any vertex leaves at least two infinite components;

- X is the cartesian product of two connected denumerable graphs of infinite diameter.

F96: [WaZ07] Let T be an infinite tree.

- If T has no 1-valent vertices and all valences are finite or countably infinite, then $D(T) = 2$.

- If T is locally finite and $D(T) = D_0 < \infty$, then T contains a finite subtree T_0 such that $D(T_0) = D_0$.

F97: [ImKlTr07] The countable random graph has distinguishing number 2.

References

[AlBout07] M. O. Albertson and D. L. Boutin, Using determining sets to distinguish Kneser graphs, *Electron. J. Combin.* 14 (2007), #R20.

[AlCol96] M. O. Albertson and K. L. Collins, Symmetry breaking in graphs, *Electron. J. Combin.* 3 (1996), #N17.

[Bab74] L. Babai, On the minimum order of graphs with given group, *Canad. Math. Bull.* 17 (1974), 467–470.

[BabGo82] L. Babai and C. D. Godsil, On the automorphism groups of almost all Cayley graphs, *Europ. J. Combin.* 3 (1982), 9–15.

[BabWa80] L. Babai and M. E. Watkins, Connectivity of infinite graphs having a transitive torsion group action, *Arch. Math.* 34 (1980), 90–96.

[Bas72] H. Bass, The degree of polynomial growth of finitely generated nilpotent groups, *Proc. London Math. Soc. (3)* 25 (1972), 603–614.

[BogCow04] B. Bogstad and L. J. Cowen, The distinguishing number of the hypercube, *Discrete Math.* 283 (2004), 29–35.

[Bouw70] I. Z. Bouwer, Vertex and edge transitive but not 1-transitive graphs, *Canad. Math. Bull.* 13 (1970), 231–237.

[Chan06] M. Chan, The maximum distinguishing number of a group, *Electron. J. Combin.* 13 (2006), #R70.

[Ch64] C.-Y. Chao, On a theorem of Sabidussi, *Proc. Amer. Math. Soc.* 15 (1964), 291–292.

[ColHovTre09] K. L. Collins, M. T. Hovey, and A. N. Trenk, Bounds on the distinguishing chromatic number, *Electron. J. Combin.* 16 (2009), #R88.

[ColTre06] K. L. Collins and A. N. Trenk, The distinguishing chromatic number, *Electron. J. Combin.* 13 (2006), #R16.

[ConTuc11] M. Conder and T. W. Tucker, Motion and distinguishing number two, *Ars Math. Contemp.* 4 (2011), 63–72.

[Fo67] J. Folkman, Regular line-symmetric graphs, *J. Combin. Theory* 3 (1967), 215–232.

[Fr37] R. Frucht, Die Gruppe des Petersen'schen Graphen und der Kantensysteme der regulären Polyheder, *Comment. Math. Helvetici* 9 (1936/37), 217–223.

[Fr38] R. Frucht, Herstellung von Graphen mit vorgegebener abstrakter Gruppe, *Compositio Math.* 6 (1938), 239–250.

[Fr49] R. Frucht, Graphs of degree 3 with given abstract group, *Canad. J. Math.* 1 (1949), 365–378.

[FrGraWa71]] R. Frucht, J. E. Graver, and M. E. Watkins, The groups of the generalized Petersen graphs, *Proc. Cambridge Phil. Soc.* 70 (1971), 211–218.

[Go81] C. D. Godsil, GRR's for non-solvable groups, in: *Algebraic Methods in Graph Theory* (Proc. Conf. Szeged 1978) L. Lovász and V. T. Sós, eds.), Colloq. Soc. János Bolyai 25, North-Holland, Amsterdam, 1981, pp. 221–239.

[GoImSeWaWo89] C. D. Godsil, W. Imrich, N. Seifter, M. E. Watkins, and W. Woess, On bounded automorphisms of infinite graphs, *Graphs and Combin.* 5 (1989), 333–338.

[GraWa88] J. E. Graver and M. E. Watkins, A characterization of finite planar primitive graphs, *Scientia* 1 (1988), 59–60.

[Gri85] R. I. Grigorchuk, Degrees of growth of finitely generated groups and the theory of invariant means, *Math. USSR-Izv* 25 (1985), 259–300.

[Hal64] R. Halin, Über unendliche Wege in Graphen, *Math. Ann* 157 (1964), 125–137.

[Hal68] R. Halin, Die Maximalzahl fremder zweiseitig unendlicher Wege in Graphen, *Math. Nachr.* 44 (1968), 119–127.

[Hal73] R. Halin, Automorphisms and endomorphisms of infinite locally finite graphs, *Abh. Math. Sem. Univ. Hamburg* 39 (1973), 251–283.

[HarPa68] F. Harary and E. M. Palmer, On the point-group and line-group of a graph, *Acta Math. Acad. Sci. Hungar.* 19 (1968), 263–269.

[He76] D. Hetzel, Über reguläre Darstellung von auflösbaren Gruppen, *Dipomarbeit, Technische Universität Berlin*, 1976.

[Ho81] D. F. Holt, A graph which is edge transitive but not arc transitive, *J. Graph Theory* 5 (1981), 201–204.

[Im69] W. Imrich, Über das lexikographische Produkt von Graphen, *Arch. Math. (Basel)* 20 (1969), 228–234.

[Im70] W. Imrich, Graphs with transitive Abelian automorphism group, in: *Combinatorial Theory and Its Applications* (Colloq. Math. Soc. János Bolyai 4 Proc. Colloq. Balatonfüred, Hungary 1969), P. Erdős, A. Renyi, and V. T. Sós eds., North-Holland, Amsterdam, 1970, pp. 651–656.

[Im72] W. Imrich, Assoziative Produkte von Graphen, *Österreich. Akad. Wiss. Math.-Natur. K. S.-B. II*, 180 (1972), 203–293.

[ImIz75] W. Imrich and H. Izbicki, Associative products of graphs, *Monatsh. Math.* 80 (1975), 277–281.

[ImKl00] W. Imrich and S. Klavžar, *Product Graphs, Structure and Recognition*, John Wiley & Sons, Inc., New York, 2000.

[ImKl06] W. Imrich and S. Klavžar, Distinguishing Cartesian powers of graphs, *J. Graph Theory* 53 (2006), 250–260.

[ImKlTr07] W. Imrich, S. Klavžar, and A. V. Trofimov, Distinguishing infinite graphs, *Electron. J. Combin.* 14 (2007), #R36.

[ImSe88] W. Imrich and N. Seifter, A note on the growth of transitive graphs, *Discrete Math.* 73 (1988/89), 111–117.

[ImWa76] W. Imrich and M. E. Watkins, On automorphisms of Cayley graphs, *Per. Math. Hung.* 7 (1976), 243–258.

[Iv87] A. V. Ivanov, On edge but not vertex transitive regular graphs, *Annals Discrete Math.* 34 (1987), 273–286.

[Ju81] H. A. Jung, A note on fragments of infinite graphs, *Combinatorica* 1 (1981), 285–288.

[Ju94] H. A. Jung, On finite fixed sets in infinite graphs, *Discrete Math.* 131 (1994), 115–125.

[JuNi94] H. A. Jung and P. Niemeyer, Decomposing ends of locally finite graphs, *Math. Nachr.* 174 (1995), 185–202.

[JuWa77a] H. A. Jung and M. E. Watkins, On the connectivities of finite and infinite graphs, *Monatsh. Math.* 83 (1977), 121–131.

[JuWa77b] H. A. Jung and M. E. Watkins, On the structure of infinite vertex-transitive graphs, *Discrete Math.* 18 (1977), 45–53.

[JuWa84] H. A. Jung and M. E. Watkins, Fragments and automorphisms of infinite graphs, *Europ. J. Combinatorics* 5 (1984), 149–162.

[JuWa89] H. A. Jung and M. E. Watkins, The connectivities of locally finite primitive graphs, *Combinatorica* 9 (1989), 261–267.

[K36] D. König, *Theorie der endlichen und unendichen Graphen*, Akad. Verlagsgesellschaft, Leipzig, 1936.

[No68] L. A. Nowitz, On the non-existence of graphs with transitive generalized dicyclic groups, *J. Combin. Theory* 4 (1968), 49–51.

[NoWa72a] L. A. Nowitz and M. E. Watkins, Graphical regular representations of non-abelian groups, I, *Canad. J. Math.* 14 (1972), 993–1008.

[NoWa72b] L. A. Nowitz and M. E. Watkins, Graphical regular representations of non-abelian groups, II, *Canad. J. Math.* 14 (1972), 1009–1018.

[PoWa95] N. Polat and M. E. Watkins, On translations of double rays in graphs, *Per. Math. Hungar.* 30 (1995), 145–154.

[Sa57] G. Sabidussi, Graphs with given group and given graph-theoretical properties, *Canad. J. Math.* 9 (1957), 515–525.

[Sa60] G. Sabidussi, Graph multiplication, *Math. Zeitschr.* 72 (1960), 446–457.

[Sa61] G. Sabidussi, The lexicographic product of graphs, *Duke Math. J.* 28 (1961), 573–578.

[Sa64] G. Sabidussii, Vertex-transitive graphs, *Monatsh. Math.* 68 (1964), 426–428.

[Sa67] G. Sabidussi, Review #2563, *Math. Rev.* 33, No. 3, March 1967.

[Se91] N. Seifter, Properties of graphs with polynomial growth, *J. Combin. Theory Ser. B* 52 (1991), 222–235.

[SeTr97] N. Seifter and V. I. Trofimov, Automorphism groups of graphs with quadratic growth, *J. Combin. Theory Ser. B* 71 (1997), 205–210.

[Sm10] S. M. Smith, Infinite primitive directed graphs, *J. Algebraic Combin.* 31 (2010), 131–141.

[SmTucWa12] S. M. Smith, T. W. Tucker, and M. E. Watkins, Distinguishability of infinite groups and graphs, *Electron. J. Combin.* 19 (2012), #P27.

[ThWa89] C. Thomassen and M. E. Watkins, Infinite vertex-transitive, edge-transitive, non 1-transitive graphs, *Proc. Amer. Math. Soc.* 105 (1989), 258–261.

[Tr85] V. I. Trofimov, Graphs with polynomial growth, *Math. USSR Sbornik* 51 (1985), No. 2, 404–417.

[Tuc11] T. W. Tucker, Distinguishing maps, *Electron. J. Combin.* 18 (2011), #50.

[Tut66] W. T. Tutte, *Connectivity in Graphs*, University of Toronto Press, Toronto, 1966.

[Wa71] M. E. Watkins, On the action of non-Abelian groups on graphs, *J. Combin. Theory* 11 (1971), 95–104.

[Wa72] M. E. Watkins, On graphical regular representations of $C_n \times Q$, in: *Graph Theory and Its Applications*, (Y. Alavi, D. R. Lick, and A. T. White, eds.) Springer-Verlag, Berlin, 1972, pp. 305–311.

[Wa74] M. E. Watkins, Graphical regular representations of alternating, symmetric, and miscellaneous small groups, *Aequat. Math.* 11 (1974), 40–50.

[Wa76] M. E. Watkins, Graphical regular representations of free products of groups, *J. Combin. Theory* 21 (1976), 47–56.

[Wa86] M. E. Watkins, Infinite paths that contain only shortest paths, *J. Combin. Theory Ser. B* 41 (1986), 341–355.

[Wa91] M. E. Watkins, Edge-transitive strips, *J. Combin. Theory Ser. B* 95 (1991), 350–372.

[WaGra04] M. E. Watkins and J. E. Graver, A characterization of infinite planar primitive graphs, *J. Combin. Theory Ser. B* 91 (2004), 87–104.

[WaZ07] M. E. Watkins and X. Zhou, Distinguishability of locally finite trees, *Electron. J. Combin.* 14 (2007), #R29.

[Weis74] R. M. Weiss, Über *s*-reguläre Graphen, *J. Combin. Theory Ser. B* 16 (1974), 229–233.

[Weic62] P. M. Weichsel, The Kronecker product of graphs, *Proc. Amer. Math. Soc.* 13 (1962), 47–52.

Section 6.2
Cayley Graphs

Brian Alspach, University of Newcastle, Australia

INTRODUCTION

There are frequent occasions for which graphs with a lot of symmetry are required. One such family of graphs is constructed using groups. These graphs are called Cayley graphs and are the subject of this section.

6.2.1   Construction and Recognition

We restrict ourselves to finite graphs, which means we use finite groups, but the basic construction is the same for infinite groups. While Cayley graphs on finite groups and Cayley graphs on infinite groups share a variety of features, there are aspects of Cayley graphs on finite groups that do not carry over to Cayley graphs on infinite groups, and vice versa.

DEFINITIONS

D1: Let $\mathcal{G}$ be a finite group with identity 1. Let $\mathcal{S}$ be a subset of $\mathcal{G}$ satisfying $1 \notin \mathcal{S}$ and $\mathcal{S} = \mathcal{S}^{-1}$; that is, $s \in \mathcal{S}$ if and only if $s^{-1} \in \mathcal{S}$. The **Cayley graph** on $\mathcal{G}$ with **connection set** $\mathcal{S}$, denoted $\mathrm{Cay}(\mathcal{G}; \mathcal{S})$, satisfies these rules:

- the vertices of $\mathrm{Cay}(\mathcal{G}; \mathcal{S})$ are the elements of $\mathcal{G}$;
- there is an edge joining $g, h \in \mathrm{Cay}(\mathcal{G}; \mathcal{S})$ if and only if $h = gs$ for some $s \in \mathcal{S}$.

We note here that it is standard to use additive notation when $\mathcal{G}$ is an abelian group and multiplicative notation for nonabelian groups. Thus, for abelian groups, we have $\mathcal{S} = -\mathcal{S}$ and $g = h + s$. Cayley graphs on the cyclic group $\mathbb{Z}_n$ are called **circulant graphs** and we use the special notation $\mathrm{Circ}(n; \mathcal{S})$.

D2: The set of all Cayley graphs on a group $\mathcal{G}$ is denoted $\mathrm{Cay}(\mathcal{G})$.

EXAMPLES

E1: The hypercube Q_n may be realized as a Cayley graph on the elementary abelian 2-group $\mathbb{Z}_2^n$ using the standard generators $e_1, e_2, \ldots, e_n$ for the connection set, where e_i has a 1 in the i-th coordinate and zeroes elsewhere. We note here that Q_n may be realized in other ways as a Cayley graph, but the realization just given is the common one.

E2: The complete graph K_n is a Cayley graph on any group $\mathcal{G}$ of order n, where the connection set is the set of non-identity elements of the group. We get the complement of K_n by using the empty set as the connection set.

E3: The complete multipartite graph $K_{m;n}$, with m parts each of cardinality n, is realizable as a circulant graph of order mn with the connection set being all the elements not congruent to zero modulo n.

E4: The graph formed on the finite field $GF(q)$, $q \equiv 1 \pmod{4}$, where the connection set is the set of quadratic residues in $GF(q)$, is called a **Paley graph**. Paley graphs have many interesting properties.

E5: The circulant graph of even order n with connection set $S = \{\pm 1, n/2\}$ is known as the **Möbius ladder** of order n.

E6: The Cayley graphs on the group $\mathbb{Z}_\ell^n$, where $\ell \geq 3$ and the connection set is the set of standard generators of the group, are of interest in computer science.

Figure 6.2.1: Two drawings of the Möbius ladder of order 8.

DEFINITIONS

D3: When $\mathcal{G}$ is a finite group and $g \in \mathcal{G}$, define g_L acting on $\mathcal{G}$ by $g_L(h) = gh$ for all $h \in \mathcal{G}$. Clearly, g_L is a permutation of the elements of $\mathcal{G}$. Define the group $\mathcal{G}_L$ by $\mathcal{G}_L = \{g_L : g \in \mathcal{G}\}$.

D4: A bijection f on the vertex set $V(G)$ of a graph G is an ***automorphism of the graph*** G if $\langle u, v \rangle$ is an edge if and only if $\langle f(u), f(v) \rangle$ is an edge.

D5: The set of all automorphisms of a graph G forms a group under function composition and is denoted $\mathrm{Aut}(G)$.

D6: A graph G is said to be ***vertex-transitive*** if $\mathrm{Aut}(G)$ acts transitively on $V(G)$.

D7: Let $\mathcal{G}$ be a transitive permutation group acting on a finite set Ω. If $\mathcal{G}$ satisfies any one of the following three equivalent conditions, then it is said to be a ***regular action***:
- the only element of $\mathcal{G}$ fixing an element of Ω is the identity permutation;
- $|\mathcal{G}| = |\Omega|$;
- for any $\omega_1, \omega_2 \in \Omega$, there is a unique element $g \in \mathcal{G}$ satisfying $\omega_1 g = \omega_2$.

FACTS

F1: Every Cayley graph is vertex-transitive.

F2: (Sabidussi [Sa58]) A graph G is a Cayley graph if and only if $\text{Aut}(G)$ contains a regular subgroup.

REMARK

R1: Sabidussi's Theorem above is the basis for all work on recognizing whether or not an arbitrary graph is a Cayley graph. It is an absolutely fundamental result.

6.2.2 Prevalence

The family of Cayley graphs provides us with a straightforward construction for vertex-transitive graphs. A natural question to pose is whether or not the family of Cayley graphs encompasses all finite vertex-transitive graphs. The Petersen graph is the smallest vertex-transitive graph that is not a Cayley graph, which suggests the topic of this section.

DEFINITION

D8: Let $\mathbb{NC}$ denote the set of integers n for which there exists a non-Cayley vertex-transitive graph of order n.

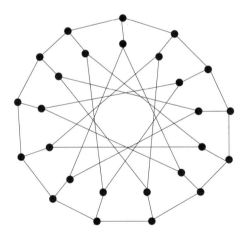

Figure 6.2.2: A non-Cayley vertex-transitive graph of order 26.

EXAMPLE

E7: It is easy to see that if $n \in \mathbb{NC}$, then any multiple of n belongs to $\mathbb{NC}$. This follows by taking the appropriate number of vertex-disjoint copies of a non-Cayley, vertex-transitive graph of order n. Thus, in order to determine $\mathbb{NC}$, it suffices to find the minimal elements belonging to $\mathbb{NC}$.

FACTS

The following results reduce the problem of trying to characterize membership in $\mathbb{NC}$ to the consideration of square-free integers.

F3: A prime power $p^e \in \mathbb{NC}$ if and only $e \geq 4$.

F4: Any positive integer, other than 12, divisible by a square is in $\mathbb{NC}$.

FACTS

F5: Let p and q be distinct primes with $p < q$. Then $pq \in \mathbb{NC}$ if and only if one of the following holds:

- p^2 divides $q - 1$;

- $q = 2p - 1 > 3$ or $q = \frac{p^2+1}{2}$;

- $q = 2^t + 1$ and either p divides $2^t - 1$ or $p = 2^{t-1} - 1$;

- $q = 2^t - 1$ and $p = 2^{t-1} + 1$; and

- $p = 7$ and $q = 11$.

F6: Let p and q be odd primes satisfying $p < q$. Then $2pq \in \mathbb{NC}$ if and only if one of the following holds:

- p^2 divides $q - 1$;

- $p \equiv 1 \pmod 4$ or $q \equiv 1 \pmod 4$;

- $p = 7$, $q = 11$;

- $p \equiv q \equiv 3 \pmod 4$, p divides $q - 1$, and p^2 does not divide $q - 1$;

- $p \equiv q \equiv 3 \pmod 4$, and $p = \frac{q+1}{4}$; and

- $p = 7$ and $q = 19$.

F7: Let p, q, r be distinct odd primes satisfying $p < q < r$. Then $pqr \in \mathbb{NC}$ if and only if at least one of pq, pr or qr is a member of $\mathbb{NC}$, or none of pq, pr and qr is a member of $\mathbb{NC}$ but one of the following holds:

- $pqr = (2^{2^t} + 1)(2^{2^{t+1}} + 1)$, for some t;

- $pqr = (2^{d\pm1} + 1)(2^d - 1)$, for some prime d;

- $pq = 2r \pm 1$ or $pq = (r + 1)/2$;

- $pq = (r^2 + 1)/2$ or $pr = (q^2 + 1)/2$;

- $pq = (r^2 - 1)/24x$ or $pr = (q^2 - 1)/24x$, where $x \in \{1, 2, 5\}$;

- $ab = 2^t + 1$ and c divides $2^t - 1$, where $\{a, b, c\} = \{p, q, r\}$;

- the largest power of p dividing $q - 1$ is p^p and the largest power of q dividing $r - 1$ is q^q;

- $q = (3p + 1)/2$ and $r = 3p + 2$, or $q = 6p - 1$ and $r = 6p + 1$;

- $q = (r - 1)/2$ and p divides $r + 1$, where $p > q$ when $p = (r + 1)/2$;

- $p = (k^{d/2} + 1)/(k + 1)$, $q = (k^{d/2} - 1)/(k - 1)$, $r = (k^{d-1} - 1)/(k - 1)$, where $k, d - 1, d/2$ are primes and $p > q$ may be the case;

- $p = (k^{(d-1)/2} + 1)/(k + 1)$, $(k^{(k-1)/2} - 1)/(k - 1)$, $r = (k^d - 1)/(k - 1)$, where $k, d, (d - 1)/2$ are primes and $p > q$ may be the case;

- $p = k^2 - k + 1$, $q = (k^5 - 1)/(k - 1)$, $r = (k^7 - 1)/(k - 1)$, where k is prime;

- $p = 3$, $q = (2^d + 1)/3$, $r = 2^d - 1$, where d is a prime;

- $p = (2^d + 1)/3$, $q = 2^d - 1$, $r = 2^{2d \pm 2} + 1$, where $d = 2^t \pm 1$ is prime;

- $p = 5$, $q = 11$ and $r = 19$; and

- $p = 7$, $q = 73$ and $r = 257$.

RESEARCH PROBLEM

RP1: Is there a number $k > 0$ such that every product of k distinct primes is in $\mathbb{NC}$? No characterization of the members of $\mathbb{NC}$ that are products of four distinct primes is known.

6.2.3 Isomorphism

Some of the most interesting and deepest work on Cayley graphs has revolved around the question of trying to determine when two Cayley graphs are isomorphic.

DEFINITIONS

D9: A Cayley graph $\mathrm{Cay}(\mathcal{G}; \mathcal{S})$ on $\mathcal{G}$ is called a **CI-graph**, if $\mathrm{Cay}(\mathcal{G}; \mathcal{S}) \cong \mathrm{Cay}(\mathcal{G}; \mathcal{S}')$ implies there exists an $\alpha \in \mathrm{Aut}(\mathcal{G})$ satisfying $\mathcal{S}' = \alpha(\mathcal{S})$.

D10: We say that a group $\mathcal{G}$ is a **CI-group** if every Cayley graph on $\mathcal{G}$ is a CI-graph.

EXAMPLE

E8: For $n = 25$, let

$$\mathcal{S} = \{1, 4, 5, 6, 9, 11, 14, 16, 19, 20, 21, 24\}$$

and
$$\mathcal{S}' = \{1, 4, 6, 9, 10, 11, 14, 15, 16, 19, 21, 24\}.$$

The two circulant graphs $\mathrm{Circ}(25; \mathcal{S})$ and $\mathrm{Circ}(25; \mathcal{S}')$ are isomorphic since both are wreath products of a 5-cycle with a 5-cycle. On the other hand, it is easy to see there is no $a \in \mathbb{Z}_{25}^*$ for which $\mathcal{S}' = a\mathcal{S}$ is satisfied. Thus, $\mathbb{Z}_{25}$ is not a CI-group.

FACTS

F8: Let G be a Cayley graph on the finite group $\mathcal{G}$. Then G is a CI-graph if and only if all regular subgroups of $\text{Aut}(G)$ isomorphic to G are conjugate in $\text{Aut}(G)$.

F9: [Li99] If $\mathcal{G}$ is a CI-group, then $\mathcal{G}$ is solvable.

F10: [Mu97] The cyclic group $\mathbb{Z}_n$ is a CI-group if and only if $n = 2^e m$, where m is odd and square-free and $e \in \{0, 1, 2\}$, or $n \in \{8, 9, 18\}$.

F11: Let p a prime. The elementary abelian p-groups $\mathbb{Z}_p^e$ are CI-groups for $1 \leq e \leq 4$. On the other hand, for $e \geq 2p - 1 + \binom{2p-1}{p}$, $\mathbb{Z}_p^e$ is not a CI-group. The latter inequality is sharp for $p = 2$.

RESEARCH PROBLEM

RP2: For an odd prime p, determine the values of e for which $\mathbb{Z}_p^e$ is a CI-group.

REMARK

R2: One motivation for classifying CI-groups is that the Cayley graphs on a CI-group may be enumerated in a straightforward way using Pólya's enumeration theorem [Br64]. The next two theorems illustrate this for circulant graphs.

DEFINITIONS

D11: Let $\mathbb{Z}_n^*$ denote the multiplicative group of units in the ring $\mathbb{Z}_n$.

D12: Let $n = p_1^{e_1} p_2^{e_2} \cdots p_t^{e_t}$ be the factorization of n as a product of distinct prime powers. For any m satisfying $0 \leq m < n$, let $m(n) = (m_1, m_2, \ldots, m_t)$ be the t-tuple satisfying $m_i \equiv m \pmod{p_i^{e_i}}$ and $0 \leq m_i < p_i^{e_i}$, $i = 1, 2, \ldots t$.

D13: Let $\mathcal{T}(n) = \{m(n) : 0 \leq m < n\}$. For a subset $R \subseteq \{1, 2, \ldots, t\}$, let $\mathcal{T}_R(n)$ be the elements of $\mathcal{T}(n)$ for which $m_i \neq 0$ if and only if $i \in T$. When the value n under discussion is clear, we use the notation $\mathcal{T}$ and $\mathcal{T}_R$.

D14: If $a = (a_1, a_2, \ldots, a_t) \in \mathbb{Z}_n^*$, then the **order-type** of a is the t-tuple $(d_1, d_2, \ldots, d_t)$, where d_i is the order of a_i in $\mathbb{Z}_{p_i^e}^*$. Note that $a \in \mathbb{Z}_n^*$ implies $a_i \neq 0$, $i = 1, 2, \ldots, t$.

D15: Let $(d_1, d_2, \ldots, d_t)$ be the order-type of some $a \in \mathbb{Z}_n^*$, and let $R \subseteq \{1, 2, \ldots, t\}$. If each d_i, $i \in R$, has the form $d_i = 2^e b$, where $e \geq 1$ and b is odd, then let $\text{lcm}^*(R) = \text{lcm}(R)/2$. In all other cases, let $\text{lcm}^*(R) = \text{lcm}(R)$, where $\text{lcm}(R)$ denotes the least common multiple of the d_i terms in the coordinates corresponding to the elements of R.

FACTS

F12: If p is an odd prime, then the number of isomorphism classes of vertex-transitive graphs of order p is

$$\frac{2}{p-1} \sum_d \Phi(d) 2^{(p-1)/2d},$$

where the summation runs over all divisors d of $(p-1)/2$ and Φ denotes the Euler totient function.

F13: If $n = p_1 p_2 \cdots p_t$ is a product of distinct primes, or if $p_1 = 4$ and $p_2, p_3, \ldots, p_t$ are distinct odd primes, then the number of isomorphism classes of circulant graphs of order n is

$$\frac{1}{\Phi(n)} \sum_{(d_1, d_2, \ldots, d_t)} \Phi(d_1)\Phi(d_2) \cdots \Phi(d_t) \prod_R 2^{|\mathcal{T}_R|/2\mathrm{lcm}^*(R)},$$

where the sum is taken over all possible order types of $a \in \mathbb{Z}_n^*$, and the product is taken over all non-empty subsets R of $\{1, 2, \ldots, t\}$ unless $p_1 = 4$, in which case $R = \emptyset$ is included and $|\mathcal{T}_\emptyset|$ is taken to be 2.

EXAMPLE

E9: We illustrate the preceding theorem for $n = 20$. The possible order-types are

$$(1,2), \ (1,2), \ (1,4), \ (2,1), \ (2,2), \ \text{and} \ (2,4)$$

There are, for example, two automorphisms of order type $(1,4)$. Since $p_1 = 4$, the term corresponding to $R = \emptyset$ appears in the product, and the contribution is 2 since $|\mathcal{T}_\emptyset| = 2$ and we consider $\mathrm{lcm}^*(\emptyset) = 1$. For $R = \{1\}$, we have $\mathrm{lcm}^*(R) = 1$ and $|\mathcal{T}_R| = 2$ making a contribution of 2 to the product. For $R = \{2\}$, we have $\mathrm{lcm}^*(R) = 2$ and $|\mathcal{T}_R| = 8$, thereby contributing 2^2. Finally, when $R = \{1, 2\}$, we have $\mathrm{lcm}^*(R) = 4$ and $|\mathcal{T}_R| = 8$ so the contribution is 2. Altogether the term in the product is 2^5. We do the same thing for all possible order-types and find that there are 336 non-isomorphic circulant graphs of order 20.

6.2.4 Subgraphs

There are interesting results and questions regarding subgraphs of Cayley graphs. Some of the results we mention hold for all vertex-transitive graphs and we state them accordingly. It is not always clear just what impact vertex-transitivity has on the existence of certain subgraphs.

DEFINITIONS

D16: A connection set $\mathcal{S}$ is said to be a ***quasi-minimal connection set*** if the elements of $\mathcal{S}$ can be ordered $s_1, s_2, \ldots, s_t$ so that
- if $|s_i| > 2$, then s_i^{-1} is either s_{i-1} or s_{i+1}, and
- if $\mathcal{S}_i$ denotes the set $\{s_1, s_2, \ldots, s_i\}$, then for each i such that $|s_i| = 2$, $\langle \mathcal{S}_i \rangle$ is a proper supergroup of $\langle \mathcal{S}_{i-1} \rangle$, and for each i such that $|s_i| > 2$ and $s_i^{-1} = s_{i-1}$, $\langle \mathcal{S}_i \rangle$ is a proper supergroup of $\langle \mathcal{S}_{i-2} \rangle$.

FACTS

F14: A graph G is said to be a ***Hamilton-connected graph*** if for any two vertices u, v of G, there is a Hamilton path whose terminal vertices are u and v. A bipartite graph G with parts A and B is said to be a ***Hamilton-laceable graph*** if for any $u \in A$ and $v \in B$, there is a Hamilton path whose terminal vertices are u and v.

F15: Let G be a connected vertex-transitive graph. If G has even order, then it has a 1-factor. If G has odd order, then $G - v$ has a 1-factor for every vertex $v \in G$.

F16: If G is a connected vertex-transitive graph of degree d, then G is d-edge-connected.

F17: [Ma71, Wa70] If G is a connected vertex-transitive graph of degree d, then γ/d, where γ denotes the vertex connectivity of G, strictly exceeds $\frac{2}{3}$. Furthermore, for every $\epsilon > 0$, there exists a connected vertex-transitive graph H of some degree d for which $\gamma/d < \epsilon + \frac{2}{3}$.

F18: If $\mathcal{S}$ is a quasi-minimal generating set of the group $\mathcal{G}$, then the Cayley graph $\text{Cay}(\mathcal{G}; \mathcal{S})$ has vertex connectivity $|\mathcal{S}|$.

F19: For every positive integer m, there exists a Paley graph containing all graphs of order m as induced subgraphs.

F20: Let G be a connected Cayley graph on a finite abelian group. If G is bipartite and has degree at least 3, then G is Hamilton-laceable. If G is not bipartite and has degree at least 3, then G is Hamilton-connected.

F21: [Wi84] Every connected Cayley graph on a group of order p^e, p a prime and $e \geq 1$, has a Hamilton cycle.

F22: [KMMMS12] Let p, q, r be distinct primes. A connected Cayley graph of any of the following orders is hamiltonian: kp with $k < 32$ and $k \neq 24$, kpq with $k \leq 5$, kp^2 with $k \leq 4$, pqr, p^3 and $2p^3$.

6.2.5 Factorization

DEFINITIONS

D17: A **1-factorization** of a graph is a partition of the edge set into 1-factors.

D18: We say the connection set $\mathcal{S}$ is a **minimal generating Cayley set** for $\mathcal{G}$ if $\mathcal{S}$ generates $\mathcal{G}$, but $\mathcal{S} - \{s, s^{-1}\}$ generates a proper subgroup for every $s \in \mathcal{S}$.

D19: A **Hamilton decomposition of a graph** G is a partition of the edge set into Hamilton cycles when the degree is even, or a partition into Hamilton cycles and a 1-factor when the degree is odd.

D20: An **isomorphic factorization of a graph** G is a partition of the edge set of G so that the subgraphs induced by the edges in each part are pairwise isomorphic.

FACTS

F23: [St85] Every connected Cayley graph on the group $\mathcal{G}$ has a 1-factorization if one of the following holds:
- $|\mathcal{G}| = 2^k$ for an integer k;
- $\mathcal{G}$ is an even order abelian group; or
- $\mathcal{G}$ is dihedral or dicyclic.

F24: Cayley graphs whose connection sets are minimal generating Cayley sets have 1-factorizations whenever the group is one of the following:
- an even order nilpotent group;
- the group contains a proper abelian normal subgroup of index 2^k; or
- the group has order $2^m p^k$ for a prime p satisfying $p > 2^m$.

F25: A cubic Cayley graph G whose automorphism group has a solvable subgroup acting transitively on the vertex set of G has a 1-factorization.

F26: [Li96, Li13] If $G = \text{Cay}(\mathcal{G}, \mathcal{S})$ is a connected Cayley graph on an abelian group $\mathcal{G}$, and if $\mathcal{S}$ is a minimal generating Cayley set, then G has a Hamilton decomposition with one small exception unsettled. Namely, when $|\mathcal{G}|$ is even, for each $s \in \mathcal{S}$, we must have that $2s$ does not lie in the subgroup generated by $\mathcal{S} - \{\pm s\}$.

F27: [Fi90] If T is any tree with n edges, then the n-dimensional cube Q_n has an isomorphic factorization by T. Furthermore, there is an isomorphic factorization so that each copy of T is an induced subgraph.

RESEARCH PROBLEM

RP3: Let $\mathcal{C}$ be one of the classes of circulant graphs, or Cayley graphs, or vertex-transitive graphs. Is it the case that for every $G \in \mathcal{C}$, whenever d divides $|E(G)|$, then there is an isomorphic factorization of G into d subgraphs?

6.2.6 Miscellaneous

Space limitations preclude discussion of several topics and we mention them briefly. We also include recommended further reading.

EMBEDDINGS

There is a long history and an extensive literature about embedding graphs on orientable and non-orientable surfaces. See Chapter 7 in this volume. The books [GrTu87, Ri74, Wh01] and a recent excellent survey [RSJTW05] provide a good starting point for this topic.

APPLICATIONS

There are a variety of meaningful applications of Cayley graphs and we mention only three. Circulant graphs appear in the study of circular chromatic number. For a recent survey see [Zh01].

Cayley graphs occur frequently in the literature on networks. A recent book on this topic is [Xu01] and a fundamental paper is [AkKr89].

Cayley graphs play a central role in the work on expanders. Two excellent references are [Al95, Lu95].

FURTHER READING

A survey on Cayley graph isomorphism is provided in [Li02]. A good general discussion about vertex-transitive graphs and Cayley graphs is [Ba95]. A good starting point for reading about $\mathbb{NC}$ is [IrPr01].

References

[AkKr89] S. Akers and B. Krishnamurthy, A group-theoretic model for symmetric interconnection networks, *IEEE Trans. Comput.* **38** (1989), 555–566.

[Al95] N. Alon, Tools from higher algebra, in *Handbook of Combinatorics Vol II*, eds. R. L. Graham, M. Grötschel and L. Lovász, MIT Press and North-Holland, 1995, 1749–1783.

[Ba95] L. Babai, Automorphism groups, isomorphism, reconstruction, in *Handbook of Combinatorics Vol II*, eds. R. L. Graham, M. Grötschel and L. Lovász, MIT Press and North-Holland, 1995, 1447–1540.

[Br64] N. G. de Bruijn, Pólya's theory of counting, Ch. 5 of *Applied Combinatorial Mathematics*, Wiley, New York, 1964.

[ChQu81] C. C. Chen and N. Quimpo, On strongly hamiltonian abelian group graphs, Combinatorial Mathematics VIII, *Lecture Notes in Mathematics* **884**, Springer-Verlag, 1981, Berlin, 23–34.

[Fi90] J. Fink, On the decomposition of n-cubes into isomorphic trees, *J. Graph Theory* **14** (1990), 405–411.

[GrTu87] J. L. Gross and T. W. Tucker, *Topological Graph Theory*, Wiley, New York, 1987.

[IrPr01] M. Iranmanesh and C. E. Praeger, On non-Cayley vertex-transitive graphs of order a product of three primes, *J. Combin. Theory Ser. B* **81** (2001), 1–19.

[KMMMS12] K. Kutnar, D. Marušič, D. Witte Morris, J. Morris, and P. Šparl: Hamiltonian cycles in Cayley graphs whose order has few prime factors, *Ars Math. Contemp.* **5** (2012), 27–71.

[Li99] C. H. Li, Finite CI-groups are soluble, *Bull. London Math. Soc.* **31** (1999), 419–423.

[Li02] C. H. Li, On isomorphisms of finite Cayley graphs — a survey, *Discrete Math.* **256** (2002), 301–334.

[Li96] J. Liu, Hamiltonian decompositions of Cayley graphs on abelian groups of odd order, *J. Combin. Theory Ser. B*, **66** (1996), 75–86.

[Li13] J. Liu, Hamiltonian decompositions of Cayley graphs on abelian groups of even order, *J. Combin. Theory Ser. B*, to appear.

[Lu95] A. Lubotzky, Cayley graphs: eigenvalues, expanders and random walks, *London Math. Soc. Lecture Note Ser.* **218** (1995), 155–189.

[Ma71] W. Mader, Eine Eigenschaft der Atome endlicher Graphen, *Arch. Math. (Basel)* **22** (1971), 333–336.

[Mu97] M. Muzychuk, On Ádám's conjecture for circulant graphs, *Discrete Math.* **167** (1997), 497–510.

[RSJTW05] R. B. Richter, J. Širáň, R. Jajcay, T. Tucker and M. Watkins, Cayley maps, *J. Combin. Theory Ser. B* **95** (2005), 189–245.

[Ri74] G. Ringel, *Map Color Theorem*, Springer-Verlag, New York, 1974.

[Sa58] G. Sabidussi, On a class of fixed-point-free graphs, *Proc. Amer. Math. Soc.* **9** (1958), 800–804.

[St85] R. Stong, On 1-factorizability of Cayley graphs, *J. Combin. Theory Ser. B* **39** (1985), 298–307.

[Wa70] M. Watkins, Connectivity of transitive graphs, *J. Combin. Theory* **8** (1970), 23–29.

[Wh01] A. White, *Graphs of Groups on Surfaces*, Mathematics Studies **188**, North-Holland, Amsterdam, 2001.

[Wi84] D. Witte, Cayley digraphs of prime-power order are hamiltonian, *J. Combin. Theory Ser. B* **40** (1984), 107–112.

[Xu01] J. Xu, *Topological Structure and Analysis of Interconnection Networks*, Kluwer, Dordrecht, 2001.

[Zh01] X. Zhu, Circular chromatic number: a survey, *Discrete Math.* **229** (2001), 371–410.

Section 6.3

Enumeration

Paul K. Stockmeyer, The College of William and Mary

INTRODUCTION

It is often important to know how many graphs there are with some desired property. Computer scientists can use such numbers in analyzing the time or space requirements of their algorithms, and chemists can make use of these numbers in organizing and cataloging lists of chemical molecules with various shapes. Indeed, any time that graphs are used to model some form of physical structure, the techniques of graphical enumeration can be extremely valuable.

Many of the techniques for counting graphs are based on the master theorem in the historic 1937 work of George Pólya. See [PoRe87] for an English translation. Frank Harary [Ha55] and others exploited this master theorem in counting simple graphs, multigraphs, digraphs, and similar graphical structures.

Tree counting began with Arthur Cayley [Ca57, Ca89], who was the first to use the word "tree" for these structures. Methods for counting trees representing chemical compounds were developed by Blair and Henze [BlHe31a, BlHe31b]. Generic tree counting methods were advanced by Pólya [PoRe87], Richard Otter [Ot48], Harary and Prins [HaPr59] and many others.

An exhaustive survey of results in graphical enumeration, far beyond what can be included here, can be found in [HaPa73]. Alternatively, if you know the first few terms of a graph-counting sequence, you can quite likely find more terms, references, and further information in the *On-Line Encyclopedia of Integer Sequences* [OEIS].

6.3.1   Counting Simple Graphs and Multigraphs

When counting graphs it is important to distinguish between the enumeration of labeled graphs and that of unlabeled graphs. Labeled graphs are relatively easy to count, usually requiring only factorials, exponentials, and binomial coefficients. Unlabeled graphs require rather sophisticated counting techniques, often utilizing permutation group theory and generating functions.

Labeled Graphs

DEFINITION

D1: A *labeled graph* is a graph with distinct labels, typically $v_1, v_2, \ldots, v_n$, assigned to its vertices. Two labeled graphs with the same set of labels are considered the same only if there is an isomorphism from one to the other that preserves the labels.

EXAMPLES

E1: Figure 6.3.1 shows the three isomorphically distinct simple graphs with 4 vertices and 3 edges. There are 4 essentially different ways to label each of the first two and 12 ways to label the third. Thus there are 20 different labeled simple graphs with 4 vertices and 3 edges. Only the last two graphs shown are connected.

Figure 6.3.1: Simple graphs with 4 vertices and 3 edges.

E2: Figure 6.3.2 shows the three isomorphically distinct loopless multigraphs that together with the graphs in Figure 6.3.1 form the six different multigraphs with 4 vertices and 3 edges. There are 6 essentially different ways to label the first and third graphs in Figure 6.3.2 and 24 ways to label the middle graph. Thus the graphs in Figures 6.3.1 and 6.3.2 represent the total of 56 labeled loopless multigraphs with 4 vertices and 3 edges.

Figure 6.3.2: Additional loopless multigraphs with 4 vertices and 3 edges.

FACTS

F1: The number of labeled simple graphs with n vertices and m edges is the binomial coefficient $\binom{\binom{n}{2}}{m}$. These numbers form sequence A084546 in [OEIS]. See Table 6.3.1.

F2: For $m > \binom{n}{2}/2$, the number of labeled simple graphs with n vertices and m edges is the same as the number of labeled simple graphs with n vertices and $\binom{n}{2} - m$ edges.

F3: The total number of labeled simple graphs with n vertices is $2^{\binom{n}{2}}$. This is sequence A006125 in [OEIS]. See Table 6.3.1.

$m \backslash n$	1	2	3	4	5	6	7	8
0	1	1	1	1	1	1	1	1
1		1	3	6	10	15	21	28
2			3	15	45	105	210	378
3			1	20	120	455	1,330	3,276
4				15	210	1,365	5,985	20,475
5				6	252	3,003	20,349	98,280
6				1	210	5,005	54,264	376,740
7					120	6,435	116,280	1,184,040
8					45	6,435	203,490	3,108,105
9					10	5,005	293,930	6,906,900
10					1	3,003	352,716	13,123,110
11						1,365	352,716	21,474,180
12						455	293,930	30,421,755
13						105	203,490	37,442,160
14						15	116,280	40,116,600
Total	1	2	8	64	1,024	32,768	2,097,152	268,435,456

Table 6.3.1: Labeled simple graphs with n vertices and m edges.

F4: [Gi56] The number $\widehat{K}_n$ of connected labeled simple graphs with n vertices can be determined from the recursive formula

$$\widehat{K}_1 = 1, \quad \text{and} \quad \widehat{K}_n = 2^{\binom{n}{2}} - \frac{1}{n} \sum_{k=1}^{n-1} k \binom{n}{k} 2^{\binom{n-k}{2}} \widehat{K}_k \quad \text{for } n > 1.$$

This is sequence A001187 in [OEIS]. See Table 6.3.2.

n	1	2	3	4	5	6	7	8
$\widehat{K}_n$	1	1	4	38	728	26,704	1,866,256	251,548,592

Table 6.3.2: Connected labeled simple graphs with n vertices.

F5: Asymptotically, most labeled simple graphs are connected. Thus the sequence $\widehat{K}_n$ satisfies

$$\widehat{K}_n \sim 2^{\binom{n}{2}}.$$

F6: The number of labeled loopless multigraphs with n vertices and m edges is the binomial coefficient $\binom{m+\binom{n}{2}-1}{m}$. When $n = 1$ this expression should be interpreted as 1 when $m = 0$ and 0 otherwise. See Table 6.3.3. These numbers form sequence A098568 in [OEIS].

$m \backslash n$	1	2	3	4	5	6	7	8
0	1	1	1	1	1	1	1	1
1		1	3	6	10	15	21	28
2		1	6	21	55	120	231	406
3		1	10	56	220	680	1,771	4,060
4		1	15	126	715	3,060	10,626	31,465
5		1	21	252	2,002	11,628	53,130	201,376
6		1	28	462	5,005	38,760	230,230	1,107,568

Table 6.3.3: Labeled loopless multigraphs with n vertices and m edges.

Unlabeled Graphs

DEFINITIONS

D2: The **symmetric group** S_n is the group of all $n!$ permutations γ acting on the set $X_n = \{1, 2, \ldots, n\}$.

D3: The **order** of a permutation group is the number of permutations it contains. The **degree** of a permutation group is the number of objects being permuted. The symmetric group S_n has order $n!$ and degree n.

D4: The **cycle index** $Z(G)$ of a permutation group G of order m and degree d is a polynomial in variables $a_1, a_2, \ldots, a_d$ given by the formula

$$Z(G) = \frac{1}{m} \sum_{\gamma \in G} \prod_{k=1}^{d} a_k^{j_k(\gamma)},$$

where $j_k(\gamma)$ is the number of cycles of length k in the permutation γ. For example, for $G = S_3 = \{(1)(2)(3),\ (123),\ (132),\ (1)(23),\ (2)(13),\ (3)(12)\}$, the symmetric group of order 6 and degree 3, the cycle index is

$$Z(G_3) = \frac{1}{6} \left(a_1^3 + 2a_3 + 3a_1 a_2 \right).$$

D5: The **pair permutation** $\gamma^{(2)}$ induced by the permutation γ acting on the set X_n is the permutation acting on unordered pairs of distinct elements of X_n defined by the rule

$$\gamma^{(2)} \left(\{x_1, x_2\} \right) = \{\gamma(x_1), \gamma(x_2)\}.$$

D6: The **symmetric pair group** $S_n^{(2)}$ induced by the symmetric group S_n is the permutation group $\{\gamma^{(2)} \mid \gamma \in S_n\}$. This group, used in counting graphs, has order $n!$ and degree $n(n-1)/2$.

FACTS

F7: The *cycle index* $Z(S_n^{(2)})$ of the symmetric pair group, used in counting graphs with n vertices, is

$$Z(S_n^{(2)}) = \frac{1}{n!} \sum_{(j)} \frac{n!}{\prod_k k^{j_k} j_k!} \prod_k a_k^{k\binom{j_k}{2}} (a_k a_{2k}^{k-1})^{j_{2k}} a_{2k+1}^{k j_{2k+1}} \prod_{r<s} a_{\mathrm{lcm}(r,s)}^{\gcd(r,s) j_r j_s},$$

where $\mathrm{lcm}(r,s)$ and $\gcd(r,s)$ are the least common multiple and greatest common divisor of r and s, respectively. The sum is taken over all partitions $(j) = j_1, j_2, \ldots, j_n$ of the integer n as an unordered sum of parts, where j_k is the number of parts of size k. For example, in the partition of 7 as $2 + 2 + 3$ we have $j_2 = 2$, $j_3 = 1$, and $j_1 = j_4 = j_5 = j_6 = j_7 = 0$. Explicit formulas for $Z(S_n^{(2)})$ for small values of n are:

$$Z(S_1^{(2)}) = 1$$

$$Z(S_2^{(2)}) = a_1$$

$$Z(S_3^{(2)}) = \frac{1}{3!}(a_1^3 + 3a_1 a_2 + 2a_3)$$

$$Z(S_4^{(2)}) = \frac{1}{4!}(a_1^6 + 9a_1^2 a_2^2 + 8a_3^2 + 6a_2 a_4)$$

$$Z(S_5^{(2)}) = \frac{1}{5!}(a_1^{10} + 10a_1^4 a_2^3 + 20a_1 a_3^3 + 15a_1^2 a_2^4 + 30a_2 a_4^2 + 20a_1 a_3 a_6 + 24a_5^2)$$

$$Z(S_6^{(2)}) = \frac{1}{6!}(a_1^{15} + 15a_1^7 a_2^4 + 40a_1^3 a_3^4 + 60a_1^3 a_2^6 + 180a_1 a_2 a_4^3 + 120a_1 a_2 a_3^2 a_6$$
$$+ 144a_5^3 + 40a_3^5 + 120a_3 a_6^2)$$

F8: [Ha55, PoRe87] Let $G_{n,m}$ denote the number of simple graphs with n vertices and m edges, and let $g_n(x)$ be the generating function for n-vertex simple graphs, so that

$$g_n(x) = \sum_{m=0}^{\binom{n}{2}} G_{n,m} x^m.$$

Pólya's enumeration theorem states that this generating function $g_n(x)$ can be obtained from the cycle index $Z(S_n^{(2)})$ by replacing each variable a_i with $1 + x^i$. See Table 6.3.4. These numbers form sequence A008406 in [OEIS].

F9: For $m > \binom{n}{2}/2$, the number of simple graphs with n vertices and m edges is the same as the number of simple graphs with n vertices and $\binom{n}{2} - m$ edges.

F10: The total number G_n of simple graphs with n vertices is obtained from the cycle index $Z(S_n^{(2)})$ by replacing each variable a_i with the number 2. See Table 6.3.4. This is sequence A000088 in [OEIS].

F11: Asymptotically, the sequence G_n satisfies $G_n \sim 2^{\binom{n}{2}}/n!$.

$m \backslash n$	1	2	3	4	5	6	7	8
0	1	1	1	1	1	1	1	1
1		1	1	1	1	1	1	1
2			1	2	2	2	2	2
3			1	3	4	5	5	5
4				2	6	9	10	11
5				1	6	15	21	24
6				1	6	21	41	56
7					4	24	65	115
8					2	24	97	221
9					1	21	131	402
10					1	15	148	663
11						9	148	980
12						5	131	1,312
13						2	97	1,557
14						1	65	1,646
Total	1	2	4	11	34	156	1,044	12,346

Table 6.3.4: Simple graphs with n vertices and m edges.

F12: [Ca71] The enumeration of connected simple graphs requires an auxiliary sequence A_n defined recursively by

$$A_1 = 1, \quad \text{and} \quad A_n = nG_n - \sum_{k=1}^{n-1} A_k \cdot G_{n-k} \quad \text{for } n > 1.$$

This sequence 1, 3 ,7, 27, 106, 681, 5972, 88963, ... is sequence A003083 in [OEIS]. The number K_n of connected simple graphs with n vertices can then be computed as

$$K_n = \frac{1}{n} \sum_{d|n} \mu(d) A_{n/d},$$

where the sum is over all divisors of n and μ is the Möbius function defined by

$$\mu(n) = \begin{cases} 1 & \text{if } n = 0 \\ 0 & \text{if } m^2 | n \text{ for some } m > 1 \\ (-1)^k & \text{if } n \text{ is the product of } k \text{ distinct primes.} \end{cases}$$

See Table 6.3.5. The sequence K_n is sequence A001349 in [OEIS].

n	1	2	3	4	5	6	7	8
K_n	1	1	2	6	21	112	853	11,117

Table 6.3.5: Connected simple graphs with n vertices.

F13: Asymptotically, most simple graphs are connected. Thus the sequence K_n satisfies $K_n \sim 2^{\binom{n}{2}}/n!$.

F14: [Ha55, PoRe87] Let $M_{n,k}$ denote the number of loopless multigraphs with n vertices and k edges, and let $m_n(x)$ be the generating function for n-vertex loopless multigraphs, so that

$$m_n(x) = \sum_{m=0}^{\binom{n}{2}} M_{n,k} x^k.$$

Pólya's enumeration theorem states that this generating function $m_n(x)$ can be obtained from the cycle index $Z(S_n^{(2)})$ by replacing each variable a_i with the infinite series $1 + x^i + x^{2i} + x^{3i} + \cdots$. See Table 6.3.6. Column $n = 3$ is sequence A001399 in [OEIS]. Column $n = 4$ is sequence A003082; column $n = 5$ is sequence A014395; and column $n = 6$ is sequence A014396.

$m \setminus n$	1	2	3	4	5	6
0	1	1	1	1	1	1
1		1	1	1	1	1
2		1	2	3	3	3
3		1	3	6	7	8
4		1	4	11	17	21
5		1	5	18	35	52
6		1	7	32	76	132
7		1	8	48	149	313
8		1	10	75	291	741
9		1	12	111	539	1,684
10		1	14	160	974	3,711

Table 6.3.6: Loopless multigraphs with n vertices and m edges.

6.3.2 Counting Digraphs and Tournaments

Labeled Digraphs

DEFINITIONS

D7: A *labeled digraph* is a digraph with distinct labels, typically $v_1, v_2, \ldots, v_n$, assigned to its vertices. Two labeled digraphs with the same set of labels are considered the same only if there is an isomorphism from one to the other that preserves the labels.

D8: A *tournament* (or *round-robin tournament*) is a digraph in which, for each pair u, v of distinct vertices, either there exists an arc from u to v or an arc from v to u but not both.

D9: A digraph is *strong* (or *strongly connected*) if for each pair u, v of vertices, there exist directed paths from u to v and from v to u. A strong tournament is also called an *irreducible* tournament.

EXAMPLES

E3: Figure 6.3.3 shows the four isomorphically distinct simple digraphs with 3 vertices and 3 arcs. The last two are tournaments. There are 6 essentially different ways to label each of the first three digraphs and 2 ways to label the fourth. Thus there are 20 different labeled simple digraphs with 3 vertices and 3 arcs. Only the last digraph is strong—an irreducible tournament.

Figure 6.3.3: The four simple digraphs with 3 vertices and 4 arcs.

E4: Figure 6.3.4 shows the four isomorphically distinct tournaments with 4 vertices. There are 24 essentially different ways to label the first and last tournaments, and 8 ways to label each of the middle two. Thus there are 64 different labeled tournaments with 4 vertices. Only the last tournament is strong.

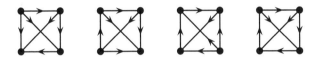

Figure 6.3.4: The four tournaments with 4 vertices.

FACTS

F15: The number of labeled simple digraphs with n vertices and m arcs is the binomial coefficient $\binom{n(n-1)}{m}$. See Table 6.3.7. These numbers form sequence A123554 in [OEIS].

$m \backslash n$	1	2	3	4	5
0	1	1	1	1	1
1		2	6	12	20
2		1	15	66	190
3			20	220	1,140
4			15	495	4,845
5			6	792	15,504
6			1	924	38,760
7				792	77,520
8				495	125,970
9				220	167,960
10				66	184,756
Total	1	4	64	4,096	1,048,576

Table 6.3.7: Labeled simple digraphs with n vertices and m arcs.

F16: For $m > n(n-1)/2$, the number of labeled simple digraphs with n vertices and m arcs is the same as the number of labeled simple digraphs with n vertices and $n(n-1) - m$ arcs.

F17: The total number of labeled simple digraphs with n vertices is $2^{n(n-1)}$. See Table 6.3.7. This is sequence A053763 in [OEIS].

F18: The number of labeled tournaments with n vertices is $2^{\binom{n}{2}}$, the same as the number of labeled simple graphs with n vertices. See Table 6.3.8. This is sequence A006125 in [OEIS].

F19: [MoMo62] The number $\widehat{S_n}$ of strong labeled tournaments with n vertices can be computed from the recursive formula

$$\widehat{S_1} = 1, \quad \text{and} \quad \widehat{S_n} = 2^{\binom{n}{2}} - \sum_{k=1}^{n-1} \binom{n}{k} 2^{\binom{n-k}{2}} \widehat{S_k} \quad \text{for } n > 1.$$

See Table 6.3.8. This is sequence A054946 in [OEIS].

n	Labeled Tournaments	Strong Labeled Tournaments
1	1	1
2	2	0
3	8	2
4	64	24
5	1,024	544
6	32,768	22,320
7	2,097,152	1,677,488
8	268,435,456	236.522,496

Table 6.3.8: Labeled tournaments and strong labeled tournaments with n vertices.

F20: Asymptotically, most labeled tournaments are strong. Thus the sequence $\widehat{S_n}$ counting strong labeled tournaments satisfies $\widehat{S_n} \sim 2^{\binom{n}{2}}$.

Unlabeled Digraphs

DEFINITIONS

D10: The **ordered pair permutation** $\gamma^{[2]}$ induced by the permutation γ acting on the set $X_n = \{1, 2, \ldots, n\}$ is the permutation acting on ordered pairs of distinct elements of X_n defined by the rule

$$\gamma^{(2)}((x_1, x_2)) = (\gamma(x_1), \gamma(x_2)).$$

D11: The **reduced ordered pair group** $S_n^{[2]}$ induced by the symmetric group S_n is the permutation group $\{\gamma^{[2]} \mid \gamma \in S_n\}$. This group, used in counting digraphs, has order $n!$ and degree $n(n-1)$.

FACTS

F21: The *cycle index* $Z(S_n^{[2]})$ of the reduced ordered pair group, used in counting digraphs with n vertices, is

$$Z(S_n^{[2]}) = \frac{1}{n!} \sum_{(j)} \frac{n!}{\prod_k k^{j_k} j_k!} \prod_k a_k^{(k-1)j_k + 2k\binom{j_k}{2}} \prod_{r<s} a_{\text{lcm}(r,s)}^{2 \gcd(r,s) j_r j_s},$$

where $\text{lcm}(r,s)$ and $\gcd(r,s)$ are the least common multiple and greatest common divisor of r and s, respectively. The sum is taken over all partitions $(j) = j_1, j_2, \ldots, j_n$ of the integer n as an unordered sum of parts, where j_k is the number of parts of size k. See Fact F7 for an example of a partition. Explicit formulas for $Z(S_n^{[2]})$ for small values of n are:

$$Z(S_1^{[2]}) = 1$$

$$Z(S_2^{[2]}) = \frac{1}{2!}(a_1^2 + a_2)$$

$$Z(S_3^{[2]}) = \frac{1}{3!}(a_1^6 + 3a_2^3 + 2a_3^2)$$

$$Z(S_4^{[2]}) = \frac{1}{4!}(a_1^{12} + 6a_1^2 a_2^5 + 8a_3^4 + 3a_2^6 + 6a_4^3)$$

$$Z(S_5^{[2]}) = \frac{1}{5!}(a_1^{20} + 10a_1^6 a_2^7 + 20a_1^2 a_3^6 + 15a_2^{10} + 30a_4^5 + 20a_2 a_3^2 a_6^2 + 24a_5^4)$$

$$Z(S_6^{[2]}) = \frac{1}{6!}(a_1^{30} + 15a_1^{12} a_2^9 + 40a_1^6 a_3^8 + 45a_1^2 a_2^{14} + 90a_1^2 a_4^7 + 120a_2^3 a_3^4 a_6^2$$
$$+ 144a_5^6 + 15a_2^{15} + 90a_2 a_4^7 + 40a_3^{10} + 120a_6^5)$$

F22: [Ha55, PoRe87] Let $D_{n,m}$ denote the number of simple digraphs with n vertices and m arcs, and let $d_n(x)$ be the generating function for n-vertex simple digraphs, so that

$$d_n(x) = \sum_{m=0}^{n(n-1)} D_{n,m} x^m.$$

Pólya's enumeration theorem states that this generating function $d_n(x)$ can be obtained from the cycle index $Z(S_n^{[2]})$ by replacing each variable a_i with $1 + x^i$. See Table 6.3.9. These numbers form sequence A052283 in [OEIS].

F23: For $m > n(n-1)/2$, the number of simple digraphs with n vertices and m arcs is the same as the number of simple digraphs with n vertices and $n(n-1) - m$ arcs.

F24: The total number D_n of simple digraphs with n vertices is obtained from the cycle index $Z(S_n^{[2]})$ by replacing each variable a_i with the number 2. See Table 6.3.9. This is sequence A000273 in [OEIS].

F25: Asymptotically, the sequence D_n satisfies $D_n \sim 2^{n(n-1)}/n!$.

F26: [Da54] The number T_n of tournaments with n vertices is given by the formula

$$T_n = \frac{1}{n!} \sum_{(j)}' \frac{n!}{\prod_k k^{j_k} j_k!} 2^{D(j)},$$

$m \backslash n$	1	2	3	4	5
0	1	1	1	1	1
1		1	1	1	1
2		1	4	5	5
3			4	13	16
4			4	27	61
5			1	38	154
6			1	48	379
7				38	707
8				27	1,155
9				13	1,490
10				5	1,670
Total	1	3	16	218	9,608

Table 6.3.9: Simple digraphs with n vertices and m arcs.

where the sum is over all partitions (j) of n into odd size parts, and where

$$D(j) = \frac{1}{2}\left(\sum_{r=1}^{n}\sum_{s=1}^{n} \gcd(r,s) j_r j_s - \sum_{k=1}^{n} j_k\right).$$

See Table 6.3.10. This is sequence A000568 in [OEIS].

F27: [Wr70] The number S_n of strong tournaments with n vertices can be determined by the recurrence relation

$$S_1 = 1, \quad \text{and} \quad S_n = T_n - \sum_{k=1}^{n-1} T_{n-k} S_k \quad \text{for } n > 1,$$

where T_n is the number of tournaments from Fact F26 above. See Table 6.3.10. Note that there are no strong tournaments with exactly two vertices. This is sequence A051337 in [OEIS].

F28: Asymptotically, most tournaments are strong. Thus the sequence S_n counting strong tournaments satisfies $S_n \sim 2^{n(n-1)}/n!$.

6.3.3 Counting Generic Trees

When counting generic trees, we must be careful to distinguish among labeled trees, rooted trees, unlabeled trees, and various other species. While labeled trees can be counted easily, unlabeled trees, both rooted and unrooted, are counted using generating functions.

n	Tournaments	Strong Tournaments
1	1	1
2	1	0
3	2	1
4	4	1
5	12	6
6	56	35
7	456	353
8	6,880	6,008
9	191,536	178,133
10	9,733,056	9,355,949
11	903,753,248	884,464,590
12	154,108,311,168	152,310,149,735

Table 6.3.10: Tournaments and strong tournaments with n vertices.

DEFINITIONS

D12: A *labeled tree* is a tree in which distinct labels, typically $v_1, v_2, \ldots, v_n$, have been assigned to the vertices. Two labeled trees with the same set of labels are considered the same only if there is an isomorphism from one to the other that preserves the labels.

D13: A *rooted tree* is a tree in which one vertex, the root, is distinguished. Two rooted trees are considered the same only if there is an isomorphism from one to the other that maps the root of the first to the root of the second.

D14: A *rooted labeled tree* is a labeled tree in which one vertex, the root, is distinguished. Two rooted labeled trees with the same set of labels are considered the same only if there is an isomorphism from one to the other that preserves the labels and maps the root of the first to the root of the second.

D15: A *reduced tree* (or *homeomorphically reduced tree*) is a tree with no vertices of degree 2. These trees are sometimes called *irreducible* trees.

EXAMPLES

E5: Figure 6.3.5 shows the three isomorphically distinct trees with 5 vertices. There are 60 essentially different ways to label each of the first two and 5 essentially different ways to label the third. Thus there are 125 different labeled trees with 5 vertices.

Figure 6.3.5: The three trees with 5 vertices.

E6:   There are 3 essentially different ways to root the first tree in Figure 6.3.5, 4 essentially different ways to root the second, and 2 essentially different ways to root the third. Thus there are 9 rooted (unlabeled) trees with 5 vertices.

E7:   Each of the 125 labeled trees discussed in Example E5 can be rooted at any of its five vertices, yielding 625 possible rooted labeled trees.

E8:   The third tree in Figure 6.3.5 is the only reduced tree with 5 vertices.

FACTS

F29: *Cayley's formula* [Ca89]: The number of labeled trees with n vertices is n^{n-2}. See Table 6.3.11. This is sequence A0000272 in [OEIS].

F30:   The number of rooted labeled trees with n vertices is n^{n-1}. See Table 6.3.11. This is sequence A000169 in [OEIS].

n	Labeled Trees	Rooted Labeled Trees
1	1	1
2	1	2
3	3	9
4	16	64
5	125	625
6	1,296	7,776
7	16,807	117,649
8	262,144	2,097,152
9	4,782,969	43,046,721
10	100,000,000	1,000,000,000
11	2,357,947,691	25,937,424,601
12	61,917,364,224	743,008,370,688
13	1,792,160,394,037	23,298,085,122,481
14	56,693,912,375,296	793,714,773,254,144
15	1,946,195,068,359,375	29,192,926,025,390,625
16	72,057,594,037,927,936	1,152,921,504,606,846,98032

Table 6.3.11: Labeled trees and rooted labeled trees with n vertices.

F31:   [Ca57] Let R_n denote the number of (unlabeled) rooted trees with n vertices, and let $r(x)$ be the generating function for rooted trees, so that

$$r(x) = \sum_{n=1}^{\infty} R_n x^n = x + x^2 + 2x^3 + 4x^4 + 9x^5 + 20x^6 + \cdots .$$

The coefficients R_n of this generating function can be determined by means of the recurrence relation

$$r(x) = x \prod_{k=1}^{\infty} (1 - x^k)^{-R_k}.$$

See Table 6.3.12. This is sequence A000081 in [OEIS].

n	Rooted Trees	Trees	Reduced Trees
1	1	1	1
2	1	1	1
3	2	1	0
4	4	2	1
5	9	3	1
6	20	6	2
7	48	11	2
8	115	23	4
9	286	47	5
10	719	106	10
11	1,842	235	14
12	4,766	551	26
13	12,486	1,301	42
14	32,973	3,159	78
15	87,811	7,741	132
16	235,381	19,320	249
17	634,847	48,629	445
18	1,721,159	123,867	842
19	4,688,676	317,955	1,561
20	12,826,228	823,065	2,988
21	35,221,832	2,144,505	5,671
22	97,055,181	5,623,756	10,981
23	268,282,855	14,828,074	21,209
24	743,724,984	39,299,897	41,472
25	2,067,174,645	104,636,890	81,181
26	5,759,636,510	279,793,450	160,176
27	16,083,734,329	751,065,460	316,749
28	45,007,066,269	2,023,443,032	629,933
29	126,186,554,308	5,469,566,585	1,256,070
30	354,426,847,597	14,830,871,802	2,515,169
31	997,171,512,998	40,330,829,030	5,049,816
32	2,809,934,352,700	109,972,410,221	10,172,638
33	7,929,819,784,355	300,628,862,480	20,543,579
34	22,409,533,673,568	823,779,631,721	41,602,425
35	63,411,730,258,053	2,262,366,343,746	84,440,886
36	179,655,930,440,464	6,226,306,037,178	171 794,492
37	509,588,049,810,620	17,169,677,490,714	350,238,175
38	1,447,023,384,581,029	47,436,313,524,262	715,497,037

Table 6.3.12: Rooted trees, trees, and reduced trees with n vertices.

F32: ***Otter's formula*** [Ot48]: Let T_n denote the number of trees with n vertices, and let $t(x)$ be the generating function for trees, so that

$$t(x) = \sum_{n=1}^{\infty} T_n x^n = x + x^2 + x^3 + 2x^4 + 3x^5 + 6x^6 + \cdots .$$

The coefficients T_n of this generating function $t(x)$ can be determined from the generating function $r(x)$ for rooted trees in Fact F31 above by using the formula

$$t(x) = r(x) - \tfrac{1}{2}\left(r^2(x) - r(x^2)\right).$$

See Table 6.3.12. This is sequence A000055 in [OEIS].

F33: Counting reduced trees requires an auxiliary sequence Q_n with generating function $q(x)$, so that

$$q(x) = \sum_{k=1}^{\infty} Q_k x^k = x + x^3 + x^4 + 2x^5 + 3x^6 + 6x^7 + 10x^8 + \cdots.$$

The coefficients Q_i of this generating function can be determined from the recurrence relation

$$q(x) = \frac{x}{1+x} \prod_{k=1}^{\infty} (1 - x^k)^{-Q_k}.$$

This is sequence A001678 in [OEIS].

F34: [HaPr59] Let H_n denote the number of reduced trees with n vertices, and let $h(x)$ be the generating function for reduced trees, so that

$$h(x) = \sum_{n=1}^{\infty} H_n x^n = x + x^2 + x^4 + x^5 + 2x^6 + 2x^7 + 4x^8 + \cdots.$$

The coefficients H_n of this generating function $h(x)$ can be determined from the auxiliary function $q(x)$ in Fact F33 above by using the formula

$$h(x) = (1+x)q(x) - \left(\frac{1+x}{2}\right) q^2(x) + \left(\frac{1-x}{2}\right) q(x^2).$$

See Table 6.3.12. Note that there are no reduced trees with exactly 3 vertices. This is sequence A000014 in [OEIS].

6.3.4   Counting Trees in Chemistry

DEFINITIONS

D16: A *1-4 tree* is a tree in which each vertex has degree 1 or 4.

D17: A *1-rooted 1-4 tree* is a 1-4 tree rooted at a vertex of degree 1.

REMARKS

R1: The 1-4 trees model many types of organic chemical molecules such as saturated hydrocarbons or alkanes. These molecules have the chemical formula $C_n H_{2n+2}$ and consist of n carbon atoms of valence 4 and $2n + 2$ hydrogen atoms of valence 1.

R2: The 1-rooted 1-4 trees model the monosubstituted hydrocarbons such as the alcohols with the chemical formula $C_n H_{2n+1} OH$ and consisting of n carbon atoms, $2n+1$ hydrogen atoms, and an OH group.

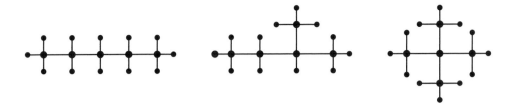

Figure 6.3.6: The three 1-4 trees with 5 vertices of degree 4.

EXAMPLES

E9: Figure 6.3.6 shows the three different 1-4 trees with 5 vertices of degree 4 and 12 vertices of degree 1.

E10: The first 1-4 tree in Figure 6.3.6 can be rooted at a vertex of degree 1 in 3 essentially different ways, the second in 4 essentially different ways, and the third in essentially only 1 way. Thus there are 8 different 1-rooted 1-4 trees with 5 vertices of degree 4.

FACTS

F35: A 1-4 tree with n vertices of degree 4 always has $2n + 2$ vertices of degree 1.

F36: [BlHe31a] Let A_n denote the number of 1-rooted 1-4 trees with n vertices of degree 4, and let $a(x)$ be the generating function for the number of 1-rooted 1-4 trees, so that

$$a(x) = \sum_{n=0}^{\infty} A_n x^n = 1 + x + x^2 + 2x^3 + 4x^4 + 8x^5 + 17x^6 + \cdots .$$

The coefficients A_n of this generating function $a(x)$ can be determined from the recurrence relation

$$a(x) = 1 + \tfrac{x}{6} \left(a^3(x) + 3a(x)a(x^2) + 2a(x^3) \right).$$

See Table 6.3.13. This is sequence A000598 in [OEIS].

F37: Counting (unrooted) 1-4 trees requires first counting 1-4 trees rooted at a vertex of degree 4. Let G_n be the number of 4-rooted 1-4 trees with n vertices of degree 4, and let $g(x)$ be the generating function for the number of 4-rooted 1-4 trees, so that

$$g(x) = \sum_{n=1}^{\infty} G_n x^n = x + x^2 + 2x^3 + 4x^4 + 9x^5 + 18x^6 + \cdots .$$

The coefficients G_n of this generating function $g(x)$ can be obtained by using the formula

$$g(x) = \tfrac{x}{24} \left(a^4(x) + 6a^2(x)a(x^2) + 8a(x)a(x^3) + 3a^2(x^2) + 6a(x^4) \right),$$

where $a(x)$ is the generating function for 1-rooted 1-4 trees from Fact F36.

n	1-Rooted 1-4 Trees (Alcohols)	1-4 Trees (Alkanes)
1	1	1
2	1	1
3	2	1
4	4	2
5	8	3
6	17	5
7	39	9
8	89	18
9	211	35
10	507	75
11	1,238	159
12	3,057	355
13	7,639	802
14	19,241	1,858
15	48,865	4,347
16	124,906	10,359
17	321,198	24,894
18	830,219	60,523
19	2,156,010	148,284
20	5,622,109	366,319
21	14,715,813	910,726
22	38,649,152	2,278,658
23	101,821,927	5,731,580
24	269,010,485	14,490,245
25	712,566,567	36,797,588
26	1,891,993,344	93,839,412
27	5,034,704,828	240,215,803
28	13,425,117,806	617,105,614
29	35,866,550,869	1,590,507,121
30	95,991,365,288	4,111,846,763
31	257,332,864,506	10,660,307,791
32	690,928,354,105	27,711,253,769
33	1,857,821,351,559	72,214,088,660
34	5,002,305,607,153	188,626,236,139
35	13,486,440,075,669	493,782,952,902
36	36,404,382,430,278	1,295,297,588,128
37	98,380,779,170,283	3,404,490,780,161
38	266,158,552,000,477	8,964,747,474,595

Table 6.3.13: 1-Rooted 1-4 trees and 1-4 trees with n vertices of degree 4.

F38: [BlHe31b] Let B_n denote the number of (unrooted) 1-4 trees with n vertices of degree 4, and let $b(x)$ be the generating function for 1-4 trees, so that

$$b(x) = \sum_{n=0}^{\infty} B_n x^n = 1 + x + x^2 + x^3 + 2x^4 + 3x^5 + 5x^6 + \cdots .$$

The coefficients B_n of this generating function $b(x)$ can be determined from the functions $a(x)$ and $g(x)$ from facts F36 and F37, respectively, by using the formula

$$b(x) = g(s) + a(x) - \tfrac{1}{2}\left(a^2(x) - a(x^2)\right).$$

See Table 6.3.13. This is sequence A000602 in [OEIS].

6.3.5   Counting Trees in Computer Science

DEFINITIONS

D18:   An *ordered tree* is recursively defined as consisting of a root vertex and a sequence $t_1, t_2, \ldots, t_m$ of $m \geq 0$ principal subtrees that are themselves ordered trees. The root vertex of an ordered tree is joined by an edge to the root of each principal subtree.

D19:   A *binary tree* consists of a root vertex and at most two principal subtrees that are themselves binary trees. Each principal subtree must be specified as either the left subtree or the right subtree.

D20:   The *children* of the root vertex of an ordered tree or a binary tree are the roots of the principal subtrees.

D21:   A *left-right tree* is a binary tree in which each vertex has either 0 or 2 children.

EXAMPLES

E11:   Figure 6.3.7 shows the 5 ordered trees with 4 vertices.

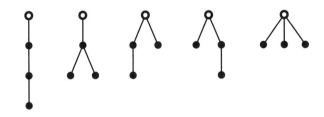

Figure 6.3.7: The 5 ordered trees with 4 vertices.

E12:   Figure 6.3.8 shows the 5 binary trees with 3 vertices.

Figure 6.3.8: The 5 binary trees with 3 vertices.

E13: Figure 6.3.9 shows the 5 left-right trees with 7 vertices.

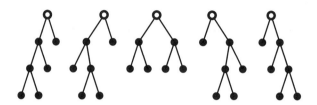

Figure 6.3.9: The 5 left-right trees with 7 vertices.

REMARKS

R3: In computer science, trees are usually drawn with the root at the top.

R4: Ordered trees are used to represent structures such as family trees, showing all descendants of a person represented by the root. The roots of the principal subtrees represent the children of the root person, in order of birth.

R5: Binary trees are some of the tree structures most easily represented in a computer. Other types of trees are often converted into binary trees for computer representation.

R6: Left-right trees are frequently used to represent arithmetic expressions, in which the leaves correspond to numbers and the other vertices represent binary operations such as $+$, $-$, $\times$, or $\div$.

FACTS

F39: The **Catalan numbers** C_n can be written as

$$C_n = \frac{1}{n+1}\binom{2n}{n} = \frac{(2n)!}{(n+1)!\,n!} \quad \text{for } n \geq 0.$$

See Table 6.3.14. This is sequence A000108 in [OEIS].

n	Catalan Number	n	Catalan Number
1	1	17	129,644,790
2	2	18	477,638,700
3	5	19	1,767,263,190
4	14	20	6,564,120,420
5	42	21	24,466,267,020
6	132	22	91,482,563,640
7	429	23	343,059,613,650
8	1,430	24	1,289,904,147,324
9	4,862	25	4,861,946,401,452
10	16,796	26	18,367,353,072,152
11	58,786	27	69,533,550,916,004
12	208,012	28	263,747,951,750,360
13	742,900	29	1,002,242,216,651,368
14	2,674,440	30	3,814,986,502,092,304
15	9,694,845	31	14,544,636,039,226,909
16	35,357,670	32	55,534,064,877,048,198

Table 6.3.14: The Catalan numbers.

F40: The number of ordered trees with n vertices is the Catalan number C_{n-1}. See Table 6.3.14.

F41: The number of binary trees with n vertices is the Catalan number C_n. See Table 6.3.14.

F42: The number of left-right trees with $2n + 1$ vertices is also C_n. See Table 6.3.14.

References

[BlHe31a] C. M. Blair and H. R. Henze, The number of structurally isomeric alcohols of the methanol series, *J. Amer. Chem. Soc.* **53** (1931), 3042–3046.

[BlHe31b] C. M. Blair and H. R. Henze, The number of isomeric hydrocarbons of the methane series, *J. Amer. Chem Soc.* **53** (1931), 3077–3085.

[Ca71] C. C. Cadogan, The Möbius function and connected graphs, *J. Combinatorial Theory, Ser. B* **11** (1971), 193–200.

[Ca57] A. Cayley, On the theory of the analytical forms called trees, *Philos. Mag.* **13** (1857), 19–30.

[Ca89] A. Cayley, A theorem on trees, *Quart. J. Math.* **23** (1889), 376–378.

[Da54] R. L. Davis, Structures of dominance relations, *Bull. Math. Biophys.* **16** (1954), 131–140.

[Ha55] F. Harary, The number of linear, directed, rooted, and connected graphs, *Trans. Amer. Math. Soc.* **78** (1955), 445–463.

[HaPa73] F. Harary and E. M. Palmer, *Graphical Enumeration*, Academic Press, 1973.

[HaPr59] F. Harary and G. Prins, The number of homeomorphically irreducible trees, and other species, *Acta Math.* **101** (1959), 141–162.

[Gi56] E. N. Gilbert, Enumeration of labeled graphs, *Canad. J. Math.* **8** (1956), 405–411.

[MoMo62] J. W. Moon and L. Moser, Almost all tournaments are irreducible, *Canad. Math. Bull.* **5** (1962), 61–65.

[OEIS] *The On-Line Encyclopedia of Integer Sequences*, published electronically at http://oeis.org.

[Ot48] R. Otter, The number of trees, *Ann. of Math.* **49** (1948), 583–599.

[PoRe87] G. Pólya and R. C. Read, *Combinatorial Enumeration of Groups, Graphs, and Chemical Compounds*, Springer-Verlag, 1987.

[Wr70] E. M. Wright, The number of irreducible tournaments, *Glasgow Math. J.* **11** (1970), 97–101.

Section 6.4
Graphs and Vector Spaces

Krishnaiyan "KT" Thulasiraman, University of Oklahoma

INTRODUCTION

Electrical circuit theory is one of the earliest applications of graph theory to a problem in physical science. The dynamic behavior of an electrical circuit is governed by three laws: Kirchhoff's voltage law, Kirchhoff's current law, and Ohm's law. Each element in a circuit is associated with two variables, namely, the current variable and the voltage variable. Kirchhoff's voltage law requires that the algebraic sum of the voltages around a circuit is zero, and Kirchhoff's current law requires that the algebraic sum of the currents across a cut is zero. Thus, circuits and cuts define a linear relationship among the voltage variables and a linear relationship among the current variables, respectively. It is for this reason that circuits, cuts, and the vector spaces associated with them have played a major role in the discovery of several fundamental properties of electrical circuits arising from the structure or the interconnection of the circuit elements. Several graph theorists and circuit theorists have immensely contributed to the development of what we may now call the structural theory of electrical circuits. The significance of the results to be presented in this section goes well beyond their application to circuit theory. They will bring out the fundamental duality that exists between circuits and cuts and the influence of this duality on the structural theory of graphs. Most of the results in this section are also relevant to the development of combinatorial optimization theory as well as matroid theory.

6.4.1 Basic Concepts and Definitions

Although the terms *node* and *oriented graph* are commonly used in electrical circuit theory, we use the terms *vertex* and *directed graph* along with all the other basic terminology of graph theory established in Chapter 1. For the sake of completeness, we begin with a review of certain basic concepts and definitions. For concepts not discussed here, the reader is referred to [GrYe06] and [ThSw92].

NOTATION: Unless otherwise specified, $G = (V, E)$ is a graph (or digraph) with n vertices, $V = \{v_1, v_2, \ldots, v_n\}$, and m edges, $E = \{e_1, e_2, \ldots e_m\}$.

NOTATION: If vertices v_i and v_j are the endpoints (or end vertices) of an edge then, when there is no ambiguity, we denote that edge by the ordered pair (v_i, v_j).

DEFINITIONS

D1: A graph is called a ***trivial graph*** if it has only one vertex and no edge. A graph with no edges is called an ***empty graph***. A graph with no vertices and hence no edges is called a ***null graph*** and will be denoted by $\emptyset$.

REMARK

R1: In this section we consider only graphs in which all edges have two distinct endpoints (i.e., no self-loops).

EXAMPLE

E1: Examples E1 through E9 in this section refer to the graph shown in Figure 6.4.1.

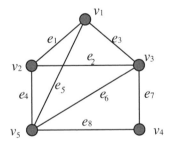

Figure 6.4.1: A graph used in several examples.

Subgraphs and Complements

DEFINITIONS

D2: A graph $G' = (V', E')$ is called a ***subgraph*** of graph $G = (V, E)$ if $V' \subseteq V, E' \subseteq E$ and V' contains all the endpoints of all the edges in E'.

D3: Each subset $E' \subseteq E$ defines a unique subgraph $G' = (V', E')$ of graph $G = (V, E)$, where V' consists of only those vertices which are the endpoints of the edges in E'. The subgraph G' is called the ***induced subgraph of G on the edge set E'***. Note that an edge-induced subgraph will not have isolated vertices.

D4: Each subset $V' \subseteq V$ defines a unique subgraph $G' = (V', E')$ of graph $G = (V, E)$, where E' consists of those edges whose endpoints are in V'. The subgraph G' is called the ***induced subgraph of G on the vertex set V'***. Note that a vertex-induced subgraph may have isolated vertices.

D5: Given a subgraph $G' = (V', E')$ of graph $G = (V, E)$, the subgraph $G'' = (V, E - E')$ is called the (***edge-***)***complement of G' in G***.

EXAMPLES

E2: For the set $E' = \{e_1, e_3, e_8\}$, the corresponding edge-induced subgraph of graph G in Figure 6.4.1 is shown in Figure 6.4.2(a). For the set $V' = \{v_1, v_2, v_4\}$, the corresponding vertex-induced subgraph of G is shown in Figure 6.4.2(b).

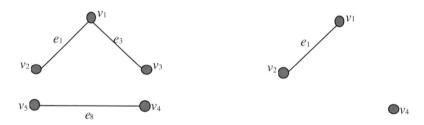

(a) An edge-induced subgraph of the graph G (b) A vertex-induced subgraph of the graph G

Figure 6.4.2: An edge-induced subgraph and a vertex-induced subgraph.

E3: The complement of the subgraph G' of Figure 6.4.3(a) in the graph G of Figure 6.4.1 is shown in Figure 6.4.3(b).

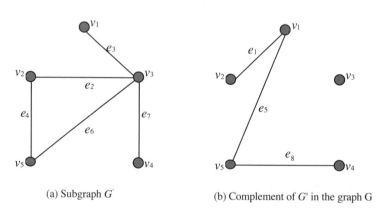

(a) Subgraph G' (b) Complement of G' in the graph G

Figure 6.4.3: A subgraph G' and its complement in G.

Components, Spanning Trees, and Cospanning Trees

DEFINITIONS

D6: A ***closed trail*** is a closed walk with no repeated edges.

TERMINOLOGY: A closed trail is also called a *circ*, which we formally state in Definition D20 and use thereafter.

D7: A *circuit* is a closed trail with no repeated vertices except the initial and terminal ones.

TERMINOLOGY: Several authors use the term *cycle* instead of circuit. In electrical circuit literature, the term circuit is commonly understood as defined in Definition D7.

D8: A graph G is **connected** if there is a path between every pair of vertices of G.

D9: A maximal connected subgraph of a graph is called a **component** of the graph. An isolated vertex is by itself considered a single component.

D10: A **tree** of a graph G is a connected subgraph containing no circuits. If a tree of a connected graph G contains all the vertices of G then it is called a **spanning tree** of G. The complement of a spanning tree T in G is called a **cospanning tree** of G.

D11: A **spanning forest** of a non-connected graph G with p components is a collection of p spanning trees, one for each component.

D12: The edges of a spanning tree T are called the **branches** of T. The edges of a cospanning tree are called the **chords** of the spanning tree.

D13: Let G be an n-vertex graph with m edges and p components. The **rank** $\rho(G)$ and **nullity** $\mu(G)$ of G are given by $\rho(G) = n - p$ and $\mu(G) = m - n + p$.

EXAMPLE

E4: A spanning tree T and the corresponding cospanning tree of the connected graph G of Figure 6.4.1 are shown in Figure 6.4.4.

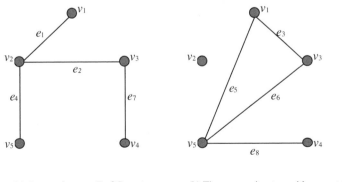

(a) A spanning tree T of G   (b) The cospanning tree with respect to T

Figure 6.4.4: A spanning tree and corresponding cospanning tree of graph G.

FACTS

F1: There is exactly one path between any two vertices of a spanning tree.

F2: A spanning tree of a connected n-vertex graph has $n-1$ branches and a cospanning tree has $m-n+1$ chords. A spanning forest of a graph having p components has $n-p$ branches and $m-n+p$ chords.

REMARK

R2: Unless stated otherwise, all graphs G considered in this section are connected.

Cuts and Cutsets

DEFINITIONS

D14: Consider a connected graph $G = (V, E)$. Let V_1 and V_2 be two disjoint subsets of V such that $V = V_1 \cup V_2$ (i.e., V_1 and V_2 form a *partition* of V). Then the set of all those edges of G having one end vertex in V_1 and the other in V_2 is called a ***cut*** of G. This cut is denoted as $\langle V_1, V_2 \rangle$. The set of edges incident on a vertex forms a cut, and is called an ***incidence set***.

D15: Removal of the edges in a cut from a connected graph G will disconnect the graph. In other words, the resulting graph will have at least two components. A cut of a connected graph is called a ***cutset*** if the removal of the edges in the cut results in a non-connected graph with *exactly* two components. Equivalently, a cutset of a connected graph is a minimal set of edges whose removal disconnects the graph.

EXAMPLE

E5: For the graph G in Figure 6.4.1, the cut $\langle V_1, V_2 \rangle$, where $V_1 = \{v_1, v_3, v_5\}$ and $V_2 = \{v_2, v_4\}$ consists of the edges e_1, e_2, e_4, e_7, and e_8, as shown in Figure 6.4.5(a). Removing these edges results in a non-connected graph with three components. So, $\langle V_1, V_2 \rangle$ is not a cutset. A cutset consisting of the edges e_4, e_5, e_6, and e_7 is shown in Figure 6.4.5(b). Removing these edges results in a non-connected graph with two components.

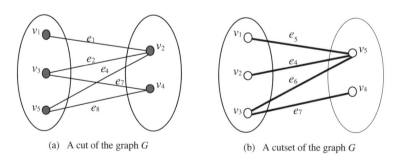

(a) A cut of the graph G (b) A cutset of the graph G

Figure 6.4.5: A cut and a cutset of the graph G.

Vector Space of a Graph under Ring Sum of Its Edge Subsets

DEFINITIONS

D16: Suppose the edge-set of a graph G is $E = \{e_1, e_2, e_3, \ldots, e_m\}$. Then each subset E' of E can be represented by a ***binary m-vector*** in which the ith component is 1 if and only if the edge e_i is in E'. For example, the binary vector $(1, 0, 0, 1, 1, 1, 0, 0)$ represents the edge subset $\{e_1, e_4, e_5, e_6\}$ of the graph G of Figure 6.4.1.

D17: The ***ring sum*** (or ***symmetric difference***) of two sets E_1 and E_2, denoted as $E_1 \oplus E_2$, is the set of those edges which belong to E_1 or to E_2 but not to both E_1 and E_2.

D18: The ***ring sum*** of two m-vectors
$$X = (x_1, x_2, x_3, \ldots, x_i, \ldots, x_m) \quad \text{and} \quad Y = (y_1, y_2, y_3, \ldots, y_i, \ldots, y_m)$$
is the vector
$$Z = (z_1, z_2, z_3, \ldots, z_i, \ldots, z_m),$$
where $z_i = x_i \otimes y_i$, and where $\otimes$ is the logical *exclusive-or* operation (i.e., $1 \otimes 0 = 1$; $0 \otimes 1 = 1$; $0 \otimes 0 = 0$; and $1 \otimes 1 = 0$).

FACT

F3: The m-vector representing the ring sum of two subsets of edges is the ring sum of the m-vectors representing these edge subsets. The set of m-vectors representing all the 2^m edge subsets of a graph G (including the null set) forms an m-dimensional ***vector space*** over GF(2), the field of integers modulo 2, under the ring sum operation $\oplus$.

NOTATION: This vector space of edge subsets of a graph G (and hence of the corresponding edge-induced subgraphs of G) is denoted by $\Psi(G)$.

REMARKS

R3: Throughout this section all vectors are assumed to be row vectors.

R4: In this section an edge subset is used to refer to the corresponding edge-induced subgraph. The vector space $\Psi(G)$ will be used to denote the vector space of all binary m-vectors as well as the vector space of all edge-induced subgraphs of G. Observe that the null set (or null graph Ø) is the 0-vector of $\Psi(G)$.

R5: In electrical engineering literature, a cut is also referred to as a ***seg*** [Re61].

R6: Proofs of most results in this section may be found in standard texts [SeRe61], [Ch71b], [De74], [ThSw92], and [SwTh81].

6.4.2 Circuit Subspace of an Undirected Graph

DEFINITIONS

D19: A graph is **even** if the degree of every vertex in the graph is even. Clearly, a circuit is an even graph.

D20: A **circ** of a graph is a closed trail. The null graph is considered as a circ.

NOTATION: The set of all circs of a graph G is denoted by $\hat{C}(G)$. In other words, $\hat{C}(G)$ is the set of all circuits and unions of edge-disjoint circuits of the graph G (including the null graph $\varnothing$).

FACTS

F4: A subgraph of a graph is a circ if and only if it is even.

F5: A circ is a circuit or union of edge-disjoint circuits. Thus, the edge set of an even graph can be partitioned into edge subsets such that each subset in the partition forms a circuit.

F6: The ring sum of any two even subgraphs of a graph is even. Thus, the set $\hat{C}(G)$ is closed under ring sum.

F7: $\hat{C}(G)$ is a subspace of the vector space $\Psi(G)$ and is called the **circuit subspace** of G.

EXAMPLE

E6: Two circs of the graph G of Figure 6.4.1 and their ring sum, which is clearly a circ, are shown in Figure 6.4.6.

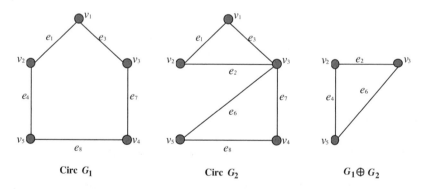

Figure 6.4.6: Two circs of the graph G and their ring sum.

REMARKS

R7: Fact F5 is attributed to Veblen [Ve31].

R8: A connected, even graph G is *eulerian*, i.e., there exists a circ that contains all the edges of G (see §4.2).

Fundamental Circuits and the Dimension of the Circuit Subspace

DEFINITION

D21: Adding a chord c to a spanning tree T of a connected graph G produces a unique circuit in G, called the ***fundamental circuit*** of G with respect to chord c.

NOTATION: If e_i is a chord of a spanning tree T, then C_i will denote the fundamental circuit with respect to e_i.

FACTS

F8: Given a connected graph G and a spanning tree T, there are $m - n + 1$ fundamental circuits, one for each chord of T.

F9: The fundamental circuit with respect to chord c contains only one chord of the spanning tree T, namely, the chord c. The chord c is not present in any other fundamental circuit with respect to T.

F10: The $(m - n + 1)$ fundamental circuits with respect to a spanning tree of a connected graph G are linearly independent in the circuit subspace $\hat{C}(G)$.

F11: If a circ of a graph G contains the chords $e_a, e_b, \ldots, e_k$, then the circ can be expressed as the ring sum of the fundamental circuits $C_a, C_b, \ldots, C_k$.

F12: The fundamental circuits with respect to a spanning tree of a connected graph G constitute a basis for the circuit subspace $\hat{C}(G)$, and hence, the dimension of $\hat{C}(G)$ is equal to $m - n + 1$, the nullity $\mu(G)$.

F13: The dimension of the circuit subspace $\hat{C}(G)$ of a graph having p components is equal to $\mu(G) = m - n + p$.

EXAMPLE

E7: The set of fundamental circuits with respect to the spanning tree $T = \{e_1, e_2, e_4, e_7\}$ of the graph shown in Figure 6.4.1 is

Chord e_3	$C_3 = \{e_3, e_1, e_2\}$
Chord e_5	$C_5 = \{e_5, e_1, e_4\}$
Chord e_6	$C_6 = \{e_6, e_2, e_4\}$
Chord e_8	$C_8 = \{e_8, e_2, e_4, e_7\}$

It can be verified that the circ $\{e_1, e_4, e_5, e_6, e_7, e_8\}$, which contains the chords e_5, e_6 and e_8, is the ring sum of the fundamental circuits C_5, C_6, and C_8. This illustrates Fact F11.

6.4.3   Cutset Subspace of an Undirected Graph

Recall from the definitions in §6.4.1 that a cutset is also a cut. Several facts that highlight the duality between cuts and circs will be presented next.

DEFINITION

D22: The collection of all cutsets and unions of edge-disjoint cutsets of a graph G is called the **cutset subspace** of G and is denoted by $\lambda(G)$. The null graph $\emptyset$ is considered a cut and hence belongs to $\lambda(G)$.

FACTS

F14: Every cut of a connected graph G is the union of some edge-disjoint cutsets of G. Thus, $\lambda(G)$ is the collection of cuts of G.

F15: The **cutset subspace** $\lambda(G)$ of a graph G is a subspace of the vector space $\Psi(G)$.

F16: The ring sum of any two cuts of a graph G is also a cut of G; i.e., $\lambda(G)$ is closed under ring sum.

EXAMPLE

E8: Consider the graph in Figure 6.4.1 and the cuts $S_1 = \langle V_1, V_2 \rangle$ and $S_2 = \langle V_3, V_4 \rangle$ in Figure 6.4.5, where $V_1 = \{v_1, v_3, v_5\}$, $V_2 = \{v_2, v_4\}$, $V_3 = \{v_1, v_2, v_3\}$, and $V_4 = \{v_4, v_5\}$. Then $S_1 = \{e_1, e_2, e_4, e_7, e_8\}$, $S_2 = \{e_4, e_5, e_6, e_7\}$, and $S_1 \oplus S_2 = \{e_1, e_2, e_5, e_6, e_8\}$.

Moreover, it can be seen that $S_1 \oplus S_2 = \langle A \cup D, B \cup C \rangle$, where

$$
\begin{aligned}
A &= V_1 \cap V_3 = \{v_1, v_3\} \\
B &= V_1 \cap V_4 = \{v_5\} \\
C &= V_2 \cap V_3 = \{v_2\} \\
D &= V_2 \cap V_4 = \{v_4\}
\end{aligned}
$$

In fact, this illustration is also the basis of the proof of Fact F16.

Fundamental Cutsets and the Dimension of the Cutset Subspace

DEFINITIONS

D23: Let T be a spanning tree of a connected graph G, and let b be a branch of T. If V_1 and V_2 are the vertex-sets of the two components of $T - b$, then we can verify that the cut $\langle V_1, V_2 \rangle$ is a cutset of G. This cutset is called the **fundamental cutset** of G with respect to the branch b of T.

NOTATION: If e_i is a branch of a spanning tree T, then S_i denotes the fundamental cutset with respect to the branch e_i.

D24: An **incidence set** of a vertex v in a graph G is the cut consisting of the set of edges of G that are incident on v.

FACTS

F17: Given a connected graph G and a spanning tree T, there are $n - 1$ fundamental cutsets, one for each branch of T.

F18: The fundamental cutset with respect to branch b of a spanning tree T contains only one branch, namely, the branch b. The branch b is not present in any other fundamental cutset with respect to T.

F19: The $n - 1$ fundamental cutsets with respect to a spanning tree of a connected n-vertex graph G are linearly independent in the cutset subspace $\lambda(G)$.

F20: If a cut of a graph G contains the branches $e_a, e_b, \ldots e_k$, then the cut can be expressed as the ring sum of the fundamental cutsets $S_a, S_b, \ldots, S_k$.

F21: The fundamental cutsets with respect to a spanning tree of a connected graph G constitute a basis for the cutset subspace $\lambda(G)$ of G, and hence the dimension of $\lambda(G)$ is equal to $n - 1$, the rank $\rho(G)$.

F22: The dimension of the cutset subspace $\lambda(G)$ of a graph having p components is equal to $\rho(G) = n - p$.

F23: The incidence sets of any $n - 1$ vertices of a connected n-vertex graph G form a basis of the cutset subspace $\lambda(G)$.

EXAMPLE

E9: For the graph shown in Figure 6.4.1, the fundamental cutsets with respect to the spanning tree $T = \{e_1, e_2, e_4, e_7\}$ are

Branch e_1	$S_1 = \{e_1, e_3, e_5\}$
Branch e_2	$S_2 = \{e_2, e_3, e_6, e_8\}$
Branch e_4	$S_4 = \{e_4, e_5, e_6, e_8)$
Branch e_7	$S_7 = \{e_7, e_8\}$

It can be verified that the cut $= \{e_1, e_2, e_4, e_7, e_8\}$ containing the branches e_1, e_2, e_4, and e_7 is the ring sum of the fundamental cutsets S_1, S_2, S_4, and S_7. This illustrates Fact F20.

6.4.4 Relationship between Circuit and Cutset Subspaces

By now it should be evident that circs and cuts are dual concepts in the sense that for each result that involves circuits or circs, there is a corresponding result involving cutsets or cuts. Facts F5 through F13 correspond to Facts F14 through F22. Spanning trees and cospanning trees provide the links between circs and cuts. This duality is further explored next.

Orthogonality of Circuit and Cutset Subspaces

DEFINITIONS

D25: The binary m-vector representing a circ is called a **circuit vector**; the binary m-vector representing a cut is called a **cut vector**; and the m-vector representing an incidence set is called an **incidence vector**.

D26: Two subspaces W' and W'' of a vector space W are **orthogonal** to each other if the inner product (or dot product) of every vector in W' with every vector in W'' is zero. Note that the zero vector belongs to every subspace.

FACTS

F24: A circuit and a cutset of a connected graph have an even number of edges in common. Hence, a circ and a cut have an even number of edges in common.

F25: The inner product of a circuit vector and a cut vector over GF(2) is zero under the ring sum operation.

F26: A subgraph of a graph G belongs to the circuit subspace of the graph if and only if it has an even number of edges in common with every subgraph in the cutset subspace of G. Equivalently, a vector is a circuit vector if and only if it is orthogonal to every cut vector.

F27: A subgraph of a graph G belongs to the cutset subspace of the graph if and only if it has an even number of edges in common with every subgraph in the circuit subspace of G. Equivalently, a vector is a cut vector if and only if it is orthogonal to every circuit vector.

F28: The circuit and cutset subspaces of a graph are orthogonal to each other.

Circ/Cut-Based Decomposition of Graphs and Subgraphs

DEFINITION

D27: Two orthogonal subspaces W' and W'' of a vector space W are **orthogonal complements** if every vector in W can be expressed as the ring sum of a vector of W' and a vector of W''. Note that the zero vector is the only vector that is in the intersection of the orthogonal complements W' and W''.

FACTS

F29: If the orthogonal subspaces W' and W'' of a vector space W are not orthogonal complements, then the dimension of their union is less than the dimension of the vector space W.

F30: [Ch71a] The circuit and the cutset subspaces of a graph are orthogonal complements if and only if the graph has an odd number of spanning forests.

F31: If the circuit and cutset subspaces of a graph are orthogonal complements, then every subgraph (including the graph itself) can be expressed as the ring sum of a circ and a cut.

F32: [Ch71b, WiMa71] Every graph can be represented as the ring sum of a circ and a cut of the graph. If the dimension of the intersection of the circuit and cutset subspaces of a graph is equal to k, then there are 2^k such representations.

EXAMPLES

E10: Consider the graph G_a in Figure 6.4.7. It can be verified that no nonempty subgraph of this graph is both a circ and a cut. So the cutset and circuit subspaces of G_a are orthogonal complements. Then the set of fundamental cutsets and fundamental circuits with respect to a spanning tree of G_a constitutes a basis of the vector space $\Psi(G)$. One such set with respect to the spanning tree formed by the edges e_1, e_2, e_3, and e_4 is as follows:

$$
\begin{aligned}
S_1 &= (1 \quad 0 \quad 0 \quad 0 \quad 1 \quad 1 \quad 0) \\
S_2 &= (0 \quad 1 \quad 0 \quad 0 \quad 1 \quad 1 \quad 0) \\
S_3 &= (0 \quad 0 \quad 1 \quad 0 \quad 1 \quad 0 \quad 1) \\
S_4 &= (0 \quad 0 \quad 0 \quad 1 \quad 0 \quad 0 \quad 1) \\
S_5 &= (1 \quad 1 \quad 1 \quad 0 \quad 1 \quad 0 \quad 0) \\
S_6 &= (1 \quad 1 \quad 0 \quad 0 \quad 0 \quad 1 \quad 0) \\
S_7 &= (0 \quad 0 \quad 1 \quad 1 \quad 0 \quad 0 \quad 1)
\end{aligned}
$$

It is easy to verify that every subgraph can be expressed as the ring sum of a circ and a cut, which illustrates Fact F31. For instance, the vector $(0\ 0\ 1\ 1\ 0\ 1\ 1)$, which represents the induced subgraph on the edge subset $\{e_3, e_4, e_6, e_7\}$, can be expressed as:

$$(0 \quad 0 \quad 1 \quad 1 \quad 0 \quad 1 \quad 1) \quad = \quad S_1 \oplus S_2 \oplus C_6 \oplus C_7$$

$$= \quad (1 \quad 1 \quad 0 \quad 0 \quad 0 \quad 0 \quad 0) \oplus (1 \quad 1 \quad 1 \quad 1 \quad 0 \quad 1 \quad 1)$$

where $(1\ 1\ 0\ 0\ 0\ 0\ 0)$ represents a cut in G_a, and $(1\ 1\ 1\ 1\ 0\ 1\ 1)$ represents a circ.

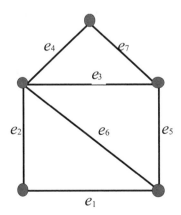

Figure 6.4.7: Graph $\boldsymbol{G_a}$ for illustration of Fact F31.

E11:   Consider the graph G_b in Figure 6.4.8. In this graph the edges e_1, e_2, e_3, and e_5 constitute a circuit as well as a cut. Hence the circuit and cutset subspaces are not orthogonal complements. This means that there is a subgraph of G_b that cannot be expressed as the ring sum of a circ and a cut. However, according to Fact F32, such a decomposition is possible for G_b. This is verified as follows:

$$(1 \ \ 1 \ \ 1 \ \ 1 \ \ 1 \ \ 1) \ = \ (1 \ \ 1 \ \ 0 \ \ 1 \ \ 0 \ \ 0) \oplus (0 \ \ 0 \ \ 1 \ \ 0 \ \ 1 \ \ 1)$$

where $(1 \ 1 \ 0 \ 1 \ 0 \ 0)$ represents the cut of edges e_1, e_2, and e_4, and $(0 \ 0 \ 1 \ 0 \ 1 \ 1)$ represents the circuit of edges e_3, e_5, and e_6 in G_b.

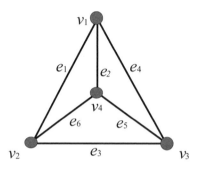

Figure 6.4.8: Graph $\boldsymbol{G_b}$ for illustration of Fact F32.

6.4.5   Circuit and Cutset Spaces in a Directed Graph

In most engineering applications of graph theory, directed graphs are encountered. But, as we shall see next, the effect of orientation is minimal in so far as the results concerning circuits and cuts are concerned. Almost all the results presented earlier in this section have their equivalents in the directed case. In fact, we can view all the results on undirected graphs presented thus far as special cases of the results to be presented next.

TERMINOLOGY: A *circuit*, *cut*, or *spanning tree* in a directed graph G is a subset of edges that constitutes a circuit, cut, or spanning tree, respectively, in the underlying graph of G.

Circuit and Cut Vectors and Matrices

DEFINITIONS

D28:   A circuit in a directed graph can be traversed in one of two directions, clockwise or counter-clockwise (relative to a plane drawing of the circuit). The traversal direction we choose is called the ***circuit orientation***.

D29: Let C be a circuit in a directed graph and $e = (v_i, v_j)$ an edge in C directed from v_i to v_j. Given an orientation of C, edge e is said to **agree with the circuit orientation** if the traversal of e specified by that orientation is from its tail v_i to its head v_j.

D30: A cut (V_a, V_b) in a directed graph can be traversed in one of two directions, from V_a to V_b or from V_b to V_a. The direction chosen is called the **cut orientation**.

D31: Given an orientation of a cut in a directed graph, an edge $e = (v_i, v_j)$ in the cut is said to **agree with the cut orientation** if the traversal of e specified by that orientation is from v_i to v_j.

D32: Let G be a directed graph with edge-set $E = \{e_1, e_2, \ldots, e_m\}$, and let C be a circuit in G with a given orientation. The **circuit vector** representing C is the m-vector $(x_1, x_2, \ldots, x_m)$, where

$$x_i = \begin{cases} 1, & \text{if edge } e_i \text{ agrees with the orientation of } C \\ -1, & \text{if edge } e_i \text{ does not agree with the orientation of } C \\ 0, & \text{if edge } e_i \text{ is not in } C \end{cases}$$

D33: Let G be a directed graph with edge-set $E = \{e_1, e_2, \ldots, e_m\}$, and let S be a cut in G with a given orientation. The **cut vector** representing S is the m-vector $(x_1, x_2, \ldots, x_m)$, where

$$x_i = \begin{cases} 1, & \text{if edge } e_i \text{ agrees with the orientation of } S \\ -1, & \text{if edge } e_i \text{ does not agree with the orientation of } S \\ 0, & \text{if edge } e_i \text{ is not in } S \end{cases}$$

D34: Let G be a directed graph with edge-set $E = \{e_1, e_2, \ldots, e_m\}$. Let $C_1, C_2, \ldots, C_t$ and $S_1, S_2, \ldots, S_r$ be the circuits and cuts of G, respectively, each with a given traversal orientation. The **circuit matrix** of G is the $t \times m$ matrix whose ith row is the circuit vector representing circuit C_i. The **cut matrix** of G is the $r \times m$ matrix whose ith row is the cut vector representing cut S_i.

Fundamental Circuit, Fundamental Cutset, and Incidence Matrices

Next, we define two special matrices corresponding to the fundamental circuits and cutsets relative to a given spanning tree in a directed graph and a third matrix corresponding to the incidence vectors of the vertices.

REMARK

R9: The definitions of these three matrices depend on how the associated circuits and cuts are oriented. The orientations of each fundamental circuit and each fundamental cut are usually chosen to agree with the defining chord and branch, respectively, and we adopt that convention here. Also, for the cut consisting of the set of edges incident on a vertex v (i.e., the incidence set of v), we assume that the orientation is away from vertex v. Accordingly, the incidence vector of vertex v is given by $(x_1, x_2, \ldots, x_m)$, where

$$x_i = \begin{cases} 1, & \text{if edge } e_i \text{ is directed } \textit{from } v \quad (v \text{ is the tail of edge } e_i) \\ -1, & \text{if edge } e_i \text{ is directed } \textit{to } v \quad (v \text{ is the head of edge } e_i) \\ 0, & \text{if edge } e_i \text{ is not incident on } v \end{cases}$$

DEFINITIONS

D35: Let T be a spanning tree of a connected directed graph. The **fundamental circuit matrix** of the graph with respect to T, denoted by $\boldsymbol{B}_f$, is the $(m - n + 1)$-rowed submatrix of the circuit matrix whose rows are the fundamental circuit vectors. Similarly, the **fundamental cutset** matrix with respect to T, denoted by Q_f, is the $(n-1)$-rowed submatrix of the cut matrix whose rows are the fundamental cutset vectors.

D36: The **incidence matrix** of a given directed graph, denoted A_c, is the n-rowed submatrix of the cut matrix whose rows are the incidence vectors of the directed graph. The submatrix of the incidence matrix containing any $n - 1$ of the incidence vectors is called a **reduced incidence matrix** and is denoted by A.

D37: A matrix of real numbers is **unimodular** if the determinant of every square submatrix of the matrix is equal to 1, -1, or 0.

EXAMPLES

E12: Consider the directed graph of Figure 6.4.9(a) below. A circuit and a cut with orientations are shown in Figures 6.4.9(b) and (c), respectively. The corresponding circuit and cut vectors are $\begin{pmatrix} 1 & -1 & -1 & 0 & 1 & 0 & 0 \end{pmatrix}$ and $\begin{pmatrix} 0 & 1 & 0 & 0 & 1 & 1 & 0 \end{pmatrix}$, respectively.

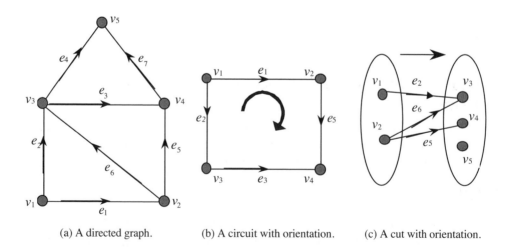

| (a) A directed graph. | (b) A circuit with orientation. | (c) A cut with orientation. |

Figure 6.4.9: A directed graph, a circuit, and a cut with orientations.

E13: Consider the spanning tree T of the graph of Figure 6.4.9(a) consisting of the edges e_1, e_2, e_3, and e_4. The fundamental circuit and the fundamental cutset matrices with respect to T, and the incidence matrix of this graph with the column i in each matrix corresponding to edge e_i are:

Fundamental Circuit Matrix:

$$
\begin{array}{c}
\text{Chord } e_5 \\
\text{Chord } e_6 \\
\text{Chord } e_7
\end{array}
\begin{pmatrix}
1 & -1 & -1 & 0 & 1 & 0 & 0 \\
1 & -1 & 0 & 0 & 0 & 1 & 0 \\
0 & 0 & 1 & -1 & 0 & 0 & 1
\end{pmatrix}
$$

Fundamental Cutset Matrix:

$$
\begin{array}{c}
\text{Branch } e_1 \\
\text{Branch } e_2 \\
\text{Branch } e_3 \\
\text{Branch } e_4
\end{array}
\begin{pmatrix}
1 & 0 & 0 & 0 & -1 & -1 & 0 \\
0 & 1 & 0 & 0 & 1 & 1 & 0 \\
0 & 0 & 1 & 0 & 1 & 0 & -1 \\
0 & 0 & 0 & 1 & 0 & 0 & 1
\end{pmatrix}
$$

Incidence Matrix:

$$
\begin{array}{c}
\text{Node } v_1 \\
\text{Node } v_2 \\
\text{Node } v_3 \\
\text{Node } v_4 \\
\text{Node } v_5
\end{array}
\begin{pmatrix}
1 & 1 & 0 & 0 & 0 & 0 & 0 \\
-1 & 0 & 0 & 0 & 1 & 1 & 0 \\
0 & -1 & 1 & 1 & 0 & -1 & 0 \\
0 & 0 & -1 & 0 & -1 & 0 & 1 \\
0 & 0 & 0 & -1 & 0 & 0 & -1
\end{pmatrix}
$$

Orthogonality and the Matrix Tree Theorem

TERMINOLOGY: A directed edge that is in both a circuit and a cut is said to have the *same relative orientation* with respect to the circuit and the cut if the edge either agrees or disagrees with the assigned orientations of both the circuit and the cut.

FACTS

F33: A circuit and a cut in a connected graph have an even number of common edges. If a circuit and a cut have $2k$ common edges, then these edges can be partitioned into two sets, each of size k, such that each edge in one set has the same relative orientation with respect to the circuit and the cut, and that each edge in the other set agrees with one of the two assigned orientations (circuit or cut) and disagrees with the other assigned orientation.

NOTATION: Let G be a directed graph and suppose that each of the circuits and cuts has been given an orientation.
(a) The collection of all circuit vectors of G and their linear combinations over the real field is denoted by $\hat{C}(G)$.
(b) The collection of all cut vectors of G and their linear combinations over the real field is denoted by $\lambda(G)$.

F34: In a directed graph every circuit vector is orthogonal to every cut vector over the real field.

F35: In a connected directed graph, every circuit vector can be expressed as a linear combination of fundamental circuit vectors with respect to a spanning tree of the graph.

The coefficients in the linear combination are 1 or -1. Similarly, every cut vector in a connected directed graph can be expressed as a linear combination of fundamental cutset vectors with respect to a spanning tree of the graph. The coefficients in the linear combination are 1 or -1.

F36: In a directed graph G, $\hat{C}(G)$ and $\lambda(G)$ are vector spaces over the real field and are orthogonal to each other. $\hat{C}(G)$ and $\lambda(G)$ are called the **circuit space** and the **cutset space**, respectively.

F37: The fundamental circuit vectors and the fundamental cutset vectors with respect to a spanning tree of a connected directed graph G form a basis of the circuit space and a basis of the cutset space, respectively. The dimension of the circuit space is equal to $m - n + p$, the nullity of G, and the dimension of the cutset space is equal to $n - p$, the rank of G, where p is the number of components of G.

F38: Any set of $n - 1$ incidence vectors of a connected directed graph forms a basis of the cutset space of the graph.

F39: The fundamental cutset and the fundamental circuit matrices of a connected directed graph are unimodular.

F40: Consider a spanning tree T of a connected directed graph G with branches $b_1, b_2, \ldots\ldots, b_{n-1}$ and chords $c_1, c_2, c_3, \ldots, c_{m-n+1}$. Suppose that the edges of G are labeled so that $e_1, e_2, \ldots, e_m = b_1, b_2, \ldots, b_{n-1}, c_1, c_2, \ldots, c_{m-n+1}$, respectively. Then the fundamental circuit matrix B_f has the form $B_f = [B_{ft}|U_{m-n+1}]$, where U_{m-n+1} is the identity matrix of size $m - n + 1$ and B_{ft} is the submatrix of B_f consisting of the columns corresponding to the branches $b_1, b_2, \ldots, b_{n-1}$ of T. Similarly, the fundamental cutset matrix Q_f has the form $Q_f = [U_{n-1}|Q_{fc}]$, where U_{n-1} is the identity matrix of size $n - 1$ and Q_{fc} is the submatrix of Q_f consisting of the columns corresponding to the chords $c_1, c_2, \ldots, c_{m-n+1}$ of T. Moreover, $Q_{fc} = -B_{ft}^t$.

F41: The columns of the cut matrix of a connected directed graph G are linearly independent if and only if they correspond to the branches of a spanning tree. Similarly, the columns of the circuit matrix are linearly independent if and only if they correspond to the chords of a cospanning tree.

F42: (*Matrix Tree Theorem*) For a connected directed graph, each cofactor of the matrix $A_c A_c^t$ equals the number of spanning trees of the graph.

EXAMPLE

E14: The matrices in Example E13 illustrate Facts F33 through F41.

REMARKS

R10: By simply replacing -1 by 1 in all the matrices defined for directed graphs, we get the corresponding matrices for undirected graphs.

R11: The rank and the nullity of the cut matrix of a connected graph are $(n - 1)$ and $(m - n + 1)$, respectively. This motivated the definitions of the rank and nullity of a graph (see Definition D13).

R12: The matrix $A_c A_c^t$ is called the *degree matrix* of the graph. It can be verified that the diagonal entry (i, i) of the degree matrix is equal to the degree of vertex v_i and the off-diagonal entry (i, j) is equal to the negative of the number of edges connecting vertex v_i and vertex v_j (regardless of the orientations of these edges). A proof of Fact F42 may be found in [ThSw92]. A weighted version of the degree matrix plays an important role in electrical circuit analysis [SwTh81].

Minty's Painting Theorem

TERMINOLOGY: Two directed edges in a circuit or cutset are said to have the ***same direction*** (relative to that circuit or cutset) if both edges agree with the same orientation of that circuit or cutset.

DEFINITIONS

D38: A *directed circuit* is a circuit whose edges all have the same direction relative to it.

D39: A *directed cutset* is a cutset whose edges all have the same direction relative to it.

D40: A *painting* of a directed graph G is a partitioning of the edges of the graph into three sets R, Y, and B and the distinguishing of one element of the set Y. We can visualize this as coloring of the edges of G with three colors, each edge being painted red, yellow, or blue, and exactly one yellow edge being colored dark yellow.

FACTS

F43: (*Painting Theorem*) [Mi66] Let G be a directed graph. For any painting of the edges of G, exactly one of the following holds:

1. There exists a circuit containing the dark yellow edge but no blue edges, in which all the yellow edges have the same direction as the dark yellow edge.

2. There exists a cutset containing the dark yellow edge but no red edges, in which all the yellow edges have the same direction as the dark yellow edge.

F44: Each edge of a directed graph is in a directed circuit or in a directed cutset, but no edge belongs to both.

REMARK

R13: Minty's painting theorem (also known as the ***arc coloring lemma***) has profound applications in electrical circuit theory. This theorem is also true for orientable matroids (see [ThSw92]). Fact F44 is a corollary of Fact F2. Other related works by Minty of considerable significance in electrical circuit theory are [Mi60, Mi61]. Some applications of the arc coloring lemma to problems in electrical circuit theory may be found in [VaCh80, ChGr76, Wo70].

6.4.6   Two Circ/Cut-Based Tripartitions of a Graph

In §6.4.4 we presented a result on the decomposition of a graph into a circ and a cut. But such circs and cuts may not be disjoint and hence they may not form a partition of the edge set of the graph. We now present two ways to partition a graph. These partitions are both tripartitions and are again based on circs and cuts.

Bicycle-Based Tripartition

DEFINITION

D41: A subgraph that is in the intersection of the circuit and cutset subspaces of an undirected graph is called a **bicycle**. That is, a bicycle is a circ as well as a cut.

EXAMPLE

E15: The edges e_1, e_2, e_3, and e_5 in the graph of Figure 6.4.8 form both a cut and a circuit.

FACT

F45: [RoRe78] Any edge e of a graph G is of one of the following types:
  1. e is in a circ that becomes a cut when e is removed from it.
  2. e is in a cut that becomes a circ when e is removed from it.
  3. e is in a bicycle.

TERMINOLOGY: The partition of the edges defined by Fact F45 is called the **bicycle-based tripartition**.

REMARK

R14: Rosenstiehl and Read [RoRe78] have proved several interesting results relating to circuits and cuts and their relationship. A proof of Fact F45 may also be found in [Pa94].

A Tripartition Based on Maximally Distant Spanning Trees

DEFINITIONS

D42: The **tree distance** $d(T_1, T_2)$ between any two spanning trees T_1 and T_2 is defined as $d(T_1, T_2) = |E(T_1) - E(T_2)| = |E(T_2) - E(T_1)|$.

D43: Two spanning trees T_1 and T_2 are **maximally distant** if $d(T_1, T_2) \geq d(T_i, T_j)$ for every pair of spanning trees T_i and T_j.

NOTATION: The maximum distance between any two spanning trees of a connected graph is denoted by d_m.

D44: Given a pair of maximally distant spanning trees T_1 and T_2 of a connected graph G. Suppose c is a common chord of T_1 and T_2. The **k-subgraph G_c** of G with respect to c is the edge-induced subgraph constructed as follows:

1. Let L_1 be the set of all the edges in the fundamental circuit with respect to T_1 defined by c.

2. Let L_2 be the union of the sets of edges in all the fundamental circuits with respect to T_2 defined by every edge in L_1.

3. Repeating the above, we can obtain a sequence of sets of edges $L_1, L_2, \ldots$ until we arrive at a set $L_{k+1} = L_k$. Then the induced subgraph on the edge set L_k is called the *k-subgraph* G_c with respect to c.

D45: The ***k-subgraph*** $\boldsymbol{G_b}$ with respect to a common branch b can be constructed in a dual manner as in Definition D44.

D46: The ***principal subgraph*** G_1 with respect to the common chords (of a pair of maximally distant spanning trees T_1 and T_2) is the union of the k-subgraphs with respect to all the common chords. The ***principal subgraph*** G_2 with respect to the common branches is the union of the k-subgraphs with respect to all the common branches.

FACTS

F46: [KiKa69] Let T_1 and T_2 form a pair of maximally distant spanning trees of a connected graph G.

1. The fundamental circuit of G with respect to T_1 or T_2 defined by a common chord of T_1 and T_2 contains no common branches of these spanning trees.

2. The fundamental cutset of G with respect to T_1 or T_2 defined by a common branch of T_1 and T_2 contains no common chords of these spanning trees.

F47: [KiKa69] Consider a graph $G = (V, E)$. Let E_1 and E_2 denote the edge-sets of the principal subgraphs G_1 and G_2, respectively, and let $E_0 = E(G) - (E_1 \cup E_2)$. Then E_0, E_1, and E_2 form a partition of the edge-set $E(G)$. The partition (E_0, E_1, E_2) is called the ***principal partition*** of G and is independent of the maximally distant trees used to construct it.

EXAMPLE

E16: It can be verified that $T_1 = \{e_2, e_3, e_4, e_7\}$ and $T_2 = \{e_1, e_3, e_5, e_6\}$ are a pair of maximally distant spanning trees for the graph in Figure 6.4.10 and that the associated principal partition is: $E_1 = \{e_6, e_7, e_8\}$, $E_2 = \{e_1, e_2, e_3\}$, and $E_0 = \{e_4, e_5\}$.

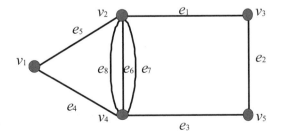

Figure 6.4.10: A graph.

REMARKS

R15: In electrical circuit analysis one is interested in solving for all the current and the voltage variables. The circuit method of analysis (also known as the loop analysis) requires solving for only $m-n+1$ independent current variables. The remaining current variables and all the voltage variables can then be determined using these $m-n+1$ independent current variables. The cutset method of analysis requires solving for only $n-1$ independent voltage variables. A question that intrigued circuit theorists for a long time was whether one could use a hybrid method of analysis involving some current variables and some voltage variables and reduce the size of the system of equations to be solved to less than both $n-1$ and $m-n+1$, the rank and nullity of the graph of the circuit. Ohtsuki, Ishizaki, and Watanabe [OhIsWa70] studied this problem and showed that d_m, the maximum distance between any two spanning trees of the graph of the circuit is, in fact, the minimum number of variables required in the hybrid method of analysis. They also showed that the variables can be determined using the principal partition of the graph. The works by Kishi and Kajitani [KiKa69] on principal partition and by Ohtsuki, Ishizaki, and Watanabe [OhIsWa70] on the hybrid method of analysis are considered landmark results in electrical circuit theory. Swamy and Thulasiraman [SwTh81] give a detailed exposition of the principal partition concept and the hybrid and other methods of circuit analysis.

R16: Lin [Li76] presented an algorithm for computing the principal partition of a graph. Bruno and Weinberg [BrWe71] extended the concept of principal partition to matroids.

6.4.7   Realization of Circuit and Cutset Spaces

In the application of graph theory to the electrical circuit synthesis problem, one encounters a certain matrix of integers modulo 2 and seeks to determine if this matrix is the cutset or the circuit matrix of an undirected graph. The complete solution to this problem was given by Tutte [Tu59]. Cederbaum [Ce58] and Gould [Go58] considered this problem before Tutte provided the solution. We now present the main result on the necessary and sufficient conditions for the realizability of a matrix of integers modulo 2 as the circuit or the cutset matrix of an undirected graph. Related results leading to this main result are also presented. Seshu and Reed [SeRe61] discuss these results in considerable detail, except for proof of the sufficiency of Tutte's realizability condition.

DEFINITIONS

D47: The graphs in Figure 6.4.11 are called **Kuratowski graphs**.

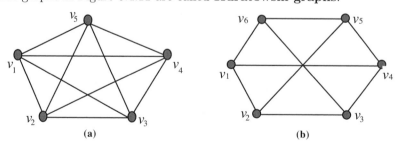

(a)   (b)

Figure 6.4.11: The two Kuratowski graphs.

D48: A matrix F of the form $F = [F' \mid U]$, where U is the identity matrix, is said to be in **normal form**.

D49: A matrix F of real integers in normal form is a **regular matrix** if for every linear combination X of the rows of M with coefficients -1, 1, and 0 we have the following:

1. The elements of X are 1, -1, and 0, or

2. There exists another such linear combination Y (with coefficients 1, -1, and 0) that has 1 and -1 for nonzero elements and these are at a (not necessarily proper) subset of the positions in which X has nonzero elements.

D50: A matrix of integers mod 2 is **regular** if the replacement of a suitable set of 1's by -1's makes it regular.

FACTS

F48: For a connected directed graph G, the fundamental cutset and fundamental circuit matrices with respect to a spanning tree T of G and the reduced incidence matrix A of G are all regular matrices.

F49: A regular matrix in normal form is unimodular.

F50: Given a regular matrix F of integers 1, -1, and 0 in normal form, the replacement of -1's by 1's will leave the ranks of the submatrices unaltered (where the rank of the derived matrix is with respect to modulo 2 arithmetic).

F51: A matrix F of integers mod 2 is regular if and only if no normal form of F contains either the matrix N_0 or its transpose, where

$$
N_0 = \begin{bmatrix} 1 & 1 & 1 & 0 \\ 1 & 1 & 0 & 1 \\ 1 & 0 & 1 & 1 \end{bmatrix}
$$

F52: A matrix F of integers mod 2 is realizable as the cutset matrix of an undirected graph if and only if it is regular and no normal form of F contains the circuit matrix of either of the two Kuratowski graphs shown in Figure 6.4.11

F53: A matrix F of integers mod 2 in normal form is realizable as the circuit matrix of an undirected graph if and only if it is regular and no normal form of F contains the cutset matrix of either of the two Kuratowski graphs shown in Figure 6.4.11

REMARKS

R17: Mayeda [Ma70] gave an alternate proof of Tutte's realizability condition, shorter than Tutte's original proof, which is 27 pages long.

R18: Early works on algorithms for constructing graphs having specified circuit or cutset matrices are in [Tu60, Tu64]. Bapeswara Rao [Ba70] defined the tree-path matrix of an undirected graph which is essentially the non-unit submatrix of the fundamental circuit matrix and presented an algorithm for constructing a graph with a prescribed tree-path matrix. This is also an algorithmic solution to the cutset and the circuit matrix realization problems. A detailed presentation of Bapeswara Rao's algorithm is given in [SwTh81].

R19:   The circuit and the cutset matrix realization problems arise in the design of multi-port resistance networks. It was in the context of this application that Cederbaum [Ce58, Ce59] encountered the realization problem. Interestingly, Bapeswara Rao [Ba70] and Boesch and Youla [BoYo65] presented circuit-theoretic approaches to the realization of a matrix as the cutset or circuit matrix of a directed graph. Details of Bapeswara Rao's algorithm based on this approach may also be found in [SwTh81].

Whitney and Kuratowski

We believe that it is appropriate to conclude this section with a reference to two classic works by Whitney [Wh33] and Kuratowski [Ku30] relating to duality. While the results of this section bring out the duality between circuits and cutsets, Whitney introduced the concept of duality between graphs. His original definition was an algebraic one (see also [ThSw92]) relating the nullity and rank of certain corresponding subgraphs of dual graphs. Definition D51 is an equivalent one.

DEFINITION

D51:   A graph G_2 is a *dual* of a graph G_1 if there is a one-to-one correspondence between their edge-sets such that a set of edges in G_2 is a circuit vector of G_2 if and only if the corresponding set of edges in G_1 is a cutset vector of G_1.

FACTS

F54:   It follows from the duality between circuits and cutsets that if G_2 is a dual of G_1, then G_1 is a dual of G_2.

F55:   [Wh33] A graph has a dual if and only if it is planar.

F56:   In another classic work, Kuratowski [Ku30] proved that a graph is planar if and only if it does not contain a subdivision of a Kuratowski graph.

REMARK

R20:   See [We01] for a proof of Kuratowski's theorem. It is quite interesting to see the role of the Kuratowski graphs in Tutte's realizability conditions for the cutset and the circuit matrix realization problems.

6.4.8   An Application: Cross-Layer Survivability in a Layered Network

DEFINITIONS

D52:   A *layered network* is a pair (G_P, G_L), where $G_P = (V_P, E_P)$, the physical graph, and $G_L = (V_L, E_L)$, the logical graph, have the same vertex set. That is, $V_P = V_L$. It is assumed that both G_P and G_L are two-edge connected.

D53: A path in G_P is called a **lightpath**.

D54: A layered network (G_P, G_L) is **survivable** if there exists a mapping of each edge $(i, j) \in E_L$ into an $i - j$ lightpath in G_P such that deletion of an edge in G_P does not cause G_L to become disconnected. Such a mapping is called a *survivable logical topology mapping*.

The following is a generalization of the concept of ear decomposition defined in §7.2.

D55: A **generalized partial circuit ear decomposition** of a connected graph G is the collection of *circuit ears* obtained by the following algorithm:

1. Let $G_C = G$.
2. Pick a circuit C of size two (called *circuit ear*) in G_C.
3. Contract the edges in C, resulting in a new G_C.
4. Repeat 1 and 2 until G_C is a single node.

D56: Some of the edges in G may not be in any circuit ear in a partial circuit ear decomposition, and such edges are called **straddling edges**.

D57: A **generalized partial cutset ear decomposition** of a connected graph G is the collection of *cutset ears* obtained by the following algorithm:

1. Let $G_S = G$.
2. Pick a cutset S of size two (called *cutset ear*) in G_S.
3. Delete the edges in S, resulting in a new G_S.
4. Repeat 1 and 2 until every cutset in G_S is a single edge.

D58: Some of the edges in G may not be in any cutset ear in a partial cutset ear decomposition, and such edges are called **isolated edges**.

FACTS

F57: [KuTh05] A layered network (G_P, G_L) is survivable if and only if there is a generalized partial circuit ear decomposition of G_L such that all the edges in a circuit ear can be mapped into edge disjoint lightpaths in G_P.

F58: [ThJaXu09] Given a generalized partial cutset ear decomposition of G_L in a layered network, let G_L^* be graph obtained by contracting the isolated edges in the decomposition. Then, the layered network (G_P, G_L^*) is survivable if and only if there is a generalized partial cutset ear decomposition of G_L^* such that all the edges in a cutset ear can be mapped into edge disjoint lightpaths in G_P.

REMARKS

R21: An IP (Internet Protocol) over a WDM (Wavelength Division Multiplexed) network is an example of a layered network. Here G_P represents the optical network and the logical graph G_L represents the network at the IP layer.

R22: The concepts of generalized partial circuit and cutset ear decompositions were introduced in [ThJaXu09] and studied further in [Th09], [ThJaXu10], [ThLiJaXu10], and [ThLiZhXu12].

References

[Ba70] V. V. Bapeswara Rao, *The Tree-Path Matrix of a Network and Its Applications*, Ph.D. Thesis, Department of Electrical Engineering, Indian Institute of Technology, Madras, India, 1970.

[BoYo65] F. T. Boesch and D. C. Youla, Synthesis of resistor n-port networks, *IEEE Trans. Circuit Theory* **12** (1965), 515–520.

[BrWe71] J. Bruno and L. Weinberg, The principal minors of a matroid, *Linear Algebra and Its Applications* **4** (71), 17–54.

[Ce58] I. Cederbaum, Matrices all of whose elements and subdeterminants are 1, -1 or 0, *J. Math. and Phys.* **36** (58), 351–361.

[Ce59] I. Cederbaum, Applications of matrix algebra to network theory, *IRE Trans. Circuit Theory* **6** (1959), 127–137.

[Ch71a] W.-K. Chen, On vector spaces associated with a graph, *SIAM J. Appl. Math.* **20** (1971), 526–529.

[Ch71b] W.-K. Chen, *Applied Graph Theory*, North Holland, Amsterdam, 1971.

[ChGr76] L. O. Chua and D. M. Greene, Graph-theoretic properties of dynamic nonlinear networks, *IEEE Trans. Circuits and Systems* **23** (1976), 292–312.

[De74] N. Deo, *Graph Theory with Applications to Engineering and Computer Science*, Prentice-Hall, 1974.

[Eu36] L. Euler, Solutios Problematis ad Geometriam Situs Pertinantis, *Academimae Petropolitanae* **8** (1736), 128–140.

[Go58] R. L. Gould, Graphs and vector spaces, *J. Math. and Phys.* **38** (1958), 193–214.

[GrYe06] J. L. Gross and J. Yellen, *Graph Theory and Its Applications*, Second Edition, CRC Press, 2006.

[KiKa69] G. Kishi and Y. Kajitani, Maximally distant trees and principal partition of a linear graph *IEEE Trans. Circuit Theory* **16** (1969), 323–330.

[Ku30] C. Kuratowski, Sur le probleme des Courbes Gauches en topologie, *Fund. Math.* **15** (1930), 271–283.

[KuTh2005] M. Kurant and P. Thiran, On survivable routing of mesh topologies in IP-over-WDM networks, *IEEE INFOCOM* 2005, 1106–1116.

[Li76] P. M. Lin, An improved algorithm for principal partition of graphs, *Proc. IEEE Intl. Symp. Circuits and Systems* (1976), 145–148.

[Ma70] W. Mayeda, A proof of Tutte's realizability condition, *IEEE Trans. Circuit Theory* **17** (1970), 506–511.

[Mi60] G. J. Minty, Monotone networks, *Proc. Roy. Soc., A* **257** (1960), 194–212.

[Mi61] G. J. Minty, Solving steady-state nonlinear networks of 'monotone' elements, *IRE. Trans. Circuit Theory* **8** (1961), 99–104.

[Mi66] G. J. Minty, On the axiomatic foundations of the theories of directed linear graphs, electrical networks and network programming, *J. Math. and Mech.* **15** (1966), 485–520.

[OhIsWa70] T. Ohtsuki, Y. Ishizaki, and H. Watanabe, Topological degrees of freedom and mixed analysis of electrical networks, *IEEE Trans. Circuit Theory* **17** (1970), 491–499.

[Pa94] K. R. Parthasarathy, *Basic Graph Theory*, Tata McGraw-Hill Publishing Company, New Delhi, India, 1994.

[Re61] M. B. Reed, The seg: a new class of subgraphs, *IEEE Trans. on Circuit Theory* **CT-8** (1961), 17–22.

[RoRe78] P. Rosenstiehl and R. C. Read, On the principal edge tripartition of a graph, *Annals of Discrete Mathematics* **3** (1978), 195–226.

[SeRe61] S. Seshu and M. B. Reed, *Linear Graphs and Electrical Networks*, Addison Wesley, 1961.

[SwTh81] M. N. S. Swamy and K. Thulasiraman, *Graphs, Networks and Algorithms*, Wiley (Interscience), 1981.

[Th09] K. Thulasiraman, Duality in Graphs and Logical Topology Survivability in Layered Networks, *India-Taiwan Conference on Discrete Mathematics*, National Taiwan University, Taipei, 2009.

[ThJaXu09] K. Thulasiraman, M. Javed, and G. Xue, Circuits/Cutsets Duality and a Unified Algorithmic Framework for Survivable Logical Topology Design in IP-over-WDM Optical Networks, *INFOCOM* 2009, Rio de Janeiro, 1026–1034.

[ThJaXu10] K. Thulasiraman, M. Javed, and G. Xue, Primal Meets Dual: A Generalized Theory of Logical Topology Survivability in IP-over-WDM Optical Networks, *2nd Intl. Conference on Comm. Syst. and Networks (COMSNETS)* 2010, 1–10.

[ThLiJaXu10] K. Thulasiraman, T. Lin, M. Javed, and G. Xue, Logical Topology Augmentation for Guaranteed Survivability under Multiple Failures in IP-over-WDM Optical Networks, *Optical Switching and Networking (OSN) Journal Special Issue of Advanced Networks and Telecom. Systems* **7** (2010), 206–214.

[ThLiZhXu12] K. Thulasiraman, T. Lin, Z. Zhou, and G. Xue, Robustness of Logical Topology Mapping Algorithms for Survivability Against Multiple Failures in an IP over WDM Optical Network, *4th Intl. Conference on Comm. Syst. and Networks (COMSNETS)*, Bengaluru, 2012.

[ThSw92] K. Thulasiraman and M. N. S. Swamy, *Graphs: Theory and Algorithms*, Wiley (Interscience), 1992.

[Tu59] W. T. Tutte, Matroids and graphs, *Trans. of the Amer. Math. Soc.* **90** (1959), 527–552.

[Tu60] W. T. Tutte, An algorithm for determining whether a given binary matroid is graphic, *Proc. Amer. Math. Soc.* **11** (1960), 905–917.

[Tu64] W. T. Tutte, From matrices to graphs, *Canad. J. Math.* **56** (1964), 108–127.

[VaCh80] J. Vandewalle and L. O. Chua, The colored branching theorem and its applications in circuit theory, *IEEE Trans. Circuits and Systems* **27** (1980), 816–825.

[Ve31] O. Veblen, *Analysis Situs*, Amer. Math. Soc., 1931.

[We01] D. B. West, *Introduction to Graph Theory*, Prentice Hall, 2001.

[Wh33] H. Whitney, Planar graphs, *Fund. Math.* **21** (1933), 73–84.

[WiMa71] T. W. Williams and L. M. Maxwell, The decomposition of a graph and the introduction of a new class of subgraphs, *SIAM J. Appl. Math.* **20** (1971), 385–389.

[Wo70] D. H. Wolaver, Proof in graph theory of the no gain property of resistor networks, *IEEE Trans. Circuits and Systems* **17** (1970), 436–437.

Section 6.5
Spectral Graph Theory

Michael Doob, University of Manitoba, Canada

INTRODUCTION

Spectral graph theory involves the investigation of the relationship of the usual (topological) properties of a graph with the (algebraic) spectral properties of various matrices associated with it. By far the most common matrix investigated has been the 0-1 adjacency matrix. The subject had its genesis with the paper by L. Collatz and U. Sinogowitz [CoSi57] in 1957. Since that time the subject has steadily grown and has shown surprising interrelationships with other mathematical areas.

Throughout this section, graphs are assumed to be simple.

6.5.1 Basic Matrix Properties

Many spectral properties of graphs follow from direct application or easy extensions of known results in matrix theory. An older but compact and useful reference is [MiMa64]. A more encyclopedic one is [Ga60].

DEFINITIONS

D1: The **adjacency matrix** of a (simple) graph G is a square matrix A (or A_G) with rows and columns corresponding to the vertices.

$$A_{i,j} = \begin{cases} 1 & \text{if } v_i \text{ and } v_j \text{ are adjacent} \\ 0 & \text{otherwise} \end{cases}$$

D2: The ***characteristic polynomial of a graph*** is the determinant $\det(xI - A)$ of its adjacency matrix.

D3: The ***eigenvalues of a graph*** are the roots to the characteristic polynomial.

D4: The ***algebraic multiplicity*** of an eigenvalue λ is the number of times it occurs as a root of the characteristic polynomial.

D5: The ***spectrum of a graph*** is the multiset of eigenvalues. For a graph with n vertices, there are n eigenvalues.

D6: The ***geometric multiplicity*** of an eigenvalue λ is the dimension of the *eigenspace* $\{x \mid Ax = \lambda x\}$.

FACTS

F1: The eigenvalues of a graph are real. This follows because the adjacency matrix is real and symmetric, which implies that it is *Hermitian*, i.e., each entry $a_{i,j}$ equals the complex conjugate of $a_{j,i}$. (The eigenvalues of any Hermitian matrix are real.)

F2: The geometric and algebraic multiplicities of each eigenvalue are equal.

F3: The eigenvalues have a corresponding set of eigenvectors that is orthonormal.

F4: The adjacency matrix A of a graph can be diagonalized, that is, there is a square matrix U (of eigenvectors) such that $UAU^T = UAU^{-1}$ is a diagonal matrix with the eigenvalues as diagonal entries.

F5: The *trace* of the adjacency matrix of a graph, that is, the sum of the eigenvalues of a graph, is 0, since the adjacency matrix is 0 on the diagonal.

F6: The spectrum of a graph is the union of the spectra of its connected components, since connected components of a graph are just blocks down the diagonal of the adjacency matrix.

F7: If r is the largest eigenvalue of a graph, then $|\lambda| \leq r$ for any eigenvalue λ of that graph. (Since the adjacency matrix has nonnegative entries and connectivity implies irreducibility, this follows from the well-known Perron–Frobenius theorem – see [MiMa64].)

F8: If a graph is connected, then the largest eigenvalue has multiplicity 1. It has an eigenvector with all entries positive. (This fact is another consequence of the Perron–Frobenius theorem.) Since the adjacency matrix is symmetric, being imprimitive is equivalent to the graph being bipartite.

F9: Let r be the largest eigenvalue of a graph. Then a graph is bipartite if and only if the number $-r$ is also an eigenvalue.

F10: Whether or not a graph is bipartite can be determined by its spectrum. (This follows immediately from Fact F9.)

F11: A graph is bipartite if and only if the spectrum is symmetric around 0, that is, λ is an eigenvalue if and only if $-\lambda$ is an eigenvalue.

REMARK

R1: Because of Fact F6, for most results in this chapter the graphs under consideration may be assumed to be connected, with no loss of generality.

EXAMPLES

The spectrum of a graph is given as a set of eigenvalues with the multiplicities as exponents (and thus, the determinant is taken as the product of the set entries).

E1: The complete graph K_n: $\{(n-1)^1, -1^{n-1}\}$.

E2: The complete bipartite graph $K_{m,n}$: $\{\sqrt{mn}^1, 0^{m+n-2}, -\sqrt{mn}^1\}$.

E3: The path with n vertices P_n: $\{2\cos(k\pi/n+1)^1, k = 1, \ldots, n\}$.

E4: The circuit with n vertices C_n: $\{2\cos(2k\pi/n)^1, k = 1, \ldots, n\}$.
Notice that these eigenvalues are not distinct. The eigenvalue 2 and, when n is even, the eigenvalue -2 are simple and all others have multiplicity 2.

E5: A **cocktail party graph** $\mathrm{CP}(n)$ is a complete graph on $2n$ vertices with a 1-factor deleted. Spectrum: $\{(2n-2)^1, 0^n, -2^{n-1}\}$.

E6: The d-dimensional hypercube Q_d: $\{(d-2k)^{\binom{d}{k}}, k = 0, \ldots, d\}$.

E7: A **wheel** W_n (with $n+1$ vertices) is the join of an n-cycle C_n and an additional vertex. Spectrum: $\{2\cos(2k\pi/n)^1, k = 1, \ldots, n-1\} \cup \{1 \pm \sqrt{1+n}\}$.

E8: The platonic graphs
- tetrahedral graph K_4: $\{3, -1^3\}$.
- cube graph Q_3: $\{3^1, 1^3, -1^3, -3^1\}$.
- octahedral graph $\mathcal{O}_3(\cong \mathrm{CP}(3))$: $\{4^1, 0^3, -2^2\}$.
- dodecahedral graph: $\{3^1, \sqrt{5}^3, 1^5, 0^4, -2^4, -\sqrt{5}^3\}$.
- icosahedral graph: $\{5^1, \sqrt{5}^3, -1^5, -\sqrt{5}^3\}$.

6.5.2   Walks and the Spectrum

Walks and the Coefficients of the Characteristic Polynomial

DEFINITIONS

D7: An **elementary figure** in a graph is a subgraph that is isomorphic to K_2 or to a cycle graph C_k.

D8: A **basic figure** is a vertex-disjoint union of elementary figures.

FACTS

F12: If A is the adjacency matrix of a graph, then $A^k{}_{i,j}$ is the number of walks of length k from the vertex v_i to v_j.

F13: If A is the adjacency matrix of a graph, then $A^k{}_{j,j}$ is the number of closed walks of length k from the vertex v_j to v_j.

F14: The trace of A^k is the total number of closed walks of length k in the graph.

F15: The trace of A^2 is twice the number of edges in the graph.

F16: The trace of A^3 is six times the number of triangles in the graph.

F17: [Sa64] If $a_n x^n + a_{n-1} x^{n-1} + \cdots + a_1 x + a_0$ is the characteristic polynomial of a graph, then $a_n = 1$ and $a_{n-1} = 0$. Also, $-a_{n-2}$ is the number of edges and $-a_{n-3}$ is twice the number of triangles in the graph. (This fact follows readily from an expansion of the determinant $\det(xI - A)$.)

F18: [Sa64] If $a_n x^n + a_{n-1} x^{n-1} + \cdots + a_1 x + a_0$ is the characteristic polynomial of a graph, then $a_k = \sum (-1)^{\text{comp}(B)} 2^{\text{circ}(B)}$, where the sum is over all basic figures with $n - k$ vertices, comp(B) is the number of connected components of B, and circ(B) is the number of circuits of B.

REMARK

R2: Fact F18 is sometimes called the *coefficients theorem*. Its interesting history is given in [CvDoSa95]. It extends the idea of Fact F17 to other coefficients of the characteristic polynomial. The coefficient of x^k will be determined by permutations with exactly k fixed points (vertices). For the other vertices, the permutations will have cycles of length 2 (corresponding to an edge in the graph) or cycles of length greater than two (corresponding to a circuit in the graph). Thus, in order to determine the coefficient of x^k we need to count all of the basic figures that have $n - k$ vertices. Furthermore, within each basic figure, a permutation corresponding to a circuit contributes 2 to the determinant (once clockwise, once counterclockwise) and an edge contributes 1 to the determinant.

Walks and the Minimal Polynomial

DEFINITIONS

D9: The *minimal polynomial* of a graph G is the monic polynomial $q(x)$ of smallest degree, such that $q(A_G) = 0$.

D10: The *eigenvalues-diameter (lower) bound* for the number of eigenvaues of a graph G is diam$(G) + 1$.

FACTS

F19: The minimal polynomial of a graph is $m(x) = \prod (x - \lambda_i)$ where the product is taken over all distinct eigenvalues.

F20: Given two vertices v_i and v_j at distance t in a graph with adjacency matrix A, we have $A_{i,j}^k = 0 \le k < t$, and $A_{i,j}^t \ne 0$.

F21: If a graph has diameter d and has m distinct eigenvalues, then $m \ge d + 1$. This substantiates the eigenvalues-diameter bound. It follows from Fact F20.

F22: The degree of the minimal polynomial is larger than the diameter of a graph.

F23: The complete graph is the only (connected) graph with exactly two distinct eigenvalues.

F24: The complete graph K_n is determined by its spectrum. (This follows from Fact F23, since the total number of eigenvalues — taking multiplicities into account — equals the number of vertices.)

EXAMPLES

E9: Note that in our previous examples, the graphs K_n, $K_{m,n}$, CP(n), P_n, C_n, and Q_n all attain the eigenvalues-diameter bound.

E10: The wheel W_n has approximately $n/2$ distinct eigenvalues and diameter 2.

RESEARCH PROBLEM

RP1: Characterizing those graphs meeting the eigenvalues-diameter bound remains an open question. Of the 31 connected graphs with 5 or fewer vertices, there are 12 that meet the bound.

Regular Graphs

DEFINITION

D11: The **Hoffman polynomial** for a regular, connected graph of degree r is the polynomial

$$h(x) = n \prod \frac{(x - \lambda_i)}{(r - \lambda_i)}$$

the product being taken over all distinct eigenvalues not equal to r.

FACTS

NOTATION: Let J denote a square matrix with every entry equal to 1.

F25: The largest eigenvalue of a regular graph of degree r is r itself. A corresponding eigenvector is $(1, 1, \ldots, 1)^T$. (This follows from the Perron–Frobenius theorem.)

F26: Any eigenvector corresponding to an eigenvalue other than r has coordinates that sum to 0. (This is because eigenvectors from different eigenvalues are orthogonal.)

F27: The multiplicity of the eigenvalue r is the number of connected components. (Each connected component contributes 1 to the multiplicity of r.)

F28: A graph is regular if and only if A and J commute.

F29: The complement of a regular graph with n vertices has an adjacency matrix equal to $J - A - I$. Hence the eigenvalues for the complement of a regular connected graph are $n - r - 1$ and $-\lambda_i - 1$, where λ_i runs over the eigenvalues of A not equal to r.

F30: [Ho63] If a regular connected graph has adjacency matrix A and Hoffman polynomial $h(x)$, then $h(A) = J$.

6.5.3   Line Graph, Root System, Eigenvalue Bounds

An early problem in spectral graph theory was bounding the eigenvalues of a graph from below. One of the basic tools for bounding eigenvalues comes from matrix theory and is called the *interlacing theorem*. Other uses for the interlacing theorem are given in [Ha95].

DEFINITIONS

D12: A **principal submatrix** of an $n \times n$ square matrix is obtained by deleting the ith row and the ith column for some $1 \leq i \leq n$.

D13: The **line graph** of a graph G, denoted $L(G)$, has the edges of G as vertices with two vertices in $L(G)$ adjacent if, as edges of G, they have an endpoint in common.

D14: The **vertex-edge incidence matrix** of a graph G has rows corresponding to its vertices and columns corresponding to its edges. An entry is 1 if the vertex corresponding to the row is incident to the edge corresponding to the column, and is 0 otherwise. It is denoted by $K(G)$ or, simply, by K.

D15: Given a graph G with n vertices and nonnegative integers $a_1, \ldots, a_n$, the **generalized line graph** $L(G; a_1, \ldots, a_n)$ is formed as follows: first, take disjoint copies of the line graph $L(G)$ and cocktail party graphs $\text{CP}(a_1), \ldots, \text{CP}(a_n)$. In addition, if a vertex in $L(G)$ corresponds to the edge joining v_i to v_j in G, then join it to all vertices in $\text{CP}(a_i)$ and $\text{CP}(a_j)$.

NOTATION: $\lambda(G)$ denotes the smallest eigenvalue of a graph G.

NOTATION: $\Lambda(G)$ denotes the largest eigenvalue of a graph G.

FACTS ABOUT INTERLACING

F31: Let A be a real symmetric matrix with eigenvalues $\lambda_1 \geq \lambda_2 \geq \cdots \geq \lambda_n$, having a principal submatrix with eigenvalues $\mu_1 \geq \mu_2 \geq \cdots \geq \mu_{n-1}$. Then

$$\lambda_1 \geq \mu_1 \geq \lambda_2 \geq \mu_2 \geq \cdots \geq \lambda_{n-1} \geq \mu_{n-1} \geq \lambda_n$$

F32: If H is an induced subgraph of G, if $\mu_1 \geq \mu_2 \geq \cdots \geq \mu_m$ are the eigenvalues of H, and if $\lambda_1 \geq \lambda_2 \geq \cdots \geq \lambda_n$ are the eigenvalues of G, then

$$\lambda_i \geq \mu_i \geq \lambda_{i+n-m} \quad \text{for } i = 1, \ldots, m$$

FACTS ABOUT THE SMALLEST EIGENVALUE $\lambda(G)$

F33: If H is an induced subgraph of G, then $\lambda(H) \leq \lambda(G)$.

F34: The least eigenvalue of a connected graph is always nonpositive. It equals zero if and only if the graph is K_1.

F35: No graph has a least eigenvalue between 0 and -1.

F36: The only connected graphs with least eigenvalue -1 are the complete graphs with two or more vertices.

F37: There are no graphs with least eigenvalue between -1 and $-\sqrt{2}$.

F38: The graph $K_{1,2}$ is the only connected graph whose least eigenvalue equals $-\sqrt{2}$.

F39: There are infinitely many connected graphs with their least eigenvalues between $-\sqrt{2}$ and -2. (This follows from Example E3.)

Line Graphs and Generalized Line Graphs

FACTS ABOUT THE LINE GRAPH $L(G)$

F40: If K is the vertex-edge incidence matrix of a graph G, then $KK^T = 2I + A(L(G))$.

F41: If K is the vertex-edge incidence matrix of a graph G, then the matrix KK^T is positive semidefinite, and hence has nonnegative eigenvalues.

F42: [Ho75] For any graph G, $\lambda(L(G)) \geq -2$.

F43: [Do70] A graph G satisfies $\lambda(L(G)) > -2$ if and only if G is a tree or G has exactly one circuit, that circuit being odd.

F44: There are infinitely many graphs G with $-2 \leq \lambda(G) < -\sqrt{2}$.

FACTS ABOUT THE GENERALIZED LINE GRAPH $\lambda(L(G; a_1, \ldots, a_n)$

The results for generalized line graphs are similar to those for line graphs. Form the matrix K' by appending columns and rows to the vertex-edge incidence matrix. For each a_i, append a_i pairs of new columns, each of which has two nonzero entries. For each pair there is a 1 in the row corresponding to v_i and a new row is added with one 1 and one -1 in the new columns. All other entries are 0. Use $K'K'^T$ as before.

F45: [CDS81] For any graph G with n vertices $\lambda(L(G; a_1, \ldots, a_n) \geq -2$.

F46: [CDS81] A graph G satisfies the lower bound $\lambda(L(G; a_1, \ldots, a_n) > -2$ if and only if $G = L(T; 1, 0, \ldots, 0)$ where T is a tree or $G = L(H)$ where H is a tree and H has exactly one cycle, that cycle being odd.

F47: [Ho75] If G is a regular connected graph of degree r with n vertices and eigenvalues $r = \lambda_1 > \lambda_2 \geq \lambda_3 \geq \cdots \geq \lambda_n$ then the eigenvalues of $L(G)$ are $\lambda_i + r - 2$, $i = 1, \ldots, n$ plus -2 of (additional) multiplicity $n(r-2)/2$.

EXAMPLES

E11:   The line graph of a complete graph is called a ***triangular graph***. From the spectrum of K_n (Example E1) we see that the spectrum of the triangular graph $L(K_n)$ is $\{2n - 4^1, n - 4^{n-1}, -2^{n(n-3)/2}\}$.

E12:   The line graph of a regular complete bipartite graph is called a ***lattice graph***. From the spectrum of $K_{n,n}$ (Example E2), it follows that the spectrum of the lattice graph $L(K_{n,n})$ is $\{2n - 2^1, n - 2^{2n-2}, -2^{(n-1)^2}\}$.

Root Systems

The converse of Fact F46 is almost true. The exact statement involves the *root systems* used in the classification of semisimple Lie algebras (found in [Ca89], for example). **Root systems** are sets of vectors used to form the columns of the matrices K' as above.

FACTS ABOUT ROOT SYSTEMS

The possible real root systems are denoted A_n, D_n, E_6, E_7, and E_8. These root systems have $n(n + 1)/2$, $n(n - 1)$, 36, 63, and 120 vectors, respectively. They also satisfy $A_{n-1} \subseteq D_n$ and $E_6 \subseteq E_7 \subseteq E_8$. One of the most beautiful results in spectral graph theory relates root systems with eigenvalues of graphs.

F48:   [GCSS76] $\lambda(H) \geq -2$ if and only if there is a matrix K' whose columns are taken from the root systems A_n, D_n, E_6, E_7, or E_8 such that $A = 2I + K'K'^T$.

F49:   [GCSS76] $H = L(G)$ where G is bipartite if and only if K' can be formed from vectors in the root system A_n.

F50:   [GCSS76] $H = L(G; a_1, \ldots, a_n)$ if and only if K' can be formed from vectors in the root system D_n.

F51:   [GCSS76] With only a finite number of exceptions, $\lambda(H) \geq -2$ implies that H is a generalized line graph.

The exceptional graphs from the last fact are those constructed from (the finite number of) vectors in E_6, E_7, and E_8 by using them as columns of a matrix K' as described in Fact F48. Using a variety of techniques (both computer-assisted and otherwise) much is known about these exceptional graphs.

F52:   If $\lambda(H) > -2$, then either H can be formed by vectors in the root system D_n, or H is one of 20 graphs with six vertices, 110 graphs with seven vertices, or 443 graphs with eight vertices.

A type of characterization of graphs with $\lambda(G) > -1 - \sqrt{2}$ by a different generalization of line graphs has been given in [WoNu95].

EXAMPLE

E13:   Let $\{e_1, \ldots, e_n\}$ be the canonical basis for $\mathbb{R}^n$. Consider the following set of vectors: $\{e_i - e_{i+1}, i = 1, \ldots, n - 1\} \cup \{e_1 + e_2, e_{n-1} + e_n\}$. In fact, these vectors are in the root system D_n. Use these vectors as columns for K' and define a graph, also denoted by D_n, from the matrix equation $A = 2I + KK'^T$. The graph then appears as in Figure 6.5.1.

Figure 6.5.1: The graph D_n.

Eigenvalue Bounds

FACTS ABOUT THE LARGEST EIGENVALUE $\Lambda(G)$
Many of the above ideas can also be used to obtain upper bounds on the eigenvalues of graphs.

F53: If H is an induced subgraph of G, then $\Lambda(H) \leq \Lambda(G)$. If H is a proper subgraph of G, then $\Lambda(H) < \Lambda(G)$.

F54: If $\bar{r}$ is the average degree and $r_{\max}$ is the maximum degree of a graph G, then $\bar{r} \leq \Lambda(G) \leq r_{\max}$. Equality is attained if and only if the graph is regular.

F55: [CvDoGu82] If $\Lambda(G) < 2$, then G is P_n, $T(1,1,n)$, $T(1,2,4)$, $T(1,2,3)$, or $T(1,2,2)$. The graph $T(i,j,k)$ is formed by taking three paths $u_0, \ldots, u_i$, $v_0, \ldots, v_j$, and $w_0, \ldots, w_k$ and identifying the vertices u_o, v_0, and w_0. It clearly has one vertex of degree three, three vertices of degree one, and all other vertices of degree two.

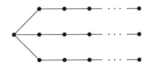

Figure 6.5.2: The graph $T(i,j,k)$.

F56: [CvDoGu82] If $\Lambda(G) = 2$ then G is C_n, $K_{1,4}$, $T(2,2,2)$, $T(3,3,1)$, $T(5,2,1)$, or D_n.

Similar results are available for graphs with $\Lambda(G) < (2 + \sqrt{5})^{1/2}$ (see [CvDoGu82]). Bounds on the second largest eigenvalue of a graph, $\lambda_2(G)$, have also been of considerable interest.

DEFINITIONS

D16: A ***Ramanujan graph*** is regular of degree k for which any eigenvalue $\lambda \neq \pm k$ satisfies $|\lambda| \leq 2\sqrt{k-1}$. In particular $\lambda_2 \leq 2\sqrt{k-1}$.

D17: The ***spectal gap*** of a graph is $\lambda_1 - \lambda_2$.

D18: If the vertices of the graph G are partitioned into $\{V, W\}$, then the ***boundary*** of V, denoted $\partial(V)$, is the number of edges with one end point in V and one in W.

D19:   A graph is an ε-*expander* if, for every partition of the vertices $\{V, W\}$ with $|V| \le |W|$, it follows that $\partial(V) \ge \varepsilon |V|$. In other words, the average outdegree from V to W is bounded from below by ε.

FACTS

F57: [LuPhSa88] Families of Ramanujan graphs with fixed degree k and a large number of vertices are difficult to construct. One such family uses the Caley graphs of groups derived from $\mathrm{PSL}_2(q)$.

F58: [LuPhSa88], [Fr91] Consider graphs $G_{k,n}$ with n vertices and regular of degree k. Then $\liminf_{n \to \infty} G_{k,n} \ge 2\sqrt{k-1}$, and for given ϵ, the probability that $G_{k,n} \le 2\sqrt{k-1} + \epsilon$ goes to 1 as $n \to \infty$.

F59: [Mu03] Ramanujan graphs make good expanders.

F60: [Al86] Good expanders have a large spectral gap.

F61: [Sm70] $\lambda_2(G) \le 0$ if and only if G is a complete multipartite graph.

F62: [Li94] $\lambda_2(G)$ is dense in the interval $\left((2 + \sqrt{5})^{1/2}, \infty\right)$ but nowhere dense in $(-\infty, \sqrt{2} - 1)$.

Graphs with $\lambda_2(G) \le \sqrt{2} - 1$ and $\lambda_2(G) \le (\sqrt{5} - 1)/2$ are described in [CvSi95], [Li94], and [Si95].

FURTHER READING

A complete discussion of the constructions of Ramanujan graphs using $\mathrm{PSL}_2(q)$ can be found in [DaSaVa03].

An excellent survey of the relationships between Ramanujan graphs and analogues of the Riemann zeta function is given in [Mu03].

A complete reference of the spectral properties of line graphs and their generalizations is [CvRoSi04].

A more general discussion on different bounds on the eigenvalues of graphs can be found in [PeRa01].

6.5.4   Distance-Regular Graphs

Eigenvalues are crucial for understanding the properties of distance-regular graphs.

DEFINITIONS

D20:   A graph of diameter d is *distance-regular* with parameters

$$\{p_{i,j}^k \mid 0 \le i, j, k \le d\}$$

if for each triple (i, j, k) and for any pair of vertices such that the distance between them is k, the number of vertices at distance i from the first and distance j from the second is $p_{i,j}^k$. Each of these numbers $p_{i,j}^k$ is independent of the particular choice of vertices.

D21: In the ***Hamming graph*** $H(d,n)$, the vertices are all d-tuples $(x_1, \ldots, x_d)$ with $1 \le x_i \le n$. Two vertices are joined if, as d-tuples, they agree in all but one coordinate. The distance between two vertices is then the number of coordinates in which, as d-tuples, they differ.

D22: In the ***Johnson graph*** $J(d,n)$, the vertices are the $\binom{n}{d}$ subsets of $\{1, \ldots, n\}$. Two vertices are joined if, as subsets, their intersection has cardinality $d-1$. The distance bewteen two vertices is k if, as subsets, they have an intersection of cardinality $d-k$.

EXAMPLES

E14: $H(d,2)$ is isomorphic to the hypercube graph Q_d. It is distance regular with diameter d.

E15: $J(2,n)$ is isomorphic to the triangular graph $L(K_n)$. It is distance regular with diameter 2.

Distance-Regular Graphs and the Hoffman Polynomial

A distance-regular graph is regular (from $p_{1,1}^0$) and connected, and so it has a Hoffman polynomial.

DEFINITIONS

D23: The lth-***order adjacency matrix*** is defined with $A_0 = I$ as the identity matrix, $A_1 = A$ as the usual adjacency matrix, and A_l as the matrix with 1 in the (i,j) position if the corresponding vertices are at distance l from each other and 0 otherwise.

D24: The lth-***order parameter matrix*** is the matrix P_k with the distance-regularity parameter $p_{j,k}^i$ in the (i,j) entry.

FACTS ABOUT THE MATRIX A_l

F63: For a distance-regular graph with diameter d, we have $A_i A_j = \sum_{k=0}^{d} p_{i,j}^k A_k$. Also, $\sum_{k=0}^{d} A_k = J$, the all-one matrix.

F64: The algebra generated by $\{A_0, \ldots, A_d\}$ is of dimension $d+1$ (since matrices A_i and A_j commute and the A_i are linearly independent). All the matrices A^k are in this algebra for $k = 0, 1, \ldots, d$.

F65: The number of distinct eigenvalues of the adjacency matrix A of a distance-regular graph of diameter d is $d+1$.

F66: The Hoffman polynomial of any distance-regular graph of diameter d is a polynomial of degree d.

F67: Any distance-regular graph meets the eigenvalues-diameter bound.

FACTS ABOUT THE PARAMETERS $p_{i,j}^k$

F68: $P_i P_j = \sum_{k=0}^{d} p_{i,j}^k P_k$, and so the commutative algebra generated by $\{P_0, \ldots, P_d\}$ is isomorphic to the one generated by $\{A_0, \ldots, A_d\}$.

F69: The minimal polynomial for A_1 and P_1 is the same, and so A_1 and P_1 have the same distinct eigenvalues.

F70: The eigenvalues of P_1 are *simple*. That is, they occur with multiplicity one.

F71: The parameters of a distance-regular graph determine the spectrum.

Strongly Regular Graphs

DEFINITION

D25: A *strongly regular graph* is a distance-regular graph of diameter 2. The parameters are (n, r, λ, μ) where n is the number of vertices, r is the degree, $\lambda = p_{1,1}^1$ and $\mu = p_{1,1}^2$. To avoid trivialities, K_n and its complement are not strongly regular.

EXAMPLES

E16: Triangular graphs: $L(K_n)$ has parameters $(n(n-1)/2, 2n-4, n-2, 4)$.

E17: Lattice graphs: $L(K_{n,n})$ has parameters $(n^2, 2n-2, n-2, 2)$.

E18: Paley graphs: $P(p^n)$ has as vertices the elements of the finite field $\mathrm{GF}(p^n)$ with two vertices adjacent if, as field elements, their difference is a quadratic residue (for this relation to be symmetric p^n must be 1 mod 4). The Paley graph has parameters $(p^n, (p^n-1)/2, (p^n-5)/4, (p^n-1)/4)$.

FACTS

The parameters of a strongly regular graph are not independent. Pick a vertex and count the number of paths of length two starting at that vertex and ending at a different one. There are $n - r - 1$ vertices at distance two from our given vertex, and each one contributes μ such paths. Also, for each of the r vertices adjacent to the given vertex, there are $r - 1 - \lambda$ choices for a second edge to get a desired path.

F72: For a strongly regular graph with parameters (n, r, λ, μ), we have $\mu(n-r-1) = r(r-1-\lambda)$.

F73: Since there are only three types of entries in A, A^2, and I, for a strongly regular graph (corresponding to equal, adjacent, and nonadjacent vertices) it's easy to recognize the Hoffman polynomial and hence the eigenvalues for a strongly regular graph. In particular, for a strongly regular graph with parameters (n, r, λ, μ), we have

$$A^2 + (\mu - \lambda)A + (\mu - r)I = \mu J.$$

F74: The eigenvalues of a strongly regular graph with parameters (n, r, λ, μ) are r and the two roots of the polynomial $x^2 + (\mu - \lambda)x + (\mu - r)$.

F75: A regular connected graph is strongly regular if and only if it has three distinct eigenvalues.

It is also easy to compute the multiplicities of the eigenvalues, since r is a simple eigenvalue, the sum of the multiplicities is n, and the trace of A is 0.

F76: For a strongly regular graph with parameters (n, r, λ, μ), the eigenvalues are $\lambda_1 = r$, $\lambda_2 = (\lambda - \mu)/2 + \Delta^{1/2}$ and $\lambda_3 = (\lambda - \mu)/2 - \Delta^{1/2}$, where $\Delta = \mu^2 - 2\mu\lambda + \lambda^2 - 4\mu - 4\lambda$. The respective multiplicities are 1, m_2, and m_3 where $m_2 + m_3 = n - 1$ and $m_2\lambda_2 + m_3\lambda_3 = -r$. Since $\lambda_2 \neq \lambda_3$, the solution for m_2 and m_3 is unique.

F77: If Δ in Fact F76 is not a square, then $m_2 = m_3$. Such a graph is called a *conference graph*.

F78: One test to see if a potential set of parameters (n, r, λ, μ) is actually attained by a graph is to see if the multiplicities m_2 and m_3 are integers.

EXAMPLE

E19: For the Paley graph, $x^2 + (\mu - \lambda)x + (\mu - r) = x^2 + x - (p^n + 1)/4$, $\{\lambda_2, \lambda_3\} = \{\frac{1}{2}(-1 \pm p^{n/2})\}$, and $m_2 = m_3 = (p^n - 1)/4$.

FURTHER READING

Further details and results concerning strongly regular graphs can be found in the excellent reference [GoRo01]. The encyclopedic reference for distance-regular graphs is [BrCoNe89]. More recent results can be found in [BrHa12].

6.5.5   Spectral Characterization

One of the earliest and continuing questions in spectral graph theory asks the following: when is a graph characterized by its spectrum? Finding pairs of nonisomorphic graphs with the same spectrum can pinpoint properties of a graph that cannot be determined spectrally.

EXAMPLES

E20: Figure 6.5.3 shows the two smallest graphs with the same spectrum, which is $\{2^1, 0^3, -2^1\}$. This example implies that the number of quadrilaterals (unlike the number of triangles) cannot be determined from the spectrum. Similarly, neither the degree sequence nor the connectivity can be determined by the spectrum.

Figure 6.5.3: The smallest pair of cospectral graphs.

E21: Figure 6.5.4 shows the two smallest connected graphs with the same spectrum. The characteristic polynomial of both graphs is $(x - 1)(x^3 - x^2 - 5x + 1)(x + 1)^2$.

Figure 6.5.4: The smallest pair of connected cospectral graphs.

E22: Figure 6.5.5 shows two cospectral trees with the smallest possible number of vertices. The spectrum is $\pm\frac{1}{2} \pm \frac{\sqrt{13}}{2}$ (all four simple) and 0 with multiplicity 4. The characteristic polynomial in this case is $x^4(x^2 + x - 3)(x^2 - x - 3)$.

Figure 6.5.5: Cospectral trees with the minimum number of vertices.

E23: Figure 6.5.6 shows a pair of strongly regular cospectral graphs with 16 vertices and spectrum $\{6^1, 2^6, -2^9\}$. Some interpretation is necessary. The graph on the left is actually drawn on the torus, that is, the vertices on the outside edges in the same row or column are identified. In the graph on the right, any pair of vertices in the same row or column are joined. The graph on the right is actually $L(K_{4,4})$ and hence is strongly regular with parameters $(16, 6, 2, 2)$. Being cospectral, the one on the left (called the *Shrikhande graph*) must be strongly regular with the same parameters.

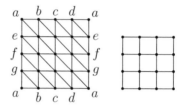

Figure 6.5.6: A pair of strongly regular cospectral graphs.

Eigenvalues and Graph Operations

One method for constructing cospectral graphs is by using various graph operations.

DEFINITIONS

D26: The **cartesian product** of two graphs G_1 and G_2, denoted $G_1 \otimes G_2$, has as vertex set all possible pairs (v_1, v_2) where v_1 is a vertex of G_1 and v_2 is a vertex of G_2. Two vertices are joined if, as ordered pairs, they are identical in one coordinate and adjacent in the other.

D27: The **coalescence** of two (disjoint) graphs G_1 and G_2 with distinguished vertices v_1 and v_2, denoted $G_1 \cdot G_2$, is formed by identifying v_1 and v_2, that is, the vertices v_1 and v_2 are replaced by a single vertex v adjacent to the same vertices in G_1 as v_1 and the same vertices in G_2 as v_2.

FACTS

F79: The straightforward extension of the cartesian product to an iterated product of more than two graphs is associative.

F80: The eigenvalues of the product graph $G_1 \otimes G_2$ are precisely all possible sums $\lambda_1 + \lambda_2$, where λ_1 is an eigenvalue of G_1 and λ_2 is an eigenvalue of G_2.

F81: Let G_1 and G_2 be a pair of cospectral nonisomorphic graphs. For $t = 0, 1, \ldots, m$, we define H_t by taking the cartesian product $G_1 \otimes \cdots \otimes G_1 \otimes G_2 \otimes \cdots \otimes G_2$ using t copies of G_1 and $m - t$ copies of G_2. Then the graphs H_t are pairwise cospectral and nonisomorphic.

F82: [Sc73] Let $P_H(x)$ be the characteristic polynomial of the graph H. Then

$$P_{G_1 \cdot G_2}(x) = P_{G_1}(x)P_{G_2 - v_2}(x) + P_{G_1 - v_1}(x)P_{G_2}(x) - xP_{G_1 - v_1}P_{G_2 - v_2}$$

F83: [Sc73] As the number of vertices gets large, the probability that a tree has a cospectral mate goes to 1.

EXAMPLES

The first example of this type of construction was given by A. J. Hoffman.

E24: (A. J. Hoffman in [Mo72]) Let G_1 and G_2 be a pair of regular cospectral nonisomorphic graphs. Define H_t by taking t copies of G_1 and $m - t$ copies of G_2 and taking the complement, for $t = 0, 1, \ldots, m$. Then the H_t are all cospectral, connected, regular, and nonisomorphic.

E25: The Hamming graph $H(d, n)$ is simply $K_n \otimes \cdots \otimes K_n$, where the number of factors is d.

REMARK

R3: The cartesian product has been generalized to the NEPS graph by D. Cvetković [CvDoSa95].

6.5.6   The Laplacian

The Laplacian is an alternative to the adjacency matrix for describing the adjacent vertices of a graph. It has many interesting properties, and, although not as much information is known about the spectral properties of the Laplacian of a graph, more recent work has indicated that there is much more to be found.

DEFINITIONS

D28: The **Laplacian** of a graph is a square matrix whose rows and columns correspond to the vertices of a graph. A diagonal entry is the degree of the corresponding vertex; an off-diagonal entry is -1 if the corresponding vertices are adjacent and 0 otherwise. In other words, $L = D - A$, where D is the diagonal matrix of degrees of the vertices and A is the usual adjacency matrix.

If the graph is regular, then $D = rI$, and the eigenvalues of A and L are obtainable from each other. Thus for regular graphs the study of the adjacency matrix and the Laplacian are identical.

D29: [Fi73] The **algebraic connectivity** of a connected graph whose Laplacian L has eigenvalues $\lambda_1 \leq \lambda_2 \leq \cdots \leq \lambda_n$ is defined to be λ_2.

FACTS

The oldest result about the Laplacian concerns the number of spanning trees of a graph. Let $\tau = \tau(G)$ be the number of spanning trees of a graph. Let $L_{i,j}$ be the matrix obtained by deleting the i-th row and j-th column from L. Also, let adj L be the adjoint of L.

F84: $\tau(G) = (-1)^{i+j} \det(L_{i,j})$.

F85: $\text{adj}(L) = \tau J$.

F86: The multiplicity of 0 as an eigenvalue of L is the number of connected components in the graph.

F87: $\prod_{i=2}^{n} \lambda_i = n\tau(G)$.

F88: Since $\lambda_2 = \lambda_3 = \cdots = \lambda_n = n$ for K_n, $\tau(K_n) = n^{n-2}$.

F89: The algebraic connectivity is positive if and only if the graph is connected.

F90: $\lambda_2(G_1 \otimes G_2) = \min\{\lambda_2(G_1), \lambda_2(G_2)\}$.

F91: [Ne00] If a graph G has diameter d, then $\lambda_2(G) \geq \frac{1}{nd}$.

EXAMPLES

E26: $\lambda_2(P_n) = 2(1 - \cos(\pi/n))$.

E27: $\lambda_2(C_n) = 2(1 - \cos(2\pi/n))$.

E28: $\lambda_2(Q_n) = 2$.

E29: $\lambda_2(K_n) = n$.

E30: $\lambda_2(K_{m,n}) = \min\{m, n\}$.

FURTHER READING

A good introduction to further properties of the Laplacian is given by B. Mohar [Mo92]. Another excellent synopsis is by M. Newman [Ne00].

References

[Al86] N. Alon, Eigenvalues and exapnders, *Combinatorica* **6** (1986), 83–96.

[BrCoNe89] A. E. Brouwer, A. M. Cohen, and A. Neumaier, *Distance-Regular Graphs*, Springer-Verlag, 1989.

[BrHa12] Andries E. Brouwer and Willem H. Haemers, *Spectra of Graphs*, Springer, 2012.

[Ca89] R. W. Carter, *Simple Groups of Lie Type*, John Wiley & Sons, 1989.

[CoSi57] L. Collatz and U. Sinogowitz, Spektren endlicher Grafen, *Ahb. Math. Sem. Univ. Hamburg* **21** (1957), 27–56.

[CvDoGu82] D. Cvetković, M. Doob, and I. Gutman, On graphs whose spectral radius does not exceed $(2 + \sqrt{5})^{1/2}$, *Ars Combiatorica* **14** (1982), 225–239.

[CDS81] D. Cvetković, M. Doob, and S. Simić, Generalized line graphs, *J. Graph Theory* **5** (1981), 385–399.

[CvDoSa95] D. Cvetković, M. Doob, and H. Sachs, *Spectra of Graphs*, Johann Ambrosius Barth, 1995.

[CvRoSi04] D. Cvetković, P. Rowlinson, and S. Simić, *Spectral Generalizations of Line Graphs*, Cambridge University Press, 2004.

[CvSi95] D. Cvetković and S. Simić, On graphs whose second largest eigenvalue does not exceed $(\sqrt{5} - 1)/2$, *Discrete Math.* **139** (1995), 213–227.

[Da95] E. R. van Dam, Regular graphs with four eigenvalues, *Lin. Alg. and Its Appl.* **226–228** (1995), 139–162.

[DaHa03] E. R. van Dam and W. H. Haemers, Which graphs are determined by their spectrum?, *Lin. Alg. and Its Appl.* **373** (2003) 241272.

[DaSaVa03] G. Davidoff, P. Sarnak, and A. Valette, *Elementary Number Theory, Group Theory, and Ramanujan Graphs* Cambridge University Press, 2003.

[Do70] M. Doob, A geometric interpretation of the least eigenvalue of a line graph, *Proc. Second Conference on Comb. Math. and Appl., Chapel Hill, NC* (1970), 126–135.

[Fi73] M. Fiedler, Algebraic connectivity of graphs, *Czechsolvak. Math. J.* **23** (1973), 298–305.

[Fr91] J. Friedman, On the second eigenvalue and random walks in random d-regular graphs, *Combinatorica* **11** (1991), 331–362.

[Ga60] F. R. Gantmacher, *The Theory of Matrices*, Vols. I, II, Chelsea, 1960.

[GCSS76] J. M. Goethals, P. Cameron, J. Seidel, and E. Shult, Line graphs, roots systems and elliptic geometry, *J. Algebra* **43** (1976), 305–327.

[Go93] C. Godsil, *Algebraic Combinatorics*, Chapman and Hall, 1993.

[GoRo01] C. Godsil and G. Royle, *Algebraic graph theory*, Springer-Verlag, 2001.

[Ha95] W. Haemers, Interlacing eigenvalues and graphs, *Lin. Alg. and Its Appl.* **226–228** (1995), 593–616.

[Ho63] A. J. Hoffman, On the polynomial of a graph, *Amer. Math. Monthly* **70** (1963), 30–36.

[Ho75] A. J. Hoffman, Eigenvalues of graphs, pp. 225–245 in *Studies in Graph Theory, II*, D. R. Fulkerson, Ed., Mathematical Assoc. of America (1970).

[Li94] J. Li, *Subdominant eigenvalues of graphs*, Ph. D. Thesis, University of Manitoba, 1994.

[LuPhSa88] A. Lubotzky, R. Phillips, and P. Sarnak, Ramanujan graphs, *Combinatorica* **8** (1988) 261–277.

[MiMa64] H. Minc and M. Marcus, *A Survey of Matrix Theory and Matrix Inequalities*, Prindle, Weber & Schmidt, 1964.

[Mo92] B. Mohar, Laplace eigenvalues of graphs—a survey, *Discrete Math* **109** (1992), 171–183.

[Mo72] A. Mowshowitz, The characteristic polynomial of a graph, *J. Comb. Theory Ser. B* **12** (1972), 177–193.

[Mu03] Ramanujan Graphs, *J. Ramanujan Math. Soc* **18** (2003) 1–20.

[Ne00] M. Newman, *The Laplacian spectrum of graphs*, Masters Thesis, University of Manitoba, 2000.

[PeRa01] M. Petrović and Z. Radosavljević, *Spectrally constrained graphs*, Faculty of Science, University of Kragujevac, 2001.

[Sa64] H. Sachs, Beziehungen zwischen den in einem Graphen enthaltenen Kreisen und seinem charakteristischen Polynomialen, *Publ. Math. Debrecen* **11** (1964), 119–134.

[Sc73] A. J. Schwenk, Almost all trees are cospectral, in *New Directions in the Theory of Graphs* (Ed: F. Harary), Academic Press, 1973, 275–307.

[Si95] S. Simić, Some notes on graphs whose second largest eigenvalue is less than $(\sqrt{5}-1)/2$, *Linear and Multilinear Algebra* **39** (1995), 59–71.

[Sm70] J. H. Smith, Some properties of the spectrum of a graph, pp. 403–406 in *Combinatorial Structures and their Applications*, Eds: R. Guy, H. Hanani, N. Sauer, and J. Schönheim, Gordon and Breach (1970).

[WoNu95] On graphs whose smallest eigenvalue is at least $-1-\sqrt{2}$, *Lin. Alg. and Its Appl.* **226–228** (1995), 577–591.

Section 6.6
Matroidal Methods in Graph Theory

James Oxley, Louisiana State University

INTRODUCTION

Every graph gives rise to a matroid so every theorem for matroids has an immediate consequence for graphs, although many of these are easy to derive directly. On the other hand, numerous results for graphs have analogs or generalizations to matroids. This link between graph theory and matroid theory is so close that the famous graph theorist W. T. Tutte (1917–2002) wrote [Tu79]: "If a theorem about graphs can be expressed in terms of edges and circuits only it probably exemplifies a more general theorem about matroids." This section provides an overview of the rich interaction between graph theory and matroid theory.

6.6.1 Matroids: Basic Definitions and Examples

The edge-sets of cycles in a graph and the minimal linearly dependent sets of columns in a matrix share many similar properties. Hassler Whitney (1907–1989) aimed to capture these similarities when he defined matroids in 1935 [Wh35].

DEFINITIONS

D1: A *matroid* M is a pair consisting of finite set $E(M)$ (the *ground set* of M) and a collection $\mathcal{C}(M)$ of nonempty incomparable subsets of $E(M)$, called *circuits*, such that if C_1 and C_2 are distinct members of $\mathcal{C}(M)$ and $e \in C_1 \cap C_2$, then there is a member C_3 of $\mathcal{C}(M)$ such that $C_3 \subseteq (C_1 \cup C_2) - \{e\}$.

NOTATION: Frequently, $E(M)$ and $\mathcal{C}(M)$ are abbreviated to E and $\mathcal{C}$.

D2: A subset of E is *dependent* if it contains a member of $\mathcal{C}$ and is *independent* otherwise.

D3: A *basis* (or *base*) is a maximal independent set.

D4: The matroid M_1 is *isomorphic* to the matroid M_2, written $M_1 \cong M_2$, if there is a 1-1 function ϕ from $E(M_1)$ onto $E(M_2)$ such that C is a circuit of M_1 if and only if $\phi(C)$ is a circuit of M_2.

NOTATION: The collections of independent sets and bases of M are denoted by $\mathcal{I}(M)$ and $\mathcal{B}(M)$, respectively.

REMARK

R1: It follows easily from the definition of a matroid that all bases are of the same cardinality.

EXAMPLES

E1: Three different classes of examples of matroids are given in Table 6.6.1.

Table 6.6.1: Examples of matroids

MATROID M	GROUND SET $E(M)$	CIRCUITS $\mathcal{C}(M)$	INDEPENDENT SETS, $\mathcal{I}(M)$	BASES $\mathcal{B}(M)$
$M(G)$, *cycle matroid* of graph G	$E(G)$, edge-set of G	edge-sets of cycles	$\{I \subseteq E(G) :$ I contains no cycle$\}$	For connected G: edge-sets of spanning trees
$M[A]$, *vector matroid* of matrix A over field F	column labels of A	minimal linearly dependent multisets of columns	$\{I \subseteq E : I$ labels a linearly independent multiset of columns$\}$	maximal linearly independent sets of columns
Uniform matroid, $U_{m,n}$ $(0 \le m \le n)$	$\{1,2,\dots,n\}$	$\{C \subseteq E :$ $|C| = m+1\}$	$\{I \subseteq E :$ $|I| \le m\}$	$\{B \subseteq E :$ $|B| = m\}$

E2:   Let M be the matroid with $E(M) = \{1, 2, \ldots, 6\}$ and $\mathcal{C}(M) = \{\{1\}, \{5, 6\},$ $\{3, 4, 5\}, \{3, 4, 6\}\}$. Then $M = M(G_1) = M(G_2)$ where G_1 and G_2 are the graphs shown in Figure 6.6.1. Also $\mathcal{B}(M) = \{\{2, 3, 4\}, \{2, 3, 5\}, \{2, 3, 6\}, \{2, 4, 5\}, \{2, 4, 6\}\}$, and $M = M[A]$ where A is the following matrix over $\mathbb{R}$.

$$
\begin{array}{cccccc}
1 & 2 & 3 & 4 & 5 & 6
\end{array}
$$
$$
\begin{bmatrix}
0 & 1 & 0 & 0 & 0 & 0 \\
0 & 0 & 1 & 0 & 1 & 1 \\
0 & 0 & 0 & 1 & 1 & 1
\end{bmatrix}
$$

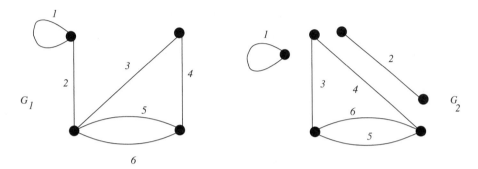

Figure 6.6.1: Graphs G_1 and G_2 yield the same matroid M.

E3:   In Table 6.6.2, several classes of matroids are defined. A matroid M is in the specified class if it satisfies the indicated condition.

Table 6.6.2: Some classes of matroids

Class	Condition
graphic	$M \cong M(G)$ for some graph G
representable over $\mathbb{F}$	$M \cong M[A]$ for some matrix A over the field $\mathbb{F}$
binary	representable over $GF(2)$, the 2-element field
ternary	representable over $GF(3)$
regular	representable over all fields

FACTS

In each of the following, M is a matroid.

F1:   The unique smallest non-graphic matroid is $U_{2,4}$.

F2:   If M is graphic, then $M \cong M(G)$ for some connected graph G.

F3:   (Whitney's 2-Isomorphism Theorem [Wh33]) Two graphs have isomorphic cycle matroids if and only if one can be obtained from the other by a sequence of the following operations: (i) choose one vertex from each of two components and identify the chosen vertices; (ii) the reverse of (i); (iii) in a graph that can be obtained from the disjoint union of two graphs G_1 and G_2 by identifying vertices u_1 and v_1 of G_1 with vertices u_2 and v_2 of G_2, **twist** the graph by identifying, instead, u_1 with v_2 and u_2 with v_1.

The following is an immediate consequence of the last fact.

F4: A 3-connected loopless graph is uniquely determined by its cycle matroid.

F5: If M is graphic, then M is regular.

F6: M is regular if and only if M can be represented over the real numbers by a ***totally unimodular matrix***, a matrix for which all subdeterminants are in $\{0, 1, -1\}$.

6.6.2   Alternative Axiom Systems

Matroids can be characterized by numerous different axiom systems. Two examples of these systems follow. Others may be found, for example, in [Ox11]. Throughout, E is a finite set and 2^E is its set of subsets.

Independent set axioms. A subset $\mathcal{I}$ of 2^E is the set of independent sets of a matroid on E if and only if

(I1) the empty set is in $\mathcal{I}$;

(I2) every subset of a member of $\mathcal{I}$ is in $\mathcal{I}$ ($\mathcal{I}$ is ***hereditary***); and

(I3) if X and Y are in $\mathcal{I}$ and Y has more elements than X, then there is an element e of $Y - X$ such that $X \cup \{e\}$ is in $\mathcal{I}$.

Basis axioms. A subset $\mathcal{B}$ of 2^E is the set of bases of a matroid on E if and only if

(B1) $\mathcal{B}$ is nonempty; and

(B2) if B_1 and B_2 are in $\mathcal{B}$ and $x \in B_1 - B_2$, then there is an element y of $B_2 - B_1$ such that $(B_1 - \{x\}) \cup \{y\} \in \mathcal{B}$.

DEFINITIONS

In all of the following, M is a matroid with ground set E.

D5: If $A \subseteq E$, all maximal independent subsets of A have the same cardinality, the ***rank***, $r(A)$ (or $r_M(A)$), of A.

D6: The ***rank $r(M)$ of the matroid*** M is the rank $r(E)$ of its ground set.

D7: A ***spanning set*** of M is a subset of E of rank $r(M)$.

D8: A ***hyperplane*** of M is a maximal nonspanning set.

D9: The ***closure*** $\mathrm{cl}(X)$ of X is $\{x \in E : r(X \cup \{x\}) = r(X)\}$.

D10: A set Y is a ***flat*** (or ***closed set***) if $\mathrm{cl}(Y) = Y$.

D11: A ***loop*** of M is an element e such that $\{e\}$ is a circuit.

D12: If $\{f, g\}$ is a circuit, then f and g are ***parallel elements***.

D13: A ***simple matroid*** (or ***combinatorial geometry***) is a matroid that has no loops and no parallel elements.

FACTS

F7: If X is a set of edges of a graph G and $G[X]$ is the subgraph of G induced by X, then $r_{M(G)}(X) = |V(G[X])| - k(G[X])$ where $k(G[X])$ is the number of components of $G[X]$.

F8: If X is a set of elements in a matroid M, then $\mathrm{cl}(X) = X \cup \{e : M \text{ has a circuit } C \text{ such that } e \in C \subseteq X \cup \{e\}\}$.

F9: For a matroid M, let $\mathcal{B}^*(M) = \{E(M) - B : B \in \mathcal{B}(M)\}$. Then $\mathcal{B}^*(M)$ is the set of bases of a matroid on $E(M)$.

EXAMPLE

E4: For the graphs G_1 and G_2 shown in Figure 6.6.1, $r(M(G_1)) = |V(G_1)| - 1 = 3$ and $r(M(G_2)) = |V(G_1)| - 3 = 3$. In each matroid, $\mathrm{cl}(\{2, 5\}) = \{1, 2, 5, 6\}$ and the last set is a flat of rank 2.

6.6.3 The Greedy Algorithm

Matroids have an important relationship to the greedy algorithm that makes them important in optimization problems. Kruskal's algorithm for finding a minimum-cost spanning tree in a connected graph G is one of the best-known efficient algorithms in graph theory. This algorithm works precisely because the spanning trees of G form the bases of a matroid.

Algorithm 6.6.1: The Greedy Algorithm for $(\mathcal{I}, w)$

Let E be a finite set and $\mathcal{I}$ be a nonempty hereditary subset of 2^E. Let w be a real-valued function on E. For $X \subseteq E$, let $w(X)$, the **weight** of X, be $\sum_{x \in X} w(x)$, and let $w(\emptyset) = 0$.

(i) Set $X_0 = \emptyset$ and $j = 0$.

(ii) If $E - X_j$ contains an element e such that $X_j \cup \{e\} \in \mathcal{I}$, choose such an element e_{j+1} of maximum weight, let $X_{j+1} = X_j \cup \{e_{j+1}\}$, and go to (iii); otherwise let $X_j = B_G$ and go to (iv).

(iii) Add 1 to j and go to (ii).

(iv) Stop.

EXAMPLE

E5: Let G be a connected graph with each edge e having a cost $c(e)$. Define $w(e) = -c(e)$. Then the greedy algorithm is just Kruskal's algorithm and the result, B_G, is the edge-set of a spanning tree of minimum cost.

FACT

F10: A nonempty hereditary set $\mathcal{I}$ of subsets of a finite set E is the set of independent sets of a matroid on E if and only if, for all real-valued weight functions w on E, the set B_G produced by the greedy algorithm is a maximal member of $\mathcal{I}$ of maximum weight.

6.6.4   Duality

Matroid theory has an attractive theory of duality that extends both the concept of a planar dual of a plane graph and the notion of orthogonality in vector spaces. This duality means that every graph gives rise to another matroid in addition to its cycle matroid.

DEFINITIONS

D14: For a matroid M, the **dual** M^* of M is the matroid on $E(M)$ having $\mathcal{B}^*(M)$ as its set of bases (see Fact F9).

D15: Circuits, bases, loops, and independent sets of M^* are called **cocircuits**, **cobases**, **coloops**, and **coindependent sets** of M.

D16: For a graph G, the matroid $(M(G))^*$ is the called the **bond matroid** of G and is denoted by $M^*(G)$.

D17: A matroid M is **cographic** if $M \cong M^*(G)$ for some graph G.

D18: A class of matroids is **closed under duality** if the dual of every member of the class is also in the class.

EXAMPLES

E6: Table 6.6.3 specifies the duals of certain types of matroids.

Table 6.6.3: Duals of some basic examples

Matroid	$M(G)$ for G plane	$U_{m,n}$	$M[I_r \vert D]$ for $r \times n$ matrix $[I_r \vert D]$
Dual	$M(G^*)$ where G^* is the dual of G^*	$U_{n-m,n}$	$M[-D^T \vert I_{n-r}]$, same order of column labels as $[I_r \vert D]$

E7: The graph G_1^* in Figure 6.6.2 is the planar dual of the graph G_1 in Figure 6.6.1. Observe that $M(G_1^*)$ is isomorphic to $M(G_1)$ under the permutation of $E(G_1)$ that interchanges 1, 3, and 4 with 2, 5, and 6, respectively.

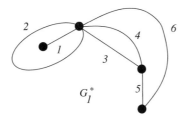

Figure 6.6.2: Graph G_1^* is the planar dual of G_1.

FACTS
For all matroids M:

F11: $(M^*)^* = M$.

F12: The rank function of M^* is $r^*(X) = |X| - r(M) + r(E - X)$.

F13: The cocircuits of M are the minimal sets having nonempty intersection with every basis of M.

F14: The cocircuits of M are the minimal nonempty sets C^* such that $|C^* \cap C| \neq 1$ for every circuit C of M.

F15: For a graph G, the circuits of $M^*(G)$ are the **bonds** or minimal edge-cuts of G. In particular, the loops of $M^*(G)$ are the isthmuses of G.

F16: A graphic matroid is cographic if and only if it is **planar** (isomorphic to the cycle matroid of a planar graph).

F17: Every row of the matrix $[I_r|D]$ is orthogonal to every row of $[-D^T|I_{n-r}]$.

F18: The following classes of matroids are closed under duality: uniform matroids, matroids representable over a fixed field $\mathbb{F}$, planar matroids, and regular matroids. The classes of graphic and cographic matroids are not closed under duality.

Table 6.6.4: Special sets and their complements in M and M^*

X	basis of M	independent set of M	circuit of M
$E - X$	basis of M^*	spanning set of M^*	hyperplane of M^*

REMARKS

R2: The fact that both the cycles and bonds of a graph are the circuits of a matroid means that cycles and bonds share many common properties.

R3: Matroids in general do not have vertices. In a 2-connected loopless graph G, the set of edges meeting a vertex is a bond of G and hence is a cocircuit of $M(G)$. Although $M(G)$ will usually have many cocircuits that do not arise in this way, in many contexts, an appropriate matroid analog of a vertex is a cocircuit.

6.6.5   Matroid Union and Its Consequences

The operation of matroid union, which was introduced by Nash-Williams [Na66], led to very straightforward proofs of two graph results whose original proofs were quite intricate.

FACTS

F19: Let $M_1, M_2, \ldots, M_n$ be matroids on a common ground set E. Then there is a matroid $M_1 \vee M_2 \vee \ldots \vee M_n$ (the **union** of $M_1, M_2, \ldots, M_n$) on E whose independent sets are all subsets of E of the form $I_1 \cup I_2 \cup \ldots \cup I_n$ such that $I_i \in \mathcal{I}(M_i)$ for all i.

F20: If M_i has rank r_i, then the rank of X in $M_1 \vee M_2 \vee \ldots \vee M_n$ is

$$\min\{\sum_{i=1}^{n} r_i(Y) + |X - Y| : Y \subseteq X\}.$$

The following covering and packing results for matroids are easily proved by taking the union of a matroid with itself multiple times, although the original proofs preceded the introduction of the operation of matroid union. The second result is the dual of the first.

F21: [Ed65] A matroid M has k disjoint bases if and only if, for every subset X of $E(M)$,

$$kr(X) + |E(M) - X| \geq kr(M).$$

F22: [Ed65] A matroid M has k independent sets whose union is $E(M)$ if and only if, for every subset X of $E(M)$,

$$kr(X) \geq |X|.$$

The last two results have the following immediate consequences for graphs.

F23: [Tu61, Na61] A connected graph G has k edge-disjoint spanning trees if and only if, for every partition π of $V(G)$, the number of edges joining vertices in different classes of the partition is at least $k(|\pi| - 1)$ where $|\pi|$ is the number of classes in π.

F24: [Tu61] The edge-set of a graph G can be partitioned into k disjoint forests if and only if, for all subsets X of $V(G)$,

$$|E(G[X])| \geq k(|X| - 1).$$

F25: [Ed65] Let G be a connected graph. Players B and C alternately tag edges of G where an edge is destroyed if it is tagged by C and made invulnerable to destruction if it is tagged by B. The goal for B is to tag all the edges of some spanning tree of G and the goal for C is to prevent B from achieving this goal (by tagging all the edges of some bond). The following are equivalent:

(i) Player C plays first and B can win against all possible strategies of C.

(ii) G has two edge-disjoint spanning trees.

(iii) For all partitions π of $V(G)$, the number of edges joining vertices in different classes of the partition is at least $2(|\pi| - 1)$.

6.6.6 Fundamental Operations

Duality is one of the three basic operations for matroids. Two other basic operations, deletion and contraction, are defined in Table 6.6.5 below along with the operation of direct sum, a special case of matroid union which generalizes the operation of direct sum of vector spaces. Each of these operations generalizes an operation for graphs.

Table 6.6.5: Three basic matroid constructions

Matroid	Ground Set	$\mathcal{C}$	$\mathcal{I}$
$M \backslash T$, the **deletion** of T from M	$E(M) - T$	$\{C \subseteq E(M) - T : C \in \mathcal{C}(M)\}$	$\{I \subseteq E(M) - T : I \in \mathcal{I}(M)\}$
M/T, the **contraction** of T from M	$E(M) - T$	minimal non-empty members of $\{C - T : C \in \mathcal{C}(M)\}$	$\{I \subseteq E(M) - T : I \cup B_T \in \mathcal{I}(M)\}$ for some B_T in $\mathcal{B}(M|T)\}$
$M_1 \oplus M_2$, **direct sum** of M_1 and M_2, $E(M_1) \cap E(M_2) = \emptyset$	$E(M_1) \cup E(M_2)$	$\mathcal{C}(M_1) \cup \mathcal{C}(M_2)$	$\{I_1 \cup I_2 : I_j \in \mathcal{I}(M_j)\}$

DEFINITIONS

D19: The matroids $M \backslash T$ and M/T are also written as $M|(E-T)$ and $M.(E-T)$ and are called the **restriction** and **contraction of M to $E-T$**.

D20: A matroid N is a **minor** of a matroid M if N can be obtained from M by a sequence of deletions and contractions. The minor N is **proper** if $N \neq M$.

D21: A graph H is a **minor** of a graph G if H can be obtained from G by a sequence of edge deletions, edge contractions, and deletions of isolated vertices.

NOTATION: For an element e of a matroid M, the matroids $M \backslash \{e\}$ and $M/\{e\}$ are frequently written as $M \backslash e$ and M/e.

EXAMPLES

E8: $M(G) \backslash e = M(G \backslash e)$, where $G \backslash e$ is the graph that is obtained from the graph G by deleting the edge e.

E9: $M(G)/e = M(G/e)$ where G/e is the graph that is obtained from the graph G by contracting the edge e, that is, by identifying the ends of e and then removing e.

E10: If a graph H is a minor of a graph G, then the cycle matroid $M(H)$ is a minor of the cycle matroid $M(G)$. To see that the converse of this fails, note that, for the graphs G_1 and G_2 in Figure 6.6.1, $M(G_1)$ is a minor of $M(G_2)$ as the two matroids are equal. But G_1 is clearly not a minor of G_2.

E11: $U_{m,n} \backslash e \cong U_{m,n-1}$ for $m \neq n$, and $U_{n,n} \backslash e \cong U_{n-1,n-1}$.

E12: $U_{m,n}/e \cong U_{m-1,n-1}$ for $m \neq 0$, and $U_{0,n}/e \cong U_{0,n-1}$.

E13: $M[A]\backslash e$ is the vector matroid of the matrix that is obtained by deleting column e from the matrix A.

E14: If e corresponds to a unit vector in A, then $M[A]/e$ is the vector matroid of the matrix obtained by deleting both the column e and the row containing the one of e.

E15: If G_1 and G_2 are vertex-disjoint graphs, then $M(G_1) \oplus M(G_2)$ is the cycle matroid of the graph that is obtained by taking the disjoint union of G_1 and G_2. Moreover, if v_1 is a vertex of G_1 and v_2 is a vertex of G_2, then $M(G_1) \oplus M(G_2)$ is also the cycle matroid of the graph that is obtained by identifying v_1 and v_2, this graph being a 1-**sum** of G_1 and G_2.

FACTS
In the following, M, M_1, and M_2 are matroids and $E(M_1) \cap E(M_2) = \emptyset$.

F26: $(M/T)^* = M\backslash T$, and $(M\backslash T)^* = M/T$. (Deletion and contraction are dual operations.)

F27: If X and Y are disjoint subsets of $E(M)$, then $M\backslash X\backslash Y = M\backslash(X \cup Y) = M\backslash Y\backslash X$, $M/X/Y = M/(X \cup Y) = M/Y/X$, and $M\backslash X/Y = M/Y\backslash X$.

F28: If $X \subseteq E(M) - T$, then $r_{M/T}(X) = r_M(X \cup T) - r_M(T)$.

F29: $M_1 \oplus M_2 = M_2 \oplus M_1$.

F30: Let N_1 and N_2 be the rank-zero matroids on $E(M_1)$ and $E(M_2)$. Then $M_1 \oplus M_2 = (M_1 \oplus N_2) \vee (N_1 \oplus M_2)$.

6.6.7   Connectedness, 2- and 3-Connectedness for Graphs and Matroids

Although connectedness for graphs does not carry over to matroids, 2-connectedness and 3-connectedness do.

DEFINITION

D22: A matroid M is 2-**connected** if, for every two distinct elements e and f of M, there is a circuit containing $\{e, f\}$.

TERMINOLOGY:
For matroids, the terms "2-connected" and "connected" are used interchangeably. Another synonym that is also used is "nonseparable".

EXAMPLE

E16: For the graphs G_1 and G_2 in Figure 6.6.1, $M(G_1) = M(G_2)$ but G_1 is a connected graph and G_2 is not. Thus, in general, for a graph G, one cannot tell from $M(G)$ whether or not G is connected.

FACTS

F31: Let G be a graph without loops or isolated vertices and assume that $|V(G)| \geq 3$. Then G is 2-connected if and only if, for every two distinct edges e and f of G, there is a cycle of G containing $\{e, f\}$.

F32: A matroid M is 2-connected if and only if M cannot be written as the direct sum of two matroids with nonempty ground sets.

F33: A matroid is 2-connected if and only if its dual is 2-connected.

F34: [Tu65] If M is 2-connected and $e \in E(M)$, then $M \backslash e$ or M/e is 2-connected.

F35: [Le64] If M is 2-connected, then M is uniquely determined by the set of circuits containing some fixed element of $E(M)$.

By combining Facts F31 and F33, one obtains the following:

F36: Let G be a graph without loops or isolated vertices and assume that $|V(G)| \geq 3$. Then G is 2-connected if and only if, for every two distinct edges e and f of G, there is a bond of G containing $\{e, f\}$.

Bounds on the number of elements

NOTATION: For a matroid M having a circuit and a cocircuit, let $c(M)$ and $c^*(M)$ be the sizes of, respectively, a largest circuit and a largest cocircuit of M. If $e \in E(M)$ and e is not a loop or a coloop, let $c_e(M)$ and $c_e^*(M)$ be the sizes of, respectively, a largest circuit of M containing e and a largest cocircuit of M containing e; and let $d_e(M)$ and $d_e^*(M)$ be the sizes of a smallest circuit of M containing e and a smallest cocircuit of M containing e.

F37: [LeOx01] Let M be a 2-connected matroid with at least two elements.

  (i) If e is an element of M, then $|E(M)| \leq (c_e(M) - 1)(c_e^*(M) - 1) + 1$.

  (ii) $|E(M)| \leq \frac{1}{2} c(M) c^*(M)$.

F38: (Length-width inequality [Le79]) Let M be a regular matroid with at least two elements and suppose $e \in E(M)$. Then

$$|E(M)| \geq (d_e(M) - 1)(d_e^*(M) - 1) + 1.$$

The next two facts for graphs are immediate consequences of Fact F37.

F39: Let u and v be distinct vertices in a 2-connected loopless graph G. Then $|E(G)|$ cannot exceed the product of the length of a longest $u - v$ path and the size of a largest bond separating u from v.

F40: [Wu97] Let G be a 2-connected loopless graph with circumference c and let c^* be the size of a largest bond. Then $|E(G)| \leq \frac{1}{2} c c^*$.

Wu [Wu00] showed that the graphs attaining equality in the last bound are certain series-parallel graphs, including cycles. Wu's bound is sometimes better and sometimes worse than the following bound of Erdős and Gallai, whose hypotheses are slightly different.

F41: [ErGa59] Let G be a simple graph with circumference c. Then

$$|E(G)| \leq \tfrac{1}{2}c(|V(G)| - 1).$$

The last bound motivated the question whether Fact F37(ii) is true for matroids. This question was answered for graphs before it was answered for all matroids. The following is a generalization of Fact F40.

F42: [NeRiUr99] Every 2-connected loopless graph with circumference c has a collection of c bonds such that every edge lies in at least two of them.

The (matroid) dual of the last result is also true.

F43: [Mc05] Every 2-connected loopless graph whose largest bond has size c^* has a family of c^* cycles so that every edge lies in at least two of them.

The last result was proved as a partial answer to the following problem of Vertigan (in [Ox01]), which remains open in general.

PROBLEM

P1: Let M be a 2-connected matroid with at least two elements. Does M have a family of $c(M)$ cocircuits such that every element is in at least two of them?

2-sums and 3-sums

As noted in Example E15, matroid direct sum generalizes the operation of 1-sum for graphs. The graph operation of 2-sum generalizes to all matroids, while 3-sum generalizes to binary matroids.

DEFINITIONS

D23: Let M_1 and M_2 be 2-connected matroids on disjoint sets, each with at least three elements. Let p_1 and p_2 be elements of M_1 and M_2, respectively. The **2-*sum of M_1 and M_2 with respect to* p_1 *and*** p_2 is the matroid $M_1 \oplus_2 M_2$ with ground set $(E(M_1) - \{p_1\}) \cup (E(M_2) - \{p_2\})$ for which the circuits are all circuits of M_1 avoiding p_1, all circuits of M_2 avoiding p_2, and all sets of the form $(C_1 - \{p_1\}) \cup (C_2 - \{p_2\})$ where C_i is a circuit of M_i containing p_i.

D24: A matroid is **3-*connected*** if it is 2-connected and cannot be written as a 2-sum.

D25: Let M_1 and M_2 be binary matroids each having at least seven elements. Suppose that $E(M_1) \cap E(M_2) = T$ where T is a 3-element circuit in each of M_1 and M_2, and that T does not contain a cocircuit of M_1 or M_2. The **3-*sum*** of M_1 and M_2 is the matroid on $(E(M_1) \cup E(M_2)) - T$ whose flats are those sets $F - T$ such that $F \cap E(M_i)$ is a flat of M_i for each i.

EXAMPLES

E17: Let G_1 and G_2 be 2-connected loopless graphs and p_i be an edge of G_i for each i. Let G be one of the two graphs that can be obtained from G_1 and G_2 by identifying p_1 with p_2 and then deleting the identified edge, that is, G is a 2-sum of the graphs G_1 and G_2. Then $M(G_1) \oplus_2 M(G_2) = M(G)$.

E18: Let G_1 and G_2 be the graphs in Figure 6.6.3, where $E(G_1) \cap E(G_2) = \{1, 2, 3\}$. Then the graph G obtained by sticking G_1 and G_2 together across the 3-cycle $\{1, 2, 3\}$ and then deleting $\{1, 2, 3\}$ is the 3-sum of the graphs G_1 and G_2. The matroid $M(G)$ is the 3-sum of the matroids $M(G_1)$ and $M(G_2)$.

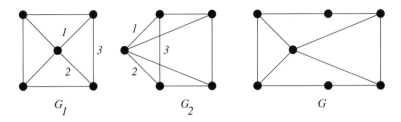

Figure 6.6.3: $M(G)$ is the 3-sum of $M(G_1)$ and $M(G_2)$.

FACTS

F44: $(M_1 \oplus_2 M_2)^* = M_1^* \oplus M_2^*$.

F45: A matroid M is 3-connected if and only if M^* is 3-connected.

F46: Let G be a graph without isolated vertices and suppose that $|V(G)| \geq 4$. Then $M(G)$ is 3-connected if and only if G is 3-connected and simple.

REMARK

R4: Tutte [Tu66] defined a notion of k-connectedness for matroids for all $k \geq 2$ that includes the cases considered above. It has the advantage of being preserved under matroid duality but the disadvantage that it departs from graph k-connectedness when $k \geq 4$. Several authors [Cu81, InWe81, Ox81] introduced the notion of vertical k-connectedness for matroids, which generalizes k-connectedness for graphs but need not be preserved under duality.

6.6.8   Graphs and Totally Unimodular Matrices

One of the most significant achievements of matroid theory is Seymour's result showing that all totally unimodular matrices are obtainable from graphs and one additional special matroid. This result leads to a polynomial-time algorithm to test whether a given matroid is totally unimodular, which is particularly useful in combinatorial optimization (see, for example, [Sc86]). Recall that a matroid is regular if and only if it can be represented by a totally unimodular matrix.

EXAMPLES

E19: Let G be a graph. Arbitrarily orient the edges of G and let D be the vertex-edge incidence matrix of the resulting directed graph. Then D is a totally unimodular matrix that represents $M(G)$ over all fields.

E20: [Bi77] Consider the matrix A over $GF(2)$ whose columns consist of the ten 5-tuples with exactly three ones. Let R_{10} be the matroid represented by A. Then $R_{10}^* \cong R_{10}$. Moreover, if e is an element of R_{10}, then $R_{10} \backslash e \cong M(K_{3,3})$ and $R_{10}/e \cong M^*(K_{3,3})$.

FACT

F47: [Se80] The class of regular matroids is the class of matroids that can be constructed by direct sums, 2-sums, and 3-sums from graphic matroids, cographic matroids, and copies of R_{10}.

6.6.9   Excluded-Minor Characterizations

The Kuratowski–Wagner Theorem [Ku30, Wa37] that a graph is planar if and only if it has no minor isomorphic to K_5 or $K_{3,3}$ has a number of extensions for graphs and matroids. The search for such results is currently the most active area of research in matroid theory.

DEFINITIONS

D26: A class of matroids is ***minor-closed*** if every minor of a member of the class is also in the class.

D27: An ***excluded minor*** of a minor-closed class is a matroid for which every proper minor is in the class yet the matroid itself is not.

EXAMPLES

E21:   The class of simple matroids is not minor-closed since it contains the cycle matroid of a 3-edge cycle but not the cycle matroid of a 2-edge cycle.

E22: The following classes of matroids are minor-closed: graphic matroids, cographic matroids, uniform matroids, matroids representable over a fixed field, regular matroids, and planar matroids.

E23: Given a finite set E of points in the plane and a collection of lines (subsets of E with at least three elements), no two of which share more than one common point, there is a matroid with ground set E whose circuits are all sets of three collinear points and all sets of four points no three of which are collinear. ***Geometric representations*** of two such matroids are shown in Figure 6.6.4, where the reader is cautioned that these diagrams are not to be interpreted as graphs. Each matroid depicted has ground set $\{1, 2, \ldots, 7\}$. On the left is the ***non-Fano matroid,*** F_7^-. It differs from the ***Fano matroid,*** F_7, on the right by the collinearity through 4, 5, and 6 in the latter. Neither of these two matroids is graphic.

E24: Table 6.6.6 specifies the collections of excluded minors for certain classes of matroids. The results in the last two rows of the table were proved in three landmark papers of Tutte [Tu58, Tu58a, Tu59]. The characterization of ternary matroids was proved independently by Bixby [Bi79] and Seymour [Se79].

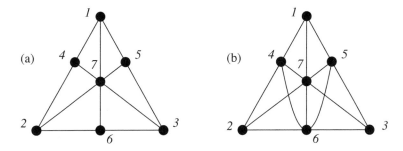

Figure 6.6.4: (a) The non-Fano matroid, F_7^-. (b) The Fano matroid, F_7.

Table 6.6.6: Excluded-minor characterizations of certain classes of matroids

Class	Excluded minors	Class	Excluded minors
uniform	$U_{0,1} \oplus U_{1,1}$	ternary	$U_{2,5}, U_{3,5}, F_7, F_7^*$
binary	$U_{2,4}$	graphic	$U_{2,4}, F_7, F_7^*, M^*(K_5), M^*(K_{3,3})$
regular	$U_{2,4}, F_7, F_7^*$	cographic	$U_{2,4}, F_7, F_7^*, M(K_5), M(K_{3,3})$

FACTS

F48: [RoSe04] For every minor-closed class of graphs, the set of excluded minors is finite.

F49: [La58] For every field $\mathbb{F}$ of characteristic 0 so, in particular, for $\mathbb{Q}$, $\mathbb{R}$, and $\mathbb{C}$, the class of matroids representable over $\mathbb{F}$ has an infinite set of excluded minors.

CONJECTURES

The following two conjectures are the main unsolved problems in matroid theory.

C1: (Rota's Conjecture [Ro71]) For all finite fields $\mathbb{F}$, there is a finite set of excluded minors for the class of $\mathbb{F}$-representable matroids.

C2: For all finite fields $\mathbb{F}$, if $\mathcal{M}$ is some minor-closed class of matroids all of which are $\mathbb{F}$-representable, then there is a finite set of excluded minors for $\mathcal{M}$.

REMARKS

R5: Fact F48 is probably the deepest result ever proved in graph theory, appearing in the twentieth paper of a very difficult series. The sixteenth paper of that series [RoSe03] proves a very powerful structure theorem for graphs that is the main tool in the proof of Fact F48. Fact F49 shows that Fact F48 does not extend to matroids. The two conjectures above propose two natural classes of matroids to which Fact F48 may be extendable.

R6: From Table 6.6.6, if $q \in \{2, 3\}$, then the set of excluded minors for the class of $GF(q)$-representable matroids is finite. Geelen, Gerards, and Kapoor [GeGeKa00] proved that the same is true for $q = 4$, there being exactly seven excluded minors in this case.

R7:   Rota's Conjecture is open for all prime powers q exceeding 4. Some recent progress on this and on Conjecture C2 has been made by Geelen and Whittle [GeWh02] and by Geelen, Gerards, and Whittle [GeGeWh02, GeGeWh06, GeGeWh07]. These three authors have been working together on Conjectures C1 and C2 for more than a decade. A survey of their work up through 2006 appears in [GeGeWh07a] and more recent progress is discussed in Chapter 14 of [Ox11]. In 2008, Geelen [Ge08] announced that they had proved the structure theorem for binary matroids that is the analog of Robertson and Seymour's main structure theorem for graphs. In 2009, Geelen [Ge09] announced that they had succeeded in using the structure theorem to prove Conjecture C2 for binary matroids, that is, for $F = GF(2)$. The papers containing these results are currently in preparation.

6.6.10   Wheels, Whirls, and the Splitter Theorem

Tutte [Tu61] identified wheels as the basic building blocks of 3-connected simple graphs. Subsequently, he generalized that result to matroids [Tu66]. The Splitter Theorem, a powerful generalization of the last result, was proved for matroids by Seymour [Se80] and, independently, for graphs by Negami [Ne82].

DEFINITIONS

D28:   For $n \geq 2$, the **wheel** $\mathcal{W}_n$ is the graph that is formed from an n-cycle C_n by adding a new vertex and joining this by a single edge (a **spoke**) to every vertex of the **rim** C_n.

D29:   For $r \geq 2$, the **rank-r whirl** $\mathcal{W}^r$ is the matroid on the edge set of $\mathcal{W}_r$ whose set of circuits consists of all the cycles of $\mathcal{W}_r$ except the rim, together with all sets of edges consisting of the rim plus a single spoke.

D30:   If M and N are matroids, then M **has an N-minor** if M has a minor isomorphic to N.

D31:   An n-**spike** with tip p is a rank-n matroid whose ground set is the union of n three-element circuits $C_1, C_2, \ldots, C_n$ all containing a common point p such that, for all $k \leq n - 1$, the union of any k of $C_1, C_2, \ldots, C_n$ has rank $k + 1$.

EXAMPLES

E25:   Figure 6.6.5 shows the graph $\mathcal{W}_3$, which is clearly isomorphic to K_4, together with geometric representations of the matroids $M(\mathcal{W}_3)$ and $\mathcal{W}^3$. The line $\{4, 5, 6\}$ in $M(\mathcal{W}_3)$ corresponds to the rim of $\mathcal{W}_3$.

E26:   Both the Fano and non-Fano matroids are examples of 3-spikes.

E27:   The unique rank-r binary spike is the vector matroid of the matrix $[I_r | I_r^c | \mathbf{1}]$ over $GF(2)$ where I_r^c is the matrix obtained by interchanging the zeros and ones in the $r \times r$ identity matrix I_r, and $\mathbf{1}$ is the column of all ones.

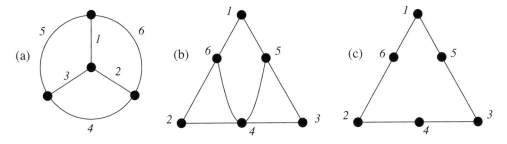

Figure 6.6.5: (a) The graph $\mathcal{W}_3$. The matroids (b) $M(\mathcal{W}_3)$ and (c) $\mathcal{W}^3$.

FACTS

F50: [Tu61] Let G be a simple 3-connected graph. Then, for every edge e of G, neither $G\backslash e$ nor G/e is both simple and 3-connected if and only if G is isomorphic to a wheel with at least three spokes.

F51: [Tu66] Let M be a 3-connected matroid. Then, for every element e of M, neither $M\backslash e$ nor M/e is 3-connected if and only if M is isomorphic to $M(\mathcal{W}_r)$ or $\mathcal{W}^r$ for some $r \geq 3$.

F52: (The Splitter Theorem [Se80]) Let M and N be 3-connected matroids such that N is a minor of M with $|E(N)| \geq 4$, and M is neither a whirl nor the cycle matroid of a wheel. Suppose that if $N \cong \mathcal{W}^2$, then M has no $\mathcal{W}^3$-minor while if $N \cong M(\mathcal{W}_3)$, then M has no $M(\mathcal{W}_4)$-minor. Then there is a sequence $M_0, M_1, \ldots, M_n$ of 3-connected matroids such that $M_0 = M$; $M_n \cong N$; and, for all i in $\{1, 2, \ldots, n\}$, M_i is a single-element deletion or a single-element contraction of M_{i-1}.

The statement above of the Splitter Theorem is a slight strengthening — due to Coullard [Co85] (see also [CoOx92]) — of Seymour's original result. The Splitter Theorem has numerous applications for both graphs and matroids. It played a key role in the proof of Fact F47 and can also be used to derive the following results, the first two of which preceded the Splitter Theorem.

F53: [Wa60] Let G be a simple 3-connected graph having no K_5-minor. Then either G has no H_8-minor or $G \cong H_8$ where H_8 is the 4-rung Möbius ladder shown in Figure 6.6.6.

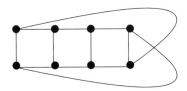

Figure 6.6.6: The 4-rung Möbius ladder, H_8.

F54: [Ha43] Let G be a simple 3-connected graph. Then G has no $K_{3,3}$-minor if and only if either G is planar or $G \cong K_5$.

F55: [Se80] Let M be a 3-connected binary matroid. Then M has no F_7-minor if and only if M is regular or $M \cong F_7^*$.

F56: Let G be a simple 3-connected graph. Then G has no $\mathcal{W}_4$-minor if and only if $G \cong \mathcal{W}_3$.

F57: [Ox89] Let G be a simple 3-connected graph. Then G has no $\mathcal{W}_5$-minor if and only if

(i) G is isomorphic to a simple 3-connected minor of one of the four graphs in Figure 6.6.7; or

(ii) for some $k \geq 3$, the graph G is obtained from $K_{3,k}$ by adding up to three edges joining distinct pairs of vertices in the 3-vertex class of the bipartition.

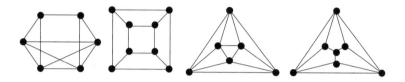

Figure 6.6.7: Four graphs with no 5-wheel minor.

The last fact and a result of Gubser [Gu93] motivated the following result on **unavoidable minors** for graphs. Fact F59 generalizes Fact F58 to matroids. Fact F60 is an immediate consequence of Fact F37(ii).

F58: [OpOxTh93] For every integer $n \geq 3$, there is an integer k such that every 3-connected simple graph with at least k edges has a minor isomorphic to $\mathcal{W}_n$ or $K_{3,n}$.

F59: [DiOpOxVe97] For every integer $n \geq 3$, there is an integer k such that every 3-connected matroid with at least k elements has a minor isomorphic to $U_{2,n}, U_{n-2,n}$, $M(K_{3,n}), M^*(K_{3,n}), M(\mathcal{W}_n), \mathcal{W}^n$, or an n-spike.

F60: For all positive integers n, a 2-connected matroid with more than $\frac{1}{2}n^2$ elements has a minor isomorphic to $U_{1,n}$ or $U_{n-1,n}$.

The next result generalizes a well-known consequence of Euler's Polyhedron Formula, namely, that if G is a simple planar graph, then $|E(G)| \leq 3|V(G)| - 6$. Fact F62 is a far-reaching matroid generalization of Fact F61 that hints at the matroid structure theorem discussed in Remark R7.

F61: [Ma67] For every integer n exceeding one, if G is a simple graph with no K_n-minor, then $|E(G)| \leq (2^n - 1)|V(G)|$.

F62: (Growth-Rate Theorem [GeKuWh09]) For a minor-closed class $\mathcal{M}$ of matroids, one of the following holds:

(i) there is a real constant c_1 such that $|E(M)| \leq c_1 r(M)$ for all simple matroids M in $\mathcal{M}$;

(ii) $\mathcal{M}$ contains the class of graphic matroids and there is a real constant c_2 such that $|E(M)| \leq c_2 (r(M))^2$ for all simple matroids M in $\mathcal{M}$;

(iii) there is a prime power q and a real constant c_3 such that $\mathcal{M}$ contains the class of $GF(q)$-representable matroids and $|E(M)| \leq c_3 q^{r(M)}$ for all simple matroids M in $\mathcal{M}$; or

(iv) $\mathcal{M}$ contains all simple rank-2-matroids.

6.6.11   Removable Circuits

A result of Mader gave conditions under which a simple k-connected graph has a cycle whose edges can be deleted without destroying k-connectedness. The natural matroid generalization of this fails for $k = 2$ even for cographic matroids. However, loose analogs of Mader's result hold for 2- and 3-connected matroids and these give new results for graphs.

DEFINITIONS

D32:   A cycle C of a k-connected graph G is **removable** if the graph obtained from G by deleting all the edges of C is k-connected.

D33:   For k in $\{2,3\}$, a circuit D of a k-connected matroid M is **removable** if $M \backslash D$ is k-connected.

FACTS

F63:   [Ma74] If G is a simple k-connected graph with minimum degree at least $k + 2$, then G has a removable cycle.

The hypothesis of Mader's result implies that $|E(G)| \geq \frac{1}{2}(k + 2)|V(G)|$. The next two facts show that imposing appropriate lower bounds on the number of elements in a matroid guarantees the existence of removable circuits.

F64:   [LeOx99] Let M be a 2-connected matroid with at least two elements and C' be a largest circuit of M. If $|E(M)| \geq 3r(M) + 3 - c(M)$, then M has a circuit C that is disjoint from C' such that $M \backslash C$ is 2-connected and $r(M \backslash C) = r(M)$. In particular, if $r(C') = r(M)$ and $|E(M)| \geq 2r(M) + 2$, then M has a removable circuit.

F65:   [LeOx99a] Let M be a 3-connected matroid with at least two elements and C' be a largest circuit of M. If

$$|E(M)| \geq \begin{cases} 3r(M) + 1 & \text{when } c(M) = r(M) + 1, \\ 4r(M) + 1 - c(M) & \text{otherwise,} \end{cases}$$

then M has a circuit C that is disjoint from C' such that $M \backslash C$ is 3-connected and $r(M \backslash C) = r(M)$.

The next two facts are obtained by applying the last two results to graphs.

F66:   Let G be a 2-connected loopless graph and C' be a largest cycle in G. If $|E(G)| \geq 3|V(G)| - c(G)$, then G has a removable cycle having no common edges with C'. In particular, if G is hamiltonian and $|E(G)| \geq 2|V(G)|$, then G has a removable cycle.

F67:   Let G be a simple 3-connected graph and C' be a largest cycle of G. Suppose that

$$|E(G)| \geq \begin{cases} 3|V(G)| - 2 & \text{if } G \text{ is hamiltonian,} \\ 4|V(G)| - 3 - c(G) & \text{otherwise.} \end{cases}$$

Then G has a cycle C that has no common edges with C' such that $G \backslash C$ is 3-connected.

F68: [GovaMc97] Let G be a 2-connected graph with minimum degree at least four. If G has no minor isomorphic to the Petersen graph, then G has two edge-disjoint removable cycles.

F69: [Mc05a] Let G be a 2-connected graph that is not a multiple edge. If G has no minor isomorphic to K_5, then G has a bond C^* such that G/C^* is 2-connected.

For 2-connected graphs, the condition that the graph is simple in Mader's result (F63) can be replaced by a higher bound on the minimum degree.

F70: [Si98] Let G be a 2-connected graph with minimum degree at least five. Then G has a removable cycle.

Example E31 below shows that the last result does not generalize to all matroids. The next result implies that it generalizes to regular and hence cographic matroids and prompts the problem as to whether it extends to binary matroids.

F71: [GoJa99] Let M be a 2-connected binary matroid in which every cocircuit has at least five elements. If M does not have minors isomorphic to both F_7 and F_7^*, then M has a removable circuit C such that $r(M \backslash C) = r(M)$.

REMARKS

R8: The last sentence of Fact F66 is easily deduced directly, but the result in the case when G is non-hamiltonian seems far less obvious.

R9: Facts F64 and F65 can also be applied to the bond matroids of, respectively, 2-connected loopless graphs and 3-connected simple graphs to give necessary conditions for such a graph G to have a bond C^* for which G/C^* is, respectively, 2-connected and loopless, or 3-connected and simple.

R10: Arthur Hobbs provided much of the impetus for the study of removable cycles by asking whether every 2-connected Eulerian graph with minimum degree at least four contains a removable cycle.

EXAMPLES

E28: [LeOx99] Consider the simple graph that is constructed as follows: begin with $K_{5,5}$ having as its two vertex classes $\{1, 2, 3, 4, 5\}$ and $\{6, 7, 8, 9, 10\}$; for every 3-element subset X of $\{1, 2, 3, 4, 5\}$ and of $\{6, 7, 8, 9, 10\}$, add two new vertices v_X and w_X each joined to all the members of X and to nothing else. Then the resulting graph G is 2-connected, having every cycle of length at least four and having every bond of size at least three. Thus $M^*(G)$ is simple and 2-connected, having every cocircuit of size at least four. But $M^*(G)$ has no removable circuit because G has no bond C^* for which G/C^* is 2-connected. Thus the generalization of Fact F63 to cographic matroids fails when $k = 2$.

E29: Jackson [Ja80] and, independently, Robertson (in [Ja80]) answered Hobbs's question negatively by producing the modified Petersen graph G_1 in Figure 6.6.8(a).

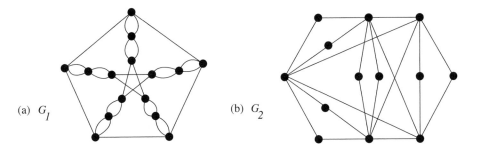

Figure 6.6.8: Neither $M(G_1)$ nor $M^*(G_2)$ has a removable circuit.

E30: [GovaMc97] For the dual problem, the graph G_2 in Figure 6.6.8(b) is 2-connected but has no bond C^* such that G_2/C^* is 2-connected. This motivated a conjecture, which McGuinness proved in Fact F69.

E31: For $r \geq 3$, the uniform matroid $U_{r,2r}$ is 2-connected, has all its cocircuits and circuits of cardinality $r + 1$, and has no removable circuits.

PROBLEMS

P2: [GoJa99] Is there an integer t such that every 2-connected binary matroid in which every cocircuit has at least t elements has a removable circuit?

P3: [GoJa99] If M is a 2-connected binary matroid in which every cocircuit has at least 5 elements, then does M have a removable circuit?

6.6.12 Minimally k-connected Graphs and Matroids

For $k \geq 2$, a k-connected graph for which no single-edge deletion is k-connected has many vertices of degree k. For k in $\{2, 3\}$, this fact has some matroid analogs that lead to new graph results.

DEFINITIONS

D34: For $k \geq 2$, a k-connected graph G is **minimally k-connected** if no single-edge deletion of G is k-connected.

D35: For k in $\{2, 3\}$, a k-connected matroid M is **minimally k-connected** if no single-element deletion of M is k-connected.

D36: Let M be a 2-connected matroid. A cocircuit C^* of M is **nonseparating** if $M \backslash C^*$ is 2-connected.

EXAMPLES

E32: If $m \geq k \geq 2$, then $K_{k,m}$ is minimally k-connected. For all $n \geq 3$, the n-spoked wheel W_n is minimally 3-connected. The cycle matroids of $K_{3,m}$ and W_n are minimally 3-connected matroids.

E33: The duals of the matroids F_7 and F_7^- are both minimally 3-connected.

FACTS FOR ARBITRARY CONNECTIVITY

F72: [Ma72] For all $k \geq 2$, every cycle of a minimally k-connected graph meets a vertex of degree k.

F73: [Ma79] For all $k \geq 2$, the number of vertices of degree k in a minimally k-connected graph G is at least

$$\frac{(k-1)|V(G)| + 2k}{2k - 1}.$$

F74: [Ox81b] For all $k \geq 2$, the number of vertices of degree k in a minimally k-connected graph G is at least $\frac{|E(G)| - |V(G)| + 1}{k - 1}$.

REMARKS

R11: Fact F73 was proved when $k = 2$ by Dirac [Di67] and Plummer [Pl68], independently, and when $k = 3$ by Halin [Ha69]. The same authors proved Fact F76 below.

R12: The bound in Fact F74, which was obtained from Fact F71 by using an elementary matroid argument, frequently sharpens the bound in Fact F72.

R13: [Ma96] The graph that is obtained from a path P of length three by adding three vertices joined to each vertex of the path is minimally 4-connected but has a cycle meeting only one vertex of degree 4. Thus Fact F72 cannot be sharpened in general although it can be improved for $k \leq 3$.

FACTS FOR SMALL CONNECTIVITY

F75: If M is a 3-connected matroid and $M = M(G)$ for some loopless graph G without isolated vertices, then a subset C^* of $E(M)$ is a nonseparating cocircuit of M if and only if C^* is the set of edges meeting at some vertex of G.

F76: [Di67, Pl68, Ha69] For k in $\{2, 3\}$, every cycle of a minimally k-connected graph meets at least two vertices of degree k.

F77: [Ox81a, Ox81b] For k in $\{2, 3\}$, let M be a minimally k-connected matroid with at least four elements. Then

  (i) every circuit of M meets at least two k-element cocircuits; and

  (ii) M has at least $\frac{r^*(M) + (k-1)}{k-1}$ k-element cocircuits.

F78: [Wu98] Let M be a minimally 3-connected binary matroid with at least four elements. Then

  (i) every circuit of M meets at least two 3-element nonseparating cocircuits; and

  (ii) M has at least $\frac{r^*(M) + 2}{2}$ 3-element nonseparating cocircuits.

F79: [Ha69a] Let G be a minimally 3-connected graph. Then $\frac{3|V(G)|}{2} \le |E(G)|$. Moreover,

$$|E(G)| \le \begin{cases} 2|V(G)| - 2 & \text{if } |V(G)| \le 6; \\ 3|V(G)| - 9 & \text{if } |V(G)| \ge 7. \end{cases}$$

The only graphs attaining equality in these bounds are $\mathcal{W}_n$ for $3 \le n \le 6$ and $K_{3,m}$ for $m \ge 4$.

F80: [Ox81a] Let M be a minimally 3-connected matroid with at least four elements. Then

$$|E(M)| \le \begin{cases} 2r(M) & \text{if } r(M) \le 5; \\ 3r(M) - 6 & \text{if } r(M) \ge 6. \end{cases}$$

The only binary matroids attaining equality in these bounds are $M(\mathcal{W}_n)$ for $3 \le n \le 6$ and $M(K_{3,m})$ for $m \ge 4$.

On combining Fact F73 and Fact F77(ii) and using a small amount of additional argument, one gets the following:

F81: [Ox81b] Let G be a minimally 3-connected graph with n_3 vertices of degree 3. Then

$$n_3 \ge \begin{cases} \frac{2|V(G)|+7}{5} & \text{when } \frac{3|V(G)|}{2} \le |E(G)| \le \frac{9|V(G)|-3}{5}; \\ \frac{|E(G)|-|V(G)|+3}{2} & \text{when } \frac{9|V(G)|-3}{5} \le |E(G)| \le 3|V(G)| - 9. \end{cases}$$

Applying Fact F77(ii) and Fact F78(ii) to cographic matroids gives the following:

F82: Let G be a graph.

(i) If G is 2-connected and loopless having no single-element contraction that is 2-connected and loopless, then the number of 2-cycles in G is at least $|V(G)|$.

(ii) Let G be a simple 3-connected graph for which no single-edge contraction is both simple and 3-connected. Then the number of 3-cycles in G such that G/C is 2-connected and loopless is at least $\frac{|V(G)|+1}{2}$.

6.6.13 Conclusion

Many areas of the interaction between graphs and matroids have not been discussed above. The most notable omission relates to the Tutte polynomial and, in particular, to colorings and flows. The interested reader is referred to the surveys in [BrOx92] and [We99]. In spite of this omission, numerous examples of this interaction are provided above. These include examples of matroid results that produce new graph results when applied to graphic or cographic matroids, graph results that have generalizations or analogs for matroids, and graph results that arise by viewing graphs from a matroid perspective. The already strong ties between matroid theory and graph theory are continuing to grow, most notably with the ongoing work of Geelen, Gerards, and Whittle generalizing Robertson and Seymour's Graph Minors Project to matroids representable over finite fields.

References

[Bi77] R. E. Bixby, Kuratowski's and Wagner's theorems for matroids, *J. Combin. Theory Ser. B* **22** (1977), 31–53.

[Bi79] R. E. Bixby, On Reid's characterization of the ternary matroids, *J. Combin. Theory Ser. B* **26** (1979), 174–204.

[BrOx92] T. Brylawski and J. Oxley, The Tutte polynomial and its applications, in *Matroid Applications* (ed. N. White), Cambridge Univ. Press, 1992, pp. 123–225.

[Co85] C. R. Coullard, *Minors of 3-connected Matroids and Adjoints of Binary Matroids*, Ph.D. thesis, Northwestern University, 1985.

[CoOx92] C. R. Coullard and J. G. Oxley, Extensions of Tutte's Wheels-and-Whirls Theorem, *J. Combin. Theory Ser. B* **56** (1992), 130–140.

[Cu81] W. H. Cunningham, On matroid connectivity, *J. Combin. Theory Ser. B* **30** (1979), 94–99.

[DiOpOxVe97] G. Ding, B. Oporowski, J. Oxley, and D. Vertigan, Unavoidable minors of large 3-connected matroids, *J. Combin. Theory Ser. B* **71** (1997), 244–299.

[Ed65] J. Edmonds, Lehman's switching game and a theorem of Tutte and Nash-Williams, *J. Res. Nat. Bur. Standards Sect. B* **69B** (1965), 67–72.

[ErGa59] P. Erdős and T. Gallai, On maximal paths and circuits of graphs, *Acta Math. Acad. Sci. Hungar.* **10** (1959), 337–356.

[Di67] G. A. Dirac, Minimally 2-connected graphs, *J. Reine Angew. Math.* **228** (1967), 204–216.

[Ge08] J. Geelen, Binary matroid minors, Fifth European Congress of Mathematics, Amsterdam, 2008.

[Ge09] J. Geelen, Well quasi ordering for binary matroids, DIMACS Workshop on Graph Coloring and Structure, Princeton, 2009.

[GeGeKa00] J. F. Geelen, A. M. H. Gerards, and A. Kapoor, The excluded minors for $GF(4)$–representable matroids, *J. Combin. Theory Ser. B* **79** (2000), 247–299.

[GeGeWh02] J. Geelen, A. M. H. Gerards, and G. Whittle, Branch-width and well-quasi-ordering in matroids and graphs, *J. Combin. Theory Ser. B* **86** (2002), 148–171.

[GeGeWh06] J. Geelen, B. Gerards, and G. Whittle, On Rota's Conjecture and excluded minors containing large projective geometries, *J. Combin. Theory Ser. B* **96** (2006), 405–425.

[GeGeWh07] J. Geelen, B. Gerards, and G. Whittle, Excluding a planar graph from $GF(q)$-representable matroids, *J. Combin. Theory Ser. B* **97** (2007), 971–998.

[GeGeWh07a] J. Geelen, B. Gerards, and G. Whittle, Towards a matroid-minor structure theory, in *Combinatorics, Complexity, and Chance*, Oxford University Press, Oxford, 2007, pp. 72–82.

[GeKuWh09] J. Geelen, J. Kung, and G. Whittle, Growth rates of minor-closed classes of matroids, *J. Combin. Theory Ser. B* **99** (2009), 420–427.

[GeWh02] J. Geelen and G. Whittle, Branch-width and Rota's conjecture, *J. Combin. Theory Ser. B* **86** (2002), 315–330.

[GoJa99] L. A. Goddyn and B. Jackson, Removable circuits in binary matroids, *Combin. Probab. Comput.* **6** (1999), 539–545.

[GovaMc97] L. A. Goddyn, J. van den Heuvel, and S. McGuinness, Removable circuits in multigraphs, *J. Combin. Theory Ser. B* **71** (1997), 130–143.

[Gu93] B. S. Gubser, Planar graphs with no 6-wheel minor, *Discrete Math.* **120** (1993), 59–73.

[Ha43] D. W. Hall, A note on primitive skew curves, *Bull. Amer. Math. Soc.* **49** (1943), 435–437.

[Ha69] R. Halin, Untersuchungen über minimale n-fach zusammenhängende Graphen, *Math. Ann.* **182** (1969), 175–188.

[Ha69a] R. Halin, Zur Theorie der n-fach zusammenhängende Graphen, *Abh. Math. Sem. Univ. Hamburg* **33** (1969), 133–164.

[InWe81] T. Inukai and L. Weinberg, Whitney conectivity of matroids, *SIAM J. Alg. Discrete Methods* **2** (1981), 311-312.

[Ja80] B. Jackson, Removable cycles in 2-connected graphs of minimum degree at least four, *J. London Math. Soc. (2)* **21** (1980), 385–392.

[Ku30] K. Kuratowski, Sur le problème des courbes gauches en topologie, *Fund. Math.* **15** (1930), 271–283.

[La58] T. Lazarson, The representation problem for independence functions, *J. London Math. Soc.* **33** (1958), 21–25.

[Le64] A. Lehman, A solution to the Shannon switching game, *J. Soc. Indust. Appl. Math.* **12** (1964), 687–725.

[Le79] A. Lehman, On the width-length inequality, *Math. Programming* **17** (1979), 403–417.

[LeOx99] M. Lemos and J. Oxley, On removable circuits in graphs and matroids, *J. Graph Theory* **30** (1999), 5166.

[LeOx99a] M. Lemos and J. Oxley, On size, circumference and circuit removal in 3-connected matroids, *Discrete Math.* **220** (1999), 145–157.

[LeOx01] M. Lemos and J. Oxley, A sharp bound on the size of a connected matroid, *Trans. Amer. Math. Soc.* **353** (2001), 4039–4056.

[Ma67] W. Mader, Homomorphieeigenschaften und mittlere Kantendichte von Graphen, *Math. Ann.* **174** (1967), 265–268.

[Ma72] W. Mader, Ecken vom Grad n in minimalen n-fach zusammenhängenden Graphen, *Arch. Math.* **23** (1972), 219–224.

[Ma74] W. Mader, Kreuzungfreie a, b-Wege in endliche Graphe, *Abh. Math. Sem. Univ. Hamburg* **42** (1974), 187–204.

[Ma79] W. Mader, Connectivity and edge-connectivity in finite graphs, in *Surveys in Combinatorics* (ed. B. Bollobás), Cambridge Univ. Press, Cambridge, 1979, pp. 66–95.

[Ma96] W. Mader, On vertices of degree n in minimally n-connected graphs and digraphs, in *Combinatorics, Paul Erdös is Eighty Vol. 2* (eds. D. Miklós, V. T. Sós, and T. Szönyi), János Bolyai Math. Soc., Budapest, 1996, pp. 423–449.

[Mc05] S. McGuinness, Circuits through cocircuits in a graph with extensions to matroids, *Combinatorica* **25** (2005), 439–450.

[Mc05a] S. McGuinness, Contractible bonds in graphs, *J. Combin. Theory Ser. B* **93** (2005), 207–249.

[Na61] C. St. J. A. Nash-Williams, Edge-disjoint spanning trees of finite graphs, *J. London Math. Soc.* **36** (1961), 445–450.

[Na66] C. St. J. A. Nash-Williams, An application of matroids to graph theory, in *Theory of Graphs* (Internat. Sympos., Rome), Dunod, Paris, 1966, pp. 263–265.

[Ne82] S. Negami, A characterization of 3-connected graphs containing a given graph, *J. Combin. Theory Ser. B* **32** (1982), 69–74.

[NeRiUr99] V. Neumann-Lara, E. Rivera-Campo, and J. Urrutia, A note on covering the edges of a graph with bonds, *Discrete Math.* **197/198** (1999), 633–636.

[OpOxTh93] B. Oporowski, J. Oxley, and R. Thomas, Typical subgraphs of 3- and 4-connected graphs, *J. Combin. Theory Ser. B* **57** (1993), 239–257.

[Ox81] J. G. Oxley, On a matroid generalization of graph connectivity, *Math. Proc. Camb. Phil. Soc.* **90** (1981), 207–214.

[Ox81a] J. G. Oxley, On matroid connectivity, *Quart. J. Math. Oxford (2)* **32** (1981), 193–208.

[Ox81b] J. G. Oxley, On connectivity in matroids and graphs, *Trans. Amer. Math. Soc.* **265** (1981), 47–58.

[Ox89] J. G. Oxley, The regular matroids with no 5-wheel minor, *J. Combin. Theory Ser. B* **46** (1989), 292–305.

[Ox01] J. G. Oxley, On the interplay between graphs and matroids, *Surveys in Combinatorics, 2001* (ed. J. W. P. Hirschfeld), London Math. Soc. Lecture Notes 288, Cambridge University Press, Cambridge, 2001, pp. 199–239.

[Ox11] J. G. Oxley, *Matroid Theory*, Second Edition, Oxford University Press, New York, 2011. (First Edition, 1992.)

[Pl68] M. D. Plummer, On minimal blocks, *Trans. Amer. Math. Soc.* **134** (1968), 85–94.

[Ro71] G.-C. Rota, Combinatorial theory, old and new, in *Proc. Internat. Cong. Math.* (Nice, Sept. 1970), Gauthier-Villars, Paris, 1971, pp. 229–233.

[RoSe03] N. Robertson and P. D. Seymour, Graph minors. XVI. Excluding a non-planar graph, *J. Combin. Theory Ser. B* **89** (2003), 43–76.

[RoSe04] N. Robertson and P. D. Seymour, Graph minors. XX. Wagner's conjecture, *J. Combin. Theory Ser. B* **92** (2004), 325–357.

[Sc86] A. Schrijver, *Theory of Linear and Integer Programming*, Wiley, Chichester, 1986.

[Se79] P. D. Seymour, Matroid representation over $GF(3)$, *J. Combin. Theory Ser. B* **26** (1979), 159–173.

[Se80] P. D. Seymour, Decomposition of regular matroids, *J. Combin. Theory Ser. B* **28** (1980), 305–359.

[Si98] P. A. Sinclair, *Strong Snarks and the Removal of Edges from Circuits in Graphs*, Ph. D. thesis, University of London, 1998.

[Tu58] W. T. Tutte, A homotopy theorem for matroids I, *Trans. Amer. Math. Soc.* **88** (1958), 144–160.

[Tu58a] W. T. Tutte, A homotopy theorem for matroids II, *Trans. Amer. Math. Soc.* **88** (1958), 161–174.

[Tu59] W. T. Tutte, Matroids and graphs, *Trans. Amer. Math. Soc.* **90** (1959), 527–552.

[Tu61] W. T. Tutte, On the problem of decomposing a graph into n connected factors, *J. London Math. Soc.* **36** (1961), 221–230.

[Tu65] W. T. Tutte, Lectures on matroids, *J. Res. Nat. Bur. Standards Sect. B* **69B** (1965), 1–47.

[Tu66] W. T. Tutte, Connectivity in matroids, *Canad. J. Math.* **18** (1966), 1301–1324.

[Tu79] W. T. Tutte, *Selected Papers of W. T. Tutte, Volume II* (eds. D. McCarthy and R. G. Stanton), Charles Babbage Research Centre, Winnipeg, 1979.

[Wa37] K. Wagner, Über eine Erweiterung eines Satzes von Kuratowski, *Deut. Math.* **2** (1937), 280–285.

[Wa60] K. Wagner, Bemerkungen zu Hadwigers Vermutung, *Math. Ann.* **141** (1960), 433–451.

[We99] D. Welsh, The Tutte polynomial, *Random Structures Algorithms* **15** (1999), 210–228.

[Wh33] H. Whitney, 2-isomorphic graphs, *Amer. J. Math.* **55** (1933), 245–254.

[Wh35] H. Whitney, On the abstract properties of linear dependence, *Amer. J. Math.* **57** (1935), 509–533.

[Wu98] H. Wu, On vertex-triads in 3-connected binary matroids, *Combin. Probab. Comput.* **7** (1998), 485–497.

[Wu97] P.-L. Wu, An upper bound on the number of edges of a 2-connected graph, *Combin. Probab. Comput.* **6** (1997), 107–113.

[Wu00] P.-L. Wu, Extremal graphs with prescribed circumference and cocircumference, *Discrete Math.* **223** (2000), 299–308.

Glossary for Chapter 6

adjacency matrix – of a simple graph G: the 0-1 matrix A_G whose rows and columns correspond to the vertices of G, with an entry being 1 if and only if the corresponding row and column vertices are adjacent.

   ___, l^{th}-order: inductively defined with $A_0 = I$ as the identity matrix, $A_1 = A$ as the usual adjacency matrix, and A_l as the matrix with 1 in the (i, j) position if the corresponding vertices are at distance l from each other and 0 otherwise.

algebraic connectivity of a graph whose Laplacian has the eigenvalues $\lambda_1 \leq \cdots \leq \lambda_n$: the eigenvalue λ_2.

almost transitive automorphism group – for an infinite graph G: a group of automorphisms of G that acts with only finitely many orbits.

almost transitive graph: a graph G whose full automorphism group $\mathcal{A}ut(G)$ is almost transitive.

s-arc: a directed walk in a graph of length s in which consective edges are distinct.

arc-transitive graph: a graph G (undirected) whose automorphism group induces a transitive group action on the set of ordered edges of G.

asymmetric graph: a graph whose automorphism group is trivial.

automorphism group – of a graph G: the set of all automorphisms of a graph, made with the operation of composition into a group, usually denoted $\mathcal{A}ut(G)$.

automorphism – of a graph: an isomorphism of the graph onto itself.

   ___, bounded – for an infinite graph G: an automorphism such that there is a uniform bound on the distances between every vertex and its image.

basic figure: a vertex-disjoint union of elementary figures.

basis – of a matroid M: a maximal set containing no circuit of M.

bicycle – in a graph: a subgraph that is both a circ and a cut.

binary vector – representing a subset E' of edges in an undirected graph: a row vector, whose i^{th} is 1 if the ith edge of the graph is in E' and is otherwise 0.

block – of objects under a permutation group action: a subset B of the set X of objects on which a permutation group $\mathcal{P}$ acts, such that for every permutation $\pi \in \mathcal{P}$, the image $\pi(B)$ either coincides with B or is disjoint from B.

   ___, nontrivial: a block other than $\emptyset$, a singleton set, or the entire set of objects on which a permutation group is acting.

bond – of a graph G: a minimal set of edges whose deletion from G increases the number of connected components.

branch$_1$ – at a vertex v of a tree T: a maximal subtree having v as a leaf.

branch$_2$ – of a spanning tree: an edge of the spanning tree.

cage: a smallest 3-valent graph with a given girth.

Catalan numbers: the sequence of numbers defined by the recursion
$$C_0 = 1, \quad C_n = C_0 C_{n-1} + C_1 C_{n-2} + \cdots + C_{n-1} C_0 \quad \text{for } n \geq 1$$

Cayley digraph – for a group $\mathcal{A}$ with *connection set* X of elements: the graph whose vertices are the elements of group $\mathcal{A}$ and such that, for each element $a \in \mathcal{A}$ and each connection $x \in X$, there is a directed edge from vertex a to vertex ax. Commonly denoted $C(\mathcal{A}, X)$.

Cayley graph$_1$: any graph isomorphic to the underlying undirected graph of a Cayley digraph.

Cayley graph$_2$: a Cayley digraph.

Cayley graph$_3$: where the connection set X is stipulated to be a generating set.

Cayley graph$_4$: where the connection set X is stipulated to be balanced, which means that $x \in X$ if and only if $x^{-1} \in X$.

characteristic polynomial – of a graph: the determinant $\det(xI - A)$ of its adjacency matrix.

chord – of a spanning tree: an edge of the cotree.

CI-graph: a Cayley graph $C(\mathcal{A}, \mathcal{X})$ such that whenever $C(\mathcal{A}, \mathcal{X}) \cong C(\mathcal{A}, \mathcal{X}')$, there exists an automorphism $\alpha \in \mathcal{A}ut(\mathcal{A})$ such that $X' = \alpha(X)$.

CI-group: a group $\mathcal{A}$ such that every Cayley graph on $\mathcal{A}$ is a CI-graph.

circ – in a graph: a circuit or union of edge-disjoint circuits of the graph.

circuit – in a graph: a subgraph isomorphic to any of the cycle graphs C_n.

    ___, **directed** – in a directed graph: a circuit in which all the edges are oriented in the same direction.

    ___, **removable** – of a k-connected matroid M: for k in $\{2, 3\}$, a circuit of M such that the deletion of C from M is k-connected.

circuit matrix – of a graph (directed or undirected): the matrix in which each row is a circuit vector, with one row for each circ in the graph.

circuit space – of a directed graph G: the set of all circuit vectors and their linear combinations over the real field, denoted by $\hat{C}(G)$.

circuit subspace – of an undirected graph G: the set of all circs of the graph, denoted by $\hat{C}(G)$.

circuit vector – of an undirected graph: the binary m-vector representing a circ of the graph.

circuit vector – in a directed graph: an m-vector representing a circ of the graph; the signs of the elements in the vector depend on the orientation assigned to each of the circuits in the circ.

circulant graph: an undirected Cayley graph on the cyclic group $\mathbb{Z}_n$.

closed under duality – of a class $\mathcal{M}$ of matroids: the dual of every member of $\mathcal{M}$ is also in $\mathcal{M}$.

closure – of a set X in a matroid M: the maximal subset of $E(M)$ that contains X and has the same rank as X.

coalescence – of two (disjoint) graphs G_1 and G_2, with distinguished vertices v_1 and v_1: the graph formed from their union by identifying the vertices v_1 and v_1; also called *amalgamation at a vertex*.

cocircuit – of a matroid M: a circuit of the dual matroid of M.

cocktail party graph $CP(n)$: the regular graph of degree $2n - 2$ with $2n$ vertices; another name for the *n-dimensional octahedral graph*.

complement of a subgraph G' – in a graph G: the graph $G'' = (V(G), E(G) - E(G'))$.

component – of a graph: a maximal connected subgraph.

conjugate subgroups – in a group $\mathcal{G}$: subgroups $\mathcal{H}_1$ and $\mathcal{H}_2$ for which there exists a subgroup $\mathcal{H}$ such that $\mathcal{H}_1 = \mathcal{H}\mathcal{H}_2\mathcal{H}^{-1}$.

connected graph: a graph in which there is a path between every pair of vertices.

connection set: see *Cayley graph*.

contraction – of a set T from a matroid M: the matroid M/T on $E(M) - T$ whose circuits are the minimal nonempty sets in $\{C - T : C \in \mathcal{C}(M)\}$.

cospanning tree – of a graph G with respect to a spanning tree T: the complement of T in G; this is generally called the *cotree* of T.

cotree – of a spanning tree T in a graph G: the *complement* of T in G.

cut matrix – of a graph (directed or undirected): the matrix in which each row is a cut vector and its number of rows is equal to the number of cuts in the graph.

cut vector – in a directed graph: the m-vector representing a cut of the graph; the signs of the elements in the vector depend on the cut orientation.

cut vector – in an undirected graph: a binary m-vector representing a cut of the graph.

cut $\langle V_1, V_2 \rangle$ – in a graph $G = (V, E)$: the set of edges with one end vertex in V_1 and the other in $V_2 = V - V_1$.

cutset$_1$ – in a connected graph G: a set of edges whose removal increases the number of components.

cutset$_2$ – in a connected graph G: a cut whose removal results in a graph with exactly two components.

___, **directed** – of a directed graph: a cut in which all the edges are oriented in the same direction.

cutset space or **cutset subspace$_1$** – of a directed graph G: the set of all cut vectors of G and their linear combinations over $GF(2)$.

cutset space or **cutset subspace$_2$** – of a directed graph G: the set of all cut vectors of G and their linear combinations over the reals.

cycle matroid – of a graph G: the matroid on the edge-set of G whose circuits are the edge-sets of the cycles of G.

degree – of a vertex: the number of edges incident on that vertex.

deletion – of a set T from a matroid M: the matroid $M \backslash T$ on $E(M) - T$ whose circuits are the circuits of M contained in $E(M) - T$.

dependent set – of a matroid M: a set containing a circuit of M.

digraph, labeled: a digraph with labels, typically $v_1, v_2, \ldots, v_n$, assigned to the vertices. Two labeled digraphs with the same set of labels are considered the same only if there is an isomorphism from one to the other that preserves the labels.

direct sum – of matroids M_1 and M_2 on disjoint sets: the matroid on $E(M_1) \cup E(M_2)$ whose circuits consist of every set that is a circuit of M_1 or of M_2.

distance-regular graph – with parameters $p_{i,j}^k$, $0 \le i, j, k \le d$: a graph such that for any pair of vertices whose distance is k, the number of vertices at distance i from the first and distance j from the second is $p_{i,j}^k$.

divisor of a graph G – with respect to any *product operation* $\natural$: either of the coordinate factors A or B, when graph G is expressed as a graph product $A \natural B$.

___, **proper**: a *divisor* of a graph other than itself and the trivial graph K_1.

doubly transitive group: a permutation group that acts transitively on ordered pairs of elements.

dual$_1$ of a graph G (*Poincare dual*): a graph obtained from a *cellular imbedding* of G on a surface, by drawing a dual vertex in each region, and then drawing a dual edge through each edge of G (see §7.6), so as to join dual vertices.

dual$_2$ of a graph G (*Whitney dual*): see Definition D51 of §6.4.

dual M^* – of a matroid M: the matroid on $E(M)$ whose set of bases consists of the set of complements of bases of M.

edge-automorphism: an edge-isomorphism from a graph to itself.

edge-group: the permutation group on the edge-set of a graph consisting of the edge-automorphisms.

edge-isomorphism: a bijection from the edge-set of one graph to the edge-set of another graph that maps every pair of adjacent edges to a pair of adjacent edges.

edge-transitive graph: a graph whose automorphism group induces a transitive group action on the edge set of the graph.

eigenvalues – of a graph: the roots of its characteristic polynomial.

eigenvalues-diameter (lower) bound – for the number of eigenvaues of a graph G: $\text{diam}(G) + 1$.

elementary abelian p-group: a group of the form $\mathbb{Z}_p^n$.

elementary figure: a subgraph isomorphic either to a K_2 or to a cycle graph C_r.

empty graph: a graph with no edges.

end – of an infinite graph: an equivalence class of rays such that no two subrays can be separated by a finite subgraph.

endomorphism – of a graph: a homomorphism of the graph to itself.

Euler totient function $\phi(n)$: the number of elements between 0 and n that are relatively prime to n.

even graph: a graph in which the degree of every vertex is even. (Such graphs are more commonly called *eulerian graphs*.)

excluded minor – of a minor-closed class of matroids: a matroid that is not in the class but has all its proper minors in the class.

factor: see *divisor*.

1-factor – of a graph: a subgraph in which every vertex has degree 1.

1-factorization – of a graph: a partitioning of its edge-set into 1-factors.

flat – of a matroid M: a maximal subset of $E(M)$ of a fixed rank.

fundamental circuit matrix – of a connected graph with respect to a spanning tree: the $(m - n + 1)$-rowed submatrix of the circuit matrix in which each row is a fundamental circuit vector with respect to the spanning tree, and will be denoted by B_f; in a directed graph, the orientation of the fundamental circuit is chosen to agree with the orientation of the chord defining the fundamental circuit.

fundamental circuit – with respect to a chord c and a spanning tree T of a graph: the unique circuit produced by adding chord c to spanning tree T.

fundamental cutset matrix – of a connected graph with respect to a spanning tree: the $(n - 1)$-rowed submatrix of the cut matrix in which each row is a fundamental cutset vector with respect to the spanning tree; in a directed graph, the orientation of a fundamental cutset is chosen to agree with the orientation of the edge defining the fundamental cutset.

fundamental cutset – of a graph with respect to an edge: the unique cutset $< V_1, V_2 >$, where V_1 and V_2 are the sets of vertices of the two trees that result when the edge is removed from the spanning tree.

generalized dicyclic group: an abstract group generated by an abelian, but not elementary abelian, subgroup $\mathcal{A}$ of index 2 and an element b of order 4 such that conjugation by b inverts every element of $\mathcal{A}$.

geodesic: a geodetic double ray.

geodetic: said of a path, ray, or double ray that contains a shortest path joining any two of its vertices.

graph product$_1$ – of two graphs: the *cartesian product*.

graph product$_2$ – of two graphs: a graph that results from applying any graph *product operation* $\natural$.

graph, labeled: a graph with labels, typically $v_1, v_2, \ldots, v_n$, assigned to the vertices. Two labeled graphs with the same set of labels are considered the same only if there is an isomorphism from one to the other that preserves the labels.

graphical regular representation of a group $\mathcal{G}$: a graph whose automorphism group is isomorphic to $\mathcal{G}$ and acts regularly on the vertex set of the graph.

growth – of an infinite graph G: $\liminf_{k\to\infty}[d(k)/a^k]$, where $d(k)$ is the number of vertices of G at distance k from a fixed vertex and $a > 1$ is a real number.

GRR: graphical regular representation.

half-transitive: vertex-transitive and edge-transitive, but not arc-transitive.

Hamilton decomposition – of a regular graph: a partition of the edge-set into Hamilton cycles (when the degree is even) or into Hamilton cycles and a 1-factor (when the degree is odd).

Hamilton-connected graph: a graph such that for any two vertices u, v, there is a Hamilton path whose terminal vertices are u and v.

Hamilton-laceable graph: a bipartite graph with parts A and B such that for any $u \in A$ and $v \in B$, there is a Hamilton path whose terminal vertices are u and v.

Hamming graph $H(d, n)$: the graph whose vertices are the d-tuples $(x_1, \ldots, x_d)$ with $1 \le x_i \le n$; two vertices are joined if, as d-tuples, they agree in all but one coordinate. (The distance between two vertices is then the number of coordinates in which, as d-tuples, they differ.)

hereditary collection – of sets: a collection $\mathcal{A}$ of sets such that every subset of a member of $\mathcal{A}$ is also in $\mathcal{A}$.

Hoffman polynomial – for an r-regular, connected graph: the polynomial $h(x) = n \prod \frac{(x-\lambda_i)}{(r-\lambda_i)}$, the product being taken over all distinct eigenvalues not equal to r.

homomorphism of general graphs G and H: a pair of mappings $f : V_G \to V_H$ and $f : E_G \to E_H$ such that the endpoint-set of each edge $e \in E_G$ is mapped onto the endpoint-set of the image edge $f(e) \in E_H$.

homomorphism of simple graphs G and H: a mapping $f : V_G \to V_H$ such that whenever the vertices u and v are adjacent in G, the vertices $f(u)$ and $f(v)$ are adjacent in H.

hyperplane – of a matroid M: a maximal subset of $E(M)$ that does not contain a basis of M.

incidence matrix$_1$ – of a graph: a matrix whose rows correspond to the vertices and whose columns correspond to the edges; the ij entry is 2 if edge j is a self-loop and vertex i is its endpoint, 1 if edge j is a proper edge and vertex i is an endpoint, and 0 otherwise.

incidence matrix$_2$ – of a graph: the n-rowed submatrix of the cut matrix in which each row is an incidence vector.

incidence set – of a vertex: the set of edges incident on that vertex.

incidence vector – for a directed graph: the cut vector representing the set of edges incident on a vertex of the graph, with the orientation of the cut chosen to be away from the vertex.

incidence vector – for an undirected graph: the binary cut vector representing the set of edges incident on a vertex of the graph.

independent set – of a matroid M: a set containing no circuit of M.

induced subgraph on an edge subset $E' \subset E(G)$: the subgraph of G with edge-set E' and vertex-set consisting of the endpoints of the edges in E'.

induced subgraph on a vertex subset $V' \subset V(G)$: the graph with vertex set V' and edge-set consisting of those edges whose endpoints are in V'.

infinite connectivity of a graph G – denoted $\kappa_\infty(G)$: the cardinality of a smallest set of vertices whose deletion leaves a graph with at least two infinite components.

isolated vertex: a vertex with degree zero.

isomorphic factorization – of a graph G: a partition of the edge set of G so that the subgraphs induced by the edges in each part are mutually isomorphic.

isomorphic matroids: matroids M_1 and M_2 for which there is a 1-1 function ϕ from $E(M_1)$ onto $E(M_2)$ such that C is a circuit of M_1 if and only if $\phi(C)$ is a circuit of M_2.

isomorphic permutation groups: a pair of isomorphic groups whose actions on their respective sets are the same, up to a bijection from one object set to the other.

isomorphism of general graphs G and H: a pair of bijections $f : V_G \to V_H$ and $f : E_G \to E_H$ such that the endpoint-set of each edge $e \in E_G$ is mapped onto the endpoint-set of the image edge $f(e) \in E_H$.

isomorphism of simple graphs G and H: a bijection $f : V_G \to V_H$ such that vertices $f(u)$ and $f(v)$ are adjacent in H if and only if vertices u and v are adjacent in G.

Johnson graph $J(d, n)$: the graph whose vertices are the $\binom{n}{d}$ subsets of $\{1, \ldots, n\}$; two vertices are joined if, as subsets, their intersection has cardinality $d-1$. (The distance bewteen two vertices is k if, as subsets, they have an intersection of cardinality $d-k$.)

Kuratowski graph: either of the two graphs in Figure 6.4.11, which characterize nonplanarity.

Laplacian matrix: a square matrix whose rows and columns correspond to the vertices of a graph, such that a diagonal entry is the degree of the corresponding vertex; an off-diagonal entry is -1 if the corresponding vertices are adjacent and 0 otherwise.

line graph of a graph G, denoted by $L(G)$: a graph whose vertex-set is the edge-set of G, with two vertices in $L(G)$ adjacent if, as edges of G, they have an endpoint in common.

   ___, **generalized** $L(G; a_1, \ldots, a_n)$ – for a graph G with n vertices and nonnegative integers $a_1, \ldots, a_n$: the graph formed by taking disjoint copies of the line graph $L(G)$ and cocktail party graphs $\mathrm{CP}(a_1), \ldots, \mathrm{CP}(a_n)$; if a vertex in $L(G)$ corresponds to the edge joining v_i to v_j in G, it is joined to all vertices in $\mathrm{CP}(a_i)$ and $\mathrm{CP}(a_j)$.

loop – of a matroid M: an element e of $E(M)$ for which $\{e\}$ is a circuit.

matroid M: a finite set $E(M)$, the ground set of M, and a collection $\mathcal{C}(M)$ of nonempty incomparable subsets of $E(M)$ called the circuits of M such that if C_1 and C_2 are distinct members of $\mathcal{C}(M)$ and $e \in C_1 \cap C_2$, then there is a member C_3 of $\mathcal{C}(M)$ such that $C_3 \subseteq (C_1 \cup C_2) - \{e\}$.

   ___, **binary**: a matroid that is isomorphic to the vector matroid of a matrix over the 2-element field $GF(2)$.

   ___, **bond** – of a graph G: the matroid on the edge-set of G whose circuits are the bonds of G.

   ___, **cographic**: a matroid that is isomorphic to the bond matroid of some graph.

   ___, **2-connected**: a matroid in which, for every two distinct elements, there is a circuit containing both.

   ___, **3-connected**: a 2-connected matroid that cannot be written as a 2-sum.

   ___, **graphic**: a matroid that is isomorphic to the cycle matroid of some graph.

   ___, **planar**: a matroid that is isomorphic to the cycle matroid of a planar graph.

   ___, **regular**: a matroid that is representable over all fields.

   ___, **representable** – over a field $\mathbb{F}$: a matroid that is isomorphic to the vector matroid of some matrix over $\mathbb{F}$.

   ___, **simple**: a matroid in which all circuits have at least three elements.

   ___, **uniform** $U_{m,n}$: for $0 \le m \le n$, the matroid on $\{1, 2, \ldots, n\}$ in which the circuits consist of all $(m+1)$-element subsets.

maximally distant trees: two spanning trees T_1 and T_2 such that $d(T_1, T_2) \geq d(T_i, T_j)$, for every pair of spanning trees T_i and T_j.

metric ray (double ray): a *ray* (*double ray*) with positive *straightness*.

metric type: describes a *ray* in an infinite graph that is an α-*essential* ray for some automorphism α.

minimally k-connected graph: a k-connected graph for which no deletion of an edge remains k-connected.

minimally k-connected matroid: for k in $\{2, 3\}$, a k-connected matroid for which no single-element deletion is k-connected.

minimum polynomial of a graph G: the monic polynomial $q(x)$ of smallest degree, such that $q(A_G) = 0$.

minor – of a graph G: a graph that can be obtained from G by a sequence of edge deletions, edge contractions, and deletions of isolated vertices.

minor – of a matroid M: a matroid that can be obtained from M by a sequence of deletions and contractions.

___, **proper** – of a matroid M: a minor of M that is not equal to M.

minor-closed – class of matroids: one in which every minor of a member of the class is also in the class.

nonseparating cocircuit – of a 2-connected matroid M: a cocircuit whose deletion from M remains 2-connected.

null graph: a graph with no vertices and hence no edges.

nullity – of a graph G having n vertices, m edges and p components: nullity is equal to $m - n + p$ and is denoted $\mu(G)$.

orientation of a cut $\langle V_1, V_2 \rangle$ – in a directed graph: the direction, either from V_1 to V_2 or from V_2 to V_1, that we choose for the cut.

orientation of a circuit – in a directed graph: the direction we choose to traverse the circuit.

orthogonal complements – of a vector space: two subspaces whose intersection is the zero vector.

orthogonal subspaces of a vector space: subspaces such that the inner product of every vector in one subspace with every vector in the other subspace is equal to zero.

painting – of a graph: a partitioning of the edges into three sets R (red), Y (yellow), and B (blue), and the distinguishing of one edge in the set Y.

Paley graph: a Cayley graph formed on the additive group of a finite field $GF(q)$, $q \equiv 1 \pmod 4$, where the connection set is the set of quadratic residues in $GF(q)$.

parallel elements e and f of a matroid M: elements such that $\{e, f\}$ is a circuit of M.

parameter matrix, l^{th}-order: the matrix P_k with the distance-regularity parameter $p^i_{j,k}$ in the (i, j) entry.

permutation group: a nonempty set $\mathcal{P}$ of permutations (on the same set X of objects), such that $\mathcal{P}$ is closed under composition and inversion.

___, **doubly transitive**: a permutation group that acts transitively on ordered pairs of objects.

___, **primitive**: a transitive permutation group whose only blocks are trivial.

___, **regular**: a permutation group that is both transitive and semiregular.

___, **semiregular**: a permutation group all of whose vertex-stabilizers are trivial.

___, **transitive**: a permutation group such that for any two objects of the set on which it acts, some permutation maps one object onto the other.

prime graph – under a given *product operation*: a graph having no proper *divisor*.

primitive graph: a graph whose automorphism group acts as a primitive permutation group on the vertex-set.

primitive group: a transitive permutation group that has no nontrivial blocks.

principal subgraphs G_1 and G_2 – of a graph G: see Definition D46 of §6.4.

product operation – on two graphs G and H: any operation ♮ such that the vertex-set $G♮H$ is the cartesian product of V_G and V_H, and such that the edge-set is determined exclusively by the adjacency relations in G and Hs.

quadratic residue – in a finite field: an element of the form x^2.

rank – of a graph G having n vertices and p components: the number of edges in the complement of a spanning forest; the rank is equal to $n - p$ and is denoted by $\rho(G)$; usually called the *cycle rank*.

rank – of a set A in a matroid: the cardinality of a maximal independent subset of A. The rank of a matroid M is the cardinality of a maximal independent subset of $E(M)$.

ray – in an infinite graph: a one-way infinite path.

___, **α-essential** – in an infinite graph: a ray that is mapped onto one of its subrays by a positive power of the automorphism α.

___, **double** – in an infinite graph: a two-way infinite path.

reduced incidence matrix – of a graph: the submatrix of the incidence matrix containing any $(n - 1)$ incidence vectors.

regular action – of a permutation group: see *permutation group, regular.*

regular matrix: see Definitions D49 and D50 of §6.4.

s-regular graph: a graph that contains at least one s-arc and whose automorphism group acts regularly on its set of s-arcs.

relatively prime graphs: graphs having no common proper divisor.

removable circuit – of a k-connected matroid M: for k in $\{2, 3\}$, a circuit of M such that the deletion of C from M is k-connected.

removable cycle – of a k-connected graph G: a cycle of G such that the deletion of the edges of C from G leaves a k-connected graph.

ring sum of two sets E_1 and E_2: the set consisting of elements that belong to E_1 or to E_2, but not to both E_1 and E_2; denoted by $E_1 \oplus E_2$.

ring sum of two vectors $(x_1, x_2, x_3, \ldots, x_i, \ldots, x_m)$ and $(y_1, y_2, y_3, \ldots, y_i, \ldots, y_m)$: the vector $Z = (z_1, z_2, z_3, \ldots, z_i, \ldots, z_m)$, where $z_i = x_i \otimes y_i$ and $\otimes$ is the logical exclusive-or operation ($1 \otimes 0 = 1$, $0 \otimes 1 = 1$, $0 \otimes 0 = 0$, and $1 \otimes 1 = 0$).

semisymmetric graph: an edge-transitive graph with constant valence (i.e., a regular graph) that is not vertex-transitive.

spanning forest – of a graph G having p components: a collection of p spanning trees, one for each component of G.

spanning set – of a matroid M: a subset of $E(M)$ containing a basis of $E(M)$.

spanning tree – of a connected graph: a tree that contains all the vertices of the graph.

spectrum of a graph: the multiset of eigenvalues; for a graph with n vertices, there are n eigenvalues.

stabilizer of a vertex u of a graph G: the subgroup of $\mathcal{A}ut$ consisting of the permutations that fix vertex u.

straightness of a ray or double ray D: the number $\liminf_{d_D(u,v) \to \infty} d(u, v)/d_D(u, v)$, where u, v are vertices of D.

strip: a connected graph G that admits a connected subgraph H and an automorphism α such that ∂H and $H - \alpha(H)$ are finite and $\alpha(H \cup \partial H) \subseteq H$.

strongly regular graph – with parameters (n, r, λ, μ): an r-regular n-vertex graph such that any pair of adjacent vertices is mutually adjacent to λ other vertices, and such that any pair of nonadjacent vertices is mutually adjacent to μ other vertices.

2-sum of matroids: for 2-connected matroids M_1 and M_2 on disjoint sets each having at least three elements, let p_i be an element of M_i; the 2-sum with respect to p_1 and p_2 is the matroid on $(E(M_1) - \{p_1\}) \cup (E(M_2) - \{p_2\})$ whose circuits are the circuits of M_1 avoiding p_1, the circuits of M_2 avoiding p_2, and all sets of the form $(C_1 - \{p_1\}) \cup (C_2 - \{p_2\})$ where C_i is a circuit of M_i containing p_i.

3-sum of matroids: for binary matroids M_1 and M_2 each having at least seven elements such that $E(M_1) \cap E(M_2)$ is a 3-element circuit T of M_1 and M_2 that does not contain a cocircuit of either matroid; the 3-sum is the matroid on $(E(M_1) \cap E(M_2)) - T$ whose flats are those sets $F - T$ such that $F \cap E(M_i)$ is a flat of M_i for each i.

symmetric difference – of two sets E_1 and E_2: the set consisting of only elements that belong to E_1 or to E_2, but not to both E_1 and E_2; denoted by $E_1 \oplus E_2$.

symmetric group S_n: the group of all permutations acting on the set $\{1, 2, \ldots, n\}$.

system of imprimitivity: collection of images of a nontrivial *block* under the action of a transitive permutation group.

ternary matroid: a matroid that is isomorphic to the vector matroid of a matrix over the 3-element field $GF(3)$.

torsion subgroup: a subgroup of an infinite group, all of whose elements have finite order.

totally unimodular matrix: a matrix over the real numbers for which the determinant of every square submatrix is in $\{0, 1, -1\}$.

tournament: a digraph in which, for each pair u, v of distinct vertices, either there exists an arc from u to v or an arc from v to u but not both.

   ___, **strong**: short for *strongly connected tournament*.

   ___, **strongly connected**: a tournament such that for each pair u, v of vertices, there exist directed paths from u to v and from v to u.

transitive action – of a permutation group: see *permutation group, transitive*.

s-transitive graph: a graph that contains at least one s-arc and whose automorpism group acts transitively on its set of s-arcs.

translation: an endomorphism of a graph that fixes no finite nonempty subset of the vertex set.

tree – in a graph: a connected subgraph of the graph containing no circuits.

   ___, **1-4**: a tree in which each vertex has degree 1 or 4.

   ___, **1-rooted 1-4**: a 1-4 tree rooted at a vertex of degree 1.

   ___, **binary**: a root vertex and at most two principal subtrees that are themselves binary trees. Each principal subtree must be specified as either the left subtree or the right subtree.

   ___, **homeomorphically reduced**: a tree with no vertices of degree 2.

   ___, **labeled**: a tree in which labels, typically $v_1, v_2, \ldots, v_n$, have been assigned to the vertices. Two labeled trees with the same set of labels are considered the same only if there is an isomorphism from one to the other that preserves the labels.

   ___, **left-right**: a binary tree in which each vertex has either 0 or 2 children.

   ___, **ordered**: a root vertex and a sequence $t_1, t_2, \ldots, t_m$ of $m \geq 0$ principal subtrees that are themselves ordered trees. The root vertex of an ordered tree is joined by an edge to the root of each principal subtree.

   ___, **reduced**: short for *tree, homeomorphically reduced*.

___, **rooted**: a tree in which one vertex, the root, is distinguished. Two rooted trees are considered the same only if there is an isomorphism from one to the other that maps the root of the first to the root of the second.

trivial graph: a graph with a single vertex and no edge.

unimodular matrix: a matrix of real numbers, the determinant of every square submatrix of which is equal to 1, -1, or 0.

union – of matroids $M_1, M_2, \ldots, M_n$ on a common set E: the matroid on E whose independent sets consist of all sets of the form $I_1 \cup I_2 \cup \ldots \cup I_n$ where I_j is an independent set of M_j for all j.

unit – of a ring: an element with a multiplicative inverse.

vector matroid of a matrix: the matroid on the set of column labels of the matrix whose circuits are the minimal linearly dependent multisets of columns.

vector space of a graph G: the set of all subsets of edges of G; also, the set of all vectors representing the subsets of edges of G; more commonly called the *edge space* of G.

vertex-edge incidence matrix – of a graph: see *incidence matrix*.

vertex-transitive graph: a graph whose automorphism group acts transitively on its vertex set.

wheel₁ W_n: for $n \geq 2$, the graph with $n+1$ vertices that is obtained by joining each vertex of an n-cycle, called the "rim", to one newly added vertex called the "hub" by an edge, called a "spoke".

wheel₂ W_n: a graph with n vertices, of which $n-1$ form a cycle (the *rim*), with the remaining vertex (the *hub*) adjacent to all the rim vertices.

whirl, $\mathcal{W}^r$: for $r \geq 2$, the matroid on the set of edges of $\mathcal{W}_r$ whose circuits are all the cycles of $\mathcal{W}_r$ except the rim along with all sets consisting of the rim plus a single spoke.

wreath product – of permutation groups G and H acting on sets S and T, resp.: a permutation group on $S \times T$ of which each element π satisfies $\pi(s, t) = (\alpha(s), \beta_{\alpha(s)}(t))$, where $\alpha \in G$ and $\beta_t \in H$ for each $t \in T$.

Chapter 7

Topological Graph Theory

Section 7.1

Graphs on Surfaces

Tomaž Pisanski, University of Ljubljana, Slovenia

Primož Potočnik, University of Ljubljana, Slovenia

INTRODUCTION

The need to imbed (draw) finite graphs on surfaces arises in various aspects of mathematics and science. Often the simplest surface in which such a graph can be imbedded is sought. Some generalizations of surfaces are briefly considered.

7.1.1   Surfaces

2-Manifolds and 2-Pseudomanifolds

DEFINITIONS

D1: The *open unit disk*, the *closed unit disk*, and the *unit half-disk* are the respective subsets

$$\{(x,y) \mid x^2 + y^2 < 1\}, \quad \{(x,y) \mid x^2 + y^2 \le 1\}, \quad \text{and } \{(x,y) \mid x \ge 0, x^2 + y^2 < 1\}$$

of the Euclidean plane, together with the inherited Euclidean topology.

D2: An *open disk*, a *closed disk*, and a *half-disk* are any topological spaces homeomorphic, respectively, to the open unit disk, the closed unit disk, or to the unit half-disk. A *disk* usually means a closed disk.

D3: A *pinched open disk* is a topological space obtained from k copies of open disks by identifying their respective centers to a single vertex, as shown in Figure 7.1.1.

Figure 7.1.1: Three disks pinched together.

D4: A *2-manifold* is a topological space in which each point has a neighborhood that is homeomorphic either to an open disk or to a half-disk.

D5: The *boundary* of a 2-manifold M is the subspace of those points in M that do not have neighborhoods homeomorphic to open disks.

D6: A *surface* is a 2-manifold, often taken in context to be connected.

D7: A *closed surface* is a compact surface without boundary.

D8: If we relax the definition of a 2-manifold to allow the neighborhoods to be homeomorphic not only to open disks or half-disks but also to pinched open disks, then the resulting topological space is called a *2-pseudomanifold*.

D9: A *pseudosurface* is a 2-pseudomanifold (usually taken to be connected). It may be obtained from a 2-manifold by successively identifying finitely many pairs of vertices.

FACTS

F1: The boundary components of a compact surface are closed curves. That is, each boundary component is homeomorphic to the unit circle.

F2: Every pseudosurface can be obtained from some 2-manifold by iteratively identifying finitely many pairs of points.

EXAMPLES

E1: The Euclidean plane is a non-compact surface.

E2: The closed disk is a compact surface with a non-empty boundary.

E3: The half-disk is a non-compact surface with a non-empty boundary.

E4: The *pinched torus* is a pseudosurface obtained from a sphere by identifying two of its points, as at the left of Figure 7.1.2.

E5: A **jellyfish pseudosurface** (also called the **spindle pseudosurface**) is obtained from two spheres by pairwise identifying some number n of points on one sphere with n points on the other, as shown at the right of Figure 7.1.2.

Figure 7.1.2: The pinched torus and a jellyfish pseudosurface.

Some Standard Surfaces

DEFINITIONS

D10: A **sphere** (usually denoted by S_0) is any surface homeomorphic to the **unit sphere** $\{(x, y, z) \mid x^2 + y^2 + z^2 = 1\}$. See Figure 7.1.3 (left).

D11: A **cylinder** (or **annulus**) is any surface which is homeomorphic to the **unit cylinder** $\{(x, y, z) \mid x^2 + y^2 = 1, -1 \leq z \leq 1\}$. See Figure 7.1.3 (right).

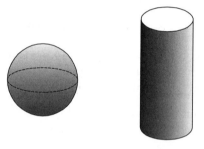

Figure 7.1.3: The sphere S_0 and the cylinder.

D12: A **projective plane** (usually denoted by N_1) is a closed surface homeomorphic to the surface obtained from the closed unit disk by identifying pairs of boundary points that are diametrically opposite relative to the center of the disk.

D13: A **Möbius band** (or **Möbius strip**) is any surface that is homeomorphic to the surface obtained from a unit square $\{(x, y) \mid -1 \leq x \leq 1, -1 \leq y \leq 1\}$ by pasting the vertical sides together with the matching $(-1, y) \rightarrow (1, -y)$. See Figure 7.1.4.

Figure 7.1.4: A Möbius band is a non-orientable surface with boundary.

D14: A *torus* (usually denoted by S_1) is a closed surface homeomorphic to the subset of the Euclidean three-dimensional space obtained by rotating a circle $\{(x, y, z) \mid (x-2)^2 + y^2 = 1, z = 0\}$ around the y-axis. See Figure 7.1.5 (left).

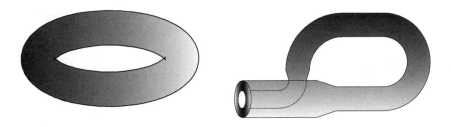

Figure 7.1.5: The torus S_1 and the Klein bottle N_2.

D15: The *Klein bottle* (usually denoted by N_2) is a closed surface homeomorphic to the surface obtained from the unit cylinder $\{(x, y, z) \mid x^2 + y^2 = 1, -1 \le z \le 1\}$ by identifying the pairs of points $\{(x, y, -1), (x, -y, 1)\}$ on the two boundary components. See Figure 7.1.5 (right).

FACTS

F3: An annulus can be obtained by excising the interior of a disk from a sphere.

F4: A Möbius band can be obtained by excising the interior of a disk from a projective plane.

F5: The torus can be obtained by identifying the pairs of points $\{(x, y, -1), (x, y, 1)\}$ on the two boundary components of the unit cylinder

$$\{(x, y, z) \mid x^2 + y^2 = 1, -1 \le z \le 1\}$$

EXAMPLES

E6: The sphere, the torus, and the Klein bottle are closed surfaces.

E7: A closed disk is a compact surface with one boundary component, while an annulus is a compact surface with two boundary components.

E8: A Möbius strip is a compact surface with one boundary component.

Surface Operations and Classification

DEFINITIONS

D16: The *connected sum* $S \# S'$ of two surfaces S and S' is obtained by excising the interior of a closed disk in each surface and then gluing the corresponding boundary curves.

D17: *Adding an orientable handle* to a surface S means forming the connected sum $S \# S_1$, as in Figure 7.1.6.

Figure 7.1.6: (left) Handle; (right) adding a handle to S_2 to obtain S_3.

D18: The *orientable surface with g handles* or the *g-torus* is the connected sum of g copies of a torus. It is denoted by S_g.

D19: *Adding a crosscap* to a surface S means forming the connected sum $S \# N_1$.

D20: The *non-orientable surface with k crosscaps*, denoted by N_k, is the connected sum of k copies of the projective plane N_1.

D21: A 2-manifold is *non-orientable* if it contains a subspace that is homeomorphic to the Möbius band. Otherwise it is *orientable*.

D22: The *genus* $\gamma(S)$ of a closed orientable surface homeomorphic to S_g is the number g of handles.

D23: The *crosscap number* or *non-orientable genus* $\tilde{\gamma}(S)$ of a closed non-orientable surface homeomorphic to N_k is k.

D24: The *Euler characteristic* $\chi(S)$ of a closed surface is defined by these formulas:

$$\chi(S_g) = 2 - 2g \qquad \text{and} \qquad \chi(N_k) = 2 - k$$

D25: A *surface with k holes* is obtained by removing the interiors of k disjoint disks from a closed surface.

D26: A simple closed curve on a surface is a *separating curve* if its excision splits the surface into two components. Otherwise it is a *non-separating curve*.

D27: A separating curve C on a surface S is a *contractible curve* if the closure of one of the components of $S - C$ is a disk.

D28: A curve C on a surface S is an *essential curve* if it is not contractible.

D29: A simple closed curve on a surface is *orientation reversing* if its regular neighborhood is a Möbius band.

FACTS

F6: An equivalent way to add a handle to an orientable surface is to remove the interiors of two disjoint disks and then to match the boundary components of a cylinder to the resulting boundary components, so as to preserve the orientation.

F7: The connected sum is well defined (up to homeomorphism of topological spaces) and is commutative and associative.

F8: A 2-manifold is orientable if and only if it is homeomorphic to a two-sided subspace of Euclidean 3-space.

F9: A closed 2-manifold is orientable if and only if it is homeomorphic to a surface in Euclidean 3-space.

F10: Classification of Closed Surfaces: Each closed surface is homeomorphic to one and only one of the following surfaces: $S_g, g \geq 0$, or $N_k, k \geq 1$.

F11: Classification of Compact Surfaces: Each compact surface with non-empty boundary components is isomorphic to a closed surface with holes. Each compact surface is completely specified by its orientability, an integer giving the genus or crosscap number, and the number b of holes.

F12: $S_g \# S_{g'} \cong S_{g+g'}$, $N_k \# N_{k'} \cong N_{k+k'}$, and $N_k \# S_g \cong N_{k+2g}$

F13: There are four mutually exclusive types of closed curves on surfaces:
- **(a)** separating and contractible
- **(b)** separating and non-contractible
- **(c)** non-separating and orientation preserving
- **(d)** non-separating and orientation reversing

EXAMPLES

E9: The sphere and the torus are orientable surfaces. Both are realizable in 3-space.

E10: Since the Klein bottle and the projective plane are non-orientable closed surfaces, it follows that they cannot be realized in 3-space.

E11: The Möbius strip is a non-orientable surface with boundary, and it can be realized in 3-space.

E12: Whereas the Jordan curve theorem asserts that every closed curve on the sphere separates the sphere, the Schönfliess theorem asserts the stronger result that every closed curve on a sphere bounds a disk.

E13: The pinched torus can be obtained by contracting ("pinching") a non-separating closed curve on a torus to a point.

7.1.2   Polygonal Complexes

DEFINITIONS

D30: A polygon is **oriented** if one of the two possible directions of traversal (i.e., clockwise or counterclockwise) of its boundary has been designated as preferred.

D31: Two topological spaces X and Y can be **pasted together** along homeomorphic subspaces by identifying the points of those subspaces under a homeomorphism.

D32: A **polygonal complex** is a structure obtained from a set of oriented polygons by pasting some of these polygons to each other and to themselves along their sides (which also results in the identification of corners). Within a polygonal complex,

- each polygon is called a **face** or a **2-cell**;
- the image of arbitrarily many polygon sides that have been pasted together is called an **edge** or a **1-cell**;
- the image of arbitrarily many polygon corners that have been pasted together is called a **vertex** or a **0-cell**.

D33: The **1-skeleton** of a polygonal complex is the graph that is formed by its vertices and edges.

D34: Each edge e of a polygonal complex is given a preferred direction of traversal, and a traversal of that edge in the **reverse direction** within a walk in the 1-skeleton is denoted e^{-1}.

D35: A polygonal complex is **consistently oriented at edge** e if within the union of the oriented boundary walks, it is not traversed twice in the same direction. Thus, a complex is consistently oriented at edge e if that edge results from a polygon side that was not pasted to another side, or if that edge results from pasting two sides together so that the traversal directions are *opposite*.

D36: A polygonal complex is **oriented** if it is consistently oriented at every edge.

D37: The **underlying topological space** of a polygonal complex is the quotient space for the union of all the polygons after all the identifications.

D38: A polygonal complex is said to **realize any topological space** that is homeomorphic to its underlying space.

D39: Occurrences of an edge e or its inverse within a walk are called **signed edges**.

D40: The **oriented boundary walk** of a face of a polygonal complex is the closed walk in the 1-skeleton that results from traversing the face boundary in the direction of orientation. (This walk is unique up to the choice of a starting/stopping vertex.)

D41: The **signed boundary walk** of a face of a polygonal complex is the list of the signed edges that occur on an oriented boundary walk of that face.

D42: The **boundary-walk specification** of a polygonal complex is a list of the signed boundary walks of the faces.

D43: The ***vertex variant*** of the boundary specification of a polygonal complex whose 1-skeleton is a simple graph gives the boundary walks as cyclic lists of vertices.

D44: A ***fundamental polygon*** for a closed surface is a polygon whose edges are pairwise identified and pasted so that the resulting polygonal complex has only one face and so that it realizes that surface.

D45: A ***specification of a fundamental polygon*** with $2n$ sides is its signed boundary walk.

D46: The standard ***fundamental polygon for the orientable surface*** S_g is specified as $a_1 b_1 a_1^{-1} b_1^{-1} a_2 b_2 a_2^{-1} b_2^{-1} \ldots a_g b_g a_g^{-1} b_g^{-1}$.

D47: The standard ***fundamental polygon for the non-orientable surface*** N_k is specified as $a_1 a_1 a_2 a_2 \ldots a_k a_k$.

FACTS

F14: A polygonal complex can be described combinatorially as the set of its signed boundary walks.

F15: A polygonal complex can realize any compact surface or pseudosurface.

F16: A polygonal complex realizes a pseudosurface or 2-manifold if and only if each side of each polygon is glued to exactly one other side; it realizes a 2-manifold if, in addition, every vertex has a topological neighborhood that is homeomorphic to a disk (this additional restriction serves to eliminate pinched disks).

EXAMPLES

E14: A ***book with n leaves***, $n \geq 3$ (or an ***n-book***) is a polygonal complex obtained by choosing a side in each of n polygons, often squares, and pasting all the chosen sides, as illustrated in Figure 7.1.7. The edge corresponding to the common side is called the ***spine*** of the book.

Figure 7.1.7: The 3-book is a polygonal complex that is not a surface.

E15: The 3-book with spine a can be specified as $\{ab_1 c_1 d_1, ab_2 c_2 d_2, ab_3 c_3 d_3\}$.

E16: The polygonal complex $\{abc, aeh^{-1}g^{-1}, bfi^{-1}e^{-1}, cgj^{-1}f^{-1}, hij\}$ is orientable but is not oriented. Reversing the orientation of the first polygon to $c^{-1}b^{-1}a^{-1}$ would make the complex oriented.

E17: The Möbius band can be specified as $\{abcd, efgb^{-1}, fid^{-1}h\}$, in which case the 1-skeleton is $K_{3,3}$.

E18: If we add a hexagon $aeh^{-1}c^{-1}g^{-1}i$ to the Möbius band specification in Example E17, the resulting polygonal complex realizes a projective plane.

E19: The polygonal complex $\{abb^{-1}a^{-1}bb^{-1}\}$ realizes the pinched torus with b as its pinch point.

E20: The standard fundamental polygon for the sphere S_0 has the form $\{a_1a_1^{-1}\}$.

7.1.3   Imbeddings

DEFINITIONS

D48: A ***topological realization of a graph*** G is obtained by first assigning to each of its edges a closed interval and then identifying endpoints of intervals according to the coincidences of the corresponding endpoints of edges of the graph.

D49: An ***immersion*** of a topological space is a continuous mapping that is locally one-to-one; that is, each point of the domain has a neighborhood that is mapped homeomorphically into the codomain.

D50: An ***imbedding*** is an immersion that is globally one-to-one.

D51: An ***imbedding of a graph*** G means an imbedding of a topological realization of G.

D52: A ***face*** of the imbedding is a connected component of the complement of the image.

D53: A ***cellular imbedding*** or ***2-cell imbedding*** of a graph into a surface is an imbedding such that the interior of each face is an open disk; thus, the complement of the image of the imbedding is a union of open disks.

D54: A ***strongly cellular imbedding*** is an imbedding such that the closure of each face is a closed disk; that is, no two points on the boundary of any face are identified.

D55: The ***minimum genus*** of a graph G (or sometimes, simply ***genus***) is the minimum of the set of integers g such that G is imbeddable in the orientable surface S_g. It is denoted by $\gamma_{\min}(G)$ or by $\gamma(G)$.

D56: A ***minimum genus imbedding*** of a graph G (or sometimes, simply ***genus imbedding***) is an imbedding of G into a closed surface of minimum genus.

D57: The ***maximum genus*** of a graph G is the maximum of the set of integers g such that G has a cellular imbedding in the orientable surface S_g. It is denoted by $\gamma_{\max}(G)$.

D58: A ***maximum genus imbedding*** of a graph is an imbedding into a closed surface of maximum genus.

D59: The ***minimum crosscap number*** of a graph G (or sometimes, simply ***crosscap number***) is the minimum of the set of integers k such that G is imbeddable in the non-orientable surface N_k. It is denoted by $\tilde{\gamma}_{\min}(G)$ or by $\tilde{\gamma}(G)$.

D60: A ***minimum crosscap imbedding*** is an imbedding into a closed non-orientable surface of minimum crosscap number.

D61: The ***maximum crosscap number*** of a graph G is the maximum of the set of integers k such that G has a cellular imbedding in the non-orientable surface N_k. It is denoted by $\tilde{\gamma}_{\max}(G)$.

D62: A ***maximum crosscap imbedding*** of a graph is 2-cell imbedding into a closed non-orientable surface of maximum crosscap number.

FACTS

F17: Every finite graph has a topological realization that can be imbedded in Euclidean 3-space.

F18: A disconnected graph has no cellular imbedding.

F19: [Yo63] Every connected graph has a minimum genus imbedding that is cellular.

F20: [PPPV87] If a connected graph is not a tree, then it has a minimum crosscap imbedding that is cellular.

F21: [Du66] Let $g' \leq g \leq g''$. If a graph admits a 2-cell imbedding in the surfaces $S_{g'}$ and $S_{g''}$ then it also admits a 2-cell imbedding in S_g.

F22: Let $k' \leq k \leq k''$. If a graph admits a 2-cell imbedding in the surfaces $N_{k'}$ and $N_{k''}$ then it also admits a 2-cell imbedding in S_k.

F23: [Wh33] Each planar 3-connected graph admits an essentially unique imbedding in the sphere. This is not generally true for imbeddings into other surfaces, not even for genus imbeddings.

F24: [Th89] The problem of determining the minimum genus of a graph is NP-hard.

F25: [Mo99] For a given graph and a fixed surface there exists a linear-time algorithm that either finds an imbedding of the graph in that surface or finds an obstruction for such an imbedding. The algorithm is not good for practical purposes since it subsumes the knowledge of all forbidden graphs for a given surface. The collection of such graphs may be quite large for a surface of moderate size genus.

F26: Euler polyhedral equation: Each cellular imbedding of a graph with v vertices, e edges, and f faces into a surface S satisfies the relation

$$v - e + f = \chi(S)$$

F27: [AuBrYo63] For any graph G, $\bar{\gamma}(G) \leq 2\gamma(G) + 1$; however, the gap may be arbitrarily large.

F28: [At68] Any graph can be imbedded in a 3-book.

F29: Every simple graph can be immersed in the plane by spacing the vertices evenly around the unit circle and joining adjacent vertices with line segments.

EXAMPLES

E21: Figure 7.1.8 shows two imbeddings of the complete graph K_4 on the torus, one non-cellular and the other cellular.

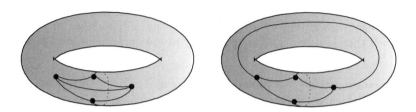

Figure 7.1.8: Two toroidal imbeddings of K_4.

E22: The vertex-variant specification $(1234)(5678)(1265)(2376)(3487)(4158)$ for the cube graph Q_3 corresponds to the following imbedding:

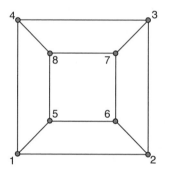

Figure 7.1.9: A standard imbedding of the cube graph Q_3 in the sphere.

E23: Two non-equivalent imbeddings of the cube graph Q_3 in the torus given by the following vertex-variant specifications

$$(123765)(341587)(234876)(126584) \qquad (148762)(123785)(326584)(567341)$$

are shown in Figure 7.1.10.

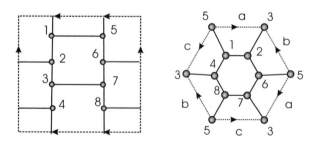

Figure 7.1.10: Two hexagonal imbeddings of Q_3 in the torus.

7.1.4   Combinatorial Descriptions of Maps

DEFINITIONS

D63: A *graph* can be defined alternatively as a combinatorial structure $\langle V, E \rangle$ with ground set S, as follows:

- The elements of the set S are called *half-edges*.

- E is a partition of S into cells of size two, such that each half-edge is paired with what amounts to the other half of the same edge. This partition is often represented as the set of orbits of an involution τ.

- V is a partition of the half-edges according to the vertex at which they are incident.

D64: A *rotation at a vertex* is a cyclic permutation on the set of half-edges at that vertex.

D65: The *surface rotation at a vertex* v of a graph imbedding is the cyclic ordering of the half-edges at v on the surface. If the surface is orientable, this ordering is taken to be consistent with the orientation.

D66: A *(global) rotation* (also called a *rotation system*) on a graph is an assignment of a rotation at each vertex. This corresponds to a permutation ρ on the set of half-edges whose orbits are the rotations at the vertices.

D67: The *(global) surface rotation of an imbedded graph* is the set of surface rotations at all the vertices.

D68: The *induced imbedding* of a global rotation ρ on a graph is an imbedding of that graph whose global surface rotation is ρ. (The face tracing algorithm below serves as proof that such an imbedding exists. It is obviously unique.)

D69: A *face tracing* for a global rotation on a graph is a list of the boundary walks of the faces of an induced imbedding.

D70: The *signature of a graph* $G = (V, E)$ is a subset $\Lambda \subseteq E_G$, whose edges are called *switches*. They represent the edges whose traversal switches the sense of orientation in an imbedding.

D71: A *generalized rotation* is a pair (ρ, Λ) composed of a global rotation and a signature.

ALGORITHM

We suppose that a global rotation ρ and an involution τ on the set of half-edges of a graph G are given as input. We want to do a face tracing. To make this easily understood, we use notation of the form e and e^{-1} for two half-edges paired by the involution τ, i.e., for the two different ends of the same edge. To each cycle of the rotation ρ, we visualize a vertex at which the half-edges within that cycle are simultaneously incident in the graph G.

Algorithm 7.1.1 Face-Tracing Algorithm

Input: half-edge list $E^{\pm}$, involution τ, rotation ρ
Output: list of all face-boundaries of the induced imbedding

{*Initialize*} Mark all half-edges *unused*
While any unused half-edges remain
    Choose next (lex order) unused half-edge y from $E^{\pm}$
    Start new cycle by writing left paren   "("
    $x := y$
    Repeat
        Write x next in current cycle
        $x := \rho(\tau(x))$ (next half-edge)
        Until $x = y$
    Close current cycle by writing right paren   ")"
    Continue with next iteration of while-loop

The algorithm for a generalized rotation is slightly more complicated, since it involves reversal of cycles. See, for example, Chapter 4 of [GrTu87].

EXAMPLES

E24: A convenient way to apply the Face-Tracing Algorithm uses a table that lists the half-edges incident at each vertex, in the cyclic order of the rotation there. For instance, this table presents an imbedding of the graph K_4 in the sphere S_0.

$$
\begin{array}{llll}
v_1. & a^+ & b^+ & c^+ \\
v_2. & a^- & e^+ & d^+ \\
v_2. & c^- & f^- & e^- \\
v_2. & b^- & d^- & f^+
\end{array}
$$

$\rho = (a^+, b^+, c^+)(a^-, e^+, d^+)(c^-, f^-, e^-)(b^-, d^-, f^+)$
$\tau = (a^+ a^-)(b^+ b^-)(c^+ c^-)(d^+ d^-)(e^+ e^-)(f^+ f^-)$

The composition permutation $\rho\tau$ has a disjoint cycle representation with four 3-cycles, which correspond to the boundary walks of the four triangular faces. Using notation that clearly associates corresponding half-edges avoids the need to write the involution.

E25: The following table presents an imbedding of K_4 in the torus S_1.

$$
\begin{array}{llll}
v_1. & a^+ & b^+ & c^+ \\
v_2. & a^- & d^+ & e^+ \\
v_2. & c^- & f^- & e^- \\
v_2. & b^- & f^+ & d^-
\end{array}
$$

The imbedding has one 4-sided face and one 8-sided face.

E26: At the left of Figure 7.1.11 is an imbedding of the dipole D_3 on the sphere. At the right are shown the three polygons of that imbedding, prior to pasting. Since D_3 is not a simple graph, the specification of that imbedding as a set of boundary walks

$$f = (b, c^{-1}), \quad g = (c, a^{-1}), \quad h = (a, b^{-1})$$

uses edges, not vertices, as does the specification by global rotation

$$\rho = u : (abc) \quad v : (a^{-1}c^{-1}b^{-1})$$

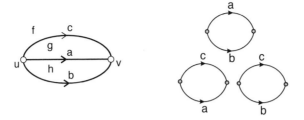

Figure 7.1.11: A spherical imbedding of the dipole D_3.

E27: At the left of Figure 7.1.12 is an imbedding of the dipole D_3 on the torus. At the right is shown the one polygon of that imbedding, prior to pasting its sides.

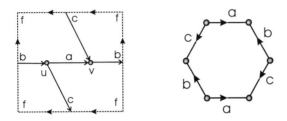

Figure 7.1.12: Toroidal imbedding of the dipole D_3.

As a set of boundary walks, it has the specification

$$f = (a, c^{-1}, b, a^{-1}, c, b^{-1})$$

and by global rotation the specification

$$\rho = u : (abc) \quad v : (a^{-1}b^{-1}c^{-1})$$

E28: At the left of Figure 7.1.13 is an imbedding of the dipole D_3 on the Klein bottle. At the right is shown the one polygon of that imbedding, prior to pasting its sides.

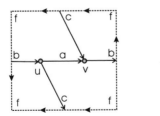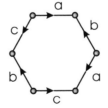

Figure 7.1.13: Klein bottle imbedding of the dipole D_3.

As a set of boundary walks, it has the specification

$$f = (a, c^{-1}, b, c^{-1}, a, b^{-1})$$

and by global generalized rotation the specification

$$\rho = u : (abc) \quad v : (a^{-1}b^{-1}c^{-1}) \quad \Lambda = \{b\}$$

FACTS

F30: The set of global rotations on a graph is in bijective correspondence with the set of oriented, cellular imbeddings of that graph.

F31: Generalized rotations correspond to cellular imbeddings into arbitrary closed surfaces.

F32: The imbedding of a graph specified by a generalized rotation is non-orientable if and only if there is a cycle in G containing an odd number of switches.

References

[Ar96] D. Archdeacon, Topological graph theory; a survey, Surveys in graph theory (San Francisco, CA, 1995) *Congr. Numer.* **115** (1996), 5–54.

[At68] G. Atneosen, *On the Embeddability of Compacta in n-Books: Intrinsic and Extrinsic Properties*, Ph. D. Thesis, Michigan State University, 1968.

[AuBrYo63] L. Auslander, T. A. Brown, and J. W. T. Youngs, The imbeddings of graphs in manifolds, *J. Math. and Mech.* **12** (1963), 629–634.

[BiWh79] N. L. Biggs and A. T. White, *Permutation Groups and Combinatorial Structures*, LMSLNS 33, Cambridge University Press, 1979.

[Du66] R. A. Duke, The genus, regional number, and Betti number of a graph, *Canad. J. Math.* **18** (1966), 817–822.

[Gr00] J. L. Gross, Topological Graph Theory, *Handbook of Discrete and Combinatorial Mathematics*, ed. by K. H. Rosen, CRC Press, 2000, 574–579.

[GrTu87] J. L. Gross and T. W. Tucker, *Topological Graph Theory*, Dover, 2001. (First Edition, Wiley-Interscience, 1987.)

[Mo99] B. Mohar, A linear time algorithm for embedding graphs in an arbitrary surface, *SIAM J. Discrete Math.* **12** (1999), 6–26.

[PPPV87] T. D. Parsons, G. Pica, T. Pisanski, and A. G. S. Ventre, Orientably simple graphs, *Math. Slovaca* **37** (1987), 391–394.

[Th89] C. Thomassen, The graph genus problem is NP-complete, *J. Algorithms* 10, (1989), pp. 568-576.

[Wh01] A. T. White, *Graphs of Groups on Surfaces*, North-Holland Mathematical Studies 188, Elsevier, 2001.

[Wh33] H. Whitney, 2-isomorphic graphs, *Amer. J. Math.* **55** (1933), 73–84.

[Yo63] J. W. T. Youngs, Minimal imbeddings and the genus of a graph, *J. Math. Mech.* **12** (1963), 303–315.

Section 7.2
Minimum Genus and Maximum Genus

Jianer Chen, Texas A&M University

INTRODUCTION

The study of graph minimum genus and maximum genus has been among the most important and interesting topics in the research in topological graph theory. Besides its theoretical importance, the study has found many impressive applications in such areas as VLSI design, computer algorithms and complexity, and computer graphics.

7.2.1   Definitions and Basic Facts

The graphs in our discussion may have multiple adjacencies or self-adjacencies. A graph with no multiple adjacencies and self-adjacencies is called a ***simple graph***. Unless stated explicitly otherwise, any graph in our discussion is assumed to be connected. An edge e in a graph is a ***cut-edge*** if removing e disconnects the graph. A graph is ***vertex-k-connected*** (resp. ***edge-k-connected***) if it remains connected after removing any $k-1$ vertices (resp. any $k-1$ edges). We sometimes shorten "vertex-k-connected" to ***k-connected***. Let C be the set of all cut-edges of a graph G. Each connected component of $G - C$ is called an ***edge-2-connected component*** of G. Clearly, each edge-2-connected component of G is either an edge-2-connected graph or a single vertex.

A theorem of Brahana [Brah32] asserts that any orientable surface is homeomorphic to the sphere with g handles, where g is called the ***genus*** of the surface. An ***embedding*** $\rho(G)$ of a graph G in an orientable surface S is a continuous one-to-one function ρ from a topological representation of the graph G into the surface S. Each connected component of $S - \rho(G)$ is called a ***face*** of the embedding $\rho(G)$. The ***genus*** of the embedding $\rho(G)$ is defined to be the genus of the surface S. An embedding is ***cellular*** if the interior of each face of the embedding is homeomorphic to a 2-dimensional open disk. Our discussion will be restricted to cellular graph embeddings.

DEFINITIONS

D1: The **minimum genus** $\gamma_{\min}(G)$ (or simply the **genus** $\gamma(G)$) of a graph G is the minimum integer g such that there exists an embedding of G into the orientable surface S_g of genus g.

D2: The **maximum genus** $\gamma_{\max}(G)$ of a graph G is the maximum integer g such that there exists an embedding of G into the orientable surface of genus g.

D3: The number $|E| - |V| + 1$ is called the **cycle rank** (or the **Betti number**) of the graph G, denoted $\beta(G)$. Intuitively, this is the number of edges remaining after the edges of a spanning tree are removed.

FACTS

F1: The genus of any embedding of a graph G is an integer between 0 and $\lfloor \beta(G)/2 \rfloor$, where $\beta(G)$ is the cycle rank of the graph G.

F2: [BHKY62, NSW71] Let $\{B_1, B_2, \cdots, B_k\}$ be the collection of edge-2-connected components of a graph G. Then

$$\gamma_{\min}(G) = \sum_{i=1}^{k} \gamma_{\min}(B_i) \qquad \text{and} \qquad \gamma_{\max}(G) = \sum_{i=1}^{k} \gamma_{\max}(B_i)$$

These results are commonly called additive properties.

F3: [**Euler Polyhedral Equation**] (for a proof, see [GrTu87]) An embedding of a graph G with vertex set V, edge set E, face set F, and genus g satisfies the relation:

$$|V| - |E| + |F| = 2 - 2g$$

REMARK

R1: According to Fact F2, in most cases we need to concentrate only on the minimum genus and maximum genus of edge-2-connected graphs.

EXAMPLE

E1: Both the minimum genus and the maximum genus of a tree are equal to 0; the minimum genus of the complete graph K_4 of four vertices is 0 while the maximum genus of K_4 is equal to 1.

Ear Decomposition

DEFINITION

D4: An **ear decomposition** $D = [P_1, P_2, \cdots, P_r]$ of a graph G is a partition of the edge set of G into an ordered collection of edge-disjoint simple paths P_1, P_2, $\cdots$, P_r such that P_1 is a simple cycle and P_i, $i \geq 2$, is a path with only its endpoints in common with $P_1 + \cdots + P_{i-1}$. Each path P_i is called an **ear**.

FACT

The class of edge-2-connected graphs has the following nice characterization.

F4: [Whit32] A graph G has an ear decomposition if and only if G is edge-2-connected.

Edge Insertion and Deletion

The operations of edge insertion and edge deletion have turned out to be important and useful in the study of graph embeddings.

DEFINITIONS

D5: Let $\rho(G)$ be an embedding of a graph G. We say a new edge e is inserted into $\rho(G)$ if the two ends of e are inserted into face corners in $\rho(G)$ to make an embedding for the graph $G + e$. The operation is called **edge insertion**.

D6: The **edge deletion** operation acts inversely to edge insertion: let $\rho(G')$ be an embedding of the graph G' and let e be an edge in G' that is not a cut-edge. If the two sides of e belong to two different faces of $\rho(G')$, then deleting e from $\rho(G')$ "merges" the two faces without changing the embedding genus; if the two sides of e belong to the same face in $\rho(G')$, then deleting e from $\rho(G')$ "splits" the face into two faces and decreases the embedding genus by 1.

FACTS

F5: If the two ends of edge e are inserted into the corners of the same face f in $\rho(G)$, then the edge e "splits" the face f into two faces and leaves the embedding genus unchanged. In this case, the two sides of the new edge e belong to two different faces in the resulting embedding for $G + e$. See Figure 7.2.1(a) for illustration.

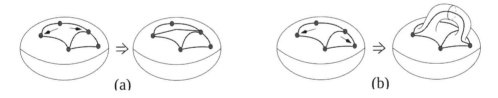

(a)   (b)

Figure 7.2.1: Inserting a new edge into an embedding.

F6: On the other hand, if the two ends of e are inserted into the corners of two different faces f_1 and f_2 in $\rho(G)$, then the edge e "merges" the faces f_1 and f_2 into a single larger face and increases the embedding genus by 1. In this case, the two sides of the new edge e belong to the same face (i.e., the new larger face) in the resulting embedding for $G + e$. Topologically, this operation can be implemented as follows: first we cut along the boundaries of the two faces f_1 and f_2 and leave two holes on the surface. Then we add a handle to the surface by pasting the two ends of a cylinder with two open ends to the boundaries of the two holes on the surface, respectively. The new edge e now runs along the new handle. See Figure 7.2.1(b) for illustration.

F7: Inserting an edge to a graph embedding can never decrease the embedding genus, and deleting an edge that is not a cut-edge from a graph embedding can never increase the embedding genus.

7.2.2   Kuratowski-Type Theorems

Any characterization of graph embeddability in terms of a finite set of forbidden subgraphs has been called a "Kuratowski-type" result, in celebration of Fact F8.

DEFINITIONS

D7: Let $e = [u, v]$ be an edge in a graph G. Denote by G/e the graph obtained from G by removing the edge e then identifying the two ends u and v. We call this operation an *edge contraction*.

D8: A graph H is a *minor* of another graph G if H can be obtained from a subgraph of G by contracting edges.

D9: A class $\mathcal{F}$ of graphs is *minor closed* if for each graph G in $\mathcal{F}$, all minors of G are also in $\mathcal{F}$.

D10: Let $\mathcal{F}$ be a minor closed graph class. A graph G is a *minimal forbidden minor* for $\mathcal{F}$ if G is not in $\mathcal{F}$ but every proper minor of G is in $\mathcal{F}$. The set of all minimal forbidden minors for the graph class $\mathcal{F}$ is called the *set of minimal forbidden minors* for $\mathcal{F}$.

FACTS

F8: [Kura30] [*Kuratowski's Theorem*] A graph G is planar (i.e., the minimum genus of G is 0) if and only if G contains no subgraphs homeomorphic to either K_5 or $K_{3,3}$.

F9: [Wagn37] A graph is planar if and only if it has neither K_5 nor $K_{3,3}$ as its minor.

F10: Every minor of a graph G can be obtained from G by a sequence of operations of edge contractions and edge deletions (we assume here that isolated vertices are automatically removed whenever they are created in the process). This follows immediately from Definition D8.

F11: The Kuratowski theorem is equivalent to the statement that a graph G is planar if and only if G has neither K_5 nor $K_{3,3}$ as its minor. Therefore, K_5 and $K_{3,3}$ are essentially the only two *forbidden minors* for the class of planar graphs.

Minimum Genus

Erdős and König [Koni36] raised the question whether there is a Kuratowski-type theorem for the class of graphs that are embeddable (not necessarily 2-cellular embeddable) in a fixed surface S. Some special cases were vastly generalized by the powerful results of Robertson and Seymour in their study of graph minor theory. The study of graph minor theory has induced significant progress in the research of graph theory. In particular, an impressive series of Robertson and Seymour's work (see [RoSe85, RoSe88, RoSe90a, RoSe90b, RoSe95]) has led to the confirmation in Fact F14 of a well-known conjecture by Wagner [Wagn37].

FACTS

F12: [GlHu78] There is a finite set $\mathcal{F}_1$ of graphs such that a graph G is embeddable in the projective plane N_1 (i.e., the non-orientable surface of crosscap number 1) if and only if G has no minor in $\mathcal{F}_1$. A complete list of these graphs can be found in [Arch81, GHW79].

F13: [ArHu89] For any non-orientable surface N, there is a finite set $\mathcal{F}_N$ of graphs such that a graph G is embeddable in the non-orientable surface N, if and only if G has no minor in $\mathcal{F}_N$.

F14: [RoSe88](Formerly known as Wagner's Conjecture) Any minor-closed class of graphs has a finite set of minimal forbidden minors.

F15: For every integer $g \geq 0$, the class of graphs of minimum genus at most g is minor closed.

F16: [RoSe90b] For every integer $g \geq 0$, the set of minimal forbidden minors for the class of graphs of minimum genus at most g is finite.

REMARKS

R2: A constructive proof for Fact F16 was developed by Mohar [Moha99]. There has been further effort to simplify the proof [Thom97b]. On the other hand, it has remained as a challenge, even for every small g such as $g = 1$, to give a good estimation on the number of graphs or the size of the graphs in the set of minimal forbidden minors in Fact F16. Seymour [Seym93] has shown that the size of the set of minimal forbidden minors for graphs of minimum genus bounded by g is bounded by

$$2^{2^{(6g+9)^9}}$$

R3: Proof that the class of planar graphs is minor closed is not difficult. Let G be a planar graph and let $\rho(G)$ be a planar embedding of G. Contracting an edge e of G on the planar embedding $\rho(G)$ can be accomplished by continuously "shrinking" the edge e on the plane until the two ends of e are identified. This gives a planar embedding of the contracted graph G/e. Moreover, by Fact F7, edge deletion does not increase embedding genus. We conclude that every minor of a planar graph is also planar, i.e., the class of planar graphs is minor closed.

Maximum Genus

We point out that a class of graphs defined in terms of maximum genus is in general not minor closed. For example, the bouquet B_2 of two self-loops (i.e., the graph with a single vertex and two self-loops) is a minor of the "dumbbell" D (i.e., the graph consisting of an edge $[u, v]$ plus two self-loops on u and v, respectively). However, it is easy to verify that $\gamma_{\max}(B_2) = 1$ while $\gamma_{\max}(D) = 0$ (see Facts F18 and F21 below).

DEFINITIONS

D11: Let G be a graph and let v be a degree-2 vertex with two neighbors u and w in G (u and w could be the same vertex). We say that a graph G' is obtained from G by **smoothing** the vertex v if G' is constructed from G by removing the vertex v then adding a new edge connecting the vertices u and w.

D12:   Two graphs G_1 and G_2 are ***homeomorphic*** if they become isomorphic after smoothing all degree-2 vertices. It is easy to see that two homeomorphic graphs have the same minimum genus and the same maximum genus.

D13:   A graph is a ***cactus*** if it can be constructed from a tree T and a subset S of vertices in T, by replacing each vertex in S by a cycle.

D14:   A ***necklace of type*** (r, s) is obtained from a cycle C_{2r+s} of $2r + s$ vertices by doubling r non-adjacent edges in C_{2r+s} (or, equivalently, by adding an extra multiple edge to each of these adjacencies), and adding a self-loop at each of the other s vertices. Figure 7.2.2 gives a type $(4, 0)$ necklace and a type $(1, 3)$ necklace.

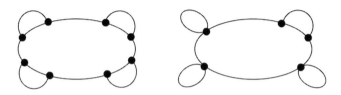

Figure 7.2.2: Left: a type $(4, 0)$ necklace; right: a type $(1, 3)$ necklace.

FACTS

F17:   An edge-2-connected graph G has maximum genus 0 if and only if G is a cycle.

F18:   A graph G has maximum genus 0 if and only if no vertex is contained in two different cycles in G.

F19:   [NSW71] A graph G has maximum genus 0 if and only if G is a cactus.

F20:   The maximum genus of any necklace is 1.

F21:   [ChGr93] An edge-2-connected graph G has maximum genus 1 if and only if G is homeomorphic to either a necklace or one of the graphs in Figure 7.2.3.

F22:   More generally, a graph G has maximum genus 1 if and only if all except one of its edge-2-connected components are either a cycle or a single vertex, and the exceptional edge-2-connected component of G is homeomorphic to either a necklace or one of the graphs in Figure 7.2.3.

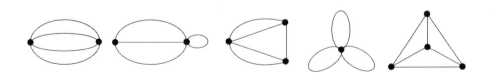

Figure 7.2.3: Graphs of maximum genus 1 that are not necklaces.

7.2.3 Planarity and Upper-Embeddability

There has been extensive research in the study on graphs of minimum genus 0 and on graphs G of maximum genus $\lfloor \beta(G)/2 \rfloor$.

DEFINITIONS

D15: A graph G whose maximum genus is $\lfloor \beta(G)/2 \rfloor$ is called an ***upper-embeddable graph***.

D16: An ear decomposition $D = [P_1, P_2, \cdots, P_r]$ of a graph G is ***3-connected*** if $r \geq 3$, $P_1 + P_2 + P_3$ is homeomorphic to the complete graph K_4, and every subgraph $P_1 + \cdots P_i$ of G, $3 \leq i \leq r$, is homeomorphic to a 3-connected graph.

D17: The edge complement $G - T$ of a spanning tree of a graph G is called a ***co-tree***. Note that the number of edges in a co-tree is exactly equal to the cycle rank $\beta(G)$ of the graph G.

D18: A component H of the co-tree $G - T$ is called an ***even component*** (resp. ***odd component***) if the number of edges in H is even (resp. odd).

D19: For a given spanning tree T of the graph G, the ***deficiency*** $\xi(G, T)$ of T is defined to be the number of odd components of the co-tree $G - T$.

D20: The ***deficiency*** $\xi(G)$ of the graph G is defined to be the minimum of $\xi(G, T)$ over all spanning trees T.

D21: A spanning tree T of G is called a ***Xuong tree*** if the deficiency $\xi(G, T)$ of T is equal to the deficiency $\xi(G)$ of the graph G.

D22: We say that two edges are ***adjacent*** if they share a common endpoint.

D23: Let A be a subset of edges of a graph G, and let $G - A$ be the graph obtained from G by deleting the edges in A. (The graph $G - A$ need not be connected.) Define $C_o(G - A)$ and $C_e(G - A)$ to be the numbers of connected components in $G - A$ with odd cycle rank and with even cycle rank, respectively, and let $\nu(G, A) = C_e(G - A) + 2C_o(G - A) - |A| - 1$. The ***Nebeský nu-invariant*** $\nu(G)$ is defined to be the maximum of $\nu(G, A)$ over all edge subsets A of the graph G.

FACTS ABOUT PLANARITY

The study of 3-connected planar graphs has played an important role in the study of planar graphs. A number of important properties of planar embeddings of 3-connected planar graphs can be derived based on a special ear-decomposition of 3-connected graphs.

F23: [Chen90] Every 3-connected graph has a 3-connected ear decomposition.

Based on Fact F23 and using induction, we can derive the following two well-known results on 3-connected planar graphs.

F24: [Whit33] There is essentially only one way to embed a 3-connected planar graph in the plane.

F25: [Tutt60] Every 3-connected planar graph has a planar embedding in the geometric plane in which every face, except the outer face, is a convex polygon.

F26: [Fary48] Every planar graph has a planar embedding in the geometric plane in which every edge is a straight line segment.

FACTS ABOUT UPPER-EMBEDDABILITY

Early study on graph upper-embeddability was focused on derivation of upper-embeddability of special graph classes. It became clearer later that most of these results could be obtained from effective characterizations of graph maximum genus. There have been a number of successful characterizations of graph maximum genus, which are now described.

F27: [Xuon79a] The maximum genus $\gamma_{\max}(G)$ of a graph G is equal to $(\beta(G) - \xi(G))/2$.

F28: [ChKa99] A spanning tree T of G can be constructed so that the co-tree $G - T$ contains at least $\gamma_{\max}(G)$ pairs of adjacent edges. Therefore, $\gamma_{\max}(G)$ is bounded by $(\beta(G) - \xi(G))/2$. This implies Fact F27.

F29: [Nebe81a] $\nu(G) = \xi(G)$. In consequence, the maximum genus $\gamma_{\max}(G)$ of the graph G is equal to $(\beta(G) - \nu(G))/2$.

F30: [NSW71, Jung78, Xuon79b, Nebe81a] A graph G is upper-embeddable if and only if $\xi(G) \leq 1$, or equivalently, if and only if G has a spanning tree T such that the co-tree $G - T$ has at most one odd component. (This result was obtained independently by a number of researchers.)

F31: [Kund74] Every edge-4-connected graph contains two edge-disjoint spanning trees.

F32: Every edge-4-connected graph is upper-embeddable.

F33: A number of other upper-embeddable graph classes have been identified, including the following:
  [Nebe81b] locally connected graphs,
  [PaXu79] cyclically edge-4-connected graphs,
  [SkNe89] k-regular vertex-transitive graphs of girth g with $k \geq 4$ or $g \geq 4$,
  [Skov91] loopless graphs of diameter 2,
  [HuLi00a] $(4k + 2)$-regular graphs and $(2k)$-regular bipartite graphs.

Readers are referred to the original papers for more detailed definitions and results.

EXAMPLES

E2: The graphs of maximum genus 1 in Figure 7.2.3 all have cycle rank 3. In consequence, all these graphs are upper-embeddable.

E3: The complete graphs K_n are upper-embeddable for all $n \geq 1$.

E4: The complete bipartite graphs $K_{n,m}$ are upper-embeddable for all $n, m \geq 1$.

REMARK

R4: The maximum genus of a graph can also be characterized in terms of ear decompositions of the graph. We refer interested readers to [ChKa99] for more details.

7.2.4 Lower Bounds

The study of lower bounds for minimum genus and maximum genus has been an interesting topic in topological graph theory. The research has led to better understanding of these important graph invariants. Moreover, the study has played an important role in the study of some well-known graph problems, such as the famous Heawood problem.

DEFINITIONS

D24: A *c-coloring of a graph* G is a function from the vertices of G to a set of c "colors" such that no two adjacent vertices are assigned the same color.

D25: The *chromatic number* of a graph G is the minimum integer c such that the graph G has a c-coloring. The *chromatic number* of a surface S is the maximum integer h such that there is a simple graph G of chromatic number h and G is embeddable in the surface S.

Lower Bounds on Minimum Genus

FACTS

F34: If a graph G with n vertices and m edges contains no cycle of size less than d, then

$$\gamma_{\min}(G) \geq \frac{(d-2)m}{(2d)} - \frac{n}{2} + 1$$

F35: If G is a simple graph with n vertices and m edges, then

$$\gamma_{\min}(G) \geq \frac{m}{6} - \frac{n}{2} + 1$$

F36: If a simple graph G of n vertices and m edges has a triangulated embedding, then the embedding is a minimum genus embedding and

$$\gamma_{\min}(G) = \frac{m}{6} - \frac{n}{2} + 1$$

F37: [RiYo68] For the complete graph K_n of n vertices, with $n \geq 3$,

$$\gamma_{\min}(K_n) = \left\lceil \frac{(n-3)(n-4)}{12} \right\rceil$$

Proving this was most of the proof of the well-known **Heawood problem**, conjectured by Heawood in 1890 [Heaw90], and confirmed by Ringel and Youngs [RiYo68].

F38: [RiYo68] **Ringel–Youngs Theorem** (formerly the Heawood Conjecture) The chromatic number of the surface S_g of genus g, for $g > 0$, is equal to

$$\left\lfloor \frac{7 + \sqrt{1 + 48g}}{2} \right\rfloor$$

This assertion for $g = 0$ is the Four-Color theorem [ApHa76].

F39: If G is a graph of chromatic number c, then $\gamma_{\min}(G) \geq (c^2 - 7c + 12)/12$.

Lower Bounds on Maximum Genus

The study of lower bounds on the maximum genus for graphs concentrates on graphs that are not upper-embeddable. In view of Fact F32, such graphs have edge-connectivity less than 4. There is no useful lower bound for maximum genus in terms of number of vertices or of cycle rank. In order to derive meaningful lower bounds for graph maximum genus, a critical structural property of necklaces must be excluded, since necklaces can have arbitrarily large cycle rank, and yet have maximum genus at most 1.

FACTS

F40: [ChGr95] A graph of any fixed maximum genus M greater than zero and arbitrarily large cycle rank can be constructed from a graph G of maximum genus $M - 1$, by replacing an edge e of G with the result of subdividing a proper edge that uniquely joins its endpoints, and splitting the necklace open at the new vertex into a "chain", so that the univalent vertices of the chain are attached where the endpoints of edge e were. Moreover, this is essentially the only way to construct graphs of very large cycle rank while keeping the maximum genus small.

F41: [CKG96] Suppose G is a simple graph of minimum degree at least 3. Then

$$\gamma_{\max}(G) \geq \left\lceil \frac{\beta(G)}{4} \right\rceil$$

This bound is tight, in the sense that there are infinitely many simple graphs G of minimum degree 3 whose maximum genus is arbitrarily close to $\beta(G)/4$ [CKG96].

F42: Let G be an edge-2-connected simple graph of minimum degree at least 3. Then

$$\gamma_{\max}(G) \geq \left\lceil \frac{\beta(G)}{3} \right\rceil$$

This was first proved by Kanchi and Chen, using Fact F27, and then by Archdeacon, Nedela, and Škoviera, using Fact F29. These results are reported in [Ar+02].

F43: [Ar+02] Let G be an edge-3-connected graph. Then $\gamma_{\max}(G) \geq \lceil \beta(G)/3 \rceil$.

F44: [CAG96] There exists an infinite class of edge-3-connected simple graphs G whose maximum genus is equal to $\lceil \beta(G)/3 \rceil$. Thus, the bound in Fact F43 is tight.

REMARKS

R5: The facts for edge-k-connected graphs just above can be translated without much trouble to k-connected graphs, since a k-connected graph is edge-k-connected.

R6: Lower bounds on graph maximum genus have been derived in terms of graph connectivity, independent number, girth, and chromatic number. See [HuLi00b, LiLi00].

7.2.5   Algorithmic Issues

Algorithms and complexity on planar graph problems have been extensively studied. In particular, there is a linear time algorithm that either constructs a planar embedding for a given graph or reports that the graph is not planar [HoTa74]. Based on graph planar embeddings, many difficult graph problems become relatively easier. For example, there is a linear time algorithm that tests the isomorphism of planar graphs [HoWo74].

Minimum Genus Algorithms

The complexity of constructing minimum genus embeddings for graphs of bounded minimum genus has been studied in the past 30 years. The first polynomial-time algorithm for the problem for graphs of minimum genus bounded by a constant g was proposed in 1979, with time complexity $O(n^{O(g)})$ [FMR79]. See [MyKo11] for recent updates and comments on this line of research.

Fellows and Langston [FeLa88] indicated that the graph minimum genus problem of graphs of bounded minimum genus can be solved in polynomial time based on Robertson and Seymour's results in graph minor theory. In fact, they showed a much stronger result that for any minor closed graph class $\mathcal{C}$, there is a polynomial time algorithm that tests the membership for the class $\mathcal{C}$. We describe their results here.

DEFINITION

D26: A graph G is an ***apex graph*** if it contains a vertex v such that $G - v$ is planar.

FACTS

F45: [RoSe95] Let H be a fixed graph. There is a polynomial-time algorithm that decides for any given graph G whether H is a minor of G.

F46: For any minor-closed graph class $\mathcal{C}$, there is a polynomial-time algorithm that tests the membership for the class $\mathcal{C}$.

F47: For any fixed integer g, there is a polynomial-time algorithm that tests whether a given graph has minimum genus at most g. The set of minimal forbidden minors for the class of graphs of minimum genus at most g is unknown (even though by Fact F16, we know such a finite set exists). An algorithm for constructing a minimum-genus embedding of graphs of bounded minimum genus was developed by Mohar.

F48: [Moha99] For each fixed integer g, there is a linear-time algorithm that, for a given graph G, either constructs an embedding of genus bounded by g for G or reports that no such embedding exists.

F49: [Thom89] The problem of deciding for a graph G and integer k, decide whether $\gamma_{\min}(G) \le k$ is NP-complete.

F50: [Thom97a] The problem of deciding for an integer k and a graph G of maximum degree 3 has minimum genus at most k is NP-complete.

F51: It is easy to test in polynomial time whether a given graph is an apex graph.

F52: [Moha98] The problem of deciding whether an apex graph has minimum genus at most k is NP-complete.

Maximum Genus Algorithms

On the other hand, the construction of graph maximum genus embeddings turns out to be easier.

FACTS

F53: [FGM88] There is a polynomial-time algorithm that constructs a maximum genus embedding for a given graph. Based on Xuong's characterization of maximum genus (Fact F27), Furst, Gross, and McGeoch construct a maximum-genus embedding for a given graph by reducing the problem to the linear matroid parity problem, which was known to be solvable in polynomial time [GaSt85].

F54: [Chen94] For any fixed integer g, there is a linear-time algorithm to decide whether a given graph has maximum genus g; if so, the algorithm constructs a maximum-genus embedding for the graph.

F55: [Chen94] For any fixed integer g, there is a linear-time isomorphism algorithm for graphs of maximum genus at most g.

F56: [GrRi91] Starting from any embedding of a graph, there is a sequence of edge deletion-then-reinsertion operations that never decreases the embedding genus and eventually leads to a maximum-genus embedding. Thus, there are no graph embeddings that are "strictly locally maximal" but not globally maximum with respect to embedding genus. On the other hand, there are graph embeddings that are not minimum genus embeddings but are "strictly locally minimal" that represent arbitrarily deep traps, in the sense that one must ascend arbitrarily higher in genus, before one reaches an embedding from which there is a monotonic descent to the global minimum genus.

References

[ApHa76] K. Appel and W. Haken, Every planar map is four-colorable, *Bull. Amer. Math. Soc.* 82 (1976), 711–712.

[Arch81] D. Archdeacon, A kuratowski theorem for the projective plane, *J. Graph Theory* 5 (1981), 243–246.

[Ar+02] D. Archdeacon, J. Chen, Y. Huang, S. P. Kanchi, D. Li, Y. Liu, R. Nedela, and M. Škoviera, Maximum genus, connectivity, and Nebeský's theorem, *Preprint*, (1994).

[ArHu89] D. Archdeacon and P. Huneke, A Kuratowski theorem for nonorientable surfaces, *J. Combin. Theory Ser. B* 46 (1989), 173–231.

[BHKY62] J. Battle, F. Harary, Y. Kodama, and J. W. T. Youngs, Additivity of the genus of a graph, *Bull. Amer. Math. Soc.* 68 (1962), 565–568.

[Brah32] H. R. Brahana, Systems of circuits of two-dimensional manifolds, *Ann. of Math.* 30 (1923), 234–243.

[Chen90] J. Chen, *The distribution of graph imbeddings on topological surfaces*, Ph.D. thesis, Department of Mathematics, Columbia University (1990).

[Chen94] J. Chen, A linear time algorithm for isomorphism of graphs of bounded average genus, *SIAM Journal on Discrete Mathematics* 7 (1994), 614–631.

[CAG96] J. Chen, D. Archdeacon, and J. L. Gross, Maximum genus and connectivity, *Discrete Mathematics* 149 (1996), 19–29.

[ChGr93] J. Chen and J. L. Gross, Kuratowski-type theorem for average genus, *J. Combinatorial Theory Ser. B* 57 (1993), 100–121.

[ChGr95] J. Chen and J. L. Gross, No lower limit points for average genus, *Graph Theory, Combinatorics, and Applications*, Y. Alavi and A. Schwenk, ed., Wiley Interscience (1995), 183–194.

[ChKa99] J. Chen and S. P. Kanchi, Graph ear decompositions and graph embeddings, *SIAM J. Discrete Math.* 12(2) (1999), 229–242.

[CKG96] J. Chen, S. P. Kanchi, and J. L. Gross, A tight lower bound on the maximum genus of a simplicial graph, *Discrete Mathematics* 156 (1996), 83–102.

[Fary48] I. Fary, On straight line representation of planar graphs, *Acta Sci. Math. (Szeged)* 11 (1948), 229–233.

[FeLa88] M. Fellows and M. Langston, Nonconstructive tools for proving polynomial-time decidability, *J. Assoc. Comput. Mach.* 35(3) (1988), 727–739.

[FMR79] L. Filotti, G. Miller, and J. Reif, On determining the genus of a graph in $O(v^{O(g)})$ steps, *Proc. 11th Annu. ACM Symp. on Theory of Computing* (1979), 27–37.

[FGM88] M. Furst, J. L. Gross, and L. A. McGeoch, Finding a maximum-genus graph imbedding, *J. Assoc. Comput. Mach.* 35(3) (1988), 523–534.

[GaSt85] H. N. Gabow and M. Stallmann, Efficient algorithms for graphic matroid intersection and parity, *Lecture Notes in Computer Science* 194 (1985), 210–220.

[GlHu78] H. Glover and J. P. Huneke, The set of irreducible graphs for the projective plane is finite, *Discrete Math.* 22 (1978), 243–256.

[GHW79] H. Glover, J. P. Huneke, and C.-S. Wang, 103 graphs that are irreducible for the projective plane, *J. Combin. Theory Ser. B* 27 (1979), 332–370.

[GKR93] J. L. Gross, E. W. Klein, and R. G. Rieper, On the average genus of a graph, *Graphs and Combinatorics* 9 (1993), 153–162.

[GrRi91] J. L. Gross and R. G. Rieper, Local extrema in genus-stratified graphs, *J. Graph Theory* 15 (1991), 159–171.

[GrTu87] J. L. Gross and T. W. Tucker, *Topological Graph Theory*, Wiley-Interscience, New York 1987.

[Heaw90] P. J. Heawood, Map-colour theorem, *Quart. J. Math.* 24 (1890), 332–338.

[HoTa74] J. Hopcroft and R. Tarjan, Efficient planarity testing, *J. Assoc. Comput. Mach.* 21 (1974), 549–568.

[HoWo74] J. Hopcroft and J. Wong, Linear time algorithm for isomorphism of planar graphs, *Proc. 6th Annu. ACM Symp. on Theory of Computing*, (1974), 172–184.

[HuLi00a] Y. Huang and Y. Liu, The classes of upper embeddable graphs with the same value of degree of vertex under modulo, *Acta Math. Sci.* 20 (2000), 251–255.

[HuLi00b] Y. Huang and Y. Liu, Maximum genus, independent number and girth, *Chinese Annals of Mathematics* 21 (2000), 77–82.

[Jung78] M. Jungerman, A characterization of upper-embeddable graphs, *Trans. Amer. Math. Soc.* 241 (1978), 401–406.

[Koni36] D. König, Theorie der endlichen und unendlichen Graphen, *Akademische Verlagsgesellschaft* (1936).

[Kund74] S. Kundu, Bounds on the number of disjoint spanning trees, *J. Combinatorial Theory Ser. B* 17 (1974), 199–203.

[Kura30] K. Kuratowski, Sur le problème des courbes gauches en topologie, *Fund. Math.* 15 (1930), 271–283.

[LiLi00] D. Li and Y. Liu, Maximum genus, girth and connectivity, *European J. Combin.* 21 (2000), 651–657.

[Moha98] B. Mohar, On the orientable genus of graphs with bounded nonorientable genus, *Discrete Math.* 182 (1998), 245–253.

[Moha99] B. Mohar, A linear time algorithm for embedding graphs in an arbitrary surface, *SIAM J. Discrete Math.* 12 (1999), 6–26.

[MyKo11] W. Myrvold and W. Kocay, Errors in graph embedding algorithms, *J. Computer and System Sciences* 77 (2011), 430–438.

[Nebe81a] L. Nebeský, A new characterization of the maximum genus of a graph, *Czechoslovak Math. J.* 31 (1981), 604–613.

[Nebe81b] L. Nebeský, Every connected, locally connected graph is upper embeddable, *Journal of Graph Theory* 5 (1981), 205–207.

[NSW71] E. Nordhaus, B. Stewart, and A. White, On the maximum genus of a graph, *J. Combin. Theory Ser. B* 11 (1971), 258–267.

[PaXu79] C. Payan and N. H. Xuong, Upper embeddability and connectivity of graphs, *Discrete Mathematics* 27 (1979), 71–80.

[RiYo68] G. Ringel and J. W. T. Youngs, Solution of the Heawood map-coloring problem, *Proc. Nat. Acad. Sci. U.S.A.* 60 (1968), 438–445.

[RoSe85] N. Robertson and P. D. Seymour, Graph minors — a survey, in *Surveys in Combinatorics 1985*, Ed. I. Anderson, Cambridge Univ. Press, Cambridge (1985), 153–171.

[RoSe04] N. Robertson and P. D. Seymour, Graph minors XX. Wagner's conjecture, *J. Combin. Theory Ser. B* 92 (2004), 325–357.

[RoSe90a] N. Robertson and P. D. Seymour, Graph minors IV. Tree-width and well-quasi-ordering, *J. Combin. Theory Ser. B* 48 (1990), 227–254.

[RoSe90b] N. Robertson and P. D. Seymour, Graph minors VIII. A Kuratowski theorem for general surfaces, *J. Combin. Theory Ser. B* 48 (1990), 255–288.

[RoSe95] N. Robertson and P. D. Seymour, Graph minors XIII. The disjoint paths problem, *J. Combin. Theory Ser. B* 63 (1995), 65–110.

[Seym93] P. D. Seymour, A bound on the excluded minors for a surface, *Preprint*, (1993).

[Skov91] M. Škoviera, The maximum genus of graphs of diameter two, *Discrete Mathematics* 87 (1991), 175–180.

[SkNe89] M. Škoviera and R. Nedela, The maximum genus of vertex-transitive graphs, *Discrete Mathematics* 78 (1989), 179–186.

[Thom89] C. Thomassen, The graph genus problem is NP-complete, *J. Algorithms* 10, (1989), 568–576.

[Thom97a] C. Thomassen, The genus problem for cubic graphs, *J. Combin. Theory Ser. B* 69, (1997), 52–58.

[Thom97b] C. Thomassen, A simpler proof of the excluded minor theorem for higher surfaces, *J. Combin. Theory Ser. B* 70, (1997), 306–311.

[Tutt60] W. Tutte, Convex representation of graphs, *Proc. London Math. Soc.* 10 (1960), 474–483.

[Wagn37] K. Wagner, Über eine Eigenschaft der ebenen Komplexe, *Math. Ann.* 114 (1937), 570–590.

[Whit32] H. Whitney, Non-separable and planar graphs, *Trans. Amer. Math. Soc.* 34 (1932), 339–362.

[Whit33] H. Whitney, A set of topological invariants for graphs, *Amer. J. Math.* 55 (1933), 231–235.

[Xuon79a] N. H. Xuong, How to determine the maximum genus of a graph, *J. Combin. Theory Ser. B* 26 (1979), 217–225.

[Xuon79b] N. H. Xuong, Upper-embeddable graphs and related topics, *J. Combin. Theory Ser. B* 26 (1979), 226–232.

Section 7.3

Genus Distributions

Jonathan L. Gross, Columbia University

INTRODUCTION

This chapter explores the natural problem of constructing a surface-by-surface inventory of the imbeddings of a fixed graph, which was introduced by Gross and Furst [GrFu87]. The present scope includes several interesting extensions of that problem.

7.3.1 Ranges and Distributions of Imbeddings

An imbedding is taken to be cellular, unless it is clear from context that a noncellular imbedding is under consideration. We regard two cellular imbeddings as "the same" if they have equivalent rotation systems. Moreover, a graph is taken to be connected unless the context implies otherwise. *Minimum genus* and *maximum genus* are presented in §7.2. We are concerned here with the entire *genus range*.

DEFINITIONS

D1: The ***minimum genus of a graph*** G, denoted $\gamma_{min}(G)$, is the smallest positive integer g such that the graph G has an imbedding in the orientable surface S_g.

D2: The ***maximum genus of a graph*** G, denoted $\gamma_{max}(G)$, is the largest integer g such that the graph G has a cellular imbedding in the orientable surface S_g.

D3: The ***genus range of a graph*** G is the integer interval $[\gamma_{min}(G), \gamma_{max}(G)]$.

D4: The jth **orientable imbedding number of a graph** G, denoted $g_j(G)$, is the number of equivalence classes of orientable imbeddings of G into the orientable surface S_j or equivalently (see §7.1), the number of rotation systems for graph G that induce an imbedding in S_j.

D5: The **genus distribution sequence** of a graph G is the sequence whose j^{th} entry is $g_j(G)$.

D6: The **genus distribution polynomial** is

$$I_G(x) = \sum_{j=0}^{\infty} g_j(G)x^j$$

D7: The **minimum crosscap number of a graph** G, also known as the **minimum non-orientable genus**, is the smallest integer k such that the graph G has an imbedding in the non-orientable surface N_k. It is denoted $\overline{\gamma}_{min}(G)$.

D8: The **maximum crosscap number of a graph** G, also known as the **maximum non-orientable genus**, is the largest integer k such that the graph G has a cellular imbedding in the no-norientable surface N_k. It is denoted $\overline{\gamma}_{max}(G)$.

D9: The **crosscap range of a graph** G is the integer interval $[\overline{\gamma}_{min}(G), \overline{\gamma}_{max}(G)]$.

D10: The jth **crosscap imbedding number of a graph** G, denoted $\overline{x}_j(G)$, is the number of equivalence classes of non-orientable imbeddings of G into the non-orientable surface N_j.

D11: The **crosscap distribution sequence** of a graph G is the sequence whose j^{th} entry is $\overline{x}_j(G)$.

D12: The **crosscap distribution polynomial** is

$$\overline{I}_G(y) = \sum_{j=1}^{\infty} \overline{x}_j(G)y^j$$

D13: The boundary walk of a face of an imbedding is called an **fb-walk**.

FACTS

F1: Let G be a graph. Then the total number of equivalence classes of orientable imbeddings equals

$$\sum_{j=0}^{\infty} g_j(G) = \prod_{v \in V(G)} [\deg(v) - 1]!$$

since the sum on the left and the product on the right both count every imbedding of G exactly once. Moreover, the polynomial evaluation $I_G(1)$ gives this same number.

F2: [Du66] [Interpolation Theorem] For every integer j within the genus range of a graph G, i.e., whenever $\gamma_{min}(G) \leq j \leq \gamma_{max}(G)$, the number $g_j(G)$ of orientable imbeddings of G is positive.

F3: [St78] For every integer j within the crosscap range of a graph G, i.e., whenever $\overline{\gamma}_{min}(G) \leq j \leq \overline{\gamma}_{max}(G)$, the number $\overline{x}_j(G)$ of non-orientable imbeddings of G is positive.

F4: Let G be a graph. Then the total number of equivalence classes of imbeddings (orientable and non-orientable) equals

$$\sum_{j=0}^{\infty} g_j(G) \; + \; \sum_{j=1}^{\infty} \overline{x}_j(G) \;\; = \;\; 2^{\beta(G)} \prod_{v \in V(G)} [\deg(v) - 1]!$$

since the sum on the left and the product on the right both count every imbedding of G exactly once. The factor of $2^{\beta(G)}$ on the right accounts for the possible choices of orientation on every edge not in a designated spanning tree for G.

EXAMPLE

E1: All of the examples of genus distributions in Table 7.1 can be calculated by considering the corresponding rotation systems (see §7.1). Consideration of symmetries expedites the calculations.

Table 7.1: Genus distributions of some familiar graphs.

graph G	$g_0(G)$	$g_1(G)$	$g_2(G)$	$g_3(G)$	$g_4(G)$	$\cdots$
K_4	2	14	0	0	0	$\cdots$
bouquet B_2	4	2	0	0	0	$\cdots$
dipole D_3	2	2	0	0	0	$\cdots$
$K_{3,3}$	0	40	24	0	0	$\cdots$
$K_2 \times C_3$	2	38	24	0	0	$\cdots$

REMARK

R1: Complementary to the graph-theoretic problems concerned with counting the imbeddings of a given graph over a range of surfaces are the map-theoretic problems of counting the maps on a given surface, taken over all possible imbedded graphs, or over all graphs with some prespecified property. (See §7.6).

7.3.2   Counting Noncellular Imbeddings

This section describes how the problem of calculating distributions of noncellular imbeddings reduces to counting cellular imbeddings. Explicit discussion of methods for achieving such a reduction are scarce in the literature.

DEFINITIONS

D14: A *semicellular graph imbedding* is an imbedding $G \to S$ whose regions are planar, but which may have more than one boundary component.

D15: A graph imbedding $G \to S$ is *strongly noncellular* if any of its regions is nonplanar.

D16: A closed curve that separates a region of a noncellular graph imbedding $G \to S$ is **boundary-separating** if there is at least one boundary component of the region on each side of the separation.

D17: A closed curve in a region of a noncellular graph imbedding $G \to S$ is **strongly noncontractible** if cutting it open and capping off the holes with disks reduces the genus of the region.

D18: Given a semicellular graph imbedding $G \to S$, the **underlying cellular imbedding** is obtained by cutting each non-cell region open along a maximal family of boundary-separating closed curves and capping the holes with disks.

D19: Given a noncellular graph imbedding $G \to S$, a **planarizing curve** for a nonplanar region is a separating closed curve such that all of the boundary components lie to one side of the separation and all of the handles lie to the other. The concept is that the resulting component with all the handles contains no part of the graph and is discarded.

D20: Given a strongly noncellular graph imbedding $G \to S$, the **underlying semicellular imbedding** is obtained by cutting each non-cell region open along a maximal family of boundary-separating closed curves and capping the holes with disks.

FACTS

F5: Every semicellular graph imbedding has an underlying cellular imbedding that is unique up to homeomorphism.

F6: The semicellular orientable imbeddings of a graph are in bijective correspondence with partitions of the regions of the underlying cellular imbedding.

F7: Every nonplanar region of a noncellular graph imbedding has a planarizing curve.

F8: Every strongly noncellular imbedding has an underlying semicellular imbedding that is unique up to homeomorphism.

F9: The strongly noncontractible imbeddings of a graph are in bijective correspondence with the set of functions from the regions of the underlying semicellular imbedding to the nonnegative integers.

F10: A strongly noncellular imbedding $G \to S_{n+k}$ of a graph in a surface can be obtained from a semicellular imbedding into $G \to S_n$ by partitioning the number k, and next selecting one face of the imbedding into S_n for each of the parts of the partition, and then increasing the genus of each selected face by the value of the associated part of the partition.

EXAMPLES

E2: The graph $K_2 \times C_3$ has six vertices, each of degree 3. Thus, by Fact F1, the total number of orientable cellular imbeddings is $64 = 2^6$. The cellular genus distribution sequence is $2, 38, 24, 0, 0, \ldots$, as given in Table 7.1.

E3:   Each of the two imbeddings of $K_2 \times C_3$ in S_0 has five faces, as shown in Figure 7.3.1. Five faces can be partitioned into four nonempty parts in five ways. (In general, one can use *Stirling subset numbers* for these partition-number calculations.) Thus, each imbedding in S_0 yields five different possible semicellular imbeddings in S_1. Thus, there are ten semicellular noncellular imbeddings of $K_2 \times C_3$ in S_1, plus 38 cellular imbeddings, as mentioned in Example E2, for a total of 48 semicellular imbeddings in S_1.

Figure 7.3.1: The graph $K_2 \times C_3$.

E4:   Each semicellular imbedding of $K_2 \times C_3$ in the surface S_2 corresponds to a partition into three parts of the five faces of a cellular imbedding in S_0 or to a partition into two parts of the three faces of a cellular imbedding into S_1. Using the Stirling numbers $\left\{ {5 \atop 3} \right\} = 25$ and $\left\{ {3 \atop 2} \right\} = 3$, and using the cellular genus distribution sequence from Example E2, we calculate that the number of semicellular (but noncellular) imbeddings into S_2 equals $2 \cdot 25 + 38 \cdot 3 = 164$. Adding in the 24 cellular imbeddings in S_2, we obtain a total of 188 semicellular imbeddings in S_2.

7.3.3   Partitioned Genus Distributions

Often a graph G is obtained by "pasting" two smaller graphs (called **amalgamands**) together, which means matching some subgraph in one of them to an isomorphic subgraph in the other. To calculate the genus distribution of G from those of the two smaller graphs, it is necessary to partition the genus distributions of the smaller graphs according to the incidence of fb-walks on the respective subgraphs that are matched together. In this subsection, we describe the partitioning for pasting on a single vertex or on a single edge.

DEFINITIONS

D21:   A **rooted graph** is a tuple $(G, x_1, x_2, \ldots, x_n)$ in which G is a graph and $x_1, \ldots, x_n$ are either vertices or edges that have been designated as **roots**.

D22:   The most frequently encountered **graph amalgamations** are of four types. Types (i) and (ii) are called **vertex amalgamations**. Types (iii) and (iv) are called **edge amalgamations**.

  (i) Let (G, u) and (G', u') be singly vertex-rooted graphs. One kind of amalgamated graph is formed from the disjoint union $G \sqcup G'$ by merging the vertices u and u'. It is usually construed to have no root.

  (ii) Let (G, u, v) and (G', s, t) be doubly vertex-rooted graphs. Another kind of amalgamated graph is formed from the disjoint union $G \sqcup G'$ by merging the vertices v and s. It is construed to have roots u and t.

(iii) Let (G, d) and (G', d') be singly edge-rooted graphs. A third kind of amalgamated graph is formed from the disjoint union $G \sqcup G'$ by either way of merging the edges d and d'.

(iv) Let (G, d, e) and (G', x, y) be doubly edge-rooted graphs. A fourth kind of amalgamated graph is formed from the disjoint union $G \sqcup G'$ by either way of merging the edges e and x. It is construed to have roots d and y.

The concept of amalgamation can be generalized to allow the roots to be arbitrary subgraphs, in which case one also specifies an isomorphism from a root of the first amalgamand to a root of the second amalgamand.

D23: Each type of root structure is associated with its own kind of **partitioned genus distribution**.

(i) Let (G, u) be a vertex-rooted graph such that root u is 2-valent. Then for each $j = \gamma_{min}(G), \ldots, \gamma_{max}(G)$, the $g_j(G)$ imbeddings $G \to S_j$ are partitioned into the $d_j(G, u)$ imbeddings such that two *different* fb-walks are incident on root-vertex u and the $s_j(G, u)$ imbeddings such that the *same* fb-walk is twice incident on u.

(ii) Let (G, b) be an edge-rooted graph such that both endpoints of root-edge e are 2-valent. Then for each $j = \gamma_{min}(G), \ldots, \gamma_{max}(G)$, the $g_j(G)$ imbeddings $G \to S_j$ are partitioned into the $d_j(G, b)$ imbeddings such that two *different* fb-walks are incident on root-edge b and the $s_j(G, b)$ imbeddings such that the *same* fb-walk is twice incident on b.

(iii) Let (G, u, v) be a doubly vertex-rooted graph such that both roots are 2-valent. Then for each $j = \gamma_{min}(G), \ldots, \gamma_{max}(G)$, the $g_j(G)$ imbeddings $G \to S_j$ are partitioned into four subtypes:

- the $dd_j(G, u)$ imbeddings such that two *different* fb-walks are incident on root-vertex u and two *different* fb-walks are incident on root-vertex v;
- the $ds_j(G, u)$ imbeddings such that two *different* fb-walks are incident on root-vertex u and the *same* fb-walk is twice incident on v;
- the $sd_j(G, u)$ imbeddings such that the *same* fb-walk is twice incident on root-vertex u and two *different* fb-walks are incident on root vertex v;
- the $ss_j(G, u)$ imbeddings such that the *same* fb-walk is twice incident on root-vertex u and the *same* fb-walk is twice incident on v.

(iv) Let (G, b, c) be a doubly edge-rooted graph such that both endpoints of both root-edges are 2-valent. Then for each $j = \gamma_{min}(G), \ldots, \gamma_{max}(G)$, the $g_j(G)$ imbeddings $G \to S_j$ are partitioned into four subtypes:

- the $dd_j(G, b, c)$ imbeddings such that two *different* fb-walks are incident on root-edge b and two *different* fb-walks are incident on c;
- the $ds_j(G, u)$ imbeddings such that two *different* fb-walks are incident on root-edge b and the *same* fb-walk is twice incident on c;
- the $sd_j(G, u)$ imbeddings such that the *same* fb-walk is twice incident on root-edge b and two *different* fb-walks are incident on root c;
- the $ss_j(G, u)$ imbeddings such that the *same* fb-walk is twice incident on root-edge u and the *same* fb-walk is twice incident on c.

Each of the names dd_j, ds_j, sd_j, and ss_j associated with the subtypes of imbeddings is called a **partial**, and the number of imbeddings of a subtype is called the **value of the corresponding partial**.

REMARK

R2: In practice, it is usually necessary to further partition double-root partials into subpartials that indicate how many fb-walks incident on one root are also incident on the other root.

EXAMPLE

E5: We consider the complete graph $\ddot{K}_4$ with two vertices u and v as roots inserted at the midpoints of two non-adjacent edges, as in Figure 7.3.2. Then we partition the

Figure 7.3.2: The doubly vertex-rooted graph $(\ddot{K}_4, u, v)$.

genus distribution $g_0(\ddot{K}_4) = 2$, $g_1(\ddot{K}_4) = 14$ as follows:

$$
\begin{array}{llll}
d_0(\ddot{K}_4, u) = 2 & & s_0(\ddot{K}_4, u) = 0 & \\
dd_0(\ddot{K}_4, u, v) = 2 & ds_0(\ddot{K}_4, u, v) = 0 & sd_0(\ddot{K}_4, u, v) = 0 & ss_0(\ddot{K}_4, u, v) = 0 \\
d_1(\ddot{K}_4, u) = 8 & & s_1(\ddot{K}_4, u) = 6 & \\
dd_1(\ddot{K}_4, u, v) = 4 & ds_1(\ddot{K}_4, u, v) = 4 & sd_0(\ddot{K}_4, u, v) = 4 & ss_0(\ddot{K}_4, u, v) = 2
\end{array}
$$

These numbers can be confirmed by using the face-tracing algorithm of the 16 imbeddings. Using symmetries reduces the required effort.

7.3.4 Graph Amalgamations

Bar Amalgamations

DEFINITIONS

D24: A *bar-amalgamation* of two disjoint vertex-rooted graphs (G, u) and (H, v) is obtained by running a new edge e between the roots u and v, as illustrated in Figure 7.3.3. Notation: $(G, u) *_{\text{bar}} (H, v)$. The isomorphism type of a bar-amalgamation depends on the choice of root-vertices in the two graphs.

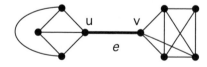

Figure 7.3.3: A bar-amalgamation of K_4 and $K_5 - e$.

D25: The *convolution* of the sequences $\langle a_i \rangle$ and $\langle b_i \rangle$ is the sequence whose k^{th} term is

$$
\sum_{i=0}^{k} a_i b_{k-i}
$$

FACTS

F11: [GrFu87] The genus distribution of a bar-amalgamation of the graphs (G, u) and (H, v) is representable as the convolution of their respective genus distributions, multiplied by a scalar equal to the product $deg(u)deg(v)$.

F12: For any two vertex-rooted graphs (G, u) and (H, v), with $\iota : u \to v$,

$$\gamma_{min}\left((G, u) *_\iota (H, v)\right) = \gamma_{min}((G, u) *_{\text{bar}} (H, v))$$

F13: For any two vertex-rooted graphs (G, u) and (H, v), with $\iota : u \to v$,

$$\gamma_{max}((G, u) *_{\text{bar}} (H, v)) \leq \gamma_{max}((G, u) *_\iota (H, v)) \leq \gamma_{max}((G, u) *_{\text{bar}} (H, v)) + 1$$

F14: The genus distribution sequence of a vertex-amalgamation $(G, u) *_\iota (H, v)$ dominates the genus distribution sequence of the bar-amalgamation $(G, u) *_{\text{bar}} (H, v)$. That is, every term of the former is at least as large as the corresponding term of the latter.

EXAMPLES

E6: The bar-amalgamation $K_4 *_{\text{bar}} K_{3,3}$ has the following genus distribution sequence:

$$9 \, (0, 80, 608, 336, 0, 0, \ldots)$$

E7: The bar-amalgamation $K_3 *_{\text{bar}} K_3$ has the genus distribution sequence

$$4, 0, 0, 0, \ldots$$

Vertex Amalgamations

Quite commonly, an interesting sequence of graphs can be specified recursively, so that after starting off with a base graph (X, s, t), subsequent graphs are obtained by iteratively amalgamating copies of some other graph (Y, u, v), as we describe in this subsection.

DEFINITIONS

D26: A **production** for an amalgamation of two doubly vertex-rooted graphs is a rule of the form

$$p_i(G, u, v) * p'_j(G', u', v') \longrightarrow \sum_\ell \alpha_\ell \, q_{i+j+k_\ell}(G * G', u, v')$$

where p_i and p'_j are types of subscripted partials and where each of the summands on the right is a subscripted partial q_{i+j+k_ℓ} preceded by the number α_ℓ of imbeddings corresponding to that subscripted partial. The genus of the surface of every imbedding must be at least $i + j$, the sum of the genera of the two imbeddings on the left of the production. The added term k_ℓ reflects the fact that the genus of some resultant imbeddings may be larger than that sum.

D27: The **doubled path** DP_n is obtained from an n-edge path by doubling every edge.

FACT

F15: The following two productions are sufficient to calculate the genus distribution of the doubled path DP_n recursively. To simplify the calculation, we have allowed the letter x to stand for either d or s.

$$xd_i(G, u, v) * dd_j(G', u', v') \longrightarrow 4xd_{i+j}(G * G', u, v') + 2xs_{i+j+1}(G * G', u, v')$$
$$xs_i(G, u, v) * dd_j(G', u', v') \longrightarrow 6xd_{i+j}(G * G', u, v')$$

EXAMPLE

E8: We use Fact F15 to calculate the partitioned genus distributions and genus distributions of the doubled paths DP_2, DP_3, and DP_4, starting from $dd(DP_1) = 1$.

$$
\begin{aligned}
xd_0(DP_1) * dd_0(DP_1) &\longrightarrow 4xd_0(DP_2) + 2xs_1(DP_2) \\
pgd(DP_2) &: \quad xd_0 = 4, \ xs_1 = 2 \\
gd(DP_2) &: \quad g_0 = 4, \ g_1 = 2 \\
xd_0(DP_2) * dd_0(DP_1) &\longrightarrow 4 \cdot 4 \cdot 1 xd_0(DP_3) + 2 \cdot 4 \cdot 1 xs_1(DP_3) \\
xs_1(DP_2) * dd_0(DP_1) &\longrightarrow 6 \cdot 2 \cdot 1 xd_1(DP_3) \\
pgd(DP_3) &: \quad xd_0 = 16, \ xd_1 = 12, \ xs_1 = 8 \\
gd(DP_3) &: \quad g_0 = 16, \ g_1 = 20 \\
xd_0(DP_3) * dd_0(DP_1) &\longrightarrow 4 \cdot 16 \cdot 1 xd_0(DP_4) + 2 \cdot 16 \cdot 1 xs_1(DP_4) \\
xd_1(DP_3) * dd_0(DP_1) &\longrightarrow 4 \cdot 12 \cdot 1 xd_1(DP_4) + 2 \cdot 12 \cdot 1 xs_2(DP_4) \\
xs_1(DP_3) * dd_0(DP_1) &\longrightarrow 6 \cdot 8 \cdot 1 xd_1(DP_4) \\
pgd(DP_4) &: \quad xd_0 = 64, \ xd_1 = 96, \ xs_1 = 32, \ xs_2 = 24 \\
gd(DP_4) &: \quad g_0 = 64, \ g_1 = 128, \ g_2 = 24
\end{aligned}
$$

7.3.5 Genus Distribution Formulas for Special Classes

Even at the outset of the program to provide explicit calculations of imbedding distributions, it was clear that a variety of techniques would be needed. Different topological and combinatorial methods seem to be needed for every class of graphs.

DEFINITIONS

D28: The **n-rung closed-end ladder** L_n is the graph obtained from the cartesian product $P_n \times K_2$ by doubling the edges $v_1 \times K_2$ and $v_n \times K_2$ at both ends of the path, as illustrated in Figure 7.3.4.

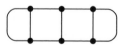

Figure 7.3.4: The 3-rung closed-end ladder L_3.

D29: The *cobblestone path* J_n is the graph obtained by doubling every edge of the n-vertex path P_n, as illustrated in Figure 7.3.5.

Figure 7.3.5: The cobblestone path J_3.

D30: The *n-bouquet* B_n is the graph with one vertex and n self-loops, as illustrated in Figure 7.3.6.

Figure 7.3.6: Some bouquets.

D31: The *dipole* D_n is the graph with two vertices, n edges, and no self-loops.

D32: An *outerplanar imbedding* of a graph G is an imbedding such that there is a face whose boundary walk contains every vertex of G.

D33: An *outerplanar graph* is a graph that has an outerplanar imbedding.

D34: A *Halin graph* is the graph that results from an ordered plane tree when a cycle is drawn in the plane through all the leaf vertices, as shown in Figure 7.3.7 in the order that they occur on a preorder traversal.

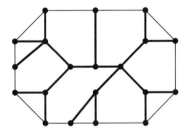

Figure 7.3.7: A Halin graph.

FACTS

F16: [FuGrSt89] The closed-end ladders have the following formula for their genus distributions:

$$
g_i\left(L_n\right) = \begin{cases} 2^{n-1+i}\binom{n+1-i}{i}\frac{2n+2-3i}{n+1-i} & \text{for } i \leq \lfloor\frac{n+1}{2}\rfloor \\ 0 & \text{otherwise} \end{cases}
$$

The following table shows the genus distributions for some of the smaller ladders.

	g_0	g_1	g_2	g_3	g_4	total
L_1	2	2	0	0	0	4
L_2	4	12	0	0	0	16
L_3	8	40	16	0	0	64
L_4	16	112	128	0	0	256
L_5	32	288	576	128	0	1024

F17: [FuGrSt89] The cobblestone paths have the following formula for their genus distributions:

$$g_i\left(J_n\right) = 3^i \cdot 4^{n-1} \cdot \binom{n-i}{i} + 2 \cdot 3^{i-1} \cdot 4^{n-1} \cdot \binom{n-i}{i-1} \quad \text{for } i \geq 0 \text{ and } n \geq 1$$

The following table shows the genus distribution for some of the smaller cobblestone paths.

	g_0	g_1	g_2	total
J_1	4	2	0	6
J_2	16	20	0	36
J_3	64	128	24	216
J_4	256	704	336	1296

F18: [GrRoTu89] The bouquets have the following formula for their orientable imbedding numbers:

$$g_j\left(B_n\right) = (n-1)! \cdot 2^{n-1} \cdot e_{n-2j+1}\left(n\right)$$

where the numbers

$$e_k\left(n\right) = \left| \left\{ \pi \in \Sigma_{2n} \;\middle|\; \begin{array}{l} \text{permutation } \pi \text{ has } k \text{ cycles, and} \\ (\exists \text{ full involution } \beta)\, [\pi = \rho_0 \circ \beta] \end{array} \right\} \right|$$

where ρ_0 is an arbitrary fixed cycle of length $2n$

are given by the formula of Jackson [Ja87]. The closed formula above for $g_j\left(B_n\right)$ leads to the following recursion.

Initial conditions:

$$g_j(B_n) = 0 \text{ for } j < 0 \text{ or } n < 0$$
$$g_j(B_0) = g_j(B_1) = 0 \text{ for } j > 0$$

$$g_j(B_2) = \begin{cases} 4 & \text{for } j = 0 \\ 2 & \text{for } j = 1 \\ 0 & \text{for } j \geq 2 \end{cases}$$

Recursion for $n > 2$:

$$(n+1)g_j(B_n) = 4(2n-1)(2n-3)(n-1)^2(n-2)g_j(B_{n-2})$$
$$+ 4(2n-1)(n-1)g_j(B_{n-1})$$

This recursion enables us to calculate numerical values.

$g_j(B_n)$	$j = 0$	1	2	Total
$n = 0$	1			1
1	1			$1! = 1$
2	4	2		$3! = 6$
3	40	80		$5! = 120$
4	672	3360	1008	$7! = 5040$
5	16128	161280	185472	$9! = 362880$

F19: Rieper [Ri90] elaborated upon the use of group characters in his analysis of the genus distribution of *dipoles*, which are graphs with two vertices and no self-loops. Andrews, Jackson, and Visentin [AnJaVi94] took a map-theoretic approach to dipole imbeddings.

F20: Stahl [St91a] calculated genus distributions for small-diameter graphs.

F21: Kwak, Kim, and Lee [KwKiLe96] took a distributional approach in studying a class of branched coverings of surfaces.

F22: Using doubly vertex-rooted and doubly edge-rooted graphs, Gross, Khan, and Poshni (see [GrKhPo10], [Gr11a], [KhPoGr10], and [PoKhGr10]) developed quadratic-time algorithms for calculating the genus distribution of chains, cycles, and twisted cycles of arbitrarily many copies of a given graph of known genus distribution.

F23: [Gr11b] and [PoKhGr11] give quadratic-time algorithms for the genus distribution of any 3-regular and 4-regular outerplanar graphs, respectively.

F24: Outerplanar graphs have treewidth 2 (see [Bo98]).

F25: [Gr12a] gives a quadratic-time algorithm for the genus distribution of any 3-regular Halin graph.

F26: Halin graphs have treewidth 3 (see [Bo98]).

F27: [GrKo12] gives a quadratic-time algorithm for any cubic series-parallel graph and, more generally, for any graph of treewidth 2 and maximum degree at most 3.

F28: [Gr12b] gives a quadratic-time algorithm for any class of graphs of fixed treewidth and bounded degree. It is not a practical algorithm.

REMARKS

R3: Ladder-like graphs played a crucial role in the solution of the Heawood map-coloring problem. (See [Ri74].) McGeoch [McG87] calculated the genus distribution of *circular ladders* and of *Möbius ladders*. Tesar [Te00] calculated the genus distribution of *Ringel ladders*.

R4: The computations of imbedding distributions of ladders and cobblestone paths were subsequently generalized by Stahl [St91a] to *linear families*.

R5: Riskin [Ri95] took a distributional approach in studying a class of polyhedral imbeddings.

R6: Stahl [St97] studied the zeroes of a class of genus polynomials.

R7: Among the important properties of bouquets to topological graph theory is that every regular graph can be derived by assigning voltages (possibly permutation voltages) to a bouquet. (See [GrTu77] or [GrTu87].)

7.3.6   Other Imbedding Distribution Calculations

Including non-orientable surfaces in the inventory requires some additional theory, partly because the possible twisting of edges complicates the recurrences one might derive. Yet another enumerative aspect of graph imbeddings regards as equivalent any two imbeddings that "look alike" when vertex and edge labels are removed.

DEFINITIONS

D35: The ***total imbedding distribution*** of a graph G is the bivariate polynomial

$$\ddot{I}_G(x,y) = I_G(x) + \overline{I}_G(y) = \sum_{k=0}^{\infty} g_k(G)x^k + \sum_{k=1}^{\infty} \overline{x}_k(G)y^k$$

D36: Given a general rotation system ρ for a graph G and a spanning tree T, the entries of the ***overlap matrix*** $M_{\rho,T} = [m_{i,j}]$ are given for all pairs of edges e_i, e_j of the co-tree $G - T$ by

$$m_{i,j} = \begin{cases} 1 & \text{if } i \neq j \text{ and } pure\,(\rho)\Big|_{T+e_i+e_j} \text{ is nonplanar} \\ 1 & \text{if } i = j \text{ and edge } i \text{ is twisted} \\ 0 & \text{otherwise} \end{cases}$$

The notation $pure(\rho)|_{T+e_i+e_j}$ means the restriction of the underlying pure part of the rotation system ρ to the subgraph $T + e_i + e_j$.

D37: The imbeddings $\iota_1 : G \to S$ and $\iota_2 : G \to S$ are ***congruent*** if there exist a graph automorphism $\alpha : G \to G$ and a surface homeomorphism $h : S \to S$ such that the diagram in Figure 7.3.8 is commutative. We write $\iota_1 \simeq \iota_2$.

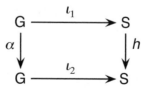

Figure 7.3.8: Commutativity condition for imbedding congruence $\iota_1 \simeq \iota_2$.

FACTS

F29: When non-orientable imbeddings of a graph G are also to be considered, the total number of imbeddings increases by a factor of $2^{\beta(G)-1}$, since each of the $\beta(G)$ edges in the complement of a spanning tree may be twisted or untwisted.

$$\ddot{I}_G(1,1) = 2^{\beta(G)} \prod_{v \in V(G)} [\deg(v) - 1]!$$

F30: [Mo89] Let G be a graph, T a spanning tree of G, and ρ a general rotation system of G. Then

$$rank\,(M_{\rho,T}) = \begin{cases} 2g & \text{if } S\,(\rho) \cong S_g \text{ (induced surface orientable)} \\ k & \text{if } S\,(\rho) \cong N_k \text{ (induced surface non-orientable)} \end{cases}$$

Thus, the genus of the surface induced by a rotation system can be calculated without doing face-tracing.

F31: [ChGrRi94] Calculating the surface type by face-tracing requires $O(n)$ time for a graph with n edges, and calculating the rank of the overlap matrix deteriorates to $O(n^2)$ time. However, regrouping the total set of imbeddings according to rank of the overlap matrix sometimes facilitates calculation of the total imbedding distribution.

F32: [ChGrRi94] Closed-end ladders have the following total imbedding distribution polynomials.

$$\ddot{I}_{L_n}(x,y) = 2^n \sum_{\substack{i_1,\cdots,i_r>0}}^{i_1+\cdots+i_r=n+1} y^{n+1-r} \prod_{h=1}^{r} \left[round\left(\frac{2^{i_h}}{3}\right) + round\left(\frac{2^{i_h+1}}{3}\right) y \right]$$
$$- I_{L_n}(y^2) + I_{L_n}(x)$$

F33: [ChGrRi94] Cobblestone paths have the following total imbedding distribution polynomials.

$$\ddot{I}_{J_n}(x,y) = 2^n \sum_{\substack{i_1,\cdots,i_r>0}}^{i_1+\cdots+i_r=n+1} 2^{n+r-1} y^{n+1-r} \prod_{h=1}^{r} \left[round\left(\frac{2^{i_h}}{3}\right) + round\left(\frac{2^{i_h+1}}{3}\right) y \right]$$
$$- I_{J_n}(y^2) + I_{J_n}(x)$$

REMARKS

R8: [MuRiWh88] counted congruence classes of imbeddings of K_n into oriented surfaces. The key to counting congruence classes was to convert the cycle index of $Aut(G)$ acting on V_G into the cycle index for the induced action on the rotation systems.

R9: [KwLe94] counted congruence classes of imbeddings into non-orientable surfaces. One of their underlying ideas is to regard an edge-twist as the voltage 1 (mod 2) and to construct the orientable double cover. Then the graph automorphisms act on the induced rotation systems.

R10: [KwSh02] developed a formula for the total imbedding distributions of bouquets.

EXAMPLES

E9: In the illustrative calculation of Figure 7.3.9, the spanning tree has edges 4, 5, and 6. Thus, the rows and columns correspond to co-tree edges 1, 2, and 3. Since the rank of the matrix is 3 and the imbedding is non-orientable, the imbedding surface must be N_3 (by Fact F30).

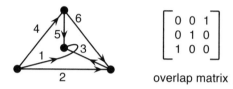

Figure 7.3.9: Sample calculation of the overlap matrix.

E10: In deriving Fact F32, [ChGrRi94] chose a tree T in the ladder graph with a path as a co-tree, as in Figure 7.3.10. This yielded a "tridiagonal" overlap matrix, which is a convenient property in rank calculations.

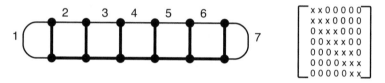

Figure 7.3.10: Ladder L_6, spanning tree, and tridiagonal overlap matrix.

E11: In deriving Fact F33, [ChGrRi94] chose a tree T in the cobblestone path again with a path as a co-tree, as in Figure 7.3.11. This again yielded a "tridiagonal" overlap matrix.

Figure 7.3.11: Cobblestone path J_5 and spanning tree.

E12: In regard to Remark R8, Figure 7.3.12 shows how the 16 different orientable imbeddings of the complete graph K_4 are partitioned into congruence classes.

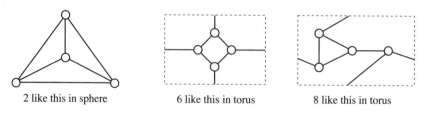

2 like this in sphere   6 like this in torus   8 like this in torus

Figure 7.3.12: Partitioning the 16 imbeddings of K_4 into congruence classes.

7.3.7 The Unimodality Problem

DEFINITIONS

D38: A sequence $\{a_m\}$ is ***unimodal*** if there exists at least one integer M such that
$$a_{m-1} \leq a_m \quad \text{for all } m \leq M \text{ and}$$
$$a_m \geq a_{m+1} \quad \text{for all } m \geq M$$

D39: A sequence $\{a_m\}$ is ***strongly unimodal*** if its convolution with any unimodal sequence yields a unimodal sequence.

FACTS

F34: A typical unimodal sequence first rises and then falls, as illustrated in Figure 7.3.13.

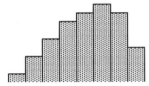

Figure 7.3.13: A unimodal sequence has no false maxima.

F35: [KeGe71] An equivalent criterion for strong unimodality is that
$$a_m^2 \geq a_{m+1} a_{m-1} \quad \text{for all } m$$

F36: The genus distributions of the closed-end ladders [FuGrSt89], cobblestone paths [FuGrSt89], and bouquets [GrRoTu89] are all strongly unimodal.

F37: [GMTW13a] Arbitrarily long chains of copies of various small graphs, including the complete graph K_4, the 4-wheel W_4, the Möbius ladder ML_4, and the circulant graph $circ(7:1,2)$, have strongly unimodal genus distributions.

F38: [GMTW13b] Iterated claw graphs have real-rooted genus polynomials, which implies that these genus polynomials are strongly unimodal.

REMARKS

R11: We observe that an imbedding of the bouquet B_n has $n+1$ faces if in the sphere S_0, $n-1$ faces if in the torus S_1, $n-3$ faces if in the surface S_2, and so on. Intuitively, this suggests that the genus distribution of the bouquet B_n might resemble the sequence of Stirling cycle numbers
$$\begin{bmatrix} 2n \\ n+1 \end{bmatrix}, \begin{bmatrix} 2n \\ n-1 \end{bmatrix}, \begin{bmatrix} 2n \\ n-3 \end{bmatrix}, \cdots$$
which is a strongly unimodal sequence.

R12: [St91a] The resemblance to Stirling numbers holds also for various graphs of small diameter, including partial suspensions of trees and of cycles.

R13: [St90] The genus distribution of the bouquet B_n is asymptotically proportional to this sequence. The proof uses group character theory.

RESEARCH PROBLEM

RP1: Decide whether the genus distribution of every graph is strongly unimodal.

7.3.8   Average Genus

DEFINITIONS

D40: The **average genus** of a graph G, denoted $\gamma_{\text{avg}}(G)$, is the average value of the genus of the imbedding surface, taken over all orientable imbeddings.

D41: The **cycle rank** of a connected graph G is the number $|E_G| - |V_G| + 1$; this is denoted $\beta(G)$ and conceptually best understood as the number of edges in the co-tree of a spanning tree for G.

D42: [GrKlRi93] A **necklace** of type (r, s) is obtained from a $2r + s$-cycle by doubling r disjoint edges and then adding a self-loop at each of the s vertices that is not an endpoint of a doubled edge.

D43: Let e be an edge of a graph. We say that we **attach an open ear** to the interior of edge e if we insert two new vertices u and v and then double the edge between them. The two new vertices are called the **ends of that open ear**.

D44: We **attach a closed ear** to the interior of edge e if we insert one new vertex w in its interior and then attach a self-loop at w. The vertex w is called the **end of that closed ear**.

D45: We say that r open ears and s closed ears are **attached serially** to the edge e if the ends of the ears are all distinct, and if no ear has an end between the two ends of an open ear.

FACTS About Average Genus

F39: [GrKlRi93] The average genus of a graph with nontrivial genus range can lie arbitrarily close to the maximum genus.

F40: [GrKlRi93] The average genus of a graph is at least as large as the average genus of any of its subgraphs.

F41: [ChGrRi95] For any 3-regular graph G,

$$\gamma_{\text{avg}}(G) \; \geq \; \frac{1}{2}\gamma_{\max}(G)$$

F42: [ChGrRi95] For any 2-connected simple graph G other than a cycle,

$$\gamma_{\text{avg}}(G) \; \geq \; \frac{1}{16}\beta(G)$$

F43: [Ch94] Isomorphism testing of graphs of bounded average genus can be achieved in linear time.

F44: [GrFu87] The average genus of the bar-amalgamation of two graphs G and H equals $\gamma_{\text{avg}}(G) + \gamma_{\text{avg}}(H)$.

F45: [ChGr92b] Let G be a 2-connected graph, and let G_+ be a graph obtained by serially attaching ears to an edge of G. Then

$$\gamma_{\text{avg}}(G) \; \leq \; \gamma_{\text{avg}}(G_+) \; \leq \; \gamma_{\text{avg}}(G) + 1$$

FACTS About Small Values of Average Genus

F46: A graph has average genus 0 if and only if at most one cycle passes through any vertex. This follows from [NRSW72].

F47: The maximum genus of a necklace is 1. This follows from [Xu79].

F48: [GrKlRi93] The average genus of any necklace of type (r, s) is

$$1 - \left(\frac{1}{2}\right)^r \left(\frac{2}{3}\right)^s$$

F49: [GrKlRi93] Each of the six smallest possible values of average genus is realizable by a necklace. Figure 7.3.14 indicates these values and shows a graph realizing each of them.

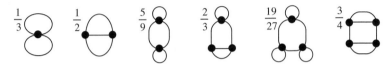

Figure 7.3.14: Realizations of the six smallest positive values of average genus.

F50: [ChGr93] Except for necklaces, there are exactly eight 2-connected graphs of average genus less than one. The bouquet B_3, the dipole D_4, and the complete graph K_4 have average genus

$$\frac{2}{3}, \quad \frac{5}{6}, \quad \frac{7}{8}$$

respectively. Figure 7.3.15 shows the other five such graphs and their average genus.

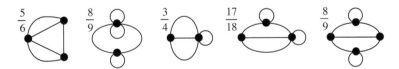

Figure 7.3.15: Five sporadic 2-connected graphs with average genus less than one.

F51: Facts F44 and F50 together yield a complete classification of all graphs of average genus less than one.

F52: [ChGr93] There are exactly three 2-connected graphs with an average genus of 1.

Figure 7.3.16: The three 2-connected graphs with average genus equal to one.

FACTS about close values of average genus

F53: [GrKlRi93] Arbitrarily many mutually nonhomeomorphic 2-connected graphs can have the same average genus.

F54: [ChGr92a] For each real number r, only finitely many 3-connected graphs have average genus less than r.

F55: [ChGr92a] For each real number r, only finitely many 2-connected simple graphs have average genus less than r.

FACTS about limit points of average genus

F56: [GrKlRi93] The number 1 is an upper limit point of the set of possible values of average genus.

F57: [ChGr92a] The set of possible values of average genus for 3-connected graphs has no limit points.

F58: [ChGr92a] The set of possible values of average genus for 2-connected simple graphs has no limit points.

F59: [ChGr95] Lower limit points of average genus do not exist.

REMARKS

R14: Fact F48 provides a means for constructing upper limit points. In fact, all limit points arise from this construction.

R15: Additional results on average genus are given by [MaSt96], [Sc99], [St95a], and [St95b].

7.3.9   Stratification of Imbeddings

Superimposing an adjacency structure on the distribution of orientable imbeddings appears to offer some insight into the problem of deciding whether two given graphs are isomorphic.

DEFINITIONS

D46: Two orientable imbeddings ι_1 and ι_2 of the same graph G are *V-adjacent* if there is a vertex v of G such that moving a single edge-end at v is sufficient to transform a rotation system representing ι_1 into a rotation system representing ι_2.

D47: Two orientable graph imbeddings ι_1 and ι_2 of the same graph G are *E-adjacent* if there is an edge e of G such that moving both edge-ends of e can transform a rotation system representing ι_1 into a rotation system representing ι_2.

D48: For any graph G, the ***stratified graph*** SG has as its vertices the orientable imbeddings of G. Its edges are the V-adjacencies and the E-adjacencies.

D49:   The induced subgraph of SG on the set of imbeddings into the surface S_j is called the jth **stratum** of SG and is denoted S_jG.

D50:   A **complete isomorphism invariant** for a graph is a graph invariant that has a different value on each isomorphism type of graph.

FACTS

F60:   The number of vertices of the jth stratum S_jG is $g_j(G)$.

F61:   [GrTu79] There may be false minima in the stratified graph, that is, local minima that are not global minima.

F62:   [GrRi91] The false minima may have arbitrarily great depth.

F63:   [GrRi91] No false maxima exist, so that it is possible to ascend from any imbedding to a maximum imbedding, even though strict ascent might not always be possible.

F64:   [GrTu95] For every vertex of the stratified graph SG, the induced subgraph on its set of neighbors in SG is a complete isomorphism invariant of the graph.

REMARKS

R16:   This is consistent with [Th89], which proves that the minimum genus problem is NP-complete.

R17:   This is consistent with [FuGrMc88], which establishes a polynomial-time algorithm for maximum genus.

R18:   [GrTu95] also demonstrated how two graphs with similar genus distributions may have markedly different imbedding strata. These findings support the plausibility of a probabilistic approach to graph isomorphism testing, based on the sampling of higher-order imbedding distribution data.

References

[AnJaVi94] G. E. Andrews, D. M. Jackson, and T. I. Visentin, A hypergeometric analysis of the genus series for a class of 2-cell embeddings in orientable surfaces, *SIAM J. Math. Anal.* **25** (1994), 243–255.

[Bo98] H. L. Bodlaender, A partial k-arboretum of graphs with bounded treewidth, *Theoretical Computer Science* **209** (1998), 1–45.

[Ch94] J. Chen, A linear-time algorithm for isomorphism of graphs of bounded average genus, *SIAM J. of Discrete Math.* **7** (1994), 614–631.

[ChGr92a] J. Chen and J. L. Gross, Limit points for average genus (I): 3-connected and 2-connected simplicial graphs, *J. Combin. Theory Ser. B* **55** (1992), 83–103.

[ChGr92b] J. Chen and J. L. Gross, Limit points for average genus (II): 2-connected non-simplicial graphs (with J. Chen), *J. Combin. Theory Ser. B* **56** (1992), 108–129.

[ChGr93] J. Chen and J. L. Gross, Kuratowski-type theorems for average genus, *J. Combin. Theory Ser. B* **57** (1993), 100–121.

[ChGrRi94] J. Chen, J. L. Gross, and R. G. Rieper, Overlap matrices and total imbedding distributions, *Discrete Math.* **128** (1994), 73–94.

[ChGr95] J. Chen and J. L. Gross, No lower limit points for average genus, *Graph Theory, Combinatorics, and Algorithms, Vol. 1 (Kalamazoo, MI, 1992)*, 183–194, Wiley-Interscience, New York, 1995.

[ChGrRi95] J. Chen, J. L. Gross, and R. G. Rieper, Lower bounds for the average genus, *J. Graph Theory* **19** (1995), 281–296.

[Du66] R. A. Duke, The genus, regional number, and Betti number of a graph, *Canad. J. Math.* **18** (1966), 817–822.

[FuGrMc88] M. L. Furst, J. L. Gross, and L. A. McGeoch, Finding a maximum genus graph imbedding, *J. Assoc. Comp. Mach.* **35** (1988), 523–534.

[FuGrSt89] M. L. Furst, J. L. Gross, and R. Statman, Genus distribution for two classes of graphs, *J. Combin. Theory Ser. B* **46** (1989), 22–36.

[Gr10] J. L. Gross, Genus distribution of graphs under surgery: adding edges and splitting vertices, *New York J. Math.* **16** (2010), 161–178.

[Gr11a] J. L. Gross, Genus distribution of graph amalgamations: Self-pasting at root-vertices, *Australasian J. Combin.* **49** (2011), 19–38.

[Gr11b] J. L. Gross, Genus distributions of cubic outerplanar graphs, *J. of Graph Algorithms and Applications* **15** (2011), 295–316.

[Gr12a] J. L. Gross, Embeddings of cubic Halin graphs: a surface-by-surface inventory, *Ars Math. Contemporanea* **7** (2013), 37–56.

[Gr12b] J. L. Gross, Embeddings of graphs of fixed treewidth and bounded degree, preprint 2012, 28 pages. Presented at AMS Annual Meeting at Boston, January 2012.

[GrFu87] J. L. Gross and M. L. Furst, Hierarchy for imbedding-distribution invariants of a graph, *J. Graph Theory* **11** (1987), 205–220.

[GrKhPo10] J. L. Gross, I. F. Khan, and M. I. Poshni, Genus distribution of graph amalgamations, I: Pasting two graphs at 2-valent roots, *Ars Combinatoria* **94** (2010), 33–53.

[GrKlRi93] J. L. Gross, E. W. Klein, and R. G. Rieper, On the average genus of a graph, *Graphs and Combinatorics* **9** (1993), 153–162.

[GrKo12] J. L. Gross and M. Kotrbčík, Genus distributions of cubic series-parallel graphs, preprint, 2012, 20 pp.

[GMTW13a] J. L. Gross, T. Mansour, T. W. Tucker, and D. G. L. Wang, Log-concavity of combinations of sequences and applications to genus distributions, preprint, 2013, 27 pp.

[GMTW13b] J. L. Gross, T. Mansour, T. W. Tucker, and D. G. L. Wang, Iterated claws have real-rooted genus polynomials, preprint, 2013, 12pp.

[GrRi91] J. L. Gross and R. G. Rieper, Local extrema in genus-stratified graphs, *J. Graph Theory* **15** (1991), 159–171.

[GrRoTu89] J. L. Gross, D. P. Robbins, and T. W. Tucker, Genus distributions for bouquets of circles, *J. Combin. Theory Ser. B* **47** (1989), 292–306.

[GrTu77] J. L. Gross and T. W. Tucker, Generating all graph coverings by permutation voltage assignments, *Discrete Math.* **18** (1977), 273–283.

[GrTu79] J. L. Gross and T. W. Tucker, Local maxima in graded graphs of imbeddings, *Ann. NY Acad. Sci.* **319** (1979), 254–257.

[GrTu87] J. L. Gross and T. W. Tucker, *Topological Graph Theory*, Dover Publications, 2001. First Edition, Wiley-Insterscience, 1987.

[GrTu95] J. L. Gross and T. W. Tucker, Stratified graphs for imbedding systems, *Discrete Math.* **143** (1995), 71–86.

[Ja87] D. M. Jackson, Counting cycles in permutations by group characters, with an application to a topological problem, *Trans. Amer. Math. Soc.* **299** (1987), 785–801.

[KeGe71] J. Keilson and H. Gerber, Some results for discrete unimodality, *J. Amer. Statist. Assoc.* **66** (1971), 386–389.

[KhPoGr10] I. Khan, M. Poshni, and J. L. Gross, Genus distribution of graph amalgamations at roots of higher degree, *Ars Mathematica Contemporanea* **3** (2010), 121–138.

[KhPoGr12] I. F. Khan, M. I. Poshni, and J. L. Gross, Genus distribution of $P_3 \times P_n$, *Discrete Math.* **312** (2012), 2863–2871.

[KwKiLe96] J. H. Kwak, S. G. Kim, and J. Lee, Distributions of regular branched prime-fold coverings of surfaces, *Discrete Math.* **156** (1996), 141–170.

[KwLe94] J. H. Kwak and J. Lee, Enumeration of graph embeddings, *Discrete Math.* **135** (1994), 129–151.

[KwSh02] J. H. Kwak and S. H. Shim, Total embedding distributions for bouquets of circles, *Discrete Math.* **248** (2002), 93–108.

[MaSt96] C. Mauk and S. Stahl, Cubic graphs whose average number of regions is small, *Discrete Math.* **159** (1996), 285–290.

[McG87] L. A. McGeoch, Ph.D. Thesis, Carnegie-Mellon University, 1987.

[Mo89] B. Mohar, An obstruction to embedding graphs in surfaces, *Discrete Math.* **78** (1989), 135–142.

[MuRiWh88] B. G. Mull, R. G. Rieper, and A. T. White, Enumerating 2-cell imbeddings of complete graphs, *Proc. Amer. Math. Soc.* **103** (1988), 321–330.

[NRSW72] E. A. Nordhaus, R. D. Ringeisen, B. M. Stewart, and A. T. White, A Kuratowski-type theorem for the maximum genus of a graph, *J. Combin. Theory B* **12** (1972) 260–267.

[PoKhGr10] M. Poshni, I. Khan, and J. L. Gross, Genus distribution of graphs under edge amalgamations, *Ars Mathematica Contemporanea* **3** (2010), 69–86.

[PoKhGr11] M. I. Poshni, I. F. Khan, and J. L. Gross, Genus distribution of 4-regular outerplanar graphs, *Electronic J. Combin.* **18** (2011) #P212, 25 pp.

[PoKhGr12] M. I. Poshni, I. F. Khan, and J. L. Gross, Genus distribution of graphs under self-edge-amalgamations, *Ars Math. Contemporanea* **5** (2012), 127–148.

[Ri90] R. G. Rieper, Ph.D. Thesis, Western Michigan University, 1990.

[Ri74] G. Ringel, *Map Color Theorem*, Springer-Verlag, 1974.

[Ri95] A. Riskin, On the enumeration of polyhedral embeddings of Cartesian products of cycles, *Ars Combinatoria* **41** (1995), 193–198.

[Sc99] M. Schultz, Random Cayley maps for groups generated by involutions, *J. Combin. Theory Ser. B* **76** (1999), 247–261.

[St78] S. Stahl, Generalized embedding schemes, *J. Graph Theory* **2** (1978), 41–52.

[St90] S. Stahl, Region distributions of graph embeddings and Stirling numbers, *Discrete Math.* **82** (1990), 57–78.

[St91a] S. Stahl, Permutation-partition pairs III: Embedding distributions of linear families of graphs, *J. Combin. Theory Ser. B* **52** (1991), 191–218.

[St91b] S. Stahl, Region distributions of some small diameter graphs, *Discrete Math.* **89** (1991), 281–299.

[St92] S. Stahl, On the number of maximum genus embeddings of almost all graphs, *Eur. J. Combin.* **13** (1992), 119–126.

[St95a] S. Stahl, Bounds for the average genus of the vertex-amalgamation of graphs, *Discrete Math.* **142** (1995), 235–245.

[St95b] S. Stahl, On the average genus of the random graph, *J. Graph Theory* **20** (1995), 1–18.

[St97] S. Stahl, On the zeros of some genus polynomials, *Canad. J. Math.* **49** (1997), 617–640.

[Te00] E. H. Tesar, Genus distribution of Ringel ladders, *Discrete Math.* **216** (2000), 235–252.

[Th89] C. Thomassen, The graph genus problem is NP-complete, *J. Algorithms* **10** (1989), 568–576.

[Xu79] N. H. Xuong, How to determine the maximum genus of a graph, *J. Combin. Theory Ser. B* **26** (1979), 217–225.

Section 7.4

Voltage Graphs

Jonathan L. Gross, Columbia University

INTRODUCTION

In the voltage graph construction, a small graph with algebraic labels (called *voltages*) on its edges specifies a large graph with global symmetries. A Cayley graph for a group can be specified by assigning group elements to the self-loops of a one-vertex graph (a *bouquet*). In this sense, voltage graphs are a generalization of Cayley graphs.

7.4.1   Regular Voltage Graphs

The usual purpose of a voltage graph is to specify an undirected graph. Accordingly, even though the voltage graph construction formally employs directions on the edges as a formal convenience, the terminology adopted concentrates on the undirected object.

DEFINITIONS

The regular voltage graph construction now described was introduced in [Gr74]. The definition of a Cayley graph here is as in §7.5. See §6.2 for an algebraic perspective.

D1: Let $G = (V, E)$ be a digraph and $\mathcal{B}$ a group. A ***regular voltage assignment*** for G in the group $\mathcal{B}$ is a function $\alpha : E \to \mathcal{B}$ that labels each edge e with a value $\alpha(e)$.

- The pair $\langle G, \alpha : E \to \mathcal{B} \rangle$ is called a ***regular voltage graph***.
- Graph G is called the ***base graph*** and group $\mathcal{B}$ is called the ***voltage group***.
- The label $\alpha(e)$ is called the ***voltage*** on edge e.

D2: The ***covering digraph*** G^α (formerly, ***derived digraph***) associated with a given regular voltage graph $\langle G = (V, E), \alpha : E \to \mathcal{B} \rangle$ is defined as follows:

- $V(G^\alpha) = V^\alpha = V \times \mathcal{B}$, the cartesian product.
- $E(G^\alpha) = E^\alpha = E \times \mathcal{B}$.
- If the edge e is directed from vertex u to vertex v in G, then the edge $e_b = (e, b)$ in G^α is from the vertex $u_b = (u, b)$ to the vertex $v_{b\alpha(e)} = (v, b\alpha(e))$.

NOTATION: Vertices and edges of the derived graph are usually specified in the subscript notation, rather than in cartesian product notation. The only standard exception to this convention is to avoid double subscripting.

TERMINOLOGY NOTE: The digraph G^α is usually called, simply, the ***covering graph*** (formerly, ***derived graph***). Moreover, its underlying (undirected) graph is also denoted G^α and is also called the ***covering graph***. Such shared terminology avoids excessively formalistic prose. In context, no ambiguity results.

D3: The ***Cayley graph*** $C(\mathcal{A}, X)$ for a group $\mathcal{A}$ with generating set X has the elements of $\mathcal{A}$ as vertices and has edges directed from a to ax for every $a \in \mathcal{A}$ and $x \in X$. We will assume that vertices are labeled by elements of $\mathcal{A}$ and that edges are labeled by elements of X. Although an involution x (i.e., an element of order 2 in the group $\mathcal{A}$) gives rise to a directed edge from a to ax and also one from ax to a, for all a, sometimes we will choose to identify these pair of edges to a single undirected edge labeled x. In §6.2, such a pair is always represented by a single edge.

EXAMPLES

E1: Figure 7.4.1 shows how a Cayley graph for the cyclic group $\mathbb{Z}_5$ is specified by assigning the elements 1 mod 5 and 2 mod 5 to the two self-loops.

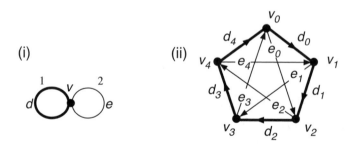

Figure 7.4.1: A voltage assignment in $\mathbb{Z}_5$ for the Cayley graph K_5.

E2: Figure 7.4.2(i) is a regular voltage graph $\langle G, \alpha : E \to \mathbb{Z}_3 \rangle$, and Figure 7.4.2(ii) is the corresponding covering graph.

Each a-edge of the covering graph G^α joins two u-vertices in G^α, because edge a of the base graph is a self-loop at vertex u. Since the voltage on edge a is 1 mod 3, each subscript increments by 1 in a traversal of an a-edge from tail to head. Since edge b of the base graph G goes from vertex u to vertex v, each of the b-edges in the covering graph G^α crosses from a u-vertex to a v-vertex. The subscripts on tail and on head are equal on all the b-edges, because edge b carries voltage 0.

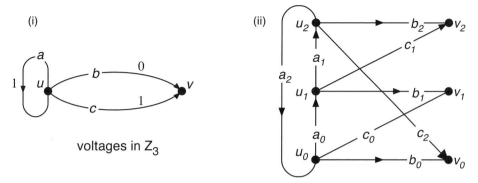

Figure 7.4.2: A regular voltage assignment and the covering graph.

TERMINOLOGY

Involutions in the voltage group. Assigning an involution x as the voltage to a self-loop e at a vertex v in the base graph causes the e-edges in the covering digraph to be paired. That is, the directed edge e_b from vertex v_b to vertex v_{bx} is paired with the directed edge e_{b+x} from v_{b+x} to v_b.

TERMINOLOGY NOTE: The term ***covering graph*** also refers to the undirected graph that is obtained by identifying these pairs of directed edges to a single edge, as one may do with (topological) Cayley graphs.

REMARK

R1:   The earliest application of voltage graphs was to construct imbeddings of large graphs on surfaces, often in connection with minimum genus or symmetric maps.

Fibers

DEFINITIONS

D4:   Let G^α be the covering graph for a regular voltage graph $\langle G = (V, E), \alpha : E \to \mathcal{B} \rangle$.

- The vertex subset $\{v\} \times \mathcal{B} = \{v_b : b \in \mathcal{B}\}$ is called the ***(vertex) fiber*** over v.

- Similarly, the edge subset $\{e\} \times \mathcal{B} = \{e_b : b \in \mathcal{B}\}$ is called the ***(edge) fiber*** over e.

D5:   Let $\langle G = (V, E), \alpha : E \to \mathcal{B} \rangle$ be a regular voltage graph. The graph mapping from the covering graph G^α to the voltage graph G given by the vertex function and edge function

$$v_b \mapsto v \quad e_b \mapsto e$$

respectively, is called the ***natural projection***. (Thus, the natural projection is given by "erasure of subscripts".)

EXAMPLE

E2, continued: In Figure 7.4.2, the subset $\{u_0, u_1, u_2\}$ of $V(G^\alpha)$ (what we were calling the "u-vertices") is the vertex fiber over u. The subset $\{b_0, b_1, b_2\}$ of $E(G^\alpha)$ (what we were calling the "b-edges") is the edge fiber over b.

FACT

F1:   It is clear from the definition of the covering graph that the vertex-set of the covering graph is partitioned into $|V|$ fibers, each with $|\mathcal{B}|$ vertices. Similarly, the edge-set of the covering graph is partitioned into $|E|$ fibers, each with $|\mathcal{B}|$ vertices.

Bouquets and Dipoles

For economy of description, it is helpful to use a base graph with as few vertices as possible.

DEFINITIONS

D6:   The ***bouquet*** B_n is the one-vertex graph with n self-loops.

D7:   The ***dipole*** D_n is the two-vertex graph with n edges joining the two vertices.

EXAMPLE

Voltage graph theory is intuitively spatial. Instead of cluttering the drawings with cumbersome labels, one uses graphic features to represent the partitions into fibers.

E3:   Figure 7.4.3 illustrates how graphic features are used. For instance, in each covering graph, a particular fiber and its corresponding voltage assignment are displayed in bold.

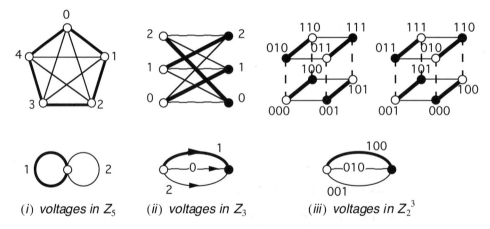

Figure 7.4.3: Three regular voltage assignments and their covering graphs.

Figure 7.4.3(i) derives K_5 with $\mathbb{Z}_5$-voltages on B_2, as in Figure 7.4.1. This time, we have "suppressed the mainscript" on the vertex fiber in the covering graph, and shown only the subscripts. Moreover, we have suppressed the directions in the covering graph. Figure 7.4.3(ii) derives $K_{3,3}$ with $\mathbb{Z}_3$-voltages on D_3. Using hollow and solid vertex graphics enables us to label all the vertices in the covering graph by their subscripts, without loss of information. Similarly, the edge graphics enable us to verify readily that the edges in each edge fiber join vertices whose labels differ by the correct amount. Figure 7.4.3(iii) derives the union of two isomorphic copies of the cube graph Q_3 with $\mathbb{Z}_2^3$-voltages on D_3. This phenomenon is examined in §7.4.2.

Additional elementary examples appear in [GrTu87] and in [GrYe06].

FACTS

F2: The Cayley graph of a group with generating set $\{x_1, \ldots, x_k\}$ is (naturally) isomorphic to the covering graph specified by the bouquet B_k with voltages $\{x_1, \ldots, x_k\}$ on its respective self-loops.

F3: The complete graph K_{2n+1} can be covering by assigning the voltages $1, 2, \ldots, n$ from the cyclic group $\mathbb{Z}_{2n+1}$ to the edges of the bouquet B_n. (This is a special instance of Fact F2.)

F4: The complete graph K_{2n} can be covering by assigning the voltages $1, 2, \ldots, n$ from the cyclic group $\mathbb{Z}_{2n}$ to the edges of the bouquet B_n, if one compresses the pairs of edges that arise from the involution $n \mod 2n$.

F5: The symmetric complete bipartite graph $K_{n,n}$ can be covering by assigning voltages $0, 1, \ldots, n-1$ in the cyclic group $\mathbb{Z}_n$ to the n edges of the dipole D_n.

F6: The d-dimensional cube graph Q_d can be covering by assigning the d elementary vectors in $\mathbb{Z}_2^d$ to the edges of the dipole D_d.

Net Voltages

DEFINITIONS

D8: A **walk in a voltage graph** is any walk, as if the voltage graph were undirected. This means that some of its edge-steps may proceed in the opposite direction from the direction on the edge it traverses.

D9: The **voltage sequence** on a walk $W = v_0, e_1, v_1, e_2, \ldots, e_n, v_n$ is the sequence of voltages $a_1, \ldots, a_n$ encountered, where $a_j = \alpha(e_j)$ or $\alpha(e_j)^{-1}$, depending on whether edge e_j is traversed in the forward or backward direction, respectively.

D10: The **net voltage on a walk** in a voltage graph is the product of the algebraic elements in its voltage sequence.

EXAMPLE

E2, continued: We return to the voltage graph of Figure 7.4.2, reproduced here for convenience.

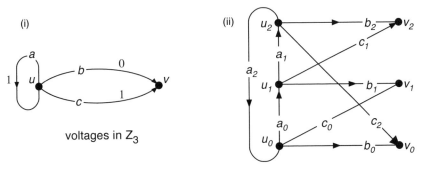

The walk $W = u, c, v, b^-, u, c, u$ has net voltage $1 + 0 + 1 = 2 \mod 3$. We observe that in the walk $u_0, c_0, v_1, b_1^-, u_1, c_1, v_2$ in the derived graph, which follows the same c, b^-, c edge pattern as walk W, the subscript increases by 2 from initial vertex to final vertex. This phenomenon is examined in §7.4.5.

7.4.2   Local Group and Natural Automorphisms

The Local Group

DEFINITION

D11: The *local group* at a vertex v of a voltage graph $\langle G = (V, E), \alpha : E \to \mathcal{B} \rangle$ is the subgroup of all elements of $\mathcal{B}$ that occur as the net voltage on a closed walk that starts and ends at vertex v. It is denoted Loc_v.

FACTS

F7: [AlGr76] If the voltage group is abelian, then the local group is the same at every vertex. If it is non-abelian, then the local group at a vertex is conjugate to the local group at any other vertex.

F8: [AlGr76] For any regular voltage graph $\langle G, \alpha : E \to \mathcal{B} \rangle$, the number of components of the derived graph equals the index $[\mathcal{B} : \mathrm{Loc}_v]$ of the local group in the voltage group.

F9: [AlGr76] The components of the derived graph are mutually isomorphic.

EXAMPLE

E3, continued: The local group for Figure 7.4.3(iii) is the subgroup of 3-tuples with evenly many 1's. This subgroup has index two in $\mathbb{Z}_2^3$. Thus, there are two components to the derived graph.

REMARKS

R2: If we select a root vertex r in the base graph, then the component of the derived graph containing the vertex r_0 (here, zero denotes the group identity, even for a non-abelian group) serves as a preferred component of the derived graph.

R3: The theory of the local group and multiple components was developed by [AlGr76] in the terminology of topological current groups and general algebra.

Natural Automorphisms

The natural action of the group on any of its Cayley graphs generalizes to a natural action of a voltage graph on the derived graph.

DEFINITION

D12: Let $\langle G, \alpha : E \to \mathcal{B} \rangle$ be a voltage graph, and let $x \in \mathcal{B}$. The *natural automorphism* $\varphi_x : G^\alpha \to G^\alpha$ is given by the rules

$$u_b \mapsto u_{xb} \quad \text{and} \quad e_b \mapsto e_{xb}$$

Thus, if edge e runs from vertex u to vertex v of the base graph, then edge e_b runs from vertex u_b to vertex $v_{xb\alpha(e)}$ in the derived graph.

FACTS

See [GrTu87] for details.

F10: The group of natural transformations is fiber preserving. That is, each vertex and edge of the derived graph is mapped to another vertex or edge, respectively, within the same fiber. (This is immediate from the definition of a natural transformation.)

F11: The group of natural transformations acts transitively on the vertices within each vertex fiber and transitively on the edges within each edge fiber.

F12: Let $\langle G, \alpha : E \to \mathcal{B} \rangle$ be a voltage graph. A component of the derived graph G^{α} that contains a vertex in the fiber over v is mapped to itself by the natural automorphism φ_x if and only if $x \in \text{Loc}_v$.

EXAMPLE

E3, continued: The natural automorphism φ_{xyz} for Figure 7.4.3(iii) maps a component of the derived graph to itself if and only if xyz has evenly many 1's.

7.4.3 Permutation Voltage Graphs

The permutation voltage graph construction of [GrTu77] uses the objects permuted by a permutation group as the subscripts, rather than using the group elements as in the regular voltage graph construction. This leads to increased generality (Fact F15).

DEFINITIONS

D13: Let $G = (V, E)$ be a digraph. A Σ_n-***permutation voltage assignment*** for G is a function $\alpha : E \to \Sigma_n$ that labels each edge with a permutation in the symmetric group.

- The pair $\langle G, \alpha \rangle$ is called a Σ_n-***permutation voltage graph.***

- Graph G is called the ***base graph*** and group Σ_n is called the ***permutation voltage group.***

- The permutation label $\alpha(e)$ is called the ***voltage*** on edge e.

D14: The *(Σ_n-permutation) **derived digraph*** G^{α} associated with a permutation voltage graph $\langle G = (V, E), \alpha : E \to \Sigma_n \rangle$ is defined as follows:

- $V(G^{\alpha}) = V^{\alpha} = V \times \{1, ..., n\}$, the cartesian product.

- $E(G^{\alpha}) = E^{\alpha} = E \times \{1, ..., n\}$.

- If the edge e is from vertex u to vertex v in G then the edge $e_j = (e, j)$ is from the vertex $u_j = (u, j)$ to the vertex $v_{\alpha(j)} = (v, \alpha(j))$.

EXAMPLE

E4: Figure 7.4.4 shows a Σ_3-permutation voltage graph and the corresponding covering digraph.

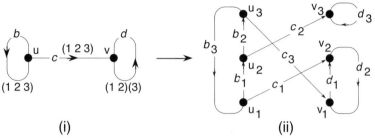

Figure 7.4.4: (i) A Σ_3-voltage graph; (ii) the covering digraph.

The edge fiber over self-loop b at vertex u of the base graph forms the single cycle $(u_1, b_1, u_2, b_2, u_3, b_3)$, because b has voltage (123). The edge fiber over self-loop d forms two disjoint cycles $(v_1, d_1 v_2, d_2)(v_3, d_3)$, because the voltage on d is $(12)(3)$. Since edge c goes from vertex u to vertex v in the base graph, each edge in the fiber over c crosses from the vertex fiber over u to the vertex fiber over v, and the vertex subscripts are permuted in accordance with the voltage (123) on edge c.

FACTS

F13: [Gr77] Every regular graph of even degree $2k$ is specifiable by assigning permutation voltages to the bouquet B_k. (See also [SiSk85].)

F14: From Bäbler's theorem [Ba38] (see Fact F6 of §5.4), it follows that every $2k$-edge-connected $(2k + 1)$-regular graph is specifiable by assigning permutation voltages to the bouquet B_k, if one permits a 1-factor to represent the fiber resulting from an involution.

F15: Any graph derivable by regular voltages is also derivable by permutation voltages. This follows from the fact that the right regular representation of any group can be imbedded in a symmetric permutation group.

REMARKS

R4: Analogous to the regular case, the vertex-set and edge-set of a Σ_n-permutation derived graph are partitioned, respectively, into $|V|$ vertex-fibers, each with n vertices, and into $|E|$ edge-fibers, each with n edges. Analogously, natural projection is by erasure of subscripts.

R5: Further elementary examples of permutation voltage graphs are given in [GrTu87] and in [GrYe06].

R6: In labeling a voltage graph drawing with permutations, one must specify whether the voltages are to be regarded as permutation voltages or as regular voltages. For permutation voltages, there are n vertices or edges in each fiber. For regular voltages, there are $n!$ vertices or edges.

R7:   In particular, it is possible to label the bouquet B_2 with permutations in the wreath product $\mathbb{Z}_n \otimes_{\text{wr}} \mathbb{Z}_2$ (with a cycle shift and a de Bruijn permutation) so that the permutation derived graph is the n-dimensional de Bruijn graph and the regular derived graph is the *wrapped butterfly* graph.

7.4.4   Representing Coverings with Voltage Graphs

Covering spaces are a topological abstraction of Riemann surfaces. In fact, every covering space of a graph can be specified by assigning voltages. The advantage of specifying a covering graph by voltages, rather than by the classical abstract descriptions, is that the derived graph over a voltage graph has every vertex and edge labeled according to its fiber, in a manner that lends itself to topological and combinatorial intuition.

Coverings and Branched Coverings of Surfaces

DEFINITIONS

D15:   Let S and $\tilde{S}$ be surfaces, and let $p : \tilde{S} \to S$ be a continuous function, such that the following condition holds:

> Every point of S has an open neighborhood U such that each component of $p^{-1}(U)$ is mapped homeomorphically by p onto U.

Then $p : \tilde{S} \to S$ is called a ***covering projection*** and the surface $\tilde{S}$ is called a ***covering space*** of S.

D16:   Let $p : \tilde{S} \to S$ be a covering projection. For each point $x \in S$, the set $p^{-1}(x)$ is called the ***fiber*** over x.

D17:   Let S and $\tilde{S}$ be closed surfaces, and let $\tilde{B}$ be a finite set of points in $\tilde{S}$ such that the restriction of the mapping $p : \tilde{S} \to S$ to $\tilde{S} - \cup \tilde{B}$ is a covering projection. Then

- The mapping $p : \tilde{S} \to S$ is called a ***branched covering***.

- The space $\tilde{S}$ is called a ***branched covering space*** of S.

- The set $\tilde{B}$ is called the ***branch set***.

- The images of points in the branch set are called ***branch points***.

EXAMPLES

E5:   The complex function e^{3ix} is a covering projection of the unit circle in the complex plane onto itself. The fiber over a point e^{ix} is the set $\{e^{ix}, e^{ix+2\pi/3}, e^{ix+4\pi/3}\}$. Moreover, the function e^{3ix} is a branched covering of the unit disk in the complex plane onto itself, in which $\{0\}$ is the only branch point.

E6:   The classical Riemann surfaces are branched coverings of the complex plane.

E7:   [A120] Every closed orientable surface is a branched covering of the sphere.

E8: Consider the unit sphere $S_0 = \{(x, y, z) \in R^3 : x^2 + y^2 + z^2 = 0\}$ and the antipodal mapping $(x, y, z) \mapsto (-x, -y, -z)$. The quotient mapping induced by the antipodal homeomorphism is a covering projection of S_0 onto the projective plane. Moreover, the restriction of this covering projection to the annular region of S_0 between the "Tropic of Cancer" and the "Tropic of Capricorn" is a covering projection of this annular region onto a Möbius band.

REMARK

R8: The branch set in any covering of manifolds has codimension 2. Thus unless a graph is imbedded in a surface, there is no branching.

Using Voltage Graph Constructions

A few basic facts serve as a guide to the use of voltage graph constructions.

FACTS

Let $\langle G = (V, E), \alpha : E \to \mathcal{B} \rangle$ be a regular voltage graph. Then the following statements hold.

F16: $|V(G^\alpha)| = |V(G)| \cdot |\mathcal{B}|$ and $|E(G^\alpha)| = |E(G)| \cdot |\mathcal{B}|$.

F17: In the fiber over a vertex $v \in V(G)$, every vertex v_b has the same degree as v.

F18: A proper coloring of the base graph can be *lifted* to a proper coloring of the voltage graph, in the following sense: every vertex in the fiber over a vertex $v \in V(G)$ is assigned the same color as v. (A graph with self-loops is considered to have no proper colorings.)

F19: [GrTu79] Let T be a spanning tree of a graph G. If a graph $\tilde{G}$ can be constructed by assigning $\mathcal{B}$-voltages to G, then it is possible to do so by completing an assignment of arbitrary voltages from $\mathcal{B}$ to the edges of T.

EXAMPLES

E9: To represent the Petersen graph as a regular covering space of the dumbbell graph, we observe that the Petersen graph has 10 vertices and 15 edges. The only nontrivial common divisor of 10 and 15 is 5, so using Fact F16, we seek a base graph with 2 vertices and 3 edges. There are four such connected graphs. In accordance with Fact F17, the base graph must be 3-valent regular, which narrows the possibilities to two graphs. One of these two, the dipole D_3, is 2-colorable. By Fact F18, it cannot be a base graph for the Petersen graph. This leaves the *dumbbell graph*, shown in Figure 7.4.5, as the only possible base graph.

We seek $\mathbb{Z}_5$-voltages on the dumbbell graph, since $\mathbb{Z}_5$ is the only group of order 5. By Fact F19, we may start by assigning the voltage 0 to the edge d. Figure 7.4.5 shows the completed assignment.

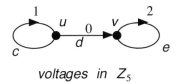

voltages in Z_5

Figure 7.4.5: A regular voltage assignment for the Petersen graph.

E10: By similar considerations, we can demonstrate that the complete graph K_4 is not nontrivially derivable as a voltage graph. It has 4 vertices and 6 edges, so the only nontrivial common divisor is 2. By the exact same progression of steps as in Example 9, we narrow the possible candidates down to the dumbbell graph, and we narrow the possible voltage groups to $\mathbb{Z}_2$.

Assigning 0 to a self-loop in the base leads to an edge fiber of self-loops. Assigning the involution 1 to both self-loops leads to a 4-vertex 6-edge graph with two double adjacencies, whose elimination yields a 4-cycle, not K_4.

Action of the Group of Covering Transformations

The subsection is confined to exploring the sense in which voltage graphs provide all possible covering graphs. Prior acquaintance with algebraic topology is helpful.

DEFINITIONS

D18: Let $p : \tilde{S} \to S$ be a covering projection. A homeomorphism h on $\tilde{S}$ is called a *covering transformation* if $ph = p$. One sometimes says that such a homeomorphism h is *fiber preserving*, since its restriction to any fiber is a permutation of that fiber.

D19: A group H of covering transformations on a covering projection $p : \tilde{S} \to S$ is said to *act freely* if no transformation in H except the identity has a fixed point in $\tilde{S}$.

D20: A group H of covering transformations on a covering projection $p : \tilde{S} \to S$ is said to *act transitively* if its restriction to each fiber acts transitively.

D21: A *regular covering projection* is a covering projection $p : \tilde{S} \to S$ such that there exists a group of freely acting covering transformations. In this case, the domain $\tilde{S}$ is called a *regular covering space* of S.

TERMINOLOGY: The phrase "covering space of a graph G" is used to describe a covering space of a topological realization of the graph G, e.g., in 3-space. It also refers to any graph $\tilde{G}$ such that there is a graph map $\tilde{G} \to G$ whose topological realization is a covering projection.

EXAMPLE

E10, continued: The three functions $e^{ix} \mapsto e^{ix}, e^{ix} \mapsto e^{ix+2\pi/3}$, and $e^{ix} \mapsto e^{ix+4\pi/3}$ form a group of covering transformations. This group acts freely, and it acts transitively on the fibers over every point of the unit circle. Thus, the complex function e^{3ix} is a regular covering projection of the unit circle onto itself.

FACTS

F20: [Gr74] Let $\langle G, \alpha \rangle$ be a regular voltage graph. Then the derived graph G^α is a regular covering graph of G.

F21: [GrTu77] Let X be a topological realization of a graph G be any graph, and let $\tilde{X}$ be a regular covering space of X. Then X is homeomorphic to the topological realization of the derived graph corresponding to a regular voltage assignment on G. ("Every regular covering space of a graph is realizable by a regular voltage assignment.")

F22: [GrTu77] Let $\langle G, \alpha \rangle$ be a permutation voltage graph. Then the derived graph G^α is a covering graph of G.

F23: [GrTu77] Let X be a topological realization of a graph G be any graph, and let $\tilde{X}$ be a covering space of X. Then X is homeomorphic to the topological realization of the derived graph corresponding to a permutation voltage assignment on G. ("Every covering space of a graph is realizable by a permutation voltage assignment.")

F24: Let $\langle G, \alpha \rangle$ be a regular voltage graph, whose voltage group $\mathcal{B}$ has order n. Then the corresponding derived graph is isomorphic to the derived graph of the Σ_n-permutation voltage graph $\langle G, \hat{\alpha} \rangle$, where $b \mapsto \hat{b}$ is the right regular representation of the group $\mathcal{B}$ embedded in Σ_n. ("Every regular voltage assignment can be represented as a permutation voltage assignment.")

F25: Let $\langle G, \alpha \rangle$ be any regular or permutation voltage graph, and let e be a directed edge of graph G. If the direction of e is reversed, and if the voltage $\alpha(e)$ is replaced by its algebraic inverse $\alpha(e)^{-1}$, then the resulting derived graph is isomorphic to the derived graph G^α.

F26: Let $p : C \to C$ be a covering projection of the unit circle onto itself, such that each point in the image is covered k times. Then $p : C \to C$ extends to a branched covering of the unit disk itself, in which 0 is the only branch point in the codomain and $\{0\}$ is the branch set in the domain.

7.4.5   The Kirchhoff Voltage Law

DEFINITIONS

D22: Let $W = v_0, e_1, v_1, e_2, ..., e_n, v_n$ be a walk in a regular voltage graph $\langle G, \alpha : E \to \mathcal{B} \rangle$, and let $a_1, \ldots, a_n$ be its voltage sequence. Let $b \in \mathcal{B}$. Then the walk

$$W_b = (v_0, b), (e_1, ba_1), (v_1, ba_1), (e_2, ba_1a_2), \ldots, (e_n, ba_1a_2 \cdots a_n), (v_n, ba_1a_2 \cdots a_n)$$

is called a **lift** of the walk W.

D23: Let $W = v_0, e_1, v_1, e_2, ..., e_k, v_k$ be a walk in a permutation voltage graph $\langle G, \alpha : E \to \Sigma_n \rangle$, and let $\eta_1, \ldots, \eta_k$ be its voltage sequence. Let $j \in \{1, \ldots, n\}$. Then the walk

$$\big(v_0, j\big), \big(e_1, \eta_1(j)\big), \big(v_1, \eta_1(j)\big), \big(e_2, \eta_2(\eta_1(j))\big),$$
$$\ldots, \big(e_k, \eta_k(\cdots(\eta_1(j)))\big), \big(v_k, \eta_k(\cdots(\eta_1(j)))\big)$$

is called a **lift** of the walk W.

D24: Let W be a closed walk in a voltage graph. If the net voltage on W is the identity of the voltage group, then we say that *the Kirchhoff voltage law (KVL)* holds on W.

FACTS

F27: [Gr74] [GrTu77] Let W be a closed walk in a voltage graph. If the Kirchhoff voltage law holds on W, then every lift W_b of W is a closed walk in the derived graph.

F28: [Gr74] Let W be a closed walk in a regular voltage graph $\langle G, \alpha : E \to \mathcal{B} \rangle$, with net voltage c. Let c have order k in the voltage group $\mathcal{B}$. Then the concatenation of the sequence of lifts $W_b, W_{bc}, W_{bc^2}, \ldots, W_{bc^{k-1}}$ is a closed walk in the derived graph G^α.

NOTATION: Under the hypotheses of Fact F28, the set of lifts of the walk W is conceptualized as partitioned into $\frac{|\mathcal{B}|}{k}$ sequences of lifts, as in the conclusion, whose concatenations are closed walks in the derived graph. This set of closed walks formed by such concatenation is denoted W^*.

F29: [GrTu77] Let W be a closed walk in a permutation voltage graph $\langle G, \alpha : E \to \Sigma_n \rangle$, with net voltage η. Let η have order k in the voltage group Σ_n. Then the concatenation of the sequence of lifts $W_j, W_{\eta(j)}, W_{\eta^2(j)}, \ldots, W_{\eta^{k-1}(j)}$ is a closed walk in the derived graph G^α.

NOTATION: Under the hypotheses of Fact F29, the set of lifts of the walk W is conceptualized as partitioned into $\frac{n}{k}$ sequences of lifts, as in the conclusion, whose concatenations are closed walks in the derived graph. This set of closed walks formed by such concatenation is denoted W^*, as in the regular case.

7.4.6 Imbedded Voltage Graphs

Imbedded voltage graphs and their duals, called current graphs, are used to specify the imbeddings of graphs on surfaces. Imbedded voltage graphs are used extensively in calculations of maximum and minimum genus of a graph (see §7.2), in calculating the minimum genus of a group (see §7.5), and in constructing regular maps on surfaces (see §7.6).

DEFINITIONS

D25: Let $\langle G, \alpha \rangle$ be a voltage graph such that the graph G is (cellularly) imbedded in a closed surface S. Then the pair $\langle G \to S, \alpha \rangle$ is called an *imbedded voltage graph*; also, $\langle G \to S, \alpha \rangle$ is called the base imbedding, and S is called the *base surface*.

D26: Let Ω be the set of closed walks of the faces of an imbedded voltage graph $\langle G \to S, \alpha \rangle$. Then the union $\tilde{\Omega}$ of the sets W^*, where $W \in \Omega$, is called the *set of lifted boundary walks*.

D27: Let $\tilde{\Omega}$ be the set of lifted boundary walks in the derived graph G^α for an imbedded voltage graph $\langle G \to S, \alpha \rangle$. The cellular 2-complex S^α that results from fitting to each closed walk in $\tilde{\Omega}$ a polygonal region (whose number of sides equals the length of that closed walk) is called the *derived surface*. The imbedding $G \to S^\alpha$ is called the *derived imbedding*.

D28: Let $\langle G \to S, \alpha \rangle$ be an imbedded voltage graph. To extend the natural projection $p : G^\alpha \to G$ to the surfaces, the natural projection p is extended from the set of lifted boundary walks in the imbedding $G^\alpha \to S^\alpha$ to the regions they bound (with branching as needed), in accordance with Fact F28. The resulting extended function is called the **natural projection**.

D29: A **monogon** is a face whose boundary walk has length equal to 1.

D30: A **digon** is a face whose boundary walk has length equal to 2.

FACTS

F30: Let $\langle G \to S, \alpha \rangle$ be an imbedded voltage graph. Then the derived surface is a closed surface, and the derived imbedding is a cellular imbedding; moreover, if the base surface S is orientable, then so is the derived surface.

F31: [GrAl74, Gr74] Let $\langle G \to S, \alpha \rangle$ be an imbedded voltage graph. If the Kirchhoff voltage law holds on the boundary walk of every face of the base imbedding, then the natural projection $p : \tilde{S} \to S$ is a covering projection. If KVL does not hold, then the natural projection is a branched covering.

EXAMPLES

E11: Figure 7.4.6 shows an imbedded ordinary voltage graph in which the base graph is the bouquet B_2, the base surface is the torus S_1, and the voltages are in the cyclic group Z_5. The derived graph is the complete graph K_5 and the derived surface is the torus S_1. There is only one base face, and KVL holds on its boundary walk. Thus, each of the derived faces has the same number of sides as the base face, and the natural projection is a covering projection.

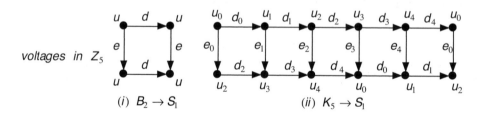

Figure 7.4.6: (i) An imbedded voltage graph; (ii) its derived imbedding.

E12: Figure 7.4.7 shows another imbedded ordinary voltage graph in which the base graph is the bouquet B_2 and the voltages are in the cyclic group Z_5, but the base surface is the sphere S_0. The derived graph is the complete graph K_5.

There are three base faces, i.e., two monogons and one digon, and KVL does not hold on any of their boundary walks; indeed, the net voltage on each boundary walk has order 5 in the group Z_5. Thus, each of the derived faces has 5 times as many sides as the base face, so there are two 5-gons and one 10-gon; the natural projection is a branched covering projection, with a branch point in each base face. Since the Euler characteristic of the derived surface is $-2 = 5 - 10 + 3$, it follows that the derived surface is S_2.

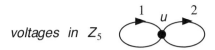

voltages in Z_5

Figure 7.4.7: An imbedded voltage graph $B_2 \to S_0$.

E13: [Gr74] Every Z-metacyclic group with presentation
$$\langle s, t \mid s^m = t^n = e, t^{-1}st = s^{-1} \rangle$$
such that m is odd and n is even is a toroidal group. Various such results on the genus of a group have been derived with the aid of voltage graphs.

7.4.7 Topological Current Graphs

The origin of voltage graphs was in the pursuit of a unified explanation (see [GrAl74] and [GrTu74]) of the 300-page Ringel–Youngs solution [RiYo68] (see also [Ri74]) to the *Heawood map-coloring problem* [He:1890], which is to calculate the chromatic number of every closed surface except the sphere. Several extensions of voltage graph theory have augmented its utility.

DEFINITIONS

D31: Let G be a digraph with vertex-set V and edge-set E, imbedded in a surface S. A ***regular current assignment*** for G in a group $\mathcal{B}$ is a function α from E to $\mathcal{B}$. The function value $\alpha(e)$ is called the ***current*** on edge e. The pair $\langle G \to S, \alpha \rangle$ is called a ***regular current graph***, and $\mathcal{B}$ is called the ***current group***.

D32: Let $\langle G = (V, E) \to S, \alpha : E \to \mathcal{B} \rangle$ be a regular current graph. Its ***dual*** is the imbedded voltage graph whose base imbedding is $G^* \to S$, the dual of the imbedding $G \to S$ (which involves reversed orientation from the primal imbedding surface, if S is orientable). For each primal directed edge $e \in E$, we define $\alpha^*(e^*) = \alpha(e)$ to be its voltage.

D33: The ***derived imbedding of a current graph*** $\langle G \to S, \alpha : E \to \mathcal{B} \rangle$ be a current graph is the derived imbedding of its dual, that is, of the imbedded voltage graph $\langle G^* \to S, \alpha^* \rangle$.

D34: Let v be a vertex in a current graph. If the net current at v is the identity of the current group, then we say that ***the Kirchhoff current law (KCL)*** holds at v. (In an abelian group, the net current is the sum of the inflowing currents. In a non-abelian group, one calculates the product in the cyclic order of the rotation at v.)

EXAMPLE

E14: In Figure 7.4.8, all three drawings shown are on tori, with the left side of the rectangle pasted to the right, and the top pasted to the bottom. The derived imbedding is pasted with a $\frac{2}{7}$ twist, so that like labels match. We observe that KVL holds on both faces of the imbedded voltage graph. In accordance with duality, KCL holds at both vertices of the corresponding current graph.

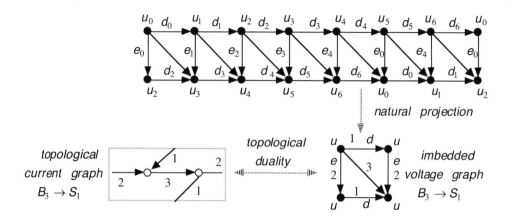

Figure 7.4.8: Deriving an imbedding from a topological current graph.

REMARKS

R9:   In Gustin's original conception [Gu63], a current graph was a 3-regular graph whose vertices were marked with instructions for traversing a small family of closed walks that doubly covered its edges, and whose edges were marked with algebraic labels. Ringel and Youngs augmented Gustin's "nomograms" (Youngs's terminology) into numerous varieties of combinatorial current graph, each with a distinct set of defining rules, whatever helped in their endeavors to solve the Heawood problem. The common feature of all varieties was that recording the algebraic elements traversed along those closed walks yielded generating rows for rotations systems (see §7.1) that specified minimum-genus imbeddings for complete graphs.

R10:   The theory of topological current graphs [GrAl74] generalized the various combinatorial current graphs referred to in Remark 9 into a single unified construction, applicable not just to complete graphs, but to a wide variety of graphs with symmetries, and it identified the underlying construction as a branched covering. The theory of regular voltage graphs [Gr74] separated the base graph from its imbedding, which facilitated a divide-and-conquer approach to constructing imbeddings, as in Examples E11 and E12. The theory of permutation voltage graphs [GrTu77] expanded the construction.

R11:   When one dualizes the natural projection of the derived imbedding onto an imbedded voltage graph, there is a natural projection of the dual of the derived graph onto the current graph, which is a folded covering in the sense of [Tu36]. The relationship of the dual derived graph to the current graph was studied by [PaPiJa80] and [GrJaPaPi82]. A different perspective on simultaneous consideration of voltages and currents is given by [Ar92].

7.4.8   Lifting Voltage Graph Mappings

In porting distributed algorithms between parallel architectures, a theoretical problem that arises is the construction of a mapping between two large symmetric graphs that would minimize the slowdown involved in emulating a computation designed for one computer architecture on a different architecture.

DEFINITIONS

D35: Let $f : G \to H$ be a graph map. The **guest** is the domain, and the **host** is the codomain. (A computation designed for the guest is to be emulated by the host.)

D36: Let $f : G \to H$ be a graph map. The **load at a vertex** v of the host is the cardinality $|f^{-1}(v)|$ of its preimage. (A processor at host vertex v is required to reproduce the computations of the processors at every guest vertex mapped to v. These must be done consecutively, so there is a delay at v proportional to the load.) The **load of the mapping** $f : G \to H$ is the maximum vertex load, taken over all vertices of the host.

D37: Let $f : G \to H$ be a graph map. The **congestion at an edge** e of the host is the cardinality $|f^{-1}(e)|$ of its preimage. (A link at host edge e is required to carry the messages of the links represented by every guest edge mapped to e. These must be done consecutively, so there is a delay at v proportional to the load.) The **congestion of the mapping** $f : G \to H$ is the maximum edge congestion, taken over all edges of the host.

FACTS

F32: [GrCh96] A graph mapping from a guest G to a host H can *sometimes* be constructed by lifting a graph mapping from a voltage graph for G to a voltage graph for H.

F33: [ArGvSi97], [ArRiSiSk97] A mapping of imbedded graphs can *sometimes* be constructed by lifting a graph mapping from an imbedded voltage graph for G to an imbedded voltage graph for H.

F34: [GrCh96] A graph mapping from a guest G to a host H that minimizes load or congestion can *sometimes* be constructed by lifting a graph mapping from a voltage graph for G to a voltage graph for H.

REMARK

R12: Other references about lifting mappings include [MaNeSk00], [MiSa00a], and [Su90].

7.4.9 Applications of Voltage Graphs

Beyond their initial uses in the construction of minimum imbeddings of graphs and maps with various forms of symmetry, voltage graphs have acquired several other uses.

REMARKS

R13:   The use of voltage graphs in the construction of graphs with special prescribed properties or in the calculation of invariants appears in [AnGa81], [ArKwLeSo00], [FeKw02], [KrPrTe97], [KwHoLeSo00], [KwLeSo99], [Le82], [MiSa97], [MiSa00b], [MiSa02], and [MiSa03].

R14:   For an extensive survey of counting covering spaces of a graph, see [KwLe04].

R15:   Among the many papers that apply voltage graphs to counting covering spaces of a graph are [EwHo93], [FeKwKiLe98], [Ho91a], [Ho91b], [Ho95a], [Ho95b], [HoKw93], [HoKwLe96], [HoKwLe99], [KiKiLe99], [KwChLe98], [KwHoLeSo00], [KwLe92], [KwLe94], [KwLe98], and [KwKiLe96].

R16:   For a study of random permutation voltages, see [AmLi02].

R17:   For connections to the Vassiliev knot invariants, see [Pl01].

R18:   For connections between voltage graphs and coding theory, see [LaPe81].

R19:   For connections between voltage graphs and biological networks, see [MoRe02].

R20:   For the applications of voltage graphs to the study of isomorphisms of graph coverings, see [Sa94] and [Sa01].

Applications of Imbedded Voltage Graphs and Topological Current Graphs

Imbedded voltage graphs and topological current graphs have often been used in connection with investigations in algebra and geometry.

REMARKS

R21:   For results on lifting imbeddings with special properties, including symmetries and the realization of minimum imbeddings for a graph, see [Mo85], [Mo86], [Mo87b], [Mo88], [NeSk97], [Pa80], [Pi80], [Pi82], and [St79].

R22:   For the use of voltage graphs in counting graph imbeddings, see [KwChLe98], [KwKiLe96], [LeKi01], [LeKi02], and [Mo87a].

R23:   Applications of voltage graphs or topological current graphs to the genus of a group (see §7.5) appear in [Gr74], [GrLo80], [JuWh80], [Pi92], [PiTu89], [Pr77], [Pr81], [PiTu92], [PiTuWi92], [Tu80], [Tu84], and [Tu83].

R24:   Voltage graphs have been used in the construction of block designs by [Al75], [BrHu87], [Ga79], and [Wh78], among others.

R25:   The use of voltage graphs in the representation of finite geometries (see §7.9) has been pioneered by A. T. White. See especially [Wh01].

References

[AbPa83] M. Abu-Sbeih and T. D. Parsons, Embeddings of bipartite graphs, *J. Graph Theory* 7 (1983), 325–334.

[Al20] J. W. Alexander, Note on Riemann spaces, *Bull. Amer. Math. Soc.* 26 (1920), 370–372.

[Al75] S. R. Alpert, Twofold triple systems and graph imbeddings, *J. Combin. Theory, Ser. A* 18 (1975), 101–107.

[AlGr76] S. R. Alpert and J. L. Gross, Components of branched coverings of current graphs, *J. Combin. Theory, Ser. B* 20 (1976), 283–303.

[AmLi02] A. Amit and N. Linial, Random graph coverings I: General theory and graph connectivity, *Combinatorica* 22 (2002), 1–18.

[AnGa91] D. Angluin and A. Gardiner, Finite common coverings of pairs of regular graphs, *J. Combin. Theory, Ser. B* 30 (1981), 184–187.

[Ar92] D. Archdeacon, The medial graph and voltage-current duality, *Discrete Math.* 104 (1992), 111–141.

[ArGvSi97] D. Archdeacon, P. Gvozdjak, and J. Širaň, Constructing and forbidding automorphisms in lifted maps, *Math. Slovaca* 47 (1997), 113–129.

[ArKwLeSo00] D. Archdeacon, J. H. Kwak, J. Lee, and M. Y. Sohn, Bipartite covering graphs, *Discrete Math.* 214 (2000), 51–63.

[ArRiSiSk94] D. Archdeacon, R. B. Richter, J. Širaň, and M. Škoviera, Branched coverings of maps and lifts of map homomorphisms, *Australas. J. Combin.* 9 (1994), 109–121.

[Ba38] F. Bäbler, Über die Zerlegung regulärer Streckencomplexe ungerader Ordnung, *Comment. Math. Helv.* 10 (1938), 275–287.

[BrHu87] N. Brand and W. C. Huffman, Constructions and topological invariants of 2-(upsilon, 3, lambda) designs with group-actions, *Eur. J. Combin.* 8 (1987), 5–26.

[EwHo93] R. Ewen and M. Hofmeister, On coverings of the complete graph with 4 vertices, *Ars Combinatoria* 35 (1993), 87–96.

[FeKwKiLe98] R. Q. Feng, J. H. Kwak, J. Kim, and J. Lee, Isomorphism classes of concrete graph coverings, *SIAM J. Discrete Math.* 11 (1998), 265–272.

[FeKw02] Y. Q. Feng and J. H. Kwak, Constructing an infinite family of cubic 1-regular graphs, *Eur. J. Combin.* 23 (2002), 559–565.

[Ga79] B. L. Garman, Voltage graph embeddings and the associated block designs, *J. Graph Theory* 3 (1979), 53–67.

[Gr74] J. L. Gross, Voltage graphs, *Discrete Math.* 9 (1974), 239–246.

[Gr77] J. L. Gross, Every regular graph of even degree is a Schreier coset graph, *J. Combin. Theory, Ser. B* 22 (1974), 227–232.

[GrAl74] J. L. Gross and S. R. Alpert, The topological theory of current graphs, *J. Combin. Theory, Ser. B* 17 (1974), 218–233.

[GrCh96] J. L. Gross and J. Chen, Algebraic specification of interconnection networks by permutation voltage graph morphisms, *Mathematical Systems Theory* 29 (1996), 451–470.

[GrJaPaPi82] J. L. Gross, B. Jackson, T. Pisanski, and T. D. Parsons, Wrapped coverings and their realizations, *Congressus Numerantium* 35 (1982), 333–347.

[GrLo80] J. L. Gross and S. J. Lomonaco, A determination of the toroidal k-metacyclic groups, *J. Graph Theory* 4 (1980), 165–172.

[GrTu74] J. L. Gross and T. W. Tucker, Quotients of complete graphs: revisiting the Heawood map-coloring problem, *Pacific J Math.* 55 (1974), 391–402.

[GrTu77] J. L. Gross and T. W. Tucker, Generating all graph coverings by permutation voltage assignments, *Discrete Math.* 18 (1977), 273–283.

[GrTu79] J. L. Gross and T. W. Tucker, Fast computations in voltage graph theory, *Ann. NY Acad. Sci.* 319 (1979), 254–257.

[GrTu87] J. L. Gross and T. W. Tucker, *Topological Graph Theory*, Dover, 2001. (First Edition, Wiley-Interscience, 1987.)

[GrYe06] J. L. Gross and J. Yellen, *Graph Theory and Its Applications*, 2nd Edition, CRC Press, 2006.

[Gu63] W. Gustin, Orientable emedding of Cayley graphs, *Bull. Amer. Math. Soc.* 69 (1963), 272–275.

[He1890] P. J. Heawood, Map-colour theorem, *Quart. J. Math.* 24 (1890), 332–338.

[Ho91a] M. Hofmeister, Concrete graph covering projections, *Ars Combinatoria* 32 (1991), 121–128.

[Ho91b] M. Hofmeister, Isomorphisms and automorphisms of graph coverings, *Discrete Math.* 98 (1991), 175–183.

[Ho95a] M. Hofmeister, Enumeration of concrete regular covering projections, *SIAM J. Discrete Math.* 8 (1995), 51–61.

[Ho95b] M. Hofmeister, Graph covering projections arising from finite vector-spaces over finite-fields, *Discrete Math.* 143 (1995), 87–97.

[HoKw93] S. P. Hong and J. H. Kwak, Regular fourfold coverings with respect to the identity automorphism, *J. Graph Theory* 17 (1993), 621–627.

[HoKwLe96] S. P. Hong, J. H. Kwak, and J. Lee, Regular graph coverings whose covering transformation groups have the isomorphism extension property, *Discrete Math.* 148 (1996), 85–105.

[HoKwLe99] S. P. Hong, J. H. Kwak, and J. Lee, Bipartite graph bundles with connected fibres, *Bull. Austral. Math. Soc.* 59 (1999), 153–161.

[JaPaPi81] B. Jackson, T. D. Parsons, and T. Pisanski, A duality theorem for graph embeddings, *J. Graph Theory* 5 (1981), 55–77.

[JuWh80] M. Jungerman and A. T. White, On the genus of finite abelian groups, *Europ. J. Combin.* 1 (1980), 243–251.

[KiKiLi99] H. K. Kim, J. H. Kim, and D. Lim, Natural isomorphism classes of cycle permutation graphs over a graph, *Graphs and Combinatorics* 15 (1999), 327–336.

[KrPrTe97] J. Kratochvil, A. Proskurowski, and J. A. Telle, Covering regular graphs, *J. Combin. Theory, Ser. B* 71 (1997), 1–16.

[KwChLe98] J. H. Kwak, J. H. Chun, and J. Lee, Enumeration of regular graph coverings having finite abelian covering transformation groups, *SIAM J. Discrete Math.* 11 (1998), 273–285.

[KwHoLeSo00] J. H. Kwak, S. P. Hong, J. Lee, and M. Y. Sohn, Isoperimetric numbers and bisection widths of double coverings of a complete graph, *Ars Combinatoria* 57 (2000), 49–64.

[KwKiLe96] J. H. Kwak, S. G. Kim, and J. Lee, Distributions of regular branched prime-fold coverings of surfaces, *Discrete Math.* 156 (1996), 141–170.

[KwKw01] J. H. Kwak and Y. S. Kwon, Characteristic polynomials of graph bundles having voltages in a dihedral group, *Linear Algebra and Its Appl.* 336 (2001), 99–118.

[KwLe92] J. H. Kwak and J. Lee, Isomorphism-classes of cycle permutation graphs, *Discrete Math.* 105 (1992), 131–142.

[KwLe94] J. H. Kwak and J. Lee, Enumeration of graph embeddings, *Discrete Math.* 135 (1994), 129–151.

[KwLe98] J. H. Kwak and J. Lee, Distribution of branched D-p-coverings of surfaces, *Discrete Math.* 183 (1998), 193–212.

[KwLe04] J. H. Kwak and J. Lee, Enumeration of graph coverings, Chapter 10 of *Studies in Topological Graph Theory*, ed. by J. L. Gross and T. W. Tucker, Cambridge University Press, 2004.

[KwLeSo99] J. H. Kwak, J. Lee, and M. Y. Sohn, Isomorphic periodic links as coverings, *J. Knot Theory and Its Ramif.* 8 (1999), 215–240.

[LaPe81] G. Lallement and D. Perrin, A graph covering construction of all the finite complete biprefix codes, *Discrete Math.* 36 (1981), 261–271.

[LeKi01] J. Lee and J. W. Kim, Enumeration of the branched mZ(P)-coverings of closed surfaces, *Eur. J. Combin.* 22 (2001), 1125–1138.

[LeKi02] J. Lee and J. W. Kim, On abelian branched coverings of the sphere, *Graphs and Combinatorics* 18 (2002), 329–342.

[Le82] F. T. Leighton, Finite common coverings of graphs, *J. Combin. Theory, Ser. B* 33 (1982), 231–238.

[MaNeSk00] A. Malnic, R. Nedela, and M. Skoviera, Lifting graph automorphisms by voltage assignments, *Eur. J. Combin.* 21 (2000), 927–947.

[MiSa97] H. Mizuno and I. Sato, Characteristic polynomials of some covers of symmetric digraphs, *Ars Combinatoria* 45 (1997), 3–12.

[MiSa00a] H. Mizuno and I. Sato, Isomorphisms of cyclic abelian covers of symmetric digraphs, *Ars Combinatoria* 54 (2000), 51–64.

[MiSa00b] H. Mizuno and I. Sato, Zeta functions of graph coverings, *J. Combin. Theory, Ser. B* 80 (2000), 247–257.

[MiSa02] H. Mizuno and I. Sato, L-functions for images of graph coverings by some operations, *Discrete Math.* 256 (2002), 335–347.

[MiSa03] H. Mizuno and I. Sato, L-functions of regular coverings of graphs, *Eur. J. Combin.* 24 (2003), 321–329.

[Mo85] B. Mohar, Akempic triangulations with 4 odd vertices, *Discrete Math.* 54 (1985), 23–29.

[Mo86] B. Mohar, A common cover of graphs and 2-cell embeddings, *J. Combin. Theory, Ser. B* 40 (1986), 94–106.

[Mo87a] B. Mohar, The enumeration of akempic triangulations, *J. Combin. Theory, Ser. B* 42 (1987), 14–23.

[Mo87b] B. Mohar, Simplicial schemes, *J. Combin. Theory, Ser. B* 42 (1987), 68–86.

[Mo88] B. Mohar, Branched-coverings, *Discrete Comput. Geom.* 3 (1988), 339–348.

[MoRe02] H. S. Mortveit and C. M. Reidys, Towards a calculus of biological networks, *Z. Phys. Chem.* 216 (2002), 235–247.

[NeSk97] R. Nedela and M. Skoviera, Regular maps from voltage assignments and exponent groups, *Eur. J. Combin.* 18 (1997), 807–823.

[Pa80] T. D. Parsons, Circulant graph imbeddings, *J. Combin. Theory, Ser. B* 29 (1980), 310–320.

[PaPiJa80] T. D. Parsons, T. Pisanski, and B. Jackson, Dual imbeddings and wrapped quasi-coverings of graphs, *Discrete Math.* 18 (1980), 43–52.

[Pi80] T. Pisanski, Genus of cartesian products of regular bipartite graphs, *J. Graph Theory* 4 (1980), 31–42.

[Pi82] T. Pisanski, Nonorientable genus of cartesian products of regular graphs, *J. Graph Theory* 6 (1982), 391–402.

[Pi92] T. Pisanski, Orientable quadrilateral embeddings of products of graphs, *Discrete Math.* 109 (1992), 203–205.

[PiTu89] T. Pisanski and T. W. Tucker, The genus of a product of a group with an abelian group, *Europ. J. Combin.* 10 (1989), 469–475.

[PiTu92] T. Pisanski and T. W. Tucker, The genus of low rank Hamiltonian groups, *Discrete Math.* 78 (1989), 157–167.

[PiTuWi92] T. Pisanski, T. W. Tucker, and D. Witte, The nonorientable genus of some metacyclic groups, *Combinatorica* 12 (1), 77–87.

[Pl01] L. Plachta, Voltage graphs, weight systems and odd symmetry, *Discrete Math.* 236 (2001), 287–313.

[Pr77] V. K. Proulx, The classification of toroidal groups, PhD Thesis, Columbia University, 1977.

[Pr81] V. K. Proulx, On the genus of symmetric-groups, *Trans. Amer. Math. Soc.* 266 (1981), 531–538.

[Ri74] G. Ringel, *Map Color Theorem*, Springer-Verlag, 1974.

[RiYo68] G. Ringel and J. W. T. Youngs, Solution of the Heawood map-coloring problem, *Proc. Natl. Acad. Sci. USA* 60 (1968), 438–445.

[Sa01] I. Sato, Isomorphisms of cyclic abelian covers of symmetric digraphs, III, *Ars Combinatoria* 61 (2001), 173–186.

[Sa94] I. Sato, Isomorphisms of some graph coverings, *Discrete Math.* 128 (1994), 317–326.

[SiSk85] J. Siran and M. Skoviera, Quotients of connected regular graphs of even degree, *J. Combin. Theory, Ser. B* 38 (1985). 214–225.

[St79] S. Stahl, Self-dual embeddings of Cayley graphs, *J. Combin. Theory, Ser. B* 27 (1979), 92–107.

[StWh76] S. Stahl and A. T. White, Genus embeddings for some tripartite graphs, *Discrete Math.* 14 (1976), 279–296.

[Su90] D. B. Surowski, Lifting map automorphisms and Macbeath theorem, *J. Combin. Theory, Ser. B* 50 (1990), 135–149.

[Tu36] A. W. Tucker, Branched and folded coverings, *Bull. Amer. Math. Soc.* 42 (1936), 859–862.

[Tu80] T. W. Tucker, Number of groups of a given genus, *Trans. Amer. Math. Soc.* 258 (1980), 167–179.

[Tu83] T. W. Tucker, Finite-groups acting on surfaces and the genus of a group, *J. Combin. Theory, Ser. B* 34 (1983), 82–98.

[Tu84] T. W. Tucker, A refined Hurwitz theorem for imbeddings of irredundant Cayley-graphs, *J. Combin. Theory, Ser. B* 36 (1984), 244–268.

[Wh78] A. T. White, Block designs and graph imbeddings, *J. Combin. Theory, Ser. B* 25 (1978), 166–183.

[Wh01] A. T. White, *Graphs of Groups on Surfaces*, North-Holland, 2001.

Section 7.5

The Genus of a Group

Thomas W. Tucker, Colgate University

INTRODUCTION

When a Cayley graph $C(\mathcal{A}, X)$ for a finite group $\mathcal{A}$ is imbedded in a surface, the face boundaries, as cycles in the Cayley graph, give relations in the generating set X, that is, words in the generators and their inverses that represent the identity element of $\mathcal{A}$. Thus, the possible imbeddings of Cayley graphs for the group $\mathcal{A}$ are closely related to the possible presentations for that group in terms of generators and relations. The smallest genus g such that some Cayley graph for the group $\mathcal{A}$ can be imbedded in the surface of genus g is called the *genus of the group* $\mathcal{A}$. White [Wh72,Wh84] first introduced the term, although Burnside [Bu11] considers a similar concept. Study of the genus of groups is closely related to questions about group actions on surfaces, regular branched coverings, and automorphisms of Riemann surfaces. For more about the genus of a group, see the survey chapter in [Tu09].

7.5.1   Symmetric Imbeddings of Cayley Graphs

If the Cayley graph $C(\mathcal{A}, X)$ is imbedded in an orientable surface S, one might ask whether the natural vertex-transitive symmetry of the Cayley graph is somehow reflected in the symmetry of its imbedding, especially for minimum genus imbeddings.

DEFINITIONS

D1: The *Cayley graph* $C(\mathcal{A}, X)$ for a group $\mathcal{A}$ with generating set X has the elements of $\mathcal{A}$ as vertices and has edges directed from a to ax for every $a \in \mathcal{A}$ and $x \in X$. We will assume that vertices are labeled by elements of $\mathcal{A}$ and that edges are labeled by elements of X.

We notice that an involution x gives rise to a directed edge from a to ax and also one from ax to a, for all a; sometimes we will choose to identify this pair of edges to a single undirected edge labeled x.

D2: The ***genus of a group*** $\mathcal{A}$, which is denoted $\gamma(\mathcal{A})$, is the smallest genus g such that some Cayley graph for the group $\mathcal{A}$ can be imbedded in the orientable surface S_g of genus g.

D3: The ***natural action*** of the group $\mathcal{A}$ on the Cayley graph $C(\mathcal{A}, X)$ is the group of automorphisms corresponding to left multiplication of the vertices of a Cayley graph $C(\mathcal{A}, X)$ by an element b of the group $\mathcal{A}$. (This respects the labeling and directing of the edges, since $(ba)x = b(ax)$.)

D4: The action of an automorphism group on a graph is ***vertex-transitive*** if it takes any vertex to any other vertex.

D5: An automorphism group acting on a graph is a ***free action*** if no element except the identity fixes any vertices of the graph.

D6: The finite group $\mathcal{A}$ ***acts on the orientable surface*** S if $\mathcal{A}$ is isomorphic to a subgroup of the group of all homeomorphisms of S.

D7: The action of a group $\mathcal{A}$ on a surface S ***preserves orientation*** if every element of the corresponding subgroup of homeomorphisms on S preserves the orientation of S.

D8: An imbedding of a Cayley graph $C(\mathcal{A}, X)$ in the orientable surface S is ***symmetric*** if the natural action of $\mathcal{A}$ on $C(\mathcal{A}, X)$ extends to an action of $\mathcal{A}$ on S.

D9: An imbedding of a Cayley graph $C(\mathcal{A}, X)$ in the orientable surface S is ***strongly symmetric*** if the natural action of $\mathcal{A}$ on $C(\mathcal{A}, X)$ extends to an orientation-preserving action of $\mathcal{A}$ on S.

D10: The ***symmetric genus*** (respectively, ***strong symmetric genus***) of the group $\mathcal{A}$, denoted $\sigma(\mathcal{A})$ (respectively, $\sigma^o(\mathcal{A})$), is the smallest g such that some Cayley graph imbeds symmetrically (respectively, strongly symmetrically) in the surface of genus g.

TERMINOLOGY: A strongly symmetric imbedding of a Cayley graph for the group $\mathcal{A}$ is also called a ***Cayley map*** for the group $\mathcal{A}$ [BiWh79, RSJTW].

FACTS

F1: The definitions immediately imply that $\gamma(\mathcal{A}) \le \sigma(\mathcal{A}) \le \sigma^o(\mathcal{A})$.

F2: The natural action of the group $\mathcal{A}$ on the Cayley graph $C(\mathcal{A}, X)$ is vertex-transitive and free.

F3: [Sa58] A graph G is a Cayley graph for the group $\mathcal{A}$ if and only if there is a group of automorphisms isomorphic to $\mathcal{A}$ acting on G, such that the action is vertex-transitive and free.

F4: [GrTu87] Any orientation-preserving automorphism of a graph imbedding must respect rotations at vertices (the cyclic ordering of edges around each vertex given by the imbedding), and the natural action of a group on a Cayley graph respects labels. Thus, if an imbedding of a Cayley graph is strongly symmetric, then it must have the same cyclic ordering of generators and their inverses at every vertex, and conversely.

F5: It is a corollary to Fact F4 that to specify a symmetric imbedding of the Cayley graph $C(\mathcal{A}, X)$, all we need do is to give a cyclic ordering of the elements of X and their inverses.

F6: The derived graph of an imbedded voltage graph for a bouquet of circles, where the assigned voltage set X generates the voltage group $\mathcal{A}$, gives a strongly symmetric imbedding of the Cayley graph $C(\mathcal{A}, X)$. Every strong symmetric imbedding of a Cayley graph can be obtained this way.

F7: [GrTu87] Any orientation-reversing automorphism of a graph imbedding reverses the rotations at vertices. If the action of the group $\mathcal{A}$ on the orientable surface S does not preserve orientation, then the set of elements of $\mathcal{A}$ that do preserve orientation form an index-two subgroup $\mathcal{B}$ of $\mathcal{A}$. Thus if an imbedding of a Cayley graph for $\mathcal{A}$ is symmetric but not strongly symmetric, there is a subgroup $\mathcal{B}$ of index two in $\mathcal{A}$ such that all vertices in $\mathcal{B}$ have the same rotation and all vertices not in $\mathcal{B}$ have the opposite rotation, and conversely.

F8: Viewing symmetric imbeddings that are not strongly symmetric as derived graphs of a small (one or two vertices) voltage graph tends to be complicated. One possibility is to begin with an imbedded voltage graph of a bouquet of circles in a non-orientable surface and hope that the derived surface is orientable. Another is to begin with a two-vertex graph imbedded in a symmetric surface having an orientation-reversing involution f that interchanges the two vertices.

F9: [GrTu87] If the voltage group $\mathcal{A}$ has an index two subgroup $\mathcal{B}$, such that all loops are assigned voltages in $\mathcal{B}$, such that all edges between the two vertices are assigned voltages not in $\mathcal{B}$, and such that e and $f(e)$ get the same voltage, then the derived graph will be a symmetric, but not strongly symmetric imbedding of a Cayley graph for $\mathcal{A}$.

F10: [Tu83] Any action of the finite group $\mathcal{A}$ on the orientable surface S comes from a symmetric imbedding of a Cayley graph $C(\mathcal{A}, X)$ in S. If the action preserves orientation, then the imbedding is strongly symmetric. Thus $\sigma(\mathcal{A})$, respectively $\sigma^o(\mathcal{A})$, is the minimal g such that $\mathcal{A}$ acts, respectively, acts preserving orientation, on a surface of genus g.

F11: If $\mathcal{B}$ is a subgroup of $\mathcal{A}$, then

$$\sigma(\mathcal{B}) \leq \sigma(\mathcal{A}) \quad \text{and} \quad \sigma^o(\mathcal{B}) \leq \sigma^o(\mathcal{A})$$

since if $\mathcal{A}$ acts on a surface, then so does $\mathcal{B}$.

F12: [Ba77] If $\mathcal{B}$ is a subgroup of $\mathcal{A}$, then any Cayley graph for $\mathcal{A}$ edge-contracts to a Cayley graph for $\mathcal{B}$. In particular, $\gamma(\mathcal{B}) \leq \gamma(\mathcal{A})$.

EXAMPLES

E1: View the standard cube as having a top and bottom face with four vertical faces. Let y denote the rotation by 90 degrees about the centers of the top and bottom faces. Let x denote the reflection that interchanges the top and bottom faces but takes each vertical face to itself. It is not hard to see that the action generated by the symmetries x and y is vertex-transitive and free, that y preserves orientation and x does not, and that $xy = yx$. Thus, the vertices and edges of the cube can be labeled to give a symmetric, but not strongly symmetric imbedding of a Cayley graph for the abelian group

$$\langle x, y : x^2 = y^4 = 1, xy = yx \rangle \cong \mathcal{Z}_2 \times \mathcal{Z}_4.$$

E2:   Let y be as in the previous example, but let z denote the rotation by 180 degrees about the midpoint of a vertical edge; then z interchanges not only the top and bottom faces, but also the vertical faces in pairs as well. It is again not hard to see that the action generated by the symmetries z and y is vertex-transitive and free, that it preserves orientation, and that $zyz = y^{-1}$. Thus, the vertices and edges of the cube can also be labeled to give a strongly symmetric imbedding of a Cayley graph for the group

$$\langle z, y : z^2 = y^4 = 1, zyz = y^{-1} \rangle$$

which is the dihedral group of order 8.

REMARKS

R1:   In both examples above, the vertical edges correspond to an involution, with the resulting pairs of edges identified. In the first example, the involution reversed orientation; it is not hard to check that the vertical edges could not be replaced by a pair of directed edges and still have the involution x respect the directions. In the second example, the vertical edges could be replaced by a pair of directed edges and still have the directions respected by the involution z. In general, pairs of edges in a symmetric imbedding of a Cayley graph corresponding to an involution that reverses orientation must be identified but need not be otherwise.

R2:   Orientation-reversing homeomorphisms of finite order, such as involutions, can be nonintuitive and subtle. One tends to think in terms of euclidean isometries, where there are two types: reflections and glides. For example, one can imagine cutting a torus in half, forming two dividing circles, and interchanging the halves by a reflection that leaves fixed the dividing circles. But it is also possible to interchange the halves by an antipodal map that also interchanges the circles. It is even possible to interchange the halves, leaving one dividing circle fixed, but rotating the other dividing circle a half turn (like a glide along the circle). See [GrTu87] for more examples.

7.5.2   Riemann–Hurwitz Equation; Hurwitz's Theorem

Given a voltage graph imbedded in a surface S, the Euler characteristic for the surface of the derived imbedding is easily calculated, in terms of the order of the voltage group, the Euler characteristic of S, and the order of the net voltages on the faces. As derived imbeddings of one-vertex imbedded voltage graphs, strongly symmetric imbeddings of Cayley graphs can be handled the same way.

DEFINITIONS

D11: Suppose that the Cayley graph $C(\mathcal{A}, X)$ has a strongly symmetric imbedding in the surface T as the derived graph for a voltage graph imbedding of the bouquet B in the surface S. Suppose that the non-identity net voltages on the faces of this imbedding are $r_1, \cdots, r_n$. Then the Euler characteristic $\chi(T)$ can be computed by the **Riemann–Hurwitz equation** (where $|\mathcal{A}|$ is the order of the group $\mathcal{A}$):

$$\chi(T) = |\mathcal{A}|(\chi(S) - \sum \left(1 - \frac{1}{r_i}\right).$$

D12: The quantity $1 - \dfrac{1}{r_i}$ is sometimes called the **deficiency of the branch point** at the center of a face (of an imbedded voltage graph) whose excess voltage has order r_i. That is because there are only $|\mathcal{A}|/r_i$ copies of that face in the derived imbedding, instead of the "expected number" $|\mathcal{A}|$ copies, which is a deficiency of $|\mathcal{A}| - \dfrac{|\mathcal{A}|}{r_i}$. (Since vertices and edges generate $|\mathcal{A}|$ copies each, this explains the Riemann–Hurwitz equation.)

D13: A similar equation holds for symmetric imbeddings that are not strongly symmetric. If the associated imbedded voltage graph still has one vertex, then the surface S is non-orientable, but the equation holds exactly as before. If the associated imbedded voltage graph has two vertices, then the **Riemann–Hurwitz equation** becomes

$$\chi(T) = \frac{|\mathcal{A}|}{2}(\chi(S) - \sum 1 - 1/r_i).$$

D14: A **triangle group** is a group with presentation

$$\langle x, r : x^p = y^q = (xy)^r = 1 \rangle$$

This is the group of isometries of the plane generated by the rotations at the vertices of a triangle with angles $\pi/p, \pi, q, \pi/r$ (the geometry of the plane is spherical, Euclidean or hyperbolic, depending on whether the angle sum is greater than, equal to, or less than π).

D15: Any *quotient group* $\mathcal{A}$ of the triangle group

$$\langle x, r : x^p = y^q = (xy)^r = 1 \rangle$$

is said to be a **(p,q,r)° group**.

D16: The **full triangle group** has presentation

$$\langle x, y, z : x^2 = y^2 = z^2 = (xy)^r = (yz)^q = (xz)^r = 1 \rangle$$

and is generated by reflections in the sides of a $\pi/p, \pi/q, \pi/r$ triangle.

D17: Any *quotient group* $\mathcal{A}$ of the full triangle group

$$\langle x, y, z : x^2 = y^2 = z^2 = (xy)^r = (yz)^q = (xz)^r = 1 \rangle$$

is said to be a **(p,q,r) group**.

D18: A (p, q, r) group $\mathcal{A}$, with generators as above, is **properly** (p, q, r) if the subgroup generated by xy and yz has index two in $\mathcal{A}$. (The index is otherwise one.)

FACTS

F13: If $\mathcal{A}$ is a $(2, q, r)^o$ group, then by the Riemann–Hurwitz equation it has a strongly symmetric imbedding in a surface of Euler characteristic

$$\chi = |\mathcal{A}| \left[2 - \left(1 - \frac{1}{2} \right) - \left(1 - \frac{1}{p} \right) - \left(1 - \frac{1}{q} \right) \right] = |\mathcal{A}| \left[\frac{1}{p} + \frac{1}{q} - \frac{1}{2} \right]$$

and genus $g = 1 - \frac{\chi}{2}$ satisfying

$$g - 1 = \frac{|\mathcal{A}|}{2} \left[\frac{1}{2} - \frac{1}{p} - \frac{1}{q} \right]$$

F14: Similarly, if $\mathcal{A}$ is properly $(2, q, r)$, then it has a symmetric imbedding in a surface of genus g satisfying

$$g - 1 = \frac{|\mathcal{A}|}{4} \left[\frac{1}{2} - \frac{1}{p} - \frac{1}{q} \right]$$

F15: (Hurwitz's Theorem [GrTu87], [Hu:1893], [Tu80]) If the group $\mathcal{A}$ has strong symmetric genus $\sigma^o(\mathcal{A}) > 1$, then

$$|\mathcal{A}| \le 84(\sigma^o(\mathcal{A}) - 1)$$

with equality if and only if $\mathcal{A}$ is $(2, 3, 7)^o$. If the symmetric genus $\sigma(\mathcal{A}) > 1$, then $|\mathcal{A}| \le 168(\sigma(\mathcal{A}) - 1)$, with equality if and only if $\mathcal{A}$ is properly $(2, 3, 7)$.

F16: [Tu80] As a corollary to Fact F15, there are only finitely many groups of a given symmetric genus or strong symmetric genus greater than one.

F17: [Tu80] (The Cayley-graph version of Hurwitz's theorem) If $\gamma(\mathcal{A}) > 1$, then

$$|\mathcal{A}| \le 168(\gamma(\mathcal{A}) - 1)$$

with equality if and only if $\mathcal{A}$ is properly $(2, 3, 7)$.

F18: [Tu80] As a corollary to Fact F17, there are only finitely many groups of a given genus greater than one.

F19: [Ba91], [Th91] (Babai–Thomassen Theorem): There are only finitely many vertex-transitive graphs of a given genus $g > 2$. In particular, there are only finitely many Cayley graphs of a given genus $g > 2$.

REMARKS

R3: The Hurwitz theorems have been stated here as an upper bound on $|\mathcal{A}|$, rather than as a lower bound on $\sigma(\mathcal{A})$ or $\gamma(\mathcal{A})$, since the traditional view was bounding the order of a group of conformal automorphisms on a Riemann surface of given genus. The proof of Hurwitz's original theorem is a brief analysis of the possibilities for the Riemann–Hurwitz equation when $\chi(T) < 0$. The same analysis can be refined to give detailed information about $\mathcal{A}$, whenever $|\mathcal{A}|$ is large compared to $\sigma(\mathcal{A})$ or $\sigma^o(\mathcal{A})$. For example, if $|\mathcal{A}| > 80(\sigma(\mathcal{A}) - 1)$, then $\mathcal{A}$ is $(2, 3, 7)^o$, properly $(2, 3, 7)$, or properly $(2, 3, 8)$. For more on refinements of Hurwitz's theorem see [GrTu87], [Tu83].

R4: The Cayley graph version of Hurwitz's theorem is unexpected. The formula for the Euler characteristic guarantees that both the valence and average face size must be small, when the number of vertices of an imbedded graph is large compared to the Euler characteristic of the imbedding surface. For a Cayley graph, this means there must be a small number of generators and many short relators. The proof is a long, exhaustive case-by-case analysis with lots of "relation chasing". This analysis can be refined to give detailed information about $\mathcal{A}$, whenever $|\mathcal{A}|$ is large, in a manner analogous to Hurwitz's theorem, although again the proofs are much harder and longer (see [GrTu87], [Tu84b]).

R5: We have stated the Riemann–Hurwitz equation in terms of a strongly symmetric imbedding of a Cayley graph for the group $\mathcal{A}$, but since all orientation-preserving group actions come from strongly symmetric imbedding, we can also view the Riemann–Hurwitz equation as holding for any group acting on the surface of genus g, preserving orientation. In that context, one might ask on which surfaces a given group acts preserving orientation. The Riemann–Hurwitz equation governs the situation for almost all genus g in the following sense [Ku87]. Given the group $\mathcal{A}$, there is a number P such that if $\mathcal{A}$ acts on the surface of genus g preserving orientation, then $g \equiv 1 \bmod(P)$. Moreover, there is such an action by $\mathcal{A}$ for all but finitely many such g. The quantity P is easily computed in the terms of the Sylow p-subgroups A_p of $\mathcal{A}$. In particular, it follows from Kulkarni's theorem that the group $\mathcal{A}$ acts preserving orientation on all but finitely many surfaces if and only if A_p is cyclic for all odd p and A does not contain the subgroup $\mathcal{Z}_2 \times \mathcal{Z}_4$.

7.5.3   Groups of Low Genus

For low genus, minimum imbeddings tend to be highly symmetric. For example, by Whitney's theorem that a 3-connected planar graph imbeds uniquely in the sphere, a Cayley graph imbedded in the sphere must be symmetrically imbedded. In addition, symmetries of the sphere and torus come from the natural geometry of the surfaces: spherical geometry for the sphere, and Euclidean geometry for the torus (viewed as the Euclidean plane rolled up by a pair of linearly independent translations).

DEFINITION

D19: A *Euclidean space group* or *Euclidean crystallographic group* is a group of isometries of the Euclidean plane that contains translations in independent directions and such that the orbit of any point under the group has no accumulation points (there is a minimum distance any point is moved by all the elements of the group not leaving the point fixed).

NOTATION: The dihedral group of order $2n$ is denoted $\mathcal{D}_n$. The symmetric group and alternating group on n symbols are denoted $\mathcal{S}_n$ and $\mathcal{A}_n$, respectively.

FACTS

Finding all groups of a given small genus has a long history.

F20: There are exactly 17 Euclidean space groups, up to isomorphism, and presentations for the groups are well-known [CxMo80].

F21: (Planar groups [Ma:1896], [GrTu87]) The groups of strong symmetric genus 0 are $\mathcal{Z}_n$, $\mathcal{D}_n$, $\mathcal{A}_4$, $\mathcal{S}_4$, and $\mathcal{A}_5$. The groups of symmetric genus 0 are these groups together with their direct products with $\mathcal{Z}_2$. In both cases, the associated group actions can be realized by automorphisms of prisms and the platonic solids. For all groups, $\gamma(\mathcal{A}) = 0$ if and only if $\sigma(\mathcal{A}) = 0$.

F22: (Toroidal groups [Ba31], [Pr77], [Tu84a], [GrTu87]) Except for three groups, $\gamma(\mathcal{A}) = 1$ if and only if $\sigma(\mathcal{A}) = 1$. Moreover, $\sigma(\mathcal{A}) = 1$ if and only if $\mathcal{A}$ is a finite quotient of one of the 17 Euclidean space groups; this yields partial presentations for all toroidal groups (see [CxMo80] or [GrTu87]). The three exceptional groups, with $\gamma(\mathcal{A}) = 1$ but $\sigma(\mathcal{A}) > 1$ have orders 24, 48, and 48 and presentations

    (a) $\mathcal{A} = \langle x, y : x^3 = y^3 = 1, xyx = yxy \rangle$,
    (b) $\mathcal{A} = \langle x, y : x^3 = y^2 = 1, xyxyxy = yxyxyx \rangle$,
    (c) $\mathcal{A} = \langle x, y : x^3 = y^2 = 1, (xyxyx^{-1})^2 = 1 \rangle$.

F23: (Genus two [Tu84c]) There is exactly one group $\mathcal{A}$ such that $\gamma(\mathcal{A}) = 2$. It has order 96 and the $(2, 3, 8)$ presentation:

$$\mathcal{A} = \langle x, y, z : x^2 = y^2 = z^2 = 1, (xy)^2 = (yz)^3 = (xz)^8 = 1, (xy)^4 z = z(xy)^4 \rangle$$

There is a sculpture on display at the Technical Museum of Slovenia showing a symmetric imbedding of $C(\mathcal{A}, \{x, y, z\}$ in the surface of genus two [GoMa07]. It appears on the cover of the journal *Ars Combinatorea Mathematica*. The generators are colored x red, y green, z yellow.

F24: (Symmetric genus 2 [MaZi95]) There are 4 groups of symmetric genus 2: the group of genus 2 and the three exceptional groups of genus 1.

F25: (Symmetric genus 3 [MaZi97]) There are 3 groups of symmetric genus 3: the proper $(2, 3, 7)$ group $PGL(2, 7)$, its $(2, 3, 7)^o$ subgroup $PSL(2, 7)$ (also known as Klein's simple group of order 168), and the proper $(2, 4, 6)$ group $\mathcal{Z}_2 \times \mathcal{Z}_2 \times \mathcal{S}_4$.

F26: (Strong symmetric low genus [MaZi00]) There are 6 groups of strong symmetric genus 2, 10 groups of strong symmetric genus 3, and 10 groups of strong symmetric genus 4.

7.5.4   Genus for Families of Groups

    Abelian groups form an interesting case study for the genus of a group. Commutators in the generators provide many possible faces of size four (quadrilaterals), but it is not easy to see how to combine them to form an all-quadrilateral imbedding. In fact, a strongly symmetric imbedding of a Cayley graph of valence greater than 4 cannot have a face of the form $xyx^{-1}y^{-1}$: the rotation would have to be $xy^{-1}x^{-1}y$ with no room for any other generators. The canonical form for the abelian group plays a key role. The simplest case, when all the factors in the canonical form are even, was part of White's original paper [Wh72] where he introduced the genus of a group. At the other extreme are groups whose minimal genus imbeddings are symmetric. For example, any proper $(2, 3, 7)$ group $\mathcal{A}$ has genus $1 + |\mathcal{A}|/168$, by the Cayley graph version of Hurwitz's theorem.

DEFINITIONS

D20: Any abelian group A can be written uniquely in the **canonical form**

$$\mathbb{Z}_{m_1} \times \mathbb{Z}_{m_2} \times \cdots \mathbb{Z}_{m_r} \quad \text{where } m_j | m_{j+1} \text{ for } j = 1 \cdots r$$

D21: The factors $\mathbb{Z}_{m_i}$ in the canonical form of an abelian group A are called the **canonical factors** of A, and the number r of factors is called the **rank of the abelian group** A.

D22: A $(2, 3, 7)^0$ group is called a **Hurwitz group**.

D23: A proper $(2, 3, 7)$ group is called a **proper Hurwitz group**.

FACTS

F27: [JuWh80] Suppose that the abelian group A does not have $\mathbb{Z}_3$ as a canonical factor and that m_1 is even if the rank $r = 3$. Then $\gamma(A) = 1 + |A|(r-2)/4$, whenever the right side of this equation is an integer.

F28: [BrSq88], [MoPiSkWh87] $\gamma(\mathbb{Z}_3 \times \mathbb{Z}_3 \times \mathbb{Z}_3) = 7$, and the minimal genus imbedding has very little symmetry, including faces of many different sizes.

F29: [PiTu89] Let C be any finite group. If A is an abelian group of rank r at least twice that of C, then in most cases $\gamma(C \times A) = 1 + |C||A|(r-2)/4$.

F30: [Mc65], [MaZi93] The strong symmetric and symmetric genus are known for all abelian groups.

F31: [Co80] For $n > 167$, the symmetric group S_n is a proper Hurwitz group. In particular, $\gamma(S_n) = 1 + n!/168$ for $n > 167$.

F32: All the 26 sporadic simple groups are $(2, p, q)$ for some p and q; and 12 of them are Hurwitz groups. All the alternating groups A_n for $n > 167$ and many of the simple groups of Lie type and of large enough dimension are Hurwitz groups [LuTa99].

F33: The symmetric genus of all the sporadic groups [CoWiWo92], of all the alternating and symmetric groups [Co85], and of many other simple groups is known. Notice that if the group A is simple, then $\sigma(A) = \sigma^o(A)$, since A has no subgroups of index two.

F34: [MaZi02] For every genus g, $\sigma^o(\mathbb{Z}_m \times D_n) = g$ for some m, n. In fact, $\sigma^o(\mathbb{Z}_3 \times D_n) = n$ if n is not divisible by 6.

REMARK

R6: Fact F34 is interesting, because it shows that the strong symmetric genus σ^o is free of gaps. For σ and γ, it is not known whether there are gaps. If $c \neq 8, 14$, there is a family of groups with $\sigma = g$ for every $g \equiv c \bmod 18$ [CoTu11]. Moreover, there is also such a family for $g \equiv 8, 14 \bmod 18$ if the prime power factorization of $g - 1$ contains no factor $p^e \equiv 5 \bmod 6$. For γ, easily constructed quadrilateral imbeddings [Wh72] show that $\gamma(\mathbb{Z}_2 \times \mathbb{Z}_{2s} \times Z_{2sm}) = ms^2 + 1$, so if g is a gap for γ, then $g - 1$ is square-free.

7.5.5 Non-Orientable Surfaces

It is possible, of course, to imbed Cayley graphs in non-orientable surfaces, and it is natural to ask about minimal non-orientable imbeddings. It is also interesting to compare minimal imbeddings in orientable and non-orientable surfaces.

DEFINITIONS

D24: If S is a non-orientable surface of Euler characteristic $\chi(S)$, then $2 - \chi(S)$ is called the **crosscap number** or **non-orientable genus** of S. (It follows from the classification of closed surfaces that every surface of crosscap number c can be obtained from the sphere by attaching c crosscaps, that is, by removing c disks from the sphere and identifying the resulting c boundary components to the boundaries of c Möbius strips.)

D25: For any closed surface S, the quantity $2 - \chi(S)$ is called the **Euler genus** of S. Thus if S is orientable of genus g, then its Euler genus is $2g$, and if S is non-orientable of crosscap number c, then its Euler genus is c.

D26: The **non-orientable genus** or **crosscap number** of the group $\mathcal{A}$, denoted $\tilde{\gamma}(\mathcal{A})$, is the smallest number c such that some Cayley graph for $\mathcal{A}$ imbeds in the non-orientable surface S of crosscap number c.

D27: The **Euler genus** of the group $\mathcal{A}$, denoted $\gamma^e(\mathcal{A})$, is the smallest number e such that some Cayley graph for $\mathcal{A}$ imbeds in a surface of Euler genus e.

D28: The **symmetric Euler genus** of the group $\mathcal{A}$, denoted $\sigma^e(\mathcal{A})$, is the smallest number e such that some Cayley graph for $\mathcal{A}$ imbeds symmetrically in a surface of Euler genus e.

D29: An imbedding for a Cayley graph $C(\mathcal{A}, X)$ in a non-orientable surface S is **symmetric** if the natural action of $\mathcal{A}$ on $C(\mathcal{A}, X)$ extends to an action on S.

D30: The **symmetric crosscap number** or **symmetric non-orientable genus**, denoted $\tilde{\sigma}(\mathcal{A})$, is the smallest number c such that some Cayley graph for $\mathcal{A}$ imbeds symmetrically in a surface of non-orientable genus c.

D31: An imbedding of a graph in a non-orientable surface can be described by assigning to each vertex a cyclic order or rotation to the set of edges incident to that vertex and assigning to each edge a **type** of 0 or 1, telling whether the edge is orientation-preserving or orientation-reversing.

FACTS

F35: An imbedding of a Cayley graph $C(\mathcal{A}, X)$ in a non-orientable surface is symmetric if and only if the rotation is the same at each vertex in terms of the directed edge labels and if every directed edge labeled by the same generator has the same type.

F36: The definitions immediately imply the following:
  (i) $\tilde{\gamma}(\mathcal{A}) \leq \tilde{\sigma}(\mathcal{A})$
  (ii) $\gamma^e(\mathcal{A}) \leq \sigma^e(\mathcal{A})$
  (iii) $\gamma^e(\mathcal{A}) = \min\{2\gamma(\mathcal{A}), \tilde{\gamma}(\mathcal{A})\}$
  (iv) $\sigma^e(\mathcal{A}) = \min\{2\sigma(\mathcal{A}), \tilde{\gamma}(\mathcal{A})\}$

F37: Any imbedding of a graph in an orientable surface can be turned into an imbedding in a non-orientable surface, decreasing the number of faces by at most 1, by changing the type of a single edge. Thus $\tilde{\gamma}(\mathcal{A}) \leq 2\gamma(\mathcal{A}) + 1$.

F38: If $\mathcal{B}$ is a subgroup of $\mathcal{A}$, then $\tilde{\gamma}(B) \leq \tilde{\gamma}(\mathcal{A})$, by Babai's theorem [Ba77].

F39: [Tu83] The group $\mathcal{A}$ has a symmetric imbedding in the non-orientable surface S if and only if $\mathcal{A}$ acts on S. In particular, if $\mathcal{B}$ is a subgroup of $\mathcal{A}$, then $\tilde{\sigma}(B) \leq \tilde{\sigma}(\mathcal{A})$.

F40: Hurwitz's theorem and its Cayley graph version apply to non-orientable surfaces.

F41: [Tu83] If the group $\mathcal{A}$ acts on the non-orientable surface S of Euler characteristic χ, then $\mathcal{Z}_2 \times \mathcal{A}$ acts on the orientable double covering of S, the surface of Euler characteristic 2χ, with the $\mathcal{A}$ factor orientation-preserving. In particular, $\sigma^o(\mathcal{A}) - 1 \leq (\tilde{\sigma}(\mathcal{A}) - 2)$.

F42: The two groups of Euler genus 1 are $\mathcal{Z}_3 \times \mathcal{Z}_3$ and its $\mathcal{Z}_2$-extension

$$\langle x, y : x^3 = y^2 = 1, [x, yxy] = 1 \rangle$$

There are no groups of symmetric Euler genus 1, since any group acting on the projective plane also acts on its orientable double covering, the sphere.   There are no groups of Euler genus 3 and one group of symmetric Euler genus 3. For $\gamma^e = 2$ or $\gamma^e = 4$, all minimal imbeddings are orientable, that is, $\gamma^e(\mathcal{A}) = 2$ if and only if $\gamma(\mathcal{A}) = 1$, and $\gamma^e(\mathcal{A}) = 4$ if and only if $\gamma(\mathcal{A}) = 2$. The only group $\mathcal{A}$ with $\sigma^e(\mathcal{A}) = 5$ is the symmetric group $\mathcal{S}_5$; it is also true that $\gamma^e(\mathcal{S}_5) = 5$ (see [Tu91], [MaZi01]).

F43: The groups with $\tilde{\gamma} = 1$ are the groups with $\gamma = 0$ together with the two groups with $\gamma^e = 1$. The groups with $\tilde{\sigma} = 1$ are the groups with $\sigma^O = 0$. It is conjectured that there are no groups with $\tilde{\gamma} = 2$. Other cases of low crosscap number are studied in [Tu91], [MaZi01].

F44: The crosscap number is known for many abelian groups.

F45: The crosscap number is known for all groups $\mathcal{Z}_m \times \mathcal{D}_n$ [EM08] .

F46: If $\mathcal{A}$ is an improper Hurwitz group, then $\tilde{\gamma}(\mathcal{A}) = 1 + |\mathcal{A}|/84$. In particular, $\tilde{\gamma}(A_n) = 1 + |\mathcal{A}|/84$ for all $n > 167$ [Co85].

R7:   There are inequalities relating all the various genus parameters to each other and to a quotient group $\mathcal{Q}$ of a given group $\mathcal{A}$ [Tu09, Tu08]. Suppose that ρ and τ are any of $\gamma, \tilde{\gamma}, \gamma^e \sigma, \sigma^o, \tilde{\sigma}, \sigma^e$ (possibly $\rho = \tau$). Define $\delta(\rho) = 1$ for $\rho = \gamma, \sigma, \sigma^o$ and $\delta(\rho) = 2$ otherwise. There is a number $m(\rho, \sigma)$ depending only on ρ and τ, such that if $\tau(\mathcal{A}) > \delta(\tau)$, then

$$\rho(\mathcal{Q}) - \delta(\rho) \leq m(\rho, \tau) \frac{|\mathcal{Q}|}{|\mathcal{A}|} (\tau(\mathcal{A}) - \delta(\tau)) .$$

In all cases, $m(\rho, \tau) < 168$. Of particular interest is the case when $\rho = \tau = \gamma$ since it is not known whether $\gamma(\mathcal{Q}) \leq \gamma(\mathcal{A})$ for any quotient $\mathcal{Q}$ of $\mathcal{A}$. Note that by the Riemann–Hurwitz equation for group actions, $m(\rho, \rho) = 1$ for $\rho = \sigma, \sigma^o, \tilde{\sigma}$, and σ^e. For more about comparisons of other genus-like parameters, see [Tu08].

References

[Ba77] L. Babai, Some applications of graph contractions, *J. Graph Theory* 1 (1977), 125–130.

[Ba91] L. Babai, Vertex-transitive graphs and vertex-transitive maps, *J. Graph Theory* 15 (1991), 587–627.

[Bk31] R. P. Baker, Cayley diagrams on the anchor ring, *Amer. J. Math.* 53 (1931), 645–669.

[BiWh79] N. Biggs and A. T. White, Permutation groups and combinatorial structures, *Math. Soc. Lect. Notes* 33, Cambridge University Press, 1979.

[BrSq88] M. G. Brin and C. C. Squier, On the genus of $\mathcal{Z}_3 \times \mathcal{Z}_3 \times \mathcal{Z}_3$, *European J. Combin.* 9 (1988), 431–443.

[Bu11] W. Burnside, *Theory of Groups of Finite Order*, Cambridge University Press, 1911.

[Co85] M. Conder, The symmetric genus of alternating and symmetric groups, *J. Combin. Theory, Ser. B* 39 (1985), 179–186.

[Co90] M. D. E. Conder, Hurwitz groups: a brief survey, *Bull. Amer. Math. Soc.* 23 (1990), 359–370.

[CoWiWo92] M. D. E. Conder, R. A. Wilson, and A. J. Woldar, The symmetric genus of sporadic groups, *Proc. Amer. Math. Soc.* 116 (1992), 653–663.

[CoTu11] M. Conder and T. W. Tucker, The spectrum of the symmetric genus, *Ars Mathematica Contemporanea* 4 (2011), 271–289.

[CxMo80] H. S. M. Coxeter and W. O. J. Moser, *Genererators and Relations for Discrete Groups* (4th ed.), Springer-Verlag, 1980.

[EM08] J. J. Etayo Gordejuela and E. Martinez, The symmetric cross-cap number of the groups $C_m \times D_n$, *Proc. Roy. Soc. Edin.* 138 (2008), 1197–1213.

[GoMa07] D. Godfrey and D. Martinez, Tucker's Group of Genus Two (sculpture), Technical Museum, Bistra, Slovenia (installed 2007).

[GrTu87] J. L. Gross and T. W. Tucker, *Topological Graph Theory*, Dover, 2001. (First Edition, Wiley-Interscience, 1987.)

[Hu:1893] A. Hurwitz, Über algebraische gebilde mit eindeutigen transformationen in sich, *Math. Ann.* 41 (1893), 403–442.

[JuWh80] M. Jungerman and A. T. White, On the genus of finite abelian groups, *Europ. J. Combin.* 1 (1980), 243–251.

[Ku87] R. S. Kulkarni, Symmetries of surfaces, *Topology* 26 (1987), 195-203.

[LuTa99] A. Luchini and M. C. Tamburini, Classical groups of large rank as Hurwitz groups, *J. Algebra* 219 (1999), 531–546.

[Mc65] C. Maclachlin, Abelian groups of automorphisms of compact Riemann surfaces, *Proc. London Math. Soc.* 15 (1965), 699–712.

[Ma:1896] H. Maschke, The representation of finite groups, *Amer. J. Math.* 18 (1896), 156–194.

[MaZi93] C. L. May and J. Zimmerman, The symmetric genus of finite abelian groups, *Illinois J. Math.* 37 (1993), 400–423.

[MaZi95] C. L. May and J. Zimmerman, Groups of small symmetric genus, *Glasgow Math. J.* 37 (1995), 115–129.

[MaZi97] C. L. May and J. Zimmerman, The groups of symmetric genus three, *Houston J. Math.* 23 (1997), 573–590.

[MaZi00] C. L. May and J. Zimmerman, Groups of small strong symmetric genus, *J. Group Theory* 3 (2000), 233–245.

[MaZi01] C. L. May and J. Zimmerman, The group of Euler characteristic-3, *Houston J. Math.* 27 (2001), 737–752.

[MaZi03] C. L. May and J. Zimmerman, There is a group of every strong symmetric genus, *Bull. London Math. Soc.*, 35 (2003), 433–439.

[MoPiSkWh85] B. Mohar, T. Pisanski, M. Skoviera, and A. T. White, The cartesian product of three triangles can be embedded into a surface of genus 7, *Discrete Math.* 56 (1985), 87–89.

[PiTu89] T. Pisanski and T. W. Tucker, The genus of a product of a group with an abelian group, *Europ. J. Combin.* 10 (1989), 469–475.

[Pr77] V. K. Proulx, The classification of toroidal groups, PhD Thesis, Columbia University, 1977.

[RSJTW] R. B. Richter, J. Širáň, R. Jajcay, T. W. Tucker, M. E. Watkins, Cayley maps, *J. Combin. Theory, Ser. B* 95 (2005), 189–245.

[Sa58] G. Sabidussi, On a class of fixed-point free graphs, *Proc. Amer. Math. Soc.* 9 (1958), 800–804.

[Th91] C. Thomassen, Tilings of the torus and the Klein bottle and vertex-transitive graphs on fixed surface, *Trans. Amer. Math. Soc.* 323 (1991), 89–105.

[Tu80] T. W. Tucker, The number of groups of a given genus, *Trans. Amer. Math. Soc.* 258 (1980), 167–179.

[Tu83] T. W. Tucker, Finite groups acting on surfaces and the genus of a group, *J. Combin. Theory, Ser. B* 34 (1983), 82–98.

[Tu84a] T. W. Tucker, On Proulx's four exceptional toroidal groups, *J. Graph Theory* 8 (1984), 29–33.

[Tu84b] T. W. Tucker, A refined Hurwitz theorem for imbeddings of irredundant Cayley graphs, *J. Combin. Theory, Ser. B* 36 (1984), 244–268.

[Tu84c] T. W. Tucker, There is one group of genus two, *J. Combin. Theory, Ser. B* 36 (1984), 269–275.

[Tu91] T. W. Tucker, Symmetric embeddings of Cayley graphs in non-orientable surfaces, pp. 1105–1120 in *Graph Theory, Combinatorics, and Applications* (Kalamazoo 1988), Wiley-Interscience, 1991.

[Tu08] T. W. Tucker, Genus parameters and sizings of groups, in *Applications of Group Theory to Combinatorics*, Koolen, Kwak, and Xu (eds.), CRC, 2008, pp 155–160.

[Tu09] T. W. Tucker, The genus of a group, in *Topics in Topological Graph Theory*, L. Beineke and R. J. Wilson (eds.), Cambridge University Press, 2009, pp 199–224.

[Wh72] A. T. White, On the genus of a group, *Trans. Amer. Math. Soc.* 173 (1972), 203–214.

[Wh84] A. T. White, *Graphs, Groups and Surfaces*, Revised Edition, North-Holland, 1984.

Section 7.6

Maps

Roman Nedela, Matej Bel University
Martin Škoviera, Comenius University

INTRODUCTION

The theory of maps is likely to be the oldest topic in this volume, going back not just to the 4-color problem posed in 1852 and to the theory of automorphic functions developed in the late 1800's, but to the Platonic solids dating from antiquity. Among the many contributors to the subject are Archimedes, Kepler, Euler, Poinsot, de Morgan, Hamilton, Dyck, Klein, Heawood, Hurwitz, Steinitz, Whitney, Tutte, Coxeter, and Grünbaum. General references on maps include [BeiWi09], [BoLi95], [BrSc95], [CoMo84], [LaZv04], [GrTu87], [JoSi87], [MoTh01], and [Wh01]. A systematic combinatorial theory of maps appears in [JoSi78], [BrSi85], and [BoLi95].

7.6.1   Maps and Polyhedral Maps

Basic notions are introduced: map and polyhedral map, duality, isomorphism, face-width. The existence and uniqueness of a map with a given underlying graph is addressed, as well as the well-known Euler–Poincaré formula, which relates combinatorial and topological invariants of a map.

DEFINITIONS

D1: A **surface** in this section is a compact, connected, 2-dimensional manifold without boundary. A surface S is determined up to homeomorphism by two invariants, its orientability and Euler characteristic. For each even integer $\chi \leq 2$, there is a unique orientable surface with Euler characteristic χ; it is a sphere with $g = (2 - \chi)/2$ handles (or, equivalently, a connected sum of g tori), and is denoted S_g. For each integer $\chi \leq 1$ there is a unique nonorientable surface with Euler characteristic χ; it is a sphere with $g = 2 - \chi$ crosscaps (or, equivalently, a connected sum of g projective planes) and is denoted N_g.

D2: The invariant g is called the **genus of the surface** S. The orientable surfaces of genus 0 and 1 are the sphere and torus, respectively, and the nonorientable surfaces of genus 1 and 2 are the projective plane and Klein bottle, respectively.

D3: A continuous mapping $f: S \to S'$ from a surface S to a surface S' is called a **branched covering** if each point $x \in S$ has a neighborhood D such that the restriction of f to D is topologically equivalent to the complex mapping $z \mapsto z^d$. If all but finitely many points of S' have precisely k preimages, the covering is said to be a **k-fold covering**. The exceptional points are called **branch points** of f. If f has no branch-points, it is called a **smooth covering**.

D4: A **map M on a surface** S is a finite cell-complex whose underlying topological space is S. The **supporting surface of a map** M is denoted S_M.

D5: The **underlying graph of a map** M is its 1-skeleton. It is denoted $G = G_M$.

D6: The **vertices and edges of a map** M are the vertices and edges, respectively, of G_M. The **faces of a map** M are the connected components of $S_M - G_M$. The boundary of each face is a closed walk in G_M.

D7: The **0-, 1-, and 2-dimensional cells of a map** M are its vertices, edges, and faces, respectively.

D8: Maps M_1 and M_2 are **isomorphic maps**, denoted $M_1 \approx M_2$, if there is a homeomorphism of the supporting surfaces that induces an isomorphism of the underlying graphs.

D9: A **map homomorphism** $f: M \to N$ is a branched covering $S_M \to S_N$ between the supporting surfaces that induces a graph epimorphism $G_M \to G_N$ between the underlying graphs.

D10: The **dual map M^*** of a map M on a surface S is a map in the same surface S, whose vertex set V^* consists of one point interior to each face of M, and whose edge set E^* consists of, for each edge e of M, an edge e^* that crosses edge e and joins the vertices of V^* that correspond to the faces incident with e.

D11: The **Petrie dual of a map** M is a map M^P with the same underlying graph as M, whose face boundaries are closed walks in G_M, such that any two consecutive edges, but not three, belong to a face of M.

D12: The **medial map M^{med}** of a map M is a map whose vertex-set is the edge-set of M, and whose edges join each pair of edges of M which are consecutive edges in a face-boundary walk of M.

D13: A **polyhedral map** M is a map whose face boundaries are simple cycles, and such that any two distinct face boundaries are either disjoint or meet in either a single edge or a single vertex.

D14: A polyhedral map M is a ***triangulation*** if the boundary of each face is a 3-cycle.

D15: The ***face-width of a map*** M, denoted $fw(M)$, is the minimum number of points $|\tau \cap G_M|$ over all noncontractible simple closed curves τ on the surface.

D16: The ***edge-width of a map*** M, denoted $ew(M)$, is the length of a shortest cycle in G_M that is noncontractible on the supporting surface.

D17: The operation of ***edge-contraction*** for a triangulation, and its inverse operation ***vertex-splitting***, are exhibited in Figure 7.6.1. After contracting an edge in a triangulation, the map may no longer be a triangulation, i.e., no longer polyhedral; this occurs if the edge is contained in a 3-cycle that is not a face boundary or if the map is the tetrahedral map.

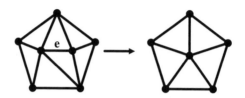

Figure 7.6.1: Edge-contraction and vertex-splitting in a triangulation.

D18: A ***minimal triangulation*** of a surface S is a triangulation such that the contraction of any edge results in a map that is no longer polyhedral.

D19: A ***k-minimal triangulation*** is a triangulation of a non-spherical surface with edge-width k such that each edge is contained in a noncontractible k-cycle. (Note that for non-spherical triangulations, minimal and 3-minimal are equivalent.)

EXAMPLES

E1: A map M in the torus, formed by a hexagonal imbedding of the 3-dimensional cube Q_3, and its dual map M^* appear in Figure 7.6.2. The torus is obtained by identifying like labeled edges on the boundary of the polygon. Neither M nor M^* is polyhedral.

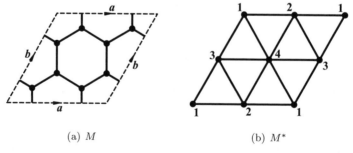

(a) M (b) M^*

Figure 7.6.2: A toroidal map and its dual.

E2: Figure 7.6.3 shows two nonisomorphic maps in the sphere that have the same 2-connected, but not 3-connected, underlying graph. The two maps are related by a *Whitney flip*. This example is relevant to Fact F8 below.

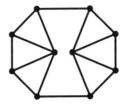

Figure 7.6.3: Maps in the sphere with the same 2-connected underlying graph.

E3: Figure 7.6.4 shows two polyhedral maps in the projective plane with isomorphic 3-connected underlying graphs. The projective plane is depicted here as a disc with antipodal boundary points identified. This example shows that the analogy to the Whitney uniqueness theorem (Fact F8) fails for projective planar graphs.

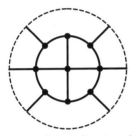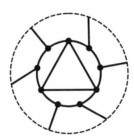

Figure 7.6.4: Maps in the projective plane with the same 3-connected underlying graph.

E4: Figure 7.6.5 shows the tetrahedral map M in the sphere and its Petrie dual. The map M^p is an imbedding of the complete graph K_4 into the projective plane known as the *hemi-cube*. Its antipodal double cover (see Remark R8) is the spherical imbedding of the cube Q_3.

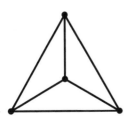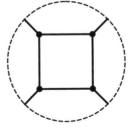

(a) M (b) M^p

Figure 7.6.5: A map and its Petrie dual.

E5: There is one minimal triangulation of the sphere (i.e., the tetrahedral map), two minimal triangulations of the projective plane (see Figure 7.6.6), 21 of the torus, and 25 of the Klein bottle.

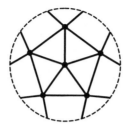 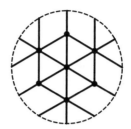

Figure 7.6.6: The two minimal triangulations of the projective plane.

REMARKS

R1: It is equivalent to regard a map as a 2-***cell imbedding*** of a graph G on a surface S, i.e., an imbedding such that the connected components of $S - G$ are 2-cells.

R2: A map may have multiple edges, self-loops, and vertices of degree 1 or 2. A polyhedral map, however, can have none of these. Moveover, in a polyhedral map, the closure of each face is topologically a closed disc.

R3: The boundary of a face may not be a simple cycle. Whether every (2-connected) graph can be imbedded on a surface in such a way that the face boundaries are (simple) cycles is not known (see the conjectures below).

R4: Face-width, introduced in [RoSe88], is a measure of local planarity, or of how dense the graph is in the surface, or of how well the graph represents the surface. It is sometimes known as *representativity*.

R5: A non-spherical map M is polyhedral if and only if $fw(M) \geq 3$ and its underlying graph is 3-connected; see [MoTh01, Proposition 5.5.12]. By Steinitz's Theorem (see Fact F22) it follows that the polyhedrality of maps generalizes the notion of a convex polyhedron.

R6: A map M on the sphere S can be drawn in the plane via stereographic projection from any point of $S - G_M$.

R7: The Petrie dual of a map M is well-defined for every M. Its faces are so-called *zig-zag polygons* of the map M. The supporting surface of the Petrie dual of a map on an orientable surface is orientable if and only if M is bipartite.

R8: For every map M on a nonorientable surface, there exists an orientable map $\tilde{M}$ and a homomorphism $\tilde{M} \to M$. Among such maps there exists a unique minimal map, the ***antipodal double cover*** of M.

R9: In the class of cubic polyhedral maps, the operation dual to edge-contraction in a triangulation is the operation of ***edge-suppression***. It consists of removal of an edge e incident with two faces and suppressing the end-vertices of e, thereby producing a smaller cubic map.

R10: The concept of map has been extended to cell-complexes whose underlying topological space is a manifold of dimension greater than 2. This includes, in particular, the boundary complex of any polytope. The generalizations, though natural and interesting, are omitted here; for more information see, e.g., [Fe76], [Ga79], and[Vi83a].

FACTS

F1: Euler–Poincaré formula: For any map M with α_0 vertices, α_1 edges, α_2 faces, and Euler characteristic $\chi(M)$,

$$\alpha_0 - \alpha_1 + \alpha_2 = \chi(M).$$

F2: Every connected graph G is the underlying graph of a map. The *rotation schemes* introduced in Definitions D31 and D32 give a systematic method for obtaining all 2-cell imbeddings of G.

F3: The underlying graph of a map M on a surface of Euler characteristic χ has cycle rank at least $2 - \chi$. A map whose underlying graph is a tree is necessarily spherical.

F4: If M is a map, then $(M^*)^* = M$, $(M^P)^P = M$ and $(M^*)^{med} = M^{med}$.

F5: If M is a map, then $fw(M^*) = fw(M)$. In particular, the dual of a polyhedral map is polyhedral.

F6: [MoTh01, Proposition 5.5.4] If M is a map, then $fw(M) = ew(M^{med})/2$.

F7: The underlying graph of a polyhedral map is 3-connected.

F8: Whitney Uniqueness Theorem [Whi32]: A 3-connected planar graph has a unique imbedding in the sphere.

F9: [Ne83] All 6-connected toroidal graphs are uniquely imbeddable in the torus. On the other hand, there are infinitely many 5-connected toroidal triangulations of the torus that are not uniquely imbeddable in the torus; see [La87].

F10: [Ne85] A 5-connected projective planar graph distinct from K_6, containing a subgraph homeomorphic to K_6, is uniquely imbeddable in the projective plane.

F11: [Th90, Corollary 5.1.7] If a 3-connected graph G has an imbedding $G \hookrightarrow S$ whose edge-width is greater than the length of any face-boundary, then G is uniquely imbeddable in S.

F12: [BaEd89] The set of minimal triangulations is finite for every fixed surface (see Example E5). In other words, for each surface, there is a finite set of triangulations from which any triangulation on that surface can be generated by vertex-splittings. In particular, every triangulation of the sphere can be obtained from the tetrahedral map by a sequence of vertex-splittings.

F13: For any $k \geq 3$, the set of k-minimal triangulations on a fixed surface is finite [MaNe95]. ([MoTh01, Theorem 5.4.1] provides another proof.)

CONJECTURES

C1: The Strong Imbedding Conjecture: Every 2-connected graph can be imbedded on a surface so that each face is bounded by a simple cycle in the graph.

C2: The Cycle Double Cover Conjecture: Every 2-connected graph contains a collection $\mathcal{C}$ of cycles such that every edge is contained in exactly two cycles of $\mathcal{C}$. The validity of the Strong Imbedding Conjecture implies validity for the Cycle Double Cover Conjecture.

C3: Grünbaum's Conjecture: If a cubic graph G has a polyhedral imbedding into an orientable surface, then G is 3-edge-colorable.

REMARKS

R11: According to Fact F2 above, every connected graph has a 2-cell imbedding into an orientable surface. Moreover, every connected graph with at least one cycle has a 2-cell imbedding into a nonorientable surface.

R12: For cubic graphs, Conjectures C1 and C2 are equivalent. It is known that the smallest counterexample to Conjecture C2 must be a non-3-edge-colorable cyclically 4-edge-connected cubic graph of girth at least 12 (see §5.1.5).

R13: An equivalent formulation of Grünbaum's Conjecture states that the dual of a triangulation of an orientable surface has a 3-edge-colorable underlying graph. Still another formulation of the conjecture states that no snark (see §5.2) admits a polyhedral imbedding into an orientable surface. In contrast, it is known that every nonorientable surface supports a polyhedral embedding of some snark [LiCh12]. For the sphere, Grünbaum's conjecture is equivalent to the Four Color Theorem. The conjecture was shown to be false for every orientable surface of genus at least 5 [Ko09], and remains open only for surfaces of genus 1, 2, 3, and 4.

7.6.2   Existence and Realizations of Polyhedral Maps

Elementary equalities hold among the basic parameters of a map. The two questions addressed in this section are, first, when are the necessary conditions also sufficient for the existence of a map with these parameters and, second, when can the map be imbedded in the Euclidean space E^3 or E^4 in such a way that the faces are plane convex polygons. The classical results for maps on the sphere are Eberhard's theorem of 1891 and Steinitz's theorem of 1922.

DEFINITIONS

D20: A map is of *map type* $\{p, q\}$ if each face has p edge incidences and each vertex has q edge incidences. (No global symmetry is implied; in fact, the automorphism group of such a map, as defined in §7.6.6, may be trivial.)

D21: The *cell-distribution vector (α-vector)* of a map M is the 3-tuple $(\alpha_0, \alpha_1, \alpha_2)$, where $\alpha_0, \alpha_1, \alpha_2$ are the numbers of vertices, edges, and faces of M, respectively.

D22: The *face-size sequence (p-sequence)* of a polyhedral map M is the sequence $\{p_i\}_{i \geq 3}$, where p_i is the number of i-gonal faces in M. The *reduced face-size sequence* is the sequence $p^* = \{p_i\}_{\substack{i \geq 3 \\ i \neq 6}}$.

D23: The ***vertex-degree sequence (v-sequence)*** of a polyhedral map M is the sequence $\{v_i\}_{i \geq 3}$, where v_i is the number of vertices of degree i in M. The ***reduced vertex-degree sequence*** is the sequence $v^* = \{v_i\}_{i \geq 4}$.

D24: For a given triple (p^*, v^*, g), let $P_6(p^*, v^*, g)$ be the set of all integers p_6 for which there is a polyhedral map of genus $g \geq 0$ with p_6 hexagonal faces, with the corresponding reduced p-sequence p^*, and with the corresponding reduced v-sequence v^*.

D25: A ***geometric realization (realization)*** of a polyhedral map M is an imbedding of M into Euclidean space E^d (without self intersection) such that each face is a plane convex polygon and that adjacent faces are not coplanar.

EXAMPLES

E6: The map M in Figure 7.6.2(b) is of type $\{3, 6\}$, with α-vector $(4, 12, 8)$. Its dual M^* is of type $\{6, 3\}$, with α-vector $(8, 12, 4)$. The maps in Figure 7.6.4 both have v-sequence $(6, 3)$, but the first has p-sequence $(0, 6, 0, 1)$, while the second has p-sequence $(1, 3, 3)$.

E7: Five polyhedral maps on the sphere and their corresponding 3-dimensional realizations appear in Figure 7.6.7.

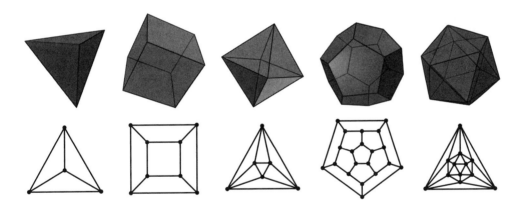

Figure 7.6.7: The Platonic solids as 3-dimensional realizations of maps.

FACTS

F14: The α-vector, the p-sequence, and the v-sequence satisfy the following elementary equalities:

$$\sum_{i \geq 1} p_i = \alpha_2, \qquad \sum_{i \geq 1} v_i = \alpha_0, \qquad \sum_{i \geq 1} i p_i = 2\alpha_1 = \sum_{i \geq 1} i v_i \qquad (7.6.1)$$

F15: For a map M on an orientable surface of genus g and $\kappa, \lambda \geq 0$ such that $\kappa + \lambda = 1$, Euler's formula implies that

$$\sum (\kappa i - 2) v_i + \sum (\lambda i - 2) p_i = 4(g - 1). \qquad (7.6.2)$$

For example, taking $\kappa = 1/3$, and M polyhedral of genus g yields

$$\sum_{i \geq 3}(6 - i)p_i + 2\sum_{i \geq 3}(3 - i)v_i = 12(1 - g). \qquad (7.6.3)$$

In particular, if M is a cubic polyhedral map on the sphere, then

$$\sum_{i \geq 3}(6 - i)p_i = 12. \qquad (7.6.4)$$

A pair of reduced sequences (p^*, v^*) is a g-**admissible pair of map sequences** if Conditions (7.6.1) and (7.6.3) are satisfied.

F16: By Equation (7.6.1), for a polyhedral map on a surface with a given p-sequence and a given reduced v-sequence the following holds

$$v_3 = \frac{1}{3}\left(\sum_{i \geq 3} ip_i - \sum_{i \geq 4} iv_i\right).$$

F17: [Eb1891] **Eberhard's Theorem**: Condition (7.6.4) above is sufficient for the existence of a cubic spherical map, in the following sense: if a sequence $\{p_i \mid i \geq 3,\ i \neq 6\}$ satisfies $\sum_{k \neq 6}(6 - k)p_k = 12$, then there exist values of p_6 such that $\{p_i \mid i \geq 3\}$ is the p-sequence of a simple polyhedral map on the sphere. For variations on Eberhard's Theorem, see [Gru70, Je93a, Je93b].

F18: Eberhard's theorem establishes that $P_6(p*, \{0\}_{i \geq 4}, 0)$ is non-empty for each $p*$ satisfying (7.4). The sets $P_6(p*, v*, 0)$ and $P_6(p*, v*, 1)$ are determined, up to a finite number of exceptions, for all admissible pairs (p^*, v^*); see [Je93a, Je93b]. There are infinitely many 0-admissible pairs (p^*, v^*) and exactly one 1-admissible pair (p^*, v^*) with $P_6(p^*, v^*, 0) = \emptyset = P_6(p^*, v^*, 1)$.

F19: [Je93b, Theorem 2] For any g-admissible pair (p^*, v^*), $g \geq 2$, the set $P_6(p^*, v^*, g)$ contains all but finitely many positive integers.

F20: [Gri83] Equation 7.6.3 for the torus (with $\kappa = 1/3$) becomes

$$2\sum(i - 3)\,v_i + \sum(i - 6)\,p_i = 0,$$

which leads to the following analogue of Eberhard's theorem for the torus. Given a sequence $\{p_i \mid i \geq 3,\ i \neq 6\}$ and a positive integer s, there is a realization in E^3 of some polyhedral map on the torus with p-sequence $\{p_i \mid i \geq 3\}$ and $\sum(i - 3)v_i = s$ if and only if $\sum_{k \neq 6}(6 - k)p_k = 2s$ and $s \geq 6$. Related results appear in [BaGrHo91].

F21: [EdEwKu82] If S is a surface with Euler characteristic χ, if $\alpha_0, \alpha_1, \alpha_2, p, q$ are positive integers such that $\alpha_0 - \alpha_1 + \alpha_2 = \chi$, and if $p\alpha_2 = 2\alpha_1 = q\alpha_0$, then there exists a map of type $\{p, q\}$ on S with α-vector $(\alpha_0, \alpha_1, \alpha_2)$, except when S is the projective plane, $\{p, q\} = \{3, 3\}$, $\alpha_0 = \alpha_2 = 2$, and $\alpha_1 = 3$.

F22: [Sti22] **Steinitz's Theorem**: Every polyhedral map on the sphere is isomorphic to the boundary complex of a 3-dimensional polytope. Thus, any polyhedral map on the sphere has a realization in E^3.

F23: [Al71, Gru67] A 3-valent polyhedral map M cannot be realized in Euclidean space of any dimension, unless the supporting surface of M is the sphere.

F24: [BrSc95] Each triangulation on the torus or on the projective plane can be realized in E^4.

F25: [BrWi93] On any nonorientable surface N_g, there exists a triangulation that cannot be realized in E^3. (When $g > 1$, it is an open question whether each triangulation of orientable genus g can be realized in E^3.)

F26: [Sti06] The vector $(\alpha_0, \alpha_1, \alpha_2)$ is the α-vector of a realization in E^3 of some polyhedral map on the sphere if and only if $\alpha_0 - \alpha_1 + \alpha_2 = 2$, $4 \leq \alpha_0 \leq 2\alpha_2 - 4$, and $4 \leq \alpha_2 \leq 2\alpha_0 - 4$.

F27: [Gri83] The vector $(\alpha_0, \alpha_1, \alpha_2)$ is the α-vector of a realization in E^3 of some polyhedral map on the torus if and only if $\alpha_0 - \alpha_1 + \alpha_2 = 0$, $\alpha_2(11 - \alpha_2)/2 \leq \alpha_0 \leq 2\alpha_2$, $\alpha_0(11 - \alpha_0)/2 \leq \alpha_2 \leq 2\alpha_0$, $2\alpha_1 - 3\alpha_0 \geq 6$, and $\alpha_1 \neq 19$.

REMARKS

R14: Cubic spherical polyhedral maps with reduced p-sequence given by $p_5 = 12$ and $p_j = 0$ for each $j \notin \{5, 6\}$, known as *fullerenes*, represent models of complex carbon molecules. They have been extensively investigated during the recent decades; see [FoMa07].

R15: If a map M with α_0 vertices and Euler characteristic $\chi(S)$ is polyhedral, then

$$\alpha_0 \geq \left\lceil \frac{7 + \sqrt{49 - 24\chi(S)}}{2} \right\rceil,$$

and this lower bound is attained for all surfaces except S_2, N_2, and N_3. By duality the same bound holds for α_2.

7.6.3 Paths and Cycles in Maps

This section covers three topics involving paths and cycles: the Lipton–Tarjan separator theorem, the existence of nonrevisiting paths in polyhedral maps, and the decomposition of maps along cycles in the graph. The third topic is related to a result of Robertson and Seymour on minors.

DEFINITIONS

D26: A path p in the underlying graph of a map M is said to be a ***nonrevisiting path*** if the intersection of p with the boundary of F is connected for each face F of M.

D27: A surface S has the ***nonrevisiting path property*** if, for every polyhedral map M on S, any two vertices of M are joined by a nonrevisiting path.

D28: A map M' is a ***minor of a map*** M if M' can be obtained from M by a sequence of edge contractions and deletions. The operations of edge deletion and edge contraction on a graph can be extended to a surface imbedding of the graph in an obvious way.

EXAMPLE

E8:   A polyhedral map on the surface S_2 that fails to have the nonrevisiting path
property appears in Figure 7.6.8 below. There is no nonrerevisiting path from x to y.
(The map is obtained by gluing along like labeled edges.)

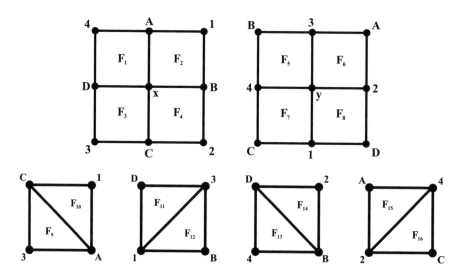

Figure 7.6.8: A map on S_2 that does not satisfy
the non-revisiting path property.

FACTS

F28:   [LiTa79] **Planar Separator Theorem**: A planar graph with n vertices has a
set of at most $2\sqrt{2n}$ vertices whose removal leaves no component with more than $2n/3$
vertices.

F29:   [AlSeTh94] Let M be a loopless map on the sphere with n vertices. Then
there is a simple closed curve τ on the surface of the sphere passing through at most
$k \leq 3\sqrt{2n}/2$ vertices (and no other points of the graph) such that each of the two open
disks bounded by τ contains at most $2n/3 - k/2$ vertices. This result slightly improves
the Lipton–Tarjan separator theorem.

F30:   [GiHuTa84] A map of genus g contains a set of at most $O(\sqrt{gn})$ vertices whose
removal leaves no component of the graph with more than $2n/3$ vertices. This generalizes
the Lipton–Tarjan theorem to maps on orientable surfaces of higher genus.

F31:   [PuVi98] For polyhedral maps, the nonrevisiting path property holds for the
sphere, torus, projective plane, and the Klein bottle. It fails for all other surfaces
except possibly the nonorientable surface of genus 3 (see [PuVi96] and Example E8).

F32:   The nonrevisiting path property holds for every polyhedral map with face-width
at least 4.

F33: [RoSe88] Let M_0 be any map on a surface S other than the sphere. There exists a constant k such that, for every map M on S with $fw(M) \geq k$, the map M_0 is a map minor of M. The following two results provide bounds for the constant k when the given M_0 contains certain sets of disjoint cycles.

F34: [Sc93] A map M on the torus with face-width w contains $\lfloor 3w/4 \rfloor$ disjoint non-contractible cycles.

F35: [BrMoRi96] For general surfaces there exist $\lfloor w/2 \rfloor$ pairwise disjoint contractible cycles in the graph of any map M, all containing a particular face, $\lfloor (w-1)/2 \rfloor$ pairwise disjoint, pairwise homotopic, surface nonseparating cycles, and $\lfloor (w-1)/8 \rfloor - 1$ pairwise disjoint, pairwise homotopic, surface separating, noncontractible cycles. (It is unknown whether any map of orientable genus $g \geq 2$ with face-width at least 3 must contain a noncontractible surface separating cycle.)

F36: [Bar88] Every polyhedral map on the torus (projective plane, Klein bottle) is isomorphic to the complex obtained by identifying the boundaries of two faces of a 3-polytope (cross identifying one face of a 3-polytope, cross identifying two faces of a 3-polytope).

F37: [Yu97] (see also [Th93]) If d is a positive integer and M is a map on S_g of face-width at least $8(d+1)(2^g - 1)$, then the underlying graph of M contains a collection of induced cycles $C_1, C_2, \ldots, C_g$ such that the distance between distinct cycles is at least d and cutting along the cycles results in a map on the sphere.

F38: [Sc91] Schrijver proved necessary and sufficient conditions (conjectured by Lovász and Seymour) for the existence of pairwise disjoint cycles $\tilde{C}_1, \ldots, \tilde{C}_k$ in the underlying graph of a map M homotopic to given closed curves $C_1, \ldots, C_k$ on the surface.

REMARK

R16: The Lipton–Tarjan separator theorem has applications to divide-and-conquer algorithms. Nonrevisiting paths arise in complexity issues for edge following linear programming algorithms like the simplex method.

7.6.4 Map Coloring

The famous problems on map coloring, the Four Color Problem and the Heawood Map Coloring Problem, stimulated considerable research in the area, and the Heawood Problem led to the birth of topological graph theory. The long-lasting effort of mathematicians to solve both problems significantly influenced development not only in graph theory and combinatorics, but also in algebra, geometry, computer science, and others. This subsection is aimed at presenting the solution of the Heawood Map Coloring Problem along with other related results.

DEFINITION

D29: The **chromatic number** $chr(S)$ of a surface S is the least number of colors sufficient to properly color the faces of any map on S. By duality, it is also the least number of colors sufficient to properly color the vertices of any map on S. In this section, coloring will mean *vertex coloring*.

FACTS

F39: [ApHa76] **Four Color Theorem:** $chr(S_0) = 4$.

F40: [Fr34] $chr(N_2) = 6$.

F41: [RiYo68] **Heawood Map Coloring Theorem:** For every surface S except the Klein bottle N_2,

$$chr(S) = \left\lfloor \frac{7 + \sqrt{49 - 24\chi(S)}}{2} \right\rfloor.$$

The right-hand side of the equation is called the ***Heawood formula***.

F42: [Di52, AlHu79] If G is a graph imbedded into a surface S other than the sphere, then $chr(G) < chr(S)$ unless G contains the complete graph of order $chr(S)$ as a subgraph.

F43: [FiMo94] There is a universal constant c such that every map M on a surface with Euler characteristic $\chi < 2$ such that $ew(M) \geq c \log(2 - \chi)$ is 6-colorable.

F44: [Th93] Any map M on S_g with $ew(M) \geq 2^{14g+6}$ is 5-colorable.

F45: [Gr59] **Grötzsch Theorem:** Every planar map of girth at least 4 is 3-colorable.

F46: [Ke1879, TsWe11] A planar triangulation is 3-colorable if and only if it is Eulerian.

F47: [HuRiSe02] There is a constant $f(g)$ such that every Eulerian triangulation M on an orientable surface of genus g with $ew(M) \geq f(g)$ is 4-colorable.

F48: [Hu95] For every positive integer g there is a constant $f(g)$ such that the following holds: If M is a map on S_g such that $ew(M) > f(g)$ and all boundary-walks are of even length, then M is 3-colorable.

F49: [Th97] For a fixed surface S, there is a polynomial time algorithm to decide whether a map on S can be 5-colored.

F50: The problem of deciding whether a map can be 3-colored is NP-complete even for maps on the sphere [GaJo79].

F51: [RSST96] On the sphere, a 4-coloring can be found in $O(n^2)$ steps.

REMARKS

R17: The problem of determining the chromatic number of the sphere appeared in an 1852 letter from Augustus de Morgan to Sir William Hamilton, and was likely due to Francis Guthrie, the brother of a student of de Morgan. A computer dependent proof of Appel and Haken [ApHa76] that four colors suffice was considerably simplified [RSST97], but still remains computer dependent.

R18: That the formula in the Heawood Map Coloring Theorem gives an upper bound on $chr(S)$ was proved by Heawood [He1890]. That there exist graphs that actually require the number of colors given by that formula is a consequence of the formula for the genus of complete graphs due to Ringel and Youngs [RiYo68].

R19: For surfaces with Euler characteristic $\chi < 0$ the problem of determining the chromatic number of a surface is equivalent to the determination of the minimum genus of the complete graph K_n [RiYo68], [GrTu87].

R20: It follows from the Heawood formula that the chromatic number of surfaces increases with the genus. In contrast, by Facts F43 and F44 maps with large edge-width or face-width have chromatic number at most six or at most five, respectively. To prove stronger upper bounds on the chromatic number one needs additional conditions on the maps in question; see Facts F46, F47, and F48.

R21: There is a quadratic-time algorithm to color a planar map with four colors [RSST97, p.27]. Since deciding the 3-colorability of a graph is NP-complete, it is widely thought to be unlikely that there exists a polynomial-time algorithm for deciding whether a map on an arbitrary surface can be 4-colored.

PROBLEMS

RP1: [Al81] Let S be any surface. Does there exist a natural number $q(S)$ such that any graph G on S contains a set A of at most $q(S)$ vertices such that $G - A$ is 4-colorable?

RP2: [GiTh97] Can the chromatic number of a triangle-free graph on a fixed surface be found in polynomial time?

EXAMPLE

E9: Figure 7.6.9(a) exhibits a triangular imbedding of K_6 in the projective plane, and Figure 7.6.9(b) exhibits a triangular imbedding of K_7 in the torus. This shows that $\chi(N_1) \geq 6$ and that $\chi(S_1) \geq 7$. In fact, $\chi(N_1) = 6$ and $\chi(S_1) = 7$, in accordance with Facts F40 and F41.

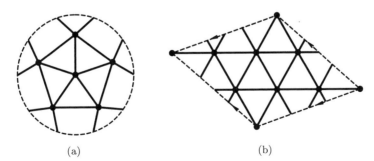

(a) (b)

Figure 7.6.9: Triangular imbeddings of K_6 into the projective plane and of K_7 into the torus.

7.6.5   Combinatorial Schemes

The definition of map in §7.6.1 as a cell complex is topological. A strictly combinatorial description, although less intuitive, is often more convenient to apply. Four such schemes are described: rotation system, signed rotation system, crystallization, and graph encoding of a map. They have been discussed in [Ed60], [Ja68], [Co75], [Wa75], [Ri77], [Sta78], [JoSi78], [Tut79], [Sta80], [Li82] and [BrSi85].

DEFINITIONS

D30: A **graph** is a triple $G = (D; V, L)$, where D is a set of **darts** (or directed edges), V is a partition of D, the **vertex-set** of G, and L is an involutory fixed-point-free permutation of D called the **dart-reversing involution**. The orbits of L are the edges of G and the incidence relation between vertices and edges of G is defined by non-empty intersection.

D31: A **rotation system** is a pair $(G; R)$, where $G = (D; V, L)$ is a connected graph, and R is a permutation of D whose system of orbits coincides with V. The cycle of R permuting the darts of a vertex $v \in V$ is called the **local rotation** at v and is denoted R_v. A rotation system $(G; R)$ with $G = (D; V, L)$ can also be represented as a triple $(D; R, L)$ without making the graph G explicit. This scheme encodes any map with G imbedded on an orientable surface with a fixed orientation [Ed60, GrTu87, JoSi78].

D32: A **signed rotation system** is a triple $(G; R, \sigma)$, where $(G; R)$ is a rotation system and σ is a mapping that assigns to each edge of G an element of $\{1, -1\}$. This scheme encodes any map with G imbedded on any closed surface including the nonorientable surfaces [Ri77, Sta78].

D33: A **switch of signed rotation system at a vertex** v is an operation transforming a signed rotation system $(G; R, \sigma)$ into $(G; R', \sigma')$, where $R'_v = R_v^{-1}$ and $R'_u = R_u$ for each vertex $u \in V - \{v\}$; and $\sigma'(e) = -\sigma(e)$ for every edge e incident with v except for loops, and $\sigma'(e) = \sigma(e)$, otherwise.

D34: The **map determined by a rotation system** $(G; R)$ is a map on an orientable surface whose faces are bounded by the cycles of RL. Regarding each cycle of RL as the boundary of a polygonal 2-cell and gluing together 2-cells along paired darts x and Lx results in an orientable surface in which G is imbedded. Conversely, the **rotation system of a map** M on an orientable surface is a pair $(G; R)$, where G is the underlying graph of M and R is the permutation induced by the chosen orientation of the supporting surface that cyclically permutes darts based at any vertex.

D35: The **map determined by a signed rotation system** $(G; R, \sigma)$ is defined as follows. Let $\tilde{L}$ and $\tilde{R}$ be permutations of $D \times \{1, -1\}$ defined by $\tilde{R}(x, i) = (R^i(x), i)$ and $\tilde{L}(x, i) = (Lx, i\sigma(x))$. The boundary walks of the map are formed from the cycles of $\tilde{R}\tilde{L}$ by ignoring the second coordinate of each pair $(x, i) \in D \times \{1, -1\}$. The cycles of $\tilde{R}\tilde{L}$ occur in pairs giving the same face boundary twice, once in each direction. The supporting surface of the map is formed by taking a representative of each pair, spanning it with a 2-cell, and gluing the boundaries of faces along the corresponding darts. Conversely, a **signed rotation system of a map** M with underlying graph G is a triple $(G; R, \sigma)$, defined as follows. For each vertex v choose a local orientation around v. Let R_v be the cyclic permutation of darts based at v induced by the chosen orientation, and set $R = \prod_v R_v$. For each edge e define $\sigma(e) = 1$ if and only if the local orientations at the end-vertices of e are consistent.

D36: The ***barycentric subdivision of a map*** M is a map M^{bar} whose vertices are the vertices of M, the centers of edges of M, and the centers of faces of M. Two vertices of M^{bar} are joined by an edge on the supporting surface of M if the corresponding cells of M are incident. The barycentric subdivision of M is a triangulation; its triangles are called ***flags*** of M. Each edge of M touches exactly four flags. A vertex of valency k in M is incident with $2k$ flags and the center of a face of size m in M is incident with $2m$ flags. Each vertex v of M^{bar} is labeled with $0, 1$, or 2, according to the dimension of the cell in M that the vertex v represents. Every edge e of M^{bar} is colored by the label missing at its end-vertices.

D37: The ***graph encoding*** of a map M is a pair $\mathcal{G} = (H, \phi)$, where H is the underlying graph of $(M^{bar})^*$, the dual of the barycentric subdivision of M, and ϕ is a proper 3-edge-coloring of H with color i assigned to the dual edge e^* if and only if the edge e of M^{bar} is colored with i. If $\mathcal{G} = (H, \phi)$ is a graph encoding of a map, then H is a cubic graph and each component of a subgraph H_1 formed by edges colored by 0 and 2 is a 4-cycle [Tut63, Ga79, Li82, Vi83a, Tut84, BrSi85, BoLi95].

D38: The ***crystallization of a map*** M is a quadruple $(F; \tau_0, \tau_1, \tau_2)$, where F is the set of flags of M (i.e., triangles of M^{bar}) and, for $i \in \{0, 1, 2\}$, τ_i is a fixed-point-free involution of F transposing two incident flags sharing an edge of M^{bar} colored i. The group $\langle \tau_0, \tau_1, \tau_2 \rangle$ is transitive on F, $(\tau_2 \tau_0)^2 = 1$, and $\tau_2 \tau_0$ is fixed-point-free. This scheme encodes a map on any closed surface.

D39: The ***map*** M ***defined by a crystallization*** $(F; \tau_0, \tau_1, \tau_2)$ is constructed as follows. First construct a cubic 3-edge-colored graph H with vertex set F by joining vertices x and y with an edge colored i whenever $y = \tau_i(x)$. Form a surface S by gluing a 2-cell to every bicolored cycle of H, which defines an imbedding of H into S. The dual map of the imbedding of H is by definition the barycentric subdivision M^{bar} of the map M defined by the crystallization $(F; \tau_0, \tau_1, \tau_2)$. To construct the ***map given by a graph encoding*** $\mathcal{G} = (H, \phi)$ we first define the associated crystallization by identifying flags with the vertices of H and setting $\tau_i(x) = y$, for $i \in \{0, 1, 2\}$, if and only if there is an edge xy in H colored by i. Then we proceed as above.

D40: The ***underlying graph*** $G = (D; V, L)$ of the map defined by a crystallization $(F; \tau_0, \tau_1, \tau_2)$ is defined by taking D to be the set of orbits of τ_2 with L sending a dart $\{x, \tau_2(x)\}$ to the dart $\{\tau_0(x), \tau_0 \tau_2(x)\}$. Each vertex of G is formed by the union of darts constituting an orbit of $\langle \tau_1, \tau_2 \rangle$ on F.

REMARKS

R22: A rotation scheme $(G; R)$ defines not just a map M on an orientable surface S but it also prescribes a global orientation of S. The scheme $(G; R^{-1})$ determines the ***mirror image of the map*** M, denoted M^{-1}, which is topologically identical to M, up to orientation. Other schemes describe maps without regard to orientation.

R23: Two rotation schemes $(G; R_1)$ and $(G; R_2)$ describe the same (isotopic) imbeddings of G if and only if $R_1 = R_2$. Similarly, two crystallizations $(F; \tau_0, \tau_1, \tau_2)$ and $(F; \tau_0', \tau_1', \tau_2')$ describe isotopic imbeddings of the underlying graph if and only if $\tau_i' = \tau_i$ for $i \in \{0, 1, 2\}$.

R24: Two signed rotation systems $(G; R, \sigma)$ and $(G; R', \sigma')$ describe the same imbedding of G if and only if there exists a sequence of switches that transforms $(G; R, \sigma)$ into $(G; R', \sigma')$.

R25: The dual of a map defined by a rotation system $(D; R, L)$ is a map determined by the system $(D; RL, L)$. The dual of a map given by a crystallization $(F; \tau_0, \tau_1, \tau_2)$ is a map defined by the crystallization $(F; \tau_2, \tau_1, \tau_0)$. If M is given as a graph encoded map $\mathcal{G} = (H, \phi)$, the dual map M^* is encoded by the same graph H with edge colors 0 and 2 interchanged.

R26: The Petrie dual of a map defined by a signed rotation system $(G; R, \sigma)$ is a map determined by the system $(G; R, -\sigma)$. The Petrie dual of a map given by a crystallization $(F; \tau_0, \tau_1, \tau_2)$ is the map defined by the crystallization $(F; \tau_0\tau_2, \tau_1, \tau_2)$.

R27: Generalizations of graph encoded maps to higher dimensions were introduced in a topological context by [Fe76], [Ga79], [Li82], [Vi83a].

R28: Lifting the restriction that $(\tau_0\tau_2)^2 = 1$ in the crystallization of a map, or that each component of H_1 in the graph encoding $\mathcal{G} = (H, \phi)$ of the map must be a 4-cycle, results in the concept of a ***hypermap*** or ***hypergraph imbedding***.

R29: Properties of maps studied within the theory of maps are combinatorial or algebraic schemes. Therefore, a map M is usually identified with a scheme describing M.

EXAMPLES

E10: Figure 7.6.10 shows the tetrahedral map on the sphere. It has twelve darts $\{1, 2, 3, 4, 5, 6\} \times \{-1, 1\}$; the dart-reversing involution L takes $(x, i) \mapsto (x, -i)$. In the diagram, the dart $(x, 1)$ is encoded as x and is represented by a directed edge, based at its origin. The dart $(x, -1) = x^{-1}$ is represented with the same edge endowed with the opposite orientation. The rotation consistent with the counterclockwise orientation of the sphere is $R = (1, 2, 3)(1^{-1}, 6, 5^{-1})(2^{-1}, 5, 4^{-1})(3^{-1}, 4, 6^{-1})$.

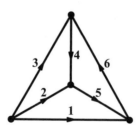

Figure 7.6.10: Tetrahedral map.

E11: Figure 7.6.11 describes an imbedding of K_4 into the projective plane. The corresponding signed rotation system is $(K_4; R, \sigma)$, with $R = (a, f^{-1}, d^{-1})(c, f, b^{-1})(d, e, c^{-1})$ (a^{-1}, b, e^{-1}) shown in the left part of the figure and the signature $\sigma \colon E(K_4) \to \{-1, 1\}$ described in the right part of the figure.

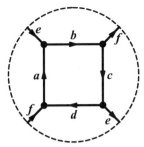 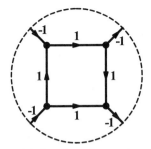

Figure 7.6.11: Map on the projective plane with its signed rotation system.

E12: The left part of Figure 7.6.12 represents a spherical map; the right part shows its graph encoding.

 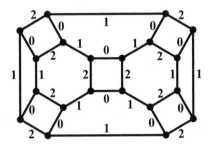

Figure 7.6.12: Map and its graph encoding.

FACTS

F52: The supporting surface of a map defined by a signed rotation system $(G; R, \sigma)$ is nonorientable if and only if G contains a cycle with an odd number of negative edges. By Definition D35 the rotation system $(\tilde{D}; \tilde{R}, \tilde{L})$, where $\tilde{D} = D \times \{1, -1\}$, $\tilde{R}(x, i) = (R^i(x), i)$ and $\tilde{L}(x, i) = (Lx, i\sigma(x))$ determines the antipodal double cover of a nonorientable map given by the signed system $(G; R, \sigma)$.

F53: The supporting surface of a map defined by a graph encoding $\mathcal{G} = (H, \phi)$ is orientable if and only if the graph H is bipartite. The same criterion can be used to determine the orientability of a map defined by a crystallization.

F54: If $(F; \tau_0, \tau_1, \tau_2)$ is a crystallization of a map on an orientable surface, then the same map (up to the choice of an orientation) is defined by the rotation system $(D; R, L)$, where D is the partition set F^+ in the bipartition of F, $L = \tau_0\tau_2$, and $R = \tau_1\tau_2$.

F55: The antipodal double cover of a nonorientable map defined by a crystallization $(F; \tau_0, \tau_1, \tau_2)$ is, up to the choice of an orientation, the map with rotation system $(F; R, L)$, where $L = \tau_0\tau_2$ and $R = \tau_1\tau_2$.

F56: The Euler characteristic of the supporting surface of a map given by a crystallization $(F; \tau_0, \tau_1, \tau_2)$ is $\chi = \alpha_0 - \alpha_1 + \alpha_2$, where α_0, α_1, and α_2 is the number of orbits of the groups $\langle \tau_1, \tau_2 \rangle$, $\langle \tau_0, \tau_2 \rangle$, and $\langle \tau_0, \tau_1 \rangle$, respectively.

7.6.6   Maps and Universal Tessellations

Every map M has a universal cover by a uniform tiling of the sphere, Euclidean plane, or the hyperbolic plane. This fact and its consequences are the subject of this section. Representation of a map as a quotient of its universal cover is closely related to groups acting as homeomorphisms on surfaces, which we also discuss in this section. Deep connections between maps, universal tessellations, Galois groups, and Grothendieck's *dessins d'enfants* are discussed in [JoSi87, JoSi96, LaZv04]. The tessellations are regarded as infinite maps, even though, by our definition, a map is a *finite* cell complex.

DEFINITIONS

D41:   The **monodromy group of an oriented map** M with rotation system $(D; R, L)$ is the subgroup $Mon^+(M) = \langle R, L \rangle$ of the symmetric group $Sym(D)$ of all permutations of D.

D42:   A **homomorphism between oriented maps** $M_1 = (D_1; R_1, L_1)$ and $M_2 = (D_2; R_2, L_2)$ is a mapping $f: M_1 \rightarrow M_2$ taking $D_1 \rightarrow D_2$ such that $fR_1 = R_2 f$ and $fL_1 = L_2 f$.

D43:   An **automorphism of an oriented map** $M = (D; R, L)$ is a permutation f of D such that $fR = Rf$ and $fL = Lf$. The automorphisms form a group $Aut^+(M)$ under composition, called the **automorphism group** of an oriented map.

D44:   The **monodromy group of a map** M with crystallization $(F; \tau_0, \tau_1, \tau_2)$ is the subgroup $Mon(M) = \langle \tau_0, \tau_1, \tau_2 \rangle$ of the symmetric group $Sym(F)$ of all permutations of F.

D45:   A **homomorphism** $f: M_1 \rightarrow M_2$ **between maps** $M = (F_1; \tau_0, \tau_1, \tau_2)$ and $N = (F; \rho_0, \rho_1, \rho_2)$ is a mapping $F_1 \rightarrow F_2$ such that $f\tau_i = \rho_i f$ for each $i \in \{0, 1, 2\}$.

D46:   An **automorphism of a map** $M = (F; \tau_0, \tau_1, \tau_2)$ is a permutation f of the set F of flags such that $\tau_i f = f\tau_i$ for each $i \in \{0, 1, 2\}$. The automorphisms form a group $Aut(M)$ under composition, called the **automorphism group** of a map.

D47:   The **tessellation** $\{p, q\}$ is the unique tessellation of the sphere or plane into regular p-gons, q incident at each vertex. This is a tiling of the sphere if $1/p + 1/q > 1/2$, of the Euclidean plane if $1/p + 1/q = 1/2$, or of the hyperbolic plane if $1/p + 1/q < 1/2$.

D48:   The **triangle group** $\Delta^+(p, q, 2)$ is the group with presentation

$$\langle x, y \mid x^p = y^2 = (xy)^q = 1 \rangle. \tag{7.6.5}$$

The group $\Delta^+ = \Delta^+(\infty, \infty, 2)$ is the **universal triangle group**.

D49:   The **extended triangle group** $\Delta(p, q, 2)$ is the group with presentation by three generators x_0, x_1, x_2 and the relations

$$\langle x_0, x_1, x_2 \mid x_0^2 = x_1^2 = x_2^2 = (x_0 x_1)^p = (x_1 x_2)^q = (x_2 x_0)^2 = 1 \rangle. \tag{7.6.6}$$

The group $\Delta = \Delta(\infty, \infty, 2)$ is the **universal extended triangle group**.

D50:   A **Belyi function** $f: S \rightarrow S_0$ is a meromorphic function from a (closed) Riemann surface S onto the Riemann sphere S_0 with at most three singular values forming a subset of $\{0, 1, \infty\}$.

D51:   A **Shabat polynomial** is a complex polynomial with at most two critical values.

D52: A *dessin d'enfant*, or briefly *dessin*, is a bipartite map with a fixed bipartition (coloring) of its vertices.

D53: The *trivial dessin* $\mathcal{I}$ is a map on the Riemann sphere whose vertices are 0 and 1 and the unique edge is formed by the unit interval $[0,1]$. By definition, 0 is colored black and 1 white. The point ∞ is the center of the single face of $\mathcal{I}$.

D54: The *universal dessin* is a dessin $\mathcal{T}$ on the extended complex upper half plane $\bar{U} = \{z \in \mathbb{C} \mid Im(z) > 0\} \cup \mathbb{Q} \cup \{\infty\}$ whose vertex-set consists of all rational numbers, with a/b is joined to c/d by an edge drawn as a hyperbolic geodesic (half-circle) if and only if $ad - bc = \pm 1$.

EXAMPLES

E13: Both maps M and M^* on the torus shown in Figure 7.6.2 are coverings of the tetrahedral map in Figure 7.6.7. The covering by M is branched at face centers and the covering by M^* is branched at vertices. Both are 2-fold coverings, that is, each non-singular point of the sphere is covered by two points of the torus.

E14: Figure 7.6.13 shows all the (hyperbolic) mirrors of reflection symmetries of the tessellation $\{6,4\}$ (or $\{4,6\}$). These lines form a *Coxeter complex*, a subdivision of the hyperbolic plane into triangles – flags of the infinite map.

Figure 7.6.13: Reflection symmetries of the hyperbolic tessellation $\{6,4\}$.

E15: Figure 7.6.14 shows three bicolored plane trees whose Belyi functions are Shabat polynomials of the form $P(x) = x^3(x-1)^2(x-a)$, where a is a root of the equation $25a^3 - 12a^2 - 24a - 16 = 0$. The trees form an orbit under the action of the universal Galois group $\Gamma(\bar{\mathbb{Q}}/\mathbb{Q})$. For more details see [LaZv04, Subsection 2.2.2.3].

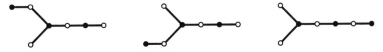

Figure 7.6.14: Conjugate trees over a cubic field.

REMARKS

R30: It is usual to identify the automorphism group of a map M with the automorphism group H of the corresponding crystallization, as done in Definition D46. To be more precise, the topological automorphism group Γ contains a normal infinite subgroup $N \vartriangleleft \Gamma$ fixing each flag of M, for which $H \cong \Gamma/N$. Another equivalent definition of H for maps with valency of each vertex at least 3 is as follows: H is isomorphic to the group of all graph automorphisms that extend to map automorphisms. In what follows the group H will be denoted $Aut(M)$.

R31: Given a map on an orientable surface, we identify its group of orientation preserving automorphisms Γ^+ with the automorphism group H^+ of the corresponding rotation system. Equivalently, H^+ consists of the automorphisms of the underlying graph which extend to orientation preserving map automorphisms. In what follows the group H^+ will be denoted $Aut^+(M)$.

R32: The subgroup $\langle x_0 x_1, x_1 x_2 \rangle \leq \Delta(p, q, 2)$ of index two is isomorphic to $\Delta^+(p, q, 2)$ for all parameters p and q.

R33: The supporting surface of a homomorphic image of an oriented map is a closed orientable surface. Let $f \colon S_g \to S_h$ be an orientation-preserving covering between orientable surfaces. If M is a (topological) map on S_g, then $f(M)$ is a map without semiedges if and only if $f(x) \neq f(Lx)$ for every dart of M. It is sometimes convenient to consider maps whose underlying graphs have semiedges; see [MaNeSk02, NeSk97]. In terms of rotation systems this means allowing the dart-reversing involution L to have fixed points.

R34: The supporting surface of a homomorphic image of a map may have a non-empty boundary. For instance, a quotient of a cycle C imbedded in the sphere by the group of order two generated by a reflection fixing the cycle point-wise is an imbedding of C into the disk. It is sometimes useful to have the set of maps closed under homomorphic images. In terms of crystallizations this means allowing the involutions τ_0, τ_1, τ_2, and $\tau_0 \tau_2$ to have fixed points; for more details see [BrSi85].

FACTS

F57: By connectivity, the monodromy group of a map is transitive on the set of flags. Similarly, the oriented monodromy group is transitive on the set of darts of an orientable map.

F58: By Fact F57, every homomorphism between maps, or between oriented maps, is surjective.

F59: The automorphism group $Aut(M)$ of a map M is the centralizer of the monodromy group $Mon(M)$ of M. Similarly, the group of orientation preserving automorphisms $Aut^+(M)$ of an oriented map M is the centralizer of the oriented monodromy group $Mon^+(M)$ of M.

F60: By Fact F59, the group $Aut(M)$ acts freely on the flags of M, and if M is orientable, then $Aut^+(M)$ acts freely on the darts of M. It follows, in turn, that $|Aut(M)| \leq |F| = 4e$, where e is the number of edges of M. Similarly, if M is an oriented map, then $|Aut^+(M)| \leq |D| = 2e$.

F61: The automorphism group of the tessellation $\{p,q\}$ is isomorphic to the extended triangle group $\Delta(p,q,2)$. The generators x_0, x_1, and x_2 (and their conjugates) correspond to reflections in the three sides of a flag, as described in Example E14; the products $x_1 x_2$, $x_0 x_2$, and $x_0 x_1$ (and their conjugates) correspond to rotations about vertices, midpoints of edges, and face centers, respectively.

F62: Every map M has a covering by a tessellation $\{p,q\}$ for some p and q. In other words, every map M is a quotient of a tessellation $\{p,q\}$ by a subgroup H_M of the group $\Delta(p,q,2)$. The subgroup H_M is determined by M up to conjugation.

F63: [Vi83a] The automorphism group $Aut(M)$ of any map M is isomorphic to the quotient $N_\Delta(H_M)/H_M$, where N_Δ denotes the normalizer and where H_M is a subgroup of Δ of index $4e$ from Fact F62. A similar statement holds for $Aut^+(M)$ and Δ^+.

F64: Conjugacy classes of torsion-free subgroups of finite index $4e$ of Δ are in a one-to-one correspondence with the isomorphism classes of maps with e edges. Furthermore, conjugacy classes of torsion-free subgroups of finite index $4e$ of $\Delta(p,q,2)$ are in a one-to-one correspondence with the isomorphism classes of maps of type $\{p,q\}$ with e edges [JoSi78, BrSi85].

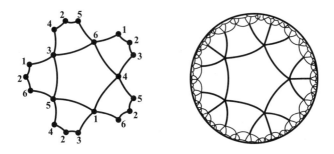

Figure 7.6.15: The regular self-dual map $\{5,5\}_3$ and its universal cover $\{5,5\}$.

F65: Every map M of type $\{p,q\}$ admits a smooth covering by the tessellation $\{p,q\}$. For example, the map on N_5 of type $\{5,5\}$ in Figure 7.6.15 is covered by the tessellation $\{5,5\}$ of the hyperbolic plane. (The map is obtained by identifying like labeled edges in the figure.)

F66: [Bi72] The automorphism group of an orientable map of genus $g > 1$ can be faithfully represented in the group of $2g \times 2g$ symplectic matrices with integral entries. From this fact it can be proved, for example, that if a prime p divides $|Aut(M)|$, then the genus of the map M is either 1, 2, or at least $(p-1)/2$.

F67: [Hu1892] *Hurwitz bound:* If a finite group Γ acts on a surface S of Euler characteristic $\chi(S) < 0$ as a group of its automorphisms, then $|\Gamma| \leq -84\chi(S)$. In contrast, surfaces of Euler characteristic $\chi \geq 0$ admit actions of groups of arbitrarily large order. The bound is satisfied with equality if and only if Γ is the automorphism group of a regular map of type $\{3,7\}$ or $\{7,3\}$ (see §7.6.7 for the definition of a regular map). This is a direct consequence of the Hurwitz bound.

F68: The Hurwitz bound is achieved for infinitely many surfaces; on the other hand, there are infinitely many surfaces where the bound is not achieved [Co90, Co10]; see also Fact F87 and Remark R40.

F69: The automorphism group of the universal dessin is the free group of rank two.

F70: Every dessin covers the trivial dessin and is covered by the universal dessin.

F71: If M is a map on an orientable surface, then there is an associated dessin $\mathcal{D}(M)$ whose black vertices are the vertices of M and white vertices are the centers of edges of M.

F72: For every map M on an orientable surface S it is possible to introduce the structure of a Riemann surface on S such that the map coverings $\mathcal{T} \overset{\alpha}{\to} \mathcal{D}(M) \overset{\beta}{\to} \mathcal{I}$ are meromorphic functions with singular values only at vertices and face centers; in particular, β is a Belyi function [LaZv04].

F73: A dessin is tree-like if and only if its Belyi function is a Shabat polynomial [LaZv04, Theorem 2.2.9.].

REMARK

R35: Fact F72 implies that the supporting surface of any map M can be assumed to carry the structure of a Riemann surface such that $Aut(M)$ acts as a group of conformal homeomorphisms [JoSi87, LaZv04]. The edges of G_M are geodesics of equal length with respect to a Riemannian metric of constant curvature (defined everywhere except perhaps at finitely many singular points located at vertices and face centers) and the angles formed by successive edges incident with a vertex are equal.

7.6.7   Highly Symmetrical Maps

Regular maps, those enjoying the greatest symmetry, are analogues of the Platonic solids on surfaces of higher genera. Cayley maps and vertex-transitive maps are less restrictive, but still very interesting classes of maps. Both regular and Cayley maps can be considered as visualizations of discrete actions of groups on surfaces, and therefore naturally appear in the investigation of symmetries of surfaces and related objects.

DEFINITIONS

D55: An oriented map M with rotation system $(D; R, L)$ is **orientably regular** if $Aut^+(M)$ acts transitively on D.

D56: An oriented map $M = (D; R, L)$ isomorphic with its mirror image $M^{-1} = (D; R^{-1}, L)$ is called a **reflexible map**; otherwise M is a **chiral map**.

D57: A map M with crystallization $(F; \tau_0, \tau_1, \tau_2)$ is a **regular map** if $Aut(M)$ acts transitively on the set of flags F.

D58: Let Γ be a finite group with generating set X such that $1 \notin X$ and $X = X^{-1}$, and let ρ be a cyclic permutation of X. A **Cayley map** $CM(\Gamma, X, \rho)$ for a group Γ is an oriented map $(D; R, L)$ with $D = \Gamma \times X$, and with $R(g, x) = (g, \rho(x))$ and $L(g, x) = (gx, x^{-1})$ for every dart $(g, x) \in D$.

D59: The **fiber transformation group** $FT(f)$ of a map homomorphism (covering) $f \colon M \to N$ is the group formed by all map automorphisms $g \in Aut(M)$ such that $fg = f$. A homomorphism (covering) $f \colon M \to N$ is a **regular covering** if $FT(f)$ acts transitively on the fiber $f^{-1}(y)$ for some flag y of N.

D60: A **skew-morphism** f of a group H is a bijection $H \to H$ satisfying the following conditions:
1. $f(1) = 1$,
2. $f(xy) = f(x)f^{\pi(x)}(y)$,

where $\pi \colon H \to \mathbb{Z}$ is an integer-valued function and $f^{\pi(x)}$ is the $\pi(x)$-th power of f in the symmetric group $Sym(H)$.

EXAMPLES

E16: Every orientably regular map on the sphere is regular. The spherical regular maps are precisely the 2-skeletons of the five **Platonic solids** (see Figure 7.6.7), which coincide with the tessellations $\{3, 3\}$, $\{3, 4\}$, $\{4, 3\}$, $\{3, 5\}$, $\{5, 3\}$, and two infinite families of non-polyhedral maps $\{p, 2\}$ and $\{2, p\}$, with $p > 0$, formed by imbedded cycles and their duals.

E17: Since every map on the projective plane has a smooth 2-fold covering by a map on the sphere (Remark R8), it follows from Example E16 that there are four regular maps on the projective plane of types $\{3, 4\}, \{4, 3\}, \{3, 5\}, \{5, 3\}$ and infinite families of types $\{p, 2\}$ and $\{2, p\}$ with $p > 0$.

E18: Regular and orientably regular maps on the torus are finite quotients of the Euclidean tessellations $\{4, 4\}$, $\{6, 3\}$ and $\{3, 6\}$. Their classification appears in [CoMo84]. There are infinitely many maps in each of these types. These include the maps $\{3, 6\}_4$ and $\{6, 3\}_4$ depicted in Figure 7.6.2 and defined in Fact F88.

E19: The **Kepler–Poinsot regular star polyhedra**, shown in Figure 7.6.16, are self-intersecting realizations of regular maps. These maps are $\{5, 5 \,|\, 3\}$ (12 pentagons on a surface of genus 4 — great dodecahedron and small stellated dodecahedron), $\{5, 3\}_{10}$ (12 pentagons on the torus — great stellated dodecahedron), and $\{3, 5\}_{10}$ (20 triangles on the torus — great icosahedron).

Figure 7.6.16: Star polyhedra.

E20: [ScWi85, ScWi86] From the history of automorphic functions come two regular maps of genus 3, the 1879 **Klein map** $\{7,3\}_8$ composed of 24 heptagons with automorphism group $PGL(2,7)$, and the 1880 **Dyck map** $\{8,3\}_6$ composed of 12 octagons (shown in dual form in Figure 7.6.17). The **Coxeter regular skew polyhedra** in E^4 also provide examples of regular maps; they are $\{4,6\,|\,3\}$, $\{6,4\,|\,3\}$, $\{4,8\,|\,3\}$, and $\{8,4\,|\,3\}$. The Klein, Dyck, and Coxeter maps all have realizations in E^3.

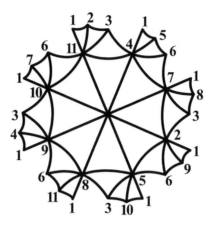

Figure 7.6.17: The dual of Dyck's map $\{8,3\}_6$.

FACTS

F74: For an orientably regular map M with e edges $|Aut^+(M)| \leq |2e| \leq |Mon^+(M)|$. A map M is orientably regular if and only if $|Aut^+(M)| = |2e| = |Mon^+(M)|$. In an orientably regular map M, both the action of $Aut^+(M)$ and the action of $Mon^+(M)$ are regular.

F75: For a map M with e edges, $|Aut(M)| \leq |4e| \leq |Mon(M)|$ and M is regular if and only if $|Aut(M)| = |4e| = |Mon(M)|$. In particular, in a regular map M both the action of $Aut(M)$ and the action of $Mon(M)$ are regular on the flags of M.

F76: An orientably regular map is regular if and only if it is reflexible.

F77: [Vi83b, Wi78] A nonorientable map is regular if and only if its antipodal double cover is regular.

F78: Regular maps with e edges are in a one-to-one correspondence with normal torsion-free subgroups of Δ of finite index $4e$. Regular maps of type $\{p,q\}$ with e edges are in a one-to-one correspondence with normal torsion-free subgroups of $\Delta(p,q,2)$ of finite index $4e$.

F79: Orientably regular maps with e edges are in a one-to-one correspondence with normal torsion-free subgroups of Δ^+ of finite index $2e$. Orientably regular maps of type $\{p,q\}$ with e edges are in a one-to-one correspondence with normal torsion-free subgroups of $\Delta^+(p,q,2)$ of finite index $2e$.

F80: If M is a regular map, then $Aut(M) \approx \Delta/N$ for some normal subgroup $N \trianglelefteq \Delta$. If M is a regular map of type $\{p,q\}$, then $Aut(M) \approx \Delta(p,q,2)/N$ for some normal subgroup $N \trianglelefteq \Delta(p,q,2)$. Similar statements hold for the class of orientably regular maps and subgroups of Δ^+ and $\Delta^+(p,q,2)$.

F81: By Fact F80, the automorphism group of every regular map can be generated by three involutions, two of which commute. The automorphism group of every orientably regular map can be generated by two elements, one of which is an involution.

F82: On each orientable surface S_g there is a regular map. It is provided, e.g., by imbedding the bouquet of g circles into S_g with a single face. However, if the underlying graphs are required to be simple, or maps to be chiral, then there are infinitely many gaps in the genus spectra in both cases [CoSiTu10].

F83: Not every nonorientable surface supports a regular map; for example, there are no regular maps on the nonorientable surfaces of genus 2 and 3. There are infinitely many nonorientable surfaces supporting no regular map [BrNeSi05, CoPoSi10].

F84: For any surface with Euler characteristic $\chi < 0$, there are at most finitely many regular maps. This follows from the Hurwitz formula stated in Fact F67 and from Fact F60.

F85: For surfaces with Euler characteristic $\chi \geq -600$ all regular and orientably regular maps have been classified [Co12a, Co12b, CoDo01].

F86: A classification of regular maps on surfaces with Euler characteristic $\chi(S) = -p$, $-p^2$, and $-3p$, and of orientably regular maps with Euler characteristic $-2p$, where p is a prime, was accomplished in [BrNeSi05, CoPoSi10, CoSiTu10, CoNeSi12].

F87: There are infinitely many regular maps of type $\{p,q\}$ for each pair (p,q) such that $1/p + 1/q \leq 1/2$. In fact, they may be chosen to have arbitrarily large face-width [Vi83b, NeSk01].

F88: Two special cases have received particular attention, the regular maps $\{p,q\}_r$ where the single relation $(\tau_0\tau_1\tau_2)^r$ has been added and the regular maps $\{p,q\,|\,m\}$ where the single relation $(\tau_0\tau_1\tau_2\tau_1)^m$ has been added to the presentation of $\Delta(p,q,2)$. Coxeter and Moser [CoMo84] have provided partial tables of parameters p,q,r and p,q,m for which a *finite* regular map with those parameters exists. Figure 7.6.15 shows the regular map $\{5,5\}_3$. It is now known exactly which of the maps $\{p,q\}_r$ are finite and which are infinite, except for one case, $(p,q,r) = (3,7,19)$; see [EdJu08, HaHo10]. The question of which maps $\{p,q\,|\,m\}$ are finite was fully answered in [EdTh97].

F89: A connected graph is the underlying graph G of some orientably regular map if and only if $Aut(G)$ contains a subgroup acting regularly on the dart-set such that the stabilizer of each vertex is cyclic [GaNeSiSk99].

F90: If a connected graph G underlies a regular map, then the map automorphism group $Aut(G)$ contains a subgroup Γ acting transitively on the dart-set of G such that the vertex-stabilizer is dihedral and the edge-stabilizer is isomorphic to the Klein four-group $\mathbb{Z}_2 \times \mathbb{Z}_2$. The existence of a subgroup $\Gamma \leq Aut(G)$ satisfying the above conditions is sufficient for G to be the underlying graph of some regular map [GaNeSiSk99].

F91: The complete graph K_n underlies an orientably regular map if and only if n is a prime power. It underlies a regular map if and only if $n = 2, 3, 4,$ or 6 [Ja83], [Wi89]. For every $n = p^k$ there are exactly $\phi(p^k - 1)/k$ non-isomorphic orientably regular imbeddings of K_n, where ϕ is the Euler function [Bi71, JaJo85].

F92: Regular and orientably regular imbeddings were classified for complete bipartite graphs [Jo10], n-dimensional cubes [CaCoDuKwNeWi11], and for some other families of graphs. In particular, for each n there exists at least one (orientably) regular imbedding of $K_{n,n}$ and Q_n, and for $n = p$ a prime number, there exists exactly one regular imbedding of $K_{p,p}$.

F93: Every Cayley map of a group Γ is vertex-transitive, with Γ acting as a group of map automorphisms by left multiplication. Numerous other useful results concerning Cayley maps can be found in [RiSiJaTuWa05].

F94: [Tu83] If a group Γ acts on an orientable surface S as a group of self-homeomorphisms, then some Cayley graph G of Γ imbeds in S, and the natural action of Γ on G (by left multiplication) extends to an action of Γ on S. In other words, there is a Cayley map $CM(\Gamma, X, \rho)$ with the supporting surface S.

F95: Biggs [BiWh79, Theorem 5.3.71] proved that if a group Γ has an automorphism ψ whose restriction on a set X of generators of Γ is ρ, then the Cayley map $CM(\Gamma, X, \rho)$ is orientably regular. In such a case the identity $\rho(x^{-1}) = (\rho(x))^{-1}$ holds. Following [SkSi92], a Cayley map satisfying this condition is called a **balanced Cayley map**. Škoviera and Širáň [SkSi92] proved that a balanced Cayley map $CM(\Gamma, X, \rho)$ is regular if and only if ρ extends to a group automorphism of Γ.

F96: Jajcay and Širáň [JaSi02] proved that a Cayley map $CM(\Gamma, X, \rho)$ is orientably regular if and only if there is a skew-morphism ψ of Γ whose restriction to X is ρ.

F97: Conder and Tucker [CoTu12] classified all orientably regular Cayley maps arising from cyclic groups.

F98: [Th91, Bab91] For each $g \geq 3$, there exist only finitely many simple vertex-transitive graphs of orientable genus g, while there are infinitely many of genus 0, 1, and 2.

F99: The double torus S_2 has the interesting property that there are only finitely many finite groups that act on S_2 as self-homeomorphisms groups, but there are infinitely many vertex-transitive graphs (in fact, Cayley graphs) with genus 2.

EXAMPLES

E21: Figure 7.6.18 represents a chiral orientably regular imbedding of the complete graph K_5 into the torus. This example belongs to a family of balanced regular Cayley maps whose underlying graph is the complete graph K_q, where q is a prime power. Let $F = F(q)$ be the additive group of the Galois field of order q, and let F^* denote the multiplicative group of the field (which is cyclic). Let μ be a primitive element of F, and let ρ be the cyclic permutation of F^* defined by the multiplication by μ. Then $CM(F, F^*, \rho)$ is a balanced orientably regular Cayley map with underlying graph the complete graph K_q.

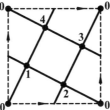

Figure 7.6.18: The complete graph K_5 imbedded as a chiral regular Cayley map of $\mathbb{Z}_5$ on the torus.

E22: Coxeter and others noticed that (orientably) regular maps frequently occur as coverings of smaller (orientably) regular maps on other surfaces. For example, the regular toroidal maps $\{3,6\}_4$ and $\{6,3\}_4$ in Figure 7.6.2 are 2-fold coverings of the tetrahedral map $\{3,3\}$ on the sphere. Constructions of families of (orientably) regular maps using coverings appear in [JoSu00], [Si01], [Vi84], [MaNeSk02], [NeSk97a], and [Wi78], among others.

E23: There are various group theoretical constructions of regular maps. In fact, every finite 2-generator group $\Gamma = \langle x, y \rangle$, where $y^2 = 1$, gives rise to an orientably regular map with $Aut^+(M) \cong \Gamma$. Such a map can be described by the rotation system $(D; R, L)$ where $D = \Gamma$, $R(g) = xg$, and $L(g) = yg$ for each dart $g \in \Gamma$ (compare with Facts F79 and F80).

E24: For every finite group $\Gamma = \langle x_0, x_1, x_2 \mid x_0^2 = x_1^2 = x_2^2 = (x_0 x_2)^2 = 1, \dots \rangle$ there is a regular map M with $Aut(M) \cong \Gamma$. It is defined by the crystallization $(\Gamma; \tau_0, \tau_1, \tau_2)$ where $\tau_i(g) = x_i g$ for each flag $g \in \Gamma$ (compare with Facts F78 and F80).

E25: The vertex-transitive maps on the sphere with simple underlying graphs, classified by [FlIm79], consist of the regular spherical maps and the boundary complexes of the Archimedean solids (semi-regular polyhedra), of the prisms, and the antiprisms. [Bab91] gave a classification of the vertex-transitive maps on the Klein bottle with simple underlying graphs.

REMARKS

R36: The automorphism group of a regular or orientably regular map acts transitively on the set of vertices, on the set of edges, and on the set of faces.

R37: If a map M is given in terms of a graph encoding $\mathcal{G} = (H, \phi)$, then the flags of M are in a bijective correspondence with the vertices of H. Therefore M is regular if and only if H is vertex-transitive. In this case, H is a cubic Cayley graph of $Aut(M)$ with generator set $\{\tau_0, \tau_1, \tau_2\}$ where each τ_i is an involution corresponding to the perfect matching of H consisting of the edges with color i.

R38: In 1994 Malle et al. [MaSaWe94] proved that every nonabelian finite simple group can be generated by two elements, one of which has order two. A stronger result due to Stein [St98] implies that the element of order two can be chosen arbitrarily. From Example E23 it follows that every nonabelian finite simple group is the automorphism group of some orientably regular map. For some classes of finite simple groups, all orientably regular maps with automorphism group in the class have been classified, for example, for the projective linear groups $PSL(2, q)$ [Sah69], for Ree groups [Jo94], or for the Suzuki groups [JoSi93].

R39: There are infinitely many nonabelian finite simple groups that do not occur as automorphism groups of regular maps. These are A_6, A_7, A_8, $S_4(3) = U_4(2)$, M_{11}, M_{22}, M_{23}, McL, further $PSL(3, q)$ and $U_3(q)$ for all prime powers q, and $PSL(4, q)$ and $U_4(q)$ for even prime powers q.

R40: Orientably regular maps of type $\{3, 7\}$ and $\{7, 3\}$ are known as ***Hurwitz maps*** because of their relationship to the Hurwitz bound (Facts F67 and F68); their automorphism groups are called ***Hurwitz groups***. Many finite simple groups are known to be Hurwitz groups while others are known not to be Hurwitz groups. For example, every sufficiently large alternating group is a Hurwitz group. Of the 26 sporadic groups 12 are Hurwitz groups; for details, see [Co90, Co10].

R41: Every group automorphism of H is a skew-morphism with constant power function $\pi(x) = 1$ for every $x \in H$. Classifying all skew-morphisms of a given group is a difficult problem in general, and is open even for cyclic groups.

R42: Let M be a Cayley orientably regular map $CM(\Gamma, X, \rho)$. Then $Aut^+(M)$ is isomorphic to a product $\Gamma\mathbb{Z}_k$ of Γ with the cyclic group $\mathbb{Z}_k$ of order k where $k = |X|$ is the valency of M. Every generator of $\mathbb{Z}_k$ gives rise to a skew-morphism of Γ. For more information on skew-morphisms of groups see [CoJaTu07] and [KoNe11].

7.6.8   Enumeration of Maps

W. T. Tutte [Tut63] pioneered map enumeration in the 1960's. He developed a machinery for deriving generating functions for several classes of rooted spherical planar maps, and as an application, derived a closed formula for the number of rooted spherical maps with given number of edges. At present, there are hundreds of results about map enumeration. In the following text we highlight results about enumeration of maps on a fixed surface by the number of edges, thus extending the classical result of Tutte for the sphere. These results can also be viewed as results about enumeration of subgroups of a given index in the universal triangle groups Δ^+ and Δ, which shows their relation to the classical results of M. Hall [Ha49] and others who enumerated subgroups of a given index in free groups. Connections between map enumeration, matrix integrals, and 2-dimensional quantum gravity are explained in [Zv97, BoFrGu02].

DEFINITIONS

D61: A ***rooted map*** is a map in which a flag has been distinguished. A ***rooted oriented map*** is an oriented map in which a dart has been distinguished. There is a one-to-one correspondence between rooted maps on orientable surfaces and rooted oriented maps; in what follows we therefore use the term rooted maps.

EXAMPLES

E26: For the sphere, the 2-connected rooted maps with four edges are shown in the first row of Figure 7.6.19. The first four of these comprise all 2-connected rooted maps with three vertices and three faces. The roots are in boldface.

E27: In the second row of Figure 7.6.19 there are the rooted near triangulations with four inner faces and a root face with two edges.

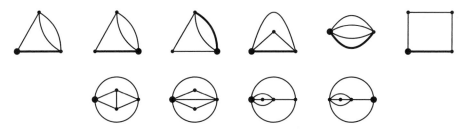

Figure 7.6.19: Counting maps on the sphere.

FACTS

F100: [Tut63] The number of rooted maps on the sphere with $n \geq 0$ edges is

$$m_0(n) = \frac{2 \cdot 3^n (2n)!}{n!(n+2)!}.$$

F101: [Tut63] The number of 2-connected rooted maps on the sphere with $n \geq 1$ edges is

$$\frac{2(3n-3)!}{n!(2n-1)!}.$$

F102: [N. Wormald] (see [GoJa83]) The number of 2-edge-connected rooted maps on the sphere with $n \geq 0$ edges is

$$\frac{2(4n+1)!}{(n+1)!(3n+2)!}.$$

F103: [BrTu64] The number of 2-connected rooted maps on the sphere with $n \geq 1$ vertices and $k \geq 2$ faces is

$$\frac{(2n+k-5)!(2k+n-5)!}{(n-1)!(k-1)!(2n-3)!(2k-3)!}.$$

F104: [Tut63] A generating function for the number of 3-connected rooted planar maps with $n \geq 3$ edges was derived by Tutte.

F105: A generating function for the number of rooted maps on a surface of genus $g > 1$ and with a given number of edges is known up to the coefficients of a polynomial of degree at most $6g - 6$ [ArGi99, WaGi13]. In particular, the formula is known for $g \leq 11$; see [Ar87] for $g = 1$, [BeCa91] for $g = 2$ and 3, [MeGi11] for $g = 4$, [WaGi13] for $g = 5$ and 6, and [WaGiMe12] for $7 \leq g \leq 11$.

F106:   A generating function that counts rooted maps of genus g by the number v of vertices and by the number f of faces is the following formal power series in two variables u and w:

$$M_g(w,u) = \sum_{v,f\geq 1} m_g(v,f)w^v u^f.$$

In [Ar85] the function $M_0(w,u)$ is determined by the following system of equations:

$$M_0(w,u) = pq(1-2p-2q),$$
$$w = p(1-p-2q),$$
$$u = q(1-2p-q).$$

F107:   [Li81, Li85] The number of isomorphism classes of oriented spherical maps with n edges is given by

$$L_0(n) = \frac{1}{2n}\left(m_0(n) + \sum_{\substack{d\mid n \\ d<n}} \phi(n/d)\binom{d+2}{2}m_0(d)\right) + \left(\frac{n-1}{4} + 2\left\lceil\frac{n}{2}\right\rceil - n\right)m_0\left(\left\lceil\frac{n}{2}-1\right\rceil\right),$$

where $m_0(n)$ is the number of rooted spherical maps and $\phi(d)$ is the Euler function.

F108:   A formula for the number of isomorphism classes of oriented maps of genus g with a given number of edges is derived in [MeNe06]. The formula is applicable provided that the numbers $m_\gamma(n)$ of rooted maps with n edges of genus $\gamma \leq g$ are known.

F109:   Rooted maps and isomorphism classes of maps and of oriented maps with a given number of edges regardless of genus are enumerated in [BrMeNe10].

F110:   In [BeCa86] it is proved that $m_g(n) \sim t_g n^{5(g-1)/2}12^n$, where t_g is a constant computable via non-linear recursions. The first two values are $t_0 = \frac{2}{\sqrt{\pi}}$ and $t_1 = \frac{1}{24}$. In [Ga93] an asymptotic pattern for the number of rooted maps with n edges is determined for numerous classes of maps, extending the previously known results.

F111:   In [RiWo95] it is proved that between n-edged maps on a fixed surface those admitting a non-trivial symmetry form an exponentially small part. It follows that almost all maps on a fixed surface are rigid.

F112:   In [DrNe11] formulas for maps regardless of genus are analyzed. In particular, it is proved that $\log A(n) \sim \frac{1}{2}\log U(n) \sim (n/2)\log n$, where $A(n)$ and $U(n)$ is, respectively, the number of reflexible and the number of all maps with n edges. It follows that for sufficiently large n, the number of reflexible maps is about the square root of the number all n-edged maps.

REMARKS

R43:   By Whitney's theorem (see Fact F8) enumeration of planar 3-connected graphs and that of planar 3-connected maps are equivalent problems.

R44:   The numbers of rooted oriented maps, and of rooted maps, with n edges are the same as the numbers of torsion-free subgroups of the group universal triangle group Δ^+ of index $2n$, and of the universal extended triangle group Δ of index $4n$, respectively.

R45: The numbers of isomorphisms classes of rooted oriented maps, and of maps, with n edges are the same as the numbers of conjugacy classes of torsion-free subgroups of the group Δ^+ of index $2n$, and of the group Δ of index $4n$, respectively.

R46: Similar statements as in Remarks 44 and 45 hold when the genus of the underlying surface is fixed.

R47: Rooted maps were also enumerated with respect to the number of vertices and faces; see for instance [WaGi13, WaGiMe12]. In particular, the generating function for maps by number of edges is $M_g(z, z)$, where $M_g(w, u)$ is the generating function for maps by number of vertices and faces from Fact F106.

R48: The "Quotient Map Method" elaborated for deriving the formula for the number of isomorphism classes of oriented maps (see Fact F107) was applied to deriving similar formulae counting nonisomorphic planar maps of several other natural classes including 2-connected maps [LiWa83], Eulerian maps, and loopless maps [LiWa04]. For more information see [Li04] and references included therein.

R49: With every map one can associate a triangulation, or a quadrangulation, for example by employing the barycentric subdivision, or the dual of the medial. Enumeration of triangulations and quadrangulations is therefore of particular interest; for a sample of results, see, for instance, [GaWo02, ViPe10].

R50: Asymptotic behavior of maps on a fixed surface was investigated in [BeCa86, BeCaRi93, Ga93, RiWo95, Li99].

PROBLEM

RP3: Derive a generating function for the number of isomorphism classes of spherical maps with a given number of edges (a map and its mirror image are regarded as isomorphic).

References

[Al81] M. O. Albertson, Open problem 2, In *The Theory and Applications of Graphs*, G. Chartrand et al., editors, Wiley 1981, p. 609.

[AlHu79] M. O. Albertson and J. P. Hutchinson, The three excluded cases of Dirac's map color theorem, *Ann. N. Y. Acad. Sci* **319** (1979), 7–17.

[AlSeTh94] N. Alon, P. Seymour, and R. Thomas, Planar separators, *SIAM J. Discrete Math.* **7** (1994), 184–193.

[Al71] A. Altshuler, Polyhedral realization in R^3 of triangulaltions of the torus and 2-manifolds in cyclic 4-polytopes, *Discrete Math.* **1** (1971), 211–238.

[ApHa76] K. Appel and W. Haken, Every planar map is four colorable, *Bull. Amer. Math. Soc.* **82** (1976), 711–712.

[Ar85] D. Arqués, Une relation fonctionnelle nouvelle sur les cartes planaires pointées, *J. Combin. Theory Ser. B* **39** (1985), 27–42.

[Ar87] D. Arqués, Relations fonctionnelles et dénombrement des cartes pointées sur le tore, *J. Combin. Theory Ser. B* **43** (1987), 253–274.

[ArGi99] D. Arqués and A. Giorgetti, Énumération des cartes pointées de genre quelconque en fonction des nombres de sommets et de faces, *J. Combin. Theory Ser. B* **77** (1999), 1–24.

[Bab91] L. Babai, Vertex-transitive graphs and vertex-transitive maps, *J. Graph Theory* **15** (1991), 587–627.

[Bar88] D. W. Barnette, Decomposition theorems for the torus, projective plane and Klein bottle, *Discrete Math.* **70** (1988), 1–16.

[BaEd89] D. Barnette and A. L. Edelson, All 2-manifolds have finitely many minimal triangulations, *Israel J. Math.* **67** (1989), 123–128.

[BaGrHo91] D. Barnette, P. Gritzmann, and R. Höhne, On valences of polyhedra, *J. Combin. Theory Ser. A* **58** (1991), 279–300.

[BeiWi09] L. W. Beineke and R. J. Wilson (editors), *Topics in Topological Graph Theory*, Cambridge Univ. Press, Cambridge, 2009.

[BeCa86] E. A. Bender and E. R. Canfield, The asymptotic number of rooted maps on a surface, *J. Comb. Theory Ser. A* **43** (1986), 244–257.

[BeCa91] E. A. Bender and E. R. Canfield, The number of rooted maps on an orientable surface, *J. Comb. Theory Ser. B* **53** (1991), 293–299.

[BeCaRi93] E. A. Bender, E. R. Canfield, and L. B. Richmond, The asymptotic number of rooted maps on a surface. II: Enumeration by vertices and faces, *J. Combin. Theory Ser. A* **63** (1993), 318–329.

[Bi71] N. Biggs, Automorphisms of embedded graphs, *J. Combin. Theory* **11** (1971), 132–138.

[Bi72] N. Biggs, The symplectic representation of map automorphisms, *Bull. London Math. Soc.* **4** (1972), 303–306.

[BiWh79] N. Biggs and A. T. White, *Permutation Groups and Combinatorial Structures*, London Math. Sot. Lect. Notes 33, Cambridge Univ. Press. Cambridge, 1979.

[BoLi95] C. P. Bonnington and C. H. C. Little, *The Foundations of Topological Graph Theory*, Springer-Verlag, New York, 1995.

[BoFrGu02] J. Bouttier, P. Di Francesco, and E. Guitter, Census of planar maps: From the one-matrix model solution to a combinatorial proof, *Nucl. Phys., B* **645** (2002), 477-499.

[BrNeSi05] A. Breda, R. Nedela, and J. Širáň, Classification of regular maps of negative prime Euler characteristic, *Trans. Amer. Math. Soc.* **357** (2005), 4175–4190.

[BrMeNe10] A. Breda, A. Mednykh, and R. Nedela, Enumeration of maps regardless of genus. Geometric approach, *Discrete Math.* **310** (2010), 1184–1203.

[BrSc95] U. Brehm and G. Schild, Realizability of the torus and the projective plane in R^4, *Israel J. Math.* **91** (1995), 249–251.

[BrSh97] U. Brehm and E. Schulte, Polyhedral maps, in *Handbook of Discrete and Computational Geometry*, 345–358, CRC Press, Boca Raton, FL, 1997.

[BrWi93] U. Brehm and J. M. Wills, Polyhedral manifolds, *Handbook of Convex Geometry*, Vol. A, B, 535–554, North-Holland, Amsterdam, 1993.

[BrTu64] W. G. Brown and W. T. Tutte, On the enumeration of rooted non-separable planar maps, *Canad. J. Math.* **16** (1964), 572–577.

[BrMoRi96] R. Brunet, B. Mohar, and R. B. Richter, Separating and nonseparating disjoint homotopic cycles in graph embeddings, *J. Combin. Theory Ser. B* **66** (1996), 201–231.

[BrSi85] R. P. Bryant and D. Singerman, Foundations of the theory of maps with boundary. *Quart. J. Math. Oxford Ser. 2* **36** (1985), 17–41.

[CaCoDuKwNeWi11] D. Catalano, M. Conder, S. F. Du, Y. S. Kwon, R. Nedela, and S. Wilson, Classification of regular embeddings of n-dimensional cubes, *J. Algebraic Combin.* **33** (2011), 215–238.

[Co90] M. D. E. Conder, Hurwitz groups: a brief survey, *Bull. Amer. Math. Soc.* **23** (1990), 359–370.

[Co10] M. Conder, An update on Hurwitz groups, *Groups, Complexity, and Cryptology* **34** (2010), 35–49.

[Co12a] M. Conder, Lists of regular maps, hypermaps and polytopes, trivalent symmetric graphs, and surface actions,
http://www.math.auckland.ac.nz/~conder/OrientableRegularMaps301.txt.

[Co12b] M. Conder, Lists of regular maps, hypermaps and polytopes, trivalent symmetric graphs, and surface actions,
http://www.math.auckland.ac.nz/~conder/ChiralMaps301.txt.

[CoDo01] M. Conder and P. Dobcsányi, Determination of all regular maps of small genus, *J. Combin. Theory Ser. B* **81** (2001), 224–242.

[CoJaTu07] M. Conder, R. Jajcay, and T. Tucker, Regular Cayley maps for finite abelian groups, *J. Algebr. Comb.* **25** (2007), 259–283.

[CoNeSi12] M. Conder, R. Nedela, and J. Širáň, Classification of regular maps of Euler characteristic $-3p$, *J. Combin. Theory Ser. B* **102** (2012), 967–981.

[CoPoSi10] M. Conder, P. Potočnik, and J. Širáň, Regular maps with almost Sylow-cyclic automorphism groups, and classification of regular maps with Euler characteristic $-p^2$, *J. Algebra* **324** (2010), 2620–2635.

[CoSiTu10] M. Conder, J. Širáň, and T. Tucker, The genera, reflexibility and simplicity of regular maps, *J. Eur. Math. Soc. (JEMS)* **12** (2010), 343–364.

[CoTu12] M. D. E. Conder and T. W. Tucker, Regular Cayley maps for cyclic groups, *Trans. Amer. Math. Soc.*, to appear.

[Co75] R. Cori, Un code pour les graphes planaires et ses applications, *Asterisk* **27** (1975).

[CoMo84] H. S. M. Coxeter and W. O. J. Moser, *Generators and Relations for Discrete Groups*, Fourth Edition, Springer-Verlag, Berlin, 1984.

[Di52] G. A. Dirac, Map colour theorems, *Canad. J. Math.* **4** (1952), 480–490.

[DrNe11] M. Drmota and R. Nedela, Asymptotic enumeration of reversible maps regardless of genus, *Ars Math. Contemp.* **5** (2012), 77–97.

[Eb1891] V. Eberhard, *Sur Morphologie der Polyeder*, Teubner, Leipzig, 1891.

[EdEwKu82] A. L. Edmonds, J. H. Ewing, and R. S. Kulkarni, Regular tessellations of surfaces and $(p, q, 2)$-triangle groups, *Ann. of Math.* **116** (1982), 113–132.

[Ed60] J. R. Edmonds, A combinatorial representation for polyhedral surfaces, *Notices Amer. Math. Soc.* **7** (1960), 646.

[EdJu08] M. Edjvet and A. Juhász, The groups $G^{m,n,p}$. *J. Algebra* **319** (2008), 248–266.

[EdTh97] M. Edjvet and R. M. Thomas, The groups $\{l, m \mid n, k\}$, *J. Pure Appl. Algebra* **114** (1997), 175–208.

[Fe76] M. Ferri, Una rappresentazione delle n-varieta topologiche triangolabili mediante grafi $(n + 1)$-colorati, *Boll. Un. Math. Ital.* **13B** (1976), 250–260.

[FiMo94] S. Fisk and B. Mohar, Coloring graphs without short non-bounding cycles, *J. Combin. Theory Ser. B* **60** (1994), 268–276.

[FlIm79] H. Fleischner and W. Imrich, Transitive planar graphs, *Math. Slovaca* **29** (1979), 97–106.

[FoMa07] P. Fowler and D. E. Manolopoulos, *An Atlas of Fullerenes*, Dover Publications, New York, 2007.

[Fr34] P. Franklin, A six color problem, *J. Mat. Phys.* **16** (1934), 363–369.

[Ga79] C. Gagliardi, A combinatorial characterization of 3-manifold crystallizations, *Boll. Un. Mat. Ital.* **16A** (1979) 441–449.

[Ga93] Z. Gao, A pattern for the asymptotic number of rooted maps on surfaces, *J. Combin. Theory Ser. A* **64** (1993), 246–264.

[GaWo02] Z. Gao and N. C. Wormald, Enumeration of rooted cubic planar maps, *Ann. Comb.* **6** (2002), 313–325.

[GaNeSiSk99] A. Gardiner, R. Nedela, J. Širáň, and M. Škoviera, Characterisation of graphs which underlie regular maps on closed surfaces, *J. London Math. Soc.* (2) **59** (1999), 100–108.

[GaJo79] M. R. Garey and D. S. Johnson, *Computers and Intractability. A Guide to the Theory of NP-completeness*, A Series of Books in the Mathematical Sciences, W. H. Freeman and Co., San Francisco, 1979.

[GiHuTa84] J. R. Gilbert, J. P. Hutchinson, and R. E. Tarjan, A separator theorem for graphs of bounded genus, *J. of Algebra* **5** (1984), 391–407.

[GiTh97] J. Gimbel and C. Thomassen, Coloring graphs with fixed genus and girth, *Trans. Amer. Math. Soc.* **349** (1997), 4555–4564.

[GoJa83] I. P. Goulden and D. M. Jackson, *Combinatorial Enumeration*, John Wiley & Sons, New York, 1983.

[Gri83] P. Gritzmann, The toroidal analogue of Eberhard's theorem, *Mathematika* **30** (1983), 274–290.

[GrTu87] J. L. Gross and T. W. Tucker, *Topological Graph Theory*, John Wiley & Sons, New York, 1987.

[Gr59] H. Grötzsch, Ein Dreifarbensatz für dreikreisfreie Netze auf der Kugel, *Wiss. Z. Martin Luther-Univ. Halle Wittenberg, Mat.-Nat. Reihe* **8** (1959), 109–120.

[Gru67] B. Grünbaum, *Convex Polytopes*, Interscience, 1967.

[Gru70] B. Grünbaum, Polytopes, graphs, and complexes. *Bull. Amer. Math. Soc.* **76** (1970), 1131–1201.

[Ha49] M. Hall, Subgroups of finite index in free groups, *Canadian J. Math.* **1** (1949), 187–190.

[HaHo10] G. Havas and D. Holt, On Coxeter's families of group presentations, *J. Algebra* **324** (2010), 1076–1082.

[He1890] P. J. Heawood, Map-colour theorem, *Quart. J. Pure Appl. Math.* **24** (1890), 332–338.

[Hu1892] A. Hurwitz, Über algebraische Gebilde mit eindeutigen Transformationen in sich, *Math. Ann.* 41 (1892), 403–442.

[Hu95] J. P. Hutchinson, Three-coloring graphs embedded on surfaces with all faces even-sided, *J. Combin. Theory Ser. B* **65** (1995), 139–155.

[HuRiSe02] J. P. Hutchinson, R. B. Richter, and P. D. Seymour, Coloring Eulerian triangulations, *J. Combin. Theory Ser. B* **84** (2002), 225–239.

[Ja68] A. Jacques, Sur le genre d'une paire de substitutions. *C. R. Acad. Sci. Paris* **367** (1968), 625–627.

[JaSi02] R. Jajcay and J. Širáň, Skew-morphisms of regular Cayley maps, *Discrete Math.* **244** (2002), 167–179.

[Ja83] L. D. James, Imbeddings of the complete graph, *Ars Combin.* **16** (1983), 57–72.

[JaJo85] L. D. James and G. A. Jones, Regular orientable imbeddings of complete graphs, *J. Combin. Theory Ser. B* **39** (1985), 353–367.

[Je93a] S. Jendrol', On face vectors and vertex vectors, *Discrete Math.* **118** (1993), 119–144.

[Je93b] S. Jendrol', On face-vectors and vertex-vectors of polyhedral maps on orientable 2-manifolds, *Math. Slovaca* **43** (1993), 393–416.

[JoSi93] G. A. Jones and S. A. Silver, Suzuki groups and surfaces, *J. London Math. Soc.* (2) **48** (1993), 117–125.

[Jo94] G. A. Jones, Ree groups and Riemann surfaces, *J. Algebra* **165** (1994), 41–62.

[Jo10] G. A. Jones, Regular embeddings of complete bipartite graphs: classification and enumeration, *Proc. Lond. Math. Soc. (3)* **101** (2010), 427–453.

[JoSi78] G. A. Jones and D. Singerman, Theory of maps on orientable surfaces, *Proc. Lond. Math. Soc.* **37** (1978), 273–301.

[JoSi87] G. A. Jones and D. Singerman, *Complex Functions: An Algebraic and Geometric Viewpoint*, Cambridge Univ. Press, Cambridge, 1987.

[JoSi96] G. A. Jones and D. Singerman, Belyĭ functions, hypermaps and Galois groups, *Bull. London Math. Soc.* **28** (1996), 561–590.

[JoSu00] G. A. Jones and D. B. Surowski, Regular cyclic coverings of the Platonic maps, *European J. Combin.* **21** (2000), 333–345.

[Ke1879] A. B. Kempe, On the geographical problem of the four colours, *Amer. J. Math.* **2** (1879), 193–200.

[Ko09] M. Kochol, Polyhedral embeddings of snarks in orientable surfaces, *Proc. Amer. Math. Soc.* **137** (2009), 1613–1619.

[KoNe11] I. Kovács and R. Nedela, Decomposition of skew-morphisms of cyclic groups, *Ars Math. Contemp.* **4** (2011), 329–349.

[LaZv04] S. K. Lando and A. K. Zvonkin, *Graphs on Surfaces and Their Applications*, Springer, Berlin, 2004.

[La87] S. Lawrencenko, An infinite set of torus triangulations of connectivity 5 whose graphs are not uniquely embeddable in the torus, *Discrete Math.* **66** (1987), 299–301.

[Li82] S. Lins, Graph-encoded maps, *J. Combin. Theory Ser. B* **32** (1982), 171–181.

[LiTa79] R. J. Lipton and R. E. Tarjan, A separator theorem for planar graphs, *SIAM J. Appl. Math.* **36** (1979), 177–189.

[Li81] V. A. Liskovets, A census of nonisomorphic planar maps, in L. Lovász and V. T. Sós, editors, *Algebraic Methods in Graph Theory*, Colloq. Math. Soc. János Bolyai **25**, 479–494, North-Holland, Amsterdam, New York, 1981.

[Li85] V. A. Liskovets, Enumeration of nonisomorphic planar maps, *Selecta Math. Sovietica* **4** (1985), 303–323.

[Li99] V. A. Liskovets, A pattern of asymptotic vertex valency distributions in planar maps, *J. Combin. Theory Ser. B* **75** (1999), 116–133.

[Li04] V. A. Liskovets, Enumerative formulae for unrooted planar maps: a pattern, *Electron. J. Comb.* **11** (2004), No.1, R88, 14 pages.

[LiWa83] V. A. Liskovets and T. R. S. Walsh, The enumeration of non-isomorphic 2-connected planar maps, *Can. J. Math.* **35** (1983), 417–435.

[LiWa04] V. A. Liskovets and T. R. S. Walsh, Enumeration of Eulerian and unicursal planar maps, *Discrete Math.* **282** (2004), 209–221.

[LiCh12] W. Liu and Y. Chen, Polyhedral embeddings of snarks with arbitrary nonorientable genera, *Electron. J. Combin.* **19** (2012), #P14.

[MaSaWe94] G. Malle, J. Saxl, and T. Weigel, Generation of classical groups. *Geom. Dedicata* **49** (1994), 85–116.

[MaNe95] A. Malnič and R. Nedela, *k*-Minimal triangulations of surfaces, *Acta Math. Univ. Comenian.* **64** (1995), 57–76.

[MaNeSk02] A. Malnič, R. Nedela, and M. Škoviera, Regular homomorphisms and regular maps, *European J. Combin.* **23** (2002), 449–461.

[MeGi11] A. Mednykh and A. Giorgetti, Enumeration of genus four maps by number of edges, *Ars Math. Contemp.* **4** (2011), 351–361.

[MeNe06] A. Mednykh and R. Nedela, Enumeration of unrooted maps of a given genus, *J. Combin. Theory Ser. B* **96** (2006), 706–729.

[MoTh01] B. Mohar and C. Thomassen, *Graphs on Surfaces*, The Johns Hopkins University Press, 2001.

[NeSk97] R. Nedela and M. Škoviera, Exponents of orientable maps, *Proc. London Math. Soc.* **75** (1997), 1–31.

[NeSk97a] R. Nedela and M. Škoviera, Regular maps from voltage assignments and exponent groups, *European J. Combin.* **18** (1997), 807–823.

[NeSk01] R. Nedela and M. Škoviera, Regular maps on surfaces with large planar width, *Europ. J. Combinatorics* **22** (2001), 243–261.

[Ne83] S. Negami, Uniqueness and faithfulness of embedding of toroidal graphs, *Discrete Math.* **44** (1983), 161–180.

[Ne85] S. Negami, Unique and faithful embeddings of projective-planar graphs, *J. Graph Theory* **9** (1985), 235–243.

[PuVi96] H. Pulapaka and A. Vince, Nonrevisiting paths on surfaces, *Discrete Comput. Geom.* **15** (1996), 352–257.

[PuVi98] H. Pulapaka and A. Vince, Nonrevisiting paths on surfaces with low genus, *Discrete Math.* **182** (1998) 267–277.

[RiWo95] L. B. Richmond and N. C. Wormald, Almost all maps are asymmetric, *J. Combin. Theory Ser. B* **63** (1995), 1–7.

[RiSiJaTuWa05] B. Richter, J. Širáň, R. Jajcay, T. W. Tucker, and M. E. Watkins, Cayley maps, *J. Combin. Theory Ser. B* **95** (2005), 189–245.

[Ri77] G. Ringel, The combinatorial map color theorem, *J. Graph Theory* **1** (1977), 141–155.

[RiYo68] G. Ringel and J. W. T. Youngs, Solution of the Heawood map-coloring problem, *Proc. Natl. Acad. Sci. U.S.A.* **60** (1968), 438–445.

[RSST96] N. Robertson, D. Sanders, P. Seymour, and R. Thomas, Efficient four-coloring planar graphs, *Proc. ACM Symp. Theory Comput.* **28** (1996), 571–575.

[RSST97] N. Robertson, D. Sanders, P. Seymour, and R. Thomas, The four-colour theorem, *J. Combin. Theory Ser. B* **70** (1997), 2–44.

[RoSe88] N. Robertson and P. Seymour, Graph minors. VII. Disjoint paths on a surface, *J. Combin. Theory Ser. B* **45** (1988), 212–254.

[Sah69] C. H. Sah, Groups related to compact Riemann surfaces, *Acta Math.* **123** (1969), 13–42.

[Sc91] A. Schrijver, Disjoint circuits of prescribed homotopies in a graph on a compact surface, *J. Combin. Theory Ser. B* **51** (1991), 127–159.

[Sc93] A. Schrijver, Graphs on the torus and geometry of numbers, *J. Combin. Theory Ser. B* **58** (1993) 147–158.

[ScWi85] E. Schulte and J. M. Wills, A polyhedral realization of Felix Klein's map $\{3,7\}_8$ on a Riemann surface of genus 3, *J. London Math. Soc.* **32** (1985), 539–547.

[ScWi86] E. Schulte and J. M. Wills, On Coxeter's regular skew polyhedra, *Discrete Math.* **60** (1986), 253–262.

[Si01] J. Širáň, Coverings of graphs and maps, orthogonality, and eigenvectors, *J. Algebraic Combin.* **14** (2001), 57–72.

[SkSi92] M. Škoviera and J. Širáň, Regular maps from Cayley graphs. I: Balanced Cayley maps, *Discrete Math.* **109** (1992), 265–276.

[Sta78] S. Stahl, Generalized embedding schemes, *J. Graph Theory* **2** (1978), 41–52.

[Sta80] S. Stahl, Permutations-partition pairs: a combinatorial generalization of graph embedding, *Trans. Amer. Math. Soc.* **259** (1980), 129–145.

[St98] A. Stein, $1\frac{1}{2}$-generation of finite simple groups, *Beiträge Algebra Geom.* **39** (1998), 349–358.

[Sti06] E. Steinitz, Über die Eulersche Polyederrelationen, *Arch. Math. Phys.* **11** (1906), 86–88.

[Sti22] E. Steinitz, Polyeder und Raumeinteilungen, *Enzykl. Math. Wiss.* **3** (1922), 1–139.

[Th90] C. Thomassen, Embeddings of graphs with no short noncontractible cycles, *J. Combin. Theory Ser. B* **48** (1990), 155–177.

[Th91] C. Thomassen, Tilings of the torus and the Klein bottle and vertex-transitive graphs on a fixed surface, *Trans. Amer. Math. Soc.* **323** (1991), 605–635.

[Th93] C. Thomassen, Five-coloring maps on surfaces, *J. Combin. Theory Ser. B* **59** (1993), 89–105.

[Th97] C. Thomassen, Color-critical graphs on a fixed surface, *J. Combin. Theory Ser. B* **70** (1997), 67–100.

[TsWe11] M.-T. Tsai and D. B. West, A new proof of 3-colorability of Eulerian triangulations, *Ars Math. Contemp.* **4** (2011), 73–77.

[Tu83] T. W. Tucker, Finite groups acting on surfaces and the genus of a group, *J. Combin. Theory Ser. B* **34** (1983), 82–98.

[Tut63] W. T. Tutte, A census of planar maps, *Canad. J. Math.* **15** (1963), 249–271.

[Tut79] W. T. Tutte, Combinatorial oriented maps, *Canad. J. Math.* **31** (1979), 986–1004.

[Tut84] W. T. Tutte, *Graph Theory*, Addison-Wesley, Menlo Park, 1984.

[ViPe10] S. Vidal and M. Petitot, Counting rooted and unrooted triangular maps, *J. Nonlinear Syst. Appl.* **1** (2010), 51–57.

[Vi83a] A. Vince, Combinatorial maps, *J. Combin. Theory Ser. B* **34** (1983), 1–21.

[Vi83b] A. Vince, Regular combinatorial maps, *J. Combin. Theory Ser. B* **35** (1983), 256–277.

[Vi84] A. Vince, Flag transitive maps, *Congr. Numer.* **45** (1984), 235–250.

[Wa75] T. R. S. Walsh, Hypermaps versus biparite maps, *J. Combin. Theory Ser. B* **18** (1975), 155–163.

[WaGi13] T. R. S. Walsh and A. Giorgetti, Efficient enumeration of rooted maps of a given orientable genus by number of faces and vertices, *Ars Math Comtemporanea* 7 (2014), 263–280.

[WaGiMe12] T. R. S. Walsh, A. Giorgetti, and A. Mednykh, Enumeration of unrooted orientable maps of arbitrary genus by number of edges and vertices, *Discrete Math.* **312** (2012), 2660–2671.

[Wh01] A. T. White, *Graphs of Groups on Surfaces. Interactions and Models*, North-Holland, Amsterdam, 2001.

[Whi32] H. Whitney, Nonseparable and planar graphs, *Trans. Amer. Math. Soc.* **34** (1932), 339–362.

[Wi78] S. Wilson, Non-orientable regular maps, *Ars Combin.* **5** (1978), 213–218.

[Wi89] S. Wilson, Cantankerous maps and rotary embeddings of K_n, *J. Combin. Theory Ser. B* **47** (1989), 262–273.

[Yu97] X. Yu, Disjoint paths, planarizing cycles, and spanning walks, *Trans. Amer. Math. Soc.* **349** (1997), 1333–1358.

[Zv97] A. Zvonkin, Matrix integrals and map enumeration: an accessible introduction, *Math. Comput. Modelling* **26** (1997), 281–304.

Section 7.7

Representativity

Dan Archdeacon, University of Vermont

INTRODUCTION

Consider the graph $C_{100} \times C_{100}$, imbedded on the torus as a grid so that all faces are quadrilaterals. For any vertex v, the graph induced by all vertices of distance at most 49 from v is imbedded exactly as if it were in the plane. In other words, it is locally planar, so it may share some properties of planar graphs. Representativity measures the extent of local planarity of an imbedded graph. Alternatively, an imbedded graph with large representativity may reveal properties of its surface.

The theory of representativity was first introduced by Robertson and Seymour [RoSe88] in their work on graph minors, although hints occur in earlier works. See [GrTu87] and [MoTh01] for some of the proofs and for general reference in topological graph theory and in representativity.

7.7.1   Basic Concepts

There are several different, but related, ways to measure the local planarity of a graph imbedded on a non-spherical surface. We present these measures and their relations.

DEFINITIONS

D1:   An ***open face*** (sometimes "open" is omitted) of a graph imbedding $G \to S$ is a connected component of $S - G$. The set of faces is denoted $F(G \to S)$, or sometimes, simply F.

D2: The union of a face f and its boundary walk is denoted $\bar{f}$ and is called a **closed face**.

D3: A **cellular** imbedding is one where each face is homeomorphic to 2-dimensional Euclidean space.

D4: In a **circular imbedding**, the boundary of every face is a simple cycle.

D5: An imbedding is a **polyhedral imbedding** if it is circular and the intersection of any two closed faces is a path. (Equivalently, we observe that each vertex has a **wheel-neighborhood** – with a possibly subdivided rim.)

D6: A cycle C on a surface S is **contractible** if it separates the surface, and if one side of the separation is a disk.

D7: If S is not a sphere, then the **interior of a contractible cycle** C, which is denoted $int(C)$, is the side that is a disk.

D8: A **k-nest** of disjoint contractible cycles is a sequence $C_1, \ldots, C_k$ such that $C_i \subset int(C_{i+1})$.

D9: The **edge-width** $ew(G)$, of a graph imbedding $G \to S$ is the (graph-theoretic) length of the shortest cycle in the graph that is non-contractible on the surface.

D10: The **face-width** $fw(G)$ of a graph imbedding $G \to S$ (alternatively the **representativity**, denoted $\rho(G)$) is the minimum cardinality $|C \cap G|$, taken over all non-contractible cycles C in S. That is, it is the smallest number k such that there exists faces $f_1, \ldots, f_k$ with a non-contractible cycle C contained in $\bar{f}_1 \cup \cdots \cup \bar{f}_k$. (The cycle C may intersect vertices, not just edges.) Informally, **width** is a synonym.

D11: An imbedded graph is a **dense imbedding** on the surface is it has large face-width, with "large" to be interpreted in context.

D12: The **medial graph** $M(G)$ of a graph imbedding $G \to S$ is the imbedded graph whose vertex set is $E(G)$, and whose edge set joins each pair of vertices representing consecutive edges in a face boundary of $G \to S$.

D13: The **radial graph** $R(G)$ is the imbedded dual of the medial graph. Equivalently, the radial graph has vertex set $V(G) \cup F(G \to S)$, with edges joining incident elements. The radial graph is also called the **vertex-face** graph, or the **vertex-face incidence graph**.

D14: The **Euler genus** $\bar{\gamma}(S)$ of a surface S is twice the number of handles if the surface is orientable, and is the number of crosscaps if the surface is non-orientable. (This definition reflects the old maxim "a handle is worth two crosscaps".)

EXAMPLES

E1: The graph in Figure 7.7.1 is drawn in the torus, where the top edge is identified with the bottom edge and the left with the right. It has edge-width 4, as shown by the cycle on the bold edges. It has face-width 2, as shown by the dotted line. All line segments in the boundary of the square are edges in the graph; the dotted lines are not.

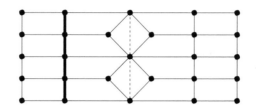

Figure 7.7.1: A graph with edge-width 4 and face-width 2.

E2: The graph in the left of Figure 7.7.2 shows K_4 imbedded in the plane. The dotted lines show its medial graph, the octahedron. On the right side is K_4 and its radial graph, the cube.

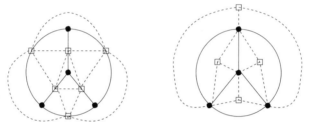

Figure 7.7.2: The medial graph and radial graph of K_4.

FACTS

F1: In any cellular imbedding, any curve in the surface is ambient isotopic (or freely homotopic, i.e., can be continuously transformed) to a closed walk in the graph. Moreover, any walk W in the graph is ambient isotopic to a curve C in the surface that meets the graph only at its vertices. The number of points in $C \cap G$ will not exceed the length of W. This gives the following two facts.

F2: For any imbedding $fw(G) \leq ew(G)$.

F3: For triangulations $fw(G) = ew(G)$.

F4: An imbedded graph M is the medial graph of some graph G if and only if M is 4-regular and the faces can be properly 2-colored.

F5: An imbedded graph R is a radial graph of some graph G if and only if R is bipartite and every face is a quadrilateral.

F6: The medial graph M of an imbedded graph G is identical to the medial graph of the dual G^*. The imbedded graph and its dual are the only two graphs whose medial and radial graphs are M and R, respectively.

F7: The face-width of an imbedded graph G is equal to half the edge-width of its radial graph, i.e., $fw(G) = ew(R(G))/2$.

F8: The face-width of an imbedded graph G is equal to the face-width of its dual G^*, i.e., $fw(G) = fw(G^*)$.

F9: An imbedding of a graph G is cellular if and only if G is connected and $fw(G) \geq 1$.

F10: The following are equivalent: 1) an imbedding of a graph G is circular, 2) G is 2-connected and $fw(G) \geq 2$, and 3) the radial graph $R(G)$ has no multiple edges.

F11: The following are equivalent: 1) an imbedding of a graph G is polyhedral, 2) G is 3-connected and $fw(G) \geq 3$, and 3) the radial graph $R(G)$ has no multi-edges and every 4-cycle bounds a face.

F12: In a polyhedral imbedding of a 3-connected graph, the face boundaries are simple chordless cycles.

F13: [RoSe88] 1) Let v be a vertex of an imbedded graph G and let $k = \lfloor (fw(G)-1)/2 \rfloor$. Then there exists a k-nest with $v \in int(C_1)$. 2) Let f be an open face of an imbedded graph G and let $k = \lfloor fw(G)/2 \rfloor - 1$. Then there exists a k-nest with $f \in int(C_1)$.

See [Mo97] for additional results on face-width.

7.7.2 Coloring Densely Imbeddable Graphs

All planar graphs are vertex-4-colorable. Nonplanar graphs may have arbitrarily high chromatic number; for instance, the chromatic number of the complete graph K_n is n. However, all imbeddings of complete graphs have many non-contractible triangles, which implies that they have edge-width 3, so complete graphs are not densely imbeddable. As an illustration of the relationship between high representativity and planarity, we consider upper bounds on the chromatic numbers of graphs with imbeddings of sufficiently high representativity.

Coloring with Few Colors

We first consider two 5-color theorems for graphs with imbeddings of high edge-width. If we allow six colors, then the required edge-width drops significantly.

FACTS

F14: [AlSt82] (combined with [Th94a]) If a graph G has a toroidal imbedding such that $ew(G) \geq 4$, then G is 5-colorable.

F15: [Th93] If a graph G admits an imbedding $G \to S_g$ such that $ew(G) \geq 2^{14g+6}$, then G is 5-colorable.

F16: [Fi78] Suppose that a graph G has exactly two vertices of odd degree, that these two vertices are adjacent, and it is imbeddable so that every face is a triangle. Then G is not 4-colorable.

F17: [FiMo94] There is a constant c such that every graph G imbeddable on a surface of Euler genus $g > 0$ with $ew(G) \geq c \log g$ is 6-colorable.

F18: [Th97] For each fixed surface S there is a polynomial-time algorithm that decides if a given graph imbedded on that surface is 5-colorable.

REMARK

R1: It is not hard to construct imbeddings on any surface with exactly two vertices of odd degree that are adjacent, as in Fact F16, so the 5-color theorem of Fact F15 is best possible.

Coloring Graphs that Quadrangulate

We next examine the chromatic number of imbedded graphs with no face a triangle, or more specifically, where every face is of even length. These generalize the classical theorem of Grötzsch, which says that every planar graph of girth at least 4 is 3-colorable.

DEFINITION

D15: A *quadrangulation* is a graph imbedding such that each face is a quadrilateral.

FACTS

F19: If every face of a planar graph G is bounded by a cycle of even length, then G is bipartite (and hence has chromatic number 2).

F20: The complete graph K_5 is a non-bipartite graph that can quadrangulate the torus S_1. By iteratively applying cut-and-paste techniques to this imbedding, we can construct quadrangulations of all of the higher surfaces S_n by non-bipartite graphs, hence, of chromatic number at least 3. Similarly, any non-orientable surface has a non-bipartite quadrangulation.

F21: [Hu95] There is a function $f(g)$ such that every graph G imbeddable on an orientable surface of genus g with every face of even size and with $ew(G) \geq f(g)$ has chromatic number at most 3.

F22: [FiMo94] There is a function $f(g)$ such that every graph G of girth at least 4 imbeddable on a non-orientable surface with g crosscaps and with $ew(G) \geq f(g)$ has chromatic number at most 4.

F23: [Yo96] Let G be a graph that has a simple quadrangulation of the projective plane. Then G has either chromatic number 2 or 4.

F24: [Yo96] There exist quadrangulations of the projective plane of arbitarily large edge-width and chromatic number 4.

F25: ([ArHuNaNeOt01] and [MoSe99]) For any non-orientable surface there exist quadrangulations of arbitrarily large edge-width and chromatic number 4.

REMARK

R2: The fact there is no quadrangulation of the projective plane with chromatic number exactly 3 is surprising. So is the contrast between orientable surfaces, where a large width implies 3-chromatic, and non-orientable surfaces, where 4-chromatic is the best possible.

Coloring Graphs That Triangulate

DEFINITIONS

D16: A *triangulation* is a graph imbedding with every face a triangle.

D17: An *Eulerian triangulation* is a triangulation of a surface such that the skeleton is Eulerian.

FACTS

F26: A graph that triangulates the plane is 3-colorable if and only if it is Eulerian.

F27: [HuRiSe02] There is a function $f(g)$ such that every Eulerian triangulation G of an orientable surface of genus g with $ew(G) \geq f(g)$ has chromatic number at most 4.

F28: [Na09] There is a function $f(g)$ such that every Eulerian triangulation G of a non-orientable surface of genus g with $ew(G) \geq f(g)$ has chromatic number at most 5.

F29: [ArHuNaNeOt01] For every non-orientable surface S and every k there exists an Eulerian triangulation G with $ew(G) \geq k$ and with chromatic number at least 5.

RESEARCH PROBLEMS

RP1: Thomassen [Th97]: Is there a surface S with chromatically 5-critical triangulations of arbitrarily large edge-width?

RP2: Albertson [Al81]: For each fixed surface S, does there exist a constant $c(S)$ such that there is a proper 4-coloring of all but $c(S)$ vertices of any imbedded graph?

RP3: For each fixed surface S, is there a polynomial-time algorithm that decides whether a graph imbedded on S is 4-colorable?

RP4: (N. Robertson) Does there exist a constant k such that each cubic graph imbedded with face-width at least k is 3-edge-colorable? Grünbaum [Gr69] conjectured that $k = 3$ suffices, but this has been disproved by Kochol [Ko09].

7.7.3 Finding Cycles, Walks, and Spanning Trees

A fundamental result by Tutte [Tu56] says that 4-connected planar graphs are Hamiltonian. We look for analogous theorems for locally planar graphs.

DEFINITIONS

D18: A *spanning walk* is a walk that visits every vertex.

D19: A *k-walk* is a spanning walk that visits no vertex more than k times.

FACTS

F30: A graph has a Hamiltonian path if and only if it contains a spanning 1-walk.

F31: If a graph G contains a k-walk, then G contains a spanning tree of maximum degree at most $k + 1$.

F32: [Ba66] Every 3-connected planar graph contains a spanning tree of maximum degree 3.

F33: [BrElGaMeRi95] Every 3-connected planar graph imbeddedable on the torus or on the Klein bottle contains a spanning tree of maximum degree 3.

F34: Results analogous to Fact F33 do not hold for surfaces of Euler genus three or more. In particular, $K_{3,n}$ quadrangulates a surface with Euler genus $(n-2)/2$ when n is even. If n is at least 8, then any spanning tree of $K_{3,n}$ contains at least one vertex of degree at least 4.

F35: [Yu97] Let G be a 3-connected graph imbedded on a surface of Euler genus g. If $fw(G) \geq 48(2^g - 1)$, then G contains a spanning 3-walk (and hence a spanning tree of maximum degree 4).

F36: [Yu97] Let G be a 4-connected graph imbedded on a surface of Euler genus g. If $fw(G) \geq 48(2^g - 1)$, then G contains a spanning 2-walk (and hence a spanning tree of maximum degree 3; see also [ElGa94]).

F37: [Yu97] Let G be a 5-connected triangulation on a surface of Euler genus g. If $fw(G) \geq 96(2^g - 1)$, then G contains a Hamiltonian cycle (and hence a spanning 1-walk and a spanning tree of maximum degree 2).

F38: [ShYu02] There is a constant c and a function $f(g)$ such that if G is a 3-connected graph on n vertices imbedded in an orientable surface of genus g with $fw(G) \geq f(g)$, then G has a cycle of length at least $cn^{\log_3(2)}$.

CONJECTURE

C1: There exists a function $f(S)$ such that every 5-connected imbedding of a graph G on a surface S with $fw(G) \geq f(S)$ is Hamiltonian.

7.7.4   Re-Imbedding Properties

Whitney [Wh33] proved that every 3-connected graph has an essentially unique imbedding in the plane, and he proved a similar theorem about imbeddings of graphs of connectivity 2 (see Fact F44). Do locally planar graphs have similar properties?

LEW-Imbeddings

DEFINITIONS

D20: A *large edge-width imbedding* or *LEW-imbedding* is an imbedding $G \to S$ whose edge-width is strictly larger than the length of the longest boundary walk.

D21: A *rooted imbedding* of a graph is an imbedding with a distinguished vertex v, an edge e incident with vertex v, and a face f incident with edge e.

FACTS

F39: [Th90] A graph has an LEW-imbedding in at most one (homeomorphism type of) surface, and if such an imbedding exists, then it is a minimum Euler genus imbedding for that graph.

F40: [Th90] In an LEW-imbedding of a 3-connected graph G, the face boundaries are the chordless cycles whose length is strictly less than $ew(G)$.

F41: [Th90] There is a polynomial-time algorithm that given a 3-connected graph G, either constructs an LEW-imbedding of G or concludes that no such imbedding exists.

F42: [BeGaRi94] For each surface S there is a constant c_S such that almost all rooted imbeddings G in S have $ew(G) \geq c_S \log(|E(G)|)$. (Thus, for a fixed surface, we can expect the edge-width of a random imbedding to be reasonably large.)

Imbeddings and Connectivity

DEFINITIONS

D22: Let G be a graph of connectivity two, and let C_1, C_2 be two subgraphs, each with a cycle, that partition the edge set of G and have only two vertices in common. Suppose that G is imbedded on a surface. Then we can replace the induced imbedding of one component, say C_2, with its mirror image, as shown in Figure 7.7.3. This is called a **Whitney 2-flip**, or more succinctly, a **2-flip**.

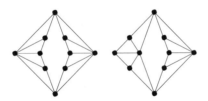

Figure 7.7.3: A Whitney 2-flip.

D23: Two imbeddings are **Whitney-similar** if one can be obtained from the other by a sequence of Whitney 2-flips.

FACTS

F43: Graphs with a cutpoint do not have unique imbeddings in the plane. If v is a cut-vertex of G incident with blocks B_1 and B_2, then B_2 can be placed in any face of an imbedding of B_1 incident with v.

F44: [Wh33] Any two imbeddings of a 2-connected graph in the plane are Whitney-similar.

F45: [Th90] If a 2-connected graph G has an LEW-imbedding on a surface S, then any other imbedding of G on S is Whitney-similar to it.

F46: [Th90] Let G be a subdivision of a 3-connected graph with an LEW-imbedding in a surface S. Then G is uniquely imbeddable in S, up to homeomorphism of pairs.

Imbeddings and Genus

We ask when an imbedding of sufficiently large face-width is a minimum genus imbedding and when it is unique.

FACTS

F47: [SeTh96] Let G be a graph imbedded on a surface of Euler genus g. If $fw(G) \geq 100 \log g / \log \log g$, then G does not imbed on any surface of smaller Euler genus.

F48: [Ar92] For each integer k there is a graph G_k that has two imbeddings on two different surfaces each of face-width at least k. (Hence in Fact F47 the bound on the face-width cannot be replaced by a constant.)

Re-Imbedding Results

DEFINITION

D24: Let G be a graph imbedded on the Klein bottle. The following *planarizing orders* measure how difficult it is to obtain a planar graph by cutting along the (image of the) graph. We define

$$ord_2(G) = \min \lfloor |C|/4 \rfloor$$

taken over all two-sided non-contractible cycles C in the Klein bottle, where $|C|$ denotes the length of C. We define

$$ord_1(G) = \min \lfloor |C_1|/4 \rfloor + \lfloor |C_2|/4 \rfloor$$

taken over all pairs of one-sided non-contractible non-homotopic cycles C_1, C_2.

FACTS

F49: [FiHuRiRo95] Let the graph G be imbeddedable in the projective plane with $fw(G) \neq 2$. Then the (orientable) genus of G is $\lfloor fw(G)/2 \rfloor$.

F50: ([RoTh91], see also [MoSc03]) Let the graph G be imbeddedable in the Klein bottle such that

$$\min\{ord_1(G), ord_2(G)\}$$

is at least four. Then this minimum is the orientable genus of G.

F51: ([RoVi90] and [Th90]) Any non-planar imbedding of a planar graph G has face-width less than or equal to 2.

F52: [MoRo98] There is a function $f(g)$ such that every 3-connected graph has at most $f(g)$ imbeddings of face-width at least 3 in a surface of Euler genus g.

F53: No such function $f(g)$ exists if we consider 2-connected graphs, or if we consider imbeddings of face-width at least 2.

For other results on re-imbedding planar graphs see [MoRo96].

CONJECTURE

RP5: [FiHuRiRo95]: For each fixed non-orientable surface S, the (orientable) genus of graphs that imbed in S can be computed in polynomial-time.

7.7.5   Minors of Imbedded Graphs

The concept of surface minors and representativity is very important in the sequence of papers by Robertson and Seymour culminating in the proof of Wagner's Conjecture. A nice background is given in [Di10]. A survey of results in this area is given in [Mo01].

Surface Minors

DEFINITIONS

D25:   A *minor of a graph* G is a graph H that can be formed by a sequence of edge deletions and edge contractions in G.

D26:   A graph property P is *hereditary under minors* if whenever G has that property, then so does every minor of G. Alternatively, one says that the class of graphs with property P is *closed under the taking of minors*.

D27:   A *surface minor of a graph imbedding* $G \to S$ is an imbedding $H \to S$ constructed by a sequence of edge deletions and edge contractions in G. (If edge e is not a loop, then the edge-contracted graph G/e is constructed by contracting e in S. For the purposes of surface minors, we consider the contraction of a loop to be equivalent to its deletion.)

FACTS

F54:   For each fixed surface S the property "G imbeds on S" is hereditary under minors. (We do not require that these imbeddings are cellular.)

F55:   For each fixed surface S the property "$fw(G) \le k$" is hereditary under surface minors.

The following results are fundamental in the theory of surface minors.

F56:   [RoSe86] For any plane imbedding of a graph G, there is a number k such that G is a surface minor of the natural plane imbedding of the Cartesian product $P_k \times P_k$. See [DiEaTaTo94] for additional references.

F57:   [RoSe88] Let $G \to S$ be a graph imbedding on a surface other than the sphere. Then there is a number k such that every graph imbedding on S with face-width at least k has the imbedding $G \to S$ as a surface minor.

Finding Imbedded Cycles

Fact F57 ensures the existence of any imbedded minor, provided that the width of the given imbedding is large enough. However, the proof is existential and does not provide explicit bounds. Such bounds are known for certain types of graphs.

DEFINITION

D28:   A set of cycles $C_1, \ldots, C_k$ in a graph G imbedded on a surface S is a *planarizing collection of cycles* if cutting along all cycles C_i simultaneously yields a connected graph imbedded in the plane.

FACTS

F58:   [Sc93] Every graph imbedded on the torus with face-width k contains $\lfloor 3k/4 \rfloor$ disjoint non-contractible cycles. (Note that any two disjoint non-contractible cycles in the torus are homotopic.)

F59:   [BrMoRi96] Every graph imbedded on a surface with face-width k contains $\lfloor (k-1)/2 \rfloor$ disjoint non-contractible homotopic cycles.

F60:   [Th93] Let S be an orientable surface of genus g and let G be a triangulation with $ew(G) \geq 8(d+1)(2^g - 1)$. Then G has a planarizing collection $C_1, \ldots, C_g$ of chordless cycles such that any two of these cycles are of distance at least d.

F61:   [Yu97] Let S be a surface with Euler genus g and let G be an imbedded graph with $fw(G) \geq 8(d+1)(2^g - 1)$. Then for some k (with $g/2 \leq k \leq g$), G has a planarizing collection $C_1, \ldots, C_k$ of chordless cycles such that any two of these cycles are of distance at least d.

7.7.6   Minor-Minimal Maps

Fact F54 asserts that for each fixed surface S, the property "$fw(G) \leq k$" is hereditary under surface minors. We look for the minor-minimal imbedded graphs with the property $fw(G) \geq k$, that is, imbedded graphs with face-width k but such that the deletion or surface contraction of any edge lowers the face-width. Any graph imbedded on S with face-width at least k must contain one of these minor-minimal imbeddings as a surface minor. The concept of minor-minimal graphs is often useful in inductive proofs.

DEFINITIONS

D29:   A *minor-minimal imbedded graph* is an imbedded graph such that the deletion or surface contraction of any edge lowers the face-width.

D30:   In a graph G, let v be a vertex adjacent to exactly three other vertices a, b, c. A $Y\Delta$-*transformation* deletes v and its three incident edges and adds three new edges ab, bc, ca.

D31:   A ΔY-*transformation* is the inverse of a $Y\Delta$-transformation.

D32:   Two graphs are $Y\Delta Y$-*equivalent* if there is a sequence of $Y\Delta$- and ΔY-transformations changing one into the other.

FACT

F62:   For every surface S and every $k \geq 1$ the number of minor-minimal maps on S with face-width k is finite.

F63:   Let G_Y be a graph and let G_Δ be formed from G_Y by a $Y\Delta$-transformation. If G_Y imbeds on a surface S, then G_Δ also imbeds on S. The converse is not necessarily true, but it is true if the 3-cycle being deleted is a face boundary. (When considering imbedded graphs, the ΔY-transformation is usually restricted to 3-cycles that bound a face.)

F64: If an imbedded G is $Y\Delta Y$-equivalent to an imbedded G', then $fw(G) = fw(G')$. Moreover if G is minor-minimal with face-width k, then so is G'.

F65: [Ra97] Any two graphs in the projective-plane that are minimal with face-width k are $Y\Delta Y$-equivalent. In particular, they have $2k^2 - k$ edges.

F66: [Ba87] There are exactly two graphs in the projective plane that are minimal with face-width 2. They are K_4 and its geometric dual.

F67: ([Ba91] and [Vi92]) There are exactly 7 minor-minimal maps in the projective plane with face-width 3.

F68: [Ba87] There are exactly 7 minor-minimal maps on the torus with face-width 2.

F69: [Hi96] There are exactly 56 minor-minimal maps on the torus with face-width 3. These fall into 7 classes under $Y\Delta Y$-equivalence.

Similarity Classes on the Torus

DEFINITION

D33: Two imbeddings are **similar imbeddings** if they are related by a sequence of operations, each a $Y\Delta$-transformation, a ΔY-transformation, or the taking of the geometric dual.

REMARK

R3: In general, the geometric dual of an imbedded graph is not necessarily $Y\Delta Y$-equivalent to the primal imbedding. Hence the number of similarity classes might be smaller than the number of $Y\Delta Y$-equivalence classes.

FACT

F70: [Sc94] For odd k there are exactly $(k^3 + 5k)/6$ similarity classes of maps on the torus with face-width k. For even k there are exactly $(k^3 + 8k)/6$ similarity classes.

Kernels

A "kernel" for a surface is an imbedding such that deleting or contracting any edge lowers the face-width in some direction, in the sense that a free homotopy class of closed curves may be regarded as a direction.

DEFINITIONS

D34: Let $G \to S$ be a graph imbedding, and let C be a curve in the surface S. We define the **μ-function** $\mu(G, C)$ as

$$\min\{|W|/2\}$$

where this minimum ranges over all closed walks W in the radial graph $R(G)$ that are freely homotopic to C. This is similar to the face-width of G, $\min\{|C' \cap G|\}$, except that we now restrict the minimum to those curves C' freely homotopic to C.

D35: An imbedded graph G is a **kernel** if for every proper imbedded minor H, there is a curve C such that $\mu(H, C) < \mu(G, C)$.

FACTS

F71: $\mu(G,C) = \mu(G^*,C)$ where G^* is the geometric dual.

F72: $\mu(G,C)$ is invariant under $Y\Delta$-exchanges.

F73: If H is a surface minor of G, $\mu(H,C) \leq \mu(G,C)$ for any curve C.

F74: ([Sc92] and [Gr94]) Suppose that G and G' are kernels on the same surface such that $\mu(G,C) = \mu(G',C)$ for all curves C. Then G and G' are similar.

References

[Al81] M. O. Albertson, Open Problem 2, p. 609 in *The Theory and Applications of Graphs*, ed. by G. Chartrand et al., Wiley, 1981.

[AlSt82] M. O. Albertson and W. R. Stromquist, Locally planar toroidal graphs are 5-colorable, *Proc. Amer. Math. Soc.* **84** (1982), 449–457.

[Ar92] D. Archdeacon, Densely embedded graphs, *J. Combin Theory Ser. B* **54** (1992), 13–36.

[ArHuNaNeOt01] D. Archdeacon, J. P. Hutchinson, A. Nakamoto, S. Negami, and K. Ota, Chromatic numbers of quadrangulations on closed surfaces, *J. Graph Theory* **37** (2001), 100-114.

[Ba66] D. W. Barnette, Trees in polyhedral graphs, *Canad. J. Math* **18** (1966), 731–736.

[Ba87] D. W. Barnette, Generating closed 2-cell embeddings in the torus and the projective plane, *Discr. Comput. Geom.* **2** (1987), 233–247.

[Ba91] D. W. Barnette, Generating projective plane polyhedral maps, *J. Combin. Theory Ser. B* **51** (1991), 277–291.

[BeGaRi94] E. A. Bender, Z. Gao, and L. B. Richmond, Almost all rooted maps have large representativity, *J. Graph Theory* **18** (1994), 545–555.

[BrElGaMeRi95] R. Brunet, M. N. Ellingham, Z. Gao, A. Metzlar, R. B. Richter, Spanning planar subgraphs of graphs in the torus and Klein bottle, *J. Combin. Theory Ser. B* **65** (1995), 7–22.

[BrMoRi96] R. Brunet, B. Mohar, and R. B. Richter, Separating and nonseparating disjoint homotopic cycles in graph embeddings, *J. Combin. Theory Ser. B* **66** (1996), 201–231.

[DiEaTaTo94] G. DiBattista, P. Eades, R. Tamassia, I. G. Tollis, Algorithms for drawing graphs: An annotated bibliography, *Comput. Geom.* **4** (1994), 235–282.

[Di10] R. Diestel, *Graph Theory, Fourth Edition*, Springer-Verlag, New York, 2010.

[ElGa94] M. N. Ellingham and Z. Gao, Spanning trees in locally planar triangulations, *J. Combin. Theory Ser. B* **61** (1994), 178–198.

[FiHuRiRo95] J. R. Fiedler, J. P. Huneke, R. B. Richter, and N. Robertson, Computing the orientable genus of projective graphs, *J. Graph Theory* **20** (1995), 297–308.

[Fi78] S. Fisk, The nonexistence of colorings, *J. Combin. Theory Ser. B* **64** (1978), 247–248.

[FiMo94] S. Fisk and B. Mohar, Coloring graphs without short non-bounding cycles, *J. Combin. Theory Ser. B* **60** (1994), 268–276.

[Gr94] M. de Graaf, Graphs and curves on surfaces, Ph.D. Thesis, University of Amsterdam, Amsterdam, 1994.

[GrTu87] J. L. Gross and T. W. Tucker, *Topological Graph Theory*, Wiley-Interscience, 1987. Dover reprint edition, 2001.

[Gr69] B. Grünbaum, Conjecture 6, p. 343 in *Recent Progress in Combinatorics*, ed. by W.T. Tutte, Academic Press, 1969.

[Ha77] G. Haggard, Edmonds' characterization of disc embeddings, *Congr. Numer.* **19** (1977), 291–302.

[Hi96] Y. Hirachi, Minor-minimal 3-representative graphs on the torus, Master's Thesis, Yokohama National University, Yokohama, 1996.

[Hu95] J. P. Hutchinson, Three-coloring graphs embedded on surfaces with all faces even-sided, *J. Combin. Theory Ser. B* **65** (1995), 139–155.

[HuRiSe02] J. P. Hutchinson, R. B. Richter, and P. D. Seymour, Coloring Eulerian triangulations, *J. Combin. Theory Ser. B* **84** (2002), 225–239.

[Ja85] F. Jaeger, A survey of the cycle double cover conjecture, in "Cycles in Graphs" (B. Alspach and C. Godsil, Eds.), *Ann. Discrete Math.* **27** (1985), 1–12.

[Ko09] M. Kochol, Polyhedral embeddings of snarks in orientable surfaces. *Proc. Amer. Math. Soc.* **137** (2009), 1613-1619.

[Mo97] B. Mohar, Face-width of embedded graphs, *Math. Slovaca* **47** (1997), 3–63.

[Mo01] B. Mohar, Graph minors and graphs on surfaces, in "Surveys in Combinatorics", J.W.P. Hirschfeld Ed., London Mathematical Society Lecture Note Series 288 (2001), pp. 145–163.

[MoRo96] B. Mohar and N. Robertson, Planar graphs on nonplanar surfaces, *J. Combin. Theory Ser. B* **68** (1996), 87–111.

[MoRo98] B. Mohar and N. Robertson, Flexibility of polyhedral embeddings of graphs in surfaces, *J. Combin. Theory Ser. B* **83** (2001), 38–57.

[MoSc03] B. Mohar and A. Schrijver, Blocking nonorientability of a surface, *J. Combin. Theory Ser. B* **87** (2003), 2–16.

[MoSe99] B. Mohar and P. D. Seymour, Coloring locally bipartite graphs on nonorientable surfaces, *J. Combin. Theory Ser. B* **84** (2002), 301–310.

[MoTh01] B. Mohar and C. Thomassen, *Graphs on Surfaces*, Johns Hopkins University Press, 2001.

[Na09] A. Nakamoto, 5-Chromatic even triangulations on surfaces, *Discrete Math.* **308** (2008), 2571–2580.

[Ra97] S. P. Randby, Minimal embeddings in the projective plane, *J. Graph Theory* **25** (1997), 153–163.

[RoSe86] N. Robertson and P. D. Seymour, Graph minors V: Excluding a planar graph, *J. Combin. Theory Ser B* **41** (1986), 92–114.

[RoSe88] N. Robertson and P. D. Seymour, Graph minors VII: Disjoint paths on a surface, *J. Combin. Theory Ser B* **48** (1990), 255–288.

[RoTh91] N. Robertson and R. Thomas, On the orientable genus of graphs embedded in the Klein bottle, *J. Graph Theory* **15** (1991), 407–419.

[RoVi90] N. Robertson and R. Vitray, Representativity of surface embeddings, pp. 293–328 in *Paths, Flows, and VLSI-Layout*, ed. by B. Korte, L. Lovász, H.J. Prömel, and A. Schrijver, Springer-Verlag, Berlin, 1990.

[Sc92] A. Schrijver, On the uniqueness of kernels, *J. Combin. Theory Ser. B* **55** (1992), 146–160.

[Sc93] A. Schrijver, Graphs on the torus and geometry of numbers, *J. Combin. Theory Ser. B* **58** (1993), 147–158.

[Sc94] A. Schrijver, Classification of minimal graphs of given face-width on the torus, *J. Combin. Theory Ser. B* **61** (1994), 217–236.

[SeTh96] P. D. Seymour and R. Thomas, Uniqueness of highly representative surface embeddings, *J. Graph Theory* **23** (1996), 337–349.

[ShYu02] L. Sheppardson and X. Yu, Long cycles in 3-connected graphs in orientable surfaces, *J. Graph Theory* **41** (2002), 80–99.

[Th99] R. Thomas, Recent excluded minor theorems for graphs, pp. 201–222 in *Surveys in Combinatorics 1999*, ed. by J. D. Lamb and D. A. Preece, Cambridge Univ. Press, 1999.

[Th90] C. Thomassen, Embeddings of graphs with no short noncontractible cycles, *J. Combin. Theory Ser. B* **48** (1990), 155–177.

[Th93] C. Thomassen, Five-coloring maps on surfaces, *J. Combin. Theory Ser. B* **59** (1993), 89–105.

[Th94a] C. Thomassen, Five-coloring graphs on the torus, *J. Combin. Theory Ser. B* **62** (1994), 11–33.

[Th94b] C. Thomassen, Grötzsch's 3-color theorem and its counterparts for the torus and the projective plane, *J. Combin. Theory Ser. B* **62** (1994), 268–279.

[Th95] C. Thomassen, Embeddings and minors, pp. 301–349 in *Handbook of Combinatorics*, Elsevier, Amsterdam, 1995.

[Th97] C. Thomassen, Color-critical graphs on a fixed surface, *J. Combin. Theory Ser. B* **70** (1997), 67–100.

[Tu56] W. T. Tutte, A theorem on planar graphs, *Trans. Amer. Math. Soc.* **82** (1956), 99–116.

[Vi92] R. Vitray, The 2- and 3-representative projective planar embeddings, *J. Combin. Theory Ser. B* **54** (1992), 1–12.

[Wh33] H. Whitney, 2-isomorphic graphs, *Amer. J. Math.* **55** (1933), 245–254.

[Yo96] D. A. Youngs, 4-Chromatic projective graphs, *J. Graph Theory* **21** (1996), 219–227.

[Yu97] X. Yu, Disjoint paths, planarizing cycles, and spanning walks, *Trans. Amer. Math. Soc.* **349** (1997), 1333–1358.

Section 7.8

Triangulations

Seiya Negami, Yokohama National University, Japan

INTRODUCTION

Triangles can be used as elementary pieces to build up a surface. Such construction of a surface generalizes to a simplicial complexin combinatorial topology. In topological graph theory, we regard the skeleton of a triangulation as a graph dividing a surface into triangles.

7.8.1   Basic Concepts

Although we are primarily interested in closed surfaces, it is worth noting that the basic concepts for triangulations on closed surfaces also work for surfaces-with-boundary, after suitable modifications. Imbeddings are implicitly taken to be cellular.

DEFINITIONS

D1: A **triangulation of a closed surface** S is an imbedding $\rho : G \to S$ of a simple graph G, such that

- each face is bounded by a 3-cycle, and

- any two faces share at most one edge.

The latter condition excludes K_3 on the sphere.

D2: A graph imbedding $G \to S$ is **triangular** if every face is 3-sided. (Some triangular imbeddings are *not* triangulations: e.g., perhaps the skeleton is not simple, or perhaps a face-boundary has a repeated edge.)

D3: The ***skeleton of an imbedding*** $\rho : G \to S$ is the image $\rho(G)$ of the imbedded graph. Informally we write G for $\rho(G)$.

D4: A graph is said to ***triangulate*** a surface if it can be imbedded on the surface as a triangulation.

D5: A ***triangulation*** $G \to S$ ***of a surface-with-boundary*** S is subject to the additional requirement that each boundary component of the surface is the image of a cycle of the skeleton.

D6: A ***Catalan triangulation*** is a triangulation $G \to S$ of a surface-with-boundary such that every vertex of G lies in a boundary component of S. (E. Catalan counted the number of such triangulations when the surface S is a disk.)

D7: The ***link in a triangulation*** $\rho : G \to S$ of a vertex v is the cycle through the neighbors of v whose edges lie on triangles incident on v. It is usually denoted by $\mathrm{lk}(v, \rho : G \to S)$, or simply by $\mathrm{lk}(v)$.

D8: The ***star neighborhood in a triangulation*** $\rho : G \to S$ of a vertex v is the wheel subgraph in G obtained by joining v to each vertex in its link. It is denoted by $\mathrm{st}(v)$.

D9: A ***clean triangulation*** is a triangulation $G \to S$ such that every 3-cycle in G bounds a face.

D10: Two triangulations $\rho_1 : G \to S$ and $\rho_2 : G \to S$ with the same vertex set are ***combinatorially equivalent triangulations*** if they have the same set of face boundary cycles. (Precisely speaking, combinatorial equivalence is for labeled triangulations.)

D11: Two triangulations $G_1 \to S$ and $G_2 \to S$ are ***isomorphic triangulations*** if there is a homeomorphism $h : S \to S$ such that $h(G_1) = G_2$.

D12: Two triangulations $G_1 \to S$ and $G_2 \to S$ are ***isotopic triangulations*** if there is a homeomorphism $h : S \to S$ with $h(G_1) = G_2$ that is isotopic to the identity mapping on S. (Roughly speaking, this means that one can be transformed continuously on the surface into the other.)

D13: An imbedding is said to be ***k-representative*** if it has *face-width* at least k. (See §7.7.)

NOTATIONS

NOTATION: As usual in topological graph theory, S_g denotes the orientable closed surface of genus g, and N_k denotes the non-orientable closed surface of crosscap number k. The Euler characteristic of a surface S is denoted by $\chi(S)$.

NOTATION: $F(G \to S)$ denotes the set of faces of a triangulation $G \to S$. However, we usually let $F(G)$ or F denote the set of faces when only one imbedding of the graph G is under consideration.

NOTATION: Each face of a triangulation $G \to S$ can be specified by listing the three vertices u, v, w at its corners. Thus, it is often identified with the triple $\{u, v, w\}$, and hence one may write $F(G) \subset \binom{V(G)}{3}$. In other contexts, a face may be denoted by its boundary cycle uvw.

EXAMPLES

E1: The 1-skeletons of the tetrahedron, the octahedron and the icosahedron all triangulate the sphere.

E2: The unique imbedding of K_7 on the torus, as shown in Figure 7.8.1, is a triangulation. (Its dual graph is the Heawood graph.)

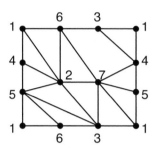

Figure 7.8.1: The complete graph K_7 on the torus.

E3: The unique imbedding of K_6 on the projective plane, as shown in Figure 7.8.2, is a triangulation. (Its dual is isomorphic to the Petersen graph.)

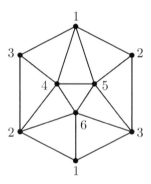

Figure 7.8.2: The complete graph K_6 on the projective plane.

E4: There exists only one 7-vertex triangulation on the torus, up to isomorphism, but there exist infinitely many up to isotopy. The skeleton is isomorphic to K_7 as a graph. Twisting it along a simple closed curve yields an infinite series of those. There exist 120 toroidal triangulations over the vertex set $\{1, \ldots, 7\}$ up to combinatorial equivalence.

E5: There exists only one 6-vertex triangulation on the projective plane, up to isomorphism, and also only one up to isotopy, since any auto-homeomorphism on the projective plane is isotopic to the identity mapping. The skeleton is isomorphic to K_6 as a graph. There exist 12 projective-planar triangulations over the vertex set $\{1, \ldots, 6\}$ up to combinatorial equivalence.

FACTS

F1: A graph with at least four vertices can triangulate the sphere if and only if it is maximal planar.

F2: A graph is isomorphic to a Catalan triangulation of the disk if and only if it is maximal outer-planar.

F3: A graph cellularly imbedded on the sphere is a triangulation if and only if its dual is 3-regular and 3-edge-connected.

F4: A graph cellularly imbedded on a closed surface other than the sphere is a triangulation if and only if its dual is 3-regular, 3-edge-connected, and 3-representative.

F5: The link of a vertex in the interior of the surface is a cycle, and the link of a vertex at the boundary is a path.

F6: Every triangulation of any closed surface has a 3-connected skeleton. More generally, a graph is 3-connected if for each vertex v all the neighbors of v lie on a cycle.

F7: The skeleton G of a triangulation on a closed surface is n-connected ($n = 4$ or 5) if and only if every cycle in G that separates the surface into two pieces, each of which includes at least one vertex, has length at least n.

F8: Every triangulation on any closed surface except the sphere is 3-representative.

F9: A triangulation on a closed surface except the sphere is 4-connected, and it is 4-representative if and only if it is clean.

F10: A triangulation with n vertices on a closed surface S with Euler characteristic $\chi(S)$ has exactly $3(n - \chi(S))$ edges and $2(n - \chi(S))$ faces.

F11: Let G be a triangulation on a closed surface S with Euler characteristic $\chi(S)$ and let V_i denote the number of vertices of degree i in G. Then we have:

$$\sum_{i \geq 3} (6 - i)V_i = 6\chi(S)$$

F12: The equivalence up to isomorphism can be rephrased combinatorially as follows; two triangulations G_1 and G_2 are isomorphic if there is a graph isomorphism $\varphi : V(G_1) \to V(G_2)$ which induces a bijection $\varphi^* : F(G_1) \to F(G_2)$ with $\varphi^*(\{u, v, w\}) = \{\varphi(u), \varphi(v), \varphi(w)\}$.

F13: Let $G_1 \to S_1$ and $G_2 \to S_2$ be triangulations, and let f_1 and f_2 be triangles in G_1 and G_2, respectively. Discard the interiors of those two faces, and paste the boundary of f_1 to the boundary of f_2, thereby producing a connected surface with an imbedded graph. That resulting imbedding is a triangulation.

EXAMPLE

E6: Let $K \to S$ and $H \to S'$ be two 6-connected triangulations on different closed surfaces. Let v be a vertex of degree $d \geq 6$ in K with link $v_1 v_2 \cdots v_d$, and choose two faces of $H \to S'$ sufficiently apart from each other. Identify these with two faces vv_1v_2 and vv_4v_5 of the imbedding $K \to S$. Then the resulting triangulation has a 5-cut $\{v, v_1, v_2, v_4, v_5\}$, but it contains no separating cycle of length less than 6. Therefore, Fact F7 does not hold for $n = 6$.

7.8.2   Constructing Triangulations

What is the minimum number of triangles needed to build up a given surface? This question must have been asked frequently, but it is difficult to answer it precisely.

Triangulations with Complete Graphs

The solution of the "Map Color Theorem" gave us the precise formulas of the genus and the nonorientable genus of K_n, namely,

$$\gamma(K_n) = \lceil (n-3)(n-4)/12 \rceil \quad \text{and} \quad \tilde{\gamma}(K_n) = \lceil (n-3)(n-4)/6 \rceil$$

The complete graph K_n triangulates a suitable surface exactly when the inside of each ceiling function becomes an integer. The constructions give us triangulations on many closed surfaces and also a hint to answer the minimum triangulation question.

DEFINITION

D14: A triangulation is said to be **tight** if

- the skeleton G is a complete graph, and

- for any partition of $V(G)$ into three nonempty subsets V_1, V_2 and V_3, there is a face $v_1 v_2 v_3 \in F(G)$ with $v_i \in V_i$.

It is **untight** otherwise.

FACTS

F14: [Ri74] The complete graph K_n over n vertices triangulates an orientable closed surface if and only if $n \equiv 0, 3, 4$ or $7 \pmod{12}$. The genus of such a surface is equal to $(n-3)(n-4)/12$.

F15: [Ri74] The complete graph K_n over n vertices triangulates a nonorientable closed surface if and only if $n \equiv 0, 1, 3$ or $4 \pmod 6$ and $n \neq 7$. The genus of such a surface is equal to $(n-3)(n-4)/6$.

F16: [Fr34] No complete graph triangulates the Klein bottle.

F17: [BrSt01] The minimum order of a complete graph that admits nonisomorphic triangulations on a nonorientable closed surface is 9; for the orientable case, the minimum is 12.

F18: [BoGrGrSi00] The complete graph K_n triangulates an orientable closed surface with bipartite duals in at least $2^{n^2/54 - O(n)}$ ways if $n \equiv 7$ or $19 \pmod{36}$ and in at least $2^{2n^2/81 - O(n)}$ ways if $n \equiv 19$ or $55 \pmod{108}$.

EXAMPLES

E7: [LaNeWh94] The complete graph K_{19} triangulates the orientable closed surface S_{20} in at least three ways.

E8: [BrSt01] The complete graph K_{10} triangulates the nonorientable closed surface N_7 in at least 14 ways.

E9: [ArBrNe95] The complete graph K_{30} triangulates the nonorientable closed surface N_{117} in at least 2 ways; they are tight and untight.

REMARK

R1: A tight triangulation is necessarily isomorphic to a complete graph as a graph. As a natural generalization of the tightness which works for general triangulations, a notion called the "looseness", has been introduced in [NeMi96] so that a tight triangulation has looseness 0. One graph may be imbedded as many triangulations on a closed surface having different loosenesses. It has been shown in [Ne05] that there is an upper bound for the difference between their maximum and minimum values, depending only on the surface.

Minimum Triangulations

Here we shall show the answer to our question on the minimum number of triangles to build up a surface. The corresponding formula is expressed below in terms of the number of vertices.

DEFINITION

D15: A *minimum triangulation* of a surface is a triangulation on the surface that has the fewest vertices (or equivalently, the fewest faces).

FACTS

F19: [JuRi80, Ri55] Let $V_{\min}(S)$ denote the order of minimum triangulations of a closed surface S. If $S \neq S_2, N_2, N_3$, then:

$$V_{\min}(S) = \left\lceil \frac{7 + \sqrt{49 - 24\chi(S)}}{2} \right\rceil$$

For the three exceptions, we have:

$$V_{\min}(S_2) = 10, \quad V_{\min}(N_2) = 8, \quad V_{\min}(N_3) = 9$$

F20: If the complete graph K_n triangulates a closed surface, then the skeleton of any minimum triangulation is isomorphic to K_n.

F21: [HaRi91] The minimum number of faces in a clean triangulation of S_2 is 24.

F22: [HaRi91] The minimum number of faces in a clean triangulation of S_g is asymptotically equal to $4g$ as $g \to \infty$.

EXAMPLES

E10: The only minimum triangulations of the sphere, the projective plane and the torus are the unique imbeddings of K_4, K_6, and K_7, respectively.

E11: There exist precisely six minimum triangulations of the Klein bottle, up to isomorphism.

Covering Constructions

Shortly after the solution of the Map Color Theorem, the theory of voltage graphs (see §7.4) provided a unified topological analysis of that solution, as a *branched covering*, and an extensive generalization of its constructive method. In fact, many triangular imbeddings of complete graphs constructed for the Map Color Theorem can now be obtained as coverings of small graphs that triangulate suitable surfaces. There are also other ways to build triangulations from triangulations.

DEFINITIONS

D16: [Gr74] A ***voltage graph*** $\langle G = (V, E), \alpha \rangle$ is a directed graph G with an assignment $\alpha : E \to \mathcal{B}$ of elements of a group $\mathcal{B}$ to its arcs. The group $\mathcal{B}$ is called the ***voltage group***.

D17: The ***net voltage on a walk*** in a graph is the product (or sum, if the voltage group is abelian) of the voltages along that walk.

D18: The ***Kirchhoff voltage law*** (abbr. KVL) holds for an imbedded voltage graph if on every face boundary walk, the net voltage equals the identity of the voltage group.

D19: The ***composition*** $G[H]$ of a graph G with a graph H is the graph with vertex set $V(G) \times V(H)$ such that (u_1, v_1) is adjacent to (u_2, v_2) whenever either u_1 is adjacent to u_2, or v_1 is adjacent to v_2 with $u_1 = u_2$. In particular, we denote $G[\overline{K}_m]$ simply by $G_{(m)}$, where $\overline{K}_m$ is the graph over m vertices with no edge.

D20: A natural projection $p : G_{(m)} \to G$ is called a ***covering with folds***. This is an m-to-1 surjective homomorphism mapping (u, v) to u for each vertex $u \in V(G)$. (There will be a more general definition of a covering with folds in other contexts.)

FACTS

F23: [Gr74] An imbedded voltage graph $\langle G \to S, \alpha \rangle$ *lifts* to an imbedding $G^{\alpha} \to S^{\alpha}$ of a covering graph of G into a branched covering of the surface S, such that branch points occur only in the interiors of the faces, with at most one branch point per face.

F24: [Gr74] Let $G \to S$ be a triangular imbedding with voltage assignment α such that the Kirchhoff voltage law holds. Then the resulting graph imbedding in the covering surface of S is also a triangular imbedding. (If the covering graph G^{α} is simple and not K_3, then the resulting imbedding is a triangulation.)

F25: [Bo82a] Let G be a triangulation on a closed surface S. If a positive integer m is not divisible by 2, 3 or 5, then $G_{(m)}$ triangulates another closed surface with the same orientability as S.

F26: [Bo82b] If a triangulation G on a closed surface S is eulerian, then $G_{(m)}$ triangulates another closed surface with the same orientability as S.

F27: [Ar92] If the complete graph K_n triangulates a closed surface S and if each prime factor of m is at least $n - 1$ except the case of $n = 4$, $m = 3$, then $K_{n(m)}$ triangulates another closed surface, where $K_{n(m)}$ stands for the n-partite graph $K_{m,...,m}$ with partite sets of size m.

EXAMPLE

E12: We observe in Figure 7.8.3 that the imbedding $B_3 \to S_1$ is a KVL triangular imbedding. Moreover, for any cyclic group $\mathbb{Z}_n$, and that for $n \geq 7$, the covering graph is simple. Accordingly, Fact F24 implies that the covering imbedding is a triangulation. For sufficiently large n, the face-width of the covering imbedding is arbitrarily large.

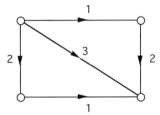

Figure 7.8.3: A KVL imbedding $B_3 \to S_1$ with voltages in arbitrary $\mathbb{Z}_n$.

7.8.3   Irreducible Triangulations

Look at one edge in a triangulation. There are two triangles incident to the edge from both sides and two wheels cover them. With such a local picture around an edge, one will guess that shrinking this edge yields another triangulation smaller than the original. Moreover, he might consider that this fact can be used for some proofs with induction on the number of vertices or edges. What is the first step of such induction? That is, what can we get, repeating this deformation as far as possible? "Irreducible triangulations" are exactly the answer. There is a strong connection between studies on irreducible triangulations and graph minor theory.

Edge Contraction

DEFINITIONS

D21: Let acb and acd be the two faces sharing the edge ac of a triangulation $G \to S$. **Contraction of the triangulation on the edge** ac is to shrink the adjacent triangles acb and acd to a path $bad = bcd$, as in Figure 7.8.4. We do not contract an edge of a triangulation unless it results in another triangulation on the surface S.

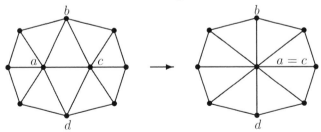

Figure 7.8.4: Edge contraction.

D22: A **vertex splitting** is the inverse operation of an edge contraction.

D23: A **contractible edge in a triangulation** $G \to S$ on a closed surface is an edge whose contraction would yield another triangulation on S.

D24: A triangulation $G \to S$ is said to be **contractible to another triangulation** $H \to S$ if it can be obtained from $G \to S$ by a sequence of edge contractions.

FACTS

F28: An edge in a triangulation G on a closed surface, except K_4 on the sphere, is contractible if and only if it is contained in exactly two cycles of length 3, which are the boundary cycles of two faces sharing the edge.

F29: [Ne94] A triangulation $G \to S$ is contractible to a triangulation $H \to S$ of the same surface if and only if the skeleton H of the latter triangulation is a minor of the graph G.

REMARK

R2: An edge contraction in a triangulation is different from that in graph minor theory. The former always decreases the number of edges by 3 at a time, while the latter does so by 1 at a time.

Classification and Finiteness in Number

DEFINITION

D25: An *irreducible triangulation* on a closed surface is one that has no contractible edge.

FACTS

F30: Every irreducible triangulation on any closed surface, except the sphere, has minimum degree at least 4.

F31: Every triangulation is contractible to an irreducible triangulation. Equivalently, it can be obtained from an irreducible triangulation by a sequence of vertex splittings.

F32: [StRa34] The only irreducible triangulation on the sphere is the tetrahedron, whose skelton is isomorphic to K_4.

F33: [Ba82] There are precisely two irreducible triangulations on the projective plane up to isomorphism, shown in Figure 7.8.3. Their skeletons are isomorphic to K_6 and $K_4 + \overline{K}_3$.

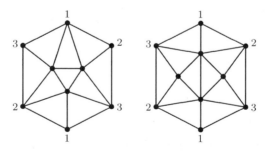

Figure 7.8.5: The two irreducible triangulations on the projective plane.

F34: [La87] There are precisely 21 irreducible triangulations on the torus up to isomorphism, which are given in Figure 7.8.6 below, where each pair of parallel sides of the rectangles should be identified. They are denoted by T1 to T21. The skeleton of T1 is isomorphic to K_7.

Figure 7.8.6: The 21 irreducible triangulations on the torus.

F35: [LaNe97, Sul06] There are precisely 25 irreducible triangulations on the Klein bottle up to isomorphism.They are classified into two classes, namely, **_handle types_** Kh1 to Kh25 in Figure 7.8.7 below and **_crosscap types_** Kc1 to Kc4 in Figure 7.8.8 below. Identify each horizontal pair of sides in parallel and each vertical pair in antiparallel to recover the handle types and identify each antipodal pair of vertices lying on the hexagons for crosscap types.

Figure 7.8.7: The 25 irreducible triangulations on N_2, handle types.

F36: [BaEd89] There are only finitely many irreducible triangulations on any closed surface, up to isomorphism.

F37: [JoWo10] Any irreducible triangulation of a closed surface S with $\chi(S) \leq 0$ has at most $13(2 - \chi(S)) - 4$ vertices.

EXAMPLES

E13: All minimum triangulations on a closed surface are irreducible, but there are irreducible triangulations that are not minimum, in general.

E14: Triangulations of closed surfaces by K_n and $K_{n(m)}$ are irreducible.

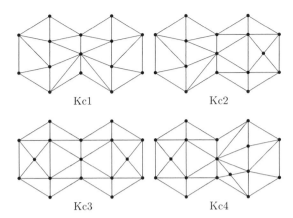

Figure 7.8.8: The 4 irreducible triangulations on N_2, crosscap types.

E15: Let $G_1 \to S_1$ and $G_2 \to S_2$ be two irreducible triangulations on closed surfaces. Form the connected sum of these triangulations by discarding the interiors of a face in each and pasting on the boundary cycles. The resulting triangulation on the surface $S_1 \# S_2$ is irreducible. In particular, all four crosscap types of irreducible triangulations on the Klein bottle are obtained from two of those on the projective plane in this way.

REMARKS

R3: No pair of skeletons for distinct irreducible triangulations on the torus are isomorphic as graphs.

R4: The only pair of isomorphic skeletons for distinct irreducible triangulations on the Klein bottle is (Kh2, Kh5).

R5: The only pair of isomorphic skeletons for distinct irreducible triangulations on the torus and the Klein bottle is (T3, Kh1).

Other Irreducibility

There are some studies on triangulations with prescribed properties which are irreducible or minimal with respect to edge contractions — those of higher representativity or with large minimum degree, for example.

DEFINITIONS

D26: An *essential cycle in a graph imbedding* on a closed surface is a cycle that bounds no 2-cell region on S.

D27: A k-*irreducible triangulation* on a closed surface is a triangulation such that each edge is contained in an essential cycle of length at least k.

D28:   A clean triangulation is *minimal* if no edge contraction results in a clean triangulation. (This is the same as 4-irreducibility.)

FACTS

F38: On any closed surface S except the sphere, every triangulation containing no essential cycle of length less than k can be obtained from some k-irreducible triangulation by a sequence of vertex splittings.

F39: [MaNe95] There are only finitely many k-irreducible triangulations on any closed surface, up to isomorphism.

F40: [MaMo92] There are only finitely many minimal clean triangulations on any closed surface, up to isomorphism.

F41: [FiMoNe94] There are exactly five minimal clean triangulations on the projective plane.

F42: [NaNe02] On any closed surface S except the sphere, every triangulation with minimum degree at least 4 is obtainable from some irreducible triangulation by splitting vertices, preserving their degrees to be at least 4, and adding octahedra to faces.

F43: [NaNe02] Every triangulation on the sphere with minimum degree at least 4 can be obtained from the octahedron in the same way as in Fact F42.

REMARKS

R6: On any closed surface S except the sphere, every irreducible triangulation is 3-irreducible.

R7: The finiteness of the number of minor-minimal k-representative imbeddings follows from Fact F39 ([MaNe95], where a k-irreducible triangulation is called a k-***minimal triangulation***). A simple proof of the same result can be found in [GaRiSe96].

7.8.4   Diagonal Flips

An edge in a triangulation can be regarded as a diagonal in a quadrilateral, and switched to the other diagonal, which is called "a diagonal flip". All sufficiently large triangulations of the same size differ from each other only by diagonal flips ([Ne94]).

DEFINITIONS

D29: A ***diagonal flip*** of an edge ac shared by triangles acb and acd in a triangulation G means to replace ac with the other diagonal bd in the quadrilateral $abcd$, as shown in Figure 7.8.9. We do not perform a diagonal flip unless the resulting skeleton would be a simple graph.

D30: Two triangulations on a surface are ***equivalent under diagonal flips*** if one can be transformed into the other by a finite sequence of diagonal flips.

D31: A ***frozen triangulation*** is a triangulation such that no edge can be flipped.

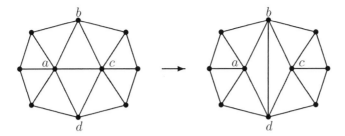

Figure 7.8.9: Diagonal flip.

D32: The **standard triangulation on the sphere** with n vertices is shown in Figure 7.8.10, and is denoted by Δ_{n-3}. The skeleton is isomorphic to $P_{n-2} + K_2$.

Figure 7.8.10: The standard triangulation Δ_4 on the sphere.

NOTATION: The result of adding Δ_m to an arbitrarily chosen face of a triangulation $G \to S$ on a closed surface is denoted by $G + \Delta_m \to S$. (All possible triangulations $G + \Delta_m \to S$ are equivalent under diagonal flips. See [Ne94].)

FACTS

F44: [Wa36] Any two triangulations on the sphere with the same number of vertices are equivalent under diagonal flips, up to isotopy.

F45: [NeWa90] Any two triangulations on the projective plane with the same number of vertices are equivalent under diagonal flips, up to isotopy.

F46: [De73] Any two triangulations on the torus with the same number of vertices are equivalent under diagonal flips, up to isomorphism.

F47: [NeWa90] Any two triangulations on the Klein bottle with the same number of vertices are equivalent under diagonal flips, up to isomorphism.

F48: [Ne94] For every closed surface S, there exists a natural number $N(S)$ such that two triangulations G_1 and G_2 with $|V(G_1)| = |V(G_2)| \geq N(S)$ are equivalent under diagonal flips, up to isomorphism.

F49: [NaOt97] For every closed surface S, there exists a natural number $\tilde{N}(S)$ such that two triangulations G_1 and G_2 with $|V(G_1)| = |V(G_2)| \geq \tilde{N}(S)$ are equivalent under diagonal flips, up to isotopy.

F50: [Ne01] Let G_1 and G_2 be two triangulations on a closed surface S with the same number n of vertices, and let $m \geq 18(n - \chi(S))$. Then any two subdivided triangulations of the form $G_1 + \Delta_m$ and $G_2 + \Delta_m$ are equivalent under diagonal flips, up to isotopy.

F51: [Ne94] A frozen triangulation is irreducible.

EXAMPLES

E16: For the sphere S_0, the projective plane N_1, the torus S_1 and the Klein bottle N_2, we have the following numbers:

$$N(S_0) = 4, \quad N(N_1) = 6, \quad N(S_1) = 7, \quad N(N_2) = 8$$

E17: The irreducible triangulations on the torus can be partitioned into five equivalence classes, as follows, under sequences of diagonal flips:

$$\{T1\}, \quad \{T2, T3, T4, T5\}, \quad \{T7\}, \quad \{T6, T8, \ldots, T20\}, \quad \{T21\}$$

E18: If K_n or $K_{n(m)}$ triangulates a closed surface, then it is a frozen triangulation.

REMARKS

R8: The problem of equivalence would be nearly trivial, and the lower bound $N(S)$ for the order of triangulations would be meaningless, if we allowed diagonal flips that resulted in non-simple skeletons. If the requirement of simpleness is removed, then there is a greedy algorithm to transform one of two triangulations into the other. See [Ne01] for details.

R9: Within the theory of diagonal flips in topological graph theory, the positions of vertices may be moved on surfaces, up to homeomorphism or isotopy. However, in computational geometry, there are studies of diagonal flips in triangulations in which the vertices have fixed positions in the plane.

R10: The bound $N(S)$ in Fact F48 is actually necessary, because there exist frozen triangulations on infinitely many surfaces, as discussed in [Ne99b]. The arguments needed to prove the theorem also work for labeled triangulations and for triangulations with boundary, with suitable modifications. See [Ne99a].

R11: The bound $\tilde{N}(S)$ in Fact F49 is large and unknown, even when S is the torus or the Klein bottle.

Estimating Bounds

We consider how many diagonal flips are necessary to transform one triangulation into another.

DEFINITION

D33: A ***pseudo-minimal triangulation*** is a triangulation such that no sequence of diagonal flips transforms it into one having a vertex of degree 3.

FACTS

F52: [Ne94] A pseudo-minimal triangulation is irreducible.

F53: [Ne94] Let $\{T_i\}$ be the set of the pseudo-minimal triangulations of a closed surface S. The precise value of $N(S)$ is equal to the minimum number N such that all the subdivisions $T_i + \Delta_{N-|V(T_i)|}$ can be transformed into one another by diagonal flips, up to isomorphism.

F54: [Ne01] Let $V_{\mathrm{pse}}(S)$ denote the maximum order taken over all the pseudo-minimal triangulations of a closed surface S with Euler characteristic $\chi(S)$. Then we have:

$$\tilde{N}(S) \le 19\,V_{\mathrm{pse}}(S) - 18\,\chi(S)$$

F55: [Ne98] Given a closed surface S, there are two constants α_1 and α_0, depending only on S, such that any two triangulations $G_1 \to S$ and $G_2 \to S$ with $n \ge N(S)$ vertices can be transformed into each other by at most $2n^2 + \alpha_1 n + \alpha_0$ diagonal flips, up to isomorphism.

F56: [MoNaOt03] Any two triangulations with n vertices on the sphere can be transformed into each other, up to isotopy, by at most $6n - 30$ diagonal flips if $n \ge 5$.

F57: [MoNa03] Any two triangulations with n vertices on the projective plane can be transformed into each other, up to isotopy, by at most $8n - 26$ diagonal flips.

F58: [GaUrWa01] Any two labeled triangulations with n vertices on the sphere can be transformed into each other, up to isotopy, by $O(n \log n)$ diagonal flips.

EXAMPLE

E19: All pseudo-minimal triangulations on the sphere, the projective plane, the torus and the Klein bottle are minimum triangulations on these surfaces.

REMARK

R12: Since there is a linear upper bound for the order of irreducible triangulations with respect to the genus of S, the upper bound for $\tilde{N}(S)$ given in Fact F54 also is linear.

Catalan Triangulations

There are some studies on diagonal flips in Catalan triangulations with the same framework as above, although the lack of interior vertices is an obstacle to the general arguments in [Ne94]. Furthermore, there has been shown an amazing method for Catalan triangulations of polygons in [SlTaTh88], in harmony with combinatorics, hyperbolic geometry and computer science.

DEFINITION

D34: A *punctured surface* is a closed surface with one hole, that is, a surface with connected boundary.

FACTS

F59: [SlTaTh88] Any two Catalan triangulations of an n-gonal disk with $n \geq 13$ can be transformed into each other by at most $2n - 10$ diagonal flips. There exists an example attaining this bound.

F60: [EdRe97] Any two Catalan triangulations of the Möbius band with the same number of vertices are equivalent under diagonal flips, up to isomorphism.

F61: [CoNa00a] Any two Catalan triangulations of the punctured torus with the same number of vertices are equivalent under diagonal flips, up to isomorphism.

F62: [CoNa00b] Any two Catalan triangulations of the punctured Klein bottle with the same number of vertices are equivalent under diagonal flips, up to isomorphism.

F63: [CoGlMaNa02] Given a puctured surface S, there exists a natural number $M(S)$ such that two Catalan triangulations G_1 and G_2 of S are equivalent under diagonal flips, up to isomorphism, if $|V(G_1)| = |V(G_2)| \geq M(S)$.

Preserving Properties

Any diagonal flip preserves the order of triangulations while an edge contraction decreases it by one. Nevertheless, the former is closely related to the latter, as Facts 51 and 52 suggest. This makes a connection of the theory of diagonal flips to graph minor theory and leads us to more general or formal arguments on conditional generating of triangulations.

DEFINITIONS

D35: A class $\mathcal{P}$ of triangulations on S is said to be **splitting-closed** if it is closed under vertex splittings.

D36: Let $\mathcal{P}$ be a class of triangulations. A triangulation is called a $\mathcal{P}$-**triangulation** (or a triangulation with property $\mathcal{P}$) if it belongs to $\mathcal{P}$.

D37: A $\mathcal{P}$-**diagonal flip** in a $\mathcal{P}$-triangulation G is a diagonal flip such that the resulting graph is also a $\mathcal{P}$-triangulation.

D38: Two $\mathcal{P}$-triangulations G_1 and G_2 are said to be $\mathcal{P}$-**equivalent** under diagonal flips if they can be transformed into each other by a finite sequence of $\mathcal{P}$-diagonal flips.

D39: A class $\mathcal{P}$ of triangulations on a closed surface S is said to be **closed under homeomorphism** if $h(G) \in \mathcal{P}$ for any member $G \in \mathcal{P}$ and for any homeomorphism $h : S \to S$.

FACTS

F64: [BrNaNe96] For any closed surface S and for any splitting-closed class $\mathcal{P}$ of triangulations on S, there exists a natural number $N_\mathcal{P}(S)$ such that if G_1 and G_2 are two $\mathcal{P}$-triangulations with $|V(G_1)| = |V(G_2)| \geq N_\mathcal{P}(S)$, then G_1 and G_2 are $\mathcal{P}$-equivalent under diagonal flips, up to isomorphism.

F65: [BrNaNe96] For any closed surface S and for any splitting-closed class $\mathcal{P}$ of triangulations on S which is closed under homeomorphism, there exists a natural number $\tilde{N}_\mathcal{P}(S)$ such that if G_1 and G_2 are two $\mathcal{P}$-triangulations with $|V(G_1)| = |V(G_2)| \geq \tilde{N}_\mathcal{P}(S)$, then G_1 and G_2 are $\mathcal{P}$-equivalent under diagonal flips, up to isotopy.

F66: [KoNaNe99] For any closed surface S except the sphere, there exists a natural number $N_4(S)$ such that two triangulations G_1 and G_2 on S with minimum degree at least 4 can be transformed into each other by a finite sequence of diagonal flips, up to isomorphism, through those triangulations if $|V(G_1)| = |V(G_2)| \geq N_4(S)$.

F67: [KoNaNe99] Two triangulations on the sphere, except the double wheels, with minimum degree at least 4 can be transformed into each other, up to isotopy, by a finite sequence of diagonal flips through those triangulations if they have the same number of vertices.

EXAMPLE

E20: The following properties are splitting-closed and closed under homeomorphism.

(i) Being k-representative.

(ii) Intersecting any non-separating simple closed curve in at least k points.

(iii) Containing at least k disjoint homotopic cycles.

(iv) Containing at least k disjoint cycles.

(v) Containing k distinct spanning trees.

REMARK

R13: The class consisting of triangulations on a closed surface with minimum degree at least 4 is not splitting-closed, and hence the meta-theorems in [BrNaNe96] cannot be used to prove the theorems in [KoNaNe99].

7.8.5   Rigidity and Flexibility

A triangulation may seem quite rigid. So one might guess that it is hardly possible for a graph that triangulates a closed surface to have another imbedding on that same surface, which is actually true for the sphere. However, the complete graph triangulates a closed surface in numerous ways. Here we shall consider many facts on the rigidity and flexibility of triangulations.

Equivalence over Imbeddings

To analyze many imbeddings of a graph, it is often useful to deal with an imbedding as a map rather than a drawing on a surface. That is, an imbedding of a graph G into a surface S is an injective continuous map $f : G \to S$ from a 1-dimensional topological space G to S.

DEFINITIONS

D40:   Two imbeddings $f_1, f_2 : G \to S$ of a graph into a surface are ***equivalent imbeddings*** if there exists a homeomorphism $h : S \to S$ with $hf_1 = f_2$.

D41:   Two imbeddings $f_1, f_2 : G \to S$ of a graph into a surface are ***congruent imbeddings*** if there exists a homeomorphism $h : S \to S$ and a graph automorphism $\sigma : G \to G$ with $hf_1 = f_2\sigma$.

D42:   An automorphism $\sigma \in \mathrm{Aut}(G)$ is called a ***symmetry*** of an imbedding $f : G \to S$ if there is a homeomorphism $h : S \to S$ with $hf = f\sigma$.

D43:   The ***symmetry group of an imbedding*** $f : G \to S$ is the subgroup $\mathrm{Sym}(f)$ in $\mathrm{Aut}(G)$ consisting of the symmetries of the imbedding.

FACTS

F68:   If a simple graph has a triangular imbedding on a closed surface, then all of its imbeddings on that surface are triangular.

F69:   Equivalent triangular imbeddings of a graph G have the same set of face boundary cycles over $V(G)$.

F70:   Congruent triangular imbeddings of a graph G correspond to isomorphic triangulations.

F71:   An imbedding $f : G \to S$ is equivalent to $f\sigma$ for any symmetry $\sigma \in \mathrm{Sym}(f)$.

F72:   An imbedding $f : G \to S$ is congruent but is not equivalent to $f\bar{\sigma}$ for any automorphism $\bar{\sigma} \in \mathrm{Aut}(G) - \mathrm{Sym}(f)$.

F73:   The number of inequivalent imbeddings of a graph G congruent to a fixed imbedding $f : G \to S$ is equal to $|\mathrm{Aut}(G)|/|\mathrm{Sym}(f)|$.

Uniqueness of Imbeddings

It has been known that the skeleton of an imbedding of sufficiently large representativity is rigid, that is, it has a unique imbedding on the surface that contains it. However, relatively simple conditions force any skeleton of a triangulation to be rigid.

DEFINITIONS

D44:   A graph is said to be ***uniquely imbeddable*** on a surface S if all of its imbeddings into S are equivalent.

D45: A *skew vertex* in a triangulation G on a closed surface S is a vertex v such that there are at least two cycles each of which contains all the neighbors of v.

FACTS

F74: The skeleton of every triangulation on the sphere is uniquely imbeddable on the sphere, up to equivalence. This is an easy consequence of the well-known fact that every 3-connected planar graph is uniquely imbeddable on the sphere.

F75: [NeNaTa97] A graph that triangulates a closed surface is uniquely imbeddable on that surface, up to equivalence, if any face has at most two skew vertices as its corners.

F76: [Ne83] The skeleton of a 4-representative triangulation on a closed surface is uniquely imbeddable on that surface, up to equivalence.

F77: [Ne83] The skeleton of a 6-connected toroidal triangulation is uniquely imbeddable on the torus, up to congruence, and also up to equivalence, with three exceptions, shown in Figure 7.8.11.

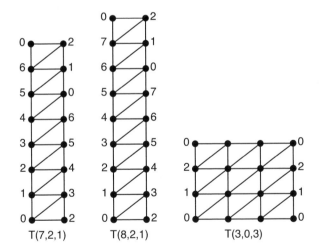

Figure 7.8.11: The three exceptional 6-connected toroidal triangulations.

F78: [Ne84a] The skeleton of a 5-connected projective-planar triangulation is uniquely imbeddable on the projective plane up to equivalence, unless it is isomorphic to K_6.

F79:   [Ne84b] The skeleton of a 6-connected Klein-bottle triangulation is uniquely imbeddable on the Klein bottle up to congruence, and likewise up to equivalence, with one exception, illustrated in Figure 7.8.12.

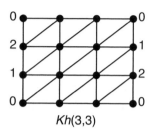

Figure 7.8.12: The exceptional 6-connected Klein-bottle triangulation.

TERMINOLOGY

In Negami's early papers [Ne84a] etc., the uniqueness means that up to congruence, and an imbedding $f : G \to S$ is said to be a ***faithful imbedding*** if $\mathrm{Sym}(f) = \mathrm{Aut}(G)$. By his terminology, a graph can be said to be ***uniquely*** and ***faithfully imbeddable*** if it is uniquely imbeddable up to equivalence in our sense. In some papers, two imbeddings are said to be ***weakly equivalent*** if they are congruent.

Re-Imbedding Structures

It is hardly possible to classify all the mechanisms that generate inequivalent imbeddings of a graph. However, there is a theory to describe the flexibility of triangulations.

DEFINITIONS

D46:   A face of a triangulation $G \to S$ is a ***panel*** if its boundary cycle bounds a face in every imbedding of G in S.

D47:   The ***panel structure*** of an imbedding $G \to S$ is a pair $(G \to S, \mathcal{P})$ in which $\mathcal{P}$ is the set of all the panels.

D48:   The panel structures of two triangulations $G_1 \to S$ and $G_2 \to S$ are said to be ***equivalent panel structures*** if the 2-simplicial complexes obtained from the skeletons G_1 and G_2 by inserting all of their panels are homeomorphic.

FACTS

F80:   [LaNe99] Two faces incident to a contractible edge of a triangulation are panels.

F81:   [NeNaTa97] A face that has at most two skew vertices at its corners is a panel.

F82:   [NeNaTa97] Two triangulations on a closed surface having equivalent panel structures admit the same number of inequivalent imbeddings on the surface.

F83:   The number of inequivalent imbeddings of a triangulation $G \to S$ (with S closed) does not exceed that of an irreducible triangulation to which $G \to S$ is contractible.

F84: [NeNaTa97] There exist only finitely many panel structures on each closed surface, up to equivalence.

F85: [La92] Every projective-planar triangulation admits exactly 1, 2, 3, 4, 6 or 12 inequivalent imbeddings on the projective plane.

F86: [Sa03] Every toroidal triangulation admits exactly 1, 2, 3, 4, 5, 6, 8, 10, 12, 14, 16, 24, 48 or 120 inequivalent imbeddings on the torus.

EXAMPLE

E21: Let $Q(S)$ denote the maximum number of inequivalent imbeddings taken over all graphs that triangulate a given closed surface S. For the sphere S_0, the projective plane N_1, the torus S_1 and the Klein bottle N_2, we have:

$$Q(S_0) = 1, \quad Q(N_1) = 12, \quad Q(S_1) = 120, \quad Q(N_2) = 36$$

The first three are attained by K_4, K_6, and K_7 on these surfaces in order while the last one is attained by the triangulation obtained from two copies of K_6 on the projective plane by pasting them along one pair of faces.

REMARK

R14: To determine the maximum number of inequivalent imbeddings taken over all triangulations on a closed surface, it suffices to investigate irreducible triangulations, by Fact F83. On the other hand, the classification of panel structures exhibits all "re-imbedding structures" and enables us to decide all possible values that appear as the number of inequivalent imbeddings of triangulations. See [NeNaTa97] for the theory of panel structures.

Imbeddings into Other Surfaces

What happens when we imbed the skeleton of a triangulation into other surfaces?

DEFINITION

D49: A graph G is said to **quadrangulate** a surface S if G can be imbedded on S so that each face is a 4-cycle.

FACTS

F87: [HoGl77] The skeleton of a triangulation on an orientable closed surface is an upper imbeddable graph. That is, it can be cellularly imbedded on a suitable orientable closed surface with one or two faces.

F88: [LaNe99] A graph triangulates both the torus and the Klein bottle if and only if it has the structure shown in Figure 7.8.13, where each triangle with ◯ may be divided into many triangles.

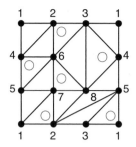 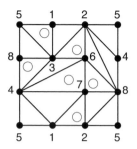

Figure 7.8.13: Triangulating both the torus and the Klein bottle.

F89: [NaNeOtSi03] For any closed surface S, there is a triangulation of the sphere whose skeleton quadrangulates the surface S.

F90: [NaNeOtSi03] No 5-connected graph triangulates the sphere and quadrangulates another orientable closed surface.

F91: [NeSu00] There is a 5-connected graph that triangulates the sphere which quadrangulates the nonorientable closed surface N_k if and only if $k = 10$ or $k \geq 12$. Such a triangulation for $k = 10$ is unique up to isomorphism.

F92: [Su03] If two closed surfaces S_1 and S_2 satisfy the relation $2\chi(S_1) - \chi(S_2) \geq 4$, then there is a graph that triangulates S_1 and quadrangulates S_2.

EXAMPLES

E22: The octahedron $K_{2,2,2}$ triangulates the sphere and quadrangulates the torus.

E23: The complete graph K_4 triangulates the sphere and quadrangulates the projective plane.

E24: The only irreducible triangulation on the torus whose skeleton also triangulates the Klein bottle is T3, and the triangulation on the Klein bottle is isomorphic to Kh1.

E25: [HaRi89] The complete graph K_n triangulates and quadrangulates two different closed surfaces if and only if $n \equiv 0, 1, 4$ or 9 (mod 12). This is exactly what Euler's formula requires.

References

[Ar92] D. Archdeacon, The medial graph and voltage-current duality, *Discrete Math.* 104 (1992) 111–141.

[ArBrNe95] J. L. Arocha, J. Bracho, and V. Neumann-Lara, Tight and untight triangulated surfaces, *J. Combin. Theory, Ser. B* 63 (1995), 185–199.

[Ba82] D. Barnette, Generating the triangulations of the projective plane, *J. Combin. Theory, Ser. B* 33 (1982), 222–230.

[BaEd89] D. W. Barnette and A. L. Edelson, All 2-manifolds have finitely many minimal triangulations, *Isr. J. Math.* 67 (1989), 123–128.

[BoGrGrSi00] C. P. Bonnington, M. J. Grannell, T. S. Griggs, and J. Širáň, Exponential families of non-isomorphic triangulations of complete graphs, *J. Combin. Theory, Ser. B* 78 (2000), 169–184.

[Bo82a] A. Bouchet, Constructing a covering triangulation by means of a nowhere-zero flow, *J. Combin. Theory, Ser. B* 32 (1982), 316–325.

[Bo82b] A. Bouchet, Constructions of covering triangulations with folds, *J. Graph Theory* 6 (1982), 57–74.

[BrSt01] J. Bracho and R. Strausz, Nonisomophic complete triangulations of a surface, *Discrete Math.* 232 (2001), 11–18.

[BrNaNe96] R. Brunet, A. Nakamoto, and S. Negami, Diagonal flips of triangulations on closed surfaces preserving specified properties, *J. Combin. Theory, Ser. B* 68 (1996), 295–309.

[CoNa00a] C. Cortés and A. Nakamoto, Diagonal flips in outer-torus triangulations, *Discrete Math.* 216 (2000), 71–83.

[CoNa00b] C. Cortés and A. Nakamoto, Diagonal flips in outer-Klein-bottle triangulations, *Discrete Math.* 222 (2000), 41–50.

[CoGlMaNa02] C. Cortés, C. Glima, A. Márquez, and A. Nakamoto, Diagonal flips in outer-triangulations of surfaces, *Discrete Math.* 254 (2002), 63–74.

[De73] A. K. Dewdney, Wagner's theorem for the torus graphs, *Discrete Math.* 4 (1973), 139–149.

[EdRe97] P. H. Edelman and V. Reiner, Catalan triangulations of Möbius band, *Graphs Combin.* 13 (1997), 231–243.

[FiNoNe94] S. Fisk, B. Mohar, and R. Nedela, Minimal locally cyclic triangulations of the projective plane, *J. Graph Theory* 18 (1994), 25–35.

[Fr34] P. Franklin, A six color problem, *J. Math. Phys.* 16 (1934), 363–369.

[GaRiSe96] Z. Gao, R. B. Richter, and P. Seymour, Irreducible triangulations of surfaces, *J. Combin. Theory, Ser. B* 68 (1996), 206–217.

[GaUrWa01] Z. Gao, J. Urrutia, and J. Wang, Diagonal flips in labelled planar triangulations, *Graphs Combin.* 17 (2001), 647–657.

[Gr74] J. L. Gross, Voltage graphs, *Discrete Math.* 9 (1974), 239–246.

[HaRi89] N. Hartsfield and G. Ringel, Minimal quadrangulations of nonorientable surfaces, *J. Combin. Theory, Ser. A* 50 (1989), 185–195.

[HaRi91] N. Hartsfield and G. Ringel, Clean triangulations, *Combinatorica* 11 (1991), 145–155.

[HoGl77] N. P. Homenko and A. D. Gluhov, Upper imbeddable graphs (Russian), *Graph Theory* (Russian), pp. 85–89, Inst. Math., Akad. Nauk Ukrain. SSR, Kiev, 1977.

[JoWo10] G. Joret and D. R. Wood, Irreducible triangulations are small, *J. Combin. Theory, Ser. B* 100 (2010), 446–455.

[JuRi80] M. Jungerman and G. Ringel, Minimal triangulations on orientable surfaces, *Acta Math.* 145 (1980), 121–154.

[KoNaNe99] H. Komuro, A. Nakamoto, and S. Negami, Diagonal flips in triangulations on closed surfaces with minimum degree at least 4, *J. Combin. Theory, Ser. B* 76 (1999), 68–92.

[La87] S. Lawrencenko, The irreducible triangulations of the torus, *Ukrain. Geom. Sb.* 30 (1987), 52–62. [In Russian; MR 89c:57002; English translation: *J. Soviet Math.* 51, No. 5 (1990), 2537–2543.]

[La92] S. Lawrencenko, The variety of triangular embeddings of a graph in the projective plane, *J. Combin. Theory, Ser. B* 54 (1992), 196–208.

[LaNe97] S. Lawrencenko and S. Negami, Irreducible triangulations of the Klein bottle, *J. Combin. Theory, Ser. B* 70 (1997), 265–291.

[LaNe99] S. Lawrencenko and S. Negami, Constructing the graphs that triangulate both the torus and the Klein bottle, *J. Combin. Theory, Ser. B* 77 (1999), 211–218.

[LaNeWh94] S. Lawrencenko, S. Negami, and A. T. White, Three nonisomorphic triangulations of an orientable surface with the same complete graph, *Discrete Math.* 135 (1994), 367–369.

[MaMo92] A. Malnič and B. Mohar, Generating locally cyclic triangulations of surfaces, *J. Combin. Theory, Ser. B* 56 (1992), 147–164.

[MaNe95] A. Malnic and R. Nedela, k-minimal triangulations of surfaces, *Acta Math. Univ. Comenian* (N.S.) 64 (1995), 57–76.

[MoNaOt03] R. Mori, A. Nakamoto, and K. Ota, Diagonal flips in Hamiltonian triangulations on the sphere, *Graphs Combin.* 19 (2003), 413–418.

[MoNa03] R. Mori and A. Nakamoto, Diagonal flips in Hamiltonian triangulations on the projective plane, *Discrete Math.* 303 (2005), 142–153.

[NaOt97] A. Nakamoto and K. Ota, Diagonal transformations of graphs and Dehn twists of surfaces, *J. Combin. Theory, Ser. B* 70 (1997), 292–300.

[NaNe02] A. Nakamoto and S. Negami, Generating triangulations on closed surfaces with minimum degree at least 4, *Discrete Math.* 244 (2002), 345–349.

[NaNeOtSi03] A. Nakamoto, S. Negami, K. Ota, and J. Širáň, Planar triangulations which quadrangulate other surfaces, *European J. Combin.* 25 (2004), 817–833.

[Ne83] S. Negami, Uniqueness and faithfulness of embedding of toroidal graphs, *Discrete Math.* 44 (1983), 161–180.

[Ne84a] S. Negami, Uniquely and faithfully embeddable projective-planar triangulations, *J. Combin. Theory, Ser. B* 36 (1984), 189–193.

[Ne84b] S. Negami, Classification of 6-regular Klein-bottlal graphs, *Res. Rep. Inf. Sci. T.I.T.* A-96 (1984).

[NeWa90] S. Negami and S. Watanabe, Diagonal transformations of triangulations on surfaces, *Tsukuba J. Math.* 14 (1990), 155–166.

[Ne94] S. Negami, Diagonal flips in triangulations of surfaces, *Discrete Math.* 135 (1994), 225–232.

[NeMi96] S. Negami and T. Midorikawa, Loosely-tightness of triangulations of closed surfaces, *Sci. Rep. Yokohama Nat. Univ., Sec. I* 43 (1996), 25–41.

[NeNaTa97] S. Negami, A. Nakamoto, and T. Tanuma, Re-embedding structures of triangulations on closed surfaces, *Sci. Rep. Yokohama Nat. Univ., Sec. I* 44 (1997), 41–55.

[Ne98] S. Negami, Diagonal flips in triangulations on closed surfaces, estimating upper bounds, *Yokohama Math. J.* 45 (1998), 113–124.

[Ne99a] S. Negami, Diagonal flips of triangulations on surfaces, a survey, *Yokohama Math. J.* 47, special issue (1999), 1–40.

[Ne99b] S. Negami, Note on frozen triangulations on closed surfaces, *Yokohama Math. J.* 47, special issue (1999), 191–202.

[Ne01] S. Negami, Diagonal flips in pseudo-triangulations on closed surfaces, *Discrete Math.* 240 (2001), 187–196.

[NeSu00] S. Negami and Y. Suzuki, 5-Connected planar triangulations quadrangulating other surfaces, *Yokohama Math. J.* 47 (2000), 187–193.

[Ne05] S. Negami, Looseness ranges of triangulations on closed surfaces, *Discrete Math.* 303 (2005), 167–174.

[Ri55] G. Ringel, Wie man die geschlossenen nichtorientierbaren Flächen in möglichst wenig Dreiecke zerlengen kann, *Math. Annalen* 130 (1955), 317–326.

[Ri74] G. Ringel, *Map Color Theorem*, Springer-Verlag, 1974.

[Sa03] A. Sasao, Panel structures of triangulations on the torus, *Discrete Math.* 303 (2005), 186–208.

[SlTaTh88] D. Sleator, R. Tarjan and W. Thurston, Rotation distance, triangulations, and hyperbolic geometry, *J. Amer. Math. Soc.* 1 (1988), 647–681.

[StRa34] E. Steinitz and H. Rademaher, *Vorlesungen über die Theorie der Polyeder*, Springer, Berlin, 1934.

[Su03] Y. Suzuki, Triangulations on closed surfaces which quadrangulate other surfaces, *Discrete Math.* 303 (2005), 234–242.

[Sul06] T. Sulanke, Note on the irreducible triangulations of the Klein bottle, *J. Combin. Theory, Ser. B* 96 (2006), 964–972.

[Wa36] K. Wagner, Bemekungen zum Vierfarbenproblem, *J. der Deut. Math., Ver.* 46, Abt. 1, (1936), 26–32.

Section 7.9

Graphs and Finite Geometries

Arthur T. White, Western Michigan University

INTRODUCTION

To every finite geometry, there are associated two distinct graphs, the Levi graph and the Menger graph. Imbeddings of these graphs into closed 2-manifolds can lead to models of the geometries. The emphasis is on imbeddings that are optimal with respect to euler characteristic, or that utilize important collineations of the geometry as map automorphisms.

The graphs in this section are simple.

7.9.1   Finite Geometries

A finite geometry might be of intrinsic interest; it might produce a useful block design; or, it might have an aesthetically pleasing model.

DEFINITIONS

D1: A *geometry* (P, L) consists of a non-empty set P called *points*, together with a non-empty collection L of subsets of P called *lines*.

D2: A *finite geometry* is a geometry (P, L) whose point-set P is finite.

D3: A (v, b, r, k, λ)-*balanced incomplete block design* (abbr. *BIBD*) is a finite geometry (P, L) with $v = |P|$ and $b = |L|$ that satisfies the following three *axioms of uniformity*:

- Every point is in exactly r lines.
- Every line consists of exactly k points.
- Every pair of points belong to exactly λ common lines.

D4: The *axiom of uniqueness* for a (v, b, r, k, λ)-BIBD is $\lambda = 1$.

D5: A *Steiner triple system* is a (v, b, r, k, λ)-BIBD with $k = 3$ and $\lambda = 1$.

D6: A $(v, b, r, k; \lambda_1, \lambda_2)$-***partially balanced incomplete block design*** is just like a BIBD, except that the points can be regarded as the vertices of a fixed strongly regular graph, in which two points are non-adjacent (or adjacent) if they belong to exactly λ_1 (or λ_2) common lines (abbr. ***PBIBD***).

D7: An (r, k)-***configuration*** is a finite geometry (P, L) that satisfies the first two axioms of uniformity, but replaces the third axiom with the following:

- Every pair of points belong to at most one common line.

D8: A ***3-configuration*** is an (r, k)-configuration such that $k = 3$.

D9: A $(v, b, r, k; 0, 1)$-***block design*** (abbr. ***BD***) is an (r, k)-configuration (P, L), where $v = |P|$ and $b = |L|$, such that no or some pairs of distinct points are not collinear ($\lambda_1 = 0$), and the other pairs of distinct points are uniquely collinear ($\lambda_2 = 1$).

D10: A ***symmetric configuration*** $(v)_k$ is a $(v, b, r, k; 0, 1)$ block design such that $r = k$, which implies $v = b$, by Fact F1 below.

D11: For each natural number $n > 1$, the n-***point geometry*** has n points, and all the 2-subsets of those points as its lines.

D12: A ***finite affine plane of order*** n is a finite geometry (P, L) that satisfies these axioms:

- Two distinct points are in a unique common line.
- For a given point not in a given line, there is a unique parallel (non-intersecting) line containing that point.
- There exist four distinct points, no three collinear.
- There exists a line having exactly n points.

D13: A ***finite projective plane of order*** n (abbr. $\Pi(n)$) is a finite geometry (P, L) that satisfies these axioms:

- Two distinct points are in a unique common line.
- Two distinct lines contain a unique common point.
- There exist four distinct points, no three collinear.
- There exists a line having exactly n + 1 points.

FACTS

F1: In any (v, b, r, k, λ)-BIBD, $(v, b, r, k; 0, 1)$-PBIBD or $(v, b, r, k; 0, 1)$-BD, the axioms of uniformity imply that $vr = bk$.

F2: A 3-configuration (P, L) exists if and only if $vr = 3b$ and $v \geq 2r + 1$. Thus, a symmetric configuration $(v)_3$ exists if and only if $v \geq 7$.

F3: For a prime power $n \geq 9$ that is not prime, there are at least two non-isomorphic projective planes of order n. There are exactly four, for $n = 9$.

F4: For $n = 2, 3, 4, 5, 7$, and 8, $\Pi(n)$ is uniquely $PG(2, n)$.

F5: Neither a $\Pi(6)$ nor a $\Pi(10)$ exists.

F6: There is no known finite projective plane $\Pi(n)$ where n is not a prime power.

F7: [BrRy49] If $n \equiv 1, 2 \pmod 4$, and if n is not a sum of two squares, then no projective plane $\Pi(n)$ exists.

F8: The affine plane $AG(2, q)$ is a resolvable $(q^2, q^2 + q, q + 1, q, 1)$-BIBD. Such planes exist for every prime power q. The resolvable feature is that the $q^2 + q$ lines partition into $q + 1$ parallel classes of q lines each, each class partitioning the point set.

F9: Every triangle in the Euclidean plane has the following four triples of concurrent lines:

- The perpendicular bisectors of the three sides meet in the ***circumcenter***.

- The altitudes meet in the ***orthocenter***.

- The internal angle bisectors meet in the ***incenter***.

- The medians meet in the ***centroid***.

If the triangle is equilateral, then all four points of concurrency coincide. This fact is used as background for Example E10.

EXAMPLES

E1: The Euclidean plane is an (infinite) affine plane.

E2: The n-point geometry is an $\left(n, \dfrac{n(n-1)}{2}, n-1, 2, 1 \right)$-BIBD.

E3: An r-regular graph is representable as an $(r, 2)$-configuration.

E4: An r-regular, k-uniform hypergraph is representable as an (r, k)-configuration.

E5: The Theorem of Pappus in Euclidean geometry states that if A, B, C are distinct points on line L and A', B', C' are three different distinct points on line $L' \neq L$, then the three points $D = AB' \cap A'B, E = AC' \cap A'C, F = BC' \cap B'C$ are collinear. This gives a 3-configuration on the nine points $\{A, B, C, A', B', C', D, E, F\}$ called the ***geometry of Pappus***, which is a $(9, 9, 3, 3; 0, 1)$-PBIBD. (See [Wh01], for example, for a diagram of the geometry of Pappus.)

E6: The Theorem of Desargues in Euclidean geometry states that if triangles ABC and $A'B'C'$ are in perspective from point P, then the three points $D = AB \cap A'B', E = AC \cap A'C', F = BC \cap B'C'$ are collinear. This gives a 3-configuration on the ten points $\{P, A, B, C, A', B', C', D, E, F\}$ called the ***geometry of Desargues***, which is a $(10, 10, 3, 3; 0, 1)$-PBIBD. (See [Wh01], for example, for a diagram of the geometry of Desargues.)

E7: For each prime power q, there is a classical finite affine plane $AG(2, q)$, which is a $(q^2, q^2 + q, q + 1, q, 1)$-BIBD with $\lambda = 1$. The affine plane $AG(2, q)$ has as its points the 1-dimensional affine subspaces of the 3-dimensional vector space over $GF(q)$, and as its lines the 2-dimensional affine subspaces. In particular, $AG(2, 2)$ is the 4-point geometry.

E8: For each prime power q, there is a classical finite projective plane $\Pi(q) = PG(2, q)$, which is a $(q^2 + q + 1, q^2 + q + 1, q + 1, q + 1, 1)$-BIBD with $\lambda = 1$. The projective plane $PG(2, q)$ has as its points the 1-dimensional vector subspaces of the 3-dimensional vector space over $GF(q)$, and as its lines the 2-dimensional vector subspaces.

E9: To every projective plane Π there corresponds an affine plane Π' (obtained by deleting one edge and all points on that edge from Π), and conversely. The affine plane obtained from $PG(2, q)$ is $AG(2, q)$.

E10: The ***Fano plane***, a familiar 3-configuration that models $PG(2, 2)$, is shown in Figure 7.9.1. It has seven points. The three sides, the three medians, and the incircle form its seven lines. This $(7, 7, 3, 3, 1)$-BIBD is the smallest non-trivial Steiner triple system.

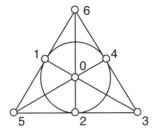

Figure 7.9.1: The Fano plane.

The line $\{0, 1, 3\}$ is called a ***perfect difference set***, since each non-identity element of $\mathbb{Z}_7$ appears uniquely as a difference of two elements in the set. It generates the other six lines, using translations by $\mathbb{Z}_7$. But note that this model has several defects (to be remedied in §7.9.3):

- The circular line $\{1, 2, 4\}$ is differently depicted. Yet $\mathbb{Z}_7$ acts transitively on the line set L, as a subgroup of the full collineation group (of order 168).

- The point 0 is distinguished by its central position, yet $\mathbb{Z}_7$ acts transitively on the point set P as well.

- The point 2 (for example) seems to be between points 3 and 5. Yet there is no concept of betweenness in this geometry.

- One cannot discern that $r = 3$ by looking at small neighborhoods of points $0, 1, 2$, and 4.

- There are three extraneous intersections of lines (i.e., of the three cevians with the circle) that have no meaning in the geometry.

E11: The projective plane $PG(2, 4)$ can be cyclically generated, using $\mathbb{Z}_{21}$, from the perfect difference set $\{0, 1, 6, 8, 18\} = L_0$. Then line $L_i = L_0 + i$, for $0 \le i \le 20$. Let $L_0' = \{0, 6, 18\}$, with $L_k' = L_0' + 3k$, for $0 \le k \le 6$. Then using the $\mathbb{Z}_7$ subgroup of $\mathbb{Z}_{21}$, we find the Fano plane $PG(2, 2)$ contained within $PG(2, 4)$.

E12: The Theorem of Desargues applies to the full projective plane $PG(2, 4)$; for example, the triangles 3 5 7 and 10 13 9 are in perspective from point 2, producing the three collinear points $19, 6, 20$ and the geometry of Desargues, contained within $PG(2, 4)$. Next, deleting line L_0 and its five points from $PG(2, 4)$ yields $AG(2, 4)$, with its five parallel classes, as shown below.

2	**7**	**9**	**19**	2	3	10	20	3	4	9	11	5	11	13	2	10	11	16	**7**
4	5	10	12	7	13	15	4	**13**	**14**	**19**	10	7	12	14	3	12	**13**	**20**	**9**
14	15	20	11	9	14	16	5	15	16	2	12	9	10	15	17	17	2	4	14
16	17	3	13	11	12	17	19	**20**	**5**	**7**	17	19	20	4	16	**19**	**3**	**5**	15

Notice that the bold (sub)lines give the three parallel classes of a Pappus configuration within $AG(2, 4)$ and hence within $PG(2, 4)$.

7.9.2   Associated Graphs

DEFINITIONS

D14: The **Menger graph** of a geometry (P, L) has the point set P as its vertex set and all edges of the form $\{p_1, p_2\}$, for $p_1, p_2 \in P$, where p_1 and p_2 are collinear; i.e., $p_1, p_2 \in l$, for some $l \in L$.

D15: The **Levi graph** of a geometry (P, L) is the bipartite graph having $P \cup L$ as its vertex set and all edges of the form $\{p, l\}$, for $p \in P, l \in L$, and $p \in l$. (Similarly, Levi graphs can encode objects and blocks of a block design, or vertices and hyperedges of a hypergraph.)

D16: A (d, g)-**cage** is a graph of minimum order among all d-regular graphs having girth g.

FACTS

F10: In general, every BIBD of order v has K_v as its Menger graph.

F11: A Levi graph represents its geometry uniquely, whereas two or more different geometries might have the same Menger graph.

F12: The Levi graph of an (r, k)-configuration has girth $g \geq 6$. The Levi graph of a symmetric configuration $(v)_k$ is k-regular.

F13: [Si66] There is an $(n + 1, 6)$-cage of order $2(n^2 + n + 1)$ if and only if there exists a finite projective plane $\Pi(n)$.

F14: The Levi graph of $PG(2, n)$ is an $(n + 1, 6)$-cage of order $2(n^2 + n + 1)$, for n a prime power.

F15: If the Menger graph of a $(v, b, r, k; 0, 1)$-BD is strongly regular, then the geometry is a partially balanced incomplete block design.

EXAMPLES

E13: In general, the Menger graph of n-point geometry is K_n, while the Levi graph of that geometry is homeomorphic to K_n (each edge of K_n is replaced by a path of length two).

E14: The projective plane $PG(2, 2)$ and the 7-point geometry are clearly different geometries, having 7 and 21 lines, respectively, yet both have Menger graph K_7.

E15: The Levi graph of $PG(2, 2)$ is the Heawood graph, the unique $(3, 6)$-cage. The Levi graph of 7-point geometry is obtained from K_7 by performing an elementary subdivision on each edge, that is, by replacing each edge with a path of length two.

E16: The Menger graph of a 3-configuration $(v, b, r, 3; 0, 1)$ is $2r$-regular of order v, and its edge set decomposes into b disjoint 3-cycles. For instance, the strongly-regular octahedral graph is the Menger graph for a $(6, 4, 2, 3; 0, 1)$-PBIBD called the "Pasch configuration". The shading in Figure 7.9.2 below depicts the C_3-decomposition.

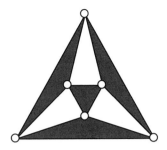

Figure 7.9.2: The Pasch configuration.

E17: The Menger graph for the geometry of Pappus is $K_{3(3)}$, the complement of the graph $3K_3$. The Menger graph for the geometry of Desargues is the complement of the Petersen graph. Since both Menger graphs are strongly regular, the corresponding incomplete block designs are both partially balanced.

7.9.3   Surface Models

DEFINITIONS

D17:   A *model for an axiom system* Σ for a finite geometry (P, L) is an interpretation of the points and lines of Σ such that each interpreted axiom in Σ is a true statement about (P, L).

D18:   A model of a geometry (P, L) is *abstract* if it specifies P as an abstract set of points and L as a collection of subsets of P. It is *concrete* if it represents P as a finite set of points of R^n, for some positive integer n, and L as a collection of locally one-dimensional subsets of R^n (often a geometric realization of a graph or of a hypergraph).

D19:   An axiom system Σ is *consistent* if no contradictions can be derived from it.

D20:   An axiom system Σ is *independent* if no axiom in Σ can be derived from the other axioms in Σ.

D21:   An axiom system Σ is *complete* if every statement in the undefined and defined terms of Σ can either be proven true or proven false, using Σ.

D22:   The *genus of a finite geometry* is the genus of its Levi graph.

FACTS
The first three facts below indicate the importance of models in the formal study of geometry. Informally, models have heuristic and pedagogical functions, and might even provide aesthetic pleasure.

F16:   An axiom system Σ is consistent if there is a model for Σ.

F17:   An axiom system Σ is independent if, for each axiom $\sigma \in \Sigma$, a model can be found satisfying $(\Sigma - \sigma) \wedge (\sim \sigma)$.

F18:   An axiom system Σ is complete, if there is a unique (up to isomorphism) model for Σ.

F19:   The genus of n-point geometry $(n > 2)$ is $\lceil (n-3)(n-4)/12 \rceil$. In consideration of Example 13, this follows from [RiYo68].

F20:   [FiWh00] The geometry of Pappus has genus 1. That of Desargues has genus 2. The Desargues model has a 3-fold rotational symmetry that fixes the point of perspectivity, the line of perspectivity, the two triangles of perspectivity, and nothing else.

F21:   [Wh95] Surface models for $PG(2,q)$, q a prime power, depend upon the residue of $q \pmod 3$:

(a) If $q \equiv 2 \pmod 3$, then $PG(2,q)$ has genus $1 + (q-2)(q^2 + q + 1)/3$; all the hyperregions are triangular.

(b) If $q \equiv 1 \pmod 3$, then $PG(2,q)$ can be modeled on the surface of genus $1 + (q-1)(q^2 + q + 1)/3$, with $q^2 + q + 1$ hyperregions pentagonal and all others triangular.

(c) If $q \equiv 0 \pmod 3$, then $PG(2,q)$ is conveniently modeled on an orientable pseudosurface of characteristic $(3 - 2q)(q^2 + q + 1)/3$, with $q^2 + q + 1$ hyperregions quadrilateral and all others triangular.

In each case, the group $\mathbb{Z}_{q^2+q+1}$ acts regularly, as a group of map automorphisms, on the point set, the line set, and on each orbit of the region set for the modified Levi graph imbedding.

F22:   Topological models for $AG(2,q)$ are obtained by deletions from the above models for $PG(2,q)$.

F23:   There is a toroidal symmetric 3-configuration $(v)_3$ on v points for all $v \geq 7$.

F24:   [Wh02] There is a 3-configuration $(3n, n^2, n, 3; 0, 1)$-PBIBD with Menger graph $K_{n,n,n}$ and genus $(n-1)(n-2)/2$ for all $n > 1$. This generalizes the Pasch configuration.

F25:   An imbedding of the Levi graph for a finite geometry on a closed orientable 2-manifold (a surface) can be readily modified to model the geometry (on the same surface) by an imbedded graph G having bichromatic dual: vertices model points, and boundaries of regions of one fixed color model lines. (The regions of the other color are hyperregions.) If the geometry is a 3-configuration, then G is its Menger graph. The process reverses, so that the Levi graph can be obtained from such an imbedding of G.

F26:   Familiar classes of strongly regular graphs include the regular complete m-partite graphs $K_{m(n)}$ (where $m, n > 1$) and the line graphs $L(K_n)$ and $L(K_{n,n})$.

F27: $L(K_{n,n}) = K_n \times K_n$, the cartesian product.

EXAMPLES

E18: An axiom system for the n-point geometry is as follows:

(i) There are exactly n points.

(ii) Two distinct points belong to a unique common line.

(iii) Each line consists of exactly two points.

This axiom system is consistent, since the complete graph K_n is a model. For $n = 4$, we have the following abstract model:

$$P = \{1, 2, 3, 4\} \qquad L = \{\{1,2\}, \{1,3\}, \{1,4\}, \{2,3\}, \{2,4\}, \{3,4\}\}$$

The graph in Figure 7.9.3 is a concrete model (which could be imbedded in R^2 or, reversing the Riemann stereographic projection, imbedded on the sphere in R^3).

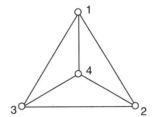

Figure 7.9.3: A concrete model for the 4-point geometry.

This axiom system is independent, for $n > 3$, since:

- The complete graph K_{n+1} shows that axiom (i) does not follow from axioms (ii) and (iii).

- The cycle graph C_n shows that axiom (ii) does not follow from axioms (i) and (iii).

- The path graph P_n (a one-line geometry) shows that axiom (iii) does not follow from axioms (i) and (ii).

This axiom system is complete, since all models are isomorphic to K_n. (The abstract and concrete models above for $n = 4$, for example, are isomorphic under the identity function on the point sets, inducing the identity function on the line sets.)

E19: Medial graph constructions serve to model geometries having Menger graphs $L(K_n)$. For $G = K_4$ as in Figure 7.9.3, we have $M(G) = L(K_4) = K_{3(2)}$, as in Figure 7.9.2. For $n > 4$, the models will be on pseudosurfaces.

E20: Surgical constructions can be employed to model geometries having Menger graphs $L(K_{n,n})$, using Fact F27 and appropriate imbeddings of K_n, for $n \equiv 1$ or 3 mod 6; these imbeddings will have bichromatic dual with one color class consisting of all triangles. These models will be on generalized pseudosurfaces.

E21: The imbedded Menger graph of Example E16 of §7.9.2 can be readily modified to obtain a spherical imbedding of the corresponding Levi graph. Thus the Pasch configuration has genus 0.

E22: Conversely, the Levi graph for the Fano plane is the Heawood graph, the unique $(3, 6)$-cage, which has genus 1. The modification of a toroidal imbedding of the Heawood graph (see Figure 7.9.4(a) below) gives a bichromatic-dual imbedding of K_7, the Menger graph for the Fano plane (see Figure 7.9.4(b) below). The seven triangular regions of either color class model the lines of this geometry. Thus the Fano plane has genus one. All the defects of the traditional model for this geometry, displayed in §7.9.1 (where the lines correspond to the unshaded regions in Figure 7.9.4(b)) are now remedied.

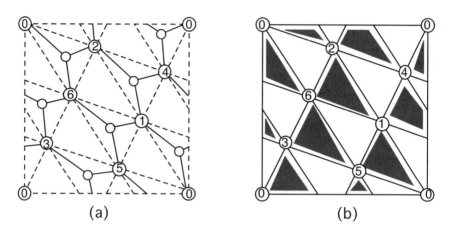

(a)   (b)

Figure 7.9.4: Levi graphs for (a) the Fano plane and (b) the Menger graph.

E23: Voltage graph constructions are useful in forming surface models of geometries having Menger graphs $K_{m(n)}$. The imbedded Menger graph of Example E17 covers the voltage graph imbedding of Figure 7.9.5, which uses the voltage group $B = \mathbb{Z}_3 \times \mathbb{Z}_3$.

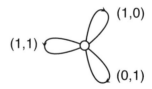

Figure 7.9.5: An imbedded bouquet with voltages in $\mathbb{Z}_3 \times \mathbb{Z}_3$; the covering graph is the Menger graph for the Pappus geometry.

References

[BrRy49] R. H. Bruck and H. J. Ryser, The nonexistence of certain finite projective planes, *Canad. J. Math.* 1 (1949), 88–93.

[FiWh00] R. M. Figueroa-Centeno and A. T. White, Topological models for classical configurations, *J. of Statistical Planning and Inference* 86 (2000), 421–434.

[RiYo68] G. Ringel and J. W. T. Youngs, Solution of the Heawood map-coloring problem, *Proc. Nat. Acad. Sci. U.S.A.* 60 (1968), 438–445.

[Si66] R. R. Singleton, On minimal graphs of maximum even girth, *J. Combin. Theory* 1 (1966), 306–332.

[Wh95] A. T. White, Efficient imbeddings of finite projective planes, *Proc. London Math. Soc.* (3) 70 (1995), 33–55.

[Wh01] A. T. White, *Graphs of Groups on Surfaces: Interactions and Models*, North-Holland Mathematical Studies 188, 2001.

[Wh02] A. T. White, Modelling finite geometries on surfaces, *Discrete Math.* 244 (2002), 479–493.

Section 7.10
Crossing Numbers

R. Bruce Richter, University of Waterloo
Gelasio Salazar, Universidad Autónoma de San Luis Potosí

INTRODUCTION

When a graph cannot be drawn in the plane without edge crossings, it is natural to seek plane representations of it in which the number of crossings is minimized. The crossing number of a graph G is the minimum number of crossings in a drawing of G in the plane. Before the turn of the century, most results in the field revolved around the estimation or calculation of the crossing number of a graph or of a family of graphs. In the last fifteen years or so we have witnessed a remarkable increase in the number of deep, general results of a structural character, giving rise to a flourishing theory of crossing numbers.

7.10.1   Drawings of Graphs and Crossing Numbers

For both theoretical and practical reasons, it is often needed or desired to generate a representation of a graph in some topological space. In an *embedding* of a graph, the most natural injective properties are satisfied, whereas in a *drawing*, some injectivity properties are relaxed.

Embeddings

DEFINITIONS

D1: An *embedding* of a graph G in a topological space X is a mapping that associates to the vertices of G distinct elements of X, and to each edge of G a homeomorphic image of $[0, 1]$, so that the endpoints of the image of an edge are precisely the elements of X representing the endvertices of the edge, and if the images of any two edges have a point in common, then this point represents a common vertex.

EXAMPLES

E1: Every finite graph that does not contain as a subgraph a subdivision of $K_{3,3}$ or K_5 can be embedded into $\mathbb{R}^2$ (Kuratowski's Theorem).

E2: Every countable graph can be embedded into $\mathbb{R}^3$, by placing the vertices on a line L and then, for each edge e, using a different plane P_e containing L to draw e.

Drawings

The preceding example justifies why most research on graph representations is focused on 2-dimensional spaces, such as the plane or compact surfaces. More often than not, an input graph cannot be embedded in a given surface. Thus we relax one condition of an embedding, and arrive at a natural concept of a drawing.

DEFINITIONS

D2: A *drawing of a graph* G in a surface S is a mapping that assigns to the vertices of G distinct elements of S, and assigns to each edge a homeomorphic image of $[0, 1]$, disjoint from the vertex points except that the endpoints of the image of an edge are precisely the elements of S representing the endvertices of the edge.

D3: A drawing of a graph G in a surface S is a *normal drawing* if
  (i) the interiors of the images of any two edges have at most one intersection point and, moreover, if such an intersection point exists then it is a *crossing* (rather than tangential);
  (ii) no point in S belongs to the interiors of the images of three distinct edges (loosely speaking, no three edges cross at a common point).

REMARKS

R1: It is an easy exercise to show that if D is any drawing of a graph G in a surface S, then there is a normal drawing D' of G in S with no more crossings than D.

R2: We follow the common practice to make no distinction between the vertices and edges of a graph and the corresponding objects (points or closed arcs) in the host topological space in an embedding or a drawing of the graph. Thus we allow ourselves, for instance, to use expressions such as "crossings between edges", instead of the unnecessarily precise "crossings between the closed arcs representing edges".

Crossing Numbers

More often than not, it is impossible to draw a graph on a surface without crossings of edges, and our attention then turns to the matter of minimizing the number of crossings over all drawings of the graph upon consideration.

DEFINITION

D4:   The **crossing number** of a graph G in a surface S, denoted $\mathrm{cr}_S(G)$, is the minimum number of crossings of edges in a drawing of G in S.

REMARKS

R3:   We are mostly concerned with the crossing number of graphs in the plane. Thus, for simplicity, whenever the host surface S is the plane, we will omit the explicit reference to S and simply write $\mathrm{cr}(G)$.

R4:   The definition of crossing number given above is the most common one, but many variants have been considered in the literature. Although we shall be mostly concerned with results around this notion of crossing number, we will also state a few results on some of these variants (see §7.10.6).

EXAMPLES

E3:   The complete bipartite graph $K_{3,3}$ has crossing number 1. It is not difficult to show, using the Jordan Curve Theorem, that $K_{3,3}$ cannot be drawn in the plane without crossings, and so $\mathrm{cr}(K_{3,3}) \geq 1$. On the other hand, $K_{3,3}$ can be drawn in the plane with exactly one crossing (see Figure 7.10.1), and so the reverse inequality $\mathrm{cr}(K_{3,3}) \leq 1$ also holds.

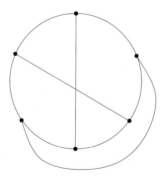

Figure 7.10.1: A drawing of $K_{3,3}$ with one crossing, showing $\mathrm{cr}(K_{3,3}) \leq 1$.

7.10.2   General Techniques and Bounds

There are few general results relating the crossing number of a graph to its number of vertices and edges. Some elementary bounds can be obtained by a trivial application of Euler's Formula, although these tend to be of little use or interest. By far the most powerful general tool known is the Crossing Lemma, which is not only of theoretical interest, but also has a surprising number of connections and applications to other branches of mathematics (see §7.10.8).

The Crossing Lemma

The following statement is perhaps the most fundamental result in the field of crossing numbers. It was conjectured by Erdős and Guy in [EG73], and proved by Ajtai, Chvátal, Newborn, and Szemerédi [ACNS82], and independently by Leighton [Le83]. Below we state the Crossing Lemma and some of its most interesting variants and refinements.

FACTS

F1: (The Crossing Lemma) There exists a positive constant c such that, for every graph G with n vertices and $m \geq 4n$ edges, $\mathrm{cr}(G) \geq c \cdot m^3/n^2$.

The current best upper and lower bounds for the *crossing constant c* (roughly 0.09 and 1/33.75) were obtained in [PT97].

F2: [PST00] Let g be a positive integer. There is a positive constant $c := c(g)$ such that if G has girth at least $2g + 1$ and $m \geq 4n$, then $\mathrm{cr}(G) \geq c \cdot m^{2+g}/n^{1+g}$.

F3: [FPT10] For every $k \in \mathbb{N}$, there is a constant $c_k > 0$ such that in every drawing of a graph with n vertices and $m \geq 3n$ edges, no two of which intersect in more than k points, there are disjoint sets of edges E_1, E_2, each of size at least $c_k m^2/n^2$, such that every edge in E_1 crosses all edges in E_2.

F4: [PRTT06] Every graph G with n vertices and m edges satisfies

$$\mathrm{cr}(G) \geq (1/31.1) \cdot m^3/n^2 - 1.06n$$

F5: [PRTT06] Every graph G with n vertices and $m \geq (103/6)n$ edges satisfies

$$\mathrm{cr}(G) \geq (1024/31827) \cdot m^3/n^2$$

Crossing Numbers, Bisection Width, and Cutwidth

The crossing number is closely related to other important graph theoretical parameters.

DEFINITIONS

D5: The ***bisection width*** $\mathrm{bw}(G)$ of a graph $G = (V, E)$ is the minimum number of edges whose removal divides G into two parts having at most $2|V|/3$ vertices each.

D6: Let $G = (V, E)$ be a graph. For each bijection $\psi : V \to \{1, 2, 3, \ldots, |V|\}$ and each $i \in \{1, 2, \ldots, |V|\}$, let $\mathrm{cw}(\psi, i)$ denote $|\{uv \in E : \psi(u) < i < \psi(v)\}|$. Then the ***cutwidth*** $\mathrm{cw}(G)$ of G is $\min_\psi \max_i \mathrm{cw}(\psi, i)$.

FACTS

F6: [PSS96] Let G be a graph with n vertices of degree $d_1, d_2, \ldots, d_n$. Then

$$\mathrm{cr}(G) + (1/16) \sum_{i=1}^{n} d_i^2 \geq (1/40)\mathrm{bw}^2(G).$$

F7: [DV03] Let G be a graph with n vertices of degree $d_1, d_2, \ldots, d_n$. Then

$$\text{cr}(G) + (1/16) \sum_{i=1}^{n} d_i^2 \geq (1/1176)\text{cw}^2(G).$$

Since $\text{cw}(G) \geq \text{bw}(G)$ for any graph G, this fact refines the previous fact.

Crossing Numbers, Immersion, and Congestion

Leighton [Le83] showed how to use *immersions* and *congestion* as a general technique to obtain bounds for crossing numbers.

DEFINITIONS

D7: Let $G_1 = (V_1, E_1)$ and $G_2 = (V_2, E_2)$ be graphs. An **immersion** of G_1 in G_2 is a pair (ϕ, ψ), where $\phi : V_1 \to V_2$ is an injection and ψ associates to each edge $e = uv$ in E_1 a path in G_2 with endpoints $\phi_1(u)$ and $\phi_1(v)$.

D8: Let $G_1 = (V_1, E_1)$ and $G_2 = (V_2, E_2)$ be graphs, and let (ϕ, ψ) be an immersion of G_1 in G_2. For each $e \in E_2$, define $\text{cg}_e(\phi, \psi) = |\{f \in E_1 \,|\, e \in \psi(f)|\}$. The **congestion of the immersion** (ϕ, ψ), denoted by $\text{cg}(\phi, \psi)$, is $\max_{e \in E_2}\{\text{cg}_e(\phi, \psi)\}$.

FACT

F8: [Le83] Let $G_1 = (V_1, E_1)$ and $G_2 = (V_2, E_2)$ be graphs, and let (ϕ_1, ψ_1) be an immersion of G_1 to G_2 with congestion $\text{cg}(\phi, \psi)$. Then,

$$\text{cr}(G_2) \geq \frac{\text{cr}(G_1)}{\text{cg}^2(\phi, \psi)} - \frac{|V_2|\Delta^2(G_2)}{2}.$$

7.10.3   Crossing Numbers of Some Families of Graphs

As with any graph-theoretical parameter, it is of natural interest to investigate the crossing number of particular (families of) graphs. Unlike most graph-theoretical parameters, the crossing numbers of perhaps the most natural families of graphs (namely, the complete graphs K_n and the complete bipartite graphs $K_{m,n}$) are not known, except for a few cases.

Complete Bipartite Graphs

It is widely accepted that the field of crossing numbers can be traced to a question posed by Turán, while he worked in a labor camp during World War II [BW10]. In our current terminology, Turán raised the question of calculating $\text{cr}(K_{m,n})$.

EXAMPLE

E4: The drawing of $K_{5,6}$ in Figure 7.10.2 has 24 crossings. It is straightforward to generalize this to a drawing of $K_{m,n}$ with $Z(m, n) := \lfloor m/2 \rfloor \lfloor (m-1)/2 \rfloor \lfloor n/2 \rfloor \lfloor (n-1)/2 \rfloor$ crossings, for any m and n. These drawings are attributed to Zarankiewicz [Za54].

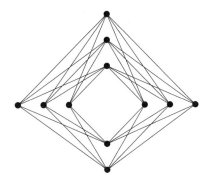

Figure 7.10.2: A drawing of $K_{5,6}$ with 24 crossings.

CONJECTURE

In [Za54], Zarankiewicz claimed to have proved that $\mathrm{cr}(K_{m,n}) = Z(m,n)$, for all positive integers m, n. Paul Kainen (1965) and Gerhard Ringel (1966) independently found a gap in Zarankiewicz's argument (see [Gu68]). This gave rise to the following conjecture, one of the most important open crossing number problems to this day.

C1: (Zarankiewicz's Conjecture) $\mathrm{cr}(K_{m,n}) = Z(m,n)$ for all positive integers m, n.

FACTS

F9: For all positive integers m and n, $\mathrm{cr}(K_{m,n}) \leq Z(m,n)$. This follows from the observation that there exist drawings of $K_{m,n}$ with exactly $Z(m,n)$ crossings.

A straightforward counting argument shows the following.

F10: (Parity Argument, [Kl70]) If $\mathrm{cr}(K_{2r-1,n}) = Z(2r-1,n)$ for some positive integers r, n, then $\mathrm{cr}(K_{2r,n}) = Z(2r,n)$.

It is not hard to prove that $\mathrm{cr}(K_{3,n}) = Z(3,n)$ for every n; for an elegant proof, see [HR90]. In [Kl70], Kleitman proved that $\mathrm{cr}(K_{5,n}) = Z(5,n)$ for every n. In [Wo93], Woodall gave computer-aided proofs showing $\mathrm{cr}(K_{7,7}) = Z(7,7)$ and $\mathrm{cr}(K_{7,9}) = Z(7,9)$. Using these facts and the Parity Argument, we have the following.

F11: $\mathrm{cr}(K_{m,n}) = Z(m,n)$ for all m, n such that $n \geq m$, $m \leq 6$, and for $(m,n) \in \{(7,7), (7,8), (7,9), (7,10), (8,8), (8,9), (8,10)\}$.

The next three statements have computer-assisted proofs, bounding $\mathrm{cr}(K_{m,n})$ using semidefinite programming techniques.

F12: [DMPRS06] $\lim_{n\to\infty} \dfrac{\mathrm{cr}(K_{7,n})}{Z(7,n)} > 0.9687$.

F13: [DPS07] $\lim_{n\to\infty} \dfrac{\mathrm{cr}(K_{8,n})}{Z(8,n)} > 0.9766$.

F14: [DPS07] $\lim_{n\to\infty} \dfrac{\mathrm{cr}(K_{9,n})}{Z(9,n)} > 0.9669$.

Complete Graphs

In [BW10], Beineke and Wilson discuss the origins of the investigation of the crossing number of the complete graphs K_n. It appears that the first to seriously devote time to produce drawings of K_n with as few crossings as possible was the British artist Anthony Hill, who eventually approached graph theorist Frank Harary with his findings, resulting in the joint paper [HH62].

EXAMPLE

E5: The drawing of K_8 in Figure 7.10.3 has 18 crossings. It is straightforward to generalize this to a drawing of K_n with $Z(n) := (1/4)\lfloor m/2 \rfloor \lfloor (m-1)/2 \rfloor \lfloor (m-2)/2 \rfloor \lfloor (m-3)/2 \rfloor$ crossings, for any positive integer n. As described in [BW10], this paradigm for drawing K_n was first devised by Hill.

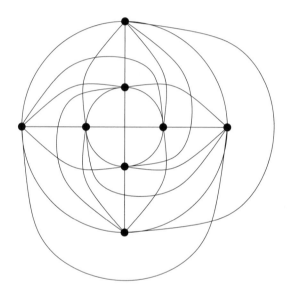

Figure 7.10.3: A drawing of K_8 with 18 crossings.

CONJECTURE

C2: [HH62] $\text{cr}(K_n) = Z(n)$, for every positive integer n.

FACTS

F15: For every positive integer n, $\text{cr}(K_n) \leq Z(n)$. This follows from the observation (see the preceding example) that there exists a drawing of K_n with exactly $Z(n)$ crossings.

In analogy with complete bipartite graphs, a straightforward counting argument shows the following.

F16: (Parity Argument) If $\text{cr}(K_{2r-1}) = Z(2r-1)$ for some integer r, then $\text{cr}(K_{2r}) = Z(2r)$.

F17: [EG73, PR07] $\mathrm{cr}(K_n) = Z(n)$ for all $n \leq 12$.

F18: [DPS07]
$$\lim_{n \to \infty} \frac{\mathrm{cr}(K_n)}{Z(n)} > 0.8594.$$

Other Families of Graphs

Other families of graphs have also received considerable attention. Special interest has been devoted to graphs with good interconnection properties (such as the n-dimensional cube Q_n and the Generalized Petersen Graphs $P(n, k)$), and to the Cartesian products of some families of graphs.

FACTS

In the next statement, the upper bound was conjectured in [EG73] to be the crossing number of Q_n. The lower and upper bounds are from [SV93] and [FDSV08], respectively.

F19:
$$\frac{1}{20}4^n - (n^2 + 1)2^{n-1} \leq \mathrm{cr}(Q_n) \leq \frac{5}{32}4^n - \left\lfloor \frac{n^2+1}{2} \right\rfloor 2^{n-2}.$$

Turning to the Generalized Petersen Graph $P(n, k)$, it is easy to verify that $P(n, 2)$ is planar if n is even or $n = 3$. It is also easy to prove that the crossing number of $P(5, 2)$ (the Petersen graph) is 2. Few exact results are known.

F20: [EHK81] For every odd integer $n \geq 7$, $\mathrm{cr}(P(n, 2)) = 3$.

F21: [RS02] For every $k \geq 3$, $\mathrm{cr}(P(3k + h, 3))$ equals $k + h$ if $h = 0$ or 2, and equals $k + 3$ if $h = 1$. The sole exception is $P(9, 3)$, whose crossing number is 2.

For $k < 3$, it is readily checked that $\mathrm{cr}(P(3k, k))$ is planar. By the previous fact, we have $\mathrm{cr}(P(3k, k)) = 2$ for $k = 3$. The following statement, proved by Fiorini and Gauci [FG03], gives $\mathrm{cr}(P(3k, k))$ for all other values of k.

F22: $\mathrm{cr}(P(3k, k)) = k$, for every $k \geq 4$.

Substantial effort has been devoted to computing the crossing numbers of the Cartesian products $C_m \square C_n$. In 1972, Harary, Kainen, and Schwenk observed that $C_m \square C_n$ can be drawn in the plane with $(m - 2)n$ crossings, and conjectured that $\mathrm{cr}(C_m \square C_n) = (m - 2)n$, for all integers m, n such that $n \geq m \geq 2$. This conjecture has been verified for small values of m (for every n). The next statement collects results from [RB78, BR80, DR95, KRS96, RS01, AR07].

F23: $\mathrm{cr}(C_m \square C_n) = (m - 2)n$ for all m, n such that $n \geq m$ and $m \leq 7$.

F24: [GS04] $\mathrm{cr}(C_m \square C_n) = (m - 2)n$ for all m, n such that $n \geq m(m + 1)$.

The crossing number of the Cartesian product of the m-star S_m and the n-path P_n was conjectured by Jendrol' and Ščerbová [JS82], and succinctly proved by Bokal [Bo07], using his novel "zip product" technique.

F25: $\mathrm{cr}(S_m \square P_n) = (n - 1)\left\lfloor \frac{m}{2} \right\rfloor \left\lfloor \frac{m-1}{2} \right\rfloor$, for all positive integers m, n.

The crossing numbers of the Cartesian products of other graphs of small order with paths, stars, or cycles have also been investigated. Good collections of results in this direction are given in [Kl01] and [DK11].

7.10.4   Crossing-Critical Graphs

As with other graph theoretical parameters, it is of natural interest to investigate those graphs that are critical, in the sense that every subgraph of the given graph has a smaller value of this parameter.

DEFINITION

D9: For a positive integer k, a graph G is k-***crossing-critical*** (or simply k-***critical***) if $\mathrm{cr}(G) \geq k$ and every proper subgraph H of G satisfies $\mathrm{cr}(H) < k$, and G is homeomorphically minimal with this property.

FACTS

F26: The only 1-critical graphs are $K_{3,3}$ and K_5. This is an immediate consequence of Kuratowski's Theorem.

A full classification of 2-critical graphs is a disproportionately harder task, as it has been known for some time that there are infinitely many 2-critical graphs. On the positive side, we have the following result.

F27: [Ri88] There are exactly eight cubic 2-critical graphs.

It is interesting to note that there exist graphs that are k-critical for some numbers $k < \mathrm{cr}(G)$, such as $C_3 \square C_3$, which is 2-crossing-critical but has crossing number 3. On the other hand, the following result from [RT95] shows that k-critical graphs cannot have an arbitrarily large crossing number.

F28: If G is k-critical, then $\mathrm{cr}(G) \leq \frac{5}{2}k + 16$.

As an immediate consequence of this, one obtains the following.

F29: Every graph G with $\mathrm{cr}(G) = k$ has an edge e with $\mathrm{cr}(G \setminus e) \geq \frac{2}{5}\mathrm{cr}(G) - \frac{37}{5}$.

In [RT95], Richter and Thomassen conjectured that this last result could be remarkably strenghtened, asking if there is a positive constant c such that every graph G with $\mathrm{cr}(G) = k$ has an edge e with $\mathrm{cr}(G \setminus e) \geq k - c\sqrt{k}$. Fox and Cs. Tóth proved a stronger version of this conjecture [FT06], showing it true, in particular, for dense graphs. Later Černý, Kynčl, and G. Tóth [CKT08] further refined this result and proved the following.

F30:   For every $\epsilon, \gamma > 0$, there is an $n_{\epsilon,\gamma}$ such that every graph G with $n \geq n_{\epsilon,\gamma}$ vertices and $m \geq n^{1+\epsilon}$ edges has a subgraph G' with m' edges such that

$$m' \leq \left(1 - \frac{\epsilon\gamma}{1224}\right) m$$

and

$$\mathrm{cr}(G') \geq (1 - \gamma)\mathrm{cr}(G).$$

In a result of a more structural character, Geelen, Richter, and Salazar [GRS02] proved that k-critical graphs have bounded tree-width (see §2.4.1). This was then strengthened by the next result, due to Hliněný [Hl03].

F31: Let k be any positive integer, and let G be a k-critical graph. Then G has path-width (see §2.4.1) at most $6(72 \log k + 248)k^3$.

It was conjectured in [RS09] that this last statement could be further refined, to show that k-critical graphs have bounded band-width. Since graphs with bounded band-width obviously have bounded degree, this conjecture got disproved by the following result by Dvořák and Mohar [DM10].

F32: For every $k \geq 171$ and every positive integer d, there exists a k-critical graph containing a vertex of degree at least d.

This last statement implies that, for any positive integer d, there exist k-critical graphs (for $k \geq 171$) containing a subgraph isomorphic to $K_{1,d}$. Following on this theme, the next result was proved in [HS10].

F33: Let k, n be positive integers with $n \geq 30k^2 + 200k$. Then no k-critical graph contains a subdivision of $K_{2,n}$.

The construction of critical graphs with prescribed properties has been a driving force behind much of the research on k-critical graphs. Particularly interesting examples of infinite families of critical graphs are the (by now) classical construction reported by Kochol in [Ko87] and the paradigm-shifting constructions given by Hliněný in [Hl08]. As an example of the knowledge gained in the quest of understanding the richness of critical graphs, we finally put forward the following mutually complementary statements.

F34: [Bo10] Let $r \in (3,6)$ be a rational number and k an integer. There exists a convex continuous function $f : (3,6) \rightarrow^+$ such that, for $k \geq f(r)$, there exists an infinite family of simple 3-connected crossing-critical graphs with average degree r and crossing number k.

F35: [HST12] For each fixed positive integer k, there are only finitely many k-crossing-critical simple graphs of average degree at least six.

7.10.5 Algorithmical Aspects

It is hardly surprising that computing crossing numbers is NP-hard. Until recently, very little was known regarding nontrivial conditions under which the crossing number can be approximated, and about restricted families of graphs computing the crossing numbers remains hard. In the last few years we have seen substantial progress on these fronts, as revealed by the collection of facts gathered below.

DEFINITIONS

D10: A graph G is a **near-planar graph** if it has an edge e such that $G \setminus e$ is planar.

D11: A graph G is an **apex graph** if it has a vertex v such that $G \setminus v$ is planar.

TERMINOLOGY: CROSSINGNUMBER is the decision problem: "Given a graph G and an integer k, is $\text{cr}(G) \leq k$?".

FACTS

F36:   [GJ83] CROSSINGNUMBER is NP-complete.

The next two statements indicate that computing the crossing number remains hard even for seriously restricted families of graphs.

F37:   [Hl06, PSS11] CROSSINGNUMBER is NP-complete, even when restricted to cubic graphs.

F38:   [CM10] CROSSINGNUMBER is NP-complete, even when restricted to near-planar graphs.

Grohe [Gr04] was the first to show that computing the crossing number is fixed-parameter tractable. His quadratic-time algorithm was later improved.

F39:   [KR07] For each fixed integer k, there is a linear-time algorithm that determines whether $cr(G) \leq k$.

F40:   [GHLS08] There is a polynomial-time constant factor approximation algorithm for the crossing number of bounded-degree graphs embeddable in the projective plane.

F41:   [HS07, CH10] Let S be any fixed orientable surface. There is a polynomial-time constant factor approximation algorithm for the crossing number of bounded-degree graphs embeddable in S.

F42:   [CHM12] There is a polynomial-time constant factor approximation algorithm for the crossing number of bounded-degree apex graphs.

F43:   [Ch11] There is a polynomial time algorithm that approximates the crossing number of a graph with n vertices and maximum degree Δ within a factor of $O(n^{9/10} \operatorname{poly}(\Delta \log n))$.

F44:   [Ca12] There is a constant $c_0 > 1$ such that, if P $\neq$ NP, then there is no c_0-approximation algorithm for the crossing number, even when restricted to cubic graphs.

7.10.6   Other Definitions of Crossing Number

While the definition of crossing number we have worked with is the one that has attracted most research, there are many more (both reasonable and interesting) ways to define the crossing number of a graph. We shall briefly review below a few alternative definitions of crossing number (closely following Székely's view in [Sz04]; see also [PT00a]), along with some interesting results that involve them. For an authoritative and comprehensive treatise on the different definitions of crossing number, we strongly recommend the survey by Schaeffer [Sc12].

DEFINITIONS

To put the definition of crossing number we have used in this chapter into the right context against other variants, we start by presenting this definition in a slightly different way. In the first four definitions below, D is a drawing of a graph G.

D12: The *standard crossing number* cr(D) of D is the number of pairs $(x, \{\alpha, \beta\})$, where x is a point of the plane and α and β are open arcs of D representing distinct edges of G such that $x \in \alpha \cap \beta$. The *standard crossing number* cr(G) of G is the minimum cr(D) taken over all drawings of G in the plane.

D13: The *pair crossing number* pcr(D) of D is the number of pairs of open arcs α and β of D representing distinct edges of G such that $\alpha \cap \beta \neq \emptyset$. The *pair crossing number* pcr(G) of G is the minimum pcr(D) taken over all drawings of G in the plane.

D14: The *odd crossing number* ocr(D) of D is the number of pairs α and β of open arcs of D representing distinct edges of G such that $|\alpha \cap \beta|$ is odd. The *odd crossing number* ocr(G) of G is the minimum ocr(D) taken over all drawings of G in the plane.

D15: The *independent odd crossing number* iocr(D) of D is the number of pairs α and β of open arcs of D representing distinct edges of G that are not incident with a common vertex and such that $|\alpha \cap \beta|$ is odd. The *independent odd crossing number* iocr(G) of G is the minimum iocr(D) taken over all drawings of G in the plane.

D16: The *minor crossing number* mcr(G) is $\min\{\text{cr}(H) \mid G$ is a minor of $H\}$.

D17: A drawing of a graph is *rectilinear* (or *geometric*) if all the edges are straight segments. The *rectilinear crossing number* $\overline{\text{cr}}(G)$ of a graph G is the minimum cr(D) taken over all rectilinear drawings of G in the plane.

FACTS

We start with some straightforward observations.

F45: For every graph G, iocr(G) $\leq$ ocr(G) $\leq$ pcr(G) $\leq$ cr(G) $\leq \overline{\text{cr}}(G)$.

Tutte introduced iocr(G) and proved the following two facts in [Tu70].

F46: If a graph G satisfies iocr(G) $= 0$, then cr(G) $= 0$.

F47: If D, D' are two drawings of a graph G, then iocr(D) $=$ iocr(D') (mod 2).

Also in [Tu70], Tutte asked whether iocr(G) $=$ cr(G) for all graphs G. Pach and Tóth [PT00] raised the (still open) important question of whether or not, for all graphs G, one has pcr(G) $=$ cr(G). They proved the following (see also [PSS07]).

F48: cr(G) $\leq 2(\text{ocr}(G))^2$.

Pelsmajer, Schaefer, and Štefankovič [PSS08] gave an elegant construction proving the existence of graphs G for which ocr(G) $<$ pcr(G) (and hence ocr(G) $<$ cr(G)), which in particular answers Tutte's question in the negative. The currently best proportionality constant separating pcr(G) and ocr(G) (for some graph G) was derived by Tóth.

F49: [To08] There exist graphs G such that $0.855 \cdot \text{pcr}(G) \geq \text{ocr}(G)$.

One strongly appealing feature of the minor crossing number (introduced in [BFM06]) is that it is minor-monotone, so that it blends well with the graph minors theory of Robertson and Seymour. A strong result involving this parameter is the following upper bound by Bokal, Fijavž, and Wood [BFW08].

F50: For every graph H there is a constant $c = c(H)$ such that every H-minor-free graph G has $\mathrm{mcr}(G) \leq c \cdot |V(G)|$.

We remark that for the standard crossing number this last fact does not hold. Indeed, in the following related result, the dependence on $|V(G)|$ and $\Delta(G)$ is best possible.

F51: [DKMW08] Let H be a fixed graph. Then every graph $G = (V, E)$ that excludes G as a minor has crossing number $O(\Delta(G) \cdot |V(G)|)$.

We also note that the general techniques and bounds in 7.10.2 were carried over to the minor crossing number in [BCSV10].

We finally mention three results involving the rectilinear crossing number, only scratching the surface of this important parameter that falls into the realm of combinatorial geometry. The next fact was proved by Bienstock and Dean [BD93].

F52: For each integer $k \geq 4$, there is a graph G_k with $\mathrm{cr}(G) = 4$ and $\overline{\mathrm{cr}}(G) > k$.

Scheinerman and Wilf [SW94] discovered the following striking connection between $\overline{\mathrm{cr}}(K_n)$ and Sylvester's Four Point Constant q_* from geometric probability [Sy84].

F53:

$$q_* = \lim_{n \to \infty} \frac{\overline{\mathrm{cr}}(K_n)}{\binom{n}{4}}.$$

Thus, the rectilinear crossing number of K_n is of interest elsewhere in mathematics. The exact value of $\overline{\mathrm{cr}}(K_n)$ is known for $n \leq 27$ and for $n = 30$ [CHLV11, ACFLS12]. It is also interesting that in general it differs from $\mathrm{cr}(K_n)$.

F54: [EG73] $\mathrm{cr}(K_n) = \overline{\mathrm{cr}}(K_n)$ if and only if $n \leq 7$ or $n = 9$. For all other values of n, $\mathrm{cr}(K_n) > \overline{\mathrm{cr}}(K_n)$.

F55: [ACFLS10, ACFLS12]

$$0.37997 < \lim_{n \to \infty} \frac{\overline{\mathrm{cr}}(K_n)}{n^4} < 0.38048.$$

7.10.7 Crossing Sequences

Any finite graph can be embedded in some (orientable or non-orientable) finite genus surface. Širáň [Si83] investigated how the crossing number of a graph increases as we move to host surfaces with smaller genus, until finally arriving at the sphere. Širáň's work is the first in a series of papers with particularly interesting and surprising results.

NOTATION: The compact orientable surface of genus h will be denoted S_h, and the compact non-orientable surface of genus k will be denoted N_k.

DEFINITIONS

D18: The *orientable crossing sequence* of G is $\mathrm{cr}_{S_0}(G), \mathrm{cr}_{S_1}(G), \mathrm{cr}_{S_2}(G) \dots$.

D19: The *non-orientable crossing sequence* of G is $\mathrm{cr}_{S_0}(G), \mathrm{cr}_{N_1}(G), \mathrm{cr}_{N_2}(G) \dots$.

D20: A sequence $a_0, a_1, a_2 \ldots$ is a **convex sequence** if $a_{i+1} - a_{i+2} \leq a_i - a_{i+1}$ for each i.

REMARKS

R5: It is trivial to see that the (orientable or non-orientable) crossing sequence of every graph is non-increasing. Moreover, if $\mathrm{cr}_{S_0}(G) \neq 0$ (that is, if G is nonplanar), then the sequence is strictly decreasing until we arrive at a (orientable or non-orientable) surface S for which $\mathrm{cr}_S(G) = 0$.

FACTS

F56: [Si83] If $a_0, a_1, \ldots, a_k$ is a convex sequence of integers such that $a_0 > a_1 > a_2 > \cdots > a_k = 0$, then there is a graph whose orientable crossing sequence is $(a_0, a_1, \ldots, a_k, 0, 0, \ldots)$. There is also a graph whose non-orientable crossing sequence is $(a_0, a_1, \ldots, a_k, 0, 0, \ldots)$.

F57: [ABS01] There exist graphs whose orientable crossing sequence is not convex.

F58: [ABS01] For all positive integers a_0, a_1 such that $a_0 > a_1$, there exists a graph whose non-orientable crossing sequence is $(a_0, a_1, 0, 0, \ldots)$.

F59: [DMS11] For all positive integers a_0, a_1 such that $a_0 > a_1$, there exists a graph whose non-orientable crossing sequence is $(a_0, a_1, 0, 0, \ldots)$.

7.10.8   Applications of Crossing Numbers

In a breakthrough paper, Székely introduced the "crossing number method" to give simple proofs of "hard" problems in combinatorial geometry, more specifically of incidence problems. At its heart it uses the Crossing Lemma for lower bounds. Below we offer a small sample of results that either use Székely's method as a crucial ingredient or make essential use of its consequent incidence results.

FACTS

F60: ([Sz97]; first proved in [ST83]): Given n points and ℓ straight lines in the plane, the number of incidences among the points and lines is $O((n\ell)^{2/3} + n + \ell)$.

TERMINOLOGY: A class Γ of curves in the plane has k *degrees of freedom and multiplicity-type s* if (i) for any k points there are at most s curves of Γ passing through all of them; and (ii) any pair of curves from Γ intersect in at most s points. If P is a finite point set and C is a finite set of curves, then $I(P, C)$ denotes the number of incidences between them.

F61: [PS98] Let P be a set of m points and let C be a set of n simples curves all lying in the plane. If C has k degrees of freedom and multiplicity-type s, then

$$I(P, C) \leq c(k, s) \left(m^{k/(2k-1)} n^{(2k-2)/(2k-1)} + m + n \right),$$

where $c(k, s)$ is a positive constant that depends on k and s.

F62: [De98] The number of $(k+1)$-sets that are possible with n points in ^{2} is at most $6.48n(k+2)^{1/3}$.

F63: Given n points and n families of concentric circles each with at most k circles in the plane, the maximal number $I(n,k)$ of incidences between the points and the circles is $O(n^{10/7}k^{5/7})$ [ATT98]. Moreover [STT02], $I(n,k)$ is $O(n^{1.4571}k^{0.6286})$ (the latter bound is better than the former when n is large enough compared to k).

We finally mention that Solymosi and Tao recently used the crossing number method to establish near-sharp Szemerédi–Trotter type bounds on the number of incidences between points and k-dimensional algebraic variaties in ^{d} for various values of k and d. The precise result is too technical to state here; we refer the reader to [ST12].

7.10.9   Suggestions for Further Reading

Pach and Tóth have a stimulating paper [PT00a] centered on crossing number problems. Many remain open and we expect they will continue to serve as a guide to important advances in the theory.

We highly recommend Beineke and Wilson's lively account [BW10] of the early history of crossing numbers.

The current authors have an earlier survey on crossing numbers [RS09]. While there is substantial overlap with this chapter, there are significant differences and we suggest the reader look there for supplementary material.

We also recommend the survey by Mutzel [Mu09], which devotes a section to an overview of the computer-assisted exact computation of crossing numbers, a topic not included in the present survey.

The survey by Schaeffer [Sc12] mentioned in Section 7.10.6 contains not only many variants of what a crossing number is, but also a very interesting, thoughtful discussion. We also mention the exhaustive (and useful) bibliography compiled and continually updated by Imrich Vrťo [Vr].

The crossing number is only one parameter that measures the nonplanarity of a graph. Although slightly outdated, the survey [Li01] is still an excellent reference on many other nonplanarity parameters.

References

[ACFLS10] B. M. Ábrego, M. Cetina, S. Fernández-Merchant, J. Leaños, and G. Salazar, 3-symmetric and 3-decomposable geometric drawings of K_n, *Discrete Appl. Math.* **158**(12) (2010), 1240–1258.

[ACFLS12] B. M. Ábrego, M. Cetina, S. Fernández-Merchant, J. Leaños, and G. Salazar, On ($\leq k$)-edges, crossings, and halving lines of geometric drawings of K_n, *Discrete Comput. Geom.* **48**(1) (2012), 192–215.

[AR07] J. Adamsson and R. B. Richter, Arrangements, circular arrangements and the crossing number of $C_7 \times C_n$, *J. Combin. Theory Ser. B* **90**(1) (2004), 21–39.

[ACNS82] M. Ajtai, V. Chvátal, M. M. Newborn, and E. Szemerédi, Crossing-free subgraphs, *Theory and Practice of Combinatorics*, 9–12, North-Holland Math. Stud. **60**, North-Holland, Amsterdam, 1982.

[ATT98] T. Akutsu, H. Tamaki, and T. Tokuyama, Distribution of distances and triangles in a point set and algorithms for computing the largest common point sets, *Discrete Comput. Geom.* **20**(3) (1998), 307–331.

[ABS01] D. Archdeacon, C. P. Bonnington, and J. Širáň, Trading crossings for handles and crosscaps, *J. Graph Theory* **38**(4) (2001), 230–243.

[BR80] L. W. Beineke and R. D. Ringeisen, On the crossing numbers of products of cycles and graphs of order four, *J. Graph Theory* **4**(2) (1980), 145–155.

[BW10] L. Beineke and R. Wilson, The early history of the brick factory problem, *Math. Intelligencer* **32**(2) (2010), 41–48.

[BD93] D. Bienstock and N. Dean, Bounds for rectilinear crossing numbers, *J. Graph Theory* **17**(3) (1993), 333–348.

[Bo07] D. Bokal, On the crossing numbers of Cartesian products with paths, *J. Combin. Theory Ser. B* **97**(3) (2007), 381–384.

[Bo10] D. Bokal, Infinite families of crossing-critical graphs with prescribed average degree and crossing number, *J. Graph Theory* **65**(2) (2010), 139–162.

[BCSV10] D. Bokal, E. Czabarka, L. A. Székely, and I. Vrťo, General lower bounds for the minor crossing number of graphs, *Discrete Comput. Geom.* **44**(2) (2010), 463–483.

[BFM06] D. Bokal, G. Fijavž, and B. Mohar, The minor crossing number, *SIAM J. Discrete Math.* **20** (2006), no. 2, 344–356.

[BFW08] D. Bokal, G. Fijavž, and D. R. Wood, The minor crossing number of graphs with an excluded minor, *Electron. J. Combin.* **15**(1) (2008), Research Paper 4, 13 pp.

[Ca12] S. Cabello, Hardness of approximation for crossing number, *Discrete and Computational Geometry* **49** (2013), 348–358. .

[CM10] S. Cabello and B. Mohar, Adding one edge to planar graphs makes crossing number hard. In: *Proc. SoCG 2010*, 68–76.

[CKT08] J. Černý, J. Kynčl, and G. Tóth, Improvement on the decay of crossing numbers, *Lecture Notes in Computer Science* **4875**, 25–30. Springer, 2008. Also: *Graphs and Combinatorics* **29** (2013), 365–371.

[CHLV11] M. Cetina, C. Hernández-Vélez, J. Leaños, and C. Villalobos, Point sets that minimize ($\leq k$)-edges, 3-decomposable drawings, and the rectilinear crossing number of K_{30}, *Discrete Math.* **311**(16) (2011), 1646–1657.

[CH10] M. Chimani and P. Hliněný, Approximating the crossing number of graphs embeddable in any orientable surface. In: *Proc. ACM-SIAM Symposium on Discrete Algorithms 2010, Austin, Texas (SODA10)*, pp. 918–927.

[CHM12] M. Chimani, P. Hliněný, and P. Mutzel, Vertex insertion approximates the crossing number of apex graphs, *European J. Combin.* **33** (2012), 326–335.

[Ch11] J. Chuzhoy, An algorithm for the graph crossing number problem. In: *Proc. STOC 2011*, pp. 303–312.

[DMPRS06] E. de Klerk, J. Maharry, D. V. Pasechnik, R. B. Richter, and G. Salazar, Improved bounds for the crossing numbers of $K_{m,n}$ and K_n. *SIAM J. Discrete Math.* **20** (2006), 189–202.

[DP12] E. de Klerk and D. V. Pasechnik, Improved lower bounds for the 2-page crossing number of $K_{m,n}$ and K_n via semidefinite programming, *SIAM Journal on Optimization*, **22**(2) (2012), 581–595.

[DPS07] E. de Klerk, D. V. Pasechnik, and A. Schrijver, Reduction of symmetric semidefinite programs using the regular -representation, *Math. Program. Ser. B*, **109**(2-3) (2007), 613–624.

[DR95] A. M. Dean and R. B. Richter, The crossing number of $C_4 \times C_4$, *J. Graph Theory* **19**(1) (1995), 125–129.

[DMS11] M. DeVos, B. Mohar, and R. Šamal, Unexpected behaviour of crossing sequences, *J. Combin. Theory Ser. B* **101** (2011), no. 6, 448–463.

[De98] T. K. Dey, Improved bounds for planar k-sets and related problems, *Discrete Comput. Geom.* **19**(3) (1998), 373–382.

[DV03] H. N. Djidjev and I. Vrťo, Crossing numbers and cutwidths, *J. Graph Algorithms Appl.* **7**(3) (2003), 245–251.

[DK11] E. Draženská and M. Klešč, On the crossing numbers of $G\square C_n$ for graphs G on six vertices. *Discuss. Math. Graph Theory* **31** (2011), 239–252.

[DKMW08] V. Dujmovic, K. Kawarabayashi, B. Mohar, and D. R. Wood, Improved upper bounds on the crossing number, SCG'08: *Proc. 24th Annual Symposium on Computational Geometry*, pp. 375-384, ACM Press, 2008.

[DM10] Z. Dvořák and B. Mohar, Crossing-critical graphs with large maximum degree, *J. Combin. Theory Ser. B* **100**(4) (2010), 413–417.

[EG70] R. B. Eggleton and R. K. Guy, The crossing number of the n-cube, *Notices Amer. Math. Soc.* **17** (1970) 757–757.

[EG73] P. Erdős and R. K. Guy, Crossing number problems, *Amer. Math. Monthly* **80** (1973), 52–58.

[EHK81] G. Exoo, F. Harary, and J. Kabell, The crossing numbers of some generalized Petersen graphs. *Math. Scand.* **48**(2) (1981), 184–188.

[FDSV08] L. Faria, C. M. H. de Figueiredo, O. Sýkora, and I. Vrťo, An improved upper bound on the crossing number of the hypercube, *J. Graph Theory* **59**(2) (2008), 145–161.

[FG03] S. Fiorini and J. B. Gauci, The crossing number of the generalized Petersen graph $P[3k, k]$. *Math. Bohem.* **128** (2003), no. 4, 337–347.

[FPT10] J. Fox, J. Pach, and Cs. D. Tóth, A bipartite strengthening of the crossing lemma, *J. Combin. Theory Ser. B* **100** (2010), 23–35.

[FT06] J. Fox and Cs. D. Tóth, On the decay of crossing numbers, *J. Combin. Theory Ser. B* **98**(1) (2008), 33–42.

[GJ83] M. R. Garey and D. S. Johnson, Crossing number is NP-complete, *SIAM J. Alg. Discr. Meth.* **4** (1983), 312–316.

[GRS02] J. F. Geelen, R. B. Richter, and G. Salazar, Embedding grids in surfaces, *European J. Combin.* **25**(6) (2004), 785–792.

[GHLS08] I. Gitler, P. Hliněný, J. Leaños, and G. Salazar, The crossing number of a projective graph is quadratic in the face-width, *Electron. J. Combin.* **15** (2008) #R46.

[GS04] L. Y. Glebsky and G. Salazar, The crossing number of $C_m \times C_n$ is as conjectured for $n \geq m(m+1)$, *J. Graph Theory* **47**(1) (2004), 53–72.

[Gr04] M. Grohe, Computing crossing numbers in quadratic time, *J. Comput. Syst. Sci.* **68** (2004), 285–302.

[GMW05] C. Gutwenger, P. Mutzel, and R. Weiskircher, Inserting an edge into a planar graph, *Algorithmica* **41** (2005), 289–308.

[Gu68] R. K. Guy, The decline and fall of Zarankiewicz's theorem. In: *Proof Techniques in Graph Theory* (ed. F. Harary), pp. 63–69. Academic Press, New York, 1969.

[HH62] F. Harary and A. Hill, On the number of crossings in a complete graph, *Proc. Edinburgh Math. Soc. (2)* **13** (1962/1963) 333–338.

[HKS72] F. Harary, P. C. Kainen, and A. J. Schwenk, Toroidal graphs with arbitrarily high crossing numbers, *Nanta Math.* **6** (1973), 58–67.

[HR90] N. Hartsfield and G. Ringel, *Pearls in Graph Theory. A Comprehensive Introduction*, Academic Press, Inc., Boston, MA, 1990.

[HST12] C. Hernández-Vélez, G. Salazar, and R. Thomas, Nested cycles in large triangulations and crossing-critical graphs, *J. Combin. Theory Ser. B* **102**(1) (2012), 86–92.

[Hl03] P. Hliněný, Crossing-number critical graphs have bounded path-width, *J. Combin. Theory Ser. B* **88**(2) (2003), 347–367.

[Hl06] P. Hliněný, Crossing number is hard for cubic graphs, *J. Comb. Theory, Ser. B* **96** (2006), 455–471.

[Hl08] P. Hliněný, New infinite families of almost-planar crossing-critical graphs, *Electron. J. Combin.* **15**(1) (2008), Research Paper 102, 12 pp.

[HS07] P. Hliněný and G. Salazar, Approximating the crossing number of toroidal graphs, *Lecture Notes in Computer Science* **4835** (2007), 148–159.

[HS10] P. Hliněný and G. Salazar, Stars and bonds in crossing-critical graphs, *J. Graph Theory* **65**(3) (2010), 198–215.

[JS82] S. Jendrol' and M. Ščerbová, On the crossing numbers of $S_m \times P_n$ and $S_m \times C_n$, *Časopis Pěst. Mat.* **107** (1982), 225–230.

[KR07] K. Kawarabayashi and B. Reed. Computing crossing number in linear time. In: *Proc. STOC 2007*, pp. 382–390.

[Kl70] D. J. Kleitman, The crossing number of $K_{5,n}$, *J. Combinatorial Theory* **9** (1970), 315–323.

[Kl01] M. Klešč, The crossing numbers of Cartesian products of paths with 5-vertex graphs, *Discrete Mathematics* **233** (2001), 353–359.

[KRS96] M. Klešč, R. B. Richter, and I. Stobert, The crossing number of $C_5 \times C_n$, *J. Graph Theory* **22**(3) (1996), 239–243.

[Ko87] M. Kochol, Construction of crossing-critical graphs, *Discrete Math.* **66**(3) (1987), 311–313.

[Le81] F. T. Leighton, New lower bound techniques for VLSI, in *Proc. 22nd. Annual IEEE Symposium on Foundations of Computer Science* (1981), 1–12.

[Le83] F. T. Leighton, *Complexity issues in VLSI: Optimal Layouts for the Shuffle-Exchange Graph and Other Networks*, MIT Press, 1983.

[Li01] A. Liebers, Planarizing graphs: a survey and annotated bibliography, *J. Graph Algorithms Appl.* **5**(1) (2001), 1–74.

[Mu09] P. Mutzel, The crossing number of graphs: Theory and computation, *Lecture Notes in Computer Science* **5760** (2009), 305–317.

[PRTT06] J. Pach, R. Radoičić, G. Tardos, and G. Tóth, Improving the crossing lemma by finding more crossings in sparse graphs, *Discrete Comput. Geom.* **36** (2006), 527–552.

[PSS96] J. Pach, F. Shahrokhi, and M. Szegedy, Applications of the crossing number, *Algorithmica* **16**(1) (1996), 111–117.

[PS98] J. Pach and M. Sharir, On the number of incidences between points and curves, *Combin. Probab. Comput.* **7** (1998), 121–127.

[PST00] J. Pach, J. Spencer, and G. Tóth. New bounds on crossing numbers, *Discrete Comput. Geom.* **24** (2000), 623–644.

[PT97] J. Pach and G. Tóth, Graphs drawn with few crossings per edge, *Combinatorica* **17** (1997), 427–439.

[PT00] J. Pach and G. Tóth, Which crossing number is it anyway? *J. Combin. Theory Ser. B* **80**(2) (2000), 225–246.

[PT00a] J. Pach and G. Tóth, Thirteen problems on crossing numbers, *Geombinatorics* **9**(4) (2000), 194–207.

[PR07] S. Pan and R. B. Richter, The crossing number of K_{11} is 100, *J. Graph Theory* **56** (2007), 128–134.

[PSS07] M. J. Pelsmajer, M. Schaefer, and D. Štefankovič, Removing even crossings, *J. Combin. Theory Ser. B* **97**(4) (2007), 489–500.

[PSS08] M. J. Pelsmajer, M. Schaefer, and D. Štefankovič, Odd crossing number and crossing number are not the same, *Discrete Comput. Geom.* **39**(1–3) (2008), 442–454.

[PSS11] M. J. Pelsmajer, M. Schaefer, and D. Štefankovič, Crossing numbers of graphs with rotation systems, *Algorithmica*, **60** (2011), 679–702.

[Ri88] R. B. Richter, Cubic graphs with crossing number two, *J. Graph Theory* **12** (1988), 363–374.

[RS01] R. B. Richter and G. Salazar, The crossing number of $C_6 \times C_n$, *Australas. J. Combin.* **23** (2001), 135–143.

[RS02] R. B. Richter and G. Salazar, The crossing number of $P(N, 3)$, *Graphs Combin.* **18**(2) (2002), 381–394.

[RS09] R. B. Richter and G. Salazar, Crossing numbers, in *Topics in Topological Graph Theory*, eds. L. W. Beineke, R. J. Wilson, J. L. Gross, and T. W. Tucker, Encyclopedia Math. Appl., **128**, 133–150. Cambridge Univ. Press, Cambridge, 2009.

[RT95] R. B. Richter and C. Thomassen, Minimal graphs with crossing number at least k, *J. Combin. Theory Ser. B* **58**(2) (1993), 217–224.

[RB78] R. B. Ringeisen and L. W. Beineke, The crossing number of $C_3 \times C_n$, *J. Combin. Theory Ser. B* **24**(2) (1978), 134–136.

[Sc12] M. Schaeffer, The graph crossing number and its variants: a survey, manuscript, 2012.

[SW94] E. R. Scheinerman and H. S. Wilf, The rectilinear crossing number of a complete graph and Sylvester's "four point problem" of geometric probability, *Amer. Math. Monthly* **101**(10) (1994), 939–943.

[Si83] J. Širáň, The crossing function of a graph, *Abh. Math. Sem. Univ. Hamburg* **53** (1983), 131–133.

[ST12] J. Solymosi and T. Tao, An incidence theorem in higher dimensions, manuscript, 2012. Available at `arXiv:1103.2926v5 [math.CO]`.

[STT02] J. Solymosi, G. Tardos, and C. D. Tóth, The k most frequent distances in the plane, *Discrete Comput. Geom.* **28** (2002), 639–648.

[ST01] J. Solymosi and Cs. D. Tóth, Different distances in the plane, *Disc. Comp. Geom.* **25**(4) (2001), 629–634.

[SST83] J. Spencer, E. Szemerédi, and W. T. Trotter, Unit distances in the Euclidean plane, in *Graph Theory and Combinatorics (Cambridge, 1983)*, 293–303. Academic Press, London, 1984.

[SV93] O. Sýkora and I. Vrťo, On crossing numbers of hypercubes and cube connected cycles, *BIT* **33**(2) (1993), 232–237.

[Sy84] J. J. Sylvester, Question 1491, *The Educational Times (London)*, April 1864.

[Sz97] L. A. Székely, Crossing numbers and hard Erdős problems in discrete geometry, *Combinatorics, Probability and Computing* **6** (1997), 353–358.

[Sz04] L. A. Székely, A successful concept for measuring non-planarity of graphs: the crossing number, *Discrete Math.* **276**(1–3) (2004), 331–352.

[ST83] E. Szemerédi and W. T. Trotter, Extremal problems in discrete geometry, *Combinatorica* **3** (1983), 381–392.

[To08] G. Tóth, Note on the pair-crossing number and the odd-crossing number, *Discrete Comput. Geom.* **39**(4) (2008), 791–799.

[Tu70] W. T. Tutte, Toward a theory of crossing numbers, *J. Combinatorial Theory* **8** (1970), 45–53.

[Vr] I. Vrťo. Crossing number of graphs: a bibliography. Available at:
ftp://ftp.ifi.savba/sk/pub/imrich/crobib.pdf.

[WT07] D. R. Wood and J. A. Telle, Planar decompositions and the crossing number of graphs with an excluded minor, *New York J. Math.* **13** (2007), 117–146.

[Wo93] D. R. Woodall, Cyclic-order graphs and Zarankiewicz's crossing-number conjecture, *J. Graph Theory* **17** (1993), 657–671.

[Za54] K. Zarankiewicz, On a problem of P. Turán concerning graphs, *Fund. Math.* **41** (1954), 137–145.

Glossary for Chapter 7

amalgamation – of two graphs G and H: forming a new graph from their disjoint union by merging a subgraph of G with an isomorphic subgraph of H.

___, **edge**: pasting two graphs together across an edge from each.

___, **vertex**: pasting two graphs together across a vertex from each.

annulus: synonym for *cylinder*.

apex graph: a graph G in which there is a vertex v such that removing v from G results in a planar graph.

Archimedean solid: a semi-regular polyhedron; it has regular polygons as faces and the same configuration of faces at each vertex.

automorphism of a map: an isomorphism of the map onto itself.

axiom system: a list of axioms for a mathematical structure.

___, **complete**: an axiom system in which every well-formed statement can either be shown to be true or be shown to be false.

___, **consistent**: an axiom system having no contradictions.

___, **independent**: a system in which no axiom is derivable from the others.

___, **model for**: an interpretation of the undefined terms so that each interpreted axiom is true.

balanced incomplete block design (abbr. BIBD): a geometry of v points (each in r lines) and b lines (each containing k points) such that each pair of points belong to λ lines.

bar-amalgamation – of two disjoint graphs G and H: the result of running a new edge e between a vertex u of G and a vertex v of H; denoted $G_u *_e H_v$.

barycenter – of a face of a cellular graph imbedding: a point in the interior of the face, corresponding to the image of the center of the geometric polygon that the face represents.

base graph – of a voltage graph construction: the graph to whose edges the voltages are assigned.

base imbedding for an *imbedded voltage graph* $\langle G \to S, \alpha \rangle$: the imbedding $\langle G \to S \rangle$ of the *base graph* in the *base surface*.

base surface for an *imbedded voltage graph* $\langle G \to S, \alpha \rangle$: the surface S in which the voltage graph is imbedded.

Belyi function: a meromorphic function from a (closed) Riemann surface S onto the Riemann sphere S_0, with at most three singular values forming a subset of $\{0, 1, \infty\}$.

Betti number $\beta(G)$ – of a graph G: the number of edges in a co-tree, which is equal to $|E_G| - |V_G| + 1$.

BIBD: see balanced incomplete block design.

bisection width $\mathrm{bw}(G)$ of a graph $G = (V, E)$: the minimum number of edges whose removal divides G into two parts having at most $2|V|/3$ vertices each.

boundary – of a 2-manifold M: the subspace of those points in M that do not have neighborhoods homeomorphic to open disks; instead, their fundamental neighborhoods are homeomorphic to half-disks.

boundary-separating closed curve – in a region of a noncellular graph imbedding: a closed curve that separates that region so that at least one boundary component of the region lies on each side of the separation.

boundary-walk specification – of a polygonal complex: a list of the signed boundary walks of the faces.

bouquet B_n: the graph with one vertex and n self-loops.

branch point – in the codomain S of a branched covering $p : \tilde{S} \to S$: a point where branching occurs.

branch set – of a branched covering $p : \tilde{S} \to S$: the discrete subset $\tilde{B} \subset \tilde{S}$ to whose complement the restriction of the *branched covering p* is a *covering projection*.

branched covering, combinatorial – of a surface S with a cellularly imbedded graph G by a surface $\tilde{S}$ with a cellularly imbedded graph $\tilde{G}$: a face-to-face, edge-to-edge, vertex-to-vertex mapping that is topologically a branched covering with every branch point occurring in the interior of some face; exemplified by the natural projection associated with an imbedded *voltage graph*.

branched covering, topological: a continuous function $p : \tilde{S} \to S$ between surfaces, whose restriction to the complement $\tilde{S} - \tilde{B}$ of a discrete subset $\tilde{B} \subset \tilde{S}$ is a *covering projection*.

branched covering space: the domain of a *branched covering $p : \tilde{S} \to S$* of a surface S.

cactus: a graph constructed from a tree T and a subset $S \subseteq V_T$ by replacing each vertex of S by a cycle.

(d, g)-**cage**: a graph of minimum order among all d-regular graphs of girth g.

canonical factors for an abelian group: the factors $\mathcal{Z}_{m_i}$ of the *canonical form*.

canonical form for an abelian group: the form $\mathcal{Z}_{m_1} \times \cdots \times \mathcal{Z}_{m_r}$, where $m_j | m_{j+1}$ for $j = 1 \ldots r$.

Cayley graph – for a group $\mathcal{A}$ and generating set X: the graph whose vertices are the elements of $\mathcal{A}$ and such that, for each element $a \in \mathcal{A}$ and each generator $x \in X$, there is a directed edge from a to ax.

Cayley map: an imbedding of a Cayley graph on a surface, possibly specified by a *rotation scheme*.

0-cell – of a polygonal complex: see *vertex of a polygonal complex*.

1-cell – of a polygonal complex: see *edge of a polygonal complex*.

2-cell – of a polygonal complex: see *face of a polygonal complex*.

2-cell imbedding – of a graph: see *cellular imbedding*.

cellular imbedding – of a graph into a surface: a graph imbedding such that the interior of each face is an open disk.

centroid of a triangle: the point common to the three medians of the triangle.

cevian of a triangle: a line from a vertex of the triangle to the opposite side.

chiral map: a map that is symmetrical, but not regular.

chromatic number of a surface S: the least number of colors sufficient to properly color the faces (or vertices) of any map on S.

circular imbedding: an imbedding where each face is bounded by a simple cycle.

circumcenter of a triangle: the point common to the three perpendicular bisectors of the sides of the triangle.

closed surface: a surface that as a topological space is compact and without boundary.

closed under minors – of a graph class: a graph class $\mathcal{C}$ such that for every graph G in $\mathcal{C}$, all minors of G are also in $\mathcal{C}$.

closed-end ladder L_n: the graph obtained from the cartesian product $P_n \times K_2$ by doubling the edges $v_1 \times K_2$ and $v_n \times K_2$ at both ends of the path.

cobblestone path J_n: the graph obtained by doubling every edge of the path P_n.

complete set of forbidden minors – for a class $\mathcal{F}$ closed under minors: a set M of minimal forbidden minors such that for every graph G that is not in $\mathcal{F}$, there exists a graph in $\mathcal{M}$ that is a minor of G.

composition of a graph G **with a graph** H: the graph obtained from copies of H corresponding to all vertices of G, by adding all possible edges between two copies corresponding to an adjacency of G.

(r, k)**-configuration**: a geometry having every point in r lines, every line consisting of k points, and each pair of points in at most one line.

___, **symmetrical**: an (r, k)-configuration such that $r = k$.

congestion at an edge e of the host of a graph mapping $f : G \to H$: the cardinality $|f^{-1}(e)|$ of its preimage; terminology for modeling the *emulation of distributed computation*.

congestion of the mapping $f : G \to H$: the maximum *congestion on any edge*, taken over all edges of H; terminology for modeling the *emulation of distributed computation*.

congruent imbeddings: two imbeddings $f_1, f_2 : G \to S$ with $hf_1 = f_2\sigma$ for some surface homeomorphism $h : S \to S$ and some graph automorphism $\sigma : G \to G$.

connected sum – of two surfaces S and S': a surface obtained by excising the interior of a closed disk in each surface and then gluing the corresponding boundary curves; denoted by $S \# S'$.

consistent orientation – of a polygonal complex: orientation of the faces such that, within a union of oriented boundary walks, none of the edges is traversed twice in the same direction.

contractible closed curve – on a surface S: a simple closed curve C on S, such that the closure of one of the components of $S - C$ is a disk.

contractible edge – in a triangulation: an edge whose contraction does not create a multiple adjacency.

contractible to the triangulation $G \to S$: said of a triangulation that can be transformed into the given triangulation $G \to S$ by a sequence of edge contractions.

convolution – of two sequences $\langle a_i \rangle$ and $\langle b_j \rangle$: a combinatorial operation that produces a sequence $\langle c_k \rangle$, with $c_k = a_0 b_k + a_1 b_{k-1} + \cdots + a_k b_0$.

co-tree: the edge complement of a spanning tree of a graph.

covering or **covering projection, combinatorial₁** – of a graph G by a graph $\tilde{G}$: an edge-to-edge, vertex-to-vertex mapping that is topologically a covering; exemplified by the natural projection associated with a *voltage graph*.

covering or **covering projection, combinatorial₂** – of a surface S with a cellularly imbedded graph G by a surface $\tilde{S}$ with a cellularly imbedded graph $\tilde{G}$: a face-to-face, edge-to-edge, vertex-to-vertex mapping that is topologically a covering; exemplified by the natural projection associated with an imbedded *voltage graph*.

covering or **covering projection, topological**: a continuous function $p : \tilde{X} \to X$ between locally arcwise connected topological spaces, in which every point of the codomain X has an open neighborhood U such that each arc-component of $p^{-1}(U)$ is mapped homeomorphically onto U by p.

___, k**-fold**: a covering projection that maps k to 1.

___, **regular** – onto a space X: a covering projection onto X, such that there exists a group of *covering transformations* that acts *freely* and *transitively* on it.

co-tree – the edge-complement of a tree.

covering space: the domain of a *covering projection* $p : \tilde{X} \to X$.

___, **regular** – of a space X: the domain of a regular covering projection onto X.

covering transformation – for a covering projection $p : \tilde{X} \to X$: an autohomeomorphism h on $\tilde{X}$ such that $ph = p$.

covering with folds: a natural projection from the composition of a graph G with $\overline{K}_m$ to the graph G.

Coxeter complex: the barycentric subdivision of the tessellation $\{p, q\}$, formed by all mirrors of reflection symmetries.

Coxeter group (of rank 3): a group with presentation by three generators ρ_0, ρ_1, ρ_2 and the relations $\rho_0^2 = \rho_1^2 = \rho_2^2 = (\rho_0\rho_1)^p = (\rho_1\rho_2)^q = (\rho_2\rho_0)^r = 1$.

k-crossing-critical graph: a graph G with $\mathrm{cr}(G) \geq k$ such that every proper subgraph H of G satisfies $\mathrm{cr}(H) < k$, and such that G is homeomorphically minimal with this property.

crosscap distribution – of a graph G: the sequence whose j^{th} entry is $\overline{\gamma}_j(G)$, starting with a (possibly empty) subsequence of zeroes, followed by the subsequence of the crosscap imbedding numbers, and then an infinite sequence of zeroes.

___, **polynomial** – of a graph G: the polynomial $\overline{I}_G(y) = \sum_{j=1}^{\infty} \overline{\gamma}_j(G)y^j$.

crosscap – on a surface: a subspace of the surface that is homeomorphic to a Möbius band.

___, **number** – of a closed nonorientable surface S or of the sphere: the integer k such that S is homeomorphic to N_k; denoted $\tilde{\gamma}(S)$; 0 for the sphere.

crosscap number – of a graph: the minimum *crosscap number* of a surface in which the graph is imbeddable.

___, **range** – of a graph G: the integer interval $[\overline{\gamma}_{min}(G), \overline{\gamma}_{max}(G)]$.

___, **symmetric** of a group $\mathcal{A}$: the smallest number c such that $\mathcal{A}$ has a Cayley graph imbedded in a nonorientable surface of crosscap number c.

crossing number of a graph G in a surface S, denoted $\mathrm{cr}_S(G)$: the minimum number of crossings of edges in a drawing of G in S.

___, **pair crossing number** of a drawing D: the number $\mathrm{pcr}(D)$ of pairs of open arcs α and β of D representing distinct edges of G such that $\alpha \cap \beta \neq \emptyset$.

___, **pair crossing number of a graph**: the minimum $\mathrm{pcr}(D)$ taken over all drawings of G in the plane.

___, **odd crossing number** of a drawing D: the number $\mathrm{ocr}(D)$ of pairs α and β of open arcs of D representing distinct edges of G such that $|\alpha \cap \beta|$ is odd.

___, **odd crossing number** of a graph G: the minimum $\mathrm{ocr}(D)$ taken over all drawings of G in the plane.

___, **minor crossing number** $\mathrm{mcr}(G)$: the minimum $\mathrm{cr}(H)$ such that G is a minor of H.

___, **rectilinear crossing number** $\overline{\mathrm{cr}}(G)$ of a graph G: minimum $\mathrm{cr}(D)$ taken over all rectilinear drawings of G in the plane.

crystallization of a map M: a quadruple $(F; \tau_0, \tau_1, \tau_2)$ that encodes a map on any closed surface. In the quadruple, F is the set of flags of M and, for $i \in \{0, 1, 2\}$, τ_i is a fixed-point-free involution of F transposing two incident flags sharing an edge of M^{bar} colored i. The group $\langle \tau_0, \tau_1, \tau_2 \rangle$ is transitive on F, $(\tau_2\tau_0)^2 = 1$, and $\tau_2\tau_0$ is fixed-point-free.

current – on a directed edge e: the value $\alpha(e)$ assigned to edge e by a *current assignment* in a group called the *current group*.

___, **assignment** – on a digraph $G = (V, E)$ imbedded in a surface S: a function α from edge set E to a group $\mathcal{B}$; used to specify a *derived digraph*; it must be stated explicitly that the algebraic values on the edges are currents, rather than voltages.

___, **group**: the group in which a *current assignment* takes its values; usually a finite group.

current graph: a pair $\langle G \to S, \alpha \rangle$ comprising an imbedded digraph and a *current assignment*; it specifies a graph imbedding.

cycle rank$_1$ – of a connected graph $G = (V, E)$: the number $|E| - |V| + 1$, which is the number of edges in a co-tree of a spanning graph; denoted $\beta(G)$ (for Betti, an Italian mathematician).

cycle rank$_2$ – of a possibly non-connected graph $G = (V, E)$: the number $|E| - |V| + c$, where c is the number of components.

cylinder: a surface homeomorphic to the unit cylinder in $\mathbb{R}^3$, i.e., to
$$\{(x, y, z) \mid x^2 + y^2 = 1, 0 \le z \le 1\}$$

dart: a directed edge.

deficiency of a branch point – of order r in a regular n-sheeted branched covering: the number $n - n/r$.

deficiency $\xi(G)$ of a graph G: the minimum value of $\xi(G, T)$, over all spanning trees T of G.

deficiency $\xi(G, T)$ of a spanning tree T – in a graph G: the number of odd components of the co-tree $G - T$.

delta-Y transformation – in the theory of triangulations: see ΔY-*transformation* under *transformation*.

dense imbedding: an imbedding of "large" face-width, where "large" is relative to the context.

derived digraph – specified by a voltage graph $\langle G = (V, E), \alpha \rangle$: the covering graph G^α associated with the specified type of voltages, i.e., *permutation* or *regular*.

derived digraph, Σ_n-permutation – for a *voltage assignment* $\alpha : E(G) \to \Sigma_n$: the digraph $G^\alpha = (V^\alpha, E^\alpha)$, with $V^\alpha = V(G) \times \{1, ..., n\}$ and $E^\alpha = E(G) \times \{1, ..., n\}$; if the edge e joins vertex u to vertex v in G, then the edge $e_j = (e, j)$ joins vertex $u_j = (u, j)$ to the vertex $v_{\alpha(j)} = (v, \alpha(j))$.

derived digraph, regular – for a *voltage assignment* $\alpha : E(G) \to \mathcal{B}$: the digraph $G^\alpha = (V^\alpha, E^\alpha)$, with $V^\alpha = V(G) \times \mathcal{B}$ and $E^\alpha = E(G) \times \mathcal{B}$; if the edge e joins vertex u to vertex v in G, then the edge $e_j = (e, j)$ joins vertex $u_j = (u, b)$ to the vertex $v_{b\alpha(e)} = (v, b\alpha(e))$.

derived graph: formally, the result of deleting directions from a *derived digraph*; informally, the derived digraph itself is also called a derived graph.

derived imbedding$_1$ – for an *imbedded voltage graph*: the imbedding $G^\alpha \to S^\alpha$ of the derived graph into the derived surface; constructed as described in §7.4.

derived imbedding$_2$ – for a *current graph*: the derived imbedding for the *voltage graph* of which it is the dual.

derived surface for an *imbedded voltage graph* $\langle G \to S, \alpha \rangle$: the cellular 2-complex S^α that results from fitting to each closed walk in the set $\hat{\Omega}$ of lifted boundary walks in the derived graph G^α a polygonal region (whose number of sides equals the length of that closed walk).

dessin d'enfant: a bipartite map with a fixed bipartition (coloring) of its vertices.

diagonal flip – in a triangulation: to switch the *diagonal* in a quadrilateral formed by two faces that meet on an edge (i.e., that edge is the "diagonal").

___, $\mathcal{P}$-**equivalent under**: two $\mathcal{P}$-triangulations that can be transformed into each other by a sequence of $\mathcal{P}$-preserving diagonal flips.

___, $\mathcal{P}$-**preserving**: a diagonal flip that preserves a specified property $\mathcal{P}$.

digon – in a graph imbedding: a two-sided face.

disk – closed, open: a topological space homeomorphic, respectively, to the closed or to the open unit disk.

drawing of a graph G in a surface S: a mapping that assigns to the vertices of G distinct elements of S, and assigns to each edge a homeomorphic image of $[0, 1]$, disjoint from the vertex points, except that the endpoints of the image of an edge are precisely the elements of S representing the endpoints of the edge.

___, **rectilinear**: a drawing in which all the edges are straight segments.

dual, topological: a concept due to Poincaré associated with an involutory mathematical property.

___, **graph** – for a cellularly imbedded graph G in any closed surface S: the graph G^* whose vertices are the *barycenters* of the faces of the imbedding $G \to S$, such that through each edge $e \in E(G)$ there is an edge that joins the dual vertex in the region on one side of the edge to the dual vertex on the other side (a self-loop if a face meets itself on edge e).

___, **map** – of a graph imbedding $G \to S$: the map corresponding to the *dual imbedding*.

___, **imbedding** – of a graph imbedding $G \to S$: the imbedding $G^* \to S$ obtained while constructing the *dual graph*.

___, **of a current graph** $\langle G = (V, E) \to S, \alpha : E \to \mathcal{B} \rangle$: the imbedded voltage graph whose base imbedding $G^* \to S$, is dual to the imbedding $G \to S$ (which involves reversing the orientation from the primal imbedding surface, if S is orientable), such that for each primal directed edge $e \in E$, the dual edge e^* has voltage $\alpha^*(e^*) = \alpha(e)$.

ear: a path attached at its end-vertices to a graph; the name was inspired by some drawings in which such paths had the shape of human ears.

ear decomposition – of a graph G: a partition of the edge set of G into an ordered collection $P_0, P_1, \cdots, P_r$, such that P_0 is a simple cycle and P_i, $i \geq 1$, is a path with only its endpoints in common with $P_0 + \cdots + P_{i-1}$.

edge contraction$_1$ – in a graph: removing the edge e then identifying the two vertices u and v; topologically, the edge is shrunk homotopically to a point.

edge contraction$_2$ – in a triangulation: topologically shrinking an edge, and then excising the two degenerate faces (digons) that result.

edge-suppression – of an edge of a cubic map: removal of an edge e incident with two faces and smoothing the end-vertices of e, thereby producing a smaller cubic map.

edge-width – $ew(G)$ of an imbedded graph G: the length of the shortest cycle in the graph that is non-contractible in the surface.

elementary subdivision – of an edge: the operation of replacing the edge by a path of length two; a special case of the PL-topological concept of *barycentric subdivision*.

emulation of distributed computation: porting a distributed algorithm from the parallel computer (*guest*) for which it is designed to a computer (*host*) with a different parallel architecture; modeled by a graph mapping from a graph model for the guest to a graph model for the host.

essential curve – on a surface: a simple closed curve that is not contractible on the surface; that is, either it does not separate the surface, or if it separates, then neither side of the separation is a disk.

essential cycle – in an imbedded graph: a cycle that bounds no cellular region of the imbedding surface.

Euclidean space group: a group of isometries of the Euclidean plane.

euler characteristic$_1$ – of a closed surface S: the integer $\chi(S)$ defined by $\chi(S) = 2-2g$ if S is homeomorphic to g-torus S_g, and $\chi(S) = 2 - k$, if S is homeomorphic to a non-orientable surface with k crosscaps N_k.

euler characteristic – of a cellular imbedding of a graph $G = (V, E)$: the alternating sum $|V| - |E| + |F|$, where F is the set of faces.

euler genus $\bar{\gamma}(S)$ – of a surface S: twice the number of handles if S is orientable, and the number of crosscaps if S is non-orientable.

euler genus of a group $\mathcal{A}$: the minimum number d such that the group $\mathcal{A}$ has a Cayley graph that imbeds in a surface of Euler genus d.

___, **symmetric** of a group $\mathcal{A}$: the smallest number d such that the group $\mathcal{A}$ has a Cayley graph imbeddable in a surface of Euler genus d.

Euler (Polyhedral) Equation (sometimes "Euler Formula") – for a cellularly imbedded graph $G \to S$: the equation $v - e + f = 2 - \gamma(S)$.

face – of a graph imbedding: a component of the complement of the image of the graph.

___, **boundary of**: the vertices and edges encountered while traversing the face boundary walk.

___, **boundary walk of**: the closed walk that encircles the face; it may have repeated edges and repeated vertices.

___, **closed**: the face and its boundary.

___, **open**: the face without its boundary.

face of a polygonal complex: a polygon used in the construction of the polygonal complex, viewed as a subspace of that complex.

face-width $fw(G \to S)$ – of a graph imbedding $G \to S$: the minimum value of the number $|C \cap G|$, taken over all non-contractible cycles C in the surface S; also called the *representativity*;

fb-walk: abbreviation for face-boundary walk.

fiber over an edge e – of a *voltage graph* $\langle G = (V, E), \alpha : E \to \mathcal{B} \rangle$: in the derived graph G^α, the edge subset $\{e\} \times \mathcal{B} = \{e_b : b \in \mathcal{B}\}$.

fiber over a vertex v – of a *voltage graph* $\langle G = (V, E), \alpha : E \to \mathcal{B} \rangle$: in the derived graph G^α, the vertex subset $\{v\} \times \mathcal{B} = \{v_b : b \in \mathcal{B}\}$.

fiber over x – where x is a point in the codomain of a *topological covering projection* $p : \tilde{X} \to X$: the set $p^{-1}(x)$.

finite geometry: a geometry whose point set is finite.

flag: an ordered triple (F_0, F_1, F_2) of pairwise incident faces of a map of dimensions 0, 1, and 2, respectively; represents a triangle in a map.

free action – of a group on a graph: an automorphism group with no fixed vertices, except by the identity automorphism.

freely acting group $\mathcal{H}$ of *covering transformations* for a covering projection $p : \tilde{X} \to X$: a group such that no transformation except the identity has a fixed point in $\tilde{X}$.

fundamental polygon – for a closed surface S: a polygon whose edges are pairwise identified and pasted so that the resulting polygonal complex has only one face and so that it is homeomorphic to the surface S.

f-**vector** – for a graph map: see *vector*.

genus distribution polynomial – of a graph: the polynomial $I_G(x) = \sum_{j=0}^{\infty} \gamma_j(G)x^j$.

genus distribution sequence – of a graph G: the sequence whose j^{th} entry is $\gamma_j(G)$, starting with a (possibly empty) subsequence of zeroes, followed by the subsequence of the orientable imbedding numbers, and then an infinite sequence of zeroes.

genus of a surface S: the number of handles for an orientable surface (and sometimes, the number of crosscaps for a nonorientable surface); the surface S_g has genus g; denoted $\gamma(S)$.

genus of a graph: the smallest genus of any orientable *minimum genus*.

genus imbedding – of a graph: an imbedding of the graph in a surface of smallest possible genus; short for *minimum genus imbedding*.

genus of a group $\mathcal{A}$: the minimum number g such that $\mathcal{A}$ has a Cayley graph that imbeds in a surface of genus g.

___, **symmetric** $\mathcal{A}$: the smallest number g such that $\mathcal{A}$ has a Cayley graph symmetrically imbedded in an orientable surface of genus g.

genus range of a graph G: the integer interval $[\gamma_{min}(G), \gamma_{max}(G)]$.

geometry: a pair (P, L) where P is a non-empty set and L is a non-empty collection of subsets of P.

___, **lines of**: elements of the set L.

___, **points of**: elements of the set P.

graph-encoded map – often abbreviated GEM: a particular system for describing a map using colored graphs.

(p, q, r)-group $\mathcal{A}$: a group with presentation
$$A = \langle x, y, z : x^2 = y^2 = z^2 = 1, (xy)^p = (yz)^q = (xz)^r = 1 \rangle$$

___, **proper**: a (p, q, r) group such that the subgroup generated by xy and yz has index two.

group action on a surface: a subgroup of the homeomorphism group of the surface.

guest graph – for a graph mapping: the domain of the mapping; terminology used when modeling the *emulation of distributed computation*; see *host graph*.

half-disk: a topological space homeomorphic to the unit half-disk.

Halin graph: a graph formed by joining pairs of consecutive leaves (in a pre-order or post-order traversal) of a plane tree, so that a cycle passes through the leaves.

hereditary property – under minors: a property such that whenever a graph has it, then so do all of its minors.

homeomorphic graphs: two graphs that become isomorphic after smoothing all their degree-2 vertices.

host graph for a graph mapping: the codomain; terminology used when modeling the *emulation of distributed computation*; see *guest graph*.

Hurwitz group: a $(2, 3, 7)^o$-group.

hypermap: a generalization of a graph imbedding to a hypergraph representation on a surface.

hyperregion – of an imbedding of a hypergraph in a surface: a component of the complement of the image of the hypergraph in the surface.

imbedded voltage graph: a pair $\langle G \to S, \alpha \rangle$, such that $\langle G, \alpha \rangle$ is a voltage graph such that S is a closed surface in which the graph G is (cellularly) imbedded.

imbedding – of a graph: an imbedding of the topological realization of the graph.

___, **2-cell**: see *cellular imbedding*.

___, **cellular**: an imbedding of a graph G on a surface S such that the components of $S \setminus G$ are open disks.

imbedding – of a topological space: an immersion which is (globally) one-to-one.

___, **congruent imbeddings**: imbeddings $f_1, f_2 : G \to S$ for which there exists a homeomorphism $h : S \to S$ and a graph automorphism $\sigma : G \to G$ with $hf_1 = f_2\sigma$.

___, **equivalent imbeddings**: two imbeddings $f_1, f_2 : G \to S$ with $hf_1 = f_2$ for some homeomorphism $h : S \to S$.

immersion – of a topological space: a continuous mapping that is locally one-to-one.

incenter of a triangle: the point common to the three internal angle bisectors of the triangle.

interior of a contractible cycle C on a surface S: the component of $S - C$ that is homeomorphic to the plane.

irreducible triangulation of N_2, crosscap-type: an irreducible triangulation on the Klein bottle that splits into two triangulations on the projective plane.

irreducible triangulation of N_2, handle-type: an irreducible triangulation of the Klein bottle that contains no separating cycle of length 3.

irreducible triangulation: a triangulation that has no contractible edge.

isomorphic triangulations: two triangulations on a surface such that there is an auto-homeomorphism on the surface mapping one skeleton onto the other.

isomorphism of maps: a homeomorphism of the respective surfaces that induces a graph isomorphism of the respective graphs.

isotopic triangulations: two triangulations on a surface one of which can be transformed continuously on the surface into the other.

kernel: an imbedded graph G such that for every proper minor H, there is a curve C with $\mu(H, C) < \mu(G, C)$.

Kepler–Poinsot regular star polyhedra: some self-intersecting realizations of regular maps.

Kirchhoff current law (KCL) – at a vertex v of a current graph: a possible condition, namely, that the net current at v is the group identity.

Kirchhoff current law (KCL) – on a current graph: a possible condition, namely, that KVL holds at every vertex.

Kirchhoff voltage law (KVL) – on a closed walk W in a voltage graph: a possible condition, namely, that the *net voltage* on W is the identity of the voltage group.

Kirchhoff voltage law (KVL) – on an imbedded voltage graph: a possible condition, namely, that KVL holds on every face boundary walk in the graph.

Klein bottle N_2: a closed nonorientable surface obtained by identifying the pairs of points $\{(x, y, -1), (x, -y, 1)\}$ on the two boundary components of the cylinder $\{(x, y, z) \mid x^2 + y^2 = 1, -1 \leq z \leq 1\}$; its crosscap number is 2.

Kuratowski's Theorem: the theorem that every non-planar graph contains a homeomorphic copy either of K_5 or of $K_{3,3}$.

large edge-width imbedding: an imbedding where the edge-width is strictly larger than the length of the longest facial cycle.

large-edge-width map: a map whose edge-width is greater than the number of edges in any face boundary.

Levi graph of a geometry: a graph whose edges join incident point/line pairs of that geometry.

LEW-imbedding: short for large edge-width imbedding.

lift of a walk W – in a *voltage graph* $\langle G, \alpha \rangle$: a walk $\widetilde{W}$ in the *derived graph* that is mapped isomorphically onto W by the *natural projection*.

line$_1$: an element of a geometry; esp. an affine line in a real Euclidean space.

line$_2$: an edge of a graph.

link of a vertex v **in a triangulation**: the cycle around v through all its neighbors.

load at a vertex v in the host of a graph mapping $f : G \to H$: the cardinality $|f^{-1}(v)|$ of its preimage; terminology for modeling the *emulation of distributed computation*.

load of the mapping $f : G \to H$: the maximum *load at a vertex*, taken over all vertices of H; terminology for modeling the *emulation of distributed computation*.

2-manifold: a topological space in which each point has a neighborhood that is homeomorphic either to an open disk or to a half-disk.

map: a cellular imbedding of a graph on a surface.

maximum crosscap imbedding – of a graph: a cellular imbedding into a closed nonorientable surface of maximum crosscap number.

maximum crosscap number of a graph G – also called the *maximum nonorientable genus*: the largest integer k such that the graph G has a cellular imbedding in the nonorientable surface N_k; denoted $\overline{\gamma}_{max}(G)$.

maximum crosscap number – of a graph: the maximum of the set of integers k such that G has a cellular imbedding in the nonorientable surface N_k; 0 if the graph is planar; denoted $\tilde{\gamma}_{\max}(G)$.

maximum genus – of a graph G: the largest integer g such that the graph G has a cellular imbedding in the orientable surface S_g; denoted $\gamma_{max}(G)$.

maximum genus imbedding – of a graph: an imbedding of the graph into a closed orientable surface of maximum genus.

medial graph – $M(G)$ of an imbedded graph G: an imbedded graph whose vertices are the edges of G and whose edges join two vertices corresponding to two consecutive edges in a face boundary of G.

Menger graph of a geometry: a graph whose edges join collinear points of a geometry.

minimal forbidden minor – for a class $\mathcal{F}$ of graphs closed under minors: a graph G that is not in $\mathcal{F}$, but such that every proper minor of G is in $\mathcal{F}$.

minimal triangulation: a simplicial polyhedral map such that the contraction of any edge results in a map that is no longer polyhedral.

minimum crosscap imbedding – of a graph: an imbedding into a closed nonorientable surface of minimum crosscap number; an imbedding in the sphere if possible.

minimum crosscap number of a graph G – also known as the *minimum nonorientable genus*: the smallest integer k such; 0 if the graph is planar; denoted $\overline{\gamma}_{min}(G)$.

minimum crosscap number – of a graph: the minimum of the set of integers k such that the graph is imbeddable in the nonorientable surface N_k; denoted $\tilde{\gamma}_{\min}(G)$ or $\tilde{\gamma}(G)$.

minimum genus – of a graph G: the minimum integer g such that the graph G has an imbedding into the orientable surface S_g of genus g; denoted $\gamma_{\min}(G)$ or $\gamma(G)$.

minimum genus imbedding – of a graph: an imbedding of the graph into a closed orientable surface of minimum possible genus.

minor of a graph G: a graph formed from G by a sequence of edge deletions and edge contractions.

minor-minimal imbedded graph: an imbedded graph such that the deletion or surface contraction of any edge lowers the face-width.

minor of a map M: a map $\overline{M}$ obtained from map M by deleting and/or contracting edges.

Möbius band: a surface obtained from a 2-×-2 square $\{(x,y) | -1 \le x \le 1, -1 \le y \le 1\}$ by pasting the vertical sides together with the matching $(-1, x) \to (1, -x)$.

monogon – in a graph imbedding: a one-sided face.

mu-invariant $\mu(G,C)$ – for a cycle C in an imbedded graph G: half the minimum length of a closed walk W, taken over all walks W in the radial graphs that are homotopic to C.

natural action of a group $\mathcal{A}$ **on the Cayley graph** $C(\mathcal{A}, X)$: left multiplication by elements of $\mathcal{A}$.

natural projection for a voltage graph $\langle G = (V, E), \alpha : E \to \mathcal{B}\rangle$: the graph mapping $G^\alpha \to G$ comprising the vertex function $v_b \mapsto v$ and the edge function $e_b \mapsto e$. (Thus, the natural projection is given by "erasure of subscripts".)

natural projection for an imbedded voltage graph $\langle G \to S, \alpha\rangle$: the extension of the natural projection $p : G^\alpha \to G$ to the surfaces, so that it maps the center of each polygon f in the derived imbedding $G^\alpha \to S^\alpha$ to the center of the region of the imbedding $G \to S$ bounded by $p(bd(f))$.

near-planar graph: a graph G that has an edge e such that $G \setminus e$ is planar.

near triangulation: a rooted map in which every nonroot face is a 3-gon.

Nebeský nu-invariant $\nu(G)$: an invariant used in calculating the crosscap number of a graph.

neighborly polyhedral map: a polyhedral map in which every pair of distinct vertices is joined by an edge.

k-nest – in an imbedding: a sequence $C_1, \ldots, C_k$ of disjoint contractible cycles such that C_i is in the interior of C_{i+1}.

net current at a vertex v of a current graph: for an abelian group, the sum of the inflowing currents; for a non-abelian group, the product of the inflowing currents in the cyclic order of the *rotation* at v.

net voltage on a walk in a voltage graph: for an abelian group, the sum of the voltages on the walk, taken in the order of traversal; for a non-abelian group, the product of the algebraic elements in its voltage sequence, in cyclic order (which is unique up to conjugacy).

non-contractible cycle – in an imbedded graph: a cycle that is non-contractible on the surface.

non-orientable 2-manifold: a 2-manifold that contains a subspace homeomorphic to the Möbius band.

non-orientable surface with k **crosscaps** N_k: a connected sum of k copies of the projective plane N_1.

non-revisiting path: a path p in the graph of a map M such that the set $p \cap F$ is connected, for each face F of M.

non-separating cycle – in an imbedded graph: a cycle whose removal separates the surface.

orientable 2-manifold: a 2-manifold which is not non-orientable.

orientation reversing curve – on a surface: a simple closed curve whose regular neighborhood is a Möbius band.

oriented boundary walk – of a face of a graph imbedding in an oriented surface: the closed walk in the 1-skeleton that results from traversing the face boundary in the direction of orientation.

oriented polygon: a polygon together with a direction (clockwise or counterclockwise) of traversal of its boundary, designated to be *preferred*.

oriented polygonal complex: polygonal complex together with a consistent orientation.

orthocenter of a triangle: the point common to the three altitudes of the triangle.

outerplanar graph: a graph that can be drawing in the plane so that every vertex lies on the exterior face.

overlap matrix – for a general rotation system ρ of a graph G and a spanning tree T: the matrix $M_{\rho,T} = [m_{i,j}]$ whose entries are given for all pairs of edges e_i, e_j of the cotree $G - T$ by

$$m_{i,j} = \begin{cases} 1 & \text{if } i \neq j \text{ and } pure(\rho)\,|_{T+e_i+e_j} \text{ is nonplanar} \\ 1 & \text{if } i = j \text{ and edge } i \text{ is twisted} \\ 0 & \text{otherwise} \end{cases}$$

The notation $pure(\rho)|_{T+e_i+e_j}$ means the restriction of the underlying pure part of the rotation system ρ to the subgraph $T + e_i + e_j$.

panel – of a triangulation: a 3-cycle in the skeleton that always bounds a face in any imbedding of the skelton graph in that same surface.

panel structure – of a triangulation: the composite structure of a triangulation and its panels.

panel structures, equivalent: two panel structures whose skeletons with all panels inserted form homeomorphic 2-complexes.

parallel lines – of a geometry: lines with no point in common.

partially balanced incomplete block system (abbr. PBIBD): a geometry of v points (each in r lines) and b lines (each containing k points), together with a strongly regular graph (whose vertices are the points of the geometry) such that two non-adjacent points belong to λ_1 lines and two adjacent points belong to λ_2 lines.

pasting topological spaces X and Y – along homeomorphic subspaces: obtaining a new topological space from the original ones by identifying the points of the homeomorphic subspaces under a homeomorphism.

permutation scheme: a particular system for describing a map using a pair of permutations.

Petrie dual of a map M: a map M^P with the same underlying graph as M, whose face boundaries are closed walks in G_M, such that any two consecutive edges, but not three, belong to a face of M.

Platonic solids: the five regular geometric solids — tetrahedron, octahedron, cube, dodecahedron, icosahedron.

pinched open disk: a topological space obtained from several copies of open disks by identifying their respective centers to a single vertex.

planar graph: a graph whose minimum genus is 0.

planarizing collection of cycles: a set $C_1, \ldots, C_g$ of cycles in an imbedded graph such that cutting along all of the C_i simultaneously yields a connected graph imbedded in the plane.

planarizing curve – for a nonplanar region of a noncellular graph imbedding: a separating closed curve such that all of the boundary components lie to one side of the separation and all of the genus lies to the other.

point$_1$: a point of Euclidean space or a topological space.

point$_2$: a vertex of a graph.

polygonal complex: roughly, a topological space obtained from a set of oriented polygons by pasting some of these polygons to each other (and to themselves) along their sides.

polyhedral imbedding: an imbedding such that the intersection of any two face boundaries is either empty or a path.

polyhedral map: a map M whose face boundaries are cycles, and such that any two distinct face boundaries are either disjoint or meet in either a single edge or vertex.

___, **weakly neighborly**: a polyhedral map for which every pair of vertices is contained on a face.

preferred direction – of the traversal of a polygon boundary: a chosen direction of traversal of sides of the polygon (clockwise or counterclockwise).

projective plane N_1: a closed surface obtained from the closed unit disk by identifying pairs of boundary points that are diametrically opposite relative to the center of the disk.

projective plane: the nonorientable surface of genus 1.

2-pseudomanifold: a topological space in which each point has a neighborhood that is homeomorphic either to an open disk, to a half-disk, or to a pinched disk.

pseudosurface: a 2-pseudomanifold, usually assumed to be connected.

punctured surface: a surface with one boundary component.

quadrangulation: a graph imbedding all of whose faces are 4-sided.

quadrilateral: a 4-sided face of a graph imbedding.

radial graph $R(G)$ of an imbedded graph G: an imbedded bipartite graph whose vertices are the vertices and faces of G, and whose edges join incident elements.

ramification point of a covering: see *branch point*.

realization – of a map: an imbedding of the map into Euclidean space $\mathbf{E}^d$ such that each face is a plane convex polygon and adjacent faces are not coplanar.

regular map: a map whose automorphism group acts transitively on the set of *flags*.

representativity $\rho(G)$ – of an imbedded graph G: same as the face-width of the imbedding.

Ringel–Youngs Theorem: the theorem that the chromatic number of the orientable surface S_g equals the Heawood number

$$\left\lfloor \frac{7 + \sqrt{1 + 48g}}{2} \right\rfloor$$

rooted graph: a graph with a distinguished vertex or edge, called the root-vertex or root-edge. Sometimes a graph has more than one root-vertex or root-edge.

rooted imbedding: an imbedding with a distinguished vertex v, an edge e incident with v, and a face f incident with e.

rooted map: a map in which a flag has been distinguished.

rotation (global) – on a graph: an assignment of a rotation at each vertex.

rotation at a vertex: a cyclic permutation on the set of half-edges incident to the vertex.

rotation system: a purely combinatorial description of an imbedding of a graph G on a surface, by giving a *rotation* at each vertex of G.

semicellular graph imbedding – of a graph G: an imbedding $G \to S$ whose regions are planar, but which may have more than one boundary component.

separating closed curve – on a surface: a simple closed curve the excision of which splits the surface into two components.

p-sequence – of a polyhedral map: the sequence $\{p_i\}$, where p_i is the number of i-gonal faces.

v-sequence – of a polyhedral map: the sequence $\{v_i\}$, where v_i is the number of vertices of degree i.

signed boundary walk – of a face of a polygonal complex: the list of the signed edges that occur on an oriented boundary walk of that face.

signed edge – in a polygonal complex: the occurrence of an oriented edge or of its reverse edge within a walk in the 1-skeleton of the polygonal complex.

similar imbeddings: two imbeddings such that one can be changed into the other by a sequence of $Y\Delta$- and ΔY-transformations and the taking of geometric duals.

simple map: a map in which each vertex has degree 3.

simplicial map: a map where each face boundary is a 3-cycle.

skeleton, or **1-skeleton** – of a polygonal complex: the graph consisting of the vertices and edges of the polygonal complex.

skew polyhedron: a realization of a polyhedral map in $\mathbb{R}^d$, for $d > 3$.

skew vertex – in a triangulation: a vertex whose skeleton has at least two different cycles that contain all of its neighbors.

smoothing – a degree-2 vertex: an operation that removes a degree-2 vertex v then adds a new edge between the two neighbors of v.

sphere S_0: a surface homeomorphic to the standard sphere $\{(x, y, z) \mid x^2 + y^2 + z^2 = 1\}$ in $\mathbb{R}^3$.

standard triangulation on the sphere: an n-vertex triangulation on the sphere whose skeleton is isomorphic to the join $P_{n-2} + K_2$.

star neighborhood of a vertex v **in a triangulation**: the wheel obtained by joining the link of v to v.

Steiner triple system: a *balanced incomplete block design* with $k = 3$ and $\lambda = 1$.

stratified graph for a graph G: a graph in which each imbedding of G is represented by a vertex, and in which each edge represents a transition between two imbeddings.

strong symmetric genus of a group $\mathcal{A}$: the smallest number g such that the group $\mathcal{A}$ has a Cayley graph with a strongly symmetric imbedding in an orientable surface of genus g.

strongly cellular imbedding – of a graph: an imbedding such that the closure of each face is a closed disk.

strongly noncellular graph imbedding $G \rightarrow S$: an imbedding with at least one nonplanar region.

strongly noncontractible closed curve – in a region of a noncellular graph imbedding: a curve such that cutting it open reduces the genus of the region.

strongly symmetric imbedding of a Cayley graph $C(\mathcal{A}, X)$ in an orientable surface S: an imbedding such that the natural action of $\mathcal{A}$ on $C(\mathcal{A}, X)$ extends to an orientation-preserving action on the surface S.

surface: a 2-manifold, usually assumed to be connected, compact, and without boundary, unless otherwise declared.

surface minor – of an imbedded graph: another imbedded graph in the same surface formed by a sequence of edge deletions and contractions in the surface.

surface with k **holes**: a surface obtained by removing the interiors of k disjoint disks from a closed surface.

symmetric imbedding of a Cayley graph $C(\mathcal{A}, X)$ in a surface S: an imbedding such that the natural action of $\mathcal{A}$ on $C(\mathcal{A}, X)$ extends to the surface S.

symmetrical map: a map with at most two orbits under the action of the automorphism group on the set of flags.

symmetry group of an imbedding $f : G \rightarrow S$: the subgroup of automorphisms that are symmetries of an imbedding f; denoted $\mathrm{Sym}(f)$.

symmetry of an imbedding $f : G \rightarrow S$: an automorphism $\sigma \in \mathrm{Aut}(G)$ such that $hf = f\sigma$ for some homeomorphism $h : S \rightarrow S$.

tessellation $\{p, q\}$: the classical tiling of the sphere, Euclidean plane, or hyperbolic plane into p-gons, of which q are incident at each vertex.

topological realization – of a graph: a topological space obtained from the graph by first assigning to each of its edges a closed interval and then identifying endpoints of intervals according to the coincidences of the corresponding endpoints of edges of the graph.

g-**torus** S_g: a connected sum of g copies of a torus; this surface is usually called the *orientable surface of genus g*.

torus S_1: the orientable surface of genus 1; a closed surface obtained by rotating a circle $\{(x, y, z) \mid (x - 2)^2 + y^2 = 1, z = 0\}$ around the y-axis.

total imbedding distribution – of a graph G: the bivariate polynomial
$$\ddot{I}_G(x, y) = I_G(x) + \overline{I}_G(y) = \sum_{j=0}^{\infty} \gamma_j(G)x^j + \sum_{j=1}^{\infty} \overline{\gamma}_j(G)y^j$$

ΔY-**transformation** – in the theory of triangulations: the operation of deleting the three edges joining three mutually adjacent vertices and inserting a new vertex with new edges to all three vertices; the inverse of a $Y\Delta$-transformation.

$Y\Delta$-**transformation** – in the theory of triangulations: a graph formed by deleting a vertex v of degree three and adding in a new 3-cycle incident with its neighbors.

transitively acting group $\mathcal{H}$ – of *covering transformations* for a *covering projection* $p : \tilde{X} \to X$: a group whose restriction to every fiber is a transitive permutation group.

triangle group: a group of isometries generated by rotation about the vertices of a triangle with angles $\pi/p, \pi/q, \pi/r$; the symmetry group of the *tessellation* of type $\{p, q\}$.

___, **full**: a group generated by the reflections in the sides of a triangle with angles $\pi/p, \pi/q, \pi/r$.

triangle$_1$: a 3-sided polygon, a figure in plane geometry; see *centroid, cevian, circumcenter, incenter, orthocenter*.

triangle$_2$: a 3-sided face of a graph imbedding.

triangle$_3$: a 3-cycle of a graph.

triangular imbedding: an imbedding that imbeds a graph with all faces 3-sided.

triangulates a surface – a possible graph property: having a triangular imbedding in some surface.

triangulation of a surface: a simplicial map where each face boundary is a 3-cycle.

___, **Catalan**: a triangulation on a surface-with-boundary whose boundary includes all vertices.

___, **clean**: a triangulation such that every 3-cycle in the skeleton bounds a face.

___, **combinatorially equivalent**: two triangulations that have the same set of face boundary cycles.

___, **eulerian**: a triangulation with each vertex of even degree.

___, **frozen**: a triangulation such that no edge can be flipped without giving the skeleton a double adjacency.

___, k-**irreducible**: a triangulation such that each edge is contained in an essential cycle of length at least k.

___, **isomorphic triangulations**: two triangulations $G_1 \to S$ and $G_2 \to S$ such that there is a homeomorphism $h : S \to S$ such that $h(G_1) = G_2$.

___, **isotopic triangulations**: two triangulations $G_1 \to S$ and $G_2 \to S$ such that there is a homeomorphism $h : S \to S$ with $h(G_1) = G_2$ that is isotopic to the identity mapping on S. (Roughly speaking, this means that one can be transformed continuously on the surface into the other.)

___, **minimal clean**: a clean triangulation minimal with respect to edge contractions.

___, **minimal** – of a surface: a triangulation on the surface having the fewest vertices.

___, k-**minimal**: the same as a k-irreducible triangulation.

___, **pseudo-minimal**: a triangulation such that no sequence of diagonal flips transforms it into one having a vertex of degree 3.

___, **tight**: a triangulation $G \to S$ such that, for any partition of $V(G)$ into three nonempty subsets V_1, V_2 and V_3, there is a face $v_1 v_2 v_3 \in F(G \to S)$ with $v_i \in V_i$.

___, **untight**: a triangulation that is not tight.

type-$\{p,q\}$ map: a map with p edges incident with each vertex and q edges incident with each face.

underlying cellular imbedding – of a semicellular graph imbedding: the imbedding obtained by cutting each non-cell region open along a maximal family of boundary-separating closed curves and capping the holes with disks.

underlying graph of a map: the 1-skeleton.

underlying semicellular imbedding – of a strongly noncellular graph imbedding $G \to S$: the imbedding obtained by cutting each non-cell region open along a maximal family of boundary-separating closed curves and then capping the holes with disks.

unimodal sequence $\{a_m\}$: a sequence such that there exists at least one integer M such that

$$a_{m-1} \leq a_m \text{ for all } m \leq M \quad \text{and} \quad a_m \geq a_{m+1} \text{ for all } m \geq M$$

uniquely imbeddable on a surface S – a possible graph property: having a unique imbedding on the surface (up to a suitable equivalence).

unit disk – closed, open: respective subsets $\{(x,y) \mid x^2 + y^2 \leq 1\}$ and $\{(x,y) \mid x^2 + y^2 < 1\}$ of the Euclidean plane together with the inherited Euclidean topology.

unit half-disk: a subset $\{(x,y) \mid x \geq 0, x^2 + y^2 < 1\}$ of the Euclidean plane together with the inherited Euclidean topology.

upper-imbeddable graph: a graph G whose maximum genus is equal to $\lfloor \beta(G)/2 \rfloor$, where $\beta(G)$ is the cycle rank of G.

f-vector – for a graph map: the triple (f_0, f_1, f_2) where f_i is the number of i-dimensional faces of the map.

vertex – of a polygonal complex: the image of arbitrarily many polygon corners that have been pasted together when building the polygonal complex.

vertex-amalgamation – of two disjoint graphs G and H: the result of identifying a vertex u of G and a vertex v of H. Notation: $G_u * H_v$.

vertex-face graph: same as the radial graph.

vertex-face incidence graph: same as the radial graph.

vertex splitting: an operation on a map inverse to *edge contraction* – a single vertex is replaced by two vertices joined by an edge.

vertex-transitive action of a group of automorphisms on a graph: a group such that for any pair of vertices, there is an automorphism taking one vertex to the other.

vertex-transitive map: a map whose automorphism group acts transitively on the set of vertices.

voltage – on a directed edge e: the value $\alpha(e)$ assigned to e by a *voltage assignment*.

voltage assignment – on a digraph $G = (V, E)$: a function α from edge-set E to a group $\mathcal{B}$; used to specify a *derived digraph*.

voltage graph: a pair $\langle G, \alpha \rangle$, where G is a digraph and $\alpha : E_G \to \mathcal{B}$ is a *voltage assignment*; an algebraic specification of a *derived graph*.

voltage group: the group in which a *voltage assignment* $\alpha : E_G \to \mathcal{B}$ takes its values.

voltage sequence on a walk $W = v_0, e_1, v_1, e_2, ..., e_n, v_n$ in a voltage graph $\langle G, \alpha \rangle$: the sequence of voltages $a_1, ..., a_n$ encountered, where $a_j = \alpha(e_j)$ or $\alpha(e_j)^{-1}$, respectively, depending on whether edge e_j is traversed in the forward or backward direction.

walk in a voltage graph $\langle G, \alpha \rangle$: a walk in G as if it were undirected, so that some of its edge-steps may proceed in the opposite direction from the assigned direction on the edge it traverses.

k-**walk**: a spanning walk that visits no vertex more than k times.

weakly neighborly polyhedral map: see *polyhedral map*.

wheel-neighborhood, having a – a possible property of a vertex v: the property that any two face boundaries containing v intersect in a path.

Whitney flip: a transformation of an imbedding of a 2-connected graph that replaces a subgraph by its mirror image.

Xuong tree T in a graph G: a spanning tree whose deficiency is equal to the deficiency of the graph G.

$Y\triangle Y$-**equivalent graphs**: two graphs such that one can be changed into the other by a sequence of $Y\triangle$- and $\triangle Y$-transformations.

$Y\triangle$-**transformation**: see *transformation*.

Chapter 8

Analytic Graph Theory

Section 8.1
Extremal Graph Theory

Béla Bollobás, University of Memphis, and Trinity College, Cambridge
Vladimir Nikiforov, University of Memphis

INTRODUCTION

Extremal graph theory is concerned with inequalities among functions of graph invariants and the structures that demonstrate that these inequalities are best possible. Accordingly, in its wide sense, it encompasses most of graph theory. Nevertheless, there is a clearly identifiable body of core extremal results: in this all-too-brief review we shall present a selection of narrowly interpreted extremal results.

Since by now there are many thousands of papers on extremal graph theory, no short survey of extremal graph theory has any hope of being complete. There is no doubt that the selection of topics in this survey, in which we shall concentrate on the basic graph invariants such as size, maximal and minimal degrees, connectivity, number of r-cliques, and independence number, strongly reflects the tastes and preferences of its authors.

CONVENTIONS AND NOTATIONS

C1: Unless explicitly stated, all graphs are assumed to be defined on the vertex set $[n] = \{1, 2, \ldots, n\}$.

C2: $G(n)$ stands for a graph with n vertices and $G(n, m)$ stands for a graph with n vertices ("of *order n*") and m edges ("of *size m*"). Thus, the statement

> "...if in $G = G(n)$ with $n \geq 3$, every vertex has degree at least $n/2$, then G is Hamiltonian..."

means that every graph of order $n > 3$ and minimal degree $n/2$ is Hamiltonian.

N1: $v(G)$ and $e(G)$ denote the numbers of vertices and edges in a graph G.

N2: $\Delta(G)$ and $\delta(G)$ stand for the maximal and minimal degrees of G.

N3: If u is a vertex of a graph G, then $\Gamma_G(u)$ is its set of neighbors, and $d_G(u) = |\Gamma_G(u)|$ is its degree. We use $d(u)$ and $\Gamma(u)$ instead of $d_G(u)$ and $\Gamma_G(u)$ when it is clear which graph G is intended.

N4: If the graphs G_1 and G_2 are disjoint graphs, then $G_1 + G_2$ denotes the *join* of G_1 and G_2 (see Subsection 1.1.3), that is, the union of G_1 and G_2 together with some new edges joining every vertex of G_1 to every vertex of G_2.

N5: We denote by $k_s(G)$ the number of copies of the complete graph K_s in the graph G.

DEFINITIONS

D1: Given graphs H and G, we say that G is H-*free* if G has no subgraph isomorphic to H.

D2: An s-*clique* is a complete subgraph on s vertices. Thus, $k_s(G)$ equals the number of s-cliques of the graph G.

D3: Given a graph G, the **clique number** $\omega(G)$ of G is the order of its largest clique.

D4: The **independence number** $\alpha(G)$ of G is the clique number of its complement.

D5: A graph is k-**connected** if the deletion of fewer than k of its vertices leaves it connected. The vertex connectivity $\kappa(G)$ of a graph G is the maximal number k such that G is k-connected.

EXAMPLES

Let us illustrate the difference between the wide and narrow interpretations of extremal graph theory with two examples.

E1: Let $f(n)$ be the minimal number of triangles needed to cover all the edges of a complete graph with n vertices. Then

$$f(n) \leq \binom{n}{2}/3$$

and equality holds iff there is a Steiner triple system of order n. This assertion is not really a result in extremal graph theory, but leads to *design theory*.

E2: Consider the statement that "every 2-connected graph with n vertices and minimal degree k contains a cycle with at least $\min\{2k, n\}$ vertices". This is clearly a result in extremal graph theory with the narrow interpretation: it expresses a relationship involving the number of vertices, the minimal degree, the connectivity, and the circumference (maximal length of a cycle).

8.1.1   Turán-Type Problems

The quintessential problem in extremal graph theory is the following question due to Turán: given $3 \leq r \leq n$, what is the maximal size graph $G(n)$ that does not contain K_r? Equivalently, what is the maximal size of a K_r-free $G(n)$?

DEFINITION

The Turán problem has been substantially generalized.

D6:   Let $\{F_n\}_{n=1}^{\infty}$ be a sequence of families of graphs, and let $\Phi(n, F_n)$ be the set of graphs $G(n)$ that are H-free for every $H \in F_n$.

- The function $ex(n, F_n)$ of n is called the ***extremal function*** of the sequence $\{F_n\}$.
- The graphs $G \in \Phi(n, F_n)$ for which $e(G) = ex(n, F_n)$ are called ***extremal graphs***.
- In this context, the families $\{F_n\}_{n=1}^{\infty}$ are called ***forbidden graphs***.
- The aim to find or estimate $ex(n, F_n)$ and to determine the extremal graphs is called a ***Turán type problem*** for the families of forbidden graphs $\{F_n\}_{n=1}^{\infty}$.

Turán's Theorem and Its Extensions

The fundamental theorem of Turán has been the driving force of extremal graph theory for more than six decades.

DEFINITIONS

D7:   Let $n \geq r \geq 2$ be integers. The ***Turán graph*** $T_r(n)$ is the complete r-partite graph whose classes are as nearly equal as possible.

D8:   The ***Turán number*** $t_r(n)$ is the size of the Turán graph $T_r(n)$.

FACTS

F1:   If $n = rs + t$ $(0 \leq t \leq r - 1)$, then $T_r(n)$ has t classes of cardinality $\lceil n/r \rceil$ and $r - t$ classes of cardinality $\lfloor n/r \rfloor$. Therefore,

$$t_2(n) = \left\lfloor \frac{n^2}{4} \right\rfloor, \quad t_r(n) = \frac{r-1}{2r}\left(n^2 - t^2\right) + \binom{t}{2}$$

$$\frac{r-1}{2r}n^2 \ \geq \ t_r(n) \ \geq \ \frac{r-1}{2r}n^2 - \frac{n}{4}$$

F2:   **Mantel** [Man07] If a graph $G = G(n)$ is K_3-free, then $e(G) \leq \lfloor n^2/4 \rfloor$.

F3:   **Turán's theorem** [Tur41] If a graph $G = G(n)$ is K_{r+1}-free, then $e(G) \leq t_r(n)$; and if $e(G) = t_r(n)$, then $G = T_r(n)$.

F4:   **Zykov** [Zyk49], **Erdős** [Erd62a] If a graph $G = G(n)$ is K_{r+1}-free, then for every $s = 2, \ldots, r$ we have $k_s(G) \leq k_s(T_r(n))$; and if $k_s(G) = k_s(T_r(n))$ for some s such that $2 \leq s \leq r$, then we have $G = T_r(n)$.

F5: Erdős [Erd70] For every K_{r+1}-free graph G, there exists an r-partite graph H with $V(H) = V(G)$, such that $d_G(u) \leq d_H(u)$ for every $u \in V(G)$. If G is not a complete r-partite graph, then H may be chosen so that $d_G(u) < d_H(u)$ for some $u \in V(G)$.

F6: Khadžiivanov [Kha77], **Fisher and Ryan** [FiRa92] If the graph G is K_{r+1}-free, then for every $s = 1, \ldots, r - 1$ we have

$$\left(\frac{k_s(G)}{\binom{r}{s}} \right)^{1/s} \geq \left(\frac{k_{s+1}(G)}{\binom{r}{s+1}} \right)^{1/(s+1)}$$

F7: Motzkin–Straus inequality [MoSt65] If a graph $G = G(n)$ is K_{r+1}-free and $x_1, \ldots, x_n$ are nonnegative numbers then

$$\sum_{v \in V(G)} \sum_{u \in \Gamma(v)} x_u x_v \leq \frac{r-1}{r} \left(\sum_{v \in V(G)} x_v \right)^2$$

F8: Bomze inequality [Bom97] If the graph $G = G(n)$ is K_{r+1}-free and if the numbers $x_1, \ldots, x_n$ are nonnegative and not all zero, then

$$\sum_{v \in V(G)} \sum_{u \in \Gamma(v)} x_u x_v + \frac{1}{2} \sum_{u \in V(G)} x_u^2 \leq \left(1 - \frac{1}{2r} \right) \left(\sum_{v \in V(G)} x_v \right)^2 \qquad (1)$$

Let $x_{u_1}, \ldots, x_{u_q}$ be the nonzero numbers in $\{x_1, \ldots, x_n\}$. Equality in (1) is attained if and only if $q = r$, $x_{u_1} = \cdots = x_{u_r}$, and the vertices $u_1, \ldots, u_r$ are an r-clique in G.

Structural Properties of the Graphs $G(n, t_r(n) + 1)$

Turán's theorem guarantees that every $G(n, t_r(n) + 1)$ contains a K_{r+1}, but further investigations revealed a lot more properties of such graphs. We present below three topics of considerable interest.

NOTATIONS

N6: $K_r(s_1, \ldots, s_r)$ denotes the complete r-partite graph with classes of size $s_1, \ldots, s_r$, respectively. The graph $K_r^+(s_1, \ldots, s_r)$ is obtained from $K_r(s_1, \ldots, s_r)$ by adding an edge to the first specified class, i.e., of order s_1.

N7: For all natural numbers n, m, r let $\delta(n, m, r)$ denote the maximal value such that every graph $G(n, m)$ has an r-clique R with $\sum \{d(u) : u \in R\} \geq \delta(n, m, r)$.

FACTS

F9: Bollobás and Thomason [BoTh81], **Erdős and Sós** [ErSo83] For $r \geq 3$ every $G(n, t_r(n) + 1)$ has a vertex u with degree $d(u) > \left(1 - 1/r - (1 + \sqrt{r})^{-1} \right) n$ and such that $\Gamma(u)$ induces more than $t_{r-1}(d(u))$ edges.

F10: Bondy [Bon83a], [Bon83b] For $r \geq 3$ every $G(n, t_r(n) + 1)$ has a vertex u of maximal degree such that $\Gamma(u)$ induces more than $t_{r-1}(d(u))$ edges.

F11: **Bollobás** [Bol99] Let $G = G(n, t_r(n) + a)$ where $a \geq 0$. Let u be a vertex of maximal degree $d(u) = n - k$. Then $e(G[\Gamma(u)]) \geq t_r(d(u))$; and the inequality is strict unless $k = \lfloor n/r \rfloor$, $d(u) = \lceil (r-1)n/r \rceil$, the set $V \backslash \Gamma(u)$ is independent, and every vertex of $\Gamma(u)$ is joined to every vertex of $V - \Gamma(u)$.

F12: **Erdős** [Erd63] For every $\varepsilon > 0$ there exists a number $c = c(\varepsilon) > 0$ such that every $G(n, \lfloor n^2/4 \rfloor + 1)$ contains a $K_2^+(c \log n, n^{1-\varepsilon})$.

F13: **Erdős and Simonovits** [ErSi73] For every q and n sufficiently large, every $G(n, t_r(n) + 1)$ contains a $K_r^+(q, \ldots, q)$.

F14: **Edwards** [Edw77], [Edw78] It $3 \leq r \leq 8$, $n > r^2$ and $m \geq t_r(n)$, then

$$\delta(n, m, r) \geq \frac{2rm}{n}$$

F15: **Faudree** [Fau92] If $r \geq 3$, $n > r^2(r-1)/4$, and $m \geq t_r(n)$, then
$$\delta(n, m, r) \geq \frac{2rm}{n} \qquad (2)$$

REMARKS

R1: In a sense, Fact F13 is best possible, since if H is a fixed graph that occurs in any graph $G = G(n, t_r(n) + 1)$ then $H \subset K_r^+(q, \ldots, q)$ if q is sufficiently large. However, Fact F12 suggests that extensions are still possible, although we are not aware of any such extension.

R2: Bollobás and Nikiforov showed that Fact F15 holds for every n.

R3: The result in Fact F16 confirms a conjecture of Bollobás and Erdős [BoEr76], and it is essentially best possible, since if G is regular, then equality holds in (2). On the other hand, if $m \leq t_{r-1}(n)$, then $\delta(n, m, r) = 0$. It is a difficult open question to determine $\delta(n, m, r)$ for $t_{r-1}(n) < m < t_r(n)$ (see, e.g., [CEV88]).

Books and Generalized Books

The study of books was initiated by Erdős in 1962 [Erd62b] and has attracted much effort since then. Nevertheless, the Turán problems about books, except for the simplest case, are largely open.

DEFINITION

D9: For $q \geq 1$, $r \geq 1$ an *r-book* is the graph $B_q^{(r)}$ consisting of q distinct $(r+1)$-cliques, sharing a common r-clique.

- The value q is called the *size of the r-book*; we write $bk^{(r)}(G)$ for the size of the largest r-book in a graph G.
- We call 2-books simply *books* and write $bk(G)$ for $bk^{(2)}(G)$.

FACTS

F16: **Dirac** [Dir63] Every $G = G(n, t_r(n) + 1)$ contains a K_{r+2} with one edge removed.

F17: Edwards [EdMS], Khadžiivanov and Nikiforov [KhNi79]

$$bk\left(G\left(n, \left\lfloor \frac{n^2}{4} \right\rfloor + 1\right)\right) > \frac{n}{6}$$

and this inequality is essentially best possible in view of the following graph. Let $n = 6k$. Partition $[n]$ into 6 sets $A_{11}, A_{12}, A_{13}, A_{21}, A_{22}, A_{23}$ with $|A_{11}| = |A_{12}| = |A_{13}| = k - 1$ and $|A_{21}| = |A_{22}| = |A_{23}| = k + 1$. For $1 \le j < k \le 3$ join every vertex of A_{ij} to every vertex of A_{ik}, and for $j = 1, 2, 3$ join every vertex of A_{1j} to every vertex of A_{2j}. The size of the resulting graph is greater than $\lfloor n^2/4 \rfloor + 1$, and its booksize is $n/6 + 1$.

F18: Erdős, Faudree, and Rousseau [EFR94] If $m > (r - 1)n^2/2r$ and if n is sufficiently large, then

$$bk^{(r)}(G(n, m)) \ge \frac{3r - 4}{8r(r + 1)} n$$

For $q \ge r > 2$, $s > 0$ define the graph G with $V(G) = [r] \times [q] \times [s]$ and join two vertices (x_1, y_1, z_1) and (x_2, y_2, z_2) if and only if $x_1 \ne x_2$ and $y_1 \ne y_2$. Setting

$$n = rqs \quad \text{and} \quad m = \left(1 - \frac{1}{q}\right)\left(1 - \frac{1}{r}\right)\frac{n^2}{2}$$

we have $G = G(n, m)$, and for $1 < k < r$,

$$bk^{(r)}(G) = \left(1 - \frac{k}{r}\right)\left(1 - \frac{k}{q}\right) n$$

F19: Erdős, Faudree and Győri [EFG95] There exists a number $c > 1/6$ such that if $G = G(n)$ and $\delta(G) > n/2$, then $bk(G) > cn$.

Vertex-Disjoint Cliques

The next two results are milestones in extremal graph theory. In particular, the proof of Hajnal–Szemerédi theorem, although somewhat simplified by Bollobás in [Bol78], is still very difficult.

DEFINITION

NOTATION: The union of s vertex-disjoint copies of a graph G is denoted by sG.

D10: Let H and G be graphs such that $v(G) = kv(H)$. If $kH \subset G$ then G is said to have an **H-factor**.

FACTS

F20: Corrádi–Hajnal theorem [CoHa63] Let n, k be natural numbers with $n \le 3k$ and $s = \lfloor n/k \rfloor$, and let $t = k - (n - ks)$. If $G = G(n)$ and $\delta(G) \ge 2k$, then G contains k vertex-disjoint cycles of length at most $s + 1$, and t of them are of length at most s. In particular, if $\delta(G) \ge 2n/3$, then G contains $\lfloor n/3 \rfloor K_3$.

F21: Hajnal–Szemerédi theorem [HaSz70] If $G = G(n)$ and $\Delta(G) \le r$, then $V(G)$ can be partitioned into $r + 1$ independent sets such that the sizes of any two sets differ by at most 1.

8.1.2   The Number of Complete Graphs

An exciting and difficult problem is to determine min $k_s(G)$ for a given value of $k_r(G)$. In spite of the few illuminating results to be presented below, the general problem remains largely unsolved.

FACTS

F22: Rademacher [Erd62b]

$$k_3\left(G\left(n, \left\lfloor \frac{n^2}{4} \right\rfloor + 1\right)\right) \geq \left\lfloor \frac{n}{2} \right\rfloor$$

and this inequality is best possible.

F23: Lovász and Simonovits [LoSi83] If $l \leq \lfloor n/2 \rfloor$, then

$$k_3\left(G\left(n, \left\lfloor \frac{n^2}{4} \right\rfloor + l\right)\right) \geq l \left\lfloor \frac{n}{2} \right\rfloor$$

and this inequality is best possible.

F24: Let $0 \leq l \leq n/2r$ and suppose the graph G is obtained by adding l disjoint edges to one of the larger classes of the Turán graph $T_r(n)$. Then $k_{r+1}(G)$ is given by

$$k_{r+1}(G) = f_r(n,l) = l \prod_{i=0}^{r-2} \left\lfloor \frac{n+i}{r} \right\rfloor$$

F25: Erdős [Erd69] For every r there exist $c = c(r) > 0$ and $n_0 = n_0(r)$, such that if $n > n_0$ and $0 < l < cn$, then

$$k_{r+1}(G(n, t_r(n) + l)) \geq f_r(n,l)$$

and this inequality is best possible.

F26: Fisher [Fis89] If $G = G(n,m)$ and $n^2/4 \leq m \leq n^2/3$ then

$$k_3(G) \geq \frac{9nm - 2n^3 - 2(n^2 - 3m)^{3/2}}{27}$$

and this is best possible up to a term of order $O(n^2)$.

F27: Nordhaus and Stewart [NoSt63], **Moon and Moser** [MoMo62] If $G = G(n)$ and $k_s(G) > 0$ then

$$(s^2 - 1)\frac{k_{s+1}(G)}{k_s(G)} \geq s^2 \frac{k_s(G)}{k_{s-1}(G)} - n$$

F28: Bollobás [Bol76] Suppose that $2 \leq s < r \leq n$. Let the function $\phi(x)$ be defined in the interval $\left[0, \binom{n}{s}\right]$ such that for every $q = s, \ldots, n$,

(i) $\phi(k_s(T_q(n))) = k_r(T_q(n))$;
(ii) ϕ is linear in the interval $[k_s(T_{q-1}(n)), k_s(T_q(n))]$.

Then

$$k_r(G) \geq \phi(k_s(G))$$

8.1.3   Erdős–Stone Theorem and Its Extensions

The fundamental theorem of Erdős and Stone has attracted the attention of researchers for more than 50 years; no doubt this will continue in the future. The theorem can be viewed as a considerable extension of Turán's theorem: slightly more than $t_r(n)$ edges in a graph of order n guarantees not only a K_{r+1} but a $K_{r+1}(q)$ for q fairly large. Equivalently, the Erdős–Stone theorem solves asymptotically the Turán problem for a fixed family of forbidden graphs.

NOTATION

N8:   For $\varepsilon > 0$ and natural $2 \leq r \leq n$, let $g(n, r, \varepsilon)$ denote the maximal number q such that every $G(n, \lceil (1 - 1/r + \varepsilon)n^2 \rceil)$ contains a $K_{r+1}(q)$ for n sufficiently large.

FACTS

F29:   Erdős–Stone theorem [ErSt46] For $\varepsilon > 0$ and $2 \leq r \leq n$, the function $g(n, r, \varepsilon)$ tends to infinity when n tends to infinity. Since $T_r(n)$ is r-chromatic, this implies the result of Erdős and Simonovits [ErSi66] that if $F = \{F_1, \ldots, F_k\}$ is a fixed family of graphs and $r + 1 = \min_i \chi(F_i) \geq 2$, then

$$ex(n, F) = \frac{r - 1}{2r}n^2 + o(n^2)$$

F30:   Bollobás and Erdős [BoEr73] There exist constants $c_1, c_2 > 0$ such that

$$c_1 \log n \leq g(n, r, \varepsilon) \leq c_2 \log n$$

F31:   Bollobás, Erdős, and Simonovits [BES76] There exists $\alpha > 0$ such that if $0 < \varepsilon < 1/r$ then

$$g(n, r, \varepsilon) \geq \frac{\alpha \log n}{r \log \frac{1}{\varepsilon}}$$

There exists $\varepsilon_r > 0$ such that if $0 < \varepsilon < \varepsilon_r$ then

$$g(n, r, \varepsilon) \leq 3\frac{\log n}{\log \frac{1}{\varepsilon}}$$

F32:   Chvátal and Szemerédi [ChSz83]

$$g(n, r, \varepsilon) \geq \frac{\log n}{500 \log \frac{1}{\varepsilon}}$$

F33:   Bollobás and Kohayakawa [BoKo94] There exists an absolute constant $\alpha > 0$ such that if

$$r \geq 2, \quad 0 < \gamma < 1, \quad 0 < \varepsilon < 1/r$$

then every graph G of sufficiently large order n with $e(G) \geq (1 - 1/r + \varepsilon)n^2$ contains a $K_{r+1}(s, m, \ldots, m, l)$ such that

$$s \geq \alpha(1 - \gamma)\frac{\log n}{r \log \frac{1}{\varepsilon}}, \quad m \geq \alpha(1 - \gamma)\frac{\log n}{\log r}, \quad l \geq \alpha\varepsilon^{1 + \gamma/2}n^\gamma$$

F34: Ishigami [Ish02] There exists an absolute constant $\beta > 0$ such that if

$$r \geq 2, \quad 0 < \gamma < 1, \quad \text{and} \quad 0 < \varepsilon < 1/r$$

then every graph G of sufficiently large order n with $e(G) \geq (1 - 1/r + \varepsilon)n^2$ contains a $K_{r+1}(s, m, \ldots, m, l)$ such that

$$s \geq \beta(1 - \gamma)\frac{\log n}{\log \frac{1}{\varepsilon}}, \quad m \geq \beta(1 - \gamma)\frac{\log n}{\log r}, \quad \text{and} \quad l \geq n^\gamma$$

The Structure of Extremal Graphs

The structure of extremal graphs is fairly well understood in the case of a fixed family of forbidden graphs. Moreover, the stability theorems of Erdős and Simonovits give useful information about the structure of a graph without forbidden subgraphs, provided the size is close to the maximum.

FACTS

F35: *Stability theorem*. Simonovits [Sim68] For every r there is some $c = c(n)$ such that if $l \leq cn$ then every K_{r+1}-free graph $G(n, t_r(n) - l)$ is r-chromatic.

F36: *Stability theorem*. Erdős [Erd68], **Simonovits** [Sim68] Let H be a graph with $\chi(F) = r + 1 \geq 3$. For every $\varepsilon > 0$ there exists $\delta > 0$ such that if G is a H-free graph of order n and $e(G) > ((r - 1)/2r - \delta)n^2$ then there is a $K_r(n_1, \ldots, n_r)$ with $n_1 + \ldots + n_r = n$ that can be obtained from G by changing fewer than εn^2 edges of G.

F37: Erdős [Erd68], **Simonovits** [Sim68] Let $F = \{F_1, \ldots, F_k\}$ be a fixed graph family with $r + 1 = \min_i \chi(F_i) \geq 3$ and suppose that F_1 has an $(r+1)$-coloring in which one of the color classes contains t vertices. Then

$$ex(n, F) = \left(\frac{(r - 1)}{2r}\right) n^2 + O(n^{2-1/t})$$

If $G \in \Phi(n, F)$ is such that $e(G) = ex(n, F)$, then $\delta(G) = (1 - 1/r)n + o(n)$ and all of the following hold:

(i) the vertices of G can be partitioned into r classes each of size $n/r + o(n)$;

(ii) each vertex is joined to at most as many vertices of its own class as to any other class. For every $\varepsilon > 0$ the number of vertices joined to at least εn vertices of their own class is $o(n)$;

(iii) there are $O(n^{2-1/t})$ edges joining vertices of the same class.

F38: Simonovits [Sim68] Let n and s be fixed integers. If n is sufficiently large, then every graph $G = G(n)$ with

$$e(G) \geq t_r(n - s - 1) + (s - 1)(n - s - 1) + \binom{s - 1}{2}$$

contains sK_{r+1}, unless $G = K_{s-1} + T_r(n - s - 1)$.

8.1.4   Zarankiewicz Problem and Related Questions

The problem of Zarankiewicz is the counterpart of Turán's theorem for bipartite graphs; this problem has turned out to be extremely difficult — it seems that even today we are very far from a satisfactory solution.

NOTATION

N9: Let $z(m, n, s, t)$ denote the largest size of an n-by-m bipartite graph not containing the complete biparatite graph $K_2(s, t)$, and set $z(n, t) = z(n, n, t, t)$.

FACTS

F39: $2ex(n, \{K_2(s, t)\}) \leq z(m, n, s, t) \leq ex(2n, \{K_2(s, t)\})$.

F40:   Kövary, Sós, and Turán [KST54] If $2 \leq s \leq m$ and $2 \leq t \leq n$, then
$$z(m, n, s, t) \leq (s - 1)^{1/t}(n - t + 1)m^{1-1/t} + (t - 1)m$$
and
$$z(n, t) \leq (t - 1)^{1/t}n^{2-1/t} + O(n)$$

F41:   Reiman [Rei58]
  (i) $z(n, 2) \leq (n/2)(1 + \sqrt{4n - 3})$;
  (ii) for every $n = q^2 + q + 1$, where q is a power of a prime,
$$z(n, 2) \leq \frac{n}{2}(1 + \sqrt{4n - 3}) = (q - 1)(q^2 + q + 1)$$

(iii) $\lim_{n \to \infty} z(n, 2)n^{-3/2} = 1$.

F42:   Erdős, Rényi, and Sós [ERS66], **Brown** [Bro66] Let q be a power of a prime. Then for the cycle $C_4 = K(2, 2)$, we have

$$\frac{1}{2}q(q + 1)^2 \leq ex(q^2 + q + 1, \{C_4\}) \leq \frac{1}{2}q(q + 1)^2 + \frac{q + 1}{2} \tag{3}$$

and
$$\lim_{n \to \infty} \frac{ex(n, \{C_4\})}{n^{3/2}} = \frac{1}{2}$$

The **Erdős–Rényi graph** giving the lower bound in (3) has for vertices the $q^2 + q + 1$ points of the projective plane $PG(2, q)$ over the field of order q, and two points (x_1, y_1, z_1) and (x_2, y_2, z_2) are joined if and only if $x_1x_2 + y_1y_2 + z_1z_2 = 0$.

F43:   Füredi [Fur83] For every natural number q,

$$ex(q^2 + q + 1, \{C_4\}) \leq \frac{1}{2}q(q + 1)^2$$

and if q is a power of a prime, then

$$ex(q^2 + q + 1, \{C_4\}) = \frac{1}{2}q(q + 1)^2$$

K_r-Free Graphs with Large Minimal Degree

In 1973 Erdős and his collaborators initiated the study of K_r-free graphs with large minimal degree. It turned out that under certain conditions, the chromatic number of such graphs is bounded, and as later investigations showed, their structure is well determined. Despite intensive efforts the general questions remain open, the most challenging of which is the conjecture that every K_3-free $G = G(n)$ with $\delta(G) > n/3$ is at most 4-chromatic.

DEFINITIONS

D11: We say that a graph G is **homomorphic to a graph** H if there exists a mapping $f : V(G) \to V(H)$ such that $(u, v) \in E(G)$ implies that $(f(u), f(v)) \in E(H)$.

D12: The **kth power of a cycle** C_n is a graph G with $V(G) = [n]$ and $(i, j) \in E(G)$ if and only if $|i - j| = 1, 2, \ldots, k$ mod n.

D13: The **square of a cycle** is its second power.

EXAMPLES

E3: **Mycielski graphs** [Myc55] Define the sequence of graphs $M_1, M_2, \ldots$ as follows: set $M_1 = K_2$; suppose that M_{s-1} is already defined, and let $V(M_{s-1}) = [n]$. Set $V(M_s) = [2n + 1]$, and let $E(M_s)$ be the union

$$E(M_{s-1}) \cup \{(i, j + n) : (i, j) \in E(M_{s-1})\} \cup \{(2n + 1, i) : n < i \le 2n\}$$

For every i, the graph M_i is K_3-free and $\chi(M_i) = i$. In particular, the graph M_2 is C_5; the graph M_3 is also known as the **Grötzsch graph**.

E4: **Andrásfai graphs** [And62] Set $A_1 = K_2$ and for every $i \ge 2$ let A_i be the complement of the $(i - 1)$th power of C_{3i-1}. For every i, the graph A_i is K_3-free and $\chi(M_i) = 3$. In particular, the graph A_2 is C_5; the graph A_3 is also known as the **Möbius ladder**.

FACTS

F44: **Andrásfai, Erdős and Sós** [AES74] If a graph $G = G(n)$ is K_{r+1}-free with minimal degree

$$\delta(G) \ge \left(1 - \frac{3}{3r - 1}\right) n$$

then G is r-chromatic.

F45: **Erdős-Hajnal-Simonovits graphs** [ErSi73] If $\varepsilon > 0$, $h > 2$ and n is sufficiently large, then there exists a K_3-free graph $G(n)$ with $\delta(G) \ge (1/3 - \varepsilon)n$ and $\chi(G) \ge h$.

F46: **Häggkvist** [Hag82] Every K_3-free graph $G = G(n)$ with $\delta(G) > 3n/8$ is homomorphic to $A_2 = C_5$ and so satisfies the inequality $\chi(G) \le 3$.

F47: **Häggkvist** [Hag82] For every natural number k, there exists a 4-chromatic, K_3-free, $10k$-regular graph of order $29k$.

F48:   Jin [Jin95] Every K_3-free $G = G(n)$ with $\delta(G) > 10n/29$ is homomorphic to the graph A_9 and so satisfies the inequality $\chi(G) \leq 3$.

F49:   Chen, Jin, and Koh [CJK97] Every K_3-free 3-chromatic $G = G(n)$ with $\delta(G) > n/3$ is homomorphic to some graph A_i. If $\chi(G) \geq 4$, then $M_3 \subset G$.

F50:   Brandt [Bra00] If $G(n)$ is a d-regular maximal K_3-free graph with $d > n/3$, then $\chi(G) \leq 4$.

F51:   Häggkvist and Jin [HaJi98] Let $G = G(n)$ be K_3-free and C_5-free. If $\delta(G) > n/4$, then G is homomorphic to C_7. The bound $n/4$ is best possible, as there is a K_3-free, C_5-free, $3k$-regular $G(12k)$ that is not homomorphic to C_7.

8.1.5   Paths and Trees

One of the most famous unsolved problems in extremal graph theory is the **Erdős–Sós conjecture**:

Every graph $G(n, \lfloor (k-1)n/2 \rfloor + 1)$ contains all trees of order k.

The conjecture is true for many types of trees and under special conditions for the graph, but the general case remains open.

NOTATION

N10: P_k denotes the path of order k.

FACTS

F52:   Erdős–Gallai theorem [ErGa59] Every $G = G(n)$ with $e(G) > (n-1)k/2$ contains a P_{k+1}. If $n = q(k-1) + 1$, then there exists a graph $G = G(n, (n-1)k/2)$ containing no P_{k+1}.

F53:   Faudree and Schelp [FaSc75] If $n = kt + r$ and $0 \leq r < t$, then

$$ex(n, P_{k+1}) = t\binom{k}{2} + \binom{r}{2}$$

and the extremal graphs are known.

F54:   Brandt and Dobson [BrDo96] Every graph $G(n, \lfloor (k-1)n/2 \rfloor + 1)$ of girth at least 5 contains all trees of order k.

F55:   Saclé and Woźniak [SaWo97] Every C_4-free graph $G(n, \lfloor (k-1)n/2 \rfloor + 1)$ contains all trees of order k.

F56:   Wang, Li and Liu [WLL00] Every graph $G(n, \lfloor (k-1)n/2 \rfloor + 1)$ whose edge-complement is of girth at least 5 contains all trees of order k.

F57:   Dobson [Dob02] Every graph $G(n, \lfloor (k-1)n/2 \rfloor + 1)$ whose edge-complement is $K(2, 4)$-free contains all trees of order k.

8.1.6   Circumference

DEFINITIONS

D14: The *circumference* of a graph G is the length of its largest cycle. It is denoted by $c(G)$.

D15: The *girth* of a graph G is the length of its smallest cycle.

NOTATIONS

N11: The set of cycle lengths of a graph G is denoted by $C(G)$.

N12: $ec(G)$ is the largest even number in $C(G)$ and $oc(G)$ is the largest odd number in $C(G)$.

FACTS

F58: Erdős and Gallai [ErGa59] If $2 \leq k \leq n$ then

$$c(G(n, \lfloor (k-1)(n-1)/2 \rfloor + 1)) \geq k$$

F59: Bollobás and Häggkvist [BoHa90] If $G = G(n)$ and $\delta(G) \geq n/k$, then $c(G) \geq \lceil n/(k-1) \rceil$ and this inequality is best possible.

F60: Egawa and Miyamoto [EgMi89] If $G = G(n)$ and if $d(u) + d(v) \geq \lceil 2n/k \rceil$ whenever u and v are two nonajacent vertices, then $c(G) \geq \lceil n/(k-1) \rceil$ and this inequality is best possible.

F61: Dirac [Dir52] If $G = G(n)$ is 2-connected with $\delta(G) = \delta \leq n/2$, then $c(G) \geq 2\delta$.

F62: Voss and Zuluaga [VoZu77] If $G = G(n)$ is a 2-connected, nonbipartite graph with $\delta(G) = \delta \leq n/2$, then

$$oc(G) \geq 2\delta - 1, \quad ec(G) \geq 2\delta$$

8.1.7   Hamiltonian Cycles

The theory of Hamiltonian graphs is one of the most popular areas of graph theory. Here we present several well-known results with an "extremal" flavor.

DEFINITIONS

D16: A graph $G = G(n)$ is said to be *Hamiltonian* if $n \in C(G)$.

D17: The *closure* of a graph $G(n)$ is obtained by successively joining every two nonadjacent vertices u and v with $d(u) + d(v) \geq n$.

FACTS

F63: Dirac's theorem [Dir52] If $G = G(n)$, $n \geq 3$, and $\delta(G) \geq n/2$, then G is Hamiltonian.

F64: Shi [Shi92], **Bollobás and Brightwell** [BoBr93] Let $G = G(n)$, and let S be the set of vertices of degree at least $n/2$. If $|S| \geq 3$, then there is a cycle in G that includes every vertex of S.

F65: Ore's theorem [Ore60] If $G = G(n)$ with $n \geq 3$, and if $d(u) + d(v) \geq n$ whenever u and v are two nonadjacent vertices, then G is Hamiltonian.

F66: Pósa's Theorem [Pos62] Let $G = G(n)$ with $n \geq 3$. If for every k with $1 \leq k < (n-1)/2$, the number of vertices of G of degree not exceeding k is less than k, and for odd n the number of vertices of degree $(n-1)/2$ does not exceed $(n-1)/2$, then G is Hamiltonian.

F67: Closure Lemma of Bondy and Chvátal [BoCv76] A graph is Hamiltonian if and only if its closure is Hamiltonian.

F68: Chvátal's theorem [Chv72] Let $G = G(n)$ with $n \geq 3$ and with vertex degrees $d(1) \leq \ldots \leq d(n)$. If for every $k \leq (n-1)/2$ either $d(k) > k$ or $d(n-k) \geq n - k$, then G is Hamiltonian.

F69: Chvátal–Erdős theorem [ChEr72] If $\alpha(G) \leq \kappa(G)$, then G is Hamiltonian.

F70: Fan and Häggkvist [FaHa94] If $G = G(n)$ and $\delta(G) \geq 5n/7$, then G contains the square of C_n.

8.1.8 Cycle Lengths

Erdős proposed the sum $\sum\{1/r : r \in C(G)\}$ as a measure of the wealth of cycle lengths in a graph G. He stated a conjecture that led to the following two results.

FACTS

F71: Gyárfás, Komlós, and Szemerédi [GKS84] There exists a number $c > 0$ such that for every graph $G = G(n, m)$ we have

$$\sum\{1/r : r \in C(G)\} \geq c \log(2m/n)$$

F72: Gyárfás, Prömel, Szemerédi, and Voigt [GPSV85] If k is sufficiently large and $2m \geq (1 + 1/k)n$, then

$$\sum\{1/r : r \in C(G)\} \geq (300 k \log k)^{-1}$$

Cycles of Consecutive Lengths

In this section we present several sufficient conditions in terms of the size and minimal degree for the existence of large intervals in the set $C(G)$ of cycle lengths.

FACTS

F73: Bondy-Simonovits theorem [BoSi74] Every graph $G(n, \lfloor 100 k n^{1/k} \rfloor + 1)$ contains the cycle C_{2l} for $k \leq 2l \leq k n^{1/k}$.

F74: Verstraëte [Ver00] Every graph $G(n, \lfloor 8(k-1)n^{1/k} \rfloor + 1)$ contains the cycle C_{2l} for $k \le 2l \le kn^{1/k}$.

F75: Fan [Fan02] If G is a graph with $\delta(G) \ge 3k$, then G contains $k+1$ cycles $C_0, C_1, \ldots, C_k$ such that

$$k + 1 < |C_0| < \ldots < |C_k|, \quad |C_i| - |C_{i-1}| = 2, (1 \le i \le k)$$

and $|C_k| - |C_{k-1}| \le 2$.

F76: Gould, Haxell, and Scott [GHS02] For every $c > 0$ there exists a constant $k = k(c)$ such that if $G = G(n)$ and $\delta(G)cn$, then G contains a cycle of order t for every even $t \in [4, ec(G) - k]$ and for every odd $t \in [k, oc(G) - k]$.

Pancyclicity and Weak Pancyclicity

In 1971 Bondy introduced the concept of pancyclicity that soon became a topic of intensive study. We present below only few of the known results.

DEFINITIONS

D18: A graph $G = G(n)$ is called **weakly pancyclic** if $C(G)$ is an interval.

D19: A graph $G = G(n)$ is called a **pancyclic graph** if $C(G) = [3, n]$.

FACTS

F77: Bondy [Bon71] If $G = G(n, \lfloor n^2/4 \rfloor)$ is Hamiltonian, then G is pancyclic unless $G = K(\lfloor n/2 \rfloor, \lceil n/2 \rceil)$.

F78: Bondy [Bon71] If $G = G(n, \lfloor n^2/4 \rfloor + 1)$, then $c(G) \ge \lfloor (n+3)/2 \rfloor$ and G is weakly pancyclic.

F79: Amar, Flandrin, Fournier, and Germa [AFFG83] If $n \ge 102$, $G = G(n)$ is Hamiltonian and $\delta(G) > 2n/5$, then G is pancyclic.

F80: Shi [Shi86] If $n > 50$, $G = G(n)$ is Hamiltonian and for every two nonadjacent vertices u and v, $d(u) + d(v) > 4n/5$, then G is pancyclic.

F81: Brandt, Faudree and Goddard [BFG98] If $\delta(G) \ge n/4 + 250$, then the graph G is weakly pancyclic unless the order of the shortest odd cycle of G is 7, in which case $C(G) = \{4, 6, 7, \ldots, c(G)\}$.

F82: Brandt, Faudree, and Goddard [BFG98] If G is a 2-connected nonbipartite graph of sufficiently large order n with $\delta(G) > 2n/7$, then G is weakly pancyclic.

F83: Brandt [Bra97] Every $G(n, \lfloor (n-1)^2/4 \rfloor + 2)$ is weakly pancyclic or bipartite.

F84: Bollobás and Thomason [BoTh99] Every graph $G(n, \lfloor n^2/4 \rfloor - n + 59)$ is weakly pancyclic or bipartite.

8.1.9 Szemerédi's Uniformity Lemma

The Uniformity Lemma of Szemerédi, whose power and versatility could hardly be overemphasized, is one of the most remarkable tools in discrete mathematics. Loosely stated, it guarantees that every dense graph has some finite rough structure, which, surprisingly often, is the basis of successful attacks on difficult combinatorial problems. The Blow-up Lemma of Komlós, Sárközy, and Szemerédi, a close relative of the Uniformity Lemma, has been used to solve a number of difficult graph embedding conjectures. For comprehensive surveys of this area see [KoSi96], [Kom99], and [Kom00].

DEFINITIONS

D20: Let $a > 0$. A **tower of** a **of length** k is the function

$$a^{a^{\cdot^{\cdot^{\cdot^a}}}}$$

where the exponentiation is done k times.

D21: A bipartite graph with classes A, B is called an ε-**uniform pair** if for every pair of vertex subsets $X \subset A$ and $Y \subset B$ with $|X| > \varepsilon |A|$ and $|Y| > \varepsilon |B|$ we have

$$\left| \frac{E(X,Y)}{|X|\,|Y|} - \frac{E(A,B)}{|A|\,|B|} \right| < \varepsilon$$

D22: A bipartite graph with classes A, B is called an (ε, δ)-**super-uniform pair** if it is ε-uniform and

$$d(u) \geq \delta |B|, \ldots, d(v) \geq \delta |A|$$

whenever every $u \in A$, $v \in B$.

FACTS

F85: Szemerédi's Uniformity Lemma [Sze76] For every $\varepsilon > 0$ there exist numbers $n_0 = n_0(\varepsilon)$ and $k_0 = k_0(\varepsilon)$ such that for every graph G of order $n > n_0$ there is a partition $V(G) = V_0 \cup V_1 \cup \ldots \cup V_k$ satisfying these criteria:
   (i) $k \leq k_0(\varepsilon)$;
   (ii) $|V_0| < \varepsilon n$, $|V_1| = \ldots = |V_k|$;
   (iii) all but εk^2 pairs (V_i, V_j) are ε-uniform.

F86: The function $k_0(\varepsilon)$ in Szemerédi's Uniformity Lemma is bounded from above by a tower of 2s of length ε^{-5}.

F87: Gowers' bound [Gow98] There exist constants $\varepsilon_0 > 0$ and $c > 0$ such that for $0 < \varepsilon \leq \varepsilon_0$, there is a graph G whose vertices cannot be partitioned according to criteria (i)–(iii) of Szemerédi's Uniformity Lemma unless k is as large as a tower of 2s of length $c\varepsilon^{-1/16}$.

F88: Blow-Up Lemma of Komlós, Sárközy, and Szemerédi [KSS97] Fix a graph R with $V(R) = [r]$. For every $\delta, \Delta > 0$ there exists $\varepsilon > 0$ such that the following holds. Fix a natural n and let $V_1, \ldots, V_r$ be r disjoint sets of size n. Define the graphs $R(n)$ and G as follows:
   (i) Set $V(R(n)) = \cup V_i$ and for every $(i,j) \in E(R)$, place all edges between V_i and V_j.
   (ii) Set $V(G)) = \cup V_i$ and for every $(i,j) \in E(R)$, place an (ε, δ)-super-uniform pair between V_i and V_j.

If $H \subset R(n)$ and $\Delta(H) \leq \Delta$ then $H \subset G$.

Applications of the Uniformity and Blow-up lemmas

The Uniformity Lemma and the Blow-up Lemma are powerful tools in graph theory. We present below only four of their applications but, in fact, many results described in other sections are also obtained applying these two lemmas.

FACTS

F89: Komlós, Sárközy, and Szemerédi [KSS98] For every $\varepsilon > 0$ and natural number k there exists a number $n_0 = n_0(k)$, such that if

$$n > n_0, \quad G = G(n), \quad \text{and} \quad \delta(G) \geq \left(1 - \frac{1}{k+1}\right) n$$

then the graph G contains the kth power of a Hamiltonian cycle.

F90: Alon and Yuster [AlYu96] For every $\varepsilon > 0$ and natural number h there exists a number $n_0 = n_0(\varepsilon, h)$, such that if H is a graph of order h with $\chi(H) = k$ and

$$n > n_0, \quad G = G(hn), \quad \text{and} \quad \delta(G) \geq \left(1 - \frac{1}{k} + \varepsilon\right) hn$$

then the graph G contains an H-factor.

F91: Komlós, Sárközy, and Szemerédi [KSS01] Let H be a graph of order h and $\chi(H) = k$. There exist numbers $c = c(H)$ and $n_0 = n_0(H)$ such that if

$$n > n_0, \quad G = G(hn), \quad \text{and} \quad \delta(G) \geq \left(1 - \frac{1}{k}\right) hn + c$$

then G contains an H-factor.

F92: Komlós, Sárközy, and Szemerédi [KSS01a] For every $\varepsilon > 0$, there exist numbers c and n_0 such that for $n > n_0$, every graph $G = G(n)$ with $\delta(G) > (1/2 + \varepsilon)n$ contains every tree T of order n such that $\Delta(T) < cn/\log n$.

8.1.10   Asymptotic Enumeration

An intriguing question is how many graphs with given properties are there. For certain natural properties like "G is K_r-free" or "G has no induced subgraph isomorphic to H" satisfactory answers have been obtained.

DEFINITIONS

D23: A **graph property** is a graph family closed under isomorphism.

D24: A graph property P is called a **monotone property** if $G \in P$ implies $H \in P$ for every subgraph of G.

D25: A graph property P is called an **hereditary property** if $G \in P$ implies $H \in P$ for every induced subgraph of G.

D26: For any graph property P, set $P^n = \{G : G \in P, v(G) = n\}$. The **logarithmic density** of P^n is the value

$$c_n(P) = (\log_2 |P^n|) / \binom{n}{2}$$

D27: Let $0 \leq s \leq r$ be integers. A graph H is called (r, s)-**colorable** if its vertices can be colored in r colors, so that the vertices colored with the ith color are a clique for $1 \leq i \leq s$, and an independent set otherwise.

D28: The **coloring number** $r(P)$ of a hereditary property P is the largest integer such that for some s, the family P contains every (r, s)-colorable graph.

NOTATION

N13: Given a hereditary property P, let $ex_{ind}(n, P)$ denote the maximal number of edges in a graph $G_0 = G(n)$ for which there is a graph $G_1 = G(n)$ with $V(G_1) = V(G_0)$ and $E(G_1) \cap E(G_0) = \emptyset$, so that every graph G with $G_1 \subseteq G \subseteq G_0 \cup G_1$ belongs to P^n.

FACTS

F93: Erdős, Kleitman, and Rothschild [EKR73] The number k_n of the K_{r+1}-free graphs of order n is given asymptotically by

$$\log_2 k_n = \left(1 - \frac{1}{r} + o(1)\right) n^2$$

F94: Erdős, Frankl, and Rödl [EFR86] Let H be a graph with $\chi(H) = r + 1$. The number h_n of the H-free graphs of order n is given asymptotically by

$$\log_2 h_n = \left(1 - \frac{1}{r} + o(1)\right) n^2$$

F95: Kolaitis, Prömel, and Rothschild [KPR87] For every n let k_n be the number of K_{r+1}-free graphs $G(n)$ and h_n the number of r-chromatic graphs $G(n)$. Then

$$\lim_{n \to \infty} \frac{k_n}{h_n} = 1$$

F96: A property P is monotone if and only if there exists some sequence of graphs $F_1, F_2, \ldots$ such that P is the collection of graphs having no subgraph isomorphic to an F_i.

F97: A property P is hereditary if and only if there exists some sequence of graphs $F_1, F_2, \ldots$ such that P is the collection of graphs having no induced subgraph isomorphic to an F_i. The coloring number of $r(P)$ is exactly the maximal r such that for some $0 \leq s \leq r$ no F_i is (r, s)-colorable.

F98: Prömel and Steger [PrSt91] There exist numbers c_0 and c_1 such that the number t_n of $G(n)$ with no induced C_4 is given by

$$t_n = (c_r + o(1)) 2^{n^2/4 + n - (\log n)/2}$$

where $r = 0, 1$, $r = n \bmod 2$.

F99: Prömel and Steger [PrSt92], [PrSt93] Fix a graph H, and let P be the hereditary property "G has no induced subgraph isomorphic to H". Then

$$\lim_{n \to \infty} ex_{ind}(n, P) \binom{n}{2}^{-1} = 1 - \frac{1}{r(P)}$$

F100: Alekseev [Ale92], **Bollobás and Thomason** [BoTh95] Let P be a hereditary property. Then
$$1 = c_1(P) \geq \ldots \geq c_n(P) \geq \ldots$$
and the limit
$$c(P) = \lim_{n \to \infty} c_n(P)$$
exists.

F101: Scheinerman and Zito [ScZi94] For every hereditary property P, one of the following is true:
  (i) for n sufficiently large $|P^n|$ is identically $0, 1$ or 2;
  (ii) $|P^n| = \Theta(1)n^k$ for some integer $k \geq 1$;
  (iii) for some $c_2 \geq c_1 > 0$, $c_1^n \leq |P^n| \leq c_2^n$;
  (iv) for some $c > 0$, $|P^n| \geq n^{cn}$.

F102: Bollobás and Thomason [BoTh97] For every hereditary property P,
$$c(P) = \lim_{n \to \infty} ex_{ind}(n, P)\binom{n}{2}^{-1} = 1 - \frac{1}{r(P)}$$

F103: Balogh, Bollobás, and Weinreich [BBW00], [BBW01], [BBW02] For every hereditary property P one of the following is true:
  (i) there exists a collection of polynomials $\{p_i(n)\}_{i=0}^k$ such that for n sufficiently large $|P^n| = \sum_{i=0}^k p_i(n)i^n$;
  (ii) for some integer $k > 1$, $|P^n| = n^{(1-1/k+o(1))n}$;
  (iii) $n^{(1+o(1))n} \leq |P^n| \leq n^{o(n^2)}$;
  (iv) for some integer $k > 1$, $|P^n| = n^{(1-1/k+o(1))n^2/2}$.

8.1.11   Graph Minors

The study of graph minors was initially motivated by the conjecture of Hadwiger that every r-chromatic graph has K_r as a minor. However, from the extremal point of view, minors happen to be of their own fascinating interest.

DEFINITIONS

D29: Let G and H be graphs. We say that H is a **_minor_** of G, and we write
$$G \succ H$$
if there are disjoint sets $W(u)$, $u \in V(H)$, such that $W(u)$ induces a connected graph in G, and for every $(u, v) \in E(G)$, there is an edge between $W(u)$ and $W(v)$.

D30: Let $\mu(H)$ be the minimal number μ such that $e(G) \geq \mu v(G)$ implies that $G \succ H$.

FACTS

F104: Mader [Mad67], [Mad68]
$$\mu(K_r) \leq 8r \log_2 r$$

F105: **Bollobás, Catlin, and Erdős** [BCE80], **de la Vega** [Fer83] For some $C > 0$,
$$\mu(K_r) \geq Cr\sqrt{\log r}$$

F106: **Kostochka** [Kos82], [Kos84], **Thomason** [Tho84]
$$\mu(K_r) = O(r\sqrt{\log r})$$

F107: **Thomason** [Tho01] There is an explicit constant $\alpha = 0.319\dots$ such that
$$\mu(K_r) = (\alpha + o(1))r\sqrt{\log r}$$

F108: **Myers and Thomason** [MyTh02] Given a graph H of order n, set
$$\gamma(H) = \min_{w} \frac{1}{n} \sum_{u \in V(H)} w(u) \quad \text{with} \quad \sum_{(u,v) \in E(H)} n^{-w(u)w(v)} = n$$
where $w(u)$ are nonnegative real numbers assigned to the vertices of H. Then
$$\mu(H) = (\alpha\gamma(H) + o(1))r\sqrt{\log r}$$

8.1.12   Ramsey–Turán Problems

Ramsey–Turán problems are in fact Turán-type problems with with restriction on the independence number. For a comprehensive survey of this topic see [SiSo01]; we present below only some of the highlights of the area.

NOTATIONS

N14: Let $F_1, \dots, F_s$ be fixed graphs. Let $RT_s(n, F_1, \dots, F_s, f(n))$ denote the maximal size of a graph $G(n)$ with $\alpha(G) \leq f(n)$ whose edges can be colored in s colors so that there is no F_i in the ith color. We write $RT(n, F_1, \dots, F_s, f(n))$ instead of $RT_s(n, F_1, \dots, F_s, f(n))$ when s is understood.

N15: Let $R(s)$ be the maximal number R such that one can color the edges of the complete graph K_R in s colors, so that there is no monochromatic triangle and so that each star is colored in at most $(r - 1)$ colors.

FACTS

F109: **Erdős graph** [Erd61] For every k there exists $\varepsilon > 0$ such that if n is sufficiently large there exists a graph $F_{n,k} = G(n)$ with girth $g(G) > k$ and independence number $\alpha(G) < n^{1-\varepsilon}$.

F110: **Erdős and Sós** [ErSo70]

$$RT(n, K_{2r+1}, o(n)) = \frac{r-1}{2r}n^2 + o(n^2)$$

The lower bound comes from the following graph: take the Turán graph $T_r(n)$ and add to each of its classes a copy of $F_{s,3}$, where s is the size of the class.

F111: Bollobás and Erdős [BoEr76a] For every $\varepsilon > 0$ and n sufficiently large, there exists a K_4-free graph $BE_n = G(n)$ with $\alpha(BE_n) \leq \varepsilon n$ and $|d(u) - n/4| < \varepsilon n$ for every $u \in V(BE_n)$. Thus,

$$RT(n, K_4, o(n)) \geq \frac{1}{8}n^2 + o(n^2)$$

F112: Szemerédi [Sze72]

$$RT(n, K_4, o(n)) \leq \frac{1}{8}n^2 + o(n^2)$$

F113: Erdős, Hajnal, Sós, and Szemerédi [EHSS83]

$$RT(n, 2r, o(n)) = \frac{3r - 5}{6r - 4}n^2 + o(n^2)$$

To prove the lower bound consider the following graph: for $l = \lceil 4n/(3r - 2) \rceil$ take $BE_l + T_{r-l}(n - l)$, and add to each of the parts of $T_{r-l}(n - l)$ a copy of $F_{s,3}$, where s is the size of the part.

F114: Erdős and Sós [ErSo70]

$$RT(n, K_3, K_3, o(n)) = \frac{1}{4}n^2 + o(n^2)$$

F115: Erdős, Hajnal, Sós, and Szemerédi [EHSS83]

$$RT_s(n, K_3, \ldots, K_3, o(n)) = \frac{R(s) - 1}{2R(s)}n^2 + o(n^2)$$

F116: Erdos, Hajnal, Simonovits, Sós, and Szemerédi [EHSSS93]

$$RT_s(n, K_3, K_4, o(n)) = \frac{1}{2}\left(1 - \frac{1}{3}\right)n^2 + o(n^2)$$

$$RT_s(n, K_3, K_5, o(n)) = \frac{1}{2}\left(1 - \frac{1}{5}\right)n^2 + o(n^2)$$

If p and q are odd integers then

$$RT_s(n, C_p, C_q, o(n)) = \frac{1}{4}n^2 + o(n^2)$$

References

[Ale92] V. E. Alekseev, Range of values of entropy of hereditary classes of graphs (Russian), *Diskret. Mat.* **4** (1992), 148–157.

[AlYu96] N. Alon and R. Yuster, *H*-factors in dense graphs, *J. Combin. Theory, Ser. B* **66** (1996), 269–282.

[AFFG83] D. Amar, E. Flandrin, I. Fournier and A. Germa, Pancyclism in Hamiltonian graphs, *Discrete Math.* **89** (1991), 111–131.

[And62] B. Andrásfai, Über ein Extremalproblem der Graphentheorie (German), *Acta Math. Acad. Sci. Hungar.* **13** (1962), 443–455.

[AES74] B. Andrásfai, P. Erdős and V. T. Sós, On the connection between chromatic number, maximal clique and minimal degree of a graph, *Discrete Math.* **8** (1974), 205–218.

[BBW00] J. Balogh, B. Bollobás, and D. Weinreich, The speed of hereditary properties of graphs. *J. Combin. Theory, Ser. B* **79** (2000), 131–156.

[BBW01] J. Balogh, B. Bollobás, and D. Weinreich, The penultimate rate of growth for graph properties, *European J. Combin.* **22** (2001), 277–289.

[BBW02] J. Balogh, B. Bollobás, and D. Weinreich, Measures on monotone properties of graphs, *Discrete Appl. Math.* **116** (2002), 17–36.

[Bol78] B. Bollobás, *Extremal Graph Theory*, Academic Press, 1978.

[Bol76] B. Bollobás, On complete subgraphs of different orders, *Math. Proc. Cambridge Philos. Soc.* **79** (1976), 19–24.

[Bol98] B. Bollobás, *Modern Graph Theory*, Graduate Texts in Mathematics, **184,** Springer-Verlag, 1998.

[Bol99] B. Bollobás, Turán's theorem and maximal degrees, *J. Combin. Theory, Ser. B* **75** (1999), 160–164.

[BoBr93] B. Bollobás and G. Brightwell, Cycles through specified vertices, *Combinatorica* **13** (1993), 147–155.

[BCE80] B. Bollobás, P. Catlin, and P. Erdős, Hadwiger's conjecture is true for almost every graph, *European J. Combin.* **1** (1980), 195–199.

[BoEr73] B. Bollobás and P. Erdős, On the structure of edge graphs, *J. London Math. Soc.* **5** (1973), 317–321.

[BoEr76] B. Bollobás and P. Erdős, On a Ramsey-Turán type problem, *J. Combin. Theory, Ser. B* **21** (1976), 166–168.

[BoEr76a] B. Bollobás and P. Erdős, Unsolved problems, *Proc. Fifth Brit. Comb. Conf. (Univ. Aberdeen, Aberdeen, 1975)*, Winnipeg, Util. Math. Publ., 678–680.

[BES76] B. Bollobás, P. Erdős, and M. Simonovits, On the structure of edge graphs II. *J. London Math. Soc.* **12** (1976), 219–224.

[BoHa90] B. Bollobás and R. Häggkvist, The circumference of a graph with a given minimal degree, pp. 97–104 in *A tribute to Paul Erdős*, Cambridge University Press, 1990.

[BoKo94] B. Bollobás and Y. Kohayakawa, An extension of the Erdös-Stone theorem, *Combinatorica* **14** (1994), 279–286.

[BoTh81] B. Bollobás and A. Thomason, Dense neighbourhoods and Turán's theorem, *J. Combin. Theory, Ser. B* **31** (1981), 111–114.

[BoTh85] B. Bollobás and A. Thomason, Random graphs of small order, *Ann. Discrete Math.*, Vol 28, pp 47–97, North-Holland, Amsterdam, 1985.

[BoTh95] B. Bollobás and A. Thomason, Projections of bodies and hereditary properties of hypergraphs, *J. London Math. Soc.* **27** (1995), 417–424.

[BoTh97] B. Bollobás and A. Thomason, Hereditary and monotone properties of graphs, *The mathematics of Paul Erdös, II,* pp 70–78, *Algorithms Comb.* **14**, Springer, Berlin, 1997.

[BoTh99] B. Bollobás and A. Thomason, Weakly pancyclic graphs, *J. Combin. Theory, Ser B* **77** (1999), 121–137.

[Bom97] I. Bomze, Evolution towards the maximum clique, *J. Global Optim.* **10** (1997), 143–164.

[Bon71] J. A. Bondy, Pancyclic graphs I, *J. Comb. Theory, Ser. B* **11** (1971), 80–84.

[Bon83a] J. A. Bondy, Large dense neighbourhoods and Turán's theorem, *J. Combin. Theory, Ser B* **34** (1983), 109–111.

[Bon83b] J. A. Bondy, Erratum: Large dense neighbourhoods and Turán's theorem, *J. Combin. Theory, Ser B* **35** (1983), 80.

[Bon95] J. A. Bondy, Basic graph theory: paths and circuits, *Handbook of Combinatorics,* Vol. **1**, Elsevier, Amsterdam, 1995, pp. 3–110.

[BoCv76] J. A. Bondy and V. Chvátal, A method in graph theory, *Discrete Math. 15* (1976), 111–135.

[BoSi74] J. A. Bondy and M. Simonovits, Cycles of even length in graphs, *J. Combin. Theory, Ser B* **16** (1974), 97–105.

[Bra97] S. Brandt, A sufficient condition for all short cycles, *Discrete Appl. Math.* **79** (1997), 63–66.

[Bra00] S. Brandt, A 4-colour problem for dense triangle-free graphs, *Discrete Math.* **251** (2002), 33–46.

[BrDo96] S. Brandt and E. Dobson, The Erdős-Sós conjecture for graphs of girth 5, *Discrete Math.* **150** (1996), 411–414.

[BFG98] S. Brandt, R. Faudree, and W. Goddard, Weakly pancyclic graphs, *J. Graph Theory* **27** (1998), 141–176.

[Bro66] W. G. Brown, On graphs that do not contain a Thomsen graph, *Canad. Math. Bull.* **9** (1966), 281–285.

[CEV88] L. Caccetta, P. Erdős, and K. Vijayan, Graphs with unavoidable subgraphs with large degrees, *J. Graph Theory* **12** (1988), 17–27.

[CJK97] C. C. Chen, G. P. Jin, and K. M. Koh, Triangle-free graphs with large degree, *Combin. Probab. Comput.* **6** (1997), 381–396.

[Chv72] V. Chvátal, On Hamilton's ideals, *J. Combin. Theory, Ser B* **12** (1972), 163–168.

[ChEr72] V. Chvátal and P. Erdős, A note on Hamiltonian circuits, *Discrete Math.* **2** (1972), 111–113.

[ChSz83] V. Chvátal and E. Szemerédi, Notes on the Erdős–Stone theorem, *Annals of Discrete Math.* **17** (1983), 207–214.

[CoHa63] K. Corradi and A. Hajnal, On the maximal number of independent circuits in a graph, *Acta Math. Acad. Sci. Hungar.* **14** (1963), 423–439.

[Dir52] G. Dirac, Some theorems on abstract graphs, *Proc. London Math. Soc.* (3) **2**, (1952). 69–81.

[Dir63] G. Dirac, Extension of Turán's theorem on graphs, *Acta Math. Acad. Sci. Hungar* **14** (1963), 417–422.

[Dob02] E. Dobson, Constructing trees in graphs whose complement has no $K_{2,s}$, *Combin. Probab. Comput.* **11** (2002), 343–347.

[Edw77] C. Edwards, The largest vertex degree sum for a triangle in a graph, *Bull. Lond. Math. Soc.*, **9** (1977), 203–208.

[Edw78] C. Edwards, Complete subgraphs with largest sum of vertex degrees, *Colloq. Math. Soc. János Bolyai* **18**, North-Holland, Amsterdam-New York, 1978, pp. 293–306.

[EdMS] C. Edwards, A lower bound for the largest number of triangles with a common edge, Manuscript.

[EgMi89] Y. Egawa and T. Miyamoto, The longest cycles in a graph G with minimum degree at least $|G|/k$, *J. Combin. Theory, Ser B* **46** (1989), 356–362.

[Erd61] P. Erdős, Graph theory and probability II, *Canad. J. Math.* **13** (1961), 346–352.

[Erd62a] P. Erdős, On the number of complete subgraphs contained in certain graphs, *Publ. Math. Inst. Hung. Acad. Sci.* **VII**, Ser. A3 (1962), 459-464.

[Erd62b] P. Erdős, On a theorem of Rademacher-Turán, *Illinois J. Math.* **6** (1962), 122–127.

[Erd63] P. Erdős, On the structure of linear graphs, *Israel J. Math.* **1** (1963), 156–160.

[Erd66] P. Erdős, Some recent results on extremal problems in graph theory (results), pp. 117–130 in *Theory of Graphs (Internat. Sympos., Rome, 1966)*, Gordon and Breach, New York; Dunod, Paris.

[Erd68] P. Erdős, On some new inequalities concerning extremal properties of graphs, pp. 77–81 in *Theory of Graphs (Proc. Colloq., Tihany, 1966)*, Academic Press, 1968.

[Erd69] P. Erdős, On the number of complete subgraphs and circuits contained in graphs, *Časopis Pěst. Mat.* **94** (1969), 290–296.

[Erd70] P. Erdős, On the graph theorem of Turán (Hungarian), *Mat. Lapok* **21** (1970), 249–251.

[EFR94] P. Erdős, R. Faudree and C. Rousseau, Extremal problems and generalized degrees, *Discrete Math.* **127** (1994), 139–152.

[EFG95] P. Erdős, R. Faudree, and E. Györi, On the book size of graphs with large minimal degree, *Studia Sci. Math. Hungar.* **30** (1995), 25–46.

[EFR86] P. Erdős, P. Frankl, and V. Rödl, The asymptotic number of graphs not containing a fixed subgraph and a problem for hypergraphs having no exponent, *Graphs Combin.* **2** (1986), 113–121.

[ErGa59] P. Erdős and T. Gallai, On maximal paths and circuits of graphs, *Acta Math. Acad. Sci. Hungar.* **10** (1959), 337–356.

[EHSS83] P. Erdős, A. Hajnal, V. Sós and E. Szemerédi, More results on Ramsey-Turán type problems, *Combinatorica* **3** (1983), 69–81.

[EHSSS93] P. Erdős, A. Hajnal, M. Simonovits, V. Sós, and E. Szemerédi, Turán-Ramsey theorems and simple asymptotically extremal structures, *Combinatorica* **13** (1993), 31–56.

[EKR73] P. Erdős, D. J. Kleitman and B. L. Rothschild, Asymptotic enumeration of K_n-free graphs, *Colloquio Internazionale sulle Teorie Combinatorie (Rome, 1973)*, Tomo II, pp. 19–27. *Atti dei Convegni Lincei, No. 17, Accad. Naz. Lincei*, Rome, 1976.

[ERS66] P. Erdős, A. Rényi, and V. Sós, On a problem of graph theory, *Studia Sci. Math. Hungar.* **1** (1966), 215–235.

[ErSi66] P. Erdős and M. Simonovits, A limit theorem in graph theory, *Studia Sci. Math. Hungar* **1** (1966), 51–57.

[ErSi73] P. Erdős and M. Simonovits, On a valence problem in extremal graph theory, *Discrete Math.*, **5** (1973), 323–334.

[ErSo70] P. Erdős and V. T. Sós, Some remarks on Ramsey's and Turán's theorem, pp. 395–404 in *Combinatorial Theory and Its Applications, II (Proc. Colloq., Balatonfüred, 1969)*, North-Holland, Amsterdam, 1970.

[ErSo83] P. Erdős and V. T. Sós, On a generalization of Turán's graph theorem, pp. 181–185 in *Studies in Pure Mathematics*, Birkhäuser, 1983.

[ErSt46] P. Erdős and A. H. Stone, On the structure of linear graphs, *Bull. Amer. Math. Soc.* **52** (1946), 1087–1091.

[Fan02] G. Fan, Distribution of cycle lengths in graphs, *J. Combin. Theory, Ser B* **84** (2002), 187–202.

[FaHa94] G. Fan and R. Häggkvist, The square of a Hamiltonian cycle, *SIAM J. Discrete Math.* **7** (1994), 203–212.

[Fau92] R. Faudree, Complete subgraphs with large degree sums, *J. Graph Theory* **16** (1992), 327–334.

[FaSc75] R. Faudree and R. H. Schelp, Path-path Ramsey-type numbers for the complete bipartite graph, *J. Combin. Theory, Ser. B* **19** (1975), 161–173.

[Fer83] W. Fernandez de la Vega, On the maximum density of graphs which have no subcontraction to K_s, *Discrete Math.* **46** (1983), 109–110.

[Fis89] D. C. Fisher, Lower bounds on the number of triangles in a graph, *J. Graph Theory* **13** (1989), 505–512.

[FiRa92] D. Fisher and J. Ryan, Bounds on the number of complete subgraphs, *Discrete Math.* **103** (1992), 313–320.

[Fur83] Z. Füredi, Graphs without quadrilaterals, *J. Combin. Theory, Ser B* **34** (1983), 187–190.

[GHS02] R. Gould, P. Haxell, and A. Scott, A note on cycle lengths in graphs, *Graphs Combin.* **18** (2002), 491–498.

[Gow98] W. Gowers, Lower bounds of tower type for Szemerédi's uniformity lemma, *Geom. Funct. Anal.* **7** (1997), 322–337.

[GKS84] A. Gyárfás, J. Komlós, and E. Szemerédi, On the distribution of cycle lengths in graphs, *J. Graph Theory* **8** (1984), 441–462.

[GPSV85] A. Gyárfás, H. Prömel, E. Szemerédi, and B. Voigt, On the sum of the reciprocals of cycle lengths in sparse graphs, *Combinatorica* **5** (1985), 41–52.

[Hag82] R. Häggkvist, Odd cycles of specified length in nonbipartite graphs, pp. 89–99 in *Graph Theory (Cambridge, 1981)*, North-Holland Math. Stud., **62**, North-Holland, 1982.

[HaJi98] R. Häggkvist and G. Jin, Graphs with odd girth at least seven and high minimal degree, *Graphs Comb.* **14** (1998), 351–362.

[HaSz70] A. Hajnal and E. Szemerédi, Proof of a conjecture of P. Erdös, pp. 601–623 in *Combinatorial Theory and Its Applications, II (Proc. Colloq., Balatonfüred, 1969)*, North-Holland, 1970.

[Ish02] Y. Ishigami, Proof of a conjecture of Bollobás and Kohayakawa on the Erdös-Stone theorem, *J. Combin. Theory, Ser B* **85** (2002), 222–254.

[Jin95] G. P. Jin, Triangle-free four-chromatic graphs, *Discrete Math.* **145** (1995), 151–170.

[Kha77] N. Khadžiivanov, Inequalities for graphs (Russian), *C. R. Acad. Sci. Bul.* **30** (1977), 793–796.

[KhNi79] N. Khadžiivanov and V. Nikiforov, A solution of a problem of Erdős on the maximum number of triangles in a graph with n vertices and $[sq(n)/4]+1$ edges, *C. R. Acad. Sci. Bull.* **32** (1979), 1315–1318 (Russian).

[KST54] . Kőváry, V. Sós, and P. Turán, On a problem of K. Zarankiewicz, *Colloq. Math.* **3** (1954), 50–57.

[KPR87] G. Kolaitis, H. Prömel, and B. Rothschild, K_{l+1}-free graphs: asymptotic structure and a $0-1$ law, *Trans. Amer. Math. Soc.* **303** (1987), 637–671.

[Kom99] J. Komlós, The blow-up lemma, *Combin. Probab. Comput.* **8** (1999), 161–176.

[Kom00] J. Komlós, Tiling Turán theorems, *Combinatorica* **20** (2000), 203–218.

[KSS97] J. Komlós, G. Sárközy, and E. Szemerédi, Blow-up lemma, *Combinatorica* **17** (1997), 109–123.

[KSS98] J. Komlós, G. Sárközy, and E. Szemerédi, Proof of the Seymour conjecture for large graphs, *Ann. Combin.* **2** (1998), 43–60.

[KSS01] J. Komlós, G. Sárközy, and E. Szemerédi, Proof of the Alon-Yuster conjecture, *Discrete Math.* **235** (2001), 255–269.

[KSS01a] J. Komlós, G. Sárközy and E. Szemerédi, Spanning trees in dense graphs, *Combin. Probab. Comput.* **10** (2001), 397–416.

[KoSi96] J. Komlós and M. Simonovits, Szemerédi's regularity lemma and its applications in graph theory, *Combinatorics, Paul Erdős is Eighty*, Vol. **2** (Keszthely, 1993), pp. 295–352, Bolyai Soc. Math. Stud., 2, János Bolyai Math. Soc., Budapest, 1996.

[Kos82] A. Kostochka, The minimum Hadwiger number for graphs with a given mean degree of vertices (Russian), *Metody Diskret. Analiz.* **38** (1982), 37–58.

[Kos84] A. Kostochka, A lower bound of the Hadwiger number of graphs by their average degree, *Combinatorica* **4** (1984), 307–316.

[LoSi83] L. Lovász and M. Simonovits, On the number of complete subgraphs of a graph II, *Studies in Pure Mathematics*, pp. 459–495, Birkhäuser, 1983.

[Mad67] W. Mader, Homomorphieeigenschaften und mittlere Kantendichte von Graphen (German), *Math. Ann.* **174** (1967), 265–268.

[Mad68] W. Mader, Homomorphiesätze für Graphen (German), *Math. Ann.* **178** (1968), 154–168.

[Man07] W. Mantel, Problem 28, soln. by H. Gouventak, W. Mantel, J. Teixeira de Mattes, F. Schuh and W. A. Wythoff, *Wiskundige Opgaven* **10** (1907), 60–61.

[MoMo62] J. Moon and L. Moser, On a problem of Turán, *Magyar Tud. Akad. Mat. Kutató Int. Közl.* 7 (1962), 283–286.

[MoSt65] T. Motzkin and E. Straus, Maxima for graphs and a new proof of a theorem of Turán, *Canad. J. Math.*, **17** (1965), 533–540.

[Myc55] J. Mycielski, Sur le coloriage des graphs (French), *Colloq. Math.* **3**, (1955), 161–162.

[MyTh02] J. Myers and A. Thomason, The extremal function for noncomplete minors, preprint.

[NoSt63] E. Nordhaus and B. Stewart, Triangles in an ordinary graph, *Canad. J. Math.* **15** (1963), 33–41.

[Ore60] O. Ore, Note on Hamilton circuits, *Amer. Math. Monthly* **67** (1960), 55.

[Pos62] L. Pósa, A theorem concerning Hamilton lines, *Magyar Tud. Akad. Mat. Kutató Int. Közl.* **7** (1962), 225–226.

[PrSt91] H. Prömel and A. Steger, Excluding induced subgraphs: quadrilaterals, *Random Structures Algorithms* **2** (1991), 55–71.

[PrSt92] H. Prömel and A. Steger, Excluding induced subgraphs III. A general asymptotic, *Random Structures Algorithms* **3** (1992), 19–31.

[PrSt93] H. Prömel and A. Steger, Excluding induced subgraphs II. Extremal graphs, *Discrete Appl. Math.* **44** (1993), 283–294.

[Rei58] I. Reiman, Über ein Problem von K. Zarankiewicz (German), *Acta. Math. Acad. Sci. Hungar.* **9** (1958), 269–273.

[SaWo97] J.-F. Saclé and M. Woźniak, The Erdős–Sós conjecture for graphs without C_4, *J. Combin. Theory, Ser B* **70** (1997), 367–372.

[ScZi94] . R. Scheinerman and J. Zito, On the size of hereditary classes of graphs *J. Combin. Theory, Ser. B* **61** (1994), 16–39.

[Shi86] R. Shi, Ore-type conditions for pancyclism of Hamiltonian graphs (Chinese), *J. Systems Sci. Math. Sci.* **11** (1991), 79–90.

[Shi92] R. Shi, 2-neighborhoods and Hamiltonian conditions, *J. Graph Theory* **16** (1992), 267–271.

[Sim68] M. Simonovits, A method for solving extremal problems in graph theory, stability problems, pp. 279–319 in *Theory of Graphs (Proc. Colloq., Tihany, 1966)*, Academic Press, 1968.

[Sim74] M. Simonovits, Extremal graph problems with symmetrical extremal graphs. Additional chromatic conditions, *Discrete Math.* **7** (1974), 349–376.

[Sim83] M. Simonovits, Extremal graph theory, pp. 161–200 in *Selected Topics in Graph Theory* (ed. L. W. Beineke and R. J. Wilson), vol. 2, Academic Press, 1983.

[Sze72] E. Szemerédi, On graphs containing no complete subgraph with 4 vertices (Hungarian), *Mat. Lapok* **23** (1972), 113–116.

[Sze76] E. Szemerédi, Regular partitions of graphs, *Problèmes Combinatoires et Théorie des Graphes (Colloques Internationaux CNRS), Orsay* **260** (1976), 399–401.

[Tho84] A. Thomason, An extremal function for contractions of graphs, *Math. Proc. Cambridge Philos. Soc.* **95** (1984), 261–265.

[Tho01] A. Thomason, The extremal function for complete minors, *J. Combin. Theory, Ser. B* **81** (2001), 318–338.

[Tur41] P. Turán, On an extremal problem in graph theory (Hugarian), *Mat. és Fiz. Lapok* **48** (1941), 436–452.

[Ver00] J. Verstraëte, On arithmetic progressions of cycle lengths in graphs, *Combin. Probab. Comput.* **9** (2000), 369–373.

[VoZu77] H.-J. Voss and C. Zuluaga, Maximale gerade und ungerade Kreise in Graphen I. (German), *Wiss. Z. Techn. Hochsch. Ilmenau* **23** (1977), 57–70.

[WLL00] M. Wang, G. Li, and A. Liu, A result of Erdös-Sós conjecture, *Ars Combin.* **55** (2000), 123–127.

[Zyk49] A. A. Zykov, On some properties of linear complexes (Russian), *Mat. Sbornik N.S.* **24(66)** (1949), 163–188.

Section 8.2

Random Graphs

Nicholas Wormald, University of Waterloo, Canada

INTRODUCTION

The field of random graphs came into its own with papers of Erdős and Rényi in 1959–61. Earlier it had surfaced mainly in probabilistic proofs, where facts about random graphs were used to prove the existence of graphs with desired properties. Then, progressively, many interesting features of the random graphs themselves were discovered. For instance, a large random graph can be relied upon (in a sense to be made precise) to have diameter 2 and to contain any arbitrary fixed subgraph. Consequently, it will contain large complete subgraphs, and thus will have large genus and will not be k-colorable for small k. A number of models of random graphs are commonly studied, including variations arising from areas such as communication networks. Several monographs have been devoted to the subject ([Bo01], [Pa85], [Ko99], [JaLuRu00], [Sp01]).

8.2.1   Random Graph Models

NOTATION: For §8.2, let n be a positive integer and let p be a real number, $0 \leq p \leq 1$, and $q = 1 - p$. An n-vertex simple graph G has vertex set $V = [n] = \{1, \ldots, n\}$. The number of edges in the complete graph K_n is $N = \binom{n}{2}$.

NOTATION: Probability and expectation are denoted by $\mathbf{P}$ and $\mathbf{E}$, respectively, and variance is denoted by $\mathbf{Var}$. With no base shown, log denotes the natural logarithm, with base e.

DEFINITIONS

Here we consider the two most common random graph models.

D1: For $0 \leq p \leq 1$, the **binomial** (or **Bernoulli**) **random graph**, denoted by $\mathcal{G}(n,p)$, is a probability space whose underlying set is the set of n-vertex graphs. The probability function is determined by specifying that the edges of K_n occur independently with probability p each. Equivalently, the probability of any given graph with m edges is defined to be

$$p^m q^{N-m}$$

D2: For $0 \leq m \leq N$, the **uniform** (or **Erdős–Rényi**) **random graph**, denoted by $\mathcal{G}(n,m)$, is the uniform probability space on those graphs with exactly n vertices and m edges. Thus, the probability of any n-vertex m-edge graph is

$$\binom{N}{m}^{-1}$$

TERMINOLOGY NOTE: To denote that G is a random graph with the probability distribution of $\mathcal{G}(n,p)$, we write $G \in \mathcal{G}(n,p)$; alternatively we write $\mathcal{G}(n,p)$ for such G.

D3: An **event** is a subset of the graphs in whichever model is under discussion. Given a graph probability space, every graph property Q defines an event in a natural way, being the set of graphs with property Q. This event is also denoted by Q.

D4: A graph property Q is **increasing** if a graph G has property Q whenever one of its spanning subgraphs has Q.

D5: A graph property Q is **convex** if G has property Q whenever $G_1 \subseteq G \subseteq G_2$ for some G_1, G_2 both having property Q and with the same vertex sets as G.

D6: **Decreasing properties** are the complements of increasing properties.

D7: **Monotone properties** are either increasing or decreasing.

EXAMPLES

E1: Let Q be the graph property *is complete*. Then for $\mathcal{G}(n,p)$, we have

$$\mathbf{P}(Q) = p^N$$

E2: Let Q be the property *vertex 1 is isolated*. For the Bernoulli graph $\mathcal{G}(n,p)$, we have

$$\mathbf{P}(Q) = q^{n-1}$$

For the Erdős–Rényi graph $\mathcal{G}(n,m)$, we have

$$\mathbf{P}(Q) = \frac{\binom{N-n+1}{m}}{\binom{N}{m}}$$

E3:　　The properties in the two previous examples are monotone and hence convex, as are the properties *G has a subgraph in $\mathcal{F}$* for any family of graphs $\mathcal{F}$, and *G has minimum degree k* for any k.

E4:　　The property *G has diameter exactly 2* is neither increasing nor decreasing, but is convex, whereas *G has a vertex of degree exactly 2* is not convex.

REMARK

R1:　　The independence of the edges in $\mathcal{G}(n,p)$ tends to simplify calculations of probability.

Asymptotics

Most of the interest in random graphs lies in the asymptotic behavior as $n \to \infty$, with $p = p(n)$ and $m = m(n)$ functions of n.

DEFINITIONS

D8:　　An event A_n holds ***asymptotically almost surely*** (a.a.s.) if $\mathbf{P}(A_n) \to 1$ as $n \to \infty$. This applies to events A_n defined on any sequence of probability spaces indexed by n, such as the Bernoulli random graph $\mathcal{G}(n,p)$ with $p = p(n)$ a function of n, or the Erdős–Rényi random graph $\mathcal{G}(n,m)$ with $m = m(n)$.

NOTATION: Suppose that $|f| < \phi g$, for some functions $f(n)$, $g(n)$, and $\phi(n)$.

- If $\phi(n)$ is bounded, then we write $f = O(g)$.

- If $\phi \to 0$ as $n \to \infty$, then we write $f = o(g)$ or alternatively, $f \ll g$ or $g \gg f$.

- If $f = O(g)$ and $g = O(f)$, then we write $f = \Theta(g)$.

- If $f(n) = (1 + o(1))g(n)$, then we write $f \sim g$.

CONVENTION:　　The appearance of $o(g)$ in a formula denotes a function f for which $f = o(g)$, and the same convention applies to $O(g)$ and $\Theta(g)$.

NOTATION: If S is a statement about a sequence of random variables involving any of these notations, rather than an event, we write "a.a.s. S" to mean that all inequalities $|f| < \phi g$ that are implicit in S hold a.a.s.

TERMINOLOGY NOTE: Our definition of $o(g)$ is nonstandard, equivalent to the usual definition, but also accommodating the a.a.s. versions. For instance, [JaLuRu00] use $f = O_C(g)$, $f = \Theta_C(g)$ and $f = o_p(g)$ for a.a.s. $f = O(g)$, a.a.s. $f = \Theta(g)$ and a.a.s. $f = o(g)$, respectively.

TERMINOLOGY NOTE: Elsewhere, the notations *a.e.* (almost every), *whp* (with high probability), or *a.s.* (almost surely) are sometimes used instead of a.a.s.

D9:　　Let $X_1, X_2, \ldots$ be random variables and $\lambda \geq 0$ constant. We say that $X = X_n$ is ***asymptotically Poisson with mean λ*** if

$$\mathbf{P}(X = k) = \frac{e^{-\lambda}\lambda^k}{k!} + o(1)$$

for all fixed integers $k \geq 0$, as $n \to \infty$. This also applies if $\lambda = \lambda(n)$ is a bounded function of n.

D10:   For a random variable X with $0 < \mathbf{Var}X < \infty$, the ***standardized variable*** is

$$\hat{X} = (X - \mathbf{E}X)/\sqrt{\mathbf{Var}X}$$

D11:   We say that $X = X_n$ is ***asymptotically normal*** if for all fixed values of a,

$$\mathbf{P}(\hat{X} \le a) = o(1) + \frac{1}{\sqrt{2\pi}} \int_{-\infty}^{a} e^{-x^2/2}\, dx \qquad \text{as } n \to \infty$$

FACTS

F1:   (See [Bo01] and [JaŁuRu00].) Let Q be a graph property and $0 \le p = p(n) \le 1$ such that $pqN \to \infty$. If $m = m(n)$ is a positive integer function, define $x = x(m,n)$ by $m = pN + x\sqrt{pqN}$.
(i) If Q is a.a.s. true in $\mathcal{G}(n,m)$ whenever x is bounded, then Q is a.a.s. true in $\mathcal{G}(n,p)$.
(ii) The converse of (i) is true if Q is convex.

F2:   (Following from [Bo79], for example.) Let $0 < p < 1$, let H be a fixed graph, and F an induced subgraph of H. Then a.a.s. for $G \in \mathcal{G}(n,p)$, every isomorphism of F with an induced subgraph of G extends to an isomorphism of H with an induced subgraph of G. It follows that for k fixed,
(i) a.a.s. every vertex in G is in a complete subgraph of size k and in an independent set of size k,
(ii) a.a.s. G has diameter 2 and is k-connected; moreover, deleting any k vertices from G leaves a graph of diameter 2.

REMARKS

R2:   Fact F1 shows that the two models under consideration share many properties. For instance, the properties in Fact F2 hold also in $\mathcal{G}(n,m)$ with $m \sim cn^2$, any constant $0 < c < 1/2$. For this reason, we usually limit ourselves to stating properties of just one model, when the corresponding property holds in the other model by Fact F1.

R3:   The topics of random graphs, and asymptotic enumeration of graphs, are intimately intertwined; many results in either area owe their existence to techniques from the other, especially with the model $\mathcal{G}(n,m)$. For example, one may derive the asymptotic number of graphs with n vertices, m edges and no triangles, by multiplying $\binom{N}{m}$ by the probability that $G \in \mathcal{G}(n,m)$ has no triangles. Such results can be equally appealing when stated in either form. For uniformity we state only the random graph form here.

8.2.2   Threshold Functions

DEFINITIONS

D12:   A ***threshold function*** for a property Q in $\mathcal{G}(n,p)$ is a function $f(n)$ such that for $G \in \mathcal{G}(n,p)$, with $p = p(n)$,

$$\mathbf{P}(G \text{ has } Q) \to \begin{cases} 0 & \text{if } p = o(f) \\ 1 & \text{if } f = o(p) \end{cases}$$

or alternatively, such that this is true with 0 and 1 interchanged.

D13: A threshold function f for a property Q is ***sharp*** if for every fixed $\epsilon > 0$,

$$\mathbf{P}(G \text{ has } Q) \to \begin{cases} 0 & \text{if } p < (1 - \epsilon)f, \\ 1 & \text{if } p > (1 + \epsilon)f \end{cases}$$

(or the same with 0 and 1 interchanged).

FACTS

F3: [BoTh87] Every monotone property has a threshold function in $\mathcal{G}(n, p)$.

F4: [FrKa96] Every monotone property with a threshold function f in $\mathcal{G}(n, p)$ such that $\log(1/f) = o(\log n)$ has a sharp threshold function.

F5: [AcFr99] For fixed $k \geq 3$, the property of being k-colorable has a sharp threshold function in $\mathcal{G}(n, p)$.

F6: [ShSp88] Let A be a property expressible in the first-order theory of graphs, that is, using variables to represent vertices, using the equality and adjacency relations, and the usual Boolean connectives, and the quantifiers $\forall$, $\exists$. For any irrational α, $0 < \alpha < 1$, in $\mathcal{G}(n, p)$ with $p = n^{-\alpha}$, A is either a.a.s. true or a.a.s. false.

F7: [ErRé60] For $G \in \mathcal{G}(n, p)$ and fixed $k \geq 1$, $(\log n)/n$ is a sharp threshold function for the minimum vertex degree of G being at least k.

F8: [BoFr85] For $G \in \mathcal{G}(n, p)$ and fixed $k \geq 1$, $(\log n)/n$ is a sharp threshold function for the property that G has k edge-disjoint Hamilton cycles.

Strengthened versions of Facts F7 and F8 are given later.

REMARKS

R4: For a given monotone property, all sharp thresholds are clearly asymptotically equal.

R5: Statements analogous to Facts F3 and F4 also hold in more general probability spaces concerning random subsets of a set (with an extra symmetry condition, in the case of Fact F4).

R6: Threshold functions are known for many properties, and there are many properties for which more accurate information is known than the mere existence of a sharp threshold.

R7: Fact F6 is an example of a zero-one law in random graphs; see [JaLuRu00] and [Sp01] for much more on this topic.

R8: Facts F7 and F8 together imply that $(\log n)/n$ is a sharp threshold function for G being k-edge-connected [ErRé61], having a matching which either is perfect (if n is even) or meets all but one vertex (if n is odd) [ErRé66], having a Hamilton cycle [KoSz83].

R9: Threshold functions are also defined in $\mathcal{G}(n, m)$: f is a threshold function for a property Q if $\mathbf{P}(G \text{ has } Q) \to 0$ for $m = o(f)$ and $\mathbf{P}(G \text{ has } Q) \to 1$ for $f = o(m)$ (or with 0 and 1 interchanged). Sharp threshold functions are then defined in the obvious way. From Fact F1, $\frac{1}{2}n \log n$ is a threshold function in $\mathcal{G}(n, m)$ for the properties mentioned in Facts F7 and F8, and is, moreover, sharp.

8.2.3   Small Subgraphs and the Degree Sequence

DEFINITIONS

D14: The **maximum density** of a graph G is

$$\mu(G) = \max\left\{\frac{|E(F)|}{|V(F)|} \,:\, F \subseteq G,\ |V(F)| > 0\right\}$$

D15: A graph G is **strictly balanced** if its maximum density is achieved uniquely by $F = G$.

FACTS ABOUT SMALL SUBGRAPHS

NOTATION: Here H is a fixed graph with at least one edge, and X_H ($\tilde{X}_H$) denotes the number of subgraphs of $G \in \mathcal{G}(n,p)$ ($G \in \mathcal{G}(n,m)$, respectively) isomorphic to H.

F9: [Bo81] Let H be strictly balanced with k vertices and $j \geq 2$ edges, and automorphism group of order a. Let $c > 0$ be fixed and $p = cn^{-k/j}$. Then X_H is asymptotically Poisson with mean c^j/a.

F10: (See [Bo01].) For arbitrary H, a threshold function for $\{X_H > 0\}$ is $n^{-1/\mu(H)}$.

F11: [Ru88] If $|E(H)| \geq 1$ then the distribution of X_H is asymptotically normal if and only if $np^{\mu(H)} \to \infty$ and $n^2(1-p) \to \infty$.

F12: [Ja94] If $|E(H)| > 1$, $m \gg \sqrt{n}$, $N - m \gg \sqrt{n}$ and $ns(m/N)^{\mu(H)} \to \infty$, then $\tilde{X}_H$ is asymptotically normal.

F13: [JaŁuRu90] For every $p = p(n) < 1$, $e^{-\Psi_H/(1-p)} \leq \mathbf{P}(X_H = 0) \leq e^{-\Theta(\Psi_H)}$ where $\Psi_H = \min\{\mathbf{E}(X_F) \,:\, F \subseteq H,\ |E(F)| > 0\}$.

F14: [OsPrTa03], [PrSt96] Let $P(n,m) = \mathbf{P}\left(\mathcal{G}(n,m)\text{ is bipartite}\right)$. For all $\epsilon > 0$,

$$\mathbf{P}(\tilde{X}_{K_3} = 0) \sim P(n,m) \quad \text{if} \quad m \geq (1+\epsilon)\frac{\sqrt{3}}{4}n^{3/2}\sqrt{\log n}$$

and $\mathbf{P}(\tilde{X}_{K_3} = 0) \gg P(n,m)$ if $n/2 \leq m \leq (1-\epsilon)\frac{\sqrt{3}}{4}n^{3/2}\sqrt{\log n}$.

F15: [Wo96] For $p = o(n^{-2/3})$,

$$\mathbf{P}(X_{K_3} = 0) \sim e^{-\frac{1}{6}p^3 n^3 + \frac{1}{4}p^5 n^4 - \frac{7}{12}p^7 n^5}$$

For $d = m/N = o(n^{-2/3})$,

$$\mathbf{P}(\tilde{X}_{K_3} = 0) \sim e^{-\frac{1}{6}d^3 n^3}$$

F16: [PrSt92] For $G \in \mathcal{G}(n, \frac{1}{2})$, $\mathbf{P}(X_H = 0) \sim \Pr(G \text{ is } k\text{-colorable})$ iff H has chromatic number $k + 1$ but contains a color critical edge (i.e., an edge whose omission from H reduces the chromatic number to k).

FACTS ABOUT THE DEGREE SEQUENCE

NOTATION: The number of vertices of degree k in a random graph is denoted D_k, and $d_1 \geq \cdots \geq d_n$ is a descending ordering of the degrees of the vertices.

F17: [ErRé61] Let k be a fixed natural number, x a fixed real, and $m = m(n) = \frac{1}{2}n(\log n + k \log \log n + x + o(1))$. In $\mathcal{G}(n,m)$, $\mathbf{P}(d_n = k+1) \to e^{-e^{-x}/k!}$, and a.a.s. $d_n = k$ or $k+1$. (Note that d_n is the minimum vertex degree.)

F18: [Bo01] Let $k = k(n)$ be a natural number, and for fixed $\epsilon > 0$ let $\epsilon n^{-3/2} \leq p = p(n) \leq 1 - \epsilon n^{-3/2}$. If $\mathbf{E}D_k \to c$ then in $\mathcal{G}(n,p)$, the random variable D_k is asymptotically Poisson with mean c. (Note that $\mathbf{E}D_k = n\binom{n-1}{k}p^k(1-p)^{n-k}$.)

F19: [BaHoJa92] If $k = k(n)$ and either (i) $np \to 0$ and $k \geq 2$, or (ii) np is bounded away from 0 and $(np)^{-1/2}|k - np| \to \infty$, then the random variable D_k is asymptotically Poisson in $\mathcal{G}(n,p)$, in the sense that the total variation distance between $D_k = D_k(n)$ and a Poisson random variable $Z = Z(n)$ tends to 0 as $n \to \infty$. (In some cases $\mathbf{E}Z(n) \neq O(1)$.)

F20: [Bo01] Suppose that $p(1-p)n \gg (\log n)^3$ and y is a fixed real number. Then for every fixed m, in $\mathcal{G}(n,p)$

$$\lim_{n \to \infty} \mathbf{P}(d_m < f(n,p,y)) = e^{-e^{-y}} \sum_{k=0}^{m-1} e^{-ky}/k!$$

where $f(n,p,y) = pn + \sqrt{2p(1-p)n \log n}\left(1 - \frac{\log \log n}{4 \log n} + \frac{y - \log(2\sqrt{\pi})}{2 \log n}\right)$.

F21: [Bo01] The random graph $\mathcal{G}(n,p)$ a.a.s. has a unique vertex of maximum degree and a unique vertex of minimum degree iff $np(1-p) \gg \log n$ iff $\mathcal{G}(n,p)$ a.a.s. has a unique vertex of maximum degree or a unique vertex of minimum degree.

F22: [McWo97, Theorem 2.6] Let $\mathcal{B}_p(n)$ be a sequence of n independent binomial variables, each $\mathrm{Binom}(n-1, p)$ (which is the distribution of the degree of any given vertex in $\mathcal{G}(n,p)$). Consider the degree sequences $\mathcal{D}_p(n)$ of $\mathcal{G}(n,p)$, and $\mathcal{D}_m(n)$ of $\mathcal{G}(n,m)$. Let A be any event defined on sequences, and assume that either $\log n/n^2 \ll p(1-p) \ll n^{-1/2}$ or $\liminf p(1-p) \log n > 2/3$. Then the following hold:

  (i) If $m = pN$ is always an integer, then the probabilities of the event A in the two models $\mathcal{D}_m(n)$, and $\mathcal{B}_p(n)$ restricted to sequences with sum $2m$, differ by $o(1)$.

  (ii) Choose p' from the normal distribution with mean p and variance $p(1-p)/(2N)$, truncated to the unit interval $(0,1)$. Then the probabilities of the event A in the two models $\mathcal{D}_p(n)$, and $\mathcal{B}_{p'}(n)$ restricted to sequences with even sum, differ by $o(1)$.

REMARKS

R10: The facts about small subgraphs concern the number of subgraphs isomorphic to a fixed graph H. Similar results on the number of *induced* copies of H were obtained by Janson (see [JaLuRu00, Chapter 6]). The property of having an induced subgraph isomorphic to H is not monotone; nevertheless there will usually be a "local threshold" near which the probability moves from $o(1)$ to $1 - o(1)$ as the edge density increases, and later a second (disappearance) local threshold, where it changes back to $o(1)$. For the existence of a copy of H vertex-disjoint from all other copies, again there are two local thresholds, with significant pioneering results in [Su90]; see also the discussion in [JaLuRu00]. Threshold results on covering every vertex by a copy of H were given by Spencer [Sp90], as part of more general results on extending all partial embeddings of k vertices of H. For threshold results on the property that every coloring of the edges of $\mathcal{G}(n,p)$ contains a monochromatic copy of a given graph G, see [RöRu95].

R11: On probabilities in the tail of the distribution of X_H and $\tilde{X}_H$, exponentially small upper bounds have been obtained but are not always sharp ([Vu01],[JaRu02]). There are a number of other papers estimating the probability of nonexistence of a given subgraph, or equivalently results on the number of graphs which do not contain the subgraph, e.g., [PrSt96a].

R12: Fact F17 (and a number of similar ones) are stated with limiting probability strictly between 0 and 1, covering the whole range. For monotone properties, this implies the results for limiting probabilities equal to 0 and 1. For example, from Fact F17 it follows that with $m = \frac{1}{2}n(\log n + k \log \log n + x)$,

$$\mathbf{P}(d_n \geq k+1) \to 1 \text{ for } x \to \infty \quad \text{and} \quad \mathbf{P}(d_n \geq k+1) \to 0 \text{ for } x \to -\infty$$

R13: Many properties of the degree sequences $\mathcal{D}_p(n)$ and $\mathcal{D}_m(n)$ of the random graphs $\mathcal{G}(n,p)$ and $\mathcal{G}(n,m)$, respectively, follow from Fact F22. It is conjectured in [McWo97] that the restriction on p can be relaxed to simply $p(1-p) \gg \log n/n^2$, which covers all p of any interest whatsoever.

8.2.4 Phase Transition

Erdős and Rényi initiated the study of the random graph $\mathcal{G}(n,m)$ as an evolving object, growing from a sparse, disconnected graph for small m to a highly connected graph for large m, and finally a complete graph when $m = N$. The biggest issue in this study has been the ***phase transition*** at $m \sim \frac{1}{2}n$, where increasing m by $o(n)$ can change the size of the largest component from a.a.s. $O(\log n)$ to a.a.s. nearly a constant times n. Here we state properties of $\mathcal{G}(n,m)$ or $\mathcal{G}(n,p)$ in a large neighborhood of this phenomenon. Most of these translate from one model to the other by Fact F1. We begin with a simple statement shown by Erdős and Rényi, the tripartite nature of which leads to the term ***double jump***.

FACT

F23: [ErRé60] Fix $c > 0$, and for $c > 1$ define $b = b(c)$ so as to satisfy $b + e^{-bc} = 1$ (see Figure 8.2.1). Let L denote the number of vertices in the largest component (called the *giant*) in $G \in \mathcal{G}(n,m)$ where $m = \lfloor cn/2 \rfloor$. Then a.a.s.

$$L = \begin{cases} O(\log n) & \text{if } c < 1, \\ \Theta(n^{2/3}) & \text{if } c = 1, \\ (b + o(1))n & \text{if } c > 1. \end{cases}$$

More precise examination by Bollobás began a revelation of details on how the phase transition takes place. The phase transition was eventually shown to have width of the order of $n^{2/3}$. So we call $m = n/2 + O(n^{2/3})$ the critical phase; before that is subcritical, and after is supercritical. Retrospectively, the significance of the "double jump" is mainly historical, as it manifests itself only when requiring m to be a fixed constant times n.

DEFINITIONS

D16: The ***excess of a graph*** with n vertices and m edges is $m - n$.

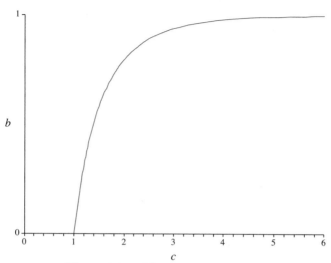

Figure 8.2.1: The growth of the giant.

D17: A connected graph is **complex** if its excess is at least 1 (so there are at least two cycles).

D18: The **k-core of a graph** is the largest subgraph with all its vertex degrees at least k. (It is quite easy to see that the k-core is unique.)

FACTS
For many of these, specific bounds on the error terms are known but unstated here.

Throughout the phase transition

F24: [Ja93] For any $m(n)$ and any $k \ll n^{2/3}$, the random graph $G(n, m)$ a.a.s has no complex component with fewer than k vertices.

F25: [Bo01, Theorem 5.15] For any $p(n)$ and any $k \gg n^{2/3}$, the random graph $G(n, p)$ a.a.s. contains no component which is a tree of order at least k.

F26: [ErRé60] If $0 < c \neq 1$ is fixed and $m \sim cn/2$, then the size of the largest tree component in $\mathcal{G}(n, m)$ is a.a.s. $(a + o(1)) \log n$ where $a = a(c) = 1/(c - 1 - \log c)$.

Figure 8.2.2 charts the size of the largest tree component.

Subcritical phase: $n/2 - m \gg n^{2/3}$

F27: [ErRé60] If $0 < c < 1$ is fixed and $m \sim cn/2$, then the probability that $\mathcal{G}(n, m)$ is a forest is asymptotic to $e^{c/2 + c^2/4}\sqrt{1 - c}$, and the expected total number of vertices belonging to cycles tends towards $c^3/(2 - 2c)$.

F28: (See [JaLuRu00].) Let $r \geq 1$ be fixed and $n^{2/3} \ll s \ll n$. In $\mathcal{G}(n, n/2 - s)$, the r largest components are a.a.s. trees of order $(1/2 + o(1))(n/s)^2 \log(s^3/n^2)$.

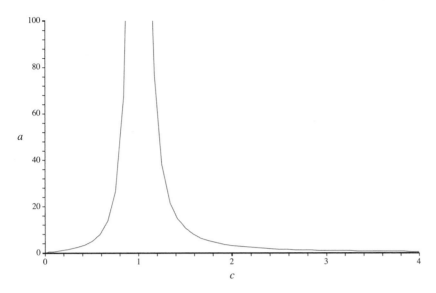

Figure 8.2.2: The size of the largest tree component.

F29: [JaLuRu00] For $m < n/2$, the probability that $\mathcal{G}(n,m)$ contains a complex component is less than $2n^2/(n-2m)^3$. (Note that this tends to 0 in the subcritical phase.)

F30: [Bo01 Corollary 5.8] For $p = c/n$, $0 < c < 1$, and any $\omega = \omega(n) \to \infty$, $G \in \mathcal{G}(n,p)$ a.a.s. has at most ω vertices in unicyclic components. Consequently, the length of the longest cycle is a.a.s. at most ω.

F31: [JaLuRu00 Section 5.4] Let $n^{2/3} \ll s \ll n$. For any $\omega = \omega(n) \to \infty$, the length ℓ of the longest cycle in $\mathcal{G}(n, n/2 - s)$ a.a.s. satisfies $\ell/\omega < n/s < \ell\omega$.

Critical phase: $m = n/2 + O(n^{2/3})$

F32: [JaLuRu00 Section 5.5] Let $m = n/2 + O(n^{2/3})$ and let r_a be the number of components of $\mathcal{G}(n,m)$ with excess a. For any $\omega = \omega(n) \to \infty$, a.a.s. $\sum_{a \geq 1} a r_a < \omega$, and the total number of vertices in complex components of $\mathcal{G}(n,m)$ is at most $\omega n^{2/3}$.

F33: [JaKnLuPi93 Theorem 5] Let $m = n/2 + O(n^{1/3})$ and fix $q \geq 1$. The probability that in $\mathcal{G}(n,m)$ there are exactly r_i components of excess i for $1 \leq i \leq q$ and none of greater excess is

$$\left(\frac{4}{3}\right)^r \sqrt{\frac{2}{3}} \frac{c_1^{r_1}}{r_1!} \frac{c_2^{r_2}}{r_2!} \cdots \frac{c_q^{r_q}}{r_q!} \frac{r!}{(2r)!} + O(n^{-1/3})$$

where $r = r_1 + 2r_2 + \cdots + q r_q$ and the c_j are (easily computed) constants. The probability that there are no components of excess 2 or more is $\sqrt{\frac{2}{3}} \cosh \sqrt{\frac{5}{18}} + O(n^{-1/3}) \approx 0.9325$.

F34: [LuPiWi94], see also [Al97] Let $m = n/2 + cn^{2/3}$ where c is constant. If (S_i, E_i) gives the size and excess of the ith component of $\mathcal{G}(n,m)$, listed so that S_i is nonincreasing with i, then the random sequence $(n^{-2/3}S_1, E_1), (n^{-2/3}S_2, E_2), \ldots$ converges in distribution to some random sequence $(X_{11}, X_{12}), (X_{21}, X_{22}), \ldots$ as $n \to \infty$.

F35: (See [JaLuRu00 Section 5.5].) If $m = n + O(n^{2/3})$ and $\omega = \omega(n) \to \infty$ then the length ℓ of the longest cycle in $\mathcal{G}(n, m)$ a.a.s. satisfies $\ell/\omega < n^{1/3} < \ell\omega$.

Supercritical phase: $m - n/2 \gg n^{2/3}$

F36: [Bo84 and Łu90] Define b so as to satisfy $b + e^{-2bm/n} = 1$. If $m - n/2 \gg n^{2/3}$ then in $\mathcal{G}(n, m)$, a.a.s. there is a complex component with $bn + O(n^{2/3})$ vertices, while every other component is not complex and has less than $n^{2/3}$ vertices.

F37: [Łu91] Let $n^{2/3} \ll s \ll n$. The longest cycle in $\mathcal{G}(n, n/2 + s)$ a.a.s. has length between $(16/3 + o(1))s^2/n$ and $(7.496 + o(1))s^2/n$.

F38: [Łu91] For any $\omega = \omega(n) \to \infty$, the length ℓ of the longest cycle in $\mathcal{G}(n, n/2 + s)$ outside the largest component a.a.s. satisfies $\ell/\omega < n/s < \ell\omega$, as does the length of the shortest cycle in the largest component.

F39: [Łu90], [JaKnLuPi93], [PiWo03] Define b as in Fact F36, $c = 2m/n$ and $t = c - cb$. By Fact F36 we may assume the largest component is unique. Let Y_1 denote the number of vertices in the 2-core of the largest component of $\mathcal{G}(n, m)$, Y_2 the number of vertices in the largest component not in the 2-core, and Y_3 the excess of the largest component. For $m = O(n)$ with $m - n/2 \gg n^{2/3}$, a.a.s. $Y_1 \sim (1-t)bn$, $Y_2 \sim tbn$ and $Y_3 \sim b(c+t-2)n/2$. Furthermore, each of the three variables is asymptotically normally distributed, and so are the numbers $Y_1 + Y_2$ (a.a.s. $\sim bn$) of vertices and $Y_1 + Y_2 + Y_3$ (a.a.s. $\sim b(c+t)n/2$) of edges in the largest component.

REMARKS

R14: Let G' denote the graph obtained by deleting the largest component from $\mathcal{G}(n, m)$ (or, if there is more than one largest component, deleting all of them). From Fact F39, it follows that for $m = O(n)$ and $m - n/2 \gg n^{2/3}$, G' a.a.s. has any particular property which is a.a.s. true for $\mathcal{G}(n', m')$ as $n' \to \infty$ when $m' \sim \frac{1}{2}tn'$. (See Theorem 5.24 in [JaLuRu00] for a more precise statement.) Here $te^{-t} = ce^{-c}$, $t \in (0, 1)$.

R15: Other interesting facts about the phase transition can be seen by viewing the random graph as a process (see Section 8.2.8).

8.2.5   Many More Properties of Random Graphs

NOTATION: The connectivity of a graph G is denoted $\kappa(G)$, the edge connectivity $\lambda(G)$, the minimum vertex degree $\delta(G)$, the independence number $\alpha(G)$, and the chromatic number $\chi(G)$.

FACTS ON CONNECTIVITY, DIAMETER, MATCHINGS, CYCLES AND PATHS

The next fact implies the threshold for the property of being k-connected, and this coincides with having minimum degree k.

F40: [ErRé61] For fixed $k \geq 0$ and $m = m(n) = \frac{1}{2}n(\log n + k \log \log n + x + o(1))$,

$$\mathbf{P}(\kappa(\mathcal{G}(n,m)) = k) \to 1 - e^{-e^{-x}/k!} \quad \text{and} \quad \text{a.a.s. } \kappa(\mathcal{G}(n,m)) = k \text{ or } k+1$$

F41: [BoTh85] For $G \in \mathcal{G}(n,p)$ and any p, a.a.s. $\kappa(G) = \lambda(G) = \delta(G)$.

F42: [Bo81a] Let p be fixed, $0 < p < 1$, and $t = \lfloor n^{1/7} \rfloor$. For $G \in \mathcal{G}(n,p)$ let $G_0 = G$ and $G_i = G_{i-1} - v_i$ ($1 \leq i \leq t$), where v_i is any member of the set S_i of vertices of minimum degree in G_{i-1}. Then a.a.s. $|S_i| = 1$, so v_i is uniquely determined, for $1 \leq i \leq t$. Also, a.a.s. $\kappa(G_i) = \delta(G_i)$ for all i, and $\delta(G_{i+1}) > \delta(G_i) + t$ for $0 \leq i < t$.

F43: [Bo81b] For constant $c > 0$ and $d = d(n) \geq 2$ an integer, we define p by $p^d n^{d-1} = \log(n^2/c)$. If $pn \gg (\log n)^3$, then for $G \in \mathcal{G}(n,p)$, a.a.s. the diameter of G is either d or $d+1$, and the probability it is d tends towards $e^{-c/2}$. Corollaries of this are: if $p^2 n - 2 \log n \to \infty$ and $n^2(1-p) \to \infty$ then a.a.s. $\mathcal{G}(n,p)$ has diameter 2; if $m < N$ and $2m^2/n^3 - \log n \to \infty$ then a.a.s. $G(n,m)$ has diameter 2.

Facts F44 and F45 discuss properties for which the obvious necessary condition, minimum degree at least 1, has the same threshold (see Fact F17). The first is the classic special case of the second.

F44: [ErRé66] For $p = p(n) = (\log n + x + o(1))/n$, the probability that $G \in \mathcal{G}(n,p)$ has a perfect matching tends to $e^{-e^{-x}}$ as $n \to \infty$ with n restricted to the even integers.

F45: [LuRu91] Let T be a tree with at $t \geq 2$ vertices. For $p = p(n) = (\log n + x + o(1))/n$, the probability that $G \in \mathcal{G}(n,p)$ has a T-factor (i.e., a spanning subgraph each of whose components is isomorphic to T) tends to $e^{-e^{-x}}$ as $n \to \infty$ with n restricted to the integers divisible by t.

More generally, one can ask for the threshold of existence of a spanning subgraph with all components isomorphic to a given graph H (assuming n is divisible by $|V(H)|$). The case when H is a triangle was an outstanding unsolved problem for many years. The threshold was shown for strictly balanced graphs by Johansson, Kahn and Vu to coincide with the threshold for the property that every vertex is contained in a copy of H. We give only the triangle case here. Their argument also extends to solve a well known problem of Shamir on perfect matchings in hypergraphs.

F46: [JoKaVu08] Restrict n to integers divisible by 3. For the property that the vertices are covered by a set of disjoint triangles, $n^{-2/3}(\log n)^{1/3}$ is a threshold function.

Fact F47 concerns a property for which the obvious necessary condition, minimum degree at least 2, has the same threshold. (See also Remark 9 and Fact F76.)

F47: [KoSz83] For $p = p(n) = (\log n + \log \log n + x + o(1))/n$, the probability that $G \in \mathcal{G}(n,p)$ has a Hamilton cycle tends to $e^{-e^{-x}}$ as $n \to \infty$.

F48: [AjKoSz81] Let $f(c)$ be the supremum of all β such that $\mathcal{G}(n, p = c/n)$ a.a.s. contains a path of length at least βn. Then $f(c) > 0$ for $c > 1$, and $\lim_{c \to \infty} f(c) = 1$.

FACTS ON INDEPENDENT SETS AND CHROMATIC NUMBER

F49: [BoEr76], [Ma76] Let $c < 1$ and suppose that $n^{-\delta} \ll p = p(n) < c$ for all $\delta > 0$. For fixed $\epsilon > 0$ the independence number $\alpha(G)$ of $G \in \mathcal{G}(n,p)$ a.a.s. satisfies $r_1 \leq \alpha(G) \leq r_2$ where
$$r_i = \lfloor 2\log_b n - 2\log_b \log_b np + 2\log_b(e/2) + 1 + (-1)^i \epsilon/p \rfloor,$$
with $b = 1/(1-p)$.

F50: [Fr90] Let $\epsilon > 0$ and $r_i = \lfloor 2p^{-1}(\log np - \log\log np + \log(e/2) + (-1)^i \epsilon) \rfloor$. For some constant C_ϵ, the independence number $\alpha(G)$ of $G \in \mathcal{G}(n,p)$ a.a.s. satisfies the inequality $r_1 \leq \alpha(G) \leq r_2$ provided that $C_\epsilon/n < p = p(n) < \log^{-2} n$.

F51: [Bo88], [Lu91] Let $c < 1$ be constant. If $1/n \ll p = p(n) < c$, then for $G \in \mathcal{G}(n,p)$ a.a.s. $\chi(G) \sim n\log\frac{1}{1-p}/(2\log np)$.

F52: [AcNa06] Let $p = d/n$ for constant $d > 0$. The chromatic number of $G \in \mathcal{G}(n,p)$ is a.a.s. either k_d or $k_d + 1$, where k_d is the smallest integer k such that $d < 2k\log k$. Moreover, if d lies in the interval $[(2k-1)\log k, 2k\log k)]$, then for $G \in \mathcal{G}(n,p)$, a.a.s. $\chi(G) = k + 1$.

F53: [Al93], [Kr00] Let $\epsilon > 0$. If $p(n) \geq n^{-1/4+\epsilon}$, then a.a.s. for $G \in \mathcal{G}(n,p)$, we have $\chi(G) \sim \chi_l(G)$, where $\chi_l(G)$ is the choice number (or list-chromatic number) of G.

F54: [PiSpWo96] For fixed $k \geq 3$, the existence of a k-core in $\mathcal{G}(n,p)$ has a sharp threshold function $p = c_k/n$ where $c_k = \inf_{\mu>0}\{\mu e^\mu/f(\mu,k)\}$, $f(\mu,k) = \sum_{i=k-1}^{\infty}\mu^i/i!$. (The k-core of a graph is the maximum subgraph of minimum degree at least k.)

F55: [Mo96] Let c_k be as above and $d_k = \sup\{d : \chi(G(n,d/n) \leq k\}$. Then $c_k \neq d_k$ for $k \geq 4$. (From Fact F54, we have $c_k \leq d_k$, since a graph with no k-core can always be $(k-1)$-colored.)

FACTS ON PLANARITY, GENUS, AND CROSSING NUMBER

F56: [ErRé60] For planarity in $\mathcal{G}(n,m)$, $m = n/2$ is a sharp threshold function. More precisely [ŁuPiWi94], there is a function f such that $0 \leq f(x) \leq 1$, $f(x) \to 0$ as $x \to -\infty$, $f(x) \to 1$ as $x \to \infty$, and $\mathbf{P}(\mathcal{G}(n,m)$ is planar$) \to f(c)$ for $m = n/2 + cn^{2/3}$ (where c is constant).

F57: [ArGr95] If $p^2(1-p)^2 \geq 8(\log n)^4/n$, the genus of $\mathcal{G}(n,p)$ is a.a.s. $(1+o(1))\frac{1}{12}pn^2$.

F58: [RöTh96] For every integer $i \geq 1$, if $n^{-i/(i+1)} \ll p \ll n^{-(i-1)/i}$, the genus of $\mathcal{G}(n,p)$ is a.a.s. $(1+o(1))(i/4(i+2))pn^2$.

F59: [SpTó02] The expected value of the crossing number of $G \in \mathcal{G}(n,p)$ (i.e., the minimum number of crossing points in a drawing of G in the plane, no three edges crossing at the same point) is $o(p^2n^4)$ if $p \sim 1/n$ and $\Theta(p^2n^4)$ if $p = c/n$ for fixed $c > 1$.

FACTS ON EIGENVAUES, AUTOMORPHISMS, AND UNLABELED GRAPHS

F60: [FüKo81] Let $\lambda_1 \geq \lambda_2 \geq \ldots \geq \lambda_n$ be the eigenvalues of the adjacency matrix of G and let $p(n) = c$. Then a.a.s. in $\mathcal{G}(n,p)$,
$$\lambda_1 \sim np \quad \text{and} \quad \max_{2 \leq i \leq n}|\lambda_i| = 2\sqrt{pqn} + O(n^{1/3}\log n)$$

F61: [Wr70] The expected number of automorphisms of $G \in \mathcal{G}(n, m)$ tends to 1 iff $\min(m, N - m)/n - (\log n)/2 \to \infty$.

The next fact follows from this and explains why random unlabeled graphs are rarely studied: for many purposes they have the same properties as random labeled graphs.

F62: [Bo01] Let Q be a property of graphs of order n, $0 < c < 1$, and suppose that $m = m(n)$ is such that $\min\left(m, N - m\right)/n - (\log n)/2 \to \infty$. Let $G \in \mathcal{G}(n, m)$ and choose an unlabeled graph H on n vertices and m edges uniformly at random. Then $\mathbf{P}(G \text{ has } Q) \to c$ iff $\mathbf{P}(H \text{ has } Q) \to c$.

8.2.6   Random Regular Graphs

There is an abundance of interesting models of random graphs besides $\mathcal{G}(n, p)$ and $\mathcal{G}(n, m)$. One of the most common is a restriction of $\mathcal{G}(n, m)$, to graphs with specified degree sequence. Random regular graphs are a special case of this.

NOTATION: For $d \geq 0$ and dn even, $\mathcal{G}_{n,d}$ is the probability space containing just the d-regular graphs on n vertices, all being equiprobable.

FACTS

The next fact comes easily from the enumeration formulae of Bender and Canfield [BeCa78] or alternatively from the model of random regular graphs given by Bollobás [Bo80]. (See the survey [Wo99], for example.)

F63: For fixed d and any fixed graph F with more edges than vertices, a random regular graph $G \in \mathcal{G}_{n,d}$ a.a.s. contains no subgraph isomorphic to F.

F64: [Bo80], [Wo81] For $d \geq 0$ and $k \geq 3$ fixed, the number of cycles of length k in a graph in $G \in \mathcal{G}_{n,d}$ is asymptotically Poisson with mean $(d - 1)^k/2k$.

F65: [KrSuVuWo01], [CoFrRe02] For $3 \leq d = d(n) \leq n - 4$, a random regular graph $G \in \mathcal{G}_{n,d}$ is a.a.s. d-connected.

F66: [BoFe82] Fix $d \geq 3$ and $\epsilon > 0$. The diameter $D(G)$ of $G \in \mathcal{G}_{n,d}$ a.a.s. satisfies the inequalities

$$1 + \lfloor \log_{d-1} n \rfloor + \left\lfloor \log_{d-1}\left(\frac{d - 2}{6d} \log n\right) \right\rfloor \; \leq \; D(G) \; \leq \; 1 + \lceil \log_{d-1}((2 + \epsilon)dn \log n) \rceil$$

F67: [McWo84] If $3 \leq d = d(n) = o(\sqrt{n})$, then the expected number of automorphisms of $G \in \mathcal{G}_{n,d}$ tends to 1 as $n \to \infty$. (c.f. Fact F61.)

F68: [KiSuVu02] If $3 \leq d = d(n) \leq n - 4$, then $G \in \mathcal{G}_{n,d}$ a.a.s. has only the trivial automorphism.

F69: [KrSuVuWo01], [CoFrReRi02] If $1 \ll d = d(n) < 0.9n$, then for $G \in \mathcal{G}_{n,d}$, a.a.s. $\alpha(G) \sim 2 \log d/\log(n/(n - d))$ and $\chi(G) \sim n/\alpha(G)$.

F70: [RoWo94] For fixed $d \geq 3$, a random regular graph $G \in \mathcal{G}_{n,d}$ a.a.s. has a Hamilton cycle and, for odd d, a.a.s. has edge chromatic number equal to d. Indeed, [KiWo01] for fixed $d \geq 3$, $G \in \mathcal{G}_{n,d}$ a.a.s. has a partition of its edge set into the edges of $\frac{d}{2}$ Hamilton cycles (for d even), or $\frac{d-1}{2}$ Hamilton cycles and a perfect matching (for d odd).

F71:   [KePéWo10] For fixed $d \geq 3$, a random regular graph $G \in \mathcal{G}_{n,d}$ asymptotically almost surely has chromatic number either k or $k-1$, where k is the smallest integer satisfying $d < 2(k-1)\log(k-1)$. If, moreover, $d > (2k-3)\log(k-1)$, then the value is k is a.a.s.

F72:   [ShWo07] A random 4-regular graph $G \in \mathcal{G}_{n,4}$ asymptotically almost surely has chromatic number 3.

F73:   [KrSuVuWo01], [CoFrRe02] For $3 \leq d = d(n) \leq n-1$, a random regular graph $G \in \mathcal{G}_{n,d}$ is a.a.s. hamiltonian.

F74:   [Mc81] For fixed $d \geq 2$ and $|y| \leq 2\sqrt{d-1}$, the proportion of eigenvalues of the adjacency matrix of $\mathcal{G}_{n,d}$ which are at most y is a.a.s.

$$o(1) + \int_{-2\sqrt{d-1}}^{y} \frac{d\sqrt{4(d-1)-x^2}}{2\pi(d^2-x^2)}\, \mathrm{d}x$$

F75:   [Fr08] For d even and fixed, the second-largest eigenvalue (in absolute value) of the adjacency matrix of $\mathcal{G}_{n,d}$ is a.a.s. at most $2\sqrt{d-1} + o(1)$.

F76:   [BrFrSuUp99, Lemma 16] The second-largest eigenvalue (in absolute value) of the adjacency matrix of $\mathcal{G}_{n,d}$ is a.a.s. $O(\sqrt{d})$ provided $d = o(\sqrt{n})$.

REMARKS

R16:   There are interesting relationships between $\mathcal{G}_{n,d}$ and other random graph models, expressed in terms of contiguity (see [Wo99] or [JaŁuRu00]).

R17:   The behavior of the size of the largest independent set, smallest dominating set, and related functions of $\mathcal{G}_{n,d}$ is not very well determined (see the bounds in [Wo99]).

R18:   There are many results on random graphs with given degree sequences, usually obtained by the same methods as for random regular graphs. For instance, random graphs with given degree sequence, all elements of which lie between 3 and $d \leq n^{0.02}$, a.a.s. have connectivity equal to minimum degree [Łu92]. Properties of the emerging giant component in random graphs with given degree sequences are studied in [MoRe98].

8.2.7   Other Random Graph Models

A vast number of other random graph models, and closely related probabilistic models, have received much attention. Not attempting a complete or balanced treatment, we give a relevant pointer for some of the main models, either to a recent or significant result or to a major source of information: random trees (these are especially relevant to the average case analysis of many algorithms) [BaBoDeFlGaGo02], [MeMo98], [Dr09]; random hypergraphs [FrJa95], [KrVu01]; random digraphs [Ka90]; random subgraphs of the cube [Ri00]; random geometric graphs [Pe03]; superposition models [Wo99], [GrJaKiWo02]; k-in, k-out models [CoFr94]; random mappings [Ko86]; random instances of k-SAT [BoBoChKiWi01]; random graphs with independent edges but of unequal probabilities [ŁuSh95], [BoJaRi07]; random maps (graphs embedded on surfaces) [GaWo00]; random planar graphs (for instance in [McStWe05]).

8.2.8 Random Graph Processes

A random graph process is a family of random graphs indexed by time. Some random graph processes are useful in proofs of facts about standard random graph models, but the following one in particular has been studied because of the interesting interpretation and extensions of threshold results which it enables.

DEFINITIONS

D19: The ***standard random graph process*** $\tilde{\mathcal{G}}(n)$ begins with no edges and adds new edges one at a time, each selected uniformly at random from those not already present. (Formally, $\tilde{\mathcal{G}}(n)$ is a Markov chain. Its state at time m (when it has m edges) is clearly equivalent to the random graph $\mathcal{G}(n, m)$, and so it is commonly represented as the sequence $\{\mathcal{G}(n, m)\}_{m \geq 0}$.)

D20: The ***hitting time*** of a graph property Q is $\min\{m : \mathcal{G}(n, m) \text{ has Q}\}$.

FACTS

As before, asymptotic statements regarding $\tilde{\mathcal{G}}(n)$ refer to the passage of n to infinity. A number of statements have been proved which show that the first edge giving a graph a certain property is a.a.s. also the first edge for which another (simpler) property holds.

F77: [Bo84] The hitting time for the property of possessing a Hamilton cycle is a.a.s. equal to the hitting time for having minimum degree at least 2.

F78: [BoTh85] The hitting time for possessing a perfect matching is a.a.s. equal to the hitting time for having minimum degree at least 1. Indeed, [BoFr85] the hitting time for possessing $\lfloor k/2 \rfloor$ edge-disjoint Hamilton cycles and, if k is odd, a matching of size $\lfloor n/2 \rfloor$ disjoint from these cycles, is a.a.s. equal to the hitting time for having minimum degree at least k.

F79: [BoTh85] For any function $k = k(n)$, the hitting time for being k-connected is a.a.s. equal to the hitting time for having minimum degree at least k.

F80: Let F_n denote the length of the first cycle to appear in $\tilde{\mathcal{G}}(n)$. Then [Ja87] for fixed $j \geq 3$, $\mathbf{P}(F_n = j) \sim p_j = \frac{1}{2} \int_0^1 t^{j-1} e^{t/2+t^2/4} \sqrt{1-t}\, dt$; $\sum_{j=3}^{\infty} p_j \sim 1$, but on the other hand [FlKnPi89] the expected value of F_n is asymptotic to $n^{1/6}$.

F81: [JaKnLuPi93] The probability that $\tilde{\mathcal{G}}(n)$ at no time contains more than one complex component is $(1 + o(1))\frac{5\pi}{18}$.

REMARK

R19: Many other random graph processes have been studied. The world-wide-web provides motivation for a number of these (see [BoRiSpTu01] and [CoFr03], for example). Some of the other random graph processes of interest include processes modeling the growth of the giant component [AlPi00], processes which randomly delete from a graph in order to find packings [AlKiSp97], simple processes for generating graphs with given maximim degree [RuWo92], and processes generating random planar graphs by adding edges at random maintaining planarity [GeScStTa08], but these are only a small representative sample.

References

[AcFr99] D. Achlioptas and E. Friedgut, A sharp threshold for k-colorability, *Random Structures Algorithms* 14 (1999), 63–70.

[AcNa06] D. Achlioptas and A. Naor, The two possible values of the chromatic number of a random graph, *Ann. of Math. (2)* 162 (2005), 1335–1351.

[AjKoSz81] M. Ajtai, J. Komlós, and E. Szemerédi, The longest path in a random graph, *Combinatorica* 1 (1981), 1–12.

[AlPi00] D. J. Aldous and B. Pittel, On a random graph with immigrating vertices: emergence of the giant component, *Random Structures and Algorithms* 17 (2000), 79–102.

[Al93] N. Alon, Restricted colorings of graphs, pp. 1–33 in *Surveys in Combinatorics 1993*, K. Walker (Ed.), Cambridge University Press, 1993.

[AlKiSp97] N. Alon, J. H. Kim, and J. Spencer, Nearly perfect matchings in regular simple hypergraphs, *Israel J. Math.* 100 (1997), 171–187.

[ArGr95] D. Archdeacon and D. A. Grable, The genus of a random graph, *Discrete Math.* 142 (1995), 21–37.

[BaBoDeFlGaGo02] C. Banderier, M. Bousquet-Mélou, A. Denise, P. Flajolet, D. Gardy and D. Gouyou-Beauchamps, Generating functions for generating trees, *Discrete Math.* 246 (2002), 29–55.

[BeCa78] E. A. Bender and E. R. Canfield, The asymptotic number of labeled graphs with given degree sequences, *J. Combin. Theory, Ser. A* 24 (1978), 296–307.

[BaHoJa92] A. D. Barbour, L. Holst, and S. Janson, *Poisson Approximation.* Clarendon Press, Oxford, 1992.

[Bo79] B. Bollobás, *Graph Theory—An introductory Course*, Graduate Texts in Mathematics, Springer-Verlag, 1979.

[Bo80] B. Bollobás, A probabilistic proof of an asymptotic formula for the number of labelled regular graphs, *Europ. J. Combinatorics* 1 (1980), 311–316.

[Bo81] B. Bollobás, Threshold functions for small subgraphs, *Math. Proc. Cambridge Phil. Soc.* 90 (1981), 197–206.

[Bo81a] B. Bollobás, Degree sequences of random graphs, *Discrete Math.* 33 (1981), 1–19.

[Bo81b] B. Bollobás, The diameter of random graphs, *Trans. Amer. Math. Soc.* 267 (1981), 41–52.

[Bo84] B. Bollobás, The evolution of sparse graphs. In *Graph Theory and Combinatorics*, B. Bollobás (Ed.), Academic Press, pp. 35–57 (1984).

[Bo88] B. Bollobás, The chromatic number of random graphs, *Combinatorica* 8 (1988), 49–56.

[Bo01] B. Bollobás, *Random Graphs*, (2nd edn.) Cambridge Studies in Advanced Mathematics, 73. Cambridge University Press, 2001.

[BoBoChKiWi01] B. Bollobás, C. Borgs, J. Chayes, J. H. Kim, and D. B. Wilson, The scaling window of the 2-SAT transition, *Random Structures and Algorithms* 18 (2001), 201–256.

[BoFe82] B. Bollobás and W. Fernandez de la Vega, The diameter of random regular graphs, *Combinatorica* 2 (1982), 125–134.

[BoFr85] B. Bollobás and A. Frieze, On matchings and Hamiltonian cycles in random graphs. In *Random Graphs '83*, M. Karoński and A. Ruciński (Eds.), North-Holland, Amsterdam, pp. 23–46 (1985).

[BoJaRi07] B. Bollobás, S. Janson, and O. Riordan, The phase transition in inhomogeneous random graphs, *Random Structures Algorithms* 31 (2007), 3–122.

[BoRiSpTu01] B. Bollobás, O. Riordan, J. Spencer, and G. Tusnády, The degree sequence of a scale-free random graph process, *Random Structures and Algorithms* 18 (2001), 279–290.

[BoTh85] B. Bollobás and A. Thomason, Random graphs of small order, pp. 47–97 in *Random Graphs '83*, M. Karoński and A. Ruciński (Eds.), North-Holland, 1985.

[BoTh87] B. Bollobás and A. Thomason, Threshold functions, *Combinatorica* 7 (1987), 35–38.

[BrFrSuUp99] A. Z. Broder, A. Frieze, S. Suen, and E. Upfal, Optimal construction of edge-disjoint paths in random graphs, *SIAM Journal on Computing* 28 (1999), 541–574.

[CoFr94] C. Cooper and A. Frieze, Hamilton cycles in a class of random directed graphs, *J. Combinatorial Theory, Ser. B* 62 (1994), 151–163.

[CoFr03] C. Cooper and A. Frieze, A general model of web graphs, *Random Structures Algorithms/* 22 (2003), 311–335.

[CoFrRe02] C. Cooper, A. Frieze, and B. Reed, Random regular graphs of non-constant degree: connectivity and Hamilton cycles, *Combinatorics, Probability and Computing* 11 (2002), 249–262.

[CoFrReRi02] C. Cooper, A. Frieze, B. Reed, and O. Riordan, Random regular graphs of non-constant degree: independence and chromatic number, *Combinatorics, Probability and Computing* 11, 323–342.

[Dr09] M. Drmota, *Random Trees*, Springer, 2009.

[ErRé60] P. Erdős and A. Rényi, On the evolution of random graphs, *Publ. Math. Inst. Hungar. Acad. Sci.* 5 (1960), 17–61.

[ErRé61] P. Erdős and A. Rényi, On the strength of connectedness of a random graph, *Acta Math. Acad. Sci. Hungar.* 12 (1961), 261–267.

[ErRé66] P. Erdős and A. Rényi, On the existence of a factor of degree one of a connected random graph, *Acta Math. Acad. Sci. Hungar.* 17 (1966), 359–368.

[FlKnPi89] P. Flajolet, D. E. Knuth, and B. Pittel, The first cycles in an evolving graph, *Discrete Math.* 75 (1989), 167–215.

[FrKa96] E. Friedgut and G. Kalai, Every monotone graph property has a sharp threshold, *Proc. Amer. Math. Soc.* 124 (1997), 2993–3002.

[Fr08] J. Friedman, A proof of Alon's second eigenvalue conjecture and related problems,
Mem. Amer. Math. Soc. 195 (2008), no. 910.

[Fr90] A. Frieze, On the independence number of random graphs. *Discrete Math.* 81
(1990), 171–175.

[FrJa95] A. Frieze and S. Janson, Perfect matchings in random *s*-uniform hypergraphs,
Random Structures and Algorithms 7 (1995), 41–57.

[FüKo81] Z. Füredi and J. Komlós, The eigenvalues of random symmetric matrices,
Combinatorica **1** (1981), 233–241.

[GaWo00] Z. Gao and N. C. Wormald, The distribution of the maximum vertex degree
in random planar maps, *J. Combinatorial Theory, Ser. A* 89 (2000), 201–230.

[GeScStTa08] S. Gerke, D. Schlatter, A. Steger, and A. Taraz, The random planar
graph process, *Random Structures Algorithms* 32 (2008), 236–261.

[GrJaKiWo02] C. Greenhill, S. Janson, J. H. Kim, and N. C. Wormald, Permutation pseudographs and contiguity, *Combinatorics, Probability and Computing* 11
(2002), 273–298.

[Ja87] S. Janson, Poisson convergence and Poisson processes with applications to random graphs, *Stochastic Processes and Their Applications* 26 (1987), 1–30.

[Ja93] S. Janson, *Orthogonal Decompositions and Functional Limit Theorems for Random Graph Statistics*, Memoirs Amer. Math. Soc. 534, Amer. Math. Soc., Providence, R.I. (1994).

[Ja94] S. Janson, Multicyclic components in a random graph process, *Random Structures and Algorithms* 4 (1993), 71–84.

[JaKnŁuPi93] S. Janson, D. E. Knuth, T. Łuczak, and B. Pittel, The birth of the giant
component, *Random Structures and Algorithms* 4 (1993) 233–358.

[JaŁuRu90] S. Janson, T. Łuczak, and A. Ruciński, An exponential bound for the
probability of nonexistence of a specified subgraph in a random graph, pp. 73–87
in *Random Graphs '83*, M. Karoński, J. Jaworski, and A. Ruciński (Eds.), Wiley,
1990.

[JaŁuRu00] S. Janson, T. Łuczak, and A. Ruciński, *Random Graphs*, Wiley, 2000.

[JaRu02] S. Janson and A. Ruciński, The infamous upper tail, *Random Structures and
Algorithms* 20 (2002), 317–342.

[JoKaVu08] A. Johansson, J. Kahn, and V. Vu, Factors in random graphs, *Random
Structures and Algorithms* 33 (2008), 1–28.

[Ka90] R. M. Karp, The transitive closure of a random digraph, *Random Structures
and Algorithms* 1 (1990), 73–93.

[KePéWo10] G. Kemkes, X. Pérez-Giménez, and N. Wormald, On the chromatic number of random *d*-regular graphs, *Advances in Mathematics* 223 (2010), 300–328.

[KiSuVu02] J. H. Kim, B. Sudakov, and V. Vu, On the asymmetry of random regular graphs and random graphs, *Random Structures and Algorithms* 21 (2002), 216–224.

[KiWo01] J. H. Kim and N. C. Wormald, Random matchings which induce Hamilton cycles, and hamiltonian decompositions of random regular graphs, *J. Combin. Theory, Ser. B* 81 (2001), 20–44.

[Ko86] V. F. Kolchin, *Random Mappings*. Translation Series in Mathematics and Engineering. Optimization Software, Inc., Publications Division, New York (1986).

[Ko99] V. F. Kolchin, *Random Graphs*. Encyclopedia of Mathematics and its Applications, 53. Cambridge University Press, 1999.

[KoSz83] J. Komlós and E. Szemerédi, Limit distributions for the existence of Hamilton circuits in a random graph, *Discrete Math.* 43 (1983), 55–63.

[Kri00] M. Krivelevich, The choice number of dense random graphs, *Combinatorics, Probability and Computing* 9 (2000), 19–26.

[KrSuVuWo01] M. Krivelevich, B. Sudakov, V. H. Vu, and N. C. Wormald, Random regular graphs of high degree, *Random Structures and Algorithms* 18 (2001), 346–363.

[KrVu01] M. Krivelevich and V. Vu, Choosability in random hypergraphs, *J. Combinatorial Theory, Ser. B* 83 (2001), 241–257.

[Łu90] T. Łuczak, Component behavior near the critical point of the random graph process, *Random Structures and Algorithms* 1 (1990), 287–310.

[Łu91] T. Łuczak, The chromatic number of random graphs, *Combinatorica* 11 (1991), 45–54.

[Łu92] T. Łuczak, Sparse random graphs with a given degree sequence, pp. 165–182 in *Random Graphs Vol. 2*, A. Frieze and T. Łuczak (Eds.), Wiley, 1992.

[ŁuRu91] T. Łuczak and A. Ruciński, Tree-matchings in graph processes, *SIAM J. Discrete Math.* 4 (1991), 107–120.

[ŁuSh95] T. Łuczak and S. Shelah, Convergence in homogeneous random graphs, *Random Structures and Algorithms* 6 (1995), 371–392.

[McStWe05] C. McDiarmid, A. Steger, and D. Welsh, Random planar graphs, *J. Combinatorial Theory, Ser. B* 93 (2005), 187–205.

[Mc81] B. D. McKay, The expected eigenvalue distribution of a large regular graph, *Linear Algebra and Its Applications* 40 (1981), 203–216.

[McWo97] B. D. McKay and N. C. Wormald, The degree sequence of a random graph. I. The models, *Random Structures and Algorithms* 11 (1997), 97–117.

[MeMo98] A. Meir and J. W. Moon, On the log-product of the subtree-sizes of random trees, *Random Structures Algorithms* 12 (1998), 197–212.

[Mo96] M. Molloy, A gap between the appearances of a k-core and a $(k+1)$-chromatic graph, *Random Structures and Algorithms* 8 (1996), 159–160.

[MoRe98] M. Molloy and B. Reed, The size of the giant component of a random graph with a given degree sequence, *Combinatorics, Probability and Computing* 7 (1998), 295–305.

[OsPrTa03] D. Osthus, H. J. Prömel, and A. Taraz, For which densities are random triangle-free graphs almost surely bipartite?, *Combinatorica* 23 (2003), 105–150.

[Pa85] E. M. Palmer, *Graphical Evolution. An Introduction to the Theory of Random Graphs*. Wiley, 1985.

[Pe03] M. D. Penrose, *Random Geometric Graphs*, volume 5 of *Oxford Studies in Probability*, Oxford University Press, 2003.

[PiSpWo96] B. Pittel, J. Spencer, and N. C. Wormald, Sudden emergence of a giant k-core in a random graph, *J. Combinatorial Theory, Ser. B* 67 (1996), 111–151.

[PrSt92] H. J. Prömel and A. Steger, The asymptotic number of graphs not containing a fixed color-critical subgraph, *Combinatorica* 12 (1992), 463–473.

[PrSt96] H. J. Prömel and A. Steger, On the asymptotic structure of sparse triangle free graphs, *J. Graph Theory* 21 (1996), 137–151.

[PrSt96a] H. J. Prömel and A. Steger, Counting H-free graphs, *Discrete Math.* 154 (1996), 311–315.

[Ri00] O. Riordan, Spanning subgraphs of random graphs, *Combinatorics, Probability and Computing* 9 (2000), 125–148.

[RoWo94] R. W. Robinson and N. C. Wormald, Almost all regular graphs are hamiltonian, *Random Structures and Algorithms* 5 (1994), 363–374.

[RöRu95] V. Rödl and A. Ruciński, Threshold functions for Ramsey properties, *J. American Math. Soc.* 8 (1995), 253–270.

[RöTh95] V. Rödl and R. Thomas, On the genus of a random graph, *Random Structures and Algorithms* 6 (1995), 1–12.

[Ru88] A. Ruciński, When are small subgraphs of a random graph normally distributed? *Probability Theory and Related Fields* 78 (1988), 1–10.

[RuWo92] A. Ruciński and N. C. Wormald, Random graph processes with degree restrictions, *Combinatorics, Probability and Computing* 1 (1992), 169–180.

[ShSp88] S. Shelah and J. Spencer, Zero-one laws for sparse random graphs, *J. Amer. Math. Soc.* 1 (1988), 97–115.

[ShWo07] L. Shi and N. Wormald, Colouring random 4-regular graphs, *Combinatorics, Probability and Computing* 16, (2007), 309–344.

[Sp90] J. Spencer, Threshold functions for extension statements, *J. Combin. Theory, Ser. A* 53 (1990), 286–305.

[Sp01] J. Spencer, *The Strange Logic of Random Graphs*. Algorithms and Combinatorics, 22. Springer-Verlag, Berlin, 2001.

[SpTó02] J. Spencer and G. Tóth, Crossing numbers of random graphs, *Random Structures and Algorithms* 21 (2002), 347–358.

[Su90] W. C. S. Suen, A correlation inequality and a Poisson limit theorem for nonoverlapping balanced subgaphs of a random graph, *Random Structures and Algorithms* 1 (1990), 231–242.

[Vu01] V. Vu, A large deviation result on the number of small subgraphs of a random graph, *Combin. Probab. Comput.* 10 (2001), 79–94.

[Wo81] N. C. Wormald, The asymptotic distribution of short cycles in random regular graphs, *J. Combinatorial Theory, Ser. B* 31 (1981), 168–182.

[Wo96] N. C. Wormald, The perturbation method for triangle-free random graphs, *Random Structures and Algorithms* 9 (1996), 253–270.

[Wo99] N. C. Wormald, Models of random regular graphs, pp. 239–298 in *Surveys in Combinatorics, 1999*, J. D. Lamb and D. A. Preece (Eds.), Cambridge University Press, 1999.

[Wr70] E. M. Wright, Graphs on unlabelled nodes with a given number of edges, *Acta Math.* 126 (1970), 1–9.

Section 8.3
Ramsey Graph Theory

Ralph J. Faudree, University of Memphis

INTRODUCTION

In any group of six people there are always either three who know each other or three mutual strangers. This same statement in the language of graph theory is that if each edge of a complete graph K_6 is colored either Red or Blue, then there is either a Red triangle (K_3) or a Blue triangle (K_3). Moreover, this conclusion is not true for K_5, so six is a minimum such number. This is a special case of a much more general observation of F. P. Ramsey [Ra30]. He observed that for all positive integers m and n, there is an integer r such that if each edge of a K_r is colored either Red or Blue, then there will be either a Red K_m or a Blue K_n. The smallest such integer r is denoted by $r(m, n)$, and is called the (m, n)-Ramsey number. Ramsey graph theory is the study of such numbers and the corresponding graphs. More generally, the number of colors is not restricted to just two, the monochromatic graphs are arbitrary — not just complete graphs, and the graph being edge-colored is not restricted to being complete.

8.3.1   Ramsey's Theorem

Ramsey's original theorem applies to general set theory and has implications to many areas of mathematics other than combinatorics and graph theory. For combinatorial results related to Ramsey's theorem, see [GaRoSp90], [Pa78], [Ne96], and survey articles [Bu74], [Bu79], [ChGr83], [GrRo87], and [Ra02]. A simplified version of Ramsey's theorem applicable to finite graphs is our starting point.

FACT

F1: (**Ramsey's Theorem** [Ra30]) Given positive integers $k, n_1, n_2, \cdots, n_k \geq 2$, there is a least positive integer $r(n_1, n_2, \cdots, n_k)$ such that, for any partition $\mathcal{C}_1, \mathcal{C}_2, \cdots \mathcal{C}_k$ of the edges of a complete graph K_p with $p \geq r(n_1, n_2, \cdots, n_k)$, there is for some i a complete subgraph K_{n_i} all of whose edges are in $\mathcal{C}_i$.

DEFINITIONS

D1: The number $r(n_1, n_2, \cdots, n_k)$ is called the **Ramsey number** for the k-tuple $(n_1, n_2, \cdots, n_k)$.

D2: The partition of the edges of a complete graph K_p into k sets is described as a coloring of the edges of K_p with k colors, or more specifically a k-**edge-coloring** of K_p.

Ramsey Numbers for Arbitrary Graphs

Ramsey's theorem implies the existence of a "monochromatic" complete subgraph in the appropriate color in any edge-coloring of a sufficiently large complete graph. Since any graph G on m vertices is isomorphic to a subgraph of K_m, an immediate consequence of Ramsey's theorem is the existence of the Ramsey numbers for arbitrary graphs.

DEFINITIONS

D3: The **(generalized) Ramsey number** $r(G_1, G_2, \cdots, G_k)$ for any collection of k graphs $\{G_1, G_2, \cdots, G_n\}$ is the least positive integer n such that for any k-edge-coloring of K_n, there is for some i a monochromatic copy of G_i in color i.

D4: Given $k \geq 2$ and graphs $G_1, G_2, \cdots, G_k$, a graph F is said to **arrow** the k-tuple $(G_1, G_2, \cdots, G_k)$ if for any k-edge-coloring of F there is for some i a monochromatic copy of G_i in the ith color. This is denoted by $F \longrightarrow (G_1, G_2, \cdots, G_n)$. Thus, the Ramsey number $r(G_1, G_2, \cdots, G_k)$ is the smallest order of a graph F such that $F \longrightarrow (G_1, G_2, \cdots, G_n)$.

D5: The **size Ramsey number** $\hat{r}(G_1, G_2, \cdots, G_k)$ is the smallest size (i.e., number of edges) of a graph F such that $F \longrightarrow (G_1, G_2, \cdots, G_n)$.

D6: A graph F is $(G_1, G_2, \cdots, G_k)$-**minimal** if $F \longrightarrow (G_1, G_2, \cdots, G_n)$, but no proper subgraph of F also arrows.

REMARKS

R1: If for each i, $G_i = K_{n_i}$, then $r(n_1, n_2, \cdots, n_k) = r(K_{n_1}, K_{n_2}, \cdots, K_{n_k})$.

R2: Classical Ramsey graph theory deals with the case when each of the required monochromatic graphs is complete, while generalized Ramsey graph theory involves the generalization to arbitrary graphs.

R3: This leads to asking questions about the structure of and the number of "different" graphs that *arrow*. Ramsey minimal graphs are considered in 8.3.6.

8.3.2   Fundamental Results

The vast majority of Ramsey graph results concern 2-colorings, so these are featured. Several useful facts are immediate consequences of the definition.

FACTS

F2:   For any pair of graphs G_1 and G_2, $r(G_1, G_2) = r(G_2, G_1)$. More generally, the order of the graphs is not important for any number of graphs.

F3:   For any graphs G_m and G_n of orders m and n, respectively, $r(G_m, G_n) \leq r(m, n)$.

F4:   For $n \geq 2$, $r(2, n) = r(n, 2) = n$.

F5:   Erdős and Szekeres [ErSz35] For $m, n \geq 3$,

$$r(m, n) \leq r(m - 1, n) + r(m, n - 1)$$

with strict inequality if both $r(m - 1, n)$ and $r(m, n - 1)$ are even. A consequence of this is

$$r(m + 1, n + 1) \leq \binom{m + n}{m}.$$

REMARKS

R4:   The Erdős–Szekeres theorem gives a finite upper bound for the Ramsey numbers of all pairs of finite graphs. There are corresponding bounds for any finite number of colors and collections of finite graphs.

R5:   To prove that the Ramsey number $r(G, H) = p$, normally two steps are taken. A proof is given to show that any 2-edge-coloring of K_p, say a Red-Blue coloring, yields either a Red G or a Blue H, and then a Red-Blue coloring of a K_{p-1} is exhibited that has neither a Red G nor a Blue H.

R6:   In the case of a 2-edge-coloring, say with Red and Blue, it is sometimes more convenient to just denote a subgraph F, which represents the subgraph induced by the Red edges. Then, the complement $\overline{F}$ of F denotes the Blue subgraph.

EXAMPLE

E1:   To show that $r(3, 3) = 6$, observe that in Figure 8.3.1 there is a Red-Blue coloring of K_5 with no K_3 in either color, and observe by the result of Erdős and Szekeres

$$r(3, 3) \leq r(2, 3) + r(3, 2) \ = \ 3 + 3 = 6.$$

8.3.3   Classical Ramsey Numbers

Determining classical Ramsey numbers is quite difficult, and the number of nontrivial classical Ramsey numbers that are known precisely is very limited. Only one nontrivial multicolor (at least three colors) classical Ramsey number is known and only nine nontrivial two color Ramsey numbers $r(m, n)$ are known, which is strong evidence of the difficulty in determining Ramsey numbers.

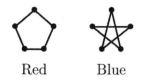

Red   Blue

Figure 8.3.1: $r(K_3, K_3) > 5$.

Ramsey Numbers for Small Graphs

Figure 8.3.2 contains the known classical Ramsey numbers $r(k, l)$ along with the best upper and lower bounds for small values of m and n. The argument k ranges from 3 through 10 and runs along the main diagonal, and the argument l runs along the top row. The exact Ramsey values are centered, upper bounds are the top entries, and lower bounds are the bottom entries. For instance, $r(3, 8) = 28$, and $r(4, 8)$ is between 56 and 84. The references for each of the exact values are also listed in the table along with references for some of the upper and lower bounds, where the leftmost column is used to indicate the argument l for the particular Ramsey number. For instance, it was established in [Ke64] that $r(3, 6) = 18$. A listing of Ramsey numbers much more complete than Figure 8.3.2 can be found in an excellent dynamic electronic paper by Radzisowski [Ra02], which is updated periodically.

FACT

F6: Greenwood and Gleason [GrGl55] $r(3, 3, 3) = 17$.

REMARK

R7: The earlier work determining classical Ramsey numbers was done by Greenwood and Gleason [GrGl55], Kéry [Ke64], and by Graver and Yackel [GrYa68]. Lower bounds were established with explicit but sophisticated colorings using algebraic techniques, and upper bounds were established using graph theory techniques. More recently Exoo, McKay, and Radzisowski, among other mathematicians, have used computational techniques, powerful algorithms, and more computing power to sharpen upper and lower bounds for larger Ramsey numbers. Numerous improvements have been made in the past few years. However, the gaps between upper and lower bounds are still enormous for even many small values of m and n.

Asymptotic Results

Considerable study of the asymptotic behavior of the Ramsey number $r(m, n)$ has not yielded sharp asymptotic results. There have been numerous improvements in the upper bound of Erdős and Szekeres (Fact F5). The lower bounds for $r(m, n)$, or in particular for $r(n, n)$, are far from the upper bounds. The only case in which sharp asymptotic results have been obtained is $m = 3$. Shearer [Sh83] proved an upper bound for $r(3, n)$, and Kim [Ki95] verified an asymptotically sharp lower bound.

In the table below each entry gives the known value or the bounds of the Ramsey number $r(k,l)$ (upper bound over lower bound). Cells in the lower-left triangle give the references establishing the corresponding result; the bold numbers run along the diagonal $k=l$.

k \ l	3	4	5	6	7	8	9	10	11	12	13	14	15
3	6	9	14	18	23	28	36	43/40	51/46	59/52	69/59	78/66	88/73
4	GreGl55	18	25	41/35	61/49	84/56	115/73	149/92	191/97	238/128	291/133	349/141	417/153
5	GreGl55	GreGl	49/43	87/58	143/80	216/101	316/125	442/143	623/159	848/185	1139/209	1461/235	1878/265
6		McRd95 / Ka65	McRd97 / Ex89b	165/102	298/111	495/129	790/169	1171/179	1804/253	2566/262	3705/317	5037/292	6911/401
7	Ke64	McRd97 / Ex93	HuZ98 / Ex93		540/205	1031/216	1713/235	2826/289	4553/415	6954/416	10581/511	15263	22116
8	GrYa68 / Ka66	M94 / Ex89	S94 / CET97			1870/282	3583/317	6090	10630	16944	27940/817	41525	63620/861
9	McZ92 / GriRb82	M94 / Ex89e	S94 / HaKr03				6588/565	12677/580	22325	39025	64871	89203	
10	GriRb82 / Ka66	M94 / Ex	Ma94 / Ex					23556/798					1265

Figure 8.3.2: The classical Ramsey numbers $r(k,l)$.

FACTS

F7: [Th88] For all $1 \leq m \leq n$, there is an absolute constant c such that

$$r(m+1, n+1) \leq e^{(-m \log n)/2n + c\sqrt{\log m}} \binom{m+n}{m}$$

and in particular

$$r(n+1, n+1) \leq n^{-1/2 + c\sqrt{\log n}} \binom{2n}{2}.$$

F8: For some constant c' and for all positive n,

$$r(n, n) \geq c' n 2^{n/2}.$$

F9: ([Sh83], [Ki95]) There are absolute constants c and c' such that

$$cn^2 / \log n \leq r(3, n) \leq c' n^2 / \log n.$$

Probabilistic techniques are used to prove the existence of colorings, but specific colorings are not exhibited.

8.3.4 Generalized Ramsey Numbers

There has been more activity and considerably more results in generalized Ramsey theory than in any other area of Ramsey graph theory. It would be impossible to survey even a fraction of the results, so we will review just a few of the highlights.

Initial Generalized Ramsey Results

FACTS

F10: [GeGy67] For positive integers $2 \leq m \leq n$,

$$r(P_m, P_n) = n + \lfloor m/2 \rfloor - 1.$$

The lower bound for $r(P_m, P_n)$ comes from the coloring determined by the graph $F = K_{n-1} \cup K_{\lfloor m/2 \rfloor - 1}$. The graph F contains no connected graph with n vertices and so no P_n, and the complement $\overline{F}$ is a bipartite graph with no P_m.

F11: [BuRo73] Let $n_1, n_2, \cdots n_k$ be positive integers with s of them being even and $k \geq 2$. Then

$$r(K_{1,n_1}, K_{1,n_2}, \cdots, K_{1,n_k}) = \sum_{i=1}^{k} (n_i - 1) + \alpha,$$

where $\alpha = 1$ if s is positive and even and $\alpha = 2$ otherwise. For $k = 2$,

$$r(K_{1,m}, K_{1,n}) = m + n - \epsilon,$$

where $\epsilon = 1$ if m and n are both even and 0 otherwise.

REMARK

R8: There are regular graphs of any order p and degree $k < p$ except when k and p are odd, and, in this case, nearly regular graphs exist. The lower bounds for the Ramsey numbers in Fact F11 depend on colorings derived directly from regular or nearly regular graphs.

Ramsey Numbers for Trees

There are many classes of pairs of trees for which the Ramsey number $r(T_m, T_n)$ is not known. However, for all such numbers that are known, $r(T_m, T_n) \le m + n - 2$.

FACT

F12: [BuRo73] When at least one of m or n is even and T_m and T_n are stars ($T_m = K_{1,m-1}$ and $T_n = K_{1,n-1}$),

$$r(T_m, T_n) = m + n - 2.$$

CONJECTURES

C1: Tree Conjecture [BuEr76] For any trees T_m and T_n with $m, n \ge 2$,

$$r(T_m, T_n) \le m + n - 2.$$

C2: Erdős–Sós Conjecture Any graph G with n vertices and at least $n(k-2)/2 + 1$ edges contains any tree T_k ($k \ge 2$) as a subgraph.

REMARKS

R9: The Erdős–Sós conjecture implies the Tree Conjecture.

R10: Personal communication has indicated that M. Ajtai, J. Komolós, M. Simonovits, and E. Szemerédi [AKSS] have proved that the Erdős–Sós conjecture is true for n sufficiently large. The paper has not been published.

Cycle Ramsey Numbers

The Ramsey numbers of cycle graphs appear to have some of the same characteristics as the Ramsey numbers of trees.

FACTS

F13: ([Ro73a], [Ro73b], [FaSc74]) If $3 \le m \le n$ with $(m,n) \ne (3,3), (4,4)$, then

$$r(C_m, C_n) = \begin{cases} 2n - 1 & \text{when } m \text{ is odd,} \\ n + \frac{m}{2} - 1 & \text{when } m \text{ and } n \text{ are even, and} \\ \max\{n + \frac{s}{2} - 1, 2m - 1\} & \text{when } m \text{ is even and } n \text{ is odd.} \end{cases}$$

For 3-edge-colorings the examples for cycle Ramsey numbers have similar properties, but determination of the numbers is much more difficult.

F14: [RoYa92] $r(C_5, C_5, C_5) = 17$.

F15: [FaScSc] $r(C_7, C_7, C_7) = 25$

F16: [Lu99] For all $n \geq 4$, $r(C_n, C_n, C_n) \leq (4 + o(1))n$.

CONJECTURE

C3: Bondy and Erdős Conjecture [BoEr73] For $n \geq 5$ and odd,
$$r(C_n, C_n, C_n) = 4n - 3.$$

EXAMPLE

E2: For $n \geq 3$ and odd, consider the 3-edge-coloring, say with Red, Blue, and Green, of a $K_{4(n-1)}$. The Red subgraph is $4K_{n-1}$, the Blue subgraph is isomorphic to $2K_{n-1,n-1}$ and contains all of the edges between the first two and the last two of the complete graphs in Red, and the remaining edges are Green and form a $K_{2(n-1),2(n-1)}$. For n odd there is no Red C_n, since no component of the Red subgraph has n vertices, and there is no Blue or Green C_n since these graphs are bipartite and have no odd cycles. Thus, $r(C_n, C_n, C_n) > 4n - 4$ for n odd.

F17: [KoSiSk05] For n odd and sufficiently large, $r(C_n, C_n, C_n) = 4n - 3$.

F18: [BeSk09] For n even and sufficiently large $r(C_n, C_n, C_n) = 2n$.

Good Results

Results of [BoEr73] and [Ch77] on complete graphs and trees (see Facts F20 and F21) motivated new lines of investigation into generalized Ramsey numbers.

DEFINITIONS

D7: [Bu81] If $\chi(G)$ is the chromatic number of G, then the ***chromatic surplus*** of G is the largest number $s = s(G)$ such that in every vertex coloring of G with $\chi(G)$ colors, every color class has at least s vertices.

D8: [Bu81] A connected graph H of order $n \geq s(G)$ is called a G-***good graph*** if
$$r(G, H) = (\chi(G) - 1)(n - 1) + s(G).$$

EXAMPLES

E3: Consider a Red-Blue coloring of a $K_{(m-1)(n-1)}$ in which the Blue graph is $m - 1$ vertex disjoint copies of a complete graph K_{n-1} ($(m - 1)K_{n-1}$) and the Red graph is the complementary graph, $K_{n_1, n_2, \cdots, n_{m-1}}$, where $n_1 = n_2 = \cdots n_{m-1} = n - 1$. There is no Blue T_n, and in fact no Blue connected graph with n vertices, and there is no Red K_m.

E4: The 2-edge-coloring of Example E3 gives the lower bound for $r(K_m, T_n)$ and also for $r(K_m, C_n)$. Moreover, there is no graph with chromatic number m in the Red graph. This coloring implies that if the chromatic number $\chi(G) = m$ and H is any connected graph of order n, then $r(G, H) > (m-1)(n-1)$.

E5: For $p = (\chi(G) - 1)(n-1) + s(G) - 1$, consider the Red-Blue edge-coloring of K_p in which the Blue graph consists of $\chi(G) - 1$ disjoint complete graphs of order $n - 1$ and one complete graph of order $s(G) - 1$, and the Red graph is the complementary graph. There is no Blue G and there is no connected graph of order n in Red for $n \geq s(G)$.

FACTS

F19: [Bu81] If H is any connected graph of order $n \geq s(G)$, then

$$r(G, H) \geq (\chi(G) - 1)(n - 1) + s(G).$$

F20: (**Bondy and Erdős** [BoEr73]) If $m \geq 3$ and $n \geq m^2 - 2$, then

$$r(K_m, C_n) = (m - 1)(n - 1) + 1.$$

It was conjectured in [EFRS78b] that Fact F20 is true for $n \geq m$ except for $n = m = 3$, and also verified for $m = 3$. It has now been verified for $m = 3, 4, 5, 6$, (see [ChS71], [YHZ99], [BJYHRZ00], and [Sc03], respectively), and in addition proved in [Ni03] for $m \geq 3$ and $n \geq 4m + 2$.

F21: (**Chvátal** [Ch77]) For integers $m, n \geq 1$,

$$r(K_m, T_n) = (m - 1)(n - 1) + 1.$$

This theorem, which can be stated as any tree T_n is K_m-good, has been generalized in many ways. The two main approaches have been to replace K_m by a graph with chromatic number m or to replace the tree T_n by a connected sparse graph.

F22: [BuFa93] A graph G satisfies $r(G, T_n) = (m-1)(n-1) + 1$ for all trees T_n of sufficiently large order n, if and only if $s(G) = 1$, and there is a $\chi(G)$-vertex coloring of G such that the graph induced by two of the color classes is a subgraph of a matching.

F23: Let G be an arbitrary graph and H a connected graph of order n. Then there are positive constants c, c_1, c_2, and α such that H is G-good if n is sufficiently large and

(i)	[BEFRS80a]	$G = K_3, n \geq 4$, and $|E(H)| \leq (17n + 1)/15$, or
(ii)	[BEFRS80a]	$G = K_m, m \geq 4, |E(H)| \leq n + cn^{2/(m-1)}$, or
(iii)	[EFRS85]	$|E(H)| \leq n + c_1 n^{\alpha}$, and $\Delta(H) \leq c_2 n^{\alpha}$, or
(iv)	[BEFRS82b]	$G = C_{2m+1}$ and $|E(H)| \leq (1 + c_3)n$, or
(v)	[BuEr83]	$G = K_3$ and $H = K_1 + C_n$ (wheel), or
(vi)	[FaRoSh91]	$G = C_{2m+1}$ and $H = K_2 + \overline{K}_{n-2}$.

REMARK

R11: A comprehensive summary of "good" results can be found in [FaRoSc92].

Small Order Graphs

Most of the generalized Ramsey numbers for very small order graphs were determined in papers by Chvátal and Harary [ChHa72], Clancy [Cl77], and Hendry [He89a]. Figure 8.3.3 pictures all of the graphs with at most five vertices that have no isolated vertices. The graphs are described using standard graphical operations such as $+, -$, and $\cup$ along with $\bullet$, where $G \bullet H$ is a graph (not unique) obtained from G and H by identifying one vertex from each graph.

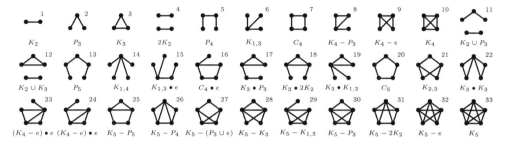

Figure 8.3.3: Graphs of order ≤ 5 without isolates.

REMARKS

R12: Figures 8.3.4 and 8.3.5 give the diagonal Ramsey numbers or the sharpest known bounds for the Ramsey numbers for all graphs of order at most five without isolated vertices. Additional information on these exact numbers and the bounds can be found in [Ra02].

R13: The Ramsey numbers for several other classes of small order graphs have been determined, and there is an excellent survey of this type of result in [Ra02].

R14: Ramsey numbers for the pair (K_3, G) where G is an arbitrary graph of order p have been determined for $p \leq 6$ in [FaRoSc80], for $p = 7$ or 8 in [Br98], and for $p = 9$ in [BrBrHa98].

R15: The diagonal Ramsey numbers $r(G, G)$ for all graphs with at most seven edges and without isolated vertices can be found in [He87].

R16: Ramsey numbers for almost all pairs (G, T) where G is a connected graph of order at most 5 and T is an arbitrary tree were calculated in [FaRoSc88].

REMARKS

R17: Figures 8.3.4 and 8.3.5 (Part 1 and Part 2), which can be found in [FaSh83a], gives the generalized Ramsey numbers for all pairs of graphs without isolated vertices and with five or less vertices. The references for the numbers in these figures appear below the diagonal, where [CH] represents [ChHa72], [C] represents [Cl77], and [H] represents [He89a], and for the single entries [h] represents [He89b], [b] represents [BoHa81], [e] represents [ExHaMe88], [y] represents [YuHe95], and [m] represents [McRa95].

	1	2	3	4	5	6	7	8	9	10	11	12	13	14	15	16	17	18
[1]	2	3	3	4	4	4	4	4	4	4	4	5	5	5	5	5	5	5
1	[2]	3	5	4	4	5	4	5	7	7	5	5	5	5	5	5	5	5
2		[3]	6	5	7	7	7	7	7	9	6	6	9	9	9	9	9	9
3			[4]	5	5	5	5	5	5	6	6	6	6	6	6	6	6	6
4				[5]	5	5	5	7	7	10	6	7	6	7	6	6	6	7
5					[6]	6	6	7	7	10	5	7	5	7	6	6	7	7
6						[7]	6	7	7	10	6	7	6	7	6	6	7	7
7	CH						[8]	7	7	10	6	7	9	9	9	9	9	9
8								[9]	10	11	6	7	9	9	9	9	9	9
0									[10]	18	7	9	13	13	13	13	13	13
10										[11]	6	7	6	6	6	6	7	6
11	–	–	–	–	–	–	–	–	–	–	[12]	7	9	9	9	9	9	9
12										—		[13]	6	7	6	6	9	9
13										—			[14]	7	9	7	9	9
14										—				[15]	6	6	9	9
15										—					[16]	6	9	9
16										—						[17]	9	9
17										—							[18]	9
18										—								[19]
19										—								
20										—								
21										—								
22			C							—								
23										—								
24										—								
25										—								
26										—								
27										—				H				
28										—								
29									h	—								
30										—								
31									h	—								
32									e	—								
33								b	m	—								

Figure 8.3.4: Generalized Ramsey numbers for small graphs - Part 1.

	19	20	21	22	23	24	25	26	27	28	29	30	31	32	33
	5	5	5	5	5	5	5	5	5	5	5	5	5	5	5
1	5	5	5	5	5	5	5	5	5	7	7	7	5	7	9
2	9	9	9	9	9	9	9	9	9	9	9	9	11	11	14
3	6	6	6	6	6	6	6	6	6	6	6	6	6	6	7
4	7	7	7	7	7	7	7	7		7	10	10	7	10	13
5	7	7	7	7	7	7	7	8	7	8	10	10	9	10	13
6	7	7	8	7	7	7	7	7	8	9	10	10	9	11	14
7	9	9	9	9	9	9	9	9	9	9	10	10	11	11	14
8	9	9	10	9	10	10	9	10	10	11	11	11	11	13	16
9	13	13	14	13	13	13	13	13	14	14	18	18	17	19	25
10	6	7	6	7	6	7	7	7	7	7	7	7	7	7	9
11	9	9	9	9	9	9	9	9	9	9	9	9	11	11	14
12	9	9	6	9	9	9	9	9	9	9	13	13	9	13	17
13	9	9	9	9	9	9	9	9	9	9	13	13	9	13	17
14	9	9	7	9	9	9	9	9	9	9	13	13	9	13	17
15	9	9	8	9	9	9	9	9	9	9	13	13	9	13	17
16	9	9	9	9	9	9	9	9	9	9	13	13	11	13	17
17	9	9	9	9	9	9	9	9	9	9	13	13	11	13	17
18	9	9	9	9	9	9	9	9	9	9	13	13	11	13	17
19	[20]	9	9	9	9	9	9	9	9	10	13	13	9	13	17
20		[21]	10	10	10	10	9	10	10	13	14	14	13	17	20
21			[22]	9	9	9	9	9	10	10	13	13	11	13	17
22				[23]	10	10	0	10	10	11	13	13	11	13	17
23					[24]	19	10	10	10	11	13	13	11	13	17
24						[25]	9	9	9	10	13	13	11	13	17
25							[26]	10	10	13	13	13	13	17	20
26								[27]	10	13	14	14	13	17	20
27									[28]	14	14	14	15	17	20-22
28										[29]	18	18	17	19	25
29											[30]	18	17	19	25-28
30												[31]	15	17	27-29
31													[32]	22	30-34
32												y		[33]	43-49
33										m					

Figure 8.3.5: Generalized Ramsey numbers for small graphs - Part 2.

Linear Bounds

By using Szemerédi's regularity lemma, [CRST83] verified the following conjecture of Erdős. Chen and Schelp [ChSc93] subsequently extended the class of "linearly bounded" graphs to a larger class of c-arrangeable graphs, and this extension has some nice applications.

CONJECTURE

C4: Erdős Conjecture [BuEr75] If G is a graph of order n with maximal degree Δ, then $r(G, G)$ has an upper bound that is linear in n.

DEFINITION

D9:   A graph G is c-***arrangeable*** if the vertices of G can be ordered in such a way that for any vertex v, each adjacency of v that succeeds v in the order has at most c adjacencies that precede v in the order.

FACTS

F24: [CRST83] if G is a graph of order n with maximal degree Δ, then $r(G, G) \leq c \cdot \Delta \cdot n$ for some positive constant c. (This verifies the Erdős Conjecture.)

F25: [ChSc93] If G is a c-arrangeable graph of order n, then there is an upper bound for $r(G, G)$ that is linear in n.

F26: [ChSc93] (Corollary) If G is a planar graph, then $r(G, G)$ has an upper bound that is linear in the order of G.

F27: [RoTh96] If G is a graph of bounded genus, then $r(G, G)$ has an upper bound that is linear in the order of G.

CONJECTURE

C5: Bounded Density Conjecture [BuEr75] If the average degree of each subgraph of a graph G of order n is at most c', then there is a constant $c = c(c')$ such that $r(G, G) \leq cn$.

8.3.5   Size Ramsey Numbers

Increased interest in the size Ramsey number $\hat{r}(G, H)$ was created in a paper by Erdős et al. [EFRS78a].

General Bounds

FACTS

F28: $|E(G)| + |E(H)| - 1 \le \hat{r}(G, H) \le \binom{r(G,H)}{2}$.

F29: [EFRS78a] For $m, n \ge 1$,
(i) $\hat{r}(K_m, K_n) = \binom{r(m,n)}{2}$, and
(ii) $\hat{r}(K_{1,m}, K_{1,n}) = m + n - 1$.

REMARKS

R18: Any graph F such that $F \longrightarrow (G, H)$ must have at least $|E(G)| + |E(H)| - 1$ edges, and the fact that $K_{r(G,H)} \longrightarrow (G, H)$ implies that $\hat{r}(G, H) \le \binom{r(G,H)}{2}$.

R19: It is natural to investigate the relationship that $\hat{r}(G, H)$ has with both $\binom{r(G,H)}{2}$ and $|E(G)| + |E(H)| - 1$. Both extreme possibilities occur.

Linear Bounds

The size Ramsey $\hat{r}(K_{1,m}, K_{1,n})$ is linear in m and n, while the number of edges in the complete Ramsey graph for the pair $(K_{1,m}, K_{1,n})$ is quadratic in m and n. Beck ([Be83], [Be90]) answered some of the questions posed in [EFRS78a] by showing that there were large classes of graphs for which the size Ramsey number has a linear property or near linear property.

FACTS

F30: [Be83], [Be90] There exist constants c and c' such that for any tree T_n of order n and maximum degree Δ and for n sufficiently large,
(i) $\hat{r}(P_n, P_n) \le cn$,
(ii) $\hat{r}(C_n, C_n) \le c'n$, and
(iii) $\hat{r}(T_n, T_n) \le \Delta \dot{n} (\log n)^{12}$.

F31: [HxKo95] For any tree T_n with maximum degree Δ, there is a constant c such that $\hat{r}(T_n, T_n) \le c \cdot \Delta \cdot n$.

REMARK

R20: The previous result was conjectured in [Be90], along with a stronger conjecture dealing with the bipartite structure of the tree T_n. The previous results led Beck and Erdős to make the following conjecture.

CONJECTURE

C6: Beck Conjecture [Be90] For a graph G of order n and bounded degree Δ, there is a constant $c = c(\Delta)$ such that $\hat{r}(G, G) \le cn$.

Bipartite Graphs

For the complete bipartite graph $K_{n,n}$, upper bounds were proved by Erdős et al. [EFRS78a] and by Nešetřil and Rödl [NeRö78], and lower bounds were proved by Erdős and Rousseau [ErRo93], but none are asymptotically sharp.

DEFINITION

D10: A star forest with s components each being a star with n edges will be denoted by $sK_{1,n}$.

FACTS

F32: [EFRS78a], [NeRö78], [ErRo93] For $n \geq 6$,

$$(1/60)n^2 2^n < \hat{r}(K_{n,n}, K_{n,n}) < (3/2)n^3 2^n.$$

F33: [BEFRS78] For positive integers m, n, s and t,

$$\hat{r}(sK_{1,m}, tK_{1,n}) = (m + n - 1)(s + t - 1).$$

REMARKS

R21: Only a limited number of precise values of size Ramsey numbers are known, since they are much more difficult to calculate than generalized Ramsey numbers. Star forests with all components equal is one class of graphs for which many numbers are known.

R22: The precise value of the size Ramsey numbers for general families of star forests is still open, but results in special cases support the following conjecture from [BEFRS78].

CONJECTURE

C7: Star Forest Conjecture [BEFRS78] Let s and t be positive integers with $m_1 \geq m_2 \cdots m_s \geq 1$ and $n_1 \geq n_2 \geq \cdots n_t \geq 1$, and let $F_1 = \cup_{i=1}^s K_{1,m_i}$ and $F_2 = \cup_{j=1}^t K_{1,n_j}$. Then,

$$\hat{r}(F_1, F_2) = \sum_{k=2}^{s+t} p_k$$

where $p_k = \max\{m_i + n_j - 1 \: : \: i + j = k\}$.

Small Order Graphs

Exact size Ramsey numbers are known for some small graphs, but even for small graphs it is sometimes difficult to calculate the number. An easier number to calculate is the restricted size Ramsey number, which is determined by restricting the "arrowing" graphs to those whose order is the Ramsey number.

DEFINITION

D11: For graphs G and H, the **restricted size Ramsey number** $r^*(G, H)$ is the minimum size graph F of order $r(G, H)$ such that $F \longrightarrow (G, H)$.

REMARK

R23: An "arrowing" graph corresponding to the size Ramsey numbers appears below the diagonal in Figure 8.3.6. These graphs are not, in general, unique. The subscript is a reference to the paper with this result. In this case [1] refers to [Bu79], [2] refers to [BEFRS78], [3] refers to [EFRS78a], [4] refers to [FaSh83b], [5] refers to [HaMi83], and [6] refers to [FaRoSh84].

$\hat{r}$		P_3	K_3	$2K_2$	P_4	$K_{1,3}$	C_4	$K_4 - P_3$	$K_4 - e$	K_4
	P_3	$3_{[2]}$	$8_{[4]}$	$4_{[2]}$	5	$4_{[2]}$	$6_{[6]}$	8	9	$18_{[4]}$
P_3	$K_{1,3}$	K_3	$15_{[3]}$	6	15	18	20	19	20	$36_{[3]}$
K_3	$K_5 - 2k_2$	K_6	$2K_3$	3	5	6	8	8	10	12
$2K_2$	C_4	$2K_3$	$3K_2$	P_4	$7_{[5]}$	8	9	15	17	35^*
P_4	$K_4 - K_2$	$K_7 - P_7$	C_5	$K_5 - P_4$	$K_{1,3}$	$5_{[2]}$	$12_{[6]}$	18	19	42^*
$K_{1,3}$	$K_{1,4}$	$K_7 - K_3$	$2K_{1,3}$	$B_5 \bullet K_2$	$K_{1,5}$	C_4	$15_{[6]}$	20	21	45^*
C_4	K_4	$K_7 - K_2$	$2C_4$	$K_5 - K_2$	$K_{2,6}$	K_6	$K_4 - P_3$	$19_{[5]}$	20	43^*
$K_4 - P_3$	$K_5 - 2K_2$	$K_7 - P_3$	$2(K_4 - P_3)$	$K_7 - P_7$	$K_7 - K_3$	$K_7 - K_2$	$K_7 - P_3$	$K_4 - e$	$39^*_{[5]}$	55^*
$K_4 - e$	$K_5 - K_2$	$K_7 - K_2$	$2(K_4 - e)$	$K_7 - P_6$	$K_7 - P_3$	K_7	$K_7 - K_2$	$K_{10} - K_4$	K_4	$153_{[3]}$
K_4	$K_7 - 3K_2$	K_9	$2K_4$	$K_{10} - 2C_5$	$K_{10} - K_3$	K_{10}	$K_{10} - P_3$	K_{11}	K_{18}	

Figure 8.3.6: Size Ramsey numbers for small order graphs.

8.3.6 Ramsey Minimal Graphs

DEFINITIONS

D12: For a pair (G, H) of graphs, $\mathcal{R}'(G, H) = \{F : F \longrightarrow (G, H)\}$. The graphs F in $\mathcal{R}'(G, H)$ are the **Ramsey graphs** for the pair (G, H).

D13: A graph F is (G, H)-**minimal** if $F \in \mathcal{R}'(G, H)$, but no proper subgraph of F is in $\mathcal{R}'$.

D14: The Ramsey minimal graphs in $\mathcal{R}'(G, H)$ will be denoted by $\mathcal{R}(G, H)$.

D15: The pair (G, H) is **Ramsey-infinite** if the number of nonisomorphic graphs in $\mathcal{R}(G, H)$ is infinite. Otherwise, the pair (G, H) is **Ramsey-finite**.

D16: A graph is 2.5-*connected* if it is 2-connected and any cutset with two vertices is independent.

EXAMPLES

E6: For any graph G, clearly $G \longrightarrow (G, K_2)$, and if $F \longrightarrow (G, K_2)$, then F must have G as a subgraph. Hence, $\mathcal{R}(G, K_2) = \{G\}$, and the pair (G, K_2) is Ramsey-finite.

E7: Observe that any 2-edge-coloring of an odd cycle C_n will have consecutive edges with the same color because of the parity of n. Thus, $C_n \longrightarrow (P_3, P_3)$ for n odd, and no proper subgraph of C_n will "arrow" (P_3, P_3). Thus, the pair (P_3, P_3) is Ramsey-infinite, and it is easy to show that $\mathcal{R}(P_3, P_3) = \{C_n \ : \ n \ \text{odd}\} \cup K_{1,3}$.

FACTS

Nešetřil and Rödl [NeRö78] initiated the Ramsey-infinite and Ramsey-finite line of investigation. Their work on forests was extended in [BEFRS81] and [BEFRS82a], but there is still no complete characterization of Ramsey-finite graphs for forests.

F34: (**Nešetřil and Rödl** [NeRö78]) The pair (G, G) is Ramsey-infinite if
(i) $\chi(G) \geq 3$,
(ii) G is 2.5-connected, or
(iii) G is a forest containing a P_4.

F35: [BEFRS81], [BEFRS82a] Let G and H be forests.

  (i) The pair (G, H) is Ramsey-infinite if either G or H has a component that is not a star.

  (ii) If G and H are star forests without isolated edges, then (G, H) is Ramsey-finite if and only if each of G and H is a single star with an odd number of edges.

  (iii) There are both Ramsey-finite and Ramsey-infinite pairs of star forests (G, H) when G and H have isolated edges.

F36: [BEFS78] For m a positive integer and G an arbitrary graph, the pair (G, mK_2) is Ramsey-finite.

F37: [Lu94] If G is a forest that is not a matching and H contains a cycle, then the pair (G, H) is Ramsey-infinite.

REMARKS

R24: One consequence of Fact F37 is that a matching $G = mK_2$ is the only graph that can be paired with any graph H to yield a Ramsey-finite pair.

R25: Fact F37 also answers a question posed in [BEFRS80b] by showing that the pair (P_3, H), and in fact the pair $(K_{1,2n}, H)$ for $n \geq 1$, is Ramsey-infinite for any graph H that is not a matching.

R26: A complete characterization of the pairs of forests that are Ramsey-finite is not known, and much less is known about pairs of graphs in general.

CONJECTURE

C8: **Ramsey-Finite Conjecture** The pair (G, H) is Ramsey-finite if and only if either
(i) G or H is a matching, or
(ii) G and H are appropriate star forests.

8.3.7   Generalizations and Variations

There are an endless number of generalizations to classical Ramsey numbers and only a few of them have been considered in this short survey. We end with a brief mention of some of the directions that have been explored.

Graphs

The induced Ramsey number leads to a stronger "arrowing" result, since the monochromatic graph must be induced. The existence of $r^*(G, H)$ was verified by Rödl in his doctoral thesis [Ro73], and was also verified independently by other mathematicians. Some general upper bounds on $r^*(G, H)$ for various graphs G and H can be found in [KoPrRo98].

DEFINITIONS

D17: The ***induced Ramsey number*** $r^*(G, H)$ is the least positive integer n such that there exists a graph F of order n such that any 2-edge-coloring (Red and Blue) of F yields an induced copy of G in Red or an induced copy of H in Blue.

D18: For bipartite graphs G and H the ***bipartite Ramsey number*** $r_b(G, H)$ is the smallest order of a bipartite graph F such that $F \longrightarrow (G, H)$.

D19: The ***connected Ramsey number*** $r_c(G, H)$ is the order of the smallest graph F such that $F \longrightarrow (G, H)$ and the graph induced by each color is also connected.

D20: For a graph G, the ***Ramsey multiplicity*** $R(G, G)$ is the minimum number of monochromatic copies of G in any 2-edge-coloring of the Ramsey graph K_n where $n = r(G, G)$.

REMARKS

R27: The existence of the bipartite Ramsey number $r_b(G, H)$ was verified by Erdős and Rado [ErRa56] and has been calculated for some basic graphs as paths, stars, and some small complete bipartite graphs. This bipartite definition can be extended to any chromatic number, not just chromatic number two.

R28: Sumner [Su78] showed that $r_c(G, H) = r(G, H)$ if neither G nor H has a bridge and each has order at least four. However, $r_c(G, H) < r(G, H)$ for some graphs with bridges, such as paths [Su78] and paths with other graphs [FaSc78].

R29: Harary and Prins [HaPr74] calculated $R(G, G)$ for small order graphs and stars, but very few Ramsey multiplicities are known.

Hypergraphs

The discussion of Ramsey theory to this point has been restricted to graphs, and nearly exclusively to 2-colorings of graphs. However, the original Ramsey theorem applied to k-uniform hypergraphs as well.

DEFINITIONS

D21: A *hypergraph* consists of a set of vertices V and a set of edges, each of which is a subset of V. A hypergraph is *k-uniform* if its edges all have cardinality k.

D22: For k-uniform hypergraphs $(G_1, G_2, \cdots, G_m)$, the *k-hypergraph Ramsey number* $r_k(G_1, G_2, \cdots, G_m)$ is the smallest integer n such that if the k-sets of a set of order n are colored with m colors, there will be for some i an isomorphic copy of G_i in color i.

FACT

F38: (**McKay, Radzisowski** [McRa91]) $r_3(4, 4) = 13$.

REMARKS

R30: The only classical hypergraph Ramsey number known is $r_3(4, 4)$.

R31: Fact F38 says that if the triples of a set with 13 elements are 2-colored, then there will be a set of order 4 with all of its subsets of order 3 having the same color, and it is not true for all colorings of the triples on a set with 12 elements.

References

[AKSS] M. Ajtai, J. Komolós, M. Simonovits, and E. Szemerédi, The Erdős Sós Conjecture, Personal communication.

[Be83] J. Beck, On size Ramsey number of paths, trees, and circuits, I, *J. Graph Theory* 7 (1983), 115–129.

[Be90] J. Beck, On size Ramsey number of paths, trees and circuits, II, *Mathematics of Ramsey Theory*, (V. Rödl, J. Nešetřil eds.), *Algorithms Combin.* 5 (1990), 34–45.

[BeSk09] F. Benevides and J. Skokan, The 3-colored Ramsey number of even cycles, *J. Comb. Theory, Ser. B* 99 (2009), 690–708.

[BoHa81] R. Bolze and H. Harborth, The Ramsey number $r(K_4 - x, K_5)$, *The Theory and Applications of Graphs* (G. Chartrand, Y. Alavi, D. L. Goldsmith, L. Lesniak-Foster, and D. R. Lick, eds.), Wiley, (1981), 108–116.

[BoEr73] J. A. Bondy and P. Erdős, Ramsey numbers for cycles in graphs, *J. Combin. Theory, Ser. B* 14 (1973), 46–54.

[Br98] G. Brinkman, All Ramsey numbers $r(K_3, G)$ for connected graphs of order 7 and 8, *Combin. Prob. and Comp.* 7 (1998), 129–140.

[BrBrHa98] S. Brandt, G. Brinkman, and T. Harmuth, All Ramsey numbers $r(K_3, G)$ for connected graphs of order 9, *Electronic J. Combin.* 5 (1998).

[Bu74] S. A. Burr, Generalized Ramsey theory for graphs – a survey, *Graphs and Combinatorics* (R. Bari and F. Harary eds.), Springer LNM 406, Berlin (1974), 52–75.

[Bu79] S. A. Burr, A survey of noncomplete Ramsey theory for graphs, *Ann. N. Y. Acad. Sci* 328 (1979), 58–75.

[Bu81] S. A. Burr, Ramsey numbers involving graphs with long suspended paths, *J. London Math. Soc. 2* 24 (1981), 405–413.

[BuEr75] S. A. Burr and P. Erdős, On the magnitude of generalized Ramsey numbers for graphs, *Infinite and Finite Sets*, I, *Colloq. Math. Soc. János Bolyai* 10 (1975), 215–240.

[BuEr76] S. A. Burr and P. Erdős, Extremal Ramsey theory for graphs, *Utilitas Math.* 9 (1976), 247–258.

[BuEr83] S. A. Burr and P. Erdős, Generalizations of a Ramsey-theoretic result of Chvátal, *J. Graph Theory* 7 (1983), 39–51.

[BEFRS78] S. A. Burr, P. Erdős, R. J. Faudree, C. C. Rousseau, and R. H. Schelp, Ramsey minimal graphs for multiple copies, *Nederl. Akad. Wetensch. Indag. Math.* 81 (1978), 187–195.

[BEFRS80a] S. A. Burr, P. Erdős, R. J. Faudree, C. C. Rousseau, and R. H. Schelp, An extremal problem in generalized Ramsey theory, *Ars Combin.* 10 (1980), 193–203.

[BEFRS80b] S. A. Burr, P. Erdős, R. J. Faudree, C. C. Rousseau, and R. H. Schelp, Ramsey minimal graphs for the pair star, connected graph, *Studia Sci. Math. Hungar.* 15 (1980), 265–273.

[BEFRS81] S. A. Burr, P. Erdős, R. J. Faudree, C. C. Rousseau, and R. H. Schelp, Ramsey minimal graphs for star-forests, *Discrete Math.* 33 (1981), 227–237.

[BEFRS82a] S. A. Burr, P. Erdős, R. J. Faudree, C. C. Rousseau, and R. H. Schelp, Ramsey minimal graphs for forests, *Discrete Math.* 38 (1982), 23–32.

[BEFRS82b] S. A. Burr, P. Erdős, R. J. Faudree, C. C. Rousseau, and R. H. Schelp, Ramsey numbers for the pair sparse graph-path or cycle, *Trans. Amer. Math. Soc.* 269 (1982), 501–512.

[BEFS78] S. A. Burr, P. Erdős, R. J. Faudree, and R. H. Schelp, A class of Ramsey-finite graphs, *Congr. Numer.* XXI (1978), 171–180.

[BJYHRZ00] B. Bollobás, C. Jayawardene, S. Yang, R. Huang, C. C. Rousseau, and M. Zhang, On a conjecture involving cycle-complete ramsey numbers, *Australasian Journal of Combinatorics* 22 (2000), 63-71.

[BuFa93] S. A. Burr and R. J. Faudree, On graphs G for which all large trees are G-good, *Graphs and Combin.* 9 (1993), 305–313.

[BuRo73] S. A. Burr and J. A. Roberts, On Ramsey numbers for stars, *Utilitas Math.* 4 (1973), 217–220.

[CET97] N. J. Calkin, P. Erdős, and C. A. Tovey, New Ramsey bounds from cyclic graphs of prime order, *SIAM J. Discrete Mathematics*, 10 (1997), 381-387.

[ChS71] G. Chartrand and S. Schuster, On the existence of specified cycles in complementary graphs, *Bull. Amer. Math. Soc.* 77 (1971), 995–998.

[ChSc93] G. Chen and R. H. Schelp, Graphs with linearly bounded Ramsey numbers, *J. Combin. Theory Ser. B* 57 (1993), 138–149.

[ChGr83] F. R. K. Chung and C. M. Grinstead, A survey of bounds for classical Ramsey numbers, *J. Graph Theory* 7 (1983), 25–38.

[Ch77] V. Chvátal, Tree-complete graph Ramsey numbers, *J. Graph Theory* 7 (1977), 93.

[ChHa72] V. Chvátal and F. Harary, Generalized Ramsey theory for graphs, *Bull. Amer. Math. Soc.* 78 (1972), 423–426.

[CRST83] V. Chvátal, V. Rödl, E. Szemerédi, and W. T. Trotter, The Ramsey number of a graph with bounded maximum degree, *J. Combin. Theory, Ser. B* 34 (1983), 239–243.

[Cl77] M. Clancy, Some small Ramsey numbers, *J. Graph Theory* 1 (1977), 89–91.

[ErRa56] P. Erdős and R. Rado, A partition calculus in set theory, *Bull. Amer. Math. Soc.* 62 (1956), 427–489.

[EFRS78a] P. Erdős, R. J. Faudree, C. C. Rousseau, and R. H. Schelp, The size Ramsey number, a new concept in generalized Ramsey theory, *Periodica Mathematica Hungarica* 9 (1978), 145–161.

[EFRS78b] P. Erdős, R. J. Faudree, C. C. Rousseau, and R. H. Schelp, On cycle-complete graph Ramsey numbers, *J. Graph Theory* 2 (1978), 53–64.

[EFRS85] P. Erdős, R. J. Faudree, C. C. Rousseau, and R. H. Schelp, Multipartite graph–sparse graph Ramsey numbers, *Combinatorica* 5 (1985), 311–318.

[ErRo93] P. Erdős and C. C. Rousseau, The size Ramsey number of a complete bipartite graph, *Discrete Math.* 113 (1993), 259–262.

[ErSz35] P. Erdős and G. Szekeres, A combinatorial problem in geometry, *Compositio Math.* 2 (1935), 463–470.

[Ex] G. Exoo, Construction available at http://ginger.indstate.edu/ge/RAMSEY, personal communication.

[Ex89a] G. Exoo, Applying optimization algorithm to Ramsey problems, *Graph Theory, Combinatorics, Algorithms, and Applications* (Y. Alavi ed.), SIAM, (1989), 175–179.

[Ex89b] G. Exoo, A lower bound for $R(5,5)$, *J. Graph Theory* 13 (1989), 97–98.

[Ex93] G. Exoo, Announcement: On the Ramsey numbers $R(4,6)$, $R(5,6)$ and $R(3,12)$, *Ars Combinatoria* 35 (1993), 85.

[Ex02] G. Exoo, Some applications of pq-graphs in graph theory, *Discuss. Math. Graph Theory* 24 (2004), 109–114.

[ExHaMe88] G. Exoo, H. Harborth, and I. Mengersen, The Ramsey number of K_4 versus $K_5 - e$, *Ars Combin.* 25A (1988), (Proc. Eleventh British Combinatorial Conf.), 277–286.

[FaRoSc80] R. J. Faudree, C. C. Rousseau, and R. H. Schelp, All triangle-graph Ramsey numbers for connected graphs of order six, *J. Graph Theory* 4 (1980), 293–300.

[FaRoSc88] R. J. Faudree, C. C. Rousseau, and R. H. Schelp, Small order graph-tree Ramsey numbers, *Discrete Math.* 72 (1988), 119–127.

[FaRoSh91] R. J. Faudree, C. C. Rousseau, and J. Sheehan, Cycle-book Ramsey numbers, *Ars Combinatoria* 31 (1991), 239–248.

[FaRoSc92] R. J. Faudree, C. C. Rousseau, and R. H. Schelp, A good idea in Ramsey theory, pp. 180–189 in *Graph Theory, Combinatorics, Algorithms, and Applications*, Soc. Indust. Appl. Math., 1992.

[FaSc74] R. J. Faudree and R. H. Schelp, All Ramsey numbers for cycles in graphs, *Discrete Math.* 8 (1974), 313–329.

[FaSc78] R. J. Faudree and R. H. Schelp, Some connected Ramsey numbers, *J. Graph Theory* 2 (1978), 119–128.

[FaScSc] R. J. Faudree, A. Schelten, and I. Schiermeyer, The Ramsey number $r(C_7, C_7, C_7)$, *Discuss. Math. Graph Theory* 23, (2003), 141–158.

[FaSh83a] R. J. Faudree and J. Sheehan, Size Ramsey numbers for small-order graphs, *J. Graph Theory* 7 (1983), 53–55.

[FaSh83b] R. J. Faudree and J. Sheehan, Size Ramsey numbers involving stars, *Discrete Math.* 46 (1983), 151–157.

[FaRoSh84] R. J. Faudree, J. Sheehan, and C. C. Rousseau, A class of size Ramsey problems involving stars, *Graph Theory and Combinatorics* (Cambridge, 1983), Academic Press, London (1984), 273–281.

[GeGy67] L. Geréncser and A. Gyárfas, On Ramsey-type problems, *Ann. Univ. Sci. Budapest Eötvös Sect. Math.* 10 (1967), 167–170.

[GrRo87] R. L. Graham and V. Rödl, Numbers in Ramsey theory, *Surveys in Combinatorics* (C. Whitehead ed.), Cambridge University Press, 1987.

[GaRoSp90] R. L. Graham, B. L. Rothschild, and J. H. Spencer, *Ramsey Theory*, John Wiley & Sons, 1990.

[GrYa68] J. E. Graver and J. Yackel, Some graph theoretic results associated with Ramsey's theorem, *J. Combin. Theory* 4 (1968), 125–175.

[GrGl55] R. E. Greenwood and A. M. Gleason, Combinatorial relations and chromatic graphs, *Canad. J. Math.* 7 (1955), 1–7.

[GrRo82] C. Grinstead and S. Roberts, On the Ramsey numbers $R(3, 8)$ and $R(3, 9)$, *J. Combin. Theory, Ser. B* 33 (1982), 27–51.

[HaKr03] H. Harborth and S. Krause, Ramsey numbers for circulant colorings, *Congressus Numerantium* 161 (2003), 139–150.

[HaMi83] F. Harary and Z. Miller, Generalized Ramsey theory VIII: the size Ramsey number of small graphs, *Studies in Pure Mathematics - To the Memory of Paul Turán* (P. Erdős, L. Alpár, G. Halász, and A. Sárközy, eds.), Birkhaüser (1983), 271–283.

[HaPr74] F. Harary and G. Prins, Generalized Ramsey theory for graphs, IV: The Ramsey multiplicity of a graph, *Networks* 4 (1974), 163–173.

[HxKo95] P. E. Haxell, and Y. Kohayakawa, The size-Ramsey number of trees, *Israel J. Math.* 89 (1995), 261–274.

[He87] G. R. T. Hendry, The diagonal Ramsey numbers for graphs with seven edges, *Utilitas Math.* 32 (1987), 11–34.

[He89a] G. Hendry, Ramsey numbers for graphs with five vertices, *J. Graph Theory* 13 (1989), 245–248.

[He89b] G. Hendry, The Ramsey numbers $r(K_2 + \overline{K}_3)$ and $r(K_1 + C_4, K_4)$, *Utilitas Math.* 35 (1989), 40–54.

[HuZh98] Y. R. Huang and K. M. Zhang, An upper bound formula for two color classical Ramsey numbers, *J. Combin. Math. Combin. Comp.* 28 (1998), 347–350.

[Ka65] J. G. Kalbfleisch, Construction of special edge-chromatic graphs, *Canadian Math. Bull.* 8 (1965), 575–584.

[Ka66] J. G. Kalbfleisch, Chromatic graphs and Ramsey's theorem, *Ph.D. Thesis*, University of Waterloo(1966).

[Ke64] G. Kéry, On a theorem of Ramsey (Hungarian), *Matematikai Lapok* 15 (1964), 204–224.

[Ki95] J. H. Kim, The Ramsey number $R(3, t)$ has order of magnitude $t^2/\log t$, *Random Structures and Algorithms* 7 (1995), 173–207.

[KoPrRo98] Y. Kohayakawa, H. J. Prömel, and V. Rödl, Induced Ramsey numbers, *Combinatorica* 18 (1998), 373–404.

[KoSiSk05] Y. Kohayakama, M. Simonovits, and J. Skokan, The 3-colored Ramsey number of odd cycles, *Proceedings of GRACO2005*, (electronic) Electronic Notes Discrete Math. 19, Elsevier, Amsterdam, (2005), 397–402.

[Lu94] T. Luczak, On Ramsey minimal graphs, *Electron. J. Combin.* 1 (1994).

[Lu99] T. Luczak, $R(C_n, C_n, C_n) \leq (4 + o(1))n$, *J. Combin. Theory, Ser. B* 75 (1999), 174–187.

[Ma94] J. Mackey, Combinatorial remedies, *Ph.D. Thesis*, University of Hawaii, 1994.

[McRa91] B. D. McKay and S. P. Radzisowski, The first classical Ramsey number for hypergraphs is computed, *Proc. of the Second Annual ACM-SIAM Symposium on Discrete Algorithms* SODA'91, San Francisco (1991), 304–308.

[McRa95] B. D. McKay and S. P. Radzisowski, $R(4, 5) = 25$, *J. Graph Theory* 19 (1995), 309–322.

[McRa97] B. D. McKay and S. P. Radzisowski, Subgraphs counting identities and Ramsey numbers, *J. Combin. Theory, Ser. B* 69 (1997), 193–209.

[McZh92] B. D. McKay and K. M. Zhang, The value of the Ramsey number $R(3, 8)$, *J. Graph Theory* 16 (1992), 99–105.

[Ne96] J. Nešetřil, Ramsey theory, Chapter 25 in *Handbook of Combinatorics, II* (R. L. Graham, M. Grótschel, and L. Lovász eds.), MIT Press (1996), 1331–1403.

[NeRö78] J. Nešetřil and V. Rödl, The structure of critical graphs, *Acta. Math. Acad. Sci. Hungar.* 32 (1978), 295–300.

[Ni03] V. Nikiforov, The cycle-complete graph Ramsey numbers, *Combin. Probab. Comput.* 14 (2005), 349–370.

[Pa78] T. D. Parsons, Ramsey graph theory, *Selected Topics in Graph Theory*, (L. W. Beineke and R. J. Wilson eds.), Academic Press (1978), 361–370.

[Ra02] S. P. Radzisowski, Small Ramsey numbers, *Electronic J. Combin.* Dynamic Surveys (2002).

[RaKr88] S. P. Radzisowski and D. L. Kreher, Search algorithms for Ramsey graphs by union of group orbits, *J. Graph Theory* 12 (1988), 59–72.

[Ra30] F. P. Ramsey, On a problem of formal logic, *Proc. London Math. Soc. (2)* 30 (1930), 264–286.

[Ro73] V. Rödl, The dimension of a graph and generalized Ramsey theorems, Thesis, Charles Univ., Praha, 1973.

[RoTh96] V. Rödl and R. Thomas, Arrangeability and clique subdivisions, *The Mathematics of Paul Erdős* II (R. L. Graham and J. Nešetřil eds.), Springer-Verlag (1996), 236–239.

[Ro73a] V. Rosta, On a Ramsey type problem of J. A. Bondy and P. Erdős, I, *J. Combin. Theory, Ser. B* 15 (1973), 94–104.

[Ro73b] V. Rosta, On a Ramsey type problem of J. A. Bondy and P. Erdős, II, *J. Combin. Theory, Ser. B* 15 (1973), 105–120.

[RoYa92] P. Rowlinson and Y. Yang, On the third Ramsey numbers of graphs with five edges, *Combin. Math. Combin. Comp.* 11 (1992), 213–222.

[Sc03] I. Schiermeyer, All cycle-complete graph Ramsey numbers $r(K_m, C_6)$, *J. Graph Theory* 44 (2003), 251–260.

[Sh83] J. B. Shearer, A note on the independence number of a triangle free-graph, *Discrete Math* 46 (1983), 83–87.

[S94] T. Spencer, Upper bounds for Ramsey numbers via linear programming, personal communication (1994).

[Su78] D. P. Sumner, The connected Ramsey number, *Discrete Math.* 22 (1978), 49–55.

[Th88] A. Thomason, An upper bound for some Ramsey numbers, *J. Graph Theory* 12 (1988), 509–517.

[YHZ99] J. Yang, R. Huang, and M. Zhang, The value of the Ramsey number $r(C_n, K_4)$ is $3(n-1)+1$ $(n \geq 4)$, *Australasian J. of Combinatorics* 20 (1999), 205–206.

[YuHe95] Y. Yuansheng, and G. R. T. Hendry, The Ramsey number $r(K_1 + C_4, K_5 - e)$, *J. Graph Theory* 19 (1995), 13–15.

Section 8.4
The Probabilistic Method

Alan Frieze and Po-Shen Loh, Carnegie Mellon University

INTRODUCTION

Even in purely deterministic settings, randomness can be artificially introduced as a powerful proof technique. This is now known as the Probabilistic Method, and it emerged in the middle of the last century. Around that time, several important results were proved by such arguments, including the myriad applications to Combinatorics that were popularized by Paul Erdős. The basic application follows the same general lines. First, one specifies a random procedure which generates a random object or identifies a random substructure according to some probability distribution. Then, one shows that the result has the desired properties with some positive probability, typically using inequalities instead of exact enumeration.

8.4.1   The First Moment Method

The *first moment method* has been used numerous times to prove results which are more difficult or perhaps impossible to prove using constructive methods.

DEFINITION

D1: The ***first moment method*** involves defining a random variable, the knowledge of whose expected value (i.e., first moment) can resolve the question of the existence of a particular structure.

EXAMPLES

E1: To prove that there is a *tournament* T with n vertices and at least $n!2^{-(n-1)}$ Hamilton paths, choose a random tournament and compute the expected number of Hamilton paths. Since there are $n!$ paths in the undirected graph, each aligned in orientation with probability $2^{-(n-1)}$, the expected number is $n!2^{-(n-1)}$. This implies that at least one tournament has this many!

E2: (Bohman, Frieze, Martin, Ruszinko, and Smyth [BoFrMaRuSm]) A coloring of the edges of a graph G is said to be k-bounded if no color is used more than k times. A subgraph H of G is said to be multi-colored if every edge has a different color. To prove that every k-bounded coloring of the edges of the hypercube graph Q_m contains a multi-colored copy of the hypercube graph Q_n, for $m \geq kn^2 2^n$, simply choose a random sub-cube and estimate the expected number of colors that appear twice or more. This is less than one for the given parameters. It follows that some Q_n does not have a color that appears twice.

E3: A *hypergraph* $H = (V, \{E_i : i = 1, 2, \ldots, m\})$ is *2-colorable* if there exists a partition of its vertex set V into two color classes $R \cup B$ such that $E_i \cap R \neq \emptyset$ and $E_i \cap B \neq \emptyset$ for $1 \leq i \leq m$. In general it is NP-hard to tell whether or not a hypergraph is 2-colorable. However, Erdős [Er63] showed that if $|E_i| \geq k$ for $1 \leq i \leq m$ and $m < 2^{k-1}$ then H is 2-colorable. Simply partition V at random and show that the expected number of edges which are mono-colored is less than one.

E4: (Spencer [Sp94]) A tournament T is said to have property S_k if for every set S of k vertices (players), there is a vertex $v = v(S) \notin S$ such that all the edges of T which join v to S are directed toward v, i.e., player v beats everyone in S. Do finite tournaments exist with property S_k? The answer is yes. To prove this, simply choose an n which satisfies

$$\binom{n}{k}(1 - 2^{-k})^{n-k} < 1 \tag{8.4.1}$$

and then randomly orient the edges of K_n. Let Z be the number of sets of k vertices for which $v(S)$ does not exist. The left hand side of (8.4.1) is the expected value of Z, and so the first moment method proves the existence of a tournament with property S_k. For large k this gives

$$n > 2^k k^2 (\ln 2)(1 + o(1))$$

and this is close to being best possible, since Szekeres has proved that if $f(k)$ is the smallest number of vertices in a tournament with property S_k, then $f(k) > ck2^k$ for some constant $c > 0$ (see Moon [Mo79]).

One can find other results like this in the book by Alon and Spencer [AlSp00].

Ramsey Numbers

DEFINITIONS

D2: A **red-blue edge-coloring** of a graph is an edge coloring in which every edge is colored either red or blue.

D3: The ***Ramsey number*** $R(k, k)$ is the smallest integer such that for $n \geq R(k, k)$, every red-blue edge-coloring of the complete graph K_n contains either an all-red K_k or an all-blue K_k.

Determining the precise values of $R(k, k)$ has proven to be extremely difficult, and $R(k, k)$ is not known exactly for any $k \geq 5$. All we have are bounds. See Section 8.4.3. One of the earliest bounds was proved by Erdős. It should be stated right away that Paul Erdős was a pioneer in the use of the probabilistic method, proving many beautiful results, as well as inspiring numerous researchers to follow in his footsteps.

FACT

F1: (Erdős [Er47]) For sufficiently large values of k,

$$R(k, k) \geq (1 - o(1)) \frac{k}{e\sqrt{2}} 2^{k/2} . \tag{8.4.2}$$

REMARKS

R1: The proof of the previous fact is quite elementary. One wants to show that if n is smaller than the right hand side of (8.4.2), then one can find an edge coloring without a mono-chromatic copy of K_k. It has proven very difficult to produce an explicit coloring that will give this result. So we proceed as follows: **we randomly color the edges of K_n and show that with positive probability this coloring will have the property we want**, which is that no K_k will be monochromatic. This proves the *existence* of such a coloring without actually explicitly constructing one. This is the essence of the probabilistic method.

R2: In the random construction above one concentrates on the random variable Z which counts the number of monochromatic K_k in the coloring. A simple calculation shows that the expected value $\mathbf{E}(Z) < 1$ and then one can use the fact that

$$\Pr(Z > 1) \leq \mathbf{E}(Z) < 1$$

to show that $\Pr(Z = 0) > 0$.

R3: The previous fact is one of the basic results in a deep and difficult theory. See Section 8.4.3 and also, for example, the books by Graham, Rothschild, and Spencer [GrRoSp90] or Nešetřil and Rödl [NeRo90].

8.4.2   Alterations

Our main example of another probabilistic proof technique concerns the possible relationship between chromatic number and girth. It would be reasonable to conjecture that graphs with large girth have small chromatic number, i.e., there is some function f such every graph with no cycle of length less than g can be properly colored with $f(g)$ colors. In spite of its appeal, it just is not true.

DEFINITION

D4: The *alteration method* is first to generate a random object and then to alter it to obtain a property we desire.

EXAMPLES

E5: Erdős [Er59] proved that for any pair of integers k, ℓ there exists a graph with girth at least k and chromatic number at least ℓ. For a probabilistic proof, let $G_{n,p}$ denote the random graph with vertex set $[n] = \{1, 2, \ldots, n\}$ and in which each of the $\binom{n}{2}$ possible edges occurs with probability p, and start with a careful choice of $p = O(1/n)$ and n sufficiently large. Erdős showed that one can delete edges and vertices to create a graph G' with n' vertices, girth at least k and no independent set of size n'/ℓ. A moment's thought will convince the reader that G' has the required property.

E6: By altering the randomly colored complete graph used to prove (8.4.2) one can show that

$$R(k,k) \geq (1 - o(1))\frac{k}{e}2^{k/2}.$$

E7: A similar, but more sophisticated alteration was used by Beck [Be78] to replace the inequality $m < 2^{k-1}$ in the question of 2-colorability of hypergraphs by $m = \Omega(2^k k^{1/3})$. Beck's proof was modified and improved so that the current best value for m is $m = \Omega(2^k(k/\ln k)^{1/2})$. This was done by Radhakrishnan and Srinivasan [RaSr00].

8.4.3   The Lovász Local Lemma

After the first moment method, perhaps the next most useful tool is the *Local Lemma*, which is the following fact.

FACT

F2: (Lovász [ErLo75]) *Symmetric* version of the Local Lemma: Given a collection of *bad* events $A_1, A_2, \ldots, A_m$, we wish to prove that there is some point in our probability space for which none of the A_i occurs. Let Γ be the corresponding *dependency graph* on the vertex set $\{1, \ldots, m\}$, defined such that for each $1 \leq i \leq m$, the event A_i is independent of the collection $\{A_j : ij \in E(\Gamma)\}$. If all $\Pr(A_i)$ are upper bounded by some real number p which satisfies $e(\Delta + 1)p < 1$, where $e \approx 2.718$ is the natural base and Δ is the maximum degree of the dependency graph Γ, then $\Pr(\bigcap_{i=1}^{m} \overline{A_i}) > 0$. In particular, the desired point in the probability space exists.

EXAMPLES

E8: The local lemma yields a slight improvement of the lower bound on $R(k, k)$. We once again randomly color, and the bad events are that a particular k-clique gets mono-colored. The computations lead to a slight improvement,

$$R(k,k) \geq (1 - o(1))\frac{k\sqrt{2}}{e}2^{k/2},\tag{8.4.3}$$

which only doubles the bound of (8.4.2).

E9: Let $H = (V, \{E_i : i = 1, 2, \ldots, m\})$ be a hypergraph in which every edge has at least k elements and suppose that each edge of H intersects at most d other edges. If $e(d + 1) < 2^{d-1}$ then H is 2-colorable. We randomly 2-color V and event A_i is defined to be {edge E_i is monochromatic}. Given the set of events $\mathcal{A} = \{A_i, i = 1, 2, \ldots, m\}$, we define a *dependency graph* Γ with vertex set $\mathcal{A}$ such that event A_i is independent of the events which are not adjacent to A_i in the graph Γ. In the present context of hypergraph coloring, $p = 2^{k-1}$ and $\Delta = d$ and so, the local lemma proves the *existence* of a coloring, i.e., proves that the hypergraph is 2-colorable.

E10: This example concerns list coloring. Here we have a graph $G = (V, E)$ and each $v \in V$ has a list of allowable colors L_v and the question is can one choose a color $c_v \in L_v$ for each $v \in V$ so that the coloring is *proper*. G is k-list colorable if $|L_v| \geq k$ for all $v \in V$ implies that a proper coloring exists. The *list chromatic number* of a graph is the minimum k such that G is k-list colorable. Here is a simple result that can be proven by one simple application of the local lemma: It is taken from *Graph Coloring and the Probabilistic Method*, by Molloy and Reed [MoRe02]. Suppose that every $v \in V$ has the following properties:

 (i) $|L_v| \geq \ell$, and

 (ii) each $c \in L_v$ appears at most $\frac{\ell}{8}$ times on the color list of a neighbor of v.

The random experiment is to choose a random color c_v independently from L_v for each $v \in V$. The collection of events is $A_{c,e}$, which denotes {color c is chosen at both ends of edge e}. The local lemma immediately implies that there is a positive probability that none of these events will occur and so a proper coloring exists under these circumstances.

E11: Here is another simple application. Suppose $G = (V, E)$, $|V| = n$ and r divides n. Let Δ denote maximum degree. Suppose that $r > 8\Delta$; then we can show that for any partition of V into sets $V_1, V_2, \ldots, V_m$, $m = n/r$ of size r, there is an independent set of G of size m which contains exactly one member of each V_i. Simply choose a random member of each V_i and use the local lemma to show that it is an independent set with positive probability. The events are defined by the edges of G. Event A_e will denote both endpoints of e are chosen.

REMARKS

R4: Sometimes a modification of the local lemma can be used, even when all the events are dependent. What is needed is some notion of being only weakly dependent. This has been called the *lopsided local lemma* and has been used by Erdős and Spencer [ErSp91] and by Albert, Frieze and Reed [AlFrRe95] to show the existence of multi-colored perfect matchings and Hamilton cycles.

R5: As described, the local lemma is non-constructive and does not yield polynomial-time algorithms for finding the objects of interest. Starting with a breakthrough by Beck [Be91], algorithmic versions have been developed by Alon [Al91], Molloy and Reed [MoRe98a], Czumaj and Scheideler [CzSc00], Salavatipour [Sa04], Moser [Mo09], and Srinivasan [Sr08]. Ultimately, Moser and Tardos [MoTa10] discovered a constructive proof producing a polynomial-time algorithm for the most common setting, where there is an underlying collection of mutually independent random variables, the events A_i are determined by various subsets of that collection, and the edge ij appears in the dependency graph Γ if and only if the subsets of variables that determine A_i and A_j overlap.

8.4.4   The Rödl Nibble

The alteration method proceeds by altering the results of a random experiment. The *Rödl nibble* takes this a step further. It was first used by Rödl [Ro85] to affirm a conjecture of Erdős and Hanani. This nibbling approach has become a powerful but technically demanding tool.

DEFINITION

D5: The **Rödl nibble** considers a random process that builds the required object of interest a little piece at a time.

FACT

F3: (Rödl [Ro85]) Let $M(n, k, \ell)$ denote the minimum size of a family of k-subsets of $[n]$ which contain every ℓ-subset of $[n]$ at least once. Then as $n \to \infty$, we have

$$M(n, k, l) = (1 + o(1))\frac{\binom{n}{\ell}}{\binom{k}{\ell}} .$$

This was generalized by Pippenger and Spencer [PiSp89] to general hypergraphs with small *co-degree*.

EXAMPLES

E12: Johansson [Jo96] used the nibble to show that the chromatic number of a triangle free graph is $O(\frac{\Delta}{\ln \Delta})$. The main idea is to randomly color a small fraction of the vertex set, update the lists of colors available at each vertex and repeat. The proof is complicated by the need to choose colors non-uniformly. Also, one needs to use the local lemma to show that with positive probability the vertex coloring has some regularity properties.

E13: Kim [Ki95] used the nibble to show that $R(3, t) = \Omega(\frac{t^2}{\ln t})$ where $R(3, t)$ is the minimum n such that every Red-Blue coloring of the complete graph K_n contains either a Red triangle or a Blue copy of K_t. This coincides with the upper bound of $O(\frac{t^2}{\ln t})$ proved earlier by Ajtai, Komlós and Szemerédi [AjKoSz80].

E14: Kahn [Ka96] used the nibble to prove that the *list chromatic index* of a graph G is $\Delta + o(\Delta)$. Here we properly color the edges of a graph G, using lists of colors for each edge.

E15: Molloy and Reed [MoRe98b] used the nibble to show that the *total chromatic number* of a graph is at most $\Delta + O(1)$. The total chromatic number is the minimum number of colors needed to color the edges *and* vertices of a graph so that no edge or vertex is incident/adjacent to an edge/vertex of the same color.

8.4.5   Bounds on Tails of Distributions

The probabilistic method often deals with events of low probability and has to use estimates for the probability of a large deviation of some random variable. The following two inequalities are widely used in probabilistic combinatorics.

FACTS

F4:   (Corollary to the Azuma–Hoeffding inequality — e.g., see [AlSp00]): Let $Z = Z(Y_1, Y_2, \ldots, Y_m)$ be a random variable, with $Y_1, Y_2, \ldots, Y_m$ independent. Suppose also that changing the value of one variable Y_i only changes the value of Z by at most one. Then for any $t > 0$ we have

$$\Pr(|Z - \mathbf{E}(Z)| \geq t) \leq 2 \exp\left\{-\frac{t^2}{2m}\right\}. \tag{8.4.4}$$

F5:   Suppose that we choose a random subset S of some set X, such that each $x \in X$ is chosen independently with probability p_x. For a collection $A_1, A_2, \ldots, A_m$ of subsets of X, we want an estimate of the probability Π that S does not contain any of the A_i. Janson [Ja90] proved an upper bound on Π which is the meat of the inequality. The lower bound

$$\Pi \geq \prod_{i=1}^{m} (1 - \Pr(A_i \subseteq X)) \tag{8.4.5}$$

follows directly from the FKG inequality of Fortuin, Kasteleyn and Ginibre [FoKaGi71].

EXAMPLES

E16:   Inequality (8.4.4) was used by Frieze, Gould, Karoński, and Pfender [FrGoKaPf02] in their paper on *graph irregularity strength*. Suppose that we weight the edges of a graph G with integers from $\{1, 2, \ldots, m\}$. The weight of a vertex is the weight of all its incident edges. A weighting is proper if every vertex has a different weight. The strength $\sigma(G)$ of a graph G is the minimum m for which a proper weighting exists. One result from [FrGoKaPf02] is that if G is r-regular and $r \leq (n/\ln n)^{1/4}$ then $s(G) \leq 1 + 10n/r$. Part of the proof involves randomly weighting each edge with a one or a two and then using (8.4.4) to bound the probability that some vertex weighting is repeated much more than its expected number of times.

E17:   Inequality (8.4.5) was used in [BoFrMaRuSm] to give a simple proof of the following result. Suppose we have k-bounded *proper* coloring of the edges of K_m and $m > 2k^{1/2}n^{3/2}$. Then there must be a multi-colored copy of K_n. We simply choose a random set, where each vertex of K_m is chosen with probability $p = 2n/m$. Then we use (8.4.5) to bound the probability that we do not choose two edges from the same color class.

REMARKS

R6:   Sometimes a related inequality due to Talagrand [Ta96] can be used in place of Inequality (8.4.4).

R7:   The interested reader can learn more about this subject and the related subject of random graphs from [AlSp00], [MoRe02], [Bo01], and [JaLuRu].

8.4.6 Dependent Random Choice

The most straightforward way to select a subset U of vertices in a graph is to independently accept each vertex at random with the same probability. Sometimes, one wishes to use U as a base for finding a desired substructure in the graph. In applications related to subgraph embedding, it can be very beneficial to find a set U for which all (or at least, many) of its r-subsets have many common neighbors, for some r of interest. It turns out that such "rich" sets can actually be constructed randomly, although not via independent choices. This robust technique, called *dependent random choice*, has now seen numerous applications, many of which are outlined in the recent survey of Fox and Sudakov [FoSu11].

DEFINITION

D6: In this section, the common neighborhood $N(S)$ of a collection of vertices S is the set of all vertices which are simultaneously adjacent to every single vertex of S.

The following result is the key lemma that appears in, or serves as a prototype for applications of dependent random choice.

FACT

F6: (As formulated in [FoSu11].) For every n, d, s, and k, every n-vertex graph with average degree d contains a subset U of at least

$$\max \left\{ \frac{d^t}{n^{t-1}} - \binom{n}{s} \left(\frac{k}{n} \right)^t : t \in \mathbb{Z}^+ \right\}$$

vertices, such that every subset $S \subset U$ of size s has $|N(S)| \geq k$.

REMARK

R8: The proof of this main technical lemma is short and flexible, and uses the alterations method introduced earlier in this chapter. Indeed, one first selects an auxiliary subset T of vertices by independently picking exactly t uniformly random vertices, with replacement, for a certain optimal t. Then, one takes $N(T)$, and removes one vertex from every problematic s-subset, producing U as the result.

EXAMPLES

E18: The Turán number of a graph F, denoted $ex(n, F)$, is the maximum number of edges in any n-vertex graph with no subgraph isomorphic to F. Erdős, Stone, and Simonovits [ErSi66, ErSt46] determined that for every fixed graph F, $ex(n, F) = \left(1 - \frac{1}{\chi(F)-1} \right) \frac{n^2}{2} + o(n^2)$, where $\chi(F)$ is the chromatic number of F, resolving this asymptotically for non-bipartite F. Erdős conjectured [Er67] that for bipartite graphs F with degeneracy r, $ex(n, F) = O(n^{2-1/r})$. Alon, Krivelevich, and Sudakov [AlKrSu03] discovered a short proof using dependent random choice that $ex(n, F) = O(n^{2-1/\Delta})$, where Δ is the maximum degree of F. In the same paper, they extended the technique to prove that $ex(n, F) = O(n^{2-1/(8r)})$.

E19: In additive combinatorics, one often considers the *sumset* $A + A$ of a subset of integers with itself, which is defined to be the collection of all possible sums of two not necessarily distinct elements of A. The sumset can have size which is linear in the size of A when, for example, A is an arithmetic progression, and much research has investigated the relationship between small sumsets and progression-like behavior. See, e.g., the book [TaVu] by Tao and Vu. The Balog–Szemerédi–Gowers theorem (see [BaSz94, Go98, SuSzVu05]) considers partial sumsets. Given two sets A and B of, say, n integers each, and a $2n$-vertex bipartite graph G with one part corresponding to A and the other corresponding to B, one can define $A +_G B$ to be the set of all sums $a + b$, where $a \in A$ and $b \in B$ form an edge in G. (The full sumset $A + B$ corresponds to the case when G is the complete bipartite graph.) The result states that for any δ and C, there are ϵ and C' (independent of n) such that whenever A and B are two n-sets, G has at least δn^2 edges, and $|A +_G B| \leq Cn$; then there exist $A' \subset A$ and $B' \subset B$ with $|A'|, |B'| \geq \epsilon n$, and the full sumset $|A' + B'| \leq C'n$. The original proof by Balog and Szemerédi produced tower-type dependencies, due to its use of the Regularity Lemma. The approach of Gowers used the philosophy of dependent random choice to establish the technical core of the proof, which was a purely graph theoretic statement.

E20: A graph homomorphism from H to G is a not necessarily injective map from the vertices of H to the vertices of G, such that whenever uv is an edge of H, the images of u and v are adjacent in G. The homomorphism density $t_H(G)$ is the probability that a uniformly random map from $V(H)$ to $V(G)$ is a homomorphism. Erdős and Simonovits [Si84] and Sidorenko [Si93] conjectured that for every m-edge bipartite graph H and any graph G,

$$t_H(G) \geq t_{K_2}(G)^m .$$

Since $t_{K_2}(G)$ is essentially the edge density of G, this conjecture can be interpreted as saying that the Erdős–Rényi random graph minimizes the number of copies of H, given a fixed number of vertices and edges. Recently, Conlon, Fox, and Sudakov [CoF0Su10] used dependent random choice to prove this conjecture for the case when H has a vertex that is adjacent to every vertex in the other part of the bipartition. From this, an approximate version of the conjecture follows for all H, in which the exponent m on the right hand side can, for example, be replaced by $m + n$, where n is the number of vertices in H.

References

[AjKoSz80] M. Ajtai, J. Komlós, and E. Szemerédi, A note on Ramsey numbers, *Journal of Combinatorial Theory A* (1980), 354–360.

[AlFrRe95] M. Albert, A. M. Frieze, and B. A. Reed, Multicoloured Hamilton cycles, *Electronic Journal of Combinatorics* 2 (1995), R10.

[Al91] N. Alon, A parallel algorithmic version of the local lemma, *Random Structures and Algorithms* 2 (1991), 367–379.

[AlSp00] N. Alon and J. Spencer, *The Probabilistic Method*, Second Edition, Wiley-Interscience, 2000.

[AlKrSu03] N. Alon, M. Krivelevich, and B. Sudakov, Turán numbers of bipartite graphs and related Ramsey-type questions, *Combin. Probab. Comput.* 12 (2003), 477–494.

[BaSz94] A. Balog and E. Szemerédi, A statistical theorem of set addition, *Combinatorica* 14 (1994), 263–268.

[Be78] J. Beck, On 3-chromatic hypergraphs, *Discrete Mathematics* 24 (1978), 127–137.

[Be91] J. Beck, An algorithmic approach to the Lovász local lemma I, *Random Structures and Algorithms* 2 (1991), 343–365.

[BoFrMaRuSm] T. Bohman, A. M. Frieze, R. Martin, M. Ruszinko, and C. Smyth, Polychromatic cliques and related questions.
http:www.math.cmu.edu/ af1p/Texfiles/AntiRamsey.pdf.

[Bo01] B. Bollobás, *Random Graphs*, second edition, Cambridge University Press, 2001.

[CoFoSu10] D. Conlon, J. Fox, and B. Sudakov, An approximate version of Sidorenko's conjecture, *Geometric and Functional Analysis* 20 (2010), 1354–1366.

[CzSc00] A. Czumaj and C. Scheideler, Coloring nonuniform hypergraphs: A new algorithmic approach to the General Lovász Local Lemma, *Random Structures and Algorithms* 17 (2000), 213–237.

[Er47] P. Erdős, Some remarks on the theory of graphs, *Bulletin of the American Mathematical Society* 53 (1947), 292–294.

[Er59] P. Erdős, Graph theory and probability, *Canadian Journal of Mathematics* 11 (1959), 34–38.

[Er63] P. Erdős, On a combinatorial problem, *Nordisk Mat. Tidskr.* 11 (1963), 5–10.

[Er67] P. Erdős, Some recent results on extremal problems in graph theory, in: *Theory of Graphs (Rome, 1966)*, Gordon and Breach, New York, 1967, 117–123.

[ErLo75] P. Erdős and L. Lovász, Problems and results on 3-chromatic hypergraphs and some related questions, pp. 609–627 in A. Hajnal, R. Rado, and V. T. Sós, eds., *Infinite and Finite Sets (to Paul Erdős on his 60th birthday) II.*, North-Holland (1975).

[ErSi66] P. Erdős and M. Simonovits, A limit theorem in graph theory, *Studia Sci. Math. Hungar.* 1 (1966), 51–57.

[ErSp91] P. Erdős and J. Spencer, Lopsided local lemma and Latin transversals, *Discrete Applied Mathematics* 30 (1991), 151–154.

[ErSt46] P. Erdős and A. Stone, On the structure of linear graphs, *Bull. Amer. Math. Soc.* 52 (1946), 1087–1091.

[FoKaGi71] C. Fortuin, P. Kasteleyn, and J. Ginibre, Correlation inequalities in some partially ordered sets, *Communications of Mathematical Physics* 22 (1971), 89–103.

[FoSu11] J. Fox and B. Sudakov, Dependent random choice, *Random Structures and Algorithms* 38 (2011), 68–99.

[FrGoKaPf02] A. M. Frieze, R. Gould, M. Karoński, and F. Pfender, On graph irregularity strength, *Journal of Graph Theory* 41 (2002), 120–137.

[Go98] W. T. Gowers, A new proof of Szemerédis theorem for arithmetic progressions of length four, *Geom. Funct. Anal.* 8 (1998), 529–551.

[GrRoSp90] R. Graham, B. Rothschild, and J. Spencer, *Ramsey Theory*, Wiley, second edition (1990).

[Ja90] S. Janson, Poisson approximation for large deviations, *Random Structures and Algorithms*, 1 (1990), 221–230.

[JaLuRu] S. Janson, T. Łuczak and A. Ruciński, *Random Graphs*, John Wiley and Sons, (2000).

[Jo96] A. Johansson, Asymptotic choice number for triangle free graphs, DIMACS Technical report (1996).

[Ka96] J. Kahn, Asymptotically good list colorings, *Journal of Combinatorial Theory: Series A* 73 (1996), 1–59.

[Ki95] J. Kim, The Ramsey number $R(3,t)$ has order of magnitude $t^2/\log t$, *Random Structures and Algorithms* 7 (1995), 173–207.

[MoRe02] M. Molloy and B. A. Reed, *Graph colouring and the probabilistic method*, Springer, 2002.

[MoRe98a] M. Molloy and B. A. Reed, Further algorithmic aspects of the local lemma, *Proceedings of the 30th Annual ACM Symposium on Theory of Computing* (1998), 524–529.

[MoRe98b] M. Molloy and B. A. Reed, A bound on the total chromatic number, *Combinatorica* 18 (1998), 241–280.

[Mo79] J. Moon, *Topics in Tournaments*, Holt, Reinhart and Winston, 1979.

[Mo09] R. Moser, A constructive proof of the Lovász local lemma, in *Proceedings of the 41st Annual ACM Symposium on Theory of Computing (STOC '09)*, 343–350.

[MoTa10] R. Moser and G. Tardos, A constructive proof of the general Lovász Local Lemma, *Journal of the ACM* 57 (2010), (2) Art. 11, 15 pp.

[NeRo90] J. Nešetřil and V. Rödl, *Mathematics of Ramsey Theory*, Springer-Verlag, 1990.

[PiSp89] N. Pippenger and J. Spencer, Asymptotic behaviour of the chromatic index for hypergraphs, *Journal of Combinatorial Theory, Series A* 51 (1989), 24–42.

[RaSr00] J. Radhakrishnan and A. Srinivasan, Improved bounds and algorithms for hypergraph two-colouring, *Random Structures and Algorithms* 16 (2000), 4–32.

[Ro85] V. Rödl, On a packing and covering problem, *European Journal of Combinatorics* 6 (1985), 69–78.

[Sa04] M. R. Salavatipour, A $(1+\epsilon)$-approximation algorithm for partitioning hypergraphs using a new algorithmic version of the Lovasz local lemma, *Random Structures and Algorithms* 25(1) (2004), 68–90.

[Si93] A. F. Sidorenko, A correlation inequality for bipartite graphs, *Graphs Combin.* 9 (1993), 201–204.

[Si84] M. Simonovits, Extremal graph problems, degenerate extremal problems and super-saturated graphs, in: *Progress in Graph Theory (Waterloo, Ont., 1982)*, Academic Press, Toronto, ON (1984), 419–437.

[Sp94] J. Spencer, *Ten Lectures on the Probabilistic Method*, SIAM Publications, second edition, 1994.

[Sr08] A. Srinivasan, Improved algorithmic versions of the Lovász Local Lemma, in *Proceedings of the Nineteenth Annual ACM-SIAM Symposium on Discrete Algorithms (SODA)*, San Francisco, California, 611–620, 2008.

[SuSzVu05] B. Sudakov, E. Szemerédi, and V. Vu, On a question of Erdős and Moser, *Duke Mathematical Journal* 129 (2005), 129–155.

[Ta96] M. Talagrand, Concentration of measures and isoperimetric inequalities in product spaces, *Publications Mathematiques de l'I.H.E.S.* 81 (1996), 73–205.

[TaVu] T. Tao and V. Vu, *Additive Combinatorics*, Cambridge University Press, 2006.

Section 8.5

Graph Limits

Bojan Mohar, Simon Fraser University, Canada, and IMFM, Slovenia

MOTIVATION

When taking limits of rational numbers, we find out the realm of all real numbers. They provide better understanding of the variety of rational numbers and turn out to be a natural notion which enables us to simplify many arguments and develop more sophisticated mathematical tools.

Discrete mathematics is traditionally about all but taking limits. However, Lovász et al. [BCL⁺08, BCL10, BCL⁺06, LS06] introduced a natural notion and developed a powerful theory of graph limits. In this theory, finite graphs are viewed as elements of a certain metric space (see Section 8.5.4). The completion of this metric space, obtained by adding all limits of "convergent" graph sequences, provides a similar boost as the real numbers provide for the rationals. What we obtain is a complete metric space and the limits themselves can be viewed as generalizations of graphs; thus they are called **graphons**. Their main use is in understanding, exploring, and manipulating very large graphs. Recent advances in mathematics, computer science, bioinformatics, life sciences and social sciences show the need to analyze large combinatorial objects, graphs being the most important among these; see [Lov09]. Emergence of the probabilistic method [AS08] and success of Szemerédi's Regularity Lemma (cf. Section 8.5.5) boosted interest in (very) large graphs as well.

The theory of graph limits emerged from three seemingly unrelated areas:

- Random graphs (see Section 8.2 or [AS08, Bol01]) and quasirandom graphs [CGW89].
- Szemerédi's Regularity Lemma [Sze78] (see also [KS96, KSSS02, SS91, Tao06]) and its algorithmic version [ADL⁺94, FK99b, FK99a].
- Computational complexity ([AFNS06, LS10, BCL⁺06]).

Proper understanding of these areas is fundamental for understanding of graph limits and their applications.

8.5.1   Graphs, Weighted Graphs, and Graphons

Graphons are generalizations of finite graphs. These objects can be used to describe large dense graphs and their limits in a similar manner as the adjacency matrices are used to represent finite graphs.

DEFINITIONS

D1:   A *graphon* is a symmetric measurable function $W : [0,1] \times [0,1] \to \mathbb{R}$. By *measurable function* we refer to the Lebesgue measure on the unit interval $[0,1]$, and by *symmetric function* we mean that $W(x,y) = W(y,x)$ for every $x, y \in [0,1]$. The graphon W is a *simple graphon* if $0 \le W(x,y) \le 1$ for every $x, y \in [0,1]$.

D2:   A map $\phi : [0,1] \to [0,1]$ is *measure-preserving* if for every measurable set $A \subseteq [0,1]$, the preimage $\phi^{-1}(A)$ is measurable and has the same measure as A.

D3:   Two graphons W and U are *isomorphic graphons* if there exist measure-preserving maps $\phi : [0,1] \to [0,1]$ and $\psi : [0,1] \to [0,1]$ such that $W(\phi(x), \phi(y)) = U(\psi(x), \psi(y))$ for every $x, y \in [0,1]$. The pair (ϕ, ψ) is said to be an *isomorphism* between graphons W and U.

D4: If G is a graph with vertices $v_1, \ldots, v_n$, then we define the *graphon corresponding to the graph* G as the graphon W_G defined as follows. If $x \in (0,1]$, we define $i = \lceil nx \rceil$ and set $u_x = v_i$ to be the ith vertex of G. For $x = 0$ we set $u_x = v_1$. Then we define

$$W_G(x,y) = \begin{cases} 1, & \text{if } u_x \text{ and } u_y \text{ are adjacent;} \\ 0, & \text{otherwise.} \end{cases}$$

D5: A *weighted graph* is a graph G together with *vertex-weights* $\alpha_v = \alpha_v(G) \in \mathbb{R}^+$ $(v \in V(G))$ and *edge-weights* $\beta_e = \beta_e(G)$ $(e \in E(G))$. The *total weight of the weighted graph* G is $\alpha_G = \sum_{v \in V(G)} \alpha_v(G)$.

NOTATION: If $S \subset V(G)$, then we write $\alpha(S) = \sum_{v \in S} \alpha_v(G)$ and $\alpha(G) = \alpha_G = \alpha(V(G))$.

D6: Let G be a weighted graph with vertices $v_1, \ldots, v_n$, vertex weights α_v $(v \in V(G))$, and edge-weights β_e $(e \in E(G))$. Set $z_0 = 0$ and define $z_i = z_{i-1} + \alpha_{v_i}/\alpha(G)$ for $i = 1, \ldots, n$. Finally, define the *graphon W_G corresponding to the weighted graph* G as follows. If $x, y \neq 0$ and $z_{i-1} < x \le z_i$ and $z_{j-1} < y \le z_j$, then

$$W_G(x,y) = \begin{cases} \beta_{v_i v_j}, & \text{if } v_i \text{ and } v_j \text{ are adjacent;} \\ 0, & \text{otherwise.} \end{cases}$$

If $x = 0$ or $y = 0$, we define $W_G(x,y)$ to be 0.

D7:   If G and H are weighted graphs, a mapping $\phi : V(G) \to V(H)$ is an ***isomorphism of weighted graphs*** if it is bijective and for any vertices $u, v \in V(G)$, we have $\alpha_{\phi(v)}(H) = \alpha_v(G)$ and $\beta_{\phi(u)\phi(v)}(H) = \beta_{uv}(G)$. An ***automorphism of a weighted graph*** H is an isomorphism of H with itself. The automorphisms of H form a group acting on $V(H)$, and they partition $V(H)$ into ***orbits of this action***.

D8:   Two vertices v, v' of a weighted graph G are ***twins*** if for every $u \in V(G)$, we have $\beta_{vu}(G) = \beta_{v'u}(G)$. If v and v' are twins, let H be the weighted graph obtained from G by replacing the vertices v and v' with a new vertex w whose weight is $\alpha_w(H) = \alpha_v(G) + \alpha_{v'}(G)$ and the edge-weights $\beta_{uw}(H)$ are equal to $\beta_{uv}(G)$ for $u \in V(G) \setminus \{v, v'\}$. Other vertex and edge-weights in H are the same as in G. The graph H is said to be obtained from G by ***merging the twins*** v and v', and G is obtained from H by ***splitting the vertex*** w into vertices v and v'.

REMARKS

R1:   Similarly as for the adjacency matrix of a graph, the graphon W_G is only defined up to isomorphism of graphons as different permutations of the vertex-set give different (but isomorphic) outcomes.

R2:   The definition of the isomorphism of graphons involves two measure-preserving maps, ϕ and ψ. Both of them are needed since a measure-preserving function does not always have a measure-preserving inverse. (An example of such a function $\phi : [0, 1] \to [0, 1]$ is given by the rule $\phi(x) = |1 - 2x|$, which is a measure-preserving map and is 2-1 almost everywhere.)

R3:   It makes sense to extend the definition of edge-weights to all pairs of vertices by setting $\beta_{uv}(G) = 0$ if $uv \notin E(G)$. This convention will be used throughout the whole section.

FACTS

F1:   For a (simple or weighted) graph G and a positive integer k, let $G[k]$ be the graph obtained from G by replacing each vertex by k twin vertices and replacing each edge uv by the complete bipartite graph $K_{k,k}$ joining the copies of v with copies of u. Then both graphs have isomorphic graphons, $W_G = W_{G[k]}$.

F2:   A weighted graph and any weighted graph obtained from it by merging twins or splitting vertices into twins have isomorphic graphons.

F3:   If vertex and edge-weights of a weighted graph G are positive integers, then there is an unweighted graph H such that $W_G = W_H$. The graph H can be obtained from G by replacing each vertex v with α_v twin vertices and replacing each edge e with β_e multiple edges joining the same pair of vertices.

8.5.2 Homomorphism Density

DEFINITIONS

D9: If F and G are simple graphs, then we denote by $\hom(F, G)$ the **number of homomorphisms** $F \to G$, i.e., the number of maps $\phi : V(F) \to V(G)$ such that for every edge $uv \in E(F)$, $\phi(u)\phi(v)$ is an edge in G.

D10: The **homomorphism density** for two simple graphs F and G, denoted by $t(F, G)$, is the normalized value of $\hom(F, G)$,

$$t(F, G) = \frac{\hom(F, G)}{|G|^{|F|}}.$$

D11: *Homomorphism density between weighted graphs:* Let $\phi : V(F) \to V(G)$ be a mapping between weighted graphs F and G. Let us set

$$\alpha_\phi = \prod_{v \in V(F)} (\alpha_{\phi(v)}(G))^{\alpha_v(F)} \qquad \text{and} \qquad \beta_\phi = \prod_{uv \in E(F)} (\beta_{\phi(u)\phi(v)}(G))^{\beta_{uv}(F)}$$

and then define

$$\hom(F, G) = \sum_{\phi : V(F) \to V(G)} \alpha_\phi \, \beta_\phi$$

and

$$t(F, G) = \frac{\hom(F, G)}{\alpha(G)^{\alpha(F)}} \tag{8.5.1}$$

where $\alpha(G)$ and $\alpha(H)$ denote the total weight of vertices of the graphs.

REMARKS

R4: The homomorphism density tells us what fraction of all maps $V(F) \to V(G)$ are homomorphisms; it can be viewed as the probability that a randomly chosen map $V(F) \to V(G)$ is a homomorphism.

R5: If $\phi : V(F) \to V(G)$ is a mapping between weighted graphs F and G, then the number of ways twins of v can be mapped to different twins of $\phi(v)$ is equal to $(\alpha_{\phi(v)}(G))^{\alpha_v(F)}$. This explains the definition of α_ϕ in Definition D11.

R6: If F and G are multigraphs (multiple edges and loops allowed), then homomorphisms also specify which edges joining u and v are mapped to which edges between $\phi(u)$ and $\phi(v)$. If $\beta_{uv}(F)$ is the multiplicity of the edge uv in F and $\beta_{\phi(u)\phi(v)}(G)$ is the edge multiplicity in G, then the number of ways to map these edges onto each other is $(\beta_{\phi(u)\phi(v)}(G))^{\beta_{uv}(F)}$ if $\phi(u)$ and $\phi(v)$ are fixed. This explains the definition of β_ϕ in Definition D11.

EXAMPLES

E1: $\hom(G, K_k)$ is the number of k-colorings of the graph G.

E2: $\hom(K_k, G)/k!$ is the number of k-cliques in G.

E3: If B is the graph with two adjacent vertices and a loop at one of them, then each homomorphism of a graph G into B is determined by the set of vertices that are mapped to the vertex without the loop. The set of these vertices is independent in G, and any independent set can arise in this way. Thus, $\hom(G, B)$ is equal to the number of independent sets in G.

FACTS

F4: Let H be a simple graph. Then the homomorphism densities into H satisfy the following submodularity property: If F and F' are simple graphs on the same vertex-set, then

$$t(F \cup F', H) + t(F \cap F', H) \geq t(F, H) + t(F', H).$$

F5: In the setting of weighted graphs, the densities and homomorphisms can be treated in the same way. Namely, if $\hat{G}$ is the weighted graph obtained from G by dividing all vertex weights by $\alpha(G)$, then $\alpha(\hat{G}) = 1$ and thus

$$t(F, G) = t(F, \hat{G}) = \hom(F, \hat{G}).$$

F6: Let F be a simple graph with vertex set $\{1, 2, \ldots, k\}$. The homomorphism density $t(F, G)$ of F into a simple graph G can be expressed as follows:

$$t(F, G) = \underbrace{\int_0^1 \int_0^1 \cdots \int_0^1}_{k} \prod_{ij \in E(F)} W_G(x_i, x_j)\, dx_1 dx_2 \ldots dx_k.$$

NOTATION: One can use the following shorter notation for multiple integrals over the k-dimensional cube $[0, 1]^k$ like the one appearing in Fact F6:

$$\iiint_{[0,1]^k} f(X)\, dX = \int_0^1 \int_0^1 \cdots \int_0^1 f(x_1, \ldots, x_k)\, dx_1 dx_2 \ldots dx_k.$$

DEFINITIONS

D12: Let F be a simple graph with vertex set $\{1, 2, \ldots, k\}$ and let W be a graphon. The **homomorphism density** for F and W is the number

$$t(F, W) = \iiint_{[0,1]^k} \prod_{ij \in E(F)} W(x_i, x_j)\, dX.$$

FACTS

F7: $t(K_1, W) = 1$ (we say that the homomorphism density is **normalized**).

F8: Homomorphism densities are **multiplicative**: For arbitrary disjoint simple graphs F, F' and every graphon W, we have

$$t(F \cup F', W) = t(F, W)\, t(F', W).$$

F9: If G is a (simple or weighted) graph, then $t(F, G) = t(F, W_G)$, where W_G is the graphon corresponding to G.

F10: If two graphons W and W' are isomorphic, then they have the same homomorphism densities, i.e., for every simple graph F, we have $t(F, W) = t(F, W')$.

8.5.3   Convergent Sequences of Graphs and Graphons

Graphons can be considered as limiting objects of sequences of finite graphs. In this section we reveal this relationship.

NOTATION: We write (X_n) or $(X_n)_{n \geq 1}$ for a sequence of objects, $X_1, X_2, X_3, \ldots$.

DEFINITIONS

D13: Let (G_n) be a sequence of (simple or weighted) graphs. We say that the sequence is **convergent** if for every simple graph F, the sequence of homomorphism densities $t(F, G_n)$ converges. If there exists a graphon W such that $\lim_{n \to \infty} t(F, G_n) = t(F, W)$ for every simple graph F, then we say that W is the **limit of the convergent graph sequence** (G_n), and we write $W = \lim_{n \to \infty} G_n$.

D14: A sequence (W_n) of graphons is **convergent** if for every simple graph F, $\lim_{n \to \infty} t(F, W_n)$ exists. A graphon W is the **limit of the convergent sequence** (W_n) **of graphons** if $\lim_{n \to \infty} t(F, G_n) = t(F, W)$ for every F; in this case we write $W = \lim_{n \to \infty} W_n$.

FACTS

F11: Every simple graphon is a limit of a convergent sequence of simple graphs.

F12: Every graphon is a limit of a convergent sequence of weighted graphs.

F13: If (G_n) is a convergent sequence of (simple or weighted) graphs, then there exists a graphon W that is the limit of the sequence, $W = \lim_{n \to \infty} G_n$.

F14: If (W_n) is a convergent sequence of graphons, then there exists a graphon W that is the limit of the sequence, $W = \lim_{n \to \infty} W_n$.

F15: The limit of a convergent sequence of graphs or graphons is determined uniquely up to isomorphisms of graphons.

8.5.4 Metric Space Topology on Graphs and Graphons

The set of graphs and the larger set of graphons can be viewed as a metric space whose topology gives the same convergent sequences as defined via homomorphism densities. This setup enables one to use analytic techniques when studying large graphs. The space of graphons turns out to be a complete metric space in the sense of Cauchy convergence.

DEFINITIONS (METRIC SPACE)

D15: A *metric on a set* X is a function $d : X \times X \to \mathbb{R}$ satisfying the following conditions:

(a) $d(x, y) = d(y, x)$ for every $x, y \in X$.

(b) $d(x, y) \geq 0$ for every $x, y \in X$.

(c) $d(x, y) = 0$ if and only if $x = y$.

(d) $d(x, y) \leq d(x, z) + d(z, y)$ for every $x, y, z \in X$.

D16: A *metric space* is a pair (X, d), where X is a set and d is a metric on X.

D17: A sequence (x_n) of elements of X is *convergent* in the metric space (X, d) if there exists $x \in X$ such that $d(x_n, x) \to 0$ as $n \to \infty$. In such a case we say that x is the *limit of the sequence* (x_n).

D18: A sequence (x_n) of elements of X is a *Cauchy sequence* in the metric space (X, d) if for every $\varepsilon > 0$ there exists an integer n_0 such that for every $m, n \geq n_0$, $d(x_n, x_m) < \varepsilon$.

D19: A metric space (X, d) is *complete* if every Cauchy sequence is convergent.

D20: Let (X, d) be a metric space. If (x_n) and (y_n) are Cauchy sequences of elements of X, then we define $\overline{d}((x_n), (y_n)) = \lim_{n \to \infty} d(x_n, y_n)$. The two Cauchy sequences (x_n) and (y_n) are said to be *equivalent sequences* if $\overline{d}((x_n), (y_n)) = 0$. This induces a metric $\overline{d}$ on the set $\overline{X}$ of equivalence classes of all Cauchy sequences. The resulting metric space $(\overline{X}, \overline{d})$ is called the *completion of the metric space* (X, d). It turns out that every metric space has a completion and that the completion is uniquely defined as the smallest complete metric space that contains (X, d) as a metric subspace. Note that every $x \in X$ can be identified with the equivalence class of the constant sequence $(x, x, x, \dots)$.

DEFINITIONS (CUT DISTANCE)

D21: *Cut distance of labeled graphs*: If G and H are two graphs of order n on the same (labeled) vertex set $V = V(G) = V(H)$, then we define the *cut distance* of G and H as

$$d_\square(G, H) = \frac{1}{n^2} \max_{S, T \subseteq V} \left| e_G(S, T) - e_H(S, T) \right|,$$

where $e_G(S,T)$ (and similarly $e_H(S,T)$) denotes the number of edges between S and T with edges whose both ends are in $S \cap T$ counted twice, i.e.,

$$e_G(S,T) = \big|\{(s,t) \mid s \in S, t \in T, st \in E(G)\}\big|.$$

D22: *Cut distance of weighted labeled graphs*: The definition of the cut distance extends to weighted graphs on the same set of vertices if the vertex weights are the same in both graphs. In that case, the normalization factor $1/n^2$ is replaced by $1/\alpha(G)^2$ (where $\alpha(G)$ is the total vertex weight of G) and define

$$e_G(S,T) = \sum_{s \in S} \sum_{t \in T} \alpha_s(G)\alpha_t(G)\beta_{st}(G).$$

The edge count $e_H(S,T)$ in H is defined in the same way. Then we set $\alpha = \alpha(G) = \alpha(H)$ and define

$$d_\square(G,H) = \frac{1}{\alpha^2} \max_{S,T \subseteq V} \big|e_G(S,T) - e_H(S,T)\big|.$$

D23: *Cut distance of unlabeled graphs*: If two graphs, G and H, have the same number of vertices that are not labeled, then one can consider all possible labelings and select those whose cut distance is the smallest:

$$\hat{\delta}_\square(G,H) = \min_{\hat{G},\hat{H}} d_\square(\hat{G},\hat{H})$$

where the minimum runs over all labelings $\hat{G}, \hat{H}$ of graphs G and H.

Fact F16 motivates the way to define the cut distance also for graphs whose numbers of vertices are different: we first blow up both graphs so that the resulting graphs have the same number of vertices, and then use Definition D23. To deal also with the weighted case and to involve a corresponding generalization of relabelings, the notion of a fractional bijection from $V(G)$ to $V(H)$ is needed.

D24: Let G and G' be weighted graphs with ***normalized vertex-weights***, i.e., $\alpha(G) = \alpha(G') = 1$. Let $V = V(G)$ and $V' = V(G')$. A function $X : V \times V' \to [0,1]$ is a ***fractional bijection*** $V \to V'$ if

$$\sum_{v' \in V'} X(v,v') = \alpha_v(G) \qquad \text{for every } v \in V, \text{ and}$$

$$\sum_{v \in V} X(v,v') = \alpha_{v'}(G') \qquad \text{for every } v' \in V'.$$

The ***transpose*** $X^T : V' \times V \to [0,1]$ of a fractional bijection X, defined by $X^T(v',v) = X(v,v')$, is a fractional bijection $V' \to V$.

D25: Having a fractional bijection X from $V(G)$ to $V(G')$, we define a weighted graph $G[X]$ on the vertex set $V \times V'$. The vertex-weights of $G[X]$ are $\alpha_{(v,v')}(G[X]) = X(v,v')$ and the edge-weights are $\beta_{(v,v')(u,u')}(G[X]) = \beta_{vu}(G)$. One can view this as splitting each vertex $v \in V$ into vertices vv' ($v' \in V'$). Similarly we define $G'[X^T]$. There is obvious bijection between their vertex sets, so we can define their cut distance $d_\square(G[X], G'[X^T])$ by Definition D22.

D26: **_Cut distance of arbitrary unlabeled weighted graphs_**: The **_cut distance_** $\delta_\square(G, G')$ is defined as

$$\delta_\square(G, G') \quad = \quad \inf_X d_\square(G[X], G'[X^T])$$

$$= \quad \inf_X \max_{S,T \subseteq V \times V'} \left| \sum_{vv' \in S} \sum_{uu' \in T} X(v, v') X(u, u') \big(\beta_{vu}(G) - \beta_{v'u'}(G') \big) \right|,$$

where the infimum is taken over all fractional bijections $X \in [0, 1]^{V \times V'}$.

D27: The **_cut norm of a graphon_** W is defined as:

$$\|W\|_\square = \sup_{A,B \subseteq [0,1]} \left| \int_A \int_B W(x, y) \, dx \, dy \right|,$$

where the supremum runs over all measurable subsets A, B of the unit interval.

D28: **_Cut distance of graphons_**: For graphons U and W, we define $d_\square(U, W) = \|U - W\|_\square$ and $\delta_\square(U, W) = \inf d_\square(\hat{U}, \hat{W})$, where the infimum is taken over all graphons $\hat{W}$ and over all graphons $\hat{U}$ that are isomorphic to W and U, respectively.

FACTS

F16: For a graph G and a positive integer k, let $G[k]$ be the graph obtained from G by replacing each vertex by k twin vertices (or equivalently, multiplying all vertex weights by k) and replacing each edge uv by the complete bipartite graph $K_{k,k}$ joining the copies of v with copies of u. Then $d_\square(G, G[k]) = 0$.

F17: For graphons U and W, we have $\delta_\square(U, W) = \inf d_\square(U, \hat{W}) = \inf d_\square(\hat{U}, W)$, where the infima are taken over all graphons $\hat{W}$ and over all graphons $\hat{U}$, respectively, that are isomorphic to W and U, respectively.

F18: If G and H are (simple or weighted) graphs, then their cut distance is the same as the cut distance between the corresponding graphons, $\delta_\square(G, H) = \delta_\square(W_G, W_H)$.

F19: The cut distance $\delta_\square(U, W)$ defines a metric on the space $\mathcal{W}$ of all graphons. In this metric, $(\mathcal{W}, \delta_\square)$ becomes a complete metric space.

F20: A sequence (G_n) of graphs (respectively, a sequence (W_n) of graphons) is convergent if and only if it is a Cauchy sequence in the metric space $(\mathcal{W}, \delta_\square)$, i.e., whenever $n, m \to \infty$, we have $\delta_\square(G_n, G_m) \to 0$ (respectively, $\delta_\square(W_n, W_m) \to 0$).

F21: The cut norm of a graphon can be expressed as

$$\|W\|_\square = \sup_{0 \leq f,g \leq 1} \left| \int_0^1 \int_0^1 W(x, y) f(x) g(y) \, dx \, dy \right|,$$

where the supremum runs over all measurable functions $f, g : [0, 1] \to [0, 1]$.

F22: For every simple graph F and arbitrary graphons U and W, we have:

$$|t(F, U) - t(F, W)| \leq |E(F)| \cdot \delta_\square(U, W).$$

8.5.5 Regularity Lemma and Approximation of Graphons by Weighted Graphs

Every graphon can be approximated by a large weighted graph. An important fact is that for a given "error of approximation", the number of vertices of such a graph is bounded, independent of the graphon to be approximated. This is made possible by applying the Szemerédi Regularity Lemma [Sze78]. This important result states that the vertices of every large enough graph can be divided into subsets of about the same size so that the edges between different subsets behave almost randomly. The Regularity Lemma has numerous applications in graph theory (see [KS96, KSSS02]). Additionally, it became an invaluable tool in number theory after it was used in the celebrated result of Green and Tao [GT08] that prime numbers contain arbitrarily long arithmetic progressions.

The basic setup for the Regularity Lemma is to partition the vertex set of a graph G into a bounded number of parts $V_1 \cup \cdots \cup V_k$ such that the edges between (almost all pairs of) different parts behave almost randomly. The formulation involves a fixed (small) positive real number $\varepsilon > 0$ and uses a notion of a random-like property given in Definition D30.

DEFINITIONS (REGULAR PARTITION)

D29: Let X, Y be disjoint sets of vertices in G. The **density of edges between X and Y** is the number

$$d(X, Y) = \frac{e_G(X, Y)}{|X|\,|Y|}$$

where $e_G(X, Y)$ denotes the number of edges of G between X and Y.

D30: Let $\varepsilon > 0$ be a real number. A pair (X, Y) of disjoint nonempty vertex-sets of G is an **ε-regular pair** if for every $X' \subseteq X$ and every $Y' \subseteq Y$ with $|X'| \geq \varepsilon|X|$ and $|Y'| \geq \varepsilon|Y|$, we have

$$|d(X', Y') - d(X, Y)| \leq \varepsilon.$$

Observe that ε plays a dual role; it is used in the lower bound on the size of vertex sets X' and Y' and is used again as the upper bound on the discrepancy between the densities.

D31: A partition $V = V_1 \cup \cdots \cup V_k$ of a set V is said to be a **partition into k parts**. Such a partition is denoted as $\mathcal{P} = \{V_1, \ldots, V_k\}$.

D32: A partition $\mathcal{P} = \{V_1, \ldots, V_k\}$ of a set V is **balanced** if $\big|\,|V_i| - |V_j|\,\big| \leq 1$ for all $i, j \in \{1, \ldots, k\}$.

D33: A partition $\mathcal{P} = \{V_1, \ldots, V_k\}$ of the vertex-set V of a graph G is **ε-regular** if

- $\mathcal{P}$ is balanced, and
- all but at most εk^2 pairs (V_i, V_j) with $1 \leq i \leq j \leq k$ are ε-regular.

D34: Given a partition $\mathcal{P} = \{V_1, \ldots, V_k\}$ of the vertex-set V of a graph G, we define a **weighted graph** $G_{\mathcal{P}}$ with vertex-set $[k] = \{1, \ldots, k\}$ as follows:

- The **weight of the vertex** $i \in [k]$ is $\alpha_i = |V_i|/|V(G)|$, and
- the **weight of the edge** ij is $\beta_{ij} = e_G(V_i, V_j)/(|V_i|\,|V_j|) = d(V_i, V_j)$.

We let $W_{\mathcal{P}}$ denote the corresponding graphon of $G_{\mathcal{P}}$. The same definition can be made if the graph G is weighted by replacing cardinalities of vertex- sets by their total weight and by defining $e_G(V_i, V_j)$ as in Definition D22.

FACTS

F23: (Regularity Lemma [Sze78]) For every $\varepsilon > 0$ and every integer L, there is an integer M such that every graph G with at least L vertices has an ε-regular partition $\mathcal{P}$ with $L \leq |\mathcal{P}| \leq M$.

F24: (Weak Regularity Lemma [FK99a]) For every $\varepsilon > 0$, every graph G has a partition $\mathcal{P}$ into at most $4^{1/\varepsilon^2}$ classes such that $d_{\square}(G, G_{\mathcal{P}}) \leq \varepsilon$.

F25: (Weak Regularity Lemma for weighted graphs [FK99a]) For every $\varepsilon > 0$, every weighted graph G has a partition $\mathcal{P}$ into at most $2^{2/\varepsilon^2}$ classes such that $d_{\square}(G, G_{\mathcal{P}}) \leq \frac{\varepsilon}{\alpha_G}\left(e_G(V, V)\right)^{1/2}$, where $V = V(G)$ and $e_G(V, V)$ is defined as in Definition D22.

F26: For every $\varepsilon > 0$, $q \geq 2^{20/\varepsilon^2}$ and every weighted graph G with edge-weights in $[0, 1]$, there exists a simple graph H of order q such that $\delta_{\square}(G, H) \leq \varepsilon$.

REMARKS

R7: The constant $M = M(\varepsilon, L)$ involved in the Szemerédi Regularity Lemma (Fact F23) is extremely large, and hence the Weak Regularity Lemma of Frieze and Kannan (Fact F24) is more appropriate for applications like parameter testing [BCL+06].

R8: The Regularity Lemma is trivial for graphs having at most M vertices since the partition into singletons is always ε-regular.

R9: If the graph is sparse, e.g., if the number of edges is less than $\frac{1}{2}\varepsilon^3 n^2/L^2$, then every balanced partition into L parts of size $\lfloor n/L \rfloor$ is ε-regular since the densities appearing in Definition D30 are smaller than ε.

R10: Numerous applications of the Regularity Lemma are presented in [KS96, KSSS02, SS91, Tao06]. Algorithmic versions of the Regularity Lemma and the algorithmically more tractable Weak Regularity Lemma appear in [AN06, ADL+94, FK99b, FK99a]. Additional applications are considered in [AFdlVKK03, AFNS06, LS10, BCL+06].

8.5.6   *W*-Random Graphs

The notion of W-random graphs, which generalize traditional notions of random graphs (cf. Section 8.2), has a close relationship with graph limits.

DEFINITIONS

D35: If W is a graphon and $n \geq 1$ is an integer, one can define the probability space $\mathcal{G}(n, W)$ of W-***random graphs*** of order n as follows. To obtain a random element of $\mathcal{G}(n, W)$, select n points $v_1, \ldots, v_n$ from the unit interval $[0, 1]$ independently at random (with the uniform probability distribution on $[0, 1]$). Then add the edge $v_i v_j$ with probability $W(v_i, v_j)$ independently for every $1 \leq i < j \leq n$. Although it is possible that some selected points v_i are equal to each other, this occurs with probability 0, so $v_1, \ldots, v_n$ are almost surely pairwise distinct; hence almost all graphs in $\mathcal{G}(n, W)$ are of order n.

A W-random graph drawn from $\mathcal{G}(n, W)$ is denoted by $G_{n,W}$. For an arbitrary fixed simple graph F, the homomorphism density $t(F, G_{n,W})$ can be considered as a random variable.

FACTS

F27: For every simple graph F and $n \geq |F|$, the following statements hold:
   (a) $\left| \mathbb{E}[t(F, G_{n,W})] - t(F, W) \right| \leq \frac{1}{n} \binom{|F|}{2}$.
   (b) $\mathrm{Var}[t(F, G_{n,W})] \leq \frac{3}{n} |F|^2$.
   (c) If $k = |F|$ and $0 < \varepsilon < 1$, then
$$\Pr\left[\left| t(F, G_{n,W}) - t(F, W) \right| > \varepsilon \right] \leq 2 \exp\left(-\frac{\varepsilon^2}{2k^2} n\right).$$

F28: The probability that a sequence of W-random graphs $(G_{n,W})$ is convergent and has limit W is equal to 1.

F29: Every graphon W is a limit of some convergent sequence of graphs. Such a sequence can be obtained, with probability 1, by taking a sequence $(G_{n,W})$ of W-random graphs.

8.5.7   Graph Parameters and Connection Matrices

An important aspect in understanding graph limits is through the study of connection matrices of homomorphism densities. They give rise to the use of algebraic tools and have application in extremal graph theory and theoretical computer science.

DEFINITIONS (GRAPH PARAMETERS)

D36: A ***graph parameter*** is a function defined on all (finite) graphs that is invariant under graph isomorphisms. We may allow multiple edges and loops. If a parameter is defined only for simple graphs, we can extend its range to allow parallel edges by defining its value in such a way that we first replace multiple edges by single edges (and delete all loops). Such a parameter is invariant under multiplying edges, and is said to be a ***simple graph parameter***.

D37: A graph parameter p is ***multiplicative*** if $p(G \cup H) = p(G)\, p(H)$ whenever G and H are disjoint graphs.

D38: A graph parameter p is ***normalized*** if $p(K_1) = 1$.

EXAMPLES

E4: Homomorphism counting function $\hom(\cdot, H)$ into a fixed weighted graph H determines a graph parameter $\hom_H$ that is multiplicative: If $F \cup F'$ is the disjoint union of two graphs, then $\hom(F \cup F', H) = \hom(F, H) \hom(F', H)$.

E5: The density function $t_H = t(\cdot, H)$ can be interpreted as the homomorphism function into the normalized weighted graph $\hat{H}$ and is therefore also multiplicative. Moreover, the homomorphism density parameter t_H is normalized.

FACT

F30: (Lovász [Lov06]) Let G and H be weighted graphs without twin vertices and with the same total weight $\alpha(G) = \alpha(H)$. Then G and H have the same densities, i.e., $t(F, G) = t(F, H)$ for every simple graph F, if and only if they are isomorphic.

DEFINITIONS (k-LABELED GRAPHS)

D39: Let $k \geq 0$ be an integer. A k-**label on a graph** G is a sequence $L = (L_1, \ldots, L_k)$ of k distinct vertices of G. Having such a k-label, we say that the vertex L_i is the i**th labeled vertex**, or the **vertex with label** i $(1 \leq i \leq k)$.

D40: A k-**labeled graph** is a pair (G, L), where G is a graph of order at least k and L is a k-label on G. We usually omit the reference to the labeling L and refer to G itself as a k-labeled graph.

D41: Two k-labeled graphs (G, L) and (G', L') are **isomorphic** if there is a graph isomorphism $\iota : G \to G'$ that preserves the labels, i.e., $\iota(L_i) = L'_i$ for every $i = 1, \ldots, k$.

D42: Let $\mathcal{L}_k$ be the set of all k-labeled graphs up to isomorphism. For $G, H \in \mathcal{L}_k$, we define the k-**sum** $G \cdot H$ as the labeled graph obtained from the disjoint union of G and H and then, for $i = 1, \ldots, k$, identify the ith labeled vertex of G with the ith labeled vertex of H to get the ith labeled vertex of the sum.

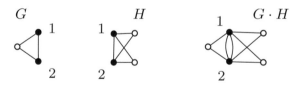

Figure 8.5.1: The 2-sum of 2-labeled graphs.

DEFINITIONS (CONNECTION MATRICES)

D43: If p is a graph parameter and $k \geq 0$, we define the k**th connection matrix** for the parameter p as the infinite matrix $M = M(p, k) \in \mathbb{R}^{\mathcal{L}_k \times \mathcal{L}_k}$ whose entries correspond to pairs of k-labeled graphs with their values being the values of the parameter on the k-sum, $M_{G,H} = p(G \cdot H)$.

EXAMPLE

E6: The 2-sum $G \cdot H$ of 2-labeled graphs G and H is illustrated in Figure 8.5.1. The black vertices are labeled as indicated. As an illustration of connection matrices, we show a part of the connection matrix $M(\chi, 2)$ for the chromatic number and $k = 2$ in Figure 8.5.2.

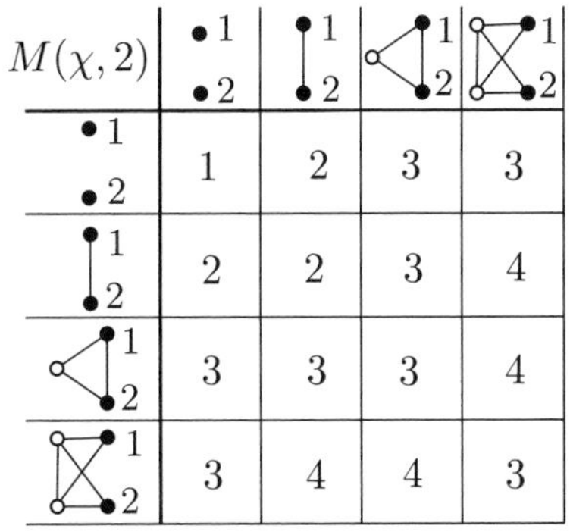

Figure 8.5.2: Part of the connection matrix $M(\chi, 2)$ for χ and $k = 2$.

NOTATION: When speaking of vectors and matrices, we will employ functional notation. Recall that B^A denotes the set of all functions $A \to B$. In this manner, a (real) *matrix indexed by a (possibly infinite) set* L is an element of $\mathbb{R}^{L \times L}$, and a *vector with entries indexed by* L is an element in $\mathbb{R}^L$. As usual, we use index notation to denote specific entries of matrices and vectors, and we apply usual notation from linear algebra to denote the matrix-matrix or matrix-vector multiplication, the transpose notation, etc.

D44: The *rank of an infinite matrix* is the supremum of ranks of its finite (principal) submatrices.

D45: An infinite matrix M is *positive semidefinite* if every finite principal submatrix of M is positive semidefinite.

D46: A graph parameter p is *reflection positive* if its connection matrices $M(p, k)$ are positive semidefinite for each $k \geq 0$.

FACTS

F31: A matrix $M \in \mathbb{R}^{\mathcal{L}_k \times \mathcal{L}_k}$ has rank at most $r < \infty$ if (and only if) there are $2r$ vectors $x_1, \ldots, x_r$ and $y_1, \ldots, y_r$ in $\mathbb{R}^{\mathcal{L}_k}$ such that

$$M = \sum_{i=1}^{r} x_i \cdot y_i^T, \tag{8.5.2}$$

where the product $F = x \cdot y^T \in \mathbb{R}^{\mathcal{L}_k \times \mathcal{L}_k}$ denotes the matrix with values $F_{G,H} = x_G y_H$.

F32: A matrix $M \in \mathbb{R}^{\mathcal{L}_k \times \mathcal{L}_k}$ is ***positive semidefinite*** if and only if it can be written in the form $M = \sum_{i \in I} x_i \cdot x_i^T$, where $x_i \in \mathbb{R}^{\mathcal{L}_k}$ for $i \in I$. If the matrix M is positive semidefinite and has finite rank r, then it is possible to choose I so that $|I| = r$.

F33: (Freedman, Lovász, and Welsh [FLW]) If $\mathrm{rank}(M(p,k))$ is finite for some $k \geq 0$, then the parameter p can be computed in polynomial time for graphs of tree-width at most k.

F34: Every graph parameter p can be considered as an element in $\mathbb{R}^{\mathcal{L}_k}$. If p is multiplicative, then $M(p,0) = p \cdot p^T$, so $M(p,0)$ is positive semidefinite with rank at most 1. For larger values of k, multiplicative parameters satisfy the following relation:

$$\mathrm{rank}(M(p,k+l)) \geq \mathrm{rank}(M(p,k)) \cdot \mathrm{rank}(M(p,l)).$$

F35: Connection matrices $M(\hom_H, k)$ and $M(t_H, k)$ corresponding to homomorphism counting are positive semidefinite. Thus, the parameters $\hom_H$ and t_H counting homomorphisms into a fixed weighted graph H are reflection positive.

F36: For every simple graph H and every k-labeled graph F, we have

$$t_H(G \cdot G) \geq t_H(F)^2.$$

F37: For every integer $k \geq 0$, the rank of the connection matrices $M(\hom_H, k)$ and $M(t_H, k)$ is exponentially bounded:

$$\mathrm{rank}(M(\hom_H, k)) \leq |H|^k.$$

F38: Let $q = \mathrm{rank}(M(\hom_H, 1))$. Then $\mathrm{rank}(M(\hom_H, k)) \geq q^k$.

F39: For every weighted graph H and $k \geq 0$, $\mathrm{rank}(M(\hom_H, k)) = \mathrm{rank}(M(t_H, k))$.

F40: (Lovász [Lov06]) If a weighted graph H has no twins and the automorphism group of H has q orbits on $V(H)$, then

$$\mathrm{rank}(M(\hom_H, k)) = q^k.$$

In particular, if H has no twins and has no nontrivial automorphisms, then

$$\mathrm{rank}(M(\hom_H, k)) = |V(H)|^k.$$

F41: (Freedman, Lovász, and Schrijver [FLS07]) Let p be a reflection positive graph parameter for which there exists a positive integer q such that $\mathrm{rank}(M(p,k)) \leq q^k$ for every $k \geq 0$. Then there exists a weighted graph H with $|V(H)| \leq q$ such that $p = \hom_H$, i.e., $p(G) = \hom(G, H)$ for every graph G.

EXAMPLES

E7: Let the parameter e be the graph parameter that counts the number of edges, i.e., $e(G) = \|G\|$. Since $e(G \cdot H) = e(G) + e(H)$, it follows that

$$M(e,k) = e \cdot \mathbf{1}^T + \mathbf{1} \cdot e^T,$$

where we interpret e as a vector in $\mathbb{R}^{\mathcal{L}_k}$, and $\mathbf{1}$ is the constant 1 vector in $\mathbb{R}^{\mathcal{L}_k}$. Thus, the rank of $M(e,k)$ is at most 2. By considering the submatrix of M on the empty graph and a graph with one edge, we see that its rank is 2, so the rank of $M(e,k)$ is 2 for every k.

E8: Suppose that for $G \in \mathcal{L}_k$, $e'(G)$ denotes the number of pairs of vertices that are adjacent in G. That is, we count the number of edges of the underlying simple graph. Then $e'(G \cdot H) = e'(G) + e'(H) - r$, where r denotes the number of pairs of the labeled vertices that are simultaneously adjacent in G and in H. For every pair of distinct labels i, j $(1 \leq i < j \leq k)$, define the vector $e_{ij} \in \mathbb{R}^{\mathcal{L}_k}$ that has value 1 for a k-labeled graph G if its ith and jth labeled vertices are adjacent, and value 0 otherwise. Then we have

$$M(e', k) = e' \cdot \mathbf{1}^T + \mathbf{1} \cdot (e')^T + \sum_{1 \leq i < j \leq k} e_{ij} \cdot e_{ij}^T.$$

Consequently, $\text{rank}(M(e', k)) \leq 2 + \binom{k}{2}$. It can be shown that the equality holds by looking at the principal submatrix of M corresponding to the k-labeled graphs having precisely the edge ij (there are $\binom{k}{2}$ of these) and two additional graphs with edges outside of the labeled part.

E9: Let $m(G)$ denote the number of perfect matchings of the k-labeled graph G. Every perfect matching of $G \cdot H$ can be decomposed into a matching in G and a matching in H. Suppose that the set of labeled vertices in G and H is $[k] = \{1, \ldots, k\}$. Let $S \subseteq [k]$ be the subset of labeled vertices that are covered by the edges in G, and let $\bar{S} = [k] \setminus S$ be the vertices covered by the edges of H. Clearly, all perfect matchings, in which S is covered by the edges in G and $\bar{S}$ is covered by the edges of H, can be obtained by taking a perfect matching in $G - \bar{S}$ and a perfect matching in $H - S$. Thus, if m_S is the parameter that counts the number of perfect matchings in $G - S$, then

$$m(G \cdot H) = \sum_{S \subseteq [k]} m_{\bar{S}}(G) \, m_S(H).$$

This shows that

$$M(m, k) = \sum_{S \subseteq [k]} m_{\bar{S}} \cdot m_S^T$$

and hence $\text{rank}(M(m, k)) \leq 2^k$. It can be shown that the equality holds.

E10: Let r be a positive integer, and let $h_r(G) = \text{hom}(G, K_r)$ denote the number of r-colorings of a graph G. The r-colorings of $G \cdot H$ can be partitioned into the following classes. For every partition π of the labeled set $[k]$ into $|\pi| \leq r$ parts, we consider all r-colorings such that the colors on the labeled set induce the partition π. Let us denote the number of such colorings of a graph G by $c_\pi(G)$. Then

$$h_r(G \cdot H) = \sum_\pi \binom{r}{|\pi|} c_\pi(G) c_\pi(H).$$

This shows that $M(h_r, k)$ is a positive semidefinite matrix whose rank is at most the number of partitions of $[k]$ into at most r parts.

8.5.8   Extremal Graph Theory and the Algebra of Quantum Graphs

The study of graph parameters can be extended to a formal setting of quantum graphs and graph algebras. This approach has applications in extremal graph theory (see Section 8.1). It is essentially the same as the theory of Razborov *flag algebras* that was independently introduced in [Raz07, Raz08].

EXAMPLE

E11: The special case of Turán's Theorem (see Section 8.1) states that every triangle-free graph H of order n contains at most $n^2/4$ edges. This can be expressed with homomorphism densities as follows: If $t_H(K_3) = 0$, then $t_H(K_2) \leq 1/2$. A more general statement (proved by Goodman [Goo59]) shows that H contains many triangles if the number of edges is larger than $n^2/4$. It can be expressed as follows:

$$t_H(K_3) \geq t_H(K_2)(2t_H(K_2) - 1) \tag{8.5.3}$$

and can be derived using multiplicativity and reflection positivity of the homomorphism density parameter t_H as follows. For $i = 1, 2$, let F_i be the 3-labeled graph on vertex set $\{1, 2, 3\}$ isomorphic to the path of length 2 in which the vertex i has degree 2. Then $t_H(F_1 \cup F_2) = t_H(K_3)$ and $t_H(F_1 \cap F_2) = t_H(K_2 \cup K_1) = t_H(K_2)$. By using Fact F4 (submodularity) on F_1 and F_2, we conclude that

$$t_H(K_3) + t_H(K_2) \geq t_H(F_1) + t_H(F_2) = 2t_H(P_3).$$

Note that P_3 can be viewed as the 1-sum $K_2 \cdot K_2$ of two copies of 1-labeled graph K_2. By Fact F36, we conclude that $t_H(P_3) \geq t_H(K_2)^2$; thus, $t_H(K_3) \geq 2t_H(P_3) - t_H(K_2) \geq 2t_H(K_2)^2 - t_H(K_2)$, which gives Inequality (8.5.3).

NOTATION: For every $k \geq 0$, let $\mathcal{L}_k$ denote the set of all k-labeled graphs, whose labeled vertices are $[k] = \{1, \ldots, k\}$.

DEFINITIONS

D47: A k-*labeled quantum graph* $\mathcal{G}$ is a formal linear combination of labeled graphs, i.e.,

$$\mathcal{G} = \sum_{H \in \mathcal{L}_k} x_H \, H,$$

where $x_H \in \mathbb{R}$ for every $H \in \mathcal{L}_k$, and only finitely many of the coefficients x_H are non-zero. The set of all k-labeled quantum graphs is a real vector space, isomorphic to a subspace of $\mathbb{R}^{\mathcal{L}_k}$.

D48: Recall that the k-*sum* of two k-labeled graphs G and H in $\mathcal{L}_k$ is the labeled graph $G \cdot H$ that is obtained from the disjoint union of G and H by identifying pairs of labeled vertices that have the same label. The sum extends to quantum graphs by the rule

$$\mathcal{G} \cdot \mathcal{H} = \left(\sum_{G \in \mathcal{L}_k} x_G \, G \right) \cdot \left(\sum_{H \in \mathcal{L}_k} y_H \, H \right) = \sum_{\substack{G \in \mathcal{L}_k \\ H \in \mathcal{L}_k}} x_G \, y_H \, G \cdot H.$$

With this notation the space of quantum graphs $\mathbb{R}^{\mathcal{L}_k}$ becomes a commutative algebra (where the k-sum has the role of the product operation in the algebra) which is called the ***algebra of k-labeled quantum graphs***. The edgeless graph on labeled vertex set $[k]$ is the identity in this algebra.

D49: If p is a graph parameter, we define its values on quantum graphs by the following rule:

$$p(\mathcal{G}) = p\left(\sum_{H \in \mathcal{L}_k} x_H \, H \right) = \sum_{H \in \mathcal{L}_k} x_H \, p(H).$$

EXAMPLE

E12: With the notation introduced above, Goodman's Theorem (8.5.3) can be stated in the form
$$t_H(K_3 - 2(K_2 \cup K_2) + K_2) \geq 0$$
where we view $K_3 - 2(K_2 \cup K_2) + K_2$ as a 0-labeled quantum graph and $\cup$ denotes the disjoint union.

FACTS

F42: A graph parameter p is reflection positive if and only if the parameter p is non-negative on squares of quantum graphs. Specifically, the connection matrix $M(p, k)$ is positive semidefinite if and only if the parameter p is non-negative on squares of k-labeled quantum graphs.

At the time of preparing this section, the topic of exploring uses of analytic techniques for homomorphism densities is very active. One of the most interesting outstanding conjectures in this area is the ***Sidorenko Conjecture***, which claims that for every bipartite graph H with m edges and every graph G,

$$t_H(G) \geq t_{K_2}(G)^m.$$

The Sidorenko Conjecture has been proved for many bipartite graphs H, and the reader is referred to [CFS10, Hat10, Lov11] for some of the known results.

References

[ADL⁺94] N. Alon, R. A. Duke, H. Lefmann, V. Rödl, and R. Yuster, The algorithmic aspects of the regularity lemma, *J. Algorithms* **16** (1994), no. 1, 80–109. MR 1251840 (94j:05112)

[AFdlVKK03] Noga Alon, W. Fernandez de la Vega, Ravi Kannan, and Marek Karpinski, Random sampling and approximation of MAX-CSPs, *J. Comput. System Sci.* **67** (2003), no. 2, 212–243. MR 2022830 (2005g:68192)

[AFNS06] Noga Alon, Eldar Fischer, Ilan Newman, and Asaf Shapira, A combinatorial characterization of the testable graph properties: it's all about regularity, STOC'06: *Proceedings of the 38th Annual ACM Symposium on Theory of Computing*, ACM, New York, 2006, pp. 251–260. MR 2277151 (2007h:68150)

[AN06] Noga Alon and Assaf Naor, Approximating the cut-norm via Grothendieck's inequality, *SIAM J. Comput.* **35** (2006), no. 4, 787–803 (electronic). MR 2203567 (2006k:68176)

[AS08] Noga Alon and Joel H. Spencer, *The Probabilistic Method, third ed.*, Wiley-Interscience Series in Discrete Mathematics and Optimization, John Wiley & Sons Inc., Hoboken, NJ, 2008, with an appendix on the life and work of Paul Erdős.

[BCL+06] Christian Borgs, Jennifer Chayes, László Lovász, Vera T. Sós, Balázs Szegedy, and Katalin Vesztergombi, Graph limits and parameter testing, STOC'06: *Proceedings of the 38th Annual ACM Symposium on Theory of Computing*, ACM, New York, 2006, pp. 261–270.

[BCL+08] C. Borgs, J. T. Chayes, L. Lovász, V. T. Sós, and K. Vesztergombi, Convergent sequences of dense graphs. I. Subgraph frequencies, metric properties and testing, *Adv. Math.* **219** (2008), no. 6, 1801–1851.

[BCL10] Christian Borgs, Jennifer Chayes, and László Lovász, Moments of two-variable functions and the uniqueness of graph limits, *Geom. Funct. Anal.* **19** (2010), no. 6, 1597–1619.

[Bol01] Béla Bollobás, *Random Graphs*, Second ed., Cambridge Studies in Advanced Mathematics, vol. 73, Cambridge University Press, 2001.

[CFS10] David Conlon, Jacob Fox, and Benny Sudakov, An approximate version of Sidorenko's conjecture, *Geom. Funct. Anal.* **20** (2010), no. 6, 1354–1366. MR 2738996 (2012h:05351)

[CGW89] F. R. K. Chung, R. L. Graham, and R. M. Wilson, Quasi-random graphs, *Combinatorica* **9** (1989), no. 4, 345–362.

[FK99a] Alan Frieze and Ravi Kannan, Quick approximation to matrices and applications, *Combinatorica* **19** (1999), no. 2, 175–220. MR 1723039 (2001i:68066)

[FK99b] _____ , A simple algorithm for constructing Szemerédi's regularity partition, *Electron. J. Combin.* **6** (1999), Research Paper 17, 7 pp. MR 1674741 (2000f:68086)

[FLS07] Michael Freedman, László Lovász, and Alexander Schrijver, Reflection positivity, rank connectivity, and homomorphism of graphs, *J. Amer. Math. Soc.* **20** (2007), no. 1, 37–51 (electronic).

[FLW] Michael Freedman, László Lovász, and Dominic Welsh, unpublished notes.

[Goo59] A. W. Goodman, On sets of acquaintances and strangers at any party, *Amer. Math. Monthly* **66** (1959), 778–783. MR 0107610 (21 #6335)

[GT08] Ben Green and Terence Tao, The primes contain arbitrarily long arithmetic progressions, *Ann. of Math.* (2) **167** (2008), no. 2, 481–547.

[Hat10] Hamed Hatami, Graph norms and Sidorenko's conjecture, *Israel J. Math.* **175** (2010), 125–150. MR 2607540 (2011m:05174)

[KS96] J. Komlós and M. Simonovits, Szemerédi's regularity lemma and its applications in graph theory, *Combinatorics, Paul Erdős is Eighty*, Vol. 2 (Keszthely, 1993), Bolyai Soc. Math. Stud., vol. 2, János Bolyai Math. Soc., Budapest, 1996, pp. 295–352.

[KSSS02] János Komlós, Ali Shokoufandeh, Miklós Simonovits, and Endre Szemerédi, The regularity lemma and its applications in graph theory, Theoretical aspects of computer science (Tehran, 2000), *Lecture Notes in Comput. Sci.*, vol. 2292, Springer, Berlin, 2002, pp. 84–112.

[Lov06] László Lovász, The rank of connection matrices and the dimension of graph algebras, *European J. Combin.* **27** (2006), no. 6, 962–970.

[Lov09] ———, Very large graphs, *Current Developments in Mathematics*, 2008, Int. Press, Somerville, MA, 2009, pp. 67–128.

[Lov11] ———, Subgraph densities in signed graphons and the local Simonovits-Sidorenko conjecture, *Electron. J. Combin.* **18** (2011), no. 1, Paper 127, 21. MR 2811096 (2012f:05158)

[LS06] László Lovász and Balázs Szegedy, Limits of dense graph sequences, *J. Combin. Theory Ser. B* **96** (2006), no. 6, 933–957.

[LS10] ———, Testing properties of graphs and functions, *Israel J. Math.* **178** (2010), 113–156. MR 2733066 (2011i:05242)

[Raz07] Alexander A. Razborov, Flag algebras, *J. Symbolic Logic* **72** (2007), no. 4, 1239–1282. MR 2371204 (2008j:03040)

[Raz08] ———, On the minimal density of triangles in graphs, *Combin. Probab. Comput.* **17** (2008), no. 4, 603–618. MR 2433944 (2009i:05118)

[SS91] Miklós Simonovits and Vera T. Sós, Szemerédi's partition and quasirandomness, *Random Structures Algorithms* **2** (1991), no. 1, 1–10.

[Sze78] Endre Szemerédi, Regular partitions of graphs, *Problèmes Combinatoires et Théorie des Graphes* (Colloq. Internat. CNRS, Univ. Orsay, Orsay, 1976), Colloq. Internat. CNRS, vol. 260, CNRS, Paris, 1978, pp. 399–401.

[Tao06] Terence Tao, Szemerédi's regularity lemma revisited, *Contrib. Discrete Math.* **1** (2006), no. 1, 8–28 (electronic).

Glossary for Chapter 8

alteration method — a probabilistic method: first generate a random object, and then alter it to obtain a property we desire.

Andrásfai graph A_i: for $i \geq 2$, the complement of the $(i-1)^{\text{th}}$ power of the cycle graph C_{3i-1}.

c-arrangeable graph: a graph G whose vertices can be ordered in such a way that for any vertex v, each neighbor of v that succeeds v in the order has at most c adjacencies that precede v in the order.

arrowing graph F **for** $(G_1, G_2, \cdots, G_n)$ – a concept in Ramsey theory: we write $F \longrightarrow$ $(G_1, G_2, \cdots, G_n)$ if for any k-edge-coloring of F there is for some i a monochromatic copy of G_i in the ith color.

asymptotically normal sequence $\{X_n\}$ – of random variables: a sequence such that for all fixed a,

$$\mathbf{P}(\hat{X} \leq a) = o(1) + \frac{1}{\sqrt{2\pi}} \int_{-\infty}^{a} e^{-x^2/2} \, dx \text{ as } n \to \infty$$

where $\hat{X} = (X - \mathbf{E}X)/\sqrt{\mathbf{Var}X}$.

asymptotically Poisson sequence $\{X_n\}$ – of random variables: a sequence such that

$$\mathbf{P}(X_n = k) = e^{-\lambda}\lambda^k/k! + o(1)$$

for all fixed integers $k \geq 0$, as $n \to \infty$. [Valid in the case that λ is any bounded function of n.]

book: a collection of K_3-graphs sharing a common edge.

___, **r-book**: a collection of $(r+1)$-cliques sharing a common K_r.

booksize – of a graph G: the number of K_3-subgraphs in the largest *book* subgraph of G.

chromatic surplus: the largest number $s = s(G)$ such that in every vertex coloring of a graph G with $\chi(G)$ colors, every color class has at least s vertices.

circumference – of a graph: the length of the longest cycle.

clique$_1$ – in a graph: a maximal complete subgraph; note that there are two different definitions.

clique$_2$ – in a graph: any complete subgraph, not necessarily maximal under inclusion; note that there are two different definitions.

clique number $\omega(G)$ – of a graph G: the number of vertices of its largest clique.

closure – of a graph: the graph obtained from $G(n)$ by successively joining all nonadjacent vertices u and v with $deg(u) + deg(v) \geq n$.

complex graph: a connected graph G with $|E(G)| > |V(G)|$.

concentration inequality: a bound on the probability that a random variable differs from its mean by a large amount; a technique within the *probabilistic method*.

2.5-connected graph: a 2-connected graph such that any cutset with two vertices is independent.

convex graph property: a property $\mathcal{P}$ that a graph G must have, if a subgraph and a supergraph of G on the same vertex set both have $\mathcal{P}$.

k-core – of a graph: the largest subgraph with all vertex degrees at least k.

decreasing graph property: a property that cannot be lost by deleting edges alone.

density$_1$ – of a graph G: the ratio $|E(G)|/|V(G)|$.

   ___, maximum – of a graph G: the maximum ratio of the number of edges divided by the number of vertices, taken over all non-null subgraphs of G.

density$_2$ – of a graph G: the ratio $|E(G)| / \binom{|V(G)|}{2}$.

excess – of a graph G: the difference $|E(G)| - |V(G)|$.

extremal function $ex(n, \mathcal{F})$ – for an integer n and a family $\mathcal{F}$ of *forbidden graphs*: the function whose value is the largest number of edges of a simple n-vertex graph that contains none of the forbidden subgraphs.

extremal graph – for an integer n and a family $\mathcal{F}$ of *forbidden graphs*: a graph with no forbidden subgraphs and with the largest possible number of edges.

H-factor – of a graph G: a collection of disjoint copies of the graph H that covers the vertices of G.

first moment method: using the expected value of a random variable to bound the probability; a technique within the *probabilistic method*.

H-free graph – where H is a graph: a graph that has no subgraph isomorphic to H.

girth – of a graph: the length of its shortest cycle.

graph parameter: a function defined on all (finite) graphs that is invariant under graph isomorphisms.

   ___, multiplicative: a graph parameter p such that $p(G \cup H) = p(G) p(H)$, whenever G and H are disjoint graphs.

   ___, normalized: a graph parameter p such that $p(K_1) = 1$.

graphon: a symmetric, measurable function $W : [0, 1] \times [0, 1] \to \mathbb{R}$.

   ___, simple: a gryphon W such that $0 \le W(x, y) \le 1$ for every $x, y \in [0, 1]$.

hereditary property $\mathcal{P}$ – of graphs: a property that is shared with every induced subgraph of a graph having $\mathcal{P}$.

homomorphism from G to H: for simple graphs, a mapping $f : V(G) \to V(H)$ such that $uv \in E(G)$ implies $f(u)f(v) \in E(H)$; for general graphs, there is also an edge mapping $f : E(G) \to E(H)$ such that if u and v are the endpoints of edge $e \in E(G)$, then $f(u)$ and $f(v)$ are the endpoints of $f(e) \in E(H)$.

hypergraph: a set of vertices V and a set of *edges*, each of which is a subset of V.

   ___, k-uniform: a hypergraph whose edges are all of size k.

k-hypergraph Ramsey number $r_k(G_1, G_2, \cdots, G_m)$: the smallest integer n such that if the k-sets of a set of order n are colored with m colors, there will be for some i an isomorphic copy of G_i in color i.

increasing graph property: a property of a simple graph that cannot be lost by adding edges alone.

independence number $\alpha(G)$ – of a graph G: the largest possible number of vertices in an independent set.

independent set – of vertices in a graph G: a set of mutually nonadjacent vertices.

Lovász Local Lemma: a tool within the *probabilistic method* for dealing with *weakly dependent events*.

minor of a graph G: any graph formed from G by a sequence of edge deletions and contractions.

monotone graph property: a property that is either increasing or decreasing.

Mycielski graphs: any of the graphs $M_1, M_2, \ldots$ in a particular sequence constructed inductively by Mycielski.

pancyclic graph: a Hamiltonian graph having cycles of all possible lengths up to its number of vertices.

phase transition – of the random graph: loosely, the range of density during which the largest component grows from very small to very large.

k^{th} **power of a cycle**: a graph G with $V(G) = \{1, \ldots, n\}$ and $(i, j) \in E(G)$ if and only if $i - j = \pm 1, \pm 2, \ldots, \pm k \bmod n$.

probabilistic method: proving the existence of an object by showing that it exists with a positive probability.

Ramsey graph for the pair (G, H): a graph F such that $F \longrightarrow (G, H)$.

Ramsey multiplicity $R(G, G)$: the minimum number of monochromatic copies of the graph G in any 2-edge-coloring of the Ramsey graph K_n where $n = r(G, G)$.

Ramsey number $r(G_1, G_2, \cdots, G_k)$: the least positive integer n such that for any k-edge-coloring of K_n there is for some i a monochromatic copy of the graph G_i in color i.

___, **bipartite** $r_b(G, H)$: smallest number of vertices of a bipartite graph F such that $F \longrightarrow (G, H)$.

___, **classical** – $r(G_1, G_2, \cdots, G_k)$ where each G_i is a complete graph K_{n_i}.

___, **connected**, denoted by $r_c(G, H)$: the least number of vertices in a graph F such that $F \longrightarrow (G, H)$ and such that the graph induced by each color is connected.

___, **diagonal**: a Ramsey number $r(G, H)$ with $G = H$.

___, **induced**, denoted by $r^*(G, H)$: the least positive integer n such that there exists a graph F of order n such that any 2-edge-coloring (Red and Blue) of F yields an induced copy of G in Red or an induced copy of H in Blue.

Ramsey-finite pair (G, H): a pair of graphs such that there are only finitely many nonisomorphic minimal graphs F with $F \longrightarrow (G, H)$.

Ramsey-infinite pair (G, H): a pair of graphs such that there are infinitely many nonisomorphic minimal graphs F with $F \longrightarrow (G, H)$.

(Ramsey) minimal graph – for $(G_1, G_2, \cdots, G_k)$: a graph F such that $F \longrightarrow (G_1, G_2, \cdots, G_n)$, but no proper subgraph of F also arrows $(G_1, G_2, \cdots, G_k)$.

random graph$_1$: a probability space whose domain is the set of n-vertex simple (labeled) graphs on the vertex-set $[n] = \{1, \dots, n\}$; the probability function is determined by specifying that the edges occur independently with probability $\frac{1}{2}$ each.

___, **binomial** $\mathcal{G}(n, p)$: a probability space whose domain is the set of n-vertex simple (labeled) graphs on the vertex-set $[n] = \{1, \dots, n\}$; the probability function is determined by specifying that the edges occur independently with probability $p = p(n)$ each; also called the *Bernoulli random graph*.

___, **uniform** $\mathcal{G}(n, m)$: the uniform probability space on those simple (labeled) graphs on the vertex-set $[n] = \{1, \dots, n\}$ with exactly m edges; also called the *Erdős-Rényi random graph*; we observe that the probability of each edge is $p(n) = m/\binom{n}{2}$.

Rödl nibble: a random process that builds the required object a little piece at a time.

sharp threshold function – regarding the binomial random graph $\mathcal{G}(n, p)$, for a graph property Q: a function $f(n)$ such that for all $\epsilon > 0$, $\mathcal{G}(n, p)$ a.a.s. does not have Q if $\frac{p(n)}{f(n)} < 1 - \epsilon$, and $\mathcal{G}(n, p)$ a.a.s. has Q if $\frac{p(n)}{f(n)} > 1 + \epsilon$ (or the same with "does not have" and "has" interchanged).

size Ramsey number: smallest number of edges of a graph F such that $F \longrightarrow (G_1, G_2, \cdots, G_n)$.

___, **restricted**: the minimum size, i.e., number of edges, of a graph F with $r(G, H)$ vertices such that $F \longrightarrow (G, H)$.

square of a cycle: the second *power* of a cycle.

star forest: a graph in which each component is a star.

star: any bipartite graph $K_{1,m}$.

strictly balanced graph: a graph whose maximum density is strictly greater than the maximum density of any of its proper subgraphs.

threshold function – for a graph property Q: regarding $\mathcal{G}(n, p)$, a function $f(n)$ such that $\mathcal{G}(n, p)$ a.a.s. does not have Q if $p = o(f)$, and $\mathcal{G}(n, p)$ a.a.s. has Q if $f = o(p)$ (or the same with "does not have" and "has" interchanged). Regarding $\mathcal{G}(n, m)$, the definition is the same but with p replaced by m.

Turán graph $T_r(n)$: the complete r-partite graph on n vertices whose classes (*partite sets*) are as nearly equal as possible.

Turán number $t_r(n)$: the number of edges of the *Turán graph* of n vertices and chromatic number r.

Turán-type problem: the problem of finding the largest possible number of edges of a graph having no graphs is a specified family of *forbidden subgraphs*.

weakly pancyclic graph: a graph having cycles of all possible lengths up to its circumference.

Chapter 9

Graphical Measurement

Section 9.1

Distance in Graphs

Gary Chartrand, Western Michigan University
Ping Zhang, Western Michigan University

INTRODUCTION

How far two objects (or sets of objects) are apart in a discrete structure is of interest, both theoretically and for its applications. Since discrete structures are naturally modeled by graphs, this leads us to studying distance in graphs. A book entirely devoted to this subject has been written (see [BuHa90]).

9.1.1   Standard Distance in Graphs

Although there is not a unique way to define the distance between two vertices in a graph, there is one definition of distance that has been used most often and is commonly accepted as the standard definition of distance.

Distance and Eccentricity

Many of the distance concepts that have been studied have their origins in maximizing distance from a vertex.

DEFINITIONS

D1: For two vertices u and v in a graph G, the **distance** $d(u, v)$ from u to v is the length (number of edges) of a shortest $u - v$ path in G.

D2: A $u - v$ path of length $d(u, v)$ is called a $u - v$ **geodesic**.

D3: For a vertex v in a connected graph G, the **eccentricity** $e(v)$ of v is the distance from v to a vertex farthest from v. That is, $e(v) = \max\{d(v, x) : x \in V(G)\}$.

EXAMPLE

E1: Each vertex in the graph G of Figure 9.1.1 is labeled with its distance from the vertex v. The distance from v to a vertex farthest from v is 3 and so $e(v) = 3$.

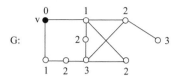

Figure 9.1.1: Distances from the vertex v to the vertices of G.

FACTS

F1: For vertices u and v in a connected graph G, $d(u, v) \geq 2$ if and only if u and v are distinct and nonadjacent.

F2: The distance d defined above is a metric, that is, d satisfies the following four properties:

1. $d(u, v) \geq 0$ for all $u, v \in V(G)$.
2. $d(u, v) = 0$ if and only if $u = v$.
3. $d(u, v) = d(v, u)$ for all $u, v \in V(G)$ [**the symmetric property**].
4. $d(u, w) \leq d(u, v) + d(v, w)$ for all $u, v, w \in V(G)$ [**the triangle inequality**].

Radius and Diameter

Two major distance parameters associated with a connected graph G are obtained by minimizing and maximizing the eccentricities of the vertices of G.

DEFINITION

D4: The minimum eccentricity among the vertices of a connected graph G is the **radius** of G, denoted by rad(G), and the maximum eccentricity is its **diameter** diam(G).

REMARK

R1: The diameter of a connected graph G also equals $\max\{d(x, y) : x, y \in V(G)\}$.

EXAMPLE

E2: Each vertex in the graph G of Figure 9.1.2 is labeled with its eccentricity. So rad$(G) = 2$ and diam$(G) = 4$.

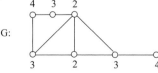

Figure 9.1.2: Eccentricities, radius, and diameter.

FACTS

F3:   For every nontrivial connected graph G, $\text{rad}(G) \leq \text{diam}(G) \leq 2\,\text{rad}(G)$. More generally, if a "distance" defined on $V(G)$ is a metric and eccentricity, radius, and diameter are defined as expected, then the inequalities $\text{rad}(G) \leq \text{diam}(G) \leq 2\,\text{rad}(G)$ always hold.

F4:   [Os73] For every pair r, d of positive integers with $r \leq d \leq 2r$, there exists a connected graph G with $\text{rad}(G) = r$ and $\text{diam}(G) = d$. Furthermore, the minimum order (number of vertices) of such a graph is $r + d$.

Center and Periphery

The radius and diameter of a connected graph G give rise to two subgraphs of G.

DEFINITIONS

D5:   A vertex v in a connected graph G is a ***central vertex*** if $e(v) = \text{rad}(G)$, while a vertex v in G is a ***peripheral vertex*** if $e(v) = \text{diam}(G)$.

D6:   The subgraph induced by the central vertices of a connected graph G is the ***center*** $C(G)$ of G and the subgraph of G induced by its peripheral vertices is its ***periphery*** $P(G)$.

EXAMPLE

E3:   Each vertex in the graph G of Figure 9.1.3 is labeled with its eccentricity. So $\text{rad}(G) = 3$ and $\text{diam}(G) = 5$. The center and periphery of G are also shown in Figure 9.1.3.

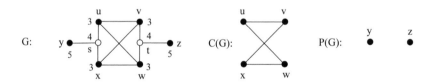

Figure 9.1.3: The center and periphery of a graph.

FACTS

F5:   [BuMiSl81] Every graph is (isomorphic to) the center of some graph.

F6:   [HaNi53] The center of every connected graph G lies in a single block of G.

F7:   [Jo69] The center of every tree either consists of a single vertex or is isomorphic to K_2. Furthermore, the center of a tree T consists of a single vertex if and only if $\text{diam}(T) = 2\,\text{rad}(T)$.

F8:   [BiSy83] A nontrivial graph G is (isomorphic to) the periphery of some graph if and only if every vertex of G has eccentricity 1 or no vertex of G has eccentricity 1.

Self-Centered Graphs

DEFINITIONS

D7: A graph G is ***self-centered*** if $C(G) = G$.

D8: An ***automorphism*** of a graph G is an isomorphism between G and itself. The set of all automorphisms of a graph G under the operation of composition forms a group called the ***automorphism group*** of G.

EXAMPLE

E4: The vertices of each of the graphs in Figure 9.1.4 are labeled with their eccentricities. Thus the graphs $K_{2,3}$ and C_5 are self-centered, while P_4 is not.

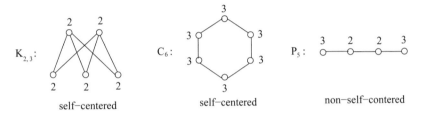

Figure 9.1.4: Self-centered graphs and a non-self-centered graph.

FACTS

F9: [Bu79] Let $n \geq 5$ and $r \geq 2$ be integers such that $n \geq 2r$. Then there exists a self-centered graph of order n, size m, and radius r if and only if

$$\lceil (nr - 2r - 1)/(r - 1) \rceil \leq m \leq (n^2 - 4nr + 5n + 4r^2 - 6r)/2.$$

If $n = 2r = 4$, then $m = 4$.

F10: [Bu79] If G is a self-centered connected graph of order n, size m, and radius 2, then $m \geq 2n - 5$.

F11: [LeWa90] For a given graph H that is not self-centered, there exists a self-centered graph G whose order exceeds the order of H by 3 and such that (1) G contains H as an induced subgraph and (2) the automorphism group of G is isomorphic to the automorphism group of H.

F12: [LeWa90] For every finite group Γ, there exists a self-centered graph whose automorphism group is isomorphic to Γ.

9.1.2   Geodetic Parameters

Geodetic Sets and Geodetic Numbers

For every connected graph G, there are sets S of vertices of G such that every vertex of G lies on a geodesic connecting two vertices of S.

DEFINITIONS

D9: A vertex w is said to **lie in** a $u - v$ path P if w is a vertex of P but $w \neq u, v$.

D10: For two vertices u and v in a connected graph G, the **closed interval** $I[u, v]$ consists of u, v, and all vertices lying in some $u - v$ geodesic of G; for $S \subseteq V(G)$, the **closed interval** $I[S]$ of S is the union of all sets $I[u, v]$ for $u, v \in S$.

D11: A set S of vertices of a connected graph G is called a **geodetic set** in G if $I[S] = V(G)$. A geodetic set of minimum cardinality is a **minimum geodetic set**. The cardinality of a minimum geodetic set is called the **geodetic number** $g(G)$.

D12: A graph F is a **minimum geodetic subgraph** if there exists a graph G containing F as an induced subgraph such that $V(F)$ is a minimum geodetic set for G.

REMARK

R2: Each vertex of a connected graph G whose neighborhood induces a complete subgraph in G belongs to every geodetic set of G. In particular, each end-vertex of G belongs to every geodetic set of G.

EXAMPLES

E5: The set $S_1 = \{x, y, z\}$ is a geodetic set of the graph G_1 in Figure 9.1.5. Since there is no 2-element geodetic set in G_1, it follows that S_1 is a minimum geodetic set of G_1 and so $g(G_1) = 3$. In G_2, the set $S_2 = \{u, v, w, t\}$ is a minimum geodetic set of G_2, so $g(G_2) = 4$. Notice that every vertex in S_2 has the property that its neighborhood induces a complete subgraph in G_2.

Figure 9.1.5: Minimum geodetic sets in graphs.

E6: The set $S = \{u, v, w, x\}$ is a minimum geodetic set of the graph G of Figure 9.1.6. Since the subgraph $\langle S \rangle$ of G induced by S is (isomorphic to) C_4, it follows that C_4 is a minimum geodetic subgraph.

Figure 9.1.6: A minimum geodetic subgraph.

FACTS

F13: [HaLoTs93] Determining the geodetic number of a graph is an NP-hard problem.

F14: [ChHaZh02] If G is a connected graph of order $n \geq 2$ and diameter d, then $g(G) \leq n - d + 1$ and this bound is sharp.

F15: [BuHaQu] Let G be a nontrivial connected graph of order n. Then (a) $g(G) = n$ if and only if $G = K_n$ and (b) $g(G) = n - 1$ if and only if $G = (K_{n_1} \cup K_{n_2} \cup \cdots \cup K_{n_r}) + K_1$, where $r \geq 2$ and $n_1, n_2, \cdots, n_r$ satisfy $n_1 + n_2 + \cdots + n_r = n - 1$.

F16: [ChHaZh02] For every three positive integers r, d, and $k \geq 2$ with $r \leq d \leq 2r$, there exists a connected graph G with $\mathrm{rad}(G) = r$, $\mathrm{diam}(G) = d$ and $g(G) = k$.

F17: [ChHaZh02] A nontrivial graph F is a minimum geodetic subgraph if and only if every vertex of F has eccentricity 1 or no vertex of F has eccentricity 1.

F18: [ChHaZh02] A nontrivial graph F is a minimum geodetic subgraph of a connected graph G if and only if F is the periphery of some connected graph H.

Convex Sets and Hull Sets

There are sets S of vertices in a connected graph with the property that every geodesic connecting two vertices of S contains only vertices of S.

DEFINITIONS

D13: A set S of vertices of a connected graph G is **convex** if $I[S] = S$ and the **convex hull** $[S]$ is the smallest convex set containing S.

D14: For a set S of vertices in a connected graph G, let $I^0[S] = S$, $I^1[S] = I[S]$, and $I^k[S] = I[I^{k-1}[S]]$ for $k \geq 2$. From some term on, this sequence is constant. The smallest nonnegative integer p for which $I^p[S] = I^{p+1}[S]$ is the **geodetic iteration number** $gin(S)$. The set $I^p[S]$ is, in fact, the convex hull $[S]$ of S. The **geodetic iteration number** of G, denoted by $gin(G)$, is given by

$$gin(G) = \max_{S \subseteq V(G)} \{gin(S)\}.$$

D15: Let S be a set of vertices of a connected graph G. If $[S] = V(G)$, then S is called a **hull set** in G. A hull set of minimum cardinality is a **minimum hull set**. The cardinality of a minimum hull set in G is called the **hull number** $h(G)$.

D16: A graph F is a ***minimum hull subgraph*** if there exists a graph G containing F as an induced subgraph such that $V(F)$ is a minimum hull set for G.

EXAMPLES

E7: In the graph G of Figure 9.1.7, let $S_1 = \{u, v, w\}$ and $S_2 = \{u, v, w, x\}$. Since $[S_1] = S_2 \neq S_1$ and $[S_2] = S_2$, it follows that S_1 is not a convex set in G; while S_2 is a convex set in G. Furthermore, S_2 is the convex hull of S_1.

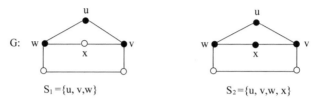

Figure 9.1.7: Convex sets in a graph.

E8: In the graph G of Figure 9.1.8, let $S = \{s, t, y\}$. Since $I[S] = \{s, t, u, v, w, x, y\}$ and $I^2[S] = V(G)$, it follows that S is a hull set of G. In fact, S is a minimum hull set and so $h(G) = 3$. Furthermore, $gin(S) = 2$.

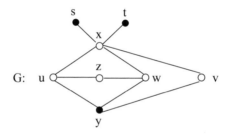

Figure 9.1.8: A minimum hull set in a graph.

REMARK

R3: Every geodetic set in a connected graph G is a hull set of G. The converse is not true in general. Thus $h(G) \leq g(G)$ for every connected graph G.

FACTS

F19: [HaNi81] Let n be the minimum order of a graph G for which $gin(G) = k$. Then $n = 1$ if $k = 0$, $n = 3$ if $k = 1$, and $n = k + 3$ if $k \geq 2$.

F20: [EvSe85] If G is a connected graph of order $n \geq 2$ and diameter d, then $h(G) \leq n - d + 1$.

F21: [EvSe85] If G is a k-connected graph of order n and diameter d, then $h(G) \leq n - k \lfloor d/2 \rfloor$.

F22: [ChHaZh00] For every pair a, b of integers with $2 \leq a \leq b$, there exists a connected graph G such that $h(G) = a$ and $g(G) = b$.

F23: [ChHaZh00] For every nontrivial connected graph G, $h(G) = h(G \times K_2)$.

F24: [ChHaZh00] A connected graph G of order $n \geq 3$ has hull number $n - 1$ if and only if $G = (K_{n_1} \cup K_{n_2} \cup \cdots \cup K_{n_r}) + K_1$ where r (≥ 2), $n_1, n_2, \cdots, n_r$ are positive integers with $n_1 + n_2 + \cdots + n_r = n - 1$.

F25: [ChHaZh00] A nontrivial graph F is a minimum hull subgraph of some connected graph if and only if every component of F is complete.

9.1.3 Total Distance and Medians of Graphs

Total Distance of a Vertex

DEFINITION

D17: The *total distance* $td(u)$ (or *distance* or *status*) of a vertex u in a connected graph G is defined by

$$td(u) = \sum_{v \in V(G)} d(u, v).$$

EXAMPLE

E9: Each vertex in the graph G of Figure 9.1.9 is labeled with its distance from the vertex u. Thus, the total distance of u is

$$td(u) = \sum_{v \in V(G)} d(u, v) = 0 + 1 + 1 + 2 + 2 + 2 + 3 + 4 + 4 + 5 + 6 + 7 = 37.$$

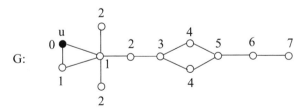

Figure 9.1.9: The total distance of a vertex u.

FACT

F26: [EnJaSn76] Let G be a connected graph of order n and size m and $v \in V(G)$. Then

$$n - 1 \leq td(v) \leq (n - 1)(n + 2)/2 - m$$

and these bounds can be attained for each m with $n - 1 \leq m \leq \binom{n}{2}$.

The Median of a Connected Graph

The center of a connected graph is not the only subgraph that's been used to describe the middle of a connected graph.

DEFINITIONS

D18: A vertex v in a connected graph G is a *median vertex* if v has the minimum total distance among the vertices of G.

D19: The *median* $M(G)$ of a connected graph G is the subgraph of G induced by its median vertices.

EXAMPLE

E10: Each vertex in the graph G of Figure 9.1.10 is labeled with its total distance. Therefore, u and v are the two median vertices of G. The median of G is also shown in Figure 9.1.10.

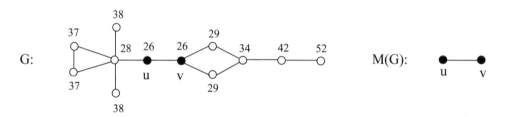

Figure 9.1.10: A graph and its median.

FACTS

F27: [Sl80] Every graph is (isomorphic to) the median of some graph.

F28: [Tr85] The median of every connected graph G lies in a single block of G.

Centers and Medians of a Connected Graph

There is no restriction on the relative locations of the center and median of a connected graph.

DEFINITION

D20: For two subgraphs F and H in a connected graph G, the *distance between F and H* is

$$d(F, H) = \min\{d(u, v) :\ u \in V(F), v \in V(H)\}.$$

FACTS

F29: [Ho89] For every two graphs G_1 and G_2 and positive integer k, there exists a connected graph G such that $C(G)$ is isomorphic to G_1, $M(G)$ is isomorphic to G_2, and $d(C(G), M(G)) = k$.

F30: [NoTi91] For every three graphs G_1, G_2, and G_3, where G_3 is isomorphic to an induced subgraph of both G_1 and G_2, there exists a connected graph G such that $C(G)$ is isomorphic to G_1, $M(G)$ is isomorphic to G_2, and $C(G) \cap M(G)$ is isomorphic to G_3.

9.1.4 Steiner Distance in Graphs

There is a generalization of the distance between two vertices in a connected graph G for any set of vertices in G that is an analogue of the Euclidean Steiner Problem which asks, for a given set S of points in the plane, the smallest network connecting the points of S.

Steiner Radius and Steiner Diameter

All of the basic distance parameters can be extended to Steiner distance.

DEFINITIONS

D21: For a nonempty set W of vertices in a connected graph G, the **Steiner distance** $sd(W)$ of W is the minimum size of a connected subgraph of G containing W. Necessarily, each such subgraph is a tree, called a **Steiner tree with respect to** W. In particular, if $W = \{u, v\}$, then $sd(W) = d(u, v)$, the ordinary distance between u and v.

D22: Let G be a connected graph of order n. For an integer k with $1 \le k \le n$, the **k-eccentricity** $e_k(v)$ of a vertex v in G is the maximum Steiner distance among all k-element sets of vertices of G containing v.

D23: The minimum k-eccentricity of a connected G is the **k-radius** $\mathrm{rad}_k(G)$ of G and the maximum k-eccentricity is its **k-diameter** $\mathrm{diam}_k(G)$.

EXAMPLES

E11: Let $S = \{u, v, x\}$ in the graph G of Figure 9.1.11. Here $sd(S) = 4$. There are several trees of size 4 containing S, one of which is the Steiner tree T of Figure 9.1.11.

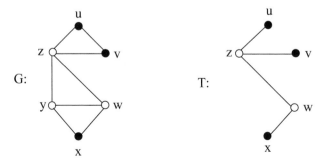

Figure 9.1.11: A graph G and a Steiner tree T.

E12: Each vertex in the graph G of Figure 9.1.12 is labeled with its 3-eccentricity so that $\text{rad}_3(G) = 4$ and $\text{diam}_3(G) = 6$.

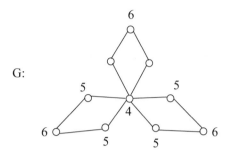

G:

Figure 9.1.12: The 3-eccentricities of the vertices of a graph.

FACTS

F31: [ChOeTiZo89] Let k and n be integers with $3 \leq k \leq n$. For every tree T of order n, $\text{diam}_{k-1}(T) = \text{rad}_k(T)$.

F32: [ChOeTiZo89] For every integer $k \geq 3$ and every tree T of order at least k,

$$\text{diam}_k(T) \leq \left(\frac{k}{k-1}\right) \text{diam}_{k-1}(T).$$

This fact is not true for graphs in general. [HeOeSw90]

F33: [HeOeSw90] If G is a connected graph of order n, then

(a) $\text{diam}_3(G) \leq \left(\frac{8}{5}\right) \text{rad}_3(G)$ if $n \geq 3$.

(b) $\text{diam}_4(G) \leq \left(\frac{10}{7}\right) \text{rad}_4(G)$ if $n \geq 4$.

F34: [HeOeSw91] For every connected graph G and every integer $k \geq 3$,

$$\text{diam}_k(G) \leq \left(\frac{k+1}{k-1}\right) \text{diam}_{k-1}(G).$$

Steiner Centers

There are a number of centers associated with Steiner distance.

DEFINITION

D24: For $k \geq 2$, a vertex v in a connected graph G is a k-*central vertex* if $e_k(v) = \text{rad}_k(G)$; the subgraph induced by the k-central vertices of G is the Steiner k-*center* of G.

FACTS

F35: [OeTi90] Let $k \geq 2$ be an integer. Every graph is (isomorphic to) the Steiner k-center of some graph.

F36: [OeTi90] Let $k \geq 3$ be an integer and T a tree. Then T is (isomorphic to) the Steiner k-center of some tree if and only if T has at most $k - 1$ end-vertices.

9.1.5 Distance in Digraphs

There is a natural definition of distance from one vertex to another in digraphs as well.

DEFINITIONS

D25: Let u and v be vertices in a digraph D. If D contains one or more directed $u - v$ paths, then the ***directed distance*** $\vec{d}(u, v)$ is the length of a shortest directed $u - v$ path in D.

D26: A digraph D is ***strong*** if D contains both a directed $u - v$ path and a directed $v - u$ path for every pair u, v of distinct vertices of D.

Radius and Diameter in Strong Digraphs

The definitions of eccentricity, radius, and diameter in a digraph are analogous to those in an undirected graph (see Definitions D3 and D4).

DEFINITION

D27: The ***eccentricity*** $e(v)$ of v in a strong digraph D is the greatest directed distance from v to a vertex of D. The minimum eccentricity among the vertices of D is its ***radius***, rad(D), and the maximum eccentricity is its ***diameter***, diam(D).

EXAMPLES

E13: There are three directed $u - v$ paths in the digraph D of Figure 9.1.13. A shortest directed $u - v$ path has length 2 and so $\vec{d}(u, v) = 2$. On the other hand, there is no directed $v - u$ path in D. In fact, there is no directed $x - u$ path in D for any vertex x ($\neq u$) of D since the indegree of u is 0. Therefore, D is not a strong digraph.

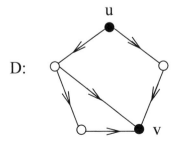

Figure 9.1.13: A digraph that is not strong.

E14: The vertices of the strong digraph D of Figure 9.1.14 are labeled by their eccentricities. Observe that $\mathrm{rad}(D) = 2$ and $\mathrm{diam}(D) = 5$. So, in general, it is not true that $\mathrm{diam}(D) \leq 2\,\mathrm{rad}(D)$.

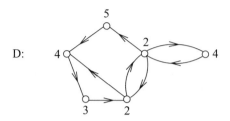

Figure 9.1.14: The eccentricities of the vertices of a strong digraph.

FACT

F37: [ChJoTi92] For every two positive integers a and b with $a \leq b$, there exists a strong digraph D with $\mathrm{rad}(D) = a$ and $\mathrm{diam}(D) = b$.

The Center of a Strong Digraph

DEFINITION

D28: The *center* $C(D)$ of a strong digraph D is the subdigraph induced by those vertices v with $e(v) = \mathrm{rad}(D)$.

EXAMPLE

E15: The strong digraph D of Figure 9.1.14 is repeated in Figure 9.1.15, where its center is also shown.

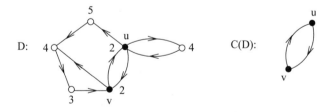

Figure 9.1.15: The center of a strong digraph.

FACT

F38: [ChJoTi92] For every oriented digraph D, there exists a strong oriented digraph whose center is (isomorphic to) D.

Strong Distance in Strong Digraphs

There is yet another reasonable way to define distance in strong digraphs, and this definition is analogous to Steiner distance in an undirected graph (see Definition D21).

DEFINITIONS

D29: For a strong oriented graph D, the **strong distance** $sd(u, v)$ from u to v is the minimum size of a strong subdigraph of D containing u and v.

D30: The **strong eccentricity** $se(v)$ of a vertex v in a strong oriented graph D is the largest strong distance from v to a vertex in D.

D31: The minimum strong eccentricity among the vertices of a strong oriented graph D is the **strong radius** $\mathrm{srad}(D)$ of D and the maximum strong eccentricity is its **strong diameter** $\mathrm{sdiam}(D)$.

EXAMPLE

E16:   In the strong digraph D_1 of Figure 9.1.16, $sd(v, w) = 3$, $sd(u, y) = 4$, and $sd(u, x) = 5$. The vertices of the digraph D_2 of Figure 9.1.16 are labeled with their strong eccentricities. Therefore, $\mathrm{srad}(D_2) = 6$ and $\mathrm{sdiam}(D_2) = 10$.

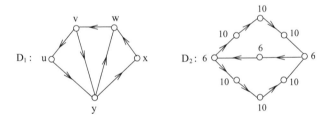

Figure 9.1.16: Strong eccentricities, strong radius, and strong diameter.

FACTS

F39: [ChErRaZh99-a] Strong distance is a metric on the vertex set of a strong oriented graph. Thus $\mathrm{srad}(D) \leq \mathrm{sdiam}(D) \leq 2\,\mathrm{srad}(D)$ for every strong oriented graph D.

F40: [ChErRaZh99-a] For every pair r, d of integers with $3 \leq r \leq d \leq 2r$, there exists a strong oriented graph D with $\mathrm{srad}(D) = r$ and $\mathrm{sdiam}(D) = d$.

F41: [ChErRaZh99-a] If D is a strong oriented graph of order $n \geq 3$, then $\mathrm{sdiam}(D) \leq \lfloor 5(n - 1)/3 \rfloor$ and this bound is sharp.

Strong Centers of Strong Digraphs

DEFINITIONS

D32:   A vertex v in a strong oriented graph D is called a **strong central vertex** if $se(v) = \mathrm{srad}(G)$, while the subgraph induced by the strong central vertices of D is the **strong center** $SC(D)$ of D.

D33:   A strong oriented graph D is called **strongly self-centered** if srad(D) = sdiam(D), that is, if D is its own strong center.

EXAMPLE

E17: The strong center $SC(D)$ of a digraph D is shown in Figure 9.1.17.

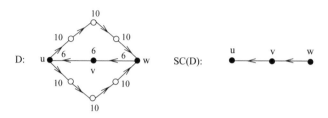

Figure 9.1.17: The strong center of a digraph.

FACTS

F42:   [ChErRaZh99-b] Every oriented graph is (isomorphic to) the strong center of some strong oriented graph.

F43:   [ChErRaZh99-b] For every integer $r \geq 3$, there exist infinitely many strongly self-centered oriented graphs of strong radius r.

References

[BiSy83] H. Bielak and M. M. Syslo, Peripheral vertices in graphs. *Studia Sci. Math. Hungar.* 18 (1983), 269–75

[Bu79] F. Buckley, Self-centered graphs with a given radius. *Congr. Numer.* 23 (1979), 211–215.

[BuHa90] F. Buckley and F. Harary, *Distance in Graphs*. Addison-Wesley, Redwood City, CA (1990).

[BuHaQu] F. Buckley, F. Harary, and L. V. Quintas, Extremal results on the geodetic number of a graph. *Sci. Ser. A Math. Sci.* 2 (1988), 17–26.

[BuMiSl81] F. Buckley, Z. Miller, and P. J. Slater, On graphs containing a given graph as center. *J. Graph Theory* 5 (1981), 427–434.

[ChErRaZh99-a] G. Chartrand, D. Erwin, M. Raines, and P. Zhang, Strong distance in strong digraphs. *J. Combin. Math. Combin. Comput.* 31 (1999), 33–44.

[ChErRaZh99-b] G. Chartrand, D. Erwin, M. Raines, and P. Zhang, On strong distance in strong oriented graphs. *Congr. Numer.* 141 (1999), 49–63.

[ChHaZh00] G. Chartrand, F. Harary, and P. Zhang, On the hull number of a graph. *Ars Combin.* 57 (2000), 129–138.

[ChHaZh02] G. Chartrand, F. Harary, and P. Zhang, On the geodetic number of a graph. *Networks.* 39 (2002), 1–6.

[ChJoTi92] G. Chartrand, G. L. Johns, and S. Tian, Directed distance in digraphs: centers and peripheries. *Congr. Numer.* 89 (1992), 89–95.

[ChOeTiZo89] G. Chartrand, O. R. Oellermann, S. Tian, and H. B. Zou, Steiner distance in graphs. *Časopis Pro Pest. Mat.* 114 (1989), 399–410.

[EnJaSn76] R. C. Entringer, D. E. Jackson, and D. E. Snyder, Distance in graphs. *Czech. Math. J.* 26 (1976), 283–296.

[EvSe85] M. G. Everett and S. B. Seidman, The hull number of a graph. *Discrete Math.* 57 (1985), 217–223.

[HaLoTs93] F. Harary, E. Loukakis, and C. Tsouros, The geodetic number of a graph. *Math. Comput. Modeling* 17 (1993), 89–95.

[HaNi81] F. Harary and J. Nieminen, Convexity in graphs. *J. Differential Geom.* 16 (1981), 185–190.

[HaNi53] F. Harary and R. Z. Norman, The dissimilarity characteristic of Husimi trees. *Ann. of Math.* 58 (1953), 134–141.

[HeOeSw90] M. A. Henning, O. R. Oellermann, and H. C. Swart, On Steiner radius and Steiner diameter of a graph. *Ars Combin.* 29 (1990), 13–19.

[HeOeSw91] M. A. Henning, O. R. Oellermann, and H. C. Swart, On vertices with maximum Steiner eccentricity in graphs. *Graph Theory, Combinatorics, Algorithms and Applications* (eds. Y. Alav, F. R. K. Chung, R. L. Graham, and D. F. Hsu), SIAM Publications (1991), 393–403.

[Ho89] K. S. Holbert, A note on graphs with distant center and median. *Recent Studies in Graph Theory.* Vishwa, Gulbarga (1989), 155–158.

[Jo69] C. Jordan, Sur les assemblages de lignes. *J. Reine Agnew. Math* 70 (1869), 185–190.

[LeWa90] S. M. Lee and P. C. Wang, On groups of automorphisms of self-centered graphs. *Bull. Math. Soc. Sci. Math. Roumanie* 34 (1990), 11–316.

[NoTi91] K. Novotny and S. Tian, On graphs with intersecting center and median. *Advances in Graph Theory.* Vishwa, Gulbarga (1991), 297–300.

[OeTi90] O. R. Oellermann and S. Tian, Steiner center in graphs. *J. Graph Theory.* 14 (1990), 585–597.

[Os73] P. A. Ostrand, Graphs with specified radius and diameter. *Discrete Math.* 4 (1973), 71–75.

[Sl80] P. J. Slater, Medians of arbitrary graphs. *J. Graph Theory* 4 (1980), 389–392.

[Tr85] M. Truszczyński, Centers and centroids of unicyclic graphs. *Math. Slovaca* 35 (1985), 223–228.

Section 9.2
Domination in Graphs

Teresa W. Haynes, East Tennessee State University
Michael A. Henning, University of Johannesburg, South Africa

INTRODUCTION

We consider sets of vertices that "are near" (dominate) all the vertices of a graph. The idea of **domination** is an area of research in graph theory that is experiencing significant growth. Its application in design and analysis of communication networks, social sciences, optimization, bioinformatics, computational complexity, and algorithm design may explain in part the increased interest.

The books by Haynes, Hedetniemi, and Slater [HaHeSl98, HaHeSl98b] deal exclusively with domination in graphs. Recent survey articles on domination in graphs can be found in [He96] and [HeLa90]. For a comprehensive bibliography of papers on dominating sets in graphs, see the reference list compiled in [HaHeSl98] that contains over 1200 entries.

9.2.1   Dominating Sets in Graphs

DEFINITIONS

D1: A set $S \subseteq V$ is a **dominating set** of a graph $G = (V, E)$ if each vertex in V is in S or is adjacent to a vertex in S.

D2: The **domination number** $\gamma(G)$ is the minimum cardinality of a dominating set of G. We refer to a minimum dominating set of a graph G as a $\gamma(G)$-**set**.

REMARK

R1: A vertex is said to dominate itself and all its neighbors.

EXAMPLE

E1: The set $S = \{2, 5, 7\}$ is a dominating set in the Petersen graph G shown in Figure 9.2.1. Since the Petersen graph is a cubic graph, each vertex dominates four vertices. Therefore no set of two vertices dominates all ten vertices in the graph, and so the set S is a dominating set of minimum cardinality (a $\gamma(G)$-set) and $\gamma(G) = 3$.

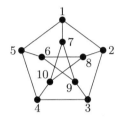

Figure 9.2.1: The Petersen graph G.

REMARK

R2: The **open neighborhood** of a vertex v, denoted $N(v)$, is the set of vertices which are adjacent to v. One can in fact show that the Petersen graph in Figure 9.2.1 has exactly ten distinct $\gamma(G)$-sets, namely, the ten open neighborhoods $N(v)$ corresponding to the ten vertices v in the graph.

Equivalent Definitions of a Dominating Set

The domination number appears in many different mathematical contexts or frameworks. We mention a few of the equivalent definitions of a dominating set:

DEFINITIONS

D3: *Vertex Set Covering Problem.*

Set $S \subseteq V$ is a **dominating set** of a graph G if each vertex in $V \setminus S$ has at least one neighbor (is covered by a vertex) in S.

D4: *Set Intersection.*

The **closed neighborhood** of a vertex v is the set $N[v] = N(v) \cup \{v\}$. Set $S \subseteq V$ is a **dominating set** if for every vertex $v \in V$, $|N[v] \cap S| \geq 1$.

D5: *Union of Neighborhoods.*

Set $S \subseteq V$ is a **dominating set** if $\bigcup_{v \in S} N[v] = V$.

D6: *Dominating Function.*

Let f be the function $f : V \to \{0, 1\}$ such that for each $v \in V$,

$$\sum_{u \in N[v]} f(u) \geq 1.$$

The vertices with the value 1 under f form a ***dominating set***.

D7: *Distance from the Set.*

Set $S \subseteq V$ is a ***dominating set*** if for every vertex $v \in V \setminus S$, $d(v, x) \leq 1$ for some vertex $x \in S$.

D8: *Integer Programming Problem.*

For a graph G of order n, the ***closed neighborhood matrix***, denoted N, is the $n \times n$ matrix formed from the adjacency matrix of G by adding 1's on the diagonal. For a set S of G, we let the n-vector $X_S = [x_i]$ be the ***characteristic vector*** of S, that is, $x_i = 1$ if $x_i \in S$ and $x_i = 0$, otherwise. Let $\bar{1}_n$ denote the column vector of all 1's. We say that S is a dominating set of G if and only if $N \cdot X_S \geq \bar{1}_n$. This leads to the integer programming formulation for the ***domination number*** $\gamma(G)$ given by:

$$\gamma(G) = \min \sum_{i=1}^{n} x_i$$

$$\text{subject to} \quad N \cdot X \geq \bar{1}_n$$

$$\text{with} \quad x_i \in \{0, 1\}.$$

Applications of Domination

The applications of domination in a wide variety of fields have surely added to its escalating popularity. For a sample of its applications, consider communication networks, facility and guard location problems, surveillance systems, and coding theory.

EXAMPLES

E2: Berge [Be73] mentions the problem of keeping a number of strategic locations under surveillance by a set of radar stations. The minimum number of radar stations needed to survey all the locations is the domination number of the associated graph.

E3: Liu [Li68] discusses the application of dominance to communications in a network, where a dominating set represents a set of cities which, acting as transmission stations, can transmit messages to every city in the network.

E4: The notion of domination is a standard one in coding theory. If one defines a graph whose vertices are the n-dimensional vectors with co-ordinates chosen from $(1, \ldots, p)$ and two vertices are adjacent if they differ in one co-ordinate, then sets of vectors which are (n, p)-covering sets, single error correcting codes, or perfect covering sets are all dominating sets of the graph with certain additional properties. See, for example, Kalfleisch, Stanton, and Horton [KaStHo71].

E5: A desirable property for a committee from a collection of people might be that every nonmember know at least one member of the committee, for ease of communication. A committee with this property is a dominating set of the acquaintance graph of the set of people.

9.2.2 Minimality Conditions

Notice that if S is a dominating set of a graph G, then so too is every superset of S. However, not every subset of S is necessarily a dominating set.

DEFINITION

D9: A *minimal dominating set* in a graph G is a dominating set that contains no dominating set as a proper subset.

EXAMPLE

E6: For the Petersen graph G of Figure 9.2.1, the sets $S_1 = N(1) = \{2, 5, 7\}$, $S_2 = \{1, 3, 6, 10\}$, and $S_3 = \{1, 2, 3, 4, 5\}$ are all minimal dominating sets. Hence, the Petersen graph contains minimal dominating sets of cardinalities 3, 4, and 5.

Early work on the topic of domination focused on properties of minimal dominating sets. We begin with two classical results of Ore [Or62].

FACTS

F1: (Ore's Theorem) [Or62] Let D be a dominating set of a graph $G = (V, E)$. Then D is a minimal dominating set of G if and only if each $v \in D$ has at least one of the following two properties: $[P_1]$: there exists a vertex $w \in V \setminus D$ such that $N(w) \cap D = \{v\}$; $[P_2]$: the vertex v is adjacent to no other vertex of D.

F2: [Or62] If $G = (V, E)$ is a graph with no isolated vertex and D is a minimal dominating set of G, then $V \setminus D$ is a dominating set of G.

F3: [BoCo79] If G is a graph with no isolated vertex, then there exists a $\gamma(G)$-set in which every vertex has property P_1.

The Domination Chain

Here we discuss a domination inequality chain.

DEFINITION

D10: A set S of vertices is said to be *independent* if no two vertices in S are adjacent.

D11: The *vertex independence number* $\alpha(G)$ is the maximum cardinality of an independent set in G.

D12: The *independent domination number* $i(G)$ is the minimum cardinality of a maximal independent set of G.

D13: While the domination number $\gamma(G)$ is the smallest cardinality of a minimal dominating set in a graph G, the *upper domination number* $\Gamma(G)$ is the maximum cardinality of a minimal dominating set in G.

D14: For any set $S \subseteq V$, the *open neighborhood* $N(S)$ is defined as $\cup_{v \in S} N(v)$ and the *closed neighborhood* $N[S] = N(S) \cup S$. A set S of vertices is *irredundant* if for every vertex $v \in S$, $N[v] \setminus N[S - \{v\}] \neq \emptyset$.

D15: The minimum cardinality of a maximal irredundant set in G is called the **irre-dundance number** of G, and is denoted $ir(G)$.

D16: The maximum cardinality of an irredundant set in G is called the **upper irre-dundance number** of G, and is denoted $IR(G)$.

EXAMPLES

E7: The tree T in Figure 9.2.2 has maximal independent sets of two sizes: $\{1, 2, 3, 6, 7, 8\}$ and $\{1, 2, 3, 5\}$. Thus, $i(T) = 4$ and $\alpha(T) = 6$.

Figure 9.2.2: A tree T with $i(T) = 4$ and $\alpha(T) = 6$.

E8: The tree T in Figure 9.2.3 has maximal irredundant sets of two different sizes: $\{2, 3, 8, 9\}$ and $\{2, 4, 6, 8, 10\}$. Thus, $ir(T) = 4$ and $IR(T) = 5$. Note that for this tree T, $\gamma(T) = 5$.

Figure 9.2.3: A tree T with ir $(T) = 4$ and IR $(T) = 5$.

FACTS

F4: [Be62] An independent set is maximal independent if and only if it is independent and dominating.

F5: [Be62] Every maximal independent set in a graph is a minimal dominating set of the graph.

F6: [CoHeMi78] A dominating set is a minimal dominating set if and only if it is dominating and irredundant.

F7: [BoCo79] Every minimal dominating set in a graph is a maximal irredundant set of the graph.

Since every maximal independent set is a dominating set, and every minimal dominating set is a maximal irredundant set, we have the following inequality chain, which was first observed by Cockayne, Hedetniemi, and Miller in 1978.

F8: [CoHeMi78] For every graph G,

$$\text{ir}\,(G) \leq \gamma(G) \leq i(G) \leq \alpha(G) \leq \Gamma(G) \leq \text{IR}\,(G).$$

F9: [CoFaPaTh81] If G is a bipartite graph, then $\alpha(G) = \Gamma(G) = \text{IR}\,(G)$.

REMARK

R3: The inequality chain, known as the **domination chain**, in Fact F8 has become one of the strongest focal points for research in domination theory; approximately 100 research papers have been published on various aspects of this sequence of inequalities. For example, Cockayne, Favaron, Mynhardt, and Puech [CoFaMyPu00] characterized trees T with $\gamma(T) = i(T)$ in terms of the sets of vertices of T which are contained in all its minimum dominating and minimum independent dominating sets. These sets were characterized by Mynhardt [My99], who used a tree pruning procedure. A simple constructive characterization of such trees is given in [DoGoHeMy06].

9.2.3   Domination Perfect Graphs

Motivated by the concept of perfect graphs in the chromatic sense, Sumner and Moore [SuMo79] defined a graph to be *domination perfect* as follows.

DEFINITION

D17:   A graph G is **domination perfect** if $\gamma(H) = i(H)$ for every induced subgraph H of G.

Building on several other results, including those by Zverovich and Zverovich [ZvZv91], Fulman [Fu93], and Topp and Volkmann [ToVo91], Zverovich and Zverovich [ZvZv95] finally provided a forbidden induced subgraph characterization of domination perfect graphs in terms of seventeen forbidden induced subgraphs.

FACT

F10:   [ZvZv95] A graph is domination perfect if and only if it does not contain one of seventeen graphs G_1–G_{17} shown in Figure 9.2.4 as an induced subgraph.

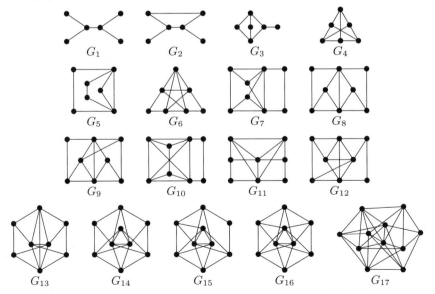

Figure 9.2.4: Minimal domination imperfect graphs G_1–G_{17}.

DEFINITION

D18: A graph G is called ***minimal domination imperfect*** if G is not domination perfect and $\gamma(H) = i(H)$, for every proper induced subgraph H of G.

REMARK

R4: The 17 subgraphs presented in the characterization of domination perfect graphs given in Fact F10 are all minimal domination imperfect graphs. Thus Fact F10 states that there are exactly 17 minimal domination imperfect graphs.

9.2.4   Bounds on the Domination Number

Since determining whether a graph has domination number at most k is NP-complete (see [GaJo79] and Chapter 1 of [HaHeSl98]), it is of interest to find bounds for this parameter.

REMARK

R5: Obviously if G is a graph of order n, then $1 \leq \gamma(G) \leq n$. Equality of the lower bound is attained if and only if $\Delta(G) = n - 1$, and equality holds for the upper bound if and only if $\Delta(G) = 0$, i.e., $G = \overline{K}_n$.

Bounds in Terms of Order and Minimum Degree

Restricting ourselves to graphs without isolated vertices, we have the following general upper bound on the domination number of a graph.

FACT

F11: [BuHeTu12] Let G be a graph of order n and size m having no isolated vertices. Then, the bound $\gamma(G) \leq an + bm$ is valid if and only if both $2a + b \geq 1$ and $b \geq 0$ hold.

REMARK

R6: Taking $(a, b) = (\frac{1}{2}, 0)$ in Fact F11, we have that the upper bound on the domination number can be improved from its order to one-half its order if we impose the condition that the graph is isolate-free. This result was first observed by Ore [Or62] and is a consequence of Facts F2 and F3.

FACT

F12: [Or62] If G is a graph of order n with no isolated vertex, then $\gamma(G) \leq n/2$.

DEFINITION

D19: The **corona** of two graphs G_1 and G_2 is the graph $G = G_1 \circ G_2$ formed from one copy of G_1 and $|V(G_1)|$ copies of G_2 where the *ith* vertex of G_1 is adjacent to every vertex in the *ith* copy of G_2. The corona $H \circ K_1$, in particular, is the graph constructed from a copy of H and for each vertex $v \in V(H)$, a new vertex v' and the pendant edge vv' are added.

FACT

F13: [FiJaKiRo85, PaXu82] If G is a graph of order n with no isolated vertex, then $\gamma(G) = n/2$ if and only if the components of G are the cycle C_4 or the corona $H \circ K_1$ for any connected graph H.

REMARK

R7: The graphs G for which $\gamma(G) = \lfloor n/2 \rfloor$ were characterized independently in [Ba-CoHaHeSh00] and [RaVo98].

Minimum Degree Two

If we restrict the minimum degree $\delta(G)$ of G to be at least two, then the upper bound in Fact F12 due to Ore on the domination number can be improved from one-half its order to two-fifths its order except for seven exceptional graphs (one of order four and six of order seven). More precisely, McCuaig and Shepherd [McSh89] defined a collection $\mathcal{B}$ of "bad" graphs, shown in Figure 9.2.5.

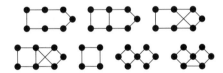

Figure 9.2.5: The family $\mathcal{B}$ of "bad" graphs.

FACTS

F14: [McSh89] If G is a connected graph of order n with $\delta(G) \geq 2$ and $G \notin \mathcal{B}$, where $\mathcal{B}$ is the collection of seven graphs shown in Figure 9.2.5, then $\gamma(G) \leq 2n/5$.

F15: [McSh89] If G is a connected graph of order $n \geq 8$ with $\delta(G) \geq 2$, then $\gamma(G) \leq 2n/5$.

DEFINITIONS

D20: A **key** $L_{4,1}$ is the graph obtained from a 4-cycle C_4 by adding a pendant edge to one of its vertices. We define a **unit** to be a graph that is isomorphic to a 5-cycle C_5 or to a key $L_{4,1}$. We call a unit a **cycle unit** or a **key unit** according to whether it is a cycle or a key, respectively. In a cycle unit, we select two vertices at distance two apart in the unit, and we call these two vertices the **link vertices** of the unit; in a key unit we call the vertex of degree one the **link vertex** of the unit.

D21:   To show that the bound of Fact F15 is sharp, McCuaig and Shepherd [McSh89] introduced a ***family $\mathcal{F}$ of graphs*** constructed as follows. Let $\mathcal{F}$ denote the family of all graphs G that are obtained from the disjoint union of at least two units, each of which is a cycle unit or a key unit, by adding edges in such a way that G is connected and every added edge joins two link vertices.

EXAMPLE

E9:   A graph in the family $\mathcal{F}$ with three cycle units and two key units is shown in Figure 9.2.6 with the link vertices indicated by the darkened vertices.

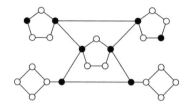

Figure 9.2.6: A graph in the family $\mathcal{F}$.

FACT

F16:   [McSh89] If $G \in \mathcal{F}$ has order n, then G is a connected graph with $\delta(G) \geq 2$ satisfying $\gamma(G) = 2n/5$.

Minimum Degree Three

If we restrict the minimum degree to be at least three, then the upper bound in Fact F15 due to McCuaig and Shepherd on the domination number can be improved from two-fifths its order to three-eighths its order.

FACT

F17:   [Re96] If G is a graph of order n and $\delta(G) \geq 3$, then $\gamma(G) \leq 3n/8$.

EXAMPLE

E10:   The two non-planar cubic graphs of order $n = 8$ (shown in Figure 9.2.7) both have domination number $3 = 3n/8$. We remark that the graph M_8 in Figure 9.2.7(a) is called the Mobius ladder on eight vertices.

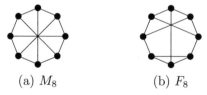

(a) M_8 (b) F_8

Figure 9.2.7: The two non-planar cubic graphs of order eight.

REMARK

R8: The ratio $3/8$ in Reed's Fact F17 is best possible. Gamble (see [McSh89], [Re96]) gave infinitely many connected graphs of minimum degree at least three with domination number exactly three-eighths their order.

EXAMPLE

E11: Let H' be any connected graph. For each vertex v of H', add a (disjoint) copy of the cubic graph F_8 shown in Figure 9.2.7(b) and identify any one of its vertices that is in a triangle with v. Let H denote the resulting graph, and let $\mathcal{H}$ denote the family of all such graphs H. When $H' = P_3$, the resulting graph H is shown in Figure 9.2.8.

Figure 9.2.8: A graph $H \in \mathcal{H}$.

REMARK

R9: Reed's Fact F17 has been generalized in several ways.

FACTS

F18: [ShSoYuHe09] If G is a graph of order n and $\delta(G) \geq 2$, then $\gamma(G) \leq (3n+|V_2|)/8$, where V_2 denotes the set of vertices of degree 2 in G.

F19: [LoRa08] If G is a graph of order n and $\delta(G) \geq 2$ with girth $g \geq 5$, then $\gamma \leq (\frac{1}{3} + \frac{2}{3g})n$.

DEFINITIONS

D22: A vertex x is called a ***bad-cut-vertex*** of G if $G - x$ contains a component, C_x, which is an induced 4-cycle and x is adjacent to at least one but at most three vertices on C_x. We let $\mathrm{bc}(G)$ denote the number of bad-cut-vertices in G.

D23: A cycle C is called a ***special-cycle*** if C is a 5-cycle in G such that if u and v are consecutive vertices on C, then at least one of u and v has degree 2 in G. We let $\mathrm{sc}(G)$ denote the maximum number of vertex disjoint special-cycles in G that contain no bad-cut-vertices.

D24: A graph is (C_4, C_5)***-free*** if it has no induced 4-cycle or 5-cycle.

FACTS

F20: [HeScYe11] If G is a connected graph of order $n \geq 14$ with $\delta(G) \geq 2$, then $\gamma(G) \leq \frac{1}{8}(3n + \mathrm{sc}(G) + \mathrm{bc}(G))$.

F21: [HeScYe11] If G is a connected graph of order $n \geq 14$ with $\delta(G) \geq 2$ that contains no special cycle and no bad-cut-vertex, then $\gamma(G) \leq 3n/8$.

F22: [HeScYe11] If G is a (C_4, C_5)-free connected graph of order $n \geq 14$ with $\delta(G) \geq 2$, then $\gamma(G) \leq 3n/8$.

F23: [HeScYe11] If G is a 2-connected graph of order $n \geq 14$ and $d_G(u) + d_G(v) \geq 5$ for every two adjacent vertices u and v, then $\gamma(G) \leq 3n/8$.

REMARKS

R10: Fact F23 can be restated as follows: If G is a 2-connected graph of order $n \geq 14$ such that the set of degree-2 vertices in G form an independent set, then $\gamma(G) \leq 3n/8$.

R11: The ratio $3/8$ in Fact F22 is best possible. Henning, Schiermeyer, and Yeo [HeScYe11] gave infinitely many (C_4, C_5)-free connected graphs of minimum degree two with domination number exactly three-eighths their order.

R12: That the bound of Fact F23 is sharp may be seen as follows. Let $k \geq 2$ be an integer, and let $\mathcal{H}$ be the family of all graphs that can be obtained from a 2-connected graph F of order $2k$ that contains a perfect matching M as follows. Replace each edge $e = uv$ in the matching M by an 8-cycle $uavbcdefu$ with two added edges, namely, be and cf. Let H denote the resulting 2-connected graph of order $n = 8k$. Then, $\gamma(H) = 3k = 3n/8$ and the set of degree-2 vertices in H form an independent set.

EXAMPLE

E12: A graph in the family $\mathcal{H}$ with $k = 4$ that is obtained from an 8-cycle F is shown in Figure 9.2.9.

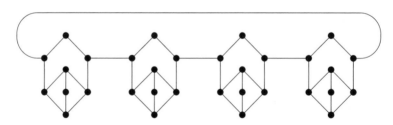

Figure 9.2.9: A graph in the family $\mathcal{H}$.

Cubic Graphs

REMARK

R13: As a special case of Fact F17, every connected cubic graph G on n vertices has domination number $\gamma(G) \leq 3n/8$. Kostochka and Stodolsky [KoSt09] proved that the two non-planar cubic graphs F_8 and G_8 shown in Figure 9.2.7 are the only connected cubic graphs that achieve the three-eighths bound in Fact F17.

DEFINITION

D25: Let $\mathcal{G}^n_{\text{cubic}}$ denote the family of all connected cubic graphs of order n.

FACT

F24: [KoSt09] If $G \in \mathcal{G}_{\text{cubic}}^n$ and $n \geq 10$, then $\gamma(G) \leq 4n/11$.

REMARK

R14: Reed [Re96] conjectured that the upper bound in Fact F17 can be improved to $\gamma(G) \leq \lceil n/3 \rceil$ if $G \in \mathcal{G}_{\text{cubic}}^n$.

FACTS

F25: Kostochka and Stodolsky [KoSt09] disproved Reed's conjecture by constructing a connected cubic graph G on 60 vertices with $\gamma(G) = 21$ and presented a sequence $\{G_k\}_{k=1}^{\infty}$ of connected cubic graphs with

$$\lim_{k \to \infty} \frac{\gamma(G_k)}{|V(G_k)|} \geq \frac{8}{23} = \frac{1}{3} + \frac{1}{69}.$$

F26: Kelmans [Ke06] constructed a smaller (with 54 vertices) counter-example to Reed's conjecture and an infinite series of 2-connected cubic graphs H_k with

$$\lim_{k \to \infty} \frac{\gamma(H_k)}{|V(H_k)|} \geq \frac{1}{3} + \frac{1}{60}.$$

F27: [Ke06], [KoSt09] The following holds.

$$0.35 = \frac{1}{3} + \frac{1}{60} \leq \sup_{G \in \mathcal{G}_{\text{cubic}}^n} \left(\lim_{n \to \infty} \frac{\gamma(G)}{n} \right) \leq \frac{4}{11} \approx 0.363636.$$

More Bounds Involving Minimum Degree

FACTS

F28: [SoXu09] If G is a graph of order n and $\delta(G) \geq 4$, then $\gamma(G) \leq 4n/11$.

F29: [XiSuCh06] If G is a graph of order n and $\delta(G) \geq 5$, then $\gamma(G) \leq 5n/14$.

F30: [CaRo85, CaRo90] For any graph G of order n and minimum degree δ,

$$\gamma(G) \leq n \left[1 - \delta \left(\frac{1}{\delta + 1} \right)^{1 + 1/\delta} \right].$$

F31: [AlSp92, Ar74, Pa75] For any graph G of order n and minimum degree δ,

$$\gamma(G) \leq n \left(\frac{1 + \ln(\delta + 1)}{\delta + 1} \right).$$

F32: [Ar74, Pa75] For a graph G of order n and minimum degree δ,

$$\gamma(G) \leq \frac{n}{\delta + 1} \sum_{j=1}^{\delta+1} \frac{1}{j}.$$

F33: [We81] Let $k = \lfloor \log_2 n - 2 \log_2(\log_2 n) + \log_2(\log_2 e) \rfloor$. Then for almost every graph G of order n, $k + 1 \leq \gamma(G) \leq k + 2$.

Bounds in Terms of Size and Degree

FACTS

F34: If G is a connected graph of size m, then $\gamma(G) \le (m+1)/2$.

F35: [Sa97] If G is a connected graph of size m with $\delta(G) \ge 2$, then $\gamma(G) \le (m+2)/3$ with equality if and only if G is a cycle of length n where $n \equiv 1 \,(\text{mod}\,3)$.

DEFINITION

D26: A graph G with m edges is called an $\frac{m}{3}$ *-graph* if $\delta(G) \ge 2$, G is connected, and $\gamma(G) > m/3$.

EXAMPLE

E13: An example of an $\frac{m}{3}$-graph is shown in Figure 9.2.10.

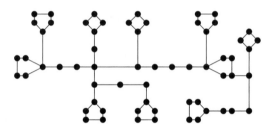

Figure 9.2.10: A graph G of size m with $\gamma(G) = (m+1)/3$.

REMARK

R15: A characterization of $\frac{m}{3}$-graphs can be found in [He99].

Bounds in Terms of Order and Maximum Degree

FACTS

F36: [Be73, WaAcSa79] For any graph G of order n with maximum degree Δ,

$$\left\lceil \frac{n}{1+\Delta} \right\rceil \le \gamma(G) \le n - \Delta.$$

F37: [FlVo90] For any graph G of order n with no isolated vertex, minimum degree δ, and maximum degree Δ,

$$\gamma(G) \le \frac{1}{2}\left(n + 1 - (\delta - 1)\frac{\Delta}{\delta} \right).$$

The following fact is an immediate consequence of Fact F37.

F38: [Pa75] For a graph G of order n with no isolated vertex and minimum degree δ,

$$\gamma(G) \le \frac{1}{2}(n + 2 - \delta).$$

Bounds in Terms of Order and Size

Vizing [Vi65] bounded the size of a graph having a given order and domination number. It follows that if a graph of a given order has sufficiently many edges, then it is guaranteed to have a dominating set of some specified order.

DEFINITION

D27:   A *minimum edge cover* in a graph is a minimum number of edges required to cover all the vertices of the graph.

FACT

F39:   [Vi65] If G is a graph of order n and size m with domination number $\gamma \geq 2$, then $m \leq \lfloor \frac{1}{2}(n-\gamma)(n-\gamma+2) \rfloor$. Furthermore, the maximum size is attained by taking the complete graph on $n - \gamma + 2$ vertices and removing a minimum edge cover and then adding $\gamma - 2$ isolated vertices.

REMARK

R16:   The graphs constructed in Fact F39 also achieve the upper bound of Fact F36; that is, such a graph G has $\Delta(G) = n - \gamma(G)$.

FACTS

F40:   [Sa91] If G is a graph of order n and size m with domination number $\gamma \geq 2$ and with $\Delta(G) \leq n - \gamma - 1$, then $m \leq \frac{1}{2}(n-\gamma)(n-\gamma+1)$.

F41:   [Be62,Vi65] For any graph G of order n and size m,

$$n - m \leq \gamma(G) \leq n + 1 - \sqrt{1 + 2m}.$$

Furthermore, $\gamma(G) = n - m$ if and only if every component of G is a star.

Bounds in Terms of Packing

DEFINITIONS

D28:   A set $S \subseteq V$ is called a *packing* in a graph $G = (V, E)$ if the vertices in S are pairwise at distance at least 3 apart in G, i.e., if $u, v \in S$, then $d_G(u, v) \geq 3$.

D29:   The *packing number* $\rho(G)$ of G is the maximum cardinality of a packing in G.

REMARK

R17:   Note that if S is a packing in G, then for each pair of vertices $u, v \in S$, $N[u] \cap N[v] = \emptyset$. Hence, the packing number provides a lower bound on $\gamma(G)$.

FACTS

F42:   For any graph G, $\rho(G) \leq \gamma(G)$.

F43:   [MeMo75] For a tree T, $\rho(T) = \gamma(T)$.

Bounds in Terms of Radius

FACT

F44:   [DePeWa10] If G is a connected graph on at least two vertices with radius r, then $\gamma(G) \geq 2r/3$.

REMARK

R18:   Equality holds in the bound of Fact F44 for cycles with orders congruent to 0 modulo 6.

9.2.5   Nordhaus–Gaddum-Type Results

The original paper by Nordhaus and Gaddum [NoGa56] in 1956 gave sharp bounds on the sum and product of the chromatic numbers of a graph and its complement. Since then such results have been given for several parameters. Fact F45 below is the first such result for domination.

FACT

F45:   [JaPa72] For any graph G of order $n \geq 2$,
  (a) $3 \leq \gamma(G) + \gamma(\overline{G}) \leq n + 1$, and
  (b) $2 \leq \gamma(G)\gamma(\overline{G}) \leq n$.

DEFINITIONS

D30:   For a pair of graphs G and H, the **cartesian product** $G \,\square\, H$ of graphs G and H is the graph with vertex set $V(G) \times V(H)$ and where two vertices are adjacent if and only if they are equal in one coordinate and adjacent in the other.

D31:   Let $G_1 \oplus G_2 \oplus G_3$ denote an edge-disjoint factoring of the complete graph.

FACTS

F46:   [PaXu82] Let G be a graph of order $n \geq 2$. Then, $\gamma(G)\gamma(\overline{G}) = n$ if and only if G is one or the complement of one of the following graphs: K_n, disjoint union of cycles of length 4 and the corona $H \circ K_1$ for any graph H, and $K_3 \,\square\, K_3$.

F47:   [JoAr95] If G is a graph of order $n \geq 2$ such that G and $\overline{G}$ have no isolated vertices, then $\gamma(G) + \gamma(\overline{G}) \leq (n + 4)/2$.

F48:   [GoHeSw92] Let $G_1 \oplus G_2 \oplus G_3 = K_n$. Then,
  (a) $\gamma(G_1) + \gamma(G_2) + \gamma(G_3) \leq 2n + 1$, and
  (b) the maximum value of the product $\gamma(G_1)\gamma(G_2)\gamma(G_3)$ is $n^3/27 + \Theta(n^2)$.

REMARK

R19:   From Fact F48, there exist constants c_1 and c_2 such that the maximum triple product always lies between $n^3/27 + c_1 n^2$ and $n^3/27 + c_2 n^2$.

9.2.6   Domination in Planar Graphs

The Dominating Set decision problem remains NP-hard even when restricted to planar graphs of maximum degree 3 (see [GaJo79]). Hence it is of interest to determine upper bounds on the domination number of a planar graph.

REMARK

R20: A tree of radius 2 and diameter 4 can have arbitrarily large domination number. So the interesting question is what happens when the diameter of a planar graph is 2 or 3. Bounding the diameter of a planar graph is a reasonable restriction to impose because planar graphs with small diameter are often important in applications (see [FeHeSe95]).

FACT

F49:   [MaSe96] The domination number of a planar graph of diameter 2 is bounded above by 3.

EXAMPLE

E14: The graph of Figure 9.2.11, constructed by MacGillivray and Seyffarth [MaSe96], shows that the bound in Fact F49 is achievable.

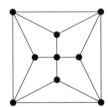

Figure 9.2.11: A planar graph of diameter 2 and domination number 3.

FACTS

F50:   [GoHe02] Every planar graph of diameter 2 has domination number at most 2 except for the graph of Figure 9.2.11, which has domination number 3.

F51:   [DoGoHe06] Every planar graph of diameter 3 and of radius 2 has domination number at most 5.

F52:   [DoGoHe06] Every planar graph of diameter 3 has domination number at most 9.

F53:   [DoGoHe06] Every sufficiently large planar graph of diameter 3 has domination number at most 6, and this bound is sharp.

F54: [GoHe02] For each orientable surface, there are finitely many graphs with diameter 2 and domination number more than 2.

F55: [GoHe02] For each orientable surface, there is a maximum domination number of graphs with diameter 3.

EXAMPLE

E15:   The sharpness of the bound in Fact F53 is shown by the graph of Figure 9.2.12, which can be made arbitrarily large by duplicating any of the vertices of degree 2. Furthermore, by adding edges joining vertices of degree 2, it is possible to construct such a planar graph with minimum degree equal to 3.

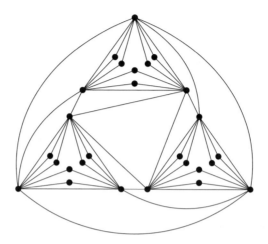

Figure 9.2.12: A planar graph with diameter 3 and domination number 6.

9.2.7   Vizing's Conjecture

One of the oldest unsolved problems in domination theory involves graph products. In 1963 Vizing [Vi63] suggested the problem of determining a lower bound on the domination number of a product graph in terms of the domination numbers of its factors G and H. Five years later he offered it as a conjecture, which remains one of the most famous open problems involving domination.

CONJECTURE

   Vizing's Conjecture [Vi68]: For any graphs G and H, $\gamma(G \,\square\, H) \geq \gamma(G)\gamma(H)$.

DEFINITION

D32:   A graph G is said to be ***decomposable*** if $\gamma(G) = k$ and $V(G)$ can be partitioned into k subsets $C_1, C_2, \ldots, C_k$ such that each of the induced subgraphs $G[C_i]$ is a complete subgraph of G.

FACTS

F56:   [BaGe79] If F is a spanning subgraph of a decomposable graph G where $\gamma(F) = \gamma(G)$, then for any graph H, $\gamma(F \,\square\, H) \geq \gamma(F)\gamma(H)$.

   Clark and Suen [ClSu00] provided the following general result.

F57:   [ClSu00] For any graphs G and H, $\gamma(G \,\square\, H) \geq \frac{1}{2}\gamma(G)\gamma(H)$.

The best general result to date is due to Suen and Tarr [SuTa10,Ta10].

F58: [SuTa10,Ta10] For any graphs G and H,

$$\gamma(G \,\square\, H) \geq \frac{1}{2}\gamma(G)\gamma(H) + \frac{1}{2}\min\{\gamma(G), \gamma(H)\}.$$

F59: [BrDoGoHaHeKlRa] Let G be a claw-free graph. Then for any graph H without isolated vertices, $\gamma(G \,\square\, H) \geq \frac{1}{2}\gamma(G)(\gamma(H) + 1)$.

F60: [AhSz09] If G is a chordal graph, then for any graph H, $\gamma(G \,\square\, H) \geq \gamma(G)\gamma(H)$.

REMARKS

R21: Since the family of graphs in Fact F56 includes trees and cycles, Vizing's Conjecture is true if one of G and H is a tree (independently proved by Jacobson and Kinch [JaKi86]) or a cycle (independently proved in [ElPa91]).

R22: For surveys on graph products and Vizing's conjecture, see [HaRa91], [HaRa95], [BrDoGoHaHeKlRa], and Hartnell and Rall's Chapter 7 of [HaHeSl98b].

9.2.8 Domination Critical Graphs

In this section we consider graphs that are critical with respect to their domination number. Criticality may be defined for different graph modifications, including vertex deletion and edge addition or deletion. For example, Brigham, Chinn, and Dutton [BrChDu88] studied graphs that are vertex domination critical in the sense that their domination number decreases when any vertex is deleted. Also, a graph may be critical in the sense that its domination number increases when any edge is deleted, as studied in [BaHaNiSu83] and [WaAc79]. However, the most attention has probably been directed to those graphs that are edge addition critical, that is, their domination number decreases when any missing edge is added (see [SuBl83]).

REMARK

R23: The addition of an edge to a graph cannot increase its domination number and can decrease it by at most one.

DEFINITION

D33: A graph G is ***domination critical*** if for every edge $e \notin E(G)$, $\gamma(G + e) = \gamma(G) - 1$. If G is a domination critical graph with $\gamma(G) = k$, then the graph G is said to be k-***critical***. Thus G is k-critical if $\gamma(G) = k$ and $\gamma(G + e) = k - 1$ for each edge $e \notin E(G)$.

FACTS

F61: The 1-critical graphs are (vacuously) K_n for $n \geq 1$.

F62: [SuBl83] A graph G is 2-critical if and only if each component of $\overline{G}$ is a star.

F63: [Su90] A disconnected graph G is 3-critical if and only if $G = A \cup B$ where either A is trivial and B is any 2-critical graph or A is complete and B is a complete graph minus a 1-factor.

F64: [SuBl83] The diameter of a connected 3-critical graph is at most 3.

F65: [SuBl83] Every connected 3-critical graph of even order has a 1-factor.

F66: [Wo90] Every connected 3-critical graph on more than six vertices has a hamiltonian path.

REMARKS

R24: To date only the 1-critical and 2-critical have been characterized. For $k > 2$, the structure of the k-critical graphs is more complex. Most of the known results concentrate primarily on the concept of 3-critical graphs.

R25: It remains an open problem to determine whether every connected 3-critical graph on more than six vertices is hamiltonian.

R26: For a survey of edge domination critical graph results, see Sumner's Chapter 16 in [HaHeSl98b].

R27: Graphs for which the domination number remains unchanged when a vertex is deleted, or an edge is deleted or added have also been studied. For a survey, see Chapter 5 of [HaHeSl98]. Note that six classes of graphs result from the effect these three graph modifications have on the domination number, that is, the changing or unchanging of the domination number. For example, the k-critical graphs form a class for which the domination number changes when an arbitrary edge is added, and a second class for this graph modification consists of the graphs where the domination number remains the same upon the addition of an arbitrary edge. Relationships among these six classes of graphs are investigated in [HaHe03].

9.2.9   Domination Parameters

As noted in [HaHa95], domination related parameters can be defined by combining domination with another graph theoretical property. We consider the parameters defined by imposing an additional constraint on the dominating set.

DEFINITION

D34: For a given graph property P, the ***conditional domination number*** $\gamma(G : P)$ is the smallest cardinality of a dominating set $S \subseteq V$ such that the induced subgraph $G[S]$ satisfies property P.

EXAMPLE

E16: Examples of conditional domination parameters.

$P1$. $G[S]$ is an independent set (independent domination [CoHe77]).

P2. $G[S]$ has no isolated vertices (total domination [CoDaHe80]).

P3. $G[S]$ is connected (connected domination [SaWa79]).

P4. $G[S]$ is a complete graph (clique domination [Ke85],[KeCo88]).

P5. $G[S]$ has a perfect matching (paired-domination [HaSl98]).

P6. $G[S]$ has a hamiltonian cycle (cycle domination [LeWi77]).

REMARKS

R28: By definition, $\gamma(G) \le \gamma(G : P)$ for any property P. With the exception of independent domination, these conditional domination parameters do not necessarily exist for all graphs. However, graphs with no isolated vertices have total and paired-dominating sets.

R29: In many cases the additional constraint on the dominating set is application driven. For example, as first suggested by Wu and Li [WuLi99], a virtual backbone in a wireless network can be represented by a connected dominating set in its underlying graph model.

R30: The generic nature of this formalization provides a method for defining new invariants by considering different properties P.

R31: Also, domination parameters have been defined by changing the method of dominating. For example, requiring that each vertex outside the dominating set has at least k neighbors in the dominating set is k-domination [FiJa85].

References

[AhSz09] R. Aharoni and T. Szabó, Vizing's conjecture for chordal graphs, *Discrete Math.* 309(6) (2009), 1766–1768.

[AlSp92] N. Alon and J. H. Spencer, *The Probablistic Method*, John Wiley and Sons, Inc., 1992.

[Ar74] V. I. Arnautov, Estimation of the exterior stability number of a graph by means of the minimal degree of its vertices (Russian), *Prikl. Mat. i Programmirovanie Vyp.* 11 (126) (1974), 3–8.

[BaCoHaHeSh00] Xu Baogen, E. J. Cockayne, T.W. Haynes, S.T. Hedetniemi, and Z. Shangchao, Extremal graphs for inequalities involving domination parameters, *Discrete Math.* 216 (2000), 1–10.

[BaGe79] A. M. Barcalkin and L. F. German, The external stability number of the Cartesian product of graphs, *Bul. Akad. Stiince RSS Moldoven* 94(1) (1979), 5–8.

[BaHaNi83] D. Bauer, F. Harary, J. Nieminen, and C. L. Suffel, Domination alteration sets in graphs, *Discrete Math.* 47 (1983), 153–161.

[Be62] C. Berge, *Theory of Graphs and its Applications*, (Methuen, London, 1962), 40–51.

[Be73] C. Berge, *Graphs and Hypergraphs*, North-Holland, Amsterdam, 1973.

[BoCo79] B. Bollobás and E. J. Cockayne, Graph-theoretic parameters concerning domination, independence, and irredundance, *J. Graph Theory* 3 (1979), 241–249.

[BrDoGoHaHeKlRa] B. Brešar, P. Dorbec, W. Goddard, B. Hartnell, M. A. Henning, S. Klavžar, and D. Rall, Vizing's conjecture: A survey and recent results, *J. Graph Theory* 69 (2012), 46–76.

[BrChDu88] R. C. Brigham, P. Z. Chinn, and R. D. Dutton, Vertex domination-critical graphs, *Networks* 18 (1988), 173–179.

[BuHeTu12] C. Bujtás, M. A. Henning, and Zs. Tuza, Transversals and domination in uniform hypergraphs, *European J. Combin.* 33 (2012), 62–71.

[CaRo85] Y. Caro and Y. Roditty, On the vertex-independence number and star decomposition of graphs, *Ars Combin.* 20 (1985), 167–180.

[CaRo90] Y. Caro and Y. Roditty, A note on the k-domination number of a graph, *Internat. J. Math. Sci.* 13 (1990), 205–206.

[ClSu00] W. E. Clark and S. Suen, An inequality related to Vizing's conjecture, *Electronic J. Combin.* 7 (2000), no. 1, Note 4, 3 pp. (electronic).

[CoDaHe80] E. J. Cockayne, R. M. Dawes, and S. T. Hedetniemi, Total domination in graphs, *Networks* 10 (1980), 211–219.

[CoFaMyPu00] E. J. Cockayne, O. Favaron, C. M. Mynhardt, and J. Puech, A characterization of (γ, i)-trees, *J. Graph Theory* 34(4) (2000), 277–292.

[CoFaPaTh81] E. Cockayne, O. Favaron, C. Payan, and A. Thomason, Contributions to the theory of domination, independence and irredundance in graphs, *Discrete Math.* 33 (1981), 249–258.

[CoHe77] E. J. Cockayne and S. T. Hedetniemi, Towards a theory of domination in graphs, *Networks* 7 (1977), 247–261.

[CoHeMi78] E. J. Cockayne, S. T. Hedetniemi, and D. J. Miller, Properties of hereditary hypergraphs and middle graphs, *Canad. Math. Bull.* 21 (1978), 461–468.

[DePeWa10] E. DeLaViña, R. Pepper, and B. Waller, Lower bounds for the domination number, *Discuss. Math. Graph Theory* 30(3) (2010), 475–487.

[DoGoHe06] M. Dorfling, W. Goddard, and M. A. Henning, Domination in planar graphs with small diameter II, *Ars Combin.* 78 (2006), 237–255.

[DoGoHeMy06] M. Dorfling, W. Goddard, M. A. Henning, and C. M. Mynhardt, Construction of trees and graphs with equal domination parameters, *Discrete Math.* 306 (2006), 2647–2654.

[ElPa91] M. El-Zahar and C. M. Pareek, Domination number of products of graphs, *Ars Combin.* 31 (1991), 223–227.

[FeHeSe95] M. Fellows, P. Hell, and K. Seyffarth, Large planar graphs with given diameter and maximum degree, *Discrete Applied Math.* 61 (1995), 133–154.

[FiJa85] J. F. Fink and M. S. Jacobson, *n*-domination in graphs, In Y. Alavi and A.J. Schwenk, editors, *Graph Theory with Applications to Algorithms and Computer Science* (Kalamazoo, MI 1984), 1985, 283–300. Wiley.

[FiJaKiRo85] J. F. Fink, M. S. Jacobson, L. F. Kinch, and J. Roberts, On graphs having domination number half their order, *Period. Math. Hungar.* 16 (1985), 287–293.

[FlVo90] P. Flach and L. Volkmann, Estimations for the domination number of a graph, *Discrete Math.* 80 (1990), 145–151.

[Fu93] J. Fulman, A note on the characterization of domination perfect graphs, *J. Graph Theory* 17 (1993), 47–51.

[GaJo79] M. R. Garey and D. S. Johnson, *Computers and Intractability: A Guide to the Theory of NP-completeness*, W.H. Freeman, San Francisco (1979).

[GoHeSw92] W. Goddard, M. A. Henning, and H. C. Swart, Some Nordhaus-Gaddum-type results, *J. Graph Theory* 16 (1992), 221–231.

[GoHe02] W. Goddard and M. A. Henning, Domination in planar graphs with small diameter, *J. Graph Theory* 40 (2002), 1–25.

[HaHa95] F. Harary and T. W. Haynes, Conditional graph theory IV: Dominating sets, *Utilitas Math.* 40 (1995), 179–192.

[HaRa91] B. Hartnell and D. F. Rall, On Vizing's conjecture, *Congr. Numer.* 82 (1991), 87–96.

[HaRa95] B. Hartnell and D. F. Rall, On Vizing's conjecture and the one-half argument, *Discuss. Math. Graph Theory* 15 (1995), 205–216.

[Ha97] T. W. Haynes, Domination in graphs: A brief overview, *J. Combin. Math. Combin. Comput.* 24 (1997), 225–237.

[HaHeSl98] T. W. Haynes, S. T. Hedetniemi, and P. J. Slater, *Fundamentals of Domination in Graphs*, Marcel Dekker, New York, 1998.

[HaHeSl98b] T. W. Haynes, S. T. Hedetniemi, and P. J. Slater, *Domination in Graphs: Advanced Topics*, Marcel Dekker, New York, 1998.

[HaHe03] T. W. Haynes and M. A. Henning, Changing and unchanging domination: a classification, *Discrete Math.* 272 (2003), 65–79.

[HaSl98] T. W. Haynes and P. J. Slater, Paired-domination in graphs, *Networks* 32 (1998), 199–206.

[HeLa90] S. T. Hedetniemi and R. C. Laskar, editors. *Topics on Domination*, Volume 48, North Holland, New York, 1990.

[He96] M. A. Henning, Domination in graphs: A survey, *In Surveys in Graph Theory* (eds. G. Chartrand and M. Jacobson), *Congress. Numer.* 116 (1996), 139–172.

[He99] M. A. Henning, A characterisation of graphs with minimum degree 2 and domination number exceeding a third their size, *J. Combin. Math. Combin. Comput.* 31 (1999), 45–64.

[HeScYe11] M. A. Henning, I. Schiermeyer, and A. Yeo, A new bound on the domination number of graphs with minimum degree two, *Electronic J. Combinatorics* 18 (2011), #P12.

[JaKi86] M. S. Jacobson and L. F. Kinch, On the domination of the products of graphs II: trees, *J. Graph Theory* 10 (1986), 97–106.

[JaPa72] F. Jaeger and C. Payan, Relations du type Nordhaus-Gaddum pour le nombre d'absorption d'un graphe simple, *C.R. Acad. Sci. Ser. A* 274 (1972), 728–730.

[JoAr95] J. P. Joseph and S. Arumugam, Domination in graphs, *Internat. J. Management Systems* 11 (1995), 177–182.

[KaStHo71] J. G. Kalfleisch, R. G. Stanton, and J. D. Horton, On covering sets and error-correcting codes, *J. Comb. Theory* 11A (1971), 233–250.

[Ke85] L. L. Kelleher, *Domination in Graphs and Its Application to Social Network Theory*, PhD thesis, Northeastern Univ., 1985.

[KeCo88] L. L. Kelleher and M. B. Cozzens, Dominating sets in social networks, *Math. Social. Sci.* 16 (1988), 267–279.

[Ke06] A. Kelmans, Counterexamples to the cubic graph domination conjecture, arXiv:math.CO/0607512 v1 20 July 2006.

[KoSt09] A. V. Kostochka and B. Y. Stodolsky, An upper bound on the domination number of n-vertex connected cubic graphs, *Discrete Math.* 309 (2009), 1142–1162.

[LeWi77] L. Lesniak-Foster and J. E. Williamson, On spanning and dominating circuits in graphs, *Canad. Math. Bull.* 20 (1977), 215–220.

[Li68] C. L. Liu, *Introduction to Combinatorial Mathematics*, McGraw-Hill, New York, 1968.

[LoRa08] C. Löwenstein and D. Rautenbach, Domination in graphs with minimum degree at least two and large girth, *Graphs Combin.* 24 (2008), 37–46.

[MaSe96] G. MacGillivray and K. Seyffarth, Domination numbers of planar graphs, *J. Graph Theory* 22 (1996), 213–229.

[McSh89] W. McCuaig and B. Shepherd, Domination in graphs with minimum degree two, *J. Graph Theory* 13 (1989), 749–762.

[MeMo75] A. Meir and J. W. Moon, Relations between packing and covering numbers of a tree, *Pacific J. Math.* 61 (1975), 225–233.

[My99] C. M. Mynhardt, Vertices contained in every minimum domination set of a tree, *J. Graph Theory* 31(3) (1999), 163–177.

[NoGa56] E. A. Nordhaus and J. W. Gaddum, On complementary graphs, *Amer. Math. Monthly* 63 (1956), 175–177.

[Pa75] C. Payan, Sur le nombre d'absorption d'un graphe simple, *Cahiers Centre Études Recherche Opér.* 17 (1975), 307–317.

[PaXu82] C. Payan and N. H. Xuong, Domination-balanced graphs, *J. Graph Theory* 6 (1982), 23–32.

[Or62] O. Ore, *Theory of Graphs, Amer. Math. Soc. Transl.* 38 (Amer. Math. Soc., Providence, RI, 1962), 206–212.

[RaVo98] B. Randerath and L. Volkmann, Characterization of graphs with equal domination and covering number, *Discrete Math.* 191 (1998), 159–169.

[Re96] B. A. Reed, Paths, stars and the number three, *Combin. Probab. Comput.* 5 (1996), 277–295.

[SaWa79] E. Sampathkumar and H. B. Walikar, The connected domination number of a graph, *J. Math. Phys. Sci.* 13 (1979), 607–613.

[Sa91] L. A. Sanchis, Maximum number of edges in connected graphs with a given domination number, *Discrete Math.* 87 (1991), 64–72.

[Sa97] L. A. Sanchis, Bounds related to domination in graphs with minimum degree two, *J. Graph Theory* 25 (1997), 139–152.

[ShSoYuHe09] E. R. Shan, M. Y. Sohn, X. D. Yuan, and M. A. Henning, Domination number in graphs with minimum degree two, *Acta Mathematica Sinica* (English Series) 25 (2009), 1253–1268.

[SoXu09] M. Y. Sohn and Y. Xudong, Domination in graphs with minimum degree four, *J. Korean Math. Soc.* 46 (2009), 759–773.

[SuTa10] S. Suen and J. Tarr, An Improved Inequality Related to Vizing's Conjecture, manuscript (2010).

[Su90] D. P. Sumner, Critical concepts in domination, *Discrete Math.* 86 (1990), 33–46.

[SuBl83] D. P. Sumner and P. Blitch, Domination critical graphs, *J. Combin. Theory Ser. B* 34 (1983), 65–76.

[SuMo79] D. P. Sumner and J. L. Moore, Domination perfect graphs, *Notice Amer. Math. Soc.* 26 (1979), A-569.

[Ta10] J. Tarr, Domination in graphs (2010). Theses and Dissertations. Paper 1786. http://scholarcommons.usf.edu/etd/1786.

[ToVo91] J. Topp and L. Volkmann, On graphs with equal domination and independent domination numbers, *Discrete Math.* 96 (1991), 75–80.

[Vi63] V. G. Vizing, The cartesian product of graphs, *Vychisl/ Sistemy* 9 (1963), 30–43.

[Vi65] V. G. Vizing, A bound on the external stability number of a graph, *Doklady A. N.* 164 (1965), 729–731.

[Vi68] V. G. Vizing, Some unsolved problems in graph theory, *Uspekhi Mat. Nauk* 23(6) (1968), 117–134.

[WaAc79] H. B. Walikar and B. D. Acharya, Domination critical graphs, *Nat. Acad. Sci. Lett.* 2 (1979), 70–72.

[WaAcSa79] H. B. Walikar, B. D. Acharya, and E. Sampathkumar, Recent developments in the theory of domination in graphs, Mehta Research Institute, Allahabad, MRI *Lecture Notes in Math.* 1 (1979).

[We81] K. Weber, Domination number for almost every graph, *Rostock. Math. Kolloq.*, 16 (1981), 31–43.

[Wo90] E. Wojcicka, Hamiltonian properties of domination-critical graphs, *J. Graph Theory* 14 (1990), 205–215.

[WuLi99] J. Wu and H. Li, On calculating connected dominating set for effficient routing in ad hoc wireless networks, In: *Proceedings of the 3rd ACM International Workshop on Discrete Algorithms and Methods for Mobile Computing and Communications* (1999), 7–14.

[XiSuCh06] H. M. Xing, L. Sun, and X. G. Chen, Domination in graphs of minimum degree five, *Graphs Combin.* 22 (2006), no. 1, 127–143.

[ZvZv91] I. E. Zverovich and V. E. Zverovich, A characterization of domination perfect graphs, *J. Graph Theory* 15 (1991), 109–114.

[ZvZv95] I. E. Zverovich and V. E. Zverovich, An induced subgraph characterization of domination perfect graphs, *J. Graph Theory* 20 (1995), 375–395.

Section 9.3
Tolerance Graphs

Martin Charles Golumbic, University of Haifa, Israel

INTRODUCTION

The study of tolerance graphs arose from a natural extension of ideas that have grown out of the now classical notion of interval graph [Be73, Be85, Go80, LeBo62]. In an interval graph, each vertex is associated with an interval on the real line with two vertices adjacent if and only if their associated intervals intersect. In 1982, Golumbic and Monma [GoMo82] suggested associating "tolerances" with each interval so that now two vertices are joined iff the length of the intersection of their associated intervals is at least the minimum of the two tolerances.

The first journal paper on tolerance graphs [GoMoTr84] set the stage for two decades of further research on global themes which focus on special families of tolerance graphs and their properties, directed graph versions, generalizations beyond intervals and restricted models. All these involve the notion of measured intersection known as *tolerance*, culminating in the book by Golumbic and Trenk [GoTr04].

In this survey, we will explore several of these themes. After introducing intersection graphs in Section 9.3.1, we will move on to the original class of (interval) tolerance graphs and their variations in Section 9.3.2. We will then describe several tolerance graph models on trees in Section 9.3.6, and finally present a more general framework called *rank-tolerance* [GoJa06] in Section 9.3.5.

9.3.1   Intersection Graphs and Their Applications

We begin with some basic definitions and results.

DEFINITION

D1: Let $\mathcal{C} = \{S_1, \ldots, S_n\}$ be a collection of sets. The ***intersection graph*** of $\mathcal{C}$ is the graph G obtained by assigning a distinct vertex v_i to each set S_i in $\mathcal{C}$ and joining two vertices by an edge precisely when their corresponding sets have a nonempty intersection, i.e., $v_i v_j \in E(G)$ if and only if $i \neq j$ and $S_i \cap S_j \neq \emptyset$.

FACT

F1: Every graph is an intersection graph. (Marczewski's Theorem [Ma45], see also [ErGoPo66, Ha94])

When the type of sets allowed in $\mathcal{C}$ is limited, interesting structured classes of graphs may result. Let $\mathcal{F}$ be a family of sets of a specified type, for example, intervals on a line, paths in a tree, circles in the plane, live areas of variables in a program flow diagram, senate subcommittee members, ideals in a commutative ring, patterns of consecutive elements in a DNA string, etc.

DEFINITION

D2: A graph G is an $\mathcal{F}$-***intersection graph*** if there exists an intersection ***representation*** $\mathcal{C} = \{S_1, \ldots, S_n\}$ of G where each S_i is in the family $\mathcal{F}$.

Among the well known classes of intersection graphs are the interval graphs, trapezoid graphs, circular-arc graphs, parallelogram graphs, chordal graphs (subtrees of a tree), permutation graphs, string graphs, segment graphs, and many others. The complements of comparability graphs (cocomparability graphs) also have an intersection graph model, known as function diagrams, as shown by Golumbic, Rotem and Urrutia [GoRoUr83]. Later, in Section 9.3.6, we will describe several types of intersection graphs on trees.

Most of the early uses of the intersection graph model have been described in the classical books [Be73, Go80, Ro76] and in some more recent books [BrLeSp99, MaPe95, McMc99, Sp03]. Yet, the scope of research in this general area has expanded significantly both from the modeling and algorithmic points of view.

Intersection graphs have become a necessary and important tool for solving real-world problems mathematically and algorithmically. Some of these applications include mobile frequency assignment [OsRo81, OsRo83], pavement deterioration analysis [GaNe81], relational databases [Go88], evolutionary trees [Wa95], physical mapping of DNA [GoGoKaSh95, GoKaSh94], container ship stowage [AvPeSh00], VLSI circuit design [DaGoPi88] and temporal reasoning notions from artificial intelligence [GoSh93, Go98, Go12].

Containment Graphs

Besides the intersection graph model, researchers have also considered a graph model based on the relation of containment.

DEFINITION

D3: The ***containment graph*** $G = (V, E)$ of a collection $\mathcal{C} = \{S_i\}$ of distinct subsets of a set $\mathbf{S}$ has vertex set $V = \{1, \ldots, n\}$ and edge set $E = \{ij \mid \text{ either } S_i \subset S_j \text{ or } S_j \subset S_i\}$. A graph with such a representation is called a ***containment graph***.

FACTS

F2: The class of containment graphs is equivalent to the class of *comparability graphs* or to the *transitively orientable (TRO)* graphs.

Golumbic and Scheinerman [GoSc89] observed the following variation on this fact.

F3: Every comparability graph can be represented as the containment graph of a collection of subtrees (substars) of a star.

Dushnik and Miller [DuMi41] characterized the containment graphs of intervals on the line as precisely those having partial order dimension 2; these are also equivalent to the permutation graphs, see [Go80, Ro76, Sp03]. Generalizing interval containment, Golumbic and Scheinerman [GoSc89] also showed the following.

F4: A graph G is the containment graph of rectilinear boxes[1] in d-space if and only if the partial order dimension of G is at most $2d$.

Using a different model, Golumbic, Rotem and Urrutia [GoRoUr83] have shown that the partial order dimension of G equals k if and only if the cocomparability graph $\overline{G}$ can be realized as the concatenation of $k - 1$ permutation diagrams.[2]

9.3.2 The Classical Model of Interval Tolerance

Tolerance graphs were introduced by Golumbic and Monma [GoMo82] and Golumbic, Monma and Trotter [GoMoTr84] to generalize some of the applications associated with interval graphs. Their motivation was the need to solve scheduling problems in which resources may be needed on an exclusive basis, such as rooms, vehicles, or support personnel, but where a measure of flexibility or tolerance would be allowed for sharing or relinquishing the resource when total exclusivity prevented a solution.

DEFINITIONS

D4: An undirected graph $G = (V, E)$ is a **tolerance graph** if there exists a collection $\mathcal{I} = \{I_v\}_{v \in V}$ of closed intervals on the real line and an assignment of positive numbers $t = \{t_v\}_{v \in V}$ such that

$$vw \in E \iff |I_v \cap I_w| \geq \min\{t_v, t_w\}. \tag{9.3.1}$$

Here $|I_u|$ denotes the length of the interval I_u. The positive number t_v is called the *tolerance* of v, and the pair $\langle \mathcal{I}, t \rangle$ is called an *interval tolerance representation* of G.

D5: A tolerance graph is said to be a **bounded tolerance graph** if it has a tolerance representation in which $t_v \leq |I_v|$ for all $v \in V$.

Tolerance graphs generalize both interval graphs and permutation graphs.

---

[1] Boxes with sides parallel to the axes.

[2] A *permutation diagram* consists of two parallel lines each labeled with a permutation of the numbers $1, \ldots, n$ and n line segments connecting the matched pairs of numbers. The intersection graph of the segments of a permutation diagram is called a *permutation graph*.

FACTS

F5: [GoMo82] A graph G is an interval graph if and only if it has a tolerance graph representation where all tolerances are equal (to some positive constant).

F6: [GoMo82] A graph G is a permutation graph if and only if it has a tolerance graph representation where the tolerance of every vertex equals the length of its interval.

F7: If we restrict the tolerances to be 1 or ∞, we obtain the class of non-partitioned interval probe graphs.[3]

F8: [GoMo82] Every bounded tolerance graph is a cocomparability graph.

F9: [La93] The bounded tolerance graphs are equivalent to the class of parallelogram graphs.[4]

F10: [GoMoTr84] Tolerance graphs are perfect and are contained in the class of weakly chordal graphs.

DEFINITION

D6: A tolerance graph is said to be a ***proper tolerance graph*** if it has a tolerance representation in which no interval properly contains another interval. A tolerance graph is said to be a ***unit tolerance graph*** if it has a tolerance representation in which each interval I_v has unit length for all $v \in V$.

For the case of intersection graphs, the proper interval graphs are equivalent to the unit interval graphs [Ro69]. This is not true for tolerance graphs.

FACTS

F11: Every interval graph is a unit tolerance graph, but not conversely.

F12: Every unit tolerance graph is a proper tolerance graph, but not conversely.

F13: Every proper tolerance graph is a bounded tolerance graph, but not conversely.

A complete hierarchy of these and other classes of perfect graphs ordered by inclusion can be found in Golumbic and Trenk [GoTr04].

In their original paper, Golumbic, Monma and Trotter [GoMoTr84] made the following conjecture:

Conjecture: *If a graph G is both a tolerance graph and a cocomparability graph, then it is a bounded tolerance graph.*

This conjecture remains open, and has become a significant challenge to the graph theory community. It became even more interesting after Langley [La93] proved that bounded tolerance graphs are equivalent to parallelogram graphs, and more recently when Mertzios, Sau and Zaks [MeSaZa09] showed a 3-dimensional geometric intersection model for tolerance graphs called the parallelepiped representation.

---

[3]$G = (V, E)$ is a *non-partitioned interval probe graph* if the vertex set can be partitioned $V = P \cup N$ into a set P of *probes* and N of *non-probes* such that N is a stable set and there exists a (fill-in) $F \subseteq N \times N$, where $G' = (V, E \cup F)$ is an interval graph (see [McWaZh98, GoLi01] and Chapter 4 of [GoTr04]).

[4]A *parallelogram graph* is the intersection graph of a set of parallelograms whose upper and lower sides are on a given pair of horizontal lines.

9.3.3   The Algorithmics of Tolerance Graphs

The computational complexity of recognizing tolerance graphs and bounded tolerance graphs had remained open for 28 years. Hayward and Shamir [HaSh04] showed that the problem is in NP, and Mertzios, Sau and Zaks [MeSaZa10] proved that it is NP-hard, thus obtaining the following:

FACTS

F14: [MeSaZa10] Recognizing tolerance graphs and bounded tolerance graphs is NP-complete.

The following related result answers Question 3.11 of [GoTr04]:

F15: [BuIs07] Recognizing bipartite tolerance graphs has linear time complexity.

Narasimhan and Manber [NaMa92] presented a polynomial time algorithm to find a maximum weighted stable set of a tolerance graph, given a tolerance representation for the graph.

F16: [NaMa92] A maximum weighted stable set of a tolerance representation can be found in time $O(n^2 \log n)$.

Coloring bounded tolerance graphs in polynomial time is an immediate consequence of their being cocomparability graphs. Narasimhan and Manber [NaMa92] use this fact (as a subroutine) to find the chromatic number of any (unbounded) tolerance graph in polynomial time, but not the coloring itself. Subsequently, Golumbic and Siani [GoSi02] gave a different algorithm to find a coloring of a tolerance graph, given a tolerance representation for it (see also Golumbic and Trenk [GoTr04]).

F17: [GoSi02] Finding a minimum coloring of a tolerance representation with at most q intervals having unbounded tolerance can be done in $O(qn + n \log n)$ time.

9.3.4   Variations of Tolerance on Intervals

A variety of "variations on the theme of tolerance" in graphs have been defined and studied over the past years. By changing the function *min* in the tolerance definition (inequality (9.3.1)) with a different binary function ϕ (for example, *max, sum, product,* etc.), we obtain a class that is called ϕ*-tolerance* graphs.

By allowing a separate left tolerance and right tolerance for each interval, various *bitolerance* graph and poset models can be obtained. Directed graph analogues to several of these models have also been defined and studied (see [GoTr04]).

By substituting a different "host" set instead of the real line and with a specified type for the subsets of that host, instead of intervals on a line, we obtain classes such as tolerance graphs of paths on a tree. Bibelnieks and Dearing [BiDe93] considered *NeST graphs*, a tolerance model on neighborhood subtrees of a continuous tree (in the plane), studied further in [HaKeMa02].

By replacing the measure of the length of an interval by some other measure μ of the intersection of the two subsets (for example, cardinality in the case of discrete sets, or number of branching nodes or maximum degree in the case of subtrees of trees), we could obtain yet other variations of tolerance graphs.

We will now survey several of these variations. For further study of tolerance graphs and related topics, we refer the reader to Golumbic and Trenk [GoTr04].

ϕ-*Tolerance* Graphs

DEFINITIONS

Let ϕ be a symmetric binary function, positive valued on positive arguments.

D7: A graph $G = (V, E)$ is a ϕ-*tolerance graph* if there is an interval representation $\mathcal{I} = \{I_v\}_{v \in V}$ with positive tolerances $t = \{t_v\}_{v \in V}$ such that

$$vw \in E \iff |I_v \cap I_w| \geq \phi\{t_v, t_w\}. \qquad (9.3.2)$$

D8: A ϕ-*tolerance chain graph* is defined to be a ϕ-*tolerance graph* which has a representation $\mathcal{I}$ consisting of a nested family of intervals (i.e., a set of intervals totally ordered by inclusion).

FACTS

The following are due to Jacobson, McMorris and Mulder [JaMcMu91].

F18: The *min*-tolerance chain graphs are equivalent to the class of threshold graphs.

F19: The *max*-tolerance chain graphs are equivalent to the class of interval graphs.

F20: The *sum*-tolerance chain graphs are equivalent to the class of coTT graphs.

We note that ϕ-tolerance chain graphs are an earlier special case of *rank-tolerance* graphs, which we will present in Section 9.3.5.

DEFINITIONS

D9: An **Archimedean function** has the property of tending to infinity whenever one of its arguments tends to infinity, that is, $\lim_{x \to \infty} \phi(x, c) = \infty$ for every fixed $c > 0$.

D10: A graph G is an **Archimedean ϕ-tolerance graph**, or more simply, an **Archimedean graph**, if G is a ϕ-tolerance graph for all Archimedean functions ϕ.

FACTS

Generalizing a known result of Jacobson, McMorris and Scheinerman [JaMcSc91] for trees, Golumbic, Jamison and Trenk [GoJaTr02] prove the following:

F21: All trees, cacti, chordless suns, and complete bipartite graphs $K_{2,k}$ are Archimedean.

In [GoJaTr02], they also prove the following:

F22: Every graph G can be represented as a ϕ_G-tolerance graph for some Archimedean polynomial ϕ_G. Moreover, there is a "universal" Archimedean function ϕ^* such that every graph G is a ϕ^*-tolerance graph.

Threshold Tolerance and coTT Graphs

DEFINITIONS

D11: An undirected graph $G = (V, E)$ is called a ***Threshold Tolerance (TT) graph*** if its vertices can be assigned positive weights $\{w_v \mid v \in V\}$ and positive tolerances $\{t_v \mid v \in V\}$ such that $xy \in E \iff w_x + w_y \geq \min\{t_x, t_y\}$.

Monma, Reed and Trotter [MoReTr88] introduced this family as a generalization of the well-known class of **threshold graphs** [ChHa77, MaPe95], namely, the special case in which all tolerances are equal to some constant (called the *threshold*).

The *complement of a threshold tolerance graph* is called a *coTT graph*. An alternate definition of coTT graphs is:

D12: A graph $G = (V, E)$ is a ***complement Threshold Tolerance (coTT) graph*** if for each vertex v there exist two positive numbers a_v and b_v such that for any pair of vertices x and y: $xy \in E \iff a_x \leq b_y$ and $a_y \leq b_x$. (See also [GoTr04] page 187.)

FACTS

F23: Threshold graphs are threshold tolerance graphs, but threshold tolerance graphs are **not** (interval) tolerance.

F24: coTT graphs are tolerance graphs, and coTT graphs are also strongly chordal.

Bounded Bitolerance Graphs and Ordered Sets

The bounded bitolerance graphs generalize bounded tolerance graphs by allowing separate left and right tolerances for each interval in the representation. We define them below in terms of their complements, which are comparability graphs.

DEFINITIONS

D13: An ordered set $P = (V, \prec)$ is a ***bounded bitolerance order*** if each vertex $v \in V$ can be assigned a closed interval $I_v = [L_v, R_v]$ and two "tolerant points" $p_v, q_v \in I_v$ satisfying $L_v < p_v$ and $q_v < R_v$, such that

$$x \prec y \iff R_x < p_y \text{ and } q_x < L_y.$$

D14: An undirected graph $G = (V, E)$ is a ***bounded bitolerance graph*** if it is the incomparability graph of a bounded bitolerance order.

D15: A ***trapezoid graph*** is the intersection graph of a set of trapezoids whose upper and lower sides are on a given pair of horizontal lines.

D16: From such a trapezoid representation $\{T_v\}$, a ***trapezoid order*** is obtained as $x \prec y \iff T_x$ lies totally to the left of T_y.

FACTS

The following are due to Langley [La93]:

F25: The bounded bitolerance orders are equivalent to the class of trapezoid orders.

F26:   The bounded bitolerance graphs are equivalent to the class of trapezoid graphs.

Trapeziod graphs were originally introduced in the context of a circuit design application [DaGoPi88]. They noted the following:

A trapezoid order can be viewed as the intersection of two interval orders $P_1 = (V, \prec_1)$ and $P_2 = (V, \prec_2)$ in which $x \prec y$ if $x \prec_1 y$ and $x \prec_2 y$, that is, the upper interval of x is disjoint and to the left of the upper interval of y, and the lower interval of x is disjoint and to the left of the lower interval of y. This property is known as having **interval dimension** *idim* 2.

9.3.5   Rank-Tolerance Graphs

Rank-tolerance graphs (to be defined formally below) were motivated by extending the notion of ϕ-tolerance chain graphs. A *rank-tolerance* representation of a graph assigns to each vertex two parameters: a *rank*, which represents the size of that vertex, and a *tolerance* which represents an allowed extent of conflict with other vertices. Two vertices will be adjacent if and only if their joint rank exceeds (or equals) their joint tolerance. By varying the coupling functions used to obtain the joint rank or joint tolerance, a variety of graph classes arise, many of which have interesting structure.

BACKGROUND

In the investigation of ϕ-tolerance graphs, as we have seen in Section 9.3.4, a particular case appeared to be of recurring interest – namely, when all the intervals in the representation are *nested* to form a *chain* under inclusion (the ϕ-*tolerance chain* graphs). In this case, the length of the intersection of two nested intervals is just the minimum of their lengths. So if we denote $r_v = |I_v|$, for chain representations we may substitute $\min(r_v, r_w)$ for $|I_v \cap I_w|$ in the inequality (9.3.2) defining ϕ-tolerance graphs. Thus, here, only the *lengths*, and not the actual intervals, play a role in determining adjacency.

Motivated by this observation, Golumbic and Jamison [GoJa06] extended and explored a large number of classes that arise when the *min* of r_v, r_w is also allowed to be replaced by a more general function ρ. That paper laid the foundation for a general theory of *rank-tolerance graphs*.

DEFINITION

Let $\mathcal{C}$ denote the class of all commutative binary operations on the positive real numbers. The functions in $\mathcal{C}$ will be thought of as combining or coupling two input tolerances (or ranks) and giving a single joint tolerance (or rank) as output.

D17:   Given two coupling functions ρ and ϕ, we say that a graph $G = (V, E)$ is a $[\![\rho; \phi]\!]$-***graph*** if there are mappings $v \to r_v$ and $v \to t_v$ from the vertex set V of G into the positive real numbers $\mathbb{R}^+$ such that for all vertices $v \neq w$ in V,

$$vw \in E \iff \rho(r_v, r_w) \geq \phi(t_v, t_w). \tag{9.3.3}$$

The r_v are called the *ranks* and the t_v are called the *tolerances*. The class of all $[\![\rho; \phi]\!]$-graphs is denoted simply by $[\![\rho; \phi]\!]$, where ρ is called the *rank function* and ϕ is called the *tolerance function*. [5]

---

[5] Our notation here $[\![\rho; \phi]\!]$-graphs differs from the original paper [GoJa06] which called them $[\phi, \rho]$ and a more recent paper [Ja12] which called them $N(\rho; \phi)$.

FACTS

In our notation, we have

F27: $[\![min;\phi]\!] \equiv \phi$-tolerance chain graphs.

The three Facts in Section 9.3.4 on tolerance chain graphs due to Jacobson, McMorris and Mulder [JaMcMu91] can now be restated as follows:

F28: $[\![min; min]\!] \equiv$ threshold graphs.

F29: $[\![min; max]\!] \equiv$ interval graphs.

F30: $[\![min; sum]\!] \equiv$ coTT graphs.

The paper by Golumbic and Jamison [GoJa06] is the primary source for rank-tolerance graphs where they prove the following:

F31: $[\![min; prod]\!] \equiv$ coTT graphs.

F32: $[\![max; sum]\!] \equiv$ TT graphs $\equiv [\![max; prod]\!]$.

Question: One interesting, unstudied class worthy of future investigation is the class $[\![sum; prod]\!]$. Jamison has shown a new, yet unpublished result that *every Threshold Tolerance graph is a $[\![sum; prod]\!]$-graph*. See also [JaSp06].

F33: If at least one of the coupling functions ρ or ϕ is moderately nice (continuous and weakly increasing), then reversing the roles of tolerance and rank leads to a representation of the complement $\overline{G}$ of G, giving us a beautiful *co-symmetry* between classes:

$$[\![\rho; \phi]\!] \equiv co[\![\phi; \rho]\!].$$

The core functions $min, max, sum, prod$ all satisfy nice properties. However, there are examples of discrete non-strict cases where this co-symmetry fails.

DEFINITIONS

D18: ϕ is ***Archimedean*** if $\lim_{x\to\infty} \phi(x, c) = \infty$ for every fixed $c > 0$.

D19: ϕ is ***dual Archimedean*** if $\lim_{x\to 0} \phi(x, c) = 0$ for every fixed $c > 0$.

The next result is one of several on reflexive classes:

FACTS

F34: If ϕ is nondecreasing, weakly increasing, and associative, then $[\![\phi; \phi]\!]$ is the class of all threshold graphs.

F35: Under any of the following three conditions,

- ρ is both Archimedean and dual Archimedean, or

- ϕ is both Archimedean and dual Archimedean, or

- ρ and ϕ are both dual Archimedean,

the class $[\![\rho; \phi]\!]$ contains all threshold graphs.

The reader is referred to [GoJa06] for further results on rank-tolerance graphs. In [Ja12], Jamison studies a more comprehensive mathematical theory of conflict-tolerance graphs.

9.3.6   Intersection and Tolerance Graphs on Trees

DEFINITIONS

D20: A *chordal graph* is one that contains no chordless cycle of size greater than or equal to 4.

D21: A *split graph* is one whose vertex set can be partitioned into a clique and an independent set.

FACTS

F36: A graph is a split graph if and only if it is chordal and its complement is chordal.

Let T be a tree and let $\{T_i\}$ be a collection of subtrees (connected subgraphs) of T. We may think of the host tree T either (1) as a continuous model of a tree embedded in the plane, thus generalizing the real line from the one-dimensional case, or (2) as a finite discrete model of a tree, namely, a connected graph of vertices and edges having no cycles, thus generalizing the path P_k from the one-dimensional case.

The distinction between these two models becomes important when measuring the *size of the intersection* of two subtrees. For example, in the continuous model (1), we might take the size of the intersection to be the length of a longest common path of the two subtrees measured along the host tree [BiDe93]. In the discrete model (2), we might count the number of common vertices or common edges [GoJa85a, GoJa85b, JaMu00, JaMu05]. We use the expressions "nonempty intersection" or "vertex intersection" to mean sharing a vertex or point of T, and "nontrivial intersection" or "edge intersection" to mean sharing an edge or otherwise measurable segment of T. In this way, edge intersection is more *tolerant* than vertex intersection.

Using this terminology, a classical result of [Bu74, Ga74, Wa78] stated the following.

F37: A graph is the *vertex intersection graph* of a set of subtrees of a tree if and only if it is a chordal graph.

McMorris and Shier [McSh83] give an analogous version for split graphs.

F38: A graph G is the vertex intersection graph of distinct induced subtrees of a star $K_{1,n}$ if and only if G is a split graph.

In contrast to these results, Golumbic and Jamison [GoJa85a] observed that the family of *edge intersection graphs* of subtrees of a tree yield all possible graphs. In fact, they proved the following variation on Marczewski's Theorem:

F39: Every graph can be represented as the edge intersection graph of substars of a star.

Two different classes of intersection graphs also arise when considering simple paths (instead of subtrees) of an arbitrary host tree T. The *path graphs*, which are the vertex intersection graphs of paths on a tree, also known as *VPT graphs*, are a subfamily of chordal graphs. However, the graphs obtained as the "edge intersection graphs of paths in a tree," called *EPT graphs*, are not necessarily chordal. The class of EPT graphs are not perfect graphs, and the recognition problem for them is NP-complete [GoJa85a], whereas the VPT graphs are perfect and can be recognized efficiently [Ga78]. See also Monma and Wei [MoWe86] and Sysło [Sy85].

Type of Interaction	Objects	Host	Graph Class
vertex intersection	subtrees	tree	chordal graphs
vertex intersection	subtrees	star	split graphs
edge intersection	subtrees	star	all graphs
vertex intersection	paths	path	interval graphs
vertex intersection	paths	tree	path graphs or VPT graphs
edge intersection	paths	tree	EPT graphs
containment	intervals	line	permutation graphs
containment	paths	tree	(open question)
containment	subtrees	star	comparability graphs

Table 9.1: Some graph classes involving trees.

FACT

F40: [GoJa85b] In the special case of the host tree having maximum vertex degree 3 (binary trees), the VPT and EPT classes are the same, i.e., deg3-VPT $\equiv$ deg3-EPT. See Table 9.1.

Thus, EPT graphs are a more *tolerant* model than VPT graphs, but they have a high algorithmic cost. Table 9.1 summarizes the subtree graph classes we have discussed here. A full treatment can be found in Chapter 11 of Golumbic and Trenk [GoTr04].

The $\langle h, s, t \rangle$ Graphs: Degree Constrained Representations

Jamison and Mulder [JaMu00, JaMu05] introduced a constant tolerance model for subtrees of a tree where degree restrictions are placed on the trees. This further generalizes VPT and EPT graphs.

DEFINITIONS

D22: An $\langle h, s, t \rangle$-representation of a graph G consists of a collection of subtrees $\{S_v \mid v \in V(G)\}$ of a tree T, such that (i) the maximum degree of T is at most h, (ii) every subtree has maximum degree at most s, and (iii) there is an edge between two vertices in G if and only if the corresponding subtrees in T have at least t vertices in common.

D23: A graph is **weakly chordal** if neither the graph nor its complement contains a chordless cycle of size greater than 4.

REMARKS

Using this notation, where ∞ denotes that no restriction is imposed, we immediately have from their definitions:

R1: interval graphs $\equiv \langle 2, 2, 1 \rangle$.

R2: EPT graphs $\equiv \langle \infty, 2, 2 \rangle$.

R3: VPT graphs or path graphs $\equiv \langle \infty, 2, 1 \rangle$.

The following results have been shown in the literature:

FACTS

F41: chordal graphs $\equiv \langle \infty, \infty, 1 \rangle$ [Bu74, Ga74, Wa78]
$\qquad\qquad\qquad\quad \equiv \langle 3, 3, 1 \rangle$ [McSc91]
$\qquad\qquad\qquad\quad \equiv \langle 3, 3, 2 \rangle$.   [JaMu05]

F42: $\langle 3, 2, 1 \rangle \equiv \langle 3, 2, 2 \rangle \equiv VPT \cap$ chordal $\equiv EPT \cap$ chordal. [GoJa85b]

F43: $\langle 4, 2, 2 \rangle \equiv EPT \cap$ weakly chordal. [GoLiSt08b]

F44: $\langle 4, 4, 2 \rangle \equiv$ weakly chordal $\cap \{K_{2,3}, \overline{4P_2}, \overline{P_2 \cup P_4}, \overline{P_6}, H_1, H_2, H_3\}$-free. [GoLiSt09]

In the paper [CoGoLiSt08], the family of $\langle 4, 3, 2 \rangle$ is characterized and a polynomial time recognition is also provided. The class $\langle 3, 3, 3 \rangle$ is studied in [JaMu00]. For further results in this area, see [GoLiSt08a], [GoLiSt08c] and [JaMu05].

References

[AvPeSh00] M. Avriel, M. Penn and N. Shpirer, Container ship stowage problem: complexity and connection to the coloring of circle graphs, *Discrete Applied Math.* 103 (2000), 271–279.

[Be73] C. Berge, *Graphs and Hypergraphs*, North-Holland, Amsterdam, 1973.

[Be85] C. Berge, *Graphs*, North-Holland, 1985.

[BiDe93] E. Bibelnieks and P. M. Dearing, Neighborhood subtree tolerance graphs, *Discrete Appl. Math.* 43 (1993), 13–26.

[BoFiIsLa95] K. P. Bogart, P. C. Fishburn, G. Isaak and L. J. Langley, Proper and unit tolerance graphs, *Discrete Appl. Math.* 60 (1995), 99–117.

[BrLeSp99] A. Brandstädt, V. B. Le and J. P. Spinrad, *Graph Classes: A Survey*, SIAM, Philadelphia, 1999.

[Bu74] P. Buneman, A characterisation of rigid circuit graphs, *Discrete Math.* 9 (1974), 205–212.

[BuIs07] A. H. Busch and G. Isaak, Recognizing bipartite tolerance graphs in linear time, in *Proc. 33rd Int'l Conference on Graph-theoretic Concepts in Computer Science* (WG'07), *Lecture Notes in Comput. Sci.* 4769, Springer-Verlag, 2007, pp. 12–20.

[ChHa77] V. Chvátal and P. L. Hammer, Aggregation of inequalities in integer programming, *Annals of Discrete Math.* 1 (1977), 145–162.

[CoGoLiSt08] E. Cohen, M. C. Golumbic, M. Lipshteyn and M. Stern What is between chordal and weakly chordal graphs? Proc. 34rd Int'l. Workshop on Graph-Theoretic Concepts in Computer Science (WG 2008), *Lecture Notes in Computer Science* 5344, Springer-Verlag, 2008, pp. 275–286.

[DaGoPi88] I. Dagan, M. C. Golumbic and R. Y. Pinter, Trapezoid graphs and their coloring, *Discrete Applied Math.* 21 (1988), 35–46.

[DuMi41] B. Dushnik and E. W. Miller, Partially ordered sets, *Amer. J. Math.* 63 (1941), 600–610.

[ErGoPo66] P. Erdős, A. W. Goodman, and L. Pósa, The representation of a graph by set intersections, *Canad. J. Math.* 18 (1966), 106–112.

[Fa83] M. Farber, Characterizations of strongly chordal graphs, *Discrete Math.* 43 (1983), 173–189.

[FoHa77] S. Földes and P. L. Hammer, Split graphs, *Congressus Numer.* 17 (1977), 311–315.

[GaNe81] E. A. Gattass and G. L. Nemhauser, An application of vertex packing to data analysis in the evaluation of pavement deterioration, *Operations Research Letters* 1 (1981), 13–17.

[Ga74] F. Gavril, The intersection graphs of subtrees in trees are exactly the chordal graphs, *J. Combin. Theory Ser. B* 16 (1974), 47–56.

[Ga78] F. Gavril, A recognition algorithm for the intersection graphs of paths in trees, *Discrete Math.* 23 (1978), 211–227.

[Go80] M. C. Golumbic, *Algorithmic Graph Theory and Perfect Graphs*, Academic Press, New York, 1980. Second edition: *Annals of Discrete Mathematics* 57, Elsevier, Amsterdam, 2004.

[Go88] M. C. Golumbic, Algorithmic aspects of intersection graphs and representation hypergraphs, *Graphs and Combinatorics* 4 (1988), 307–321.

[Go98] M. C. Golumbic, Reasoning about time, in *Mathematical Aspects of Artificial Intelligence*, F. Hoffman, ed., American Mathematical Society, *Proc. Symposia in Applied Math.*, vol. 55 (1988), 19–53.

[Go12] M. C. Golumbic, Perspectives on reasoning about time, in *Ubiquitous Display Environments*, A. Krger and T. Kuflik, eds., Springer Verlag, 2012, pp. 53–70.

[GoJa85a] M. C. Golumbic and R. E. Jamison, The edge intersection graphs of paths in a tree, *Journal of Combinatorial Theory, Series B* 38 (1985), 8–22.

[GoJa85b] M. C. Golumbic and R. E. Jamison, Edge and vertex intersection of paths in a tree, *Discrete Mathematics* 55 (1985), 151–159.

[GoJa06] M. C. Golumbic and R. E. Jamison, Rank-tolerance graph classes, *J. Graph Theory* 52 (2006), 317–340.

[GoJaTr02] M. C. Golumbic, R. E. Jamison and A. N. Trenk, Archimedian ϕ-tolerance graphs, *J. Graph Theory* 41 (2002), 179–194.

[GoGoKaSh95] P. W. Goldberg, M. C. Golumbic, H. Kaplan and R. Shamir, Four strikes against physical mapping of DNA, *J. Comput. Biology* 3 (1995), 139–152.

[GoKaSh94] M. C. Golumbic, H. Kaplan and R. Shamir, On the complexity of DNA physical mapping, *Advances in Applied Math.* 15 (1994), 251–261.

[GoLi01] M. C. Golumbic and M. Lipshteyn, On the hierarchy of tolerance, probe and interval graphs, *Congressus Numerantium* 153 (2001), 97–106.

[GoLiSt08a] M. C. Golumbic, M. Lipshteyn and M. Stern, The k-edge intersection graphs of paths in a tree, *Discrete Applied Math.* 156 (2008), 451–461.

[GoLiSt08b] M. C. Golumbic, M. Lipshteyn and M. Stern, Representing edge intersection graphs of paths on degree 4 trees, *Discrete Math.* 308 (2008), 1381–1387.

[GoLiSt08c] M. C. Golumbic, M. Lipshteyn and M. Stern, Equivalences and the complete hierarchy of intersection graphs of paths in a tree, *Discrete Applied Math.* 156 (2008), 3203–3215.

[GoLiSt09] M. C. Golumbic, M. Lipshteyn and M. Stern, Intersection models of weakly chordal graphs, *Discrete Applied Math.* 157 (2009), 2031–2047.

[GoMo82] M. C. Golumbic and C. L. Monma, A generalization of interval graphs with tolerances, *Congressus Numer.* 35 (1982), 321–331.

[GoMoTr84] M. C. Golumbic, C. L. Monma, and W. T. Trotter, Tolerance graphs, *Discrete Applied Math.* 9 (1984), 157–170.

[GoRoUr83] M. C. Golumbic, D. Rotem and J. Urrutia, Comparability graphs and intersection graphs, *Discrete Math.* 43 (1983) 37–46.

[GoSc89] M. C. Golumbic and E. Scheinerman, Containment graphs, posets, and related classes of graphs, *Ann. N.Y. Acad. Sci.* 555 (1989) 192–204.

[GoSh93] M. C. Golumbic and R. Shamir, Complexity and algorithms for reasoning about time, *J. Assoc. for Comput. Mach.* 40 (1993), 1108–1133.

[GoSi02] M. C. Golumbic and A. Siani, Coloring algorithms for tolerance graphs: reasoning and scheduling with interval constraints, *Lecture Notes in Comput. Sci.* 2385, Springer-Verlag, 2002, pp. 196–207.

[GoTr04] M. C. Golumbic and A. N. Trenk, *Tolerance Graphs*, Cambridge University Press, 2004.

[Ha94] F. Harary, *Graph Theory*, Perseus reprint, 1994. (First Edition, Addison-Wesley, 1969.)

[HaKeMa02] R. B. Hayward, P. E. Kearney and A. Malton, NeST graphs, *Discrete Applied Math.* 121 (2002), 139–153.

[HaSh04] R. B. Hayward and Ron Shamir, A note on tolerance graph recognition, *Discrete Applied Math.* 143 (2004), 307–311.

[JaLeLe84] M. S. Jacobson, J. Lehel and L. Lesniak, ϕ-threshold and ϕ-tolerance chain graphs, *Discrete Applied Math.* 44 (1984), 191–203.

[JaMcMu91] M. S. Jacobson, F. R. McMorris and H. M. Mulder, An introduction to tolerance intersection graphs, in *Proc. Sixth Int. Conf. on Theory and Applications of Graphs*, Y. Alavi, G. Chartrand, O. Oellermann and A. Schwenk, eds., 1991, pp. 705–724.

[JaMcSc91] M. S. Jacobson, F. R. McMorris and E. R. Scheinerman, General results on tolerance intersection graphs, *J. Graph Theory* 15 (1991), 573–577.

[Ja12] R. E. Jamison, Towards a comprehensive theory of conflict-tolerance graphs, *Discrete Applied Mathematics* 160 (2012), 2742–2751.

[JaMu00] R. E. Jamison and H. M. Mulder, Tolerance intersection graphs on binary trees with constant tolerance 3, *Discrete Math.* 215 (2000), 115–131.

[JaMu05] R. E. Jamison and H. M. Mulder, Constant tolerance intersection graphs of subtrees of a tree, *Discrete Math.* 290 (2005), 27–46.

[JaSp06] Robert E. Jamison and Alan Sprague, Symmetry of extended representations of mix graphs and sum-product graphs, *Congr. Numer.* 182 (2006), 111–128.

[La93] L. Langley, Interval tolerance orders and dimension, Ph.D. Thesis, Dartmouth College, 1993.

[LeBo62] C. Lekkerkerker and D. Boland, Representation of finite graphs by a set of intervals on the real line, *Fund. Math.* 51 (1962), 45–64.

[MaPe95] N. V. R. Mahadev and U. N. Peled, *Threshold Graphs and Related Topics*, North-Holland, Amsterdam, 1995.

[Ma45] E. (Szpilrajn-) Marczewski, Sur deux propriétés des classes d'ensembles, *Fund. Math.* 33 (1945), 303–307.

[McMc99] T. A. McKee and F. R. McMorris, *Topics in Intersection Graph Theory*, SIAM, Philadelphia, 1999.

[McWaZh98] F. R. McMorris, C. Wang and P. Zhang, On probe interval graphs, *Discrete Applied Math.* 88 (1998), 315–324.

[McSc91] F. R. McMorris and E. M. Scheinerman, Connectivity threshold for random chordal graphs. *Graphs and Combin.* 7 (1991), 177–181.

[McSh83] F. R. McMorris and D. R. Shier, Representing chordal graphs on $K_{1,n}$, *Comment. Math. Univ. Carolin.* 24 (1983), 489–494.

[MeSaZa09] G. B. Mertzios, I. Sau and S. Zaks, A new intersection model and improved algorithms for tolerance graphs, *SIAM Journal on Discrete Mathematics* 23 (2009), 1800–1813.

[MeSaZa10] G. B. Mertzios, I. Sau and S. Zaks, The recognition of tolerance and bounded tolerance graphs, *SIAM J. Comput.* 40 (2010), 1234–1257.

[MoReTr88] C. L. Monma, B. Reed and W. T. Trotter, Threshold tolerance graphs, *J. of Graph Theory* 12 (1988), 343–362.

[MoWe86] C. L. Monma and V. K. Wei, Intersection graphs of paths in a tree, *J. Combin. Theory B* 41 (1986), 141–181.

[NaMa92] G. Narasimhan and R. Manber, Stability number and chromatic number of tolerance graphs, *Discrete Applied Math.* 36 (1981) 47–56.

[OsRo81] R. J. Opsut and F. S. Roberts, On the fleet maintenance, mobile radio frequency, task assignment, and traffic phasing problems, in *The Theory and Applications of Graphs*, G. Chartrand et al. (eds.), Wiley, New York, 1981, pp. 479–492.

[OsRo83] R. J. Opsut and F. S. Roberts, I-colorings, I-phasings, and I-intersection assignments for graphs and their applications, *Networks* 13 (1983), 327–345.

[Ro69] F. S. Roberts, Indifference graphs, in *Proof Techniques in Graph Theory*, F. Harary, ed., Academic Press, New York, 1969, pp. 139–146.

[Ro76] F. S. Roberts, *Discrete Mathematical Models with Applications to Social, Biological and Environmental Problems*, Prentice Hall, 1976.

[Sp03] J. P. Spinrad, *Efficient Graph Representations*, Fields Institute Monographs 19, American Mathematical Society, Providence, 2003.

[Sy85] M. M. Sysło, Triangulated edge intersection graphs of paths in a tree, *Discrete Math.* 55 (1985), 217–220.

[Wa78] J. R. Walter, Representations of chordal graphs as subtrees of a tree, *J. Graph Theory* 2 (1978), 265–267.

[Wa95] M. S. Waterman, *Introduction to Computational Biology*, Chapman Hall, London, 1995.

Section 9.4
Bandwidth

Robert C. Brigham, University of Central Florida

INTRODUCTION

Harper [Ha64] discusses a coding problem in which the integers $1, 2, \ldots, 2^n$ form the code words, each assigned to a vertex of the n-dimensional hypercube. If code words i and j are assigned to adjacent vertices, then Δ_{ij} is defined to be $|i - j|$. The paper determines the minimum value of $\sum \Delta_{ij}$ over all possible assignments. In concluding remarks Harper says "Another problem, as yet not solved, is this: how to number the vertices of an n-cube so that $max\Delta_{ij}$ ... is minimized." This latter problem is precisely that of determining the bandwidth of a hypercube, and this is the first known reference in graph theoretic terms.

[ChChDeGi82] and [LaWi99], each with extensive bibliographies, provide comprehensive surveys of bandwidth. Further results can be found in [Ch88] and [Mi91]. All graphs discussed in this section are assumed to be simple and finite.

9.4.1   Fundamentals

The Bandwidth Concept

EXAMPLE

E1: Figures 9.4.1 and 9.4.2 show different vertex labelings of the three-dimensional hypercube Q_3 and the corresponding adjacency matrices.

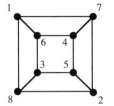

Figure 9.4.1: Hypercube Q_3 with labeling and associated adjacency matrix.

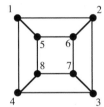

Figure 9.4.2: A second labeling of Q_3 and associated adjacency matrix.

For the labeling of Figure 9.4.1, all ones in the adjacency matrix lie in the seven diagonals above and below the main diagonal. Since the matrix is symmetric, we can restrict our attention to the diagonals in the upper triangular portion. The matrix corresponding to the labeling of Figure 9.4.2 has ones in only the four ("upper") diagonals closest to the main diagonal. The bandwidth of a graph corresponds to the minimum number of such diagonals, taken over all possible labelings. For Q_3 this number is four. Thus the labeling of Figure 9.4.2 yields the minimum. This concept is formalized as follows.

DEFINITIONS

NOTATION: An edge with endpoints (end vertices) u and v is denoted uv.

D1: A **proper numbering** of G is a bijection $f : V \to \{1, 2, \ldots, n\}$.

D2: Let f be a proper numbering of a graph G. The **bandwidth of** f, denoted $B_f(G)$, is given by $B_f(G) = max\{|f(u) - f(v)| : uv \in E\}$.

D3: The **bandwidth** of G is $B(G) = min\{B_f(G) : f$ is a proper numbering of $G\}$.

D4: A **bandwidth numbering** of G is a proper numbering f such that $B(G) = B_f(G)$ (i.e., a proper numbering that achieves $B(G)$).

Applications

The introduction mentions an application of bandwidth related to coding theory. [LaWi99] includes a survey, along with references, of other applications.

EFFICIENT MATRIX STORAGE

Storing the entire upper triangular portion of the adjacency matrix is one of several computer representations of a graph. However, only the $B(G)$ diagonals above the main diagonal need be stored. If $B(G)$ is small, this can represent a significant savings.

VLSI LAYOUT

The ***placement problem*** for modules of a VLSI design is the location of the modules on a two-dimensional grid so that certain criteria are met. Modules that must communicate with each other should be as close as possible. A simplified model of the geometry can be given by a graph G whose vertices correspond to modules and whose edges correspond to wires between modules. Then $B(G)$ represents the maximum distance between communicating modules.

INTERCONNECTION NETWORKS

An ***interconnection*** or ***parallel computation network*** is a collection of processors with links between them. This can be modeled by a graph G where the vertices represent the processors and edges correspond to the links. Sometimes it is desirable to simulate the network represented by G on a second network modeled by graph H. This can be done by a one-to-one mapping $f : V(G) \to V(H)$, where processor u in G is simulated by processor $f(u)$ in H, and link uv in G is simulated by a shortest path between $f(u)$ and $f(v)$ in H. If t is the communication time for a link uv in G, then the corresponding time in H is dt where d is the distance between $f(u)$ and $f(v)$ in H. If $t = 1$ and H is a path, the greatest possible delay in the simulation is $B(G)$.

BINARY CONSTRAINT SATISFACTION PROBLEM

A ***binary constraint satisfaction problem*** involves a collection of variables, a set of possible values for each, and constraints between them. The problem is to assign to each variable a permissible value such that all constraints are satisfied. The associated ***constraint graph*** G has vertices representing the variables, with an edge between two vertices if the corresponding variables share a nontrivial constraint. If $B(G)$ is small, the problem may be more easily solvable than otherwise since then it might be possible to deal with only a small number of variables at a time.

Algorithms

DEFINITIIONS

D5: The ***bandwidth decision problem*** is the problem which accepts as input an arbitrary graph G and an arbitrary integer K and returns "YES" if $B(G) \le K$ and "NO" otherwise.

D6: For a given fixed positive integer k, the ***bandwidth-k decision problem*** is the problem which accepts as input an arbitrary graph G and returns "YES" if $B(G) \le k$ and "NO" otherwise.

D7: A ***polynomial algorithm*** for graphs is one whose execution time is bounded by a polynomial in some parameter of the problem, often the number of vertices.

D8: An ***NP-complete problem*** is a problem having a "YES" or "NO" answer that can be solved nondeterministically in polynomial time, and all other such problems can be transformed to it in polynomial time. Such problems are generally accepted as being computationally difficult.

FACTS

F1:   [Pa76] The bandwidth decision problem is NP-complete.

F2:   [GaGrJoKn78] The bandwidth decision problem is NP-complete for trees with maximum degree three.

F3:   [Sa80] The bandwidth-k problem is solvable in polynomial time for any fixed positive integer k.

REMARKS

R1:   The important distinction between the bandwidth decision problem and the bandwidth-k decision problem is that the integer K in the former is an input variable while the integer k in the latter is fixed for all graphs and does not appear as an input.

R2:   In view of Facts F1 and F2 above, it is highly unlikely that a polynomial algorithm can be found for computing the bandwidth of all graphs or even for trees with maximum degree three. Sometimes, though, it is possible to restate an NP-complete problem, by limiting its generality, so that the revised problem becomes polynomial. This has been done for bandwidth, as illustrated in Fact F3.

R3:   Many approximation algorithms have been developed, some dealing with a matrix equivalent of the bandwidth problem, and a listing of several of them is given in [LaWi99]. Further surveys and references are in [Ev79, HeGr79, Sm85]. While it certainly would be advantageous to determine bandwidth exactly, approximate values remain useful in practical applications, since any reduction in the number of nonzero diagonals of the adjacency matrix provides benefit.

R4:   Apparently the first attempt to develop an approximate algorithm was reported in [AlMa65], and its effective use was limited to small matrices. The first algorithm to receive wide acceptance is discussed in [Ro68]. [CuMc69] describes an algorithm which took center stage during the 1970's, despite several limitations. [GiPoSt76] presents a greatly improved version. Details can be found in [ChChDeGi82].

R5:   More recent work on both approximate algorithms for general graphs and exact algorithms for specific classes of graphs includes [GoOp90, KaSh96, HaMa97, Ya98, BlKoRaVe00, Fe00, Gu01, KlTa01, CaMaPr02, KrSt02].

9.4.2   Elementary Results

The Bandwidth of Some Common Families of Graphs

DEFINITIONS

D9:   The **complete k-partite graph** $K_{n_1,n_2,\ldots,n_k}$ is the graph whose vertex set is partitioned into sets A_i of n_i vertices, $1 \le i \le k$, with two vertices adjacent if and only if they are in distinct sets.

D10: The n-**dimensional hypercube** Q_n is the graph having 2^n vertices, each labeled with a distinct n-digit binary sequence, and two vertices are adjacent if and only if their labels differ in exactly one position.

FACTS

F4: $B(P_n) = 1$, where P_n is the path having n vertices.

F5: $B(C_n) = 2$, where C_n is the cycle having n vertices.

F6: $B(K_n) = n - 1$, where K_n is the complete graph having n vertices.

F7: [Ei79] Let $n_1 \geq n_2 \geq \cdots \geq n_k$ be positive integers. Then $B(K_{n_1,n_2,\ldots,n_k}) = |V(K_{n_1,n_2,\ldots,n_k})| - \lceil (n_1 + 1)/2 \rceil$. Thus, $B(K_{n_1,n_2}) = \lceil n_1/2 \rceil + n_2 - 1$ [Ch70].

F8: [Ha66] $B(Q_n) = \sum_{k=0}^{n-1} \binom{k}{\lfloor k/2 \rfloor}$.

EXAMPLE

E2: Figure 9.4.2 shows a bandwidth numbering of Q_3 and Figure 9.4.3 presents bandwidth numberings for C_7 and $K_{5,3}$.

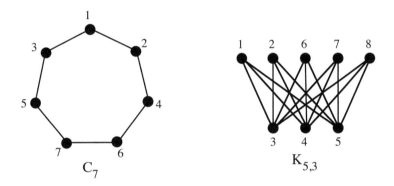

Figure 9.4.3: Bandwidth numberings of C_7 and $K_{5,3}$.

A Few Basic Relations

NOTATION: The *minimum degree* and *maximum degree* of the vertices in a graph G are denoted $\delta_{\min}(G)$ and $\delta_{\max}(G)$, respectively.

TERMINOLOGY NOTE: Several authors use $\delta(G)$ and $\Delta(G)$ instead.

FACTS

F9: [ChDeGiKo75] $B(G) \geq \lceil \delta_{\max}(G)/2 \rceil$.

F10: If H is a subgraph of G, then $B(H) \leq B(G)$.

F11: If graph G has components $G_1, G_2, \ldots, G_k$, then $B(G) = max\{B(G_1), B(G_2), \ldots, B(G_k)\}$.

F12: [ChDeGiKo75] If G is a nonplanar graph, then $B(G) \geq 4$.

On the Bandwidth of Trees

DEFINITIONS

D11: An (edge) **subdivision** of edge $e = uv$ in graph G is the graph obtained from G by replacing e by the path $\langle u, w, v \rangle$ where w is a new vertex of degree two. A **refinement** of G is a graph obtained from G by a finite number of subdivisions.

D12: The **complete k-ary tree** $T_{k,d}$ **of depth** d is the rooted tree in which all vertices at level $d - 1$ or less have exactly k children, and all vertices at level d are leaves.

FACTS

F13: [ChDeGiKo75] For any tree T, $B(T) \leq \lfloor |V(T)|/2 \rfloor$. Equality holds if and only if $|V(T)|$ is even and T is the *star* $K_{1,|V(T)|-1}$.

F14: [WaYa95, AnKaGe96] Let T be a tree with k univalent vertices. Then $B(T) \leq \lceil k/2 \rceil$.

F15: [Sm95] Let $T_{k,d}$ be the complete k-ary tree of depth d. Then we have $B(T_{k,d}) = \lceil k(k^d - 1)/(2d(k - 1)) \rceil$.

F16: [Ch88] If tree T contains a refinement of the complete binary tree $T_{2,d}$, then $B(T) \geq \lceil d/2 \rceil$.

EXAMPLE

E3: Figure 9.4.4 shows bandwidth numberings of two trees, one for which $B(T) < \lfloor n/2 \rfloor$ and $K_{1,7}$, for which equality occurs.

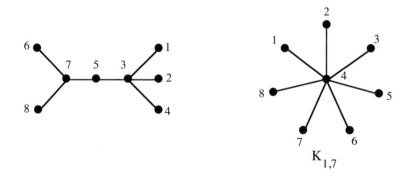

Figure 9.4.4: Bandwidth numberings of two trees.

Alternative Interpretations of Bandwidth

Three alternative interpretations of bandwidth are shown below. Others are given in [Li00].

DEFINITIONS

D13: The kth *power* of graph G, denoted G^k, is the graph having the same vertex set as G and an edge between two vertices if and only if the distance between them is at most k in G.

D14: The *complementary numbering f_c of proper numbering f* of G is defined by $f_c(v) = n + 1 - f(v)$ for each vertex v of G.

FACTS

F17: [ChChDeGi82] For a real symmetric matrix M, let m_{ij} be the value in position (i, j). Consider the problem of finding a symmetric permutation of the rows and columns of M such that the maximum of $|i - j|$, taken over all pairs (i, j) for which m_{ij} is nonzero, is minimized. This problem's equivalence to the bandwidth problem follows by replacing each nonzero entry of M by 1 and considering the resulting matrix as an adjacency matrix of a graph.

F18: [ChChDeGi82] G has bandwidth k if and only if k is the smallest integer such that G can be embedded in P_n^k where P_n^k is the kth power of the path P_n on n vertices.

F19: $B_{f_c}(G) = B_f(G)$ so the complementary numbering of a bandwidth numbering also is a bandwidth numbering.

9.4.3 Bounds on Bandwidth

Two General Bounds

FACTS

F20: [Ha66] For $S \subseteq V$, let ∂S be the subset of S with at least one neighbor outside of S. Then $B(G) \geq \max_k \min\{|\partial S| : |S| = k\}$.

F21: [Ch80-a] For $S \subseteq V$, let ΔS be the subset of edges of G with exactly one endpoint in S. Then $B \geq \max_k \min \left\{ \frac{(1 + 8\Delta|S|)^{1/2} - 1}{2} : |S| = k \right\}$.

Subdivisions, Mergers, Contractions, and Edge Additions

DEFINITION

D15: The *merger of two vertices u and v of graph* G is the graph, denoted $G|_{u,v}$, obtained from G by identifying u and v and then eliminating any loops and duplicate edges. If $e = uv$, the merger $G|_{u,v}$ is called a *contraction* of G along e and denoted $G|_e$.

FACTS

F22: [ChOp86] If H is obtained from G by a subdivision of an edge, then

$$B(H) \geq \lceil (3B(G) - 1)/4 \rceil$$

and this result is sharp.

F23: [ChOp86] For any graph G and vertices $u, v \in V(G)$,

$$B(G) - 1 \leq B(G|_{u,v}) \leq 2B(G)$$

and both bounds are sharp.

F24: [Ch80-a, ChOp86] For any graph G and edge $e \in E(G)$,

$$B(G) - 1 \leq B(G|_e) \leq \lceil [3B(G) - 1]/2 \rceil$$

and both bounds are sharp.

F25: [WaWeYa95] Let $B(G) = b$ and $g(b, |V(G)|)$ be the maximum possible value of $B(G + e)$. Then

$$g(b, |V(G)|) = \begin{cases} b + 1 & \text{if } |V(G)| \leq 3b + 4 \\ \lceil (|V(G)| - 1)/3 \rceil & \text{if } 3b + 5 \leq |V(G)| \leq 6b - 2 \\ 2b & \text{if } |V(G)| \geq 6b - 1 \end{cases}$$

EXAMPLE

E4: Figure 9.4.5 illustrates Fact F25 by showing a graph G having bandwidth 2 and a corresponding $G + e$ having bandwidth 4, both shown with bandwidth numberings.

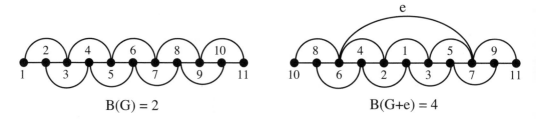

Figure 9.4.5: A graph whose bandwidth doubles when an edge is added.

REMARK

R6: Fact F25 gives a complete solution to a question originally posed by Erdős: whether $B(G + e) - B(G) \leq 1$, where $G + e$ is a graph obtained from G by adding an edge e not originally in G. This was first shown not to be the case in [Ch80-a].

Nordhaus–Gaddum Types of Bounds

DEFINITIONS

D16: The *complement of graph* G, denoted $\overline{G}$, is the graph with the same vertex set as G and $e \in \overline{G}$ if and only if $e \notin G$.

D17: A property P of a graph holds for *almost all graphs* if the ratio of the number of n-vertex graphs possessing P divided by the number of n-vertex graphs approaches one as n approaches infinity.

FACTS

F26: [ChErChGr81] For any graph G, $|V(G)| - 2 \leq B(G) + B(\overline{G})$.

F27: [ChErChGr81] There is a positive constant c_1 such that $B(G) + B(\overline{G}) \leq 2|V(G)| - c_1 \log |V(G)|$ for any graph.

F28: [ChErChGr81] There is a positive constant c_2 such that $2|V(G)| - c_1 \log |V(G)| \leq B(G) + B(\overline{G})$ for almost all graphs.

F29: [FuWe01] Let $f(n) = \max\{B(G) + B(\overline{G}) : G \text{ an } n\text{-vertex graph}\}$. Then

$$2n - \lceil (4 + 2\sqrt{2}) \log_2 n \rceil \leq f(n) \leq 2n - 4 \log_2 n + o(\log n)$$

Other Bounds

FACTS

F30: [ChDeGiKo75] Let G be a graph and G^k be its kth power. Then $B(G^k) \leq kB(G)$.

F31: [Ch80-a] For graph G, $B(G) \leq |V(G)| - 3$ if and only if $\overline{G}$ contains a P_4.

F32: [Ch80-a] Let G be a graph such that $d(x, y) \leq 2$ for every pair of vertices x, y. Then $B(G) = |V(G)| - 2$ if and only if every component of $\overline{G}$ is a vertex, a $K_{1,n}$, or a K_3.

9.4.4   On the Bandwidth of Combinations of Graphs

Two or more graphs can be combined in a variety of ways to form a new graph, and information about the bandwidth of the new graph often can be gleaned from the bandwidths of the original graphs.

Cartesian Product

DEFINITION

D18: The *Cartesian product* of graphs G and H, denoted $G \times H$, is the graph where $V(G \times H) = V(G) \times V(H)$ and $(g_1, h_1)(g_2, h_2) \in E(G \times H)$ if and only if either (i) $g_1 = g_2$ and $h_1 h_2 \in E(H)$ or (ii) $h_1 = h_2$ and $g_1 g_2 \in E(G)$.

FACTS

F33: [Ch75, ChDeGiKo75] For graphs G and H,

$$B(G \times H) \leq \min\{|V(H)|B(G), |V(G)|B(H)\}.$$

F34: [Ch75] For paths P_m and P_n, where $\max\{m, n\} \geq 2$, $B(P_m \times P_n) = \min\{m, n\}$.

F35: If $m \geq 2$ and $n \geq 3$, then $B(P_m \times C_n) = \min\{2m, n\}$.

EXAMPLE

E5: Figure 9.4.6 shows a bandwidth numbering of $P_3 \times C_7$ which, by Fact F35, has bandwidth 6.

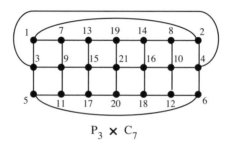

$$P_3 \times C_7$$

Figure 9.4.6: A bandwidth numbering of $P_3 \times C_7$.

Sum of Two Graphs

DEFINITION

D19: The *sum* of graphs G and H, denoted $G + H$, is the graph with $V(G + H) = V(G) \cup V(H)$ and $E(G + H) = E(G) \cup E(H) \cup \{uv : u \in V(G) \text{ and } v \in V(H)\}$.

FACTS

F36: [LiWiWa91] Let G and H be graphs such that $|V(G)| \geq |V(H)|$ and $B(G) < \lceil |V(G)|/2 \rceil$. Then $B(G + H) = \lceil |V(G)|/2 \rceil + |V(H)| - 1$.

F37: [LiWiWa91] Let G and H be graphs such that $|V(G)| \geq |V(H)|$ and $B(G) > \lceil |V(G)|/2 \rceil$. Then $|V(G)|/2 + |V(H)| - 1 \leq B(G+H) \leq \min\{B(G) + |V(H)|, \max\{B(H) + |V(G)|, \lceil |V(H)|/2 \rceil + |V(G)| - 1\}\}$.

F38: [LiWiWa91] For paths P_n and P_m with $n \geq m$, $B(P_n + P_m) = \lceil n/2 \rceil + m - 1$.

F39: [LiWiWa91] For cycles C_n and C_m with $n \geq m$,

$$B(C_n + C_m) = \begin{cases} \lceil n/2 \rceil + m - 1 & \text{if } n \geq 5 \\ 5 & \text{if } n \leq 4 \text{ and } m = 3 \\ 6 & \text{if } n = m = 4. \end{cases}$$

EXAMPLE

E6: Figure 9.4.7 shows a bandwidth numbering of $C_5 + C_3$ which, by Fact F39, has bandwidth 5.

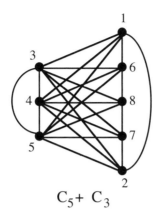

Figure 9.4.7: A bandwidth numbering of $C_5 + C_3$.

Corona and Composition

DEFINITIONS

D20: The **corona** of graphs G and H, denoted $G \circ H$, is the graph constructed from one copy of G and $|V(G)|$ copies of H, one associated with each vertex of G. If $v \in V(G)$ and H_v is the copy of H associated with v, there are the additional edges vh for every $h \in V(H_v)$.

D21: The **composition** of graphs G and H, denoted $G(H)$, is the graph where $V(G(H)) = V(G) \times V(H)$ and $(g_1, h_1)(g_2, h_2) \in E(G(H))$ if and only if either (i) $g_1 g_2 \in E(G)$ or (ii) $g_1 = g_2$ and $h_1 h_2 \in E(H)$.

FACTS

F40: [Ch80-b] For graphs G and H, $B(G \circ H) \le B(G)(|V(H)| + 1)$, and this bound is sharp.

F41: [Ch80-b] For graphs G and H, $B(G(H)) \le (B(G) + 1)|V(H)| - 1$, and this bound is sharp.

Strong Product and Tensor Product

DEFINITIONS

D22: The **strong product** of graphs G and H, denoted $G(Sp)H$, is the graph with $V(G(Sp)H) = V(G) \times V(H)$ and $(g_1, h_1)(g_2, h_2) \in E(G(Sp)H)$ if and only if either $g_1 g_2 \in E(G)$ and $h_1 h_2 \in E(H)$ or $g_1 = g_2$ and $h_1 h_2 \in E(H)$ or $h_1 = h_2$ and $g_1 g_2 \in E(G)$.

D23:   The **tensor product** of graphs G and H, denoted $G(Tp)H$, is the graph with $V(G(Tp)H) = V(G) \times V(H)$ and $(g_1, h_1)(g_2, h_2) \in E(G(Tp)H)$ if and only if $g_1 g_2 \in E(G)$ and $h_1 h_2 \in E(H)$.

FACTS

F42:   [LaWi95] (a) If $m \geq n \geq 2$, $B(P_m(S_p)P_n) = n + 1$.

(b) If $m \geq 3$ and $n \geq 2$, $B(C_m(S_p)P_n) = \begin{cases} m+2 & \text{if } n \geq \lfloor n/2 \rfloor + 1 \\ 2n+1 & \text{otherwise.} \end{cases}$

(c) If $m \geq n$, $B(C_m(S_p)C_n) = 2n + 2$.

F43:   [LaWi97-b]

$$B(C_m(T_p)C_n) = \begin{cases} n+1 & \text{if } m \geq n \geq 4 \text{ and } m, n \text{ even} \\ \min\{n+1, 2m+1\} & \text{if } m \text{ odd}, n \text{ even} \\ 2n+1 & \text{if } m \geq n \geq 3 \text{ and } m, n \text{ odd} \end{cases}$$

9.4.5   Bandwidth and Its Relationship to Other Invariants

Many bounds for bandwidth in terms of other graphical invariants have been found. Several are listed in the previously mentioned survey papers and in [BrDu85, BrDu91].

NOTATION: Throughout this subsection, G is a graph with $V = V(G)$, $E = E(G)$, $B = B(G)$, $\delta_{\max} = \delta_{\max}(G)$, and $\delta_{\min} = \delta_{\min}(G)$.

Vertex Degree

DEFINITION

D24:   The **degree sequence** of graph G is a listing of the degrees of the vertices of G, usually in monotonic order.

FACTS

F44:   [Ch70] If the graph G has the degree sequence $d_1 \leq d_2 \leq \cdots \leq d_n$, then

$$B \geq \max_j \{d_j - \lfloor (j-1)/2 \rfloor, d_j/2\}.$$

Setting $j = 1$ yields $B \geq \delta_{\min}$.

F45:   [ChChDeGi82] If G contains no copies of K_3, $B \geq \lfloor (3\delta_{\min} - 1)/2 \rfloor$.

EXAMPLE

E7:   Two graphs that show Fact F45 is sharp are the 3-dimensional hypercube Q_3 and the graph along with the bandwidth numbering shown in Figure 9.4.8. Both examples have no K_3, $\delta_{\min} = 3$, and bandwidth $B = \lfloor (3(3) - 1)/2 \rfloor = 4$.

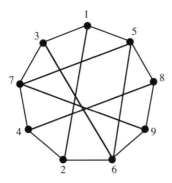

Figure 9.4.8: Bandwidth numbering of a graph with no K_3, $\delta_{\min} = 3$, $B = 4$.

Number of Vertices and Edges for Arbitrary Graphs

FACTS

F46: [DuBr89] $B \leq (|E| + 1)/2$.

F47: [BrDu85] $B \geq \left[2|V| - 1 - \sqrt{(2|V| - 1)^2 - 8|E|} \right] /2$.

F48: [LaWi97-a] If G is connected, then $B \geq |V| - s$, where s is the largest integer such that $s(s - 1) \leq |V|(|V| - 1) - 2|E|$, and the bound is sharp.

F49: [DuBr89] If $B \geq |V|/2$, then $|E| \geq |V|(|V| - 1)/[2(|V| - B)]$.

F50: [DuBr89] If $B \geq |V|/2$, then $|E| \geq (2\lfloor |V|/2 \rfloor - 1) \left[|V|/(|V| - 2) \right]^{B - \lfloor |V|/2 \rfloor}$.

F51: [AlLiMcEr92] Let $B = \lceil (1 - \epsilon)|V| \rceil$ with $0 < \epsilon < 1$. Then there are positive constants c_1 (which depends on ϵ) and c_2 such that $c_1 |V|/\epsilon \leq m(|V|, B) \leq c_2 (\log (2/\epsilon)|V|/\epsilon$ where $m(|V|, B)$ is the minimum possible number of edges in a graph with $|V|$ vertices and bandwidth B.

Number of Vertices and Edges for Graphs with no K_3

FACTS

F52: [ChTr84] Let $t(n, B)$ be the maximum number of edges that an n-vertex graph having no K_3 and bandwidth at most B can have. Then $\left(2 - \sqrt{2} \right) nB \leq t(n, B) \leq \frac{5 + \sqrt{3}}{11} nB$.

F53: [BrCaDuFiVi00] Let G be a bipartite graph with partite set sizes m and n, $m \leq n$, bandwidth B, and $(\lceil m/2 \rceil + 1)B \leq n \leq (m + 1)B - 1$. Then $|E| \leq 2mB - 2m - 3 + \lceil (n + 1)/B \rceil + \lfloor (n + 1)/B \rfloor$, and this bound is sharp.

F54: [BrCaDuFiVi00] Let G be a bipartite graph with partite set sizes m and n, $m \leq n = (t + 1)B + \lfloor \alpha B \rfloor$ where α is a fixed constant such that $0 \leq \alpha < 1$, and $B \geq (m + t + 4\lfloor m/t \rfloor - 5/2)/(1 - \alpha)$. Then $|E| \leq 2mB - \lfloor m/t \rfloor (2m - t\lfloor m/t \rfloor - t)$, and this bound is sharp.

Radius and Diameter

DEFINITIONS

D25: The **radius** of graph G, denoted $rad(G)$, is the smallest number r such that there is a vertex u of G with distance at most r from every other vertex of G.

D26: The **diameter** of graph G, denoted $diam(G)$, is the maximum distance between any two vertices of G.

FACTS

F55: [ChChDeGi82] For any graph G, $B \leq \delta_{max}(\delta_{max} - 1)^{rad(G)-1}$.

F56: [Ch70, ChDeGiKo75] For any graph G,

$$\lceil (|V| - 1)/diam(G) \rceil \leq B \leq |V| - diam(G)$$

F57: [ChSe89] For any graph G, $B \geq \max\{(|V(G')| - 1)/diam(G')\}$ where the maximum is taken over all connected subgraphs G' of G that have at least two vertices.

REMARK

R7: Paths and cycles achieve the lower bound of Fact F56. Figure 9.4.9 shows a bandwidth numbering of a graph having $|V| = 9$, diameter $= 4$, and $B = 5$, so the graph achieves the upper bound.

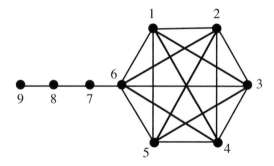

Figure 9.4.9: Bandwidth numbering of a graph with $|V| = 9$, $d = 4$, $B = 5$.

Vertex and Edge Chromatic Number

DEFINITIONS

D27: The (vertex) **chromatic number** of graph G, denoted $\chi(G)$, is the smallest number k such that there is a function $f : V(G) \to \{1, 2, \ldots, k\}$ with the property that, if uv is an edge, then $f(u) \neq f(v)$.

D28: The **edge-chromatic number** of graph G, denoted $\chi'(G)$, is the smallest number k such that there is a function $f : E(G) \to \{1, 2, \ldots, k\}$ with the property that, if edges e_1 and e_2 share a common vertex, then $f(e_1) \neq f(e_2)$.

FACTS

F58: [ChDeGiKo75] For any graph G, $B \geq \chi(G) - 1$.

F59: [BrDu85] For any graph G, $B \geq \chi'(G)/2$.

Vertex Independence and Vertex Cover Numbers

DEFINITIONS

D29: The **vertex independence number** of graph G, denoted $ind(G)$, is the largest cardinality of a set of vertices which induces a graph with no edges.

D30: The **vertex cover number** of graph G, denoted $\alpha_0(G)$, is the smallest cardinality of a set of vertices such that every edge is incident to at least one of the vertices in the set.

FACTS

F60: [Ch70, ChDeGiKo75] For any graph G,

$$\lceil |V|/ind(G) \rceil - 1 \leq B \leq |V| - \lfloor ind(G)/2 \rfloor - 1$$

F61: [De76] For any graph G, $B \geq \alpha_0(G)/ind(G)$.

Girth, Vertex Arboricity, and Thickness

DEFINITIONS

D31: The **girth** of graph G, denoted $girth(G)$, is the size of a smallest induced cycle of G.

D32: The **vertex arboricity** of graph G, denoted $arbor(G)$, is the minimum number of subsets into which $V(G)$ can be partitioned such that the vertices of each subset induce an acyclic subgraph.

D33: The **thickness** of graph G, denoted $thick(G)$, is the smallest number of planar subgraphs of G whose union is G.

FACTS

F62: [BrDu91] If G is not a forest, then $B \geq (girth(G) - 1)(arbor(G) - 2) + 2$.

F63: [BrDu91] If G is not a forest, then

$$B \geq [(girth(G) - 1)|V|/(2 \cdot ind(G)] - girth(G) + 2$$

F64: [BrDu85] For any graph G, $thick(G) \leq \max(B/2, 1)$.

9.4.6   Related Concepts

The study of bandwidth has spawned investigations into a variety of related ideas.

Bandsize

Bandsize is discussed briefly in [LaWi99].

DEFINITION

D34: Let f be a proper numbering of a graph G. The **bandsize of** f, denoted $bs_f(G)$, is the number of distinct edge differences produced by f. The **bandsize of** G is given by $bs(G) = \min\{bs_f(G) : f$ is a proper numbering of $G\}$.

FACT

F65: [ErHeWi89] For any graph G, $B(G) \geq bs(G)$.

Edgesum (Bandwidth Sum)

The edgesum first appeared in [Ha64]. Edgesums are discussed in [Se70, Io74, Io76, ChChDeGi82, Ch88, YaWa95, YuHu95, YuHu96, LaWi99].

DEFINITION

D35: Let f be a proper numbering of G. The **edgesum generated by** f is $s_f(G) = \sum_{uv \in E(G)} |f(u) - f(v)|$. The **edgesum of** G is given by $s(G) = \min\{s_f(G) : f$ is a proper numbering of $G\}$.

FACTS

F66: For the n-dimensional hypercube Q_n, $s(Q_n) = 2^{n-1}(2^n - 1)$.

F67: Like bandwidth, the edgesum decision problem is NP-complete.

Cyclic Bandwidth

Cyclic bandwidth is discussed in [Li94, Li97, HaKaRi99, LaShCh02].

DEFINITION

D36: Let f be a proper numbering of G. The **cyclic bandwidth of** f is $B_{cf}(G) = \max\{\|f(u) - f(v)\|_c : uv \in E\}$ where $\|x\|_c = \min\{|x|, n - |x|\}$. The **cyclic bandwidth** of G is given by $B_c(G) = \min\{B_{cf}(G) : f$ is a proper numbering of $G\}$.

Edge-Bandwidth

Edge-bandwidth is introduced and several results are presented in [JiMuShWe99].

DEFINITIONS

D37: An **edge-numbering** f of a graph G is a bijection from $E(G)$ to the set of integers.

D38: Let f be an edge-numbering of G. The ***edge-bandwidth of*** f is $B'_f(G) = \max\{|f(e_1) - f(e_2)| : \text{edges } e_1 \text{ and } e_2 \text{ adjacent in } G\}$. The ***edge-bandwidth of graph*** G is given by $B'(G) = \min\{B'_f(G) : f \text{ an edge numbering of } G\}$.

D39: The ***line graph*** of a graph G is the graph $L(G)$ such that $V(L(G)) = E(G)$ and two vertices in $L(G)$ are adjacent if and only if the corresponding edges are adjacent in G.

FACTS

F68: For any graph G, $B'(G) = B(L(G))$.

F69: For any graph G, $B(G) \le B'(G)$ and, if G is a forest, $B'(G) \le 2B(G)$.

EXAMPLE

E8: Figure 9.4.10 shows an edge-bandwidth numbering of a graph G and a bandwidth numbering of the line graph $L(G)$. Notice that the edge-bandwidth numbering of an edge of G is identical to the bandwidth numbering of the corresponding vertex in $L(G)$. It is not difficult to see that $B(L(G)) = 3$.

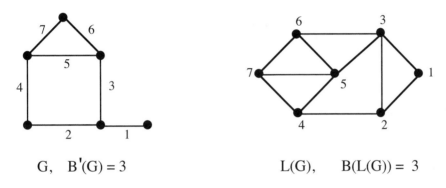

G, B'(G) = 3   L(G), B(L(G)) = 3

Figure 9.4.10: $B(L(G)) = B'(G) = 3$.

Profile

The *profile* of a graph is discussed in [LiYu94, LaWi99].

DEFINITIONS

D40: Let f be a proper numbering of a graph G, and let vertex $v \in V(G)$. The ***profile width*** is $w_f(v) = \max_{x \in N[v]}(f(v) - f(x))$ where $N[v]$ is the closed neighborhood of v.

D41: Let f be a proper numbering of a graph G. The ***profile of*** f is $P_f(G) = \sum_{v \in V} w_f(v)$. The ***profile of*** G, denoted $P(G)$, is given by $P(G) = \min\{P_f(G) : f \text{ is a proper numbering of } G\}$.

FACTS

F70: $P(P_n) = n - 1$.

F71: $P(C_n) = 2n - 3$.

F72: If $m \le n$, then $P(K_{m,n}) = mn + m(m-1)/2$.

Cutwidth

References for cutwidth include [GuSu82, Le82, Ch85, MaPaSu85, Ya85, Ch88].

DEFINITION

D42: Let f be a proper numbering of a graph G. The **cutwidth of** f is $c_f(G) = \max_i |\{vw \in E(G) : f(v) \leq i < f(w)\}|$. The **cutwidth of** G is $cutw(G) = \min\{c_f(G) : f$ is a proper numbering of $G\}$.

FACTS

F73: $cutw(P_n) = 1$.

F74: $cutw(C_n) = 2$.

F75: $cutw(K_n) = \lfloor n^2/4 \rfloor$.

F76: $cutw(K_{1,n}) = \lfloor n/2 \rfloor$.

F77: $cutw(T_{k,d}) = \lceil (d-1)(kt-1)/2 \rceil + 1$ if $d \geq 3$.

Topological Bandwidth

Some references for topological bandwidth are [Ch80-a, MaPaSu85, Ch88].

DEFINITION

D43: The **topological bandwidth** of graph G, denoted $B^*(G)$, is given by $B^*(G) = \min\{B(G') : G'$ is a refinement of $G\}$.

FACTS

F78: For any graph G, $B^*(G) \leq cutw(G)$.

F79: For any tree T, $B^*(T) \leq cutw(T) \leq B^*(T) + \log_2 B^*(T) + 2$.

F80: If G is P_n, C_n, or $K_{1,n}$, then $B^*(G) = cutw(G)$.

F81: $B^*(K_n) = n - 1 < cutw(K_n)$.

Additive Bandwidth

The proper numbering of Q_3 given in Section 9.4.1 corresponds to an adjacency matrix with ones on all but one diagonal above the main diagonal, including the diagonal farthest away. Since $B(Q_3) = 4$, this is far from a bandwidth numbering. However, consider the main *contradiagonal* (running from the lower left corner of the adjacency matrix to the upper right). All ones are on the main contradiagonal and the two contradiagonals above and below it. This concept, recognized and elucidated in [BaRuSl92], is the basis for a second type of bandwidth called *additive bandwidth*. Since the ones of the adjacency matrix are all on contradiagonals within two of the main contradiagonal, the additive bandwidth for Q_3 is at most two, and it is easy to see that equality holds.

DEFINITION

D44: Let f be a proper numbering of a graph G. The ***additive bandwidth of*** f is $B_f^+(G) = max\{|f(u) + f(v) - (n+1)| : uv \in E\}$. The ***additive bandwidth of*** G, denoted $B^+(G)$, is given by $B^+(G) = min\{B_f^+(G) : f$ is a proper numbering of $G\}$.

REMARKS

R8: The expression $|f(u) + f(v) - (n+1)|$ indicates, in the adjacency matrix, the number of contradiagonals (perhaps zero) from the main contradiagonal which contains the one corresponding to edge uv, and the summation involved motivated the name "additive bandwidth."

R9: Many of the investigations which have been made into bandwidth have been repeated for this new concept. However, while it is believed that the corresponding decision problem is NP-complete, this had not been proven nor have any algorithms yet been developed.

R10: Since $B^+(Q_3) = 2 < B(Q_3) = 4$, we see that additive bandwidth can be smaller than bandwidth. The graphs $P_m \times P_n$ represent an infinite family for which this is true by the same factor of two. In fact, this factor of two is best possible. On the other hand, additive bandwidth can be arbitrarily larger than bandwidth (by Fact F34 and Fact F82 below).

R11: In addition to the three results given below, the values of $B^+(G)$ for other families of graphs have been determined, as have several relationships between it and other invariants. The complete k-ary tree has proven difficult, and only partial results are available for it. We have seen that adding an edge to a graph can double its original bandwidth. In fact, the addition of an edge can triple the original additive bandwidth [BrCaViWiYe03]. Additional references for additive bandwidth are [Ha93, HaCaCa94, VoBr94, BrCaRoVi96, Ca96, DuBr97].

FACTS

F82: [BaRuSl92] $B^+(P_m \times P_n) = \lceil min\{m, n\}/2 \rceil$.

F83: [BaRuSl92] If $B^+(G) \geq 1$, then $B(G) \leq 2B^+(G)$.

F84: [BaRuBrCaSlVi95] $B(tK_3) = 2$ and $B^+(tK_3) = t$.

References

[AlLiMcEr92] Y. Alavi, J. Liu, J. McCanna, and P. Erdős, On the minimum size of graphs with a given bandwidth, *Bulletin of the Institute of Combinatorics and its Applications* 6 (1992), 22–32.

[AlMa65] G. G. Alway and D. W. Martin, An algorithm for reducing the bandwidth of a matrix of symmetrical configuration, *The Computer Journal* 8 (1965), 264–272.

[AnKaGe96] K. Ando, A. Kaneko, and S. Gervacio, The bandwidth of a tree with k leaves is at most $\lceil k/2 \rceil$, Discrete Mathematics 150 (1996), 403–406.

[BaRuBrCaSlVi95] M. E. Bascuñán, S. Ruiz, R. C. Brigham, R. M. Caron, P. J. Slater, and R. P. Vitray, On the additive bandwidth of graphs, Journal of Combinatorial Mathematics and Combinatorial Computing 18 (1995), 129–144.

[BaRuSl92] M. E. Bascuñán, S. Ruiz, and P. J. Slater, The additive bandwidth of grids and complete bipartite graphs, Congressus Numerantium 88 (1992), 245–254.

[BlKoRaVe00] A. Blum, G. Konjevod, R. Ravi, and S. Vempala, Semi-definite relaxations for minimum bandwidth and other vertex-ordering problems, Theoretical Computer Science 235 (2000), 25–42.

[BrCaDuFiVi00] R. C. Brigham, J. R. Carrington, R. D. Dutton, J. Fiedler, and R. P. Vitray, An extremal bandwidth problem for bipartite graphs, Journal of Graph Theory 35 (2000), 278–289.

[BrCaRoVi96] R. C. Brigham, J. R. Carrington, D. G. Rogers, and R. P. Vitray, On the additive bandwidth of the complete k-ary tree, Congressus Numerantium 118 (1996), 209–214.

[BrCaViWiYe03] R. C. Brigham, J. R. Carrington, R. P. Vitray, D. J. Williams, and J. Yellen, Change in additive bandwidth when an edge is added, Ars Combinatoria 68 (2003) 283–317.

[BrDu85] R. C. Brigham and R. D. Dutton, A compilation of relations between graph invariants, Networks 15 (1985), 73–107.

[BrDu91] R. C. Brigham and R. D. Dutton, A compilation of relations between graph invariants–supplement I, Networks 21 (1991), 421–455.

[Ca96] R. M. Caron, Free additive bandwidth of a graph, Congressus Numerantium 121 (1996), 49–57.

[CaMaPr02] A. Caprara, F. Malucelli, and D. Pretolani, On bandwidth-2 graphs, Discrete Applied Mathematics 117 (2002), 1–13.

[Ch70] V. Chvátal, A remark on a problem of Harary, Czechoslovak Mathematical Journal 20 (1970), 109–111.

[Ch75] J. Chvátalová, Optimal labeling of a product of two paths, Discrete Mathematics 11 (1975), 249–253.

[Ch80-a] J. Chvátalová, On the bandwidth problem for graphs, Ph.D. dissertation, Department of Combinatorics and Optimization, University of Waterloo, Ontario (1980).

[Ch80-b] P. Z. Chinn, The bandwidth of the corona and composition of two graphs, Department of Mathematics, Humboldt State University, Arcata, California (1980).

[Ch85] F. R. K. Chung, On the cutwidth and the topological bandwidth of a tree, SIAM Journal on Algebraic and Discrete Methods 6 (1985), 268–277.

[Ch88] F. R. K. Chung, Labelings of graphs, Selected Topics in Graph Theory 3, Academic Press Limited, San Diego, CA (1988), 151–168.

[ChChDeGi82] P. Z. Chinn, J. Chvátalová, A. K. Dewdney, and N. E. Gibbs, The bandwidth problem for graphs and matrices–a survey, *Journal of Graph Theory* 6 (1982), 223–254.

[ChDeGiKo75] J. Chvátalová, A. K. Dewdney, N. E. Gibbs, and R. R. Korfhage, The bandwidth problem for graphs: a collection of recent results, Research Report 24, Department of Computer Science, University of Western Ontario, London, Ontario (1975).

[ChErChGr81] P. Z. Chinn, P. Erdős, F. R. K. Chung, and R. L. Graham, On the bandwidth of a graph and its complement, *The Theory and Applications of Graphs*, G. Chartrand, Ed., Wiley, New York (1981), 243–253.

[ChOp86] J. Chvátalová and J. Opatrný, The bandwidth problem and operations on graphs, *Discrete Mathematics* 61 (1986), 141–150.

[ChSe89] F. R. K. Chung and P. D. Seymour, Graphs with small bandwidth and cutwidth, *Discrete Mathematics* 75 (1989), 113–119.

[ChTr84] F. R. K. Chung and W. T Trotter, Jr., Triangle-free graphs with restricted bandwidth, *Progress in Graph Theory*, Academic Press, Toronto, Ontario (1984), 175–190.

[CuMc69] E. Cuthill and J. McKee, Reducing the bandwidth of sparse symmetric matrices, *Proc. 24th National Conference of the ACM* (1969), 157–172.

[De76] A. K. Dewdney, The bandwidth of a graph — some recent results, *Congressus Numerantium* 17 (1976), 273–288.

[DuBr89] R. D. Dutton and R. C. Brigham, On the size of graphs of a given bandwidth, *Discrete Mathematics* 76 (1989), 191–195.

[DuBr97] R. D. Dutton and R. C. Brigham, Invariant relations involving the additive bandwidth, *Journal of Combinatorial Mathematics and Combinatorial Computing* 23 (1997), 77–85.

[Ei79] P. G. Eitner, The bandwidth of the complete multipartite graph, presented at the *Toledo Symposium on Applications of Graph Theory* (1979).

[ErHeWi89] P. Erdős, P Hell, and P. Winkler, Bandwidth versus bandsize, *Annals of Discrete Mathematics* 41 (1989), 117–129.

[Ev79] G. C. Everstine, A comparison of three resequencing algorithms for the reduction of matrix profile and wavefront, *International Journal for Numerical Methods in Engineering* 14 (1979), 837–853.

[Fe00] U. Feige, Approximating the bandwidth via volume respecting embeddings, *Journal of Computer and System Sciences* 60 (2000), 510–539.

[FuWe01] Z. Füredi and D. B. West, Ramsey theory and bandwidth of graphs, *Graphs and Combinatorics* 17 (2001), 463–471.

[GaGrJoKn78] M. R. Garey, R. L. Graham, D. S. Johnson, and D. E. Knuth, Complexity results for bandwidth minimization, *SIAM Journal on Applied Mathematics* 34 (1978), 477–495.

[GiPoSt76] N. E. Gibbs, W. G. Poole, Jr., and P. K. Stockmeyer, An algorithm for reducing the bandwidth and profile of a sparse matrix, *SIAM Journal on Numerical Analysis* 13 (1976), 236–250.

[GoOp90] C. GowriSankaran and J. Opatrný, New bandwidth reduction algorithms, *Congressus Numerantium* 76 (1990), 77–88.

[Gu01] A. Gupta, Improved bandwidth approximation for trees and chordal graphs, *Journal of Algorithms* 40 (2001), 24–36.

[GuSu82] E. M. Gurari and I. H. Sudborough, Improved dynamic programming algorithms for the bandwidth minimization problem and the min cut linear arrangement problem, Technical Report, Department of Electrical Engineering and Computer Science, Northwestern University, Evanston, IL (1982).

[Ha64] L. H. Harper, Optimal assignment of numbers to vertices, *Journal of SIAM* 12 (1964), 131–135.

[Ha66] L. H. Harper, Optimal numberings and isoperimetric problems on graphs, *Journal of Combinatorial Theory* 1 (1966), 385–393.

[Ha93] F. W. Hackett, The additive bandwidth of a union of star graphs, Master's report, Department of Mathematics, University of Central Florida (1993).

[HaCaCa94] F. W. Hackett, R. M. Caron, and J. R. Carrington, The additive bandwidth of a union of stars, *Congressus Numerantium* 101 (1994), 155–160.

[HaKaRi99] F. Harary, P. C. Kainen, and A. Riskin, Every graph of cyclic bandwidth 3 is toroidal, *Bulletin of the Institute of Combinatorics and Its Applications* 27 (1999), 81–84.

[HaMa97] J. Haralambides and F. Makedon, Approximation algorithms for the bandwidth minimization problem for a large class of trees, *Theory of Computing Systems* 30 (1997), 67–90.

[HeGr79] G. Hein and E. Groten, On the use of bandwidth and profile reduction in combination solutions of satellite altimetry, *Acta Geodaetia, Geophys Montanistica Acad. Sci. Hung.* 14 (1979), 59–69.

[Io74] M. A. Iordanskiĭ, Minimal numerations of the vertices of trees, *Soviet Mathematics Doklady* 15 (1974), 1311–1315.

[Io76] M. A. Iordanskiĭ, Minimal numerations of the vertices of trees, *Problemy Kibernetiki* 31 (1976), 109–132.

[JiMuShWe99] T. Jiang, D. Mubayi, A. Shastri, and D. B. West, Edge-bandwidth of graphs, *SIAM Journal on Discrete Mathematics* 12 (1999), 307–316.

[KaSh96] H. Kaplan and R. Shamir, Pathwidth, bandwidth, and completion problems to proper integral graphs with small cliques, *SIAM Journal on Computing* 25 (1996), 540–561.

[KlTa01] T. Kloks and R. B. Tan, Bandwidth and topological bandwidth of graphs with few P_4's, *Discrete Applied Mathematics* 115 (1997), 117–133.

[KrSt02] D. Kratsch and L. Stewart, Approximating bandwidth by mixing layouts of interval graphs, *SIAM Journal on Discrete Mathematics* 15 (2002), 435–449.

[LaShCh02] P. C. B. Lam, W. C. Shiu, and W. H. Chan, Characterization of graphs with equal bandwidth and cyclic bandwidth, *Discrete Mathematics* 242 (2002), 283–289.

[LaWi95] Y. Lai and K. Williams, Bandwidth of the strong product of paths and cycles, *Congressus Numerantium* 109 (1995), 123–128.

[LaWi97-a] Y. Lai and K. Williams, Some bounds on bandwidth, edgesum, and profile of graphs, *Congressus Numerantium* 125 (1997), 25–31.

[LaWi97-b] Y. Lai and K. Williams, On bandwidth for the tensor product of paths and cycles, *Discrete Applied Mathematics* 73 (1997), 133–141.

[LaWi99] Y. Lai and K. Williams, A survey of solved problems and applications on bandwidth, edgesum, and profile of graphs, *Journal of Graph Theory* 31 (1999), 75–94.

[Le82] T. Lengauer, Upper and lower bounds on the complexity of the min-cut linear arrangement problem on trees, *SIAM Journal on Algebraic and Discrete Methods* 3 (1982), 99–113.

[Li94] Y. Lin, The cyclic bandwidth problem, *Systems Science and Mathematical Sciences* 7 (1994), 282–288.

[Li97] Y. Lin, Minimum bandwidth problem for embedding graphs in cycles, *Networks* 29 (1997), 135–140.

[Li00] Y. Lin, On characterizations of graph bandwidth, *OR Transactions* 4 (2000), 1–6.

[LiWiWa91] J. Liu, K. Williams, and J. F. Wang, Bandwidth for the sum of two graphs, *Congressus Numerantium* 82 (1991), 79–85.

[LiYu94] Y. X. Lin and J. J. Yuan, Minimum profile of grid networks, *Systems Science and Mathematical Sciences* 7 (1994), 56–66.

[MaPaSu85] F. S. Makedon, C. H. Papadimitriou, and I. H. Sudborough, Topological bandwidth, *SIAM Journal on Algebraic and Discrete Methods* 6 (1985), 418–444.

[Mi91] Z. Miller, Graph layouts, *Applications of Discrete Mathematics*, J. G. Michaels and K. H. Rosen (Editors), McGraw-Hill, New York (1991), 365–393.

[Pa76] C. H. Papadimitriou, The NP-completeness of the bandwidth minimization problem, *Computing* 16 (1976), 263–270.

[Ro68] R. Rosen, Matrix bandwidth minimization, *Proc. 23rd National Conference of the ACM*, Brandon Systems, Princeton, New Jersey (1968), 585–595.

[Sa80] J. B. Saxe, Dynamic-programming algorithms for recognizing small-bandwidth graphs in polynomial time, *SIAM Journal on Algebraic and Discrete Methods* 1 (1980), 363–369.

[Se70] M. A. Seĭdvasser, The optimal numbering of the vertices of a tree, *Diskretnyĭ Analiz* 17 (1970), 56–74.

[Sm85] W. F Smyth, Algorithms for the reduction of matrix bandwidth and profile, *Journal of Computational and Applied Mathematics* 12-13 (1985), 551–561.

[Sm95] L. Smithline, Bandwidth of the complete k-ary tree, *Discrete Mathematics* 142 (1995), 203–212.

[VoBr94] M. P. Vogt and R. C. Brigham, On the additive bandwidth of simple trees, *Congressus Numerantium* 103 (1994), 155–160.

[WaWeYa95] J. F. Wang, D. B. West, and B. Yao, Maximum bandwidth under edge addition, *Journal of Graph Theory* 20 (1995), 87–90.

[WaYa95] J. F. Wang and B. Yao, On upper bounds of bandwidths of trees, *Acta Mathematicae Applicatae Sinica (English Series)* 11 (1995), 152–159.

[Ya85] M. Yannakakis, A polynomial algorithm for the min-cut linear arrangement of trees, *Journal of the Association for Computing Machinery* 32 (1985), 950–988.

[Ya98] J. Yan, Algorithm aspects of the bandwidth problem on P_4-sparse graphs, *Tamsui Oxford Journal of Mathematical Sciences* 14 (1998), 11–18.

[YaWa95] B. Yao and J. F. Wang, On bandwidth sums of graphs, *Acta Mathematicae Applicatae Sinica (English Series)* 11 (1995), 69–78.

[YuHu95] J. Yuan and Q. Huang, A note on the bandwidth sum of complete multipartite graphs, *Journal of Mathematical Study* 28 (1995), 19–22.

[YuHu96] J. Yuan and Q. Huang, Some lower bounds of bandwidth sum of graphs with applications, *Mathematica Applicata* 9 (1996), 536–538.

Section 9.5

Pursuit-Evasion Problems

Richard B. Borie, University of Alabama
Sven Koenig, University of Southern California
Craig A. Tovey, Georgia Institute of Technology

INTRODUCTION

In *pursuit-evasion* problems, a team of mobile pursuers (or searchers) attempts to capture one or more mobile evaders (fugitives, intruders) within a graph. For example, the pursuers may represent soldiers, policemen, or robots. The evaders might be terrorists, criminals, lost children, or even a poisonous gas. The graph may represent a road map, building floor plan, cave system, pipe network, etc. Many distinct variations of pursuit-evasion problems can be formulated by specifying the rules of movement for the pursuers and for the evaders, the knowledge each opponent has about the other, the rules of capture, the kind of graph, and the objective function. Typical objectives include minimizing the number of pursuers, the distance travelled by pursuers, or the elapsed time until capture. Because finiteness of the latter two objectives is equivalent to optimization of the first objective, the complexity of the latter two problems is bounded below by that of the first. However, optimization of the first objective is usually $\mathcal{NP}$-hard for general graphs; hence the bulk of the graph-theoretic pursuit-evasion literature focuses on that objective.

The following subsections discuss several of the most-studied pursuit-evasion variations. Other surveys on pursuit-evasion include [Bi91], [FoPe96], [Al04], [Ha07], [FoTh08], [ChHoIs11], and [BoYa11]. Now there is also an entire book on this topic [BoNo11]. Throughout this section, except where explicitly specified otherwise, G will denote a connected undirected graph or multigraph, possibly with loops.

9.5.1   Sweeping and Edge Search

FACT

F1:   Parsons [Pa78] describes the original pursuit-evasion problem. The following definitions are adapted from [Pa78].

DEFINITIONS

D1:   Consider an embedding of G in 3D space such that each vertex resides at a distinct location and no two edges intersect except at a common endpoint. (For every G such an embedding exists.) Let k denote the number of pursuers, and let $P = (P_1, \ldots, P_k)$ where each $P_j : [0, \infty) \to G$ is a continuous function. Then P is a **sweep strategy** for G if for every continuous function $E : [0, \infty) \to G$, there exists some pursuer $j \in \{1, \ldots, k\}$ and time t such that $P_j(t) = E(t)$. Here $P_j(t)$ denotes the location within graph G of pursuer j at time t, $E(t)$ denotes the location of an evader at time t, and capture occurs when $P_j(t) = E(t)$.

D2:   The **sweep number** of G, denoted $\mathrm{sw}(G)$, is the smallest k such that a sweep strategy $P = (P_1, \ldots, P_k)$ exists. The **sweep problem** on G is to determine $\mathrm{sw}(G)$, and G is k-**sweepable** if $\mathrm{sw}(G) \leq k$.

FACTS

F2:   Petrov [Pe82] independently develops another pursuit-evasion model, based on a system of differential equations. See [Pe82] for details.

F3:   Golovach [Go89] describes yet another formulation for pursuit-evasion. The following definitions are adapted from [Go89].

DEFINITIONS

D3:   An **edge search operation** is one of the following: $p(x)$ = place a pursuer at vertex x; $r(x)$ = remove a pursuer from vertex x; and $s(e, x, y)$ = slide a pursuer along edge e from endpoint x to other endpoint y. (We may write $s(e, x, y)$ as $s(x, y)$ if only one edge (x, y) exists, or as $s(e)$ if the sliding direction is forced or inconsequential.)

D4:   Initially every edge of G is *contaminated* (might contain an evader). An edge $e = (x, y)$ becomes *clear* if a pursuer slides along e from x to y while either (i) another pursuer resides at x or (ii) every other edge incident to x is clear. If ever any unoccupied vertex x is incident to a contaminated edge, then any clear edges incident to x immediately become recontaminated. (So if a pursuer slides from x to y while neither (i) nor (ii) holds, then edge (x, y) does not become clear.) An **edge search strategy** for G is any sequence of edge search operations that ends with every edge of G being simultaneously clear.

D5:   The **edge search number** of G, denoted $\mathrm{es}(G)$, is the smallest number of pursuers needed to implement any edge search strategy. The **edge search problem** on G is to determine $\mathrm{es}(G)$, and G is k-**edge-searchable** if $\mathrm{es}(G) \leq k$.

FACT

F4:   Golovach [Go89] shows that the formulations of Parsons, Petrov, and Golovach
are all equivalent problems. Therefore $sw(G) = es(G)$ for every G.

EXAMPLES

E1:   The graph in Figure 9.5.1(a) has edge search number 2. Here is a strategy that
clears this graph using 2 pursuers: p(1), p(1), s(a), s(b), s(c), s(d,1,2), s(e), s(f), s(3,2),
s(g), s(h,2,4), s(i), s(j), s(5,4), s(k), s(l,4,6), s(m), s(n).

E2:   The graph in Figure 9.5.1(b) has edge search number 3. Here is a strategy
that clears this graph using 3 pursuers: p(1), p(2), p(2), s(a,2,1), s(b,1,2), s(c,2,1),
s(d), s(e,1,3), s(3,2), s(g), s(h,2,4), s(4,3), s(j), s(k,3,5), s(5,4), s(m), s(n,4,6), s(o,6,5),
s(p,5,6), s(q).

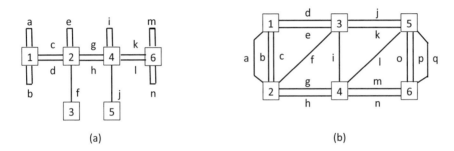

(a)   (b)

Figure 9.5.1: Typical graphs with edge search numbers 2 and 3.

FACTS

F5:   The edge search number $es(G) = 1$ if and only if G is a simple path. So $es(G) = 1$
if and only if G contains neither a cycle nor a vertex of degree 3 or more. Equivalently,
$es(G) = 1$ if and only if G contains neither of these two minimal forbidden minors: a
loop with one vertex and one edge, or a star with three edges.

F6:   Megiddo et al. [MeHaGaJoPa88] show that $es(G) \leq 2$ if and only if G contains none of the minimal forbidden minors illustrated in Figure 9.5.2. This paper also
provides a structural characterization for the 2-edge-searchable graphs, similar to the
graph in Figure 9.5.1(a).

F7:   Megiddo et al. [MeHaGaJoPa88] show that if G is biconnected, then $es(G) \leq 3$
if and only if G contains none of the minimal forbidden minors illustrated in Figure 9.5.3. This paper also provides a structural characterization for the biconnected
3-edge-searchable graphs, similar to the graph in Figure 9.5.1(b), and also a more general characterization for the non-biconnected 3-edge-searchable graphs.

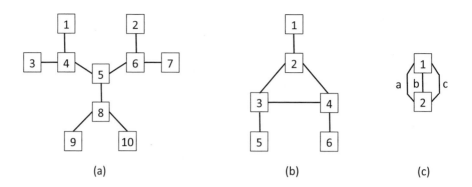

Figure 9.5.2: Forbidden minors for graphs with edge search number ≤ 2.

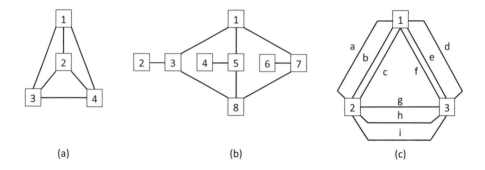

Figure 9.5.3: Forbidden minors for biconnected graphs with edge search number ≤ 3.

EXAMPLES

E3:   The graph in Figure 9.5.2(a) has edge search number 3: p(1), s(1,4), p(3), s(3,4), r(4), s(4,5), p(2), s(2,6), p(7), s(7,6), s(6,5), s(5,8), s(5,8), s(8,9), s(8,10). However, if edge (8,10) is removed, the resulting graph has edge search number 2: p(1), s(1,4), p(3), s(3,4), s(4,5), s(4,5), s(5,8), s(8,9), r(9), s(5,6), p(2), s(2,6), s(6,7).

E4:   The graph in Figure 9.5.2(b) has edge search number 3: p(1), s(1,2), p(2), s(2,3), s(2,4), p(5), s(5,3), s(3,4), s(4,6). However, if edge (4,6) is removed, the resulting graph has edge search number 2: p(1), s(1,2), p(2), s(2,3), s(2,4), s(4,3), s(3,5).

E5:   The graph in Figure 9.5.2(c) has edge search number 3: p(1), p(1), p(1), s(a), s(b,1,2), s(c). However, if edge c is removed, the resulting graph has edge search number 2: p(1), p(1), s(a), s(b).

E6:   The graph K_4 in Figure 9.5.3(a) has edge search number 4: p(1), p(1), p(1), s(1,2), s(1,3), s(1,4), p(3), s(3,2), s(2,4), s(4,3). However, if edge (3,4) is removed, the resulting graph has edge search number 3: p(1), p(1), p(1), s(1,2), s(1,3), s(1,4), s(3,2), s(2,4).

E7:   The graph in Figure 9.5.3(b) has edge search number 4: p(1), p(1), p(1), s(1,3), s(1,5), s(1,7), p(2), s(2,3), r(3), s(3,8), p(4), s(4,5), s(5,8), s(8,7), s(7,6). However, if edge (6,7) is removed, the resulting graph has edge search number 3: p(2), s(2,3), p(3), s(3,1), s(3,8), p(1), s(1,7), s(7,8), s(1,5), s(8,5), s(5,4).

E8: The graph in Figure 9.5.3(c) has edge search number 4: p(1), p(1), p(1), p(1), s(a), s(b,1,2), s(c,1,2), s(g), s(h,2,3), s(i,2,3), s(d,3,1), s(e), s(f). However, if edge i is removed, the resulting graph has edge search number 3: p(1), p(1), s(a), p(2), s(b,2,1), s(c,1,2), s(g), s(h,2,3), s(d,3,1), s(e,1,3), s(f).

FACTS

F8: Megiddo et al. [MeHaGaJoPa88] prove that the decision version of the edge search problem is $\mathcal{NP}$-complete for arbitrary graphs G.

F9: For every $n \geq 4$, es$(K_n) = n$, where K_n denotes a complete graph (or clique) with n vertices.

F10: Megiddo et al. [MeHaGaJoPa88] present a linear-time algorithm for computing es(G) when G is a tree.

F11: There exist polynomial-time algorithms for computing es(G) when G is a split graph, an interval graph, or a cograph.

REMARK

R1: It currently remains unresolved whether or not polynomial-time algorithms exist for computing es(G) when G is a permutation graph, an outerplanar graph, a series-parallel graph, or a planar graph.

FACTS

F12: Parsons [Pa78] shows that if G is a tree and $k \geq 2$, then es$(G) \geq k$ if and only if G has a vertex v with degree $d \geq 3$ such that splitting v into d vertices each having degree 1 yields a forest in which at least three trees have edge search number at least $k - 1$.

F13: Let T_k denote a smallest tree such that es$(T_k) = k$. Then T_1 has a single edge, T_2 is a star with three edges, and T_3 is the tree shown in Figure 9.5.2(a). In general for $k \geq 2$, T_k may be formed from three copies of T_{k-1} by choosing one leaf from each copy of T_{k-1} and fusing together these three vertices.

F14: Let m_k denote the number of edges in T_k. Then $m_1 = 1$, $m_2 = 3$, and $m_3 = 9$. In general for $k \geq 2$, it follows that $m_k = 3m_{k-1}$, so $m_k = 3^{k-1}$. Hence if T is any tree with m edges, then es$(T) \leq 1 + \log_3 m$.

F15: LaPaugh [La93] shows that recontamination is not useful for edge search. That is, es(G) pursuers can always clear G using an edge search strategy in which no clear edge ever becomes recontaminated.

REMARK

R2: LaPaugh's result that recontamination is not useful applies only to edge search but not to sweeping. This is because edge search permits arbitrary removal and placement of a pursuer, which essentially allows pursuers to jump between any vertices of the graph. However, sweeping requires each pursuer to move continuously through the graph, and therefore may require a pursuer to traverse (and unintentionally clear) a contaminated edge.

EXAMPLE

E9:　The graph in Figure 9.5.4 illustrates that recontamination is sometimes useful for sweeping. First note that this graph has edge search number 3 as follows: p(1), p(1), p(1), s(a), s(b,1,2), s(c,1,2), r(2), r(2), p(3), p(3), s(f), s(g), s(3,4), s(4,2), s(j), s(e), s(2,5), s(h), s(5,7), r(7), p(6), s(i), r(6), r(6), p(7), p(7), s(l), s(m,7,8), s(n). However, this solution requires jumping to avoid recontamination. That is, rather than removing two pursuers from vertex 2 and placing them on vertex 3, instead let these two pursuers slide along edges d and g to reach vertex 3. Then edge d is temporarily cleared, but later it becomes recontaminated when both pursuers depart from vertex 4. Because sweeping only permits moving along edges (no jumping), the graph in Figure 9.5.4 has sweep number 3, but every sweep strategy with 3 pursuers requires recontamination to occur.

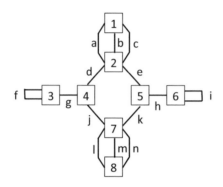

Figure 9.5.4: Recontamination can be useful for sweeping.

DEFINITIONS

D6:　A search strategy is **monotonic** if recontamination does not occur.

D7:　A search strategy is **internal** if no pursuer is ever removed from a vertex (so jumping does not occur).

D8:　A search strategy is **connected** if the set of clear edges always induces a connected subgraph.

FACT

F16:　Barriere et al. [BaFrSaTh03] provide inequalities that show the relationships between the numbers of pursuers needed to clear a graph when one or more of these constraints (m = monotonic, i = internal, c = connected) are required during edge search. In particular, $es(G) = m(G) = i(G) \leq mi(G) \leq c(G) = ic(G) \leq mc(G) = mic(G)$.

9.5.2 Node Search and Mixed Search

FACT

F17: Kirousis and Papadimitriou [KiPa85], [KiPa86] introduce a variation of pursuit-evasion that lacks sliding and that has a novel rule for clearing edges (capturing evaders). The following definitions are adapted from [KiPa86].

DEFINITIONS

D9: A *node search operation* is one of the following: $p(x) = $ place a pursuer at vertex x, and $r(x) = $ remove a pursuer from vertex x.

D10: Initially every edge of G is *contaminated*. An edge $e = (x, y)$ becomes *clear* if pursuers simultaneously occupy both endpoint vertices x and y. As previously stated with edge search, if ever an unoccupied vertex x is incident to a contaminated edge, then all clear edges incident to x become recontaminated. A *node search strategy* for G is any sequence of node search operations that ends with every edge of G being simultaneously clear.

D11: The *node search number* of G, denoted ns(G), is the smallest number of pursuers needed to implement any node search strategy. The *node search problem* on G is to determine ns(G), and G is k-*node-searchable* if ns(G) $\leq k$.

FACTS

F18: Kirousis and Papadimitriou [KiPa86] show that recontamination is not useful for node search. That is, ns(G) pursuers can always clear G using a node search strategy in which no clear edge ever becomes recontaminated.

F19: Bienstock and Seymour [BiSe91] unify edge search and node search into a more general framework called mixed search. The following definitions are adapted from [BiSe91].

DEFINITIONS

D12: *Mixed search operations* are the same as edge search operations: $p(x) = $ place a pursuer at vertex x; $r(x) = $ remove a pursuer from vertex x; and $s(e, x, y) = $ slide a pursuer along edge e from endpoint x to other endpoint y.

D13: Initially every edge of G is *contaminated*. As with edge search, edge $e = (x, y)$ becomes *clear* if a pursuer slides along e from x to y while either (i) another pursuer resides at x or (ii) every other edge incident to x is clear. Also, as with node search, edge $e = (x, y)$ becomes *clear* if pursuers simultaneously occupy both endpoint vertices x and y. Recontamination may occur the same as with edge search and node search. A *mixed search strategy* for G is any sequence of mixed search operations that ends with every edge of G being simultaneously clear.

D14: The *mixed search number* of G, denoted ms(G), is the smallest number of pursuers needed to implement any mixed search strategy. The *mixed search problem* on G is to determine ms(G), and G is k-*mixed-searchable* if ms(G) $\leq k$.

FACTS

F20:   Bienstock and Seymour [BiSe91] show that recontamination is not useful for mixed search. Thus ms(G) pursuers can always clear G using a mixed search strategy in which no clear edge ever becomes recontaminated.

F21:   For any G, construct G^e and G^n by replacing each edge of G with two edges in series or with two edges in parallel, respectively. Bienstock and Seymour [BiSe91] show that es(G) = ms(G^e) and ns(G) = ms(G^n), so edge search and node search both reduce to mixed search. Therefore the recontamination result for mixed graphs in [BiSe91] implies the previous recontamination results for edge search in [La93] and for node search in [KiPa86].

F22:   [KiPa86] and [BiSe91] provide inequalities that show the relationships between the edge search, node search, and mixed search numbers. Combining those inequalities yields that max{es(G), ns(G)} $- 1 \leq$ ms(G) $\leq$ min{es(G), ns(G)}, so these three parameter values are always within one of each other.

EXAMPLES

E10:   If G is a path with at least one edge then es(G)=1, ns(G)=2, and ms(G)=1.

E11:   If G is a loop with one vertex and one edge then es(G)=2, ns(G)=1, and ms(G)=1. Here is a (trivial) node search strategy that requires only 1 pursuer: p(1).

E12:   If G is a cycle with two vertices and two edges then es(G)=ns(G)=ms(G)=2.

E13:   If G is a cycle with at least three edges then es(G)=2, ns(G)=3, and ms(G)=2.

E14:   If G is a star with at least three edges then es(G)=ns(G)=ms(G)=2.

E15:   If G is the graph in Figure 9.5.2(c) then es(G)=3, ns(G)=2, and ms(G)=2. Here is a node search strategy that requires only 2 pursuers: p(1), p(2).

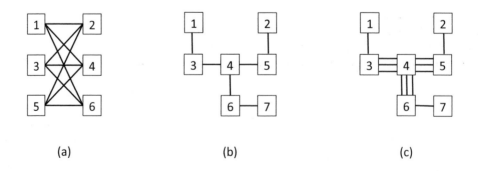

Figure 9.5.5: Examples for edge search, node search, and mixed search.

E16:   If G is K_4 then es(G)=ns(G)=ms(G)=4.

E17:   If G is the graph in Figure 9.5.5(a) then es(G)=5, ns(G)=4, and ms(G)=4. Here is a node search strategy that requires only 4 pursuers: p(1), p(3), p(5), p(2), r(2), p(4), r(4), p(6).

E18:   If G is the graph in Figure 9.5.5(b) then es(G)=2, ns(G)=3, and ms(G)=2.

E19:   If G is the graph in Figure 9.5.5(c) then es(G)=3, ns(G)=3, and ms(G)=2. Here is a mixed search strategy that requires only 2 pursuers: p(4), p(3), s(3,1), r(1), p(5), s(5,2), r(2), p(6), s(6,7).

FACTS

F23:   There exist polynomial-time algorithms for computing ns(G) and ms(G) when G is a split graph, an interval graph, a cograph, or a permutation graph.

F24:   There exist polynomial-time algorithms for computing ns(G) when G is a tree, an outerplanar graph, or a series-parallel graph.

REMARKS

R3:   It currently remains unresolved whether or not polynomial-time algorithms exist for computing ms(G) when G is a tree, an outerplanar graph, or a series-parallel graph.

R4:   It currently remains unresolved whether or not polynomial-time algorithms exist for computing ns(G) and ms(G) when G is a planar graph.

FACTS

F25:   Kirousis and Papadimitriou [KiPa86] prove that the decision version of the node search problem is $\mathcal{NP}$-complete for arbitrary graphs G.

F26:   Bienstock and Seymour [BiSe91] prove that the decision version of the mixed search problem is $\mathcal{NP}$-complete for arbitrary graphs G.

F27:   Kirousis and Papadimitriou [KiPa86] show that for every G, ns(G) is exactly one plus the *vertex separation* of G. Subsequently, Kinnersley [Ki92] showed that the vertex separation of G always equals the *pathwidth* of G; hence ns(G) is exactly one plus the pathwidth of G.

TERMINOLOGY NOTE: See Section 2.4 of this *Handbook* for a definition of pathwidth.

FACTS

F28:   Suppose the evader is *visible*, that is, the evader's position is always known to the pursuers. In this situation Seymour and Thomas [SeTh93] show that the fewest pursuers needed to implement a node search strategy is exactly one plus the *treewidth* of G.

F29:   Suppose instead that the (invisible) evader is *lazy*, that is, the evader can only move immediately before a pursuer is placed on the vertex where it resides (so that if the evader did not move it would be captured). In this situation Dendris et al. [DeKiTh97] show that the fewest pursuers needed to implement a node search strategy is again exactly one plus the *treewidth* of G.

TERMINOLOGY NOTE: See Section 2.4 of this *Handbook* for a definition of treewidth.

9.5.3 Cops-and-Robbers

The cops-and-robbers problem differs from the previously considered pursuit-evasion problems in several significant ways: both the cops (pursuers) and the robber (evader) must reside only at vertices, the cops and robber take alternating turns; everybody's location is visible to everyone else.

FACT

F30:   Nowakowski and Winkler [NoWi83] and Quilliot [Qu83] each independently originate the cops-and-robbers problem. However, each only considers the special case when there is only one cop.

F31:   Aigner and Fromme [AiFr84] extend the cops-and-robbers problem to permit multiple cops. The following definitions are adapted from [AiFr84].

DEFINITIONS

D15:   A *cops-and-robbers game* on G proceeds as follows. There are two players, C (a team of k cops) and R (a robber). C begins by placing each of the k cops at any vertex of G. (C is permitted to place more than one cop at the same location.) Next, R places the robber at any vertex. The players continue alternating turns. On C's turns, each cop either remains at its present location or moves to an adjacent vertex (so multiple cops may move simultaneously). Similarly, on R's turns, the robber either remains at its present location or moves to an adjacent vertex. Both C and R always know the locations of all participants. A cop *captures* the robber if the cop resides at the same vertex as the robber, and in this case player C wins the game. Player C has a *winning strategy* if no matter what choices R makes, player C can eventually win the game. Otherwise, if the robber can indefinitely avoid capture no matter what choices C makes, then player R wins the game.

D16:   The *cop number* of G, denoted $c(G)$, is the smallest number of cops k needed for player C to win the cops-and-robbers game on G. The *cops-and-robbers problem* on G is to determine $c(G)$, and G is *k-cop-winnable* if $c(G) \leq k$.

EXAMPLES

E20:   If G is a tree, then $c(G) = 1$. Player C's winning strategy is for the cop to move toward the robber along the shortest path that connects them.

E21:   If G is a complete graph (or clique), then $c(G) = 1$.

E22:   If G is a cycle with at least four edges, then $c(G) = 2$.

E23:   If G is a complete bipartite graph $K_{p,q}$ with $p \geq 2$ and $q \geq 2$, then $c(G) = 2$. Player C initially places one cop on each side of the bipartition.

E24:   If G is the graph shown in Figure 9.5.6(a) (the 3-cube), then $c(G) = 2$. If C places cops at vertices $\{1, 8\}$ then a robber placed at any vertex can be captured immediately. But if only one cop is available, then no matter where it is placed, the robber can always escape to a vertex that is not adjacent to the cop's location.

E25: If G is the graph shown in Figure 9.5.6(b) (Petersen's graph), then $c(G) = 3$. If C places cops at vertices $\{2, 5, 6\}$ then a robber placed at any vertex can be captured immediately. But if only two cops are available, then no matter where they are placed, the robber can always escape to a vertex that is not adjacent to either cop's location.

E26: If G is a p-by-q grid graph with $p \geq 2$ and $q \geq 2$, then $c(G) = 2$. (Figure 9.5.6(c) illustrates a 4-by-4 grid graph.) Cop 1 moves toward the row of the robber, but if already on the same row, then cop 1 moves toward the column of the robber. Cop 2 moves toward the column of the robber, but if already on the same column, then cop 2 moves toward the row of the robber.

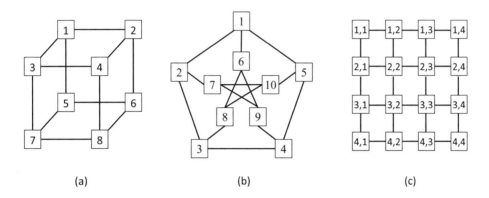

(a) (b) (c)

Figure 9.5.6: Examples for the cops-and-robbers problem.

DEFINITIONS

D17: Let N[v] denote the *closed neighborhood* of vertex v. Then v is a **corner vertex** if there exists some other vertex u with N[v] $\subseteq$ N[u].

D18: G is **dismantlable** if there exists a sequence of removing corner vertices that ends when only one vertex remains. Such a sequence is called an **elimination ordering**.

FACTS

F32: Nowakowski and Winkler [NoWi83] and Quilliot [Qu83] each developed a characterization of the graphs G with $c(G) = 1$, as follows: $c(G) = 1$ if and only if G is dismantlable.

F33: If G is a chordal graph, then $c(G) = 1$. (A chordal graph always has a *simplicial* vertex v such that N[v] is a clique, and any such v is a corner vertex.)

F34: Clarke [Cl02] shows that if G is an outerplanar graph then $c(G) \leq 2$.

F35: Theis [Th11] shows that if G is a series-parallel graph then $c(G) \leq 2$.

F36: Aigner and Fromme [AiFr84] show that if G is a planar graph then $c(G) \leq 3$.

F37:   Schroeder [Sc01] shows that if G is a toroidal graph (can be embedded in a torus) then $c(G) \leq 4$.

F38:   Joret et al. [JoKaTh10] show that if G is a treewidth-k graph then $c(G) \leq \lfloor \frac{k}{2} \rfloor + 1$.

F39:   Frankl [Fr87] shows that if G is a d-cube then $c(G) = \lceil \frac{d+1}{2} \rceil$.

F40:   Aigner and Fromme [AiFr84] show that if G is a graph with *girth* (length of smallest cycle) ≥ 5, then $c(G) \geq$ the minimum degree of any vertex in G.

F41:   Fomin et al. [FoGoKr08] prove that the cops-and-robbers problem is $\mathcal{NP}$-hard for arbitrary graphs G.

F42:   Goldstein and Reingold [GoRe95] show that if the initial location of each cop is specified as part of the problem instance, then this variation of the cops-and-robbers problem is $EXPTIME$-complete.

F43:   Chung et al. [ChHoIs11] give a pseudo-polynomial-time dynamic programming algorithm for solving the cops-and-robbers problem when the number of cops is fixed. The algorithm's running time is $O(n^{2k+2})$, where n is the number of vertices in G and k is the number of cops.

F44:   Llewellyn et al. [LlToTr89] show that it is $\mathcal{NP}$-complete to determine whether k cops can capture an infinitely fast robber if the cops are placed sequentially but cannot move once placed.

REMARKS

R5:   The complexity status of the decision version of the (standard) cops-and-robbers problem currently remains unresolved. Is it $\mathcal{NP}$-complete? Is it $EXPTIME$-complete?

R6:   Another currently unresolved question is known as Meyniel's conjecture: Is $c(G)$ in $O(\sqrt{n})$ for connected graphs G? Chiniforooshan [Ch08] shows that $c(G)$ is in $O(n/\log n)$, which is currently the best known bound. (Here again n denotes the number of vertices of G.)

9.5.4   Additional Variations

The previous subsections have discussed some of the best-known pursuit-evasion problems such as sweeping, edge search, node search, mixed search, and cops-and-robbers. The current subsection discusses some of the many possible additional variations that can be constructed by increasing or restricting the capabilities of the pursuers and/or the evader, and/or by modifying the kind of graph structure through which the pursuers and evaders move.

FACT

F45:   Nowakowski [No93], Dyer [Dy04], Barat [Ba06], Alspach et al [AlDyHaYa07], and Yang and Cao [YaCa07-a] [YaCa07-b] examine the sweeping, edge search, node search, and mixed search problems when G is a directed graph or multidigraph. Different variants occur depending on which participants must obey the specified edge directions.

DEFINITIONS

D19: In *directed sweeping*, both the pursuers and the evader must obey the specified edge directions.

D20: In *undirected sweeping*, both the pursuers and the evader may ignore the edge directions (so they can traverse each edge in either direction).

D21: In *weak sweeping*, the pursuers must obey the edge directions, but the evader may ignore these directions.

D22: In *strong sweeping*, the evader must obey the edge directions, but the pursuers may ignore these directions.

FACT

F46: Let $d(G)$, $u(G)$, $w(G)$, and $s(G)$ denote the minimum number of pursuers needed to capture an evader using directed, undirected, weak, or strong sweeping, respectively. Dyer [Dy04] shows these inequalities: $s(G) \leq \min\{d(G), u(G)\}$ and $\max\{d(G), u(G)\} \leq w(G)$.

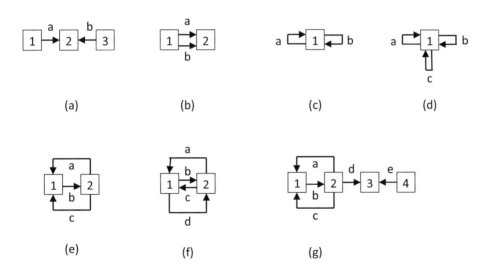

Figure 9.5.7: Directed graph examples.

EXAMPLES

E27: If G is a directed path then $d(G)=1$, $u(G)=1$, $w(G)=1$, and $s(G)=1$.

E28: If G is the graph in Figure 9.5.7(a) then $d(G)=2$, $u(G)=1$, $w(G)=2$, and $s(G)=1$.

E29: If G is the graph in Figure 9.5.7(b) then $d(G)=2$, $u(G)=2$, $w(G)=2$, and $s(G)=1$.

E30: If G is the graph in Figure 9.5.7(c) then $d(G)=2$, $u(G)=2$, $w(G)=2$, and $s(G)=1$. For strong sweeping, one pursuer can clear G by starting at vertex 1 and then traversing each edge in a backward direction.

E31: If G is the graph in Figure 9.5.7(d) then d(G)=2, u(G)=2, w(G)=2, and s(G)=2.

E32: If G is the graph in Figure 9.5.7(e) then d(G)=2, u(G)=3, w(G)=3, and s(G)=1. For strong sweeping, one pursuer can clear G by starting at vertex 1 and then traversing edges a, b, c each in a backward direction.

E33: If G is the graph in Figure 9.5.7(f) then d(G)=2, u(G)=3, w(G)=3, and s(G)=2.

E34: If G is the graph in Figure 9.5.7(g) then d(G)=3, u(G)=3, w(G)=4, and s(G)=1.

FACTS

F47: Dyer [Dy04] shows that s(G) = 1 if and only if each strongly connected component of G is either a single vertex or a cycle or a subdivision (homeomorphism) of one of the graphs shown in Figure 9.5.7(c) or (e).

F48: Gottlob et al. [GoLeSc03] examine pursuit-evasion when G in a hypergraph. This variation is known as *robber-and-marshals*. The robber (evader) resides in a vertex, and each marshal (pursuer) resides in a hyperedge. Capture occurs when any marshal occupies a hyperedge that is incident to the robber's vertex.

F49: Barriere et al. [BaFlFrSa02], Kolling and Carpin [KoCa08] [KoCa10], Daniel et al. [DaBoKoTo10], and Borie et al. [BoToKo11] consider pursuit-evasion on graphs in which each vertex and each edge has a specified *width*. These vertex and edge widths represent the number of pursuers needed to guard or clear each vertex or edge. The latter two papers also consider graphs where each edge may have a specified length.

F50: Kolling and Carpin [KoCa08] [KoCa10] define a variation of node search known as the Graph-Clear problem, and also present a polynomial-time algorithm for Graph-Clear on trees.

F51: Fomin and Golovach [FoGo00] [FoHeTe05], Daniel et al. [DaBoKoTo10], and Borie et al. [BoToKo11] consider variations of pursuit-evasion with different objectives such as minimizing the elapsed time until capture, or the total distance travelled by all the pursuers, or the sum of the times that each pursuer is present in the graph.

F52: Daniel et al. [DaBoKoTo10] present a pseudo-polynomial-time heuristic algorithm called ESP for edge search on series-parallel graphs.

F53: Borie et al. [BoToKo11] develop polynomial-time and pseudo-polynomial-time algorithms, and also $\mathcal{NP}$-completeness and strong $\mathcal{NP}$-completeness results, for several variations of sweeping on an assortment of graph classes.

EXAMPLES

Each vertex and edge in the graphs of Figure 9.5.8 is labeled with both its name and its width. Each edge in these graphs has length 1, and each pursuer travels at speed 1.

E35: The graph in Figure 9.5.8(a) can be cleared with 2 pursuers: Both pursuers start at vertex u and depart u simultaneously. In parallel, one pursuer clears edge a, and one pursuer clears edge b. Finally both pursuers arrive at v simultaneously. The total distance travelled is 2, and the elapsed time is 1.

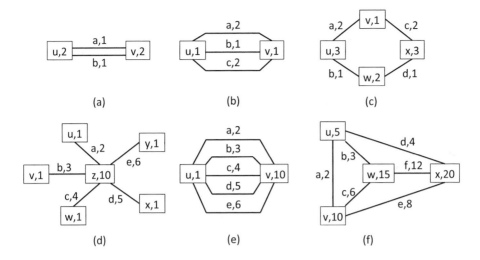

Figure 9.5.8: Examples for sweeping with specified vertex and edge widths.

E36: The graph in Figure 9.5.8(b) can be cleared with 3 pursuers: All pursuers start at vertex u. One pursuer guards u while the other two pursuers depart from u, clear edge a, and arrive at v. One pursuer guards v while the other pursuer departs from v, clears b, and arrives at u. Finally two pursuers depart from u, clear c, and arrive at v. The total distance travelled is 5, and the elapsed time is 3. [Alternatively, if 4 pursuers are available, this graph can be cleared in elapsed time 2. Or, if 5 pursuers are available, the graph can be cleared in elapsed time 1.]

E37: The graph in Figure 9.5.8(c) can be cleared with 3 pursuers: All pursuers start at vertex u and depart u simultaneously; two pursuers clear edge a, while the third pursuer clears edge b. When the first two pursuers arrive at v, one guards v while the other travels through the graph toward w. Two pursuers arrive at w simultaneously and clear w. Next these two pursuers depart w simultaneously; one clears d while the other travels through the graph toward v. When two pursuers reside at v, they both depart v and clear c. Finally all three pursuers arrive at x simultaneously. The total distance travelled is 10, and the elapsed time is 6. [Alternatively, if 4 pursuers are available, this graph can be cleared with total distance 6 and elapsed time 2, as follows: In parallel, two pursuers travel from u to x along edges a and c, while another pursuer travels from u to x along edges b and d. The fourth pursuer remains stationary at w.]

E38: The graph in Figure 9.5.8(d) can be cleared with 10 pursuers: Initially, 2 pursuers reside at vertex u, 3 pursuers reside at vertex v, and 5 pursuers reside at vertex x. In parallel, these 10 pursuers clear edges a, b, and d, respectively, and all pursuers arrive at vertex z simultaneously. Next, again in parallel, 4 pursuers clear edge c, while the other 6 pursuers clear edge e. The total distance travelled is 20, and the elapsed time is 2. [Alternatively, if 20 pursuers are available, this graph can be cleared in elapsed time 1.]

E39: The graph in Figure 9.5.8(e) can be cleared with 11 pursuers: All pursuers start at vertex u, and one pursuer remains stationary to guard u. In parallel, 2 pursuers clear edge a, 3 pursuers clear edge b, and 5 pursuers clear edge d. These 10 pursuers arrive at vertex v simultaneously. Next, again in parallel, 4 pursuers clear edge c, while 6 pursuers clear edge e. The total distance travelled is 20, and the elapsed time is 2. [Alternatively, if 20 pursuers are available, this graph can be cleared in elapsed time 1.]

E40:   The graph in Figure 9.5.8(f) can be cleared with 25 pursuers: 3 pursuers start at vertex u, 18 pursuers start at vertex x, and 4 pursuers start in the center of edge d. These last 4 pursuers clear edge d; 2 pursuers travel toward u, and 2 pursuers travel toward x. When they arrive, the 5 pursuers at u clear u, and the 20 pursuers at x clear x. Next, 5 pursuers depart from u; 2 pursuers clear edge a, and 3 pursuers clear edge b. Simultaneously, 20 pursuers depart from x; 8 pursuers clear edge e, and 12 pursuers clear edge f. When they arrive, the 10 pursuers at v clear v, and the 15 pursuers at w clear w. Finally, in parallel, 6 pursuers depart v along edge c, and 6 pursuers depart w along c. Eventually these 12 pursuers will meet in the middle of edge c. The total distance travelled is 35, and the elapsed time is 2.

FACTS

F54:   Sugihara and Suzuki [SuSu89], Dawes [Da92], Neufeld [Ne96], Tanaka [Ta96], Dumitrescu et al. [DuKoSuZy08], and Munteanu and Borie [MuBo10] study sweeping on p-by-q grid graphs such that the evader becomes visible to any pursuer that occupies the same row or column, the pursuers can communicate information such as the evader's position, and the ratio of the speeds of the evader and pursuers is fixed or bounded.

F55:   Hahn and MacGillivray [HaMa06], [Ha07] describe an exponential-time algorithm for solving the cops-and-robbers problem on directed graphs when the number of cops is fixed.

F56:   Goldstein and Reingold [GoRe95] show that the cops-and-robbers problem is $EXPTIME$-complete for directed graphs.

F57:   The literature considers many additional variants of pursuit-evasion, some of which we briefly mention here:

- The pursuers and/or evader must begin at specified locations within the graph.

- A pursuer can see and/or capture the evader if within distance at most ϵ.

- Pursuers are non-uniform (different speeds, visibility, and/or capture capabilities).

- Evaders can capture/destroy pursuers.

- Evaders move randomly rather than adversarially, for example, via a Markovian random walk.

- The pursuers' strategy is randomized rather than deterministic.

- The pursuers' strategy yields capture with high probability rather than guaranteed capture.

- The pursuers' strategy performs well in average-case rather than worst-case.

- The algorithm that produces the pursuers' strategy is approximate or heuristic/experimental rather than optimal.

- The size and/or structure of the graph is not known in advance.

- The environment is a geometric space rather than a graph, for example, a circle, convex polygon, non-convex polygon, or any other 2D or 3D region (bounded or unbounded, possibly containing holes or obstacles).

References

[AiFr84] M. Aigner and M. Fromme, A game of cops and robbers, *Discrete Applied Mathematics* **8** (1984), 1–12.

[Al04] B. Alspach, Searching and sweeping graphs: a brief survey, *Matematiche* **59** (2004), 5–37.

[AlDyHaYa07] B. Alspach, D. Dyer, D. Hanson, and B. Yang, Lower bounds on edge searching, in *Lecture Notes in Computer Science* **4614** (2007), Springer, 516–527.

[Ba06] J. Barat, Directed path-width and monotonicity in digraph searching, *Graphs and Combinatorics* **22** (2006), 161–172.

[BaFlFrSa02] L. Barriere, P. Flocchini, P. Fraigniaud, and N. Santoro, Capture of an intruder by mobile agents, in *Proceedings of the 14th Annual ACM Symposium on Parallel Algorithms and Architectures* (2002), 200–209.

[BaFrSaTh03] L. Barriere, P. Fraigniaud, N. Santoro, and D. Thilikos, Searching is not jumping, *Graph-Theoretic Concepts in Computer Science* **2880** (2003), 34–45.

[Bi91] D. Bienstock, Graph searching, path-width, tree-width and related problems (a survey), *DIMACS Series in Discrete Mathematics and Theoretical Computer Science* **5** (1991), 33–49.

[BiSe91] D. Bienstock and P. Seymour, Monotonicity in graph searching, *Journal of Algorithms* **12** (1991), 239–245.

[Bo98] H. Bodlaender, A partial *k*-arboretum of graphs with bounded treewidth, *Theoretical Computer Science* **209** (1998), 1–45.

[BoKl96] H. Bodlaender and T. Kloks, Efficient and constructive algorithms for the pathwidth and treewidth of graphs, *Journal on Algorithms* **21** (1996), 358–402.

[BoKlKr95] H. Bodlaender, T. Kloks, and D. Kratsch, Treewidth and pathwidth of permutation graphs, *SIAM Journal on Discrete Mathematics* **8** (1995), 606–616.

[BoMo93] H. Bodlaender and R. Möhring, The pathwidth and treewidth of cographs, *SIAM Journal on Discrete Mathematics* **6** (1993), 181–188.

[BoTh04] H. Bodlaender and D. Thilikos, Computing small search numbers in linear time, in *Proceedings of the International Workshop on Parameterized and Exact Computation* (2004), 37–48.

[BoNo11] A. Bonato and R. Nowakowski, *The Game of Cops and Robbers on Graphs*, American Mathematical Society, 2011.

[BoYa11] A. Bonato and B. Yang, Graph searching and related problems, manuscript
(2011), to appear in *Handbook of Combinatorial Optimization*.

[BoToKo11] R. Borie, C. Tovey, and S. Koenig, Algorithms and complexity results for
graph-based pursuit-evasion, *Autonomous Robots* **31** (2011), 317–332.

[Ch08] E. Chiniforooshan, A better bound for the cop number of general graphs, *Journal
of Graph Theory* **58** (2008), 45–48.

[ChHoIs11] T. Chung, G. Hollinger, and V. Isler, Search and pursuit-evasion in mobile
robotics: a survey, *Autonomous Robots* **31** (2011), 299–316.

[Cl02] N. Clarke, Constrained cops and robber, Ph.D. Dissertation, Dalhousie University, 2002.

[DaBoKoTo10] K. Daniel, R. Borie, S Koenig, and C. Tovey, ESP: Pursuit evasion
on series-parallel graphs, in *Proceedings of 9th International Conference on Autonomous Agents and Multiagent Systems* (2010), 1519–1520.

[Da92] R. Dawes, Some pursuit-evasion problems on grids, *Information Processing Letters* **43** (1992), 241–247.

[DeKiTh97] N. Dendris, L. Kirousis, and D. Thilikos, Fugitive-search games on graphs
and related parameters, *Theoretical Computer Science* **172** (1997), 233–254.

[DuKoSuZy08] A. Dumitrescu, H. Kok, I. Suzuki, and P. Zylinski, Vision-based pursuit-
evasion in a grid, in *Proceedings of 11th Scandinavian Workshop on Algorithm
Theory* (2008), 53–64.

[Dy04] D. Dyer, Sweeping graphs and digraphs, Ph.D. Dissertation, Department of
Mathematics, Simon Fraser University, 2004.

[ElSuTu94] J. Ellis, I. Sudborough, and J. Turner, The vertex separation and search
number of a graph, *Information and Computation* **113** (1994), 50–74.

[ElWa08] J. Ellis and R. Warren, Lower bounds on the pathwidth of some grid-like
graphs, *Discrete Applied Mathematics* **156** (2008), 545–555.

[FoGo00] F. Fomin and P. Golovach, Graph searching and interval completion, *SIAM
Journal on Discrete Mathematics* **13** (2000), 454–464.

[FoGoKr08] F. Fomin, P. Golovach, and J. Kratochvil, On tractability of cops and
robbers game, in *Proceedings of 5th IFIP International Conference on Theoretical
Computer Science* (2008), 171–185.

[FoPe96] F. Fomin and N. Petrov, Pursuit-evasion and search problems on graphs,
in *Proceedings of 27th Southeastern International Conference on Combinatorics,
Graph Theory and Computing*, in *Congressus Numerantium* **122** (1996), 47–58.

[FoHeTe05] F. Fomin, P. Heggernes, and J. Telle, Graph searching, elimination trees,
and a generalization of bandwidth, *Algorithmica* **41** (2005), 73–87.

[FoTh08] F. Fomin and D. Thilikos, An annotated bibliography on guaranteed graph
searching, *Theoretical Computer Science* **399** (2008), 236–245.

[Fr87] P. Frankl, On a pursuit game on Cayley graphs, *Combinatorica* **7** (1987), 289–295.

[GoRe95] A. Goldstein and E. Reingold, The complexity of pursuit on a graph, *Theoretical Computer Science* **143** (1995), 93–112.

[Go89] P. Golovach, A topological invariant in pursuit problems, *Differentsial'nye Uravneniya (Differential Equations)* **25** (1989), 923–929.

[GoPeFo00] P. Golovach, N. Petrov, and F. Fomin, Search in graphs, in *Proceedings of the Steklov Institute of Mathematics* (2000), S90–S103.

[GoLeSc03] G. Gottlob, N. Leone, and F. Scarcello, Robbers, marshals, and guards: game theoretic and logical characterizations of hypertree width, *Journal of Computer and System Sciences* **66** (2003), 775-808.

[Gu93] J. Gustedt, On the path width of chordal graphs, *Discrete Applied Mathematics* **45** (1993), 233–248.

[Ha07] G. Hahn, Cops, robbers and graphs, *Tatra Mountains Mathematical Publications* **36** (2007), 163–176.

[HaMa06] G. Hahn and G. MacGillivray, A note on k-cop, l-robber games on graphs, *Discrete Mathematics* **306** (2006), 2492–2497.

[Ha87] Y. Hamidoune, On a pursuit game on Cayley digraphs, *European Journal of Combinatorics* **8** (1987), 285–289.

[JoKaTh10] G. Joret, M. Kaminski, and D. Theis, The cops and robbers game on graphs with forbidden (induced) subgraphs, *Contributions to Discrete Mathematics* **5** (2010), 40–51.

[Ki92] N. Kinnersley, The vertex separation number of a graph equals its path-width, *Information Processing Letters* **42** (1992), 345–350.

[KiPa85] L. Kirousis and C. H. Papadimitriou, Interval graphs and searching, *Discrete Mathematics* **55** (1985), 181–184.

[KiPa86] M. Kirousis and C. Papadimitriou, Searching and pebbling, *Theoretical Computer Science* **47** (1986), 205–218.

[KoCa08] A. Kolling and S. Carpin, Multi-robot surveillance: an improved algorithm for the GRAPH-CLEAR problem, in *Proceedings of the 2008 IEEE International Conference on Robotics and Automation* (2008), 2360–2365.

[KoCa10] A. Kolling and S. Carpin, Pursuit-evasion on trees by robot teams, *IEEE Transactions on Robotics* **26** (2010), 32–47.

[La93] A. LaPaugh, Recontamination does not help to search a graph, *Journal of the ACM* **40** (1993), 224–245.

[LlToTr89] D. Llewellyn, C. Tovey, and M. Trick, Local optimization on graphs, *Discrete Applied Mathematics* **3** (1989), 157–178.

[MeHaGaJoPa88] N. Megiddo, S. Hakimi, M. Garey, D. Johnson, and C. Papadimitriou, The complexity of searching a graph, *Journal of the ACM* **35** (1988), 18–44.

[MuBo10] B. Munteanu and R. Borie, Variations of the vision-based pursuit-evasion problem on a grid, in *Proceedings of the International Conference on Foundations of Computer Science* (2010), 139–144.

[Ne96] S. Neufeld, A pursuit-evasion problem on a grid, *Information Processing Letters* **58** (1996), 5–9.

[No93] R. Nowakowski, Search and sweep numbers of finite directed acyclic graphs, *Discrete Applied Mathematics* **41** (1993), 1–11.

[NoWi83] R. Nowakowski and P. Winkler, Vertex-to-vertex pursuit in a graph, *Discrete Mathematics* **43** (1983), 235–239.

[Pa78] T. Parsons, Pursuit-evasion in a graph, in *Theory and Applications of Graphs, Lecture Notes in Mathematics* **642** (1978), Springer, Berlin, 426–441.

[Pe82] N. Petrov, A problem of pursuit in the absence of information on the pursued, *Differentsial'nye Uravneniya (Differential Equations)* **18** (1982), 1345–1352.

[Qu83] A. Quilliot, Problemes de jeux, de point fixe, de connectivite et de representation sur des graphes, des ensembles ordonnes et des hypergraphes, These d'Etat, Universite de Paris VI, 1983.

[Sc01] B. Schroeder, The copnumber of a graph is bounded by $\lfloor \frac{3}{2} \, \mathrm{genus}(G) \rfloor + 3$, *Categorical Perspectives* (2001), 243–263.

[SeTh93] P. Seymour and R. Thomas, Graph searching and a min-max theorem for tree-width, *Journal of Combinatorial Theory Series B* **58** (1993), 22–33.

[SuSu89] K. Sugihara and I. Suzuki, Optimal algorithms for a pursuit-evasion problem in grids, *SIAM Journal on Discrete Mathematics* **2** (1989), 126–143.

[Ta96] K. Tanaka, An improved strategy for a pursuit-evasion problem on grids, manuscript (1996).

[Th11] D. Theis, The cops and robber game on series-parallel graphs, manuscript (2011), to appear in *Graphs and Combinatorics*.

[Th00] D. Thilikos, Algorithms and obstructions for linear-width and related search parameters, *Discrete Applied Mathematics* **105** (2000), 239–271.

[YaCa07-a] B. Yang and Y. Cao, Directed searching digraphs: monotonicity and complexity, in *Lecture Notes in Computer Science* **4484** (2007), Springer, 136–147.

[YaCa07-b] B. Yang and Y. Cao, Digraph strong searching: monotonicity and complexity, in *Lecture Notes in Computer Science* **4508** (2007), Springer, 37–46.

Glossary for Chapter 9

additive bandwidth $B^+(G)$ – of graph G:

$$B^+(G) = min\left\{B_f^+(G) \mid f \text{ is proper numbering of } G\right\}$$

additive bandwidth $B_f^+(G)$ **of proper numbering** f – of graph G:

$$B_f^+(G) = max\left\{|f(u) + f(v) - (n+1)| \mid uv \in E(G)\right\}$$

almost all graphs – property P holds for: if the ratio of the number of n-vertex graphs possessing P divided by the number of n-vertex graphs approaches one as n approaches infinity.

Archimedean ϕ if $\lim_{x\to\infty} \phi(x, c) = \infty$ for every fixed $c > 0$

___, **dual** if $\lim_{x\to 0} \phi(x, c) = 0$ for every fixed $c > 0$

Archimedean function: a function ϕ such that for every $c > 0$, $\lim_{x\to\infty} \phi(x, c) = \infty$ and $\lim_{x\to\infty} \phi(c, x) = \infty$.

Archimedean graph: if G is a ϕ-tolerance graph for all Archimedean functions ϕ.

asteroidal triple: three vertices in a graph such that, for any two of them, there is a path containing those two but no neighor of the third.

automorphism group: the set of all automorphisms of a graph under the operation of composition.

automorphism: an isomorphism between a graph and itself.

bandsize$_1$ – of proper numbering f in a graph G: the number of distinct edge differences produced by f; denoted $bs_f(G)$.

bandsize$_2$ – of graph G: $bs(G) = min\{bs_f(G) : f \text{ is proper numbering of } G\}$.

bandwidth$_1$ $B_f(G)$ – of proper numbering f in a graph G:

$$B_f(G) = max\left\{|f(u) - f(v)| \mid uv \in E(G)\right\}$$

bandwidth$_2$ $B(G)$ – of graph G: $B(G) = min\{B_f(G) : f \text{ is aproper numbering of } G\}$.

bandwidth decision problem: the problem which has answer "YES" if $B(G) \le K$ and "NO" otherwise when presented with given graph G and positive integer K.

bandwidth numbering – of graph G: proper numbering f such that $B(G) = B_f(G)$.

bandwidth-k decision problem: for fixed integer k, the problem which has answer "YES" if $B(G) \le k$ and "NO" otherwise when presented with given graph G.

bounded bitolerance graph: a graph that is the incomparability graph of a bounded bitolerance order.

bounded bitolerance order: an ordered set $P = (V, \prec)$ if each vertex $v \in V$ can be assigned a closed interval $I_v = [L_v, R_v]$ and two "tolerant points" $p_v, q_v \in I_v$ satisfying $L_v < p_v$ and $q_v < R_v$, such that $x \prec y \iff R_x < p_y$ and $q_x < L_y$.

cartesian product $G \square H$ **or** $G \times H$ – of graphs G and H: a graph where $V(G \times H) = V(G) \times V(H)$ and $(g_1, h_1)(g_2, h_2) \in E(G \square H)$ if and only if either (i) $g_1 = g_2$ and $h_1 h_2 \in E(H)$ or (ii) $h_1 = h_2$ and $g_1 g_2 \in E(G)$.

center – of a connected graph: the subgraph induced by the central vertices.

___, **k-**: the subgraph induced by the k-central vertices.

central vertex – in a connected graph: a vertex whose eccentricity equals the radius of the graph.

___, **k-**: a vertex whose k-eccentricity is the k-radius.

chordal graph: a graph containing no chordless cycle of size greater than or equal to 4.

chromatic number $\chi(G)$ – of graph G: the smallest number k such that there is a function $f : V(G) \to \{1, 2, \ldots, k\}$ with the property that, if uv is an edge, then $f(u) \neq f(v)$.

closed interval$_1$ – between two vertices: a set consisting of these two vertices and all vertices lying in some geodesic between them.

closed interval$_2$ – of a set: the union of all closed intervals between every pair of vertices in the set.

closed neighborhood$_1$ – of a vertex v: $N(v) \cup \{v\}$, where $N(v)$ is the open neighborhood of v; denoted $N[v]$.

closed neighborhood$_2$ – of a set S: $\cup_{v \in S} N[v]$, where $N[v]$ is the closed neighborhood of v; denoted $N[S]$.

competition graph: an intersection graph of the family of outsets of the vertices in some digraph.

___, **p-**: a p-intersection graph of the family of outsets of the vertices of some digraph.

___, **ϕ-tolerance**: a graph that is the ϕ-tolerance intersection graph of the family of out-sets of the vertices of some digraph.

complement – of graph $G = (V, E)$: a graph $\overline{G}$ with $V(\overline{G}) = V$ and vertices u and v are adjacent in $\overline{G}$ if and only if they are not adjacent in G.

complement Threshold Tolerance graph: if for each vertex v there exist two positive numbers a_v and b_v such that for any pair of vertices x and y: $xy \in E \iff a_x \leq b_y$ and $a_y \leq b_x$.

complementary numbering f_c – of proper numbering f: $f_c(v) = n + 1 - f(v)$ for each vertex v of G.

complete k-ary tree $T_{k,d}$ **of depth** d: a rooted tree in which all vertices at level $d-1$ or less have exactly k children, and all vertices at level d are leaves.

complete k-partite graph $K_{n_1, n_2, \ldots, n_k}$: a graph whose vertex-set is partitioned into sets A_i of n_i vertices, $1 \leq i \leq k$, with two vertices adjacent if and only if they are in distinct sets.

composition $G(H)$ – of graphs G and H: a graph where $V(G(H)) = V(G) \times V(H)$ and $(g_1, h_1)(g_2, h_2) \in E(G(H))$ if and only if either (i) $g_1 g_2 \in E(G)$ or (ii) $g_1 = g_2$ and $h_1 h_2 \in E(H)$.

containment graph – of a collection $\{S_i\}$ of distinct subsets of a set **S**: a graph with vertex set $\{1, \ldots, n\}$ and edge set $\{ij \mid$ either $S_i \subset S_j$ or $S_j \subset S_i\}$.

contraction $G|_e$ – of a graph G along edge e: a merger where $e = uv$.

convex hull – of a set: the smallest convex set containing a given set.

convex set: a set of vertices in a graph whose closed interval is itself.

cop number $\mathbf{c}(G)$: the smallest number of cops k needed for player C to win the cops-and-robbers game on G.

cops-and-robbers game: a two-player game with players C and R, as follows. First player C places k cops at any vertices of G. Next player R places a robber at any vertex. The players continue alternating turns. On C's turns, each cop either remains at its present location or moves to an adjacent vertex. On R's turns, the robber either remains at its present location or moves to an adjacent vertex. Both C and R always know the locations of all participants. Player C wins the game if any cop ever resides at the same vertex as the robber.

corona – of two graphs G_1 and G_2: the graph formed from one copy of G_1 and $|V(G_1)|$ copies of G_2 where the ith vertex of G_1 is adjacent to every vertex in the ith copy of G_2.

cutwidth$_1$ – of proper numbering f in a graph G:

$$c_f(G) = \max_i |\{vw \in E(G) : f(v) \le i < f(w)\}|.$$

cutwidth$_2$ – of a graph G: $cutw(G) = \min\{c_f(G) : f \text{ is a proper numbering of } G\}$.

cyclic bandwidth$_1$ $B_{cf}(G)$ – of proper numbering f in a graph G:

$$B_{cf}(G) = \max\{\|f(u) - f(v)\|_c : uv \in E\}, \text{ where } \|x\|_c = \min\{|x|, n - |x|\}.$$

cyclic bandwidth$_2$ $B_c(G)$ – of graph G:

$$B_c(G) = \min\{B_{cf}(G) : f \text{ is proper numbering of } G\}.$$

daisy: a graph that can be constructed from $k \ge 2$ disjoint cycles by identifying a set of k vertices, one from each cycle, into one vertex.

decomposable graph: a graph G whose vertex-set can be partitioned into $\gamma(G)$ subsets $C_1, C_2, \ldots, C_{\gamma(G)}$ such that each of the induced subgraphs $G[C_i]$ is a complete subgraph of G.

degree sequence – of a graph G: a listing of the degrees of the n vertices of G, usually in monotonic order.

diameter – of a graph G: the maximum distance between any two vertices of G; denoted $diam(G)$.

___, **k-**: the maximum k-eccentricity among the vertices in a graph.

directed distance from u to v: the length of a shortest directed $u - v$ path.

distance$_1$ – between two vertices: the length of a shortest path between these two vertices.

distance$_2$ – between two subgraphs: the minimum distance between a vertex in one subgraph and a vertex in the other subgraph.

dominating function – of a graph G: a function $f : V \to \{0, 1\}$ such that for each $v \in V$, $\sum_{u \in N[v]} f(u) \ge 1$.

dominating set – of a graph G: a set $S \subset V$ such that every vertex in V is either in S or adjacent to a vertex in S.

___, **clique** – of a graph G: a dominating set S of G such that $G[S]$ is a clique.

___, **connected** – of a graph G: a dominating set S of G such that $G[S]$ is connected.

___, **cycle** – of a graph G: a dominating set S of G such that $G[S]$ has a hamiltonian cycle.

___, **minimal**: a dominating set that contains no dominating set as a proper subset.

___, **paired** – of a graph G: a dominating set S of G such that the induced subgraph $G[S]$ has a perfect matching.

___, **total** – of a graph G: a set $S \subseteq V$ such that every vertex of V is adjacent to a vertex of S.

domination critical graph: a graph G with the property that adding an arbitrary edge not in G results in a graph with domination number less than the domination number of G.

___, **k-γ**: a domination critical graph with domination number k.

domination number – of a graph G: the minimum cardinality of a dominating set of G.

___, **clique** – of a graph G: the minimum cardinality of a clique dominating set of G.

___, **connected** – of a graph G: the minimum cardinality of a connected dominating set of G.

___, **cycle** – of a graph G: the minimum cardinality of a cycle dominating set of G.

___, **independent** – of a graph G: the minimum cardinality of a maximal independent set. (A maximal independent set must be a dominating set.)

___, **paired** – of a graph G: the minimum cardinality of a paired dominating set of G.

___, **total** – of a graph G: the minimum cardinality of a total dominating set of G.

___, **upper** – of a graph G: the maximum cardinality of a minimal dominating set of G.

domination perfect graph: a graph G having the property that $\gamma(H) = i(H)$ for every induced subgraph H of G.

dual Archimedean ϕ if $\lim_{x \to 0} \phi(x, c) = 0$ for every fixed $c > 0$.

eccentricity – of a vertex v: the distance from vertex v to a vertex farthest from it.

___, **k-**: the maximum Steiner distance among all k-element sets of vertices containing the vertex.

edge-bandwidth$_1$ – of edge-numbering f of a graph G: $B'_f(G) = \max\{|f(e_1) - f(e_2)| :$ edges e_1 and e_2 adjacent in $G\}$.

edge-bandwidth$_2$ – of a graph G: $B'(G) = \min\{B'_f(G) : f$ an edge numbering of $G\}$.

edge chromatic number $\chi_1(G)$ – of a graph G: the smallest number k such that there is a function $f : E(G) \to \{1, 2, \ldots, k\}$ with the property that, if edges e_1 and e_2 share a common vertex, then $f(e_1) \neq f(e_2)$.

edge clique cover – of a graph G: a family $\mathcal{E} = \{C_1, \ldots, C_k\}$ of complete subgraphs of G such that every edge of G is in at least one of $E(C_1), \ldots, E(C_k)$. (Elsewhere, a clique is required to be a maximal complete subgraph.)

___, **p-** – of a graph G: a family $\{V_1, \ldots, V_m\}$ of not necessarily distinct subsets of $V(G)$ such that, for every set $\{i_1, \ldots, i_p\}$ of p distinct subscripts, $T = V_{i_1} \cap \cdots \cap V_{i_p}$ induces a complete subgraph of G, and such that the collection of sets of the form T is an edge clique cover of G.

___, **ϕ-T-edge** – of a graph G: a family $\{V_1, \ldots, V_n\}$ such that $v_i v_j \in E(G)$ if and only if at least $\phi(t_i, t_j)$ of the sets V_k contain both v_i and v_j.

edge cover – of a graph G: a set of edges covering all the vertices of G.

edge-numbering f – of a graph G: a bijection $f : E(G) \to \{1, 2, \ldots, |E(G)|\}$.

edge search number $es(G)$: the smallest number of pursuers needed to implement an edge search strategy on G.

edge search operation: any of $p(x) =$ place a pursuer at vertex x, $r(x) =$ remove a pursuer from vertex x, or $s(e, x, y) =$ slide a pursuer along edge e from endpoint x to other endpoint y.

edge search strategy: any sequence of edge search operations that ends with every edge of G being simultaneously clear, as follows. Initially every edge of G is contaminated. An edge $e = (x, y)$ becomes clear if a pursuer slides along e from x to y while either (i) another pursuer resides at x or (ii) every other edge incident to x is clear. If ever an unoccupied vertex x is incident to a contaminated edge, then any clear edges incident to x immediately become recontaminated.

edgesum$_1$ $s_f(G)$ – generated by proper numbering f of a graph G:
$$s_f(G) = \sum_{uv \in E(G)} |f(u) - f(v)|.$$
edgesum$_2$ $s(G)$ – of a graph G: $s(G) = \min\{s_f(G) : f \text{ is a proper numbering of } G\}$.

geodesic: a shortest path between two vertices.

geodetic number: the minimum cardinality of a geodetic set in a graph.

geodetic set: a set of vertices of a graph whose closed interval is the vertex-set of the graph.

girth – of a graph G: the size of a smallest induced cycle of G.

hull number – of a graph G: the minimum cardinality of a hull set in G.

hull set: a set of vertices of a graph whose convex hull is the vertex-set of the graph.

hypercube – n-dimensional: the graph having 2^n vertices, each labeled with a distinct n-digit binary sequence, and two vertices adjacent if and only if their labels differ in exactly one position.

independent set – of vertices: a set of vertices in which no two vertices are adjacent.

intersection graph– of a collection $\mathcal{C} = \{S_1, \ldots, S_n\}$ of sets: a graph G obtained by assigning a distinct vertex v_i to each set S_i in $\mathcal{C}$ and joining two vertices by an edge precisely when their corresponding sets have a nonempty intersection.

interval graph: an intersection graph of a family of intervals of the real line.

irredundance number – of a graph G: the minimum cardinality of a maximal irredundant set of G.

___, **upper** – of a graph G: the maximum cardinality of an irredundant set of G.

irredundant set – of vertices: a set S of vertices such that for every vertex $v \in S$, $N[v] - N[S - \{v\}] \neq \emptyset$.

key: the graph obtained by joining with an edge a vertex in C_m to an end-vertex of P_n; denoted $L_{m,n}$.

line graph $L(G)$ – of a graph G: a graph with vertex-set equal to the edges of G and two vertices are adjacent if and only if the corresponding edges are adjacent in G.

link edge: an edge joining two link vertices.

link vertex$_1$ – in a type (a) unit: two non-adjacent vertices in that unit.

link vertex$_2$ – in a type (b) unit: the vertex of degree 1 in that unit.

$\frac{m}{3}$**-graph**: a graph with m edges that satisfies the three conditions: (i) $\delta(G) \geq 2$, (ii) G is connected, and (iii) $\gamma(G) > m/3$.

median vertex: a vertex whose total distance is minimum among the vertices in a graph.

median subgraph: the subgraph induced by the median vertices.

merger $G|_{u,v}$ – of vertices u and v of a graph G: a graph obtained from G by identifying u and v and then eliminating any loops and duplicate edges.

$\frac{2}{5}$**-minimal graph**: an n-vertex graph that is edge-minimal with respect to satisfying the three conditions: (i) $\delta(G) \geq 2$, (ii) G is connected, and (iii) $\gamma(G) \geq 2n/5$.

minimum hull subgraph: an induced subgraph whose vertex set is a hull set of minimum cardinality for some graph.

mixed search number ms(G): the smallest number of pursuers needed to implement a mixed search strategy on G.

mixed search operation: same as edge search operation.

mixed search strategy: any sequence of mixed search operations that ends with every edge of G being simultaneously clear, as follows. Initially every edge of G is contaminated. As with edge search, edge $e = (x, y)$ becomes clear if a pursuer slides along e from x to y while either (i) another pursuer resides at x or (ii) every other edge incident to x is clear. Also, as with node search, edge $e = (x, y)$ becomes clear if pursuers simultaneously occupy both endpoint vertices x and y. Recontamination may occur the same as with edge search and node search.

node search number ns(G): the smallest number of pursuers needed to implement a node search strategy on G.

node search operation: either $p(x)$ = place a pursuer at vertex x, or $r(x)$ = remove a pursuer from vertex x.

node search strategy: any sequence of node search operations that ends with every edge of G being simultaneously clear, as follows. Initially every edge of G is contaminated. An edge $e = (x, y)$ becomes clear if pursuers simultaneously occupy both endpoint vertices x and y. If ever an unoccupied vertex x is incident to a contaminated edge, then all clear edges incident to x become recontaminated.

NP-complete – problem: a problem having a "YES" or "NO" answer that can be solved nondeterministically in polynomial time, and all other such problems can be transformed into it in a polynomial time.

open neighborhood$_1$ – of a vertex v: the set of vertices that are adjacent to v; denoted $N(v)$.

open neighborhood$_2$ – of a set S: $\cup_{v \in S} N(v)$, where $N(v)$ is the open neighborhood of vertex v; denoted $N(S)$.

out-set – of a vertex v in a digraph: the set of vertices x such that (v, x) is an arc in the digraph.

packing – in a graph G: a set S of vertices such that each pair of vertices in S are at a distance at least 3 apart in G.

packing number – of a graph G: the maximum cardinality of a packing in G.

perfect elimination ordering – of a graph G: an ordering $\langle v_1, \ldots, v_n \rangle$ of all the vertices of G such that, for each $i \in \{1, \ldots, n\}$, v_i is a simplicial vertex of the subgraph induced on the vertex subset $\{v_i, v_{i+1}, \ldots, v_n\}$.

peripheral vertex: a vertex in a connected graph whose eccentricity equals the diameter of the graph.

periphery: the subgraph induced by the peripheral vertices.

polynomial algorithm: an algorithm whose execution time is bounded by a polynomial in some parameter of the problem, often the number of vertices for graph problems.

power – k^{th} of a graph G: the graph having the same vertex-set as G and an edge between two vertices if and only if the distance between them is at most k in G.

profile$_1$ $P_f(G)$ – of proper numbering f in a graph G: $P_f(G) = \sum_{v \in V} w_f(v)$.

profile$_2$ $P(G)$ – of a graph G: $P(G) = \min\{P_f(G) : f \text{ is a proper numbering of } G\}$.

profile width $w_f(v)$ – for vertex $v \in V(G)$ and proper numbering f of a graph G: $w_f(v) = \max_{x \in N[v]}(f(v) - f(x))$ where $N[v]$ is the closed neighborhood of v.

proper numbering – of a graph G: a bijection $f : V(G) \to \{1, 2, \ldots, |V(G)|\}$.

proper tolerance graph: a graph having a tolerance representation in which no interval properly contains another interval.

pursuit-evasion: a kind of problem in which a team of mobile pursuers attempts to capture one or more mobile evaders within a graph or other environment.

radius $rad(G)$ – of a graph G: the smallest number r such that there is a vertex u of G with distance at most r from every other vertex of G; equivalently, the minimum eccentricity among the vertices of a connected graph.

___, k-: the minimum k-eccentricity among the vertices in a graph.

refinement – of a graph G: a graph obtained from G by a finite number of subdivisions.

self-centered graph: a graph whose center is itself.

simplicial vertex: a vertex whose neighbors induce a complete subgraph.

split graph: a graph whose vertex set can be partitioned into a clique and an independent set.

Steiner distance – of a set of vertices in a graph: the minimum size of a connected subgraph containing the set.

strong diameter – of a strong digraph: the maximum strong eccentricity among the vertices of the strong digraph.

strong eccentricity – of a vertex in a strong digraph: the greatest strong distance from the vertex to a vertex in the strong digraph.

strong radius – of a strong digraph: the minimum strong eccentricity among the vertices of the strong digraph.

strong center – of a strong digraph: the subdigraph induced by the strong central vertices.

strong central vertex – in a strong digraph: a vertex whose strong eccentricity is the strong radius.

strong digraph: synonym for strongly connected digraph.

strong (Steiner) distance – between two vertices in a strong digraph: the minimum size of a strong subdigraph containing these two vertices.

strong product $G(Sp)H$ – of graphs G and H: a graph with $V(G(Sp)H) = V(G) \times V(H)$ and $(g_1, h_1)(g_2, h_2) \in E(G(Sp)H)$ if and only if either $g_1 g_2 \in E(G)$ and $h_1 h_2 \in E(H)$ or $g_1 = g_2$ and $h_1 h_2 \in E(H)$ or $h_1 = h_2$ and $g_1 g_2 \in E(G)$.

strongly connected digraph: a digraph containing both a directed $u - v$ path and a directed $v - u$ path for every pair u, v of vertices in the digraph.

strongly self-centered digraph: a strong digraph whose strong center is itself.

subdivision – of edge uv: a graph obtained by replacing uv with path $\langle u, w, v \rangle$ where w is a new vertex of degree two.

subtree graph: an intersection graph of a familiy of subtrees of a tree.

sum $G + H$ – of graphs G and H: a graph with $V(G + H) = V(G) \cup V(H)$ and $E(G + H) = E(G) \cup E(H) \cup \{uv : u \in V(G) \text{ and } v \in V(H)\}$.

sweep number $sw(G)$: the smallest k such that a sweep strategy $P = (P_1, \ldots, P_k)$ exists for G.

sweep strategy: a tuple $P = (P_1, \ldots, P_k)$ of k functions specified as follows. Embed G in 3D space such that each vertex resides at a distinct location and no two edges intersect except at a common endpoint. Let k denote the number of pursuers, and let $P = (P_1, \ldots, P_k)$ where each $P_j : [0, \infty) \to G$ is a continuous function. Then P is a sweep strategy for G if for every continuous function $E : [0, \infty) \to G$, there exists some pursuer $j \in \{1, \ldots, k\}$ and time t such that $P_j(t) = E(t)$.

tensor product $G(Tp)H$ – of graphs G and H: a graph with $V(G(Tp)H) = V(G) \times V(H)$ and $(g_1, h_1)(g_2, h_2) \in E(G(Tp)H)$ if and only if $g_1 g_2 \in E(G)$ and $h_1 h_2 \in E(H)$.

thickness $thick(G)$ – of a graph G: a smallest number of planar subgraphs of G whose union is G.

Threshold Tolerance graph: $G = (V, E)$ if its vertices can be assigned positive weights $\{w_v \mid v \in V\}$ and positive tolerances $\{t_v \mid v \in V\}$ such that $xy \in E \iff w_x + w_y \geq \min\{t_x, t_y\}$.

___, **complement**: if for each vertex v there exists two positive numbers a_v and b_v such that for any pair of vertices x and y: $xy \in E \iff a_x \leq b_y$ and $a_y \leq b_x$.

ϕ-tolerance chain graph: a ϕ-tolerance graph which has a representation $\mathcal{I}$ consisting of a nested family of intervals.

tolerance graph: $G = (V, E)$ if there exists a collection $\mathcal{I} = \{I_v\}_{v \in V}$ of closed intervals on the real line and an assignment of positive numbers $t = \{t_v\}_{v \in V}$ such that $vw \in E \iff |I_v \cap I_w| \geq \min\{t_v, t_w\}$. Here $|I_u|$ denotes the length of the interval I_u. The positive number t_v is called the *tolerance* of v, and the pair $\langle \mathcal{I}, t \rangle$ is called an *interval tolerance representation* of G.

___, **bounded**: a tolerance graph $G = (V, E)$ that has a tolerance representation $\langle \mathcal{I}, t \rangle$ in which $t_v \leq |I_v|$ for all $v \in V$.

___, **ϕ-**: a tolerance graph $G = (V, E)$ such that for a symmetric binary function ϕ, positive valued on positive arguments, there is an interval representation $\mathcal{I} = \{I_v\}_{v \in V}$ with positive tolerances $t = \{t_v\}_{v \in V}$ such that $vw \in E \iff |I_v \cap I_w| \geq \phi\{t_v, t_w\}$.

___, **proper**: a tolerance graph having a tolerance representation in which no interval properly contains another interval.

___, **unit**: a tolerance graph such that it has a tolerance representation in which each interval I_v has unit length for each vertex v.

topological bandwidth $B^*(G)$ – of graph G:

$$B^*(G) = \min\{B(G') : G' \text{ is a refinement of } G\}.$$

total distance of a vertex: the sum of the distances from the vertex to all other vertices.

transitive orientation – of a graph G: an assignment of directions to the edges such that the resulting binary relation is transitive.

trapezoid graph: the intersection graph of a set of trapezoids whose upper and lower sides are on a given pair of horizontal lines.

type-(a) unit: a graph that is isomorphic to a cycle C_5.

type-(b) unit: a graph that is isomorphic to a key $L_{4,1}$.

unit tolerance graph: if it has a tolerance representation in which each interval I_v has unit length for each vertex v.

vertex arboricity – of a graph G: the minimum number of subsets into which $V(G)$ can be partitioned such that the vertices of each subset induce an acyclic subgraph.

vertex cover number – of a graph G: the smallest cardinality of a set of vertices such that every edge is incident to at least one of the vertices in the set.

vertex independence number – of a graph G: the largest cardinality of a set of vertices which induces a graph with no edges.

vertex independence number – of a graph G: the maximum cardinality of a maximal independent set of G.

weakly chordal graph: if neither the graph nor its complement contains a chordless cycle of size greater than 4.

Chapter 10

Graphs in Computer Science

Section 10.1

Searching

Harold N. Gabow, University of Colorado

INTRODUCTION

A **search** of a graph is a methodical exploration of all the vertices and edges. It must run in "linear time," i.e., in one pass (or a small number of passes) over the graph. Even with this restriction, a surprisingly large number of fundamental graph properties can be tested and identified.

This section examines the two most important search methods. *Breadth-first search* gives an efficient way to compute distances. *Depth-first search* is useful for checking many basic connectivity properties, for checking planarity, and also for data flow analysis for compilers. A treatment of at least some aspects of both these methods can be found in almost any algorithms text (some recent ones are [BrBr96, CLRS01, GoTa02, HSR98, Se02, We99]).

All the algorithms of this section (except for §10.1.7) run in linear time or very close to it. Since it takes linear time just to read the graph, the algorithms are essentially as efficient as possible (they are "asymptotically optimal").

NOTATION: Throughout this chapter, the number of vertices and edges of a graph $G = (V, E)$ are denoted n and m, respectively. Time bounds for algorithms are given using asymptotic notation, e.g., $O(n)$ denotes a quantity that, for sufficiently large values of n, is at most cn, for some constant c that is independent of n.

Convention: In all algorithms, we assume that the graph G is given as an adjacency list representation. If G is undirected, this means that each vertex has a list of all its neighbors. The list can be sequentially allocated or linked. If G is directed, then each vertex has a list of all its out-neighbors.

10.1.1 Breadth-First Search

The **breadth-first search** method (abbr. **bfs**) finds shortest paths from a given vertex of a graph to other vertices. It generalizes to Dijkstra's algorithm, which allows numerical (nonnegative) edge-lengths. Throughout this section, the given graph G can be directed or undirected.

DEFINITIONS

D1: A **length function** on a graph specifies the numerical length of each edge. Each edge is assumed to have length one, unless there is an explicitly declared length function.

D2: The **distance** from vertex u to vertex v in a graph, denoted $d(u, v)$, is the length of a shortest path from u to v.

D3: The **diameter** of a graph is the maximum value of $d(u, v)$ for $u \neq v$.

D4: A **shortest-path tree** T from a vertex s is a tree, rooted at s, that contains all the vertices that are reachable from s. The path in T from s to any vertex x is a shortest path in G, i.e., it has length $d(s, x)$.

EXAMPLE

E1: Figure 10.1.1 gives a shortest path tree from vertex s.

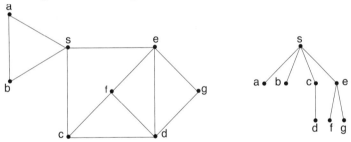

Figure 10.1.1: Undirected graph and shortest path tree.

E2: The *small-world phenomenon* [Mi67, Kl00] occurs when relatively sparse graphs have low diameter. Studies have shown that the graphs of movie actors, neural connections in the *c.~elegans* worm, and the electric power grid of the western United States all exhibit the small-world phenomenon. The world-wide web is believed to have this structure too.

E3: For several decades, mathematicians have computed their *Erdős number* as their distance from the prolific mathematician Paul Erdős, in the graph where an edge joins two mathematicians who have coauthored a paper.

E4: The premise of the *6 Degrees of Kevin Bacon* game is that the graph whose vertices are movie actors and whose edges join two actors appearing in the same movie has diameter at most 6.

E5: In computer and communications networks, a message is typically broadcast from one site s to all others by passing it down a shortest path tree from s.

E6: To solve a puzzle like Sam Lloyd's "15 puzzle" [St96], we can represent each position by a vertex. A directed edge (i, j) exists if we can legally move from i to j. We seek a shortest path from the initial position to a winning position.

Ordered Trees

DEFINITIONS

D5: In a rooted tree, a vertex x is an ***ancestor*** of a vertex y, and y is a ***descendant*** of x, if there is a path from x to y whose edges all go from parent to child. By convention x is an ancestor and descendant of itself (e.g., in the tree of Figure 10.1.1 vertex e has 3 descendants).

D6: Vertex x is a ***proper ancestor (descendant)*** of vertex y if it is an ancestor (descendant) and $x \neq y$.

D7: An ***ordered tree*** is a rooted tree in which the children of each vertex are linearly ordered. In a plane drawing of such a tree, left-to-right order gives the order of the children. (The leftmost child is first.)

D8: Vertex x is *to the left of* vertex y if some vertex has children c and d, with c to the left of d, c an ancestor of x and d an ancestor of y.

D9: In a graph G, a ***breadth-first tree*** T from a vertex s contains the vertices that are reachable from s. It is an ordered tree, rooted at s. If x is a vertex at depth δ in the tree T, then the children of x in T are the vertices of G that are adjacent in G to x, but not adjacent (in G) to any vertex in T at depth less than δ, or to any vertex at depth δ in T that is at the left of x.

FACTS

F1: Any breadth-first tree is a shortest-path tree.

F2: A high level bfs algorithm is given below as Algorithm 10.1.1. It constructs a breadth-first tree. It starts from s, finds the vertices at distance 1 from s, then the vertices at distance 2, etc.

Algorithm 10.1.1: Breadth-first Search

> *Input*:   directed or undirected graph $G = (V, E)$, vertex s.
> *Output*: breadth-first tree T from s.
> $V_i = \{$all vertices at distance i from $s\}$
> $V_0 = \{s\}$
> make s the root of T
> $i = 0$
> while $V_i \neq \emptyset$ do construct V_{i+1}
> $V_{i+1} = \emptyset$
>   for each vertex $v \in V_i$ do
>   "scan" v
>   for each edge (v, w) do
>   if $w \notin \bigcup_j V_j$ then
>   make w the next child of v in T
>   add w to V_{i+1}
> $i = i + 1$

F3: The high-level algorithm can be implemented to run in total time $O(n+m)$. The main data structure is a queue of vertices that have been added to T, but whose children in T have not been computed.

F4: In general we verify that an algorithm takes time $O(n+m)$ by checking that it spends constant time (i.e., $O(1)$ time) on each vertex and edge of G.

F5: Not every shortest path tree is a breadth-first tree (e.g., the tree of Figure 10.1.1). This does not cause any problems in applications.

F6: The diameter can be found by doing a breadth-first search from each vertex.

F7: Dijkstra's algorithm computes a shortest path tree from s in a graph with a nonnegative length function. It generalizes breadth-first search. Like bfs it finds the set V_d of all vertices at distance d from s, for increasing values of d. An appropriate data structure implements the algorithm in time $O(m + n \log n)$ [FrTa87, CLRS01].

10.1.2 Depth-First Search

Depth-first search (abbr. *dfs*) was investigated in the 19th century as a strategy for exploring a maze [Lu82, Tarr95]. The fundamental properties of the depth-first search tree were discovered by Hopcroft and Tarjan [HoTa73a, Ta72]. Tarjan also developed many other elegant and efficient dfs algorithms (see §10.1.6). The idea of depth-first search is to scan repeatedly an edge incident to the most recently discovered vertex that still has unscanned edges.

DEFINITIONS

D10: Two vertices in a tree are *related vertices* if one is an ancestor of the other.

D11: In an undirected graph $G = (V, E)$, a *depth-first tree* (abbr. *dfs tree*) from a vertex s is a tree subgraph T, rooted at s, that contains all the vertices of G that are reachable from s.

- Edges of $E(T)$ and $E(G) - E(T)$ are called *tree edges* and *nontree edges*, respectively.
- Each nontree edge is also called a *back edge*.

The crucial property is that the two endpoints of each back edge are related.

D12: In an undirected graph G, a *depth-first spanning forest* is a collection of depth-first trees, one for each connected component of G. Each vertex of G belongs to exactly one tree of the forest.

D13: Let $G = (V, E)$ be a directed graph where every vertex is reachable from a designated vertex s. A *depth-first tree* from s is an ordered tree in G, rooted at s that contains all vertices V. Each edge of T is called a *tree edge*. Each *nontree edge* $(x, y) \in E - T$ can be classified into one of three types:

- A *back edge* has y an ancestor of x.
- A *forward edge* has y a descendant of x.
- A *cross edge* joins two unrelated vertices.

The crucial property is that each cross edge (x, y) has x to the right of y.

D14: Let G be a directed graph, in which we no longer assume that some vertex can reach all others. A ***depth-first forest*** is an ordered collection of trees in G so that each vertex of G belongs to exactly one tree. The edges of G are classified into the 4 types of edges in Definition 13 with one additional possibility:

- A ***cross edge*** can join 2 vertices in different trees as long as it goes from right to left (i.e., from a higher numbered tree to a lower numbered tree).

EXAMPLES

E7: Figure 10.1.2 illustrates a depth-first search of an undirected graph. In drawings of depth-first spanning trees, tree edges are solid and nontree edges are dashed. There can be many depth-first trees with the same root. For instance, the tree edge $(5, 6)$ could be replaced by $(5, 7)$.

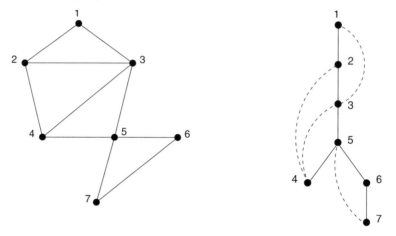

Figure 10.1.2: Undirected graph and depth-first spanning tree.

E8: Figure 10.1.3 illustrates a depth-first search of a directed graph. There is 1 forward edge, 2 back edges and 2 cross edges.

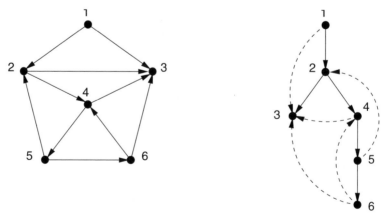

Figure 10.1.3: Directed graph and depth-first spanning tree.

FACTS

F8: Any vertex s of an undirected graph has a depth-first tree from s. Any vertex s of a directed graph has a depth-first tree of the subgraph induced by the vertices reachable from s. A high level algorithm to find such a tree is the following.

Algorithm 10.1.2: Depth-First Search

> *Input:*   directed or undirected graph $G = (V, E)$, vertex s
> *Output:*   depth-first tree T from s
> make s the root of T
> DFS(s)
> procedure DFS(s)
> vertex v is ***discovered*** at this point
> for each edge (v, w) do
> >   edge (v, w) is ***scanned*** (***from*** v) at this point
> >   if w has not been discovered then
> >   make w the next child of v
> >   DFS(w)
> >   vertex v is ***finished*** at this point

F9: The procedure DFS is recursive, i.e., it calls itself. The overhead for a recursive call is $O(1)$. Algorithm 10.1.2 uses linear time, $O(n + m)$.

F10: If scanning edge (v, w) from the vertex v results in the discovery of the vertex w, then (v, w) is a tree edge.

F11: Suppose that the graph G is undirected. For the tree T produced by Algorithm 10.1.2 to be a valid depth-first tree, any edge $(v, w) \in E - T$ must have v and w related vertices. Why does T have this property? By symmetry suppose v gets discovered before w. Then w will either be made a child of v (like edge $(3, 5)$ in Figure 10.1.2) or a nonchild descendant of v (like edge $(3, 4)$ in Figure 10.1.2).

F12: Suppose that the graph G is directed. For T to be a valid depth-first tree, any edge (v, w) must be one of the 4 possible types. Why does T have this property? First suppose v gets discovered before w. In that case w will be a descendant of v and (v, w) will be a tree or forward edge (as in Fact F11). Next suppose v is discovered after w. Then either v descends from w or v is to the right of w. In the former case (v, w) is a back edge and in the latter case (v, w) is a cross edge.

F13: Algorithm 10.1.2 can be extended to a procedure that constructs a depth-first forest F: The procedure starts with $F = \emptyset$. It repeatedly chooses a vertex $s \notin F$, uses $DFS(s)$ to grow a depth-first tree T from s, and adds T to F.

F14: Algorithm 10.1.2 uses linear time. (For directed graphs a point to note is that a vertex w gets added to only 1 tree of F. This is because once discovered, vertex w remains "discovered" throughout the whole procedure.)

REMARKS

R1: We can test whether an undirected graph is connected in linear time, by using a depth-first search. The trees of a depth-first search spanning forest give the connected components.

R2: We can test whether all vertices of a directed graph are reachable from a vertex s in linear time, by a depth-first search.

Discovery Order

DEFINITIONS

D15: *Discovery order* is a numbering of the vertices from 1 to n in the order they are discovered. This is also called the *preorder* of the dfs tree.

D16: In *finish time order* the vertices are numbered from 1 to n by increasing finish time. This is the *postorder* of the dfs tree.

FACTS

F15: Most algorithms based on the depth-first search tree use discovery order. These algorithms identify each vertex v with its discovery number, also called v. This is how the vertices are named in Figure 10.1.3.

F16: In discovery order, the descendants of a vertex v are numbered consecutively, with v first, followed by all its proper descendants. This gives a quick way to test if a given vertex w descends from another given vertex v: Let v have d descendants. w is a descendant of v exactly when $v \leq w < v + d$. This method can be implemented to run in $O(1)$ (i.e., constant) time.

REMARKS

R3: The power of depth-first search comes from its simplification of the edge structure — the absence of cross edges in undirected graphs, and the absence of left-to-right edges in directed graphs. Depth-first search algorithms work by propagating information up or down the dfs tree(s).

R4: Many simple properties of graphs can be analyzed without using the full power of depth-first search. The algorithm always works with a path in the dfs tree, rather than with the entire dfs tree. The algorithm propagates information along the path.

R5: As a simple example of Remark 4 we give a procedure that shows an undirected graph with minimum degree δ has a path of length $> \delta$: execute $DFS(s)$ (for any s), stopping at the first vertex t that becomes finished. The portion of tree T constructed by this procedure is a path from s to t of length $> \delta$. The reason is that all of t's neighbors must be in the path for t to be finished.

R6: Sections 10.1.3–10.1.5 deal with simpler graph properties that can be handled by the path view of depth-first search. Section 10.1.6 covers deeper properties whose algorithms require the full power of the depth-first search tree. Section 10.1.7 deals with both views of depth-first search.

10.1.3   Topological Order

Topological order is the fundamental property of directed acyclic graphs. In conjunction with dynamic programming, topological order leads to efficient algorithms for many fundamental properties of directed acyclic graphs — even properties that are NP-complete in general graphs.

DEFINITIONS

D17: A *dag* is a directed acyclic graph, i.e., it has no cycles.

D18: A *source* of a dag is a vertex with indegree 0, and a *sink* is a vertex with outdegree 0.

D19: A *topological numbering* (topological order, topological sort) of a directed graph assigns an integer to each vertex so that each edge is directed from lower number to higher number.

EXAMPLES

E9: The dag of Figure 10.1.4 has source a and sink f. Alphabetic order is a valid topological ordering. In general a dag has many topological numberings. In this figure 12 are possible.

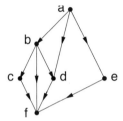

Figure 10.1.4: Dag and topological order.

E10: A dag can always be drawn so that all edges are directed downwards, as in Figure 10.1.4. Topological numbers guide the vertical placement of the vertices. This principle is useful in algorithms for drawing graphs (see Section 10.3).

E11: Prerequisite graphs in a university department are dags: if course X is a prerequisite to course Y, then an arrow is drawn from X to Y. There cannot be a cycle, else no one could graduate! The course numbering is a topological numbering: a prerequisite to a course always has a lower number.

E12: A combinational circuit is a collection of logic gates and interconnecting wires, with no feedback. The no-feedback property makes it a dag.

E13: A graph of program dependencies is a dag (assuming no recursion is allowed). For instance, the dependency graph specified by a `makefile` is a dag. The `make` utility always ensures that a file's timestamp is no later than the timestamp of any dependent file. Thus the timestamps form a topological numbering.

E14: The formulas in a spreadsheet depend on one another, and this dependence relation is a dag. When the value of a cell is changed, the values of dependent cells are recalculated in topological order.

E15: In ecology, a *food web* is a graph whose vertices are the species of an ecosystem. An arrow is drawn from one species to all the other species it preys upon. This model is commonly assumed to be a dag, to disallow cycles in the food chain.

FACTS

F17: Every dag has one or more sources and one or more sinks. This can be seen by examining a path of maximal length. The first (last) vertex must be a source (sink), since otherwise the path could be extended at the beginning (end).

F18: A graph with a topological numbering is a dag. To see this observe that topological numbers increase along a path, so a path cannot return to its starting vertex. Thus no cycle exists.

F19: Any dag has a topological numbering. To construct such a numbering with lowest number 1, assign the lowest number to a source s. Then proceed recursively on the dag $G - s$, using lowest number 2.

F20: One can similarly construct a topological numbering by repeatedly numbering a sink s with the highest number, and proceeding recursively on dag $G - s$.

F21: The strategy of Fact F20 can be implemented efficiently by depth-first search. The reason is that as we grow a depth-first path in a dag, the first vertex to become finished is a sink. More succinctly, we can grow a depth-first path until a sink is reached. This gives the following high-level algorithm.

Algorithm 10.1.3: Topological Numbering (High Level)

Input: dag $G = (V, E)$
Output: topological numbering of G: vertex v has number $I[v]$
repeat until G has no vertices:
            grow a dfs path P until a sink s is reached
            set $I[s] = n$, decrease n by 1 and delete s from P & G

To make this algorithm efficient, each iteration grows the dfs-path P by starting with the previous P and extending it, if possible.

F22: A lower level implementation of Algorithm 10.1.3 runs in linear time. The idea is to use array $I[1..n]$ for 2 purposes:

$$I[v] = \begin{cases} 0 & \text{if } v \text{ has never been in } P \\ t & \text{if } v \text{ has been deleted and assigned topological number } t \end{cases}$$

Algorithm 10.1.4: Topological Numbering (Lower Level)

Input: dag $G = (V, E)$
Output: topological numbering of G: vertex v has number $I[v]$
$num = n$;
for each vertex v do $I[v] = 0$
for each vertex v do if $I[v] = 0$ then $DFS(v)$
procedure $DFS(v)$
for each edge (v, w) do
        if $I[v] = 0$ then $DFS(v)$
v is now a sink in the high level algorithm
$I[v] = num$; decrease num by 1
v is now deleted in the high level algorithm

F23: Algorithm 10.1.4 runs in linear time. It spends $O(1)$ time on each vertex and edge.

EXAMPLE

E16: Figure 10.1.5 illustrates how the algorithm numbers the dag of Figure 10.1.4.

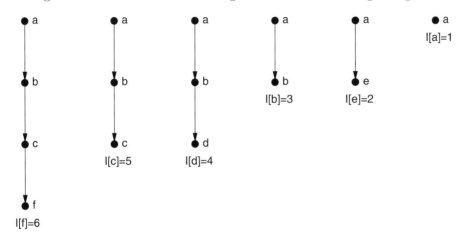

Figure 10.1.5: Execution of topological numbering algorithm.

FACTS

F24: Listing the vertices in order of decreasing finish time (Definition D16) is a valid topological order.

F25: Tarjan's algorithm for topological order [Ta74b, CLRS01] is based on Fact F24. Algorithm 10.1.4 is a reinterpretation of Tarjan's algorithm.

To illustrate this, Figure 10.1.6 shows a dfs tree for Figure 10.1.4. Each vertex is labelled by its name and finish number. Subtracting each finish number from 7 gives the topological number of Figure 10.1.5.

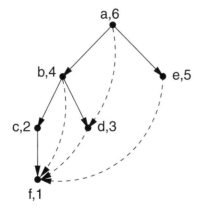

Figure 10.1.6: Topologically numbering by finish times.

F26: Another linear-time topological numbering algorithm [Kn73] works by repeatedly deleting a source. The algorithm maintains a queue of sources, as well as the in-degree of each vertex. If the in-degrees are not initially available this algorithm can do more work than Algorithm 10.1.4, since it makes two passes over the graph.

F27: Dag algorithms often propagate information from higher topological numbers to lower, either after scanning each edge (v, w) or at the end of $DFS(v)$. Propagating information in the opposite direction is also possible.

F28: As an example suppose each edge e of a dag G has a real-valued length $\ell[e]$. We can find the longest path in G in linear time. The idea is to set $d[v]$ to the length of a longest path starting at v. These values $d[v]$ can be computed in reverse topological order, using the recurrence

$$d[v] = \max 0, \ell[v, w] + d[w](v, w) \in E$$

It is easy to modify `DFS` to calculate these values.

The algorithm can recover the longest path from the $d[\ ]$ values in a second pass. The second pass can be faster if the first pass stores a pointer for each vertex indicating its successor on its longest path. Longest paths are useful in critical path scheduling. Finding the longest path in a general graph is NP-complete.

F29: Similar algorithms can be used to calculate the longest path from s to t, shortest paths from a vertex s, etc.

F30: More generally Fact F28 illustrates how the technique of dynamic programming can be used to solve problems on dags. Dynamic programming is based on similar recurrences [CLRS01].

EXAMPLE

E17: Figure 10.1.7 illustrates how the algorithm finds a longest path in a dag. Edges are labelled with their length, and vertices are labelled with their $d[\ ]$ values. The longest path corresponds to the largest $d[\ ]$ value, which is 5; it is the upper path from source to sink.

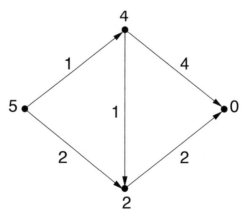

Figure 10.1.7: Illustrating a longest path algorithm.

10.1.4   Connectivity Properties

Depth-first search is the method of choice to calculate low order connectivity information. This section surveys notions of 1- and 2-connectivity. It starts with 1-connectivity of directed graphs, and it then examines 2-connectivity of undirected graphs. These connectivity algorithms are originally due to Tarjan [Ta72]. This section follows the path-based development of [Ga00], which simplifies the algorithms to eliminate the depth-first spanning tree.

Strong Components of a Directed Graph

In this section, $G = (V, E)$ is a directed graph.

DEFINITIONS

D20: For two vertices u and v, a *uv-path* is a path starting at u and ending at v.

D21: A directed graph $G = (V, E)$ is **strongly connected** if for every two distinct vertices u and v, there is a uv-path and a vu-path.

D22: In general, a directed graph will not be strongly connected. But the vertices can be partitioned into blocks that are strongly connected, according to this definition: two vertices u & v are in the same **strong component** (SC) if and only if they can reach each other, i.e., there is a uv-path and a vu-path. This defines a partition of V since it is an equivalence relation.

D23: For any directed graph G, contracting each SC to a vertex gives the **strong component graph** (SC graph) (also called the *condensation* of G).

D24: A **tournament** is a directed graph G such that each pair of vertices is joined by exactly one edge. This models a round robin tournament, where edge (x, y) represents the fact that player x beat player y.

FACTS

F31: Let C be a cycle in a graph G. All vertices of C are in the same SC. Contracting the vertices of cycle C to a single vertex yields a graph with the same SC graph as G.

F32: The SC graph is always a dag. This follows from Fact F31.

F33: A topological numbering of the SC graph of a tournament gives a ranking of the players. To see why, note that if player x is in an SC with lower topological number than y, then the tournament contains the edge (x, y) (not (y, x)). Thus SC number 1 contains the players that are unequivocally in the top tier — they all beat all other players. SC number 2 contains the 2nd tier players — they all beat all other players except those in tier 1, etc.

F34: All the vertices on a cycle belong to the same SC. In fact the SC graph is formed by repeatedly contracting cycles, until no cycle remains.

F35: A sink s is a vertex of the SC graph. In fact the SC's are $\{s\}$ and the SC's of $G - s$.

F36: Facts F34 and F35 justify the following high-level algorithm for finding the SC graph. It repeatedly contracts a cycle or deletes a sink.

F37: Next we present a linear-time depth-first search algorithm for finding the strong components and the SC graph of a given directed graph.

Algorithm 10.1.5: Strong Components

Input: directed graph $G = (V, E)$
Output: strong components of G
repeat until G has no vertices:
      grow a dfs path P until a sink or a cycle is found
          sink s: mark $\{s\}$ as an SC & delete s from P & G
          cycle C: contract the vertices of C

Like Algorithm 10.1.3, for efficiency each iteration grows P by starting with the previous P and extending it, if possible.

F38: The algorithm has a low-level implementation that finds the SC graph in linear time [Ga00]. Sinks are deleted similar to Algorithm 10.1.3. Cycles are contracted using a stack to represent P and another stack to give the boundaries of contracted vertices in P.

F39: The algorithm discovers each SC as a sink of the SC graph. So the SC's can be numbered in topological order by the method of Algorithm 10.1.3.

F40: The first linear-time algorithm for strong components is due to Tarjan [Ta72]. It computes a value called *lowpoint(v)* for each vertex v. *lowpoint(v)* is the lowest-numbered vertex (in preorder) in v's SC that is reachable from v by a path of (0 or more) tree edges followed by a back or cross edge (*lowpoint(v)* equals v if no smaller numbered vertex can be reached). The vertices with *lowpoint(v)* = v are the "roots" of the strong components.

F41: A third linear-time strong component algorithm is due to Sharir [Sh81] and Kosaraju (unpublished; see also [CLRS01]). It does a depth-first search, followed by a second depth-first search on the reverse graph. This makes good sense — the first search discovers which vertices can reach which others, and the second search discovers which vertices can be reached by which others.

EXAMPLES

E18: Figure 10.1.8 shows a directed graph, its three strong components, and its SC graph. Each strong component is strongly connected.

An elementary misperception is that a strongly connected graph has a Hamiltonian cycle. The component $\{2, 4, 5, 6\}$ illustrates that this is not always true.

E19: Figure 10.1.9 gives a dfs tree of Figure 10.1.8. (To better illustrate the algorithm a different dfs from Figure 10.1.8 is used.) Each vertex is labelled by its preorder number followed by its lowpoint value.

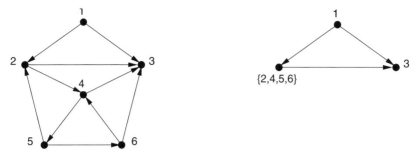

Figure 10.1.8: Strong components of a directed graph.

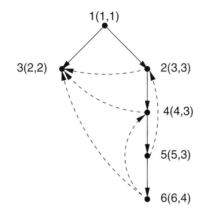

Figure 10.1.9: Execution of strong component algorithm.

E20: Suppose we number the vertices of an arbitrary directed graph by topologically numbering the SC graph, and then listing first the vertices in SC number 1, then the vertices in SC number 2, etc. The adjacency matrix of the graph with new vertex numbers is upper block triangular. This is because no edge goes from a higher numbered SC to a lower numbered SC. For instance, Figure 10.1.10 gives the adjacency matrix. It is upper triangular except for the block corresponding to SC $\{b, d, e\}$.

	a	b	d	e	c
a	0	1	1	1	1
b	0	0	0	1	1
d	0	1	0	0	1
e	0	0	1	0	1
c	0	0	0	0	0

Figure 10.1.10: Upper block triangular adjacency matrix.

E21: Example E20 shows how the SC graph is used to speed up operations on sparse matrices like Gaussian elimination, matrix inversion, finding eigenvalues, etc. The given matrix M is interpreted as a directed graph, with m_{ij} corresponding to edge (i, j). The adjacency matrix of Example E20 is constructed, and the 1 for each edge (i, j) is replaced by the value m_{ij}. The resulting block upper triangular matrix has less fill-in for Gaussian elimination and nice properties for other matrix operations [Ha69].

E22: Figure10.1.11 below illustrates the execution of the algorithm on the graph of Figure 10.1.8.

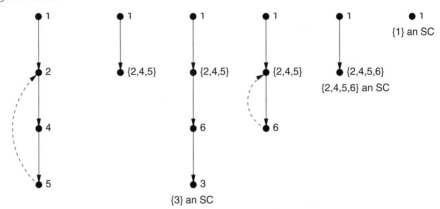

Figure 10.1.11: Execution of strong component algorithm.

E23: Figure 10.1.12 below shows a tournament and its SC graph. Player a is first, players b, d, e are in the 2nd tier, and player c is last.

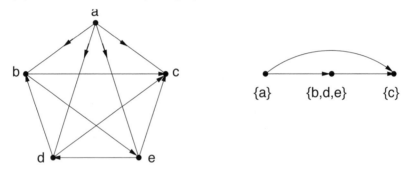

Figure 10.1.12: Tournament and its SC graph.

E24: A Markov chain is *irreducible* if the graph of its (nonzero) transition probabilities is strongly connected.

REMARK

R7: The algorithm of Fact F41 is very simple to code and is covered in many textbooks. It can be appreciably slower than the other two algorithms, because it makes two passes over the graph and has larger memory requirement.

Bridges and Cutpoints of an Undirected Graph

In this section $G = (V, E)$ is a connected undirected graph.

DEFINITIONS

D25: A vertex v is an *cutpoint (articulation point)* if $G - v$ is not connected. A graph is a *biconnected graph* if it has no cutpoint.

D26: A *biconnected component* is a maximal subgraph that has no cutpoint.

D27: An edge e is a *bridge* if $G - e$ is not connected. An edge is a bridge if and only if it's not in any cycle. A graph is *bridgeless* if it has no bridges.

D28: Let B be the set of all bridges of G. The *bridge components* (BCs) of G are the connected components of $G - B$. Equivalently a BC is the induced subgraph on a maximal set of vertices, any of which can reach any other without crossing a bridge.

D29: Contracting each BC to a vertex gives a tree, the *bridge tree*.

D30: An *orientation* of an undirected graph assigns a unique direction to each edge.

D31: A *perfect matching* of an undirected graph G is a spanning subgraph in which every vertex has degree exactly 1.

EXAMPLES

E25: Figure 10.1.13 shows a graph with 3 bridges, 6 cutpoints, and 7 biconnected components. It illustrates that an end of a bridge is a cutpoint unless it has degree one. However, a cutpoint need not be the end of a bridge.

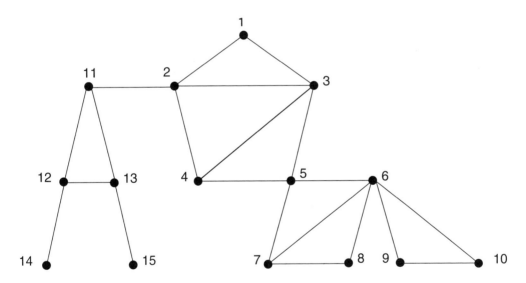

Figure 10.1.13: Undirected graph with bridges and cutpoints.

E26: If a communications network (e.g., Internet) has a bridge, that link's failure disables communication, i.e., there are sites that cannot send messages to each other. If the network has an articulation point, that site's failure also disables communication.

FACTS

F42: All vertices on a cycle are in the same BC. In fact the bridge tree is formed by repeatedly contracting cycles.

F43: A vertex x of degree ≤ 1 is a vertex of the bridge tree. In fact the BC's are $\{x\}$ and the BC's of $G - x$.

F44: Facts F42 and F43 justify the following high-level algorithm for finding the bridges and bridge tree. It has a linear-time implementation almost identical to Algorithm 10.1.5, the strong component algorithm. We call the last vertex x of a dfs path a **dead end** if x has degree ≤ 1.

Algorithm 10.1.6: Bridges

Input: connected undirected graph $G = (V, E)$
Output: bridge components and bridges of G
repeat until G has no vertices:
      grow a dfs path P until a cycle is found or a dead end is reached
          cycle C: contract the vertices of C
          dead end x: mark $\{x\}$ as a BC
                if x has degree 1, then mark its edge as a bridge (of G)

F45: A similar linear-time algorithm finds the cutpoints and biconnected components of an undirected graph [Ga00].

F46: The original linear-time dfs algorithm of Hopcroft and Tarjan for cutpoints and biconnected components [Ta72] is based on the idea of lowpoints (recall Fact F40). Start with a dfs tree T. Assume that the vertices are numbered in discovery order and that each vertex is identified with its discovery number. Define

$$lowpoint(v) = \min\{v\} \cup \{w : \text{some back edge goes from a descendant of } v \text{ to } w\}.$$

Hopcroft and Tarjan proved that G is biconnected if and only if

  (i) vertex 1 has exactly one child (which must be vertex 2);

  (ii) $lowpoint(2) = 1$;

  (iii) each vertex $w > 2$ has $lowpoint(w) < v$, where v is the parent of w.

The cutpoints have a similar characterization.

Lowpoint is easy to compute in a bottom-up pass over T, since

$$lowpoint(v) = \min\{v\} \cup \{lowpoint(w) :\ w \text{ a child of } v\} \cup \{w :(v,w)\text{a back edge}\}.$$

EXAMPLES

E27: Figure 10.1.14 below illustrates the execution of the bridges algorithm on the graph of Figure 10.1.13.

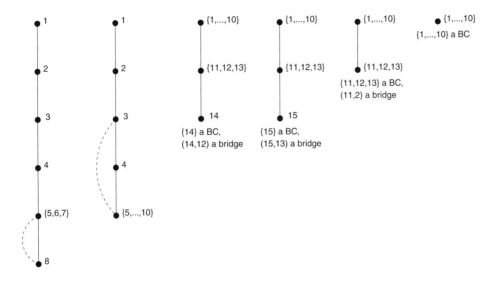

Figure 10.1.14: Execution of bridge algorithm.

E28: Figure 10.1.15 below illustrates **Robbins's Theorem** that a connected undirected graph has a strongly connected orientation if and only if it is bridgeless [Ro39]. If one of the horizontal edges is deleted, making the other a bridge, then the graph has no strongly connected orientation.

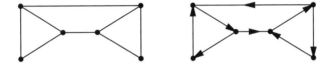

Figure 10.1.15: Undirected graph and strongly connected orientation.

E29: Kotzig's Theorem [Ko59] states that a unique perfect matching must contain a bridge of G. Figure 10.1.16 shows a graph with a unique perfect matching — matched edges are drawn heavy. Note that deleting the bridge of the matching gives another graph with a unique perfect matching. This idea can be used to efficiently find a unique perfect matching or show it does not exist [GaKaTa01].

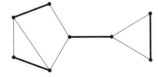

Figure 10.1.16: Graph with a unique perfect matching.

E30: Whitney's Flipping Theorem asserts that a graph is planar if and only if each biconnected component is planar [Wh32a].

10.1.5   DFS as a Proof Technique

In addition to being a powerful algorithmic tool, depth-first search can be used to easily prove many theorems of graph theory. (It's a handy way to remember the theorems too!) This subsection gives several examples.

DEFINITIONS

D32: A *mixed graph* G can have both directed and undirected edges.

D33: A mixed graph G is *traversable* if every ordered pair of vertices u, v has a uv-path with all its directed edges pointing in the forward direction. (Traversability is equivalent to connectedness if G is undirected and to strong connectedness if G is directed.)

D34: A *bridge* in a mixed graph is an undirected edge that is a bridge of G when edge directions are ignored.

D35: An *orientation* of a mixed graph assigns a unique direction to each undirected edge.

EXAMPLES

E31: Robbins's Theorem can be proved using the high-level bridge algorithm (Algorithm 10.1.6) and the strong components algorithm (Algorithm 10.1.5). When the BC algorithm is executed on a bridgeless graph G, it ends with G contracted to a single vertex. But if the SC algorithm ends with the entire graph contracted to a single vertex, then the initial graph is strongly connected. So orient the given undirected graph G to make the execution of the SC algorithm on the orientation mimic the execution of the BC algorithm on G. To do this orient edges that extend the dfs path or cause contractions (in the BC algorithm) so they do the same in the SC algorithm.

This is illustrated in Figure 10.1.17, which shows how a depth-first search executed on the undirected graph of Figure 10.1.15 gives the orientation shown in that figure. Enlarged hollow vertices are contractions of original vertices.

Figure 10.1.17: dfs proof of Robbins's Theorem.

E32: The same approach proves a generalization of Robbins's Theorem by Boesch and Tindell [BoTi80] that a traversable graph has a strongly connected orientation if and only if it has no bridge. It can be proved using Algorithm 10.1.5 with the sink rule replaced by a rule for a "1-sink," i.e., a vertex with no leaving directed edge and only one incident undirected edge.

E33: Kotzig's Theorem can be proved by dfs [Ga79]. We illustrate by proving a simple special case: a bipartite graph with a unique perfect matching has a vertex of degree one. The idea is to grow a dfs path P two edges at a time, repeatedly adding an unmatched edge (x, y) and the matched edge containing y. When the path cannot be extended, the last vertex y has degree 1. If not, a back edge from y creates an even length cycle, whose edges yield another perfect matching, as shown in Figure 10.1.18 below.

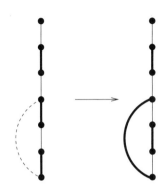

Figure 10.1.18: dfs proof of Kotzig's Theorem.

A linear-time dfs algorithm for testing if a perfect matching is unique is given in [GaKaTa01].

E34: *Rédei's Theorem* [Re34] states that any tournament has a Hamiltonian path, i.e., a simple path through all its vertices. This is easy to see by dfs: listing the vertices in order of decreasing finish time gives a Hamiltonian path.

10.1.6   More Graph Properties

The basic properties of depth-first search were developed by Hopcroft and Tarjan as stepping-stones to their goal of an efficient planarity algorithm. This subsection starts by surveying the high-level principles of the planarity algorithm. It then surveys other important properties that can be decided by efficient dfs algorithms. The depth-first tree plays a central role in all these algorithms.

Planarity Testing

The first complete linear-time algorithm to decide whether or not a graph is planar is due to Hopcroft and Tarjan. This property has obvious applications to graph drawing, circuit layout, etc. This section gives the high-level depth-first approach.

DEFINITIONS

D36: Let G be a biconnected graph with a cycle C. The edge set $E - E(C)$ can be partitioned into a family of subgraphs called ***segments*** as follows:

(i) An edge not in C that joins 2 vertices of C is a segment.
(ii) The remaining segments each consist of a connected component of $G - V(C)$, plus all edges joining that component to C.

D37: Two segments S, T of a cycle C in a graph **_interlace_** either if $|V(S) \cap V(T) \cap V(C)| \geq 3$, or if there are 4 distinct vertices u, v, w, x that occur along cycle C (not necessarily consecutively) in that order such that $u, w \in S$ and $v, x \in T$.

EXAMPLE

E35: Figure 10.1.19 below shows a cycle C (dotted) with 5 segments. Segments S_1 and S_2 interlace, and S_4 interlaces with both S_3 and S_5.

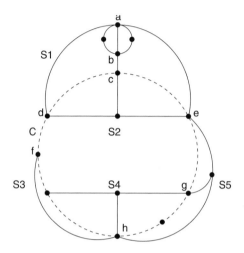

Figure 10.1.19: Planar graph with interlacing segments.

FACTS

F47: By Whitney's Flipping Theorem (E30), one can test planarity by treating each biconnected component separately.

F48: The graph theoretic approach used by Hopcroft and Tarjan is the following theorem of Auslander and Parter [AuPa61]: a biconnected graph G is planar if and only if

 (a) $C \cup S$ is planar for every segment S;

 (b) the segments can be partitioned into two families such that no two segments in the same family interlace.

The necessity of both (a) and (b) is clear. An outline of a complete proof of this theorem is given in [Ev79].

F49: Here is the overall structure of the algorithm of Hopcroft and Tarjan [HoTa74] which decides in linear time whether or not a graph is planar. Each biconnected component is processed separately.

  A depth-first spanning tree of the component is found.

  A cycle C is chosen, consisting of a path in the dfs tree plus one back edge.

  Then segments are found:

   (i) each back edge that joins two vertices of C is a segment;

(ii) each remaining segment S is determined by a vertex $w \notin C$ whose parent is in C. The edges of S are those edges with at least one endpoint descending from w. (Specifically, this amounts to the tree edge joining w to its parent, plus all edges of the subtree rooted at w, plus all back edges that join two descendants of w or join a descendant of w with a vertex of C.)

The algorithm processes each segment S recursively, checking that $C \cup S$ is planar and S can be added to an imbedding of all subgraphs processed so far. (The latter uses the interlacing criterion.)

F50: A number of additional ideas are used to achieve linear time. The *lowpoint* values (Fact F46) are used to guide the construction of cycles C. In fact the "second lowpoint" is also used. A second depth-first search is done for cycle generation. The planarity algorithm is intricate, but is very fast in practice.

Triconnectivity

Hopcroft and Tarjan show how to find the triconnected components in linear time [HoTa73b]. Like their planarity algorithm the approach is based on segments.

DEFINITIONS

D38: An undirected graph is a **triconnected graph** if it is connected and remains so whenever any two or fewer vertices are deleted.

D39: Two vertices in a biconnected graph form a **separation pair** if deleting them leaves a disconnected graph.

There is a natural definition of the *triconnected components* of a graph.

EXAMPLE

E36: In Figure 10.1.19 above there are 5 separation pairs: a, b; a, c; d, e; e, f; and g, h.

FACTS

F51: The following characterization of the separation pairs is easy to prove. Let G be a biconnected graph with a cycle C. Let a, b be a separation pair. Then a and b either both belong to C or both belong to a common segment. Moreover, suppose a and b both belong to C. Then either

(a) some segment S has $V(S) \cap V(C) = \{a, b\} \subset V(S)$; or
(b) $C - \{a, b\}$ consists of two nonempty paths, and no segment contains a vertex of both paths.

(The symbol "$\subset$" denotes proper set containment.)

F52: The triconnectivity algorithm applies the characterization of Fact F51 recursively. Hopcroft and Tarjan's triconnectivity algorithm shares algorithmic ideas with their planarity algorithm.

F53: Another useful fact is that the two vertices of a separation pair are related (Definition D10).

Ear Decomposition and *st*-numbering

DEFINITIONS

D40: An *open ear decomposition* of an undirected graph is a partition of the edges into a simple cycle P_0 and simple paths $P_1, \ldots, P_k$ such that for each $i > 0$, P_i is joined to previous paths only at its (2 distinct) ends, i.e., $V(P_i) \cap V(\cup_{j<i} P_j)$ consists of the 2 ends of P_i. (The concept, but not the name, is due to Whitney.)

D41: Let (s, t) be any edge of a biconnected graph. An *st-numbering* numbers the vertices from 1 to n so that s is numbered 1, t is numbered n, and every other vertex has both a higher-numbered neighbor and a lower-numbered neighbor.

EXAMPLE

E37: Figure 10.1.20 shows an ear decomposition consisting of cycle P_0 and simple paths $P_1, \ldots, P_6$. The 15 vertices are numbered in an *st*-numbering (corresponding to the ear decomposition).

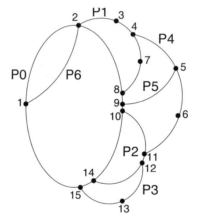

Figure 10.1.20: Ear decomposition and *st*-numbering of a biconnected graph.

FACTS

F54: Whitney [Wh32b] proved that an undirected graph is biconnected if and only if it has an open ear decomposition.

F55: An algorithm based on *lowpoint* values can be used to find an open ear decomposition of a biconnected graph in linear time (*pathfinder* in [EvTa76], although the term "ear decomposition" is not used).

F56: An open ear decomposition with $(s, t) \in P_0$ can be used to give an *st*-numbering in linear time [EvTa76].

REMARKS

R8: *st*-numbering is the basis of the linear-time planarity algorithm of Lempel, Even and Cederbaum [LeEvCe67]. It constructs a planar imbedding by repeatedly adding a vertex. More precisely it starts with an imbedding of one vertex and its incident edges. Then it repeatedly adds all edges incident to the next vertex, updating the imbedding. The vertices are added in *st*-order.

R9: Ear decomposition is closely related to depth-first search. An open ear decomposition can be found efficiently on parallel computers with large numbers of processors; the same cannot be said for doing a depth-first search. Efficient parallel algorithms for bi- and triconnectivity and planarity are based on ear decomposition [Ra93].

Reducibility

DEFINITIONS

D42: A *(program) flow graph* is a directed graph with a distinguished vertex r, the *start vertex*, that can reach every vertex.

D43: A flow graph is *reducible* if it can be transformed into the single vertex r by a sequence of operations of the following type:

if (v, w) is the only edge entering w and if $w \neq r$, then contract edge (v, w) to vertex v.

(The contraction operation discards parallel edges and self-loops.)

D44: We define a problem in data structures that arises in many dfs algorithms (and other contexts). A universe of n elements is given. The problem is to maintain a partition $\mathcal{P}$ of this universe into sets.

Initially each element forms a singleton set of $\mathcal{P}$.

Partition $\mathcal{P}$ is updated by the operation $union(A, B)$, which replaces two sets A and B of $\mathcal{P}$ by their union $A \cup B$.

A second operation $find(x)$ computes the name of the set currently containing element x.

The *set-merging problem* is to process a sequence of m intermixed *union* and *find* operations.

EXAMPLE

E38: Figure 10.1.21 shows an irreducible flow graph. In fact, this flow graph gives a

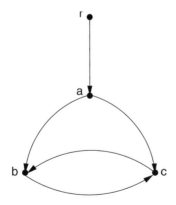

Figure 10.1.21: Irreducible flow graph.

forbidden subgraph characterization of reduciblity: a flow graph is reducible if and only if it does not contain a subgraph consisting of 4 vertices r, a, b, c (where r and a may coincide but otherwise the vertices are distinct) joined by vertex disjoint paths from r to a, a to b, a to c, b to c and c to b.

REMARK

R10:   Flow graphs model the structure of computer programs. Any program without goto's has a reducible flow graph. Many methods for code optimization (e.g., eliminating common subexpressions, identifying active variables, finding useless definitions, etc.) depend on the graph being reducible [AhSeUl86].

FACTS

The starting point of a linear-time reducibility algorithm of Tarjan [Ta74a] for flow graphs is a reformulation of reducibility. It produces the sequence of contractions that reduce it to the start vertex. For each vertex w in a dfs tree of a flow graph, we define

$$I(w) = v : \text{there is a simple } vw\text{-path ending with a back edge (to } w)$$

F57:   [Ta74a] A flow graph is reducible if and only if every set $I(w)$ consists only of descendants of w.

F58:   [Ta74a] Assume that the vertices of the flow graph are indexed by preorder number. Let w be the largest vertex (in preorder) with an entering back edge. Suppose that $I(w)$ consists only of descendants of w. If we contract the vertices of $I(w)$ into vertex w, then the new graph is reducible if and only if the original was.

F59:   Fact F58 specifies a sequence of contractions that substantiate the reducibility of a graph. The sets $I(w)$ can be computed simply by scanning edges in the backwards direction, starting at w. If a nondescendant of w is ever reached, then the graph is not reducible. The efficient descendance test of Fact F16 is used.

F60:   The contractions performed by the algorithm change the vertex set of the graph. At all times the vertices of the current graph form a partition of the original vertex set. This partition is manipulated by the *union* and *find* operations (Definition D44).

F61:   The best known algorithm for set-merging is based on the so-called weighted union and path compression rules. Tarjan showed that this algorithm solves the set-merging problem in time $O(m\alpha(m, n))$. Here α is an inverse of Ackermann's function and is very slowly growing [Ta75, CLRS01].

F62:   Gabow and Tarjan [GaTa85] showed that a special case of the set-merging problem can be solved in linear time. Using this special case algorithm makes the reducibility algorithm run in linear time.

F63:   Suppose the vertices are numbered in discovery order. If $v < w$ then any vw-path contains a common ancestor of v and w [Ta72]. This property holds in both directed and undirected graphs.

EXAMPLE

E39:   Figure 10.1.22 shows a depth-first spanning tree of a flow graph with the vertices labelled by discovery number. Any path from vertex 5 to vertex 8 passes through one or both of the common ancestors of 5 and 8, i.e., vertices 1 and 2.

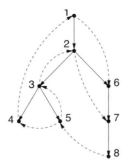

Figure 10.1.22: Depth-first search with preorder numbers.

Two Directed Spanning Trees

DEFINITIONS

D45: In a flow graph, an *r-tree* is a directed spanning tree rooted at start vertex r.

D46: Consider a dfs tree with root r. Edge (v, w) is a *(directed) bridge* in a flow graph if every rw-path includes (v, w).

EXAMPLE

E40: In Figure 10.1.21 edge (r, a) is a bridge. Duplicating it gives a graph with 2 edge-disjoint r-trees (the directed paths r, a, b, c and r, a, c, b).

FACTS

Consider a dfs tree with root r. For any vertex w define

$$I(w) = \{v : \text{ there is a simple } vw\text{-path containing only descendants of } w\}$$

Clearly the path of this definition ends in a back edge to w (unless $v = w$). Note the similarity with Fact F57.

F64: A flow graph has two edge-disjoint r-trees if and only if each vertex $v \neq r$ has two edge-disjoint rv-paths (recall Definition D20). This is a special case of *Edmonds's Branching Theorem*, which is the same statement generalized from 2 to any $k \geq 2$ [Ed72].

F65: A flow graph has 2 edge-disjoint r-trees if and only if there are no bridges.

F66: Tarjan presents a linear-time algorithm to find 2 edge-disjoint r-trees if they exist [Ta74a]. More generally in an arbitrary flow graph the algorithm finds 2 r-trees that contain the fewest possible number of common edges. The more general problem is solved by identifying the bridges and duplicating each of them. This gives a graph with 2 edge-disjoint r-trees (Fact F65).

F67: In terms of dfs, an edge (v, w) is a bridge if and only if (v, w) is a tree edge and is the only edge entering $I(w)$. Tarjan's algorithm identifies the bridges using techniques similar to the reducibility algorithm. Computing the trees is more involved.

F68: The algorithm performs set-merging to keep track of the contracted vertices, as in the reducibility algorithm. As in that algorithm the data structure of [GaTa85] is used to achieve linear time.

Dominators

DEFINITIONS

D47: In a flow graph with start vertex r, vertex v **dominates** vertex $w \neq v$ if every rw-path contains v.

D48: The **immediate dominator** of w, denoted $idom(w)$, is a vertex v that dominates w such that every other dominator of w dominates v.

D49: The **dominator tree** is a tree T whose nodes are the vertices of G. The root of T is the start vertex r. The parent of a vertex $v \neq r$ is $idom(v)$.

D50: The **internal vertices** of a path are all its vertices except the endpoints.

D51: Consider a dfs tree with root r. For every vertex $w \neq r$, the **semidominator** of w, $sdom(w)$, is defined by

$$sdom(w) = \min\{v : \text{some } vw\text{-path has all its internal vertices} > w\}.$$

EXAMPLE

E41: Figure 10.1.23 shows the dominator tree for the graph of Figure 10.1.22. Note that vertex 2 does not dominate 3 because of path $1, 6, 7, 8, 5, 3$. The start vertex 1 is the semidominator of every vertex except two: $sdom(7) = 2$, $sdom(8) = 7$. Although vertex 1 is the immediate dominator of 7 it is not the semidominator of 7.

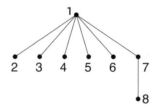

Figure 10.1.23: Dominator tree for Figure 10.1.22.

FACTS

F69: The basic properties of dominance are due to Lowry and Medlock [LoMe69]: Every vertex except r has a unique immediate dominator. This justifies the notion of dominator tree. A vertex v dominates w if and only if v is a proper ancestor of w in the dominator tree.

F70: Lengauer and Tarjan [LeTa79] give an efficient algorithm to find the dominator tree T. It is a refinement of an earlier dfs algorithm of Tarjan [Ta74b].

F71: For any vertex w, $sdom(w)$ is a proper ancestor of w. This follows from Fact F63.

FACTS ABOUT SEMIDOMINATORS
Semidominators are useful because of the next two facts proved by Lengauer and Tarjan:

F72: Take any vertex $w \neq r$. Let u be a vertex with minimum value $sdom(u)$ among all vertices in the tree path from $sdom(w)$ to w, excluding $sdom(w)$. Then

$$idom(w) = \begin{cases} sdom(w) & \text{if sdom(w) = sdom(u)} \\ idom(u) & \text{otherwise} \end{cases}$$

F73: Semidominators can be computed by a recursive definition:

$$sdom(w) \quad = \quad \min\{v : (v,w)\text{an edge}\} \cup$$
$$\{sdom(u) : u > w \text{ and some edge goes from a descendant of } u \text{ to } w\}.$$

(Note the similarity with *lowpoint* in Fact F46.)

F74: The algorithm of Lengauer and Tarjan [LeTa79] computes semidominators using Fact F73 in a backwards pass (i.e., w is decreasing). Then it computes immediate dominators using Fact F72 in a forwards pass.

F75: The time for the algorithm is $O(m\alpha(m, n))$. An implementation of this algorithm in linear time is presented in [AlHaLaTh99].

10.1.7 Approximation Algorithms

Finding small spanning subgraphs with prespecified connectivity properties is usually a difficult (NP-hard) problem. For example, finding a bridgeless spanning subgraph with the fewest possible number of edges is NP-hard. (The reason is that this subgraph contains n edges if and only if there is a Hamiltonian cycle.)

Depth-first search has been used to design good approximation algorithms for such difficult problems. Here the goal is to find a subgraph that has all the desired properties except that instead of having the fewest possible number of edges, it is within a small constant factor of this goal. This section surveys the use of depth-first search in approximation algorithms for connectivity properties. Other dfs approximation algorithms are surveyed in [Kh97].

DEFINITIONS

D52: Consider an optimization problem that seeks to find a smallest feasible solution OPT. An α-***approximation algorithm*** is a polynomial-time algorithm that is guaranteed to find a solution of size at most $\alpha|OPT|$ [CLRS01]. For the graph problems of this section, the size of the solution is the number of edges.

D53: The ***smallest bridgeless spanning subgraph*** of a connected bridgeless undirected graph is a bridgeless spanning subgraph with the minimum possible number of edges.

D54: An undirected graph is k-***edge connected*** if it is connected and remains so when any set of fewer than k edges is deleted. This concept makes good sense for a multigraph. A k-*ECSS* is a k-edge connected spanning subgraph; the graph is assumed to be k-edge connected. So a bridgeless spanning subgraph is a 2-ECSS. A *smallest k-ECSS* has the fewest possible number of edges. From now on instead of "smallest bridgeless spanning subgraph," we use the shorter equivalent phrasing, "smallest 2-ECSS."

ALGORITHM

Approximation algorithms for the smallest 2-ECSS are our first concern. A 2-approximation can be designed from Algorithm 10.1.6 in a straightforward way. Khuller and Vishkin [KhVi94] were the first to go beyond this. They presented an elegant dfs algorithm based on a "tree carving" using the dfs tree. The following modification of Algorithm 10.1.6 is a path-based reinterpretation of their algorithm.

Algorithm 10.1.7: Smallest 2-ECSS Approximation

Input: bridgeless undirected graph $G = (V, E)$
Output: edge set $F \subseteq E$, a 3/2-approximation to the smallest 2-ECSS
$F = \emptyset$
repeat until G has 1 vertex:
       grow a dfs path P until its endpoint x has all neighbors belonging to P
       let y be the neighbor of x closest to the start of P
       let C be the cycle formed by edge (x, y) edges of P
       add all edges of C to F
       contract the vertices of C

EXAMPLE

E42:   Figure 10.1.24 below gives a sample execution of the algorithm. The given graph on top has a Hamiltonian cycle, so the smallest 2-ECSS has n edges.

The algorithm grows the depth-first path of solid edges shown in the middle, starting from r. It then adds the dashed edges.

A typical edge addition is illustrated in the bottom graph, where the enlarged hollow vertex is the contraction of the last vertices on the path.

As n approaches ∞, the algorithm's solution approaches $3n/2$ edges: n solid edges and $n/2$ dashed edges. So the approximation ratio approaches $3/2$.

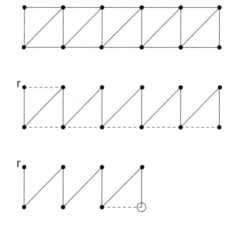

Figure 10.1.24: Smallest 2-ECSS approximation algorithm.

FACTS

Good approximation algorithms require good lower bounds on the size of the optimum solution. We analyze this algorithm using 2 lower bounds.

F76: The **Degree Lower Bound** says that any 2-ECSS has at least n edges. This results from the fact that every vertex must have degree at least 2.

F77: The **Carving Lower Bound** says that if Algorithm 10.1.7 contracts c cycles, then any 2-ECSS has at least $2c$ edges [KhVi94]. To see this let x be an endpoint of P giving a contraction in the algorithm. Any 2-ECSS contains ≥ 2 edges leaving x. These edges disappear in the contraction operation. So we can repeat this argument for every contraction, getting a lower bound of $2c$ edges.

F78: Algorithm 10.1.7 is a 3/2 approximation. This follows because the edge set F consists of $n-1$ edges from paths P and c edges that cause contractions. If OPT is the edge set of a 2-ECSS, then $|OPT| > n$ (Degree Lower Bound) and $|OPT|/2 \geq c$ (Carving Lower Bound). Thus $|F| < 3|OPT|/2$.

F79: Vempala and Vetta [VeVe00] present a 4/3-approximation algorithm for the smallest 2-ECSS. Their algorithm is based on the idea of doing a depth-first search of objects of the graph, specifically cycles and paths. It uses the **Matching Lower Bound**: Any 2-ECSS has at least as many edges as a smallest spanning subgraph where every vertex has degree ≥ 2. Vempala and Vetta give a similar 4/3-approximation algorithm for the smallest biconnected subgraph of a biconnected graph.

F80: Jothi, Raghavachari & Varadrajan [JoRaVa03] use a stronger version of the Matching Lower Bound in a dfs algorithm that achieves performance ratio 5/4 for the smallest 2-ECSS. Vetta [Ve01] uses a version of the Matching Lower Bound in a dfs algorithm that approximates the smallest strongly connected subgraph of a strongly connected graph to within a factor 3/2.

F81: The Carving Lower Bound extends to k-ECSS: If Algorithm 10.1.7 contracts c cycles then any k-ECSS has at least kc edges [KhVi94]. This can be proved by simply changing the "2"'s to k's in Fact F77.

F82: Gabow [Ga02] gives a dfs algorithm that is a 3/2-approximation for the smallest 3-ECSS of a multigraph. It uses the above dfs approach of [KhVi94] for 2-ECSS, the Carving Lower Bound, and ear decomposition.

F83: Khuller and Raghavachari [KhRa96] present the first approximation algorithm that achieves ratio < 2 for the smallest k-ECSS of a multigraph. It boosts the edge-connectivity of the solution graph in steps of 2. Each of these steps is a slight variant of the above algorithm of [KhVi94]. The analysis is based on a refinement of the Carving Lower Bound. Gabow [Ga03] improves the analysis to show it is a 1.61-approximation.

References

[AhSeUl86] A. V. Aho, R. Sethi, and J. D. Ullman, *Compilers: Principles, Techniques and Tools*, Addison-Wesley, 1986.

[AlHaLaTh99] S. Alstrup, D. Harel, P. W. Lauridsen, and M. Thorup, Dominators in linear time, *SIAM J. Comput.* 28 (1999), 2117–2132.

[AuPa61] L. Auslander and S. V. Parter, On imbedding graphs in the plane, *J. Math. and Mech.* 10 (1961), 517–523.

[BoTi80] F. Boesch and R. Tindell, Robbins's theorem for mixed multigraphs, *Amer. Math. Monthly* 87 (1980), 716–719.

[BrBr96] G. Brassard and P. Bratley, *Fundamentals of Algorithmics*, Prentice-Hall, 1996.

[CLRS01] T. H. Cormen, C. E. Leiserson, R. L. Rivest, and C. Stein, *Introduction to Algorithms*, Second Edition, McGraw-Hill, 2001.

[Ed72] J. Edmonds, Edge-disjoint branchings, pp. 91–96 in *Combinatorial Algorithms*, ed. by R. Rustin, Algorithmics Press, New York (1972).

[Ev79] S. Even, *Graph Algorithms*, Computer Science Press, Baltimore, 1979.

[EvTa76] S. Even and R. E. Tarjan, Computing an *st*-numbering, *Theoret. Comp. Sci.* 2 (1976), 339–344.

[FrTa87] M. L. Fredman and R. E. Tarjan, Fibonacci heaps and their uses in improved network optimization algorithms, *J. ACM* 34 (1987), 596–615.

[Ga79] H. N. Gabow, Algorithmic proofs of two relations between connectivity and the 1-factors of a graph, *Disc. Math.* 26 (1979), 33–40.

[Ga00] H. N. Gabow, Path-based depth-first search for strong and biconnected components, *Inf. Proc. Letters* 74 (2000), 107–114.

[Ga02] H. N. Gabow, An ear decomposition approach to approximating the smallest 3-edge connected spanning subgraph of a multigraph, *Proc. 13th Annual ACM-SIAM Symp. on Disc. Algorithms* (2002), 84–93.

[Ga03] H. N. Gabow, Better performance bounds for finding the smallest *k*-edge connected spanning subgraph of a multigraph, *Proc. 14th Annual ACM-SIAM Symp. on Disc. Algorithms* (2003), 460–469.

[GaTa85] H. N. Gabow and R. E. Tarjan, A linear-time algorithm for a special case of disjoint set union, *J. Comp. and Sys. Sci.* 30 (1985), 209–221.

[GaKaTa01] H. N. Gabow, H. Kaplan, and R. E. Tarjan, Unique maximum matching algorithms, *J. Algorithms* 40 (2001), 159–183.

[GoTa02] M. T. Goodrich and R. Tamassia, *Algorithm Design: Foundations, Analysis and Internet Examples*, John Wiley & Sons, 2002.

[Ha69] F. Harary, *Graph Theory*, Addison-Wesley, Reading, MA, 1969.

[HoTa73a] J. Hopcroft and R. E. Tarjan, Efficient algorithms for graph manipulation, *Comm. ACM* 16 (1973), 372–378.

[HoTa73b] J. E. Hopcroft and R. E. Tarjan, Dividing a graph into triconnected components, *SIAM J. Comput.* 2 (1973), 135–158.

[HoTa74] J. Hopcroft and R. Tarjan, Efficient planarity testing, *J. ACM* 21 (1974), 549–568.

[HSR98] E. Horowitz, S. Sahni, and S. Rajasekaran, *Computer Algorithms*, Computer-Science Press, Baltimore, 1998.

[JoRaVa03] R. Jothi, B. Raghavachari, and S. Varadarajan, A 5/4-approximation algorithm for minimum 2-edge-connectivity, *Proc. 14th Annual ACM-SIAM Symp. on Disc. Algorithms* (2003) 725–734.

[Kh97] S. Khuller, Approximation algorithms for finding highly connected subgraphs, in *Appoximation Algorithms for NP-hard Problems*, ed. by D. S. Hochbaum, PWS Publishing, 1997.

[KhRa96] S. Khuller and B. Raghavachari, Improved approximation algorithms for uniform connectivity problems, *J. Algorithms* 21 (1996), 434–450.

[KhVi94] S. Khuller and U. Vishkin, Biconnectivity approximations and graph carvings, *J. ACM* 41 (1994), 214–235.

[Kl00] J. Kleinberg, The small-world phenomenon: An algorithmic perspective, *Proc. 32nd Annual ACM Symp. on Th. Comput.* (2000), 163–170.

[Kn73] D. E. Knuth, *The Art of Computer Programming, Vol. 1: Fundamental Algorithms*, Second Edition, Addison-Wesley, 1973.

[Ko59] A. Kotzig, On the theory of finite graphs with a linear factor I, *Mat.-Fyz. Časopis Slovensk. Akad. Vied* 9 (1959), 73–91.

[LeEvCe67] A. Lempel, S. Even, and I. Cederbaum, An algorithm for planarity testing of graphs, pp. 215–232 in *Theory of Graphs: Int. Symp*, ed. by P. Rosenstiehl, Gordon and Breach, 1967.

[LeTa79] T. Lengauer and R. E. Tarjan, A fast algorithm for finding dominators in a flowgraph, *ACM Trans. on Prog. Lang. and Sys.* 1 (1979), 121–141.

[LoMe69] E. S. Lowry and C. W. Medlock, Object code optimization, *C. ACM* 12 (1969), 13–21.

[Lu82] E. Lucas, *Récreations Mathématiques*, Paris, 1882.

[Mi67] S. Milgram, The small world problem, *Psychology Today* 1 (1967), 60–67.

[Ra93] V. Ramachandran, Parallel open ear decomposition with applications to graph biconnectivity and triconnectivity, in *Synthesis of Parallel Algorithms*, ed. by J. H. Reif, Morgan Kaufmann, 1993.

[Re34] L. Rédei, Ein kombinatorischer Satz., *Acta Litt. Sci. Szeged* 7 (1934), 39–43.

[Ro39] H. E. Robbins, A theorem on graphs, with an application to a problem in traffic control, *Amer. Math. Monthly* 46 (1939), 281–283.

[Se02] R. Sedgewick, *Algorithms in C++, Part 5: Graph Algorithms*, Addison-Wesley, 2002.

[Sh81] M. Sharir, A strong-connectivity algorithm and its application in data flow analysis, *Comp. and Math. with Applications* 7 (1981), 67–72.

[St96] I. Stewart, "Cows in the maze," pp. 116-118 in Mathematical Recreations, *Sci. American*, Dec. 1996.

[Ta72] R. E. Tarjan, Depth-first search and linear graph algorithms, *SIAM J. Comput.* 1 (1972), 146–160.

[Ta74a] R. E. Tarjan, Testing flow graph reducibility, *J. Comput. Sys. Sci.* 9 (1974), 355–365.

[Ta74b] R. E. Tarjan, Finding dominators in directed graphs, *SIAM J. Comput.* 3 (1974), 62–89.

[Ta75] R. E. Tarjan, Efficiency of a good but not linear set union algorithm, *J. ACM* 22 (1975), 215–225.

[Ta76] R. E. Tarjan, Edge-disjoint spanning trees and depth-first search, *Acta Inf.* 6 (1976), 171–185.

[Tarr95] G. Tarry, Le probléme des labyrinthes, *Nouvelles Ann. de Math.* 14 (1895), 187.

[VeVe00] S. Vempala and A. Vetta, Factor 4/3 approximations for minimum 2-connected subgraphs, pp. 262–273 in *Appoximation Algorithms for Combinatorial Optimization*, ed. by K. Jansen and S. Khuller, Lecture Notes in Computer Science 1931, Springer-Verlag, 2000.

[Ve01] A. Vetta, Approximating the minimum strongly connected subgraph via a matching lower bound, *Proc. 12th Annual ACM-SIAM Symp. on Disc. Algorithms* (2001), 417–426.

[We99] M. A. Weiss, *Data Structures and Algorithm Analysis in C++*, Second Edition, Addison-Wesley, 1999.

[Wh32a] H. Whitney, Non-separable and planar graphs, *Trans. Amer. Math. Soc.* 34 (1932), 339–362.

[Wh32b] H. Whitney, Congruent graphs and the connectivity of graphs, *Amer. J. Math.* 54 (1932), 150–168.

Section 10.2
Dynamic Graph Algorithms

Camil Demetrescu, University of Rome La Sapienza, Italy
Irene Finocchi, University of Rome La Sapienza, Italy
Giuseppe F. Italiano, University of Rome Tor Vergata, Italy

INTRODUCTION

In many applications of graph algorithms, including social networks, communication networks, transportation networks, VLSI design, graphics, and assembly planning, graphs are subject to discrete changes, such as additions or deletions of edges or vertices, or edge cost changes. In the last two decades there has been a growing interest in such dynamically changing graphs, and a whole body of algorithms and data structures for dynamic graphs has been discovered. This section of the *Handbook* is intended as an overview of this field.

10.2.1   Basic Terminology

In a typical dynamic graph problem, one would like to answer queries on graphs that are changing dynamically. For instance, while a graph is undergoing dynamic changes, one might be interested to know whether the graph is connected, or which is the shortest path between any two vertices.

DEFINITIONS

D1: An ***update on a graph*** is an operation that inserts or deletes edges or vertices of the graph or changes attributes associated with edges or vertices, such as cost or color.

D2: A ***dynamic graph*** is a graph that is undergoing a sequence of updates.

REMARK

R1:   The goal of a dynamic graph algorithm is to update efficiently the solution of a problem after dynamic changes, rather than having to recompute it from scratch each time. Given their powerful versatility, it is not surprising that dynamic algorithms and dynamic data structures are often more difficult to design and to analyze than their static counterparts.

DEFINITIONS

We can classify dynamic graph problems according to the types of updates allowed.

D3:   A dynamic graph problem is said to be *fully dynamic* if the update operations include unrestricted insertions and deletions of edges or vertices.

D4:   A dynamic graph problem is said to be *partially dynamic* if only one type of update, either insertions or deletions, is allowed.

D5:   A dynamic graph problem is said to be *incremental* if only insertions are allowed.

D6:   A dynamic graph problem is said to be *decremental* if only deletions are allowed.

REMARKS

R2:   In the first part of this work we will present the main algorithmic techniques used to solve dynamic problems on *undirected* graphs. To illustrate those techniques, we will focus particularly on dynamic minimum spanning trees and on connectivity problems.

R3:   In the second part of this work we will deal with dynamic problems on *directed* graphs, and we will investigate as paradigmatic problems the dynamic maintenance of transitive closure and shortest paths.

R4:   Interestingly enough, dynamic problems on directed graphs seem much harder to solve than their counterparts on undirected graphs, and they require completely different techniques and tools.

10.2.2   Dynamic Problems on Undirected Graphs

This part considers fully dynamic algorithms for undirected graphs. These algorithms maintain efficiently some property of a graph that is undergoing structural changes defined by insertion and deletion of edges, and/or updates of edge costs. To check the graph property throughout a sequence of these updates, the algorithms must be prepared to answer queries on the graph property efficiently.

EXAMPLES

E1:   The *fully dynamic minimum spanning tree* problem consists of maintaining a minimum spanning forest of a graph during insertions of edges, deletions of edges, and edge cost changes.

E2:   A *fully dynamic connectivity* algorithm must be able to insert edges, delete edges, and answer a query on whether the graph is connected, or whether two vertices are in the same connected component.

REMARKS

R5: The goal of a dynamic algorithm is to minimize the amount of recomputation required after each update.

R6: All the dynamic algorithms that we describe are able to maintain dynamically the graph property at a cost (per update operation) which is significantly smaller than the cost of recomputing the graph property from scratch.

In this part, first we present general techniques and tools used in designing dynamic algorithms on undirected graphs, and then we survey the fastest algorithms for solving two of the most fundamental graph problems: connectivity and minimum spanning trees.

General Techniques for Undirected Graphs

Many of the algorithms proposed in the literature use the same general techniques, and hence we begin by describing these techniques. As a common theme, most of these techniques use some sort of graph decomposition, and they partition either the vertices or the edges of the graph to be maintained. Moreover, data structures that maintain properties of dynamically changing trees are often used as building blocks by many dynamic graph algorithms. The basic update operations are edge insertions and edge deletions. Many properties of dynamically changing trees have been considered in the literature.

EXAMPLES

E3: The basic query operation is tree membership: while the forest of trees is dynamically changing, we would like to know at any time which tree contains a given vertex, or whether two vertices are in the same tree. Dynamic tree membership is a special case of dynamic connectivity in undirected graphs, and indeed in the following we will see that some of the data structures presented here for trees are useful for solving the more general problem on graphs.

E4: Other properties have also been considered: the parent of a vertex, the least common ancestor of two vertices, and the center or the diameter of a tree [AlHoDeTh97, AlHoTh00, SlTa83]. When costs are associated either to vertices or to edges, one could also ask what is the minimum or maximum cost in a given path.

In what follows, we first present three different data structures that maintain properties of dynamically changing trees: topology trees, ET trees, and top trees. Next, we discuss techniques that can be applied on general undirected graphs: clustering, sparsification, and randomization. In the course of the presentation, we also highlight how these techniques have been applied to solve the fully dynamic connectivity and/or minimum spanning tree problems, and which update and query bounds can be achieved when they are deployed.

Topology Trees

Topology trees have been introduced by Frederickson [Fr85] to maintain dynamic trees upon insertions and deletions of edges.

DEFINITIONS

D7: Given a tree T of a forest, a ***cluster*** is a connected subgraph of T.

D8: The ***cardinality*** of a cluster is the number of its vertices.

D9: The ***external degree*** of a cluster is the number of tree edges incident to it.

ASSUMPTION

In order to illustrate the solution proposed by Frederickson [Fr85, Fr97], we assume that the tree T has maximum vertex degree 3: this is without loss of generality, since a standard transformation can be applied if this is not the case [Ha69].

DEFINITION

D10: A ***restricted partition*** of a tree T is a partition of its vertex set V into clusters of external degree ≤ 3 and cardinality ≤ 2 such that:

(1) Each cluster of external degree 3 has cardinality 1.

(2) Each cluster of external degree < 3 has cardinality at most 2.

(3) No two adjacent clusters can be combined and still satisfy the above.

EXAMPLE

E5: A restricted partition of order 2 of a tree T is shown in Figure 10.2.1.

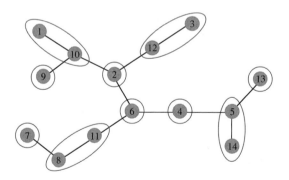

Figure 10.2.1: A restricted partition of order 2 of a tree T.

REMARKS

R7: There can be several restricted partitions for a given tree T, based upon different choices of the vertices to be unioned. For instance, vertex 8 in Figure 10.2.1 could be unioned with vertex 7, instead of vertex 11, and the partition would still be valid.

R8: Because of clause (3), the restricted partition implements a cluster-forming scheme according to a locally greedy heuristic, which does not always obtain the minimum number of clusters, but which has the advantage of requiring only local adjustments during updates.

DEFINITIONS

D11: A *restricted multi-level partition* consists of a collection of restricted partitions satisfying the following:

(1) The clusters at level 0 (known as *basic clusters*) contain one vertex each.

(2) The clusters at level $\ell \geq 1$ form a restricted partition with respect to the tree obtained after shrinking all the clusters at level $\ell - 1$.

(3) There is exactly one vertex cluster at the topmost level.

D12: A *topology tree* is a hierarchical representation of a tree T based on multi-level partitions. Each level of the topology tree partitions the vertices of T into clusters. Clusters at level 0 contain one vertex each. Clusters at level $\ell \geq 1$ form a restricted partition of order 2 of the vertices of the tree T' obtained by shrinking each cluster at level $\ell - 1$ into a single vertex.

EXAMPLE

E6: An example of topology tree, together with the restricted partitions used to obtain its levels, is given in Figure 10.2.2.

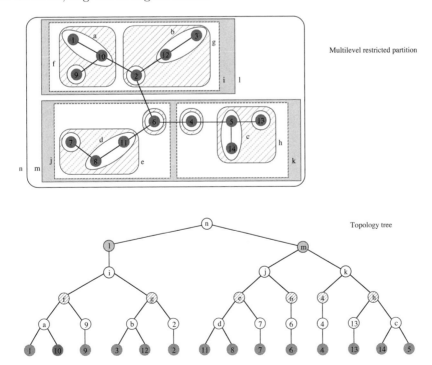

Figure 10.2.2: Restricted partitions and topology tree of a tree T.

APPROACH

Edge deletion. We sketch how to update the clusters of a restricted partition when an edge e is deleted from a tree T. First, removing e splits T into two trees, say T_1 and T_2, which inherit all of the clusters of T, possibly with the following exceptions.

- If e is entirely contained in a cluster, this cluster is no longer connected and therefore must be split. After the split, we must check whether each of the two resulting clusters is adjacent to a cluster of tree degree at most 2, and if these two adjacent clusters together have cardinality ≤ 2. If so, we combine these two clusters in order to maintain condition (3) in Definition D10.

- If e is between two clusters, then no split is needed. However, since the tree degree of the clusters containing the endpoints of e has been decreased, we must check if each cluster should be combined with an adjacent cluster, again because of condition (3) in Definition D10.

Edge insertion. Similar local manipulations can be applied to restore invariants (1) – (3) in Definition D10 in case of edge insertions.

Construction of the topology tree. The levels of the topology tree are built in a bottom up fashion by repeatedly applying the locally greedy heuristic.

Update of the topology tree. Each level can be updated upon insertions and deletions of edges in tree T by applying a few locally greedy adjustments similar to the ones described before. In particular, a constant number of basic clusters (corresponding to leaves in the topology tree) are examined: the changes in these basic clusters percolate up in the topology tree, possibly causing vertex clusters to be regrouped in different ways.

FACTS

F1: The number of nodes at each level of the topology tree is a constant fraction of that at the previous level, and thus the number of levels is $O(\log n)$ (see [Fr85, Fr97]).

F2: The property that only a constant amount of work has to be done on $O(\log n)$ topology tree nodes implies a logarithmic bound on the update time.

F3: (Frederickson's Theorem) [Fr85] The update of a topology tree because of an edge insertion or deletion can be supported in $O(\log n)$ time.

ET Trees

ET trees have been introduced by Henzinger and King [HeKi99] to work on dynamic forests whose vertices are associated with weighted or unweighted keys. Updates allow it to cut arbitrary edges, to insert edges linking different trees of the forest, and to add or remove the weighted key associated to a vertex. Supported queries are the following:

- `Connected`(u, v): tells whether vertices u and v are in the same tree.
- `Size`(v): returns the number of vertices in the tree that contains v.
- `Minkey`(v): returns a key of minimum weight in the tree that contains v; if keys are unweighted, an arbitrary key is returned.

DEFINITIONS

D13: An ***Euler tour*** of a tree T is a maximal closed walk over the graph obtained by replacing each edge of T by two directed edges with opposite direction. The walk traverses each edge exactly once; hence, if T has n vertices, the Euler tour has length $2n - 2$ (see Figure 10.2.3).

D14: An **ET tree** is a *balanced binary tree* (the number of nodes in the left and right subtrees of each node differs by at most one) over some Euler tour around T. Namely, leaves of the balanced binary tree are the nodes of the Euler tour, in the same order in which they appear (see Figure 10.2.3).

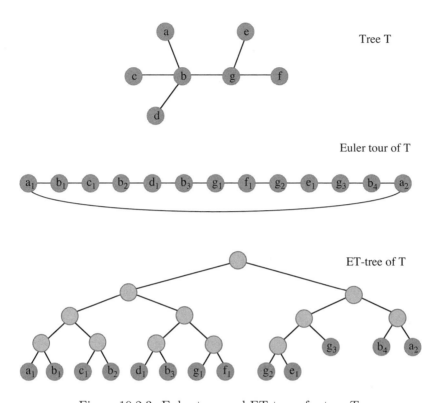

Figure 10.2.3: Euler tour and ET tree of a tree T.

REMARK

R9: Although each vertex of T may occur several times in the Euler tour (an arbitrary occurrence is marked as *representative* of the vertex), an ET tree has $O(n)$ nodes.

APPROACH

Edge insertion and deletion. If trees in the forest are linked or cut, a constant number of splits and concatenation allows reconstruction of the new Euler tour(s); the ET tree(s) can then be rebalanced by affecting only $O(\log n)$ nodes.

Connectivity queries. The query Connected(u, v) can be easily supported in $O(\log n)$ time by finding the roots of the ET trees containing u and v and checking if they coincide.

Size and Minkey queries. To support Size and Minkey queries, each node q of the ET tree maintains two additional values: the number $s(q)$ of representatives below it and the minimum weight key $k(q)$ attached to a representative below it. Such values can be maintained in $O(\log n)$ time per update, which allows answering queries of the form Size(v) and Minkey(v) in $O(\log n)$ time for any vertex v of the forest: the root r of the ET tree containing v is found and values $s(r)$ and $k(r)$ are returned, respectively. See [HeKi99] for additional details of the method.

FACT

F4: Both updates and queries can be supported in $O(\log n)$ time using ET trees (see [HeKi99]).

Top Trees

Top trees have been introduced by Alstrup et al. [AlHoDeTh97] to maintain efficiently information about paths in trees, such as, e.g., the maximum weight on the path between any pair of vertices in a tree. The basic idea is taken from Frederickson's topology trees, but instead of partitioning vertices, top trees work by partitioning edges: the same vertex can then appear in more than one cluster.

DEFINITIONS

D15: Similarly to [Fr85, Fr97], a ***cluster*** is a connected subtree of tree T, with the additional constraint that at most two vertices, called ***boundary vertices***, have edges out of the subtree.

D16: Two clusters are said to be ***neighbors*** if their intersection contains exactly one vertex.

D17: A ***top tree*** of T is a binary tree such that:
- The leaves and the internal nodes represent edges and clusters of T, respectively.
- The subtree represented by an internal node is the union of the subtrees represented by its two children, which must be neighbors.
- The root represents the entire tree T.
- The height is $O(\log n)$.

EXAMPLE

E7: We refer to Figure 10.2.4 for an example of a top tree.

APPROACH

Top trees can be maintained under edge insert and delete operations in tree T by making use of two basic `Merge` and `Split` operations.

Merge. It takes two top trees whose roots are neighbor clusters and joins them to form a unique top tree.

Split. This is the reverse operation, deleting the root of a given top tree.

Edge insertion and deletion. The implementation of an edge insertion/deletion starts with a sequence of `Split` of all ancestor clusters of edges whose boundary changes and finishes with a sequence of `Merge`. Since an end-point v of an edge has to be a boundary vertex of the edge if v is not a leaf, each edge insert/delete can change the boundary of at most two edges, excluding the edge being inserted/deleted.

FACT

F5: [AlHoDeTh97] For a dynamic forest we can maintain top trees of height $O(\log n)$ supporting edge insertions and deletions with a sequence of $O(\log n)$ `Split` and `Merge`. The sequence itself is identified in $O(\log n)$ time.

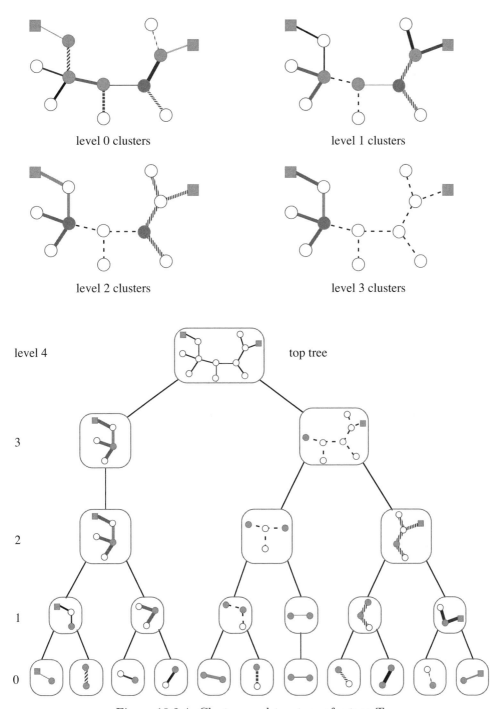

Figure 10.2.4: Clusters and top tree of a tree T.

REMARKS

R10: Top trees are typically used by attaching extra information to their nodes. A careful choice of the extra information makes it possible to maintain easily path properties of trees, such as the maximum weight of an edge in the unique path between any two vertices. See [AlHoDeTh97, AlHoTh00, HoDeTh01] for sample applications.

R11: Top trees are a natural generalization of standard balanced binary trees over dynamic collections of lists that may be concatenated and split, where each node of the balanced binary tree represents a segment of a list. In the terminology of top trees, this is just a special case of a cluster.

Clustering

The clustering technique of [Fr85] is based upon partitioning the graph into a suitable collection of *clusters*, such that each update involves only a small number of such clusters.

REMARKS

R12: Typically, the decomposition defined by the clusters is applied recursively, and the information about the subgraphs is combined with the topology trees described above.

R13: A refinement of the clustering technique appears in the idea of *ambivalent data structures* [Fr97], in which edges can belong to multiple groups, only one of which is actually selected depending on the topology of the given spanning tree.

EXAMPLE

E8: We briefly describe the application of clustering to the problem of maintaining a minimum spanning forest [Fr85]. Let $G = (V, E)$ be a graph with a designated spanning tree S. Clustering is used for partitioning the vertex set V into subtrees connected in S, so that each subtree is only adjacent to a few other subtrees. A topology tree is then used for representing a recursive partition of the tree S. Finally, a generalization of topology trees, called *2-dimensional topology trees*, is formed from pairs of nodes in the topology tree and allows it to maintain information about the edges in $E \setminus S$ [Fr85].

FACTS

F6: Fully dynamic algorithms based only on a single level of clustering obtain typically time bounds of the order of $O(m^{2/3})$ (see for instance [GaIt92, Ra95]).

F7: When the partition can be applied recursively, better $O(m^{1/2})$ time bounds can be achieved by using 2-dimensional topology trees (see, for instance, [Fr85, Fr97]).

F8: (Frederickson's theorem) [Fr85] The minimum spanning forest of an undirected graph can be maintained in time $O(m^{1/2})$ per update, where m is the current number of edges in the graph.

REMARKS

R14: See [Fr85, Fr97] for details about Frederickson's algorithm. With the same technique, an $O(m^{1/2})$ time bound can be obtained also for fully dynamic connectivity and 2-edge connectivity [Fr85, Fr97].

R15: The type of clustering used can be very problem-dependent, however, which makes this technique difficult to be used as a black box.

Sparsification

Sparsification is a general technique due to Eppstein et al. [EpGaItNi97] that can be used as a black box (without having to know the internal details), in order to design and dynamize graph algorithms. It is a divide-and-conquer technique that allows it to reduce the dependence on the number of edges in a graph, so that the time bounds for maintaining some property of the graph match the times for computing in sparse graphs. More precisely, when the technique is applicable, it speeds up a $T(n, m)$ time bound for a graph with n vertices and m edges to $T(n, O(n))$, i.e., to the time needed if the graph were sparse. For instance, if $T(n, m) = O(m^{1/2})$, we get a better bound of $O(n^{1/2})$. The technique itself is quite simple. A key concept is the notion of certificate.

DEFINITIONS

D18: For any graph property P and graph G, a ***certificate*** for G is a graph G' such that G has property P if and only if G' has the property.

D19: A subgraph on n vertices is ***sparse*** if it has $O(n)$ edges.

D20: A time bound $T(n)$ is ***well-behaved*** if, for some $c < 1$, $T(n/2) < cT(n)$. Well-behavedness eliminates strange situations in which a time bound fluctuates wildly with n. For instance, all polynomials are well-behaved.

APPROACH

Let G be a graph with m edges and n vertices. We partition the edges of G into a collection of $O(m/n)$ sparse subgraphs. The information relevant for each subgraph can be summarized in a sparse certificate. Certificates are then merged in pairs, producing larger subgraphs which are made sparse by again computing their certificate. The result is a balanced binary tree in which each node is represented by a sparse certificate. Each update involves $O(\log(m/n))$ graphs with $O(n)$ edges each, instead of one graph with m edges.

NOTATION

In the present context, $\log x$ stands for $\max(1, \log_2 x)$, so that $\log(m/n)$ is never smaller than 1, even if $m < 2n$.

REMARKS

There exist two variants of sparsification.

R16: The first variant is used in situations where no previous fully dynamic algorithm is known. A static algorithm is used for recomputing a sparse certificate in each tree node affected by an edge update. If the certificates can be found in time $O(m + n)$, this variant gives time bounds of $O(n)$ per update.

R17: In the second variant, certificates are maintained using a dynamic data structure. For this to work, a *stability* property of certificates is needed, to ensure that a small change in the input graph does not lead to a large change in the certificates. (We refer the interested reader to [EpGaItNi97] for a precise definition of stability.) This variant transforms time bounds of the form $O(m^p)$ into $O(n^p)$.

FACTS

F9: [EpGaItNi97] Let P be a property for which we can find sparse certificates in time $f(n, m)$ for some well-behaved f, and such that we can construct a data structure for testing property P in time $g(n, m)$ which can answer queries in time $q(n, m)$. Then there is a fully dynamic data structure for testing whether a graph has property P, for which edge insertions and deletions can be performed in time $O(f(n, O(n))) + g(n, O(n))$, and for which the query time is $q(n, O(n))$.

F10: [EpGaItNi97] Let P be a property for which stable sparse certificates can be maintained in time $f(n, m)$ per update, where f is well-behaved, and for which there is a data structure for property P with update time $g(n, m)$ and query time $q(n, m)$. Then P can be maintained in time $O(f(n, O(n))) + g(n, O(n))$ per update, with query time $q(n, O(n))$.

REMARKS

R18: Basically, the first version of sparsification (Fact F9) can be used to dynamize static algorithms, in which case we only need to *compute* efficiently *sparse* certificates, while the second version (Fact F10) can be used to speed up existing fully dynamic algorithms, in which case we need to *maintain* efficiently *stable sparse* certificates.

R19: Sparsification applies to a wide variety of dynamic graph problems, including minimum spanning forests, edge-connectivity, and vertex-connectivity. As an example, for the fully dynamic minimum spanning tree problem, it reduces the update time from $O(m^{1/2})$ [Fr85, Fr97] to $O(n^{1/2})$ [EpGaItNi97].

R20: Since sparsification works on top of a given algorithm, we need not know the internal details of this algorithm. Consequently, it can be applied orthogonally to other data structuring techniques: in a large number of situations both clustering and sparsification have been combined to produce an efficient dynamic graph algorithm.

Randomization

Clustering and sparsification allow one to design efficient deterministic algorithms for fully dynamic problems. The last technique we present in this section is due to Henzinger and King [HeKi99]; it achieves faster update times for some problems by exploiting the power of randomization.

APPROACH

We sketch how the randomization technique works, taking the fully dynamic connectivity problem as an example. In a graph $G = (V, E)$ to be maintained dynamically, the edges of a spanning forest F are called *tree edges*, and the edges in $E \setminus F$ are called *non-tree edges*.

Maintaining spanning forests. Trees in the spanning forests are maintained using the Euler tours data structure (ET trees) described above, which allows one to obtain logarithmic updates and queries within the forest.

Random sampling. A key idea behind the technique of Henzinger and King is the following: when e is deleted from a tree T, use random sampling among the non-tree edges incident to T, in order to find quickly a replacement edge for e, if any.

Graph decomposition. The second key idea is to combine randomization with a suitable graph decomposition. We maintain an edge decomposition of the current graph G into $O(\log n)$ edge disjoint subgraphs $G_i = (V, E_i)$. These subgraphs are hierarchically ordered. The lower levels contain tightly connected portions of G (i.e., dense edge cuts), while the higher levels contain loosely connected portions of G (i.e., sparse cuts). For each level i, a spanning forest for the graph defined by all the edges in levels i or below is also maintained.

REMARKS

R21: Note that the hard operation is the deletion of a tree edge: indeed, a spanning forest is easily maintained throughout edge insertions, and deleting a non-tree edge does not change the forest.

R22: The goal is an update time of $O(\log^3 n)$: after an edge deletion, in the quest for a replacement edge, we can afford a number of sampled edges of $O(\log^2 n)$. However, if the candidate set of edge e is a small fraction of all non-tree edges which are adjacent to T, it is unlikely to find a replacement edge for e among this small sample. If we found no candidate among the sampled edges, we must check explicitly all the non-tree edges adjacent to T. After random sampling has failed to produce a replacement edge, we need to perform this check explicitly; otherwise we would not be guaranteed to provide correct answers to the queries.

R23: Since there might be numerous edges adjacent to T, this explicit check could be an expensive operation, so it should be made a low probability event for the randomized algorithm. This can produce pathological updates, however, since deleting all edges in a relatively small candidate set, reinserting them, deleting them again, and so on will almost surely produce many of those unfortunate events.

R24: The graph decomposition is used to prevent the undesirable behavior described above. If a spanning forest edge e is deleted from a tree at some level i, random sampling is used to quickly find a replacement for e at that level. If random sampling succeeds, the tree is reconnected at level i. If random sampling fails, the edges that can replace e in level i form with high probability a sparse cut. These edges are moved to level $i + 1$ and the same procedure is applied recursively on level $i + 1$.

FACT

F11: (Henzinger and King's Theorem) [HeKi99] Let G be a graph with n vertices and m_0 edges, subject to edge deletions only. A spanning forest of G can be maintained in $O(\log^3 n)$ expected amortized time per deletion, if there are at least $\Omega(m_0)$ deletions. The time per query is $O(\log n)$.

Connectivity

We now give a high level description of the fastest deterministic algorithm for the fully dynamic connectivity problem in undirected graphs [HoDeTh01]: the algorithm answers connectivity queries in $O(\log n/\log\log n)$ worst-case running time while supporting edge insertions and deletions in $O(\log^2 n)$ amortized time. Like the randomized algorithm in [HeKi99], the deterministic algorithm in [HoDeTh01] maintains a spanning forest F of the dynamically changing graph G.

FACTS

F12: Let e be a tree edge of forest F, and let T be the tree of F containing it. When e is deleted, the two trees T_1 and T_2 obtained from T after the deletion of e can be reconnected if and only if there is a non-tree edge in G with one endpoint in T_1 and the other endpoint in T_2. We call such an edge a *replacement edge* for e. In other words, if there is a replacement edge for e, then T is reconnected via this replacement edge; otherwise, the deletion of e creates a new connected component in G.

F13: To accommodate systematic search for replacement edges, the algorithm associates to each edge e a level $\ell(e)$ and, based on edge levels, maintains a set of sub-forests of the spanning forest F: for each level i, forest F_i is the sub-forest induced by tree edges of level $\geq i$.

F14: $F = F_0 \supseteq F_1 \supseteq F_2 \supseteq \ldots \supseteq F_L$, where L denotes the maximum edge level.

F15: Initially, all edges have level 0; levels are then progressively increased, but never decreased. The changes of edge levels are accomplished so as to maintain the following invariants, which obviously hold at the beginning.

INVARIANTS

Invariant (1): F is a maximum spanning forest of G if edge levels are regarded as weights.

Invariant (2): The number of nodes in each tree of F_i is at most $n/2^i$.

REMARKS

R25: Invariant (1) should be interpreted as follows. Let (u, v) be a non-tree edge of level $\ell(u, v)$, and let $u \cdots v$ be the unique path between u and v in F (such a path exists since F is a spanning forest of G). Let e be any edge in $u \cdots v$ and let $\ell(e)$ be its level. Due to invariant (1), $\ell(e) \geq \ell(u, v)$. Since this holds for each edge in the path, and by construction $F_{\ell(u,v)}$ contains all the tree edges of level $\geq \ell(u, v)$, the entire path is contained in $F_{\ell(u,v)}$, i.e., u and v are connected in $F_{\ell(u,v)}$.

R26: Invariant (2) implies that the maximum number of levels is $L \leq \lfloor \log_2 n \rfloor$.

FACTS

F16: When a new edge is inserted, it is given level 0. Its level can be then increased at most $\lfloor \log_2 n \rfloor$ times as a consequence of edge deletions.

F17: When a tree edge $e = (v, w)$ of level $\ell(e)$ is deleted, the algorithm looks for a replacement edge at the highest possible level, if any. Due to invariant (1), such a replacement edge has level $\ell \leq \ell(e)$. Hence, a replacement subroutine `Replace((u, w), ℓ(e))` is called with parameters e and $\ell(e)$. We now sketch the operations performed by this subroutine.

F18: `Replace((u, w), ℓ)` finds a replacement edge of the highest level $\leq \ell$, if any. If such a replacement does not exist in level ℓ, we have two cases: if $\ell > 0$, we recurse on level $\ell - 1$; otherwise, $\ell = 0$, and we can conclude that the deletion of (v, w) disconnects v and w in G.

F19: During the search at level ℓ, suitably chosen tree and non-tree edges may be promoted at higher levels as follows. Let T_v and T_w be the trees of forest F_ℓ obtained after deleting (v, w) and let, w.l.o.g., T_v be smaller than T_w. Then T_v contains at most $n/2^{\ell+1}$ vertices, since $T_v \cup T_w \cup \{(v, w)\}$ was a tree at level ℓ and due to invariant (2). Thus, edges in T_v of level ℓ can be promoted at level $\ell+1$ by maintaining the invariants. Non-tree edges incident to T_v are finally visited one by one: if an edge does connect T_v and T_w, a replacement edge has been found and the search stops; otherwise its level is increased by 1.

F20: We maintain an ET-tree, as described before, for each tree of each forest. Consequently, all the basic operations needed to implement edge insertions and deletions can be supported in $O(\log n)$ time.

F21: [HoDeTh01] A dynamic graph G with n vertices can be maintained upon insertions and deletions of edges using $O(\log^2 n)$ amortized time per update and answering connectivity queries in $O(\log n / \log \log n)$ worst-case running time.

REMARKS

R27: In addition to inserting and deleting edges from a forest, ET-trees must also support operations such as finding the tree of a forest that contains a given vertex, computing the size of a tree, and, more importantly, finding tree edges of level ℓ in T_v and non-tree edges of level ℓ incident to T_v. This can be done by augmenting the ET-trees with a constant amount of information per node: see [HoDeTh01] for details.

R28: Using an amortization argument based on level changes, the claimed $O(\log^2 n)$ bound on the update time can be finally proved. Namely, inserting an edge costs $O(\log n)$, as well as increasing its level. Since this can happen $O(\log n)$ times, the total amortized insertion cost, inclusive of level increases, is $O(\log^2 n)$. With respect to edge deletions, cutting and linking $O(\log n)$ forest has a total cost $O(\log^2 n)$; moreover, there are $O(\log n)$ recursive calls to `Replace`, each of cost $O(\log n)$ plus the cost amortized over level increases. The ET-trees over $F_0 = F$ allows answering connectivity queries in $O(\log n)$ worst-case time. As shown in [HoDeTh01], this can be reduced to $O(\log n / \log \log n)$ by using a $\Theta(\log n)$-ary version of ET-trees.

Minimum Spanning Trees

A few simple changes to the previous connectivity algorithm are sufficient to maintain a minimum spanning forest of a weighted undirected graph upon deletions of edges [HoDeTh01]. A general reduction from [HeKi01] can then be applied to make the deletions-only algorithm fully dynamic.

Decremental Minimum Spanning Tree

APPROACH

In addition to starting from a *minimum* spanning forest, the only change concerns the function `Replace`, which should be implemented so as to consider candidate replacement edges of level ℓ in order of increasing weight, and not in arbitrary order. To do so, the ET-trees can be augmented so that each node maintains the minimum weight of a non-tree edge incident to the Euler tour segment below it. All the operations can still be supported in $O(\log n)$ time, yielding the same time bounds as for connectivity.

We now discuss the correctness of the algorithm. In particular, function `Replace` returns a replacement edge of minimum weight on the highest possible level: it is not immediate that such a replacement edge has the minimum weight among all levels. This can be proved by first showing that the following invariant, proved in [HoDeTh01], is maintained by the algorithm.

INVARIANT

Invariant (3): Every cycle $\mathcal{C}$ has a non-tree edge of maximum weight and minimum level among all the edges in $\mathcal{C}$.

FACTS

F22: Invariant (3) can be used to prove that, among all the replacement edges, the lightest edge is on the maximum level. Let e_1 and e_2 be two replacement edges with $w(e_1) < w(e_2)$, and let $\mathcal{C}_i$ be the cycle induced by e_i in F, $i = 1, 2$. Since F is a minimum spanning forest, e_i has maximum weight among all the edges in $\mathcal{C}_i$. In particular, since by hypothesis $w(e_1) < w(e_2)$, e_2 is also the heaviest edge in cycle $\mathcal{C} = (\mathcal{C}_1 \cup \mathcal{C}_2) \setminus (\mathcal{C}_1 \cap \mathcal{C}_2)$. Thanks to Invariant (3), e_2 has minimum level in $\mathcal{C}$, proving that $\ell(e_2) \leq \ell(e_1)$. Thus, considering non-tree edges from higher to lower levels is correct.

F23: [HoDeTh01] There exists a deletions-only minimum spanning forest algorithm that can be initialized on a graph with n vertices and m edges and supports any sequence of edge deletions in $O(m \log^2 n)$ total time.

Fully Dynamic Minimum Spanning Tree

The reduction used to obtain a fully dynamic algorithm is a slight generalization of the construction proposed by Henzinger and King [HeKi01] and works as follows.

FACT

F24: [HeKi01, HoDeTh01] Suppose we have a deletions-only minimum spanning tree algorithm that, for any k and l, can be initialized on a graph with k vertices and l edges and supports any sequence of $\Omega(l)$ deletions in total time $O(l \cdot t(k, l))$, where t is a non-decreasing function. Then there exists a fully dynamic minimum spanning tree algorithm for a graph with n nodes starting with no edges, that, for m edges, supports updates in time

$$O\left(\log^3 n + \sum_{i=1}^{3+\log_2 m} \sum_{j=1}^{i} t(min\{n, 2^j\}, 2^j) \right)$$

REMARKS

R29: See [HeKi01] and [HoDeTh01] for a description of the construction that proves Fact F24.

R30: From Fact F23 we get $t(k, l) = O(\log^2 k)$. Hence, by combining Fact F23 and Fact F24, we get the claimed result.

FACT

F25: [HoDeTh01] There exists a fully dynamic minimum spanning forest algorithm that, for a graph with n vertices, starting with no edges, maintains a minimum spanning forest in $O(\log^4 n)$ amortized time per edge insertion or deletion.

10.2.3   Dynamic Problems on Directed Graphs

In this part we survey the known results for dynamic problems on directed graphs. In particular, we focus on two of the most fundamental problems: transitive closure and shortest paths. These problems play a crucial role in many applications, including network optimization and routing, traffic information systems, databases, compilers, garbage collection, interactive verification systems, industrial robotics, dataflow analysis, and document formatting.

We first present general techniques and tools used in designing dynamic path problems on directed graphs, and then we address the results for dynamic transitive closure and dynamic shortest paths. In the first problem, the goal is to maintain reachability information in a directed graph subject to insertions and deletions of edges. The fastest known algorithm supports graph updates in quadratic time and reachability queries in constant time [DeIt08]. In the second problem, we wish to maintain information about shortest paths in a directed graph subject to insertion and deletion of edges, or updates of edge weights. Similarly to dynamic transitive closure, this can be done in near-quadratic time per update and optimal time per query [DeIt04, Th04].

General Techniques for Directed Graphs

In this subsection we discuss the main techniques used to solve dynamic path problems on directed graphs. We first address combinatorial and algebraic properties, and then we consider some efficient data structures, which are used as building blocks in designing dynamic algorithms for transitive closure and shortest paths.

Path Problems and Kleene Closures

Path problems such as transitive closure and shortest paths are tightly related to matrix sum and matrix multiplication over a closed semiring (see [CoLeRiSt01] for more details).

NOTATION: The usual sum and multiplication operations over Boolean matrices are denoted by $+$ and $\cdot$, respectively.

NOTATION: Given two real-valued matrices A and B, $C = A \odot B$ is the matrix product such that $C[x, y] = \min_{1 \leq z \leq n}\{A[x, z] + B[z, y]\}$, and $D = A \oplus B$ is the matrix sum such that $D[x, y] = \min\{A[x, y], B[x, y]\}$.

NOTATION: We also denote by AB the product $A \odot B$ and by $AB[x, y]$ entry (x, y) of matrix AB.

FACTS

F26: Let $G = (V, E)$ be a directed graph and let $TC(G)$ be the (reflexive) transitive closure of G. If X is the Boolean adjacency matrix of G, then the Boolean adjacency matrix of $TC(G)$ is the Kleene closure of X on the $\{+, \cdot, 0, 1\}$ Boolean semiring:

$$X^* = \sum_{i=0}^{n-1} X^i$$

F27: Let $G = (V, E)$ be a weighted directed graph with no negative-length cycles. If X is a weight matrix such that $X[x, y]$ is the weight of edge (x, y) in G, then the distance matrix of G is the Kleene closure of X on the $\{\oplus, \odot, \mathcal{R}\}$ semiring:

$$X^* = \bigoplus_{i=0}^{n-1} X^i$$

The next two facts recall two well-known methods for computing the Kleene closure X^* of an $n \times n$ matrix X.

Logarithmic Decomposition. A simple method to compute X^*, based on repeated squaring, requires $O(n^\mu \cdot \log n)$ worst-case time, where $O(n^\mu)$ is the time required for computing the product of two matrices over a closed semiring.

F28: This method performs $\log_2 n$ sums and products of the form $X_{i+1} = X_i + X_i^2$, where $X = X_0$ and $X^* = X_{\log_2 n}$.

Recursive Decomposition. Another method, due to Munro [Mu71], is based on a Divide and Conquer strategy and computes X^* in $O(n^\mu)$ worst-case time.

F29: Munro observed that, if we partition a matrix X into four submatrices A, B, D, C of size $n/2 \times n/2$ (considered in clockwise order), and the closure X^* similarly into four submatrices E, F, H, G of size $n/2 \times n/2$, then X^* is definable recursively according to the following equations:

$$E = (A + BD^*C)^*$$
$$F = EBD^*$$
$$G = D^*CE$$
$$H = D^* + D^*CEBD^*$$

Surprisingly, using this decomposition the cost of computing X^* starting from X is asymptotically the same as the cost of multiplying two matrices over a closed semiring.

Locally Shortest Paths

Some combinatorial properties of shortest paths in directed graphs have been recently discovered by Demetrescu and Italiano [DeIt04]. In particular, we consider shortest paths as a special case of a broader class of paths called *locally shortest paths*. To characterize how locally shortest paths change in a fully dynamic graph, we consider the notions of *historical path* and *locally historical path*.

DEFINITIONS

D21: A path π in a graph is *locally shortest* if every proper subpath of π is a shortest path.

D22: A ***historical path*** is a path that has been a shortest path at some point during the sequence of updates, and none of its edges has been updated since then.

Using this notion we can define a superset of locally shortest paths that are called *locally historical paths*.

D23: A path π in a graph is *locally historical* if every proper subpath of π is a historical path.

REMARKS

R31: As an alternative equivalent definition, a path π_{xy} is locally shortest in a graph if every edge (u, v) in π_{xy} satisfies the relation $d_{xu} + w_{uv} + d_{vy} = w(\pi_{xy})$, where d_{xy} denotes the distance between vertex x and vertex y in the graph, w_{uv} is the weight of edge (u, v), and $w(\pi_{xy})$ is the weight of π_{xy}.

R32: It is not difficult to prove that the amortized number of locally shortest paths that may change due to an edge weight update is $O(n^2)$ if updates are partially dynamic, i.e., increase-only or decrease-only.

FACTS

F30: [DeIt04] If we denote by SP, LSP, HP, and LHP, respectively, the sets of shortest paths, locally shortest paths, historical paths, and locally historical paths in a graph, then at any time the following inclusions hold: $SP \subseteq LSP \subseteq LHP$ and $SP \subseteq HP \subseteq LHP$.

F31: [DeIt04] Let G be a graph subject to a sequence of update operations. If at any time throughout the sequence of updates there are at most $O(z)$ historical paths between each pair of vertices, then the amortized number of paths that become locally historical at each update is $O(zn^2)$.

REMARKS

R33: Locally historical paths exhibit strong combinatorial properties in graphs subject to (fully) dynamic updates. In particular, it is possible to prove that the number of paths that become locally historical in a graph at each edge weight update depends on the number of historical paths in the graph.

R34: To keep changes in locally historical paths small, it is then desirable to have as few historical paths as possible. Indeed, it is possible to transform every update sequence into a slightly longer equivalent sequence that generates only a few historical paths. In particular, there exists a simple *smoothing* strategy that, given any update sequence Σ of length k, produces an operationally equivalent sequence $F(\Sigma)$ of length $O(k \log k)$ that yields only $O(\log k)$ historical paths between each pair of vertices in the graph. We refer the interested reader to [DeIt04] for a detailed description of this smoothing strategy. According to Fact F31, this technique implies that only $O(n^2 \log k)$ locally historical paths change at each edge weight update in the smoothed sequence $F(\Sigma)$.

R35: As elaborated in [DeIt04], locally historical paths can be maintained very efficiently. Since by Fact F30 locally historical paths include shortest paths, this yields the fastest known algorithm for fully dynamic all pairs shortest paths.

Long Paths Property

If we pick a subset S of vertices at random from a graph G, then a sufficiently long path will intersect S with high probability. This can be very useful in finding a long path by using short searches. This property has been rediscovered many times and it has been exploited to design efficient algorithms for transitive closure and shortest paths (see e.g., [DeIt06, Ki99, UlYa91, Zw98]).

FACT

F32: (Ullman and Yannakakis [UlYa91]) Let $S \subseteq V$ be a set of vertices chosen uniformly at random. Then the probability that a given simple path has a sequence of more than $(cn \log n)/|S|$ vertices, none of which is from S, for any $c > 0$, is, for sufficiently large n, bounded by $2^{-\alpha c}$ for some positive α.

REMARK

R36: As shown in [Zw98], it is possible to choose set S deterministically by a reduction to a hitting set problem [Ch79, Lo75]. A similar technique has also been used in [Ki99].

Reachability Trees

A special tree data structure has been widely used to solve dynamic path problems on directed graphs. The first appearance of this tool dates back to 1981, when Even and Shiloach showed how to maintain a breadth-first tree of an undirected graph under any sequence of edge deletions [EvSh81]; they used this as a kernel for decremental connectivity on undirected graphs. Later on, Henzinger and King [HeKi99] showed how to adapt this data structure to fully dynamic transitive closure in directed graphs. King [Ki99] designed an extension of this tree data structure to weighted directed graphs for solving fully dynamic all pairs shortest paths.

PROBLEM

In the unweighted directed version, the goal is to maintain information about breadth-first search (BFS) on a directed graph G undergoing deletions of edges. In particular, in the context of dynamic path problems, we are interested in maintaining BFS trees of depth up to d, with $d \leq n$. Given a directed graph $G = (V, E)$ and a vertex $r \in V$, we would like to support any intermixed sequence of the following operations:

Delete(x, y): delete edge (x, y) from G.

Level(u): return the level of vertex u in the BFS tree of depth d rooted at r (return $+\infty$ if u is not reachable from r within distance d).

FACT

F33: [Ki99] Maintaining BFS levels up to depth d from a given root requires $O(md)$ time in the worst case throughout any sequence of edge deletions in a directed graph with m initial edges.

REMARKS

R37: Fact F33 means that maintaining BFS levels requires d times the time needed for constructing them. Since $d \leq n$, we obtain a total bound of $O(mn)$ if there are no limits on the depth of the BFS levels.

R38: As was shown in [HeKi99, Ki99], it is possible to extend the BFS data structure presented in this section to deal with weighted directed graphs. In this case, a shortest path tree is maintained in place of BFS levels: after each edge deletion or edge weight increase, the tree is reconnected by essentially mimicking Dijkstra's algorithm rather than BFS. Details can be found in [Ki99].

Matrix Data Structures

We now consider matrix data structures for keeping information about paths in dynamic directed graphs. As we have seen above (Path Problems and Kleene Closures), Kleene closures can be constructed by evaluating polynomials over matrices. It is therefore natural to consider data structures for maintaining polynomials of matrices subject to updates of entries, like the one introduced in [DeIt08].

PROBLEM

In the case of Boolean matrices, the problem can be stated as follows. Let P be a polynomial over $n \times n$ Boolean matrices with constant degree, constant number of terms, and variables $X_1 \ldots X_k$. We wish to maintain a data structure for P subject to any intermixed sequence of update and query operations of the following kind:

`SetRow`$(i, \Delta X, X_b)$: sets to one the entries in the i-th row of variable X_b of polynomial P corresponding to one-valued entries in the i-th row of matrix ΔX.

`SetCol`$(i, \Delta X, X_b)$: sets to one the entries in the i-th column of variable X_b of polynomial P corresponding to one-valued entries in the i-th column of matrix ΔX.

`Reset`$(\Delta X, X_b)$: resets to zero the entries of variable X_b of polynomial P corresponding to one-valued entries in matrix ΔX.

`Lookup`(): returns the maintained value of P.

We add to the previous four operations a further update operation especially designed for maintaining path problems:

`LazySet`$(\Delta X, X_b)$: sets to 1 the entries of variable X_b of P corresponding to one-valued entries in matrix ΔX. However, the maintained value of P might not be immediately affected by this operation.

REMARK

R39: Let C_P be the correct value of P that we would have by recomputing it from scratch after each update, and let M_P be the actual value that we maintain. If no `LazySet` operation is ever performed, then always $M_P = C_P$. Otherwise, M_P is not necessarily equal to C_P, and we guarantee the following weaker property on M_P: if $C_P[u, v]$ flips from 0 to 1 due to a `SetRow`/`SetCol` operation on a variable X_b, then $M_P[u, v]$ flips from 0 to 1 as well. This means that `SetRow` and `SetCol` always correctly reveal new 1's in the maintained value of P, possibly taking into account the 1's inserted through previous `LazySet` operations. This property is crucial for dynamic path problems.

FACTS

F34: [DeIt08] Let P be a polynomial with constant degree of matrices over the Boolean semiring. Any `SetRow`, `SetCol`, `LazySet`, and `Reset` operation on a polynomial P can be supported in $O(n^2)$ amortized time. `Lookup` queries are answered in optimal time.

F35: [DeIt06] Let P be a polynomial with constant degree of matrices over the $\{\min, +\}$ semiring. Any `SetRow`, `SetCol`, `LazySet`, and `Reset` operation on variables of P can be supported in $O(D \cdot n^2)$ amortized time, where D is the maximum number of different values assumed by entries of variables during the sequence of operations. `Lookup` queries are answered in optimal time.

Dynamic Transitive Closure

In this subsection we survey the best known algorithms for fully dynamic transitive closure. Given a directed graph G with n vertices and m edges, the problem consists of supporting any intermixed sequence of operations of the following kind:
`Insert`(u, v): insert edge (u, v) in G;
`Delete`(u, v): delete edge (u, v) from G;
`Query`(x, y): answer a reachability query by returning "yes" if there is a path from vertex x to vertex y in G, and "no" otherwise.

FACTS

F36: A simple-minded solution to this problem consists of maintaining the graph under insertions and deletions, searching if y is reachable from x at any query operation. This yields $O(1)$ time per update (`Insert` and `Delete`), and $O(m)$ time per query, where m is the current number of edges in the maintained graph.

F37: Another simple-minded solution would be to maintain the Kleene closure of the adjacency matrix of the graph, rebuilding it from scratch after each update operation. Using the recursive decomposition of Munro [Mu71] discussed in the section on Path Problems and Kleene Closures and fast matrix multiplication, this takes constant time per reachability query and $O(n^\omega)$ time per update, where $\omega < 2.3727$ is the current best exponent for matrix multiplication [CoWi90, Va12].

REMARKS

R40: Despite many years of research in this topic, no better solution to this problem was known until 1995, when Henzinger and King [HeKi99] proposed a randomized Monte Carlo algorithm with one-sided error supporting a query time of $O(n/\log n)$ and an amortized update time of $O(n\hat{m}^{0.58}\log^2 n)$, where $\hat{m}$ is the average number of edges in the graph throughout the whole update sequence. Since $\hat{m}$ can be as high as $O(n^2)$, their update time is $O(n^{2.16}\log^2 n)$.

R41: Khanna, Motwani and Wilson [KhMoWi96] proved that, when a lookahead of $\Theta(n^{0.18})$ in the updates is permitted, a deterministic update bound of $O(n^{2.18})$ can be achieved.

R42: King and Sagert [KiSa02] showed how to support queries in $O(1)$ time and updates in $O(n^{2.26})$ time for general directed graphs and $O(n^2)$ time for directed acyclic graphs; their algorithm is randomized with one-sided error. These bounds were further improved by King [Ki99], who exhibited a deterministic algorithm on general digraphs with $O(1)$ query time and $O(n^2 \log n)$ amortized time per update operations, where updates are insertions of a set of edges incident to the same vertex and deletions of an arbitrary subset of edges.

R43: Using a completely different approach, Demetrescu and Italiano [DeIt08] obtained a deterministic fully dynamic algorithm that achieves $O(n^2)$ amortized time per update for general directed graphs. Sankowski [Sa04] showed how to make the amortized bound worst-case.

R44: We note that each update might change a portion of the transitive closure as large as $\Omega(n^2)$. Thus, if the transitive closure has to be maintained explicitly after each update so that queries can be answered with one lookup, $O(n^2)$ is the best update bound one could hope for.

R45: By combining in a novel way techniques of Italiano [It86, It88], King [Ki99], King and Thorup [KiTh01] and Frigioni et al. [FrMiZa01], Roditty [Ro08] showed how to reduce from $O(n^3)$ to $O(mn)$ the preprocessing time required by the algorithm of Demetrescu and Italiano [DeIt08].

R46: If one is willing to pay more for queries, Demetrescu and Italiano [DeIt05] showed how to break the $O(n^2)$ barrier on the single-operation complexity of fully dynamic transitive closure: building on a previous path counting technique introduced by King and Sagert [KiSa02], they devised a randomized algorithm with one-sided error for directed acyclic graphs that achieves $O(n^{1.575})$ worst-case time per update and $O(n^{0.575})$ worst-case time per query.

R47: The bounds of Demetrescu and Italiano [DeIt05] were extended to general directed graphs by Sankowski [Sa04]. Sankowski [Sa04] also showed how to achieve $O(n^{1.495})$ worst-case time both per update and per query.

R48: Further trade-offs between queries and updates were given by Roditty and Zwick [RoZw08]. They presented one deterministic algorithm, with amortized update time of $O(m\sqrt{n})$ and worst-case query time of $O(\sqrt{n})$, and one randomized algorithm with amortized update time of $O(m^{0.58}n)$ and worst-case query time of $O(m^{0.43})$. Both algorithms are competitive for sparse graphs.

King's $O(n^2 \log n)$ Update Algorithm

King [Ki99] devised the first deterministic near-quadratic update algorithm for fully dynamic transitive closure. The algorithm is based on the tree data structure considered in §10.2.4 (Reachability Trees) and on the logarithmic decomposition discussed in §10.2.4 (Path Problems and Kleene Closures). It maintains explicitly the transitive closure of a graph G in $O(n^2 \log n)$ amortized time per update, and supports inserting and deleting several edges of the graph with just one operation. Insertion of a bunch of edges incident to a vertex and deletion of any subset of edges in the graph require asymptotically the same time of inserting/deleting just one edge.

APPROACH

The algorithm maintains $\log n + 1$ levels: level i, $0 \le i \le \log n$, maintains a graph G_i whose edges represent paths of length up to 2^i in the original graph G. Thus, $G_0 = G$ and $G_{\log n}$ is the transitive closure of G.

FACTS

F38: Each level i is built on top of the previous level $i-1$ by keeping two trees of depth ≤ 2 rooted at each vertex v of G: an out-tree $OUT_i(v)$ maintaining vertices reachable from v by traversing at most two edges in G_{i-1}, and an in-tree $IN_i(v)$ maintaining vertices that reach v by traversing at most two edges in G_{i-1}. An edge (x, y) will be in G_i if and only if $x \in IN_i(v)$ and $y \in OUT_i(v)$ for some v.

F39: The $2 \log n$ trees $IN_i(v)$ and $OUT_i(v)$ are maintained with instances of the BFS tree data structure considered in the section on Reachability Trees.

F40: To update the levels after an insertion of edges around a vertex v in G, the algorithm simply rebuilds $IN_i(v)$ and $OUT_i(v)$ for each i, $1 \le i \le \log n$, while other trees are not touched. This means that some trees might not be up to date after an insertion operation. Nevertheless, any path in G is represented in at least the in/out trees rooted at the latest updated vertex in the path, so the reachability information is correctly maintained. This idea is the key ingredient of King's algorithm.

F41: When an edge is deleted from G_i, it is also deleted from any data structures $IN_i(v)$ and $OUT_i(v)$ that contain it. For details, see [Ki99].

Demetrescu and Italiano's $O(n^2)$ Update Algorithm

The algorithm by Demetrescu and Italiano [DeIt08] is based on the matrix data structure considered in the section on Matrix Data Structures and on the recursive decomposition discussed in the section on Path Problems and Kleene Closures. It maintains explicitly the transitive closure of a graph in $O(n^2)$ amortized time per update, supporting the same generalized update operations of King's algorithm, i.e., insertion of a bunch of edges incident to a vertex and deletion of any subset of edges in the graph with just one operation. This is the best known update bound for fully dynamic transitive closure with constant query time.

APPROACH

The algorithm maintains the Kleene closure X^* of the $n \times n$ adjacency matrix X of the graph as the sum of two matrices X_1 and X_2.

NOTATION

Let V_1 be the subset of vertices of the graph corresponding to the first half of indices of X, and let V_2 contain the remaining vertices.

FACTS

F42: Both matrices X_1 and X_2 are defined according to Munro's equations given in the section on Path Problems and Kleene Closures, but in such a way that paths appearing due to an insertion of edges around a vertex in V_1 are correctly recorded in X_1, while paths that appear due to an insertion of edges around a vertex in V_2 are correctly recorded in X_2. Thus, neither X_1 nor X_2 encodes complete information about X^*, but their sum does.

F43: In more detail, assuming that X is decomposed in sub-matrices A, B, C, D as explained in the section on Path Problems and Kleene Closures, and that X_1, and X_2 are similarly decomposed in sub-matrices E_1, F_1, G_1, H_1 and E_2, F_2, G_2, H_2, the algorithm maintains X_1 and X_2 with the following 8 polynomials using the data structure discussed in the section on Matrix Data Structures:

$$
\begin{aligned}
Q &= A + BP^2C & E_2 &= E_1BH_2^2CE_1 \\
F_1 &= E_1^2BP & F_2 &= E_1BH_2^2 \\
G_1 &= PCE_1^2 & G_2 &= H_2^2CE_1 \\
H_1 &= PCE_1^2BP & R &= D + CE_1^2B
\end{aligned}
$$

where $P = D^*$, $E_1 = Q^*$, and $H_2 = R^*$ are Kleene closures maintained recursively as smaller instances of the problem of size $n/2 \times n/2$.

F44: To support an insertion of edges around a vertex in V_1, strict updates are performed on polynomials Q, F_1, G_1, and H_1 using `SetRow` and `SetCol`, while E_2, F_2, G_2, and R are updated with `LazySet`.

F45: Insertions around V_2 are performed symmetrically, while deletions are supported via `Reset` operations on each polynomial in the recursive decomposition.

F46: Finally, P, E_1, and H_2 are updated recursively. The low-level details of the method appear in [DeIt08].

Dynamic Shortest Paths

In this subsection we survey the best known algorithms for fully dynamic all pairs shortest paths (in short APSP). Given a weighted directed graph G with n vertices and m edges, the problem consists of supporting any intermixed sequence of operations of the following kind:
`Update`(u, v, w): updates the weight of edge (u, v) in G to the new value w (if $w = +\infty$ this corresponds to edge deletion);
`Query`(x, y): returns the distance from vertex x to vertex y in G, or $+\infty$ if no path between them exists.

NOTATION
In the the following, we use $\widetilde{O}(f(n))$ to denote $O(f(n) \operatorname{polylog}(n))$.

REMARKS

R49: The dynamic maintenance of shortest paths has a remarkably long history, as the first papers date back to 35 years ago [Lo67, Mu67, Ro68]. Since then, many dynamic shortest paths algorithms have been proposed (see, e.g., [EvGa85, FrMaNa98, FrMaNa00, RaRe96a, RaRe96b, Ro85]), but their running times in the worst case were comparable to recomputing APSP from scratch.

R50: The first dynamic shortest path algorithms which are provably faster than re-computing APSP from scratch only worked on graphs with small integer weights.

R51: In particular, Ausiello et al. [AuItMaNa91] proposed a decrease-only shortest path algorithm for directed graphs having positive integer weights less than C: the amortized running time of their algorithm is $O(Cn \log n)$ per edge insertion.

R52: Henzinger et al. [HeKiRaSu97] designed a fully dynamic algorithm for APSP on planar graphs with integer weights, with a running time of $O(n^{4/3} \log(nC))$ per operation.

R53: This bound has been improved by Fakcharoemphol and Rao in [FaRa06], who designed a fully dynamic algorithm for single-source shortest paths in planar directed graphs that supports both queries and edge weight updates in $O(n^{4/5} \log^{13/5} n)$ amortized time per edge operation.

R54: The first big step on general graphs and integer weights was made by King [Ki99], who presented a fully dynamic algorithm for maintaining all pairs shortest paths in directed graphs with positive integer weights less than C: the running time of her algorithm is $O(n^{2.5} \sqrt{C \log n})$ per update.

R55: Demetrescu and Italiano [DeIt06] gave the first algorithm for fully dynamic APSP on general directed graphs with real weights assuming that each edge weight can attain a limited number S of different *real* values throughout the sequence of updates. In particular, the algorithm supports each update in $O(n^{2.5} \sqrt{S \log^3 n})$ amortized time and each query in $O(1)$ worst-case time.

R56: The same authors discovered the first algorithm that solves the fully dynamic all pairs shortest paths problem in its generality [DeIt04]. The algorithm maintains explicitly information about shortest paths, supporting any edge weight update in $O(n^2 \log^3 n)$ amortized time per operation in directed graphs with non-negative real edge weights. Distance queries are answered with one lookup and actual shortest paths can be reconstructed in optimal time.

R57: Using the same approach as Demetrescu and Italiano [DeIt04], but with a different smoothing strategy, Thorup [Th04] showed how to improve slightly to $O(n^2(\log n + \log^2((m+n)/n)))$ the amortized bound per update for the fully dynamic APSP problem, while still maintaining constant query times. Thorup's algorithm works with negative weights as well.

R58: We note that each update might change a portion of the distance matrix as large as $\Omega(n^2)$. Thus, if the distance matrix has to be maintained explicitly after each update so that queries can be answered with one lookup, $O(n^2)$ is the best update bound one could hope for.

R59: The currently best worst-case bound per update for the fully dynamic APSP problem is much higher than known amortized bounds. Indeed Thorup [Th05] has shown a worst-case bound per update of $\widetilde{O}(n^{2.75})$.

R60: In the special case of unweighted graphs, Roditty and Zwick [RoZw11] have shown how to support updates in $\widetilde{O}(m\sqrt{n})$ amortized time and queries in $O(n^{3/4})$ worst-case time. Those bounds are competitive for sparse graphs.

R61: Other deletions-only algorithms for APSP, in the simpler case of unweighted graphs, are presented in [BaHaSe07].

R62: As shown by Sankowski [Sa05], dynamic shortest distances in unweighted graphs can be maintained in $O(n^{1.932})$ randomized time per update and $O(n^{1.288})$ randomized time per query.

King's $O(n^{2.5}\sqrt{C \log n})$ Update Algorithm

The dynamic shortest paths algorithm by King [Ki99] is based on the long paths property discussed in the section on Long Paths Property and on the tree data structure of the section on Reachability Trees. Similarly to the transitive closure algorithms previously described, generalized update operations are supported within the same bounds, i.e., insertion (or weight decrease) of a bunch of edges incident to a vertex, and deletion (or weight increase) of any subset of edges in the graph with just one operation.

APPROACH

The main idea of the algorithm is to maintain dynamically all pairs shortest paths up to a distance d, and to recompute longer shortest paths from scratch at each update by stitching together shortest paths of length $\leq d$. For the sake of simplicity, we only consider the case of unweighted graphs: an extension to deal with positive integer weights less than C is described in [Ki99].

FACTS

F47: To maintain shortest paths up to distance d, similarly to the transitive closure algorithm by King described in §10.2.5, the algorithm keeps a pair of in/out shortest paths trees $IN(v)$ and $OUT(v)$ of depth $\leq d$ rooted at each vertex v. Trees $IN(v)$ and $OUT(v)$ are maintained with the decremental data structure mentioned in §10.2.4 (Reachability Trees). It is easy to prove that, if the distance d_{xy} between any pair of vertices x and y is at most d, then d_{xy} is equal to the minimum of $d_{xv} + d_{vy}$ over all vertices v such that $x \in IN(v)$ and $y \in OUT(v)$. To support updates, insertions of edges around a vertex v are handled by rebuilding only $IN(v)$ and $OUT(v)$, while edge deletions are performed via operations on any trees that contain them. The amortized cost of such updates is $O(n^2 d)$ per operation.

F48: To maintain shortest paths longer than d, the algorithm exploits the long paths property of Fact F30: in particular, it hinges on the observation that, if H is a random subset of $\Theta((n \log n)/d)$ vertices in the graph, then the probability of finding more than d consecutive vertices in a path, none of which is from H, is very small. Thus, if we look at vertices in H as "hubs," then any shortest path from x to y of length $\geq d$ can be obtained by stitching together shortest subpaths of length $\leq d$ that first go from x to a vertex in H, then jump between vertices in H, and eventually reach y from a vertex in H. This can be done by first computing shortest paths only between vertices in H using any cubic-time static all-pairs shortest paths algorithm, and then by extending them at both endpoints with shortest paths of length $\leq d$ to reach all other vertices. This stitching operation requires $O(n^2|H|) = O((n^3 \log n)/d)$ time.

F49: Choosing $d = \sqrt{n \log n}$ yields an $O(n^{2.5}\sqrt{\log n})$ amortized update time. As mentioned in the section on the Long Paths Property, since H can be computed deterministically, the algorithm can be derandomized. For further details, see [Ki99].

Demetrescu and Italiano's $O(n^2 \log^3 n)$ Update Algorithm

Demetrescu and Italiano [DeIt04] devised the first deterministic near-quadratic update algorithm for fully dynamic all-pairs shortest paths. This algorithm is also the first solution to the problem in its generality. It is based on the notions of locally shortest path, locally historical path, and historical paths in a graph subject to a sequence of updates, as discussed in the section on Locally Shortest Paths.

APPROACH

The main idea is to maintain dynamically the locally historical paths of the graph in a data structure. Since by Fact F30 shortest paths are locally historical, this guarantees that information about shortest paths is maintained as well.

FACTS

F50:   To support an edge weight update operation, the algorithm implements the smoothing strategy mentioned in the section on Locally Shortest Paths and works in two phases. It first removes from the data structure all maintained paths that contain the updated edge: this is correct since historical paths, in view of their definition, are immediately invalidated as soon as they are touched by an update. This means that also locally historical paths that contain them are invalidated and have to be removed from the data structure. As a second phase, the algorithm runs an all-pairs modification of Dijkstra's algorithm [Di59], where at each step a shortest path with minimum weight is extracted from a priority queue and it is combined with existing historical paths to form new locally historical paths. At the end of this phase, paths that become locally historical after the update are correctly inserted in the data structure.

F51:   The update algorithm spends $O(\log n)$ time for each of the $O(zn^2)$ new locally historical path (see Fact F31). Since the smoothing strategy lets $z = O(\log n)$ and increases the length of the sequence of updates by an additional $O(\log n)$ factor, this yields $O(n^2 \log^3 n)$ amortized time per update. For further details, see [DeIt04].

10.2.4   Research Issues

In this work we have surveyed the algorithmic techniques underlying the fastest known dynamic graph algorithms for several problems, both on undirected and on directed graphs. Most of the algorithms that we have presented achieve bounds that are close to optimum. In particular, we have presented fully dynamic algorithms with polylogarithmic amortized time bounds for connectivity and minimum spanning trees [HoDeTh01] on undirected graphs. It remains an interesting open problem to show whether polylogarithmic update bounds can be achieved also in the worst case: we recall that for both problems the current best worst-case bound is $O(\sqrt{n})$ per update, and it is obtained with the sparsification technique [EpGaItNi97] described in Section 10.2.2.

For directed graphs, we have shown how to achieve constant-time query bounds and nearly-quadratic update bounds for transitive closure and all pairs shortest paths. These bounds are close to optimal in the sense that one update can make as many as $\Omega(n^2)$ changes to the transitive closure and to the all-pairs shortest paths matrices. While the quadratic bounds for dynamic transitive closure are worst-case, the nearly quadratic bounds for dynamic shortest paths are amortized, and the best worst-case update bound in this case is only $\widetilde{O}(n^{2.75})$. Can this worst-case bound be improved? Furthermore, if the problem is just to maintain reachability or shortest paths between two fixed vertices s and t, no solution better than the static is known. If one is willing to pay more for queries, Demetrescu and Italiano [DeIt05] have shown how to break the $O(n^2)$ barrier on the single-operation complexity of fully dynamic transitive closure for directed acyclic graphs. It remains an interesting open problem to show whether effective query/update tradeoffs can be achieved for shortest paths problems in general graphs.

Finally, dynamic algorithms for other fundamental problems such as matching and flow problems deserve further investigation.

Further Information

Research on dynamic graph algorithms is published in many computer science journals, including *Algorithmica*, *Journal of ACM*, *ACM Transactions on Algorithms*, *Journal of Algorithms*, *Journal of Computer and System Science*, *SIAM Journal on Computing* and *Theoretical Computer Science*. Work on this area is published also in the proceedings of general theoretical computer science conferences, such as the *ACM Symposium on Theory of Computing* (STOC), the *IEEE Symposium on Foundations of Computer Science* (FOCS) and the *International Colloquium on Automata, Languages and Programming* (ICALP). More specialized conferences devoted exclusively to algorithms are the *ACM–SIAM Symposium on Discrete Algorithms* (SODA), and the *European Symposium on Algorithms* (ESA).

Acknowledgments

This work has been supported in part by the Italian Ministry of University and Scientific Research (Project "ALGODeep: Algorithmic Challenges for Data-intensive Processing on Emerging Computing Platforms").

References

[AlHoDeTh97] S. Alstrup, J. Holm, K. de Lichtenberg, and M. Thorup, Minimizing diameters of dynamic trees, *Proc. 24th Int. Colloquium on Automata, Languages and Programming (ICALP 97)* (1997), LNCS 1256, 270–280.

[AlHoTh00] S. Alstrup, J. Holm, and M. Thorup, Maintaining center and median in dynamic trees, *Proc. 7th Scandinavian Workshop on Algorithm Theory (SWAT 00)* (2000), 46–56.

[AuItMaNa91] G. Ausiello, G. F. Italiano, A. Marchetti-Spaccamela, and U. Nanni, Incremental algorithms for minimal length paths, *J. of Algorithms* 12(4) (1991), 615–638.

[BaHaSe07] S. Baswana, R. Hariharan, and S. Sen, Improved decremental algorithms for transitive closure and all-pairs shortest paths, *J. Algorithms* 62(2) (2007), 74–92.

[Ch79] V. Chvátal, A greedy heuristic for the set-covering problem, *Mathematics of Operations Research* 4(3) (1979), 233–235.

[CoWi90] D. Coppersmith and S. Winograd, Matrix multiplication via arithmetic progressions, *J. of Symbolic Computation* 9 (1990), 251–280.

[CoLeRiSt01] T. H. Cormen, C. E. Leiserson, R. L. Rivest, and C. Stein, *Introduction to Algorithms*, Second Edition, MIT Press, 2001.

[DeIt04] C. Demetrescu and G. F. Italiano, A new approach to dynamic all pairs shortest paths, *J. Assoc. Comput. Mach.* 51(6) (2004), 968–992.

[DeIt05] C. Demetrescu and G. F. Italiano, Trade-offs for fully dynamic transitive closure on DAGs: breaking through the $O(n^2)$ barrier. *J. Assoc. Comput. Mach.* 52(2) (2005), 147–156.

[DeIt06] C. Demetrescu and G. F. Italiano, Fully dynamic all pairs shortest paths with real edge weights, *J. Comput. Syst. Sci.* 72(5) (2006), 813–837.

[DeIt08] C. Demetrescu and G. F. Italiano, Mantaining dynamic matrices for fully dynamic transitive closure, *Algorithmica* 51(4) (2008), 387–427.

[Di59] E. W. Dijkstra, A note on two problems in connection with graphs, *Numerische Mathematik* 1 (1959), 269–271.

[EpGaItNi97] D. Eppstein, Z. Galil, G. F. Italiano, and A. Nissenzweig, Sparsification – A technique for speeding up dynamic graph algorithms, *J. Assoc. Comput. Mach.* 44 (1997), 669–696.

[EvGa85] S. Even and H. Gazit, Updating distances in dynamic graphs, *Methods of Operations Research* 49 (1985), 371–387.

[EvSh81] S. Even and Y. Shiloach, An on-line edge deletion problem, *J. Assoc. Comput. Mach.* 28 (1981), 1–4.

[FaRa06] J. Fakcharoemphol and S. Rao, Planar graphs, negative weight edges, shortest paths, and near linear time, *J. Comput. Syst. Sci.* 72(5) (2006), 868–889.

[Fr85] G. N. Frederickson, Data structures for on-line updating of minimum spanning trees, *SIAM J. Comput.* 14 (1985), 781–798.

[Fr97] G. N. Frederickson, Ambivalent data structures for dynamic 2-edge-connectivity and k smallest spanning trees, *SIAM J. Comput.* 26(2) (1997), 484–538.

[FrMiZa01] D. Frigioni, T. Miller, and C. Zaroliagis, An experimental study of dynamic algorithms for transitive closure, *ACM J. Experimental Algorithmics* 6 (2001).

[FrMaNa98] D. Frigioni, A. Marchetti-Spaccamela, and U. Nanni, Semi-dynamic algorithms for maintaining single source shortest paths trees, *Algorithmica* 22(3) (1998), 250–274.

[FrMaNa00] D. Frigioni, A. Marchetti-Spaccamela, and U. Nanni, Fully dynamic algorithms for maintaining shortest paths trees, *J. of Algorithms* 34 (2000), 251–281.

[GaIt92] Z. Galil and G. F. Italiano, fully dynamic algorithms for 2-edge connectivity, *SIAM J. Comput.* 21 (1992), 1047–1069.

[Ha69] F. Harary, *Graph Theory*, Addison-Wesley, 1969.

[HeKi01] M. R. Henzinger and V. King, Maintaining minimum spanning forests in dynamic graphs, *SIAM J. Comput.* 31(2) (2001), 364–374.

[HeKi99] M. R. Henzinger and V. King, Randomized fully dynamic graph algorithms with polylogarithmic time per operation, *J. Assoc. Comput. Mach.* 46(4) (1999), 502–536.

[HeKiRaSu97] M. R. Henzinger, P. Klein, S. Rao, and S. Subramanian, Faster shortest-path algorithms for planar graphs, *J. of Computer and System Sciences* 55(1) (1997), 3–23.

[HoDeTh01] J. Holm, K. de Lichtenberg, and M. Thorup, Poly-logarithmic deterministic fully dynamic algorithms for connectivity, minimum spanning tree, 2-edge, and biconnectivity, *J. Assoc. Comput. Mach.* 48(4) (2001), 723–760.

[It86] G. F. Italiano, Amortized efficiency of a path retrieval data structure, *Theoretical Computer Science* 48(2–3) (1986), 273–281.

[It88] G. F. Italiano, Finding paths and deleting edges in directed acyclic graphs, *Information Processing Letters*, 28(1) (1988), 5–11.

[KhMoWi96] S. Khanna, R. Motwani, and R. H. Wilson, On certificates and lookahead on dynamic graph problems, *Algorithmica* 21(4) (1998), 377–394.

[Ki99] V. King, Fully dynamic algorithms for maintaining all-pairs shortest paths and transitive closure in digraphs, *Proc. 40th Symposium on Foundations of Computer Science (FOCS 99)* (1999).

[KiTh01] V. King and M. Thorup, A space saving trick for directed dynamic transitive closure and shortest path algorithms, *Proc. 7th International Conference on Computing and Combinatorics (COCOON 2001)* (2001), 268–277.

[KiSa02] V. King and G. Sagert, A fully dynamic algorithm for maintaining the transitive closure, *J. Comput. Syst. Sci.* 65(1) (2002), 150–167.

[Lo67] P. Loubal, A network evaluation procedure, *Highway Research Record 205* (1967), 96–109.

[Lo75] L. Lovász, On the ratio of optimal integral and fractional covers, *Discrete Mathematics* 13 (1975), 383–390.

[Mu71] I. Munro, Efficient determination of the transitive closure of a directed graph, *Information Processing Letters* 1(2) (1971), 56–58.

[Mu67] J. Murchland, The effect of increasing or decreasing the length of a single arc on all shortest distances in a graph, Technical report, LBS-TNT-26, London Business School, Transport Network Theory Unit, London, UK, 1967.

[RaRe96a] G. Ramalingam and T. Reps, An incremental algorithm for a generalization of the shortest path problem, *J. of Algorithms* 21 (1996), 267–305.

[RaRe96b] G. Ramalingam and T. Reps, On the computational complexity of dynamic graph problems, *Theoretical Computer Science* 158 (1996), 233–277.

[Ra95] M. Rauch, Fully dynamic biconnectivity in graphs, *Algorithmica* 13 (1995), 503–538.

[Ro68] V. Rodionov, The parametric problem of shortest distances, *U.S.S.R. Computational Math. and Math. Phys.* 8(5) (1968), 336–343.

[Ro08] L. Roditty, A faster and simpler fully dynamic transitive closure, *ACM Trans. on Algorithms* 4(1) (2008), 16

[RoZw08] L. Roditty and U. Zwick, Improved dynamic reachability algorithms for directed graphs, *SIAM J. Comput.* 37(5) (2008), 1455–1471.

[RoZw11] L. Roditty and U. Zwick, On dynamic shortest paths problems, *Algorithmica* 61(2) (2011) 389–401.

[Ro85] H. Rohnert, A dynamization of the all-pairs least cost problem, *Proc. 2nd Annual Symposium on Theoretical Aspects of Computer Science (STACS 85), LNCS 182* (1985), 279–286.

[Sa04] P. Sankowski, Dynamic transitive closure via dynamic matrix inverse, *Proc. 45th IEEE Symposium on Foundations of Computer Science (FOCS 04)* (2004), 509–517.

[Sa05] P. Sankowski, Subquadratic algorithm for dynamic shortest distances, *Proc. 11th International Conference on Computing and Combinatorics (COCOON 2005)* (2005), 461–470.

[SlTa83] D. D. Sleator and R. E. Tarjan, A data structure for dynamic trees, *J. Comp. Syst. Sci.* 24 (1983), 362–381.

[Th04] M. Thorup, fully dynamic all-pairs shortest paths: faster and allowing negative cycles, *Proc. 9th Scandinavian Workshop on Algorithm Theory (SWAT 2004)* (2004), 384-396.

[Th05] M. Thorup, Worst-case update times for fully dynamic all-pairs shortest paths, *Proc. of the 37th Symposium on Theory of Computing (STOC 2005)* (2005), 112–119.

[UlYa91] J. D. Ullman and M. Yannakakis, High-probability parallel transitive-closure algorithms, *SIAM J. on Computing* 20(1) (1991), 100–125.

[Va12] V. Vassilevska Williams, Multiplying matrices faster than Coppersmith-Winograd. *Proc. of the 44th Symposium on Theory of Computing (STOC 2012)* (2012), 887–898.

[Zw98] U. Zwick, All pairs shortest paths in weighted directed graphs — exact and almost exact algorithms, *Proc. of the 39th IEEE Annual Symposium on Foundations of Computer Science (FOCS'98)* (1998), 310–319.

Section 10.3
Drawings of Graphs

Emilio Di Giacomo, University of Perugia, Italy
Giuseppe Liotta, University of Perugia, Italy
Roberto Tamassia, Brown University

INTRODUCTION

Research on graph drawing has been conducted within several diverse areas, including discrete mathematics (topological graph theory, geometric graph theory, order theory), algorithmics (graph algorithms, data structures, computational geometry, VLSI), and human-computer interaction (visual languages, graphical user interfaces, software visualization). In this section, we overview two different aspects of the current research in graph drawing: the study of the graph theoretic properties of families of geometric representations of graphs and the algorithmic issues involved in computing a drawing of a graph that satisfies a given set of geometric constraints.

10.3.1 Types of Graphs and Drawings

Graph drawing concerns geometric representations of graphs, and it has important applications to key computer technologies such as software engineering, database systems, visual interfaces, and computer-aided design.

Types of Graphs

First, we define some terminology on graphs pertinent to graph drawing. Throughout this section, let n and m be the number of graph vertices and edges, respectively, and let d be the maximum vertex degree (i.e., number of incident edges).

DEFINITIONS

D1:   A *degree-k graph* is a graph with maximum degree $d \leq k$.

D2:   A *transitive edge* of a digraph is an edge (u, v) such that there is a directed path from u to v not containing edge (u, v).

D3:   A *reduced digraph* is a digraph without transitive edges.

D4:   A *source vertex of a digraph* is a vertex without incoming edges.

D5:   A *sink vertex of a digraph* (also called a *target*) is a vertex without outgoing edges.

D6:   An st-*digraph* (also called a *bipolar digraph*) is an acyclic digraph with exactly one source and one sink, which are joined by an edge.

D7:   A *biconnected graph* is a 2-connected graph; that is, any two vertices are joined by two vertex-disjoint paths.

D8:   A *triconnected graph* is a 3-connected graph; that is, any two vertices are joined by three (pairwise) vertex-disjoint paths.

D9:   A *rooted tree* is a directed tree with a distinguished vertex, called the *root*, such that each vertex lies on a directed path to the root. (We observe that this reverses the usual convention.)

D10:   A *binary tree* is a rooted tree such that each vertex has at most two incoming edges.

D11:   A *ternary tree* is a rooted tree such that each vertex has at most three incoming edges.

D12:   A *layered (di)graph* is a (di)graph whose vertices are partitioned into sets, called layers. A rooted tree can be viewed as a layered digraph where the layers are sets of vertices at the same distance from the root.

D13:   A *k-layered (di)graph* layered (di)graph has k layers.

Types of Drawings

In a drawing of a graph, vertices are represented by points (or by geometric figures such as circles or rectangles) and edges are represented by curves such that any two edges intersect at most in a finite number of points. The following definitions are relative to drawings in the plane, which are the main subject of this section.

DEFINITIONS

D14:   In a *polyline drawing*, each edge is a polygonal chain (see Figure 10.3.1(a)).

D15:   In a *straight-line drawing*, each edge is a straight-line segment (see Figure 10.3.1(b)).

D16: In an *orthogonal drawing*, each edge is a chain of horizontal and vertical segments (see Figure 10.3.1(c)).

D17: A *bend in a polyline drawing* is a point where two segments belonging to the same edge meet (see Figure 10.3.1(a)).

D18: An *orthogonal representation* of an orthogonal drawing is in terms of the bends along each edge and the angles around each vertex.

D19: A *crossing* is a point of a graph drawing where two edges intersect (see Figure 10.3.1(b)).

D20: A *grid drawing* is a polyline drawing such that the vertices, crossings, and bends all have integer coordinates.

D21: In a *planar drawing*, no two edges cross (see Figure 10.3.1(d)).

D22: A *planar (di)graph* is a (di)graph that admits a planar drawing.

D23: An *imbedded (di)graph* is a planar (di)graph with a prespecified topological imbedding (i.e., set of faces), which must be preserved in the drawing.

D24: In an *upward drawing* of a digraph, each edge is monotonically nondecreasing in the vertical direction (see Figure 10.3.1(d)).

D25: An *upward planar digraph* admits an upward planar drawing.

D26: In a *layered drawing* of a layered graph (also called a *hierarchical drawing*), the vertices in the same layer all lie on the same horizontal line.

D27: A *face* is a region of a planar drawing, and the unbounded region is called the *external face*.

D28: An *outerplanar (di)graph* is a planar (di)graph that admits a planar drawing with all vertices on the boundary of the external face.

D29: A *series-parallel digraph* is a planar digraph with a single source s and a single sink t recursively defined as follows: (i) a single edge (s, t) is a series-parallel digraph. Given two series-parallel digraphs G' and G'' with sources s' and s'', respectively, and sinks t' and t'', respectively, (ii) the digraph obtained by identifying t' with s'' is a series-parallel digraph; (iii) the digraph obtained by identifying s' with s'' and t' with t'' is a series-parallel digraph.

D30: A *series-parallel graph* is the underlying undirected graph of a series-parallel digraph.

D31: A *convex drawing* is a planar straight-line drawing of a graph such that the boundary of each face is a convex polygon.

D32: A *visibility drawing* of a graph is based on a geometrically visible relation; e.g., the vertices might be drawn as horizontal segments, and the edges associated with vertically visible segments.

D33:   A **dominance drawing** is an upward drawing of an acyclic digraph such that there exists a directed path from vertex u to vertex v if and only if $x(u) \leq x(v)$ and $y(u) \leq y(v)$, where $x(\cdot)$ and $y(\cdot)$ denote the coordinates of a vertex.

D34:   An **hv-*drawing*** is an upward orthogonal straight-line drawing of a binary tree such that the drawings of the subtrees of each node are separated by a horizontal or vertical line.

EXAMPLE

E1:   In Figure 10.3.1 the first three drawings are of the complete bipartite graph $K_{3,3}$.

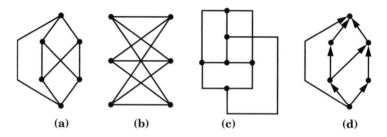

(a)   (b)   (c)   (d)

Figure 10.3.1: Drawings: (a) polyline; (b) straight-line; (c) orthogonal; (d) planar upward.

REMARKS

R1:   Polyline drawings provide great flexibility since they can approximate drawings with curved edges. However, edges with more than two or three bends may be difficult to "follow" for the eye. Also, a system that supports editing of polyline drawings is more complicated than one limited to straight-line drawings. Hence, depending on the application, polyline or straight-line drawings may be preferred.

R2:   If vertices are represented by points, then orthogonal drawings exist only for graphs of maximum vertex degree 4.

R3:   The series-parallel (di)graphs defined above are often called **two-terminal series-parallel (di)graphs**. Throughout this section series-parallel (di)graphs have no multiple edges.

10.3.2   Combinatorics of Some Geometric Graphs

Geometric graphs [PS90, OBS92] are straight-line drawings. Geometric graphs are often studied under the assumption that they satisfy some additional geometric constraints. For example, many papers study the structure of geometric graphs that avoid forbidden edge configurations, such as the *k-quasi planar graphs* where no k mutually crossing edges are allowed (see, e.g. [AT07, AAP+97, FPS11, PRTT06, DDLM12b]), *k-planar graphs*, where no edge can be crossed more than k times (see, e.g. [BKRS01, FM07, Kor08, HELP12]), the ACL_α graphs where any two crossing edges cannot form an angle smaller than a given value α (see, e.g. [AFT11, ACD+11, ABS11a, ABS11b, AFK+12, DDE+12, DDLM11, DEL10, DEL11, DGMW11, DDEL11, vK11] and [DL12] for a survey).

In this section we shall, however, concentrate on another broad family of geometric graphs, namely, the *proximity graphs*. Results about some other types of geometric graphs are described in Section 10.3.7. A proximity graph is a geometric graph such that pairs of adjacent vertices are represented as pairs of points that are deemed to be "sufficiently close," while pairs of non-adjacent vertices are "relatively far" from each other. At a first, broad approximation, the definition of closeness in a proximity drawing can be either based on the concept of *proximity region* or based on a *global proximity* measure. In a proximity region based proximity graph two or more vertices are adjacent if and only if some suitably defined region that describes the neighborhood of these vertices contains at most k other vertices, for a given integer value $k \geq 0$. Global proximity, in turn, gives rise to geometric graphs where the overall sum of the lengths of the edges is minimized. See, e.g., the survey by Jaromczyk and Toussaint [JT92] for extensive lists of different definitions of proximity.

It is worth observing that the problem of analyzing the combinatorial properties of a given type of proximity graph naturally raises the question of the characterization of those graphs which admit the given type of straight-line drawing. This, in turn, leads to the investigation of the design of efficient algorithms for computing such a drawing when one exists. Although these questions are far from being resolved in general, many partial answers have appeared in the literature. See, e.g., [Lio] for an up-to-date survey of these questions. We report below some basic facts and results.

Delaunay Triangulations

DEFINITIONS

D35: A *Delaunay triangulation* is a planar straight-line drawing with all internal faces triangles and such that two adjacent vertices u and v are adjacent if and only if all disks having $\overline{u,v}$ as a chord do not contain any other vertices in their interior (the disks are assumed to be open sets).

D36: A planar triangulated graph is *Delaunay drawable* if it admits a drawing that is a Delaunay triangulation.

D37: A *Voronoi diagram* is the dual graph of a Delaunay triangulation.

FACTS

F1: All Delaunay drawable triangulations are 1-tough and have perfect matchings [Dil90b].

F2: All maximal outerplanar graphs are Delaunay drawable [Dil90a].

F3: Any triangulation without chords or non-facial triangles is Delaunay drawable [DS96].

REMARKS

R4: Di Battista and Vismara [DV96] give a characterization based on a non-linear system of equations involving the angles in the triangulation.

R5: Liotta and Meijer have studied the combinatorial properties of a Voronoi diagram. In particular, a characterization of Voronoi drawable trees can be found in [LM03].

β-drawings and Rectangle of Influence Drawings

DEFINITIONS

D38:   In 1985, Kirkpatrick and Radke [KR85, Rad88] introduced a family of proximity regions called β-**neighborhoods**, denoted by $R[u, v, \beta]$ and defined as follows:

1. For $\beta = 0$, $R[u, v, \beta]$ is the line segment $\overline{uv}$.
2. For $0 < \beta < 1$, $R[u, v, \beta]$ is the intersection of the two closed disks of radius $d(u, v)/(2\beta)$ passing through both u and v.
3. For $1 \leq \beta < \infty$, $R[u, v, \beta]$ is the intersection of the two closed disks of radius $\beta d(u, v)/2$ and centered on the line through u and v.
4. For $\beta = \infty$, $R[u, v, \beta]$ is the closed infinite strip perpendicular to the line segment $\overline{uv}$.

D39:   For a given real value $\beta \geq 0$, a β-**drawing** is a geometric graph where two vertices u and v are adjacent if and only if $R[u, v, \beta]$ does not contain any vertices other than u and v in its interior. A **Gabriel graph** is a β-drawing for $\beta = 1$.

D40:   For a given real value $\beta \geq 0$, a **weak β-drawing** is a geometric graph where if two vertices u and v are adjacent then $R[u, v, \beta]$ does not contain any vertices other than u and v in its interior.

D41:   The **rectangle of influence graph** is a geometric graph such that there exists an edge (u, v) if and only if the axis-aligned rectangle having u and v at opposite corners does not contain any other vertices (the rectangle is assumed to be an open set in some papers and a closed set in some other papers).

D42:   A **weak rectangle of influence drawing** is a geometric graph where if two vertices u and v are adjacent then the axis-aligned rectangle having u and v at opposite corners does not contain any other vertices.

FACTS

F4:   Families of graphs that admit a β-drawing for different values of β are studied in [BDLL95, BLL96, LS93, LL97, IR07, SIR08].   Weak β-drawings are studied in [DLW06, LL97, PV04]

F5:   Different families of graphs that admit a rectangle of influence drawing are described in [LLMW98].   Weak rectangle of influence drawings are studied in [BBM99, MMN09, AB12].

Minimum Spanning Trees

DEFINITIONS

D43:   A **minimum spanning tree of a set P of points** is a connected, straight-line drawing that has P as vertex set and minimizes the total edge length.

D44:   A tree T is **drawable as a minimum spanning tree** if there exists a set P of points such that the minimum spanning tree of P is isomorphic to T. (The problem is that whatever plane locations are assigned to vertices of the tree T, perhaps the image of T itself is not the minimum spanning tree for those vertex locations.)

FACTS

The problem of testing whether a tree can be drawn as a Euclidean minimum spanning tree in the plane is essentially solved. The 3-dimensional counterpart of the problem is not yet solved.

F6: Monma and Suri [MS92] show that each tree with maximum vertex degree at most five can be drawn as a minimum spanning tree of some set of vertices. There is a linear time (real RAM) algorithm. No tree with maximum degree greater than six can be drawn as a minimum spanning tree.

F7: Eades and Whitesides [EW96] show that it is NP-hard to decide whether trees of maximum degree equal to six can be drawn as minimum spanning trees.

F8: No trees with maximum degree greater than twelve can be drawn as a Euclidean minimum spanning tree in 3D-space, while all trees with vertex degree at most nine are drawable [LD95].

F9: King [Kin06] improves this last result by showing that all trees whose vertices have vertex degree at most ten can be realized as a Euclidean minimum spanning tree in 3-dimensional space.

One of the most challenging questions in the seminal paper by Monma and Suri [MS92] was about the area required by a minimum weight drawing of a tree. Namely, the construction by Monma and Suri used a grid of size $O(2^{n^2}) \times O(2^{n^2})$ and the authors conjectured an exponential lower bound for minimum weight drawings of trees with maximum vertex degree five (i.e., the existence of a tree T with n vertices such that any minimum weight drawing of T requires area at least $c^n \times c^n$ for some constant $c > 1$).

F10: The above long standing conjecture by Monma and Suri was recently proved to be correct by Angelini et al. [ABC+11], who describe a tree T with n vertices having maximum degree five such that in any minimum weight drawing of T the ratio between the longest and the shortest edge is $2^{\Omega(n)}$, which implies that the drawing requires exponential area.

F11: Frati and Kaufmann [FK11] proved that the exponential area lower bound of minimum weight drawings of trees does not hold for maximum vertex degree smaller than five.

F12: The area bound for a complete binary tree has been further reduced in [DDLM10, DDLM12a].

Minimum Weight Triangulations

DEFINITIONS

D45: A **triangulation** T of a set P of points on the plane is a straight-line drawing whose vertices are the elements in P and all internal faces are triangles.

D46: T is a **minimum weight triangulation** if it is a triangulation of P that minimizes the total edge length.

D47: A *minimum weight drawing* of a planar triangulated graph G is a straight-line drawing Γ of G with the additional property that Γ is a minimum weight triangulation of the points representing the vertices.

D48: If a graph admits a minimum weight drawing it is called *minimum weight drawable*; otherwise, it is called *minimum weight forbidden*.

FACTS

Little is known about the problem of constructing a minimum weight drawing of a planar triangulation.

F13: The problem of computing a Euclidean minimum weight triangulation of a set of points in the plane is NP-hard [MR08].

F14: All maximal outerplanar triangulations are minimum weight drawable and a linear time (real RAM) drawing algorithm for computing a minimum weight drawing of these graphs is also known [LL96].

This naturally leads us to investigate the internal structure of minimum weight drawable triangulations.

F15: In [LL02] Lenhart and Liotta examine the ***endoskeleton*** of a triangulation: that is, the subgraph induced by the internal vertices of the triangulation. They construct skeletons that cannot appear in any minimum weight drawable triangulation, skeletons that do appear in minimum weight drawable triangulations, and skeletons that guarantee minimum weight drawability.

F16: Wang, Chin, and Yang [WCY00] also focus on the minimum weight drawability of triangulations and show examples of triangulations with acyclic skeletons that do not admit a minimum weight drawing.

F17: There exists an infinite class of minimum weight drawable triangulations that cannot be realized as Delaunay triangulations (that is, for any triangulation T of the class, there does not exist a set P of points such that the Delaunay triangulation of P is isomorphic to T) [LL02] .

It is worth remarking that the study of the geometric differences between the minimum-weight and Delaunay triangulations of a given set of points in order to compute good approximations of the former has a long tradition (see, e.g., [Kir80, LK96, MZ79]); little is known about the combinatorial difference between Delaunay triangulations and minimum-weight triangulations.

Open Problems

P1. Give a complete combinatorial characterization of Delaunay drawable triangulations.

P2. Let T be a tree with maximum vertex degree at most twelve. Is there a polynomial time algorithm to decide whether T can be drawn as a Euclidean minimum spanning tree in 3D-space? If so, compute such a drawing.

P3. Define new families of minimum weight drawable triangulations. For example, characterize the class of triangulations with acyclic skeleton that admit a minimum weight drawing.

P4. Investigate the combinatorial relationship between minimum weight and Delaunay drawable triangulations. Are there any Delaunay drawable and minimum weight forbidden triangulations?

P5. Further study the combinatorial structure of proximity graphs. For example, characterize the family of Gabriel drawable triangulations, that is, the family of those triangulations that admit a straight-line drawing where the angles of each triangular face are less than $\pi/2$.

10.3.3   Properties of Drawings and Bounds

For various classes of graphs and drawing types, many universal/existential upper and lower bounds for specific drawing properties have been discovered. Such bounds typically exhibit tradeoffs between drawing properties. A universal bound applies to all the graphs of a given class. An existential bound applies to infinitely many graphs of the class. Whenever we give bounds on the area or edge length, we assume that the drawing is constrained by some resolution rule that prevents it from being arbitrarily scaled down, reduced by an arbitrary scaling (e.g., requiring a grid drawing, or stipulating a minimum unit distance between any two vertices).

Properties of Drawings

In computing graph drawings, we would like to take into account a variety of properties. For example, planarity and the display of symmetries are highly desirable in visualization applications. Or we may want to display trees and acyclic digraphs with upward drawings. In general, to avoid wasting valuable space on a page or a computer screen, it is important to keep the area of the drawing small. Moreover, it is typically desirable to maximize the angular resolution and to minimize the other measures.

DEFINITIONS

D49: The *crossing number χ of a drawing* is its total number of edge-crossings.

D50: The *area of a drawing* is the area of its convex hull.

D51: The *total edge length of a drawing* is the sum of the lengths of the edges.

D52: The *number of bends of a polyline drawing* is the total number of bends on the edges of a drawing.

D53: The *maximum number of bends of a polyline drawing* is the maximum number of bends on any edge.

D54: The *angular resolution ρ in a polyline drawing* is the smallest angle formed by any two edges or segments of edges, incident on the same vertex or bend.

D55: The *aspect ratio of a drawing* is the ratio of the longest side to the shortest side of the smallest rectangle with horizontal and vertical sides covering the drawing.

EXAMPLES

The need to satisfy different drawing properties at the same time leads to formalizing many graph drawing problems as multi-objective optimization problems (e.g., construct a drawing with minimum area and minimum number of crossings), so that tradeoffs are inherent in solving them.

E2:   Figure 10.3.2(a–b) below shows two drawings of K_4, the complete graph on four vertices. The drawing of part (a) is planar, while the drawing of part (b) "maximizes symmetries." It can be shown that no drawing of K_4 is optimal with respect to both criteria, i.e., the maximum number of symmetries cannot be achieved by a planar drawing.

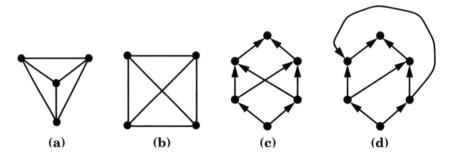

      (a)   **(b)**   **(c)**   **(d)**

Figure 10.3.2: (a–b) Tradeoff between planarity and symmetry in drawing K_4. (c–d) Tradeoff between planarity and upwardness in drawing an acyclic digraph G.

E3:   Figure 10.3.2(c–d) shows two drawings of the same acyclic digraph G. The drawing of part (c) is upward, while the drawing of part (d) is planar. It can be shown that there is no drawing of G that is both planar and upward.

Bounds on the Area

Tables 10.3.1–10.3.3 below summarize selected universal upper bounds and existential lower bounds on the area of drawings of graphs. In the tables, a is an arbitrary constant $0 \leq a < 1$, b and c are fixed constants $1 < b < c$, and ϵ is an arbitrary positive constant. The abbreviations "PSL" and "PSLg" are used for "planar straight-line" and "planar straight-line grid," respectively.

In general, the effect of bends on the area requirement is dual. On one hand, bends occupy space and hence negatively affect the area. On the other hand, bends may help in routing edges without using additional space.

Table 10.3.1: **Universal upper and existential lower bounds on the area of trees.**

	CLASS OF GRAPHS	DRAWING TYPE	AREA	
1	Fibonacci trees	strictly upward PSLg	$\Omega(n)$	$O(n)$
2	AVL trees	strictly upward PSLg	$\Omega(n)$	$O(n)$
3	balanced binary trees	strictly upward PSLg	$\Omega(n)$	$O(n)$
4	binary trees	PSLg	$\Omega(n)$	$O(n)$
5	binary trees	upward order preserving PSLg	$\Omega(n \log n)$	$O(n \log n)$
6	binary trees	order preserving PSLog	$\Omega(n)$	$O(n^{1.5})$
7	binary trees	upward order preserving PSLog	$\Omega(n^2)$	$O(n^2)$
8	binary trees	upward planar orthogonal grid	$\Omega(n \log \log n)$	$O(n \log \log n)$
9	binary trees	planar polyline grid order preserving	$\Omega(n)$	$O(n \log \log n)$
10	binary trees	upward planar polyline grid order preserving	$\Omega(n \log n)$	$O(n \log n)$
11	ternary trees	PSLog	$\Omega(n)$	$O(n^{1.631})$
12	ternary trees	order preserving PSLog	$\Omega(n^2)$	$O(n^2)$
13	ternary trees	upward planar orthogonal grid	$\Omega(n \log n)$	$O(n \log n)$
14	ternary trees	planar orthog grid order preserving	$\Omega(n)$	$O(n)$
15	degree-$O(n^{\frac{a}{2}})$ rooted trees	PSLg	$\Omega(n)$	$O(n)$
16	degree-$O(n^a)$ rooted trees	upward planar polyline grid	$\Omega(n)$	$O(n)$
17	rooted trees	PSLg	$\Omega(n)$	$O(n \log n)$
18	rooted trees	upward PSLg	$\Omega(n)$	$O(n \log n)$
19	rooted trees	strictly upward order preserving PSLg	$\Omega(n)$	$O(n^{1+\epsilon})$

Table 10.3.2: **Universal upper and existential lower bounds on the area of directed trees.**

1	directed binary trees	strictly upward PSLg	$\Omega(n \log n)$	$O(n \log n)$
2	directed binary trees	strictly upward order preserving PSLg	$\Omega(b^n)$	$O(c^n)$
3	directed binary trees	strictly upward order preserving planar polyline grid	$\Omega(n^2)$	$O(n^2)$
4	directed trees	strictly upward PSLg	$\Omega(n \log n)$	$O(n \log n)$
5	directed trees	strictly upward order preserving PSLg	$\Omega(b^n)$	$O(c^n)$
6	directed trees	strictly upward order preserving planar polyline grid	$\Omega(n^2)$	$O(n^2)$

FACTS

F18: Linear or almost-linear bounds on the area can be achieved for several families of trees; typically superlinear bounds are associated with order preserving drawings (Table 10.3.1).

F19: No linear area lower bounds exist for upward drawings of directed trees (Table 10.3.2). Exponential lower bounds are known for the drawing conventions of rows 2 and 5 of Table 10.3.2.

F20: Almost linear area can be achieved for undirected outerplanar graphs (rows 1–3 of Table 10.3.3) and for non-planar drawings of degree-4 graphs (row 18 of Table 10.3.3). Subquadratic area upper bounds and/or superlinear lower bounds are known for undirected SP-graphs (rows 6–7 of Table 10.3.3). Exponential lower bounds are known for upward drawings of directed acyclic graphs (rows 8 and 20 of Table 10.3.3).

F21: A quadratic area bound is achieved only at the expense of a linear number of bends (row 22 of Table 10.3.3).

See Table 10.3.6 below for tradeoffs between area and aspect ratio in drawings of trees.

Bounds on the Angular Resolution

Table 10.3.4 below summarizes selected universal lower bounds and existential upper bounds on the angular resolution of drawings of graphs. Here c is a fixed constant with $c > 1$.

Table 10.3.3: **Universal upper and existential lower bounds on the area of planar graphs and digraphs.**

	CLASS OF GRAPHS	DRAWING TYPE	AREA	
1	outerplanar graphs	outerplanar SLg	$\Omega(n)$	$O(n^{1.48})$
2	degree-d outerplanar graphs	outerplanar SLg	$\Omega(n)$	$O(d \log n)$
3	outerplanar graphs	outerplanar polyline grid	$\Omega(n)$	$O(n \log n)$
4	directed outerplanar graphs	upward outerplanar SLg	$\Omega(b^n)$	
5	directed outerplanar graphs	upward outerplanar polyline grid	$\Omega(n^2)$	
6	series-parallel graphs	PSLg	$\Omega(n2^{\sqrt{\log n}})$	$O(n^2)$
7	series-parallel graphs	planar polyline grid	$\Omega(n2^{\sqrt{\log n}})$	$O(n^{3/2})$
8	series-parallel digraphs	upward planar embedding preserving SLg	$\Omega(b^n)$	$O(c^n)$
9	series-parallel digraphs	upward PSLg	$\Omega(n^2)$	$O(n^2)$
13	planar graphs	planar polyline grid	$\Omega(n^2)$	$O(n^2)$
14	planar graphs	PSL (with angular resolution $\geq \rho$)	$\Omega(c^{\rho n})$	
15	planar graphs	PSLg	$\Omega(n^2)$	$O(n^2)$
16	triconnected planar graphs	PSL convex grid	$\Omega(n^2)$	$O(n^2)$
17	planar graphs	planar orthog grid	$\Omega(n^2)$	$O(n^2)$
18	planar degree-4 graphs	orthogonal grid	$\Omega(n \log n)$	$O(n \log^2 n)$
19	general graphs	polyline grid	$\Omega(n + \chi)$	$O((n + \chi)^2)$
20	up planar digraphs	upward PSLg	$\Omega(b^n)$	$O(c^n)$
21	reduced planar st-digraphs	upward PSLg dominance	$\Omega(n^2)$	$O(n^2)$
22	up planar digraphs	up planar grid polyline	$\Omega(n^2)$	$O(n^2)$

Table 10.3.4: **Universal lower bounds and existential upper bounds on angular resolution.**

CLASS OF GRAPHS	DRAWING TYPE	ANGULAR RESOLUTION	
outerplanar graph	planar straight-line	$\Omega(\frac{1}{d})$	$O(\frac{1}{d})$
series-parallel graph	planar straight-line	$\Omega(\frac{1}{d^2})$	$O(\frac{1}{d})$
general graph	straight-line	$\Omega(\frac{1}{d^2})$	$O(\frac{\log d}{d^2})$
planar graph	straight-line	$\Omega(\frac{1}{d})$	$O(\frac{1}{d})$
planar graph	planar straight-line	$\Omega(\frac{1}{c^d})$	$O(\sqrt{\frac{\log d}{d^3}})$
planar graph	planar straight-line	$\Omega(\frac{1}{n^2})$	$O(\frac{1}{n})$
planar graph	planar polyline	$\Omega(\frac{1}{d})$	$O(\frac{1}{d})$

Bounds on the Number of Bends

Table 10.3.5 summarizes selected universal upper bounds and existential lower bounds on the total and maximum number of bends in orthogonal drawings. Some bounds are stated for $n \geq 5$ or $n \geq 7$ because the maximum number of bends is at least 2 for K_4 and at least 3 for the skeleton graph of an octahedron, in any planar orthogonal drawing.

Table 10.3.5: **Orthogonal drawings: universal upper bounds and existential lower bounds on the number of bends. Notes:** † $n \geq 7$; ‡ $n \geq 5$.

CLASS OF GRAPHS	DRAWING TYPE	TOTAL # BENDS		MAX # BENDS	
deg-4 †	orthog	$\geq n$	$\leq 2n + 2$	≥ 2	≤ 2
planar deg-4 †	orthog planar	$\geq 2n - 2$	$\leq 2n + 2$	≥ 2	≤ 2
imbedded deg-4	orthog planar	$\geq 2n - 2$	$\leq \frac{12}{5}n + 2$	≥ 3	≤ 3
biconnected imbedded deg-4	orthog planar	$\geq 2n - 2$	$\leq 2n + 2$	≥ 3	≤ 3
triconnected imbedded deg-4 †	orthog planar	$\geq \frac{4}{3}(n-1)+2$	$\leq \frac{3}{2}n + 4$	≥ 2	≤ 2
imbedded deg-3 ‡	orthog planar	$\geq \frac{1}{2}n + 1$	$\leq \frac{1}{2}n + 1$	≥ 1	≤ 1

Tradeoff Between Area and Aspect Ratio

The ability to construct area-efficient drawings is essential in practical visualization applications, where screen space is at a premium. However, achieving small area is not enough: e.g., a drawing with high aspect ratio may not be conveniently placed on a workstation screen, even if it has modest area. Hence, it is important to keep the aspect ratio small. Ideally, one would like to obtain small area for any given aspect ratio in a wide range. This would provide graphical user interfaces with the flexibility of fitting drawings into arbitrarily shaped windows. A variety of tradeoffs for the area and aspect ratio arise even when drawing graphs with a simple structure, such as trees.

Table 10.3.6 below summarizes selected universal bounds that can be simultaneously achieved on the area and the aspect ratio of various types of drawings of trees.

Table 10.3.6: **Trees: universal upper bounds simultaneously achievable for area and aspect ratio.**

CLASS OF GRAPHS	DRAWING TYPE	AREA	ASPECT RATIO
binary tree	PSLg	$O(n)$	$[O(1), O(n^\epsilon)]$
binary tree	PSLg order preserving	$O(n \log n)$	$[O(1), O(n/\log n)]$
binary tree	PSLg order preserving	$O(n \log \log n)$	$O\left(\frac{n \log \log n}{\log^2 n}\right)$
binary tree	PSLog	$O(n \log \log n)$	$\left[O(1), O\left(\frac{n \log \log n}{\log^2 n}\right)\right]$
binary tree	up planar orthog grid	$O(n \log \log n)$	$O(n \log \log n / \log^2 n)$
binary tree	upward PSLog	$O(n \log n)$	$[O(1), O(n/\log n)]$
deg-4 tree	orthog grid	$O(n)$	$O(1)$
deg-4 tree	orthog grid, leaves on convex hull	$O(n \log n)$	$O(1)$
rooted deg-$O(n^a)$ tree	upward planar polyline grid	$O(n)$	$[O(1), O(n^\epsilon)]$
rooted tree	upward PSL layered grid	$O(n^2)$	$O(1)$
rooted tree	upward PSLg	$O(n \log n)$	$O(n/\log n)$

In Table 10.3.6, a is an arbitrary constant with $0 \le a < 1$, while ϵ is an arbitrary positive constant. The abbreviation "PSLog" is used for "planar straight-line orthgonal grid," that is, a straight-line grid drawing where the edges are either horizontal or vertical straight-line segments. Only for a few cases there exist algorithms that guarantee efficient area performance and that can accept any user-specified aspect ratio in a given range. For such cases the aspect ratio in Table 10.3.6 is given as an interval.

REMARKS

R6: While upward planar straight-line drawings are the most natural way of visualizing rooted trees, the existing drawing techniques are unsatisfactory with respect to either the area requirement or the aspect ratio. The situation is similar for orthogonal drawings.

R7:   Regarding polyline drawings, linear area can be achieved with a prescribed aspect ratio. However, experiments show that this is done at the expense of a somehow aesthetically unappealing drawing.

R8:   For nonupward drawings of trees, linear area and optimal aspect ratio are possible for planar orthogonal drawings, and a small (logarithmic) amount of extra area is needed if the leaves are constrained to be on the convex hull of the drawing (e.g., pins on the boundary of a VLSI circuit). However, the nonupward drawing methods do not seem to yield aesthetically pleasing drawings, and are suited more for VLSI layout than for visualization applications.

Tradeoff between Area and Angular Resolution

Table 10.3.7 summarizes selected universal bounds that can be simultaneously achieved on the area and the angular resolution of drawings of graphs. Here b and c are fixed constants, $b > 1$ and $c > 1$. Universal lower bounds on the angular resolution exist that depend only on the degree of the graph. Also, substantially better bounds can be achieved by drawing a planar graph with bends or in a nonplanar way.

Table 10.3.7: **Universal upper bounds for area and lower bounds for angular resolution, simultaneously achievable.**

CLASS OF GRAPHS	DRAWING TYPE	AREA	ANGULAR RESOLUTION
trees	planar straight-line	$O(n^8)$	$\Omega(\frac{1}{d})$
planar graph	straight-line	$O(d^6 n)$	$\Omega(\frac{1}{d^2})$
planar graph	straight-line	$O(d^3 n)$	$\Omega(\frac{1}{d})$
planar graph	planar straight-line grid	$O(n^2)$	$\Omega(\frac{1}{n^2})$
planar graph	planar straight-line	$O(b^n)$	$\Omega(\frac{1}{c^d})$
planar graph	planar polyline grid	$O(n^2)$	$\Omega(\frac{1}{d})$

FACTS

F22:   Any unordered tree has a planar straight-line drawing with perfect angular resolution.[1]

F23:   There are ordered trees that require exponential area for any planar straight-line drawing having perfect angular resolution.

Ordered preserving planar drawings of trees with perfect angular resolution and polynomial area can be obtained in the Lombardi drawing style.

---

[1] A drawing has perfect angular resolution if for every vertex v the angle formed by any two consecutive edges around v is $\frac{2\pi}{d(v)}$, where $d(v)$ is the degree of v.

Open Problems

P6. Determine the area requirement of planar straight-line orthogonal drawings of binary and ternary trees. There are currently wide gaps between the known upper and lower bounds (Table 10.3.1 rows 6 and 11).

P7. Determine the area requirement of (upward) planar straight-line drawings of trees. There is currently an $O(\log n)$ gap between the known upper and lower bounds (Table 10.3.1 rows 17 and 18).

P8. Determine the area requirement of strictly upward planar order preserving straight-line drawings of rooted trees (Table 10.3.1 row 19).

P9. Determine the area requirement of outerplanar straight-line grid drawings of outerplanar graphs. There is currently an $O(n^{0.48})$ gap between the known upper and lower bounds (Table 10.3.3 row 1).

P10. Determine the area requirement of planar straight-line grid drawings of series-parallel graphs. In particular it would be interesting to prove a subquadratic upper bound (Table 10.3.3 row 6).

P11. Determine the area requirement of orthogonal (or, more generally, polyline) non-planar drawings of planar graphs. There is currently an $O(\log n)$ gap between the known upper and lower bounds (Table 10.3.3 row 18).

P12. Close the gap between the $\Omega(\frac{1}{d^2})$ universal lower bound and the $O(\frac{\log d}{d^2})$ existential upper bound on the angular resolution of straight-line drawings of general graphs (Table 10.3.4).

P13. Close the gap between the $\Omega(\frac{1}{c^d})$ universal lower bound and the $O(\sqrt{\frac{\log d}{d^3}})$ existential upper bound on the angular resolution of planar straight-line drawings of planar graphs (Table 10.3.4).

P14. Determine the best possible aspect ratio and area simultaneously achievable for (upward) planar straight-line and orthogonal drawings of trees (Table 10.3.6).

10.3.4 Complexity of Graph Drawing Problems

Tables 10.3.8–10.3.11 below summarize selected results on the time complexity of some fundamental graph drawing problems.

It is interesting that apparently similar problems exhibit very different time complexities. For example, while planarity testing can be done in linear time, upward planarity testing is NP-hard. Note that, as illustrated in Figure 10.3.1(c–d), planarity and acyclicity are necessary but not sufficient conditions for upward planarity. While many efficient algorithms exist for constructing drawings of trees and planar graphs with good universal area bounds, exact area minimization for most types of drawings is NP-hard, even for trees.

Table 10.3.8: **Time complexity of some fundamental graph drawing problems: general graphs and digraphs.**

CLASS OF GRAPHS	PROBLEM	TIME COMPLEXITY	
general graph	minimize crossings	NP-hard	
2-layered graph	minimize crossings in layered drawing with preassigned order on one layer	NP-hard	
general graph	maximum planar subgraph	NP-hard	
general graph	test the existence of a drawing where each edge is crossed at most once	NP-hard	
general graph	planarity testing and computing a planar imbedding	$\Omega(n)$	$O(n)$
general graph	maximal planar subgraph	$\Omega(n + m)$	$O(n + m)$
general digraph	upward planarity testing	NP-hard	
imbedded digraph	upward planarity testing	$\Omega(n)$	$O(n^2)$
biconnected series-parallel digraph	upward planarity testing	$\Omega(n)$	$O(n^4)$
biconnected outerplanar	upward planarity testing	$\Omega(n)$	$O(n^2)$
biconnected bipartite	upward planarity testing	$\Omega(n)$	$O(n)$
single-source digraph	upward planarity testing	$\Omega(n)$	$O(n)$
general graph	draw as the intersection graph of a set of unit diameter disks in the plane	NP-hard	

Table 10.3.9: **Time complexity of some fundamental graph drawing problems: planar graphs and digraphs.**

CLASS OF GRAPHS	PROBLEM	TIME COMPLEXITY	
planar graph	planar straight-line drawing with prescribed edge lengths	NP-hard	
planar graph	planar straight-line drawing with max angular resolution	NP-hard	
imbedded graph	test the existence of a planar st-line drawing with prescribed angles betw pairs of consecutive edges incident on a vertex	NP-hard	
maximal planar graph	test the existence of a planar st-line drawing with prescribed angles betw pairs of consecutive edges incident on a vertex	$\Omega(n)$	$O(n)$
planar graph	planar st-line grid drawing with $O(n^2)$ area and $O(1/n^2)$ angular resolution	$\Omega(n)$	$O(n)$
planar graph	planar polyline drawing with $O(n^2)$ area, $O(n)$ bends, and $O(1/d)$ angular resolutions	$\Omega(n)$	$O(n)$
triconn planar graph	planar straight-line convex grid drawing with $O(n^2)$ area and $O(1/n^2)$ angular resolution	$\Omega(n)$	$O(n)$
triconn planar graph	planar st-line strictly convex drawing	$\Omega(n)$	$O(n)$
reduced planar st-digraph	up planar grid st-line dominance drawing with min area	$\Omega(n)$	$O(n)$
upward planar digraph	up planar polyline grid drawing with $O(n^2)$ area & $O(n)$ bends	$\Omega(n)$	$O(n)$

Table 10.3.10: **Time complexity of some fundamental graph drawing problems: planar graphs and digraphs.**

CLASS OF GRAPHS	PROBLEM	TIME COMPLEXITY	
planar deg-4 graph	planar orthogonal grid drawing with min number of bends	NP-hard	
planar biconnected deg-3 graph	planar orthog grid drawing with min # bends and $O(n^2)$ area	$\Omega(n)$	$O(n^5 \log n)$
planar biconnected deg-4 series-parallel graph	planar orthog grid drawing with min # bends and $O(n^2)$ area	$\Omega(n)$	$O(n^4)$
planar biconnected deg-3 series-parallel graph	planar orthog grid drawing with min # bends and $O(n^2)$ area	$\Omega(n)$	$O(n^3)$
imbedded deg-4 graph	planar orthog grid drawing with min # bends and $O(n^2)$ area	$\Omega(n)$	$O(n^{\frac{3}{2}})$
planar deg-4 graph	planar orthog grid drawing with $O(n^2)$ area and $O(n)$ bends	$\Omega(n)$	$O(n)$
imbedded deg-4 graph	test the existence of a PSLog drawing with rectangular faces	$\Omega(n)$	$O(\frac{n^{1.5}}{\log n})$
planar deg-3 graph	test the existence of a PSLog drawing with rectangular faces	$\Omega(n)$	$O(n)$
deg-3 series-parallel graph	test the existence of a planar orthog grid drawing with no bends	$\Omega(n)$	$O(n)$
planar orthogonal representation	planar orthog grid drawing with minimum area	NP-hard	

Table 10.3.11: **Time complexity of some fundamental graph drawing problems: trees.**

CLASS OF GRAPHS	PROBLEM	TIME COMPLEXITY	
tree	draw as the Euclidean min spanning tree of a set of points in the plane	NP-hard	
degree-4 tree	minimize area in planar orthogonal grid drawing	NP-hard	
degree-4 tree	minimize total/maximum edge length in planar orthogonal grid drawing	NP-hard	
rooted tree	minimize area in a planar st-line up layered grid drawing that displays symmetries and isom's of subtrees	NP-hard	
rooted tree	minimize area in a planar straight-line up layered drawing that displays symmetries and isom's of subtrees	$\Omega(n)$	$O(n^k)$, $k \geq 1$
binary tree	minimize area in hv-drawing	$\Omega(n)$	$O(n\sqrt{n \log n})$
rooted tree	planar straight-line up layered grid drawing with $O(n^2)$ area	$\Omega(n)$	$O(n)$
rooted tree	planar polyline up grid drawing with $O(n)$ area	$\Omega(n)$	$O(n)$

1260 Chapter 10. Graphs in Computer Science

Open Problems

P15. Reduce the time complexity of upward planarity testing for imbedded digraphs (currently $O(n^2)$), or prove a superlinear lower bound (Table 10.3.8).

P16. Reduce the time complexity of upward planarity testing for series-parallel and outerplanar digraphs (currently $O(n^4)$ and $O(n^2)$, respectively), or prove a superlinear lower bound (Table 10.3.8).

P17. Reduce the time complexity of bend minimization for planar orthogonal drawings of imbedded graphs (currently $O(n^{3/2})$), or prove a superlinear lower bound (Table 10.3.9).

P18. Reduce the time complexity of bend minimization for planar orthogonal drawings of several families of graphs (deg-3 graphs, deg-3 and deg-4 series-parallel graphs, imbedded deg-4 graphs) (Table 10.3.9).

P19. Determine the time complexity of testing the existence a planar orthogonal straight-line grid drawing with rectangular faces for planar deg-4 graphs (Table 10.3.10).

P20. Reduce the time complexity of testing the existence a planar orthogonal grid drawing with no bends for deg-4 series-parallel graphs. The current bound is $O(n^3)$ deriving from bend minimization (Table 10.3.10).

10.3.5 Example of a Graph Drawing Algorithm

In this subsection we outline the algorithm by one of the authors [Tam87] for computing, for an imbedded degree-4 graph G, a planar orthogonal grid drawing with minimum number of bends and using $O(n^2)$ area (see Table 10.3.7). This algorithm is the core of a practical drawing algorithm for general graphs (see §10.3.6 and Figure 10.3.3(d)).

Graph Drawing Algorithm The algorithm consists of two main phases:
1. Computation of an orthogonal representation for G, where only the bends and the angles of the orthogonal drawing are defined.
2. Assignment of integer lengths to the segments of the orthogonal representation.

REMARKS

R9: Phase 1 uses a transformation into a network flow problem (Figure 10.3.3(a–c)), where each unit of flow is associated with a right angle in the orthogonal drawing. Hence, angles are viewed as a commodity that is produced by the vertices, transported across faces by the edges through their bends, and eventually consumed by the faces.

R10: From the imbedded graph G we construct a flow network N as follows. The nodes of network N are the vertices and faces of G. Let $\deg(f)$ denote the number of edges of the circuit bounding face f. Each vertex v supplies $\sigma(v) = 4$ units of flow, and each face f consumes $\tau(f)$ units of flow, where

$$\tau(f) = \begin{cases} 2\deg(f) - 4 & \text{if } f \text{ is an internal face} \\ 2\deg(f) + 4 & \text{if } f \text{ is the external face} \end{cases}$$

By Euler's formula, $\sum_v \sigma(v) = \sum_f \tau(f)$, i.e., the total supply is equal to the total consumption.

R11: Network N has two types of arcs:

- arcs of the type (v, f), where f is a face incident on vertex v; the flow in (v, f) represents the angle at vertex v in face f, and has lower bound 1, upper bound 4, and cost 0;

- arcs of the type (f, g), where face f shares an edge e with face g; the flow in (f, g) represents the number of bends along edge e with the right angle inside face f, and has lower bound 0, upper bound $+\infty$, and cost 1.

R12:   The conservation of flow at the vertices expresses the fact that the sum of the angles around a vertex is equal to 2π. The conservation of flow at the faces expresses the fact that the sum of the angles at the vertices and bends of an internal face is equal to $\pi(p-2)$, where p is the number of such angles. For the external face, the above sum is equal to $\pi(p+2)$.

R13:   It can be shown that every feasible flow ϕ in network N corresponds to an admissible orthogonal representation for graph G, whose number of bends is equal to the cost of flow ϕ. Hence, an orthogonal representation for G with the minimum number of bends can be computed from a minimum cost flow in G. This flow can be constructed in $O(n^2 \log n)$ time with standard flow-augmentation methods.

R14:   Phase 2 uses a simple compaction strategy derived from VLSI layout, where the lengths of the horizontal and vertical segments are computed independently after a preliminary refinement of the orthogonal representation that decomposes each face into rectangles. The resulting drawing is shown in Figure 10.3.3(d).

10.3.6   Techniques for Drawing Graphs

In this subsection we outline some of the most successful techniques that have been devised for drawing general graphs.

Planarization

The planarization approach is motivated by the availability of many efficient and well-analyzed drawing algorithms for planar graphs (see Table 10.3.9). If the graph is non-planar, it is transformed into a planar graph by means of a preliminary planarization step that replaces each crossing with a fictitious vertex. The planarization approach consists of two main steps: in the first step a maximal planar subgraph G' of the input graph G is computed; in the second step, all the edges of G that are not in G' are added to G' and the crossings formed by each added edge are replaced with dummy vertices. Clearly when adding an edge one wants to produce as few crossings as possible. The two optimization problems arising in the two steps of the planarization approach, i.e., the maximum planar subgraph problem and the edge insertion problem, are NP-hard. Hence, existing planarization algorithms use heuristics. The best available heuristic for the maximum planar subgraph problem is described in [JM96]. This method has a solid theoretical foundation in polyhedral combinatorics, and achieves good results in practice. A sophisticated algorithm for edge insertion (that insert each edge minimizing the number of crossings over all possible imbeddings of the planar subgraph) is described in [GMW05]. See also [BCG$^+$] for more references.

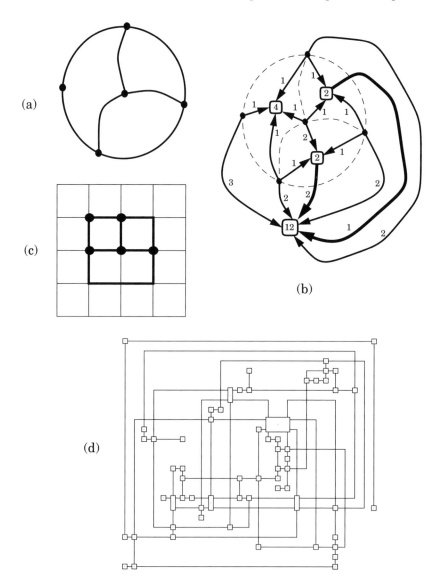

Figure 10.3.3: (a) Imbedded graph G. (b) Minimum cost flow in network N: the flow is shown next to each arc; arcs with zero flow are omitted; arcs with unit cost are drawn with thick lines; a face f is represented by a box labeled with $\tau(f)$. (c) Planar orthogonal grid drawing of G with minimum number of bends. (d) Orthogonal grid drawing of a nonplanar graph produced by a drawing method for general graphs based on the algorithm of this subsection.

A successful drawing algorithm based on the planarization approach and a bend-minimization method [Tam87] is described in [TDB88] (Figure 10.3.3(d) was generated by this algorithm). It has been widely used in software visualization systems.

Layering

The layering approach for constructing polyline drawings of directed graphs transforms the digraph into a layered digraph and then constructs a layered drawing. A typical algorithm based on the layering approach consists of the following main steps:

1. Assign each vertex to a layer, with the goal of maximizing the number of edges oriented upward.
2. Insert fictitious vertices along the edges that cross layers, so that each edge in the resulting digraph connects vertices in consecutive layers. (The fictitious vertices will be displayed as bends in the final drawing.)
3. Permute the vertices on each layer with the goal of minimizing crossings.
4. Adjust the positions of the vertices in each layer with the goal of distributing the vertices uniformly and minimizing the number of bends.

Most of the subproblems involved in the various steps are NP-hard; hence heuristics must be used. The layering approach was pioneered by Sugiyama et al. [STT81] and since then, a lot of research has been devoted to all optimization problems defined in each of the four steps above (see, e.g., [ELS93, GKNV93, TNB04, HN02, JM97, Nag05, NY04, EK86, EW94, MSM99, BWZ10, BK02, BBBH10, CGMW11, CGMW10]). See also [HN] for more references.

Physical Simulation

This approach uses a physical model where the vertices and edges of the graph are viewed as objects subject to various forces. Starting from an initial random configuration, the physical system evolves into a final configuration of minimum energy, which yields the drawing. Rather than solving a system of differential equations, the evolution of the system is usually simulated using numerical methods (e.g., at each step, the forces are computed and corresponding incremental displacements of the vertices are performed). Drawing algorithms based on the physical simulation approach are often able to detect and display symmetries in the graph. However, their running time is typically high. The physical simulation approach was pioneered in [Ead84, KS80]. Sophisticated developments and applications include [DH96, EH00, FR91, KK89, HK02, GGK04, GKN05, BP07]. See also [Kob] for additional references.

10.3.7 Selected Topics

This subsection presents an overview of selected areas of graph drawing that have attracted increasing attention.

Point-Set Embeddings and Universal Point Sets

In the "typical" graph drawing setting, algorithms compute a layout of the graph by suitably choosing vertex locations in order to satisfy drawing requirements or to optimize aesthetic criteria. In some applications, however, one wants to lay out a graph onto preassigned vertex locations. A point-set embedding of a graph G to a set of points S is a drawing of G where the vertices are represented as points of S.

DEFINITIONS

D56: Let $G = (V, E)$ be a graph. A k-**coloring** of G ($k > 0$) is a partition of V into k sets $\{V_1, V_2, \ldots, V_k\}$. A graph G with a k-coloring is called a k-**colored graph**. A 1-colored graph is simply called a graph.

D57: Let S be a set of points in the plane. A k-**coloring** of S ($k > 0$) is a partition of S into k sets $\{S_1, S_2, \ldots, S_k\}$. A set of points S with a k-coloring is called a k-**colored point set**. A 1-colored set of points is simply called a point set.

D58: Let G be a k-colored graph and let S be a k colored point set. G and S are **compatible** if $|V_i| \leq |S_i|$ for every $i = 1, 2, \ldots, k$.

D59: Let $G = (V, E)$ be a k-colored graph, let S be a k-colored point set, and let $\mathcal{D}$ be a type of drawing (planar, straight-line, orthogonal, etc.). A k-**colored point-set embedding of G to S of type $\mathcal{D}$** is a drawing Γ of G of type $\mathcal{D}$ where each vertex $v \in V_i$ is represented in Γ by a point of S_i. A 1-colored point-set embedding of G to S of type $\mathcal{D}$ is simply called a point-set embedding of G to S of type $\mathcal{D}$. Also, when the type of drawing is clear from the context we omit the specification "of type" $\mathcal{D}$.

D60: Let $\mathcal{F}$ be a family of k-colored graphs and let $\mathcal{D}$ be a type of drawing. A k-colored point set S is **universal** for $\mathcal{F}$ for drawings of type $\mathcal{D}$ if every graph $G \in \mathcal{F}$ admits a k-colored point-set embedding to S of type $\mathcal{D}$.

D61: An h-**bend drawing** ($h > 0$) is a polyline drawing where each edge has at most h bends.

REMARKS

R15: The definition of k-colored graph given above allows adjacent vertices to have the same color.

R16: In the literature about point-set embeddability different versions of the problem have been investigated that are specializations of the general definition given above. Typically, the number of points considered is equal to the number of vertices of the input graph. Concerning the number of colors k, typical values considered are 1 or n. In the first case any vertex can be mapped to any point, while in the second one the coloring defines a one-to-one mapping between the vertices and the points. Values of k between 1 and n define a mapping of groups of vertices to group of points. Drawings are often required to be planar. Straight-line, polyline, and orthogonal drawings are those most considered. Straight-line and polyline have been studied also in the upward version. When studying polyline point-set embeddings one typically wants to keep the number of bends per edge as small as possible.

R17: Universal point sets have been typically studied in the version with $k = 1$. When studying universal point sets one is typically interested in the size of a universal point set for a family and for a type of drawing.

FACTS

F24: Deciding whether an n-vertex planar graph admits a planar straight-line point-set embedding on a point set of size n is NP-complete [Cab06].

F25: An n-vertex planar graph admits a planar straight-line point-set embedding on every point set of size n in general position if and only if it is outerplanar [GMPP91, CU96].

F26: A planar straight-line point-set embedding of an n-vertex tree on any point set of size n in general position can be computed in $\Theta(n \log n)$ [BMS97].

F27: A planar straight-line point-set embedding of an n-vertex outerplanar graph on any point set of size n in general position can be computed in $O(n \log^3 n)$ [Bos02].

F28: A planar 2-bend point-set embedding of an n-vertex planar graph on any point set of size n can be computed in $O(n \log n)$ [KW02].

F29: There exists an n-vertex planar graph G and a point set S of size n such that in every polyline point-set embedding of G on S there exists at least one edge with two bends [KW02].

F30: A planar n-colored $O(n)$-bend point-set embedding of an n-vertex n-colored planar graph on any n-colored point set of size n in general position can be computed in $O(n^2)$ [PW01].

F31: There exists an n-vertex planar graph G and a point set S of size n such that in every polyline n-colored point-set embedding of G on S, almost surely there exist $\Omega(n)$ edges with $\Omega(n)$ bends [PW01, BDL08, Gor12].

F32: For every $1 < k \leq n$, a planar k-colored $O(n)$-bend point-set embedding of an n-vertex planar graph on any k-colored point set of size n can be computed in $O(n^2 \log n)$ [BDL08].

F33: For every $1 < k \leq n$, there exists an n-vertex k-colored planar graph G and a k-colored point set S of size n such that in every polyline k-colored point-set embedding of G on S, there exist $\Omega(n)$ edges with $\Omega(n)$ bends [BDL08].

F34: A planar 2-colored 5-bend point-set embedding of an n-vertex outerplanar graph on any 2-colored point set of size n can be computed in $O(n \log n)$ time [DDL$^+$08].

F35: For every positive integer $h > 0$, there exists an n-vertex 3-colored planar graph G (with $n = \Theta(h^2)$) and a 3-colored point set S of size n such that in every polyline 3-colored point-set embedding of G on S, there exists at least one edge with more than h bends [DDL$^+$08].

F36: Deciding whether an n-vertex planar graph admits a planar orthogonal point-set embedding on a point set of size n is NP-complete [KKRW10].

F37: Deciding whether an n-vertex n-colored perfect matching admits a planar n-colored orthogonal grid point-set embedding on an n-colored point set of size n is NP-complete [KKRW10].

F38: Deciding whether an n-vertex n-colored perfect matching admits a planar n-colored orthogonal grid point-set embedding on an n-colored point set of size n is NP-complete [KKRW10].

F39: It is possible to decide in $O(n^3)$ time whether an n-vertex n-colored perfect matching admits a planar n-colored orthogonal point-set embedding on an n-colored point set of size n [KKRW10].

F40: An upward planar 1-bend point-set embedding of an n-vertex series-parallel digraph on any point set of size n can be computed in $O(n \log n)$ [DDLW06].

F41: An upward planar 2-bend point-set embedding of an n-vertex planar digraph on any point set of size n can be computed in $O(n \log n)$ [GLMS07].

F42: No biconnected digraph with $n \geq 4$ vertices admits an upward planar straight-line point-set embedding on every convex point set of size n [BDD$^+$10].

F43: For every $k \geq 3$, there exists a $3k$-vertex upward planar digraph that admits an upward planar straight-line point-set embedding on every convex point set of size $3k$ but not on every point set in general position of size $3k$. [AFG$^+$11].

F44: For every odd $n \geq 5$, there exists an n-vertex directed tree T and a convex point set S of size n such that T does not admit an upward planar straight-line point-set embedding on S. [BDD$^+$10].

F45: Every n-vertex directed caterpillar admits an upward straight-line point-set embedding on every convex point set of size n [AFG$^+$11].

F46: Every universal point set for straight-line planar drawings of n-vertex planar graphs has size at least $1.235n$ [Kur04].

F47: There exists a universal point set of size $\frac{8}{9}n^2$ for planar straight-line drawings of n-vertex planar graphs [Bra08].

F48: There exists a universal point set of size $O(n(\frac{\log n}{\log \log n})^2)$ for planar straight-line drawings of n-vertex simply-nested planar graphs [ADK$^+$12].

F49: There exists a universal point set of size n for planar 1-bend drawings of n-vertex planar graphs [ELLW10].

F50: There exists a universal point set of size $10n - 18$ for planar 3-bend drawings of n-vertex planar graphs, when bends are required to be drawn at points of the point set [DEL$^+$12].

F51: There exists a universal point set of size $O(n \log n)$ for planar 2-bend drawings of n-vertex planar graphs, when bends are required to be drawn at points of the point set [DEL$^+$12].

F52: There exists a universal point set of size $O(\frac{n^2}{\log n})$ for planar 1-bend drawings of n-vertex planar graphs, when bends are required to be drawn at points of the point set [DEL$^+$12].

F53: For every $1 \leq k \leq n$, there exists a universal point set of size $kn - k^2 + 1$ for planar $(3k + 7)$-bend drawings of n-vertex k-colored planar graphs [DLT10].

F54: For every $1 \leq k \leq n$, there exists a universal point set of size $O(kn^2) \times O(kn^2)$ for planar $(6k + 5)$-bend drawings of n-vertex k-colored planar graphs [DLT10].

F55: There exists a universal point set of size $n^2 - 2n + 2$ for planar 1-bend orthogonal drawings of n-vertex degree-4 trees [DFF+12].

F56: There exists a universal point set of size $4n$ for planar geodesic orthogonal[2] drawings of n-vertex degree-4 trees [DFF+12].

F57: There exists a universal point set of size n for planar geodesic orthogonal drawings of n-vertex degree-3 trees [DFF+12].

F58: There exists a universal point set of size $O(n^2)$ for upward planar straight-line drawings of n-vertex directed paths [Mch12].

Simultaneous Embeddings

In some application areas one has different types of relations defined over a set of elements. For example, in social network analysis different relationships among the same set of people can be studied. Analogously in biology, different algorithms can produce different phylogenetic trees on the same set of organisms. In terms of graphs we have two (or more) graphs with the same vertex set and different edge sets. In this situation it may be desirable to simultaneously visualize two (or more) drawings of the two (or more) graphs under consideration. A simultaneous embedding of k graphs ($k > 1$) consists of k drawings (one for each graph) such that every vertex is represented by the same point in all drawings. The typical requirement for a simultaneous embedding is that each single drawing is planar, while the union of all drawings may not be planar. We report below basic definitions and facts. See also [BKR] for more references.

DEFINITIONS

D62: Let $G_i = (V, E_i)$ ($i = 1, 2, \ldots, k$) be a set of k planar graphs. A ***simultaneous embedding*** of $G_1, G_2, \ldots G_k$ is a set of k drawings Γ_i of G_i such that each vertex v is represented by the same point of the plane in all Γ_i.

D63: A ***simultaneous embedding with fixed edges*** is a simultaneous embedding where each edge $e \in E_i \cap E_j$ ($1 \leq i, j \leq k$) is drawn in the same way (i.e., by the same curve) in Γ_i and Γ_j.

D64: A ***simultaneous geometric embedding*** is a simultaneous embedding such that every Γ_i ($1 \leq i \leq k$) is a straight-line drawing.

REMARKS

R18: A simultaneous geometric embedding is also a simultaneous embedding with fixed edges, and a simultaneous embedding with fixed edges is also a simultaneous embedding.

R19: Most of the research about simultaneous embeddings concentrated on studying whether two (or more) graphs from some families can always be simultaneous embedded or not.

---

[2]In a geodesic orthogonal drawing the length of each edge is equal to the Manhattan distance of its endpoints.

FACTS

F59:   There exist a planar graph and an outerplanar graph that do not admit a simultaneous geometric embedding. [BCD+07]

F60:   There exist a planar graph and a path that do not admit a simultaneous geometric embedding. [BCD+07]

F61:   There exist a planar graph and a matching that do not admit a simultaneous geometric embedding. [CvKL+10]

F62:   There exist two outerplanar graphs that do not admit a simultaneous geometric embedding. [BCD+07]

F63:   There exists two trees that do not admit a simultaneous geometric embedding. [GKV09]

F64:   There exists a tree and a path that do not admit a simultaneous geometric embedding. [AGKN12]

F65:   There exist three paths that do not admit a simultaneous geometric embedding. [BCD+07]

F66:   Every pair consisting of a tree and a matching both with n vertices admits a simultaneous geometric embedding. [CvKL+10]

F67:   Every pair of caterpillars[3] with n vertices admits a simultaneous geometric embedding on an integer grid of size $3n \times 3n$. [BCD+07]

F68:   Every pair consisting of a caterpillar and a path both with n vertices admits a simultaneous geometric embedding on an integer grid of size $n \times 2n$. [BCD+07]

F69:   Every pair of paths with n vertices admits a simultaneous geometric embedding on an integer grid of size $n \times n$. [BCD+07]

F70:   Every pair of cycles with n vertices admits a simultaneous geometric embedding on an integer grid of size $4n \times 4n$. [BCD+07]

F71:   For any fixed $k > 1$, deciding whether k planar graphs admit a simultaneous geometric embedding is NP-hard. [EBGJ+08]

F72:   For any fixed $k > 13$, deciding whether k planar graphs, each with a fixed imbedding, admit a simultaneous geometric embedding is NP-complete. [ADF11]

F73:   There exist two outerplanar graphs that do not admit a simultaneous embedding with fixed edges. [Fra07]

F74:   Every pair consisting of a tree and a path both with n vertices admits a simultaneous embedding with fixed edges on an integer grid of size $O(n) \times O(n^2)$. The drawing of the path is a straight-line drawing, while in the drawing of the tree each edge has at most one bend. [EK05]

---

[3]A caterpillar is a tree such that after removing all leaves we are left with a path.

F75: Every pair consisting of an outerplanar graph and a path both with n vertices admits a simultaneous embedding with fixed edges on an integer grid of size $O(n) \times O(n^2)$. The drawing of the path is a straight-line drawing, while in the drawing of the outerplanar graph each edge has at most one bend. [DL07]

F76: Every pair consisting of an outerplanar graph and a cycle both with n vertices admits a simultaneous embedding with fixed edges on an integer grid of size $O(n^2) \times O(n^2)$. In both drawings each edge has at most one bend. [DL07]

F77: Every pair consisting of a planar graph and a tree both with n vertices admits a simultaneous embedding with fixed edges. [Fra07]

F78: For any fixed $k > 2$, deciding whether k planar graphs admit a simultaneous embedding with fixed edges is NP-complete. [GJP+06]

F79: It is possible to decide in $O(n)$ time whether two n-vertex planar graphs admit a simultaneous embedding with fixed edges if one of them has a fixed imbedding. [ADF+10]

F80: It is possible to decide in $O(n)$ time whether two n-vertex planar graphs admit a simultaneous embedding with fixed edges if their intersection is a biconnected graph. [HJL10]

F81: It is possible to decide in $O(n^3)$ time whether three n-vertex planar graphs admit a simultaneous embedding with fixed edges if each of them has a fixed imbedding. [ADF11]

F82: For any fixed $k > 14$, deciding whether k planar graphs, each with a fixed imbedding, admit a simultaneous embedding with fixed edges is NP-complete. [ADF11]

F83: Every pair of planar graphs with n vertices admits a simultaneous embedding on an integer grid of size $O(n^2) \times O(n^2)$. In both drawings each edge has at most two bends. [DL07]

F84: Every pair of trees with n vertices admits a simultaneous embedding on an integer grid of size $O(n^2) \times O(n^2)$. In both drawings each edge has at most one bend. [EK05]

Lombardi Drawings

Inspired by the work of the American artist Mark Lombardi, Duncan et al. introduced and studied Lombardi drawings of graphs. In a Lombardi drawing of a graph edges are drawn with circular arcs and perfect angular resolution, i.e., consecutive edges are equiangularly spaced around each vertex.

DEFINITIONS

D65: A ***Lombardi drawing*** of a graph G is a drawing of G in the plane in which vertices are represented as points, edges are represented as line segments or circular arcs between their endpoints, and every vertex has perfect angular resolution.

D66: A ***circular Lombardi drawing*** is a Lombardi drawing where all vertices lie on a circle.

D67: A *k-circular Lombardi drawing* is a Lombardi drawing where all vertices lie on k concentric circles ($k > 1$).

D68: A *k-Lombardi drawing* ($k > 1$) of a graph G is a drawing of G in the plane in which vertices are represented as points, edges are represented as a sequence of line segments or circular arcs between their endpoints, and every vertex has perfect angular resolution. A k-Lombardi drawing is *smooth* if every edge is continuously differentiable, i.e., no edge in the drawing has a sharp bend. A non-smooth k-Lombardi drawing is said to be *pointed*.

FACTS

F85: Every ordered tree with n vertices admits a planar order preserving Lombardi drawing in area $O(n^3)$. [DEG$^+$12b]

F86: Every bipartite d-regular graph with n vertices admits a circular Lombardi drawing that can be constructed in $O(dn \log d)$ time. [DEG$^+$12b]

F87: Every $4k$-regular graph with n vertices admits a circular Lombardi drawing that can be constructed in $O(kn \log k)$ time. [DEG$^+$12b]

F88: Every d-regular graph of odd degree with n vertices admits a circular Lombardi drawing if and only if has a perfect matching. When such a drawing exists it can be constructed in $O(dn^{1.5})$ time. [DEG$^+$12b]

F89: Every 3-regular bridgeless graph with n vertices admits a circular Lombardi drawing that can be constructed in $O(n \log^3 n \log \log n)$ time. [DEG$^+$12b]

F90: When $d \equiv 2 \pmod 4$, it is NP-complete to test whether a d-regular graph has a circular Lombardi drawing. [DEG$^+$12b]

F91: Every outerplanar series-parallel graph admits a Lombardi drawing. [DEG$^+$12b]

F92: There exist planar graphs that do not admit planar Lombardi drawings. [DEG$^+$12b]

F93: Every Halin graph with n vertices admits a planar Lombardi drawing that can be computed in $O(n)$ time. [DEG$^+$12b]

F94: There exist graphs that do not admit Lombardi drawings. [DEG$^+$12a]

F95: Every graph has a smooth 2-Lombardi drawing. [DEG$^+$12a]

F96: Every planar graph of maximum degree 3 has a smooth planar 2-Lombardi drawing. [DEG$^+$12a]

F97: Every planar graph has a smooth planar 3-Lombardi drawing. [DEG$^+$12a]

F98: Every planar graph has a pointed planar 2-Lombardi drawing. [DEG$^+$12a]

Drawings with Large Crossing Angles

One of the most desirable properties when drawing a graph is planarity. Planar drawings are easily readable and aesthetically pleasant. This is not only suggested by intuition but also confirmed by experiments with end-users [PCA02, Pur00, WPCM02]. Unfortunately, since not all graphs are planar, crossings are often unavoidable. Some recent cognitive experiments [Hua07, HHE08], however, showed that if edges cross forming "large" angles, then crossings do not inhibit human performance in reading the graph. Based on this experiment an emerging research line in graph drawing studies non-planar drawings of graphs with large angles at crossings. We report below basic definitions and facts. See also [DL12] for more references.

DEFINITIONS

D69: A **RAC drawing (Right Angle Crossing drawing)** is a drawing of a graph where any two crossing edges form a $\frac{\pi}{2}$ angle.

D70: A 2-**layer RAC drawing** is a straight-line drawing of a bipartite graph such that each vertex partition set is drawn on a distinct horizontal line.

D71: An ACE_α **drawing** is a drawing of a graph where any two crossing edges form an angle that is exactly α $(0 < \alpha < \frac{\pi}{2})$.

D72: An ACL_α **drawing** is a drawing of a graph where any two crossing edges form an angle that is at least α $(0 < \alpha < \frac{\pi}{2})$.

FACTS

F99: Every straight-line RAC drawing with n vertices has at most $4n - 10$ edges. Also, for any $k \geq 3$ there exists a graph with $n = 3k - 5$ vertices and $4n - 10$ edges that admits a straight-line RAC drawing [DEL11].

F100: Every straight-line RAC drawing with n vertices and $4n - 10$ edges is 1-planar.[4] Also, for any $k \geq 0$ there exists a 1-planar graph with $n = 8 + 4k$ vertices and $4n - 10$ edges that does not admit a RAC drawing [EL12].

F101: Deciding whether a graph admits a straight-line RAC drawing is NP-hard. [ABS11b]

F102: Deciding whether a digraph admits a straight-line upward RAC drawing is NP-hard. [ACD$^+$11]

F103: There exists a $O(n)$-time algorithm to decide whether an n-vertex bipartite graph admits a 2-layer RAC drawing. If the drawing exists it can be computed in $O(n)$ time. [DDEL11]

F104: For any given $k > 0$, deciding whether a bipartite graph has a subgraph with at least k edges that admits a 2-layer RAC drawing is NP-hard. [DDEL11]

F105: A complete bipartite graph K_{n_1,n_2} $(n_1 \leq n_2)$ admits a straight-line RAC drawing if either $n_1 \leq 2$, or $n_1 = 3$ and $n_2 \leq 4$. [DEL10]

---

[4] A 1-planar graph is a graph that admits a drawing where each edge is crossed at most once.

F106: There exists a planar graph G such that any straight-line RAC drawing of G requires $\Omega(n^2)$ area. [ACD$^+$11]

F107: A RAC drawing with n vertices and at most one bend per edge has at most $6.5n - 13$ edges. [AFK$^+$12]

F108: A RAC drawing with n vertices and at most two bends per edge has at most $74.2n$ edges. [AFK$^+$12]

F109: Every graph admits a RAC drawing with at most three bends per edge and $O(n^4)$ area. [DEL11]

F110: Every graph admits a RAC drawing with at most four bends per edge and $O(n^3)$ area. [DDLM11]

F111: Every n-vertex graph with maximum vertex degree six admits a RAC drawing with at most two bends per edge and $O(n^2)$ area. [ACD$^+$11]

F112: Every n-vertex graph with maximum vertex degree three admits a RAC drawing with at most one bend per edge and $O(n^2)$ area. [ACD$^+$11]

F113: A straight-line ACL_α drawing with n vertices has at most $\frac{\pi}{\alpha}(3n - 6)$ edges. [DGMW11]

F114: A straight-line ACE_α drawing with n vertices has at most $3(3n - 6)$ edges. [AFT11]

F115: An ACE_α drawing with n vertices and at most one bend per edge has at most $27n$ edges. [AFT11]

F116: An ACE_α drawing with n vertices and at most two bends per edge has at most $477n$ edges. [AFT11]

F117: Every graph with n vertices admits an ACL_α drawing with at most one bend per edge and area $O(n^2)$. [DDLM11]

REMARK

R20: We conclude this section by mentioning that recently RAC simultaneous embeddings, i.e., simultaneous embedding where crossings between edges of the two different graphs form right angles, have been introduced and studied by Argyriou et al. [ABKS12]

Drawings with Few Slopes

In a paper by Dujmović et al. [DESW07] a new aesthetic requirement is introduced for straight-line drawings of graphs: the minimization of the number of distinct segment slopes used in the drawing. The idea is that using few slopes, especially if chosen from a "nice" set of slopes, should increase the readability. As an example, consider orthogonal drawings where only two "nice" slopes are used: the horizontal one and the vertical one. The study of straight-line drawings using few slopes is related to the study of the slope number of a graph introduced by Wade and Chu [WC94]. The slope number of a graph G is the minimum number of segment slopes in any straight-line drawing of G. It is immediate to see that the slope number of G is at least $\lceil \frac{d}{2} \rceil$ and at most m, where d is the maximum vertex degree and m is the number of edges of G. More recently, the study of polyline drawings of graphs with the minimum number of slopes has also been considered.

DEFINITIONS

D73: Let Γ be a polyline drawing of a graph G. Let s be a segment of Γ and let ℓ be the straight-line containing s. The **slope** of s is the angle swept from the X-axis in the anticlockwise direction to ℓ (and is thus a value in $[0, \pi)$).

D74: Let G be a graph. The **slope number** of G is the minimum number of slopes needed to construct any straight-line drawing of G.

D75: Let G be a planar graph. The **planar slope number** of G is the minimum number of slopes needed to construct any planar straight-line drawing of G.

D76: Let G be a plane graph, i.e., a planar graph with a fixed planar imbedding. The **plane slope number** of G is the minimum number of slopes needed to construct any planar straight-line drawing of G that preserves the given planar imbedding.

FACTS

F118: The plane slope number of every tree with maximum degree Δ is $\lceil \frac{\Delta}{2} \rceil$. [DESW07]

F119: The plane slope number of every maximal outerplanar graph with n vertices is at most n. [DESW07]

F120: The (outer)planar slope number of every outerplanar graph with maximum vertex degree d vertices is at most $d - 1$. For every $d \geq 4$ there exists an outerplanar graph of maximum vertex degree d whose (outer)planar slope number is at least $d - 1$. [KMW12]

F121: The plane slope number of every n-vertex plane 2-tree is at most $2n - 3$. For every $n \geq 3$ there exists an n-vertex plane 2-tree whose plane slope number is at least $2n - 3$. [DESW07]

F122: The planar slope number of every n-vertex 2-tree is at most $\frac{3}{2}n$. [DESW07]

F123: The plane slope number of every n-vertex plane 3-tree is at most $2n - 2$. For every $n \geq 4$ there exists an n-vertex plane 3-tree whose plane slope number is at least $2n - 2$. [DESW07]

F124: The planar slope number of every n-vertex planar 3-tree of maximum vertex degree d is at most d^5. [JJK$^+$]

F125: The plane slope number of every n-vertex plane graph is at most $2n - 10$. [DESW07]

F126: For every $d \geq 5$ there exist graphs with maximum vertex degree d whose slope number is arbitrarily large. [PP06, DSW07]

F127: The slope number of every graph with maximum vertex degree 3 is at most 4. [MP12]

F128: The planar slope number of every planar graph with maximum vertex degree d is at most $2^{O(d)}$. [KPP11]

F129: Every planar graph with maximum vertex degree d admits a polyline drawing with at most 1 bend per edge using at most $2d$ slopes. [KPP11]

F130: Every planar graph with maximum vertex degree $d \geq 3$ admits a polyline drawing with at most 2 bends per edge using at most $\lceil \frac{d}{2} \rceil$ slopes. The only exception is the graph formed by the edges of an octahedron, which is 4-regular, but requires 3 slopes. [KPP11]

Witness Proximity and Approximate Proximity Drawings

In a witness proximity drawing we look at a set of points that represent the vertices and at a set of points that play the role of the witnesses. The existence/absence of an edge in the drawing depends on the location of the witness points (the set of witness points and the set of points representing the vertices of the graph in a drawing may not coincide).

DEFINITIONS

D77: In a *positive witness proximity drawing* Γ two vertices u,v are adjacent if and only if the proximity region of u and v contains at least one vertex that belongs to the set of witness points.

D78: In a *negative witness proximity drawing* u and v are adjacent if and only if their region of influence does not contain any of the witness points (it may, however, contain other vertices of the graph that are not witnesses).

FACT

F131: *Witness proximity* has been introduced and studied in a series of papers by Aronov, Dulieu, and Hurtado [ADH, ADH11a, ADH11b].

Proximity drawability imposes severe restrictions on the families of the representable graphs. For example, a tree can be realized as a minimum spanning tree only if its vertex degree is at most six. In order to overcome restrictions on the combinatorial structure of the drawable graphs, recent papers study straight-line drawings of graphs that are "good approximations" of proximity drawings.

DEFINITIONS

D79: A $(1 + \varepsilon)$-*EMST drawing* is a planar straight-line drawing of a tree such that, for any fixed $\varepsilon > 0$, the distance between any two vertices is at least $\frac{1}{1+\varepsilon}$ the length of the longest edge in the path connecting them.

D80: Let D be a disk with center c and radius r, and let ε_1 and ε_2 be two nonnegative real numbers. The ε_1-*shrunk disk of D* is the disk centered at c and having radius $\frac{r}{1+\varepsilon_1}$; the ε_2-*expanded disk of D* is the disk centered at c and having radius $(1 + \varepsilon_2)r$. An $(\varepsilon_1, \varepsilon_2)$-*proximity drawing* is a planar straight-line drawing where the proximity region of two adjacent vertices is defined by using ε_1-shrunk disks, while the region of influence of two nonadjacent vertices uses ε_2-expanded disks.

FACTS

F132: in [DDLM10, DDLM12a] it is shown that every tree T has a $(1 + \varepsilon)$-EMST drawing for any given $\varepsilon > 0$ and that this drawing can be computed in linear time in the real RAM model of computation.

F133: In [EGK$^+$12] it is proved that one can arbitrarily approximate a proximity drawing of any planar graph for some of the most studied definitions of proximity. Namely, it is shown that for any positive values of $\varepsilon_1, \varepsilon_2$ an embedded planar graph admits an $(\varepsilon_1, \varepsilon_2)$-Gabriel drawing and an $(\varepsilon_1, \varepsilon_2)$-Delaunay drawing and an $(\varepsilon_1, \varepsilon_2)$-$\beta$-drawing ($1 \leq \beta \leq \infty$) that preserve the given embedding.

REMARKS

R21: The previous results are, in a sense, tight since for each of the above types of proximity rules there are embedded planar graphs that do not have an embedding preserving $(\varepsilon_1, \varepsilon_2)$-proximity drawing with either $\varepsilon_1 = 0$ or $\varepsilon_2 = 0$.

R22: Approximate proximity drawings generalize weak proximity drawings. Namely, an $(\varepsilon_1, \varepsilon_2)$-proximity drawing is a weak proximity drawing if $\varepsilon_1 = 0$ and $\varepsilon_2 = \infty$.

Cluster Planarity

When the graph to be displayed consists of hundreds or thousands of vertices, a complete visualization of the data is typically not effective for the user. To handle large graphs and their visualizations, a lot of attention has been devoted in the last decade to the study of graphs whose vertices are organized into a hierarchy of clusters. Making clusters with the vertices helps in the analysis of complex relational data, since it makes it possible to explore large graphs at different levels of detail by collapsing or expanding the clusters.

DEFINITIONS

D81: A *clustered graph* $C = (G, T)$, also called *c-graph*, consists of an undirected graph G and a rooted tree T called an *inclusion tree* of C, such that: (i) the leaves of T are the vertices of G; (ii) each internal node μ of T has at least two children and represents a *cluster* $V(\mu)$ of the vertices of G that are leaves of the subtree rooted at μ.

D82: A *c-planar drawing* (resp. an *oc-planar drawing*) of a c-graph C is a visualization of C in the plane such that each vertex v of G is drawn as a point $p(v)$, each edge of G is drawn as a simple Jordan arc, and each node μ of T is drawn as a simple closed region $R(\mu)$ according to the following rules: (1) $R(\mu)$ contains the drawing of $G(\mu)$. (2) If $V(\mu) \subset V(\nu)$, $R(\nu)$ contains $R(\mu)$. (3) If $v \notin V(\mu)$, $p(v)$ is outside $R(\mu)$. (4) There is no *edge crossing*, i.e., any two edges of G never cross. (5) There is no *edge-region crossing*, i.e., there is no edge of G that crosses the boundary of a region $R(\mu)$ twice.

D83: A c-graph is *c-planar* if it admits a c-planar drawing

FACTS

F134:   The problem of testing whether a c-graph is c-planar was first introduced in a paper by Feng, Cohen, and Eades [FCE95b] that inspired and motivated a sequence of papers on this topic. Feng et al. [FCE95b] describe a quadratic-time c-planarity testing algorithm for clustered graphs where each cluster induces a connected subgraph.

F135:   Linear-time testing algorithms for the same class of clustered graphs as the one studied by Feng, Cohen, and Eades are described in [CDF$^+$08, Dah98, DKM06].

REMARK

R23:   Feng et al. leave as open the problem of testing a c-graph for c-planarity when clusters can induce non-connected subgraphs; the time complexity of this problem is still unknown.

FACTS

F136:   Special cases for which polynomial-time testing algorithms exist have been described in the literature, such as [DF09, JKK$^+$09].

F137:   The relationship between planarity and c-planarity has been studied in [CW06] and a planarization algorithm for c-graphs that are not c-planar is described in [DDM02].

F138:   Assuming that a clustered graph C is c-planar, several algorithms and bounds are known for constructing c-planar drawings of C. [DDF09, EFN99, NK07]

F139:   Every c-planar graph admits a straight-line convex drawing [EFLN06], even if the shape of each cluster is fixed in advance [AFK11]. Straight-line convex drawings might require exponential area [FCE95a].

F140:   Extensions of the studies above to the case of a clustered graph with clusters that can partially intersect can be found in [DGL08, OS07].

10.3.8   Sources and Related Material

Several books devoted to graph drawing have been published [Kam89, DETT99, JM03, KW01, NR04, Sug02]. The proceedings of the annual Symposium on *Graph Drawing* are published by Springer-Verlag in the *Lecture Notes in Computer Science* series (volumes 7034, 6502, 5849, 5417, 4875, 4372, 3843, 3383, 2912, 2528, 2265, 1984, 1731, 1547, 1353, 1190, 1027, 894).

REMARKS

R24:   Special issues devoted to graph drawing have appeared in *Algorithmica* (vol. 16, no. 1, 1996), *Computational Geometry: Theory and Applications* (vol. 9, no. 1–2, 1998), the *Journal of Visual Languages and Computing* (vol. 6, 1995), *International Journal of Foundations of Computer Science* (vol. 17, no. 5, 2006), and the *Journal of Graph Algorithms and Applications* (vol. 3, no. 4, 1999; vol. 4, no. 3, 2000; vol. 6, no. 1, 2002; vol. 6, no. 3, 2002; vol. 7, no. 4, 2003; vol. 8, no. 2, 2004; vol. 9, no. 1, 2005; vol. 9, no. 3, 2005; vol. 11, no. 2, 2007; vol. 12, no. 1, 2008; vol. 13, no. 3, 2009; vol. 14, no. 1, 2010; vol. 15, no. 1, 2011; vol. 16, no. 1, 2012).

R25: Sites with pointers to graph drawing resources and tools include the WWW page maintained by Tamassia (`http://www.cs.brown.edu/people/rt/gd.html`), the WWW page maintained by Brandes (`http://graphdrawing.org/`), the Graph drawing e-print archive (`http://gdea.informatik.uni-koeln.de/`), and the Graph-Archive (`http://www.graph-archive.org/doku.php`).

References

[AAP⁺97] Pankaj K. Agarwal, Boris Aronov, János Pach, Richard Pollack, and Micha Sharir, *Quasi-planar graphs have a linear number of edges*, Combinatorica **17** (1997), no. 1, 1–9.

[AB12] Soroush Alamdari and Therese Biedl, *Planar open rectangle-of-influence drawings with non-aligned frames*, Graph Drawing (Marc van Kreveld and Bettina Speckmann, eds.), Lecture Notes in Computer Science, vol. 7034, Springer, Berlin/Heidelberg, 2012, 10.1007/978-3-642-25878-7_3, pp. 14–25.

[ABC⁺11] Patrizio Angelini, Till Bruckdorfer, Marco Chiesa, Fabrizio Frati, Michael Kaufmann, and Claudio Squarcella, *On the area requirements of Euclidean minimum spanning trees*, Algorithms and Data Structures (Frank Dehne, John Iacono, and Jrg-Rdiger Sack, eds.), Lecture Notes in Computer Science, vol. 6844, Springer, Berlin/Heidelberg, 2011, 10.1007/978-3-642-22300-6_3, pp. 25–36.

[ABKS12] Evmorfia Argyriou, Michael Bekos, Michael Kaufmann, and Antonios Symvonis, *Geometric RAC simultaneous drawings of graphs*, Computing and Combinatorics, Lecture Notes in Computer Science, Springer, Berlin/Heidelberg, 2012, To appear.

[ABS11a] Evmorfia Argyriou, Michael Bekos, and Antonios Symvonis, *Maximizing the total resolution of graphs*, Graph Drawing (Ulrik Brandes and Sabine Cornelsen, eds.), Lecture Notes in Computer Science, vol. 6502, Springer Berlin/Heidelberg, 2011, 10.1007/978-3-642-18469-7_6, pp. 62–67.

[ABS11b] _____, *The straight-line RAC drawing problem is NP-hard*, SOFSEM 2011: Theory and Practice of Computer Science (Ivana Cern, Tibor Gyimthy, Juraj Hromkovic, Keith Jefferey, Rastislav Krlovic, Marko Vukolic, and Stefan Wolf, eds.), Lecture Notes in Computer Science, vol. 6543, Springer, Berlin/Heidelberg, 2011, 10.1007/978-3-642-18381-2_6, pp. 74–85.

[ACD⁺11] Patrizio Angelini, Luca Cittadini, Giuseppe Di Battista, Walter Didimo, Fabrizio Frati, Michael Kaufmann, and Antonios Symvonis, *On the perspectives opened by right angle crossing drawings*, J. Graph Algorithms Appl. **15** (2011), no. 1, 53–78.

[ADF⁺10] Patrizio Angelini, Giuseppe Di Battista, Fabrizio Frati, Vít Jelínek, Jan Kratochvíl, Maurizio Patrignani, and Ignaz Rutter, *Testing planarity of partially embedded graphs*, Proceedings of the Twenty-First Annual ACM-SIAM Symposium on Discrete Algorithms, SODA 2010, Austin, Texas,

USA, January 17-19, 2010 (Moses Charikar, ed.), SIAM, 2010, pp. 202–221.

[ADF11]   Patrizio Angelini, Giuseppe Di Battista, and Fabrizio Frati, *Simultaneous embedding of embedded planar graphs*, Algorithms and Computation (Takao Asano, Shin-ichi Nakano, Yoshio Okamoto, and Osamu Watanabe, eds.), Lecture Notes in Computer Science, vol. 7074, Springer, Berlin/Heidelberg, 2011, 10.1007/978-3-642-25591-5_29, pp. 271–280.

[ADH]   Boris Aronov, Muriel Dulieu, and Ferran Hurtado, *Witness gabriel graphs*, Computational Geometry, to appear.

[ADH11a]   ―――, *Witness (delaunay) graphs*, Computational Geometry **44** (2011), no. 67, 329–344.

[ADH11b]   Boris Aronov, Muriel Dulieu, and Ferran Hurtado, *Witness rectangle graphs*, Algorithms and Data Structures (Frank Dehne, John Iacono, and Jrg-Rdiger Sack, eds.), Lecture Notes in Computer Science, vol. 6844, Springer, Berlin/Heidelberg, 2011, 10.1007/978-3-642-22300-6_7, pp. 73–85.

[ADK+12]   Patrizio Angelini, Giuseppe Di Battista, Michael Kaufmann, Tamara Mchedlidze, Vincenzo Roselli, and Claudio Squarcella, *Small point sets for simply-nested planar graphs*, Graph Drawing (Marc van Kreveld and Bettina Speckmann, eds.), Lecture Notes in Computer Science, vol. 7034, Springer, Berlin/Heidelberg, 2012, 10.1007/978-3-642-25878-7_8, pp. 75–85.

[AFG+11]   Patrizio Angelini, Fabrizio Frati, Markus Geyer, Michael Kaufmann, Tamara Mchedlidze, and Antonios Symvonis, *Upward geometric graph embeddings into point sets*, Graph Drawing (Ulrik Brandes and Sabine Cornelsen, eds.), Lecture Notes in Computer Science, vol. 6502, Springer, Berlin/Heidelberg, 2011, 10.1007/978-3-642-18469-7_3, pp. 25–37.

[AFK11]   Patrizio Angelini, Fabrizio Frati, and Michael Kaufmann, *Straight-line rectangular drawings of clustered graphs*, Discrete & Computational Geometry **45** (2011), 88–140, 10.1007/s00454-010-9302-z.

[AFK+12]   Karin Arikushi, Radoslav Fulek, Balázs Keszegh, Filip Morić, and Csaba D. Tóth, *Graphs that admit right angle crossing drawings*, Computational Geometry **45** (2012), no. 4, 169–177.

[AFT11]   Eyal Ackerman, Radoslav Fulek, and Csaba Tóth, *On the size of graphs that admit polyline drawings with few bends and crossing angles*, Graph Drawing (Ulrik Brandes and Sabine Cornelsen, eds.), Lecture Notes in Computer Science, vol. 6502, Springer, Berlin/Heidelberg, 2011, 10.1007/978-3-642-18469-7_1, pp. 1–12.

[AGKN12]   Patrizio Angelini, Markus Geyer, Michael Kaufmann, and Daniel Neuwirth, *On a tree and a path with no geometric simultaneous embedding*, J. Graph Algorithms Appl. **16** (2012), no. 1, 37–83.

[AT07]   Eyal Ackerman and Gábor Tardos, *On the maximum number of edges in quasi-planar graphs*, J. Comb. Theory, Ser. A **114** (2007), no. 3, 563–571.

[BBBH10] Christian Bachmaier, Franz Brandenburg, Wolfgang Brunner, and Ferdinand Hübner, *A global k-level crossing reduction algorithm*, WALCOM: Algorithms and Computation (Md. Rahman and Satoshi Fujita, eds.), Lecture Notes in Computer Science, vol. 5942, Springer, Berlin/Heidelberg, 2010, 10.1007/978-3-642-11440-3_7, pp. 70–81.

[BBM99] Therese Biedl, Anna Bretscher, and Henk Meijer, *Rectangle of influence drawings of graphs without filled 3-cycles*, Graph Drawing (Jan Kratochvyl, ed.), Lecture Notes in Computer Science, vol. 1731, Springer Berlin / Heidelberg, 1999, 10.1007/3-540-46648-7_37, pp. 359–368.

[BCD+07] Peter Brass, Eowyn Cenek, Cristian A. Duncan, Alon Efrat, Cesim Erten, Dan P. Ismailescu, Stephen G. Kobourov, Anna Lubiw, and Joseph S.B. Mitchell, *On simultaneous planar graph embeddings*, Computational Geometry **36** (2007), no. 2, 117–130.

[BCG+] Christoph Buchheim, Markus Chimani, Carsten Gutwenger, Michael Jünger, and Petra Mutzel, *Handbook of graph drawing and visualization*, ch. Crossings and planarization, CRC Press, To appear.

[BDD+10] Carla Binucci, Emilio Di Giacomo, Walter Didimo, Alejandro Estrella-Balderrama, Fabrizio Frati, Stephen G. Kobourov, and Giuseppe Liotta, *Upward straight-line embeddings of directed graphs into point sets*, Computational Geometry **43** (2010), no. 2, 219–232.

[BDL08] Melanie Badent, Emilio Di Giacomo, and Giuseppe Liotta, *Drawing colored graphs on colored points*, Theoretical Computer Science **408** (2008), no. 2-3, 129–142.

[BDLL95] Prosenjit Bose, Giuseppe Di Battista, William Lenhart, and Giuseppe Liotta, *Proximity constraints and representable trees*, Graph Drawing (Roberto Tamassia and Ioannis Tollis, eds.), Lecture Notes in Computer Science, vol. 894, Springer, Berlin/Heidelberg, 1995, 10.1007/3-540-58950-3_389, pp. 340–351.

[BK02] Ulrik Brandes and Boris Köpf, *Fast and simple horizontal coordinate assignment*, Graph Drawing (Petra Mutzel, Michael Jnger, and Sebastian Leipert, eds.), Lecture Notes in Computer Science, vol. 2265, Springer, Berlin/Heidelberg, 2002, 10.1007/3-540-45848-4_3, pp. 33–36.

[BKR] Thomas Bläsius, Stephen G. Kobourov, and Ignaz Rutter, *Handbook of graph drawing and visualization*, ch. Simultaneous drawings, CRC Press, To appear.

[BKRS01] Oleg V. Borodin, Alexandr V. Kostochka, André Raspaud, and Eric Sopena, *Acyclic colouring of 1-planar graphs*, Discrete Applied Mathematics **114** (2001), no. 1–3, 29–41.

[BLL96] Prosenjit Bose, William Lenhart, and Giuseppe Liotta, *Characterizing proximity trees*, Algorithmica **16** (1996), 83–110, 10.1007/BF02086609.

[BMS97] Prosenjit Bose, Michael McAllister, and Jack Snoeyink, *Optimal algorithms to embed trees in a point set*, Journal of Graph Algorithms and Applications **2** (1997), no. 1, 1–15.

[Bos02]   Prosenjit Bose, *On embedding an outer-planar graph on a point set*, Computational Geometry: Theory and Applications **23** (2002), 303–312.

[BP07]   Ulrik Brandes and Christian Pich, *Eigensolver methods for progressive multidimensional scaling of large data*, Graph Drawing (Michael Kaufmann and Dorothea Wagner, eds.), Lecture Notes in Computer Science, vol. 4372, Springer, Berlin/Heidelberg, 2007, 10.1007/978-3-540-70904-6_6, pp. 42–53.

[Bra08]   Franz J. Brandenburg, *Drawing planar graphs on $\frac{8}{9}n^2$ area*, Electronic Notes in Discrete Mathematics **31** (2008), 37–40.

[BWZ10]   Christoph Buchheim, Angelika Wiegele, and Lanbo Zheng, *Exact algorithms for the quadratic linear ordering problem*, INFORMS Journal on Computing **22** (2010), no. 1, 168–177.

[Cab06]   Sergio Cabello, *Planar embeddability of the vertices of a graph using a fixed point set is np-hard*, J. Graph Algorithms Appl. **10** (2006), no. 2, 353–363.

[CDF$^+$08] Pier Francesco Cortese, Giuseppe Di Battista, Fabrizio Frati, Maurizio Patrignani, and Maurizio Pizzonia, *C-planarity of c-connected clustered graphs*, J. Graph Algorithms Appl. **12** (2008), no. 2, 225–262.

[CGMW10]   Markus Chimani, Carsten Gutwenger, Petra Mutzel, and Hoi-Ming Wong, *Layer-free upward crossing minimization*, J. Exp. Algorithmics **15** (2010), 2.2:2.1–2.2:2.27.

[CGMW11]   Markus Chimani, Carsten Gutwenger, Petra Mutzel, and Hoi-Ming Wong, *Upward planarization layout*, J. Graph Algorithms Appl. **15** (2011), no. 1, 127–155.

[CU96]   Netzahualcoyotl Castañeda and Jorge Urrutia, *Straight line embeddings of planar graphs on point sets*, Canadian Conference on Computational Geometry (CCCG '96) (F. Fiala, E. Kranakis, and J.-R. Sack, eds.), 1996, pp. 312–318.

[CvKL$^+$10] Sergio Cabello, Marc van Kreveld, Giuseppe Liotta, Henk Meijer, Bettina Speckmann, and Kevin Verbeek, *Geometric simultaneous embeddings of a graph and a matching*, Graph Drawing (David Eppstein and Emden Gansner, eds.), Lecture Notes in Computer Science, vol. 5849, Springer, Berlin/Heidelberg, 2010, 10.1007/978-3-642-11805-0_18, pp. 183–194.

[CW06]   Sabine Cornelsen and Dorothea Wagner, *Completely connected clustered graphs*, Journal of Discrete Algorithms **4** (2006), no. 2, 313–323.

[Dah98]   Elias Dahlhaus, *A linear time algorithm to recognize clustered planar graphs and its parallelization*, LATIN'98: Theoretical Informatics (Cludio Lucchesi and Arnaldo Moura, eds.), Lecture Notes in Computer Science, vol. 1380, Springer, Berlin/Heidelberg, 1998, 10.1007/BFb0054325, pp. 239–248.

[DDE$^+$12] Emilio Di Giacomo, Walter Didimo, Peter Eades, Seok-Hee Hong, and Giuseppe Liotta, *Bounds on the crossing resolution of complete geometric graphs*, Discrete Applied Mathematics **160** (2012), no. 12, 132–139.

[DDEL11] Emilio Di Giacomo, Walter Didimo, Peter Eades, and Giuseppe Liotta, *2-layer right angle crossing drawings*, Combinatorial Algorithms (Costas Iliopoulos and William Smyth, eds.), Lecture Notes in Computer Science, vol. 7056, Springer, Berlin/Heidelberg, 2011, 10.1007/978-3-642-25011-8_13, pp. 156–169.

[DDF09] Giuseppe Di Battista, Guido Drovandi, and Fabrizio Frati, *How to draw a clustered tree*, Journal of Discrete Algorithms **7** (2009), no. 4, 479–499.

[DDL⁺08] Emilio Di Giacomo, Walter Didimo, Giuseppe Liotta, Henk Meijer, Francesco Trotta, and Stephen K. Wismath, *k-colored point-set embeddability of outerplanar graphs*, J. Graph Algorithms Appl. **12** (2008), no. 1, 29–49.

[DDLM10] Emilio Di Giacomo, Walter Didimo, Giuseppe Liotta, and Henk Meijer, *Drawing a tree as a minimum spanning tree approximation*, Algorithms and Computation (Otfried Cheong, Kyung-Yong Chwa, and Kunsoo Park, eds.), Lecture Notes in Computer Science, vol. 6507, Springer, Berlin/Heidelberg, 2010, 10.1007/978-3-642-17514-5_6, pp. 61–72.

[DDLM11] _____, *Area, curve complexity, and crossing resolution of non-planar graph drawings*, Theory of Computing Systems **49** (2011), 565–575, 10.1007/s00224-010-9275-6.

[DDLM12a] Emilio Di Giacomo, Walter Didimo, Giuseppe Liotta, and Henk Meijer, *Drawing a tree as a minimum spanning tree approximation*, Journal of Computer and System Sciences **78** (2012), no. 2, 491–503.

[DDLM12b] Emilio Di Giacomo, Walter Didimo, Giuseppe Liotta, and Fabrizio Montecchiani, *h-quasi planar drawings of bounded treewidth graphs in linear area*, Graph-Theoretic Concepts in Computer Science, Lecture Notes in Computer Science, Springer, Berlin/Heidelberg, 2012, To appear.

[DDLW06] Emilio Di Giacomo, Walter Didimo, Giuseppe Liotta, and Stephen K. Wismath, *Book embeddability of series-parallel digraphs*, Algorithmica **45** (2006), no. 4, 531–547.

[DDM02] Giuseppe Di Battista, Walter Didimo, and Alessandro Marcandalli, *Planarization of clustered graphs*, Graph Drawing (Petra Mutzel, Michael Jünger, and Sebastian Leipert, eds.), Lecture Notes in Computer Science, vol. 2265, Springer, Berlin/Heidelberg, 2002, 10.1007/3-540-45848-4_5, pp. 113–117.

[DEG⁺12a] Christian Duncan, David Eppstein, Michael Goodrich, Stephen G. Kobourov, and Maarten Löffler, *Planar and poly-arc lombardi drawings*, Graph Drawing (Marc van Kreveld and Bettina Speckmann, eds.), Lecture Notes in Computer Science, vol. 7034, Springer, Berlin/Heidelberg, 2012, 10.1007/978-3-642-25878-7_30, pp. 308–319.

[DEG⁺12b] Christian A. Duncan, David Eppstein, Michael T. Goodrich, Stephen G. Kobourov, and Martin Nöllenburg, *Lombardi drawings of graphs*, J. Graph Algorithms Appl. **16** (2012), no. 1, 85–108.

[DEL10]   Walter Didimo, Peter Eades, and Giuseppe Liotta, *A characterization of complete bipartite RAC graphs*, Inf. Process. Lett. **110** (2010), no. 16, 687–691.

[DEL11]   ———, *Drawing graphs with right angle crossings*, Theoretical Computer Science **412** (2011), no. 39, 5156–5166.

[DEL⁺12]   Vida Dujmović, William Evans, Sylvain Lazard, William Lenhart, Giuseppe Liotta, David Rappaport, and Stephen Wismath, *On point-sets that support planar graphs*, Graph Drawing (Marc van Kreveld and Bettina Speckmann, eds.), Lecture Notes in Computer Science, vol. 7034, Springer, Berlin/Heidelberg, 2012, 10.1007/978-3-642-25878-7_7, pp. 64–74.

[DESW07]   Vida Dujmović, David Eppstein, Matthew Suderman, and David R. Wood, *Drawings of planar graphs with few slopes and segments*, Computational Geometry **38** (2007), no. 3, 194–212.

[DETT99]   Giuseppe Di Battista, Peter Eades, Roberto Tamassia, and Ioannis G. Tollis, *Graph drawing. algorithms for the visualization of graphs*, Prentice Hall, 1999.

[DF09]   Giuseppe Di Battista and Fabrizio Frati, *Efficient c-planarity testing for embedded flat clustered graphs with small faces*, J. Graph Algorithms Appl. **13** (2009), no. 3, 349–378.

[DFF⁺12]   Emilio Di Giacomo, Fabrizio Frati, Radoslav Fulek, Luca Grilli, and Marcus Krug, *Orthogeodesic point-set embedding of trees*, Graph Drawing (Marc van Kreveld and Bettina Speckmann, eds.), Lecture Notes in Computer Science, vol. 7034, Springer, Berlin/Heidelberg, 2012, 10.1007/978-3-642-25878-7_6, pp. 52–63.

[DGL08]   Walter Didimo, Francesco Giordano, and Giuseppe Liotta, *Overlapping cluster planarity*, J. Graph Algorithms Appl. **12** (2008), no. 3, 267–291.

[DGMW11]   Vida Dujmović, Joachim Gudmundsson, Pat Morin, and Thomas Wolle, *Notes on large angle crossing graphs*, Chicago Journal of Theoretical Computer Science **2011** (2011), no. 4.

[DH96]   Ron Davidson and David Harel, *Drawing graphs nicely using simulated annealing*, ACM Trans. Graph. **15** (1996), no. 4, 301–331.

[Dil90a]   Michael B. Dillencourt, *Realizability of Delaunay triangulations*, Information Processing Letters **33** (1990), no. 6, 283–287.

[Dil90b]   Michael B. Dillencourt, *Toughness and Delaunay triangulations*, Discrete & Computational Geometry **5** (1990), 575–601, 10.1007/BF02187810.

[DKM06]   Elias Dahlhaus, Karsten Klein, and Petra Mutzel, *Planarity testing for c-connected clustered graphs*, Tech. Report SYS-1/06, LSXI, University of Dortmund, 2006.

[DL07]   Emilio Di Giacomo and Giuseppe Liotta, *Simultaneous embedding of outerplanar graphs, paths, and cycles*, Int. J. Comput. Geometry Appl. **17** (2007), no. 2, 139–160.

[DL12]   Walter Didimo and Giuseppe Liotta, *Thirty essays on geometric graph theory*, ch. The crossing angle resolution in Graph Drawing, Springer, 2012.

[DLT10]   Emilio Di Giacomo, Giuseppe Liotta, and Francesco Trotta, *Drawing colored graphs with constrained vertex positions and few bends per edge*, Algorithmica **57** (2010), 796–818, 10.1007/s00453-008-9255-2.

[DLW06]   Giuseppe Di Battista, Giuseppe Liotta, and Sue H. Whitesides, *The strength of weak proximity*, Journal of Discrete Algorithms **4** (2006), no. 3, 384–400.

[DS96]   Michael B. Dillencourt and Warren D. Smith, *Graph-theoretical conditions for inscribability and delaunay realizability*, Discrete Mathematics **161** (1996), no. 13, 63–77.

[DSW07]   Vida Dujmović, Matthew Suderman, and David R. Wood, *Graph drawings with few slopes*, Computational Geometry **38** (2007), no. 3, 181–193.

[DV96]   Giuseppe Di Battista and Luca Vismara, *Angles of planar triangular graphs*, SIAM J. Discret. Math. **9** (1996), no. 3, 349–359.

[Ead84]   Peter Eades, *A heuristic for graph drawing*, Congressus Numerantium **42** (1984), 149–160.

[EBGJ+08] Alejandro Estrella-Balderrama, Elisabeth Gassner, Michael Jünger, Merijam Percan, Marcus Schaefer, and Michael Schulz, *Simultaneous geometric graph embeddings*, Graph Drawing (Seok-Hee Hong, Takao Nishizeki, and Wu Quan, eds.), Lecture Notes in Computer Science, vol. 4875, Springer, Berlin/Heidelberg, 2008, 10.1007/978-3-540-77537-9_28, pp. 280–290.

[EFLN06]   Peter Eades, Qingwen Feng, Xuemin Lin, and Hiroshi Nagamochi, *Straight-line drawing algorithms for hierarchical graphs and clustered graphs*, Algorithmica **44** (2006), 1–32, 10.1007/s00453-004-1144-8.

[EFN99]   Peter Eades, Qing-Wen Feng, and Hiroshi Nagamochi, *Drawing clustered graphs on an orthogonal grid*, J. Graph Algorithms Appl. **3** (1999), no. 4, 3–29.

[EGK+12]   William Evans, Emden Gansner, Michael Kaufmann, Giuseppe Liotta, Henk Meijer, and Andreas Spillner, *Approximate proximity drawings*, Graph Drawing (Marc van Kreveld and Bettina Speckmann, eds.), Lecture Notes in Computer Science, vol. 7034, Springer, Berlin/Heidelberg, 2012, 10.1007/978-3-642-25878-7_17, pp. 166–178.

[EH00]   Peter Eades and Mao Lin Huang, *Navigating clustered graphs using force-directed methods*, J. Graph Algorithms Appl. **4** (2000), no. 3, 157–181.

[EK86]   Peter Eades and David Kelly, *Heuristics for drawing 2-layered networks*, Ars Comb. **21** (1986), 89–98.

[EK05]   Cesim Erten and Stephen G. Kobourov, *Simultaneous embedding of planar graphs with few bends*, J. Graph Algorithms Appl. **9** (2005), no. 3, 347–364.

[EL12]   Peter Eades and Giuseppe Liotta, *Right angle crossing graphs and 1-planarity*, Graph Drawing (Marc van Kreveld and Bettina Speck- mann, eds.), Lecture Notes in Computer Science, vol. 7034, Springer, Berlin/Heidelberg, 2012, 10.1007/978-3-642-25878-7_15, pp. 148–153.

[ELLW10]   Hazel Everett, Sylvain Lazard, Giuseppe Liotta, and Stephen Wismath, *Universal sets of n points for one-bend drawings of planar graphs with n vertices*, Discrete & Computational Geometry **43** (2010), 272–288, 10.1007/s00454-009-9149-3.

[ELS93]   Peter Eades, Xuemin Lin, and W.F. Smyth, *A fast and effective heuristic for the feedback arc set problem*, Information Processing Letters **47** (1993), no. 6, 319–323.

[EW94]   Peter Eades and Nicholas C. Wormald, *Edge crossings in drawings of bi- partite graphs*, Algorithmica **11** (1994), no. 4, 379–403.

[EW96]   Peter Eades and Sue Whitesides, *The realization problem for Euclidean minimum spanning trees is NP-hard*, Algorithmica **16** (1996), 60–82, 10.1007/BF02086608.

[FCE95a]   Qing Feng, Robert Cohen, and Peter Eades, *How to draw a planar clustered graph*, Computing and Combinatorics (Ding-Zhu Du and Ming Li, eds.), Lecture Notes in Computer Science, vol. 959, Springer, Berlin/Heidelberg, 1995, 10.1007/BFb0030816, pp. 21–30.

[FCE95b]   Qing-Wen Feng, Robert Cohen, and Peter Eades, *Planarity for clustered graphs*, Algorithms ESA '95 (Paul Spirakis, ed.), Lecture Notes in Com- puter Science, vol. 979, Springer, Berli /Heidelberg, 1995, 10.1007/3-540- 60313-1_145, pp. 213–226.

[FK11]   Fabrizio Frati and Michael Kaufmann, *Polynomial area bounds for MST embeddings of trees*, Computational Geometry **44** (2011), no. 9, 529–543.

[FM07]   Igor Fabrici and Tomás Madaras, *The structure of 1-planar graphs*, Dis- crete Mathematics **307** (2007), no. 7–8, 854–865.

[FPS11]   Jacob Fox, János Pach, and Andrew Suk, *The number of edges in k-quasi- planar graphs*, CoRR **abs/1112.2361** (2011).

[FR91]   Thomas M. J. Fruchterman and Edward M. Reingold, *Graph drawing by force-directed placement*, Softw., Pract. Exper. **21** (1991), no. 11, 1129– 1164.

[Fra07]   Fabrizio Frati, *Embedding graphs simultaneously with fixed edges*, Graph Drawing (Michael Kaufmann and Dorothea Wagner, eds.), Lecture Notes in Computer Science, vol. 4372, Springer, Berlin/Heidelberg, 2007, 10.1007/978-3-540-70904-6_12, pp. 108–113.

[GGK04]   Pawel Gajer, Michael T. Goodrich, and Stephen G. Kobourov, *A multi- dimensional approach to force-directed layouts of large graphs*, Comput. Geom. **29** (2004), no. 1, 3–18.

[GJP$^+$06] Elisabeth Gassner, Michael Jünger, Merijam Percan, Marcus Schaefer, and Michael Schulz, *Simultaneous graph embeddings with fixed edges*, Graph-Theoretic Concepts in Computer Science (Fedor Fomin, ed.), Lecture Notes in Computer Science, vol. 4271, Springer, Berlin/Heidelberg, 2006, 10.1007/11917496_29, pp. 325–335.

[GKN05] Emden R. Gansner, Yehuda Koren, and Stephen North, *Graph drawing by stress majorization*, Graph Drawing (János Pach, ed.), Lecture Notes in Computer Science, vol. 3383, Springer, Berlin/Heidelberg, 2005, 10.1007/978-3-540-31843-9_25, pp. 239–250.

[GKNV93] Emden R. Gansner, Eleftherios Koutsofios, Stephen C. North, and Kiem-Phong Vo, *A technique for drawing directed graphs*, Software Engineering, IEEE Transactions on **19** (1993), no. 3, 214–230.

[GKV09] Markus Geyer, Michael Kaufmann, and Imrich Vrt'o, *Two trees which are self-intersecting when drawn simultaneously*, Discrete Mathematics **309** (2009), no. 7, 1909–1916.

[GLMS07] Francesco Giordano, Giuseppe Liotta, Tamara Mchedlidze, and Antonios Symvonis, *Computing upward topological book embeddings of upward planar digraphs*, Algorithms and Computation (Takeshi Tokuyama, ed.), Lecture Notes in Computer Science, vol. 4835, Springer, Berlin/Heidelberg, 2007, 10.1007/978-3-540-77120-3_17, pp. 172–183.

[GMPP91] Peter Gritzmann, Bojan Mohar, János Pach, and Richard Pollack, *Embedding a planar triangulation with vertices at specified points*, Amer. Math. Monthly **98** (1991), no. 2, 165–166.

[GMW05] Carsten Gutwenger, Petra Mutzel, and René Weiskircher, *Inserting an edge into a planar graph*, Algorithmica **41** (2005), no. 4, 289–308.

[Gor12] Taylor Gordon, *Simultaneous embeddings with vertices mapping to pre-specified points*, Computing and Combinatorics, Lecture Notes in Computer Science, Springer, Berlin/Heidelberg, 2012, To appear.

[HELP12] Seok-Hee Hong, Peter Eades, Giuseppe Liotta, and Sheung-Hung Poon, *Fáry's theorem for 1-planar graphs*, Computing and Combinatorics, Lecture Notes in Computer Science, Springer, Berlin/Heidelberg, 2012, To appear.

[HHE08] Weidong Huang, Seok-Hee Hong, and Peter Eades, *Effects of crossing angles*, PacificVis, 2008, pp. 41–46.

[HJL10] Bernhard Haeupler, Krishnam Jampani, and Anna Lubiw, *Testing simultaneous planarity when the common graph is 2-connected*, Algorithms and Computation (Otfried Cheong, Kyung-Yong Chwa, and Kunsoo Park, eds.), Lecture Notes in Computer Science, vol. 6507, Springer, Berlin/Heidelberg, 2010, 10.1007/978-3-642-17514-5_35, pp. 410–421.

[HK02] David Harel and Yehuda Koren, *A fast multi-scale method for drawing large graphs*, J. Graph Algorithms Appl. **6** (2002), no. 3, 179–202.

[HN] Patrick Healy and Nikola S. Nikolov, *Handbook of graph drawing and visualization*, ch. Hierarchical drawing algorithms, CRC Press, To appear.

[HN02]   Patrick Healy and Nikola Nikolov, *A branch-and-cut approach to the directed acyclic graph layering problem*, Graph Drawing (Michael Goodrich and Stephen Kobourov, eds.), Lecture Notes in Computer Science, vol. 2528, Springer, Berlin/Heidelberg, 2002, 10.1007/3-540-36151-0_10, pp. 235–256.

[Hua07]   Weidong Huang, *Using eye tracking to investigate graph layout effects*, APVIS, 2007, pp. 97–100.

[IR07]   Mohammad Tanvir Irfan and Md. Saidur Rahman, *Computing β-drawings of 2-outerplane graphs*, Workshop on Algorithms and Computation 2007 (M. Kaykobad and Md. Saidur Rahman, eds.), Bangladesh Academy of Sciences (BAS), 2007, pp. 46–61.

[JJK+]   Vít Jelínek, Eva Jelínková, Jan Kratochvíl, Bernard Lidický, Marek Tesař, and Tomáš Vyskočil, *The planar slope number of planar partial 3-trees of bounded degree*, Graphs and Combinatorics, 1–25, 10.1007/s00373-012-1157-z.

[JKK+09]   Eva Jelínková, Jan Kára, Jan Kratochvíl, Martin Pergel, Ondřej Suchý, and Tomáš Vyskočil, *Clustered planarity: Small clusters in cycles and eulerian graphs*, J. Graph Algorithms Appl. **13** (2009), no. 3, 379–422.

[JM96]   Michael Jünger and Petra Mutzel, *Maximum planar subgraphs and nice embeddings: Practical layout tools*, Algorithmica **16** (1996), no. 1, 33–59.

[JM97]   _____, *2-layer straightline crossing minimization: Performance of exact and heuristic algorithms*, J. Graph Algorithms Appl. **1** (1997).

[JM03]   Michael Jünger and Petra Mutzel (eds.), *Graph drawing software*, Springer, 2003.

[JT92]   Jerzy W. Jaromczyk and Godfried T. Toussaint, *Relative neighborhood graphs and their relatives*, Proceedings of the IEEE **80** (1992), no. 9, 1502–1517.

[Kam89]   Tomihisa Kamada, *Visualizing abstract objects and relations*, World Scientific, 1989.

[Kin06]   James A. King, *Realization of degree 10 minimum spanning trees in 3-space*, Proceedings of the 18th Annual Canadian Conference on Computational Geometry, CCCG 2006, 2006.

[Kir80]   David G. Kirkpatrick, *A note on Delaunay and optimal triangulations*, Information Processing Letters **10** (1980), no. 3, 127–128.

[KK89]   Tomihisa Kamada and Satoru Kawai, *An algorithm for drawing general undirected graphs*, Information Processing Letters **31** (1989), no. 1, 7–15.

[KKRW10]   Bastian Katz, Marcus Krug, Ignaz Rutter, and Alexander Wolff, *Manhattan-geodesic embedding of planar graphs*, Graph Drawing (David Eppstein and Emden Gansner, eds.), Lecture Notes in Computer Science, vol. 5849, Springer, Berlin/Heidelberg, 2010, 10.1007/978-3-642-11805-0_21, pp. 207–218.

[KMW12] Kolja Knauer, Piotr Micek, and Bartosz Walczak, *Outerplanar graph drawings with few slopes*, Computing and Combinatorics, Lecture Notes in Computer Science, Springer, Berlin/Heidelberg, 2012, To appear.

[Kob] Stephen G. Kobourov, *Handbook of graph drawing and visualization*, ch. Force-directed drawing algorithms, CRC Press, To appear.

[Kor08] Vladimir P. Korzhik, *Minimal non-1-planar graphs*, Discrete Mathematics **308** (2008), no. 7, 1319–1327.

[KPP11] Balázs Keszegh, János Pach, and Dömötör Pálvölgyi, *Drawing planar graphs of bounded degree with few slopes*, Graph Drawing (Ulrik Brandes and Sabine Cornelsen, eds.), Lecture Notes in Computer Science, vol. 6502, Springer, Berlin/Heidelberg, 2011, 10.1007/978-3-642-18469-7_27, pp. 293–304.

[KR85] David G. Kirkpatrick and John D. Radke, *A framework for computational morphology*, Computational Geometry (Godfried T. Toussaint, ed.), North-Holland, Amsterdam, Netherlands, 1985, pp. 217–248.

[KS80] Joseph B. Kruskal and Judith B. Seery, *Designing network diagrams*, Proc. First General Conference on Social Graphics, U. S. Dept. of the Census, 1980, pp. 22–50.

[Kur04] Maciej Kurowski, *A 1.235 lower bound on the number of points needed to draw all n-vertex planar graphs*, Information Processing Letters **92** (2004), no. 2, 95–98.

[KW01] Michael Kaufmann and Dorothea Wagner (eds.), *Drawing graphs, methods and models*, Lecture Notes in Computer Science, vol. 2025, Springer, 2001.

[KW02] Michael Kaufmann and Roland Wiese, *Embedding vertices at points: Few bends suffice for planar graphs*, Journal of Graph Algorithms and Applications **6** (2002), no. 1, 115–129.

[LD95] Giuseppe Liotta and Giuseppe Di Battista, *Computing proximity drawings of trees in the 3-dimensional space*, Algorithms and Data Structures (Selim Akl, Frank Dehne, Jrg-Rdiger Sack, and Nicola Santoro, eds.), Lecture Notes in Computer Science, vol. 955, Springer, Berlin/Heidelberg, 1995, 10.1007/3-540-60220-8_66, pp. 239–250.

[Lio] Giuseppe Liotta, *Handbook of graph drawing and visualization*, ch. Proximity drawings, CRC Press, To appear.

[LK96] Christos Levcopoulos and Drago Krznaric, *Tight lower bounds for minimum weight triangulation heuristics*, Information Processing Letters **57** (1996), no. 3, 129–135.

[LL96] William Lenhart and Giuseppe Liotta, *Drawing outerplanar minimum weight triangulations*, Information Processing Letters **57** (1996), no. 5, 253–260.

[LL97] William Lenhart and Giuseppe Liotta, *Proximity drawings of outerplanar graphs*, Graph Drawing (Stephen North, ed.), Lecture Notes in Computer Science, vol. 1190, Springer, Berlin/Heidelberg, 1997, 10.1007/3-540-62495-3_55, pp. 286–302.

[LL02]   William Lenhart and Giuseppe Liotta, *The drawability problem for minimum weight triangulations*, Theoretical Computer Science **270** (2002), no. 12, 261–286.

[LLMW98]   Giuseppe Liotta, Anna Lubiw, Henk Meijer, and Sue H. Whitesides, *The rectangle of influence drawability problem*, Computational Geometry **10** (1998), no. 1, 1–22.

[LM03]   Giuseppe Liotta and Henk Meijer, *Voronoi drawings of trees*, Computational Geometry **24** (2003), no. 3, 147–178.

[LS93]   Anna Lubiw and Nora Sleumer, *Maximal outerplanar graphs are relative neighborhood graphs*, Proc. 5th Canad. Conf. Comput. Geom., 1993, pp. 198–203.

[Mch12]   Tamara Mchedlidze, *Upward planar embedding of a n-vertex oriented path into $O(n^2)$ points*, 28th European Workshop on Computational Geometry (EuroCG 2012), 2012, pp. 141–144.

[MMN09]   Kazuyuki Miura, Tetsuya Matsuno, and Takao Nishizeki, *Open rectangle-of-influence drawings of inner triangulated plane graphs*, Discrete & Computational Geometry **41** (2009), 643–670, 10.1007/s00454-008-9098-2.

[MP12]   Padmini Mukkamala and Dömötör Pálvölgyi, *Drawing cubic graphs with the four basic slopes*, Graph Drawing (Marc van Kreveld and Bettina Speckmann, eds.), Lecture Notes in Computer Science, vol. 7034, Springer, Berlin/Heidelberg, 2012, 10.1007/978-3-642-25878-7_25, pp. 254–265.

[MR08]   Wolfgang Mulzer and Günter Rote, *Minimum-weight triangulation is np-hard*, J. ACM **55** (2008), no. 2, 11:1–11:29.

[MS92]   Clyde Monma and Subhash Suri, *Transitions in geometric minimum spanning trees*, Discrete & Computational Geometry **8** (1992), 265–293, 10.1007/BF02293049.

[MSM99]   Christian Matuszewski, Robby Schönfeld, and Paul Molitor, *Using sifting for k-layer straightline crossing minimization*, Graph Drawing (Jan Kratochvyl, ed.), Lecture Notes in Computer Science, vol. 1731, Springer, Berlin/Heidelberg, 1999, 10.1007/3-540-46648-7_22, pp. 217–224.

[MZ79]   Glenn K. Manacher and Albert L. Zobrist, *Neither the greedy nor the delaunay triangulation of a planar point set approximates the optimal triangulation*, Information Processing Letters **9** (1979), no. 1, 31–34.

[Nag05]   Hiroshi Nagamochi, *On the one-sided crossing minimization in a bipartite graph with large degrees*, Theoretical Computer Science **332** (2005), no. 13, 417–446.

[NK07]   Hiroshi Nagamochi and Katsutoshi Kuroya, *Drawing c-planar biconnected clustered graphs*, Discrete Applied Mathematics **155** (2007), no. 9, 1155–1174.

[NR04]   Takao Nishizeki and Md Saidur Rahman, *Planar graph drawing*, World Scientific, 2004.

[NY04]   Hiroshi Nagamochi and Nobuyasu Yamada, *Counting edge crossings in a 2-layered drawing*, Information Processing Letters **91** (2004), no. 5, 221–225.

[OBS92]   Atsuyuki Okabe, Barry Boots, and Kokichi Sugihara, *Spatial tessellations: Concepts and applications of Voronoi diagrams*, John Wiley & Sons, Chichester, UK, 1992.

[OS07]   Hiroki Omote and Kozo Sugiyama, *Force-directed drawing method for intersecting clustered graphs*, APVIS'07, 2007, pp. 85–92.

[PCA02]   Helen C. Purchase, David A. Carrington, and Jo-Anne Allder, *Empirical evaluation of aesthetics-based graph layout*, Empirical Software Engineering **7** (2002), no. 3, 233–255.

[PP06]   János Pach and Dömötör Pálvölgyi, *Bounded-degree graphs can have arbitrarily large slope numbers*, Electr. J. Comb. **13** (2006), no. 1.

[PRTT06]   János Pach, Rados Radoicic, Gábor Tardos, and Géza Tóth, *Improving the crossing lemma by finding more crossings in sparse graphs*, Discrete & Computational Geometry **36** (2006), no. 4, 527–552.

[PS90]   Franco P. Preparata and Michael I. Shamos, *Computational geometry: An introduction*, 3rd ed., Springer-Verlag, October 1990.

[Pur00]   Helen C. Purchase, *Effective information visualisation: a study of graph drawing aesthetics and algorithms*, Interacting with Computers **13** (2000), no. 2, 147–162.

[PV04]   Paolo Penna and Paola Vocca, *Proximity drawings in polynomial area and volume*, Computational Geometry **29** (2004), no. 2, 91–116.

[PW01]   János Pach and Rephael Wenger, *Embedding planar graphs at fixed vertex locations*, Graphs and Combinatorics **17** (2001), 717–728.

[Rad88]   John D. Radke, *On the shape of a set of points*, Computational Morphology (Godfried T. Toussaint, ed.), North-Holland, Amsterdam, Netherlands, 1988, pp. 105–136.

[SIR08]   Md. Samee, Mohammad Irfan, and Md. Rahman, *Computing β-drawings of 2-outerplane graphs in linear time*, WALCOM: Algorithms and Computation (Shin-ichi Nakano and Md. Rahman, eds.), Lecture Notes in Computer Science, vol. 4921, Springer, Berlin/Heidelberg, 2008, 10.1007/978-3-540-77891-2_8, pp. 81–87.

[STT81]   Kozo Sugiyama, Shojiro Tagawa, and Mitsuhiko Toda, *Methods for visual understanding of hierarchical system structures*, Systems, Man and Cybernetics, IEEE Transactions on **11** (1981), no. 2, 109–125.

[Sug02]   Kozo Sugiyama, *Graph drawing and applications for software and knowledge engineers*, World Scientific, 2002.

[Tam87]   Roberto Tamassia, *On embedding a graph in the grid with the minimum number of bends*, SIAM J. Comput. **16** (1987), no. 3, 421–444.

[TDB88]   Roberto Tamassia, Giuseppe Di Battista, and Carlo Batini, *Automatic graph drawing and readability of diagrams*, Systems, Man and Cybernetics, IEEE Transactions on **18** (1988), no. 1, 61–79.

[TNB04]   Alexandre Tarassov, Nikola Nikolov, and Jürgen Branke, *A heuristic for minimum-width graph layering with consideration of dummy nodes*, Experimental and Efficient Algorithms (Celso Ribeiro and Simone Martins, eds.), Lecture Notes in Computer Science, vol. 3059, Springer, Berlin/Heidelberg, 2004, 10.1007/978-3-540-24838-5_42, pp. 570–583.

[vK11]   Marc van Kreveld, *The quality ratio of RAC drawings and planar drawings of planar graphs*, Graph Drawing (Ulrik Brandes and Sabine Cornelsen, eds.), Lecture Notes in Computer Science, vol. 6502, Springer, Berlin/Heidelberg, 2011, 10.1007/978-3-642-18469-7_34, pp. 371–376.

[WC94]   Greg A. Wade and Jiang-Hsing Chu, *Drawability of complete graphs using a minimal slope set*, The Computer Journal **37** (1994), no. 2, 139–142.

[WCY00]   Cao An Wang, Francis Y. Chin, and Boting Yang, *Triangulations without minimum-weight drawing*, Information Processing Letters **74** (2000), no. 56, 183–189.

[WPCM02]   Colin Ware, Helen C. Purchase, Linda Colpoys, and Matthew McGill, *Cognitive measurements of graph aesthetics*, Information Visualization **1** (2002), no. 2, 103–110.

Section 10.4
Algorithms on Recursively Constructed Graphs

Richard B. Borie, University of Alabama
R. Gary Parker, Georgia Institute of Technology
Craig A. Tovey, Georgia Institute of Technology

INTRODUCTION

In this section we demonstrate algorithms for the recursively defined classes of trees, series-parallel graphs, treewidth-k graphs, cographs, cliquewidth-k graphs, and k-HB graphs. For convenience, definitions of these graph classes, which appear in §2.4 (among others given there), are repeated here. Our emphasis in this section is on the solution *technique* for each graph class and less so on the variety of problems solvable *within* each class per se. Accordingly, for each graph class, we will consider the problem of finding an *independent set* in a given graph (§5.3); where it is natural to do so, some additional problems are solved as well. For a much broader accounting using this same format and one that includes a host of other problems such as *clique, dominating set, vertex coloring, matching, hamiltonian cycle, hamiltonian path*, and others, the reader is directed to [BoPaTo08].

DEFINITIONS

D1: A ***recursively constructed graph class*** is defined by a set (usually finite) of primitive or ***base graphs***, in addition to one or more operations (called ***composition rules***) that compose larger graphs from smaller subgraphs. Each operation involves fusing specific vertices from each subgraph or adding new edges between specific vertices from each subgraph.

D2:   Each graph in a recursive class has a corresponding ***decomposition tree*** that shows how to build it from base graphs.

D3:   For a graph G, a set S of vertices is an ***independent set*** if no two vertices in S are adjacent.

Algorithm Design Strategy

Efficient algorithms for problems restricted to recursively constructed graph classes typically employ a *dynamic programming* approach as follows: first solve the problem on the base graphs defined for the given class; then combine the solutions for subgraphs into a solution for a larger graph that is formed by the specific composition rules that govern construction of members in the class. A linear-time algorithm is achieved by determining a finite number of equivalence classes that correspond to each node in a member graph's decomposition tree. The number of such equivalence classes is constant with respect to the size of the input graph, but may depend upon a parameter (k) associated with the class. A polynomial algorithm can often be created if the number of equivalence classes required for the problem grows only polynomially with input graph size. Also key is that a graph's decomposition tree be given or be computable efficiently.

NOTATION: In the descriptions of the algorithms in this section, we use $G.x$ to denote an attribute of a given graph G. When we write $G.x = $ *maximum-cardinality independent set*, then $G.x$ carries with it two pieces of information: the *size* of a maximum-cardinality independent set, and *one particular instance* of such a set. Moreover, these two pieces of information are carried forward in computations and assignments involving $G.x$.

REMARK

R1:   When the simplest version of a problem (cardinality or existence) can be solved using a dynamic programming approach, then other more complicated versions (involving vertex or edge weights, counting, bottleneck, min-max, etc.) can generally also be routinely solved. Following, we begin with the simple recursive class of trees.

10.4.1   Algorithms on Trees

DEFINITION

D4:   The graph with a single vertex r (and no edges) is a ***tree*** with root r (the sole base graph). Let (G, r) denote a tree with root r. Then $(G_1, r_1) \oplus (G_2, r_2)$ is a tree formed by taking the disjoint union of G_1 and G_2 and adding an edge (r_1, r_2). The root of this new tree is $r = r_1$.

TERMINOLOGY NOTE: Technically, the pairs (G, r) in Definition D4 denote *rooted trees*. However, the specification of distinguished vertices r_1 and r_2 (and hence r) is relevant here only as a vehicle in the recursive construction.

NOTATION: Given any tree (or subtree) G, the designated root is denoted by $root[G]$.

Maximum-Cardinality Independent Set in a Tree

FACTS

F1: Any independent set of vertices either includes $root[G]$ or does not. Whether or not independent sets of two trees G_1, G_2 can be combined into an independent set of the tree $G_1 \oplus G_2$ depends only on these inclusions.

NOTATION: The following notation is used in the description of Algorithm 10.4.1 below.

- $G.a$ = max-cardinality independent set that includes $root[G]$.
- $G.b$ = max-cardinality independent set that excludes $root[G]$.
- $G.c$ = max-cardinality independent set.

F2: The following multiplication table suffices to describe all possible types of outcomes from the composition $G_1 \oplus G_2$. In the table, rather than displaying $G.a$ and $G.b$ we simply specify a and b, respectively. The row by column product assumes the convention where subgraph (i.e., subtree) G_1 is on the left and subtree G_2 is on the right.

$\oplus$	**a**	**b**
a		a
b	b	b

REMARK

R2: The values for $G.a$, $G.b$, and $G.c$ are known trivially for the base graph – a single-vertex tree (which is its own root). For composed graphs, the values may be computed via $O(1)$ additions and comparisons across the outcomes in the table. There is only one product producing a possible member of $G.a$ while $G.b$ may be produced from a pair of possible products; the maximum of these yields the desired $G.b$. The final step in Algorithm 10.4.1 computes $G.c$, which, at the root of the decomposition tree, is the solution.

Algorithm 10.4.1: Maximum-Cardinality Independent Set in a Tree

Input: Tree $G = (V, E)$.
Output: $G.c$ (size and instance of a max-cardinality independent set in G).
    If $|V| = 1$
        $G.a \leftarrow 1$
        $G.b \leftarrow 0$
    Else
        If $G = G_1 \oplus G_2$
            $G.a \leftarrow G_1.a + G_2.b$
            $G.b \leftarrow \max \{G_1.b + G_2.a, G_1.b + G_2.b\}$
    $G.c \leftarrow \max \{G.a, G.b\}$

COMPUTATIONAL NOTE: The decomposition tree for trees is easy to determine and, accordingly, can be assumed to be part of the instance.

Maximum-Weight Independent Set in a Tree

Here we assume that each vertex in the graph is assigned a weight. Switching to a weighted version of the independent set problem is straightforward.

NOTATION: The notation used in the Algorithm 10.4.2 is as follows:

- $G.d$ = max-weight independent set containing $root[G]$.

- $G.e$ = max-weight independent set without $root[G]$.

- $G.f$ = max-weight independent set.

Algorithm 10.4.2: Maximum-Weight Independent Set in a Tree

Input: Tree $G = (V, E)$.
Output: $G.f$ (weight and instance of a max-weight independent set in G).
    If $|V| = 1$
        $G.d \leftarrow \text{weight}(root[G])$
        $G.e \leftarrow 0$
    Else
        If $G = G_1 \oplus G_2$
            $G.d \leftarrow G_1.d + G_2.e$
            $G.e \leftarrow \max \{G_1.e + G_2.d, G_1.e + G_2.e\}$
    $G.f \leftarrow \max \{G.d, G.e\}$

EXAMPLE

E1:   Consider the tree T shown in Figure 10.4.1. Vertices are labeled $t, u, \ldots, z$ and beside each label is a vertex weight. Algorithms 10.4.1 and 10.4.2 are applied and the computations are summarized by the listing on the right. The 6-tuples aligned with each composed subgraph, G_k, correspond to values $[G.a, G.b, G.c, G.d, G.e, G.f]$. The maximum cardinality and maximum weight of any independent set is $G.c = 5$ and $G.f = 16$, respectively; these are read from the computation for G_{13}. Standard backtracking can be applied to determine that the explicit solutions are sets $\{t, v, w, y, z\}$ and $\{u, y, z\}$, respectively.

REMARKS

R3:   Many other problems such as variations of the vertex cover, dominating set, matching, or longest path problems could have been selected to represent the basic computation on trees (cf. [BoPaTo92], [BoPaTo08]). Note that some problems such as minimum bandwidth [GaGrJoKn78] are $\mathcal{NP}$-complete on trees.

R4:   A number of problems are trivial when restricted to trees. For example, any tree with at least 2 vertices has maximum clique size of 2 and chromatic number 2; otherwise these are both 1. Also, no tree can contain a hamiltonian cycle, and a tree has a hamiltonian path if and only if it is a path.

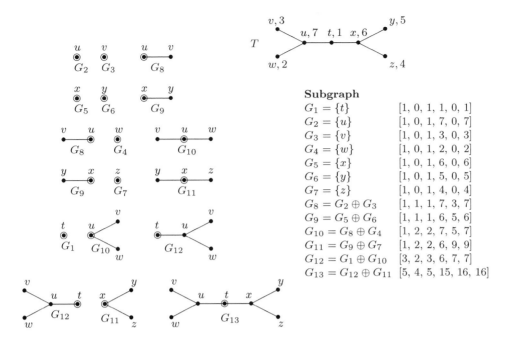

Figure 10.4.1: Max-cardinality and max-weight independent sets in a tree.

R5: The definition and number of equivalence classes that are required to solve a given problem depend on both the graph class and the problem to be solved. The effect of graph class is demonstrated throughout in subsequent subsections of this section. To provide a sense of dependence on the problem, let the distance in connected graph $G = (V, E)$ between $U \subset V$ and $W \subset V$ be the shortest (edge) length of a path from any $u \in U$ to any $w \in W$, and let a κ-*independent set* be a subset of V containing no two distinct vertices at distance κ or less. Finding a maximum-cardinality κ-independent set requires $\kappa + 2$ equivalence classes $G.c$ and $G.i : i = 0 \ldots \kappa$, where: $G.c$ is the maximum-cardinality κ-independent sets; $G.\kappa$ denotes maximum-cardinality κ-independent sets at distance at least κ to the root; $G.i : i = 0 \ldots \kappa - 1$ denotes maximum-cardinality κ-independent sets at distance i to the root; multiplication table entry $G_1.i, G_2.j = \min \{i, j + 1\}$ if $i + j \geq \kappa$ and null otherwise. Note that the resulting algorithm may be superlinear if κ is not $O(1)$.

10.4.2 Algorithms on Series-Parallel Graphs

Every series-parallel graph can be composed from single edges using only the three composition rules given in Definition D5 below. For illustrations of these rules, see §2.4.1.

DEFINITION

D5: A *series-parallel graph* with distinguished **terminals** l and r is denoted (G, l, r) and is defined recursively as follows:

- (base graph) The graph consisting of a single edge (v_1, v_2) is a series-parallel graph (G, l, r) with $l = v_1$ and $r = v_2$.

- The **series operation** $(G_1, l_1, r_1) \odot_s (G_2, l_2, r_2)$ forms a series-parallel graph by identifying r_1 with l_2. The terminals of the new graph are l_1 and r_2.

- The **parallel operation** $(G_1, l_1, r_1) \odot_p (G_2, l_2, r_2)$ forms a series-parallel graph by identifying l_1 with l_2 and r_1 with r_2. The terminals of the new graph are l_1 and r_1.

- The **jackknife operation** $(G_1, l_1, r_1) \odot_j (G_2, l_2, r_2)$ forms a series-parallel graph by identifying r_1 with l_2; the new terminals are l_1 and r_1.

REMARK

R6:   Series-parallel graphs are recognizable, and their decomposition trees can be constructed in linear time (see §2.4.3). This leads to fast, often linear-time, dynamic programming algorithms for many problems when instances are confined to series-parallel graphs.

Maximum-Cardinality Independent Set in a Series-Parallel Graph

In the case of trees, only a single point of composition involving constituent subtrees, the root vertex, was relevant; a series-parallel graph G has two such points, its terminal vertices. These are denoted $left[G]$ and $right[G]$ in the following description.

NOTATION: Algorithm 10.4.3 uses the following notation:

- $G.a$ = max cardinality independent set containing both $left[G]$ and $right[G]$.

- $G.b$ = max cardinality independent set with $left[G]$ but not $right[G]$.

- $G.c$ = max cardinality independent set with $right[G]$ but not $left[G]$.

- $G.d$ = max cardinality independent set with neither $left[G]$ nor $right[G]$.

- $G.e$ = max cardinality independent set.

Multiplication Tables for Series, Parallel, and Jacknife Operations

$\odot_s$	a	b	c	d
a	a	b		
b			a	b
c	c	d		
d			c	d

$\odot_p$	a	b	c	d
a	a			
b		b		
c			c	
d				d

$\odot_j$	a	b	c	d
a	a	a		
b			b	b
c	c	c		
d			d	d

Algorithm 10.4.3: Maximum-Cardinality Independent Set in a Series-Parallel Graph

Input: Series-parallel graph $G = (V, E)$.

Output: $G.e$ (size and instance of a max-cardinality independent set in G).

If $|E| = 1$

$\qquad [G.a, G.b, G.c, G.d] \leftarrow [-\infty, 1, 1, 0]$

Else

$\qquad$ If $G = G_1 \odot_s G_2$

$\qquad\qquad G.a \leftarrow \max\{G_1.a + G_2.a - 1, G_1.b + G_2.c\}$

$\qquad\qquad G.b \leftarrow \max\{G_1.a + G_2.b - 1, G_1.b + G_2.d\}$

$\qquad\qquad G.c \leftarrow \max\{G_1.c + G_2.a - 1, G_1.d + G_2.c\}$

$\qquad\qquad G.d \leftarrow \max\{G_1.c + G_2.b - 1, G_1.d + G_2.d\}$

$\qquad$ Else

$\qquad\qquad$ If $G = G_1 \odot_p G_2$

$\qquad\qquad\qquad G.a \leftarrow G_1.a + G_2.a - 2$

$\qquad\qquad\qquad G.b \leftarrow G_1.b + G_2.b - 1$

$\qquad\qquad\qquad G.c \leftarrow G_1.c + G_2.c - 1$

$\qquad\qquad\qquad G.d \leftarrow G_1.d + G_2.d$

$\qquad\qquad$ Else

$\qquad\qquad\qquad$ If $G = G_1 \odot_j G_2$

$\qquad\qquad\qquad\qquad G.a \leftarrow G_1.a + \max\{G_2.a, G_2.b\} - 1$

$\qquad\qquad\qquad\qquad G.b \leftarrow G_1.b + \max\{G_2.c, G_2.d\}$

$\qquad\qquad\qquad\qquad G.c \leftarrow G_1.c + \max\{G_2.a, G_2.b\} - 1$

$\qquad\qquad\qquad\qquad G.d \leftarrow G_1.d + \max\{G_2.c, G_2.d\}$

$\quad G.e \leftarrow \max\{G.a, G.b, G.c, G.d\}$

COMPUTATIONAL NOTE: Subtraction of values 2 and 1 in the respective computational expressions above avoids multiple counting when terminal vertices are fused.

EXAMPLE

E2:   Algorithm 10.4.3 is demonstrated on the series-parallel graph G given to the left in Figure 10.4.2. Vertices are labeled in G as shown and to the right, the explicit computation is summarized. The 4-tuples exhibit values for $G.a, G.b, G.c$, and $G.d$. From the last computation, it follows that either $G.b$ or $G.d$ produces an optimum. In the first case, the set is $\{z, v, w\}$, while in the second, we have $\{y, v, w\}$.

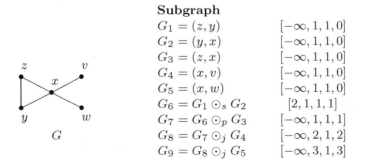

Subgraph	
$G_1 = (z, y)$	$[-\infty, 1, 1, 0]$
$G_2 = (y, x)$	$[-\infty, 1, 1, 0]$
$G_3 = (z, x)$	$[-\infty, 1, 1, 0]$
$G_4 = (x, v)$	$[-\infty, 1, 1, 0]$
$G_5 = (x, w)$	$[-\infty, 1, 1, 0]$
$G_6 = G_1 \odot_s G_2$	$[2, 1, 1, 1]$
$G_7 = G_6 \odot_p G_3$	$[-\infty, 1, 1, 1]$
$G_8 = G_7 \odot_j G_4$	$[-\infty, 2, 1, 2]$
$G_9 = G_8 \odot_j G_5$	$[-\infty, 3, 1, 3]$

Figure 10.4.2: Max-cardinality of an independent set in a series-parallel graph.

FACTS

F3: Other problems solvable in linear time on series-parallel graphs include variations of clique, dominating set, matching, hamiltonian path, and hamiltonian cycle. Indeed, a series-parallel graph can have at most one hamiltonian cycle (cf. [Sy83]).

F4: A series-parallel graph has chromatic number 3 if it is not bipartite; otherwise it has chromatic number 2 (because it has at least one edge).

REMARKS

R7: Following Fact F3, solving the *traveling salesman problem* (cf. §4.6) on series-parallel graphs reduces to deciding hamiltonicity.

R8: Following Fact F4, the chromatic number for a series-parallel graph can be determined in linear time by using depth-first search to simply test for the existence of an odd cycle.

R9: In some problem settings, the jackknife operation can simply be neglected. For example, a hamiltonian graph must be *2-connected* (i.e., no *cut-vertices*) but the jackknife operation destroys this property; hence, the jackknife operation is not relevant in this case. On the other hand, if the aim is deciding the existence of a hamiltonian path, the jackknife operation is relevant.

R10: Solutions to numerous other problems on series-parallel graphs follow the machinery demonstrated by Algorithm 10.4.3. In addition to the references already cited in Remark R2, a good basic source dealing with problems such as *vertex cover, maximum eulerian subgraph, Steiner subgraph, edge-covering*, etc. is [Ri85].

10.4.3 Algorithms on Treewidth-k Graphs

DEFINITIONS

D6: A ***tree-decomposition*** of a graph $G = (V, E)$ is a pair $(\{X_i \mid i \in I\}, T)$, where $\{X_i \mid i \in I\}$ is a family of subsets of V, and T is a tree with vertex set I such that:

- $\bigcup_{i \in I} X_i = V$,

- for all edges $(x, y) \in E$ there is an element $i \in I$ with $x, y \in X_i$, and

- for all triples $i, j, k \in I$, if j is on the path from i to k in T, then $X_i \bigcap X_k \subseteq X_j$.

D7: The ***width*** of a given tree-decomposition is measured as max $_{i \in I}\{|X_i| - 1\}$.

D8: The ***treewidth*** of a graph G is the minimum width taken over all tree-decompositions of G.

D9: A graph G is a ***treewidth-k graph*** if it has treewidth no greater than k.

FACT

F5: Every treewidth-k graph has a tree-decomposition T such that T is a rooted binary tree [Sc89]. We write

$$(G, X) = (G_1, X_1) \otimes (G_2, X_2)$$

where $X \subseteq V$ is the set of vertices of G associated with $root[T]$, and graphs G_1 and G_2 have tree-decompositions given by the left and right subtrees of T. This is enough to produce linear-time dynamic programming algorithms for many problems on treewidth-k graphs, because each $|X| \leq k + 1$.

Maximum-Cardinality Independent Set in a Treewidth-k Graph

NOTATION: For a graph G, let binary tree T be a tree-decomposition of G and let $X \subseteq V$ be the set of vertices of G associated with $root[T]$. Also,

- $G[S]$ = max-cardinality independent set that contains $S \subseteq X$ but not $X - S$.

- $G.max$ = max-cardinality independent set.

Algorithm 10.4.4: Maximum-Cardinality Independent Set in a Treewidth-k Graph

Input: Treewidth-k graph $G = (V, E)$.
Output: $G.max$ (size and instance of a max-cardinality independent set in G).
    If $X = V$
        For all $S \subseteq X$
            If S contains two adjacent vertices
                $G[S] \leftarrow -\infty$
            Else
                $G[S] \leftarrow |S|$
    Else
        If $(G, X) = (G_1, X_1) \otimes (G_2, X_2)$
            For all $S \subseteq X$
                If S contains two adjacent vertices
                    $G[S] \leftarrow -\infty$
                Else
                    $G[S] \leftarrow \max\{G_1[S_1] + G_2[S_2] - |S_1 \cap S| - |S_2 \cap S| + |S| :$
                          $S_1 \subseteq X_1, S_2 \subseteq X_2, S_1 \cap X = S \cap X_1, S_2 \cap X = S \cap X_2\}$
     $G.max \leftarrow \max\{G[S] : S \subseteq X\}$

EXAMPLE

E3:   Algorithm 10.4.4 is demonstrated on the treewidth-2 graph G shown in Figure 10.4.3. T is a binary rooted tree-decomposition of G, and each 8-tuple exhibits values for $G[S]$ for each $S \subseteq X$. However, only values larger than $-\infty$ within each 8-tuple are shown as the computation progresses. The maximum independent set has size 4, and the explicit solution is $\{a, c, e, g\}$.

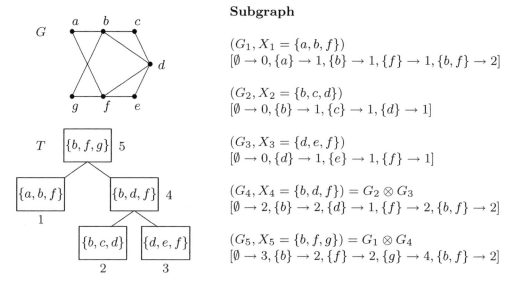

Subgraph

$(G_1, X_1 = \{a, b, f\})$
$[\emptyset \to 0, \{a\} \to 1, \{b\} \to 1, \{f\} \to 1, \{b, f\} \to 2]$

$(G_2, X_2 = \{b, c, d\})$
$[\emptyset \to 0, \{b\} \to 1, \{c\} \to 1, \{d\} \to 1]$

$(G_3, X_3 = \{d, e, f\})$
$[\emptyset \to 0, \{d\} \to 1, \{e\} \to 1, \{f\} \to 1]$

$(G_4, X_4 = \{b, d, f\}) = G_2 \otimes G_3$
$[\emptyset \to 2, \{b\} \to 2, \{d\} \to 1, \{f\} \to 2, \{b, f\} \to 2]$

$(G_5, X_5 = \{b, f, g\}) = G_1 \otimes G_4$
$[\emptyset \to 3, \{b\} \to 2, \{f\} \to 2, \{g\} \to 4, \{b, f\} \to 2]$

Figure 10.4.3: Maximum-cardinality independent set in a treewidth-2 graph.

REMARK

R11:   Many other problems including variations of independent set, dominating set, clique, m-vertex coloring (for arbitrary, fixed m), matching, and hamiltonian cycle/path problems can be solved in linear time for treewidth-k graphs (cf. [BoPaTo92]).

Monadic Second-Order Logic Expressions for a Graph

NOTATION: Let variables v_i denote a vertex with domain V, e_i denote an edge with domain E, V_i denote a vertex set with domain 2^V (subsets of V), and E_i denote an edge set with domain 2^E.

DEFINITION

D10:   *Monadic second-order logic* (MSOL) for a graph $G = (V, E)$ is a *predicate calculus language* in which predicates are constructed recursively as follows:
- MSOL contains *primitive* predicates such as $v_i = v_j$, $v_i \in V_j$, $e_i \in E_j$, and Incident(v_i, e_j).
- If P and Q are MSOL predicates then each of $(\neg P)$, $(P \wedge Q)$, and $(P \vee Q)$ is also a MSOL predicate.
- If P is a MSOL predicate and x is any variable, then $(\exists x)(P)$ and $(\forall x)(P)$ are also MSOL predicates.

EXAMPLE

E4:   Some simple MSOL predicates are listed below.

$P \to Q \Leftrightarrow \neg P \vee Q$
$P \leftrightarrow Q \Leftrightarrow (P \to Q) \wedge (Q \to P)$
$e_i = e_j \Leftrightarrow (\forall v_1)\,(\text{Incident}(v_1, e_i) \leftrightarrow \text{Incident}(v_1, e_j))$
$\text{Adjacent}(v_i, v_j) \Leftrightarrow \neg(v_i = v_j) \wedge (\exists e_1)\,(\text{Incident}(v_i, e_1) \wedge \text{Incident}(v_j, e_1))$

MSOL-Expressible Graph Problems

Many important graph problems can be expressed in MSOL (cf. [Co90], [ArLaSe91], [BoPaTo91], [BoPaTo92], [CoMo93]). Here is a sampling of several such problems.

$\text{IndependentSet}(V_1) \Leftrightarrow (\forall v_2) (\forall v_3) ((v_2 \in V_1 \wedge v_3 \in V_1) \rightarrow \neg \text{Adjacent}(v_2, v_3))$

$\text{Clique}(V_1) \Leftrightarrow (\forall v_2) (\forall v_3) ((v_2 \in V_1 \wedge v_3 \in V_1) \rightarrow \text{Adjacent}(v_2, v_3))$

$\text{DominatingSet}(V_1) \Leftrightarrow (\forall v_2) (v_2 \in V_1 \vee (\exists v_3) (v_3 \in V_1 \wedge \text{Adjacent}(v_2, v_3)))$

$\text{VertexColorable}_m(V_1, \ldots, V_m) \Leftrightarrow (\forall v_0) (v_0 \in V_1 \vee \ldots \vee v_0 \in V_m)$
$\qquad\qquad \wedge \text{IndependentSet}(V_1) \wedge \ldots \wedge \text{IndependentSet}(V_m)$

$\text{Matching}(E_1) \Leftrightarrow (\forall e_2) (\forall e_3) ((e_2 \in E_1 \wedge e_3 \in E_1 \wedge \neg (e_2 = e_3)) \rightarrow$
$\qquad\qquad \neg (\exists v_4) (\text{Incident}(v_4, e_2) \wedge \text{Incident}(v_4, e_3)))$

$\text{Connected}(E_1) \Leftrightarrow (\forall V_2) (\forall V_3) (\neg (\exists v_4) (v_4 \in V_2) \vee \neg (\exists v_5) (v_5 \in V_3) \vee$
$\qquad\qquad (\exists v_6) (\neg (v_6 \in V_2) \wedge \neg (v_6 \in V_3)) \vee$
$\qquad\qquad (\exists e_7) (\exists v_8) (\exists v_9) (e_7 \in E_1 \wedge v_8 \in V_2 \wedge v_9 \in V_3 \wedge$
$\qquad\qquad \text{Incident}(v_8, e_7) \wedge \text{Incident}(v_9, e_7)))$

$\text{HamCycle}(E_1) \Leftrightarrow \text{Connected}(E_1) \wedge (\forall v_2) (\exists e_3) (\exists e_4) (e_3 \in E_1 \wedge e_4 \in E_1 \wedge$
$\qquad\qquad \neg (e_3 = e_4) \wedge \text{Incident}(v_2, e_3) \wedge \text{Incident}(v_2, e_4) \wedge$
$\qquad\qquad (\forall e_5) ((e_5 \in E_1 \wedge \text{Incident}(v_2, e_5)) \rightarrow (e_5 = e_3 \vee e_5 = e_4)))$

$\text{HamPath}(E_1) \Leftrightarrow \text{Connected}(E_1) \wedge (\forall v_2) (\exists e_3) (\exists e_4) (e_3 \in E_1 \wedge e_4 \in E_1 \wedge$
$\qquad\qquad \text{Incident}(v_2, e_3) \wedge \text{Incident}(v_2, e_4) \wedge$
$\qquad\qquad (\forall e_5) ((e_5 \in E_1 \wedge \text{Incident}(v_2, e_5)) \rightarrow (e_5 = e_3 \vee e_5 = e_4)) \wedge$
$\qquad\qquad (\exists v_6) (\exists e_7) (\forall e_8) ((e_8 \in E_1 \wedge \text{Incident}(v_6, e_8)) \rightarrow e_8 = e_7))$

FACTS

F6: Every MSOL-expressible problem can be solved in linear time for treewidth-k graphs [Co90], [ArLaSe91], [BoPaTo92], [CoMo93]. Moreover, this is the case for many variations of each MSOL problem, including existence, minimum or maximum cardinality, minimum or maximum total weight, minimum-maximal or maximum-minimal sets, bottleneck weight, and counting.

F7: Once a problem is expressed in MSOL, a linear-time dynamic programming algorithm can be created mechanically [BoPaTo92].

F8: The *chromatic number* problem (§5.1) for treewidth-k graphs is solvable in linear time, because every treewidth-k graph possesses a vertex coloring with at most $k + 1$ colors.

F9: For some problems, a MSOL expression cannot be written and a linear-time algorithm cannot be found. In these cases it may still be possible to develop a linear-time algorithm via an extension to MSOL [BoPaTo92], or to develop a polynomial-time algorithm. Polynomial time is achieved by constructing a polynomial-size data structure that corresponds to each node in the tree decomposition (see Remark R1; also see [Bo95] and [BoPaTo08]).

REMARK

R12: The literature contains hundreds of linear-time algorithms for problems on trees, series-parallel graphs, treewidth-k graphs, and related classes. Many of these algorithms are predicted by the results cited in Fact F6. For example, all of the linear-time algorithms given in [BoPaTo08] for problems on trees, series-parallel graphs, and treewidth-k graphs are predicted by these results.

Two Open Problems

The hidden constant in the running time of a mechanically created algorithm can grow superexponentially with the number of quantifiers $\exists$ and $\forall$ present in the formula, so Fact F7 is impractical for complex formulas. *Ad hoc* methods often suffice to design an equivalent linear-time dynamic programming algorithm with a small hidden constant, but a computationally practical algorithm remains elusive. However, progress has been made for certain kinds of problems and graph classes [Kl98], [SaHuTaOg00], [Ka01], [KlMoSc02].

Open Problem 1. Determine an optimally efficient algorithm to create a linear time dynamic programming algorithm given an MSOL expression.

Open Problem 2. Determine a procedure that given an MSOL expression produces a linear-time dynamic programming algorithm with minimum hidden constant.

REMARKS

R13: Chromatic index can be solved on treewidth-k graphs in polynomial time, by constructing a polynomial-size data structure that corresponds to each node in the tree decomposition (cf. [Bo90]). More recently, a linear-time algorithm has been developed for this chromatic index problem (cf. [ZhNaNi96]).

R14: Algorithms on treewidth-k graphs can be adapted to solve the same problems on related classes such as Halin graphs, partial k-trees, bandwidth-k graphs, pathwidth-k graphs, branchwidth-k graphs, and k-terminal graphs (cf. [Wi87], [WiHe88]).

10.4.4   Algorithms on Cographs

One can produce linear-time dynamic programming algorithms for problems related to cographs. Here, in addition to the maximum-cardinality independent-set problem, we present algorithms for several other problems.

Algorithm 10.4.5: Maximum-Cardinality Independent Set in a Cograph

Input: Cograph $G = (V, E)$.
Output: $G.i$ (size and instance of a max-cardinality independent set in G).
  If $|V| = 1$
     $G.i \leftarrow 1$
  Else
     If $G = G_1 \cup G_2$
        $G.i \leftarrow G_1.i + G_2.i$
     Else
        If $G = G_1 \times G_2$
           $G.i \leftarrow \max\{G_1.i, G_2.i\}$

DEFINITIONS

D11: A *clique* in a graph is a maximal set of pairwise adjacent vertices (§5.3).

D12: The *chromatic number* of a graph G is the minimum number of colors that can be used to color the vertices of G so that no two adjacent vertices get the same color (§5.1).

D13: A *dominating set* in a graph G is a subset S of vertices such that every vertex in G is either in S or adjacent to a vertex in S (§9.2).

D14: A *matching* in a graph is a set of edges no two of which have an endpoint in common.

D15: A *cograph* is defined recursively as follows:

- (base graph) A graph with a single vertex is a cograph.

- If G_1 and G_2 are cographs, then the disjoint union $G_1 \cup G_2$ is a cograph.

- If G_1 and G_2 are cographs, then the cross-product $G_1 \times G_2$ is a cograph, which is formed by taking the union of G_1 and G_2 and adding all edges (v_1, v_2) where v_1 is in G_1 and v_2 is in G_2.

Three More Algorithms for Cographs

Algorithm 10.4.6 solves both the *maximum-clique* and *chromatic number* problems in a cograph.

Algorithm 10.4.6: Maximum-Clique and Chromatic Number in a Cograph

Input: Cograph $G = (V, E)$.
Output: $G.c$ (maximum clique in G, which is equal to chromatic number of G).

    If $|V| = 1$
        $G.c \leftarrow 1$
    Else
        If $G = G_1 \cup G_2$
            $G.c \leftarrow \max\{G_1.c, G_2.c\}$
        Else
            If $G = G_1 \times G_2$
                $G.c \leftarrow G_1.c + G_2.c$

Algorithm 10.4.7: Minimum-Cardinality Dominating Set in a Cograph

Input: Cograph $G = (V, E)$.
Output: $G.d$ (size and instance of a min-cardinality dominating set in G).

    If $|V| = 1$
        $G.d \leftarrow 1$
    Else
        If $G = G_1 \cup G_2$
            $G.d \leftarrow G_1.d + G_2.d$
        Else
            If $G = G_1 \times G_2$
                $G.d \leftarrow \min\{G_1.d, G_2.d, 2\}$

Algorithm 10.4.8: Maximum-Cardinality Matching in a Cograph

Input: Cograph $G = (V, E)$.
Output: $G.m$ (size and instance of a max-cardinality matching in G).
    If $|V| = 1$
        $G.m \leftarrow 0$
    Else
        If $G = G_1 \cup G_2$
            $G.m \leftarrow G_1.m + G_2.m$
        Else
            If $G = G_1 \times G_2$
                $G.m \leftarrow \min\{G_1.m + |V_2|,\ G_2.m + |V_1|,\ \lfloor(|V_1| + |V_2|)/2\rfloor\}$

REMARKS

R15: The right-hand side of the final assignment in Algorithm 10.4.8 is obtained by simplifying the more straightforward but less efficient formula below, wherein k denotes the number of matching edges with one endpoint in each of the subgraphs G_1 and G_2.

$$\max_{0 \le k \le \min\{|V_1|,|V_2|\}} \left\{ k + \min\left\{G_1.m, \lfloor(|V_1| - k)/2\rfloor\right\} + \min\left\{G_2.m, \lfloor(|V_2| - k)/2\rfloor\right\} \right\}$$

R16: The hamiltonian cycle and hamiltonian path problems can be solved in linear time on cographs. Weighted versions of the independent set, clique, and dominating set problems are also solvable in linear time on cographs by extending Algorithms 10.4.5 through 10.4.7. However, weighted versions of the matching and hamiltonian problems do not appear to be solvable in linear time on cographs (although they are solvable in polynomial time). Intuitively, the reason is that the cross product operation adds too many edges, where each edge potentially has a different weight.

EXAMPLE

E5: Algorithms 10.4.5 through 10.4.8 are demonstrated on the cograph G shown in Figure 10.4.4. T denotes the tree decomposition of G, and each 4-tuple exhibits values for $G.i, G.c, G.d$, and $G.m$. The maximum independent set has size 3, for example $\{a, b, c\}$. The maximum clique has size 5, given by $\{c, d, e, g, h\}$. The minimum dominating set has size 2, for example $\{a, f\}$. The maximum matching has size 4, for example $\{(a, f), (b, g), (c, h), (d, e)\}$.

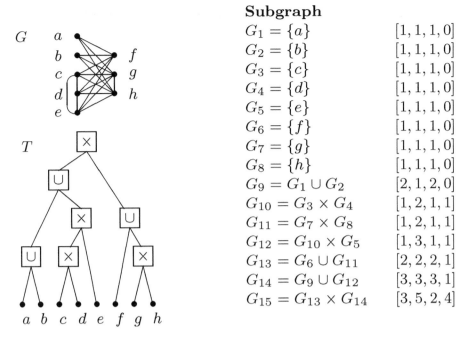

Subgraph	
$G_1 = \{a\}$	$[1,1,1,0]$
$G_2 = \{b\}$	$[1,1,1,0]$
$G_3 = \{c\}$	$[1,1,1,0]$
$G_4 = \{d\}$	$[1,1,1,0]$
$G_5 = \{e\}$	$[1,1,1,0]$
$G_6 = \{f\}$	$[1,1,1,0]$
$G_7 = \{g\}$	$[1,1,1,0]$
$G_8 = \{h\}$	$[1,1,1,0]$
$G_9 = G_1 \cup G_2$	$[2,1,2,0]$
$G_{10} = G_3 \times G_4$	$[1,2,1,1]$
$G_{11} = G_7 \times G_8$	$[1,2,1,1]$
$G_{12} = G_{10} \times G_5$	$[1,3,1,1]$
$G_{13} = G_6 \cup G_{11}$	$[2,2,2,1]$
$G_{14} = G_9 \cup G_{12}$	$[3,3,3,1]$
$G_{15} = G_{13} \times G_{14}$	$[3,5,2,4]$

Figure 10.4.4: Illustrating Algorithms 10.4.5 through 10.4.8.

10.4.5 Algorithms on Cliquewidth-k Graphs

DEFINITION

D16: Let $[k]$ denote the set of integers $\{1, 2, \ldots, k\}$. A **cliquewidth-k graph** is defined recursively as follows:

- (base graph) Any graph G with $V(G) = \{v\}$ and $l(v) \in [k]$ is a cliquewidth-k graph.

- If G_1 and G_2 are cliquewidth-k graphs and $i, j \in [k]$ with $i \neq j$, then:
    - The disjoint union $G_1 \cup G_2$ is a cliquewidth-k graph.
    - The graph $(G_1)_{i \times j}$ is a cliquewidth-k graph, where $(G_1)_{i \times j}$ is formed from G_1 by adding all edges (v_1, v_2) such that $l(v_1) = i$ and $l(v_2) = j$.
    - The graph $(G_1)_{i \to j}$ is a cliquewidth-k graph, where $(G_1)_{i \to j}$ is formed from G_1 by switching all vertices with label i to label j.

Maximum-Cardinality Independent Set of a Cliquewidth-k Graph

NOTATION: Algorithm 10.4.9 below uses the following notation:

- $G[S]$ = max-cardinality independent set that contains only labels from $S \subseteq [k]$.

- $G.max$ = max-cardinality independent set.

Algorithm 10.4.9: Maximum-Cardinality Independent Set in a Cliquewidth-k Graph

Input: Cliquewidth-k graph $G = (V, E)$.
Output: $G.max$ (size and instance of a max-cardinality independent set in G).
    If $V = \{v\}$
       For all $S \subseteq [k]$
          If $l(v) \in S$
             $G[S] \leftarrow 1$
          Else
             $G[S] \leftarrow 0$
   Else
      If $G = G_1 \cup G_2$
         For all $S \subseteq [k]$
            $G[S] \leftarrow G_1[S] + G_2[S]$
      Else
         If $G = (G_1)_{i \times j}$
           For all $S \subseteq [k]$
              $G[S] \leftarrow \max\{G_1[S - \{i\}],\ G_1[S - \{j\}]\}$
         Else
            If $G = (G_1)_{i \rightarrow j}$
              For all $S \subseteq [k]$
                 If $j \in S$
                    $G[S] \leftarrow G_1[S \cup \{i\}]$
                 Else
                    $G[S] \leftarrow G_1[S - \{i\}]$
     $G.max \leftarrow \max\{G[S] : S \subseteq [k]\}$

EXAMPLE

E6:   Algorithm 10.4.9 is demonstrated on the cliquewidth-3 graph G shown in Figure 10.4.5. T denotes the decomposition tree of G. Each 8-tuple consists of $G[\emptyset]$, $G[\{1\}]$, $G[\{2\}]$, $G[\{3\}]$, $G[\{1,2\}]$, $G[\{1,3\}]$, $G[\{2,3\}]$, and $G[\{1,2,3\}]$. The maximum independent set has size 3, given by either $\{a, c, e\}$ or $\{b, d, f\}$.

A Subset of the MSOL Expressions for a Graph

Many problems including variations of independent set, dominating set, clique, and m-vertex colorability (for any fixed m) can be solved in linear time on cliquewidth-k graphs, provided that a decomposition tree is known. These problems are all expressible using a certain subset of the MSOL expressions.

DEFINITION

D17:   The **MSOL′ set** of expressions for a graph $G = (V, E)$ is the subset of MSOL expressions restricted to variables v_i with domain V, e_i with domain E, and V_i with domain 2^V. The MSOL′ set contains primitive predicates such as $v_i = v_j$, Incident(v_i, e_j), and $v_i \in V_j$. MSOL′ permits the logical operators ($\neg$, $\wedge$, $\vee$) and quantifiers ($\exists$, $\forall$). Thus, MSOL′ is the same as MSOL without the edge-set variables E_i and without primitive predicates such as $e_i \in E_j$ that refer to edge-set variables.

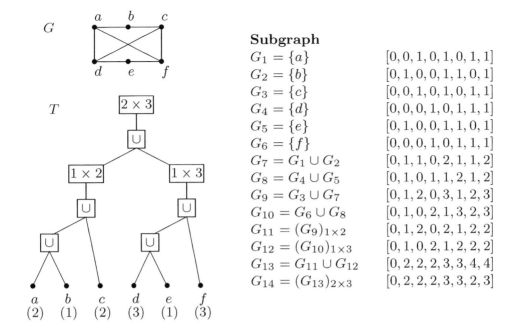

Figure 10.4.5: Maximum-cardinality independent set in a cliquewidth-3 graph.

FACT

F10: Every MSOL'-expressible problem can be solved in linear time on any class of cliquewidth-k graphs [CoMaRo00], provided that either there exists a linear time decomposition algorithm for the class (as for cographs) or a decomposition tree is provided as part of the input. This statement holds for variations of each MSOL' problem that involve existence, optimum cardinality or total weight, counting the number of solutions, etc. Once a problem is expressed in MSOL', a linear-time dynamic programming algorithm can be created mechanically.

REMARKS

R17: Observe that the MSOL expressions given in §10.4.3 for IndependentSet(V_1), Clique(V_1), DominatingSet(V_1), and VertexColorable$_m$(V_1, ..., V_m), are also MSOL' expressions. However, the MSOL expressions given for Matching(E_1), Connected(E_1), HamCycle(E_1), and HamPath(E_1) are not in MSOL'.

R18: Some problems such as variations of matching and hamiltonicity do not appear to be expressible in MSOL', and it is not known whether these problems can be solved in linear time on cliquewidth-k graphs. However, such problems can often be solved in polynomial time, given the decomposition tree. Polynomial time is achieved by constructing a polynomial-size data structure that corresponds to each node in the tree decomposition (cf. [Wa94], [EsGuWa01]).

R19: Algorithms on cliquewidth-k graphs can be adapted to solve the same problems on related classes such as k-NLC graphs [Wa94].

10.4.6   Algorithms on k-HB Graphs

DEFINITION

D18:   ***k-HB (homogeneous balanced) graphs*** are graphs for which there is a particular $O(n^{k+2})$-time top-down decomposition algorithm that constructs a pseudo-cliquewidth-$(k+2^k)$ balanced decomposition. (Also see §2.4.1 and [Jo03], [BoJoRaSp04].)

FACTS

F11:   Every k-HB graph can be composed from single vertices using only the operation $G = G_1 \times_{B,h} G_2$. Here G_1 and G_2 denote child subgraphs, each $|V_i| \leq 2 \cdot |V|/3$, $B = (V_B, E_B)$ is a bipartite graph with $V_B = Z_1 \cup Z_2$ and $E_B \subseteq Z_1 \times Z_2$, $|Z_1| \leq k$, $|Z_2| \leq 2^k$, $h: V \to V_B$ is a mapping with each $h(V_i) \subseteq Z_i$, and $(x, y) \in E$ iff $(h(x), h(y)) \in E_B$ for all $x \in V_1$ and $y \in V_2$.

F12:   This k-HB decomposition leads to polynomial-time dynamic programming algorithms for many problems on k-HB graphs, using recursion (top-down) rather than dynamic programming (bottom-up). Each algorithm's running time is polynomial because at each node of the decomposition it evaluates $O(1)$ parameters, each of which produces $O(1)$ recursive calls on smaller subproblems. Also, the decomposition has $O(\log |V|)$ height, hence $|V|^{O(1)}$ nodes. [Jo03], [BoJoRaSp04]

Maximum-Cardinality Independent Set in a k-HB Graph

NOTATION: Algorithm 10.4.10 below uses the following notation:

- $G[S]$ = max-cardinality independent set that contains only vertices in $S \subseteq V$.

- $G.indep$ = max-cardinality independent set.

Algorithm 10.4.10: Maximum-Cardinality Independent Set in a k-HB Graph

Input: k-HB graph $G = (V, E)$.
Output: $G.indep$ (size and instance of a max-cardinality independent set in G).
  If $|V| = 1$
      $G[S] \leftarrow |S|$
  Else
      If $G = G_1 \times_{B,h} G_2$
          $G[S] \leftarrow \max\{G_1[T] + G_2[U] : X \subseteq h(V_1),\ Y \subseteq h(V_2),$
              $(X \times Y) \cap E_B = \emptyset,\ T = S \cap h^{-1}(X),\ U = S \cap h^{-1}(Y)\}$
  $G.indep = G[V]$

EXAMPLE

E7:   We demonstrate Algorithm 10.4.10 on the 2-HB graph G shown in Figure 10.4.6. Note that $G = G_1 \times_{B,h} G_2$, where G_1, G_2, B, and h are as shown. The top-level computations are summarized on the right. The maximum independent set has size 4, and the explicit solution is $\{r, t, w, y\}$.

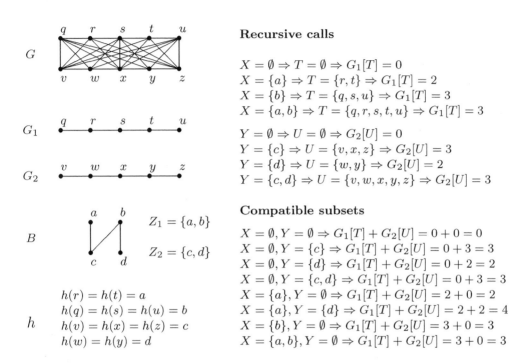

Recursive calls

$X = \emptyset \Rightarrow T = \emptyset \Rightarrow G_1[T] = 0$
$X = \{a\} \Rightarrow T = \{r, t\} \Rightarrow G_1[T] = 2$
$X = \{b\} \Rightarrow T = \{q, s, u\} \Rightarrow G_1[T] = 3$
$X = \{a, b\} \Rightarrow T = \{q, r, s, t, u\} \Rightarrow G_1[T] = 3$

$Y = \emptyset \Rightarrow U = \emptyset \Rightarrow G_2[U] = 0$
$Y = \{c\} \Rightarrow U = \{v, x, z\} \Rightarrow G_2[U] = 3$
$Y = \{d\} \Rightarrow U = \{w, y\} \Rightarrow G_2[U] = 2$
$Y = \{c, d\} \Rightarrow U = \{v, w, x, y, z\} \Rightarrow G_2[U] = 3$

$Z_1 = \{a, b\}$

$Z_2 = \{c, d\}$

$h(r) = h(t) = a$
$h(q) = h(s) = h(u) = b$
$h(v) = h(x) = h(z) = c$
$h(w) = h(y) = d$

Compatible subsets

$X = \emptyset, Y = \emptyset \Rightarrow G_1[T] + G_2[U] = 0 + 0 = 0$
$X = \emptyset, Y = \{c\} \Rightarrow G_1[T] + G_2[U] = 0 + 3 = 3$
$X = \emptyset, Y = \{d\} \Rightarrow G_1[T] + G_2[U] = 0 + 2 = 2$
$X = \emptyset, Y = \{c, d\} \Rightarrow G_1[T] + G_2[U] = 0 + 3 = 3$
$X = \{a\}, Y = \emptyset \Rightarrow G_1[T] + G_2[U] = 2 + 0 = 2$
$X = \{a\}, Y = \{d\} \Rightarrow G_1[T] + G_2[U] = 2 + 2 = 4$
$X = \{b\}, Y = \emptyset \Rightarrow G_1[T] + G_2[U] = 3 + 0 = 3$
$X = \{a, b\}, Y = \emptyset \Rightarrow G_1[T] + G_2[U] = 3 + 0 = 3$

Figure 10.4.6: Maximum-cardinality independent set in a 2-HB graph.

REMARK

R20:   The maximum-cardinality clique and m-vertex colorability problems can also be solved in polynomial time on k-HB graphs. However, the chromatic number, dominating set, and hamiltonian problems are not known to be solvable in polynomial time on k-HB graphs. Maximum matching is of course solvable in polynomial time on k-HB graphs, but it is not known whether this can be done more efficiently than for arbitrary graphs.

A Subset of the MSOL′ Expressions

Most problems that are known to be solvable in polynomial time for k-HB graphs are expressible in a particular predicate language whose expressions form a subset of the MSOL′ expressions.

DEFINITION

D19: MSOL″ for a graph $G = (V, E)$ denotes a subset of MSOL′ restricted to variables v_i with domain V, and variables V_i with domain 2^V. MSOL″ contains primitive predicates such as Adjacent(v_i, v_j) and $v_i \in V_j$. MSOL″ permits the logical operators $(\neg, \wedge, \vee)$ and quantifiers $(\exists, \forall)$. However, these primitives and connectors cannot be combined in any arbitrary way; rather every MSOL″ expression must possess the following format.

$$(\exists V_1) \ldots (\exists V_m) \, ((\forall v_1) F_0(v_1 \in V_1, \ldots, v_1 \in V_m)$$
$$\wedge (\forall v_2)(\forall v_3) \, (\text{Adjacent}(v_2, v_3) \rightarrow \wedge_{1 \le i \le j \le m} F_{ij}(v_2 \in V_i, v_3 \in V_j))$$
$$\wedge (\forall v_4)(\forall v_5)(\neg \, \text{Adjacent}(v_4, v_5) \rightarrow \wedge_{1 \le i \le j \le m} F'_{ij}(v_4 \in V_i, v_5 \in V_j)))$$

REMARK

R21: In Definition D19, each F_0, each F_{ij}, and each F'_{ij} is an arbitrary formula that combines the indicated primitive predicates using operators $\neg$, $\wedge$, and $\vee$. If any of these formulas is identically true, it may be omitted.

EXAMPLE

E8: The MSOL expressions for IndependentSet, Clique, and VertexColorable$_m$ can be rewritten as equivalent MSOL″ expressions as shown below. However, other MSOL expressions such as DominatingSet do not appear to be expressible in MSOL″.

IndependentSet $\Leftrightarrow (\exists V_1)(\forall v_2)(\forall v_3) \, (\text{Adjacent}(v_2, v_3) \rightarrow \neg(v_2 \in V_1 \wedge v_3 \in V_1))$
Clique $\Leftrightarrow (\exists V_1)(\forall v_4)(\forall v_5)(\neg \, \text{Adjacent}(v_4, v_5) \rightarrow \neg(v_4 \in V_1 \wedge v_5 \in V_1))$
VertexColorable$_m \Leftrightarrow (\exists V_1) \ldots (\exists V_m)((\forall v_1)(v_1 \in V_1 \vee \ldots \vee v_1 \in V_m)$
$$\wedge (\forall v_2)(\forall v_3) \, (\text{Adjacent}(v_2, v_3) \rightarrow \wedge_{1 \le i \le m} \neg(v_2 \in V_i \wedge v_3 \in V_i)))$$

FACT

F13: Every MSOL″-expressible problem can be solved in polynomial time when the input graph is restricted to any class of k-HB graphs [Jo03], [BoJoRaSp04]. This includes every cliquewidth-k graph, even if its decomposition tree is not provided as part of the input. Once a problem is expressed in MSOL″, the polynomial-time recursive algorithm can be created mechanically.

References

[Ar85] S. Arnborg, Efficient algorithms for combinatorial problems on graphs with bounded decomposibility — a survey, *BIT* **25** (1985), 2–23.

[ArCoPrSe93] S. Arnborg, B. Courcelle, A. Proskurowski, D. Seese, An algebraic theory of graph reduction, *Journal of the ACM* **40** (1993), 1134–1164.

[ArLaSe91] S. Arnborg, J. Lagergren, D. Seese, Easy problems for tree-decomposable graphs, *Journal of Algorithms* **12** (1991), 308–340.

[ArPr89] S. Arnborg, A. Proskurowski, Linear time algorithms for $\mathcal{NP}$-hard problems restricted to partial k-trees, *Discrete Applied Mathematics* **23** (1989), 11–24.

[BeLaWo87] M. W. Bern, E. L. Lawler, A. L. Wong, Linear time computation of optimal subgraphs of decomposable graphs, *Journal of Algorithms* **8** (1987), 216–235.

[Bo87] H. L. Bodlaender, Dynamic programming on graphs with bounded tree-width, Ph.D. Dissertation, Massachusetts Institute of Technology, 1987; extended abstract in *Proceedings of ICALP* (1988).

[Bo90] H. L. Bodlaender, Polynomial algorithms for graph isomorphism and chromatic index on partial k-trees, *Journal of Algorithms* **11** (1990), 631–643.

[Bo95] R. B. Borie, Generation of polynomial-time algorithms for some optimization problems on tree-decomposable graphs, *Algorithmica* **14** (1995), 123–137.

[BoJoRaSp04] R. B. Borie, J. L. Johnson, V. Raghavan, J. Spinrad, Robust algorithms for some problems on graphs of bounded clique-width, manuscript, presented at *SIAM Conference on Discrete Mathematics* (2004).

[BoPaTo91] R. Borie, R. G. Parker, C. A. Tovey, Algorithms for recognition of regular properties and decomposition of recursive graph families, *Annals of Operations Research* **33** (1991), 127–149.

[BoPaTo92] R. B. Borie, R. G. Parker, C. A. Tovey, Automatic generation of linear-time algorithms from predicate calculus descriptions of problems on recursively constructed graph families, *Algorithmica* **7** (1992), 555–581.

[BoPaTo08] R. B. Borie, R. G. Parker, C. A. Tovey, Solving problems on recursively constructed graphs, *ACM Computing Surveys* **41** (2008), Article 4, 1–51.

[BrLeSp99] A. Brandstadt, V. B. Lee, J. P. Spinrad, *Graph Classes: A Survey*, SIAM monograph, 1999.

[CoLeBu81] D. G. Corneil, H. Lerchs, L. S. Burlingham, Complement reducible graphs, *Discrete Applied Mathematics* **3** (1981), 163–174.

[CoPeSt85] D. G. Corneil, Y. Perl, L. K. Stewart, A linear recognition algorithm for cographs, *SIAM Journal on Computing* **14** (1985), 926–934.

[Co90] B. Courcelle, The monadic second-order logic of graphs I: Recognizable sets of finite graphs, *Information and Computation* **85** (1990), 12–75.

[CoMaRo00] B. Courcelle, J. A. Makowsky, U. Rotics, Linear time solvable optimization problems on graphs of bounded clique width, *Theory of Computing Systems* **33** (2000), 125–150.

[CoMo93] B. Courcelle, M. Mosbah, Monadic second-order evaluations on tree-decomposable graphs, *Theoretical Computer Science* **109** (1993), 49–82.

[CoOl00] B. Courcelle, S. Olariu, Upper bounds to the clique-width of graphs, *Discrete Applied Mathematics* **101** (2000), 77–114.

[De97] B. De Fluiter, Algorithms for graphs of small treewidth, Ph.D. Dissertation, University of Utrecht, 1997.

[Ed65-a] J. Edmonds, Paths, trees, and flowers, *Canadian Journal of Mathematics* **17** (1965), 449–467.

[Ed65-b] J. Edmonds, Maximum matching and polyhedron of 0,1 vertices, *J. Research National Bureau of Standards* **69B** (1965), 125–130.

[ElCo88] E. El-Mallah, C. Colbourn, Partial k-tree algorithms, *Congressus Numerantium* **64** (1988), 105–119.

[EsGuWa01] W. Espelage, F. Gurski, E. Wanke, How to solve $\mathcal{NP}$-hard graph problems on clique-width bounded graphs in polynomial time, *Lecture Notes in Computer Science* **2204** (2001), 117–128.

[GaGrJoKn78] M. Garey, R. Graham, D. Johnson, D. Knuth, Complexity results for bandwidth minimization, *SIAM Journal on Applied Mathematics* **34** (1978), 477–495.

[HaHeLaPeWi87] E. Hare, S. Hedetniemi, R. Laskar, K. Peters, T. Wimer, Linear-time computability of combinatorial problems on generalized series-parallel graphs, *Discrete Algorithms and Complexity* **14** (1987), 437–457.

[JaOl95] B. Jamison, S. Olariu, Linear time optimization algorithms for P4-sparse graphs, *Discrete Applied Mathematics* **61** (1995), 155–175.

[Jo03] J. Johnson, Polynomial time recognition and optimization algorithms on special classes of graphs, Ph.D. Dissertation, Computer Science, Vanderbilt University, 2003.

[Ka01] I. T. Kassios, Translating Borie-Parker-Tovey calculus into mutumorphisms, manuscript, 2001.

[Kl98] N. Klarlund, Mona and Fido: the logic-automaton connection in practice, *Lecture Notes in Computer Science* **1414** (1998), 311–326.

[KlMoSc02] N. Klarlund, A. Molle, M. I. Schwartzbach, MONA implementation secrets, *International Journal of Foundations of Computer Science* **13** (2002), 571–586.

[Ri85] M. B. Richey, Combinatorial optimization on series-parallel graphs: algorithms and complexity, Ph.D. Dissertation, Georgia Institute of Technology, 1985.

[SaHuTaOg00] I. Sasano, Z. Hu, M. Takeichi, M. Ogawa, Make it practical: a generic linear-time algorithm for solving maximum-weightsum problems, *ACM SIGPLAN Notices* **35** (2000), 137–149.

[Sc87] P. Scheffler, Linear-time algorithms for $\mathcal{NP}$-complete problems restricted to partial k-trees, Technical report R-MATH-03/87, *IMATH*, Berlin, 1987.

[Sc89] P. Scheffler, Die Baumweite von Graphen als ein Maβ für die Kompliziertheit algorithmischer Probleme, Ph.D. Dissertation, Akademie der Wissenschaften der DDR, 1989.

[ScSe86] P. Scheffler, D. Seese, Graphs of bounded tree-width and linear-time algorithms for $\mathcal{NP}$-complete problems, *Proceedings of the Bilateral Seminar*, Samarkand, 1986.

[ScSe88] P. Scheffler, D. Seese, A combinatorial and logical approach to linear-time computability, *Lecture Notes in Computer Science* **378** (1988), 379–380.

[Sp03] J. P. Spinrad, *Efficient Graph Representations*, Fields Institute Monographs, American Mathematical Society, 2003.

[Sy83] M. Syslo, $\mathcal{NP}$-complete problems on some tree-structured graphs: a review, *Proceedings of 9th Workshop on Graph-theoretic Concepts in Computer Science* (1983), 342–353.

[TaNiSa82] K. Takamizawa, T. Nishizeki, N. Saito, Linear-time computability of combinatorial problems on series-parallel graphs, *Journal of the ACM* **29** (1982), 623–641.

[Wa94] E. Wanke, k-NLC graphs and polynomial algorithms, *Discrete Applied Mathematics* **54** (1994), 251–266; later revised with new co-author F. Gurski.

[Wi87] T. V. Wimer, Linear algorithms on k-terminal recursive graphs, Ph.D. Dissertation, Clemson University, 1987.

[WiHe88] T. V. Wimer, S. T. Hedetniemi, k-terminal recursive families of graphs, *Congressus Numerantium* **63** (1988), 161–176.

[WiHeLa85] T. V. Wimer, S. T. Hedetniemi, R. Laskar, A methodology for constructing linear graph algorithms, *Congressus Numerantium* **50** (1985), 43–60.

[ZhNaNi96] X. Zhou, S. Nakano, T. Nishizeki, Edge-coloring partial k-trees, *Journal of Algorithms* **21** (1996), 598–617.

Section 10.5

Fuzzy Graphs

John N. Mordeson, Creighton University
D. S. Malik, Creighton University

INTRODUCTION

In 1965, L. A. Zadeh introduced the concept of a fuzzy subset of a set as a way for representing uncertainty, [Za65]. Let S denote a set. A *fuzzy subset* of S is a function of S into the closed interval $[0, 1]$. His seminal paper described fuzzy set theory and consequently fuzzy logic. The purpose of Zadeh's paper was to develop a theory which could deal with ambiguity and imprecision of certain classes or sets in human thinking, particularly in the domains of pattern recognition, communication of information, and abstraction. This theory proposed making the grade of membership of an element in a subset of a universal a value in the closed interval $[0, 1]$.

Zadeh's ideas have found applications in engineering, computer science, artificial intelligence, decision analysis, pattern recognition, management science, operation research, robotics, and other areas. All areas in mathematics have been touched by fuzzy set theory. In this paper, we concentrate on both theoretical and applied issues concerning fuzzy graphs.

Kauffman was the first to define a fuzzy graph, [Ka73]. However, the cornerstone papers for the development of graph theory were by Rosenfeld [Ro75] and by Yeh and Bang, [YeBa75]. In [Ro75], Rosenfeld presented fuzzy analogs of several basic graph-theoretic concepts, including subgraphs, paths and connectedness, cliques, bridges and cutnodes, forests and trees. The paper [YeBa75] was partly motivated from the desire to extend the existing theoretical techniques at the time to data graphs directly rather than to threshold graphs

---

10.5.1 Definitions and Basic Properties

Let S be a set and μ a fuzzy subset of S.

DEFINITIONS

D1: The set of all fuzzy subsets of S is called the **fuzzy power set** of S and is denoted by $\mathcal{FP}(S)$.

D2: Let $t \in [0,1]$. Then μ^t is defined to be the set $\{x \in S \mid \mu(x) \geq t\}$.

D3: The set μ^t is called a **level set** or the t-level set of μ.

D4: The set $\{x \in S \mid \mu(x) > 0\}$ is called the **support** of μ and is denoted by $\text{Supp}(\mu)$.

TERMINOLOGY: Throughout we use the notation $\vee$ for supremum and $\wedge$ for infimum.

D5: Let μ, ν be fuzzy subsets of S. Then we write $\mu \subseteq \nu$ if $\mu(x) \leq \nu(x)$ for all $x \in S$ and we write $\mu \subset \nu$ if $\mu \subseteq \nu$ and there exists $x \in S$ such that $\mu(x) < \nu(x)$.

D6: We define the fuzzy subsets $\mu \cup \nu$ and $\mu \cap \nu$ of S by for all $x \in S, (\mu \cup \nu)(x) = \mu(x) \vee \nu(x)$ and $(\mu \cap \nu)(x) = \mu(x) \wedge \nu(x)$.

D7: The fuzzy subset μ^c of S is defined by for all $x \in S, \mu^c(x) = 1 - \mu(x)$.

TERMINOLOGY: We consider $\mu \cup \nu$ and $\mu \cap \nu$ to be the **union** and **intersection** of μ and ν, respectively, while μ^c is considered to be the **complement** of μ in S.

D8: A **fuzzy relation** of a set S into a set T is a fuzzy subset of the set of ordered pairs $S \times T$.

D9: If ρ is a fuzzy relation of S into T and σ is a fuzzy relation of T into a set W, then the **composition** of ρ with σ, written $\rho \circ \sigma$, is defined by for all $(x, w) \in S \times W, (\rho \circ \sigma)(x, w) = \vee\{\rho(x, y) \wedge \sigma(y, w) \mid y \in T\}$.

D10: If ρ is a fuzzy relation of a set S into S, we say that ρ is a fuzzy relation on S and define $\rho^2 = \rho \circ \rho$. Let ρ be a fuzzy relation on S and n any positive integer greater than or equal to 2. We let $\rho^1 = \rho$ and define ρ^n recursively by $\rho^n = \rho \circ \rho^{n-1}$, where ρ^{n-1} is assumed to have been defined. We define the fuzzy relation on ρ^∞ on S by for all $x, y \in S, \rho^\infty(x, y) = \vee\{\rho^k(x, y) \mid k = 1, 2, ...\}$.

D11: A ***graph*** is a pair (V, R), where V is a set and R is a relation on V. The elements of V are thought of as vertices of the graph and the elements of R are thought of as the edges.

D12: A ***fuzzy graph*** $G = (V, \mu, \rho)$ is a triple, where V is a nonempty set, μ is a fuzzy subset of V, and ρ is a fuzzy relation on V such that for all $x, y \in V, \rho(x, y) \le \mu(x) \wedge \mu(y)$.

FACT

F1: If (V, μ, ρ) is a fuzzy graph, then (μ^t, ρ^t) is a graph for all $t \in [0, 1]$ and in fact is a subgraph of $(V, \text{Supp}(\rho))$.

DEFINITIONS

D13: Let $G = (V, \mu, \rho)$ be a fuzzy graph. Then a fuzzy graph $H = (V, \nu, \tau)$ is called a ***partial fuzzy subgraph*** of G if $\nu \subseteq \mu$ and $\tau \subseteq \rho$.

D14: The fuzzy graph $H = (P, \nu, \tau)$ is called a ***fuzzy subgraph*** of G induced by P if $P \subseteq V, \nu(x) = \mu(x)$ for all $x \in P$ and $\tau(x, y) = \rho(x, y)$ for all $x, y \in P$.

TERMINOLOGY: If the set of vertices V is understood, we sometimes write $G = (\mu, \rho)$ for a fuzzy graph. It can be shown that if (ν, τ) is a partial fuzzy subgraph of (μ, ρ), then $\tau^\infty \subseteq \rho^\infty$.

D15: Let $G = (\mu, \rho)$ be a fuzzy graph. Then a partial fuzzy subgraph of (ν, τ) of G is said to ***span*** G if $\mu = \nu$. In this case, we call (ν, τ) a ***spanning fuzzy subgraph*** of (μ, ρ).

10.5.2   Paths and Connectedness

Let $G = (V, \mu, \rho)$ be a fuzzy graph.

DEFINITIONS

D16: A ***path*** P in G is a sequence of distinct vertices $x_0, x_1, ..., x_n$ (except possibly x_0 and x_n) such that $\rho(x_{i-1}, x_i) > 0, i = 1, ..., n$. Here n is called the ***length*** of the path. The consecutive pairs are called the ***edges*** of the path.

D17: The ***diameter*** of $x, y \in V$, written $\text{diam}(x, y)$, is the length of the longest path joining x to y.

D18: Let P be a path. The ***strength*** of P is defined to be $\wedge_{i=1}^n \rho(x_{i-1}, x_i)$. In words, the strength of a path is defined to be the weight of the weakest edge. We call P a ***cycle*** if $x_0 = x_n$ and $n \ge 3$.

D19: Two vertices that are joined by a path are called ***connected vertices***.

FACT

F2: The notion of connectedness is an equivalence relation.

DEFINITIONS

D20: The equivalence classes of vertices under the connected relation are called **connected components** of the given fuzzy graph. They are just its maximal connected partial fuzzy subgraphs.

D21: A strongest path joining any two vertices x, y has strength $\rho^\infty(x, y)$. We sometimes refer to this as the **strength of connectedness** between vertices.

D22: Let $G = (\mu, \rho)$ be a fuzzy graph, let x, y be two distinct vertices, and let G' be the partial fuzzy subgraph of G obtained by deleting the edge (x, y). That is, $G' = (\mu, \rho')$, where $\rho'(x, y) = 0$ and $\rho' = \rho$ for all other pairs. We call (x, y) a **bridge** in G if $\rho'^\infty(u, v) < \rho^\infty(u, v)$ for some u, v in V. In words, if deleting the edge (x, y) reduces the strength of connectedness between some pair of vertices.

FACTS

F3: Let x and y be two vertices in G. Then (x, y) is a bridge if and only if there exist vertices u, v such that (x, y) is an edge of every strongest path from u to v.

F4: Let (μ, ρ) be a fuzzy graph. Then the following statements are equivalent.
(1) (x, y) is a bridge.
(2) $\rho'^\infty(x, y) < \rho(x, y)$.
(3) (x, y) is not the edge of any weakest cycle.

DEFINITION

D23: Let w be any vertex and let G' be the partial fuzzy subgraph of G obtained by deleting the vertex w. That is, $G' = (\mu', \rho')$ is the partial fuzzy subgraph of G such that $\mu'(w) = 0, \mu = \mu'$ for all other vertices, $\rho'(w, z) = 0$ for all vertices z, and $\rho' = \rho$ for all other edges. We call w a **cutvertex** in G if $\rho'^\infty(u, v) < \rho^\infty(u, v)$ for some u, v in V such that $u \neq w \neq v$. In words, if deleting the vertex w reduces the strength of connectedness between some other pair of vertices.

FACT

F5: A vertex w is a cutvertex if and only if there exist vertices u, v distinct from w such that w is on every strongest path from u to v.

DEFINITION

D24: A fuzzy graph G is called **nonseparable** or a **block** if it has no cutvertices.

REMARK

R1: Although in a fuzzy graph, a block may have bridges, this cannot happen for crisp graphs.

10.5.3   Forests and Trees

DEFINITIONS

D25: A crisp graph that has no cycles is called **acyclic** or a **forest**.

D26: A connected forest is a **tree**.

D27: A fuzzy graph is called a *forest* if the graph consisting of its nonzero edges is a forest, and a *tree* if this graph is also connected.

D28: If $G = (\mu, \rho)$ is a fuzzy graph, we call G a **fuzzy forest** if it has a partial fuzzy spanning subgraph $F = (\mu, \tau)$ which is a forest, where for all edges (x, y) not in F, i.e., such that $\tau(x, y) = 0$, we have $\rho(x, y) < \tau^{\infty}(x, y)$. In words, if (x, y) is in G, but is not in F, there is a path in F between x and y whose strength is greater than $\rho(x, y)$. Clearly, a forest is a fuzzy forest.

FACTS

F6: A fuzzy graph G is a fuzzy forest if and only if in any cycle of G there is an edge (x, y) such that $\rho(x, y) < \rho'^{\infty}(x, y)$, where $G' = (\mu, \rho')$ is the partial fuzzy subgraph obtained by the deletion of the edge (x, y) from G.

F7: Let G be a fuzzy graph. If there is at most one strongest path between any two vertices of G, then G is a fuzzy forest.

REMARK

R2: Note that the converse of Fact F7 does not hold.

DEFINITION

D29: Let $G = (\mu, \rho)$ be a fuzzy graph. We call G a **tree** if and only if $(\mathrm{Supp}(\mu), \mathrm{Supp}(\rho))$ is a tree and we call G a **cycle** if and only if $(\mathrm{Supp}(\mu), \mathrm{Supp}(\rho))$ is a cycle.

FACTS

F8: Let $G = (\mu, \rho)$ be a cycle. Then G is a fuzzy cycle if and only if G is not a fuzzy tree.

F9: Let $G = (\mu, \rho)$ be a fuzzy graph. If there exists $t \in (0, 1]$ such that $(\mathrm{Supp}(\mu), \rho^t)$ is a tree, then G is a fuzzy tree. Conversely, if G is a cycle and G is a fuzzy tree, then there exists $t \in (0, 1]$ such that $(\mathrm{Supp}(\mu), \rho^t)$ is a tree.

We next give some properties of fuzzy graphs.

F10: Let $G = (\mu, \rho)$ be a fuzzy graph such that $(\mathrm{Supp}(\mu), \mathrm{Supp}(\rho))$ is a cycle. Then a vertex is a cutvertex of G if and only if it is a common vertex of two bridges.

F11: If w is a common vertex of at least two bridges, then w is a cutvertex.

F12:   If (u, v) is a bridge, then $\rho^\infty(u, v) = \rho(u, v)$.

F13:   If $G = (\mu, \rho)$ is a fuzzy tree and $(\text{Supp}(\mu), \text{Supp}(\rho))$ is not a tree, then there exists at least one edge (u, v) in $\text{Supp}(\rho)$ for which $\rho(u, v) < \rho^\infty(u, v)$.

DEFINITION

D30:   A ***complete fuzzy graph*** is a fuzzy graph $G = (\mu, \rho)$ such that $\rho(u, v) = \mu(u) \wedge \mu(v)$ for all $u, v \in V$.

FACTS

F14:   If $G = (\mu, \rho)$ is a fuzzy tree, then G is not complete.

F15:   If G is a fuzzy tree, then the internal vertices of F are cutvertices of G.

F16:   Let $G = (\mu, \rho)$ be a fuzzy graph. Then G is a fuzzy tree if and only if the following conditions are equivalent for all $u, v \in V$:
(1) (u, v) is a bridge.
(2) $\rho^\infty(u, v) = \rho(u, v)$.

F17:   A fuzzy graph is a fuzzy tree if and only if it has a unique maximum fuzzy spanning tree.

REMARK

R3:   If G is a fuzzy graph which is not a fuzzy tree and T is the maximum fuzzy spanning tree of G, then there is at least one edge in T which is not a bridge and edges not in T are not bridges of G.

FACT

F18:   If $G = (\mu, \rho)$ is a fuzzy graph with $\text{Supp}(\mu) = V$ and $|V| = p$, then G has at most $p - 1$ bridges.

REMARK

R4:   It follows that if $G = (\mu, \rho)$ is a fuzzy graph with T a maximum fuzzy spanning tree of G, then end vertices of T are not cutvertices of G. Also, every fuzzy graph has at least two vertices which are not cutvertices.

10.5.4   Fuzzy Cut Sets

DEFINITIONS

D31:   Let $G = (\mu, \rho)$ be a fuzzy graph. Let $x \in V$ and let $t \in [0, 1]$. Define the fuzzy subset x_t of V by $\forall y \in V, x_t(y) = 0$ if $y \neq x$ and $x_t(y) = t$ if $y = x$. Then x_t is called a ***fuzzy singleton*** in V. If $(x, y) \in V \times V$, then $(x, y)_{\rho(x,y)}$ denotes a fuzzy singleton in $V \times V$.

D32: Let $G = (\mu, \rho)$ be a fuzzy graph and let E be a subset of Supp(ρ).
(1) $\{(x,y)_{\rho(x,y)}|(x,y) \in E\}$ is called a ***cut set*** of (μ, ρ) if E is a cut set of (Supp(μ),Supp(ρ)).
(2) $\{(x,y)_{\rho(x,y)}|(x,y) \in E\}$ is called a ***fuzzy cut set*** of (μ, ρ) if $\exists u, v \in$ Supp(μ) such that $\rho'^\infty(u,v) < \rho^\infty(u,v)$, where ρ' is the fuzzy subset of $V \times V$ defined by $\rho' = \rho$ on Supp(ρ) and $\rho'(x,y) = 0 \forall (x,y) \in E$.

When E is a singleton set, a cut set is called a *bridge* and a fuzzy cut set a *fuzzy bridge*.

FACTS

F19: Let $G = (\mu, \rho)$ be a fuzzy graph. Let $V = \{v_1, ..., v_n\}$ and $C = \{(v_1, v_2), (v_2, v_3), ..., (v_{n-1}, v_n), (v_n, v_1)\}, n \geq 3$.
(1) Suppose that $C \subseteq$ Supp(ρ) and that $\forall (v_j, v_k) \in$ Supp(ρ)$\backslash C, \rho(v_j, v_k) < \vee\{\rho(v_i, v_{i+1})|$ $i = 1, ..., n\}$, where $v_{n+1} = v_1$. Then either ρ is a constant function on C or G has fuzzy bridge.
(2) Suppose that $\emptyset \neq$ Supp(ρ) $\subset C$. Then G has a bridge.

F20: Let $G = (\mu, \rho)$ be a fuzzy graph. Suppose that the dimension of the cycle space of (Supp(μ), Supp(ρ)) is 1. Then G does not have a fuzzy bridge if and only if G is a cycle and ρ is a constant function.

One can find a discussion of fuzzy chords, fuzzy cotrees, and fuzzy twigs in [MoNa00].

10.5.5   Fuzzy 1-Chain with Boundary 0, Fuzzy Coboundary, and Fuzzy Cocycles

DEFINITION

D33: Let $(x,y) \in V \times V$. Then (x,y) is called ***exceptional*** in G if and only if there exists a cycle $C \subseteq V \times V$ such that $(x,y) \in C$ and (x,y) is unique with respect to $\rho(x,y) = \wedge\{\rho(u,v)|(u,v) \in C\}$. Let $E = \{(x,y) \in V \times V|(x,y)$ is exceptional$\}$. Let ρ_E be a fuzzy subset of $V \times V$ defined by $\rho_E = \rho$ on $V \times V \backslash E$ and $\rho_E(x,y) = 0 \forall (x,y) \in E$.

FACT

F21: Let $S_\rho = \{(x,y)_t|(x,y) \in$ Supp(ρ)$, t \in (0,1]\} \cup \{0_t|t \in (0,1]\}$. Let addition of elements of Supp(ρ) be a formal addition modulo 2, i.e., $\forall (x,y), (u,v) \in$ Supp(ρ), we write $(x,y) + (u,v)$ if $(x,y) \neq (u,v)$ and $(x,y) + (u,v) = 0$ if $(x,y) = (u,v)$. Then $\forall (x,y)_t, (u,v)_s \in S_\rho, (x,y)_t + (u,v)_s = ((x,y) + (u,v))_{t \wedge s}$. Also, $\forall (x,y)_t \in S_\rho, (x,y)_t + 0_s = (x,y)_{t \wedge s} = 0_s + (x,y)_t$ and $0_t + 0_s = 0_{t \wedge s}$. Clearly, $(S_\rho, +)$ is a commutative semigroup with identity 0_1.

DEFINITION

D34: Let $\mathcal{S}$ be a set of fuzzy singletons of a set W. Then foot($\mathcal{S}$) = $\{w \in W|w_t \in \mathcal{S}\}$.

EXAMPLE

E1: Since $\mathbb{Z}_2 = \{0, 1\}$ is a field of integers modulo $2, 1 + 1 = 0$. We have that

$$\sum \epsilon_i(x_i, y_i)_{\rho(x_i, y_i)} + \sum \epsilon'_i(x_i, y_i)_{\rho(x_i, y_i)}$$
$$= \sum (\epsilon_i + \epsilon'_i)(x_i, y_i)_{\rho(x_i, y_i)}, \epsilon_i(x_i, y_i)_{\rho(x_i y_i)}$$
$$= (x_i, y_i)_{\rho(x_i, y_i)}$$

if $\epsilon_i = 1$ and $(x_i, y_i)_{\rho(x_i, y_i)} = 0_{\rho(x_i y_i)}$ if $\epsilon_i = 0, \epsilon_i, \epsilon'_i \in \mathbb{Z}_2$. We have that

$$\sum \epsilon_i(x_i, y_i)_{\rho(x_i, y_i)} = \left(\sum \epsilon_i(x_i, y_i)\right)_m,$$

where $m = \wedge_i\{\rho(x_i, y_i)\}$.

DEFINITIONS

D35: Let $G = (\mu, \rho)$ be a fuzzy graph. Then
(1) $\sum \epsilon_i(x_i, y_i)_{\rho(x_i, y_i)}$ is a 1-***chain with boundary*** 0 in G, where $(x_i, y_i) \in \text{Supp}(\rho)$ if and only if $\sum \epsilon_i(x_i, y_i)$ is a 1-chain with boundary 0 in $(\text{Supp}(\mu), \text{Supp}(\rho))$;
(2) $\sum \epsilon_i(x_i, y_i)_{\rho(x_i, y_i)}$ is ***fuzzy*** 1-***chain with boundary*** 0 in (μ, ρ), where $(x_i, y_i) \in \text{Supp}(\rho_E)$ if and only if $\sum \epsilon_i(x_i, y_i)$ is a 1-chain with boundary 0 in $(\text{Supp}(\mu), \text{Supp}(\rho_E))$.

D36: A (fuzzy) 1-chain with boundary 0 in G is called a ***(fuzzy) cycle vector***.

D37: Let $G = (\mu, \rho)$ be a fuzzy graph. Then
(1) $\sum \epsilon_i(x_i, y_i)_{\rho(x_i, y_i)}$ is a ***coboundary*** of G, where $(x_i, y_i) \in \text{Supp}(\rho)$ if and only if $\sum \epsilon_i(x_i, y_i)$ is a coboundary of $(\text{Supp}(\mu), \text{Supp}(\rho))$;
(2) $\sum \epsilon_i(x_i, y_i)_{\rho(x_i, y_i)}$ is ***fuzzy coboundary*** of G, where $(x_i, y_i) \in \text{Supp}(\rho_E)$ if and only if $\sum \epsilon_i(x_i, y_i)$ is a coboundary of $(\text{Supp}(\mu), \text{Supp}(\rho_E))$.

D38: $S' \subseteq S_\rho$ is called a ***(fuzzy) cocycle*** of G if and only if $\text{foot}(S')$ is a cocycle of $((\text{Supp}(\mu), \text{Supp}(\rho_E)))$ $(\text{Supp}(\mu), \text{Supp}(\rho))$.

D39: Let $G = (\mu, \rho)$ be a fuzzy graph. Then
(1) the set of all (fuzzy) cycle vectors of G is called the ***(fuzzy) cycle set*** of G;
(2) the set of all (fuzzy) coboundaries of G is called the ***(fuzzy) cocycle set*** of G.

TERMINOLOGY: Let $\mathcal{CS}(\mu, \rho), \mathcal{FCS}(\mu, \rho), \mathcal{CoS}(\mu, \rho)$, and $\mathcal{FCoS}(\mu, \rho)$ denote the cycle set, the fuzzy cycle set, the cocycle set, and the fuzzy cocycle set of the fuzzy graph $G = (\mu, \rho)$, respectively. When the fuzzy graph (μ, ρ) is understood, we sometimes write $\mathcal{CS}, \mathcal{FCS}, \mathcal{CoS}$, and $\mathcal{FCoS}$.

REMARKS

R5: It follows that $\mathcal{CS}, \mathcal{FCS}, \mathcal{CoS}$, and $\mathcal{FCoS}$ are not necessarily vector spaces over $\mathbb{Z}_2$. The details can be found in [MoNa00].

R6: The concepts of (fuzzy) twigs and (fuzzy) chords have results similar to their counterparts in the crisp case.

R7:   Clearly, $\mathcal{CS}, \mathcal{FCS}, \mathcal{CoS}$, and $\mathcal{FCoS}$ are subsets of $S_\rho = \{e_t | e \in \mathrm{Supp}(\rho), t \in (0,1]\} \cup \{0_t | t \in (0,1]\}$. Let $\mathcal{S}$ be a subset S_ρ. We let $\langle \mathcal{S} \rangle$ denote the intersection of all subsemigroups of S_ρ which contain $\mathcal{S}$. Then $\langle \mathcal{S} \rangle$ is the smallest subsemigroup of S_ρ which contains $\mathcal{S}$. Let $\mathcal{S}^+ = \{(e_1)_{t_1} + ... + (e_n)_{t_n} | (e_i)_{t_i} \in \mathcal{S}, i = 1, ..., n; n \in \mathbb{N}\}$, where $\mathbb{N}$ denotes the set of positive integers. Then $\mathcal{S}^+$ is a subsemigroup of S_ρ.

FACTS

F22: $\langle \mathcal{CS} \rangle = (\mathcal{CS})^+ = \mathcal{CS} \cup \{e_a + 0_b | e_a \in \mathcal{CS}, 0_b \in (\mathcal{CS})^+\}$. $\langle \mathcal{CS} \rangle$ has 0_m as its identity where $m = \vee\{b | 0_b \in (\mathcal{CS})^+\}$.

F23: $\langle \mathcal{FCS} \rangle = (\mathcal{FCS})^+ = \mathcal{FCS} \cup \{e_a + 0_b | e_a \in \mathcal{FCS}, 0_b \in (FCS)^+\}$. $\langle \mathcal{FCS} \rangle$ has 0_m as its identity where $m = \vee\{b | 0_b \in (\mathcal{FCS})^+\}$.

F24: $\langle \mathcal{CoS} \rangle = (\mathcal{CoS})^+ = \mathcal{CoS} \cup \{e_a + 0_b | e_a \in \mathcal{CoS}, 0_b \in (\mathcal{CoS})^+\}$. $\langle \mathcal{CoS} \rangle$ has 0_m as its identity where $m = \vee\{b | 0_b \in (\mathcal{CoS})^+\}$.

F25: $\langle \mathcal{FCoS} \rangle = (\mathcal{FCoS})^+ = \mathcal{FCoS} \cup \{e_a + 0_b | e_a \in \mathcal{FCoS}, 0_b \in (\mathcal{FCoS})^+\}$. $\langle \mathcal{FCoS} \rangle$ has 0_m as its identity where $m = \vee\{b | 0_b \in (\mathcal{FCoS})^+\}$.

DEFINITION

D40: Let $G = (\mu, \rho)$ be a fuzzy graph. Then the **cycle rank** of G, written $m(\mu, \rho)$, is defined to be $m(\mu, \rho) = \vee\{\sum_{i=1}^n t_i | (e_i)_{t_i} \in \mathcal{CS}, i = 1, ..., n, \{e_1, ..., e_n\}$ is a basis for $\mathrm{foot}(\mathcal{CS})\}$. The **fuzzy cycle rank** of G, written $fm(\mu, \rho)$, is defined to be the cycle rank of (μ, ρ_E). If $\{e_1, ..., e_n\}$ is a basis for $\mathrm{foot}(\mathcal{CS})$ such that $m(\mu, \rho) = \sum_{i=1}^n t_i$, where $(e_i)_{t_i} \in \mathcal{CS}, i = 1, ..., n$, then $\{e_1, ..., e_n\}$ is called a **cycle basis** of $\langle \mathcal{CS} \rangle$. If $\{e_1, ..., e_n\}$ is a basis for $\mathrm{foot}(\mathcal{FCS}(\mu, \rho))$ such that $fm(\mu, \rho) = \sum_{i=1}^n t_i$, where $(e_i)_{t_i} \in \mathcal{FCS}(\mu, \rho), i = 1, ..., n$, then $\{e_1, ..., e_n\}$ is called a **fuzzy cycle** basis of $\langle \mathcal{CS}(\mu, \rho) \rangle$.

FACTS

F26: Let $\{e_1, ..., e_n\}$ be a cycle basis of $\langle \mathcal{CS} \rangle$. Then for all $e_t \in \mathcal{CS}$, there is a reordering of $e_1, ..., e_n$ such that $e_t = (e_1)_{t_1} + ... + (e_m)_{t_m}, m \leq n$, where $t_i = \rho(e_i), i = 1, ..., m$.

F27: Let $\{e_1, ..., e_n\}$ be a fuzzy cycle basis of $\langle \mathcal{CS}(\mu, \rho) \rangle$. Then for all $e_t \in \mathcal{FCS}(\mu, \rho)$, there is a reordering of $e_1, ..., e_n$ such that $e_t = (e_1)_{t_1} + ... + (e_m)_{t_m}, m \leq n$, where $t_i = \rho(e_i), i = 1, ..., m$.

DEFINITION

D41: Let $G = (\mu, \rho)$ be a fuzzy graph. Then the **cocycle rank** of G, written $m_c(\mu, \rho)$, is defined as follows:

$$m_c(\mu, \rho) = \vee\{\sum_{i=1}^n t_i | (e_i)_{t_i} \in \mathcal{CoS}, i = 1, ..., n, \{e_1, ..., e_n\} \textit{ is a basis for foot}(\mathcal{CoS})\}.$$

The **fuzzy cocycle rank** of G, written $fm_c(\mu, \rho)$, is defined to be the cocycle rank of (μ, ρ_E). If $\{e_1, ..., e_n\}$ is a basis for $\mathrm{foot}(\mathcal{CoS})$ such that $m_c(\mu, \rho) = \sum_{i=1}^n t_i$, where $(e_i)_{t_i} \in \mathcal{CoS}, i = 1, ..., n$, then $\{e_1, ..., e_n\}$ is called a **cocycle basis** of $\langle \mathcal{CoS} \rangle$. If $\{e_1, ..., e_n\}$ is a basis for foot $\mathcal{FCoS}(\mu, \rho)$ such that $fm(\mu, \rho) = \sum_{i=1}^n t_i$, where $(e_i)_{t_i} \in \mathcal{FC} \rtimes \mathcal{S}(\mu, \rho), i = 1, ..., n$, then $\{e_1, ..., e_n\}$ is called a **fuzzy cocycle basis** of $\langle \mathcal{CS}(\mu, \rho) \rangle$.

FACTS

F28: Let $\{e_1, ..., e_n\}$ be a cocycle basis of $\langle \mathcal{C}o\mathcal{S} \rangle$. Then for all $e_t \in \mathcal{C}o\mathcal{S}$, there is a reordering of $e_1, ..., e_n$ such that $e_t = (e_1)_{t_1} + ... + (e_m)_{t_m}, m \leq n$, where $t_i = \rho(e_i), i = 1, ..., m$.

F29: Let $\{e_1, ..., e_n\}$ be a fuzzy cocycle basis of $\langle \mathcal{C}\mathcal{S}(\mu, \rho) \rangle$. Then for all $e_t \in \mathcal{F}\mathcal{C}o\mathcal{S}$, there is a reordering of $e_1, ..., e_n$ such that $e_t = (e_1)_{t_1} + ... + (e_m)_{t_m}, m \leq n$, where $t_i = \rho(e_i), i = 1, ..., m$.

10.5.6   Fuzzy Line Graphs

The line graph, $L(G)$, of a (crisp) graph G is the intersection graph of the set of edges of G. Hence the vertices of $L(G)$ are the edges of G with two vertices of $L(G)$ adjacent whenever the corresponding edges of G are. We present the notion of a fuzzy line graph. Let $G = (V, X)$ and $G' = (V', X')$ be graphs. If μ is a fuzzy subset of V and ρ is a fuzzy subset of $V \times V$ such that (μ, ρ) is a fuzzy graph with $\text{Supp}(\rho) \subseteq X$, we call (μ, ρ) a *partial fuzzy subgraph* of G.

DEFINITIONS

D42: Let (μ, ρ) and (μ', ρ') be partial fuzzy subgraphs of G and G', respectively. Let f be a one-to-one function of V onto V'. Then
(1) f is called a *(weak) vertex-isomorphism* of (μ, ρ) onto (μ', ρ') if and only if $\forall v \in V, (\mu(v) \leq \mu'(f(v))$ and $\text{Supp}(\mu') = f(\text{Supp}(\mu)))$ $\mu(v) = \mu'(f(v))$;
(2) f is called a *(weak) line-isomorphism* of (μ, ρ) onto (μ', ρ') if and only if $\forall (u, v) \in X, (\rho(u, v) \leq \rho'(f(u), f(v))$ and $\text{Supp}(\rho') = \{(f(u), f(v)) | (u, v) \in \text{Supp}(\rho)\})$ $\rho(u, v) = \rho'(f(u), f(v))$.

D43: If f is a (weak) vertex-isomorphism and a (weak) line-isomorphism of (μ, ρ) onto (μ', ρ'), then f is called a *(weak) isomorphism* of (μ, ρ) onto (μ', ρ'). If (μ, ρ) is isomorphic to (μ', ρ'), then we write $(\mu, \rho) \simeq (\mu', \rho')$.

D44: Let $G = (V, X)$ be a graph, where $V = \{v_1, ..., v_n\}$. Let $S_i = \{v_i, x_{i1}, ..., x_{iq_i}\}$, where $x_{ij} \in X$ and x_{ij} has v_i as a vertex, $j = 1, ..., q_i; i = 1, ..., n$. Let $S = \{S_1, ..., S_n\}$. Let $T = \{(S_i, S_j) | S_i S_j \in S, S_i \cap S_j \neq \emptyset, i \neq j\}$. Then $\mathcal{I}(S) = (S, T)$ is an intersection graph and $G \simeq \mathcal{I}(S)$. Any partial fuzzy subgraph (ι, γ) of $I(S)$ with $\text{Supp}(\gamma) = T$ is called a *fuzzy intersection graph*.

D45: Let (μ, ρ) be a partial fuzzy subgraph of G. Let $I(S)$ be the intersection graph described above. Define the fuzzy subsets ι, γ of S and T, respectively, as follows:
$\forall S_i \in S, \iota(S_i) = \mu(v_i)$;
$\forall (S_i, S_j) \in T, \gamma(S_i, S_j) = \rho(v_i, v_j)$.

FACT

F30: Let (μ, ρ) be a partial fuzzy subgraph of G. Then
(1) (ι, γ) is a partial fuzzy subgraph of $\mathcal{I}(S)$;
(2) $(\mu, \rho) \simeq (\iota, \gamma)$.

REMARK

R8: Let $\mathcal{I}(S)$ be the intersection graph of (V, X). Let (i, γ) be the fuzzy intersection graph of $\mathcal{I}(S)$ as defined above. We call (ι, γ) the *fuzzy intersection graph* of (μ, ρ). Fact F30 shows that any fuzzy graph is isomorphic to a fuzzy intersection graph.

The line graph $L(G)$ of G is by definition the intersection graph $\mathcal{I}(X)$. That is, $L(G) = (Z, W)$, where $Z = \{\{x\} \cup \{u_x, v_x\} | x \in X, u_x, v_x \in V, x = (u_x, v_x)\}$ and $W = \{(S_x, S_y) | S_x \cap S_y \neq \emptyset, x, y \in X, x \neq y\}$ and where $S_x = \{\{x\} \cup \{u_x, v_x\}, x \in X$.

DEFINITION

D46: Let (μ, ρ) be a partial fuzzy subgraph of G. Define the fuzzy subsets λ, ω of Z, W, respectively, as follows:
$\forall S_x \in Z, \lambda(S_x) = \rho(x);$
$\forall (S_x, S_y) \in W, \omega(S_x, S_y) = \rho(x) \wedge \rho(y).$

FACT

F31: (λ, ω) is a fuzzy subgraph of $L(G)$ and is called the *fuzzy line graph* corresponding to (μ, ρ).

Every cutpoint of $L(G)$ is a bridge of G which is not an endline, and conversely, [MoNa00, p. 42]. It is shown in [MoNa00] that the relationship between cutpoints in $L(G)$ and bridges in G does not carry over to the fuzzy case.

REMARK

R9: Let (μ, ρ) and (μ', ρ') be partial fuzzy subgraphs of G and G', respectively. If f is a weak isomorphism of (μ, ρ) onto (μ', ρ'), then it can be shown that f is an isomorphism of $(\text{Supp}(\mu), \text{Supp}(\rho))$ onto $(\text{Supp}(\mu'), \text{Supp}(\rho'))$. If (λ, ω) is the fuzzy line graph of (μ, ρ), then it can also be shown that $(\text{Supp}(\lambda), \text{Supp}(\omega))$ is the fuzzy line graph of $(\text{Supp}(\mu), \text{Supp}(\rho))$.

FACTS

F32: Let (λ, ω) be the fuzzy line graph corresponding to (μ, ρ). Suppose that $(\text{Supp}(\mu), \text{Supp}(\rho))$ is connected. Then
(1) there is a weak isomorphism of (μ, ρ) onto (λ, ω) if and only if $(\text{Supp}(\mu), \text{Supp}(\rho))$ is a cycle and μ and ρ are constant functions on $\text{Supp}(\mu)$ and $\text{Supp}(\rho)$, respectively, taking on the same value;
(2) if f is a weak isomorphism of (μ, ρ) onto (λ, ω), then f is an isomorphism.

F33: Let (μ, ρ) and (μ', ρ') be partial fuzzy subgraphs of G and G', respectively, such that $(\text{Supp}(\mu), \text{Supp}(\rho))$ and $(\text{Supp}(\mu'), \text{Supp}(\rho'))$ are connected. Let (λ, ω) and (λ', ω') be the line graphs corresponding to (μ, ρ) and (μ', ρ'), respectively. Suppose that it is not the case that one of $(\text{Supp}(\mu), \text{Supp}(\rho))$ and $(\text{Supp}(\mu'), \text{Supp}(\rho'))$ is K_3 and the other is $K_{1,3}$. If $(\lambda, \omega) \simeq (\lambda', \omega')$, then (μ, ρ) and (μ', ρ') are line isomorphic.

F34: Let (τ, ν) be a partial fuzzy subgraph of $L(G)$. Then (τ, ν) is a fuzzy line graph of some partial fuzzy subgraph of G if and only if $\forall (S_x, S_y) \in W, \nu(S_x, S_y) = \tau(x) \wedge \tau(y)$.

F35: (μ, ρ) is a fuzzy line graph if and only if $(\text{Supp}(\mu), \text{Supp}(\rho))$ is a line graph and $\forall (u, v) \in \text{Supp}(\mu), \rho(u, v) = \mu(u) \wedge \mu(v)$.

10.5.7   Fuzzy Interval Graphs

Intersection graphs and in particular interval graphs are used extensively in mathematical modeling. Applications in archaeology, developmental psychology, ecological modeling, mathematical sociology, and organization theory are cited in [Ro76]. These disciplines all have components that are ambiguously defined, require subjective evaluation, or are satisfied to differing degrees. They are active areas of applications of fuzzy methods. It is therefore worthwhile to study the extent that intersection graph results can be extended using fuzzy set theory.

The intersection graph of a family (possibly with repeated members) of sets is a graph with a vertex representing each member of the family and an edge connecting two vertices if and only if the two sets have nonempty intersection. Generally, loops are suppressed. If the family is composed of intervals or is the edge set of a hypergraph, then the intersection graph is called an *interval graph* or a *line graph*, respectively.

A fuzzy analog of Marczewski's theorem [Ma45] shows that every fuzzy graph without loops is the intersection graph of some family of fuzzy subsets. However, the natural generalization of the Fulkerson and Gross characterization [FuGr65] of interval graphs fails. However, a natural generalization of the Gilmore and Hoffman characterization [GiHo64] holds.

Let $G = (V, \mu, \rho)$ be a fuzzy graph.

DEFINITION

D47: A *fuzzy digraph* is a triple $D = (V, \mu, \delta)$, where μ is a fuzzy subset of V and δ is a fuzzy subset of $V \times V$ such that $\delta(x, y) \leq \mu(x) \wedge \mu(y)$ for all $x, y \in V$. We note that δ need not be symmetric.

REMARK

R10: A fuzzy graph (fuzzy digraph) can be represented by an adjacency matrix, where the rows and columns are indexed by the vertex set V and the x, y entry is $\rho(x, y)$ $(\delta(x, y))$. Vertex strength can be indicated by adding a column indexed by μ and letting the x, μ entry be $\mu(x)$.

DEFINITIONS

D48:   For a family $\mathcal{F}$ of fuzzy subsets, we define the t *level family* of $\mathcal{F}$ to be $\mathcal{F}^t = \{\alpha^t | \alpha \in \mathcal{F}\}$.

D49:   Let α be a fuzzy subset of V. The *height* of α is $h(\alpha) = \vee\{\alpha(x) | x \in V\}$.

We construct a sequence of crisp level graphs in order to see how a fuzzy subset's structure changes between various levels. Theorems characterizing a fuzzy property in terms of level set properties are significant in that such theorems demonstrate the extent to which the crisp theory can be generalized. To formalize this sequence of graphs, we define the notion of fundamental sequence.

D50:   The *fundamental sequence* of a fuzzy graph $G = (\mu, \rho)$ is defined to be the ordered set

$$\mathrm{fs}(G) = \{\mu(x) > 0 | x \in V\} \cup \{\rho(x, y) > 0 | x, y \in V\},$$

where we use the decreasing order inherited from the closed real interval $[0, 1]$.

REMARK

R11:   The first element listed in fs(G) is the maximal vertex strength while the last element is the minimal nonzero edge strength.

DEFINITION

D51:   Let $\mathcal{F} = \{\alpha_1, ..., \alpha_n\}$ be a finite set of fuzzy subsets of a set V. The *fuzzy intersection graph* of $\mathcal{F}$ is the fuzzy graph Int($\mathcal{F}$) = (μ, ρ), where $\mu : \mathcal{F} \to [0, 1]$ is defined by $\mu(\alpha_i) = h(\alpha_i), i = 1, ..., n$ and $\rho : \mathcal{F} \times \mathcal{F} \to [0, 1]$ is defined by

$$\rho(\alpha_i, \alpha_j) = \begin{cases} h(\alpha_i \cap \alpha_j) \text{ if } i \neq j, \\ \\ 0 \text{ if } i = j, \end{cases} \qquad \text{for } i, j = 1, ..., n.$$

REMARKS

R12: $\mathcal{F}$ in the previous definition is considered a crisp set of vertices. The notion of a fuzzy intersection graph is different here than in the previous definition and so a different notation is used. Recall that every graph $G = (V, X)$ is an intersection graph. For all $x \in V$, let S_x denote the union of $\{x\}$ with the set of all edges incident with x. It follows that G is isomorphic to the intersection graph of $\{S_x | x \in V\}$.

R13:   If $\mathcal{F} = \{\alpha_1, ..., \alpha_n\}$ is a family of fuzzy subsets of a set V and $t \in [0, 1]$, then Int($\mathcal{F}^t$) = (Int($\mathcal{F}$))^t. The graph Int($\mathcal{F}^t$) has a vertex set representing $\alpha_i \in \mathcal{F}$ if and only if $h(\alpha_i) \geq t$. The set $\{(\alpha_i)^t, (\alpha_j)^t\}$ is an edge of Int($\mathcal{F}^t$) if and only if $i \neq j$ and $h(\alpha_i \cap \alpha_j) \geq t$. These conditions also characterize the graph (Int($\mathcal{F}$))^t. In particular, if $\mathcal{F}$ is a family of crisp subsets of V, then the fuzzy intersection graph and the crisp intersection graph definitions coincide.

We next state the fuzzy analog of Marczewski's theorem [Ma45].

FACT

F36:   If $G = (\mu, \rho)$ is a fuzzy graph (without loops), then for some family of fuzzy subsets of $\mathcal{F}, G = $ Int($\mathcal{F}$).

The families of sets most often considered in connection with intersection graphs are families of intervals of a linearly ordered set. This class of interval graphs is central to many applications.

In both the crisp and fuzzy cases, distinct families of sets can have the same intersection graph. In particular, the intersection properties of a finite family of real intervals (fuzzy numbers) can be characterized by a family of intervals (fuzzy intervals) defined on a finite set. Therefore, as is common in interval theory, we restrict our attention to intervals (fuzzy intervals) with finite support.

We next generalize two characterizations of crisp interval graphs. Both make use of relationships between the finite number of points which define the intervals and the cliques of the corresponding interval graph.

DEFINITION

D52: A *clique* of a graph is a maximal (with respect to inclusion) complete subgraph.

REMARK

R14: If a vertex z is not a member of a clique K, then there exists $x \in K$ such that (x, z) is not an edge of the graph.

DEFINITION

D53: Let V be a linearly ordered set.
(1) A *fuzzy interval* $\mathcal{I}$ on V is a normal, convex fuzzy subset of V. That is, there exists an $x \in V$ with $\mathcal{I}(x) = 1$ and the ordering $w \le y \le z$ implies that $\mathcal{I}(y) \ge \mathcal{I}(w) \wedge \mathcal{I}(z)$.
(2) A *fuzzy number* is a fuzzy interval.
(3) A *fuzzy interval graph* is the fuzzy intersection graph of a finite family of fuzzy intervals.

REMARK

R15: Let $G = \mathrm{Int}(\mathcal{F})$ be a fuzzy interval graph. It follows easily that for all $t \in (0, 1]$, the level graph G^t is an interval graph. The converse of this result is not true [MoNa00, Example 2.12, p. 47].

The Fulkerson and Gross characterization makes use of a correspondence between the set of points on which the family of intervals is defined and the set of cliques of the corresponding interval graph. We provide natural generalizations of the crisp definitions. It follows from fuzzy graphs that the relationship holds only in one direction.

DEFINITION

D54: Let $G = (\mu, \rho)$ be a fuzzy graph.
(1) We say that a fuzzy subset $\mathcal{K}$ defines a fuzzy clique of G if for each $t \in (0, 1], \mathcal{K}^t$ induces a clique of G^t.
(2) We associate with G a vertex clique incidence matrix where the rows are indexed by the domain of μ, the columns are indexed by the family of all cliques of G, and the $x, \mathcal{K}$ entry is $\mathcal{K}(x)$.

FACTS

F37: Suppose that G is a fuzzy graph with $\mathrm{fs}(G) = \{r_1, ..., r_n\}$ and let $\mathcal{K}$ be a fuzzy clique of G. The level sets of $\mathcal{K}$ define a sequence $\mathcal{K}^{r_1} \subseteq ... \subseteq \mathcal{K}^{r_n}$, where each $\mathcal{K}^{r_i}$ is a clique of G^{r_i}. Conversely, any sequence $K_1 \subseteq ... \subseteq K_n$, where each K_i is a clique of G^{r_i} defines a fuzzy clique $\mathcal{K}$, where $\mathcal{K}(x) = \vee \{r_i | x \in K_i\}$. Therefore, K is a clique of the t-level graph G^t if and only if $K = \mathcal{K}^t$ for some fuzzy clique $\mathcal{K}$.

We now state the fuzzy analog of Fulkerson and Gross.

F38: Let $G = (V, \rho)$ be a fuzzy graph. Then the row of any vertex clique incidence matrix of G defines a family of fuzzy subsets $\mathcal{F}$ for which $G = \mathrm{Int}(\mathcal{F})$. Further, if there exists an ordering of the fuzzy cliques of G such that each row of the vertex clique incidence matrix is convex, then G is a fuzzy interval graph.

F39:   The converse of Fact F38 is not true [MoNa00, Example 2.13, p. 50].

TERMINOLOGY: Let $G = (V, X)$ be a graph and D be a directed graph. We use the notation (x, y) for an edge in G and $\langle x, y \rangle$ for a directed edge in D.

DEFINITIONS

D55:   A graph is called **chordal** or *triangulated* if each cycle with $n \geq 4$ vertices has a chord, i.e., there exist integers $j \neq 0$ or $k \neq n$ with $0 \leq j < k - 1 \leq n$ and $(x_j, x_k) \in X$.

D56:   An **orientation** of a graph $G = (V, X)$ is a directed graph $G_A = (V, A)$ that has G as its underlying graph. That is, $(x, y) \in X$ implies that $\langle x, y \rangle \in A$ or $\langle y, x \rangle \in A$, but not both.

D57:   A graph G is **transitively orientable** if there exists an orientation of G for which $\langle u, v \rangle \in A$ and $\langle v, w \rangle \in A$ implies $\langle u, w \rangle \in A$.

D58:   The **complement** of a graph G, denoted by G^c, is the graph with vertex set V and edge set consisting of those edges which are not in X. For a fuzzy graph $G = (\mu, \rho)$, we let $G^c = (\mu, 1 - \rho)$.

D59:   A **cycle of length** n in a fuzzy graph is a sequence of distinct vertices $x_0, x_1, ..., x_n$ such that $\rho(x_0, x_n) > 0$ and if $1 \leq i \leq n$, then $\rho(x_{i-1}, x_i) > 0$. A fuzzy graph $G = (\mu, \rho)$ is **chordal** if for each cycle with $n \geq 4$,

1. there exist integers $j \neq 0$ or $k \neq n$ such that $0 \leq j < k - 1 \leq n$ and

2. $\rho(x_j, x_k) \geq \wedge \{\rho(x_{i-1}, x_i) | i = 1, 2, ..., n\} \wedge \rho(x_0, x_n)$.

REMARK

R16:   It is easily shown that a fuzzy graph $G = (\mu, \rho)$ is chordal if and only if for each $t \in (0, 1]$ the t-level graph of G is chordal.

FACT

F40:   If G is a fuzzy interval graph, then G is chordal.

DEFINITION

D60:   Let $G = (\mu, \rho)$ be a fuzzy graph with $\text{fs}(G) = \{r_1, ..., r_n\}$ and let A be an orientation of G^{r_n}. Then the **orientation** of G by A is the fuzzy digraph $G_A = (\mu, \rho_A)$, where

$$\rho_A(\langle x, y \rangle) = \begin{cases} \rho(x, y) \text{ if } \langle x, y \rangle \in A, \\ 0 \text{ if } \langle x, y \rangle \notin A. \end{cases}$$

The fuzzy graph G is called **transitively orientable** if there exists an orientation which is transitive, i.e., $\rho_A(\langle x, y \rangle) \wedge \rho_A(\langle y, z \rangle) \leq \rho_A(\langle x, z \rangle)$.

REMARK

R17: Let G be a fuzzy graph. The t level graph of G_A has arc set $\{\langle x, y\rangle | \rho_A(\langle x, y\rangle) \geq t\}$. Therefore an orientation of a fuzzy graph induces consistent orientations on each member of the fundamental sequence of cut level graphs. Conversely, it is possible to have a sequence of transitively oriented subgraphs $G_1 \subseteq G_2 \subseteq G_3$, where the transitive orientation of G_2 does not induce a transitive orientation of G_1, and the transitive orientation of G_2 cannot be extended to a transitive orientation of G_3.

FACTS

F41: Suppose that $G = \text{Int}(F)$ is a fuzzy interval graph. Then there exists an orientation A that induces a transitive orientation of G^c.

We next state the fuzzy analog of the Gilmore and Hoffman characterization.

F42: A fuzzy graph $G = (\mu, \rho)$ is a fuzzy interval graph if and only if the following conditions hold:
(1) for all $x \in \text{Supp}(\mu) = V$, $\mu(x) = 1$,
(2) each fuzzy subgraph of G induced by four vertices is chordal,
(3) G^c is transitively orientable.

10.5.8   Operations on Fuzzy Graphs

By a partial fuzzy subgraph of a graph $G = (V, X)$, we mean a partial fuzzy subgraph of (χ_V, χ_X), where χ_V and χ_X denote the characteristic functions of V and X, respectively. We denote the edge between vertices u, v by uv rather than (u, v) in this section. Let (μ_i, ρ_i) be a partial fuzzy subgraph of the graph $G_i = (V_i, X_i), i = 1, 2$. The operations of Cartesian product, composition, union, and join on (μ_1, ρ_1) and (μ_2, ρ_2) are given in [MoNa00]. If the graph G is formed from G_1 and G_2 by one of the these operations, necessary and sufficient conditions are given in [MoNa00] for an arbitrary partial fuzzy subgraph of G to also be formed by the same operation from partial fuzzy subgraphs of G_1 and G_2. Recall that the Cartesian product $G = G_1 \times G_2$ of graphs $G_1 = (V_1, X_1)$ and $G_2 = (V_2, X_2)$ is given by $V = V_1 \times V_2$ and

$$X = \{(u, u_2)(u, v_2) | u \in V_1, u_2 v_2 \in X_2\} \cup \{(u_1, w)(v_1 w) | w \in V_2, u_1 v_1 \in X_1\}.$$

Let μ_i be a fuzzy subset of V_i and ρ_i be a fuzzy subset of $X_i, i = 1, 2$.

DEFINITION

D61: Define the fuzzy subsets $\mu_1 \times \mu_2$ of V and $\rho_1 \rho_2$ of X as follows:

(1) $\forall (u_1, u_2) \in V$, $(\mu_1 \times \mu_2)(u_1, u_2) = \mu_1(u_1) \wedge \mu_2(u_2)$,
(2) $\forall u \in V_1$, $\forall u_2 v_2 \in X_2$, $\rho_1 \rho_2((u, u_2)(u, v_2)) = \mu_1(u) \wedge \rho_2(u_2 v_2)$,
(3) $\forall w \in V_2$, $\forall u_1 v_1 \in X_1$, $\rho_1 \rho_2((u_1, w)(v_1, w)) = \mu_2(w) \wedge \rho_1(u_1 v_1)$.

It follows easily that if G is the Cartesian product of graphs G_1 and G_2 and (μ_i, ρ_i) is a partial fuzzy subgraph of $G_i, i = 1, 2$, then $(\mu_1 \times \mu_2, \rho_1 \rho_2)$ is a partial subgraph of G.

FACT

F43: Suppose that G is a Cartesian product of two graphs G_1 and G_2. Let (μ, ρ) be a partial fuzzy subgraph of G. Then (μ, ρ) is a Cartesian product of a partial fuzzy subgraph of G_1 and a partial fuzzy subgraph of G_2 if and only if the following three equations have solutions for x_i, y_j, z_{jk}, and w_{ih}, where $V_1 = \{v_{11}, v_{12}, ..., v_{1n}\}$ and $V_2 = \{v_{21}, v_{22}, ..., v_{2m}\}$:

(1) $x_i \wedge y_j = \mu(v_{1i}, v_{2j}), i = 1, ..., n; j = 1, ..., m;$

(2) $x_i \wedge z_{jk} = \rho((v_{1i}, v_{2j})(v_{1i}, v_{2k})), i = 1, ..., n; j, k$ such that $v_{2j}v_{2k} \in X_2;$

(3) $y_j \wedge w_{ik} = \rho((v_{1i}, v_{2j})(v_{1h}, v_{2j})), j = 1, ..., m; i, h$ such that $v_{1i}v_{1h} \in X_1.$

REMARK

R18: Definitions and results concerning the composition of graphs and partial fuzzy subgraphs follow along the same lines as for the Cartesian product. The results with examples can be found in [MoNa00]. The results for the union and join of graphs and partial fuzzy subgraphs follow differently. It can be shown, for example, that if G is a union of two subgraphs G_1 and G_2, then every partial fuzzy subgraph (μ, ρ) is a union of a partial fuzzy subgraph of G_1 and a partial fuzzy subgraph of G_2. It can also be shown that if G is the join of two subgraphs G_1 and G_2, then every strong partial fuzzy subgraph (μ, ρ) of G is a join of a strong partial fuzzy subgraph of G_1 and a strong partial fuzzy subgraph of G_2. Definitions, proofs, and examples can be found in [MoNa00].

10.5.9 Clusters

In graph theory, there are several ways of defining "clusters" of vertices. One approach is to call a subset C of V a *cluster of order* k if the following two conditions hold:

(a) for all vertices x, y in $C, d(x, y) \leq k;$

(b) for all vertices $z \notin C, d(z, w) > k$ for some $w \in C,$

where $d(u, v)$ is length of a shortest path between two vertices u, v.

A 1-cluster is called a *clique*; it is a maximal complete subgraph. That is, a maximal subgraph in which each pair of vertices is joined by an edge. At the other extreme, if we let $k \to \infty$, a k-cluster becomes a connected component, that is, a maximal subgraph in which each pair of vertices is joined by a path (of any length).

These ideas can be generalized to fuzzy graphs as follows: In $G = (\mu, \rho)$, we call $C \subseteq V$ a *fuzzy cluster of order* k if

$$\wedge \{\rho^k(x, y) | x, y \in C\} > \vee \{\wedge(\rho^k(w, z) | w \in C\} | z \notin C\}.$$

Note that C is an ordinary subset of V, not a fuzzy subset. If G is an ordinary graph, we have $\rho^k(a, b) = 0$ or 1 for all a, b. Hence this definition reduces to

(1) $\rho^k(x, y) = 1$ for all x, y in $C,$

(2) $\rho^k(w, z) = 0$ for all $z \notin C$ and some $w \in C.$

Property (1) implies that for all x, y in C, there exists a path of length $\leq k$ between x and y and property (2) implies that for all $z \notin C$ and some $w \in C$, there does not exist a path of length $\leq k$. This is the same as the definition of a cluster of order k.

In fact, the k-clusters obtained using this definition are just ordinary cliques in graphs obtained by thresholding the k-th power of the given fuzzy graph. Indeed, let C be a fuzzy k-cluster, and let $\wedge\{\rho^k(x, y)|x, y \in C\} = t$. If we threshold ρ^k (and μ) at t, we obtain an ordinary graph in which C is now an ordinary clique.

10.5.10   Application to Cluster Analysis

The usual graph theoretical approaches to cluster analysis involve first obtaining a threshold graph from a fuzzy graph and then applying various techniques to obtain clusters as maximal components under different connectivity considerations. These methods have a common weakness, namely, the weight of edges is not treated fairly in that any weight greater (less) than the threshold is treated as 1(0). We discuss an extension of these techniques to fuzzy graphs. It turns out that the fuzzy graph approach can be more powerful.

The following table provides a summary of various graph theoretical techniques for cluster analysis. For cluster procedures (1), (2), and (3), the cluster independence can be considered disjoint while that of cluster procedure (4) is limited overlap and that of (5) is considerable overlap. The extent of chaining is high, moderate, low, low, and none for cluster procedures (1)–(5), respectively.

	Cluster procedure	Graph theoretical interpretation of clusters
(1)	Single linkage	Maximal connected subgraphs
(2)	k-linkage	Maximal connected subgraphs of minimum degree
(3)	k-edge connectivity	Maximal k-edge connected subgraph
(4)	k-vertex connectivity	Maximal k-vertex connected subgraph and cliques on k or less vertices
(5)	Complete linkage	Cliques

DEFINITION

D62: Let $G = (V, \rho)$ be a fuzzy graph. A ***cluster of type*** k $(k = 1, 2, 3, 4)$ is defined by the following conditions:
(1) maximal ϵ-connected subgraphs for some $0 < \epsilon \leq 1$.
(2) maximal τ-degree connected subgraphs.
(3) maximal τ-edge connected subgraphs.
(4) maximal τ-vertex connected subgraphs.

Hierarchial cluster analysis is a method of generating a set of classifications of a finite set of objects based on some measure of similarity between a pair of objects. It follows from the previous definition that clusters of type (1), (2), and (3) are hierarchial with different ϵ and τ, whereas clusters of type (4) are not due to the fact τ-vertex components need not be disjoint.

It is also easily seen that all clusters of type (1) can be determined by the single-linkage procedure. The difference between the two procedures lies in the fact that ϵ-connected subgraphs can be obtained directly from M_{ρ^∞} by at most $n - 1$ matrix multiplications, where n is the rank of M_G, whereas in the single-linkage procedure, it is necessary to obtain as many threshold graphs as the number of distinct values in the graph.

The output of hierarchial clustering is called a *dendogram*, which is a directed tree that describes the process of generating clusters.

It is shown in [MoNa00] that not all clusters of types 2, 3, and 4 are obtainable by procedures of k-linkage, k-edge connectivity, and k-vertex connectivity, respectively.

FACTS

F44: The τ-degree connectivity procedure for the construction of clusters is more powerful than the k-linkage procedure.

F45: The τ-edge connectivity procedure for the construction of clusters is more powerful than the k-edge connectivity procedure.

Single Linkage

An important result in hierarchical clustering is the equivalence between the single linkage and connected components of a fuzzy graph. That is, the following four methods generate the same partition:

(1) the single linkage method,
(2) connected components of an undirected graph,
(3) transitive closure of a reflexive and symmetric fuzzy relation,
(4) the maximal spanning tree of a weighted graph.

The formal statement of the above result can be found in [Mi90, Proposition 6.1, p. 161] along with its proof and pertinent algorithms.

10.5.11   Fuzzy Graphs in Database Theory

In the classical relational database theory, in order to design good databases (no data redundancy, no update anomalies) one has to know additional information called functional dependencies, which say that some values determine other values. This notion can be generalized for fuzzy relations. Certain kinds of decompositions of fuzzy relational databases can be obtained using level cuts [Ki90, Ki91].

DEFINITION

D63: Let $U = \{A_1, ..., A_n\}$ be a set of attributes and V a set of values. Let $\mathrm{DOM}(A_i)$ be a nonempty subset of $V, i = 1, ..., n$. Let R be a subset of the Cartesian cross product $\times_{i=1}^n \mathrm{DOM}(A_i)$. For all $t = (t_1, ..., t_n) \in R$ and $A_i \in U$, let $t[A_i] = t_i, i = 1, ..., n$. Let $X, Y \in \mathcal{P}(U)$, the power set of U. Then X is said to functionally determine Y in R if for two elements in R, the Y values are equal whenever the X values are equal. Formally, $\forall t, t' \in V^n, \forall X, Y \in U$,

$$t \in R, t' \in R, t[X] = t'[X] \Rightarrow t[Y] = t'[Y],$$

where $t(X) = (t_{i_1}, ..., t_{i_k})$ for $X = \{A_{i_1}, ..., A_{i_k}\}$.

We now fuzzify these ideas. Let ρ be a fuzzy subset of $\times_{i=1}^{n} \mathrm{DOM}(A_i)$. We replace $\Rightarrow$ with the implication $\rightarrow$, where $\forall a, b \in [0, 1]$,

$$a \rightarrow b = \begin{cases} 1 \text{ if } a \leq b, \\ 1 - (a - b) \text{ otherwise.} \end{cases}$$

Then we get that the truth value of the fuzzy relation ρ satisfies a given functional dependency $X \rightarrow Y$ for $X, Y \in \mathcal{P}(U) : \forall t_1, t_2 \in V^n$,

$$\mu(X, Y) = 1 - \vee\{\rho(t_1) \wedge \rho(t_2) \mid t_1[X] = t_2[X], \text{ but } t_1[Y] \neq t_2[Y]\}.$$

For $X, Y \in P(U)$, let $XY = X \cup Y$. The following properties are easily shown.

A1 If $Y \subseteq X$, then $\mu(X, Y) = 1$;
A2 $\mu(X, Y) \wedge \mu(Y, Z) \leq \mu(X, Z)$;
A3 $\mu(X, Y) \leq \mu(XZ, YZ)$.

If τ is a fuzzy relation on $\mathcal{P}(U)$, the smallest fuzzy relation τ^+ on $\mathcal{P}(U)$ which satisfies $A1, A2, A3$ is called the *closure* of τ. It can be shown $\tau^+ = \tau^{++}$.

We now associate with τ a fuzzy graph $G_\tau = (\omega, \rho)$ as follows: The vertices are ordered pairs (X, Y) with $\omega(X, Y) = \tau(X, Y)$ and edges are ordered pairs of vertices of the form $((X, Y), (X, Z))$ with $\rho((X, Y), (X, Z)) = \tau(Y, Z)$. An algorithm in [Ki90] and [Ki91] can be found that gives τ^+ by modifying step by step the labels of the graph.

10.5.12 Strengthening and Weakening Members of a Group

A fuzzy directed graph can be utilized to characterize the role played by an individual member in a group that a class of group members having relationship with any given member has no sharply defined boundary. The theory of graphs is an important tool in the study of the group structure [Ha72]. A strengthening member of a group is one whose presence causes the graph corresponding to the group to be more highly connected than that obtained when he is absent, while a weakening member is one whose presence causes the graph to belong to a weaker category of connectedness. The graph can be used to study problems concerning redundancies, liaison persons, cliques, structural balance, and so forth.

In many cases, however, the mere presence or absence of a relation is not adequate to represent group structure. There may be different strengths of the relations between individuals. There may even be situations in which it is fuzzy rather than well-defined whether or not an arbitrary individual has a relationship with a given member, that is, a class of group members being in relationship with any given member does not have a sharply defined boundary. In such cases, fuzzy graphs become a more relevant model [TaNi76].

10.5.13 Network Analysis of Fuzzy Graphs

Posing problems on networks serves as a means for visualizing a problem and for developing a better understanding of the problem. It also has certain computational advantages. It is easier for a decision maker to draw a picture of what he wants than it is to write down constraints. There are a wide variety of network type problems such as location, transportation, flow, reliability, and shortest path. These models often deal with deterministic data and a single objective. Fuzzy counterparts can be found in [Kl91, MaMo00].

The most basic network problem is the shortest path problem. The fuzzy shortest path problem was first analyzed in [DuPr80]. The fuzzy shortest path can be found, but it may not correspond to an actual path in the network. Generally fuzziness is introduced into the network through arc capacities, arc lengths, or vertex restrictions. To circumvent the problem of the fuzzy shortest path distance not corresponding to an actual path, a different category of fuzzy path problems is considered. Rather than viewing each arc as a fuzzy number, let each possible arc length and path length be a fuzzy set. Then each arc has a membership grade in each fuzzy set corresponding to a length. Then through a DP recursion, it is possible to find the shortest fuzzy path length. This length will be a fuzzy number that may not correspond to an actual path. however, each value in the fuzzy set with positive membership grade will correspond to a path in the network. Results on decision trees can be found in [Ad80].

In [Ko92], a way is shown for modeling various quantitatively featurable functional capabilities of computer, communication, and similar networks. It is shown that functional capability of distributed hierarchical multicomponent systems (networks) can be described by the directed rooted tree model according to fuzzy graph ideas. The problem of optimizing the overall functional capability with a given component (vertex) set is introduced. The fuzzy formulae based on the algebraic connectives which are to be optimized are constructed.

Imprecise observations or possible perturbations mean that capacities and flows in a network may well be better represented by intervals or fuzzy numbers than crisp quantities. In [Di01] analogues of the MFMCT and Karp–Edwards algorithm for networks with fuzzy capacities and flows are derived. The principal difference between fuzzified and traditional crisp versions is that although the maximum fuzzy flow corresponds to a minimum fuzzy capacity, the latter may incorporate a number of network cuts. The preliminary results are for interval-valued flows and capacities which, in themselves, provide robustness estimates for flows in an uncertain environment.

Many types of document networks exist such as bibliographic databases containing scientific publications, social networking services, as well as databases of datasets used in scientific endeavors. However, the prime example of a document network is the World Wide Web (WWW). Each of these databases possesses several distinct relationships among documents and between documents and semantic tags or indices that classify documents appropriately. For instance, documents in the WWW are related via a hyperlink network, while documents in bibliographic databases can be related to semantic tags such as keywords used to describe their content. Given these relations, we can compute distance functions (typically via co-occurrence measures) among documents and/or semantic tags, thus creating associative, weighted networks between these items – which denote stronger or weaker co-associations.

10.5.14   Intuitionistic Fuzzy Graphs and Group Decision-Making

DEFINITION

D64: Let E_1 and E_2 be sets and let $G \subseteq E_1 \times E_2$. Let $\mu_G, \nu_G : E_1 \times E_2 \to [0,1]$ be such that $0 \leq \mu_G(x,y) + \nu_G(x,y) \leq 1$ for all $(x,y) \in E_1 \times E_2$. Then the set

$$G^* = \{\langle (x,y), \mu_G(x,y), \nu_G(x,y)\rangle | (x,y) \in E_1 \times E_2\}$$

is called an ***intutionistic fuzzy graph***. (The functions μ_G and ν_G are interpreted as the degree of membership and nonmembership, respectively, of the element (x,y) in the set G.)

In group decision-making, a set of experts in a given field is involved in a decision process concerning the selection of the best alternatives among a set of predefined ones. Each expert is asked to evaluate at least a subset of the alternatives in terms of its performance with respect to predefined criterion. The expert evaluations are expressed as a pair of numeric values, interpreted in the intuitionistic fuzzy framework. These numbers express a "positive" and a "negative" evaluation, respectively. Intuitionistic fuzzy graphs can be constructed for each expert. A suitable operation is performed on these intuitionistic fuzzy graphs to obtain a single intuitionistic fuzzy graph which provides an aggregation of the experts' opinions [At99, AtPaYaAt03].

References

[Ad80] J. H. Adamo, Fuzzy decision tress, *Fuzzy Sets and Systems* 4 (1980), 207–219.

[At99] K. Atanassov, *Intuitionistic Fuzzy Sets*, Springer Physica-Verlag, Berlin, 1999.

[AtPaYaAt03] K. Atanassov, G. Pasi, R. Yager, and V. Atanassova, Intuitionistic fuzzy graph interpretations of multi-person multi-criteria decision making, www.eusflat.org/proceedings/EUSFLAT_2003papers/09/Atanassov/pdf.

[ChKo82] S. Chanas and W. Kolodziejczynk, Maximum flow in a network with fuzzy arc capacities, *Fuzzy Sets and Systems* 8 (1982), 165–173.

[DeVeVi85] M. Delgado, J. L. Verdegay, and M. A. Vila, On fuzzy tree definition, *European J. Operational Res.* 22 (1985), 243–249.

[Di01] P. Diamond, A fuzzy max-flow min-cut theorem, *Fuzzy Sets and Systems* 119 (2001), 139–148.

[DuPr80] D. Dubois and H. Prade, *Fuzzy Sets and Systems*, Academic Press, New York 1980.

[FuGr65] D. R. Fulkerson and O. A. Gross, Incidence matrices and interval graphs, *Pacific J. Math.* 5 (1965), 835–855.

[GiHo64] P. C. Gilmore and A. J. Hoffman, A characterization of comparability graphs and interval graphs, *Canad. J. Math.* 16 (1964), 539–548.

[Ha72] F. Harary, *Graph Theory*, Addison Wesley, third printing, October 1972.

[Ka73] A. Kauffmann, *Introduction to the Theory of Fuzzy Sets*, Vol. 1, Academic Press, Inc., Orlando, Florida, 1973.

[Ki90] A. Kiss, An application of fuzzy graphs in database theory, *PU. M. A. Ser. A*, Vol. 1 (1990), 337–342.

[Ki91] A. Kiss, λ-decomposition of fuzzy relational databases, *Annales Univ. Sci. Budapest, Sect. Comp. 12* (1991), 133–148.

[Kl91] C. M. Klein, Fuzzy shortest paths, *Fuzzy Sets and Systems*, 39 (1991), 27-41.

[Ko92] L. T. Koczy, Fuzzy graphs in the evaluation and optimization of networks, *Fuzzy Sets and Systems* 46 (1992), 307–319.

[MaMo00] D. S. Malik and J. N. Mordeson, Fuzzy Discrete Structures, *Studies in Fuzziness and Soft Computing* 58 Physica-Verlag, 2000.

[Ma45] E. Marczewski, Sur deux Proprieties des Classes d'ensembles, *Fund. Math.* 33 (1945), 303–307.

[Ma70] D. W. Matula, Cluster analysis via graph theoretic techniques, *Proc. of Louisiana Conf. on Combinatorics, Graph Theory, and Computing* (1970), 199–212.

[Mi90] S. Miyamoto, *Fuzzy Sets in Information Retieval and Cluster Analysis, Theory and Decision Library*, Series D: System Theory, Knowledge Engineering and Problem Solving, Kluwer Academic Publishers, 1990.

[MoNa00] J. N. Mordeson and P. S. Nair, Fuzzy Graphs and Fuzzy Hypergraphs, *Studies in Fuzziness and Soft Computing* 46 Physica-Verlag, 2000.

[Ro76] F. Roberts, *Discrete Mathematical Models*, Prentice Hall, Englewood Cliffs, New Jersey, 1976.

[Ro75] A. Rosenfeld, Fuzzy graphs, In: L. A. Zadeh, K. S. Fu, and M. Shimura, Eds., *Fuzzy Sets and Their Applications*, Academic Press, New York, 77–95, 1975.

[TaNi76] E. Takeda and T. Nishida, An application of fuzzy graph to the problem concerning group structure, *J. Operations Res. Soc. Japan* 19 (1976), 217–227.

[YeBa75] R. T. Yeh and S. Y. Bang, Fuzzy graphs, fuzzy relations, and their applications to cluster analysis, In: L. A. Zadeh, K. S. Fu, and M. Shimura, Eds., *Fuzzy Sets and Their Applications*, Academic Press, New York, 125–149, 1975.

[Za65] L. A. Zadeh, Fuzzy sets, *Inform. and Control* 8 (1965), 338–353.

Section 10.6
Expander Graphs

Mike Krebs, California State University, Los Angeles
Anthony Shaheen, California State University, Los Angeles

INTRODUCTION

Roughly speaking, expander graphs are large, sparse, pseudorandom graphs. They enjoy a remarkable range of applications (particularly in computer science), including switching networks, derandomization, error-correcting codes, cryptographic hash functions, and much, much more. Moreover, dozens of other branches of mathematics connect in some way to the theory; the list includes, among others, functional analysis, analytic number theory, ergodic theory, combinatorics, random walk theory, operator algebras, representation theory, and geometric group theory. To quote Tao [Ta12], "it is quite remarkable that a single problem—namely the construction of expander graphs—is so deeply connected with such a rich and diverse array of mathematical topics. (Perhaps this is because so many of these fields are all grappling with aspects of a single general problem in mathematics, namely when to determine whether a given mathematical object or process of interest 'behaves pseudorandomly' . . .)."

10.6.1   Foundational Definitions and Results

Isoperimetric Constants and Expander Families

DEFINITIONS

D1: Let X be a graph with vertex set V, and let $S \subset V$. The **boundary** of S, denoted ∂S, is the set of all edges in X incident to both a vertex in S and a vertex not in S.

D2: Let X be a finite graph with a vertex set V. The **isoperimetric constant** of X, denoted $h(X)$, is the minimum, over all subsets S of V with $|S| \leq n/2$, of $|\partial S|/|S|$, where $|S|$ denotes the number of vertices in S, and $|\partial S|$ denotes the number of edges in ∂S. A set S that achieves this minimum is called an **isoperimetric set**.

The invariant $h(X)$ is sometimes also called the **expansion constant** of X, the **edge expansion constant** of X, the **Cheeger constant** of X, or (for regular graphs) the **conductance** of X. It is sometimes denoted $i(X)$.

Equivalently, one can remove the restriction on S and replace $|S|$ with $\min\{|S|, |S^c|\}$.

D3: Let (X_n) be a sequence of finite regular graphs, each with the same degree. We say (X_n) is an **expander family** if $|X_n| \to \infty$ as $n \to \infty$, and there exists $\epsilon > 0$ such that $h(X_n) \geq \epsilon$ for all n. (Here $|X_n|$ denotes the order of the graph X_n.)

While we sometimes speak of "expander graphs," the more apropos term is "expander family," because generally speaking, no graph can be a straight-out expander in and of itself.

EXAMPLES

E1: We have that $h(X) = 0$ if and only if X is not connected.

E2: Regard the cycle graph C_n as a Cayley graph on the group $\mathbb{Z}_n$ of integers modulo n with generating set $\{\pm 1\}$. Let $\lfloor x \rfloor$ denote the largest integer less than or equal to x. Then $S = \{1, \ldots, \lfloor n/2 \rfloor\}$ is an isoperimetric set, with $|\partial S| = 2$. Figure 10.6.1 illustrates the situation for C_6. Hence $h(C_n) = 4/n$ if n is even, and $h(C_n) = 4/(n-1)$ if n is odd. It follows that (C_n) is not an expander family. So d-regular expander families do not exist for $d < 3$.

Figure 10.6.1: The cycle graph C_6, with an isoperimetric set as hollow dots and its boundary as dashed line segments.

Relationship to Graph Spectra

DEFINITIONS

D4: Let X be a finite graph with vertex set V. Let $L^2(X)$ be the set of all complex-valued functions on V. The **adjacency operator** of X is the linear mapping $A : L^2(X) \to L^2(X)$ defined by $(Af)(v) = \sum f(w)$, where the sum is over all vertices w adjacent to v. We use the notation $\lambda_0(X), \lambda_1(X), \ldots, \lambda_{n-1}(X)$ for the eigenvalues of A, with the convention that $\lambda_0(X) \geq \lambda_1(X) \geq \cdots \geq \lambda_{n-1}(X)$. We frequently write λ_j instead of $\lambda_j(X)$ when the graph X is understood.

D5: Let X be a finite d-regular graph. The quantity $d - \lambda_1$ is the **spectral gap** of X.

FACTS

F1: Let X be a finite d-regular graph. For $j = 0, 1, \ldots, n-1$, we have $|\lambda_j| \leq d$. (See, for example, [GoRo01] or [KrSh11] for a proof.)

F2: Let X be a finite d-regular graph. Then

$$\frac{d - \lambda_1}{2} \leq h(X) \leq \sqrt{(d + \lambda_1)(d - \lambda_1)}.$$

The inequality $(d - \lambda_1)/2 \leq h(X)$ follows from Rayleigh–Ritz theorem of linear algebra; see [GoRo01] or [KrSh11]. See [Mo89] for a proof of the right-hand inequality.

A weaker inequality, $h(X) \leq \sqrt{2d(d - \lambda_1)}$, is due to Alon, Milman, Tanner, and Dodziuk. Computing or estimating the isoperimetric constant directly from its definition can be quite difficult. Fact F2 allows us to obtain information about it indirectly via eigenvalues, for which one can employ algebraic tools such as the Rayleigh–Ritz theorem. Fact F2 also induces a connection between expander graphs and random walk theory, as λ_1 plays a prominent role in the latter.

F3: As an immediate corollary of the preceding fact, we see that a sequence of d-regular graphs forms an expander family if and only if the spectral gaps are uniformly bounded away from zero. In other words, let (X_n) be a sequence of d-regular graphs with $|X_n| \to \infty$. Then (X_n) is an expander family if and only if there exists $\epsilon > 0$ such that $d - \lambda_1(X_n) \geq \epsilon$ for all n.

In lieu of the adjacency operator, one sometimes studies the eigenvalues of the Laplacian operator Δ instead. For a d-regular graph, we have $\Delta = dI - A$, where I denotes the identity operator.

Ramanujan Graphs and the Alon–Boppana Theorem

DEFINITIONS

D6: Let X be a finite regular graph of order n. We define

$$\lambda(X) := \begin{cases} \max\{|\lambda_1(X)|, |\lambda_{n-1}(X)|\} & \text{if } X \text{ is nonbipartite} \\ \max\{|\lambda_1(X)|, |\lambda_{n-2}(X)|\} & \text{if } X \text{ is bipartite}. \end{cases}$$

For finite nonbipartite d-regular graphs, the largest eigenvalue d can be thought of as a trivial eigenvalue. For finite bipartite d-regular graphs, the extremal eigenvalues d and $-d$ are trivial. So $\lambda(X)$ more or less captures the second-largest non-trivial eigenvalue of X.

D7: Let X be a finite d-regular graph. We say X is ***Ramanujan*** if $\lambda(X) \leq 2\sqrt{d-1}$.

Some authors define X to be Ramanujan when $\lambda_1(X) \leq 2\sqrt{d-1}$.
Definition D7 is motivated by the Alon–Boppana theorem (Fact F5).

EXAMPLE

E3: For $n \geq 3$, the complete graph K_n is nonbipartite with eigenvalues $n - 1$ (with multiplicity one) and -1 and is therefore Ramanujan.

FACTS

F4: Suppose (X_n) is a sequence of finite d-regular graphs, where $d \geq 3$ and $|X_n| \to \infty$. Using Fact F3, a quick computation shows that if each X_n is Ramanujan, then (X_n) is an expander family.

F5: Let (X_n) be a sequence of finite d-regular graphs with $|X_n| \to \infty$. Then

$$\liminf \lambda(X_n) \geq 2\sqrt{d-1}.$$

As discussed in [Mu03], this theorem is attributed to Alon and Boppana but is also independently due to Serre.

Group Representations and Kazhdan Constants

DEFINITIONS

D8: Let G be a finite group, and let $\Gamma \subset G$. For any finite-dimensional unitary representation $\pi : G \to \mathrm{GL}(V)$, define $\kappa(G, \Gamma, \pi)$ to be the minimum, over all unit vectors $v \in V$ and all $\gamma \in \Gamma$, of $\|\pi(\gamma)v - v\|$. (Note that a minimum exists by compactness of the unit sphere in V.) We define the **Kazhdan constant** $\kappa(G, \Gamma)$ to be the minimum, over all nontrivial irreducible unitary representations π of G, of $\kappa(G, \Gamma, \pi)$. (Note that a minimum exists because there are only finitely many irreducible representations of G, and because compatible inner products are unique up to a scalar constant—see [KrSh11] for details.)

Many variations of Definition D8 are possible. The following variant is sometimes also called the Kazhdan constant. Recall that for a finite group G, the right regular representation $R : L^2(G) \to L^2(G)$ is defined by $(R(\gamma)f)(g) = f(g\gamma)$, where $L^2(G)$ denotes the set of all complex-valued functions on G, with standard inner product $\langle f, g \rangle = \sum_{x \in G} f(x)\overline{g(x)}$.

D9: Let G be a finite group, and let $\Gamma \subset G$. Let $L_0^2(G)$ denote the set of complex-valued functions on G orthogonal to a nonzero constant function. Let $\hat{R}$ be the restriction of the right regular representation to $L_0^2(G)$. Define $\hat{\kappa}(G, \Gamma) := \kappa(G, \Gamma, \hat{R})$.

D10: Let G be a group, and let $\Gamma \subset G$. We denote by $\mathrm{Cay}(G, \Gamma)$ the **Cayley graph** of G with respect to Γ, that is, the graph with vertex set G so that x and y are adjacent if and only if $y^{-1}x \in \Gamma$. Recall that we say Γ is **symmetric** if $\gamma \in \Gamma$ implies that $\gamma^{-1} \in \Gamma$. Recall also that $\mathrm{Cay}(G, \Gamma)$ is an undirected graph if and only if Γ is symmetric.

FACTS

F6: Let G be a finite group, and let $\Gamma \subset G$. Then $\kappa(G, \Gamma) \geq \hat{\kappa}(G, \Gamma) \geq \frac{\kappa(G,\Gamma)}{\sqrt{d}}$.

F7: Let G be a a a finite group, and let Γ be a symmetric subset of G. Let $\pi_1, \ldots, \pi_k$ be a complete set of inequivalent irreducible representations of G. For $j = 1, \ldots, k$, let $M_{\pi_j} = \sum_{\gamma \in \Gamma} \pi_j(\gamma)$. Let A be the adjacency operator of $\mathrm{Cay}(G, \Gamma)$. Then A is unitarily equivalent to a direct sum of the operators M_{π_j}, where for each $j = 1, \ldots, k$, the number of times M_{π_j} appears in this direct sum equals the dimension of π_j.

Fact F7, first observed in [Lo75], provides a crucial link that allows one to apply the theory of group representations to the analysis of Cayley graphs; the next theorem follows from this link and is of particular interest in the theory of expander graphs.

F8: Let G be a finite group, Γ be a symmetric subset of G, and $d = |\Gamma|$. Let $X = \mathrm{Cay}(G, \Gamma)$. Then

$$2\sqrt{h(X)} \geq \hat{\kappa}(G, \Gamma) \geq \sqrt{\frac{2(d - \lambda_1(X))}{d}}.$$

For proofs of Facts F6 and F8, see [KrSh11] or [Lu94] or [LuWe93]. Putting them together with Fact F3, we get the following.

F9: Let (G_n) be a sequence of finite groups with $|G_n| \to \infty$. Let d be a positive integer. For each n, let Γ_n be a symmetric subset of G_n with $|\Gamma_n| = d$. Then $(\text{Cay}(G_n, \Gamma_n))$ is an expander family if and only if there exists a positive real number ϵ such that $\kappa(G_n, \Gamma_n) \geq \epsilon$ for all n.

In other words, for a sequence of Cayley graphs, the isoperimetric constants are bounded away from zero if and only if the spectral gaps are bounded away from zero if and only if the Kazhdan constants are bounded away from zero.

10.6.2   Major Results and Open Problems

Existence and Construction of Expander Families

DEFINITION
In the following definition, graphs will be undirected, but multiple edges and loops are allowed. If v is a vertex in a graph and e is an edge incident to v, then we denote by $e(v)$ the other endpoint of e.

D11: Let X, Y be finite regular graphs, where the degree of X equals the order of Y. Let V_X, V_Y be the vertex sets of X and Y, respectively. Let E_X, E_Y be the edge sets of X and Y, respectively; treat multiple edges as distinct elements. For each $v \in V_X$, let $E_v = \{e \in E_X \mid e \text{ is incident to } v\}$, and let $L_v : V_Y \to E_v$ be a bijective function. Define the **zig-zag product** $X \circledz Y$ to be the graph with vertex set $V_X \times V_Y$ so that the number of edges between (x_1, y_1) and (x_2, y_2) equals the number of ordered pairs $(z_1, z_2) \in E_Y \times E_Y$ such that y_1 is an endpoint of z_1; y_2 is an endpoint of z_2; and $L_{x_1}(z_1(y_1)) = L_{x_2}(z_2(y_2))$.

Many variations of this definition are possible; see, for example, [ASW08], [BeTa11], and [RVW02].

FACTS

F10: Expander families exist.

Fact F10 is originally due to Pinsker [Pi73], who used a probabilistic argument. Margulis gave the first explicit construction of an expander family in [Ma73]. His construction relies on forming quotients of infinite discrete groups with Kazhdan's Property T. Fact F13 implies that for any integer $d \geq 3$, a d-regular expander family exists.

For any d-regular graph T, let $\mu(T) = \lambda(T)/d$. (Dividing by d effectively replaces the adjacency operator with the Markov chain transition operator.)

F11: Let X be a finite d_X-regular nonbipartite graph, and let Y be a finite d_Y-regular nonbipartite graph such that d_X equals the order of Y. Then

$$\mu(X \circledz Y) \leq \mu(X) + \mu(Y) + \mu(Y)^2.$$

Fact F11 is independent of the choice of L.

F12: Recall that for a graph T with adjacency operator A, we denote by T^2 the graph with adjacency operator A^2. Let W be a nonbipartite d-regular graph with order d^4 such that $\mu(W) \leq 1/5$. (Such a graph W indeed exists.) Recursively define a sequence (W_n) by $W_1 = W^2$ and $W_{n+1} = W_n^2 \textcircled{z} W$. It then follows from Facts F3 and F11 that (W_n) is an expander family.

Facts F11 and F12 are due to [RVW02]. Other constructions and stronger estimates can be found in that paper as well. The significance of the zig-zag product approach is that the proofs involve nothing more than elementary linear algebra, in contrast to previous constructions of expander families, where the proofs relied on deep results in analytic number theory. A related graph product, the (balanced) replacement product, is used in [ASW08] to construct an expander family; that proof is purely combinatorial, as it bypasses eigenvalues altogether and instead uses the definition of the isoperimetric constant directly.

Ramanujan Graphs and Zeta Functions

DEFINITIONS

D12: Let X be a graph. Let $C = (v_0, e_0, v_1, \ldots, e_{n-1}, v_n = v_0)$ be a closed walk in X. Then C **backtracks** if $e_i = e_{i+1}$ for some i with $0 \leq i < n - 1$. Otherwise, we say that C is **backtrackless**. (In other words, C is backtrackless if no edge is traversed twice consecutively.) We say C has a **tail** if $e_0 = e_{n-1}$. Otherwise, we say that C is **tailless**. We say a closed walk P is **prime** if it is backtrackless, is tailless, and cannot be written as a k-fold concatenation of a closed walk C with itself.

D13: Let X be a finite regular graph. The **Ihara zeta function** $\zeta_X(u)$ of X is defined to be the product, over all equivalence classes of prime walks, of the complex-valued functions $\left(1 - u^{\ell(P)}\right)^{-1}$. Here $\ell(P)$ denotes the length of the walk P, and two walks are said to be equivalent if one is obtained from the other by cyclic permutation. That is, the walks $(v_0, e_0, v_1, \ldots, e_{n-1}, v_n), (v_1, e_1, \ldots, e_{n-1}, v_n, e_0, v_1), \ldots, (v_{n-1}, e_{n-1}, v_n, e_0, v_1, \ldots, e_{n-2}, v_{n-1})$ are all equivalent.

The Ihara zeta function satisfies many functional relationships like those of other zeta functions. The book [Te11] provides a thorough treatment.

FACTS

F13: Let q be a prime power. Then there exists a family of $(q+1)$-regular Ramanujan graphs.

Fact F13 is due to Lubotzky, Phillips, and Sarnak [LPS88] in the case where q is an odd prime. (Similar results were obtained by Margulis [Ma88].) These "LPS graphs" are Cayley graphs on the groups $\mathrm{PGL}(2, \mathbb{Z}_p)$ and $\mathrm{PSL}(2, \mathbb{Z}_p)$ for certain primes p. They provide an explicit construction of regular graphs with arbitrarily large girth and chromatic numbers; Erdős had previously nonconstructively demonstrated the existence of such graphs. Chiu [Ch92] has a construction of Ramanujan graphs for the case $q = 2$, and Morgenstern [Mo94] has one for arbitrary prime powers q.

Open problem: For any integer $d \geq 3$ not equal to one plus a prime power, determine whether a family of d-regular Ramanujan graphs exists.

F14: The Ihara zeta function of a finite regular graph X satisfies the Riemann hypothesis if and only if X is Ramanujan. That is, for a finite $(q+1)$-regular graph X, we have that X is Ramanujan if and only if $0 < \text{Re}(s) < 1$ and $\zeta_X(q^{-s}) = 0$ implies that $\text{Re}(s) = 1/2$.

See, for example, [Te99] for a proof of Fact F14.

Group Structure and Expansion

DEFINITIONS

D14: Let (X_n) be a sequence of finite graphs. For each n, let V_n be the vertex set of X_n. We say (X_n) *has logarithmic diameter* if $\text{diam}(X_n) = O(\log |V_n|)$.

D15: Let (G_n) be a sequence of finite groups. We say (G_n) *has logarithmic diameter* if there exists a positive integer d and symmetric subsets $\Gamma_n \subset G_n$ with $|\Gamma_n| = d$ for all n such that $\text{diam}(\text{Cay}(G_n, \Gamma_n)) = O(\log |G_n|)$. We say (G_n) *yields an expander family* if there exists a positive integer d and symmetric subsets $\Gamma_n \subset G_n$ with $|\Gamma_n| = d$ for all n such that $(\text{Cay}(G_n, \Gamma_n))$ is an expander family.

FACTS

F15: Let (X_n) be an expander family. Then (X_n) has logarithmic diameter.

The proof of Fact F15 follows quickly from the definitions; [KrSh11], for example, works out the details.

F16: Let (G_n) be a sequence of finite groups. For each n, let Q_n be a homomorphic image of G_n. Suppose that $|Q_n| \to \infty$. If (G_n) yields an expander family, then (Q_n) yields an expander family.

For a subgroup H of a group G, denote by $[G : H]$ the index of H in G.

F17: Let (G_n) be a sequence of finite groups. For each n, let H_n be a subgroup of G_n. Suppose that there exists a positive integer M such that for all n, we have $[G_n : H_n] \leq M$. If (G_n) yields an expander family, then (H_n) yields an expander family.

See [KrSh11] for proofs of Facts F16 and F17 directly from the definition of isoperimetric constant and [KrSh11] or [LuWe93] for proofs using Kazhdan constants.

For a group G, recall that the derived subgroup $G^{(m)}$ is defined recursively by taking $G^{(0)} = G$ and defining $G^{(m+1)}$ to be the commutator subgroup of $G^{(m)}$. For a solvable group G, denote by $\ell(G)$ the derived length of G. The following fact is an immediate consequence of Facts F16 and F17 as well as Example E4.

F18: Suppose (G_n) is a sequence of finite groups that yields an expander family. Then for all positive integers m, the sequence $(|G_n^{(m)}|)$ is bounded. In particular, if each G_n is solvable, then the sequence $(\ell(G_n))$ must be unbounded.

We remark that there exists a sequence of solvable groups (indeed, p-groups, hence nilpotent groups) that yields an expander family [LuWe93].

Other group-theoretic obstructions to expansion are discussed in [Lu94] and [LuWe93].

F19: Let (G_n) be a sequence of finite nonabelian simple groups with $|G_n| \to \infty$. Then (G_n) yields an expander family.

Fact F19 represents the combined work of many authors. An overview of all cases except that of Suzuki groups can be found in [KLN06]. This final case is dealt with in [BGT11], which draws on techniques developed in [BoGa08].

Open problem: Find necessary and sufficient conditions for a sequence of groups to yield an expander family.

EXAMPLES

E4: Let G be an abelian group, Γ a symmetric subset of G, and $d = |\Gamma|$. Then

$$\mathrm{diam}(\mathrm{Cay}(G, \Gamma)) \geq |G|^{1/d} - 1.$$

It follows that no sequence (G_n) of abelian groups with $|G_n| \to \infty$ has logarithmic diameter. Hence by Fact F15, no sequence of abelian groups yields an expander family. See [KrSh11] for a detailed proof.

E5: Let D_n denote the dihedral group of order $2n$. Then D_n contains a cyclic group of order n as a subgroup of order 2. So by Fact F17 and Example E4, we see that (D_n) does not yield an expander family.

E6: For each positive integer n, let V_n be the set of all ordered pairs of the form (s, k), where s is a string of 0s and 1s of length n, and k is an element of the integers modulo n. Define the *cube-connected cycle graph* CCC_n to be the cubic graph with vertex set V_n so that (s, k) is adjacent to $(s, k \pm 1)$ and to (s', k), where s' is the string identical to s except in in position k. So CCC_n is something like a hypercube, but with each vertex replaced by an n-cycle. See Figure 10.6.2 for a picture of CCC_3. Then (CCC_n) has logarithmic diameter but is not an expander family. Let G_n be the wreath product of the cyclic group of order 2 with the cyclic group of order n. Then CCC_n can be realized as a Cayley graph on G_n. Hence (G_n) has logarithmic diameter. However, because G_n admits the cyclic group of order n as a homomorphic image, by Fact F16 and Example E4, we see that (G_n) does not yield an expander family. So the converse of Fact F15 fails. See [KrSh11] for a more thorough discussion of this example.

Figure 10.6.2: A cube-connected cycle graph.

E7: Let $\gamma = (1, \dots, n)$ and $\tau = (1, 2)$ be elements of the symmetric group S_n on n letters. The *bubblesort graphs* are defined to be the graphs $\mathrm{Cay}(S_n, \{\gamma, \gamma^{-1}, \tau\})$. The bubblesort graphs do not form an expander family (see [KrSh11] or [Lu94]). However, in [Ka07], Kassabov proves that the symmetric groups yield an expander family. Therefore group structure does not determine expansion.

Random Graphs and Expansion

DEFINITIONS

D16: Let X be a graph, and let A be a nonempty set of vertices of X. The **vertex boundary** of A, denoted $\partial'(A)$, is the set of all vertices of distance 1 from A.

D17: Let X be a finite graph of order n. Let V be the vertex set of X. We say X is an ϵ-**expander** if for every nonempty subset A of V, we have $|\partial' A| \geq \epsilon(1 - |A|/n)|A|$.

FACTS

The following fact essentially shows that for $d \geq 3$, a random sequence of d-regular graphs will form an expander family.

F20: [Bo88] Let $d \geq 3$ be an integer, and let $0 < \alpha < 1$ be a real number such that $2^{4/d} < (1 - \alpha)^{1-\alpha}(1 + \alpha)^{1+\alpha}$. Then as the order n goes to infinity, the probability that $h(X) \geq (1 - \alpha)d/2$ for a randomly chosen d-regular graph X of order n goes to 1.

F21: [Fr91] Let d be a positive integer, and let ϵ be a positive real number. Then as the order n goes to infinity, the probability that $\lambda_1(X) \geq 2\sqrt{d-1} + \epsilon$ for a randomly chosen d-regular graph X of order n goes to 1.

Fact F21 had been conjectured by Alon. Sarnak [Sa04] summarizes its implication by noting, "So the random graph is asymptotically Ramanujan."

F22: [AlRo94] Let $\epsilon < 1$ be a positive real number. Then there exists $\delta > 0$ such that for any positive integer n and any group G of order n and any randomly chosen set Γ of at least $\delta \log_2(n)$ elements of G, the probability that $\mathrm{Cay}(G, \Gamma)$ is an ϵ-expander goes to 1 as $n \to \infty$.

In the facts above, the papers cited make precise the meaning of "randomly chosen."

Computer-generated statistical evidence seems to indicate that slightly more than half of bipartite (and slightly more than one-fourth of nonbipartite) regular graphs are Ramanujan. Based on their data, Miller and Novikoff [MiNo08] conjecture that the distribution of second-largest eigenvalues approaches a Tracy–Widom distribution as the number of vertices goes to infinity.

10.6.3 Other Surveys and General Sources

Readers looking for other general surveys on expanders will find no shortage of material. In only two pages, [Sa04] provides a remarkably thorough overview. The award-winning survey [HLW06] goes into considerably more depth, with an especially rich discussion of applications in computer science. Another survey, [Lu12], stands out for its discussion of connections to number theory, group theory, and geometry. Terrence Tao's blog post [Ta12] on the subject is well worth a read. The book [Lu94] is an essential guide for many in the field. The authors' text [KrSh11] presents some elementary aspects of expander graphs in a way that is meant to be accessible to advanced undergraduates or beginning graduate students and contains many ideas for student research projects such as REUs. Xiao's undergraduate thesis [Xi03] deserves special mention, particularly for its discussion of various applications in computer science. Relationships between graph eigenvalues and other graph invariants are explored in detail in [Ch97]. Murty's survey article [Mu03] on Ramanujan graphs goes into more detail on that subject.

References

[AlRo94] N. Alon and Y. Roichman, *Random Cayley graphs and expanders*, Rand. Struct. Alg. **5** (1994), no. 2, 271–284.

[ASW08] N. Alon, O. Schwartz, and A. Shapira, *An elementary construction of constant-degree expanders*, Combin. Probab. Comput. **17** (2008), no. 3, 319–327.

[BeTa11] A. Ben-Aroya and A. Ta-Shma, *A combinatorial construction of almost-Ramanujan graphs using the zig-zag product*, SIAM J. Comput. **40** (2011), no. 2, 267–290.

[Bo88] B. Bollobás, *The isoperimetric number of random regular graphs*, Eur. J. Combin. **9** (1988), no. 3, 241–244.

[BoGa08] J. Bourgain and A. Gamburd, *Uniform expansion bounds for Cayley graphs of* $SL_2(\mathbb{F}_p)$, Ann. of Math. (2) **167** (2008), no. 2, 625–642.

[BGT11] E. Breuillard, B. Green, and T. Tao, *Suzuki groups as expanders*, Groups Geom. Dyn. **5** (2011), no. 2, 281–299.

[Ch92] P. Chiu, *Cubic Ramanujan graphs*, Combinatorica **12** (1992), 275–285.

[Ch97] F. Chung, *Spectral Graph Theory*, American Mathematical Society, 1997.

[Fr91] J. Friedman, *On the second eigenvalue and random walks in random d-regular graphs*, Combinatorica **11** (1991), 331–362.

[GoRo01] C. Godsil and G. Royle, *Algebraic Graph Theory*, Springer, 2001.

[HLW06] S. Hoory, N. Linial, and A. Wigderson, *Expander graphs and their applications*, Bull. Amer. Math. Soc. **43** (2006), no. 4, 439–561.

[Ka07] M. Kassabov, *Symmetric groups and expander graphs*, Invent. Math. **170** (2007), no. 2, 327–354.

[KLN06] M. Kassabov, A. Lubotzky, and N. Nikolov, *Finite simple groups as expanders*, Proceedings of the National Academy of Sciences of the United States of America **103** (2006), no. 16, 6116–6119.

[KrSh11] M. Krebs and A. Shaheen, *Expander Families and Cayley Graphs: A Beginner's Guide*, Oxford University Press, USA, 2011.

[Lo75] L. Lovász, *Spectra of graphs with transitive groups*, Period. Math. Hungar. **6** (1975), no. 2, 191–195.

[Lu94] A. Lubotzky, *Discrete Groups, Expanding Graphs, and Invariant Measures*, Birkhauser Verlag, 1994.

[Lu12] A. Lubotzky, *Expander graphs in pure and applied mathematics*, Bull. Amer. Math. Soc. (N.S.) **49** (2012), no. 1, 113–162.

[LPS88] A. Lubotzky, R. Phillips, and P. Sarnak, *Ramanujan graphs*, Combinatorica **8** (1988), no. 3, 261–277.

[LuWe93] A. Lubotzky and B. Weiss, *Groups and expanders*, Expanding Graphs, DI-MACS Series in Discrete Mathematics and Theoretical Computer Science, vol. 10, American Mathematical Society, 1993, pp. 95–109.

[Ma73] G. Margulis, *Explicit constructions of expanders*, Problemy Peredači Informacii **9** (1973), no. 4, 71–80.

[Ma88] G. Margulis, *Explicit group-theoretical constructions of combinatorial schemes and their application to the design of expanders and concentrators*, Prob. Inform. Trans. **24** (1988), no. 1, 39–46.

[MiNo08] S. Miller and T. Novikoff, *The distribution of the largest nontrivial eigenvalues in families of random regular graphs*, Experiment. Math. **17** (2008), no. 2, 231–244.

[Mo89] B. Mohar, *Isoperimetric numbers of graphs*, Journal of Combinatorial Theory, Series B **47** (1989), 274–291.

[Mo94] M. Morgenstern, *Existence and explicit constructions of q+1 regular Ramanujan graphs for every prime power q*, J. Comb. Theory, Ser. B **62** (1994), 44–62.

[Mu03] M. Ram Murty, *Ramanujan graphs*, J. Ramanujan Math. Soc. **18** (2003), 33–52.

[Pi73] M. Pinsker, *On the complexity of a concentrator*, 7th International Teletraffic Conference, Stockholm, June 1973, pp. 318/1–318/4.

[RVW02] O. Reingold, S. Vadhan, and A. Wigderson, *Entropy waves, the zig-zag graph product, and new constant-degree expanders*, Ann. Math. 155 (2002), no. 1, 157–187.

[Sa04] P. Sarnak, *What is . . . an expander?* Notices of the AMS **51** (2004), no. 7, 762–763.

[Ta12] T. Tao, unpublished, available at terrytao.wordpress.com/2011/12/02/245b-notes-1-basic-theory-of-expander-graphs

[Te99] A. Terras, *Fourier analysis on finite groups and applications*, Cambridge University Press, 1999.

[Te11] A. Terras, *Zeta functions of graphs*, Cambridge University Press, 2011.

[Xi03] D. Xiao, *The Evolution of Expander Graphs*, Honor's Thesis, Harvard University, April 2003.

Section 10.7
Visibility Graphs

Alice M. Dean, Skidmore College
Joan P. Hutchinson, Macalester College

INTRODUCTION

Visibility graphs are often studied within the field of graph drawing, itself a sub-discipline of computational geometry. Good overviews of these fields can be found in [DiEaTaTo99, GoOR04, DeOR11], and the websites www.graphdrawing.org and www.cs.smith.edu/~orourke/ are resources for additional material. Visibility graphs are studied for their theoretical, algorithmic, and applied interest. Most generally a visibility layout is formed by a collection of objects placed in an ambient space with prescribed visibility between objects. The vertices of a visibility graph correspond to objects and its edges correspond to visibilities between two objects. Here we specialize to polygonal shapes in $\mathbb{R}^n$ and unobstructed line-of-sight visibility, typically requiring both object placement and visibility parallel to one or more axes in $\mathbb{R}^n$. Often our objects are axis-aligned line segments, rectangles, boxes, etc. Other types of visibility graphs involve more complex configurations (see §10.7.7 and [DeHaMo03]) and are studied, for example, in motion planning problems [DeOR11, Str05], VLSI design and two-layer routing [Ul84].

10.7.1   Bar-Visibility Graphs

The first study of rectilinear objects in $\mathbb{R}^2$ with vertical and horizontal visibility is due to Garey, Johnson, and So [GaJoSo76] in which circuit board elements are laid out in a piecewise linear fashion and errors in fabrication are detected using an algorithm that employs graph coloring results. Possible erroneous short circuits are detected with at most twelve tests; for an exposition and improvement to at most four tests, see [Hu93]. Theoretically the question of representing planar graphs with horizontal segments and vertical visibility was posed by Melnikov [Me81] and by de Fraysseix and Rosenstiehl [deFRo81]. Initial results were obtained by Duchet et al. [DuHaLaMe83] and by Thomassen [Th84], and a comprehensive history can be found in [DiEaTaTo99].

DEFINITIONS

D1: A *bar-visibility* (or *BV*) layout of a graph G is a representation of G in the plane by disjoint, closed, horizontal line segments (called *bars*) in which each vertex corresponds to a bar and two vertices are adjacent if and only if there is an unobstructed, non-degenerate vertical visibility band between the corresponding bars [Wi85]. (Non-degenerate visibility means visibility through a positive-width band.) A *bar-visibility graph* (or *BVG*) is a graph with a BV layout.

D2: A planar graph has a *strong visibility* layout if its vertices can be represented by disjoint, closed, horizontal line segments and two vertices are adjacent if and only if there is unobstructed vertical visibility for edges in which visibility may be along a (width-0) line.

D3: In the settings of both bar-visibility and strong visibility, a planar graph is said to have a *weak visibility* layout if it can be represented by disjoint, closed, horizontal line segments for vertices and with vertical visibility between the bars for each pair of adjacent vertices. (Thus there may be visibilities that do not correspond to edges.)

REMARKS

R1: A BV layout is also known as an *ε-visibility layout* [TaTo86].

R2: A BV layout induces a plane embedding of G by placing each vertex on its corresponding bar and drawing edges between pairs of vertices whose bars have vertical visibility; see also Fact F3 and Example E2.

R3: A subgraph of a BVG may not be a BVG; see Example E1 and Fact F2.

R4: When disjoint horizontal line segments are laid out, the resulting BVG is the same, regardless of whether the segments are closed, open, or neither, but these choices can make a difference in the graph of the strong visibility layout. As is customary [TaTo86], we choose always to use closed line segments.

R5: It is an artifact of the evolution of terminology that the definitions of strong and weak visibility do not describe complementary properties of visibility layouts.

EXAMPLE

E1: Figure 10.7.1 shows a bar-visibility graph G and a layout of G. The layout is not a strong layout of G because, for instance, there is a degenerate visibility line between bars 2 and 7, but the vertices are not adjacent. If we form a subgraph G' by deleting the edge (1,6), we obtain the smallest planar graph that is not a bar-visibility graph, namely, the bipartite graph $K_{2,3}$ plus three pendant edges. However, G' is a weak bar-visibility graph since it is a subgraph of a bar-visibility graph.

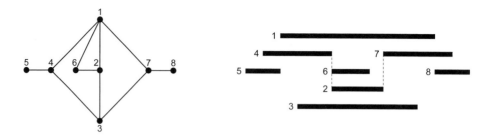

Figure 10.7.1: A graph and its BV layout.

FACTS

F1: Any set S of horizontal segments in $\mathbb{R}^2$ determines a BVG G of which it is a BV layout. S also determines a graph G' of which it is a strong BV layout, and G is a subgraph of G'. Every subgraph of G and of G' has a weak visibility layout using the segments of S and the appropriate model of bar-visibility or strong visibility.

F2: Wismath [Wi85] and Tamassia and Tollis [TaTo86] showed that a planar graph is a BVG if and only if it can be drawn in the plane with all cut-vertices on a common face (and without loss of generality that face is the infinite, exterior face). [TaTo86] gives a linear-time algorithm to determine whether a graph is a BVG. An alternative layout algorithm follows from giving the graph a *canonical ordering* [deFHuPaPo90, Ka96, Nu99].

F3: Every BVG (and every graph with a strong visibility layout) has a planar drawing with edges that are polylines with at most two bends (i.e., consisting of at most three contiguous straight-line segments) [LuMaWo87, DiTa88, DiEaTaTo99]; see also E2.

EXAMPLE

E2: A plane drawing of the graph induced by a BV layout can be obtained by vertically thickening each bar to become a rectangle and placing a vertex in its center. A polyline edge with at most two bends is produced by choosing a visibility line and drawing radial lines from each of its endpoints to the vertex in the center of the thickened bar; see Fig. 10.7.2.

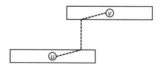

Figure 10.7.2: A BV layout induces a 2-bend polyline embedding.

FACTS

F4: Tamassia and Tollis [TaTo86] show that every maximal planar graph has a strong visibility layout that can be determined in linear time, and every 4-connected planar graph has one that can be determined in polynomial time. They also provide a characterization of graphs G that have a strong visibility layout in terms of extensions of G that have a "strong st-numbering."

F5: Andreae [An92] proved that deciding whether a planar graph has a strong visibility layout is an NP-complete problem.

F6: In the settings of bar-visibility [TaTo86] and strong visibility [DuHaLaMe83], a graph has a weak visibility layout if and only if it is planar. Thomassen [Th06] proved that a countably infinite graph is a weak visibility graph if and only if it is planar.

DEFINITIONS

D4: A visibility layout is *collinear* if there are two bars with endpoints having the same x-coordinate; otherwise it is *noncollinear*.

D5: A graph G is a *unit bar-visibility graph* if it has a BV layout in which all bars have equal length.

D6: A digraph D is a *directed bar-visibility graph* if there is a BV layout of the underlying, undirected graph G such that for each arc (u, v) of D, the visibility of u to v is in the positive y direction.

D7: A planar graph is a *near-triangulation* if it has a plane embedding with all finite, interior faces 3-sided.

D8: A *caterpillar* is a tree containing a path with the property that every vertex is at distance at most one from the path.

FACTS

F7: Every noncollinear visibility layout is both a BV layout and a strong BV layout of the same graph, but the bar-visibility graph and the strong visibility graph induced by a collinear layout may be different. In [TaTo86] it is shown that every graph with a strong visibility layout also has a BV layout. The 4-cycle C_4 has a strong visibility layout but not a noncollinear visibility layout [LuMaWo87, TaTo86], and the complete bipartite graph $K_{2,4}$ has a BV layout but not a strong visibility layout [TaTo86]. Hence the noncollinear BVGs form a strict subclass of the graphs with strong visibility layouts, which in turn form a strict subclass of the BVGs.

F8: Luccio et al. [LuMaWo87] proved that a planar graph has a noncollinear visibility layout if and only if after possibly duplicating some edges it is a near-triangulation.

F9: A tree is a unit BVG if and only if it is a subdivision of a caterpillar with maximum degree 3 [DeVe03]. In [DeGeHu05] triangulated polygons with unit-length BV layouts are characterized. [ChHuKeSh06] studies BVGs in which all bars have length between $1/k$ and 1 given integer $k \geq 1$, and they characterize all trees that have such a layout.

F10: A digraph D is a directed bar-visibility graph if and only if D^* is planar and acyclic where D^* is formed by adding two vertices, a *source* s and a *sink* t, an arc (s, v) to every vertex v with indegree 0, an arc (w, t) from every vertex w with outdegree 0, and the arc (s, t) [Wi85, TaTo86].

REMARKS

R6: In [KiWi96] bar-visibility layouts of edge-weighted graphs are studied with the requirement that the width of a visibility band be proportional to the weight of the edge to which it corresponds.

R7:   From a BV layout of G one obtains a *tessellation* representation in $\mathbb{R}^2$ with an axis-aligned rectangular tile representing each vertex, edge, and face of the induced plane embedding of G. This is obtained essentially by expanding each visibility band to a maximal band. Then the vertices and faces are represented by line segments or degenerate tiles. Such tessellations have been studied in the context of undirected and directed graphs; see [RoTa86, TaTo89, MoRo98, DiEaTaTo99].

R8:   BV layouts have also been studied and characterized on the surface of a cylinder, a Möbius band, and the projective plane [TaTo91, De01, Hu05].

RESEARCH PROBLEMS

RP1:   Is there a characterization of planar graphs that have BV layouts with all bars of unit length or a characterization with all bars lying within the interval $[1/k, 1]$?

10.7.2   Rectangle-Visibility Graphs

Another visibility model that derives from the rectilinear polygon layouts studied in [GaJoSo76] is the class of rectangle-visibility graphs, in which vertices are represented as axis-parallel rectangles in the plane and edges are induced by both vertical and horizontal visibilities. The first studies of these occurred in [Wi89, BoDeHuSh97, DeHu97, HuShVi99].

DEFINITIONS

D9:   A *rectangle-visibility (or RV)* layout of a graph G is a representation of G in the plane by interior-disjoint axis-parallel rectangles in which each vertex corresponds to a rectangle and two vertices are adjacent if and only if there is an unobstructed, non-degenerate vertical or horizontal visibility band between the corresponding rectangles. A *rectangle-visibility graph* (or *RVG*) is a graph with an RV layout.

D10:   A *unit rectangle-visibility graph* (or *unit RVG)* is a graph with an RV layout using unit squares.

D11:   For integer $k \geq 1$, a graph has *thickness* k if k is the minimum for which the graph's edges can be partitioned into k planar graphs.

D12:   A *k-tree* is either a k-clique or a graph formed from a smaller k-tree T by adding a vertex adjacent precisely to the vertices of a k-clique within T. A *partial k-tree* is a subgraph of a k-tree.

REMARKS

R9:   In [Wi89] RVGs are called *2-box-representable graphs*.

R10:   The analogous definitions of *strong, weak, collinear* and *noncollinear* visibility apply to RV layouts as to BV layouts.

R11:   The horizontal and vertical visibilities of an RV layout induce a partition of edges of the corresponding RVG into two (planar) BVG subgraphs; hence the thickness of an RVG is at most two [GaJoSo76].

EXAMPLES

E3: Figure 10.7.3 shows an RV layout of the nonplanar graph $K_{4,4}$ and the partition of its edges, induced by the horizontal and vertical visibilities, into two planar subgraphs, drawn with straight lines.

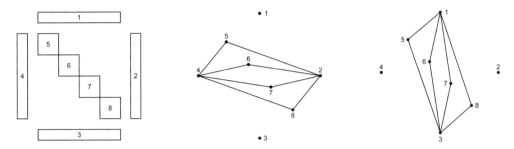

Figure 10.7.3: An RV layout of $K_{4,4}$ and the subgraphs induced by its horizontal and vertical visibilities.

E4: Figure 10.7.4 shows a layout of K_8 as an RVG, and the two induced planar graph drawings with polylines for edges.

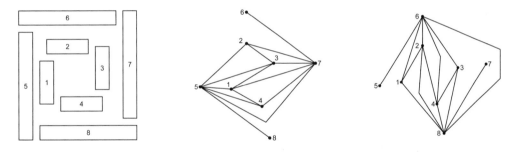

Figure 10.7.4: An RV layout of K_8 and the subgraphs induced by its horizontal and vertical visibilities.

FACTS

F11: Every planar graph is an RVG [Wi89].

F12: The complete graph K_n is an RVG if and only if $n \leq 8$ [DeHu97, HuShVi99].

F13: Dean and Hutchinson [DeHu97] showed that for integers $p \leq q$ the complete bipartite graph $K_{p,q}$ is an RVG if and only if $p \leq 4$, although $K_{5,q}$ has thickness two for $5 \leq q \leq 12$ [BeHaMo64]. From Fig. 10.7.3 it is clear how to lay out $K_{4,q}$ for $q > 4$. They show that $K_{5,5}$ is a weak RVG, and $K_{p,q}$ is a noncollinear RVG if and only if $p \leq 2$ or $p = 3$ and $q \leq 4$; hence the noncollinear RVGs form a strict subclass of the RVGs.

F14: Bose et al. [BoDeHuSh97] and Biedl [Bi11] proved that for $k = 1, 2$ all partial k-trees are noncollinear RVGs. (Note that a 1-tree is a tree, and a 2-tree is also called a ***series-parallel graph***.) For $1 \leq k \leq 4$ every k-tree is an RVG [BoDeHuSh97]. Since 5-trees can have thickness 3, 5-trees need not be RVGs.

F15: Every graph that is a union of two forests of caterpillars is a noncollinear RVG and consequently so are graphs with maximum degree 3 [BoDeHuSh97].

F16:   A tree is a unit RVG if and only if it is the union of two forests of subdivided caterpillars, each having maximum degree 3 [DeElHaPa08]. Results on representing 3-trees as unit RVGs are in [DeHu10].

F17: Every graph with maximum degree four is a weak RVG, and every graph whose vertices of degree four or more are mutually at distance at least two apart is a weak RVG [BoDeHuSh97].

F18:   An RVG on $n \geq 5$ vertices has at most $6n - 20$ edges, and for each $n \geq 8$ there is a (connected) RVG with $6n - 20$ edges. There are thickness-2 graphs that are not RVGs (since a thickness-2 graph may have as many as $6n - 12$ edges) [HuShVi99]. A bipartite RVG on $n \geq 4$ vertices has at most $4n - 12$ edges [DeHu97].

F19: A unit RVG on $n \geq 1$ vertices has at most $6n - 4\lceil \sqrt{n} \rceil + 1$ edges, and for each $n \geq 64$ there is a unit RVG with at least $6n - 12\lfloor \sqrt{n} \rfloor + 6$ edges. A bipartite unit RVG with $n \geq 7$ vertices has at most $4n - 2\lceil \sqrt{n} \rceil + 5$ edges [DeElHaPa08].

F20: Mansfield [Ma83] proved that it is an NP-complete problem to determine whether a graph has thickness two; Shermer [Sh96a] proved the same for determining whether a graph is an RVG.

REMARK

R12: An RV layout has a natural orientation with each visibility directed in the positive x or y direction. An RVG inherits an orientation by directing each edge according to that of its visibility band, as do the two underlying BVGs of horizontal and vertical visibilities.

DEFINITION

D13: A planar, acyclic digraph is an *st-graph* if it has a single source s and a single sink t.

FACTS

F21:   Streinu and Whitesides initiated a more detailed study of *topological RVGs* [StrWh03]. Let $\mathcal{R}$ be a set of interior-disjoint, axis-aligned rectangles in $\mathbb{R}^2$ with visibilities directed as in Remark R12. To this add a *frame* of two wide rectangles N and S at the top and the bottom of the layout and two tall rectangles W and E at the left and the right that span the layout of $\mathcal{R}$ and so that there is no $N - E$, $E - S$, $S - W$, or $W - N$ visibility. The *topological rectangle visibility graph of* $\mathcal{R}$ or $TRVG(\mathcal{R})$ is a pair (D_V, D_H) of digraphs, where D_V records the upwardly directed, cyclically ordered vertical visibilities of $\mathcal{R} \cup \{S, N\}$ including multiplicities, and similarly D_H records the left-to-right directed, cyclically ordered visibilities of $\mathcal{R} \cup \{W, E\}$. A *pseudo TRVG* is a TRVG with digraphs (D_V, D_H) that are both directed, acyclic planar *st*-graphs whose underlying graph is 2-connected. In [StrWh03] they characterize TRVGs in terms of a pseudo TRVG (D_V, D_H) and conditions on paths within D_V and D_H. More precisely, given a pseudo TRVG (D_V, D_H), necessary and sufficient conditions are given for the existence of $\mathcal{R}$ for which $TRVG(\mathcal{R}) = (D_V, D_H)$. There is a quadratic time algorithm to recognize and lay out a TRVG specified by a set $\mathcal{R}$ and the related pair (D_V, D_H).

REMARKS

R13: Since an RV layout induces two BV layouts of the horizontal and vertical visibilities, it induces a drawing of its RVG as the union of two planar graphs with identical vertex locations and edges that are polylines, each with at most two bends; see Fig. 10.7.2. An RV layout and related polyline drawing is used in [Bi11] to show that series-parallel graphs (or partial 2-trees) can be drawn in the plane in small area. For more information on *orthogonal* polyline graph drawing in the plane, see Section 10.3.

R14: See also Remark R18 on *2-box or rectangle intersection graphs*.

RESEARCH PROBLEMS

RP2: A graph G is *doubly-linear* [HuShVi99] (or equivalently G has *geometric thickness two* [DiEpHi98]) if it can be drawn as the union of two straight-edged planar graphs with identical vertex locations. All known examples of RVGs are doubly linear (as compared with Remark R13), and doubly-linear graphs have at most $6n - 18$ edges [HuShVi99], higher than the bound of Fact F18. Is there a doubly-linear graph that is not an RVG?

RP3: Is it an NP-complete problem to determine if a graph is an RVG with all rectangles squares (or unit squares)? (See also Fact F26.)

The next research problems were raised by T. Shermer [Sh96b].

RP4: The noncollinear RVGs form a subclass of the strong RVGs—is it a strict subclass? (See also Fact F13.) Is there a containment relation between the classes of strong RVGs and RVGs, and if so, is it strict?

RP5: Is there a graph that is the union of three caterpillar forests that is not an RVG?

RP6: If G has maximum degree 4, is it an RVG? If it has maximum degree 5, is it a weak RVG?

RP7: For $n > 0$, let $c(n)$ be the maximum such that every graph with n vertices and crossing number at most $c(n)$ is an RVG. What can be said about the function $c(n)$?

RP8: How much information can be dropped from that given for a TRVG and still have a polynomial time recognition algorithm for that class of RVGs [StrWh03]?

10.7.3 Visibility Representations in $\mathbb{R}^3$

Many early papers from this and the previous subsection were derived from the McGill University Bellairs Research Institute's Workshop on Visibility Representations, organized by S.H. Whitesides and J.P. Hutchinson in 1993. There the model of using axis-aligned rectangles in $\mathbb{R}^3$, placed parallel to the $z = 0$ plane, and with visibility parallel to the z-axis, was first studied. This is one of several models of visibility in $\mathbb{R}^3$.

In this subsection through §10.7.6 visibility bands are always non-degenerate, i.e., they must have positive width.

DEFINITIONS

D14: A 3D *visibility representation* of a graph G in $\mathbb{R}^3$ is a layout of the vertices of G with disjoint, (x, y)-axis-aligned rectangles, lying perpendicular to the (vertical) z-axis. Edges are represented with vertical visibility so that two vertices in G are adjacent if and only if there is a closed cylinder of non-zero radius within an unobstructed vertical visibility corridor between the corresponding rectangles [BoEvFeLu94, BoEvFeHo98]. This visibility is also known as *z-visibility*, and the represented graphs are called **VR-representable graphs** and **ZPR** graphs [BoEvFeLu94, FeMe99].

D15: A graph has a *weak* 3D *visibility representation* if it is a subgraph of a graph with a 3D visibility representation.

D16: A 3D *visibility representation by unit squares, polygons, or unit discs* is a 3D representation in which unit squares, polygons, or unit discs, respectively, are used instead of rectangles.

D17: A class of objects is *universal* if for every graph there is a 3D visibility representation of the graph by objects from that class [AlGoWh98].

D18: The *multipartite number* of graphs with 3D visibility representations is the maximum integer k for which every complete k-partite graph has a 3D visibility representation. Equivalently the multipartite number is the maximum k for which every k-partite graph (or k-colorable) graph has a weak 3D visibility representation [Sto08].

EXAMPLE

E5: Figure 10.7.5(a) shows a 3D layout of K_6 using unit squares; in the figure darker squares are placed behind lighter squares. In Figure 10.7.5(b) a translucent square is added on top to achieve a layout of K_7. Both layouts are based on figures in [FeHoWh96].

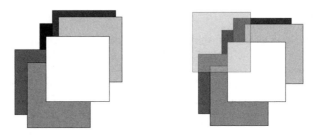

Figure 10.7.5: 3D layouts of K_6 and K_7 using unit squares.

FACTS

F22: Every planar graph is 3D representable, and the set of 3D representable graphs is not closed under graph minors [BoEvFeLu94].

F23: K_n is 3D representable for all $n \leq 22$ [BoEvFeHo98], and no K_n is 3D representable for $n \geq 51$ [FeHoWh96, Sto09].

F24: $K_{p,q}$ is $3D$ representable for all positive p and q. Bounds are also known for tripartite graphs and for complete bipartite graphs minus a perfect matching. Every K_n is $3D$ representable by unit discs [BoEvFeLu94].

F25: The multipartite number for $3D$ representable graphs is 8; see §10.7.4 and [Sto08] for related definitions and results.

F26: Fekete, Houle and Whitesides [FeHoWh96] prove that K_n is $3D$ representable by unit squares if and only if $n \leq 7$. They also show that it is an NP-complete problem to recognize graphs that have a $3D$ visibility representation by squares [FeHoWh97].

F27: Fekete and Meijer [FeMe99] give bounds on the sizes of possible complete graphs when the number of different sizes (congruent by translations) and shapes (congruent by translations and rotations) for $3D$ representation is considered. For example, if one shape of rectangle is allowed, K_n, $n \leq 12$, is $3D$ representable and no K_n with $n > 14$ is so representable. The same bounds (12 and 14) exist for $3D$ representations of K_n when two different sizes of rectangles are allowed. All bounds of 55 in Table 1 in [FeMe99] have now been reduced to 50 by results of Stola [Sto09].

F28: Alt, Godau and Whitesides [AlGoWh98] show that there is no k for which the class of polygons with at most k sides is universal. Every graph with n vertices has a $3D$ representation by polygons, each having at most $2n$ sides, and there is a quadratic time algorithm for constructing such a representation from a graph.

REMARKS

R15: There is extensive literature on $3D$ visibility representations of complete graphs using regular k-gons; see, for example, [Sto10] where the biggest representable K_n is bounded by $O(k^4)$.

R16: Cobos et al. [CoDaHuMa96] have initiated a study of graph representations using hyper-rectangles in $\mathbb{R}^n$ with orthogonal visibility. Every graph can be so represented, and a fundamental problem is the determination of the minimum n which allows for such a representation for different families of graphs.

RESEARCH PROBLEMS

RP9: In [FeMe99] there are many instances of distinct upper and lower bounds for complete graphs representable by different shapes and sizes of rectangles, as in Fact F27. What is the best possible bound in each case?

RP10: What is the best bound on the size of K_n that has a $3D$ representation by regular k-gons?

10.7.4   Box-Visibility Graphs

Another natural model in $\mathbb{R}^3$ represents the vertices of a graph by axis-parallel 3-dimensional boxes (called ***boxes*** or ***3-boxes***) and edges by visibility in directions parallel to each of the three axes.

DEFINITIONS

D19: A ***box-visibility graph*** has vertices represented by boxes in $\mathbb{R}^3$ and edges represented by unobstructed visibility through a positive-radius, axis-parallel cylinder [BoJoMiOR94]; such a representation is also called a ***box-visibility representation*** or ***BR*** [FeMe99].

D20: K_n is said to have a ***BR by unit cubes*** if it has a BR with each box a unit cube [FeMe99].

D21: The ***multipartite number*** for box-visibility graphs is the maximum integer k for which every complete k-partite graph is a box-visibility graph [Sto08].

EXAMPLE

E6: Figure 10.7.6, based on a figure in [FeMe99], shows a layout of K_8 using unit cubes in $\mathbb{R}^3$. The layout is shown from three perspectives, with the second and third views obtained by rotating the preceding view $90°$ about the vertical axis, to aid in seeing the visibility cylinders between each pair of cubes. The white cubes at top and bottom see all the others vertically.

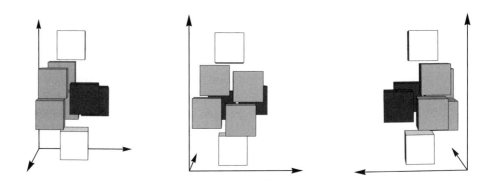

Figure 10.7.6: Three views of a layout of K_8 in $\mathbb{R}^3$ using unit cubes.

FACTS

F29: K_n, $n \leq 56$, is a box-visibility graph; this bound with $n \leq 42$ was first shown in [BoJoMiOR94] and the better bound in [FeMe99]. There is no such representation of K_n with $n \geq 169$. The latter bound with $n \geq 184$ was shown in [FeMe99]; the better bound follows from the work of [Sto09].

F30: K_n has a BR by unit cubes for $n \leq 8$, and for $n \geq 10$ no K_n has such a box representation [FeMe99].

F31: The multipartite number k for box-visibility graphs is bounded by $22 \leq k \leq 42$ [Sto08].

REMARKS

R17: Results of [FeMe99] are obtained by studying forbidden configurations in related partially ordered sets, and the authors obtain additional results by specifying shapes and sizes of representing boxes. Romanik [Rom97] also uses posets to determine that directed RVGs and digraphs with $3D$ visibility representations have infinite dimension.

R18: Let a *d-box* be a d-dimensional rectilinear box in $\mathbb{R}^d$ or equivalently the Cartesian product of closed line segments, one within each of the d axes [Th86]. The *intersection graph* of a set of d-boxes in $\mathbb{R}^d$ has a vertex for each d-box and an edge joining two vertices precisely when the corresponding d-boxes intersect. Intersection graphs of 1-boxes are interval graphs. Early work on d-box intersection graphs is found in [Rob69, Sc84]. For $d = 2$ these are also called *rectangle intersection graphs* and graphs with *boxicity 2*. Kratochvil [Kr94] proved that determining if a graph is a 2-box intersection graph is NP-complete. For $d = 3$ these graphs are also called *box intersection graphs* and graphs with *boxicity 3*. Thomassen [Th86] proved that every planar graph is a 3-box intersection graph. When vertices of graphs are represented by orthogonal polyhedra in $\mathbb{R}^3$, every K_n is the intersection graph of such polyhedra [Wh84].

RESEARCH PROBLEMS

RP11: Obtain better or best possible bounds on the size of K_n that is box-representable and also best bounds for representation by unit cubes.

RP12: Determine the multipartite number of box-visibility graphs.

10.7.5   Bar k-Visibility Graphs

Since bar-visibility graphs are planar, they are limited in their applicability for modeling more complex systems, like multi-layer circuits, that may correspond to non-planar graphs. One way to broaden the visibility model is to permit the *opacity* of the bars to vary, in other words, to permit visibility through one or more bars. Bar k-visibility graphs, which permit visibility through k bars, were introduced by Dean et al. [DeEvGeLa07] and further studied by Hartke et al. [HaVaWe07] and by Felsner and Massow [FeMa08].

DEFINITION

D22: Given a layout in the plane of disjoint horizontal line segments (called *bars*) and an integer $k \geq 0$, a *k-visibility band* between two bars is a non-degenerate vertical band between the two bars that passes through at most k intermediate bars. If a k-visibility band passes through no intermediate bars, then the two bars it joins have *direct* visibility; otherwise they have *indirect* visibility. A graph G is a *bar k-visibility* graph if G can be represented by such a layout, so that vertices correspond to bars and two vertices are adjacent if and only if there is a k-visibility band between the corresponding bars. A graph is a *weak bar k-visibility* graph if it is a subgraph of a bar k-visibility graph.

REMARK

R19: A bar-visibility graph is the same as a bar 0-visibility graph, and an interval graph with n vertices is the same as a bar k-visibility graph with $k \geq n - 2$.

EXAMPLE

E7: Figure 10.7.7 shows a bar 1-visibility layout of the (non-planar) complete graph K_8. Note that, for example, bar 1 has direct visibility only to bars 2 and 8, while its visibilities to bars 3–7 are all indirect visibilities with intermediate bar 2 in each case.

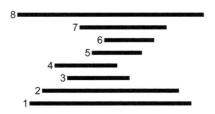

Figure 10.7.7: A bar 1-visibility layout of K_8.

FACTS

F32: A bar k-visibility graph with $n \geq 2k+3$ vertices has at most $(k+1)(3n-4k-6)$ edges [DeEvGeLa07, HaVaWe07], and this bound is tight.

F33: K_{4k+4} is the largest complete bar k-visibility graph [DeEvGeLa07, HaVaWe07]; in particular, K_8 is the largest complete bar 1-visibility graph (see Fig. 10.7.7).

F34: The chromatic number $\chi(G)$ of a bar k-visibility graph G is at most $6k+6$ [DeEvGeLa07].

F35: If G is a bar 1-visibility graph, then its thickness $\Theta(G)$ is at most 4. If G is a bar k-visibility graph with $k \geq 2$, then $\Theta(G) \leq 2k(9k-1)$. There are thickness-2 graphs that are not weak bar 1-visibility graphs [DeEvGeLa07], but there are bar 1-visibility graphs with thickness 3 [FeMa08].

F36: For every $k \geq 0$ there are bar k-visibility graphs that are not bar $(k+1)$-visibility graphs, and there are bar $(k+1)$-visibility graphs that are not bar k-visibility graphs [HaVaWe07].

F37: If a graph G contains a triangle-free, non-planar induced subgraph, then G is not a bar k-visibility graph for any k [HaVaWe07].

F38: If G is a connected, d-regular bar k-visibility graph with $d \leq 2k+2$, then G is a complete graph [HaVaWe07].

RESEARCH PROBLEMS

RP13: (See Fact F35.) What is the tight bound on the thickness of bar 1-visibility graphs? More generally, what is the tight bound on the thickness of bar k-visibility graphs? Is it a quadratic bound in k, like the one given in [DeEvGeLa07], or is it linear in k?

RP14: Dean et al. [DeEvGeLa07] also give bounds on the chromatic number of bar k-visibility graphs, but they do not claim that these are tight bounds. What is the tight bound on the chromatic number of bar k-visibility graphs?

RP15: Are there forbidden induced subgraphs for bar k-visibility graphs besides the triangle-free, nonplanar graphs [HaVaWe07]?

10.7.6   Bar Visibility Number

Bar k-visibility graphs broaden the visibility model by permitting visibility to pass through one or more bars. Another way to extend the model is to permit vertices to be represented by more than one bar. This model was first introduced by Chang et al. [ChHuJaLe04]; results for directed graphs were introduced by Axenovich et al. in [AxBeHuWe11].

DEFINITION

D23:   For $t \geq 1$ a t-**bar representation** (or t-**bar layout**) of a graph G is a bar-visibility layout in which each vertex of G corresponds to a maximum of t bars, and two vertices of G are adjacent if and only if there is an unobstructed, non-degenerate vertical visibility band between two bars that correspond to the two vertices. The **bar visibility number** $b(G)$ is the minimum t such that G has a t-bar representation.

D24:   For $t \geq 1$ a (directed) t-**bar representation** of a digraph D is a t-bar visibility layout of the underlying undirected graph such that there is an arc from vertex u to vertex v if and only if the visibility band from u to v is directed in the positive y direction; see Definition D6. The (directed) **bar visibility number** $b(D)$ is the minimum t such that D has a directed t-bar representation.

EXAMPLE

E8:   Figure 10.7.8 (modified from a figure in [DePhRa12]) shows a 2-bar layout of the complete graph K_9 (which has thickness 3). The layout can be viewed as a directed 2-bar layout of a tournament on nine vertices. The figure also demonstrates that K_9 has a 2-bar layout using all bars of unit length.

Figure 10.7.8: A unit 2-bar visibility layout of K_9.

REMARKS

R20:   A bar-visibility graph is the same as a 1-bar visibility graph.

R21:   For every pair of graphs G and H, $b(G \cup H) \leq b(G) + b(H)$, and the same is true for directed bar visibility number.

R22:   If G is the underlying graph of a directed graph D, then $b(G) \leq b(D)$.

FACTS

F39:   The bar visibility number of the complete graph K_n is $\lceil n/6 \rceil$ [ChHuJaLe04], and the bar visibility number of the complete bipartite graph $K_{p,q}$ is $\left\lceil \frac{pq+4}{2p+2q} \right\rceil$ [Ca06, ChHuJaLe04].

F40:   If G has n vertices, then $b(G) \leq \lceil n/6 \rceil + 2$ [ChHuJaLe04]. If D is a digraph with n vertices, then $b(D) \leq (n + 10)/3$. If the underlying graph G of D has maximum degree Δ, then $b(G) \leq b(D) \leq \lceil (\Delta(G) + 1)/2 \rceil$ [AxBeHuWe11].

F41:   If G has n vertices and e edges, then $b(G) \geq \lceil \frac{e+6}{3n} \rceil$, and if G is triangle-free, then $b(G) \geq \lceil \frac{e+4}{2n} \rceil$ [ChHuJaLe04]. It follows that these lower bounds also hold for directed bar visibility number.

F42:   It follows from a result of Wismath [Wi85, Wi89] that if G is planar, then $b(G) \leq 2$ [ChHuJaLe04]. Furthermore, every planar graph has a 2-bar representation in which every vertex that is not a cut-vertex is assigned only one bar [ChHuJaLe04].

F43:   If D is a planar digraph, then $b(D) \leq 4$. If D is also triangle-free, then $b(D) \leq 3$, and this bound is sharp. If D is outerplanar or has girth ≥ 6, then $b(D) \leq 2$ [AxBeHuWe11].

F44:   If D is a transitive tournament on n vertices, then $b(D) \leq 7n/24 + 2\sqrt{n \log n}$; for n sufficiently large, $b(D) < 3n/14 + 42$ [AxBeHuWe11].

RESEARCH PROBLEMS

RP16:   What is the bar visibility number of the d-dimensional hypercube Q_d? Does it equal the lower bound $\lceil (n + 1)/4 \rceil$?

RP17:   Is the directed bar visibility of a graph at most twice the bar visibility number of the underlying graph [AxBeHuWe11]?

RP18:   Is there a planar digraph D with $b(D) = 4$ [AxBeHuWe11]?

RP19:   If D is a transitive tournament on n vertices, what is the best upper bound on $b(D)$? Asymptotically what is the best bound on $b(D)$?

10.7.7   Other Visibility Graphs

We conclude with a brief mention of other types of visibility graphs (especially those with names similar to terms in this subsection) but do not include comprehensive references on fundamental and latest results.

Three well-studied visibility graphs are defined below. More information can be found in O'Rourke [OR93] and in Chapter 28 of Goodwin and O'Rourke [GoOR04].

DEFINITIONS

D25: The **vertex visibility graph** or **polygon vertex visibility graph** has vertices represented by the vertices of a simple polygon in the plane and edges by straight-line visibility within the polygon between polygon vertices.

D26: Given a set of disjoint segments in the plane, the ***endpoint visibility graph*** or ***segment endpoint visibility graph*** has vertices represented by the endpoints of the segments and edges by straight-line visibility between endpoints.

D27: The ***segment visibility graph*** or ***(whole) segment visibility graph*** has vertices represented by disjoint line segments in the plane and edges by (any) straight-line visibility between corresponding segments.

In the third definition, when the segments are constrained to be horizontal and visibility to be vertical, then we are in the case of bar- or strong visibility graphs.

All three concepts are studied for graph characterization, recognition, and algorithmic reconstruction. The chromatic number, clique number, Hamiltonicity, and size constraints of these graphs are studied, among other structural and algorithmic properties.

RESEARCH PROBLEMS

RP20: Find necessary and sufficient conditions to determine when a graph is the vertex visibility graph of a polygon [DeOR11].

RP21: A collection of open problems from 2003 is available in [BrEpGoKo04].

References

[AlGoWh98] H. Alt, M. Godau, and S. Whitesides, Universal 3-dimensional visibility representations for graphs, *Comp. Geom.* 9 (1998), 111–125.

[An92] T. Andreae, Some results on visibility graphs, *Discrete Appl. Math.* 40 (1992), 5–17.

[AxBeHuWe11] M. Axenovich, A. Beveridge, J. Hutchinson, and D. West, Visibility number of directed graphs, preprint (2011).

[BeHaMo64] L. Beineke, F. Harary, and J. Moon, On the thickness of the complete bipartite graph, *Math. Proc. Camb. Philos. Soc.* 60 (1964), 1–5.

[Bi11] T. Biedl, Small drawings of outerplanar graphs, series-parallel graphs, and other planar graphs *Discrete Comput. Geom.* 45 (2011), 141–160.

[BoDeHuSh97] P. Bose, A. Dean, J. Hutchinson, and T. Shermer, On rectangle visibility graphs, in *Graph Drawing (GD96)*, S. North (ed.), *Lecture Notes Comput. Sci.* Vol. 1190, pp. 25–44, Springer-Verlag, 1997.

[BoEvFeLu94] P. Bose, H. Everett, S. Fekete, A. Lubiw, H. Meijer, K. Romanik, T. Shermer, and S. Whitesides, On a visibility representation for graphs in three dimensions, in *Snapshots of Computational and Discrete Geometry*, D. Avis and P. Bose (eds.), Vol. 3, pp. 2–25, McGill Univ. School Comput. Sci. tech. report SOCS-94.50, 1994.

[BoEvFeHo98] P. Bose, H. Everett, S. Fekete, M. Houle, A. Lubiw, H. Meijer, K. Romanik, G. Rote, T. Shermer, S. Whitesides, and C. Zelle, A visibility representation for graphs in three dimensions, *J. Graph Algorithms Appl.* 2 (1998), 1–16.

[BoJoMiOR94] P. Bose, A. Josefczyk, J. Miller, and J. O'Rourke, K_{42} is a box visibility graph, in *Snapshots of Computational and Discrete Geometry*, D. Avis and P. Bose (eds.), Vol. 3, pp. 88–91, McGill Univ. School Comput. Sci. tech. report SOCS-94.50, 1994.

[BrEpGoKo04] F. Brandenburg, D. Eppstein, M. Goodrich, S. Kobourov, G. Liotta, and P. Mutzel, Selected open problems in graph drawing, in *Graph Drawing (GD 2003)*, G. Liotta (ed.), *Lecture Notes Comput. Sci.* Vol. 2912, pp. 515–539, Springer-Verlag, 2004.

[Ca06] W. Cao, Some problems in structural graph theory, Ph.D. thesis, U. Illinois (2006).

[ChHuJaLe04] Y Chang, J. Hutchinson, M. Jacobson, J Lehel, and D. West, The bar visibility number of a graph, *SIAM J. Discrete Math.* 18 (2004), 462–471.

[ChHuKeSh06] G. Chen, J. Hutchinson, K. Keating, and J. Shen, Characterization of $[1, k]$-bar visibility trees, *Electron. J. Comb.* 13 (2006), R90.

[CoDaHuMa96] F. Cobos, J. Dana, F. Hurtado, A. Marquez, and F. Mateos, On a visibility representation of graphs, in *Graph Drawing (GD95)*, F. Brandenburg (ed.), *Lecture Notes Comput. Sci.* Vol. 1027, pp. 152–161, Springer-Verlag, 1996.

[De01] A. Dean, A layout algorithm for bar-visibility graphs on the Möbius band, in *Graph Drawing (GD 2000)*, J. Marks (ed.), *Lecture Notes Comput. Sci.* Vol. 1984, pp. 350–359, Springer-Verlag, 2001.

[DeElHaPa08] A. Dean, J. Ellis-Monaghan, S. Hamilton, and G. Pangborn, Unit rectangle visibility graphs, *Electron. J. Comb.* 15 (2008), R79.

[DeEvGeLa07] A. Dean, W. Evans, E. Gethner, J. Laison, M. Safari, and W. Trotter, Bar k-visibility graphs, *J. Graph Algorithms Appl.* 11 (2007), 45–59.

[DeGeHu05] A. Dean, E. Gethner, and J. Hutchinson, Unit bar-visibility layouts of triangulated polygons, in *Graph Drawing (GD 2004)*, J. Pach (ed.), *Lecture Notes Comput. Sci.* Vol. 3383, pp. 111–121, Springer-Verlag, 2005.

[DeHu97] A. Dean and J. Hutchinson, Rectangle-visibility representations of bipartite graphs, *Discrete Appl. Math.* 75 (1997), 9–25.

[DeHu10] A. Dean and J. Hutchinson, Representing 3-trees as unit rectangle-visibility Graphs, *Congr. Numer.* 203 (2010), 139–160.

[DePhRa12] A. Dean, M. Philley, and N. Rangelov, k-transparent and m-bar unit bar visibility graphs, *Skidmore College tech. report* 2012.

[DeVe03] A. Dean and N. Veytsel, Unit bar-visibility graphs, *Congr. Numer.* 160 (2003), 161–175.

[deFRo81] H. de Fraysseix and P. Rosenstiehl, Problem at the Seminaire du Lundi, Paris, 1981.

[deFHuPaPo90] H. de Fraysseix, J. Pach, and R. Pollack, How to draw a planar graph on a grid, *Combinatorica* 10 (1990), 41–51.

[DeOR11] S. Devadoss and J. O'Rourke, *Discrete and Computational Geometry*, Princeton Univ. Press, Princeton, NJ, 2011.

[DeHaMo03] M. Develin, S.G. Hartke, and D.P. Moulton, A general notion of visibility graphs, *Discrete Comput. Geom.* 29 (2003), 511–524.

[DiEaTaTo99] G. Di Battista, P. Eades, R. Tamassia, and I. Tollis, *Graph Drawing*, Prentice Hall, Upper Saddle River, NJ, 1999.

[DiTa88] G. Di Battista and R. Tamassia, Algorithms for plane representations of acyclic digraphs, *Theoret. Comput. Sci.* 61 (1988), 175–198.

[DiEpHi98] M. Dillencourt, D. Eppstein, and D. Hirschberg, Geometric thickness of complete graphs, in *Graph Drawing (GD98)*, S. Whitesides (ed.), *Lecture Notes Comput. Sci.* Vol. 1547, pp. 102–110, Springer-Verlag, 1998.

[DuHaLaMe83] P. Duchet, Y. Hamidoune, M. Las Vergnas, and H. Meyniel, Representing a planar graph by vertical lines joining different levels, *Discrete Math.* 46 (1983), 319–321.

[FeHoWh96] S. Fekete, M. Houle, and S. Whitesides, New results on a visibility representation of graphs in 3D, in *Graph Drawing (GD95)*, F. Brandenburg (ed.), *Lecture Notes Comput. Sci.* Vol. 1027, pp. 234–241, Springer-Verlag, 1996.

[FeHoWh97] S. Fekete, M. Houle, and S. Whitesides, The wobbly logic engine: proving hardness of non-rigid geometric graph representation problems, in *Graph Drawing (GD97)*, G. Di Battista (ed.), *Lecture Notes Comput. Sci.* Vol. 1353, pp. 272–283, Springer-Verlag, 1997.

[FeMe99] S. Fekete and H. Meijer, Rectangle and box visibility graphs in 3D, *Int. J. Comput. Geom. Appl.* 9 (1999), 1–27.

[FeMa08] S. Felsner and M. Massow, Parameters of bar k-visibility graphs, *J. Graph Algorithms Appl.* 12 (2008), 5–27.

[FoKaKa97] U. Fößmeier, G. Kant, and M. Kaufmann, 2-visibility drawings of planar graphs, in *Graph Drawing (GD96)*, S. North (ed.), *Lecture Notes Comput. Sci.* Vol. 1190, pp. 155–168, Springer-Verlag, 1997.

[GaJoSo76] M. Garey, D. Johnson, and H. So, An application of graph coloring to printed circuit testing, *IEEE Trans. Circuits Syst.* 23 (1976), 591–599.

[GoOR04] J.E. Goodman and J. O'Rourke, eds., *Handbook of Discrete and Computational Geometry (2nd ed.)*, Chapman & Hall/CRC Press LLC, Boca Raton, FL, 2004.

[HaVaWe07] S. Hartke, J. Vandenbussche, and P. Wenger, Further results on bar k-visibility graphs, *SIAM J. Discrete Math.* 21 (2007), 523–531.

[Hu93] J. Hutchinson, Coloring ordinary maps, maps of empires, and maps of the Moon, *Math. Mag.* 66 (1993), 211–226.

[Hu05] J. Hutchinson, A note on rectilinear and polar visibility graphs, *Discrete Appl. Math.* 148 (2005), 263–272.

[HuShVi99] J. Hutchinson, T. Shermer, and A. Vince, On representations of some thickness-two graphs, *Comp. Geom.* 13 (1999), 161–171.

[Ka96] G. Kant, Drawing planar graphs using the canonical ordering, *Algorithmica* 16 (1996), 4–32.

[KiWi96] D. Kirkpatrick and S. Wismath, Determining bar-representability for ordered weighted graphs, *Comp. Geom.* 6 (1996), 99–122.

[Kr94] J. Kratochvil, A special planar satisfiability problem and a consequence of its NP-completeness, *Discrete Appl. Math* 52 (1994), 233–252.

[LuMaWo87] F. Luccio, S. Mazzone, and C. Wong, A note on visibility graphs, *Discrete Math.* 64 (1987), 209–219.

[Ma83] A. Mansfield, Determining the thickness of graphs is NP-hard, *Math. Proc. Camb. Philos. Soc.* 93 (1983), 9–23.

[Me81] L.A. Melnikov, Problem at the Sixth Hung. Coll. on Combinatorics, Eger, 1981.

[MoRo98] B. Mohar and P. Rosenstiehl, Tessellation and visibility representations of maps on the torus, *Discrete Comput. Geom.* 19 (1998), 249–263.

[Nu99] J. Nummenmaa, Constructing compact rectilinear planar layouts using canonical representation of planar graphs, *Theoret. Comput. Sci.* 99 (1992), 213–230.

[OR93] J. O'Rourke, Computational Geometry Column 18, *Int. J. Comput. Geom. Appl.* 3 (1993), 107-113; also *ACM SIGACT News* 24 (1993), 20–25.

[Rob69] F. Roberts, On the boxicity and cubicity of a graph, in *Recent Progress in Combinatorics*, W. Tutte (ed.), Academic Press, NY (1969), 301–310.

[Rom97] K. Romanik, Directed rectangle-visibility graphs have unbounded dimension, *Discrete Appl. Math.* 73 (1997), 35–39.

[RoTa86] P. Rosenstiehl and R. Tarjan, Rectilinear planar layouts and bipolar orientations of planar graphs, *Discrete Comput. Geom.* 1 (1986), 343–353.

[Sc84] E. Scheinerman, Intersection classes and multiple intersection parameters, Ph.D. thesis, Princeton Univ. (1984).

[Sh96a] T. Shermer, On rectangle visibility graphs. III. External visibility and complexity, in *Proc. 8th Canad. Conf. Comput. Geom.* (1996), 234–239.

[Sh96b] T. Shermer, personal communication, 1996.

[Sto08] J. Stola, Colorability in orthogonal graph drawing, in *Graph Drawing (GD 2007)*, S.-H. Hong, T. Nishizeki, and W. Quan (eds.), *Lecture Notes Comput. Sci.* Vol. 4875, pp. 327–338, Springer-Verlag, 2008.

[Sto09] J. Stola, Unimaximal sequences of pairs of rectangle visibility drawing, in *Graph Drawing (GD 2008)*, I. Tollis and M. Patrignani (eds.), *Lecture Notes Comput. Sci.* Vol. 5417, pp. 61–66, Springer-Verlag, 2009.

[Sto10] J. Stola, 3D visibility representations by regular polygons, in *Graph Drawing (GD 2008)*, D. Eppstein and E. Gansner (eds.), *Lecture Notes Comput. Sci.* Vol. 5849, pp. 323–333, Springer-Verlag, 2010.

[Str05] I. Streinu, Pseudo-triangulations, rigidity and motion planning, *Discrete Comput. Geom.* 34 (2005), 587–635; also *Discrete Comput. Geom.* 35 (2006), 358.

[StrWh03] I. Streinu and S. Whitesides, Rectangle visibility graphs: characterization, construction, and compaction, in *Proc. 20th STACS 2003*, H. Alt and M. Habib (eds.), *Lecture Notes Comput. Sci.* Vol. 2607, pp. 26–37, Springer-Verlag, 2003.

[TaTo86] R. Tamassia and I. Tollis, A unified approach to visibility representations of planar graphs, *Discrete Comput. Geom.* 1 (1986), 321–341.

[TaTo89] R. Tamassia and I. Tollis, Tessellation representations of planar graphs, in *Proc. 27th Allerton Conf. Commun. Control Comput.* (1989), 48–57.

[TaTo91] R. Tamassia and I. Tollis, Representations of graphs on a cylinder, *SIAM J. Discrete Math.* 4 (1991), 139–149.

[Th84] C. Thomassen, Plane representations of graphs, in *Progress in Graph Theory*, J.A. Bondy and U.S.R. Murty (eds.), Academic Press, NY (1984), 43–69.

[Th86] C. Thomassen, Interval representations of planar graphs, *J. Comb. Theory B* 40 (1986), 9–20.

[Th06] C. Thomassen, Rectangular and visibility representations of infinite planar graphs, *J. Graph Th.* 52 (2006), 257–265.

[Ul84] J. Ullman, *Computational Aspects of VLSI*, Computer Science Press, Rockville, MD, 1984.

[Wh84] A. White, *Graphs, Groups and Surfaces*, revised ed., North-Holland, Amsterdam, 1984.

[Wi85] S. Wismath, Characterizing bar line-of-sight graphs, *Proc. 1st Ann. Symp. Comput. Geom.*, ACM (1985), 147–152.

[Wi89] S. Wismath, Bar-representable visibility graphs and a related network flow problem, Ph.D. thesis, Depart. Comput. Sci., Univ. B. C., tech. report 89–24 (1989).

Glossary for Chapter 10

$3D$ **visibility representation of a graph** G: a set of disjoint, (x, y)-axis aligned
   rectangles lying perpendicular to the z-axis in $\mathbb{R}^3$, each corresponding to a vertex of
   G, with two vertices adjacent if and only if there is unobstructed, vertical visibility
   through a non-degenerate cylinder between the corresponding rectangles.

adjacency operator: the linear mapping A, from the set of all complex-valued functions on the vertex set of a graph to itself, so that the value of Af at a vertex v
   equals the sum of the values of f over all neighbors of v.

ancestor – of a vertex y in a rooted tree: a vertex x that lies on the unique path from
   the root to y.

___, **proper** – of a vertex y in a rooted tree: an ancestor other than vertex y itself.

angular resolution ρ – in a polyline drawing: the smallest angle formed by two edges,
   or segments of edges, incident on the same vertex or bend.

α-**approximation algorithm** – for minimization problems: a polynomial-time algorithm that is guaranteed to find a solution of size at most α times the minimum.

area – of a drawing: the area of the convex hull of the drawing.

articulation point: see *cutpoint*.

aspect ratio – of a drawing: the ratio of the longest to the shortest side of the smallest
   rectangle with horizontal and vertical sides covering the drawing.

back edge$_1$ – for a dfs-tree in a directed graph: an edge directed from descendant to
   ancestor.

back edge$_2$ – for a dfs-tree in an undirected graph: any non-tree edge.

backtracks: goes back and forth along an edge, as in a walk.

bar k-visibility graph: a graph whose vertices can be represented by horizontal bars in
   the plane, with two vertices adjacent if and only if there is a non-degenerate vertical
   band joining the corresponding bars and passing through at most k intermediate
   bars.

bar-visibility graph (or BVG): a graph whose vertices can be represented by horizontal bars in the plane, with two vertices adjacent if and only if there is an unobstructed, non-degenerate vertical band joining the corresponding bars. The graph is
   also known as an ϵ-*visibility graph*.

bar visibility number $b(G)$: the minimum t for which G has a t-bar representation.

bend – in a polyline drawing: a point where two segments belonging to the same edge
   meet.

bfs: abbreviation for *breadth-first search*.

biconnected component – of an undirected graph: a maximal set of edges that has
   no cutpoint.

biconnected graph: an undirected graph with no cutpoint.

breadth-first search (bfs): a graph search method that finds shortest paths.

breadth-first tree: an ordered tree in which the children of a vertex x are the vertices discovered from x in a breadth-first search.

bridge: an edge whose removal reduces the strength of connectedness between some pair of vertices.

bridge$_1$ – in an undirected graph: a cutedge, i.e., an edge whose removal disconnects the graph.

bridge$_2$ – in a mixed graph: an edge whose removal disconnects the underlying undirected graph.

bridge$_3$ – in a flow graph with root r: an edge (v, w) in a flow graph that belongs to every rw-path.

bridgeless graph: an undirected graph with no bridges.

bridge component (BC) – of a graph: a connected component of the graph that results when all the bridges are deleted.

bridge tree: the tree formed by contracting every bridge component of a connected graph.

bridgeless spanning subgraph, smallest: a bridgeless spanning subgraph of a connected bridgeless undirected graph that has the minimum possible number of edges.

boundary: the set of edges connecting a given set of vertices to its complement.

box-visibility graph: a graph whose vertices can be represented by axis-aligned, 3-dimensional boxes in $\mathbb{R}^3$, with two vertices adjacent if and only if there is an unobstructed, non-degenerate, axis-parallel cylinder joining the corresponding boxes.

cardinality – of a cluster: the number of its vertices.

Cartesian product, join and union: operations on fuzzy graphs.

caterpillar: a tree containing a path with every vertex at distance at most one from the path.

Cayley graph: a graph with vertex set G for a given group G and symmetric subset Γ of G, where x and y are adjacent iff $y^{-1}x \in \Gamma$.

certificate – for a graph property P and a graph G: a graph G' such that G has property P if and only if G' has the property.

chordal or triangulated graph: if each cycle with $n \geq 4$ vertices has a chord.

chromatic index – of a graph: the minimum number of colors needed to color all edges such that edges with a common endpoint receive different colors.

chromatic number – of a graph: the minimum number of colors needed to color all vertices such that adjacent vertices receive different colors.

clique: a maximal set of vertices that are pairwise adjacent; sometimes maximality is not required.

cliquewidth-k graph – defined recursively (with $[k]$ denoting the set $\{1, 2, \ldots, k\}$):
- Any graph G with $V(G) = \{v\}$ and $l(v) \in [k]$ is a cliquewidth-k graph.
- If G_1 and G_2 are cliquewidth-k graphs and $i, j \in [k]$, then

  (1) the disjoint union $G_1 \cup G_2$ is a cliquewidth-k graph.

  (2) the graph $(G_1)_{i \times j}$ is a cliquewidth-k graph, where $(G_1)_{i \times j}$ is formed from G_1 by adding all edges (v_1, v_2) such that $l(v_1) = i$ and $l(v_2) = j$.

  (3) the graph $(G_1)_{i \to j}$ is a cliquewidth-k graph, where $(G_1)_{i \to j}$ is formed from G_1 by switching all vertices with label i to label j.

cluster – in dynamic graph algorithms: a connected subgraph, subject to various additional problem-specific conditions.

clustering technique – used in the design of dynamic algorithms: a technique based on partitioning the graph into a suitable collection of connected subgraphs called clusters, such that each update involves only a small number of such clusters.

cograph – defined recursively:
- A graph with a single vertex is a cograph.
- If G_1 and G_2 are cographs, then the the disjoint union $G_1 \cup G_2$ is a cograph.
- If G_1 and G_2 are cographs, then the cross-product $G_1 \times G_2$ is a cograph, which is formed by taking the union of G_1 and G_2 and adding all edges (v_1, v_2) where v_1 is in G_1 and v_2 is in G_2.

collinear visibility layout: a bar-visibility (resp., rectangle-visibility) graph layout that contains two bars (resp., rectangles) having endpoints with the same x-coordinate (resp., having collinear sides). If a visibility layout contains no such pair of bars (resp., rectangles), it is *noncollinear.*

complete: a fuzzy graph (μ, ρ) is complete if $\rho(x, y) = \mu(x) \wedge \mu(y)$.

composition: if ρ is a fuzzy relation of S into T and σ is a fuzzy relation of T into W, the composition $\rho \circ \sigma$ is defined by $\rho \circ \sigma(x, w) = \vee\{\rho(x, y) \wedge \sigma(y, w) \mid y \in T\}$. If $S = T, \rho^2 = \rho \circ \rho, \rho^n = \rho^{n-1} \circ \rho$, and $\rho^\infty = \vee\{\rho^k(x, y) \mid k = 1, 2, \ldots\}$.

connected fuzzy graph: if there is a path between every pair of vertices.

convex drawing: a planar straight-line drawing such that the boundary of each face is a convex polygon.

cross edge – for a *dfs-tree* in a directed graph: a nontree edge joining two unrelated vertices.

crossing: a point where two edges intersect.

cutedge: an edge whose removal disconnects a graph.

cutpoint: a vertex whose removal disconnects a graph.

cutvertex: a vertex whose removal reduces the degree of connectedness between some other pair of vertices.

cycle set, fuzzy cycle set, the cocycle set, and the fuzzy cocycle set: sets in a fuzzy graph that are not necessarily vector spaces over the integers modulo 2.

dag: acronym for "directed acyclic graph," i.e., a directed graph with no (directed) cycle.

Delaunay drawable graph: a planar triangulated graph that admits a drawing that is a Delaunay triangulation.

Delaunay triangulation: a planar straight-line drawing with all internal faces triangles, and such that three vertices form a face if and only if their convex hull does not contain any other vertex of the triangulation.

depth-first search (dfs): a graph search method that iteratively scans an edge incident to the most recently discovered vertex that still has unscanned edges.

depth-first spanning forest₁ – in an undirected graph: a collection of depth-first trees, one for each connected component of the graph.

depth-first spanning forest₂ – in a directed graph: a collection of depth-first trees containing every vertex once, with all cross edges joining two different trees directed from right to left.

depth-first tree (dfs-tree) – in a graph: an ordered tree in which the children of a vertex x are the vertices discovered from x in a depth-first search.

descendant – of a vertex y in a rooted tree: a vertex x such that y lies on the unique path from the root to x.

___, proper – of a vertex y in a rooted tree: a descendant other than vertex y itself.

dfs: abbreviation for *depth-first search.*

diameter – of a graph: the maximum distance between two vertices.

directed bar-visibility graph: a BVG representable such that for each arc (u, v) of the digraph, the visibility of u to v is in the positive y direction.

discovery – of a vertex: when a search reaches that vertex for the first time.

discovery order – induced by a graph search algorithm: a numbering of the vertices from 1 to $|V|$, in the order the vertices are discovered in the search, i.e., the preorder of the search tree.

distance – from vertex u to vertex v: length of a shortest uv-path.

dominance drawing: an upward drawing of an acyclic digraph, such that there exists a directed path from vertex u to vertex v if and only if $x(u) \leq x(v)$ and $y(u) \leq y(v)$, where $x(\cdot)$ and $y(\cdot)$ denote the coordinates of a vertex.

dominating set – of a graph: a set of vertices such that every vertex is either in this set or has a neighbor in this set.

domination of a vertex in a flow graph: vertex v dominates vertex $w \neq v$ if every rw-path contains v.

dominator tree – for a flow graph: a tree that represents all the dominance relations.

dynamic graph problem: a problem concerned with efficiently answering queries regarding whether a dynamic graph has the specified property.

___, **connectivity**: answering a query whether the graph is connected, or whether two vertices are in the same connected component.

___, **decremental**: a partially dynamic problem in which only deletions are allowed.

___, **fully**: a problem in which the update operations include unrestricted insertions and deletions of edges or vertices.

___, **incremental**: a partially dynamic problem in which only insertions are allowed.

___, **minimum spanning tree**: the problem of maintaining a minimum spanning forest of a graph during insertions of edges, deletions of edges, and edge-cost changes.

___, **partially**: a problem in which only one type of update, either insertions or deletions, is allowed.

dynamic graph: a graph that is undergoing a sequence of updates; formally, a graph-valued variable.

dynamic programming: evaluation of a recursive formula in such a way as to avoid repeated computations.

k-ECSS: a k-edge-connected spanning subgraph of a k-edge-connected graph.

___, **smallest**: a k-ECSS that has the minimum possible number of edges.

endpoint visibility graph: a graph whose vertices can be represented by the vertices of a simple polygon in the plane and edges by straight-line visibility between endpoints.

ET tree – for a tree T: a dynamic balanced binary tree whose leaves are the sequence of vertex occurrences in an Euler tour of T.

Euler tour – of a tree T: in dynamic graph algorithms, a closed walk over the digraph obtained by replacing each edge of T by two directed edges with opposite direction, such that the walk traverses each edge exactly once.

expander family: a sequence of finite regular graphs, each with the same degree, so that the number of vertices goes to infinity but the isoperimetric constant is uniformly bounded away from zero.

external degree – of a cluster: the number of tree edges.

face – of a drawing: a region of the plane defined by a planar drawing.

finish-time – at a vertex: when a search leaves that vertex for the last time.

finish-time order – induced by a graph search: a numbering of the vertices from 1 to $|V|$, in the order they are finished in a search, i.e., the postorder of the search tree.

flexibility – of a network: an application of fuzzy graphs is to model information networks. A measure of flexibility of a network is $(m - n)/n(n - 2)$, where m is the numbers of edges and n is the number of nodes.

flow graph: a directed graph with a distinguished root vertex r that can reach every vertex.

forest: a fuzzy graph if the graph consisting of its nonzero edges is a forest.

forward edge – for a dfs-tree in a directed graph: a nontree edge from ancestor to descendant.

fuzzy clique: a fuzzy subgraph $\mathcal{K}$ of a fuzzy graph G such that the level subgraphs $\mathcal{K}^t$ induce a clique of G^t.

fuzzy cluster – of order k: a subset C of the set of vertices in a fuzzy graph (μ, ρ) such that $\wedge\{\rho^k(x, y) \mid x, y \in C\} > \vee\{\wedge\{\rho^k(w, z) \mid w \in C\} \mid z \notin C\}$. k-clusters are just ordinary cliques in graphs by thresholding the k-th

fuzzy forest: a fuzzy graph (μ, ρ) if it has a partial fuzzy spanning subgraph $F = (\mu, \tau)$ which is a forest, where for all edges (x, y) such that $\tau(x, y) = 0$, $\rho(x, y) < \tau^\infty(x, y)$, i.e., if there is a path in F between x and y whose strength is greater than $\rho(x, y)$.

fuzzy graph: let V be a set, μ a fuzzy subset of V, and ρ a fuzzy relation on V. The pair (μ, ρ) is a fuzzy graph if $\rho(x, y) \leq \mu(x) \wedge \mu(y)$.

fuzzy intersection graph: Let $G = (V, X)$ be a graph, where $V = \{v_1, \ldots, v_n\}$. Let $S_i = \{v_i, x_{i1}, \ldots, x_{iq_i}\}$, where $x_{ij} \in X$ and x_{ij} has v_i as a vertex, $j = 1, \ldots, q_i; i = 1, \ldots, n$. Let $S = \{S_1, \ldots, S_n\}$. Let $T = \{(S_i, S_j) \mid S_i S_j \in S, S_i \cap S_j \neq \emptyset, i \neq j\}$. Then $\mathcal{I}(S) = (S, T)$ is an intersection graph and $G \simeq \mathcal{I}(S)$. Any partial fuzzy subgraph (ι, γ) of $\mathcal{I}(S)$ with $\text{Supp}(\gamma) = T$ is called a *fuzzy intersection graph*.

fuzzy interval graph: the fuzzy intersection graph of a finite family of fuzzy intervals.

fuzzy line graph: a partial fuzzy subgraph (λ, ω) of a fuzzy graph (μ, ρ) such that for all $S_x \in Z, \lambda(S_x) = \rho(x)$ and for all $(S_x, S_y) \in W, \omega(S_x, S_y) = \rho(x) \wedge \rho(y)$, where $Z = \{\{x\} \cup \{u_x, v_x\} \mid x \in X, u_x, v_x \in V, x = (u_x, v_x)\}$ and $W = \{(S_x, S_y) \mid S_x \cap S_y \neq \emptyset, x, y \in X, x \neq y\}$ and where $S_x = \{\{x\} \cup \{u_x, v_x\}, x \in X$. $L(G) = (Z, W)$ is the intersection graph $\mathcal{I}(X)$.

fuzzy relation: a fuzzy subset of $S \times T$, where S and T are sets.

fuzzy set operations: basic set operations for fuzzy subsets of a set.

fuzzy tree: a connected fuzzy forest.

grid drawing: a polyline drawing such that vertices, crossings, and bends have integer coordinates.

hamiltonian cycle – in a graph: a simple cycle that includes every vertex.

hamiltonian path – in a graph: a simple path that includes every vertex.

historical shortest path – in a dynamic graph: a path that has been a shortest path at some point during a sequence of updates of the graph, and such that none of its edges has been updated since then.

hv-drawing: an upward orthogonal straight-line drawing of a binary tree such that the drawings of the subtrees of each node are separated by a horizontal or vertical line.

Ihara zeta function: the product over all prime walks P modulo cyclic permutation of $\left(1 - u^{\text{length of } P}\right)^{-1}$, where u is an indeterminate.

imbedded (di)graph drawing problem: a planar (di)graph with a prespecified topological imbedding (i.e., set of faces) that must be preserved in the drawing.

immediate dominator – of a vertex w in a flow graph: the unique dominating vertex v such that every other dominator of w dominates v.

independent set – in a graph: a set of vertices that are pairwise non-adjacent.

interlacing segments – of a cycle C: two segments S, T such that either $|V(S) \cap V(T) \cap V(C)| \geq 3$, or there are four distinct vertices u, v, w, x, which occur in that cyclic order along C, such that $u, w \in S$ and $v, x \in T$.

internal vertex – of path: a vertex that is not an endpoint of the path.

in-tree: a tree representing paths of a graph that lead to a given vertex.

irreducible Markov chain: a Markov chain whose transition graph remains strongly connected after all null-probable edges are deleted.

isoperimetric constant: the minimum ratio, over all subsets containing no more than half the vertices of a finite graph, of the number of edges in the boundary to the number of vertices in the set.

Kazhdan constant: the best possible lower bound, given a finite group G and a subset Γ of G, for the distance a unit vector in a nontrivial irreducible unitary representation space of G is moved by some element of Γ.

layer of vertices – in a rooted graph or digraph: a vertex subset comprising all the vertices at a given distance from the root.

layered drawing: a drawing of a rooted graph such that vertices at the same distance from the root lie on the same horizontal line.

left of a vertex y – in an ordered tree: a vertex x such that some common ancestor of x and y has children c and d, with c preceding d as siblings.

length function – on a graph: an assignment of numerical lengths to the edges, usually nonnegative numbers.

level set for a fuzzy subset μ of a set S and $t \in [0, 1]$: the set $\mu^t = \{x \in S \mid \mu(x) \geq t\}$, which is also called the *t-level set* of μ.

locally historical path π if every proper subpath of π is a historical path.

locally shortest path π if every proper subpath of π is a shortest path.

matching, perfect matching – in a graph: a spanning subgraph such that every vertex has degree exactly 1.

matching – in a graph: a set of edges that share no common endpoints.

mixed graph: a graph in which directed and undirected edges both occur.

monadic second-order logic (MSOL): a type of logic in which each variable may represent an individual element (vertex or edge) or a set of elements (vertex set or edge set); see Definition D10 of §10.4.

___, **MSOL′** – for a graph $G = (V, E)$: the subset of MSOL expressions restricted to variables v_i with domain V, e_i with domain E, and V_i with domain 2^V.

multipartite number: the maximum k for which every complete k-partite graph has a $3D$ visibility representation.

near-triangulation: a graph with a plane embedding having all finite, interior faces 3-sided.

nonseparable or a block: a fuzzy graph that has no cutvertices. A block may have bridges, but this is not the case in ordinary graphs.

open ear decomposition – of an undirected graph: a partition of the edges into a simple cycle P_0 and simple paths $P_1, \ldots, P_k$ such that for each $i > 0$, P_i is joined to P_0 and previous paths only at its (2 distinct) ends.

ordered tree: a rooted tree in which the children of each vertex are linearly ordered.

orientation – of a graph: assignment of a unique direction to each undirected edge.

orthogonal drawing: a drawing in which each edge is a chain of horizontal and vertical segments.

orthogonal representation: a representation of orthogonal drawing in terms of bends along each edge and angles around each vertex.

out-tree: a tree representing paths of a graph that originate from a given vertex.

partial fuzzy subgraph: a fuzzy graph (ν, τ) is a partial fuzzy subgraph of a fuzzy graph (μ, ρ) if $\nu \subseteq \mu$ and $\tau \subseteq \rho$.

path: a sequence of distinct vertices $x_0, x_1, \ldots, x_n$ (except possibly x_0 and x_n) with the edges (x_{i-1}, x_i) having positive weight.

uv**-path**: a path starting at vertex u and ending at vertex v.

planar drawing: a drawing in which no two edges cross.

polyline drawing: a drawing in which each edge is a polygonal chain.

postorder – of an ordered tree: the finish-time order of a depth first search of the tree itself.

potentially uniform path – of a graph G: a path such that every proper subpath is a historical shortest path.

predicate calculus: a type of logic in which predicates have arguments and expressions are built using various operators ($\neg$, $\wedge$, $\vee$), and in which the quantifiers ($\forall$, $\exists$).

preorder – of an ordered tree: the discovery order of a depth first search of the tree itself.

prime closed walk: a closed walk that is backtrackless, tailless, and not expressable as a repeated concatenation of a closed walk with itself.

proximity drawing: a drawing of a graph based on a geometric proximity relation.

Ramanujan graph: a d-regular graph such that every non-trivial eigenvalue λ satisfies $|\lambda| \leq 2\sqrt{d-1}$.

recursively constructed graph class: defined by a set (usually finite) of primitive or *base graphs*, in addition to one or more operations that compose larger graphs from smaller subgraphs; each operation involves either fusing specific vertices from each subgraph or adding new edges between specific vertices from each subgraph.

rectangle-visibility graph (or RVG): a graph whose vertices can be represented by axis-parallel rectangles in the plane, with two vertices adjacent if and only if there is an unobstructed, non-degenerate, horizontal or vertical band joining the corresponding rectangles.

reducible flow graph: a flow graph that can be transformed into its root vertex r by a sequence of reduction operations; that is, if e is the only edge entering a vertex $w \neq r$, then contract edge e (and its other endpoint) into the vertex v.

related vertices – in a rooted tree: two vertices such that one is an ancestor of the other.

restricted multi-level partition: a partition that consists of a collection of restricted partitions satisfying the following:

(1) The clusters at level 0 (known as *basic clusters*) contain one vertex each.

(2) The clusters at level $\ell \geq 1$ form a restricted partition with respect to the tree obtained after shrinking all the clusters at level $\ell - 1$.

(3) There is exactly one vertex cluster at the topmost level.

restricted partition – of a tree T: a partition of its vertex set V into clusters of external degree ≤ 3 and cardinality ≤ 2 such that:

(1) Each cluster of external degree 3 has cardinality 1.

(2) Each cluster of external degree < 3 has cardinality at most 2.

(3) No two adjacent clusters can be combined and still satisfy the above.

scanning an edge: the work done by a graph searching algorithm when it traverses the edge.

searching a graph: a methodical (linear-time) exploration of all the vertices and edges of a graph.

segment – of a cycle C in a biconnected graph G: either (i) a chord of C (i.e., an edge not in C that joins two vertices of C); or (ii) a connected component of the graph $G - V(C)$, plus all the edges of G joining that component to C.

segment visibility graph: a graph whose vertices can be represented by disjoint line segments in the plane and edges by straight-line visibility between corresponding segments.

semidominator: a useful intermediate concept in computing dominators, defined in terms of a depth-first search tree.

separation pair: two vertices in a biconnected graph whose removal disconnects the graph.

series-parallel graph – a recursively defined graph: see Definition D5 of §10.4.

set-merging problem: the problem of maintaining a partition of a given universe subject to a sequence of *union* and *find* operations.

shortest-path tree – in a rooted graph: a tree in which the path from the root r to any vertex v is a shortest rv-path.

sink – in a directed graph: a vertex with outdegree 0.

source – in a directed graph: a vertex with indegree 0.

spanning fuzzy subgraph: a partial fuzzy subgraph (ν, τ) of a fuzzy graph (μ, ρ) spans (μ, ρ) if $\mu = \nu$.

sparse graph: a graph with at most $O(|V|)$ edges.

sparsification technique: a technique for speeding up dynamic graph algorithms, which when applicable, transforms a time bound of $T(|V|, |E|)$ into $O(T(|V|, |V|))$.

spectral gap: the distance between the degree and second-largest eigenvalue of a finite regular graph.

start vertex: the distinguished vertex of a flow graph.

straight-line drawing: a drawing in which each edge is a straight-line segment.

strength – of a path: the weight of the weakest edge in the path.

strength – of connectedness between vertices x, y: $\rho^{\infty}(x, y)$.

strong component (SC) – of a directed graph: a maximal subgraph in which any two vertices are reachable from each other.

strong component graph (SC graph) – for a directed graph: the result of contracting every strong component to a vertex; also called the *condensation graph*.

strong visibility graph: a graph whose vertices can be represented by horizontal bars in the plane, with two vertices adjacent if and only if there is an unobstructed, vertical band, possibly with width zero, joining the corresponding bars.

strongly connected digraph: a digraph in which every vertex can reach every other vertex by a directed path.

support: for μ a fuzzy subset of a set S, the support of μ is the set $\text{Supp}(\mu) = \{x \in S \mid \mu(x) > 0\}$.

***t*-bar representation** – of a graph (resp., digraph): a bar-visibility representation in which each vertex corresponds to at most t bars, such that two vertices are adjacent if and only if there is an unobstructed, non-degenerate vertical band joining two bars corresponding to the two vertices (resp., such that there is an arc (u, v) if and only if there is such a band with the visibility of u to v in the positive y direction).

tail: an edge traversed in both the first and last step of a closed walk.

top tree – in dynamic graph algorithms: a tree that describes a hierarchical partition of the edges of another tree, well suited to maintaining path information.

topological numbering (topological order, topological sort) – of an acyclic directed graph: assignment of an integer to each vertex so that each edge is directed from a lower number to a higher number.

topology tree: in dynamic graph algorithms, a hierarchical representation of a tree T into clusters.

tournament: a directed simple graph such that each pair of vertices is joined by exactly one edge.

traversable mixed graph: a mixed graph in which every vertex can reach every other by a path with all its directed edges pointing in the forward direction.

tree: a connected graph with no cycles and sometimes with a designated *root* used to describe recursive constructions.

  ___, **drawable as a minimum spanning tree**: a tree T such that there exists a set P of points (especially in $\mathbb{R}2$ or $\mathbb{R}3$) such that the minimum spanning tree of P (using Euclidean metric distances) is isomorphic to T.

  ___, **rooted, recursively defined**: the graph whose only vertex is the root r (and no edges); or the result of joining the roots of two disjoint trees with a new edge.

tree edge: edge of a spanning tree in a graph.

treewidth – of a graph G: the minimum width taken over all tree-decompositions of G; measures how closely the graph resembles a tree.

treewidth-k graph: a graph whose treewidth is no greater than k.

triconnected graph: an undirected connected graph that remains connected whenever any two or fewer vertices are deleted.

uniform path – of a graph G: a path such that every proper subpath is a shortest path.

unit bar-visibility graph: a BVG representable with all bars having equal length.

unit rectangle-visibility graph: an RVG represented by unit squares.

universal class – of objects: a class for which every graph has a $3D$ visibility representation with the objects as vertices.

update – on a graph: an operation that inserts or deletes edges or vertices of the graph or changes attributes associated with edges or vertices, such as cost or color.

upward drawing: a drawing of a digraph such that each edge is monotonically nondecreasing in the vertical direction.

upward planar digraph: a digraph that admits an upward planar drawing.

vertex cover – of a graph: a set of vertices such that every edge is incident on at least one vertex in the set.

vertex visibility graph: a graph whose vertices can be represented by the vertices of a simple polygon in the plane and edges by straight-line visibility within the polygon between polygon vertices.

visibility drawing: a drawing of a graph based on a geometric visibility relation; e.g., the vertices might be drawn as horizontal segments, and the edges associated with vertically visible segments.

Voronoi diagram: the dual graph of a Delaunay triangulation.

weak visibility graph: a graph whose vertices can be represented by horizontal bars in the plane, so that if two vertices are adjacent then there is an unobstructed vertical band, possibly with width zero, joining the corresponding bars.

well-behaved time bound $T(n)$ if $T(n/2) < cT(n)$ for some $c < 1$.

zig-zag product: a certain graph product formed by two regular graphs X and Y where the degree of X equals the order of Y and the edges of X are labeled at each vertex by vertices of Y.

Chapter 11

Networks and Flows

Section 11.1
Maximum Flows

Clifford Stein, Columbia University

INTRODUCTION

The *Maximum Flow Problem* is one of the basic problems in combinatorial optimization. It models a large variety of problems in a diverse set of application areas including data flowing through a communications network, power flowing through an electrical network, liquids flowing through pipes and parts flowing through a factory. It has also served as a prototypical problem in algorithm design, and many useful and powerful ideas were first introduced in the context of maximum flows. This section will describe maximum flows and some generalizations. The book by Ahuja, Magnanti and Orlin [AhMaOr93] is an excellent reference and includes a large number of applications. Other texts and surveys with significant coverage of maximum flows include [CoCuPuSc98], [Ev79], [La76], [PaSt82], [Ta83], and [GoTaTa90].

11.1.1   The Basic Maximum Flow Problem

Informally, in a maximum flow problem, we wish to send as much stuff as possible from one place in a network to another, while limiting the amount of stuff sent through an arc by the capacity of that arc.

DEFINITIONS

D1: An *s-t (flow) network* $G = (V, E, s, t, cap)$ is a directed graph with vertex-set V and arc-set E, two distinguished vertices, a *source* s and a *sink* t, and a nonnegative *capacity function* $cap : E \to \mathbb{N}$. We adopt the convention, without loss of generality, that if arc $(v, w) \in E$ then the reverse arc $(w, v) \notin E$. (See Subsection 11.1.4.)

D2: A *(feasible) flow* is a function $f : E \to \mathbb{R}$ which obeys three types of constraints:

capacity constraints: $f(v, w) \le cap(v, w)$, for each arc $(v, w) \in E$.

conservation constraints:

$\sum_{(w,v) \in E} f(w, v) = \sum_{(v,w) \in E} f(v, w)$ for each vertex $v \in V - \{s, t\}$.

nonnegativity constraints: $f(v, w) \ge 0$, for each arc $(v, w) \in E$.

D3: The *value of a flow* f, denoted $val(f)$ (or $|f|$), is the total flow into the sink, i.e.,

$$val(f) = \sum_{(v,t) \in E} f(v, t)$$

D4: In the *maximum flow problem*, we are given a flow network $G = (V, E, s, t, cap)$ and wish to find a flow f of maximum value.

EXAMPLE

E1: An example of a flow network appears in Figure 11.1.1. A maximum flow for the network appears in Figure 11.1.2. The numbers on the arcs are capacities; the flows appear in parentheses. It is straightforward to verify that the three properties of a flow are satisfied.

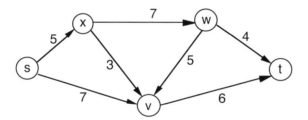

Figure 11.1.1: An *s-t* network.

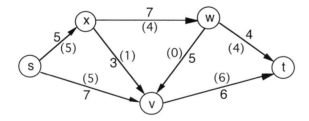

Figure 11.1.2: A maximum flow of value 10.

FACTS

F1: The flow conservation properties imply that the value of a flow is also equal to total flow out of the source, i.e.,

$$val(f) = \sum_{(v,t) \in E} f(v, t) = \sum_{(s,v) \in E} f(s, v)$$

F2: Although a flow is defined as a real-valued function, an integer-valued maximum flow always exists (since the capacity function is integer-valued).

11.1.2   Minimum Cuts and Duality

An important and *dual concept* related to maximum flows is that of *minimum cuts*.

Cuts in a Network

An *s-t cut* combines the concepts of *partition-cut* (§6.4) and an *(s|t)-disconnecting set of edges* (§4.1).

DEFINITIONS

D5: Let $G = (V, E, s, t, cap)$ be an *s-t* network, and let S and T form a partition of V such that source $s \in S$ and sink $t \in T$. Then the set of all arcs that have one endpoint in set S and the other endpoint in set T is called an *s-t* **cut** of network G and is denoted $\langle S, T \rangle$. An arc (v, w) is a **forward arc** of the cut if $v \in S$ and $w \in T$, and (v, w) is a **backward arc** if $v \in T$ and $w \in S$.

D6: The **capacity of an** *s-t* **cut** $\langle S, T \rangle$, denoted $cap\langle S, T \rangle$, is the sum of the capacities of the forward arcs of the cut in the forward direction, i.e.,

$$cap\langle S, T \rangle = \sum_{(v,w) \in E : v \in S, w \in T} cap(v, w)$$

D7: A **minimum** *s-t* **cut** in a flow network G is a cut of minimum value, that is, $\min\{cap\langle S, T \rangle : \langle S, T \rangle \text{ is an } s\text{-}t \text{ cut}\}$.

EXAMPLES

E2: The *s-t* cut $\langle \{s, x, v\}, \{w, t\} \rangle = \{(x, w), (v, t)\}$ appears in Figure 11.1.3. The capacity of the cut $cap\langle S, T \rangle = cap(x, w) + cap(v, t) = 13$. Notice that (w, v) is a backward arc and hence its capacity is not included in the capacity of the cut.

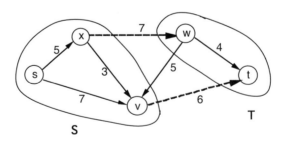

Figure 11.1.3: The *s-t* cut $\langle \{s, x, v\}, \{w, t\} \rangle = \{(x, w), (v, t)\}$ has capacity 13.

E3: A minimum cut of capacity 10 appears in Figure 11.1.4.

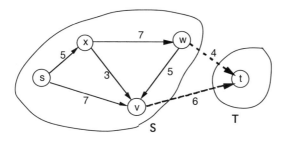

Figure 11.1.4: A minimum cut of capacity 10.

Weak Duality

DEFINITION

D8: Given a flow f and an *s-t* cut $\langle S, T \rangle$, the ***flow across cut*** $\langle S, T \rangle$, denoted $f\langle S, T \rangle$, is the sum of the flows on the forward arcs minus the sum of the flows on the backward arcs, i.e.,

$$f\langle S, T \rangle = \sum_{(v,w)\in E:v\in S, w\in T} f(v, w) - \sum_{(v,w)\in E:v\in T, w\in S} f(v, w).$$

FACTS

F3: Flow conservation implies that for a given flow f and *any* cut $\langle S, T \rangle$, $val(f) = f\langle S, T \rangle$.

F4: (*Weak Duality*) Let f be any flow in an *s-t* network G, and let $\langle S, T \rangle$ be any *s-t* cut. Then

$$val(f) \leq cap\langle S, T \rangle$$

F5: Let f be a flow in an *s-t* network G and $\langle S, T \rangle$ an *s-t* cut, and suppose that $val(f) = cap\langle S, T \rangle$. Then flow f is a maximum flow in network G, and $\langle S, T \rangle$ a minimum *s-t* cut. (This is an immediate consequence of weak duality [Fact F4].)

EXAMPLE

E4: The flow f (in parentheses) and *s-t* cut $\langle S, T \rangle$ shown in Figure 11.1.5 illustrate Fact F3. In particular, $val(f) = 6$, and the flow across the cut, $f\langle S, T \rangle$, equals $f(x, w) + f(v, t) - f(w, v) = 2 + 5 - 1 = 6$.

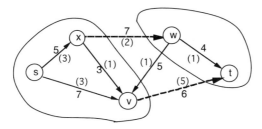

Figure 11.1.5: $val(f) = f\langle S, T \rangle = 6$.

11.1.3   Max-Flow Min-Cut Theorem

The relationship between the maximum-flow problem and its dual, the minimum-cut problem, is an example of *strong max-min duality* that occurs between certain optimization problems and their dual problems. Two other instances are König's theorem (§11.3), which states that the size of a minimum vertex cover in a bipartite graph equals the size of a maximum matching, and Menger's theorem (§4.1), which relates the local connectivity between two vertices of any graph and the number of internally disjoint paths between them.

The Residual Network and Flow-Augmenting Paths

Algorithms that iteratively increase the flow in a network via *flow-augmenting paths* often use an associated digraph, called a *residual network*, for finding these paths more easily.

DEFINITIONS

D9: Let f be the current flow in a network $G = (V, E, s, t, cap)$. An arc $(v, w) \in E$ is *increasable* if $f(v, w) < cap(v, w)$ and is *reducible* if $f(v, w) > 0$.

NOTATION: Let I denote the set of all increasable arcs, and let R be the set of all reducible arcs. (Of course, in general, $I \cap R \neq \emptyset$.)

D10: Given a flow f in a network $G = (V, E, s, t, cap)$, the *residual network* $G_f = (V, E_f, s, t, r_f)$ has vertex-set V, and the arc-set E_f is constructed from network G as follows: for each arc $(v, w) \in E$, if arc $(v, w) \in I$, then create an arc (v, w) in G_f, and label it with a *residual capacity* $r_f(v, w) = cap(v, w) - f(v, w)$; if arc $(v, w) \in R$, then create an arc (w, v) in G_f, and label it $r_f(w, v) = f(v, w)$.

D11: Given a flow f in a network $G = (V, E, s, t, cap)$, a *flow-augmenting path* P for network G is a directed s-t path in the residual network G_f. The *capacity of flow-augmenting path* P, denoted Δ_P, is given by $\Delta_P = \min_{(v,w) \in E_f} r_f(v, w)$.

REMARK

R1: It follows from the definitions that the capacity Δ_P is always positive.

FACTS

F6: [Flow Augmentation] Let f be a flow in a network $G = (V, E, s, t, cap)$, and let P be a flow-augmenting path with capacity Δ_P. Let f_P be defined as follows:

$$f_P(v, w) = \begin{cases} f(v, w) + \Delta_P & \text{if } (v, w) \in E(P) \cap I \\ f(v, w) - \Delta_P & \text{if } (w, v) \in E(P) \text{ and } (v, w) \in R \\ f(v, w) & \text{otherwise} \end{cases}$$

Then f_P is a feasible flow in network G and $val(f_P) = val(f) + \Delta_P$.

F7: [Characterization of Maximum Flow] Let f be flow in a network G. Then f is a maximum flow if and only if there does not exist an f-augmenting path in G.

F8: Max-Flow Min-Cut [FuDa55, ElFeSh56, FoFu56] For a given network, the value of a maximum flow is equal to the capacity of a minimum cut.

EXAMPLES

E5: The current flow f in the network G shown at the top in Figure 11.1.6 has value 9. The corresponding residual network G_f is shown at the bottom.

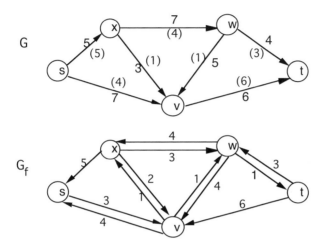

Figure 11.1.6: A network G with flow f and its residual network G_f.

There are exactly two directed paths from s to t in the residual network, each corresponding to a different flow-augmenting path for increasing the flow. One of them is the directed path $s \to v \to w \to t$ in the network G_f. The arc (v, w) in path P corresponds to the reducible arc (w, v) in network G. Notice that $\Delta_P = 1$.

E6: If the flow f given in Figure 11.1.6 above is augmented by the flow-augmenting path P in Example E5, then the resulting flow is as shown in Figure 11.1.7. Observe that the one reducible arc, (w, v), of the flow-augmenting path P reduces the flow on that arc by 1.

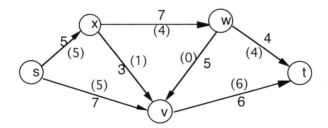

Figure 11.1.7: Increasing the flow in Example E5 by 1.

A simple application of Fact F5 shows that f_P is a maximum flow. Also straightforward to show is the non-existence of a flow-augmenting path corresponding to f_P.

11.1.4   Algorithms for Maximum Flow

There are three popular approaches to compute a maximum flow in polynomial time. The first uses flow-augmenting paths, the second uses a *push-relabel* method, and the third uses *linear programming*. We discuss the first two methods briefly.

Ford–Fulkerson Algorithm

The first published maximum-flow algorithm, due to Ford and Fulkerson [FoFu62], is essentially a *greedy* method: we iteratively push flow along flow-augmenting paths from source to sink.

Algorithm 11.1.1: Ford–Fulkerson Algorithm

Input: a flow network $G = (V, E, s, t, cap)$
Output: a maximum flow f

    Initialize $f(v, w) = 0$ for all $(v, w) \in E$.
    Calculate residual network G_f.
    While an augmenting path in G_f exists
        Let P be an augmenting path in G_f with capacity Δ_P.
        Obtain increased flow f_P using flow-augmenting path P (Fact F6).
        $f := f_P$
        Update residual network G_f

COMPUTATIONAL NOTE: The efficiency of the Ford–Fulkerson algorithm depends on which augmenting path is chosen, and on the data structures used to facilitate the computation. If an arbitrary augmenting path is chosen, the algorithm may not run in polynomial time [Za72]. However, many natural choices of a path lead to a polynomial time algorithm. The first such algorithm, due to Edmonds and Karp [EdKa72], and Dinic [Di70], always chooses the shortest augmenting path, where the length of a path is defined to be the number of arcs on the path. This algorithm runs in $O(|E||V|)$ iterations of the main loop. Each iteration requires a breadth-first search and some updates of flow variables, which can be done in $O(|E|)$ time, and hence the algorithm runs in $O(|E|^2|V|)$ time. Many further improvements are possible with data structures such as dynamic trees [SlTa83] or by augmenting flow on several shortest paths simultaneously [Ka74]. (These are known as blocking flows.) The current fastest running time using this approach is due to Goldberg and Rao [GoRa98].

Preflow-Push Algorithms

An alternative approach to computing a maximum flow, called a *preflow-push* algorithm, was introduced by Goldberg [Go87] and Goldberg and Tarjan [GoTa88]. It uses a *push-relabel* strategy, which pushes flow over individual arcs, rather than paths. To do so, it allows flow to "accumulate" at some vertices, creating an *excess* at those vertices. The *push* operation selects one of these *active* vertices and tries to remove its excess by pushing flow to neighbors that are "closer" to the sink. The *relabel* operation maintains distance labels that help keep track of these neighbors.

DEFINITIONS

D12: A **preflow** is a relaxed version of a flow, a function $f : E \to R^+$ which obeys three types of constraints:

capacity constraints: $f(v, w) \le cap(v, w)$, for each arc $(v, w) \in E$.

relaxed conservation constraints:

$\sum_{(w,v) \in E} f(w, v) - \sum_{(v,w) \in E} f(v, w) \ge 0$ for each vertex $v \in V - \{s, t\}$.

nonnegativity constraints: $f(v, w) \ge 0$, for each arc $(v, w) \in E$.

D13: Let f be a preflow in an s-t network $G = (V, E, s, t, cap)$. The **excess** at vertex v, denoted $e(v)$, is given by $e(v) = \sum_{(w,v) \in E} f(w, v) - \sum_{(v,w) \in E} f(v, w)$. A vertex v with $e(v) > 0$ is called an **active** vertex.

TERMINOLOGY: Given a preflow f, the sets of *increasable* arcs, I, and *reducible* arcs, R, are defined as in the case when f is a feasible flow (Definition D9). Similarly, the residual network $G_f = (V, E_f, s, t, r_f)$ is defined in the same way as before (Definition D10).

D14: Let f be a preflow in an s-t network $G = (V, E, s, t, cap)$. A **distance function** $h : V \to N$ satisfies:

$$h(s) = |V|, \quad h(t) = 0, \quad \text{and} \quad h(v) \le h(w) + 1$$

for each arc $(v, w) \in E_f$ in the residual graph G_f.

D15: For a given preflow f and distance function h, an arc (v, w) in the residual network G_f is **admissible** if $h(v) = h(w) + 1$.

Algorithm 11.1.2: Preflow-Push Algorithm

Input: a flow network $G = (V, E, s, t, cap)$
Output: a maximum flow f

{Initialization}
$f(v, w) := 0$ for all $(v, w) \in E$.
$h(v) := 0$ for all $v \in V$
$h(s) := |V|$
$f(s, w) := cap(s, w)$ for all arcs $(s, w) \in E$.
Compute residual network G_f and excesses e.
While there are active vertices
    Select an active vertex v.
    If G_f contains an admissible arc (v, w)
        {Push flow on arc (v, w).}
        $\Delta := \min\{e(v), r_f(v, w)\}$
        $e(v) := e(v) - \Delta$
        $e(w) := e(w) + \Delta$
        If arc $(v, w) \in I$
            $f(v, w) := f(v, w) + \Delta$
        Else
            $f(v, w) := f(v, w) - \Delta$
    Else {Relabel v.}
        $h(v) := \min\{h(w) : (v, w) \in E_f \text{ and } r_f(v, w) > 0\}$
    Update residual network G_f.

COMPUTATIONAL NOTE: The preflow-push algorithm terminates in $O(|V|^2|E|)$ time using simple data structures. More careful selection of operations and careful use of data structures leads to algorithms with running times of $O(|V|^3)$ or $O(|V||E|\log(|V|^2/|E|))$ [GoTa88]. The fastest preflow-push algorithm is slightly faster than these and is due to King, Rao and Tarjan [KiRaTa94].

COMPUTATIONAL NOTE: In practice a good implementation of a preflow-push algorithm seems to be faster than a good implementation of an augmenting path algorithm. Two heuristics are essential ingredients in implementing a push relabel algorithm well. First, a backwards breadth-first search of the residual graph is performed periodically, in order to update distance labels. Second, the *gap heuristic*, which quickly identifies vertices that must be on the sink side of the minimum *s-t* cut, is employed [ChGo97].

11.1.5   Variants and Extensions of Maximum Flow

We briefly mention some variations and extensions of the basic maximum-flow problem. For more extensive coverage, see, e.g., [AhMaOr93, EvMi92].

FACTS

F9: The convention that if arc $(v, w) \in E$ then the reverse arc $(w, v) \notin E$ is without loss of generality. Any maximum flow algorithm can easily be extended to handle this case. Alternatively, arc (v, w) can be converted to two arcs (v, x) and (x, w) each with capacity $cap(v, w)$, and arc (w, v) can be converted to two arcs (w, y) and (y, v) each with capacity $cap(w, v)$.

F10: [multiple-source multiple-sink] Suppose that we have a flow network G with multiple sources $\{s_1, \ldots, s_k\}$ and multiple sinks, $\{t_1, \ldots, t_\ell\}$. We can still find a maximum flow in this network by the following transformation. We create a new supersource s and supersink t, and add an arc (s, s_i) with $cap(s, s_i) = \infty$ for each source s_i, and an arc (t_i, t) with $cap(t_i, t) = \infty$ for each sink t_i. A maximum flow in this network is easily interpreted as a maximum flow for the multiple-source multiple-sink problem.

F11: Maximum flow can be used to find a maximum matching in a bipartite graph. (Matchings are discussed in §11.3.)

F12: If each arc has a cost, then we obtain the *minimum-cost flow problem* (see §11.2).

F13: [flow on undirected edges] The network G can contain undirected as well as directed edges. An undirected edge (v, w) with capacity $cap(v, w)$ is understood to be an edge that can carry up to $cap(v, w)$ units of flow in either direction. In any flow, it is only necessary that flow be carried in one direction or the other, as flows in opposing directions cancel each other out. For example, an undirected edge with flow $f(v, w) = 3$ and $f(w, v) = 2$ is equivalent to an edge with $f(v, w) = 1$ and no flow in the (w, v) direction. We can therefore convert an undirected edge to two oppositely directed arcs, each with capacity $cap(v, w)$.

F14: [lower bounds on flow] The network G can also contain lower bounds on the flow over an arc. It is still possible to find a maximum flow in such a graph, providing that one exists (see, e.g., [AhMaOr93, EvMi92]).

Multicommodity Flow

Perhaps the most important extension of maximum flow is the extension to the case of multiple commodities.

DEFINITIONS

D16: A **commodity** i is a triple (s_i, t_i, d_i) where s is a source, t is a sink, and d is a **demand**, or amount of flow to be routed.

D17: A **multicommodity flow network** $G = (V, E, K, cap)$ is a directed graph with vertex set V and arc-set E, commodity set K, and a nonnegative **capacity** function $cap : E \rightarrow N$. We adopt the convention that if arc $(v, w) \in E$ then the reverse arc $(w, v) \notin E$. The commodities are indexed by the integers $1, 2, \ldots, k$.

D18: A **multicommodity flow** in a multicommodity flow network $G = (V, E, K, cap)$ is a set of $k = |K|$ functions $f_i : E \rightarrow \mathbb{R}^+$ satisfying the following conditions:

joint capacity constraints: $\sum_{i=1}^{k} f_i(v, w) \leq cap(v, w)$, for each arc $(v, w) \in E$.

conservation constraints:
$\sum_{(w,v) \in E} f_i(w, v) = \sum_{(v,w) \in E} f_i(v, w)$ for each $v \in V - \{s_i, t_i\}$ and $i = 1, \ldots, k$, and
$\sum_{(w,t_i) \in E} f_i(w, t_i) - \sum_{(t_i,w) \in E} f_i(t_i, w) = d_i$ for $i = 1, \ldots, k$.

nonnegativity constraints: $f_i(v, w) \geq 0$, for each arc $(v, w) \in E$ and $i = 1, \ldots, k$.

Variants of Multicommodity Flow Problems

DEFINITIONS

D19: In the **feasible multicommodity flow problem**, we are given a multicommodity flow network G, and wish to know if a multicommodity flow exists. We call such a flow a **feasible multicommodity flow**.

D20: In the **concurrent flow problem**, we are given a multicommodity flow network G, and we wish to compute the maximum value z for which there is a feasible multicommodity flow in the network with all demands multiplied by z.

D21: In the **maximum multicommodity flow problem**, we are given a multicommodity flow network, except that for each commodity, we are *not* given a demand. We wish to find, for each commodity i, a flow f_i of value (demand) $val(f_i)$ such that $\sum_{i=1}^{k} val(f_i)$ is maximized.

COMPUTATIONAL NOTE: If we do not require that the flows be integral (even though capacities and demands are integral), then all the above multicommodity flow problems can be solved in polynomial time via linear programming. More efficient combinatorial algorithms that compute approximately optimal solutions also exist [LeMaPlStTaTr95, Yo95, and GaKo98].

COMPUTATIONAL NOTE: If we require that all flows be integral, then all the above problems are NP-hard. The degree to which we can approximate them varies from problem to problem. (See [Va01] for a survey.) Note that integral multicommodity flow generalizes the *disjoint-paths problem*.

D22: In an ***unsplittable flow problem***, we have the additional restriction that each commodity must be routed on one path. All variants of this problem are NP-hard, but constant factor approximation algorithms exist for single-source multiple-sink variants [Kl96, KoSt97, DiGaGo98].

References

[AhMaOr93] R. K. Ahuja, T. L. Magnanti, and J. B. Orlin, *Network Flows: Theory, Algorithms, and Applications*, Prentice Hall, 1993.

[CoCuPuSc98] W. J. Cook, W. H. Cunningham, W. R. Pulleyblank, and A. Schrijver, *Combinatorial Optimization*, John Wiley and Sons, New York, 1998.

[ChGo97] B. V. Cherkassky and A. V. Goldberg, On implementing the push-relabel method for the maximum flow problem, *Algorithmica* 19(4) (1997), 390–410.

[Di70] E. A. Dinic, Algorithm for solution of a problem of maximum flow in networks with power estimation, *Soviet Math. Dokl.* 11 (1970), 1277–1280.

[DiGaGo98] Y. Dinitz, N. Garg, and M. Goemans, On the single source unsplittable flow problem, In *Proceedings of the 39th Annual Symposium on Foundations of Computer Science* (1998), 290–299.

[EdKa72] J. Edmonds and R. M. Karp, Theoretical improvements in the algorithmic efficiency for network flow problems, *Journal of the ACM* 19 (1972), 248–264.

[ElFeSh56] P. Elias, A. Feinstein, and C. E. Shannon, Note on maximum flow through a network, *IRE Transactions on Information Theory IT-2* 19 (1956), 117–119.

[Ev79] S. Even, *Graph Algorithms*, Computer Science Press, 1979.

[EvMi92] J. R. Evans and E. Minieka, *Optimization Algorithms for Networks and Graphs*, Dekker, 1992.

[FoFu56] L. R. Ford, Jr. and D. R. Fulkerson, Maximal flow through a network, *Canadian J. of Math.* 8 (1956), 399–404.

[FoFu62] L. R. Ford, Jr. and D. R. Fulkerson, *Flows in Networks*, Princeton University Press, 1962.

[FuDa55] D. R. Fulkerson and G. B. Dantzig, Computation of maximum flow in networks, *Naval Research Logistics Quarterly* 2 (1955), 277–283.

[GaKo98] N. Garg and J. Konemann, Faster and simpler algorithms for multicommodity flow and other fractional packing problems, In *Proceedings of the 39th Annual Symposium on Foundations of Computer Science* (1998), 300–309.

[Go87] A. V. Goldberg, *Efficient graph algorithms for sequential and parallel computers*, PhD thesis, MIT, Cambridge, MA, January 1987.

[GoRa98] A. V. Goldberg and S. Rao, Beyond the flow decomposition barrier. *Journal of the ACM* 45 (1998), 783–797.

[GoTa88] A. V. Goldberg and Robert E. Tarjan, A new approach to the maximum flow problem, *Journal of the ACM* 35 (1988), 921–940.

[GoTaTa90] A. V. Goldberg, É. Tardos, and R. E. Tarjan, Network flow algorithms, In B. Korte, L. Lovász, H. J. Prömel, and A. Schrijver, editors, *Paths, Flows, and VLSI-Layout*, pp. 101–164, Springer-Verlag, 1990.

[Ka74] A. V. Karzanov, Determining the maximal flow in a network by the method of preflows, *Soviet Math. Dokl.* 15 (1974), 434–437.

[KiRaTa94] V. King, S. Rao, and R. E. Tarjan, A faster deterministic maximum flow algorithm, *Journal of Algorithms* 17 (1994), 447–474.

[Kl96] J. M. Kleinberg, Single-source unsplittable flow, In *Proceedings of the 37th Annual Symposium on Foundations of Computer Science* (1996), 68–77.

[KoSt97] S. G. Kolliopoulos and C. Stein, Improved approximation algorithms for unsplittable flow problems, In *Proceedings of the 38th Annual Symposium on Foundations of Computer Science* (1997), 426–436.

[La76] E. L. Lawler, *Combinatorial Optimization: Networks and Matroids*, Holt, Rinehart and Winston, 1976.

[LeMaPlStTaTr95] T. Leighton, F. Makedon, S. Plotkin, C. Stein, É. Tardos, and S. Tragoudas, Fast approximation algorithms for multicommodity flow problems, *Journal of Computer and System Sciences* 50 (1995), 228–243.

[PaSt82] C. H. Papadimitriou and K. Steiglitz, *Combinatorial Optimization: Algorithms and Complexity*, Prentice-Hall, 1982.

[SlTa83] D. Sleator and R. E. Tarjan, A data structure for dynamic trees, *Journal of Computer and System Sciences* 26 (1983), 362–391.

[Ta83] R. E. Tarjan, *Data Structures and Network Algorithms*, SIAM, 1983.

[Va01] V. Vazirani, *Approximation Algorithms*, Springer-Verlag, 2001.

[Yo95] N. Young, Randomized rounding without solving the linear program, In *Proceedings of the 6th ACM-SIAM Symposium on Discrete Algorithms* (1995), 170–178.

[Za72] N. Zadeh, Theoretical efficiency of the Edmonds-Karp algorithm for computing maximal flows, *Journal of the ACM* 19 (1972), 184–192.

Section 11.2
Minimum Cost Flows

Lisa Fleischer, Dartmouth College

INTRODUCTION

Minimum cost flows are a powerful and useful network flow model that is distinguished by supply nodes, demand nodes, and linear flow costs on the edges of a directed network. They are used to model complex problems occurring in transportation, transshipment, manufacturing, telecommunications, graph drawing, human resources, statistics, numerical algebra, physics, and many other engineering disciplines. Minimum cost flows generalize many other network problems. They lie on the tractable side of a boundary between computable and intractable problems: while there exist efficient algorithms to find integer minimum cost flows, most generalizations of integer minimum cost flow problems are NP-hard.

The book *Network Flows* by Ahuja, Magnanti, and Orlin provides thorough coverage of minimum cost flows, applications, and related topics [AhMaOr93]. Other recent texts and surveys with significant coverage of minimum cost flows include [CoCuPuSc98, GoTaTa90, IwMcSh00, Sc03].

11.2.1   The Basic Model and Definitions

DEFINITIONS

D1: A *(standard) flow network* $G = (V, A, cap, c, b)$ is a directed graph with vertex-set V, arc-set A, a nonnegative capacity function $cap : A \to N$, a linear cost function $c : A \to Z$, and an integral supply vector $b : V \to Z$ that satisfies $\sum_{w \in V} b(w) = 0$.

D2: An *s-t **flow network*** (or ***single-source single-sink network***) is a flow network $G = (V, A, cap, c, b)$ that contains two distinguished vertices s and t such that $b(v) = 0$ for all $v \in V - \{s, t\}$ and $b(s) = -b(t) > 0$.

D3: An ***extended flow network*** $G' = (V', A', cap')$ of $G = (V, A, cap, c, b)$ is an *s-t* network with vertex-set $V' = V \cup \{s, t\}$, arc-set $A' = A \cup \{(s, v) | b(v) > 0\} \cup \{(w, t) | b(w) < 0\}$ and capacity function cap' defined by

$$cap'(v, w) = \begin{cases} cap(v, w), & \text{if } (v, w) \in A \\ b(v), & \text{if } v = s \\ -b(v), & \text{if } w = t \end{cases}$$

D4: A ***transshipment network*** is a flow network in which all arcs have infinite capacity.

D5: A ***(standard) flow*** (also called a ***(standard) feasible flow***) is a function $f : A \to Z$ that satisfies

capacity constraints: $f(v, w) \le cap(v, w)$ for all $(v, w) \in A$,

nonnegativity constraints: $f(v, w) \ge 0$ for all $(v, w) \in A$,

flow conservation constraints: $\sum_w [f(v, w) - f(w, v)] = b(v)$ for each $v \in V$.

D6: A flow f in an *s-t* flow network $G = (V, A, cap, c, b)$ is called an *s-t **flow*** and is said to have ***volume*** $b(s)$.

D7: A ***minimum cost flow*** is a flow f with minimum $c^T f$ value among all flows.

D8: A ***circulation*** is a flow for the supply vector $b \equiv \mathbf{0}$. A ***minimum cost circulation*** is a circulation f with minimum $c^T f$ value among all circulations.

NOTATION: Sometimes we use the subscript notation f_{uv}, c_{uv}, cap_{uv} or b_v instead of $f(u, v)$, $c(u, v)$, $cap(u, v)$, or $b(v)$, respectively.

REMARK

R1: In the definitions above, we can extend the definition of the capacity function cap to *all* ordered pairs in $V \times V$ by defining $cap(v, w) = 0$ for $(v, w) \notin A$. This has the effect of extending any flow function f to all such ordered pairs as well. This extended view of the capacity and flow functions is notationally convenient for interpreting expressions like the flow conservation constraints given above and for various other expressions appearing later in this section.

ASSUMPTIONS

A1: G has no parallel arcs and no oppositely directed pairs of arcs.

A2: No arc in G has negative cost.

EXAMPLE

E1: An example of a flow network appears on the left in Figure 11.2.1. The supply at each node is indicated in brackets. In parentheses at each arc is the arc cost followed by the arc capacity. A feasible flow in this network is indicated by the italic numerals in the figure on the right.

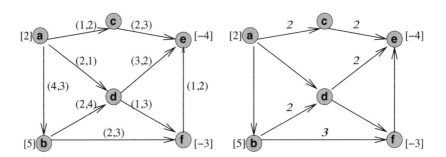

Figure 11.2.1: An example of a flow network and a feasible flow.

NOTATION: The following notation is used throughout this section:

(a) $n = |V|$ and $m = |A|$ (number of vertices and arcs, respectively, in the network).

(b) $C = \max\limits_{e \in A} |c(e)|$ (largest arc cost).

(c) $U = \max\limits_{e \in A} \{cap(e) | cap(e) < \infty\}$ (largest finite capacity).

(d) $B = \max\limits_{v \in V} |b(v)|$ (largest magnitude of a supply [or demand]).

(e) For an arc subset $F \subseteq A$, $c(F) \doteq \sum\limits_{e \in F} c(e)$.

(f) $M = \sum\limits_{e} c(e) cap(e)$.

(g) $\mathcal{S}(n, m, C)$ denotes the time complexity of computing single source shortest paths from a fixed vertex in a directed network with n vertices, m arcs, and arc lengths (costs) bounded by C.

FACTS

F1: The quantity M is an upper bound on the cost of a flow.

F2: Assumption 1 is without loss of generality and is made for notational convenience only. Alternatively, we can remove a parallel or opposite arc (v, w) by introducing a new node z and replacing (v, w) with (v, z) and (z, w), where (v, z) has the same cost and capacity as (v, w) and (z, w) has 0 cost and infinite capacity.

F3: Assumption 2 is without loss of generality. If arc $(v, w) \in A$ has negative cost, it can be removed by **saturating** the arc: modify functions cap, c and b by setting $cap(w, v) = cap(v, w)$, $c(w, v) = -c(v, w)$, $b(v) = b(v) - cap(v, w)$, $b(w) = b(w) + cap(v, w)$. If the resulting problem yields a flow with a value of $f'(w, v)$ on arc (w, v), then the value of $f(v, w) = cap(v, w) - f'(w, v)$.

F4: A minimum cost flow may also be defined in an undirected graph. By replacing each undirected arc by two oppositely directed arcs of the same cost and capacity, the undirected problem may be solved in the directed graph. If f' is the flow in the directed graph, the flow f in the undirected graph is obtained as follows: for all ordered pairs (v, w),

$$f(v, w) = \max\{0, f'(v, w) - f'(w, v)\}.$$

F5: A (feasible) flow in flow network G can be found by finding a maximum s-t flow in the extended flow network G'. (See the multiple-source multiple-sink extension in Subsection 11.1.5.)

F6: The minimum cost flow problem in G is equivalent to the minimum cost circulation problem in the network obtained by adding an infinite capacity arc (t, s) to G' with cost $-mC$.

F7: The following graph optimization problems are all special cases of minimum cost flows. The directed versions are described, but the undirected versions can be treated similarly.

Maximum flows: for $b_s = \beta = -b_t$ and $b_v = 0$ for $v \in V \setminus \{s, t\}$, find a flow that maximizes β. Using

$$b \equiv \mathbf{0}, \quad c \equiv \mathbf{0}, \quad A \leftarrow A \cup \{(t, s)\}, \quad c_{ts} = -1, \quad \text{and} \quad cap_{ts} = \infty,$$

a minimum cost circulation in $G = (V, A, cap, c, b)$ is a maximum flow in the original network. (See §11.1.)

Single-source shortest paths: for a given vertex s, find the shortest path using arc lengths l from s to every other node. Using

$$b_s = |V| - 1, \quad b_v = -1, \quad cap \equiv \infty, \quad \text{and} \quad c \equiv l,$$

a minimum cost flow in this graph describes a shortest path tree by taking all arcs with positive flow. (See §10.1.)

k arc-disjoint s-t paths: given vertices s and t, arc cost function c, and an integer k, find k paths from s to t such that no two share an arc. Using

$$b_s = k = -b_t, \quad b_v = 0, \; v \in V \setminus \{s, t\}, \quad cap \equiv 1, \quad \text{and} \quad c \text{ as given},$$

the minimum cost flow yields a solution to this problem by taking all arcs with positive flow. (See §4.1, Menger's theorems.)

Maximum-weight bipartite matching: given a bipartite graph $G = (V_1 \cup V_2, A)$ with arc weights w, find a maximum weight subset of arcs such that no two share an end point. Using

$$b_v = 1, \; \forall v \in V_1, \quad b_v = -1, \; \forall v \in V_2, \quad c = -w, \quad \text{and} \quad cap \equiv 1,$$

a minimum cost flow yields a maximum weight matching by taking all arcs with positive flow. (See §11.3.)

Minimum cost transshipment: a minimum cost flow in a transshipment network.

Residual Networks

DEFINITIONS

D9: Given a flow f in a network G, let

$$F = \{e \in A | f_e < cap_e\} \text{ and } B = \{e \in A | f_e > 0\}.$$

Note that F and B may intersect. The **residual network** of G (with respect to) f, denoted G_f, is the network that contains for each arc $e = (v, w) \in F$ a **forward arc** (v, w) with **residual capacity** $cap_e - f_e$ and cost c_e, and for each arc $e = (v, w) \in B$ a **backward arc** (w, v) with residual capacity f_e and cost $-c_e$. Denote the arc-set of G_f by A_f.

D10: A set $S \subseteq A_f$ of arcs is **augmenting** if each arc in S has positive residual capacity. An **augmenting path** is a path whose set of arcs is augmenting.

D11: A flow f in a network G is augmented by u on arc set $S \subseteq A_f$ as follows: for each forward arc $(v, w) \in S$, set $f(v, w) = f(v, w) + u$, and for each backward arc $(v, w) \in S$, set $f(w, v) = f(w, v) - u$.

D12: An arc set $S \subseteq A_f$ is **saturated** if f is augmented on S by a quantity equal to the minimum capacity of an arc in S.

FACT

F8: Let f be a flow in G and f' be a circulation in G_f. Then $f + f'$ defined as

$$(f + f')(v, w) = f(v, w) + f'(v, w) - f'(w, v)$$

is a flow in G.

EXAMPLE

E2: Figure 11.2.2 shows the residual network of the flow appearing in Figure 11.2.1.

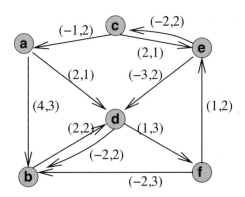

Figure 11.2.2: The residual network of the flow in Figure 11.2.1.

11.2.2 Optimality Conditions

DEFINITION

D13: Given a labeling $\pi : V \to Z$, the **reduced cost** vector c^π is defined as $c^\pi_{vw} := c_{vw} + \pi_v - \pi_w$.

FACTS

F9: If f is a minimum cost flow, then G_f has no negative cost cycles.

F10: If there exists a minimum cost flow, then there exists one such that the set of arcs in $F \cap B$ forms a forest.

F11: If G_f has no negative cost cycles, then any $\pi : V \to Z$ obtained as shortest path distance labels from a selected vertex r by applying Dijkstra's algorithm (see §10.1) satisfies $c^\pi_{vw} \geq 0$ for all $(v, w) \in G_f$.

F12: For any flow f and labeling π, we have

(i) $c^\pi(\mathcal{C}) = c(\mathcal{C})$ for any cycle $\mathcal{C}$ in G.

(ii) $c^\pi(\mathcal{P}) = c(\mathcal{P}) + \pi_s - \pi_t$ for any s to t path $\mathcal{P}$ in G.

(iii) $\sum_{v,w} c^\pi_{vw} f_{vw} = \sum_{v,w} c_{vw} f_{vw} + \sum_v \pi_v b_v$.

Thus f minimizes $\sum_{v,w} c_{vw} f_{vw}$ if and only if f minimizes $\sum_{v,w} c^\pi_{vw} f_{vw}$.

F13: (Reduced-cost optimality conditions) If there is a labeling $\pi : V \to Z$ such that $c^\pi_{vw} \geq 0$ for all $(v, w) \in G_f$, then f is a minimum cost flow.

F14: (Complementary slackness) If there is a labeling $\pi : V \to Z$ such that

$$c^\pi_{vw} < 0 \;\Rightarrow\; f_{vw} = cap_{vw}$$
$$c^\pi_{vw} > 0 \;\Rightarrow\; f_{vw} = 0$$
$$c^\pi_{vw} = 0 \;\Rightarrow\; 0 \leq f_{vw} \leq cap_{vw}$$

then f is a minimum cost flow. In this case, f and π are **complementary**.

REMARKS

R2: There is an economic interpretation of the reduced-cost optimality conditions: if $c(v, w)$ is the cost of shipping one unit from v to w and π_v is the sales price of one unit at v, then the reduced cost for (v, w) is the cost associated with buying an item at v, shipping it to w, and selling it at w. If this is negative, it is worth doing, and thus the arc should be saturated. If it is positive, it is not worth doing, and thus the arc should not carry flow.

TERMINOLOGY: Consistent with the economic interpretation of Fact F13, the vertex labels π are often called **dual prices**.

R3: Fact F14 is a restatement of Fact F13.

EXAMPLE

E3: Figure 11.2.3 highlights a negative cost cycle in the residual network appearing in
Figure 11.2.2. On the left of Figure 11.2.4 is the new flow obtained by saturating this
cycle, and on the right is the new residual graph with node labels in italics. The node
labels and the residual network together show that the flow on the left is a minimum
cost flow in the network in Figure 11.2.1.

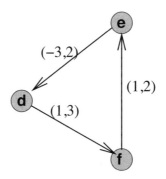

Figure 11.2.3: A negative cost cycle.

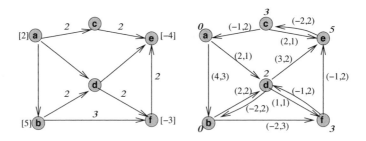

Figure 11.2.4: A minimum cost flow, residual network, and node labels.

A Basic Cycle-Canceling Algorithm

The correctness of the following simple algorithm for finding minimum cost flows rests
on Facts F9, F11, and F13.

Algorithm 11.2.1: Basic Cycle Canceling

Input: a flow network G.
Output: a minimum cost flow f

    Find a feasible flow f and compute the residual network G_f.
    While G_f contains a negative cost cycle
        Find a negative cost cycle $\mathcal{C}$.
        Augment flow f by saturating the arc-set of cycle $\mathcal{C}$.
        Compute G_f.
    Return flow f.

FACTS

F15: If *cap* is integral, then the initial feasible flow is integral, and each augmentation modifies the existing flow by integral amounts. If in addition c is integral, each augmentation decreases the cost of the initial flow by at least -1.

F16: Negative cycles in G_f can be found using the algorithm of Bellman and Ford [Be58, Fo56]. (See also the references suggested in the introduction to this section.)

F17: There is a minimum cost flow in G if and only if G has no negative cost cycles with infinite capacity. In this case, the cycle-canceling algorithm finds an integer minimum cost flow after at most M augmentations.

COMPUTATIONAL NOTE:

(a) Zadeh [Za73] demonstrated by a family of examples that the basic cycle-canceling algorithm may require $O(U)$ augmentations.

(b)) A modified cycle-canceling algorithm that simultaneously augments on a set of node-disjoint cycles that decrease the cost of the flow the most finds a minimum cost flow after at most $O(m \log(nCU))$ augmentations [BaTa89].

(c) A modified algorithm that augments on a *minimum-mean cycle* (a cycle $\mathcal{C}$ that minimizes $c(\mathcal{C})/|\mathcal{C}|$) can be implemented to find a minimum cost flow after at most $O(mn \log(nC))$ augmentations [GoTa89].

REMARK

R4: Further discussion of useful ways to select cycles and comparisons with dual algorithmic approaches that cancel cuts can be found in [IwMcSh00].

11.2.3 The Dual Problem

Frequently, problems arise that can be solved using minimum cost flow methods, but their equivalence to a minimum cost flow problem may not be immediately obvious. Certainly, recognizing such problems is important. The *linear programming dual* of the minimum cost flow problem is one such type of problem.

DEFINITIONS

D14: The **dual** of the minimum cost flow problem (DualMCF) defined on flow network G is

$$\text{Minimize} \sum_{v,w} cap_{vw} z_{vw} + \sum_{v} \pi_v b_v$$

Subject to

$$c_{vw} + \pi_v - \pi_w + z_{vw} \geq 0 \quad \text{for all } (v, w) \in A$$
$$z_{vw} \geq 0 \quad \text{for all } (v, w) \in A$$

D15: The pair of vectors z and π are **dual feasible** if they satisfy the constraints of DualMCF.

FACTS

F18: Given an arbitrary vector $\pi \in \mathbb{R}^n$, the pair (π, z^π) is dual feasible if $z^\pi \in \mathbb{R}^m$ is defined by $z_{vw}^\pi = \max\{0, -c_{vw}^\pi\}$.

F19:
$$\min_{z,\pi \text{ dual feasible}} \left\{ \sum_{v,w} cap_{vw} z_{vw} + \sum_v \pi_v b_v \right\} + \min_{f \text{ a flow}} \left\{ \sum_{v,w} c_{vw} f_{vw} \right\} = 0.$$

F20: The vectors z and π are optimal for DualMCF if and only if there exists a flow $\hat{f}$ with $\sum_{v,w} cap_{vw} z_{vw} + \sum_{v,w} c_{vw}^\pi \hat{f}_{vw} = 0$.

F21: If flow $\hat{f}$ satisfying the conditions in Fact F20 exists, it can be found by fixing $\hat{f}$ on all arcs (v, w) with $c_{vw}^\pi \neq 0$ to $\hat{f}_{vw} = -\frac{cap_{vw} z_{vw}}{c_{vw}^\pi}$, $\hat{b}_v = b_v - \sum_w \hat{f}_{vw} + \sum_w \hat{f}_{wv}$, and setting $\hat{f}$ on all remaining arcs to the value of a feasible flow for $\hat{b}$ in the network $\hat{G} = (V, \hat{A})$ with $\hat{A} = \{(v, w) | c_{vw}^\pi = 0\}$.

11.2.4 Algorithms for Minimum Cost Flow

There are a wide variety of algorithms designed to solve minimum cost flow problems. Some algorithms, like *network simplex*, are based on linear programming. The cycle-canceling algorithm of Subsection 11.2.2 can be interpreted as a primal algorithm: it starts with a feasible flow, and improves this solution until the complementary dual solution (in the sense of Fact F14) is feasible. In this subsection, we discuss two algorithms: a (primal-dual) augmenting-path algorithm that maintains primal and dual solutions that satisfy complementary slackness and works to make both solutions feasible; and a *push-relabel* algorithm based on the *preflow-push* maximum flow algorithm in §11.1. In the process we describe two powerful ideas that help algorithms achieve polynomial run-time: *capacity scaling* and *cost scaling*.

The first algorithm is simpler to describe and discuss if we assume capacities are infinite and G is a *complete* digraph (each pair of nodes has two oppositely directed arcs between them). Thus, we begin by discussing the relation between the transshipment problem and the minimum cost flow problem.

A Transshipment Problem Associated with a Minimum Cost Flow Problem

DEFINITIONS

D16: Let $G = (V, A, cap, c, b)$ be a flow network with nonnegative costs. The ***associated transshipment network*** is the network obtained by replacing each arc $e = (v, w)$ by three arcs (v, x_e), (y_e, x_e), and (y_e, w) having infinite capacity and with costs c_e, 0, and 0, respectively. Also, the supplies at the new nodes x_e and y_e are defined to be $b(x_e) = -cap_e$ and $b(y_e) = cap_e$.

D17: The ***completion*** of a transshipment network $G = (V, A, cap, c, b)$ is the complete transshipment network obtained by adding all missing arcs and giving each of them infinite capacity and arc cost $M + 1$.

FACTS

F22: Let G be a flow network and let $\hat{G}$ be its associated transshipment network. Then G has a feasible flow f if and only if $\hat{G}$ has a feasible flow $\hat{f}$ such that for each arc $e = (v, w)$, $\hat{f}_{vx_e} = f_e$, $\hat{f}_{y_e x_e} = cap_e - f_e$, and $\hat{f}_{y_e w} = f_e$. Moreover, the cost of $\hat{f}$ equals the cost of f.

F23: Let G be a flow network and let G^* be the completion of its associated transshipment network. If the minimum cost transshipment uses an arc with cost $M + 1$, then the original minimum cost flow problem is not feasible.

F24: Let G be a flow network and let G^* be the completion of its associated transshipment network. Any algorithm that finds a minimum cost transshipment in G^* yields a solution to the minimum cost flow problem for G.

A Primal-Dual Algorithm

Edmonds and Karp gave the first polynomial time algorithm to find a minimum cost flow [EdKa72]. Their algorithm introduces the idea of capacity scaling. As mentioned above, the algorithm presented here is a simplified version designed for the transshipment problem on a complete transshipment network.

DEFINITION

D18: Given a flow f and a vertex v, the **excess** at v, denoted excess(v), is the initial supply at v minus the net flow out of v, i.e.,

$$\text{excess}(v) = b(v) - \sum_{w \in V} [f(v, w) - f(w, v)].$$

NOTATION: For a length (or cost) function c, a flow f, and vertex labels π, dist(s, t) denotes the shortest (or least cost) augmenting path from v to w using c^π. By Fact F12, this is the same as the shortest (or least cost) augmenting path using c.

Algorithm 11.2.2: Capacity-Scaling Algorithm

Input: a complete transshipment network $G = (V, A, b, c)$
Output: a minimum cost transshipment f and optimal dual prices π.

Initialize $\pi(v) = 0$ for all $v \in V$; $f(v, w) = 0$ for all $(v, w) \in A$; $\Delta = 2^{\lfloor \log_2 B \rfloor}$.
While $\Delta \geq 1$
    While $\max_{s \in V}\{\text{excess}(s)\} \geq \Delta$ and $\min_{t \in V}\{\text{excess}(t)\} \leq -\Delta$
        Compute dist(s, v) for all $v \in V$.
        Augment flow by Δ on a least cost s-t augmenting path.
        Update dual price $\pi(v) = \pi(v) + \text{dist}(s, v)$ for all $v \in V$.
    $\Delta := \Delta/2$.
Return transshipment f and dual prices π.

TERMINOLOGY: A Δ'-**phase** is the set of algorithmic steps while $\Delta = \Delta'$.

FACTS

F25: Throughout Algorithm 11.2.2, f is nonnegative, and after each augmentation and label update, f and π are complementary.

F26: When Algorithm 11.2.2 terminates, f is feasible. Thus, since f and π are complementary, f is minimum cost.

F27: The number of augmentations in a Δ-phase is at most $n - 1$.

COMPUTATIONAL NOTE: The capacity-scaling algorithm finds a minimum cost transshipment in $O(n(\log B)\mathcal{S}(n, m, C))$ time. Moreover, if G is a standard flow network with n vertices and m edges, then the number of vertices in the completion of its associated transshipment network (described in Definitions D16 and D17) equals $n + m$. Thus, the number of augmentations necessary to find a minimum cost flow in that complete transshipment network (which, by Fact F24, finds one in the original network G) equals $m + n - 1$, which is $O(m)$. Since the complexity of a shortest-path computation is unaffected by vertices of degree 2 (added in the transformation in Definition D16), it follows that the run-time complexity of the capacity-scaling algorithm for solving a minimum cost flow problem is $O(m(\log B)\mathcal{S}(n, m, C))$.

COMPUTATIONAL NOTE: Algorithm 11.2.2 can be easily modified to work directly with a network with finite capacities. To ensure that the final f and π are complementary (so that the final f is optimal), the following condition must be maintained throughout the algorithm: $c^\pi(v, w) \geq 0$ if residual capacity of (v, w) is at least Δ. In the modified algorithm, the search for augmenting paths in a Δ-phase is restricted to arcs with residual capacity at least Δ. To maintain the relaxed complementary slackness, immediately after Δ is decreased, each arc with negative reduced cost and residual capacity at least $\Delta/2$ is saturated.

A Push-Relabel Algorithm

The push-relabel method was initially introduced for maximum flows (see the preflow-push algorithm in §11.1). We describe here a modification for minimum cost flows that runs in polynomial time. It relies on cost scaling and is due to Goldberg and Tarjan [GoTa89].

DEFINITIONS

D19: The pair (f, π) defined on $A \times V$ is $\boldsymbol{\epsilon\text{-}optimal}$ if f is nonnegative and $c^\pi(v, w) \geq -\epsilon$ for all arcs with positive residual capacity.

D20: Arc (v, w) is $\boldsymbol{admissible}$ if it has positive residual capacity and $c^\pi(v, w) < 0$.

D21: Given a current flow f in a network G, the operation $\boldsymbol{push}(v, w)$ assigns a flow of $\min\{cap(v, w), f(v, w) + \text{excess}(v)\}$ to arc(u, v).

D22: The operation push(v, w) is $\boldsymbol{saturating}$ if the resulting flow on arc (v, w) equals $cap(v, w)$; otherwise it is $\boldsymbol{non\text{-}saturating}$.

Algorithm 11.2.3: Push-Relabel Algorithm

Input: a flow network $G = (V, A, u, b, c)$
Output: a minimum cost flow f, and optimal dual prices π.

    Initialize $\pi(v) = 0$ for all $v \in V$; $f(v, w) = 0$ for all $(v, w) \in A$; $\epsilon = C$.
    While $\epsilon \geq \frac{1}{n}$
        Saturate each arc (v, w) satisfying $c^\pi(v, w) < 0$.
        While $\max_{v \in V}\{\text{excess}(v)\} > 0$,
            Select v with $\text{excess}(v) > 0$.
            If there is w with (v, w) admissible   {then push(v, w)}
                $f(v, w) := \min\{cap(v, w),\ f(v, w) + \text{excess}(v)\}$
            Else   {relabel(v)}
                $\pi(v) := \pi(v) - \epsilon$
      $\epsilon := \epsilon/2$
    Return flow f and dual prices π.

TERMINOLOGY: An ϵ'-***phase*** is the set of algorithmic steps while $\epsilon = \epsilon'$.

FACTS

F28: If f is nonnegative and $\pi(v) = 0$ for all $v \in V$, then (f, π) is C-optimal. If f is a flow and (f, π) is ϵ-optimal for $\epsilon < \frac{1}{n}$, then f is a minimum cost flow.

F29: After saturation at the start of the ϵ-phase, the pair (f, π) is ϵ-optimal, and it remains ϵ-optimal throughout the phase. At the end of each ϵ-phase, f is a flow. Thus at the end of the last phase, f is a minimum cost flow.

F30: The network of admissible arcs is acyclic throughout the algorithm.

F31: π is monotone non-increasing throughout the algorithm.

F32: There are at most $3n$ relabels per vertex in an ϵ-phase.

F33: There are $O(n)$ saturating pushes per arc in an ϵ-phase.

COMPUTATIONAL NOTE: The run-time complexity of the algorithm depends on how v is selected and, for a given v, how w is selected. Goldberg and Tarjan [GoTa87], and independently Bertsekas and Eckstein [BeEc88], suggest selecting v to be the first vertex in a topological order implied by the graph of admissible arcs, and then selecting w by cycling through the adjacency list of v. They show that using such protocols, the number of non-saturating pushes is $O(n^3)$ per ϵ-phase. Using more sophisticated data structures, Goldberg and Tarjan [GoTa87] show that an implementation of this algorithm runs in $O(nm \log(n^2/m) \log(nC))$ time. Coupled with heuristic improvements, such as set relabels, *price buckets*, *flow fixing*, and *push look-ahead*, this algorithm has been shown to be experimentally competitive [Go97].

Strongly Polynomial Algorithms

There are numerous algorithms to solve the minimum cost flow problem, and detailed descriptions and comparisons already exist in [AhMaOr93, IwMcSh00, Sc03]. Both algorithms presented here have run-time complexities that depend on the size of numbers in the input data. The first *strongly polynomial time algorithm* (an algorithm with a run-time complexity that does not depend on the size of the numbers in the input data) to solve minimum cost flows is due to Tardos [Ta85] and introduces the idea of fixing flows on arcs. Most subsequent strongly polynomial time algorithms fix either arc flows or vertex labels. The minimum-mean cycle-canceling algorithm of Goldberg and Tarjan mentioned in Subsection 11.2.2 can also be made to run in strongly polynomial time [GoTa89]. The fastest strongly polynomial algorithm is due to Orlin [Or88], is based on capacity scaling and arc-flow fixing, and has run-time complexity $O(m \log n \mathcal{S}(n, m, C))$.

11.2.5   Extensions to Minimum Cost Flow

Convex Cost Flows

Separable convex flow costs model some natural phenomena not captured by linear costs, for example traffic flows and matrix balancing.

DEFINITIONS

D23: Given separable convex functions $c_e : \mathbb{R} \to \mathbb{R}$ for all $e \in A$, and a flow network G, the **minimum convex cost flow** is a flow that minimizes $\sum_{e \in A} c_e(f_e)$ over all feasible flows.

D24: Convex costs in the **discrete model** are defined by piecewise linear curves, where each cost curve c_e is defined by at most p linear segments. The linear segments meet at **breakpoints**. Convex costs in the **continuous model** are defined in functional form.

FACT

F34: In the discrete model, breakpoints of the cost curve can be assumed to be integers. This problem can be solved in polynomial time by transforming the convex cost flow problem into a minimum cost flow problem.

COMPUTATIONAL NOTE: In the continuous model, there is an issue of error introduced due to approximation of the values of the continuous function. This can be handled either by restricting the search for optimal flows to integral optimal flows, or by obtaining a close to optimal solution, within a tolerance dictated by the accuracy of the oracle for the continuous function. In this sense, the continuous problem can be solved using techniques from nonlinear programming, or by extending various algorithms for minimum linear cost flows (e.g., capacity-scaling augmenting-path algorithm [Mi86]; minimum-mean cycle-canceling algorithm [KaMc97]; cost-scaling algorithm [BePoTs97]).

Flows Over Time

Flows over time model problems where time plays a crucial role, such as transportation and telecommunications. They are also called *dynamic flows*.

DEFINITIONS

D25: A *flow-over-time network* is a flow network $G_\tau = (V, A, cap, \tau, b)$ where each arc $(v, w) \in A$ has an associated *transit time* τ_{vw}. The transit time τ_{vw} represents the amount of time that elapses between when flow enters arc (v, w) at v and when the same flow arrives at w.

D26: A *flow over time* x on G_τ with time horizon T is a collection of Lebesgue-measurable functions $x_e : [0, T) \to \mathbb{R}$ where $x_e(\theta)$ is the rate of flow (per time unit) entering arc e at time θ. For notational convenience, define $x_e(\theta) = 0$ for all θ outside the interval $[0, T)$. A flow over time satisfies the following conditions:

nonnegativity constraints: $x_{vw}(\theta) \geq 0$ for all $(v, w) \in A$ and $\theta \in [0, T]$.

flow conservation: The flow entering arc (w, v) at time θ arrives at v at time $\theta + \tau_{wv}$, i.e.,

$$\int_0^\xi \left(\sum_{w \in V} x_{vw}(\theta) - \sum_{w \in V} x_{wv}(\theta - \tau_{wv}) \right) d\theta = 0$$

for all $\xi \in [0, T)$, and $v \in V$ with $b_v = 0$.

time horizon: There is no flow after time T: $x_e(\theta) = 0$ for $\theta \in [T - \tau_e, T)$; and at time T, no flow should remain in the network, i.e.,

$$\int_0^T \left(\sum_{w \in V} x_{vw}(\theta) - \sum_{w \in V} x_{wv}(\theta - \tau_{wv}) \right) d\theta = b_v, \quad \text{for all } v \in V$$

(flow rate) capacity constraints: $x_e(\theta) \leq cap_e$, for all $\theta \in [0, T)$ and $e \in A$.

D27: In the setting with costs, the cost of a flow over time x is defined as

$$c(x) := \sum_{e \in A} \int_0^T c_e x_e(\theta) \, d\theta$$

D28: A *maximum s-t flow over time* with time horizon T is a flow over time for which $b_v = 0$ for all $v \in V - \{s, t\}$ and b_s is maximum among all such flows over time with time horizon T.

D29: A *path flow* (or *cycle flow*) f in a flow network G is a flow on a set of arcs S that forms a path (or cycle) such that $f(e) = \nu$ on $e \in S$ and $f(e) = 0$ on $e \notin S$. In this case, the *volume* of the path flow (or cycle flow) is ν.

D30: Let f be a standard flow in a network $G = (V, A, cap, c, b)$, and let Γ be a set of path flows and cycle flows in G. The set Γ is a *flow decomposition* of the flow f if for each arc $(v, w) \in A$, and for each $\gamma \in \Gamma$, $\gamma(v, w) > 0$ only if $f(v, w) > 0$, and the flow $f(v, w) = \sum_{\gamma \in \Gamma} \gamma(v, w)$.

NOTATION: For a given flow-over-time network G_τ, the standard flow network obtained by ignoring transit times τ is denoted G.

NOTATION: For a path or cycle flow $\gamma \in \Gamma$, denote by $\nu(\gamma)$ the volume of γ, and denote by $P(\gamma)$ the path of arcs corresponding to γ. If Γ is a flow decomposition of flow in the standard flow network obtained from G_τ, let $\tau(\gamma)$ denote the sum of the transit times of arcs in $P(\gamma)$.

FACTS

F35: For any standard flow f in $G = (V, A, cap, c, b)$, there exists a flow decomposition Γ of flows on simple paths and cycles, where $|\Gamma| \leq m$.

F36: [FoFu58] Let G be an s-t flow network and let $\hat{G}$ be the augmented network obtained by adding an infinite capacity arc (t, s) with cost $c(t, s) = -T$. Suppose f is a minimum cost circulation in $\hat{G}$ and Γ is a flow decomposition of f restricted to G. Then $\tau(\gamma) \leq T$ for all path flows $\gamma \in \Gamma$. Define a flow over time as follows: for each path flow $\gamma \in \Gamma$, send flow along $P(\gamma)$ by inserting flow into the first arc in $P(\gamma)$ at rate $\nu(\gamma)$ from time 0 until time $T - \tau(\gamma)$. This flow will arrive at the end of $P(\gamma)$ in the interval $[\tau(\gamma), T]$. Ford and Fulkerson [FoFu58] showed that this flow over time is a maximum s-t flow over time with time horizon T.

F37: Unlike the case with a standard multiple-source multiple-sink, maximum flow problem (Subsection 11.1.5), there is no simple transformation of a multiple-source, multiple-sink flow-over-time problem into an s-t flow-over-time problem. However, there is a combinatorial algorithm that solves this problem in polynomial time [HoTa00].

F38: Finding a minimum cost flow-over-time with time horizon T is NP-hard [KlWo95], but, for any fixed $\epsilon > 0$, an integral solution with time horizon $T(1 + \epsilon)$ and cost at most the cost of the minimum cost flow-over-time with time horizon T can be found in polynomial time (where the run time complexity depends linearly on ϵ^{-2}) [FlSk03].

Flows with Losses and Gains

Flows with losses and gains model flow problems where leakage or loss may occur. Some examples of this are financial transactions, shipping, conversion of raw materials into products, and machine loading. These flows with losses and gains are also called *generalized flows*.

DEFINITIONS

D31: A *gain network* $\bar{G} = (V, A, cap, \gamma, c, s, t)$ is a network $G = (V, A, cap, c, b)$ with positive-valued gain function $\gamma : E \to \mathbb{R}^+$ and supply function $b_v = 0$ for all $v \in V - \{s, t\}$. The *gain factor* $\gamma(e) > 0$ for arc e enforces that for each unit of flow that enters the arc, $\gamma(e)$ units exit. If $\gamma(e) < 1$, the arc e is *lossy*; if $\gamma(e) > 1$, the arc e is *gainy*. In general, the term gains is used to denote the gain functions of both lossy and gainy arcs. For standard network flows, the gain factor of every arc is one.

D32:   A *flow with gains* is a function $f : A \to \mathbb{R}$ that satisfies:

capacity constraints: $f(v, w) \leq cap(v, w)$ for all $(v, w) \in A$,

nonnegativity constraints: $f(v, w) \geq 0$ for all $(v, w) \in A$,

flow conservation constraints:

$$\sum_{w \in V} [f(v, w) - f(w, v)\gamma(w, v)] = 0, \quad \text{for each } v \in V - \{s, t\}.$$

D33:   A *maximum flow with gains* is a flow with gains that maximizes the amount of flow reaching t given an unlimited supply at s.

D34:   A *minimum cost maximum flow with gains* is a maximum flow with gains that minimizes $\sum_{e \in A} c(e)g(e)$.

FACT

F39:   There are combinatorial optimality conditions for flows with losses and gains that generalize the optimality conditions for standard flows described in §11.2.2 [On67, Tr77].

COMPUTATIONAL NOTE: Maximum and minimum cost flows with gains can be solved by linear programming. The versions in which f is restricted to be integral are NP-hard. Combinatorial, polynomial algorithms for maximum flow with gains are based on *high capacity* paths (paths that send the most to the sink) [GoPlTa91], or on high gain paths [GoJiOr97]. A combinatorial, polynomial time algorithm for minimum cost flow with gains is based on a generalization of the minimum-mean cycle-canceling algorithm for minimum cost flow [Wa02]. It is an interesting open question if there exists a strongly polynomial algorithm for this problem.

References

[AhMaOr93] R. K. Ahuja, T. L. Magnanti, and J. B. Orlin, *Network Flows: Theory, Algorithms, and Applications*. Prentice Hall, Englewood Cliffs, NJ, 1993.

[BaTa89] F. Barahona and É. Tardos, Note on Weintraub's minimum-cost circulation algorithm, *SIAM J. Comput.* 18 no. 3 (1989), 579–583.

[Be58] R. E. Bellman, On a routing problem, *Quart. Appl. Math.* 16 (1958), 87–90.

[BeEc88] D. P. Bertsekas and J. Eckstein, Dual coordinate step methods for linear network flow problems, *Math. Programming* 42 (Ser. B) no. 2 (1988), 203–243.

[BePoTs97] D. P. Bertsekas, L. C. Polymenakos, and P. Tseng, An ϵ-relaxation method for separable convex cost network flow problems, *SIAM J. Optim.* 7, no. 3 (1997), 853–870.

[CoCuPuSc98] W. J. Cook, W. H. Cunningham, W. R. Pulleyblank, and A. Schrijver, *Combinatorial Optimization*, John Wiley and Sons, New York, 1998.

[EdKa72] J. Edmonds and R. M. Karp, Theoretical improvements in algorithmic efficiency for network flow problems, *J. ACM* 19 (1972), 248–264.

[FlSk03] L. Fleischer and M. Skutella, Minimum cost flows over time without intermediate storage. *Proceedings of the 14th Annual ACM-SIAM Symposium on Discrete Algorithms* (2003) 66–75.

[Fo56] L. R. Ford, Jr., Network flow theory, The Rand Corp., P-923, August 1956.

[FoFu58] L. R. Ford, Jr. and D. R. Fulkerson, Constructing maximal dynamic flows from static flows, *Operations Res.* 6 (1958), 419–433.

[Go97] A. V. Goldberg, An efficient implementation of a scaling minimum-cost flow algorithm, *J. Algorithms* 22, no. 1 (1997), 1–29.

[GoPlTa91] A. V. Goldberg, S. A. Plotkin, and É. Tardos, Combinatorial algorithms for the generalized circulation problem, *Math. Oper. Res.* 16, no. 2 (1991), 351–381.

[GoTa89] A. V. Goldberg and R. E. Tarjan, Finding minimum-cost circulations by canceling negative cycles, *J. Assoc. Comput. Mach.* 36, no. 4 (1989), 873–886.

[GoTa87] A. V. Goldberg and R. E. Tarjan, Finding minimum-cost circulations by successive approximation, *Math. Oper. Res.* 15, no. 3 (1990), 430–466.

[GoTaTa88] A. V. Goldberg, É. Tardos, and R. E. Tarjan, Network flow algorithms, *Paths, Flows, and VLSI-layout* (Bonn, 1988), 101–164.

[GoJiOr97] D. Goldfarb, Z. Y. Jin, and J. B. Orlin, Polynomial-time highest-gain augmenting path algorithms for the generalized circulation problem, *Math. Oper. Res.* 22, no. 4 (1997), 793–802.

[HoTa00] B. Hoppe and É. Tardos, The quickest transshipment problem, *Math. Oper. Res.* 25, no. 1 (2000), 36–62.

[IwMcSh00] S. Iwata, S. T. McCormick, and M. Shigeno, Relaxed most negative cycle and most positive cut canceling algorithms for minimum cost flow, *Math. Oper. Res.* 25, no. 1 (2000), 76–104.

[KaMc97] A. V. Karzanov and S. T. McCormick, Polynomial methods for separable convex optimization in unimodular linear spaces with applications, *SIAM J. Comput.* 26, no. 4 (1997), 1245–1275.

[KlWo95] B. Klinz and G. J. Woeginger, Minimum cost dynamic flows: the series-parallel case, *Integer programming and combinatorial optimization* (Copenhagen, 1995) 329–343, Lecture Notes in Comput. Sci., 920, Springer, Berlin, 1995.

[Mi86] M. Minoux, Solving integer minimum cost flows with separable convex cost objective polynomially. Netflow at Pisa (Pisa, 1983), *Math. Programming Stud.* No. 26 (1986), 237–239.

[On67] K. Onaga, Optimum flows in general communication networks, *J. Franklin Inst.* 283 (1967) 308–327.

[Or88] J. B. Orlin, A faster strongly polynomial minimum cost flow algorithm, *Oper. Res.* 41 (1993) 338–350.

[Sc03] A. Schrijver. *Combinatorial Optimization*, Springer-Verlag, Berlin, 2003.

[Ta85] É. Tardos, A strongly polynomial minimum cost circulation algorithm, *Combinatorica* 5, no. 3 (1985), 247–255.

[Tr77] K. Truemper, On max flows with gains and pure min-cost flows, *SIAM J. Appl. Math.* 32, no. 2 (1977), 450–456.

[Wa02] K. D. Wayne, A polynomial combinatorial algorithm for generalized minimum cost flow, *Math. Oper. Res.* 27, no. 3 (2002), 445–459.

[Za73] N. Zadeh, A bad network problem for the simplex method and other minimum cost flow algorithms, *Math. Program.* 5 (1973), 255–266.

Section 11.3
Matchings and Assignments

Jay Sethuraman, Columbia University
Douglas Shier, Clemson University

INTRODUCTION

In an undirected graph, the maximum matching problem requires finding a set of nonadjacent edges having the largest total size or largest total weight. This graph optimization problem arises in a diverse number of applications, often involving the optimal pairing of a set of objects.

11.3.1   Matchings

Matchings are defined on undirected graphs, in which the edges can be weighted. Matchings are useful in a wide variety of applications, such as vehicle and crew scheduling, sensor location, snowplowing streets, scheduling on parallel machines, among others.

Basic Terminology

DEFINITIONS

D1: Let $G = (V, E)$ be an undirected graph with vertex set V and edge set E. Each edge $e \in E$ has an associated **weight** w_e.

D2: A **matching** in $G = (V, E)$ is a set $M \subseteq E$ of pairwise nonadjacent edges.

D3: A **vertex cover** in G is a set C of vertices such that every edge in G is incident on at least one vertex in C.

D4: A **perfect matching** in $G = (V, E)$ is a matching M in which each vertex of V is incident on exactly one edge of M.

TERMINOLOGY: A perfect matching of G is also called a **1-factor** of G; see §5.3.

D5: The **size** (**cardinality**) of a matching M is the number of edges in M, written $|M|$. The **weight** of a matching M is $wt(M) = \sum_{e \in M} w_e$.

D6: A **maximum-size matching** of G is a matching M having the largest size $|M|$.

D7: A **maximum-weight matching** of G is a matching M having the largest weight $wt(M)$.

D8: Relative to a matching M in $G = (V, E)$, edges $e \in M$ are **matched** edges, while edges $e \in E - M$ are **free** edges. Vertex v is **matched** if it is incident on a matched edge; otherwise vertex v is **free** (or **unmatched**).

D9: Every matched vertex v has a **mate**, the other endpoint of the matched edge incident on v.

D10: With respect to a matching M, the **weight** $wt(P)$ of path P is the sum of the weights of the free edges in P minus the sum of the weights of the matched edges in P.

D11: An **alternating** path has edges that are alternately free and matched. An **augmenting** path is an alternating path that starts at a free vertex and ends at another free vertex.

NOTATION: Throughout this section, edges are represented as ordered pairs of vertices, and when discussing matchings, paths are represented as edge sets.

EXAMPLES

E1: Figure 11.3.1 shows a graph G together with the matching $M_1 = \{(2, 3), (4, 5)\}$ of size 2; the matched edges are highlighted. The mate of vertex 2 is vertex 3, and the mate of vertex 5 is vertex 4. Relative to matching M_1, vertices 1 and 6 are free vertices, and an augmenting path P from 1 to 6 is given by $P = \{(1, 2), (2, 3), (3, 4), (4, 5), (5, 6)\}$. The matching M_1 is not a maximum-size matching; the matching $M_2 = \{(1, 2), (3, 4), (5, 6)\}$ of size 3 is a perfect matching, and so is also a maximum-size matching.

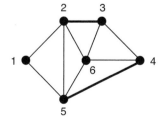

Figure 11.3.1: A matching in a graph.

E2: Figure 11.3.2 shows a graph G with 6 vertices. The matching $M = \{(2,4),(3,5)\}$ displayed is a maximum-size matching, of size 2. This graph G does not have a perfect matching since vertices 1 and 5 are only adjacent to vertex 3.

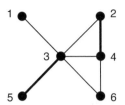

Figure 11.3.2: A maximum-size matching that is not perfect.

E3: In the weighted graph G of Figure 11.3.3 below, the weight w_e is shown next to each edge e. The weight of matching $M = \{(1,2),(3,5)\}$ is $wt(M) = 7$. Relative to this matching, the path $P = \{(1,2),(2,5),(3,5),(3,6)\}$ is an alternating path with $wt(P) = 7 + 1 - 2 - 5 = 1$. The path $\{(1,4),(1,2),(2,3),(3,5),(5,6)\}$ is an augmenting path, joining the free vertices 4 and 6.

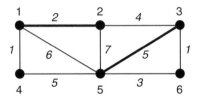

Figure 11.3.3: A matching in a weighted graph.

Some Fundamental Results

FACTS

F1: If M is a matching of $G = (V, E)$, then the number of matched vertices is $2|M|$ and the number of free vertices is $|V| - 2|M|$.

F2: If M is any matching in G, then $|M| \le \left\lfloor \frac{|V|}{2} \right\rfloor$.

F3: (Weak Duality) The size of any vertex cover of G is an upper bound on the size of any matching in G. Consequently the minimum size of a vertex cover is at least as large as the maximum size of a matching.

F4: Every augmenting path has an odd number of edges.

F5: If M is a matching and P is an augmenting path with respect to M, then the symmetric difference $M \triangle P$ is a matching of size $|M| + 1$.

NOTE: The symmetric difference $M \triangle P$ is taken with respect to the *edge sets* defining M and P.

F6: (Augmenting Path Theorem) M is a maximum-size matching if and only if there is no augmenting path with respect to M. (See [Pe1891, Be57, NoRa59].)

F7: If M is a matching and P is an augmenting path with respect to M, then $wt(M \Delta P)$ $= wt(M) + wt(P)$.

F8: Suppose M is a matching having maximum weight among all matchings of a fixed size k. If P is an augmenting path of maximum weight with respect to M, then $M \Delta P$ is a maximum-weight matching among all matchings of size $k+1$.

F9: Let M_i be a maximum-weight matching among all matchings of a fixed size i, $i = 1, 2 \ldots, k$, and let P_i be a maximum-weight augmenting path with respect to M_i. Then $wt(P_1) \geq wt(P_2) \geq \cdots \geq wt(P_k)$.

F10: An immediate consequence of Facts F7 and F8 is a weighted-matching analogue of the Augmenting Path Theorem: A matching M is of maximum weight if and only if the weight of every augmenting path relative to M is nonpositive.

EXAMPLES

E4: In Figure 11.3.1, the path $P = \{(1,2),(2,3),(3,4),(4,5),(5,6)\}$ is augmenting with respect to the matching $M_1 = \{(2,3),(4,5)\}$. As guaranteed by Fact F4, path P has an odd number of edges. The new matching $M_2 = M_1 \Delta P = \{(1,2),(3,4),(5,6)\}$ has size one greater than M_1, and is a maximum-size matching. There are other maximum-size matchings, such as $\{(1,2),(3,6),(4,5)\}$ and $\{(1,5),(2,6),(3,4)\}$.

E5: In Figure 11.3.2, the set $S = \{3,4\}$ is a vertex cover of G. Thus, by Fact F3, the size of any matching M satisfies $|M| \leq 2 = |S|$. On the other hand, $S = \{2,3,4,5\}$ is a (minimum cardinality) vertex cover of the graph in Figure 11.3.1, yet a maximum-size matching M for this graph satisfies $|M| = 3 < |S|$.

E6: Figure 11.3.4(a) shows a matching M_1 of size 1, with $wt(M_1) = 7$. Since edge $(2,5)$ has maximum weight among all edges, M_1 is a maximum-weight matching of size 1. Relative to M_1, the augmenting path $P_1 = \{(1,5),(2,5),(2,3)\}$ has weight $wt(P_1) = 6 + 4 - 7 = 3$, whereas the augmenting path $P_2 = \{(3,6)\}$ has weight 1. It can be verified that P_1 is a maximum-weight augmenting path relative to M_1. Illustrating Fact F8, $M_2 = M_1 \Delta P_1 = \{(1,5),(2,3)\}$ is a maximum-weight matching of size 2, with $wt(M_2) = wt(M_1) + wt(P_1) = 10$ (see Figure 11.3.4(b)). Relative to M_2 there are several augmenting paths between the free vertices 4 and 6:

$$Q_1 = \{(1,4),(1,5),(5,6)\}, \qquad wt(Q_1) = 1 + 3 - 6 = -2,$$
$$Q_2 = \{(1,4),(1,5),(2,5),(2,3),(3,6)\}, \quad wt(Q_2) = 1 + 7 + 1 - 6 - 4 = -1,$$
$$Q_3 = \{(4,5),(1,5),(1,2),(2,3),(3,6)\}, \quad wt(Q_3) = 5 + 2 + 1 - 6 - 4 = -2.$$

Path Q_2 is a maximum-weight augmenting path and so (by Fact F8) $M_3 = M_2 \Delta Q_2 = \{(1,4),(2,5),(3,6)\}$ is a maximum-weight matching of size 3, with $wt(M_3) = 9$. All augmenting paths relative to M_2 have negative weight, and so (by Fact F9) M_2 is a maximum-weight matching in G.

Figure 11.3.4: Maximum-weight matchings of sizes 1 and 2.

REMARKS

R1: Fact F6 was obtained independently by C. Berge [Be57] and also by R. Z. Norman and M. O. Rabin [NoRa59]. This result can also be found in an 1891 paper of J. Petersen [Pe1891].

R2: An historical perspective on the theory of matchings is provided in [Pl92].

R3: Plummer [Pl93] describes a number of variations on the standard matching problem, together with their computational complexity.

11.3.2   Matchings in Bipartite Graphs

Bipartite graphs arise in a number of applications (such as in assigning personnel to jobs or tracking objects over time). See the surveys [AhMaOr93, AhMaOrRe95, Ge95] as well as the text [LoPl86] for additional applications. This section describes properties and algorithms for maximum-size and maximum-weight matchings in bipartite graphs.

DEFINITIONS

D12: Let $G = (X \cup Y, E)$ be a bipartite graph with edge weights w_e.

D13: If $S \subseteq X$ then $\Gamma(S) = \{y \in Y \mid (x, y) \in E \text{ for some } x \in S\}$ is the set of vertices in Y adjacent to some vertex of S.

D14: A *complete* (or *X-saturating*) *matching* of $G = (X \cup Y, E)$ is a matching M in which each vertex of X is incident on an edge of M. Such a matching is also called an *assignment* from X to Y.

APPLICATIONS

A1: A drug company is testing n antibiotics on n volunteer patients in a hospital. Some patients have known allergic reactions to certain of these antibiotics. To determine if there is a feasible assignment of the n different antibiotics to n different patients, construct the bipartite graph $G = (X \cup Y, E)$, where X is the set of antibiotics and Y is the set of patients. An edge $(i, j) \in E$ exists when patient j is *not* allergic to antibiotic i. A complete matching of G is then sought.

A2: An important preprocessing step in solving large sparse systems of linear equations involves rearranging the given $n \times n$ coefficient matrix A using row and column permutations. The objective is to place the maximum number of nonzero coefficients on the diagonal of the permuted matrix [Du81]. This can be viewed as a maximum-size matching problem on a bipartite graph $G = (X \cup Y, E)$. The set X contains the n row indices of A and the set Y contains the n column indices of A. An edge $(i, j) \in E$ exists when $a_{ij} \neq 0$.

A3: There are n applicants to be assigned to n jobs, with each job being filled with exactly one applicant. The weight w_{ij} measures the suitability (or productivity) of applicant i for job j. Finding a valid assignment (matching) achieving the best overall suitability is a weighted matching problem on the bipartite graph $G = (X \cup Y, E)$, where X is the set of applicants and Y is the set of jobs.

A4: The movements of n objects (such as submarines or missiles) are to be followed over time. The locations of the group of objects are known at two distinct times, though without identification of the individual objects. Suppose $X = \{x_1, x_2, \ldots, x_n\}$ and $Y = \{y_1, y_2, \ldots, y_n\}$ represent the spatial coordinates of the objects detected at times t and $t + \Delta t$. If Δt is sufficiently small, then the Euclidean distance between a given object's position at these two times should be relatively small. To aid in identifying the objects (as well as their velocities and directions of travel), a pairing between set X and set Y is desired that minimizes the overall sum of Euclidean distances. This can be formulated as a maximum-weight matching problem on the complete bipartite graph $G = (X \cup Y, E)$, where edge (i, j) indicates pairing position x_i with position y_j. The weight of this edge is the negative of the Euclidean distance between x_i and y_j. A maximum-weight matching of size n in G then provides an optimal (minimum distance) pairing of observations at the two times t and $t + \Delta t$.

FACTS

F11: (König's Theorem) For a bipartite graph G, the maximum size of a matching in G is the minimum cardinality of a vertex cover in G. Thus for bipartite graphs the general inequality stated in Fact F3 can always be satisfied as an equality. (See [Bo90].)

F12: (Hall's Theorem) $G = (X \cup Y, E)$ has a complete matching if and only if $|\Gamma(S)| \geq |S|$ holds for every $S \subseteq X$. In words, a complete matching exists precisely when every set of vertices in X is adjacent to at least an equal number of vertices in Y. (See [Bo90, Gr04].)

F13: Suppose there exists some k such that $deg(x) \geq k \geq deg(y)$ holds in $G = (X \cup Y, E)$ for all $x \in X$ and $y \in Y$. Then G has a complete matching. (See [Gr04].)

EXAMPLES

E7: In the bipartite graph of Figure 11.3.5, $S = \{1, 3, b, d\}$ is a vertex cover of minimum cardinality, and $M = \{(1, a), (3, c), (4, b), (5, d)\}$ is a maximum-size matching. As guaranteed by König's Theorem, $|M| = |S|$. Also, by choosing $A = \{2, 4, 5\}$ we have $\Gamma(A) = \{b, d\}$. Since $|\Gamma(A)| < |A|$ holds, Hall's Theorem shows that there is no complete matching with respect to the set $X = \{1, 2, 3, 4, 5\}$. In fact, the maximum matching M above has size $4 < 5$.

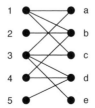

Figure 11.3.5: Covers and matchings in a bipartite graph.

E8: In the chessboard of Figure 11.3.6 we are to place non-taking rooks at certain allowable positions, those marked with an X. For example, we can place rooks at the *independent* positions $(1, 3), (2, 4), (4, 1)$: no two selected positions are in the same row or column. It turns out that three is the maximum number of rooks that can be so placed with regard to the allowable positions X. Also, notice that row 2, row 4, and column 3 are three *lines* in the chessboard containing all X entries; in fact, no fewer number of lines suffice. Here the maximum number of non-taking rooks among the X entries equals the minimum number of lines containing all the X entries. This is a manifestation of König's Theorem, obtained by constructing the bipartite graph $G = (X \cup Y, E)$ where X contains the rows $\{1, 2, 3, 4\}$ and Y contains the columns $\{1, 2, 3, 4, 5\}$; edge $(i, j) \in E$ indicates an X in row i and column j. In this context, independent positions correspond to a matching and covering lines correspond to a vertex cover in G.

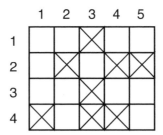

Figure 11.3.6: A chessboard with allowable X entries.

REMARKS

R4: König's Theorem and Hall's Theorem can be derived from the Max-Flow Min-Cut Theorem of §11.1.

R5: Maximum-size matching problems in bipartite graphs can be formulated as maximum flow problems in unit capacity networks and solved using maximum flow algorithms (§11.1).

R6: Maximum-weight matching problems in bipartite graphs can be formulated as minimum cost flow problems in two-terminal flow networks and solved using minimum cost flow algorithms (§11.2).

Bipartite Maximum-Size Matching Algorithm

Algorithm 11.3.1, based on Fact F6, produces a maximum-size matching of the bipartite graph $G = (X \cup Y, E)$. Each iteration involves a modified breadth-first search of G, starting with all free vertices in set X. The vertices of G are structured into levels that alternate between free and matched edges. Algorithm 11.3.1 can be implemented to run in $O(nm)$ time on a graph with n vertices and m edges; see [PaSt82].

Algorithm 11.3.1: Bipartite Maximum-Size Matching

Input: Bipartite graph $G = (X \cup Y, E)$.
Output: Maximum-size matching M.

$M := \emptyset$
$DONE := $ FALSE
While NOT $DONE$
   Let $FREE$ consist of all the free vertices of G.
   $S_X := X \cap FREE$
   $SEEN := \emptyset$
   $STILL_LOOKING := $ TRUE
   While $STILL\_LOOKING$ {for an augmenting path}
     $S_Y := \{y \mid y \notin SEEN \text{ and } (x, y) \in E,\ x \in S_X\}$
     If $S_Y \cap FREE \neq \emptyset$ {an augmenting path exists}
       Construct an augmenting path P to y^*.   [⋆]
       $M := M \Delta P$
       $STILL_LOOKING := $ FALSE
     Else {continue looking for an augmenting path}
       $SEEN := SEEN \cup S_Y$
       $S_X := \{x \mid (y, x) \in M,\ y \in S_Y\}$
       If $S_X = \emptyset$
         $STILL_LOOKING := $ FALSE
         $DONE := $ TRUE

REMARK

R7: The augmenting path at Step [⋆] is constructed in reverse, starting at the free Y-vertex y^*. Choose a vertex $x \in S_X$ (adjacent to y^*) by which y^* was defined to be an element of S_Y. Then choose the vertex $y \in S_Y$ that is matched to x in M. Vertices from X and Y are alternately chosen in this way until an x is chosen from the initial S_X, which means that it is a free vertex.

EXAMPLE

E9: Algorithm 11.3.1 can be used to find a maximum-size matching in the bipartite graph of Figure 11.3.7 below. We begin with the matching $M = \{(1, a), (2, b)\}$ of size 2, shown in Figure 11.3.7(a). At the next iteration, $S_X = \{3, 4\}$ and $S_Y = \{a, b\}$. Since both vertices of S_Y are matched, the algorithm continues with $S_X = \{1, 2\}$ and $S_Y = \{c\}$. Since $c \in S_Y$ is free, with augmenting path $P = \{(3, a), (a, 1), (1, c)\}$, the new matching produced is $M = \{(1, c), (2, b), (3, a)\}$; see Figure 11.3.7(b). The next iteration produces $S_X = \{4\}$, $S_Y = \{b\}$; $S_X = \{2\}$, $S_Y = \{a, c\}$; finally $S_X = \{1, 3\}$, $S_Y = \emptyset$, $S_X = \emptyset$. No further augmenting paths are found, and Algorithm 11.3.1 terminates with the maximum-size matching $M = \{(1, c), (2, b), (3, a)\}$.

Figure 11.3.7: Maximum-size matching in a bipartite graph.

Bipartite Maximum-Weight Matching Algorithm

Algorithm 11.3.2, based on Facts F8–F10, produces a maximum-weight matching of $G = (X \cup Y, E)$. Each iteration finds a maximum-weight augmenting path relative to the current matching M. The algorithm terminates when the path has nonpositive weight. A straightforward implementation of Algorithm 11.3.2 runs in $O(n^2 m)$ time on a graph with n vertices and m edges.

NOTATION: The tentative largest weight of an alternating path from a free vertex in X to vertex j is maintained using the label $d(j)$.

Algorithm 11.3.2: Bipartite Maximum-Weight Matching

Input: Bipartite graph $G = (X \cup Y, E)$ with edge weights w_e.
Output: Maximum-weight matching M.

  $M := \emptyset$
  $DONE :=$ FALSE
  While NOT $DONE$
    Let S_X consist of all the free vertices of X.
    Let $d(j) := 0$ for $j \in S_X$ and $d(j) := -\infty$ otherwise.
    While $S_X \neq \emptyset$
      $S_Y := \emptyset$
      For each edge $(x, y) \in E - M$ with $x \in S_X$
        If $d(x) + w_{xy} > d(y)$
          $d(y) := d(x) + w_{xy}$
          $S_Y := S_Y \cup \{y\}$
      $S_X := \emptyset$
      For each edge $(y, x) \in M$ with $y \in S_Y$
        If $d(y) - w_{yx} > d(x)$
          $d(x) := d(y) - w_{yx}$
          $S_X := S_X \cup \{x\}$
    Let y be a free vertex with maximum label $d(y)$
     and let P be the associated path.
    If $d(y) > 0$
      $M := M \Delta P$
    Else
      $DONE :=$ TRUE

EXAMPLE

E10: Algorithm 11.3.2 can be used to find a maximum-weight matching in the bipartite graph of Figure 11.3.8. If we begin with the empty matching, then the first iteration yields the augmenting path $P_1 = \{(3, a)\}$, with $wt(P_1) = 6$, and the maximum-weight matching (of size 1) $M = \{(3, a)\}$, with $wt(M) = 6$; see Figure 11.3.8(a). The next iteration starts with $S_X = \{1, 2\}$. The labels on vertices a, b, c are then updated to $d(a) = 4$, $d(b) = 4$, $d(c) = 5$, so $S_Y = \{a, b, c\}$. Using the matched edge $(a, 3)$, vertex 3 has its label updated to $d(3) = -2$ and $S_X = \{3\}$. No further updates occur, and the free vertex c with maximum label $d(c) = 5$ is selected. This label corresponds to the augmenting path $P_2 = \{(2, c)\}$, with $wt(P_2) = 5$. The new matching is $M = \{(2, c), (3, a)\}$, with $wt(M) = 11$; see Figure 11.3.8(b). At the next iteration, $S_X = \{1\}$ and vertices a, b receive updated labels $d(a) = 4$, $d(b) = 1$. Subsequent updates produce $d(3) = -2$, $d(c) = 3$, $d(2) = -2$, $d(b) = 2$. Finally, the free vertex b is selected with $d(b) = 2$, corresponding to the augmenting path $P_3 = \{(1, a), (a, 3), (3, c), (c, 2), (2, b)\}$, with $wt(P_3) = 2$. This gives the maximum-weight matching $M = \{(1, a), (2, b), (3, c)\}$, with $wt(M) = 13$, shown in Figure 11.3.8(c). As predicted by Fact F9, the weights of the augmenting paths are nonincreasing: $wt(P_1) \geq wt(P_2) \geq wt(P_3)$.

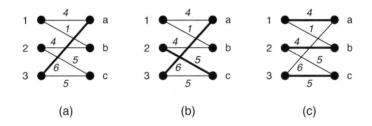

(a) (b) (c)

Figure 11.3.8: Maximum-weight matchings of sizes 1, 2, and 3.

11.3.3 Matchings in Nonbipartite Graphs

This section discusses matchings in more general (nonbipartite) graphs. Algorithms for constructing maximum-size and maximum-weight matchings are considerably more intricate than for bipartite graphs. The important new concept is that of a *blossom*.

DEFINITIONS

D15: Suppose P is an alternating path from a free vertex s in graph $G = (V, E)$. Then a vertex v on P is **even** if the subpath P_{sv} of P joining s to v has even length; it is **odd** if P_{sv} has odd length.

D16: Suppose P is an alternating path from a free vertex s to an even vertex v and edge $(v, w) \in E$ joins v to another even vertex w on P. Then $P \cup \{(v, w)\}$ contains a unique cycle, called a **blossom.**

D17: A **shrunken blossom** results when a blossom B is collapsed into a single vertex b, whereby any edge (x, y) with $x \notin B$ and $y \in B$ is transformed into the edge (x, b). The reverse of this process gives an **expanded blossom.**

FACTS

F14: A blossom B has odd length $2k + 1$ and contains k matched edges, for some $k \geq 1$.

F15: A bipartite graph contains no blossoms.

F16: (Edmonds's Theorem) [Ed65a] Suppose graph G^B is formed from G by collapsing blossom B. Then G^B contains an augmenting path if and only if G does.

F17: (General Maximum-Size Matching) Algorithm 11.3.3, based on Fact F6, produces a maximum-size matching of G. At each iteration, a forest of trees is grown, rooted at the free vertices of G, in order to identify an augmenting path. As encountered, blossoms B are shrunk, with the search continued in the resulting graph G^B.

Algorithm 11.3.3: General Maximum-Size Matching

Input: Graph $G = (V, E)$.
Output: Maximum-size matching M.

  $M := \emptyset$
  $DONE := \text{FALSE}$
  While NOT $DONE$
    Mark all free vertices as even.
    Mark all matched vertices as unreached.
    Mark all free edges as unexamined.
    While there are unexamined edges and no augmenting path is found
      Let (v, w) be an unexamined edge.
      Mark (v, w) as examined.
      {*Case 1*}
      If v is even and w is unreached
        Mark w as odd and its mate z as even.
        Extend the forest by adding (v, w) and matched edge (w, z).
      {*Case 2*}
      If v and w are even and they belong to different subtrees
        An augmenting path has been found.
      {*Case 3*}
      If v and w are even and they belong to the same subtree
        A blossom B is found.
        Shrink B to an even vertex b.
    If an augmenting path P has been found
      $M := M \triangle P$
    Else
      $DONE := \text{TRUE}$

NOTATION: Throughout we let $n = |V|$ and $m = |E|$.

F18: Algorithm 11.3.3 was initially proposed by Edmonds [Ed65a] with a time bound of $O(n^4)$. An improved implementation of Algorithm 11.3.3 runs in $O(nm)$ time; see [Ta83, Ge95].

F19: Maximum-size matchings in nonbipartite graphs can also be found using the algorithm of Gabow [Ga76], which runs in $O(n^3)$ time, and the algorithm of Micali and Vazirani [MiVa80], which runs in $O(m\sqrt{n})$ time.

F20: More complicated algorithms are required for solving weighted-matching problems in general graphs. The first such algorithm, also involving blossoms, was developed by Edmonds [Ed65b] and has a time bound of $O(n^4)$.

F21: Improved algorithms exist for the weighted-matching problem, with running times $O(n^3)$ and $O(nm \log n)$, respectively; see [AhMaOr93, Ge95]. An $O(nm + n^2 \log n)$ algorithm is given by Gabow [Ga90]. An efficient implementation of the Edmonds blossom algorithm is given in [Ko09].

EXAMPLES

E11: In Figure 11.3.9(a), $P = \{(1,2),(2,3),(3,4),(4,5)\}$ is an alternating but not augmenting path, with respect to the matching $M = \{(2,3),(4,5)\}$. Relative to path P, vertices $1,3,5$ are even while vertices $2,4$ are odd. Since $(5,3)$ is an edge joining two even vertices on P, the blossom $B = \{(3,4),(4,5),(5,3)\}$ is formed. On the other hand, $Q = \{(1,2),(2,3),(3,5),(5,4),(4,6)\}$ is an augmenting path relative to M so that $M \Delta P = \{(1,2),(3,5),(4,6)\}$ is a matching of larger size—in fact a matching of maximum size. Notice that relative to path Q, vertices $1,3,4$ are even while vertices $2,5,6$ are odd.

E12: Shrinking the blossom B relative to path P in Figure 11.3.9(a) produces the graph G^B shown in Figure 11.3.9(b). The path $P^B = \{(1,2),(2,b),(b,6)\}$ is now augmenting in G^B. By expanding P^B so that $(2,3)$ remains matched and $(4,6)$ remains free, the augmenting path $Q = \{(1,2),(2,3),(3,5),(5,4),(4,6)\}$ in G is obtained.

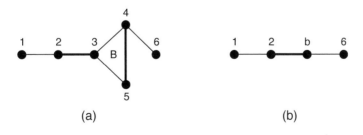

(a)   (b)

Figure 11.3.9: Shrinking blossom B.

E13: Algorithm 11.3.3 can be applied to the nonbipartite graph shown in Figure 11.3.10(a). Suppose the matching $M = \{(3,4),(6,8)\}$ of size 2 is already available.

Iteration 1: The free vertices $1,2,5,7$ are marked as even, and the matched vertices $3,4,6,8$ are marked as unreached. The initial forest consists of the isolated vertices $1,2,5,7$.

• If the free edge $(2,3)$ is examined then Case 1 applies, so vertex 3 is marked odd and vertex 4 even; the free edge $(2,3)$ and the matched edge $(3,4)$ are added to the forest.

• If the free edge $(7,4)$ is next examined then Case 2 applies, and the augmenting path $P = \{(2,3),(3,4),(4,7)\}$ is found. Using P the new matching $M = \{(2,3),(4,7),(6,8)\}$ of size 3 is obtained; see Figure 11.3.10(b).

Iteration 2: The forest is initialized with the free (even) vertices $1,5$.

• If the free edge $(1,2)$ is examined then Case 1 applies, so vertex 2 is marked odd and vertex 3 even; edges $(1,2)$ and $(2,3)$ are added to the forest.

• Examining in turn the free edges $(3,4)$ and $(7,6)$ makes $4,6$ odd vertices and $7,8$ even. Edges $(3,4),(4,7),(7,6),(6,8)$ are then added to the subtree rooted at 1.

• If edge $(8,7)$ is examined, then Case 3 applies, and the blossom $B = \{(7,6),(6,8),(8,7)\}$ is detected and shrunk; Figure 11.3.10(c) shows the resulting G^B. The current subtree rooted at 1 now becomes $\{(1,2),(2,3),(3,4),(4,b)\}$.

• If the free edge $(b,5)$ is examined, then Case 2 applies and the augmenting path $\{(1,2),(2,3),(3,4),(4,b),(b,5)\}$ is found in G^B. The corresponding augmenting path in G is $P = \{(1,2),(2,3),(3,4),(4,7),(7,8),(8,6),(6,5)\}$. Forming $M\Delta P$ produces the new matching $\{(1,2),(3,4),(5,6),(7,8)\}$, which is a maximum-size matching; see Figure 11.3.10(d).

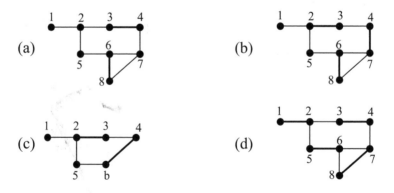

(a) (b) (c) (d)

Figure 11.3.10: Illustrating Algorithm 11.3.3.

APPLICATIONS

A5: Pairs of pilots are to be assigned to aircraft serving international routes. Pilots i and j are considered compatible if they are fluent in a common language and have comparable flight training. Form the graph G whose vertices represent pilots and whose edges represent compatible pairs of pilots. The problem of flying the largest number of aircraft with compatible pilots is then a maximum-size matching problem on G.

A6: Bus drivers are hired to work two four-hour shifts each day. Union rules require a certain minimum amount of time between the shifts that a driver can work. There are also costs associated with transporting the driver between the ending location of the first shift and the starting location of the second shift. The problem of optimally combining pairs of shifts that satisfy union regulations and incur minimum total cost can be formulated as a maximum-weight matching problem. Namely, define the graph G with vertices representing each shift that must be covered and edges between pairs of compatible shifts (satisfying union regulations). The weight of edge (i, j) is the negative of the cost of assigning a single driver to shifts i and j. It is convenient also to add edges (i, i) to G to represent the possibility of needing a part-time driver to cover a single shift; edge (i, i) is given a sufficiently large negative weight to discourage single-shift assignments unless absolutely necessary. A maximum-weight perfect matching in G then provides a minimum-cost pairing of shifts for the bus drivers.

11.3.4   Stable Matchings

The stable matching problem, first discussed by D. Gale and L. S. Shapley [GaSh62], is a fundamental model for assignment problems with preferences (e.g., the assignment of medical residents to hospitals in the United States).

DEFINITIONS

D18: Given two disjoint sets X and Y with $|X| = |Y|$, let G be the complete bipartite graph on the vertex set $X \cup Y$. For each $x \in X$, let $\succ_x$ be an ordering of Y and for each $y \in Y$, let $\succ_y$ be an ordering of X.

D19: Let M be a perfect matching in G. For any $v \in X \cup Y$, $M(v) = w$ if $(v, w) \in M$.

D20: Relative to a perfect matching M of G, a pair $(x, y) \in X \times Y$ is a **blocking pair** if $y \succ_x M(x)$ **and** $x \succ_y M(y)$.

D21: A perfect matching M of G is **stable** if there are no blocking pairs relative to M.

EXAMPLES

E14: Figure 11.3.11 shows an instance of the stable matching problem with $X = \{1, 2, 3, 4\}$ and $Y = \{a, b, c, d\}$, along with the corresponding ordering for each member of X and Y. The matching $\{(1, a), (2, b), (3, c), (4, d)\}$ is not stable: $(2, a)$ is a blocking pair for this matching because $a >_2 b$ and $2 >_a 1$.

1	b	d	a	c
2	c	a	d	b
3	b	c	a	d
4	d	a	c	b

a	2	1	4	3
b	4	3	1	2
c	1	4	3	2
d	2	1	4	3

Figure 11.3.11: An instance of the stable matching problem.

E15: The matching $\{(1,d),(2,c),(3,b),(4,a)\}$ is stable. To verify this, observe that 2 and 3 are matched with their top choice, so they cannot be part of a blocking pair. The only potential blocking pair involving 1 is $(1,b)$, but b prefers 3 to 1; finally, the only potential blocking pair involving 4 is $(4,d)$, but d prefers 1 to 4.

E16: The matching $\{(1,d),(2,a),(3,b),(4,c)\}$ is also stable. That this is the only other stable matching can be verified by an exhaustive search.

Gale–Shapley Algorithm

Algorithm 11.3.4, discovered by Gale and Shapley, produces a stable matching for any complete bipartite graph on $X \cup Y$ with $|X| = |Y|$. Each iteration involves a free vertex x "proposing" to its most-preferred choice $y \in Y$ that has not yet rejected it. The algorithm is such that each vertex in $X \cup Y$ is either *tentatively matched* or *free*; vertices in X—the *active* side—may go from one of these states to the other, but once a vertex in Y—the *passive* side—becomes *tentatively matched*, it will stay so until the end (although its partner may change during the course of the algorithm). From the description of the algorithm, it is not obvious that the algorithm terminates (and it does so with a stable matching); furthermore, as stated, the algorithm needs to make a choice whenever there are multiple free vertices in X. However, it can be shown that the algorithm terminates with a stable matching, and that the outcome is the same regardless of the order in which the free vertices in X are examined.

Algorithm 11.3.4: Stable Matching

Input: Complete bipartite graph $G = (X \cup Y, X \times Y)$, with $|X| = |Y|$.
Output: Stable matching M.

$M := \emptyset$
$FREE := X \cup Y$
for each $x \in X$, $REJECT(x) = \emptyset$
While $X \cap FREE \neq \emptyset$
  pick any $x \in FREE \cap X$
  let y be the greatest (according to $\succ_x$) in $Y \setminus REJECT(x)$.
  If $y \in FREE$
    Add (x,y) to M
    $FREE := FREE \setminus \{x,y\}$
  Else
    If $x >_y M(y)$ then
      $FREE := (FREE \setminus x) \cup M(y)$
      $REJECT(M(y)) := REJECT(M(y)) \cup \{y\}$
      $M := [M \cup (x,y)] \setminus (M(y),y)$
    Else
      $REJECT(x) := REJECT(x) \cup \{y\}$

FACTS

Proofs of the following facts about the GS algorithm may be found in the first chapter of [GuIr89].

F22: Given any instance of the stable matching problem, the Gale–Shapley (GS) algorithm terminates with a stable matching. In particular, there is always at least one stable matching.

F23: The outcome of the GS algorithm does not depend on the order in which the free vertices in X are examined.

F24: Say that x and y are **stable partners** if $(x, y) \in M$ for *some* stable matching M. In the stable matching computed by the GS algorithm, *every* $x \in X$ is matched with their *most-preferred* stable partner. Consequently, the GS algorithm finds an optimal stable matching for X.

F25: In the stable matching computed by the GS algorithm, *every* $y \in Y$ is matched with their *least-preferred* stable partner. Consequently, the GS algorithm finds the worst stable matching for Y.

F26: By reversing the roles of X and Y, we can obtain a symmetric algorithm that computes an optimal stable matching for Y.

F27: If the same stable matching is produced by applying the GS algorithm as prescribed in Facts F24 and F26, then the given instance has a unique stable matching.

F28: Let M and M' be two distinct stable matchings, and suppose that x and y are matched to each other in M. Then exactly one of the following three possibilities holds: (i) x and y are matched to each other in M'; (ii) x prefers $M'(x)$ to y and y prefers x to $M'(y)$; (iii) x prefers y to $M'(x)$ and y prefers $M'(y)$ to x.

F29: Suppose M and M' are two distinct stable matchings. Define the assignment $\mu(x) = \max_{\succ_x} \{M(x), M'(x)\}$. That is, assign to x the better of the two stable partners that x has (in M and M'). Then μ is a *stable matching*. (A priori it is not even clear that μ is a matching.) Similarly, define $\nu(y) = \min_{\succ_y} \{M(y), M'(y)\}$. That is, assign to y the worse of the two partners that y has (in M and M'). It can be shown that ν is a stable matching and that ν and μ are identical.

F30: Given any stable matchings $M_1, M_2, \ldots, M_t$ and any $k \in \{1, 2, \ldots, t\}$, let $\mu_k(x)$ be the assignment that gives each x its kth best partner (among the t stable partners, with multiplicity). Similarly, let ν_k assign to each y the kth worst partner among its t stable partners in $M_1, M_2, \ldots, M_t$. Then μ_k and ν_k are both stable matchings; moreover $\mu_k = \nu_k$.

EXAMPLES

E17: Applying the GS algorithm to the instance of Figure 11.3.11 results in the following. First, 1 proposes to b and is tentatively accepted; 2 proposes to c and is tentatively accepted; 3 proposes to b, who accepts 3 and rejects 1; 1 then proposes to d and is tentatively accepted; 4 proposes to d and is rejected; 4 then proposes to a and is tentatively accepted. At this point none of the members of X is free and the algorithm terminates with the matching $\{(1, d), (2, c), (3, b), (4, a)\}$. The symmetric execution of the GS algorithm with X and Y reversed results in the matching mentioned in Example E16.

E18:   Figure 11.3.12 shows an instance of the stable matching problem with $X = \{1, 2, 3, 4\}$ and $Y = \{a, b, c, d\}$, along with the corresponding ordering for each member of X and Y. Consider the stable matchings $M = \{(1, b), (2, d), (3, a), (4, c)\}$ and $M' = \{(1, c), (2, a), (3, d), (4, b)\}$. With $x = 2$ and $y = d$, we see that x prefers M' to M, whereas d prefers M to M', consistent with case (ii) of Fact F28. The matching μ of Fact F29 is $\{(1, b), (2, a), (3, d), (4, c)\}$, which is yet another stable matching.

1	a	b	c	d
2	b	a	d	c
3	c	d	a	b
4	d	c	b	a

a	4	3	2	1
b	3	4	1	2
c	2	1	4	3
d	1	2	3	4

Figure 11.3.12: An instance with 10 stable matchings.

APPLICATIONS AND EXTENSIONS

A7:   The stable matching problem as formulated requires $|X| = |Y|$ and a complete bipartite graph. These assumptions can be relaxed—one could formulate the stable matching problem on an arbitrary bipartite graph. In this case, each $x \in X$ has an ordering of $\Gamma(x)$, the set of vertices it is adjacent to; similarly, each $y \in Y$ has an ordering of $\Gamma(y)$, the set of vertices it is adjacent to. One can think of $\Gamma(v)$ as the set of all vertices in the other set that are *acceptable* partners for v. A stable matching M may not be perfect any longer. The only requirement for M to be stable is that no pair of vertices that are not matched with each other (in M) prefer each other to their respective partners (in M). It can be shown that the vertices that are unmatched in *any* stable matching remain unmatched in *all* stable matchings.

A8:   Most applications of the stable matching model involve entities that are allowed to have multiple partners. For example, in the assignment of schools to students, each student is typically assigned to a single school, but schools are allowed to admit many students. This generalization can easily be accommodated in the stable matching model by allowing each $y \in Y$ to have an integer quota $q_y \geq 1$. A pair (x, y) blocks the matching M if x prefers y to $M(x)$ and y prefers x to *at least* one student in the set of students $M(y)$. As Gale and Shapley observed [GaSh62], one can reduce this to the classical one-to-one matching problem by making q_y copies of y, and by letting each copy have the same preference ordering over X as the original y. Each $x \in X$ now has a preference ordering over the copies in a consistent way: if $y >_x y'$ then any copy of y is preferred by x to any copy of y'; also the copies of a given y are always ranked $1, 2, \ldots, q_y$. It is easy to verify that any stable matching of the expanded instance is stable in the original instance and vice versa.

A9:   The stable matching problem can be posed on a general graph with an even number of vertices. In this model, each vertex has an ordering over the other vertices, and the objective is to find a matching with no blocking pair. Unlike the bipartite case, a stable matching may not exist. However, it is possible to decide in polynomial time whether or not a given instance admits a stable matching, and much is known about the structure of such instances. See Gusfield and Irving [GuIr89] for additional details.

A10: The most celebrated application of the stable matching model is the National Resident Matching Program (NRMP) in the United States which assigns medical interns (or residents) to hospitals. Roth [Ro84] discusses the early history of this program and observes that the assignment algorithm used by this centralized matching program was in fact equivalent to the hospital-optimal version of the GS algorithm. It is astonishing that this algorithm had been in use from 1951, predating the seminal paper of Gale and Shapley by more than a decade. Roth and Peranson [RoPe99] led the effort to redesign the NRMP matching algorithm, which is now a variant of the resident-optimal GS algorithm.

A11: A relatively recent application of the stable matching algorithm occurs in the assignment of school students to public schools in various US cities that have active school choice programs. In New York City, a variant of the student-optimal GS algorithm is used to assign students to specialized high schools. Students submit their preferences to a central authority, and school preferences are determined by various factors such as grades, test-scores, etc. An important new element in some districts is that school preferences have indifferences (or ties) and the algorithm has to be enhanced to accommodate such preferences. See Abdulkadiroglu et al. [AbPaRo09] for additional details.

REMARKS

R8: The stable matching model was introduced by Gale and Shapley [GaSh62]. The books by Gusfield and Irving [GuIr89] and by Knuth and Goldstein [KnGo97] deal with algorithmic and structural aspects of the stable matching problem and their generalizations. For an overview of applications, especially in economics, see Roth and Sotomayor [RoSo91]. Subramanian [Su94] and Feder [Fe95] apply a fixed-point approach to the study of stable matchings and discuss applications to circuit complexity. Fleiner [Fl03] discusses substantive generalizations of the basic model that are amenable to a fixed-point approach.

R9: The stable matching problem can be formulated as an integer programming problem. Vande Vate [Va89] and Rothblum [Ro92] discuss linear programming formulations of the stable matching problem that are integral; these results have been generalized by Teo and Sethuraman [TeSe98], who give further insight into the structure of fractional stable matchings.

References

[AbPaRo09] A. Abdulkadiroglu, P. A. Pathak, and A. E. Roth, Strategyproofness versus efficiency in matching with indifferences: Redesigning the NYC high school match. *American Economic Review* **99** (2009), 1954–1978.

[AhMaOr93] R. K. Ahuja, T. L. Magnanti, and J. B. Orlin, *Network Flows: Theory, Algorithms, and Applications*, Prentice-Hall, 1993.

[AhMaOrRe95] R. K. Ahuja, T. L. Magnanti, J. B. Orlin, and M. R. Reddy, Applications of network optimization, in M. Ball, T. Magnanti, C. Monma, and G. Nemhauser (Eds.), *Network Models*, North-Holland, 1995, 1–83.

[Be57] C. Berge, Two theorems in graph theory. *Proc. Natl. Acad. Sci.* (U.S.A.) **43** (1957), 842–844.

[Bo90] K. P. Bogart, *Introductory Combinatorics*, Harcourt Brace Jovanovich, 1990.

[Du81] I. S. Duff, On algorithms for obtaining a maximum transversal. *ACM Transactions on Mathematical Software* **7** (1981), 315–330.

[Ed65a] J. Edmonds, Paths, trees, and flowers. *Canadian Journal of Mathematics* **17** (1965), 449–467.

[Ed65b] J. Edmonds, Maximum matching and a polyhedron with 0, 1-vertices. *Journal of Research of the National Bureau of Standards* **B-69** (1965), 125–130.

[Fe95] T. Feder, Stable networks and product graphs, *Memoirs of the American Mathematical Society* **555**, 1995.

[Fl03] T. Fleiner, A fixed-point approach to stable matchings and some applications. *Math. Oper. Res.* **28** (2003), 103–126.

[Ga76] H. N. Gabow, An efficient implementation of Edmonds' algorithm for maximum matching on graphs. *Journal of the ACM* **23** (1976), 221–234.

[Ga90] H. N. Gabow, Data structures for weighted matchings and nearest common ancestors with linking. *Proceedings of the 1st Annual ACM-SIAM Symposium on Discrete Algorithms*, 1990, 434–443.

[GaSh62] D. Gale and L. S. Shapley, College admissions and the stability of marriage. *American Mathematical Monthly* **69** (1962), 9–15.

[Ge95] A. M. H. Gerards, Matching, in M. Ball, T. Magnanti, C. Monma, and G. Nemhauser (Eds.), *Network Models*, North-Holland, 1995, 135–224.

[Gr04] R. P. Grimaldi, *Discrete and Combinatorial Mathematics*, Fifth Edition, Pearson, 2004.

[GuIr89] D. Gusfield and R. W. Irving, *The Stable Marriage Problem: Structure and Algorithms*, MIT Press, 1989.

[KnGo97] D. E. Knuth and M. Goldstein, *Stable Marriage and Its Relation to Other Combinatorial Problems: An Introduction to the Mathematical Analysis of Algorithms*, American Mathematical Society, 1997.

[Ko09] V. Kolmogorov, Blossom V: A new implementation of a minimum cost perfect matching algorithm. *Mathematical Programming Computation* **1** (2009), 43–67.

[LoPl86] L. Lovász and M. D. Plummer, *Matching Theory*, North-Holland, 1986.

[MiVa80] S. Micali and V. V. Vazirani, An $O(\sqrt{|V|} \cdot |E|)$ algorithm for finding maximum matching in general graphs. *Proceedings of the 21st Annual Symposium on Foundations of Computer Science*, 1980, 17–27.

[NoRa59] R. Z. Norman and M. O. Rabin, An algorithm for a minimum cover of a graph. *Proceedings of the American Mathematics Society* **10** (1959), 315–319.

[PaSt82] C. H. Papadimitriou and K. Steiglitz, *Combinatorial Optimization*, Prentice-Hall, 1982.

[Pe1891] J. Petersen, Die Theorie der regulären graphs. *Acta Mathematica* **15** (1891), 193–220.

[Pl92] M. D. Plummer, Matching theory—a sampler: from Dénes König to the present. *Discrete Mathematics* **100** (1992), 177–219.

[Pl93] M. D. Plummer, Matching and vertex packing: how "hard" are they?, in *Quo Vadis, Graph Theory?*, J. Gimbel, J. W. Kennedy, and L. V. Quintas (Eds.), *Annals of Discrete Mathematics* **55**, North-Holland, 1993, 275–312.

[Ro84] A. E. Roth, The evolution of the labor market for medical interns and residents: A case study in game theory. *Journal of Political Economy* **92** (1984), 991–1016.

[RoPe99] A. E. Roth and E. Peranson, The redesign of the matching market for American physicians: Some engineering aspects of economic design. *American Economic Review* **89** (1999), 748–780.

[RoSo91] A. E. Roth and M. Sotomayor, *Two-Sided Matching: A Study in Game-Theoretic Modeling and Analysis*, Cambridge University Press, 1991.

[Ro92] U. G. Rothblum, Characterization of stable matchings as extreme points of a polytope. *Mathematical Programming* **54** (1992), 57–67.

[Su94] A. Subramanian, A new approach to stable matching problems. *SIAM Journal on Computing* **23** (1994), 671–700.

[Ta83] R. E. Tarjan, *Data Structures and Network Algorithms*, SIAM, 1983.

[TeSe98] C. P. Teo and J. Sethuraman, The geometry of fractional stable matchings and its applications. *Mathematics of Operations Research* **23** (1998), 874–891.

[Va89] J. H. Vande Vate, Linear programming brings marital bliss. *Operations Research Letters* **8** (1989), 147–153.

Section 11.4
Graph Pebbling

Glenn Hurlbert, Arizona State University

INTRODUCTION

Graph Pebbling is a network optimization model for the transportation of resources that are consumed in transit. Electricity, heat, or other energy may dissipate as it moves from one location to another, oil tankers may use up some of the oil they transport, or information may be lost as it travels through its medium. The central problem in this model asks whether discrete pebbles from one set of vertices can be moved to another while pebbles are lost in the process. A typical question asks how many pebbles are necessary to guarantee that, from any configuration of that many pebbles, one can move a pebble to any particular vertex. This section will describe this question and other variations of it, and will present the main results and applications in the theory. Good surveys of the subject can be found in [Hu05, Hu12, HurlGPP].

All graphs considered are simple and connected.

11.4.1 Solvability

Here we develop the notion of moving from one configuration of pebbles to another via pebbling steps.

NOTATION

The set of nonnegative integers is denoted by $\mathcal{N}$. We use $n = n(G)$ to denote the number of vertices of a graph G. When H is a subgraph of G, we write $G - H$ to denote the graph having vertices $V(G - H) = V(G)$ and edges $E(G - H) = E(G) - E(H)$. The eccentricity at vertex r, diameter, girth, connectivity, and domination number of a graph G are written $\mathsf{ecc}_G(r)$, $\mathsf{diam}(G)$, $\mathsf{gir}(G)$, $\kappa(G)$, and $\mathsf{dom}(G)$, respectively, while

$\text{dist}_G(u,v)$ denotes the distance between vertices u and v in G (we may write $\text{ecc}(r)$ and $\text{dist}(u,v)$ when G is understood). Also, the minimum degree of G is denoted $\delta(G)$ and we write lg for the base 2 logarithm.

DEFINITIONS

D1: A *configuration* C on a graph G is a function $C : V(G) \to \mathcal{N}$. The value $C(v)$ signifies the number of pebbles at vertex v. We also write $C(S) = \sum_{v \in S} C(v)$ for a subset $S \subseteq V(G)$ of vertices.

D2: For an edge $\{u,v\} \in E(G)$, if u has at least two pebbles on it, then a *pebbling step from u to v* removes two pebbles from u and places one pebble on v. That is, if C is the original configuration, then the resulting configuration C' has $C'(u) = C(u) - 2$, $C'(v) = C(v) + 1$, and $C'(x) = C(x)$ for all $x \in V(G) - \{u,v\}$.

D3: A pebbling step from u to v is r-*greedy* if $\text{dist}(v,r) < \text{dist}(u,r)$. It is r-*semigreedy* if $\text{dist}(v,r) \le \text{dist}(u,r)$.

D4: We say that a configuration C on G is r-*solvable* if it is possible from C to place a pebble on r via pebbling steps. It is r-*unsolvable* otherwise.

D5: More generally, for a configuration D, we say that C is D-*solvable* if it is possible to perform pebbling steps from C to arrive at another configuration C' for which $C'(v) \ge D(v)$ for all $v \in V(G)$. It is D-*unsolvable* otherwise. We denote by $G(S)$ the directed subgraph of G induced by a set S of pebbling steps.

D6: We say that a configuration C on G is k-*fold* r-*solvable* if it is possible from C to place k pebbles on r via pebbling steps.

NOTE: The k-fold r-solvability of C is the specific instance of D-solvability for which D has k pebbles on r and none elsewhere.

D7: The *size* $|C|$ of a configuration C on a graph G is the total number of pebbles on G; i.e., $|C| = \sum_{v \in V(G)} C(v)$.

D8: For a graph G and a particular *root* vertex r, the *rooted pebbling number* $\pi(G,r)$ is defined to be the minimum number t so that every configuration C on G of size t is r-solvable.

D9: A sequence of paths $\mathcal{P} = (P[1], \dots, P[h])$ is a *maximum r-path partition* of a rooted tree (T,r) if $\mathcal{P}$ forms a partition of $E(T)$, r is a leaf of $P[1]$, $T_i = \cup_{j=1}^{i} P[j]$ is a tree for all $1 \le i \le h$, and $P[i]$ is a maximum length path in $T - T_{i-1}$, among all such paths with one endpoint in T_{i-1}, for all $1 \le i \le h$.

D10: We define the function $f(T,r) = \sum_{i=1}^{h} 2^{l_i} - h + 1$, where $(l_1, \dots, l_h)$ is the sequence of lengths $l_i = \text{diam}(P[i])$ in a maximum r-path partition $\mathcal{P}$ of a rooted tree (T,r). Also, set $f_k(T,r) = f(T,r) + (k-1)2^{l_1}$.

D11: A *thread* in a graph G is a subpath of G whose vertices have degree two in G.

EXAMPLES

E1: Figure 11.4.1 shows two r-unsolvable configurations (of maximum size, right) on the path with 7 vertices.

Figure 11.4.1: Two r-unsolvable configurations on the path P_7.

E2: Figure 11.4.2 shows a maximum sized r-unsolvable configuration on a tree.

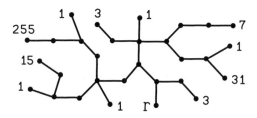

Figure 11.4.2: An r-unsolvable configuration on a tree.

E3: Figure 11.4.3 shows an r-solvable configuration on the 4-cycle with pendant edge.

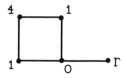

Figure 11.4.3: An r-solvable configuration.

FACTS

F1: If H is a connected, spanning subgraph of a graph G then $\pi(H,r) \geq \pi(G,r)$ for every root vertex r.

F2: Every graph G on n vertices has rooted pebbling number $\pi(G,r) \geq n$ for every root vertex r.

F3: The complete graph K_n on n vertices has rooted pebbling number $\pi(K_n,r) = n$ for every root vertex r.

F4: Every graph G on n vertices has rooted pebbling number $\pi(G,r) \geq 2^{\mathrm{ecc}_G(r)}$ for every root vertex r.

F5: The path P_n on n vertices has rooted pebbling number $\pi(P_n, r) = 2^{n-1}$ when r is one of its leaves.

F6: Every graph G on n vertices has rooted pebbling number $\pi(G, r) \leq (n-1)(2^{\mathrm{ecc}_G(r)} - 1) + 1$ for every root vertex r.

F7: [Ch89] If (T, r) is a tree with root r then $\pi(T, r) \geq f(T, r)$.

F8: [**No-Cycle Lemma**] [BuChCrMiWe08] If a configuration C is D-solvable then there exists a D-solution S for which $G(S)$ is acyclic.

F9: [**Squishing Lemma**] [BuChCrMiWe08] For every root vertex r of a graph G there is a maximum-sized r-unsolvable configuration such that, on each thread not containing r, all pebbles sit on one vertex or two adjacent vertices.

Weight Functions

Weight functions can be used to provide upper bounds on rooted pebbling numbers of graphs.

DEFINITIONS

D12: For a tree T rooted at a vertex r we define the ***parent*** of vertex $v \in V(T) - \{r\}$ to be the unique neighbor v^+ of v for which $\mathrm{dist}(v^+, r) = \mathrm{dist}(v, r) - 1$. We say also that v is a ***child*** of v^+.

D13: We say that a rooted subtree (T, r) of (G, r) is an r-***strategy*** if associated with it is a ***weight function*** $\mathsf{w} : V(G) \to \mathcal{N}$ having the properties that $\mathsf{w}(v) = 0$ for all $v \notin V(T)$ and $\mathsf{w}(v^+) \geq 2\mathsf{w}(v)$ for every vertex $v \neq r$. The r-strategy T is ***basic*** if equality holds for all such $v \in V(T)$.

D14: For a rooted graph (G, r) with r-strategy (T, w), we say that the ***weight of a vertex*** v is $\mathsf{w}(v)$ when $v \in T$ and 0 otherwise, and define the ***weight of a configuration*** C on G to be

$$\mathsf{w}(C) = \sum_{v \in V(G)} C(v)\mathsf{w}(v).$$

NOTATION: We denote by $\mathbf{J}_r$ the configuration on any rooted graph (G, r) having no pebbles on r and one pebble on every other vertex. Furthermore, let wt denote the weight function for any breadth-first search spanning tree (G, r), where $wt(r) = 1$; that is, $wt(v) = 2^{-\mathrm{dist}(v,r)}$ for all $v \in V(G)$.

D15: For a rooted graph (G, r) on n vertices, let $\mathcal{C}$ be the set of all r-unsolvable configurations on G, viewed as points in $\mathcal{N}^{n-1}$: each $C \in \mathcal{C}$ is identified with the coordinates $(C(v_2), \ldots, C(v_n))$, where $V(G) = \{r, v_2, \ldots, v_n\}$. The convex hull of $\mathcal{C}$ is called the r-***unsolvability polytope*** of G, denoted $\mathbf{U}(G, r)$. Define the r-***strategy polytope*** $\mathbf{T}(G, r)$ by the set of linear inequalities given by the Weight Function Lemma over all r-strategies.

FACTS

F10:   If C is a configuration on the rooted graph (G, r) and C' is the configuration obtained from C after a pebbling step from u to v then, for any r-strategy (T, w) of (G, r) containing the edge $\{u, v\}$, we have $\mathsf{w}(C') \le \mathsf{w}(C)$, with equality if and only if $\mathsf{w}(v) = 2\mathsf{w}(u)$ (when $\mathsf{w} = \mathit{wt}$ this means that the step is greedy).

F11:   If C is an r-solvable configuration on G then $\mathit{wt}(C) \ge 1$.

F12:   A configuration C on a path rooted at a leaf r is r-solvable if and only if $\mathit{wt}(C) \ge 1$.

F13:   Every r-strategy is a conic combination of basic r-strategies; that is, for every r-strategy (T, w) of a rooted graph (G, r), there are basic r-strategies $(T_1, \mathsf{w}_1), \dots, (T_h, \mathsf{w}_h)$ of (G, r) and nonnegative coefficients $\alpha_1, \dots, \alpha_h$ so that, for all $v \in v(G)$, we have $\mathsf{w}(v) = \sum_{i=1}^{h} \mathsf{w}_i(v)$.

F14:   [**Weight Function Lemma**] [Hu10] Let (T, w) be an r-strategy of the rooted graph (G, r) and suppose that C is an r-unsolvable configuration on G. Then $\mathsf{w}(C) \le \mathsf{w}(\mathbf{J}_r)$.

EXAMPLES

E4:   Figure 11.4.4 displays the upper bound for the rooted tree (T, r) of Figure 11.4.2 given by a basic r-strategy: $\pi(T, r) = 320$.

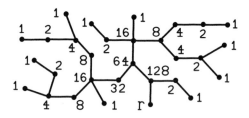

Figure 11.4.4: A rooted tree with a basic r-strategy.

E5:   Figure 11.4.5 displays the lower and upper bounds for the rooted cycle (C_7, r) given by an r-unsolvable configuration and basic r-strategy, respectively: $\pi(C_7, r) = 15$.

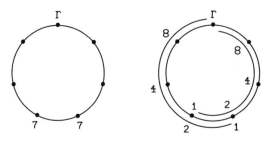

Figure 11.4.5: A rooted cycle with (left) its maximum r-unsolvable configuration and (right) two basic r-strategies.

E6: Figure 11.4.6 displays the upper bound for a rooted Petersen graph (P, r) given by three basic r-strategies: $\pi(P, r) = 10$.

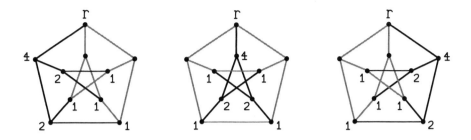

Figure 11.4.6: Three basic r-strategies of a rooted Petersen graph.

FACTS

F15: [Ch89] If (T, r) is a tree with root r then $\pi(T, r) \geq f(T, r)$ (and hence $\pi(T, r) = f(T, r)$).

F16: Every rooted graph (G, r) has rooted pebbling number $\pi(G, r) \leq f(T, r)$ for any breadth-first search spanning tree T of G rooted at r.

F17: [CuHiHuMo09] Let G be a graph in which each of its blocks is a clique, and suppose that T is a breadth-first search spanning tree of G rooted at r. Then $\pi(G, r) = \pi(T, r)$.

F18: [PaSnVo95] For every root vertex r in the cycle C_n we have $\pi(C_{2k}, r) = 2^k$ for all $k \geq 2$ and $\pi(C_{2k+1}, r) = \lceil (2^{k+2} - 1)/3 \rceil$ for all $k \geq 1$.

F19: [Hu10] For a polytope $\mathbf{P}$ of configurations on a rooted graph (G, r) define $z_{\mathbf{P}}(G, r) = \max_{C \in \mathbf{P}} |C|$ and $\pi_{\mathbf{P}}(G, r) = \lfloor z_{\mathbf{P}}(G, r) \rfloor + 1$. The Weight Function Lemma implies that $\mathbf{U}(G, r) \subseteq \mathbf{T}(G, r)$, and hence $\pi(G, r) = \pi_{\mathbf{U}}(G, r) \leq \pi_{\mathbf{T}}(G, r)$.

F20: [**Uniform Covering Lemma**] [Hu10] Let G be a graph on n vertices. If some collection of r-strategies $\{(T_i, \mathbf{w}_i)\}_{i=1}^{k}$ has the property that there is a constant c such that, for every $v \in V(G) - \{r\}$, we have $\sum_{i=1}^{k} \mathbf{w}_i(v) = c$, then $\pi(G, r) = n$.

F21: [Ch89] The d-dimensional cube Q^d has $\pi(Q^d, r) = n(Q^d) = 2^d$ for every root vertex r.

RESEARCH PROBLEMS

RP1: Is there a characterization for r-solvable configurations on trees rooted at r?

RP2: Is $\pi_{\mathbf{T}}(G, r) \leq 2\pi(G, r)$ for every rooted graph (G, r)?

RP3: Find larger classes of strategies than those arising from trees.

Complexity

Here we discuss questions such as how long it takes to decide if a particular configuration C on a graph G is D-solvable, or to calculate $\pi(G, r)$ for a rooted graph (G, r).

DEFINITIONS

D16: A graph G is a ***split*** graph if its vertices can be partitioned into a clique K and an independent set I.

D17: Let $\mathcal{H}$ be a hypergraph with vertices $V(\mathcal{H})$ and edges $E(\mathcal{H}) = \{e_1, \dots, e_k\}$. For a given t define the ***pebbling graph*** $G = G_t(\mathcal{H})$ as follows. The vertices of G are given by $V(G) = V(\mathcal{H}) \cup E(\mathcal{H}) \cup \{u_1, \dots, u_k\} \cup \{r, w_1, \dots, w_t\}$. The edges of G include $ve_i \in E(G)$ for every $v \in e_i$, as well as the paths $w_t u_i e_i$ for every $i \leq k$ and the path $r w_1 \cdots w_t$.

D18: We define **SOLVABLE** to be the problem of deciding, for configurations C and D on a graph G, if C is D-solvable.

D19: We define **UPPERBOUND** to be the problem of deciding, for given k and configuration D on a graph G, if $\pi(G, D) \leq k$.

FACTS

F22: [MiCl06] The configuration C is D-***solvable on*** G if and only if there is a nonnegative integral solution to the system $\{C(u) + \sum_{v \in V}(x_{v,u} - 2x_{u,v}) \geq D(u)$ for all $u \in V\}$. Hence **SOLVABLE** $\in$ NP.

F23: [HuKi05] Let $\mathcal{H}$ be a 4-uniform hypergraph on 2^{t+2} vertices with pebbling graph $G = G_t(\mathcal{H})$. Define the configuration C on G by $C(v) = 2$ for all $v \in V(\mathcal{H})$ and $C(v) = 0$ otherwise. Then C is r-solvable if and only if $\mathcal{H}$ has a perfect matching. Hence **SOLVABLE** is NP-complete.

F24: [CuLeSiTa12] When restricted to the class of diameter two graphs, **SOLVABLE** remains NP-complete.

F25: [CuDiLe12] When restricted to the class of planar graphs, **SOLVABLE** remains NP-complete.

F26: [CuDiLe12] When restricted to the class of diameter two planar graphs, **SOLVABLE** $\in$ P.

F27: [MiCl06] **UPPERBOUND** is complete for the class of decision problems computable in polynomial time by a co-NP machine equipped with an NP-complete oracle (Π_2^P-complete).

F28: [BuChCrMiWe08] If (T, r) is a rooted tree then $\pi(T, r)$ can be calculated in linear time. Moreover, for any configuration C, in linear time, we can find an r-solution or determine that none exists.

RESEARCH PROBLEMS

RP4: Is r-**SOLVABLE** $\in$ P when restricted to the class of cubes?

RP5: Is r-**SOLVABLE** $\in$ P when restricted to the class of split graphs?

11.4.2   Pebbling Numbers

We turn our attention now to configurations that solve every possible root.

DEFINITIONS

D20: We say that a configuration C on G is **(k-fold) solvable** if it is (k-fold) r-solvable for every vertex r.

D21: The **pebbling number** $\pi(G)$ is defined to be the minimum number t so that every configuration C on G of size t is solvable.

D22: For two graphs G_1 and G_2, define the **cartesian product** $G_1 \square G_2$ to be the graph with vertex set $V(G_1 \square G_2) = \{(v_1, v_2) | v_1 \in V(G_1), v_2 \in V(G_2)\}$ and edge set $E(G_1 \square G_2) = \{\{(v_1, v_2), (w_1, w_2)\} | (v_1 = w_1 \text{ and } (v_2, w_2) \in E(G_2)) \text{ or } (v_2 = w_2 \text{ and } (v_1, w_1) \in E(G_1))\}$. We write $\Pi_{i=1}^{k} G_i$ to mean $G_1 \square \ldots \square G_k$ and set $G^k = \Pi_{i=1}^{k} G$.

D23: The **support** $s(C)$ of a configuration C on G is the set of vertices that have a pebble of C; i.e., $s(C) = \{v \in V(G) \mid C(v) > 0\}$. The size of the support is denoted $\sigma(C) = |s(C)|$.

D24: A graph G has the **2-pebbling property** if every configuration C of size at least $2\pi(G) - \sigma(C) + 1$ is 2-fold solvable. A **Lemke** graph is any graph that does not have the 2-pebbling property; the smallest of these is called **the Lemke graph**.

D25: For $n = 2k(+1)$, the **sun** S_n is the split graph with perfect matching joining $I = I_k$ to $K = K_k$ (and one extra leaf when n is odd).

D26: For $m \geq 2t + 1$ the **Kneser** graph $K(m, t)$ has as vertices all t-subsets of $\{1, 2, \ldots, m\}$ and edges between every pair of disjoint sets. For example $K_n = K(n, 1)$ and $P = K(5, 2)$.

EXAMPLES

E7: The complete graph K_n on n vertices has $\pi(K_n) = n$.

E8: The complete graph K_n has the 2-pebbling property because the maximum number of pebbles that can be placed on σ vertices without having two vertices with at least two pebbles or one vertex with at least four pebbles is $\sigma + 2$, which is stricly less than $2n - \sigma + 1$.

E9: The smallest graph without the 2-pebbling property is the Lemke graph L, shown in Figure 11.4.7. We have $\pi(L) = 8$ and $|C| = 12 = 2(8) - 5 + 1$, but C cannot place two pebbles on r.

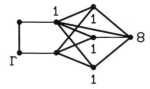

Figure 11.4.7: The Lemke graph with a 2-fold r-unsolvable configuration.

FACTS

F29: [Ch89] The path P_n on n vertices has $\pi(P_n) = 2^{n-1}$. More generally, let r^* be a leaf of a longest path in a tree T. Then $\pi(T) = \pi(T, r^*)$.

F30: [Hu99] The Petersen graph P has $\pi(P) = 10$.

F31: [PaSnVo95] The cycle C_n on $n \geq 3$ vertices has $\pi(C_{2k}) = 2^k$ for all $k \geq 2$ and $\pi(C_{2k+1}) = \lceil (2^{k+2} - 1)/3 \rceil$ for all $k \geq 1$.

F32: [AlGuHu12] If G is a diameter 3 split graph then $\pi(G)$ is given as follows. Let x be the number of cut vertices of G and, for a vertex r, define $\delta^*(G, r)$ to be the minimum degree of a vertex at maximum distance from r.

1. If x ≥ 2 then
$$\pi(G) = n + \text{x} + 2.$$

2. If x $= 1$ then
$$\pi(G) = \begin{cases} n + 5 - \delta^* & \text{if } r \text{ is a leaf with } \text{ecc}(r) = 3 \text{ and } \delta^* = \delta^*(G,r) \leq 4; \\ n + 1 & \text{otherwise.} \end{cases}$$

3. If x $= 0$ then
$$\pi(G) = \begin{cases} n + 4 - \delta^* & \text{if there is a cone vertex } r \text{ with } \deg(r) = 2, \\ & \quad \text{ecc}(r) = 3, \text{ and } \delta^* = \delta^*(G,r) \leq 3; \\ n + 1 & \text{if no such } r \text{ exists and } G \text{ is Pereyra;} \\ n & \text{otherwise.} \end{cases}$$

F33: [Ch89] The d-dimensional cube Q^d has $\pi(Q^d) = 2^d$. More generally, let $G = \Pi_{i=1}^{k} P_{l_i+1}$ be the cartesian product of k paths of lengths $l_i = diam(P_{l+i+1})$, with $l = \sum_{i=1}^{k} l_i$. Then $\pi(G) = 2^l$.

F34: [FoSn00, He03] If G_1 and G_2 are both cycles then $\pi(G_1 \square G_2) \leq \pi(G_1)\pi(G_2)$.

F35: [Ch89, FoSn00] If G_1 and G_2 are both trees then $\pi(G_1 \square G_2) \leq \pi(G_1)\pi(G_2)$.

F36: [Ch89, Mo92, He08] If G is a tree, cycle, complete graph, or complete bipartite graph and H has the 2-pebbling property then $\pi(G \square H) \leq \pi(G)\pi(H)$.

F37: [Wa01] None of the graphs in Figure 11.4.8 has the 2-pebbling property.

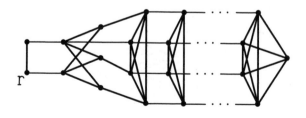

Figure 11.4.8: An infinite family of Lemke graphs.

F38: [GaYi12] If G is a bipartite graph with largest part size $s \geq 15$ and minimum degree at least $\lceil \frac{s+1}{2} \rceil$ then G is Class 0 and has the 2-pebbling property.

F39: [CzHuKiTr02] If G_1 and G_2 are connected graphs on n vertices that satisfy $\delta(G_i) \geq k$ and $k \geq 2^{12n/k+15}$, then $\pi(G_1 \square G_2) \leq \pi(G_1)\pi(G_2)$.

F40: If G is a graph on n vertices with $\mathsf{diam}(G) = d$ then $e_\pi(G) \geq d/\lg n$.

RESEARCH PROBLEMS

RP6: Does every bipartite graph have the 2-pebbling property?

RP7: [**Graham's Conjecture**] Every pair of graphs G_1 and G_2 satisfy $\pi(G_1 \square G_2) \leq \pi(G_1)\pi(G_2)$.

RP8: Is $\pi(L^2) = 64$?

Diameter, Connectivity, and Class 0

In this subsection we study graphs having smallest possible pebbling number.

DEFINITIONS

D27: A graph G is of **Class 0** if $\pi(G) = n$.

D28: The kth **graph power** $G^{(k)}$ of a graph G is formed from G by adding edges between every pair of vertices of distance at most k in G.

D29: The **pyramid** is any graph on 6 vertices isomorphic to the union of the 6-cycle (r, a, p, c, q, b) and the (inner) triangle (a, b, c). A **near-pyramid** is a pyramid minus one of the edges of its inner triangle (a, b, c).

D30: A graph G is **pyramidal** if it contains an induced (near-) pyramid, having 6-cycle C and inner (near-) triangle K, and can be drawn in the plane so that

  1. the edges of K are drawn in the interior of the region bounded by C and

  2. every other edge of G can be drawn inside the convex hull of exactly one of the sets $\{r, a, b\}$, $\{p, a, c\}$, $\{q, b, c\}$, or $\{a, b, c\}$.

D31: Define the **pebbling exponent** $e_\pi(G)$ of a graph G to be the minimum k such that $G^{(k)}$ is Class 0.

EXAMPLES

E10: Complete graphs, balanced complete bipartite graphs, cubes, and the Petersen graph are all Class 0.

E11: If u is a cut vertex of G then, for vertices r and v in different components of $G - u$, the configuration C with $C(r, u, v) = (0, 0, 3)$ and $C(w) = 1$ otherwise is r-unsolvable of size n.

E12: Let G be pyramidal with 6-cycle (r, a, p, c, q, b) and inner (near-) triangle (a, b, c). Then the configuration C with $C(r, a, p, c, q, b) = (0, 0, 3, 0, 3, 0)$ and $C(v) = 1$ otherwise is r-unsolvable of size n.

FACTS

F41: [ChGo03] Set $d = \mathsf{diam}(G)$. Then

1. $\pi(G) \leq (n - d)(2^d - 1) + 1$,

2. $\pi(G) \leq (n + \lfloor \frac{n-1}{d} \rfloor - 1)2^{d-1} - n + 2$, and

3. $\pi(G) \leq 2^{d-1}(n + 2\mathsf{dom}(G)) - \mathsf{dom}(G) + 1$.

The inequalities in parts 1 and 2 are sharp, and the coefficient of 2 in part 3 can be reduced to 1 in the case of perfect domination.

F42: If G is Class 0 then $\kappa(G) \geq 2$.

F43: [BlCzFuHeHuSc12] If G is Class 0 with n vertices and e edges then $e \geq \lfloor 3n/2 \rfloor$.

F44: [PaSnVo95] If G is a graph with n vertices and e edges and $e \geq \binom{n-1}{2} + 2$ then G is Class 0. Because the complete graph K_{n-1} plus a pendant edge has a cut vertex, this result is tight.

F45: [PaSnVo95] If $\mathsf{diam}(G) = 2$ then $\pi(G) \leq n(G) + 1$.

F46: [ClHoHu97] If $\mathsf{diam}(G) = 2$ and $\kappa(G) \geq 2$ then $\pi(G) = n + 1$ if and only if G is pyramidal.

F47: [ClHoHu97] If $\mathsf{diam}(G) = 2$ and $\kappa(G) \geq 3$ then G is Class 0.

F48: [CzHuKiTr02] There is a function $k(d) \leq 2^{2d+3}$ such that if G is a graph with $\mathsf{diam}(G) = d$ and $\kappa(G) \geq k(d)$ then G is of Class 0. Moreover, $k(d) \geq 2^d/d$.

F49: [CzHuKiTr02] For any constant $c > 0$ there is an integer t_0 such that, for $t > t_0$, $s \geq c(t/\lg_2 t)^{1/2}$ and $m = 2t + s$, we have $\kappa(K(m, t)) \geq 2^{2d+3}$, where $d = \mathsf{diam}(K(m, t))$; hence $K(m, t)$ is Class 0.

F50: [CzHuKiTr02] Let $G \in \mathcal{G}(n, p)$ be a random graph on n vertices with edge probability p and let $d = \mathsf{diam}(G)$. If $p \gg (n \lg_2 n)^{1/d}/n$ then $\Pr[\kappa(G) \geq 2^{2d+3}] \to 1$ as $n \to \infty$; hence $\Pr[G$ is Class 0$] \to 1$ as $n \to \infty$.

F51: [AlGuHu12] If G is a split graph with $\delta(G) \geq 3$ then G is Class 0.

F52: [Hu10] The pebbling exponent of the cycle satisfies

$$\frac{n/2}{\lg n} \leq e_\pi(C_n) \leq \frac{n/2}{\lg n - \lg \lg n}.$$

F53: [PoStYe] If $\mathsf{diam}(G) = 3$ then $\pi(G) \leq \lfloor 3n/2 \rfloor + 2$, which is best possible, as shown by the sun S_n.

F54: [PoStYe] If $\mathsf{diam}(G) = 4$ then $\pi(G) \leq 3n/2 + c$, for some constant c.

F55: [Po12] If $\mathsf{diam}(G) = d$ then $\pi(G) \leq (2^{\lceil d/2 \rceil} - 1)n/\lceil d/2 \rceil + c$, for some constant c.

F56: [CzHu03] There is a constant c so that if $\delta(G_i) > cn/\lg n$ for $i \in \{1, 2\}$ then $G_1 \square G_2$ is Class 0.

F57: [CzHu06] Let $g_0(n)$ denote the maximum number g such that there exists a Class 0 graph G on at most n vertices with finite $\mathsf{gir}(G) \geq g$. Then for all $n \geq 3$ we have

$$\left\lfloor \sqrt{(\lg_2 n)/2 + 1/4} - 1/2 \right\rfloor \leq g_0(n) \leq 1 + 2\lg_2 n.$$

RESEARCH PROBLEMS

RP9: Find infinitely many Class 0 graphs with n vertices and at most $3n/2 + o(n)$ edges.

RP10: Decide if $K(m, t)$ is Class 0 for all $m = 2t + s$ with $s \in O((t/\lg_2 t)^{1/2})$.

RP11: Find the smallest $k(d)$ such that G is Class 0 for every diameter d graph G with $\kappa(G) \geq k(d)$.

Complexity

Calculating $\pi(G)$ and $\pi(G, r)$ are polynomially equivalent, but it may be possible to calculate $\pi(G)$ faster than by calculating $\pi(G, r)$ for every r.

DEFINITION

D32: We define **PEBBLINGNUMBER** to be the problem of deciding if $\pi(G) \leq k$.

FACTS

F58: [HeHeHu13] Calculating $\pi(G)$ when G is a diameter two graph can be done in $O(n^4)$ time.

F59: [AlGuHu12] Calculating $\pi(G)$ when G is a split graph can be done in $O(n^\beta)$ time, where $\omega \cong 2.376$ is the exponent of matrix multiplication and $\beta = 2\omega/(\omega + 1) \cong 1.41$.

RESEARCH PROBLEM

RP12: Is **PEBBLINGNUMBER** $\in$ P when restricted to interval graphs of fixed diameter?

11.4.3   Optimal Pebbling

While pebbling can be thought of as a worst-case scenario — we give an adversary enough pebbles so that we can solve the graph no matter how he arranges them — optimal pebbling can be considered a best-case scenario — we place few pebbles carefully so as to solve the graph.

DEFINITIONS

D33: The *optimal pebbling number* $\pi^*(G)$ is the minimum number t for which there exists a solvable configuration of size t.

D34: Let C be a configuration on G and suppose that $\deg(v) = 2$ and $C(v) \geq 3$. A *smoothing move* at v removes two pebbles from v and adds one pebble to each of its neighbors. A *smooth* configuration has no smoothing move available; that is, C is smooth if $C(v) \leq 2$ whenever v has degree 2.

D35: For $S \subseteq V(G)$, the operation of *collapsing* S forms a new graph H in which S is replaced by a single vertex that is adjacent to all the neighbors of vertices of S that are in $V - S$. (Note that S need not be connected.)

EXAMPLES

E13: Figure 11.4.9 displays the upper bound of $\pi^*(P_8) \leq \lceil 2(8)/3 \rceil$.

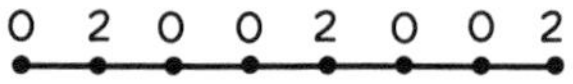

$$0 \quad 2 \quad 0 \quad 0 \quad 2 \quad 0 \quad 0 \quad 2$$

Figure 11.4.9: A minimum solvable configuration on the path P_8.

E14: The configuration with 2 pebbles on a single vertex can reach any other vertex of the complete graph, and so $\pi^*(K_n) = 2$ for all n.

FACTS

F60: Every graph G satisfies $\pi^*(G) \leq 2\mathrm{dom}(G)$.

F61: [**Smoothing Lemma**] [BuChCrMiWe08] If G has at least 3 vertices then G has a smooth minimum solvable configuration with no pebbles on leaves.

F62: [**Collapsing Lemma**] [BuChCrMiWe08] If H is obtained from G by collapsing sets of vertices then $\pi^*(G) \geq \pi^*(H)$.

F63: [BuChCrMiWe08] Every graph G satisfies $\pi^*(G) \leq \lceil 2n/3 \rceil$, with equality for paths and cycles.

F64: [Mo98] The d-cube has $(4/3)^d \leq \pi^*(Q^d) \leq (4/3)^{d+O(\lg d)}$.

F65: [FuSh00, HeHeHu11] For all graphs G and H we have $\pi^*(G \square H) \leq \pi^*(G)\pi^*(H)$.

F66: [BuChCrMiWe08] If G has n vertices and $\delta(G) = k$, then $\pi^*(G) \leq \frac{4n}{k+1}$.

F67: [BuChCrMiWe08] For all $t \geq 1$, $k = 3t$ and $n \geq k+3$, there is a graph G with n vertices, $\delta(G) = k$ and $\pi^*(G) \geq (2.4 - \frac{24}{5k+15} - o(1))\frac{n}{k+1}$.

Complexity

DEFINITION

D36: We define **OPTIMALPEBBLINGNUMBER** to be the problem of deciding if $\pi^*(G) \le k$.

FACT

F68: [MiCl06] The problem OPTIMALPEBBLINGNUMBER is NP-complete.

RESEARCH PROBLEMS

RP13: Is there a graph G with $\pi^*(G) \ge 3n(G)/(\delta(G)+1)$?

RP14: Does $\delta(G) \ge 3$ imply that $\pi^*(G) \le \lceil n(G)/2 \rceil$?

11.4.4   Thresholds

The probabilistic model of pebbling studies the typical case; that is, small configurations are usually unsolvable and large configurations are usually solvable — at roughly how many pebbles is the transition?

We assume that all sequences $\mathcal{G} = (G_1, \ldots, G_k, \ldots)$ of graphs considered have an increasing number of vertices $n = n_k = n(G_k)$.

NOTATION

The sequences of complete graphs, stars, paths, cycles, and cubes are denoted $\mathcal{K}$, $\mathcal{S}$, $\mathcal{P}$, $\mathcal{C}$, and $\mathcal{Q}$, respectively. The sequence of graph products is written $\mathcal{G} \square \mathcal{H} = (G_1 \square H_1, \ldots, G_k \square H_k, \ldots)$, with $\mathcal{G}^2 = \mathcal{G} \square \mathcal{G}$. For sets of functions A and B on the integers we write $A \lesssim B$ to mean that $a \in O(b)$ for every $a \in A, b \in B$.

DEFINITION

D37: Let $C_k : [n] \to \mathcal{N}$ denote a configuration on $V(G_k)$ and, for a function $h : \mathcal{N} \to \mathcal{N}$ and fixed $n = n_k$, define the uniform probability space $X_{n,h}$ of all configurations C_k of size $h = h(n)$. Denote by P_n^+ the probability that C_n is solvable on G_k and let $t : \mathcal{N} \to \mathcal{N}$ be any function. We say that t is a **pebbling threshold** for $\mathcal{G}$, and write $\tau(\mathcal{G}) = \Theta(t)$, if $P_n^+ \to 0$ whenever $h(n) \ll t(n)$ and $P_n^+ \to 1$ whenever $h(n) \gg t(n)$.

EXAMPLE

E15: Solvability on K_k is equivalent to the labelled version of Feller's Birthday Problem (also hashing collisions in computer science). Thus $\tau(\mathcal{K}) = \Theta(\sqrt{n})$.

FACTS

F69: [BeBrCzHu03] Every graph sequence $\mathcal{G}$ has nonempty threshold $\tau(\mathcal{G})$.

F70: [CzEaHuKa02] Every graph sequence $\mathcal{G}$ satisfies $\tau(\mathcal{K}) \lesssim \tau(\mathcal{G}) \lesssim \tau(\mathcal{P})$.

F71: [CzEaHuKa02, BeBrCzHu03, GoJaSaWi04, CzHu08] For every constant $c > 1$, we have $\tau(\mathcal{P}) \subseteq \Omega\left(n2^{\sqrt{\lg n}/c}\right) \cap O\left(n2^{c\sqrt{\lg n}}\right)$.

F72: [BeHu08] The sequence of squares of cliques have threshold $\tau(\mathcal{K}^2) = \Theta(\sqrt{n})$.

F73: [Al03, CzWa03] For all $\epsilon > 0$ the sequence of cubes has threshold $\tau(\mathcal{Q}) \in \Omega(n^{1-\epsilon}) \cap O(n/(\lg \lg n)^{1-\epsilon})$.

F74: If $\mathcal{G}$ is a sequence of graphs of bounded diameter then $\tau(\mathcal{G}) \subseteq O(n)$.

F75: [CzHu08] Let t_1 and t_2 be functions satisfying $\tau(\mathcal{K}) \lesssim t_1 \ll t_2 \lesssim \Theta(n)$. Then there is some graph sequence $\mathcal{G}$ such that $t_1 \lesssim \tau(\mathcal{G}) \lesssim t_2$.

F76: [BjHo12] There exist graph sequences $\mathcal{G} = (G_1, \ldots, G_k, \ldots)$ and $\mathcal{H} = (H_1, \ldots, H_k, \ldots)$ such that $\pi(G_k) < \pi(H_k)$ for all k but $\tau(\mathcal{H}) \lesssim \tau(\mathcal{G})$.

F77: [CzHu06] Suppose that $t \in \tau(\mathcal{P})$ and $s \in \tau(\mathcal{P}^2)$. Then $s(n) \in O\left(t\left(\sqrt{n}\right)^2\right)$.

F78: [CzHu03] Define $\mathbf{G}(n, \delta)$ to be the set of all connected graphs on n vertices having minimum degree at least $\delta = \delta(n)$. Let $\mathcal{G}_\delta = \{G_1, \ldots, G_k, \ldots\}$ denote any sequence of graphs with each $G_k \in \mathbf{G}(k, \delta)$. For every function $n^{1/2} \ll \delta = \delta(n) \leq n - 1$, $\tau(\mathcal{G}_\delta) \subseteq O(n^{3/2}/\delta)$. In particular, if in addition $\delta \in \Omega(n)$ then $\tau(\mathcal{G}_\delta) = \Theta(n^{1/2})$.

REMARKS

R1: Note the need to rescale threshold functions of products of graph sequences in terms of the new number of vertices $n(G_k \square H_k) = n(G_k) \square n(H_k)$; for example, in Fact F72 we have $\sqrt{n^2} = \sqrt{n}^2$.

RESEARCH PROBLEMS

RP15: Determine $\tau(\mathcal{P})$.

RP16: Determine $\tau(\mathcal{Q})$.

RP17: Extend Fact F75 to the range $\Omega(n) \cap \tau(\mathcal{P})$.

RP18: Suppose that $\mathcal{G}$ is any graph sequence, $t \in \tau(\mathcal{G})$ and $s \in \tau(\mathcal{G}^2)$. Is it true that $s(n) \in O\left(t\left(\sqrt{n}\right)^2\right)$?

11.4.5   Other Variations

Here we present a few variations on the pebbling theme and a taste of the main results for each.

NOTATION

We write kI_v for the configuration with k pebbles on v and 0 elsewhere, and J for the configuration with 1 pebble on each vertex. Also $k\mathscr{I}$ denotes the set of all such kI_v, and

$\mathscr{C}_t$ is the set of all configurations of size t. Denote by $\mathscr{M}(G)$ the set of all configurations corresponding to dominating sets in a graph G; that is, the configuration corresponding to a dominating set has one pebble on each of its vertices and none elsewhere. Next, consider the set of all induced paths on $d+1$ vertices in G and write $\mathscr{P}_d^+(G)$ for those configurations on such paths with two pebbles on one leaf of the path, no pebbles on the other leaf, and one pebble on all other vertices of the path. Finally, for $\mathbf{d} = \langle d_1, \ldots, d_m \rangle$ let $P^{\mathbf{d}}$ denote the graph $P_{d_1+1} \square \cdots \square P_{d_m+1}$.

DEFINITIONS

D38: Let G be a **weighted graph** with edge weights $\mathsf{w} : E(G) \to \mathcal{N}$. For an edge $\{u, v\} \in E(G)$, if u has at least $\mathsf{w}(uv)$ pebbles on it, then a **weighted pebbling step from u to v** removes $\mathsf{w}(uv)$ pebbles from u and places one pebble on v. The corresponding **weighted pebbling number** $\pi(G_{\mathsf{w}})$ is defined to be the minimum number t so that every configuration of size t solves any r via weighted pebbling steps.

D39: For a graph G and set of configurations $\mathscr{D}$ on G the **(optimal) pebbling number** $\pi(G, \mathscr{D})$ (resp. $\pi^*(G, \mathscr{D})$) is the minimum t for which every (resp. some) $C \in \mathscr{C}_t$ is D-solvable for every $D \in \mathscr{D}$. The k-**fold (optimal) pebbling number** $\pi_k(G) = \pi(G, k\mathscr{I})$ (resp. $\pi_k^*(G) = \pi^*(G, k\mathscr{I})$).

D40: For a graph G the **fractional (optimal) pebbling number** is defined to be $\hat{\pi}(G) = \lim_{k \to \infty} \pi_k(G)/k$ (resp. $\hat{\pi}^*(G) = \lim_{k \to \infty} \pi_k^*(G)/k$).

D41: The **cover pebbling number** of a graph G is defined to be $\pi(G, J)$. A configuration D is **positive** if $D(v) > 0$ for every vertex v. For positive D on G we define the function $s(G, D) = \max_v \sum_u D(u) 2^{\mathrm{dist}(u,v)}$.

D42: For a set of configurations $\mathscr{D}$ the configuration C is **weakly $\mathscr{D}$-solvable** if C solves some $D \in \mathscr{D}$. The **target pebbling number** $\pi^-(G, \mathscr{D})$ is the minimum t for which every $C \in \mathscr{C}_t$ is weakly D-solvable.

D43: The **domination target pebbling number** of G is defined to be $\pi^-(G, \mathscr{M})$.

D44: For given d the **distance pebbling number** $\vec{\pi}_d(G)$ of a graph G is defined to be the minimum t such that, for every size t configuration, there is some pebble that can move to a vertex at distance d from where it started; in other words, $\vec{\pi}_d(G) = \pi^-(G, \mathscr{P}_d^+)$.

D45: A **rubbling step** on a graph G is either a pebbling step or a strict rubbling step. A **strict rubbling step** takes one pebble from each of two neighbors u and w of a vertex v and places one pebble on v. The **(optimal) rubbling number** of G, denoted $\rho(G)$ (resp. $\rho^*(G)$), is defined to be the minimum t so that every (resp. some) configuration of size t can solve any root vertex r via rubbling steps.

EXAMPLES

E16: For a positive configuration D on the complete graph we have $s = s(K_n, D) = 2|D| - \min D$. If $D(v) = \min D$ then the configuration that places $s - 1$ pebbles on v and none elsewhere cannot solve D.

E17: For a vertex $v \in V(G)$, the configuration that places no pebbles on v and all its neighbors and one pebble on every other vertex cannot solve any dominating set of G.

E18: The pigeonhole principle implies that $\vec{\pi}_1(G) = n(G) + 1$ for every graph G.

E19: Every configuration of two pebbles on K_n solves every vertex via rubbling steps.

FACTS

F79: [Ch89] Given $\mathbf{d} = \langle d_1, \ldots, d_m \rangle$ we represent the vertices of $P^{\mathbf{d}}$ by coordinates $\langle v_1, \ldots, v_m \rangle$ with each $0 \le v_i \le d_i$ and denote by $\mathbf{e}_i = \langle 0, \ldots, 1, \ldots, 0 \rangle$ the ith standard basis vector. For any $\mathbf{w} = \langle w_1, \ldots, w_m \rangle$ define the weight function $\mathbf{w}(\mathbf{uv}) = w_i$ when $|\mathbf{u} - \mathbf{v}| = \mathbf{e}_i$ and write $\mathbf{w}^{\mathbf{d}} = \prod_{i=1}^{m} w_i^{d_i}$. Then $\pi(P_{\mathbf{w}}^{\mathbf{d}}) = \mathbf{w}^{\mathbf{d}}$.

F80: [Ch89] For every root r of a tree T we have $\pi_k(T, r) = f_k(T, r)$ for all k.

F81: [HeHeHu13] If G is a diameter two graph with n vertices and m edges then $\pi_k(G) \le \pi(G) + 4(k-1)$. Furthermore, from any configuration of size at least $\pi(G) + 4(k-1)$, k pebbles can be placed on any root vertex r in at most $6n + \min\{3t, m\}$ steps.

F82: [HeHeHu13, HoMaOkZu11] Every graph G satisfies $\hat{\pi}(G) = 2^{\text{diam}(G)}$.

F83: [HeHeHu13] If G is a complete graph, cycle, tree, or has $\pi(G) = 2^{\text{diam}(G)}$ then $\pi(G, \mathscr{C}_k) = \pi_k(G)$.

F84: [HeHeHu13] For all $d \ge 0$, $n \ge 1$, and $k \ge 1$ we have $\hat{\pi}^*(K_n) = 2n/(n+1)$, $\hat{\pi}^*(P_n) = (n+2)/3$, $\hat{\pi}^*(C_{2k}) = k2^{k+1}/3(2^k - 1)$, $\hat{\pi}^*(C_{2k+1}) = (2k+1)(2^{k-1})/(3(2^{k-1}) - 1)$, $\hat{\pi}^*(Q^d) = (4/3)^d$, and $\hat{\pi}^*(P) = 5/2$.

F85: [**Stacking Theorem**] [Sj05] Every positive configuration D on a graph G has $\pi(G, D) = s(G, D)$.

F86: [GaGoTeVuWaYe08] The complete r-partite graph $K = K_{s_1, \ldots, s_t}$ has domination target pebbling number $\pi^-(K, \mathscr{M}) = 3$ if every $s_i = 2$ and $\max_i s_i$.

F87: [GaGoTeVuWaYe08] For the path P_n on n vertices, $\pi^-(P_n, \mathscr{M}) = 2(2^n - 2^{n \bmod 3})/7 + \left\lfloor \frac{n \bmod 3}{2} \right\rfloor$.

F88: [Kn12] The distance pebbling number of the cycle is $\vec{\pi}_d(C_n) = (2^d - 1)\lfloor n/d \rfloor + 2^{n \bmod d}$.

F89: [BeSi09] If G has n vertices and diameter d then $\rho(G) \le (n - d + 1)(2^{d-1} - 1)$.

F90: [KaSi] If G has n vertices and diameter 2 then $\rho(G) \le \sqrt{2n-1} + 5$. Furthermore, for all n there is a diameter 2 graph G with $\rho(G) \ge \lfloor \sqrt{2n-1} \rfloor + 2$.

F91: [KaSi] If G has n vertices and diameter d then $\lceil (d+2)/2 \rceil \le \rho^*(G) \le \lceil (n+1)/2 \rceil$.

REMARK

R2: Note that the Stacking Theorem implies that the computational complexity of calculating pebbling numbers applies only to nonpositive target configurations.

RESEARCH PROBLEMS

RP19: Is it true that $\pi(G, \mathscr{C}_k) = \pi_k(G)$ for every graph G?

RP20: Is there a constant c such that $\pi^*(G) \le c\hat{\pi}^*(G)$ for every graph G?

11.4.6 Applications

Graph pebbling arose as a method to prove a conjecture of Erdős and Lemke in combinatorial number theory. It has since produced a more general result in combinatorial group theory and another in *p*-adic diophantine equations.

FACTS

F92: [Ch89, ElHu05] Fact F79 implies that, if $g_1, \ldots, g_n$ is a sequence of elements of an abelian group $\mathbf{G}$ of size n, then there is a nonempty subsequence $(g_k)_{k \in K}$ such that $\sum_{k \in K} a_k = 0_{\mathbf{G}}$ and $\sum_{k \in K} 1/|g_k| \leq 1$, where $|g|$ denotes the order of the element g in $\mathbf{G}$ and $0_{\mathbf{G}}$ is the identity element in $\mathbf{G}$.

F93: [Kn12] Write $n = 2^t m$, where m is odd. Define $d = 1$ for odd n ($t = 0$) and $d = t + 2$ for even n ($t > 0$). If $s \geq \vec{\pi}(C_n, d)$ then, for all integer coefficients $a_1, \ldots, a_s$, the additive form $F(\mathbf{x}) = \sum_{i=1}^s a_i x_i^n$ has a nontrivial (not all zero) solution to $F(\mathbf{x}) = 0$ in the 2-adic integers.

REMARK

R3: The pebbling steps studied here have a cost in the loss of pebbles. Various no-cost rules for pebbling steps have been studied for years and have found applications in a wide array of areas. One version, dubbed *black and white* pebbling, was applied to computational complexity theory in studying time-space tradeoffs, as well as to optimal register allocation for compilers. Connections have been made also to pursuit and evasion games and graph searching. Another (*black* pebbling) is used to reorder large sparse matrices to minimize in-core storage during an out-of-core Cholesky factorization scheme. A third version yields results in computational geometry in the rigidity of graphs, matroids, and other structures.

RESEARCH PROBLEM

RP21: Prove that Fact F92 holds for all groups $\mathbf{G}$.

References

[AlGuHu12] L. Alcón, M. Gutierrez, and G. Hurlbert, Pebbling in split graphs, *preprint*, 2012.

[Al03] N. Alon, *personal communication*, 2003.

[BeBrCzHu03] A. Bekmetjev, G. Brightwell, A. Czygrinow, and G. Hurlbert, Thresholds for families of multisets, with an application to graph pebbling, *Discrete Math.* 269 (2003), 21–34.

[BeCu09] A. Bekmetjev and C. Cusack, Pebbling algorithms in diameter two graphs, *SIAM J. Discrete Math.* 23 (2009), 634–646.

[BeHu08] A. Bekmetjev and G. Hurlbert, The pebbling threshold of the square of cliques, *Discrete Math.* 308 (2008), 4306–4314.

[BeSi09] C. Belford and N. Sieben, Rubbling and optimal rubbling of graphs, *Discrete Math.* 309 (2009), 3436–3446.

[BjHo12] J. Björklund and C. Holmgren, Counterexamples to a monotonicity conjecture for the threshold pebbling number, *Discrete Math.* 312 (2012), 2401–2405.

[BlCzFuHeHuSc12] A. Blasiak, A. Czygrinow, A. Fu, D. Herscovici, G. Hurlbert, and J. R. Schmitt, Sparse graphs with small pebbling number, *preprint*, 2012.

[BlSc08] A. Blasiak and J. Schmitt, Degree sum conditions in graph pebbling, *Austral. J. Combin.* 42 (2008), 83–90.

[Bo02] J. Boyle, Thresholds for random distributions on graph sequences with applications to pebbling, *Discrete Math.* 259 (2002), 59–69.

[Bu06] B. Bukh, Maximum pebbling number of graphs of diameter three, *J. Graph Th.* 52 (2006), 353–357.

[BuChCrMiWe08] D. Bunde, E. Chambers, D. Cranston, K. Milans, and D. West, Pebbling and optimally pebbling in graphs, *J. Graph Theory* 57 (2008), 215–238.

[Ch89] F. R. K. Chung, Pebbling in hypercubes, *SIAM J. Discrete Math.* 2 (1989), 467–472.

[ChGo03] M. Chan and A. Godbole, Improved pebbling bounds, *Discrete Math.* 308 (2003), 2301–2306.

[ClHoHu97] T. Clarke, R. Hochberg and G. Hurlbert, Pebbling in diameter two graphs and products of paths, *J. Graph Th.* 25 (1997), 119–128.

[CrCuFeHuPuSzTu05] B. Crull, T. Cundif, P. Feltman, G. Hurlbert, L. Pudwell, Z. Szaniszlo, and Z. Tuza, The cover pebbling number of graphs, *Discrete Math.* 296 (2005), 15–23.

[CuDiLe12] C. Cusack, L. Dion and T. Lewis, The complexity of pebbling reachability in planar graphs, *preprint*, 2012.

[CuHiHuMo09] D. Curtis, T. Hines, G. Hurlbert, and T. Moyer, Pebbling graphs by their blocks, *Integers: Elec. J. Combin. Number Theory*, 9:#G02 (2009), 411–422.

[CuLeSiTa12] C. Cusack, T. Lewis, D. Simpson, and S. Taggart, The complexity of pebbling in diameter two graphs, *SIAM J. Discrete Math.* 26 (2012), 919–928.

[CzEaHuKa02] A. Czygrinow, N. Eaton, G. Hurlbert, and P. M. Kayll, On pebbling threshold functions for graph sequences, *Discrete Math.* 247 (2002), 93–105.

[CzHu03] A. Czygrinow and G. Hurlbert, Pebbling in dense graphs, *Austral. J. Combin.* 29 (2003), 201–208.

[CzHu06] A. Czygrinow and G. Hurlbert, Girth, pebbling and grid thresholds, *SIAM J. Discrete Math.* 20 (2006), 1–10.

[CzHu08] A. Czygrinow and G. Hurlbert, On the pebbling threshold of paths and the pebbling threshold spectrum, *Discrete Math.* 308 (2008), 3297–3307.

[CzHuKiTr02] A. Czygrinow, G. Hurlbert, H. Kierstead, and W. T. Trotter, A note on graph pebbling, *Graphs and Combin.* 18 (2002), 219–225.

[CzWa03] A. Czygrinow and M. Wagner, *unpublished*, 2003.

[ElHu05] S. Elledge and G. Hurlbert, An application of graph pebbling to zero-sum sequences in abelian groups, *Integers: Elec. J. Combin. Number Theory*, 5(1):#A17 (2005), 10 pp.

[FeKi01] R. Feng and J. Y. Kim, Graham's pebbling conjecture on product of complete bipartite graphs, *Sci. China Ser. A* 44 (2001), 817–822.

[FoSn00] J. A. Foster and H. S. Snevily, The 2-pebbling property and a conjecture of Graham's, *Graphs and Combin.* 16 (2000), 231–244.

[FrWy05] T. Friedman and C. Wyels, Optimal pebbling of paths and cycles, `arXiv:math/0506076`, 2005.

[FuSh00] H. L. Fu and C. L. Shiue, The optimal pebbling number of the complete m-ary tree, *Discrete Math.* 222 (2000), 89–100.

[FuSh02] H. L. Fu and C. L. Shiue, The optimal pebbling number of the caterpillar, *Taiwanese J. Math.* 13(2A) (2009), 419–429.

[GaGoTeVuWaYe08] J. Gardner, A. Godbole, A. Teguia, A. Vuong, N. Watson, and C. Yerger, Domination cover pebbling: graph families, *J. Combin. Math. Combin. Comput.* 64 (2008), 255–271.

[GaYi12] Z. Gao and J. Yin, The 2-pebbling property of bipartite graphs, *preprint*, 2012.

[GiLeTa80] J. Gilbert, T. Lengauer, and R. Tarjan, The pebbling problem is complete in polynomial space, *SIAM J. Comput.* 9 (1980), 513–525.

[GoJaSaWi04] A. Godbole, M. Jablonski, J. Salzman, and A. Wierman, An improved upper bound for the pebbling threshold of the n-path, *Discrete Math.* 275 (2004), 367–373.

[GuSh96] Y. Gurevich and S. Shelah, On finite rigid structures, *J. Symbolic Logic* 61 (1996), 549–562.

[He03] D. Herscovici, Graham's pebbling conjecture on products of cycles, *J. Graph Theory* 42 (2003), 141–154.

[He08] D. Herscovici, Graham's pebbling conjecture on products of many cycles, *Discrete Math.* 308 (2008), 6501–6512.

[He10] D. Herscovici, On graph pebbling numbers and Graham's conjecture, *Graph Theory Notes of New York* (2010), LIX:15–21.

[HeHeHu11] D. Herscovici, B. Hester, and G. Hurlbert, Optimal pebbling in products of graphs, *Austral. J. Combin.* 50 (2011), 3–24.

[HeHeHu12] D. Herscovici, B. Hester, and G. Hurlbert, Generalizations of Grahams pebbling conjecture, *Discrete Math.* 312 (2012), 2286–2293.

[HeHeHu13] D. Herscovici, B. Hester, and G. Hurlbert, t-Pebbling and extensions, *Graphs and Combinatorics*. To appear.

[HeHi98] D. Herscovici and A. Higgins, The pebbling number of $C_5 \square C_5$, *Discrete Math.* 187 (1998), 123–135.

[HoMaOkZu11] M. Hoffmann, J. Matousek, Y. Okamoto and P. Zumstein, The t-pebbling number is eventually linear in t, *Electron. J. Combin.* 18(1) #153 (2011), 4 pp.

[HoPaVa77] J. Hopcroft, W. Paul, and L. Valiant, On time versus space, *J. Assoc. Comput. Mach.* 24 (1977), 332–337.

[Hu99] G. Hurlbert, A survey of graph pebbling, *Congr. Numer.* 139 (1999), 41–64.

[Hu05] G. H. Hurlbert, Recent progress in graph pebbling, *Graph Theory Notes of New York* (2005), XLIX:25–37.

[Hu10] G. Hurlbert, A linear optimization technique for graph pebbling, *Preprints of the Centre de Recerca Matematica*, 988 (2010), 39 pp.

[Hu12] G. H. Hurlbert, General graph pebbling, *Discrete Appl. Math.* To appear.

[HurlGPP] G. Hurlbert, *The Graph Pebbling Page*, `mingus.la.asu.edu/~hurlbert/pebbling/pebb.html`.

[HuKi05] G. Hurlbert and H. Kierstead, Graph pebbling complexity and fractional pebbling, *unpublished*, 2005.

[HuMu06] G. Hurlbert and B. Munyan, Cover pebbling hypercubes, *Bull. Inst. Combin. Appl.* 47 (2006), 71–76.

[KaSi] G. Y. Katona and N. Sieben, Bounds on the rubbling and optimal rubbling numbers of graphs, *Graphs and Combin.* To appear.

[KiPi86] L. M. Kirousis and C. H. Papadimitriou, Searching and pebbling, *Theoret. Comput. Sci.* 47 (1986), 205–218.

[Kl85] M. Klawe, The complexity of pebbling for two classes of graphs, in *Graph Theory with Applications to Algorithms and Computer Science*, Y. Alavi, G. Chartrand, and L. Lesniak (Eds.), 475–487, Wiley, New York, 1985.

[Kn12] M. P. Knapp, 2-adic zeros of diagonal forms and distance pebbling of graphs, *preprint*, 2012.

[Li87] J. W. H. Liu, An application of generalized tree pebbling to sparse matrix factorization, *SIAM J. Algebraic Discrete Methods* 8 (1987), 375–395.

[MiCl06] K. Milans and B. Clark, The complexity of graph pebbling, *SIAM J. Discrete Math.* 20 (2006), 769–798.

[Mo92] D. Moews, Pebbling graphs, *J. Combin. Th. (Ser. B)* 55 (1992), 244–252.

[Mo98] D. Moews, Optimally pebbling hypercubes and powers. *Discrete Math.* 190 (1998), 271–276.

[Pa76] T. D. Parsons, Pursuit-evasion in a graph. In Y. Alani and D. R. Lick, editors, *Theory and Applications of Graphs*, 426–441, Springer, Berlin, 1976.

[PaHe70] M. S. Paterson and C. E. Hewitt, Comparative schematology. In J. Dennis, editor, *Proj. MAC Conf. on Concurrent Systems and Parallel Computation*, 119–127, Assoc. Computing Machinery, New York, 1970.

[PaSnVo95] L. Pachter, H. S. Snevily, and B. Voxman, On pebbling graphs, *Congr. Numer.* 107 (1995), 65–80.

[Po12] L. Postle, Pebbling graphs of fixed diameter, *preprint*, 2012.

[PoStYe] L. Postle, N. Streib and C. Yerger, Pebbling graphs of diameter three and four, *J. Graph Theory*. To appear.

[Se75] R. Sethi, Complete register allocation problems, *SIAM J. Comput.* 4 (1975), 226–248.

[Sj05] J. Sjostrand, The cover pebbling theorem, *Electron. J. Combin.*, 12:#22 (2005), 5 pp.

[StTh09] I. Streinu and L. Theran, Sparse hypergraphs and pebble game algorithms, *European J. Combin.* 30 (2009), 1944–1964.

[Wa01] S. Wang, Pebbling and Graham's conjecture, *Discrete Math.* 226 (2001), 431–438.

[YeZhZh12a] Y. Ye, P. Zhang, and Y. Zhang, Pebbling number of squares of odd cycles, *Discrete Math.* 312 (2012), 3174–3178.

[YeZhZh12b] Y. Ye, P. Zhang, and Y. Zhang, The pebbling number of squares of even cycles, *Discrete Math.* 312 (2012), 3203–3211.

Glossary for Chapter 11

alternating path – relative to a matching: a path whose edges alternate between free and matched.

augmenting path$_1$ – relative to a matching: an alternating path that starts at one free vertex and ends at another free vertex.

augmenting path$_2$ P – in a flow network: a directed path from s to t in the residual network.

___, **capacity of** – in a residual network $G_f = (V, E_f, s, t, u_f)$: denoted Δ_P and given by $\Delta_P = \min_{(v,w)\in E_f} r_f(v, w)$.

backward arc (v, w) – across a cut $\langle S, T\rangle$: when $v \in T$ and $w \in S$.

bandwidth – of a communication facility: capacity of a communication facility in bits per second (bps).

blossom: an odd length cycle formed by joining two even vertices of an alternating path, rooted at a free vertex.

cartesian product of two graphs $G_1 \square G_2$: the graph with vertex set $V(G_1 \square G_2) = \{(v_1, v_2) \mid v_1 \in V(G_1), v_2 \in V(G_2)\}$ and edge set $E(G_1 \square G_2) = \{\{(v_1, v_2), (w_1, w_2)\} \mid (v_1 = w_1 \text{ and } (v_2, w_2) \in E(G_2)) \text{ or } (v_2 = w_2 \text{ and } (v_1, w_1) \in E(G_1))\}$.

circulation – in a cost-flow network $G = (V, A, cap, c, b)$: a flow for the supply vector $b \equiv \mathbf{0}$.

Class 0: the set of graphs having pebbling number equal to its number of vertices.

complete graph K_n: the graph on n vertices with an edge between every pair of vertices.

configuration on a graph G: a function from the vertices of G to the nonnegative numbers that indicates how many pebbles are on each vertex.

___, **capacity of**: the number of pebbles in the configuration, i.e., the sum of the function values of the configuration.

cost-flow network $G = (V, A, cap, c, b)$: a directed graph with vertex-set V, arc-set A, a nonnegative capacity function $cap : A \to N$, a linear cost function $c : A \to Z$, and an integral supply vector $b : V \to Z$ that satisfies $\sum_{w\in V} b(w) = 0$.

___, **s-t**: a flow network $G = (V, A, cap, c, b)$ that contains two distinguished vertices s and t such that $b(v) = 0$ for all $v \in V - \{s, t\}$ and $b(s) = -b(t) > 0$.

___, **extended**: an s-t network $G' = (V', A', cap')$ of $G = (V, A, cap, c, b)$ with vertex-set $V' = V \cup \{s, t\}$, arc-set $A' = A \cup \{(s, v)|b(v) > 0\} \cup \{(w, t)|b(w) < 0\}$, and capacity function cap' defined by

$$cap'(v, w) = \begin{cases} cap(v, w), & \text{if } (v, w) \in A; \\ b(v), & \text{if } v = s; \\ -b(w), & \text{if } w = t \end{cases}$$

cover pebbling number of a graph G: equals $\pi(G, J)$, where J is the configuration with one pebble on each vertex.

cube graph: see *d-dimensional cube graph*.

s-t cut $\langle S, T \rangle$ – corresponding to a partition (S, T) of V: the set of arcs that have one endpoint in one of the sets and the other endpoint in the other set.

___, **capacity of**: the sum of the capacities of the arcs crossing the cut in the forward direction, i.e., $cap\langle S, T \rangle = \sum\limits_{(v,w) \in E : v \in S, w \in T} cap(v, w)$.

___, **minimum**: a cut of minimum value, i.e., $\min\{cap\langle S, T \rangle : \langle S, T \rangle \text{ is an } s\text{-}t \text{ cut}\}$.

cycle graph C_n: the path graph P_n with an extra edge joining the last vertex to the first.

d-dimensional cube graph Q_d: the graph on the 2^d vertices labeled by all binary d-tuples, with an edge between pairs of vertices whose labels differ in exactly one coordinate.

distance pebbling number $\vec{\pi}_d(G)$: the minimum t such that, for every size t configuration, there is some pebble that can move to a vertex at distance d from where it started.

dual – of a minimum-cost flow problem: see §11.2, Definition D14.

even vertex – relative to an alternating path: a vertex that is an even distance (number of edges) from the root of the path.

excess $e(v)$ – at vertex v: $e(v) = \sum\limits_{(w,v) \in E} f(w, v) - \sum\limits_{(v,w) \in E} f(v, w)$.

exponent: see *pebbling exponent*.

facility: a transmission medium (e.g., fibre-optic or wireless) installed on an edge for sending data, voice or video signals.

flow₁ f – in an s-t flow network $G = (V, E, s, t, cap)$: a function $f : E \to \mathbb{N}$ which obeys three types of constraints:

capacity constraints: $f(v, w) \leq cap(v, w)$, for each arc $(v, w) \in E$.

conservation constraints:

$\sum_{(w,v) \in E} f(w, v) = \sum_{(v,w) \in E} f(v, w)$ for each $v \in V - \{s, t\}$.

nonnegativity constraints: $f(v, w) \geq 0$, for each arc $(v, w) \in E$.

flow₂ f – in a cost-flow network $G = (V, A, cap, c, b)$: a function $f : A \to \mathbb{Z}$ that satisfies

capacity constraints: $f(v, w) \leq cap(v, w)$ for all $(v, w) \in A$,

flow conservation constraints: $\sum\limits_{w}[f(v, w) - f(w, v)] = b(v)$ for each $v \in V$,

nonnegativity constraints: $f(v, w) \geq 0$ for all $(v, w) \in A$.

___, **minimum**: a flow f with minimum $c^T f$ value among all flows.

___, **value of**: the total flow into the sink, i.e., $val(f) = \sum\limits_{(v,t) \in E} f(v, t)$.

flow across cut $\langle S, T \rangle$: the flow crossing the cut in the forward direction minus the flow crossing the cut in the backward direction, i.e.,

$$f\langle S, T \rangle = \sum_{(v,w) \in E : v \in S, w \in T} f(v, w) - \sum_{(v,w) \in E : v \in T, w \in S} f(v, w).$$

s-t flow network $G = (V, E, s, t, cap)$: a directed graph with vertex set V and arc-set E, two distinguished vertices, a *source* s and a *sink* t, and a nonnegative *capacity* function $cap : E \to \mathbb{N}$.

flow with gains – in a gain network $\bar{G} = (V, A, cap, \gamma, c, s, t)$: a function $f : A \to \mathbb{R}$ that satisfies:

capacity constraints: $f(v, w) \leq cap(v, w)$ for all $(v, w) \in A$,

nonnegativity constraints: $f(v, w) \geq 0$ for all $(v, w) \in A$,

flow conservation constraints:

$\sum_{w \in V}[f(v, w) - f(w, v)\gamma(w, v)] = 0$, for each $v \in V - \{s, t\}$.

___, **maximum**: a flow with gains that maximizes the amount of flow reaching t given an unlimited supply at s.

___, **minimum-cost maximum-**: a maximum flow with gains that minimizes $\sum_{e \in A} c(e)g(e)$.

flow-over-time network: a flow network $G_\tau = (V, A, cap, \tau, b)$, where each arc $(v, w) \in A$ has an associated *transit time* τ_{vw}. The transit time τ_{vw} represents the amount of time that elapses between when flow enters arc (v, w) at v and when the same flow arrives at w.

flow-over-time: see §11.2, Definition D26.

k-**fold** r-**solvable configuration**: a configuration from which it is possible to place at least k pebbles on vertex r via pebbling steps.

forward arc (v, w) – across a cut $\langle S, T \rangle$: when $v \in S$ and $w \in T$.

free edges – of a matching M: the edges of the graph not in M.

free vertices – of a graph G: the vertices of G not incident on a matched edge.

gain network $\bar{G} = (V, A, cap, \gamma, c, s, t)$: a network $G = (V, A, cap, c, b)$ with positive-valued *gain function* $\gamma : E \to \mathbb{R}^+$ and supply function $b_v = 0$ for all $v \in V - \{s, t\}$; the *gain factor* $\gamma(e) > 0$ for arc e enforces that for each unit of flow that enters the arc, $\gamma(e)$ units exit. For standard network flows, the gain factor of every arc is one.

graph power $G^{(k)}$: the graph formed from G by adding edges between every pair of vertices of distance at most k in G.

increasable arc (v, w) – in network G for a given flow: $f(v, w) < cap(v, w)$.

Kneser graph $K(m, t)$: the graph having as vertices all t-subsets of $\{1, 2, \ldots, m\}$, with edges between every pair of disjoint sets.

Lemke graph: a graph that does not have the 2-pebbling property is *a* Lemke graph; *the* Lemke graph is the smallest among them.

matched edges – of a matching M: the edges of M.

matched vertices – of a graph G: the vertices of G incident on a matched edge.

matching – in a graph G: a set of pairwise nonadjacent edges.

___, **complete** – of a bipartite graph $G = (X \cup Y, E)$: a matching that meets each vertex of X; also called X-*saturating*.

___, **maximum-size**: a matching M having the largest size $|M|$.

___, **maximum-weight**: a matching M having the largest weight $wt(M)$.

___, **perfect** – of a graph G: a matching that meets each vertex of G exactly once.

___, **size of**: the number of edges in the matching.

___, **weight of**: the sum of the weights of edges in the matching.

maximum flow problem: given a flow network $G = (V, E, s, t, cap)$, find a flow of maximum value.

maximum multicommodity flow problem: for each commodity i, find a flow f_i of value (demand) $val(f_i)$ such that $\sum_{i=1}^k val(f_i)$ is maximized.

maximum r-**path partition of a tree** T: a sequence of paths $P[1], \ldots, P[h]$ that partitions the edges of T so that r is a leaf of $P[1]$, $T_i = \cup_{j=1}^i P[j]$ is a tree for all $1 \le i \le h$, and $P[i]$ is a maximum length path in $T - T_{i-1}$, among all such paths with one endpoint in T_{i-1}, for all $1 \le i \le h$.

multicommodity flow – in a multicommodity flow network $G = (V, E, K, u)$: a set of $k = |K|$ functions $f_i : E \to \mathbb{R}^+$ satisfying the following conditions:

joint capacity constraints: $\sum_{i=1}^k f_i(v, w) \le cap(v, w)$, for each arc $(v, w) \in E$.

conservation constraints: $\sum_{(w,v) \in E} f_i(w, v) = \sum_{(v,w) \in E} f_i(v, w)$ for each vertex $v \in V - \{s_i, t_i\}$ and $i = 1, \ldots, k$, and $\sum_{(w,t_i) \in E} f_i(w, t_i) - \sum_{(t_i,w) \in E} f_i(t_i, w) = d_i$ for each $i = 1, \ldots, k$.

nonnegativity constraints: $f_i(v, w) \geq 0$, for each arc $(v, w) \in E$ and each
$i = 1, \ldots, k$.

multicommodity flow network $G = (V, E, K, u)$: a directed graph with vertex-set
V and arc-set E, commodity set K, and a nonnegative *capacity* function $u : E \to N$.
We adopt the convention that if arc $(v, w) \in E$ then the reverse arc $(w, v) \notin E$. The
commodities are indexed by the integers $1, 2, \ldots, k$.

near pyramid graph: a pyramid graph minus one of the edges of its inner triangle
(a, b, c).

odd vertex – relative to an alternating path: a vertex that is an odd distance (number
of edges) from the root of the path.

optimal pebbling number $\pi^*(G)$: the minimum size of a solvable configuration on
G.

parent of a vertex in a rooted tree: the unique neighbor that is closer to the root
of the tree.

path graph P_n: the graph with a sequence of n vertices and $n - 1$ edges joining
consecutive pairs of vertices.

pebbling exponent $e_\pi(G)$ **of a graph** G: the minimum k such that the graph power
$G^{(k)}$ is Class 0.

pebbling number $\pi(G)$ **of a graph** G: the maximum rooted pebbling number $\pi(G, r)$
over all root vertices r.

2-pebbling property of a graph G: the property that every configuration C of size
more than $2\pi(G) - |\sigma(C)|$ is 2-fold r-solvable for all r.

pebbling step: the removal of two pebbles from one vertex, along with the placement
of one pebble on an adjacent vertex.

pebbling threshold of a graph sequence $G_1, G_2, \ldots$: a function $t : \mathcal{N} \to \mathcal{N}$ such
that a random configuration on G_n of asymptotically more (less) pebbles than $t(n)$
is almost surely solvable (unsolvable).

power: see *graph power*.

preflow: a relaxed version of a flow, a function $f : E \to Z^+$ which obeys three types
of constraints:
capacity constraints: $f(v, w) \leq cap(v, w)$, for each arc $(v, w) \in E$.
relaxed conservation constraints:
$$\sum_{(w,v)\in E} f(w, v) - \sum_{(v,w)\in E} f(v, w) \geq 0 \text{ for each vertex } v \in V - \{s, t\}.$$
nonnegativity constraints: $f(v, w) \geq 0$, for each arc $(v, w) \in E$.

pyramid graph: the graph on 6 vertices formed by the union of the 6-cycle (r, a, p, c, q, b)
and the (inner) triangle (a, b, c).

pyramidal graph: a graph G that contains an induced (near-) pyramid, having 6-cycle
C and inner (near-) triangle K, and can be drawn in the plane so that the edges
of K are drawn in the interior of the region bounded by C and every other edge of
G can be drawn inside the convex hull of exactly one of the sets $\{r, a, b\}$, $\{p, a, c\}$,
$\{q, b, c\}$, or $\{a, b, c\}$.

reducible arc (v, w) in network G for a given flow: $f(v, w) > 0$.

residual capacity $r_f(v, w)$ – of arc (v, w) in residual network G_f for a given flow or
preflow f: see §10.1, Definition D10.

residual network$_1$ – for a maximum-flow network: see §11.1, Definition D10.

residual network$_2$ – for a cost-flow network: see §11.2, Definition D9.

rooted pebbling number $\pi(G, D)$ **of a graph** G: the minimum number t of pebbles
such that every size t configuration is D-solvable.

shrunked blossom: obtained by collapsing a blossom into a single vertex.

solvable configuration: a configuration that is r-solvable for every vertex r.

D-solvable configuration: a configuration from which it is possible via pebbling steps to reach another configuration having at least as many pebbles on each vertex as the configuration D.

r-solvable configuration: a configuration from which it is possible to place a pebble on vertex r via pebbling steps.

split graph: a graph whose vertices can be partitioned into a clique and an independent set.

Steiner tree problem: given a weighted graph in which a subset of vertices are identified as terminals, find a minimum-weight connected subgraph that includes all the terminals.

strategy: a weight function on a rooted tree with the property that, for every nonroot vertex v, the parent of v has weight at least twice that of v.

support $\sigma(C)$ of a configuration C on a graph G: the set of vertices of G that have at least one pebble.

switch – in a communication network: node equipment for routing and processing communication traffic.

thread: a set of degree two vertices that form a path in a graph.

threshold: see *pebbling threshold*.

transshipment network: a cost-flow network $G = (V, A, cap, c, b)$ in which all arcs have infinite capacity.

    ___, **associated**: the transshipment network obtained by replacing each arc $e = (v, w)$ by three arcs (v, x_e), (y_e, x_e), and (y_e, w) having infinite capacity and with costs c_e, 0, and 0, respectively; the supplies at the new nodes x_e and y_e are defined to be $b(x_e) = -cap_e$ and $b(y_e) = cap_e$.

    ___, **completion of**: the complete transshipment network obtained by adding all missing arcs and giving each of them infinite capacity and arc cost $M + 1$, where $M = \sum_e c(e)cap(e)$.

unsplittable flow problem: multicommodity flow problem with the additional restriction that each commodity must be routed on one path.

vertex cover – of a graph G: a set of vertices incident on all edges of G.

weight of a path P – relative to a matching M: the sum of the weights of the free edges in P minus the sum of the weights of the matched edges in P, denoted $wt(P)$.

Chapter 12

Communication Networks

Section 12.1
Complex Networks

Anthony Bonato, Ryerson University
Fan Chung, University of California, San Diego

INTRODUCTION

The study of complex networks analyzes graph-theoretical properties arising in real-world networks, ranging from technological, social, and biological. Web pages and their links, protein-protein interaction networks, and on-line social networks such as Facebook and LinkedIn are some of the commonly studied examples of such networks.

Never before have we confronted graphs of not only such tremendous sizes but also extraordinary richness and complexity, both at a theoretical and a practical level. Numerous problems arise. For example, what are basic structures of such large networks? How do they evolve? What are the underlying principles that dictate their behavior? How are subgraphs (that we observe) related to the large (and often incomplete) host graphs? What are the main graph invariants that capture the myriad properties of such large graphs?

To deal with these questions, graph theory comes into play. There have been a great many advances in the field over the past thirty years in combinatorial, probabilistic, and spectral methods. Still, the traditional random graphs mostly consider the same degree distribution for all nodes or edges while real-world graphs are uneven and clustered. The classical algebraic and analytic methods are efficient in dealing with highly symmetric structures, whereas real graphs are quite the opposite. Guided by examples of complex networks, many new and challenging directions in graph theory emerge. Here we include several selected topics that have been developing. This article is based in part on the surveys [BoTi12, Ch10], and further references can be found in [Bo08, ChLu04b].

12.1.1 Examples of Complex Networks

Nowadays we are surrounded by various information networks that are of prohibitively large size. Dealing with graphs arising from these networks, the usual graph parameters such as the exact number of nodes are no longer as important. Instead, only partial or locally available information can be obtained. This leads to a whole range of new graph parameters and problems, which are motivated by examples of complex networks. Some of these complex networks we will mention here. Our main focus in this survey will be on the web graph and on-line social networks.

EXAMPLES

E1: The *web graph* has nodes consisting of web pages, and edges corresponding to links between them. The web graph may be viewed as directed or undirected, depending on the context.

E2: The *collaboration graph* has nodes as co-authors and two authors are connected by an edge if they have written a joint paper together. The collaboration graph (according to the Math Reviews database) has 401,000 nodes and 676,000 edges. The reader is referred to the website of Grossman [Gr23] for many interesting properties of the collaboration graph.

E3: *On-line social networks* have nodes consisting of users on some social networking site such as Facebook, and edges consisting of friendship links between them. Twitter may be viewed as a directed graph, where users follow each other, but following may not be reciprocal.

E4: *Protein-protein interaction networks* have nodes consisting of proteins in a living cell, with two proteins joined if they share some biochemical interaction. For a survey of protein-protein interaction networks, see [Pr05].

E5: Other important examples of complex networks are *router graphs* (nodes are routers and edges correspond to physical connections between them), *call graphs* (nodes are phone numbers and directed edges correspond to calls placed between them), *citation graphs* (nodes are academic papers in a given discipline and there are directed edges between cited papers).

12.1.2 Properties of Complex Networks

Complex networks arise is diverse arenas but have a completely unexpected coherence. The prevailing characteristics of the complex graphs are *large-scale, small-world, and possessing power-law degree distribution*, which we will now describe. Throughout, graphs are taken as finite and undirected, unless otherwise stated.

DEFINITIONS

D1: The order and size of complex networks vary considerably, but all of them are *large-scale*.

The web graph has over a trillion nodes, with billions of pages appearing and disappearing each day. Facebook [Fa12] has over 955 million users, and over 70 billion friendship links. Some of the nodes of Twitter corresponding to well-known celebrities including Lady Gaga and Justin Bieber have out-degree over 26 million [Tw12].

Because of the large-scale property of complex networks, an examination of almost all nodes, called a **sweep** or **crawl**, is costly and rarely done, and only performed off-line. However, on-line and local computation should ideally be computed in constant time or cost (independent of n, the number of nodes), or of order $O(\log n)$ or $O(\log \log n)$.

D2: Complex graphs are often **sparse**, which means that the graphs have a linear number of edges (that is, $|E(G) \leq c|V(G)|$, for some small constant c. In fact, the constant is usually less than 10 in most examples.) Some networks, such as on-line social networks, however, tend to be dense graphs.

In extremal graph theory and random graph theory, dense graphs are quite well understood, partly due to Szemeredi's regularity lemma. To deal with certain complex graphs with billions of nodes such as the web graph, research has focused primarily on the study of sparse graphs.

D3: The **small world phenomenon** can be regarded as a combination of *small distances* and *clustering*. Namely, there is a short path joining any two nodes and if two nodes share a common neighbor, they are more likely to be adjacent.

There are two different notions of "small distance." The **diameter** of a connected graph G is the maximum distance between a pair of nodes, and is written as $\mathrm{diam}(G)$. The **average distance**, denoted by $\mathrm{adist}(G)$, is defined by

$$\mathrm{adist}(G) = \sum_{u,v \in F} \frac{d(u,v)}{|F|},$$

where F is the set of pairs of distinct nodes u, v of G with the property that the distance $d(u,v)$ between u and v is finite. The directed analogue of this parameter, where distance refers to shortest directed paths (that is, with no back edges), is denoted by $\mathrm{adist}_d(G)$.

FACT

F1: As evidence of the small world property for the web graph, in [AlJeBa99] it was reported that some subgraph of the web graph in 1999 has average distance 19, while some other values were reported earlier in [Br-etal00]. The collaboration graph of Math Reviews has diameter 23 and average distance 7.64 (see [Gr23]). This can be phrased as, "eight degrees of separation" among mathematicians. In some complex networks such as on-line social networks and citations networks, distances have been observed to decrease over time; see [LeKlFa05]. In a recent study of 700 million users of Facebook in [Ba-etal12], the average distance of that network was given as 4.74.

There are also a number of graph parameters associated with clustering. A direct way is to define the **clustering coefficient** of G to be

$$\frac{2}{n} \sum_{x \in V(G)} \frac{|E(N(v))|}{\deg(v)(\deg(v) - 1)},$$

where $E(N(v))$ denotes the number of edges in the subgraph induced by the neighborhood of v in G.

DEFINITIONS

D4: A useful isoperimetric graph invariant related to clustering is the **Cheeger constant**, which can be defined as the maximum **Cheeger ratios** $h(S)$ over all subsets S of nodes in G. The **Cheeger ratio**, written $h(S)$, is defined by

$$h(S) = \frac{|E(S, \bar{S})|}{\mathrm{vol}(S)}, \quad \text{and} \quad h_G = \max_{\substack{S \subseteq V(G) \\ \mathrm{vol}(S) \le \mathrm{vol}(G)/2}} h(S)$$

where we define the **volume** $\mathrm{vol}(S) = \sum_{v \in S} \deg(v)$, $\mathrm{vol}(G) = \mathrm{vol}(V(G))$ and $E(A, B)$ denotes the set of edges with one endpoint in A and one endpoint in B. We note that in the literature, the **isoperimetric number** $i(S) = \frac{|E(S, \bar{S})|}{|S|}$ ignores the uneven degree distributions of complex graphs and is therefore less effective for capturing clustering in complex networks.

D5: In a graph G on n nodes, let $N_k = N_k(n)$ denote the number of nodes of degree k. The degree distribution of G follows a **power law** in some range of k if N_k is proportional to k^{-b}, for a fixed **exponent** $b > 2$. If G is directed, then we may consider (possibly different) power laws for the in- and out-degree distributions by defining $N_{k,G}^{in}$ and $N_{k,G}^{out}$, respectively, in the obvious way.

Power law degree distributions are sometimes called **heavy-tailed distributions**, since the real-valued function $f(k) = k^{-\beta}$ exhibits a polynomial (rather than exponential) decay to 0 as k tends to ∞. If G possesses a power law degree distribution, then we simply say G is a **power law graph**.

FACTS

F2: By taking logarithms, power laws can be expressed as

$$\log(N_{k,G}) \sim \log(t) - \beta \log(k).$$

Hence, in the log-log plot, we obtain a straight line with slope $-\beta$.

F3: Based on their crawl of the domain of Notre Dame University, Albert, Barabási, and Jeong [AlJeBa99] claimed that the web graph exhibited a power law in-degree distribution, with $\beta = 2.1$. An independent crawl corroborating the findings of [AlJeBa99] was reported in [Ku-etal99], which considered 40 million web pages from 1997 data. The exponent of $\beta = 2.1$ was further corroborated by a larger crawl of the entire web (including 200 million web pages) reported in Broder et al. [Br-etal00]. There were several other reports concerning slightly different power law exponents $\beta = 2.45$ and $\beta = 2.72$ for Web graphs [AlJeBa99, Br-etal00]. Kumar, Novak, and Tomkins [KuNoTo06] studied the evolution of Flickr and Yahoo!360, and found that these networks exhibit power-law degree distributions. Power law degree distributions for both the in- and out-degree distributions were documented in Flickr, YouTube, LiveJournal, and Orkut [Mi-etal07], as well as in Twitter [Ja-etal07, Kw-etal10].

DEFINITION

D6:   In a graph G, let $0 = \lambda_0 \leq \lambda_1 \cdots \leq \lambda_{n-1} \leq 2$ denote the eigenvalues of the **normalized Laplacian** $\mathcal{L} = I - D^{-1/2}AD^{-1/2}$ where A is the adjacency matrix and D is the diagonal degree matrix of G. The eigenvalue λ_1 is intimately related to the Cheeger constant h_G by the **Cheeger inequality** (see [Ch97]):

$$2h_G \geq \lambda_1 \geq \frac{h_G^2}{2}.$$

FACTS

F4:   In 1999 Faloutsos et al. [FaFaFa99] made an experimental study of an autonomous systems graph, finding a power law distribution for the highest eigenvalues of the adjacency matrix.

F5:   Mihail and Papadimitriou [MiPa02] showed that the largest eigenvalues of the adjacency power law graphs are themselves distributed according to a power law. This phenomenon is sometimes referred to as the **eigenvalue power law**. In the other direction, the eigenvalues of the normalized Laplacian of a random graph with general degree distribution satisfy the **semi-circle law**.

F6:   Social networks often organize into separate clusters in which the intra-cluster links are significantly higher than the number of inter-cluster links. In particular, social networks contain communities (characteristic of social organization), where tightly knit groups correspond to the clusters [NePa03]. As a result, it is reported in [Es06] that social networks, unlike other complex networks such as the web graph, possess bad spectral expansion properties realized by small spectral gaps in their adjacency matrices.

12.1.3   Random Graphs with General Degree Distributions

To describe a complex graph (or indeed, any graph), we wish to use as few parameters as possible. This is exactly what graph models are supposed to do. For example, the Erdős–Reńyi random graph model $G(n, p)$ uses only two parameters to describe a family of quite complicated graphs by selecting each pair of nodes to be an edge with probability p independently. We note that a random graph in $G(n, p)$ has the same expected degree at every node, and therefore, $G(n, p)$ does not capture some basic behaviors of complex networks. Still, the methods and approaches of classical random graph theory are instrumental in the study of general random graphs.

DEFINITION

D7:   The $G(\mathbf{w})$ model for random graphs with expected degree sequence $\mathbf{w}$ is defined by selecting an edge between v_i and v_j independently with probability p_{ij} where p_{ij} is proportional to the product $w_i w_j$. For example, $G(n, p)$ is a special case of $G(\mathbf{w})$ by taking $\mathbf{w}$ to be $(pn, pn, \ldots, pn)$. The $G(\mathbf{w})$ with a power law distribution $\mathbf{w}$ is called a **power law random graph**.

FACT

F7: [ChLu02] Suppose that a random power law graph G in $G(\mathbf{w})$ has n nodes and expected degree sequence $\mathbf{w}$ following a power law with exponent $2 < \beta < 3$. Let G have average degree $d > 1$ and maximum degree m satisfy

$$\log m \gg \frac{\log n}{\log \log n}.$$

Then, for all values of $\beta > 2$, asymptotically almost surely the graph G is connected and the diameter satisfies

$$\mathrm{diam}(G) = \Theta(\log n),$$

and the average distance satisfies

$$\mathrm{adist}(G) \leq (2 + o(1)) \frac{\log \log n}{\log(1/(\beta - 2))}.$$

Many probabilistic and spectral properties of power law random graphs have been studied (see [ChLu04b] for details).

DEFINITION

D8: An important random graph model for general degree sequences is the *configuration model*, which is a spin-off of random regular graphs. One way to define *random regular graphs* $\mathcal{G}_k$ of degree k on n nodes is to consider all possible matchings in a complete graph K_{kn}. Note that a matching is a maximum set of vertex-disjoint edges. Each matching is chosen with equal probability. We then obtain a random k-regular graph by partitioning the nodes into subsets of size k. Each k-subset then is associated with a node in a random regular graph $\mathcal{G}_k$. Although such a random regular graph might contain loops (that is, an edge having both endpoints the same node), the probability of such an event is of a lower order and can be controlled. Now, instead of partitioning the set of nodes of the large graph into equal parts, we choose a random matching of a complete graph on $\sum_i d_i$ nodes which are partitioned into subsets of sizes $d_1, d_2, \ldots, d_n$. Then we form the random graph by associating each edge in the matching with an edge between associated nodes. Clearly, in the configuration model, there are nontrivial dependencies among the edges.

FACTS

F8: A useful tool in a configuration model is a result of Molloy and Reed [MoRe95, MoRe98]. For a random graph with $(\gamma_i + o(1))n$ nodes of degree i, where γ_i are non-negative values which sum to 1 and n is the number of nodes, the giant component emerges when $Q = \sum_{i \geq 1} i(i - 2)\gamma_i > 0$, provided that the maximum degree is less than $n^{1/4-\epsilon}$ (where $\epsilon > 0$) and some "smoothness" conditions are satisfied. Also, there is asymptotically almost surely no giant component when $Q = \sum_{i \geq 1} i(i - 2)\gamma_i < 0$ and the maximum degree is less than $n^{1/8-\epsilon}$.

F9:   [AiChLu00] The evolution of a random power law graph with exponent β is as follows:

1. When $\beta > \beta_0 = 3.47875\ldots$, the random graph asymptotically almost surely has no giant component where the value $\beta_0 = 3.47875\ldots$ is a solution to

$$\zeta(\beta - 2) - 2\zeta(\beta - 1) = 0.$$

   When $\beta < \beta_0 = 3.47875\ldots$, there is asymptotically almost surely a unique giant component.

2. When $2 < \beta < \beta_0 = 3.47875\ldots$, the second largest component is asymptotically almost surely of size $\Theta(\log n)$. For any $2 \leq x < \Theta(\log n)$, there is asymptotically almost surely a component of size x.

3. When $\beta = 2$, asymptotically almost surely the second largest component is of size $\Theta(\frac{\log n}{\log \log n})$. For any $2 \leq x < \Theta(\frac{\log n}{\log \log n})$, there is asymptotically almost surely a component of size x.

4. When $1 < \beta < 2$, the second largest component is asymptotically almost surely of size $\Theta(1)$. The graph is asymptotically almost surely not connected.

5. When $0 < \beta < 1$, the graph is asymptotically almost surely connected.

6. When $\beta = \beta_0 = 3.47875\ldots$, the situation is complicated. It is similar to the double jump of the random graph $\mathcal{G}(n, p)$ with $p = \frac{1}{n}$. For $\beta = 1$, there is a nontrivial probability for either case that the graph is connected or disconnected.

12.1.4   On-Line Models of Complex Networks

Models of complex networks are usually (but not always) stochastic, and graphs evolve over time. As there are a large number of models, our survey is far from exhaustive. We focus on three models which have applications to modelling the web graph and on-line social networks. Other models which have been rigorously analyzed are **competition-based preferential attachment models** [Be-etal04], **forest fire models** [LeKlFa05], **ranking models** [FoFlMe06, JaPr09, LuPr06], **growth-deletion models** [ChLu04b], **Kronecker graphs** [Le-etal05], **Multiplicative Attribute Graph model** [KiLe12], **planar power law graphs** [FrTs12], and models based on **algorithmic game theory** [Bo-etal12].

Preferential Attachment Model

Arguably the most commonly studied complex network models are ones incorporating some form of preferential attachment. The first evolving graph model explicitly designed to model the web graph was given by Barabási and Albert [BaAl99], and this model is now referred to as the **preferential attachment**. A rigorous analysis of a preferential attachment model for the case of $\beta = 3$ was given in [Bo-etal01]. Here we give a general definition for a preferential attachment model for $\beta > 2$ introduced and analyzed in [AiChLu01].

DEFINITION

D9:   The **preferential attachment model** $G(p)$ has one parameter $0 < p < 1$ and involves two possible operations:

- *Vertex-step*: Add a new node v, and add an edge uv from v by randomly and independently choosing u in proportion to the degree of u in the current graph.
- *Edge-step*: Add a new edge rs by independently choosing nodes r and s with probability proportional to their degrees.

We start with an initial graph G_0 (which can usually be taken to be the graph formed by one node having one loop). At time t, the graph G_t is formed by modifying G_{t-1} as follows: with probability p, take a vertex-step, otherwise take an edge-step.

FACT

F10: [AiChLu01] For the preferential attachment model $G(p)$, asymptotically almost surely the number of nodes with degree k at time t is

$$M_k t + O(2\sqrt{k^3 t \ln(t)}),$$

where $M_1 = \frac{2p}{4-p}$ and $M_k = \frac{2p}{4-p}\frac{\Gamma(k)\Gamma(1+\frac{2}{2-p})}{\Gamma(k+1+\frac{2}{2-p})} = O(k^{-(2+\frac{p}{2-p})})$, for $k \geq 2$. In other words, asymptotically almost surely the graphs generated by $G(p)$ have a power law degree distribution with the exponent $\beta = 2 + \frac{p}{2-p}$.

The Copying Model

The copying model was introduced in Kleinberg et al. [Kl-etal99].

DEFINITION

D10: The ***copying model*** has parameters $p \in (0,1)$, $d \in \mathbb{N}^+$, and an initial directed graph H with constant out-degree d. Assume that the graph G_t at time t has constant out-degree d. At time $t+1$, we add a new node v_{t+1} and the d out-link of v_{t+1} is generated as follows: We choose a "prototype" node u_t from the set of all existing nodes at random. For each of the d out-neighbours w of u_t with probability p, add a directed edge (v_{t+1}, z), where z is chosen uniformly at random from $V(G_t)$, and with the remaining probability $1-p$ add the directed edge (v_{t+1}, w).

FACTS

F11: [Ku-etal00] If $k > 0$, then in the copying model with parameter p, asymptotically almost surely

$$\frac{N_{k,t}^{in}}{t} = \Theta\left(k^{-\frac{2-p}{1-p}}\right).$$

In particular, the copying model asymptotically almost surely generates directed graphs G_t whose in-degree distribution follows a power law with exponent

$$\beta = \frac{2-p}{1-p} \in (2, \infty).$$

F12: [Ku-etal00] Let $N_{t,i,j}$ denote the expected number of distinct $K_{i,j}$'s which are subgraphs of G_t. Asymptotically almost surely in the copying model with constant outdegree d, for $i \leq \log t$,

$$N_{t,i,j} = \Omega(t \exp(-i)).$$

This abundance of bipartite cliques, which model community structure, is provably absent from the preferential attachment model.

Iterated Local Transitivity Model

We describe a recent model for on-line social networks based on transitivity in social networks (that is, friends of friends are often friends). The model is deterministic and simple to describe, but leads to complex behaviour over time.

DEFINITIONS

D11: In the ***Iterated Local Transitivity*** model [Bo-etal11], given some initial graph as a starting point, nodes are repeatedly added over time which clone each node, so that the new nodes form an independent set. The only parameter of the model is the initial graph G_0, which is any fixed finite connected graph. Assume that for a fixed $t \geq 0$, the graph G_t has been constructed. To form G_{t+1}, for each node $x \in V(G_t)$, add its ***clone*** x', such that x' is joined to x and all of its neighbors at time t. Note that the set of new nodes at time $t + 1$ forms an independent set of cardinality $|V(G_t)|$.

D12: We write $\deg_t(x)$ for the degree of a node at time t, n_t for the order of G_t, and e_t for its number of edges. Define the ***volume*** of G_t by

$$\text{vol}(G_t) = \sum_{x \in V(G_t)} \deg_t(x) = 2e_t.$$

FACT

F13: [Bo-etal11] For $t > 0$, the average degree of G_t equals

$$\left(\frac{3}{2}\right)^t \left(\frac{\text{vol}(G_0)}{n_0} + 2\right) - 2.$$

Note that the average degree tends to infinity with n; that is, the model generates graphs satisfying a **densification power law**. In [LeKlFa05], densification power laws were reported in several real-world networks such as the physics citation graph and the internet graph at the level of autonomous systems.

DEFINITION

D13: Define the ***Wiener index*** of G_t as

$$W(G_t) = \frac{1}{2} \sum_{x,y \in V(G_t)} d(x,y).$$

FACTS

F14: [Bo-etal11] For $t > 0$,

$$L(G_t) = \frac{4^t \left(W(G_0) + (e_0 + n_0)\left(1 - \left(\frac{3}{4}\right)^t\right)\right)}{4^t n_0^2 - 2^t n_0}.$$

Note that the average distance of G_t is bounded above by $\text{diam}(G_0) + 1$ (in fact, by $\text{diam}(G_0)$ in all cases except cliques). Further, for many initial graphs G_0 (such as large cycles) the average distance decreases.

F15: [Bo-etal11] If we let $n_t = n$ (so $t \sim \log_2 n$), then

$$C(G_t) = n^{\log_2(7/8)+o(1)}.$$

In contrast, for a random graph $G(n, p)$ with comparable average degree

$$pn = \Theta((3/2)^{\log_2 n}) = \Theta(n^{\log_2(3/2)}).$$

F16: [Bo-etal11] Define $\lambda(G_t)$ to be the **spectral gap** of the normalized Laplacian of G_t. For $t \geq 1$, $\lambda(G_t) > \frac{1}{2}$. This represents a drastic departure from the good expansion found in random graphs, where $\lambda = o(1)$; see [Ch97].

12.1.5   Geometric Models for Complex Networks

In **geometric random graph models**, nodes are identified with points in some metric space $\mathcal{S}$, and edges are determined via a mixture of probabilistic rules and distance between nodes in $\mathcal{S}$. Complex networks do not usually live in physical space, but live in some **feature space**, where nodes are associated with vectors of features, and nodes with similar features are more likely to be adjacent. For example, we may view the web graph as residing in **topic-space**, where pages are close if they have common topics. On-line social networks are embedded in **social-space** (sometimes called **Blau space**; see [Mi83]), where nodes with similar social attributes are close.

Several geometric models for complex networks have been proposed, such as **geometric preferential attachment** [FlFrVe07], **random dot product graphs** [ScYo08], and **spatially preferred preferential attachment** [Ai-etal09, CoFrPr12]. We focus on the geometric Protean model which was introduced as a model of an on-line social network, in particular, as it generates dense graphs.

DEFINITION

D14: [BoJaPr12] The **geometric Protean model**, written GEO-P(α, β, m, p), produces a sequence $(G_t : t \geq 0)$ of undirected graphs on n nodes, where t denotes time. We write $G_t = (V_t, E_t)$. There are four parameters: the *attachment strength* $\alpha \in (0, 1)$, the *density parameter* $\beta \in (0, 1 - \alpha)$, the *dimension* $m \in \mathbb{N}$, and the *link probability* $p \in (0, 1]$. Each node $v \in V_t$ has rank $r(v, t) \in [n] = \{1, 2, \ldots, n\}$. The rank function $r(\cdot, t) : V_t \to [n]$ is a bijection for all t, so every node has a unique rank. The highest ranked node has rank equal to 1; the lowest ranked node has rank n. The initialization and update of the ranking is done so that the node added at time t obtains an initial rank R_t, which is randomly chosen from $[n]$ according to a prescribed distribution. Ranks of all nodes are adjusted accordingly. Let S be the unit hypercube in $\mathbb{R}^m$, with the torus metric $d(\cdot, \cdot)$ derived from the L_∞ metric. More precisely, for any two points x and y in $\mathbb{R}^m$, their distance is given by

$$d(x, y) = \min\{||x - y + u||_\infty : u \in \{-1, 0, 1\}^m\}.$$

To initialize the model, let $G_0 = (V_0, E_0)$ be any graph on n nodes that are chosen from S. We define the *influence region* of node v at time $t \geq 0$, written $R(v, t)$, to be the ball around v with volume

$$|R(v, t)| = r(v, t)^{-\alpha} n^{-\beta}.$$

For $t \geq 1$, we form G_t from G_{t-1} according to the following rules.

1. Add a new node v that is chosen uniformly at random from S. Next, independently, for each node $u \in V_{t-1}$ such that $v \in R(u, t-1)$, an edge vu is created with probability p. Note that the probability that u receives an edge is proportional to $p \, r(u, t-1)^{-\alpha}$. The negative exponent guarantees that nodes with higher ranks ($r(u, t-1)$ close to 1) are more likely to receive new edges than lower ranks.

2. Choose uniformly at random a node $u \in V_{t-1}$, and delete u and all edges incident to u.

3. Vertex v obtains an initial rank $r(v, t) = R_t$, which is randomly chosen from $[n]$ according to a prescribed distribution.

4. Update the ranking function $r(\cdot, t) : V_t \to [n]$.

FACTS

F17: Since the geometric Protean process is an ergodic Markov chain, it will converge to a stationary distribution.

F18: [BoJaPr12] The geometric Protean model asymptotically almost surely generates power law graphs. More precisely, let $N_k = N_k(n, p, \alpha, \beta)$ denote the number of nodes of degree k, and $N_{\geq k} = \sum_{l \geq k} N_l$. Let $\alpha \in (0, 1)$, $\beta \in (0, 1 - \alpha)$, $m \in \mathbb{N}$, $p \in (0, 1]$, and

$$n^{1-\alpha-\beta} \log^{1/2} n \leq k \leq n^{1-\alpha/2-\beta} \log^{-2\alpha-1} n.$$

Then asymptotically almost surely GEO-P(α, β, m, p) satisfies

$$N_{\geq k} = \left(1 + O(\log^{-1/3} n)\right) \frac{\alpha}{\alpha + 1} p^{1/\alpha} n^{(1-\beta)/\alpha} k^{-1/\alpha}.$$

F19: [BoJaPr12] Asymptotically almost surely the average degree of GEO-P(α, β, m, p) is

$$d = (1 + o(1)) \frac{p}{1 - \alpha} n^{1-\alpha-\beta}.$$

Note that the average degree tends to infinity with n; that is, the model generates graphs satisfying a **densification power law**.

F20: [BoJaPr12] Let $\alpha \in (0, 1)$, $\beta \in (0, 1 - \alpha)$, $m \in \mathbb{N}$, and $p \in (0, 1]$. Then asymptotically almost surely the diameter D of GEO-P(α, β, m, p) satisfies

$$D = O(n^{\frac{\beta}{(1-\alpha)m}} \log^{\frac{2\alpha}{(1-\alpha)m}} n). \tag{12.1.1}$$

F21: We note that in a geometric model where regions of influence have constant volume and possess the same average degree as the geo-protean model, the diameter is $\Theta(n^{\frac{\alpha+\beta}{m}})$. This is a larger diameter than in the geometric Protean model. If $m = C \log n$, for some constant $C > 0$, then by (12.1.1) asymptotically almost surely we obtain a diameter bounded above by a constant.

F22: The dimensions of social space quantify user traits such as interests or geography; for instance, nodes representing users from the same city or in the same profession would likely be closer in social space. The **Logarithmic Dimension Hypothesis** [BoJaPr12] conjectures that the dimension m of an on-line social network is best fit by about $\log n$, where n is the number of users in the on-line social network. The motivation for the conjecture comes from both the geometric protean and multiplicative attribute models (see [KiLe12] for more on the latter model).

12.1.6 Percolation in a General Host Graph

Most information networks that we observe are subgraphs of certain host graphs for which only partial and incomplete information is available. A natural question is to deduce properties of the host graph from its random subgraphs and vice versa. It is of interest to understand the connections between a graph and its subgraphs. What graph invariants of the host graph are preserved in its subgraphs? Under what conditions can we predict the behavior of the host graph from some samples of subgraphs?

DEFINITION

D15: One way to deal with a random subgraph of a given graph G is a type of (bond) ***percolation*** problem. For a positive value $p \leq 1$, we consider G_p, which is formed by retaining each edge independently with probability p and discarding the edge with probability $1 - p$. A fundamental problem of interest is to determine the ***critical probability*** for which G_p contains a giant connected component. As an example, if the host graph is the complete graph K_n, G_p is the same as the random graph model $G(n, p)$.

FACTS

F23: In the application of epidemics, the host graph is taken to be a **contact graph**, consisting of edges formed by pairs of people with possible contact. The question of interest is to determine the critical probability which corresponds to the problem of finding the epidemic threshold for the spreading of diseases.

F24: Percolation problems have long been studied in theoretical physics, especially for the case of the host graph taken to be the lattice graphs $\mathbb{Z}^k$ or special families of Cayley graphs [Bo-etal05, Bo-etal05a, Bo-etal06, MaPa02], and dense graphs [Bp-etal10].

F25: For a general host graph G, it was shown in [ChLu02] that under some (mild) conditions depending on its spectral gap and higher moments of its degree sequence of G, for any $\epsilon > 0$, if $p > (1 + \epsilon)/\tilde{d}$, then asymptotically almost surely the percolated subgraph G_p has a giant component. In the other direction, if $p < (1 - \epsilon)/\tilde{d}$, then asymptotically almost surely the percolated subgraph G_p contains no giant component. We note that the second order average degree $\tilde{d}$ is $\tilde{d} = \sum_v d_v^2 / (\sum_v d_v)$, where d_v denotes the degree of v. The volume of the giant component in G_p is given in [ChLu06].

F26: Suppose the host graph G has a large spectral gap λ, defined by

$$\lambda = \min\{\lambda_1, 2 - \lambda_{n-1}\},$$

where $0 = \lambda_0 \leq \lambda_1 \leq \ldots \leq \lambda_{n-1}$ are eigenvalues of the normalized Laplacian of G. It was shown in [ChHo07] that the random subgraph H in G_p preserves the spectral gap as follows:

$$\lambda_H = \lambda - O\left(\sqrt{\frac{\log n}{pd_{\min}}} + \frac{(\log n)^{3/2}}{pd_{\min}(\log \log n)^{3/2}}\right).$$

F27: In a paper by Liben-Nowell and Kleinberg [LiKl08] it was observed that the tree-like subgraphs derived from some chain-letter data seem to have relatively large diameter. In the study of the Erdős–Rényi graph model $G(n, p)$, it was shown that the diameter of a random spanning tree is of order $\sqrt{n}$, in contrast to the fact that the diameter of the host graph K_n is 1. Aldous [Al90] proved that in a regular graph G with a certain spectral bound σ, the expected diameter of a random spanning tree T of G, denoted by diam(T), has expected value satisfying

$$\frac{c\sigma\sqrt{n}}{\log n} \leq \mathrm{E}(\mathrm{diam}(T)) \leq \frac{c\sqrt{n}\log n}{\sqrt{\sigma}}$$

for some absolute constant c. In [ChHoLu12], it was shown that for a general host graph G, with high probability the diameter of a random spanning tree of G is between $c\sqrt{n}$ and $c'\sqrt{n}\log n$, where c and c' depend on the spectral gap of G and the ratio of the moments of the degree sequence.

12.1.7   PageRank for Ranking Nodes

The notion of PageRank, first introduced by Brin and Page [BrPa98], forms the basis for Google's web search algorithms. Furthermore, PageRank can be used to measure the global importance of nodes, and capture quantitative correlations between pairs or subsets of nodes. Roughly speaking, PageRank is the stationary distribution of some random walk on a given directed graph. In other words, the PageRank value of a node is the probability that, at any given moment, a random surfer is visiting this node following a random walk whose diffusion is under control (by a specified constant α which we will define).

DEFINITIONS

D16: In a directed graph G without vertices with zero outdegree, the **transition probability matrix** P for a typical random walk on G is defined as

$$P(u,v) = \left\{ \begin{array}{ll} 1/\deg^+(u) & \text{if } (u,v) \in E(G) \\ 0 & \text{otherwise} \end{array} \right.$$

where $\deg^+(u)$ denotes the out-degree of vertex u.

D17: For a **teleportation constant** $\alpha \in (0,1]$, the **PageRank matrix** of G is defined by

$$P_\alpha = \alpha J_n/n + (1-\alpha)P$$

where J_n is the all 1's matrix of size $n \times n$ and n is the number of vertices in G.

FACTS

F28: [BrPa98] For a teleportation constant α, the PageRank matrix P_α is **stochastic** (that is, its row sums are all 1) and hence is a transition probability matrix of a Markov chain. The teleportation constant α is sometimes called the **damping** or **diffusion constant**.

F29: [BrPa98] The PageRank Markov chain with transition probability matrix P_α converges to a stationary distribution $\mathbf{pr}_\alpha$ which is called the **PageRank vector** (with teleportation constant α) and the entry associated with vertex v is the **PageRank** of the v in G.

DEFINITION

D18: [Be06, JeWi03] A useful generalization of PageRank is defined as follows. Given a digraph G, consider a typical random walk on G with the transition probability matrix defined by $W = D^{-1}A$, where D is the diagonal degree matrix and A is the adjacency matrix. Personalized PageRank vectors are based on random walks with two governing parameters: a **seed vector** $\mathbf{s}$, representing a probability distribution over $V(G)$, and a teleportation constant α, controlling the rate of diffusion. The **personalized PageRank** $\mathrm{pr}(\alpha, \mathbf{s})$ is defined to be the solution to the following recurrence relation:

$$\mathrm{pr}(\alpha, \mathbf{s}) = \alpha \mathbf{s} + (1 - \alpha)\mathrm{pr}(\alpha, \mathbf{s})W$$

where the PageRank is taken to be a row vector.

FACTS

F30: The original PageRank $\mathbf{pr}_\alpha$ is the special case where the seed vector $\mathbf{s}$ is the uniform distribution.

F31: Personalized PageRank has found a number of applications, such as finding graph clusters and drawing graphs [ChTs10], devising nearly linear time local partitioning algorithms [AnChLa06, AnCh07], and to graph sparsification and graph partitioning [ChZh10].

F32: PageRank has lead to the discovery of a number of ranking algorithms in various complex networks (beyond web search) such as **TwitterRank** [We-etal10], **Social-PageRank** [Ba-etal07], and **ProteinRank** [Fr07].

F33: Because of the close connection of PageRank with random walks, there are very efficient and robust algorithms for computing and approximating PageRank [AnChLa06, Be06, JeWi03].

F34: The usage of PageRank leads to numerous applications including the basic problem of finding a "good" cut in a graph with small Cheeger ratio. Some of the most widely used approximation algorithms are spectral partitioning algorithms with their performance guaranteed by the **Cheeger inequality**:

$$2h_G \geq \lambda_1 \geq \frac{h_f^2}{2} \geq \frac{h_G^2}{2},$$

where h_f is the minimum Cheeger ratio among subsets which are initial segments in the order determined by the eigenvector f associated with the spectral gap λ. Instead of using eigenvector (which requires expensive computation), one can use PageRank vector to derive a local partitioning algorithm [AnChLa06], based on the **local Cheeger inequality** for a subset S of vertices in a graph G:

$$h_S \geq \lambda_S \geq \frac{h_g^2}{8 \log \mathrm{vol}(S)} \geq \frac{h_S^2}{8 \log \mathrm{vol}(S)},$$

where λ_S is the Dirichlet eigenvalue of the induced subgraph on S, h_S is the local Cheeger constant of S defined by $h_S = \min_{T \subseteq S} h(T)$, h_g is the minimum Cheeger ratio over initial segments determined by PageRank vectors g with the seed as a vertex in S, and α is appropriately chosen depending only on the volume of S.

12.1.8   Network Games

In morning traffic, drivers choose the most convenient way to get to work without necessarily paying attention to the consequences of their decisions on others. The Internet network can be viewed as a similar macrocosm which functions neither by the control of a central authority nor by coordinated rules. The basic motivation for each individual can only be deduced by greed and selfishness. Every player chooses the most convenient route and use strategies to maximize possible payoff. In other words, we face a combination of **game theory** and graph theory for dealing with large networks both in quantitative analysis and algorithm design. There has been a great deal of progress in algorithmic game theory; see [Ni-etal07]. Of special interest are games on networks that involve graph theory as well as internet economics.

DEFINITION

D19: Classical **chromatic graph theory** can be examined from the perspective of game theory. The **chromatic number** of a graph G, written $\chi(G)$, is the minimum number of colors needed to properly color the nodes of G so that adjacent nodes have different colors. Suppose there is a payoff of 1 unit for each player (represented by a node) if its color is different from all its neighbors. A proper coloring is a **Nash equilibrium**, while no player has an incentive to change their color.

FACT

F35: Kearns et al. [KeSuMo06] conducted an experimental study of several coloring games on specified networks. A multiple round model of graph coloring games was analyzed in [ChChJa08].

DEFINITIONS

D20: The analysis of **selfish routing** arises naturally in network management. The **"price of anarchy"** refers to the ratio of the average travel time of the decentralized selfish routing versus the coordinated routing respecting the collective welfare.

D21: **Braess' paradox** is the counter-intuitive observation that, if the travelers were behaving selfishly it was possible to improve everyone's travel time by removing roads.

FACTS

F36: Valiant and Roughgarden [VaRo10] showed that in a random graph $G(n, p)$ with edge density $p \geq n^{-1/2+\epsilon}$, Braess' paradox occurs asymptotically almost surely. Chung and Young proved that Braess' paradox occurs in a sparse random graph with $p \geq c \log n/n$. In fact, it was shown [ChYoZh12] that Braess' paradox is ubiquitous in expander graphs.

F37: There has been extensive research done on selfish routing [RoTa02]. The reader is referred to several surveys and some recent books on this topic [Ro06].

References

[AdBuAd03] L.A. Adamic, O. Buyukkokten, E. Adar, A social network caught in the web, *First Monday* **8** (2003).

[Ah-etal07] Y. Ahn, S. Han, H. Kwak, S. Moon, H. Jeong, Analysis of topological characteristics of huge on-line social networking services, In: *Proceedings of the 16th International Conference on World Wide Web*, 2007.

[AiChLu00] W. Aiello, F.R.K. Chung, L. Lu, A random graph model for massive graphs, In: *Proceedings of the Thirty-Second Annual ACM Symposium on Theory of Computing*, 2000.

[AiChLu01] W. Aiello, F.R.K. Chung, L. Lu, Random evolution of massive graphs, In: *Proceedings of FOCS 2001*.

[Ai-etal09] W. Aiello, A. Bonato, C. Cooper, J. Janssen, P. Prałat, A spatial web graph model with local influence regions, *Internet Mathematics* **5** (2009), 175–196.

[Al90] D. Aldous, The random walk construction of uniform spanning trees, *SIAM J. Discrete Math.* **3** (1990), 450–465.

[AlJeBa99] R. Albert, H. Jeong, A. Barabási, Diameter of the world-wide web, *Nature* **401** (1999), 130.

[AnChLa06] R. Andersen, F.R.K. Chung, K. Lang, Local graph partitioning using pagerank vectors, In: *Proceedings of FOCS'06*, 2006.

[AnCh07] R. Andersen, F.R.K. Chung, Detecting sharp drops in PageRank and a simplified local partitioning algorithm theory and applications of models of computation, In: *Proceedings of TAMC'07*, 2007.

[Ba-etal12] L. Backstrom, P. Boldi, M. Rosa, J. Ugander, S. Vigna, Four degrees of separation, Preprint 2012.

[Ba-etal07] S. Bao, G. Xue, X. Wu, Y. Yu, B. Fei, Z. Su, Optimizing web search using social annotations, In *Proc. of 16th Int. World Wide Web Conference*, 2007.

[BaAl99] A. Barabási, R. Albert, Emergence of scaling in random networks, *Science* **286** (1999), 509–512.

[Be-etal04] N. Berger, C. Borgs, J.T. Chayes, R. D'Souza, R.D. Kleinberg, Competition-induced preferential attachment, In: *Proceedings of the 31st International Colloquium on Automata, Languages and Programming (ICALP)*, 2004.

[Be06] P. Berkhin, Bookmark-coloring approach to personalized pagerank computing, *Internet Math.*, **3**, (2006), 41–62.

[Bo01] B. Bollobás, *Random graphs, Second edition*, Cambridge Studies in Advanced Mathematics **73**, Cambridge University Press, Cambridge, 2001.

[Bp-etal10] B. Bollobás, C. Borgs, J. Chayes, O. Riordan, Percolation on dense graph sequences, *Ann. Probab.* **38** (2010), 150–183.

[Bo-etal01] B. Bollobás, O. Riordan, J. Spencer, G. Tusnády, The degree sequence of a scale-free random graph process, *Random Structures and Algorithms* **18** (2001), 279–290.

[BoRi04] B. Bollobás, O. Riordan, The diameter of a scale-free random graph, *Combinatorica* **24** (2004), 5–34.

[Bo08] A. Bonato, *A Course on the Web Graph*, American Mathematical Society Graduate Studies Series in Mathematics, Providence, Rhode Island, 2008.

[Bo-etal11] A. Bonato, N. Hadi, P. Horn, P. Prałat, C. Wang, Models of on-line social networks, *Internet Mathematics* **6** (2011), 285–313.

[BoJaPr12] A. Bonato, J. Janssen, P. Prałat, Geometric protean graphs, *Internet Mathematics* **8** (2012), 2–28.

[BoTi12] A. Bonato, A. Tian, Complex networks and social networks, In: *Social Networks*, editor E. Kranakis, Springer, Mathematics in Industry series, 2012.

[Bo-etal05] C. Borgs, J. Chayes, R. van der Hofstad, G. Slade, J. Spencer, Random subgraphs of finite graphs: I. The scaling window under the triangle condition, *Random Structures and Algorithms*, **27** (2005), 137–184.

[Bo-etal05a] C. Borgs, J. Chayes, R. van der Hofstad, G. Slade, J. Spencer, Random subgraphs of finite graphs: II. The lace expansion and the triangle condition, *Annals of Probability*, **33** (2005) 1886–1944.

[Bo-etal06] C. Borgs, J. Chayes, R. van der Hofstad, G. Slade, J. Spencer, Random subgraphs of finite graphs: III. The phase transition for the n-cube, *Combinatorica*, **26** (2006), 395–410.

[Bo-etal12] C. Borgs, J.T. Chayes, J. Ding, B. Lucier, The hitchhiker's guide to affiliation networks: A game-theoretic approach, In: *Proceedings of the 2nd Symposium on Innovations in Computer Science*, 2011.

[BrPa98] S. Brin, L. Page, Anatomy of a large-scale hypertextual web search engine, In: *Proceedings of the 7th International World Wide Web Conference*, 1998.

[Br-etal00] A. Broder, R. Kumar, F. Maghoul, P. Raghavan, S. Rajagopalan, R. Stata, A. Tomkins, J. Wiener, Graph structure in the web, *Computer Networks* **33** (2000), 309–320.

[ChChJa08] K. Chaudhuri, F.R.K. Chung, M.S. Jamall, A network game, In: *Proceedings of WINE*, 2008.

[Ch97] F.R.K. Chung, *Spectral Graph Theory*, American Mathematical Society, Providence, Rhode Island, 1997.

[Ch10] F. Chung, Graph theory in the information age, *Notices of AMS*, **56**, no. 6, (2010), 726–732.

[ChLu02] F.R.K. Chung, L. Lu, Connected components in random graphs with given degree sequences, *Annals of Combinatorics* **6** (2002), 125–145.

[ChLu03] F.R.K. Chung, L. Lu, The average distances in random graphs with given expected degrees, *Internet Mathematics* **1** (2003), 91–113.

[ChLu04a] F.R.K. Chung, L. Lu, Coupling on-line and on-line analyses for random power law graphs, *Internet Mathematics* **1** (2004), 409–461.

[ChLu04b] F.R.K. Chung, L. Lu, *Complex Graphs and Networks*, American Mathematical Society, U.S.A., 2004.

[ChLu06] F.R.K. Chung, L. Lu, The volume of the giant component of a random graph with given expected degrees, *SIAM J. Discrete Math.* **20** (2006), 395–411.

[ChHo07] F.R.K. Chung, P. Horn, The spectral gap of a random subgraph of a graph, *Internet Mathematics* **4** (2007), 225–244.

[ChHoLu12] F.R.K. Chung, P. Horn, L. Lu, Diameter of random spanning trees in a given graph, *Journal of Graph Theory,* **69** (2012), 223–240.

[ChTs10] F.R.K. Chung, A. Tsiatas, Finding and visualizing graph clusters using PageRank optimization, In: *Proceedings of WAW'10*, 2010.

[ChYo10] F.R.K. Chung, S. Young, Braess's paradox in large sparse graphs, In: *Proceedings of WINE 2010*.

[ChYoZh12] F.R.K. Chung, S. Young, W. Zhao, Braess's paradox in expander graphs, Random Structures and Algorithms, to appear.

[ChZh10] F.R.K. Chung, W. Zhao, PageRank algorithm with applications to edge ranking and graph sparsification, In: *Proceedings of WAW'10*, 2010.

[CoFrPr12] C. Cooper, A. Frieze, P. Prałat, Some typical properties of the Spatial Preferred Attachment model, In: *Proceedings of WAW'12*, 2012.

[ErRe59] P. Erdős, A. Rényi, On random graphs I, *Publicationes Mathematicae Debrecen* **6** (1959), 290–297.

[ErRe60] P. Erdős, A. Rényi, On the evolution of random graphs, *Publ. Math. Inst. Hungar. Acad. Sci.* **5** (1960), 17–61.

[Es06] E. Estrada, Spectral scaling and good expansion properties in complex networks, *Europhys. Lett.* **73** (2006), 649–655.

[Fa12] Facebook: statistics. Accessed September 21, 2012. `http://www.facebook.com/press/info.php?statistics`.

[FaFaFa99] M. Faloutsos, P. Faloutsos, C. Faloutsos, Michalis Faloutsos, On power-law relationships of the Internet topology, In: *Proceedings of SIGCOMM*, 1999.

[FlFrVe07] A. Flaxman, A. Frieze, J. Vera, A geometric preferential attachment model of networks, *Internet Mathematics* **3** (2007), 187–205.

[FoFlMe06] S. Fortunato, A. Flammini, F. Menczer, Scale-free network growth by ranking, *Phys. Rev. Lett.* **96** 218701 (2006).

[Fr07] V. Freschi, Protein function prediction from interaction networks using random walk ranking algorithm, In: *Proceedings of the IEEE-BIBE07 7th International Symposium on Bioinformatics and Bioengineering*, 2007.

[FrTs12] A. Frieze, C. E. Tsourakakis, On certain properties of random Apollonian networks, In: *Proceedings of WAW'12*, 2012.

[Gr23] J. Grossman, The Erdős number project. Accessed September 21, 2012. `http://www.oakland.edu/enp/`.

[JaPr09] J. Janssen, P. Prałat, Protean graphs with a variety of ranking schemes, *Theoretical Computer Science* **410** (2009), 5491–5504.

[Ja-etal07] A. Java, X. Song, T. Finin, B. Tseng, Why we twitter: understanding microblogging usage and communities, In: *Proceedings of the Joint 9th WEBKDD and 1st SNA-KDD Workshop 2007*, 2007.

[JeWi03] G. Jeh, J. Widom, Scaling personalized Web search, In: *Proceedings of the 12th International Conference on World Wide Web*, 2003.

[KiLe12] M. Kim, J. Leskovec, Multiplicative attribute graph model of real-world networks, In: *Proceedings of the 7th Workshop on Algorithms and Models for the Web Graph (WAW 2010)*, 2010.

[KeSuMo06] M. Kearns, S. Suri, N. Montfort, An experimental study of the coloring problem on human subject networks, *Science*, **322** (2006), 824–827.

[Kl99] J. M. Kleinberg, Authoritative sources in a hyperlinked environment, *Journal for the Association of Computing Machinery* **46** (1999), 604–632.

[Kl00] J. Kleinberg, The small-world phenomenon: An algorithmic perspective, In: *Proceedings of the 32nd ACM Symposium on Theory of Computing*, 2000.

[Kl-etal99] J. Kleinberg, S. R. Kumar, P. Raghavan, S. Rajagopalan, A. Tomkins, The web as a graph: Measurements, models and methods, In: *Proceedings of the International Conference on Combinatorics and Computing*, 1999.

[Ku-etal99] R. Kumar, P. Raghavan, S. Rajagopalan, A. Tomkins, Trawling the web for emerging cyber-communities, In: *Proceedings of the 8th WWW Conference*, 1999.

[Ku-etal00] R. Kumar, P. Raghavan, S. Rajagopalan, D. Sivakumar, A. Tomkins, E. Upfal, Stochastic models for the web graph, In: *Proceedings of the 41th IEEE Symposium on Foundations of Computer Science*, 2000.

[KuNoTo06] R. Kumar, J. Novak, A. Tomkins, Structure and evolution of on-line social networks, In: *Proceedings of the 12th ACM SIGKDD International Conference on Knowledge Discovery and Data Mining*, 2006.

[Kw-etal10] H. Kwak, C. Lee, H. Park, S. Moon, What is Twitter, a social network or a news media? In: *Proceedings of the 19th International World Wide Web Conference*, 2010.

[LeKlFa05] J. Leskovec, J. Kleinberg, C. Faloutsos, Graphs over time: densification laws, shrinking diameters and possible explanations, In: *Proceedings of the 13th ACM SIGKDD International Conference on Knowledge Discovery and Data Mining*, 2005.

[Le-etal05] J. Leskovec, D. Chakrabarti, J. Kleinberg, C. Faloutsos, Realistic, mathematically tractable graph generation and evolution, using Kronecker multiplication, In: *Proceedings of European Conference on Principles and Practice of Knowledge Discovery in Databases*, 2005.

[Li-etal05] D. Liben-Nowell, J. Novak, R. Kumar, P. Raghavan, A. Tomkins, Geographic routing in social networks, In: *Proceedings of the National Academy of Sciences* **102** (2005), 11623–11628.

[LiKl08] D. Liben-Nowell, J. Kleinberg, Tracing information flow on a global scale using Internet, *PNAS* **105** (2008), 4633–4638.

[LuPr06] T. Łuczak, P. Prałat, Protean graphs, *Internet Mathematics* **3** (2006), 21–40.

[MaPa02] C. Malon, I. Pak, Percolation on finite Cayley graphs, *Lecture Notes in Comput. Sci.* **2483**, Springer, Berlin (2002), 91–104.

[MiPa02] M. Mihail, C.H. Papadimitriou, On the eigenvalue power law, In: *Proceedings of 6th Int. Workshop on Randomization and Approximation Techniques*, 2002.

[Mi83] M. Miller, Ecology of affiliation, *American Sociological Review* **48** (1983), 519–532.

[Mi-etal07] A. Mislove, M. Marcon, K. Gummadi, P. Druschel, B. Bhattacharjee, Measurement and analysis of on-line social networks, In: *Proceedings of the 7th ACM SIGCOMM Conference on Internet Measurement*, 2007.

[Mi04] M. Mitzenmacher, A brief history of generative models for power law and log-normal distributions, *Internet Mathematics* **1** (2004), 226–251.

[MoRe95] M. Molloy, B. Reed, A critical point for random graphs with a given degree sequence. *Random Structures and Algorithms* **6** (1995), 161–179.

[MoRe98] M. Molloy, B. Reed, The size of the giant component of a random graph with a given degree sequence, *Combin. Probab. Comput.* **7** (1998), 295–305.

[NePa03] M.E.J. Newman, J. Park, Why social networks are different from other types of networks, *Phys. Rev. E* **68** 036122 (2003).

[Ni-etal07] N. Nisan, T. Roughgarden, E. Tardos, V.V. Vazirani, *Algorithmic Game Theory*, Cambridge University Press, 2007.

[Pr05] N. Przulj, Graph theory analysis of protein-protein interactions, In: *Knowledge Discovery in Proteomics*, (I. Jurisica and D. Wigle eds.), CRC Press, 2005.

[Ro06] T. Roughgarden, *Selfish Routing and the Price of Anarchy*, MIT Press, 2006.

[RoTa02] T. Roughgarden, E. Tardos, How bad is selfish routing? *Journal of the ACM*, **49** (2002), 236–259.

[ScYo08] E.R. Scheinerman, S. Young, Directed random dot product graphs, *Internet Mathematics* **5** (2008) 91–111.

[Tw12] Twitaholic. Accessed September 21, 2012. `http://twitaholic.com/`.

[VaRo10] G. Valiant, T. Roughgarden, Braess's paradox in large random graphs, *Random Structures and Algorithms* **37** (2010), 495–515.

[WaDoNe02] D.J. Watts, P.S. Dodds, M.E.J. Newman, Identity and search in social networks, *Science* **296** (2002), 1302–1305.

[WaStr98] D.J. Watts, S.H. Strogatz, Collective dynamics of 'small-world' networks, *Nature* **393** (1998), 440–442.

[We-etal10] J. Weng, E. Lim, J. Jiang, Q. He, TwitterRank: finding topic-sensitive
    influential twitterers, In: *Proceedings of the 3rd ACM International Conference on
    Web Search and Data Mining*, 2010.

[Wi12] Wikipedia: List of social networking websites. Accessed September 21, 2012.
    `http://en.wikipedia.org/wiki/List of social networking websites`.

Section 12.2
Broadcasting and Gossiping

Hovhannes A. Harutyunyan, Concordia University, Canada
Arthur L. Liestman, Simon Fraser University, Canada
Joseph G. Peters, Simon Fraser University, Canada
Dana Richards, George Mason University

INTRODUCTION

A communication network can be modeled by a connected graph with vertices representing nodes of the network and edges representing communication links between them. A low level operation in a communication network is the transmission of a message from one node to an adjacent node. A collection of such transmissions can be combined to achieve higher level goals such as broadcasting, in which information originating at one node in the network must be distributed to all other nodes of the network. A protocol specifies the transmissions and the order in which they are made to achieve a higher level goal and can be described by a labeled subgraph. We consider two broad categories of problems — finding the best network for a given high level goal and finding the best protocol for a given goal and network.

We assume that the nodes of the network are synchronized and that transmitting a message from a node to its neighbor takes one unit of time. A protocol can be specified by a sequence of sets, each containing the transmissions made during a particular time unit. We represent this in a graph by assigning labels to each edge indicating the time(s) during which that edge transmits a message. In this model, a message originating at vertex u is received by a vertex v if there is a path from u to v with increasing edge labels.

We assume that every vertex can send a message to at most one neighbor during a given time unit. In the labeled graph, this means that no two edges incident on a single vertex have the same label. We will discuss variants of this model at the end of this section.

12.2.1   Broadcasting

In this subsection, we consider broadcasting. We begin by looking in more detail at the notion of broadcast time in this model, giving some basic definitions and simple facts.

DEFINITIONS

D1: *Broadcasting* is the goal of sending a single message from a particular vertex (called the *originator*) to all of the other vertices in the graph.

D2: A *broadcast protocol for originator (source)* s is represented by a graph $G = (V, E)$ on n vertices such that every edge e is labeled with at most one positive integer i and the labels satisfy the following constraints:

- the set of labels on the edges incident on any vertex are disjoint,
- there is exactly one path with increasing labels from s to each of the other vertices in the graph.

D3: A vertex v is *informed at time* t by a broadcast protocol for originator s if the last edge in the path from s to v is labeled with t. The *completion time* of a broadcast protocol for originator s is the least integer t such that every vertex v in the graph is informed by time t.

D4: A *broadcast protocol for a graph* G is a collection of broadcast protocols for each originator $s \in V(G)$.

D5: The *broadcast time of a broadcast protocol for a graph* G is the maximum completion time for any originator $s \in V(G)$.

D6: The *broadcast time* $b(G)$ *of a graph* G is the minimum broadcast time for any broadcast protocol for G.

D7: Let $b(n)$ be the minimum $b(G)$ over all graphs G with n vertices.

EXAMPLE

E1: Figure 12.2.1 shows examples of two broadcast protocols for a given originator s.

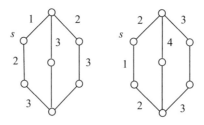

Figure 12.2.1: Two broadcast protocols for originator s.

REMARK

R1: To broadcast quickly, the labeled edges incident on s should have labels $1, 2, \ldots, i$ and for each other vertex v, one of its incident edges is labeled with the time t at which v is informed, and the other labeled edges incident on v should be labeled $t+1, t+2, \ldots, t+j$.

FACTS

F1: Due to the restrictions on the labels, at most 2^t vertices can be informed at time $t > 0$ by any broadcast protocol for any originator.

F2: $b(n) = \lceil \log n \rceil$. The previous fact implies $b(n) \geq \lceil \log n \rceil$. Protocols showing $b(n) \leq \lceil \log n \rceil$ are discussed next.

Minimum Broadcast Trees

Here we discuss those trees which correspond to minimum time broadcast protocols.

DEFINITIONS

D8: A **minimum broadcast tree** is a rooted tree on n vertices with root s for which there exists a broadcast protocol for originator s with completion time $\lceil \log n \rceil$.

D9: The **binomial tree** on 2^k vertices is a rooted tree defined recursively as follows. A single vertex is a binomial tree on 2^0 vertices. A binomial tree on 2^k vertices is formed from two binomial trees on 2^{k-1} vertices by joining their roots with an edge and making one of them the root of the binomial tree on 2^k vertices.

EXAMPLE

E2: Figure 12.2.2 shows a binomial tree and two of its subtrees that are minimum broadcast trees.

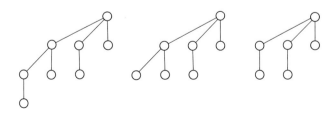

Figure 12.2.2: Binomial tree (left) and two subtrees.

FACTS

F3: The binomial tree on 2^k vertices is a minimum broadcast tree [Prosk].

F4: Every n vertex connected subtree, $2^{k-1} < n \le 2^k$, of the binomial tree on 2^k vertices that includes the root is a minimum broadcast tree and these subtrees are the only minimum broadcast trees on n vertices [Prosk].

$B(n)$ and Minimum Broadcast Graphs

One research goal has been to determine the graphs on n vertices with the fewest edges that allow minimum time broadcasting from each vertex.

DEFINITIONS

D10: A graph $G = (V, E)$ on n vertices is a **broadcast graph** if there is a broadcast protocol with completion time $\lceil \log n \rceil$ for each originator $v \in V$. In other words, for every $v \in V$ there is a spanning subgraph of G that is a minimum broadcast tree rooted at v. A broadcast graph on n vertices is a **minimum broadcast graph** (MBG) if the number of its edges is the minimum over all broadcast graphs on n vertices. The number of edges in an MBG on n vertices is denoted $B(n)$.

D11: A k-**dimensional binary hypercube** $\mathcal{H}_k$ is a k-regular vertex-transitive bipartite graph with $n = 2^k$ vertices and $k \cdot 2^{k-1}$ edges. Each vertex is labeled with a different binary string $x_1 x_2 \cdots x_k$, and there is an edge between two vertices if and only if their labels differ in exactly one bit position $1 \le \ell \le k$. Such vertices are called dimension ℓ neighbors and the edge connecting them is a dimension ℓ edge. The set of 2^{k-1} dimension ℓ edges is a perfect matching in $\mathcal{H}_k$. Thus, $\mathcal{H}_k$ can be defined recursively as two copies of $\mathcal{H}_{k-1}$ joined by a perfect matching corresponding to dimension k.

D12: The **Cayley graph** $\mathcal{D}_k$ is a $(k-1)$-regular vertex-transitive bipartite graph with $n = 2^k - 2$ vertices and $(k - 1) \cdot (2^{k-1} - 1)$ edges. The vertices are labeled with the integers mod $2^k - 2$. There is an edge between two vertices with labels i and j if and only if $(i + j) \pmod{2^k - 2} = 2^\ell - 1$ for some $1 \le \ell \le k - 1$. Such vertices are called dimension ℓ neighbors and the edge connecting them is a dimension ℓ edge. The set of $2^{k-1} - 1$ dimension ℓ edges is a perfect matching in $\mathcal{D}_k$.

This definition is specifically for Cayley graphs based on dihedral groups.

D13: The **recursive circulant graph** $\mathcal{G}_{n,d}$, $d \ge 2$ has n vertices labeled with the integers mod n and an edge between two vertices with labels i and j if and only if $i + d^\ell = j \pmod{n}$ for some $0 \le \ell \le \lceil \log_d n \rceil - 1$.

D14: The **Knödel graph** $\mathcal{W}_{n,\Delta}$ is a Δ-regular bipartite graph, $2 \le \Delta \le \lfloor \log n \rfloor$, with n vertices, n even, and $\frac{n\Delta}{2}$ edges. The vertices are labeled (i, j), $i = 1, 2$, $0 \le j \le \frac{n}{2} - 1$. There is a dimension ℓ edge, $0 \le \ell \le \Delta - 1$, between every pair of vertices $(1, j)$ and $(2, j + 2^\ell - 1 \pmod{\frac{n}{2}})$, $0 \le j \le \Delta - 1$. The set of $\frac{n}{2}$ dimension ℓ edges is a perfect matching in $\mathcal{W}_{n,\Delta}$.

EXAMPLES

E3: Figure 12.2.3 shows three non-isomorphic MBGs on 16 vertices.

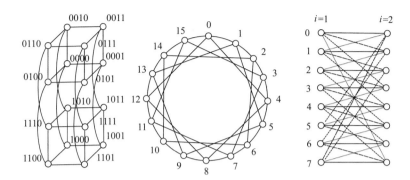

Figure 12.2.3: $\mathcal{H}_4$ (left), $\mathcal{G}_{16,4}$ (center), $\mathcal{W}_{16,4}$ (right).

E4: Figure 12.2.4 shows two isomorphic MBGs on 14 vertices.

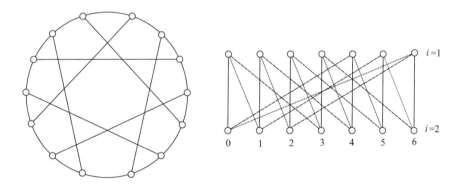

Figure 12.2.4: Heywood graph $\mathcal{D}_4$ (left), $\mathcal{W}_{14,3}$ (right).

FACTS

F5: The set $\{\mathcal{H}_k \mid k = 1, 2, 3, \ldots\}$ of binary hypercubes is an infinite family of
MBGs [FHMP]. $\mathcal{H}_k$ is a broadcast graph because there is a spanning subgraph isomorphic to a binomial tree on 2^k vertices rooted at each vertex. One broadcast protocol is
to use **dimension order**: each vertex that is informed at time t informs its dimension
$t + 1, t + 2, \ldots, k$ neighbors during time steps $t + 1, t + 2, \ldots, k$, respectively. $\mathcal{H}_k$ is an
MBG because the number of informed vertices must double during each of k time units
to inform all 2^k vertices and this requires that the originator has at least k neighbors.
Any vertex can be the originator, so the minimum number of edges is $k \cdot 2^{k-1}$.

F6: The set $\{\mathcal{G}_{2^k,4} \mid k = 2, 3, 4, \ldots\}$ of recursive circulant graphs is an infinite family of
MBGs. $\mathcal{G}_{2^k,4}$ is isomorphic to $\mathcal{H}_k$ for $k = 2$ and non-isomorphic to $\mathcal{H}_k$ for $k \geq 3$ [FR98].

F7: The set $\{\mathcal{W}_{2^k,k} \mid k = 2, 3, \ldots\}$ of Knödel graphs is an infinite family of MBGs. $\mathcal{W}_{2^k,k}$ is isomorphic to $\mathcal{H}_k$ for $k = 2, 3$ and non-isomorphic for $k \geq 4$ [FR98, FP01]. It is isomorphic to $\mathcal{G}_{2^k,4}$ for $k = 2$ and non-isomorphic for $k \geq 3$ [FR98].

F8: If G_1 and G_2 are MBGs on 2^k vertices, then the graph that results from adding any perfect matching between the vertices of G_1 and G_2 is an MBG on 2^{k+1} vertices. This construction can be applied recursively with different matchings at each stage to construct many non-isomorphic MBGs [FR98].

F9: The set $\{\mathcal{D}_k \mid k = 2, 3, 4, \ldots\}$ of Cayley graphs is an infinite family of MBGs with $2^k - 2$ vertices and $(k - 1) \cdot (2^{k-1} - 1)$ edges [DFF91, KH90]. One broadcast protocol is a dimension order protocol similar to the hypercube protocol except that there are $k - 1$ dimensions instead of k. The originator and its dimension 1 neighbor are idle during the kth time step because they have no uninformed neighbors after $k - 1$ time steps; every other informed vertex informs its dimension 1 neighbor during the kth time step. $\mathcal{D}_k$ is an MBG because any originator with degree less than $k - 1$ cannot inform $2^k - 3$ other vertices in k time units, so every vertex must have degree at least $k - 1$.

F10: The Cayley graph $\mathcal{D}_k$ is isomorphic to the Knödel graph $\mathcal{W}_{2^k-2,k-1}$ for $k \geq 3$ [FR98, HMP97], so $\{\mathcal{W}_{2^k-2,k-1} \mid k = 3, 4, \ldots\}$ is an infinite family of MBGs.

F11: Table 12.1 shows the known values of $B(n)$ for small n, indicated by asterisks, and the best upper bounds currently known for other small n. Some of the graphs that verify these values are the results of ad hoc constructions, while others are the results of construction methods that have produced several MBGs. The contents of this table are based on [BFP92, BHLP92b, DFF91, FHMP, KH90, Labahn, MS94, MitHed, Sac96, VenWen, WenVen95, WenVen, XiaWan, ZZ01].

n	$B(n)$	n	$B(n)$	n	$B(n)$	n	$B(n)$	n	$B(n)$	n	$B(n)$	n	$B(n)$
1	0*	10	12*	19	25*	28	48*	37	57	46	82	55	111
2	1*	11	13*	20	26*	29	52*	38	57	47	83	56	111
3	2*	12	15*	21	28*	30	60*	39	60	48	83	57	126
4	4*	13	18*	22	31*	31	65*	40	60	49	94	58	121*
5	5*	14	21*	23	34	32	80*	41	65	50	95	59	124*
6	6*	15	24*	24	36	33	48	42	66	51	100	60	130*
7	8*	16	32*	25	40	34	49	43	71	52	99	61	136*
8	12*	17	22*	26	42*	35	51	44	72	53	107	62	155*
9	10*	18	23*	27	44*	36	52	45	81	54	108	63	162*

Table 12.1: Upper bounds on $B(n)$ for small n. Values that are known to be optimal are indicated by asterisks.

F12: $B(n) \in \Theta(L(n)n)$, where $L(n)$ is the number of consecutive leading 1's in the binary representation of $n - 1$ [GP91].

F13: $B(n) \leq (n \lceil \log n \rceil)/2$ [Far].

F14: $B(n) \leq (n \lfloor \log n \rfloor)/2$ for even values of n [FR98].

F15: $B(n)$ is monotonically non-decreasing for any n in the first quarter of the range between any two consecutive powers of 2. More precisely, $B(n) \leq B(n+1)$ for $2^{m-1}+1 \leq n \leq 2^{m-1} + 2^{m-3} - 1$ [HL03].

RESEARCH PROBLEMS

RP1: Is $B(n)$ monotonically non-decreasing between two consecutive powers of 2? This question has only been solved for the first quarter of each range (see **F15**).

RP2: In a broadcast graph on $n = 2^k - 1$ vertices, every vertex must have degree at least $k - 1$ and if the originator has degree $k - 1$ then it must have at least one neighbor of degree k. This gives a lower bound $B(2^k - 1) \geq \left\lceil \frac{1}{2}(n(k-1) + \left\lceil \frac{n}{k+1} \right\rceil) \right\rceil$. MBGs matching this lower bound are known for $k \leq 7$ [BHLP92b, FHMP, H08, Labahn]. The question is open for $k > 7$. A good starting point to explore this problem would be the values of k that are one less than a prime because the lower bound formula is exact for these values by Fermat's little theorem (i.e., the ceilings disappear).

RP3: $B(n)$ is only known for a few infinite families of graphs and some small values of n. The general problems of determining $B(n)$ and of finding more infinite families are open. MBGs on $n = 2^k$ and $n = 2^k - 2$ vertices are k-regular and $(k-1)$-regular graphs, respectively. For most other values of n the MBGs will not be regular and the difficulty of proving upper and lower bounds seems to increase the farther that n is from a power of 2. MBGs for $n = 2^k - 1$ vertices are known for small k (see **RP2**). General lower bounds for $n = 2^k - l$, $l = 3, 4, 5, 6$, $k \geq 4$ were proved in [Sac96] and MBGs matchings these lower bounds are known for $k = 4, 5, 6$. It is unknown if there are MBGs matching these bounds for $k \geq 7$. The degree of any originator in an MBG on $n = 2^k - 2^p - r$ vertices is at least $k - p$ [GarVac] and this gives a general lower bound on $B(n)$.

Construction of Sparse Broadcast Graphs

As MBGs have proven to be difficult to construct, numerous methods have been proposed to construct sparse broadcast graphs that have a small number of edges. Almost all of these methods are combinations and variations of a few techniques.

DEFINITIONS

D15: Given a broadcast protocol in a broadcast graph G on n vertices, a vertex u is **idle at time** $t \leq \lceil \log n \rceil$ if and only if u is aware of the message at (the beginning of) time step t and u does not communicate with any of its neighbors during time step t.

D16: A subset of vertices C in a broadcast graph G is a **solid 1-cover** if and only if C is a vertex cover of G, and for each $u \notin C$, there is a broadcast protocol for u such that at least one neighbor of u is idle at some time during the broadcast.

D17: Let $G = (V(G), E(G))$ and $H = (V(H), E(H))$ be two graphs. The **compound of G into H relative to a set** $S \subseteq V(G)$, denoted $G_S[H]$, is obtained by replacing each vertex x of H by a graph G_x isomorphic to G and adding a matching between two sets S_x and S_y if x and y are adjacent in H. More precisely, the matching between S_x and S_y connects each vertex in S_x with its copy in S_y. For any vertex $u \in S$, we use H_u to denote the graph isomorphic to H that interconnects the copies of $u \in V(G)$.

EXAMPLES

E5: Figure 12.2.5 shows a broadcast protocol for C_6 (left) in which the vertex to the left of the originator is idle during time step 3, and two copies of C_6 joined by a partial matching between their solid covers of size 3 (right).

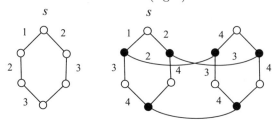

Figure 12.2.5: Partial matching of two copies of C_6.

E6: Figure 12.2.6 shows a solid 1-cover of the Heywood graph.

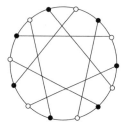

Figure 12.2.6: Solid 1-cover of the Heywood graph $\mathcal{D}_4$.

FACTS

F16: If $G = (V, E)$ is a broadcast graph on n vertices and C is a solid 1-cover of G, then $B(2n) \leq 2|E| + |C|$ [BFP92, KH90]. Construct G' by joining two isomorphic copies of G, G_1 and G_2, with a perfect matching between their solid covers C_1 and C_2. Let $u_1 \in V(G_1)$ be the originator of the broadcast. If $u_1 \in C_1$, then u_1 sends the message to its copy u_2, and then u_1 and u_2 broadcast in G_1 and G_2, respectively. If $u_1 \notin C_1$, then u_1 initiates a broadcast in G_1 and whenever a vertex $v_1 \in C_1$ is informed, it first informs its copy v_2 and then both copies of v continue to broadcast in their copies of G. C_2 is a solid cover so u_2 can be informed by an idle vertex. See Figure 12.2.5 for an example.

F17: Let $G = (V(G), E(G))$ and $H = (V(H), E(H))$ be broadcast graphs on n_1 and n_2 vertices, respectively, and let C be a solid 1-cover of G. If $\lceil \log n_1 n_2 \rceil = \lceil \log n_1 \rceil + \lceil \log n_2 \rceil$, then $G_C[H]$ is a broadcast graph and therefore $B(n_1 n_2) \leq n_2 |E(G)| + |C||E(H)|$ [BFP92]. A specific case is seen in **F16**, where H is $\mathcal{H}_2$.

F18: $B(2^k - 2^p) \leq \frac{2^k - 2^p}{4}(2k - p - 1)$ for any $1 \leq p \leq k - 2$ and $k \geq 4$ [KH90]. This also follows from **F17**.

F19: Vertex deletion was introduced in [XiaWan]. If G is a broadcast graph with n vertices, $2^{k-1} + 1 < n \leq 2^k$, e edges, and a vertex v of degree d, then $B(n-1) \leq e + \frac{1}{2}d(d-3)$. Construct G' on $n-1$ vertices by deleting v and adding enough edges to form a clique among the former neighbors of v. Each broadcast protocol for G can be modified so that the vertex u that informed v in G informs the former neighbors of v in the same order that v informed them in G.

REMARKS

R2: Vertex addition was introduced in [BHLP92b]. In this construction method, a vertex is added to a broadcast graph on $n-1$ vertices and is connected to some of the original vertices, yielding a broadcast graph on n vertices. This method is used in [BHLP92b, H08, HL12].

R3: Solid 1-covers are vertex covers for which the broadcast protocols have enough idle vertices to "cover" all paths of length 1 (i.e., edges). They can be generalized to **solid h-covers** which cover all paths of length h. Solid h-covers with $h > 1$ are used in [BFP92, KH90, MS94].

R4: m-way splits were introduced in [ChaLie] (for $m = 5, 6, 7$) and generalized in [BFP92]. The construction is similar to graph compounding and uses a graph G that has a perfect matching with certain broadcast properties, and graphs that have **even adjacency splits**. (A graph on ℓ vertices has an even adjacency split if it has two dominating sets of size $\lceil \frac{\ell}{2} \rceil$ and $\lfloor \frac{\ell}{2} \rfloor$.) The method produces sparse broadcast graphs on $n = mi + j$ vertices for which i is not a power of 2, $0 \le j < m$, and $\lceil \log(mi + j) \rceil = \lceil \log m \rceil + \lceil \log i \rceil$. This method produces broadcast graphs for more values of n than solid h-covers, which require that $n = mi$ for $m, i \in \mathbb{N}$.

R5: Several other construction methods have been developed. Most are combinations of matchings, partial matchings, compounding, m-way splits, vertex deletion, vertex addition, and operations on hypercubes. The methods in [Di99, VenWen, WenVen95] are similar to the solid covers method with vertex deletion and give similar results. The methods in [Che, GarVac] use solid covers, hypercubes, and vertex deletion. Matchings were used in the first construction methods [Far] and early uses of solid covers, vertex addition, and vertex deletion appear in [BHLP92a]. The methods in [HL99] include compounding, merging vertices, and deleting edges based on a broadcast graph construction using binomial trees.

R6: The constructions in [GP91] use hypercubes and generalized Fibonacci numbers to produce asymptotic bounds on $B(n)$.

Bounded Degree Broadcast Graphs

If we restrict our attention to graphs of given maximum degree, the time required to broadcast increases. Researchers have developed efficient broadcast protocols for some bounded degree graphs that are of interest to network designers.

DEFINITIONS

D18: The **k-dimensional cube-connected cycles graph** $\mathcal{CCC}_k$ is derived from $\mathcal{H}_k$ by replacing each vertex $x = x_1 x_2 \ldots x_k$ of $\mathcal{H}_k$ by a cycle of length k with vertices labeled (i, x), $0 \le i \le k - 1$. Each of the k vertices on a cycle that replaces x inherits one of the k edges that were incident to x in $\mathcal{H}_n$. In particular, vertex (i, x) is connected to its neighbors $(i + 1, x)$ and $(i - 1, x)$ on its cycle and to vertex $(i, x_1 x_2 \cdots \overline{x_{i+1}} \cdots x_k)$ where $\overline{x_{i+1}}$ is the complement of the binary value x_{i+1}.

D19: The **k-dimensional (wrapped) butterfly graph** $\mathcal{BF}_k$ is derived in a similar way to $\mathcal{CCC}_k$. $\mathcal{BF}_k$ has the same vertex set as $\mathcal{CCC}_k$. The difference is that $\mathcal{BF}_k$ has two edges corresponding to each former hypercube edge instead of one: (i, x) is connected to $(i + 1, x_1 x_2 \cdots \overline{x_{i+1}} \cdots x_k)$ and $(i - 1, x_1 x_2 \cdots \overline{x_i} \cdots x_k)$.

D20: The *k-dimensional binary shuffle-exchange graph* $\mathcal{SE}_k$ has 2^k vertices with the same labels as the vertices of $\mathcal{H}_k$. Each vertex $x_1 x_2 \cdots x_k$ is connected to its *shuffle neighbor* $x_2 x_3 \cdots x_k x_1$ and its *unshuffle neighbor* $x_k x_1 \cdots x_{k-1}$, and to its *exchange neighbor* $x_1 x_2 \cdots \overline{x_k}$. Parallel edges and loops are removed.

D21: The *k-dimensional binary de Bruijn graph* $\mathcal{UB}_k$ has the same vertex set as $\mathcal{H}_k$ and $\mathcal{SE}_k$. As in $\mathcal{SE}_k$, each vertex $x_1 x_2 \cdots x_k$ is connected to its shuffle neighbor $x_2 x_3 \cdots x_k x_1$ and its unshuffle neighbor $x_k x_1 \cdots x_{k-1}$. It is also connected to its *shuffle-exchange neighbor* $x_2 x_3 \cdots x_k \overline{x_1}$ and its *unshuffle-exchange neighbor* $x_k x_1 \cdots \overline{x_{k-1}}$. Parallel edges and loops are removed.

D22: The *de Bruijn graph* $\mathcal{UB}_{d,k}$ **with degree** $2d$ **and diameter** k has d^k vertices with labels that are strings of length k over the alphabet $\{0, 1, \ldots, d-1\}$. There is an edge between vertex $x_1, x_2, \ldots, x_k$ and each vertex $x_2, x_3, \ldots, x_k, \lambda$ and $\lambda, x_1, x_2, \ldots, x_{k-1}$ with $\lambda \in \{0, 1, \ldots, d-1\}$. $\mathcal{UB}_{2,k}$ is the special case $\mathcal{UB}_k$ in **D21**.

D23: Let $b(n, \Delta)$ be minimum broadcast time over all graphs on n vertices with maximum degree Δ.

EXAMPLE

E7: Figure 12.2.7 shows $\mathcal{CCC}_3$, $\mathcal{BF}_3$, $\mathcal{SE}_3$, and $\mathcal{UB}_3$.

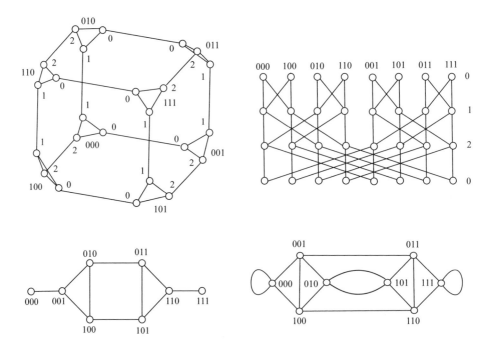

Figure 12.2.7: $\mathcal{CCC}_3$ (top left), $\mathcal{BF}_3$ (top right), $\mathcal{SE}_3$ (bottom left), and $\mathcal{UB}_3$ (bottom right).

Table 12.2 lists some parameters of hypercubes and bounded-degree approximations of hypercubes.

	vertices	edges	degree	diameter
$\mathcal{H}_k$	2^k	$k \cdot 2^{k-1}$	regular k	k
$\mathcal{CCC}_k$	$k2^k$	$\frac{3}{2} \cdot k2^k$	regular 3	$2k - 1 + \max(1, \lfloor \frac{k-2}{2} \rfloor)$
$\mathcal{BF}_k$	$k2^k$	$2 \cdot k2^k$	regular 4	$\lfloor \frac{3k}{2} \rfloor$
$\mathcal{SE}_k$	2^k	$\frac{3}{2} \cdot 2^k - 3 + (k \pmod 2)$	maximum 3	$2k - 1$
$\mathcal{UB}_k$	2^k	$2 \cdot 2^k - 3$	maximum 4	k

Table 12.2: Some parameters of hypercube-derived graphs.

FACTS

F20: Let a_t^Δ be the maximum number of vertices that can be informed after t time units in any graph with maximum degree Δ. There is one informed vertex at time $t = 0$ (the originator), and $a_t^\Delta \leq 2 \cdot a_{t-1}^\Delta$ for $t \geq 1$ (see **F1**). Furthermore, at time $t + \Delta - 1$, $t \geq 1$, all vertices that were informed by time t have no more uninformed neighbors, so $a_{t+\Delta}^\Delta \leq a_{t+\Delta-1}^\Delta + (a_{t+\Delta-1}^\Delta - a_t^\Delta) = 2a_{t+\Delta-1} - a_t^\Delta$. Based on this recurrence, $1.440 \log_2 n - 1.769$ is a lower bound on $b(n, 3)$ and $1.137 \log_2 n - 0.637$ is a lower bound on $b(n, 4)$ [LP88]. Closed form lower bounds are not known for $\Delta > 4$, but asymptotically $b(n, \Delta) > p(n, \Delta)$ where $p(n, \Delta) \approx (1 + \frac{\log_2 e}{2\Delta})$ [BHLP92a].

F21: $b(\mathcal{CCC}_k) = \lceil \frac{5k}{2} \rceil - 1$ [LP88].

F22: $1.7417k \leq b(\mathcal{BF}_k) \leq 2k - 1$. The upper bound may be improved to $2k - \frac{1}{2} \log \log k + c$ for some constant c and a sufficiently large k [KMPS94].

F23: $b(\mathcal{SE}_k) = 2k - 1$ [HJM93].

F24: $1.3171k \leq b(\mathcal{UB}_k) \leq \frac{3}{2}(k + 1)$. The upper bound is from [BP88] and the lower bound is from [KMPS94].

F25: $b(n, \Delta) < p(n, \Delta)$ where $p(n, \Delta) \approx (1 + \frac{0.415}{\Delta}) \log_2 n$ [BHLP92a]. This bound is obtained using graph compounds $\mathcal{H}[\mathcal{UB}_{d,k}]$ where $\mathcal{H}$ is a hypercube and $\mathcal{UB}_{d,k}$ is a de Bruijn graph. Better upper bounds for particular values of Δ can be obtained by compounding different graphs in de Bruijn graphs [BHLP92a]. See **F20** for a lower bound on $p(n, \Delta)$.

12.2.2 Gossiping

In this subsection, we turn our attention to gossiping — a related problem that was first investigated before broadcasting. Results on broadcasting provide lower bounds on gossiping and graphs constructed for broadcasting are also useful for gossiping.

DEFINITIONS

D24: *Gossiping* is the goal of sending a unique message from each vertex to all of the other vertices in the graph.

D25: A *gossip protocol* is represented by a graph $G = (V, E)$ on n vertices such that every edge e is labeled with a set of positive integers c_e and the labels satisfy the following constraints:

- the sets of labels on any two edges incident on a vertex are disjoint,
- there is a *transmission path* between any two vertices u and v; in particular a path of edges $(e_1, e_2, \ldots, e_k)$ such that there exist elements of the corresponding label sets $t_i \in c_{e_i}$ such $t_1 < t_2 < \cdots < t_k$. Such a transmission path "ends" at time t_k.

D26: A vertex v is *informed at time t by a gossip protocol* if there exists a transmission path from every other vertex to v that ends at or before time t.

D27: The *completion time* of a gossip protocol is the least integer t such that every vertex v in the graph is informed at time t.

D28: The *gossip time* $g(G)$ of a graph G is the minimum completion time of any gossip protocol for G.

EXAMPLE

E8: Figure 12.2.8 shows an example of a gossip protocol for a unicyclic graph.

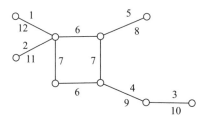

Figure 12.2.8: A gossip protocol.

REMARK

R7: Implicit in the above definitions is a basic model of communication. This model allows "calls" between two vertices such that all messages currently known to both vertices are exchanged. The model allows this unbounded exchange of information to occur in one time unit. Further, the model allows calls on disjoint edges, e_1 and e_2, to occur during the same time unit t, when $t \in c_{e_1}$ and $t \in c_{e_2}$; each time unit is a "round." When these assumptions are found to be untenable, researchers have augmented the basic model in a variety of ways that are too numerous to consider within the bounds of this paper.

FACTS

F26: Clearly $g(G) \geq b(G)$ since from a gossip protocol a broadcast protocol for each vertex of G can be inferred that has a completion time which is at most the completion time of the gossip protocol.

F27: Given a gossip protocol, the time-reversed protocol derived from it is also a valid gossip protocol. In particular, if for each edge e we replace each $t \in c_e$ with $g(G) - t + 1$, transmission paths still exist for all pairs of vertices [T71].

Optimal Gossip Graphs

Here, we discuss the structure of gossiping in terms of the number of calls made without regard to the time needed to make the calls.

DEFINITIONS

D29: The *size of a gossip protocol* for a graph G is the total number of calls used. In particular, the size is $\sum_{e \in E(G)} |c_e|$. Let $f(G)$ be the minimum size of any gossip protocol for G. A protocol of size $f(G)$ is an *optimal gossip protocol* for G.

D30: The *optimal gossip size* $f(n)$ is the minimum $f(G)$ over all graphs with n vertices.

REMARKS

R8: As remarked above, the basic model of communication allows calls to occur "at the same time." For the purpose of understanding $f(G)$ this is de-emphasized; it is as if parallel calls have been linearized. Bandwidth consumed by a protocol is measured by this quantity, not time.

FACTS

F28: $f(n+1) \leq f(n) + 2$ for $n \geq 1$ [T71]. This follows from the following observation: Given a graph G with $f(G) = f(n)$, add an edge from any vertex v of G to a new vertex v' to form a new graph G'. A valid gossip protocol for G' can be obtained by adding a call between v and v' both before and after a valid gossip protocol for G. This observation can be used to recursively construct gossip protocols from protocols for smaller graphs.

F29: $f(G) \leq 2n - 3$ for any (connected) graph G. This follows from the previous fact, building a protocol recursively using a trivial protocol for the subgraph of two adjacent vertices as the basis. For some graphs G with n vertices we have $f(G) > f(n)$; for example, the graph which is a path of four vertices requires, by exhaustive analysis, 5 calls, while $f(n) = 4$.

F30: $f(n) = 2n - 4$ for $n \geq 4$, while $f(1) = 0$, $f(2) = 1$, and $f(3) = 2$. Various researchers independently proved this fact using very different proof techniques [BS72, HMS72, T71]. Other proofs appeared later as corollaries to stronger theorems. The difficulty was the lower bound; the upper bound follows from the next fact.

F31: $f(G) = 2n - 4$ for any graph G which contains C_4 as a subgraph; C_4 is a cycle of four edges. Note that $f(C_4) = 4$. This fact follows from the above facts, building a protocol recursively using a protocol for the C_4 subgraph as the basis. An example of such a protocol was seen in Figure 12.2.8.

F32: $f(G) = 2n - 4$ if and only if the graph G contains C_4 as a subgraph. This resisted proof for many years even though it was widely conjectured. (The difficulty is that a C_4 subgraph is necessary even though the protocol in the previous fact can be replaced by a fundamentally different protocol.) Eventually two independent proofs appeared [B81, KS80]. Later proofs used deeper results about "information flow" [K96, L95].

Minimum Gossip Graphs

In this subsection, we return our focus to minimum time gossiping and consider the graphs on n vertices with the fewest edges that allow minimum time gossiping.

DEFINITIONS

D31: Recall that $g(G)$ is the gossip time for G, the number of rounds necessary to gossip in G. The **minimum gossip time** $g(n)$ is the minimum $g(G)$ over all graphs with n vertices.

D32: Let $M_n = \{G \mid G \text{ has } n \text{ vertices and } g(G) = g(n)\}$. A **minimum gossip graph** G (MGG) is a member of M_n, for some n, and the number of its edges is the minimum over all graphs in M_n. The number of edges in an MGG on n vertices is denoted $\boldsymbol{G(n)}$.

FACTS

F33: As noted above $g(G) \geq b(G)$ for every G, so it follows that $g(n) \geq b(n)$.

F34: $g(n) = \lceil \log_2 n \rceil$ if n is even and $g(n) = \lceil \log_2 n \rceil + 1$ if n is odd. The lower bound follows from the bound on $b(n)$ and the fact that with odd n some node does not participate in the first round. The upper bounds come from constructions [K75]. This result was anticipated in earlier literature where the problem was posed in a different context, e.g. [B50].

F35: It follows from the above that $G(n) \geq B(n)$ for even n [F00].

F36: For any tree T with n vertices $g(T) \geq 2\lceil \log_2 n \rceil - 1$ [L86].

F37: For all $n = 2^k$, $k \geq 1$, $G(n) = \frac{n}{2} \log_2 n$. In particular, the set $\{\mathcal{H}_k \mid k = 1, 2, 3, \ldots\}$ of binary hypercubes is an infinite family of MGGs for such n [L93]. $\mathcal{H}_k$ has a gossip protocol that makes all calls across each dimension at the same time and proceeds in dimension order; it will broadcast from each originator in the same manner as was discussed for broadcasting. For $n = 16$, all MGGs have been characterized; they are all formed from $\mathcal{H}_4$ by perturbing C_4 subgraphs [LP91]. Further, when $k \geq 2$ the families $\mathcal{W}_{2^k,k}$ and $\mathcal{G}_{2^k,4}$ are also MGGs [FR98].

F38: For all $n = 2^k - 2$, $k \geq 3$, $G(n) = \frac{n}{2}(\lceil \log_2 n \rceil - 1)$. In particular, the set $\{\mathcal{D}_k \mid k = 3, 4, 5, \ldots\}$ of Cayley graphs is an infinite family of MGGs for such n [DFF91, L93]. One gossip protocol is a dimension order protocol similar to the hypercube protocol except there are $k - 1$ dimensions instead of k; after calling across all dimensions the first dimension is repeated.

F39: For all $n = 2^k - 4$, $k \geq 6$, $G(n) = \frac{n}{2}(\lceil \log_2 n \rceil - 1)$. In particular, the infinite family $\mathcal{W}_{n,k-1}$ are MGGs for such n [FR98, L93].

F40: $G(n) \leq (n \lfloor \log n \rfloor)/2$ for even values of n [BHLP97].

F41: Table 12.3 shows the known exact values of $G(n)$ for small n, indicated by asterisks, and the best upper bounds currently known for other small n. These are apparently the only known values, beyond the families in the above facts. Some of the graphs that verify these values are the results of ad hoc constructions, while others are the results of construction methods that have produced several MBGs. The contents of this table are based on [F00, FL00, FR98, L93].

n	$G(n)$	n	$G(n)$	n	$G(n)$	n	$G(n)$	n	$G(n)$	n	$G(n)$	n	$G(n)$	n	$G(n)$
1	0*	5	5*	9	9*	13	17	17	20	21	27	25	32	29	45
2	1*	6	6*	10	13*	14	21*	18	25	22	36	26	52	30	60*
3	2*	7	7*	11	11*	15	19*	19	22	23	29	27	34	31	47
4	4*	8	12*	12	18*	16	32*	20	28	24	36*	28	56*	32	80*

Table 12.3: Upper bounds on $B(n)$ for small n. Values that are known to be optimal are indicated by asterisks.

12.2.3 Other Variations of Broadcasting and Gossiping

Many variations of broadcasting and gossiping have been investigated. Most of these involve changes to the model that we have been using, considering digraphs, hypergraphs, multiple originators, multiple messages from a single source, random transmissions, or restricted protocols. The model described here is somewhat unrealistic for gossiping, as it assumes that combined messages can be sent in a single transmission in constant time. Many of the other papers on gossiping use different models that account for the size of the combined messages sent.

For more information on these and other variations, see [FL94, HHL88, HKMP96, HKPRU05].

References

[BS72] B. Baker , R. Shostak, Gossips and telephones, *Discrete Math.* 2 (1972), 191–193.

[B50] A. Bavelas, Communication patterns in task-oriented groups, *J. Acous. Soc. Am.* 22(6) (1950), 725–730.

[BFP92] J.-C. Bermond, P. Fraigniaud, and J.G. Peters, Antepenultimate broadcasting, *Networks* 26 (1995), 125–137.

[BHLP97] J.-C. Bermond, H.A. Harutyunyan, A.L. Liestman, and S. Pérennes, A note on the dimensionality of modified Knödel graphs, *Int. J. Found. Comput. Sci.* 8(2) (1997), 109–116.

[BHLP92a] J.-C. Bermond, P. Hell, A.L. Liestman, and J.G. Peters, Broadcasting in bounded degree graphs, *SIAM J. Discrete Math.* (5) (1992), 10–24.

[BHLP92b] J.-C. Bermond, P. Hell, A. L. Liestman, and J.G. Peters, Sparse broadcast graphs, *Discrete Appl. Math.* 36 (1992), 97–130.

[BP88] J.-C. Bermond and C. Peyrat, Broadcasting in de Bruijn networks. In: *Proc. 19th SE Conference on Combinatorics, Graph Theory and Computing*, Congr. Numer. (1988), 283–292.

[B81] R. Bumby, A Problem with telephones, *SIAM J. Alg. Discrete Math.* 2 (1981), 13–18.

[ChaLie] S. C. Chau and A. L. Liestman, Constructing minimal broadcast networks, *J. Comb. Inf. Syst. Sci.* 10 (1985), 110–122.

[Che] X. Chen, An upper bound for the broadcast function $B(n)$, *Chinese J. Comput.* 13 (1990), 605-611.

[R94] J. de Rumeur, *Communications dans les réseaux de processeurs*, Collection Etudes et Recherches en Informatique, Masson, Paris, 1994.

[DFF91] M. J. Dinneen, M. R. Fellows, and V. Faber, Algebraic constructions of efficient broadcast networks, In *Applied Algebra, Algebraic Algorithms and Error Correcting Codes 9*. Lecture Notes in Computer Science 539 (1991), 152–158.

[Di99] M. J. Dinneen, J. A. Ventura, M. C. Wilson, and G. Zakeri, Compound constructions of broadcast networks, *Discrete Math.* 93 (1999), 205–232.

[Far] A. M. Farley, Minimal broadcast networks, *Networks* 9 (1979), 313–332.

[FHMP] A. M. Farley, S. Hedetniemi, S. Mitchell, and A. Proskurowski, Minimum broadcast graphs, *Discrete Math.* 25 (1979), 189-193.

[F00] G. Fertin, A study of minimum gossip graphs, *Discrete Math.* 215 (2000), 33–57.

[FL00] G. Fertin and R. Labahn, Compounding of gossip graphs, *Networks* 36(2) (2000), 126–137.

[FR98] G. Fertin and A. Raspaud, Families of graphs having broadcasting and gossiping properties. In: *Proc. 24th Int. Workshop on Graph-Theoretic Concepts in Computer Science (WG '98)*, Smolenice, Lecture Notes in Computer Science 1517, Springer-Verlag, 1998, 63–77.

[FL94] P. Fraigniaud and E. Lazard, Methods and problems of communication in usual networks, *Discrete Appl. Math.* 53 (1994), 79–133.

[FP01] P. Fraigniaud and J. G. Peters, Minimum linear gossip graphs and maximal linear (Δ, k)-gossip graphs, *Networks* 38 (2001), 150–162.

[GarVac] L. Gargano and U. Vaccaro, On the construction of minimal broadcast networks, *Networks* 19 (1989), 673-689.

[GP91] M. Grigni and D. Peleg, Tight bounds on minimum broadcast networks, *SIAM J. Discrete Math.* 4(2) (1991), 207–222.

[HMS72] A. Hajnal, E. C. Milner, and E. Szemeredi, A cure for the telephone disease, *Canad. Math. Bull.* 15(3) (1972), 447–450.

[H08] H. A. Harutyunyan, An efficient vertex addition method for broadcast networks, *Internet Math.* 5(3) (2008), 211–225.

[HL99] H. A. Harutyunyan and A. L. Liestman, More broadcast graphs, *Discrete Appl. Math.* 98 (1999), 81–102.

[HL03] H. A. Harutyunyan and A. L. Liestman, On the monotonicity of the broadcast function, *Discrete Math.* 262(1-3) (2003), 149–157.

[HL12] H. A. Harutyunyan and A. L. Liestman, Upper bounds on the broadcast function using minimum dominating sets, *Discrete Math.* 312(20) (2012), 2992–2996.

[HHL88] S. M. Hedetniemi, S. T. Hedetniemi, and A. L. Liestman, A survey of gossiping and broadcasting in communication networks, *Networks* 18 (1988), 319–349.

[HMP97] M.-C. Heydemann, N. Marlin and S. Pérennes, Complete rotations in Cayley graphs, *Eur. J. Comb.* 22(2) (2001), 179–196.

[HJM93] J. Hromkovič, C. D. Jeschke, and B. Monien, Optimal algorithms for dissemination of information in some interconnection networks, *Algorithmica* 10 (1993), 24–40.

[HKMP96] J. Hromkovič, R. Klasing, B. Monien, and R. Peine, Dissemination of information in interconnection networks (broadcasting & gossiping). In F. Hsu and D.-Z. Du, editors, *Combinatorial Network Theory*. Kluwer Academic Publishers, 1996, 125–212.

[HKPRU05] J. Hromkovič, R. Klasing, A. Pelc, P. Ružiča, and W. Unger, Dissemination of information in communication networks: part 1. broadcasting, gossiping, leader election, and fault-tolerance, Springer Monograph, Springer-Verlag, 2005.

[KH90] L. H. Khachatrian and O. S. Haroutunian, Construction of new classes of minimal broadcast networks. In *Proc. 3rd International Colloquium on Coding Theory, Dilijan, Armenia*, 1990, 69–77.

[KMPS94] R. Klasing, B. Monien, R. Peine, and E. A. Stöhr, Broadcasting in butterfly and de Bruijn networks, *Discrete Appl. Math.* 53(1-3) (1994), 183–197.

[KS80] D. J. Kleitman and J. B. Shearer, Further gossip problems, *Discrete Math.* 30 (1980), 151–156.

[K75] W. Knödel, New gossips and telephones, *Discrete Math.* 13 (1975), 95.

[K96] D. W. Krumme, Reordered gossip schemes, *Discrete Math.* 156 (1996), 113–140.

[Labahn] R. Labahn, A minimum broadcast graph on 63 vertices, *Discrete Appl. Math.* 53 (1994), 247–250.

[L95] R. Labahn, Kernels of minimum size gossip schemes, *Discrete Math* 143 (1995), 99–139.

[L93] R. Labahn, Some minimum gossip graphs, *Networks* 23 (1993), 333–341.

[L86] R. Labahn, The telephone problem for trees, *Elektron. Info. u Kybernet.* 22 (1986), 475–485.

[LP91] R. Labahn and C. Pietsch, Characterizing minimum gossip graphs on 16 vertices, *Technical Report*, Univ. Bonn (1991).

[LP88] A. L. Liestman and J.G. Peters, Broadcast networks of bounded degree, *SIAM J. Discrete Math.* 1(4) (1988), 531–540.

[MS94] M. Mahéo and J.-F. Saclé, Some minimum broadcast graphs, Technical Report 685, LRI, Université de Paris-Sud, 1991. (Some of the results of this report have since been published with the same title in *Discrete Appl. Math.* 53 (1994), 275–285.)

[MitHed] S. Mitchell and S. Hedetniemi, A census of minimum broadcast graphs, *J. Comb. Inf. Syst. Sci.* 5 (1980), 141–151.

[Prosk] A. Proskurowski, Minimum broadcast trees, *IEEE Trans. Comput.* 30 (1981), 363–366.

[Sac96] J.-F. Saclé, Lower bounds for the size in four families of minimum broadcast graphs, *Discrete Math.* 150 (1996), 359–369.

[T71] R. Tijdeman, On a telephone problem, *Nieuw Arch. Wiskd.* 19(3) (1971), 188–192.

[VenWen] J. A. Ventura and X. Weng, A new method for constructing minimal broadcast networks, *Networks* 23 (1993), 481–497.

[WenVen95] X. Weng and J. A. Ventura, A doubling procedure for constructing minimal broadcast networks, *Telecommunication Systems* 3 (1995), 259–293.

[WenVen] X. Weng and J. A. Ventura, Constructing optimal broadcast networks, Working Paper 91-132, Dept. of Industrial and Management Systems Engineering, Pennsylvania State Univ., 1991.

[XiaWan] J. Xiao and X. Wang, A research on minimum broadcast graphs, *Chinese J. Comput.* 11 (1988), 99–105.

[ZZ01] J.-G. Zhou and K.-M. Zhang, A minimum broadcast graph on 26 vertices, *Appl. Math. Lett.* 14 (2001), 1023–1026.

Section 12.3
Communication Network Design Models

Prakash Mirchandani, University of Pittsburgh
David Simchi-Levi, Massachusetts Institute of Technology

INTRODUCTION

Telecommunication network design (along with logistics and supply chain configuration, electricity distribution, and road infrastructure development) represents a major application of graph theory. Although telecommunication network design problems come in many different flavors that correspond to different telecommunication application contexts, the common theme across all such problems is that we need to connect a set of locations (for example, customers, computers, cities, or communication switches) using transmission links (which may or may not be capacitated) in order to satisfy the demand (consisting of voice, video, or data traffic) between pairs of locations at minimum cost. Specific application contexts may warrant additional constraints. While simple to state, network design problems are challenging from the modeling, algorithmic, and computational perspectives, and even the simplest network design problems belong to the class of NP-hard problems. Many classical graph theory problems, for example, the Steiner Tree Problem, the Traveling Salesman Problem, and the two-connected problem, are special cases of the general network design problem.

In this chapter, we first present a general network design model and then consider a number of its special cases. Since the vast scope of network design prevents us from being comprehensive, our objective is to introduce some central network design models, along with their important structural properties and solution algorithms.

12.3.1   General Network Design Model

Preliminaries

DEFINITIONS

D1:   A *network* is a graph $G = (N, E)$, where N is the set of nodes (or vertices) and E is the set of edges. The nodes denote geographical locations where communication demands originate or terminate, or locations where routing hardware and software are installed. The edges denote potential transmission links. Throughout this section, $|N| = n$ and $|E| = m$.

D2:   A *commodity*, denoted by k, refers to a communication *demand* of value d_k (measured in *bits per second (bps)*) between the *origin node* $O(k) \in N$ and *destination node* $D(k) \in N$. Let $\mathcal{K}$ denote the set of all commodities and $K = |\mathcal{K}|$.

D3:   A *communication facility*, or simply a *facility*, is a transmission medium (e.g., a copper cable, a fiber-optic cable, or a cellular tower-to-satellite connection) installed on a network edge. In our discussion, we assume that the facilities are *undirected facilities*, that is, if a facility of capacity C is installed on an edge, then it permits a total flow of C on the edge in each direction. (Facilities that permit flow only in one direction are said to be *directed facilities*.)

D4:   If the facility allows unlimited commodity flow, it is said to be a *uncapacitated facility*, otherwise it is said to be a *capacitated facility*. The capacity measured in *bps* of a facility is referred to as its *bandwidth*.

NOTATION: Let $\mathcal{L} = \{1, 2, \ldots, L\}$ denote the set of *facility levels* ordered by *facility capacity*; the capacity of the facility level i is C_i, $i = 1, 2 \ldots, L$, where $C_1 \leq C_2 \leq \ldots \leq C_L$.

NOTATION: The (nonnegative) *fixed cost* of installing a facility of capacity C_l on the edge $\{i, j\} \in E$ is a_{ij}^l for $l = 1, 2, \ldots, L$, with $a_{ij}^l \leq a_{ij}^{l+1}$. The (nonnegative) *variable cost* of sending one unit of commodity k from node i to node j on edge $\{i, j\}$ is denoted b_{ij}^k.

D5:   A *switch* routes or processes the data signal and can be installed on each node in a subset $M \subseteq N$. Each switch has capacity T, and for each node $i \in M$, t_i denotes the *switch installation cost*.

D6:   The *network design problem* is to install capacity, install switches, and route the commodities at minimum cost.

D7:   A *(linear) mixed-integer program* (MIP) is a problem consisting of a linear multi-variable function to be minimized or maximized subject to linear inequalities involving those variables, where some of the variables are restricted to integer values. An *integer program* (IP) is one in which all variables are restricted to be integer-valued, and a *linear program* (LP) is one in which no variables are restricted to be integer (i.e., all variables are *continuous variables*).

NOTATION: The optimal value of any mathematical program P defined above (LP, IP, or MIP) is denoted z_P.

SUMMARY OF NOTATION
- Network $G = (N, E)$, n nodes and m edges.
- Set $\mathcal{K}$ of commodities, $k = 1, 2, \ldots, K$.
- Each commodity k has origin $O(k)$, destination $D(k)$, demand d_k bps, and a variable cost b_{ij}^k of sending one unit from node i to j on edge $\{i, j\}$.
- Set $\mathcal{L}$ of facility levels, $l = 1, 2, \ldots, L$.
- Each facility level l has capacity C_l and fixed installation cost a_{ij}^l on edge $\{i, j\}$.
- Each node $i \in M \subseteq N$ has a switch installation cost t_i.

General Edge-Based Flow Model

The network design problem defined above can be modeled by the mixed integer program shown below.

NOTATION: The **edge design** decision variables, upper bounds, any restrictions on the network topology are defined as follows:
- The continuous decision variable f_{ij}^k denotes the **flow** of commodity k on edge $\{i, j\}$ from i to j.
- The integer decision variable u_{ij}^l denotes the number of facilities of level l to install on edge $\{i, j\}$, with $u_{ij}^l \leq \mu_{ij}^l$, the upper bound.
- The binary decision variable v_i equals one if a switch is installed on node $i \in M$ and 0 otherwise.
- For each facility level l, U^l is a set specifying any topological restrictions on the design (for example, U^1 might specify that the chosen edge design variables u_{ij}^1 define a cycle). The m-component vector of u_{ij}^l's is denoted u^l.

General Model: Edge-Flow [GM:EF]

$$\text{Minimize} \left\{ \sum_{\{i,j\} \in E, l \in \mathcal{L}} a_{ij}^l u_{ij}^l + \sum_{\{i,j\} \in E, k \in \mathcal{K}} b_{ij}^k f_{ij}^k + \sum_{i \in M} t_i v_i \right\}$$

subject to:

$$\sum_{j \in N : \{i,j\} \in E} f_{ji}^k - \sum_{j \in N : \{i,j\} \in E} f_{ij}^k = \begin{cases} -d_k & \text{if } i = O(k) \\ d_k & \text{if } i = D(k) \\ 0 & \text{otherwise} \end{cases} \quad \forall i \in N, \forall k \in K \quad \text{(FC)}$$

$$\sum_{k \in \mathcal{K}} f_{ij}^k \leq \sum_{l \in \mathcal{L}} C_l u_{ij}^l \quad \forall \{i, j\} \in E \quad \text{(EF2)}$$

$$\sum_{k \in \mathcal{K}} f_{ji}^k \leq \sum_{l \in \mathcal{L}} C_l u_{ij}^l \quad \forall \{i, j\} \in E \quad \text{(EF3)}$$

$$\sum_{k \in \mathcal{K}} \sum_{j \in N : \{i,j\} \in E} f_{ji}^k \leq T v_i \quad \forall i \in M \quad \text{(EF4)}$$

$$u^l \in U^l \quad \forall \, l \in \mathcal{L} \quad \text{(EF5)}$$

$$f_{ij}^k \geq 0 \quad \forall \{i, j\} \in E, \forall k \in \mathcal{K} \quad \text{(EF6)}$$

$$0 \leq u_{ij}^l \leq \mu_{ij}^l \text{ and integer} \quad \forall \{i, j\} \in E, \forall l \in \mathcal{L} \quad \text{(EF7)}$$

$$v_i = 0 \text{ or } 1 \quad \forall \, i \in M \quad \text{(EF8)}$$

REMARKS

R1:   The constraints (FC) are the usual *flow conservation* constraints for each commodity at each node. Constraints (EF2) and (EF3) limit the total flow of all commodities in each direction of an edge by its installed capacity, and (EF4) constrains the total flow through nodes in M. Constraints (EF5) restrict the topology of the chosen design. Constraints (EF6) through (EF8) define the nonnegativity and integrality requirements.

R2:   This formulation assumes that the initial capacity of the edges is zero. By modifying the right hand side of constraints (EF2) and (EF3), we can model situations where edges have some positive level of initial capacity.

R3:   Constraints (EF2) and (EF3) limit the total flow of all commodities on an edge; hence, we refer to them as the *bundle* capacity constraints. In addition, some applications limit the flow for *individual* commodities on each edge. For an edge $\{i, j\}$, if C_{ij}^k denotes the maximum allowable flow of commodity k on the edge from i to j, then the *individual* capacity constraints are of the form:

$$f_{ij}^k \le C_{ij}^k \quad \forall k \in \mathcal{K}, \ \forall \{i, j\} \in E.$$

R4:   Formulation [GM:EF] allows fractional flows, although some applications require integral flows. The formulation also allows a commodity to flow over multiple origin-destination paths, called a **bifurcated flow**. Some applications might require *non-bifurcated* flow where each commodity is required to flow on only one origin-destination path.

R5:   Parallel edges in the graph can refer to different transmission technologies, for example, copper cables, fiber-optic cables, or wireless transmission.

R6:   In subsequent models that have the flow conservation constraints, we simply write "Flow Conservation Constraints (FC)" instead of rewriting those constraints each time.

General Path-Flow Model

Instead of using the edge-based flow variables as defined in formulation [GM:EF] above, an alternate approach for modeling the network design problem is to use *path-based* flow variables. We replace each edge $\{i, j\}$ in the original network by two oppositely directed arcs (i, j) and (j, i) having the same costs as edge $\{i, j\}$.

NOTATION: The relevant notation for the model given below is as follows:

- For each commodity $k = 1, 2, \ldots, K$, P_k denotes the set of directed paths from origin $O(k)$ to destination $D(k)$.

- The flow (in bps) of commodity k on path $p \in P_k$ is denoted g_p^k.

- The cost of sending one unit of flow of commodity k from origin $O(k)$ to destination $D(k)$ on path p is denoted $c_p^k = \sum_{(i,j) \in p} b_{ij}^k$.

General Model: Path-Flow [GM:PF]

$$\text{Minimize} \left\{ \sum_{\{i,j\}\in E, l\in \mathcal{L}} a_{ij}^l u_{ij}^l + \sum_{k\in \mathcal{K}} \sum_{p\in P_k} c_p^k g_p^k + \sum_{i\in M} t_i v_i \right\}$$

subject to:

$$\sum_{p\in P_k} g_p^k = d_k \quad \forall k \in \mathcal{K}$$

$$\sum_{k\in \mathcal{K}} \sum_{p\in P_k : (i,j)\in p} g_p^k \leq \sum_{l\in \mathcal{L}} C_l u_{ij}^l \quad \text{and}$$

$$\sum_{k\in \mathcal{K}} \sum_{p\in P_k : (j,i)\in p} g_p^k \leq \sum_{l\in \mathcal{L}} C_l u_{ij}^l \quad \forall \{i,j\} \in E$$

$$\sum_{k\in \mathcal{K}} \sum_{p\in P_k : i\in p} g_p^k \leq T v_i \quad \forall i \in M$$

$$u^l \in U^l \quad \forall l \in \mathcal{L}$$

$$g_p^k \geq 0 \quad \forall \{i,j\} \in E, \ \forall k \in \mathcal{K}$$

$$0 \leq u_{ij}^l \leq \mu_{ij}^l \text{ and integer} \quad \forall \{i,j\} \in E, \ \forall l \in \mathcal{L}$$

$$v_i = 0 \text{ or } 1 \quad \forall i \in M$$

EXAMPLE

E1: Suppose there are seven nodes ($n = 7$), two facility levels (L=2) with capacities $C_1 = 1$ and $C_2 = 10$, $U^1 = U^2 = \emptyset$, and the internodal demand is as follows.

	1	2	3	4	5	6	7
1	0	0	0	0	0	0	0
2	10	0	0	0	0	0	0
3	10	10	0	0	0	0	0
4	20	0	30	0	0	0	0
5	1	0	0	0	0	0	0
6	1	0	0	0	0	0	0
7	1	0	0	0	0	0	0

In the Figure 12.3.1 below, which gives a feasible solution for this situation, each thick line represents ten units of capacity and each thin line represents one unit of capacity.

REMARKS

R7: Nodes 1, 2, 3, and 4 in Figure 12.3.1, represented by squares, are *hub* nodes; they are typically 2-edge connected by high capacity facilities and form the *backbone network*. Nodes 5, 6, and 7, represented by circles, are *end-office* or *terminal* nodes. Each end-office node is uniquely assigned to a hub. Each subset of end-office nodes assigned to the same hub forms the **local access network**, which is usually a tree.

R8: Because configuring the entire telecommunication network simultaneously is computationally difficult, a decomposition approach that separates the full problem into the backbone network and the local access network design problems can be used.

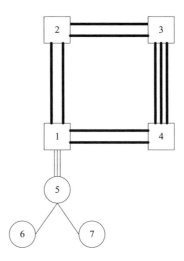

Figure 12.3.1: Example showing a capacitated network.

12.3.2   Uncapacitated Network Design

In certain applications, we can assume that the edge facilities and nodes switches are uncapacitated. Three such scenarios are:

(1) If $\sum_{k \in \mathcal{K}} d_k \leq C_1$, then installing the lowest capacity facility on an edge permits us to send the maximum possible flow on it.

(2) In the telecommunications setting, because fiber-optic cables have high transmission capacity, their capacity may be practically unlimited for some applications.

(3) We may be initially interested in designing the topology of the network only. (A later analysis, if necessary, will determine the edge capacities and flow routes.)

DEFINITIONS

D8:   The **uncapacitated network design model** (UND) is the network design model that results from the assumption that the edge facilities and node switches are uncapacitated.

D9:   A **linear programming relaxation** (LPR) of an integer or mixed-integer program is the linear program that results from relaxing the integrality requirement.

D10:   Let IP1 and IP2 denote two minimization integer or mixed-integer programs for a discrete optimization problem. If LP1 and LP2, respectively, denote their linear programming relaxations, then the LP1 relaxation is said to be *at least as strong* as the LP2 relaxation if $z_{LP1} \geq z_{LP2}$.

D11:   The **integrality gap** for a minimization integer (or mixed-integer) program IP is the quantity $(z_{IP} - z_{LPR})/z_{LPR}$.

Uncapacitated Network Design [UND]

For any of the three scenarios mentioned above, we are able to make certain assumptions about our design problem. The last of the three listed below is an additional assumption that we make for the purpose of this presentation.

ASSUMPTIONS

- The design will use only (the least cost) facility 1.
- We can scale all the commodity demands d_k to one, and correspondingly, we can scale the variable flow costs to $b_{ij}^k \leftarrow b_{ij}^k d_k \ \forall k \in \mathcal{K}, \forall \{i,j\} \in E$, and set $C_1 = C$.
- We assume $U^l = \emptyset$ for all $l \in \mathcal{L}$, and we suppress the facility index l.

$$\text{Minimize} \left\{ \sum_{\{i,j\}\in E} a_{ij} u_{ij} + \sum_{\{i,j\}\in E, k\in\mathcal{K}} b_{ij}^k f_{ij}^k \right\}$$

subject to:

Flow Conservation Constraints (FC)

$$\sum_{k\in\mathcal{K}} f_{ij}^k \leq Cu_{ij} \quad \text{and} \tag{UND1}$$

$$\sum_{k\in\mathcal{K}} f_{ji}^k \leq Cu_{ij} \quad \forall\{i,j\} \in E \tag{UND2}$$

$$f_{ij}^k \geq 0 \quad \forall\{i,j\} \in E, \ \forall k \in \mathcal{K} \tag{UND3}$$

$$u_{ij} = 0 \text{ or } 1 \quad \forall\{i,j\} \in E \tag{UND4}$$

FACTS

F1: When all the flow costs are zero, $K = n - 1, O(k) = k$, and $D(k) = n$ for all $k \in \mathcal{K}$, the above formulation models the (polynomially solvable) minimum spanning tree problem.

F2: If the values of the design variables u_{ij} are known, then solving problem UND amounts to finding K shortest paths, one for each commodity k using cost b_{ij}^k on the directed graph $G(N, A)$, where A is the set of directed arcs $(i, j), (j, i)$ for which $u_{ij} = 1, \{i, j\} \in E$.

F3: Replacing the aggregate *"forcing"* constraints (UND1) and (UND2) by the disaggregate constraints

$$f_{ij}^k \leq u_{ij} \quad \text{and} \tag{UND5}$$

$$f_{ji}^k \leq u_{ij} \quad \forall\{i,j\} \in E, \ \forall k \in \mathcal{K} \tag{UND6}$$

results in an equivalent integer program, [UNDStr]. However, the linear programming relaxation obtained by eliminating the integrality restriction on the u_{ij} variables for formulation [UNDStr] is at least as strong as the linear programming relaxation of formulation [UND].

F4: Suppose the costs satisfy the mild flow-cost conditions:

(a) $b_{ij}^k = b_{ji}^k = b_{ij}$ for all $k \in \mathcal{K}$ and for all edges $\{i, j\} \in E$;

(b) $b_{ij} + b_{ji} \geq 0$ for all $\{i, j\} \in E$.

In this case [BaMaWo89], if two commodities k_1 and k_2 share the same origin or the same destination, then there exists an optimal [UND] solution in which k_1 and k_2 flow in the same direction on every edge, that is,

$$f_{ij}^{k_1} + f_{ji}^{k_2} \leq u_{ij} \quad \forall k_1, k_2 \ \left(O(k_1) = O(k_2) \text{ or } D(k_1) = D(k_2)\right), \ \forall \{i, j\} \in E \quad \text{(UND7)}$$

F5:   Balakrishnan et al. [BaMaWo89] develop a very effective dual ascent method that approximately solves the dual of formulation [UNDStr] and includes constraints (UND7). Using local search (*add–drop*) heuristics on an initial solution provided by the dual heuristic results in low performance gaps on large-scale, randomly generated problems.

F6:   The results in [BaMaWo89] indicate that the integrality gap of formulation [UND-Str] with constraints (UND7) is, on average, small for randomly generated problems. [BaMaMi98] conduct a theoretical analysis of the uncapacitated network design model and develop heuristics with a worst-case integrality gap of $\sqrt{K}$.

EXAMPLES

E2:   Consider a complete graph on three nodes $1, 2, 3$ with three commodities defined by $O(1) = O(2) = 1, O(3) = 2, D(1) = 2, D(2) = D(3) = 3$. Let the fixed costs equal one and the flow costs equal zero for all three edges, i.e., $a_{12} = a_{13} = a_{23} = 1$ and $b_{12} = b_{13} = b_{23} = 0$. Setting $u_{12} = u_{13} = u_{23} = 1/3$ and $f_{12}^1 = f_{13}^2 = f_{23}^3 = 1$ gives a feasible solution of cost 1 to the linear programming relaxation of formulation [UND]. However, this solution is not feasible to the linear programming relaxation of formulation [UNDStr]. The optimal solution to the linear programming relaxation of formulation [UNDStr] costs $3/2$ ($u_{12} = u_{13} = u_{23} = 1/2, f_{12}^1 = f_{13}^1 = f_{32}^1 = f_{13}^2 = f_{12}^2 = f_{23}^2 = f_{23}^3 = f_{21}^3 = f_{13}^3 = 1/2$), which shows, in this case, that the relaxation of [UNDstr] is stronger than the relaxation of [UND].

E3:   Constraints (UND7) strengthen the linear programming relaxation of formulation [UNDStr]. Consider a 3-node, 2-commodity network with $O(1) = O(2) = 1, D(1) = 2, D(2) = 3$. As in Example E2, $a_{12} = a_{13} = a_{23} = 1$, and $b_{12} = b_{13} = b_{23} = 0$. The solution to the linear programming relaxation of the UNDStr model sets all three design variables to $1/2$; if the model is enhanced by adding constraints (UND7), the linear programming relaxation obtains the optimal solution by setting $u_{12} = u_{13} = 1$.

Multi-Level Network Design

Another uncapacitated network design that occurs in many telecommunication and transportation settings is one in which the nodes are classified into a hierarchy of groups based on their importance, and different grades of facilities are available. The more critical nodes need to be connected using higher grade (level) facilities. The **Multi-Level Network Design** (MLND) model considers topological network design applications in such hierarchical settings.

DEFINITIONS

D12:   In a **multi-level network design** (MLND) problem, the facility levels are $l = 1, 2, \ldots, L$, the node-set N is partitioned into $N_1, N_2, \ldots, N_L$ non-empty levels (groups), and we assume that the higher indexed node groups are more critical than the lower indexed ones and the facility levels are indexed in increasing order of grade (and expense). The objective is to assign a facility level to each selected design edge subject to the constraint that every pair of nodes $i \in N_{l'}$ and $j \in N_{l''}$ can communicate along a path that uses only facilities at level at least $\min(l', l'')$. We assume that there are no flow costs and that the facilities are uncapacitated.

NOTATION: Installing a level-l facility on each edge $\{i, j\}$ of the network costs a_{ij}^l, where $a_{ij}^1 \le a_{ij}^2 \le \cdots \le a_{ij}^L$.

D13: The MLND problem has a ***proportional cost structure*** if the ratio $(a_{ij}^l / a_{ij}^{l-1})$, $l = 2, \ldots, L$, is the same for all edges.

D14: The ***Steiner tree problem***: Given a weighted graph in which a subset of vertices is identified as terminals, find a minimum-weight connected subgraph that includes all the terminals. In an optimal solution, the non-terminal nodes are called ***Steiner nodes***.

Multi-Level Network Design Model [MLND]

In the formulation of the MLND problem shown below [BaMaMi94a], the variable u_{ij}^l equals 1 if a level-l facility is installed on edge $\{i, j\}$ and 0 otherwise. We also define commodities $k = 1, 2 \ldots, n - 1$ such that for each commodity k, $O(k) = n, D(k) = k$, and $d_k = 1 \ \forall k$.

$$\text{Minimize} \ \sum_{\{i,j\} \in E} \sum_{l=1}^{L} a_{ij}^l u_{ij}^l$$

subject to:

Flow Conservation Constraints (FC)

$$f_{ij}^k \le \sum_{l \le l' \le L} u_{ij}^{l'} \quad \text{and}$$

$$f_{ji}^k \le \sum_{l \le l' \le L} u_{ij}^{l'} \quad \forall \{i, j\} \in E, \ \forall k : D(k) \in N_l$$

$$f_{ij}^k \ge 0 \quad \forall \{i, j\} \in E, \ \forall k \in \mathcal{K}$$

$$u_{ij}^l = 0 \text{ or } 1 \quad \forall \{i, j\} \in E, \ \forall l = 1, 2, \ldots, L$$

FACTS

F7: Since the edge costs are nonnegative, the edges chosen by the optimal MLND solution define a tree.

F8: For each $l, l = 2, 3, \ldots, L$, the subtree in the optimal solution defined by facilities at level l is embedded in the (sub) tree defined by facilities at level $l - 1$.

F9: When $L = 2$ and $a_{ij}^1 = 0$, the MLND problem is equivalent to the Steiner tree problem with N_2 defining the terminal nodes and N_1 defining the Steiner nodes.

F10: Since the MLND problem generalizes the Steiner problem, the MLND problem is NP-hard. The problem continues to be NP-hard when $L = 2, |N_2| = 2$ and (i) the costs are proportional or (ii) $a_{ij}^2 = 1$ and $a_{ij}^1 = 0$ or 1 for all $\{i, j\} \in E$ [Or91].

MLND Composite Heuristic

The following composite heuristic [BaMaMi94a] for $L = 2$ takes the better of two heuristic values to develop a worst-case performance bound for the heuristic solution value relative to the optimal solution value.

Step 1. Minimum Spanning Tree (Forward) heuristic: Treat the level-1 nodes as level-2 nodes, and find the minimum spanning tree in $G(N, E)$ using costs a_{ij}^2. Set $u_{i,j}^2 = 1$ on all edges of this tree to get a feasible solution.

Step 2. Steiner Overlay (Backward) heuristic: Find the Minimum Spanning Tree T_{MLND} spanning all nodes using costs a_{ij}^1. Using the incremental costs $a_{ij}^2 - a_{ij}^1$ for all edges in the minimum spanning tree, and a_{ij}^2 otherwise, solve a Steiner tree problem with N_2 as the terminal nodes and N_1 as the Steiner nodes. Install level-2 facilities on the edges of the Steiner tree and level-1 facilities on the remaining edges of T_{MLND} if they do not create a cycle.

COMPUTATIONAL NOTE: If $\rho < 2$ denotes the worst-case performance ratio of the Steiner tree solution method used in *Step 2*, then the worst-case performance ratio of the MLND composite heuristic is $\dfrac{4}{4 - \rho}$ for proportional costs and $\rho + 1$ for general costs.

COMPUTATIONAL NOTE: Although, for simplicity, we have described the heuristic for $L = 2$, it generalizes to an exponential run-time recursive heuristic for the general MLND problem (the composite heuristic iterates $O(2^L)$ times). For the case when $L = 3, \rho = 1$ and costs are proportional, this generalized recursive heuristic provides a worst-case performance ratio of 1.52241 [Mi96].

REMARKS

R9: As with the uncapacitated network design problem, we can "directize" the edges of the MLND model (i.e., replace each undirected edge by two oppositely directed arcs). A solution approach based on the dual of this directized formulation solves large, randomly generated test cases, producing solutions that are within 0.9 percent of the optimal solution for $L = 2$ and within 6 percent of the optimal solution for $L = 5$.

R10: The MLND model can be generalized by imposing more restrictive connectivity requirements. For example, we may require that the subgraph defined by the l-level facilities be k_l-connected for a pre-specified integer $k_l, \forall l$. The extant literature has not addressed this generalization even when $L = 2$, $k_1 = 1$, and $k_2 = 2$.

12.3.3   Survivable Network Design (SND)

Communication networks designed purely from a cost minimization perspective to satisfy commodity demand tend to be sparse (due to economies of scale), and an edge (representing a transmission facility) or a node (representing a communication switch) failure can lead to interruptions in communication service. Therefore, network designers build in redundancy by providing alternate communication paths in the network, so that the network can continue to satisfy communication demands even after a failure. The degree of network redundancy depends on the trade-off between network cost and the importance of maintaining the required connections between pairs of nodes. Since the probability of simultaneous failure of two or more elements (edges or nodes) is very small, it is generally assumed for network planning purposes that only one failure occurs at a time.

Uncapacitated Survivable Network Design [SNDUnc]

The mathematical program given below is based on the concept of an *edge-cut*, a set of edges whose deletion disconnects the network. Edge-cuts and their relation to connectivity and internally disjoint paths (Menger's theorem) are discussed in §4.1 and §4.7. The connection between edge-cuts and the algebraic structure of a graph is presented in §6.4, and their role in finding maximum flows in networks is discussed in §10.1.

DEFINITIONS

D15: A network is said to be a ***survivable network*** if it can continue to satisfy demand even when one of its edges or nodes fails.

D16: The ***connectivity requirement*** between nodes $i, j \in N, i \neq j$, denoted by a nonnegative integer r_{ij}, is the minimum number of edge-disjoint paths needed between i and j.

D17: When the maximum connectivity requirement is no more than two, the problem is said to be a ***low connectivity survivable*** [LCS] network design problem.

D18: Let $G = (N, E)$ be a network and let S be a proper nonempty subset of the node-set N. The ***(edge-)cut*** defined by S, denoted $\langle S, N \setminus S \rangle$, is the set of edges defined by

$$\langle S, N \setminus S \rangle = \{\{i, j\} \in E \mid i \in S \text{ and } j \in N \setminus S\}$$

NOTATION: For node subsets A and B, let $[A, B]$ denote the set of node-pairs given by
$$[A, B] \doteq \{i, j \mid i \in A \text{ and } j \in B\}$$

REMARKS

R11: Since the graph is undirected, we can assume that the connectivity requirements are symmetric, that is, $r_{ij} = r_{ji}$.

R12: When both i and j are backbone nodes, r_{ij} is at least two, and when either of them is a local access node, r_{ij} is typically one. In addition, $r_{ij} = 0$ if either i or j is an optional (Steiner) node that the network can, but is not required to, use.

Cut-Based Formulation of SND [SND-CUT]

If $S \subset N$, the basic cut formulation of the survivable network design (SND) problem is given below.

$$\text{Minimize} \quad \sum_{\{i,j\} \in E} a_{ij} u_{ij}$$

subject to:

$$\sum_{\{i,j\} \in \langle S, N \setminus S \rangle} u_{ij} \geq \max_{i,j \in [S, N \setminus S]} \{r_{ij}\} \quad \text{for all proper nonempty } S \subset N$$

$$u_{ij} = 0 \text{ or } 1 \quad \text{for all } \{i, j\} \in E$$

FACTS

F11: If there exist ***node levels*** $r_i, i \in N$ such that $r_{ij} = \min(r_i, r_j)$, then an optimal solution to SND has at most one 2-connected component.

F12:   For a slightly more general connectivity requirement function, using a *primal-dual* solution approach, [WiGoMiVa95] develop a worst-case performance bound of $2R$, where R denotes the maximum connectivity level; [GoGoPlShTaWi94] improve this bound to $2(1 + \frac{1}{2} + \frac{1}{3} + \ldots + \frac{1}{R})$.

EXAMPLE

E4:   Figure 12.3.2 shows an example of a survivable network with $R = 3$ and with two 2-connected components.

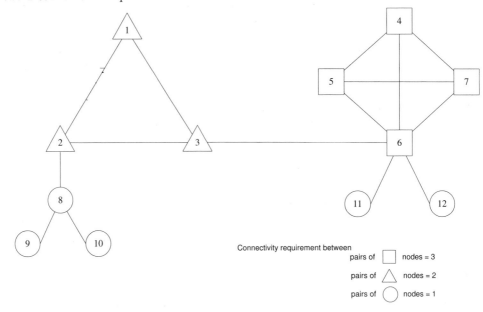

Figure 12.3.2: Solution to an SND problem.

SND Iterative Rounding Heuristic

[Ja01] develops a linear programming based iterative rounding heuristic, presented below, that has a (tight) worst-case performance bound of two relative to the linear programming relaxation value of formulation [SND-CUT]. The two-step strategy repeats until the heuristic finds a feasible integer SND solution.

Step 1. Find an optimal basic solution $u_{ij}, \{i,j\} \in E$ to the linear programming relaxation of formulation [SND-CUT]. (Comment: Except for the first iteration, some of the u_{ij} variables would have been set to one in Step 2.)

Step 2. For all edges $\{i,j\}$ with $u_{ij} \geq \frac{1}{2}$, fix u_{ij} to one.

FACTS

F13:   Using insightful arguments, Jain shows that in any optimal basic solution to the formulation [SND-CUT], at least one u_{ij} variable is at least $\frac{1}{2}$. Rounding this edge to one at most doubles its contribution to the solution value. Jain departs from traditional rounding heuristics that solve one linear program and then simultaneously round all fractional solutions suitably. By setting only those variables with value at least a half to one and resolving the linear program, he finds the "valuable" edges in each iteration, and provides the first known SND heuristic with a constant worst-case performance bound.

F14: For specific connectivity requirements, or for specific cost parameters, or when defined over special classes of graphs, the SND problem is polynomially solvable. For example, when $r_{ij} = 1$ for all pairs of nodes i and j in N, the SND problem simplifies to the minimum spanning tree problem, and when $r_{ij} = 1$ for a specific pair of nodes i and j in N amd 0 otherwise, the SND problem is the shortest path problem and is thus polynomially solvable. The SND problem defined on *Halin* and on *series-parallel* graphs is polynomially solvable for several situations, such as when all connectivity requirements are even, or are equal. See [KeMa05] for an excellent survey of these as well as other results relating to the SND problem.

F15: [DiKeMa08] provide a complete linear description of the integer polytope and show that the SND problem on a subclass of series-parallel graphs, including *outerplanar* graphs, is polynomially solvable when the connectivity requirements are either one or two.

NOTATION: Let $G = (N, E)$ be a network with requirements r_{ij} as defined above, and let $A, B \subset N$. Let $[A, B]_{r \geq 2}$ denote the set of node-pairs i, j whose connectivity requirement is at least 2. That is, $[A, B]_{r \geq 2} = \{i, j \mid i \in A, j \in B \text{ and } r_{ij} \geq 2\}$.

FACT

F16: The following constraint models the redundancy requirement that the network contain at least two node-disjoint paths between every pair of nodes i and j for which $r_{ij} \geq 2$.

$$\sum_{\{i,j\} \in E \,:\, i,j \in [S, \, (N \setminus \{S \cup \{z\}\})]} u_{ij} \geq 1$$

$\forall z \in N, \quad$ for all nonempty $S \not\subseteq N \setminus \{z\} \; : \; [S, \, (N \setminus \{S \cup \{z\}\})]_{r \geq 2} \neq \emptyset$

Flow-Based Formulation of SND [SND-FLOW]

By using flow variables, we can model the SND problem as a special case of the general network design model.

NOTATION: For every pair of nodes with $r_{ij} > 0$, define a commodity k such that $d_k = r_{ij}$, $O(k) = i$, $D(k) = j$, and set $L = 1$.

$$\text{Minimize} \quad \sum_{\{i,j\} \in E} a_{ij} u_{ij}$$

subject to:

Flow Conservation Constraints (FC)

$f_{ij}^k \leq u_{ij} \quad$ and

$f_{ji}^k \leq u_{ij} \quad \forall \{i, j\} \in E, \forall k \in \mathcal{K}$

$f_{ij}^k \geq 0 \quad \forall \{i, j\} \in E, \forall k \in \mathcal{K}$

$u_{ij} = 0 \text{ or } 1 \quad \forall \{i, j\} \in E$

REMARKS

R13: Using a novel connectivity upgrade strategy, [BaMiNa09] develop several classes of valid inequalities to strengthen this flow-based formulation and test their computational effectiveness for the case when the SND solution has at most one 2-connected component.

R14: [MaRa05] develop another flow-based formulation, doubling the number of commodities with connectivity requirements at least two and show that this formulation is stronger than formulation [SND-FLOW] when some connectivity requirements are at least one.

Survivable Network Design: Bounded Cycles

A limitation of both the SND formulations discussed above is that they allow long cycles. Thus, in an extreme case, the solution to an LCS network design problem might be a minimum cost Hamiltonian cycle through all the nodes. In such a solution, any edge failure requires the rerouting of the affected demands using long alternate paths. To prevent long cycles, we can impose the condition that every chosen edge belong to at least one cycle with length bounded by a specified constant.

NOTATION: We use the following notation to impose this requirement:

- For each edge $\{i,j\} \in E$, Y_{ij} denotes the set of cycles that contain $\{i,j\}$ and satisfy the length bound.

- For edge $\{i,j\} \in E$ and cycle $C \in Y_{ij}$, y_{ij}^C denotes a binary variable that is one if cycle C is included in the solution and zero otherwise. The following (exponentially sized) set of constraints imposes the bounded cycles condition.

$$\sum_{C \in Y_{ij}} y_{ij}^C \geq u_{ij} \quad \forall \{i,j\} \in E$$

$$\sum_{C \in Y_{ij}, \{i',j'\} \in C} y_{ij}^C \leq u_{i'j'} \quad \forall \{i,j\} \in E, \ \{i',j'\} \in E \setminus \{\{i,j\}\}$$

$$y_{ij}^C = 0 \text{ or } 1 \quad \forall \{i,j\} \in E, \ C \in Y_{ij}$$

REMARKS

R15: Fortz et al. [FoLaMa00] design a *branch-and-cut* approach for the design of minimum-cost bounded-cycles networks that contain two node-disjoint paths between every pair of nodes.

R16: Another way of making a network robust to failures is to limit the number of edges (also called the number of *hops*) in the path used to satisfy the demand for each commodity (e.g., ([BaAl92], [Go98]).

12.3.4   Capacitated Network Design

In our discussion so far, we have not considered the capacity constraints appearing in the general models of §12.3.1. Many practical applications require activation of these constraints; such capacitated situations result in very hard optimization problems even when there are no node capacities and the edge capacity levels may be chosen from a continuous range (see, for example, [Mi81]). We describe three models below in each of which the capacities are available at discrete levels.

Network Loading Problem [NLP]

In telecommunication settings, we often encounter situations where we have L different types of facilities available such that their capacities are *modular*, that is, the capacity of a level-$(l+1)$ facility is a multiple of the capacity of a level-(l) facility, $l = 1, 2, \ldots, L-1$. Moreover, we can install (load) any number of facilities on an edge. We refer to this problem as the ***Network Loading Problem (NLP)***. The number of different types of facilities is often small (less than five) in practical situations. For example, in some telecommunications settings, *T1* and *T3* facilities may be the available transmission facilities. A T3 facility has 28 times the capacity of a T1 line, but costs less than 28 times the cost of a T1 line. This results in economies of scale in the fixed costs structure for any edge. Figure 12.3.3 below depicts an illustrative cost structure when we have two facilities and the capacity of the higher level facility is 12 times the capacity of the lower level facility. Note that, in this example, the break-even point is eight.

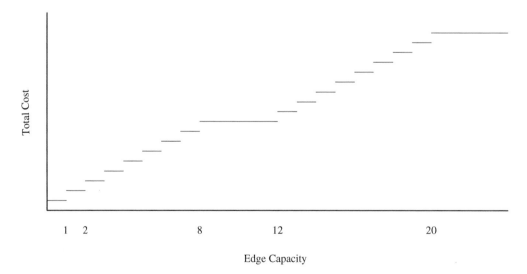

Figure 12.3.3: Cost structure for a network loading problem.

The following formulation models the Network Loading Problem when $L = 2$ with $C_1 = 1$ and $C_2 = C$.

$$\text{Minimize } \left\{ \sum_{\{i,j\}\in E} (a_{ij}^1 u_{ij}^1 + a_{ij}^2 u_{ij}^2) + \sum_{\{i,j\}\in E,\ k\in\mathcal{K}} b_{ij}^k f_{ij}^k \right\}$$

subject to:

$$\text{Flow Conservation Constraints (FC)}$$

$$\sum_{k \in \mathcal{K}} f_{ij}^k \; \leq \; u_{ij}^1 + C u_{ij}^2 \quad \text{and}$$

$$\sum_{k \in \mathcal{K}} f_{ji}^k \; \leq \; u_{ij}^1 + C u_{ij}^2 \quad \forall \{i,j\} \in E$$

$$f_{ij}^k \geq 0 \qquad \forall \{i,j\} \in E, \; \forall k \in \mathcal{K}$$

$$0 \leq u_{ij}^l \leq \mu_{ij}^l \text{ and integer} \qquad \forall \{i,j\} \in E, l = 1, 2$$

EXAMPLE

E5: Consider the single-commodity case (i.e., $K = 1$) with $L = 1, C_1 = C$. If d denotes the commodity demand, intuition suggests that the solution will use at most two paths: one carrying $(\lceil d/C \rceil - 1)C$ units of flow, and the other carrying $d - (\lceil d/C \rceil - 1)C$ units of flow. The following example ([Mi89]) (shown in Figure 12.3.4 below) shows that this intuition is not correct. In this example, the commodity origin is node 1, the commodity destination is node 4, $d = 3$ and $C = 2$.

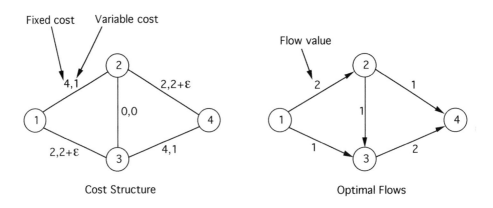

Figure 12.3.4: An optimal solution with three flow paths.

FACTS

F17: [MaMiVa95] and [BiGu95] develop and implement a polyhedral approach for solving the network loading problem, and [MaMiVa92] provide a complete polyhedral description of two of its core subproblems.

F18: The single-commodity network loading problem with one facility type and flow costs, or with two facility types and zero flow costs is NP-hard ([ChGiSa98]). [At02] develops a complete linear description for the single facility, single commodity cut-set polytope.

Capacitated Concentrator Location [CCL]

The design of local access networks (recall that local access networks are frequently trees) requires connecting the end-user nodes (terminals) on the local access tree to consolidation points, called *concentrators* on the tree. These concentrators compress the data signal and transmit it directly to the backbone network hub to which the local access tree is connected.

DEFINITION

D19: In the ***capacitated concentrator location*** problem (**CCL**), we are given the local access tree $G(N, E)$ with node-set N and edge-set E, the demands d_i at the terminals $i \in N$, a set $M \subseteq N$ of possible sites for locating concentrators, the cost t_j associated with installing a concentrator of capacity T at node $j \in M$, and the cost a_{ij} of connecting terminal i to concentrator j. We want to determine the minimum cost location of the concentrators and assign each terminal to exactly one concentrator. The assignment should not violate the concentrator capacity constraint, nor the *contiguity* constraint: if terminal i is assigned to concentrator j, then all terminals on the unique path in G, P_{ij}, from node i to concentrator j are also assigned to the concentrator at node j.

NOTATION: Let v_j equal one if a concentrator is located at node $j \in M$ and 0 otherwise, and let u_{ij} equal one if terminal $i \in N$ is served by a concentrator at site $j \in M$.

Capacitated Concentrator Location Model [CCL]

The (CCL) problem can be formulated as the following 0-1 integer linear program:

$$\text{Minimize} \left\{ \sum_{i \in N} \sum_{j \in M} a_{ij} u_{ij} + \sum_{j \in M} t_j v_j \right\}$$

$$\sum_{j \in M} u_{ij} = 1 \qquad \forall i \in N \tag{CCL1}$$

$$\sum_{i \in N} d_i x_{ij} \leq T v_j \qquad \forall j \in M \tag{CCL2}$$

$$u_{ij} \leq v_j \qquad \forall i \in N, \, \forall j \in M \tag{CCL3}$$

$$u_{i'j} \geq u_{ij} \qquad \forall i' \in P_{ij} \tag{CCL4}$$

$$u_{ij} = 0 \text{ or } 1 \qquad \forall i \in N, \forall j \in M \tag{CCL5}$$

$$v_j = 0 \text{ or } 1 \qquad \forall j \in M \tag{CCL6}$$

REMARKS

R17: In formulation [CCL], the first set of constraints, CCL1, ensures that each terminal is connected to exactly one concentrator, and the CCL2 models the concentrator capacity constraint. The CCL3 constraints ensure that terminal i can be assigned to a concentrator only if it has been installed. The CCL4 constraints model the contiguity requirement.

R18:   This problem can be considered to be a generalization of the bin-packing problem but with an additional cost incurred for making the item and bin assignments and the contiguity constraint on the items.

R19:   Another problem that arises in local access networks is the *capacitated minimum spanning tree* problem (AlGa88], [Ga91]). In this problem, there are $n - 1$ commodities and one of the nodes of $G(N, E)$, say node n, is the destination node for all commodities. The origin node for commodity k is k, $k = 1, 2, \ldots, n - 1$ and its demand is d_k. The cost of using edge $\{i, j\}$ is a_{ij} and all flow costs equal zero. We need to select the least cost subset of E such that (i) the subset of edges form a spanning tree of $G(N, E)$, and (ii) the sum of the demands of the nodes included in each subtree formed by deleting all the chosen edges incident to node n does not exceed a pre-specified capacity limit. A number of heuristic and optimization based approaches have been developed for this problem.

Survivable Network Design (Capacitated)

When we are designing capacitated networks, one way of improving survivability is to have two edge-disjoint paths, a working path and a backup path from $O(k)$ to $D(k)$, each with a dedicated capacity of d_k for meeting the demand for commodity k. While this approach, called *1+1 Diverse Path Protection*, provides the necessary survivability instantaneously, it does so at a high cost because it more than doubles the capacity (the backup path is typically longer than the working path) in the network. Therefore, network designers have devised a number of other ways of imposing the survivability condition in capacitated networks. We discuss two of these approaches. The first uses *self-healing rings* (SHRs) (which may still provide dedicated protection capacity against a failure) and the second one limits the amount of disrupted flow by using a *diversification* and *reservation* strategy.

DEFINITIONS

D20:   *Self-Healing Rings Approach.* **Self-healing rings** (SHRs) are cycles in the network formed by groups of nodes. Different SHRs may share edges (and thus be connected to each other); together the SHRs cover all the demand nodes. A switching device (called Add-Drop Multiplexer) is placed at the nodes that connect two SHRs and allows the signal to be transferred between the SHRs. Each edge in an SHR permits the signal to flow in both directions. Hence, each pair of nodes in an SHR is connected by two edge and two node disjoint paths. Therefore, any signal flowing through a ring is protected against a single edge or a single node failure on that ring. A number of different design problems arise when SHRs are used (see, for example, [SoWySeLaGeFo98]).

D21:   *Diversification and Reservation Approach.* **Diversification** splits the flow of commodity k such that no more than a fraction δ_k flows through any edge or node (except $O(k)$ or $D(k)$). **Reservation** reserves enough spare capacity in the network such that it can reroute at least a fraction of ρ_k of commodity k if an edge or a node fails.

EXAMPLE

E6:   [PaJoAlGrWe96] Figure 12.3.5 illustrates the difference between diversification and reservation. For this example, $K = 1, O(1) = 1, D(1) = 4$, and $d_1 = 2$.

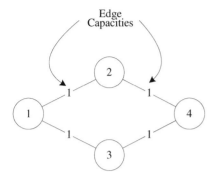

Diversification: Normal State

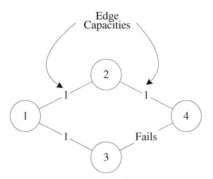

Diversification: Failure State.
Failure of any edge still allows 1/2
the original demand to be satisfied

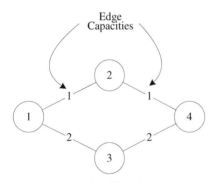

Reservation: Normal State
Working Path: 1-3-4
Backup Path: 1-2-4

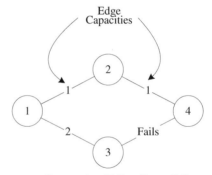

Reservation: Failure State. Failure
of any edge on the working path still
allows 1/2 the original demand to
be satisfied

Figure 12.3.5: Two ways of enhancing network survivability.

Diversification and Reservation Model [DR]

In the model given below, all flow costs are zero.

NOTATION: The model uses the following notation:

- $\theta \in \Theta$ denotes the operating state of the network, where $\theta = 0$ denotes the normal operating state (when all edges and nodes are operational).
- The state when node i, $i \in N$, breaks down is denoted $\theta = i$, and $\theta = \{i, j\}$, $\{i, j\} \in E$, denotes the state when edge $\{i, j\}$ breaks down.
- $G(\theta) = (N(\theta), E(\theta))$, where $N(\theta)$ and $E(\theta)$ are the sets of nodes and edges, respectively, that are still operating under state θ.
- $P_k(\theta)$ denotes the set of feasible paths from $O(k)$ to $D(k)$ under operating state θ.

$$\text{Minimize} \quad \sum_{\{i,j\}\in E} \sum_{l=1}^{L} a_{ij}^l u_{ij}^l$$

subject to:

$$\sum_{k\in\mathcal{K}} \sum_{p\in P_k(\theta):\{i,j\}\in p} g_p^k(\theta) \leq \sum_{l\in\mathcal{L}} C_l u_{ij}^l \qquad \forall\, \{i,j\} \in E,\ \forall\theta\in\Theta \qquad\qquad \text{(DR1)}$$

$$\sum_{p\in P_k(0)} g_p^k(0) = d_k \qquad \forall k \in \mathcal{K} \qquad\qquad \text{(DR2)}$$

$$\sum_{p\in P_k(\theta)} g_p^k(\theta) = \rho_k d_k \qquad \forall k \in \mathcal{K},\ \forall\theta \in \Theta \setminus \{0\} \qquad\qquad \text{(DR3)}$$

$$\sum_{p\in P_k(0):i\in p} g_p^k(0) \leq \delta_k d_k \qquad \forall k \in \mathcal{K},\ \forall i \in N \setminus \{O(k), D(k)\} \qquad\qquad \text{(DR4)}$$

$$g_p^k(0) \leq \delta_k d_k \qquad \forall k \in \mathcal{K}, p = \{O(k), D(k)\} \qquad\qquad \text{(DR5)}$$

$$g_p^k(\theta) \geq 0 \qquad \forall\, \{i,j\} \in E, k \in \mathcal{K},\ \forall\theta\in\Theta \qquad\qquad \text{(DR6)}$$

$$u_{ij}^l \geq 0 \text{ and integer} \qquad \forall\, \{i,j\} \in E, l \in \mathcal{L} \qquad\qquad \text{(DR7)}$$

REMARK

R20: In the formulation above, (DR1) are the capacity constraints. Constraints (DR2) ensure that the full demand of each commodity is routed under normal operating conditions (no failures), and constraints (DR3) ensure that at least a fraction ρ_k of commodity k is routed under all other operating states. The next set of constraints (DR4) ensures that no node (and hence no edge other than the direct edge $\{O(k), D(k)\}$) carries a flow of more than δ_k. Constraints (DR5) ensure this diversification for direct edges. The remaining constraints are nonnegativity and integrality constraints. See [StDa94] and [AlGrJoPaWe98] for a discussion of cutting plane approaches for solving such models.

p-Cycle Protection Model [PP]

DEFINITIONS

D22: *Directed version of G.* Given an undirected graph $G(N, E)$, let the corresponding *directed graph* be $G_{Dir}(N, A)$ where A, the set of directed arcs, is $\{(i,j), (j,i) : i, j \in E\}$.

D23: *p-cycle.* A *p-cycle* (or, a pre-configured protection cycle) is a simple, directed cycle in a directed graph $G_{Dir}(N, A)$ with at least three arcs.

EXAMPLE

E7: Installing sufficient capacity on a p-cycle permits rerouting the flow in case of an edge failure. In Figure 12.3.6, the p-cycle is the cycle defined by $\{1, 2, 4, 5, 7, 6\}$. In case an edge corresponding to a p-cycle arc, say edge $\{1, 2\}$ fails, the flow from node 1 to node 2 is rerouted on the path $1 - 6 - 7 - 5 - 4 - 2$ and the flow from node 2 to node 1 is is rerouted on the path $2 - 4 - 5 - 7 - 6 - 1$. If a chord of the p-cycle, say edge $\{1, 7\}$ fails, the flow from node 1 to node 7 is is rerouted on the path $1 - 2 - 4 - 5 - 7$ and the flow from node 7 to node 1 is is rerouted on the path $7 - 6 - 1$. Note that the same p-cycle can protect against the (non-simultaneous) failure of mupltiple edges.

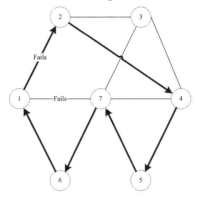

Figure 12.3.6: Directed p-cycle for re-routing flows.

NOTATION: Model [PP] [AtRa08] uses the following notation:
- Let Ψ denote the set of p-cycles of $G_{Dir}(N, A)$ with at least three arcs.
- Let z_ψ denote the amount of capacity reserved for p-cycle $\psi \in \Psi$.
- Let $\alpha_{ij}^\psi = 1$ if arc (i, j) belongs to p-cycle $\psi \in \Psi$ and 0 otherwise.
- Let $\beta_{ij}^\psi = 1$ if edge $\{i, j\}$ is a chord of p-cycle $\psi \in \Psi$ and 0 otherwise.

$$\text{Minimize} \left\{ \sum_{\{i,j\} \in E, l \in \mathcal{L}} a_{ij}^l u_{ij}^l + \sum_{\{i,j\} \in E, k \in \mathcal{K}} b_{ij}^k f_{ij}^k \right\}$$

subject to:

Flow Conservation Constraints (FC)

$$\sum_{k \in \mathcal{K}} f_{ij}^k \leq \sum_{\psi \in \Psi} \alpha_{ij}^\psi z_\psi + \sum_{\psi \in \Psi} \beta_{ij}^\psi z_\psi \qquad \forall \{i, j\} \in E \tag{PP1}$$

$$\sum_{k \in \mathcal{K}} f_{ji}^k \leq \sum_{\psi \in \Psi} \alpha_{ji}^\psi z_\psi + \sum_{\psi \in \Psi} \beta_{ij}^\psi z_\psi \qquad \forall \{i, j\} \in E \tag{PP2}$$

$$\sum_{k \in \mathcal{K}} f_{ij}^k + \sum_{\psi \in \Psi} \alpha_{ij}^\psi z_\psi \leq \sum_{l \in \mathcal{L}} C_l u_{ij}^l \qquad \forall \{i, j\} \in E \tag{PP3}$$

$$\sum_{k \in \mathcal{K}} f_{ji}^k + \sum_{\psi \in \Psi} \alpha_{ij}^\psi z_\psi \leq \sum_{l \in \mathcal{L}} C_l u_{ij}^l \qquad \forall \{i, j\} \in E \tag{PP4}$$

$$f_{ij}^k \geq 0 \qquad \forall \{i, j\} \in E, \ \forall k \in \mathcal{K} \tag{PP5}$$

$$z_\psi \geq 0 \qquad \forall \psi \in \Psi \tag{PP6}$$

$$0 \leq u_{ij}^l \leq \mu_{ij}^l \text{ and integer} \qquad \forall \{i, j\} \in E, \ \forall l \in \mathcal{L} \tag{PP7}$$

REMARK

R21: The objective function in this formulation minimizes the sum of the variable flow and the capacity installation costs. In addition to the flow conservation constraints, the formulation's constraints (PP3) and (PP4) ensure that flow can be rerouted on the associated p-cycles if an edge failure occurs. Constraints (PP5) and (PP6) ensure that the installed capacity on an edge exceeds the sum of the commodity flows and the reserved capacity on the p-cycles associated with that edge. [AtRa08] develop a cutting plane approach for solving the capacitated survivable network design problem using p-cycles.

References

[AlGa88] K. Altinkemer and B. Gavish, Heuristics with Constant Error Guarantees for Topological Design of Local Access Tree Networks, *Management Science* **34** (1988), 331–341.

[AlGrJoPaWe98] D. Alevras, M. Grötschel, P. Jonas, U. Paul, and R. Wessäly, Survivable Mobile Phone Network Architectures: Models and Solution Methods, *IEEE Communications Magazine* **36** (1998), 88–93.

[AlGrWe98] D. Alevras, M. Grötschel, and R. Wessäly, Cost-Efficient Network Synthesis from Leased Lines, *Annals of Operations Research* **76** (1998),1–20.

[ArHaMa90] G. J. R. Araque, L. Hall, and T. Magnanti, Capacitated Trees, Capacitated Routing and Associated Polyhedra, *C.O.R.E. Discussion Paper*, 1990.

[At02] A. Atamtürk, On Capacitated Network Design Cut-set Polyhedra, *Mathematical Programming* **92** (2002), 425–437.

[AtRa08] A. Atamtürk and D. Rajan, Partition Inequalities for Capacitated Survivable Network Design Based on Directed p-cycles, *Discrete Optimization* **5** (2008), 415–433.

[BaAl92] A. Balakrishnan and K. Altinkemer, Using a Hop-constrained Model to Generate Alternative Communication Network Designs, *INFORMS Journal on Computing* **4** (1992), 192–205.

[BaMaMi94a] A. Balakrishnan, T. L. Magnanti, and P. Mirchandani, A Dual-based Algorithm for Multi-level Network Design, *Management Science* **40** (1994), 567–581.

[BaMaMi94b] A. Balakrishnan, T. L. Magnanti, and P. Mirchandani, Modeling and Heuristic Worst-case Performance Analysis of the Two-Level Network Design Problem, *Management Science* **40** (1994), 846–867.

[BaMaMi97] A. Balakrishnan, T. L. Magnanti, and P. Mirchandani, Network Design, In *Annotated Bibliographies in Combinatorial Optimization*, M. Dell'Amico, F. Maffioli, and S. Martello (editors), John Wiley and Sons, New York (1997), 311–334.

[BaMaMi98] A. Balakrishnan, T. L. Magnanti, and P. Mirchandani, Designing Hierarchical Survivable Networks, *Operations Research* **46** (1998), 116–136.

[BaMaWo89] A. Balakrishnan, T. L. Magnanti, and R. T. Wong, A Dual-Ascent Procedure for Large-Scale Uncapacitated Network Design, *Operations Research* **37** (1989), 716–740.

[BaMiNa09] A. Balakrishnan, P. Mirchandani, and H. P. Natarajan, Connectivity Upgrade Models for Survivable Network Design, *Operations Research* **57** (2009), 170–186.

[BiChGuTs96] D. Bienstock, S. Chopra, O. Günlük, and C. Tsai, Minimum Cost Capacity Installation for Multi-Commodity Network Flows, *Mathematical Programming* **81** (1996), 177–199.

[BiDeSi94] D. Bienstock, Q. Deng, and D. Simchi-Levi, *A Branch-and-cut Algorithm for the Capacitated Minimum Spanning Tree Problem*, 1994.

[BiGu95] D. Bienstock and O. Günlük, Computational Experience with a Difficult Mixed-integer Multicommodity Flow Problem, *Mathematical Programming* **68** (1995), 213–237.

[ChGiSa98] S. Chopra, I. Gilboa, and T. Sastry, Source Sink Flows with Capacity Installation in Batches, *Discrete Applied Mathematics* **85** (1998), 165–192.

[DiKeMa08] M. Didi Biha, H. Kerivin, and A. R. Mahjoub, On the Polytope of the (1-2)-Survivable Network Design Problem, *SIAM Journal on Discrete Mathematics* **22** (2008), 1640–1666.

[EsWi66] L. R. Esau and K. C. Williams, On Teleprocessing System Design, *IBM Systems Journal* **5** (1966), 142–147.

[FoLaMa00] B. Fortz, M. Labbé, and F. Maffioli, Solving the Two-Connected Network with Bounded Meshes Problem, *Operations Research* **48** (2000), 866–877.

[Ga85] B. Gavish, Augmented Lagrangian Based Algorithms for Centralized Network Design, *IEEE Trans. Commun.* **33** (1985), 1247–1257.

[Ga91] B. Gavish, Topological Design of Telecommunication Networks — Local Access Design Methods, *Annals of Operations Research* **33** (1991), 17–71.

[Go98] L. Gouveia, Using Variable Redefinition for Computing Lower Bounds for Minimum Spanning and Steiner Trees with Hop Constraints, *INFORMS Journal on Computing* **10** (1998), 180–188.

[GoBe93] M. X. Goemans and D. J. Bertsimas, Survivable Networks, Linear Programming Relaxations and the Parsimonious Property, *Mathematical Programming* **60** (1993), 145–166.

[GoGoPlShTaWi94] M. X. Goemans, A. Goldberg, S. Plotkin, D. Shmoys, E. Tardos, and D. P. Williamson, Approximation Algorithms for Network Design Problems, *SODA*, (1994), 223–232.

[GrMoSt96] M. Grötschel, C. L. Monma, and M. Stoer, Design of Survivable Networks, In *Network Models* of *Handbooks in Operations Research and Management Science*, M. O. Ball, T. L. Magnanti, C. L. Monma, and G. L. Nemhauser (editors), Elsevier Science, The Netherlands (1995), 617–672.

[Ja01] K. Jain, A Factor-2 Approximation Algorithm for the Generalized Steiner Network Problem, *Combinatorica* **21** (2001), 39–60.

[KeMa05] H. Kerivin and A. R. Mahjoub, Design of Survivable Networks: A Survey, *Networks* **46** (2005), 1–21.

[MaMiVa92] T. L. Magnanti, P. Mirchandani, and R. Vachani, The Convex Hull of Two Core Capacitated Network Design Problems, *Mathematical Programming* **60** (1995), 233–250.

[MaMiVa95] T. L. Magnanti, P. Mirchandani, and R. Vachani, Modeling and Solving the Two-facility Capacitated Network Loading Problem, *Operations Research* **43** (1995), 142–157.

[MaRa05] T. L. Magnanti, S. Raghavan, Strong Formulations for Network Design Problems with Connectivity Requirements, *Networks* **45** (2005), 61–79.

[Mi81] M. Minoux, Optimum Synthesis of a Network with Non-simultaneous Multi-commodity Flow Requirements, In *Studies of Graphs and Discrete Programming*, P. Hansen (editor), North-Holland (1981), 269–277.

[Mi00] P. Mirchandani, Projections of the Network Loading Problem, *European Journal of Operational Research* **122** (2000), 534–560.

[Mi89] P. Mirchandani, Polyhedral Structure of a Capacitated Network Design Problem with an Application to the Telecommunication Industry, Unpublished PhD Dissertation, MIT, Cambridge, MA, 1989.

[Mi96] P. Mirchandani, The Multi-Tier Tree Problem, *INFORMS Journal on Computing* **8** (1996), 202–218.

[NeWo88] G. L. Nemhauser and L. A. Wolsey, *Integer Programming*, John Wiley & Sons, New York, 1988.

[Or91] J. Orlin. *Personal Communication*, 1991.

[PaJoAlGrWe96] U. Paul, P. Jonas, D. Alveras, M. Grötschel, and R. Wessäly, *Survivable Mobile Phone Network Architectures: Models and Solution Methods*, Preprint SC 96-48, Konrad-Zuse-Centrum für Informationstechnik, Berlin, 1996.

[RaMa97] S. Raghavan and T. L. Magnanti, Network Connectivity, In *Annotated Bibliographies in Combinatorial Optimization*, M. Dell'Amico, F. Maffioli, and S. Martello (editors), John Wiley and Sons, New York (1997) 335–354.

[SoWySeLaGeFo98] P. Soriano, C. Wynants, R. Séguin, M. Labbé, M. Gendreau, and B. Fortz, Designing and Dimensioning of Survivable SDH/SONET Networks, In *Telecommunications Network Planning*, B. Sansò and P. Soriano (editors), Kluwer Academic Publishers, The Netherlands (1998), 147–168.

[StDa94] M. Stoer and G. Dahl, A Polyhedral Approach to Multicommodity Survivable Network Design, *Numerische Mathematik* **68** (1994), 149–167.

[WiGoMiVa95] D. P. Williamson, M. X. Goemans, M. Mihail, and V. V. Vazirani, A Primal-Dual Approximation Algorithm for Generalized Steiner Network Problems, *Combinatorica* **15** (1995), 435–454.

[Wo84] R. T. Wong, A Dual Ascent Approach for Steiner Tree Problems on a Directed Graph, *Mathematical Programming* **28** (1984), 271–287.

Section 12.4
Network Science for Graph Theorists

David C. Arney, West Point
Steven B. Horton, West Point

INTRODUCTION

We start by making a distinction that graphs are elements of pure or abstract mathematics and networks are modeling tools in applied mathematics. There is insight into these structures obtained by thinking about their roles in light of the above statement.

One way to differentiate a network from a graph is in the complexity of the definition. In graph theory, a mathematically rigorous, elegant, and simple definition of the graph structure is sought and used: a graph is a set of nodes and and a set of edges [ChZh05, Di97, We01]. In network science, a more complex, multi-perspective, multi-element definition is sought and used for a network. To show some of that complexity, a common way to define a network is to establish its components (its graph – nodes and edges, processes, and data/attributes associated with nodes and edges), its properties (dynamic, layered), and its application (social, propagating, organizational). This definition requires more than an adjacency matrix to meet all of these definitional elements. A network also typically has some kind of database of attributes and/or a process algorithm. Another approach often used to define a network is to use the concept of a mathematical graph (its nodal-link structure) with its topological features and then classify the various types of graphs that occur (random, scale free, small world, scale rich, etc.). A foundational research management report on network science [CoNS05] defined networks using a layered architecture of network roles — physical, communication, informational, biological, and social/cognitive — that connect the layers together to produce the overall network framework. See Figure 12.4.1 for a schematic of that architecture.

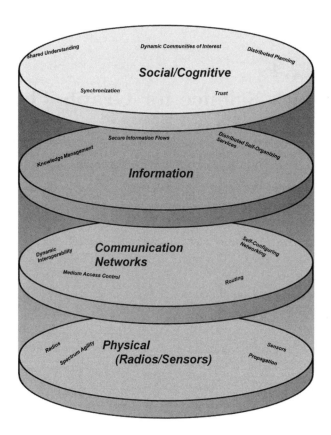

Figure 12.4.1: The layered elements of a network.

These approaches and elements of the various definitions show both the power and complexity of networks as a modeling paradigm. Network Science analyzes the nodes and edges with their associated attributes (weights, direction, roles, capabilities) along with the additional element that the network itself has a systematic process to meet its mission or purpose. Like graphs, networks are often represented by matrices (adjacency, weighted) and can possess specialized topological classifications [ChLu06]. However, another distinction is the dynamical nature of networks. Understanding and modeling the changing spatial-temporal behavior of the network structure is a significant component of network science [B03]. This element of dynamics perhaps produces the most significant challenges for the network scientist. Choices in representing and measuring the changing elements of the network can significantly influence the nature and effectiveness of a network model [NBW06].

12.4.1 Network Measures and Properties

No matter what definition or framework is used, one of the basic steps in network modeling and analysis is to determine, calculate, or measure basic network properties. Most networks are sufficiently complex that simply relying on visualization can produce counter-intuitive conceptions of the network capabilities and performance or worse yet complete misunderstanding of the properties such as the relative importance or roles of certain network nodes and edges. Precisely computing and measuring well-defined properties counter those potential visual misperceptions and this in turn improves and strengthens network modeling and analysis [BE05].

Since terminology is important in mathematics and science, especially to expedite progress in an emerging discipline, network science gives thanks to its well-established graph theoretic foundation. Many of the measures developed in network science (often from graph theoretic roots) are very precise with careful definitions which contain helpful normalizations and clear notations.

In this section, we define and discuss several mathematical measures for networks and, in particular, try to provide insights and definitions for basic network properties. Network properties can be classified many ways: structural (nodal and link), process, or data; local, global, or regional; discrete or continuous; dynamic or static. We attempt to sort out these classifications and use them in our network measures schema.

Structural Measures: Centrality

Let's consider the underlying system used to measure structural (geometrically related) properties of networks. These basic structural measures are important aspects of networks as they make up the bulk of the measures in the literature [VCLC08]. We will discuss the most basic elements such as degree, closeness, betweenness, and eigenvector centralities. For these network measures, centrality refers to the geometric center or the level of importance (as in playing a central role). Unfortunately, even this broad definition may be neither sufficient nor enlightening, since network scientists have pronounced most network measures as centralities. Since these definitions are precise, we will introduce them using graph theoretic notation. All graphs in this section are undirected and simple.

DEFINITION

D1: Given a graph $G = (V, E)$ and a node v, the ***degree centrality*** of v is simply its degree, *i.e.*, $C_D(v) = \deg(v)$.

REMARK

R1: Degree centrality provides some measure of the importance of each node.

DEFINITIONS

D2: Given a node v, its ***farness*** is $F(v) = \sum_{u \in V - v} d(u, v)$.

D3: Given a node v, its ***closeness centrality*** is $C_C(v) = 1/F(v)$.

REMARKS

R2: Closeness centrality provides some measure of how fast information can spread to the rest of a network from a given node.

R3:   The given definition is not meaningful for unconnected graphs; the literature contains many modifications and alternative definitions to address this problem.

DEFINITIONS

D4: Given a graph $G = (V, E)$ and distinct nodes s and t, we define σ_{st} to be the total number of shortest paths in G from s to t.

D5: Given a graph $G = (V, E)$ and distinct nodes s, t, and v, we define $\sigma_{st}(v)$ to be the total number of shortest paths in G from s to t that pass through v.

D6: Given a graph $G = (V, E)$ and a node v, the ***betweenness centrality*** of v is

$$C_B(v) = \sum_{s \neq v \neq t} \frac{\sigma_{st}(v)}{\sigma_{st}}$$

REMARKS

R4: The betweenness centrality of a node v measures the sum of the fraction of shortest $s - t$ paths that pass through v.

R5:   There are several ways this metric can be normalized to provide a means of comparing C_B across graphs of differing order.

DEFINITION

D7:   Given a graph $G = (V, E)$, we can compute the ***eigenvector centrality*** by solving the system $x_v = \frac{1}{\lambda} \sum_{u \in V} a_{uv} x_u$ where $x = [x_1, x_2, x_3, ..., x_n]$ is a vector and x_i is the eigenvector centrality of vertex i.

Here the a_{ij} values are the entries in the adjacency matrix A.

REMARK

R6: While there can be several values of λ that admit solutions for the vector x in the above definition, the standard approach is to use the largest eigenvalue of A for λ as this results in an x that has all positive components.

12.4.2   Other Structural Measures

We look at the concept of structural balance within local regions of a network to determine network properties that are not nodal or edge-based centralities. The idea is to determine if analyzing the regional structural relationships can provide measures for valuable properties of the network. We will do so by looking at 2 nodes at a time (dyads), 3 nodes (triads), and other size subnetworks (groups). We also present the process of finding structurally related nodes by looking at clustering algorithms.

Reciprocity on Dyads

In this section we consider directed networks. One basic question for dyads is: if one node is linked to another, is that relationship reciprocated by a link back? In more precise terms, if there is a link from node u to node v, is there a link from node v back to node u? This characteristic is only relevant to directed networks. The idea is to aggregate the results for all the individual nodes to determine statistically the overall symmetry level of the network [WaFa94]. A major motivation for studying this property is to understand mutual relationships in social networks. The overall data analysis for a directed network of n nodes with k links is based on its *density* $\frac{k}{n(n-1)}$. There are $D = \binom{n}{2}$ possible dyads that can be classified as mutual (M) (reciprocated), asymmetric (A) (nonreciprocated), or null (N) (not present) with $M+A+N = D$. There are several classic examples of calculations for a measure called the Index of Mutuality (I) based on whether there are fixed or free choices for the number of outbound arcs at each node. A fixed number of links could come from a survey on a network of friends where the survey asks each person (node) in the network to list his or her top d friends for fixed d. Free choice links come from similar survey data where there is no restriction on the number of friends that can be listed. By calculating the random independent probability that a dyad is reciprocated for a fixed choice of d friends, an expected value of M can be calculated.

DEFINITIONS

D8: For a fixed-choice social network, the ***index of mutuality***

$$I = \frac{2(n-1)M - nd^2}{nd(n-1-d)}$$

D9: For a free-choice social network, the ***index of mutuality***

$$I = \frac{2(n-1)^2 M - L^2 + L_2}{L(n-1)^2 - L^2 + L_2}$$

where L is the sum of the number of friends chosen and L_2 is the sum of the squares of the number of friends chosen.

REMARK

R7: A convenient feature of this measure for both fixed and free choice is that the values of I are benchmarked and normalized by $-\infty < I < 1$, where the maximum reciprocity level is given when $I = 1$ (meaning $uv \in E \iff vu \in E$) and reciprocity less than that expected by random ties gives $I < 0$.

The following two examples show a network with low Mutuality Index under free choice and a network with high Mutuality Index under fixed choice.

EXAMPLES

E1: The graph in Figure 12.4.2 shows a free choice directed social network that results in $I = -0.087$.

E2: The graph in Figure 12.4.3 shows a fixed choice directed social network that results in $I = 0.72$.

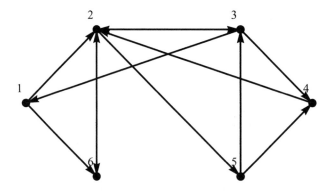

Figure 12.4.2: A free-choice social network with $I = -0.087$.

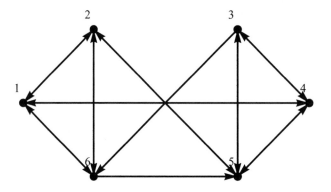

Figure 12.4.3: A fixed-choice social network with $I = 0.72$.

The Mutuality Index is very helpful in comparing networks or judging networks to known norms for reciprocity. Network modelers can use the reciprocity measure I to predict other behaviors in a social network, such as effective communication or decision making.

Transitivity on Triads

We now turn our attention to transitivity. Here we seek to understand the extent to which $uv, vw \in E \implies uw \in E$. For social networks, the adage is "the friend of my friend is also my friend." We seek to measure the level of transitivity in the entire network by finding the portion of transitive triads to the total number of triads in the network. This is equivalent to finding the proportion of closed triads in the network and can be found for both directed and undirected networks. One such measure is the network's *clustering coefficient* [WaFa94].

DEFINITION

D10: A network *clustering coefficient* C is the fraction of paths of length two that are closed.

$$C = \frac{\text{the number of closed paths of length 2}}{\text{the number of paths of length 2}}$$

REMARK

R8: If $C = 1$, the network has perfect or maximal transitivity, or $uv, vw \in E(G) \Rightarrow uw \in E(G)$. $C = 0$ equates to no transitivity in the network, or $uv, vw \in E(G) \Rightarrow uw \notin E(G)$. Many familiar network structures, such as trees and rectangular lattices, have $C = 0$.

Another measure is the local clustering coefficient which measures transitivity as a local property.

DEFINITION

D11: The *local clustering coefficient* $C_l(v)$ is the fraction of pairs of neighbors of a vertex v that are themselves neighbors.

$$C_l(v) = \frac{\sum_{i,j \in N(v), i<j} a_{ij}}{\binom{|N(v)|}{2}}$$

Here a_{ij} are the entires in the adjacent matrix A, and $N(v)$ denotes the open neighborhood of v.

REMARKS

R9: Nodes with lower than average values of C_l lead to structural holes in the network, and statistical analysis of the values of C_l in a network can provide additional insight into its structure.

R10: Some network scientists define a clustering coefficient that is the mean of the local cluster coefficients. Nodes with less than 2 neighbors are typically ignored for this purpose. This value is different than the clustering coefficient C defined above.

Balance on Triads

In some networks, the node connections are signed (positive or negative). Obvious examples of signed connections in a social network are friendship (positive) and animosity (negative). These networks are called signed networks and some signed network structures exhibit balance and others do not. Lack of balance may lead to network instability or poor network performance. Situations that lead to balance are "the friend of my friend is my friend" and "the enemy of my enemy is my friend." But when nodes have situations where "my enemy and my friend are friends," the network exhibits imbalance. Network modelers often seek to measure the level of balance in the entire network by finding the fraction of triads that are balanced.

REMARK

R11: Analysis of cycles and other structural components can be performed on signed networks, both directed and undirected, to determine conditions for balance and imbalance. Another known result is that when a network is in balance it can be divided into 2 friendly clusters (all positive connections within the cluster) with no positive friendship connections between the clusters (negative connections between nodes in the different clusters).

E3: Figure 12.4.4 shows a network with signed edges. The triad $\{3, 5, 6\}$ is an example of a balanced triad; this represents three mutual friends. The triad $\{1, 2, 4\}$ is also balanced; here friends 2 and 4 are both enemies of 1. The triad $\{4, 6, 7\}$ is unbalanced; 7 is friends with both 4 and 6 but the latter two are mutual enemies.

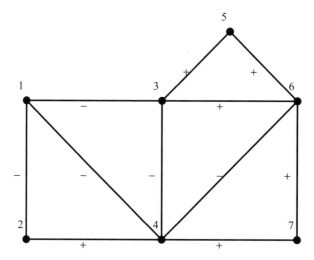

Figure 12.4.4: A network with signed edges.

Communities and Partitions

Finding structural properties by identifying natural groups (subnetworks) or dividing the network into communities or partitions are network science objectives. We distinguish grouping or clustering (not all nodes are included) from community detecting and partitioning where we seek to place all nodes in the network. Partitions differ from communities since the number and/or size of the partitions are fixed in partitioning and not set a priori in community detecting. Graph partitioning algorithms often find the best division of the network given the set conditions of number or size of the partitions. The goal in community detection is to understand the structure of the network even though that sometimes results in a situation where no viable communities are discovered.

REMARK

R12: The usual goal of clustering is to find groups such that the densities of connections within the groups are substantially more than the densities of connections between groups or to the nodes outside the group.

Large cliques (complete subgraphs) are not all that common in social networks. It is more common that groups, even cohesive ones, form at best near-cliques. Since some members of the group may not be acquainted with everyone in the group, but with almost everyone, a new construct was needed for "near-cliques," giving rise to the notion of a k-plex.

DEFINITION

D12: A k-*plex* of order n is a subgraph of n nodes such that each node is adjacent to at least $n - k$ of the others.

REMARK

R13: A k-plex with $k = 1$ is a clique.

DEFINITION

D13: A *k-core* is a subgraph of n nodes such that each node is adjacent to k other nodes in the induced subgraph.

REMARK

R14: A k-core of n nodes is an $(n-k)$-plex on n nodes.

In general, k-cores are easy to find through iteration. Start by removing all nodes of degree less than k, along with edges incident with those nodes. Repeat until all nodes have degree k or higher. The remaining network is either empty (in which case the original network has no k-core), or the remaining network consists of k-core(s).

EXAMPLE

E4: Figure 12.4.5 shows examples of cliques, k-plexes, and k-cores. The subgraph induced by $\{1, 2, 3, 4\}$ is a 4-clique. The subgraph induced by $\{1, 2, 3, 4, 7, 8, 9\}$ is a 3-plex on 7 nodes or, equivalently, a 4-core on 7 nodes.

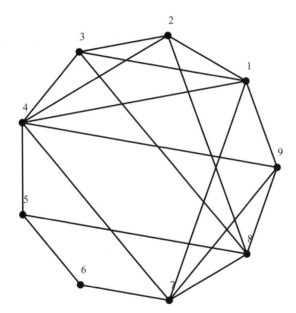

Figure 12.4.5: A graph containing a 4-clique and a 3-plex on 7 nodes.

We will not cover the details of partitioning algorithms since partitioning is a mainstay of graph theory covered in other sections. Partitioning algorithms can be computationally demanding and complex in their implementation. Community detection is also challenging and we merely give general descriptions of these algorithms. One community detection algorithm attempts to find sets of nodes with high modularity scores (a measure of the extent that similar nodes are connected to one another). Determining modularity can also be demanding from a computational standpoint. Another algorithm for community detection is hierarchical clustering, which is an agglomerative method. The main idea of this method is to form groups with high similarity and then iteratively combine groups to form larger ones.

12.4.3   Attribute or Data Measures

Similarity

In this section we look at the similarity of the attributes of nodes, not just their geometrical or structural similarities. However, we start with a similarity measure involving common neighbors of a node [N10].

DEFINITION

D14: *Cosine similarity* is a measure of commonality of neighbors. It is computed by

$$CS_{uv} = \frac{n_{uv}}{\sqrt{\deg(u)\deg(v)}}$$

where n_{uv} is the number of common neighbors of u and v that have a specified attribute.

REMARKS

R15:   The specified property can either be structural, or a statistical attribute from the network's database. For an example of the latter, we might be interested in cosine similarity related to friends who smoke cigarettes in a social network. In this case, n_{uv} would count smokers who are friends of both u and v.

R16:   An example of a purely structural application of this measure is to let n_{uv} count common neighbors (of all types). In this case, the measure is

$$CS_{uv} = \frac{|\,N(u) \cap N(v)\,|}{\sqrt{\deg(u)\deg(v)}}$$

R17:   In the case of the previous remark, if $CS_{uv} = 1$, then u and v have exactly the same neighbors, *i.e.*, $N(u) = N(v)$. If $CS_{uv} = 0$, then u and v have no common neighbors, *i.e.*, $N(u) \cap N(v) = \emptyset$.

Homophily

A basic notion in network science is homophily — the principle that nodes tend to be similar to their neighbors. The adage "birds of a feather flock together" describes the idea. One measure of homophily is called *modularity*.

DEFINITION

D15: *Modularity* is a measure of homophily. It is computed by

$$Q = \frac{1}{m}(L - E(L)).$$

Here m is the number of edges in the network, L is the number of "like" edges, and $E(L)$ is the expected number of like edges.

REMARKS

R18: Vertices are categorized based on structural (related to the underlying graph) or other (related to the database mentioned in the introduction, or some other source) properties at the discretion of the modeler. An edge between two vertices that are sufficiently similar, based on the vertex categorization, are said to be "like." $E(L)$ is computed assuming edges exist independently of the properties that create like edges.

R19: It is clear that $-1 \leq Q \leq 1$. Relatively high values of Q show high homophily.

R20: For social networks the tendency of people to form friendships with others similar in some characteristic(s) to themselves is termed *selection*. When people modify their behaviors/characteristics to bring themselves in alignment with their friend it is called *social influence*.

12.4.4 Process Measures

We will be less precise for process measures which are more like utility or performance functions. Modelers often attempt to design process measures (utility functions) in models that are simply well-behaved. That is, they are generally quantitative and consistent. However, when we have human preference involved even these basic properties are difficult to require because human-based preferences are not easily represented by rationality-based utility function measures. Other process measures relate to the flow of something (physical, informational, cognitive, emotional) through the network.

12.4.5 Modeling with Networks

It should be noted that constructing and calculating effective measures can be difficult for modelers. Finding the "right" measure could be called a facet of the art of modeling [EK10]. Some commonly used measures have very little value and often serve to obscure rather than enlighten. In organizational models, this is referred to as measurement inversion. For example, some traditional measures are merely what organizations find immediately measurable from already available data, even if they are low value and not compatible with the real property that is under investigation. Additional modeling ineffectiveness occurs when modelers ignore high value or more precise measurements simply because they seem more difficult to calculate.

12.4.6 Process Dynamics

Equally important, yet often not considered, are the network processes and their measures related to network dynamics. As Barabási pointed out in 2003, "we must move beyond structure and topology and start focusing on the dynamics that take place along the links [B03]." Some researchers, especially those modeling resource or information systems would classify the process as "flow". However, these processes are often more complex than flow, which connotes direction and continuity, but operates on an unchanging underlying structure.

12.4.7   Structural Classification

Two well-known and much studied classes of complex networks are scale-free networks and small-world networks. Both are characterized by similar special structural features: inverse power-law degree distributions for the former and short path lengths and high clustering coefficients for the latter. These networks provide models for many network applications to include the internet, power grids, and airline travel and route maps.

The small-world network is characterized by the properties that the shortest path lengths between nodes are proportional to the logarithm of the number of nodes in the network and the presence of many near-cliques and highly connected hubs [W99]. For a large social network, this results in strangers being connected by only a few links of acquaintances. In [W03], Duncan Watts describes his work on small-world networks and the application of small-world network theory to sociology. In pop culture, the trivia game "Six Degrees of Kevin Bacon" is based on the small-world network concept. In this game, movie buffs challenge each other to find the shortest path between an arbitrary actor and Hollywood actor Kevin Bacon through a network where vertices are actors and edges between two actors are present when they acted together in a movie. A similar phenomenon exists in mathematics, where research mathematicians try to connect each other to prolific research mathematician Paul Erdős through a co-authorship network.

The defining property of a scale-free network is its degree distribution following an inverse power law or Pareto distribution:

$$F(k) \sim k^{-p}$$

where $F(k)$ is the fraction of nodes with k connections and the parameter p is often determined to be between 2 and 3 [BA99]. Modelers have conjectured that these networks form through a phenomenon called preferred attachment, where the common phrase used to describe this is "the rich get richer." In this concept, new nodes are more likely to attach to already popular or high density nodes and therefore build a heavier tail in the statistical distribution of node degrees than expected from random connections. Barabási and Alberts studied this generative model for web pages on the internet. Other researchers, such as West and Grigolini [WG11], have used this model for many network studies and cultural phenomena.

12.4.8   Future Directions

Network Science has had tremendous growth and development as a research area and discipline over the last decade. The NRC study of 2005 provided network science a baseline framework, along with attention from research funding agencies [CoNS05]. Several factors, including the rising importance of cybersecurity in the National Defense Strategy of the United States and other nations may accelerate this trend.

References

[B03] A. Barabási, *Linked: The New Science of Networks*, Perseus, 2002.

[BA99] A. Barabási and R. Albert, Emergence of Scaling in Random Networks. *Science* 1999 October 15 **286** (5439): 509–512.

[BE05] U. Brandes and T. Erlebach, *Network Analysis: Methodological Foundations*, Springer, 2005.

[ChZh05] G. Chartrand and P. Zhang, *Introduction to Graph Theory*, McGraw Hill, 2005.

[ChLu06] F. Chung and L. Lu, *Complex Graphs and Networks*, American Mathematical Society, 2006.

[CoNS05] Committee on Network Science for Future Army Applications, *Network Science*, National Academies Press, 2005.

[Di97] R. Diestel, *Graph Theory*, Springer, 1997.

[EK10] D. Easley and J. Kleinberg, *Networks, Crowds, and Markets: Reasoning about a Highly Connected World*, Cambridge University Press, 2010.

[N10] M. E. J. Newman, *Networks: An Introduction*, Oxford, 2010.

[NBW06] M. Newman, A. Barabási, D. Watts, eds., *The Structure and Dynamics of Networks*, Princeton University Press, 2006.

[VCLC08] T. Valente, K. Coronges, C. Lakon, and E. Costenbader, How Correlated Are Network Centrality Measures, *Connect* (Tor). 2008 January 1; 28(1): 16–26.

[WaFa94] S. Wasserman and K. Faust, *Social Network Analysis: Models and Applications*, Cambridge University Press, 1994.

[W03] D. Watts, *Six Degrees: The Science of the Connected Age*, W. W. Norton, 2003.

[W99] D. Watts, *Small Worlds: The Dynamics of Networks Between Order and Randomness*, Princeton University Press, 1999.

[WG11] B. West and P. Grigolini, *Complex Webs: Anticipating the Improbable*, Cambridge University Press, 2011.

[We01] D. B. West, *Introduction to Graph Theory*, Second Edition, Prentice-Hall, 2001. (First Edition, 1996.)

Glossary for Chapter 12

adjacency matrix: given a graph G on n nodes, the $n \times n$ matrix with a 1 in entry (i, j) if there is an edge from nodes i to j.

average distance: for a graph G, the term

$$\mathrm{adist}(G) = \sum_{u,v \in F} \frac{d(u, v)}{|F|},$$

where F is the set of pairs of distinct nodes u, v of G with the property that the distance $d(u, v)$ between u and v is finite. The directed analogue of this parameter, where distance refers to shortest directed paths, is denoted by $\mathrm{adist}_d(G)$.

$b(n)$: minimum broadcast time of any graph with n vertices.

$b(n, \Delta)$: minimum broadcast time over all graphs on n vertices with maximum degree Δ.

$B(n)$: the number of edges in an MBG on n vertices.

backbone network: a two-connected network containing high capacity edges and nodes.

bandwidth: capacity of a communications facility in bits per second (bps).

betweenness centrality: a measure of the extent to which a given vertex is on a shortest path between all pairs of other vertices.

binomial tree: a recursively defined rooted tree on 2^k vertices. A binomial tree on 2^k vertices is formed from two binomial trees on 2^{k-1} vertices by joining their roots with an edge.

Braess' paradox: the counter-intuitive observation that, if the travelers were behaving selfishly it was possible to improve everyone's travel time by removing roads.

broadcast graph: a graph on n vertices for which there is a broadcast protocol with completion time $\lceil \log n \rceil$ for each vertex of the graph.

broadcast protocol – for originator: a group of message transmissions that deliver a message from the originator to all of the other vertices in the graph.

broadcast protocol – for a graph: a collection of broadcast protocols, one for each originator.

broadcast time – of a broadcast protocol for a graph: maximum completion time for any vertex of the graph.

broadcast time – of a graph: minimum broadcast time for any broadcast protocol for the graph.

broadcasting: sending a single message from a particular vertex (the originator) to all of the other vertices in the graph.

capacitated facility: a facility with limited bandwidth.

Cayley graph $\mathcal{D}_k$: a graph with $2^k - 2$ vertices labeled with the integers mod $2^k - 2$. There is an edge between two vertices with labels i and j if and only if $(i + j)$ $(\mathrm{mod}\ 2^k - 2) = 2^\ell - 1$ for some $1 \le \ell \le k - 1$.

Cheeger ratio: given a subset S of nodes in a graph G, the terms

$$h(S) = \frac{|E(S, \bar{S})|}{\text{vol}(S)}, \quad \text{and} \quad h_G = \max_{\substack{S \subseteq V(G) \\ \text{vol}(S) \leq \text{vol}(G)/2}} h(S)$$

where we define the **volume** $\text{vol}(S) = \sum_{v \in S} \deg(v)$, $\text{vol}(G) = \text{vol}(V(G))$ and $E(A, B)$ denotes the set of edges with one endpoint in A and one endpoint in B.

closeness centrality: a measure of how close a vertex is to all of the other vertices in a connected graph.

clustering coefficient – of a graph G: the term

$$\frac{2}{n} \sum_{x \in V(G)} \frac{|E(N(v))|}{\deg(v)(\deg(v) - 1)},$$

where $E(N(v))$ denotes the number of edges in the subgraph induced by the neighborhood of v in G. (See §12.1.)

clustering coefficient: a measure of the extent to which the network exhibits transitivity. (See §12.4.)

collaboration graph: the graph whose nodes are co-authors and two authors are connected by an edge if they have written a joint paper together.

completion time – of a broadcast protocol: the time at which the last vertex of a graph is informed of the message being broadcast.

completion time – of a gossip protocol: the time at which the last vertex of a graph is informed of all of the messages.

complex network model: a stochastic or deterministic model which generates graphs which simulate one or more properties of complex networks. Examples include the preferential attachment and copying models.

compound – of G into H relative to a set $S \subseteq V(G)$: the graph obtained by replacing each vertex x of H by a graph G_x isomorphic to G and adding a matching between two sets S_x and S_y if x and y are adjacent in H.

cosine similarity: a measure of commonality between neighbors.

degree centrality: the degree of a vertex.

de Bruijn graph $\mathcal{UB}_{d,k}$ **with degree** $2d$ **and diameter** k: a graph with d^k vertices with labels that are strings of length k over the alphabet $\{0, 1, \ldots, d-1\}$. There is an edge between vertex $x_1, x_2, \ldots, x_k$ and each vertex $x_2, x_3, \ldots, x_k, \lambda$ and $\lambda, x_1, x_2, \ldots, x_{k-1}$ with $\lambda \in \{0, 1, \ldots, d-1\}$.

degree matrix: given a graph G, the diagonal matrix with the degree of node i in the (i, i) entry.

diversification: splitting of flow between an origin and destination so that no more than a pre-specified fraction flows on any one edge or node (except the origin or the destination node).

eigenvector centrality: a centrality measure based on eigenvectors of a graph's adjacency matrix.

facility: a transmission medium (e.g., fibre-optic or wireless) installed on an edge for sending data, voice or video signals.

$G(n)$: the number of edges in an MGG on n vertices.

gossip protocol: a group of message transmissions that complete gossiping.

gossip time $g(G)$ – of a graph G: the minimum completion time of any gossip protocol for G.

gossiping: sending a unique message from each vertex to all of the other vertices in the graph.

idle: a vertex is idle at a given time in a broadcast protocol if it has learned the message but does not communicate with any of its neighbors at that time.

informed by a broadcast protocol: a vertex is informed at any given time in a broadcast protocol if it has received the message being broadcast.

informed by a gossip protocol: a vertex is informed at any given time in a gossip protocol if it has learned all of the messages.

integrality gap: for the minimization case, the difference between the optimal objective function value of an integer program and its linear programming relaxation expressed as a fraction of the optimal objective function value of the linear programming relaxation.

index of mutuality: a measure of the extent to which reciprocated arcs appear in a directed graph.

k-**core**: a way to specify a "near-clique" subgraph.

k-**dimensional binary de Bruijn graph** $\mathcal{UB}_k$: a graph with 2^k vertices, each labeled with a different binary string of length k. Each vertex $x_1x_2\cdots x_k$ is connected to $x_2x_3\cdots x_kx_1$, $x_kx_1\cdots x_{k-1}$, $x_2x_3\cdots x_k\overline{x_1}$ and $x_kx_1\cdots\overline{x_{k-1}}$.

k-**dimensional binary hypercube** $\mathcal{H}_k$: a graph with 2^k vertices, each labeled with a different binary string of length k. There is an edge between two vertices if and only if their labels differ in exactly one bit position.

k-**dimensional binary shuffle-exchange graph** $\mathcal{SE}_k$: a graph with 2^k vertices, each labeled with a different binary string of length k. Each vertex $x_1x_2\cdots x_k$ is connected to $x_2x_3\cdots x_kx_1$, $x_kx_1\cdots x_{k-1}$, and $x_1x_2\cdots\overline{x_k}$.

k-**dimensional cube-connected cycles graph** $\mathcal{CCC}_k$: a graph derived from the k-dimensional binary hypercube $\mathcal{H}_k$ by replacing each vertex of $\mathcal{H}_k$ by a cycle of length k. There are edges between each vertex and its two neighbors on its cycle. There is one edge between each pair of cycles that correspond to neighbors in the hypercube.

k-**dimensional (wrapped) butterfly graph** $\mathcal{BF}_k$: a graph derived from the k-dimensional binary hypercube $\mathcal{H}_k$ by replacing each vertex of $\mathcal{H}_k$ by a cycle of length k. There are edges between each vertex and its two neighbors on its cycle. There are two edges between each pair of cycles that correspond to neighbors in the hypercube.

k-**plex**: another way to specify a "near-clique" subgraph.

Knödel graph $\mathcal{W}_{n,\Delta}$: a Δ-regular bipartite graph with n vertices. The vertices are labeled (i,j), $i = 1, 2$, $0 \le j \le \frac{n}{2} - 1$. There is an edge between every pair of vertices $(1,j)$ and $(2, j + 2^\ell - 1 \pmod{\frac{n}{2}})$, $0 \le j \le \Delta - 1$, for each $0 \le \ell \le \Delta - 1$.

linear programming relaxation: the linear program obtained by dropping the integrality requirement of an integer or a mixed-integer program.

local access network: a network, usually a tree, for transmitting the data signal from the backbone network to the end user.

Logarithmic Dimension Hypothesis: the conjecture that the dimension of an on-line social network is best fit by about $\log n$, where n is the number of users in the on-line social network.

minimum broadcast graph (MBG): a broadcast graph on n vertices with the minimum number of edges over all broadcast graphs on n vertices.

minimum broadcast tree: a rooted tree on n vertices for which there exists a broadcast protocol for the root as originator with completion time $\lceil \log n \rceil$.

minimum gossip graph (MGG): a graph for which there is a minimum time gossip protocol and which has the minimum number of edges of any such graph.

minimum gossip time $g(n)$: the minimum gossip time over all graphs with n vertices.
modularity: a measure of homophily of a network.
normalized Laplacian matrix: the matrix $\mathcal{L} = I - D^{-1/2}AD^{-1/2}$ where A is the adjacency matrix and D is the diagonal degree matrix.
on-line social networks: graphs whose nodes consisting of users on some social networking site such as Facebook, and edges consisting of friendship links between them.
optimal gossip protocol for a graph: a gossip protocol with the minimum completion time of any protocol for the graph.
optimal gossip size $f(n)$: the minimum $f(G)$ over all graphs with n vertices.
originator: initial sender of a broadcast message.
p-cycle: a simple, directed cycle in a directed graph with at at least three arcs — used in an approach for designing survivable networks.
PageRank matrix: for a graph G and teleportation constant $\alpha \in (0,1]$, the matrix

$$P_\alpha = \alpha J_n/n + (1-\alpha)P$$

where J_n is the all 1's matrix of size $n \times n$ and n is the number of vertices in G.
PageRank vector: the stationary distribution of the PageRank Markov chain.
price of anarchy: the ratio of the average travel time of the decentralized selfish routing versus the coordinated routing respecting the collective welfare.
power law graph: a graph G on n nodes such that for some range of degrees k, N_k is proportional to k^{-b}, for a fixed exponent $b > 2$, where N_k is the number of nodes of degree k. If G is directed, then we may consider (possibly different) power laws for the in- and out-degree distributions by defining $N_{k,G}^{in}$ and $N_{k,G}^{out}$, respectively, in the obvious way.
protein-protein interaction networks: graphs whose have nodes consisting of proteins in a living cell, with two proteins joined if they share some biochemical interaction.
recursive circulant graph $\mathcal{G}_{n,d}$ for $d \geq 2$: an n vertex graph with vertices labeled with the integers mod n and an edge between two vertices with labels i and j if and only if $i + d^\ell = j \pmod{n}$ for some $0 \leq \ell \leq \lceil \log_d n \rceil - 1$.
reservation: reserving spare capacity in a survivable network for re-routing flow in case of an edge failure.
size of a gossip protocol for a graph: the total number of calls used in the gossip protocol.
small world phenomenon: a combination of small distances and clustering. Namely, there is a short path joining any two nodes and if two nodes share a common neighbor, they are more likely to be adjacent.
solid 1-cover: a subset of the vertices in a broadcast graph that is a vertex cover of that graph, and for each originator not in that subset there is a broadcast protocol such that at least one neighbor of the originator is idle at some time during the broadcast.
survivable network: a network in which traffic disrupted by a node or edge failure can be rerouted using spare capacity.
switch: node equipment for routing and processing communication traffic.
two-level network: a network where edges can have two types of facilities (say, fiber-optic or wireless).
uncapacitated facility: a facility with practically unlimited bandwidth relative to the demand.

web graph: the graph whose nodes consisting of web pages, and edges corresponding to links between them.

Wiener index: for a graph of G, the term

$$W(G) = \frac{1}{2} \sum_{x,y \in V(G_t)} d(x,y).$$

Chapter 13

Natural Science & Processes

Section 13.1
Chemical Graph Theory

Ernesto Estrada, University of Strathclyde, Scotland
Danail Bonchev, Virginia Commonwealth University

INTRODUCTION

Chemical graph theory (CGT) is a branch of mathematical chemistry which deals with the nontrivial applications of graph theory to solve molecular problems. In general, a graph is used to represent a molecule by considering the atoms as the vertices of the graph and the molecular bonds as the edges. Then, the main goal of CGT is to use algebraic invariants to reduce the topological structure of a molecule to a single number which characterizes either the energy of the molecule as a whole or its orbitals, its molecular branching, structural fragments, and its electronic structures, among others. These graph theoretic invariants are expected to correlate with physical observable measures by experiments in a way that theoretical predictions can be used to gain chemical insights even for not yet existing molecules. In this brief review we shall present a selection of results in some of the most relevant areas of CGT.

13.1.1   Basic Definitions

DEFINITIONS

D1: A ***molecular graph*** $G = (V, E)$ is a simple graph having $n = |V|$ nodes and $m = |E|$ edges. The ***nodes*** $v_i \in V$ represent non-hydrogen atoms and the ***edges*** $(v_i, v_j) \in E$ represent covalent bonds between the corresponding atoms. In particular, hydrocarbons are formed only by carbon and hydrogen atoms and their molecular graphs represent the carbon skeleton of the molecule.

D2: An ***alternant conjugated hydrocarbon*** is a hydrocarbon with alternant multiple (double and/or triple) and single bonds, such as the molecular graph is bipartite and the edges of the graph represent $C = C$ and $= C - C =$ or $C \equiv C$ and $\equiv C - C \equiv$ bonds only.

13.1.2   Molecular Energy

FACTS

F1: In the Hückel Molecular Orbital (HMO) method for conjugated hydrocarbons the energy of the jth molecular orbital of the so-called π-electrons is related to the graph spectra by

$$\lambda_j = \frac{\alpha - E_\pi(j)}{\beta},$$

where λ_j is an eigenvalue of the adjacency matrix of the hydrogen-depleted graph representing the conjugated hydrocarbon and α, β are empirical parameters [CoOlMa78, GrGuTr77, Ku06, Ya78].

F2: The total π (molecular) energy is given by

$$E_\pi = \alpha n_e + \beta \sum_{j=1}^{n} g_j \lambda_j + \beta E,$$

where n_e is the number of π-electrons in the molecule and g_j is the occupation number of the jth molecular orbital.

F3: For neutral conjugated systems in their ground state [Gu05],

$$f(n) = \begin{cases} 2 \displaystyle\sum_{j=1}^{n/2} \lambda_j & \text{if } n \text{ is even,} \\ 2 \displaystyle\sum_{j=1}^{(n+1)/2} \lambda_j + \lambda_{(j+1)/2} & \text{if } n \text{ is odd.} \end{cases}$$

REMARKS

R1: In most of the conjugated molecules studied by HMO n is an even number. In such cases, E can be expressed as $E = \sum_{j=1}^{n} |\lambda_j|$.

R2: The concept of graph energy is defined for any graph as $E = \sum_{j=1}^{n} |\lambda_j|$ [Ni07]. In this case this term is not related to any "physical" energy but the index can be considered as a graph-theoretic invariant.

R3: $\beta < 0$, then in representing the energy of molecular orbitals $\varepsilon_j = \alpha + \beta \lambda_j$ it is assumed that the largest eigenvalue represents the minimum energy, then the second largest, and so forth [CoOlMa78, GrGuTr77, Ku06, Ya78].

R4: Because an alternant conjugated hydrocarbon has a bipartite molecular graph: $\lambda_j = -\lambda_{n-j+1}$ for all $j = 1, 2, \ldots, n$.

EXAMPLE

E1: The molecule of 1,3-butadiene is a conjugated hydrocarbon whose molecular graph is the path graph with four nodes P_4. The energy of the four molecular orbitals in 1,3-butadiene, expressed by the eigenvalues of the adjacency matrix are -1.618, -0.618, 0.618, 1.618. The total energy of the molecule is -4.472.

FACTS

F4: Let G be a graph with n vertices and m edges. Then [Mc71],

$$\sqrt{2m + n(n-1)(det\mathbf{A})^{n/2}} \le E \le \sqrt{mn}.$$

F5: Let G be a graph with m edges. Then, $2\sqrt{m} \le E \le 2m$.

F6: Let G be a graph with n vertices. Then, $E \ge 2\sqrt{n-1}$, where the equality holds if G is the star graph with n vertices.

F7: [KoMo01] $E \le 2m/n + \sqrt{(n-1)(2m - 4m^2/n^2)}$ where the equality holds if and only if G is K_n, $\frac{n}{2}K_2$, or a strongly regular graph with two eigenvalues having absolute value

$$\sqrt{\frac{2m - (2m/n)^2}{n-1}}.$$

F8: Let G be a graph with n vertices. Then [KoMo01],

$$E \le \frac{n}{2}(\sqrt{n} + 1),$$

where the equality holds if and only if G is a strongly regular graph with parameters

$$(n, (n + \sqrt{n})/2, (n + 2\sqrt{n})/4, (n + 2\sqrt{n})/4).$$

F9: Let G be a bipartite graph with n vertices and m edges. Then [KoMo03],

$$E \le 4m/n + \sqrt{(n-2)(2m - 8m^2/n^2)}.$$

F10: For all sufficiently large n, there is a graph G of order n such that [Ni07]

$$E \ge \frac{n}{2}(\sqrt{n} - n^{1/10}).$$

13.1.3 Graph Nullity and Zero-Energy States

DEFINITION

D3: The ***nullity*** of a (molecular) graph, denoted by $\eta = \eta(G)$, is the algebraic multiplicity of the number zero in the spectrum of the adjacency matrix of the (molecular) graph.

REMARKS

R5: An alternant unsaturated conjugated hydrocarbon with $\eta = 0$ is predicted to have a stable, closed-shell, electron configuration. Otherwise, the respective molecule is predicted to have an unstable, open-shell, electron configuration.

R6: If n is even, then η is either zero or it is an even positive integer.

EXAMPLE

E2: The molecule of 1,3-cyclobutadiene is a conjugated hydrocarbon whose molecular graph is the cycle graph with four nodes c_4. The energy of the four molecular orbitals in 1,3-butadiene is $E_1 = \alpha - 2|\beta|$, $E_2 = \alpha + 0|\beta|$, $E_3 = \alpha + 0|\beta|$ and $E_4 = 4(\alpha + 2|\beta|)$. The nullity of this graph is $\eta = 2$ and the first orbital is occupied by a pair of electrons while the two zero-energy states have one electron each. The total π-energy is $E_\pi = 4(\alpha - |\beta|)$.

FACTS

F11: Let P_n, C_n and K_n be the path, cycle and complete graph with n vertices, respectively. Then [BoGu09],

   i) $\eta(P_n) = 0$ if n is even and $\eta(P_n) = 1$ if n is odd.

   ii) $\eta(C_n) = 2$ if $n \equiv 0 \ (mod \ 4)$ or *zero* otherwise.

   iii) $\eta(K_1) = 1$ and $\eta(K_{n>1}) = 0$.

F12: [CvGu72] Let $M = M(G)$ be the size of the maximum matching of a graph, i.e., the maximum number of mutually non-adjacent edges of G. Let T be a tree with $n \geq 1$ vertices. Then, $\eta(T) = n - 2M$.

F13: [CvGuTr72] Let G be a bipartite graph with $n \geq$ vertices and no cycle of length $4s(s = 1, 2, \ldots)$, then $\eta(G) = n - 2M$.

REMARK

R7: The nullity of benzenoid graphs, which may contain cylces of length $4s$, is also given by $\eta(G) = n - 2M$ [Gu83, FaJoSa05].

FACTS

F14: [Lo50] Let G be a bipartite graph with incidence matrix $\mathbf{B}$, $\eta(G) = n - 2r(\mathbf{B})$, where $r(\mathbf{B})$ is the rank of $\mathbf{B}$.

F15: [ChLi07] Let G be a graph with n vertices and at least one cycle,

$$\eta(G) = \begin{cases} n - 2g(G) + 2 & g(G) \equiv 0 \ (mod \ 4), \\ n - 2g(G) & \text{otherwise} \end{cases}$$

where $g(G)$ is the *girth* (length of miminal cycle) of the graph.

F16: [ChLi07] If there is a path of length $d(p, q)$ between the vertices p and q of G

$$\eta(G) = \begin{cases} n - d(p, q) & \text{if } d(p, q) \text{ is even,} \\ n - d(p, q) - 1 & \text{otherwise.} \end{cases}$$

F17: [ChLi07] Let G be a simple connected graph of diameter D

$$\eta(G) = \begin{cases} n - D & \text{if } D \text{ is even,} \\ n - D - 1 & \text{otherwise.} \end{cases}$$

F18: [ChLi07] Let G be a simple connected graph on n vertices having K_p as a subgraph, where $2 \leq p \leq n$. Then,

$$\eta(G) \leq n - p.$$

13.1.4   Graph-Based Molecular Descriptors

DEFINITIONS

D4: A *graph-based molecular descriptor*, commonly known as *topological index* (TI), is a graph-theoretic invariant characterizing numerically the topological structure of a molecule [DeBa00].

D5: The *Wiener index* of a (molecular) graph is a TI defined by

$$W = \sum_{i<j} d_{ij}$$

where d_{ij} is the *shortest-path distance* between the vertices i and j [Wi47].

D6: The *Hosoya index* of a (molecular) graph is a TI defined by

$$H = \sum_{i=0}^{n/2} P(G,i)$$

where $P(G,i)$ is the *number of selections of i mutually nonadjacent edges* in the graph. By definition $P(G,0) = 1$ and $P(G,1) = m$ [Ho71].

D7: The *Zagreb indices* of a (molecular) graph are TIs defined by [GuTr72]

$$M_1 = \sum_{j=1}^{n} (\delta_j)^2,$$

$$M_2 = \sum_{i,j \epsilon E} \delta_i \delta_j.$$

D8: The *Randić index* of a (molecular) graph is a TI defined by [Ra75]

$$\chi = \sum_{i,j \epsilon E} (\delta_i \delta_j)^{-1/2}.$$

D9: Let $k = 0, 1, 2, 3, \ldots$ be the number of adjacent vertices of degrees $\delta_i, \delta_j, \delta_l, \ldots$ in graph G. Then [KiHaMuRa75, KiHa76], the *Kier and Hall molecular connectivity index* is defined as

$$k_\chi = \sum_{i,j,l,\ldots} (\delta_i, \delta_j, \delta_l, \ldots)^{-1/2}$$

where the summation is taken over all subgraphs of size k, and the null term is the sum of all the vertex degrees (the total adjacency of G).

D10: Let $s_i = \sum\limits_{j=1}^{n} d_{ij}$ be the *distance sum for the vertex i* in a (molecular) graph. The *Balaban index* is a TI defined by

$$J = \frac{m}{C+1} \sum_{i,j \epsilon E}^{m} (s_i s_j)^{-1/2},$$

where $C = m - n + 1$ is the *cyclomatic number* of the graph [Ba82].

D11: The ***atom-bond connectivity index*** of a (molecular) graph is a TI defined by [EsToRoGu98, ?]

$$ABC = \sum_{i,j \epsilon E} \sqrt{\frac{\delta_i + \delta_j - 2}{\delta_i \delta_j}}.$$

D12: Let G be a connected graph with adjacency matrix $\mathbf{A}$ and let $\mathbf{D}$ be a diagonal matrix of vertex degrees of G. The ***Laplacian matrix*** of the graph is defined as $\mathbf{L} = \mathbf{D} - \mathbf{A}$.

D13: Let G be a connected graph with Laplacian matrix $\mathbf{L}$ and let p and q be two vertices of G. The ***resistance distance*** between p **and** q is defined by [KlRa93]

$$\Omega_{pq} = \mathbf{L}_{pp}^{\dagger} + \mathbf{L}_{qq}^{\dagger} - 2\mathbf{L}_{pq}^{\dagger}$$

where $\mathbf{L}_{pq}^{\dagger}$ is the *p, q-entry of the Moore-Penrose pseudo-inverse* of the Laplacian matrix.

D14: The ***Kirchhoff index*** of a (molecular) graph is a TI defined by [KlRa93]

$$Kf = \sum_{i<j} \Omega_{ij}$$

REMARKS

R8: The Wiener number has been modified to describe the basic topology of infinite polymeric macromolecules and named *Wiener infinite*, W_∞ [BoMeKa92]:

$$W_\infty = \lim_{lim \to \infty} \frac{an^3 + bn^2 + cn + d}{m\left[\dfrac{n(n-1)}{2}\right]}$$

R9: The Randić index has been generalized to [BoErSa99]

$$\chi^t = \sum_{i,j \epsilon E} (\delta_i \delta_j)^t,$$

and a few mathematical results exist for the different values of t [LiSh08].

FACTS

F19: Let T_n be a tree with n vertices, then [EnJaSn76, BoTr77]

$$W(S_n) < W(T_n) < W(P_n),$$

where $W(S_n) = (n-1)^2$ and $W(P_n) = \binom{n+1}{3}$.

F20: Let T_n be a tree with n vertices and let $0 = \mu_1 < \mu_2 \leq \ldots \leq \mu_n$ be the eigenvalues of the Laplacian matrix of the tree. Then [Me90, Mo91, DoEnGu01],

$$W(T_n) = n \sum_{j=2}^{n} (\mu_j)^{-1}.$$

F21: Let H_k be a hexagonal chain with $k \geq 1$ linearly fused hexagons, then [ShLa97]

$$W(H_k) = \frac{1}{3}(16k^3 + 36k^2 + 26k + 3).$$

F22: Let T be a tree on n vertices. Let for an edge $e = (x, y)$ define $n_1(e) = |\{v|v\epsilon V(T), d(v, x|T) < d(v, y|T)$ and $n_2(e) = |\{v|v\epsilon V(T), d(v, y|T) < d(v, x|T)$. Then [Wi47, GuPo86, DoEnGu01],

$$W(T) = \sum_{e\epsilon E(T)} n_1(e)n_2(e)$$

REMARK

R10: This is the manner in which Wiener introduced his index in 1947.

FACTS

F23: Let T be a tree. Let the bipartite sets of its vertices be of cardinality $|V_a|$ and $|V_b|$. Then [BoGuPo87], $W(T)$ is odd if and only if both $|V_a|$ and $|V_b|$ are odd. If $|V_a|$ or/and $|V_b|$ is even, $W(G)$ is even.

F24: Let $m \geq 2$. Let $T_1, T_2, \ldots, T_m$ be trees with disjoint vertex sets and orders $n_1, n_2, \ldots, n_m$. Let $i = 1, 2, \ldots, m, w_i\epsilon V(T_i)$. Let T be a tree on $n \geq 3$ vertices, obtained by joining a new vertex u to each of the vertices $w_1, w_2, \ldots, w_m$. Then [CaRoRo85, DoEnGu01],

$$W(T) = \sum_{i=1}^{m} [W(T_i) + (n - n_i)d(w_i|T_i) - n_i^2] + n(n - 1)$$

F25: Let T be a tree on n vertices. Let v and u be vertices on a pendant edge. Then [DoGu94],

$$W(T) = \frac{1}{4}[n^2(n - 1) - \sum_{(u,v)\epsilon E(T)} [d(v|T) - d(u|T)]^2].$$

F26: Let T be a tree on n vertices. Let $\deg(v)$ be the degree of vertex v. Then [KlMiPlTr92, DoGu94, Gu94],

$$W(T) = \frac{1}{4}[n(n - 1) + \sum_{v\epsilon V(T)} \deg(v)d(v|T)]$$

F27: Let T be a tree on n vertices and u branching points. Then [DoGr77],

$$W(T) = \binom{n + 1}{3} - \sum_{u} \sum_{1 \leq i < j < k \leq m} n_i n_j n_k.$$

F28: Let T be a tree on n vertices and let $L(T)$ be its line graph. Then [Bu81],

$$W(L(T)) = W(T) - \binom{n}{2}$$

F29: Let W_∞ be the Wiener infinite index, N_1 and C_1 the number of atoms and cycles in the monomeric unit, and d the distance between two neighboring monomeric units in the polymer graph. Then [BaBaBo01],

$$W_\infty = \frac{d}{3(N_1 + C_1)}.$$

F30: Let N, R_g^2, and W be the number of atoms of a polymer whose macromolecule contains no atomic rings, the mean-square radius of gyration of the polymer, and the Wiener number of the polymer graph. Let also b be the length of the covalent bond connecting two monomeric units, let c be the number of polymer chains in a unit volume, and let ξ be the friction coefficient. Then [BoMaDe02],

$$R_g^2 = \frac{b^2}{N^2}W; \eta_0 = \frac{cb^2\xi}{6N^2}W.$$

F31: Let g be the Zimm–Stockmayer branching ratio of a branched macromolecule containing no atomic rings. Let also W, W_{lin}, and R_g^2, $R_{g,lin}^2$ be the Wiener indices and the mean-square radius of gyration of the branched and linear polymer graph with the same molecular weight. Then [BoMaDe02],

$$g = \frac{R_g^2}{R_{g,lin}^2} = \frac{w}{w_{lin}}.$$

F32: Let subgraphs G_i cover upon a vertex u'. Let also $d(u \epsilon G)$ and $d(u_i \epsilon G_i)$ be the distance numbers of the common vertex u in graph G and its ith component G_i. Then [PoBo86],

$$W(G) = \sum_i W(G_i) + nd(u \ \epsilon \ G) - \sum_i n_i d(u_i \ \epsilon \ G_i).$$

F33: Let I be the number of isomorphic components G', which cover to form graph H, and let each of the G's have n' vertices. Let also $W(G')$, $W(H)$, and $d(u|G')$ be the Wiener number of G' and H, and the distance number of vertex u in G'. Then [PoBo90],

$$W(H) = I.W(G') + (n' - 1).I(I - 1).d(u|G').$$

F34: Let graphs G_1 and G_2 have n_1 and n_2 vertices, and let the graphs be linked by a bridge $\{uv\}$. Then,

$$W(H) = W(G_1) + W(G_2) + n_1 n_2 + n_2 d(u|G_1) + n_1 d(v|G_2).$$

F35: Let an edge $\{uv\}$ be divided by an inserted vertex x. Let also the total distance of vertex x in the graph H obtained by $d(xH)$, and the number of geodesics containing vertex s, which are enlarged due to the division of the edge be $b(s)$. Then,

$$W(H) = W(G) + d(x|H) + [\sum_{s \epsilon G} b(s)]/2.$$

F36: Let the edge considered in Fact F35 be a bridge. Let also the number of vertices in the two subgraphs G_1 and G_2 be n_1 and n_2, and let $u \epsilon G_1$ and $v \epsilon G_2$. Then,

$$W(H) = W(G) + n_1 n_2 + [d(u|G) + d(v|G) + n_1 + n_2]2.$$

F37: Let a subgraph of n_1 vertices be transferred from a terminal vertex u to another terminal vertex v. Let also the distance numbers of u and v be denoted by $d(u|G)$ and $d(v|G)$. Then [PoBo86],

$$\Delta W = n_1[d(u|G) - d(v|G)].$$

F38: Let a subgraph be transferred from a terminal vertex u to another terminal vertex v. Let also the length of the shortest path uv be L, the position of the branches located between u and v be i, and the number of vertices in these intermediate branches i, located symmetrically with respect to u and v, be $n_{v,i}$ and $n_{u,i}$. Then [PoBo90],

$$\Delta W = \sum i[(L - 2i)(n_{u,i} - n_{v,i})].$$

F39: Let T_n be a tree with n vertices and let F_n be the nth Fibonacci number. Then,

$$n \leq Z(T_n) \leq F_n + 1$$

where the lower bound is obtained for S_n and the upper bound is obtained for P_n [Gu77].

F40: Let G be a graph with k components $G_1, G_2, \ldots, g_k$. Then [GuPo86],

$$Z(G) = \prod_{i=1}^{k} Z(G_i).$$

F41: Let G be a graph, let $pq \epsilon E$ be an edge, and $p \in V$ be a vertex of G [GuPo86]. Then,

   i) $Z(G) = Z(G - pq) + Z(G - \{p, q\})$.

   ii) $Z(G) = Z(G - p) + \sum_{p,q \epsilon E} Z(G - \{p, q\})$.

F42: Let G be a graph and let $pq \epsilon E$. Then [WaYeYa10],

$$Z(G) \geq Z(G - pq).$$

F43: Let G be a graph with $|P_i|$ paths of length i, $|P_i| = m$, and $|C_3|$ triangles. Then [BrKeMeRu05],

   i) $M_1 = 2m + 2|P_2|$,

   ii) $M_2 = m + 2|P_2| + |P_3| + 3|C_3|$.

F44: Let G be a connected graph with n vertices and m edges. Then [De98],

$$M_1 \leq m\left(\frac{2m}{n-1} + n - 2\right),$$

with equality if and only if the graph is S_n or K_n.

F45: [DaGu04] Let G be a graph with n vertices. Then

$$0 \le M_2 \le \frac{1}{2}n(n-1)^3,$$

where the upper bound is obtained for the complete graph and the lower one for the empty graph.

F46: [DaGu04] Let G be a connected graph with n vertices, m edges, and minimum degree δ_{min}. Then

$$M_2 \le 2m^2 - (n-1)m\delta_{min} + \frac{1}{2}(\delta_{min} - 1)m\left(\frac{2m}{n-1} + n - 2\right),$$

with equality if and only if the graph is S_n or K_n.

F47: The Randić index is bounded as [CaGuHaPa03, LiSh08]

$$\sqrt{n-1} \le \chi \le \frac{n}{2}$$

where the lower bound is reached for the star S_n and the upper bound is attained for any regular network with n nodes indistinct of its degree.

F48: Let T_n be a chemical tree ($\delta_{min} \le 4$) with n vertices and $n_1 \ge 3$ pendant vertices. Then,

$$\chi(T_n) \le \frac{n}{2} + n_1\left(\frac{1}{\sqrt{2}} + \frac{1}{\sqrt{6}} - \frac{7}{6}\right),$$

with equality if and only if the tree is $T(3,2)$ [HaMe03].

F49: [Es10] Let $\mathbf{k} = \left[\delta_1^{-1/2}\ \delta_2^{-1/2} \dots \delta_n^{-1/2}\right]^T$. Then

$$\chi = \frac{1}{2}(n - \mathbf{k}^T \mathbf{L} \mathbf{k}).$$

F50: Let G be a connected graph with n vertices, m edges and let λ_1 be the largest eigenvalue of the adjacency matrix of G. Then [FaMaSa93, CaHa04],

i) $\lambda_1 \ge \dfrac{m}{\chi}$,

ii) $\chi + \lambda_1 \ge 2\sqrt{n-1}$ ($n \ge 3$),

iii) $\chi . \lambda_1 \ge n - 1 (n \ge 3)$.

F51: Let G be a connected graph with n vertices. Then [DoGu10]

$$J(P_n) \le J(G) \le J(K_n),$$

where

$$J(P_n) = (n-1)\sum_{i=1}^{n-1}(s_i s_{i+1})^{-1/2}, \quad s_i = \frac{(n-i+1)(i-1)i}{2} + \frac{(i-1)i}{2}$$

and

$$J(K_n) = \frac{n^2(n-1)}{2(n^2 - 3n + 4)}.$$

F52: [Da10] Let G be a connected graph with m edges and let δ_{max} be the maximum vertex degree. Then,

$$ABC \geq \frac{2^{7/4} m \sqrt{\delta_{max} - 1}}{\delta_{max}^{3/4} \left(\sqrt{\delta_{max}} + \sqrt{2} \right)}$$

when equality is attained for the path graph with n vertices.

F53: [ChGu11, DaGuFu11] Among graphs with n vertices the complete graph has the greatest ABC index and this maximal-ABC graph is unique.

F54: [ChGu11, DaGuFu11] The smallest ABC index for a connected graph with n vertices must be a tree and this minimal-ABC tree need not be unique.

F55: [FuGrVu09] Among trees with n vertices, the star has the greatest ABC index and this maximal-ABC tree is unique.

REMARK

R11: The trees with vertices for which the ABC index is minimum are not known.

FACTS

F56: Let $\mathbf{L}(G - u)$ be the matrix resulting from removing the uth row and column of the Laplacian and let $\mathbf{L}(G - u - v)$ be the matrix resulting from removing both the uth and v^{th} rows and columns of $\mathbf{L}$. The resistance distance can be calculated as [BaGuXi03]:

$$\Omega(u, v) = \frac{det\ \mathbf{L}(G - u - v)}{det\ \mathbf{L}(G - u)}.$$

F57: Let $U_k(u)$ be the uth entry of the kth orthonormal eigenvector associated to the Laplacian eigenvalue μ_k, which has been ordered as $0 = \mu_1 < \mu_2 \leq \ldots \leq \mu_n$. Then [XiGu03],

$$\Omega(u, v) = \sum_{k=2}^{n} \frac{1}{\mu_k} [U_k(u) - U_k(v)]^2.$$

F58: Let the resistance matrix $\mathbf{\Omega}$ be the matrix containing the resistance distance between every pair of vertices in a graph. Then [GoBoSa08],

$$\mathbf{\Omega} = |\mathbf{1}\rangle diag\{[\mathbf{L} + (1/n)\mathbf{J}]^{-1}\}^T + diag[\mathbf{L} + (1/n)\mathbf{J}]^{-1}\langle \mathbf{1}| - 2(\mathbf{L} + (1/n)\mathbf{J})^{-1}$$

where $\mathbf{J} = |\mathbf{1}\rangle\langle\mathbf{1}|$ is an all-ones matrix.

F59: Let G be a connected graph with n vertices; the Kirchhoff index is given by

$$Kf(G) = nTr \int_0^\infty \left(e^{-t\mathbf{L}} - \frac{1}{n}\mathbf{1}\mathbf{1}^T \right) dt,$$

where $\mathbf{1}$ is an all-ones column vector [GoBoSa08].

F60: Let G be a connected graph with $n \geq 3$ vertices, m edges, and let δ_{max} be the maximum vertex degree. Then [ZhTr08],

$$Kf(G) \geq \frac{n}{1 + \delta_{max}} + \frac{n(n-2)^2}{2m - 1 - \delta_{max}}.$$

F61: Let G be a connected graph with $n \geq 2$ vertices, m edges, and let δ_{min} and δ_{max} be the minimum and maximum vertex degree, respectively. Let $0 = \mu_1 < \mu_2 \leq \ldots \leq \mu_n$ be the eigenvalues of the Laplacian matrix. Then [ZhTr09],

$$\frac{n}{\delta_{max}} \sum_{j=2}^{n} \frac{1}{\mu_j} \leq Kf(G) \leq \frac{n}{\delta_{min}} \sum_{j=2}^{n} \frac{1}{\mu_j},$$

with equalities at both sides if and only if it is regular.

F62: Let G be a connected bipartite graph with $n \geq 2$ vertices, and let δ_{max} be the maximum vertex degree. Then [ZhTr09],

$$Kf(G) \geq \frac{n(2n-3)}{\delta_{max}},$$

with equality if and only if G is $K_{\frac{n}{2},\frac{n}{2}}$.

13.1.5   Walk-Based Molecular Parameters

DEFINITIONS

D15: A **walk** of length k is a sequence of (not necessarily distinct) nodes v_0, v_1, ..., v_{k-1}, v_k such that for each $i = 1, 2, \ldots, k$ there is a link from v_{i-1} to v_i. A walk is a **closed walk** if $v_0 = v_k$. The number of edges in the walk is called the **length of the walk**.

D16: The **vector** $w = [\mu_1, \mu_2, \ldots, \mu_k]$, where μ_j is the *number of closed walks of length j* or j^{th} **spectral moment of the adjacency matrix** in the graph and $k < \infty$, represents a molecular descriptor, such as a molecular property A can be expressed as

$$A = \sum_j b_j \mu_j + \alpha,$$

where b_j and α are **empirical coefficients** [GuTr72, JiTaHo84, BoKi92, Es08b].

REMARK

R12: Every μ_j can be expressed in terms of subgraphs, which allows us to express a molecular property as a combination of fragmental molecular contributions.

DEFINITIONS

D17: The **weighted sum** of all closed walks starting at a given node represents an atomic descriptor, subgraph centrality, for the corresponding atom in a molecule [Es00, EsRo05],

$$EE_p = \sum_{k=0}^{\infty} \frac{(A^k)_{pp}}{k!} = (e^A)_{pp},$$

where e^A is a **matrix function** that can be defined using the following Taylor series:

$$e^A = I + A + \frac{A^2}{2!} + \frac{A^3}{3!} + \ldots + \frac{A^k}{k!} + \ldots$$

D18: The *sum of subgraph centralities* of all atoms in a molecule is a molecular descriptor called the Estrada index of the graph [Es00, EsRo05, DeGuRa07],

$$EE(G) = \sum_{p=1}^{n} EE_p.$$

D19: The *subgraph centrality* and *Estrada index* have the following spectral representations [Es00, EsRo05, DeGuRa07]:

$$EE_p = \sum_{j=1}^{n} [\varphi_j(p)]^2 e^{\lambda j}, \qquad (13.1.1)$$

$$EE(G) = \sum_{j=1}^{n} e\lambda j. \qquad (13.1.2)$$

REMARK

R13: The Estrada index of a molecular graph in which every edge is weighted by the parameter $\beta = (kT)^{-1}$, where T is the temperature and k is the Boltzmann constant, represents the electronic partition function of a molecule as defined by $Z_e = \sum_{j=1}^{n} e^{-\beta \epsilon_j}$ [EsHa07].

DEFINITIONS

D20: The *probability* that the system is found in a particular state can be obtained by considering a Maxwell–Boltzmann distribution [EsHa07]:

$$p_j = \frac{e^{\beta \lambda_j}}{\sum_j e^{\beta \lambda_j}} = \frac{e^{\beta \lambda_j}}{EE(G, \beta)}.$$

D21: The *enthalpy* $H(G)$ and *Helmholtz free energy* $F(G)$ of the graph are, respectively, [EsHa07]

$$H(G, \beta) = -\sum_{j=1}^{n} \lambda_j p_j, \qquad (13.1.3)$$

$$F(G, \beta) = \beta^{-1} \ln EE. \qquad (13.1.4)$$

FACTS

F63: [EsHi10] The Estrada index can be obtained as $EE = tr(e^{\beta \mathbf{A}})$, where tr is the trace and

$$exp(\mathbf{A}) = \sum_{k=0}^{\infty} \frac{\mathbf{A}^k}{k!}.$$

F64: [DeGuRa07] The Estrada index of a network G of size n is bounded as

$$n < EE(G) < e^{n-1} + \frac{n-1}{e},$$

where the lower bound is obtained for the graph having n nodes and no links and the upper bound is attained for the complete graph K_n.

F65: Let T_n be a tree with n vertices, then [De09, DeRaGu09]

$$EE(s_n) > EE(T_n) > EE(P_n),$$

where $EE(S_n) = n - 2 + 2\cosh(\sqrt{n-1})$, and $EE(P_n) = \sum\limits_{r-1}^{n} e^{2\cos(2r\pi/(n+1))}$.

F66: [DeRaGu09] Let G be a graph with n vertices and m edges. Then

$$\sqrt{n^2 + 4m} \le EE(G_n) \le n - 1 + e^{\sqrt{2m}}.$$

F67: [BeBo10] Let G be a graph with n vertices. Let δ_j be the degree of the jth vertex and let $a, b \in \mathbb{R}$ be such that the spectrum of $\mathbf{A}$ is contained in $[a, b]$. Then,

$$\sum_{j=1}^{n} \frac{b^2 e^{\frac{\delta_j}{b}} + \delta_j e^{-b}}{b^2 + \delta_j} \le EE \le \sum_{j=1}^{n} \frac{a^2 e^{\frac{\delta_j}{a}} + \delta_j e^{-a}}{a^2 + \delta_j}.$$

F68: [BeBo10] Let G be a graph with n vertices. Let δ_j be the degree of the jth vertex and let $a = 1 - n$ and $b = n - 1$. Then

$$\frac{(n-1)^2 e^{\frac{1}{n-1}} + e^{1-n}}{n-1} \le EE \le \frac{n-1}{e} \cdot \frac{n-1+e^n}{n-2}.$$

F69: [EjFiLuZo07] Let G be a regular graph with n nodes of degree $d = q + 1$. Then,

$$EE(G, \beta) = n \left[\frac{q+1}{2\pi} \int_{-2\sqrt{q}}^{2\sqrt{q}} e^{\beta s} \frac{\sqrt{4q - s^2}}{(q+1)^2 - s^2} ds + \frac{1}{n} \sum_{\gamma} \sum_{k=1}^{\infty} \frac{i(\gamma)}{2^{kl(\gamma)/2}} I_{kl(\gamma)}(2\sqrt{q}\beta) \right],$$

where γ runs over all (oriented) primitive geodesics in the network, $l(\gamma)$ is the length of γ, and $I_m(z)$ is the Bessel function of the first kind

$$I_m(z) = \sum_{r=0}^{\infty} \frac{(z/2)^{n+2r}}{r!(n+r)!}.$$

F70: [EsHa07] The electronic parameters are bounded as:

i) $0 \le S(G, \beta) \le \beta \ln n,$

ii) $-\beta(n-1) \le H(G, \beta) \le 0,$

iii) $-\beta(n-1) \le F(G, \beta) \le -\beta \ln n.$

The lower bounds are obtained for the complete graph as $n \to \infty$ and the upper bounds are reached for the null graph with n nodes.

13.1.6   Vibrational Analysis of Graphs

DEFINITIONS

D22: A *ball-spring graph* is a graph in which every node is a ball of mass m and every link is a spring with the spring constant $m\omega^2$ connecting two balls. The ball-spring graph is submerged into a thermal bath at the temperature T, such that the balls oscillate under thermal disturbances.

D23: The *coordinates chosen to describe a configuration of the system are* $x_i, i = 1, 2, \ldots, n$, each of which indicates the fluctuation of the ball i from its equilibrium point $x_i = 0$.

D24: The *ball-spring graphs* are described by any of the following Hamiltonians:

$$H_A = \sum_i \left(\frac{p_i^2}{2m} + \frac{Km\omega^2}{2} x_i^2 \right) - \frac{m\omega^2}{2} \sum_{i,j} x_i A_{ij} x_j,$$

$$H_L = \sum_i \frac{p_i^2}{2m} + \frac{m\omega^2}{2} \sum_{i,j} x_i L_{ij} x_j,$$

where p_i is the **momentum** of the node i, K is a *constant* satisfying $K \geq max_i k_i$, and k_i is the *degree* of the node i [EsHaBe12].

D25: A *classical vibrational scenario* is one in which the momenta p_i and the coordinates x_i are independent variables. A *quantum vibrational scenario* is one in which the momenta p_j and the coordinates x_i are not independent variables but they are operators that satisfy the commutation relation $[x_i, p_j] = i\hbar\delta_{ij}$, where $i = \sqrt{-1}, \hbar$ is the **Dirac constant**, and δ_{ij} is the **Dirac delta function** [EsHaBe12].

FACT

F71: The mean displacement of node i in the classical vibrational scenario is given by any of the following expressions depending on the Hamiltonian selected [EsHaBe12]:

$$\Delta x_i = \sqrt{\langle x_i^2 \rangle} = \frac{1}{\beta m K \omega^2} \left[(I - A/K)^{-1} \right]_{ii},$$

$$\Delta x_i = \sqrt{\langle x_i^2 \rangle} = \frac{1}{\beta m \omega^2} \left[(L^\dagger) \right]_{ii},$$

where $L^\dagger$ is the Moore–Penrose generalized inverse of the Laplacian.

REMARK

R14: By obviating the physical constants the mean atomic displacements in the classical picture are given by the diagonal entries of the resolvent of the adjacency matrix or of the pseudoinverse of the Laplacian, respectively. The last expression was also investigated in [BaAtEr97, EsHa07].

FACT

F72: The mean displacement of node i in the quantum vibrational scenario is given by any of the following expressions depending on the Hamiltonian selected [EsHaBe12]:

$$\Delta x_i = \sqrt{\langle x_i^2 \rangle} = e^{-\beta\hbar\Omega}\left(\exp\left[\frac{\beta\hbar\omega^2}{2\Omega}A\right]\right)_{ii},$$

$$\Delta x_i = \sqrt{\langle x_i^2 \rangle} = \lim_{\Omega\to 0}\left(\exp\left[\frac{\beta\hbar\omega^2}{2\Omega}L\right]\right)_{pq},$$

$$= 1 + \lim_{\Omega\to 0} O_{2p}O_{2q}\exp\left[\frac{\beta\hbar\omega^2}{2\Omega}\mu_2\right],$$

where μ_2 is the second eigenvalue of the Laplacian matrix and $\Omega = \sqrt{K/m\omega}$.

REMARK

R15: The displacement correlation between a pair of nodes $\langle x_i x_j \rangle$ is given by the (i,j)-entry of the corresponding matrix [EsHaBe12].

FACTS

F73: The resistance distance between a pair of nodes in a graph can be expressed in terms of the node displacements due to small vibrations/oscillations as follows [EsHa07]:

$$\Omega_{ij} = \left[(\Delta x_i)^2 + (\Delta x_j)^2 - \langle x_i x_j \rangle - \langle x_j x_i \rangle\right] = \langle (x_i - x_j)^2 \rangle.$$

F74: The sum of resistance distances for a given node in a graph,

$$R_i = \sum_{j=1}^{n}(L_{ii}^\dagger + L_{jj}^\dagger - 2L_{ij}^\dagger),$$

is related to the node displacements as [EsHa07]

$$R_i = n(\Delta x_i)^2 + \sum_{i=1}^{n}(\Delta x_i)^2 = n\left[(\Delta x_i)^2 + \overline{(\Delta x)^2}\right].$$

F75: The potential energy of the vibrations in a graph is given by [EsHa07]

$$\langle V(\overrightarrow{x}) \rangle = \frac{1}{2n}\sum_{i=1}^{n} k_i R_i - \frac{1}{2n}\sum_{i,j\epsilon E}(R_i + R_j - n\Omega_{ij}).$$

References

[Ba82] A. T. Balaban, Highly discriminating distance-based topological index, *Chem. Phys. Lett.* 89 (1982), 399–404.

[BaAtEr97] I. Bahar, A. R. Atilgan, and B. Erman, Direct evaluation of thermal fluctuations in proteins using a single-parameter harmonic potential, *Folding Des.* 2 (1997), 173–181.

[BaBaBo01] T.-S. Balaban, A. T. Balaban, and D. Bonchev, A topological approach to the predicting of properties of infinite polymers. VI. Rational formulas for normalized the Wiener index and a comparison with index J, *J. Mol. Structure (Theochem)* 535 (2001), 81–92.

[BaGuXi03] R. Bapat, I. Gutman, and W. Xiao, A simple method for computing resistance distance, *Zeitschr. Naturfors. A* 58 (2003), 494–498.

[BeBo10] M. Benzi, and P. Boito, Quadrature rule-based bounds for functions of adjacency matrices, *Lin. Algebra Appl.* 433 (2010), 637–652.

[BoErSa99] B. Bollobás, P. Erdős, and A. Sarkar, Extremal graphs for weights, *Discr. Math.* 200 (1999), 5–19.

[BoGu09] B. Borovićanin and I. Gutman, Nullity of graphs, pp. 107-122 in *Applications of Graph Spectra*, D. Cvetković and I. Gutman (Eds.), Math. Inst. SANU, 2009.

[BoGuPo87] D. Bonchev, I. Gutman, and O. Polansky, Parity of the distance numbers and Wiener numbers of bipartite graphs. *MATCH Commun. Math. Comput. Chem.* 22 (1987), 209–214.

[BoKi92] D. Bonchev and L. B. Kier, Topological atomic indices and the electronic charges in alkanes, *J. Math. Chem.* 9 (1992), 75–85.

[BoMaDe02] D. Bonchev, E. Markel, and A. Dekmezian, Long-chain branch polymer dimensions: application of topology to the Zimm-Stockmayer model, *Polymer*, 43 (2002), 203–222.

[BoMeKa92] D. Bonchev, O. Mekenyan, and V. Kamenska, A topological approach to the modeling of polymer properties (The TEMPO Method), *J. Math. Chem.* 11 (1992), 107–132.

[BoTr77] D. Bonchev and N. Trinajstić Information theory, distance matrix and molecular branching, *J. Chem. Phys.* 67 (1977), 4517–4533.

[BrKeMeRu05] J. Braun, A. Kerber, M. Meringer, and C. Rücker, Similarity of molecular descriptors: the equivalence of Zagreb indices and walk counts. *MATCH: Commun. Math. Comput. Chem.* 54 (2005), 163–176.

[Bu81] F. Buckley, Mean distance in line graphs, *Congr. Numer.* 32 (1981), 153–162.

[CaGuHaPa03] G. Caporossi, I. Gutman, P. Hansen, and L. Pavlović, Graphs with maximum connectivity index, *Comp. Biol. Chem.* 27 (2003), 85–90.

[CaHa04] G. Caporossi and P. Hansen, Variable neighborhood search for extremal graphs. 5. Three ways to automate finding conjectures, *Discr. Math.* 276 (2004), 81–94.

[CaRoRo85] E. R. Canfield, R. W. Robinson, and D. H. Rouvray, Determination of the Wiener molecular branching index for the general tree, *J. Comput. Chem.* 6 (1985), 598–609.

[ChGu11] J. Chen and X. Guo, Extreme atombond connectivity index of graphs, *MATCH: Comm. Math. Comput. Chem.* 65 (2011), 713–722.

[ChLi07] B. Cheng and B. Liu, On the nullity of graphs, *Electron. J. Linear Algebra* 16 (2007), 60–67.

[CoOlMa78] C. A. Coulson, B. O'Leary, and R. B. Mallion, *Hückel theory for organic chemists.* Academic Press, London, 1978.

[CvGu72] D. M. Cvetković and I. Gutman, The algebraic multiplicity of the number zero in the spectrum of a bipartite graph, *Mat. Vesnik* 9 (1972), 141–150.

[CvGuTr72] D. M. Cvetković, I. Gutman, and N. Trinajstić, Graph theory and molecular orbitals II, *Croat. Chem. Acta* 44 (1972) 195–201.

[Da10] K. C. Das, Atombond connectivity index of graphs, *Discr. Appl. Math.* 158 (2010), 1181–1188.

[DaGu04] K. Ch. Das and I. Gutman, Some properties of the second Zagreb index, *MATCH: Comm. Math. Comput. Chem.* 52 (2004), 103-112.

[DaGuFu11] K. C. Das, I. Gutman, and B. Furtula, On atombond connectivity index, *Chem. Phys. Lett.* 511 (2011), 452–454.

[De09] H. Deng, A proof of a conjecture on the Estrada index, *MATCH: Comm. Math. Comput. Chem.* 62 (2009), 599.

[De98] D. de Caen, An upper bound on the sum of squares of degrees in a graph, *Discr. Math.* 185 (1998), 245–248.

[DeBa00] J. Devillers and A. T. Balaban (Eds.) *Topological indices and related descriptors in QSAR and QSPAR.* CRC, 2000.

[DeGuRa07] J. A. de la Peña, I. Gutman, and J. Rada, Estimating the Estrada index, *Lin. Algebra Appl.* 427 (2007), 70–76.

[DeRaGu09] H. Deng, S. Radenković, and I. Gutman, The Estrada index, pp. 124-140 122 in *Applications of Graph Spectra*, D. Cvetković and I. Gutman (Eds.), Math. Inst. SANU, 2009.

[DoEnGu01] A. Dobrynin, R. Entringer, and I. Gutman, Wiener index of trees: theory and applications, *Acta Appl. Math.* 66 (2001), 211–249.

[DoGu10] H. Dong and X. Guo, Character of graphs with extremal Balaban index, *MATCH: Comm. Math. Comput. Chem.* 63 (2010), 799–812.

[DoGr77] J. K. Doyle and J. E. Graver, Mean distance in a graph, *Discrete Math.* 7 (1977), 147–154.

[DoGu94] A. A. Dobrynin and I. Gutman, On a graph invariant related to the sum of all distances in a graph, *Publ. Inst. Math. (Beograd)* 56 (1994), 18–22.

[EjFiLuZo07] V. Ejov, J. A. Filar, S. K. Lucas, and P. Zograf, Clustering of spectra and fractals of regular graphs, *J. Math. Anal. Appl.* 333 (2007), 236–246.

[EnJaSn76] R. C. Entringer, D. E. Jackson, and D. A. Snyder, Distance in graphs, *Czech. Math. J.* 26 (1976), 283–296.

[Es00] E. Estrada, Characterization of 3D molecular structure, *Chem. Phys. Lett.* 319 (2000), 713–718.

[Es08b] E. Estrada, Quantum-chemical foundations of the topological sub-structural molecular design, *J. Phys. Chem. A,* 112 (2008), 5208–5217.

[Es10] E. Estrada, Randic index, irregularity and complex biomolecular networks, *Acta Chim. Slov.* 57 (2010), 597–603.

[EsHa07] E. Estrada and N. Hatano, Statistical-mechanical approach to subgraph centrality in complex networks, *Chem. Phys. Lett.* 439 (2007), 247–251.

[EsHaBe12] E. Estrada, N. Hatano, and M. Benzi, The physics of communicability in complex networks, *Phys. Rep.* 514 (2012), 89–119.

[EsHi10] E. Estrada and D. J. Higham, Network properties revealed through matrix functions, *SIAM Rev.* 52 (2010), 696–714.

[EsRo05] E. Estrada and J. A. Rodríguez-Velázquez, Subgraph centrality in complex networks, *Phys. Rev. E* 71 (2005), 056103.

[EsToRoGu98] E. Estrada, L. Torres, L. Rodrguez, and I. Gutman, An atombond connectivity index: Modelling the enthalpy of formation of alkanes, *Indian J. Chem.* 37A (1998), 849–855.

[FaJoSa05] S. Fajtlowicz, P. E. John, and H. Sachs, On maximum matchings and eigenvalues of benzenoid graphs, *Croat. Chem. Acta* 78 (2005), 195–201.

[FaMaSa93] O. Favaron, M. Mahéo, and J.-F. Saclé, Some eigenvalue properties in graphs (conjectures of GraffitiII), *Discr. Math.* 111 (1993), 197–220.

[FuGrVu09] B. Furtula, A. Graovac, and D. Vukićević, Atombond connectivity index of trees, *Discr. Appl. Math.* 157 (2009), 2828–2835.

[GoBoSa08] A. Ghosh, S. Boyd, and A. Saberi, Minimizing effective resistance of a graph, *SIAM Rev.* 50 (2008), 37–66.

[GrGuTr77] A. Graovac, I. Gutman, and N. Trinajstić, *Topological approach to the chemistry of conjugated molecules.* Springer-Verlag, Berlin, 1977.

[Gu83] I. Gutman, Characteristic and matching polynomials of benzenoid hydrocarbons, *J. Chem. Soc., Faraday Trans.* II 79 (1983), 337–345.

[Gu05] I. Gutman, Topology and stability of conjugated hydrocarbons: The dependence of total - electron energy on molecular topology, *J. Serbian Chem. Soc.* 70 (2005), 441–456.

[Gu77] I. Gutman, Acyclic systems with extremal Hückel-electron energy, *Theor. Chem. Acc.* 45 (1977), 79–87.

[Gu94] I. Gutman, Selected properties of the Schultz molecular topological index, *J. Chem. Inf. Comput. Sci.* 34 (1994), 1087–1089.

[GuPo86] I. Gutman and O. E. Polansky, *Mathematical concepts in organic chemistry.* Springer-Verlag, Berlin, 1986.

[GuTr72] I. Gutman and N. Trinajstić, Graph theory and molecular orbitals. Total-electron energy of alternant hydrocarbons, *Chem. Phys. Lett.* 17 (1972), 535–538.

[HaMe03] P. Hansen and H. Mélot, Variable neighborhood search for extremal graphs. 6. Analyzing bounds for the connectivity index, *J. Chem. Inf. Comput. Sci.* 43 (2003), 1–14.

[Ho71] H. Hosoya, Topological index. A newly proposed quantity characterizing the topological nature of structural isomers of saturated hydrocarbons, *Bull. Chem. Soc. Japan* 44 (1971), 2332–2339.

[JiTaHo84] Y. Jiang, A Tang, and R. Hoffmann, Evaluation of moments and their application in Hückel molecular orbital theory, *Theor. Chem. Acc.* 66 (1984), 183–192.

[KiHaMuRa75] L. B. Kier, L. H. Hall, W. J. Murray, and M. Randić, Molecular connectivity. Part 1. Relationship to nonspecific local anesthesia, *J. Pharma. Sci.* 64 (1975), 1971–74.

[KiHa76] L. B. Kier and L. H. Hall, *Molecular connectivity in chemistry and drug research*, Academic Press, New York (1976) p. 257.

[KlMiPlTr92] D. J. Klein, Z. Mihalić, D. Plavić, and N. Trinajstić, Molecular topological index: A relation with the Wiener index, *J. Chem. Inf. Comput. Sci.* 32 (1992), 304–305.

[KlRa93] D. J. Klein and M. Randić, Resistance distance, *J. Math. Chem.* 12 (1993), 81–95.

[KoMo01] J. H. Koolen and V. Moulton, Maximal energy graphs, *Adv. Appl. Math.* 26 (2001), 47–52.

[KoMo03] J. H. Koolen and V. Moulton, Maximal energy bipartite graphs, *Graphs and Combinatorics*, 19 (2003), 131–135.

[Ku06] W. Kutzelnigg, What I like about Hückel theory, *J. Comp. Chem.* 28 (2006), 25–34.

[LiSh08] X. Li and Y. Shi, A survey on the Randić index, *MATCH: Comm. Math. Comp. Chem.* 59 (2008), 127–156.

[Lo50] H. C. Longuet Higgins, Some studies in molecular orbital theory. I. Resonance structures and molecular orbitals in unsaturated hydrocarbons, *J. Chem. Phys.* 18 (1950), 265.

[Mc71] B. J. McClelland, Properties of the latent roots of a matrix: The estimation of n electron energies, *J. Chem. Phys.* 54 (1971), 640.

[Me90] R. Merris, The distance spectrum of a tree. *J. Graph Theory* 14 (1990), 365–369.

[Mo91] B. Mohar, Eigenvalues, diameter, and mean distance in graphs, *Graphs and Combinatorics* 7 (1991), 53–64.

[Ni07] V. Nikiforov, The energy of graphs and matrices, *J. Math. Anal. Appl.* 326 (2007), 1472–1475.

[PoBo86] O. E. Polansky and D. Bonchev, The Wiener number of graphs. I. General theory and changes due to graph operations, *MATCH Commun. Math. Comput. Chem.* 21 (1986), 133–186.

[PoBo90] O. E. Polansky and D. Bonchev, The Wiener number of graphs. II. Transfer graphs and some of their metric properties, *MATCH Commun. Math. Comput. Chem.* 25 (1990), 3–40.

[Ra75] M. Randić, Characterization of molecular branching, *J. Am. Chem. Soc.* 97 (1975), 6609–6615.

[ShLa97] W. C. Shiu and P. C. B. Lam, The Wiener number of the hexagonal net, *Discr. Appl. Math.* 73 (1997), 101–111.

[WaYeYa10] B. Wang, C. Ye, and L. Yan, On the Hosoya index of graphs, *Appl. Math.-A J. Chin. Univ.* 25 (2010), 155–161.

[Wi47] H. Wiener, Structural determination of paraffin boiling points, *J. Am. Chem. Soc.* 69 (1947), 17–20.

[XiGu03] W. Xiao and I. Gutman, Resistance distance and Laplacian spectrum. *Theor. Chem. Acc.* 110 (2003), 284–289.

[Ya78] K. Yates, *Hückel molecular orbital theory*, Academic Press, New York, 1978.

[ZhTr08] B. Zhou and N. Trinajstić, A note on Kirchhoff index, *Chem. Phys. Lett.* 455 (2008), 120–123.

[ZhTr09] B. Zhou and N. Trinajstić, On resistance-distance and Kirchhoff index, *J. Math. Chem.* 46 (2009), 283–289.

Section 13.2
Ties between Graph Theory and Biology

Jacek Blazewicz, Poznan University of Technology, Poland
Marta Kasprzak, Poznan University of Technology, Poland
Nikos Vlassis, LCSB, University of Luxembourg

INTRODUCTION

The last decades brought us a new scientific area of computational biology, which is placed at the junction of biology (especially molecular biology), computer science and mathematics. Its aim is to solve real-world problems arising in biology with the use of mathematical models and methods, and tools from computer science. Molecular biology, due to its rapid progress, yields more and more experimental data, possible to be processed on computers only. Efficient processing and advisable analysis must be accompanied by well suited models and methods; the ones coming from graph theory frequently appeared to be most useful. Here, the most interesting and breakthrough approaches of computational biology tied with graph theory are characterized.

13.2.1 Biological Primer

Except for the very first moments, where collected data were small enough for non-automated treatment, processing and analysis of biological data were done on computers with the use of algorithms. Huge amounts of data produced by a new generation of specialized laboratory equipment need efficient computational approaches. Many of them refer to known models from graph theory. In the following subsections these issues are discussed. Before that, let us introduce basic notions of molecular biology used in this section.

Deoxyribonucleic acid (DNA for short) carries genetic information and exists in the form of a double helix, i.e., two twisted chains/strands of *nucleotides* joined together. Each nucleotide is composed of a nitrogenous base, a saccharide (deoxyribose) and a phosphoric acid. Nucleotides differ only in their nitrogenous bases. There are four bases: adenine (A), cytosine (C), guanine (G), and thymine (T). Their order in a DNA chain codes genetic information, symbolically written as a sequence of the letters A, C, G, T. Bases from the two DNA strands are joined by double or triple hydrogen bonds. Against adenine always thymine stands, against cytosine — guanine. This property was named the *complementarity*. The DNA strands are complementary, i.e., knowing a sequence of bases from a fragment of one strand, one can always determine a sequence of bases from the corresponding fragment of the other strand. A length of a DNA chain is expressed in nucleotides or in bases (in base pairs when we have double-stranded DNA). The length of a human genome is about $3 \cdot 10^9$ base pairs. A short fragment of a DNA chain (formerly up to 20 nucleotides, now understood as of length smaller than 100) is called an *oligonucleotide*.

A *transcription factor* (TF) is a protein that binds to specific short DNA sequences (typically 5–15 bp long), thereby controlling the transcription of genetic information from DNA to mRNA [La97]. TFs are the largest set of regulatory proteins in mammalian cells, covering about 10% of all known proteins of mammals, including humans [KuSiJe10]. In a mammalian genome the number of binding sites for a given TF could be on the order of hundreds or thousands. A TF attaches to one or more binding sites on the genome and helps initiate a gene transcription program—see Figure 13.2.1 for a cartoon view. As such, TFs are vital for many important cellular processes. A fundamental problem, which we will discuss in Section 13.2.3, is the characterization and the extraction from data of the DNA binding specificities of a given transcription factor.

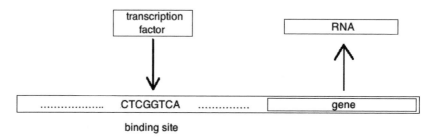

Figure 13.2.1: A cartoon view of gene transcription regulation: A transcription factor attaches to a binding site on the DNA and helps initiate a gene transcription program.

DNA sequencing, which is the process of determining a sequence of bases of a DNA fragment, is the first stage of discovering genetic information. This process is based on the property of complementarity of DNA duplexes. Later, the read DNA fragments are combined into the whole chromosome, or genome, by assembling and mapping algorithms (Figure 13.2.2). Nowadays, next-generation sequencing approaches produce, at relatively low cost, huge volumes of data composed of *reads*, being recognized short DNA or RNA fragments. For these approaches, the step of mapping often is not necessary. One of the former approaches, *sequencing by hybridization* (SBH), became the inspiration for defining a series of graph classes (characterized in the next subsection); thus we pay more attention to it here. What is more, the algorithmic part of SBH is basically very similar to current algorithmic approaches solving the assembling problem.

The aim of the *SBH experiment in its standard version* is to detect all oligonucleotides of a given length l (usually 8–12 nucleotides) composing a part of a DNA chain of a known length n (a few hundreds of nucleotides). For this purpose, the oligonucleotide library is generated, which consists of all possible single-stranded DNA fragments of length l (i.e., their number is 4^l). Next, the library is exposed to the reaction of hybridization with many copies of the studied DNA. In order to operate on that great number of molecules, microarray technology is used (Figure 13.2.3).

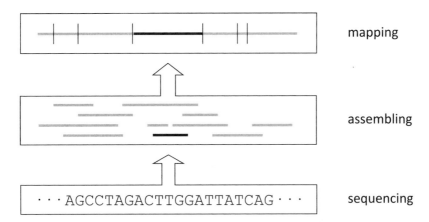

Figure 13.2.2: Discovering genetic information as a three-stage process. DNA fragments read at the lowest level, in SBH or next-generation sequencing experiments, become input data for the assembling stage. At the mapping stage the assembled sequences are ordered within a chromosome or a genome.

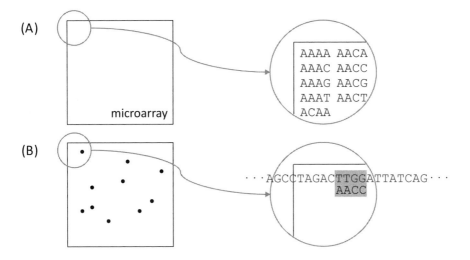

Figure 13.2.3: Sequencing by hybridization involving microarray technology. (A) A microarray containing a complete library of oligonucleotides of length 4. Every oligonucleotide (in many copies) is immobilized to the microarray at known coordinates. (B) During the reaction of hybridization, clones of the studied DNA join oligonucleotides being (reversely) complementary to their subchains. Due to fluorescent labeling of the clones, the image of the microarray gives the information about the spectrum contents.

During the *hybridization* process, oligonucleotides from the library join the DNA chains containing subchains complementary to them (to form a duplex). As a result, one can select oligonucleotides composing the studied DNA, and the oligonucleotides written as words of equal length over the alphabet {A,C,G,T} make a set called a *spectrum* (for details see, e.g., [BaSm88, LyFlKh⁺88, DrLaBr⁺89]). The computational phase of the sequencing process consists in the reconstruction of an original DNA sequence on the basis of the spectrum.

The hybridization experiment executed without any errors provides an *ideal spectrum*, that is, containing only all substrings of length l of the original sequence of the known length n (see Example E1). However, experiments usually end with several errors in spectra. There are two general types of *errors*: *negative* ones, i.e., oligonucleotides missing in the spectrum (the words which are parts of the original sequence but are not present in the set), and *positive* ones, which are erroneous oligonucleotides (the words which are not parts of the original sequence but are present in the set). In the literature these errors are alternatively named false negatives and false positives, respectively.

EXAMPLE

E1:   Suppose the original sequence to be found is CAGTCAGAGTA, $n = 11$. In the hybridization experiment one can use, for example, the complete library of oligonucleotides of length $l = 4$, composed of the following $4^4 = 256$ oligonucleotides: {AAAA, AAAC, AAAG, AAAT, AACA, ..., TTTG, TTTT}. As a result of the experiment performed without errors one obtains the ideal spectrum for this sequence, containing all four-letter substrings of the original sequence: {AGAG, AGTA, AGTC, CAGA, CAGT, GAGT, GTCA, TCAG}. The reconstruction of the sequence consists in finding such an order of the spectrum elements, where each pair of neighboring elements overlaps on $l - 1 = 3$ letters (i.e., for every pair the suffix of $l - 1$ letters of the predecessor is the same as the prefix of the successor). Two possible solutions for this instance are shown in Figure 13.2.4.

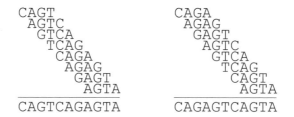

Figure 13.2.4: Two possible solutions of the sequencing problem for the ideal spectrum {AGAG, AGTA, AGTC, CAGA, CAGT, GAGT, GTCA, TCAG}.

In the standard approach to SBH, the complete library contains oligonucleotides of the same length but of compositions differing in the ratio of C/G nucleotides to A/T ones. However, DNA duplexes of C/G rich oligonucleotides are more stable than those containing more A/T nucleotides, and the hybridization conditions (e.g., temperature) should differ for these two groups. The constant temperature kept during the hybridization reaction (i.e., not taking into account C/G content) causes a number of hybridization errors in the resulting spectra. In the earliest studies a simple equation was used to calculate melting temperatures of oligonucleotide duplexes, assuming 4 degrees for C/G pairs and 2 degrees for A/T pairs. Although simple, this rule gives a useful estimation (for more advanced rules see, e.g., [XiSaBu⁺98]). Therefore, to equalize melting temperatures of all oligonucleotides in the library, the ones composed mainly of A/T nucleotides should be longer than the ones composed mainly of C/G. This fact justifies a new approach to DNA sequencing by hybridization using *isothermic oligonucleotide libraries* [BlFoKa⁺04]. Such a library is composed of all possible oligonucleotides of a constant (for the library) melting temperature of oligonucleotide duplexes, but of different lengths. The hybridization with isothermic libraries should result in spectra with a lower number of experimental errors, and then the computational phase of the DNA sequencing would be much more effective. But, to ensure that every possible DNA sequence is covered by such a library, two isothermic oligonucleotide libraries differing by 2 degrees must be used.

We refer the interested reader to [Wa95, Pe00] for a broader treatment of algorithmic and graph-theory issues in computational biology.

13.2.2 Graphs in Sequencing and Mapping

Computational biology, where molecular biology and computer science meet, gave a fresh view on problems of both sides due to its interdisciplinary nature. The biological side was supplied by models and methods from computer science, which allow us to solve biological problems efficiently. Many of them were based on graph theory. The computer science side also gained a lot from this junction; among others, new classes of biologically inspired graphs useful for problem modeling were defined and analyzed.

Graph theory appeared to have wide application in modeling biological problems, following the combinatorial nature of nucleic acids. Especially the recognition of one-dimensional DNA and RNA structures well fits this kind of model. Small DNA/RNA particles are then represented as vertices in a directed graph, overlap of strings corresponding to their nucleotide chains is expressed by arcs, and a solution path corresponds to a resulting sequence of nucleotides.

Interval Graphs in DNA Mapping

Historically the first connection between graph theory and molecular biology happened in 1959, when Benzer proposed a method for identifying the topology of genetic information in a bacteriophage [Be59]. His experiment led to the detection of its linear structure and, as a byproduct, to the invention of interval graphs.

DEFINITION

D1: *Interval graphs* represent the relation of interval overlapping in a linear space. Two vertices, corresponding to intervals, are joined by an edge in an undirected graph if the intervals intersect.

FACTS

F1: If a graph belongs to the class of interval graphs, the problem of verifying its Hamiltonicity becomes polynomially solvable [Ke85].

F2: Combinatorial problems, which are in general NP-complete but easy in the case of interval graphs, include vertex coloring, independent set, and clique problem.

REMARK

R1: Since 1959 interval graphs were used in modeling problems from different areas, where a linear arrangement is looked for, for example, in computational biology for mapping by hybridization to unique probes.

The problem of mapping by hybridization aims at ordering long DNA fragments without referring to their nucleotide sequences, unlike the sequencing or assembling stages.

FACT

F3: In the process of mapping, the hybridization experiment provides a binary matrix of information about pairwise hybridization of longer fragments against a set of short (possibly unique) probes. In the case of errorless experiment, when all appearances of probes within fragments have been detected and there is no false information, the matrix possesses the **consecutive 1s property** in rows, i.e., the columns can be ordered in such a way that in every row all 1s are consecutive.

EXAMPLE

E2: Figure 13.2.5 visualizes exemplary hybridization data and the resulting interval graph. This kind of data is enough to find a solution in polynomial time, which is a Hamiltonian path in the graph.

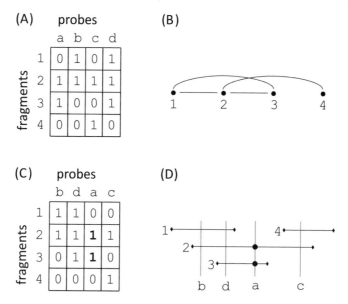

Figure 13.2.5: Example for mapping by hybridization to unique probes.
(A) Result of hybridization has the property, and to find the correct order of its columns, one can model this problem with the use of an interval graph. (B) Vertices correspond to fragments and edges join fragments having at least one probe common. Every Hamiltonian path in the graph represents a possible ordering. (C) One of possible orderings of the columns. (D) The solution, which is an order in the set of fragments, can be immediately read from the rows of the matrix with all 1s consecutive. Two distinguished points correspond to two cells of the matrix in bold.

Adjoint Graphs in DNA Sequencing

Coming back to DNA sequencing, the story began with two papers of Lysov et al. [LyFlKh+88] and Pevzner [Pe89]. The algorithms presented there had been modeled with the use of two fundamental problems of graph theory. The approach by Lysov and co-authors refered to the *Hamiltonian path* problem (Algorithm 13.2.1), while Pevzner used the problem of searching for the *Eulerian trail* (Algorithm 13.2.2).

EXAMPLE

E3: In Figure 13.2.6 graphs modeling the SBH problem in its standard version with constant-length oligonucleotide libraries are presented. They are constructed for the errorless spectrum from Example E1 by Algorithms 13.2.1 and 13.2.2.

Algorithm 13.2.1: Algorithm of Lysov et al.

Input: an ideal spectrum S composed of words of length l.
Output: a Hamiltonian path in G corresponding to an order within S.

Build a directed graph G:
    For every $o_i \in S$ add vertex v_i labeled by o_i.
    For every pair of vertices v_i and v_j
        If suffix of length $l-1$ of o_i is equal to prefix of length $l-1$ of o_j
            Add arc (v_i, v_j).
Find a Hamiltonian path P in G.
Return P.

Algorithm 13.2.2: Pevzner's Algorithm

Input: an ideal spectrum S composed of words of length l.
Output: an Eulerian trail in G corresponding to an order within S.

Build a directed graph G:
    For every $o \in S$
        Add, if not yet present, vertices v_i and v_j labeled by prefix of length
            $l-1$ of o and suffix of length $l-1$ of o, respectively.
        Add arc (v_i, v_j).
Find an Eulerian trail P in G.
Return P.

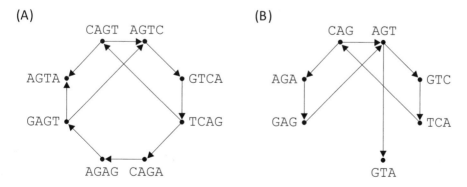

Figure 13.2.6: (A) The graph from the method of Lysov et al., where every Hamiltonian path corresponds to a possible solution. (B) The graph from Pevzner's method with solutions represented by Eulerian trails.

REMARKS

R2: This interesting transformation of the graph, in which the Hamiltonian path is looked for (the strongly NP-hard problem in general), to the graph, in which the Eulerian trail is searched, which changed the computational complexity of the way the solution is produced, was not commented on by Pevzner.

R3: The class of graphs for which such transformation is possible (*labeled graphs*) was widely examined in [BlHeKo⁺99, BlFoKa⁺02]. The graphs built on the base of the ideal spectrum according to Lysov's manner, called **DNA graphs**, belong to this class — they are labeled graphs over the four-letter alphabet.

R4: The class of Pevzner's graphs does not belong as a whole to the class of labeled/DNA graphs. This is because there is not a 1-1 correspondence between the presence of arcs in such graphs and the overlaps of vertex labels (see Figure 13.2.6B, where arc (GAG,AGA) is not present).

DEFINITIONS

D2: *Directed de Bruijn graphs* constitute a class of labeled graphs which are complete with respect to the size of the alphabet and the length of labels [Br46]. For an alphabet of size α and labels of constant length l, de Bruijn graph $B(\alpha, l)$ has α^l vertices, every one labeled by a different word over the alphabet. Arcs are present between all pairs of vertices u and v satisfying the identity of the $(l-1)$-length suffix of label of u and the $(l-1)$-length prefix of label of v.

D3: *DNA graphs* are vertex-induced subgraphs of directed de Bruijn graphs with $\alpha = 4$. *Pevzner's graphs* are subgraphs of DNA graphs.

D4: As defined in [Be73] in the context of *directed graphs*, the *adjoint* $G = (V, A)$ of a graph $H = (U, V)$ is a 1-graph whose vertices represent arcs of H, and which has an arc from x to y if the head of the arc in H corresponding to x is the tail of the arc corresponding to y.

D5: A *directed line graph* is an adjoint G of a 1-graph H.

FACTS

F4: For directed graphs, the Hamiltonian path/cycle problem is easily solvable in an adjoint by transforming the adjoint into its original graph and then by searching for an Eulerian trail/closed trail within it. The existence of an Eulerian trail/closed trail in the original directed graph is a necessary and sufficient condition of the existence of a Hamiltonian path/cycle in its adjoint [BlHeKo⁺99].

F5: The graph constructed by the method of Lysov et al. is the directed line graph of the graph by Pevzner for the same spectrum. For that pair of graphs, the two problems of looking for a Hamiltonian path and an Eulerian trail are equivalent.

REMARK

R5: What is interesting, this transformation does not work for undirected graphs [Be81], unlike the statements present in [La76, GaJo79]. See, for example, the undirected graph in Figure 13.2.7A, and its line graph in Figure 13.2.7B. The first one does not possess any Eulerian trail while the other has a Hamiltonian path. This is why further work utilizing this transformation was restricted to directed graphs.

(A)   (B)

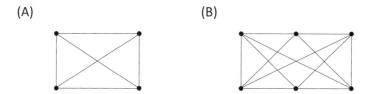

Figure 13.2.7: The equivalence between the problems of Eulerian trail and Hamiltonian
                path in undirected graphs and their line graphs, respectively, is not
                valid. (A) A graph without any Eulerian trail. (B) Its line graph
                having a Hamiltonian path.

Quasi-Adjoint Graphs in DNA Sequencing

The problem of isothermic DNA sequencing by hybridization without errors in the
spectrum was solved in [BlKa06]. The polynomial-time algorithm presented there used
a graph model, in which vertices represented oligonucleotides and arcs were added in
a similar way as in DNA graphs, but with special rules concerning vertices with labels
being substrings of others. (The substrings are possible for two isothermic libraries
differing in the melting temperatures by two degrees, which are used together in the
isothermic approach.) In order to make such graph a directed line graph, some arcs
must be deleted and some "temporary" arcs must be added. After the transformation
of this graph into an original graph, a modified algorithm searching for the Eulerian trail
was applied. This work, which proved that the equivalence of the Hamiltonian path and
Eulerian trail in the two kinds of graphs holds true also for non-adjoints, became strong
motivation for looking for a wider class of graphs with this property. In [BlKaLe$^+$08]
such a class was defined.

DEFINITIONS

D6: A directed graph is a **quasi-adjoint graph** if, for any two vertices x and y, the
following property holds:

$$N^+(x) \cap N^+(y) \neq \emptyset \quad \Rightarrow \quad \begin{array}{ll} N^+(x) = N^+(y) & \vee \\ N^+(x) \subset N^+(y) & \vee \\ N^+(y) \subset N^+(x), \end{array}$$

where $N^+(x)$ is the set of immediate successors of vertex x.

FACTS

F6: For the class of quasi-adjoint graphs, the Hamiltonian cycle problem (HCP) is
solvable in polynomial-time [BlKaLe$^+$08]. The exact algorithm solving this problem
used the transformation between graphs.

F7: The class of quasi-adjoint graphs is a generalization of, among others, adjoints
and the graphs modeling the problem of isothermic SBH without errors in experimental
data.

REMARKS

R6:   One step toward a further extension of this class was made in [BlKa12], where
an algorithm for removing some additional superfluous arcs from a graph was examined. The algorithm takes as the input any directed 1-graph, and by solving a series
of perfect matchings over its subgraphs reduces arcs, which are guaranteed to not compose any Hamiltonian cycle in the graph. The graphs on the output of the algorithm,
named ***reduced-by-matching graphs***, in the best case become quasi-adjoint graphs,
but mostly they remain "hard" for the problem. However, even hard, the reduced graph
becomes an easier instance for some exact or heuristic algorithm solving HCP. It was
confirmed in tests, which also showed that the algorithm is especially efficient for graphs
with average outdegree equal to 2 (HCP is NP-hard for such graphs). Those graphs after the reduction had the mean outdegree close to 1, what made them easy (on average)
for HCP algorithms.

R7:   In [BlKa12] a systematization of several classes of digraphs referring to directed
line graphs was provided, together with the proof of its correctness. In Figure 13.2.8
this relationship is shown. For most of the classes, polynomial-time algorithms solving
HCP exist.

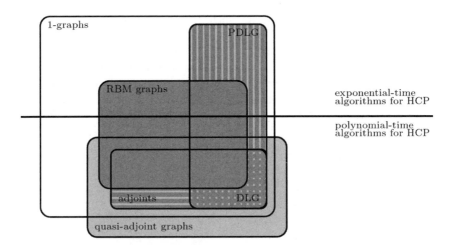

Figure 13.2.8: [BlKa12] The relationship between a few digraph classes with reference
            to HCP solvability. DLG stands for directed line graphs, PDLG for
            partial directed line graphs [ApFr07], and RBM graphs for reduced-
            by-matching graphs. DNA graphs lie within the DLG area.

Labeled Graphs in DNA Assembling

Another field of computational biology, DNA sequence assembling, also intensively uses
models and methods of graph theory. Sequence assembling is the next step, after sequencing, in the reconstruction of a genomic fragment. The output of the sequencing
process (e.g., SBH) becomes the input for sequence assembling and the latter is somehow
similar to the former but on a much larger scale. Now the input sequences generally have

different lengths, misreadings (insertions, deletions, and substitutions of nucleotides), and come from any of the two strands of a fragment of a DNA helix (with no hint from which one). Huge amount of erroneous and incomplete data make the problem well known for its high complexity. It is strongly NP-hard even in the case of data without errors and derived from one DNA strand (compare with the shortest common superstring problem [GaJo79]).

FACTS

F8: The goal of DNA sequence assembling is to compose input fragments into one resulting sequence (or a series of disjoint sequences — contigs) in a proper order. This problem can be modeled as a graph theoretic problem, the searching for a Hamiltonian path (or a series of paths in disjoint components) in a certain digraph (e.g., [KeMy95]).

F9: Unlike in sequencing, due to inexact matchings allowed between fragments, such graphs do not possess the useful property of being adjoints or quasi-adjoint graphs.

REMARK

R8: The inexact matchings are the result of sequencing errors (insertions, deletions, substitutions of nucleotides) and researchers sought ways to overcome this disadvantage. One successful idea, applied next in many assembling algorithms (e.g., [PeTaWa01]), was to decompose the input sequences into series of shorter overlapping constant-length oligonucleotides and to build the graph with only exact matchings allowed, as in Pevzner's approach. Erroneous information present in the input data was then either corrected by some heuristic procedures or ignored. Finally, the Eulerian trail was searched for. Novel next-generation sequencing approaches, like 454 sequencing, Illumina's one or SOLiD, well suit to this model thanks to a small guaranteed experimental error rate.

FACT

F10: The *assembling graphs* constructed with the decomposition of input sequences are — as Pevzner's graphs — subgraphs (but not vertex-induced) of de Bruijn graphs over a four-letter alphabet.

With the progress in sequencing technology, the form of the data change and theoretical models have to adjust accordingly. Paired-end sequencing protocol has been theoretically and experimentally proved to be much more efficient for de novo sequencing than standard single-read approach. In paired-end sequencing, instead of a set of single fragments at the input, one gets a set of paired fragments placed close to each other in the original genome (within some predefined approximate distance). Theoretical models proposed for this kind of data usually skipped this additional information at the beginning, built the assembling graphs as defined above and used the pairs just to guide the process of searching for a path. They did not lead to a simplification of the graph. One of the latest papers brought a new concept of so-called *paired de Bruijn graphs* [MePhCh+11], a modification of the assembling graphs incorporating the information about paired fragments.

REMARKS

R9: Recall that graphs from [MePhCh+11], similarly to other decomposition-based assembling graphs, are not in fact de Bruijn graphs due to the incompleteness of both the vertex and arc sets.

R10: In [MePhCh+11] the input pairs of fragments, distant by a number of nucleotides called the insert size, are decomposed into pairs of oligonucleotides: "left" oligonucleotides in the pairs come from the decomposition of the "left" fragments, and the corresponding "right" oligonucleotide lies at the same position as the "left" one but within the corresponding "right" fragment. In the graph, arcs correspond to the pairs of oligonucleotides and vertices to pairs of their prefixes and suffixes. Such graphs, thanks to the more restricted rule of joining arcs, are indeed less tangled than the regular decomposition-based assembling graphs. See Example E4.

EXAMPLE

E4: For the original sequence TATTTATTACGTACG and for insert size equal to 1 (a theoretical assumption since in fact the insert sizes differ to a large extent in one biological experiment), the paired fragments may be as follows: TATTT+TTACG, ATTTA+TACGT, TTTAT+ACGTA, TTATT+CGTAC, TATTA+GTACG. After their decomposition into pairs of oligonucleotides of length $l = 4$, one obtains: TATT+TTAC, ATTT+TACG, TTTA+ACGT, TTAT+CGTA, TATT+GTAC, ATTA+TACG. The graph from [MePhCh+11] is shown in Figure 13.2.9A. The regular decomposition-based assembling graph constructed without the information about pairing is presented in Figure 13.2.9B.

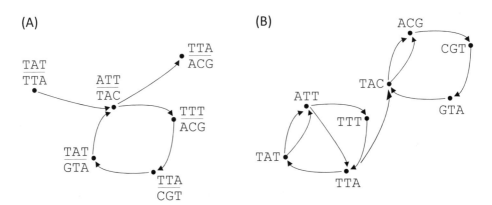

Figure 13.2.9: Decomposition-based assembling graphs constructed for the sequence TATTTATTACGTACG. (A) The graph with paired labels from the method of Medvedev et al. (B) The regular decomposition-based assembling graph.

We see that there is a number of graph classes inspired by biological problems concerning the reconstruction of one-dimensional DNA or RNA structures. What is more, these graphs are not only nice theoretical models of real-world problems, but — first of all — they led to polynomial-time solutions of these problems. The class most known beyond the area of computational biology (and the oldest) is the one of interval graphs. The new class of quasi-adjoint graphs also may become significant for the non-biological community, since it enables polynomial-time exact solution of the Hamiltonian cycle problem, commonly used in modeling real-world problems.

13.2.3 Probabilistics and Graphs

In the previous subsections, we encountered several graphs where vertices correspond to short DNA sequences, and edges reflect some notion of vertex similarity (such as overlap of the corresponding vertex sequences). Here we will briefly describe a different use of graphs in biology, in particular in the context of transcription factor (TF) binding (see the introductory section for some background information). Here, the vertices of the graph correspond to single positions on the DNA, and the edges of the graph capture a certain notion of *stochastic dependence* between nucleotides at different positions.

Graphical Models of TF Binding

The dominant model for TF binding in the literature is the *position weight matrix* (PWM) [DuEdKr⁺98, St00]. This is a probabilistic model that assumes that all DNA positions in a binding site are mutually independent given the binding event (See Figure 13.2.10 for a cartoon view). However, it has been recognized that more expressive models need to be developed, in particular models that can capture higher-order dependencies among nucleotides at different positions on the genome [BaBePh⁺09, StZh10].

Figure 13.2.10: A 'logo' of a PWM model for a binding site of length seven, where the height of a letter in some position of the binding site reflects the probability under that model that the corresponding nucleotide appears in that position.

DEFINITIONS

D7: A *probabilistic graphical model* is a directed or undirected graph, where each vertex defines a random variable and the edges define stochastic (in)dependence relations between these variables.

D8: A *Markov network* is an undirected probabilistic graphical model.

D9: A *Markov tree* is a cycle-free Markov network.

D10: A *mixture of trees* is a probabilistic model that is a convex combination of a (typically small) set of Markov trees [BaElFr⁺03].

D11: An *ensemble of trees* is a probabilistic model that is a mixture over *all* spanning trees of the complete graph formed by all variables [MeJa06].

FACTS

F11: In a Markov network, a local Markov property holds among variables: A variable is conditionally independent of all other variables given its neighbors in the graph [KoFr09].

F12: The probability of a vector $\mathbf{x}$ under a Markov network with strictly positive distribution is given by a log-linear function:

$$p(\mathbf{x}) = \frac{1}{Z} \exp\left(\sum_{c \in \mathcal{C}} \theta_c^\top \phi_c(\mathbf{x}_c) \right), \tag{13.2.1}$$

where $\mathcal{C}$ is a set of cliques in the graph, and ϕ_c is a function that maps the subset of variables involved in clique c to the real numbers. The model is parametrized by a set of parameters θ_c for each clique c, and the term Z is a normalizer that ensures that $p(\mathbf{x})$ is a proper distribution, that is, $\sum_{\mathbf{x}} p(\mathbf{x}) = 1$.

F13: The probability of a vector $\mathbf{x}$ under a Markov tree model T is given by

$$p(\mathbf{x}|T;\theta) = \prod_{e_{ij} \in T} \frac{\theta_{ij}(x_i, x_j)}{\theta_i(x_i)\theta_j(x_j)} \prod_{i=1}^{d} \theta_i(x_i), \tag{13.2.2}$$

where e_{ij} denotes the edge that is incident to the vertices i and j in tree T, and the parameters $\theta = \{\theta_i, \theta_{ij}\}$ encode all single and pairwise marginals of the distribution of $\mathbf{x}$ under the model.

F14: The probability of a vector $\mathbf{x}$ under a mixture of trees model is given by

$$p(\mathbf{x}) = \sum_{T \in \mathcal{T}} p(T)\, p(\mathbf{x}|T;\theta_T), \tag{13.2.3}$$

where $\mathcal{T}$ is a fixed set of trees, and $p(T)$ defines a probability distribution over the set of trees.

F15: The probability of a vector $\mathbf{x}$ under an ensemble of trees model is given by

$$p(\mathbf{x}) = \sum_{T \in \mathcal{S}} p(T;\beta)\, p(\mathbf{x}|T;\theta), \tag{13.2.4}$$

where $\mathcal{S}$ is the set of all spanning trees in the complete graph formed by the set of variables x_i, for $i = 1, \ldots, d$, and the probability of a tree $p(T;\beta)$ is given by

$$p(T;\beta) = \frac{1}{Z} \prod_{(i,j) \in T} \beta_{ij}, \tag{13.2.5}$$

for parameters $\beta = \{\beta_{ij} \geq 0\}$, for each edge (i,j) of the graph, with $\beta_{ij} = \beta_{ji}$. Here, $Z = \sum_{T \in \mathcal{S}} \prod_{(i,j) \in T} \beta_{ij}$ is a normalizer that ensures that $p(\mathbf{x})$ sums to one over all $\mathbf{x}$.

F16: The probability $p(\mathbf{x})$ in an ensemble of trees model (13.2.4), as well as the normalizer Z in (13.2.5), can be computed analytically via the *weighted matrix tree theorem* [Mo70]. This theorem states that the weighted sum of all spanning trees of a given graph equals any minor of order $d - 1$ of the weighted graph Laplacian L:

$$L_{ij} = \begin{cases} -\beta_{ij} & \text{if } i \neq j \\ \sum_k \beta_{ik} & \text{if } i = j. \end{cases} \tag{13.2.6}$$

REMARKS

R11: Comparing the ensemble of trees model (13.2.4) with the mixture of trees model (13.2.3), we note a few differences. The first is that we are now using a large (super-exponential) number of spanning trees (the number of spanning trees of a complete graph of order d is d^{d-2}, according to Cayley's formula). The second is that each spanning tree T defines a Markov tree model $p(\mathbf{x}|T; \theta)$ that is parametrized by the same parameters θ; in other words, all trees share the same parameters $\theta = \{\theta_i, \theta_{ij}\}$. The last crucial point is that the mixing distribution $p(T; \beta)$ is parametrized by an extra set of parameters β, which play a special role in this model.

R12: In the context of TF binding, the random vector $\mathbf{x} = (x_1, \ldots, x_d)$ corresponds to a length-d genomic sequence (a TF binding site), a vertex of a Markov network corresponds to a position on the DNA site where the given TF binds to, and an edge captures stochastic dependencies between different nucleotides at different positions on the binding site.

R13: Markov networks and Markov trees have been used as models of TF binding [BaElFr+03, ShLuSe08]. As shown in [BaElFr+03], a mixture of trees provides better predictions over a single Markov tree in general, and better predictions than the classical PWM model.

EXAMPLE

E5: Figure 13.2.11 shows a hypothetical example of a binding site of length 8 on the DNA, and a graphical model that can be used as a model of TF binding. In this example, the TF binding model is a Markov tree with 8 vertices and 7 edges. Each vertex indexes the corresponding DNA position on the binding site (e.g., vertex 1 denotes the left-most position in the binding site), and edges capture independencies of nucleotides at different positions on the site according to the local Markov property of the graph. For example, the neighborhood structure of vertex 2 in the graph implies that, for observed values of nucleotides at positions 3 and 5, position 2 is stochastically independent of all other positions.

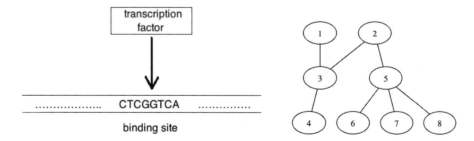

Figure 13.2.11: An example of a hypothetical TF binding site on the DNA of length eight (on the left), and an undirected graphical model (a Markov tree) that can be used as a model of TF binding (on the right). The particular values of the nucleotides shown on the left can be viewed as realizations of the corresponding random variables, one per DNA position.

Learning Graphical Models

For any choice of TF binding model, a fundamental question is how to *learn* the model i.e., estimate its structure as well as its parameters (typically via maximum likelihood), from measured data. The recent technological advances in biology permit the simultaneous measurement of thousands of candidate binding regions on the DNA [BaBePh$^+$09]. The measured data, after some preprocessing, are in the form of short DNA sequences that serve as binding sites of a given TF, and from this set of sequences a model can be learned.

EXAMPLES

E6: In Figure 13.2.6 graphs modeling the SBH problem in its standard version with constant-length oligonucleotide libraries are presented. They are constructed for the errorless spectrum from Example E1 by Algorithms 13.2.1 and 13.2.2.

E7: In Figure 13.2.6 graphs modeling the SBH problem in its standard version with constant-length oligonucleotide libraries are presented. They are constructed for the errorless spectrum from Example E1 by Algorithms 13.2.1 and 13.2.2.

FACTS

F17: Learning a probabilistic graphical model from data by likelihood maximization is in general NP-hard [KoFr09].

F18: A Markov tree (13.2.2) can be learned in polynomial time by casting the problem as a maximum weight spanning tree problem, for appropriate weights computed from the observed data [ChLi68].

F19: Learning Markov networks with higher tree-width is NP-hard [Sr03].

REMARKS

R14: The NP-hardness of the problem can be understood by noticing that in the case of general Markov networks (13.2.1) the normalizer Z involves a summation over an exponentially large set of all combinations of values for each variable. In the context of our TF binding problem, a brute-force computation of Z would require enumerating a set involving $O(4^d)$ elements, for a binding set of length d. Even for modest values of d, this can be an expensive computation.

R15: Learning a mixture model is also a difficult problem, but complexity results are scarcer [AnHsHu$^+$12]. The complexity of learning an ensemble of trees model (13.2.4) is not known, but we conjecture that the problem is NP-hard to solve to global optimality. In practice, local optimization techniques are used, such as the Expectation-Maximization (EM) algorithm [DeLaRu77], but these techniques offer little or no optimality guarantees. Better results can be obtained by taking a greedy approximation in a Hilbert space, an approach that also offers theoretical guarantees [LiBa00, VlLi02].

13.2.4   Hypergraphs in Biology

In this section we briefly discuss possible applications of *hypergraphs* in biology. We mainly follow Klamt et al. [KlHaTh09], who provide a readable review on the topic. The need for hypergraph modeling in biology stems from the fact that biological processes often involve more than two participating elements. A typical example is a biochemical reaction such as $A + B \rightarrow C + D$ that involves four molecules. Hypergraphs can naturally capture such multi-partner relations and therefore they can model a large class of problems in biology, such as protein-protein networks, gene regulation, biochemical reaction sets, and others [KlHaTh09, Ma11, FlMaYe^{+}12].

DEFINITIONS

D12:   An **undirected hypergraph** H is a pair $H = (V, E)$, where V is a set of vertices, and E is a set of nonempty subsets of V called **hyperedges** [Be89].

D13:   A **directed hypergraph** H is a pair $H = (V, E)$, where V is a set of vertices, and E is a set of hyperedges, where each hyperedge is assigned a direction.

D14:   A **transversal** of a hypergraph (V, E) is a subset of V that has nonempty intersection with every edge in E. A hypergraph transversal T is called a **minimal transversal** if no proper subset of T is a transversal.

EXAMPLES

E8:   A protein-protein interaction network can be modeled by an undirected hypergraph, where the vertices of the hypergraph represent proteins, and the hyperedges represent protein 'complexes' in which two or more proteins are physically connected to each other [KlHaTh09].

E9:   A chemical reaction $A + B \rightarrow C + D$ can be modeled by a (trivial) directed hypergraph $H = (V, E)$, where $V = \{A, B, C, D\}$ is the set of participating molecules, and E consists of only one hyperedge $\{A, B\} \rightarrow \{C, D\}$.

E10:   Figure 13.2.12 shows an example of a small reaction network and its corresponding directed hypergraph. If all reactions were reversible, the hypergraph would have been undirected.

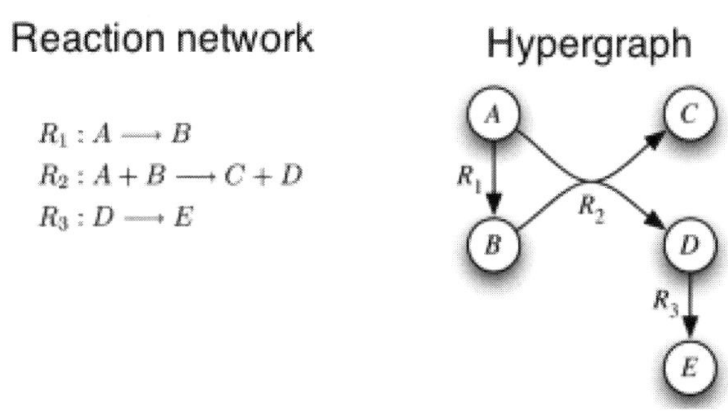

Figure 13.2.12: A reaction network can be represented by a directed hypergraph in which vertices represent molecules and hyperedges represent reactions. Adapted from [KlHaTh09].

E11: The influence of a change of the concentration of a metabolite A to another metabolite B in a metabolic reaction network can be characterized by the shortest hyperpath between vertices A and B in the corresponding reaction hypergraph.

REMARKS

R16: Several tractable problems in graph theory become intractable when moving to hypergraphs (e.g., bipartite hypergraph testing) [GaJo79, Be89].

R17: In a biological context, a minimal transversal of a hypergraph could define a set of distinct representative molecules, for instance a set of essential proteins in a protein-protein interaction network, or it could characterize the most economical way to 'disrupt' a biological system such as a cancer cell: In the case of a metabolic network involving a set of reactions modeled via a hypergraph, a minimal transversal of the hypergraph would index the smallest set of molecules the perturbation of which could have an effect on every single reaction in the network.

R18: A faithful model of a biological system often requires imposing additional semantics on its objects, as in the case of *stoichiometry* where each biochemical reaction includes also the molecularities of the involved species [FlMaYe$^+$12], in which case a hypergraph model must be annotated with additional semantics. Accordingly, several classical graph concepts when generalized to hypergraphs, such as hypercircuits or hypercycles, may need to be carefully redefined in a biological context [Ze00].

ACKNOWLEDGMENTS

The research has been partially supported by grant No. DEC-2011/01/B/ST6/07021 from the National Science Centre, Poland. A part of this section is an extended version of results published in *Fundamenta Informaticae* journal. A part of Section 13.2.3 is based on joint work of the last author with Anke Wienecke-Baldacchino and Merja Heinäniemi.

References

[AnHsHu$^+$12] A. Anandkumar, D. Hsu, F. Huang, and S. Kakade, Learning high-dimensional mixtures of graphical models, arXiv:1203.0697v2, 2012.

[ApFr07] N. Apollonio and P. G. Franciosa, A characterization of partial directed line graphs, *Discrete Mathematics* 307 (2007), 2598–2614.

[BaBePh$^+$09] G. Badis, M. F. Berger, A. A. Philippakis, S. Talukder, A. R. Gehrke, S. A. Jaeger, E. T. Chan, G. Metzler, A. Vedenko, X. Chen, H. Kuznetsov, C. Wang, D. Coburn, D. E. Newburger, Q. Morris, T. R. Hughes, and M. L. Bulyk, Diversity and complexity in DNA recognition by transcription factors, *Science* 324 (5935) (2009), 1720–1723.

[BaSm88] W. Bains and G. C. Smith, A novel method for nucleic acid sequence determination, *Journal of Theoretical Biology* 135 (1988), 303–307.

[BaElFr⁺03] Y. Barash, G. Elidan, N. Friedman, and T. Kaplan, Modeling dependencies in protein-DNA binding sites, in *Proceedings of the seventh annual international conference on research in computational molecular biology*, RECOMB'03, 28–37, ACM, New York, 2003.

[Be59] S. Benzer, On the topology of the genetic fine structure, *Proceedings of the National Academy of Sciences of USA* 45 (1959), 1607–1620.

[Be73] C. Berge, *Graphs and Hypergraphs*, North-Holland Publishing Company, London, 1973.

[Be89] C. Berge, *Hypergraphs: Combinatorics on Finite Sets*, Elsevier, 1989.

[Be81] A. A. Bertossi, The edge Hamiltonian path problem is NP-complete, *Information Processing Letters* 13 (1981), 157–159.

[BlFoKa⁺02] J. Blazewicz, P. Formanowicz, M. Kasprzak, and D. Kobler, On the recognition of de Bruijn graphs and their induced subgraphs, *Discrete Mathematics* 245 (2002), 81–92.

[BlFoKa⁺04] J. Blazewicz, P. Formanowicz, M. Kasprzak, and W. T. Markiewicz, Sequencing by hybridization with isothermic oligonucleotide libraries, *Discrete Applied Mathematics* 145 (2004), 40–51.

[BlHeKo⁺99] J. Blazewicz, A. Hertz, D. Kobler, and D. de Werra, On some properties of DNA graphs, *Discrete Applied Mathematics* 98 (1999), 1–19.

[BlKa06] J. Blazewicz and M. Kasprzak, Computational complexity of isothermic DNA sequencing by hybridization, *Discrete Applied Mathematics* 154 (2006), 718–729.

[BlKa12] J. Blazewicz and M. Kasprzak, Reduced-by-matching graphs: toward simplifying Hamiltonian circuit problem, *Fundamenta Informaticae* 118 (2012), 225–244.

[BlKaLe⁺08] J. Blazewicz, M. Kasprzak, B. Leroy-Beaulieu, and D. de Werra, Finding Hamiltonian circuits in quasi-adjoint graphs, *Discrete Applied Mathematics* 156 (2008), 2573–2580.

[Bo98] B. Bollobas, *Modern Graph Theory*, Springer, 1998.

[ChLi68] C. Chow and C. Liu, Approximating discrete probability distributions with dependence trees, *IEEE Transactions on Information Theory* 14(3) (1968), 462–467.

[Br46] N. G. de Bruijn, A combinatorial problem, *Koninklijke Nederlandse Akademie v. Wetenschappen Proceedings* 49 (1946), 758–764.

[DeLaRu77] A. P. Dempster, N. M. Laird, and D. B. Rubin, Maximum likelihood from incomplete data via the EM algorithm, *J. Roy. Statist. Soc. B* 39 (1977), 1–38.

[DrLaBr⁺89] R. Drmanac, I. Labat, I. Brukner, and R. Crkvenjakov, Sequencing of megabase plus DNA by hybridization: theory of the method, *Genomics* 4 (1989), 114–128.

[DuEdKr⁺98] R. Durbin, S. Eddy, A. Krogh, and G. Mitchison, *Biological Sequence Analysis: Probabilistic Models of Proteins and Nucleic Acids*, Cambridge University Press, 1998.

[FlMaYe⁺12] R. Fleming, C. Maes, Y. Ye, M. Saunders, and B. Palsson, A variational principle for computing nonequilibrium fluxes and potentials in genome-scale biochemical networks, *Journal of Theoretical Biology* 292 (2012), 71–77.

[GaJo79] M. R. Garey and D. S. Johnson, *Computers and Intractability. A Guide to the Theory of NP-Completeness*, W.H. Freeman and Company, San Francisco, 1979.

[KeMy95] J. D. Kececioglu and E. W. Myers, Combinatorial algorithms for DNA sequence assembly, *Algorithmica* 13 (1995), 7–51.

[Ke85] J. M. Keil, Finding Hamiltonian circuits in interval graphs, *Information Processing Letters* 20 (1985), 201–206.

[KlHaTh09] S. Klamt, U.-U. Haus, and F. Theis, Hypergraphs and cellular networks, *PLoS Comput Biology* 5(5):e1000385, 2009.

[KoFr09] D. Koller and N. Friedman, *Probabilistic Graphical Models: Principles and Techniques*, MIT Press, 2009.

[KuSiJe10] V. A. Kuznetsov, O. Singh, and P. Jenjaroenpun, Statistics of protein-DNA binding and the total number of binding sites for a transcription factor in the mammalian genome, *BMC Genomics* 11(Suppl 1):S12, 2010.

[La97] D. S. Latchman, Transcription factors: an overview, *The International Journal of Biochemistry Cell Biology* 29(12) (1997), 1305–1312.

[La76] E. Lawler, *Combinatorial Optimization: Networks and Matroids*, Holt, Rinehart and Winston, New York, 1976.

[LiBa00] J. Q. Li and A. R. Barron, Mixture density estimation, in S. A. Solla, T. K. Leen, and K.-R. Müller, editors, *Advances in Neural Information Processing Systems 12*, MIT Press, 2000.

[LyFlKh⁺88] Yu. P. Lysov, V. L. Florentiev, A. A. Khorlin, K. R. Khrapko, V. V. Shik, and A. D. Mirzabekov, Determination of the nucleotide sequence of DNA using hybridization with oligonucleotides. A new method, *Doklady Akademii Nauk SSSR* 303 (1988), 1508–1511.

[Ma11] A. Marchetti-Spaccamela, Structures and hyperstructures in metabolic networks, in P. Kolman and J. Kratochvl, editors, *Graph-Theoretic Concepts in Computer Science*, volume 6986 of *Lecture Notes in Computer Science*, 1–4, Springer, Berlin/Heidelberg, 2011.

[MePhCh⁺11] P. Medvedev, S. Pham, M. Chaisson, G. Tesler, and P. Pevzner, Paired de Bruijn graphs: a novel approach for incorporating mate pair information into genome assemblers, *Lecture Notes in Computer Science* 6577 (2011), 238–251.

[MeJa06] M. Meilă and T. Jaakkola, Tractable bayesian learning of tree belief networks, *Statistics and Computing* 16(1) (2006), 77–92.

[Mo70] J. Moon, *Counting labelled trees*, Canadian Mathematical Congress, Montreal, 1970.

[Pe89] P. A. Pevzner, *l*-tuple DNA sequencing: computer analysis, *Journal of Biomolecular Structure and Dynamics* 7 (1989), 63–73.

[Pe00] P. A. Pevzner, *Computational Molecular Biology: an Algorithmic Approach*, MIT Press, Cambridge, 2000.

[PeTaWa01] P. A. Pevzner, H. Tang, and M. S. Waterman, An Eulerian path approach to DNA fragment assembly, *Proceedings of the National Academy of Sciences of the USA* 98 (2001), 9748–9753.

[ShLuSe08] E. Sharon, S. Lubliner, and E. Segal, A feature-based approach to modeling protein-DNA interactions, *PLoS Comput Biol* 4(8):e1000154, 2008.

[Sr03] N. Srebro, Maximum likelihood bounded tree-width Markov networks, *Artificial Intelligence* 143 (2003), 123–138.

[St00] G. D. Stormo, DNA binding sites: representation and discovery, *Bioinformatics* 16(1) (2000), 16–23.

[StZh10] G. D. Stormo and Y. Zhao, Determining the specificity of protein-DNA interactions, *Nature Reviews Genetics* 11(11) (2010), 751–760.

[VlLi02] N. Vlassis and A. Likas, A greedy EM algorithm for Gaussian mixture learning, *Neural Processing Letters* 15(1) (2002), 77–87.

[Wa95] M. S. Waterman, *Introduction to Computational Biology. Maps, Sequences, and Genomes*, Chapman & Hall, London, 1995.

[XiSaBu$^+$98] T. Xia, J. SantaLucia Jr., M. E. Burkard, R. Kierzek, S. J. Schroeder, X. Jiao, C. Cox, and D. H. Turner, Thermodynamic parameters for an expanded nearest-neighbor model for formation of RNA duplexes with Watson-Crick base pairs, *Biochemistry* 37 (1998), 14719–14735.

[Ze00] A. Zeigarnik, On hypercycles and hypercircuits in hypergraphs, *Discrete Mathematical Chemistry (DIMACS Series in Discrete Mathematics and Theoretical Computer Science)* 51 (2000), 377–383.

Glossary for Chapter 13

ABC index - atom-bond connectivity index: the sum of the square root of edge
weights for the graph, where the edge weights are defined as the edge degree divided
by the product of vertex degrees of the pair of vertices forming the edge.

adjoint $G = (V, A)$ - of a digraph $H = (U, V)$: a directed 1-graph whose vertices
represent arcs of H and which has an arc from x to y if the head of the arc in H
corresponding to x is the tail of the arc corresponding to y.

alternant conjugate molecule: a molecule in which multiple bonds (double or triple)
alternate with single ones.

assembling graph: a directed graph used for modeling instances of the DNA assembling problem, constructed with decomposition of input sequences into equal-length
subsequences. It is a subgraph (but not vertex-induced) of a de Bruijn graph over a
four-letter alphabet.

Balaban index: an analog of the Randić index in which the vertex degrees are replaced
by the total distances of the graph vertices, with a normalizing coefficient including
the number of graph edges and cycles of a molecule.

benzenoid molecule: a molecule formed by fused hexagonal rings.

consecutive 1s property in rows: when columns of a matrix can be ordered in such
a way that in every row all 1s are consecutive.

cycle graph: a graph in which every node has degree two.

de Bruijn graph: a labeled graph, which is complete with respect to the size of the
alphabet and the length of labels. For an alphabet of size α and labels of constant
length l, de Bruijn graph $B(\alpha, l)$ has α^l vertices, every one labeled by a different
word over the alphabet.

directed hypergraph: a hypergraph where each hyperedge is assigned a direction.

directed line graph: an adjoint of a directed 1-graph.

displacement correlation: refers to the correlation function of the displacements of
two atoms (vertices) in a molecule.

DNA graph: a labeled graph with labels over a four-letter alphabet.

edge degree: number of edges adjacent to a given edge.

ensemble of trees: a probabilistic model that is a mixture over all spanning trees of
the complete graph formed by a set of variables.

Estrada index: the sum of the exponential of the eigenvalues of the adjacency matrix,
i.e., the trace of the exponential of the adjacency matrix.

graph diameter: the length of the largest shortest-path distance in a graph.

graph energy: the sum of the absolute values of graph eigenvalues.

graph invariant: a characterization of a graph which does not depend on the labeling
of vertices or edges.

graph nullity: the multiplicity of the zero eigenvalue of the adjacency matrix, i.e., the
number of times eigenvalue zero occurs in the spectrum of the adjacency matrix.

Hosoya index: the total number of selections of k mutually nonadjacent edges with k (max) equal to half of the number of vertices in the graph.

hydrocarbon: a molecule formed only by carbon and hydrogen.

hypergraph transversal: a subset of the set of vertices V of a hypergraph (V, E) that has nonempty intersection with every edge in E.

incidence matrix: of a graph: a matrix whose rows correspond to vertices and its columns to edges of the graph and the i, j entry is one or zero if the ith vertex is incident with the jth edge or not, respectively.

interval graph: a graph that represents the relation of interval overlapping in a linear space. Two vertices, corresponding to intervals, are joined by an edge in an undirected graph if the intervals intersect.

Kirchhoff index: the sum of the resistances of the graph.

labeled graph: a directed graph whose vertices can be labeled over an alphabet in such a way that labels are unique and of the same length l, and the suffix of length $l - 1$ of a label of vertex x is equal to the prefix of length $l - 1$ of a label of vertex y if and only if there is the arc (x, y) in the graph.

Laplacian matrix: a square symmetric matrix with diagonal entries equal to the degree of the corresponding vertex and out-diagonal equal to -1 or zero depending if the corresponding vertices are connected or not, respectively.

Markov network: an undirected probabilistic graphical model.

Markov tree: a cycle-free Markov network.

matching - of a graph: the number of mutually non-adjacent edges in the graph.

mean displacement - of an atom (vertex): refers to the oscillations of an atom from its equilibrium position due to thermal fluctuations.

molecular descriptor: a quantitative characteristic of a molecule based on its structure or composition.

molecular graph: simple graph with non-hydrogen atoms as nodes and covalent bonds between them representing links.

molecular Hamiltonian: the operator representing the energy of the electrons and atomic nuclei in a molecule.

paired 'de Bruijn' graph: a directed graph being a modification of the assembling graph, which incorporates information about paired input sequences (see §13.2).

path graph: a graph formed by vertices a degree two except two nodes of degree one.

Position Weight Matrix (PWM): a probabilistic model for TF binding that assumes that positions in a binding site are mutually stochastically independent.

probabilistic graphical model: a directed or undirected graph, where each vertex defines a random variable and the edges define stochastic (in)dependence relations between these variables.

quasi-adjoint graph: a directed graph with the following property satisfied for all pairs of vertices x and y: either their sets of immediate successors are disjoint, or equal, or one of the sets is contained in the other.

Randić index: the sum of the inverse-square root of the product of degrees of all pairs of vertices in the graph.

reduced-by-matching graph: a directed graph being a product of an algorithm reducing arcs not composing any Hamiltonian cycle in a digraph, via a series of perfect matchings solved for its subgraphs (see §13.2).

resistance distance: a distance between any pair of vertices of the graph, determined by the Kirchhoff rules for electrical sets.

shortest path: a path having the least number of edges among all paths connecting two vertices.

star graph: a tree consisting of a node with degree $n-1$ and $n-1$ nodes with degree one.

stoichiometry: a directed hypergraph that models a set of biochemical reactions, where each reaction corresponds to a hyperedge that is annotated by the molecularities of the involved chemicals in the reaction.

strongly regular graph: a regular graph in which every two adjacent vertices and every two non-adjacent vertices have an integer number of common neighbors.

subgraph centrality - of a vertex: the corresponding diagonal entry of the exponential of the adjacency matrix.

topological index: a graph invariant characterizing numerically the topological structure of a molecule.

transcription factor (TF): a protein that binds to DNA and regulates the activity of genes.

transcription factor binding site: a site on the DNA where a transcription factor binds to.

undirected hypergraph: a pair $H = (V, E)$, where V is a set of vertices, and E is a set of non-empty subsets of V (hyperedges).

weighted matrix tree theorem: a theorem that states that the weighted sum of all spanning trees of a given graph with d nodes equals any minor of order $d-1$ of the weighted graph Laplacian.

Wiener index: the sum of the shortest-path distances between all pairs of vertices in the graph.

Zagreb indices: a pair of topological indices based on sums of squared vertex degrees or the product of vertex degrees of adjacent vertices.

Index